THE
DIGITAL
SIGNAL
PROCESSING
HANDBOOK

The Electrical Engineering Handbook Series

Series Editor
Richard C. Dorf
University of California, Davis

Titles Included in the Series

The Electrical Engineering Handbook, Richard C. Dorf
The Biomedical Engineering Handbook, Joseph D. Bronzino
The Circuits and Filters Handbook, Wai-Kai Chen
The Transforms and Applications Handbook, Alexander D. Poularikas
The Control Handbook, William S. Levine
The Electronics Handbook, Jerry C. Whitaker
The Industrial Electronics Handbook, J. David Irwin
The Communications Handbook, Jerry D. Gibson
The Mobile Communications Handbook, Jerry D. Gibson

THE
DIGITAL
SIGNAL
PROCESSING
HANDBOOK

EDITED BY

VIJAY K. MADISETTI
DOUGLAS B. WILLIAMS

 **CRC PRESS**

 **IEEE PRESS**

A CRC Handbook Published in Cooperation with IEEE Press

The cover design background by Vijay K. Madisetti and Douglas B. Williams consists of a two-dimensional "sinc" function superimposed on a fractal image. The fractal was taken from the Mandelbrot set around (x, y) = (–0.48, –0.61) and was generated using MandelZot 4.01 (© Dave Platt). MATLAB 5.0 (© The MathWorks, Inc.) was used to plot the sinc function.

Acquiring Editor:	Jerry Papke
Project Editor:	Carol Whitehead
Marketing Manager:	Jane Stark
Cover design:	Dawn Boyd
PrePress:	Carlos Esser
Manufacturing:	Carol Royal

Library of Congress Cataloging-in-Publication Data

The digital signal processing handbook / editors, Vijay K. Madisetti,
    Douglas B. Williams.
        p.   cm. -- (Electrical engineering handbook series)
    Includes bibliographical references and index.
    ISBN 0-8493-8572-5 (alk. paper)
    1. Signal processing--Digital techniques--Handbooks, manuals, etc.
I. Madisetti, V. (Vijay)   II. Williams, Douglas B. (Douglas
Bennett), 1962–   . III. Series.
TK5102.9.D534   1997
621.382'2--DC21

97-21848
CIP

TK
5102.9
D534
1998

Preface

Digital Signal Processing (DSP) is concerned with the theoretical and practical aspects of representing information bearing signals in digital form and with using computers or special purpose digital hardware either to extract that information or to transform the signals in useful ways. Areas where digital signal processing has made a significant impact include telecommunications, man-machine communications, computer engineering, multimedia applications, medical technology, radar and sonar, seismic data analysis, and remote sensing, to name just a few.

During the first fifteen years of its existence, the field of DSP saw advancements in the basic theory of discrete-time signals and processing tools. This work included such topics as fast algorithms, A/D and D/A conversion, and digital filter design. The past fifteen years has seen an ever quickening growth of DSP in application areas such as speech and acoustics, video, radar, and telecommunications. Much of this interest in using DSP has been spurred on by developments in computer hardware and DSP microprocessors. This Handbook is an attempt to capture the entire range of DSP: from theory to applications — from algorithms to hardware.

Given the widespread use of DSP, a need developed for an authoritative reference, written by some of the top experts in the world, that provides information on both theoretical and practical issues suitable for a broad audience — ranging from professionals in electrical engineering, computer science, and related engineering fields, to managers involved in design and marketing, and to graduate students and scholars in the field. Given the large number of excellent introductory texts in DSP, it was also important to focus on topics that are useful to the engineer or scholar without overemphasizing those aspects that are already widely accessible. In short, we wished to create a Handbook that was relevant to the needs of the engineering community.

A task of this magnitude was only possible through the cooperation of many of the foremost DSP researchers and practitioners. This collaboration, over the past three years, has resulted in a Handbook containing a comprehensive range of DSP topics presented with a clarity of vision and a depth of coverage that is expected to inform, educate, and fascinate the reader. Indeed, many of the articles, written by leaders in their fields, embody unique visions and perceptions that enable a quick, yet thorough, exposure to knowledge garnered over years of development.

As with other CRC Press handbooks, we have attempted to provide a balance between essential information, background material, technical details, and introduction to relevant standards and software. The Handbook pays equal attention to theory, practice, and application areas. The DSP Handbook can be used in a number of ways. While simply picking it up and reading it is a good way to learn a lot about this exciting field, it is anticipated that most readers will look up a topic of interest and then read the applicable chapters. As such, each chapter has been written to stand alone and give an overview of its subject matter while providing key references for those interested in learning more. The Handbook can also be used as a reference book for graduate classes, or as supporting material for continuing education courses in the DSP area. Industrial organizations may wish to provide the Handbook with their products to enhance their value by providing a standard and up-to-date reference book.

We have been very impressed with the quality of the Handbook which is due entirely to the contributions of all the authors, and we would like to thank them all. The Advisory Board was instrumental in helping to choose subjects and leaders for all the sections. Being experts in their fields, the section leaders provided the vision and fleshed out the contents for their sections.

Finally, the authors produced the necessary content for this Handbook. To them fell the challenging task of writing for such a broad audience, and they excelled at their jobs.

In addition to these technical contributors, we wish to thank a number of outstanding individuals whose administrative skills made this project possible. Without the outstanding organizational skills of Elaine

M. Gibson, this Handbook may never have been finished. Not only did Elaine manage the paperwork, but she had the unenviable task of reminding authors about deadlines and pushing them to finish. We also thank a number of individuals associated with the CRC Press Handbook Series over a period of time, especially Joel Claypool, Dick Dorf, Kristen Maus, Jerry Papke, Ron Powers, Suzanne Lassandro, and Carol Whitehead.

We welcome you to this Handbook, and hope you find it worth your interest.

Vijay K. Madisetti and Douglas B. Williams
Center for Signal and Image Processing
School of Electrical and Computer Engineering
Georgia Institute of Technology
Atlanta, Georgia

Advisory Board

Editors

Vijay K. Madisetti is an Associate Professor in the School of Electrical and Computer Engineering at Georgia Institute of Technology in Atlanta. He teaches undergraduate and graduate courses in signal processing and computer engineering, and is affiliated with the Center for Signal and Image Processing (CSIP) and the Microelectronics Research Center (MiRC) on campus. He received his B. Tech (honors) from the Indian Institute of Technology (IIT), Kharagpur, in 1984, and his Ph.D. from the University of California at Berkeley, in 1989, in electrical engineering and computer sciences.

Dr. Madisetti is active professionally in the area of signal processing, having served as an Associate Editor of the *IEEE Transactions on Circuits and Systems II*, the *International Journal in Computer Simulation*, and the *Journal of VLSI Signal Processing*. He has authored, co-authored, or edited six books in the areas of signal processing and computer engineering, including *VLSI Digital Signal Processors* (IEEE Press, 1995), *Quick-Turnaround ASIC Design in VHDL* (Kluwer, 1996), and a CD-ROM tutorial on VHDL (IEEE Standards Press, 1997). He serves as the IEEE Press Signal Processing Society liaison, and is counselor to Georgia Tech's IEEE Student Chapter, which is one of the largest in the world with over 600 members in 1996. Currently, he is serving as the Technical Director of DARPA's RASSP Education and Facilitation program, a multi-university/industry effort to develop a new digital systems design education curriculum.

Dr. Madisetti is a frequent consultant to industry and the U.S. government, and also serves as the President and CEO of VP Technologies, Inc., Marietta, GA., a corporation that specializes in rapid prototyping, virtual prototyping, and design of embedded digital systems. Dr. Madisetti's home page URL is at http://www.ee.gatech.edu/users/215/index.html, and he can be reached at vkm@ee.gatech.edu.

Editors

Douglas B. Williams received the B.S.E.E. degree (summa cum laude), the M.S. degree, and the Ph.D. degree, in electrical and computer engineering from Rice University, Houston, Texas in 1984, 1987, and 1989, respectively. In 1989, he joined the faculty of the School of Electrical and Computer Engineering at the Georgia Institute of Technology, Atlanta, Georgia, where he is currently an Associate Professor. There he is also affiliated with the Center for Signal and Image Processing (CSIP) and teaches courses in signal processing and telecommunications.

Dr. Williams has served as an Associate Editor of the *IEEE Transactions on Signal Processing* and was on the conference committee for the 1996 International Conference on Acoustics, Speech, and Signal Processing that was held in Atlanta. He is currently the faculty counselor for Georgia Tech's student chapter of the IEEE Signal Processing Society. He is a member of the Tau Beta Pi, Eta Kappa Nu, and Phi Beta Kappa honor societies.

Dr. Williams's current research interests are in statistical signal processing with emphasis on radar signal processing, communications systems, and chaotic time-series analysis. More information on his activities may be found on his home page at http://dogbert.ee.gatech.edu/users/276. He can also be reached at dbw@ee.gatech.edu.

Contributors

Kenzo Akagiri
Sony Corporation
Tokyo, Japan

Osama Al-Shaykh
University of California
Berkeley, California

Joseph Arrowood
Georgia Institute of Technology
Atlanta, Georgia

K. Balemarthy
Georgia Institute of Technology
Atlanta, Georgia

Victor A. N. Barroso
Instituto Superior Técnico
Instituto de Sistemas e Robótica
Lisboa, Portugal

Kurt Baudendistel
Momentum Data Systems
St. Louis, Missouri

Jan Biemond
Delft University of Technology
Delft, The Netherlands

Bruce W. Bomar
University of Tennessee
Space Institute
Tullahoma, Tennessee

Kevin M. Buckley
Villanova University
Villanova, Pennsyvania

C. Sidney Burrus
Rice University
Houston, Texas

Richard V. Cox
AT&T Labs - Research
Florham Park, New Jersey

Kevin M. Cuomo
Massachusetts Institute of
  Technology
Lincoln Laboratory
Lexington, Massachusetts

Grant A. Davidson
Dolby Laboratories, Inc.
San Francisco, California

J. Debardelaben
Georgia Institute of Technology
Atlanta, Georgia

R. D. DeGroat
The University of Texas at Dallas
Richardson, Texas

Gerard de Haan
Philips Research Laboratories
Eindhoven, The Netherlands

Ricardo L. de Queiroz
Advanced Color Imaging
Xerox Corporation
Webster, New York

N. Desai
Georgia Institute of Technology
Atlanta, Georgia

Zhi Ding
Auburn University
Auburn, Alabama

Petar M. Djurić
State University of New York
Stony Brook, New York

John F. Doherty
Pennsylvania State University
University Park, Pennsylvania

Sean Dorward
Bell Laboratories
Lucent Technologies
Murray Hill, New Jersey

Scott C. Douglas
University of Utah
Salt Lake City, Utah

E. M. Dowling
The University of Texas at Dallas
Richardson, Texas

P. Duhamel
Signal Processing Department
École Nationale Supérieure des
  Télécommunications (ENST)
Paris, France

Lan-Rong Dung
Georgia Institute of Technology
Atlanta, Georgia

T. Egolf
Georgia Institute of Technology
Atlanta, Georgia

S. Famorzadeh
Georgia Institute of Technology
Atlanta, Georgia

Kevin R. Farrell
T-Netix/SpeakEZ
Englewood, Colorado

Ephraim Feig
IBM Corporation
T.J. Watson Research Center
Yorktown Heights, New York

Ruggero E. H. Franich
AEA Technology
Automation and Vision Systems
Culham Laboratory
Abingdon
Oxfordshire
United Kingdom

Daniel R. Fuhrmann
Washington University
St. Louis, Missouri

Sadaoki Furui
Tokyo Institute of Technology
Tokyo, Japan

W. Gehrke
Philips Semiconductors
Hamburg, Germany

Jan J. Gerbrands
Delft University of Technology
Delft, The Netherlands

Georgios B. Giannakis
University of Virginia
Charlottesville, Virginia

Egemen Gönen
Globalstar
San Jose, California

Martin Haardt
Siemens AG
Mobile Radio Networks
Munich, Germany

Joseph L. Hall
Bell Laboratories
Lucent Technologies
Murray Hill, New Jersey

Emile A. Hendriks
Delft University of Technology
Delft, The Netherlands

Cormac Herley
Hewlett Packard Laboratories
Palo Alto, California

Gabor T. Herman
University of Pennsylvania
Philadelphia, Pennsylvania

Alfred Hero
University of Michigan
Ann Arbor, Michigan

R. Hezar
Georgia Institute of Technology
Atlanta, Georgia

Steven H. Isabelle
Massachusetts Institute of
  Technology
Cambridge, Massachusetts

Nikil Jayant
Bell Laboratories
Lucent Technologies
Murray Hill, New Jersey

W. Kenneth Jenkins
University of Illinois at
  Urbana-Champaign
Urbana; Illinois

James D. Johnston
AT&T Research Labs
Murray Hill, New Jersey

B. H. Juang
Bell Laboratories
Lucent Technologies
Murray Hill, New Jersey

Yong-kyu Jung
Georgia Institute of Technology
Atlanta, Georgia

Thomas Kailath
Stanford University
Stanford, California

Ton Kalker
Philips Research Laboratories
Eindhoven, The Netherlands

Lina J. Karam
Arizona State University
Tempe, Arizona

M. Katakura
Sony Corporation
Kanagawa, Japan

Aggelos K. Katsaggelos
Northwestern University
Evanston, Illinois

Mostafa Kaveh
University of Minnesota
Minneapolis, Minnesota

A. Kavipurapu
Georgia Institute of Technology
Atlanta, Georgia

Steven M. Kay
University of Rhode Island
Kingston, Rhode Island

M. Khan
Georgia Institute of Technology
Atlanta, Georgia

M. Kohut
Sony Corporation
Culver City, California

Stephen Kosonocky
IBM Corporation
T.J. Watson Research Center
Yorktown Heights, New York

Jelena Kovačević
Bell Laboratories
Lucent Technologies
Murray Hill, New Jersey

Reginald L. Lagendijk
Delft University of Technology
Delft, The Netherlands

Vic Larson
Science Applications International
  Corporation
Arlington, Virginia

B. P. Lathi
Emeritus Professor
California State University
Sacramento, California

D. A. Linebarger
The University of Texas at Dallas
Richardson, Texas

Vijay K. Madisetti
Georgia Institute of Technology
Atlanta, Georgia

Richard J. Mammone
Rutgers University
Piscataway, New Jersey

Petros Maragos
Georgia Institute of Technology
Atlanta, Georgia

Daniel F. Marshall
Massachusetts Institute of
  Technology
Lincoln Laboratory
Lexington, Massachusetts

Cherian P. Mathews
University of West Florida
Pensacola, Florida

James H. McClellan
Georgia Institute of Technology
Atlanta, Georgia

Jerry M. Mendel
University of Southern California
Los Angeles, California

Russell M. Mersereau
Georgia Institute of Technology
Atlanta, Georgia

José M. F. Moura
Carnegie Mellon University
Pittsburgh, Pennsylvania

Kambiz Nayebi
Sharif University
Tehran, Iran

Ralph Neff
University of California
Berkeley, California

Arye Nehorai
The University of Illinois
  at Chicago
Chicago, Illinois

Masayuki Nishiguchi
Sony Corporation
Tokyo, Japan

Peter Noll
Technical University of Berlin
Berlin, West Germany

Joseph Olive
Bell Laboratories
Lucent Technologies
Murray Hill, New Jersey

Alan V. Oppenheim
Massachusetts Institute of
  Technology
Cambridge, Massachusetts

Eytan Paldi
Haifa, Israel

C. B. Papadias
Stanford University
Stanford, California

Panos Papamichalis
Texas Instruments
Dallas, Texas

A. Paulraj
Stanford University
Stanford, California

Athina P. Petropulu
Drexel University
Philadelphia, Pennsylvania

M. Pettigrew
Georgia Institute of Technology
Atlanta, Georgia

P. Pirsch
Laboratorium für
  Informationstechnologie
University of Hannover
Hannover, Germany

Christine Podilchuk
Bell Laboratories
Lucent Technologies
Murray Hill, New Jersey

K. Venkatesh Prasad
Ford Motor Company
Scientific Research Laboratory
Dearborn, Michigan

Schuyler R. Quackenbush
AT&T Research Labs
Murray Hill, New Jersey

Lawrence R. Rabiner
AT&T Labs - Research
Florham Park, New Jersey

Ravi P. Ramachandran
Rowan University
Glassboro, New Jersey

Javier Ramos
E.T.S.I. Communications
Polytechnic University of Madrid
Madrid, Spain

Tor A. Ramstad
Norwegian University of Science
  and Technology
  (NTNU)
Trondheim, Norway

Tami Randolph
Georgia Institute of Technology
Atlanta, Georgia

Stanley J. Reeves
Auburn University
Auburn, Alabama

Aaron E. Rosenberg
AT&T Labs - Research
Florham Park, New Jersey

Markus Rupp
Bell Laboratories
Lucent Technologies
Holmdel, New Jersey

E. Saito
Sony Corporation
Kanagawa, Japan

Ali H. Sayed
University of California
Los Angeles, California

Juergen Schroeter
AT&T Labs - Research
Florham Park, New Jersey

Ivan W. Selesnick
Polytechnic University
Brooklyn, New York

John Shore
Entropic Research Laboratory, Inc.
Washington, D. C.

Andrew C. Singer
Sanders, A Lockheed Martin
  Company
Manchester, New Hampshire

Deepen Sinha
Bell Laboratories
Lucent Technologies
Murray Hill, New Jersey

Mark J. T. Smith
Georgia Institute of Technology
Atlanta, Georgia

Iraj Sodagar
David Sarnoff Research Center
Princeton, New Jersey

M. Mohan Sondhi
Bell Laboratories
Lucent Technologies
Murray Hill, New Jersey

Richard Sproat
Bell Laboratories
Lucent Technologies
Murray Hill, New Jersey

Clay Stewart
Science Applications International
  Corporation
Arlington, Virginia

P. Stoica
Uppsala University
Uppsala, Sweden

A. C. Surendran
Bell Laboratories
Lucent Technologies
Murray Hill, New Jersey

David Taubman
Hewlett Packard
Palo Alto, California

A. Murat Tekalp
University of Rochester
Rochester, New York

Charles W. Therrien
Naval Postgraduate School
Monterey, California

K. Tsutsui
Sony Corporation
Tokyo, Japan

Jitendra K. Tugnait
Auburn University
Auburn, Alabama

Kou-Hu Tzou
Hyundai Network Systems
Herndon, Virginia

Barry Van Veen
University of Wisconsin
Madison, Wisconsin

Lucas J. van Vliet
Delft University of Technology
Delft, The Netherlands

M. Vetterlli
Communication Systems Division
École Polytechnique Fédérale de
  Lausanne (EPFL)
Lausanne, Switzerland
and
University of California at Berkeley
Berkeley, California

M. Viberg
Chalmers University of Technology
Gothenburg, Sweden

Hong Wang
Syracuse University
Syracuse, New York

Douglas B. Williams
Georgia Institute of Technology
Atlanta, Georgia

Geoffrey A. Williamson
Illinois Institute of Technology
Chicago, Illinois

M. Wong
McMaster University
Hamilton, Ontario
Canada

Gregory W. Wornell
Massachusetts Institute of
  Technology
Cambridge, Massachusetts

Q. Wu
CELWAVE
Corvallis, Oregon

Peter Xiao
NeoParadigm Labs, Inc.
San Jose, California

Andrew E. Yagle
University of Michigan
Ann Arbor, Michigan

H. Yamauchi
Sony Corporation
Kanagawa, Japan

Ian T. Young
Delft University of Technology
Delft, The Netherlands

Avideh Zakhor
University of California
Berkeley, California

Jun Zhang
University of Wisconsin
Milwaukee, Wisconsin

Xiaoyu Zhang
Rutgers University
Piscataway, New Jersey

Michael D. Zoltowski
Purdue University
West Lafayette, Indiana

Contents

SECTION III   Fast Algorithms and Structures

SECTION IV   Digital Filtering

SECTION V   Statistical Signal Processing

SECTION VI Adaptive Filtering

SECTION VII Inverse Problems and Signal Reconstruction

SECTION VIII   Time Frequency and Multirate Signal Processing

SECTION IX   Digital Audio Communications

SECTION X   Speech Processing

SECTION XI Image and Video Processing

SECTION XII Sensor Array Processing

SECTION XIII Nonlinear and Fractal Signal Processing

SECTION XIV   DSP Software and Hardware

To our families

THE
DIGITAL
SIGNAL
PROCESSING
HANDBOOK

I

Signals and Systems

Vijay K. Madisetti
Georgia Institute of Technology
Douglas B. Williams
Georgia Institute of Technology

T HE STUDY OF "SIGNALS AND SYSTEMS" has formed a cornerstone for the development of digital signal processing and is crucial for all of the topics discussed in this Handbook. While the reader is assumed to be familiar with the basics of signals and systems, a small portion is reviewed in this chapter with an emphasis on the transition from continuous time to discrete time. The reader wishing more background may find in it any of the many fine textbooks in this area, for example [1]-[6].

In the chapter "Fourier Series, Fourier Transforms, and the DFT" by W. Kenneth Jenkins, many important Fourier transform concepts in continuous and discrete time are presented. The discrete Fourier transform (DFT), which forms the backbone of modern digital signal processing as its most common signal analysis tool, is also described, together with an introduction to the fast Fourier transform algorithms.

In "Ordinary Linear Differential and Difference Equations", the author, B.P. Lathi, presents a detailed tutorial of differential and difference equations and their solutions. Because these equations are the most common structures for both implementing and modelling systems, this background is necessary for the understanding of many of the later topics in this Handbook. Of particular interest are a number of solved examples that illustrate the solutions to these formulations.

While most software based on workstations and PCs is executed in single or double precision arithmetic, practical realizations for some high throughput DSP applications must be implemented in fixed point arithmetic. These low cost implementations are still of interest to a wide community in the consumer electronics arena. The chapter "Finite Wordlength Effects" by Bruce W. Bomar describes basic number representations, fixed and floating point errors, roundoff noise, and practical considerations for realizations of digital signal processing applications, with a special emphasis on filtering.

References

[1] Jackson, L.B., *Signals, Systems, and Transforms,* Addison-Wesley, Reading, MA, 1991.
[2] Kamen, E.W. and Heck, B.S., *Fundamentals of Signals and Systems Using MATLAB,* Prentice-Hall, Upper Saddle River, NJ, 1997.
[3] Oppenheim, A.V. and Willsky, A.S., with Nawab, S.H., *Signals and Systems,* 2nd Ed., Prentice-Hall, Upper Saddle River, NJ, 1997.
[4] Strum, R.D. and Kirk, D.E., *Contemporary Linear Systems Using MATLAB,* PWS Publishing, Boston, MA, 1994.
[5] Proakis, J.G. and Manolakis, D.G., *Introduction to Digital Signal Processing,* Macmillan, New York; Collier Macmillan, London, 1988.
[6] Oppenheim, A.V. and Schafer, R.W., *Discrete Time Signal Processing,* Prentice-Hall, Englewood Cliffs, NJ, 1989.

1

Fourier Series, Fourier Transforms, and the DFT

W. Kenneth Jenkins
University of Illinois,
Urbana-Champaign

1.1   Introduction

Fourier methods are commonly used for signal analysis and system design in modern telecommunications, radar, and image processing systems. Classical Fourier methods such as the Fourier series and the Fourier integral are used for continuous time (CT) signals and systems, i.e., systems in which a characteristic signal, $s(t)$, is defined at all values of t on the continuum $-\infty < t < \infty$. A more recently developed set of Fourier methods, including the discrete time Fourier transform (DTFT) and the discrete Fourier transform (DFT), are extensions of basic Fourier concepts that apply to discrete time (DT) signals. A characteristic DT signal, $s[n]$, is defined only for values of n where n is an integer in the range $-\infty < n < \infty$. The following discussion presents basic concepts and outlines important properties for both the CT and DT classes of Fourier methods, with a particular emphasis on the relationships between these two classes. The class of DT Fourier methods is particularly useful as a basis for digital signal processing (DSP) because it extends the theory of classical Fourier analysis to DT signals and leads to many effective algorithms that can be directly implemented on general computers or special purpose DSP devices.

The relationship between the CT and the DT domains is characterized by the operations of sampling and reconstruction. If $s_a(t)$ denotes a signal $s(t)$ that has been uniformly sampled every T seconds, then the mathematical representation of $s_a(t)$ is given by

$$s_a(t) = \sum_{n=-\infty}^{\infty} s(t)\delta(t - nT) \tag{1.1}$$

where $\delta(t)$ is a CT impulse function defined to be zero for all $t \neq 0$, undefined at $t = 0$, and has unit area when integrated from $t = -\infty$ to $t = +\infty$. Because the only places at which the product $s(t)\delta(t - nT)$ is not identically equal to zero are at the sampling instances, $s(t)$ in (1.1) can be replaced with $s(nT)$ without changing the overall meaning of the expression. Hence, an alternate expression for $s_a(t)$ that is often useful in Fourier analysis is given by

$$s_a(t) = \sum_{n=-\infty}^{\infty} s(nT)\delta(t - nT) \tag{1.2}$$

The CT sampling model $s_a(t)$ consists of a sequence of CT impulse functions uniformly spaced at intervals of T seconds and weighted by the values of the signal $s(t)$ at the sampling instants, as depicted in Fig. 1.1. Note that $s_a(t)$ is not defined at the sampling instants because the CT impulse function itself is not defined at $t = 0$. However, the values of $s(t)$ at the sampling instants are imbedded as "area under the curve" of $s_a(t)$, and as such represent a useful mathematical model of the sampling process. In the DT domain the sampling model is simply the sequence defined by taking the values of $s(t)$ at the sampling instants, i.e.,

$$s[n] = s(t)|_{t=nT} \tag{1.3}$$

In contrast to $s_a(t)$, which is not defined at the sampling instants, $s[n]$ is well defined at the sampling instants, as illustrated in Fig. 1.2. Thus, it is now clear that $s_a(t)$ and $s[n]$ are different but equivalent models of the sampling process in the CT and DT domains, respectively. They are both useful for signal analysis in their corresponding domains. Their equivalence is established by the fact that they have equal spectra in the Fourier domain, and that the underlying CT signal from which $s_a(t)$ and $s[n]$ are derived can be recovered from either sampling representation, provided a sufficiently large sampling rate is used in the sampling operation (see below).

1.2 Fourier Series Representation of Continuous Time Periodic Signals

It is convenient to begin this discussion with the classical Fourier series representation of a periodic time domain signal, and then derive the Fourier integral from this representation by finding the limit of the Fourier coefficient representation as the period goes to infinity. The conditions under which a periodic signal $s(t)$ can be expanded in a Fourier series are known as the Dirichet conditions. They require that in each period $s(t)$ has a finite number of discontinuities, a finite number of maxima and minima, and that $s(t)$ satisfies the following absolute convergence criterion [1]:

$$\int_{-T/2}^{T/2} |s(t)|\, dt < \infty \tag{1.4}$$

It is assumed in the following discussion that these basic conditions are satisfied by all functions that will be represented by a Fourier series.

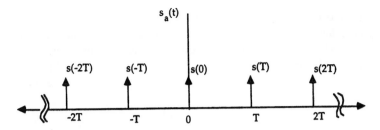

FIGURE 4.1   CT model of a sampled CT signal.

FIGURE 1.1   CT model of a sampled CT signal.

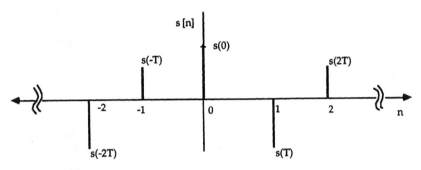

FIGURE 1.2   DT model of a sampled CT signal.

1.2.1   Exponential Fourier Series

If a CT signal $s(t)$ is periodic with a period T, then the classical complex Fourier series representation of $s(t)$ is given by

$$s(t) = \sum_{n=-\infty}^{\infty} a_n e^{jn\omega_0 t} \tag{1.5a}$$

where $\omega_0 = 2\pi/T$, and where the a_n are the complex Fourier coefficients given by

$$a_n = (1/T) \int_{-T/2}^{T/2} s(t) e^{-jn\omega_0 t} \, dt \tag{1.5b}$$

It is well known that for every value of t where $s(t)$ is continuous, the right-hand side of (1.5a) converges to $s(t)$. At values of t where $s(t)$ has a finite jump discontinuity, the right-hand side of (1.5a) converges to the average of $s(t^-)$ and $s(t^+)$, where $s(t^-) \equiv \lim_{\epsilon \to 0} s(t - \epsilon)$ and $s(t^+) \equiv \lim_{\epsilon \to 0} s(t + \epsilon)$.

For example, the Fourier series expansion of the sawtooth waveform illustrated in Fig. 1.3 is characterized by $T = 2\pi$, $\omega_0 = 1$, $a_0 = 0$, and $a_n = a_{-n} = A\cos(n\pi)/(jn\pi)$ for $n = 1, 2, \ldots$. The coefficients of the exponential Fourier series represented by (1.5b) can be interpreted as the spectral representation of $s(t)$, because the a_n-th coefficient represents the contribution of the $(n\omega_0)$-th frequency to the total signal $s(t)$. Because the a_n are complex valued, the Fourier domain representation has both a magnitude and a phase spectrum. For example, the magnitude of the a_n is plotted in Fig. 1.4 for the sawtooth waveform of Fig. 1.3. The fact that the a_n constitute a discrete set is consistent with the fact that a periodic signal has a "line spectrum," i.e., the spectrum contains only integer multiples of the fundamental frequency ω_0. Therefore, the equation pair given by (1.5a) and (1.5b) can be interpreted as a transform pair that is similar to the CT Fourier transform for periodic signals. This leads to the observation that the classical Fourier series can be interpreted as a special transform that provides a one-to-one invertible mapping between the discrete-spectral domain and the CT domain. The next section shows how the periodicity constraint can be removed to produce the more general classical CT Fourier transform, which applies equally well to periodic and aperiodic time domain waveforms.

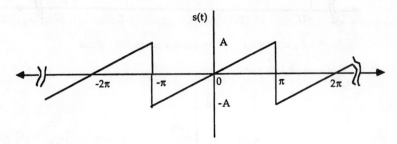

FIGURE 1.3   Periodic CT signal used in Fourier series example.

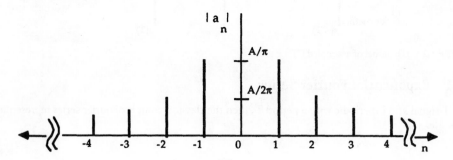

FIGURE 1.4   Magnitude of the Fourier coefficients for example of Figure 1.3.

1.2.2   The Trigonometric Fourier Series

Although Fourier series expansions exist for complex periodic signals, and Fourier theory can be generalized to the case of complex signals, the theory and results are more easily expressed for real-valued signals. The following discussion assumes that the signal $s(t)$ is real-valued for the sake of simplifying the discussion. However, all results are valid for complex signals, although the details of the theory will become somewhat more complicated.

For real-valued signals $s(t)$, it is possible to manipulate the complex exponential form of the Fourier series into a trigonometric form that contains $\sin(\omega_0 t)$ and $\cos(\omega_0 t)$ terms with corresponding real-valued coefficients [1]. The trigonometric form of the Fourier series for a real-valued signal $s(t)$ is given by

$$s(t) = \sum_{n=0}^{\infty} b_n \cos(n\omega_0 t) + \sum_{n=1}^{\infty} c_n \sin(n\omega_0 t) \qquad (1.6a)$$

where $\omega_0 = 2\pi/T$. The b_n and c_n are real-valued Fourier coefficients determined by

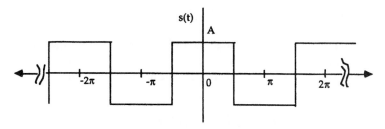

FIGURE 1.5   Periodic CT signal used in Fourier series example 2.

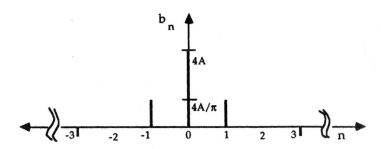

FIGURE 1.6   Fourier coefficients for example of Figure 1.5.

$$b_0 = (1/T) \int_{-T/2}^{T/2} s(t)\, dt$$

$$b_n = (2/T) \int_{-T/2}^{T/2} s(t) \cos(n\omega_0 t)\, dt, \qquad n = 1, 2, \ldots, \qquad (1.6b)$$

$$c_n = (2/T) \int_{-T/2}^{T/2} s(t) \sin(n\omega_0 t)\, dt, \qquad n = 1, 2, \ldots,$$

An arbitrary real-valued signal $s(t)$ can be expressed as a sum of even and odd components, $s(t) = s_{\text{even}}(t) + s_{\text{odd}}(t)$, where $s_{\text{even}}(t) = s_{\text{even}}(-t)$ and $s_{\text{odd}}(t) = -s_{\text{odd}}(-t)$, and where $s_{\text{even}}(t) = [s(t) + s(-t)]/2$ and $s_{\text{odd}}(t) = [s(t) - s(-t)]/2$. For the trigonometric Fourier series, it can be shown that $s_{\text{even}}(t)$ is represented by the (even) cosine terms in the infinite series, $s_{\text{odd}}(t)$ is represented by the (odd) sine terms, and b_0 is the DC level of the signal. Therefore, if it can be determined by inspection that a signal has DC level, or if it is even or odd, then the correct form of the trigonometric series can be chosen to simplify the analysis. For example, it is easily seen that the signal shown in Fig. 1.5 is an even signal with a zero DC level. Therefore it can be accurately represented by the cosine series with $b_n = 2A \sin(\pi n/2)/(\pi n/2), n = 1, 2, \ldots$, as illustrated in Fig. 1.6. In contrast, note that the sawtooth waveform used in the previous example is an odd signal with zero DC level; thus, it can be completely specified by the sine terms of the trigonometric series. This result can be demonstrated by pairing each positive frequency component from the exponential series with its conjugate partner, i.e., $c_n = \sin(n\omega_0 t) = a_n e^{jn\omega_0 t} + a_{-n} e^{-jn\omega_0 t}$, whereby it is found that $c_n = 2A \cos(n\pi)/(n\pi)$ for this example. In general it is found that $a_n = (b_n - jc_n)/2$ for $n = 1, 2, \ldots, a_0 = b_0$, and $a_{-n} = a_n^*$. The trigonometric Fourier series is common in the signal processing literature because it replaces complex coefficients with real ones and often results in a simpler and more intuitive interpretation of the results.

1.2.3   Convergence of the Fourier Series

The Fourier series representation of a periodic signal is an approximation that exhibits mean squared convergence to the true signal. If $s(t)$ is a periodic signal of period T, and $s'(t)$ denotes the Fourier series approximation of $s(t)$, then $s(t)$ and $s'(t)$ are equal in the mean square sense if

$$\text{MSE} = \int_{-T/2}^{T/2} |s(t) - s(t)'|^2 \, dt = 0 \tag{1.7}$$

Even with (1.7) satisfied, mean square error (MSE) convergence does not mean that $s(t) = s'(t)$ at every value of t. In particular, it is known that at values of t, where $s(t)$ is discontinuous, the Fourier series converges to the average of the limiting values to the left and right of the discontinuity. For example, if t_0 is a point of discontinuity, then $s'(t_0) = [s(t_0^-) + s(t_0^+)]/2$, where $s(t_0^-)$ and $s(t_0^+)$ were defined previously. (Note that at points of continuity, this condition is also satisfied by the definition of continuity.) Because the Dirichet conditions require that $s(t)$ have at most a finite number of points of discontinuity in one period, the set S_t, defined as all values of t within one period where $s(t) \neq s'(t)$, contains a finite number of points, and S_t is a set of measure zero in the formal mathematical sense. Therefore, $s(t)$ and its Fourier series expansion $s'(t)$ are *equal almost everywhere*, and $s(t)$ can be considered identical to $s'(t)$ for the analysis of most practical engineering problems.

Convergence almost everywhere is satisfied only in the limit as an infinite number of terms are included in the Fourier series expansion. If the infinite series expansion of the Fourier series is truncated to a finite number of terms, as it must be in practical applications, then the approximation will exhibit an oscillatory behavior around the discontinuity, known as the Gibbs phenomenon [1]. Let $s'_N(t)$ denote a truncated Fourier series approximation of $s(t)$, where only the terms in (1.5a) from $n = -N$ to $n = N$ are included if the complex Fourier series representation is used, or where only the terms in (1.6a) from $n = 0$ to $n = N$ are included if the trigonometric form of the Fourier series is used. It is well known that in the vicinity of a discontinuity at t_0 the Gibbs phenomenon causes $s'_N(t)$ to be a poor approximation to $s(t)$. The peak magnitude of the Gibbs oscillation is 13% of the size of the jump discontinuity $s(t_0^-) - s(t_0^+)$ regardless of the number of terms used in the approximation. As N increases, the region that contains the oscillation becomes more concentrated in the neighborhood of the discontinuity, until, in the limit as N approaches infinity, the Gibbs oscillation is squeezed into a single point of mismatch at t_0.

If $s'(t)$ is replaced by $s'_N(t)$ in (1.7), it is important to understand the behavior of the error MSE_N as a function of N, where

$$\text{MSE}_N = \int_{-T/2}^{T/2} |s(t) - s'_N(t)|^2 \, dt \tag{1.8}$$

An important property of the Fourier series is that the exponential basis functions $e^{jn\omega_0 t}$ (or $\sin(n\omega_0 t)$ and $\cos(n\omega_0 t)$ for the trigonometric form) for $n = 0, \pm 1, \pm 2, \ldots$ (or $n = 0, 1, 2, \ldots$ for the trigonometric form) constitute an orthonormal set, i.e., $t_{nk} = 1$ for $n = k$, and $t_{nk} = 0$ for $n \neq k$, where

$$t_{nk} = (1/T) \int_{-T/2}^{T/2} (e^{-jn\omega_0 t})(e^{jk\omega_0 t}) \, dt \tag{1.9}$$

As terms are added to the Fourier series expansion, the orthogonality of the basis functions guarantees that the error decreases in the mean square sense, i.e., that MSE_N monotonically decreases as N is increased. Therefore, a practitioner can proceed with the confidence that when applying Fourier series analysis more terms are always better than fewer in terms of the accuracy of the signal representations.

1.3   The Classical Fourier Transform for Continuous Time Signals

The periodicity constraint imposed on the Fourier series representation can be removed by taking the limits of (1.5a) and (1.5b) as the period T is increased to infinity. Some mathematical preliminaries are required so that the results will be well defined after the limit is taken. It is convenient to remove the $(1/T)$ factor in front of the integral by multiplying (1.5b) through by T, and then replacing $T a_n$ by a'_n in both (1.5a) and (1.5b). Because $\omega_0 = 2\pi/T$, as T increases to infinity, ω_0 becomes infinitesimally small, a condition that is denoted by replacing ω_0 with $\Delta\omega$. The factor $(1/T)$ in (1.5a) becomes $(\Delta\omega/2\pi)$. With these algebraic manipulations and changes in notation (1.5a) and (1.5b) take on the following form prior to taking the limit:

$$s(t) \;=\; (1/2\pi) \sum_{n=-\infty}^{\infty} a'_n e^{jn\Delta\omega t} \, \Delta\omega \tag{1.10a}$$

$$a'_n \;=\; \int_{-T/2}^{T/2} s(t) e^{-jn\Delta\omega t} \, dt \tag{1.10b}$$

The final step in obtaining the CT Fourier transform is to take the limit of both (1.10a) and (1.10b) as $T \to \infty$. In the limit the infinite summation in (1.10a) becomes an integral, $\Delta\omega$ becomes $d\omega$, $n\Delta\omega$ becomes ω, and a'_n becomes the CT Fourier transform of $s(t)$, denoted by $S(j\omega)$. The result is summarized by the following transform pair, which is known throughout most of the engineering literature as the classical CT Fourier transform (CTFT):

$$s(t) \;=\; (1/2\pi) \int_{-\infty}^{\infty} S(j\omega) e^{j\omega t} \, d\omega \tag{1.11a}$$

$$S(j\omega) \;=\; \int_{-\infty}^{\infty} s(t) e^{-j\omega t} \, dt \tag{1.11b}$$

Often (1.11a) is called the Fourier integral and (1.11b) is simply called the Fourier transform. The relationship $S(j\omega) = \mathcal{F}\{s(t)\}$ denotes the Fourier transformation of $s(t)$, where $\mathcal{F}\{\cdot\}$ is a symbolic notation for the Fourier transform operator, and where ω becomes the continuous frequency variable after the periodicity constraint is removed. A transform pair $s(t) \leftrightarrow S(j\omega)$ represents a one-to-one invertible mapping as long as $s(t)$ satisfies conditions which guarantee that the Fourier integral converges.

From (1.11a) it is easily seen that $\mathcal{F}\{\delta(t-t_0)\} = e^{-j\omega t_0}$, and from (1.11b) that $\mathcal{F}^{-1}\{2\pi\delta(\omega-\omega_0)\} = e^{j\omega_0 t}$, so that $\delta(t-t_0) \leftrightarrow e^{-j\omega t_0}$ and $e^{j\omega_0 t} \leftrightarrow 2\pi\delta(\omega-\omega_0)$ are valid Fourier transform pairs. Using these relationships it is easy to establish the Fourier transforms of $\cos(\omega_0 t)$ and $\sin(\omega_0 t)$, as well as many other useful waveforms that are encountered in common signal analysis problems. A number of such transforms are shown in Table 1.1.

The CTFT is useful in the analysis and design of CT systems, i.e., systems that process CT signals. Fourier analysis is particularly applicable to the design of CT filters which are characterized by Fourier magnitude and phase spectra, i.e., by $|H(j\omega)|$ and $\arg H(j\omega)$, where $H(j\omega)$ is commonly called the frequency response of the filter. For example, an **ideal transmission channel** is one which passes a signal without distorting it. The signal may be scaled by a real constant A and delayed by a fixed time increment t_0, implying that the impulse response of an ideal channel is $A\delta(t - t_0)$, and its corresponding frequency response is $Ae^{-j\omega t_0}$. Hence, the frequency response of an ideal channel is specified by constant amplitude for all frequencies, and a phase characteristic which is linear function given by ωt_0.

TABLE 1.1   Some Basic CTFT Pairs

Signal	Fourier Transform	Fourier Series Coefficients (if periodic)
$\sum\limits_{k=-\infty}^{+\infty} a_k e^{jk\omega_0 t}$	$2\pi \sum\limits_{k=-\infty}^{+\infty} a_k \delta(\omega_k \omega_0)$	a_k
$e^{j\omega_0 t}$	$2\pi \delta(\omega + \omega_0)$	$a_1 = 1$ $a_k = 0,$ otherwise
$\cos \omega_0 t$	$\pi [\delta(\omega - \omega_0) + \delta(\omega + \omega_0)]$	$a_1 = a_{-1} = \frac{1}{2}$ $a_k = 0,$ otherwise
$\sin \omega_0 t$	$\frac{\pi}{j} [\delta(\omega - \omega_0) - \delta(\omega + \omega_0)]$	$a_1 = -a_{-1} = \frac{1}{2j}$ $a_k = 0,$ otherwise
$x(t) = 1$	$2\pi \delta(\omega)$	$a_0 = 1,\quad a_k = 0,\quad k \neq 0$ $\left(\text{has this Fourier series representation for} \atop \text{any choice of } T_0 > 0 \right)$
Periodic square wave $x(t) = \begin{cases} 1, & \|t\| < T_1 \\ 0, & T_1 < \|t\| \leq \frac{T_0}{2} \end{cases}$ and $x(t + T_0) = x(t)$	$\sum\limits_{k=-\infty}^{+\infty} \frac{2 \sin k\omega_0 T_1}{k} \delta(\omega_k \omega_0)$	$\frac{\omega_0 T_1}{\pi} \operatorname{sinc}\left(\frac{k\omega_0 T_1}{\pi}\right) = \frac{\sin k\omega_0 T_1}{k\pi}$
$\sum\limits_{n=-\infty}^{+\infty} \delta(t - nT)$	$\frac{2\pi}{T} \sum\limits_{k=-\infty}^{+\infty} \delta\left(\omega - \frac{2\pi k}{T}\right)$	$a_k = \frac{1}{T}$ for all k
$x(t) = \begin{cases} 1, & \|t\| < T_1 \\ 0, & \|t\| > T_1 \end{cases}$	$2T_1 \operatorname{sinc}\left(\frac{\omega T_1}{\pi}\right) = \frac{2 \sin \omega T_1}{\omega}$	—
$\frac{W}{\pi} \operatorname{sinc}\left(\frac{Wt}{\pi}\right) = \frac{\sin Wt}{\pi t}$	$X(\omega) = \begin{cases} 1, & \|\omega\| < W \\ 0, & \|\omega\| > W \end{cases}$	—
$\delta(t)$	1	—
$u(t)$	$\frac{1}{j\omega} + \pi \delta(\omega)$	—
$\delta(t - t_0)$	$e^{j\omega t_0}$	—
$e^{-at} u(t), \operatorname{Re}\{a\} > 0$	$\frac{1}{a + j\omega}$	—
$te^{-at} u(t), \operatorname{Re}\{a\} > 0$	$\frac{1}{(a + j\omega)^2}$	—
$\frac{t^{n-1}}{(n-1)!} e^{-at} u(t),$ $\operatorname{Re}\{a\} > 0$	$\frac{1}{(a + j\omega)^n}$	—

Source: A. V. Oppenheim et al., *Signals and Systems*, ©1983. Reprinted by permission of Prentice-Hall, Inc., Upper Saddle River, NJ.

1.3.1 Properties of the Continuous Time Fourier Transform

The CTFT has many properties that make it useful for the analysis and design of linear CT systems. Some of the more useful properties are stated below. A more complete list of the CTFT properties is given in Table 1.2. Proofs of these properties can be found in [2] and [3]. In the following discussion $\mathcal{F}\{\cdot\}$ denotes the Fourier transform operation, $\mathcal{F}^{-1}\{\cdot\}$ denotes the inverse Fourier transform operation, and $*$ denotes

TABLE 1.2 Properties of the CTFT

Name	If $\mathcal{F} f(t) = F(j\omega)$, then
Definition	$f(j\omega) = \displaystyle\int_{-\infty}^{\infty} f(t) e_{j\omega t}\, dt$
	$f(t) = \dfrac{1}{2\pi} \displaystyle\int_{-\infty}^{\infty} F(j\omega) e^{j\omega t}\, d\omega$
Superposition	$\mathcal{F}[a f_1(t) + b f_2(t)] = a F_1(j\omega) + b F_2(j\omega)$
Simplification if:	
(a) $f(t)$ is even	$F(j\omega) = 2 \displaystyle\int_0^{\infty} f(t) \cos \omega t\, dt$
(b) $f(t)$ is odd	$F(j\omega) = 2j \displaystyle\int_0^{\infty} f(t) \sin \omega t\, dt$
Negative t	$\mathcal{F} f(-t) = F^*(j\omega)$
Scaling:	
(a) Time	$\mathcal{F} f(at) = \dfrac{1}{|a|} F\left(\dfrac{j\omega}{a}\right)$
(b) Magnitude	$\mathcal{F} a f(t) = a F(j\omega)$
Differentiation	$\mathcal{F}\left[\dfrac{d^n}{dt^n} f(t)\right] = (j\omega)^n F(j\omega)$
Integration	$\mathcal{F}\left[\displaystyle\int_{-\infty}^{t} f(x)\, dx\right] = \frac{1}{j\omega} F(j\omega) + \pi F(0)\delta(\omega)$
Time shifting	$\mathcal{F} f(t - a) = F(j\omega) e^{j\omega a}$
Modulation	$\mathcal{F} f(t) e^{j\omega_0 t} = F[j(\omega - \omega_0)]$
	$\{\mathcal{F} f(t) \cos \omega_0 t = \frac{1}{2} F[j(\omega - \omega_0)] + F[j(\omega + \omega_0)]\}$
	$\{\mathcal{F} f(t) \sin \omega_0 t = \frac{1}{2} j[F[j(\omega - \omega_0)] - F[j(\omega + \omega_0)]]\}$
Time convolution	$\mathcal{F}^{-1}[F_1(j\omega) F_2(j\omega)] = \displaystyle\int_{-\infty}^{\infty} f_1(\tau) f_2(\tau) f_2(t_\tau)\, d\tau$
Frequency convolution	$\mathcal{F}[f_1(t) f_2(t)] = \dfrac{1}{2\pi} \displaystyle\int_{-\infty}^{\infty} F_1(j\lambda) F_2[j(\omega_\lambda)]\, d\lambda$

Source: M. E. VanValkenburg, *Network Analysis*, (3rd edition), ©1974. Reprinted by permission of Prentice-Hall, Inc. Upper Saddle River, NJ.

the convolution operation defined as

$$f_1(t) * f_2(t) = \int_{-\infty}^{\infty} f_1(t - \tau) f_2(\tau)\, d\tau$$

1. Linearity (superposition): $\mathcal{F}\{a f_1(t) + b f_2(t)\} = a\mathcal{F}\{f_1(t)\} + b\mathcal{F}\{f_2(t)\}$
   (a and b, complex constants)
2. Time shifting: $\mathcal{F}\{f(t - t_0)\} = e^{-j\omega t_0} \mathcal{F}\{f(t)\}$
3. Frequency shifting: $e^{j\omega_0 t} f(t) = \mathcal{F}^{-1}\{F(j(\omega - \omega_0))\}$
4. Time domain convolution: $\mathcal{F}\{f_1(t) * f_2(t)\} = \mathcal{F}\{f_1(t)\}\mathcal{F}\{f_2(t)\}$
5. Frequency domain convolution: $\mathcal{F}\{f_1(t) f_2(t)\} = (1/2\pi)\mathcal{F}\{f_1(t)\} * \mathcal{F}\{f_2(t)\}$

1-10

6. Time differentiation: $-j\omega F(j\omega) = \mathcal{F}\{d(f(t))/dt\}$
7. Time integration: $\mathcal{F}\{\int_{-\infty}^{t} f(\tau)\,d\tau\} = (1/j\omega)F(j\omega) + \pi F(0)\delta(\omega)$

The above properties are particularly useful in CT system analysis and design, especially when the system characteristics are easily specified in the frequency domain, as in linear filtering. Note that properties 1, 6, and 7 are useful for solving differential or integral equations. Property 4 provides the basis for many signal processing algorithms because many systems can be specified directly by their impulse or frequency response. Property 3 is particularly useful in analyzing communication systems in which different modulation formats are commonly used to shift spectral energy to frequency bands that are appropriate for the application.

1.3.2   Fourier Spectrum of the Continuous Time Sampling Model

Because the CT sampling model $s_a(t)$, given in (1.1), is in its own right a CT signal, it is appropriate to apply the CTFT to obtain an expression for the spectrum of the sampled signal:

$$\mathcal{F}\{s_a(t)\} = \mathcal{F}\left\{\sum_{n=-\infty}^{\infty} s(t)\delta(t - nT)\right\} = \sum_{n=-\infty}^{\infty} s(nT)e^{-j\omega Tn} \tag{1.12}$$

Because the expression on the right-hand side of (1.12) is a function of $e^{j\omega T}$ it is customary to denote the transform as $F(e^{j\omega T}) = \mathcal{F}\{s_a(t)\}$. Later in the chapter this result is compared to the result of operating on the DT sampling model, namely $s[n]$, with the DT Fourier transform to illustrate that the two sampling models have the same spectrum.

1.3.3   Fourier Transform of Periodic Continuous Time Signals

We saw earlier that a periodic CT signal can be expressed in terms of its Fourier series. The CTFT can then be applied to the Fourier series representation of $s(t)$ to produce a mathematical expression for the "line spectrum" characteristic of periodic signals.

$$\mathcal{F}\{s(t)\} = \mathcal{F}\left\{\sum_{n=-\infty}^{\infty} a_n e^{jn\omega_0 t}\right\} = 2\pi \sum_{n=-\infty}^{\infty} a_n \delta(\omega - n\omega_0) \tag{1.13}$$

The spectrum is shown pictorially in Fig. 1.7. Note the similarity between the spectral representation of Fig. 1.7 and the plot of the Fourier coefficients in Fig. 1.4, which was heuristically interpreted as a "line spectrum". Figures 1.4 and 1.7 are different but equivalent representations of the Fourier spectrum. Note that Fig. 1.4 is a DT representation of the spectrum, while Fig. 1.7 is a CT model of the same spectrum.

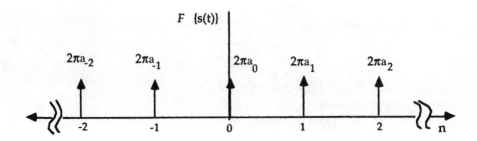

FIGURE 1.7   Spectrum of the Fourier series representation of $s(t)$.

1.3.4   The Generalized Complex Fourier Transform

The CTFT characterized by (1.11a) and (1.11b) can be generalized by considering the variable $j\omega$ to be the special case of $u = \sigma + j\omega$ with $\sigma = 0$, writing (1.11) in terms of u, and interpreting u as a complex frequency variable. The resulting complex Fourier transform pair is given by (1.14a) and (1.14b)

$$s(t) = (1/2\pi j) \int_{\sigma-j\infty}^{\sigma+j\infty} S(u)e^{jut} \, du \tag{1.14a}$$

$$S(u) = \int_{-\infty}^{\infty} s(t)e^{-jut} \, dt \tag{1.14b}$$

The set of all values of u for which the integral of (1.14b) converges is called the region of convergence (ROC). Because the transform $S(u)$ is defined only for values of u within the ROC, the path of integration in (1.14a) must be defined by σ so that the entire path lies within the ROC. In some literature this transform pair is called the **bilateral Laplace transform** because it is the same result obtained by including both the negative and positive portions of the time axis in the classical Laplace transform integral. [Note that in (1.14) the complex frequency variable was denoted by u rather than by the more common s, in order to avoid confusion with earlier uses of $s(\cdot)$ as signal notation.] The complex Fourier transform (bilateral Laplace transform) is not often used in solving practical problems, but its significance lies in the fact that it is the most general form that represents the point at which Fourier and Laplace transform concepts become the same. Identifying this connection reinforces the notion that Fourier and Laplace transform concepts are similar because they are derived by placing different constraints on the same general form.

1.4   The Discrete Time Fourier Transform

The discrete time Fourier transform (DTFT) can be obtained by using the DT sampling model and considering the relationship obtained in (1.12) to be the definition of the DTFT. Letting $T = 1$ so that the sampling period is removed from the equations and the frequency variable is replaced with a normalized frequency $\omega' = \omega T$, the DTFT pair is defined in (1.15). Note that in order to simplify notation it is not customary to distinguish between ω and ω', but rather to rely on the context of the discussion to determine whether ω refers to the normalized ($T = 1$) or the unnormalized ($T \neq 1$) frequency variable.

$$S(e^{j\omega'}) = \sum_{n=-\infty}^{\infty} s[n]e^{-j\omega'n} \tag{1.15a}$$

$$s[n] = (1/2\pi) \int_{-\pi}^{\pi} S(e^{j\omega'})e^{jn\omega'} \, d\omega' \tag{1.15b}$$

The spectrum $S(e^{j\omega'})$ is periodic in ω' with period 2π. The fundamental period in the range $-\pi < \omega' \leq \pi$, sometimes referred to as the baseband, is the useful frequency range of the DT system because frequency components in this range can be represented unambiguously in sampled form (without aliasing error). In much of the signal processing literature the explicit primed notation is omitted from the frequency variable. However, the explicit primed notation will be used throughout this section because the potential exists for confusion when so many related Fourier concepts are discussed within the same framework.

By comparing (1.12) and (1.15a), and noting that $\omega' = \omega T$, it is established that

$$\mathcal{F}\{s_a(t)\} = \text{DTFT}\{s[n]\} \tag{1.16}$$

where $s[n] = s(t)_{t=nT}$. This demonstrates that the spectrum of $s_a(t)$, as calculated by the CT Fourier transform is identical to the spectrum of $s[n]$ as calculated by the DTFT. Therefore, although $s_a(t)$ and

$s[n]$ are quite different sampling models, they are equivalent in the sense that they have the same Fourier domain representation.

A list of common DTFT pairs is presented in Table 1.3. Just as the CT Fourier transform is useful in CT signal system analysis and design, the DTFT is equally useful in the same capacity for DT systems. It is indeed fortuitous that Fourier transform theory can be extended in this way to apply to DT systems.

TABLE 1.3   Some Basic DTFT Pairs

Sequence	Fourier Transform
1. $\delta[n]$	1
2. $\delta[n - n_0]$	$e^{-j\omega n_0}$
3. $1 \quad (-\infty < n < \infty)$	$\displaystyle\sum_{k=-\infty}^{\infty} 2\pi\delta(\omega + 2\pi k)$
4. $a^n u[n] \qquad (|a| < 1)$	$\dfrac{1}{1 - ae^{-j\omega}}$
5. $u[n]$	$\dfrac{1}{1 - e^{-j\omega}} + \displaystyle\sum_{k=-\infty}^{\infty} \pi\delta(\omega + 2\pi k)$
6. $(n + 1)a^n u[n] \qquad (|a| < 1)$	$\dfrac{1}{(1 - ae^{-j\omega})^2}$
7. $\dfrac{r^2 \sin\omega_p(n+1)}{\sin\omega_p} u[n] \qquad (|r| < 1)$	$\dfrac{1}{1 - 2r\cos\omega_p e^{-j\omega} + r^2 e^{j2\omega}}$
8. $\dfrac{\sin\omega_c n}{\pi n}$	$Xe^{j\omega} = \begin{cases} 1, & |\omega| < \omega_c \\ 0, & \omega_c < |\omega| \leq \pi \end{cases}$
9. $x[n] - \begin{cases} 1, & 0 \leq n \leq M \\ 0, & \text{otherwise} \end{cases}$	$\dfrac{\sin[\omega(M+1)/2]}{\sin(\omega/2)} e^{-j\omega M/2}$
10. $e^{j\omega_0 n}$	$\displaystyle\sum_{k=-\infty}^{\infty} 2\pi\delta(\omega - \omega_0 + 2\pi k)$
11. $\cos(\omega_0 n + \phi)$	$\pi \displaystyle\sum_{k=-\infty}^{\infty} [e^{j\phi}\delta(\omega - \omega_0 + 2\pi k) + e^{-j\phi}\delta(\omega + \omega_0 + 2\pi k)]$

Source: A. V. Oppenheim and R. W. Schafer, *Discrete-Time Signal Processing*, © 1989. Reprinted by permission of Prentice-Hall, Inc., Upper Saddle River, NJ.

In the same way that the CT Fourier transform was found to be a special case of the complex Fourier transform (or bilateral Laplace transform), the DTFT is a special case of the bilateral z-transform with $z = e^{j\omega' t}$. The more general bilateral z-transform is given by

$$S(z) = \sum_{n=-\infty}^{\infty} s[n]z^{-n} \tag{1.17a}$$

$$s[n] = (1/2\pi j)\int_C S(z)z^{n-1}\, dz \tag{1.17b}$$

where **C** is a counterclockwise contour of integration which is a closed path completely contained within the region of convergence of $S(z)$. Recall that the DTFT was obtained by taking the CT Fourier transform of the CT sampling model represented by $s_a(t)$. Similarly, the bilateral z-transform results by taking the bilateral Laplace transform of $s_a(t)$. If the lower limit on the summation of (1.17a) is taken to be $n = 0$, then (1.17a) and (1.17b) become the one-sided z-transform, which is the DT equivalent of the one-sided LT for CT signals. The hierarchical relationship among these various concepts for DT systems is discussed later in this chapter, where it will be shown that the family structure of the DT family tree is identical to that of the CT family. For every CT transform in the CT world there is an analogous DT transform in the DT world, and vice versa.

1.4.1 Properties of the Discrete Time Fourier Transform

Because the DTFT is a close relative of the classical CT Fourier transform it should come as no surprise that many properties of the DTFT are similar to those presented for the CT Fourier transform in the previous section. In fact, for many of the properties presented earlier an analogous property exists for the DTFT. The following list parallels the list that was presented in the previous section for the CT Fourier transform, to the extent that the same property exists. A more complete list of DTFT pairs is given in Table 1.4. (Note that the primed notation on ω' is dropped in the following to simplify the notation, and to be consistent with standard usage.)

TABLE 1.4   Properties of the DTFT

Sequence	Fourier Transform
$x[n]$	$X(e^{j\omega})$
$y[n]$	$Y(e^{j\omega})$
1. $ax[n] + by[n]$	$aX(e^{j\omega}) + bY(e^{j\omega})$
2. $x[n - n_d]$ (n_d an integer)	$e^{-j\omega n_d} X(e^{j\omega})$
3. $e^{j\omega_0 n} x[n]$	$X(e^{j(\omega - \omega_0)})$
4. $x[-n]$	$X(e^{-j\omega})$ if $x[n]$ is real $X^*(e^{j\omega})$
5. $nx[n]$	$j\dfrac{dX(e^{j\omega})}{d\omega}$
6. $x[n] * y[n]$	$X(e^{j\omega})Y(e^{j\omega})$
7. $x[n]y[n]$	$\dfrac{1}{2\pi}\displaystyle\int_{-x}^{x} X(e^{j\theta})Y(e^{j(\omega-\theta)})\, d\theta$

Parseval's Theorem

8. $\displaystyle\sum_{n=-\infty}^{\infty} |x[n]|^2 = \frac{1}{2\pi}\int_{-\pi}^{\pi} |X(e^{j\omega})|^2\, d\omega$

9. $\displaystyle\sum_{n=-\infty}^{\infty} x[n]y^*[n] = \frac{1}{2\pi}\inf_{-\pi}^{\pi} X(e^{j\omega})Y^*(e^{j\omega})\, d\omega$

Source: A. V. Oppenheim and R. W. Schafer, *Discrete-Time Signal Processing*, © 1989. Reprinted by permission of Prentice-Hall, Inc., Upper Saddle River, NJ.

1. Linearity (superposition): $\text{DTFT}\{af_1[n] + bf_2[n]\} = a\text{DTFT}\{f_1[n]\} + b\text{DTFT}\{f_2[n]\}$
   (a and b, complex constants)

2. Index shifting: $\mathrm{DTFT}\{f[n - n_0]\} = e^{-j\omega n_0}\mathrm{DTFT}\{f[n]\}$

3. Frequency shifting: $e^{j\omega_0 n}f[n] = \mathrm{DTFT}^{-1}\{F(e^{j(\omega-\omega_0)})\}$

4. Time domain convolution: $\mathrm{DTFT}\{f_1[n] * f_2[n]\} = \mathrm{DTFT}\{f_1[n]\}\mathrm{DTFT}\{f_2[n]\}$

5. Frequency domain convolution: $\mathrm{DTFT}\{f_1[n]f_2[n]\} = (1/2\pi)\mathrm{DTFT}\{f_1[n]\} * \mathrm{DTFT}\{f_2[n]\}$

6. Frequency differentiation: $nf[n] = \mathrm{DTFT}^{-1}\{dF(e^{j\omega})/d\omega\}$

Note that the time-differentiation and time-integration properties of the CTFT do not have analogous counterparts in the DTFT because time domain differentiation and integration are not defined for DT signals. When working with DT systems practitioners must often manipulate difference equations in the frequency domain. For this purpose property 1 and property 2 are very important. As with the CTFT, property 4 is very important for DT systems because it allows engineers to work with the frequency response of the system, in order to achieve proper shaping of the input spectrum or to achieve frequency selective filtering for noise reduction or signal detection. Also, property 3 is useful for the analysis of modulation and filtering operations common in both analog and digital communication systems.

The DTFT is defined so that the time domain is discrete and the frequency domain is continuous. This is in contrast to the CTFT that is defined to have continuous time and continuous frequency domains. The mathematical dual of the DTFT also exists, which is a transform pair that has a continuous time domain and a discrete frequency domain. In fact, the dual concept is really the same as the Fourier series for periodic CT signals presented earlier in the chapter, as represented by (1.5a) and (1.5b). However, the classical Fourier series arises from the assumption that the CT signal is inherently periodic, as opposed to the time domain becoming periodic by virtue of sampling the spectrum of a continuous frequency (aperiodic time) function [8]. The dual of the DTFT, the discrete frequency Fourier transform (DFFT), has been formulated and its properties tabulated as an interesting and useful transform in its own right [5]. Although the DFFT is similar in concept to the classical CT Fourier series, the formal properties of the DFFT [5] serve to clarify the effects of frequency domain sampling and time domain aliasing. These effects are obscured in the classical treatment of the CT Fourier series because the emphasis is on the inherent "line spectrum" that results from time domain periodicity. The DFFT is useful for the analysis and design of digital filters that are produced by frequency sampling techniques.

1.4.2   Relationship between the Continuous and Discrete Time Spectra

Because DT signals often originate by sampling CT signals, it is important to develop the relationship between the original spectrum of the CT signal and the spectrum of the DT signal that results. First, the CTFT is applied to the CT sampling model, and the properties listed above are used to produce the following result:

$$
\begin{aligned}
\mathcal{F}\{s_a(t)\} &= \mathcal{F}\left\{s(t)\sum_{n=-\infty}^{\infty}\delta(t - nT)\right\} \\
&= (1/2\pi)S(j\omega) * \mathcal{F}\left\{\sum_{n=-\infty}^{\infty}\delta(t - nT)\right\}
\end{aligned}
\tag{1.18}
$$

In this section it is important to distinguish between ω and ω', so the explicit primed notation is used in the following discussion where needed for clarification. Because the sampling function (summation of shifted impulses) on the right-hand side of the above equation is periodic with period T it can be replaced with a CT Fourier series expansion as follows:

$$
S(e^{j\omega T}) = \mathcal{F}\{s_a(t)\} = (1/2\pi)S(j\omega) * \mathcal{F}\left\{\sum_{n=-\infty}^{\infty}(1/T)e^{j(2\pi/T)nt}\right\}
$$

Applying the frequency domain convolution property of the CTFT yields

$$S(e^{j\omega T}) = (1/2\pi) \sum_{n=-\infty}^{\infty} S(j\omega) * (2\pi/T)\delta(\omega - (2\pi/T)n)$$

The result is

$$S(e^{j\omega T}) = (1/T) \sum_{n=-\infty}^{\infty} S(j[\omega - (2\pi/T)n]) = (1/T) \sum_{n=-\infty}^{\infty} S(j[\omega - n\omega_s]) \qquad (1.19a)$$

where $\omega_s = (2\pi/T)$ is the sampling frequency expressed in radians per second. An alternate form for the expression of (1.19a) is

$$S(e^{j\omega'}) = (1/T) \sum_{n=-\infty}^{\infty} S(j[(\omega' - n2\pi)/T]) \qquad (1.19b)$$

where $\omega' = \omega T$ is the normalized DT frequency axis expressed in radians. Note that $S(e^{j\omega T}) = S(e^{j\omega'})$ consists of an infinite number of replicas of the CT spectrum $S(j\omega)$, positioned at intervals of $(2\pi/T)$ on the ω axis (or at intervals of 2π on the ω' axis), as illustrated in Fig. 1.8. Note that if $S(j\omega)$ is band limited with a bandwidth ω_c, and if T is chosen sufficiently small so that $\omega_s > 2\omega_c$, then the DT spectrum is a copy of $S(j\omega)$ (scaled by $1/T$) in the baseband. The limiting case of $\omega_s = 2\omega_c$ is called the Nyquist sampling frequency. Whenever a CT signal is sampled at or above the Nyquist rate, no aliasing distortion occurs (i.e., the baseband spectrum does not overlap with the higher-order replicas) and the CT signal can be exactly recovered from its samples by extracting the baseband spectrum of $S(e^{j\omega'})$ with an ideal low-pass filter that recovers the original CT spectrum by removing all spectral replicas outside the baseband and scaling the baseband by a factor of T.

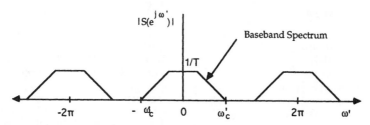

FIGURE 1.8   Illustration of the relationship between the CT and DT spectra.

1.5 The Discrete Fourier Transform

To obtain the discrete Fourier transform (DFT) the continuous frequency domain of the DTFT is sampled at N points uniformly spaced around the unit circle in the z-plane, i.e., at the points $\omega_k = (2\pi k/N)$, $k = 0, 1, \ldots, N-1$. The result is the DFT pair defined by (1.20a) and (1.20b). The signal $s[n]$ is either a finite length sequence of length N, or it is a periodic sequence with period N.

$$S[k] = \sum_{n=0}^{N-1} s[n]e^{-j2\pi kn/N} \qquad k = 0, 1, \ldots, N-1 \qquad (1.20a)$$

$$s[n] = (1/N)\sum_{k=0}^{N-1} S[k]e^{j2\pi kn/N} \qquad n = 0, 1, \ldots, N-1 \qquad (1.20b)$$

Regardless of whether $s[n]$ is a finite length or periodic sequence, the DFT treats the N samples of $s[n]$ as though they are one period of a periodic sequence. This is an important feature of the DFT, and one that must be handled properly in signal processing to prevent the introduction of artifacts. Important properties of the DFT are summarized in Table 1.5. The notation $((k))_N$ denotes k modulo N, and $R_N[n]$ is a rectangular window such that $R_N[n] = 1$ for $n = 0, \ldots, N-1$, and $R_N[n] = 0$ for $n < 0$ and $n \geq N$. The transform relationship given by (1.20a) and (1.20b) is also valid when $s[n]$ and $S[k]$ are periodic sequences, each of period N. In this case n and k are permitted to range over the complete set of real integers, and $S[k]$ is referred to as the discrete Fourier series (DFS). The DFS is developed by some authors as a distinct transform pair in its own right [6]. Whether the DFT and the DFS are considered identical or distinct is not very important in this discussion. The important point to be emphasized here is that the DFT treats $s[n]$ as though it were a single period of a periodic sequence, and all signal processing done with the DFT will inherit the consequences of this assumed periodicity.

TABLE 1.5 Properties of the DFT

Finite-Length Sequence (Length N)	N-Point DFT (Length N)
1. $x[n]$	$X[k]$
2. $x_1[n], x_2[n]$	$X_1[k], X_2[k]$
3. $ax_1[n] + bx_2[n]$	$aX_1[k] + bX_2[k]$
4. $X[n]$	$Nx[((-k))_N]$
5. $x[((n_m))_N]$	$W_N^{km}X[k]$
6. $W_N^{-ln}x[n]$	$X[((k-l))_N]$
7. $\displaystyle\sum_{m=0}^{N-1} x_1(m)x_2[((n_m))_N]$	$X_1[k]X_2[k]$
8. $x_1[n]x_2[n]$	$\dfrac{1}{N}\displaystyle\sum_{l=0}^{N-1} X_1(l)X_2[((k-l)_N]$
9. $x^*[n]$	$X^*[((-k))_N]$
10. $x^*[((-n))_N]$	$X^*[k]$
11. $\mathrm{Re}\{x[n]\}$	$X_{ep}[k] = \frac{1}{2}\{X[((k))_N] + K^*[((-k))_N]\}$
12. $j\mathrm{Im}\{x[n]\}$	$X_{op}[k] = \frac{1}{2}\{X[((k))_N] - X^*[((-k))_N]\}$
13. $x_{ep}[n] = \frac{1}{2}\{x[n] + x^*[((-n))_N]\}$	$\mathrm{Re}\{X[k]\}$
14. $x_{op}[n] = \frac{1}{2}\{x[n] - x^*[((-n))_N]\}$	$j\mathrm{Im}\{X[k]\}$
Properties 15–17 apply only when $x[n]$ is real	
15. Symmetry properties	$\begin{cases} X[k] &= X^*[((-k))_N] \\ \mathrm{Re}\{X[k]\} &= \mathrm{Re}\{X[((-k))_N]\} \\ \mathrm{Im}\{X[k]\} &= -\mathrm{Im}\{((-k))_N]\} \\ |X[k]| &= |X[((-k))_N]| \\ \not\angle\{X[k]\} &= -\not\angle\{X[((-k))_N]\} \end{cases}$
16. $x_{ep}[n] = \frac{1}{2}\{x[n] + x[((-n))_N]\}$	$\mathrm{Re}\{X[k]\}$
17. $x_{op}[n] = \frac{1}{2}\{x[n] - x[((-n))_N]\}$	$j\mathrm{Im}\{X[k]\}$

Source: A. V. Oppenheim and R. W. Schafer, *Discrete-Time Signal Processing*, © 1989. Reprinted by permission of Prentice-Hall, Inc., Upper Saddle River, NJ.

1.5.1 Properties of the Discrete Fourier Series

Most of the properties listed in Table 1.5 for the DFT are similar to those of the z-transform and the DTFT, although some important differences exist. For example, property 5 (time-shifting property), holds for *circular* shifts of the finite length sequence $s[n]$, which is consistent with the notion that the DFT treats $s[n]$ as one period of a periodic sequence. Also, the multiplication of two DFTs results in the **circular convolution** of the corresponding DT sequences, as specified by property 7. This latter property is quite different from the **linear convolution** property of the DTFT. Circular convolution is the result of the assumed periodicity discussed in the previous paragraph. Circular convolution is simply a linear convolution of the periodic extensions of the finite sequences being convolved, in which each of the finite sequences of length N defines the structure of one period of the periodic extensions.

For example, suppose one wishes to implement a digital filter with finite impulse response (FIR) $h[n]$. The output $y(n)$ in response to input $s[n]$ is given by

$$y[n] = \sum_{k=0}^{N-1} h[k]s[n-k] \tag{1.21}$$

where $y(n)$ is obtained by transforming $h[n]$ and $s[n]$ into $H[k]$ and $S[k]$ using the DFT, multiplying the transforms point-wise to obtain $Y[k] = H[k]S[k]$, and then using the inverse DFT to obtain $y[n] = \text{DFT}^{-1}\{Y[k]\}$. If $s[n]$ is a finite sequence of length M, then the results of the circular convolution implemented by the DFT will correspond to the desired linear convolution if the block length of the DFT, N_{DFT}, is chosen sufficiently large so that $N_{\text{DFT}} \geq N + M - 1$ and both $h[n]$ and $s[n]$ are padded with zeroes to form blocks of length N_{DFT}.

1.5.2 Fourier Block Processing in Real-Time Filtering Applications

In some practical applications either the value of M is too large for the memory available, or $s[n]$ may not actually be finite in length, but rather a continual stream of data samples that must be processed by a filter at real-time rates. Two well-known algorithms are available that partition $s[n]$ into smaller blocks and process the individual blocks with a smaller-length DFT: (1) overlap-save partitioning and (2) overlap-add partitioning. Each of these algorithms is summarized below.

Overlap-Save Processing

In this algorithm N_{DFT} is chosen to be some convenient value with $N_{\text{DFT}} > N$. The signal $s[n]$ is partitioned into blocks which are of length N_{DFT} and which overlap by $N - 1$ data points. Hence, the kth block is $s_k[n] = s[n + k(N_{\text{DFT}} - N + 1)], n = 0, \ldots, N_{\text{DFT}} - 1$. The filter impulse response is augmented with $N_{\text{DFT}} - N$ zeroes to produce

$$h_{\text{pad}}[n] = \left[\begin{array}{ll} h[n], & n = 0, \ldots, N - 1 \\ 0, & n = N, \ldots, N_{\text{DFT}} - 1 \end{array} \right] \tag{1.22}$$

The DFT is then used to obtain $Y_{\text{pad}}[n] = \text{DFT}\{h_{\text{pad}}[n]\} \cdot \text{DFT}\{s_k[n]\}$, and $y_{\text{pad}}[n] = \text{IDFT}\{Y_{\text{pad}}[n]\}$. From the $y_{\text{pad}}[n]$ array the values that correctly correspond to the linear convolution are saved; values that are erroneous due to wrap-around error caused by the circular convolution of the DFT are discarded. The kth block of the filtered output is obtained by

$$y_k[n] = \left[\begin{array}{ll} y_{\text{pad}}[n], & n = N - 1, \ldots, N_{\text{DFT}} - 1 \\ 0, & n = 0, \ldots, N - 2 \end{array} \right] \tag{1.23}$$

For the overlap-save algorithm, each time a block is processed there are $N_{\text{DFT}} - N + 1$ points saved and $N - 1$ points discarded. Each block moves forward by $N_{\text{DFT}} - N + 1$ data points and overlaps the previous block by $N - 1$ points.

Overlap-Add Processing

This algorithm is similar to the previous one except that the kth input block is defined as

$$s_k[n] = \left[\begin{array}{ll} s[n + kL], & n = 0, \ldots, L - 1 \\ 0, & n = L, \ldots, N_{\mathrm{DFT}} - 1 \end{array} \right] \tag{1.24}$$

where $L = N_{\mathrm{DFT}} - N + 1$. The filter function $h_{\mathrm{pad}}[n]$ is augmented with zeroes, as before, to create $h_{\mathrm{pad}}[n]$, and the DFT processing is executed as before. In each block $y_{\mathrm{pad}}[n]$ that is obtained at the output the first $N - 1$ points are erroneous, the last $N - 1$ points are erroneous, and the middle $N_{\mathrm{DFT}} - 2(N - 1)$ points correctly correspond to the linear convolution. However, if the last $N - 1$ points from block k are overlapped with the first $N - 1$ points from block $k + 1$ and added pairwise, correct results corresponding to linear convolution are also obtained from these positions. Hence, after this addition the number of correct points produced per block is $N_{\mathrm{DFT}} - N + 1$, which is the same as that for the overlap-save algorithm. The overlap-add algorithm requires approximately the same amount of computation as the overlap-save algorithm, although the addition of the overlapping portions of blocks is extra. This feature, together with the added delay of waiting for the next block to be finished before the previous one is complete, has resulted in more popularity for the overlap-save algorithm in practical applications.

Block filtering algorithms make it possible to efficiently filter continual data streams in real time because the fast Fourier transform (FFT) algorithm can be used to implement the DFT, thereby minimizing the total computation time and permitting reasonably high overall data rates. However, block filtering generates data in bursts, i.e., a delay occurs during which no filtered data appear, and then an entire block is suddenly generated. In real-time systems buffering must be used. The block algorithms are particularly effective for filtering very long sequences of data that are prerecorded on magnetic tape or disk.

1.5.3   Fast Fourier Transform Algorithms

The DFT is typically implemented in practice with one of the common forms of the FFT algorithm. The FFT is not a Fourier transform in its own right, but simply a computationally efficient algorithm that reduces the complexity of computing the DFT from order $\{N^2\}$ to order $\{N \log_2 N\}$. When N is large, the computational savings provided by the FFT algorithm is so great that the FFT makes real-time DFT analysis practical in many situations that would be entirely impractical without it. Fast Fourier transform algorithms abound, including decimation-in-time (D-I-T) algorithms, decimation-in-frequency (D-I-F) algorithms, bit-reversed algorithms, normally ordered algorithms, mixed-radix algorithms (for block lengths that are not powers of 2), prime factor algorithms, and Winograd algorithms [7]. The D-I-T and the D-I-F radix-2 FFT algorithms are the most widely used in practice. Detailed discussions of various FFT algorithms can be found in [3, 6, 7], and [10].

The FFT is easily understood by examining the simple example of $N = 8$. The FFT algorithm can be developed in numerous ways, all of which deal with a nested decomposition of the summation operator of (1.20a). The development presented here is called an **algebraic development** of the FFT because it follows straightforward algebraic manipulation. First, the summation indices (k, n) in (1.20a) are expressed as explicit binary integers, $k = k_2 4 + k_1 2 + k_0$ and $n = n_2 4 + n_1 2 + n_0$, where k_i and n_i are bits that take on the values of either 0 or 1. If these expressions are substituted into (1.20a), all terms in the exponent that contain the factor $N = 8$ can be deleted because $e^{-j2\pi l} = 1$ for any integer l. Upon deleting such terms and regrouping the remaining terms, the product nk can be expressed in either of two ways:

$$nk = (4k_0)n_2 + (4k_1 + 2k_0)n_1 + (4k_2 + 2k_1 + k_0)n_0 \tag{1.25a}$$

$$nk = (4n_0)k_2 + (4n_1 + 2n_0)k_1 + (4n_2 + 2n_1 + n_0)k_0 \tag{1.25b}$$

Substituting (1.25a) into (1.20a) leads to the D-I-T FFT, whereas substituting (1.25b) leads to the D-I-F FFT. Only the D-I-T FFT is discussed further here. The D-I-F and various related forms are treated in detail in [6].

The D-I-T FFT decomposes into $\log_2 N$ stages of computation, plus a stage of bit reversal,

$$x_1[k_0, n_1, n_0] = \sum_{n_2=0}^{1} s[n_2, n_1, n_0] W_8^{4k_0 n_2} \qquad \text{(stage 1)} \qquad (1.26a)$$

$$x_2[k_0, k_1, n_0] = \sum_{n_1=0}^{1} x[k_0, n_1, n_0] W_8^{(4k_1+2k_0)n_2} \qquad \text{(stage 2)} \qquad (1.26b)$$

$$x_3[k_0, k_1, k_2] = \sum_{n_0=0}^{1} x[k_0, k_1, n_0] W_8^{(4k_2+2k_1+k_0)n_0} \qquad \text{(stage 3)} \qquad (1.26c)$$

$$S[k_2, k_1, k_0] = x_3[k_0, k_1, k_2] \qquad \text{(bit reversal)} \qquad (1.26d)$$

In each summation above one of the n_i is summed out of the expression, while at the same time a new k_i is introduced. The notation is chosen to reflect this. For example, in stage 3, n_0 is summed out, k_2 is introduced as a new variable, and n_0 is replaced by k_2 in the result. The last operation, called bit reversal, is necessary to correctly locate the frequency samples $X[k]$ in the memory. It is easy to show that if the samples are paired correctly, an **in-place computation** can be done by a sequence of butterfly operations. The term "in-place" means that each time a butterfly is to be computed, a pair of data samples is read from memory, and the new data pair produced by the butterfly calculation is written back into the memory locations where the original pair was stored, thereby overwriting the original data. An in-place algorithm is designed so that each data pair is needed for only one butterfly, and thus the new results can be immediately stored on top of the old in order to minimize memory requirements.

For example, in stage 3 the $k = 6$ and $k = 7$ samples should be paired, yielding a "butterfly" computation that requires one complex multiply, one complex add, and one subtract:

$$x_3(1, 1, 0) = x_2(1, 1, 0) + W_8^3 x_2(1, 1, 1) \qquad (1.27a)$$

$$x_3(1, 1, 1) = x_2(1, 1, 0) - W_8^3 x_2(1, 1, 1) \qquad (1.27b)$$

Samples $x_2(6)$ and $x_2(7)$ are read from the memory, the butterfly is executed on the pair, and $x_3(6)$ and $x_3(7)$ are written back to the memory, overwriting the original values of $x_2(6)$ and $x_2(7)$. In general, $N/2$ butterflies are found in each stage and there are $\log_2 N$ stages, so the total number of butterflies is $(N/2)\log_2 N$. Because one complex multiplication per butterfly is the maximum, the total number of multiplications is bounded by $(N/2)\log_2 N$ (some of the multiplies involve factors of unity and should not be counted).

Figure 1.9 shows the signal flow graph of the D-I-T FFT for $N = 8$. This algorithm is referred to as an in-place FFT with normally ordered input samples and bit-reversed outputs. Minor variations that include both bit-reversed inputs and normally ordered outputs and non-in-place algorithms with normally ordered inputs and outputs are possible. Also, when N is not a power of 2, a mixed-radix algorithm can be used to reduce computation. The mixed-radix FFT is most efficient when N is highly composite, i.e., $N = p_1^{r_1} p_2^{r_2} \cdots p_L^{r_L}$, where the p^i are small prime numbers and the r^i are positive integers. It can be shown that the order of complexity of the mixed radix FFT is order $\{N(r_1(p_1-1)+r_2(p_2-1)+\cdots+r_L(p_L-1))\}$. Because of the lack of uniformity of structure among stages, this algorithm has not received much attention for hardware implementation. However, the mixed-radix FFT is often used in software applications, especially for processing data recorded in laboratory experiments in which it is not convenient to restrict the block lengths to be powers of 2. Many advanced FFT algorithms, such as higher-radix forms, the mixed-

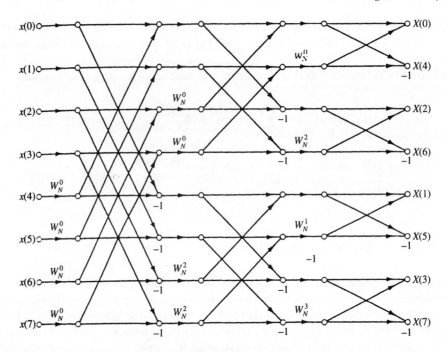

FIGURE 1.9   D-I-T FFT algorithm with normally ordered inputs and bit-reversed outputs.

radix form, the prime-factor algorithm, and the Winograd algorithm are described in [9]. Algorithms specialized for real-valued data reduce the computational cost by a factor of two. A radix-2 D-I-T FFT program, written in C language, is listed in Table 1.6.

1.6   Family Tree of Fourier Transforms

It is now possible to illustrate the functional relationships among the various forms of Fourier transforms that have been discussed in the previous sections. The family tree of CT Fourier transform is shown in Fig. 1.10, where the most general, and consequently the most powerful, Fourier transform is the classical complex Fourier transform (or equivalently, the bilateral Laplace transform). Note that the complex Fourier transform is identical to the bilateral Laplace transform, and it is at this level that the classical Laplace transform and Fourier transform techniques become identical. Each special member of the CT Fourier family is obtained by impressing certain constraints on the general form, thereby producing special transforms that are simpler and more useful in practical problems where the constraints are met.

The analogous family of DT Fourier techniques is presented in Fig. 1.11, in which the bilateral z-transform is analogous to the complex Fourier transform, the unilateral z-transform is analogous to the classical (one-sided) Laplace transform, the DTFT is analogous to the classical Fourier (CT) transform, and the DFT is analogous to the classical (CT) Fourier series.

1.7   Selected Applications of Fourier Methods

1.7.1   Fast Fourier Transform in Spectral Analysis

An FFT program is often used to perform spectral analysis on signals that are sampled and recorded as part of laboratory experiments, or in certain types of data acquisition systems. Several issues must be addressed when spectral analysis is performed on (sampled) analog waveforms that are observed over a finite interval of time.

TABLE 1.6 An In-Place D-I-T FFT Program in C Language

```
/*******************************************************
*     fft: in-place radix-2 DFT of a complex input
*
*     input:
*        n:     length of FFT: must be a power of two
*        m:     n = 2**m
*     input/output:
*        x:     float array of length n with real part of data
*        y:     float array of length n with image part of data
*******************************************************/
fft(n,m,x,y)
tnt    n,m;
float  x[ ], y[ ]:
{
       int    i,j,k,nl,n2:
       float  c,s,e,a,t1,t2;

       j = 0;               /*BIT-REVERSE */
       n2 = n/2;
       for (i=1; 1 < n-1; i++)          /*bit-reverse counter */
       {
        nl = n1/2;
        while ( j >= n1)
         {
         j = j - n1;
         nl = n1/2;
         }
        j = j + nl;
        if (i < j)              /*swap data */
         {
         t1 = x[i]; x[i] = x[j]; x[j] = t1;
         t1 = y[i]; y[i] = y[j]; y[j] = t1;
         }
        }
       n1 = 0; n2 = 1;         /* FFT */
       for (i = 0; i < m; i++)         /*state loop */
       {
        n1 = n2;  n2 = n2 + n2;
        e = -6.283185307179586/n2;
        a = 0.0;

        for (j=0; j < n1; j++)          /*flight loop */
         {
         c = cos(a); s=sin (a);
         a = a + e;

         for (k=j; k < n; k=k+n2)      /*butterfly loop */
          {
          t1 = c*x[k+n1] - s*y[k+n1];
          t2 = s*x[k+n1] + c*y[k+n1];
          x[k+n1] = x[k] - t1;
          y[k+n1] = y[k] - t2;
          x[k] = x[k] + t1;
          y[k] = y[k] + t2;
          }
         }
        }
        return;
}
```

Windowing

The FFT treats the block of data as though it were one period of a periodic sequence. If the underlying waveform is not periodic, then harmonic distortion may occur because the periodic waveform created by the FFT may have sharp discontinuities at the boundaries of the blocks. This effect is minimized by removing the mean of the data (it can always be reinserted) and by windowing the data so the ends of the block are smoothly tapered to zero. A good rule of thumb is to taper 10% of the data on each end of the block using either a cosine taper or one of the other common windows shown in Table 1.7. An alternate

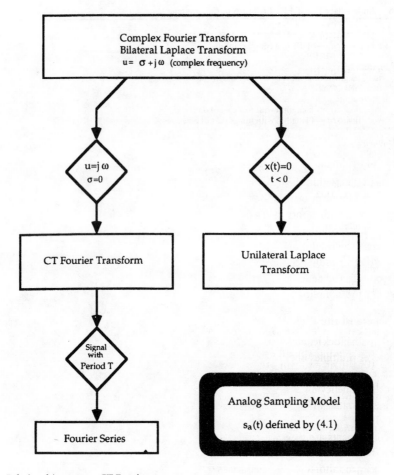

FIGURE 1.10   Relationships among CT Fourier concepts.

interpretation of this phenomenon is that the finite length observation has already windowed the true waveform with a rectangular window that has large spectral sidelobes (see Table 1.7). Hence, applying an additional window results in a more desirable window that minimizes frequency domain distortion.

TABLE 1.7   Common Window Functions

Name	Function	Peak Side-Lobe Amplitude (dB)	Mainlobe Width	Minimum Stopband Attenuation (dB)
Rectangular	$\omega(n) = 1.\quad 0 \le n \le N-1$	-13	$4\pi/N$	-21
Bartlett	$\omega(n) = \begin{cases} 2/N, & 0 \le n \le (N-1)/2 \\ 22n/N, & (N-1)/2 \le n \le N-1 \end{cases}$	-25	$8\pi/N$	-25
Hanning	$\omega(n) = (1/2)[1 - \cos(2\pi n/N)]$ $0 \le n \le N-1$	-31 -43	$8\pi/N$ $8\pi/N$	-44 -53
Hamming	$\omega(n) = 0.54 - 0.46\cos(2\pi n/N),$ $0 \le n \le N-1$	-43	$8\pi/N$	-53
Backman	$\omega(n) = 0.42 - 0.5\cos(2\pi n/N)$ $+ 0.08\cos(4\pi n/N),\quad 0 \le n \le N-1$	-57	$12\pi/N$	-74

Zero Padding

An improved spectral analysis is achieved if the block length of the FFT is increased. This can be done by taking more samples within the observation interval, increasing the length of the observation interval, or augmenting the original data set with zeroes. First, it must be understood that the finite observation interval results in a fundamental limit on the spectral resolution, even before the signals are sampled. The CT rectangular window has a $(\sin x)/x$ spectrum, which is convolved with the true spectrum of the analog signal. Therefore, the frequency resolution is limited by the width of the mainlobe in the $(\sin x)/x$ spectrum, which is inversely proportional to the length of the observation interval. Sampling causes a certain degree of aliasing, although this effect can be minimized by sampling at a high enough rate. Therefore, lengthening the observation interval increases the fundamental resolution limit, while taking more samples within the observation interval minimizes aliasing distortion and provides a better definition (more sample points) on the underlying spectrum.

Padding the data with zeroes and computing a longer FFT does give more frequency domain points (improved spectral resolution), but it does not improve the fundamental limit, nor does it alter the effects of aliasing error. The resolution limits are established by the observation interval and the sampling rate. No amount of zero padding can improve these basic limits. However, zero padding is a useful tool for providing more spectral definition, i.e., it allows a better view of the (distorted) spectrum that results once the observation and sampling effects have occurred.

Leakage and the Picket Fence Effect

An FFT with block length N can accurately resolve only frequencies $\omega_k = (2\pi/N)k, k = 0, \ldots, N-1$ that are integer multiples of the fundamental $\omega_1 = (2\pi/N)$. An analog waveform that is sampled and subjected to spectral analysis may have frequency components between the harmonics. For example, a component at frequency $\omega_{k+1/2} = (2\pi/N)(k + 1/2)$ will appear scattered throughout the spectrum. The effect is illustrated in Fig. 1.12 for a sinusoid that is observed through a rectangular window and then sampled at N points. The **picket fence effect** means that not all frequencies can be seen by the FFT. Harmonic components are seen accurately, but other components "slip through the picket fence" while their energy is "leaked" into the harmonics. These effects produce artifacts in the spectral domain that must be carefully monitored to assure that an accurate spectrum is obtained from FFT processing.

1.7.2 Finite Impulse Response Digital Filter Design

A common method for designing FIR digital filters is by use of windowing and FFT analysis. In general, window designs can be carried out with the aid of a hand calculator and a table of well-known window functions. Let $h[n]$ be the impulse response that corresponds to some desired frequency response, $H(e^{j\omega})$. If $H(e^{j\omega})$ has sharp discontinuities, such as the low-pass example shown in Fig. 1.13, then $h[n]$ will represent an infinite impulse response (IIR) function. The objective is to time limit $h[n]$ in such a way as to not distort $H(e^{j\omega})$ any more than necessary. If $h[n]$ is simply truncated, a ripple (Gibbs phenomenon) occurs around the discontinuities in the spectrum, resulting in a distorted filter (Fig. 1.13).

Suppose that $w[n]$ is a window function that time limits $h[n]$ to create an FIR approximation, $h'[n]$; i.e., $h'[n] = w[n]h[n]$. Then if $W(e^{j\omega})$ is the DTFT of $w[n]$, $h'[n]$ will have a Fourier transform given by $H'(e^{j\omega}) = W(e^{j\omega}) * H(e^{j\omega})$, where $*$ denotes convolution. Thus, the ripples in $H'(e^{j\omega})$ result from the sidelobes of $W(e^{j\omega})$. Ideally, $W(e^{j\omega})$ should be similar to an impulse so that $H'(e^{j\omega})$ is approximately equal to $H(e^{j\omega})$.

Special Case. Let $h[n] = \cos n\omega_0$, for all n. Then $h[n] = w[n]\cos n\omega_0$, and

$$H'(e^{j\omega}) = (1/2)W(e^{j(\omega+\omega_0)}) + (1/2)W(e^{j(\omega-\omega_0)}) \tag{1.28}$$

as illustrated in Fig. 1.14. For this simple class, the center frequency of the bandpass is controlled by ω_0, and both the shape of the bandpass and the sidelobe structure are strictly determined by the choice of the

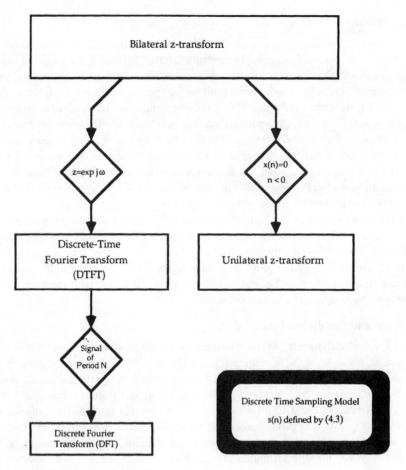

FIGURE 1.11   Relationships among DT concepts.

window. While this simple class of FIRs does not allow for very flexible designs, it is a simple technique for determining quite useful low-pass, bandpass, and high-pass FIRs.

    General Case.   Specify an ideal frequency response, $H(e^{j\omega})$, and choose samples at selected values of ω. Use a long inverse FFT of length N' to find $h'[n]$, an approximation to $h[n]$, where if N is the desired length of the final filter, then $N' \gg N$. Then use a carefully selected window to truncate $h'[n]$ to obtain $h[n]$ by letting $h[n] = \omega[n]h'[n]$. Finally, use an FFT of length N' to find $H'(e^{j\omega})$. If $H'(e^{j\omega})$ is a satisfactory approximation to $H(e^{j\omega})$, the design is finished. If not, choose a new $H(e^{j\omega})$ or a new $w[n]$ and repeat. Throughout the design procedure it is important to choose $N' = kN$, with k an integer that is typically in the range of 4 to 10. Because this design technique is a trial and error procedure, the quality of the result depends to some degree on the skill and experience of the designer. Table 1.7 lists several well-known window functions that are often useful for this type of FIR filter design procedure.

1.7.3   Fourier Analysis of Ideal and Practical Digital-to-Analog Conversion

From the relationship characterized by (1.19b) and illustrated in Fig. 1.8, CT signal $s(t)$ can be recovered from its samples by passing $s_a(t)$ through an ideal lowpass filter that extracts only the baseband spectrum. The ideal lowpass filter, shown in Fig. 1.15, is a zero-phase CT filter whose magnitude response is a constant of value T in the range $-\pi < \omega' \leq \pi$, and zero elsewhere. The impulse response of this "reconstruction filter" is given by $h(t) = T\,\text{sinc}((\pi/T)t)$, where $\text{sinc}\,x = (\sin x)/x$. The reconstruction

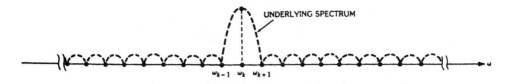

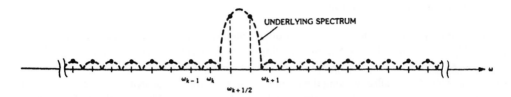

FIGURE 1.12   Illustration of leakage and the picket-fence effects.

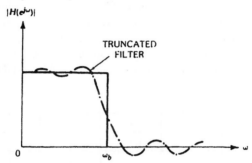

FIGURE 1.13   Gibbs effect in a low-pass filter caused by truncating the impulse response.

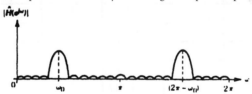

FIGURE 1.14   Design of a simple bandpass FIR filter by windowing.

can be expressed as $s(t) = h(t) * s_a(t)$, which, after some mathematical manipulation, yields the following classical reconstruction formula

$$s(t) = \sum_{n=-\infty}^{\infty} s(nT)\operatorname{sinc}((\pi/T)(t - nT)) \tag{1.29}$$

Note that the signal $s(t)$ is exactly recovered from its samples only if an infinite number of terms is included in the summation of (1.29). However, good approximation of $s(t)$ can be obtained with only a finite number of terms if the lowpass reconstruction filter $h(t)$ is modified to have a finite interval of support, i.e., if $h(t)$ is nonzero only over a finite time interval. The reconstruction formula of (1.29) is an important result in that it represents the inverse of the sampling operation. By this means Fourier transform theory establishes that as long as CT signals are sampled at a sufficiently high rate, the information content contained in $s(t)$ can be represented and processed in either a CT or DT format. Fourier sampling and reconstruction theory provides the theoretical mechanism for translation between one format or the other without loss of information.

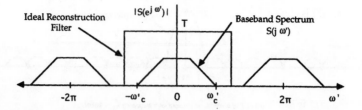

FIGURE 1.15   Illustration of ideal reconstruction.

A CT signal $s(t)$ can be perfectly recovered from its samples using (1.29) as long as the original sampling rate was high enough to satisfy the Nyquist sampling criterion, i.e., $\omega_s > 2\omega_B$. If the sampling rate does not satisfy the Nyquist criterion the adjacent periods of the analog spectrum will overlap, causing a distorted spectrum. This effect, called **aliasing distortion**, is rather serious because it cannot be corrected easily once it has occurred. In general, an analog signal should always be prefiltered with an CT low-pass filter prior to sampling so that aliasing distortion does not occur.

Figure 1.16 shows the frequency response of a fifth-order elliptic analog low-pass filter that meets industry standards for prefiltering speech signals. These signals are subsequently sampled at an 8-kHz sampling rate and transmitted digitally across telephone channels. The band-pass ripple is less than ± 0.01 dB from DC up to the frequency 3.4 kHz (too small to be seen in Fig. 1.16), and the stopband rejection reaches at least -32.0 dB at 4.6 kHz and remains below this level throughout the stopband.

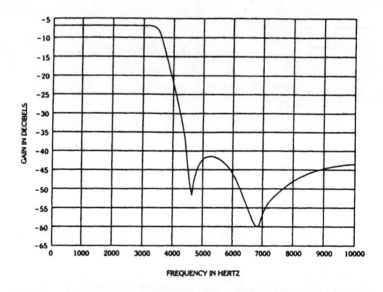

FIGURE 1.16   A fifth-order elliptic analog anti-aliasing filter used in the telecommunications industry with an 8-kHz sampling rate.

Most practical systems use digital-to-analog converters for reconstruction, which results in a staircase approximation to the true analog signal, i.e.,

$$\hat{s}(t) = \sum_{n=-\infty}^{\infty} s(nT)\{u(t - nT) - u[t - (n + 1)]\}, \qquad (1.30)$$

where $\hat{s}(t)$ denotes the reconstructed approximation to $s(t)$, and $u(t)$ denotes a CT unit step function.

The approximation $\hat{s}(t)$ is equivalent to a result obtained by using an approximate reconstruction filter of the form

$$H_a(j\omega) = 2Te^{-j\omega T/2}\sin c(\omega T/2) \tag{1.31}$$

The approximation $\hat{s}(t)$ is said to contain "$\sin x/x$ distortion," which occurs because $H_a(j\omega)$ is not an ideal low-pass filter. $H_a(j\omega)$ distorts the signal by causing a droop near the passband edge, as well as by passing high-frequency distortion terms which "leak" through the sidelobes of $H_a(j\omega)$. Therefore, a practical digital to analog converter is normally followed by an analog postfilter

$$H_p(j\omega) = \left[\begin{array}{ll} H_a^{-1}(j\omega), & 0 \le |\omega| < \pi/T \\ 0, & \omega \text{ otherwise} \end{array}\right] \tag{1.32}$$

which compensates for the distortion and produces the correct $\hat{s}(t)$, i.e., the correctly constructed CT output. Unfortunately, the postfilter $H_p(j\omega)$ cannot be implemented perfectly, and, therefore, the actual reconstructed signal always contains some distortion in practice that arises from errors in approximating the ideal postfilter. Figure 1.17 shows a digital processor, complete with analog-to-digital and digital-to-analog converters, and the accompanying analog pre- and postfilters necessary for proper operation.

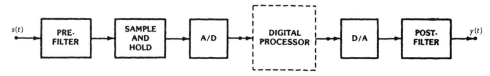

FIGURE 1.17   Analog pre- and postfilters required at the analog to digital and digital to analog interfaces.

1.8   Summary

This chapter presented many different Fourier transform concepts for both continuous time (CT) and discrete time (DT) signals and systems. Emphasis was placed on illustrating how these various forms of the Fourier transform relate to one another, and how they are all derived from more general complex transforms, the complex Fourier (or bilateral Laplace) transform for CT, and the bilateral z-transform for DT. It was shown that many of these transforms have similar properties which are inherited from their parent forms, and that a parallel hierarchy exists among Fourier transform concepts in the CT and the DT worlds. Both CT and DT sampling models were introduced as a means of representing sampled signals in these two different "worlds," and it was shown that the models are equivalent by virtue of having the same Fourier spectra when transformed into the Fourier domain with the appropriate Fourier transform. It was shown how Fourier analysis properly characterizes the relationship between the spectra of a CT signal and its DT counterpart obtained by sampling. The classical reconstruction formula was obtained as an outgrowth of this analysis. Finally, the discrete Fourier transform (DFT), the backbone for much of modern digital signal processing, was obtained from more classical forms of the Fourier transform by simultaneously discretizing the time and frequency domains. The DFT, together with the remarkable computational efficiency provided by the fast Fourier transform (FFT) algorithm, has contributed to the resounding success that engineers and scientists have experienced in applying digital signal processing to many practical scientific problems.

References

[1] VanValkenburg, M.E., *Network Analysis*, 3rd ed., Englewood Cliffs, NJ: Prentice-Hall, 1974.

[2] Oppenheim, A.V., Willsky, A.S., and Young, I.T., *Signals and Systems*, Englewood Cliffs, NJ: Prentice-Hall, 1983.

[3] Bracewell, R.N., *The Fourier Transform*, 2nd ed., New York: McGraw-Hill, 1986.

[4] Oppenheim, A.V. and Schafer, R.W., *Discrete-Time Signal Processing*, Englewood Cliffs, NJ: Prentice-Hall, 1989.

[5] Jenkins, W.K. and Desai, M.D., The discrete-frequency Fourier transform, *IEEE Trans. Circuits Syst.*, vol. CAS-33, no. 7, pp. 732–734, July 1986.

[6] Oppenheim, A.V. and Schafer, R.W., *Digital Signal Processing*, Englewood Cliffs, NJ: Prentice-Hall, 1975.

[7] Blahut, R.E., *Fast Algorithms for Digital Signal Processing*, Reading, MA: Addison-Wesley, 1985.

[8] Deller, J.R., Jr., Tom, Dick, and Mary discover the DFT, *IEEE Signal Processing Mag.*, vol. 11, no. 2, pp. 36–50, Apr. 1994.

[9] Burrus, C.S. and Parks, T.W., *DFT/FFT and Convolution Algorithms*, New York: John Wiley and Sons, 1985.

[10] Brigham, E.O., *The Fast Fourier Transform*, Englewood Cliffs, NJ: Prentice-Hall, 1974.

2

Ordinary Linear Differential and Difference Equations

B.P. Lathi
California State University, Sacramento

2.1 Differential Equations

A function containing variables and their derivatives is called a *differential expression*, and an equation involving differential expressions is called a *differential equation*. A differential equation is an *ordinary* differential equation if it contains only one independent variable; it is a *partial* differential equation if it contains more than one independent variable. We shall deal here only with ordinary differential equations.

In the mathematical texts, the independent variable is generally x, which can be anything such as time, distance, velocity, pressure, and so on. In most of the applications in control systems, the independent variable is time. For this reason we shall use here independent variable t for time, although it can stand for any other variable as well.

The following equation

$$\left(\frac{d^2 y}{dt^2}\right)^4 + 3\frac{dy}{dt} + 5y^2(t) = \sin t$$

is an ordinary differential equation of second *order* because the highest derivative is of the second order. An nth-order differential equation is *linear* if it is of the form

$$a_n(t)\frac{d^n y}{dt^n} + a_{n-1}(t)\frac{d^{n-1}y}{dt^{n-1}} + \cdots + a_1(t)\frac{dy}{dt}$$
$$+ a_0(t)y(t) = r(t) \tag{2.1}$$

where the coefficients $a_i(t)$ are not functions of $y(t)$. If these coefficients (a_i) are constants, the equation is linear with *constant coefficients*. Many engineering (as well as nonengineering) systems can be modeled by these equations. Systems modeled by these equations are known as *linear time-invariant* (LTI) systems. In this chapter we shall deal exclusively with linear differential equations with constant coefficients. Certain other forms of differential equations are dealt with elsewhere in this volume.

0-8493-8572-5/98/$0.00+$.50
© 1998 by CRC Press LLC

Role of Auxiliary Conditions in Solution of Differential Equations

We now show that a differential equation does not, in general, have a unique solution unless some additional constraints (or conditions) on the solution are known. This fact should not come as a surprise. A function $y(t)$ has a unique derivative dy/dt, but for a given derivative dy/dt there are infinite possible functions $y(t)$. If we are given dy/dt, it is impossible to determine $y(t)$ uniquely unless an additional piece of information about $y(t)$ is given. For example, the solution of a differential equation

$$\frac{dy}{dt} = 2 \tag{2.2}$$

obtained by integrating both sides of the equation is

$$y(t) = 2t + c \tag{2.3}$$

for any value of c. Equation 2.2 specifies a function whose slope is 2 for all t. Any straight line with a slope of 2 satisfies this equation. Clearly the solution is not unique, but if we place an additional constraint on the solution $y(t)$, then we specify a unique solution.

For example, suppose we require that $y(0) = 5$; then out of all the possible solutions available, only one function has a slope of 2 and an intercept with the vertical axis at 5. By setting $t = 0$ in Equation 2.3 and substituting $y(0) = 5$ in the same equation, we obtain $y(0) = 5 = c$ and

$$y(t) = 2t + 5$$

which is the unique solution satisfying both Equation 2.2 and the constraint $y(0) = 5$.

In conclusion, differentiation is an irreversible operation during which certain information is lost. To reverse this operation, one piece of information about $y(t)$ must be provided to restore the original $y(t)$. Using a similar argument, we can show that, given d^2y/dt^2, we can determine $y(t)$ uniquely only if two additional pieces of information (constraints) about $y(t)$ are given. In general, to determine $y(t)$ uniquely from its nth derivative, we need n additional pieces of information (constraints) about $y(t)$. These constraints are also called *auxiliary conditions*. When these conditions are given at $t = 0$, they are called *initial conditions*.

We discuss here two systematic procedures for solving linear differential equations of the form in Eq. 2.1. The first method is the *classical method*, which is relatively simple, but restricted to a certain class of inputs. The second method (the convolution method) is general and is applicable to all types of inputs. A third method (Laplace transform) is discussed elsewhere in this volume. Both the methods discussed here are classified as *time-domain* methods because with these methods we are able to solve the above equation directly, using t as the independent variable. The method of Laplace transform (also known as the *frequency-domain* method), on the other hand, requires transformation of variable t into a frequency variable s.

In engineering applications, the form of linear differential equation that occurs most commonly is given by

$$\frac{d^n y}{dt^n} + a_{n-1}\frac{d^{n-1}y}{dt^{n-1}} + \cdots + a_1\frac{dy}{dt} + a_0 y(t)$$
$$= b_m\frac{d^m f}{dt^m} + b_{m-1}\frac{d^{m-1}f}{dt^{m-1}} + \cdots + b_1\frac{df}{dt} + b_0 f(t) \tag{2.4a}$$

where all the coefficients a_i and b_i are constants. Using operational notation D to represent d/dt, this equation can be expressed as

$$(D^n + a_{n-1}D^{n-1} + \cdots + a_1 D + a_0)y(t)$$
$$= (b_m D^m + b_{m-1}D^{m-1} + \cdots + b_1 D + b_0)f(t) \tag{2.4b}$$

or

$$Q(D)y(t) = P(D)f(t) \tag{2.4c}$$

where the polynomials $Q(D)$ and $P(D)$, respectively, are

$$
\begin{aligned}
Q(D) &= D^n + a_{n-1}D^{n-1} + \cdots + a_1 D + a_0 \\
P(D) &= b_m D^m + b_{m-1}D^{m-1} + \cdots + b_1 D + b_0
\end{aligned}
$$

Observe that this equation is of the form of Eq. 2.1, where $r(t)$ is in the form of a linear combination of $f(t)$ and its derivatives. In this equation, $y(t)$ represents an output variable, and $f(t)$ represents an input variable of an LTI system. Theoretically, the powers m and n in the above equations can take on any value. Practical noise considerations, however, require [1] $m \le n$.

2.1.1   Classical Solution

When $f(t) \equiv 0$, Eq. 2.4 is known as the *homogeneous* (or complementary) equation. We shall first solve the homogeneous equation. Let the solution of the homogeneous equation be $y_c(t)$, that is,

$$Q(D)y_c(t) = 0$$

or

$$(D^n + a_{n-1}D^{n-1} + \cdots + a_1 D + a_0)y_c(t) = 0$$

We first show that if $y_p(t)$ is the solution of Eq. 2.4, then $y_c(t) + y_p(t)$ is also its solution. This follows from the fact that

$$Q(D)y_c(t) = 0$$

If $y_p(t)$ is the solution of Eq. 2.4, then

$$Q(D)y_p(t) = P(D)f(t)$$

Addition of these two equations yields

$$Q(D)\left[y_c(t) + y_p(t)\right] = P(D)f(t)$$

Thus, $y_c(t) + y_p(t)$ satisfies Eq. 2.4 and therefore is the general solution of Eq. 2.4. We call $y_c(t)$ the *complementary* solution and $y_p(t)$ the *particular* solution. In system analysis parlance, these components are called the *natural* response and *the forced* response, respectively.

Complementary Solution (The Natural Response)

The complementary solution $y_c(t)$ is the solution of

$$Q(D)y_c(t) = 0 \tag{2.5a}$$

or

$$\left(D^n + a_{n-1}D^{n-1} + \cdots + a_1 D + a_0\right)y_c(t) = 0 \tag{2.5b}$$

A solution to this equation can be found in a systematic and formal way. However, we will take a short cut by using heuristic reasoning. Equation 2.5b shows that a linear combination of $y_c(t)$ and its n

successive derivatives is zero, not at *some* values of t, but for all t. This is possible *if and only if* $y_c(t)$ and all its n successive derivatives are of the same form. Otherwise their sum can never add to zero for all values of t. We know that only an exponential function $e^{\lambda t}$ has this property. So let us assume that

$$y_c(t) = ce^{\lambda t}$$

is a solution to Eq. 2.5b. Now

$$Dy_c(t) \quad = \quad \frac{dy_c}{dt} = c\lambda e^{\lambda t}$$

$$D^2 y_c(t) \quad = \quad \frac{d^2 y_c}{dt^2} = c\lambda^2 e^{\lambda t}$$

$$\ldots\ldots \qquad \ldots \qquad \ldots\ldots$$

$$D^n y_c(t) \quad = \quad \frac{d^n y_c}{dt^n} = c\lambda^n e^{\lambda t}$$

Substituting these results in Eq. 2.5b, we obtain

$$c\left(\lambda^n + a_{n-1}\lambda^{n-1} + \cdots + a_1\lambda + a_0\right)e^{\lambda t} = 0$$

For a nontrivial solution of this equation,

$$\lambda^n + a_{n-1}\lambda^{n-1} + \cdots + a_1\lambda + a_0 = 0 \tag{2.6a}$$

This result means that $ce^{\lambda t}$ is indeed a solution of Eq. 2.5 provided that λ satisfies Eq. 2.6a. Note that the polynomial in Eq. 2.6a is identical to the polynomial $Q(D)$ in Eq. 2.5b, with λ replacing D. Therefore, Eq. 2.6a can be expressed as

$$Q(\lambda) = 0 \tag{2.6b}$$

When $Q(\lambda)$ is expressed in factorized form, Eq. 2.6b can be represented as

$$Q(\lambda) = (\lambda - \lambda_1)(\lambda - \lambda_2)\cdots(\lambda - \lambda_n) = 0 \tag{2.6c}$$

Clearly λ has n solutions: $\lambda_1, \lambda_2, \ldots, \lambda_n$. Consequently, Eq. 2.5 has n possible solutions: $c_1 e^{\lambda_1 t}$, $c_2 e^{\lambda_2 t}$, $\ldots, c_n e^{\lambda_n t}$, with $c_1, c_2, \ldots, c_n$ as arbitrary constants. We can readily show that a general solution is given by the sum of these n solutions,[1] so that

$$y_c(t) = c_1 e^{\lambda_1 t} + c_2 e^{\lambda_2 t} + \cdots + c_n e^{\lambda_n t} \tag{2.7}$$

---

[1] To prove this fact, assume that $y_1(t)$, $y_2(t)$, $\ldots$, $y_n(t)$ are all solutions of Eq. 2.5. Then

$$Q(D)y_1(t) \quad = \quad 0$$
$$Q(D)y_2(t) \quad = \quad 0$$
$$\ldots\ldots \qquad \ldots \qquad \ldots\ldots$$
$$Q(D)y_n(t) \quad = \quad 0$$

Multiplying these equations by $c_1, c_2, \ldots, c_n$, respectively, and adding them together yields

$$Q(D)\left[c_1 y_1(t) + c_2 y_2(t) + \cdots + c_n y_n(t)\right] = 0$$

This result shows that $c_1 y_1(t) + c_2 y_2(t) + \cdots + c_n y_n(t)$ is also a solution of the homogeneous Eq. 2.5.

where $c_1, c_2, \ldots, c_n$ are arbitrary constants determined by n constraints (the auxiliary conditions) on the solution.

The polynomial $Q(\lambda)$ is known as the *characteristic polynomial*. The equation

$$Q(\lambda) = 0 \tag{2.8}$$

is called the *characteristic* or auxiliary equation. From Eq. 2.6c, it is clear that $\lambda_1, \lambda_2, \ldots, \lambda_n$ are the roots of the characteristic equation; consequently, they are called the *characteristic roots*. The terms *characteristic values, eigenvalues*, and *natural frequencies* are also used for characteristic roots.[2] The exponentials $e^{\lambda_i t}$ ($i = 1, 2, \ldots, n$) in the complementary solution are the *characteristic modes* (also known as *modes* or *natural modes*). There is a characteristic mode for each characteristic root, and the *complementary solution is a linear combination of the characteristic modes*.

Repeated Roots

The solution of Eq. 2.5 as given in Eq. 2.7 assumes that the n characteristic roots $\lambda_1, \lambda_2, \ldots, \lambda_n$ are distinct. If there are repeated roots (same root occurring more than once), the form of the solution is modified slightly. By direct substitution we can show that the solution of the equation

$$(D - \lambda)^2 y_c(t) = 0$$

is given by

$$y_c(t) = (c_1 + c_2 t)e^{\lambda t}$$

In this case the root λ repeats twice. Observe that the characteristic modes in this case are $e^{\lambda t}$ and $te^{\lambda t}$. Continuing this pattern, we can show that for the differential equation

$$(D - \lambda)^r y_c(t) = 0 \tag{2.9}$$

the characteristic modes are $e^{\lambda t}, te^{\lambda t}, t^2 e^{\lambda t}, \ldots, t^{r-1} e^{\lambda t}$, and the solution is

$$y_c(t) = \left(c_1 + c_2 t + \cdots + c_r t^{r-1} \right) e^{\lambda t} \tag{2.10}$$

Consequently, for a characteristic polynomial

$$Q(\lambda) = (\lambda - \lambda_1)^r (\lambda - \lambda_{r+1}) \cdots (\lambda - \lambda_n)$$

the characteristic modes are $e^{\lambda_1 t}, te^{\lambda_1 t}, \ldots, t^{r-1} e^{\lambda t}, e^{\lambda_{r+1} t}, \ldots, e^{\lambda_n t}$. and the complementary solution is

$$y_c(t) = (c_1 + c_2 t + \cdots + c_r t^{r-1})e^{\lambda_1 t} + c_{r+1} e^{\lambda_{r+1} t} + \cdots + c_n e^{\lambda_n t}$$

Particular Solution (The Forced Response): Method of Undetermined Coefficients

The particular solution $y_p(t)$ is the solution of

$$Q(D)y_p(t) = P(D)f(t) \tag{2.11}$$

It is a relatively simple task to determine $y_p(t)$ when the input $f(t)$ is such that it yields only a finite number of independent derivatives. Inputs having the form $e^{\zeta t}$ or t^r fall into this category. For example, $e^{\zeta t}$ has only one independent derivative; the repeated differentiation of $e^{\zeta t}$ yields the same form, that is, $e^{\zeta t}$. Similarly, the repeated differentiation of t^r yields only r independent derivatives. The particular solution

---

[2] The term *eigenvalue* is German for characteristic value.

to such an input can be expressed as a linear combination of the input and its independent derivatives. Consider, for example, the input $f(t) = at^2 + bt + c$. The successive derivatives of this input are $2at + b$ and $2a$. In this case, the input has only two independent derivatives. Therefore the particular solution can be assumed to be a linear combination of $f(t)$ and its two derivatives. The suitable form for $y_p(t)$ in this case is therefore

$$y_p(t) = \beta_2 t^2 + \beta_1 t + \beta_0$$

The undetermined coefficients β_0, β_1, and β_2 are determined by substituting this expression for $y_p(t)$ in Eq. 2.11 and then equating coefficients of similar terms on both sides of the resulting expression.

Although this method can be used only for inputs with a finite number of derivatives, this class of inputs includes a wide variety of the most commonly encountered signals in practice. Table 2.1 shows a variety of such inputs and the form of the particular solution corresponding to each input. We shall demonstrate this procedure with an example.

TABLE 2.1

Input $f(t)$	Forced Response
1. $e^{\zeta t}$ $\zeta \neq \lambda_i$ ($i = 1, 2, \ldots, n$)	$\beta e^{\zeta t}$
2. $e^{\zeta t}$ $\zeta = \lambda_i$	$\beta t e^{\zeta t}$
3. k (a constant)	β (a constant)
4. $\cos(\omega t + \theta)$	$\beta \cos(\omega t + \phi)$
5. $\left(t^r + \alpha_{r-1} t^{r-1} + \cdots + \alpha_1 t + \alpha_0\right) e^{\zeta t}$	$(\beta_r t^r + \beta_{r-1} t^{r-1} + \cdots + \beta_1 t + \beta_0) e^{\zeta t}$

Note: By definition, $y_p(t)$ cannot have any characteristic mode terms. If any term $p(t)$ shown in the right-hand column for the particular solution is also a characteristic mode, the correct form of the forced response must be modified to $t^i p(t)$, where i is the smallest possible integer that can be used and still can prevent $t^i p(t)$ from having characteristic mode term. For example, when the input is $e^{\zeta t}$, the forced response (right-hand column) has the form $\beta e^{\zeta t}$. But if $e^{\zeta t}$ happens to be a characteristic mode, the correct form of the particular solution is $\beta t e^{\zeta t}$ (see Pair 2). If $t e^{\zeta t}$ also happens to be characteristic mode, the correct form of the particular solution is $\beta t^2 e^{\zeta t}$, and so on.

EXAMPLE 2.1:

Solve the differential equation

$$\left(D^2 + 3D + 2\right) y(t) = Df(t) \tag{2.12}$$

if the input

$$f(t) = t^2 + 5t + 3$$

and the initial conditions are $y(0^+) = 2$ and $\dot{y}(0^+) = 3$.

The characteristic polynomial is

$$\lambda^2 + 3\lambda + 2 = (\lambda + 1)(\lambda + 2)$$

Therefore the characteristic modes are e^{-t} and e^{-2t}. The complementary solution is a linear combination of these modes, so that

$$y_c(t) = c_1 e^{-t} + c_2 e^{-2t} \qquad t \geq 0$$

Here the arbitrary constants c_1 and c_2 must be determined from the given initial conditions.

The particular solution to the input $t^2 + 5t + 3$ is found from Table 2.1 (Pair 5 with $\zeta = 0$) to be

$$y_p(t) = \beta_2 t^2 + \beta_1 t + \beta_0$$

Moreover, $y_p(t)$ satisfies Eq. 2.11, that is,

$$\left(D^2 + 3D + 2\right) y_p(t) = Df(t) \tag{2.13}$$

Now

$$Dy_p(t) \;=\; \frac{d}{dt}\left(\beta_2 t^2 + \beta_1 t + \beta_0\right) = 2\beta_2 t + \beta_1$$

$$D^2 y_p(t) \;=\; \frac{d^2}{dt^2}\left(\beta_2 t^2 + \beta_1 t + \beta_0\right) = 2\beta_2$$

and

$$Df(t) = \frac{d}{dt}\left[t^2 + 5t + 3\right] = 2t + 5$$

Substituting these results in Eq. 2.13 yields

$$2\beta_2 + 3(2\beta_2 t + \beta_1) + 2(\beta_2 t^2 + \beta_1 t + \beta_0) = 2t + 5$$

or

$$2\beta_2 t^2 + (2\beta_1 + 6\beta_2)t + (2\beta_0 + 3\beta_1 + 2\beta_2) = 2t + 5$$

Equating coefficients of similar powers on both sides of this expression yields

$$2\beta_2 \;=\; 0$$
$$2\beta_1 + 6\beta_2 \;=\; 2$$
$$2\beta_0 + 3\beta_1 + 2\beta_2 \;=\; 5$$

Solving these three equations for their unknowns, we obtain $\beta_0 = 1$, $\beta_1 = 1$, and $\beta_2 = 0$. Therefore,

$$y_p(t) = t + 1 \qquad t > 0$$

The total solution $y(t)$ is the sum of the complementary and particular solutions. Therefore,

$$y(t) \;=\; y_c(t) + y_p(t)$$
$$\;=\; c_1 e^{-t} + c_2 e^{-2t} + t + 1 \qquad t > 0$$

so that

$$\dot{y}(t) \;=\; -c_1 e^{-t} - 2c_2 e^{-2t} + 1$$

Setting $t = 0$ and substituting the given initial conditions $y(0) = 2$ and $\dot{y}(0) = 3$ in these equations, we have

$$2 \;=\; c_1 + c_2 + 1$$
$$3 \;=\; -c_1 - 2c_2 + 1$$

The solution to these two simultaneous equations is $c_1 = 4$ and $c_2 = -3$. Therefore,

$$y(t) = 4e^{-t} - 3e^{-2t} + t + 1 \qquad t \geq 0$$

The Exponential Input $e^{\zeta t}$

The exponential signal is the most important signal in the study of LTI systems. Interestingly, the particular solution for an exponential input signal turns out to be very simple. From Table 2.1 we see that the particular solution for the input $e^{\zeta t}$ has the form $\beta e^{\zeta t}$. We now show that $\beta = Q(\zeta)/P(\zeta)$.[3] To determine the constant β, we substitute $y_p(t) = \beta e^{\zeta t}$ in Eq. 2.11, which gives us

$$Q(D)\left[\beta e^{\zeta t}\right] = P(D)e^{\zeta t} \tag{2.14a}$$

Now observe that

$$
\begin{aligned}
De^{\zeta t} &= \frac{d}{dt}\left(e^{\zeta t}\right) = \zeta e^{\zeta t} \\
D^2 e^{\zeta t} &= \frac{d^2}{dt^2}\left(e^{\zeta t}\right) = \zeta^2 e^{\zeta t} \\
\cdots\cdots \quad \cdots \quad &\cdots\cdots \\
D^r e^{\zeta t} &= \zeta^r e^{\zeta t}
\end{aligned}
$$

Consequently,

$$Q(D)e^{\zeta t} = Q(\zeta)e^{\zeta t} \qquad \text{and} \qquad P(D)e^{\zeta t} = P(\zeta)e^{\zeta t}$$

Therefore, Eq. 2.14a becomes

$$\beta Q(\zeta)e^{\zeta t} = P(\zeta)e^{\zeta t} \tag{2.14b}$$

and

$$\beta = \frac{P(\zeta)}{Q(\zeta)}$$

Thus, for the input $f(t) = e^{\zeta t}$, the particular solution is given by

$$y_p(t) = H(\zeta)e^{\zeta t} \qquad t > 0 \tag{2.15a}$$

where

$$H(\zeta) = \frac{P(\zeta)}{Q(\zeta)} \tag{2.15b}$$

This is an interesting and significant result. It states that for an exponential input $e^{\zeta t}$ the particular solution $y_p(t)$ is the same exponential multiplied by $H(\zeta) = P(\zeta)/Q(\zeta)$. The total solution $y(t)$ to an exponential input $e^{\zeta t}$ is then given by

$$y(t) = \sum_{j=1}^{n} c_j e^{\lambda_j t} + H(\zeta)e^{\zeta t}$$

where the arbitrary constants $c_1, c_2, \ldots, c_n$ are determined from auxiliary conditions.

---

[3] This is true only if ζ is not a characteristic root.

Recall that the exponential signal includes a large variety of signals, such as a constant ($\zeta = 0$), a sinusoid ($\zeta = \pm j\omega$), and an exponentially growing or decaying sinusoid ($\zeta = \sigma \pm j\omega$). Let us consider the forced response for some of these cases.

The Constant Input f(t) = C

Because $C = Ce^{0t}$, the constant input is a special case of the exponential input $Ce^{\zeta t}$ with $\zeta = 0$. The particular solution to this input is then given by

$$
\begin{aligned}
y_p(t) &= CH(\zeta)e^{\zeta t} \quad \text{with} \quad \zeta = 0 \\
&= CH(0)
\end{aligned}
\tag{2.16}
$$

The Complex Exponential Input e$^{j\omega t}$

Here $\zeta = j\omega$, and

$$
y_p(t) = H(j\omega)e^{j\omega t}
\tag{2.17}
$$

The Sinusoidal Input f(t) = cos ω_0t

We know that the particular solution for the input $e^{\pm j\omega t}$ is $H(\pm j\omega)e^{\pm j\omega t}$. Since $\cos \omega t = (e^{j\omega t} + e^{-j\omega t})/2$, the particular solution to $\cos \omega t$ is

$$
y_p(t) = \frac{1}{2}\left[H(j\omega)e^{j\omega t} + H(-j\omega)e^{-j\omega t} \right]
$$

Because the two terms on the right-hand side are conjugates,

$$
y_p(t) = \mathrm{Re}\left[H(j\omega)e^{j\omega t} \right]
$$

But

$$
H(j\omega) = |H(j\omega)|e^{j\angle H(j\omega)}
$$

so that

$$
\begin{aligned}
y_p(t) &= \mathrm{Re}\left\{ |H(j\omega)|e^{j[\omega t + \angle H(j\omega)]} \right\} \\
&= |H(j\omega)| \cos\left[\omega t + \angle H(j\omega)\right]
\end{aligned}
\tag{2.18}
$$

This result can be generalized for the input $f(t) = \cos(\omega t + \theta)$. The particular solution in this case is

$$
y_p(t) = |H(j\omega)| \cos\left[\omega t + \theta + \angle H(j\omega)\right]
\tag{2.19}
$$

EXAMPLE 2.2:

Solve Eq. 2.12 for the following inputs:
(a) $10e^{-3t}$ (b) 5   (c) e^{-2t} (d) $10 \cos(3t + 30°)$.
The initial conditions are $y(0^+) = 2$, $\dot{y}(0^+) = 3$.
The complementary solution for this case is already found in Example 2.1 as

$$y_c(t) = c_1 e^{-t} + c_2 e^{-2t} \qquad t \geq 0$$

For the exponential input $f(t) = e^{\zeta t}$, the particular solution, as found in Eq. 2.15 is $H(\zeta)e^{\zeta t}$, where

$$H(\zeta) = \frac{P(\zeta)}{Q(\zeta)} = \frac{\zeta}{\zeta^2 + 3\zeta + 2}$$

(a) For input $f(t) = 10e^{-3t}$, $\zeta = -3$, and

$$
\begin{aligned}
y_p(t) &= 10H(-3)e^{-3t} \\
&= 10 \left[\frac{-3}{(-3)^2 + 3(-3) + 2} \right] e^{-3t} \\
&= -15e^{-3t} \qquad t > 0
\end{aligned}
$$

The total solution (the sum of the complementary and particular solutions) is

$$y(t) = c_1 e^{-t} + c_2 e^{-2t} - 15e^{-3t} \qquad t \geq 0$$

and

$$\dot{y}(t) = -c_1 e^{-t} - 2c_2 e^{-2t} + 45e^{-3t} \qquad t \geq 0$$

The initial conditions are $y(0^+) = 2$ and $\dot{y}(0^+) = 3$. Setting $t = 0$ in the above equations and substituting the initial conditions yields

$$c_1 + c_2 - 15 = 2 \qquad \text{and} \qquad -c_1 - 2c_2 + 45 = 3$$

Solution of these equations yields $c_1 = -8$ and $c_2 = 25$. Therefore,

$$y(t) = -8e^{-t} + 25e^{-2t} - 15e^{-3t} \qquad t \geq 0$$

(b) For input $f(t) = 5 = 5e^{0t}$, $\zeta = 0$, and

$$y_p(t) = 5H(0) = 0 \qquad t > 0$$

The complete solution is $y(t) = y_c(t) + y_p(t) = c_1 e^{-t} + c_2 e^{-2t}$. We then substitute the initial conditions to determine c_1 and c_2 as explained in Part a.

(c) Here $\zeta = -2$, which is also a characteristic root. Hence (see Pair 2, Table 2.1, or the comment at the bottom of the table),

$$y_p(t) = \beta t e^{-2t}$$

To find β, we substitute $y_p(t)$ in Eq. 2.11, giving us

$$\left(D^2 + 3D + 2 \right) y_p(t) = D f(t)$$

or

$$\left(D^2 + 3D + 2 \right) \left[\beta t e^{-2t} \right] = D e^{-2t}$$

But

$$D\left[\beta t e^{-2t}\right] = \beta(1-2t)e^{-2t}$$

$$D^2\left[\beta t e^{-2t}\right] = 4\beta(t-1)e^{-2t}$$

$$De^{-2t} = -2e^{-2t}$$

Consequently,

$$\beta(4t-4+3-6t+2t)e^{-2t} = -2e^{-2t}$$

or

$$-\beta e^{-2t} = -2e^{-2t}$$

This means that $\beta = 2$, so that

$$y_p(t) = 2te^{-2t}$$

The complete solution is $y(t) = y_c(t) + y_p(t) = c_1 e^{-t} + c_2 e^{-2t} + 2te^{-2t}$. We then substitute the initial conditions to determine c_1 and c_2 as explained in Part a.

(d) For the input $f(t) = 10\cos(3t + 30°)$, the particular solution (see Eq. 2.19) is

$$y_p(t) = 10|H(j3)|\cos\left[3t + 30° + \angle H(j3)\right]$$

where

$$H(j3) = \frac{P(j3)}{Q(j3)} = \frac{j3}{(j3)^2 + 3(j3) + 2}$$

$$= \frac{j3}{-7 + j9} = \frac{27 - j21}{130} = 0.263e^{-j37.9°}$$

Therefore,

$$|H(j3)| = 0.263, \qquad \angle H(j3) = -37.9°$$

and

$$y_p(t) = 10(0.263)\cos(3t + 30° - 37.9°)$$
$$= 2.63\cos(3t - 7.9°)$$

The complete solution is $y(t) = y_c(t) + y_p(t) = c_1 e^{-t} + c_2 e^{-2t} + 2.63\cos(3t - 7.9°)$. We then substitute the initial conditions to determine c_1 and c_2 as explained in Part a.

2.1.2   Method of Convolution

In this method, the input $f(t)$ is expressed as a sum of impulses. The solution is then obtained as a sum of the solutions to all the impulse components. The method exploits the superposition property of the linear differential equations. From the sampling (or sifting) property of the impulse function, we have

$$f(t) = \int_0^t f(x)\delta(t-x)\,dx \qquad t \geq 0 \tag{2.20}$$

The right-hand side expresses $f(t)$ as a sum (integral) of impulse components. Let the solution of Eq. 2.4 be $y(t) = h(t)$ when $f(t) = \delta(t)$ and all the initial conditions are zero. Then use of the linearity property yields the solution of Eq. 2.4 to input $f(t)$ as

$$y(t) = \int_0^t f(x)h(t-x)\,dx \tag{2.21}$$

For this solution to be general, we must add a complementary solution. Thus, the general solution is given by

$$y(t) = \sum_{j=1}^{n} c_j e^{\lambda_j t} + \int_0^t f(x)h(t-x)\,dx \tag{2.22}$$

The first term on the right-hand side consists of a linear combination of natural modes and should be appropriately modified for repeated roots. For the integral on the right-hand side, the lower limit 0 is understood to be 0^- in order to ensure that impulses, if any, in the input $f(t)$ at the origin are accounted for. The integral on the right-hand side of (2.22) is well known in the literature as the *convolution integral*. The function $h(t)$ appearing in the integral is the solution of Eq. 2.4 for the impulsive input $[f(t) = \delta(t)]$. It can be shown that [3]

$$h(t) = P(D)[y_o(t)u(t)] \tag{2.23}$$

where $y_o(t)$ is a linear combination of the characteristic modes subject to initial conditions

$$\begin{aligned} y_o^{(n-1)}(0) &= 1 \\ y_o(0) = y_o^{(1)}(0) &= \cdots = y_o^{(n-2)}(0) = 0 \end{aligned} \tag{2.24}$$

The function $u(t)$ appearing on the right-hand side of Eq. 2.23 represents the unit step function, which is unity for $t \geq 0$ and is 0 for $t < 0$.

The right-hand side of Eq. 2.23 is a linear combination of the derivatives of $y_o(t)u(t)$. Evaluating these derivatives is clumsy and inconvenient because of the presence of $u(t)$. The derivatives will generate an impulse and its derivatives at the origin [recall that $\frac{d}{dt}u(t) = \delta(t)$]. Fortunately when $m \leq n$ in Eq. 2.4, the solution simplifies to

$$h(t) = b_n\delta(t) + [P(D)y_o(t)]u(t) \tag{2.25}$$

EXAMPLE 2.3:

Solve Example 2.2, Part a using the method of convolution.

We first determine $h(t)$. The characteristic modes for this case, as found in Example 2.1, are e^{-t} and e^{-2t}. Since $y_o(t)$ is a linear combination of the characteristic modes

$$y_o(t) = K_1 e^{-t} + K_2 e^{-2t} \qquad t \geq 0$$

Therefore,

$$\dot{y}_o(t) = -K_1 e^{-t} - 2K_2 e^{-2t} \qquad t \geq 0$$

The initial conditions according to Eq. 2.24 are $\dot{y}_o(0) = 1$ and $y_o(0) = 0$. Setting $t = 0$ in the above equations and using the initial conditions, we obtain

$$K_1 + K_2 = 0 \qquad \text{and} \qquad -K_1 - 2K_2 = 1$$

Solution of these equations yields $K_1 = 1$ and $K_2 = -1$. Therefore,

$$y_o(t) = e^{-t} - e^{-2t}$$

Also in this case the polynomial $P(D) = D$ is of the first-order, and $b_2 = 0$. Therefore, from Eq. 2.25

$$\begin{aligned} h(t) &= [P(D)y_o(t)]u(t) = [Dy_o(t)]u(t) \\ &= \left[\frac{d}{dt}(e^{-t} - e^{-2t})\right]u(t) \\ &= (-e^{-t} + 2e^{-2t})u(t) \end{aligned}$$

and

$$\int_0^t f(x)h(t-x)\,dx = \int_0^t 10e^{-3x}[-e^{-(t-x)}$$
$$+ 2e^{-2(t-x)}]\,dx$$
$$= -5e^{-t} + 20e^{-2t} - 15e^{-3t}$$

The total solution is obtained by adding the complementary solution $y_c(t) = c_1 e^{-t} + c_2 e^{-2t}$ to this component. Therefore,

$$y(t) = c_1 e^{-t} + c_2 e^{-2t} - 5e^{-t} + 20e^{-2t} - 15e^{-3t}$$

Setting the conditions $y(0^+) = 2$ and $y(0^+) = 3$ in this equation (and its derivative), we obtain $c_1 = -3$, $c_2 = 5$ so that

$$y(t) = -8e^{-t} + 25e^{-2t} - 15e^{-3t} \qquad t \geq 0$$

which is identical to the solution found by the classical method.

Assessment of the Convolution Method

The convolution method is more laborious compared to the classical method. However, in system analysis, its advantages outweigh the extra work. The classical method has a serious drawback because it yields the total response, which cannot be separated into components arising from the internal conditions and the external input. In the study of systems it is important to be able to express the system response to an input $f(t)$ as an explicit function of $f(t)$. This is not possible in the classical method. Moreover, the classical method is restricted to a certain class of inputs; it cannot be applied to any input.[4]

If we must solve a particular linear differential equation or find a response of a particular LTI system, the classical method may be the best. In the theoretical study of linear systems, however, it is practically useless. General discussion of differential equations can be found in numerous texts on the subject [1].

2.2   Difference Equations

The development of difference equations is parallel to that of differential equations. We consider here only linear difference equations with constant coefficients. An nth-order difference equation can be expressed in two different forms; the first form uses delay terms such as $y[k-1]$, $y[k-2]$, $f[k-1]$, $f[k-2]$, ..., etc., and the alternative form uses advance terms such as $y[k+1]$, $y[k+2]$, ..., etc. Both forms are useful. We start here with a general nth-order difference equation, using advance operator form

$$y[k+n] + a_{n-1}y[k+n-1] + \cdots + a_1 y[k+1] + a_0 y[k]$$
$$= b_m f[k+m] + b_{m-1}f[k+m-1] + \cdots$$
$$+ b_1 f[k+1] + b_0 f[k] \qquad (2.26)$$

---

[4]Another minor problem is that because the classical method yields total response, the auxiliary conditions must be on the total response, which exists only for $t \geq 0^+$. In practice we are most likely to know the conditions at $t = 0^-$ (before the input is applied). Therefore, we need to derive a new set of auxiliary conditions at $t = 0^+$ from the known conditions at $t = 0^-$. The convolution method can handle both kinds of initial conditions. If the conditions are given at $t = 0^-$, we apply these conditions only to $y_c(t)$ because by its definition the convolution integral is 0 at $t = 0^-$.

Causality Condition

The left-hand side of Eq. 2.26 consists of values of $y[k]$ at instants $k+n$, $k+n-1$, $k+n-2$, and so on. The right-hand side of Eq. 2.26 consists of the input at instants $k+m$, $k+m-1$, $k+m-2$, and so on. For a causal equation, the solution cannot depend on future input values. This shows that when the equation is in the advance operator form of Eq. 2.26, causality requires $m \leq n$. For a general causal case, $m = n$, and Eq. 2.26 becomes

$$y[k+n] + a_{n-1}y[k+n-1] + \cdots + a_1 y[k+1] + a_0 y[k]$$
$$= b_n f[k+n] + b_{n-1}f[k+n-1] + \cdots$$
$$+ b_1 f[k+1] + b_0 f[k] \qquad (2.27a)$$

where some of the coefficients on both sides can be zero. However, the coefficient of $y[k+n]$ is normalized to unity. Eq. 2.27a is valid for all values of k. Therefore, the equation is still valid if we replace k by $k-n$ throughout the equation. This yields the alternative form (the delay operator form) of Eq. 2.27a

$$y[k] + a_{n-1}y[k-1] + \cdots + a_1 y[k-n+1] + a_0 y[k-n]$$
$$= b_n f[k] + b_{n-1}f[k-1] + \cdots$$
$$+ b_1 f[k-n+1] + b_0 f[k-n] \qquad (2.27b)$$

We designate the form of Eq. 2.27a the *advance operator form*, and the form of Eq. 2.27b the *delay operator form*.

2.2.1   Initial Conditions and Iterative Solution

Equation 2.27b can be expressed as

$$y[k] = -a_{n-1}y[k-1] - a_{n-2}y[k-2] - \cdots$$
$$- a_0 y[k-n] + b_n f[k] + b_{n-1}f[k-1] + \cdots$$
$$+ b_0 f[k-n] \qquad (2.27c)$$

This equation shows that $y[k]$, the solution at the kth instant, is computed from $2n + 1$ pieces of information. These are the past n values of $y[k]$: $y[k-1]$, $y[k-2]$, ..., $y[k-n]$ and the present and past n values of the input: $f[k]$, $f[k-1]$, $f[k-2]$, ..., $f[k-n]$. If the input $f[k]$ is known for $k = 0, 1, 2, \ldots$, then the values of $y[k]$ for $k = 0, 1, 2, \ldots$ can be computed from the $2n$ initial conditions $y[-1]$, $y[-2]$, ..., $y[-n]$ and $f[-1]$, $f[-2]$, ..., $f[-n]$. If the input is causal, that is, if $f[k] = 0$ for $k < 0$, then $f[-1] = f[-2] = \ldots = f[-n] = 0$, and we need only n initial conditions $y[-1]$, $y[-2]$, ..., $y[-n]$. This allows us to compute iteratively or recursively the values $y[0]$, $y[1]$, $y[2]$, $y[3]$, ..., and so on.[5] For instance, to find $y[0]$ we set $k = 0$ in Eq. 2.27c. The left-hand side is $y[0]$, and the right-hand side contains terms $y[-1]$, $y[-2]$, ..., $y[-n]$, and the inputs $f[0]$, $f[-1]$, $f[-2]$, ..., $f[-n]$.

---

[5]For this reason Eq. 2.27 is called a *recursive difference equation*. However, in Eq. 2.27 if $a_0 = a_1 = a_2 = \cdots = a_{n-1} = 0$, then it follows from Eq. 2.27c that determination of the present value of $y[k]$ does not require the past values $y[k-1]$, $y[k-2]$, ..., etc. For this reason when $a_i = 0$, $(i = 0, 1, \ldots, n-1)$, the difference Eq. 2.27 is *nonrecursive*. This classification is important in designing and realizing digital filters. In this discussion, however, this classification is not important. The analysis techniques developed here apply to general recursive and nonrecursive equations. Observe that a nonrecursive equation is a special case of recursive equation with $a_0 = a_1 = \ldots = a_{n-1} = 0$.

Therefore, to begin with, we must know the n initial conditions $y[-1]$, $y[-2]$, ..., $y[-n]$. Knowing these conditions and the input $f[k]$, we can iteratively find the response $y[0]$, $y[1]$, $y[2]$, ..., and so on. The following example demonstrates this procedure. This method basically reflects the manner in which a computer would solve a difference equation, given the input and initial conditions.

EXAMPLE 2.4:

Solve iteratively

$$y[k] - 0.5y[k-1] = f[k] \tag{2.28a}$$

with initial condition $y[-1] = 16$ and the input $f[k] = k^2$ (starting at $k = 0$). This equation can be expressed as

$$y[k] = 0.5y[k-1] + f[k] \tag{2.28b}$$

If we set $k = 0$ in this equation, we obtain

$$\begin{aligned} y[0] &= 0.5y[-1] + f[0] \\ &= 0.5(16) + 0 = 8 \end{aligned}$$

Now, setting $k = 1$ in Eq. 2.28b and using the value $y[0] = 8$ (computed in the first step) and $f[1] = (1)^2 = 1$, we obtain

$$y[1] = 0.5(8) + (1)^2 = 5$$

Next, setting $k = 2$ in Eq. 2.28b and using the value $y[1] = 5$ (computed in the previous step) and $f[2] = (2)^2$, we obtain

$$y[2] = 0.5(5) + (2)^2 = 6.5$$

Continuing in this way iteratively, we obtain

$$y[3] = 0.5(6.5) + (3)^2 = 12.25$$
$$y[4] = 0.5(12.25) + (4)^2 = 22.125$$

. . . .

This iterative solution procedure is available only for difference equations; it cannot be applied to differential equations. Despite the many uses of this method, a closed-form solution of a difference equation is far more useful in the study of system behavior and its dependence on the input and the various system parameters. For this reason we shall develop a systematic procedure to obtain a closed-form solution of Eq. 2.27.

Operational Notation

In difference equations it is convenient to use operational notation similar to that used in differential equations for the sake of compactness and convenience. For differential equations, we use the operator D to denote the operation of differentiation. For difference equations, we use the operator E to denote the operation for advancing the sequence by one time interval. Thus,

$$Ef[k] \equiv f[k+1]$$
$$E^2 f[k] \equiv f[k+2]$$
$$\cdots\cdots \quad \cdots \quad \cdots\cdots$$
$$E^n f[k] \equiv f[k+n] \tag{2.29}$$

A general nth-order difference Eq. 2.27a can be expressed as

$$(E^n + a_{n-1}E^{n-1} + \cdots + a_1 E + a_0)y[k]$$
$$= (b_n E^n + b_{n-1}E^{n-1} + \cdots + b_1 E + b_0)f[k] \tag{2.30a}$$

or

$$Q[E]y[k] = P[E]f[k] \tag{2.30b}$$

where $Q[E]$ and $P[E]$ are nth-order polynomial operators, respectively,

$$Q[E] = E^n + a_{n-1}E^{n-1} + \cdots + a_1 E + a_0 \tag{2.31a}$$
$$P[E] = b_n E^n + b_{n-1}E^{n-1} + \cdots + b_1 E + b_0 \tag{2.31b}$$

2.2.2 Classical Solution

Following the discussion of differential equations, we can show that if $y_p[k]$ is a solution of Eq. 2.27 or Eq. 2.30, that is,

$$Q[E]y_p[k] = P[E]f[k] \tag{2.32}$$

then $y_p[k] + y_c[k]$ is also a solution of Eq. 2.30, where $y_c[k]$ is a solution of the homogeneous equation

$$Q[E]y_c[k] = 0 \tag{2.33}$$

As before, we call $y_p[k]$ the particular solution and $y_c[k]$ the complementary solution.

Complementary Solution (The Natural Response)

By definition

$$Q[E]y_c[k] = 0 \tag{2.33a}$$

or

$$(E^n + a_{n-1}E^{n-1} + \cdots + a_1 E + a_0)y_c[k] = 0 \tag{2.33b}$$

or

$$y_c[k+n] + a_{n-1}y_c[k+n-1] + \cdots + a_1 y_c[k+1]$$
$$+ a_0 y_c[k] = 0 \tag{2.33c}$$

We can solve this equation systematically, but even a cursory examination of this equation points to its solution. This equation states that a linear combination of $y_c[k]$ and delayed $y_c[k]$ is zero *not for some values of k, but for all k*. This is possible *if and only if* $y_c[k]$ and delayed $y_c[k]$ have the same form. Only an exponential function γ^k has this property as seen from the equation

$$\gamma^{k-m} = \gamma^{-m}\gamma^k$$

This shows that the delayed γ^k is a constant times γ^k. Therefore, the solution of Eq. 2.33 must be of the form

$$y_c[k] = c\gamma^k \qquad (2.34)$$

To determine c and γ, we substitute this solution in Eq. 2.33. From Eq. 2.34, we have

$$
\begin{aligned}
Ey_c[k] &= y_c[k+1] = c\gamma^{k+1} = (c\gamma)\gamma^k \\
E^2 y_c[k] &= y_c[k+2] = c\gamma^{k+2} = (c\gamma^2)\gamma^k \\
\cdots \quad \cdots & \quad \dots\dots\dots\dots\dots \\
E^n y_c[k] &= y_c[k+n] = c\gamma^{k+n} = (c\gamma^n)\gamma^k
\end{aligned}
\qquad (2.35)
$$

Substitution of this in Eq. 2.33 yields

$$c(\gamma^n + a_{n-1}\gamma^{n-1} + \cdots + a_1\gamma + a_0)\gamma^k = 0 \qquad (2.36)$$

For a nontrivial solution of this equation

$$(\gamma^n + a_{n-1}\gamma^{n-1} + \cdots + a_1\gamma + a_0) = 0 \qquad (2.37a)$$

or

$$Q[\gamma] = 0 \qquad (2.37b)$$

Our solution $c\gamma^k$ [Eq. 2.34] is correct, provided that γ satisfies Eq. 2.37. Now, $Q[\gamma]$ is an nth-order polynomial and can be expressed in the factorized form (assuming all distinct roots):

$$(\gamma - \gamma_1)(\gamma - \gamma_2)\cdots(\gamma - \gamma_n) = 0 \qquad (2.37c)$$

Clearly γ has n solutions $\gamma_1, \gamma_2, \cdots, \gamma_n$ and, therefore, Eq. 2.33 also has n solutions $c_1\gamma_1^k, c_2\gamma_2^k, \cdots, c_n\gamma_n^k$. In such a case we have shown that the general solution is a linear combination of the n solutions. Thus,

$$y_c[k] = c_1\gamma_1^k + c_2\gamma_2^k + \cdots + c_n\gamma_n^k \qquad (2.38)$$

where $\gamma_1, \gamma_2, \cdots, \gamma_n$ are the roots of Eq. 2.37 and $c_1, c_2, \ldots, c_n$ are arbitrary constants determined from n auxiliary conditions. The polynomial $Q[\gamma]$ is called the *characteristic polynomial*, and

$$Q[\gamma] = 0 \qquad (2.39)$$

is the *characteristic equation*. Moreover, $\gamma_1, \gamma_2, \cdots, \gamma_n$, the roots of the characteristic equation, are called *characteristic roots* or *characteristic values* (also *eigenvalues*). The exponentials $\gamma_i^k (i = 1, 2, \ldots, n)$ are the *characteristic* modes or *natural* modes. A characteristic mode corresponds to each characteristic root, and the complementary solution is a linear combination of the characteristic modes of the system.

Repeated Roots

For repeated roots, the form of characteristic modes is modified. It can be shown by direct substitution that if a root γ repeats r times (root of multiplicity r), the characteristic modes corresponding to this root are $\gamma^k, k\gamma^k, k^2\gamma^k, \ldots, k^{r-1}\gamma^k$. Thus, if the characteristic equation is

$$Q[\gamma] = (\gamma - \gamma_1)^r (\gamma - \gamma_{r+1})(\gamma - \gamma_{r+2}) \cdots$$
$$(\gamma - \gamma_n) \tag{2.40}$$

the complementary solution is

$$
\begin{aligned}
y_c[k] &= (c_1 + c_2 k + c_3 k^2 + \cdots + c_r k^{r-1})\gamma_1^k \\
&\quad + c_{r+1}\gamma_{r+1}^k + c_{r+2}\gamma_{r+2}^k + \cdots \\
&\quad + c_n \gamma_n^k
\end{aligned}
\tag{2.41}
$$

Particular Solution

The particular solution $y_p[k]$ is the solution of

$$Q[E]y_p[k] = P[E]f[k] \tag{2.42}$$

We shall find the particular solution using the method of undetermined coefficients, the same method used for differential equations. Table 2.2 lists the inputs and the corresponding forms of solution with undetermined coefficients. These coefficients can be determined by substituting $y_p[k]$ in Eq. 2.42 and equating the coefficients of similar terms.

TABLE 2.2

Input $f[k]$	Forced Response $y_p[k]$
1. $\quad r^k \quad r \neq \gamma_i \ (i = 1, 2, \cdots, n)$	βr^k
2. $\quad r^k \quad r = \gamma_i$	$\beta k r^k$
3. $\quad \cos(\Omega k + \theta)$	$\beta \cos(\Omega k + \phi)$
4. $\quad \left(\displaystyle\sum_{i=0}^{m} \alpha_i k^i \right) r^k$	$\left(\displaystyle\sum_{i=0}^{m} \beta_i k^i \right) r^k$

Note: By definition, $y_p[k]$ cannot have any characteristic mode terms. If any term $p[k]$ shown in the right-hand column for the particular solution should also be a characteristic mode, the correct form of the particular solution must be modified to $k^i p[k]$, where i is the smallest integer that will prevent $k^i p[k]$ from having a characteristic mode term. For example, when the input is r^k, the particular solution in the right-hand column is of the form cr^k. But if r^k happens to be a natural mode, the correct form of the particular solution is $\beta k r^k$ (see Pair 2).

EXAMPLE 2.5:

Solve

$$(E^2 - 5E + 6)y[k] = (E - 5)f[k] \tag{2.43}$$

if the input $f[k] = (3k + 5)u[k]$ and the auxiliary conditions are $y[0] = 4$, $y[1] = 13$.

The characteristic equation is

$$\gamma^2 - 5\gamma + 6 = (\gamma - 2)(\gamma - 3) = 0$$

Therefore, the complementary solution is

$$y_c[k] = c_1(2)^k + c_2(3)^k$$

To find the form of $y_p[k]$ we use Table 2.2, Pair 4 with $r = 1$, $m = 1$. This yields

$$y_p[k] = \beta_1 k + \beta_0$$

Therefore,

$$
\begin{aligned}
y_p[k+1] &= \beta_1(k+1) + \beta_0 = \beta_1 k + \beta_1 + \beta_0 \\
y_p[k+2] &= \beta_1(k+2) + \beta_0 = \beta_1 k + 2\beta_1 + \beta_0
\end{aligned}
$$

Also,

$$f[k] = 3k + 5$$

and

$$f[k+1] = 3(k+1) + 5 = 3k + 8$$

Substitution of the above results in Eq. 2.43 yields

$$\beta_1 k + 2\beta_1 + \beta_0 - 5(\beta_1 k + \beta_1 + \beta_0) + 6(\beta_1 k + \beta_0)$$
$$= 3k + 8 - 5(3k + 5)$$

or

$$2\beta_1 k - 3\beta_1 + 2\beta_0 = -12k - 17$$

Comparison of similar terms on two sides yields

$$
\left.
\begin{aligned}
2\beta_1 &= -12 \\
-3\beta_1 + 2\beta_0 &= -17
\end{aligned}
\right\}
\implies
\begin{aligned}
\beta_1 &= -6 \\
\beta_2 &= -\tfrac{35}{2}
\end{aligned}
$$

This means

$$y_p[k] = -6k - \frac{35}{2}$$

The total response is

$$
\begin{aligned}
y[k] &= y_c[k] + y_p[k] \\
&= c_1(2)^k + c_2(3)^k - 6k - \frac{35}{2} \quad k \geq 0
\end{aligned}
\tag{2.44}
$$

To determine arbitrary constants c_1 and c_2 we set $k = 0$ and 1 and substitute the auxiliary conditions $y[0] = 4$, $y[1] = 13$ to obtain

$$
\left.
\begin{aligned}
4 &= c_1 + c_2 - \tfrac{35}{2} \\
13 &= 2c_1 + 3c_2 - \tfrac{47}{2}
\end{aligned}
\right\}
\implies
\begin{aligned}
c_1 &= 28 \\
c_2 &= \tfrac{-13}{2}
\end{aligned}
$$

Therefore,

$$y_c[k] = 28(2)^k - \tfrac{13}{2}(3)^k \tag{2.45}$$

and

$$y[k] = \underbrace{28(2)^k - \frac{13}{2}(3)^k}_{y_c[k]} \underbrace{- 6k - \frac{35}{2}}_{y_p[k]} \tag{2.46}$$

A Comment on Auxiliary Conditions

This method requires auxiliary conditions $y[0], y[1], \ldots, y[n-1]$ because the total solution is valid only for $k \geq 0$. But if we are given the initial conditions $y[-1], y[-2], \ldots, y[-n]$, we can derive the conditions $y[0], y[1], \ldots, y[n-1]$ using the iterative procedure discussed earlier.

Exponential Input

As in the case of differential equations, we can show that for the equation

$$Q[E]y[k] = P[E]f[k] \tag{2.47}$$

the particular solution for the exponential input $f[k] = r^k$ is given by

$$y_p[k] = H[r]r^k \qquad r \neq \gamma_i \tag{2.48}$$

where

$$H[r] = \frac{P[r]}{Q[r]} \tag{2.49}$$

The proof follows from the fact that if the input $f[k] = r^k$, then from Table 2.2 (Pair 4), $y_p[k] = \beta r^k$. Therefore,

$$E^i f[k] = f[k+i] = r^{k+i} = r^i r^k \quad \text{and} \quad P[E]f[k] = P[r]r^k$$

$$E^j y_p[k] = \beta r^{k+j} = \beta r^j r^k \quad \text{and} \quad Q[E]y[k] = \beta Q[r]r^k$$

so that Eq. 2.47 reduces to

$$\beta Q[r]r^k = P[r]r^k$$

which yields $\beta = P[r]/Q[r] = H[r]$.

This result is valid only if r is not a characteristic root. If r is a characteristic root, the particular solution is $\beta k r^k$ where β is determined by substituting $y_p[k]$ in Eq. 2.47 and equating coefficients of similar terms on the two sides. Observe that the exponential r^k includes a wide variety of signals such as a constant C, a sinusoid $\cos(\Omega k + \theta)$, and an exponentially growing or decaying sinusoid $|\gamma|^k \cos(\Omega k + \theta)$.

A Constant Input $f(k) = C$

This is a special case of exponential Cr^k with $r = 1$. Therefore, from Eq. 2.48 we have

$$y_p[k] = C\frac{P[1]}{Q[1]}(1)^k = CH[1] \tag{2.50}$$

A Sinusoidal Input

The input $e^{j\Omega k}$ is an exponential r^k with $r = e^{j\Omega}$. Hence,

$$y_p[k] = H[e^{j\Omega}]e^{j\Omega k} = \frac{P[e^{j\Omega}]}{Q[e^{j\Omega}]}e^{j\Omega k}$$

Similarly for the input $e^{-j\Omega k}$

$$y_p[k] = H[e^{-j\Omega}]e^{-j\Omega k}$$

Consequently, if the input

$$
\begin{aligned}
f[k] &= \cos \Omega k = \frac{1}{2}(e^{j\Omega k} + e^{-j\Omega k}) \\
y_p[k] &= \frac{1}{2}\left\{ H[e^{j\Omega}]e^{j\Omega k} + H[e^{-j\Omega}]e^{-j\Omega k} \right\}
\end{aligned}
$$

Since the two terms on the right-hand side are conjugates

$$y_p[k] = \text{Re}\left\{H[e^{j\Omega}]e^{j\Omega k}\right\}$$

If

$$H[e^{j\Omega}] = |H[e^{j\Omega}]|e^{j\angle H[e^{j\Omega}]}$$

then

$$
\begin{aligned}
y_p[k] &= \text{Re}\left\{\left|H[e^{j\Omega}]\right|e^{j(\Omega k+\angle H[e^{j\Omega}])}\right\} \\
&= |H[e^{j\Omega}]|\cos\left(\Omega k + \angle H[e^{j\Omega}]\right)
\end{aligned}
\tag{2.51}
$$

Using a similar argument, we can show that for the input

$$
\begin{aligned}
f[k] &= \cos\left(\Omega k + \theta\right) \\
y_p[k] &= |H[e^{j\Omega}]|\cos\left(\Omega k + \theta + \angle H[e^{j\Omega}]\right)
\end{aligned}
\tag{2.52}
$$

EXAMPLE 2.6:

Solve

$$(E^2 - 3E + 2)y[k] = (E + 2)f[k]$$

for $f[k] = (3)^k u[k]$ and the auxiliary conditions $y[0] = 2$, $y[1] = 1$.

In this case

$$H[r] = \frac{P[r]}{Q[r]} = \frac{r+2}{r^2 - 3r + 2}$$

and the particular solution to input $(3)^k u[k]$ is $H[3](3)^k$; that is,

$$y_p[k] = \frac{3+2}{(3)^2 - 3(3) + 2}(3)^k = \frac{5}{2}(3)^k$$

The characteristic polynomial is $(\gamma^2 - 3\gamma + 2) = (\gamma - 1)(\gamma - 2)$. The characteristic roots are 1 and 2. Hence, the complementary solution is $y_c[k] = c_1 + c_2(2)^k$ and the total solution is

$$y[k] = c_1(1)^k + c_2(2)^k + \frac{5}{2}(3)^k$$

Setting $k = 0$ and 1 in this equation and substituting auxiliary conditions yields

$$2 = c_1 + c_2 + \frac{5}{2} \quad \text{and} \quad 1 = c_1 + 2c_2 + \frac{15}{2}$$

Solution of these two simultaneous equations yields $c_1 = 5.5$, $c_2 = -5$. Therefore,

$$y[k] = 5.5 - 6(2)^k + \frac{5}{2}(3)^k \qquad k \geq 0$$

2.2.3 Method of Convolution

In this method, the input $f[k]$ is expressed as a sum of impulses. The solution is then obtained as a sum of the solutions to all the impulse components. The method exploits the superposition property of the linear difference equations. A discrete-time unit impulse function $\delta[k]$ is defined as

$$\delta[k] = \begin{cases} 1 & k = 0 \\ 0 & k \neq 0 \end{cases} \tag{2.53}$$

Hence, an arbitrary signal $f[k]$ can be expressed in terms of impulse and delayed impulse functions as

$$f[k] = f[0]\delta[k] + f[1]\delta[k-1] + f[2]\delta[k-2] + \cdots$$
$$+ f[k]\delta[0] + \cdots \qquad k \geq 0 \tag{2.54}$$

The right-hand side expresses $f[k]$ as a sum of impulse components. If $h[k]$ is the solution of Eq. 2.30 to the impulse input $f[k] = \delta[k]$, then the solution to input $\delta[k-m]$ is $h[k-m]$. This follows from the fact that because of constant coefficients, Eq. 2.30 has time invariance property. Also, because Eq. 2.30 is linear, its solution is the sum of the solutions to each of the impulse components of $f[k]$ on the right-hand side of Eq. 2.54. Therefore,

$$y[k] = f[0]h[k] + f[1]h[k-1] + f[2]h[k-2] + \cdots$$
$$+ f[k]h[0] + f[k+1]h[-1] + \cdots$$

All practical systems with time as the independent variable are causal, that is $h[k] = 0$ for $k < 0$. Hence, all the terms on the right-hand side beyond $f[k]h[0]$ are zero. Thus,

$$\begin{aligned} y[k] &= f[0]h[k] + f[1]h[k-1] + f[2]h[k-2] + \cdots \\ &\quad + f[k]h[0] \\ &= \sum_{m=0}^{k} f[m]h[k-m] \end{aligned} \tag{2.55}$$

The first term on the right-hand side consists of a linear combination of natural modes and should be appropriately modified for repeated roots. The general solution is obtained by adding a complementary solution to the above solution. Therefore, the general solution is given by

$$y[k] = \sum_{j=1}^{n} c_j \gamma_j^k + \sum_{m=0}^{k} f[m]h[k-m] \tag{2.56}$$

The last sum on the right-hand side is known as the *convolution sum* of $f[k]$ and $h[k]$.

The function $h[k]$ appearing in Eq. 2.56 is the solution of Eq. 2.30 for the impulsive input ($f[k] = \delta[k]$) when all initial conditions are zero, that is, $h[-1] = h[-2] = \cdots = h[-n] = 0$. It can be shown that [3] $h[k]$ contains an impulse and a linear combination of characteristic modes as

$$h[k] = \frac{b_0}{a_0}\delta[k] + A_1\gamma_1^k + A_2\gamma_2^k + \cdots + A_n\gamma_n^k \tag{2.57}$$

where the unknown constants A_i are determined from n values of $h[k]$ obtained by solving the equation $Q[E]h[k] = P[E]\delta[k]$ iteratively.

EXAMPLE 2.7:

Solve Example 2.5 using convolution method. In other words solve

$$(E^2 - 3E + 2)y[k] = (E+2)f[k]$$

for $f[k] = (3)^k u[k]$ and the auxiliary conditions $y[0] = 2$, $y[1] = 1$.

The unit impulse solution $h[k]$ is given by Eq. 2.57. In this case $a_0 = 2$ and $b_0 = 2$. Therefore,

$$h[k] = \delta[k] + A_1(1)^k + A_2(2)^k \tag{2.58}$$

To determine the two unknown constants A_1 and A_2 in Eq. 2.58, we need two values of $h[k]$, for instance $h[0]$ and $h[1]$. These can be determined iteratively by observing that $h[k]$ is the solution of $(E^2 - 3E + 2)h[k] = (E + 2)\delta[k]$, that is,

$$h[k+2] - 3h[k+1] + 2h[k] = \delta[k+1] + 2\delta[k] \tag{2.59}$$

subject to initial conditions $h[-1] = h[-2] = 0$. We now determine $h[0]$ and $h[1]$ iteratively from Eq. 2.59. Setting $k = -2$ in this equation yields

$$h[0] - 3(0) + 2(0) = 0 + 0 \implies h[0] = 0$$

Next, setting $k = -1$ in Eq. 2.59 and using $h[0] = 0$, we obtain

$$h[1] - 3(0) + 2(0) = 1 + 0 \implies h[1] = 1$$

Setting $k = 0$ and 1 in Eq. 2.58 and substituting $h[0] = 0$, $h[1] = 1$ yields

$$0 = 1 + A_1 + A_2 \qquad \text{and} \qquad 1 = A_1 + 2A_2$$

Solution of these two equations yields $A_1 = -3$ and $A_2 = 2$. Therefore,

$$h[k] = \delta[k] - 3 + 2(2)^k$$

and from Eq. 2.56

$$
\begin{aligned}
y[k] &= c_1 + c_2(2)^k + \sum_{m=0}^{k}(3)^m [\delta[k-m] - 3 + 2(2)^{k-m}] \\
&= c_1 + c_2(2)^k + 1.5 - 4(2)^k + 2.5(3)^k
\end{aligned}
$$

The sums in the above expression are found by using the geometric progression sum formula

$$\sum_{m=0}^{k} r^m = \frac{r^{k+1} - 1}{r - 1} \qquad r \neq 1$$

Setting $k = 0$ and 1 and substituting the given auxiliary conditions $y[0] = 2$, $y[1] = 1$, we obtain

$$2 = c_1 + c_2 + 1.5 - 4 + 2.5 \quad \text{and} \quad 1 = c_1 + 2c_2 + 1.5 - 8 + 7.5$$

Solution of these equations yields $c_1 = 4$ and $c_2 = -2$. Therefore,

$$y[k] = 5.5 - 6(2)^k + 2.5(3)^k$$

which confirms the result obtained by the classical method.

Assessment of the Classical Method

The earlier remarks concerning the classical method for solving differential equations also apply to difference equations. General discussion of difference equations can be found in texts on the subject [2].

References

[1] Birkhoff, G. and Rota, G.C., *Ordinary Differential Equations*, 3rd ed., John Wiley & Sons, New York, 1978.

[2] Goldberg, S., *Introduction to Difference Equations*, John Wiley & Sons, New York, 1958.

[3] Lathi, B.P., *Signal Processing and Linear Systems*, Berkeley-Cambridge Press, Carmichael, CA, 1998.

3

Finite Wordlength Effects

Bruce W. Bomar
University of Tennessee
Space Institute

3.1   Introduction

Practical digital filters must be implemented with finite precision numbers and arithmetic. As a result, both the filter coefficients and the filter input and output signals are in discrete form. This leads to four types of finite wordlength effects.

Discretization (quantization) of the filter coefficients has the effect of perturbing the location of the filter poles and zeroes. As a result, the actual filter response differs slightly from the ideal response. This *deterministic* frequency response error is referred to as **coefficient quantization error**.

The use of finite precision arithmetic makes it necessary to quantize filter calculations by rounding or truncation. **Roundoff noise** is that error in the filter output that results from rounding or truncating calculations within the filter. As the name implies, this error looks like low-level noise at the filter output.

Quantization of the filter calculations also renders the filter slightly nonlinear. For large signals this nonlinearity is negligible and roundoff noise is the major concern. However, for recursive filters with a zero or constant input, this nonlinearity can cause spurious oscillations called **limit cycles**.

With fixed-point arithmetic it is possible for filter calculations to overflow. The term **overflow oscillation**, sometimes also called **adder overflow limit cycle**, refers to a high-level oscillation that can exist in an otherwise stable filter due to the nonlinearity associated with the overflow of internal filter calculations.

In this chapter, we examine each of these finite wordlength effects. Both fixed-point and floating-point number representations are considered.

0-8493-8572-5/98/$0.00+$.50
© 1998 by CRC Press LLC

3.2   Number Representation

In digital signal processing, $(B + 1)$-bit fixed-point numbers are usually represented as two's-complement signed fractions in the format

$$b_0 \cdot b_{-1} b_{-2} \cdots b_{-B}$$

The number represented is then

$$X = -b_0 + b_{-1} 2^{-1} + b_{-2} 2^{-2} + \cdots + b_{-B} 2^{-B} \tag{3.1}$$

where b_0 is the sign bit and the number range is $-1 \leq X < 1$. The advantage of this representation is that the product of two numbers in the range from -1 to 1 is another number in the same range.

Floating-point numbers are represented as

$$X = (-1)^s m 2^c \tag{3.2}$$

where s is the *sign bit*, m is the **mantissa**, and c is the *characteristic* or *exponent*. To make the representation of a number unique, the mantissa is *normalized* so that $0.5 \leq m < 1$.

Although floating-point numbers are always represented in the form of (3.2), the way in which this representation is actually *stored* in a machine may differ. Since $m \geq 0.5$, it is not necessary to store the 2^{-1}-weight bit of m, which is always set. Therefore, in practice numbers are usually stored as

$$X = (-1)^s (0.5 + f) 2^c \tag{3.3}$$

where f is an unsigned fraction, $0 \leq f < 0.5$.

Most floating-point processors now use the IEEE Standard 754 32-bit floating-point format for storing numbers. According to this standard the exponent is stored as an unsigned integer p where

$$p = c + 126 \tag{3.4}$$

Therefore, a number is stored as

$$X = (-1)^s (0.5 + f) 2^{p-126} \tag{3.5}$$

where s is the sign bit, f is a 23-b unsigned fraction in the range $0 \leq f < 0.5$, and p is an 8-b unsigned integer in the range $0 \leq p \leq 255$. The total number of bits is $1 + 23 + 8 = 32$. For example, in IEEE format 3/4 is written $(-1)^0 (0.5 + 0.25) 2^0$ so $s = 0$, $p = 126$, and $f = 0.25$. The value $X = 0$ is a unique case and is represented by all bits zero (i.e., $s = 0$, $f = 0$, and $p = 0$). Although the 2^{-1}-weight mantissa bit is not actually stored, it does exist so the mantissa has 24 b plus a sign bit.

3.3   Fixed-Point Quantization Errors

In fixed-point arithmetic, a multiply doubles the number of significant bits. For example, the product of the two 5-b numbers 0.0011 and 0.1001 is the 10-b number 00.000 110 11. The extra bit to the left of the decimal point can be discarded without introducing any error. However, the least significant four of the remaining bits must ultimately be discarded by some form of quantization so that the result can be stored to 5 b for use in other calculations. In the example above this results in 0.0010 (quantization by rounding) or 0.0001 (quantization by truncating). When a sum of products calculation is performed, the quantization can be performed either after each multiply or after all products have been summed with double-length precision.

We will examine three types of fixed-point quantization—rounding, truncation, and magnitude truncation. If X is an exact value, then the rounded value will be denoted $Q_r(X)$, the truncated value $Q_t(X)$, and the magnitude truncated value $Q_{mt}(X)$. If the quantized value has B bits to the right of the decimal point, the quantization step size is

$$\Delta = 2^{-B} \tag{3.6}$$

Since rounding selects the quantized value nearest the unquantized value, it gives a value which is never more than $\pm\Delta/2$ away from the exact value. If we denote the rounding error by

$$\epsilon_r = Q_r(X) - X \tag{3.7}$$

then

$$-\frac{\Delta}{2} \leq \epsilon_r \leq \frac{\Delta}{2} \tag{3.8}$$

Truncation simply discards the low-order bits, giving a quantized value that is always less than or equal to the exact value so

$$-\Delta < \epsilon_t \leq 0 \tag{3.9}$$

Magnitude truncation chooses the nearest quantized value that has a magnitude less than or equal to the exact value so

$$-\Delta < \epsilon_{mt} < \Delta \tag{3.10}$$

The error resulting from quantization can be modeled as a random variable uniformly distributed over the appropriate error range. Therefore, calculations with roundoff error can be considered error-free calculations that have been corrupted by additive white noise. The mean of this noise for rounding is

$$m_{\epsilon_r} = E\{\epsilon_r\} = \frac{1}{\Delta} \int_{-\Delta/2}^{\Delta/2} \epsilon_r \, d\epsilon_r = 0 \tag{3.11}$$

where $E\{\}$ represents the operation of taking the expected value of a random variable. Similarly, the variance of the noise for rounding is

$$\sigma_{\epsilon_r}^2 = E\{(\epsilon_r - m_{\epsilon_r})^2\} = \frac{1}{\Delta} \int_{-\Delta/2}^{\Delta/2} (\epsilon_r - m_{\epsilon_r})^2 \, d\epsilon_r = \frac{\Delta^2}{12} \tag{3.12}$$

Likewise, for truncation,

$$
\begin{aligned}
m_{\epsilon_t} &= E\{\epsilon_t\} = -\frac{\Delta}{2} \\
\sigma_{\epsilon_t}^2 &= E\{(\epsilon_t - m_{\epsilon_t})^2\} = \frac{\Delta^2}{12}
\end{aligned}
\tag{3.13}
$$

and, for magnitude truncation

$$
\begin{aligned}
m_{\epsilon_{mt}} &= E\{\epsilon_{mt}\} = 0 \\
\sigma_{\epsilon_{mt}}^2 &= E\{(\epsilon_{mt} - m_{\epsilon_{mt}})^2\} = \frac{\Delta^2}{3}
\end{aligned}
\tag{3.14}
$$

3.4 Floating-Point Quantization Errors

With floating-point arithmetic it is necessary to quantize after both multiplications and additions. The addition quantization arises because, prior to addition, the mantissa of the smaller number in the sum is shifted right until the exponent of both numbers is the same. In general, this gives a sum mantissa that is too long and so must be quantized.

We will assume that quantization in floating-point arithmetic is performed by rounding. Because of the exponent in floating-point arithmetic, it is the relative error that is important. The relative error is defined as

$$\varepsilon_r = \frac{Q_r(X) - X}{X} = \frac{\epsilon_r}{X} \tag{3.15}$$

Since $X = (-1)^s m 2^c$, $Q_r(X) = (-1)^s Q_r(m) 2^c$ and

$$\varepsilon_r = \frac{Q_r(m) - m}{m} = \frac{\epsilon}{m} \tag{3.16}$$

If the quantized mantissa has B bits to the right of the decimal point, $|\epsilon| < \Delta/2$ where, as before, $\Delta = 2^{-B}$. Therefore, since $0.5 \le m < 1$,

$$|\varepsilon_r| < \Delta \tag{3.17}$$

If we assume that ϵ is uniformly distributed over the range from $-\Delta/2$ to $\Delta/2$ and m is uniformly distributed over 0.5 to 1,

$$
\begin{aligned}
m_{\varepsilon_r} &= E\left\{\frac{\epsilon}{m}\right\} = 0 \\
\sigma_{\varepsilon_r}^2 &= E\left\{\left(\frac{\epsilon}{m}\right)^2\right\} = \frac{2}{\Delta} \int_{1/2}^{1} \int_{-\Delta/2}^{\Delta/2} \frac{\epsilon^2}{m^2}\, d\epsilon\, dm \\
&= \frac{\Delta^2}{6} = (0.167) 2^{-2B} \tag{3.18}
\end{aligned}
$$

In practice, the distribution of m is not exactly uniform. Actual measurements of roundoff noise in [1] suggested that

$$\sigma_{\varepsilon_r}^2 \approx 0.23\Delta^2 \tag{3.19}$$

while a detailed theoretical and experimental analysis in [2] determined

$$\sigma_{\varepsilon_r}^2 \approx 0.18\Delta^2 \tag{3.20}$$

From (3.15) we can represent a quantized floating-point value in terms of the unquantized value and the random variable ε_r using

$$Q_r(X) = X(1 + \varepsilon_r) \tag{3.21}$$

Therefore, the finite-precision product $X_1 X_2$ and the sum $X_1 + X_2$ can be written

$$fl(X_1 X_2) = X_1 X_2 (1 + \varepsilon_r) \tag{3.22}$$

and

$$fl(X_1 + X_2) = (X_1 + X_2)(1 + \varepsilon_r) \tag{3.23}$$

where ε_r is zero-mean with the variance of (3.20).

3.5   Roundoff Noise

To determine the roundoff noise at the output of a digital filter we will assume that the noise due to a quantization is stationary, white, and uncorrelated with the filter input, output, and internal variables. This assumption is good if the filter input changes from sample to sample in a sufficiently complex manner. It is not valid for zero or constant inputs for which the effects of rounding are analyzed from a limit cycle perspective.

To satisfy the assumption of a sufficiently complex input, roundoff noise in digital filters is often calculated for the case of a zero-mean white noise filter input signal $x(n)$ of variance σ_x^2. This simplifies calculation of the output roundoff noise because expected values of the form $E\{x(n)x(n-k)\}$ are zero for $k \neq 0$ and give σ_x^2 when $k = 0$. This approach to analysis has been found to give estimates of the output roundoff noise that are close to the noise actually observed for other input signals.

Another assumption that will be made in calculating roundoff noise is that the product of two quantization errors is zero. To justify this assumption, consider the case of a 16-b fixed-point processor. In this case a quantization error is of the order 2^{-15}, while the product of two quantization errors is of the order 2^{-30}, which is negligible by comparison.

If a linear system with impulse response $g(n)$ is excited by white noise with mean m_x and variance σ_x^2, the output is noise of mean [3, pp.788–790]

$$m_y = m_x \sum_{n=-\infty}^{\infty} g(n) \tag{3.24}$$

and variance

$$\sigma_y^2 = \sigma_x^2 \sum_{n=-\infty}^{\infty} g^2(n) \tag{3.25}$$

Therefore, if $g(n)$ is the impulse response from the point where a roundoff takes place to the filter output, the contribution of that roundoff to the variance (mean-square value) of the output roundoff noise is given by (3.25) with σ_x^2 replaced with the variance of the roundoff. If there is more than one source of roundoff error in the filter, it is assumed that the errors are uncorrelated so the output noise variance is simply the sum of the contributions from each source.

3.5.1 Roundoff Noise in FIR Filters

The simplest case to analyze is a finite impulse response (FIR) filter realized via the convolution summation

$$y(n) = \sum_{k=0}^{N-1} h(k)x(n-k) \tag{3.26}$$

When fixed-point arithmetic is used and quantization is performed after each multiply, the result of the N multiplies is N-times the quantization noise of a single multiply. For example, rounding after each multiply gives, from (3.6) and (3.12), an output noise variance of

$$\sigma_o^2 = N \frac{2^{-2B}}{12} \tag{3.27}$$

Virtually all digital signal processor integrated circuits contain one or more double-length accumulator registers which permit the sum-of-products in (3.26) to be accumulated without quantization. In this case only a single quantization is necessary following the summation and

$$\sigma_o^2 = \frac{2^{-2B}}{12} \tag{3.28}$$

For the floating-point roundoff noise case we will consider (3.26) for $N = 4$ and then generalize the result to other values of N. The finite-precision output can be written as the exact output plus an error term $e(n)$. Thus,

$$
\begin{aligned}
y(n) + e(n) = \ & (\{[h(0)x(n)[1+\varepsilon_1(n)] \\
& + h(1)x(n-1)[1+\varepsilon_2(n)]][1+\varepsilon_3(n)] \\
& + h(2)x(n-2)[1+\varepsilon_4(n)]\}\{1+\varepsilon_5(n)\} \\
& + h(3)x(n-3)[1+\varepsilon_6(n)])[1+\varepsilon_7(n)]
\end{aligned} \tag{3.29}
$$

In (3.29), $\varepsilon_1(n)$ represents the error in the first product, $\varepsilon_2(n)$ the error in the second product, $\varepsilon_3(n)$ the error in the first addition, etc. Notice that it has been assumed that the products are summed in the order implied by the summation of (3.26).

Expanding (3.29), ignoring products of error terms, and recognizing $y(n)$ gives

$$
\begin{aligned}
e(n) \;=\; & h(0)x(n)[\varepsilon_1(n) + \varepsilon_3(n) + \varepsilon_5(n) + \varepsilon_7(n)] \\
& + h(1)x(n-1)[\varepsilon_2(n) + \varepsilon_3(n) + \varepsilon_5(n) + \varepsilon_7(n)] \\
& + h(2)x(n-2)[\varepsilon_4(n) + \varepsilon_5(n) + \varepsilon_7(n)] \\
& + h(3)x(n-3)[\varepsilon_6(n) + \varepsilon_7(n)]
\end{aligned}
\tag{3.30}
$$

Assuming that the input is white noise of variance σ_x^2 so that $E\{x(n)x(n-k)\}$ is zero for $k \neq 0$, and assuming that the errors are uncorrelated,

$$
E\{e^2(n)\} = [4h^2(0) + 4h^2(1) + 3h^2(2) + 2h^2(3)]\sigma_x^2\sigma_{\varepsilon_r}^2
\tag{3.31}
$$

In general, for any N,

$$
\sigma_o^2 = E\{e^2(n)\} = \left[Nh^2(0) + \sum_{k=1}^{N-1}(N+1-k)h^2(k) \right]\sigma_x^2\sigma_{\varepsilon_r}^2
\tag{3.32}
$$

Notice that if the order of summation of the product terms in the convolution summation is changed, then the order in which the $h(k)$'s appear in (3.32) changes. If the order is changed so that the $h(k)$ with smallest magnitude is first, followed by the next smallest, etc., then the roundoff noise variance is minimized. However, performing the convolution summation in nonsequential order greatly complicates data indexing and so may not be worth the reduction obtained in roundoff noise.

3.5.2   Roundoff Noise in Fixed-Point IIR Filters

To determine the roundoff noise of a fixed-point infinite impulse response (IIR) filter realization, consider a causal first-order filter with impulse response

$$
h(n) = a^n u(n)
\tag{3.33}
$$

realized by the difference equation

$$
y(n) = ay(n-1) + x(n)
\tag{3.34}
$$

Due to roundoff error, the output actually obtained is

$$
\hat{y}(n) = Q\{ay(n-1) + x(n)\} = ay(n-1) + x(n) + e(n)
\tag{3.35}
$$

where $e(n)$ is a random roundoff noise sequence. Since $e(n)$ is injected at the same point as the input, it propagates through a system with impulse response $h(n)$. Therefore, for fixed-point arithmetic with rounding, the output roundoff noise variance from (3.6), (3.12), (3.25), and (3.33) is

$$
\sigma_o^2 = \frac{\Delta^2}{12} \sum_{n=-\infty}^{\infty} h^2(n) = \frac{\Delta^2}{12} \sum_{n=0}^{\infty} a^{2n} = \frac{2^{-2B}}{12}\frac{1}{1-a^2}
\tag{3.36}
$$

With fixed-point arithmetic there is the possibility of overflow following addition. To avoid overflow it is necessary to restrict the input signal amplitude. This can be accomplished by either placing a *scaling* multiplier at the filter input or by simply limiting the maximum input signal amplitude. Consider the case of the first-order filter of (3.34). The transfer function of this filter is

$$H(e^{j\omega}) = \frac{Y(e^{j\omega})}{X(e^{j\omega})} = \frac{1}{e^{j\omega} - a} \tag{3.37}$$

so

$$|H(e^{j\omega})|^2 = \frac{1}{1 + a^2 - 2a\cos(\omega)} \tag{3.38}$$

and

$$|H(e^{j\omega})|_{\max} = \frac{1}{1 - |a|} \tag{3.39}$$

The peak gain of the filter is $1/(1 - |a|)$ so limiting input signal amplitudes to $|x(n)| \leq 1 - |a|$ will make overflows unlikely.

An expression for the output roundoff noise-to-signal ratio can easily be obtained for the case where the filter input is white noise, uniformly distributed over the interval from $-(1 - |a|)$ to $(1 - |a|)$ [4, 5]. In this case

$$\sigma_x^2 = \frac{1}{2(1 - |a|)} \int_{-(1-|a|)}^{1-|a|} x^2 \, dx = \frac{1}{3}(1 - |a|)^2 \tag{3.40}$$

so, from (3.25),

$$\sigma_y^2 = \frac{1}{3} \frac{(1 - |a|)^2}{1 - a^2} \tag{3.41}$$

Combining (3.36) and (3.41) then gives

$$\frac{\sigma_o^2}{\sigma_y^2} = \left(\frac{2^{-2B}}{12} \frac{1}{1 - a^2} \right) \left(3 \frac{1 - a^2}{(1 - |a|)^2} \right) = \frac{2^{-2B}}{12} \frac{3}{(1 - |a|)^2} \tag{3.42}$$

Notice that the noise-to-signal ratio increases without bound as $|a| \to 1$.

Similar results can be obtained for the case of the causal second-order filter realized by the difference equation

$$y(n) = 2r\cos(\theta)y(n - 1) - r^2 y(n - 2) + x(n) \tag{3.43}$$

This filter has complex-conjugate poles at $re^{\pm j\theta}$ and impulse response

$$h(n) = \frac{1}{\sin(\theta)} r^n \sin[(n + 1)\theta] u(n) \tag{3.44}$$

Due to roundoff error, the output actually obtained is

$$\hat{y}(n) = 2r\cos(\theta)y(n - 1) - r^2 y(n - 2) + x(n) + e(n) \tag{3.45}$$

There are two noise sources contributing to $e(n)$ if quantization is performed after each multiply, and there is one noise source if quantization is performed after summation. Since

$$\sum_{n=-\infty}^{\infty} h^2(n) = \frac{1 + r^2}{1 - r^2} \frac{1}{(1 + r^2)^2 - 4r^2 \cos^2(\theta)} \tag{3.46}$$

the output roundoff noise is

$$\sigma_o^2 = \nu \frac{2^{-2B}}{12} \frac{1 + r^2}{1 - r^2} \frac{1}{(1 + r^2)^2 - 4r^2 \cos^2(\theta)} \tag{3.47}$$

where $\nu = 1$ for quantization after summation, and $\nu = 2$ for quantization after each multiply.

To obtain an output noise-to-signal ratio we note that

$$H(e^{j\omega}) = \frac{1}{1 - 2r\cos(\theta)e^{-j\omega} + r^2 e^{-j2\omega}} \tag{3.48}$$

and, using the approach of [6],

$$|H(e^{j\omega})|^2_{max} = \frac{1}{4r^2 \left\{ \left[\text{sat}\left(\frac{1+r^2}{2r}\cos(\theta) \right) - \frac{1+r^2}{2r}\cos(\theta) \right]^2 + \left[\frac{1-r^2}{2r}\sin(\theta) \right]^2 \right\}} \tag{3.49}$$

where

$$\text{sat}(\mu) = \begin{cases} 1 & \mu > 1 \\ \mu & -1 \le \mu \le 1 \\ -1 & \mu < -1 \end{cases} \tag{3.50}$$

Following the same approach as for the first-order case then gives

$$\frac{\sigma_o^2}{\sigma_y^2} = \nu \frac{2^{-2B}}{12} \frac{1+r^2}{1-r^2} \frac{3}{(1+r^2)^2 - 4r^2\cos^2(\theta)}$$

$$\times \frac{1}{4r^2 \left\{ \left[\text{sat}\left(\frac{1+r^2}{2r}\cos(\theta) \right) - \frac{1+r^2}{2r}\cos(\theta) \right]^2 + \left[\frac{1-r^2}{2r}\sin(\theta) \right]^2 \right\}} \tag{3.51}$$

Figure 3.1 is a contour plot showing the noise-to-signal ratio of (3.51) for $\nu = 1$ in units of the noise variance of a single quantization, $2^{-2B}/12$. The plot is symmetrical about $\theta = 90°$, so only the range from $0°$ to $90°$ is shown. Notice that as $r \to 1$, the roundoff noise increases without bound. Also notice that the noise increases as $\theta \to 0°$.

It is possible to design state-space filter realizations that minimize fixed-point roundoff noise [7] – [10]. Depending on the transfer function being realized, these structures may provide a roundoff noise level

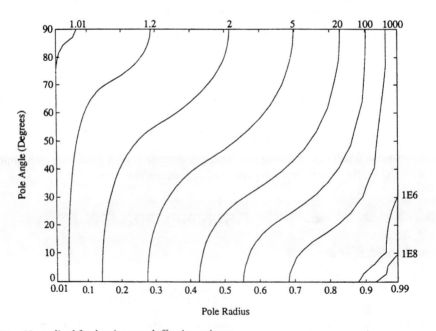

FIGURE 3.1   Normalized fixed-point roundoff noise variance.

that is orders-of-magnitude lower than for a nonoptimal realization. The price paid for this reduction in roundoff noise is an increase in the number of computations required to implement the filter. For an Nth-order filter the increase is from roughly $2N$ multiplies for a direct form realization to roughly $(N + 1)^2$ for an optimal realization. However, if the filter is realized by the parallel or cascade connection of first- and second-order optimal subfilters, the increase is only to about $4N$ multiplies. Furthermore, near-optimal realizations exist that increase the number of multiplies to only about $3N$ [10].

3.5.3 Roundoff Noise in Floating-Point IIR Filters

For floating-point arithmetic it is first necessary to determine the injected noise variance of each quantization. For the first-order filter this is done by writing the computed output as

$$y(n) + e(n) = [ay(n-1)(1 + \varepsilon_1(n)) + x(n)](1 + \varepsilon_2(n)) \tag{3.52}$$

where $\varepsilon_1(n)$ represents the error due to the multiplication and $\varepsilon_2(n)$ represents the error due to the addition. Neglecting the product of errors, (3.52) becomes

$$
\begin{aligned}
y(n) + e(n) \;\approx\; & ay(n-1) + x(n) + ay(n-1)\varepsilon_1(n) \\
& + ay(n-1)\varepsilon_2(n) + x(n)\varepsilon_2(n)
\end{aligned} \tag{3.53}
$$

Comparing (3.34) and (3.53), it is clear that

$$e(n) = ay(n-1)\varepsilon_1(n) + ay(n-1)\varepsilon_2(n) + x(n)\varepsilon_2(n) \tag{3.54}$$

Taking the expected value of $e^2(n)$ to obtain the injected noise variance then gives

$$
\begin{aligned}
E\{e^2(n)\} \;=\; & a^2 E\{y^2(n-1)\}E\{\varepsilon_1^2(n)\} + a^2 E\{y^2(n-1)\}E\{\varepsilon_2^2(n)\} \\
& + E\{x^2(n)\}E\{\varepsilon_2^2(n)\} + E\{x(n)y(n-1)\}E\{\varepsilon_2^2(n)\}
\end{aligned} \tag{3.55}
$$

To carry this further it is necessary to know something about the input. If we assume the input is zero-mean white noise with variance σ_x^2, then $E\{x^2(n)\} = \sigma_x^2$ and the input is uncorrelated with past values of the output so $E\{x(n)y(n-1)\} = 0$ giving

$$E\{e^2(n)\} = 2a^2\sigma_y^2\sigma_{\varepsilon_r}^2 + \sigma_x^2\sigma_{\varepsilon_r}^2 \tag{3.56}$$

and

$$
\begin{aligned}
\sigma_o^2 \;=\; & \left(2a^2\sigma_y^2\sigma_{\varepsilon_r}^2 + \sigma_x^2\sigma_{\varepsilon_r}^2\right) \sum_{n=-\infty}^{\infty} h^2(n) \\
\;=\; & \frac{2a^2\sigma_y^2 + \sigma_x^2}{1 - a^2}\sigma_{\varepsilon_r}^2
\end{aligned} \tag{3.57}
$$

However,

$$\sigma_y^2 = \sigma_x^2 \sum_{n=-\infty}^{\infty} h^2(n) = \frac{\sigma_x^2}{1 - a^2} \tag{3.58}$$

so

$$\sigma_o^2 = \frac{1 + a^2}{(1 - a^2)^2}\sigma_{\varepsilon_r}^2\sigma_x^2 = \frac{1 + a^2}{1 - a^2}\sigma_{\varepsilon_r}^2\sigma_y^2 \tag{3.59}$$

and the output roundoff noise-to-signal ratio is

$$\frac{\sigma_o^2}{\sigma_y^2} = \frac{1 + a^2}{1 - a^2}\sigma_{\varepsilon_r}^2 \tag{3.60}$$

Similar results can be obtained for the second-order filter of (3.43) by writing

$$y(n) + e(n) = ([2r\cos(\theta)y(n-1)(1+\varepsilon_1(n)) - r^2 y(n-2)(1+\varepsilon_2(n))]$$
$$\times [1 + \varepsilon_3(n)] + x(n))(1 + \varepsilon_4(n)) \tag{3.61}$$

Expanding with the same assumptions as before gives

$$e(n) \approx 2r\cos(\theta)y(n-1)[\varepsilon_1(n) + \varepsilon_3(n) + \varepsilon_4(n)]$$
$$- r^2 y(n-2)[\varepsilon_2(n) + \varepsilon_3(n) + \varepsilon_4(n)] + x(n)\varepsilon_4(n) \tag{3.62}$$

and

$$E\{e^2(n)\} = 4r^2 \cos^2(\theta)\sigma_y^2 3\sigma_{\varepsilon_r}^2 + r^2 \sigma_y^2 3\sigma_{\varepsilon_r}^2$$
$$+ \sigma_x^2 \sigma_{\varepsilon_r}^2 - 8r^3 \cos(\theta)\sigma_{\varepsilon_r}^2 E\{y(n-1)y(n-2)\} \tag{3.63}$$

However,

$$E\{y(n-1)y(n-2)\}$$
$$= E\{[2r\cos(\theta)y(n-2) - r^2 y(n-3) + x(n-1)]y(n-2)\}$$
$$= 2r\cos(\theta)E\{y^2(n-2)\} - r^2 E\{y(n-2)y(n-3)\}$$
$$= 2r\cos(\theta)E\{y^2(n-2)\} - r^2 E\{y(n-1)y(n-2)\}$$
$$= \frac{2r\cos(\theta)}{1+r^2}\sigma_y^2 \tag{3.64}$$

so

$$E\{e^2(n)\} = \sigma_{\varepsilon_r}^2 \sigma_x^2 + \left[3r^4 + 12r^2 \cos^2(\theta) - \frac{16r^4 \cos^2(\theta)}{1+r^2}\right]\sigma_{\varepsilon_r}^2 \sigma_y^2 \tag{3.65}$$

and

$$\sigma_o^2 = E\{e^2(n)\} \sum_{n=-\infty}^{\infty} h^2(n)$$
$$= \xi\left[\sigma_{\varepsilon_r}^2 \sigma_x^2 + \left[3r^4 + 12r^2 \cos^2(\theta) - \frac{16r^4 \cos^2(\theta)}{1+r^2}\right]\sigma_{\varepsilon_r}^2 \sigma_y^2\right] \tag{3.66}$$

where from (3.46),

$$\xi = \sum_{n=-\infty}^{\infty} h^2(n) = \frac{1+r^2}{1-r^2} \frac{1}{(1+r^2)^2 - 4r^2 \cos^2(\theta)} \tag{3.67}$$

Since $\sigma_y^2 = \xi\sigma_x^2$, the output roundoff noise-to-signal ratio is then

$$\frac{\sigma_o^2}{\sigma_y^2} = \xi\left[1 + \xi\left[3r^4 + 12r^2 \cos^2(\theta) - \frac{16r^4 \cos^2(\theta)}{1+r^2}\right]\right]\sigma_{\varepsilon_r}^2 \tag{3.68}$$

Figure 3.2 is a contour plot showing the noise-to-signal ratio of (3.68) in units of the noise variance of a single quantization $\sigma_{\varepsilon_r}^2$. The plot is symmetrical about $\theta = 90°$, so only the range from $0°$ to $90°$ is shown. Notice the similarity of this plot to that of Fig. 3.1 for the fixed-point case. It has been observed that filter structures generally have very similar fixed-point and floating-point roundoff characteristics [2]. Therefore, the techniques of [7] – [10], which were developed for the fixed-point case, can also be used to design low-noise floating-point filter realizations. Furthermore, since it is not necessary to scale the floating-point realization, the low-noise realizations need not require significantly more computation than the direct form realization.

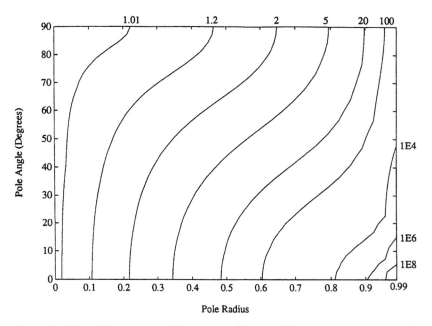

FIGURE 3.2   Normalized floating-point roundoff noise variance.

3.6   Limit Cycles

A limit cycle, sometimes referred to as a **multiplier roundoff limit cycle**, is a low-level oscillation that can exist in an otherwise stable filter as a result of the nonlinearity associated with rounding (or truncating) internal filter calculations [11]. Limit cycles require recursion to exist and do not occur in nonrecursive FIR filters.

As an example of a limit cycle, consider the second-order filter realized by

$$y(n) = Q_r \left\{ \frac{7}{8} y(n-1) - \frac{5}{8} y(n-2) + x(n) \right\} \tag{3.69}$$

where $Q_r\{\}$ represents quantization by rounding. This is stable filter with poles at $0.4375 \pm j0.6585$. Consider the implementation of this filter with 4-b (3-b and a sign bit) two's complement fixed-point arithmetic, zero initial conditions ($y(-1) = y(-2) = 0$), and an input sequence $x(n) = \frac{3}{8}\delta(n)$, where $\delta(n)$ is the unit impulse or unit sample. The following sequence is obtained;

$$
\begin{aligned}
y(0) &= Q_r \left\{ \frac{3}{8} \right\} = \frac{3}{8} \\[2mm]
y(1) &= Q_r \left\{ \frac{21}{64} \right\} = \frac{3}{8} \\[2mm]
y(2) &= Q_r \left\{ \frac{3}{32} \right\} = \frac{1}{8} \\[2mm]
y(3) &= Q_r \left\{ -\frac{1}{8} \right\} = -\frac{1}{8} \\[2mm]
y(4) &= Q_r \left\{ -\frac{3}{16} \right\} = -\frac{1}{8} \\[2mm]
y(5) &= Q_r \left\{ -\frac{1}{32} \right\} = 0
\end{aligned}
$$

$$y(6) = Q_r \left\{ \frac{5}{64} \right\} = \frac{1}{8} \qquad\qquad (3.70)$$

$$y(7) = Q_r \left\{ \frac{7}{64} \right\} = \frac{1}{8}$$

$$y(8) = Q_r \left\{ \frac{1}{32} \right\} = 0$$

$$y(9) = Q_r \left\{ -\frac{5}{64} \right\} = -\frac{1}{8}$$

$$y(10) = Q_r \left\{ -\frac{7}{64} \right\} = -\frac{1}{8}$$

$$y(11) = Q_r \left\{ -\frac{1}{32} \right\} = 0$$

$$y(12) = Q_r \left\{ \frac{5}{64} \right\} = \frac{1}{8}$$

$$\vdots$$

Notice that while the input is zero except for the first sample, the output oscillates with amplitude 1/8 and period 6.

Limit cycles are primarily of concern in fixed-point recursive filters. As long as floating-point filters are realized as the parallel or cascade connection of first- and second-order subfilters, limit cycles will generally not be a problem since limit cycles are practically not observable in first- and second-order systems implemented with 32-b floating-point arithmetic [12]. It has been shown that such systems must have an extremely small margin of stability for limit cycles to exist at anything other than underflow levels, which are at an amplitude of less than 10^{-38} [12].

There are at least three ways of dealing with limit cycles when fixed-point arithmetic is used. One is to determine a bound on the maximum limit cycle amplitude, expressed as an integral number of quantization steps [13]. It is then possible to choose a word length that makes the limit cycle amplitude acceptably low. Alternately, limit cycles can be prevented by randomly rounding calculations up or down [14]. However, this approach is complicated to implement. The third approach is to properly choose the filter realization structure and then quantize the filter calculations using magnitude truncation [15, 16]. This approach has the disadvantage of producing more roundoff noise than truncation or rounding [see (3.12)–(3.14)].

3.7 Overflow Oscillations

With fixed-point arithmetic it is possible for filter calculations to overflow. This happens when two numbers of the same sign add to give a value having magnitude greater than one. Since numbers with magnitude greater than one are not representable, the result overflows. For example, the two's complement numbers 0.101 (5/8) and 0.100 (4/8) add to give 1.001 which is the two's complement representation of $-7/8$.

The overflow characteristic of two's complement arithmetic can be represented as $R\{\}$ where

$$R\{X\} = \begin{cases} X - 2 & X \geq 1 \\ X & -1 \leq X < 1 \\ X + 2 & X < -1 \end{cases} \qquad\qquad (3.71)$$

For the example just considered, $R\{9/8\} = -7/8$.

An overflow oscillation, sometimes also referred to as an *adder overflow limit cycle*, is a high-level oscillation that can exist in an otherwise stable fixed-point filter due to the gross nonlinearity associated with the overflow of internal filter calculations [17]. Like limit cycles, overflow oscillations require recursion to exist and do not occur in nonrecursive FIR filters. Overflow oscillations also do not occur with floating-point arithmetic due to the virtual impossibility of overflow.

As an example of an overflow oscillation, once again consider the filter of (3.69) with 4-b fixed-point two's complement arithmetic and with the two's complement overflow characteristic of (3.71):

$$y(n) = Q_r \left\{ R \left[\frac{7}{8} y(n-1) - \frac{5}{8} y(n-2) + x(n) \right] \right\} \tag{3.72}$$

In this case we apply the input

$$\begin{aligned} x(n) &= -\frac{3}{4} \delta(n) - \frac{5}{8} \delta(n-1) \\ &= \left\{ -\frac{3}{4}, -\frac{5}{8}, 0, 0, \cdots \right\}, \end{aligned} \tag{3.73}$$

giving the output sequence

$$\begin{aligned} y(0) &= Q_r \left\{ R \left[-\frac{3}{4} \right] \right\} = Q_r \left\{ -\frac{3}{4} \right\} = -\frac{3}{4} \\ y(1) &= Q_r \left\{ R \left[-\frac{41}{32} \right] \right\} = Q_r \left\{ \frac{23}{32} \right\} = \frac{3}{4} \\ y(2) &= Q_r \left\{ R \left[\frac{9}{8} \right] \right\} = Q_r \left\{ -\frac{7}{8} \right\} = -\frac{7}{8} \\ y(3) &= Q_r \left\{ R \left[-\frac{79}{64} \right] \right\} = Q_r \left\{ \frac{49}{64} \right\} = \frac{3}{4} \\ y(4) &= Q_r \left\{ R \left[\frac{77}{64} \right] \right\} = Q_r \left\{ -\frac{51}{64} \right\} = -\frac{3}{4} \\ y(5) &= Q_r \left\{ R \left[-\frac{9}{8} \right] \right\} = Q_r \left\{ \frac{7}{8} \right\} = \frac{7}{8} \\ y(6) &= Q_r \left\{ R \left[\frac{79}{64} \right] \right\} = Q_r \left\{ -\frac{49}{64} \right\} = -\frac{3}{4} \\ y(7) &= Q_r \left\{ R \left[-\frac{77}{64} \right] \right\} = Q_r \left\{ \frac{51}{64} \right\} = \frac{3}{4} \\ y(8) &= Q_r \left\{ R \left[\frac{9}{8} \right] \right\} = Q_r \left\{ -\frac{7}{8} \right\} = -\frac{7}{8} \\ &\vdots \end{aligned} \tag{3.74}$$

This is a large-scale oscillation with nearly full-scale amplitude.

There are several ways to prevent overflow oscillations in fixed-point filter realizations. The most obvious is to scale the filter calculations so as to render overflow impossible. However, this may unacceptably restrict the filter dynamic range. Another method is to force completed sums-of-products to saturate at ± 1, rather than overflowing [18, 19]. It is important to saturate only the completed sum, since intermediate overflows in two's complement arithmetic do not affect the accuracy of the final result. Most fixed-point digital signal processors provide for automatic saturation of completed sums if their *saturation arithmetic* feature is enabled. Yet another way to avoid overflow oscillations is to use a filter structure for which any internal filter transient is guaranteed to decay to zero [20]. Such structures are desirable anyway, since they tend to have low roundoff noise and be insensitive to coefficient quantization [21].

3.8   Coefficient Quantization Error

Each filter structure has its own finite, generally nonuniform grids of realizable pole and zero locations when the filter coefficients are quantized to a finite word length. In general the pole and zero locations desired in filter do not correspond exactly to the realizable locations. The error in filter performance (usually measured in terms of a frequency response error) resulting from the placement of the poles and zeroes at the nonideal but realizable locations is referred to as coefficient quantization error.

Consider the second-order filter with complex-conjugate poles

$$
\begin{aligned}
\lambda &= r e^{\pm j\theta} \\
&= \lambda_r \pm j\lambda_i \\
&= r\cos(\theta) \pm jr\sin(\theta)
\end{aligned}
\tag{3.75}
$$

and transfer function

$$
H(z) = \frac{1}{1 - 2r\cos(\theta)z^{-1} + r^2 z^{-2}}
\tag{3.76}
$$

realized by the difference equation

$$
y(n) = 2r\cos(\theta)y(n-1) - r^2 y(n-2) + x(n)
\tag{3.77}
$$

Figure 3.3 from [5] shows that quantizing the difference equation coefficients results in a nonuniform grid of realizable pole locations in the z plane. The grid is defined by the intersection of vertical lines corresponding to quantization of $2\lambda_r$ and concentric circles corresponding to quantization of $-r^2$. The sparseness of realizable pole locations near $z = \pm 1$ will result in a large coefficient quantization error for poles in this region.

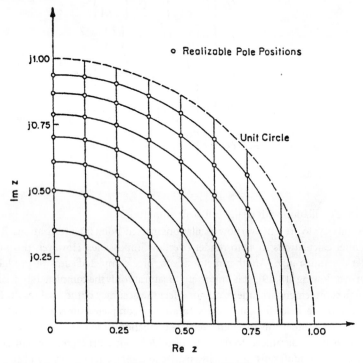

FIGURE 3.3   Realizable pole locations for the difference equation of (3.76).

Figure 3.4 gives an alternative structure to (3.77) for realizing the transfer function of (3.76). Notice that quantizing the coefficients of this structure corresponds to quantizing λ_r and λ_i. As shown in Fig. 3.5 from [5], this results in a uniform grid of realizable pole locations. Therefore, large coefficient quantization errors are avoided for all pole locations.

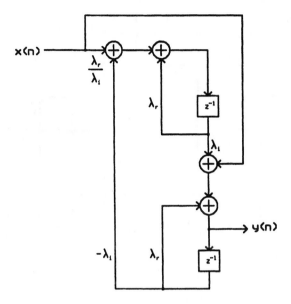

FIGURE 3.4   Alternate realization structure.

It is well established that filter structures with low roundoff noise tend to be robust to coefficient quantization, and visa versa [22]– [24]. For this reason, the uniform grid structure of Fig. 3.4 is also popular because of its low roundoff noise. Likewise, the low-noise realizations of [7]– [10] can be expected to be relatively insensitive to coefficient quantization, and digital wave filters and lattice filters that are derived from low-sensitivity analog structures tend to have not only low coefficient sensitivity, but also low roundoff noise [25, 26].

It is well known that in a high-order polynomial with clustered roots, the root location is a very sensitive function of the polynomial coefficients. Therefore, filter poles and zeros can be much more accurately controlled if higher order filters are realized by breaking them up into the parallel or cascade connection of first- and second-order subfilters. One exception to this rule is the case of linear-phase FIR filters in which the symmetry of the polynomial coefficients and the spacing of the filter zeros around the unit circle usually permits an acceptable direct realization using the convolution summation.

Given a filter structure it is necessary to assign the ideal pole and zero locations to the realizable locations. This is generally done by simply rounding or truncating the filter coefficients to the available number of bits, or by assigning the ideal pole and zero locations to the nearest realizable locations. A more complicated alternative is to consider the original filter design problem as a problem in discrete optimization, and choose the realizable pole and zero locations that give the best approximation to the desired filter response [27]– [30].

3.9   Realization Considerations

Linear-phase FIR digital filters can generally be implemented with acceptable coefficient quantization sensitivity using the direct convolution sum method. When implemented in this way on a digital signal

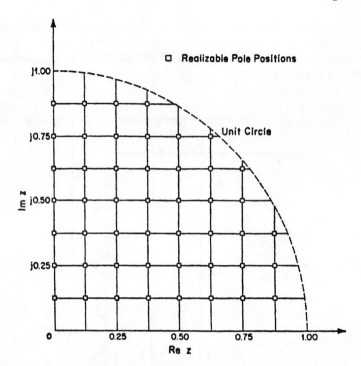

FIGURE 3.5   Realizable pole locations for the alternate realization structure.

processor, fixed-point arithmetic is not only acceptable but may actually be preferable to floating-point arithmetic. Virtually all fixed-point digital signal processors accumulate a sum of products in a double-length accumulator. This means that only a single quantization is necessary to compute an output. Floating-point arithmetic, on the other hand, requires a quantization after every multiply and after every add in the convolution summation. With 32-b floating-point arithmetic these quantizations introduce a small enough error to be insignificant for many applications.

When realizing IIR filters, either a parallel or cascade connection of first- and second-order subfilters is almost always preferable to a high-order direct-form realization. With the availability of very low-cost floating-point digital signal processors, like the Texas Instruments TMS320C32, it is highly recommended that floating-point arithmetic be used for IIR filters. Floating-point arithmetic simultaneously eliminates most concerns regarding scaling, limit cycles, and overflow oscillations. Regardless of the arithmetic employed, a low roundoff noise structure should be used for the second-order sections. Good choices are given in [2] and [10]. Recall that realizations with low fixed-point roundoff noise also have low floating-point roundoff noise. The use of a low roundoff noise structure for the second-order sections also tends to give a realization with low coefficient quantization sensitivity. First-order sections are not as critical in determining the roundoff noise and coefficient sensitivity of a realization, and so can generally be implemented with a simple direct form structure.

References

[1] Weinstein, C. and Oppenheim, A.V., A comparison of roundoff noise in floating-point and fixed-point digital filter realizations, *Proc. IEEE,* 57, 1181–1183, June 1969.
[2] Smith, L.M., Bomar, B.W., Joseph, R.D., and Yang, G.C., Floating-point roundoff noise analysis of second-order state-space digital filter structures, *IEEE Trans. Circuits Syst. II,* 39, 90–98, Feb. 1992.

[3] Proakis, G.J. and Manolakis, D.J., *Introduction to Digital Signal Processing*, New York, Macmillan, 1988.

[4] Oppenheim, A.V. and Schafer, R.W., *Digital Signal Processing*, Englewood Cliffs, NJ, Prentice-Hall, 1975.

[5] Oppenheim, A.V. and Weinstein, C.J., Effects of finite register length in digital filtering and the fast Fourier transform, *Proc. IEEE*, 60, 957–976, Aug. 1972.

[6] Bomar, B.W. and Joseph, R.D., Calculation of L_∞ norms for scaling second-order state-space digital filter sections, *IEEE Trans. Circuits Syst.*, CAS-34, 983–984, Aug. 1987.

[7] Mullis, C.T. and Roberts, R.A., Synthesis of minimum roundoff noise fixed-point digital filters, *IEEE Trans. Circuits Syst.*, CAS-23, 551–562, Sept. 1976.

[8] Jackson, L.B., Lindgren, A.G., and Kim, Y., Optimal synthesis of second-order state-space structures for digital filters, *IEEE Trans. Circuits Syst.*, CAS-26, 149–153, Mar. 1979.

[9] Barnes, C.W., On the design of optimal state-space realizations of second-order digital filters, *IEEE Trans. Circuits Syst.*, CAS-31, 602–608, July 1984.

[10] Bomar, B.W., New second-order state-space structures for realizing low roundoff noise digital filters, *IEEE Trans. Acoust., Speech, Signal Processing*, ASSP-33, 106–110, Feb. 1985.

[11] Parker, S.R. and Hess, S.F., Limit-cycle oscillations in digital filters, *IEEE Trans. Circuit Theory*, CT-18, 687–697, Nov. 1971.

[12] Bauer, P.H., Limit cycle bounds for floating-point implementations of second-order recursive digital filters, *IEEE Trans. Circuits Syst. II*, 40, 493–501, Aug. 1993.

[13] Green, B.D. and Turner, L.E., New limit cycle bounds for digital filters, *IEEE Trans. Circuits Syst.*, 35, 365–374, Apr. 1988.

[14] Buttner, M., A novel approach to eliminate limit cycles in digital filters with a minimum increase in the quantization noise, in *Proc. 1976 IEEE Int. Symp. Circuits Syst.*, Apr. 1976, pp. 291–294.

[15] Diniz, P.S.R. and Antoniou, A., More economical state-space digital filter structures which are free of constant-input limit cycles, *IEEE Trans. Acoust., Speech, Signal Processing*, ASSP-34, 807–815, Aug. 1986.

[16] Bomar, B.W., Low-roundoff-noise limit-cycle-free implementation of recursive transfer functions on a fixed-point digital signal processor, *IEEE Trans. Industr. Electron.*, 41, 70–78, Feb. 1994.

[17] Ebert, P.M., Mazo, J.E. and Taylor, M.G., Overflow oscillations in digital filters, *Bell Syst. Tech. J.*, 48. 2999–3020, Nov. 1969.

[18] Willson, A.N., Jr., Limit cycles due to adder overflow in digital filters, *IEEE Trans. Circuit Theory*, CT-19, 342–346, July 1972.

[19] Ritzerfield, J.H.F., A condition for the overflow stability of second-order digital filters that is satisfied by all scaled state-space structures using saturation, *IEEE Trans. Circuits Syst.*, 36, 1049–1057, Aug. 1989.

[20] Mills, W.T., Mullis, C.T., and Roberts, R.A., Digital filter realizations without overflow oscillations, *IEEE Trans. Acoust., Speech, Signal Processing*, ASSP-26, 334–338, Aug. 1978.

[21] Bomar, B.W., On the design of second-order state-space digital filter sections, *IEEE Trans. Circuits Syst.*, 36, 542–552, Apr. 1989.

[22] Jackson, L.B., Roundoff noise bounds derived from coefficient sensitivities for digital filters, *IEEE Trans. Circuits Syst.*, CAS-23, 481–485, Aug. 1976.

[23] Rao, D.B.V., Analysis of coefficient quantization errors in state-space digital filters, *IEEE Trans. Acoust., Speech, Signal Processing*, ASSP-34, 131–139, Feb. 1986.

[24] Thiele, L., On the sensitivity of linear state-space systems, *IEEE Trans. Circuits Syst.*, CAS-33, 502–510, May 1986.

[25] Antoniou, A., *Digital Filters: Analysis and Design*, New York, McGraw-Hill, 1979.

[26] Lim, Y.C., On the synthesis of IIR digital filters derived from single channel AR lattice network, *IEEE Trans. Acoust., Speech, Signal Processing*, ASSP-32, 741–749, Aug. 1984.

[27] Avenhaus, E., On the design of digital filters with coefficients of limited wordlength, *IEEE Trans. Audio Electroacoust.*, AU-20, 206–212, Aug. 1972.

[28] Suk, M. and Mitra, S.K., Computer-aided design of digital filters with finite wordlengths, *IEEE Trans. Audio Electroacoust.*, AU-20, 356–363, Dec. 1972.

[29] Charalambous, C. and Best, M.J., Optimization of recursive digital filters with finite wordlengths, *IEEE Trans. Acoust., Speech, Signal Processing*, ASSP-22, 424–431, Dec. 1979.

[30] Lim, Y.C., Design of discrete-coefficient-value linear-phase FIR filters with optimum normalized peak ripple magnitude, *IEEE Trans. Circuits Syst.*, 37, 1480–1486, Dec. 1990.

Signal Representation and Quantization

Jelena Kovačević
Bell Laboratories, Lucent Technologies

Christine Podilchuk
Bell Laboratories, Lucent Technologies

S AMPLING THEOREMS CAN BE TRACED to the original paper by Whittaker in 1915 on interpolation. He proved the exactness of a method for interpolating between the samples from a function. Nyquist then presented the sampling theory for sampled telephone signals in 1928 establishing for the first time the term Nyquist frequency. Shannon in 1948 and Kotel'nikov in 1933 wrote additional treatises on this topic [1]-[4].

Extensions from one-dimensional to multidimensional sampling can be traced to papers by Bracewell in 1956, and to Miyakawa in 1959. Multidimensional Fourier analysis, however, can be traced back to papers by Germain and Navier in the early 18th and 19th centuries [5]-[7].

In this section, the first chapter, "On Multidimensional Sampling" by Kalker presents a thorough discussion of the techniques that are currently used and their underlying theory.

Of related interest is structure of the conversion process from the analog domain to the digital domain, and the chapter by Kosonocky and Xiao presents a thorough survey of the various architectures for analog-to-digital conversion.

Finally, the process of quantization of discrete samples is discussed in the chapter by Ramachandran. This discussion considers the accuracy issues arising due to quantization, in addition to other related topics.

References

[1] Whittaker, E. T., *Proc. R. Soc. Edinburgh* 35: 181-194, 1915.

[2] Nyquist, H., Certain topics in telegraph transmission theory, *Trans. AIEE* 47: 617-644, 1928.

[3] Shannon, C. E., A mathematical theory of communication, *Bell System Technical Journal* 27:379-423, 1948.

[4] Sullivan, W. et al., *The Early Years of Radio Astronomy,* Cambridge University Press, Cambridge, England, 1984.

[5] Bracewell, R. N., Two-dimensional aerial smoothing in radio astronomy, *Aust. J. Phys.* 9:197-314, 1956.

[6] Miyakawa, K., Sampling theory of stationary stochastic variables in multidimensional space, *J. Inst. Elec. Commun.* (Japan), 421-427, 1959.

[7] Bracewell, R. N., *Two-Dimensional Imaging,* Prentice-Hall, Englewood Cliffs, NJ, 1995.

4

On Multidimensional Sampling

Ton Kalker
*Philips Research Laboratories,
Eindhoven*

This chapter gives an overview of the most relevant facts of sampling theory, paying particular attention to the multidimensional aspect of the problem. It is shown that sampling theory formulated in a multidimensional setting provides insight to the supposedly simpler situation of one-dimensional sampling.

4.1 Introduction

The signals we encounter in the physical reality around us almost invariably have a continuous domain of definition. We like to model a speech signal as continuous function of amplitudes, where the domain of definition is a (finite) length interval of real numbers. A video signal is most naturally viewed as continuous function of luminance (chrominance) values, where the domain of definition is some volume in space-time.

In modern electronic systems we deal with many (in essence) continuous signals in a digital fashion. This means that we do not deal with these signals directly, but only with *sampled* versions of it: we only

0-8493-8572-5/98/$0.00+$.50
© 1998 by CRC Press LLC

retain the values of these signals at a discrete set of points. Moreover, due to the inherently finite precision arithmetic capabilities of digital systems, we only record an approximated (quantized) value at every point of the sampling set. If we define sampling as the process of restricting a signal to a discrete set, explicitly without quantization of the sampled values, we can describe the contribution of this chapter as a study of the relation between continuous signals and their sampled versions.

Many textbooks start this topic by only considering sampling in the one-dimensional case. Digressions into the multidimensional case are usually made in later and more advanced sections. In this chapter we will start from the outset with the multidimensional case. It will be argued that this is the most natural setting, and that this approach will even lead to greater understanding of the one-dimensional case.

I will assume that not every reader is familiar with the concept of a *lattice*. As lattices are the most basic kind of sets onto which to sample signals, this chapter will start with a crash course on lattices in Section 4.2. After this the real work starts in Section 4.3 with an overview of the sampling theory for continuous functions. The central theme of this section is the intimate relationship between sampling and the discrete space-time Fourier transform (DSFT). In Section 4.4 we consider simultaneous sampling in both spatial and frequency domain. The central theme in this section is the relationship with the discrete fourier transform (DFT). We continue with a digression on cascaded sampling (Section 4.5), and with some useful results on changing variables (Section 4.6). We end with an application of sampling theory to HDTV-to-SDTV conversion. The proofs (or hints to it) of the stated result can be found in the Appendix.

We end this introduction with some conventions. We will refer to a signal as a function, defined on some appropriate domain. As all of our functions are in principle multidimensional, we will lighten the burden of notation by suppressing the multidimensional character of variables involved wherever possible. In particular we will use $f(x)$ to denote a function $f(x_1, \cdots, x_n)$ on some continuous domain (say $\mathbb{R}^n$). Similarly we will use $f(k)$ to denote a function $f(k_1, \cdots, k_n)$ on some discrete domain (say $\mathbb{Z}^n$). By abuse of terminology we will refer to a function defined on a continuous domain as a continuous function and to a function on discrete domain as discrete function.

4.2 Lattices

Although *sampling* of a function can in principle be done with respect to any set of points (*nonuniform sampling*), the most common form of sampling is done with respect to sets of points which have a certain algebraic structure and are known as *lattices*. They are the object of study in this section.

4.2.1 Definition

Formally, the definition of a lattice is given as

DEFINITION 4.1

A (sub)lattice $\mathcal{L}$ of $\mathbb{C}^n$ ($\mathbb{R}^n$, $\mathbb{Z}^n$) is a set of points satisfying that

1. There is a shortest nonzero element,
2. If $\lambda_1, \lambda_2 \in \mathcal{L}$, then $a\lambda_1 + b\lambda_2 \in \mathcal{L}$ for all integers a and b, and
3. $\mathcal{L}$ contains n linearly independent elements.

This definition may seem to make lattices rather abstract objects, but they can be made more tangible by representing them by *generating matrices*. Namely, one can show that every lattice $\mathcal{L}$ contains a set of linearly independent points $\{\lambda_1, \cdots, \lambda_n\}$ such that every other point $\lambda \in \mathcal{L}$ is an integer linear combination $\sum_{i=1}^{n} a_i \lambda_i$. Arranging such a set in a matrix $L = [\lambda_1, \cdots, \lambda_n]$ yields a generating matrix L of $\mathcal{L}$. It has the property that every $\lambda \in \mathcal{L}$ can be written as $\lambda = Lk$, where $k \in \mathbb{Z}^n$ is an integer vector. At this point it is important to note that there is no such thing as *the* generating matrix L of a lattice $\mathcal{L}$. Defining a unimodular matrix U as an integer matrix with $|\det(U)| = 1$, every other generating matrix is of the form LU, and every such matrix is a generating matrix. However, this also shows that the determinant of a generating matrix is determined up to a sign.

DEFINITION 4.2

Let $\mathcal{L}$ be a lattice and let L be a generating matrix of $\mathcal{L}$. Then the *determinant* of $\mathcal{L}$ is defined by

$$\det(\mathcal{L}) = |\det(L)|.$$

In case the dimension is 1 ($n = 1$), every lattice is given as all the integer multiples of a single scalar. This scalar is unique up to a sign, and by convention one usually defines the positive scalar as *the sampling period T* (for time).

$$\mathcal{L}_T = \{nT : n \in \mathbb{Z}\} \subset \mathbb{C}, \mathbb{R}, \mathbb{Z} \tag{4.1}$$

In case the dimension is 2 ($n = 2$) it is no longer possible to single out a natural candidate as *the* generating matrix for a lattice. As an example consider the lattice $\mathcal{L}$ generated by the matrix (see also Fig. 4.1)

$$L_1 = \begin{bmatrix} \sqrt{3} & \sqrt{3} \\ -1 & 1 \end{bmatrix}.$$

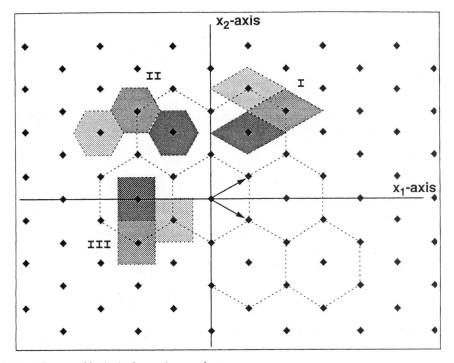

FIGURE 4.1 A hexagonal lattice in the continuous plane.

There is no reason to consider the matrix L_1 as *the* generating matrix of the lattice $\mathcal{L}$, and in fact the matrix

$$L_2 = \begin{bmatrix} \sqrt{3} & 2\sqrt{3} \\ 1 & 0 \end{bmatrix}$$

is just as valid a generating matrix as L_1.

4.2.2 Fundamental Domains and Cosets

Each lattice $\mathcal{L}$ can be used to partition its embedding space into so-called *fundamental domains*. The importance of the concept of fundamental domains lies in their ability to define $\mathcal{L}$-periodic functions, i.e., functions $f(x)$ for which $f(x) = f(x + \lambda)$ for every $\lambda \in \mathcal{L}$. Knowing a $\mathcal{L}$-periodic function $f(x)$ on a fundamental domain is sufficient to know the complete function. Periodic functions will emerge naturally when we come to speak about sampling of continuous functions.

Let $\mathcal{L} \subset \mathcal{D}$ be a lattice, where $\mathcal{D}$ is either a lattice $\mathcal{M} \subset \mathbb{R}^n$ or the space $\mathbb{R}^n$ itself. Let L be a generating matrix of $\mathcal{L}$, and let P be an arbitrary subset of $\mathcal{D}$. With every $p \in P$ we can associate a translated version or *coset* $p + \mathcal{L}$ of $\mathcal{L}$. The set of cosets is referred to as the *coset group* of $\mathcal{L}$ with respect to $\mathcal{D}$ and is denoted by the expression $\mathcal{D}/\mathcal{L}$. A fundamental domain is defined as a subset $P \subset \mathcal{D}$ which intersects every coset in exactly one point.

DEFINITION 4.3

The set P is called a fundamental domain of the lattice $\mathcal{L}$ in $\mathcal{D}$ if and only if

1. $p \neq q$ implies $p + \mathcal{L} \neq q + \mathcal{L}$, and
2. $\bigcup_{p \in P} p + \mathcal{L} = \mathcal{D}$.

A fundamental domain is not a uniquely defined object. For example, the shaded areas in Fig. 4.1 show three possibilities for the choice of a fundamental domain. Although the shapes may differ, their volume is defined by the lattice $\mathcal{L}$.

THEOREM 4.1   *Let P be a fundamental domain of the lattice $\mathcal{L}$ in $\mathcal{D}$, and assume that P is measurable, i.e., that its volume is defined.*

1. *If $\mathcal{D} = \mathbb{R}^n$, then the volume of P is given by*

$$vol(P) = \det(\mathcal{L}) .$$

2. *If $\mathcal{D} = \mathcal{M}$, and if Q is a fundamental domain of $\mathcal{L}$ in $\mathbb{R}^n$, then $Q \cap \mathcal{M}$ is a fundamental domain of $\mathcal{L}$ in $\mathcal{M}$.*
3. *If $\mathcal{D} = \mathcal{M}$, then the number of points in P is given by*

$$\#(P) = \det(\mathcal{L})/\det(\mathcal{M}).$$

This number is referred to as the index of $\mathcal{L}$ in $\mathcal{M}$, and is denoted by the symbol $\iota(\mathcal{L}, \mathcal{M})$.

As a consequence of assertion 1 of this theorem, all the shaded areas in Fig. 4.1, being fundamental domains of the same hexagonal lattice, have a volume equal to $2\sqrt{3}$.

4.2.3 Reciprocal Lattices

For any lattice $\mathcal{L}$ there exists a *reciprocal* lattice $\mathcal{L}^*$ as defined below. Reciprocal lattices appear in the theory of Fourier transforms of sampled continuous functions (see Section 4.3).

DEFINITION 4.4   Let $\mathcal{L}$ be a lattice. Its reciprocal lattice $\mathcal{L}^*$ is defined by

$$\mathcal{L}^* = \{\lambda^* : \langle \lambda^*, \lambda \rangle \in \mathbb{Z} \ \forall \lambda \in \mathcal{L}\} \,,$$

where $\langle \lambda^*, \lambda \rangle$ denotes the usual inner product $\sum_i \lambda_i^* \lambda_i$.

This notion of reciprocal lattice is made more tangible by the observation that the reciprocal lattice of $[L]$ is the lattice $[L^{-1}]$, where $[M]$ denotes the lattice generated by a matrix M. In particular $\det(\mathcal{M}^*) = \det(\mathcal{M})^{-1}$. For example, the reciprocal lattice of the lattice of Fig. 4.1 is generated by the matrix

$$\frac{1}{2\sqrt{3}} \begin{bmatrix} 1 & 1 \\ -\sqrt{3} & \sqrt{3} \end{bmatrix}$$

This lattice is very similar to the original lattice: it differs by a rotation by $\pi/2$, and a scaling factor of $1/2\sqrt{3}$. In particular, the volume of a fundamental domain of $\mathcal{L}^*$ is equal to $1/2\sqrt{3}$.

An important property of reciprocal lattices is that subset inclusions are reversed. To be precise, the inclusion $\mathcal{M} \subset \mathcal{L}$ holds if and only if $\mathcal{L}^* \subset \mathcal{M}^*$. Using some elementary math it follows that the coset groups $\mathcal{L}/\mathcal{M}$ and $\mathcal{M}^*/\mathcal{L}^*$ have the same number of elements.

4.3 Sampling of Continuous Functions

In this section we will give the main results on the theory of sampled continuous functions. It will be shown that there is a strong relationship between sampling in the spatial domain and *periodizing* in the frequency domain. In order to state this result this section starts with a short overview of multidimensional Fourier transforms. This allows us to formulate the main result (Theorem 4.3), which states very informally that sampling in the spatial domain is equivalent to periodizing in the frequency domain.

4.3.1 The Continuous Space-Time Fourier Transform

Let $f(x)$ be a *nice*[1] function defined on the continuous domain $\mathbb{R}^n$. Let its continuous space-time Fourier transform[2] (CSFT) $F(v)$ be defined by

$$F(v) = \mathcal{F}(f)(v) = \int_{\mathbb{R}^n} e^{-2\pi i \langle x, v \rangle} f(x) \, dx \tag{4.2}$$

with inverse transform given by

$$f(x) = \mathcal{F}^{-1}(F)(x) = \int_{\mathbb{R}^n} e^{2\pi i \langle x, v \rangle} F(v) \, dv \,. \tag{4.3}$$

Forgetting many technicalities, the CSFT has the following basic properties:

---

[1] Nice means in this context that all sums, integrals, Fourier transforms, etc. involving the function exist and are finite.
[2] Contrary to the conventional wisdom, we choose to exclude the factor 2π from the frequency term $\omega = 2\pi v$. This has the advantage that the Fourier transform is orthogonal, without any need for normalizing factors.

- The CSFT is an *isometry*, i.e., it preserves inner products.

$$\langle f, g \rangle = \langle \mathcal{F}(f), \mathcal{F}(g) \rangle .$$

- The CSFT of the point-wise multiplication of two functions is the convolution of the two separate CSFTs.

$$\mathcal{F}(f \cdot g) = \mathcal{F}(f) * \mathcal{F}(g) .$$

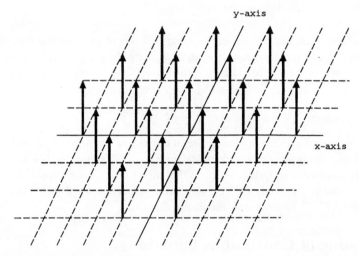

FIGURE 4.2   Lattice comb for the quincunx lattice.

A special class of functions[3] is the class of *lattice combs* (Fig. 4.2 illustrates the lattice comb of the quincunx lattice generated by the matrix $\begin{bmatrix} 1 & -1 \\ 1 & 1 \end{bmatrix}$). If $\mathcal{L}$ is a lattice, the lattice comb $\text{Ш}_{\mathcal{L}}$ is a set of δ functions with support on $\mathcal{L}$ and is formally defined by

$$\text{Ш}_{\mathcal{L}}(x) = \sum_{\lambda \in \mathcal{L}} \delta_{\lambda}(x) . \tag{4.4}$$

The following theorem states the most important facts about lattice combs.

THEOREM 4.2   *With notations as above we have the following properties:*

$$\text{Ш}_{\mathcal{L}}(x) \quad = \quad \frac{1}{\det(\mathcal{L})} \sum_{\lambda^* \in \mathcal{L}^*} e^{-2\pi i \langle x, \lambda^* \rangle} \tag{4.5}$$

$$\mathcal{F}(\text{Ш}_{\mathcal{L}})(v) \quad = \quad \sum_{\lambda \in \mathcal{L}} e^{-2\pi i \langle \lambda, v \rangle}$$

$$\quad = \quad \det(\mathcal{L}^*) \, \text{Ш}_{\mathcal{L}^*}(v) . \tag{4.6}$$

The last equation says that the CSFT of a lattice comb is the lattice comb of the reciprocal lattice, up to a constant.

---

[3] Actually distributions.

4.3.2 The Discrete Space-Time Fourier Transform

The CSFT is a functional on continuous functions. We also need a similar functional on (multidimensional) sequences. This functional will be the discrete space-time Fourier transform (DSFT). In this section we will only state the definition. The properties of this functional and its relation to the CSFT will be highlighted in the next section. So let $\mathcal{L}$ be a lattice and let P^* be a fundamental domain of the reciprocal lattice $\mathcal{L}^*$. Let $\tilde{f}(x) = \Sigma_{\mathcal{L}}(f)(x)$ be the sampled version of f, and let $\tilde{F}(v) = \Pi_{\mathcal{L}^*}(F)(v)$ be the periodized version of $F(v)$. Then we define the forward and backward discrete space-time Fourier transform (DSFT) by

$$\tilde{\mathcal{F}}(\tilde{f})(v) = \sum_{x \in \mathcal{L}} e^{-2\pi i \langle x, v \rangle} \tilde{f}(x) , \tag{4.7}$$

and

$$\tilde{\mathcal{F}}^{-1}(\tilde{F})(v) = \det(\mathcal{L}) \int_{P^*} e^{2\pi i \langle x, v \rangle} \tilde{F}(v) dv , \tag{4.8}$$

respectively.

Note that the function $\tilde{\mathcal{F}}(\tilde{f})(v)$ is a $\mathcal{L}^*$-periodic function. This implies that the formula for the inverse DSFT is independent of the choice of the fundamental domain P^*.

4.3.3 Sampling and Periodizing

One of the most important issues in the sampling of functions concerns the relationship between the CSFT of the original function and the DSFT of a sampled version. In this section we will state the main theorem (Theorem 4.3) of sampling theory.

Before continuing we need two definitions. If $f(x)$ is a function and $\mathcal{L} \subset \mathbb{R}^n$ is a lattice, *sampling $f(x)$ on $\mathcal{L}$* is defined by

$$\Sigma_{\mathcal{L}}(f)(x) = \begin{cases} f(x) & \text{if } x \in \mathcal{L} \\ 0 & \text{if } x \notin \mathcal{L} . \end{cases} \tag{4.9}$$

The above definition has to be read carefully: sampling a function $f(x)$ on a lattice means that we modify $f(x)$ by putting all its values outside of the lattice to 0. It *does not* mean that we forget how the lattice is embedded in the continuous domain. For example, when we sample a one-dimensional continuous function $f(x)$ on the set of even numbers, the down sampled function $f_s(k)$ is not defined by $f_s(k) = f(2k)$, but by $f_s(k) = f(k)$ when k is even, and 0 otherwise.

Closely related to the sampling operator is the periodizing operator $\Pi_{\mathcal{L}}$, which modifies a function $f(x)$ such that it becomes $\mathcal{L}$-periodic. This operator is defined by

$$\Pi_{\mathcal{L}}(f)(x) = \det(\mathcal{L}) \sum_{\lambda \in \mathcal{L}} f(x - \lambda) \tag{4.10}$$

Clearly $\Pi_{\mathcal{L}}(f)(x)$ is $\mathcal{L}$-periodic, i.e., $\Pi_{\mathcal{L}}(f)(x) = \Pi_{\mathcal{L}}(f)(x - \lambda)$ for all $\lambda \in \mathcal{L}$. With these tools at our disposal we are now in a position to formulate the main theorem of sampling theory.

THEOREM 4.3 *With definitions and notations as above, consider the following diagram:*

$$
\begin{array}{ccc}
f & \xrightarrow{\mathcal{F}} & F \\
\downarrow \Sigma_{\mathcal{L}} & & \downarrow \Pi_{\mathcal{L}^*} \\
\tilde{f} & \xrightarrow{\tilde{\mathcal{F}}} & \tilde{F}
\end{array}
$$

The following assertions hold:

1. *The above diagram commutes,[4] i.e., whichever way we take to go from top left to bottom right, the result is the same. Informally this can be formulated as saying that first sampling and taking the DSFT is the same as first taking the CSFT and then periodizing.*

2. $\sqrt{\det(\mathcal{L})}\,\tilde{\mathcal{F}}$ *(and, therefore,* $\sqrt{\det(\mathcal{L}^*)}\,\tilde{\mathcal{F}}^{-1}$*) is an isometry with respect to the inner products*

$$\langle \tilde{f}, \tilde{g} \rangle_{\mathcal{L}} = \sum_{\lambda \in \mathcal{L}} \tilde{f}^{\dagger}(\lambda)\tilde{g}(\lambda)$$

and

$$\langle \tilde{F}, \tilde{G} \rangle_{P*} = \int_{P*} \tilde{F}^{\dagger}(\nu)\tilde{G}(\nu)d\nu \,,$$

respectively.

PROOF 4.1   The proof relies heavily on the property of lattice combs and can be found in the Appendix.

This theorem has many important consequences, the best known of which is the Shannon sampling theorem. This theorem says that a function can be retrieved from a sampled version if the support of its CSFT is contained within a fundamental domain of the reciprocal lattice. Given the above theorem this result is immediate: we only need to verify that a function $F(\nu)$ can be retrieved from $\Pi_{\mathcal{L}^*}(F)$ by restriction to a fundamental domain when $F(\nu)$ has sufficiently restricted support.

THEOREM 4.4   *(Shannon) Let $\mathcal{L}$ be a lattice, and let $f(x)$ be a continuous function with CSFT $F(\nu)$. Let $\tilde{f} = \Sigma_{\mathcal{L}}(f)$. The function $f(x)$ can be retrieved from $\tilde{f}(\lambda)$ if and only if the support of $F(\nu)$ is contained in some fundamental domain P^* of the reciprocal lattice $\mathcal{L}^*$. In that case we can retrieve $f(x)$ from $\tilde{f}(\lambda)$ with the formula*

$$f(x) = \sum_{\lambda \in \mathcal{L}} f(\lambda)\mathrm{Int}(x - \lambda) \,,$$

where

$$\mathrm{Int}(x) = \det(\mathcal{L})\int_{P*} e^{2\pi i \langle x,\nu \rangle} \, d\nu \,.$$

PROOF 4.2   We only need to prove the interpolation formula.

$$
\begin{aligned}
f(x) &= \int_{P*} e^{2\pi i \langle x,\nu \rangle} F(\nu)\,d\nu \\
&= \det(\mathcal{L})\sum_{\lambda \in \mathcal{L}} f(\lambda)\int_{P*} e^{2\pi i \langle x-\lambda,\nu \rangle}\,d\nu \\
&= \sum_{\lambda \in \mathcal{L}} f(\lambda)\mathrm{Int}(x - \lambda) \,.
\end{aligned}
\tag{4.11}
$$

We end this section with an example showing all the aspects of Theorem 4.3.

---

[4]Commuting diagrams are a common mathematical tool to describe that certain sequences of function applications are equivalent.

EXAMPLE 4.1:

Let $\mathcal{L} \subset \mathbb{Z}^2$ be the *quincunx* sampling lattice generated by the matrix $L = \frac{1}{2}\begin{bmatrix} 1 & -1 \\ 1 & 1 \end{bmatrix}$. Let

$$f(x_1, x_2) = \text{sinc}(x_1 - x_2)\text{sinc}(x_1 + x_2) .$$

A simple computation shows that CSFT $F(\nu_1, \nu_2)$ of $f(x_1, x_2)$ is given by

$$F(\nu_1, \nu_2) = \frac{1}{2} X_\Lambda(\nu_1, \nu_2) ,$$

where Λ is the set $\Lambda = \{(\nu_1, \nu_2) : |\nu_1| + |\nu_2| \leq 1\}$. Observing that $\mathcal{L}^*$ is generated by $\begin{bmatrix} 1 & -1 \\ 1 & 1 \end{bmatrix}$, we find that the periodized function $\Pi_{\mathcal{L}^*}(F)$ is constant with value 1.

Sampling $f(x)$ on the quincunx lattice yields the function $\tilde{f}(\lambda)$

$$\tilde{f}(\lambda_1, \lambda_2) = \begin{cases} 1 & \text{if } (\lambda_1, \lambda_2) = (0, 0) \\ 0 & \text{if } (\lambda_1, \lambda_2) \neq (0, 0) . \end{cases}$$

It is now trivial to check that $\tilde{\mathcal{F}}(\tilde{f}) = \tilde{F}$, as predicted by Theorem 4.3. Moreover, as

$$\left\| \tilde{f} \right\|_2^2 = \sum_{\lambda \in \mathcal{L}} \delta_0(\lambda)^2 = 1$$

and

$$\left\| \tilde{F} \right\|_2^2 = \int_\Lambda d\nu = 1/2 ,$$

it follows that $\left\| \tilde{\mathcal{F}} \right\|$ and $\left\| \tilde{f} \right\|$ differ by a factor of $\sqrt{2} = \sqrt{\det(\mathcal{L}^*)}$, again as predicted by Theorem 4.3.

4.4 From Infinite Sequences to Finite Sequences

In the previous section we considered sampling in the spatial domain and saw that this was equivalent to periodizing in the frequency domain. One obvious question now arises: what happens if we sample the DSFT of a (spatially) sampled function? In this section we will answer this question and show that sampling in both spatial and frequency domains simultaneously is closely related to properties of the discrete Fourier transform (DFT).

4.4.1 The Discrete Fourier Transform

The discrete Fourier transform (DFT) is a frequency transform on finite sequences. In a multidimensional context the DFT is best defined by assuming two lattices $\mathcal{L}$ and $\mathcal{M}, \mathcal{M} \subset \mathcal{L} \subset \mathbb{R}^n$. Let P be a fundamental domain of $\mathcal{L}$ in $\mathcal{M}$, and let P^* be a fundamental domain of $\mathcal{M}^*$ in $\mathcal{L}^*$ (recall that lattice inclusions invert when going over to the reciprocal domain [Section 4.2]). Note that both P and P^* have the same number points, viz. $\#(P) = \#(P^*) = \iota(\mathcal{L}^*, \mathcal{M}^*) = \iota(\mathcal{M}, \mathcal{L})$. Let $\hat{f}(p), p \in P$ be a finite sequence over P. The DFT $\hat{\mathcal{F}}$ is now defined as functional which maps sequences $\hat{f}$ to sequences $\hat{F}$ over P^*. The formal definitions of $\hat{\mathcal{F}}$ and $\hat{\mathcal{F}}^{-1}$ are as follows.

DEFINITION 4.5

$$\hat{\mathcal{F}}(\hat{f})(p^*) = \frac{1}{\det(\mathcal{M})} \sum_{p \in P} e^{-2\pi i \langle p, p^* \rangle} \hat{f}(p) \tag{4.12}$$

$$\hat{\mathcal{F}}^{-1}(\hat{F})(p) = \frac{1}{\det(\mathcal{L}^*)} \sum_{p^* \in P^*} e^{2\pi i \langle p, p^* \rangle} \hat{F}(p^*) . \tag{4.13}$$

It is obvious that the conventional one-dimensional DFT is a special case of the more general multidimensional DFT defined above. The next example makes this more explicit.

EXAMPLE 4.2:

Let $\mathcal{M} \subset \mathcal{L} \subset \mathbb{R}$ be defined by $\mathcal{M} = \mathbb{Z}$ for some positive integer p, and let $\mathcal{L} = \frac{1}{p}\mathbb{Z}$. One easily checks that the set P and P^* can be chosen as $\{0/p, \cdots, (p-1)/p\}$ and $\{0, \cdots, p-1\}$, respectively. If x_n and X_m are the values of $\hat{f}$ on $n/p \in P$ and of $\hat{F}$ on $m \in P^*$, respectively, then the functionals $\hat{\mathcal{F}}$ and $\hat{\mathcal{F}}^{-1}$ are defined in the (x_n, X_m) domain as

$$X_m = \sum_{n=0}^{p-1} e^{-\frac{2\pi i n m}{p}} x_n, \tag{4.14}$$

$$x_n = \frac{1}{p} \sum_{m=0}^{p-1} e^{\frac{2\pi i n m}{p}} X_m. \tag{4.15}$$

This is, of course, nothing else but the usual definition of the one-dimensional DFT on finite sequences of length p.

The following example shows the general DFT at work in a two-dimensional setting.

EXAMPLE 4.3:

(Example 4.1 continued) Continuing Example 4.1, we choose the lattice $\mathcal{M} = \mathbb{Z}^2$ as the periodizing lattice. We can then choose

$$P = \{p_0, p_1\} = \left\{(0, 0), \left(\frac{1}{2}, \frac{1}{2}\right)\right\}$$

and

$$P^* = \{p_0^*, p_1^*\} = \{(0, 0), (1, 0)\} .$$

The functional $\hat{\mathcal{F}}$ is then given by

$$\begin{aligned}
X_0 &= x_0 e^{-2\pi i \langle p_0, p_0^* \rangle} + x_1 e^{-2\pi i \langle p_1, p_0^* \rangle} \\
&= x_0 + x_1 \\
X_1 &= x_0 e^{-2\pi i \langle p_0, p_1^* \rangle} + x_1 e^{-2\pi i \langle p_1, p_1^* \rangle} \\
&= x_0 - x_1 ,
\end{aligned}$$

and the functional $\hat{\mathcal{F}}^{-1}$ by

$$\begin{aligned}
x_0 &= \frac{1}{2}\left(X_0 e^{-2\pi i \langle p_0, p_0^* \rangle} + X_1 e^{-2\pi i \langle p_0, p_1^* \rangle}\right) \\
&= \frac{1}{2}(X_0 + X_1) \\
x_1 &= \frac{1}{2}\left(X_0 e^{-2\pi i \langle p_1, p_0^* \rangle} + X_1 e^{-2\pi i \langle p_1, p_1^* \rangle}\right) \\
&= \frac{1}{2}(X_0 - X_1) .
\end{aligned}$$

4.4.2   Combined Spatial and Frequency Sampling

We start with setting up the context of the problem. So let $f(x)$ be a nice continuous function on $\mathbb{R}^n$ and let $\mathcal{M}$ and $\mathcal{L}$ be two lattices such that $\mathcal{M} \subset \mathcal{L} \subset \mathbb{R}^n$. Sampling $f(x)$ on $\mathcal{L}$ and periodizing on $\mathcal{M}$ we construct a function $\hat{f}(x)$ that has support on $\mathcal{L}$ and is $\mathcal{M}$-periodic. In formula:

$$\hat{f}(x) = \begin{cases} \det(\mathcal{M}) \sum_{\mu \in \mathcal{M}} f(x - \mu) & \text{if } x \in \mathcal{L} \\ 0 & \text{if } x \notin \mathcal{L}. \end{cases}$$

A similar definition can be given for the function $\hat{F}(\nu)$, which is obtained from the CSFT $F(\nu)$ of $f(x)$ by periodizing on $\mathcal{L}^*$ and sampling on $\mathcal{M}^*$.

One easily verifies that $\hat{f}(x)$ is completely specified by its values on a (finite) fundamental domain P of $\mathcal{M}$ in $\mathcal{L}$. Similarly $\hat{F}(\nu)$ is completely specified by its values on a fundamental domain P^* of $\mathcal{L}^*$ in $\mathcal{M}^*$. Now we are in a position to extend the commutative diagram of Theorem 4.3.

THEOREM 4.5   *With notations and definitions as above, consider the following extensions of the diagram of Theorem 4.3:*

$$\begin{array}{ccc} f & \xrightarrow{\mathcal{F}} & F \\ \downarrow \Sigma_{\mathcal{L}} & & \downarrow \Pi_{\mathcal{L}^*} \\ \tilde{f} & \xrightarrow{\tilde{\mathcal{F}}} & \tilde{F} \\ \downarrow \Pi_{\mathcal{M}} & & \downarrow \Sigma_{\mathcal{M}^*} \\ \hat{f} & \xrightarrow{\hat{\mathcal{F}}} & \hat{F} \end{array}$$

The following assertions hold:

1. *The above diagram commutes;*
2. *The functionals $\sqrt{\det(\mathcal{L})}\sqrt{\det(\mathcal{M})}\hat{\mathcal{F}}$ and $\sqrt{\det(\mathcal{L}^*)}\sqrt{\det(\mathcal{M}^*)}\hat{\mathcal{F}}^{-1}$ are isometries with respect to the inner products*

$$\langle \hat{f}, \hat{g} \rangle_P = \sum_{p \in P} \hat{f}^\dagger(p)\hat{g}(p)$$

and

$$\langle \hat{F}, \hat{G} \rangle_{P^*} = \sum_{p^* \in P^*} \hat{F}^\dagger(p^*)\hat{G}(p^*).$$

PROOF 4.3   See Appendix.

The theorem above says that sampling the Fourier transform of a sampled function amounts to periodizing that sampled version. In this process only a finite number of data points in both the spatial and the frequency domain are sufficient to specify the resulting functions. Moreover, the CSFT can be pushed down to a DFT to provide for a one-to-one orthogonal correspondence between the two domains.

We close this section with two examples.

EXAMPLE 4.4:

(Example 4.2 continued) The formulas for the DFT obtained in Example 4.2 are *not orthonormal*. According to Theorem 4.5 above we have to multiply the forward transform with $\sqrt{\det(\mathcal{L})\det(\mathcal{M})} = \frac{1}{\sqrt{p}}$

and the backward transform with the inverse of this number to obtain orthonormal versions of the DFT. This result in the following well-known formulas for the orthonormal one-dimensional DFT.

$$X_m = \frac{1}{\sqrt{p}} \sum_{n=0}^{p-1} e^{-\frac{2\pi i n m}{p}} x_n , \tag{4.16}$$

$$x_n = \frac{1}{\sqrt{p}} \sum_{m=0}^{p-1} e^{\frac{2\pi i n m}{p}} X_m . \tag{4.17}$$

EXAMPLE 4.5:

(Example 4.3 continued) With $\mathcal{L}$, $\mathcal{M}$, $f(x)$, P and P^* as in Example 4.3, we find that the periodized sampled function $\hat{f}$ is represented by the pair $(1, 0)$, and that the periodized sampled CSFT $\hat{F}$ of F is represented by the pair $(1, 1)$. Using the formulas for the DFT of Example 4.3 is now easy to verify that $\hat{\mathcal{F}}(\{1, 0\}) = \{1, 1\}$ and $\hat{\mathcal{F}}^{-1}(\{1, 1\}) = \{1, 0\}$, as predicted by Theorem 4.5.

4.5 Lattice Chains

In the previous section we considered the sampling of continuous functions. In this section we will consider the sampling of discrete functions. The necessity of studying this topic comes from the fact that very often the sampling of a continuous function $f(x)$ is done in steps: $f(x)$ is first sampled to a fine grid $\mathcal{L}_1$, and subsequently sampled to a coarser grid $\mathcal{L}_2$, $\mathcal{L}_2 \subset \mathcal{L}_1$. Letting $\tilde{f}^{(i)} = \Sigma_{\mathcal{L}_i}(f)$ and letting $\tilde{F}^{(i)}$ be the corresponding DFST, a natural question is whether we can obtain $\tilde{F}^{(2)}$ directly from $\tilde{F}^{(1)}$, without having to go back to CSFT of $f(x)$. This question is addressed in the following theorem and answered affirmatively.

THEOREM 4.6 *With notation as above, and letting P^* be a fundamental domain of $\mathcal{L}_1^*$ in $\mathcal{L}_2^*$, we have the following result.*

$$\tilde{F}^{(2)}(\nu) = \frac{1}{\#(P^*)} \sum_{p^* \in P^*} \tilde{F}^{(1)}(\nu - p^*) .$$

PROOF 4.4 See Appendix.

The above result has a natural interpretation. The function $\tilde{F}^{(1)}$ is by construction $\mathcal{L}_1^*$-periodic. The function $\tilde{F}^{(2)}$ has more symmetries as it is $\mathcal{L}_2^*$-periodic. The above theorem can be phrased as saying that $\tilde{F}^{(2)}$ is obtained from $\tilde{F}^{(1)}$ by periodizing (and thereby enlarging the set of symmetries) and averaging (dividing by $\#(P^*)$). The following example shows an application of Theorem 4.6 in the one-dimensional case.

EXAMPLE 4.6:

Let $f(x) = \text{sinc}(x/2)$. Let $\mathcal{L}_1 = \mathbb{Z}$ be the lattice of integers and let $\mathcal{L}_2 = 2\mathbb{Z}$ be the lattice of even integers. Let as before $\tilde{F}^{(i)}(x)$ denote the sampled versions of $f(x)$. Then one easily computes that

$$\tilde{F}^{(1)}(\nu) = 2 \sum_{\lambda^* \in \mathbb{Z}} X_{[-1/4; 1/4]}(\nu - \lambda^*) ,$$
$$\tilde{F}^{(2)}(\nu) = 1 ,$$

where X_A denotes the characteristic function of a set A.

Using Theorem 4.6 above we can also compute $\tilde{F}^{(2)}(v)$ directly from $\tilde{F}^{(1)}(v)$. We proceed as follows. Computing the reciprocal lattices we find $\mathcal{L}_1^* = \mathbb{Z}$ and $\mathcal{L}_2^* = \frac{1}{2}\mathbb{Z}$. We find two shifted versions of $\mathcal{L}_1^*$ within $\mathcal{L}_2^*$, viz. $\mathcal{L}_1^*$ and $\frac{1}{2} + \mathcal{L}_1^*$. Picking an arbitrary point in each coset, say 0 and $\frac{1}{2}$ respectively, we find

$$\begin{aligned} \tilde{F}^{(2)}(v) &= \frac{1}{2}\left(\tilde{F}^{(1)}(v) + \tilde{F}^{(1)}\left(v - \frac{1}{2}\right)\right) \\ &= 1 \end{aligned}$$

4.6 Change of Variables

Consider the case of a one-dimensional continuous function $f(x)$. It is not always the case that $f(x)$ has a nice form, suitable for direct mathematical treatment. In such a situation a change of variables can sometimes help out. If A is an invertible linear transformation on $\mathbb{R}^n$, it might be more convenient to work with the variable $y = Ax$. Substituting $x = A^{-1}y$ we formally define the *change of variable* functional $f(x) \rightarrow f^A(x)$ by

$$f^A(x) = f\left(A^{-1}x\right).$$

A similar approach can be used for discrete functions. Instead of using a linear transform A on some continuous domain, we need in this case an *isomorphism* $A : \mathcal{L}_1 \rightarrow \mathcal{L}_2$ between two lattices $\mathcal{L}_1$ and $\mathcal{L}_2$. If $\tilde{f}(k)$ is a discrete function on $\mathcal{L}_1$, a change of variables by A yields a discrete function on $\mathcal{L}_2$ defined by

$$\tilde{f}^A(k) = \tilde{f}\left(A^{-1}k\right).$$

A typical example for a change of variables on discrete functions is the following. Let the lattice $\mathcal{L}_1 = 2\mathbb{Z}$, let $\mathcal{L}_2 = \mathbb{Z}$ and define $A : \mathcal{L}_1 \rightarrow \mathcal{L}_2$ by $2k \rightarrow k$. Given a function $f(x)$ on $\mathbb{R}$, downsampling it to $\mathcal{L}_1$ and changing variables with A, yield a discrete function $\tilde{f}(k)$ on $\mathbb{Z}$ defined by $\tilde{f}(k) = f(2k)$. In many textbooks this function $\tilde{f}(k)$ is referred to as the downsampled version of $f(x)$, but our analysis shows that it is better to view the discrete function $\tilde{f}(k)$ as the result of *two* consecutive operations: downsampling and change of variables.

The following two theorems address the question of how the CSFT and DSFT behave under a change of variables for the continuous and discrete case, respectively.

THEOREM 4.7

Let A be an invertible linear transform on $\mathbb{R}^n$, and let $f(x)$ be a function on $\mathbb{R}^n$. Then the CSFT of $f^A(x)$ is given by

$$\mathcal{F}\left(f^A\right) = |\det(A)|\mathcal{F}(f)^{A^{-t}}.$$

PROOF 4.5 See Appendix.

THEOREM 4.8 *Let $A : \mathcal{L}_1 \rightarrow \mathcal{L}_2$ be an isomorphism of lattices, and let $\tilde{f}(k)$ be a function on $\mathcal{L}_1$. Then the DSFT of $\tilde{f}^A(k)$ is given by*

$$\tilde{\mathcal{F}}\left(\tilde{f}^A\right) = \tilde{\mathcal{F}}\left(\tilde{f}\right)^{A^{-t}}.$$

PROOF 4.6 See Appendix.

Note that in the assertion of Theorem 4.7 a factor $|\det(A)|$ is present, which is lacking in the assertion of Theorem 4.8. The last theorem of this section addresses the situation in which a function is extended by zero-padding to a larger domain.

THEOREM 4.9

Let $\mathcal{L}$, $\mathcal{L} \subset \mathcal{D}$ be a lattice, where $\mathcal{D}$ is either a lattice $\mathcal{M}$ or the ambient space $\mathbb{R}^n$. Let $\tilde{f}(\lambda)$ be a function on $\mathcal{L}$. Define the $\mathcal{D}$-extension $\tilde{f}_{\mathcal{D}}$ of $\tilde{f}$ by

$$\tilde{f}_{\mathcal{D}}(x) = \begin{cases} \tilde{f}(x) & \text{if } x \in \mathcal{L} \\ 0 & \text{otherwise.} \end{cases}$$

Define $\Phi(\nu)$ by

$$\Phi(\nu) = \begin{cases} \mathcal{F}\left(\tilde{f}_{\mathcal{D}}\right)(\nu) & \text{if } \mathcal{D} = \mathbb{R}^n \\ \tilde{\mathcal{F}}\left(\tilde{f}_{\mathcal{D}}\right)(\nu) & \text{if } \mathcal{D} = \mathcal{M}, \end{cases}$$

i.e., $\Phi(\nu)$ is the appropriate Fourier transform of $\tilde{f}_{\mathcal{D}}$. Then the equality $\Phi(\nu) = \tilde{\mathcal{F}}(\tilde{f})(\nu)$ holds.

Informally, the above theorem says that the Fourier transform of an extended function is equal to the Fourier transform of the function itself, i.e., extending a function does not change the Fourier transform. We will now apply the three theorems above in two examples.

EXAMPLE 4.7:

Let $A : \mathbb{Z}^n \to \mathbb{R}^n$ be a nonsingular linear mapping, and let $\mathcal{L} = [A]$ be the lattice generated by A. Let $f(x)$ be a continuous function on $\mathbb{R}^n$, and let $g = f^{A^{-1}}$. Define a discrete function $\tilde{g}(m)$ on $\mathbb{Z}^n$ by the rule[5]

$$\tilde{g}(m) = f(Am) .$$

The question is how the Fourier transforms of $f(x)$ and $\tilde{g}(k)$ are related. To answer this question we define $\tilde{f}(\lambda)$ to be the sampled version $\Sigma_{\mathcal{L}}(f)(\lambda)$ of $f(x)$. The following commutative diagram results.

$$
\begin{array}{ccc}
(\mathbb{R}^n, g) & \stackrel{A^{-1}}{\longleftarrow} & (\mathbb{R}^n, f) \\
\downarrow \Sigma_{\mathbb{Z}^n} & & \downarrow \Sigma_{\mathcal{L}} \\
(\mathbb{Z}^n, \tilde{g}) & \stackrel{A^{-1}}{\longleftarrow} & (\mathcal{L}, \tilde{f})
\end{array}
$$

Tracing the diagram from top right to bottom right to bottom left we find

$$
\begin{aligned}
\tilde{\mathcal{F}}(\tilde{g})(\nu) &= (\mathcal{F}(\tilde{f}))^{A^t}(\nu) \\
&= \det\left(\mathcal{L}^*\right) \sum_{\lambda^* \in \mathcal{L}^*} \left(\mathcal{F}(f)^{A^t}(\nu - \lambda^*)\right) \\
&= \frac{1}{\det(A)} \sum_{\lambda^* \in \mathcal{L}^*} \mathcal{F}(f)\left(A^{-t}\nu - \lambda^*\right) ,
\end{aligned}
$$

---

[5]This is a common situation when we have to sample a continuous function (on points of the form An) and store it in some rectangular storage space (with addresses n).

where we have used Theorem 4.8 and Theorem 4.3 in the first and second steps, respectively. Of course we should find the same result tracing the diagram from top right to top left to bottom left.

$$
\begin{aligned}
\tilde{\mathcal{F}}(\tilde{g})(v) &= \sum_{k \in \mathbb{Z}^n} \mathcal{F}(g)(v - k) \\
&= \sum_{k \in \mathbb{Z}^n} \mathcal{F}\left(f^{A^{-1}}\right)(v - k) \\
&= \frac{1}{\det(A)} \sum_{k \in \mathbb{Z}^n} \mathcal{F}(f)^{A^t}(v - k) \\
&= \frac{1}{\det(A)} \sum_{k \in \mathbb{Z}^n} \mathcal{F}(f)\left(A^{-t}v - A^{-t}k\right) \\
&= \frac{1}{\det(A)} \sum_{\lambda^* \in \mathcal{L}^*} \mathcal{F}(f)\left(A^{-t}v - \lambda^*\right) ,
\end{aligned}
$$

where we have first applied Theorem 4.3, followed by an application of Theorem 4.8. As one sees, both calculations end up with the same result.

EXAMPLE 4.8:

Let $\mathcal{L}_1$ and $\mathcal{L}_2$ be two lattices. Let $A : \mathcal{L}_1 \to \mathcal{L}_2$ be a nonsingular linear mapping, and let $\tilde{f}$ be a function on $\mathcal{L}_1$. Let $\mathcal{L}_3$ be the lattice generated by A, $\mathcal{L}_3 = [A] \subset \mathcal{L}_2$. Define $\tilde{g}$ on $\mathcal{L}_2$ by

$$
\tilde{g}(\lambda_2) = \begin{cases} \tilde{f}(\lambda_1) & \text{if } \lambda_2 = A\lambda_1 \\ 0 & \text{otherwise.} \end{cases}
$$

The question is to find an expression for the DSFT of $\tilde{g}$. To this end we define $\tilde{h}$ on $\mathcal{L}_3$ by $\tilde{h} = \tilde{f}^A$. The following diagram results.

$$
\left(\mathcal{L}_1, \tilde{f}\right) \xrightarrow{A} \left(\mathcal{L}_3, \tilde{h}\right) \xrightarrow{\text{extension}} (\mathcal{L}_2, \tilde{g})
$$

For the DSFT of $\tilde{g}$ we find

$$
\begin{aligned}
\tilde{\mathcal{F}}(\tilde{g})(v) &= \tilde{\mathcal{F}}\left(\tilde{h}\right)(v) \\
&= \tilde{\mathcal{F}}\left(\tilde{f}^A\right)(v) \\
&= \tilde{\mathcal{F}}\left(\tilde{f}\right)^{A^{-t}}(v) \\
&= \tilde{\mathcal{F}}(\tilde{f})(A^t v) ,
\end{aligned}
$$

where we have used Theorem 4.9 and Theorem 4.8 in the first and second step, respectively.

4.7 An Extended Example: HDTV-to-SDTV Conversion

This section will introduce an application of sampling theory as it occurs in the problem of interlaced high definition television (HDTV) to interlaced standard definition television (SDTV) conversion. This problem exists because an HDTV broadcast can at present only be viewed by a minority of people. Most people can only view SDTV broadcast. As broadcasters like their programs to be viewed by as many

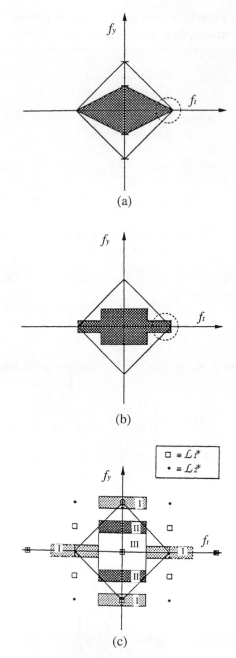

(a)

(b)

(c)

FIGURE 4.3   HDTV-to-SDTV conversion in the frequency domain.

customers as possible, they are interested in (preferably inexpensive) schemes which can convert HDTV in SDTV. In this section we present an approach to this conversion problem as has been suggested in [1].

In order to keep the notational burden low, our television signal will be one-dimensional. This leaves us with a spatial axis, referred to as the y-axis (y for vertical), and a time axis, referred to as the t-axis.

An interlaced television signal is constructed by sampling a continuous luminance signal with at times kT, but only even lines for even k and only the odd lines for odd k. Choosing T to be 1 in some unit of time, and recalling that we assume one-dimensional images, we may model an interlaced HDTV signal as a luminance signal sampled at the quincunx lattice $\mathcal{L}_2$ generated by the matrix

$$\begin{bmatrix} 1 & -1 \\ 1 & 1 \end{bmatrix}.$$

In order to prevent alias distortion, i.e., in order to prevent that frequencies overlap after sampling, the continuous luminance signal has to be sufficiently band limited. An often-used pass band region is given by the diamond in Fig. 4.3(c).

An SDTV interlaced signal has half the vertical resolution of the HDTV signal, but the same temporal resolution, and we may model this as the sampling of the continuous luminance signal on the skew quincunx lattice $\mathcal{L}_1$ generated by the matrix

$$\begin{bmatrix} 1 & -1 \\ 2 & 2 \end{bmatrix}.$$

Note that the lattice $\mathcal{L}_1$ is not a sublattice of the $\mathcal{L}_2$. This has the consequence that the extraction of an SDTV signal from an HDTV signal is not simply a question of subsampling the HDTV signal; interpolation is needed to compute the values of the luminance signal at the missing points. In the frequency domain this is equivalent to restricting the pass band region of the HDTV signal to a smaller pass band region, such that no alias occurs when the interpolated signal is sampled to the SDTV lattice.

Figure 4.3(a) gives a possible solution. The SDTV pass band region is chosen as the skew diamond region within the HDTV pass band (the outer diamond). This solution has several disadvantages. One disadvantage is the fact that the realization of this diamond pass band region can only be realized by nonseparable filters, and, therefore, that it is expensive. A second disadvantage is the temporal attenuation at maximum temporal frequency, which may introduce visible artifacts for moving video.

As argued in [1], the best compromise between vertical resolution and temporal attenuation at maximum temporal frequency is given by a pass band of the form as given in Fig. 4.3(b). This pass band can even be realized cheaply.

Following [1] we note that the temporal information at maximum frequency (region I on the f_t-axis in Fig. 4.3(c)) is repeated at maximal vertical frequency (region I on the f_y-axis in Fig. 4.3(c)). This is simply a consequence of the fact that the DSFT of the HDTV signal is $\mathcal{L}_2^*$-periodic. We can retain this information by using an appropriately chosen *vertical high pass* filter. In a practical implementation this implies that (after temporal low-pass filtering) we extract from the HDTV signal a base-band signal using a vertical low-pass filter (the rectangle III in Fig. 4.3(c)) and a *temporal* band using a vertical high-pass filter. The temporal band is now modulated to position II in Fig. 4.3(c) by multiplying the sample at position $(2k, t)$ with $(-1)^k$.

The base band and the temporal band are now merged and sampled to the SDTV lattice. Due to this last sampling operation, region II is repeated at its original position I in frequency space: this follows immediately from computing the reciprocal SDTV quincunx lattice.

This proves (as first shown in [1]) that a high quality HDTV-to-SDTV conversion can be achieved using only separable filters.

4.8 Conclusions

We have presented the basic facts of multidimensional sampling theory. Particular attention has been paid to the interaction of the different kinds of Fourier transforms, the sampling operator, and the periodizing operator. Every basic result is accompanied by one or more examples. An application of the theory to a format conversion problem has been presented.

References

[1] Albani, L., Mian, G. and Rizzi, A., A new intra-frame solution for HDTV-to-SDTV down-conversion, in *HDTV–1995 International Workshop and the Evolution of Television,* 1995.

[2] Cassels, J., *An Introduction to the Geometry of Numbers.* Springer-Verlag, Berlin, 1971.

[3] Hungerford, T., *Algebra, Graduate Texts in Mathematics,* vol. 73. Springer-Verlag, New York, 1974.

[4] Dudgeon, D.E. and Mersereau, R.M., *Multidimensional Digital Signal Processing.* Signal Processing Series, Prentice-Hall, Englewood Cliffs, NJ, 1984.

[5] Dubois, E., The sampling and reconstruction of time-varying imagery with application in video systems, *Proc. IEEE,* 73: 502–522, April, 1985.

[6] Viscito, E. and Allebach, J., The analysis and design of multidimensional FIR perfect reconstruction filter banks for arbitrary sampling lattices, *IEEE Trans. Circuits Syst.,* 38: 29–42, January, 1991.

[7] Chen, T. and Vaidyanathan, P., Recent developments in multidimensional multirate systems, *IEEE Trans. Circuits Syst. Video Technol.,* 3: 116–137, April, 1993.

[8] Vetterli, M. and Kovačević, J., *Wavelets and Subband Coding.* Signal Processing Series, Prentice-Hall, Englewood Cliffs, NJ, 1995.

[9] Jerri, A., The Shannon sampling theorem – its various extensions and applications: A tutorial review, *Proc. IEEE,* pp. 1565–1596, November, 1977.

Appendix

A.1 Proof of Theorem 4.3

PROOF 4.7   We first observe that

$$
\begin{aligned}
\Sigma_{\mathcal{L}}(f) &= f \cdot \amalg_{\mathcal{L}}, \\
\Pi_{\mathcal{L}}(F) &= F * \amalg_{\mathcal{L}^*}.
\end{aligned}
$$

It follows immediately that $\mathcal{F}(\Sigma_{\mathcal{L}}(f)) = \Pi_{\mathcal{L}^*}(\mathcal{F}(f))$. To prove the first assertion of this theorem, it suffices to verify that $\tilde{\mathcal{F}}(\tilde{f}) = \tilde{F}$.

$$
\begin{aligned}
\tilde{F}(v) &= \mathcal{F}(f \cdot \amalg_{\mathcal{L}})(v) \\
&= \int_{\mathbb{R}^n} \sum_{\lambda \in \mathcal{L}} e^{-2\pi i \langle x, v \rangle} f(x) \delta_{\lambda}(x) dx \\
&= \sum_{\lambda \in \mathcal{L}} e^{-2\pi i \langle \lambda, v \rangle} f(\lambda) \\
&= \tilde{\mathcal{F}}(\tilde{f}).
\end{aligned}
$$

The second assertion of the theorem, viz. the isometry property of the DSFT, follows from

$$
\begin{aligned}
\langle \tilde{F}, \tilde{G} \rangle_{P*} &= \frac{1}{\det(\mathcal{L})^2} \int_{P*} \langle \amalg_{\mathcal{L}^*} * F, \amalg_{\mathcal{L}^*} * G \rangle_{P*} \\
&= \frac{1}{\det(\mathcal{L})^2} \int_{P*} \left(\sum_{\lambda_1^* \in \mathcal{L}^*} F(v - \lambda_1^*) \right) \left(\sum_{\lambda_1^* \in \mathcal{L}^*} G(v - \lambda_2^*) \right) dv
\end{aligned}
$$

$$
\begin{aligned}
&= \frac{1}{\det(\mathcal{L})^2} \int_{\mathbb{R}^n} F(v) \left(\sum_{\lambda^* \in \mathcal{L}^*} G(v - \lambda^*) \right) dv \\
&= \frac{1}{\det(\mathcal{L})} \langle F, \tilde{G} \rangle \\
&= \frac{1}{\det(\mathcal{L})} \langle f, \tilde{g} \rangle \\
&= \frac{1}{\det(\mathcal{L})} \langle \tilde{f}, \tilde{g} \rangle_{\mathcal{L}}.
\end{aligned}
$$

A.2 Proof of Theorem 4.5

PROOF 4.8 Similar to the proof of Theorem 4.3, to prove the first assertion it suffices to show that $\tilde{\mathcal{F}}(\hat{f}) = \hat{F}$.

$$
\begin{aligned}
\tilde{\mathcal{F}}(\hat{f})(v) &= \sum_{\lambda \in \mathcal{L}} e^{-2\pi i \langle \lambda, v \rangle} \hat{f}(\lambda) \\
&= \left(\sum_{\mu \in \mathcal{M}} e^{-2\pi i \langle \mu, v \rangle} \right) \left(\sum_{p \in P} e^{-2\pi i \langle p, v \rangle} \hat{f}(p) \right) \\
&= \frac{1}{\det(\mathcal{M})} \, \amalg_{\mathcal{M}^*} \cdot \left(\sum_{p \in P} e^{-2\pi i \langle p, v \rangle} \hat{f}(p) \right) \\
&= \amalg_{\mathcal{M}^*} \cdot \hat{\mathcal{F}}(\hat{f})(v).
\end{aligned}
$$

The isometry property of the DFT follows from

$$
\begin{aligned}
\langle \hat{f}, \hat{g} \rangle_P &= \sum_{p \in P} \hat{f}^\dagger(p) \hat{g}(p) \\
&= \det(\mathcal{M})^2 \sum_{p \in P} \left(\sum_{\mu_1 \in \mathcal{M}} \tilde{f}^\dagger(p - \mu_1) \right) \left(\sum_{\mu_2 \in \mathcal{M}} \tilde{g}(p - \mu_2) \right) \\
&= \det(\mathcal{M})^2 \sum_{\lambda \in \mathcal{L}} \tilde{f}^\dagger(\lambda) \left(\sum_{\mu \in \mathcal{M}} \tilde{g}(\lambda - \mu) \right) \\
&= \det(\mathcal{M}) \langle \tilde{f}, \hat{g} \rangle_{\mathcal{L}} \\
&= \det(\mathcal{M})^2 \langle f, \amalg_{\mathcal{L}} \cdot (\amalg_{\mathcal{M}} * g) \rangle \\
&= \frac{\det(\mathcal{M})}{\det(\mathcal{L})} \langle F, \amalg_{\mathcal{L}^*} * (\amalg_{\mathcal{M}^*} \cdot G) \rangle \\
&= \frac{\det(\mathcal{M})}{\det(\mathcal{L})} \langle F, \amalg_{\mathcal{M}^*} \cdot (\amalg_{\mathcal{L}^*} * G) \rangle \\
&= \det(\mathcal{M}) \det(\mathcal{L}) \langle \hat{F}, \hat{G} \rangle_{P^*}.
\end{aligned}
$$

The last step in this derivation follows from reversing the other steps, replacing the spatial functions f and g by their frequency domain counterparts F and G.

A.3 Proof of Theorem 4.6

PROOF 4.9

$$
\begin{aligned}
\tilde{F}^{(2)}(v) &= \frac{1}{\det(\mathcal{L}_2)} \sum_{\lambda_2^* \in \mathcal{L}_2^*} F(v - \lambda_2^*) \\
&= \frac{1}{\det(\mathcal{L}_2)} \sum_{p^* \in P^*} \sum_{\lambda_1^* \in \mathcal{L}_1^*} F(v - p^* - \lambda_1^*) \\
&= \frac{\det(\mathcal{L}_1)}{\det(\mathcal{L}_2)} \sum_{p^* \in P^*} \tilde{F}^{(1)}(v - p^*) \\
&= \frac{1}{\iota(\mathcal{L}_2, \mathcal{L}_1)} \sum_{p^* \in P^*} \tilde{F}^{(1)}(v - p^*) \\
&= \frac{1}{\#(P^*)} \sum_{p^* \in P^*} \tilde{F}^{(1)}(v - p^*).
\end{aligned}
$$

A.4 Proof of Theorem 4.7

PROOF 4.10

$$
\begin{aligned}
\mathcal{F}(f^A)(v) &= \int_{\mathbb{R}^n} e^{-2\pi i \langle x, v \rangle} f^A(x) dx \\
&= \int_{\mathbb{R}^n} e^{-2\pi i \langle x, v \rangle} f(A^{-1}x) dx \\
&= |\det(A)| \int_{\mathbb{R}^n} e^{-2\pi i \langle Ay, v \rangle} f(y) dy \\
&= |\det(A)| \int_{\mathbb{R}^n} e^{-2\pi i \langle y, A^t v \rangle} f(y) dy \\
&= |\det(A)| F(A^t v) \\
&= |\det(A)| F^{A^{-t}}(v).
\end{aligned}
$$

A.5 Proof of Theorem 4.8

PROOF 4.11

$$
\begin{aligned}
\tilde{\mathcal{F}}(\tilde{f}^A)(v) &= \sum_{\lambda_2 \in \mathcal{L}_2} e^{-2\pi i \langle \lambda_2, v \rangle} \tilde{f}^A(\lambda_2) \\
&= \sum_{\lambda_2 \in \mathcal{L}_2} e^{-2\pi i \langle \lambda_2, v \rangle} \tilde{f}(A^{-1}\lambda_2) \\
&= \sum_{\lambda_1 \in \mathcal{L}_1} e^{-2\pi i \langle A\lambda_1, v \rangle} \tilde{f}(\lambda_1) \\
&= \sum_{\lambda_1 \in \mathcal{L}_1} e^{-2\pi i \langle \lambda_1, A^t v \rangle} \tilde{f}(\lambda_1) \\
&= \tilde{\mathcal{F}}(\tilde{f})^{A^{-t}}(v).
\end{aligned}
$$

Glossary of Symbols and Expressions

$\mathbb{Z}^n$	n-dimensional integer space
$\mathbb{R}^n$	n-dimensional real space
$\mathbb{C}^n$	n-dimensional complex space
CSFT	Continuous space-time Fourier transform
DSFT	Discrete space-time Fourier transform
DFT	Discrete Fourier transform
$\mathcal{L}, \mathcal{M}$	Sampling lattice
λ, μ	Elements of lattice $\mathcal{L}, \mathcal{M}$
λ^*, μ^*	Elements of reciprocal lattice $\mathcal{L}^*, \mathcal{M}^*$
$[L]$	Lattice generated by matrix L
$\#(A)$	Number of points of set A
$\mathrm{vol}(A)$	Volume (measure) of set A
$\det(\mathcal{L})$	Determinant of lattice $\mathcal{L}$
$\iota(\mathcal{M}, \mathcal{L})$	Index of lattice $\mathcal{M}$ w.r.t. lattice $\mathcal{L}$
$\mathcal{L}/\mathcal{M}$	Coset group of lattice $\mathcal{M}$ w.r.t. lattice $\mathcal{L}$
$\mathcal{L}^*$	Reciprocal lattice of $\mathcal{L}$
$\amalg_{\mathcal{L}}$	Lattice comb
P	Fundamental domain
$\|\alpha\|_2$	L_2-norm of α
α^t	Hermitian transpose of α
$\langle \alpha, \beta \rangle_{\mathcal{N}}$	Inner products of α and β with respects to $\mathcal{N}$-norm
$\alpha^{\dagger}$	Complex conjugate of α
$\alpha \cdot \beta$	Point-wise multiplication
$\alpha * \beta$	Convolution
$f^A(x)$	Change of variables $f(A^{-1}x)$
X_A	Characteristic function of set A
$\mathcal{F}$	Continuous space-time Fourier transform
$\tilde{\mathcal{F}}$	Discrete space-time Fourier transform
$\hat{\mathcal{F}}$	Discrete Fourier transform
$\Sigma_{\mathcal{L}}$	Sampling operator
$\Pi_{\mathcal{L}}$	Periodizing operator

$$\mathrm{sinc}(x) \quad \begin{cases} \sin(\pi x)/\pi x & \text{if } x \neq 0 \\ 1 & \text{if } x = 0 \end{cases}$$

5

Analog-to-Digital Conversion Architectures

Stephen Kosonocky
IBM Corporation
T.J. Watson Research Center

Peter Xiao
NeoParadigm Labs, Inc.

5.1 Introduction

Digital signal processing methods fundamentally require that signals are quantized at discrete time instances and represented as a sequence of words consisting of 1's and 0's. In nature, signals are usually nonquantized and continuously varied with time. Natural signals such as air pressure waves as a result of speech are converted by a transducer to a proportional analog electrical signal. Consequently, it is necessary to perform a conversion of the analog electrical signal to a digital representation or vice versa if an analog output is desired. The number of quantization levels used to represent the analog signal and the rate at which it is sampled is a function of the desired accuracy, bandwidth that is required, and the cost of the system. Figure 5.1 shows the basic elements of a digital signal processing system. The analog signal is

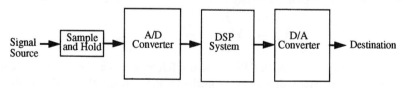

FIGURE 5.1   Digital signal processing system.

first converted to a discrete time signal by a sample and hold circuit. The output of the sample and hold is then applied to an analog-to-digital converter (A/D) circuit where the sampled analog signal is converted to a digitally coded signal. The digital signal is then applied to the digital signal processing (DSP) system where the desired DSP algorithm is performed. Depending on the application, the output of the DSP system can be used directly in digital form or converted back to an analog signal by a digital-to-analog converter (D/A). A digital filtering application may produce an analog signal as its output, whereas a speech recognition system may pass the digital output of the DSP system to a computer system for further processing. This section will describe basic converter terminology and a sample of common architectures for both conventional Nyquist rate converters and oversampled delta-sigma converters.

5.2 Fundamentals of A/D and D/A Conversion

The analog signal can be given as either a voltage signal or current signal, depending on the signal source. Figure 5.2 shows the ideal transfer characteristics for a 3-bit A/D conversion. The output of the converter

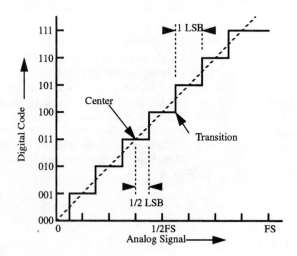

FIGURE 5.2   Ideal transfer characteristics for an A/D converter.

is an n-bit digital code given as,

$$D = \frac{A_{sig}}{FS} = \frac{b_n}{2^n} + \frac{b_{n-1}}{2^{n-1}} + \ldots + \frac{b_1}{2^1} \tag{5.1}$$

where A_{sig} is the analog signal, FS is the analog full scale level, and b_n is a digital value of either 0 or 1. As shown in the figure, each digital code represents a quantized analog level. The width of the quantized region is one least-significant bit (LSB) and the ideal response line passes through the center of each quantized region. The converse D/A operation can be represented as viewing the digital code in Fig. 5.2 as the input and the analog signal as the output. An n-bit D/A converter transfer equation is given as

$$A_{sig} = FS \left(\frac{b_n}{2^n} + \frac{b_{n-1}}{2^{n-1}} + \ldots + \frac{b_1}{2^1} \right) \tag{5.2}$$

where A_{sig} is the analog output signal, FS is the analog full scale level and b_n is a binary coefficient.

The *resolution* of a converter is defined as the smallest distinct change that can be resolved (produced) at an analog input (output) for an A/D (D/A) converter. This can be expressed as

$$\Delta A_{sig} = \frac{FS}{2^N} \tag{5.3}$$

where ΔA_{sig} is the smallest reproducible analog signal for an N-bit converter with full scale analog signal of FS.

The *accuracy* of a converter, often referred to also as *relative accuracy,* is the worst-case error between the actual and the ideal converter output after gain and offset errors are removed [1]. This can be quantified as the number of equivalent bits of resolution or as a fraction of an LSB.

The *conversion rate* specifies the rate at which a digital code (analog signal) can be accurately converted into an analog signal (digital code). Accuracy is often expressed as a function of conversion rate and the two are closely linked. The conversion rate is often an underlying factor in choosing the converter architecture. The speed and accuracy of analog components are a limiting factor. Sensitive analog operations can either be done in parallel, at the expense of accuracy, or cyclicly reused to allow high accuracy with lower conversion speeds.

5.2.1   Nonideal A/D and D/A Converters

Actual A/D and D/A converters exhibit deviations from the ideal characteristics shown in Fig. 5.2. Integration of a complete converter on a single monolithic circuit or as a macro within a very large scale integration (VLSI) DSP system presents formidable design challenges. Converter architectures and design trade-offs are most often dictated by the fabrication process and available device types. Device parameters such as voltage threshold, physical dimensions, etc. vary across a semiconductor die. These variations can manifest themselves into errors. The following terms are used to describe converter nonideal behavior:

1. *Offset error,* described in Fig. 5.3, is a d.c. error between the actual response with the ideal response. This can usually be removed by trimming techniques.

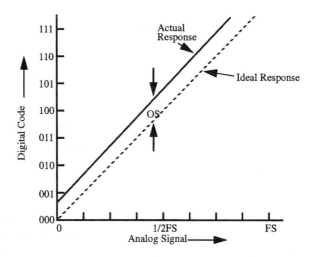

FIGURE 5.3   Offset error.

2. *Gain error* is defined as an error in the slope of the transfer characteristic shown in Fig. 5.4, which can also usually be removed by trimming techniques.
3. *Integral nonlinearity* is the measure of worst-case deviation from an ideal line drawn between the full scale analog signal and zero. This is shown in Fig. 5.5 as a monotonic nonlinearity.
4. *Differential nonlinearity* is the measure of nonuniform step sizes between adjacent steps in a converter. This is usually specified as a fraction of an LSB.
5. *Monotonicity* in a converter specifies that the output will increase with an increasing input. Certain converter architectures can guarantee monotonicity for a specified number of bits of resolution. A nonmonotonic transfer characteristic is detailed in Fig. 5.6.

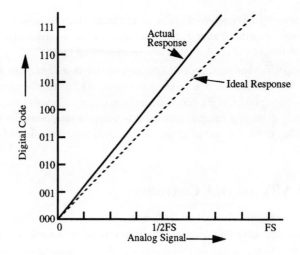

FIGURE 5.4   Gain error.

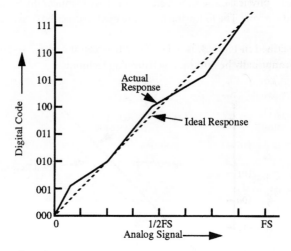

FIGURE 5.5   Monotonic nonlinearity.

6. *Settling time* for D/A converters refers to the time taken from a change of the digital code to the point at which the analog output settles within some tolerance around the final value.

7. *Glitches* can occur during changes in the output at major transitions, i.e., at 1 MSB, 1/2 MSB, 1/4 MSB. During large changes, switching time delays between internal signal paths can cause a spike in the output.

The choice of converter architecture can greatly affect the relative weight of each of these errors. Data converters are often designed for low cost implementation in standard digital processes, i.e., digital CMOS, which often do not have well-controlled resistors or capacitors. Absolute values of these devices can vary by as much as ± 20% under typical process tolerances. Post-fabrication trimming techniques can be used to compensate for process variations, but at the expense of added cost and complexity to the manufacturing process. As will be shown, various architectural techniques can be used to allow high speed or highly accurate data conversion with such variations of process parameters.

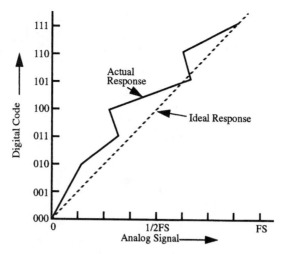

FIGURE 5.6   Nonmonotonic nonlinearity.

5.3   Digital-to-Analog Converter Architecture

The digital-to-analog (D/A) converter, also known as a DAC, decodes a digital word into a discrete analog level. Depending on the application, this can be either a voltage or current. Figure 5.7 shows a high level block diagram of a D/A converter. A binary word is latched and decoded and drives a set of switches that control a scaling network. A basic analog scaling network can be based on voltage scaling, current scaling, or charge scaling [1, 2]. The scaling network scales the appropriate analog level from the analog reference circuit and applies it to the output driver. A simple serial string of identical resistors between a reference voltage and ground can be used as a voltage scaling network. Switches can be used to tap voltages off the resistors and apply them to the output driver. Current scaling approaches are based on switched scaled current sources. Charge scaling is achieved by applying a reference voltage to a capacitor divider using scaled capacitors where the total capacitance value is determined by the digital code [1]. Choice of the architecture depends on the available components in the target technology, conversion rate, and resolution. Detailed description of these trade-offs and designs can be found in the references [1]–[5].

5.4   Analog-to-Digital Converter Architectures

The analog-to-digital (A/D) converter, also known as an ADC, encodes an analog signal into a digital word. Conventional converters work by sampling the time varying analog signal at a sufficient rate to fully resolve the highest frequency components. According to the sampling theorem, the minimum sampling rate is twice the frequency of the highest frequency contained in the signal source. The sampling rate requirement thus becomes the major deterministic factor in choosing a proper converter architecture. Certain architectures exploit parallelism to achieve high speed operation on the order of 100's of MHz, and others which can be used for high accuracy 16-bit resolution for signals with maximum frequencies on the order of 10's of KHz.

5.4.1   Flash A/D

The flash A/D, also known as a parallel A/D, is the highest speed architecture for A/D conversion since maximum parallelism is used. Figure 5.8 shows a block diagram of a 3-bit flash A/D converter. A flash converter requires $2^n - 1$ analog comparators, $2^n - 1$ reference voltages, and a digital encoder. The reference voltages are required to be evenly spaced between 0.5 LSB above the most negative signal and 1.5

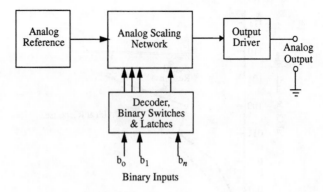

FIGURE 5.7   Basic D/A converter block diagram.

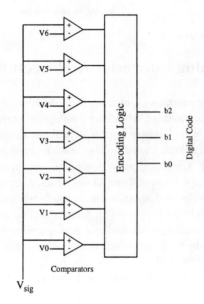

FIGURE 5.8   3-bit flash A/D converter.

LSB below the most positive signal and spaced 1 LSB apart. Each reference voltage is applied to the negative input of a comparator and the analog signal voltage is applied simultaneously to all the comparators. A thermometer code results at the output of the comparators which is converted to a digital word by encoding logic. The speed of the converter is limited by the time delay through a comparator and the encoding logic. This speed is gained at the expense of accuracy, which is limited by the ability to generate evenly spaced reference voltages and the precision of the comparators. Each analog comparator must be precisely matched in order to achieve acceptable performance at a given resolution. For these reasons, flash A/D converters are typically used only for very high speed low resolution applications.

5.4.2   Successive Approximation A/D Converter

A successive approximation A/D converter is formed creating a feedback loop around a D/A converter. Figure 5.9 shows a block diagram for an 8-bit successive approximation A/D. The operation of the converter works by initializing the successive approximation register (SAR) to a value where all bits are set to 0 except the MSB which is set to 1. This represents the mid-level value. The analog signal is applied to a sample-and-

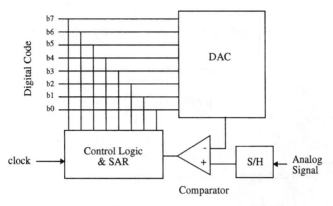

FIGURE 5.9   8-bit successive approximation A/D converter.

hold (S/H) circuit, and on the first clock cycle the DAC converts the digital code stored in the SAR into an analog signal. The comparator is used to determine whether the analog signal is greater or less than the mid level, and control logic determines whether to leave the MSB set to 1 or to change it back to 0. The process is repeated on the next clock cycle, but instead the next MSB is tested. For an n-bit converter n clock cycles are required to fully quantize each sample-and-hold signal. The speed of the successive approximation converter is largely limited by the speed of the DAC and the time delay through the comparator. This type of converter is widely used for medium speed and medium accuracy applications. The resolution is limited by the DAC converter and the comparator.

5.4.3   Pipelined A/D Converter

A pipelined A/D converter achieves high-speed conversion and high accuracy at the expense of latency in the conversion process. A pipelined A/D converter block diagram is shown in Fig. 5.10. The conversion

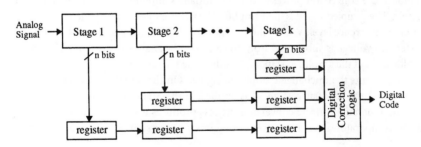

FIGURE 5.10   Pipelined A/D converter.

process is broken into multiple stages where, at each stage, a partial conversion is done and the converted bits are shifted down the pipeline in digital registers. Figure 5.11 shows the detail of a single pipeline stage. The analog signal is applied to a sample-and-hold circuit and the output is applied to an n-bit flash ADC where n is less then the total desired resolution. The outputs of the ADC are connected directly to a DAC, and the output of the DAC is subtracted from the original analog signal stored in the S/H to produce a residual signal. The residual signal is then amplified by 2^n so that it will vary within the entire full scale range of the next stage and is transferred on the next clock cycle. At this point the first stage begins conversion on the next analog sample. The maximum conversion rate is determined by the time delay through a single stage. Pipelining allows high resolution conversion without the need for many comparators. An 8-bit converter can be ideally constructed with $k = 4$ stages with $n = 2$ bits of resolution

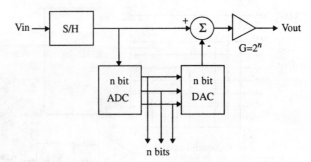

FIGURE 5.11   Diagram of single pipelined A/D converter stage.

per stage, requiring only 12 total comparators. This can be contrasted with an 8-bit flash converter requiring 255 comparators. Each pipeline stage adds an additional cycle of latency before the final code is converted. Pipelined converters also accommodate digital correction schemes for errors generated in the analog circuitry. Digital correction can be achieved by using higher resolution ADC and DAC circuits in each stage than required so that errors in the preceding stage can be detected and corrected digitally [5]. Auto calibration can also be achieved by adding additional stages after the required stages to convert errors in the DAC values and storing these digitally to be added to the final result [6].

5.4.4   Cyclic A/D Converter

Cyclic A/D converters, also known as algorithmic converters, trade off conversion speed for high accuracy without the need for calibration or device trimming. Figure 5.12 shows a block diagram of a cyclic A/D converter [5]. Here the same analog components are cyclicly reused for conversion of each bit for each analog sample. The conversion process works by initially sampling the input signal by setting switch S1 appropriately. The sampled signal is then amplified by a factor of two and applied to a comparator where it is compared to a reference level, Vref. If the voltage exceeds the reference level, a bit value of 1 is produced and the reference voltage is subtracted from the amplified signal by control of switch S2 to produce the residual voltage V_e. If the amplified signal is less than the reference voltage, Vref, the comparator outputs a 0, and V_e represents the unchanged amplified signal. On the remaining cycles for the sample, switch S1 changes so that the residual voltage V_e is applied to the S/H circuit. The cycle is repeated for each remaining bit. Operation on the conversion process produces a serial stream of digital bit values from output of the comparator. An n-bit converter requires n conversion cycles for each sampled signal.

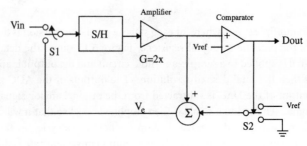

FIGURE 5.12   Block diagram of a cyclic A/D converter.

5.5 Delta-Sigma Oversampling Converter

The oversampling delta-sigma A/D converter was first proposed 30 years ago [7], while it only became popular after the maturity of the VLSI digital technology. With the advancement of semi-conductor technology, an increasing portion of signal processing tasks have been shifted from the usual analog domain to digital domain. For digital systems to interact with analog signal sources, such as voice, data, and video, the role of analog-to-digital interface is essential. In voice data processing and communication, an accurate digital form is often desired to represent the voice. Due to the large demand of these systems, the cost must be kept at a minimum. All these requirements call upon a need to implement monolithic high resolution analog-to-digital interfaces in economical semiconductor technology. However, with the increasing complexity of integration and a trend of reducing supply voltage, the accuracy of device components and analog signal dynamic range deteriorate. It becomes more difficult to realize high resolution conversions by conventional Nyquist rate converter architecture.

Compared to Nyquist rate converters, the oversampling converters use coarse analog components at the front end and employ more digital signal processing in the later stages. High resolution conversions are achieved by trading off speed and digital signal processing complexity, both of which can be easily realized in modern VLSI technology.

The oversampling A/D converter and Nyquist rate converter are compared in Fig. 5.13. A nonover-sampled A/D converter has an anti-aliasing lowpass filter in the front. The anti-aliasing filter attenuates high-frequency components buried in the analog input and prevents them from being aliased into the signal frequency band. Because the converter is sampled at the Nyquist rate, which is twice the input signal bandwidth, the anti-aliasing filter's transition band must be very narrow and its stop-band must have enough suppression of the out-of-band noise. This requirement makes the filter very complex and adds to the complexity that a nonoversampled A/D already has.

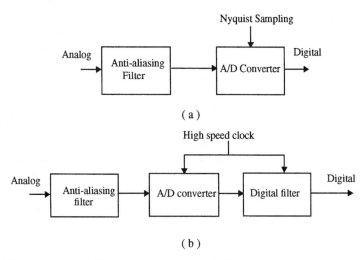

FIGURE 5.13   (a) Nonoversampled A/D converter. (b) oversampled A/D converter.

In comparison, an oversampled delta-sigma A/D converter, as shown in Fig. 5.13(b), is sampled at a higher rate than the input Nyquist rate. A simple first-order lowpass filter is sufficient to attenuate the noise components at the sampling frequency region to avoid the noise aliasing. This is because only the noise components close to the sampling frequency can be aliased back into the signal band. This arrangement simplifies the design and implementation of the filter. The complexity of the A/D itself is much simpler than the nonoversampled A/D converters as we will see later. The only extra complexity in the oversampled

A/D converters is that more digital signal processing is required after the A/D conversion. But this becomes less and less an issue with the advancement of the VLSI technology. In the following sections, we will explain the conversion principle and various architectures of the oversampling delta-sigma converter.

5.5.1 Delta-Sigma A/D Converter Architecture

Delta-Sigma Oversampling A/D Converter Principle

The structure of a first-order delta-sigma converter is shown in Fig. 5.14. The input signal is sampled

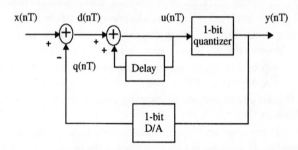

FIGURE 5.14   The modulator of a first-order delta-sigma converter. T is the sampling period and n is the index.

at a frequency $f_s (T = 1/f_s)$. A feedback signal from a 1-bit D/A converter is subtracted from the input and the residue signal is accumulated by an integrator. The output of the integrator is quantized to generate a 1-bit digital stream. This digital output sets the sign of the feedback. If the digital output is **1**, it feeds back a large negative signal to subtract from the input signal. The net effect of the feedback loop is to keep the output of the integrator small so that the output digits always track the amplitudes of the input signal.

The resolution of an A/D converter is determined by the quantization noise generated in the process. Even though a delta-sigma converter only has an 1-bit quantizer, much higher resolution is achieved by employing the noise shaping mechanism to move the noise out of the signal band and later blocking it using a lowpass digital filter.

Quantization is a nonlinear process and the feedback mechanism makes the noise highly dependent on the input signal spectrum. Rigorous treatment of this noise component in a delta-sigma converter can be found in the literature [8]. Useful information can still be obtained by linearizing the quantization process. The noise component is approximated by white additive noise uniformly distributed up to half of the sampling frequency. This approximation is valid because over a long period of time, the input to the quantizer will spread over a large number of values and appear to be quasi-random, so the noise introduced is quasi-random as well. Similar to a nonoversampled A/D converter, the rms value of the noise is $e_{rms}^2 = \frac{\Delta^2}{12}$, where Δ is the quantization step. When the quantizer is sampled at f_s, the noise power is sampled into a frequency band: $0 \le f < f_s/2$ and its spectral density is

$$Q(f) = \sqrt{2} \cdot e_{rms} \tag{5.4}$$

where f is normalized to f_{-s}.

The delta-sigma converter can be generalized as shown in Fig. 5.15. The forward path is modeled by transfer function $B(z)$ plus the noise, and the feedback path can be modeled by $C(z)$. The system output and input transfer function is governed by

$$Y(z) = \frac{B(z) \cdot X(z) + Q}{1 + B(z) \cdot C(z)} \tag{5.5}$$

To achieve high-resolution A/D conversion, the system needs to convert the input signal within a specified frequency bandwidth and minimize the noise component in that band. One method is to pass the signal

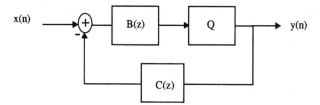

FIGURE 5.15   General feedback system.

component and block the noise component. This can be expressed as

$$Y(z) = X(z) + H_{ns}(z) \cdot Q , \qquad (5.6)$$

where the input $X(z)$ passes through the system, but the quantization noise is modified by a noise-shaping function $H_{ns}(z)$.

Comparing Eq. 5.5 to Eq. 5.6, to achieve the noise-shaping effect, the system in Fig. 5.15 needs to have the following property:

$$C(z) = 1 - \frac{1}{B(z)} \qquad (5.7)$$

$$B(z) = \frac{1}{H_{ns}(z)}$$

Now, we can see the delta-sigma A/D converter shown in Fig. 5.14 as a noise-shaping data converter. The transfer function of the integrator in the forward pass is $\frac{1}{1-z^{-1}}$; the D/A converter in the feedback path is equivalent to a delay element and its transfer function is z^{-1}. They satisfy the relation required by a noise-shaping converter in Eq. 5.7. Therefore, its noise-shaping function $H_{ns}(z)$ is

$$H_{ns}(z) = \frac{1}{B(z)} = 1 - z^{-1} \qquad (5.8)$$

which is a highpass filtering function. The amplitude of its response is

$$|H_{ns}(z)| = |1 - z^{-1}| = 2\sin(\pi f) \qquad (5.9)$$

where f is the normalized frequency with respect to f_s. This function is plotted in Fig. 5.16. As shown in the figure, the noise is evenly distributed across the frequency, before applying the noise shaping function. The noise power in the signal band is the area of a region highlighted by the grey color underneath the flat line. After applying the noise-shaping function, the noise in the signal band is suppressed to a much lower level and the total noise power left (dark grey region) is much smaller than the original noise power. The high-frequency noise portion will be filtered by the digital filter. Therefore, the signal-to-noise ratio of the converter is greatly enhanced.

Quantitatively, the noise power left in the signal band is the integration of its spectrum up to signal bandwidth f_b as

$$N^2 = \int_0^{f_b/f_s} \left(|H_{ns}(z)|^2 Q^2 \right) df = \frac{2\Delta^2}{3 f_s} \int_0^{f_b/f_s} [\sin(\pi f)]^2 df \qquad (5.10)$$

where Q^2 is substituted for the noise spectral density in Eq. 5.4. In a delta-sigma converter the signal bandwidth is significantly lower than the sampling frequency. The resulting integration is

$$N_q^2 = \frac{2\pi^2 \Delta^2}{9} \left(\frac{f_b}{f_s} \right)^3 . \qquad (5.11)$$

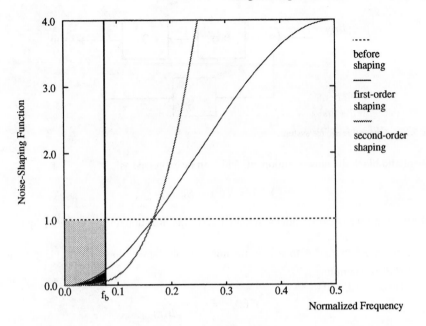

FIGURE 5.16   Plot of noise-shaping effect of the delta-sigma modulator comparing the noise power left within the baseband f_h. The noise (cross-hatched region) of a first-order modulator is much less than the noise before shaping (shaded region). Noise from the second-order shaping is even less.

For a sine wave input, the maximum signal amplitude is $\frac{\Delta}{2}$ and its average power is $\frac{\Delta^2}{8}$. This gives a peak signal-to-noise ratio (SNR) as

$$\frac{S^2}{N^2} = \frac{9}{16\pi^2} \left(\frac{f_s}{f_b} \right)^3 . \tag{5.12}$$

We can see that the peak SNR is only a function of the frequency ratio $\frac{f_s}{f_b}$. The faster the converter is sampled, the higher the resolution can be achieved. The expression in Eq. 5.12 can be transformed into

$$SNR = 10 \log_{10} \frac{S^2}{N^2} = 20 \log_{10} \left(\frac{3}{\sqrt{2\pi}} \right) + 9 \log_2 M (dB) , \tag{5.13}$$

where M is an important parameter called the oversampling ratio, defined as the ratio of the sampling frequency over the Nyquist sampling frequency $2 f_b$. From this expression, we can see that we can get $9\,dB$ of increase in SNR for every doubling of the sampling frequency. This corresponds to 1.5 bits. For example, if $M = 128$, we have 11.5 bits more resolution than sampling at the Nyquist rate. This method allows a high resolution A/D conversion by using only a one-bit quantizer.

We can see that higher resolution is achieved by trading off the input signal bandwidth. In order to get 1.5 more bits, the bandwidth has to be cut by a half in this structure. To have a more favorable resolution and bandwidth trade-off, we can go to higher order delta-sigma converters.

Higher-Order Single-Stage Converters

In the first-order delta-sigma converter, the noise-shaping function is $H_{ns}(z) = 1 - z^{-1}$. Higher order converters can allow the noise-shaping function go up to Lth power, given as

$$H_{ns}(z) = \left(1 - z^{-1} \right)^L , \tag{5.14}$$

where L is an integer greater than one. Thus, the magnitude of this noise-shaping function is

$$|H_{ns}(z)| = \left|\left(1 - z^{-1}\right)^L\right| = [2\sin(\pi f)]^L . \tag{5.15}$$

This function is also plotted in Fig. 5.16 for $L = 2$. As seen in the figure, more noise from the signal band is blocked than with the first-order function. Integrating Eq. 5.14 over the signal band allows calculation of the SNR of an Lth order delta-sigma converter as

$$\frac{S^2}{N^2} = \frac{3(2L + 1)}{2^{2L+2} \cdot \pi^{2L}} \cdot \left(\frac{f_s}{f_b}\right)^{2L+1} , \tag{5.16}$$

which is equivalent to

$$SNR = 20\log_{10} = \left(\frac{\sqrt{3(2L + 1)/2}}{\pi^L}\right) + 3(2L + 1)\log_2 M (dB) , \tag{5.17}$$

where M is the oversampling ratio. For every doubling of the sampling frequency, the SNR is increased by $3(2L + 1)dB$, i.e., $L + 0.5$ bits more resolution. For example, $L = 2$ adds 2.5 bits and $L = 3$ adds 3.5

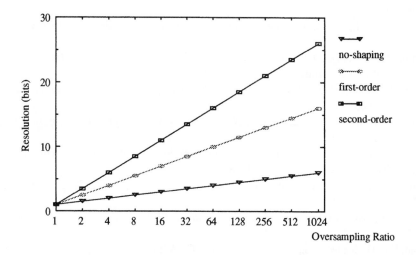

FIGURE 5.17 A plot of the resolution vs. oversampling ratio for different types of delta-sigma converters and Nyquist sampling converter.

bits of resolution. Therefore, compared to the first-order system, by employing a higher order delta-sigma converter architecture, the same resolution can be achieved with a lower sampling frequency, or a higher input bandwidth can be allowed at the same resolution with the same sampling frequency. Figure 5.17 shows a plot of Eq. 5.17 comparing resolution vs. oversampling ratio for different order delta-sigma converters.

A second-order delta-sigma converter can be realized as shown in Fig. 5.18 with two integrators. Higher order converters can be similarly constructed. However, when the order of the converter is greater than two, special care must be taken to insure the converter stability [9]. More zeroes are introduced in the transfer function of the forward path to suppress the signal swing after the integrators.

Other methods can be used to improve the resolution of the delta-sigma converter. A first-order and a second-order converter can be cascaded to achieve the same performance as a third-order converter, but with better stability over the frequency range [10]. A multi-bit quantizer can also be used to replace the

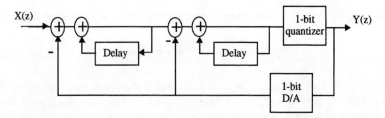

FIGURE 5.18   Block diagram of a second order D-S modulator.

1-bit quantizer in the architecture presented here [11]. This improves the resolution at the same sampling speed. Interested readers are referred to reference articles.

In an oversampling converter, the digital decimation filter is also an integral part. Only after the decimation filter is the resolution of the converter realized. The design of decimation filters are discussed in other sections of this book and can also be found in the reference article by Candy [12].

References

[1] Grebene, A.B., *Bipolar and MOS Analog Integrated Circuit Design,* John Wiley & Sons, New York, 1984.

[2] Sheingold, D.H., Ed., *Analog-Digital Conversion Handbook,* Prentice-Hall, Englewood Cliffs, NJ, 1986.

[3] Toumazou, C., Lidgey F.J., and Haigh, D.G., eds., *Analogue IC Design: The Current-Mode Approach,* Peter Peregrinus Ltd., London, 1990.

[4] Gray, P.R., Hodges, D.A., Broderson, R.W., eds., *Analog MOS Integrated Circuits,* IEEE Press, New York, 1980.

[5] Gray, P.R., Wooley, B.A., Broderson, R.W., eds., *Analog MOS Integrated Circuits, II,* IEEE Press, New York, 1989.

[6] Lee, S.H, Song B.S, Digital-domain calibration of multistep analog-to-digital converters, *IEEE J. Solid-State Circuits,* 27: (12) 1679–1688, Dec., 1992.

[7] Inose, H. and Yasuda, Y., A unity bit coding method by negative feedback, *Proc. IEEE,* 51: 1524–1535, Nov., 1963.

[8] Gray, R.M., Oversampled sigma-delta modulation, *IEEE Trans. Commun.,* 35: 481–489, May, 1987.

[9] Chao, K.C-H., Nadeem, S., Lee, W.L., Sodini, C.G., A higher order topology for interpolative modulators for oversampled A/D converters, *IEEE Trans. Circuits and Syst.,* CAS-37: 309–318, March, 1990.

[10] Matsuya, Y., Uchimura, K., Iwata, A., Kobayashi, T., Ishikawa, M., and Yoshitoma, T., A 16-bit oversampling A-to-D conversion technology using triple-integration noise shaping, *IEEE J. Solid-State Circuits,* SC-22: 921–929, Dec., 1987.

[11] Larson, L.E., Cataltepe, T., and Temes, G.C., Multibit oversampled $\Sigma - \Delta$ A/D converter with digital error correction, *Electron. Lett.,* 24: 1051–1052, Aug., 1988.

[12] Candy, J.C., Decimation for sigma delta modulation, *IEEE Trans. Commun.,* COM-24: 72–76, Jan., 1986.

6

Quantization of Discrete Time Signals

Ravi P. Ramachandran
Rowan University

6.1 Introduction

Signals are usually classified into four categories. A continuous time signal $x(t)$ has the field of real numbers **R** as its domain in that t can assume any real value. If the range of $x(t)$ (values that $x(t)$ can assume) is also **R**, then $x(t)$ is said to be a continuous time, continuous amplitude signal. If the range of $x(t)$ is the set of integers **Z**, then $x(t)$ is said to be a continuous time, discrete amplitude signal. In contrast, a discrete time signal $x(n)$ has **Z** as its domain. A discrete time, continuous amplitude signal has **R** as its range. A discrete time, discrete amplitude signal has **Z** as its range. Here, the focus is on discrete time signals. Quantization is the process of approximating any discrete time, continuous amplitude signal into one of a finite set of discrete time, continuous amplitude signals based on a particular distortion or distance measure. This approximation is merely signal compression in that an infinite set of possible signals is converted into a finite set. The next step of encoding maps the finite set of discrete time, continuous amplitude signals into a finite set of discrete time, discrete amplitude signals.

A signal $x(n)$ is quantized one block at a time in that p (almost always consecutive) samples are taken as a vector **x** and approximated by a vector **y**. The signal or data vectors **x** of dimension p (derived from $x(n)$) are in the vector space $\mathbf{R}^p$ over the field of real numbers **R**. Vector quantization is achieved by mapping the infinite number of vectors in $\mathbf{R}^p$ to a finite set of vectors in $\mathbf{R}^p$. There is an inherent compression of the data vectors. This finite set of vectors in $\mathbf{R}^p$ is encoded into another finite set of vectors in a vector space of dimension q over a finite field (a field consisting of a finite set of numbers). For communication applications, the finite field is the binary field (0, 1). Therefore, the original vector **x** is converted or compressed into a bit stream either for transmission over a channel or for storage purposes. This compression is necessary due to channel bandwidth or storage capacity constraints in a system.

The purpose of this chapter is to describe the basic definition and properties of vector quantization, introduce the practical aspects of design and implementation, and relate important issues. Note that two excellent review articles [1, 2] give much insight into the subject. The outline of the article is as follows. The basic concepts are elaborated on in Section 6.2. Design algorithms for scalar and vector quantizers are described in Section 6.3. A design example is also provided. The practical issues are discussed in Section 6.4. The multistage and split manifestations of vector quantizers are described in Section 6.5. In Section 6.6, two applications of vector quantization in speech processing are discussed.

6.2   Basic Definitions and Concepts

In this section, we will elaborate on the definitions of a vector and scalar quantizer, discuss some commonly used distance measures, and examine the optimality criteria for quantizer design.

6.2.1   Quantizer and Encoder Definitions

A quantizer, Q, is mathematically defined as a mapping [3] $Q : \mathbf{R}^p \rightarrow C$. This means that the p-dimensional vectors in the vector space $\mathbf{R}^p$ are mapped into a finite collection C of vectors that are also in $\mathbf{R}^p$. This collection C is called the codebook and the number of vectors in the codebook, N, is known as the codebook size. The entries of the codebook are known as codewords or codevectors. If $p = 1$, we have a scalar quantizer (SQ). If $p > 1$, we have a vector quantizer (VQ).

A quantizer is completely specified by p, C and a set of disjoint regions in $\mathbf{R}^p$ which dictate the actual mapping. Suppose C has N entries $\mathbf{y}_1$, $\mathbf{y}_2$, $\cdots$, $\mathbf{y}_N$. For each codevector, $\mathbf{y}_i$, there exists a region, R_i, such that any input vector $\mathbf{x} \in R_i$ gets mapped or quantized to $\mathbf{y}_i$. The region R_i is called a Voronoi region [3, 4] and is defined to be the set of all $\mathbf{x} \in \mathbf{R}^p$ that are quantized to $\mathbf{y}_i$. The properties of Voronoi regions are as follows:

1. Voronoi regions are convex subsets of $\mathbf{R}^p$.
2. $\bigcup_{i=1}^{N} R_i = \mathbf{R}^p$.
3. $R_i \cap R_j$ is the null set for $i \neq j$.

It is seen that the quantizer mapping is nonlinear and many to one and hence noninvertible.

Encoding the codevectors $\mathbf{y}_i$ is important for communications. The encoder, E, is mathematically defined as a mapping $E : C \rightarrow C_B$. Every vector $\mathbf{y}_i \in C$ is mapped into a vector $\mathbf{t}_i \in C_B$ where $\mathbf{t}_i$ belongs to a vector space of dimension $q = \lceil \log_2 N \rceil$ over the binary field $(0, 1)$. The encoder mapping is one to one and invertible. The size of C_B is also N. As a simple example, suppose C contains four vectors of dimension p, namely, $(\mathbf{y}_1, \mathbf{y}_2, \mathbf{y}_3, \mathbf{y}_4)$. The corresponding mapped vectors in C_B are $\mathbf{t}_1 = [0\,0]$, $\mathbf{t}_2 = [0\,1]$, $\mathbf{t}_3 = [1\,0]$ and $\mathbf{t}_4 = [1\,1]$. The decoder D described by $D : C_B \rightarrow C$ performs the inverse operation of the encoder.

A block diagram of quantization and encoding for communications applications is shown in Fig. 6.1. Given that the final aim is to transmit and reproduce $\mathbf{x}$, the two sources of error are due to quantization and channel. The quantization error is $\mathbf{x} - \mathbf{y}_i$ and is heavily dealt with in this article. The channel introduces errors that transform $\mathbf{t}_i$ into $\mathbf{t}_j$ thereby reproducing $\mathbf{y}_j$ instead of $\mathbf{y}_i$ after decoding. Channel errors are ignored for the purposes of this article.

6.2.2   Distortion Measure

A distortion or distance measure between two vectors $\mathbf{x} = [x_1 \, x_2 \, x_3 \, \cdots \, x_p]^T \in \mathbf{R}^p$ and $\mathbf{y} = [y_1 \, y_2 \, y_3 \, \cdots \, y_p]^T \in \mathbf{R}^p$ where the superscript T denotes transposition is symbolically given by $d(\mathbf{x}, \mathbf{y})$.

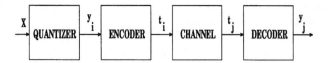

FIGURE 6.1 Block diagram of quantization and encoding for communication systems.

Most distortion measures satisfy three properties given by:

1. Positivity: $d(\mathbf{x}, \mathbf{y})$ is a real number greater than or equal to zero with equality if and only if $\mathbf{x} = \mathbf{y}$
2. Symmetry: $d(\mathbf{x}, \mathbf{y}) = d(\mathbf{y}, \mathbf{x})$
3. Triangle inequality: $d(\mathbf{x}, \mathbf{z}) \leq d(\mathbf{x}, \mathbf{y}) + d(\mathbf{y}, \mathbf{z})$

To qualify as a valid measure for quantizer design, only the property of positivity needs to be satisfied. The choice of a distance measure is dictated by the specific application and computational considerations. We continue by giving some examples of distortion measures.

EXAMPLE 6.1: The L_r Distance

The L_r distance is given by

$$d(\mathbf{x}, \mathbf{y}) = \sum_{i=1}^{p} |x_i - y_i|^r \tag{6.1}$$

This is a computationally simple measure to evaluate. The three properties of positivity, symmetry, and the triangle inequality are satisfied. When $r = 2$, the squared Euclidean distance emerges and is very often used in quantizer design. When $r = 1$, we get the absolute distance. If $r = \infty$, it can be shown that [2]

$$\lim_{r \to \infty} d(\mathbf{x}, \mathbf{y})^{1/r} = \max_i |x_i - y_i| \tag{6.2}$$

This is the maximum absolute distance taken over all vector components.

EXAMPLE 6.2: The Weighted L_2 Distance

The weighted L_2 distance is given by:

$$d(\mathbf{x}, \mathbf{y}) = (\mathbf{x} - \mathbf{y})^T \mathbf{W}(\mathbf{x} - \mathbf{y}) \tag{6.3}$$

where $\mathbf{W}$ is the matrix of weights. For positivity, $\mathbf{W}$ must be positive-definite. If $\mathbf{W}$ is a constant matrix, the three properties of positivity, symmetry, and the triangle inequality are satisfied. In some applications, $\mathbf{W}$ is a function of $\mathbf{x}$. In such cases, only the positivity of $d(\mathbf{x}, \mathbf{y})$ is guaranteed to hold. As a particular case, if $\mathbf{W}$ is the inverse of the covariance matrix of $\mathbf{x}$, we get the Mahalanobis distance [2]. Other examples of weighting matrices will be given when we discuss the applications of quantization.

6.2.3 Optimality Criteria

There are two necessary conditions for a quantizer to be optimal [2, 3]. As before, the codebook C has N entries $\mathbf{y}_1$, $\mathbf{y}_2$, $\cdots$, $\mathbf{y}_N$ and each codevector $\mathbf{y}_i$ is associated with a Voronoi region R_i. The first condition known as the nearest neighbor rule states that a quantizer maps any input vector $\mathbf{x}$ to the codevector closest to it. Mathematically speaking, $\mathbf{x}$ is mapped to $\mathbf{y}_i$ if and only if $d(\mathbf{x}, \mathbf{y}_i) \leq d(\mathbf{x}, \mathbf{y}_j) \; \forall j \neq i$. This enables us to more precisely define a Voronoi region as:

$$R_i = \left\{ \mathbf{x} \in \mathbf{R}^P : d\left(\mathbf{x}, \mathbf{y}_i\right) \leq d\left(\mathbf{x}, \mathbf{y}_j\right) \; \forall j \neq i \right\} \tag{6.4}$$

The second condition specifies the calculation of the codevector $\mathbf{y}_i$ given a Voronoi region R_i. The codevector $\mathbf{y}_i$ is computed to minimize the average distortion in R_i which is denoted by D_i where:

$$D_i = E\left[d\left(\mathbf{x}, \mathbf{y}_i\right) \mid \mathbf{x} \in R_i\right] \tag{6.5}$$

6.3   Design Algorithms

Quantizer design algorithms are formulated to find the codewords and the Voronoi regions so as to minimize the overall average distortion D given by:

$$D = E[d(\mathbf{x}, \mathbf{y})] \tag{6.6}$$

If the probability density $p(\mathbf{x})$ of the data $\mathbf{x}$ is known, the average distortion is [2, 3]

$$D = \int d(\mathbf{x}, \mathbf{y}) p(\mathbf{x}) d\mathbf{x} \tag{6.7}$$

$$= \sum_{i=1}^{N} \int_{R_i} d\left(\mathbf{x}, \mathbf{y}_i\right) p(\mathbf{x}) d\mathbf{x} \tag{6.8}$$

Note that the nearest neighbor rule has been used to get the final expression for D. If the probability density is not known, an empirical estimate is obtained by computing many sampled data vectors. This is called training data, or a training set, and is denoted by $T = \{\mathbf{x}_1, \mathbf{x}_2, \mathbf{x}_3, \cdots \mathbf{x}_M\}$ where M is the number of vectors in the training set. In this case, the average distortion is

$$D = \frac{1}{M} \sum_{k=1}^{M} d\left(\mathbf{x}_k, \mathbf{y}\right) \tag{6.9}$$

$$= \frac{1}{M} \sum_{i=1}^{N} \sum_{\mathbf{x}_k \in R_i} d\left(\mathbf{x}_k, \mathbf{y}_i\right) \tag{6.10}$$

Again, the nearest neighbor rule has been used to get the final expression for D.

6.3.1   Lloyd-Max Quantizers

The Lloyd-Max method is used to design scalar quantizers and assumes that the probability density of the scalar data $p(x)$ is known [5, 6]. Let the codewords be denoted by $y_1, y_2, \cdots, y_N$. For each codeword y_i, the Voronoi region is a continuous interval $R_i = (v_i, v_{i+1}]$. Note that $v_1 = -\infty$ and $v_{N+1} = \infty$. The average distortion is

$$D = \sum_{i=1}^{N} \int_{v_i}^{v_{i+1}} d\left(x, y_i\right) p(x) dx \tag{6.11}$$

Setting the partial derivatives of D with respect to v_i and y_i to zero gives the optimal Voronoi regions and codewords.

In the particular case when $d(x, y_i) = (x - y_i)^2$, it can be shown that [5] the optimal solution is

$$v_i = \frac{y_i + y_{i+1}}{2} \tag{6.12}$$

for $2 \le i \le N$ and

$$y_i = \frac{\int_{v_i}^{v_{i+1}} x p(x) dx}{\int_{v_i}^{v_{i+1}} p(x) dx} \tag{6.13}$$

for $1 \le i \le N$. The overall iterative algorithm is

1. Start with an initial codebook and compute the resulting average distortion.
2. Solve for v_i.
3. Solve for y_i.
4. Compute the resulting average distortion.
5. If the average distortion decreases by a small amount that is less than a given threshold, the design terminates. Otherwise, go back to Step 2.

The extension of the Lloyd-Max algorithm for designing vector quantizers has been considered [7]. One practical difficulty is whether the multidimensional probability density function $p(\mathbf{x})$ is known or must be estimated. Even if this is circumvented, finding the multidimensional shape of the convex Voronoi regions is extremely difficult and practically impossible for dimensions greater than 5 [7]. Therefore, the Lloyd-Max approach cannot be extended to multidimensions and methods have been configured to design a VQ from training data. We will now elaborate on one such algorithm.

6.3.2 Linde-Buzo-Gray Algorithm

The input to the Linde-Buzo-Gray (LBG) algorithm [7] is a training set $T = \{\mathbf{x}_1, \mathbf{x}_2, \mathbf{x}_3, \cdots \mathbf{x}_M\} \in \mathbf{R}^p$ having M vectors, a distance measure $d(\mathbf{x}, \mathbf{y})$, and the desired size of the codebook N. From these inputs, the codewords $\mathbf{y}_i$ are iteratively calculated. The probability density $p(\mathbf{x})$ is not explicitly considered and the training set serves as an empirical estimate of $p(\mathbf{x})$. The Voronoi regions are now expressed as:

$$R_i = \left\{ \mathbf{x}_k \in T : d\left(\mathbf{x}_k, \mathbf{y}_i\right) \le d\left(\mathbf{x}_k, \mathbf{y}_j\right) \ \forall j \ne i \right\} \tag{6.14}$$

Once the vectors in R_i are known, the corresponding codevector $\mathbf{y}_i$ is found to minimize the average distortion in R_i as given by

$$D_i = \frac{1}{M_i} \sum_{\mathbf{x}_k \in R_i} d\left(\mathbf{x}_k, \mathbf{y}_i\right) \tag{6.15}$$

where M_i is the number of vectors in R_i. In terms of D_i, the overall average distortion D is

$$D = \sum_{i=1}^{N} \frac{M_i}{M} D_i \tag{6.16}$$

Explicit expressions for $\mathbf{y}_i$ depend on $d(\mathbf{x}, \mathbf{y}_i)$ and two examples are given. For the L_1 distance,

$$\mathbf{y}_i = \text{median } [\mathbf{x}_k \in R_i] \tag{6.17}$$

For the weighted L_2 distance in which the matrix of weights $\mathbf{W}$ is constant,

$$\mathbf{y}_i = \frac{1}{M_i} \sum_{\mathbf{x}_k \in R_i} \mathbf{x}_k \tag{6.18}$$

which is merely the average of the training vectors in R_i. The overall methodology to get a codebook of size N is

1. Start with an initial codebook and compute the resulting average distortion.
2. Find R_i.
3. Solve for y_i.
4. Compute the resulting average distortion.
5. If the average distortion decreases by a small amount that is less than a given threshold, the design terminates. Otherwise, go back to Step 2.

If N is a power of 2 (necessary for coding), a growing algorithm starting with a codebook of size 1 is formulated as follows:

1. Find codebook of size 1.
2. Find initial codebook of double the size by doing a binary split of each codevector. For a binary split, one codevector is split into two by small perturbations.
3. Invoke the methodology presented earlier of iteratively finding the Voronoi regions and codevectors to get the optimal codebook.
4. If the codebook of the desired size is obtained, the design stops. Otherwise, go back to Step 2 in which the codebook size is doubled.

Note that with the growing algorithm, a locally optimal codebook is obtained. Also, scalar quantizer design can also be performed.

Here, we present a numerical example in which $p = 2$, $M = 4$, $N = 2$, $T = \{x_1 = [0\ 0], x_2 = [0\ 1], x_3 = [1\ 0], x_4 = [1\ 1]\}$, and $d(x, y) = (x - y)^T (x - y)$. The codebook of size 1 is $y_1 = [0.5\ 0.5]$. We will invoke the LBG algorithm twice, each time using a different binary split. For the first run:

1. Binary split: $y_1 = [0.51\ 0.5]$ and $y_2 = [0.49\ 0.5]$.
2. Iteration 1
   (a) $R_1 = \{x_3, x_4\}$ and $R_2 = \{x_1, x_2\}$.
   (b) $y_1 = [1\ 0.5]$ and $y_2 = [0\ 0.5]$.
   (c) Average distortion: $D = 0.25[(0.5)^2 + (0.5)^2 + (0.5)^2 + (0.5)^2] = 0.25$.
3. Iteration 2
   (a) $R_1 = \{x_3, x_4\}$ and $R_2 = \{x_1, x_2\}$.
   (b) $y_1 = [1\ 0.5]$ and $y_2 = [0\ 0.5]$.
   (c) Average distortion: $D = 0.25[(0.5)^2 + (0.5)^2 + (0.5)^2 + (0.5)^2] = 0.25$.
4. No change in average distortion, the design terminates.

For the second run:

1. Binary split: $y_1 = [0.5\ 0.51]$ and $y_2 = [0.5\ 0.49]$.
2. Iteration 1
   (a) $R_1 = \{x_2, x_4\}$ and $R_2 = \{x_1, x_3\}$.
   (b) $y_1 = [0.5\ 1]$ and $y_2 = [0.5\ 0]$.
   (c) Average distortion: $D = 0.25[(0.5)^2 + (0.5)^2 + (0.5)^2 + (0.5)^2] = 0.25$.
3. Iteration 2
   (a) $R_1 = \{x_2, x_4\}$ and $R_2 = \{x_1, x_3\}$.

      (b) $\mathbf{y}_1 = [0.5\ 1]$ and $\mathbf{y}_2 = [0.5\ 0]$.

      (c) Average distortion: $D = 0.25[(0.5)^2 + (0.5)^2 + (0.5)^2 + (0.5)^2] = 0.25$.

    4. No change in average distortion, the design terminates.

The two codebooks are equally good locally optimal solutions that yield the same average distortion. The initial condition as determined by the binary split influences the final solution.

6.4 Practical Issues

When using quantizers in a real environment, there are many practical issues that must be considered to make the operation feasible. First we enumerate the practical issues and then discuss them in more detail. Note that the issues listed below are interrelated.

1. Parameter set
2. Distortion measure
3. Dimension
4. Codebook storage
5. Search complexity
6. Quantizer type
7. Robustness to different inputs
8. Gathering of training data

A parameter set and distortion measure are jointly configured to represent and compress information in a meaningful manner that is highly relevant to the particular application. This concept is best illustrated with an example. Consider linear predictive (LP) analysis [8] of speech that is performed by the autocorrelation method. The resulting minimum phase nonrecursive filter

$$A(z) = 1 - \sum_{k=1}^{p} a_k z^{-k} \tag{6.19}$$

removes the near-sample redundancies in the speech. The filter $1/A(z)$ describes the spectral envelope of the speech. The information regarding the spectral envelope as contained in the LP filter coefficients a_k must be compressed (quantized) and coded for transmission. This is done in predictive speech coders [9]. There are other parameter sets that have a one-to-one correspondence to the set a_k. An equivalent parameter set that can be interpreted in terms of the spectral envelope is desired. The line spectral frequencies (LSFs) [10, 11] have been found to be the most useful.

The distortion measure is significant for meaningful quantization of the information and must be mathematically tractable. Continuing the above example, the LSFs must be quantized such that the spectral distortion between the spectral envelopes they represent is minimized. Mathematical tractability implies that the computation involved for (1) finding the codevectors given the Voronoi regions (as part of the design procedure) and (2) quantizing an input vector with the least distortion given a codebook is small. The L_1, L_2, and weighted L_2 distortions are mathematically feasible. For quantizing LSFs, the L_2 and weighted L_2 distortions are often used [12, 13, 14]. More details on LSF quantization will be provided in a forthcoming section on applications. At this point, a general description is provided just to illustrate the issues of selecting a parameter set and a distortion measure.

The issues of dimension, codebook storage, and search complexity are all related to computational considerations. A higher dimension leads to an increase in the memory requirement for storing the codebook and in the number of arithmetic operations for quantizing a vector given a codebook (search complexity). The dimension is also very important in capturing the essence of the information to be quantized. For example, if speech is sampled at 8 kHz, the spectral envelope consists of 3 to 4 formants (vocal tract resonances) which must be adequately captured. By using LSFs, a dimension of 10 to 12 suffices for capturing the formant information. Although a higher dimension leads to a better description of the fine details of the spectral envelope, this detail is not crucial for speech coders. Moreover, this higher dimension imposes more of a computational burden. The codebook storage requirement depends on the codebook size N. Obviously, a smaller value of N imposes less of a memory requirement. Also for coding, the number of bits to be transmitted should be minimized, thereby diminishing the memory requirement. The search complexity is directly related to the codebook size and dimension. However, it is also influenced by the type of distortion measure.

The type of quantizer (scalar or vector) is dictated by computational considerations and the robustness issue (discussed later). Consider the case when a total of 12 bits are used for quantization, the dimension is 6, and the L_2 distance measure is utilized. For a VQ, there is one codebook consisting of $2^{12} = 4096$ codevectors each having 6 components. A total of $4096 \times 6 = 24576$ numbers need to be stored. Computing the L_2 distance between an input vector and one codevector requires 6 multiplications and 11 additions. Therefore, searching the entire codebook requires $6 \times 4096 = 24576$ multiplications and $11 \times 4096 = 45056$ additions. For an SQ, there are six codebooks, one for each dimension. Each codebook requires 2 bits or $2^2 = 4$ codewords. The overall codebook size is $4 \times 6 = 24$. Hence, a total of 24 numbers needs to be stored. Consider the first component of an input vector. Four multiplications and four additions are required to find the best codeword. Hence, for all 6 components, 24 multiplications and 24 additions are needed to complete the search. The storage and search complexity are always much less for an SQ.

The quantizer type is also closely related to the robustness issue. A quantizer is said to be robust to different test input vectors if it can maintain the same performance for a large variety of inputs. The performance of a quantizer is measured as the average distortion resulting from the quantization of a set of test inputs. A VQ takes advantage of the multidimensional probability density of the data as empirically estimated by the training set. An SQ does not consider the correlations among the vector components as a separate design is performed for each component based on the probability density of that component. For test data having a similar density to the training data, a VQ will outperform an SQ given the same overall codebook size. However, for test data having a density that is different from that of the training data, an SQ will outperform a VQ given the same overall codebook size. This is because an SQ can accomplish a better coverage of a multidimensional space. Consider the example in Fig. 6.2. The vector space is of two dimensions ($p = 2$). The component x_1 lies in the range 0 to $x_1(max)$ and x_2 lies between 0 and $x_2(max)$. The multidimensional probability density function (pdf) $p(x_1, x_2)$ is shown as the region ABCD in Fig. 6.2. The training data will represent this pdf and can be used to design a vector and scalar quantizer of the same overall codebook size. The VQ will perform better for test data vectors in the region ABCD. Due to the individual ranges of the values of x_1 and x_2, the SQ will cover the larger space OKLM. Therefore, the SQ will perform better for test data vectors in OKLM but outside ABCD. An SQ is more robust in that it performs better for data with a density different from that of the training set. However, a VQ is preferable if the test data is known to have a density that resembles that of the training set.

In practice, the true multidimensional pdf of the data is not known as the data may emanate from many different conditions. For example, LSFs are obtained from speech material derived from many environmental conditions (like different telephones and noise backgrounds). Although getting a training set that is representative of all possible conditions gives the best estimate of the multidimensional pdf, it is impossible to configure such a set in practice. A versatile training set contributes to the robustness of the VQ but increases the time needed to accomplish the design.

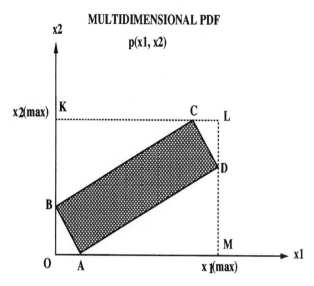

FIGURE 6.2   Example of a multidimensional probability density for explanation of the robustness issue.

6.5   Specific Manifestations

Thus far, we have considered the implementation of a VQ as being a one-step quantization of **x**. This is known as full VQ and is definitely the optimal way to do quantization. However, in applications such as LSF coding, quantizers between 25 and 30 bits are used. This leads to a prohibitive codebook size and search complexity. Two suboptimal approaches are now described that use multiple codebooks to alleviate the memory and search complexity requirements.

6.5.1   Multistage VQ

In multistage VQ consisting of R stages [3], there are R quantizers, Q_1, Q_2, $\cdots$, Q_R. The corresponding codebooks are denoted as C_1, C_2, $\cdots$, C_R. The sizes of these codebooks are N_1, N_2, $\cdots$, N_R. The overall codebook size is $N = N_1 + N_2 + \cdots + N_R$. The entries of the ith codebook C_i are $y_1^{(i)}, y_2^{(i)}, \cdots, y_{N_i}^{(i)}$. Figure 6.3 shows a block diagram of the entire system.

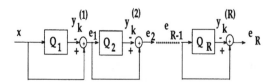

FIGURE 6.3   Multistage vector quantization.

The procedure for multistage VQ is as follows. The input **x** is first quantized by Q_1 to $\mathbf{y}_k^{(1)}$. The quantization error is $\mathbf{e}_1 = \mathbf{x} - \mathbf{y}_k^{(1)}$, which is in turn quantized by Q_2 to $\mathbf{y}_k^{(2)}$. The quantization error at the second stage is $\mathbf{e}_2 = \mathbf{e}_1 - \mathbf{y}_k^{(2)}$. This error is quantized at the third stage. The process repeats and at the Rth stage, $\mathbf{e}_{R-1}$ is quantized by Q_R to $\mathbf{y}_k^{(R)}$ such that the quantization error is $\mathbf{e}_R$. The original vector **x** is quantized to $\mathbf{y} = \mathbf{y}_k^{(1)} + \mathbf{y}_k^{(2)} + \cdots + \mathbf{y}_k^{(R)}$. The overall quantization error is $\mathbf{x} - \mathbf{y} = \mathbf{e}_R$.

The reduction in the memory requirement and search complexity is best illustrated by a simple example. A full VQ of 30 bits will have one codebook of 2^{30} codevectors (cannot be used in practice). An equivalent multistage VQ of $R = 3$ stages will have three 10-bit codebooks C_1, C_2, and C_3. The total number of codevectors to be stored is 3×2^{10}, which is practically feasible. It follows that the search complexity is also drastically reduced over that of a full VQ.

The simplest way to train a multistage VQ is to perform sequential training of the codebooks. We start with a training set $T = \{\mathbf{x}_1, \mathbf{x}_2, \mathbf{x}_3, \cdots \mathbf{x}_M\} \in \mathbf{R}^p$ to get C_1. The entire set T is quantized by Q_1 to get a training set for the next stage. The codebook C_2 is designed from this new training set. This procedure is repeated so that all the R codebooks are designed. A joint design procedure for multistage VQ has been recently developed in [15] but is outside the scope of this article.

6.5.2 Split VQ

In split VQ [3], $\mathbf{x} = [x_1\, x_2\, x_3 \, \cdots \, x_p]^T \in \mathbf{R}^p$ is split or partitioned into R subvectors of smaller dimension as $\mathbf{x} = [\mathbf{x}^{(1)}\, \mathbf{x}^{(2)}\mathbf{x}^{(3)} \, \cdots \, \mathbf{x}^{(R)}]^T$. The ith subvector $\mathbf{x}^{(i)}$ has dimension d_i. Therefore, $p = d_1 + d_2 + \cdots + d_R$. Specifically,

$$\mathbf{x}^{(1)} = [x_1\, x_2\, \cdots \, x_{d_1}]^T \tag{6.20}$$

$$\mathbf{x}^{(2)} = [x_{d_1+1}\, x_{d_1+2}\, \cdots \, x_{d_1+d_2}]^T \tag{6.21}$$

$$\mathbf{x}^{(3)} = [x_{d_1+d_2+1}\, x_{d_1+d_2+2}\, \cdots \, x_{d_1+d_2+d_3}]^T \tag{6.22}$$

and so forth.

There are R quantizers, one for each subvector. The subvectors $\mathbf{x}^{(i)}$ are individually quantized to $\mathbf{y}_k^{(i)}$ so that the full vector $\mathbf{x}$ is quantized to $\mathbf{y} = [\mathbf{y}_k^{(1)}\, \mathbf{y}_k^{(2)}\mathbf{y}_k^{(3)} \, \cdots \, \mathbf{y}_k^{(R)}]^T \in \mathbf{R}^p$. The quantizers are designed using the appropriate subvectors in the training set T. The extreme case of a split VQ is when $R = p$. Then, $d_1 = d_2 = \cdots = d_p = 1$ and we get a scalar quantizer.

The reduction in the memory requirement and search complexity is again illustrated by a similar example as for multistage VQ. Suppose the dimension $p = 10$. A full VQ of 30 bits will have one codebook of 2^{30} codevectors. An equivalent split VQ of $R = 3$ splits uses subvectors of dimensions $d_1 = 3$, $d_2 = 3$, and $d_3 = 4$. For each subvector, there will be a 10-bit codebook having 2^{10} codevectors.

Finally, note that split VQ is feasible if the distortion measure is separable in that

$$d(\mathbf{x}, \mathbf{y}) = \sum_{i=1}^{R} d\left(\mathbf{x}^{(i)}, \mathbf{y}_k^{(i)}\right) \tag{6.23}$$

This property is true for the L_r distance and for the weighted L_2 distance if the matrix of weights $\mathbf{W}$ is diagonal.

6.6 Applications

In this article, two applications of quantization are discussed. One is in the area of speech coding and the other is in speaker identification. Both are based on LP analysis of speech [8] as performed by the autocorrelation method. As mentioned earlier, the predictor coefficients, a_k, describe a minimum phase nonrecursive LP filter $A(z)$ as given by Eq. (6.19). We recall that the filter $1/A(z)$ describes the spectral envelope of the speech, which in turn gives information about the formants.

6.6.1 Predictive Speech Coding

In predictive speech coders, the predictor coefficients (or a transformation thereof) must be quantized. The main aim is to preserve the spectral envelope as described by $1/A(z)$ and, in particular, preserve the formants. The coefficients a_k are transformed into an LSF vector **f**. The LSFs are more clearly related to the spectral envelope in that (1) the spectral sensitivity is local to a change in a particular frequency and (2) the closeness of two adjacent LSFs indicates a formant.

Ideally, LSFs should be quantized to minimize the spectral distortion (SD) given by

$$\text{SD} = \sqrt{\frac{1}{B} \int_R \left[10 \log \left(|A_q \left(e^{j2\pi f} \right)|^2 / |A \left(e^{j2\pi f} \right)|^2 \right) \right]^2 \, df} \qquad (6.24)$$

where $A(.)$ refers to the original LP filter, $A_q(.)$ refers to the quantized LP filter, B is the bandwidth of interest, and R is the frequency range of interest. The SD is not a mathematically tractable measure and is also not separable if split VQ is to be used. A weighted L_2 measure is used in which **W** is diagonal and the ith diagonal element is $w(i)$ is given by [14]:

$$w(i) = \frac{1}{f_i - f_{i-1}} + \frac{1}{f_{i+1} - f_i} \qquad (6.25)$$

where $\mathbf{f} = [f_1 \ f_2 \ f_3 \ \cdots \ f_p]^T \in \mathbf{R}^p$, f_0 is taken to be zero, and f_{p+1} is taken to be the highest digital frequency (π or 0.5 if normalized). Regarding this distance measure, note the following:

1. The LSFs are ordered ($f_{i+1} > f_i$) if and only if the LP filter $A(z)$ is minimum phase. This guarantees that $w(i) > 0$.
2. The weight $w(i)$ is high if two adjacent LSFs are close to each other. Therefore, more weight is given to regions in the spectrum having formants.
3. The weights are dependent on the input vector **f**. This makes the computation of the codevectors using the LBG algorithm different from the case when the weights are constant. However, for finding the codevector given a Voronoi region, the average of the training vectors in the region is taken so that the ordering property is preserved.
4. Mathematical tractability and separability of the distance measure are obvious.

A quantizer can be designed from a training set of LSFs using the weighted L_2 distance. Consider LSFs obtained from speech that is lowpass filtered to 3400 Hz and sampled at 8 kHz. If there are additional highpass or bandpass filtering effects, some of the LSFs tend to migrate [16]. Therefore, a VQ trained solely on one filtering condition will not be robust to test data derived from other filtering conditions [16]. The solution in [16] to robustize a VQ is to configure a training set consisting of two main components. First, LSFs from different filtering conditions are gathered to provide a reasonable empirical estimate of the multidimensional pdf. Second, a uniformly distributed set of vectors provides for coverage of the multidimensional space (similar to what is accomplished by an SQ). Finally, multistage or split LSF quantizers are used for practical feasibility [13, 15, 16].

6.6.2 Speaker Identification

Speaker recognition is the task of identifying a speaker by his or her voice. Systems performing speaker recognition operate in different modes. A closed set mode is the situation of identifying a particular speaker as one in a finite set of reference speakers [17]. In an open set system, a speaker is either identified as belonging to a finite set or is deemed not to be a member of the set [17]. For speaker verification, the claim of a speaker to be one in a finite set is either accepted or rejected [18]. Speaker recognition can either be done as a text-dependent or text-independent task. The difference is that in the former case, the speaker is constrained as to what must be said, while in the latter case no constraints are imposed. In this article, we focus on the closed set, text-independent mode. The overall system will have three components, namely, (1) LP analysis for parameterizing the spectral envelope, (2) feature extraction for ensuring speaker discrimination, and (3) classifier for making a decision. The input to the system will be a speech signal. The output will be a decision regarding the identity of the speaker.

After LP analysis of speech is carried out, the LP predictor coefficients, a_k, are converted into the LP cepstrum. The cepstrum is a popular feature as it provides for good speaker discrimination. Also, the cepstrum lends itself to the L_2 or weighted L_2 distance that is simple and yet reflective of the log spectral distortion between two LP filters [19]. To achieve good speaker discrimination, the formants must be captured. Hence, a dimension of 12 is usually used.

The cepstrum is used to develop a VQ classifier [20] as shown in Fig. 6.4. For each speaker enrolled in the system, a training set is established from utterances spoken by that speaker. From the training set,

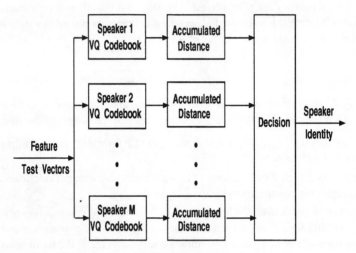

FIGURE 6.4   A VQ based classifier for speaker identification.

a VQ codebook is designed that serves as a speaker model. The VQ codebook represents a portion of the multidimensional space that is characteristic of the feature or cepstral vectors for a particular speaker. Good discrimination is achieved if the codebooks show little or no overlap as illustrated in Fig. 6.5 for the case of three speakers. Usually, a small codebook size of 64 or 128 codevectors is sufficient [21]. Even if there are 50 speakers enrolled, the memory requirement is feasible for real-time applications. An SQ is of no use because the correlations among the vector components are crucial for speaker discrimination. For the same reason, multistage or split VQ is also of no use. Moreover, full VQ can easily be used given the relatively smaller codebook size as compared to coding.

Given a random speech utterance, the testing procedure for identifying a speaker is as follows (see Fig. 6.4). First, the S test feature (cepstrum) vectors are computed. Consider the first vector. It is quantized by the codebook for speaker 1 and the resulting minimum L_2 or weighted L_2 distance is recorded. This quantization is done for all S vectors and the resulting minimum distances are accumulated (added up) to get an overall score for speaker 1. In this manner, an overall score is computed for all the speakers. The identified speaker is the one with the least overall score. Note that with the small codebook sizes, the search complexity is practically feasible. In fact, the overall score for the different speakers can be obtained in parallel. The performance measure for a speaker identification system is the identification success rate, which is the number of test utterances for which the speaker is identified correctly divided by the total number of test utterances.

The robustness issue is of great significance and emerges when the cepstral vectors derived from certain test speech material have not been considered in the training phase. This phenomenon of a full VQ not being robust to a variety of test inputs has been mentioned earlier and has been encountered in our discussion on LSF coding. The use of different training and testing conditions degrades performance since the components of the cepstrum vectors (such as LSFs) tend to migrate. Unlike LSF coding, appending

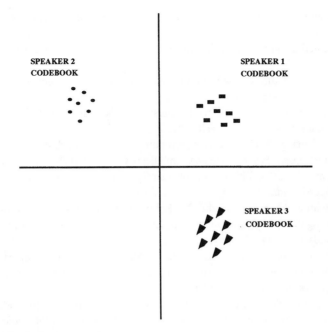

FIGURE 6.5   VQ codebooks for three speakers.

the training set with a uniformly distributed set of vectors to accomplish coverage of a large space will not work as there will be much overlap among the codebooks of different speakers. The focus of the research is to develop more robust features that show little variation as the speech material changes [22, 23].

6.7 Summary

This article has presented a tutorial description of quantization. Starting from the basic definition and properties of vector and scalar quantization, design algorithms are described. Many practical aspects of design and implementation (such as distortion measure, memory, search complexity, and robustness) are discussed. These practical aspects are interrelated. Two important applications of vector quantization in speech processing are discussed in which these practical aspects play an important role.

References

[1] Gray, R.M., Vector quantization, *IEEE Acoust. Speech Sig. Proc.*, 1, 4–29, Apr. 1984.

[2] Makhoul, J., Roucos, S., and Gish, H., Vector quantization in speech coding, *Proc. IEEE*, 73, 1551–1588, Nov. 1985.

[3] Gersho, A. and Gray, R.M., *Vector Quantization and Signal Compression*, Kluwer Academic Publishers, 1991.

[4] Gersho, A., Asymptotically optimal block quantization, *IEEE Trans. Infor. Theory*, IT-25, 373–380, July 1979.

[5] Jayant, N.S. and Noll, P., *Digital Coding of Waveforms, Principles and Applications to Speech and Video*, Prentice-Hall, Englewood Cliffs, NJ, 1984.

[6] Max, J., Quantizing for minimum distortion, *IEEE Trans. Infor. Theory*, 7–12, Mar. 1960.

[7] Linde, Y., Buzo, A., and Gray, R.M., An algorithm for vector quantizer design, *IEEE Trans. Comm.*, COM-28, 84–95, Jan. 1980.

[8] Rabiner, L.R. and Schafer, R.W., *Digital Processing of Speech Signals*, Prentice-Hall, Englewood Cliffs, NJ, 1978.

[9] Atal, B.S., Predictive coding of speech at low bit rates, *IEEE Trans. Comm.*, COM-30, 600–614, Apr. 1982.

[10] Itakura, F., Line spectrum representation of linear predictor coefficients of speech signals, *J. Acoust. Soc. Amer.*, 57, S35(A), 1975.

[11] Wakita, H., Linear prediction voice synthesizers: Line spectrum pairs (LSP) is the newest of several techniques, *Speech Technol.*, Fall 1981.

[12] Soong, F.K. and Juang, B.-H., Line spectrum pair (LSP) and speech data compression, *IEEE Int. Conf. Acoust. Speech Signal Processing*, San Diego, CA, pp. 1.10.1–1.10.4, March 1984.

[13] Paliwal, K.K. and Atal, B.S., Efficient vector quantization of LPC parameters at 24 bits/frame, *IEEE Trans. Speech Audio Processing*, 1, 3–14, Jan. 1993.

[14] Laroia, R., Phamdo, N., and Farvardin, N., Robust and efficient quantization of speech LSP parameters using structured vector quantizers, *IEEE Intl. Conf. Acoust. Speech Signal Processing*, Toronto, Canada, 641–644, May 1991.

[15] LeBlanc, W.P., Cuperman, V., Bhattacharya, B., and Mahmoud, S.A., Efficient search and design procedures for robust multi-stage VQ of LPC parameters for 4 kb/s speech coding, *IEEE Trans. Speech Audio Processing*, 1, 373–385, Oct. 1993.

[16] Ramachandran, R.P., Sondhi, M.M., Seshadri, N., and Atal, B.S., A two codebook format for robust quantization of line spectral frequencies, *IEEE Trans. Speech Audio Processing*, 3, 157–168, May 1995.

[17] Doddington, G.R., Speaker recognition—identifying people by their voices, *Proc. IEEE*, 73, 1651–1664, Nov. 1985.

[18] Furui, S., Cepstral analysis technique for automatic speaker verification, *IEEE Trans. Acoust. Speech Sig. Proc.*, ASSP-29, 254–272, Apr. 1981.

[19] Rabiner, L.R. and Juang, B.-H., *Fundamentals of Speech Recognition*, Prentice-Hall, Englewood Cliffs, NJ, 1993.

[20] Rosenberg, A.E. and Soong, F.K., Evaluation of a vector quantization talker recognition system in text independent and text dependent modes, *Comp. Speech Lang.*, 22, 143–157, 1987.

[21] Farrell, K.R., Mammone, R.J., and Assaleh, K.T., Speaker recognition using neural networks versus conventional classifiers, *IEEE Trans. Speech Audio Processing*, 2, 194–205, Jan. 1994.

[22] Assaleh, K.T. and Mammone, R.J., New LP-derived features for speaker identification, *IEEE Trans. Speech Audio Processing*, 2, 630–638, Oct. 1994.

[23] Zilovic, M.S., Ramachandran, R.P., and Mammone, R.J., Speaker identification based on the use of robust cepstral features derived from pole-zero transfer functions, accepted in *IEEE Trans. Speech Audio Processing*.

Fast Algorithms and Structures

P. Duhamel
École Nationale Supérieure des Télécommunications (ENST)

T HE FIELD OF DIGITAL SIGNAL PROCESSING grew rapidly and achieved its current prominence primarily through the discovery of efficient algorithms for computing various transforms (mainly the Fourier transforms) in the 1970s. In addition to fast Fourier transforms (FFTs), discrete cosine transforms (DCTs) have also gained importance owing to their performance being very close to the statistically optimum Karhunen Loeve transform.

Transforms, convolutions, and matrix-vector operations form the basic tools utilized by the signal processing community, and this section reviews and presents the state of art in these areas of increasing importance.

The chapter by Duhamel and Vetterli, "Fast Fourier Transforms: A Tutorial Review and a State of the Art", presents a thorough discussion of this important transform. Selesnick and Burrus present an excellent survey of filtering and convolution techniques in the chapter "Fast Convolution and Filtering".

One approach to understanding the time and space complexities of signal processing algorithms is through the use of quantitative complexity theory, and Feig's "Complexity Theory of Transforms in Signal Processing" applies quantitative measures to the computation of transforms. Finally, Yagle presents a comprehensive discussion of matrix computations in signal processing in "Fast Matrix Computations".

7

Fast Fourier Transforms: A Tutorial Review and a State of the Art[1]

P. Duhamel
ENST, Paris

M. Vetterli
*EPFL, Lausanne
and University of California,
Berkeley*

---

[1]Reprinted from *Signal Processing* 19:259-299, 1990 with kind permission from Elsevier Science-NL, Sara Burgerhartstraat 25, 1055 KV Amsterdam, The Netherlands.

The publication of the Cooley-Tukey fast Fourier transform (FFT) algorithm in 1965 has opened a new area in digital signal processing by reducing the order of complexity of some crucial computational tasks such as Fourier transform and convolution from N^2 to $N \log_2 N$, where N is the problem size. The development of the major algorithms (Cooley-Tukey and split-radix FFT, prime factor algorithm and Winograd fast Fourier transform) is reviewed. Then, an attempt is made to indicate the state of the art on the subject, showing the standing of research, open problems, and implementations.

7.1   Introduction

Linear filtering and Fourier transforms are among the most fundamental operations in digital signal processing. However, their wide use makes their computational requirements a heavy burden in most applications. Direct computation of both convolution and discrete Fourier transform (DFT) requires on the order of N^2 operations where N is the filter length or the transform size. The breakthrough of the Cooley-Tukey FFT comes from the fact that it brings the complexity down to an order of $N \log_2 N$ operations. Because of the convolution property of the DFT, this result applies to the convolution as well. Therefore, fast Fourier transform algorithms have played a key role in the widespread use of digital signal processing in a variety of applications such as telecommunications, medical electronics, seismic processing, radar or radio astronomy to name but a few.

Among the numerous further developments that followed Cooley and Tukey's original contribution, the fast Fourier transform introduced in 1976 by Winograd [54] stands out for achieving a new theoretical reduction in the order of the multiplicative complexity. Interestingly, the Winograd algorithm uses convolutions to compute DFTs, an approach which is just the converse of the conventional method of computing convolutions by means of DFTs. What might look like a paradox at first sight actually shows the deep interrelationship that exists between convolutions and Fourier transforms.

Recently, the Cooley-Tukey type algorithms have emerged again, not only because implementations of the Winograd algorithm have been disappointing, but also due to some recent developments leading to the so-called split-radix algorithm [27]. Attractive features of this algorithm are both its low arithmetic complexity and its relatively simple structure.

Both the introduction of digital signal processors and the availability of large scale integration has influenced algorithm design. While in the sixties and early seventies, multiplication counts alone were taken into account, it is now understood that the number of addition and memory accesses in software and the communication costs in hardware are at least as important.

The purpose of this chapter is first to look back at 20 years of developments since the Cooley-Tukey paper. Among the abundance of literature (a bibliography of more than 2500 titles has been published [33]), we will try to highlight only the key ideas. Then, we will attempt to describe the state of the art on the subject. It seems to be an appropriate time to do so, since on the one hand, the algorithms have now reached a certain maturity, and on the other hand, theoretical results on complexity allow us to evaluate how far we are from optimum solutions. Furthermore, on some issues, open questions will be indicated.

Let us point out that in this chapter we shall concentrate strictly on the computation of the discrete Fourier transform, and not discuss applications. However, the tools that will be developed may be useful in other cases. For example, the polynomial products explained in Section 7.5.1 can immediately be applied to the derivation of fast running FIR algorithms [73, 81].

The chapter is organized as follows.

Section 7.2 presents the history of the ideas on fast Fourier transforms, from Gauss to the splitradix algorithm.

Section 7.3 shows the basic technique that underlies all algorithms, namely the divide and conquer approach, showing that it always improves the performance of a Fourier transform algorithm.

Section 7.4 considers Fourier transforms with twiddle factors, that is, the classic Cooley-Tukey type schemes and the split-radix algorithm. These twiddle factors are unavoidable when the transform length is composite with non-coprime factors. When the factors are coprime, the divide and conquer scheme can be made such that twiddle factors do not appear.

This is the basis of Section 7.5, which then presents Rader's algorithm for Fourier transforms of prime lengths, and Winograd's method for computing convolutions. With these results established, Section 7.5 proceeds to describe both the prime factor algorithm (PFA) and the Winograd Fourier transform (WFTA).

Section 7.6 presents a comprehensive and critical survey of the body of algorithms introduced thus far, then shows the theoretical limits of the complexity of Fourier transforms, thus indicating the gaps that are left between theory and practical algorithms.

Structural issues of various FFT algorithms are discussed in Section 7.7.

Section 7.8 treats some other cases of interest, like transforms on special sequences (real or symmetric) and related transforms, while Section 7.9 is specifically devoted to the treatment of multidimensional transforms.

Finally, Section 7.10 outlines some of the important issues of implementations. Considerations on software for general purpose computers, digital signal processors, and vector processors are made. Then, hardware implementations are addressed. Some of the open questions when implementing FFT algorithms are indicated.

The presentation we have chosen here is constructive, with the aim of motivating the "tricks" that are used. Sometimes, a shorter but "plug-in" like presentation could have been chosen, but we avoided it because we desired to insist on the mechanisms underlying all these algorithms. We have also chosen to avoid the use of some mathematical tools, such as tensor products (that are very useful when deriving some of the FFT algorithms) in order to be more widely readable.

Note that concerning arithmetic complexities, all sections will refer to synthetic tables giving the computational complexities of the various algorithms for which software is available. In a few cases, slightly better figures can be obtained, and this will be indicated.

For more convenience, the references are separated between books and papers, the latter being further classified corresponding to subject matters (1-D FFT algorithms, related ones, multidimensional transforms and implementations).

7.2 A Historical Perspective

The development of the fast Fourier transform will be surveyed below because, on the one hand, its history abounds in interesting events, and on the other hand, the important steps correspond to parts of algorithms that will be detailed later.

A first subsection describes the pre-Cooley-Tukey area, recalling that algorithms can get lost by lack of use, or, more precisely, when they come too early to be of immediate practical use. The developments following the Cooley-Tukey algorithm are then described up to the most recent solutions. Another subsection is concerned with the steps that lead to the Winograd and to the prime factor algorithm, and finally, an attempt is made to briefly describe the current state of the art.

7.2.1 From Gauss to the Cooley-Tukey FFT

While the publication of a fast algorithm for the DFT by Cooley and Tukey [25] in 1965 is certainly a turning point in the literature on the subject, the divide and conquer approach itself dates back to Gauss as noted in a well-documented analysis by Heideman et al. [34]. Nevertheless, Gauss's work on FFTs in the early 19th century (around 1805) remained largely unnoticed because it was only published in Latin and this after his death.

Gauss used the divide and conquer approach in the same way as Cooley and Tukey have published it later in order to evaluate trigonometric series, but his work predates even Fourier's work on harmonic analysis (1807)! Note that his algorithm is quite general, since it is explained for transforms on sequences with lengths equal to any composite integer.

During the 19th century, efficient methods for evaluating Fourier series appeared independently at least three times [33], but were restricted on lengths and number of resulting points. In 1903, Runge derived an algorithm for lengths equal to powers of 2 which was generalized to powers of 3 as well and used in the forties. Runge's work was thus quite well known, but nevertheless disappeared after the war.

Another important result useful in the most recent FFT algorithms is another type of divide and conquer approach, where the initial problem of length $N_1 \cdot N_2$ is divided into subproblems of lengths N_1 and N_2 without any additional operations, N_1 and N_2 being coprime.

This result dates back to the work of Good [32] who obtained this result by simple index mappings. Nevertheless, the full implication of this result will only appear later, when efficient methods will be derived for the evaluation of small, prime length DFTs. This mapping itself can be seen as an application of the Chinese remainder theorem (CRT), which dates back to 100 years A.D.! [10]–[18].

Then, in 1965, appeared a brief article by Cooley and Tukey, entitled "An algorithm for the machine calculation of complex Fourier series" [25], which reduces the order of the number of operations from N^2 to $N \log_2(N)$ for a length $N = 2^n$ DFT.

This turned out to be a milestone in the literature on fast transforms, and was credited [14, 15] with the tremendous increase of interest in DSP beginning in the seventies. The algorithm is suited for DFTs on any composite length, and is thus of the type that Gauss had derived almost 150 years before. Note that all algorithms published in-between were more restrictive on the transform length [34].

Looking back at this brief history, one may wonder why all previous algorithms had disappeared or remained unnoticed, whereas the Cooley-Tukey algorithm had such a tremendous success. A possible explanation is that the growing interest in the theoretical aspects of digital signal processing was motivated by technical improvements in semiconductor technology. And, of course, this was not a one-way street.

The availability of reasonable computing power produced a situation where such an algorithm would suddenly allow numerous new applications. Considering this history, one may wonder how many other algorithms or ideas are just sleeping in some notebook or obscure publication.

The two types of divide and conquer approaches cited above produced two main classes of algorithms. For the sake of clarity, we will now skip the chronological order and consider the evolution of each class separately.

7.2.2 Development of the Twiddle Factor FFT

When the initial DFT is divided into sublengths which are not coprime, the divide and conquer approach as proposed by Cooley and Tukey leads to auxiliary complex multiplications, initially named twiddle factors, which cannot be avoided in this case.

While Cooley-Tukey's algorithm is suited for any composite length, and explained in [25] in a general form, the authors gave an example with $N = 2^n$, thus deriving what is now called a radix-2 decimation in time (DIT) algorithm (the input sequence is divided into decimated subsequences having different phases). Later, it was often falsely assumed that the initial Cooley-Tukey FFT was a DIT radix-2 algorithm only.

A number of subsequent papers presented refinements of the original algorithm, with the aim of increasing its usefulness.

The following refinements were concerned:

- with the structure of the algorithm: it was emphasized that a dual approach leads to "decimation in frequency" (DIF) algorithms,

- or with the efficiency of the algorithm, measured in terms of arithmetic operations: Bergland showed that higher radices, for example radix-8, could be more efficient, [21]

– or with the extension of the applicability of the algorithm: Bergland [60], again, showed that the FFT could be specialized to real input data, and Singleton gave a mixed radix FFT suitable for arbitrary composite lengths.

While these contributions all improved the initial algorithm in some sense (fewer operations and/or easier implementations), actually no new idea was suggested.

Interestingly, in these very early papers, all the concerns guiding the recent work were already here: arithmetic complexity, but also different structures and even real-data algorithms.

In 1968, Yavne [58] presented a little-known paper that sets a record: his algorithm requires the least known number of multiplications, as well as additions for length-2^n FFTs, and this both for real and complex input data. Note that this record still holds, at least for practical algorithms. The same number of operations was obtained later on by other (simpler) algorithms, but due to Yavne's cryptic style, few researchers were able to use his ideas at the time of publication.

Since twiddle factors lead to most computations in classical FFTs, Rader and Brenner [44], perhaps motivated by the appearance of the Winograd Fourier transform which possesses the same characteristic, proposed an algorithm that replaces all complex multiplications by either real or imaginary ones, thus substantially reducing the number of multiplications required by the algorithm. This reduction in the number of multiplications was obtained at the cost of an increase in the number of additions, and a greater sensitivity to roundoff noise. Hence, further developments of these "real factor" FFTs appeared in [24, 42], reducing these problems. Bruun [22] also proposed an original scheme particularly suited for real data. Note that these various schemes only work for radix-2 approaches.

It took more than 15 years to see again algorithms for length-2^n FFTs that take as few operations as Yavne's algorithm. In 1984, four papers appeared or were submitted almost simultaneously [27, 40, 46, 51] and presented so-called "split-radix" algorithms. The basic idea is simply to use a different radix for the even part of the transform (radix-2) and for the odd part (radix-4). The resulting algorithms have a relatively simple structure and are well adapted to real and symmetric data while achieving the minimum known number of operations for FFTs on power of 2 lengths.

7.2.3   FFTs Without Twiddle Factors

While the divide and conquer approach used in the Cooley-Tukey algorithm can be understood as a "false" mono- to multi-dimensional mapping (this will be detailed later), Good's mapping, which can be used when the factors of the transform lengths are coprime, is a true mono- to multi-dimensional mapping, thus having the advantage of not producing any twiddle factor.

Its drawback, at first sight, is that it requires efficiently computable DFTs on lengths that are coprime: For example, a DFT of length 240 will be decomposed as $240 = 16 \cdot 3 \cdot 5$, and a DFT of length 1008 will be decomposed in a number of DFTs of lengths 16, 9, and 7. This method thus requires a set of (relatively) small-length DFTs that seemed at first difficult to compute in less than N_i^2 operations. In 1968, however, Rader [43] showed how to map a DFT of length N, N prime, into a circular convolution of length $N - 1$. However, the whole material to establish the new algorithms was not ready yet, and it took Winograd's work on complexity theory, in particular on the number of multiplications required for computing polynomial products or convolutions [55] in order to use Good's and Rader's results efficiently.

All these results were considered as curiosities when they were first published, but their combination, first done by Winograd and then by Kolba and Parks [39] raised a lot of interest in that class of algorithms. Their overall organization is as follows:

After mapping the DFT into a true multidimensional DFT by Good's method and using the fast convolution schemes in order to evaluate the prime length DFTs, a first algorithm makes use of the intimate structure of these convolution schemes to obtain a nesting of the various multiplications. This algorithm is known as the Winograd Fourier transform algorithm (WFTA) [54], an algorithm requiring the least known number of multiplications among practical algorithms for moderate lengths DFTs. If the nesting is not used, and the multi-dimensional DFT is performed by the row-column method, the resulting algorithm is known as the prime factor algorithm (PFA) [39], which, while using more multiplications, has less additions and a better structure than the WFTA.

From the above explanations, one can see that these two algorithms, introduced in 1976 and 1977, respectively, require more mathematics to be understood [19]. This is why it took some effort to translate the theoretical results, especially concerning the WFTA, into actual computer code.

It is even our opinion that what will remain mostly of the WFTA are the theoretical results, since although a beautiful result in complexity theory, the WFTA did not meet its expectations once implemented, thus leading to a more critical evaluation of what "complexity" meant in the context of real life computers [41, 108, 109].

The result of this new look at complexity was an evaluation of the number of additions and data transfers as well (and no longer only of multiplications). Furthermore, it turned out recently that the theoretical knowledge brought by these approaches could give a new understanding of FFTs with twiddle factors as well.

7.2.4   Multi-Dimensional DFTs

Due to the large amount of computations they require, the multi-dimensional DFTs as such (with common factors in the different dimensions, which was not the case in the multi-dimensional translation of a mono-dimensional problem by PFA) were also carefully considered.

The two most interesting approaches are certainly the vector radix FFT (a direct approach to the multi-dimensional problem in a Cooley-Tukey mood) proposed in 1975 by Rivard [91] and the polynomial transform solution of Nussbaumer and Quandalle [87, 88] in 1978.

Both algorithms substantially reduce the complexity over traditional row-column computational schemes.

7.2.5   State of the Art

From a theoretical point of view, the complexity issue of the discrete Fourier transform has reached a certain maturity. Note that Gauss, in his time, did not even count the number of operations necessary in his algorithm. In particular, Winograd's work on DFTs whose lengths have coprime factors both sets lower bounds (on the number of multiplications) and gives algorithms to achieve these [35, 55], although they are not always practical ones. Similar work was done for length-2^n DFTs, showing the linear multiplicative complexity of the algorithm [28, 35, 105] but also the lack of practical algorithms achieving this minimum (due to the tremendous increase in the number of additions [35]).

Considering implementations, the situation is of course more involved since many more parameters have to be taken into account than just the number of operations.

Nevertheless, it seems that both the radix-4 and the split-radix algorithm are quite popular for lengths which are powers of 2, while the PFA, thanks to its better structure and easier implementation, wins over the WFTA for lengths having coprime factors.

Recently, however, new questions have come up because in software on the one hand, new processors may require different solutions (vector processors, signal processors), and on the other hand, the advent of VLSI for hardware implementations sets new constraints (desire for simple structures, high cost of multiplications vs. additions).

7.3   Motivation (or: why dividing is also conquering)

This section is devoted to the method that underlies all fast algorithms for DFT, that is the "divide and conquer" approach.

The discrete Fourier transform is basically a matrix-vector product. Calling $(x_0, x_1, \ldots, x_{N-1})^T$ the vector of the input samples,

$$(X_0, X_1, \ldots, X_{N-1})^T$$

the vector of transform values and W_N the primitive Nth root of unity ($W_N = e^{-j2\pi/N}$) the DFT can be written as

$$
\begin{bmatrix} X_0 \\ X_1 \\ X_2 \\ \vdots \\ X_{N-1} \end{bmatrix} =
\begin{bmatrix}
1 & 1 & 1 & 1 & \cdots & 1 \\
1 & W_N & W_N^2 & W_N^3 & \cdots & W_N^{N-1} \\
1 & W_N^2 & W_N^4 & W_N^6 & \cdots & W_N^{2(N-1)} \\
\vdots & \vdots & \vdots & \vdots & & \vdots \\
1 & W_N^{N-1} & W_N^{2(N-1)} & \cdots & \cdots & W_N^{(N-1)(N-1)}
\end{bmatrix}
\times
\begin{bmatrix} x_0 \\ x_1 \\ x_2 \\ x_3 \\ \vdots \\ x_{N-1} \end{bmatrix}
\tag{7.1}
$$

The direct evaluation of the matrix-vector product in (7.1) requires of the order of N^2 complex multiplications and additions (we assume here that all signals are complex for simplicity).

The idea of the "divide and conquer" approach is to map the original problem into several subproblems in such a way that the following inequality is satisfied:

$$\sum \text{cost(subproblems)} + \text{cost(mapping)}$$
$$< \text{cost(original problem)}.
\tag{7.2}$$

But the real power of the method is that, often, the division can be applied recursively to the subproblems as well, thus leading to a reduction of the order of complexity.

Specifically, let us have a careful look at the DFT transform in (7.3) and its relationship with the z-transform of the sequence $\{x_n\}$ as given in (7.4).

$$X_k = \sum_{i=0}^{N-1} x_i W_N^{ik}, \quad k = 0, \ldots, N-1,
\tag{7.3}$$

$$X(z) = \sum_{i=0}^{N-1} x_i z^{-i}.
\tag{7.4}$$

$\{X_k\}$ and $\{x_i\}$ form a transform pair, and it is easily seen that X_k is the evaluation of $X(z)$ at point $z = W_N^{-k}$:

$$X_k = X(z)_{z=W_N^{-k}}.
\tag{7.5}$$

Furthermore, due to the sampled nature of $\{x_n\}$, $\{X_k\}$ is periodic, and vice versa: since $\{X_k\}$ is sampled, $\{x_n\}$ must also be periodic.

From a physical point of view, this means that both sequences $\{x_n\}$ and $\{X_k\}$ are repeated indefinitely with period N. This has a number of consequences as far as fast algorithms are concerned.

All fast algorithms are based on a divide and conquer strategy; we have seen this in Section 7.2. But how shall we divide the problem (with the purpose of conquering it)?

The most natural way is, of course, to consider subsets of the initial sequence, take the DFT of these subsequences, and reconstruct the DFT of the initial sequence from these intermediate results.

Let I_l, $l = 0, \ldots, r - 1$ be the partition of $\{0, 1, \ldots, N - 1\}$ defining the r different subsets of the input sequence. Equation (7.4) can now be rewritten as

$$X(z) = \sum_{i=0}^{N-1} x_i z^{-i} = \sum_{l=0}^{r-1} \sum_{i \in I_l} x_i z^{-i} , \tag{7.6}$$

and, normalizing the powers of z with respect to some x_{0l} in each subset I_l:

$$X(z) = \sum_{l=0}^{r-1} z^{-i_{0l}} \sum_{i \in I_l} x_i z^{-i+i_{0l}} . \tag{7.7}$$

From the considerations above, we want the replacement of z by W_N^{-k} in the innermost sum of (7.7) to define an element of the DFT of $\{x_i | i \in I_l\}$. Of course, this will be possible only if the subset $\{x_i | i \in I_l\}$, possibly permuted, has been chosen in such a way that it has the same kind of periodicity as the initial sequence. In what follows, we show that the three main classes of FFT algorithms can all be casted into the form given by (7.7).

- In some cases, the second sum will also involve elements having the same periodicity, hence will define DFTs as well. This corresponds to the case of Good's mapping: all the subsets I_l, have the same number of elements $m = N/r$ and $(m, r) = 1$.
- If this is not the case, (7.7) will define one step of an FFT with twiddle factors: when the subsets I_l all have the same number of elements, (7.7) defines one step of a radix-r FFT.
- If $r = 3$, one of the subsets having $N/2$ elements, and the other ones having $N/4$ elements, (7.7) is the basis of a split-radix algorithm.

Furthermore, it is already possible to show from (7.7) that the divide and conquer approach will always improve the efficiency of the computation.

To make this evaluation easier, let us suppose that all subsets I_l, have the same number of elements, say N_1. If $N = N_1 \cdot N_2$, $r = N_2$, each of the innermost sums of (7.7) can be computed with N_1^2 multiplications, which gives a total of $N_2 N_1^2$, when taking into account the requirement that the sum over $i \in I_l$ defines a DFT. The outer sum will need $r = N_2$ multiplications per output point, that is $N_2 \cdot N$ for the whole sum.

Hence, the total number of multiplications needed to compute (7.7) is

$$N_2 \cdot N + N_2 \cdot N_1^2 \;=\; N_1 \cdot N_2(N_1 + N_2) < N_1^2 \cdot N_2^2$$
$$\text{if } N_1, N_2 > 2 , \tag{7.8}$$

which shows clearly that the divide and conquer approach, as given in (7.7), has reduced the number of multiplications needed to compute the DFT.

Of course, when taking into account that, even if the outermost sum of (7.7) is not already in the form of a DFT, it can be rearranged into a DFT plus some so-called twiddle-factors, this mapping is always even more favorable than is shown by (7.8), especially for small N_1, N_2 (for example, the length-2 DFT is simply a sum and difference).

Obviously, if N is highly composite, the division can be applied again to the subproblems, which results in a number of operations generally several orders of magnitude better than the direct matrix vector product.

The important point in (7.2) is that two costs appear explicitly in the divide and conquer scheme: the cost of the mapping (which can be zero when looking at the number of operations only) and the cost of the subproblems. Thus, different types of divide and conquer methods attempt to find various balancing schemes between the mapping and the subproblem costs. In the radix-2 algorithm, for example, the subproblems end up being quite trivial (only sum and differences), while the mapping requires twiddle factors that lead to a large number of multiplications. On the contrary, in the prime factor algorithm, the mapping requires no arithmetic operation (only permutations), while the small DFTs that appear as subproblems will lead to substantial costs since their lengths are coprime.

7.4 FFTs with Twiddle Factors

The divide and conquer approach reintroduced by Cooley and Tukey [25] can be used for any composite length N but has the specificity of always introducing twiddle factors. It turns out that when the factors of N are not coprime (for example if $N = 2^n$), these twiddle factors cannot be avoided at all. This section will be devoted to the different algorithms in that class.

The difference between the various algorithms will consist in the fact that more or fewer of these twiddle factors will turn out to be trivial multiplications, such as $1, -1, j, -j$.

7.4.1 The Cooley-Tukey Mapping

Let us assume that the length of the transform is composite: $N = N_1 \cdot N_2$.

As we have seen in Section 7.3, we want to partition $\{x_i | i = 0, \ldots, N - 1\}$ into different subsets $\{x_i | i \in I_l\}$ in such a way that the periodicities of the involved subsequences are compatible with the periodicity of the input sequence, on the one hand, and allow to define DFTs of reduced lengths on the other hand.

Hence, it is natural to consider decimated versions of the initial sequence:

$$
\begin{aligned}
I_{n_1} &= \{n_2 N_1 + n_1\}, \\
&\quad n_1 = 0, \ldots, N_1 - 1, \quad n_2 = 0, \ldots, N_2 - 1,
\end{aligned}
\tag{7.9}
$$

which, introduced in (7.6), gives

$$
X(z) = \sum_{n_1=0}^{N_1-1} \sum_{n_2=0}^{N_2-1} x_{n_2 N_1 + n_1} z^{-(n_2 N_1 + n_1)},
\tag{7.10}
$$

and, after normalizing with respect to the first element of each subset,

$$
\begin{aligned}
X(z) &= \sum_{n_1=0}^{N_1-1} z^{-n_1} \sum_{n_2=0}^{N_2-1} x_{n_2 N_1 + n_1} z^{-n_2 N_1}, \\
X_k &= X(z)\big|_{z = W_N^{-k}} \\
&= \sum_{n_1=0}^{N_1-1} W_N^{n_1 k} \sum_{n_2=0}^{N_2-1} x_{n_2 N_1 + n_1} W_N^{n_2 N_1 k}.
\end{aligned}
\tag{7.11}
$$

Using the fact that

$$
W_N^{i N_1} = e^{-j 2\pi N_1 i / N} = e^{-j 2\pi / N_2} = W_{N_2}^i,
\tag{7.12}
$$

(7.11) can be rewritten as

$$
X_k = \sum_{n_1=0}^{N_1-1} W_N^{n_1 k} \sum_{n_2=0}^{N_2-1} x_{n_2 N_1 + n_1} W_{N_2}^{n_2 k}.
\tag{7.13}
$$

Equation (7.13) is now nearly in its final form, since the right-hand sum corresponds to N_1 DFTs of length N_2, which allows the reduction of arithmetic complexity to be achieved by reiterating the process. Nevertheless, the structure of the CooleyTukey FFT is not fully given yet.

Call $Y_{n_1,k}$ the kth output of the n_1th such DFT:

$$Y_{n_1,k} = \sum_{n_2=0}^{N_2-1} x_{n_2 N_1 + n_1} W_{N_2}^{n_2 k} . \tag{7.14}$$

Note that in $Y_{n_1,k}$, k can be taken modulo N_2, because

$$W_{N_2}^k = W_{N_2}^{N_2 + k'} = W_{N_2}^{N_2} \cdot W_{N_2}^{k'} = W_{N_2}^{k'} . \tag{7.15}$$

With this notation, X_k becomes

$$X_k = \sum_{n_1=0}^{N_1-1} Y_{n_1,k} W_N^{n_1 k} . \tag{7.16}$$

At this point, we can notice that all the X_k for ks being congruent modulo N_2 are obtained from the same group of N_1 outputs of $Y_{n_1,k}$. Thus, we express k as

$$k = k_1 N_2 + k_2$$
$$k_1 = 0, \dots, N_1 - 1, \quad k_2 = 0, \dots, N_2 - 1 . \tag{7.17}$$

Obviously, $Y_{n_1,k}$ is equal to Y_{n_1,k_2} since k can be taken modulo N_2 in this case [see (7.12) and (7.15)]. Thus, we rewrite (7.16) as

$$X_{k_1 N_2 + k_2} = \sum_{n_1=0}^{N_1-1} Y_{n_1,k_2} W_N^{n_1 (k_1 N_2 + k_2)} , \tag{7.18}$$

which can be reduced, using (7.12), to

$$X_{k_1 N_2 + k_2} = \sum_{n_1=0}^{N_1-1} Y_{n_1,k_2} W_N^{n_1 k_2} W_{N_1}^{n_1 k_1} \tag{7.19}$$

Calling Y'_{n_1,k_2} the result of the first multiplication (by the twiddle factors) in (7.19) we get

$$Y'_{n_1,k_2} = Y_{n_1,k_2} W_N^{n_1 k_2} . \tag{7.20}$$

We see that the values of $X_{k_1 N_2 + k_2}$ are obtained from N_2 DFTs of length N_1 applied on Y'_{n_1,k_2}:

$$X_{k_1 N_2 + k_2} = \sum_{n_1=0}^{N_1-1} Y'_{n_1,k_2} W_{N_1}^{n_1 k_1} . \tag{7.21}$$

We recapitulate the important steps that led to (7.21). First, we evaluated N_1 DFTs of length N_2 in (7.14). Then, N multiplications by the twiddle factors were performed in (7.20). Finally, N_2 DFTs of length N_1 led to the final result (7.21).

A way of looking at the change of variables performed in (7.9) and (7.17) is to say that the one-dimensional vector x_i has been mapped into a two-dimensional vector x_{n_1,n_2} having N_1 lines and N_2 columns. The computation of the DFT is then divided into N_1 DFTs on the lines of the vector x_{n_1,n_2}, a point by point multiplication with the twiddle factors and finally N_2 DFTs on the columns of the preceding result.

Until recently, this was the usual presentation of FFT algorithms, by the so-called "index mappings" [4, 23]. In fact, (7.9) and (7.17), taken together, are often referred to as the "Cooley-Tukey mapping" or "common factor mapping." However, the problem with the two-dimensional interpretation is that it does not include all algorithms (like the split-radix algorithm that will be seen later). Thus, while this interpretation helps the understanding of some of the algorithms, it hinders the comprehension of others. In our presentation, we tried to enhance the role of the periodicities of the problem, which result from the initial choice of the subsets.

Nevertheless, we illustrate pictorially a length-15 DFT using the two-dimensional view with $N_1 = 3$, $N_2 = 5$ (see Fig. 7.1), together with the Cooley-Tukey mapping in Fig. 7.2, to allow a precise comparison with Good's mapping that leads to the other class of FFTs: the FFTs without twiddle factors. Note that for the case where N_1 and N_2 are coprime, the Good's mapping will be more efficient as shown in the next section, and thus this example is for illustration and comparison purpose only. Because of the twiddle factors in (7.20), one cannot interchange the order of DFTs once the input mapping has been chosen. Thus, in Fig. 7.2(a), one has to begin with the DFTs on the rows of the matrix. Choosing $N_1 = 5$, $N_2 = 3$ would lead to the matrix of Fig. 7.2(b), which is obviously different from just transposing the matrix of Fig. 7.2(a). This shows again that the mapping does not lead to a true two-dimensional transform (in that case, the order of row and column would not have any importance) .

7.4.2 Radix-2 and Radix-4 Algorithms

The algorithms suited for lengths equal to powers of 2 (or 4) are quite popular since sequences of such lengths are frequent in signal processing (they make full use of the addressing capabilities of computers or DSP systems).

We assume first that $N = 2^n$. Choosing $N_1 = 2$ and $N_2 = 2^{n-1} = N/2$ in (7.9) and (7.10) divides the input sequence into the sequence of even- and odd-numbered samples, which is the reason why this approach is called "decimation in time" (DIT). Both sequences are decimated versions, with different phases, of the original sequence. Following (7.17), the output consists of $N/2$ blocks of 2 values. Actually, in this simple case, it is easy to rewrite (7.14) and (7.21) exhaustively:

$$X_{k_2} = \sum_{n_2=0}^{N/2-1} x_{2n_2} W_{N/2}^{n_2 k_2}$$
$$+ W_N^{k_2} \sum_{n_2=0}^{N/2-1} x_{2n_2+1} W_{N/2}^{n_2 k_2} , \qquad (7.22a)$$

$$X_{N/2+k_2} = \sum_{n_2=0}^{N/2-1} x_{2n_2} W_{N/2}^{n_2 k_2}$$
$$- W_N^{k_2} \sum_{n_2=0}^{N/2-1} x_{2n_2+1} W_{N/2}^{n_2 k_2} . \qquad (7.22b)$$

Thus, X_m and $X_{N/2+m}$ are obtained by 2-point DFTs on the outputs of the length-$N/2$ DFTs of the even- and odd-numbered sequences, one of which is weighted by twiddle factors. The structure made by a sum and difference followed (or preceded) by a twiddle factor is generally called a "butterfly." The DIT radix-2 algorithm is schematically shown in Fig. 7.3.

Its implementation can now be done in several different ways. The most natural one is to reorder the input data such that the samples of which the DFT has to be taken lie in subsequent locations. This results in the bit-reversed input, in-order output decimation in time algorithm. Another possibility is to selectively compute the DFTs over the input sequence (taking only the even- and odd-numbered samples), and perform an in-place computation. The output will now be in bit-reversed order. Other implementation schemes can lead to constant permutations between the stages (constant geometry algorithm [15]).

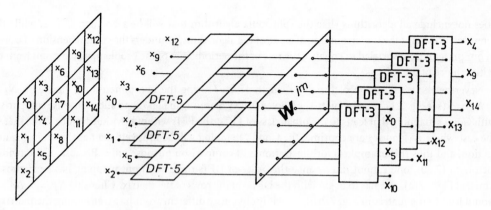

FIGURE 7.1   2-D view of the length-15 Cooley-Tukey FFT.

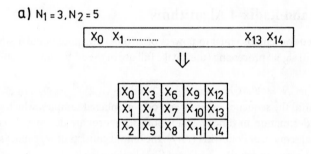

a) $N_1 = 3, N_2 = 5$

| X_0 | X_1 | .... | | X_{13} | X_{14} |

⇓

X_0	X_3	X_6	X_9	X_{12}
X_1	X_4	X_7	X_{10}	X_{13}
X_2	X_5	X_8	X_{11}	X_{14}

b) $N_1 = 5, N_2 = 3$

X_0	X_5	X_{10}
X_1	X_6	X_{11}
X_2	X_7	X_{12}
X_3	X_8	X_{13}
X_4	X_9	X_{14}

FIGURE 7.2   Cooley-Tukey mapping. (a) $N_1 = 3$, $N_2 = 5$; (b) $N_1 = 5$, $N_2 = 3$.

If we reverse the role of N_1 and N_2, we get the decimation in frequency (DIF) version of the algorithm. Inserting $N_1 = N/2$ and $N_2 = 2$ into (7.9), (7.10) leads to [again from (7.14) and (7.21)]

$$X_{2k_1} = \sum_{n_1=0}^{N/2-1} W_{N/2}^{n_1 k_1} \left(x_{n_1} + x_{N/2+n_1} \right) , \qquad (7.23a)$$

$$X_{2k_1+1} = \sum_{n_1=0}^{N/2-1} W_{N/2}^{n_1 k_1} W_N^{n_1} \left(x_{n_1} - x_{N/2+n_1} \right) , \qquad (7.23b)$$

This first step of a DIF algorithm is represented in Fig. 7.5(a), while a schematic representation of the full DIF algorithm is given in Fig. 7.4. The duality between division in time and division in frequency is obvious, since one can be obtained from the other by interchanging the role of $\{x_i\}$ and $\{X_k\}$.

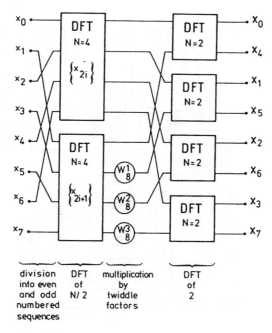

FIGURE 7.3   Decimation in time radix-2 FFT.

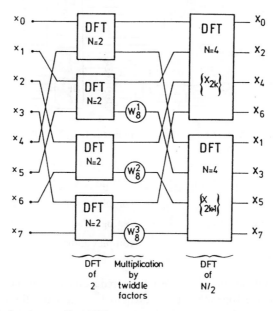

FIGURE 7.4   Decimation in frequency radix-2 FFT.

Let us now consider the computational complexity of the radix-2 algorithm (which is the same for the DIF and DIT version because of the duality indicated above). From (7.22) or (7.23), one sees that a DFT of length N has been replaced by two DFTs of length $N/2$, and this at the cost of $N/2$ complex multiplications as well as N complex additions. Iterating the scheme $\log_2 N - 1$ times in order to obtain trivial transforms (of length 2) leads to the following order of magnitude of the number of operations:

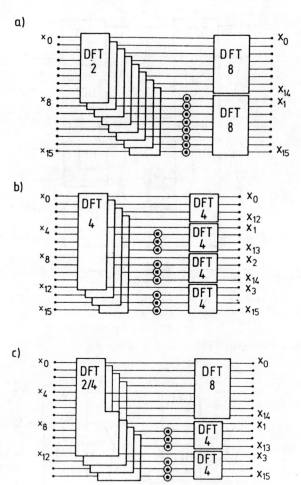

FIGURE 7.5   Comparison of various DIF algorithms for the length-16 DFT. (a) Radix-2; (b) radix-4; (c) split-radix.

$$O_M\left[\text{DFT}_{\text{radix-2}}\right] \;\approx\; N/2\left(\log_2 N - 1\right)$$

$$\text{complex multiplications,} \tag{7.24a}$$

$$O_A\left[\text{DFT}_{\text{radix-2}}\right] \;\approx\; N\left(\log_2 N - 1\right)$$

$$\text{complex additions.} \tag{7.24b}$$

A closer look at the twiddle factors will enable us to still reduce these numbers. For comparison purposes, we will count the number of real operations that are required, provided that the multiplication of a complex number x by W_N^i is done using three real multiplications and three real additions [12]. Furthermore, if i is a multiple of $N/4$, no arithmetic operation is required, and only two real multiplications and additions are required if i is an odd multiple of $N/8$. Taking into account these simplifications results in the following total number of operations [12]:

$$M\left[\text{DFT}_{\text{radix-2}}\right] \;=\; 3N/2\log_2 N - 5N + 8\,, \tag{7.25a}$$

$$A\left[\text{DFT}_{\text{radix-2}}\right] \;=\; 7N/2\log_2 N - 5N + 8\,. \tag{7.25b}$$

Nevertheless, it should be noticed that these numbers are obtained by the implementation of four different butterflies (one general plus three special cases), which reduces the regularity of the programs. An evaluation of the number of real operations for other number of special butterflies is given in [4], together with the number of operations obtained with the usual 4-mult, 2-adds complex multiplication algorithm.

Another case of interest appears when N is a power of 4. Taking $N_1 = 4$ and $N_2 = N/4$, (7.13) reduces the length-N DFT into 4 DFTs of length $N/4$, about $3N/4$ multiplications by twiddle factors, and $N/4$ DFTs of length 4. The interest of this case lies in the fact that the length-4 DFTs do not cost any multiplication (only 16 real additions). Since there are $\log_4 N - 1$ stages and the first set of twiddle factors (corresponding to $n_1 = 0$ in (7.20)) is trivial, the number of complex multiplications is about

$$O_M \left[\text{DFT}_{\text{radix-4}} \right] \approx 3N/4 \left(\log_4 N - 1 \right) . \tag{7.26}$$

Comparing (7.26) to (7.24a) shows that the number of multiplications can be reduced with this radix-4 approach by about a factor of 3/4. Actually, a detailed operation count using the simplifications indicated above gives the following result [12]:

$$M \left[\text{DFT}_{\text{radix-4}} \right] = 9N/8 \log_2 N - 43N/12 + 16/3 , \tag{7.27a}$$

$$A \left[\text{DFT}_{\text{radix-4}} \right] = 25N/8 \log_2 N - 43N/12 + 16/3 . \tag{7.27b}$$

Nevertheless, these operation counts are obtained at the cost of using six different butterflies in the programming of the FFT. Slight additional gains can be obtained when going to even higher radices (like 8 or 16) and using the best possible algorithms for the small DFTs. Since programs with a regular structure are generally more compact, one often uses recursively the same decomposition at each stage, thus leading to full radix-2 or radix-4 programs, but when the length is not a power of the radix (for example 128 for a radix-4 algorithm), one can use smaller radices towards the end of the decomposition. A length-256 DFT could use two stages of radix-8 decomposition, and finish with one stage of radix-4. This approach is called the "mixed-radix" approach [45] and achieves low arithmetic complexity while allowing flexible transform length (not restricted to powers of 2, for example), at the cost of a more involved implementation.

7.4.3 Split-Radix Algorithm

As already noted in Section 7.2, the lowest known number of both multiplications and additions for length-2^n algorithms was obtained as early as 1968 and was again achieved recently by new algorithms. Their power was to show explicitly that the improvement over fixed- or mixed-radix algorithms can be obtained by using a radix-2 and a radix-4 simultaneously on different parts of the transform. This allowed the emergence of new compact and computationally efficient programs to compute the length-2^n DFT.

Below, we will try to motivate (*a posteriori*!) the split-radix approach and give the derivation of the algorithm as well as its computational complexity.

When looking at the DIF radix-2 algorithm given in (7.23), one notices immediately that the even indexed outputs X_{2k_1} are obtained without any further multiplicative cost from the DFT of a length-$N/2$ sequence, which is not so well-done in the radix-4 algorithm for example, since relative to that length-$N/2$ sequence, the radix-4 behaves like a radix-2 algorithm. This lacks logical sense because it is well-known that the radix-4 is better than the radix-2 approach.

From that observation, one can derive a first rule: the even samples of a DIF decomposition X_{2k} should be computed separately from the other ones, with the same algorithm (recursively) as the DFT of the original sequence (see [53] for more details).

However, as far as the odd indexed outputs X_{2k+1} are concerned, no general simple rule can be established, except that a radix-4 will be more efficient than a radix-2, since it allows computation of the samples through two $N/4$ DFTs instead of a single $N/2$ DFT for a radix-2, and this at the same multiplicative cost, which will allow the cost of the recursions to grow more slowly. Tests showed that computing the odd indexed output through radices higher than 4 was inefficient.

The first recursion of the corresponding "split-radix" algorithm (the radix is split in two parts) is obtained by modifying (7.23) accordingly:

$$X_{2k_1} = \sum_{n_1=0}^{N/2-1} W_{N/2}^{n_1 k_1} \left(x_{n_1} + x_{N/2+n_1} \right) , \tag{7.28a}$$

$$X_{4k_1+1} = \sum_{n_1=0}^{N/4-1} W_{N/4}^{n_1 k_1} W_N^{n_1} \left[\left(x_{n_1} - x_{N/2+n_1} \right) + j \left(x_{n_1+N/4} - x_{n_1+3N/4} \right) \right] , \tag{7.28b}$$

$$X_{4k_1+3} = \sum_{n_1=0}^{N/4-1} W_{N/4}^{n_1 k_1} W_N^{3n} \left[\left(x_{n_1} + x_{N/2+n_1} \right) - j \left(x_{n_1+N/4} - x_{n_1+3N/4} \right) \right] . \tag{7.28c}$$

The above approach is a DIF SRFFT, and is compared in Fig. 7.5 with the radix-2 and radix-4 algorithms. The corresponding DIT version, being dual, considers separately the subsets $\{x_{2i}\}$, $\{x_{4i+1}\}$ and $\{x_{4i+3}\}$ of the initial sequence.

Taking $I_0 = \{2i\}$, $I_1 = \{4i + 1\}$, $I_2 = \{4i + 3\}$ and normalizing with respect to the first element of the set in (7.7) leads to

$$X_k = \sum_{I_0} x_{2i} W_N^{k(2i)} + W_N^k \sum_{I_1} x_{4i+1} W_N^{k(4i+1)-k} + W_N^{3k} \sum_{I_2} x_{4i+3} W_N^{k(4i+3)-3k} , \tag{7.29}$$

which can be explicitly decomposed in order to make the redundancy between the computation of X_k, $X_{k+N/4}$, $X_{k+N/2}$ and $X_{k+3N/4}$ more apparent:

$$X_k = \sum_{i=0}^{N/2-1} x_{2i} W_{N/2}^{ik} + W_N^k \sum_{i=0}^{N/4-1} x_{4i+1} W_{N/4}^{ik} + W_N^{3k} \sum_{i=0}^{N/4-1} x_{4i+3} W_{N/4}^{ik} , \tag{7.30a}$$

$$X_{k+N/4} = \sum_{i=0}^{N/2-1} x_{2i} W_{N/2}^{ik} + j W_N^k \sum_{i=0}^{N/4-1} x_{4i+1} W_{N/4}^{ik} - j W_N^{3k} \sum_{i=0}^{N/4-1} x_{4i+3} W_{N/4}^{ik} , \tag{7.30b}$$

$$X_{k+N/2} = \sum_{i=0}^{N/2-1} x_{2i} W_{N/2}^{ik} - W_N^k \sum_{i=0}^{N/4-1} x_{4i+1} W_{N/4}^{ik} - W_N^{3k} \sum_{i=0}^{N/4-1} x_{4i+3} W_{N/4}^{ik} , \tag{7.30c}$$

$$X_{k+3N/4} = \sum_{i=0}^{N/2-1} x_{2i} W_{N/2}^{ik} - j W_N^k \sum_{i=0}^{N/4-1} x_{4i+1} W_{N/4}^{ik} + j W_N^{3k} \sum_{i=0}^{N/4-1} x_{4i+3} W_{N/4}^{ik} . \tag{7.30d}$$

The resulting algorithms have the minimum known number of operations (multiplications plus additions) as well as the minimum number of multiplications among practical algorithms for lengths which are powers of 2. The number of operations can be checked as being equal to

$$M \left[\text{DFT}_{\text{split-radix}} \right] = N \log_2 N - 3N + 4 , \tag{7.31a}$$

$$A \left[\text{DFT}_{\text{split-radix}} \right] = 3N \log_2 N - 3N + 4 , \tag{7.31b}$$

These numbers of operations can be obtained with only four different building blocks (with a complexity slightly lower than the one of a radix-4 butterfly), and are compared with the other algorithms in Tables 7.1 and 7.2.

TABLE 7.1 Number of Non-Trivial Real Multiplications for Various FFTs on Complex Data

N	Radix 2	Radix 4	SRFFT	PFA	Winograd
16	24	20	20		
30				100	68
32	88		68		
60				200	136
64	264	208	196		
120				460	276
128	712		516		
240				1100	632
256	1800	1392	1284		
504				2524	1572
512	4360		3076		
1008				5804	3548
1024	10248	7856	7172		
2048	23560		16388		
2520				17660	9492

TABLE 7.2 Number of Real Additions for Various FFTs on Complex Data

N	Radix 2	Radix 4	SRFFT	PFA	Winograd
16	152	148	148		
30				384	384
32	408		388		
60				888	888
64	1032	976	964		
120				2076	2076
128	2504		2308		
240				4812	5016
256	5896	5488	5380		
504				13388	14540
512	13566		12292		
1008				29548	34668
1024	30728	28336	27652		
2048	68616		61444		
2520				84076	99628

Of course, due to the asymmetry in the decomposition, the structure of the algorithm is slightly more involved than for fixed-radix algorithms. Nevertheless, the resulting programs remain fairly simple [113] and can be highly optimized. Furthermore, this approach is well suited for applying FFTs on real data. It allows an in-place, butterfly style implementation to be performed [65, 77].

The power of this algorithm comes from the fact that it provides the lowest known number of operations for computing length-2^n FFTs, while being implemented with compact programs. We shall see later that there are some arguments tending to show that it is actually the best possible compromise.

Note that the number of multiplications in (7.31a) is equal to the one obtained with the so-called "real-factor" algorithms [24, 44]. In that approach, a linear combination of the data, using additions only, is made such that all twiddle factors are either pure real or pure imaginary. Thus, a multiplication of a complex number by a twiddle factor requires only two real multiplications. However, the real factor algorithms are quite costly in terms of additions, and are numerically ill-conditioned (division by small constants).

7.4.4 Remarks on FFTs with Twiddle Factors

The Cooley-Tukey mapping in (7.9) and (7.17) is generally applicable, and actually the only possible mapping when the factors on N are not coprime. While we have paid particular attention to the case $N = 2^n$, similar algorithms exist for $N = p^m$ (p an arbitrary prime). However, one of the elegances of the length-2^n algorithms comes from the fact that the small DFTs (lengths 2 and 4) are multiplication-free, a fact that does not hold for other radices like 3 or 5, for instance. Note, however, that it is possible, for radix-3, either to completely remove the multiplication inside the butterfly by a change of base [26], at the cost of a few multiplications and additions, or to merge it with the twiddle factor [49] in the case where the implementation is based on the 4-mult 2-add complex multiplication scheme. It was also recently shown that, as soon as a radix p^2 algorithm was more efficient than a radix-p algorithm, a split-radix p/p^2 was more efficient than both of them [53]. However, unlike the 2^n case, efficient implementations for these p^n split-radix algorithms have not yet been reported. More efficient mixed radix algorithms also remain to be found (initial results are given in [40]).

7.5 FFTs Based on Costless Mono- to Multidimensional Mapping

The divide and conquer strategy, as explained in Section 7.3, has few requirements for feasibility: N needs only to be composite, and the whole DFT is computed from DFTs on a number of points which is a factor of N (this is required for the redundancy in the computation of (7.11) to be apparent). This requirement allows the expression of the innermost sum of (7.11) as a DFT, provided that the subsets I_1, have been chosen in such a way that $x_i, i \in I_1$, is periodic. But, when N factors into relatively prime factors, say $N = N_1 \cdot N_2, (N_1, N_2) = 1$, a very simple property will allow a stronger requirement to be fulfilled:

Starting from any point of the sequence x_i, you can take as a first subset with compatible periodicity either $\{x_{i+N_1 \cdot n_2} | n_2 = 1, \ldots, N_2 - 1\}$ or, equivalently $\{x_{i+N_2 \cdot n_1} | n_1 = 1, \ldots, N_1 - 1\}$, and both subsets only have one common point x_i (by compatible, it is meant that the periodicity of the subsets divides the periodicity of the set). This allows a rearrangement of the input (periodic) vector into a matrix with a periodicity in both dimensions (rows and columns), both periodicities being compatible with the initial one (see Fig. 7.6).

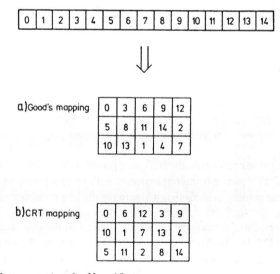

FIGURE 7.6   The prime factor mappings for $N = 15$.

7.5.1 Basic Tools

FFTs without twiddle factors are all based on the same mapping, which is explained in the next section ("The Mapping of Good"). This mapping turns the original transform into sets of small DFTs, the lengths of which are coprime. It is therefore necessary to find efficient ways of computing these short-length DFTs. The section "DFT Computation as a Convolution" explains how to turn them into cyclic convolutions for which efficient algorithms are described in the Section "Computation of the Cyclic Convolution."

The Mapping of Good [32]

Performing the selection of subsets described in the introduction of Section 7.5 for any index i is equivalent to writing i as

$$\begin{aligned}
i &= \langle n_1 \cdot N_2 + n_2 \cdot N_1 \rangle_N\,, \\
n_1 &= 1, \ldots, N_1 - 1\,, \quad n_2 = 1, \ldots, N_2 - 1\,, \\
N &= N_1 N_2\,,
\end{aligned} \tag{7.32}$$

and, since N_1 and N_2 are coprime, this mapping is easily seen to be one to one. (It is obvious from the right-hand side of (7.32) that all congruences modulo N_1 are obtained for a given congruence modulo N_2, and vice versa.)

This mapping is another arrangement of the "Chinese Remainder Theorem" (CRT) mapping, which can be explained as follows on index k.

The CRT states that if we know the residue of some number k modulo two relatively prime numbers N_1 and N_2, it is possible to reconstruct $\langle k \rangle_{N_1 N_2}$ as follows:

Let $\langle k \rangle_{N_1} = k_1$ and $\langle k \rangle_{N_2} = k_2$. Then the value of $k \mod N (N = N_1 \cdot N_2)$ can be found by

$$k = \langle N_1 t_1 k_2 + N_2 t_2 k_1 \rangle_N\,, \tag{7.33}$$

t_1 being the multiplicative inverse of $N_1 \mod N_2$, that is $\langle t_1, N_1 \rangle_{N_2} = 1$, and t_2 the multiplicative inverse of $N_2 \mod N_1$ [these inverses always exist, since N_1 and N_2 are coprime: $(N_1, N_2) = 1$].

Taking into account these two mappings in the definition of the DFT (7.3) leads to

$$X_{N_1 t_1 k_2 + N_2 t_2 k_1} = \sum_{n_1=0}^{N_1-1} \sum_{n_2=0}^{N_2-1} x_{n_1 N_2 + n_2 N_1} W_N^{(n_1 N_2 + N_1 n_2)(N_1 t_1 k_2 + N_2 t_2 k_1)}\,, \tag{7.34}$$

but

$$W_N^{N_2} = W_{N_1} \tag{7.35}$$

and

$$W_{N_1}^{N_2 t_2} = W_{N_1}^{\langle N_2 t_2 \rangle N_1} = W_{N_1}\,, \tag{7.36}$$

which implies

$$X_{N_1 t_1 k_2 + N_2 t_2 k_1} = \sum_{n_1=0}^{N_1-1} \sum_{n_2=0}^{N_2-1} x_{n_1 N_2 + n_2 N_1} W_{N_1}^{n_1 k_2} W_{N_2}^{n_2 k_2}\,, \tag{7.37}$$

which, with

$$x'_{n_1, n_2} = x_{n_1 N_2 + n_2 N_1}$$

and

$$X'_{k_1, k_2} = X_{N_1 t_1 k_2 + N_2 t_2 k_1}\,,$$

leads to a formulation of the initial DFT into a true bidimensional transform:

$$X'_{k_1 k_2} = \sum_{n_1=0}^{N_1-1} \sum_{n_2=0}^{N_2-1} x'_{n_1 n_2} W_{N_1}^{n_1 k_1} W_{N_2}^{n_2 k_2} \tag{7.38}$$

An illustration of the prime factor mapping is given in Fig. 7.6(a) for the length $N = 15 = 3 \cdot 5$, and Fig. 7.6(b) provides the CRT mapping. Note that these mappings, which were provided for a factorization of N into two coprime numbers, easily generalizes to more factors, and that reversing the roles of N_1, and N_2 results in a transposition of the matrices of Fig. 7.6.

DFT Computation as a Convolution

With the aid of Good's mapping, the DFT computation is now reduced to that of a multidimensional DFT, with the characteristic that the lengths along each dimension are coprime. Furthermore, supposing that these lengths are small is quite reasonable, since Good's mapping can provide a full multi-dimensional factorization when N is highly composite.

The question is now to find the best way of computing this M-D DFT and these small-length DFTs. A first step in that direction was obtained by Rader [43], who showed that a DFT of prime length could be obtained as the result of a cyclic convolution: Let us rewrite (7.1) for a prime length $N = 5$:

$$\begin{bmatrix} X_0 \\ X_1 \\ X_2 \\ X_3 \\ X_4 \end{bmatrix} = \begin{bmatrix} 1 & 1 & 1 & 1 & 1 \\ 1 & W_5^1 & W_5^2 & W_5^3 & W_5^4 \\ 1 & W_5^2 & W_5^4 & W_5^1 & W_5^3 \\ 1 & W_5^3 & W_5^1 & W_5^4 & W_5^2 \\ 1 & W_5^4 & W_5^3 & W_5^2 & W_5^1 \end{bmatrix} \begin{bmatrix} x_0 \\ x_1 \\ x_2 \\ x_3 \\ x_4 \end{bmatrix} . \tag{7.39}$$

Obviously, removing the first column and first row of the matrix will not change the problem, since they do not involve any multiplication. Furthermore, careful examination of the remaining part of the matrix shows that each column and each row involves every possible power of W_5, which is the first condition to be met for this part of the DFT to become a cyclic convolution. Let us now permute the last two rows and last two columns of the reduced matrix:

$$\begin{bmatrix} X'_1 \\ X'_2 \\ X'_4 \\ X'_3 \end{bmatrix} = \begin{bmatrix} W_5^1 & W_5^2 & W_5^4 & W_5^3 \\ W_5^2 & W_5^4 & W_5^3 & W_5^1 \\ W_5^4 & W_5^3 & W_5^1 & W_5^2 \\ W_5^3 & W_5^1 & W_5^2 & W_5^4 \end{bmatrix} \begin{bmatrix} x_1 \\ x_2 \\ x_4 \\ x_3 \end{bmatrix} . \tag{7.40}$$

Equation (7.40) is then a cyclic correlation (or a convolution with the reversed sequence).

It turns out that this a general result.

It is well-known in number theory that the set of numbers lower than a prime p admits some primitive elements g such that the successive powers of g modulo p generate all the elements of the set. In the example above, $p = 5$, $g = 2$, and we observe that

$$g^0 = 1, \quad g^1 = 2, \quad g^2 = 4, \quad g^3 = 8 = 3 \quad (\text{mod } 5) .$$

The above result (7.40) is only the writing of the DFT in terms of the successive powers of W_p^g:

$$X'_k = \sum_{i=1}^{p-1} x_i W_p^{ik} , \quad k = 1, \ldots, p-1 , \tag{7.41}$$

$$\langle ik \rangle_p = \langle \langle i \rangle_p \cdot \langle k \rangle_p \rangle_p = \langle \langle g^{u_i} \rangle_p \langle g^{v_k} \rangle_p \rangle_p ,$$

$$X'_{g^{v_i}} = \sum_{u_i=0}^{p-2} x_{g^{u_i}} \cdot \left(W_p^g \right)^{u_i + v_i} , \quad v_i = 0, \ldots, p-2 , \tag{7.42}$$

and the length-p DFT turns out to be a length $(p-1)$ cyclic correlation:

$$\{X'_g\} = \{x_g\} * \{W^g_p\} \, . \tag{7.43}$$

Computation of the Cyclic Convolution

Of course (7.42) has changed the problem, but it is not solved yet. And in fact, Rader's result was considered as a curiosity up to the moment when Winograd [55] obtained some new results on the computation of cyclic convolution.

And, again, this was obtained by application of the CRT. In fact, the CRT, as explained in (7.33), (7.34) can be rewritten in the polynomial domain: if we know the residues of some polynomial $K(z)$ modulo two mutually prime polynomials

$$\langle K(z) \rangle_{P_1(z)} = K_1(z) \, ,$$
$$(P_1(z), \; P_2(z)) = 1 \, , \tag{7.44}$$
$$\langle K(z) \rangle_{P_2(z)} = K_2(z) \, ,$$

we shall be able to obtain

$$K(z) \bmod P_1(z) \cdot P_2(z) = P(z)$$

by a procedure similar to that of (7.33).

This fact will be used twice in order to obtain Winograd's method of computing cyclic convolutions:

A first application of the CRT is the breaking of the cyclic convolution into a set of polynomial products. For more convenience, let us first state (7.43) in polynomial notation:

$$X'(z) = x'(z) \cdot w(z) \bmod \left(z^{p-1} - 1 \right) \, . \tag{7.45}$$

Now, since $p-1$ is not prime (it is at least even), $z^{p-1} - 1$ can be factorized at least as

$$z^{p-1} - 1 = \left(z^{(p-1)/2} + 1 \right) \left(z^{(p-1)/2} - 1 \right) \, , \tag{7.46}$$

and possibly further, depending on the value of p. These polynomial factors are known and named cyclotomic polynomials $\varphi_q(z)$. They provide the full factorization of any $z^N - 1$:

$$z^N - 1 = \prod_{q|N} \varphi_q(z) \, . \tag{7.47}$$

A useful property of these cyclotomic polynomials is that the roots of $\varphi_q(z)$ are all the qth primitive roots of unity, hence degree $\{\varphi_q(z)\} = \varphi(q)$, which is by definition the number of integers lower than q and coprime with it. Namely, if $w_q = e^{-j2\pi/q}$, the roots of $\varphi_q(z)$ are $\{W^r_q | (r, q) = 1\}$.

As an example, for $p = 5$, $z^{p-1} - 1 = z^4 - 1$,

$$\begin{aligned} z^4 - 1 &= \varphi_1(z) \cdot \varphi_2(z) \cdot \varphi_4(z) \\ &= (z - 1)(z + 1)(z^2 + 1) \, . \end{aligned}$$

The first use of the CRT to compute the cyclic convolution (7.45) is then as follows:

1. compute

$$x'_q(z) = x'(z) \bmod \varphi_q(z) \, ,$$
$$q | p - 1$$
$$w'_q(z) = w(z) \bmod \varphi_q(z) \, ,$$

2. then obtain

$$X'_q(z) = x'_q(z) \cdot w'_q(z) \bmod \varphi_q(z)$$

3. reconstruct $X'(z) \bmod z^{p-1} - 1$ from the polynomials $X'_q(z)$ using the CRT.

Let us apply this procedure to our simple example:

$$
\begin{aligned}
x'(z) &= x_1 + x_2 z + x_4 z^2 + x_3 z^3 , \\
w(z) &= W_5^1 + W_5^2 z + W_5^4 z^2 + W_5^3 z^3 .
\end{aligned}
$$

Step 1.

$$
\begin{aligned}
w_4(z) &= w(z) \bmod \varphi_4(z) \\
&= \left(W_5^1 - W_5^4\right) + \left(W_5^2 - W_5^3\right) z , \\
w_2(z) &= w(z) \bmod \varphi_2(z) \\
&= \left(W_5^1 + W_5^4 - W_5^2 - W_5^3\right) , \\
w_1(z) &= w(z) \bmod \varphi_1(z) \\
&= \left(W_5^1 + W_5^4 + W_5^2 + W_5^3\right) \quad [= -1] , \\
x'_4(z) &= (x_1 - x_4) + (x_2 - x_3) z , \\
x'_2(z) &= (x_1 + x_4 - x_2 - x_3) , \\
x'_1(z) &= (x_1 + x_4 + x_2 + x_3) .
\end{aligned}
$$

Step 2.

$$
\begin{aligned}
X'_4(z) &= x'_4(z) \cdot w_4(z) \bmod \varphi_4(z) , \\
X'_2(z) &= x'_2(z) \cdot w_2(z) \bmod \varphi_2(z) , \\
X'_1(z) &= x'_1(z) \cdot w_1(z) \bmod \varphi_1(z) ,
\end{aligned}
$$

Step 3.

$$
\begin{aligned}
X'(z) &= \left[X'_1(z)(1+z)/2 + X'_2(z)(1-z)/2\right] \\
&\quad \times \left(1+z^2\right)/2 + X'_4(z)\left(1-z^2\right)/2 .
\end{aligned}
$$

Note that all the coefficients of $W_q(z)$ are either real or purely imaginary. This is a general property due to the symmetries of the successive powers of W_p.

The only missing tool needed to complete the procedure now is the algorithm to compute the polynomial products modulo the cyclotomic factors. Of course, a straightforward polynomial product followed by a reduction modulo $\varphi_q(z)$ would be applicable, but a much more efficient algorithm can be obtained by a second application of the CRT in the field of polynomials.

It is already well-known that knowing the values of an Nth degree polynomial at $N+1$ different points can provide the value of the same polynomial anywhere else by Lagrange interpolation. The CRT provides an analogous way of obtaining its coefficients.

Let us first recall the equation to be solved:

$$
X'_q(z) = x'_q(z) \cdot w_q(z) \bmod \varphi_q(z) , \tag{7.48}
$$

with

$$
\deg \varphi_q(z) = \varphi(q) .
$$

Since $\varphi_q(z)$ is irreducible, the CRT cannot be used directly. Instead, we choose to evaluate the product $X''_q(z) = x'_q(z) \cdot w_q(z)$ modulo an auxiliary polynomial $A(z)$ of degree greater than the degree of the product. This auxiliary polynomial will be chosen to be fully factorizable. The CRT hence applies, providing

$$X_q''(z) = x_q'(z) \cdot w_q(z) \,,$$

since the mod $A(z)$ is totally artificial, and the reduction modulo $\varphi_q(z)$ will be performed afterwards.

The procedure is then as follows.

Let us evaluate both $x_q'(z)$ and $w_q(z)$ modulo a number of different monomials of the form

$$(z - a_i) \,, \quad i = 1, \ldots, 2\varphi(q) - 1.$$

Then compute

$$X_q''(a_i) = x_q'(a_i) w_q(a_i), \quad i = 1, \ldots, 2\varphi(q) - 1 \,. \tag{7.49}$$

The CRT then provides a way of obtaining

$$X_q''(z) \bmod A(z) \,, \tag{7.50}$$

with

$$A(z) = \prod_{i=1}^{2\varphi(q)-1} (z - a_i) \,,$$

which is equal to $X_q''(z)$ itself, since

$$\deg X_q''(z) = 2\varphi(q) - 2 \,. \tag{7.51}$$

Reduction of $X_q''(z) \bmod \varphi_z(z)$ will then provide the desired result.

In practical cases, the points $\{a_i\}$ will be chosen in such a way that the evaluation of $w_q'(a_i)$ involves only additions (i.e.: $a_i = 0, \pm 1, \ldots$).

This limits the degree of the polynomials whose products can be computed by this method. Other suboptimal methods exist [12], but are nevertheless based on the same kind of approach [the "dot products" (7.49) become polynomial products of lower degree, but the overall structure remains identical].

All this seems fairly complicated, but results in extremely efficient algorithms that have a low number of operations. The full derivation of our example ($p = 5$) then provides the following algorithm:

5 point DFT:

$$
\begin{aligned}
u &= 2\pi/5 \\
t_1 &= x_1 + x_4, \quad t_2 = x_2 + x_3 \,, \quad \text{(reduction modulo } z^2 - 1) \\
t_3 &= x_1 - x_4, \quad t_4 = x_3 - x_2 \,, \quad \text{(reduction modulo } z^2 + 1) \\
t_5 &= t_1 + t_2 \quad \text{(reduction modulo } z - 1) \,, \\
t_6 &= t_1 - t_2 \quad \text{(reduction modulo } z + 1) \,, \\
m_1 &= [(\cos u + \cos 2u)/2] t_5 \,, \quad \left(X_1'(z) = x_1'(z) \cdot w_1(z) \bmod \varphi_1(z) \right) \\
m_2 &= [(\cos u - \cos 2u)/2] t_6 \,, \quad \left(X_2'(z) = x_2'(z) \cdot w_2(z) \bmod \varphi_2(z) \right) \\
&\quad \text{polynomial product modulo } z^2 + 1 \,, \quad \left(X_4'(z) = x_4'(z) \cdot w_4(z) \bmod \varphi_u(z) \; \text{:)} \right) \\
m_3 &= -j(\sin u)(t_3 + t_4) \,, \\
m_4 &= -j(\sin u + \sin 2u) t_4 \,, \\
m_5 &= j(\sin u - \sin 2u) t_3 \,, \\
s_1 &= m_3 - m_4 \,, \\
s_2 &= m_3 + m_5 \,,
\end{aligned}
$$

(reconstruction following Step 3, the 1/2 terms have been included into the polynomial products:)

$$
\begin{aligned}
s_3 &= x_0 + m_1 , \\
s_4 &= s_3 + m_2 , \\
s_5 &= s_3 - m_2 , \\
X_0 &= x_0 + t_5 , \\
X_1 &= s_4 + s_1 , \\
X_2 &= s_5 + s_2 , \\
X_3 &= s_5 - s_2 , \\
X_4 &= s_4 - s_1 ,
\end{aligned}
$$

When applied to complex data, this algorithm requires 10 real multiplications and 34 real additions vs. 48 real multiplications and 88 real additions for a straightforward algorithm (matrix-vector product).

In matrix form, and slightly changed, this algorithm may be written as follows:

$$
\left(X'_0, X'_1, \ldots, X'_4 \right)^T = C \cdot D \cdot B \cdot (x_0, x_1, \ldots, x_4)^T , \tag{7.52}
$$

with

$$
C = \begin{bmatrix}
1 & 0 & 0 & 0 & 0 & 0 \\
1 & 1 & 1 & 1 & -1 & 0 \\
1 & 1 & -1 & 1 & 0 & 1 \\
1 & 1 & -1 & -1 & 0 & -1 \\
1 & 1 & 1 & -1 & 1 & 0
\end{bmatrix} ,
$$

$$
D = \operatorname{diag} \left[1, ((\cos u + \cos 2u)/2 - 1) , (\cos u - \cos 2u)/2 , -j \sin u , \right.
$$
$$
\left. - j (\sin u + \sin 2u) , j (\sin u - \sin 2u) \right] ,
$$

$$
B = \begin{bmatrix}
1 & 1 & 1 & 1 & 1 \\
0 & 1 & 1 & 1 & 1 \\
0 & 1 & -1 & -1 & 1 \\
0 & 1 & -1 & 1 & -1 \\
0 & 0 & -1 & 1 & 0 \\
0 & 1 & 0 & 0 & 1
\end{bmatrix} .
$$

By construction, D is a diagonal matrix, where all multiplications are grouped, while C and B only involve additions (they correspond to the reductions and reconstructions in the applications of the CRT).

It is easily seen that this structure is a general property of the short-length DFTs based on CRT: all multiplications are "nested" at the center of the algorithms. By construction, also, D has dimension M_p, which is the number of multiplications required for computing the DFT, some of them being trivial (at least one, needed for the computation of X_0). In fact, using such a formulation, we have $M_p \geq p$. This notation looks awkward, at first glance (why include trivial multiplications in the total number?), but Section 7.5.3 will show that it is necessary in order to evaluate the number of multiplications in the Winograd FFT.

It can also be proven that the methods explained in this section are essentially the only ways of obtaining FFTs with the minimum number of multiplications. In fact, this gives the optimum structure, mathematically speaking. These methods always provide a number of multiplications lower than twice the length of the DFT:

$$
M_{N_1} < 2 N_1 .
$$

This shows the linear complexity of the DFT in this case.

7.5.2 Prime Factor Algorithms [95]

Let us now come back to the initial problem of this section: the computation of the bidimensional transform given in (7.38). Rearranging the data in matrix form, of size $N_1 N_2$, and F_1 (resp. F_2) denoting the Fourier matrix of size N_1 (resp. N_2), results in the following notation, often used in the context of image processing:

$$X = F_1 x F_2^T . \tag{7.53}$$

Performing the FFT algorithm separately along each dimension results in the so-called prime factor algorithm (PFA).

To summarize, PFA makes use of Good's mapping (Section "The Mapping of Good") to convert the length $N_1 \cdot N_2$ 1-D DFT into a size $N_1 \times N_2$ 2-D DFT, and then computes this 2-D DFT in a row-column fashion, using the most efficient algorithms along each dimension.

Of course, this applies recursively to more than two factors, the constraints being that they must be mutually coprime. Nevertheless, this constraint implies the availability of a whole set of efficient small DFTs ($N_i = 2, 3, 4, 5, 7, 8, 16$ is already sufficient to provide a dense set of feasible lengths).

A graphical display of PFA for length $N = 15$ is given in Fig. 7.7. Since there are N_2 applications of length N_1 FFT and N_1, applications of length N_2 FFTs, the computational costs are as follows:

$$
\begin{aligned}
M_{N_1 N_2} &= N_1 M_2 + N_2 M_1 , \\
A_{N_1 N_2} &= N_1 A_2 + N_2 A_1 ,
\end{aligned}
\tag{7.54}
$$

or, equivalently, the number of operations to be performed per output point is the sum of the individual number of operations in each short algorithm: let m_N and a_N be these reduced numbers

$$
\begin{aligned}
m_{N_1 N_2 N_3 N_4} &= m_{N_1} + m_{N_2} + m_{N_3} + m_{N_4} , \\
a_{N_1 N_2 N_3 N_4} &= a_{N_1} + a_{N_2} + a_{N_3} + a_{N_4} .
\end{aligned}
\tag{7.55}
$$

An evaluation of these figures is provided in Tables 7.1 and 7.2.

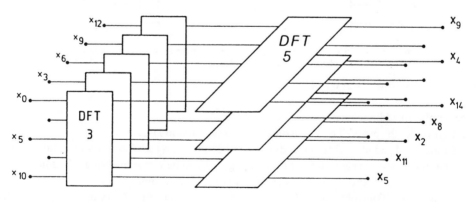

FIGURE 7.7 Schematic view of PFA for $N = 15$.

7.5.3 Winograd's Fourier Transform Algorithm (WFTA) [56]

Winograd's FFT makes full use of all the tools explained in Section 7.5.1.

Good's mapping is used to convert the length $N_1 \cdot N_2$ 1-D DFT into a length $N_1 \times N_2$ 2-D DFT, and the intimate structure of the small-length algorithms is used to nest all the multiplications at the center of the overall algorithm as follows.

Reporting (7.52) into (7.53) results in

$$X = C_1 D_1 B_1 x B_2^T D_2 C_2^T .\qquad (7.56)$$

Since C and B do not involve any multiplication, the matrix $(B_1 x B_2^T)$ is obtained by only adding properly chosen input elements. The resulting matrix now has to be multiplied on the left and on the right by diagonal matrices D_1 and D_2, of respective dimensions M_1 and M_2. Let M_1' and M_2' be the numbers of trivial multiplications involved.

Premultiplying by the diagonal matrix D_1 multiplies each row by some constant, while postmultiplying does it for each column. Merging both multiplications leads to a total number of

$$M_{N_1 N_2} = M_{N_1} \cdot M_{N_2}\qquad (7.57)$$

out of which $M_{N_1}' \cdot M_{N_2}'$ are trivial.

Pre- and postmultiplying by C_1 and C_2^T will then complete the algorithm.

A graphical display of WFTA for length $N = 15$ is given in Fig. 7.8, which clearly shows that this algorithm cannot be performed in place.

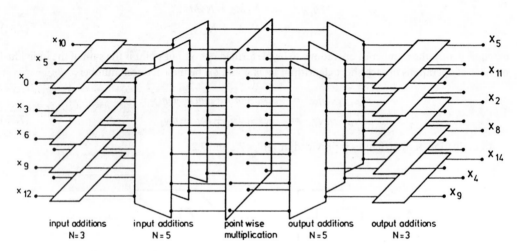

FIGURE 7.8   Schematic view of WFTA for $N = 15$.

The number of additions is more intricate to obtain.

Let us consider the pictorial representation of (7.56) as given in Fig. 7.8.

Let C_1 involve A_1^1 additions (output additions) and B_1 involve A_2^1 additions (input additions). (Which means that there exists an algorithm for multiplying C_1 by some vector involving A_1^1 additions. This is different from the number of ± 1s in the matrix—see the $p = 5$ example.)

Under these conditions, obtaining $x B_2$ will cost $A_2^2 \cdot N_1$ additions, $B_1(x B_2^T)$ will cost $A_1^2 \cdot M_2$ additions, $C_1(D_1 B_1 x B_2^T)$ will cost $A_1^1 \cdot M_2$ additions and $(C_1 D_1 B_1 x B_2^T)C_2$ will cost $A_2^1 \cdot N_1$ additions, which gives a total of

$$A_{N_1 N_2} = N_1 A_2 + M_2 A_1 .\qquad (7.58)$$

This formula is not symmetric in N_1 and N_2. Hence, it is possible to interchange N_1 and N_2, which does not change the number of multiplications. This is used to minimize the number of additions.

Since $M_2 \geq N_2$, it is clear that WFTA will always require at least as many additions as PFA, while it will always need fewer multiplications, as long as optimum short length DFTs are used. The demonstration is as follows.

Let

$$M_1 = N_1 + \varepsilon_1, \quad M_2 = N_2 + \varepsilon_2 ,$$
$$M_{\text{PFA}} = N_1 M_2 + N_2 M_1$$
$$= 2N_1 N_2 + N_1 \varepsilon_2 + N_2 \varepsilon_1 ,$$
$$M_{\text{WFTA}} = M_1 \cdot M_2$$
$$= N_1 N_2 + \varepsilon_1 \varepsilon_2 + N_1 \varepsilon_2 + N_2 \varepsilon_1 .$$

Since ε_1 and ε_2 are strictly smaller than N_1 and N_2 in optimum short-length DFTs, we have, as a result

$$M_{\text{WFTA}} < M_{\text{PFA}} .$$

Note that this result is not true if suboptimal short-length FFTs are used. The numbers of operations to be performed per output point [to be compared with (7.55)] are as follows in the WFTA:

$$m_{N_1 N_2} = m_{N_1} \cdot M_{N_2} , \quad a_{N_1 N_2} = a_{N_2} + m_{N_2} a_{N_1} . \tag{7.59}$$

These numbers are given in Tables 7.1 and 7.2.

Note that the number of additions in the WFTA was reduced later by Nussbaumer [12] with a scheme called "split nesting," leading to the algorithm with the least known number of operations (multiplications + additions).

7.5.4 Other Members of This Class [38]

PFA and WFTA are seen to be both described by the following equation:

$$X = C_1 D_1 B_1 x B_2^T D_2 C_2^T . \tag{7.60}$$

Each of them is obtained by different ordering of the matrix products.

— The PFA multiplies $(C_1 D_1 B_1)x$ first, and then the result is postmultiplied by $(B_2^T D_2 C_2^T)$.
— The WFTA starts with $B_1 x B_2^T$, then $(D_1 \times D_2)$, then C_1 and finally C_2^T.

Nevertheless, these are not the only ways of obtaining X : C and B can be factorized as two matrices each, to fully describe the way the algorithms are implemented. Taking this fact into account allows a great number of different algorithms to be obtained. Johnson and Burrus [38] systematically investigated this whole class of algorithms, obtaining interesting results, such as

— some WFTA-type algorithms, with reduced number of additions.
— algorithms with lower number of multiplications than both PFA and WFTA in the case where the short-length algorithms are not optimum.

7.5.5 Remarks on FFTs Without Twiddle Factors

It is easily seen that members of this class of algorithms differ fundamentally from FFTs with twiddle factors.

Both classes of algorithms are based on a divide and conquer strategy, but the mapping used to eliminate the twiddle factors introduced strong constraints on the type of lengths that were possible with Good's mapping.

Due to those constraints, the elaboration of efficient FFTs based on Good's mapping required considerable work on the structure of the short FFTs. This resulted in a better understanding of the mathematical structure of the problem, and a better idea of what was feasible and what was not.

This new understanding has been applied to the study of FFTs with twiddle factors. In this study, issues, such as optimality, distance (in cost) of the practical algorithms from the best possible ones and the structural properties of the algorithms, have been prominent in the recent evolution of the field of algorithms.

7.6 State of the Art

FFT algorithms have now reached a great maturity, at least in the 1-D case, and it is now possible to make strong statements about what eventual improvements are feasible and what are not.

In fact, lower bounds on the number of multiplications necessary to compute a DFT of given length can be obtained by using the techniques described in Section 7.5.1.

7.6.1 Multiplicative Complexity

Let us first consider the FFTs with lengths that are powers of two.

Winograd [57] was first able to obtain a lower bound on the number of complex multiplications necessary to compute length 2^n DFTs. This work was then refined in [28], which provided realizable lower bounds, with the following multiplicative complexity:

$$\mu_c \left[\text{DFT } 2^n \right] = 2^{n+1} - 2n^2 + 4n - 8 . \tag{7.61}$$

This means that there will never exist any algorithm computing a length 2^n DFT with a lower number of non-trivial complex multiplications than the one in (7.61).

Furthermore, since the demonstration is constructive [28], this optimum algorithm is known. Unfortunately, it is of no practical use for lengths greater than 64 (it involves much too many additions).

The lower part of Fig. 7.9 shows the variation of this lower bound and of the number of complex multiplications required by some practical algorithms (radix 2, radix 4, SRFT). It is clearly seen that SRFFT follows this lower bound up to $N = 64$, and is fairly close for $N = 128$. Divergence is quite fast afterwards.

It is also possible to obtain a realizable lower bound on the number of real multiplications [35, 36].

$$\mu_r \left[\text{DFT } 2^n \right] = 2^{n+2} - 2n^2 - 2n + 4 . \tag{7.62}$$

The variation of this bound, together with that of the number of real multiplications required by some practical algorithms is provided on the upper part of Fig. 7.9. Once again, this realizable lower bound is of no practical use above a certain limit. But, this time, the limit is much lower: SRFFT, together with radix 4, meets the lower bound on the number of real multiplications up to $N = 16$, which is also the last point where one can use an optimal polynomial product algorithm (modulo $u^2 + 1$) which is still practical. ($N = 32$ would require an optimal product modulo $u^4 + 1$ that requires a large number of additions).

It was also shown [31, 76] that all of the three following algorithms: optimum algorithm minimizing complex multiplications, optimum algorithm minimizing real multiplications and SRFFT, had exactly the same structure. They performed the decomposition into polynomial products exactly in the same manner, and they differ only in the way the polynomial products are computed.

Another interesting remark is as follows: the same number of multiplications as in SRFFT could also be obtained by so-called "real factor radix-2 FFTs" [24, 42, 44] (which were, on another respect, somewhat numerically ill-conditioned and needed about 20% more additions). They were obtained by making use of some computational trick to replace the complex twiddle factors by purely real or purely imaginary ones. Now, the question is: is it possible to do the same kind of thing with radix 4, or even SRFFT? Such a result would provide algorithms with still fewer operations. The knowledge of the lower bound tells us that it is impossible because, for some points ($N = 16$, for example) this would produce an algorithm

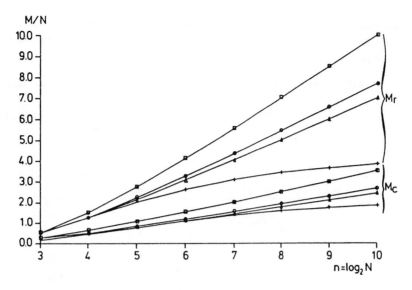

□ : radix 2
o : radix 4
▲ : split.radix
+ : lower bound

FIGURE 7.9 Number of non-trivial real or complex multiplications per output point.

with better performance than the lower bound. The challenge of eventually improving SRFFT is now as follows:

Comparison of SRFFT with $\mu_c[\text{DFT } 2^n]$ tells us that no algorithm using complex multiplications will be able to improve significantly SRFFT for lengths < 512. Furthermore, the trick allowing real factor algorithms to be obtained cannot be applied to radices greater than 2 (or at least not in the same manner).

The above discussion thus shows that there remain very few approaches (yet unknown) that could eventually improve the best known length 2^n FFT.

And what is the situation for FFTs based on Good's mapping?

Realizable lower bounds are not so easily obtained. For a given length $N = \prod N_i$, they involve a fairly complicated number theoretic function [8], and simple analytical expressions cannot be obtained. Nevertheless, programs can be written to compute $\mu_r\{\text{DFTN}_N\}$, and are given in [36]. Table 7.3 provides numerical values for a number of lengths of interest.

Careful examination of Table 7.3 provides a number of interesting conclusions.

First, one can see that, for comparable lengths (since SRFFT and WFTA cannot exist for the same lengths), a classification depending on the efficiency is as follows: WFTA always requires the lowest number of multiplications, followed by PFA, and followed by SRFFT, all fixed or mixed radix FFTs being next. Nevertheless, none of these algorithms attains the lower bound, except for very small lengths.

Another remark is that the number of multiplications required by WFTA is always smaller than the lower bound for the corresponding length that is a power of 2. This means, on the one hand, that transform lengths for which Good's mapping can be applied are well suited for a reduction in the number of multiplications, and on the other hand, that they are very efficiently computed by WFTA, from this point of view.

And this states the problem of the relative efficiencies of these algorithms: How close are they to their respective lower bound?

The last column of Table 7.3 shows that the relative efficiency of SRFFT decreases almost linearly with the length (it requires about twice the minimum number of multiplications for $N = 2048$), while the relative efficiency of WFTA remains almost constant for all the lengths of interest (it would not be the same result for much greater N). Lower bounds for Winograd-type lengths are also seen to be smaller than for the corresponding power of 2 lengths.

All these considerations result in the following conclusion: lengths for which Good's mapping is applicable allow a greater reduction of the number of multiplications (which is due directly to the mathematical structure of the problem). And, furthermore, they allow a greater relative efficiency of the actual algorithms vs. the lower bounds (and this is due indirectly to the mathematical structure).

TABLE 7.3   Practical Algorithms vs. Lower Bounds (Number of Non-Trivial Real Multiplications for FFTs on Real Data)

N	SRFFT	WFTA	Lower bound (L.B.)	SRFT L.B.	WFTA L.B.
16	20		20	1	
30		68	56		1.21
32	68		64	1.06	
60		136	112		1.21
64	196		168	1.16	
120		276	240		1.15
128	516		396	1.3	
240		632	548		1.15
256	1284		876	1.47	
504		1572	1320		1.19
512	3076		1864	1.64	
1008		3548	2844		1.25
1024	7172		3872	1.85	
2048	16388		7876	2.08	
2520		9492	7440		1.27

7.6.2   Additive Complexity

Nevertheless, the situation is not the same as regards the number of additions.

Most of the work on optimality was concerned with the number of multiplications. Concerning the number of additions, one can distinguish between additions due to the complex multiplications and the ones due to the butterflies. For the case $N = 2^n$, it was shown in [106, 110] that the latter number, which is achieved in actual algorithms, is also the optimum. Differences between the various algorithms is thus only due to varying numbers of complex multiplications. As a conclusion, one can see that the only way to decrease the number of additions is to decrease the number of true complex multiplications (which is close to the lower bound).

Figure 7.10 gives the variation of the total number of operations (multiplications plus additions) for these algorithms, showing that SRFFT has the lowest operation count. Furthermore, its more regular structure results in faster implementations.

Note that all the numbers given here concern the initial versions of SRFFT, PFA, and WFTA, for which FORTRAN programs are available. It is nevertheless possible to improve the number of additions in WFTA by using the so-called split nesting technique [12] (which is used in Fig. 7.10), and the number of multiplications of PFA by using small-length FFTs with scaled output [12], resulting in an overall scaled DFT.

As a conclusion, one can realize that we now have practical algorithms (mainly WFTA and SRFFT) that follow the mathematical structure of the problem of computing the DFT with the minimum number of multiplications, as well as a knowledge of their degree of suboptimality.

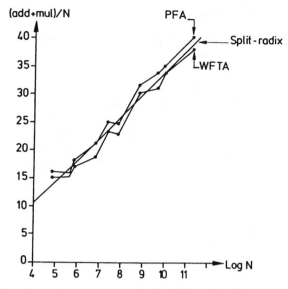

FIGURE 7.10 Total number of operations per output point for different algorithms.

7.7 Structural Considerations

This section is devoted to some points that are important in the comparison of different FFT algorithms, namely easy obtention of inverse FFT, in-place computation, regularity of the algorithm, quantization noise and parallelization, all of which are related to the structure of the algorithms.

7.7.1 Inverse FFT

FFTs are often used regardless of their "frequency" interpretation for computing FIR filtering in blocks, which achieves a reduction in arithmetic complexity compared to the direct algorithm. In that case, the forward FFT has to be followed, after pointwise multiplication of the result, by an inverse FFT. It is of course possible to rewrite a program along the same lines as the forward one, or to reorder the outputs of a forward FFT. A simpler way of computing an inverse FFT by using a forward FFT program is given (or reminded) in [99], where it is shown that, if CALL FFT (XR, Xl, N) computes a forward FFT of the sequence { $XR(i) + jXI(i)|i = 0, \ldots, N - 1$}, CALL FFT(XI, XR, N) will compute an inverse FFT of the same sequence, whatever the algorithm is. Thus, all FFT algorithms on complex data are equivalent in that sense.

7.7.2 In-Place Computation

Another point in the comparison of algorithms is the memory requirement: most algorithms (Cooley-Tukey, SRFFT, PFA) allow in-place computation (no auxiliary storage of size depending on N is necessary), while WFTA does not. And this may be a drawback for WFTA when applied to rather large sequences.

Cooley-Tukey and split-radix FFTs also allow rather compact programs [4, 113], the size of which is independent of the length of the FFT to be computed.

On the contrary, PFA and WFTA will require longer and longer programs when the upper limit on the possible lengths is increased: an 8-module program ($n = 2, 4, 8, 16, 3, 5, 7, 9$) allows obtaining a rather dense set of lengths up to $N = 5040$ only. Longer transforms can only be obtained either by the use of rather "exotic" modules that can be found in [37], or by some kind of mixture between Cooley-Tukey FFT (or SRFFT) and PFA.

7.7.3   Regularity, Parallelism

Regularity has been discussed for nearly all algorithms when they were described. Let us recall here that Cooley-Tukey FFT (CTFFT) is very regular (based on repetitive use of a few modules). SRFFT follows (repetitive use of very few modules in a slightly more involved manner). Then, PFA requires repetitive use (more intricate than CTFFT) of more modules, and finally WFTA requires some combining of parts of these modules, which means that, even if it has some regularity, this regularity is more hidden.

Let us point out also that the regularity of an algorithm cannot really be seen from its flowgraph. The equations describing the algorithm, as given in (7.13) or (7.38) do not fully define the implementations, which is partially done in the flowgraph. The reordering of the nodes of a flowgraph may provide a more regular one (the classical radix 2 and 4 CTFFT can be reordered into a constant geometry algorithm. See also [30] for SRFFT).

Parallelization of CTFFT and SRFFT is fairly easy, since the small modules are applied on sets of data that are separable and contiguous, while it is slightly more difficult with PFA, where the data required by each module are not in contiguous locations.

Finally, let us point out that mathematical tools such as tensor products can be used to work on the structure of the FFT algorithms [50, 101], since the structure of the algorithm reflects the mathematical structure of the underlying problem.

7.7.4   Quantization Noise

Roundoff noise generated by finite precision operations inside the FFT algorithm is also of importance. Of course, fixed point implementations of CTFFT for lengths 2^n were studied first, and it was shown that the error-to-signal ratio of the FFT process increases as $\sqrt{N}$ (which means 1/2 bit per stage) [117]. SRFFT and radix-4 algorithms were also reported to generate less roundoff than radix-2 [102].

Although the WFTA requires fewer multiplications than the CTFFT (hence has less noise sources), it was soon recognized that proper scaling was difficult to include in the algorithm, and that the resulting noise-to-signal ratio was higher. It is usually thought that two more bits are necessary for representing data in the WFTA to give an error of the same order as CTFFT (at least for practical lengths). A floating point analysis of PFA is provided in [104].

7.8   Particular Cases and Related Transforms

The previous sections have been devoted exclusively to the computation of the matrix-vector product involving the Fourier matrix. In particular, no assumption has been made on the input or output vector. In the following subsections, restrictions will be put on these vectors, showing how the previously described algorithms can be applied when the input is, e.g., real valued, or when only a part of the output is desired. Then, transforms closely related to the DFT will be discussed as well.

7.8.1   DFT Algorithms for Real Data

Very often in applications, the vector to be transformed is made up of real data. The transformed vector then has an hermitian symmetry, that is,

$$X_{N-k} = X_k^*, \tag{7.63}$$

as can be seen from the definition of the DFT. Thus, X_0 is real, and when N is even, $X_{N/2}$ is real as well. That is, the N input values map to 2 real and $N/2 - 1$ complex conjugate values when N is even, or 1 real and $(N - 1)/2$ complex conjugate values when N is odd (which leaves the number of free variables unchanged).

This redundancy in both input and output vectors can be exploited in the FFT algorithms in order to reduce the complexity and storage by a factor of 2. That the complexity should be half can be shown by the following argument. If one takes a real DFT of the real and imaginary parts of a complex vector separately, then $2N$ additions are sufficient in order to obtain the result of the complex DFT [3]. Therefore, the goal is to obtain a real DFT that uses half as many multiplications and less than half as many additions. If one could do better, then it would improve the complex FFT as well by the above construction.

For example, take the DIF SRFFT algorithm (7.28). First, X_{2k} requires a half length DFT on real data, and thus the algorithm can be reiterated. Then, because of the hermitian symmetry property (7.63):

$$X_{4k+1} = X^*_{4(N/4-k-1)+3} , \tag{7.64}$$

and therefore (7.28c) is redundant and only one DFT of size $N/4$ on complex data needs to be evaluated for (7.28b). Counting operations, this algorithm requires exactly half as many multiplications and slightly less than half as many additions as its complex counterpart, or [30]

$$M\left(\text{R-DFT}(2^m)\right) = 2^{n-1}(n-3) + 2 , \tag{7.65}$$
$$A\left(\text{R-DFT}(2^m)\right) = 2^{n-1}(3n-5) + 4 . \tag{7.66}$$

Thus, the goal for the real DFT stated earlier has been achieved. Similar algorithms have been developed for radix-2 and radix-4 FFTs as well. Note that even if DIF algorithms are more easily explained, it turns out that DIT ones have a better structure when applied to real data [29, 65, 77].

In the PFA case, one has to evaluate a multidimensional DFT on real input. Because the PFA is a row-column algorithm, data become hermitian after the first 1-D FFTs, hence an accounting has to be made of the real and conjugate parts so as to divide the complexity by 2 [77]. Finally, in the WFTA case, the input addition matrix and the diagonal matrix are real, and the output addition matrix has complex conjugate rows, showing again the saving of 50% when the input is real. Note, however, that these algorithms generally have a more involved structure than their complex counterparts (especially in the PFA and WFTA case). Some algorithms have been developed which are inherently "real," like the real factor FFTs [22, 44] or the FFCT algorithm [51], and do not require substantial changes for real input.

A closely related question is how to transform (or actually back transform) data that possess hermitian symmetry. An actual algorithm is best derived by using the transposition principle: since the Fourier transform is unitary, its inverse is equal to its hermitian transpose, and the required algorithm can be obtained simply by transposing the flow graph of the forward transform (or by transposing the matrix factorization of the algorithm). Simple graph theoretic arguments show that both the multiplicative and additive complexity are exactly conserved.

Assume next that the input is real and that only the real (or imaginary) part of the output is desired. This corresponds to what has been called a cosine (or sine) DFT, and obviously, a cosine and a sine DFT on a real vector can be taken altogether at the cost of a single real DFT. When only a cosine DFT has to be computed, it turns out that algorithms can be derived so that only half the complexity of a real DFT (that is, the quarter of a complex DFT) is required [30, 52], and the same holds for the sine DFT as well [52]. Note that the above two cases correspond to DFTs on real and symmetric (or antisymmetric) vectors.

7.8.2 DFT Pruning

In practice, it may happen that only a small number of the DFT outputs are necessary, or that only a few inputs are different from zero. Typical cases appear in spectral analysis, interpolation, and fast convolution applications. Then, computing a full FFT algorithm can be wasteful, and advantage should be taken of the inputs and outputs that can be discarded.

We will not discuss "approximate" methods which are based on filtering and sampling rate changes [2, pp. 317-319] but only consider "exact" methods. One such algorithm is due to Goertzel [68] which is based on the complex resonator idea. It is very efficient if only a few outputs of the FFT are required. A direct approach to the problem consists in pruning the flowgraph of the complete FFT so as to disregard redundant paths (corresponding to zero inputs or unwanted outputs). As an inspection of a flowgraph quickly shows, the achievable gains are not spectacular, mainly because of the fact that data communication is not local (since all arithmetic improvements in the FFT over the DFT are achieved through data shuffling).

More complex methods are therefore necessary in order to achieve the gains one would expect. Such methods lead to an order of $N \log_2 K$ operations, where N is the transform size and K the number of active inputs or outputs [48]. Reference [78] also provides a method combining Goertzel's method with shorter FFT algorithms. Note that the problems of input and output pruning are dual, and that algorithms for one problem can be applied to the other by transposition.

7.8.3   Related Transforms

Two transforms which are intimately related to the DFT are the discrete Hartley transform (DHT) [61, 62] and the discrete cosine transform (DCT) [1, 59]. The former has been proposed as an alternative for the real DFT and the latter is widely used in image processing.

The DHT is defined by

$$X_k = \sum_{n=0}^{N-1} x_n (\cos(2\pi nk/N) + \sin(2\pi nk/N)) \tag{7.67}$$

and is self-inverse, provided that X_0 is further weighted by $1/\sqrt{2}$. Initial claims for the DHT were

— improved arithmetic efficiency. This was soon recognized to be false, when compared to the real DFT. The structures of both programs are very similar and their arithmetic complexities are equivalent (DHTs actually require slightly more additions than real-valued FFTs).

— self-inverse property. It has been explained above that the inverse real DFT on hermitian data has exactly the same complexity as the real DFT (by transposition). If the transposed algorithm is not available, it can be found in [65] how to compute the inverse of a real DFT with a real DFT with only a minor increase in additive complexity.

Therefore, there is no computational gain in using a DHT, and only a minor structural gain if an inverse real DFT cannot be used.

The DCT, on the other hand, has found numerous applications in image and video processing. This has led to the proposal of several fast algorithms for its computation [51, 64, 70, 72]. The DCT is defined by

$$X_k = \sum_{n=0}^{N-1} x_n \cos(2\pi (2k+1)n/4N) . \tag{7.68}$$

A scale factor of $1/\sqrt{2}$ for X_0 has been left out in (7.68), mainly because the above transform appears as a subproblem in a length-$4N$ real DFT [51]. From this, the multiplicative complexity of the DCT can be related to that of the real DFT as [69]

$$\mu(\text{DCT}(N)) = (\mu(\text{real-DFT}(4N)) - \mu(\text{real-DFT}(2N)))/2 . \tag{7.69}$$

Practical algorithms for the DCT depend, as expected, on the transform length.

— N odd: the DCT can be mapped through permutations and sign changes only into a same length real DFT [69].

— N even: the DCT can be mapped into a same length real DFT plus $N/2$ rotations [51]. This is not the optimal algorithm [69, 100] but, however, a very practical one.

Other sinusoidal transforms [71], like the discrete sine transform (DST), can be mapped into DCTs as well, with permutations and sign changes only. The main point of this paragraph is that DHTs, DCTs, and other related sinusoidal transforms can be mapped into DFTs, and therefore one can resort to the vast and mature body of knowledge that exists for DFTs. It is worth noting that so far, for all sinusoidal transforms that have been considered, a mapping into a DFT has always produced an algorithm that is at least as efficient as any direct factorization. And if an improvement is ever achieved with a direct factorization, then it could be used to improve the DFT as well. This is the main reason why establishing equivalences between computational problems is fruitful, since it allows improvement of the whole class when any member can be improved.

Figure 7.11 shows the various ways the different transforms are related: starting from any transform with the best-known number of operations, you may obtain by following the appropriate arrows the corresponding transform for which the minimum number of operations will be obtained as well.

1) a	Complex DFT 2^n		2 real DFT's 2^n $+ 2^{n+1} - 4$ additions
b	Real DFT 2^n		1 real DFT 2^{n-1} + 1 complex DFT 2^{n-2} $+ (3.2^{n-2} - 4)$ multiplications $+ (2^n + 3.2^{n-2} - n)$ additions
2) a	Real DFT 2^n		1 real DFT 2^{n-1} + 2 DCT's 2^{n-2} $+ 3.2^{n-1} - 2$ additions
b	DCT 2^n		1 real DFT 2^n $+ (3.2^{n-1} - 2)$ multiplications $+ (3.2^{n-1} - 3)$ additions
3) a	Complex DFT 2^n		1 odd DFT 2^{n-1} + 1 complex DFT 2^{n-1} $+ 2^{n+1}$ additions
b	Odd DFT 2^{n-1}		2 complex DFT's 2^{n-2} $+ 2(3.2^{n-2} - 4)$ multiplications $+ (2^n + 3.2^{n-1} - 8)$ additions
4) a	Real DFT 2^n		1 DHT 2^n $- 2$ additions
b	DHT 2^n		1 real DFT 2^n $+ 2$ additions
5)	Complex DFT $2^n \text{x} 2^n$		3.2^{n-1} odd DFT 2^{n-1} + 1 complex DFT $2^{n-1} \text{x} 2^{n-1}$ $+ n.2^n$ additions
6) a	Real DFT 2^n		1 real symmetric DFT 2^n + 1 real antisymmetric DFT 2^n $+ (6n+10).4^{n-1}$ additions
b	Real symm DFT 2^n		1 real symmetric DFT 2^{n-1} + 1 inverse real DFT $+ 3(2^{n-3} - 1) + 1$ multiplications $+ (3n-4).2^{n-3} + 1$ additions

FIGURE 7.11   (a). Consistency of the split-radix based algorithms. Path showing the connections between the various transforms.

7.9   Multidimensional Transforms

We have already seen in Sections 7.4 and 7.5 that both types of divide and conquer strategies resulted in a multi-dimensional transform with some particularities: in the case of the Cooley-Tukey mapping, some "twiddle factors" operations had to be performed between the treatment of both dimensions, while in the Good's mapping, the resulting array had dimensions that were coprime.

Here, we shall concentrate on true 2-D FFTs with the same size along each dimension (generalization to more dimensions is usually straightforward).

Another characteristic of the 2-D case is the large memory size required to store the data. It is therefore important to work in-place. As a consequence, in-place programs performing FFTs on real data are also more important in the 2-D case, due to this memory size problem. Furthermore, the required memory is often so large that the data are stored in mass memory and brought into core memory when required, by rows or columns. Hence, an important parameter when evaluating 2-D FFT algorithms is the amount of memory calls required for performing the algorithm.

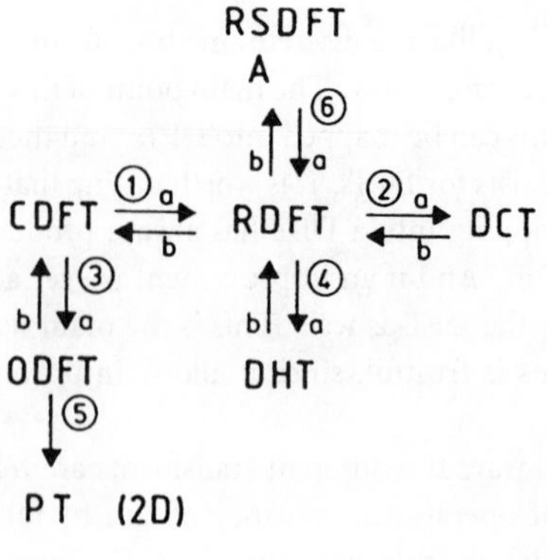

FIGURE 7.11 (b). Consistency of the split-radix based algorithms. Weighting of each connection in terms of real operations.

The 2-D DFT to be computed is defined as follows:

$$X_{k,r} = \sum_{i=0}^{N-1} \sum_{j=0}^{N-1} x_{i,j} W_N^{ik+jr}, \quad k, r = 0, \ldots, N-1. \tag{7.70}$$

The methods for computing this transform are distributed in four classes: row-column algorithms, vector-radix algorithms, nested algorithms, and polynomial transform algorithms. Among them, only the vector-radix and the polynomial transform were specifically designed for the 2-D case. We shall only give the basic principles underlying these algorithms and refer to the literature for more details.

7.9.1   Row-Column Algorithms

Since the DFT is separable in each dimension, the 2-D transform given in (7.70) can be performed in two steps, as was explained for the PFA.

— First compute N FFTs on the columns of the data.

— Then compute N FFTs on the rows of the intermediate result.

Nevertheless, when considering 2-D transforms, one should not forget that the size of the data becomes huge quickly: a length 1024×1024 DFT requires 10^6 words of storage, and the matrix is therefore stored in mass memory. But, in that case, accessing a single data is not more costly than reading the whole block in which it is stored. An important parameter is then the number of memory accesses required for computing the 2-D FFT.

This is why the row-column FFT is often performed as shown in Fig. 7.12, by performing a matrix transposition between the FFTs on the columns and the FFTs on the rows, in order to allow an access to the data by blocks. Row-column algorithms are very easily implemented and only require efficient 1-D FFTs, as described before, together with a matrix transposition algorithm (for which an efficient algorithm [84] was proposed). Note, however, that the access problem tends to be reduced with the availability of huge core memories.

7.9.2   Vector-Radix Algorithms

A computationally more efficient way of performing the 2-D FFT is a direct approach to the multi-dimensional problem: the vector-radix (VR) algorithm [85, 91, 92].

They can easily be understood through an example: the radix-2 DIT VRFFT.

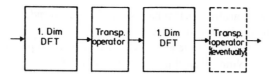

FIGURE 7.12 Row-column implementation of the 2-D FFT.

This algorithm is based on the following decomposition:

$$
X_{k,r} = \sum_{i=0}^{N/2-1} \sum_{j=0}^{N/2-1} x_{2i,2j} W_{N/2}^{ik+jr} + W_N^k \sum_{i=0}^{N/2-1} \sum_{j=0}^{N/2-1} x_{2i+1,2j} W_{N/2}^{ik+jr}
$$

$$
+ W_N^r \sum_{i=0}^{N/2-1} \sum_{j=0}^{N/2-1} x_{2i,2j+1} W_{N/2}^{ik+jr} + W_N^{k+r} \sum_{i=0}^{N/2-1} \sum_{j=0}^{N/2-1} x_{2i+1,2j+1} W_{N/2}^{ik+jr} , \quad (7.71)
$$

and the redundancy in the computation of $X_{k,r}$, $X_{k+N/2,r}$, $X_{k,r+N/2}$ and $X_{k+N/2,r+N/2}$ leads to simplifications which allow reduction of the arithmetic complexity.

This is the same approach as was used in the Cooley-Tukey FFTs, the decomposition being applied to both indices altogether.

Of course, higher radix decompositions or split radix decompositions are also feasible [86], the main difference being that the vector-radix SRFFT, as derived in [86], although being more efficient than the one in [90], is not the algorithm with the lowest arithmetic complexity in that class: For the 2-D case, the best algorithm is not only a mixture of radices 2 and 4.

Figure 7.13 shows what kind of decompositions are performed in the various algorithms. Due to the fact that the VR algorithms are true generalizations of the Cooley-Tukey approach, it is easy to realize that they will be obtained by repetitive use of small blocks of the same type (the "butterflies", by extension). Figure 7.14 provides the basic butterfly for a vector radix-2 FFT, as derived by (7.71). It should be clear, also, from Fig. 7.13 that the complexity of these butterflies increases very quickly with the radix: a radix-2 butterfly involves 4 inputs (it is a 2×2 DFT followed by some "twiddle factors"), while VR4 and VSR butterflies involve 16 inputs.

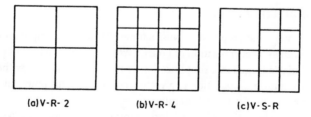

FIGURE 7.13 Decomposition performed in various vector radix algorithms.

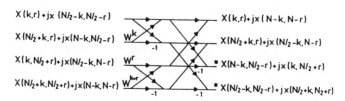

FIGURE 7.14 General vector-radix 2 butterfly.

Note also that the only VR algorithms that have seriously been considered all apply to lengths that are powers of 2, although other radices are of course feasible.

The number of read/write cycles of the whole set of data needed to perform the various FFTs of this class, compared to the row-column algorithm, can be found in [86].

7.9.3 Nested Algorithms

They are based on the remark that the nesting property used in Winograd's algorithm, as explained in Section 7.5.3, is not bound to the fact that the lengths are coprime (this requirement was only needed for Good's mapping). Hence, if the length of the DFT allows the corresponding 1-D DFT to be of a nested type (product of mutually prime factors), it is possible to nest further the multiplications, so that the overall 2-D algorithm is also nested.

The number of multiplications thus obtained are very low (see Table 7.4), but the main problem deals with memory requirements: WFTA is not performed in-place, and since all multiplications are nested, it requires the availability of a number of memory locations equal to the number of multiplications involved in the algorithms. For a length 1008×1008 FFT, this amounts to about $6 \cdot 10^6$ locations. This restricts the practical usefulness of these algorithms to small or medium length DFTs.

TABLE 7.4 Number of Non-Trivial Real Multiplications Per Output Point for Various 2-D FFTs on Real Data

$N \times N$ (WFTA)	$N \times N$ (Others)	R.C.	VR2	VR4	VSR	WFTA	P.T.
	2×2	0	0		0		0
	4×4	0	0	0	0		0
	8×8	0.5	0.375		0.375		0.375
	16×16	1.25	1.25	0.844	0.844		0.844
30×30	32×32	2.125	2.062		1.43	1.435	1.336
	64×64	3.0625	3.094	2.109	2.02		1.834
120×120	128×128	4.031	4.172		2.655	1.4375	2.333
240×240	256×256	5.015	5.273	3.48	3.28	1.82	2.833
504×504	512×512	6.008	6.386		3.92	2.47	3.33
1008×1008	1024×1024	7.004	7.506	4.878	4.56	3.12	3.83

7.9.4 Polynomial Transform

Polynomial transforms were first proposed by Nussbaumer [74] for the computation of 2-D cyclic convolutions. They can be seen as a generalization of Fourier transforms in the field of polynomials. Working in the field of polynomials resulted in a simplification of the multiplications by the root of unity, which was changed from a complex multiplication to a vector reordering. This powerful approach was applied in [87, 88] to the computation of 2-D DFTs as follows.

Let us consider the case where $N = 2^n$, which is the most common case. The 2-D DFT of (7.70) can be represented by the following three polynomial equations:

$$X_i(z) = \sum_{j=0}^{N-1} x_{i,j} \cdot z^j , \tag{7.72a}$$

$$\bar{X}_k(z) = \sum_{i=0}^{N-1} X_i(z) W_N^{ik} \bmod \left(z^N - 1 \right) , \tag{7.72b}$$

$$X_{k,r} = \bar{X}_k(z) \bmod \left(z - W_N^r \right) . \tag{7.72c}$$

This set of equations can be interpreted as follows: (7.72a) writes each row of the data as a polynomial, (7.72b) computes explicitly the DFTs on the columns, while (7.72c) computes the DFTs on the rows as a polynomial reduction [it is merely the equivalent of (7.5)]. Note that the modulo operation in (7.72b) is not necessary (no polynomial involved has a degree greater than N), but it will allow a divide and conquer strategy on (7.72c).

In fact, since $(z^N - 1) = (z^{N/2} - 1)(z^{N/2} + 1)$, the set of two equations (7.72b), (7.72c) can be separated into two cases, depending on the parity of r:

$$\bar{X}_k^1(z) = \sum_{i=0}^{N-1} X_i(z) W_N^{ik} \bmod \left(z^{N/2} - 1 \right), \tag{7.73a}$$

$$X_{k,2r} = \bar{X}_k^1(z) \bmod \left(z - W_N^{2r} \right), \tag{7.73b}$$

$$\bar{X}_k^2(z) = \sum_{i=0}^{N-1} X_i(z) W_N^{ik} \bmod \left(z^{N/2} + 1 \right), \tag{7.74a}$$

$$X_{k,2r+1} = \bar{X}_k^2(z) \bmod \left(z - W_N^{2r+1} \right). \tag{7.74b}$$

Equation (7.73) is still of the same type as the initial one, hence the same procedure as the one being derived will apply. Let us now concentrate on (7.74) which is now recognized to be the key aspect of the problem.

Since $(2r + 1, N) = 1$, the permutation $(2r + 1) \cdot k \pmod N$ maps all values of k, and replacing k with $(2r + 1) \cdot k$ in (7.73a) will merely result in a reordering of the outputs:

$$\bar{X}_{k(2r+1)}^2(z) = \sum_{i=0}^{N-1} X_i(z) W_N^{(2r+1)ik} \bmod \left(z^{N/2} + 1 \right), \tag{7.75a}$$

$$X_{k(2r+1),2r+1} = \bar{X}_{k(2r+1)}^2(z) \bmod \left(z - W_N^{2r+1} \right). \tag{7.75b}$$

and, since $z = W_N^{2r+1}$ in (7.75b), we can replace W_N^{2r+1} by z in (7.75a):

$$\bar{X}_{k(2r+1)}^2(z) = \sum_{i=0}^{N-1} X_i(z) z^{ik} \bmod \left(Z^{N/2} + 1 \right), \tag{7.76}$$

which is exactly a polynomial transform, as defined in [74]. This polynomial transform can be computed using an FFT-type algorithm, without multiplications, and with only $N^2/2 \log_2 N$ additions.

$X_{k,2r+1}$ will now be obtained by application of (7.75b). $\bar{X}^2(z)$ being computed mod $(z^{N/2} + 1)$ is of degree $N/2 - 1$. For each k, (7.75b) will then correspond to the reduction of one polynomial modulo the odd powers of W_N. From (7.5), this is seen to be the computation of the odd outputs of a length N DFT, which is sometimes called an odd DFT.

The terms $X_{k,2r+1}$ are seen to be obtained by one reduction mod $(z^{N/2} + 1)$ (7.74), one polynomial transform of N terms mod $Z^{N/2} + 1$ (7.76) and N odd DFTs. This procedure is then iterated on the terms $X_{2k+1,2r}$, by using exactly the same algorithm, the role of k and r being interchanged. $X_{2k,2r}$ is exactly a length $N/2 \times N/2$ DFT, on which the same algorithm is recursively applied.

In the first version of the polynomial transform computation of the 2-D FFT, the odd DFT was computed by a real-factor algorithm, resulting in an excess in the number of additions required.

As seen in Tables 7.4 and 7.5, where the number of multiplications and additions for the various 2-D FFT algorithms are given, the polynomial transform approach results in the algorithm requiring the lowest arithmetic complexity, when counting multiplications and additions altogether. The addition counts given in Table 7.5 are updates of the previous ones, assuming that the odd DFTs are computed by a split-radix algorithm.

TABLE 7.5   Number of Real Additions Per Output Point for Various 2-D FFTs on Real Data

$N \times N$ (WFTA)	$N \times N$ (Others)	R.C.	VR2	VR4	VSR	WFTA	P.T.
	2×2	2.	2.		2.		2.
	4×4	3.25	3.25	3.25	3.25		3.25
	8×8	5.56	5.43		5.43		5.43
	16×16	8.26	8.14	7.86	7.86		7.86
30×30	32×32	11.13	11.06		10.43	12.98	10.34
	64×64	14.06	14.09	13.11	13.02		12.83
120×120	128×128	17.03	17.17		15.65	17.48	15.33
240×240	256×256	20.01	20.27	18.48	17.67	22.79	17.83
504×504	512×512	23.00	23.38		20.92	34.42	20.33
1008×1008	1024×1024	26.00	26.5	23.88	23.56	45.30	22.83

Note that the same kind of performance was obtained by Auslander et al. [82, 83] with a similar approach which, while more sophisticated, gave a better insight on the mathematical structure of this problem. Polynomial transforms were also applied to the computation of 2-D DCT [52, 79].

7.9.5   Discussion

A number of conclusions can be stated by considering Tables 7.4 and 7.5, keeping the principles of the various methods in mind.

VR2 is more complicated to implement than row-column algorithms, and requires more operations for lengths ≥ 32. Therefore, it should not be considered. Note that this result holds only because efficient and compact 1-D FFTs, such as SRFFT, have been developed.

The row-column algorithm is the one allowing the easiest implementation, while having a reasonable arithmetic complexity. Furthermore, it is easily parallelized, and simplifications can be found for the reorderings (bit reversal, and matrix transposition [66]), allowing one of them to be free in nearly any kind of implementation. WFTA has a huge number of additions (twice the number required for the other algorithms for $N = 1024$), requires huge memory, has a difficult implementation, but requires the least multiplications. Nevertheless, we think that, in today's implementations, this advantage will in general not outweigh its drawbacks.

VSR is difficult to implement, and will certainly seldom defeat VR4, except in very special cases (huge memory available and N very large).

VR4 is a good compromise between structural and arithmetic complexity. When row-column algorithms are not fast enough, we think it is the next choice to be considered.

Polynomial transforms have the greatest possibilities: lowest arithmetic complexity, possibility of in-place computation, but very little work was done on the best way of implementing them. It was even reported to be slower than VR2 [103]. Nevertheless, it is our belief that looking for efficient implementations of polynomial transform based FFTs is worth the trouble. The precise understanding of the link between VR algorithms and polynomial transforms may be a useful guide for this work.

7.10   Implementation Issues

It is by now well recognized that there is a strong interaction between the algorithm and its implementation. For example, regularity, as discussed before, will only pay off if it is closely matched by the target architecture.

This is the reason why we will discuss in the sequel different types of implementations. Note that very often, the difference in computational complexity between algorithms is not large enough to differentiate between the efficiency of the algorithm and the quality of the implementation.

7.10.1   General Purpose Computers

FFT algorithms are built by repetitive use of basic building blocks. Hence, any improvement (even small) in these building blocks will pay in the overall performance. In the Cooley-Tukey or the split-radix case, the building blocks are small and thus easily optimizable, and the effect of improvements will be relatively more important than in the PFA/WFTA case where the blocks are larger.

When monitoring the amount of time spent in various elementary ftoating point operations, it is interesting to note that more time is spent in load/store operations than in actual arithmetic computations [30, 107, 109] (this is due to the fact that memory access times are comparable to ALU cycle times on current machines). Therefore, the locality of the algorithm is of paramount importance. This is why the PFA and WFTA do not meet the performance expected from their computational complexity only.

On another side, this drawback of PFA is compensated by the fact that only a few coefficients have to be stored. On the contrary, classical FFTs must store a large table of sine and cosine values, calculate them as needed, or update them with resulting roundoff errors.

Note that special automatic code generation techniques have been developed in order to produce efficient code for often used programs like the FFT. They are based on a "de-looping" technique that produces loop free code from a given piece of code [107]. While this can produce unreasonably large code for large transforms, it can be applied successfully to sub-transforms as well.

7.10.2   Digital Signal Processors

Digital signal processors (DSPs) strongly favor multiply/accumulate based algorithms. Unfortunately, this is not matched by any of the fast FFT algorithms (where sums of products have been changed to fewer but less regular computations). Nevertheless, DSPs now take into account some of the FFT requirements, like modulo counters and bit-reversed addressing. If the modulo counter is general, it will help the implementation of all FFT algorithms, but it is often restricted to the Cooley-Tukey/SRFFT case only (modulo a power of 2) for which efficient timings are provided on nearly all available machines by manufacturers, at least for small to medium lengths.

7.10.3   Vector and Multi-Processors

Implementations of Fourier transforms on vectorized computers must deal with two interconnected problems [93]. First, the vector (the size of data that can be processed at the maximal rate) has to be full as often as possible. Then, the loading of the vector should be made from data available inside the cache memory (as in general purpose computers) in order to save time. The usual hardware design parameters will, in general, favor length-2^m FFT implementations. For example, a radix-4 FFT was reported to be efficiently realized on a commercial vector processor [93].

In the multi-processor case, the performance will be dependent on the number and power of the processing nodes but also strongly on the available interconnection network. Because the FFT algorithms are deterministic, the resource allocation problem can be solved off-line. Typical configurations include arithmetic units specialized for butterfly operations [98], arrays with attached shuffle networks, and pipelines of arithmetic units with intermediate storage and reordering [17]. Obviously, these schemes will often favor classical Cooley-Tukey algorithms because of their high regularity. However, SRFFT or PFA implementations have not been reported yet, but could be promising in high speed applications.

7.10.4   VLSI

The discussion of partially dedicated multi-processors leads naturally to fully dedicated hardware structures like the ones that can be realized in very large scale integration (VLSI) [9, 11]. As a measure of efficiency, both chip area (A) and time (T) between two successive DFT computations (set-up times are neglected since only throughput is of interest) are of importance. Asymptotic lower bounds for the product $A \cdot T^2$ have been reported for the FFT [116] and lead to

$$\Omega_{AT2}(\text{DFT } (N)) = N^2 \log^2(N) , \tag{7.77}$$

that is, no circuit will achieve a better behavior than (7.77) for large N. Interestingly, this lower bound is achieved by several algorithms, notably the algorithms based on shuffle-exchange networks and the ones based on square grids [96, 114]. The trouble with these optimal schemes is that they outperform more traditional ones, like the cascade connection with variable delay [98] (which is asymptotically suboptimal), only for extremely large $N s$ and are therefore not relevant in practice [96].

Dedicated chips for the FFT computation are therefore often based on some traditional algorithm which is then efficiently mapped into a layout. Examples include chips for image processing with small size DCTs [115] as well as wafer scale integration for larger transforms. Note that the cost is dominated both by the number of multiplications (which outweigh additions in VLSI) and the cost of communication. While the former figure is available from traditional complexity theory, the latter one is not yet well studied and depends strongly on the structure of the algorithm as discussed in Section 7.7. Also, dedicated arithmetic units suited for the FFT problem have been devised, like the butterfly unit [98] or the CORDIC unit [94, 97] and contribute substantially to the quality of the overall design. But, similarly to the software case, the realization of an efficient VLSI implementation is still more an art than a mere technique.

7.11   Conclusion

The purpose of this paper has been threefold: a tutorial presentation of classic and recent results, a review of the state of the art, and a statement of open problems and directions.

After a brief history of the FFT development, we have shown by simple arguments, that the fundamental technique used in all fast Fourier transforms algorithms, namely the divide and conquer approach, will always improve the computational efficiency.

Then, a tutorial presentation of all known FFT algorithms has been made. A simple notation, showing how various algorithms perform various divisions of the input into periodic subsets, was used as the basis for a unified presentation of Cooley-Tukey, split-radix, prime factor, and Winograd fast Fourier transforms algorithms. From this presentation, it is clear that Cooley-Tukey and split-radix algorithms are instances of one family of FFT algorithms, namely FFTs with twiddle factors.

The other family is based on a divide and conquer scheme (Good's mapping) which is costless (computationally speaking). The necessary tools for computing the short-length FFTs which then appear were derived constructively and led to the presentation of the PFA and of the WFTA.

These practical algorithms were then compared to the best possible ones, leading to an evaluation of their suboptimality. Structural considerations and special cases were addressed next. In particular, it was shown that recently proposed alternative transforms like the Hartley transform do not show any advantage when compared to real valued FFTs.

Special attention was then paid to multidimensional transforms, where several open problems remain. Finally, implementation issues were outlined, indicating that most computational structures implicitly favor classical algorithms. Therefore, there is room for improvements if one is able to develop architectures that match more recent and powerful algorithms.

Acknowledgments

The authors would like to thank Prof. M. Kunt for inviting them to write this paper, as well as for his patience. Prof. C. S. Burrus, Dr. J. Cooley, Dr. M. T. Heideman, and Prof. H. J. Nussbaumer are also thanked for fruitful interactions on the subject of this paper. We are indebted to J. S. White, J. C. Bic, and P. Gole for their careful reading of the manuscript.

References

Books

[1] Ahmed, N. and Rao, K.R., *Orthogonal Transforms for Digital Signal Processing*, Springer, Berlin, 1975.

[2] Blahut, R.E., *Fast Algorithms for Digital Signal Processing*, Addison-Wesley, Reading, MA, 1986.

[3] Brigham, E.O., *The Fast Fourier Transform*, Prentice-Hall, Englewood Cliffs, NJ, 1974.

[4] Burrus, C.S. and Parks, T.W., *DFT/FFT and Convolution Algorithms*, John Wiley & Sons, New York, 1985.

[5] Burrus, C.S., Efficient Fourier transform and convolution algorithms, in: J.S. Lim and A.V. Oppenheim, Eds., *Advanced Topics in Digital Signal Processing*, Prentice-Hall, Englewood Cliffs, NJ, 1988.

[6] Digital Signal Processing Committee, Ed., *Selected Papers in Digital Signal Processing, II*, IEEE Press, New York, 1975.

[7] Digital Signal Processing Committee, Ed., *Programs for Digital Signal Processing*, IEEE Press, New York, 1979.

[8] Heideman, M.T., *Multiplicative Complexity, Convolution and the DFT*, Springer, Berlin, 1988.

[9] Kung, S.Y., Whitehouse, H.J. and Kailath, T., Eds., *VLSI and Modern Signal Processing*, Prentice-Hall, Englewood Cliffs, NJ, 1985.

[10] McClellan, J.H. and Rader, C.M., *Number Theory in Digital Signal Processing*, Prentice-Hall, Englewood Cliffs, NJ, 1979.

[11] Mead, C. and Conway, L., *Introduction to VLSI*, AddisonWesley, Reading, MA, 1980.

[12] Nussbaumer, H.J., *Fast Fourier Transform and Convolution Algorithms*, Springer, Berlin, 1982.

[13] Oppenheim, A.V., Ed., *Papers on Digital Signal Processing*, MIT Press, Cambridge, MA, 1969.

[14] Oppenheim, A.V. and Schafer, R.W., *Digital Signal Processing*, Prentice-Hall, Englewood Cliffs, NJ, 1975.

[15] Rabiner, L.R. and Rader, C.M., Ed., *Digital Signal Processing*, IEEE Press, New York, 1972.

[16] Rabiner, L.R. and Gold, B., *Theory and Application of Digital Signal Processing*, Prentice-Hall, Englewood Cliffs, NJ, 1975.

[17] Schwartzlander, E.E., *VLSI Signal Processing Systems*, Kluwer Academic Publishers, Dordrecht, 1986.

[18] Soderstrand, M.A., Jenkins, W.K., Jullien, G.A., and Taylor, F.J., Eds., *Residue Number System Arithmetic: Modern Applications in Digital Signal Processing*, IEEE Press, New York, 1986.

[19] Winograd, S., *Arithmetic Complexity of Computations*, SIAM CBMS-NSF Series, No. 33, SIAM, Philadelphia, 1980.

1-D FFT algorithms

[20] Agarwal, R.C. and Burrus, C.S., Fast one-dimensional digital convolution by multi-dimensional techniques, *IEEE Trans. Acoust. Speech Signal Process.*, ASSP-22(1), 1–10, Feb. 1974.

[21] Bergland, G.D., A fast Fourier transform algorithm using base 8 iterations, *Math. Comp.*, 22(2), 275–279, April 1968 (reprinted in [13]).

[22] Bruun, G., z-Transform DFT filters and FFTs, *IEEE Trans. Acoust. Speech Signal Process.*, ASSP-26(1), 56–63, Feb. 1978.

[23] Burrus, C.S., Index mappings for multidimensional formulation of the DFT and convolution, *IEEE Trans. Acoust. Speech Signal Process.*, ASSP-25(3), 239–242, June 1977.

[24] Cho, K.M. and Temes, G.C., Real-factor FFT algorithms, *Proc. ICASSP 78*, Tulsa, OK, 634–637, April 1978.

[25] Cooley, J.W. and Tukey, J.W., An algorithm for the machine calculation of complex Fourier series, *Math. Comp.*, 19, 297–301, April 1965.

[26] Dubois, P. and Venetsanopoulos, A.N., A new algorithm for the radix-3 FFT, *IEEE Trans. Acoust. Speech Signal Process.*, ASSP-26, 222–225, June 1978.

[27] Duhamel, P. and Hollmann, H., Split-radix FFT algorithm, *Electron. Lett.*, 20(1), 14–16, 5 January 1984.

[28] Duhamel, P. and Hollmann, H., Existence of a 2^n FFT algorithm with a number of multiplications lower than 2^{n+1}, *Electron. Lett.*, 20(17), 690–692, August 1984.

[29] Duhamel, P., Un algorithme de transformation de Fourier rapide à double base, *Annales des Telecommunications*, 40(9-10), 481–494, September 1985.

[30] Duhamel, P., Implementation of "split-radix" FFT algorithms for complex, real and real-symmetric data, *IEEE Trans. Acoust. Speech Signal Process.*, ASSP-34(2), 285–295, April 1986.

[31] Duhamel, P., Algorithmes de transformés discrètes rapides pour convolution cyclique et de convolution cyclique pour transformés rapides, Thèse de doctorat d'état, Université Paris XI, Sept. 1986.

[32] Good, I.J., The interaction algorithm and practical Fourier analysis, *J. Roy. Statist. Soc. Ser. B*, B-20, 361–372, 1958, B-22, 372–375, 1960.

[33] Heideman, M.T. and Burrus, C.S., A bibliography of fast transform and convolution algorithms II, Technical Report No. 8402, Rice University, 24 February 1984.

[34] Heideman, M.T., Johnson, D.H., and Burrus, C.S., Gauss and the history of the FFT, *IEEE Acoust. Speech Signal Process. Magazine*, 1(4), 14–21, Oct. 1984.

[35] Heideman, M.T. and Burrus, C.S., On the number of multiplications necessary to compute a length-2^n DFT, *IEEE Trans. Acoust. Speech Signal Process.*, ASSP-34(1), 91–95, Feb. 1986.

[36] Heideman, M.T., Application of multiplicative complexity theory to convolution and the discrete Fourier transform, PhD Thesis, Dept. of Elec. and Comp. Eng., Rice Univ., April 1986.

[37] Johnson, H.W. and Burrus, C.S., Large DFT modules: 11, 13, 17, 19, and 25, Tech. Report 8105, Dept. of Elec. Eng., Rice Univ., Houston, TX, December 1981.

[38] Johnson, H.W. and Burrus, C.S., The design of optimal DFT algorithms using dynamic programming, *IEEE Trans. Acoust. Speech Signal Process.*, ASSP-31(2), 378–387, 1983.

[39] Kolba, D.P. and Parks, T.W., A prime factor algorithm using high-speed convolution, *IEEE Trans. Acoust. Speech Signal Process.*, ASSP-25, 281–294, Aug. 1977.

[40] Martens, J.B., Recursive cyclotomic factorization—A new algorithm for calculating the discrete Fourier transform, *IEEE Trans. Acoust. Speech Signal Process.*, ASSP32(4), 750–761, Aug. 1984.

[41] Nussbaumer, H.J., Efficient algorithms for signal processing, *Second European Signal Processing Conference*, EUSIPC0-83, Erlangen, September 1983.

[42] Preuss, R.D., Very fast computation of the radix-2 discrete Fourier transform, *IEEE Trans. Acoust. Speech Signal Process.*, ASSP-30, 595–607, Aug. 1982.

[43] Rader, C.M., Discrete Fourier transforms when the number of data samples is prime, *Proc. IEEE*, 56, 1107–1008, 1968.

[44] Rader, C.M. and Brenner, N.M., A new principle for fast Fourier transformation, *IEEE Trans. Acoust. Speech Signal Process.*, ASSP-24, 264–265, June 1976.

[45] Singleton, R., An algorithm for computing the mixed radix fast Fourier transform, *IEEE Trans. Audio Electroacoust.*, AU-17, 93–103, June 1969 (reprinted in [13]).

[46] Stasinski, R., Asymmetric fast Fourier transform for real and complex data, *IEEE Trans. Acoust. Speech Signal Process.*, submitted.

[47] Stasinski, R., Easy generation of small-N discrete Fourier transform algorithms, *IEE Proc.*, 133, Pt. G, 3, 133–139, June 1986.

[48] Stasinski, R., FFT pruning. A new approach, *Proc. Eusipco 86*, 267–270, 1986.

[49] Suzuki, Y., Sone, T., and Kido, K., A new FFT algorithm of radix 3, 6, and 12, *IEEE Trans. Acoust. Speech Signal Process.*, ASSP-34(2), 380–383, April 1986.

[50] Temperton, C., Self-sorting mixed-radix fast Fourier transforms, *J. Comput. Phys.*, 52(1), 1–23, Oct. 1983.

[51] Vetterli, M. and Nussbaumer, H.J., Simple FFT and DCT algorithms with reduced number of operations, *Signal Process.*, 6(4), 267–278, Aug. 1984.

[52] Vetterli, M. and Nussbaumer, H.J., Algorithmes de transformé de Fourier et cosinus mono et bi-dimensionnels, *Annales des Télécommunications*, Tome 40, 9-10, 466–476, Sept.-Oct. 1985.

[53] Vetterli, M. and Duhamel, P., Split-radix algorithms for length-p^m DFTs, *IEEE Trans. Acoust. Speech Signal Process.*, ASSP-37(1), 57–64, Jan. 1989.

[54] Winograd, S., On computing the discrete Fourier transform, *Proc. Nat. Acad. Sci. USA*, 73, 1005–1006, April 1976.

[55] Winograd, S., Some bilinear forms whose multiplicative complexity depends on the field of constants, *Math. Systems Theory*, 10(2), 169–180, 1977 (reprinted in [10]).

[56] Winograd, S., On computing the DFT, *Math. Comp.*, 32(1), 175–199, Jan. 1978 (reprinted in [10]).

[57] Winograd, S., On the multiplicative complexity of the discrete Fourier transform, *Adv. in Math.*, 32(2), 83–117, May 1979.

[58] Yavne, R., An economical method for calculating the discrete Fourier transform, *AFIPS Proc.*, 33, 115–125, Fall Joint Computer Conf., Washington, 1968.

Related algorithms

[59] Ahmed, N., Natarajan, T., and Rao, K.R., Discrete cosine transform, *IEEE Trans. Comput.*, C-23, 88–93, Jan. 1974.

[60] Bergland, G.D., A radix-eight fast Fourier transform subroutine for real-valued series, *IEEE Trans. Audio Electroacoust.*, 17(1), 138–144, June 1969.

[61] Bracewell, R.N., Discrete Hartley transform, *J. Opt. Soc. Amer.*, 73(12), 1832–1835, Dec. 1983.

[62] Bracewell, R.N., The fast Hartley transform, *Proc. IEEE*, 22(8), 1010–1018, Aug. 84.

[63] Burrus, C.S., Unscrambling for fast DFT algorithms, *IEEE Trans. Acoust. Speech Signal Process.*, ASSP-36(7), 1086–1087, July, 1988.

[64] Chen, W.-H., Smith, C.H. and Fralick, S.C., A fast computational algorithm for the discrete cosine transform, *IEEE Trans. Comm.*, COM-25, 1004–1009, Sept. 1977.

[65] Duhamel, P. and Vetterli, M., Improved Fourier and Hartley transform algorithms. Application to cyclic convolution of real data, *IEEE Trans. Acoust. Speech Signal Process.*, ASSP-35(6), 818–824, June 1987.

[66] Duhamel, P. and Prado, J., A connection between bitreverse and matrix transpose. Hardware and software consequences, *Proc. IEEE Acoust. Speech Signal Process.*, 1403–1406.

[67] Evans, D.M., An improved digit reversal permutation algorithm for the fast Fourier and Hartley transforms, *IEEE Trans. Acoust. Speech Signal Process.*, ASSP-35(8), 1120–1125, Aug. 87.

[68] Goertzel, G., An algorithm for the evaluation of finite Fourier series, *Am. Math. Monthly*, 65(1), 34–35, Jan. 1958.

[69] Heideman, M.T., Computation of an odd-length DCT from a real-valued DFT of the same length, *IEEE Trans. Acoust. Speech Signal Process.*, submitted.

[70] Hou, H.S., A fast recursive algorithm for computing the discrete Fourier transform, *IEEE Trans. Acoust. Speech Signal Process.*, ASSP-35(10), 1455–1461, Oct. 1987.

[71] Jain, A.K., A sinusoidal family of unitary transforms, *IEEE Trans. PAMI*, 1(4), 356–365, Oct. 1979.

[72] Lee, B.G., A new algorithm to compute the discrete cosine transform, *IEEE Trans. Acoust. Speech Signal Process.*, ASSP-32, 1243–1245, Dec. 1984.

[73] Mou, Z.J. and Duhamel, P., Fast FIR filtering: algorithms and implementations, *Signal Process.*, 13(4), 377–384, Dec. 1987.

[74] Nussbaumer, H.J., Digital filtering using polynomial transforms, *Electron. Lett.*, 13(13), 386–386, June 1977.

[75] Polge, R.J., Bhaganan, B.K. and Carswell, J.M., Fast computational algorithms for bit-reversal, *IEEE Trans. Comput.*, 23(1), 1–9, Jan. 1974.

[76] Duhamel, P., Algorithms meeting the lower bounds on the multiplicative complexity of length-2^n DFTs and their connection with practical algorithms, *IEEE Trans. Acoust. Speech Signal Process.*, Sept. 1990.

[77] Sorensen, H.V., Jones, D.L., Heideman, M.T., and Burrus, C.S., Real-valued fast Fourier transform algorithms, *IEEE Trans. Acoust. Speech Signal Process.*, ASSP-35(6), 849–863, June 1987.

[78] Sorensen, H.V., Burrus, C.S., and Jones, D.L., A new efficient algorithm for computing a few DFT points, *Proc. 1988 IEEE Internat. Symp. on CAS*, 1915–1918, 1988.

[79] Vetterli, M., Fast 2-D discrete cosine transform, *Proc. 1985 IEEE Internat. Conf. Acoust. Speech Signal Process.*, Tampa, 1538–1541, March 1985.

[80] Vetterli, M., Analysis, synthesis and computational complexity of digital filter banks, PhD Thesis, Ecole Polytechnique Federale de Lausanne, Switzerland, April 1986.

[81] Vetterli, M., Running FIR and IIR filtering using multirate filter banks, *IEEE Trans. Acoust. Speech Signal Process.*, ASSP-36(5), 730–738, May 1988.

Multi-dimensional transforms

[82] Auslander, L., Feig, E., and Winograd, S., New algorithms for the multidimensional Fourier transform, *IEEE Trans. Acoust. Speech Signal Process.*, ASSP-31(2), 338–403, April 1983.

[83] Auslander, L., Feig, E., and Winograd, S., Abelian semisimple algebras and algorithms for the discrete Fourier transform, *Adv. Applied Math.*, 5, 31–55, 1984.

[84] Eklundh, J.O., A fast computer method for matrix transposing, *IEEE Trans. Comput.*, 21(7), 801–803, July 1972 (reprinted in [6]).

[85] Mersereau, R.M. and Speake, T.C., A unified treatment of Cooley-Tukey algorithms for the evaluation of the multidimensional DFT, *IEEE Trans. Acoust. Speech Signal Process.*, 22(5), 320–325, Oct. 1981.

[86] Mou, Z.J. and Duhamel, P., In-place butterfly-style FFT of 2-D real sequences, *IEEE Trans. Acoust. Speech Signal Process.*, ASSP-36(10), 1642–1650, Oct. 1988.

[87] Nussbaumer, H.J. and Quandalle, P., Computation of convolutions and discrete Fourier transforms by polynomial transforms, *IBM J. Res. Develop.*, 22, 134–144, 1978.

[88] Nussbaumer, H.J. and Quandalle, P., Fast computation of discrete Fourier transforms using polynomial transforms, *IEEE Trans. Acoust. Speech Signal Process.*, ASSP-27, 169–181, 1979.

[89] Pease, M.C., An adaptation of the fast Fourier transform for parallel processing, *J. Assoc. Comput. Mach.*, 15(2), 252–264, April 1968.

[90] Pei, S.C. and Wu, J.L., Split-vector radix 2-D fast Fourier transform, *IEEE Trans. Circuits Systems*, 34(1), 978–980, Aug. 1987.

[91] Rivard, G.E., Algorithm for direct fast Fourier transform of bivariant functions, *1975 Annual Meeting of the Optical Society of America*, Boston, MA, Oct. 1975.

[92] Rivard, G.E., Direct fast Fourier transform of bivariant functions, *IEEE Trans. Acoust. Speech Signal Process.*, 25(3), 250–252, June 1977.

Implementations

[93] Agarwal, R.C. and Cooley, J.W., Fourier transform and convolution subroutines for the IBM 3090 Vector Facility, *IBM J. Res. Develop.*, 30(2), 145–162, March 1986.

[94] Ahmed, H., Delosme, J.M. and Morf, M., Highly concurrent computing structures for matrix arithmetic and signal processing, *IEEE Trans. Comput.*, 15(1), 65–82, Jan. 1982.

[95] Burrus, C.S. and Eschenbacher, P.W., An in-place, in-order prime factor FFT algorithm, *IEEE Trans. Acoust. Speech Signal Process.*, ASSP-29(4), 806–817, Aug. 1981.

[96] Card, H.C., VLSI computations: from physics to algorithms, *Integration*, 5, 247–273, 1987.

[97] Despain, A.M., Fourier transform computers using CORDIC iterations, *IEEE Trans. Comput.*, 23(10), 993–1001, Oct. 1974.

[98] Despain, A.M., Very fast Fourier transform algorithms hardware for implementation, *IEEE Trans. Comput.*, 28(5), 333–341, May 1979.

[99] Duhamel, P., Piron, B., and Etcheto, J.M., On computing the inverse DFT, *IEEE Trans. Acoust. Speech Signal Process.*, ASSP-36(2), 285–286, Feb. 1988.

[100] Duhamel, P. and H'mida, H., New 2^n DCT algorithms suitable for VLSI implementation, *Proc. IEEE Internat. Conf. Acoust. Speech Signal Process.*, 1805–1809, 1987.

[101] Johnson, J., Johnson, R., Rodriguez, D., and Tolimieri, R., A methodology for designing, modifying, and implementing Fourier transform algorithms on various architectures, preliminary draft, Sept. 1988 (to be submitted).

[102] Elterich, A. and Stammler, W., Error analysis and resulting structural improvements for fixed point FFT's, *Proc. IEEE Internat. Conf. Acoust. Speech Signal Process.*, 1419–1422, April 1988.

[103] Lhomme, B., Morgenstern, J., and Quandalle, P., Implantation de transformés de Fourier de dimension 2^n, *Techniques et Science Informatiques*, 4(2), 324–328, 1985.

[104] Manson, D.C. and Liu, B., Floating point roundoff error in the prime factor FFT, *IEEE Trans. Acoust. Speech Signal Process.*, 29(4), 877–882, Aug. 1981.

[105] Mescheder, B., On the number of active *-operations needed to compute the DFT, *Acta Inform.*, 13, 383–408, May 1980.

[106] Morgenstern, J., The linear complexity of computation, *Assoc. Comput. Mach.*, 22(2), 184–194, April 1975.

[107] Morris, L.R., Automatic generation of time efficient digital signal processing software, *IEEE Trans. Acoust. Speech Signal Process.*, ASSP-25, 74–78, Feb. 1977.

[108] Morris, L.R., A comparative study of time efficient FFT and WFTA programs for general purpose computers, *IEEE Trans. Acoust. Speech Signal Process.*, ASSP26, 141–150, April 1978.

[109] Nawab H. and McClellan, J.H., Bounds on the minimum number of data transfers in WFTA and FFT programs, *IEEE Trans. Acoust. Speech Signal Process.*, ASSP-27, 394–398, Aug. 1979.

[110] Pan, V.Y., The additive and logical complexities of linear and bilinear arithmetic algorithms, *J. Algorithms*, 4(1), 1–34, March 1983.

[111] Rothweiler, J.H., Implementation of the in-order prime factor transform for variable sizes, *IEEE Trans. Acoust. Speech Signal Process.*, ASSP-30(1), 105–107, Feb. 1982.

[112] Silverman, H.F., An introduction to programming the Winograd Fourier transform algorithm, *IEEE Trans. Acoust. Speech Signal Process.*, ASSP-25(2), 152–165, April 1977, with corrections in: *IEEE Trans. Acoust Speech Signal Process.*, ASSP-26(3), 268, June 1978, and in ASSP-26(5), 482, Oct. 1978.

[113] Sorensen, H.V., Heideman, M.T., and Burrus, C.S., On computing the split-radix FFT, *IEEE Trans. Acoust. Speech Signal Process.*, ASSP-34(1), 152–156, Feb. 1986.

[114] Thompson, C.D., Fourier transforms in VLSI, *IEEE Trans. Comput.*, 32(11), 1047–1057, Nov. 1983.

[115] Vetterli, M. and Ligtenberg, A., A discrete Fourier-cosine transform chip, *IEEE J. Selected Areas in Communications,* Special Issue on VLSI in Telecommunications, SAC-4(1), 49–61, Jan. 1986.

[116] Vuillemin, J., A combinatorial limit to the computing power of VLSI circuits, *Proc. 21st Symp. Foundations of Comput. Sci., IEEE Comp. Soc.,* 294–300, Oct. 1980.

[117] Welch, P.D., A fixed-point fast Fourier transform error analysis, *IEEE Trans. Audio Electro.,* 15(2), 70–73, June 1969, (reprinted in [13] and [15]).

Software

FORTRAN (or DSP) code can be found in the following references.

[7] contains a set of classical FFT algorithms.

[111] contains a prime factor FFT program.

[4] contains a set of classical programs and considerations on program optimization, as well as TMS 32010 code.

[113] contains a compact split-radix Fortran program.

[29] contains a speed-optimized split-radix FFT.

[77] contains a set of real-valued FFTs with twiddle factors.

[65] contains a split-radix real valued FFT, as well as a Hartley transform program.

[112] as well as [7] contains a Winograd Fourier transform Fortran program.

[66], [67] and [75] contain improved bit-reversal algorithms.

8

Fast Convolution and Filtering

Ivan W. Selesnick
Polytechnic University

C. Sidney Burrus
Rice University

8.1 Introduction

One of the first applications of the Cooley-Tukey fast Fourier transform (FFT) algorithm was to implement convolution faster than the usual direct method [13, 25, 30]. Finite impulse response (FIR) digital filters and convolution are defined by

$$y(n) = \sum_{k=0}^{L-1} h(k) \, x(n-k) \tag{8.1}$$

where, for an FIR filter, $x(n)$ is a length-N sequence of numbers considered to be the input signal, $h(n)$ is a length-L sequence of numbers considered to be the filter coefficients, and $y(n)$ is the filtered output. Examination of this equation shows that the output signal $y(n)$ must be a length-$(N + L - 1)$ sequence of numbers, and the direct calculation of this output requires NL multiplications and approximately NL additions (actually, $(N - 1)(L - 1)$). If the signal and filter length are both length-N, we say the arithmetic complexity is of order N^2, $O(N^2)$. Our goal is calculate this convolution or filtering faster than directly implementing (8.1). The most common way to achieve "fast convolution" is to section or block the signal and use the FFT on these blocks to take advantage of the efficiency of the FFT. Clearly, one disadvantage of this technique is an inherent delay of one block length.

0-8493-8572-5/98/$0.00+$.50
© 1998 by CRC Press LLC

Indeed, this approach is so common as to be almost synonymous with fast convolution. The problem is to implement on-going, noncyclic convolution with the finite-length, cyclic convolution that the FFT gives. An answer was quickly found in a clever organization of piecing together blocks of data using what is now called the *overlap-add* method and the *overlap-save* method. These two methods convolve length-L blocks using one length-L FFT, L complex multiplications, and one length-L inverse FFT [22].

Later this was generalized to arbitrary length blocks or sections to give block convolution and block recursion [5]. By allowing the block lengths to be even shorter than one word (bits and bytes!) we come up with an interesting implementation called distributed arithmetic that requires no explicit multiplications [7, 34].

Another approach for improving the efficiency of convolution and recursion uses fast algorithms other than the traditional FFT. One possibility is to use a transform based on number-theoretic roots of unity rather than the usual complex roots of unity [17]. This gives rise to number-theoretic transforms that require no multiplications and no trigonometric functions. Still another method applies Winograd's fast algorithms directly to convolution rather than through the Fourier transform. Finally, we remark that some filters $h(n)$ require fewer arithmetic operations because of their structure.

8.2   Overlap-Add and Overlap-Save Methods for Fast Convolution

If one implements convolution by use of the FFT, then it is cyclic convolution that is obtained. In order to use the FFT, zeros are appended to the signal or filter sequence until they are both the same length. If the FFT of the signal $x(n)$ is term-by-term multiplied by the FFT of the filter $h(n)$, the result is the FFT of the output $y(n)$. However, the length of $y(n)$ obtained by an inverse FFT is the same as the length of the input. Because the DFT or FFT is a periodic transform, the convolution implemented using this FFT approach is cyclic convolution, which means the output of (8.1) is wrapped or aliased. The tail of $y(n)$ is added to it head — but that is not usually what is wanted for filtering or normal convolution and correlation. This aliasing, the effects of cyclic convolution, can be overcome by appending zeros to both $x(n)$ and $h(n)$ until their lengths are $N + L - 1$ and by then using the FFT. The part of the output that is aliased is zero and the result of the cyclic convolution is exactly the same as noncyclic convolution. The cost is taking the FFT of lengthened sequences — sequences for which about half the numbers are zero. Now that we can do noncyclic convolution with the FFT, how do we account for the effects of sectioning the input and output into blocks?

8.2.1   Overlap-Add

Because convolution is linear, the output of a long sequence can be calculated by simply summing the outputs of each block of the input. What is complicated is that the output blocks are longer than the input. This is dealt with by overlapping the tail of the output from the previous block with the beginning of the output from the present block. In other words, if the block length is N and it is greater than the filter length L, the output from the second block will overlap the tail of the output from the first block and they will simply be added. Hence the name: *overlap-add*. Figure 8.1 illustrates why the overlap-add method works, for $N = 10, L = 5$.

Combining the overlap-add organization with use of the FFT yields a very efficient algorithm for calculating convolution that is faster than direct calculation for lengths above 20 to 50. This cross-over point depends on the computer being used and the overhead needed by use of the FFTs.

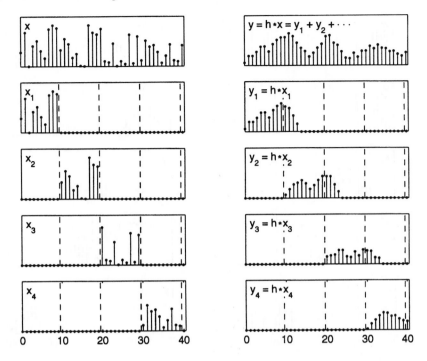

FIGURE 8.1 Overlap-add algorithm. The sequence $y(n)$ is the result of convolving $x(n)$ with an FIR filter $h(n)$ of length 5. In this example, $h(n) = 0.2$ for $n = 0, \ldots, 4$. The block length is 10, the overlap is 4. As illustrated in the figure, $x(n) = x_1(n) + x_2(n) + \cdots$ and $y(n) = y_1(n) + y_2(n) + \cdots$ where $y_i(n)$ is the result of convolving $x_i(n)$ with the filter $h(n)$.

8.2.2 Overlap-Save

A slightly different organization of the above approach is also often used for high-speed convolution. Rather than sectioning the input and then calculating the output from overlapped outputs from these individual input blocks, we will section the output and then use whatever part of the input contributes to that output block. In other words, to calculate the values in a particular output block, a section of length $N + L - 1$ from the input will be needed. The strategy is to save the part of the first input block that contributes to the second output block and use it in that calculation. It turns out that exactly the same amount of arithmetic and storage are used by these two approaches. Because it is the input that is now overlapped and, therefore, must be saved, this second approach is called *overlap-save*.

This method has also been called *overlap-discard* in [12] because, rather than adding the overlapping output blocks, the overlapping portion of the output blocks are discarded. As illustrated in Fig. 8.2, both the head and the tail of the output blocks are discarded. It may appear in Fig. 8.2 that an FFT of length 18 is needed. However, with the use of the FFT (to get cyclic convolution), the head and the tail overlap, so the FFT length is 14. (In practice, block lengths are generally chosen so that the FFT length $N + L - 1$ is a power of 2).

8.2.3 Use of the Overlap Methods

Because the efficiency of the FFT is $O(N \log(N))$, the efficiency of the overlap methods for convolution increases with length. To use the FFT for convolution will require one length-N forward FFT, N complex multiplications, and one length-N inverse FFT. The FFT of the filter is done once and stored rather than done repeatedly for each block. For short lengths, direct convolution will be more efficient. The exact length of filter where the efficiency cross-over occurs depends on the computer and software being used.

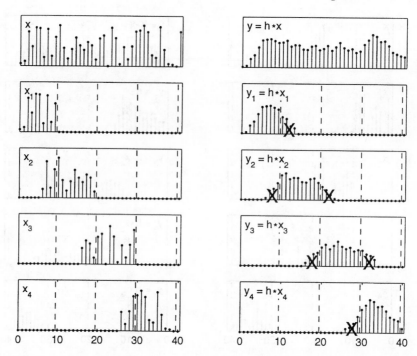

FIGURE 8.2 Overlap-save algorithm. The sequence $y(n)$ is the result of convolving $x(n)$ with an FIR filter $h(n)$ of length 5. In this example, $h(n) = 0.2$ for $n = 0, \ldots, 4$. The block length is 10, the overlap is 4. As illustrated in the figure, the sequence $y(n)$ is obtained, block by block, from the appropriate block of $y_i(n)$, where $y_i(n)$ is the result of convolving $x_i(n)$ with the filter $h(n)$.

If it is determined that the FFT is potentially faster than direct convolution, the next question is what block length to use. Here, there is a compromise between the improved efficiency of long FFTs and the fact you are processing a lot of appended zeros that contribute nothing to the output. An empirical plot of multiplication (and, perhaps, additions) per output point vs. block length will have a minimum that may be several times the filter length. This is an important parameter that should be optimized for each implementation. Remember that this increased block length may improve efficiency but it adds a delay and requires memory for storage.

8.3 Block Convolution

The operation of a finite impulse response (FIR) filter is described by a finite convolution as

$$y(n) = \sum_{k=0}^{L-1} h(k) x(n - k) \tag{8.2}$$

where $x(n)$ is causal, $h(n)$ is causal and of length L, and the time index n goes from zero to infinity or some large value. With a change of index variables this becomes

$$y(n) = \sum_{k=0}^{n} h(n - k) x(k) \tag{8.3}$$

which can be expressed as a matrix operation by

$$
\begin{bmatrix} y_0 \\ y_1 \\ y_2 \\ \vdots \end{bmatrix} = \begin{bmatrix} h_0 & 0 & 0 & \cdots & 0 \\ h_1 & h_0 & 0 & & \\ h_2 & h_1 & h_0 & & \\ \vdots & & & \vdots & \end{bmatrix} \begin{bmatrix} x_0 \\ x_1 \\ x_2 \\ \vdots \end{bmatrix} . \tag{8.4}
$$

The H matrix of impulse response values is partitioned into N by N square submatrices and the X and Y vectors are partitioned into length-N blocks or sections. This is illustrated for $N = 3$ by

$$
H_0 = \begin{bmatrix} h_0 & 0 & 0 \\ h_1 & h_0 & 0 \\ h_2 & h_1 & h_0 \end{bmatrix}, \quad H_1 = \begin{bmatrix} h_3 & h_2 & h_1 \\ h_4 & h_3 & h_2 \\ h_5 & h_4 & h_3 \end{bmatrix}, \quad \text{etc.} \tag{8.5}
$$

$$
\underline{x}_0 = \begin{bmatrix} x_0 \\ x_1 \\ x_2 \end{bmatrix}, \quad \underline{x}_1 = \begin{bmatrix} x_3 \\ x_4 \\ x_5 \end{bmatrix}, \quad \underline{y}_0 = \begin{bmatrix} y_0 \\ y_1 \\ y_2 \end{bmatrix}, \quad \text{etc.} \tag{8.6}
$$

Substituting these definitions into (8.4) gives

$$
\begin{bmatrix} \underline{y}_0 \\ \underline{y}_1 \\ \underline{y}_2 \\ \vdots \end{bmatrix} = \begin{bmatrix} H_0 & 0 & 0 & \cdots & 0 \\ H_1 & H_0 & 0 & & \\ H_2 & H_1 & H_0 & & \\ \vdots & & & \vdots & \end{bmatrix} \begin{bmatrix} \underline{x}_0 \\ \underline{x}_1 \\ \underline{x}_2 \\ \vdots \end{bmatrix} \tag{8.7}
$$

The general expression for the nth output block is

$$
\underline{y}_n = \sum_{k=0}^{n} H_{n-k} \, \underline{x}_k \tag{8.8}
$$

which is a vector or block convolution. Since the matrix-vector multiplication within the block convolution is itself a convolution, (8.9) is a sort of convolution of convolutions and the finite length matrix-vector multiplication can be carried out using the FFT or other fast convolution methods.

The equation for one output block can be written as the product

$$
\underline{y}_2 = \begin{bmatrix} H_2 & H_1 & H_0 \end{bmatrix} \begin{bmatrix} \underline{x}_0 \\ \underline{x}_1 \\ \underline{x}_2 \end{bmatrix} \tag{8.9}
$$

and the effects of one input block can be written

$$
\begin{bmatrix} H_0 \\ H_1 \\ H_2 \end{bmatrix} \underline{x}_1 = \begin{bmatrix} \underline{y}_0 \\ \underline{y}_1 \\ \underline{y}_2 \end{bmatrix} . \tag{8.10}
$$

These are generalized statements of overlap-save and overlap-add [11, 30]. The block length can be longer, shorter, or equal to the filter length.

8.3.1 Block Recursion

Although less well known, infinite impulse response (IIR) filters can be implemented with block processing [5, 6]. The block form of an IIR filter is developed in much the same way as the block convolution implementation of the FIR filter. The general constant coefficient difference equation which describes an IIR filter with recursive coefficients a_l, convolution coefficients b_k, input signal $x(n)$, and output signal $y(n)$ is given by

$$y(n) = \sum_{l=1}^{N-1} a_l\, y_{n-l} + \sum_{k=0}^{M-1} b_k\, x_{n-k} \tag{8.11}$$

using both functional notation and subscripts, depending on which is easier and clearer. The impulse response $h(n)$ is

$$h(n) = \sum_{l=1}^{N-1} a_l\, h(n-l) + \sum_{k=0}^{M-1} b_k\, \delta(n-k) \tag{8.12}$$

which, for $N = 4$, can be written in matrix operator form

$$\begin{bmatrix} 1 & 0 & 0 & \cdots & 0 \\ a_1 & 1 & 0 \\ a_2 & a_1 & 1 \\ a_3 & a_2 & a_1 \\ 0 & a_3 & a_2 \\ & \vdots & & \vdots \end{bmatrix} \begin{bmatrix} h_0 \\ h_1 \\ h_2 \\ h_3 \\ h_4 \\ \vdots \end{bmatrix} = \begin{bmatrix} b_0 \\ b_1 \\ b_2 \\ b_3 \\ 0 \\ \vdots \end{bmatrix}$$

In terms of smaller submatrices and blocks, this becomes

$$\begin{bmatrix} A_0 & 0 & 0 & \cdots & 0 \\ A_1 & A_0 & 0 \\ 0 & A_1 & A_0 \\ & \vdots & & \vdots \end{bmatrix} \begin{bmatrix} \underline{h}_0 \\ \underline{h}_1 \\ \underline{h}_2 \\ \vdots \end{bmatrix} = \begin{bmatrix} \underline{b}_0 \\ \underline{b}_1 \\ 0 \\ \vdots \end{bmatrix} \tag{8.13}$$

for blocks of dimension two. From this formulation, a block recursive equation can be written that will generate the impulse response block by block.

$$A_0\, \underline{h}_n + A_1\, \underline{h}_{n-1} = 0 \qquad \text{for } n \geq 2 \tag{8.14}$$

or

$$\underline{h}_n = -A_0^{-1} A_1\, \underline{h}_{n-1} = K\, \underline{h}_{n-1} \qquad \text{for } n \geq 2 \tag{8.15}$$

with initial conditions given by

$$\underline{h}_1 = -A_0^{-1} A_1 A_0^{-1}\, \underline{b}_0 + A_0^{-1}\, \underline{b}_1 \tag{8.16}$$

Next, we develop the recursive formulation for a general input as described by the scalar difference equation (8.12) and in matrix operator form by

$$\begin{bmatrix} 1 & 0 & 0 & \cdots & 0 \\ a_1 & 1 & 0 \\ a_2 & a_1 & 1 \\ a_3 & a_2 & a_1 \\ 0 & a_3 & a_2 \\ & \vdots & & \vdots \end{bmatrix} \begin{bmatrix} y_0 \\ y_1 \\ y_2 \\ y_3 \\ y_4 \\ \vdots \end{bmatrix} = \begin{bmatrix} b_0 & 0 & 0 & \cdots & 0 \\ b_1 & b_0 & 0 \\ b_2 & b_1 & b_0 \\ 0 & b_2 & b_1 \\ 0 & 0 & b_2 \\ & \vdots & & \vdots \end{bmatrix} \begin{bmatrix} x_0 \\ x_1 \\ x_2 \\ x_3 \\ x_4 \\ \vdots \end{bmatrix} \tag{8.17}$$

which, after substituting the definitions of the submatrices and assuming the block length is larger than the order of the numerator or denominator, becomes

$$
\begin{bmatrix}
A_0 & 0 & 0 & \cdots & 0 \\
A_1 & A_0 & 0 & & \\
0 & A_1 & A_0 & & \\
& & & & \vdots \\
\vdots & & & \vdots &
\end{bmatrix}
\begin{bmatrix}
\underline{y}_0 \\
\underline{y}_1 \\
\underline{y}_2 \\
\vdots
\end{bmatrix}
=
\begin{bmatrix}
B_0 & 0 & 0 & \cdots & 0 \\
B_1 & B_0 & 0 & & \\
0 & B_1 & B_0 & & \\
& & & & \vdots \\
\vdots & & & \vdots &
\end{bmatrix}
\begin{bmatrix}
\underline{x}_0 \\
\underline{x}_1 \\
\underline{x}_2 \\
\vdots
\end{bmatrix}.
\tag{8.18}
$$

From the partitioned rows of (8.19), one can write the block recursive relation

$$
A_0\, \underline{y}_{n+1} + A_1\, \underline{y}_n = B_0\, \underline{x}_{n+1} + B_1\, \underline{x}_n
\tag{8.19}
$$

Solving for $\underline{y}_{n+1}$ gives

$$
\underline{y}_{n+1} = -A_0^{-1} A_1\, \underline{y}_n + A_0^{-1} B_0\, \underline{x}_{n+1} + A_0^{-1} B_1\, \underline{x}_n
\tag{8.20}
$$

$$
\underline{y}_{n+1} = K\, \underline{y}_n + H_0\, \underline{x}_{n+1} + \tilde{H}_1\, \underline{x}_n
\tag{8.21}
$$

which is a first order vector difference equation [5, 6]. This is the fundamental block recursive algorithm that implements the original scalar difference equation in (8.12). It has several important characteristics.

1. The block recursive formulation is similar to a state variable equation but the states are blocks or sections of the output [6].

2. If the block length were shorter than the denominator, the vector difference equation would be higher than first order. There would be a nonzero A_2. If the block length were shorter than the numerator, there would be a nonzero B_2 and a higher order block convolution operation. If the block length were one, the order of the vector equation would be the same as the scalar equation. They would be the same equation.

3. The actual arithmetic that goes into the calculation of the output is partly recursive and partly convolution. The longer the block, the more the output is calculated by convolution, and the more arithmetic is required.

4. There are several ways of using the FFT in the calculation of the various matrix products in (8.20). Each has some arithmetic advantage for various forms and orders of the original equation. It is also possible to implement some of the operations using rectangular transforms, number theoretic transforms, distributed arithmetic, or other efficient convolution algorithms [6, 36].

8.4   Short and Medium Length Convolution

For the cyclic convolution of short sequences ($n \leq 10$) and medium length sequences ($n \leq 100$), special algorithms are available. For short lengths, algorithms that require the minimum number of multiplications possible have been developed by Winograd [8, 17, 35]. However, for longer lengths Winograd's algorithms, based on his theory of multiplicative complexity, require a large number of additions and become cumbersome to implement. Nesting algorithms, such as the Agarwal-Cooley and split-nesting algorithm, are methods that combine short convolutions. By nesting Winograd's short convolution algorithms, efficient medium length convolution algorithms can thereby be obtained.

In the following section we give a matrix description of these algorithms and of the Toom-Cook algorithm. Descriptions based on polynomials can be found in [4, 8, 19, 21, 24]. The presentation that follows relies upon the notions of similarity transformations, companion matrices, and Kronecker products. With them, the algorithms are described in a manner that brings out their structure and differences. It is found that when companion matrices are used to describe cyclic convolution, the algorithms block-diagonalize the cyclic shift matrix.

8.4.1   The Toom-Cook Method

A basic technique in fast algorithms for convolution is interpolation: two polynomials are evaluated at some common points, these values are multiplied, and by computing the polynomial interpolating these products, the product of the two original polynomials is determined [4, 19, 21, 31]. This interpolation method is often called the Toom-Cook method and can be described by a bilinear form. Let $n = 2$,

$$
\begin{aligned}
X(s) &= x_0 + x_1 s + x_2 s^2 \\
H(s) &= h_0 + h_1 s + h_2 s^2 \\
Y(s) &= y_0 + y_1 s + y_2 s^2 + y_3 s^3 + y_4 s^4 .
\end{aligned}
$$

The linear convolution of x and h can be represented by a matrix-vector product $y = Hx$,

$$
\begin{bmatrix} y_0 \\ y_1 \\ y_2 \\ y_3 \\ y_4 \end{bmatrix} =
\begin{bmatrix} h_0 & & \\ h_1 & h_0 & \\ h_2 & h_1 & h_0 \\ & h_2 & h_1 \\ & & h_2 \end{bmatrix}
\begin{bmatrix} x_0 \\ x_1 \\ x_2 \end{bmatrix}
$$

or as a polynomial product $Y(s) = H(s)X(s)$. In the former case, the linear convolution matrix can be written as $h_0 H_0 + h_1 H_1 + h_2 H_2$ where the meaning of H_k is clear. In the later case, one obtains the expression

$$
y = C\{Ah * Ax\} \tag{8.22}
$$

where $*$ denotes point-by-point multiplication. The terms Ah and Ax are the values of $H(s)$ and $X(s)$ at some points $i_1, \ldots i_{2n-1}$ ($n = 2$). The point-by-point multiplication gives the values $Y(i_1), \ldots, Y(i_{2n-1})$. The operation of C obtains the coefficients of $Y(s)$ from its values at the point $i_1, \ldots i_{2n-1}$. Equation (8.22) is a bilinear form and it implies that

$$
H_k = C \operatorname{diag}(Ae_k) A
$$

where e_k is the kth standard basis vector. (Ae_k is the kth column of A). However, A and C do not need to be Vandermonde matrices as suggested above. As long as A and C are matrices such that $H_k = C \operatorname{diag}(Ae_k)A$, then the linear convolution of x and h is given by the bilinear form $y = C\{Ah * Ax\}$. More generally, as long as A, B, and C are matrices satisfying $H_k = C \operatorname{diag}(Be_k)A$, then $y = C\{Bh * Ax\}$ computes the linear convolution of h and x. For convenience, if $C\{Bh * Ax\}$ computes the n point linear convolution of h and x (both h and x are n point sequences), then we say "(A, B, C) describes a bilinear form for n point linear convolution."

EXAMPLE 8.1:

(A, A, C) describes a 2-point linear convolution where

$$
A = \begin{bmatrix} 1 & 0 \\ 1 & 1 \\ 0 & 1 \end{bmatrix} \quad \text{and} \quad C = \begin{bmatrix} 1 & 0 & 0 \\ 0 & 1 & 0 \\ -1 & -1 & 1 \end{bmatrix} . \tag{8.23}
$$

8.4.2   Cyclic Convolution

The cyclic convolution of x and h can be represented by a matrix-vector product

$$
\begin{bmatrix} y_0 \\ y_1 \\ y_2 \end{bmatrix} =
\begin{bmatrix} h_0 & h_2 & h_1 \\ h_1 & h_0 & h_2 \\ h_2 & h_1 & h_0 \end{bmatrix}
\begin{bmatrix} x_0 \\ x_1 \\ x_2 \end{bmatrix}
$$

or as the remainder of a polynomial product after division by $s^n - 1$, denoted by $Y(s) = \langle H(s)X(s)\rangle_{s^n-1}$. In the former case, the cyclic convolution matrix can be written as $h_0 I + h_1 S_2 + h_2 S_2^2$ where S_n is the cyclic shift matrix,

$$
S_n = \begin{bmatrix} & & & 1 \\ 1 & & & \\ & \ddots & & \\ & & 1 & \end{bmatrix}.
$$

It will be useful to make a more general statement.

The companion matrix of a monic polynomial, $M(s) = m_0 + m_1 s + \cdots + m_{n-1}s^{n-1} + s^n$ is given by

$$
C_M = \begin{bmatrix} & & & -m_0 \\ 1 & & & -m_1 \\ & \ddots & & \vdots \\ & & 1 & -m_{n-1} \end{bmatrix}.
$$

Its usefulness in the following discussion comes from the following relation, which permits a matrix formulation of convolution:

$$
Y(s) = \langle H(s)X(s)\rangle_{M(s)} \quad \Longleftrightarrow \quad y = \left(\sum_{k=0}^{n-1} h_k C_M^k\right) x \tag{8.24}
$$

where x, h, and y are the vectors of coefficients and C_M is the companion matrix of $M(s)$. In (8.24), we say y is the convolution of x and h with respect to $M(s)$. In the case of cyclic convolution, $M(s) = s^n - 1$ and C_{s^n-1} is the cyclic shift matrix, S_n.

Similarity transformations can be used to interpret the action of some convolution algorithms. If $C_M = T^{-1}QT$ for some matrix T (C_M and Q are similar, denoted $C_M \sim Q$), then (8.24) becomes

$$
y = T^{-1} \left(\sum_{k=0}^{n-1} h_k Q^k\right) Tx .
$$

That is, by employing the similarity transformation given by T in this way, the action of S_n^k is replaced by that of Q^k. Many cyclic convolution algorithms can be understood, in part, by understanding the manipulations made to S_n and the resulting new matrix Q. If the transformation T is to be useful, it must satisfy two requirements: (1) Tx must be simple to compute, and (2) Q must have some advantageous structure. For example, by the convolution property of the DFT, the DFT matrix F diagonalizes S_n and, therefore, it diagonalizes every circulant matrix. In this case, Tx can be computed by an FFT and the structure of Q is the simplest possible: a diagonal.

8.4.3 Winograd Short Convolution Algorithm

The Winograd algorithm [35] can be described using the notation above. Suppose $M(s)$ can be factored as $M(s) = M_1(s)M_2(s)$ where $M_1(s)$ and $M_2(s)$ have no common roots, then $C_M \sim (C_{M_1} \oplus C_{M_2})$ where $\oplus$ denotes the matrix direct sum. Using this similarity and recalling (8.24), the original convolution can be decomposed into two disjoint convolutions. This is a statement of the Chinese remainder theorem for polynomials expressed in matrix notation. In the case of cyclic convolution, $s^n - 1$ can be written as the product of *cyclotomic* polynomials — polynomials whose coefficients are small integers. Denoting the dth cyclotomic polynomial by $\Phi_d(s)$, one has $s^n - 1 = \prod_{d|n} \Phi_d(s)$. Therefore, S_n can be transformed to a block diagonal matrix,

$$S_n \sim \begin{bmatrix} C_{\Phi_1} & & & \\ & C_{\Phi_d} & & \\ & & \ddots & \\ & & & C_{\Phi_n} \end{bmatrix} = \left(\bigoplus_{d|n} C_{\Phi_d} \right). \tag{8.25}$$

The symbol $\oplus$ denotes the matrix direct sum (diagonal concatenation). Each matrix on the diagonal is the companion matrix of a cyclotomic polynomial.

EXAMPLE 8.2:

$$
\begin{aligned}
s^{15} - 1 &= \Phi_1(s)\Phi_3(s)\Phi_5(s)\Phi_{15}(s) \\
&= (s-1)(s^2+s+1)(s^4+s^3+s^2+s+1)(s^8-s^7+s^5-s^4+s^3-s+1)
\end{aligned}
$$

$$
S_{15} = T^{-1}
\begin{bmatrix}
1 & & & & & & & & & & & & & & \\
 & -1 & & & & & & & & & & & & & \\
 & 1 & -1 & & & & & & & & & & & & \\
 & & & & -1 & & & & & & & & & & \\
 & & & 1 & -1 & & & & & & & & & & \\
 & & & & 1 & -1 & & & & & & & & & \\
 & & & & & 1 & -1 & & & & & & & & \\
 & & & & & & & & & & -1 & & & & \\
 & & & & & & & 1 & & & 1 & & & & \\
 & & & & & & & & 1 & & & & & & \\
 & & & & & & & & & 1 & & & -1 & & \\
 & & & & & & & & & & 1 & & 1 & & \\
 & & & & & & & & & & & 1 & & -1 & \\
 & & & & & & & & & & & & 1 & & \\
 & & & & & & & & & & & & & 1 & 1 \\
\end{bmatrix}
T. \tag{8.26}
$$

Each block represents a convolution with respect to a cyclotomic polynomial, or a "cyclotomic convolution." When n has several prime divisors the similarity transformation T becomes quite complicated. However, when n is a prime power, the transformation is very structured, as described in [29].

As in the previous section, we can write a bilinear form for cyclotomic convolution. Let d be any positive integer and let $X(s)$ and $H(s)$ be polynomials of degree $\phi(d) - 1$ where $\phi(\cdot)$ is the Euler totient function. If A, B, and C are matrices satisfying $(C_{\Phi_d})^k = C \operatorname{diag}(Be_k)A$ for $0 \le k \le \phi(d) - 1$, then the coefficients of $Y(s) = \langle X(s)H(s) \rangle_{\Phi_d(s)}$ are given by $y = C\{Bh * Ax\}$. As above, for such A, B, and C, we say "(A, B, C) describes a bilinear form for $\Phi_d(s)$ convolution."

But since $\langle X(s)H(s) \rangle_{\Phi_d(s)}$ can be found by computing the product of $X(s)$ and $H(s)$ and reducing the result, a cyclotomic convolution algorithm can always be derived by following a linear convolution algorithm by the appropriate reduction operation: If G is the appropriate reduction matrix and if (A, B, C) describes a bilinear form for a $\phi(d)$ point *linear* convolution, then (A, B, GC) describes a bilinear form for $\Phi_d(s)$ convolution. That is, $y = GC\{Bh * Ax\}$ computes the coefficients of $\langle X(s)H(s) \rangle_{\Phi_d(s)}$.

EXAMPLE 8.3:

A bilinear form for $\Phi_3(s)$ convolution is described by (A, A, GC) where A and C are given in (8.23) and G is given by

$$
G = \begin{bmatrix} 1 & 0 & -1 \\ 0 & 1 & -1 \end{bmatrix}.
$$

The Winograd short cyclic convolution algorithm decomposes the convolution into smaller (cyclotomic) ones, and can be described as follows. If (A_d, B_d, C_d) describes a bilinear form for $\Phi_d(s)$ convolution, then a bilinear form for cyclic convolution is provided by

$$A = \left(\oplus_{d|n} A_d \right) T \qquad B = \left(\oplus_{d|n} B_d \right) T \qquad C = T^{-1} \left(\oplus_{d|n} C_d \right) .$$

The matrix T decomposes the problem into disjoint parts, and T^{-1} recombines the results.

8.4.4 The Agarwal-Cooley Algorithm

The Agarwal-Cooley [3] algorithm uses a similarity of another form. Namely, when $n = n_1 n_2$, and $(n_1, n_2) = 1$

$$S_n = P^t \left(S_{n_1} \otimes S_{n_2} \right) P \tag{8.27}$$

where $\otimes$ denotes the Kronecker product and P is a permutation matrix. The permutation is $k \rightarrow \langle k \rangle_{n_1} + n_1 \langle k \rangle_{n_2}$. This converts a one-dimensional cyclic convolution of length n into a two-dimensional one of length n_1 along one dimension and length n_2 along the second. Then an n_1-point and an n_2-point cyclic convolution algorithm can be combined to obtain an n-point algorithm.

8.4.5 The Split-Nesting Algorithm

The split-nesting algorithm [21] combines the structures of the Winograd and Agarwal-Cooley methods, so that S_n is transformed to a block diagonal matrix as in (8.25),

$$S_n \sim \bigoplus_{d|n} \Psi(d) . \tag{8.28}$$

Here $\Psi(d) = \bigotimes_{p|d, p \in \mathcal{P}} C_{\Phi_{H_d(p)}}$ where $H_d(p)$ is the highest power of p dividing d, and $\mathcal{P}$ is the set of primes. An example clarifies this decomposition.

EXAMPLE 8.4:

$$S_{45} = P^t R^{-1} \begin{bmatrix} 1 & & & & & \\ & C_{\Phi_3} & & & & \\ & & C_{\Phi_9} & & & \\ & & & C_{\Phi_5} & & \\ & & & & C_{\Phi_3} \otimes C_{\Phi_5} & \\ & & & & & C_{\Phi_9} \otimes C_{\Phi_5} \end{bmatrix} R P \tag{8.29}$$

where P is the same permutation matrix of (8.27), and R is a matrix described in [29].

In the split-nesting algorithm, each matrix along the diagonal represents a multidimensional cyclotomic convolution rather than a one-dimensional one. To obtain a bilinear form for the split-nesting method, bilinear forms for one-dimensional convolutions can be combined to obtain bilinear forms for multi-dimensional cyclotomic convolution. This is readily explained by an example.

EXAMPLE 8.5:

A 45-point circular convolution algorithm:

$$y = P^t R^{-1} C \{ B R P h * A R P x \} \tag{8.30}$$

where

$$
\begin{aligned}
A &= 1 \oplus A_3 \oplus A_9 \oplus A_5 \oplus (A_3 \otimes A_5) \oplus (A_9 \otimes A_5) \\
B &= 1 \oplus B_3 \oplus B_9 \oplus B_5 \oplus (B_3 \otimes B_5) \oplus (B_9 \otimes B_5) \\
C &= 1 \oplus C_3 \oplus C_9 \oplus C_5 \oplus (C_3 \otimes C_5) \oplus (C_9 \otimes C_5)
\end{aligned}
$$

and where $(A_{p^i}, B_{p^i}, C_{p^i})$ describes a bilinear form for $\Phi_{p^i}(s)$ convolution.

Split-nesting (1) requires a simpler similarity transformation than the Winograd algorithm and (2) decomposes cyclic convolution into several disjoint multidimensional convolutions. For these reasons, for medium lengths, split-nesting can be more efficient than the Winograd convolution algorithm, even though it does not achieve the minimum number of multiplications. An explicit matrix description of the similarity transformation is provided in [29].

8.5 Multirate Methods for Running Convolution

While fast FIR filtering, based on block processing and the FFT, is computationally efficient, for real-time processing it has three drawbacks: (1) A delay is incurred; (2) the multiply-accumulate structure of the convolutional sum, a command for which DSPs are optimized, is lost; and (3) extra memory and communication (data transfer) time is needed. For real-time applications, this has motivated the development of alternative methods for convolution that partially retain the FIR filtering structure [18, 33].

In the z-domain, the running convolution of x and h is described by a polynomial product

$$
Y(z) = H(z)X(z) \tag{8.31}
$$

where $X(z)$ and $Y(z)$ are of infinite degree, and $H(z)$ is of finite degree. Let us write the polynomials as follows

$$
\begin{aligned}
X(z) &= X_0\left(z^2\right) + z^{-1}X_1\left(z^2\right) \tag{8.32} \\
Y(z) &= Y_0\left(z^2\right) + z^{-1}Y_1\left(z^2\right) \tag{8.33} \\
H(z) &= H_0\left(z^2\right) + z^{-1}H_1\left(z^2\right) \tag{8.34}
\end{aligned}
$$

where

$$
X_0(z) = \sum_{i=0}^{\infty} x_{2i} z^{-i} \qquad X_1(z) = \sum_{i=0}^{\infty} x_{2i+1} z^{-i}
$$

and Y_0, Y_1, H_0, H_1 are similarly defined. (These are known as polyphase components, although that is not important here). The polynomial product (8.31) can then be written as

$$
Y_0\left(z^2\right) + z^{-1}Y_1\left(z^2\right) = \left(H_0\left(z^2\right) + z^{-1}H_1\left(z^2\right)\right)\left(X_0\left(z^2\right) + z^{-1}X_1\left(z^2\right)\right) \tag{8.35}
$$

or in matrix form as

$$
\begin{bmatrix} Y_0 \\ Y_1 \end{bmatrix} = \begin{bmatrix} H_0 & z^{-2}H_1 \\ H_1 & H_0 \end{bmatrix} \begin{bmatrix} X_0 \\ X_1 \end{bmatrix} \tag{8.36}
$$

where $Y_0 = Y_0(z^2)$, etc.

The general form of (8.34) is given by

$$
X(z) = \sum_{k=0}^{N-1} z^{-1} X_k(z^N)
$$

where

$$X_k(z) = \sum_i x_{Ni+k} z^{-i}$$

and similarly for H and Y. For clarity, $N = 2$ is used in this exposition.

Note that the right hand side of (8.35) is a product of two polynomials of degree N, where the coefficients are themselves polynomials, either of finite degree (H_i), or of infinite degree (X_i). Accordingly, the Toom-Cook algorithm described previously can be employed, in which case the sums and products become polynomial sums and products. The essential key is that the polynomial products are themselves equivalent to FIR filtering, with shorter filters.

A Toom-Cook algorithm for carrying out (8.35) is given by

$$\begin{bmatrix} Y_0 \\ Y_1 \end{bmatrix} = C \left\{ A \begin{bmatrix} H_0 \\ H_1 \end{bmatrix} * A \begin{bmatrix} X_0 \\ X_1 \end{bmatrix} \right\}$$

where

$$A = \begin{bmatrix} 1 & 0 \\ 1 & 1 \\ 0 & 1 \end{bmatrix} \qquad C = \begin{bmatrix} 1 & 0 & z^{-2} \\ -1 & 1 & -1 \end{bmatrix}.$$

This Toom-Cook algorithm yields the multirate filter bank structure shown in Fig. 8.3. The outputs of the two downsamplers, on the left side of the structure shown in the figure, are $X_0(z)$ and $X_1(z)$. The outputs of the two upsamplers, on the right side of the structure, are $Y_0(z^2)$ and $Y_1(z^2)$. Note that the three filters H_0, $H_0 + H_1$, and H_1 operate at half the sampling rate. The right-most operation shown in Fig. 8.3 is not an arithmetic addition — it is a merging of the two sequences, $Y_0(z^2)$ and $z^{-1}Y_1(z^2)$, by interleaving. The arithmetic overhead is 1 "input" addition and 3 "output" additions per 2 samples; that is a total of 2 additions per sample.

If the original filter $H(z)$ is of length L and operates at the rate f_s, then the structure in Fig. 8.3 is an implementation of $H(z)$ that employs three filters of length $L/2$, each operating at the rate $\frac{1}{2} f_s$.

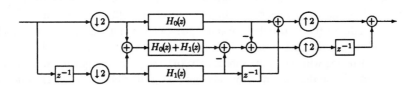

FIGURE 8.3 Filter structure based on a two-point convolution algorithm. Let H_0 be the even coefficients of a filter H, let H_1 be the odd coefficients. The structure implements the filter H using three half-length filters, each running at half the rate of H.

The convolutional sum for $H(z)$, when implemented directly, requires L multiplications per output point and $L - 1$ additions per output point. Per output point, the structure in Fig. 8.3 requires $\frac{3}{4} L$ multiplications and $2 + \frac{3}{2}(L/2 - 1) = \frac{3}{4} L + \frac{1}{2}$ additions.

The decomposition can be repeatedly applied to each of the three filters; however, the benefit diminishes for small L, and quantization errors may accumulate. Table 8.1 gives the number of multiplications needed to implement a length 32 FIR filter, using various levels of decomposition.

Other short linear convolution algorithms can be obtained from existing ones by a technique known as transposition. The transposed form of a short convolution algorithm has the same arithmetic complexity, but in a different arrangement. It was observed in [18] that the transposed forms generally have more input additions and fewer output additions. Consequently, the transposed forms should be more robust to quantization noise.

TABLE 8.1   Computation of Running Convolution

Method	Subsampling	Delay	Mult./Point
1 32-pt. FIR filter	1	0	32
3 16-pt. FIR filters	2	1	24
9 8-pt. FIR filters	4	3	18
27 4-pt. FIR filters	8	7	13.5
81 2-pt. FIR filters	16	15	10.125
243 1-pt. mults.	32	31	7.59

Based on repeated application of two-point convolution structure in Fig. 8.3. (From [33].)

Various short-length convolution algorithms that are appropriate for this approach are provided in [18]. Also addressed is the issue of when to stop successive decompositions — and the problem of finding the best way to combine small-length filters, depending on various criteria. In particular, it is noted that DSPs generally perform a multiply-accumulate (MAC) operation in a single clock cycle, in which case a MAC should be considered a single operation.

It appears that this approach is amenable to (1) efficient multiprocessor implementations due to their inherent parallelism, and (2) efficient VLSI realization, since the implementation requires only local communication, instead of global exchange of data as in the case of FFT-based algorithms.

In [33], the following is noted. The mapping of long convolutions into small, subsampled convolutions is attractive in hardware (VLSI), software (signal processors), and multiprocessor implementations since the basic building blocks remain convolutions which can be computed efficiently once small enough.

8.6   Convolution in Subbands

Maximally decimated perfect reconstruction filter banks have been used for a variety of applications where processing in subbands is advantageous. Such filter banks can be regarded as generalizations of the short-time Fourier transform, and it turns out that the convolution theorem can be extended to them [23, 32]. In other words, the convolution of two signals can be found by directly convolving the subband signals and combining the results. In [23], both uniform and nonuniform decimation ratios are considered for orthonormal and biorthonormal filter banks. In [32], the results of [23] are generalized.

The advantage of this method is that the subband signals can be quantized based on the signal variance in each subband and other perceptual considerations, as in traditional subband coding. Instead of quantizing $x(n)$ and then convolving with $g(n)$, the subbands $x_k(n)$ and $g_k(n)$ are quantized, and the results are added. When quantizing in the subbands, the subband energy distribution can be exploited and bits can be allocated to subbands accordingly. For a fixed bit rate, this approach increases the accuracy of the overall convolution — that is, this approach offers a coding gain.

In [23] an optimal bit allocation formula and the optimized coding gain is derived for orthogonal filter banks. The contribution to coding gain comes partly from the nonuniformity of the signal spectrum and partly from the nonuniformity of the filter spectrum. When the filter impulse response is taken to be the unit impulse $\delta(n)$, the formulas for the bit allocation and coding gain reduce to those for traditional subband and transform coding.

The efficiency that is gained from subband convolution comes from the ability to use a fewer number of bits to achieve a given level of accuracy. In addition, in [23], low sensitivity filter structures are derived from the subband convolution theorem and examined.

8.7   Distributed Arithmetic

Rather than grouping the individual scalar data values in a discrete-time signal into blocks, the scalar values can be partitioned into groups of bits. Because multiplication of integers, multiplication of polynomials, and discrete-time convolution are the same operations, the bit-level description of multiplication can be mixed with the convolution of the signal processing. The resulting structure is called distributed arithmetic [7, 34].

8.7.1 Multiplication is Convolution

To simplify the presentation, we will assume the data and coefficients to be positive integers with simple binary coding and the problem of carrying will be omitted. Assume the product of two B-bit words is desired

$$y = ax \tag{8.37}$$

where

$$a = \sum_{i=0}^{B-1} a_i 2^i \text{ and } x = \sum_{i=0}^{B-1} a_j 2^j \tag{8.38}$$

with $a_i, x_j \in \{0, 1\}$. This gives

$$y = \sum_i a_i 2^i \sum_j x_j 2^j \tag{8.39}$$

which, with a change of variables $k = i + j$, becomes

$$y = \sum_k \sum_i a_i x_{k-i} 2^k . \tag{8.40}$$

Using the binary description of y as

$$y = \sum_k y_k 2^k \tag{8.41}$$

we have for the binary coefficients

$$y_k = \sum_i a_i x_{k-i} \tag{8.42}$$

as a convolution of the binary coefficients for a and x. We see that multiplying two numbers is the same as convolving their coefficient representation any base. Multiplication is convolution.

8.7.2 Convolution is Two Dimensional

Consider the following convolution of number strings (FIR filtering)

$$y(n) = \sum_\ell a(\ell) x(n - \ell) . \tag{8.43}$$

Using the binary representation of the coefficients and data, we have

$$y(n) = \sum_\ell \sum_i a_i(\ell) 2^i \sum_j x_j(n - \ell) 2^j \tag{8.44}$$

$$y(n) = \sum_\ell \sum_i \sum_i a_i(\ell) x_j(n - \ell) 2^{i+j} \tag{8.45}$$

which after changing variables, $k = i + j$, becomes

$$y(n) = \sum_k \sum_\ell \sum_i a_i(\ell) x_{k-i}(n - \ell) 2^k . \tag{8.46}$$

A one-dimensional convolution of numbers is a two-dimensional convolution of the binary (or other base) representations of the numbers.

8.7.3 Distributed Arithmetic by Table Lookup

The usual way that distributed arithmetic convolution is calculated does the arithmetic in a special concentrated algorithm or piece of hardware. We are now going to reorder the very general description in (8.46) to allow some of the operations to be precomputed and stored in a lookup table. The arithmetic will then be distributed with the convolution itself.

If (8.46) is summed over the index i, we have

$$y(n) = \sum_j \sum_\ell a(\ell) x_j(n - \ell) 2^j. \tag{8.47}$$

Each sum over ℓ convolves the word string $a(n)$ with the bit string $x_j(n)$ to produce a partial product which is then shifted and added by the sum over j to give $y(n)$.

If (8.47) is summed over ℓ to form a table which can be addressed by the binary numbers $x_j(n)$, we have

$$y(n) = \sum_j f(x_j(n), x_j(n - 1), \cdots) 2^j \tag{8.48}$$

where

$$f(x_j(n), x_j(n - 1), \cdots) = \sum_\ell a(\ell) x_j(n - \ell) \tag{8.49}$$

The numbers $a(i)$ are the coefficients of the filter, which as usual is assumed to be fixed. Consider a filter of length L. This function $f()$ is a function of L binary variables and, therefore, takes on 2^L possible values. The function is determined by the filter, $a(i)$. For example, if $L = 3$, the table (function values) would contain eight values:

$$0, \ a(0), \ a(1), \ a(2), \ (a(0) + a(1)), \ (a(1) + a(2)), \ (a(0) + a(2)), \ (a(0) + a(1) + a(2)) \tag{8.50}$$

and if the words were stored as B bits, they would require $2^L B$ bits of memory.

There are extensions and modifications of this basic idea to allow a very flexible trade of memory for logic. The idea is to precompute as much as possible, store it in a table, and fetch it when needed. The two extremes of this are on one hand to compute all possible outputs and simply fetch them using the input as an address. The other extreme is the usual system which simply stores the coefficients and computes what is needed as needed.

This table lookup is illustrated in Fig. 8.4 where the blocks represent 4 b words, where the least significant bit of each of the four most recent data words form the address for the table lookup from memory. After 4 b shifts and accumulates, the output word $y(n)$ is available, using no multiplications.

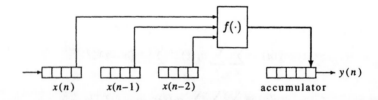

FIGURE 8.4   Distributed arithmetic by Table Lookup. In this example, a sequence $x(n)$ is filtered with a length 3 FIR filter. The wordlength for $x(n)$ is 4 b. The function $f(\cdot)$ is a function of three binary variables, and can be implemented by table lookup. The bits of $x(n)$ are shifted, bit by bit, through the input registers. Accordingly, the bits of $y(n)$ are shifted through the accumulator — after 4 b shifts, a new output $y(n)$ becomes available.

Distributed arithmetic with table lookup can be used with FIR and IIR filters and can be arranged in direct, transpose, cascade, parallel, etc. structures. It can be organized for serial or parallel calculations or for combinations of the two. Because most microprocessors or DSP chips do not have appropriate instructions or architectures for distributed arithmetic, it is best suited for special purpose VLSI design and in those cases, it can be extremely fast.

An alternative realization of these ideas can be developed using a form of periodically time varying system that is oversampled [10].

8.8   Fast Convolution by Number Theoretic Transforms

If one performs all calculations in a finite field or ring of integers rather than the usual infinite field of real or complex numbers, a very efficient type of Fourier transform can be formulated that requires no floating point operations — it supports exact convolution with finite precision arithmetic [1, 2, 17, 26]. This is particularly interesting because a digital computer is a finite machine and arithmetic over finite systems fits it perfectly. In the following, all arithmetic operations are performed modulo some integer M, called the modulus. A bit of number theory can be found in [17, 20, 28].

8.8.1   Number Theoretic Transforms

Here we look at the conditions placed on a general linear transform in order for it to support cyclic convolution. The form of a linear transformation of a length-N sequence of number is given by

$$X(k) = \sum_{n=0}^{N-1} t(n, k) \, x(n) \bmod M \tag{8.51}$$

for $k = 0, 1, \cdots, (N-1)$. The definition of cyclic convolution of two sequences in Z_M is given by

$$y(n) = \sum_{m=0}^{N-1} x(m) \, h(n-m) \bmod M \tag{8.52}$$

for $n = 0, 1, \cdots, (N-1)$ where all indices are evaluated modulo N. We would like to find the properties of the transformation such that it will support cyclic convolution. This means that if $X(k)$, $H(k)$, and $Y(k)$ are the transforms of $x(n)$, $h(n)$, and $y(n)$ respectively, then

$$Y(k) = X(k) \, H(k) \, . \tag{8.53}$$

The conditions are derived by taking the transform defined in (8.1) of both sides of Eq. (8.52) which gives the form for our general linear transform (8.51) as

$$X(k) = \sum_{n=0}^{N-1} \alpha^{nk} \, x(n) \tag{8.54}$$

where α is a *root of order* N, which means that N is the smallest integer such that $\alpha^N = 1$.

THEOREM 8.1   *The transform (8.11) supports cyclic convolution if and only if α is a root of order N and $N^{-1} \bmod M$ is defined.*

This is discussed in [1, 2]. This transform supports N-point cyclic convolution only if a particular relationship between the modulus M and the data length N is satisfied. The following theorem describes that relationship.

THEOREM 8.2   *The transform (8.11) supports N-point cyclic convolution if and only if*

$$N \mid O(M) \tag{8.55}$$

where

$$O(M) = gcd\{p_1 - 1, p_2 - 1, \cdots, p_l - 1\} \tag{8.56}$$

and the prime factorization of M is

$$M = p_1^{r_1} p_2^{r_2} \cdots p_l^{r_l}. \tag{8.57}$$

Equivalently, N must divide $p_i - 1$ for every prime p_i dividing M. This theorem is a more useful form of Theorem 8.1. Notice that $N_{max} = O(M)$.

One needs to find appropriate N, M, and α such that

- N should be appropriate for a fast algorithm and handle the desired sequence lengths.
- M should allow the desired dynamic range of the signals and should allow simple modular arithmetic.
- α should allow a simple multiplication for $\alpha^{nk} x(n)$.

We see that if M is even, it has a factor of 2 and, therefore, $O(M) = N_{max} = 1$ which implies M should be odd. If M is prime the $O(M) = M - 1$ which is as large as could be expected in a field of M integers. For $M = 2^k - 1$, let k be a composite $k = pq$ where p is prime. Then $2^p - 1$ divides $2^{pq} - 1$ and the maximum possible length of the transform will be governed by the length possible for $2^p - 1$. Therefore, only the prime k need be considered interesting. Numbers of this form are know as Mersenne numbers and have been used by Rader [26]. For Mersenne number transforms, it can be shown that transforms of length at least $2p$ exist and the corresponding $\alpha = -2$. Mersenne number transforms are not of as much interest because $2p$ is not highly composite and, therefore, we do not have FFT-type algorithms.

For $M = 2^k + 1$ and k odd, 3 divides $2^k + 1$ and the maximum possible transform length is 2. Thus, we consider only even k. Let $k = s2^t$, where s is an odd integer. Then 2^{2^t} divides $2^{s2^t} + 1$ and the length of the possible transform will be governed by the length possible for $2^{2^t} + 1$. Therefore, integers of the form $M = 2^{2^t} + 1$ are of interest. These numbers are known as Fermat numbers [26]. Fermat numbers are prime for $0 \le t \le 4$ and are composite for all $t \ge 5$.

Since Fermat numbers up to F_4 are prime, $O(F_t) = 2^b$ where $b = 2^t$ and $t \le 4$, we can have a Fermat number transform for any length $N = 2^m$ where $m \le b$. For these Fermat primes the integer $\alpha = 3$ is of order $N = 2^b$ allowing the largest possible transform length. The integer $\alpha = 2$ is of order $N = 2b = 2^{t+1}$. Then all multiplications by powers of α are bit shifts — which is particularly attractive because in (8.54), the data values are multiplied by powers of α.

Table 8.2 gives possible parameters for various Fermat number moduli.

TABLE 8.2   Fermat Number Transform Moduli

t	b	$M = F_t$	N_2	$N_{\sqrt{2}}$	N_{max}	α for N_{max}
3	8	$2^8 + 1$	16	32	256	3
4	16	$2^{16} + 1$	32	64	65536	3
5	32	$2^{32} + 1$	64	128	128	$\sqrt{2}$
6	64	$2^{64} + 1$	128	256	256	$\sqrt{2}$

This table gives values of N for the two most important values of α which are 2 and $\sqrt{2}$. The second column gives the approximate number of bits in the number representation. The third column gives the Fermat number modulus, the fourth is the maximum convolution length for $\alpha = 2$, the fifth is the maximum length for $\alpha = \sqrt{2}$, the sixth is the maximum length for any α, and the seventh is the α for that maximum length. Remember that the first two rows have a Fermat number modulus which is prime and the second two rows have a composite Fermat number as modulus. Note the differences.

The number theoretic transform itself seems to be very difficult to interpret or use directly. It seems to be useful only as a means for high-speed convolution where it has remarkable characteristics. The books, articles, and presentations that discuss NTT and related topics are [4, 17, 21]. A recent book discusses NT in a signal processing context [14].

8.9 Polynomial-Based Methods

The use of polynomials in representing elements of a digital sequence and in representing the convolution operation has led to the development of a family of algorithms based on the fast polynomial transform [4, 16, 21]. These algorithms are especially useful for two-dimensional convolution. The Chinese remainder theorem for polynomials (CRT), which is central to Winograd's short convolution algorithm, is also conveniently described in polynomial notation. An interesting approach combines the use of the polynomial-based methods with the number theoretic approach to convolution (NTTs), wherein the elements of a sequence are taken to lie in a finite field [9, 15]. In [15] the CRT is extended to the case of a ring of polynomials with coefficients from a finite ring of integers. It removes the limitations on both word length and sequence length of NNTs and serves as a link between the two methods (CRT and NNT). The new result so obtained, which specializes to both the NNTs and the CRT for polynomials, has been called the AICE-CRT (the American-Indian-Chinese extension of the CRT). A complex version has also been derived.

8.10 Special Low-Multiply Filter Structures

In the use of convolution for digital filtering, the convolution operation can be simplified, if the filter $h(n)$ is chosen appropriately.

Some filter structures are especially simple to implement. Some examples are:

- A simple implementation of the recursive running sum (RRS) is based on the factorization
$$\sum_{k=0}^{L-1} z^k = (z^L + 1)/(z - 1).$$

- If the transfer function $H(z)$ of the filter possesses a root at $z = -1$ of multiplicity K, the factor $(z + 1)/2$ can be extracted from the transfer function. The factor $(z + 1)/2$ can be implemented very simply.

- This idea is extended in prefiltering and IFIR filtering techniques — a filter is implemented as a cascade of two filters: one with a crude response that is simple to implement, another that makes up for it, but requires the usual implementation complexity. The overall response satisfies specifications and can be implemented with reduced complexity.

- The maximally flat symmetric FIR filter can be implemented without multiplications using the De Casteljau algorithm [27].

In summary, a filter can often be designed so that the convolution operation can be performed with less computational complexity and/or at a faster rate. Much work has focused on methods that take into account implementation complexity during the approximation phase of the filter design process. (See the chapter on digital filter design).

References

[1] Agarwal, R.C. and Burrus, C.S., Fast convolution using Fermat number transforms with applications to digital filtering, *IEEE Trans. Acoustics Speech Signal Process.*, ASSP-22(2):87–97, April, 1974. Reprinted in [17].

[2] Agarwal, R.C. and Burrus, C.S., Number theoretic transforms to implement fast digital convolution, *Proc. IEEE*, 63(4):550–560, April, 1975. (Also in IEEE Press DSP Reprints II, 1979).

[3] Agarwal, R.C. and Cooley, J.W., New algorithms for digital convolution, *IEEE Trans. Acoustics Speech Signal Process.*, 25(5):392–410, October, 1977.

[4] Blahut, R.E. *Fast Algorithms for Digital Signal Processing*, Addison-Wesley, Reading, MA, 1985.

[5] Burrus, C.S., Block implementation of digital filters, *IEEE Trans. Circuit Theory*, CT-18(6):697–701, November, 1971.

[6] Burrus, C.S., Block realization of digital filters, *IEEE Trans. Audio Electroacoust.*, AU-20(4):230–235, October, 1972.

[7] Burrus, C.S., Digital filter structures described by distributed arithmetic, *IEEE Trans. Circuits Syst.*, CAS-24(12):674–680, December, 1977.

[8] Burrus, C.S., Efficient Fourier transform and convolution algorithms, in Jae S. Lim and Alan V. Oppenheim, Eds., *Advanced Topics in Signal Processing*, Prentice-Hall, Englewood Cliffs, NJ, 1988.

[9] Garg, H.K., Ko, C.C., Lin, K.Y., and Liu, H., On algorithms for digital signal processing of sequences, *Circuits Syst. Signal Process.*, 15(4):437–452, 1996.

[10] Ghanekar, S.P., Tantaratana, S., and Franks, L.E., A class of high-precision multiplier-free FIR filter realizations with periodically time-varying coefficients, *IEEE Trans. Signal Process.*, 43(4):822–830, 1995.

[11] Gold, B. and Rader, C.M., *Digital Processing of Signals*, McGraw-Hill, New York, 1969.

[12] Harris, F.J., Time domain signal processing with the DFT, in D. F. Elliot, ed., *Handbook of Digital Signal Processing*, ch. 8, 633–699, Academic Press, NY, 1987.

[13] Helms, H.D., Fast Fourier transform method of computing difference equations and simulating filters, *IEEE Trans. Audio Electroacoust.*, AU-15:85–90, June, 1967.

[14] Krishna, H., Krishna, B., Lin, K.-Y, and Sun, J.-D., *Computational Number Theory and Digital Signal Processing*, CRC Press, Boca Raton, FL, 1994.

[15] Lin, K.Y., Krishna, H., and Krishna, B., Rings, fields the Chinese remainder theorem and an American-Indian-Chinese extension, part I: Theory. *IEEE Trans. Circuits Syst. II*, 41(10):641–655, 1994.

[16] Loh, A.M. and Siu, W.-C., Improved fast polynomial transform algorithm for cyclic convolutions, *Circuits Syst. Signal Process.*, 14(5):603–614, 1995.

[17] McClellan, J.H. and Rader, C.M., *Number Theory in Digital Signal Processing*, Prentice-Hall, Englewood Cliffs, NJ, 1979.

[18] Mou, Z.-J. and Duhamel, P., Short-length FIR filters and their use in fast nonrecursive filtering, *IEEE Trans. Signal Process.*, 39(6):1322–1332, June, 1991.

[19] Myers, D.G., *Digital Signal Processing: Efficient Convolution and Fourier Transform Techniques*, Prentice-Hall, Englewood Cliffs, NJ, 1990.

[20] Niven, I. and Zuckerman, H.S., *An Introduction to the Theory of Numbers*, 4th ed., John Wiley & Sons, New York, 1980.

[21] Nussbaumer, H.J., *Fast Fourier Transform and Convolution Algorithms*, Springer-Verlag, New York, 1982.

[22] Oppenheim, A.V. and Schafer, R.W., *Discrete-Time Signal Processing*, Prentice-Hall, Englewood Cliffs, NJ, 1989.

[23] Phoong, S-.M. and Vaidyanathan, P.P., One- and two-level filter-bank convolvers, *IEEE Trans. Signal Process.*, 43(1):116–133, January, 1995.

[24] Proakis, J.G., Rader, C.M., Ling, F., and Nikias, C.L., *Advanced Digital Signal Processing*, Macmillan, New York, 1992.

[25] Rabiner, L.R. and Gold, B., *Theory and Application of Digital Signal Processing*, Prentice-Hall, Englewood Cliffs, NJ, 1975.

[26] Rader, C.M., Discrete convolution via Mersenne transforms, *IEEE Trans. Comput.*, 21(12):1269–1273, December, 1972.

[27] Samadi, S., Cooklev, T., Nishihara, A., and Fujii, N., Multiplierless structure for maximally flat linear phase FIR filters, *Electron. Lett.*, 29(2):184–185, Jan. 21, 1993.

[28] Schroeder, M.R., *Number Theory in Science and Communication*, 2nd ed., Springer-Verlag, Berlin, 1984, 1986.

[29] Selesnick, I.W. and Burrus, C.S., Automatic generation of prime length FFT programs, *IEEE Trans. Signal Process.*, 44(1):14–24, January, 1996.

[30] Stockham, T.G., High speed convolution and correlation, in *AFIPS Conf. Proc.*, vol. 28, pp. 229–233, Spring Joint Computer Conference, 1966.

[31] Tolimieri, R., An, M., and Lu, C., *Algorithms for Discrete Fourier Transform and Convolution*, Springer-Verlag, New York, 1989.

[32] Vaidyanathan, P.P, Orthonormal and biorthonormal filter banks as convolvers, and convolutional coding gain, *IEEE Trans. Signal Process.*, 41(6):2110–2129, June, 1993.

[33] Vetterli, M., Running FIR and IIR filtering using multirate filter banks, *IEEE Trans. Acoust. Speech Signal Process.*, 36(5):730–738, May, 1988.

[34] White, S.A., Applications of distributed arithmetic to digital signal processing, *IEEE ASSP Mag.*, 6(3):4–19, July, 1989.

[35] Winograd, S., *Arithmetic Complexity of Computations*, SIAM, 1980.

[36] Zalcstein, Y., A note on fast cyclic convolution, *IEEE Trans. Comput.*, 20:665–666, June, 1971.

9

Complexity Theory of Transforms in Signal Processing

Ephraim Feig
IBM Corporation
T.J. Watson Research Center

9.1 Introduction

Complexity theory of computation attempts to determine how "inherently" difficult are certain tasks. For example, how inherently complex is the task of computing an inner product of two vectors of length N? Certainly one can compute the inner product $\sum_{j=1}^{N} x_j y_j$ by computing the N products $x_j y_j$ and then summing them. But can one compute this inner product with fewer than N multiplications? The answer is no, but the proof of this assertion is no trivial matter. One first abstracts and defines the notions of the algorithm and its components (such as addition and multiplication); then a theorem is proven that any algorithm for computing a bilinear form which uses K multiplications can be transformed to a quadratic algorithm (some algorithm of a very special form, which uses no divisions, and whose multiplications only compute quadratic forms) which uses at most K multiplications [20]; and finally a proof by induction on the length N of the summands in the inner product is made to obtain the lower bound result [6, 13, 22, 25]. We will not present the details here; we just want to let the reader know that the process for even proving what seems to be an intuitive result is quite complex.

Consider next the more complex task of computing the product of an N point vector by an $M \times N$ matrix. This corresponds to the task of computing M separate inner products of N-point vectors. It is tempting to jump to the conclusion that this task requires MN multiplications. But we should not jump to fast conclusions. First, the M inner products are separate, but not independent (the term is used loosely, and not in any linear algebra sense). After all, the second factor in the M inner products is always the same. It turns out [6, 22, 25] that, indeed, our intuition this time is correct again. And the proof is really not much more difficult than the proof for the complexity result for inner products. In fact, once the general machinery is built, the proof is a slight extension of the previous case. So far intuition proved accurate.

In complexity theory one learns early on to be skeptical of intuitions. An early surprising result in complexity theory — and to date still one of its most remarkable — contradicts the intuitive guess that computing the product of two 2×2 matrices requires 8 multiplications. Remarkably, Strassen [21] has shown that it can be done with 7 multiplication. His algorithm is very nonintuitive; I am not aware of any good algebraic explanation for it except for the assertion that the mathematical identities which define the algorithm indeed are valid. It can also be shown [15] that 7 is the minimum number of multiplications required for the task.

The consequences of Strassen's algorithm for general matrix multiplication tasks are profound. The task of computing the product of two 4×4 matrices with real entries can be viewed as a task of computing two 2×2 matrices whose entries are themselves 2×2 matrices. Each of the 7 multiplications in Strassen's algorithm now become matrix multiplications requiring 7 real multiplications plus a bunch of additions; and each addition in Strassen's algorithm becomes an addition of 2×2 matrices, which can be done with 4 real additions. This process of obtaining algorithms for large problems, which are built up of smaller ones in a structures manner, is called the "nesting" procedure [25]. It is a very powerful tool in both complexity theory and algorithm design. It is a special form of recursion.

The set of $N \times N$ matrices form a noncommutative algebra. A branch of complexity theory called "multiplicative complexity theory" is quite well established for certain relatively few algebras, and wide open for the rest. In this theory complexity is measured by the number of "essential multiplications." Given an algebra over a field F, an algorithm is a sequence of arithmetic operations in the algebra. A multiplication is called essential if neither factor is an element in F. If one of the factors in a multiplication is an element in F, the operation is called a scaling.

Consider an algebra of dimension N over a field F, with basis $b_1, \ldots, b_N$. An algorithm for computing the product of two elements $\sum_{j=1}^{N} f_j b_j$ and $\sum_{j=1}^{N} g_j b_j$ with $f_j, g_j \in F$ is called bilinear, if every multiplication in the algorithm is of the form $L_1(f_1, \ldots, f_N) * L_2(g_1, \ldots, g_N)$, where L_1 and L_2 are linear forms and $*$ is the product in the algebra, and it uses no divisions. Because none of the arithmetic operations in bilinear algorithms rely on the commutative nature of the underlying field, these algorithms can be used to build recursively via the nesting process algorithms for noncommutative algebras of increasingly large dimensions, which are built from the smaller algebras via the tensor product. For example, the algebra of 4×4 matrices (over some field F; I will stop adding this necessary assumption, as it will be obvious from content) is isomorphic to the tensor product of the algebra of 2×2 matrices with itself. Likewise, the algebra of 16×16 matrices is isomorphic to the tensor product of the algebra of 4×4 matrices with itself. And this proceeds to higher and higher dimensions.

Suppose we have a bilinear algorithm for computing the product in an algebra T_1 of dimension D, which uses M multiplications and A additions (including subtractions) and S scalings. The algebra $T_2 = T_1 \otimes T_1$ has dimension D^2. By the nesting procedure we can obtain an algorithm for computing the product in T_2 which uses M multiplications of elements in T_1, A additions of elements in T_1, and S scalings of elements in T_1. Each multiplication in T_1 requires M multiplications, A additions, and S scalings; each addition in T_1 requires D additions; and each scaling in T_1 requires D scalings. Hence, the total computational requirements for this new algorithm is M^2 multiplications, $A(M + D)$ additions and $S(M + D)$ scalings. If the nesting procedure is continued to yield an algorithm for the product in the D^4 dimensional algebra $T_4 = T_2 \otimes T_2$, then its computational requirements would be M^4 multiplications, $A(M + D)(M^2 + D^2)$ additions and $S(M + D)(M^2 + D^2)$ scalings. One more iteration would yield an algorithm for the D^8 dimensional algebra $T_8 = T_4 \otimes T_4$, which uses M^8 multiplications, $A(M + D)(M^2 + D^2)(M^4 + D^4)$ additions, M^8 multiplications, and $S(M + D)(M^2 + D^2)(M^4 + D^4)$ scalings. The general pattern should be apparent by now. We see that the growth of the number of operations (the high order term, that is) is governed by M and not by A or S. A major goal of complexity theory is the understanding of computational requirements as problem sizes increase, and nesting is the natural way of building algorithms for larger and larger problems. We see one reason why counting multiplications (as opposed to all arithmetic operations) became so important in complexity theory. (Historically, in the early days multiplications were indeed much more expensive than additions.)

Algebras of polynomials are important in signal processing; filtering can be viewed as polynomial multiplications. The product of two polynomials of degrees d_1 and d_2 can be computed with $d_1 + d_2 - 1$ multiplications. Furthermore, it is rather easy to prove (a straightforward dimension argument) that this is the minimal number of multiplications necessary for this computation. Algorithms which compute these products with these numbers of multiplications (so-called optimal algorithms) are obtained using Lagrange interpolation techniques. For even moderate values of d_j, they use inordinately many additions and scalings. Indeed, they use $(d_1 + d_2 - 3)(d_1 + d_2 - 2)$ additions, and a half as many scalings. So these algorithms are not very practical, but they are of theoretical interest. Also of interest is the asymptotic complexity of polynomial products. They can be computed by embedding them in cyclic convolutions of sizes at most twice as long. Using FFT techniques, these can be achieved with order $D \log D$ arithmetic operations, where D is the maximum of the degrees. With optimal algorithms, while the number of (essential) multiplications is linear, the total number of operations is quadratic. If nesting is used, then the asymptotic behavior of the number of multiplications is also quadratic.

Convolution algebras are derived from algebras of polynomials. Given a polynomial $P(u)$ of degree D, one can define an algebra of dimension D whose entries are all polynomials of degree less than D, with addition defined in the standard way, and multiplication is modulo $P(u)$. Such algebras are called convolution algebras. For polynomials $P(u) = u^D - 1$, the algebras are cyclic convolutions of dimension D. For polynomials $P(u) = u^D + 1$, these algebras are called signed-cyclic convolutions. The product of two polynomials modulo $P(u)$ can be obtained from the product of the two polynomials without any extra essential multiplications. Hence, if the degree of $P(u)$ is D, then the product modulo $P(u)$ can be done with $2D - 1$ multiplications. But can it be done with fewer multiplications?

Whereas complexity theory has huge gaps in almost all areas, it has triumphed in convolution algebras. The minimum number of multiplications required to compute a product in an algebra is called the multiplicative complexity of the algebra. The multiplicative complexity of convolution algebras (over infinite fields) is completely determined [22]. If P(u) factors (over the base field; the role of the field will be discussed in greater detail soon) to a product of k irreducible polynomials, then the multiplicative complexity of the algebra is $2D - k$. So if $P(u)$ is irreducible, then the answer to the question in the previous paragraph is no. Otherwise, it is yes.

The above complexity result for convolution algebras is a sharp bound. It is a lower bound in that every algorithm for computing the product in the algebra requires at least $2D - k$ multiplications, where k is the number of factors of the defining polynomial $P(u)$. It is also an upper bound, in that there are algorithms which actually achieve it. Let us factor $P(u) = \prod P_j(u)$ into a product of irreducible polynomials (here we see the role of the field; more about this soon). Then the convolution algebra modulo $P(u)$ is isomorphic to a direct sum of algebras modulo $P_j(u)$; the isomorphism is via the Chinese remainder theorem. The multiplicative complexity of the direct summands are $2d_j - 1$, where d_j are the degrees of $P_j(u)$; these are sharp bounds. The algorithm for the algebra modulo $P(u)$ is derived from these smaller algorithms; because of the isomorphism, putting them all together requires no extra multiplications. The proof that this is a lower bound, first given by Winograd [23], is quite complicated.

The above result is an example of a "direct sum theorem." If an algebra is decomposable to a direct sum of subalgebras, then clearly the multiplicative complexity of the algebra is less than or equal to the sum of the multiplicative complexities of the summands. In some (relatively rare) circumstances equality can be shown. The example of convolution algebras is such a case. The results for convolution algebras are very strong. Winograd has shown that every minimal algorithm for computing products in a convolution algebra is bilinear and is a direct sum algorithm. The latter means that the algorithm actually computes a minimal algorithm for each direct summand and then combines these results without any extra essential multiplications to yield the product in the algebra itself.

Things get interesting when we start considering algebras which are tensor products of convolution algebras (these are called multi-dimensional convolution algebras). A simple example already is enlightening. Consider the algebra C of polynomial multiplications modulo $u^2 + 1$ over the rationals Q; this algebra is called the Gaussian rationals. The polynomial $u^2 + 1$ is irreducible over Q (the algebra is a field), so by the previous result, its multiplicative complexity is 3. The nesting procedure would yield an algorithm the product in $C \otimes C$ which uses 9 multiplications. But it can in fact be computed with 6 multiplications. The reason is due to an old theorem, probably due to Kroeneker (though I cannot find the original proof); the reference I like best is Adrian Albert's book [1]. The theorem asserts that the tensor product of fields is isomorphic to a direct sum of fields, and the proof of the theorem is actually a construction of this isomorphsim. For our example, the theorem yields that the tensor product $C \otimes C$ is isomorphic to a direct sum of two copies of C. The product in $C \otimes C$ can, therefore, be computed by computing separately the product in each of the two direct summands, each with 3 multiplications, and the final result can be obtained without any more essential multiplications. The explicit isomorphism was presented to the complexity theory community by Winograd [22]. Since the example is sufficiently simple to work out, and the results so fundamental to much of our later discussions, we will present it here explicitly.

Consider A, the polynomial ring modulo $u^2 + 1$ over the Q. This is a field of dimension 2 over Q, and it has the matrix representation (called its regular representation) given by

$$\rho(a + bu) \;=\; \begin{pmatrix} a & -b \\ b & a \end{pmatrix}. \tag{9.1}$$

While for all $b \neq 0$ the matrix above is not diagonalizable over Q, the field (algebra) is diagonalizable over the complexes. Namely,

$$\begin{pmatrix} 1 & i \\ 1 & -i \end{pmatrix} \begin{pmatrix} a & -b \\ b & a \end{pmatrix} \begin{pmatrix} 1 & i \\ 1 & -i \end{pmatrix}^{-1} \;=\; \begin{pmatrix} a+ib & 0 \\ 0 & a-ib \end{pmatrix}. \tag{9.2}$$

The elements 1 and i of A correspond (in the regular representation) in the tensor algebra $A \otimes A$ to the matrices

$$\rho(1) \;=\; \begin{pmatrix} 1 & 0 \\ 0 & 1 \end{pmatrix} \tag{9.3}$$

and

$$\rho(i) \;=\; \begin{pmatrix} 0 & -1 \\ 1 & 0 \end{pmatrix}, \tag{9.4}$$

respectively. Hence, the 4×4 matrix

$$R \;=\; \begin{pmatrix} \rho(1) & \rho(i) \\ \rho(1) & \rho(-i) \end{pmatrix} \tag{9.5}$$

diagonalizes the algebra $A \otimes A$. Explicitly, we can compute

$$\begin{pmatrix} 1 & 0 & 0 & -1 \\ 0 & 1 & 1 & 0 \\ 1 & 0 & 0 & 1 \\ 0 & 1 & -1 & 0 \end{pmatrix} \begin{pmatrix} x_0 & -x_1 & -x_2 & -x_3 \\ x_1 & x_0 & -x_3 & x_2 \\ x_2 & -x_3 & x_0 & -x_1 \\ x_3 & x_2 & x_1 & x_0 \end{pmatrix}$$

$$\begin{pmatrix} 1 & 0 & 0 & -1 \\ 0 & 1 & 1 & 0 \\ 1 & 0 & 0 & 1 \\ 0 & 1 & -1 & 0 \end{pmatrix}^{-1} \;=\; \begin{pmatrix} y_0 & -y_1 & 0 & 0 \\ y_1 & y_0 & 0 & 0 \\ 0 & 0 & y_2 & -y_2 \\ 0 & 0 & y_3 & y_3 \end{pmatrix}, \tag{9.6}$$

where $y_0 = x_0 - x_3$, $y_1 = x_1 + x_2$, $y_2 = x_0 + x_3$ and $y_3 = x_1 - x_2$. A simple way to derive this is by setting X_0 to be the top left 2×2 minor of the matrix with x_j entries in the above equation, X_1 to be its bottom left 2×2 minor, and observing that

$$R \begin{pmatrix} X_0 & -X_1 \\ X_1 & X_0 \end{pmatrix} R^{-1} \;=\; \begin{pmatrix} \rho(1)X_0 + \rho(i)X_1 & \\ & \rho(0)X_0 - \rho(i)X_1 \end{pmatrix}. \tag{9.7}$$

The algorithmic implications are straightforward. The product in $A \otimes A$ can be computed with fewer multiplications than the nesting process would yield. Straightforward extensions of the above construction yield recipes for obtaining minimal algorithms for products in algebras which are tensor products of convolution algebras. The example also highlights the role of the base field. The complexity of A as an algebra over Q is 3; the complexity of A as an algebra over the complexes is 2, as over the complexes this algebra diagonalizes.

Historically, multiplicative complexity theory generalized in two ways (and in various combinations of the two). The first addressed the question: what happens when one of the factors in the product is not an arbitrary element but a fixed element not in the basefield? The second addressed: what is the complexity of semidirect systems — those in which several products are to be computed, and one factor is arbitrary but fixed, while the others are arbitrary? Computing an arbitrary product in an n-dimensional algebra can be thought of (via the regular representation) as computing a product of a matrix $A(X)$ times a vector Y, where the entries in the matrix $A(X)$ are linear combinations of n indeterminates $x_1, \ldots, x_n$ and y is a vector of n indeterminates $y_1, \ldots, y_n$. When one factor is a fixed element in an extension field, the entries in $A(X)$ are now entries in some extension field of the basefield which may have algebraic relations. For example, consider

$$G = \begin{pmatrix} \gamma(1,8) & -\gamma(3,8) \\ \gamma(3,8) & \gamma(1,8) \end{pmatrix} \quad (9.8)$$

where $\gamma(m,n) = \cos(2\pi m/n)$. The complex numbers $\gamma(1,8)$ and $\gamma(3,8)$ are linearly independent over Q, but they satisfy the algebraic relation $\gamma(1,8) / \gamma(3,8) = \sqrt{2}$. This algebraic relation gives a relation of the two numbers to the rationals, namely $\gamma(1,8)^2 / \gamma(3,8)^2 = 2$. Now this is not a linear relation; linear independence over Q has complexity ramifications. But this algebraic relation also has algorithmic ramifications. The linear independence implies that the multiplicative complexity of multiplying an arbitrary vector by G is 3. But because of the algebraic relation, it is not true (as is the case for quadratic extensions by indeterminates) that all minimal algorithms for this product are quadratic. A nonquadratic minimal algorithm is given via the factorization

$$G = \begin{pmatrix} \gamma(1,8) & 0 \\ 0 & \gamma(1,8) \end{pmatrix} \begin{pmatrix} 1 & 1-\sqrt{2} \\ \sqrt{2}-1 & 1 \end{pmatrix}. \quad (9.9)$$

As for computing the product of G and k distinct vectors, theory has it that the multiplicative complexity is 3k [5]. In other words, a direct sum theorem hold for this case. This result, and its generalization, due to Auslander and Winograd [5], is very deep; its proof is very complicated. But it yields great rewards.

The multiplicative complexity of all DFTs and DCTs are established using this result. The key to obtaining multiplicative complexity results for DFTs and DCTs is to find the appropriate block diagonalizations that transform these linear operators to such direct sums, and then to invoke this fundamental theorem. We will next cite this theorem, and then describe explicitly how we apply it to DFTs and DCTs.

Fundamental Theorem (Auslander-Winograd): *Let P_j be polynomials of degrees d_j, respectively, over a field ϕ. Let F_j denote polynomials of degree $d_j - 1$ with complex coefficients (that is, they are complex numbers). For non-negative integers k_j, let $T(k_j, F_j, P_j)$ denote the task of computing k_j products of arbitrary polynomials by F_j modulo P_j. Let $\sum_j T(k_j, F_j, P_j)$ denote the task of simultaneously computing all of these products. If the coefficients span a vector space of dimension $\sum_j d_j$ over ϕ, then the multiplicative complexity of $\sum_j T(k_j, F_j, P_j)$ is $\sum_j k_j(2d_j - 1)$. In other words, if the dimension assumption holds, then so does the direct sum theorem for this case.*

Multiplicative complexity results for DFTs and DCTs assert that their computation is linear in the size of the input. The measure is number of nonrational multiplications. More specifically, in all cases (arbitrary input sizes, arbitrary dimensions), the number of nonrational multiplications necessary for computing these transforms is always less than twice the size of the input. The exact numbers are interesting, but more important is the algebraic structure of the transforms which lead to these numbers. This is what will be emphasized in the remainder of this chapter. Some special cases will be discussed in greater detail; general results will be reviewed rather briefly.

The following notation will be convenient. If A, B are matrices with real entries, and R, S are invertible rational matrices such that $A = RBS$, then we will say that A is rationally equivalent (or more plainly, equivalent) to B and write $A \approx B$. The multiplicative complexity of A is the same as that of B.

9.2 One-Dimensional DFTs

We will build up the theory for the DFT in stages. The one-dimensional DFT on input size N is a linear operator whose matrix is given by $F_N = \left(w^{jk} \right)$, where $w = e^{2\pi i/N}$, and j, k index the rows and columns of the matrix, respectively. The first row and first column of F_N have all entries equal to 1, so the multiplicative complexity of F_N are the same as that of its "core" C_N, its minor comprising its last $N - 1$ rows and $N - 1$ columns. The first results were for one-dimensional DFTs on input sizes which are prime [24]. For p a prime integer, the set of integers between 0 and $p - 1$ form a cyclic group under multiplication modulo p. It was shown by Rader [19] that there exist permutations of the rows and columns of the core C_N that bring it to the cyclic convolution $w^{g^j + k}$, where g is any generator of the cyclic group described above. Using the decomposition for cyclic convolutions described above, we decompose the core to a direct sum of convolutions modulo the irreducible factors of $u^{p-1} - 1$. This decomposition into cyclotomic polynomials is well known [18]. There are $\tau(p - 1)$ irreducible factors, where $\tau(n)$ is the number of positive divisors of the positive integer n. One direct summand is the 1×1 matrix corresponding to the factor $u - 1$, and its entry is -1 (in particular, rational). Also, the coefficients of the other polynomials comprising the direct summands are all linearly independent over Q, hence the fundamental theorem (in its weakest form) applies. It yields that the multiplicative complexity of F_p for p a prime is $2p - \tau(p - 1) - 3$.

Next is the case for $N = p^k$ where p is an odd prime and the integer k is greater than 1. The group of units comprising those integers between 0 and $p - 1$ which are relatively prime to p, and under multiplication modulo p, is of order $p^k - p^{k-1}$. A Rader-like permutation [24] brings the sub-core, whose rows and columns are indexed by the entries in this group of units, to a cyclic convolution. The group of units, when multiplied by p, forms an orbit of order $p^{k-1} - p^{k-2}$ (p elements in the group of units map to the same element in the orbit), and the Rader-like permutations induces a permutation on the orbit, which yields cyclic convolutions of the sizes of the orbit. This proceeds until the final orbit of size $p - 1$. These cyclic convolutions are decomposed via the Chinese remainder theorem, and (after much cancellation and rearrangement) it can be shown that the core C_N in this case reduces to k direct summands, each of which is a semi-direct sum of $j(p - 1)(p^{k-j} - p^{k-j-1})$ dimensional convolutions modulo irreducible polynomials, $j = 1, 2, \ldots, k$. Also, the dimension of the coefficients of the polynomials is precisely $\sum_{j=1}^{k}(p - 1)(p^{k-j} - p^{k-j-1})$. These are precisely the conditions sufficient to invoke the fundamental theorem. This algebraic decomposition yields minimal algorithms. When one adds all these up, the numerical result is that the multiplicative complexity for the DFT on p^k points where p is an odd prime and k a positive integer, is $2p^k - k - 2 - \frac{k^2+k}{2}\tau(p - 1)$.

The case of the one dimensional DFT on $N = 2^n$ points is most familiar. In this case,

$$F_N = P_N \begin{pmatrix} F_{N/2} & \\ & G_{N/2} \end{pmatrix} R_N \qquad (9.10)$$

where P_N is the permutation matrix which rearranges the output to even entries followed by odd entries, R_N is a rational matrix for computing the so-called "butterfly additions," and $G_{N/2} = D_{N/2}F_{N/2}$, where $D_{N/2}$ is a diagonal matrix whose entries are the so-called "twiddle factors." This leads to the classical divide-and-conquer algorithm called the FFT. For our purposes, $G_{N/2}$ is equivalent to a direct sum of two polynomial products modulo u^{2^j} $j = 0, \ldots, n - 3$. It is routine to proceed inductively, and then show that the hypothesis of the fundamental theorem are satisfied. Without details, the final result is that the complexity of the DFT on $N = 2^n$ points is $2^{n+1} - n^2 - n - 2$. Again, the complexity is below $2N$.

For the general one-dimensional DFT case, we start with the equivalence $F_{mn} \approx F_m \otimes F_n$, whenever m and n are relatively prime, and where $\otimes$ denotes the tensor product. If m and n are of the forms p^k for some prime p and positive integer k, then from above, both F_m and F_n are equivalent to direct sums of polynomial products modulo irreducible polynomials. Applying the theorem of Kroeneker/Albert, which states that the tensor product of algebraic extension fields is isomorphic to a direct sum of fields, we have that F_{mn} is, therefore, equivalent to a direct sum of polynomial products modulo irreducible polynomials. When one follows the construction suggested by the theorem and counts the dimensionality of the coefficients, one can show that this direct sum system satisfies the hypothesis of the fundamental theorem. This argument extends to the general one-dimensional case of F_N where $N = \prod_j p_j^{k_j}$ with p_j distinct primes.

9.3 Multidimensional DFTs

The k-dimensional DFT on $N_1, \ldots, N_k$ points is equivalent to the tensor product $F_{N_1} \otimes \cdots \otimes F_{N_k}$. Directly from the theorem of Kroeneker/Albert, this is equivalent to a direct sum of polynomial products modulo irreducible polynomials. It can be shown that this system satisfies the hypothesis of the fundamental theorem so that complexity results can be directly invoked for the general multidimensional DFT. Details can be found in [4]. More interesting than the general case are some special cases with unique properties.

The k-dimensional DFT on $p, \ldots, p$ points, where p is an odd prime, is quite remarkable. The core of this transform is a cyclic convolution modulo $u^{p^k-1} - 1$. The core of the matrix corresponding to $F_p \otimes \cdots \otimes F_p$, which is the entire matrix minus its first row and column, can be brought into this large cyclic convolution by a permutation derived from a generator of the group of units of the field with p^k elements. The details are in [2]. Even more remarkably, this large cyclic convolution is equivalent to a direct sum of $p + 1$ copies of the same cyclic convolution obtainable from the core of the one-dimensional DFT on p points. In other words, the k-dimensional DFT on $p, \ldots, p$ points, where p is an odd prime, is equivalent to a direct sum of $p + 1$ copies of the one-dimensional DFT on p points. In particular, its multiplicative complexity is $(p + 1)(2p - \tau(p - 1) - 3)$.

Another particularly interesting case is the k-dimensional DFT on $N, \ldots, N$ points, where $N = 2^k$. This transform is equivalent to the k-fold tensor product $F_N \otimes \cdots \otimes F_N$, and we have seen above the recursive decomposition of F_N to a direct sum of $F_{N/2}$ and $G_{N/2}$. The semi-simple Abelian construction [3, 8] yields that $F_{N/2} \otimes G_{N/2}$ is equivalent to $N/2$ copies of $G_{N/2}$, and likewise that $F_{N/2} \otimes G_{N/2}$ is equivalent to $N/2$ copies of $G_{N/2}$. Hence, F_N and F_N is equivalent to $3N/2$ copies of $G_{N/2}$ plus $F_{N/2} \otimes F_{N/2}$. This leads recursively to a complete decomposition of the two-dimensional DFT to a direct sum of polynomial products modulo irreducible polynomials (of the form $u^{2^m} + 1$ in this case). The extensions to arbitrary dimensions are quite detailed but straightforward.

9.4 One-Dimensional DCTs

As in the case of DFTs, DCTs are also all equivalent to direct sums of polynomial multiplications modulo irreducible polynomials and satisfy the hypothesis of the fundamental theorem. In fact, some instances are easier to handle. A fast way to see the structure of the DCT is by relating it to the DFT. Let C_N denote the the one-dimensional DCT on N points; recall we defined F_N to be the one-dimensional DFT on N points.

It can be shown [14] that F_{4N} is equivalent to a direct sum of two copies of C_N plus one copy of F_{2N}. This is sufficient to yield complexity results for all one-dimensional DCTs. But for some special cases, direct derivations are more revealing. For example, when $N = 2^k$, C_N is equivalent to a direct sum of polynomial products modulo $u^{2^j} + 1$, for $j = 1, \ldots, k - 1$. This is a much simpler form than the corresponding one for the DFT on 2^k points. It is then straightforward to check that this direct sum system satisfies the hypothesis of the fundamental theorem, and then that the multiplicative complexity of C_{2^k} is $2^{k+1} - n - 2$.

Another (not so) special case is when N is an odd integer. Then C_N is equivalent to F_N, from which complexity results follow directly. Another useful result is that, as in the case of the DFT, C_{pq} is equivalent to $C_p \otimes C_q$ where p and q are relatively prime [26]. We can then use the theorem of Kroeneker/Albert [10] to build direct sum structures for DCTs of composites given direct sums of the various components.

9.5 Multidimensional DCTs

Here too, once the one-dimensional DCT structures are known, their extensions to multidimensions via tensor products, utilizing the theorem of Kroeneker/Albert, is straightforward. This leads to the appropriate direct sum structures, proving that the coefficients satisfy the hypothesis of the fundamental theorem does require some careful applications of elementary number theory. This is done in [10].

A most interesting special case is multidimensional DCT on input sizes which are powers of 2 in each dimension. If the input is k dimensional with size $2^{j_1} \times \ldots \times 2^{j_k}$, and $j_1 \leq j_i$, $i = 2, \ldots, k$, then the multidimensional DCT is equivalent to $2^{j_2} \times \ldots \times 2^{j_k}$ copies of the one-dimensional DCT on 2^{j_1} points [11]. This is a much more straightforward result than the corresponding one for multidimensional DFTs.

9.6 Nonstandard Models and Problems

DCTs have become popular because of their role in compression. In such roles, the DCT is usually followed by quantization. Therefore, in such applications, one need not actually compute the DCT but a scaled version of it, and then absorb the scaling into the quantization step. For the one-dimensional case this means that one can replace the computation of a product by C with a product by a matrix DC, where D is diagonal. It turns out [9, 16] that for propitious choices of D, the computation of the product by DC is easier than that by C. The question naturally arises—what is the minimum number of steps required to compute a product of the form DC, where D can be any diagonal matrix? Our ability to answer such a question is very limited. All we can say today is that if we can compute a scaled DCT on N points with m multiplications, then certainly we can compute a DCT on N multiplications with $m + N$ points. Since we know the complexity of DCTs, this gives a lower bound on the complexity of scaled DCTs. For example, the one-dimensional DCT on 8 points (the most popular applied case) requires 12 multiplications. (The reader may see the number 11 in the literature; this is for the case of the "unnormalized DCT" in which the DC component is scaled. The unnormalized DCT is not orthogonal.) Suppose a scaled DCT on 8 points can be done with m multiplications. Then $8 + m \geq 12$, or $m \geq 4$. An algorithm for the scaled DCT on 8 points which uses 5 multiplications is known [9, 16]. It is an open question whether one can actually do it in 4 multiplications or not. Similarly, the two-dimensional DCT on 8×8 points can be done with 54 multiplications [9, 12], and theory says that at least 24 are needed [11]. The gap is very wide, and I know of stronger results as of this writing.

Machines whose primitive operations are fused multiply-accumulate are becoming very popular, especially in the higher end workstation arena. Here a single cycle can yield a result of the form $ab + c$ for arbitrary floating point numbers a, b, c; we call such an operation a "mutiply/add." Lower bounds are obviously bounded below by lower bounds for number of multiplications and also for lower bounds on number of additions. The latter is a wide open subject. A simple yet instructive example involves multiplications of a 4×4 Hadamard matrix. It is well known that, in general, multiplication by an

$N \times N$ Hadamard matrix, where N is a power of 2, can be done with $N \, log_2 N$ additions. Recently it was shown [7] that the 4×4 case can be done with 7 multiply/add operations [7]. This result has not been extended, and it may in fact be rather hard to extend except in most trivial (and uninteresting) ways.

Upper bounds of DFTs have been obtained. It was shown in [17] that a complex DFT on $N = 2^k$ points can be done with $\frac{8}{3}Nk - \frac{16}{9}N + 2 - \frac{2}{9}(-1)^k$ real multiply/adds. For real input, an upper bound of $\frac{4}{3}Nk - \frac{17}{9}N + 3 - \frac{2}{9}(-1)^k$ real multiply/adds was given. These were later improved slightly using the results of the Hadamard transform computation. Similar multidimensional results were also obtained.

In the past several years new, more powerful, processors have been introduced. Sun and HP have incorporated new vector instructions. Intel has introduced its aggressive Intel's MMX architecture. And new MSPs (multimedia signal processors) from Philips, Samsung, and Chromatic are pushing similar designs even more aggressively. These will lead to new models of computation. Astounding (though probably not surprising) upper bounds will be announced; lower bounds are sure to continue to baffle.

References

[1] Albert, A., *Structure of Algebras*, AMS Colloqium Publications, Vol. 21, 1939.

[2] Auslander, L., Feig, E., and Winograd, S., New algorithms for the multidimensional discrete Fourier transform, *IEEE Trans. Accoust. Speech Signal Process.*, ASSP-31(2): 388–403, Apr., 1983.

[3] Auslander, L., Feig, E., and Winograd, S., Abelian semi-simple algebras and algorithms for the discrete Fourier transform, *Adv. Appl. Math.*, 5: 31–55, Mar., 1984.

[4] Auslander, L., Feig, E., and Winograd, S., The multiplicative complexity of the discrete Fourier transform, *Adv. Appl. Math.*, 5: 87–109, Mar., 1984.

[5] Auslander, L. and Winograd, S., The multiplicative complexity of certain semilinear systems defined by polynomials, *Adv. Appl. Math.*, 1(3): 257–299, 1980.

[6] Brocket, R.W. and Dobkin, D., On the optimal evaluation of a set of bilinear forms, *Linear Algebra Appl.*, 19(3): 207–235, 1978.

[7] Coppersmith, D., Feig, E., and Linzer, E., Hadamard transforms on multiply/add architectures, *IEEE Trans. Signal Processing*, 46(4): 969–970, Apr., 1994.

[8] Feig, E., New algorithms for the 2-dimensional discrete Fourier transform, IBM RC 8897 (No. 39031), June, 1981.

[9] Feig, E., A fast scaled DCT algorithm, Proc. SPIE-SPSE, Santa Clara, CA, Feb. 11–16, 1990.

[10] Feig, E. and Linzer, E., The multiplicative complexity of discrete cosine transforms, *Adv. Appl. Math.*, 13: 494–503, 1992.

[11] Feig, E. and Winograd, S., On the multiplicative complexity of discrete cosine transforms, *IEEE Trans. Inf. Theory*, 38(4): 1387–1391, July, 1992.

[12] Feig, E. and Winograd, S., Fast algorithms for the discrete cosine transform, *IEEE Trans. Signal Processing*, 40:(9) Sept., 1992.

[13] Fiduccia C.M., and Zalcstein, Y., Algebras having linear multiplicative complexities, *J. ACM*, 24(2): 311–331, 1977.

[14] Heideman, M.T., *Multiplicative Complexity, Convolution, and the DFT*, Springer-Verlag, New York, 1988.

[15] Hopcroft, J. and Kerr, L., On minimizing the number of multiplications necessary for matrix multiplication, *SIAM J. Appl. Math.*, 20: 30–36, 1971.

[16] Arai, Y., Agui, T., and Nakajima, M., A fast DCT-SQ scheme for images, *Trans. IEICE*, E-71(11): 1095–1097, Nov., 1988.

[17] Linzer, E. and Feig, E., Modified FFTs for fused multiply-add architectures, *Math. Comput.*, 60(201): 347–361, Jan., 1993.

[18] Niven, I. and Zuckerman, H.S., *An Introduction to the Theory of Numbers,* John Wiley & Sons, New York, 1980.

[19] Rader, C.M., Discrete Fourier transforms when the number of data samples is prime, *Proc. IEEE,* 56(6): 1107–1108, June, 1968.

[20] Strassen, V., Vermeidung con divisionen, *J. Reine Angew. Math.,* 264: 184–202, 1973.

[21] Strassen, V., Gaussian elimination is not optimal, *Numer. Math.,* 13: 354–356, 1969.

[22] Winograd, S., On the number of multiplications necessary to compute certain functions, *Commun Pure Appl. Math.,* No. 23, 165–179, 1970.

[23] Winograd, S., Some bilinear forms whose multiplicative complexity depends on the field of constants, *Math. Syst. Theory,* 10(2): 169–180, 1977.

[24] Winograd, S., On the multiplicative complexity of the discrete Fourier transform, *Adv. Math.,* 32(2): 83–117, May, 1979.

[25] Winograd, S., Arithmetic Complexity of Computations, CBMS-NSF Regional Conference Series in Applied Math, 1980.

[26] Yang, P.P.N. and Narasimha, M.J., Prime Factor Decomposition of the Discrete Cosine Transform and its Hardware Realization, *Proc. IEEE ICASSP,* 1985.

10

Fast Matrix Computations

Andrew E. Yagle
University of Michigan

10.1 Introduction

This chapter presents two major approaches to fast matrix multiplication. We restrict our attention to matrix multiplication, excluding matrix addition and matrix inversion, since matrix addition admits no fast algorithm structure (save for the obvious parallelization), and matrix inversion (i.e., solution of large linear systems of equations) is generally performed by iterative algorithms that require repeated matrix-matrix or matrix-vector multiplications. Hence, matrix multiplication is the real problem of interest.

We present two major approaches to fast matrix multiplication. The first is the divide-and-conquer strategy made possible by Strassen's [1] remarkable reformulation of non-commutative 2×2 matrix multiplication. We also present the APA (arbitrary precision approximation) algorithms, which improve on Strassen's result at the price of approximation, and a recent result that reformulates matrix multiplication as convolution and applies number theoretic transforms. The second approach is to use a wavelet basis to sparsify the representation of Calderon-Zygmund operators as matrices. Since electromagnetic Green's functions are Calderon-Zygmund operators, this has proven to be useful in solving integral equations in electromagnetics. The sparsified matrix representation is used in an iterative algorithm to solve the linear system of equations associated with the integral equations, greatly reducing the computation. We also present some new insights that make the wavelet-induced sparsification seem less mysterious.

10.2 Divide-and-Conquer Fast Matrix Multiplication

10.2.1 Strassen Algorithm

It is not obvious that there should be any way to perform matrix multiplication other than using the definition of matrix multiplication, for which multiplying two $N \times N$ matrices requires N^3 multiplications and additions (N for each of the N^2 elements of the resulting matrix). However, in 1969 Strassen [1]

0-8493-8572-5/98/$0.00+$.50
© 1998 by CRC Press LLC

made the remarkable observation that the product of two 2×2 matrices

$$\begin{bmatrix} a_{1,1} & a_{1,2} \\ a_{2,1} & a_{2,2} \end{bmatrix} \begin{bmatrix} b_{1,1} & b_{1,2} \\ b_{2,1} & b_{2,2} \end{bmatrix} = \begin{bmatrix} c_{1,1} & c_{1,2} \\ c_{2,1} & c_{2,2} \end{bmatrix} \tag{10.1}$$

may be computed using only seven multiplications (fewer than the obvious eight), as

$$
\begin{aligned}
m_1 &= (a_{1,2} - a_{2,2})(b_{2,1} + b_{2,2}); \quad m_3 = (a_{1,1} - a_{2,1})(b_{1,1} + b_{1,2}) \\
m_2 &= (a_{1,1} + a_{2,2})(b_{1,1} + b_{2,2}) \\
m_4 &= (a_{1,1} + a_{1,2})b_{2,2}; \quad m_7 = (a_{2,1} + a_{2,2})b_{1,1} \\
m_5 &= a_{1,1}(b_{1,2} - b_{2,2}); \quad m_6 = a_{2,2}(b_{2,1} - b_{1,1}) \\
c_{1,1} &= m_1 + m_2 - m_4 + m_6; \quad c_{1,2} = m_4 + m_5 \\
c_{2,2} &= m_2 - m_3 + m_5 - m_7; \quad c_{2,1} = m_6 + m_7
\end{aligned} \tag{10.2}
$$

A vital feature of (10.2) is that it is *non-commutative*, i.e., it does not depend on the commutative property of multiplication. This can be seen easily by noting that each of the m_i are the product of a linear combination of the elements of A by a linear combination of the elements of B, in that order, so that it is never necessary to use, say $a_{2,2}b_{2,1} = b_{2,1}a_{2,2}$. We note there exist commutative algorithms for 2×2 matrix multiplication that require even fewer operations, but they are of little practical use.

The significance of noncommutativity is that the noncommutative algorithm (10.2) may be applied as is to *block* matrices. That is, if the $a_{i,j}$, $b_{i,j}$ and $c_{i,j}$ in (10.1) and (10.2) are replaced by block matrices, (10.2) is still true. Since matrix multiplication can be subdivided into block submatrix operations (i.e. (10.1) is still true if $a_{i,j}$, $b_{i,j}$ and $c_{i,j}$ are replaced by block matrices), this immediately leads to a divide-and-conquer fast algorithm.

10.2.2 Divide-and-Conquer

To see this, consider the $2^n \times 2^n$ matrix multiplication $AB = C$, where A, B, C are all $2^n \times 2^n$ matrices. Using the usual definition, this requires $(2^n)^3 = 8^n$ multiplications and additions. But if A, B, C are subdivided into $2^{n-1} \times 2^{n-1}$ blocks $a_{i,j}, b_{i,j}, c_{i,j}$, then $AB = C$ becomes (10.1), which can be implemented with (10.2) since (10.2) does not require the products of subblocks of A and B to commute. Thus the $2^n \times 2^n$ matrix multiplication $AB = C$ can actually be implemented using only seven matrix multiplications of $2^{n-1} \times 2^{n-1}$ subblocks of A and B. And these subblock multiplications can in turn be broken down by using (10.2) to implement them as well. The end result is that the $2^n \times 2^n$ matrix multiplication $AB = C$ can be implemented using only 7^n multiplications, instead of 8^n.

The computational savings grow as the matrix size increases. For $n = 5$ (32×32 matrices) the savings is about 50%. For $n = 12$ (4096×4096 matrices) the savings is about 80%. The savings as a fraction can be made arbitrarily close to unity by taking sufficiently large matrices. Another way of looking at this is to note that $N \times N$ matrix multiplication requires $O(N^{\log_2 7}) = O(N^{2.807}) < N^3$ multiplications using Strassen.

Of course we are not limited to subdividing into $2 \times 2 = 4$ subblocks. Fast non-commutative algorithms for 3×3 matrix multiplication requiring only $23 < 3^3 = 27$ multiplications were found by exhaustive search in [2] and [3]; 23 is now known to be optimal. Repeatedly subdividing $AB = C$ into $3 \times 3 = 9$ subblocks computes a $3^n \times 3^n$ matrix multiplication in $23^n < 27^n$ multiplications; $N \times N$ matrix multiplication requires $O(N^{\log_3 23}) = O(N^{2.854})$ multiplications, so this is not quite as good as using (10.2). A fast noncommutative algorithm for 5×5 matrix multiplication requiring only $102 < 5^3 = 125$ multiplications was found in [4]; this also seems to be optimal. Using this algorithm, $N \times N$ matrix multiplication requires $O(N^{\log_5 102}) = O(N^{2.874})$ multiplications, so this is even worse. Of course, the idea is to write $N = 2^a 3^b 5^c$ for some a, b, c and subdivide into $2 \times 2 = 4$ subblocks a

times, then subdivide into $3 \times 3 = 9$ subblocks b times, etc. The total number of multiplications is then $7^a 23^b 102^c < 8^a 27^b 125^c = N^3$.

Note that we have not mentioned additions. Readers familiar with nesting fast convolution algorithms will know why; now we review why reducing multiplications is much more important than reducing additions when nesting algorithms. The reason is that at each nesting stage (reversing the divide-and-conquer to build up algorithms for multiplying large matrices from (10.2)), each scalar addition is replaced by a matrix addition (which requires N^2 additions for $N \times N$ matrices), and each scalar multiplication is replaced by a matrix multiplication (which requires N^3 multiplications *and additions* for $N \times N$ matrices). Although we are reducing N^3 to about $N^{2.8}$, it is clear that each multiplication will produce more multiplications *and additions* as we nest than each addition. So reducing the number of multiplications from eight to seven in (10.2) is well worth the extra additions incurred. In fact, the number of additions is also $O(N^{2.807})$.

The design of these base algorithms has been based on the theory of bilinear and trilinear forms. The review paper [5] and book [6] of Pan are good introductions to this theory. We note that reducing the exponent of N in $N \times N$ matrix multiplication is an area of active research. This exponent has been reduced to below 2.5; a known lower bound is two. However, the resulting algorithms are too complicated to be useful.

10.2.3 Arbitrary Precision Approximation (APA) Algorithms

APA algorithms are noncommutative algorithms for 2×2 and 3×3 matrix multiplication that require even fewer multiplications than the Strassen-type algorithms, but at the price of requiring longer word lengths. Proposed by Bini [7], the APA algorithm for multiplying two 2×2 matrices is this:

$$
\begin{aligned}
p_1 &= (a_{2,1} + \epsilon a_{1,2})(b_{2,1} + \epsilon b_{1,2}) ; \\
p_2 &= (-a_{2,1} + \epsilon a_{1,1})(b_{1,1} + \epsilon b_{1,2}) \\
p_3 &= (a_{2,2} - \epsilon a_{1,2})(b_{2,1} + \epsilon b_{2,2}) ; \\
p_4 &= a_{2,1}(b_{1,1} - b_{2,1}) ; \\
p_5 &= (a_{2,1} + a_{2,2})b_{2,1} \\
c_{1,1} &= (p_1 + p_2 + p_4)/\epsilon - \epsilon(a_{1,1} + a_{1,2})b_{1,2} ; \\
c_{2,1} &= p_4 + p_5; \\
c_{2,2} &= (p_1 + p_3 - p_5)/\epsilon - \epsilon a_{1,2}(b_{1,2} - b_{2,2}) .
\end{aligned}
\tag{10.3}
$$

If we now let $\epsilon \to 0$, the second terms in (10.3) become negligible next to the first terms, and so they need not be computed. Hence, three of the four elements of $C = AB$ may be computed using only five multiplications. $c_{1,2}$ may be computed using a sixth multiplication, so that, in fact, two 2×2 matrices may be multiplied to arbitrary accuracy using only six multiplications. The APA 3×3 matrix multiplication algorithm requires 21 multiplications. Note that APA algorithms improve on the exact Strassen-type algorithms ($6 < 7$, $21 < 23$).

The APA algorithms are often described as being numerically unstable, due to roundoff error as $\epsilon \to 0$. We believe that an electrical engineering perspective on these algorithms puts them in a light different from that of the mathematical perspective. In fixed point implementation, the computation $AB = C$ can be scaled to operations on integers, and the p_i can be bounded. Then it is easy to set ϵ a sufficiently small (negative) power of two to ensure that the second terms in (10.3) do not overlap the first terms, provided that the wordlength is long enough. Thus, the reputation for instability is undeserved. However, the requirement of large wordlengths to be multiplied seems also to have escaped notice; this may be a more serious problem in some architectures.

The divide-and-conquer and resulting nesting of APA algorithms work the same way as for the Strassen-type algorithms. $N \times N$ matrix multiplication using (10.3) requires $O(N^{\log_2(6)}) = O(N^{2.585})$ multiplications, which improves on the $O(N^{2.807})$ multiplications using (10.2). But the wordlengths are longer.

A design methodology for fast matrix multiplication algorithms by grouping terms has been proposed in a series of papers by Pan (see References [5] and [6]). While this has proven quite fruitful, the methodology of grouping terms becomes somewhat ad hoc.

10.2.4 Number Theoretic Transform (NTT) Based Algorithms

An approach similar in flavor to the APA algorithms, but more flexible, has been taken recently in [8]. First, matrix multiplication is reformulated as a linear convolution, which can be implemented as the multiplication of two polynomials using the z-transform. Second, the variable z is scaled, producing a scaled convolution, which is then made cyclic. This aliases some quantities, but they are separated by a power of the scaling factor. Third, the scaled convolution is computed using pseudo-number-theoretic transforms. Finally, the various components of the product matrix are read off of the convolution, using the fact that the elements of the product matrix are bounded. This can be done without error if the scaling factor is sufficiently large.

This approach yields algorithms that require the same number of multiplications or fewer as APA for 2×2 and 3×3 matrices. The multiplicands are again sums of scaled matrix elements as in APA. However, the design methodology is quite simple and straightforward, and the reason why the fast algorithm exists is now clear, unlike the APA algorithms. Also, the integer computations inherent in this formulation make possible the engineering insights into APA noted above.

We reformulate the product of two $N \times N$ matrices as the linear convolution of a sequence of length N^2 and a sparse sequence of length $N^3 - N + 1$. This results in a sequence of length $N^3 + N^2 - N$, from which elements of the product matrix may be obtained. For convenience, we write the linear convolution as the product of two polynomials. This result (of [8]) seems to be new, although a similar result is briefly noted in ([3], p. 197). Define

$$a_{i,j} = a_{i+jN}; \quad b_{i,j} = b_{N-1-i+jN}; \quad 0 \le i, j \le N-1$$

$$\left(\sum_{i=0}^{N-1} \sum_{j=0}^{N-1} a_{i+jN} x^{i+jN} \right) \left(\sum_{i=0}^{N-1} \sum_{j=0}^{N-1} b_{N-1-i+jN} x^{N(N-1-i+jN)} \right)$$

$$= \sum_{i=0}^{N^3+N^2-N-1} c_i x^i$$

$$c_{i,j} = c_{N^2-N+i+jN^2}; \quad 0 \le i, j \le N-1. \tag{10.4}$$

Note that coefficients of all three polynomials are read off of the matrices A, B, C column-by-column (each column of B is reversed), and the result is noncommutative. For example, the 2×2 matrix multiplication (10.1) becomes

$$\left(a_{1,1} + a_{2,1}x + a_{1,2}x^2 + a_{2,2}x^3\right) \left(b_{2,1} + b_{1,1}x^2 + b_{2,2}x^4 + b_{1,2}x^6\right)$$
$$= * + *x + c_{1,1}x^2 + c_{2,1}x^3 + *x^4 + *x^5 + c_{1,2}x^6 + c_{2,2}x^7 + *x^8 + *x^9, \tag{10.5}$$

where $*$ denotes an irrelevant quantity. In (10.5) substitute $x = sz$ and take the result $mod(z^6 - 1)$. This gives

$$\left(a_{1,1} + a_{2,1}sz + a_{1,2}s^2z^2 + a_{2,2}s^3z^3\right) \left((b_{2,1} + b_{1,2}s^6) + b_{1,1}s^2z^2 + b_{2,2}s^4z^4\right)$$
$$= (* + c_{1,2}s^6) + (*s + c_{2,2}s^7)z + (c_{1,1}s^2 + *s^8)z^2$$
$$+ (c_{2,1}s^3 + *s^9)z^3 + *z^4 + *z^5; \quad mod(z^6 - 1) \tag{10.6}$$

If $|c_{i,j}|, |*| < s^6$ then the $*$ and $c_{i,j}$ may be separated without error, since both are known to be integers. If s is a power of two, $c_{0,1}$ may be obtained by discarding the $6 \log_2 s$ least significant bits in the binary representation of $* + c_{0,1}s^6$. The polynomial multiplication $mod(z^6 - 1)$ can be computed using number-theoretic transforms [9] using six multiplications. Hence, 2×2 matrix multiplication requires six multiplications. Similarly, 3×3 matrices may be multiplied using 21 multiplications. Note these are the same numbers required by the APA algorithms, quantities multiplied are again sums of scaled matrix elements, and results are again sums in which one quantity is partitioned from another quantity which is of no interest.

However, this approach is more flexible than the APA approach (see [8]). As an extreme case, setting $z = 1$ in (10.5) computes a 2×2 matrix multiplication using ONE (very long wordlength) multiplication! For example, using $s = 100$

$$\begin{bmatrix} 2 & 4 \\ 3 & 5 \end{bmatrix} \begin{bmatrix} 9 & 8 \\ 7 & 6 \end{bmatrix} = \begin{bmatrix} 46 & 40 \\ 62 & 54 \end{bmatrix} \tag{10.7}$$

becomes the single scalar multiplication

$$(5, 040, 302)(8, 000, 600, 090, 007) = 40, 325, \mathbf{440}, 634, \mathbf{862}, \mathbf{462}, 114 . \tag{10.8}$$

This is useful in optical computing architectures for multiplying large numbers.

10.3 Wavelet-Based Matrix Sparsification

10.3.1 Overview

A common application of solving large linear systems of equations is the solution of integral equations arising in, say, electromagnetics. The integral equation is transformed into a linear system of equations using Galerkin's method, so that entries in the matrix and vectors of knowns and unknowns are coefficients of basis functions used to represent the continuous functions in the integral equation. Intelligent selection of the basis functions results in a sparse (mostly zero entries) system matrix. The sparse linear system of unknowns is then usually solved using an iterative algorithm, which is where the sparseness becomes an advantage (iterative algorithms require repeated multiplication of the system matrix by the current approximation to the vector of unknowns).

Recently, wavelets have been recognized as a good choice of basis function for a wide variety of applications, especially in electromagnetics. This is true because in electromagnetics the kernel of the integral equation is a 2-D or 3-D Green's function for the wave equation, and these are Calderon-Zygmund operators. Using wavelets as basis functions makes the matrix representation of the kernel drop off rapidly away from the main diagonal, more rapidly than discretization of the integral equation would produce.

Here we quickly review the wavelet transform as a representation of continuous functions and show how it sparsifies Calderon-Zygmund integral operators. We also provide some insight into why this happens and present some alternatives that make the sparsification less mysterious. We present our results in terms of continuous (integral) operators, rather than discrete matrices, since this is the proper presentation for applications, and also since similar results can be obtained for the explicitly discrete case.

10.3.2 The Wavelet Transform

We will not attempt to present even an overview of the rich subject of wavelets. The reader is urged to consult the many papers and textbooks (e.g., [10]) now being published on the subject. Instead, we restrict our attention to aspects of wavelets essential to sparsification of matrix operator representations.

The wavelet transform of an L^2 function $f(x)$ is defined as

$$f_i(n) = 2^{i/2} \int_{-\infty}^{\infty} f(x)\psi(2^i x - n)dx; \quad f(x) = \sum_i \sum_n f_i(n)\psi(2^i x - n)2^{i/2} \tag{10.9}$$

where $\{\psi(2^i x - n), i, n \in Z\}$ is a complete orthonormal basis for L^2. That is L^2 (the space of square-integrable functions) is spanned by dilations (scaling) and translations of a wavelet basis function $\psi(x)$. Constructing this $\psi(x)$ is nontrivial, but has been done extensively in the literature.

Since the summations must be truncated to finite intervals in practice, we define the wavelet scaling function $\phi(x)$ whose translations on a given scale span the space spanned by the wavelet basis function $\psi(x)$ at all translations and at scales coarser than the given scale. Then we can write

$$f(x) = 2^{I/2} \sum_n c_I(n) \phi(2^I x - n) + \sum_{i=I}^{\infty} \sum_n f_i(n) \psi(2^i x - n) 2^{i/2}$$

$$c_I(n) = 2^{I/2} \int_{-\infty}^{\infty} f(x) \phi(2^I x - n) dx \tag{10.10}$$

So the projection $c_I(n)$ of $f(x)$ on the scaling function $\phi(x)$ at scale I replaces the projections $f_i(n)$ on the basis function $\psi(x)$ on scales coarser (smaller) than I. The scaling function $\phi(x)$ is orthogonal to its translations but (unlike the basis function $\psi(x)$) is not orthogonal between scales. Truncating the summation at the upper end approximates $f(x)$ at the resolution defined by the finest (largest) scale i; this is somewhat analogous to truncating Fourier series expansions and neglecting high-frequency components.

We also define the 2-D wavelet transform of $f(x, y)$ as

$$f_{i,j}(m, n) = 2^{i/2} 2^{j/2} \int_{-\infty}^{\infty} \int_{-\infty}^{\infty} f(x, y) \psi(2^i x - m) \psi(2^j y - n) dx \, dy$$

$$f(x, y) = \sum_{i,j,m,n} f_{i,j}(m, n) \psi(2^i x - m) \psi(2^j y - n) 2^{i/2} 2^{i/2} \tag{10.11}$$

However, it is more convenient to use the 2-D counterpart of (10.10), which is

$$c_I(m, n) = 2^I \int_{-\infty}^{\infty} \int_{-\infty}^{\infty} f(x, y) \phi(2^I x - m) \phi(2^I y - n) dx \, dy$$

$$f_i^1(m, n) = 2^i \int_{-\infty}^{\infty} \int_{-\infty}^{\infty} f(x, y) \phi(2^i x - m) \psi(2^i y - n) dx \, dy$$

$$f_i^2(m, n) = 2^i \int_{-\infty}^{\infty} \int_{-\infty}^{\infty} f(x, y) \psi(2^i x - m) \phi(2^i y - n) dx \, dy$$

$$f_i^3(m, n) = 2^i \int_{-\infty}^{\infty} \int_{-\infty}^{\infty} f(x, y) \psi(2^i x - m) \psi(2^i y - n) dx \, dy$$

$$f(x, y) = \sum_{m,n} c_I(m, n) \phi(2^I x - m) \phi(2^I y - n) 2^I$$

$$+ \sum_{i=I}^{\infty} \sum_{m,n} f_i^1(m, n) \phi(2^i x - m) \psi(2^i y - n) 2^i$$

$$+ \sum_{i=I}^{\infty} \sum_{m,n} f_i^2(m, n) \psi(2^i x - m) \phi(2^i y - n) 2^i$$

$$+ \sum_{i=I}^{\infty} \sum_{m,n} f_i^3(m, n) \psi(2^i x - m) \psi(2^i y - n) 2^i . \tag{10.12}$$

Once again the projection $c_I(m, n)$ on the scaling function at scale I replaces all projections on the basis functions on scales coarser than M.

Some examples of wavelet scaling and basis functions:

| Scaling | pulse | B-spline | sinc | softsinc | Daubechies |
| Wavelet | Haar | Battle-Lemarie | Paley-Littlewood | Meyer | Daubechies |

An important property of the wavelet basis function $\psi(x)$ is that its first k moments can be made zero, for any integer k [10]:

$$\int_{-\infty}^{\infty} x^i \psi(x)dx = 0, \quad i = 0 \ldots k \tag{10.13}$$

10.3.3 Wavelet Representations of Integral Operators

We wish to use wavelets to sparsify the L^2 integral operator $K(x, y)$ in

$$g(x) = \int_{-\infty}^{\infty} K(x, y) f(y)dy \tag{10.14}$$

A common situation: (10.14) is an integral equation with known kernel $K(x, y)$ and known $g(x)$ in which the goal is to compute an unknown function $f(y)$. Often the kernel $K(x, y)$ is the Green's function (spatial impulse response) relating observed wave field or signal $g(x)$ to unknown source field or signal $f(y)$.

For example, the Green's function for Laplace's equation in free space is

$$G(r) = -\frac{1}{2\pi} \log r \quad (2D); \quad \frac{1}{4\pi r} \quad (3D) \tag{10.15}$$

where r is the distance separating the points of source and observation. Now consider a line source in an infinite 2-D homogeneous medium, with observations made along the same line. The observed field strength $g(x)$ at position x is

$$g(x) = -\frac{1}{2\pi} \int_{-\infty}^{\infty} \log |x - y| f(y)dy \tag{10.16}$$

where $f(y)$ is the source strength at position y.

Using Galerkin's method, we expand $f(y)$ and $g(x)$ as in (10.9) and $K(x, y)$ as in (10.11). Using the orthogonality of the basis functions yields

$$\sum_j \sum_n K_{i,j}(m, n) f_j(n) = g_i(m) \tag{10.17}$$

Expanding $f(y)$ and $g(x)$ as in (10.10) and $K(x, y)$ as in (10.12) leads to another system of equations which is difficult notationally to write out in general, but can clearly be done in individual applications. We note here that the entries in the system matrix in this latter case can be rapidly generated using the fast wavelet algorithm of Mallat (see [10]).

The point of using wavelets is as follows. $K(x, y)$ is a *Calderon-Zygmund* operator if

$$|\frac{\partial^k}{\partial x^k} K(x, y)| + |\frac{\partial^k}{\partial y^k} K(x, y)| \le \frac{C_k}{|x - y|^{k+1}} \tag{10.18}$$

for some $k \ge 1$. Note in particular that the Green's functions in (10.15) are Calderon-Zygmund operators. Then the representation (10.12) of $K(x, y)$ has the property [11]

$$|f_i^1(m, n)| + |f_i^2(m, n)| + |f_i^3(m, n)| \le \frac{C_k}{1 + |m - n|^{k+1}}, \quad |m - n| > 2k \tag{10.19}$$

if the wavelet basis function $\psi(x)$ has its first k moments zero (10.13).

This means that using wavelets satisfying (10.13) *sparsifies* the matrix representation of the kernel $K(x, y)$. For example, a direct discretization of the 3-D Green's function in (10.15) decays as $1/|m - n|$ as one moves away from the main diagonal $m = n$ in its matrix representation. However, using wavelets, we can attain the much faster decay rate $1/(1 + |m - n|^{k+1})$ far away from the main diagonal. By neglecting matrix entries less than some threshold (typically 1% of the largest entry) a sparse and mostly banded matrix is obtained. This greatly speeds up the following matrix computations:

1. Multiplication by the matrix for solving the forward problem of computing the response to a given excitation (as in (10.16));
2. Fast solution of the linear system of equations for solving the inverse problem of reconstructing the source from a measured response (solving (10.16) as an integral equation). This is typically performed using an iterative algorithm such as conjugate gradient method. Sparsification is essential for convergence in a reasonable time.

A typical sparsified matrix from an electromagnetics application is shown in Figure 6 of [12]. Battle-Lemarie wavelet basis functions were used to sparsify the Galerkin method matrix in an integral equation for planar dielectric millimeter-wave waveguides and a 1% threshold applied (see [12] for details). Note that the matrix is not only sparse but (mostly) banded.

10.3.4   Heuristic Interpretation of Wavelet Sparsification

Why does this sparsification happen? Considerable insight can be gained using (10.13). Let $\hat{\psi}(\omega)$ be the Fourier transform of the wavelet basis function $\psi(x)$. Since the first k moments of $\psi(x)$ are zero by (10.13) we can expand $\hat{\psi}(\omega)$ in a power series around $\omega = 0$:

$$\hat{\psi}(\omega) \approx \omega^k; \quad |\omega| << 1 . \tag{10.20}$$

This shows that for small $|\omega|$ taking the wavelet transform of $f(x)$ is roughly equivalent to taking the kth derivative of $f(x)$. This can be confirmed that many wavelet basis functions bear a striking resemblance to the impulse responses of regularized differentiators. Since $K(x, y)$ is assumed a Calderon-Zygmund operator, its kth derivatives in x and y drop off as $1/|x - y|^{k+1}$. Thus, it is not surprising that the wavelet transform of $K(x, y)$, which is roughly taking k^{th} derivatives, should drop off as $1/|m - n|^{k+1}$. Of course there is more to it, but this is why it happens.

It is not surprising that $K(x, y)$ can be sparsified by taking advantage of its derivatives being small. To see a more direct way of accomplishing this, apply integration by parts to (10.14) and take the partial derivative with respect to x. This gives

$$\frac{dg(x)}{dx} = -\int_{-\infty}^{\infty} \left(\frac{\partial}{\partial x} \frac{\partial}{\partial y} K(x, y) \right) \left(\int_{-\infty}^{y} f(y')dy' \right) dy \tag{10.21}$$

which will likely sparsify a smooth $K(x, y)$. Of course, higher derivatives can be used until a condition like (10.18) is reached. The operations of integrating $f(y)$ and $\frac{\partial^k g}{\partial x^k}$ (to get $g(x)$) k times can be accomplished using $nk << n^2$ additions, so considerable savings can result. This is different from using wavelets, but in the same spirit.

References

[1] Strassen, V., Gaussian elimination is not optimal, *Numer. Math.*, 13: 354–356, 1969.
[2] Landerman, J.D., A noncommutative algorithm for multiplying 3 × 3 matrices using 23 multiplications, *Bull. Am. Math. Soc.*, 82: 127–128, 1976.

[3] Johnson, R.W. and McLoughlin, A.M., Noncommutative bilinear algorithms for 3 × 3 matrix multiplication, *SIAM J. Comput.,* 15: 595–603, 1976.

[4] Makarov, O.M., A noncommutative algorithm for multiplying 5 × 5 matrices using 102 multiplications, *Inform. Proc. Lett.,* 23: 115–117, 1986.

[5] Pan, V., How can we speed up matrix multiplication? *SIAM Rev.,* 26(3): 393–415, 1984.

[6] Pan, V., *How Can We Multiply Matrices Faster?,* Springer-Verlag, New York, 1984.

[7] Bini, D., Capovani, M., Lotti, G. and Romani, F., $O(n^{2.7799})$ complexity for matrix multiplication, *Inform. Proc. Lett.,* 8: 234–235, 1979.

[8] Yagle, A.E., Fast algorithms for matrix multiplication using pseudo number theoretic transforms, *IEEE Trans. Signal Process.,* 43: 71–76, 1995.

[9] Nussbaumer, H.J., *Fast Fourier Transforms and Convolution Algorithms,* Springer-Verlag, Berlin, 1982.

[10] Daubechies, I., *Ten Lectures on Wavelets,* SIAM, Philadelphia, PA, 1992.

[11] Beylkin, G., Coifman, R. and Rokhlin, V., Fast wavelet transforms and numerical algorithms I, *Comm. Pure Appl. Math.,* 44: 141–183, 1991.

[12] Sabetfakhri, K. and Katehi, L.P.B., Analysis of integrated millimeter wave and submillimeter wave waveguides using orthonormal wavelet expansions, *IEEE Trans. Microwave Theor. Technol.,* 42: 2412–2422, 1994.

IV

Digital Filtering

Lina J. Karam
Arizona State University
James H. McClellan
Georgia Institute of Technology

D IGITAL FILTERING is one of the most important functions in digital signal processing, and this single chapter not only provides a thorough coverage of conventional topics such as FIR and IIR filtering, it also presents material on design methods and new research directions that have not been widely available in the open literature.

Karam and McClellan present an "Introduction to Digital Filtering", followed by Karam's "Steps in Filter Design". A comprehensive coverage of FIR and IIR classical filter design follows in "FIR Design Methods" by Selesnick, Burrus, Karam, and McClellan, and "IIR Design Methods" by Karam, Selesnick and Burrus, respectively. Unique to this chapter is the special discussion of "Software Tools" for filtering by McClellan.

The various topics covered in this section are integrated together very well, ensuring a coherent and authoritative coverage of the filtering area.

11

Digital Filtering

Lina J. Karam
Arizona State University

James H. McClellan
Georgia Intitute of Technology

Ivan W. Selesnick
Polytechnic University

C. Sidney Burrus
Rice University

11.1 Introduction

Digital filters are widely used in processing digital signals of many diverse applications, including speech processing and data communications, image and video processing, sonar, radar, seismic and oil exploration, and consumer electronics. One class of digital filters, the linear shift-invariant (LSI) type, are the most frequently used because they are simple to analyze, design, and implement. This chapter treats the LSI case only; other filter types, such as adaptive filters, require quite different design methodologies.

An LSI digital filter can be uniquely identified in the time/space domain by its impulse response $h(n)$ (where n is an integer index). Alternatively, the LSI digital filter can be uniquely characterized in the frequency domain by its frequency response $H(\omega)$ (where ω is a real-valued frequency variable in radians), which is also the Discrete-Time Fourier Transform (DTFT) of the sequence $h(n)$. LSI digital filters are of two main types: Finite-duration Impulse Response (FIR) filters for which the impulse response $h(n)$ is non-zero for only a finite number of samples, and Infinite-duration Impulse Response (IIR) filters for which $h(n)$ has an infinite number of non-zero samples. In the FIR case, the samples of the sequence $h(n)$ are commonly referred to as the filter coefficients; for the IIR case, the filter coefficients include feedback terms in a difference equation.

Digital filter design has been extensively addressed within the last 25 years. The design and realization of digital filters involve a blend of theory, applications, and technologies. For most applications, it is desirable to design frequency-selective filters which alter or pass unchanged different frequency components. In this case, the desired design specifications are given in the frequency domain by specifying a desired frequency response $D(f)$. Note that $D(f)$ is, in general, complex valued, consisting of a desired magnitude response $|D(f)|$ and a desired phase response $\angle D(f)$. One of the most important problems is the design of a highly frequency-selective filter with sharp cutoff edges (short transition bands). However, ideal sharp edges correspond mathematically to discontinuities and cannot be realized in practice. Therefore, the filter design problem consists in finding an implementable filter whose order is low and whose frequency response $H(f)$ best approximates the specified ideal magnitude and phase responses which are given as the desired design specifications or constraints.

The design of digital filters is typically done by performing the following steps:

1. Convert the desired design constraints into precise specifications of the desired magnitude and phase responses, designed filter type (FIR or IIR), filter order, error tolerance, or criteria.

2. Approximate the design specifications (of Step 1) by finding the implementable FIR or IIR filter such that the obtained filter frequency response best meets the design specs according to a mathematical error criterion.

3. Realize the filter using the digital technology most suitable for the considered application.

While Step 2 is performed using mathematical optimization and approximation methods, Step 1 is highly dependent on the application and the detail provided by the user. Step 3 depends on the technology or software used to build the filter.

Nowadays, the optimization needed in Step 2 is usually done with computer software that implements sophisticated numerical optimization routines. In addition, these design packages usually have a convenient graphical user interface to aid in the conversion of specs needed in Step 1. With such software, a filter design can be carried out quickly so that many designs can be tried in the process of getting the best filter. Since most filter design techniques involve the trade-off among competing parameters, the software can also incorporate design rules that allow the user to predict the order needed for certain specs without actually designing the filter, for example.

This chapter is organized as follows. Section 11.2 provides a discussion of Steps 1 and 3, including creating the design specifications, selecting the filter type and order, specifying the error tolerances and criteria, and realizing the designed filter. Step 2 is treated in Sections 11.3 and 11.4. Section 11.3 describes the classical FIR and IIR design methods. Section 11.4 presents nonclassic and more recently developed design methods with added efficiency and/or flexibility. Finally, Section 11.5 gives examples of some of the currently available software design tools and describes the characteristics that a user can expect from such tools.

11.2   Steps in Filter Design

Lina J. Karam

The general filter design problem can be briefly stated as follows. Given some ideal frequency response, $D(\omega)$, find a realizable IIR or FIR digital filter whose frequency response, $H(\omega)$, approximates $D(\omega)$. The realizable filter is found by optimizing some measure of the filter's performance, e.g., minimizing the filter order (IIR) or the filter length (FIR), or minimizing the width of the transition bands, or reducing the passband error and/or stopband error. Setting up the specifications for the general filter design problem will define these parameters and show which trade-offs are possible.

11.2.1   Creating the Design Specifications

Since the frequency response of a digital filter is always periodic in the frequency variable ω with a period of 2π, the design specifications need only be specified for one period; usually, over the frequency region $[-\pi, \pi]$. Furthermore, when the frequency response is conjugate-symmetric (i.e., $D^*(\omega) = D(-\omega)$), then it is sufficient to specify the response only on the positive frequency interval $[0, \pi]$. The conjugate-symmetric case is the most common, because it corresponds to filters with real coefficients.

The simplest case is that of an ideal low-pass digital filter with zero phase, whose frequency response can be expressed as:

$$D(\omega) = \begin{cases} 1, & |\omega| < \omega_c \\ 0, & \omega_c < |\omega| < \pi \end{cases} \tag{11.1}$$

where ω_c is the cutoff frequency corresponding to the location of a sharp cutoff edge, as shown in Fig. 11.1(a). In this case, the frequency response, $D(\omega)$, is real-valued and, therefore, corresponds also to the magnitude response of the filter (since the phase is zero). Ideal frequency responses of other commonly used frequency-selective filters are shown in Fig. 11.1.

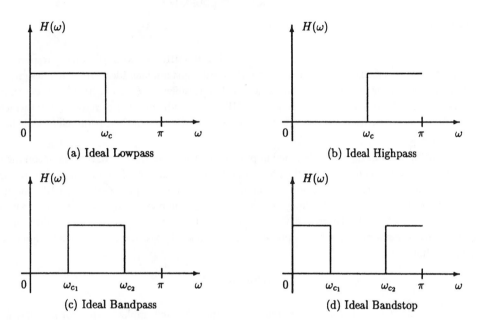

FIGURE 11.1   Common ideal digital filter types.

These ideal filters have frequency responses with sharp cutoff edges (discontinuities) and cannot be implemented directly. They must be approximated with a realizable system—the sharp cutoff edges need to be replaced with *transition bands* in which the designed frequency response would change smoothly in going from one band to the other. So, *design templates* need to be provided where the sharp cutoff edges are replaced with non-zero width transition bands located around the ideal cutoff edges. A typical design template for a lowpass filter is shown in Fig. 11.2, where:

- ω_p is the passband cutoff frequency.
- ω_s is the stopband cutoff frequency. The cutoff frequency ω_c is usually taken to be midway between the passband and stopband cutoff frequencies.
- The open interval (ω_p, ω_s) is the transition band of width $\Delta\omega_t = \omega_s - \omega_p$. In the common design methods, no design specifications are given in the transition bands which are therefore commonly known as "don't care bands." However, it is usually desirable to have the frequency response change smoothly (i.e., no fluctuations or overshoots) in the transition bands; this requirement might not be satisfied by a design method that places no design constraints on the frequency response in the transition bands.
- δ_p is known as the passband ripple and is the maximum allowable error in the passband.
- δ_s is known as the stopband ripple and is the maximum allowable error in the stopband.

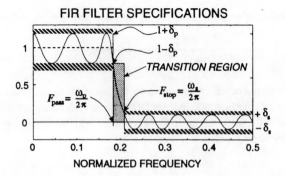

FIGURE 11.2   Design template for a lowpass filter.

The objective of filter design then is to find a realizable FIR or IIR filter whose frequency response $H(\omega)$ approximates the specified design constraints given by the design template. Ideally, the filter design process would make each of the following parameters as small as possible: δ_p, δ_s, $\Delta\omega_t$, IIR filter order (number of poles of $H(z)$ which is a rational function) or FIR filter length (number of zeros of $H(z)$ which is a finite polynomial). Practically, the filter design process minimizes one of these parameters while holding the others fixed.

Traditionally, many of the filters designed in practice are specified in terms of constraints on the magnitude response and no constraints on the phase response other than those imposed implicitly by stability and/or causality requirements (e.g., poles inside unit circle in the complex Z-plane for IIR, and linear-phase for FIR [1]). More recently, design methods that include phase design specifications have been presented [2, 3, 4, 5]. In this latter case, two design templates must be provided, one for the magnitude response and another for the (passband) phase response. An ideal phase response is most likely a constant slope phase function:

$$\angle D(\omega) = -M\omega$$

The parameter M is equivalent to the desired delay of the filter (in samples). An error template for the phase would be a tolerance about the desired phase, e.g., δ_ϕ would denote the maximum allowable phase ripple, so that we require

$$|\angle H(\omega) - \angle D(\omega)| < \delta_\phi$$

11.2.2   Specs Derived from Analog Filtering

Often, the desired design specifications are not given directly in the digital domain. Instead, an equivalent analog filtering operation is desired but is to be performed using an embedded digital filter. Figure 11.3 shows a standard system for processing continuous-time (-space) signals using a digital filter. The analog input signal is first transformed into a digital signal through an analog-to-digital (A/D) conversion operation; then, filtering is carried out using a digital filter; finally, the filtered digital output is converted back to the analog domain using a digital-to-analog (D/A) converter. For this system, if the sampling period T_s of the A/D and D/A converters is chosen appropriately to avoid aliasing of the input spectrum, the overall system (consisting of the A/D converter, the digital filter, and the D/A converter) behaves as an equivalent *analog* filter. In this case, the frequency response $H_a(\Omega)$ of the equivalent analog filter is related to the frequency response $H(\omega)$ of the digital filter through a simple linear scaling relation between the digital frequency ω and the analog frequency Ω. This linear scaling relation is given by

$$\omega = \Omega T_s \tag{11.2}$$

leading to the following expression of the analog $H_a(\Omega)$ in terms of the digital $H(\omega)$:

$$H_a(\Omega) = \begin{cases} H(\Omega T_s), & |\Omega| < \dfrac{\pi}{T_s} \\ \\ 0, & |\Omega| \geq \dfrac{\pi}{T_s} \end{cases} \tag{11.3}$$

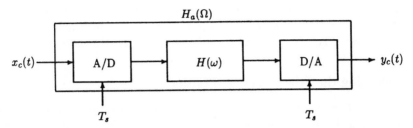

FIGURE 11.3   Standard system for processing analog signals using a digital (discrete-time) filter.

Equivalently, $H(\omega)$ can also be expressed in terms of $H_a(\Omega)$ as follows:

$$H(\omega) = H_a(\omega/T_s), \qquad |\omega| < \pi. \tag{11.4}$$

A typical filter design problem corresponding to this system is to design the digital filter such that the overall equivalent analog filter best approximates some ideal analog specifications. So, if we are given the desired analog specifications of the overall analog system, these can be turned into specifications for the desired digital filter by using Eq. (11.4). Then, a digital filter $H(\omega)$ can be designed to approximate the derived desired digital specifications. Finally, the resulting analog frequency response of the overall system can be found using Eq. (11.3), for example, to compare with the ideal analog response.

11.2.3 Specifying an Error Measure

An error measure is needed to assess how much the designed filter $H(\omega)$ deviates from the desired filter $D(\omega)$. Defining the pointwise error $E(\omega)$ as:

$$E(\omega) = [D(\omega) - H(\omega)], \tag{11.5}$$

we must reduce $E(\omega)$ to a scalar error measure (also called an error norm). With a correctly chosen norm, there are many possible optimization algorithms that will compute the best filter parameters to minimize the chosen error norm. The following error norms are the most commonly used in filter design:

- *Mean Squared Error (MSE)* or L_2 *norm*

$$E_2 = \left[\frac{1}{2\pi} \int_B |E(\omega)|^2 \, d\omega \right]^{1/2} \tag{11.6}$$

- L_p *norm* which is a generalization of the L_2 norm and where p is a non-zero integer

$$E_p = \left[\frac{1}{2\pi} \int_B |E(\omega)|^p \, d\omega \right]^{1/p} \tag{11.7}$$

- *Chebyshev* or L_∞ *norm*

$$E_\infty = \max_{\omega \in B} |E(\omega)| \tag{11.8}$$

The Chebyshev error norm limits the worst case deviation from the ideal specifications.

In the above definitions, $|\cdot|$ denotes the complex error magnitude and B is the frequency region of interest over which the error norm is to be minimized. The frequency subset $B \subset [-\pi, \pi)$ is taken to be the union of the desired passbands and stopbands.

A more selective control of the approximation accuracy can be achieved by introducing a weighting function $W(\omega)$ in Eq. (11.5) as follows:

$$E(\omega) = W(\omega)[D(\omega) - H(\omega)]. \tag{11.9}$$

The weighting function $W(\omega)$ must be a real, strictly positive and continuous function on B. It can force a better match over selected regions or frequency points relative to other regions in B. Alternatively, note that Eq. (11.5) reduces to Eq. (11.9) if we replace $D(\omega)$ with $W(\omega)D(\omega)$ and $H(\omega)$ with $W(\omega)H(\omega)$.

11.2.4　Selecting the Filter Type and Order

As mentioned in Section 11.1, there are two main types of filters, namely FIR and IIR. These differ in their characteristics and in the way they are designed. Since the design algorithm depends strongly on the choice of IIR vs. FIR filter, the designer should make this decision as early as possible. Although the desired frequency response specifications can be approximated with either type of filter, deciding which of the two filter types to use depends on many factors including the implementation hardware, as well as the magnitude and phase characteristics of the resulting filter. To aid in this decision, the main characteristics of FIR and IIR filters are discussed below.

11.2.4.1　FIR Characteristics

1. The impulse response $h(n)$ has a finite length, i.e., $h(n)$ is non-zero only for a finite range of indices n. For a general N-length FIR system, $h(n) \neq 0$ only for $N_1 \leq n \leq N_2 = (N_1 + N - 1)$. When $N_1 \geq 0$, the filter is also causal.

2. The FIR frequency response $H(\omega)$ is *a finite-degree polynomial* in $e^{j\omega}$ of the form

$$H(\omega) = \sum_{n=N_1}^{N_2} h(n)(e^{j\omega})^{-n} \tag{11.10}$$

   where N_1 and N_2 are (negative or positive) integers corresponding to the indices of the first and last samples of $h(n)$, respectively. The N impulse response samples are the free parameters of the design procedure. This form is general enough to represent non-causal filters such as zero-phase filters.

3. Designing an FIR filter consists in finding the polynomial $H(\omega)$ that best approximates the design specifications. This is done by computing the "optimal" (relative to some criteria) impulse response samples $\{h(n)\}_{n=N_1}^{N_2}$, which correspond to the unknown coefficients of the polynomial $H(\omega)$. The impulse response length N is usually fixed, but it could also be considered as a free parameter to be optimized. Procedures for designing FIR filters are given in Sections 11.3.1 and 11.4.1.

4. The filter transfer function, denoted by $H(z)$, is the z-transform of $h(n)$ and is useful for studying the stability of the system. For FIR filters, $H(z)$ is a finite-degree polynomial in the complex variable z and is given by

$$H(z) = H(e^{j\omega})|_{e^{j\omega}=z} = \sum_{n=N_1}^{N_2} h(n)z^{-n}. \tag{11.11}$$

It follows that the function $H(z)$ has no poles except possibly at 0 or ∞, i.e., it cannot be infinite for any point z with $0 < |z| < \infty$. It has only zeros (points z at which $H(z) = 0$). Therefore, an FIR filter is *always stable*.

5. FIR filters allow the design of *causal linear-phase* systems which are very important and widely used in practice. In fact, in many signal processing applications, such as speech and image processing, it is desirable to pass some portion of the signal frequency band with minimal distortion. For that purpose, linear-phase systems are particularly desirable since the effect of the linear-phase is a pure time delay. For a more detailed discussion of linear-phase systems, the reader is referred to [1].

6. Because the impulse response is of finite length, FIR filters are realized using the *convolution* operation [1] which can be implemented directly in the time/space domain, or in terms of the FFT in the frequency domain. More details about the implementation will be given in Section 11.2.6.

7. Since FIR filters have no feedback loops, they are relatively insensitive to round-off noise. Noise due to coefficient quantization can be a problem for very long filters, but can be mitigated by avoiding the direct-form structures, and using special structures such as the cascade form for implementation.

8. FIR filters with very long impulse responses ($N \approx 500$) might be required to meet certain design specifications, e.g., high accuracy and/or short transition bands. Longer filters lead to an increased complexity for both design and implementation. They require significant computing time to optimize all the parameters $h(n)$, and also many operations per second in the actual filter implementation.

9. The trade-off among the filter design parameters has been determined empirically for some types of FIR designs. The following simple (approximate) formula shows the relationship among the ripples, bandedges, and filter length (N) for one method, the Parks-McClellan algorithm (Section on page **11**-21):

$$(N - 1)\Delta\omega \approx \frac{-20 \log_{10} \sqrt{\delta_p \delta_s} - 13}{2.324}$$

where $\Delta\omega = \omega_s - \omega_p$ is the transition width. This formula allows the designer to predict the value of N that will be needed to satisfy specs given for $\{\omega_p, \omega_s, \delta_p, \delta_s\}$. Other design formulas are given in Section 11.3.1.

11.2.4.2 IIR Characteristics

1. The impulse response $h(n)$ has an infinite number of non-zero samples (infinite length). As an example, for a general IIR filter, $h(n) \neq 0$ only for $N_o \leq n \leq \infty$, where N_o is a non-negative integer (commonly, N_o is taken to be 0; in this case, the filter is said to be *causal*).

2. The frequency response $H(\omega)$ is a *rational function*, i.e., a ratio of two finite-degree polynomials in $e^{j\omega}$ of the form

$$H(\omega) = \frac{B(\omega)}{A(\omega)} = e^{-j\omega N_o} \frac{\sum_{k=0}^{M} b_k e^{-j\omega k}}{\sum_{k=0}^{N} a_k e^{-j\omega k}} \tag{11.12}$$

where N_o is an integer constant. The order of an IIR filter is equal to N, which is the degree of the denominator in Eq. (11.12); usually the degree of the numerator M is no greater than N. The order N also determines the number of previous output samples that need to be stored and then fed back to compute the current output sample. Therefore, IIR systems are also known as *feedback systems*. The filter coefficients $\{b_n\}$ and $\{a_n\}$ in Eq. (11.12) correspond to the unknown (free) parameters of the design.

3. Designing an IIR filter amounts to finding the rational function $H(\omega)$ that best approximates the design specifications. In the frequency domain, this is done by computing the "optimal" (relative to some criteria) coefficients $\{b_n\}$ and $\{a_n\}$ in Eq. (11.12) for the rational function $H(\omega)$. The filter order N is usually fixed, but can also be considered as a free parameter to be optimized. Procedures for designing IIR filters are given in Sections 11.3.2 and 11.4.2.

4. As mentioned previously in Section on page **11**-6, the filter transfer function, denoted by $H(z)$, is the z-transform of $h(n)$ and is useful for studying the stability of the system. In the context of LSI filters, stability implies that a bounded input to the filter will always result in a bounded output. For IIR filters, $H(z)$ is a rational function in the complex variable z and is given by

$$H(z) = H\left(e^{j\omega}\right)|_{e^{j\omega}=z} = z^{-N_o} \frac{\sum_{k=0}^{M} b_k z^{-k}}{\sum_{k=0}^{N} a_k z^{-k}} \tag{11.13}$$

The roots of the denominator polynomial are poles of the function $H(z)$, i.e., $H(z)$ is infinite at points z with $0 \leq |z| < \infty$. Stability then requires that no poles lie on the Unit Circle (U.C.) ($|z| = 1$) in the z-plane. Causality and stability require that the poles lie inside the U.C. in the z-plane. So, it is possible to obtain a resulting IIR filter that is unstable. Also, coefficient quantization noise might severely affect the response of the filter and its stability by disturbing the poles locations and by driving some of the poles closer to or onto the U.C.

5. It is not possible to design *causal linear-phase* IIR filters. The resulting IIR causal realizable filters must have a non-linear phase response. Forward-backward filtering can be used as an implementation to approximate a zero-phase response [1].

6. Because the impulse response is infinitely long, convolution can no longer be used to implement the IIR filters. Instead, IIR filters are efficiently implemented using *feedback difference equations* as described in Section 11.2.6.

7. The noise characteristics of an IIR filter can be a major consideration when doing an implementation, especially in fixed-point arithmetic. Coefficient quantization degrades the actual filter response from that designed by high-precision software. More critical is round-off noise sensitivity which can be amplified by the feedback loops in the filter.

8. Compared to FIR filters, IIR filters can achieve the desired design specifications with a relatively low order (as few as 4 to 6 poles). So, fewer unknown parameters need to be computed and stored, which might lead to a lower design and implementation complexity. However, the phase response of IIR filters is never linear, which leads to the use of all-pass filters to compensate the group delay, and thus raises the order of the filter and the complexity of the design process.

9. IIR filters are commonly designed by using closed-form design formulas corresponding to classical filter types. While for FIR filters the length-estimating formulas are only approximate, the order-estimating formulas for IIR filters are exact since they are derived from the mathematical properties of the classical prototypes. These formulas are very useful to obtain the IIR filter order needed to satisfy the desired design specifications.

11.2.5 Designing the Filter

After the designed filter type (FIR or IIR) is specified, a suitable design procedure can be selected depending on the chosen filter type. Popular design procedures are based on computing the unknown filter parameters by optimizing one of the error criteria indicated in Section 11.2.3.

For FIR filters, the two main classical methods are the *windowing method* [1] and the *Parks-McClellan (Remez) algorithm* [6]. The windowing method minimizes the MSE when a rectangular window (corresponding to pure truncation of the ideal impulse response) is used at the expense of possible large overshoots near the band edges and large ripples in the resulting frequency response. It is suboptimal when other general windows are used. However, the edge overshoot, transition width, and ripple height can be controlled by using different types of windows as described in Section on page **11**-12. The Parks-McClellan (Remez) algorithm minimizes the Chebyshev (L_∞) error norm resulting in optimal equiripple designs. However, the original Parks-McClellan algorithm is restricted to the design of linear-phase filters with a symmetric magnitude response. An extension of this algorithm that allows the design of optimal FIR filters with arbitrary magnitude and phase specifications has been presented by Karam and McClellan in [2, 3]. Linear-programming-based [4, 7] and Constrained least square [8] optimization methods also have been presented to allow the inclusion of additional important design constraints. These and other FIR design procedures are described in Sections 11.3.1 and 11.4.1.

While the design of FIR filters is typically performed directly in the digital domain, IIR filters are commonly designed by transforming the digital design specifications into analog design specifications and performing the filter design in the analog domain. The resulting analog filter is then transformed into a digital filter using a suitable transformation. One important classical IIR design method is the *Bilinear Transformation method*. Digital-only IIR design methods have also been presented. A description of IIR design procedures is given in Sections 11.3.2 and 11.4.2.

11.2.6 Realizing the Designed Filter

Realizing the designed digital filter corresponds to computing the output of the filter in response to any given input. For LSI filters, this is simplified by the fact that the input and output signals are related through a simple convolution operation in the time/space domain. If $x(n)$ is the input, $y(n)$ the corresponding output, and $h(n)$ the impulse response of the LSI filter, then this relation is given by

$$y(n) = h(n) * x(n) = \sum_{k=N_1}^{N_2} h(k)x(n-k), \qquad (11.14)$$

where N_1 and N_2 are the indices of the first and last non-zero samples of $h(n)$. In the frequency (Fourier transform) domain, the convolution relation (11.14) corresponds to a multiplication of the respective Fourier transforms:

$$Y(\omega) = H(\omega)X(\omega) \qquad (11.15)$$

where $X(\omega)$, $H(\omega)$, and $Y(\omega)$ are the DTFT of $x(n)$, $h(n)$, and $y(n)$, respectively. The variable ω in Eq. (11.15) is continuous and, therefore, Eq. (11.15) cannot be implemented in practice. An implementable version of Eq. (11.15) is obtained by using the Discrete Fourier Transform (DFT), which is a sampled version of the DTFT and which consists of samples of the DTFT evaluated at the points $\omega = (2\pi k/N_{\text{DFT}})$, $k = 0, \ldots, (N-1)$. N_{DFT} is the size of the DFT and corresponds to the number of sample points within the period 2π. It is a known fact that the time/space digital signal can be exactly recovered from its DFT if N_{DFT} is chosen to be greater than or equal to the length of the time/space signal. Using the DFT, Eq. (11.15) becomes

$$Y(k) = H(k)X(k), \qquad k = 0, \ldots, N_{\text{DFT}} \qquad (11.16)$$

where $N_{\text{DFT}} \geq \max\{\text{length of } x(n) + \text{length of } h(n) - 1\}$ in order to perform the pointwise multiplication. The DFT can be computed very efficiently using the Fast Fourier Transform (FFT) algorithm.

11.2.6.1 Realizing FIR Filters

For FIR filters, the impulse response has a finite length and, therefore, N_1 and N_2 in Eq. (11.14) are finite. Also, in this case, a finite-size DFT is sufficient to exactly represent $h(n)$ ($N_{\text{DFT}} \geq (N_2 - N_1 + 1)$). Consequently, for finite-length input signals $x(n)$, Eq. (11.14) or Eq. (11.16) can be directly used to realize the designed FIR filter in software or hardware. Commonly, the FIR filter coefficients $h(n)$ (or the DFT values if Eq. (11.16) is used) are quantized to the precision of the processor or chip, stored, and used as in Eq. (11.14) to realize the designed FIR filter. While for Eq. (11.14) the storage can be fixed to the size of $h(n)$ and is independent of the input, the size of the DFTs in Eq. (11.16) and, therefore, the needed storage vary with the size of the input signal. To overcome this problem and to handle the processing of large-size signals, block-based convolution (also known as sectioned or high-speed convolution) is used where the input signal is divided into blocks (sections) of fixed equal size; then, the convolution of each input block with $h(n)$ is computed using Eq. (11.16) with $X(k)$ being, in this case, the DFT of the considered block; the computed block convolutions are finally properly combined to lead the final output $y(n)$. Two popular ways of performing block convolutions are [1, 9] (1) overlap-add and (2) overlap-save.

11.2.6.2 Realizing IIR Filters

For IIR filters, the impulse response has *infinite* length and, therefore, the summation in Eq. (11.14) involves an infinite number of terms (N_1 and/or N_2 infinite). This makes Eq. (11.14) not suitable for realizing IIR filters. Similarly, the direct realization of Eq. (11.16) would require computing the infinite-length DFT $H(k)$, which is not possible. These problems are overcome by using *feedback difference equations* to realize the designed IIR filters. In fact, using Eq. (11.15) with $H(\omega)$ replaced by Eq. (11.12), we get

$$Y(\omega) = e^{-j\omega N_o} \frac{\sum_{k=0}^{M} b_k e^{-j\omega k}}{\sum_{k=0}^{N} a_k e^{-j\omega k}} X(\omega). \tag{11.17}$$

For simplicity and without loss of generality, assume $N_o = 0$; we can rewrite Eq. (11.17) as:

$$\sum_{k=0}^{N} a_k e^{-j\omega k} Y(\omega) = \sum_{k=0}^{M} b_k e^{-j\omega k} X(\omega). \tag{11.18}$$

Taking the inverse DTFT of both sides of Eq. (11.18) and noting that multiplication by $e^{-j\omega k}$ corresponds to a shift by k in the time/space domain, we obtain the input-output relation of the system in the time/space domain:

$$\sum_{k=0}^{N} a_k y(n-k) = \sum_{n=0}^{M} b_k x(n-k) \tag{11.19}$$

The difference equation (11.19) can be rearranged leading a recursive (feedback) input-output relation. For instance, in order to compute the right-sided output sequence $y(n)$, for $n \geq n_o$ (n_o integer constant), Eq. (11.19) can be rewritten as:

$$y(n) = \sum_{n=0}^{M} \frac{b_k}{a_o} x(n-k) - \sum_{k=1}^{N} \frac{a_k}{a_o} y(n-k). \tag{11.20}$$

where a_o is commonly taken to be 1, without loss of generality, since it can be integrated into the parameters b_k and a_k. Realizing Eq. (11.20) requires that N initial output values, $y(n_o-1), \dots y(n_o-N)$, be specified. For LSI filters, initial rest conditions are required: if the input $x(n) = 0$ for $n < n_o$, then set $y(n) = 0$ for $n < n_o$.

For a left-sided output sequence, Eq. (11.19) can be rearranged as follows:

$$y(\underbrace{n-N}_{m}) = \sum_{k=0}^{M} \frac{b_k}{a_N} x(n-k) - \sum_{k=0}^{N-1} \frac{a_k}{a_N} y(n-k). \tag{11.21}$$

So, Eq. (11.21) can be used to compute $y(m)$, $m \leq n_o$, by setting $n = m + N$ and specifying the N initial values $y(n_o + 1), \dots, y(n_o + N)$.

The feedback difference equations (11.20) and (11.21) are simple to implement in software or hardware. The MATLAB[TM1] software command $\mathbf{y} = \textbf{filter(b,a,x)}$ implements Eq. (11.20). In hardware, typical DSP chips implement low-order filters ($N = 1$ or $N = 2$); the low-order filters can be combined together (in cascade and/or parallel) to produce the desired higher-order filters (see Section 11.5). To implement the filter in hardware, the difference equations (or, equivalently, the rational frequency response) are represented by *structures*, which are flow graphs describing the algorithm, that is to be implemented, in terms of basic building blocks [1, Chap. 6]. The basic building blocks include adders, multipliers, branch points, and delay elements.

11.2.6.3 Quantization: Finite Wordlength Effect

In the design step, the filter coefficients are usually computed with a very high precision. In practice, these coefficients can be implemented with finite wordlength only. Since the design algorithm yields coefficients computed to the highest precision available (e.g., double-precision floating-point), the filter coefficients must be quantized to the internal format of the DSP. In addition, fixed-point chips are widely used since they generally provide higher processing speed at lower cost than do the floating point systems. In the case of a fixed-point DSP, this quantization also requires scaling of the coefficients to a predetermined maximum value. The quantization and/or truncation of the coefficients will generally cause the frequency response of the implemented filter to deviate from the designed filter frequency response. The deviation from the desired specifications will depend on the chosen filter type and on the structure used to implement the filter. For IIR filters, the quantization of the coefficients might turn a stable filter into an unstable one. Other effects are due to the fact that arithmetic operations performed on finite wordlength numbers generally result in numbers with larger wordlengths, which then need to be quantized or truncated to the allowable precision. Therefore, it is important to specify the required minimum wordlength that can be tolerated. As indicated in Section 11.5, very few design algorithms perform the optimization of quantized coefficients. Studies of the different wordlength effects has resulted in "rules of thumb" for the design and realization of a system such that the desired properties can be achieved with reduced errors and expense. A detailed study of the wordlength effects and the characterization of the resulting errors can be found in Sections 6.7 through 6.10 of [1] and Sections 7.5 through 7.7 of [9].

11.3 Classical Filter Design Methods

The methods described in this section are magnitude-only approximation methods, i.e., the desired phase response is assumed to be constant or linear and is not included in the design. These classical methods mainly design frequency-selective filters with real-valued coefficients $h(n)$.

Methods for the design of filters with general specifications [2, 4, 10] have been developed more recently and are presented in Section 11.4.

11.3.1 FIR Design Methods

Ivan W. Selesnick, C. Sidney Burrus,
Lina J. Karam, and
James H. McClellan

The classical FIR design methods are mainly concerned with the design of linear-phase FIR filters with real-valued coefficients $h(n)$. These filters are of four possible types [1, 11]. The properties of the four types of linear-phase filters are summarized in Table 11.1 and illustrated in Fig. 11.4.

---

[1]MATLAB   is a trademark of The Mathworks, Inc.

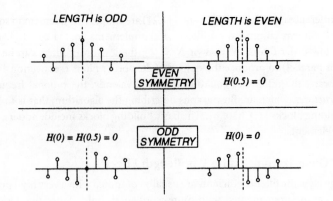

FIGURE 11.4 Examples of impulse responses corresponding to the four types of linear-phase filters. $H(f)$ is the corresponding frequency response, where the normalized frequency variable $f = \omega/2\pi$.

TABLE 11.1 Summary of the Four Types of Linear-Phase FIR Filters

	Odd length (N)	Even length (N)
Even symmetry	**Type I**	**Type II**
$h(\alpha + n) = h(\alpha - n)$	$\displaystyle\sum_{k=0}^{\frac{1}{2}[N-1]} a(k)\cos(\omega k)$	$\displaystyle\sum_{k=1}^{\frac{1}{2}N} b(k)\cos(\omega[k - \tfrac{1}{2}])$
$\alpha = \frac{N-1}{2}$	$a(0) = h(\frac{N-1}{2})$	zero at $\omega = \pi$
$\beta = 0$	$a(k) = 2h(\frac{N-1}{2} - k)$	$b(k) = 2h(\frac{N}{2} - k)$
		$\displaystyle\cos(\tfrac{1}{2}\omega) \sum_{k=0}^{\frac{1}{2}N-1} \hat{b}(k)\cos(\omega k)$
Odd symmetry	**Type III**	**Type IV**
$h(\alpha + n) = -h(\alpha - n)$	$\displaystyle\sum_{k=1}^{\frac{1}{2}[N-1]} c(k)\sin(\omega k)$	$\displaystyle\sum_{k=1}^{\frac{1}{2}N} d(k)\sin(\omega[k - \tfrac{1}{2}])$
$\alpha = \frac{N-1}{2}$	zeros at $\omega = 0, \pi$	zero at $\omega = 0$
$\beta = \frac{\pi}{2}$	$c(k) = 2h(\frac{N-1}{2} - k)$	$d(k) = 2h(\frac{N}{2} - k)$
	$h(\frac{N-1}{2}) = 0$	
	$\displaystyle\sin(\omega) \sum_{k=0}^{\alpha-1} \hat{c}(k)\cos(\omega k)$	$\displaystyle\sin(\tfrac{1}{2}\omega) \sum_{k=0}^{\frac{1}{2}N-1} \hat{d}(k)\cos(\omega k)$

11.3.1.1 Design by Windowing

The Fourier relationship between the impulse response and $H(\omega)$ suggests that $h(n)$ can be obtained via

$$h(n) = \frac{1}{2\pi} \int_{-\pi}^{\pi} D(\omega)e^{j\omega n}\, d\omega \tag{11.22}$$

where $D(\omega)$ is the desired frequency response. However, these Fourier series coefficients are usually infinitely supported. The windowing technique proposes that the infinitely supported Fourier series be truncated and multiplied by an appropriate function (a "window") to obtain an FIR filter. For the design of odd length symmetric filters, it is appropriate that $D(\omega)$ be a real-valued even function — then $h(n)$ is real and $h(n) = h(-n)$. A casual filter is obtained by then shifting $h(n)$.

Steps in Window Filter Design

1. Create the ideal impulse response, using the inverse DTFT to obtain $h_d[n]$:

$$h_d[n] = \frac{1}{2\pi} \int_{-\pi}^{\pi} D(\omega) \, e^{j\omega n} \, d\omega$$

where $D(\omega)$ is the *ideal* frequency response. For example, $D(\omega)$ might be the ideal LPF.

2. Note: If the length of the window is N, then the "ideal" frequency response *must contain a linear phase term*. For example, the ideal LPF would be specified as:

$$D(\omega) = \begin{cases} 1 \cdot e^{-j\omega(N-1)/2} & -\omega_c \leq \omega \leq +\omega_c \\ 0 & \omega_c \leq |\omega| < \pi \end{cases}$$

This allows both even-length and odd-length filters to be designed.

3. Create the FIR filter coefficients by multiplying by the window:

$$h[n] = w[n] \cdot h_d[n] \qquad n = 0, 1, \ldots, N-1$$

4. In the frequency domain, this windowing operation results in a convolution of the ideal frequency response with the Fourier transform of the window, $W(\omega)$.

$$H(\omega) = \frac{1}{2\pi} \int_{-\pi}^{\pi} D(\theta) \, W(\omega - \theta) \, d\theta$$

Note that this convolution is periodic with period 2π.

5. *Transition Width:* The result is that the ideal frequency response is smeared by the convolution, so the actual frequency response has a smooth roll-off from the passband to the stopband.

6. *Passband and Stopband Deviations:* In addition, all windows have sidelobes in their Fourier transforms, so the convolution gives rise to ripples in the frequency response of the FIR filter.

Examples of commonly used windows and their transforms are shown in Fig. 11.5. Windowed filter design examples are shown in Fig. 11.6.

    Window Selection   Let $D(\omega)$ be the response of an ideal lowpass filter with cut-off frequency ω_c, illustrated in Fig. 11.7. The Fourier series of $D(\omega)$ are samples of the **sinc** function:

$$\text{sinc}(n) = \begin{cases} \dfrac{\omega_c}{\pi} \dfrac{sin(\omega_c n)}{\omega_c n} & n \neq 0 \\ \dfrac{\omega_c}{\pi} & n = 0. \end{cases} \qquad (11.23)$$

Simple truncation of the sinc function samples is generally not found to be acceptable because the frequency responses of filters so obtained have large errors near the cut-off frequency. Moreover, as the filter length is increased, the size of this error does not diminish to zero (although the square error does). This is known as Gibbs phenomenon. Figure 11.8 illustrates a filter obtained by truncating the sinc function.

To overcome this problem, the windowing technique obtains $h(n)$ by multiplying the sinc function by a "window" that is tapered near its endpoints:

$$h(n) = w(n) \cdot \text{sinc}(n). \qquad (11.24)$$

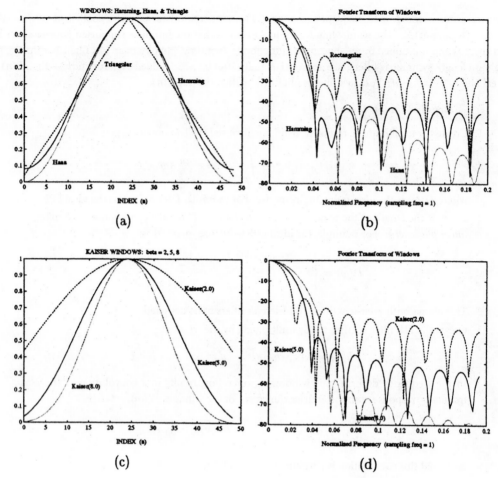

FIGURE 11.5   Common windows and their Fourier transforms. The window length is $N = 49$.

The generalized cosine windows and the Bartlett (triangular) window are examples of well-known windows. A useful window function has a frequency response that has a narrow mainlobe, a small relative peak sidelobe height, and good sidelobe roll-off. Roughly, the width of the mainlobe affects the width of the transition band of $H(\omega)$, while the relative height of the sidelobes affects the size of the ripples in $H(\omega)$. These cannot be made arbitrarily good at the same time. There is a trade-off between mainlobe width and relative sidelobe height. Some windows, such as the Kaiser window [12], provide a parameter that can be varied to control this trade-off.

One approach to window design computes the window sequence that has most of its energy in a given frequency band, say $[-B, B]$. Specifically, the problem is formulated as follows. Find $w(n)$ of specified finite support that maximizes

$$\lambda = \frac{\int_{-B}^{B} |W(\omega)|^2 d\omega}{\int_{-\pi}^{\pi} |W(\omega)|^2 d\omega} \tag{11.25}$$

where $W(\omega)$ is the Fourier transform of $w(n)$. The solution is a particular discrete prolate spheroidal (DPS) sequence [13], that can be normalized so that $W(0) = 1$. The solution to this problem was traditionally

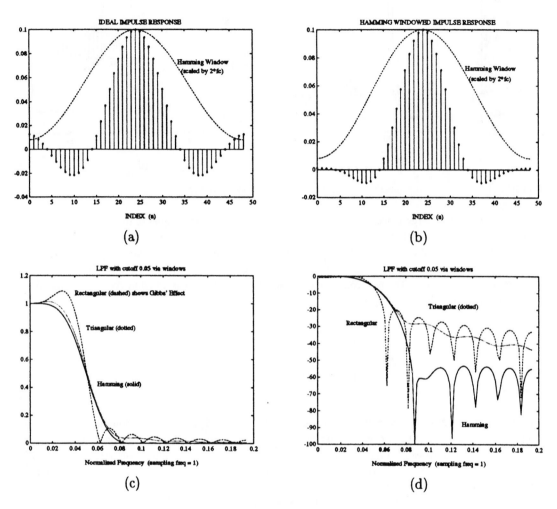

FIGURE 11.6 Examples of windowed filter design. The window length is $N = 49$.

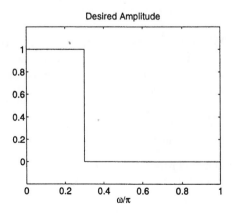

FIGURE 11.7 Ideal lowpass filter, $\omega_c = 0.3\pi$.

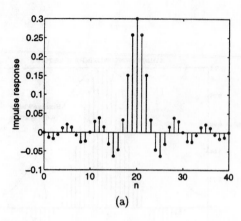
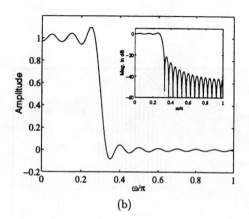

(a)   (b)

FIGURE 11.8   Lowpass filter obtained by sinc function truncation, $\omega_c = 0.3\pi$.

found by finding the largest eigenvector[2] of a matrix whose entries are samples of the sinc function [13]. However, that eigenvalue problem is numerically ill conditioned — the eigenvalues cluster around to 0 and 1. Recently, an alternative eigenvalue problem has become more widely known, that has exactly the same eigenvectors as the first eigenvalue problem (but different eigenvalues), and is numerically well conditioned [14, 15, 16]. The well conditioned eigenvalue problem is described by $\mathbf{A}\mathbf{v} = \theta\mathbf{v}$ where $\mathbf{A}$ is tridiagonal and has the following form:

$$
\mathbf{A}_{i,j} = \begin{cases}
\frac{1}{2}i(N-i) & j = i-1 \\
\left(\frac{N-1}{2} - i\right)^2 \cos B & j = i \\
\frac{1}{2}(i+1)(N-1-i) & j = i+1 \\
0 & |j-i| > 1
\end{cases}
\tag{11.26}
$$

for $i, j = 0, \ldots, N-1$. Again, the eigenvector with the largest eigenvalue is the sought solution. The advantage of $\mathbf{A}$ in Eq. (11.26) over the first eigenvalue problem is twofold: (1) The eigenvalues of $\mathbf{A}$ in Eq. (11.26) are well spread (so that the computation of its eigenvectors is numerically well conditioned); (2) The matrix $\mathbf{A}$ in Eq. (11.26) is tridiagonal, facilitating the computation of the largest eigenvector via the power method.

By varying the bandwidth, B, a family of DPS windows is obtained. By design, these windows are optimal in the sense of energy concentration. They have good mainlobe width and relative peak sidelobe height characteristics. However, it turns out that the sidelobe roll-off of the DPS windows is relatively poor, as noted in [16].

The Kaiser [12] and Saramäki [17, 18] windows were originally developed in order to avoid the numerically ill conditioning of the first matrix eigenvalue problem described above. They approximate the prolate spheroidal sequence, and do not require the solution to an eigenvalue problem. Kaiser's approximation to the prolate spheroidal window [12] is given by

$$
w(n) = \frac{I_0(\beta\sqrt{1 - (n-M)^2/M^2})}{I_0(\beta)} \qquad \text{for } n = 0, 1, \ldots N-1
\tag{11.27}
$$

where $M = \frac{1}{2}(N-1)$, β is an adjustable parameter, and $I_o(x)$ is the modified zero-th-order Bessel function of the first kind. The window in Eq. (11.27) is known as the Kaiser window of length N. For an odd-length

---

[2]The eigenvector with the largest eigenvalue.

window, the midpoint M is an integer. The parameter β controls the tradeoff between the mainlobe width and the peak sidelobe level — it should be chosen to lie between 0 and 10 for useful windows. High values of β produce filters having high stopband attenuation, but wide transition widths. The relationship between β and the ripple height in the stopband (or passband) is illustrated in Fig. 11.9 and is given by:

$$\beta = \begin{cases} 0 & \text{ATT} < 21 \\ 0.5842\,(\text{ATT} - 21)^{0.4} + 0.07886(\text{ATT} - 21) & 21 \leq \text{ATT} \leq 50 \\ 0.1102(\text{ATT} - 8.7) & 50 < \text{ATT} \end{cases} \tag{11.28}$$

where $\text{ATT} = -20\log_{10}\delta_s$ is the ripple height in dB.

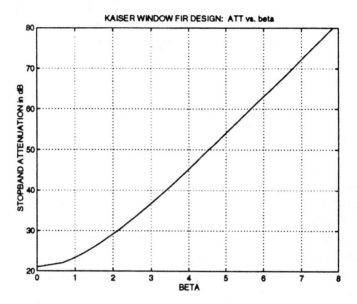

FIGURE 11.9 Kaiser window: stopband attenuation vs. β.

For lowpass FIR filter design, the following design formula helps the designer to estimate the Kaiser window length N in terms of the desired maximum passband and stopband error δ,[3] and transition width $\Delta F = (\omega_p - \omega_s)/2\pi$:

$$N \approx \frac{-20\log_{10}(\delta) - 7.95}{14.357\Delta F} + 1 \tag{11.29}$$

Examples of filter designs using the Kaiser window are shown in Fig. 11.10.

A second approach to window design minimizes the relative peak sidelobe height. The solution is the Dolph-Chebyshev window [17, 19], all the sidelobes of which have equal height. Saramäki has described a family of transitional windows that combine the optimality properties of the DPS window and the Dolph-Chebyshev window. He has found that the transitional window yields better results than both the DPS window and the Dolph-Chebyshev window, in terms of attenuation vs. transition width [17].

An extensive list and analysis of windows is given in [19]. In addition, the use of nonsymmetric windows for the design of fractional delay filters has been discussed in [20, 21].

---

[3] For Kaiser window designs, $\delta = \delta_p = \delta_s$.

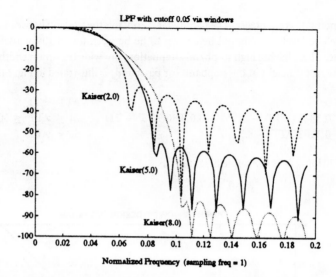

FIGURE 11.10 Frequency responses (log scale) of filters designed using the Kaiser window with selected values for the parameter β. Note the tradeoff between mainlobe width and sidelobes height.

Remarks

- The technique is conceptually and computationally simple.
- Using the window method, it is not possible to weight the passband and stopband differently. The ripple sizes in each band will be approximately the same. But requirements are often more strict in the stopband.
- It is difficult to specify the band edges and maximum ripple size precisely.
- The technique is not suitable for arbitrary desired responses.
- The use of windows for filter design is generally considered suboptimal because they do not solve a clear optimization problem, but see [22].

11.3.1.2 Optimal Square Error Design

The formulation is as follows. Given a filter length N, a desired amplitude function $D(\omega)$, and a non-negative function $W(\omega)$, find the symmetric filter that minimizes the weighted integral square error (or "L_2 error"), defined by

$$||E(\omega)||_2 = \left(\frac{1}{\pi} \int_0^\pi W(\omega)\,(A(\omega) - D(\omega))^2\,d\omega \right)^{\frac{1}{2}}. \tag{11.30}$$

For simplicity, symmetric odd-length filters[4] will be discussed here, in which case $A(\omega)$ can be written as

$$A(\omega) = \frac{1}{\sqrt{2}}a(0) + \sum_{n=1}^{M} a(n)\cos n\omega \tag{11.31}$$

---

[4]To treat the four linear phase types together, see Eqs. (11.51) through (11.55) in the sequel. Then, $||E(\omega)||_2$ becomes $\left(\frac{1}{\pi} \int_0^\pi \overline{W}(\omega)(A(\omega) - \overline{D}(\omega))^2\,d\omega \right)^{\frac{1}{2}}$ where $\overline{W}(\omega) = W(\omega)Q^2(\omega)$ and $\overline{D}(\omega) = D(\omega)/Q^2(\omega)$ and $A(\omega)$ is as in Eq. (11.31).

where $N = 2M + 1$ and where the impulse response coefficients $h(n)$ are related to the cosine coefficients $a(n)$ by

$$
h(n) = \begin{cases}
\frac{1}{2}a(M-n) & \text{for } 0 \le n \le M-1 \\
\frac{1}{\sqrt{2}}a(0) & \text{for } n = M \\
\frac{1}{2}a(n-M) & \text{for } M+1 \le n \le N-1 \\
0 & \text{otherwise.}
\end{cases}
\tag{11.32}
$$

The nonstandard choice of $\frac{1}{\sqrt{2}}$ here simplifies the notation below.

The coefficients $\mathbf{a} = (a(0), \dots, a(M))^t$ are found by solving the linear system

$$
\mathbf{Ra} = \mathbf{c}
\tag{11.33}
$$

where the elements of the vector $\mathbf{c}$ are given by

$$
c_0 = \frac{\sqrt{2}}{\pi} \int_0^\pi W(\omega) D(\omega)\, d\omega
\tag{11.34}
$$

$$
c_k = \frac{2}{\pi} \int_0^\pi W(\omega) D(\omega) \cos k\omega\, d\omega
\tag{11.35}
$$

and the elements of the matrix $\mathbf{R}$ are given by

$$
R_{0,0} = \frac{1}{\pi} \int_0^\pi W(\omega)\, d\omega
\tag{11.36}
$$

$$
R_{0,k} = R_{k,0} = \frac{\sqrt{2}}{\pi} \int_0^\pi W(\omega) \cos k\omega\, d\omega
\tag{11.37}
$$

$$
R_{k,l} = R_{l,k} = \frac{2}{\pi} \int_0^\pi W(\omega) \cos k\omega \cos l\omega\, d\omega
\tag{11.38}
$$

for $l, k = 1, \dots, M$. Often it is desirable that the coefficients satisfy some linear constraints, say $\mathbf{Ga} = \mathbf{b}$. Then the solution, found with the use of Lagrange multipliers, is given by the linear system

$$
\begin{bmatrix} \mathbf{R} & \mathbf{G}^t \\ \mathbf{G} & \mathbf{0} \end{bmatrix}
\begin{bmatrix} \mathbf{a} \\ \boldsymbol{\mu} \end{bmatrix} =
\begin{bmatrix} \mathbf{c} \\ \mathbf{b} \end{bmatrix}
\tag{11.39}
$$

the solution of which is easily verified to be given by

$$
\boldsymbol{\mu} = \left(\mathbf{G} \mathbf{R}^{-1} \mathbf{G}^t \right)^{-1} (\mathbf{G} \mathbf{R}^{-1} \mathbf{c} - \mathbf{b}) \qquad \mathbf{a} = \mathbf{R}^{-1}(\mathbf{c} - \mathbf{G}^t \boldsymbol{\mu})
\tag{11.40}
$$

where $\boldsymbol{\mu}$ are the Lagrange multipliers.

In the unweighted case ($W(\omega) = 1$) the solution is given by a simpler system:

$$
\begin{bmatrix} \mathbf{I}_{M+1} & \mathbf{G}^t \\ \mathbf{G} & \mathbf{0} \end{bmatrix}
\begin{bmatrix} \mathbf{a} \\ \boldsymbol{\mu} \end{bmatrix} =
\begin{bmatrix} \mathbf{c} \\ \mathbf{b} \end{bmatrix}.
\tag{11.41}
$$

In Eq. (11.41), $\mathbf{I}_{M+1}$ is the $(M+1)$ by $(M+1)$ identity matrix. It is interesting to note that in the unweighted case, the least square filter minimizes a worst case *pointwise* error in the time domain over a set of bounded energy input signals [23].

In the unweighted case with no constraint, the solution becomes: $\mathbf{a} = \mathbf{c}$. This is equivalent to truncation of the Fourier series coefficients (the "rectangular window" method). This simple solution is due to the orthogonality of the basis functions $\{\frac{1}{\sqrt{2}}, \cos \omega, \cos 2\omega, \dots\}$ when $W(\omega) = 1$. In general, whenever the basis functions are orthogonal, then the solution takes this simple form.

   Discrete Squares Error   When $D(\omega)$ is simple, the integrals above can be found analytically. Otherwise, entries of $\mathbf{R}$ and $\mathbf{b}$ can be found numerically. Define a dense uniform grid of frequencies over $[0, \pi)$ as $\omega_i = i\pi/L$ for $i = 0, \ldots, L - 1$ and for some large L (say $L \approx 10M$). Let $\mathbf{d}$ be the vector given by $\mathbf{d}_i = D(\omega_i)$ and $\mathbf{C}$ be the L by $M + 1$ matrix of cosine terms: $\mathbf{C}_{i,0} = \frac{1}{\sqrt{2}}$, $\mathbf{C}_{i,k} = \cos k\omega_i$ for $k = 1, \ldots, M$. ($\mathbf{C}$ has many more rows than columns.) Let $\mathbf{W}$ be the diagonal weighting matrix $diag\{W(\omega_i)\}$. Then

$$\mathbf{R} \approx \frac{2}{L\pi}\mathbf{C}^t\mathbf{W}\mathbf{C} \qquad \mathbf{c} \approx \frac{2}{L\pi}\mathbf{C}^t\mathbf{W}\mathbf{d}. \tag{11.42}$$

Using these numerical approximations for $\mathbf{R}$ and $\mathbf{c}$ is equivalent to minimizing the discrete squares error,

$$\sum_{i=0}^{L-1} W(\omega_i)\,(D(\omega_i) - A(\omega_i))^2 \tag{11.43}$$

that approximates the integral square error. In this way, an FIR filter can be obtained easily, whose response approximates an arbitrary $D(\omega)$ with an arbitrary $W(\omega)$. This makes the least squares error approach very useful. It should be noted that the minimization of Eq. (11.43) is most naturally formulated as the least squares solution to an over-determined linear system of equations, an approach described in [11]. The solution is the same, however.

   Transition Regions   As an example, the least squares design of a length $N = 2M + 1$ symmetric lowpass filter according to the desired response and weight functions

$$D(\omega) = \begin{cases} 1 & \omega \in [0, \omega_p] \\ 0 & \omega \in [\omega_s, \pi] \end{cases} \qquad W(\omega) = \begin{cases} K_p & \omega \in [0, \omega_p] \\ 0 & \omega \in [\omega_p, \omega_s] \\ K_s & \omega \in [\omega_s, \pi] \end{cases} \tag{11.44}$$

is developed. For this $D(\omega)$ and $W(\omega)$, the vector $\mathbf{c}$ in Eq. (11.33) is given by

$$c_0 = \frac{\sqrt{2}K_p\omega_p}{\pi} \qquad c_k = \frac{2K_p \sin(k\omega_p)}{k\pi} \qquad 1 \le k \le M \tag{11.45}$$

and the matrix $\mathbf{R}$ is given by

$$\mathbf{R} = \mathbf{T}(\text{toeplitz}(\mathbf{p}, \mathbf{p}) + \text{hankel}(\mathbf{p}, \mathbf{q}))\mathbf{T} \tag{11.46}$$

where the matrix $\mathbf{T}$ is the identity matrix everywhere except for $T_{0,0}$, which is $\frac{1}{\sqrt{2}}$. The vectors $\mathbf{p}$ and $\mathbf{q}$ are given by

$$p_0 = \frac{K_p\omega_p + K_s(\pi - \omega_s)}{\pi} \tag{11.47}$$

$$p_k = \frac{K_p \sin(k\omega_p) - K_s \sin(k\omega_s)}{k\pi} \qquad 1 \le k \le M \tag{11.48}$$

$$q_k = \frac{K_p \sin((k + M)\omega_p) - K_s \sin((k + M)\omega_s)}{(k + M)\pi} \qquad 0 \le k \le M. \tag{11.49}$$

The matrix toeplitz $(\mathbf{p}, \mathbf{p})$ is a symmetric matrix with constant diagonals, the first row and column of which is $\mathbf{p}$. The matrix hankel $(\mathbf{p}, \mathbf{q})$ is a symmetric matrix with constant anti-diagonals, the first column of which is $\mathbf{p}$, the last row of which is $\mathbf{q}$. The structure of the matrix $\mathbf{R}$ makes possible the efficient solution of $\mathbf{R}\mathbf{a} = \mathbf{b}$ [24]. Because the error is weighted by zero in the transition band $[\omega_p, \omega_s]$, the Gibbs phenomenon is eliminated: the peak error diminishes to zero as the filter length is increased. Figure 11.11 illustrates an example.

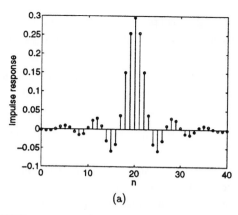
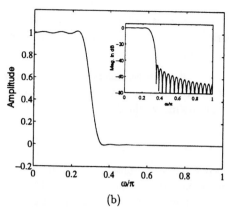

(a)   (b)

FIGURE 11.11   Weighted least squares example. $N = 41, \omega_p = 0.25\pi, \omega_s = 0.35\pi, K = 4$.

Other Least Squares Approaches   Another approach modifies the discontinuous ideal lowpass response of Fig. 11.7 so that a fractional order spline is used to continuously connect the passband and stopband [25]. In this case, with uniform error weighting, (1) a simple closed form expression for the least squares error solution is available, and (2) Gibbs phenomenon is eliminated. The use of spline transition regions also facilitates the design of multiband filters by combining various lowpass filters [26]. In that case, a least squares error multiband filter can be obtained via closed form expressions, where the transition region widths can be independently specified.

Similar expressions can be derived for the even length filter and the odd symmetric filters. It should also be noted that the least squares error approach is directly applicable to the design of nonsymmetric FIR filters, complex-valued FIR filters, and two-dimensional FIR filters.

In addition, another approach to filter design according to a square error criterion, produces filters known as *eigenfilters* [27]. This approach gives the filter coefficients as the largest eigenvalue of a matrix that is readily constructed.

Remarks

- Optimal with respect to square error criterion.
- Simple, non-iterative method.
- Analytic solutions sometimes possible, otherwise solution is obtained via solution to linear system of equations.
- Allows the use of a frequency dependent weighting function.
- Suitable for arbitrary $D(\omega)$ and $W(\omega)$.
- Easy to include arbitrary linear constraints.
- Does not allow direct control of maximum ripple size.

11.3.1.3   Equiripple Optimal Chebyshev Filter Design

The minimization of the Chebyshev norm is useful because it permits the user to explicitly specify band-edges and relative error sizes in each band. Furthermore, the designed equiripple FIR filters have the smallest transition width among all FIR filters with the same deviation.

Linear phase FIR filters that minimize a Chebyshev error criterion can be obtained with the Remez exchange algorithm [28, 29] or by linear programming techniques [30]. Both these methods are iterative numerical procedures and are applicable to arbitrary desired frequency response amplitudes.

Remez Exchange (Parks-McClellan)   Parks and McClellan proposed the use of the Remez algorithm for FIR filter design and made programs available [29, 31, 6]. Many texts describe the Parks-McClellan (PM) algorithm in detail [1, 11].

Problem Formulation   Given a filter length, N, a desired (real-valued) amplitude function, $D(\omega)$, and a non-negative weighting function, $W(\omega)$, find the symmetric (or antisymmetric) filter that minimizes the weighted Chebyshev error, defined by

$$||E(\omega)||_\infty = \max_{\omega \in B} |W(\omega)(A(\omega) - D(\omega))| \tag{11.50}$$

where B is a closed subset of $[0, \pi]$. Both $D(\omega)$ and $W(\omega)$ should be continuous over B. The solution to this problem is called the best weighted Chebyshev approximation to $D(\omega)$ over B.

To treat each of the four linear phase cases together, note that in each case, the amplitude $A(\omega)$ can be written as [32]:

$$A(\omega) = Q(\omega)P(\omega) \tag{11.51}$$

where $P(\omega)$ is a cosine polynomial (Table 11.1). By expressing $A(\omega)$ in this way, the weighted error function in each of the four cases can be written as:

$$E(\omega) = W(\omega)[A(\omega) - D(\omega)] \tag{11.52}$$

$$= W(\omega)Q(\omega)\left[P(\omega) - \frac{D(\omega)}{Q(\omega)}\right]. \tag{11.53}$$

Therefore, an equivalent problem is the minimization of

$$||E(\omega)||_\infty = \max_{\omega \in \overline{B}} |\overline{W}(\omega)(P(\omega) - \overline{D}(\omega))| \tag{11.54}$$

where

$$\overline{W}(\omega) = W(\omega)Q(\omega) \qquad \overline{D}(\omega) = \frac{D(\omega)}{Q(\omega)} \qquad P(\omega) = \sum_{k=0}^{r-1} a(k)\cos k\omega \tag{11.55}$$

and $\overline{B} = B - \{\text{endpoints where } Q(\omega) = 0\}$.

The Remez exchange algorithm, for computing the best Chebyshev solution, uses the alternation theorem. This theorem characterizes the best Chebyshev solution.

Alternation Theorem   If $P(\omega)$ is given by Eq. (11.55), then a necessary and sufficient condition that $P(\omega)$ be the unique minimizer of Eq. (11.54) is that there exist in $\overline{B}$ at least $r + 1$ extremal points $\omega_1, \ldots, \omega_{r+1}$ (in order: $\omega_1 < \omega_2 < \cdots < \omega_{r+1}$), such that

$$E(\omega_i) = c \cdot (-1)^i ||E(\omega)||_\infty \quad \text{for} \quad i = 1, \ldots, r+1 \tag{11.56}$$

where c is either 1 or -1.

The alternation theorem states that $|E(\omega)|$ attains its maximum value at a minimum of $r + 1$ points, and that the weighted error function alternates sign on at least $r + 1$ of those points. Consequently, the weighted error functions of best Chebyshev solutions exhibit an *equiripple* behavior.

For lowpass filter design via the PM algorithm, the functions $D(\omega)$ and $W(\omega)$ in Eq. (11.44) are usually used. For lowpass filters so obtained, the deviations δ_p and δ_s satisfy the relation $\delta_p/\delta_s = K_s/K_p$. For example, consider the design of a real symmetric lowpass filter of length $N = 41$. Then $Q(\omega) = 1$ and $r = (N + 1)/2 = 21$. With the desired amplitude and weight function, Eq. (11.44), with $K = 4$ and $\omega_p = 0.25\pi$, $\omega_s = 0.35\pi$, the best Chebyshev solution and its weighted error function are illustrated in Fig. 11.12. The maximum errors in the passband and stopband are $\delta_p = 0.0178$ and $\delta_s = 0.0714$, respectively. The circular marks in Fig. 11.12(c) indicate the extremal points of the alternation theorem.

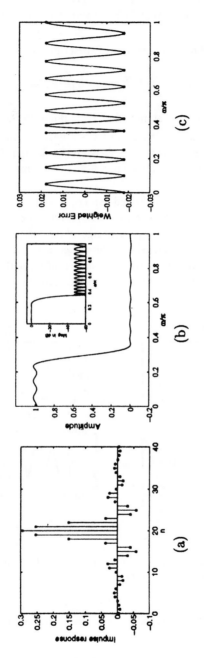

FIGURE 11.12 Equiripple lowpass filter obtained via the PM algorithm. $N = 41$, $\omega_p = 0.25\pi$, $\omega_s = 0.35\pi$, $\delta_p/\delta_s = 4$.

To elaborate on the alternation theorem, consider the design of a length 21 lowpass filter and a length 41 bandpass filter. Several optimal Chebyshev filters are illustrated in Figs. 11.13 through 11.16. It can be verified by inspection that each of the filters illustrated in Figs. 11.13 through 11.16 is Chebyshev optimal, by verifying that the alternation theorem is satisfied. In each case, a set of $r + 1$ extremal points, which satisfies the necessary and sufficient conditions of the alternation theorem, is indicated by circular marks in Figs. 11.13 through 11.16.

Several remarks regarding the weighted error function of a best Chebyshev solution are worth noting.

1. $E(\omega)$ may have local minima and maxima in $\overline{B}$ at which $|E(\omega)|$ does not attain its maximum value. See Fig. 11.14.

2. $|E(\omega)|$ may attain its maximum value at more than $r + 1$ points in $\overline{B}$. See Fig. 11.15.

3. If there exists in $\overline{B}$ s ordered points $\omega_1, \ldots, \omega_s$, with $s > r+1$, at which $|E(\omega_i)| = ||E(\omega)||_\infty$ (i.e., there are more than $r + 1$ extremal points), then it is possible that $E(\omega_i) = E(\omega_{i+1})$ for some i. See Fig. 11.16. This is rare and, for lowpass filter design, impossible.

Figure 11.14 illustrates two filters that possess "scaled-extra ripples" (ripples of non-maximal size [30]). Figure 11.15 illustrates two maximal ripple filters. Maximal ripple filters are a subset of optimal Chebyshev filters that occur for special values of ω_p, ω_s, etc. (The first algorithms for equiripple filter design produced only maximal ripple filters [33, 34]). Figure 11.16 illustrates a filter that possesses two scaled-extra ripples and one extra ripple of maximal size. These extra ripples have no bearing on the alternation theorem. The set of $r + 1$ points, indicated in Fig. 11.16, is a set that satisfies the alternation theorem; therefore, the filter is optimal in the Chebyshev sense.

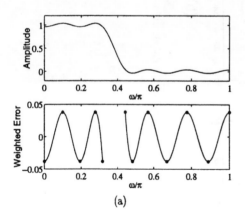
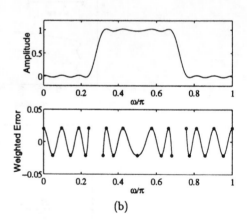

(a)   (b)

FIGURE 11.13   Parks-McClellan example. (a) Lowpass: $N = 21, \omega_p = 0.3161\pi, \omega_s = 0.4444\pi$. (b) Bandpass: $N = 41, \omega_1 = 0.2415\pi, \omega_2 = 0.3189\pi, \omega_3 = 0.6811\pi, \omega_4 = 0.7585\pi$.

Remez Algorithm   To understand the Remez exchange algorithm, first note that Eq. (11.56) can be written as

$$\sum_{k=0}^{r-1} a(k) \cos k\omega_i - \frac{(-1)^i \delta}{\overline{W}(\omega_i)} = \overline{D}(\omega_i) \quad \text{for} \quad i = 1, \ldots, r + 1. \tag{11.57}$$

where δ represents $||E(\omega)||_\infty$, and consider the following. If the set of extremal points in the alternation theorem were known in advance, then the solution could be found by solving the system of Eq. (11.57).

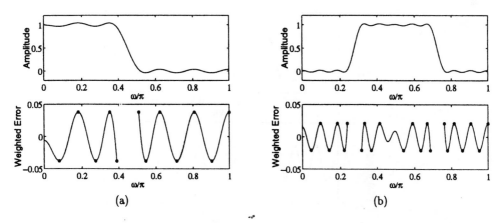

FIGURE 11.14 Parks-McClellan example. (a) Lowpass: $N = 21, \omega_p = 0.3889\pi, \omega_s = 0.5082\pi$. (b) Bandpass: $N = 41, \omega_1 = 0.2378\pi, \omega_2 = 0.3132\pi, \omega_3 = 0.6870\pi, \omega_4 = 0.7621\pi$.

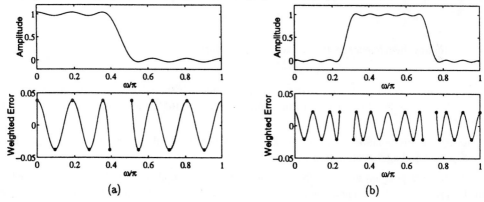

FIGURE 11.15 Parks-McClellan example. Lowpass: $N = 21, \omega_p = 0.3919\pi, \omega_s = 0.5103\pi$. Bandpass: $N = 41 \; \omega_1 = 0.2370\pi, \omega_2 = 0.3115\pi, \omega_3 = 0.6885\pi, \omega_4 = 0.7630\pi$.

The system in Eq. (11.57) represents an interpolation problem, which in matrix form becomes

$$
\begin{bmatrix}
1 & \cos\omega_1 & \cdots & \cos(r-1)\omega_1 & 1/\overline{W}(\omega_1) \\
1 & \cos\omega_2 & \cdots & \cos(r-1)\omega_2 & -1/\overline{W}(\omega_2) \\
\vdots & & & & \vdots \\
1 & \cos\omega_{r+1} & \cdots & \cos(r-1)\omega_{r+1} & (-1)^r/\overline{W}(\omega_{r+1})
\end{bmatrix}
\begin{bmatrix}
a(0) \\
a(1) \\
\vdots \\
a(r-1) \\
\delta
\end{bmatrix}
$$
$$
=
\begin{bmatrix}
\overline{D}(\omega_1) \\
\overline{D}(\omega_2) \\
\vdots \\
\overline{D}(\omega_{r+1})
\end{bmatrix}
\tag{11.58}
$$

to which there is a unique solution. Therefore, the problem becomes one of finding the correct set of points over which to solve the interpolation problem in Eq. (11.57).

The Remez exchange algorithm proceeds by iteratively

1. solving the interpolation problem in Eq. (11.58) over a specified set of $r + 1$ points (a reference set), and

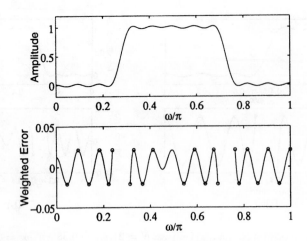

FIGURE 11.16   Parks-McClellan example. $N = 41$, $\omega_1 = 0.2374\pi$, $\omega_2 = 0.3126\pi$, $\omega_3 = 0.6876\pi$, $\omega_4 = 0.7624\pi$.

2. updating the reference set (by an exchange procedure).

The initial reference set can be taken to be $r + 1$ points uniformly spaced over $\overline{B}$. Convergence is achieved when $\|E(\omega)\|_\infty - |\delta| < \epsilon$, where ϵ is a small number (such as 10^{-6}) indicating the numerical accuracy desired.

During the interpolation step, the solution to Eq. (11.58) is facilitated by the use of a closed form solution for δ and interpolation formulas [29].

After the interpolation step is performed, the reference set is updated as follows. The weighted error function is computed, and a new reference set $\omega_1, \ldots, \omega_{r+1}$ is found such that: (1) The current weighted error function $E(\omega)$ alternates sign on the new reference set, (2) $|E(\omega_i)| \geq |\delta|$ for each point ω_i of the new reference set and (3) $|E(\omega_i)| > |\delta|$ for at least one point ω_i of the new reference set. Generally, the new reference set is found by taking the set of local minima and maxima of $E(\omega)$ that exceed the current value of δ, and taking a subset of this set that satisfies the alternation property. Figure 11.17 illustrates the operation of the Parks-McClellan algorithm.

Design Rules for Lowpass Filters   [12, 35, 36, 37] While the PM algorithm is applicable for the approximation of arbitrary responses $D(\omega)$, the lowpass case has received particular attention. In the design of lowpass filters via the PM algorithm, there are five parameters of interest: the filter length N, the passband and stopband edges ω_p and ω_s, and the maximum error in the passband and stopband δ_p and δ_s. Their values are not independent — any four determines the fifth. Formulas for predicting the required filter length for a given set of specifications make this clear. Kaiser developed the following approximate relation for estimating the equiripple FIR filter length for meeting the specifications,

$$N \approx \frac{-20\log_{10}(\sqrt{\delta_p \delta_s}) - 13}{14.6 \Delta F} + 1 \tag{11.59}$$

where $\Delta F = (\omega_s - \omega_p)/(2\pi)$. Defining the filter attenuation ATT to be $-20\log_{10}(\sqrt{\delta_p \delta_s})$, and comparing Eq. (11.29) with Eq. (11.59), it can be seen that the optimal Chebyshev design results in filters with about 5 dB more attenuation than the windowed designed filters when the same specs are used for the other design parameters (N and ΔF). Figure 11.18 compares window-based designs with Chebyshev (Parks-McClellan)-based designs.

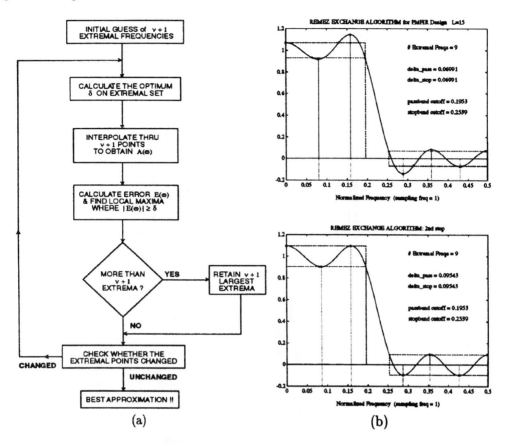

FIGURE 11.17 Operation of the Parks-McClellan algorithm. (a) Block Diagram. (b) Exchange steps. Extremal points constituting the current extremal set are shown as solid circles; extremal points selected to form the new extremal set are shown as solid squares.

Herrmann et al. gave a somewhat more accurate design formula for the optimal Chebyshev FIR filter design [37]:

$$N \approx \frac{D_\infty(\delta_p, \delta_s) - f(\delta_p, \delta_s)(\Delta F)^2}{\Delta F} + 1 \tag{11.60}$$

where

$$D_\infty(\delta_p, \delta_s) = 0.005309(\log_{10}^2 \delta_p + 0.07114 \log_{10} \delta_p - 0.4761) \log_{10} \delta_s$$
$$- (0.00266 \log_{10}^2 \delta_p + 0.5941 \log_{10} \delta_p + 0.4278),$$

$$f(\delta_p, \delta_s) = 11.01217 + 0.51244(\log_{10} \delta_p - \log_{10} \delta_s). \tag{11.61}$$

These formulas assume that $\delta_s < \delta_p$. If otherwise, then interchange δ_p and δ_s. Equation (11.60) is the one used in the Matlab implementation (**remezord**() function) as part of the Matlab Signal Processing toolbox.

To use the PM algorithm for lowpass filter design, the user specifies N, ω_p, ω_s, δ_p/δ_s. The PM algorithm can be modified so that the user specifies other parameter sets [38]. For example, with one modification, the user specifies N, ω_p, δ_p, δ_s; or similarly, N, ω_s, δ_p, δ_s. With a second modification, the user specifies N, ω_p, ω_s, δ_p; or similarly, N, ω_p, ω_s, δ_s.

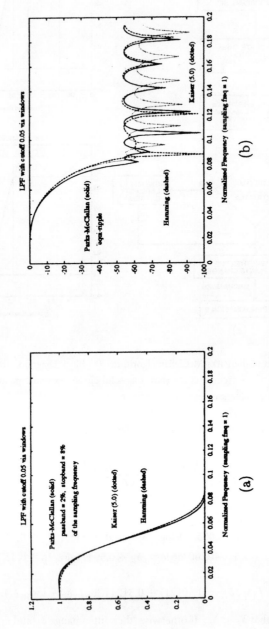

FIGURE 11.18 Comparison of window designs with optimal Chebyshev (Parks-McClellan) designs. The window length is $N = 49$. (a) Frequency response of designed filter using linear scale. (b) Frequency response of designed filter using log (dB) scale.

Note that Eq. (11.59) states that the filter length N and the transition width ΔF are inversely proportional. This is in contrast to the relation for maximally flat symmetric filters. For equiripple filters with fixed δ_p and δ_s, ΔF diminishes like $1/N$; while for maximally flat filters, ΔF diminishes like $1/\sqrt{N}$.

Remarks

- Optimal with respect to Chebyshev norm.
- Explicit control of band edges and relative ripple sizes.
- Efficient algorithm, always converges.
- Allows the use of a frequency dependent weighting function.
- Suitable for arbitrary $D(\omega)$ and $W(\omega)$.
- Does not allow arbitrary linear constraints.

Summary of Optimal Chebyshev Linear Phase FIR Filter Design

1. The desired frequency response can be written as

$$D(\omega) = A(\omega)\, e^{-j(\alpha\omega + \beta)}$$

where $\alpha = (N-1)/2$ always, and $\beta = 0$ for filters with even symmetry. Since $A(\omega)$ is a real-valued function, the Chebyshev approximation is applied to $A(\omega)$ and the linear phase comes for free. However, the delay will be proportional to the designed filter length.

2. The mathematical theory of *Chebyshev Approximation* is applied. In this type of optimization, the *maximum value of the error is minimized*, as opposed to the error energy as in least squares. Minimizing the maximum error is consistent with the desire to keep the passband and stopband deviations as small as possible. (Recall that least squares suffers from the Gibbs effect). However, minimization of the maximum error does not permit the use of derivatives to find the optimal solution.

3. The *Alternation Theorem* gives the necessary and sufficient conditions for the optimum in terms of equal-height ripples in the (weighted) error function.

4. The *Remez exchange algorithm* will compute the optimal approximation by searching for the locations of the peaks in the error function. This algorithm is iterative.

5. The inputs to the algorithm are the filter length, N, the locations of the passband, and stopband cutoff frequencies: ω_p and ω_s, and a weight function to weight the error in the passband and stopband differently.

6. The Chebyshev approximation problem can also be reformulated as a linear program. This is useful if additional linear design constraints need to be included.

7. *Transition Width* is minimized among all FIR filters with the same deviations.

8. *Passband and Stopband Deviations:* The response is equiripple, it does not fall off away from the transition region. Compared to the Kaiser window design, the optimal Chebyshev FIR design gives about 5 dB more attenuation (where attenuation is given by $-20\log_{10}\delta$ and δ is the stopband or passband error) for the same specs on all other filter design parameters.

    Linear Programming   Often it is desirable that an FIR filter be designed to minimize the Chebyshev error subject to linear constraints that the Parks-McClellan algorithm does not allow. An example described by Rabiner and Gold includes time domain constraints — in that example [30], the oscillatory behavior of the step response of a lowpass filter is included in the design formulation.

    Another example comes from a communication application [39] — given $h_1(n)$, design $h_2(n)$ so that $h(n) = (h_1 * h_2)(n)$ is an Mth band filter [i.e., $h(Mn) = 0$ for all $n \neq 0$ and $M \neq 0$]. Such constraints are linear in $h_1(n)$. [In the special case that $h_1(n) = \delta(n)$, $h_2(n)$ is itself an Mth band filter, and is often used for interpolation.]

Linear programming formulations of approximation problems (and optimization problems in general) are very attractive because well-developed algorithms exist (namely the simplex algorithm and more recently, interior point methods) for solving such problems. Although linear programming requires significantly more computation than the methods described above, for many problems it is a very rapid and viable technique [7]. Furthermore, this approach is very flexible — it allows arbitrary linear equality and *inequality* constraints.

The problem of minimizing the weighted Chebyshev error $W(\omega)(A(\omega) - D(\omega))$ where $A(\omega)$ is given by $Q(\omega) \sum_{k=0}^{r-1} a(k) \cos k\omega$ can be formulated as a linear program as follows:

$$\text{minimize} \quad \delta \tag{11.62}$$

subject to

$$A(\omega) - \frac{\delta}{W(\omega)} \leq D(\omega) \tag{11.63}$$

$$-A(\omega) - \frac{\delta}{W(\omega)} \leq -D(\omega). \tag{11.64}$$

The variables are $a(0), \ldots, a(r-1)$ and δ. The cost function and the constraints are linear functions of the variables, hence the formulation is that of a linear program.

Remarks

- Optimal with respect to chosen criteria.
- Easy to include arbitrary linear constraints.
- Criteria limited to linear programming formulation.
- High computational cost.

11.3.2 IIR Design Methods

Lina J. Karam,
Ivan W. Selesnick, and C. Sidney Burrus

The objective in IIR filter design is to find a rational function $H(\omega)$ [as in Eq. (11.12)] that approximates the ideal specifications according to some design criteria.

The approximation of an arbitrary specified frequency response is more difficult for IIR filters than is so for FIR filters. This is due to the nonlinear dependence of $H(\omega)$ on the filter coefficients in the IIR case. However, for the ideal lowpass response, there exist analytic techniques to directly obtain IIR filters. These techniques are based on converting analog filters into IIR digital filters. One such popular IIR design method is the *Bilinear Transformation Method* [1, 11]. Other types of frequency-selective filters (shown in Fig. 11.1) can be obtained from the designed lowpass prototype using additional frequency transformations [1, Chap. 7].

Direct "discrete-time" iterative IIR design methods have also been proposed (see Section 11.4.2). While these methods can be used to approximate general magnitude responses (i.e., not restricted to the design of the standard frequency-selective filters), they are iterative and slower than the traditional "continuous-time/space" based approaches that make use of simple and efficient closed-form design formulas.

11.3.2.1 Bilinear Transformation Method

The traditional IIR design approaches reduce the "discrete-time/space" (digital) filter design problem into a "continuous-time/space" (analog) filter design problem, which can be solved using well-developed and relatively simple design procedures based on closed-form design formulas. Then, a *transformation* is used to map the designed analog filter into a digital filter meeting the desired specifications.

Let $H(z)$ denote the transfer function of a digital filter [i.e., $H(z)$ is the Z-transform of the filter impulse response $h(n)$] and let $H_a(s)$ denote the transfer function of an analog filter [i.e., $H_a(s)$ is the Laplace transform of the continuous-time filter impulse response $h(t)$]. The bilinear transformation is a mapping between the complex variables s and z and is given by:

$$s = K \left(\frac{1 - z^{-1}}{1 + z^{-1}} \right) \tag{11.65}$$

where K is a design parameter. Replacing s by Eq. (11.65) in $H_a(s)$, the analog filter with transfer function $H_a(s)$ can be converted into a digital filter whose transfer function is equal to

$$H(z) = H_a(s) \big|_{s=K\left(\frac{1-z^{-1}}{1+z^{-1}}\right)} \tag{11.66}$$

Alternatively, the mapping can be used to convert a digital filter into an analog filter by expressing z in function of s.

Note that the analog frequency variable Ω corresponds to the imaginary part of s (i.e., $s = \sigma + j\Omega$), while the digital frequency variable ω (in radians) corresponds to the angle (phase) of z (i.e., $z = re^{j\omega}$). The bilinear transformation (11.65) was constructed such that it satisfies the following important properties:

1. The left-half plane (LHP) of the s-plane maps into the inside of the unit circle in the z-plane. As a result, a stable and causal analog filter will always result in a stable and causal digital filter.

2. The $j\Omega$ axis (imaginary axis) in the s-plane maps into the U.C. in the z-plane (i.e, $z = e^{j\omega}$). This results in a direct relationship between the continuous-time frequency Ω and the discrete-time frequency ω. Replacing z by $e^{j\omega}$ (unit circle) in Eq. (11.65), we obtain the following relation:

$$\Omega = K \tan(\omega/2) \tag{11.67}$$

or, equivalently,

$$\omega = 2 \arctan(\Omega/K) \tag{11.68}$$

The design parameter K can be used to map one specific frequency point in the analog domain to a selected frequency point in the digital domain, and to control the location of the designed filter cutoff frequency. Equations (11.67) and (11.68) are non-linear, resulting in a warping of the frequency axis as the filter frequency response is transformed from one domain to another. This follows from the fact that the bilinear transformation maps [via Eq. (11.67) or Eq. (11.68)] the entire $j\Omega$ axis, i.e., $-\infty \leq \Omega \leq \infty$, onto one period $-\pi \leq \omega \leq \pi$ (which corresponds to one revolution of the unit circle in the z-plane).

The bilinear transformation design procedure can be summarized as follows:

1. Transform the digital frequency domain specifications to the analog domain using Eq. (11.67). The frequency domain specs are given typically in terms of magnitude response specs as shown in Fig. 11.2. After the transformation, the digital magnitude response specs are converted into specs on the analog magnitude response.

2. Design a *stable and causal* analog filter with transfer function $H_a(s)$ such that $|H_a(s = j\Omega)|$ approximates the derived analog specs. This is typically done by using one of the classical frequency-selective analog filters whose magnitude responses are given in terms of closed-form formulas; the parameters in the closed-form formulas (e.g., needed analog filter order,

analog cutoff frequency) can then be computed to meet the desired analog specs. Typical analog prototypes include Butterworth, Chebyshev, and Elliptic filters; the characteristics of these filters are discussed in Section on page **11**-32. The closed-form formulas give only the magnitude response $|H_a(j\Omega)|$ of the analog filter and, therefore, do not uniquely specify the complete frequency response (or corresponding transfer function) which also should include a phase response. From all the filters having magnitude response $|H_a(j\Omega)|$, we need to select the filter that is stable and, if needed, causal. Using the fact that the computed magnitude-squared response $|H_a(j\Omega)|^2 = |H_a(s)|^2$, for $s = j\Omega$, and that $|H_a(s)|^2 = H_a(s)H_a^*(-s^*)$, where s^* denotes the complex conjugate of s, the system function $H_a(s)$ of the desired stable and causal filter is obtained by selecting the poles of $|H_a(j\Omega)|^2$ lying in the LHP of the s-plane [11].

3. Obtain the transfer function $H(z)$ for the digital filter by applying the bilinear transformation (11.65) to $H_a(s)$. The design parameter K can be fixed or chosen to map one analog frequency point Ω (e.g., the passband or stopband cutoff) into a desired digital frequency point ω.

4. The frequency response $H(\omega)$ of the resulting stable digital filter can be obtained from the transfer function $H(z)$ by replacing z by $e^{j\omega}$; i.e.,

$$H(\omega) = H(z)|_{z=e^{j\omega}} \tag{11.69}$$

11.3.2.2 Classical IIR Filter Types

The four standard classical analog filter types are known as (1) Butterworth, (2) Chebyshev I, (3) Chebyshev II, and (4) Elliptic [1, 11]. The characteristics of these analog filters are described briefly below.

Digital versions of these filters are obtained via the *bilinear transformation* [1, 11], and examples are illustrated in Fig. 11.19.

Butterworth The magnitude-squared function of an Nth order Butterworth lowpass filter is given by

$$|H_a(j\Omega)|^2 = \frac{1}{1 + (\Omega/\Omega_c)^{2N}}. \tag{11.70}$$

where Ω_c is the cutoff frequency.

The Butterworth filter is optimal according to a flatness criterion. For a specified filter order and cut-off frequency, the magnitude response of the Butterworth filter is the solution that attains the maximum number of derivatives equal to 0 at $\Omega = 0$ and ∞ ($\omega = 0$ and π for the digital filter). This magnitude response is maximally flat in the passband [i.e., the first $(2N - 1)$ derivatives of $|H_a(j\Omega)|^2$ are zero at $\Omega = 0$], and it decreases monotonically in the passband and stopband. Note that $|H_a(\Omega = 0)| = 1$ and $|H_a(\Omega = \Omega_c)| = 1/\sqrt{2}$, for all N. Also, as the filter order N increases, the transition width decreases, yielding a sharper cutoff edge.

The Butterworth filter has the poorest frequency selectivity compared to the Chebyshev and Elliptic filters, but it is the simplest to design.

Chebyshev: Types I and II If the filter specs are given in terms of passband and stopband ripples (as shown in Fig. 11.2), then these specs are exceeded for a Butterworth filter because of the monotonic behavior of the magnitude response. The specs can be met more efficiently with a lower-order filter if the error is distributed uniformly over the passband or the stopband or (best) both. This can be accomplished by choosing an approximating filter with an *equiripple* behavior.

The magnitude response of a Type I Chebyshev filter is equiripple in the passband and monotonic in the stopband. The magnitude-squared response is given by

$$|H_a(j\Omega)|^2 = \frac{1}{1 + \epsilon^2 T_N^2(\Omega/\Omega_c)}, \tag{11.71}$$

where $T_N(x)$ is the Nth degree Chebyshev polynomial in x, ϵ is a parameter specified by the allowable passband ripple, Ω_c is the filter cutoff frequency, and N is the filter order. The Type I Chebyshev filter is optimal according to a Chebyshev criterion in the passband and a flatness criterion in the stopband. For a specified filter order and passband edge, the magnitude response of this filter attains the minimum Chebyshev error in the passband and the maximum number of vanishing derivatives at $\Omega = \infty$ ($\omega = \pi$ for the digital filter).

Note that $|H_a(\jmath\Omega)|^2$ ripples between 1 and $1/(1+\epsilon^2)$ in the passband ($0 \leq |\Omega| \leq \Omega_c$) since $0 \leq T_N^2(x) \leq 1$ for $0 \leq x \leq 1$. For $x > 1$, $T_N^2(x)$ increases monotonically; so, $|H_a(\jmath\Omega)|^2$ decreases monotonically in the stopband ($\Omega > \Omega_c$).

From Eq. (11.71), three parameters are required to specify the filter: ϵ, Ω_c, and N. In a typical design, ϵ is specified by the allowable passband ripple δ_p by solving

$$\frac{1}{1+\epsilon^2} = (1 - \delta_p)^2. \tag{11.72}$$

Ω_c is specified by the desired passband cutoff frequency, and N is then chosen so that the stopband specs are met.

A similar treatment can be made for Chebyshev II filters (also called inverse Chebyshev). The Type II Chebyshev filter has a magnitude response that is monotonic in the passband and equiripple in the stopband. It can be obtained from the Type I Chebyshev filter by replacing $\epsilon^2 T_N^2(\Omega/\Omega_c)$ in Eq. (11.71) by $[\epsilon^2 T_N^2(\Omega_c/\Omega)]^{-1}$, resulting in the following magnitude-squared function:

$$|H_a(\jmath\Omega)|^2 = \frac{1}{1 + \left[\epsilon^2 T_N^2(\Omega_c/\Omega)\right]^{-1}}. \tag{11.73}$$

For the Chebyshev II filter, the parameter ϵ is determined by the allowable stopband ripple δ_s as follows:

$$\frac{\epsilon^2}{1+\epsilon^2} = (1 - \delta_s)^2. \tag{11.74}$$

The order N is determined so that the passband specs are met.

The Chebyshev filter is so called because the Chebyshev polynomials are used in the formula.

Elliptic The magnitude response of an Elliptic filter is equiripple in both the passband and stopband. It is optimal according to a weighted Chebyshev criterion. For a specified filter order and band edges, the magnitude response of the Elliptic filter attains the minimum weighted Chebyshev error. In addition, for a given order N, the transition width is minimized among all filters with the same passband and stopband deviations.

The magnitude-squared response of an Elliptic filter is given by:

$$|H_a(\jmath\Omega)|^2 = \frac{1}{1 + \epsilon^2 E_N^2(\Omega)}, \tag{11.75}$$

where $E_N(\Omega)$ is a Jacobian elliptic function [11]. Elliptic filters are so called because elliptic functions are used in the formula.

Remarks Note that, for these four filter types, the approximation is in the magnitude and no phase approximation is achieved. Also note that each of these filter types has a symmetric FIR counterpart.

The four types of IIR filters shown in Fig. 11.19 are usually obtained from analog prototypes via the bilinear transformation (BLT), as described in Section on page **11**-30. The analog filter $H(s)$ is designed to approximate the ideal lowpass filter over the imaginary axis. The BLT maps the imaginary axis to the unit circle $|z| = 1$, and is given by the change of variables, $s = K\frac{z-1}{z+1}$. This mapping preserves the optimality of the four classical filter types. Another method for obtaining IIR digital filters from analog prototypes is the impulse-invariant method [11]. In this method, the impulse response of a digital filter is obtained

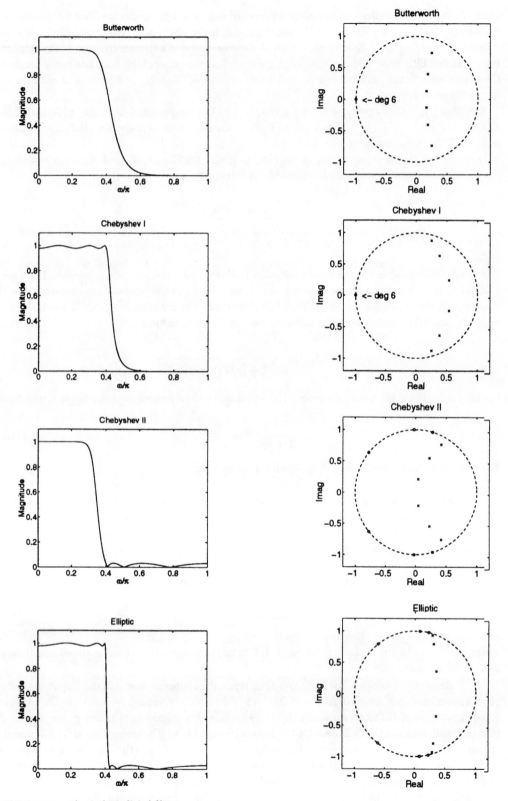

FIGURE 11.19 Classical IIR digital filters.

by sampling the continuous-time/space impulse response of the analog prototype. However, the impulse invariance method usually results in aliasing distortion and is appropriate only for bandlimited filters. For this reason, the bilinear transformation method is usually preferred.

Note that, for the four analog prototypes described above, the numerator degree of the designed digital IIR filter equals the denominator degree.[5] For the design of digital IIR filters with unequal numerator and denominator degree, analytic techniques are available only for special cases (see Section 11.4.2). For other cases, iterative numerical methods are required.

Highpass, bandpass, and band-reject filters can also be obtained from analog prototypes (or from the digital versions) by appropriate frequency transformations [11]. Those transformations are generally useful only when the IIR filter has equal degree numerator and denominator, which is the case for the digital versions of the classical analog prototypes.

A fifth IIR filter for which closed form expressions are readily available is the all-pole filter that possesses a maximally flat group delay at $\omega = 0$. In this case, no magnitude approximation is achieved. It should be noted that this filter is not obtained directly from the analog equivalent, the Bessel filter (the BLT does not preserve the maximally flat group delay characteristic). Instead, it can be derived directly in the digital domain [40]. For a specified filter order and DC group delay, the group delay of this filter attains the maximal number of vanishing derivatives at $\omega = 0$. The particularly simple formula for $H(z)$ is

$$H(z) = \frac{\sum_{k=0}^{N} a_k}{\sum_{k=0}^{N} a_k z^{-k}} \quad \text{where} \quad a_k = (-1)^k \binom{N}{k} \frac{(2\tau)_k}{(2\tau + N + 1)_k} \tag{11.76}$$

where τ is the DC group delay, and the pochhammer symbol $(x)_k$ denotes the rising factorial: $(x) \cdot (x + 1) \cdot (x + 2) \cdots (x + k - 1)$. An example is shown in Fig. 11.20, where it is evident that the magnitude response makes a poor lowpass filter. However, such a filter (1) can be cascaded with a symmetric FIR filter that improves the magnitude without affecting its phase linearity [41], and (2) is useful for fractional delay allpass filters as described in Section 11.4.2.2.

11.3.2.3 Comments and Generalizations

The design of IIR digital filters by transformation of classical analog prototypes is attractive because formulas exist for these filters. Unfortunately, digital filters so obtained necessarily possess an equal number of poles and zeros away from the origin. For some specifications, it is desired that the numerator and denominator degrees not be restricted to be equal.

Several authors have addressed the design and the advantages of IIR filters with unequal numerator and denominator degrees [42, 43, 44, 45, 46, 47, 48]. In [46, 49], Saramäki finds that the classical Elliptic and Chebyshev filter types are seldom the best choice. In [42] Jackson improves the Martinez/Parks algorithm and notes that, for equiripple filters, the use of just two poles "is often the most attractive compromise between computational complexity and other performance measures of interest."

Generally, the design of recursive digital filters having unequal denominator and numerator degrees requires the use of iterative numerical methods. However, for some special cases, formulas are available. For example, a digital generalization of the classical Butterworth filter can be obtained with the formulas given in [50]. Figure 11.21 illustrates an example. It is evident from the figure, that some zeros of the filter contribute to the shaping of the passband. The zeros at $z = -1$ produce a flat behavior at $\omega = \pi$, while the remaining zeros, together with the poles, produce a flat behavior at $\omega = 0$. The specified cut-off frequency determines the way in which the zeros are split between the $z = -1$ and the passband.

---

[5] Possibly, however, a single pole is located at $z = 0$, in which case their degrees differ by one.

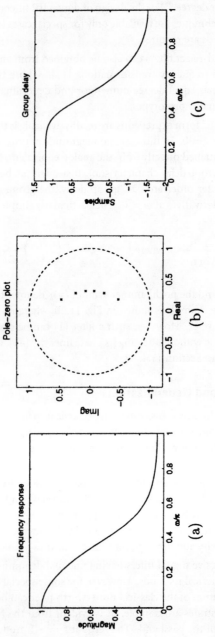

FIGURE 11.20 Maximally flat delay IIR filter, $N = 6$, $\tau = 1.2$.

To illustrate the effect of various numerator and denominator degrees, examine a set of filters for which (1) the sum of the numerator degree and the denominator degree is constant, say 20, and (2) the cut-off frequency is constant, say $\omega_c = 0.6\pi$. By varying the number of poles from 0 to 10 in steps of 2 (so that the number of zeros is decreased from 20 to 10 in steps of 2), the filters shown in Fig. 11.22 are obtained.

Figure 11.22 also shows the negative reciprocal of the slope of the magnitude response at the cut-off frequency — this indicates the width of the transition band. Notice that, for this example, as the number of poles and zeros become more equal, the transition becomes sharper. It is interesting to note that the improvement is greatest when the number of poles is increased from 0 to 2. When implementation issues are taken into consideration, the filters with two or four poles appear to attain a good trade-off between performance and implementation complexity.

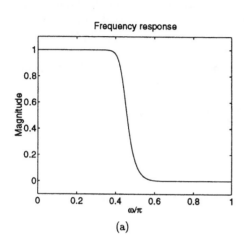
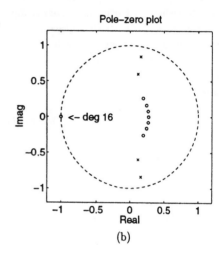

(a)   (b)

FIGURE 11.21   Generalized Butterworth filter.

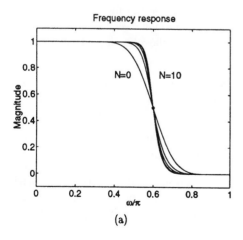
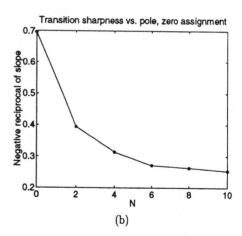

(a)   (b)

FIGURE 11.22   The filters for which the cut-off frequency is $\omega_o = 0.6\pi$, and for which the sum of the number of poles and the number of zeros is 20. N denotes the number of poles.

11.4   Other Developments in Digital Filter Design

11.4.1   FIR Filter Design

Ivan W. Selesnick, C. Sidney Burrus,
Lina J. Karam, and
James H. McClellan

11.4.1.1   Maximally Flat Real Symmetric FIR Filters

By requiring the derivatives of the amplitude function $A(\omega)$ to satisfy derivative constraints at $\omega = 0$ and $\omega = \pi$, a lowpass filter is obtained having a very flat monotone response, see Fig. 11.23. The resulting design is very simple, efficient implementations of such filters exist [51, 52], and the filters have been found to be useful when used together [53] or in conjunction with other filters [54]. Such filters preserve the input signal around $\omega = 0$ very well, and achieve very high attenuation in the stopband. The transition between the passband and stopband is wide, however. This design problem was introduced by Herrmann [55] and is formulated as follows.

Given $N = 2M + 1$ and K ($1 \leq K \leq M$), find a symmetric filter of length N such that the amplitude response, given by

$$A(\omega) = h(M) + 2 \sum_{n=1}^{M} h(M - n) \cos n\omega \tag{11.77}$$

satisfies the following constraints:

1. $A(\omega = 0) = 1$

2. $\dfrac{\partial^{2i}}{\partial^{2i}\omega} A(\omega = 0) = 0$ for $i = 1, 2, \ldots, M - K$.

3. $\dfrac{\partial^{2i}}{\partial^{2i}\omega} A(\omega = \pi) = 0$ for $i = 0, 1, \ldots, K - 1$.

The odd indexed derivatives of $A(\omega)$ are automatically zero at $\omega = 0$, so they do not need to be specified. The solution has the property that $A^{(i)}(\omega = 0) = 0$ for $i = 1, \ldots, 2(M - K) + 1$ and $A^{(i)}(\omega = \pi) = 0$ for $i = 1, \ldots, 2K - 1$. These equations are linear in the unknown filter coefficients; however, they are ill-conditioned. Fortunately, the solution can be written in closed form in several ways [55, 56].

It is convenient to use the transformation $x = \frac{1}{2}(1 - \cos \omega)$, then the solution can be written [55] as

$$A(x) = (1 - x)^K \sum_{n=0}^{M-K} d(n) x^n \tag{11.78}$$

where

$$d(n) = \binom{K - 1 + n}{n} = \frac{(K - 1 + n)!}{(K - 1)! \, n!}. \tag{11.79}$$

The transfer function has $2K$ zeros at $z = -1$, and these are the only stopband zeros. The zeros not lying at $z = -1$ can be found by computing the roots of $\sum_{n=0}^{M-K} d(n) x^n$ and mapping them back to the z domain via $z = 1 - 2x \pm \sqrt{(2x - 1) - 1}$. This equation is understood by writing $\cos \omega$ as $\frac{1}{2}(e^{j\omega} + e^{-j\omega})$ and, in turn, as $\frac{1}{2}(z + \frac{1}{z})$.

For the special case $2K = M + 1$, the polynomial $A(x)$ in Eq. (11.78) has become famous for its role in Daubechies' construction of compactly supported orthogonal wavelets [57].

Given a desired cut-off frequency and transition width, design formulas have been found [55, 58] that give approximate values for N and K. In particular, Kaiser reported that the filter length is approximately

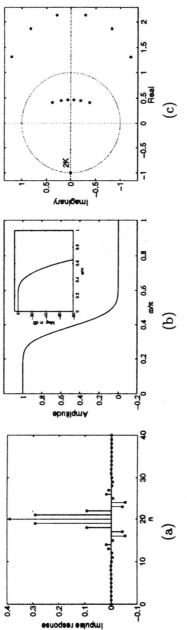

FIGURE 11.23 Maximally flat filter, $N = 41$, $K = 14$.

inversely proportional to the *square* of the transition width: $M \approx \left(\frac{\pi}{\omega_b - \omega_a}\right)^2$ where ω_b is that frequency at which $A(\omega) = 0.05$ and ω_a is that frequency at which $A(\omega) = 0.95$. Accordingly, halving the width of the transition band requires increasing the filter length by roughly a factor of four.

Because the filter has $2K$ zeros at $z = -1$ the number of multiplications can be reduced by extracting the factor $\left(\frac{1+z^{-1}}{2}\right)^{2K}$ as is indicated in Eq. (11.78). (This factor can be implemented without multiplications.) The large dynamic range of $d(n)$ can be avoided by using the structure suggested by Vaidyanathan [52] that uses the observation $d(n) = \frac{K+n-1}{n} d(n-1)$. A multiplierless implementation based on the De Casteljau algorithm is described in [51].

The formulas above permit only an approximate specification of the cut-off frequency — the only parameters the user controls is N and K. For $N = 21$, Fig. 11.24 illustrates the filters obtained by letting $K = 5$ and $K = 6$. Call them $h_1(n)$ and $h_2(n)$. To obtain a maximally flat symmetric filter having a half-magnitude frequency[6] ω_o between those of h_1 and h_2, a weighted average of h_1 and h_2 can be used [59, 60]. The desired filter is $h(n) = c \cdot h_1(n) + (1-c) \cdot h_2(n)$ where $c = (0.5 - H_2(\omega_o))/(H_1(\omega_o) - H_2(\omega_o))$. For $\omega_o = 0.56\pi$, the response of the new filter $h(n)$ is shown as a dashed line in Fig. 11.24.

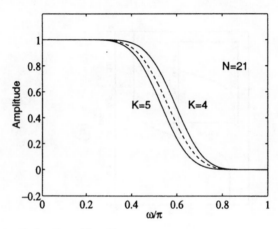

FIGURE 11.24   Three maximally flat filters, $N = 21$.

Remarks

- Extremely good at $\omega = 0$ and $\omega = \pi$.
- Simple design.
- Efficient implementations.
- Smooth impulse response.
- Wide transition.

11.4.1.2   The Affine Filter Structure

It is frequently useful to employ the structure shown in Fig. 11.25, the transfer function of which is

$$H(z) = H_1(z)H_2(z) + H_3(z). \tag{11.80}$$

---

[6]The half-magnitude frequency ω_o is that frequency such that $A(\omega_o) = \frac{1}{2}$.

In many cases, $H_2(z)$ and $H_3(z)$ are already known or determined, and it is desired that $H_1(z)$ be designed so that the overall transfer function approximates a desired transfer function $D(z)$ according to some chosen criteria.

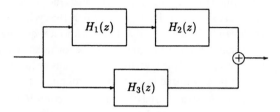

FIGURE 11.25   Affine filter structure.

Note that (1) if h_1, h_2, and h_3 are each symmetric, (2) if $h_1 * h_2$ has the same type of symmetry as h_3, and (3) if $h_1 * h_2$ and h_3 are of the same length, then the filter Eq. (11.80) is itself symmetric. In this case, designing $H_1(z)$ by minimizing either the weighted square error or the weighted Chebyshev error is particularly straightforward. An equivalent problem is obtained as follows, having a modified desired function and a modified weighting function.

Let the amplitudes of the filters be $A_1(\omega)$, $A_2(\omega)$, and $A_3(\omega)$, where $A_1(\omega) = Q(\omega)P(\omega)$ and $P(\omega)$ is a cosine polynomial as in Table 11.1. Then $A(\omega) = Q(\omega)P(\omega)A_2(\omega) + A_3(\omega)$. First consider the design via the Chebyshev norm:

$$||E(\omega)||_\infty = \max_\omega |\overline{W}(\omega)(P(\omega) - \overline{D}(\omega))| \qquad (11.81)$$

where

$$\overline{W}(\omega) = W(\omega)Q(\omega)A_2(\omega) \qquad \overline{D}(\omega) = \frac{D(\omega) - A_3(\omega)}{Q(\omega)A_2(\omega)}. \qquad (11.82)$$

The minimization of Eq. (11.81) can be accomplished by the Parks-McClellan algorithm or by linear programming if it is required that additional linear constraints be satisfied.

For the least squares error:

$$||E(\omega)||_2 = \left(\frac{1}{\pi}\int_0^\pi \overline{W}(\omega)\left(P(\omega) - \overline{D}(\omega)\right)^2 d\omega\right)^{\frac{1}{2}} \qquad (11.83)$$

where

$$\overline{W}(\omega) = W(\omega)(Q(\omega)A_2(\omega))^2 \qquad \overline{D}(\omega) = \frac{D(\omega) - A_3(\omega)}{(Q(\omega)A_2(\omega))^2}. \qquad (11.84)$$

The minimization of Eq. (11.83) can be accomplished by solving the linear system Eq. (11.33), or Eq. (11.39) if it is required that additional linear constraints be satisfied.

In some design problems, the form of Eq. (11.80) is useful because it describes a parameterization (or constraint) where $H_1(z)$ represents the available degrees of freedom [61, 62, 63].

Prefilters   In addition, the design of filters having low implementation complexity often employs the structure in Fig. 11.25. One strategy is to choose transfer functions $H_2(z)$, $H_3(z)$, having very low implementation complexity — such filters may have crude frequency responses, but they can often be implemented without multipliers and few additions. $H_1(z)$ is then designed so that the overall transfer function meets the specified requirements.

This approach, introduced in [64], is often called "prefiltering," especially when $H_3(z) = 0$. In this case, $H_2(z)$ is the prefilter. Prefilters are filters having (1) very low implementation complexity, but (2) imperfect frequency responses. In this case, $H_1(z)$ is sometimes called an equalizer. In [64], it is shown that this approach provides benefits in (1) reduced computational complexity, (2) reduced sensitivity to coefficient quantization, and (3) reduced roundoff noise. For narrowband filters, this approach gives a particularly good reduction in implementation complexity.

One class of prefilters [64, 65] is obtained by combining recursive running sum (RRS) building blocks.[7] The RRS filter is simple to implement and has all its zeros equally spaced on the unit-circle (except at $z = 1$). Other prefilters are obtained from cyclotomic polynomials [66] — all the roots of which lie on the U.C. Because all the coefficients are simple small integers [the first 105 cyclotomic polynomials (CPs) have coefficients in $\{-1, 0, 1\}$], CPs can be implemented as filters without requiring multipliers. In [67], it is shown that the problem of designing prefilters from CPs can be formulated as an optimization problem with linear objective functions by applying the logarithm to the transfer function of the CP prefilter. The design problem is then solved in [67] by mixed integer linear programming.

IFIR Filters   Another useful structure has the transfer function $H_1(z^M)H_2(z)$ [54]. The impulse response of $H_1(z^M)$ is sparse, so arithmetic complexity is reduced. A time domain interpretation emerges by considering the convolution of $h_1(Mn)$ and $h_2(n)$. $h_2(n)$ fills in, or interpolates, the gaps in $h_1(Mn)$. This structure is particularly well suited for efficient implementations of narrow band lowpass filters. For other frequency responses, the generalization is *masking*, see for example [17].

11.4.1.3 Nonsymmetric or Nonlinear Phase FIR Filter Design

Although the requirement that an FIR filter be real and symmetric simplifies the filter approximation problem, it is sometimes more restrictive than is desirable. The following scenarios motivate the consideration of nonsymmetric and/or non-linear phase FIR filters:

1. In some cases, phase linearity is of little importance and it is more important that the delay be low. Recall that the group delay of a symmetric filter is half its filter order. This delay is higher than necessary. In other cases, exactly linear phase is not required, but some degree of phase linearity is desired. It is then desirable to sacrifice exactly linear phase in exchange for delay reduction and/or delay control. The desired constant delay can be specified by explicitly including the phase or desired group delay as part of the design specifications as indicated in the following subsection on optimal design of FIR filters. The resulting designed nearly linear-phase filter has a conjugate symmetric frequency response and a real-valued, nonsymmetric, impulse response (See Design Examples at the end of the subsection on optimal design of FIR filters).

2. Sometimes it is required that $H(\omega)$ approximate a desired nonsymmetric or nonlinear phase frequency response $D(\omega)$.[8] Examples include equalizer design [68], fractional delay filter design [21], and seismic migration filter design [2].

In each case, the additional degrees of freedom that are made available by giving up symmetry or phase linearity can be used to improve the phase and/or magnitude response.

Approaches to the design of nonsymmetric and/or non-linear phase FIR filters fall roughly into at least three categories:

1. General complex approximation (see "Optimal Design of FIR Filters with Arbitrary Magnitude and Phase, below). Given an arbitrary desired frequency response $D(\omega)$, the best Chebyshev, or least square, approximation is found. For the Chebyshev criterion, the approximation is significantly more difficult in the general complex case than in the *real symmetric* case.

---

[7]Based on the factorization $\sum_{k=0}^{L-1} z^k = (z^L - 1)/(z - 1)$, the RRS filter is a recursive implementation of the running sum.

[8]Note that the frequency response of a filter can be symmetric with a *nonlinear* phase (e.g., seismic migration filters designed in the next section).

Recently, several algorithms have been presented for designing general filters in the Chebyshev sense [2, 3, 4, 5, 69, 141, 143].

2. Design of minimum-phase filters by spectral factorization of square magnitude approximation [70]. This is a very effective technique, and it can be used in conjunction with the maximally flat, least square, and Chebyshev criterion.

3. The simultaneous approximation of magnitude and group delay. There is little theory to facilitate the solution to this nonlinear problem, but see [71, 72, 73, 74, 75, 142] and "Delay Variation of Maximally Flat FIR Filters" on page **11**-54.

11.4.1.4 Optimal Design of FIR Filters with Arbitrary Magnitude and Phase

As indicated before, the *alternation theorem* [76] is at the basis of the Parks-McClellan (second Remez exchange) algorithm described in Section 11.3.1. Karam and McClellan recently extended the alternation theorem from the real-only to the general complex case [2]. As a result, they derived an efficient multiple-exchange algorithm [3, 10] for the design of optimal FIR filters with arbitrary magnitude and phase specifications approximated in the Chebyshev sense. Both causal and non-causal filters with complex or real-valued impulse responses can be designed. In addition, the Karam-McClellan algorithm *exactly* reduces to the classic Parks-McClellan (second Remez exchange) algorithm when real-only or imaginary-only filters are designed and is, therefore, a true generalization of the classic Remez algorithm to the complex case. A version of the Karam-McClellan algorithm (**cremez**) is currently available as part of the Signal Processing Toolbox in MATLAB^{TM} (Version 5).

Problem Formulation   The complex FIR filter design problem may be stated as follows.

Let $D(\omega)$ be the desired magnitude and phase of the filter frequency response defined on a compact frequency subset $B \subset [-\pi, \pi)$. $D(\omega)$ is to be approximated by an FIR filter having a frequency response $H(\omega)$ and an impulse response $h_n, n = N_1, \ldots, N_2$, of length $N = N_2 - N_1 + 1$. The filter design problem consists in finding the filter coefficients $\{h_n\}$ that will minimize the Chebyshev error norm

$$\|E(\omega)\| = \max_{\omega \in B}\{|D(\omega) - H(\omega)|\}, \tag{11.85}$$

where

$$H(\omega) = \sum_{n=N_1}^{N_2} h_n e^{-j\omega n} \tag{11.86}$$

The error norm (11.85) can include a real, strictly positive, and continuous weighting function $W(\omega)$ on B by simply replacing $D(\omega)$ with $W(\omega)D(\omega)$ and $H(\omega)$ with $W(\omega)H(\omega)$.

Note that this formulation will handle both causal filters ($N_1 \geq 0$) and noncausal filters ($N_1 < 0$). Although some authors [77] have reported an ill-conditioned behavior when using Eq. (11.86), the error (11.85) can be rewritten so that the problem is well-posed by removing a linear phase term due to N_1. This new problem, with a guaranteed unique optimal solution, results by rewriting $D(\omega)$ and $H(\omega)$ with respect to a linear phase term as

$$D(\omega) = e^{-j\frac{N_1+N_2}{2}\omega} A(\omega) \tag{11.87}$$

and

$$H(\omega) = e^{-j\frac{N_1+N_2}{2}\omega} H_{nc}(\omega). \tag{11.88}$$

The linear phase $e^{-j\frac{N_1+N_2}{2}\omega}$ does not affect the magnitude of the error (11.85); so the design problem works with the following equivalent expression for the error magnitude:

$$|E(\omega)| = |A(\omega) - H_{nc}(\omega)|. \tag{11.89}$$

The function $H_{nc}(\omega)$ can be expressed as a linear combination of real basis functions satisfying the Haar Condition [2, 78]:

$$H_{nc}(\omega) = \begin{cases} \sum_{k=0}^{(N-1)/2} \alpha_k \cos k\omega + \sum_{k=1}^{(N-1)/2} \beta_k \sin k\omega, & N \text{ odd} \\ \\ \sum_{k=0}^{(N-2)/2} [\alpha_k \cos(k+\frac{1}{2})\omega + \beta_k \sin(k+\frac{1}{2})\omega], & N \text{ even} \end{cases} \quad (11.90)$$

The Haar condition [76, 79], which is satisfied by the cos() and sin() basis functions, guarantees that the optimal solution is unique and that the set of extremal points of the optimal error function, $E_o(\omega)$, consists of at least $n + 1$ points, where n is the number of approximating basis functions.

The parameters $\{\alpha_k, \beta_k\}$ in Eq. (11.90) are the complex coefficients that need to be determined such that $H_{nc}(\omega)$ best approximates $A(\omega)$. The filter coefficients $\{h_n\}$ can be very easily obtained from $\{\alpha_k, \beta_k\}$ [78]. Usually, the number of approximating basis functions in Eq. (11.90) is $n = N$, but this number is reduced by half when $A(\omega)$ is symmetric (all $\{\beta_k\}$ are equal to 0), or antisymmetric (all $\{\alpha_k\}$ are equal to 0).

The Design Algorithm　A main strategy in Chebyshev approximation is to work on sparse finite subsets, B_s, of the desired frequency set B and relate the optimal error on B_s to the optimal error on B. The norm of the optimal error on B_s will always be a lower bound to the error norm on B [79]. If $\|E_s\|$ denotes the optimal error norm on the sparse set B_s, and $\|E_o\|$ the optimal error norm on B, the design problem on B is solved by finding the subset B_s on which $\|E_s\|$ is maximal and equal to its upper bound $\|E_o\|$. This could be done by iteratively constructing new subsets B_s with monotonically increasing error norms $\|E_s\|$. For that purpose, two main issues must be addressed in developing the approximation algorithm:

1. Finding an efficient way to compute the best approximation $H_s(\omega)$ on a given subset B_s of r points ($r \geq n + 1$).
2. Devising a simple strategy to construct a new subset B_s where the optimal error norm $\|E_s\|$ is guaranteed to increase.

While in the real case it is sufficient to consider subsets containing $r = n + 1$ points, the minimal subset size r is not known _a priori_ in the complex case. The fundamental theorem of complex Chebyshev approximation tells us that r can take any value between $n + 1$ and $2n + 1$. It is desirable, whenever possible, to keep the size of the subsets, B_s, small since the computational complexity increases with the size of B_s. The case where $r = n + 1$ points is important because, in that case, it was shown [2] that the best approximation on a subset of $n + 1$ points can be simply computed by solving a linear system of equations. So, the first issue is directly resolved.

In addition, by exploiting the alternation property[9] of the complex optimal error on B_s efficient multi-point exchange rules can be derived and the second issue is easily resolved. These exchange rules were derived in [2, 78] resulting in the very efficient complex Remez algorithm which iteratively constructs best approximations on subsets of $n + 1$ points with monotonically increasing error norms $\|E_s\|$.

The complex Remez algorithm terminates when finding the set B_s having the largest error norm ($\|E_s\| = |\delta|$) among all subsets consisting of exactly $n + 1$ points. This complex Remez multiple-exchange algorithm converges to the optimal Chebyshev solution on B when the optimal error $E_o(\omega)$ satisfies an alternating property [78]. Otherwise, the computed solution is optimal over a reduced set $B' \subset B$. In this latter case, the maximal error norm $|\delta|$ over the sets of $n + 1$ points is strictly less than, but usually very close to, the upper bound $\|E_o\|$. To compute the optimum over B, subsets consisting of more than $n + 1$

---

[9]_Alternation_ in the complex case corresponds to a phase shift of π when going from one extremal point to the next in sequence.

($r > n + 1$) need to be considered. Such sets are constructed by the second stage of the new algorithm presented in [3, 10], starting with the solution generated by the initial complex Remez stage.

When $r > n + 1$, both issues mentioned above are much harder to resolve. In particular, a simple and efficient point-exchange strategy, where the size of B_s is kept minimal and constant, does not seem possible when $r > n + 1$. The approach in [3, 10] is to use a second ascent stage for constructing a sequence of best approximations on subsets of r points ($r > n + 1$) with monotonically increasing error norms (ascent strategy). The algorithm starts with the best approximation on subsets of $n + 1$ points (minimum possible size) using the very efficient complex Remez algorithm [2] and then continues constructing the sequence of best approximations with increasing error norms on subsets B_s of more than $n + 1$ points by means of a second stage. Since the continuous domain B is represented by a dense set of discrete points, the proposed design algorithm must yield an approximation of maximum norm in a finite number of iterations since there is a finite number of distinct subsets B_s containing r ($n + 1 \leq r \leq 2n + 1$) points in the discrete set B.

A detailed block diagram of the design algorithm is shown in Fig. 11.26. The two stages of the new algorithm have the same basic ascent structure. They both consist of the two main steps shown in Fig. 11.26, and they only differ in the way these steps are implemented.

A detailed block diagram of the complex Remez stage (Stage 1) is also shown in Fig. 11.27. Note that when $D(\omega)$ is real-valued, δ will also be real and, therefore, the real phase-rotated error $E_r(\omega)$ is equal to $\pm E(\omega)$. In this case, the presented algorithm reduces to the Parks-McClellan algorithm as modified by McCallig [80] for approximating general real-valued frequency responses in the Chebyshev sense. Moreover, for many problems, the resulting initial approximation computed by the complex Remez method is the optimal Chebyshev solution and, thus, the second stage of the algorithm does not need to execute. Even when the resulting initial solution is not optimal, it has been observed that the computed deviation $|\delta|$ is very close to the optimal error norm $\|E_o\|$ (its upper bound).

As indicated above, the second stage is invoked only when the complex Remez stage (Stage 1) results in a subset optimal solution. In this case, the initial set B_s of Stage 2 is formed by taking the set of all local maxima of the error corresponding to the final solution computed by Stage 1. The resulting $B_s \subset B$ would then contain r points, where $n + 1 < r \leq 2n + 1$. The best approximation on the constructed subset, B_s, is computed by means of a generalized descent method [10, 78] suitably adapted for minimizing the nondifferentiable Chebyshev error norm. The total number of ascent iterations is independent of the method used for computing the best solution $H_s(\omega)$ on B_s. Then, the new sets, B_s, are constructed by locating and adding the new local maxima of the error on B to the current subset, B_s, and by removing from B_s those points where the error magnitude is relatively small. So, the size of the constructed subsets varies up and down. The algorithm terminates when all the extremal points of $E(\omega)$ are in B_s.

It should be noted that each iteration of Stage 2 includes descent iterations, which we will refer to as descent steps.[10] An observation in relation to the complexity of the two stages of the algorithm is in order. The initial complex Remez stage is extremely efficient and does not produce any significant overhead. However, one iteration of the second stage includes several descent steps, each one having higher computational complexity than the initial complex Remez stage. For convenience, the term *major iterations* will be used to refer to the iterations of the second stage. From the discussion above, it follows that the initial complex Remez stage is comparable to one step in a major iteration and can thus be regarded as an initialization step in the first major iteration.

An interesting analogy of the proposed two-stage algorithm with the first and second algorithms of Remez can be made. It should be noted that both Remez algorithms can be used for solving real one-dimensional Chebyshev approximation problems satisfying the Haar condition. The two real Remez algorithms involve the solution of a sequence of discrete problems [81]: at each iteration, a finite discrete

---

[10]The simplex method of linear programming could also be used for the descent steps.

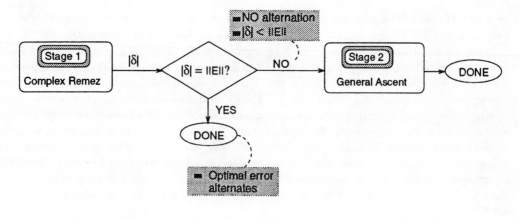

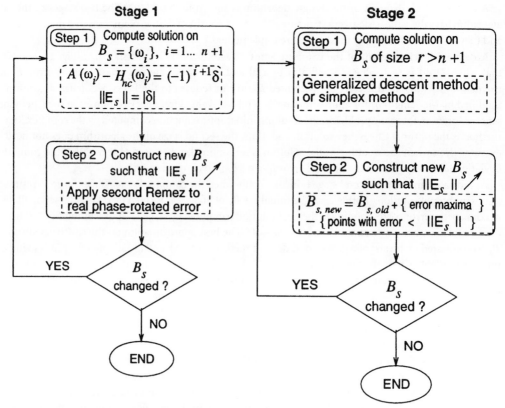

FIGURE 11.26 Block diagram of the Karam-McClellan design algorithm. $|\delta|$ is the maximal optimal deviation on the sets B_s consisting of $n + 1$ points in B. $\|E\|$ is the Chebyshev error norm on B.

subset, B_s, is defined and the best Chebyshev approximation is computed on B_s. In the second algorithm of Remez, the successive subsets B_s contain exactly $n + 1$ points: an initial subset of $n + 1$ points is replaced by $n + 1$ local maxima of the current real error function. In the first algorithm of Remez, the initial point set contains at least $n + 1$ points, and these points are supplemented at each iteration by the global maximum of the current approximation error. As shown in [2], the complex Remez stage (Stage 1) of the new proposed algorithm is a generalization of the second Remez algorithm to the complex case and reduces to it when real-valued or pure imaginary functions are approximated. On the other hand, the

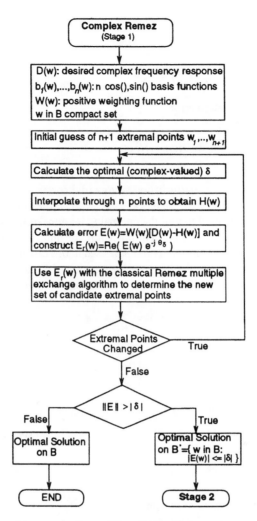

FIGURE 11.27   Block diagram of the complex Remez (Stage 1) algorithm.

second stage of the proposed algorithm can be compared to the first Remez algorithm in that the size of the constructed subsets B_s is variable and is greater than $n + 1$, except at the initial iteration. A main difference between the second stage and the first Remez algorithm is that the second stage is based on a multiple-exchange strategy while the first algorithm of Remez is a single-exchange method.

Descent Steps   In what follows, we describe the generalized descent method and the simplex method which can be used in Step 1 of Stage 2 to compute the optimal Chebyshev solution on the discrete set of points B_s. The descent method presented in this section is based on the work of Demjanov–Malozemov [82, 83] and Wolfe [84], and is suitably adapted for minimizing the nondifferentiable Chebyshev error norm.

Let $D(\omega)$ be the function that is to be approximated on B_s, and let $H_{s,0}(\omega)$ be an initial approximation given by the basis coefficient vector

$$\mathbf{c}_0 = [c_{01}, c_{02}, \ldots, c_{0n}]^T \tag{11.91}$$

whose elements are the n (complex or real) coefficients associated with the cos() and/or sin() basis functions $\{\phi_i\}_{i=0}^n$. The superscript T in Eq. (11.91) refers to the transpose operation. The descent method iteratively generates a sequence $\{\mathbf{c}_k\}$ of basis coefficient vectors, $\{\mathbf{d}_k\}$ of perturbation vectors, and $\{t_k\}$ of positive scalars such that

$$\mathbf{c}_{k+1} = \mathbf{c}_k + t_k \mathbf{d}_k \tag{11.92}$$

and

$$\|E_{s,k+1}(\omega)\| \leq \|E_{s,k}(\omega)\| \quad \text{for } \omega \in B_s \tag{11.93}$$

where $E_{s,k}(\omega)$ is the approximation error

$$E_{s,k}(\omega) = D(\omega) - H_{s,k}(\omega) = D(\omega) - \sum_{i=1}^{n} c_{ki} \phi_i(\omega) \tag{11.94}$$

and k is the iteration number. The perturbation vectors $\{\mathbf{d}_k\}$ correspond to descent directions and $\{t_k\}$ must be chosen so that $\|E_k(\omega)\|$ would significantly decrease at the next iteration. Once $\mathbf{d}_k$ is chosen, a line search method could be used to find the optimal t_k for a maximum decrease of $\|E_{s,k}(\omega)\|$ along the direction $\mathbf{d}_k$. Alternatively, a more efficient procedure for finding the best t_k was presented in [83, pp. 109–112]. Standard gradient techniques cannot be used in this case for generating the directions $\{\mathbf{d}_k\}$ since the Chebyshev error norm is a nondifferentiable function of the coefficient vector $\mathbf{c}$.

With r denoting the number of points in B_s, the Chebyshev approximation problem can be reformulated as the minimization of the function

$$\varphi(\mathbf{c}) = \max_{i \in (1,\dots,r)} e_i(\mathbf{c}) \tag{11.95}$$

where

$$e_i(\mathbf{c}) = \left| D(\omega_i) - \Phi_i^T \mathbf{c} \right|^2 \tag{11.96}$$

and

$$\Phi_i = [\phi_1(\omega_i), \phi_2(\omega_i), \dots, \phi_n(\omega_i)]^T . \tag{11.97}$$

Each $e_i(\mathbf{c})$ is a convex differentiable function with a complex gradient vector g_i given by

$$g_i = \frac{\partial e_i(\mathbf{c})}{\partial \mathbf{c}} = -2\overline{\Phi}_i E_i \tag{11.98}$$

where $\overline{\Phi}_i$ is the complex conjugate of Φ_i, and $E_i = D(\omega_i) - \Phi_i^T \mathbf{c}$. Note that g_i is a vector in the n-dimensional complex space Z_n which is isomorphic to the $2n$-dimensional real Euclidean space R_{2n}. A point $z = (z_1, \dots, z_n) \in Z_n$, with complex coordinates $z_j = \alpha_j + j\beta_j$, corresponds to the point $z = (\alpha_1, \dots, \alpha_n, \beta_1, \dots, \beta_n) \in R_{2n}$. In what follows, g_i refers to the real vector in R_{2n}.

For a given coefficient vector $\mathbf{c}$, consider the set of extremal indices $I_e(\mathbf{c})$ defined as

$$I_e(\mathbf{c}) = \{i \in (1, \dots, r) : e_i(\mathbf{c}) = \varphi(\mathbf{c})\}. \tag{11.99}$$

In other words, $I_e(\mathbf{c})$ contains every index i (corresponding to the ith point ω_i in B_s) for which $E(\omega)$ attains its maximum on B_s. Letting

$$G(\mathbf{c}) = \{g_i : i \in I_e(\mathbf{c})\}, \tag{11.100}$$

consider the convex hull $G_c(\mathbf{c})$ of $G(\mathbf{c})$. $G_c(\mathbf{c})$ is a polyhedron in R_{2n} and there is a unique point $g_{min} \in G_c(\mathbf{c})$ having minimum Euclidean norm [85]. The following gradient characterization results for $\varphi(\mathbf{c})$ [82, 85]

$$\nabla\varphi(\mathbf{c}) = g_{min} \tag{11.101}$$

and $-g_{min}$ is the direction of steepest descent at $\mathbf{c}$. Note that $\nabla\varphi(\mathbf{c})$ depends only on the set of extremal points represented by $I_e(\mathbf{c})$. So, the problem of finding the steepest descent direction reduces to the problem of finding the point of smallest norm in the convex hull of a given finite point set. An algorithm especially designed for that calculation has been presented by Wolfe [84]. The filter coefficient vector $\mathbf{c}_o$ minimizes $\varphi(\mathbf{c})$, and therefore the approximation error norm $\|E_s\|$, if and only if

$$\nabla\varphi(\mathbf{c}_o) = 0 \tag{11.102}$$

or, equivalently [see Eq. (11.101)],

$$0 \in G_c(\mathbf{c}_o). \tag{11.103}$$

Using Eq. (11.98), it can be shown that the optimality condition (11.103) reduces to the Kolmogoroff optimality criterion for Chebyshev approximation [86, p. 21].

While a direct generalization of the steepest descent method does not in general lead to convergence [82, 85], successive approximation and conjugate subgradient methods based on Eq. (11.101) have been developed for minimizing nondifferentiable functions [83, 85, 87]. The descent method presented in this section is based on the techniques presented in [83] and [84]. It is suitably adapted for solving the Chebyshev approximation problem, which was reformulated as Eqs. (11.95 through 11.97), and, consequently, for solving the filter design problem. Before describing the steps of the proposed descent method, some new definitions are needed. Define

$$I_{e,\epsilon}(\mathbf{c}) = \{i \in (1, \ldots, r) : \varphi(\mathbf{c}) - e_i(\mathbf{c}) \leq \epsilon\}, \qquad \epsilon \geq 0 \tag{11.104}$$

and

$$G_\epsilon(\mathbf{c}) = \{g_i : i \in I_{e,\epsilon}(\mathbf{c})\}. \tag{11.105}$$

Also, let $G_{c,\epsilon}(\mathbf{c})$ denote the convex hull of $G_\epsilon(\mathbf{c})$ and $g_{min,\epsilon}$ the point in $G_{c,\epsilon}(\mathbf{c})$ nearest to the origin. Clearly, $I_{e,0}(\mathbf{c}) = I_e(\mathbf{c})$, $G_0(\mathbf{c}) = G(\mathbf{c})$, $G_{c,0}(\mathbf{c}) = G_c(\mathbf{c})$, and $g_{min,0} = g_{min}$.

The basic steps of the descent algorithm can now be summarized as follows:

1. **Set initial parameters.** Fix two parameters $\epsilon_0 > 0$ and $\rho_0 > 0$, and take an initial approximation $\mathbf{c}_0$ on the desired set B_s, i.e., $\phi_{s,0}(x) = \sum_{i=1}^n c_{0i}\phi_i(x)$. Suggested values for ϵ_0 and ρ_0 are $\epsilon_0 = 0.012$ and $\rho_0 = 1.0$. Since the passage from $\mathbf{c}_k$ to $\mathbf{c}_{k+1}$ ($k = 0, 1, \ldots$) is effected the same way, suppose that the kth approximation $\mathbf{c}_k$ is already computed.

2. **Set current approximation and accuracy.** Set $\mathbf{c} = \mathbf{c}_k$, $\epsilon = \epsilon_0/2^k$, and $\rho = \rho_0/2^k$.

3. **Compute the ϵ-gradient, $g_{min,\epsilon}$.** Find the point $g_{min,\epsilon}$ of $G_{c,\epsilon}(\mathbf{c})$ nearest to the origin using the technique by Wolfe [84].

4. **Check accuracy of current approximation.** If $\|g_{min,\epsilon}\| \leq \rho$, go to Step 8.

5. **Compute the ϵ-steepest descent direction $\mathbf{d}_k$**

$$\mathbf{d}_k = -\frac{g_{min,\epsilon}}{\|g_{min,\epsilon}\|} \tag{11.106}$$

6. **Determine the best step size t_k.** Consider the ray

$$\mathbf{c}(t) = \mathbf{c} + t\mathbf{d}_k \tag{11.107}$$

and determine $t_k \geq 0$ such that

$$\varphi(\mathbf{c}(t_k)) = \min_{t \geq 0} \varphi(\mathbf{c}(t)) \tag{11.108}$$

7. **Refine approximation accuracy.** Set $\mathbf{c} = \mathbf{c}(t_k)$ and repeat from Step 3.

8. **Compute generalized gradient, g_{min}.** The technique by Wolfe [84] is used to find the point g_{min} of $G_c(\mathbf{c}_k)$ nearest to the origin (see also [83, Appendix IV]).

9. **Check stopping criteria.** If $g_{min} \equiv 0$, then $\mathbf{c}$ is the vector of the coefficients of the best approximation $H_s(\omega)$ of the function $D(\omega)$ on $B_s = \{\omega_i : i = 1, \ldots, r\}$ and the algorithm terminates.

10. **Update approximation and repeat with higher accuracy.** The approximation $\mathbf{c}_{k+1}$ is now given by

$$\mathbf{c}_{k+1} = \mathbf{c}. \tag{11.109}$$

Return to Step 2.

This successive approximation descent method is guaranteed to converge, as shown in [83].

Descent via the Simplex Method [4, 88]   Other general optimization techniques (e.g., the simplex method of linear programming [4, 88]) can also be used instead of the descent method in the second stage of the proposed algorithm. The advantage of the linear-programming method over the generalized descent method is that additional linear constraints can be incorporated into the design problem.

Using the real rotation theorem [11, p. 122]

$$|z| = \max_{-\pi \leq \theta < \pi} \text{Re}\{ze^{j\theta}\}, \qquad \text{where } z \text{ complex,} \tag{11.110}$$

the complex filter design problem on the frequency set B_s can be restated as the following linear approximation problem: find the optimal length-N impulse response $\mathbf{h}^* = [h_{N_1} \ldots h_{N_2}]^*$ such that

$$\delta(\mathbf{h}^*) = \min_{\mathbf{h}} \delta(\mathbf{h}) \tag{11.111}$$

where

$$\delta(\mathbf{h}) = \max_{\omega \in B_s} \max_{-\pi \leq \theta < \pi} (\text{Re}\{E(\omega)e^{j\theta}\})$$

$$E(\omega) = D(\omega) - H(\omega) \qquad [\text{refer to Eq. (11.86)}].$$

This problem can, in turn, be formulated as a linear program by defining

$$\mathbf{u} = \left[h_{N_1}^r, \ldots, h_{N_2}^r, h_{N_1}^i, \ldots, h_{N_1}^i, \delta \right]$$

$$\mathbf{k} = [0, \ldots, 0, 0, \ldots, 0, 1]$$

where $h_n^r = \text{Re}\{h_n\}$, $h_n^i = \text{Re}\{h_n\}$, and $\delta = \mathbf{u}\mathbf{k}^T$. The resulting linear program becomes

$$\min_{\mathbf{u}} \mathbf{u}\mathbf{k}^T \tag{11.112}$$

subject to $\text{Re}\{E(\omega)e^{j\theta}\} \leq \delta$, for all $\omega \in B_s$ and $\theta \in [-\pi, \pi]$. Alternatively, the dual linear program can be formulated and solved [4, 88].

Design Examples   In the following design examples, the filter specifications are given in terms of the normalized frequency $f = \omega/2\pi$.

Low Delay Filters with Nearly Linear Phase   In many signal processing applications, linear-phase systems are particularly desirable because the effect of *exact* linear-phase is a perfect delay. While exactly linear-phase causal filters exhibit a constant group delay, the delay they introduce is proportional to the filter length N and is always equal to $(N - 1)/2$. This delay may be unacceptably large, especially when using filters having a high degree of selectivity (sharp cutoff edges). Furthermore, in real-time applications (e.g., real-time speech and video processing), selective filters are required to have a constant group delay that is as small as possible. Minimum-phase FIR filters cause less delay, but introduce phase distortion which may have a severe effect on the shape of the processed signal. Chen and Parks [89] observed that the desired group delay which gives the minimum error deviation can be smaller than that of an exactly linear-phase filter of the same length.

Complex approximation can be used to design filters that have less delay than the exactly linear-phase filter of the same length, and which have approximately a constant group delay in the passband. The resulting complex filters are called "nearly linear-phase" and are obtained by defining the desired linear-phase frequency response to be:

$$D(\omega) = \begin{cases} e^{-j\,\tau_i\,\omega}, & \omega \in i^{\text{th}} \text{ passband} \\ 0, & \omega \in \text{stopbands} \end{cases} \qquad (11.113)$$

where τ_i is the desired group delay in the ith passband. Since the phase term is explicitly included in the approximation problem, the desired delay is fixed (τ_i) and is not determined by the FIR filter length N. Moreover, increasing N does not increase the group delay, but potentially leads to a better approximation of the desired constant delay. The definition of $D(\omega)$ should be conjugate symmetric. Then the frequency response of the optimal Chebyshev filter approximating Eq. (11.113) will also be conjugate symmetric [79, p. 27] and, therefore, the approximating filter coefficients will be real-valued.

Figure 11.28 shows the properties of a reduced delay, length-32, FIR filter designed with the following desired specifications:

$$\text{Desired:} \quad D(f) = \begin{cases} e^{-j(2\pi f)12.5} & \text{if } 0 \le |f| \le 0.06 \\ 0 & \text{if } 0.12 \le |f| \le 0.5 \end{cases}$$

$$\text{Weight:} \quad W(f) = \begin{cases} 1 & \text{if } |f| \le 0.06 \\ 10 & \text{if } 0.12 \le |f| \le 0.5 \end{cases}$$

The term "reduced delay" refers to the fact that the desired group delay ($\tau = 12.5$) is set to be smaller than $(N-1)/2 = 15.5$, which is the delay of an exactly linear-phase filter. The complex Remez stage of the multiple-exchange algorithm converges to the optimal solution in 11 exchange steps. The resulting optimal filter has a Chebyshev error norm $\|E_{opt}\| = |\delta| = 0.0439$. The FIR filter's group delay (Fig. 11.28(b)) corresponds to the nearly linear-phase characteristic in the passband. Note that the optimal error (Fig. 11.28(c)) assumes its maximum value at $N+2$ extremal points with an alternating phase shift of π. This alternation can be clearly seen from the plot of the real phase-rotated error $E_r(f)$ in Fig. 11.28(c).

Real-Valued or Exactly Linear-Phase Filters The real-valued filter design problem corresponds to the case where the function $A(\omega)$, given by Eq. (11.87), reduces to a real-valued function. In this case, the initial complex Remez stage (Stage 1 described above) always converges to the unique optimal solution of the desired function $D(\omega)$ on the specified frequency bands B.

Figure 11.29 shows the characteristics of the exactly linear-phase filter corresponding to the nearly linear-phase filter shown in Fig. 11.28; i.e., the same design specifications were used except that the delay is set to be equal to $(N-1)/2 = 15.5$ in the passband, because $N = 32$ in this example. The complex Remez stage converged to the optimal solution in eight exchange steps. The resulting optimal filter has an optimal error norm $\|E_{opt}\| = |\delta| = 0.04956$ which is *larger* than the corresponding reduced delay, nearly linear-phase filter of Fig. 11.28, which indicates that a nearly linear-phase filter not only can have less group delay but also a reduced Chebyshev error norm. Note that although the phase specifications were explicitly included as part of the approximation problem, the exactly linear-phase optimal solution was obtained as expected.

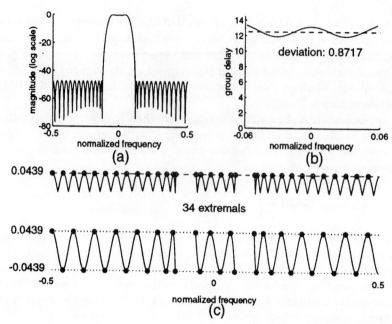

FIGURE 11.28 Example 1 — nearly linear-phase filter, $N = 32$. (a) FIR filter magnitude response in dB. (b) FIR filter (solid) and desired filter (dashed) passband group delays. (c) Magnitude of the weighted error (top), and real phase-rotated weighted error (bottom).

Seismic Migration Filters   The objective of seismic migration is to define the boundaries of the earth layers [90, 91]. For this purpose, downward propagating waves are initiated by acoustic sources at the earth. Then the migration procedure starts with the wave field measured at the earth's surface and computes the wave field values at all desired depths. This extrapolation operation is performed using a digital space-time filter whose frequency response approximates

$$D(k_x, \omega) = D_A\left(\frac{k_x}{\Delta x}, \frac{\omega}{\Delta t}\right) = \exp\left\{ j\frac{\Delta z}{\Delta x}\sqrt{\frac{\Delta x^2}{\Delta t^2}\frac{\omega^2}{v^2} - k_x^2} \right\} \tag{11.114}$$

where k_x is the spatial frequency and ω is the temporal frequency. The extrapolation in depth is usually done for a fixed frequency ω_o and a fixed velocity v_o. The process is repeated for other frequency and velocity values using frequency- and velocity-dependent migration filters. Therefore, for a fixed ratio ω_o/v_o, $D(k_x, \omega_o) = D(k_x)$ is a 1-D migration filter with a cutoff frequency equal to $\alpha = (\Delta x/\Delta t)(\omega_o/v_o)$. The objective of the migration filter design problem is to approximate the ideal frequency response $D(k_x)$. Since $D(k_x)$ is a complex-valued even-symmetric function, it can be approximated by an N-length FIR digital filter whose frequency response is given by [92]

$$H(k_x) = h_o + 2 \sum_{n=1}^{(N+1)/2} h_n \cos nk_x \tag{11.115}$$

where the filter coefficients h_n are complex-valued. Note that, even if a symmetry constraint is not imposed on the FIR filter, the resulting optimal filter will be even-symmetric when an odd-length filter is used to approximate $D(k_x)$. This property follows directly from Chebyshev approximation theory [79, p. 27]. The approximation of $D(k_x)$ needs mostly to be accurate in the region $|k_x| < |\alpha|$ (passband) which corresponds to the wavenumbers (k_x) for which waves are propagating [90, 93]. The evanescent region $|k_x| > |\alpha|$ (stopband) will contain little or no energy.

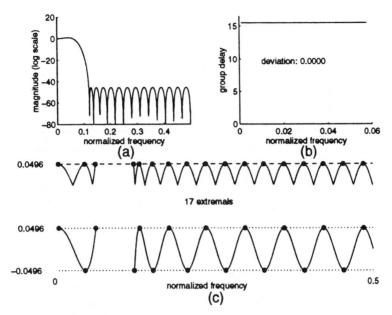

FIGURE 11.29 Example 2 — exactly linear-phase lowpass filter, $N = 32$. (a) FIR filter magnitude response in dB. (b) FIR filter (solid) and desired filter (dashed) passband group delays. (c) Magnitude of the weighted error (top), and real phase-rotated weighted error (bottom).

Figure 11.30 displays the properties of the optimal, length-31, seismic migration filter that approximates the following specifications:

$$\text{Desired:} \quad D(f) = \begin{cases} e^{j\sqrt{(2)^2 - (2\pi f)^2}} & \text{if } |f| \leq 2\sin(75^o)/2\pi \\ 0 & \text{if } 2/2\pi \leq |f| \leq 0.5 \end{cases}$$

$$\text{Weight:} \quad W(f) = \begin{cases} 500 & \text{if } |f| \leq 2\sin(75^o)/2\pi \\ 1 & \text{if } 2/2\pi \leq |f| \leq 0.5 \end{cases}$$

For this design example, the starting impulse response index is $N_1 = -15$. The large passband weighting is used to force an almost perfect match in the passband. In fact, for migration filters, the approximation need only be accurate in the passband as long as the stopband magnitude deviation is not larger than unity [93, p. 361]. The optimal solution was obtained in five major iterations. Since $D(f)$ is symmetric with respect to $f = 0$, the resulting optimal filter is also symmetric with an optimal error $\|E_o\| = 0.9755$. The lower bound $|\delta|$, which is computed by the complex Remez stage, is 0.9704.

11.4.1.5 Design of Minimum-Phase FIR Filters

A minimum-phase FIR lowpass filter whose magnitude response is optimal in the Chebyshev sense is most conveniently designed by first designing a symmetric FIR filter [63, 70]. By modifying an equiripple symmetric filter, a new filter can be obtained, the amplitude of which is nonnegative. That filter can then be spectrally factored to obtain a filter whose magnitude is equiripple. For example, see Fig. 11.31.

The top row in Fig. 11.31 illustrates a filter $h(n)$ obtained using the Parks-McClellan program. Let δ_1 and δ_2 denote the deviations from 1 and 0 in the passband and stopband, respectively. By adding δ_2 to $h(15)$, and then scaling $h(n)$ appropriately, the filter illustrated in the second row of Fig. 11.31 is obtained. The amplitude response of that filter is nonnegative and the zeros of that filter lying on the unit circle are double zeros. That being the case, that filter can be spectrally factored to obtain the filter shown in the

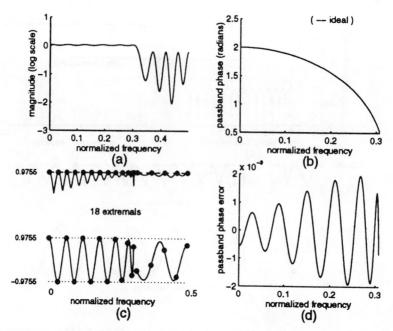

FIGURE 11.30   Example 3 — seismic migration filter, $N = 31$. (a) FIR filter magnitude response in dB. (b) FIR filter (solid) and desired filter (dashed) passband phase responses. (c) Magnitude of the weighted error (top), and real phase-rotated weighted error (bottom). (d) Phase error in passband.

third row of Fig. 11.31. The new nonsymmetric filter has a smaller delay, although its phase is nonlinear. Note that the frequency response magnitude of the nonsymmetric filter is the square root of that of the filter from which it was obtained. Denote the deviations from 1 and 0 in the passband and stopband of the nonsymmetric filter by δ_p and δ_s. Then, the following relationship holds [63]:

$$\delta_1 = \frac{4\delta_p}{2 + 2\delta_p^2 - \delta_s^2} \tag{11.116}$$

$$\delta_2 = \frac{\delta_s^2}{2 + 2\delta_p^2 - \delta_s^2} \tag{11.117}$$

Given specifications for δ_p and δ_s, Eqs. (11.116) and (11.117) give the appropriate values to guide the design of the prototype symmetric FIR filter.

   This method can also be used for the Chebyshev design of minimum-phase bandpass filters, as long as the stopband error in each stopband is equally weighted. If this is not the case, then the Remez exchange algorithm, used in the Parks-McClellan algorithm, can be modified so that it produces symmetric filters whose amplitude functions are nonnegative. The appropriate modification is simple: the interpolation equations in the Remez algorithm of the form $A(\omega_i) = -\delta$ are to be replaced by interpolation equations of the form $A(\omega_i) = 0$. The resulting symmetric FIR filters can then be spectrally factored. This modification makes possible the design of equiripple minimum-phase FIR bandpass filters where the stopband ripple in each band does not have to be the same. If a least squares error criterion is used, then symmetric filters with non-negative amplitudes can be obtained by using a constrained least squares method.

11.4.1.6   Delay Variation of Maximally Flat FIR Filters

   Consider the problem of giving up exactly linear-phase for approximately linear-phase in return for a smaller delay. This problem was also considered in Section on page **11**-43. By subjecting the frequency

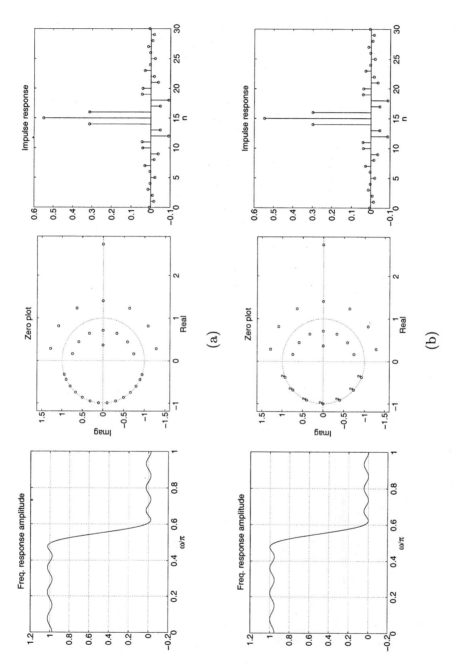

FIGURE 11.31 Design of a minimum-phase FIR filter whose magnitude response is optimal in the Chebyshev sense.

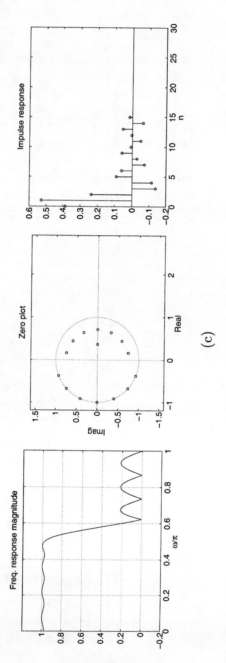

FIGURE 11.31   Design of a minimum-phase FIR filter whose magnitude response is optimal in the Chebyshev sense *(continued)*.

response magnitude and the group delay (individually) to differing numbers of flatness constraints, a family of nonsymmetric lowpass maximally flat FIR filters is obtained [94]. This approach is appropriate when:

1. Exactly linear phase is not required.
2. Some degree of phase linearity is desired.
3. A maximally flat frequency response is desired.

The resulting filters can be made to have approximately linear phase in the passband and a smaller group delay at $\omega = 0$, in comparison to a symmetric filter of equal length.

Problem Formulation Let $F(\omega)$ denote the square magnitude response: $F(\omega) = |H(\omega)|^2$, let $G(\omega)$ denote the group delay: $G(\omega) = -\frac{\partial}{\partial \omega} \angle H(\omega)$. Given the flatness parameters K, L, M, (with $K > 0, M \geq 0, L \leq M$), find N filter coefficients $h(0), \ldots, h(N-1)$ such that:

1. $N = K + L + M + 1$.
2. $F(0) = 1$.
3. $H(z)$ has a root at $z = -1$ of order K.
4. $F^{(2i)}(0) = 0$ for $i = 1, \ldots, M$.
5. $G^{(2i)}(0) = 0$ for $i = 1, \ldots, L$.

The odd indexed derivatives of $F(\omega)$ and $G(\omega)$ are automatically zero at $\omega = 0$, so they do not need to be specified. Linear-phase filters and minimum-phase filters result from the special cases $L = M$ and $L = 0$, respectively.

This problem gives rise to nonlinear equations. Consequently, the existence of multiple solutions should not be surprising and, indeed, that is true here. It is informative to construct a table indicating the number of solutions as a function of K, L, and M. It turns out that the number of solutions is independent of K. The number of solutions as a function of L and M is indicated in Table 11.2 for the first few L and M. Many solutions have complex coefficients or possess frequency response magnitudes that are unacceptable between 0 and π. For this reason, it is useful to tabulate the number of *real* solutions possessing *monotonic*

TABLE 11.2 Total Number of Solutions

		L							
		0	1	2	3	4	5	6	7
	0	1							
	1	2	3						
	2	4	4	5					
M	3	8	6	6	7				
	4	16	8	8	8	9			
	5	32	16	10	10	10	11		
	6	64	26	12	12	12	12	13	
	7	128	48	24	14	14	14	14	15

responses, as is done in Table 11.3. From Table 11.3, two distinct regions emerge. Define two regions in the (L, M) plane. Define region I as all pairs (L, M) for which

$$\lfloor \frac{M-1}{2} \rfloor \leq L \leq M.$$

Define region II as all pairs (L, M) for which

$$0 \leq L \leq \lfloor \frac{M-1}{2} \rfloor - 1.$$

See Table 11.4. It turns out that for (L, M) in region I, all the variables in the problem formulation, except $G(0)$, are linearly related and can be eliminated, yielding a polynomial in $G(0)$; the details are given in [94]. For region II, no similarly simple technique is yet available (except for $L = 0$).

TABLE 11.3   Number of Real Monotonic Solutions, Not Counting Time-Reversals

		0	1	2	3	4	5	6	7
					L				
	0	1							
	1	1	1						
	2	1	1	1					
	3	2	1	1	1				
M	4	2	1	1	1	1			
	5	4	2	1	1	1	1		
	6	4	2	1	1	1	1	1	
	7	8	4	2	1	1	1	1	1

TABLE 11.4   Regions I and II

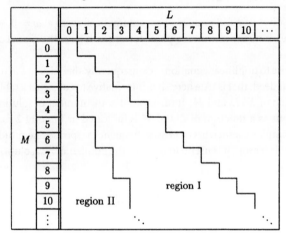

Design Examples   Figures 11.32 and 11.33 illustrate four different FIR filters of length 13 for which $K + L + M = 12$. Each of these filters has 6 zeros at $z = -1$ ($K = 6$) and 6 zeros contributing to the flatness of the passband at $z = 1$ ($L + M = 6$). The four filters shown were obtained using the four values $L = 0, 1, 2, 3$.

When $L = 3$, $M = 3$, the symmetric filter shown in Fig. 11.32 is obtained. This filter is most easily obtained using formulas for maximally flat symmetric filters [55]. When $L = 0$, $M = 6$, the minimum-phase filter shown in Fig. 11.33 is obtained. This filter is most easily obtained by spectrally factoring a length 25 maximally flat symmetric filter. The other two filters shown ($L = 2$, $M = 4$ and $L = 1$, $M = 5$) cannot be obtained using the formulas of Herrmann. They provide a compromise solution.

Observe that for the filters shown, the way in which the passband zeros are split between the interior of the unit circle and its exterior is given by the values L and M. For real monotonic solutions in region I,

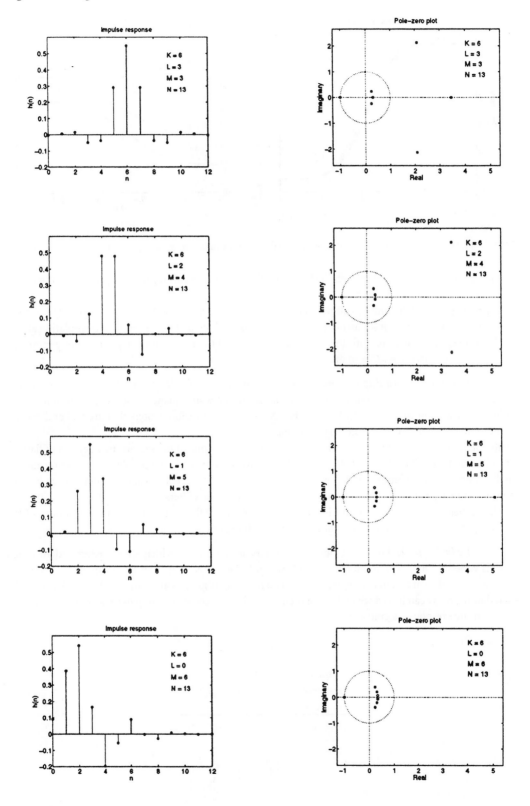

FIGURE 11.32 A selection of nonlinear-phase maximally flat filters of length 13 (for which $K + L + M = 12$). For each filter shown, the zero at $z = -1$ is of multiplicity 6.

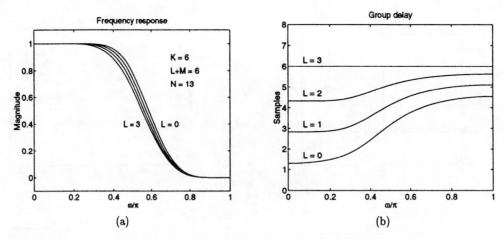

FIGURE 11.33   The magnitude responses and group delays of the filters shown in Fig. 11.32.

this is true in general — even though the location of these zeros in this regard was not part of the way in which the problem was formulated.

It may be observed that the cut-off frequencies of the four filters in Fig. 11.32 are unequal. This is to be expected because the cut-off frequency (denoted ω_o) was not included in the problem formulation above. In the problem formulation, both the cut-off frequency and the DC group delay can be only indirectly controlled by specifying K, L, and M.

Continuously Tuning ω_o ***and*** $G(0)$ To understand the relationship between ω_o, $G(0)$ and K, L, M, it is useful to consider ω_o and $G(0)$ as coordinates in a plane. Then each solution can be indicated by a point in the ω_o-$G(0)$ plane. For $N = 13$, those region I filters that are real and possess monotonic responses appear as the vertices in Fig. 11.34.

To obtain filters of length 13 for which $(\omega_o, G(0))$ lie within one of the sectors, two degrees of flatness must be given up. (Then $K + L + M + 3 = N$, in contrast to item 1 in the problem formulation above.) In this way arbitrary (noninteger) DC group delays and cut-off frequencies can be achieved exactly. This is ideally suited for applications requiring fractional delay lowpass filters.

The flatness parameters of a point in the ω_o-$G(0)$ plane are the (component-wise) minimum of the flatness parameters of the vertices of the sector in which the point lies [94].

Reducing the Delay   To design a set of filters of length 13 for which $\omega_o = 0.636\pi$ and for which $G(0)$ is varied from 3.5 to 6 in increments of 0.5, Fig. 11.34 is used to determine the appropriate flatness parameters — they are tabulated in Table 11.5. The resulting responses are shown in Fig. 11.35. It can be seen that the delay can be reduced while maintaining relatively constant group delay around $\omega = 0$, with no magnitude response degradation.

TABLE 11.5   The Flatness Parameters for the Filters Shown in Fig. 11.35.

N	ω_o/π	$G(0)$	K	L	M
		3.5	3	2	5
		4	3	2	5
		4.5	4	2	4
13	0.636	5	3	3	4
		5.5	3	3	4
		6	4	3	3

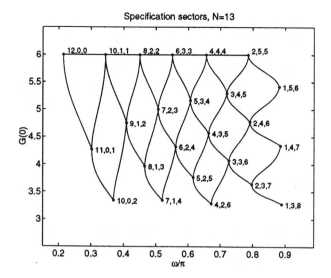

FIGURE 11.34   Specification sectors in the ω_o-$G(0)$ plane for length 13 filters in region I. The vertices are points at which $K + L + M + 1 = 13$. The three integers by each vertex are the flatness parameters (K, L, M).

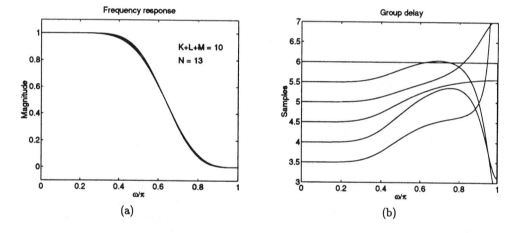

FIGURE 11.35   Length 13 filters obtained by giving up two degrees of flatness and by specifying that the cut-off frequency be 0.636π — and that the specified DC group delay be varied from 3.5 to 6.

11.4.1.7   Combining Criteria in FIR Filter Design

Ivan W. Selesnick and C. Sidney Burrus

Savitzky-Golay Filters   The Savitzky-Golay filters are one example where two of the above described criteria are combined. The two criteria that are combined in the Savitzky-Golay filter are (1) maximally flat behavior (Section on page **11**-38) and (2) least squares error (Section on page **11**-18). Interestingly, the Savitzky-Golay filters illustrate an equivalence between digital lowpass filtering and the smoothing of noisy data by polynomials [63, 95, 96]. As a consequence of this equivalence, Savitzky-Golay filters can be obtained by two different derivations. Both derivations assume that a sequence $x(n)$ is available, where $x(n)$ is composed of an unknown sequence of interest $s(n)$, corrupted by an additive zero-mean white noise sequence $r(n)$: $x(n) = s(n) + r(n)$. The problem is the estimation of $s(n)$ from

$x(n)$ in a way that minimizes the distortion suffered by $s(n)$. Two approaches yield the Savitzky-Golay filters: (1) polynomial smoothing and (2) moment preserving maximal noise reduction.

Polynomial Smoothing Suppose a set of $N = 2M + 1$ contiguous samples of $x(n)$, centered around n_0, can be well approximated by a degree L polynomial in the least squares sense. Then an estimate of $s(n_0)$ is given by $p(n_0)$ where $p(n)$ is the degree L polynomial that minimizes

$$\sum_{k=-M}^{M} (p(n_o + k) - x(n_o + k))^2 .\tag{11.118}$$

It turns out that the estimate of $s(n_0)$ provided by $p(n_0)$ can be written as

$$p(n_0) = (h * x)(n_0)\tag{11.119}$$

where $h(n)$ is the Savitzky-Golay filter of length $N = 2M + 1$ and smoothing parameter L. Therefore, the smoothing of noisy data by polynomials is equivalent to lowpass FIR filtering. Assuming L is odd, with $L = 2K + 1$, $h(n)$ can be written [63] as

$$h(n) = \begin{cases} C_K \frac{1}{n} q_{2K+1}(n) & n = \pm 1, \ldots, \pm M \\ C_K q'_{2K+1}(0) & n = 0 \end{cases}\tag{11.120}$$

where

$$C_K = (-1)^K \frac{(2K+1)!}{(K!)^2} \prod_{k=-K}^{K} \frac{1}{2M + 2k + 1}\tag{11.121}$$

and the polynomials q_l are generated via the recurrence

$$q_0(n) = 1 \qquad q_1(n) = n\tag{11.122}$$

$$q_{l+1}(n) = \frac{2l + 1}{l + 1} n\, q_l(n) - \frac{l(2M + 1 + l)(2M + 1 - l)}{4(l + 1)} q_{l-1}(n).\tag{11.123}$$

$q'_l(n)$ denotes the derivative of $q_l(n)$.

The impulse response (shifted so that it is casual) and frequency response amplitude of a length 41, $L = 13$, Savitzky-Golay filter is shown in Fig. 11.36. As is evident from the figure, Savitzky-Golay filters have poor stopband attenuation — however, they are optimal according to the criteria by which they are designed.

Moment Preserving Maximal Noise Reduction Consider again the problem of estimating $s(n)$ from $x(n)$ via FIR filtering.

$$\begin{aligned} y(n) &= (h_1 * x)(n) & (11.124)\\ &= (h_1 * s)(n) + (h_1 * r)(n) & (11.125)\\ &= y_1(n) + e_r(n) & (11.126) \end{aligned}$$

where $y_1(n) = (h_1 * s)(n)$ and $e_r(n) = (h_1 * r)(n)$. Consider designing $h_1(n)$ by minimizing the variance of $e_r(n)$, $\sigma^2(n) = E[e_r^2(n)]$. Because $\sigma^2(n)$ is proportional to $\|h_1\|_2^2 = \sum_{n=-M}^{M} h_1^2(n)$, the filter minimizing $\sigma^2(n)$ is the zero filter, $h_1(n) \equiv 0$. However, the zero filter also eliminates $s(n)$. A more useful approach requires that $h_1(n)$ preserve the moments of $s(n)$ up to a specified order L. Define the lth moment:

$$m_l[s] = \sum_{n=-M}^{M} n^l s(n).\tag{11.127}$$

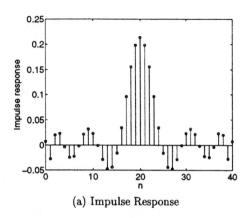

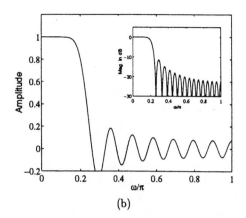

(a) Impulse Response

(b)

FIGURE 11.36 Savitzky-Golay filter, $N = 41$, $L = 13$, ($K = 6$). (a) Impulse response. (b) Magnitude response.

The requirement that $m_l[y_1] = m_l[s]$ for $l = 0, \ldots, L$, is equivalent to the requirement that $m_0[h_1] = 1$ and $m_l[h_1] = 0$ for $l = 1, \ldots, L$. The filter $h_1(n)$ is then obtained by the problem formulation

$$\text{minimize} \quad ||h_1||_2^2 \tag{11.128}$$

subject to

$$m_0[h_1] = 1 \tag{11.129}$$

$$m_l[h_1] = 0 \quad \text{for } l = 1, \ldots, L. \tag{11.130}$$

As shown in [63, 96], the solution $h_1(n)$ is the Savitzky-Golay filter [Eq. (11.120)].

It should be noted that the problem formulated in Eqs. (11.128) through (11.130) is equivalent to the least squares approach, as described in Section on page **11**-40: minimize Eq. (11.30) with $D(\omega) = 0$, $W(\omega) = 1$ subject to the constraints

$$A(\omega = 0) = 1 \tag{11.131}$$

$$A^{(i)}(\omega = 0) = 0 \quad \text{for} \quad i = 1, \ldots, L. \tag{11.132}$$

(These derivative constraints can be expressed as $\mathbf{Ga} = \mathbf{b}$). As such, the solution to Eq. (11.41) is the Savitzky-Golay filter [Eq. (11.120)] — however, with the constraints (11.131, 11.132), the resulting linear system (11.41) is numerically ill-conditioned. Fortunately, the explicit solution (11.120) eliminates the need to solve ill-conditioned equations.

Structure for Symmetric FIR Filter Having Flat Passband Define the transfer function $G(z) = z^{-M} - H(z)$, where $H(z) = \sum_{n=0}^{2M+1} h(n)z^{-n}$ and $h(n)$ is the length $N = 2M + 1$ Savitzky-Golay filter in Eq. (11.120), shifted so that it is casual, as in Fig. 11.36. The filter $G(z)$ is a highpass filter that satisfies derivative constraints at $\omega = 0$. It follows that $G(z)$ possesses a zero at $z = 1$ of order $2K + 2$, and so can be expressed as $G(z) = (-1)^{K+1} \left(\frac{1-z^{-1}}{2} \right)^{2K+2} H_1(z)$. Accordingly,[11] the transfer function of a symmetric filter of length $N = 2M + 1$, satisfying Eqs. (11.131 and 11.132), can be written as

$$H(z) = z^{-M} - (-1)^{K+1} \left(\frac{1-z^{-1}}{2} \right)^{2K+2} H_1(z) \tag{11.133}$$

---

[11]Note that $-1 \cdot \left(\frac{1-z^{-1}}{2} \right)^2 \Big|_{z=e^{j\omega}} = e^{-j\omega} \left(\frac{1-\cos\omega}{2} \right)$, so the amplitude response of $-1 \cdot \left(\frac{1-z^{-1}}{2} \right)^2$ is $\frac{1-\cos\omega}{2}$.

where $H_1(z)$ is a symmetric filter of length $N - 2K - 2 = 2(M - K) - 1$. The amplitude response of $H(z)$ is

$$A(\omega) = 1 - \left(\frac{1 - \cos\omega}{2}\right)^{K+1} A_1(\omega) \tag{11.134}$$

where $A_1(\omega)$ is the amplitude response of $H_1(z)$. Equation (11.133) structurally imposes the desired derivative constraints (11.131, 11.132) with $L = 2K + 1$, and reduces the implementation complexity by extracting the multiplierless factor $\left(\frac{1-z^{-1}}{2}\right)^{2K+2}$. In addition, this structure possesses good passband sensitivity properties with respect to coefficient quantization [97].

Equation (11.133) is a special case of the affine form (11.80). Accordingly, as discussed in Section on page **11**-40, $h_1(n)$ in Eq. (11.133) could be obtained by minimizing Eq. (11.83), with suitably defined $\overline{D}(\omega)$ and $\overline{W}(\omega)$. Although this is unnecessary for the design of Savitzky-Golay filters, it is useful for the design of other symmetric filters for which $A(\omega)$ is flat at $\omega = 0$, for example, the design of such filters in the least squares sense with various $W(\omega)$ and $D(\omega)$, or the design of such filters according to the Chebyshev norm.

Remarks

- Solution to two optimal smoothing techniques: (1) polynomial smoothing and (2) moment preserving maximal noise reduction.
- Explicit formulas for solution.
- Excellent at $\omega = 0$.
- Polynomial assumption for $s(n)$.
- Poor stopband attenuation.

Flat Passband, Chebyshev Stopband   The use of a filter having a very flat passband is desirable because it minimizes the distortion of low frequency signals. However, in the removal of high frequency noise from a low frequency signal by lowpass filtering, it is often desirable that the stopband attenuation be greater than that offered by a Savitzky-Golay filter. One approach [98] minimizes the weighted Chebyshev error, subject to the derivative constraints (11.131, 11.132) imposed at $\omega = 0$. As discussed above, the form (11.133) facilitates the design and implementation of such filters. To describe this approach [97], let the desired amplitude and weight function be as in Eq. (11.44). For the form (11.133), $A_2(\omega)$ and $A_3(\omega)$ in Section on page **11**-40 are given by $A_2(\omega) = -\left(\frac{1-\cos\omega}{2}\right)^K$ and $A_3(\omega) = 1$. $H_1(z)$ can then be designed by minimizing Eq. (11.81) via the Parks-McClellan algorithm. Passband monotonicity, which is sometimes desired, can be ensured by setting $K_p = 0$ in Eq. (11.44) [99]. Then the passband is shaped by the derivative constraints at $\omega = 0$ that are structurally imposed by Eq. (11.133).

Figure 11.37 illustrates a length 41 symmetric filter, whose passband is monotonic. The filter shown was obtained with $K = 6$ and

$$D(\omega) = 0 \quad \omega \in [\omega_s, \pi] \qquad W(\omega) = \begin{cases} 0 & \omega \in [0, \omega_s] \\ 1 & \omega \in [\omega_s, \pi] \end{cases} \tag{11.135}$$

where $\omega_s = 0.3387\pi$. Because $W(\omega)$ is positive only in the stopband, ω_p is not part of the problem formulation.

Bandpass Filters   To design bandpass filters having very flat passbands, one specifies a passband frequency, ω_p, where one wishes to impose flatness constraints. The appropriate form is $H(z) = z^{-(N-1)/2} + H_1(z)H_2(z)$ with

$$H_2(z) = \left(\frac{1 - 2(\cos\omega_p)z^{-1} + z^{-2}}{4}\right)^K \tag{11.136}$$

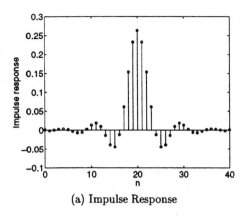
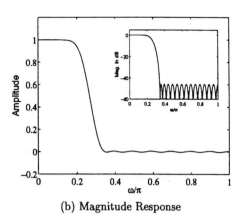

(a) Impulse Response   (b) Magnitude Response

FIGURE 11.37   Lowpass FIR filter designed via minimization of stopband Chebyshev error subject to derivative constraints at $\omega = 0$.

where N is odd, and $H_1(z)$ is a filter whose impulse response is symmetric and of length $N - 2K$. The overall frequency response amplitude $A(\omega)$ is given by

$$A(\omega) = 1 + (-1)^K \left(\frac{\cos \omega_p - \cos \omega}{2} \right)^K A_1(\omega). \tag{11.137}$$

As above, $H_1(z)$ can be found via the Parks-McClellan algorithm. Monotonicity of the passband on either side of ω_p can be ensured by weighting the passband by 0, and by taking K to be even. The filter of length 41 illustrated in Fig. 11.38 was obtained by minimizing the Chebyshev error with $\omega_p = 0.25\pi$, $K = 8$, and

$$D(\omega) = 0 \qquad W(\omega) = \begin{cases} 1 & \omega \in [0, \omega_1] \\ 0 & \omega \in [\omega_1, \omega_2] \\ 1 & \omega \in [\omega_2, \pi] \end{cases} \tag{11.138}$$

where $\omega_1 = 0.1104\pi$ and $\omega_2 = 0.3889\pi$.

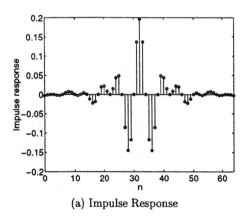
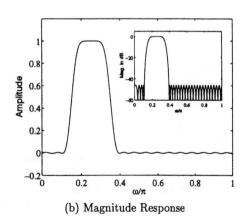

(a) Impulse Response   (b) Magnitude Response

FIGURE 11.38   Bandpass FIR filter designed via minimization of stopband Chebyshev error subject to derivative constraints at $\omega = 0.25\pi$.

Constrained Least Square   The constrained least square approach to filter design provides a compromise between the square error and Chebyshev criteria. This approach produces least square error and best Chebyshev filters as special cases, and is motivated by an observation made by Adams [100]. Least square filter design is based on the assumption that the size of the peak error can be ignored. Likewise, filter design according to the Chebyshev norm assumes the integral square error is irrelevant. In practice, however, both of these criteria are often important. Furthermore, the peak error of a least square filter can be reduced with only a slight increase in the square error. Similarly, the square error of an equiripple filter can be reduced with only a slight increase in the Chebyshev error [100, 8]. In Adams' terminology, both equiripple filters and least square filters are *inefficient*.

Problem Formulation   Suppose the following are given: the filter length N, the desired response $D(\omega)$, a lower bound function $L(\omega)$, and an upper bound function $U(\omega)$, where $D(\omega)$, $L(\omega)$, and $U(\omega)$ satisfy

1. $L(\omega) \leq D(\omega)$
2. $U(\omega) \geq D(\omega)$
3. $U(\omega) > L(\omega)$.

Find the filter of length N that minimizes

$$||E||_2^2 = \frac{1}{\pi} \int_0^\pi W(\omega)(A(\omega) - D(\omega))^2 \, d\omega \qquad (11.139)$$

such that (1) the local maxima of $A(\omega)$ do not exceed $U(\omega)$ and (2) the local minima of $A(\omega)$ do not fall below $L(\omega)$.

Design Examples   Figure 11.39 illustrates two length 41 filters obtained by minimizing Eq. (11.139), subject to the bound constraints, where

$$D(\omega) = \begin{cases} 1 & \omega \in [0, \omega_c] \\ 0 & \omega \in (\omega_c, \pi] \end{cases} \qquad (11.140)$$

$$W(\omega) = \begin{cases} 1 & \omega \in [0, \omega_c] \\ 20 & \omega \in (\omega_c, \pi] \end{cases} \qquad (11.141)$$

$$L(\omega) = \begin{cases} 1 - \delta_p & \omega \in [0, \omega_c] \\ -\delta_s & \omega \in (\omega_c, \pi] \end{cases} \qquad (11.142)$$

$$U(\omega) = \begin{cases} 1 + \delta_p & \omega \in [0, \omega_c] \\ \delta_s & \omega \in (\omega_c, \pi] \end{cases} \qquad (11.143)$$

and where $\omega_c = 0.3\pi$. For the filter on the left of the figure, $\delta_p = \delta_s = 0.0178 = 10^{-35/20}$; for the filter on the right of the figure, $\delta_p = \delta_s = 0.0032 = 10^{-50/20}$. The extremal points of $A(\omega)$ lie within the upper and lower bound functions. Note that the filter on the right is an equiripple filter — it could have been obtained with the PM algorithm, given the appropriate parameter values.

This approach is not a quadratic program (QP) because the domain of the constraints are not explicit. Two observations regarding this formulation and example should be noted:

1. For a fixed length, the maximum ripple size can be made arbitrarily small. When the specified values δ_p and δ_s are small enough, the solution is an equiripple filter. As the constraints are made more strict, the transition width of the solution becomes wider. The width of the transition automatically increases as appropriate.

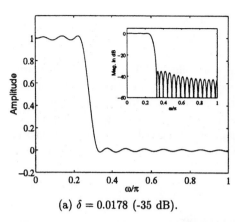
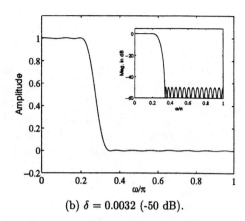

(a) $\delta = 0.0178$ (-35 dB).   (b) $\delta = 0.0032$ (-50 dB).

FIGURE 11.39   Lowpass filter design via bound constrained least squares.

2. As the example illustrates, it is not necessary to use a "don't care" band, e.g., it is not necessary
    to exclude from the square error a region around the discontinuity of the ideal lowpass filter.
    The problem formulation, however, does not preclude the use of a zero-weighted transition
    band.

Quadratic Programming Approach   Some lowpass filter specifications require that $A(\omega)$ lie
within $U(\omega)$ and $L(\omega)$ for all $\omega \in [0, \omega_p] \cup [\omega_s, \pi]$ for given bandedges ω_p and ω_s. While the approach described above ensures that the local maxima and minima of $A(\omega)$ lie below $U(\omega)$ and above
$L(\omega)$, respectively, it does not ensure that this is true at the given bandedges ω_p and ω_s. This is because ω_p
and ω_s are not generally extremal points of $A(\omega)$. The approach described above can be modified so that
bandedge constraints are satisfied; however, it should be recognized that in this case, a quadratic program
(QP) formulation is possible.

Adams formulates the constrained least square filter design problem as a QP and describes algorithms
for solving the relevant QP in [100, 101]. The design of a lowpass filter, for example, can be formulated as
a QP as follows.

QP Formulation   Suppose the following are given: the filter length, N, the bandedges, ω_p and
ω_s, and maximum allowable deviations, δ_p and δ_s. Find the filter that minimizes the square error:

$$||E||_2^2 = \frac{1}{\pi} \int_0^\pi W(\omega) \left(A(\omega) - D(\omega) \right)^2 \, d\omega \qquad (11.144)$$

such that

$$L(\omega) \leq A(\omega) \leq U(\omega) \quad \omega \in [0, \omega_p] \cup [\omega_s, \pi]. \qquad (11.145)$$

where

$$D(\omega) = \begin{cases} 1 & \omega \in [0, \omega_p] \\ 0 & \omega \in [\omega_s, \pi] \end{cases} \qquad (11.146)$$

$$W(\omega) = \begin{cases} K_p & \omega \in [0, \omega_p] \\ 0 & \omega \in [\omega_p, \omega_s] \\ K_s & \omega \in [\omega_s, \pi] \end{cases} \qquad (11.147)$$

$$L(\omega) = \begin{cases} 1 - \delta_p & \omega \in [0, \omega_p] \\ -\delta_s & \omega \in [\omega_s, \pi] \end{cases} \qquad (11.148)$$

$$U(\omega) \;\; = \;\; \begin{cases} 1 + \delta_p & \omega \in [0, \omega_p] \\ \delta_s & \omega \in [\omega_s, \pi] \end{cases} \tag{11.149}$$

This is a QP because the constraints are linear inequality constraints and the cost function is a quadratic function of the variables. The QP formulation is useful because it is very general and flexible. For example, it can be used for arbitrary $D(\omega)$, $W(\omega)$ and arbitrary constraint functions.

Note, however, that for a fixed filter length and a fixed δ_p and δ_s (each less than 0.5), it is not possible to obtain an arbitrarily narrow transition band. Therefore, if the band edges ω_p and ω_s are taken to be too close together, then the quadratic program has no solution. Similarly, for a fixed ω_p and ω_s, if δ_p and δ_s are taken too small, then there is again no solution.

> *Remarks*

- Compromise between square error and Chebyshev criterion.
- Two options: formulation without bandedge constraints or as a QP.
- QP allows (requires) bandedge constraints, but may have no solution.
- Formulation without bandedge constraints can satisfy arbitrarily strict bound constraints.
- QP is well formulated for arbitrary $D(\omega)$ and $W(\omega)$.
- QP is well formulated for the inclusion of arbitrary linear constraints.

11.4.2 IIR Filter Design

Ivan W. Selesnick and C. Sidney Burrus

11.4.2.1 Numerical Methods for Magnitude-Only IIR Design

Numerical methods for magnitude only approximation for IIR filters generally proceed by constructing a noncausal symmetric IIR filter whose amplitude response is nonnegative. Equivalently, a rational function is found, the numerator and denominator of which are both symmetric polynomials of odd degree, with two properties: (1) all zeros lying on the U.C. $|z| = 1$ have even multiplicity and (2) no poles lie on the U.C. A spectral factorization then yields a stable casual digital filter.

The differential correction algorithm for Chebyshev approximation by rational functions, and variations thereof, have been applied to IIR filter design [102, 103, 104, 105, 106]. This algorithm is guaranteed to converge to an optimal solution, and is suitable for arbitrary desired magnitude responses. However, (1) it does not utilize the characterization theorem (see [28] for a characterization theorem for rational Chebyshev approximation), and (2) it proceeds by solving a sequence of (semi-infinite) linear programs. Therefore, it can be slow and computationally intensive.

A Remez algorithm for rational Chebyshev approximation [28] is applicable to IIR filter design, but it is not guaranteed to converge. Deczky's numerical optimization program [107] is also applicable to this problem, as are other optimization methods. It should be noted that general optimization methods can be used for IIR filter design according to a variety of criteria, but the following aspects make it a challenge: (1) initialization, (2) local optimal (nonglobal) solutions, and (3) ensuring the filter's stability.

11.4.2.2 Allpass (Phase-Only) IIR Filter Design

An allpass filter is a filter with a frequency response $H(\omega)$ for which $|H(\omega)| = 1$ for all frequencies ω. The only FIR allpass filter is the trivial delay $h(n) = \delta(n - k)$. IIR allpass filters, on the other hand, must have a transfer function of the form

$$H(z) = \frac{z^N P(z^{-1})}{P(z)} \tag{11.150}$$

where $P(z)$ is a degree N polynomial in z. The problem is the design of the polynomial $P(z)$ so that the phase, or group delay, of $H(z)$ approximates a desired function. The form (11.150) structurally imposes the allpass property of $H(z)$.

The design of digital allpass filters has received much attention, for (1) low complexity structures with low roundoff noise behavior are available for allpass filters [108, 109] and (2) they are useful components in a variety of applications. Indeed, while the traditional application of allpass filters is phase equalization [68, 107], their uses in fractional delay design [21], multirate filtering, filterbanks, notch filtering, recursive phase splitters, and other applications have also been described [63, 110]. Of particular recent interest has been the design of frequency selective filters realizable as a parallel combination of two allpasses,

$$H(z) = \frac{1}{2} \left[A_1(z) + A_2(z) \right]. \tag{11.151}$$

It is interesting to note that digital filters, obtained from the classical analog (Butterworth, Chebyshev, and elliptic) prototypes via the bilinear transformation, can be realized as allpass sums [109, 111, 112]. As allpass sums, such filters can be realized with low complexity structures that are robust to finite precision effects [109]. More importantly, the allpass sum is a generalization of the classical transfer functions that is honored with a number of benefits. Certainly, examples have been given where the utility of allpass sums is well illustrated [113, 114]. Specifically, when some degree of phase linearity is desired, nonclassical filters of the form (11.151) can be designed that achieve superior results with respect to implementation complexity, delay, and phase linearity.

The desired degree of phase linearity can, in fact, be structurally incorporated. If one of the allpass branches in an allpass sum contains only delay elements, then the allpass sum exhibits approximately linear phase in the passbands [115, 116]. The frequency selectivity is then obtained by appropriately designing the remaining allpass branch. Interestingly, by varying the number of delay elements used and the degrees of $A_1(z)$ and $A_2(z)$, the phase linearity can be affected. Simultaneous approximation of the phase and magnitude is a difficult problem in general, so the ability to structurally incorporate this aspect of the approximation problem is most useful.

While general procedures for allpass design [117, 118, 119, 120, 121, 122] are applicable to the design of frequency selective allpass sums, several publications have addressed, in addition to the general problem, the details specific to allpass sums [63, 123, 124, 125]. Of particular interest are the recently described iterative Remez-like exchange algorithms for the design of allpass filters and allpass sums according to the Chebyshev criterion [113, 114, 126, 127].

A simple procedure for obtaining a fractional delay allpass filter uses the maximally flat delay all-pole filter (11.76). By using the denominator of that IIR filter for $P(z)$ in Eq. (11.150), a fractional delay filter is obtained [21]. The group delay of the allpass filter is $2\tau + N$ where τ is that of the all-pole filter used and N is the filter order.

11.4.2.3 Magnitude and Phase Approximation

The optimal frequency domain design of an IIR filter where both the magnitude and the phase are specified, is more difficult than the approximation of one alone. One of the difficulties lies in the choice of the phase function. If the chosen phase function is inconsistent with a stable filter, then the best approximation according to a chosen norm may be unstable. In that case, additional stability constraints must be made explicit. Nevertheless, several numerical methods have been described for the approximation of both magnitude and phase. Let $D(e^{j\omega})$ denote the complex valued desired frequency response.

The minimization of the weighed integral square error

$$\int_0^\pi W(\omega) \left| \frac{B(e^{j\omega})}{A(e^{j\omega})} - D(e^{j\omega}) \right|^2 d\omega \tag{11.152}$$

is a nonlinear optimization problem. If a good initial solution is known, and if the phase of $D(e^{j\omega})$ is chosen appropriately, then Newton's method, or other optimization algorithms, can be successfully used [107, 128]. A modified minimization problem, that comes from the observation that $B/A \approx D \rightarrow B \approx DA$ is the minimization of the weighted *equation error* [11]

$$\int_0^\pi W(\omega)|B(e^{j\omega}) - D(e^{j\omega})A(e^{j\omega})|^2 d\omega \qquad (11.153)$$

which is linear in the filter coefficients. There is a family of iterative methods [129] based on iteratively min-imizing the weighted equation error, or a variation thereof, with a weighting function that is appropriately modified from one iteration to the next.

The minimization of the complex Chebyshev error has also been addressed by several authors. The Ellacott-Williams algorithm for complex Chebyshev approximation by rational functions, and variations thereof, have been applied to this problem [130]. This algorithm calls for the solution to a sequence of complex polynomial Chebyshev problems, and is guaranteed to converge to a local minimum.

Structure Based Methods   Several approaches to the problem of magnitude and phase approx-imation, or magnitude and group delay approximation, use a combination of filters. There are at least three such approaches.

1. One approach cascades (1) a magnitude optimal IIR filters and (2) an allpass filter [107]. The allpass filter is designed to equalize the phase.
2. A second approach cascades (1) a phase optimal IIR filter and (2) a symmetric FIR filter [41]. The FIR filter is designed to equalize the magnitude.
3. A third approach employs a parallel combination of allpass filters. Their phases can be de-signed so that their combined frequency response is selective and has approximately linear phase [113].

11.4.2.4 Time-Domain Approximation

Another approach is based on knowledge of the time domain behavior of the filter sought. Prony's method [11] obtains filter coefficients of an IIR filter that has specified impulse response val-ues $h(0), \ldots, h(K-1)$, where K is the total number of degrees of freedom in the filter coefficients. To obtain an IIR filter whose impulse response approximates desired values $d(0), \ldots, d(L-1)$, where $L > K$, an equation error approach can be minimized, as above, by solving a linear system. The true square error, a nonlinear function of the coefficients, can be minimized by iterative methods [131]. As above, initialization, local-minima, and stability can make this problem difficult.

A more general problem is the requirement that the filter approximately reproduce other input-output data. In those cases, where the sought filter is given only by input-output data, the problem is the *identification* of the system. The problem of designing an IIR filter that reproduces observed input-output data is an important modeling problem in system and control theory, some methods for which can be used for filter design [129].

11.4.2.5 Model Order Reduction

Model order reduction (MOR) techniques, developed largely in the control theory literature, are generally noniterative linear algebraic techniques. Given a transfer function, these techniques produce a second transfer function of specified (lower) degree that approximates the given transfer function. Suppose input-output data of an unknown system is available. One two-step modeling approach proceeds by first constructing a high order model that well reproduces the observed input-output data and, second, obtains a lower order model by reducing the order of the high-order model. Two common methods for MOR are (1) balanced model truncation [132] and (2) optimal Hankel norm MOR [133]. These methods, developed for both continuous and discrete time, produce stable models for which the numerator and denominator degrees are equal.

MOR has been applied to filter design in [134, 135, 136, 137]. One approach [134] begins with a high order FIR filter (obtained by any technique), and uses MOR to obtain a lower order IIR filter, that approximates the FIR filter. As noted above, the phase of the FIR filter used can be important. MOR techniques can yield different results when applied to minimum, maximum, and linear phase FIR filters [134].

11.5   Software Tools

James H. McClellan

Over the past 30 years, many design algorithms have been introduced for optimizing the characteristics of frequency-selective digital filters. Most of these algorithms now rely on numerical optimization, especially when the number of filter coefficients is large. Many sophisticated computer optimization methods have been programmed and distributed for widespread use in the DSP engineering community. Since it is challenging to learn the details of every one of these methods and to understand subtleties of various methods, a designer must now rely on software packages that contain a subset of the available methods. With the proliferation of DSP boards for PCs, the manufacturers have been eager to place design tools in the hands of their users so that the complete design process can be accomplished with one piece of software. This software includes the filter design and optimization, followed by a filter implementation stage. The steps in the design process include:

1. Filter specification via a graphical user interface.
2. Filter design via numerical optimization algorithms. This includes the order estimation stage where the filter specifications are used to compute a predicted filter length (FIR) or number of poles (IIR).
3. Coefficient formatting for the DSP board. Since the design algorithm yields coefficients computed to the highest precision available (e.g., double-precision floating-point), the filter coefficients must be quantized to the internal format of the DSP. In the extreme case of a fixed-point DSP, this quantization also requires scaling of the coefficients to a predetermined maximum value.
4. Optimization of the quantized coefficients. Very few design algorithms perform this step. Given the type of arithmetic in the DSP and the structure for the filter, search algorithms can be programmed to find the best filter; however, it is easier to use some "rules of thumb" that are based on approximations.
5. Downloading the coefficients. If the DSP board is attached to a host computer, then the filter coefficients must be loaded to the DSP and the filtering program started.

11.5.1   Filter Design: Graphical User Interface (GUI)

Operating systems and application programs based on windowing systems have interface building tools that provide an easy way to unify many algorithms under one view. This view concentrates on the filter specifications, so the designer can set up the problem once and then try many different approaches. If the view is a graphical rendition of the tolerance scheme, then the designer can also see the difference between the actual frequency response and the template. Buttons or menu choices can be given for all the different algorithms and parameters available.

With such a GUI, the human is placed *in the filter design loop*. It has always been necessary for the human to be in the loop because filter design is the art of trading off many competing objectives. The filter design programs will optimize a mathematical criterion such as minimum L_p error, but that result might not exactly meet all the expectations of the designer. For example, trades between the length of an FIR implementation and the order of an IIR implementation can only be done by designing the individual filters and then comparing the order vs. length in a proposed implementation.

One implementation of the GUI approach to filter design can be found in a recent version of the MATLAB™ software.[12] The screen shot in Fig. 11.40 shows the GUI window presented by `sptool`, which is the graphical tool for various signal processing operations, including filter design, in MATLAB version 5.0. In this case, the filter being designed is a length-23 FIR filter optimized for minimum Chebyshev error via the Parks-McClellan method for FIR design. The filter order was estimated from the ripples and band edges, but in this case N is too small. The simultaneous graphical view of both the specifications and the actual frequency response makes it clear that the designed filter does meet the desired specifications.

In the MATLAB GUI, the user interface contains two types of controls: display modes and filter design specifications. The display mode buttons are located across the top of the window and are self-explanatory. The filter design specification fields and menus are at the left side of the window. Figure 11.41 shows these in more detail. Previously, we listed the different parameters needed to define the filter specifications: band edges, ripple heights, etc. In the GUI, we see that each of these has an entry. The available design methods come from the pop-up menu that is presently set to "Elliptic" in Fig. 11.41. The design method must be chosen from the list given in Fig. 11.41. The shape of the desired magnitude response must also be chosen from four types; in Fig. 11.41, the type is set to "Bandpass", but the other choices are given in the list "Desired Magnitude." This elliptic bandpass filter is shown in Fig. 11.44.

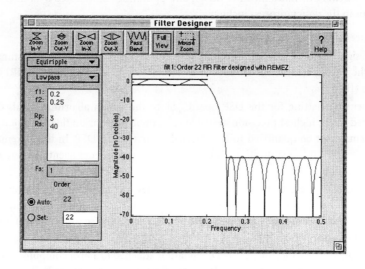

FIGURE 11.40   Screen shot from the MATLAB filter design tool called `sptool`. The equiripple filter was designed by the MATLAB function `remez`.

11.5.1.1   Band Edges and Ripples

An open box is provided so the user can enter numerical values for the parameters that define the boundaries of the tolerance scheme. In the bandpass case, four band edges are needed, as well as the

---

[12]MATLAB is a trademark of the The Mathworks, Inc. The screen shots were made with permission of The Mathworks, Inc.

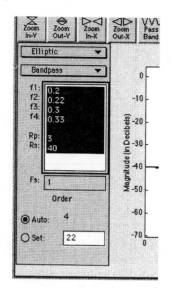

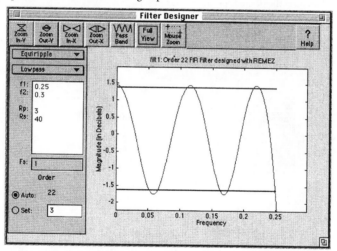

FIGURE 11.41 Pop-up menu choices for filter design options.

FIGURE 11.42 Expanded view of the passband of the lowpass filter from Fig. 11.40.

desired ripple heights for the passband and the two stopbands. The band edges are denoted by f1, f2, f3, and f4 in Fig. 11.41; the ripple heights (in dB) by Rp and Rs. A value of $R_s = 40$ dB is taken to mean 40 dB of attenuation in both stopbands, i.e., $|\delta_s| \leq 0.01$. For the elliptic filter design, the ripples cannot be different in the two stopbands. The passband specification is the difference between the positive-going ripples at 1 and the negative-going ripples at $1 - \delta_p$.

$$R_p = -20 \log_{10} \left(1 - \delta_p\right)$$

In the FIR case, the specification for R_p can be confusing because it is the total ripple which is the difference between the positive-going ripples at $1 + \delta_p$ and the negative-going ripples at $1 - \delta_p$:

$$R_p = 20 \log_{10}(1 + \delta_p) - 20 \log_{10} \left(1 - \delta_p\right)$$

In Fig. 11.42, the value 3 dB is the same as $\delta_p \approx 0.171$. As the expanded view of the passband in Fig. 11.42 shows, the ripples are not expected to be symmetric on a logarithmic scale. This expanded view for the FIR filter from Fig. 11.40 was obtained by pressing the ⎰Pass Band⎱ button at the top.

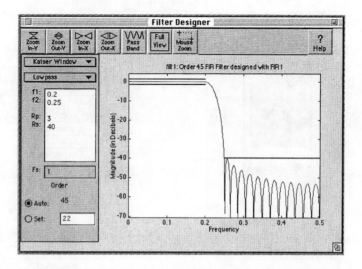

FIGURE 11.43   Length-47 FIR filter designed by the Kaiser window method. The order was estimated to be 46, and in this case the filter does meet the desired specifications.

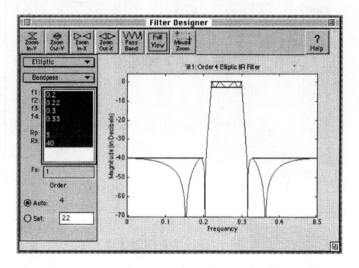

FIGURE 11.44   Eight-pole elliptic bandpass filter. The order was calculated to be four, but the filter exceeds the desired specifications by quite a bit.

11.5.1.2   Graphical Manipulation of the Specification Template

With the graphical view of the filter specifications, it is possible to use a pointing device such as a mouse to "grab" the specifications and move them around. This has the advantage that the relative placement of band edges can be visualized while the movement is taking place. In the MATLAB GUI, the filter is quickly redesigned every time the mouse is released, so the user also gets immediate feedback on how close the filter approximation can be to the new specification. Order estimation is also done instantaneously, so the designer can develop some intuition concerning tradeoffs such as transition width vs. filter order.

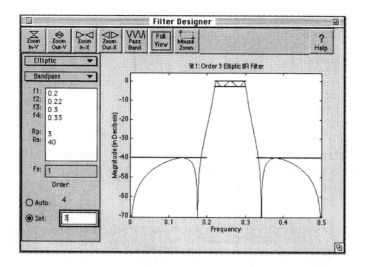

FIGURE 11.45 Six-pole elliptic bandpass filter. The order was set at three, which is too small to meet the desired specifications.

11.5.1.3 Frequency Scaling

The field for `Fs` is useful when the filter specifications come from the "analog world", and are expressed in Hertz with the sampling frequency given separately. Then the sampling frequency can be specified, and the horizontal axis is labeled and scaled in terms of `Fs`. Since the design is only carried out for $0 \leq \omega \leq \pi$, the highest frequency on the horizontal axis will be $F_s/2$. When $Fs = 1$, we say that the frequency is *normalized* and the numbers on the horizontal axis can be interpreted as a percentage of the sampling frequency, i.e., a value of 0.2 means 20% of F_s.

11.5.1.4 Automatic Order Estimation

Perhaps the most important feature of a software filter design package is its use of *design rules*. Since the design problem is always trying to trade off among the parameters of the specification, it is useful to be able to predict what the result will be without actually carrying out the design. A typical design formula involves the band edges, the desired ripples and the filter order. For example, a simple *approximate* formula [12, 37] for FIR filters designed by the Remez exchange method is:

$$N(\omega_s - \omega_p) = \frac{-20 \log_{10} \sqrt{\delta_p \delta_s} - 13}{2.324} \tag{11.154}$$

Most often the desired filter is specified by { ω_p, ω_s, δ_p, δ_s }, so the design formula can be used to predict the filter order. Since most algorithms must work with a fixed number of parameters (determined by N) in doing optimization, this step is necessary before an iterative numerical optimization can be done.

The MATLAB GUI allows the user to turn on this *order-estimating* feature, so that an estimate of the filter order is calculated automatically whenever the filter specifications change. In the case of the FIR filters, the order-estimating formulae are only approximate—being derived from an empirical study of the parameters taken over many different designs. In some cases, the length N obtained is not large enough, and when the filter is designed it will fail to meet the desired specifications (see Fig. 11.40). On the other hand, the Kaiser window design in Fig. 11.43 does meet the specifications, even though its length (47) was also estimated from an approximate formula [12] similar to Eq. (11.154).

For the IIR case, however, the formulas are exact because they are derived from the mathematical properties of the Chebyshev polynomials or elliptic functions that define the classical filter types. Typically, the band edges and the bilinear transformation define several simultaneous nonlinear equations that must be satisfied, but these can be solved in succession to get an order N that is guaranteed to work. The filter in Fig. 11.44 shows the case where the order estimate was used for the bandpass design and the filter meets the specifications; but in Fig. 11.45 the filter order was set to 3, which gave a sixth-order bandpass that fails to meet the specifications because its transition regions are too wide.

11.5.2   Filter Implementation

Another type of filter design tool ties in the filter's implementation with the design. Many DSP board vendors offer software products that perform filter design and then download the filter information to a DSP to process the data stream. Representative of this type of design is the DFDP-4/plus software[13] shown in the screen shots of Figs. 11.46 through 11.51.

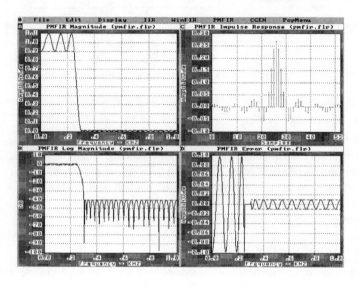

FIGURE 11.46   Length-57 FIR filter designed by the Parks-McClellan method, using the ASPI DFDP-4/plus software.

Similar to the MATLAB software, DFDP-4 can do the specification and design of the filter coefficients. In fact, it possesses an even wider range of filter design methods that includes filter banks and other special structures. It can design FIR filters based on the window method and the Parks-McClellan algorithm (an example is shown in Fig. 11.46). For the IIR problem, the classical filter types (Butterworth, Chebyshev, and Elliptic) are provided; Fig. 11.47 shows an elliptic bandpass filter. In addition to the standard lowpass, highpass, and bandpass filter shapes, DFDP-4 can also handle the multiband case as well as filters with an arbitrary desired magnitude (as in Fig. 11.51). When designing IIR filters, the phase response presents a difficulty because it is not linear or close to linear. The screen shot in Fig. 11.47 shows the phase response in the lower left-hand panel and the group delay in the upper right-hand. The wide variation in the group delay, which is the derivative of the phase, indicates that the phase is far from linear. DFDP-4 provides an algorithm to optimize the group delay, which is a useful feature to compensate the phase response of an elliptic filter by using several all-pass sections to flatten the group delay.

---

[13]DFDP is a trademark of Atlanta Signal Processors, Inc. The screen shots were made with permission of Atlanta Signal Processors, Inc.

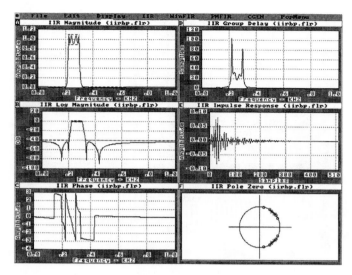

FIGURE 11.47 Eighth-order IIR bandpass elliptic filter designed using DFDP-4.

In DFDP-4, the filter design stage is specified by entering the band edges and the desired ripples in dialog boxes until all the parameters are filled in for that type of design. Conflicts among the specifications can be resolved at this point before the design algorithm is invoked. For some designs such as the arbitrary magnitude design, the specification can involve many parameters to properly define the desired magnitude.

The filter design stage is followed by an implementation stage in which DFDP-4 produces the appropriate filter coefficients for either a fixed-point or floating-point implementation, targeted to a specific DSP microprocessor. The filter coefficients can be quantized over a range from 4 to 24 bits, as shown in Fig. 11.50. The filter's frequency response would then be checked after quantization to compare with the designed filter and the original specifications. In the FIR case, coefficient quantization is the primary step needed prior to generating code for the DSP microprocessor, since the preferred implementation on a DSP is direct form. Internal wordlength scaling is also needed if a fixed-point implementation is being done. Once the wordlength is chosen, DFDP-4 will generate the entire assembly language program needed for the TMS-320 processor used on the boards supported by ASPI. As shown in Fig. 11.48, there are a variety of supported processors, and even within a given processor family, the user can choose options such as "time optimization," "size optimization," etc. In Fig. 11.48, the choice of "11" dictates a filter implementation on a TMS 320-C30, with ASM30 assembly language calls, and size optimization. The filter coefficients are taken from the file called PMFIR.FLT, and the assembly code is written to the file PMFIR.S31.

11.5.2.1 Cascade of Second-Order Sections

In the IIR case, the implementation is often done with a cascade of second-order sections. The numerator and denominator of the transfer function $H(z)$ must first be factored as:

$$H(z) = \frac{B(z)}{A(z)} = \frac{G \prod_{i=1}^{M} \left(1 - z_i z^{-1}\right)}{\prod_{i=1}^{N} \left(1 - p_i z^{-1}\right)} \tag{11.155}$$

where p_i and z_i are the poles and zeros of the filter. In the screen shot of Fig. 11.47 we see that the poles and zeros of the eighth-order elliptic bandpass filter are displayed to the user. The second-order sections are obtained by grouping together two poles and two zeros to create each second-order section; conjugate pairs must be kept together if the filter coefficients are going to be real.

$$H(z) = \frac{B(z)}{A(z)} = \prod_{k=1}^{N/2} \frac{\beta_{0k} + \beta_{1k} z^{-1} + \beta_{2k} z^{-2}}{1 + \alpha_{1k} z^{-1} + \alpha_{2k} z^{-2}} \tag{11.156}$$

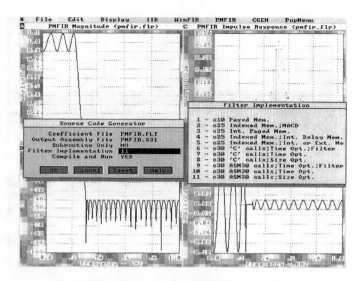

FIGURE 11.48　Code generation for an FIR filter using DFDP-4.

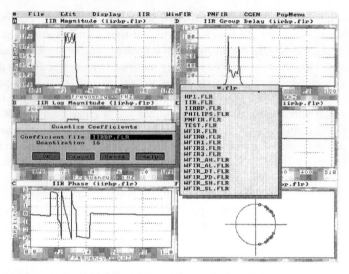

FIGURE 11.49　Eighth-order IIR bandpass elliptic filter with quantized coefficients.

Each second-order factor defines a recursive difference equation with two feedback terms, α_{1k} and α_{2k}. The product of all the sections is implemented as a cascade of the individual second-order feedback filters. This implementation has the advantage that the overall filter response is relatively insensitive to coefficient quantization and round-off noise when compared to a direct form structure. Therefore, the cascaded second-order sections provide a robust implementation, especially for IIR filters with poles very close to the unit circle.

Clearly, there are many different ways to pair the poles and zeros when defining the second-order sections. Furthermore, there are many different orderings for the cascade, and each one will produce different noise gains through the filter. Sections with a pole pair close to the U.C. will be extremely narrowband with a very high gain at one frequency. The rules of thumb originally developed by Jackson [138] give good orderings depending on the nature of the input signal—wideband vs. narrowband. This choice can be seen in Fig. 11.51 where the `section ordering` slot is set to `NARROWBAND`.

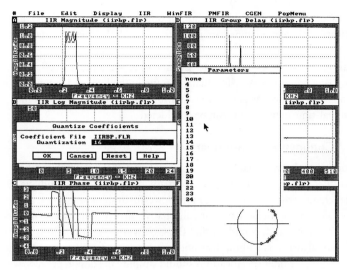

FIGURE 11.50   Eighth-order IIR bandpass elliptic filter. Saving 16-bit coefficients.

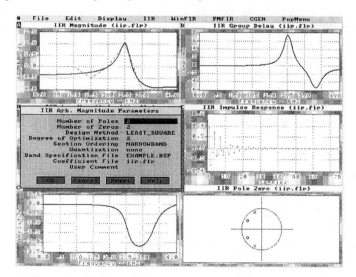

FIGURE 11.51   Arbitrary magnitude IIR filter.

11.5.2.2   Scaling for Fixed-Point

A second consideration when ordering the second-order sections is the problem of scaling to avoid overflow. This issue only arises when the IIR filter is targeted to a fixed-point DSP microprocessor. Since the gain of individual sections may vary widely, the fixed-point data might overflow beyond the maximum value allowed by the wordlength. To combat this problem, multipliers (or shifters that multiply by a power of two) can be inserted in-between the cascaded sections to guard against overflow. However, dividing by two will shift bits off the lower end of the fixed-point word, thereby introducing more round-off noise. The value of the scaling factor can be approximated via a worst-case analysis that prevents overflow entirely, or a mean square method that reduces the likelihood of overflow depending on the input signal characteristics.

Proper treatment of the scaling problem requires that it be solved in conjunction with the ordering of sections for minimal round-off noise. Similar "rules of thumb" can be employed to get a good (if not optimal) implementation that simultaneously addresses ordering, pole-zero pairing, and scaling [138]. The theoretical problem of optimizing the implementation for word length and noise performance is rarely done because it is such a difficult problem, and not one for which an efficient solution has been found. Thus, most software tools rely on approximations to perform the implementation and code-generation steps quickly.

Once the transfer function is factored into second-order sections, the code-generation phase creates the assembly language program that will actually execute in the DSP and downloads it to the DSP board. Coefficient quantization is done as part of the assembly code generation. With the program loaded into the DSP, tests on real-time data streams can be conducted.

11.5.2.3 Comments and Summary

The two design tools presented here are representative of the capabilities that one should expect in a state of the art filter design package. There are many software design products available and most of them have similar characteristics, but may be more powerful in some respects, e.g., more design algorithm choices, different DSP microprocessor support, alternative display options, etc. A user can choose a design tool with these criteria in mind, confident that the GUI will make it relatively easy to use the powerful mathematical design algorithms without learning the idiosyncrasies of each method. The uniform view of the GUI as managing the filter specifications should simplify the design process, while allowing the best possible filters to be designed through trial and comparison.

One limiting aspect of the GUI filter design tool is that it can easily do magnitude approximation, but only for the standard cases of bandpass and multiband filters. It is easy to envision, however, that the GUI could support graphical user entry of the specifications by having the user draw the desired magnitude. Then other magnitude shapes could be supported, as in DFDP-4. Another extension would be to provide a graphical input for the desired phase response, or group delay, in addition to the magnitude specification. Although a great majority of filter designs are done for the bandpass case, there has been a recent surge of interest in having the flexibility to do simultaneous magnitude and phase approximation. With the development of better general magnitude and phase design methods, the filter design packages now offer this capability.

References

[1] Oppenheim, A.V. and Schafer, R.W. *Discrete-Time Signal Processing,* Prentice-Hall, Englewood Cliffs, NJ, 1989.

[2] Karam, L.J. and McClellan, J.H. Complex Chebyshev approximation for FIR filter design, *IEEE Trans. Circuits Sys. II,* 42, 207–216, March 1995.

[3] Karam, L.J. and McClellan, J.H. Design of optimal digital FIR filters with arbitrary magnitude and phase responses, *Proc. IEEE ISCAS,* 1996.

[4] Burnside, D. and Parks, T.W. Optimal design of FIR filters with the complex Chebyshev error criteria, *IEEE Trans. Signal Processing,* 43, 605–616, March 1995.

[5] Preuss, K. On the design of FIR filters by complex Chebyshev approximation, *IEEE Trans. Acoust., Speech, Signal Processing,* 37, 702–712, May 1989.

[6] Parks, T.W. and McClellan, J.H. Chebyshev approximation for nonrecursive digital filters with linear phase, *IEEE Trans. Circuit Theory,* CT-19, 189–194, March 1972.

[7] Steiglitz, K., Parks, T.W., and Kaiser, J.F. METEOR: A constraint-based FIR filter design program, *IEEE Trans. Signal Processing,* 40, 1901–1909, Aug. 1992.

[8] Selesnick, I.W., Lang, M., and Burrus, C.S. Constrained least square design of FIR filters without specified transition bands, *IEEE Trans. Signal Processing,* 44, 1879–1892, Aug. 1996.

[9] Proakis, J.G. and Manolakis, D.G. *Digital Signal Processing: Principles, Algorithms, and Applications,* Prentice-Hall, Englewood Cliffs, NJ, 1996.

[10] Karam, L.J. and McClellan, J.H. Optimal digital FIR filters design, June 1996, submitted to *IEEE Trans. Signal Processing.*

[11] Parks, T.W. and Burrus, C.S. *Digital Filter Design,* John Wiley & Sons, New York, 1987.

[12] Kaiser, J.F. Nonrecursive digital filter design using the I_o − sinh window function, *Proc. IEEE Intl. Symp. Circuits Systems (ISCAS),* 20–23, Apr. 1974.

[13] Slepian, D. Prolate spheroidal wave functions, Fourier analysis and uncertainty, *Bell Syst. Tech. J.,* 57, May 1978.

[14] Gruenbacher, D.M. and Hummels, D.R. A simple algorithm for generating discrete prolate spheroidal sequences, *IEEE Trans. Signal Processing,* 42, 3276–3278, Nov. 1994.

[15] Percival, D.B. and Walden, A.T. *Spectral Analysis for Physical Applications: Multitaper and Conventional Univariate Techniques,* Cambridge University Press, 1993.

[16] Verma, T., Bilbao, S., and Meng, T.H.Y. The digital prolate spheroidal window, *Proc. IEEE Intl. Conf. Acoust., Speech, Signal Processing (ICASSP),* 1351–1354, May 1996.

[17] Saramäki, T. Finite impulse resonse filter design, in *Handbook For Digital Signal Processing,* Mitra, S.K. and Kaiser, J.F. Eds., John Wiley & Sons, New York, 1993, chap. 4, pp. 155–277.

[18] Saramäki, T. Adjustable windows for the design of FIR filters—a tutorial, *Proc. Mediter. Electrotech. Conf., 6th, Ljubljana, Yugoslavia,* 28–33, 1991.

[19] Elliot, D.F. *Handbook of Digital Signal Processing,* Academic Press, New York, 1987.

[20] Cain, G.D., Yardim, A., and Henry, P. Offset windowing for FIR fractional-sample delay, *Proc. IEEE Intl. Conf. Acoust., Speech, Signal Processing (ICASSP),* Detroit, 1276–1279, May 9-12, 1995.

[21] Laakso, T.I., Välimäki, V., Karjalainen, M., and Laine, U.K. Splitting the unit delay, *IEEE Signal Processing Mag.,* 13, 30–60, Jan. 1996.

[22] Gopinath, R.A. Thoughts on least square-error optimal windows, *IEEE Trans. Signal Processing,* 44, 984–987, Apr. 1996.

[23] Weisburn, E.A., Parks, T.W., and Shenoy, R.G. Error criteria for filter design, *Proc. IEEE Intl. Conf. Acoust., Speech, Signal Processing (ICASSP),* 565–568, Apr. 1994.

[24] Merchant, G.A. and Parks, T.W. Efficient solution of a Toeplitz-plus-Hankel coefficient matrix system of equations, *IEEE Trans. Acoust., Speech, Signal Proc.,* 30, 40–44, Feb. 1982.

[25] Burrus, C.S., Soewito, A.W. and Gopinath, R.A. Least squared error FIR filter design with transition bands, *IEEE Trans. Signal Processing,* 40, 1327–1340, June 1992.

[26] Burrus, C.S. Multiband least squares FIR filter design, *IEEE Trans. Signal Processing,* 43, 412–421, Feb. 1995.

[27] Vaidyanathan, P.P. and Nguyen, T.Q. Eigenfilters: a new approach to least-squares FIR filter design and applications including nyquist filters, *IEEE Trans. Circuits Syst.,* 34, 11–23, Jan. 1987.

[28] Powel, M.J.D. *Approximation Theory and Methods,* Cambridge University Press, New York, 1981.

[29] Rabiner, L.R., McClellan, J.H., and Parks, T.W. FIR digital filter design techniques using weighted Chebyshev approximation, *Proc. IEEE,* 63, 595–610, Apr. 1975.

[30] Rabiner, L.R. and Gold, B. *Theory and Application of Digital Signal Processing,* Prentice-Hall, Englewood Cliffs, NJ, 1975.

[31] McClellan, J.H., Parks, T.W., and Rabiner, L.R. A computer program for designing optimum FIR linear phase digital filters, *IEEE Trans. Audio Electroacoust.,* 21, 506–526, Dec. 1973.

[32] McClellan, J.H. On the Design of One-Dimensional and Two-Dimensional FIR Digital Filters, Ph.D. thesis, Rice University, April 1973.

[33] Herrmann, O. Design of nonrecursive filters with linear phase, *Electron. Lett.,* 6, 328–329, May 28 1970.

[34] Hofstetter, E., Oppenheim, A., and Siegel, J. A new technique for the design of nonrecursive digital filters, *Proc. Fifth Annu. Princeton Conf. Information Sci. Syst.,* 64–72, Oct. 1971.

[35] Parks, T.W. and McClellan, J.H. On the transition region width of finite impulse-response digital filters, *IEEE Trans. Audio Electroacoust.,* 21, 1–4, Feb. 1973.

[36] Rabiner, L.R. Approximate design relationships for lowpass FIR digital filters, *IEEE Trans. Audio Electroacoust.*, 21, 456–460, Oct. 1973.

[37] Herrmann, O., Rabiner, L.R., and Chan, D.S.K. Practical design rules for optimum finite impulse response lowpass digital filters, *Bell Sys. Tech. J.*, 52, 769–799, 1973.

[38] Selesnick, I.W. and Burrus, C.S. Exchange algorithms that complement the Parks-McClellan algorithm for linear phase FIR filter design, *IEEE Trans. Circuits Syst. II*, 44(2), 137–143, Feb. 1997.

[39] de Saint-Martin, F.M. and Siohan, P. Design of optimal linear-phase transmitter and receiver filters for digital systems, *Proc. IEEE Intl. Symp. Circuit Sys. (ISCAS)*, 885–888, April 30-May 3 1995.

[40] Thiran, J.P. Recursive digital filters with maximally flat group delay, *IEEE Trans. Circuit Theory*, 18, 659–664, Nov. 1971.

[41] Saramäki, T. and Neuvo, Y. Digital filters with equiripple magnitude and group delay, *IEEE Trans. Acoust., Speech, Signal Processing*, 32, 1194–1200, Dec. 1984.

[42] Jackson, L.B. An improved Martinez/Parks algorithm for IIR design with unequal numbers of poles and zeros, *IEEE Trans. Signal Processing*, 42, 1234–1238, May 1994.

[43] Liang, J. and Figueiredo, R.J.P.D. An efficient iterative algorithm for designing optimal recursive digital filters, *IEEE Trans. Acoust., Speech, Signal Proc.*, 31, 1110–1120, Oct. 1983.

[44] Martinez, H.G. and Parks, T.W. Design of recursive digital filters with optimum magnitude and attenuation poles on the unit circle, *IEEE Trans. Acoust., Speech, Signal Processing*, 26, 150–156, Apr. 1978.

[45] Saramäki, T. Design of optimum wideband recursive digital filters, *Proc. IEEE Intl. Symp. Circuits Systems (ISCAS)*, 503–506, 1982.

[46] Saramäki, T. Design of digital filters with maximally flat passband and equiripple stopband magnitude, *Intl. J. Circuit Theory Applications*, 13, 269–286, Apr. 1985.

[47] Unbehauen, R. On the design of recursive digital low-pass filters with maximally flat pass-band and Chebyshev stop-band attenuation, *Proc. IEEE Intl. Symp. Circuits Sys. (ISCAS)*, 528–531, 1981.

[48] Zhang, X. and Iwakura, H. Design of IIR digital filters based on eigenvalue problem, *IEEE Trans. Signal Processing*, 44, 1325–1333, June 1996.

[49] Saramäki, T. Design of optimum recursive digital filters with zeros on the unit circle, *IEEE Trans. Acoust., Speech, Signal Processing*, 31, 450–458, Apr. 1983.

[50] Selesnick, I.W. and Burrus, C.S. Generalized digital Butterworth filter design, *Proc. IEEE Intl. Conf. Acoust., Speech, Signal Processing (ICASSP)*, (Atlanta), 1367–1370, May 7-10 1996.

[51] Samadi, S., Cooklev, T., Nishihara, A., and Fujii, N. Multiplierless structure for maximally flat linear phase FIR filters, *Electron. Lett.*, 29, 184–185, Jan. 21 1993.

[52] Vaidyanathan, P.P. On maximally-flat linear-phase FIR filters, *IEEE Trans. Circuits Sys.*, 31, 830–832, Sep. 1984.

[53] Vaidyanathan, P.P. Efficient and multiplierless design of FIR filters with very sharp cutoff via maximally flat building blocks, *IEEE Trans. Circuits Sys.*, 32, 236–244, March 1985.

[54] Neuvo, Y., Dong, C.-Y., and Mitra, S.K. Interpolated finite impulse response filters, *IEEE Trans. Acoust., Speech, Signal Processing*, 32, 563–570, June 1984.

[55] Herrmann, O. On the approximation problem in nonrecursive digital filter design, *IEEE Trans. Circuit Theory*, 18, 411–413, May 1971.

[56] Rajagopal, L.R. and Roy, S.C.D. Design of maximally-flat FIR filters using the Bernstein polynomial, *IEEE Trans. Circuits Sys.*, 34, 1587–1590, Dec. 1987.

[57] Daubechies, I. *Ten Lectures On Wavelets*, SIAM, 1992.

[58] Kaiser, J.F. Design subroutine (MXFLAT) for symmetric FIR low pass digital filters with maximally-flat pass and stop bands, in *Programs for Digital Signal Processing*, I.A.S. Digital Signal Processing Committee, Ed., IEEE Press, New York, 1979, chap 5.3, pp. 5.3–1 – 5.3–6.

[59] Jinaga, B.C. and Roy, S.C.D. Coefficients of maximally flat low and high pass nonrecursive digital filters with specified cutoff frequency, *Signal Processing*, 9, 121–124, Sep. 1985.

[60] Thajchayapong, P., Puangpool, M., and Banjongjit, S. Maximally flat FIR filter with prescribed cutoff frequency, *Electron. Lett.*, 16, 514–515, Jun 19 1980.

[61] Rabenstein, R. Design of FIR digital filters with flatness constraints for the error function, *Circuits, Systems, and Signal Processing*, 13(1), 77–97, 1993.

[62] Schüssler, H.W. and Steffen, P. An approach for designing systems with prescribed behavior at distinct frequencies regarding additional constraints, *Proc. IEEE Intl. Conf. Acoust., Speech, Signal Processing (ICASSP)*, 1985.

[63] Schüssler, H.W. and Steffen, P. Some advanced topics in filter design, in *Advanced Topics in Signal Processing*, Lim, J.S. and Oppenheim, A.V. Eds., Prentice-Hall, Englewood Cliffs, NJ, 1988, chap 8, pp. 416–491.

[64] Adams, J.W. and Willson, A.N., Jr., A new approach to FIR digital filter with fewer multipliers and reduced sensitivity, *IEEE Trans. Circuits Sys.*, 30, 277–283, May 1983.

[65] Adams, J.W. and Willson, A.N., Jr., Some efficient prefilter structures, *IEEE Trans. Circuits Sys.*, 31, 260–266, March 1984.

[66] Hartnett, R.J. and Boudreaux-Bartels, G.F. On the use of cyclotomic polynomials prefilters for efficient FIR filter deisgn, *IEEE Trans. on Signal Processing*, 41, 1766–1779, May 1993.

[67] Oh, W.J. and Lee, Y.H. Design of efficient FIR filters with cyclotomic polynomial prefilters using mixed integer linear programming, *Proc. IEEE Intl. Conf. Acoust., Speech, Signal Processing (ICASSP)*, 1287–1290, May 1996.

[68] Lang, M. Optimal weighted phase equalization according to the l_∞-norm, *Signal Processing*, 27, 87–98, Apr. 1992.

[69] Leeb, F. and Henk, T. Simultaneous amplitude and phase approximation for FIR filters, *Intl. J. Circuit Theory Applications*, 17, 363–374, July 1989.

[70] Herrmann, O. and Schüssler, H.W. Design of nonrecursive filters with minimum phase, *Electron. Lett.*, 6, 329–330, May 28 1970.

[71] Baher, H. FIR digital filters with simultaneous conditions on amplitude and delay, *Electron. Lett.*, 18, 296–297, April 1 1982.

[72] Calvagno, G., Cortelazzo, G.M., and Mian, G.A. A technique for multiple criterion approximation of FIR filters in magnitude and group delay, *IEEE Trans. Signal Processing*, 43, 393–400, Feb. 1995.

[73] Rhodes, J.D. and Fahmy, M.I.F. Digital filters with maximally flat amplitude and delay characteristics, *Intl. J. Circuit Theory Applications*, 2, 3–11, March 1974.

[74] Sullivan, J.L. and Adams, J.W. A new nonlinear optimization algorithm for asymmetric FIR digital filters, *Proc. IEEE Intl. Symp. Circuits and Systems (ISCAS)*, 541–544, May-June 1994.

[75] Scanlan, S.O. and Baher, H. Filters with maximally flat amplitude and controlled delay responses, *IEEE Trans. on Circuits and Systems*, 23, 270–278, May 1976.

[76] Rice, J.R. *The Approximation of Functions*, Addison-Wesley, Reading, MA, 1969.

[77] Alkhairy, A.S., Christian, K.S., and Lim, J.S. Design and characterization of optimal FIR filters with arbitrary phase, *IEEE Trans. Signal Processing*, 41, 559–572, Feb. 1993.

[78] Karam, L.J. Design of Complex Digital FIR Filters in the Chebyshev sense, Ph.D. thesis, Georgia Institute of Technology, March 1995.

[79] Meinardus, G. *Approximation of Functions: Theory and Numerical Methods*, Springer-Verlag, New York, 1967.

[80] McCallig, M.T. Design of digital FIR filters with complex conjugate pulse responses, *IEEE Trans. Circuit Sys.*, CAS-25, 1103–1105, Dec. 1978.

[81] Cheney, E.W. *Introduction to Approximation Theory*, McGraw-Hill, New York, 1966.

[82] Demjanov, V.F. Algorithms for some minimax problems, *J. Comp. Sys. Sci.*, 2, 342–380, 1968.

[83] Demjanov, V.F and Malozemov, V.N. *Introduction To Minimax*. John Wiley & Sons, New York, 1974.

[84] Wolfe, P. Finding the nearest point in a polytope, *Mathematical Programming*, 11, 128–149, 1976.

[85] Wolfe, P. A method of conjugate subgradients for minimizing nondifferentiable functions, *Mathematical Programming Study*, 3, 145–173, 1975.

[86] Lorentz, G.G. *Approximation of Functions*, Holt, Rinehart and Winston, New York, 1966.

[87] Feuer, A. Minimizing well-behaved functions, *12th Annual Allerton Conference on Circuit and System Theory*, Oct. 1974.

[88] Watson, G.A. The calculation of best restricted approximations, *SIAM J. Num. Anal.*, 11, 693–699, Sept. 1974.

[89] Chen, X. and Parks, T.W. Design of FIR filters in the complex domain, *IEEE Trans. Acoust., Speech, Signal Processing*, ASSP-35, 144–153, Feb. 1987.

[90] Harris, D.B. Design and Implementaion of Rational 2-D Digital Filters, Ph.D. thesis, Massachusetts Institute of Technology, Nov. 1979.

[91] Claerbout, J. *Fundamentals of Geophysical Data Processing*, McGraw-Hill, New York, 1976.

[92] Hale, D. 3-D depth migration via McClellan transformations, *Geophysics*, 56, 1778–1785, Nov. 1991.

[93] Dudgeon, D.E. and Mersereau, R.M *Multidimensional Digital Signal Processing*, Prentice-Hall, Englewood Cliffs, NJ, 1984.

[94] Selesnick, I.W. *New Techniques for Digital Filter Design*, Ph.D. thesis, Rice University, 1996.

[95] Orfanidis, S.J. *Introduction to Signal Processing*, Prentice-Hall, Englewood Cliffs, NJ, 1996.

[96] Steffen, P. On digital smoothing filters: A brief review of closed form solutions and two new filter approaches, *Circuits, Systems, and Signal Processing*, 5(2), 187–210, 1986.

[97] Vaidyanathan, P.P. Optimal design of linear-phase FIR digital filters with very flat passbands and equiripple stopbands, *IEEE Trans. Circuits Sys.*, 32, 904–916, Sep. 1985.

[98] Kaiser, J.F. and Steiglitz, K. Design of FIR filters with flatness constraints, *Proc. IEEE Intl. Conf. Acoust., Speech, Signal Processing (ICASSP)*, 197–200, 1983.

[99] Selesnick, I.W. and Burrus, C.S. Exchange algorithms for the design of linear phase FIR filters and differentiators having flat monotonic passbands and equiripple stopbands, *IEEE Trans. Circuits Sys. II*, 43, 671–675, Sep. 1996.

[100] Adams, J.W. FIR digital filters with least squares stop bands subject to peak-gain constraints, *IEEE Trans. Circuits Sys.*, 39, 376–388, Apr. 1991.

[101] Adams, J.W., Sullivan, J.L., Hashemi, R., Ghadimi, R., Franklin, J., and Tucker, B. New approaches to constrained optimization of digital filters, *Proc. IEEE Intl. Symp. Circuits Systems (ISCAS)*, 80–83, May 1993.

[102] Barrodale, I., Powell, M.J.D., and Roberts, F.D.K. The differential correction algorithm for rational L_∞-approximation, *SIAM J. Numer. Anal.*, 9, 493–504, Sep. 1972.

[103] Crosara, S. and Mian, G.A. A note on the design of IIR filters by the differential-correction algorithm, *IEEE Trans. Circuits Sys.*, 30, 898–903, Dec. 1983.

[104] Dudgeon, D.E. Recursive filter design using differential correction, *IEEE Trans. Acoust., Speech, Signal Proc.*, 22, 443–448, Dec. 1974.

[105] Kaufman, E.H., Jr., Leeming, D.J., and Taylor, G.D. A combined Remes-differential correction algorithm for rational approximation, *Mathematics of Computation*, 32, 233–242, Jan. 1978.

[106] Rabiner, L.R., Graham, N.Y., and Helms, H.D. Linear programming design of IIR digital filters with arbitrary magnitude function, *IEEE Trans. on Acoust., Speech, Signal Proc.*, 22, 117–123, Apr. 1974.

[107] Deczky, A.G. Synthesis of recursive digital filters using the minimum p-error criterion, *IEEE Trans. Audio Electroacoust.*, 20, 257–263, Oct. 1972.

[108] Renfors, M. and Zigouris, E. Signal processor implementation of digital all-pass filters, *IEEE Trans. Acoust., Speech, Signal Processing,* 36, 714–729, May 1988.

[109] Vaidyanathan, P.P., Mitra, S.K., and Neuvo, Y. A new approach to the realization of low-sensitivity IIR digital filters, *IEEE Trans. Acoust., Speech, Signal Processing,* 34, 350–361, Apr. 1986.

[110] Regalia, P.A., Mitra, S.K., and Vaidyanathan, P.P. The digital all-pass filter: a versatile signal processing building block, *Proc. IEEE,* 76, 19–37, Jan. 1988.

[111] Vaidyanathan, P.P., Regalia, P.A., and Mitra, S.K. Design of doubly-complementary IIR digital filters using a single complex allpass filter, with multirate applications, *IEEE Trans. Circuits Sys.,* 34, 378–389, Apr. 1987.

[112] Vaidyanathan, P.P. *Multirate Systems and Filter Banks,* Prentice-Hall, Englewood Cliffs, NJ, 1993.

[113] Gerken, M., Schüßler, H.W., and Steffen, P. On the design of digital filters consisting of a parallel connection of allpass sections and delay elements, *Archiv für Electronik und Übertragungstechnik (AEÜ),* 49, 1–11, Jan. 1995.

[114] Jaworski, B. and Saramäki, T. Linear phase IIR filters composed of two parallel allpass sections, *Proc. IEEE Intl. Symp. Circuits Sys. (ISCAS),* (London), 537–540, May 30-June 2 1994.

[115] Kim, C.W. and Ansari, R. Approximately linear phase IIR filters using allpass sections, in *Proc. IEEE Intl. Symp. Circuits Sys. (ISCAS),* San Jose, 661–664, May 5-7 1986.

[116] Renfors, M. and Saramäki, T. A class of approximately linear phase digital filters composed of allpass subfilters, *Proc. IEEE Intl. Symp. Circuits Sys. (ISCAS),* San Jose, 678–681, May 5-7 1986.

[117] Chen, C.-K. and Lee, J.-H. Design of digital all-pass filters using a weighted least squares approach, *IEEE Trans. Circuits Sys. II,* 41, 346–351, May 1994.

[118] Kidambi, S.S. Weighted least-squares design of recursive allpass filters, *IEEE Trans. Signal Processing,* 44, 1553–1556, June 1996.

[119] Lang, M. and Laakso, T. Simple and robust method for the design of allpass filters using least-squares phase error criterion, *IEEE Trans. Circuits Sys. II,* 41, 40–48, Jan. 1994.

[120] Nguyen, T.Q., Laakso, T.I., and Koilpillai, R.D. Eigenfilter approach for the design of allpass filters approximating a given phase response, *IEEE Trans. Signal Processing,* 42, 2257–2263, Sep. 1994.

[121] Pei, S.-C. and Shyu, J.-J. Eigenfilter design of 1-D and 2-D IIR digital all-pass filters, *IEEE Trans. Signal Processing,* 42, 966–968, Apr. 1994.

[122] Schüßler, H.W. and Steffan, P. On the design of allpasses with prescribed group delay, *Proc. IEEE Intl. Conf. Acoust., Speech, Signal Processing (ICASSP),* Albuquerque, 1313–1316, April 3-6 1990.

[123] Anderson, M.S. and Lawson, S.S. Direct design of approximately linear phase (ALP) 2-D IIR digital filters, *Electron. Lett.,* 29, 804–805, April 29 1993.

[124] Ansari, R. and Liu, B. A class of low-noise computationally efficient recursive digital filters with applications to sampling rate alterations, *IEEE Trans. Acoust., Speech, Signal Processing,* 33, 90–97, Feb. 1985.

[125] Saramäki, T. On the design of digital filters as a sum of two all-pass filters, *IEEE Trans. Circuits Sys.,* 32, 1191–1193, Nov. 1985.

[126] Lang, M. Allpass filter design and applications, in *Proc. IEEE Intl. Conf. Acoust., Speech, Signal Processing (ICASSP),* Detroit, 1264–1267, May 9-12 1995.

[127] Schüssler, H.W. and Weith, J. On the design of recursive Hilbert-transformers, *Proc. IEEE Intl. Conf. Acoust., Speech, Signal Processing (ICASSP),* Dallas, 876–879, April 6-9 1987.

[128] Steiglitz, K. Computer-aided design of recursive digital filters, *IEEE Trans. Audio Electroacoust.,* 18, 123–129, 1970.

[129] Shaw, A.K. Optimal design of digital IIR filters by model-fitting frequency response data, *IEEE Trans. Circuits Sys. II*, 42, 702–710, Nov. 1995.

[130] Chen, X. and Parks, T.W. Design of IIR filters in the complex domain, *Proc. IEEE Intl. Conf. Acoust., Speech, Signal Processing (ICASSP)*, 1443–1446, 1988.

[131] Therrian, C.W. and Velasco, C.H. An iterative Prony method for ARMA signal modeling, *IEEE Trans. Signal Processing*, 43, 358–361, Jan. 1995.

[132] Pernebo, L. and Silverman, L.M. Model reduction via balanced state space representations, *IEEE Trans. Automatic Control*, 27, 382–387, Apr. 1982.

[133] Glover, K. All optimal Hankel-norm approximations of linear multivariable systems and their l^∞-error bounds, *Int. J. Control*, 39(6), 1115–1193, 1984.

[134] Beliczynski, B., Kale, I., and Cain, G.D. Approximation of FIR by IIR digital filters: an algorithm based on balanced model reduction, *IEEE Trans. Signal Processing*, 40, 532–542, March 1992.

[135] Chen, B.-S., Peng, S.-C., and Chiou, B.-W. IIR filter design via optimal Hankel-norm approximation, *IEE Proc., Part G*, 139, 586–590, Oct. 1992.

[136] Rudko, M. A note on the approximation of FIR by IIR digital filters: an algorithm based on balanced model reduction, *IEEE Trans. Signal Processing*, 43, 314–316, Jan. 1995.

[137] Tufan, E. and Tavsanoglu, V. Design of two-channel IIR PRQMF banks based on the approximation of FIR filters, *Electron. Lett.*, 32, 641–642, March 28, 1996.

[138] Jackson, L.B. *Digital Filters and Signal Processing (3rd ed.) with MATLAB Exercises*, Kluwer Academic Publishers, Amsterdam, 1996.

[139] Committee, I.D. Ed., *Selected Papers In Digital Signal Processing, II*, IEEE Press, New York, 1976.

[140] Rabiner, L.R. and Rader, C.M. Eds., *Digital Signal Processing*, IEEE Press, New York, 1972.

[141] Potchinkov, A. and Reemtsen, R., The design of FIR filters in the complex plane by convex optimization, *Signal Processing*, 46, 127–146, 1995.

[142] Potchinkov, A. and Reemtsen, R., The simultaneous approximation of magnitude and phase by FIR digital filters, I and II, *Int. J. Circuit Theory Appl.*, 25, 167–197, 1997.

[143] Lang, M.C., Design of nonlinear phase FIR digital filters using quadratic programming, in *Proc. IEEE Int. Conf. Acoust., Speech, Signal Processing*, Munich, Vol. 3:2169–2172, April 1997.

Statistical Signal Processing

Georgios B. Giannakis
University of Virgina

S TATISTICAL SIGNAL PROCESSING deals with random signals, their acquisition, their properties,
their transformation by system operators, and their characterization in the time and frequency domains. The goal is to extract pertinent information about the underlying mechanisms that generate

them or transform them. The area is grounded in the theories of signals and systems, random variables and stochastic processes, detection and estimation, and mathematical statistics. Random signals are temporal or spatial and can be derived from man-made (e.g., binary communication signals) or natural (e.g., thermal noise in a sensory array) sources. They can be continuous or discrete in their amplitude or index, but no exact expression describes their evolution. Signals are often described statistically when the engineer has incomplete knowledge about their description or origin. In these cases, statistical descriptors are used to characterize one's degree of knowledge (or ignorance) about the randomness. Especially interesting are those signals (e.g., stationary and ergodic) that can be described using deterministic quantities computable from finite data records. Applications of statistical signal processing algorithms to random signals are omnipresent in science and engineering in such areas as speech, seismic, imaging, sonar, radar, sensor arrays, communications, controls, manufacturing, atmospheric sciences, econometrics, and medicine, just to name a few. This chapter deals with the fundamentals of statistical signal processing, including some interesting topics that deviate from traditional assumptions. The focus is on discrete index random signals (i.e., time series) with possibly continuous-valued amplitudes. The reason is twofold: measurements are often made in discrete fashion (e.g., monthly temperature data); and continuously recorded signals (e.g., speech data) are often sampled for parsimonious representation and efficient processing by computers.

The first chapter of the section, written by Charles Therrien, reviews definitions, characterization, and estimation problems entailing random signals. The important notions outlined are stationarity, independence, ergodicity, and Gaussianity. The basic operations involve correlations, spectral densities, and linear time-invariant transformations. Stationarity reflects invariance of a signal's statistical description with index shifts. Absence (or presence) of relationships among samples of a signal at different points is conveyed by the notion of (in)dependence, which provides information about the signal's dynamical behavior and memory as it evolves in time or space. Ergodicity allows computation of statistical descriptors from finite data records. In increasing order of computational complexity, descriptors include the mean (or average) value of the signal, the autocorrelation, and higher than second-order correlations which reflect relations among two or more signal samples. Complete statistical characterization of random signals is provided by probability density and distribution functions. Gaussianity describes probabilistically a particular distribution of signal values which is characterized completely by its first- and second-order statistics. It is often encountered in practice because, thanks to the central limit theorem, averaging a sufficient number of random signal values (an operation often performed by, e.g., narrowband filtering) yields outputs which are (at least approximately) distributed according to the Gaussian probability law. Frequency-domain statistical descriptors inherit all the merits of deterministic Fourier transforms and can be computed efficiently using the fast Fourier transform. The standard tool here is the power spectral density which describes how average power (or signal variance) is distributed across frequencies; but polyspectral densities are also important for capturing distributions of higher-order signal moments across frequencies. Random input signals passing through linear systems yield random outputs. Input-output auto- and cross-correlations and spectra characterize not only the random signals themselves but also the transformation induced by the underlying system.

Many random signals as well as systems with random inputs and outputs possess finite degrees of freedom and can thus be modeled using finite parameters. Depending on *a priori* knowledge, one estimates parameters from a given data record, treating them either as random or deterministic. Various approaches become available by adopting different figures of merit (estimation criteria). Those outlined in this chapter include the maximum likelihood, minimum variance, and least-squares criteria for deterministic parameters. Random parameters are estimated using the maximum *a posteriori* and Bayes criteria. Unbiasedness, consistency, and efficiency are important properties of estimators which, together with performance bounds and computational complexity, guide the engineer to select the proper criterion and estimation algorithm.

While estimation algorithms seek values in the continuum of a parameter set, the need arises often in signal processing to classify parameters or waveforms as one or another of prespecified classes. Decision making with two classes is sought frequently in practice, including as a special case the simpler problem of detecting the presence or absence of an information-bearing signal observed in noise. Such signal detection and classification problems along with the associated theory and practice of hypotheses testing is the subject of the second chapter written by Alfred Hero. The resulting strategies are designed to minimize the average number of decision errors. Additional performance measures include receiver operating characteristics, signal-to-noise ratios, probabilities of detection (or correct classification), false alarm (or misclassification) rates, and likelihood ratios. Both temporal and spatio-temporal signals are considered, focusing on linear single- and multi-variate Gaussian models. Trade-offs include complexity versus optimality, off-line versus real time processing, and separate versus simultaneous detection and estimation for signal models containing unknown parameters.

Parametric and nonparametric methods are described in the third chapter, written by Petar Djurić and Steven Kay, for the basic problem of spectral estimation. Estimates of the power spectral density have been used over the last century and continue to be of interest in numerous applications involving retrieval of hidden periodicities, signal modeling, and time series analysis problems. Starting with the periodogram (normalized square magnitude of the data Fourier transform), its modifications with smoothing windows, and moving on to the more recent minimum variance and multiple window approaches, the nonparametric methods described here constitute the first step used to characterize the spectral content of stationary stochastic signals. Factors dictating the designer's choice include computational complexity, bias-variance, and resolution trade-offs. For data adequately described by a parametric model, such as the auto-regressive (AR), moving-average (MA), or ARMA model, spectral analysis reduces to estimating the model parameters. Such a data reduction step achieved by modeling offers parsimony and increases resolution and accuracy, provided that the model and its order (number of parameters) fit well the available time series. Processes containing harmonic tones (frequencies) have line spectra, and the task of estimating frequencies appears in diverse applications in science and engineering. The methods presented here include both the traditional periodogram as well as modern subspace approaches such as the MUSIC and its derivatives.

Estimation from discrete-time observations is the theme of the next chapter, written by Jerry Mendel. The unifying viewpoint treats both parameter and waveform (or signal) estimation from the perspective of minimizing the averaged square error between observations and input-output or state variable signal models. Starting from the traditional linear least-squares formulation, the exposition includes weighted and recursive forms, their properties, and optimality conditions for estimating deterministic parameters as well as their minimum mean-square error and maximum *a posteriori* counterparts for estimating random parameters. Waveform estimation, on the other hand, includes not only input-output signals but also state space vectors in linear and nonlinear state variable models. Prediction, smoothing, and the celebrated Kalman filtering problems are outlined in this framework and relationships are highlighted with the Wiener filtering formulation. Nonlinear least-squares and iterative minimization schemes are discussed for problems where the desired parameters are nonlinearly related with the data. Nonlinear equations can often be linearized, and the extended Kalman filter is described briefly for estimating nonlinear state variable models. Minimizing the mean-square error criterion leads to the basic orthogonality principle which appears in both parameter and waveform estimation problems. Generally speaking, the mean-square error criterion possesses rather universal optimality when the underlying models are linear and the random data involved are Gaussian distributed.

Before accessing applicability and optimality of estimation algorithms in real life applications, models need to be checked for linearity, and the random signals involved need to tested for Gaussianity and stationarity. Performance bounds and parameter confidence intervals must also be derived in order to evaluate the fit of the model. Finally, diagnostic tools for model falsification are needed to validate that the chosen model represents faithfully the underlying physical system. These important issues are discussed in the chapter written by Jitendra Tugnait. Stationarity, Gaussianity, and linearity tests are presented in a hypothesis-testing framework relying upon second- and higher-order statistics of the data. Tests are also described for estimating the number of parameters (or degrees of freedom) necessary for parsimonious modeling. Model validation is accomplished by checking for whiteness and independence of the error processes formed by subtracting model data from measured data. Tests may declare signal or noise data as non-Gaussian and/or nonstationary. The non-Gaussian models outlined here include the generalized Gaussian, Middleton's class, and the stable noise distribution models.

As for nonstationary signals and time-varying systems, detection and estimation tasks become more challenging and solutions are not possible in the most general case. However, structured nonstationarities such as those entailing periodic and almost periodic variations in their statistical descriptors are tractable. The resulting random signals are called (almost) cyclostationary and their analysis is the theme of the final chapter in this section, which I have written. The exposition starts with motivation and background material including links between cyclostationary signals and multivariate stationary processes, time-frequency representations, and multirate operators. Examples of cyclostationary signals and cyclostationarity-inducing operations are also described along with applications to signal processing and communication problems with emphasis on signal separation and channel equalization.

Modern theoretical directions in the field appear toward non-Gaussian, nonstationary, and nonlinear signal models. Advanced statistical signal processing tools (algorithms, software, and hardware) are of interest in current applications such as manufacturing, biomedicine, multimedia services, and wireless communications. Scientists and engineers will continue to search and exploit determinism in signals that they create or encounter, and find it convenient to model, as random.

12

Overview of Statistical Signal Processing

Charles W. Therrien
Naval Postgraduate School

12.1 Discrete Random Signals

12.1.1 Random Signals and Sequences

A discrete random signal or *sequence* $x[n]$ is a signal such that for any choice of the independent variable, say $n = n_\circ$, $x[n_\circ]$ is a random variable. If the parameter n represents time, as is usually the case, the random sequence is sometimes referred to as a "time series." The independent variable could represent another quantity, however, such as a spatial index in a uniform linear array.

The underlying model that represents the random sequence is known as a *random process* or a stochastic process. Figure 12.1 shows some examples of random signals. The noise signal of (a) can take on any real value, while the binary data sequence (b) (in the absence of noise) takes on only two discrete values ($+1$ and -1). The examples in (c) and (d) are interesting because, while they satisfy the definition of a random signal, their evolution (in time) is known forever once a few values of the process are observed. In the case of the sinusoid, its amplitude and/or phase may be random variables, but its future values can be determined from any two consecutive values of the signal. In the case of the constant voltage, its value is a random variable, but any one sample of the signal specifies the signal for all time. Such random signals are called *predictable* and form a set of processes distinct from those such as (a) and (b), which are called *regular*. Predictable random processes can always be estimated perfectly (i.e., with zero error) from a linear combination of past values of the process.

The fundamental statistical characterization of a random process is through the joint probability distribution or density function of its samples. For purposes of this chapter it is sufficient to work with the

0-8493-8572-5/98/$0.00+$.50

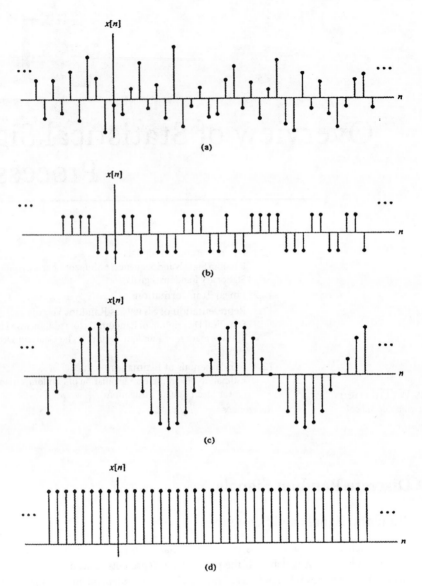

FIGURE 12.1 Examples of random sequences. (a) Noise, (b) binary data, (c) random sinusoid, (d) constant random voltage. (From Therrien, C.W., *Discrete Random Signals and Statistical Signal Processing*, Prentice-Hall, Englewood Cliffs, NJ, 1992. With permission).

density function, using impulses to formally represent any discrete probability values.[1] To characterize the signal completely, it must be possible to form the joint density of any set of samples of the process as shown in Fig. 12.2. If this density function is independent of where the samples are taken in the process as long as the spacing is the same, then the process is said to be *stationary* in the strict sense (see Fig. 12.2).

---

[1] For example, the probability density for a sample of the binary random signal of Fig. 12.1(b) taking on values of ±1 would be written as $f_{x[n]}(x_n) = P\delta_c(x_n - 1) + (1 - P)\delta_c(x_n + 1)$ where P is the probability of a positive value (+1) and $\delta_c(x_n)$ is the continuous impulse function.

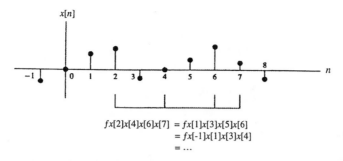

FIGURE 12.2   Stationary random process. Any set of samples with the same spacing has the same density function. (From Therrien, C.W., *Discrete Random Signals and Statistical Signal Processing*, Prentice-Hall, Englewood Cliffs, NJ, 1992. With permission).

There are three main cases that occur in signal processing where a complete statistical characterization of the random signal is possible. These are:

1. When the samples of the signal are *independent*. In that case the joint density for any set of samples can be written as a *product* of the density functions for the individual samples. If the samples have mean zero, this type of process is known as a *strictly white* process.
2. When the conditional density for the samples $f_{x[n]|x[n-1],x[n-2],\ldots}$ depends only on the *previous* sample $x[n-1]$ (or on the previous k samples). This type of process is known as a Markov process (or a kth order Markov process).
3. When the samples of the process are jointly Gaussian. This is called a *Gaussian* random process and occurs frequently in real life, for example when the random sequence is a sampled version of noise (see [1] for a more complete discussion).

Complete statistical analysis of signals is possible in cases when the nature of the problem allows one of the above models to apply. In a great many cases, however, there is incomplete knowledge of the statistical distribution of the signals; nevertheless, a very useful analysis can still be carried out using only certain statistical moments of the signal.

For a real-valued random signal the first and second order moments are the mean and correlation function defined respectively by

$$m_x[n] \stackrel{\text{def}}{=} \mathcal{E}\{x[n]\} \tag{12.1}$$

and

$$R_x[n_1, n_0] \stackrel{\text{def}}{=} \mathcal{E}\{x[n_1]x[n_0]\} \tag{12.2}$$

Higher order moments, say of order 3 and 4, are defined in an analogous way:

$$M_x^{(3)}[n_0, n_1, n_2] \stackrel{\text{def}}{=} \mathcal{E}\{x[n_0]x[n_1]x[n_2]\} \tag{12.3}$$

$$M_x^{(4)}[n_0, n_1, n_2, n_3] \stackrel{\text{def}}{=} \mathcal{E}\{x[n_0]x[n_1]x[n_2]x[n_3]\} \tag{12.4}$$

More general moments can be represented by expressions like

$$\mathcal{E}\left\{x^{k_0}[n_0] \cdot x^{k_1}[n_1] \ldots x^{k_L}[n_L]\right\}$$

for various selections of the powers $k_0, k_1, \ldots, k_L$ and values of L. Moments are usually not known *a priori* but must be *estimated* from data. In the case of a stationary random process it is useful if the moment computed from the signal average defined as

$$\left\langle x^{k_0}[n]x^{k_1}[n+l_1]\ldots x^{k_L}[n+l_L]\right\rangle \overset{\text{def}}{=} \lim_{N\to\infty} \frac{1}{2N+1}$$

$$\sum_{n=-N}^{N} x^{k_0}[n]x^{k_1}[n+l_1]\ldots x^{k_L}[n+l_L] \tag{12.5}$$

satisfies the property

$$\left\langle x^{k_0}[n]x^{k_1}[n+l_1]\ldots x^{k_L}[n+l_L]\right\rangle \doteq \mathcal{E}\left\{x^{k_0}[n]x^{k_1}[n+l_1]\ldots x^{k_L}[n+l_L]\right\} \tag{12.6}$$

where the notation "$\doteq$" means that the *event*

$$\left\langle x^{k_0}[n]x^{k_1}[n+l_1]\ldots x^{k_L}[n+l_L]\right\rangle = \mathcal{E}\left\{x^{k_0}[n]x^{k_1}[n+l_1]\ldots x^{k_L}[n+l_L]\right\}$$

has probability one. If (12.6) is satisfied for all L, all choices of the spacings $l_1, l_2, \ldots, l_L$, and all choices of the powers $k_0, k_1, \ldots, k_L$, then the process is said to be *strictly ergodic*. A random process that satisfies only the condition

$$\langle x[n]\rangle \doteq \mathcal{E}\{x[n]\} \tag{12.7}$$

is said to be "ergodic in the mean" while one that satisfies

$$\langle x[n]x[n+l]\rangle \doteq \mathcal{E}\{x[n]x[n+l]\} \tag{12.8}$$

is said to be "ergodic in correlation." These last two conditions are sufficient for many analyses.

Ergodicity implies that statistical moments can be estimated from a single realization of a random process, which is sometimes all that is available in a practical application. A noise process such as that depicted in Fig. 12.1(a) is typically an ergodic process while the battery voltage depicted in (d) is not. (Averaging in time will produce only the value of the signal in the given realization, not the mean of the distribution from which the random signal was drawn.)

12.1.2 Moment Characterization of Stationary Random Signals

Moments and Cumulants

The moments of a stationary random process are independent of the time index n. Thus, the *mean* is a constant and can be defined by

$$m_x \overset{\text{def}}{=} \mathcal{E}\{x[n]\} \tag{12.9}$$

The *(auto)correlation* function depends on only the time difference or *lag l* between the two signal samples and can be defined as

$$R_x[l] \overset{\text{def}}{=} \mathcal{E}\left\{x[n]x^*[n-l]\right\} \tag{12.10}$$

The *covariance* function is likewise defined as

$$C_x[l] \overset{\text{def}}{=} \mathcal{E}\left\{(x[n]-m_x)(x[n-l]-m_x)^*\right\} \tag{12.11}$$

and satisfies the relation

$$R_x[l] = C_x[l] + |m_x|^2 \tag{12.12}$$

If a random signal is not strictly stationary, but its mean is constant and its correlation function depends only on l (not n), then the process is called *wide-sense stationary*. The specific values $R_x[0] = \mathcal{E}\left\{|x[n]|^2\right\}$ and $C_x[0] = \mathcal{E}\left\{|x[n]-m_x|^2\right\}$ represent the *power* and the *variance* of the signal respectively.

An example of an almost trivial but fundamental correlation function is that of a *white noise* process. A white noise process is any process having mean zero and uncorrelated samples; that is, $R_x[l] = 0$ for $l \neq 0$. A white noise process thus has a correlation function of the form

$$R_x[l] = C_x[l] = \sigma_o^2 \delta[l]$$

where $\delta[l]$ is the unit sample function (discrete time impulse).

The need for the complex conjugate in the definitions (12.10) and (12.11) above for a *complex* random signal deserves just a bit of discussion. Complex random signals typically arise as representations of bandlimited signals at baseband. The stationarity assumption for such signals implies certain requirements of symmetry among the correlation and cross-correlation functions for the real and imaginary parts of the signal. In particular, if the signal is written as

$$x[n] = x_r[n] + j x_i[n]$$

then it can be shown [1] that the correlation functions for the real and imaginary parts of the signal are equal and are given by

$$R_{x_r}[l] = R_{x_i}[l] = \tfrac{1}{2} \operatorname{Re}[R_x[l]] \tag{12.13}$$

while the cross-correlation functions between the real and imaginary parts (see (12.21) for definition of cross-correlation) are given by

$$R_{x_r x_i}[l] = -R_{x_i x_r}[l] = -\tfrac{1}{2} \operatorname{Im}[R_x[l]] \tag{12.14}$$

Therefore, for stationary complex random signals, all of the correlation information is contained in the correlation function as defined by (12.10). In fact, similar arguments can be used to show that for these complex processes, if the conjugate is dropped from the second term, then $\mathcal{E}\{x[n]x[n-l]\}$ is *identically zero* for all values of l. Thus, the conjugate used in the definition is really *essential*.

The correlation (or covariance) function has two defining properties; the function is

1. Conjugate symmetric:

$$R_x[l] = R_x^*[-l] \tag{12.15}$$

and

2. Positive semidefinite:

$$\sum_{n_1=-\infty}^{\infty} \sum_{n_0=-\infty}^{\infty} a^*[n_1] R_x[n_1 - n_0] a[n_0] \geq 0 \tag{12.16}$$

for *any* sequence $a[n]$

These properties follow easily from the definitions [1]. The positive semidefinite property can be shown to imply that

$$R_x[0] \geq |R_x[l]| \quad l \neq 0$$

Note, however, that this is a *derived* property and not a fundamental defining property for the correlation function.

Higher order moments and cumulants are also used sometimes in modern signal processing. The third and fourth order moments for a stationary process are usually written as

$$M_x^{(3)}[l_1, l_2] = \mathcal{E}\left\{x^*[n]x[n+l_1]x[n+l_2]\right\} \tag{12.17}$$

$$M_x^{(4)}[l_1, l_2, l_3] = \mathcal{E}\left\{x^*[n]x^*[n+l_1]x[n+l_2]x[n+l_3]\right\} \tag{12.18}$$

while for a *zero-mean* random process the third and fourth order cumulants are given by

$$C_x^{(3)}[l_1, l_2] = \mathcal{E}\left\{x^*[n]x[n+l_1]x[n+l_2]\right\} \tag{12.19}$$

$$
\begin{aligned}
C_x^{(4)}[l_1, l_2, l_3] = \; & \mathcal{E}\left\{x^*[n]x^*[n+l_1]x[n+l_2]x[n+l_3]\right\} \\
& - C_x^{(2)}[l_2]C_x^{(2)}[l_3 - l_1] - C_x^{(2)}[l_3]C_x^{(2)}[l_2 - l_1]
\end{aligned}
\tag{12.20a}
$$

$$\text{(complex random process)}$$

$$
\begin{aligned}
C_x^{(4)}[l_1, l_2, l_3] = \; & \mathcal{E}\left\{x[n]x[n+l_1]x[n+l_2]x[n+l_3]\right\} - C_x^{(2)}[l_1]C_x^{(2)}[l_3 - l_2] \\
& - C_x^{(2)}[l_2]C_x^{(2)}[l_3 - l_1] - C_x^{(2)}[l_3]C_x^{(2)}[l_2 - l_1]
\end{aligned}
\tag{12.20b}
$$

$$\text{(real random process)}$$

where $C_x^{(2)}[l] = \mathcal{E}\left\{x[n]x^*[n+l]\right\}$ is the second order cumulant, identical to the covariance function. It should be noted that unlike the second order moments, the definition of these statistics for a complex random process is not standard, so alternate definitions to (12.17) to (12.20) are sometimes encountered.

For most analyses, cumulants are preferred to moments because the cumulants of order three and higher for a Gaussian process are identically zero. Thus, signal processing methods based on higher order cumulants have the advantage of being "blind" to any form of Gaussian noise.

For real-valued signals these higher order cumulants have many regions of symmetry. The symmetry regions for the third order cumulant are shown in Fig. 12.3. Symmetry regions for the third order cumulant of complex signals consist of only the half planes defined by $C_x^{(3)}[l_1, l_2] = C_x^{(3)}[l_2, l_1]$.

Cross-moments between two or more random signals are also of utility. For two jointly stationary[2] random signals x and y the cross-correlation and cross-covariance functions are defined by

$$R_{xy}[l] = \mathcal{E}\left\{x[n]y^*[n-l]\right\} \tag{12.21}$$

and

$$C_{xy}[l] = \mathcal{E}\left\{(x[n] - m_x)(y[n-l] - m_y)^*\right\} \tag{12.22}$$

and satisfy the relation

$$R_{xy}[l] = C_{xy}[l] + m_x m_y^* \tag{12.23}$$

These cross-moment functions have no particular properties except that $R_{xy}[l] = R_{yx}^*[-l]$ and $C_{xy}[l] = C_{yx}^*[-l]$. Higher order cross-moments and cumulants can be defined in an analogous way to (12.17) and (12.20) and are also encountered in some applications.

Frequency and Transform Domain Characterization

Random signals can be characterized in the frequency domain as well as in the signal domain. The *power spectral density function* is defined by the Fourier transform of the correlation function

$$S_x(e^{j\omega}) = \sum_{l=-\infty}^{\infty} R_x[l]e^{-j\omega l} \tag{12.24}$$

---

[2]Two signals are said to be *jointly stationary* (in the wide sense) if each of the signals is itself wide sense stationary, and the cross-correlation is a function of only the time difference or lag l.

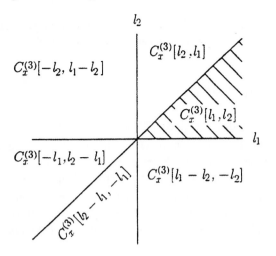

FIGURE 12.3 Regions of symmetry for the third order cumulant of real-valued signals. (From Therrien, C.W., *Discrete Random Signals and Statistical Signal Processing,* Prentice-Hall, Englewood Cliffs, NJ, 1992. With permission).

with inverse transform

$$R_x[l] = \frac{1}{2\pi} \int_{-\pi}^{\pi} S_x(e^{J\omega}) e^{J\omega l} d\omega \tag{12.25}$$

The name "power spectral density" comes from the fact that

$$\text{Avg. Power} = \mathcal{E}\left\{|x[n]|^2\right\} = R_x[0] = \frac{1}{2\pi} \int_{-\pi}^{\pi} S_x(e^{J\omega}) d\omega$$

which follows directly from (12.10) and (12.25). Since the power spectral density may contain both continuous and discrete components (see Fig. 12.4), its general form is

$$S_x(e^{J\omega}) = S_x'(e^{J\omega}) + \sum_i 2\pi P_i \delta_c(e^{J\omega} - e^{J\omega_i}) \tag{12.26}$$

The term $S_x'(e^{J\omega})$ represents the *continuous* part of the spectrum while the sum of weighted impulses represents the discrete part or "lines" in the spectrum. Impulses or lines arise from periodic or almost periodic random signals such those of Fig. 12.1(c) and (d).

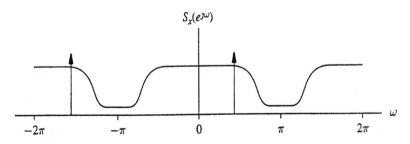

FIGURE 12.4 Typical power density spectrum for a complex random process showing continuous and discrete components. (From Therrien, C.W., *Discrete Random Signals and Statistical Signal Processing,* Prentice-Hall, Englewood Cliffs, NJ, 1992. With permission).

The two defining properties for the correlation function ((12.15) and (12.16)) are manifested as two corresponding properties of the power spectral density function, namely:

1. $S_x(e^{J\omega})$ is *real*.
2. $S_x(e^{J\omega})$ is *nonnegative*: $S_x(e^{J\omega}) \geq 0$.

In addition, for *real-valued* random signals $S_x(e^{J\omega})$ is an *even* function of frequency.

The white noise process, introduced in the previous subsection (Moment and Cumulants), has a power spectral density function that is a constant $S_x(e^{J\omega}) = \sigma_o^2$. The term "white" refers to the fact that the spectrum, like that of ideal white light, is flat and represents all frequencies in equal proportions.

The multidimensional Fourier transforms of the cumulants are also of considerable importance and are referred to generically as cumulant spectra, higher order spectra, or *polyspectra*. For the third and fourth order cumulants, these higher order spectra are called the *bispectrum* and *trispectrum*, respectively, and are defined by

$$B_x(\omega_1, \omega_2) = \sum_{l_1=-\infty}^{\infty} \sum_{l_2=-\infty}^{\infty} C_x^{(3)}[l_1, l_2] e^{-J(\omega_1 l_1 + \omega_2 l_2)} \tag{12.27}$$

and

$$T_x(\omega_1, \omega_2, \omega_3) = \sum_{l_1=-\infty}^{\infty} \sum_{l_2=-\infty}^{\infty} \sum_{l_3=-\infty}^{\infty} C_x^{(4)}[l_1, l_2, l_3] e^{-J(\omega_1 l_1 + \omega_2 l_2 + \omega_3 l_3)} \tag{12.28}$$

These quantities have many regions of symmetry. The regions of symmetry of the bispectrum of a real-valued signal are shown in Fig. 12.5. For a complex signal there is only symmetry between half planes.

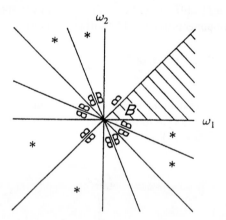

FIGURE 12.5   Regions of symmetry for the bispectrum of a real-valued signal. (From Therrien, C.W., *Discrete Random Signals and Statistical Signal Processing*, Prentice-Hall, Englewood Cliffs, NJ, 1992. With permission).

Higher order processes whose cumulants are proportional to the unit sample function and whose higher order spectra are therefore constant are sometimes called *higher order white noise processes*. For a "strictly white" process (see the discussion in Random Signals and Sequences), the cumulants of all orders are impulses, and thus the polyspectra of all orders are constant functions of frequency.

Cross-power spectral density functions are also defined as Fourier transforms of the corresponding cross-correlation functions, for example

$$S_{xy}(e^{J\omega}) = \sum_{l=-\infty}^{\infty} R_{xy}[l] e^{-J\omega l} \tag{12.29}$$

Since the cross-correlation function has no particular properties, the cross-power spectral density function will also have no distinctive properties; it is complex-valued in general. The cross-spectral density evaluated at a particular point in frequency can be interpreted as a measure of the correlation that exists between components of the two processes at the chosen frequency. The normalized cross-spectrum

$$\Gamma_{xy}(e^{J\omega}) \stackrel{\text{def}}{=} \frac{S_{xy}(e^{J\omega})}{\sqrt{S_x(e^{J\omega})}\sqrt{S_y(e^{J\omega})}} \tag{12.30}$$

is called the *coherence function* and its squared magnitude

$$|\Gamma_{xy}(e^{J\omega})|^2 = \frac{|S_{xy}(e^{J\omega})|^2}{S_x(e^{J\omega}) S_y(e^{J\omega})} \tag{12.31}$$

is called the *magnitude squared coherence* (MSC). The MSC is often used instead of $|S_{xy}(e^{J\omega})|$ and has the convenient property

$$0 \le |\Gamma_{xy}(e^{J\omega})|^2 \le 1 \tag{12.32}$$

Random signals can also be characterized in the z (transform) domain. In particular, the z-transform of the correlation and cross-correlation functions is needed in many analyses such as in the design of filters for random signals. For the correlation function the quantity

$$S_x(z) = \sum_{l=-\infty}^{\infty} R_x[l] z^{-l} \tag{12.33}$$

is known as the *complex spectral density function*. It has the basic symmetry property

$$S_x(z) = S_x^*(1/z^*) \tag{12.34}$$

and is real and nonnegative on the unit circle. For real-valued random processes (12.34) can be expressed as

$$S_x(z) = S_x(z^{-1})$$

however, expressing the property in this way sometimes hides the function's true features. For a *rational*[3] complex spectral density function, (12.34) implies that for any root of the numerator or denominator, say at location z_o, there is a corresponding root at the conjugate reciprocal position, $1/z_o^*$. This also implies that zeros on the unit circle occur in even multiplicities. (Poles are not allowed to occur *on* the unit circle.) In addition, since a real-valued random process has real coefficients in the polynomials that define $S_x(z)$, the complex roots of such processes occur in conjugate pairs. Therefore, for real-valued processes, poles or zeros not on the real axis occur in groups of four:

$$z_o, \quad 1/z_o, \quad z_o^*, \quad 1/z_o^*$$

---

[3]The term *rational* is used to describe functions of the form $S_x(z) = N(z)/D(z)$ where $N(z)$ and $D(z)$ are polynomials.

The correlation function can be obtained from the inverse transform

$$R_x[l] = \frac{1}{2\pi j} \oint_C S_x(z) z^{l-1} dz \qquad (12.35)$$

which involves a contour integral in the region of convergence of the transform. Because of the symmetry the region of convergence is always an *annular* region of the form

$$a < |z| < \frac{1}{a}$$

and the integral can be evaluated using the method of residues [1].

The complex cross-spectral density function $S_{xy}(z)$ and its inverse are defined by equations analogous to (12.33) and (12.35). Again, because the cross-correlation function has no special properties, none are imparted to the complex cross-spectral density function.

A simple but useful real correlation function has the exponential form[4]

$$R_x[l] = \sigma^2 \rho^{|l|} \qquad (12.36)$$

This function and its corresponding power spectral density

$$S_x(e^{j\omega}) = \frac{\sigma^2 \left(1 - \rho^2\right)}{1 + \rho^2 - 2\rho \cos \omega} \qquad (12.37)$$

are illustrated in Fig. 12.6. The corresponding complex spectral density function can be expressed as

$$S_x(z) = \frac{\sigma^2 \left(1 - \rho^2\right)}{-\rho z + \left(1 + \rho^2\right) - \rho z^{-1}} \qquad (12.38)$$

This function has one pair of real axis poles at $z = \rho$ and $z = 1/\rho$ and the region of convergence lies *between* the two poles.

12.2   Linear Transformations

Linear shift-invariant systems can be represented in the signal domain by their impulse response sequence $h[n]$. If a random process $x[n]$ is applied to the linear system, the output $y[n]$ is given by the convolution

$$y[n] = \sum_{k=-\infty}^{\infty} h[k] x[n-k] \qquad (12.39)$$

If $x[n]$ is stationary, then $y[n]$ will also be stationary. Taking expectations on both sides of the equation yields

$$\mathcal{E}\{y[n]\} = \sum_{k=-\infty}^{\infty} h[k] \mathcal{E}\{x[n-k]\}$$

or

$$m_y = m_x \cdot \sum_{k=-\infty}^{\infty} h[k] \qquad (12.40)$$

---

[4]A complex version of this correlation function can be found in [1].

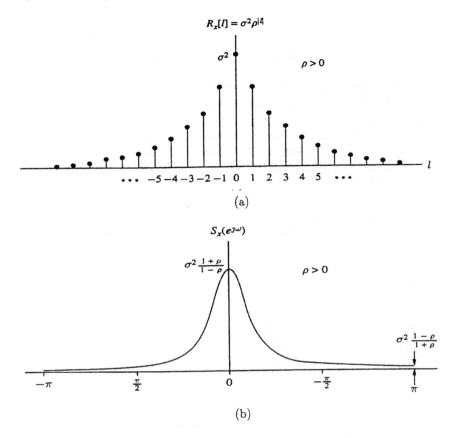

FIGURE 12.6   Real exponential correlation function and corresponding power spectral density ($\rho > 0$). (a) Correlation function, (b) power spectral density function. (From Therrien, C.W., *Discrete Random Signals and Statistical Signal Processing*, Prentice-Hall, Englewood Cliffs, NJ, 1992. With permission).

The output correlation function can be computed by the following steps. Multiplying (12.39) on both sides by $y^*[n - l]$ and taking the expectation yields

$$\mathcal{E}\left\{y[n]y^*[n - l]\right\} = \sum_{k=-\infty}^{\infty} h[k]\mathcal{E}\left\{x[n - k]y^*[n - l]\right\}$$

or

$$R_y[l] = \sum_{k=-\infty}^{\infty} h[k]R_{xy}[l - k]$$

which will be written as

$$R_y[l] = h[l] * R_{xy}[l] \tag{12.41}$$

where "$*$" denotes *convolution* of the sequences. Multiplying (12.39) by $x^*[n - l]$ and performing similar steps yields

$$R_{yx}[l] = h[l] * R_x[l] \tag{12.42}$$

Conjugating terms and noting that $R_{xy}[l] = R_{yx}^*[-l]$ and $R_x[l] = R_x^*[-l]$ permits (12.42) to be written as

$$R_{xy}[l] = h^*[-l] * R_x[l] \tag{12.43}$$

Then combining (12.41) and (12.43) yields

$$R_y[l] = h[l] * h^*[-l] * R_x[l] \tag{12.44}$$

In other words, the output correlation function is obtained as a double convolution of the input correlation function with the impulse response and the reversed conjugated impulse response. It can easily be shown that the covariance and cross-covariance functions also satisfy the relations (12.41) through (12.44).

By using the Fourier and z-transform relations and the last four equations it is easy to derive expressions for the results of a linear transformation in the frequency and transform domains. The complete set of relations is listed in Table 12.1; those for the output process are the ones most frequently used and appear in the last row of the table.

TABLE 12.1   Linear Transformation Relations

System defined by: $y[n] = h[n] * x[n]$		
$R_{yx}[l] = h[l] * R_x[l]$	$S_{yx}(e^{j\omega}) = H(e^{j\omega})S_x(e^{j\omega})$	$S_{yx}(z) = H(z)S_x(z)$
$R_{xy}[l] = h^*[-l] * R_x[l]$	$S_{xy}(e^{j\omega}) = H^*(e^{j\omega})S_x(e^{j\omega})$	$S_{xy}(z) = H^*(1/z^*)S_x(z)$
$R_y[l] = h[l] * R_{xy}[l]$	$S_y(e^{j\omega}) = H(e^{j\omega})S_{xy}(e^{j\omega})$	$S_y(z) = H(z)S_{xy}(z)$
$R_y[l] = h[l] * h^*[-l] * R_x[l]$	$S_y(e^{j\omega}) = |H(e^{j\omega})|^2 S_x(e^{j\omega})$	$S_y(z) = H(z)H^*(1/z^*)S_x(z)$

Note: For real $h[n]$, $H^*(1/z^*) = H(z^{-1})$.
(From Therrien, C.W., *Discrete Random Signals and Statistical Signal Processing*, Prentice-Hall, Englewood Cliffs, NJ, 1992. With permission).

As an example of the use of linear transformations, consider the simple first-order causal system described by the difference equation

$$y[n] = \rho\, y[n-1] + x[n]$$

The system has an impulse response given by $h[n] = \rho^n u[n]$ where $u[n]$ is the unit step function, and a transfer function given by

$$H(z) = \frac{1}{1 - \rho z^{-1}}$$

If the input is a white noise process with $S_x(z) = \sigma_o^2$, and all signals are real, then the output complex spectral density function is (see Table 12.1)

$$S_y(z) = H(z)H(z^{-1})S_x(z) = \frac{\sigma_o^2}{(1 - \rho z^{-1})(1 - \rho z)}$$

This is identical in form to (12.38) with $\sigma_o^2 = \sigma^2(1 - \rho^2)$. It follows that the correlation function and power spectral density function of the output also have the forms (12.36) and (12.37). (This could be shown directly by applying the other relations in the table.) Thus, a process with exponential correlation function can be obtained by driving a first-order filter with white noise.

The higher order moments and cumulants of the output of a linear system can also be computed from the corresponding input quantities, although the formulas are more complicated. For the third- and fourth-order cumulants the formulas are

$$C_y^{(3)}[l_1, l_2] = \sum_{k_0=-\infty}^{\infty} \sum_{k_1=-\infty}^{\infty} \sum_{k_2=-\infty}^{\infty} C_x^{(3)}[l_1 - k_1 + k_0, l_2 - k_2 + k_0]h[k_2]h[k_1]h^*[k_0] \tag{12.45}$$

and

$$C_y^{(4)}[l_1, l_2, l_3] =$$
$$\sum_{k_0=-\infty}^{\infty} \sum_{k_1=-\infty}^{\infty} \sum_{k_2=-\infty}^{\infty} \sum_{k_3=-\infty}^{\infty} C_x^{(4)}[l_1 - k_1 + k_0, l_2 - k_2 + k_0, l_3 - k_3 + k_0]$$
$$\cdot h[k_3]h[k_2]h^*[k_1]h^*[k_0] \tag{12.46}$$

These formulas can be interpreted as a sequence of convolutions with the filter impulse response in various directions (see [1]).

The corresponding frequency domain expressions are relatively simpler since they contains only products of terms. The expressions for the bispectrum and trispectrum are given by

$$B_y(\omega_1, \omega_2) = H^*\left(e^{J(\omega_1+\omega_2)}\right) H(e^{J\omega_1})H(e^{J\omega_2})B_x(\omega_1, \omega_2) \tag{12.47}$$

and

$$T_y(\omega_1, \omega_2, \omega_3) = H^*\left(e^{J(\omega_1+\omega_2+\omega_3)}\right) H^*\left(e^{-J\omega_1}\right) H(e^{J\omega_2})H(e^{J\omega_3})T_x(\omega_1, \omega_2, \omega_3) \tag{12.48}$$

Unlike the power spectral density function, these higher order spectra are affected by the *phase* of the linear system. For example, the phase of the output bispectrum is given by

$$\angle B_y(\omega_1, \omega_2) = -\angle H\left(e^{J(\omega_1+\omega_2)}\right) + \angle H(e^{J\omega_1}) + \angle H(e^{J\omega_2}) + \angle B_x(\omega_1, \omega_2)$$

Using higher order statistics it is possible to identify both the magnitude *and* phase of a linear system, while with second-order statistics it is possible to identify only the *magnitude*.

12.3 Representation of Signals as Random Vectors

12.3.1 Statistical Description of Random Vectors

For discrete signals or sequences it is often useful to define a vector $\boldsymbol{x}$ as shown in Fig. 12.7 consisting of

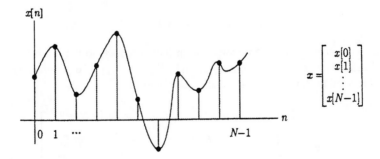

FIGURE 12.7 Representation of a random sequence as a random vector. (From Therrien, C.W., *Discrete Random Signals and Statistical Signal Processing*, Prentice-Hall, Englewood Cliffs, NJ, 1992. With permission).

N consecutive values of the signal and to denote the joint density of these N values by $f_{\boldsymbol{x}}(\mathbf{x})$. The latter quantity is referred to as the probability density function of the random vector $\boldsymbol{x}$. Consider the case of a real-valued signal first. If $\mathbf{x}^\circ$ denotes a particular value of the random vector

$$\mathbf{x}^\circ = \begin{bmatrix} x_0^\circ \\ x_1^\circ \\ \vdots \\ x_{N-1}^\circ \end{bmatrix}$$

and if we multiply it by small increments in each of the components, the expression

$$f_{\boldsymbol{x}}(\mathbf{x}^\circ)\Delta x_0 \Delta x_1 \ldots \Delta x_{N-1}$$

represents the *probability* that the signal (i.e., the random vector $\boldsymbol{x}$) lies in a small region of the vector space described by

$$x_0^\circ < x[0] \le x_0^\circ + \Delta x_0, \quad \ldots, \quad x_{N-1}^\circ < x[N-1] \le x_{N-1}^\circ + \Delta x_{N-1} \tag{12.49}$$

For a complex-valued random signal, $\boldsymbol{x}$ has complex components and $f_{\boldsymbol{x}}(\mathbf{x})$ represents the joint density between the $2N$ real and imaginary parts of the components of $\boldsymbol{x}$. Conditional and joint densities for random vectors are defined in a corresponding way [1] and have interpretations that are analogous to those for scalar random variables.

12.3.2 Moments

The first and second moment properties for random vectors are of great importance and are represented as follows. The *mean vector* is defined by

$$\boldsymbol{m_x} \overset{\text{def}}{=} \mathcal{E}\{\boldsymbol{x}\} = \begin{bmatrix} m_0 \\ m_1 \\ \vdots \\ m_{N-1} \end{bmatrix} \tag{12.50}$$

where $m_i = \mathcal{E}\{x[i]\}$ for $i = 0, 1, \ldots N - 1$. In the usual case of a stationary signal, all of the m_i have the same value (frequently zero).

The *correlation matrix* is defined by

$$\boldsymbol{R_x} \overset{\text{def}}{=} \mathcal{E}\{\boldsymbol{x}\,\boldsymbol{x}^{*T}\} \tag{12.51}$$

Note that this is an *outer* product of vectors, not an *inner* product, so the result is an $N \times N$ square matrix with the element in row i and column j given by $\mathcal{E}\{x[i]x^*[j]\}$. For a *stationary* random process, $\mathcal{E}\{x[i]x^*[j]\}$ is equal to $R_x[i-j]$, so the matrix has the form

$$\boldsymbol{R_x} = \begin{bmatrix} R_x[0] & R_x[-1] & \cdots & R_x[-N+1] \\ R_x[1] & R_x[0] & \ddots & \vdots \\ \vdots & \ddots & \ddots & R_x[-1] \\ R_x[N-1] & \cdots & R_x[1] & R_x[0] \end{bmatrix} \tag{12.52}$$

The matrix is Hermitian symmetric and Toeplitz (all elements on each diagonal are equal). The Hermitian symmetry property follows from the basic definition (12.51) and is true for *all* correlation matrices; the Toeplitz property appears only for correlation matrices of stationary random processes.

The *covariance matrix* is defined as

$$C_x = \mathcal{E}\left\{(x - m_x)(x - m_x)^{*T}\right\} \tag{12.53}$$

and satisfies the relation

$$R_x = C_x + m_x m_x^{*T} \tag{12.54}$$

The covariance matrix is thus the correlation matrix of the vector with the mean removed.

The symmetry properties for stationary complex random signals discussed in conjunction with the correlation function (in the previous section) have implications for the correlation matrices for the real and imaginary parts of the signal. These relations are summarized in Table 12.2. In addition, the symmetry

TABLE 12.2   Relations for the Complex Correlation Matrix of a Stationary Random Signal

Complex correlation matrix	$R_x = \mathcal{E}\left\{xx^{*T}\right\} = 2R_x^E + j2R_x^O$
Correlation matrices for components	$R_x^E = \mathcal{E}\left\{x_r x_r^T\right\} = \mathcal{E}\left\{x_i x_i^T\right\}$ $R_x^O = -\mathcal{E}\left\{x_r x_i^T\right\} = \mathcal{E}\left\{x_i x_r^T\right\}$

properties can be shown to imply that

$$\mathcal{E}\left\{xx^T\right\} = 0 \qquad \text{(for a complex random vector)} \tag{12.55}$$

hence, the importance of the *Hermitian (conjugate)* transpose in (12.51).

The correlation and covariance matrices of any random vector are always positive semidefinite, that is

$$a^{*T} R_x a \geq 0$$

(and $a^{*T} C_x a \geq 0$) for *any* vector a. The correlation matrix for a regular random process is, in fact, strictly positive definite ($>$ rather than $\geq$) while that for a predictable random process is just positive semidefinite if the size is sufficiently large.

Cross-correlation and cross-covariance matrices for two random signals or two random vectors x and y can also be defined as

$$R_{xy} = \mathcal{E}\left\{xy^{*T}\right\} \tag{12.56}$$

and

$$C_{xy} = \mathcal{E}\left\{(x - m_x)(y - m_y)^{*T}\right\} \tag{12.57}$$

These matrices have no particular properties and are not even square if x and y have different sizes. They exhibit a Toeplitz-like structure, however (all terms on the same diagonal are equal), if the two random processes are jointly stationary.

12.3.3   Linear Transformation of Random Vectors

When a vector y is defined by a linear transformation

$$y = Ax \tag{12.58}$$

the mean of y is given by $\mathcal{E}\{y\} = A\mathcal{E}\{x\}$ or

$$m_y = Am_x \tag{12.59}$$

while the correlation matrix is given by $\mathcal{E}\left\{\boldsymbol{y}\boldsymbol{y}^{*T}\right\} = \boldsymbol{A}\,\mathcal{E}\left\{\boldsymbol{x}\boldsymbol{x}^{*T}\right\}\boldsymbol{A}^{*T}$ or

$$\boldsymbol{R}_{\boldsymbol{y}} = \boldsymbol{A}\boldsymbol{R}_{\boldsymbol{x}}\boldsymbol{A}^{*T} \tag{12.60}$$

From these last two equations and (12.54) it can be shown that the covariance matrix transforms in a similar manner, i.e.,

$$\boldsymbol{C}_{\boldsymbol{y}} = \boldsymbol{A}\boldsymbol{C}_{\boldsymbol{x}}\boldsymbol{A}^{*T} \tag{12.61}$$

Transformations that result in random vectors with uncorrelated components are of special interest. Strictly speaking, the term "uncorrelated" applies to the *covariance* matrix. That is, if a random vector has uncorrelated components, its covariance matrix is *diagonal*. Following common practice, however, we will assume the mean is zero and discuss the methods using the correlation matrix. If the mean is nonzero, then the resulting vectors will have components that are said to be *orthogonal* rather than uncorrelated.

Since correlation matrices are Hermitian symmetric and positive semidefinite, their eigenvalues are nonnegative and eigenvectors are orthogonal (see e.g., [2, 3]). Any correlation matrix can, therefore, be factored as

$$\boldsymbol{R}_{\boldsymbol{x}} = \boldsymbol{E}\boldsymbol{\Lambda}\boldsymbol{E}^{*T} \tag{12.62}$$

where $\boldsymbol{E}$ is a unitary matrix $(\boldsymbol{E}^{*T}\boldsymbol{E} = \boldsymbol{I})$ whose columns are the eigenvectors and $\boldsymbol{\Lambda}$ is a diagonal matrix whose elements are the eigenvalues. Since the inverse of a unitary matrix is its Hermitian transpose, the last equation can be rewritten as

$$\boldsymbol{\Lambda} = \boldsymbol{E}^{*T}\boldsymbol{R}_{\boldsymbol{x}}\boldsymbol{E}$$

Comparing this with (12.60) above you can observe that if $\boldsymbol{y}$ is defined by

$$\boldsymbol{y} = \boldsymbol{E}^{*T}\boldsymbol{x} \tag{12.63}$$

then $\boldsymbol{R}_{\boldsymbol{y}}$ will be equal to $\boldsymbol{\Lambda}$, a diagonal matrix. Since $\boldsymbol{R}_{\boldsymbol{y}}$ is diagonal, the components of $\boldsymbol{y}$ are uncorrelated $(\mathcal{E}\left\{y_i y_j^*\right\} = 0,\ i \neq j)$. Thus, one way to produce a vector with uncorrelated components is to apply the eigenvector transformation (12.63).

Another way to produce a vector with uncorrelated components involves triangular decomposition of the correlation matrix. Matrices that satisfy certain conditions of their principal minors [4] can be factored into a product of a lower triangular and an upper triangular matrix. (This is called *LU* decomposition). Correlation matrices always satisfy the needed conditions, and since they are Hermitian symmetric, they can be written as a unique product

$$\boldsymbol{R}_{\boldsymbol{x}} = \boldsymbol{L}\boldsymbol{D}\boldsymbol{L}^{*T} \tag{12.64}$$

where $\boldsymbol{L}$ is a lower triangular matrix with ones on the diagonal and $\boldsymbol{D}$ is a diagonal matrix. The product $\boldsymbol{D}\boldsymbol{L}^{*T}$ is the upper triangular matrix $\boldsymbol{U}$ in the *LU* decomposition.

Equation 12.64 can be written as

$$\boldsymbol{D} = \boldsymbol{L}^{-1}\boldsymbol{R}_{\boldsymbol{x}}(\boldsymbol{L}^{-1})^{*T} \tag{12.65}$$

where it can be shown that $\boldsymbol{L}^{-1}$ is of the same form as $\boldsymbol{L}$ (i.e., lower triangular with ones on the diagonal). From (12.65) and (12.60) it can be recognized that $\boldsymbol{D}$ is the correlation matrix for a random vector $\boldsymbol{y}$ defined by

$$\boldsymbol{y} = \boldsymbol{L}^{-1}\boldsymbol{x} \tag{12.66}$$

Since $\boldsymbol{D}$ is a diagonal matrix, the components of $\boldsymbol{y}$ are seen to be uncorrelated.

The two transformations described above correspond to two fundamentally different ways of decorrelating a signal. The eigenvector transformation represents the signal in terms of an orthogonal set of basis functions (the eigenvectors) and has important geometric interpretations (see the following section and [1, Chapter 2]). It is also the basis for modern subspace methods of spectrum analysis and array processing. The transformation defined by the triangular decomposition has the advantage that it can be implemented by a *causal* linear filter. Thus, it has important practical applications. It is the transformation that naturally arises in the very important area of signal processing known as linear predictive filtering.

12.3.4 The Gaussian Density Function

One of the cases mentioned in the first section in which a complete statistical description of a random process is possible is the Gaussian case. When a random signal is Gaussian the density function for the random vector x representing that signal is specified in terms of just the mean vector and covariance matrix. For a *real* random signal this density function has the form

$$f_x(x) \;=\; \frac{1}{(2\pi)^{\frac{N}{2}}|C_x|^{\frac{1}{2}}}e^{-\frac{1}{2}(x-m_x)^T C_x^{-1}(x-m_x)} \tag{12.67}$$

(real random vector)

where N is the dimension of x. For a complex random signal the Gaussian density has the slightly different form

$$f_x(x) \;=\; \frac{1}{\pi^N |C_x|}e^{-(x-m_x)^{*T} C_x^{-1}(x-m_x)} \tag{12.68}$$

(complex random vector)

The reason for the different expression is that the density function for a complex random vector is actually the joint density between the real and imaginary parts. Because of the symmetry properties existing for the correlation or covariance matrix of a complex stationary random process (see Table 12.2), it is possible to express this joint density in the simpler form (12.68).

For the case $N = 2$ the contours of the density function defined by

$$f_x(x) = \text{const.} \tag{12.69}$$

are ellipses centered about the mean vector which can easily be drawn as in Fig. 12.8. The orientation and

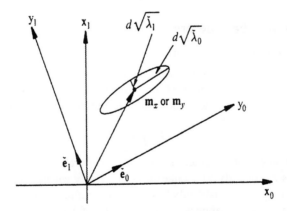

FIGURE 12.8 Typical contour of a Gaussian density function. (From Therrien, C.W., *Discrete Random Signals and Statistical Signal Processing*, Prentice-Hall, Englewood Cliffs, NJ, 1992. With permission).

eccentricity depend on the correlation between components of the random vector. It is straightforward to show that in general the contours of the Gaussian density for any value of N are *ellipsoids*. These are known as *concentration ellipsoids* (because they represent regions where the data is concentrated) and are useful in representing the signal from a geometric point of view. To show this property notice that (12.69) defining the contours simply implies that the quadratic form in the exponent of either (12.67) or (12.68) satisfies the condition

$$(x - m_x)^{*T} C_x^{-1}(x - m_x) = d \tag{12.70}$$

where C is a positive constant. Using the eigenvector decomposition given in (12.62) this quadratic form can be rewritten as

$$(\mathbf{x} - \mathbf{m}_x)^{*T} C_x^{-1} (\mathbf{x} - \mathbf{m}_x) = (\mathbf{x} - \mathbf{m}_x)^{*T} \check{\mathbf{E}} \check{\mathbf{\Lambda}}^{-1} \check{\mathbf{E}}^{*T} (\mathbf{x} - \mathbf{m}_x)$$

$$= (\mathbf{y} - \mathbf{m}_y)^{*T} \check{\mathbf{\Lambda}}^{-1} (\mathbf{y} - \mathbf{m}_y) = d \qquad (12.71)$$

where $\mathbf{y} = \check{\mathbf{E}}^{*T} \mathbf{x}$ and "hats" have been added to the variables to indicate that they pertain to the covariance matrix rather than the correlation matrix. Since $\check{\mathbf{\Lambda}}^{-1}$ is diagonal, this last expression can be written in expanded form as

$$\frac{|y_0 - m_{y_0}|^2}{\check{\lambda}_0} + \frac{|y_1 - m_{y_1}|^2}{\check{\lambda}_1} + \cdots + \frac{|y_{N-1} - m_{y_{N-1}}|^2}{\check{\lambda}_{N-1}} = d \qquad (12.72)$$

which is the equation of an N-dimensional ellipsoid with center at $\mathbf{m}_y$. The transformation $\mathbf{y} = \check{\mathbf{E}}^{*T} \mathbf{x}$ represents a rotation of the coordinate system to one aligned with the eigenvectors, which are *parallel to the axes of the ellipsoid*. The sizes of the axes are proportional to the square roots of the eigenvalues.

12.4 Fundamentals of Estimation

Problems of statistical estimation deal with deriving values for quantities that cannot be observed or measured directly from quantities that *can* be observed. These problems may deal with finding the parameters for a model of a signal from direct measurements of that signal or for estimating one signal from another (say a clean transmitted signal from a received noisy and distorted one). The former type of problem is known as *parameter estimation* while the latter is called *random variable estimation*. Although the two types of problems have much in common, it is convenient to consider them separately here.

12.4.1 Estimation of Parameters

Maximum Likelihood Estimation

The problem of parameter estimation can be thought of as having observations of a random variable described by a probability density function of some known form, but with a parameter whose value is unknown. In many problems both the random variable and the parameter are vector quantities. The density function is thus denoted by

$$f_{\mathbf{x};\boldsymbol{\theta}}(\mathbf{x}; \boldsymbol{\theta}) \qquad (12.73)$$

For a given fixed set of observations, say $\mathbf{x} = \mathbf{x}^\circ$, the *maximum likelihood estimate* of the parameter is the value $\boldsymbol{\theta}$ that maximizes $f_{\mathbf{x};\boldsymbol{\theta}}(\mathbf{x}^\circ; \boldsymbol{\theta})$. This value or "estimate" of $\boldsymbol{\theta}$ will be denoted by $\hat{\boldsymbol{\theta}}_{ml}$.

To see why this may be a good estimate and why it is called "maximum likelihood," consider the following simple example involving a scalar parameter. It is desired to estimate the mean m of a Gaussian density function

$$f_{x;m}(x; m) = \frac{1}{\sqrt{2\pi}} e^{-\frac{(x-m)^2}{2}}$$

using a single observation x°. (The variance is assumed known and equal to one.) The density function is depicted in Fig. 12.9 for several proposed values of the mean. It is obvious that if the true value of the mean were any of m_1, m_2, or m_4, the given observation x° would be unlikely. The choice m_3, however, makes the given observation *most* likely (i.e., it maximizes the quantity $f_{x;m}(x^\circ; m)$) and is, therefore, the maximum likelihood estimate.

When $f_{\mathbf{x};\boldsymbol{\theta}}(\mathbf{x}; \boldsymbol{\theta})$ of (12.73) is viewed *as a function of* $\boldsymbol{\theta}$ rather than as a function of $\mathbf{x}$, it is known as the *likelihood function*. Thus, the maximum likelihood estimate of a parameter maximizes the likelihood

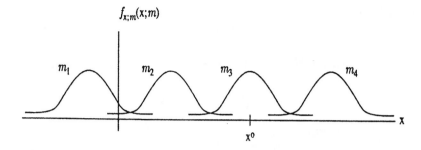

$f_{x;m}(x;m)$

m_1 m_2 m_3 m_4

x^0

x

FIGURE 12.9 Maximum likelihood estimation of the mean. (From Therrien, C.W., *Discrete Random Signals and Statistical Signal Processing*, Prentice-Hall, Englewood Cliffs, NJ, 1992. With permission).

function. If the likelihood function is continuous, and the maximum does not occur at a boundary, then the maximum likelihood estimate for a real scalar parameter θ can be found through either of the necessary conditions

$$\left.\frac{\partial f_{x;\theta}(x;\theta)}{\partial\theta}\right|_{\theta=\hat{\theta}_{ml}} = 0 \qquad (a)$$

$$\left.\frac{\partial \ln f_{x;\theta}(x;\theta)}{\partial\theta}\right|_{\theta=\hat{\theta}_{ml}} = 0 \quad (b)$$

(12.74)

These equations are sometimes referred to as the *likelihood equation* and the *log likelihood equation*, respectively. Equation 12.74(b) follows from (12.74)(a) because the logarithm is a strictly monotonic increasing function; this form is convenient in many problems where the likelihood function involves an exponential. For vector parameters $\boldsymbol{\theta}$ the corresponding likelihood and log likelihood equations involve setting the gradient or vector of derivatives with respect to *all* components of the parameter to zero. (See [1] for procedures involving both real and complex vector parameters.)

Properties of Estimates

The maximum likelihood estimate for a parameter $\boldsymbol{\theta}$ can be written as

$$\hat{\boldsymbol{\theta}}_{ml}(x) = \text{argmax} \ f_{x;\theta}(x;\boldsymbol{\theta}) \tag{12.75}$$

It is clear that the estimate is a function of the observations. Moreover, since the observations are a random vector, the estimate is itself a random variable and has a mean, covariance, density function, and so on. Not all estimates are maximum likelihood estimates. For example, the estimate for the mean of a Gaussian random signal, given by (12.82) is a maximum likelihood estimate, while the following estimate for the variance (where $\hat{m}_x$ is given by (12.82)) is not:

$$\hat{\sigma}_x^2 = \frac{1}{N}\sum_{n=0}^{N-1}(x[n] - \hat{m}_x)^2$$

All estimates are functions of the observations, however, so it is useful to denote a general estimate by

$$\hat{\boldsymbol{\theta}}_N = \hat{\boldsymbol{\theta}}_N(x)$$

where the subscript N denotes the number of observations (dimension of x), and examine its statistical properties.

Among the properties of estimates that are most useful are the following:

1. An estimate $\hat{\boldsymbol{\theta}}_N$ is *unbiased* if

$$\mathcal{E}\left\{\hat{\boldsymbol{\theta}}_N\right\} = \boldsymbol{\theta}$$

Otherwise the estimate is *biased* with bias $\boldsymbol{b}(\boldsymbol{\theta}) = \mathcal{E}\left\{\hat{\boldsymbol{\theta}}_N\right\} - \boldsymbol{\theta}$. An estimate is *asymptotically unbiased* if

$$\lim_{N\to\infty} \mathcal{E}\left\{\hat{\boldsymbol{\theta}}_N\right\} = \boldsymbol{\theta}$$

2. An estimate is $\hat{\boldsymbol{\theta}}_N$ is *consistent* if

$$\lim_{N\to\infty} \Pr[|\hat{\boldsymbol{\theta}}_N - \boldsymbol{\theta}| < \epsilon] = 1$$

for any arbitrarily small number ϵ. The sequence of estimates $\{\hat{\boldsymbol{\theta}}_N\}$ is then said to *converge in probability* to the parameter $\boldsymbol{\theta}$.

3. An estimate $\hat{\boldsymbol{\theta}}$ is said to be *efficient* with respect to another estimate $\hat{\boldsymbol{\theta}}'$ if the difference of their covariance matrices $\boldsymbol{C}_{\hat{\boldsymbol{\theta}}'} - \boldsymbol{C}_{\hat{\boldsymbol{\theta}}}$ is positive definite (written $\boldsymbol{C}_{\hat{\boldsymbol{\theta}}'} > \boldsymbol{C}_{\hat{\boldsymbol{\theta}}}$). This implies that the variance of every component of $\hat{\boldsymbol{\theta}}$ must be smaller than the variance of the corresponding component of $\hat{\boldsymbol{\theta}}'$. If $\hat{\boldsymbol{\theta}}_N$ is unbiased and efficient with respect to $\hat{\boldsymbol{\theta}}_{N-1}$ for all N, then $\boldsymbol{\theta}_N$ is a consistent estimate.

The last statement needs a little more explanation, which can best be given for the case of a scalar estimate. For a scalar estimate, (3) is a statement about its variance. The Tchebycheff inequality (see e.g., [5]) states that

$$\Pr[|\hat{\theta}_N - \theta| \geq \epsilon] \leq \frac{\text{Var}\left[\hat{\theta}_N\right]}{\epsilon^2}$$

Thus, if the variance of $\hat{\theta}_N$ decreases with N, the probability is that $|\hat{\theta}_N - \theta| \geq \epsilon$ approaches zero as $N \to \infty$. In other words, the probability is that $|\hat{\theta}_N - \theta| < \epsilon$ approaches one. This last property is illustrated in Fig. 12.10.

The variance of any unbiased estimate can be bounded with a powerful result known as the Cramér-Rao inequality. For the case of a scalar parameter the Cramér-Rao bound has the form

$$\text{Var}\left[\hat{\theta}\right] \geq \frac{1}{\mathcal{E}\left\{\left(\frac{\partial \ln f_{\boldsymbol{x};\theta}(\boldsymbol{x};\theta)}{\partial\theta}\right)^2\right\}} = \frac{1}{-\mathcal{E}\left\{\frac{\partial^2 \ln f_{\boldsymbol{x};\theta}(\boldsymbol{x};\theta)}{\partial\theta^2}\right\}} \tag{12.76}$$

where equality occurs *if and only if*

$$\hat{\theta}(\boldsymbol{x}) - \theta = K(\theta) \cdot \frac{\partial \ln f_{\boldsymbol{x};\theta}(\boldsymbol{x};\theta)}{\partial\theta}$$

The two alternate expressions on the right-hand side are valid as long as the partial derivatives exist and are absolutely integrable.

The general form of the Cramér-Rao bound for vector parameters is usually written as

$$\boldsymbol{C}_{\hat{\boldsymbol{\theta}}} \geq \boldsymbol{J}^{-1} \tag{12.77}$$

meaning that the difference matrix $\boldsymbol{C}_{\hat{\boldsymbol{\theta}}} - \boldsymbol{J}^{-1}$ is positive semidefinite. The bounding matrix on the right-hand side of (12.77) is the inverse of the *Fisher information matrix* defined by

$$\boldsymbol{J}(\boldsymbol{\theta}) = \mathcal{E}\left\{\boldsymbol{s}(\boldsymbol{x};\boldsymbol{\theta})\boldsymbol{s}^T(\boldsymbol{x};\boldsymbol{\theta})\right\} \tag{12.78}$$

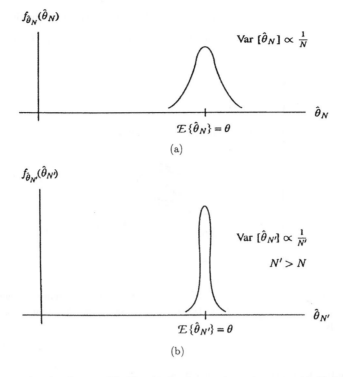

FIGURE 12.10 Density function for an unbiased estimate whose variance decreases with N. (a) Density function of the estimate $\hat{\theta}_N$; (b) density function of the estimate $\hat{\theta}_{N'}$ with $N' > N$. (From Therrien, C.W., *Discrete Random Signals and Statistical Signal Processing*, Prentice-Hall, Englewood Cliffs, NJ, 1992. With permission).

where $s(x; \theta)$ is a vector whose ith component is the derivative of $\ln f_{x;\theta}(x; \theta)$ with respect to $\hat{\theta}_i$, the ith component of θ. Equation 12.77 implies that the variance of $\hat{\theta}_i$ is bounded by

$$\text{Var}\left[\hat{\theta}_i\right] \geq j_{ii}^{(-1)} \tag{12.79}$$

where $j_{ii}^{(-1)}$ is the ith diagonal element of the inverse Fisher information matrix. The bound (12.77) is satisfied with equality if and only if the estimate satisfies an equation of the form

$$\hat{\theta}(x) - \theta = K(\theta)s(x; \theta) \tag{12.80}$$

In this case K is uniquely defined by

$$K(\theta) = J^{-1}(\theta) \tag{12.81}$$

(see [1]). An estimate satisfying the bound with equality is known as a *minimum-variance* estimate. It can be shown that if an unbiased minimum variance exists and the maximum likelihood estimate does not occur at a boundary, then the maximum likelihood estimate *is* that minimum-variance estimate.

An interpretation of the Cramér-Rao bound in terms of concentration ellipsoids is given in Fig. 12.11. If the *deviation* in the estimate is defined as

$$\delta(x; \theta) \stackrel{\text{def}}{=} \hat{\theta}(x) - \theta$$

then the bias of the estimate $b(\theta)$ is the mean deviation (i.e., its expected value). The concentration ellipse for the deviation with covariance $C_{\hat{\theta}}$ is shown in the figure where C is an arbitrary positive constant. The minimum deviation covariance of the Cramér-Rao bound is represented by the smaller ellipse with covariance J^{-1}. Geometrically the bound states that the J^{-1} ellipsoid lies entirely within the $C_{\hat{\theta}}$ ellipsoid. In the best case (when $\hat{\theta}$ is the maximum likelihood estimate), the two ellipsoids coincide.

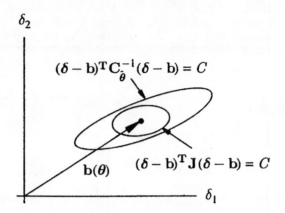

FIGURE 12.11   Concentration ellipses for the deviation of the estimate of a vector parameter; geometric interpretation of the Cramér-Rao bound. (From Therrien, C.W., *Discrete Random Signals and Statistical Signal Processing*, Prentice-Hall, Englewood Cliffs, NJ, 1992. With permission).

Estimates for Moments of Discrete Random Signals

Some of the most important parameters for random signals are their mean, correlation (or covariance) functions, and perhaps higher order statistics. Some common forms of these estimates and some of their statistical properties are cited here.

Given N time samples of a random signal, an estimate for the mean can be formed as

$$\hat{m}_x = \frac{1}{N} \sum_{n=0}^{N} x[n] \tag{12.82}$$

This estimate, known as the *sample mean*, is unbiased and efficient and, therefore, a consistent estimate. An expression for the variance of the estimate is not difficult to derive in terms of the covariance function for the process (see [1]).

The correlation function is usually estimated by one of the two formulas

$$\hat{R}_x[l] = \frac{1}{N-l} \sum_{n=0}^{N-1-l} x[n+l]x^*[n]; \quad 0 \le l < N \tag{12.83}$$

or

$$\hat{R}_x[l] = \frac{1}{N} \sum_{n=0}^{N-1-l} x[n+l]x^*[n]; \quad 0 \le l < N \tag{12.84}$$

Values of the correlation for negative lags are computed via the relation $\hat{R}_x[-l] = \hat{R}_x^*[l]$. Equation 12.83 generates an unbiased estimate while (12.84) is only asymptotically unbiased. Explicit expressions for the variance of these estimates can be derived in the Gaussian case [1]; these expressions have been found to be good approximations for non-Gaussian cases as well when $l \ll N$. Both estimates are found to be efficient and therefore consistent, but (12.83) has one serious defect: it cannot be guaranteed to be a positive semidefinite sequence. Therefore, (12.84) is usually the preferred estimate for the correlation function.

Estimates for the cross-correlation function of two sequences x and y can be generated by substituting $y^*[n]$ for $x^*[n]$ in either of the above two equations. The cross-correlation estimates have similar statistical properties. Since the cross-correlation function is not symmetric, however, explicit values for negative lags have to be computed by interchanging x and y in these equations and using the relation $\hat{R}_{xy}[-l] = \hat{R}_{yx}^*[l]$.

Estimates for higher-order moments and cumulants can be obtained by an analogous procedure. For example, after removal of the mean, an estimate for the third-order cumulant is given by

$$\hat{C}_x^{(3)}[l_1, l_2] = \frac{1}{N} \sum_{n=n_I}^{n_F} x^*[n]x[n+l_1]x[n+l_2] \tag{12.85}$$

$$0 \le l_2 \le l_1 \le N - 1$$

where

$$n_I = \max(0, -l_1, -l_2), \quad n_F = \min(N-1, N-1-l_1, N-1-l_2)$$

Estimates of the correlation *matrix* can be found by using (12.84) and generating a corresponding Toeplitz correlation matrix. It is not advisable to use (12.83) in this procedure because of the possibility of generating an estimate for the correlation matrix which is not positive semidefinite. In many applications, however, estimates for the correlation matrix are generated directly from products of suitably arranged matrices of the data. These methods, such as the autocorrelation method, the covariance method, and the modified covariance method can be found in many references dealing with signal modeling and statistical signal processing (again, see [1]).

12.4.2 Estimation of Random Variables

The problems of filtering and prediction for random signals and many others can be treated in the general context of random variable estimation. A typical problem is illustrated in Fig. 12.12, where the signal, treated as random variable y, is to be estimated from related observations $x_1, x_2, \ldots, x_N$. The form of the estimate is

$$\hat{y} = \hat{y}(x) = \phi(x_1, x_2, \ldots, x_N) \tag{12.86}$$

where the function ϕ in general is nonlinear.

A framework for this problem is provided by the procedure known as Bayes estimation. Here one seeks to minimize the *risk* defined by

$$\mathcal{R} = \mathcal{E}\{\mathcal{C}(y, \hat{y})\} \tag{12.87}$$

where $\mathcal{C}$ is known as the *cost* of the estimate and depends on both y and $\hat{y}$.

Two special cases that are commonly considered are shown in Fig. 12.13. In both cases the cost depends on the difference $y - \hat{y}$, otherwise known as the *error*. In the case depicted by Fig. 12.13(a) the cost is a quadratic function and the risk becomes equal to the mean-square error, while in case (b) the cost of any error that is not less than an arbitrarily small amount ϵ is uniform and equal to 1. It can be shown that the optimal estimate in both of these cases depends only on the conditional density function $f_{y|x}$ and is given by

$$\hat{y}_{ms}(x) = \int_{-\infty}^{\infty} y\, f_{y|x}(y\,|x)dy \tag{12.88}$$

for the mean-square case and

$$\hat{y}_{MAP}(x) = \operatorname{argmax}\, f_{y|x}(y\,|x) \tag{12.89}$$

for the uniform cost function. Notice that (12.88) is the *mean* of the conditional density, while (12.89) is the *maximum* of the conditional density function. The latter is called the *maximum a posteriori (MAP)* estimate because it maximizes the posterior density (i.e., the density *after* knowing the observations).

In general, the mean-square and the MAP estimates are different nonlinear functions of the observations. The choice of estimates depends on the particular problem at hand. In the case of zero-mean jointly

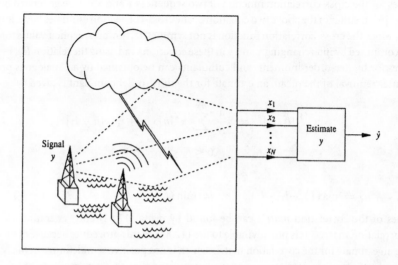

FIGURE 12.12   Estimation of a random variable y from related observations $x_1, x_2, \ldots, x_N$. (From Therrien, C.W., *Discrete Random Signals and Statistical Signal Processing,* Prentice-Hall, Englewood Cliffs, NJ, 1992. With permission).

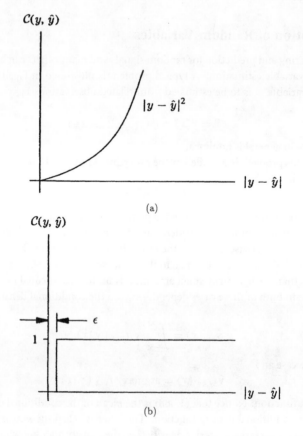

FIGURE 12.13   Cost functions for Bayes estimation: (a) mean-square, (b) uniform. (From Therrien, C.W., *Discrete Random Signals and Statistical Signal Processing,* Prentice-Hall, Englewood Cliffs, NJ, 1992. With permission).

Gaussian random variables, however, both estimates become identical and are *linear* functions of the observations. This is particularly useful for signal processing applications since the estimates can be implemented with linear filters. Signal processing considerations thus motivate finding an estimate that is linear, regardless of the distribution of the random variables. Such a linear estimate can easily be found if the mean-square error criterion is used. Further, the resulting estimate depends only upon second moment statistics of the signals. This topic, which is taken up below, is of fundamental importance to most areas of modern statistical signal processing.

12.4.3 Linear Mean-Square Estimation

The problem to be addressed here is an estimation problem of the type depicted in Fig. 12.12, where the estimate is assumed to be of the form

$$\hat{y} = \boldsymbol{a}^{*T}\boldsymbol{x} \tag{12.90}$$

where $\boldsymbol{x}$ is the vector of observations $x_1, x_2, \ldots, x_N$ and $\boldsymbol{a} = [a_1 a_2 \cdots a_N]^T$ is a vector of weighting coefficients chosen to minimize the mean-square error

$$\mathcal{E}\left\{|y - \hat{y}|^2\right\} \tag{12.91}$$

If the mean of $\boldsymbol{x}$ and y are not zero, the mean-square error can be further reduced by using an estimate of the form

$$\hat{y} = \boldsymbol{a}^{*T}\boldsymbol{x} + b$$

where it can be shown that the constant b should be chosen as $b = m_y - \boldsymbol{a}^{*T}\boldsymbol{m}_x$. Since this estimate is not "linear" in the strict mathematical sense, however, it is traditional to consider only the form represented by (12.90). To obtain the best estimate in practice one should remove the mean of the variables before considering the estimation problem. This is equivalent to using covariances instead of correlations in the solution that appears below and then adding m_y back onto the resulting linear estimate.

The solution to the linear mean-square estimation problem is based on a key idea called the *orthogonality principle*, which can be stated as follows:

THEOREM 12.1 *(Orthogonality) Let $\varepsilon = y - \hat{y}$ be the error in estimation. Then $\boldsymbol{a}$ minimizes the mean-square error $\sigma_\varepsilon^2 = \mathcal{E}\left\{|y - \hat{y}|^2\right\}$ if $\boldsymbol{a}$ is chosen such that $\mathcal{E}\left\{x_i\varepsilon^*\right\} = \mathcal{E}\left\{\varepsilon x_i^*\right\} = 0, \quad i = 1, 2, \ldots, N$. Further, the minimum mean-square error is given by $\sigma_\varepsilon^2 = \mathcal{E}\left\{y\varepsilon^*\right\} = \mathcal{E}\left\{\varepsilon y^*\right\}$.*

The condition $\mathcal{E}\left\{x_i\varepsilon^*\right\} = \mathcal{E}\left\{\varepsilon x_i^*\right\} = 0$ is known as *orthogonality* of the random variables ε and x_i. While the proof of this theorem is not difficult, a geometric argument is perhaps more illuminating.

Let us consider the case for $N = 2$. If the random variables x_1, x_2, y, and ε are thought of as "vectors" in an abstract vector space,[5] then the linear form of the estimate

$$\hat{y} = a_1^* x_1 + a_2^* x_2$$

implies that $\hat{y}$ lies in a plane defined by x_1 and x_2 (see Fig. 12.14). The random variable represented by the "vector" y is in general not restricted to this plane and the error ε is the difference between the two vectors, extending from the tip of $\hat{y}$ to the tip of y. From the illustration, it is clear that the magnitude of the error is minimized when ε is made orthogonal to the plane. This is equivalent to requiring that ε be orthogonal to both x_1 and x_2, which is what the orthogonality principle states.

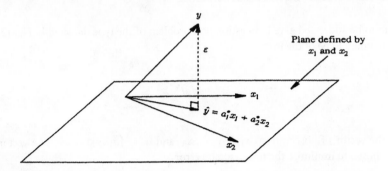

FIGURE 12.14 Vector space interpretation of linear mean-square estimation. (From Therrien, C.W., *Discrete Random Signals and Statistical Signal Processing*, Prentice-Hall, Englewood Cliffs, NJ, 1992. With permission).

In order to define orthogonality in an abstract vector space, a suitable inner product between vectors is needed. It can be shown that the operation $\mathcal{E}\left\{x_i \varepsilon^*\right\}$ satisfies the necessary conditions to serve as this definition of inner product. With this form of inner product, it is also clear from Fig. 12.14 that the magnitude of the squared error $\sigma_\varepsilon^2 = \mathcal{E}\left\{|\varepsilon|^2\right\}$ is equal to the inner product $\mathcal{E}\left\{y\varepsilon^*\right\}$.

The orthogonality principle can be used to easily derive the equations needed to find the linear mean-square estimate. The theorem requires that

$$\mathcal{E}\left\{\boldsymbol{x}\varepsilon^*\right\} = \mathcal{E}\left\{\boldsymbol{x}(y - \boldsymbol{a}^{*T}\boldsymbol{x})^*\right\} = \mathcal{E}\left\{\boldsymbol{x}(y^* - \boldsymbol{x}^{*T}\boldsymbol{a})\right\} = 0 \tag{12.92}$$

Taking the expectation then yields

$$\boxed{\mathbf{R_x a = r_{xy}}} \tag{12.93}$$

where

$$\boldsymbol{R_x} = \mathcal{E}\left\{\boldsymbol{x}\boldsymbol{x}^{*T}\right\} \tag{12.94}$$

---

[5] Hopefully, readers will not be confused by the use of the word "vector" in a new context here. The term "vector" does *not* refer to the array of observations $x = [x_1 x_2 \cdots x_N]^T$. Rather the term is used to refer to objects in some abstract space, which in this case are random variables. Mathematically a *vector space* $\mathcal{V}$ is a set of elements $u, v, \ldots$ such that if $u \in \mathcal{V}$ and $v \in \mathcal{V}$ then there is a unique element

$$u + v \in \mathcal{V}$$

called the *sum*. Further, if c is an element from an associated field such as the field of real or complex numbers, then the *scalar product*

$$c \cdot u$$

with certain associative and distributive properties, is also an element of $\mathcal{V}$ (see e.g., [6]). Random variables with the associated field of complex numbers satisfy these conditions.

is the correlation matrix of the observations and

$$r_{xy} = \mathcal{E}\left\{xy^*\right\} \tag{12.95}$$

is the cross-correlation between the x and the random variable y to be estimated. The coefficients **a** to produce the optimal estimate are thus obtained by solving the *linear* equations (12.93).

The minimum mean-square error follows directly from the second part of the theorem. That is,

$$\sigma_\varepsilon^2 = \mathcal{E}\left\{y\varepsilon^*\right\} = \mathcal{E}\left\{y(y^* - x^{*T}a)\right\} = \sigma_y^2 - r_{xy}^{*T}a \tag{12.96}$$

Note, as anticipated, the calculations in (12.93) and (12.96) involve only second moment statistics.

It seems appropriate to conclude this section with a simple example involving the ideas developed here. Consider the estimation of a stationary random signal in additive noise. The observations are given by

$$x[n] = s[n] + \eta[n]$$

where s is the desired signal and η is the additive noise. If the signal and noise are assumed uncorrelated and the signal and/or noise has mean zero, then the correlation function for the observations is given by

$$R_x[l] = R_s[l] + R_\eta[l] \tag{12.97}$$

where R_s is the signal correlation function and R_η is the noise correlation function. The cross-correlation between the signal and observation sequence is given by

$$R_{sx}[l] = \mathcal{E}\left\{s[n]x^*[n-l]\right\} = R_s[l] \tag{12.98}$$

Suppose the optimal estimator is to be implemented as a linear FIR filter with impulse response sequence $h[0], h[1], \ldots h[N-1]$. Then the estimate is generated by the expression

$$\hat{s}[n] = h[0]x[n] + h[1]x[n-1] + \cdots + h[N-1]x[n-N+1]$$

which is of the form (12.90). If the mean-square error $\mathcal{E}\left\{|\varepsilon[n]|^2\right\}$ is to be minimized, where $\varepsilon[n] = s[n] - \hat{s}[n]$, then this is a problem in linear mean-square estimation.

The correlation matrix and cross-correlation vector needed in (12.93) are easily formulated using the expressions for the corresponding correlation functions. To illustrate, for the case $N = 3$ these equations take the form

$$\begin{bmatrix} R_x[0] & R_x[-1] & R_x[-2] \\ R_x[1] & R_x[0] & R_x[-1] \\ R_x[2] & R_x[1] & R_x[0] \end{bmatrix} \begin{bmatrix} h[0] \\ h[1] \\ h[2] \end{bmatrix} = \begin{bmatrix} R_{sx}[0] \\ R_{sx}[1] \\ R_{sx}[2] \end{bmatrix} \tag{12.99}$$

where R_x and R_{sx} are computed from (12.97) and (12.98). The Eq. (12.96) for the mean-square error becomes

$$\sigma_\varepsilon^2 = R_s[0] - \sum_{l=0}^{N-1} h^*[l]R_{sx}[l] \tag{12.100}$$

The Eq. (12.99), which is a special form of (12.93) for optimal filtering problems, is known as the *Wiener-Hopf* equation. It appears in various forms for many important problems including those of signal modeling and linear prediction discussed elsewhere in this Handbook.

References

[1] Therrien, C.W., *Discrete Random Signals and Statistical Signal Processing*, Prentice-Hall, Englewood Cliffs, New Jersey, 1992.

[2] Wilkinson, J.H., *The Algebraic Eigenvalue Problem*, Oxford University Press, New York, 1965.

[3] Strang, G., *Linear Algebra and Its Applications*, 3rd ed., Harcourt Brace Jovanovich, San Diego, 1976.

[4] Golub, G.H. and Van Loan, C.F., *Matrix Computations*, 2nd ed. The Johns Hopkins University Press, Baltimore, Maryland, 1989.

[5] Papoulis, A., *Probability, Random Variables, and Stochastic Processes*, 3rd ed., McGraw-Hill, New York, 1991.

[6] Birkhoff, G. and MacLane, S., *A Survey of Modern Algebra*, 4th ed., Macmillan, New York, 1977.

13

Signal Detection and Classification

Alfred Hero
University of Michigan

13.1 Introduction

Detection and classification arise in signal processing problems whenever a decision is to be made among a finite number of hypotheses concerning an observed waveform. Signal detection algorithms decide whether the waveform consists of "noise alone" or "signal masked by noise." Signal classification algorithms decide whether a detected signal belongs to one or another of prespecified classes of signals. The objective of signal detection and classification theory is to specify systematic strategies for designing algorithms which minimize the average number of decision errors. This theory is grounded in the mathematical discipline of statistical decision theory where detection and classification are respectively called binary and M-ary *hypothesis testing* [1, 2]. However, signal processing engineers must also contend with the exceedingly large size of signal processing datasets, the absence of reliable and tractible signal models, the associated requirement of fast algorithms, and the requirement for real-time imbedding of unsupervised algorithms into specialized software or hardware. While ad hoc statistical detection algorithms were implemented by engineers before 1950, the systematic development of signal detection theory was first undertaken by radar and radio engineers in the early 1950s [3, 4].

This chapter provides a brief and limited overview of some of the theory and practice of signal detection and classification. The focus will be on the Gaussian observation model. For more details and examples see the cited references.

0-8493-8572-5/98/$0.00+$.50

13.2   Signal Detection

Assume that for some physical measurement a sensor produces an output waveform $x = \{x(t) : t \in [0, T]\}$ over a time interval $[0, T]$. Assume that the waveform may have been produced by ambient noise alone or by an impinging signal of known form plus the noise. These two possibilities are called the *null hypothesis* H and the *alternative hypothesis* K, respectively, and are commonly written in the compact notation:

$$H \quad : \quad x = \text{noise alone}$$
$$K \quad : \quad x = \text{signal} + \text{noise}.$$

The hypotheses H and K are called *simple hypotheses* when the statistical distributions of x under H and K involve no unknown parameters such as signal amplitude, signal phase, or noise power. When the statistical distribution of x under a hypothesis depends on unknown (*nuisance*) parameters the hypothesis is called a *composite hypothesis*.

   To decide between the null and alternative hypotheses one might apply a high threshold to the sensor output x and make a decision that the signal is present if and only if the threshold is exceeded at some time within $[0, T]$. The engineer is then faced with the practical question of where to set the threshold so as to ensure that the number of decision errors is small. There are two types of error possible: the error of missing the signal (decide H under K (signal is present)) and the error of false alarm (decide K under H (no signal is present)). There is always a compromise between choosing a high threshold to make the average number of false alarms small versus choosing a low threshold to make the average number of misses small. To quantify this compromise it becomes necessary to specify the statistical distribution of x under each of the hypotheses H and K.

13.2.1   The ROC Curve

Let the aforementioned threshold be denoted γ. Define the K decision region $\mathcal{R}_K = \{x : x(t) > \gamma$, for some $t \in [0, T]\}$. This region is also called the *critical region* and simply specifies the conditions

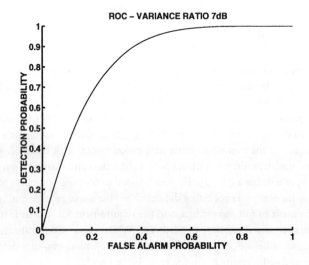

FIGURE 13.1   The receiver operating characteristic (ROC) curve describes the tradeoff between maximizing the power P_D and minimizing the probability of false alarm P_{FA} of a test between two hypotheses H and K. Shown is the ROC curve of the LRT (energy detector) which tests between $H : x = complex\ Gaussian\ random\ variable\ with\ variance\ \sigma^2 = 1$, vs. $K : x = complex\ Gaussian\ random\ variable\ with\ variance\ \sigma^2 = 5$ (7dB variance ratio).

on x for which the detector declares the signal to be present. Since the detector makes mutually exclusive binary decisions, the critical region completely specifies the operation of the detector. The probabilities of false alarm and miss are functions of γ given by $P_{FA} = P(\mathcal{R}_K|H)$ and $P_M = 1 - P(\mathcal{R}_K|K)$ where $P(A|H)$ and $P(A|K)$ denote the probabilities of arbitrary event A under hypothesis H and hypothesis K, respectively. The probability of correct detection $P_D = P(\mathcal{R}_K|K)$ is commonly called the *power* of the detector and P_{FA} is called the *level* of the detector.

The plot of the pair $P_{FA} = P_{FA}(\gamma)$ and $P_D = P_D(\gamma)$ over the range of thresholds $-\infty < \gamma < \infty$ produces a curve called the receiver operating characteristic (ROC) which completely describes the error rate of the detector as a function of γ (Fig. 13.1). Good detectors have ROC curves which have desirable properties such as concavity (negative curvature), monotone increase in P_D as P_{FA} increases, high slope of P_D at the point $(P_{FA}, P_D) = (0, 0)$, etc. [5]. For the energy detection example shown in Fig. 13.1 it is evident that an increase in the rate of correct detections P_D can be bought only at the expense of increasing the rate of false alarms P_{FA}. Simply stated, the job of the signal processing engineer is to find ways to test between K and H which push the ROC curve towards the upper left corner of Fig. 13.1 where P_D is high for low P_{FA}: this is the regime of P_D and P_{FA} where reliable signal detection can occur.

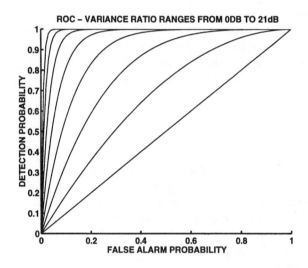

FIGURE 13.2   Eight members of the family of ROC curves for the LRT (energy detector) which tests between $H : x = $ *complex Gaussian random variable with variance* $\sigma^2 = 1$, vs. composite $K : x = $ *complex Gaussian random variable with variance* $\sigma^2 > 1$. ROC curves shown are indexed over a range [0dB, 21dB] of variance ratios in equal 3dB increments. ROC curves approach a step function as variance ratio increases.

13.2.2 Detector Design Strategies

When the signal waveform and the noise statistics are fully known, the hypotheses are simple, and an optimal detector exists which has a ROC curve that upper bounds the ROC of any other detector, i.e., it has the highest possible power P_D for any fixed level P_{FA}. This optimal detector is called the most powerful (MP) test and is specified by the ubiquitous likelihood ratio test described below. In the more common case where the signal and/or noise are described by unknown parameters, at least one hypothesis is composite, and a detector has different ROC curves for different values of the parameters (see Fig. 13.2). Unfortunately, there seldom exists a uniformly most powerful detector whose ROC curves remain upper bounds for the entire range of unknown parameters. Therefore, for composite hypotheses other design strategies must generally be adopted to ensure reliable detection performance. There are a wide range of different strategies available including Bayesian detection [5] and hypothesis testing [6], min-max hypothesis testing [2], CFAR detection [7], unbiased hypothesis testing [1], invariant hypothesis testing [8, 9], sequential detection [10], simultaneous detection and estimation [11], and nonparametric detection [12]. Detailed discussion of these strategies is outside the scope of this chapter. However, all of these strategies have a common link: their application produces one form or another of the *likelihood ratio test*.

13.2.3   Likelihood Ratio Test

Here we introduce an unknown parameter θ to simplify the upcoming discussion on composite hypothesis testing. Define the probability density of the measurement x as $f(x|\theta)$ where θ belongs to a parameter space Θ. It is assumed that $f(x|\theta)$ is a known function of x and θ. We can now state the detection problem as the problem of testing between

$$H \quad : \quad x \sim f(x|\theta), \quad \theta \in \Theta_H \tag{13.1}$$

$$K \quad : \quad x \sim f(x|\theta), \quad \theta \in \Theta_K , \tag{13.2}$$

where Θ_H and Θ_K are nonempty sets which partition the parameter space into two regions. Note it is essential that Θ_H and Θ_K be *disjoint* ($\Theta_H \cap \Theta_K = \emptyset$) so as to remove any ambiguity on the decisions, and *exhaustive* ($\Theta_H \cup \Theta_K = \Theta$) to ensure that all states of nature in Θ are accounted for. Let a detector be specified by a critical region $\mathcal{R}_K$. Then for any pair of parameters $\theta_H \in \Theta_H$ and $\theta_K \in \Theta_K$ the level and power of the detector can be computed by integrating the probability density $f(x|\theta)$ over $\mathcal{R}_K$

$$P_{FA} = \int_{x \in \mathcal{R}_K} f(x|\theta_H)dx, \tag{13.3}$$

and

$$P_D = \int_{x \in \mathcal{R}_K} f(x|\theta_K)dx. \tag{13.4}$$

The hypotheses (13.1) and (13.2) are simple when $\Theta = \{\theta_H, \theta_K\}$ consists of only two values and $\Theta_H = \{\theta_H\}$ and $\Theta_K = \{\theta_K\}$ are point sets. For simple hypotheses the Neyman-Pearson Lemma [1] states that there exists a most powerful test which maximizes P_D subject to the constraint that $P_{FA} \leq \alpha$, where α is a prespecified maximum level of false alarm. This test takes the form of a threshold test known as the likelihood ratio test (LRT)

$$L(x) \overset{\text{def}}{=} \frac{f(x|\theta_K)}{f(x|\theta_H)} \underset{H}{\overset{K}{\underset{<}{>}}} \eta, \tag{13.5}$$

where η is a threshold which is determined by the constraint $P_{FA} = \alpha$

$$\int_{\eta}^{\infty} g(l|\theta_H)dl = \alpha. \tag{13.6}$$

Here $g(l|\theta_H)$ is the probability density function of the likelihood ratio statistic $L(x)$ when $\theta = \theta_H$. It must also be mentioned that if the density $g(l|\theta_H)$ contains delta functions a simple randomization [1] of the LRT may be required to meet the false alarm constraint (13.6).

The test statistic $L(x)$ is a measure of the strength of the evidence provided by x that the probability density $f(x|\theta_K)$ produced x as opposed to the probability density $f(x|\theta_H)$. Similarly, the threshold η represents the detector designer's prior level of "reasonable doubt" about the sufficiency of the evidence — only above a level η is the evidence sufficient for rejecting H.

When θ takes on more than two values at least one of the hypotheses (13.1) or (13.2) are composite, and the Neyman Pearson lemma no longer applies. A popular but ad hoc alternative which enjoys some asymptotic optimality properties is to implement the *generalized likelihood ratio test* (GLRT):

$$L_g(x) \overset{\text{def}}{=} \frac{\max_{\theta_K \in \Theta_K} f(x|\theta_K)}{\max_{\theta_H \in \Theta_H} f(x|\theta_H)} \overset{K}{\underset{H}{\overset{>}{\underset{<}{}}}} \eta \tag{13.7}$$

where, if feasible, the threshold η is set to attain a specified level of P_{FA}. The GLRT can be interpreted as a LRT which is based on the *most likely* values of the unknown parameters θ_H and θ_K, i.e., the values which maximize the *likelihood functions* $f(x|\theta_H)$ and $f(x|\theta_K)$, respectively.

13.3   Signal Classification

When, based on a noisy observed waveform x, one must decide among a number of possible signal waveforms $s_1, \ldots, s_p$, $p > 1$, we have a *p-ary signal classification problem*. Denoting $f(x|\theta_i)$ the density function of x when signal s_i is present, the classification problem can be stated as the problem of testing between the p hypotheses

$$H_1 \quad : \quad x \sim f(x|\theta_1), \theta_1 \in \Theta_1$$

$$\vdots \quad \vdots \quad \vdots$$

$$H_p \quad : \quad x \sim f(x|\theta_p), \theta_p \in \Theta_p$$

where Θ_i is a space of unknowns which parameterize the signal s_i. As before, it is essential that the hypotheses be disjoint, which is necessary for $\{f(x|\theta_i)\}_{i=1}^p$ to be distinct functions of x for all $\theta_i \in \Theta_i$, $i = 1, \ldots, p$, and that they be exhaustive, which ensures that the true density of x is included in one of the hypotheses. Similarly to the case of detection, a classifier is specified by a partition of the space of observations x into p disjoint decision regions $\mathcal{R}_{H_1}, \ldots, \mathcal{R}_{H_p}$. Only $p - 1$ of these decision regions are needed to specify the operation of the classifier. The performance of a signal classifier is characterized by its set of p *misclassification* probabilities $P_{M_1} = 1 - P(x \in \mathcal{R}_{H_1}|H_1), \ldots, P_{M_p} = P(x \in \mathcal{R}_{H_p}|H_p)$. Unlike the case of detection ($p = 2$), even for simple hypotheses, where $\Theta_i = \{\theta_i\}$ consists of a single point, $i = 1, \ldots, p$, optimal p-ary classifiers that uniformly minimize all P_{M_i}'s do not exist. However, classifiers can be designed to minimize other weaker criteria such as average misclassification probability $\frac{1}{p} \sum_{i=1}^p P_{M_i}$ [5], worst case misclassification probability $\max_i P_{M_i}$ [2], Bayes posterior misclassification probability [12], and others.

The maximum likelihood (ML) classifier is a popular classification technique which is closely related to maximum likelihood parameter estimation. This classifier is specified by the rule

$$\text{decide } H_j \text{ if and only if } \max_{\theta_j \in \Theta_j} f(x|\theta_j) \geq \max_k \max_{\theta_k \in \Theta_k} f(x|\theta_k), \quad j = 1, \ldots, p. \tag{13.8}$$

When the hypotheses $H_1, \ldots, H_p$ are simple, the ML classifier takes the simpler form:

$$\text{decide } H_j \text{ if and only if } f_j(x) \geq \max_k f_k(x), \quad j = 1, \ldots, p$$

where $f_k = f(x|\theta_k)$ denotes the known density function of x under H_k. For this simple case it can be shown that the ML classifier is an optimal decision rule which minimizes the total misclassification error probability, as measured by the average $\frac{1}{p} \sum_{i=1}^p P_{M_i}$. In some cases a weighted average $\frac{1}{p} \sum_{i=1}^p \beta_i P_{M_i}$ is a more appropriate measure of total misclassification error, e.g., when β_i is the prior probability of H_i, $i = 1, \ldots, p$, $\sum_{i=1}^p \beta_i = 1$. For this latter case, the optimal classifier is given by the *maximum a posteriori* (MAP) decision rule [5, 13]

$$\text{decide } H_j \text{ if and only if } f_j(x)\beta_j \geq \max_k f_k(x)\beta_k, \quad j = 1, \ldots, p.$$

13.4   The Linear Multivariate Gaussian Model

Assume that $\mathbf{X}$ is an $m \times n$ matrix of complex valued Gaussian random variables which obeys the following linear model [9, 14]

$$\mathbf{X} = \mathbf{ASB} + \mathbf{W} \tag{13.9}$$

where $\mathbf{A}$, $\mathbf{S}$, and $\mathbf{B}$ are rectangular $m \times q$, $q \times p$, and $p \times n$ complex matrices, and $\mathbf{W}$ is an $m \times n$ matrix whose n columns are i.i.d. zero mean circular complex Gaussian vectors each with positive definite covariance matrix $\mathbf{R}_w$. We will assume that $n \geq m$. This model is very general, and, as will be seen in subsequent sections, covers many signal processing applications.

A few comments about random matrices are now in order. If $\mathbf{Z}$ is an $m \times n$ random matrix the mean, $E[\mathbf{Z}]$, of $\mathbf{Z}$ is defined as the $m \times n$ matrix of means of the elements of $\mathbf{Z}$, and the covariance matrix is defined as the $mn \times mn$ covariance matrix of the $mn \times 1$ vector, vec[$\mathbf{Z}$], formed by stacking columns of $\mathbf{Z}$. When the columns of $\mathbf{Z}$ are uncorrelated and each have the same $m \times m$ covariance matrix $\mathbf{R}$, the covariance of $\mathbf{Z}$ is block diagonal:

$$\text{cov}[\mathbf{Z}] = \mathbf{R} \otimes \mathbf{I}_n. \tag{13.10}$$

where $\mathbf{I}_n$ is the $n \times n$ identity matrix. For $p \times q$ matrix $\mathbf{C}$ and $r \times s$ matrix $\mathbf{D}$ the notation $\mathbf{C} \otimes \mathbf{D}$ denotes the Kronecker product which is the following $pr \times qs$ matrix:

$$\mathbf{C} \otimes \mathbf{D} = \begin{bmatrix} \mathbf{C}\,d_{11} & \mathbf{C}\,d_{12} & \ldots & \mathbf{C}\,d_{1s} \\ \mathbf{C}\,d_{21} & \mathbf{C}\,d_{22} & \ldots & \mathbf{C}\,d_{2s} \\ \vdots & \vdots & \vdots & \vdots \\ \mathbf{C}\,d_{r1} & \mathbf{C}\,d_{r2} & \ldots & \mathbf{C}\,d_{rs} \end{bmatrix}. \tag{13.11}$$

The density function of $\mathbf{X}$ has the form [14]

$$f(\mathbf{X}; \theta) = \frac{1}{\pi^{mn} |\mathbf{R}_w|^n} \exp\left(-\text{tr}\left\{[\mathbf{X} - \mathbf{ASB}][\mathbf{X} - \mathbf{ASB}]^H \mathbf{R}_w^{-1}\right\}\right), \tag{13.12}$$

where $|\mathbf{C}|$ is the determinant and tr$\{\mathbf{D}\}$ is the trace of square matrices $\mathbf{C}$ and $\mathbf{D}$, respectively. For convenience we will use the shorthand notation

$$\mathbf{X} \sim \mathcal{N}_{mn}(\mathbf{ASB}, \mathbf{R}_w \otimes \mathbf{I}_n)$$

which is to be read as $\mathbf{X}$ is distributed as an $m \times n$ complex Gaussian random matrix with mean $\mathbf{ASB}$, and covariance $\mathbf{R}_w \otimes \mathbf{I}_n$,

In the examples presented in the next section, several distributions associated with the complex Gaussian distribution will be seen to govern the various test statistics. The complex noncentral chi-square distribution with p degrees of freedom and vector of noncentrality parameters $(\rho, \underline{d})$ plays a very important role here. This is defined as the distribution of the random variable $\chi^2(\rho, \underline{d}) \stackrel{\text{def}}{=} \sum_{i=1}^{p} d_i |z_i|^2 + \rho$ where the z_i's are independent univariate complex Gaussian random variables with zero mean and unit variance and where ρ is scalar and $\underline{d}$ is a (row) vector of positive scalars. The complex noncentral chi-square distribution is closely related to the real noncentral chi-square distribution with $2p$ degrees of freedom and noncentrality parameters $(\rho, \text{diag}([\underline{d}, \underline{d}]))$ defined in [14]. The case of $\rho = 0$ and $\underline{d} = [1, \ldots, 1]$ corresponds to the standard (central) complex chi-square distribution. For derivations and details on this and other related distributions see [14].

13.5   Temporal Signals in Gaussian Noise

Consider the time-sampled superposed signal model

$$x(t_i) = \sum_{j=1}^{p} s_j b_j(t_i) + w(t_i), \quad i = 1, \ldots, n,$$

where here we interpret t_i as time; but it could also be space or other domain. The temporal signal waveforms $\underline{b}_j = [b_j(t_1), \ldots, b_j(t_n)]^T$, $j = 1, \ldots, p$, are assumed to be linearly independent where $p \leq n$. The scalar s_j is a time-independent complex gain applied to the jth signal waveform. The noise $w(t)$ is complex Gaussian with zero mean and correlation function $r_w(t, \tau) = E[w(t)w^*(\tau)]$. By concatenating the samples into a column vector $\underline{x} = [x(t_1), \ldots, x(t_n)]^T$ the above model is equivalent to:

$$\underline{x} = \mathbf{B}\underline{s} + \underline{w}, \tag{13.13}$$

where $\mathbf{B} = [\underline{b}_1, \ldots, \underline{b}_p]$, $\underline{s} = [s_1, \ldots, s_p]^T$. Therefore, the density function (13.12) applies to the vector $x = \underline{x}^T$ with $\mathbf{R}_w = \text{cov}(\underline{w})$, $m = q = 1$, and $\mathbf{A} = 1$.

13.5.1   Signal Detection: Known Gains

For known gain factors s_i, known signal waveforms $\underline{b}_i$, and known noise covariance $\mathbf{R}_w$, the LRT (13.5) is the most powerful signal detector for deciding between the simple hypotheses $H : \underline{x} \sim \mathcal{N}_n(0, \mathbf{R}_w)$ vs. $K : \underline{x} \sim \mathcal{N}_n(\mathbf{B}\underline{s}, \mathbf{R}_w)$. The LRT has the form

$$L(x) = \exp\left(-2 * \text{Re}\left\{\underline{x}^H \mathbf{R}_w^{-1} \mathbf{B}\underline{s}\right\} + \underline{s}^H \mathbf{B}^H \mathbf{R}_w^{-1} \mathbf{B}\underline{s}\right) \underset{H}{\overset{K}{\gtrless}} \eta. \tag{13.14}$$

This test is equivalent to a linear detector with critical region $\mathcal{R}_K = \{x : T(x) > \gamma\}$ where

$$T(x) = \text{Re}\left\{\underline{x}^H \mathbf{R}_w^{-1} \underline{s}_c\right\}$$

and $\underline{s}_c = \mathbf{B}\underline{s} = \sum_{j=1}^{p} s_j \underline{b}_j$ is the observed compound signal component.

Under both hypotheses H and K the test statistic T is Gaussian distributed with common variance but different means. It is easily shown that the ROC curve is monotonically increasing in the *detectability index* $\rho = \underline{s}_c^H \mathbf{R}_w^{-1} \underline{s}_c$. It is interesting to note that when the noise is white, $\mathbf{R}_w = \sigma^2 \mathbf{I}_n$ and the ROC curve depends on the form of the signals only through the signal-to-noise ratio (SNR) $\rho = \frac{\|\underline{s}_c\|^2}{\sigma^2}$. In this special case the linear detector can be written in the form of a correlator detector

$$T(x) = \text{Re}\left\{\sum_{i=1}^{n} s_c^*(t_i) x(t_i)\right\} \underset{H}{\overset{K}{\gtrless}} \gamma$$

where $s_c(t) = \sum_{j=1}^{p} s_j b_j(t)$. When the sampling times t_i are equispaced, e.g., $t_i = i$, the correlator takes the form of a matched filter

$$T(x) = \text{Re}\left\{\sum_{i=1}^{n} h(n - i) x(i)\right\} \underset{H}{\overset{K}{\gtrless}} \gamma,$$

where $h(i) = s_c^*(-i)$. Block diagrams for the correlator and matched filter implementations of the LRT are shown in Figs. 13.3 and 13.4.

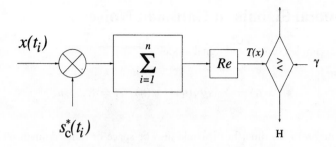

FIGURE 13.3   The correlator implementation of the most powerful LRT for signal component $s_C(t_i)$ in additive Gaussian white noise. For nonwhite noise a prewhitening transformation must be performed on $x(t_i)$ and $s_C(t_i)$ prior to implementation of correlator detector.

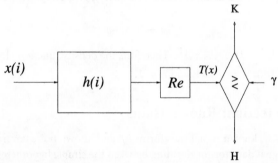

FIGURE 13.4   The matched filter implementation of the most powerful LRT for signal component $s_C(i)$ in additive Gaussian white noise. Matched filter impulse response is $h(i) = s_C^*(-i)$. For nonwhite noise a prewhitening transformation must be performed on $x(i)$ and $s_C(i)$ prior to implementation of matched filter detector.

13.5.2   Signal Detection: Unknown Gains

When the gains s_j are unknown the alternative hypothesis K is composite, the critical region $\mathcal{R}_K$ depends on the true gains for $p > 1$, and no most powerful test for $H : \underline{x} \sim \mathcal{N}_n(0, \mathbf{R}_w)$ vs. $K : \underline{x} \sim \mathcal{N}_n(\mathbf{B}\underline{s}, \mathbf{R}_w)$ exists. However, the GLRT (13.7) can easily be derived by maximizing the likelihood ratio for known gains (13.14) over $\underline{s}$. Recalling from least squares theory that $\min_{\underline{s}} (\underline{x} - \mathbf{B}\underline{s})^H \mathbf{R}_w^{-1}(\underline{x} - \mathbf{B}\underline{s}) = \underline{x}^H \mathbf{R}_w^{-1}\underline{x} - \underline{x}^H \mathbf{R}_w^{-1}\mathbf{B}[\mathbf{B}^H \mathbf{R}_w^{-1}\mathbf{B}]^{-1}\mathbf{B}^H \mathbf{R}_w^{-1}\underline{x}$ the GLRT can be shown to take the form

$$T_g(x) = \underline{x}^H \mathbf{R}_w^{-1}\mathbf{B}[\mathbf{B}^H \mathbf{R}_w^{-1}\mathbf{B}]^{-1}\mathbf{B}^H \mathbf{R}_w^{-1}\underline{x} \underset{H}{\overset{K}{\underset{<}{>}}} \gamma.$$

A more intuitive form for the GLRT can be obtained by expressing T_g in terms of the prewhitened observations $\underline{\tilde{x}} = \mathbf{R}_w^{-\frac{1}{2}}\underline{x}$ and prewhitened signal waveform matrix $\tilde{\mathbf{B}} = \mathbf{R}_w^{-\frac{1}{2}}\mathbf{B}$, where $\mathbf{R}_w^{-\frac{1}{2}}$ is the right Cholesky factor of $\mathbf{R}_w^{-1}$

$$T_g(x) = \|\tilde{\mathbf{B}}[\tilde{\mathbf{B}}^H \tilde{\mathbf{B}}]^{-1}\tilde{\mathbf{B}}^H \underline{\tilde{x}}\|^2. \tag{13.15}$$

$\tilde{\mathbf{B}}[\tilde{\mathbf{B}}^H \tilde{\mathbf{B}}]^{-1}\tilde{\mathbf{B}}^H$ is the idempotent $n \times n$ matrix which projects onto column space of the prewhitened signal waveform matrix $\tilde{\mathbf{B}}$ (whitened signal subspace). Thus, the GLRT decides that some linear combination of the signal waveforms $\underline{b}_1, \ldots, \underline{b}_p$ is present only if the energy of the component of x lying in the whitened signal subspace is sufficiently large.

Under the null hypothesis the test statistic T_g is distributed as a complex central chi-square random variable with p degrees of freedom, while under the alternative hypothesis T_g is noncentral chi-square with noncentrality parameter vector $(\underline{s}^H \mathbf{B}^H \mathbf{R}_w^{-1} \mathbf{B} \underline{s}, 1)$. The ROC curve is indexed by the number of signals p and the noncentrality parameter but is not expressible in closed form for $p > 1$.

13.5.3 Signal Detection: Random Gains

In some cases a random Gaussian model for the gains may be more appropriate than the unknown gain model considered above. When the p-dimensional gain vector $\underline{s}$ is multivariate normal with zero mean and $p \times p$ covariance matrix $\mathbf{R}_s$ the compound signal component $\underline{s}_c = \mathbf{B}\underline{s}$ is an n-dimensional random Gaussian vector with zero mean and rank p covariance matrix $\mathbf{B}\mathbf{R}_s\mathbf{B}^H$. A standard assumption is that the gains and the additive noise are statistically independent. The detection problem can then be stated as testing the two simple hypotheses $H : \underline{x} \sim \mathcal{N}_n(0, \mathbf{R}_w)$ vs. $K : \underline{x} \sim \mathcal{N}_n(0, \mathbf{B}\mathbf{R}_s\mathbf{B}^H + \mathbf{R}_w)$. It can be shown that the most powerful LRT has the form

$$T(x) = \sum_{i=1}^{p} \left(\frac{\lambda_i}{1 + \lambda_i} \right) |\underline{v}_i^* \mathbf{R}_w^{-\frac{1}{2}} \underline{x}|^2 \underset{H}{\overset{K}{\underset{<}{>}}} \gamma, \tag{13.16}$$

where $\{\lambda_i\}_{i=1}^{p}$ are the nonzero eigenvalues of the matrix $\mathbf{R}_w^{-\frac{1}{2}} \mathbf{B}\mathbf{R}_s\mathbf{B}^H \mathbf{R}_w^{-\frac{H}{2}}$ and $\{\underline{v}_i\}_{i=1}^{p}$ are the associated eigenvectors. Under H the test statistic $T(x)$ is distributed as complex noncentral chi-square with p degrees of freedom and noncentrality parameter vector $(0, \underline{d}_H)$ where $\underline{d}_H = [\lambda_1/(1 + \lambda_1), \dots, \lambda_p/(1 + \lambda_p)]$. Under the alternative hypothesis T is also distributed as noncentral complex chi-square, however, with noncentrality vector $(0, \underline{d}_K)$ where $\underline{d}_K$ are the nonzero eigenvalues of $\mathbf{B}\mathbf{R}_s\mathbf{B}^H$. The ROC is not available in closed form for $p > 1$.

13.5.4 Signal Detection: Single Signal

We obtain a unification of the GLRT for unknown gain and the LRT for random gain in the case of a single impinging signal waveform: $\mathbf{B} = \underline{b}_1$, $p = 1$. In this case the test statistic T_g in (13.15) and T in (13.16) reduce to the identical form and we get the same detector structure

$$\frac{|\underline{x}^H \mathbf{R}_w^{-1} \underline{b}_1|^2}{\underline{b}_1^H \mathbf{R}_w^{-1} \underline{b}_1} \underset{H}{\overset{K}{\underset{<}{>}}} \eta,$$

This establishes that the GLRT is uniformly most powerful over all values of the gain parameter s_1 for $p = 1$. Note that even though the form of the unknown parameter GLRT and the random parameter LRT are identical for this case, their ROC curves and their thresholds γ will be different since the underlying observation models are not the same. When the noise is white the test simply compares the magnitude squared of the complex correlator output $\sum_{i=1}^{n} b_1^*(t_i)x(t_i)$ to a threshold γ.

13.6 Spatio-Temporal Signals

Consider the general spatio-temporal model

$$\underline{x}(t_i) = \sum_{j=1}^{q} \underline{a}_j \sum_{k=1}^{p} s_{jk} b_k(t_i) + \underline{w}(t_i), \quad i = 1, \dots, n.$$

This model applies to a wide range of applications in narrowband array processing and has been thoroughly studied in the context of signal detection in [14]. The m-element vector $\underline{x}(t_i)$ is a snapshot at time t_i of the m-element array response to p impinging signals arriving from q different directions. The vector $\underline{a}_j$ is a known *steering vector* which is the complex response of the array to signal energy arriving from the jth direction. From this direction the array receives the superposition $\sum_{k=1}^{p} s_{jk}\underline{b}_k$ of p known time varying signal waveforms $\underline{b}_k = [b_k(t_1), \ldots, b_k(t_n)]^T$, $k = 1, \ldots, p$. The presence of the superposition accounts for both direct and multipath arrivals and allows for more signal sources than directions of arrivals when $p > q$. The complex Gaussian noise vectors $\underline{w}(t_i)$ are spatially correlated with spatial covariance $\mathrm{cov}[\underline{w}(t_i)] = \mathbf{R}_w$ but are temporally uncorrelated $\mathrm{cov}[\underline{w}(t_i), \underline{w}(t_j)] = 0$, $i \neq j$.

By arranging the n column vectors $\{\underline{x}(t_i)\}_{i=1}^{n}$ in an $m \times n$ matrix $\mathbf{X}$ we obtain the equivalent matrix model

$$\mathbf{X} = \mathbf{ASB}^H + \mathbf{W},$$

where $\mathbf{S} = (s_{ij})$ is a $q \times p$ matrix whose rows are vectors of signal gain factors for each different direction of arrival, $\mathbf{A} = [\underline{a}_1, \ldots, \underline{a}_q]$ is an $m \times q$ matrix whose columns are steering vectors for different directions of arrival, and $\mathbf{B} = [\underline{b}_1, \ldots, \underline{b}_p]^T$ is a $p \times n$ matrix whose rows are different signal waveforms. To avoid singular detection it is assumed that $\mathbf{A}$ is of rank q, $q \leq m$, and that $\mathbf{B}$ is of rank p, $p \leq n$. We consider only a few applications of this model here. For many others see [14].

13.6.1   Detection: Known Gains and Known Spatial Covariance

First we assume the gain matrix $\mathbf{S}$ and the spatial covariance $\mathbf{R}_w$ are known. This case is only relevant when one knows the direct path and multipath geometry of the propagation medium ($\mathbf{S}$), the spatial distribution of the ambient (possibly coherent) noise ($\mathbf{R}_w$), the q directions of the impinging superposed signals ($\mathbf{A}$), and the p signal waveforms ($\mathbf{B}$). Here, the detection problem is stated in terms of the simple hypotheses $H : \mathbf{X} \sim \mathcal{N}_{nm}(0, \mathbf{R}_w \otimes \mathbf{I}_n)$ vs. $K : \mathbf{X} \sim \mathcal{N}_{nm}(\mathbf{ASB}, \mathbf{R}_w \otimes \mathbf{I}_n)$. For this case, the LRT (13.5) is the most powerful test and, using (13.12), has the form

$$T(x) = \mathrm{Re}\left(\mathrm{tr}\left\{\mathbf{A}^H \mathbf{R}_w^{-1} \mathbf{X} \mathbf{B}^H \mathbf{S}^H\right\}\right) \underset{H}{\overset{K}{\underset{<}{\gtrless}}} \gamma.$$

Since the test statistic is Gaussian under H and K the ROC curve is of similar form to the ROC for detection of temporal signals with known gains.

Identifying the quantities $\tilde{\mathbf{X}} = \mathbf{R}_w^{-\frac{1}{2}}\mathbf{X}$ and $\tilde{\mathbf{A}} = \mathbf{R}_w^{-\frac{1}{2}}\mathbf{A}$ as the spatially whitened measurement matrix and spatially whitened array response matrix, respectively, the test statistic T can be interpreted as a multivariate spatiotemporal correlator detector. In particular, when there is only one signal impinging on the array from a single direction then $p = q = 1$, $\tilde{\mathbf{A}} = \underline{\tilde{a}}$ a column vector, $\mathbf{B} = \underline{b}^T$ a row vector, $\mathbf{S} = s$ a complex scalar, and the test statistic becomes

$$
\begin{aligned}
T(x) &= \mathrm{Re}\left\{\underline{\tilde{a}}^H \cdot_s \tilde{\mathbf{X}} \cdot_t \underline{b}^* s^*\right\} \\
&= \mathrm{Re}\left\{s^* \sum_{j=1}^{m} \tilde{a}_j^* \sum_{i=1}^{n} b^*(t_i)\tilde{x}_j(t_i)\right\}.
\end{aligned}
$$

In the above the multiplication notation $\cdot_s$ and $\cdot_t$ is used to simply emphasize the respective matrix multiplication operations (correlation) which occur over the spatial domain and the time domain. It can be shown that the ROC curve monotonically increases in the detectability index $\rho = n\underline{a}^H \mathbf{R}_w^{-1}\underline{a} \cdot \|s\underline{b}\|^2$.

13.6.2   Detection: Unknown Gains and Unknown Spatial Covariance

By assuming the gain matrix $\mathbf{S}$ and $\mathbf{R}_w$ to be unknown, the detection problem becomes one of testing for noise alone against noise plus p coherent signal waveforms, where the waveforms lie in the subspace formed by all linear combinations of the rows of $\mathbf{B}$ but are otherwise unknown. This gives a composite null and alternative hypothesis for which the generalized likelihood ratio test can be derived by maximizing the known-gain likelihood ratio over the gain matrix **S**. The result is the GLRT [14]

$$T_g(x) = \frac{\left| \mathbf{A}^H \hat{\mathbf{R}}_K^{-1} \mathbf{A} \right|}{\left| \mathbf{A}^H \hat{\mathbf{R}}_H^{-1} \mathbf{A} \right|} \begin{array}{c} K \\ \gtrless \\ H \end{array} \gamma,$$

where $|\cdot|$ denotes the determinant, $\hat{\mathbf{R}}_H = \frac{1}{n}\mathbf{X}\mathbf{X}^H$ is a sample estimate of the spatial covariance matrix using all of the snapshots, and $\hat{\mathbf{R}}_K = \frac{1}{n}\mathbf{X}[\mathbf{I}_n - \mathbf{B}^H[\mathbf{B}\mathbf{B}^H]^{-1}\mathbf{B}]\mathbf{X}^H$ is the sample estimate using only those components of the snapshots lying outside of the row space of the signal waveform matrix **B**. To gain insight into the test statistic T_g consider the asymptotic convergence of T_g as the number of snapshots n goes to infinity. By the strong law $\hat{\mathbf{R}}_K$ converges to the covariance matrix of $\mathbf{X}[\mathbf{I}_n - \mathbf{B}^H[\mathbf{B}\mathbf{B}^H]^{-1}\mathbf{B}]$. Since $\mathbf{I}_n - \mathbf{B}^H[\mathbf{B}\mathbf{B}^H]^{-1}\mathbf{B}$ annihilates the signal component **ASB**, this covariance is the same quantity **R**, $\mathbf{R} \le \mathbf{R}_w$, under both H and K. On the other hand, $\hat{\mathbf{R}}_H$ converges to $\mathbf{R}_w$ under H while it converges to $\mathbf{R}_w + \mathbf{ASBB}^H\mathbf{S}^H\mathbf{A}^H$ under K. Hence when strong signals are present T_g tends to take on very large values near the quantity $\left(|\mathbf{A}^H\mathbf{R}^{-1}\mathbf{A}|\right) / \left(|\mathbf{A}^H[\mathbf{R}_w + \mathbf{ASBB}^H\mathbf{S}^H\mathbf{A}^H]^{-1}\mathbf{A}|\right) \gg 1$.

The distribution of T_g under H (K) can be derived in terms of the distribution of a sum of central (noncentral) complex beta random variables. See [14] for discussion of performance and algorithms for data recursive computation of T_g. Generalizations of this GLRT exist which incorporate nonzero mean [14, 15].

13.7 Signal Classification

Typical classification problems arising in signal processing are: classifying an individual signal waveform out of a set of possible linearly independent waveforms, classifying the presence of a particular set of signals as opposed to other sets of signals, classifying among specific linear combinations of signals, and classifying the number of signals present. The problem of classification of the number of signals, also known as the order selection problem, is treated elsewhere in this Handbook. While the Gaussian spatiotemporal model could be treated in analogous fashion, for concreteness we focus on the case of the temporal signal model (13.13).

13.7.1 Classifying Individual Signals

Here it is of interest to decide which one of the p-scaled signal waveforms $s_1\underline{b}_1, \ldots, s_p\underline{b}_p$ are present in the observations $\underline{x} = [x(t_1), \ldots x(t_n)]^T$. Denote by H_k the hypothesis that $\underline{x} = s_k\underline{b}_k + \underline{w}$. Signal classification can then be stated as the problem of testing between the following simple hypotheses

$$H_1 \quad : \quad \underline{x} = s_1\underline{b}_1 + \underline{w}$$

$$\vdots \quad \vdots \quad \vdots$$

$$H_p \quad : \quad \underline{x} = s_p\underline{b}_p + \underline{w}$$

For known gain factors s_k, known signal waveforms $\underline{b}_k$, and known noise covariance $\mathbf{R}_w$, these hypotheses are simple, the density function $f(x|s_k, \underline{b}_k) = \mathcal{N}_n(s_k\underline{b}_k, \mathbf{R}_w)$ under H_k involves no unknown parameters, and the maximum likelihood classifier (13.8) reduces to the decision rule

$$\text{decide } H_j \text{ if and only if } j = \text{argmin}_{k=1,\ldots,p}(\underline{x} - s_k\underline{b}_k)^H \mathbf{R}_w^{-1}(\underline{x} - s_k\underline{b}_k) . \tag{13.17}$$

Thus, the classifier chooses the most likely signal as that signal $s_j\underline{b}_j$ which has minimum normalized distance from the observed waveform $\underline{x}$. The classifier can also be interpreted as a *minimum distance classifier* which chooses the signal which minimizes the Euclidean distance $\|\tilde{\underline{x}} - s_k \tilde{\underline{b}}_k\|$ between the prewhitened signal $\tilde{\underline{b}}_k = \mathbf{R}_w^{-\frac{1}{2}} \underline{b}_k$ and the prewhitened measurement $\tilde{\underline{x}} = \mathbf{R}_w^{-\frac{1}{2}} \underline{x}$.

Written in the minimum normalized distance form, the ML classifier appears to involve nonlinear statistics. However, an obvious simplification of (13.17) reveals that the ML classifier actually only requires computing linear functions of $\underline{x}$

$$\text{decide } H_j \text{ if and only if } j = \text{argmax}_{k=1,\dots,p} \left\{ \text{Re} \left(\underline{x}^H \mathbf{R}_w^{-1} \underline{b}_k \ s_k \right) - \tfrac{1}{2} |s_k|^2 \ \underline{b}_k^H \mathbf{R}_w^{-1} \underline{b}_k \right\}.$$

Note that this linear reduction only occurs when the covariances $\mathbf{R}_w$ are identical under each H_k, $k = 1, \dots, p$. In this case the ML classifier can be implemented using prewhitening filters followed by a bank of correlators or matched filters, an offset adjustment, and a maximum selector (Fig. 13.5).

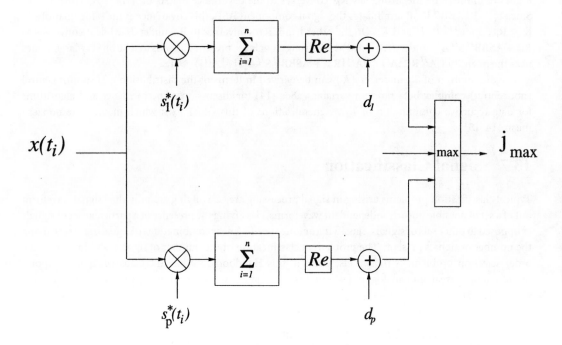

FIGURE 13.5 The ML classifier for classifying presence of one of p signals $s_j(t_i) \stackrel{\text{def}}{=} s_j \underline{b}_j(t_i)$, $j = 1, \dots, p$, under additive Gaussian white noise. $d_j = -\tfrac{1}{2} |s_j|^2 \ \|\underline{b}_j\|^2$ is an offset and j_{max} is the index of correlator output which is maximum. For nonwhite noise a prewhitening transformation must be performed on $x(t_i)$ and the $b_j(t_i)$'s prior to implementation of ML classifier.

An additional simplification occurs when the noise is white, $\mathbf{R}_w = \mathbf{I}_n$, and all signal energies $|s_k|^2 \|\underline{b}_k^H\|^2$ are identical: the classifier chooses the most likely signal as that signal $b_j(t_i)s_j$ which is maximally correlated with the measurement x:

$$\text{decide } H_j \text{ if and only if } j = \text{argmax}_{k=1,\dots,p} \text{Re} \left(s_k \sum_{i=1}^n b_k^*(t_i) x(t_i) \right).$$

The decision regions $\mathcal{R}_{H_k} = \{x : \text{decide } H_k\}$ induced by (13.17) are piecewise linear regions, known as Voronoi cells $\mathcal{V}_k$, centered at each of the prewhitened signals $s_k \tilde{\underline{b}}_k$. The misclassification error probabilities $P_{M_k} = 1 - P(x \in \mathcal{R}_{H_k}|H_k) = 1 - \int_{x \in \mathcal{V}_k} f(x|H_k)dx$ must generally be computed by integrating complex multivariate Gaussian densities $f(x|H_k) = \mathcal{N}_n(s_k \underline{b}_k, \mathbf{R}_w)$ over these regions. In the case of orthogonal signals $\underline{b}_i \mathbf{R}_w^{-1} \underline{b}_j = 0$, $i \neq j$, this integration reduces to a single integral of a univariate $\mathcal{N}_1(\rho_k, \rho_k)$ density function times the product of $p - 1$ univariate $\mathcal{N}_1(0, \rho_i)$ cumulative distribution functions, $i = 1, \ldots, p, i \neq k$, where $\rho_k = \underline{b}_k^H \mathbf{R}_w^{-1} \underline{b}_k$. Even for this case no general closed form expressions for P_{M_k} is available. However, analytical lower bounds on P_{M_k} and on average misclassification probability $\frac{1}{p} \sum_{k=1}^{p} P_{M_k}$ can be used to qualitatively assess classifer performance [12].

13.7.2 Classifying Presence of Multiple Signals

We conclude by treating the problem where the signal component of the observation is the linear combination of one of J hypothesized subsets $\mathcal{S}_k$, $k = 1, \ldots, J$, of the signal waveforms $\underline{b}_1, \ldots, \underline{b}_p$. Assume that subset $\mathcal{S}_k$ contains p_k signals and that the $\mathcal{S}_k$, $k = 1, \ldots, J$, are disjoint, i.e., they do not contain any signals in common. Define the $n \times p_k$ matrix $\mathbf{B}_k$ whose columns are formed from the subset $\mathcal{S}_k$. We can now state the classification problem as testing between the J composite hypotheses

$$H_1 \quad : \quad \underline{x} = \mathbf{B}_1 \underline{s}_1 + \underline{w}, \quad \underline{s}_1 \in \mathbb{C}^{p_1}$$

$$\vdots \quad \vdots \quad \vdots$$

$$H_J \quad : \quad \underline{x} = \mathbf{B}_J \underline{s}_J + \underline{w}, \quad \underline{s}_J \in \mathbb{C}^{p_J}$$

where $\underline{s}_k$ is a column vector of p_k unknown complex gains.

The density function under H_k, $f(x|\underline{s}_k, \mathbf{B}_k) = \mathcal{N}_n(\mathbf{B}_k \underline{s}_k, \mathbf{R}_w)$, is a function of unknown parameters $\underline{s}_k$ and, therefore, the ML classifier (13.8) involves finding the largest among maximized likelihoods $\max_{\underline{s}_k} f(x|\underline{s}_k, \mathbf{B}_k), k = 1, \ldots, J$. This yields the following form for the ML classifier:

$$\text{decide } H_j \text{ if and only if } j = \text{argmin}_{k=1,\ldots,J} \left(\underline{x} - \mathbf{B}_k \hat{\underline{s}}_k\right)^H \mathbf{R}_w^{-1} \left(\underline{x} - \mathbf{B}_k \hat{\underline{s}}_k\right), \tag{13.18}$$

where $\hat{\underline{s}}_k = \left[\mathbf{B}_k^H \mathbf{R}_w^{-1} \mathbf{B}_k\right]^{-1} \mathbf{B}_k^H \mathbf{R}_w^{-1} \underline{x}$ is the maximum likelihood gain vector estimate. The decision regions are once again piecewise linear but with Voronoi cells having centers at the least squares estimates of the hypothesized signal components $\mathbf{B}_k \hat{\underline{s}}_k$, $k = 1, \ldots, J$.

Similarly to the case of noncomposite hypotheses considered in the previous subsection, a simplification of (13.18) is possible

$$\text{decide } H_j \text{ if and only if } j = \text{argmax}_{k=1,\ldots,J} \underline{x}^H \mathbf{R}_w^{-1} \mathbf{B}_k \left[\mathbf{B}_k^H \mathbf{R}_w^{-1} \mathbf{B}_k\right]^{-1} \mathbf{B}_k^H \mathbf{R}_w^{-1} \underline{x}$$

Defining the prewhitened versions $\tilde{\underline{x}} = \mathbf{R}_w^{-\frac{1}{2}} \underline{x}$ and $\tilde{\mathbf{B}}_k = \mathbf{R}_w^{-\frac{1}{2}} \mathbf{B}_k$ of the observations and the kth signal matrix, the ML classifier is seen to decide that the linear combination of the p_j signals in H_j is present when the length $\|\tilde{\mathbf{B}}_j [\tilde{\mathbf{B}}_j^H \tilde{\mathbf{B}}_j]^{-1} \tilde{\mathbf{B}}_j^H] \tilde{\underline{x}}\|$ of the projection of $\tilde{\underline{x}}$ onto the jth signal space (colspan$\{\tilde{\mathbf{B}}_j\}$) is greatest. This classifer can be implemented as a bank of p *adaptive* matched filters each matched to one of the least squares estimates $\tilde{\mathbf{B}}_k \hat{\underline{s}}_k$, $k = 1, \ldots, p$, of the prewhitened signal component. Under any H_i the quantities $\underline{x}^H \mathbf{R}_w^{-1} \mathbf{B}_k [\mathbf{B}_k^H \mathbf{R}_w^{-1} \mathbf{B}_k]^{-1} \mathbf{R}_w^{-1} \underline{x}, k = 1, \ldots J$, are distributed as complex noncentral chi-square with p_k degrees of freedom. For the special case of orthogonal prewhitened signals $\underline{b}_i \mathbf{R}_w^{-1} \underline{b}_j = 0, i \neq j$, these variables are also statistically independent and P_{M_i} can be computed as a one-dimensional integral of a univariate noncentral chi-square density times the product of $J - 1$ univariate noncentral chi-square cumulative distribution functions.

References

[1] Lehmann, E.L., *Testing Statistical Hypotheses*, John Wiley & Sons, New York, 1959.

[2] Ferguson, T.S., *Mathematical Statistics — A Decision Theoretic Approach*, Academic Press, Orlando, FL, 1967.

[3] Middleton, D., *An Introduction to Statistical Communication Theory*, Peninsula Publishing, Los Altos, CA (reprint of 1960 McGraw-Hill edition), 1987.

[4] Davenport, W. and Root, W., *An Introduction to the Theory of Random Signals and Noise*, IEEE Press, New York (reprint of 1958 McGraw-Hill edition), 1987.

[5] Van-Trees, H.L., *Detection, Estimation, and Modulation Theory: Part I*, John Wiley & Sons, New York, 1968.

[6] Blackwell, D. and Girshik, M.A., *Theory of Games and Statistical Decisions*, John Wiley & Sons, New York, 1954.

[7] Helstrom, C., *Elements of Signal Detection and Estimation*, Prentice-Hall, Englewood Cliffs, NJ, 1995.

[8] Scharf, L.L., *Statistical Signal Processing: Detection, Estimation, and Time Series Analysis*, Addison-Wesley, Reading, MA, 1991.

[9] Siegmund, D., *Sequential Analysis: Tests and Confidence Intervals*, Springer-Verlag, New York, 1985.

[10] Baygun, B. and Hero, A.O., Optimal simultaneous detection and estimation under a false alarm constraint, *IEEE Trans. Inform. Theory*, 41(3): 688–703, 1995.

[11] Kassam, S. and Thomas, J., *Nonparametric Detection — Theory and Applications*, Dowden, Hutchinson, and Ross, 1980.

[12] Fukunaga, K., *Statistical Pattern Recognition*, 2nd ed., Academic Press, San Diego, CA, 1990.

[13] Kelly, E.J. and Forsythe, K.M., Adaptive Detection and Parameter Estimation for Multidimensional Signal Models, Technical Report 848, M.I.T. Lincoln Laboratory, April, 1989.

[14] Muirhead, R.J., *Aspects of Multivariate Statistical Theory*, John Wiley & Sons, New York, 1982.

[15] Kariya, T. and Sinha, B.K., *Robustness of Statistical Tests*, Academic Press, San Diego, 1989.

14

Spectrum Estimation and Modeling

Petar M. Djurić
State University of New York
at Stony Brook

Steven M. Kay
University of Rhode Island

14.1 Introduction

The main objective of spectrum estimation is the determination of the power spectrum density (PSD) of a random process. The PSD is a function that plays a fundamental role in the analysis of stationary random processes in that it quantifies the distribution of total power as a function of frequency. The estimation of the PSD is based on a set of observed data samples from the process. A necessary assumption is that the random process is at least wide sense stationary, that is, its first and second order statistics do not change with time. The estimated PSD provides information about the structure of the random process which can then be used for refined modeling, prediction, or filtering of the observed process.

Spectrum estimation has a long history with beginnings in ancient times [17]. The first significant discoveries that laid the grounds for later developments, however, were made in the early years of the eighteenth century. They include one of the most important advances in the history of mathematics, Fourier's theory. According to this theory, an arbitrary function can be represented by an infinite summation of sine and cosine functions. Later came the Sturm-Liouville spectral theory of differential equations, which was followed by the spectral representations in quantum and classical physics developed by John von Neuman and Norbert Wiener, respectively. The statistical theory of spectrum estimation started practically in 1949 when Tukey introduced a numerical method for computation of spectra from empirical data. A very important milestone for further development of the field was the reinvention of the fast Fourier transform (FFT) in 1965, which is an efficient algorithm for computation of the discrete Fourier transform. Shortly thereafter came the work of John Burg, who proposed a fundamentally new approach to spectrum estimation based on the principle of maximum entropy. In the past three decades his work was followed up by many researchers who have developed numerous new spectrum estimation procedures and applied them to various physical processes from diverse scientific fields. Today, spectrum estimation is a vital scientific discipline which plays a major role in many applied sciences such as radar, speech processing, underwater acoustics, biomedical signal processing, sonar, seismology, vibration analysis, control theory, and econometrics.

14.2   Important Notions and Definitions

14.2.1   Random Processes

The objects of interest of spectrum estimation are random processes. They represent time fluctuations of a certain quantity which cannot be fully described by deterministic functions. The voltage waveform of a speech signal, the bit stream of zeros and ones of a communication message, or the daily variations of the stock market index are examples of random processes. Formally, a random process is defined as a collection of random variables indexed by time. (The family of random variables may also be indexed by a different variable, for example space, but here we will consider only random *time* processes.) The index set is infinite and may be continuous or discrete. If the index set is continuous, the random process is known as a continuous-time random process, and if the set is discrete, it is known as a discrete-time random process. The speech waveform is an example of a continuous random process and the sequence of zeros and ones of a communication message, a discrete one. We shall focus only on discrete-time processes where the index set is the set of integers.

A random process can be viewed as a collection of a possibly infinite number of functions, also called realizations. We shall denote the collection of realizations by $\{\tilde{x}[n]\}$ and an observed realization of it by $\{x[n]\}$. For fixed n, $\{\tilde{x}[n]\}$ represents a random variable, also denoted as $\tilde{x}[n]$, and $x[n]$ is the n-th sample of the realization $\{x[n]\}$. If the samples $x[n]$ are real, the random process is real, and if they are complex, the random process is complex. In the discussion to follow, we assume that $\{\tilde{x}[n]\}$ is a *complex* random process.

The random process $\{\tilde{x}[n]\}$ is fully described if for any set of time indices $n_1, n_2, ..., n_m$, the joint probability density function of $\tilde{x}[n_1], \tilde{x}[n_2], ...,$ and $\tilde{x}[n_m]$ is given. If the statistical properties of the process do not change with time, the random process is called stationary. This is always the case if for any choice of random variables $\tilde{x}[n_1], \tilde{x}[n_2], ...,$ and $\tilde{x}[n_m]$, their joint probability density function is identical to the joint probability density function of the random variables $\tilde{x}[n_1 + k], \tilde{x}[n_2 + k], ...,$ and $\tilde{x}[n_m + k]$ for any k. Then we call the random process strictly stationary. For example, if the samples of the random process are independent and identically distributed random variables, it is straightforward to show that the process is strictly stationary. Strict stationarity, however, is a very severe requirement and is relaxed by introducing the concept of wide-sense stationarity. A random process is wide-sense stationary if the following two conditions are met:

$$E\left(\tilde{x}[n]\right) = \mu \tag{14.1}$$

and

$$
\begin{aligned}
r[n, n+k] &= E\left(\tilde{x}^*[n]\tilde{x}[n+k]\right) \\
&= r[k]
\end{aligned}
\tag{14.2}
$$

where $E(\cdot)$ is the expectation operator, $\tilde{x}^*[n]$ is the complex conjugate of $\tilde{x}[n]$, and $\{r[k]\}$ is the autocorrelation function of the process. Thus, if the process is wide-sense stationary, its mean value μ is constant over time, and the autocorrelation function depends only on the lag k between the random variables. For example, if we consider the random process

$$\tilde{x}[n] = a\cos(2\pi f_0 n + \tilde{\theta}) \tag{14.3}$$

where the amplitude a and the frequency f_0 are constants, and the phase $\tilde{\theta}$ is a random variable that is uniformly distributed over the interval $(-\pi, \pi)$, one can show that

$$E(\tilde{x}[n]) = 0 \qquad (14.4)$$

and

$$
\begin{aligned}
r[n, n+k] &= E\left(\tilde{x}^*[n]\tilde{x}[n+k]\right) \\
&= \frac{a^2}{2}\cos(2\pi f_0 k) .
\end{aligned}
\qquad (14.5)
$$

Thus, Eq. (14.3) represents a wide-sense stationary random process.

14.2.2   Spectra of Deterministic Signals

Before we define the concept of spectrum of a random process, it will be useful to review the analogous concept for deterministic signals, which are signals whose future values can be exactly determined without any uncertainty. Besides their description in the time domain, the deterministic signals have a very useful representation in terms of superposition of sinusoids with various frequencies, which is given by the discrete-time Fourier transform (DTFT). If the observed signal is $\{g[n]\}$ and it is not periodic, its DTFT is the complex valued function $G(f)$ defined by

$$G(f) = \sum_{n=-\infty}^{\infty} g[n]e^{-j2\pi fn} \qquad (14.6)$$

where $j = \sqrt{-1}$, f is the normalized frequency, $0 \leq f < 1$, and $e^{j2\pi fn}$ is the complex exponential given by

$$e^{j2\pi fn} = \cos(2\pi fn) + j\sin(2\pi fn) . \qquad (14.7)$$

The sum in Eq. (14.6) converges uniformly to a continuous function of the frequency f if

$$\sum_{n=-\infty}^{\infty} |g[n]| < \infty . \qquad (14.8)$$

The signal $\{g[n]\}$ can be determined from $G(f)$ by the inverse DTFT defined by

$$g[n] = \int_0^1 G(f)e^{j2\pi fn}df \qquad (14.9)$$

which means that the signal $\{g[n]\}$ can be represented in terms of complex exponentials whose frequencies span the continuous interval $[0,1)$.

The complex function $G(f)$ can be alternatively expressed as

$$G(f) = |G(f)|e^{j\phi(f)} \qquad (14.10)$$

where $|G(f)|$ is called the amplitude spectrum of $\{g[n]\}$, and $\phi(f)$ the phase spectrum of $\{g[n]\}$. For example, if the signal $\{g[n]\}$ is given by

$$g[n] = \begin{cases} 1, & n = 1 \\ 0, & n \neq 1 \end{cases} \qquad (14.11)$$

then

$$G(f) = e^{-j2\pi f} \qquad (14.12)$$

and the amplitude and phase spectra are

$$
\begin{aligned}
|G(f)| &= 1, & 0 \le f < 1 \\
\phi(f) &= -2\pi f, & 0 \le f < 1.
\end{aligned}
\tag{14.13}
$$

The total energy of the signal is given by

$$
\mathcal{E} = \sum_{n=-\infty}^{\infty} |g[n]|^2
\tag{14.14}
$$

and according to Parseval's theorem, it can also be obtained from the amplitude spectrum of the signal, i.e.,

$$
\sum_{n=-\infty}^{\infty} |g[n]|^2 = \int_0^1 |G(f)|^2 df .
\tag{14.15}
$$

From Eq. (14.15), we deduce that $|G(f)|^2 df$ is the contribution to the total energy of the signal from the frequency band $(f, f + df)$. Therefore, we say that $|G(f)|^2$ represents the energy density spectrum of the signal $\{g[n]\}$.

When $\{g[n]\}$ is periodic with period N, that is

$$
g(n) = g(n + N)
\tag{14.16}
$$

for all n, and where N is the period of $\{g[n]\}$, we use the discrete Fourier transform (DFT) to express $\{g[n]\}$ in the frequency domain, that is,

$$
G(f_k) = \sum_{n=0}^{N-1} g[n]e^{-j2\pi f_k n}, \quad f_k = \frac{k}{N}, \quad k \in \{0, 1, \cdots, N-1\}.
\tag{14.17}
$$

Note that the frequency here takes values from a discrete set. The inverse DFT is defined by

$$
g[n] = \frac{1}{N} \sum_{k=0}^{N-1} G(f_k)e^{j2\pi f_k n}, \quad f_k = \frac{k}{N}.
\tag{14.18}
$$

Now Parseval's relation becomes

$$
\sum_{n=0}^{N-1} |g[n]|^2 = \frac{1}{N} \sum_{k=0}^{N-1} |G(f_k)|^2, \quad f_k = \frac{k}{N}
\tag{14.19}
$$

where the two sides are the total energy of the signal in one period. If we define the average power of the discrete-time signal by

$$
\mathcal{P} = \frac{1}{N} \sum_{n=0}^{N-1} |g[n]|^2
\tag{14.20}
$$

then from Eq. (14.19)

$$
\mathcal{P} = \frac{1}{N^2} \sum_{k=0}^{N-1} |G(f_k)|^2, \quad f_k = \frac{k}{N}.
\tag{14.21}
$$

Thus, $|G(f_k)|^2/N^2$ is the contribution to the total power from the term with frequency f_k, and so it represents the power spectrum "density" of $\{g[n]\}$. For example, if the periodic signal in one period is defined by

$$
g[n] = \begin{cases} 1, & n = 0 \\ 0, & n = 1, 2, \cdots, N-1 \end{cases}
\tag{14.22}
$$

its PSD $P(f_k)$ is

$$P(f_k) = \frac{1}{N^2}, \quad f_k = \frac{k}{N}, \quad k \in \{0, 1, \cdots, N-1\}. \tag{14.23}$$

Again, note that the PSD is defined for a discrete set of frequencies.

In summary, the spectra of deterministic aperiodic signals are energy densities defined on the continuous set of frequencies $\mathcal{C}_f = [0, 1)$. On the other hand, the spectra of periodic signals are power densities defined on the discrete set of frequencies $\mathcal{D}_f = \{0, 1/N, 2/N, \cdots, (N-1)/N\}$, where N is the period of the signal.

14.2.3 Spectra of Random Processes

Suppose that we observe one realization of the random process $\{\tilde{x}[n]\}$, or $\{x[n]\}$. From the definition of the DTFT and the assumption of wide-sense stationarity of $\{\tilde{x}[n]\}$, it is obvious that we cannot use the DTFT to obtain $X(f)$ from $\{x[n]\}$ because Eq. (14.8) does not hold when we replace $g[n]$ by $x[n]$. And indeed, if $\{x[n]\}$ is a realization of a wide-sense stationary process, its energy is infinite. Its power, however, is finite as was the case with the periodic signals. So if we observe $\{x[n]\}$ from $-N$ to N, $\{x[n]\}_{-N}^{N}$, and assume that outside this interval the samples $x[n]$ are equal to zero, we can find its DTFT, $X_N(f)$ from

$$X_N(f) = \sum_{n=-N}^{N} x[n]e^{-j2\pi fn}. \tag{14.24}$$

Then according to Eq. (14.15), $|X_N(f)|^2 df$ represents the energy of the truncated realization that is contributed by the components whose frequencies are between f and $f + df$. The power due to these components is given by

$$\frac{|X_N(f)|^2 df}{2N+1} \tag{14.25}$$

and $|X_N(f)|^2/(2N+1)$ can be interpreted as power density. If we let $N \to \infty$, under suitable conditions [15],

$$\lim_{N \to \infty} \frac{|X_N(f)|^2}{2N+1} \tag{14.26}$$

is finite for all f, and this is then the PSD of $\{x[n]\}$. We would prefer to find, however, the PSD of $\{\tilde{x}[n]\}$, which we define as

$$P(f) = \lim_{N \to \infty} E\left\{ \frac{|\tilde{X}_N(f)|^2}{2N+1} \right\} \tag{14.27}$$

where $\tilde{X}_N(f)$ is the DTFT of $\{\tilde{x}[n]\}_{-N}^{N}$. Clearly, $P(f)df$ is interpreted as the *average* contribution to the total power from the components of $\{\tilde{x}[n]\}$ whose frequencies are between f and $f + df$.

There is a very important relationship between the PSD of a wide-sense stationary random process and its autocorrelation function. By Wold's theorem, which is the analogue of Wiener-Khintchine theorem for continuous-time random processes, the PSD in Eq. (14.27) is the DTFT of the autocorrelation function of the process [15], that is,

$$P(f) = \sum_{k=-\infty}^{\infty} r[k]e^{-j2\pi fk} \tag{14.28}$$

where $r[k]$ is defined by Eq. (14.2).

For all practical purposes, there are three different types of $P(f)$ [15]. If $P(f)$ is an absolutely continuous function of f, the random process has a purely continuous spectrum. If $P(f)$ is identically equal to zero for all f except for frequencies $f = f_k, k = 1, 2, \ldots$, where it is infinite, the random process has a line spectrum. In this case, a useful representation of the spectrum is given by the Dirac δ-functions,

$$P(f) = \sum_k P_k \delta(f - f_k) \tag{14.29}$$

where P_k is the power associated with the k line component. Finally, the spectrum of a random process may be mixed if it is a combination of a continuous and line spectra. Then $P(f)$ is a superposition of a continuous function of f and δ-functions.

14.3　The Problem of Power Spectrum Estimation

The problem of power spectrum estimation can be stated as follows: Given a set of N samples $\{x[0], x[1], \ldots, x[N-1]\}$ of a realization of the random process $\{\tilde{x}[n]\}$, denoted also by $\{x[n]\}_0^{N-1}$, estimate the PSD of the random process, $P(f)$. Obviously this task amounts to estimation of a function and is distinct from the typical problem in elementary statistics where the goal is to estimate a finite set of parameters.

Spectrum estimation methods can be classified into two categories: nonparametric and parametric. The nonparametric approaches do not assume any specific parametric model for the PSD. They are based solely on the estimate of the autocorrelation sequence of the random process from the observed data. For the parametric approaches on the other hand, we first postulate a model for the process of interest, where the model is described by a small number of parameters. Based on the model, the PSD of the process can be expressed in terms of the model parameters. Then the PSD estimate is obtained by substituting the estimated parameters of the model in the expression for the PSD. For example, if a random process $\{\tilde{x}[n]\}$ can be modeled by

$$\tilde{x}[n] = -a\tilde{x}[n] + \tilde{w}[n] \tag{14.30}$$

where a is an unknown parameter and $\{\tilde{w}[n]\}$ is a zero mean wide-sense stationary random process whose random variables are uncorrelated and with the same variance σ^2, it can be shown that the PSD of $\{\tilde{x}[n]\}$ is

$$P(f) = \frac{\sigma^2}{|1 + ae^{-j2\pi f}|^2}. \tag{14.31}$$

Thus, to find $P(f)$ it is sufficient to estimate a and σ^2.

The performance of a PSD estimator is evaluated by several measures of goodness. One is the bias of the estimator defined by

$$b(f) = E\left(\hat{P}(f) - P(f)\right) \tag{14.32}$$

where $\hat{P}(f)$ and $P(f)$ are the estimated and true PSD, respectively. If the bias $b(f)$ is identically equal to zero for all f, the estimator is said to be unbiased, which means that on average it yields the true PSD. Among the unbiased estimators, we search for the one that has minimal variability. The variability is measured by the variance of the estimator

$$v(f) = E\left([\hat{P}(f) - E(P(\hat{f}))]^2\right). \tag{14.33}$$

A measure that combines the bias and the variance is the relative mean square error given by [15]

$$v(f) = \frac{v(f) + b(f)^2}{P(f)}. \tag{14.34}$$

The variability of a PSD estimator is also measured by the normalized variance [8]

$$\psi(f) = \frac{v(f)}{E^2(\hat{P}(f))}. \tag{14.35}$$

Finally, another important metric for comparison is the resolution of the PSD estimators. It corresponds to the ability of the estimator to provide the fine details of the PSD of the random process. For example if the PSD of the random process has two peaks at frequencies f_1 and f_2, then the resolution of the estimator would be measured by the minimum separation of f_1 and f_2 for which the estimator still reproduces two peaks at f_1 and f_2.

14.4 Nonparametric Spectrum Estimation

When the method for PSD estimation is not based on any assumptions about the generation of the observed samples other than wide-sense stationarity, then it is termed a nonparametric estimator. According to Eq. (14.28), $P(f)$ can be obtained by first estimating the autocorrelation sequence from the observed samples $x[0], x[1], \cdots, x[N-1]$, and then applying the DTFT to these estimates. One estimator of the autocorrelation is given by

$$\hat{r}[k] = \frac{1}{N} \sum_{n=0}^{N-1-k} x^*[n]x[n+k], \quad 0 \le k \le N-1 . \tag{14.36}$$

The estimates of $\hat{r}[k]$ for $-N < k < 0$ are obtained from the identity

$$\hat{r}[-k] = \hat{r}^*[k] \tag{14.37}$$

and those for $|k| \ge N$ are set equal to zero. This estimator, although biased, has been preferred over others. An important reason for favoring it is that it always yields nonnegative estimates of the PSD, which is not the case with the unbiased estimator.

Many nonparametric estimators rely on using Eq. (14.36) and then transform the obtained autocorrelation sequence to estimate the PSD. Other nonparametric methods, however, operate directly on the observed data.

14.4.1 Periodogram

The periodogram was introduced by Schuster in 1898 when he was searching for hidden periodicities while studying sunspot data [19]. To find the periodogram of the data $\{x[n]\}_0^{N-1}$, first we determine the autocorrelation sequence $r[k]$ for $-(N-1) \le k \le N-1$ and then take the DTFT, i.e.,

$$\hat{P}_{\text{PER}}(f) = \sum_{k=-N+1}^{N-1} \hat{r}[k]e^{-j2\pi fk} . \tag{14.38}$$

It is more convenient to write the periodogram directly in terms of the observed samples $x[n]$. It is then defined as

$$\hat{P}_{\text{PER}}(f) = \frac{1}{N} \left| \sum_{n=0}^{N-1} x[n]e^{-j2\pi fn} \right|^2 . \tag{14.39}$$

Thus, the periodogram is proportional to the squared magnitude of the DTFT of the observed data. In practice, the periodogram is calculated by applying the FFT, which computes it at a discrete set of frequencies $\mathcal{D}_f = \{f_k : f_k = k/N, \ k = 0, 1, 2, \cdots, (N-1)\}$. The periodogram is then expressed by

$$\hat{P}_{\text{PER}}(f_k) = \frac{1}{N} \left| \sum_{n=0}^{N-1} x[n]e^{-j2\pi kn/N} \right|^2 , \qquad f_k \in \mathcal{D}_f . \tag{14.40}$$

To allow for finer frequency spacing in the computed periodogram, we define a zero padded sequence according to

$$x'[n] = \begin{cases} x[n], & n = 0, 1, \cdots, N-1 \\ 0, & n = N, N+1, \cdots, N' \end{cases} . \tag{14.41}$$

Then we specify the new set of frequencies $\mathcal{D}'_f = \{f_k : f_k = k/N', \ k \in \{0, 1, 2, \cdots, (N'-1)\}\}$, and obtain

$$\hat{P}_{\text{PER}}(f_k) = \frac{1}{N} \left| \sum_{n=0}^{N-1} x[n] e^{-j2\pi kn/N'} \right|^2 , \quad f_k \in \mathcal{D}'_f . \tag{14.42}$$

A general property of good estimators is that they yield better estimates when the number of observed data samples increases. Theoretically, if the number of data samples tends to infinity, the estimates should converge to the true values of the estimated parameters. So, in the case of a PSD estimator, as we get more and more data samples, it is desirable that the estimated PSD tends to the true value of the PSD. In other words, if for finite number of data samples the estimator is biased, the bias should tend to zero as $N \to \infty$ as should the variance of the estimate. If this is indeed the case, the estimator is called consistent. Although the periodogram is asymptotically unbiased, it can be shown that it is not a consistent estimator. For example, if $\{\tilde{x}[n]\}$ is real zero-mean white Gaussian noise, which is a process whose random variables are independent, Gaussian, and identically distributed with variance σ^2, the variance of $\hat{P}_{\text{PER}}(f)$ is equal to σ^4 regardless of the length N of the observed data sequence [12]. The performance of the periodogram does not improve as N gets larger because as N increases, so does the number of parameters that are estimated, $P(f_0), P(f_1), ..., P(f_{N-1})$. In general, for the variance of the periodogram, we can write [12]

$$\text{var}(\hat{P}_{\text{PER}}(f)) \simeq P^2(f) \tag{14.43}$$

where $P(f)$ is the true PSD.

Interesting insight can be gained if one writes the periodogram as follows

$$\begin{aligned} \hat{P}_{\text{PER}}(f) &= \frac{1}{N} \left| \sum_{n=0}^{N-1} x[n] e^{-j2\pi fn} \right|^2 \\ &= \frac{1}{N} \left| \sum_{n=-\infty}^{\infty} x[n] w_R[n] e^{-j2\pi fn} \right|^2 \end{aligned} \tag{14.44}$$

where $w_R[n]$ is a rectangular window defined by

$$w_R[n] = \begin{cases} 1, & n \in \{0, 1, \cdots, N-1\} \\ 0, & \text{otherwise} \end{cases} . \tag{14.45}$$

Thus, we can regard the finite data record used for estimating the PSD as being obtained by multiplying the whole realization of the random process by a rectangular window. Then it is not difficult to show that the expected value of the periodogram is given by [8]

$$E\left\{ \hat{P}_{\text{PER}}(f) \right\} = \frac{1}{N} \int_0^1 |W_R(f-\xi)|^2 P(\xi) d\xi \tag{14.46}$$

where $W_R(f)$ is the DTFT of the rectangular window. Hence, the mean value of the periodogram is a smeared version of the true PSD. Since the implementation of the periodogram as defined in Eq. (14.44) implies the use of a rectangular window, a question arises as to whether we could use a window of different shape to reduce the variance of the periodogram. The answer is yes, and indeed many windows have been proposed which weight the data samples in the middle of the observed data more than those towards the ends of the observed data. Some frequently used alternatives to the rectangular window are the windows of Bartlett, Hanning, Hamming, and Blackman. The magnitude of the DTFT of a window provides two important characteristics about it. One is the width of the window's mainlobe and the other is the strength of its sidelobes. A narrow mainlobe allows for a better resolution, and low sidelobes improve the smoothing of the estimated spectrum. Unfortunately, the narrower its mainlobe, the higher the sidelobes, which is a typical trade-off in spectrum estimation. It turns out that the rectangular window allows for the best resolution but has the largest sidelobes.

14.4.2 The Bartlett Method

One approach to reduce the variance of the periodogram is to subdivide the observed data record into K nonoverlapping segments, find the periodogram of each segment, and finally evaluate the average of the so-obtained periodograms. This spectrum estimator, also known as the Bartlett's estimator, has variance that is smaller than the variance of the periodogram.

Suppose that the number of data samples N is equal to KL, where K is the number of segments and L is their length. If the i-th segment is denoted by $\{x_i[n]\}_0^{L-1}, i = 1, 2 \cdots, K$, where

$$x_i[n] = x[n + (i - 1)L], \quad n \in \{0, 1, \cdots, L - 1\} \tag{14.47}$$

and its periodogram by

$$\hat{P}_{\text{PER}}^{(i)}(f) = \frac{1}{L} \left| \sum_{n=0}^{L-1} x_i[n] e^{-j2\pi fn} \right|^2 \tag{14.48}$$

then the Bartlett spectrum estimator is

$$\hat{P}_{\text{B}}(f) = \frac{1}{K} \sum_{i=1}^{K} \hat{P}_{\text{PER}}^{(i)}(f) . \tag{14.49}$$

This estimator is consistent and its variance compared to the variance of the periodogram is reduced by a factor of K. This reduction, however, is paid by a decrease in resolution. The Bartlett estimator has a resolution K times less than that of the periodogram. Thus, this estimator allows for a straightforward trading of resolution for variance.

14.4.3 The Welch Method

The Welch method is another estimator that exploits the periodogram. It is based on the same idea as Bartlett's approach of splitting the data into segments and finding the average of their periodograms. The difference is that the segments are overlapped, where the overlaps are usually 50% or 75% large, and the data within a segment are windowed. Let the length of the segments be L, the i-th segment be denoted again by $\{x_i[n]\}_0^{L-1}$, and the offset of successive sequences by D samples. Then

$$N = L + D(K - 1) \tag{14.50}$$

where N is the total number of observed samples and K the total number of sequences. Note that if there is no overlap, $K = N/L$, and if there is 50% overlap, $K = 2N/L - 1$. The i-th sequence is defined by

$$x_i[n] = x[n + (i - 1)D], \quad n \in \{0, 1, \cdots, L - 1\} \tag{14.51}$$

where $i = 1, 2, \cdots, K$, and its periodogram by

$$\hat{P}_{\text{M}}^{(i)}(f) = \frac{1}{L} \left| \sum_{n=0}^{L-1} w[n] x_i[n] e^{-j2\pi fn} \right|^2 . \tag{14.52}$$

Here $\hat{P}_M^{(i)}(f)$ is the modified periodogram of the data because the samples $x[n]$ are weighted by a non-rectangular window $w[n]$. The Welch spectrum estimate is then given by

$$\hat{P}_B(f) = \frac{1}{K} \sum_{i=1}^{K} \hat{P}_M^{(i)}(f) . \tag{14.53}$$

By permitting overlap of sequences, we can form more segments than in the case of Bartlett's method. Also, if we keep the same number of segments, the overlap allows for longer segments. The increased number of segments reduces the variance of the estimator, and the longer segments improve its resolution. Thus, with the Welch method we can trade reduction in variance for improvement in resolution in many more ways than with the Bartlett method. It can be shown that if the overlap is 50%, the variance of the Welch estimator is approximately 9/16 of the variance of the Bartlett estimator [8].

14.4.4 Blackman-Tukey Method

The periodogram can be expressed in terms of the estimated autocorrelation lags as

$$\hat{P}_{\text{PER}}(f) = \sum_{k=-(N-1)}^{N-1} \hat{r}[k]e^{-j2\pi fk} . \tag{14.54}$$

where

$$\hat{r}[k] = \begin{cases} \frac{1}{N} \sum_{n=0}^{N-1-k} x^*[n]x[n+k], & k = 0, 1, \cdots, N-1 \\ \hat{r}^*[-k], & k = -(N-1), -(N-2), \cdots, -1 \end{cases} . \tag{14.55}$$

From Eqs. (14.54) and (14.55) we see that the estimated autocorrelation lags are given the same weight in the periodogram regardless of the difference of their variances. From Eq. (14.55), however, it is obvious that the autocorrelations with smaller lags will be estimated more accurately than the ones with lags close to N because of the different number of terms that are used in the summation. For example, $\hat{r}[N-1]$ has only the term $x^*[0]x[n-1]$ compared to the N terms used in the computation of $\hat{r}[0]$. Therefore, the large variance of the periodogram can be ascribed to the large weight given to the poor autocorrelation estimates used in its evaluation.

Blackman and Tukey proposed to weight the autocorrelation sequence so that the autocorrelations with higher lags are weighted less [3]. Their estimator is given by

$$\hat{P}_{\text{BT}}(f) = \sum_{k=-(N-1)}^{N-1} w[k]\hat{r}[k]e^{-j2\pi fk} \tag{14.56}$$

where the window $w[k]$ is real nonnegative, symmetric, and nonincreasing with $|k|$, that is

1. $0 \le w[k] \le w[0] = 1$,
2. $w[-k] = w[k]$, and
3. $w[k] = 0$, $M < |k|, \ M \le N - 1$.

$\tag{14.57}$

Note that the symmetry property of $w[k]$ ensures that the spectrum is real.

The Blackman-Tukey estimator can be expressed in the frequency domain by the convolution

$$\hat{P}_{\text{BT}}(f) = \int_0^1 W(f - \xi)\hat{P}_{\text{PER}}(\xi)d\xi . \tag{14.58}$$

From Eq. (14.58) we deduce that the window's DTFT should satisfy

$$W(f) \geq 0, \quad f \in [0, 1) \tag{14.59}$$

so that the spectrum is guaranteed to be a nonnegative function, i.e.,

$$\hat{P}_{\text{BT}}(f) \geq 0, \quad |f| \leq \frac{1}{2}. \tag{14.60}$$

The bias, the variance, and the resolution of the Blackman-Tukey method depend on the applied window. For example, if the window is triangular (Bartlett),

$$w_B[k] = \begin{cases} \frac{M-|k|}{M}, & |k| \leq M \\ 0, & \text{otherwise} \end{cases} \tag{14.61}$$

and if $N \gg M \gg 1$, the variance of the Blackman-Tukey estimator is [12]

$$\text{var}(\hat{P}_{\text{BT}}(f)) \simeq \frac{2M}{3N} P^2(f) \tag{14.62}$$

where $P(f)$ is the true spectrum of the process. Compared to Eq. (14.43), it is clear that the variance of this estimator may be significantly smaller than the variance of the periodogram. However, as M decreases, so does the resolution of the Blackman-Tukey estimator.

14.4.5 Minimum Variance Spectrum Estimator

The periodogram [Eq. (14.44)] can also be written as

$$\begin{aligned} \hat{P}_{\text{PER}}(f) &= \frac{1}{N} \left| \mathbf{e}^H(f)\mathbf{x} \right|^2 \\ &= N \left| \mathbf{h}^H(f)\mathbf{x} \right|^2 \end{aligned} \tag{14.63}$$

where $\mathbf{e}(f)$ is an $N \times 1$ vector defined by

$$\mathbf{e}(f) = [1 \ e^{j2\pi f} \ e^{j4\pi f} \ \cdots \ e^{j2(N-1)\pi f}]^T \tag{14.64}$$

and $\mathbf{h}(f) = \mathbf{e}(f)/N$ with H denoting complex conjugate transpose. We could interpret $\mathbf{h}(f)$ as the impulse response of a finite impulse response (FIR) filter. It is easy to show that $\mathbf{h}(f)$ is a bandpass filter centered at f with a bandwidth of approximately $1/N$. Then starting with Eq. (14.63) we can show that the value of the periodogram at frequency f can be obtained by squaring the magnitude of the filter output at $N-1$. Such filters inherently exist for all the frequencies where the periodogram is evaluated, and they all have the same bandwidth. Thus, the periodogram may be viewed as a bank of FIR filters with equal bandwidths.

Capon proposed a spectrum estimator for processing large seismic arrays which, like the periodogram, can be interpreted as a bank of filters [5]. The width of these filters, however, is data dependent and optimized to minimize their response to components outside the band of interest. If the impulse response of the filter centered at f_0 is $\mathbf{h}(f_0)$, then it is desired to minimize

$$\rho = \int_0^1 |H(f)|^2 P(f) df \tag{14.65}$$

subject to the constraint

$$H(f_0) = 1 \tag{14.66}$$

where $H(f)$ is the DTFT of $\mathbf{h}(f_0)$. This is a constrained minimization problem, and the solution provides the optimal impulse response. When the solutions are used to determine the PSD of the observed data, we obtain the minimum variance (MV) spectrum estimator

$$\hat{P}_{MV}(f) = \frac{N}{\mathbf{e}^H(f)\hat{\mathbf{R}}^{-1}\mathbf{e}(f)} \tag{14.67}$$

where $\hat{\mathbf{R}}^{-1}$ is the inverse matrix of the $N \times N$ autocorrelation matrix $\hat{\mathbf{R}}$ defined by

$$\hat{\mathbf{R}} = \begin{bmatrix} \hat{r}[0] & \hat{r}[-1] & \hat{r}[-2] & \cdots & \hat{r}[-N+1] \\ \hat{r}[1] & \hat{r}[0] & \hat{r}[-1] & \cdots & \hat{r}[-N+2] \\ \vdots & \vdots & \vdots & \vdots & \vdots \\ \hat{r}[N-1] & \hat{r}[N-2] & \hat{r}[N-3] & \cdots & \hat{r}[0] \end{bmatrix}. \tag{14.68}$$

The length of the FIR filter does not have to be N, especially if we want to avoid the use of the unreliable estimates of $r[k]$. If the length of the filter's response is $p < N$, then the vector $\mathbf{e}(f)$, the autocorrelation matrix $\hat{\mathbf{R}}$, and the spectrum estimate $\hat{P}_{MV}(f)$ are defined by Eqs. (14.64), (14.68), and (14.67), respectively, with N replaced by p [12].

The MV estimator has better resolution than the periodogram and the Blackman-Tukey estimator. The resolution and the variance of the MV estimator depend on the choice of the filter length p. If p is large, the bandwidth of the filter is small, which allows for better resolution. A larger p, however, requires more autocorrelation lags in the autocorrelation matrix $\hat{\mathbf{R}}$, which increases the variance of the estimated spectrum. Again, we have a trade-off between resolution and variance.

14.4.6 Multiwindow Spectrum Estimator

Many efforts have been made to improve the performance of the periodogram by multiplying the data with a nonrectangular window. The introduction of such windows has been more or less ad hoc, although they have been constructed to have narrow mainlobes and low sidelobes. By contrast, Thomson has proposed a spectrum estimation method that also involves use of windows but is derived from fundamental principles. The method is based on the approximate solution of a Fredholm equation using an eigenexpansion [21]. The method amounts to applying multiple windows to the data, where the windows are discrete prolate spheroidal (Slepian) sequences. These sequences are orthogonal and their Fourier transforms have the maximum energy concentration in a given bandwidth W.

The multiwindow (MW) spectrum estimator is given by [21]

$$\hat{P}_{MW}(f) = \frac{1}{m} \sum_{i=0}^{m-1} \hat{P}_i(f) \tag{14.69}$$

where the $\hat{P}_i(f)$ is the i-th eigenspectrum defined by

$$\hat{P}_i(f) = \frac{1}{\lambda_i} \left| \sum_{n=0}^{N-1} x[n]w_i[n]e^{-j2\pi fn} \right|^2 \tag{14.70}$$

with $w_i[n]$ being the i-th Slepian sequence, λ_i the i-th Slepian eigenvalue, and W the analysis bandwidth. The steps for obtaining $\hat{P}_{MW}(f)$ are [22]

1. Selection of the analysis bandwidth W whose typical values are between $1.5/N$ and $20/N$. The number of windows m depends on the selected W, and is given by $\lfloor 2NW \rfloor$, where $\lfloor x \rfloor$ denotes the largest integer less than or equal to x. The spectrum estimator has a resolution equal to W.

2. Evaluation of the m eigenspectra according to Eq. (14.70), where the Slepian sequences and eigenvalues satisfy

$$\mathbf{C}\mathbf{w}_i = \lambda_i \mathbf{w}_i . \tag{14.71}$$

with the elements of the matrix $\mathbf{C}$ being given by

$$c_{mn} = \frac{\sin(2\pi W(m-n))}{\pi(m-n)}, \quad m, n = 1, 2, \cdots, N . \tag{14.72}$$

In the evaluation of the eigenspectra, only the Slepian sequences that correspond to the m largest eigenvalues of $\mathbf{C}$ are used.

3. Computation of the average spectrum according to Eq. (14.69). If the spectrum is mixed, that is, the observed data contain harmonics, the MW method uses a likelihood ratio test to determine if harmonics are present. If the test shows that there is a harmonic around the frequency f_0, the spectrum is reshaped by adding an impulse at f_0 followed by correction of the "local" spectrum for the inclusion of the impulse. For details, see [9, 21].

The MW method is consistent, and its variance for fixed W tends to zero as $1/N$ when $N \to \infty$. The variance, however, as well as the bias and the resolution, depend on the bandwidth W.

14.5 Parametric Spectrum Estimation

A philosophically different approach to spectrum estimation of a random process is the parametric one, which is based on the assumption that the process can be described by a parametric model. Based on the model, the spectrum of the process can then be expressed in terms of the parameters of the model. The approach thus consists of three steps: (1) selection of an appropriate parametric model (usually based on *a priori* knowledge about the process), (2) estimation of the model parameters, and (3) computation of the spectrum using the so-obtained parameters. In the literature the parametric spectrum estimation methods are known as high-resolution methods because they can achieve better resolution than the nonparametric methods.

The most frequently used models in the literature are the autoregressive (AR), the moving average (MA), the autoregressive moving average (ARMA), and the sum of harmonics (complex sinusoids) embedded in noise. With the AR model we assume that the observed data have been generated by a system whose input-output difference equation is given by

$$x[n] = -\sum_{k=1}^{p} a_k x[n-k] + e[n] \tag{14.73}$$

where $x[n]$ is the observed output of the system, $e[n]$ is the unobserved input of the system, and the a_k's are its coefficients. The input $e[n]$ is a zero mean white noise process with unknown variance σ^2, and p is the order of the system. This model is usually abbreviated as AR(p). The MA model is given by

$$x[n] = \sum_{k=0}^{q} b_k e[n-k] \tag{14.74}$$

where the b_k's denote the MA parameters, $e[n]$ is a zero mean white noise process with unknown variance σ^2, and q is the order of the model. The first MA coefficient b_0 is set usually to be $b_0 = 1$, and the model is denoted by MA(q). The ARMA model combines the AR and MA models and is described by

$$x[n] = -\sum_{k=1}^{p} a_k x[n-k] + \sum_{k=0}^{q} b_k e[n-k] . \tag{14.75}$$

Since the AR and MA orders are p and q, respectively, the model in Eq. (14.75) is referred to as ARMA (p, q). Finally, the model of complex sinusoids in noise is

$$x[n] = \sum_{i=1}^{m} A_i e^{j2\pi f_i n} + e[n], \quad n = 0, 1, \cdots, N - 1 \tag{14.76}$$

where m is the number of complex sinusoids, A_i and f_i are the complex amplitude and frequency of the i-th complex sinusoid, respectively, and $e[n]$ is a sample of a noise process, which is not necessarily white. Frequently, we assume that the samples $e[n]$ are generated by a certain parametric probability distribution whose parameters are unknown, or $e[n]$ itself is modeled as an AR, MA, or ARMA process.

14.5.1 Spectrum Estimation Based on Autoregressive Models

When the model of $x[n]$ is AR(p), the PSD of the process is given by

$$P_{AR}(f) = \frac{\sigma^2}{|1 + \sum_{k=1}^{p} a_k e^{-j2\pi fk}|^2}. \tag{14.77}$$

Thus, to find $P_{AR}(f)$ we need the estimates of the AR coefficients a_k and the noise variance σ^2.

If we multiply the two sides of Eq. (14.73) by $x^*[n - k]$, $k \geq 0$, and take their expectations, we obtain

$$E(x[n]x^*[n - k]) = -\sum_{l=1}^{p} a_l E(x[n - l]x^*[n - k]) + E(e[n]x^*[n - k]) \tag{14.78}$$

or

$$r[k] = \begin{cases} -\sum_{l=1}^{p} a_l r[k - l], & k > 0 \\ -\sum_{l=1}^{p} a_l r[k - l] + \sigma^2, & k = 0 \end{cases}. \tag{14.79}$$

The expressions in Eq. (14.79) are known as the Yule-Walker equations. To estimate the p unknown AR coefficients from Eq. (14.79), we need at least p equations as well as the estimates of the appropriate autocorrelations. The set of equations that requires the estimation of the minimum number of correlation lags is

$$\hat{\mathbf{R}}\mathbf{a} = -\hat{\mathbf{r}} \tag{14.80}$$

where $\hat{\mathbf{R}}$ is the $p \times p$ matrix

$$\hat{\mathbf{R}} = \begin{bmatrix} \hat{r}[0] & \hat{r}[-1] & \hat{r}[-2] & \cdots & \hat{r}[-p+1] \\ \hat{r}[1] & \hat{r}[0] & \hat{r}[-1] & \cdots & \hat{r}[-p+2] \\ \vdots & \vdots & \vdots & \vdots & \vdots \\ \hat{r}[p-1] & \hat{r}[p-2] & \hat{r}[p-3] & \cdots & \hat{r}[0] \end{bmatrix} \tag{14.81}$$

and

$$\hat{\mathbf{r}} = [\hat{r}[1]\ \hat{r}[2]\ \cdots\ \hat{r}[p]]^T. \tag{14.82}$$

The parameters $\mathbf{a}$ are estimated by

$$\hat{\mathbf{a}} = -\hat{\mathbf{R}}^{-1}\hat{\mathbf{r}} \tag{14.83}$$

and the noise variance is found from

$$\hat{\sigma}^2 = \hat{r}[0] + \sum_{k=1}^{p} a_k \hat{r}^*[k]. \tag{14.84}$$

The PSD estimate is obtained when $\hat{\mathbf{a}}$ and $\hat{\sigma}^2$ are substituted in Eq. (14.77). This approach for estimating the AR parameters is known in the literature as the autocorrelation method.

Many other AR estimation procedures have been proposed including the maximum likelihood method, the covariance method, and the Burg method [12]. Burg's work in the late sixties has a special place in the history of spectrum estimation because it kindled the interest in this field. Burg showed that the AR model provides an extrapolation of a known autocorrelation sequence $r[k]$, $|k| \leq p$, for $|k|$ beyond p so that the spectrum corresponding to the extrapolated sequence is the flattest of all spectra consistent with the $2p + 1$ known autocorrelations [4].

An important issue in finding the AR PSD is the order of the assumed AR model. There exist several model order selection procedures, but the most widely used are the Information Criterion A (AIC) due to Akaike [2] and the Information Criterion B (BIC), also known as the Minimum Description Length (MDL) principle, of Rissanen [16] and Schwarz [20]. According to the AIC criterion, the best model is the one that minimizes the function $AIC(k)$ over k defined by

$$AIC(k) = N \log \hat{\sigma}_k^2 + 2k \tag{14.85}$$

where k is the model order, and $\hat{\sigma}_k^2$ is the estimated noise variance of that model. Similarly, the MDL criterion chooses the order which minimizes the function $MDL(k)$ defined by

$$MDL(k) = N \log \hat{\sigma}_k^2 + k \log N \tag{14.86}$$

where N is the number of observed data samples. It is important to emphasize that the MDL rule can be derived if, as a criterion for model selection, we use the maximum *a posteriori* principle. It has been found that the AIC is an inconsistent criterion whereas the MDL rule is consistent. Consistency here means that the probability of choosing the correct model order tends to one as $N \to \infty$.

The AR-based spectrum estimation methods show very good performance if the processes are narrow-band and have sharp peaks in their spectra. Also, many good results have been reported when they are applied to short data records.

14.5.2 Spectrum Estimation Based on Moving Average Models

The PSD of a moving average process is given by

$$P_{\mathrm{MA}}(f) = \sigma^2 |1 + \sum_{k=1}^{q} b_k e^{-j2\pi f k}|^2 . \tag{14.87}$$

It is not difficult to show that the $r[k]$'s for $|k| > q$ of an MA(q) process are identically equal to zero, and that Eq. (14.87) can be expressed also as

$$P_{\mathrm{MA}}(f) = \sum_{k=-q}^{q} r[k] e^{-j2\pi f k} . \tag{14.88}$$

Thus, to find $\hat{P}_{\mathrm{MA}}(f)$ it would be sufficient to estimate the autocorrelations $r[k]$ and use the found estimates in Eq. (14.88). Obviously, this estimate would be identical to $\hat{P}_{\mathrm{BT}}(f)$ when the applied window is rectangular and of length $2q + 1$.

A different approach is to find the estimates of the unknown MA coefficients and σ^2 and use them in Eq. (14.87). The equations of the MA coefficients are nonlinear, which makes their estimation difficult. Durbin has proposed an approximate procedure that is based on a high order AR approximation of the MA process. First the data are modeled by an AR model of order L, where $L >> q$. Its coefficients are estimated from Eq. (14.83) and $\hat{\sigma}^2$ according to Eq. (14.84). Then the sequence $1, \hat{a}_1, \hat{a}_2, \cdots, \hat{a}_L$ is fitted with an AR(q) model, whose parameters are also estimated using the autocorrelation method. The estimated coefficients $\hat{b}_1, \hat{b}_2, \cdots, \hat{b}_q$ are subsequently substituted in Eq. (14.87) together with $\hat{\sigma}^2$.

Good results with MA models are obtained when the PSD of the process is characterized by broad peaks and sharp nulls. The MA models should not be used for processes with narrowband features.

14.5.3   Spectrum Estimation Based on Autoregressive Moving Average Models

The PSD of a process that is represented by the ARMA model is given by

$$P_{\text{ARMA}}(f) = \sigma^2 \frac{|1 + \sum_{k=1}^{q} b_k e^{-j2\pi f k}|^2}{|1 + \sum_{k=1}^{p} a_k e^{-j2\pi f k}|^2}. \tag{14.89}$$

The ML estimates of the ARMA coefficients are difficult to obtain, so we usually resort to methods that yield suboptimal estimates. For example, we can first estimate the AR coefficients based on the equation,

$$\begin{bmatrix} \hat{r}[q] & \hat{r}[q-1] & \cdots & \hat{r}[q-p+1] \\ \hat{r}[q+1] & \hat{r}[q] & \cdots & \hat{r}[q-p+2] \\ \vdots & \vdots & \vdots & \vdots \\ \hat{r}[M-1] & \hat{r}[M-2] & \cdots & \hat{r}[M-p] \end{bmatrix} \begin{bmatrix} a_1 \\ a_2 \\ \vdots \\ a_p \end{bmatrix} + \begin{bmatrix} \epsilon_{q+1} \\ \epsilon_{q+2} \\ \vdots \\ \epsilon_M \end{bmatrix} = - \begin{bmatrix} \hat{r}[q+1] \\ \hat{r}[q+2] \\ \vdots \\ \hat{r}[M] \end{bmatrix} \tag{14.90}$$

or

$$\hat{\mathbf{R}}\mathbf{a} + \boldsymbol{\epsilon} = -\hat{\mathbf{r}} \tag{14.91}$$

where ϵ_i is a term that models the errors in the Yule-Walker equations due to the estimation errors of the autocorrelation lags, and $M \geq p + q$. From Eq. (14.91), we can find the least squares estimates of $\mathbf{a}$ by

$$\hat{\mathbf{a}} = - \left(\hat{\mathbf{R}}^H \hat{\mathbf{R}} \right)^{-1} \hat{\mathbf{R}}^H \hat{\mathbf{r}}. \tag{14.92}$$

This procedure is known as the least-squares modified Yule-Walker equation method.   Once the AR coefficients are estimated, we can filter the observed data

$$y[n] = x[n] + \sum_{k=1}^{p} \hat{a}_k x[n-k] \tag{14.93}$$

and obtain a sequence that is approximately modeled by an MA(q) model. From the data $y[n]$ we can estimate the MA PSD by Eq. (14.88) and obtain the PSD estimate of the data $x[n]$

$$\hat{P}_{\text{ARMA}}(f) = \frac{\hat{P}_{\text{MA}}(f)}{|1 + \sum_{k=1}^{p} \hat{a}_k e^{-j2\pi f k}|^2} \tag{14.94}$$

or estimate the parameters $b_1, b_2, ..., b_q$ and σ^2 by Durbin's method, for example, and then use

$$\hat{P}_{\text{ARMA}}(f) = \hat{\sigma}^2 \frac{|1 + \sum_{k=1}^{q} \hat{b}_k e^{-j2\pi f k}|^2}{|1 + \sum_{k=1}^{p} \hat{a}_k e^{-j2\pi f k}|^2}. \tag{14.95}$$

The ARMA model has an advantage over the AR and MA models because it can better fit spectra with nulls and peaks. Its disadvantage is that it is more difficult to estimate its parameters than the parameters of the AR and MA models.

14.5.4 Pisarenko Harmonic Decomposition Method

Let the observed data represent m complex sinusoids in noise, i.e.,

$$x[n] = \sum_{i=1}^{m} A_i e^{j2\pi f_i n} + e[n], \quad n = 0, 1, \cdots, N-1 \tag{14.96}$$

where f_i is the frequency of the i-th complex sinusoid, A_i is the complex amplitude of the i-th sinusoid,

$$A_i = |A_i| e^{j\phi_i} \tag{14.97}$$

with ϕ_i being a random phase of the i-th complex sinusoid, and $e[n]$ is a sample of a zero mean white noise. The PSD of the process is a sum of the continuous spectrum of the noise and a set of impulses with area $|A_i|^2$ at the frequencies f_i, or

$$P(f) = \sum_{i=1}^{m} |A_i|^2 \delta(f - f_i) + P_e(f) \tag{14.98}$$

where $P_e(f)$ is the PSD of the noise process.

Pisarenko studied the model in Eq. (14.96) and found that the frequencies of the sinusoids can be obtained from the eigenvector corresponding to the smallest eigenvalue of the autocorrelation matrix. His method, known as Pisarenko harmonic decomposition (PHD), led to important insights and stimulated further work which resulted in many new procedures known today as "signal and noise subspace" methods.

When the noise $\{\tilde{e}[n]\}$ is zero mean white with variance σ^2, the autocorrelation of $\{\tilde{x}[n]\}$ can be written as

$$r[k] = \sum_{i=1}^{m} |A_i|^2 e^{j2\pi f_i k} + \sigma^2 \delta[k] \tag{14.99}$$

or the autocorrelation matrix can be represented by

$$\mathbf{R} = \sum_{i=1}^{m} |A_i|^2 \mathbf{e}_i \mathbf{e}_i^H + \sigma^2 \mathbf{I} \tag{14.100}$$

where

$$\mathbf{e}_i = \left[1 \; e^{j2\pi f_i} \; e^{j4\pi f_i} \; e^{j2\pi(N-1)f_i} \right]^T \tag{14.101}$$

and $\mathbf{I}$ is the identity matrix. It is seen that the autocorrelation matrix $\mathbf{R}$ is composed of the sum of signal and noise autocorrelation matrices

$$\mathbf{R} = \mathbf{R}_s + \sigma^2 \mathbf{I} \tag{14.102}$$

where

$$\mathbf{R}_s = \mathbf{E}\mathbf{P}\mathbf{E}^H \tag{14.103}$$

for

$$\mathbf{E} = [\mathbf{e}_1 \; \mathbf{e}_2 \; \cdots \; \mathbf{e}_m] \tag{14.104}$$

and $\mathbf{P}$ a diagonal matrix

$$\mathbf{P} = \text{diag} \left\{ |A_1|^2, \; |A_2|^2, \; \cdots, \; |A_m|^2 \right\}. \tag{14.105}$$

If the matrix $\mathbf{R}_s$ is $M \times M$, where $M \geq m$, its rank will be equal to the number of complex sinusoids m. Another important representation of the autocorrelation matrix $\mathbf{R}$ is via its eigenvalues and eigenvectors, i.e.,

$$\mathbf{R} = \sum_{i=1}^{m} (\lambda_i + \sigma^2) \mathbf{v}_i \mathbf{v}_i^H + \sum_{i=m+1}^{M} \sigma^2 \mathbf{v}_i \mathbf{v}_i^H \tag{14.106}$$

where the λ_i's, $i = 1, 2, \cdots, m$, are the nonzero eigenvalues of $\mathbf{R}_s$. Let the eigenvalues of $\mathbf{R}$ be arranged in decreasing order so that $\lambda_1 \geq \lambda_2 \geq \cdots \geq \lambda_M$, and let $\mathbf{v}_i$ be the eigenvector corresponding to λ_i. The space spanned by the eigenvectors $\mathbf{v}_i$, $i = 1, 2, \cdots, m$, is called the signal subspace, and the space spanned by $\mathbf{v}_i$, $i = m+1, m+2, \cdots, M$, the noise subspace. Since the set of eigenvectors are orthonormal, that is

$$\mathbf{v}_i^H \mathbf{v}_l = \begin{cases} 1, & i = l \\ 0, & i \neq l \end{cases} \tag{14.107}$$

the two subspaces are orthogonal. In other words if $\mathbf{s}$ is in the signal subspace, and $\mathbf{z}$ is in the noise subspace, then $\mathbf{s}^H \mathbf{z} = 0$.

Now suppose that the matrix $\mathbf{R}$ is $(m+1) \times (m+1)$. Pisarenko observed that the noise variance corresponds to the smallest eigenvalue of $\mathbf{R}$ and that the frequencies of the complex sinusoids can be estimated by using the orthogonality of the signal and noise subspaces, that is,

$$\mathbf{e}_i^H \mathbf{v}_{m+1} = 0, \quad i = 1, 2, \cdots, m . \tag{14.108}$$

We can estimate the f_i's by forming the pseudospectrum

$$\hat{P}_{\text{PHD}}(f) = \frac{1}{|\mathbf{e}^H(f)\mathbf{v}_{m+1}|^2} \tag{14.109}$$

which should theoretically be infinite at the frequencies f_i. In practice, however, the pseudospectrum does not exhibit peaks exactly at these frequencies because $\mathbf{R}$ is not known and, instead, is estimated from finite data records.

The PSD estimate in Eq. (14.109) does not include information about the power of the noise and the complex sinusoids. The powers, however, can easily be obtained by using Eq. (14.98). First note that $P_e(f) = \sigma^2$, and $\hat{\sigma}^2 = \lambda_{m+1}$. Second, the frequencies f_i are determined from the pseudospectrum Eq. (14.109), so it remains to find the powers of the complex sinusoids $P_i = |A_i|^2$. This can readily be accomplished by using the set of m linear equations

$$\begin{bmatrix} |\hat{\mathbf{e}}_1^H \mathbf{v}_1|^2 & |\hat{\mathbf{e}}_2^H \mathbf{v}_1|^2 & \cdots & |\hat{\mathbf{e}}_m^H \mathbf{v}_1|^2 \\ |\hat{\mathbf{e}}_1^H \mathbf{v}_2|^2 & |\hat{\mathbf{e}}_2^H \mathbf{v}_2|^2 & \cdots & |\hat{\mathbf{e}}_m^H \mathbf{v}_2|^2 \\ \vdots & \vdots & \vdots & \vdots \\ |\hat{\mathbf{e}}_1^H \mathbf{v}_m|^2 & |\hat{\mathbf{e}}_2^H \mathbf{v}_m|^2 & \cdots & |\hat{\mathbf{e}}_m^H \mathbf{v}_m|^2 \end{bmatrix} \begin{bmatrix} P_1 \\ P_2 \\ \vdots \\ P_m \end{bmatrix} = \begin{bmatrix} \lambda_1 - \hat{\sigma}^2 \\ \lambda_2 - \hat{\sigma}^2 \\ \vdots \\ \lambda_m - \hat{\sigma}^2 \end{bmatrix} \tag{14.110}$$

where

$$\hat{\mathbf{e}}_i = \begin{bmatrix} 1 & e^{j2\pi \hat{f}_i} & e^{j4\pi \hat{f}_i} & \cdots & e^{j2\pi(N-1)\hat{f}_i} \end{bmatrix}^T . \tag{14.111}$$

In summary, Pisarenko's method consists of four steps:

1. Estimate the $(m+1) \times (m+1)$ autocorrelation matrix $\mathbf{R}$ (provided it is known that the number of complex sinusoids is m).

2. Evaluate the minimum eigenvalue λ_{m+1} and the eigenvectors of $\hat{\mathbf{R}}$.

3. Set the white noise power to $\hat{\sigma}^2 = \lambda_{m+1}$, estimate the frequencies of the complex sinusoids from the peak locations of $\hat{P}_{\text{PHD}}(f)$ in Eq. (14.109), and compute their powers from Eq. (14.110).

4. Substitute the estimated parameters in Eq. (14.98).

Pisarenko's method is not used frequently in practice because its performance is much poorer than the performance of some other signal and noise subspace based methods developed later.

14.5.5   Multiple Signal Classification (MUSIC)

A procedure very similar to Pisarenko's is the MUltiple SIgnal Classification (MUSIC) method, which was proposed in the late 1970's by Schmidt [18]. Suppose again that the process $\{\tilde{x}[n]\}$ is described by m complex sinusoids in white noise. If we form an $M \times M$ autocorrelation matrix $\mathbf{R}$, find its eigenvalues and eigenvectors and rank them as before, then as mentioned in the previous subsection, its m eigenvectors corresponding to the m largest eigenvalues span the signal subspace, and the remaining eigenvectors, the noise subspace. According to MUSIC, we estimate the noise variance from the $M - m$ smallest eigenvalues of $\hat{\mathbf{R}}$

$$\hat{\sigma}^2 = \frac{1}{M-m} \sum_{i=m+1}^{M} \lambda_i \tag{14.112}$$

and the frequencies from the peak locations of the pseudospectrum

$$\hat{P}_{MU}(f) = \frac{1}{\sum_{i=m+1}^{M} |\mathbf{e}(f)^H \mathbf{v}_i|^2}. \tag{14.113}$$

It should be noted that there are other ways of estimating the f_i's. Finally the powers of the complex sinusoids are determined from Eq. (14.110), and all the estimated parameters substituted in Eq. (14.98).

MUSIC has better performance than Pisarenko's method because of the introduced averaging via the extra noise eigenvectors. The averaging reduces the statistical fluctuations present in Pisarenko's pseudospectrum, which arise due to the errors in estimating the autocorrelation matrix. These fluctuations can further be reduced by applying the Eigenvector method [11], which is a modification of MUSIC and whose pseudospectrum is given by

$$\hat{P}_{EV}(f) = \frac{1}{\sum_{i=m+1}^{M} |\frac{1}{\lambda_i}\mathbf{e}(f)^H \mathbf{v}_i|^2}. \tag{14.114}$$

Pisarenko's method, MUSIC, and its variants exploit the noise subspace to estimate the unknown parameters of the random process. There are, however, approaches that estimate the unknown parameters from vectors that lie in the signal subspace. The main idea there is to form a reduced rank autocorrelation matrix which is an estimate of the signal autocorrelation matrix. Since this estimate is formed from the m principal eigenvectors and eigenvalues, the methods based on them are called principal component spectrum estimation methods [8, 12]. Once the signal autocorrelation matrix is obtained, the frequencies of the complex sinusoids are found, followed by estimation of the remaining unknown parameters of the model.

14.6   Recent Developments

Spectrum estimation continues to attract the attention of many researchers. The answers to many interesting questions are still unknown, and many problems still need better solutions. The field of spectrum estimation is constantly enriched with new theoretical findings and a wide range of results obtained from examinations of various physical processes. In addition, new concepts are being introduced that provide tools for improved processing of the observed signals and that allow for a better understanding. Many new developments are driven by the need to solve specific problems that arise in applications, such as in sonar and communications.

Recently, for example, the notion of canonical autoregressive decomposition has been introduced [14]. It is a parametric approach for estimation of mixed spectra where the continuous part of the spectrum is modeled by an AR model. Another development is related to Bayesian spectrum estimation. Jaynes has introduced it in [10] and some interesting results for spectra of harmonics in white Gaussian noise have been reported in [7]. A Bayesian spectrum estimate is based on

$$\hat{P}_{BA}(f) = \int_{\Theta} P(f, \boldsymbol{\theta}) f(\boldsymbol{\theta} | \{x[n]\}_0^{N-1}) d\boldsymbol{\theta} \qquad (14.115)$$

where $P(f, \boldsymbol{\theta})$ is the theoretical parametric spectrum, $\boldsymbol{\theta}$ denotes the parameters of the process, Θ is the parameter space, and $f(\boldsymbol{\theta} | \{x[n]\}_0^{N-1})$ is the *a posteriori* probability density function of the process parameters. Therefore, the Bayesian spectrum estimate is defined as the expected value of the theoretical spectrum over the joint posterior density function of the model parameters.

The processes that we have addressed here are wide-sense stationary. The stationarity assumption, however, is often a mathematical abstraction and only an approximation in practice. Many physical processes are actually nonstationary and their spectra change with time. In biomedicine, speech analysis, and sonar, for example, it is typical to observe signals whose power during some time intervals is concentrated at high frequencies and, shortly thereafter, at low or middle frequencies. In such cases it is desirable to describe the PSD of the process at every instant of time, which is possible if we assume that the spectrum of the process changes smoothly over time. Such description requires a combination of the time- and frequency-domain concepts of signal processing into a single framework [6]. So there is an important distinction between the PSD estimation methods discussed here and the time-frequency representation approaches. The former provide the PSD of the process for all times, whereas the latter yield the local PSD's at every instant of time. This area of research is well developed but still far from complete. Although many theories have been proposed and developed, including evolutionary spectra [15], the Wigner-Wille method [13], and the kernel choice approach [1], time-varying spectrum analysis has remained a challenging and fascinating area of research.

References

[1] Amin, M.G., Time-frequency spectrum analysis and estimation for nonstationary random processes, in *Time-Frequency Signal Analysis*, B. Boashash, Ed., pp. 208–232, Longman Cheshire, 1992.

[2] Akaike, H., A new look at the statistical model identification, *IEEE Trans. Automatic Control*, Vol. AC-19, pp. 716–723, 1974.

[3] Blackman, R.B. and Tukey, J.W., *The Measurement of Power Spectra from the Point of View of Communications Engineering*, Dover Publications, New York, 1958.

[4] Burg, J.P., Maximum Entropy Spectral Analysis, Ph.D. dissertation, Stanford University, 1975.

[5] Capon, J., High-resolution frequency-wavenumber spectrum analysis, *Proc. IEEE*, Vol. 57, pp. 1408–1418, 1969.

[6] Cohen, L., *Time-Frequency Analysis*, Prentice Hall, Englewood Cliffs, NJ, 1995.

[7] Djurić, P.M. and Li, H.-T., Bayesian spectrum estimation of harmonic signals, *Signal Process. Lett.*, Vol. 2, pp. 213–215, 1995.

[8] Hayes, M.S., *Statistical Digital Signal Processing and Modeling*, John Wiley & Sons, New York, 1996.

[9] Haykin, S., *Advances in Spectrum Analysis and Array Processing*, Prentice Hall, Englewood Cliffs, NJ, 1991.

[10] Jaynes, E.T., Bayesian spectrum and chirp analysis, in *Maximum Entropy and Bayesian Spectral Analysis and Estimation Problems*, C. R. Smith and G. J. Erickson, Eds., pp. 1–37, D. Reidel, Dordrecht, Holland, 1987.

[11] Johnson, D.H. and DeGraaf, S.R., Improving the resolution of bearing in passive sonar arrays by eigenvalue analysis, *IEEE Trans. Acoustics, Speech, Signal Process.*, Vol. ASSP-30, pp. 638–647, 1982.

[12] Kay, S.M., *Modern Spectral Estimation*, Prentice Hall, Englewood Cliffs, NJ, 1988.

[13] Martin, W. and Flandrin, P., Wigner-Ville spectral analysis of nonstationary processes, *IEEE Trans. Acoustics, Speech, Signal Process.*, Vol. 33, pp. 1461–1470, 1985.

[14] Nagesha, V. and Kay, S.M., Spectral analysis based on the canonical autoregressive decomposition, *IEEE Trans. Signal Process.*, Vol. SP-44, pp. 1719–1733, 1996.

[15] Priestley, M.B., *Spectral Analysis and Time Series*, Academic Press, New York, 1981.

[16] Rissanen, J., Modeling by shortest data description, *Automatica*, Vol. 14, pp. 465–471, 1978.

[17] Robinson, E.A., A historical perspective of spectrum estimation, *Proc. IEEE*, Vol. 70, pp. 885–907, 1982.

[18] Schmidt, R., Multiple emitter location and signal parameter estimation, *Proc. RADC Spectrum Estimation Workshop*, pp. 243–258, 1979.

[19] Schuster, A., On the investigation on hidden periodicities with application to a supposed 26-day period of meteorological phenomena, *Terrestrial Magnetism*, Vol. 3, pp. 13–41, 1898.

[20] Schwarz, G., Estimating the dimension of the model, *Annals Statist.*, Vol. 6, pp. 461–464, 1978.

[21] Thomson, D.J., Spectrum estimation and harmonic analysis, *Proc. IEEE*, Vol. 70, pp. 1055–1096, 1982.

[22] Thomson, D.J., Quadratic-inverse spectrum estimates: applications to paleoclimatology, *Phil. Trans. R. Soc. London*, A, Vol. 332, pp. 539–597, 1990.

15

Estimation Theory and Algorithms: From Gauss to Wiener to Kalman

Jerry M. Mendel
University of Southern California

15.1 Introduction

Estimation is one of four modeling problems. The other three are representation (how something should be modeled), measurement (which physical quantities should be measured and how they should be measured), and validation (demonstrating confidence in the model). Estimation, which fits in between the problems of measurement and validation, deals with the determination of those physical quantities that cannot be measured from those that can be measured. We shall cover a wide range of estimation techniques including weighted least squares, best linear unbiased, maximum-likelihood, mean-squared, and maximum-*a posteriori*. These techniques are for parameter or state estimation or a combination of the two, as applied to either linear or nonlinear models.

The discrete-time viewpoint is emphasized in this chapter because: (1) much real data is collected in a digitized manner, so it is in a form ready to be processed by discrete-time estimation algorithms; and (2) the mathematics associated with discrete-time estimation theory is simpler than with continuous-time estimation theory. We view (discrete-time) estimation theory as the extension of classical signal processing to the design of discrete-time (digital) filters that process uncertain data in a optimal manner. Estimation theory can, therefore, be viewed as a natural adjunct to digital signal processing theory. Mendel [12] is the primary reference for all the material in this chapter.

Estimation algorithms process data and, as such, must be implemented on a digital computer. Our computation philosophy is, whenever possible, leave it to the experts. Many of our chapter's algorithms can be used with MATLAB^{TM} and appropriate toolboxes (MATLAB is a registered trademark of The MathWorks, Inc.). See [12] for specific connections between MATLAB^{TM} and toolbox M-files and the algorithms of this chapter.

The main model that we shall direct our attention to is linear in the unknown parameters, namely

$$\mathbf{Z}(k) = \mathbf{H}(k)\theta + \mathbf{V}(k) .$$ (15.1)

In this model, which we refer to as a "generic linear model," $\mathbf{Z}(k) = $ col $(z(k), z(k-1), \ldots, z(k-N+1))$, which is $N \times 1$, is called the measurement vector. Its elements are $z(j) = \mathbf{h}'(j)\theta + v(j)$; θ which is $n \times 1$, is called the parameter vector, and contains the unknown deterministic or random parameters that will be estimated using one or more of this chapter's techniques; $\mathbf{H}(k)$, which is $N \times n$, is called the observation matrix; and, $\mathbf{V}(k)$, which is $N \times 1$, is called the measurement noise vector. By convention, the argument "k" of $\mathbf{Z}(k)$, $\mathbf{H}(k)$, and $\mathbf{V}(k)$ denotes the fact that the last measurement used to construct (15.1) is the kth.

Examples of problems that can be cast into the form of the generic linear model are: identifying the impulse response coefficients in the convolutional summation model for a linear time-invariant system from noisy output measurements; identifying the coefficients of a linear time-invariant finite-difference equation model for a dynamical system from noisy output measurements; function approximation; state estimation; estimating parameters of a nonlinear model using a linearized version of that model; deconvolution; and identifying the coefficients in a discretized Volterra series representation of a nonlinear system.

The following estimation notation is used throughout this chapter: $\hat{\theta}(k)$ denotes an estimate of θ and $\tilde{\theta}(k)$ denotes the error in estimation, i.e., $\tilde{\theta}(k) = \theta - \hat{\theta}(k)$. The generic linear model is the starting point for the derivation of many classical parameter estimation techniques, and the estimation model for $\mathbf{Z}(k)$ is $\hat{\mathbf{Z}}(k) = \mathbf{H}(k)\hat{\theta}(k)$. In the rest of this chapter we develop specific structures for $\hat{\theta}(k)$. These structures are referred to as estimators. Estimates are obtained whenever data are processed by an estimator.

15.2   Least-Squares Estimation

The method of least squares dates back to Karl Gauss around 1795 and is the cornerstone for most estimation theory. The weighted least-squares estimator (WLSE), $\hat{\theta}_{WLS}(k)$, is obtained by minimizing the objective function $J[\hat{\theta}(k)] = \tilde{\mathbf{Z}}'(k)\mathbf{W}(k)\tilde{\mathbf{Z}}(k)$, where [using (15.1)] $\tilde{\mathbf{Z}}(k) = \mathbf{Z}(k) - \hat{\mathbf{Z}}(k) = \mathbf{H}(k)\tilde{\theta}(k) + \mathbf{V}(k)$, and weighting matrix $\mathbf{W}(k)$ must be symmetric and positive definite. This weighting matrix can be used to weight recent measurements more (or less) heavily than past measurements. If $\mathbf{W}(k) = c\mathbf{I}$, so that all measurements are weighted the same, then weighted least-squares reduces to least squares, in which case, we obtain $\hat{\theta}_{LS}(k)$. Setting $dJ[\hat{\theta}(k)]/d\hat{\theta}(k) = \mathbf{0}$, we find that:

$$\hat{\theta}_{WLS}(k) = \left[\mathbf{H}'(k)\mathbf{W}(k)\mathbf{H}(k)\right]^{-1}\mathbf{H}'(k)\mathbf{W}(k)\mathbf{Z}(k)$$ (15.2)

and, consequently,

$$\hat{\theta}_{LS}(k) = \left[\mathbf{H}'(k)\mathbf{H}(k)\right]^{-1}\mathbf{H}'(k)\mathbf{Z}(k)$$ (15.3)

Note, also, that $J[\hat{\theta}_{WLS}(k)] = \mathbf{Z}'(k)\mathbf{W}(k)\mathbf{Z}(k) - \hat{\theta}'_{WLS}(k)\mathbf{H}'(k)\mathbf{W}(k)\mathbf{H}(k)\hat{\theta}_{WLS}(k)$.

Matrix $\mathbf{H}'(k)\mathbf{W}(k)\mathbf{H}(k)$ must be nonsingular for its inverse in (15.2) to exist. This is true if $\mathbf{W}(k)$ is positive definite, as assumed, and $\mathbf{H}(k)$ is of maximum rank. We know that $\hat{\theta}_{WLS}(k)$ minimizes $J[\hat{\theta}_{WLS}(k)]$ because $d^2 J[\hat{\theta}(k)]/d\hat{\theta}^2(k) = 2\mathbf{H}'(k)\mathbf{W}(k)\mathbf{H}(k) > 0$, since $\mathbf{H}'(k)\mathbf{W}(k)\mathbf{H}(k)$ is invertible. Estimator $\hat{\theta}_{WLS}(k)$ processes the measurements $\mathbf{Z}(k)$ linearly; hence, it is referred to as a *linear estimator*. In practice, we do not compute $\hat{\theta}_{\mathrm{WLS}}(k)$ using (15.2), because computing the inverse of $\mathbf{H}'(k)\mathbf{W}(k)\mathbf{H}(k)$ is fraught with numerical difficulties. Instead, the so-called *normal equations* $[\mathbf{H}'(k)\mathbf{W}(k)\mathbf{H}(k)]\hat{\theta}_{\mathrm{WLS}}(k) = \mathbf{H}'(k)\mathbf{W}(k)\mathbf{Z}(k)$ are solved using stable algorithms from numerical linear algebra (e.g., [3] indicating that one approach to solving the normal equations is to convert the original least squares problem into an equivalent, easy-to-solve problem using orthogonal transformations such as Householder or Givens transformations). Note, also, that (15.2) and (15.3) apply to the estimation of either deterministic or random parameters, because nowhere in the derivation of $\hat{\theta}_{\mathrm{WLS}}(k)$ did we have to assume that θ was or was not random. Finally, note that WLSEs may not be invariant under changes of scale. One way to circumvent this difficulty is to use normalized data.

Least-squares estimates can also be computed using the singular-value decomposition (SVD) of matrix $\mathbf{H}(k)$. This computation is valid for both the overdetermined ($N < n$) and underdetermined ($N > n$) situations and for the situation when $\mathbf{H}(k)$ may or may not be of full rank. The SVD of $K \times M$ matrix $\mathbf{A}$ is:

$$\mathbf{U}'\mathbf{A}\mathbf{V} = \left[\begin{array}{c|c} \Sigma & 0 \\ \hline 0 & 0 \end{array} \right] \tag{15.4}$$

where $\mathbf{U}$ and $\mathbf{V}$ are unitary matrices, $\Sigma = \mathrm{diag}\,(\sigma_1, \sigma_2, \ldots, \sigma_r)$, and $\sigma_1 \geq \sigma_2 \geq \ldots \geq \sigma_r > 0$. The σ_i's are the singular values of $\mathbf{A}$, and r is the rank of $\mathbf{A}$. Let the SVD of $\mathbf{H}(k)$ be given by (15.4). Even if $\mathbf{H}(k)$ is not of maximum rank, then

$$\hat{\theta}_{LS}(k) = \mathbf{V} \left[\begin{array}{c|c} \Sigma^{-1} & 0 \\ \hline 0 & 0 \end{array} \right] \mathbf{U}'\mathbf{Z}(k) \tag{15.5}$$

where $\Sigma^{-1} = \mathrm{diag}\,(\sigma_1^{-1}\sigma_2^{-1}, \ldots, \sigma_r^{-1})$ and r is the rank of $\mathbf{H}(k)$. Additionally, in the overdetermined case,

$$\hat{\theta}_{LS}(k) = \sum_{i=1}^{r} \frac{\mathbf{v}_i(k)}{\sigma_i^2(k)} \mathbf{v}_i'(k)\mathbf{H}'(k)\mathbf{Z}(k) \tag{15.6}$$

Similar formulas exist for computing $\hat{\theta}_{\mathrm{WLS}}(k)$.

Equations (15.2) and (15.3) are batch equations, because they process all of the measurements at one time. These formulas can be made recursive in time by using simple vector and matrix partitioning techniques. The information form of the recursive WLSE is:

$$\hat{\theta}_{\mathrm{WLS}}(k+1) = \hat{\theta}_{\mathrm{WLS}}(k) + \mathbf{K_W}(k+1)[z(k+1) - \mathbf{h}'(k+1)\hat{\theta}_{\mathrm{WLS}}(k)] \tag{15.7}$$

$$\mathbf{K_W}(k+1) = \mathbf{P}(k+1)\mathbf{h}(k+1)w(k+1) \tag{15.8}$$

$$\mathbf{P}^{-1}(k+1) = \mathbf{P}^{-1}(k) + \mathbf{h}(k+1)w(k+1)\mathbf{h}'(k+1) \tag{15.9}$$

Equations (15.8) and (15.9) require the inversion of $n \times n$ matrix $\mathbf{P}$. If n is large, then this will be a costly computation. Applying a matrix inversion lemma to (15.9), one obtains the following alternative covariance form of the recursive WLSE: Equation (15.7), and

$$\mathbf{K_W}(k+1) = \mathbf{P}(k)\mathbf{h}(k+1) \left[\mathbf{h}'(k+1)\mathbf{P}(k)\mathbf{h}(k+1) + \frac{1}{w(k+1)} \right]^{-1} \tag{15.10}$$

$$\mathbf{P}(k+1) = \left[\mathbf{I} - \mathbf{K_W}(k+1)\mathbf{h}'(k+1) \right] \mathbf{P}(k) \tag{15.11}$$

Equations (15.7)–(15.9) or (15.7), (15.10), and (15.11), are initialized by $\hat{\theta}_{\mathrm{WLS}}(n)$ and $\mathbf{P}^{-1}(n)$, where $\mathbf{P}(n) = [\mathbf{H}'(n)\mathbf{W}(n)\mathbf{H}(n)]^{-1}$, and are used for $k = n, n+1, \ldots, N-1$.

Equation (15.7) can be expressed as

$$\hat{\theta}_{\mathrm{WLS}}(k+1) = \left[\mathbf{I} - \mathbf{K_W}(k+1)\mathbf{h}'(k+1) \right] \hat{\theta}_{\mathrm{WLS}}(k) + \mathbf{K_W}(k+1)z(k+1) \tag{15.12}$$

which demonstrates that the recursive WLSE is a time-varying digital filter that is excited by random inputs (i.e., the measurements), one whose plant matrix $[\mathbf{I} - \mathbf{K_W}(k+1)\mathbf{h}'(k+1)]$ may itself be random because

$\mathbf{K_W}(k + 1)$ and $\mathbf{h}(k + 1)$ may be random, depending upon the specific application. The random natures of these matrices make the analysis of this filter exceedingly difficult.

Two recursions are present in the recursive WLSEs. The first is the vector recursion for $\hat{\theta}_{\mathrm{WLS}}$ given by (15.7). Clearly, $\hat{\theta}_{\mathrm{WLS}}(k + 1)$ cannot be computed from this expression until measurement $z(k + 1)$ is available. The second is the matrix recursion for either $\mathbf{P}^{-1}$ given by (15.9) or $\mathbf{P}$ given by (15.11). Observe that values for these matrices can be precomputed *before* measurements are made. A digital computer implementation of (15.7)–(15.9) is $\mathbf{P}^{-1}(k + 1) \to \mathbf{P}(k + 1) \to \mathbf{K_W}(k + 1) \to \hat{\theta}_{\mathrm{WLS}}(k + 1)$, whereas for (15.7), (15.10), and (15.11), it is $\mathbf{P}(k) \to \mathbf{K_W}(k + 1) \to \hat{\theta}_{\mathrm{WLS}}(k + 1) \to \mathbf{P}(k + 1)$. Finally, the recursive WLSEs can even be used for $k = 0, 1, \ldots, N - 1$. Often $z(0) = 0$, or there is no measurement made at $k = 0$, so that we can set $z(0) = 0$. In this case we can set $w(0) = 0$, and the recursive WLSEs can be initialized by setting $\hat{\theta}_{\mathrm{WLS}}(0) = \mathbf{0}$ and $\mathbf{P}(0)$ to a diagonal matrix of very large numbers. This is very commonly done in practice. Fast fixed-order recursive least-squares algorithms that are based on the Givens rotation [3] and can be implemented using systolic arrays are described in [5] and the references therein.

15.3 Properties of Estimators

How do we know whether or not the results obtained from the WLSE, or for that matter any estimator, are good? To answer this question, we must make use of the fact that all estimators represent transformations of random data; hence, $\hat{\theta}(k)$ is itself random, so that its properties must be studied from a statistical viewpoint. This fact, and its consequences, which seem so obvious to us today, are due to the eminent statistician R.A. Fischer.

It is common to distinguish between small-sample and large-sample properties of estimators. The term "sample" refers to the number of measurements used to obtain $\hat{\theta}$, i.e., the dimension of $\mathbf{Z}$. The phrase "small sample" means any number of measurements (e.g., 1, 2, 100, 10^4, or even an infinite number), whereas the phrase "large sample" means "an infinite number of measurements." Large-sample properties are also referred to as asymptotic properties. If an estimator possesses as small-sample property, it also possesses the associated large-sample property; but the converse is not always true. Although large sample means an infinite number of measurements, estimators begin to enjoy large-sample properties for much fewer than an infinite number of measurements. How few usually depends on the dimension of θ, n, the memory of the estimators, and in general on the underlying, albeit unknown, probability density function.

A thorough study into $\hat{\theta}$ would mean determining its probability density function $p(\hat{\theta})$. Usually, it is too difficult to obtain $p(\hat{\theta})$ for most estimators (unless $\hat{\theta}$ is multivariate Gaussian); thus, it is customary to emphasize the first-and second-order statistics of $\hat{\theta}$ (or its associated error $\tilde{\theta} = \theta - \hat{\theta}$), the mean and the covariance.

Small-sample properties of an estimator are unbiasedness and efficiency. An estimator is unbiased if its mean value is tracking the unknown parameter at every value of time, i.e., the mean value of the estimation error is zero at every value of time. Dispersion about the mean is measured by error variance. Efficiency is related to how small the error variance will be. Associated with efficiency is the very famous Cramer-Rao inequality (Fisher information matrix, in the case of a vector of parameters) which places a lower bound on the error variance, a bound that does not depend on a particular estimator.

Large-sample properties of an estimator are asymptotic unbiasedness, consistency, asymptotic normality, and asymptotic efficiency. Asymptotic unbiasedness and efficiency are limiting forms of their small sample counterparts, unbiasedness and efficiency. The importance of an estimator being asymptotically normal (Gaussian) is that its entire probabilistic description is then known, and it can be entirely characterized just by its asymptotic first- and second-order statistics. Consistency is a form of convergence of $\hat{\theta}(k)$ to θ; it is synonymous with convergence in probability. One of the reasons for the importance of consistency in estimation theory is that any continuous function of a consistent estimator is itself a consistent estimator, i.e., "consistency carries over." It is also possible to examine other types of stochastic convergence for estimators, such as mean-squared convergence and convergence with probability 1. A general carry-over property does not exist for these two types of convergence; it must be established case-by case (e.g., [11]).

Generally speaking, it is very difficult to establish small sample or large sample properties for least-squares estimators, except in the very special case when $\mathbf{H}(k)$ and $\mathbf{V}(k)$ are statistically independent. While this condition is satisfied in the application of identifying an impulse response, it is violated in the important application of identifying the coefficients in a finite difference equation, as well as in many other important engineering applications. Many large sample properties of LSEs are determined by establishing that the LSE is equivalent to another estimator for which it is known that the large sample property holds true. We pursue this below.

Least-squares estimators require no assumptions about the statistical nature of the generic model. Consequently, the formula for the WLSE is easy to derive. The price paid for not making assumptions about the statistical nature of the generic linear model is great difficulty in establishing small or large sample properties of the resulting estimator.

15.4 Best Linear Unbiased Estimation

Our second estimator is both unbiased and efficient by design, and is a linear function of measurements $\mathbf{Z}(k)$. It is called a best linear unbiased estimator (BLUE), $\hat{\theta}_{\mathrm{BLU}}(k)$. As in the derivation of the WLSE, we begin with our generic linear model; but, now we make two assumptions about this model, namely: (1) $\mathbf{H}(k)$ must be deterministic, and (2) $\mathbf{V}(k)$ must be zero mean with positive definite known covariance matrix $\mathbf{R}(k)$. The derivation of the BLUE is more complicated than the derivation of the WLSE because of the design constraints; however, its performance analysis is much easier because we build good performance into its design.

We begin by assuming the following linear structure for $\hat{\theta}_{\mathrm{BLU}}(k)$, $\hat{\theta}_{\mathrm{BLU}}(k) = \mathbf{F}(k)\mathbf{Z}(k)$. Matrix $\mathbf{F}(k)$ is designed such that: (1) $\hat{\theta}_{\mathrm{BLU}}(k)$ is an unbiased estimator of θ, and (2) the error variance for each of the n parameters is minimized. In this way, $\hat{\theta}_{\mathrm{BLU}}(k)$ will be unbiased and efficient (within the class of linear estimators) by design. The resulting BLUE estimator is:

$$\hat{\theta}_{\mathrm{BLU}}(k) = [\mathbf{H}'(k)\mathbf{R}^{-1}(k)\mathbf{H}(k)]\mathbf{H}'(k)\mathbf{R}^{-1}(k)\mathbf{Z}(k) \tag{15.13}$$

A very remarkable connection exists between the BLUE and WLSE, namely, the BLUE of θ is the special case of the WLSE of θ when $\mathbf{W}(k) = \mathbf{R}^{-1}(k)$. Consequently, all results obtained in our section above for $\hat{\theta}_{\mathrm{WLS}}(k)$ can be applied to $\hat{\theta}_{\mathrm{BLU}}(k)$ by setting $\mathbf{W}(k) = \mathbf{R}^{-1}(k)$. Matrix $\mathbf{R}^{-1}(k)$ weights the contributions of precise measurements heavily and deemphasizes the contributions of imprecise measurements. The best linear unbiased estimation design technique has led to a weighting matrix that is quite sensible.

If $\mathbf{H}(k)$ is deterministic and $\mathbf{R}(k) = \sigma_v^2 \mathbf{I}$, then $\hat{\theta}_{\mathrm{BLU}}(k) = \hat{\theta}_{LS}(k)$. This result, known as the Gauss-Markov theorem, is important because we have connected two seemingly different estimators, one of which, $\hat{\theta}_{\mathrm{BLU}}(k)$, has the properties of unbiasedness and minimum variance by design; hence, in this case $\hat{\theta}_{LS}(k)$ inherits these properties.

In a recursive WLSE, matrix $\mathbf{P}(k)$ has no special meaning. In a recursive BLUE [which is obtained by substituting $\mathbf{W}(k) = \mathbf{R}^{-1}(k)$ into (15.7)–(15.9), or (15.7), (15.10) and (15.11)], matrix $\mathbf{P}(k)$ is the covariance matrix for the error between θ and $\hat{\theta}_{\mathrm{BLU}}(k)$, i.e., $\mathbf{P}(k) = [\mathbf{H}'(k)\mathbf{R}^{-1}(k)\mathbf{H}(k)]^{-1} = \mathrm{cov}\,[\tilde{\theta}_{\mathrm{BLU}}(k)]$. Hence, every time $\mathbf{P}(k)$ is calculated in the recursive BLUE, we obtain a quantitative measure of how well we are estimating θ.

Recall that we stated that WLSEs may change in numerical value under changes in scale. BLUEs are invariant under changes in scale. This is accomplished automatically by setting $\mathbf{W}(k) = \mathbf{R}^{-1}(k)$ in the WLSE.

The fact that $\mathbf{H}(k)$ must be deterministic severely limits the applicability of BLUEs in engineering applications.

15.5   Maximum-Likelihood Estimation

Probability is associated with a forward experiment in which the probability model, $p(\mathbf{Z}(k)|\theta)$, is specified, including values for the parameters, θ, in that model (e.g., mean and variance in a Gaussian density function), and data (i.e., realizations) are generated using this model. Likelihood, $l(\theta|\mathbf{Z}(k))$, is proportional to probability. In likelihood, the data is given as well as the nature of the probability model; but the parameters of the probability model are not specified. They must be determined from the given data. Likelihood is, therefore, associated with an inverse experiment.

The maximum-likelihood method is based on the relatively simple idea that different (statistical) populations generate different samples and that any given sample (i.e., set of data) is more likely to have come from some populations than from others.

In order to determine the maximum-likelihood estimate (MLE) of deterministic θ, $\hat{\theta}_{ML}$, we need to determine a formula for the likelihood function and then maximize that function. Because likelihood is proportional to probability, we need to know the entire joint probability density function of the measurements in order to determine a formula for the likelihood function. This, of course, is much more information about $\mathbf{Z}(k)$ than was required in the derivation of the BLUE. In fact, it is the most information that we can ever expect to know about the measurements. The price we pay for knowing so much information about $\mathbf{Z}(k)$ is complexity in maximizing the likelihood function. Generally, mathematical programming must be used in order to determine $\hat{\theta}_{ML}$.

Maximum-likelihood estimates are very popular and widely used because they enjoy very good large sample properties. They are consistent, asymptotically Gaussian with mean θ and covariance matrix $\frac{1}{N}\mathbf{J}^{-1}$, in which $\mathbf{J}$ is the Fisher information matrix, and are asymptotically efficient. Functions of maximum-likelihood estimates are themselves maximum-likelihood estimates, i.e., if $\mathbf{g}(\theta)$ is a vector function mapping θ into an interval in r-dimensional Euclidean space, then $\mathbf{g}(\hat{\theta}_{ML})$ is a MLE of $\mathbf{g}(\theta)$. This "invariance" property is usually not enjoyed by WLSEs or BLUEs.

In one special case it is very easy to compute $\hat{\theta}_{ML}$, i.e., for our generic linear model in which $\mathbf{H}(k)$ is deterministic and $\mathbf{V}(k)$ is Gaussian. In this case $\hat{\theta}_{ML} = \hat{\theta}_{BLU}$. These estimators are: unbiased, because $\hat{\theta}_{BLU}$ is unbiased; efficient (within the class of linear estimators), because $\hat{\theta}_{BLU}$ is efficient; consistent, because $\hat{\theta}_{ML}$ is consistent; and, Gaussian, because they depend linearly on $\mathbf{Z}(k)$, which is Gaussian. If, in addition, $\mathbf{R}(k) = \sigma_v^2\mathbf{I}$, then $\hat{\theta}_{ML}(k) = \hat{\theta}_{BLU}(k) = \hat{\theta}_{LS}(k)$, and these estimators are unbiased, efficient (within the class of linear estimators), consistent, and Gaussian.

The method of maximum-likelihood is limited to deterministic parameters. In the case of random parameters, we can still use the WLSE or the BLUE, or, if additional information is available, we can use either a mean-squared or maximum-*a posteriori* estimator, as described below. The former does not use statistical information about the random parameters, whereas the latter does.

15.6   Mean-Squared Estimation of Random Parameters

Given measurements $\mathbf{z}(1), \mathbf{z}(2), \ldots, \mathbf{z}(k)$, the mean-squared estimator (MSE) of random θ, $\hat{\theta}_{MS}(k) = \phi[\mathbf{z}(i), i = 1, 2, \ldots, k]$, minimizes the mean-squared error $J[\tilde{\theta}_{MS}(k)] = \mathbf{E}\{\tilde{\theta}'_{MS}(k)\tilde{\theta}_{MS}(k)\}$ [where $\tilde{\theta}_{MS}(k) = \theta - \hat{\theta}_{MS}(k)$]. The function $\phi[\mathbf{z}(i), i = 1, 2, \ldots, k]$ may be nonlinear or linear. Its exact structure is determined by minimizing $J[\tilde{\theta}_{MS}(k)]$.

The solution to this mean-squared estimation problem, which is known as the fundamental theorem of estimation theory is:

$$\hat{\theta}_{MS}(k) = \mathbf{E}\{\theta|\mathbf{Z}(k)\} \tag{15.14}$$

As it stands, (15.14) is not terribly useful for computing $\hat{\theta}_{MS}(k)$. In general, we must first compute $p[\theta|\mathbf{Z}(k)]$ and then perform the requisite number of integrations of $\theta p[\theta|\mathbf{Z}(k)]$ to obtain $\hat{\theta}_{MS}(k)$. It is useful to separate this computation into two major cases; (1) θ and $\mathbf{Z}(k)$ are jointly Gaussian — the Gaussian case, and (2) θ and $\mathbf{Z}(k)$ are not jointly Gaussian — the non-Gaussian case.

When θ and $\mathbf{Z}(k)$ are jointly Gaussian, the estimator that minimizes the mean-squared error is

$$\hat{\theta}_{MS}(k) = \mathbf{m}_\theta + \mathbf{P}_{\theta z}(k)\mathbf{P}_z^{-1}(k)\left[\mathbf{Z}(k) - \mathbf{m}_z(k)\right] \tag{15.15}$$

where $\mathbf{m}_\theta$ is the mean of θ, $\mathbf{m}_z(k)$ is the mean of $\mathbf{Z}(k)$, $\mathbf{P}_z(k)$ is the covariance matrix of $\mathbf{Z}(k)$, and $\mathbf{P}_{\theta z}(k)$ is the cross-covariance between θ and $\mathbf{Z}(k)$. Of course, to compute $\hat{\theta}_{MS}(k)$ using (15.15), we must somehow know all of these statistics, and we must be sure that θ and $\mathbf{Z}(k)$ are jointly Gaussian. For the generic linear model, $\mathbf{Z}(k) = \mathbf{H}(k)\theta + \mathbf{V}(k)$, in which $\mathbf{H}(k)$ is deterministic, $\mathbf{V}(k)$ is Gaussian noise with known invertible covariance matrix $\mathbf{R}(k)$, θ is Gaussian with mean $\mathbf{m}_\theta$ and covariance matrix $\mathbf{P}_\theta$, and, θ and $\mathbf{V}(k)$ are statistically independent, then θ and $\mathbf{Z}(k)$ are jointly Gaussian, and, (15.15) becomes

$$\hat{\theta}_{MS}(k) = \mathbf{m}_\theta + \mathbf{P}_\theta \mathbf{H}'(k)\left[\mathbf{H}(k)\mathbf{P}_\theta\mathbf{H}'(k) + \mathbf{R}(k)\right]^{-1}\left[\mathbf{Z}(k) - \mathbf{H}(k)\mathbf{m}_\theta\right] \tag{15.16}$$

where error-covariance matrix $\mathbf{P}_{MS}(k)$, which is associated with $\hat{\theta}_{MS}(k)$, is

$$\begin{aligned}
\mathbf{P}_{MS}(k) &= \mathbf{P}_\theta - \mathbf{P}_\theta\mathbf{H}'(k)\left[\mathbf{H}(k)\mathbf{P}_\theta\mathbf{H}'(k) + \mathbf{R}(k)\right]^{-1}\mathbf{H}(k)\mathbf{P}_\theta \\
&= \left[\mathbf{P}_\theta^{-1} + \mathbf{H}'(k)\mathbf{R}^{-1}(k)\mathbf{H}(k)\right]^{-1}.
\end{aligned} \tag{15.17}$$

Using (15.17) in (15.16), $\hat{\theta}_{MS}(k)$ can be reexpressed as

$$\hat{\theta}_{MS}(k) = \mathbf{m}_\theta + \mathbf{P}_{MS}(k)\mathbf{H}'(k)\mathbf{R}^{-1}(k)\left[\mathbf{Z}(k) - \mathbf{H}(k)\mathbf{m}_\theta\right] \tag{15.18}$$

Suppose θ and $\mathbf{Z}(k)$ are not jointly Gaussian and that we know $\mathbf{m}_\theta, \mathbf{m}_z(k), \mathbf{P}_z(k)$, and $\mathbf{P}_{\theta z}(k)$. In this case, the estimator that is constrained to be an affine transformation of $\mathbf{Z}(k)$ and that minimizes the mean-squared error is also given by (15.15).

We now know the answer to the following important question: When is the linear (affine) mean-squared estimator the same as the mean-squared estimator? The answer is when θ and $\mathbf{Z}(k)$ are jointly Gaussian. If θ and $\mathbf{Z}(k)$ are not jointly Gaussian, then $\hat{\theta}_{MS}(k) = \mathbf{E}\{\theta|\mathbf{Z}(k)\}$, which, in general, is a nonlinear function of measurements $\mathbf{Z}(k)$, i.e., it is a nonlinear estimator.

Associated with mean-squared estimation theory is the orthogonality principle: Suppose $\mathbf{f}[\mathbf{Z}(k)]$ is any function of the data $\mathbf{Z}(k)$; then the error in the mean-squared estimator is orthogonal to $\mathbf{f}[\mathbf{Z}(k)]$ in the sense that $\mathbf{E}\{[\theta - \hat{\theta}_{MS}(k)]\mathbf{f}'[\mathbf{Z}(k)]\} = \mathbf{0}$. A frequently encountered special case of this occurs when $\mathbf{f}[\mathbf{Z}(k)] = \hat{\theta}_{MS}(k)$, in which case $\mathbf{E}\{\tilde{\theta}_{MS}(k)\tilde{\theta}'_{MS}(k)\} = 0$.

When θ and $\mathbf{Z}(k)$ are jointly Gaussian, $\hat{\theta}_{MS}(k)$ in (15.15) has the following properties: (1) it is unbiased; (2) each of its components has the smallest error variance; (3) it is a "linear" (affine) estimator; (4) it is unique; and, (5) both $\hat{\theta}_{MS}(k)$ and $\tilde{\theta}_{MS}(k)$ are multivariate Gaussian, which means that these quantities are completely characterized by their first- and second-order statistics. Tremendous simplifications occur when θ and $\mathbf{Z}(k)$ are jointly Gaussian!

Many of the results presented in this section are applicable to objective functions other than the mean-squared objective function. See the supplementary material at the end of Lesson 13 in [12] for discussions on a wide number of objective functions that lead to $\mathbf{E}\{\theta|\mathbf{Z}(k)\}$ as the optimal estimator of θ, as well as discussions on a full-blown nonlinear estimator of θ.

There is a connection between the BLUE and the MSE. The connection requires a slightly different BLUE, one that incorporates the *a priori* statistical information about random θ. To do this, we treat $\mathbf{m}_\theta$ as an additional measurement that is augmented to $\mathbf{Z}(k)$. The additional measurement equation is obtained by adding and subtracting θ in the identity $\mathbf{m}_\theta = \mathbf{m}_\theta$, i.e., $\mathbf{m}_\theta = \theta + (\mathbf{m}_\theta - \theta)$. Quantity $(\mathbf{m}_\theta - \theta)$ is now treated as zero-mean measurement noise with covariance $\mathbf{P}_\theta$. The augmented linear model is

$$\begin{pmatrix} \mathbf{Z}(k) \\ \hline \mathbf{m}_\theta \end{pmatrix} = \begin{pmatrix} \mathbf{H}(k) \\ \hline \mathbf{I} \end{pmatrix}\theta + \begin{pmatrix} \mathbf{V}(k) \\ \hline \mathbf{m}_\theta - \theta \end{pmatrix} \tag{15.19}$$

Let the BLUE estimator for this augmented model be denoted $\hat{\theta}^a_{\text{BLU}}(k)$. Then it is always true that $\hat{\theta}_{MS}(k) = \hat{\theta}^a_{\text{BLU}}(k)$. Note that the weighted least-squares objective function that is associated with $\hat{\theta}^a_{\text{BLU}}(k)$ is $J_a[\hat{\theta}^a(k)] = [\mathbf{m}_\theta - \hat{\theta}^a(k)]'\mathbf{P}_\theta^{-1}[\mathbf{m}_\theta - \hat{\theta}^a(k)] + \tilde{\mathbf{Z}}'(k)\mathbf{R}^{-1}(k)\tilde{\mathbf{Z}}(k)$.

15.7   Maximum A Posteriori Estimation of Random Parameters

Maximum *a posteriori* (MAP) estimation is also known as Bayesian estimation. Recall Bayes's rule: $p(\theta|\mathbf{Z}(k)) = p(\mathbf{Z}(k)|\theta) p(\theta)/p(\mathbf{Z}(k))$ in which density function $p(\theta|\mathbf{Z}(k))$ is known as the *a posteriori* (or posterior) conditional density function, and $p(\theta)$ is the prior density function for θ. Observe that $p(\theta|\mathbf{Z}(k))$ is related to likelihood function $l\{\theta|\mathbf{Z}(k)\}$, because $l\{\theta|\mathbf{Z}(k)\} \propto p(\mathbf{Z}(k)|\theta)$. Additionally, because $p(\mathbf{Z}(k))$ does not depend on θ, $p(\theta|\mathbf{Z}(k)) \propto p(\mathbf{Z}(k)|\theta) p(\theta)$. In MAP estimation, values of θ are found that maximize $p(\mathbf{Z}(k)|\theta) p(\theta)$. Obtaining a MAP estimate involves specifying both $p(\mathbf{Z}(k)|\theta)$ and $p(\theta)$ and finding the value of θ that maximizes $p(\theta|\mathbf{Z}(k))$. It is the knowledge of the *a priori* probability model for θ, $p(\theta)$, that distinguishes the problem formulation for MAP estimation from MS estimation.

If $\theta_1, \theta_2, \ldots, \theta_n$ are uniformly distributed, then $p(\theta|\mathbf{Z}(k)) \propto p(\mathbf{Z}(k)|\theta)$, and the MAP estimator of θ equals the ML estimator of θ. Generally, MAP estimates are quite different from ML estimates. For example, the invariance property of MLEs usually does not carry over to MAP estimates. One reason for this can be seen from the formula $p(\theta|\mathbf{Z}(k)) \propto p(\mathbf{Z}(k)|\theta) p(\theta)$. Suppose, for example, that $\phi = \mathbf{g}(\theta)$ and we want to determine $\hat{\phi}_{\text{MAP}}$ by first computing $\hat{\theta}_{\text{MAP}}$. Because $p(\theta)$ depends on the Jacobian matrix of $\mathbf{g}^{-1}(\phi)$, $\hat{\phi}_{\text{MAP}} \neq \mathbf{g}(\hat{\theta}_{\text{MAP}})$. Usually $\hat{\theta}_{\text{MAP}}$ and $\hat{\theta}_{ML}(k)$ are asymptotically identical to one another since in the large sample case the knowledge of the observations tends to swamp the knowledge of the prior distribution [10].

Generally speaking, optimization must be used to compute $\hat{\theta}_{\text{MAP}}(k)$. In the special but important case, when $\mathbf{Z}(k)$ and θ are jointly Gaussian, then $\hat{\theta}_{\text{MAP}}(k) = \hat{\theta}_{MS}(k)$. This result is true regardless of the nature of the model relating θ to $\mathbf{Z}(k)$. Of course, in order to use it, we must first establish that $\mathbf{Z}(k)$ and θ are jointly Gaussian. Except for the generic linear model, this is very difficult to do.

When $\mathbf{H}(k)$ is deterministic, $\mathbf{V}(k)$ is white Gaussian noise with known covariance matrix $\mathbf{R}(k)$, and θ is multivariate Gaussian with known mean $\mathbf{m}_\theta$ and covariance $\mathbf{P}_\theta$, $\hat{\theta}_{\text{MAP}}(k) = \hat{\theta}_{\text{BLU}}^a(k)$; hence, for the generic linear Gaussian model, MS, MAP, and BLUE estimates of θ are all the same, i.e., $\hat{\theta}_{MS}(k) = \hat{\theta}_{\text{BLU}}^a(k) = \hat{\theta}_{\text{MAP}}(k)$.

15.8   The Basic State-Variable Model

In the rest of this chapter we shall describe a variety of mean-squared state estimators for a linear, (possibly) time-varying, discrete-time, dynamical system, which we refer to as the basic state-variable model. This system is characterized by $n \times 1$ state vector $\mathbf{x}(k)$ and $m \times 1$ measurement vector $\mathbf{z}(k)$, and is:

$$\mathbf{x}(k + 1) = \Phi(k + 1, k)\mathbf{x}(k) + \Gamma(k + 1, k)\mathbf{w}(k) + \Psi(k + 1, k)\mathbf{u}(k) \tag{15.20}$$

$$\mathbf{z}(k + 1) = \mathbf{H}(k + 1)\mathbf{x}(k + 1) + \mathbf{v}(k + 1) \tag{15.21}$$

where $k = 0, 1, \ldots$. In this model $\mathbf{w}(k)$ and $\mathbf{v}(k)$ are $p \times 1$ and $m \times 1$ mutually uncorrelated (possibly nonstationary) jointly Gaussian white noise sequences; i.e., $\mathbf{E}\{\mathbf{w}(i)\mathbf{w}'(j)\} = \mathbf{Q}(i)\delta_{ij}$, $\mathbf{E}\{\mathbf{v}(i)\mathbf{v}'(j)\} = \mathbf{R}(i)\delta_{ij}$, and $\mathbf{E}\{\mathbf{w}(i)\mathbf{v}'(j)\} = \mathbf{S} = \mathbf{0}$, for all i and j. Covariance matrix $\mathbf{Q}(i)$ is positive semidefinite and $\mathbf{R}(i)$ is positive definite [so that $\mathbf{R}^{-1}(i)$ exists]. Additionally, $\mathbf{u}(k)$ is an $l \times 1$ vector of known system inputs, and initial state vector $\mathbf{x}(0)$ is multivariate Gaussian, with mean $\mathbf{m_x}(0)$ and covariance $\mathbf{P_x}(0)$, and $\mathbf{x}(0)$ is not correlated with $\mathbf{w}(k)$ and $\mathbf{v}(k)$. The dimensions of matrices Φ, Γ, Ψ, $\mathbf{H}$, $\mathbf{Q}$, and $\mathbf{R}$ are $n \times n, n \times p, n \times l, m \times n, p \times p$, and $m \times m$, respectively. The double arguments in matrices Φ, Γ, and Ψ may not always be necessary, in which case we replace $(k + 1, k)$ by k.

Disturbance $\mathbf{w}(k)$ is often used to model disturbance forces acting on the system, errors in modeling the system, or errors due to actuators in the translation of the known input, $\mathbf{u}(k)$, into physical signals. Vector $\mathbf{v}(k)$ is often used to model errors in measurements made by sensing instruments, or unavoidable disturbances that act directly on the sensors.

Not all systems are described by this basic model. In general, $\mathbf{w}(k)$ and $\mathbf{v}(k)$ may be correlated, some measurements may be made so accurate that, for all practical purposes, they are "perfect" (i.e., no measurement noise is associated with them), and either $\mathbf{w}(k)$ or $\mathbf{v}(k)$, or both, may be nonzero mean or colored noise processes. How to handle these situations is described in Lesson 22 of [12].

When $\mathbf{x}(0)$ and $\{\mathbf{w}(k), k = 0, 1, \ldots\}$ are jointly Gaussian, then $\{\mathbf{x}(k), k = 0, 1, \ldots\}$ is a Gauss-Markov sequence. Note that if $\mathbf{x}(0)$ and $\mathbf{w}(k)$ are individually Gaussian and statistically independent, they will be jointly Gaussian. Consequently, the mean and covariance of the state vector completely characterize it. Let $\mathbf{m}_{\mathbf{X}}(k)$ denote the mean of $\mathbf{x}(k)$. For our basic state-variable model, $\mathbf{m}_{\mathbf{X}}(k)$ can be computed from the vector recursive equation

$$\mathbf{m}_{\mathbf{X}}(k + 1) = \Phi(k + 1, k)\mathbf{m}_{\mathbf{X}}(k) + \Psi(k + 1, k)\mathbf{u}(k) \tag{15.22}$$

where $k = 0, 1, \ldots$, and $\mathbf{m}_{\mathbf{X}}(0)$ initializes (15.22). Let $\mathbf{P}_{\mathbf{X}}(k)$ denote the covariance matrix of $\mathbf{x}(k)$. For our basic state-variable model, $\mathbf{P}_{\mathbf{X}}(k)$ can be computed from the matrix recursive equation

$$\mathbf{P}_{\mathbf{X}}(k + 1) = \Phi(k + 1, k)\mathbf{P}_{\mathbf{X}}(k)\Phi'(k + 1, k) + \Gamma(k + 1, k)\mathbf{Q}(k)\Gamma'(k + 1, k) \tag{15.23}$$

where $k = 0, 1, \ldots$, and $\mathbf{P}_{\mathbf{X}}(0)$ initializes (15.23). Equations (15.22) and (15.23) are easily programmed for a digital computer.

For our basic state-variable model, when $\mathbf{x}(0)$, $\mathbf{w}(k)$, and $\mathbf{v}(k)$ are jointly Gaussian, then $\{\mathbf{z}(k), k = 1, 2, \ldots\}$ is Gaussian, and

$$\mathbf{m}_{\mathbf{Z}}(k + 1) = \mathbf{H}(k + 1)\mathbf{m}_{\mathbf{X}}(k + 1) \tag{15.24}$$

and

$$\mathbf{P}_{\mathbf{Z}}(k + 1) = \mathbf{H}(k + 1)\mathbf{P}_{\mathbf{X}}(k + 1)\mathbf{H}'(k + 1) + \mathbf{R}(k + 1) \tag{15.25}$$

where $\mathbf{m}_{\mathbf{X}}(k + 1)$ and $\mathbf{P}_{\mathbf{X}}(k + 1)$ are computed from (15.22) and (15.23), respectively.

For our basic state-variable model to be stationary, it must be time-invariant, and the probability density functions of $\mathbf{w}(k)$ and $\mathbf{v}(k)$ must be the same for all values of time. Because $\mathbf{w}(k)$ and $\mathbf{v}(k)$ are zero-mean and Gaussian, this means that $\mathbf{Q}(k)$ must equal the constant matrix $\mathbf{Q}$ and $\mathbf{R}(k)$ must equal the constant matrix $\mathbf{R}$. Additionally, either $\mathbf{x}(0) = \mathbf{0}$ or $\Phi(k, 0)\mathbf{x}(0) \approx \mathbf{0}$ when $k > k_0$; in both cases $\mathbf{x}(k)$ will be in its steady-state regime, so stationarity is possible.

If the basic state-variable model is time-invariant and stationary and if Φ is associated with an asymptotically stable system (i.e., one whose poles all lie within the unit circle), then [1] matrix $\mathbf{P}_{\mathbf{X}}(k)$ reaches a limiting (steady-state) solution $\bar{\mathbf{P}}_{\mathbf{X}}$ and $\bar{\mathbf{P}}_{\mathbf{X}}$ is the solution of the following steady-state version of (15.23): $\bar{\mathbf{P}}_{\mathbf{X}} = \Phi\bar{\mathbf{P}}_{\mathbf{X}}\Phi' + \Gamma\mathbf{Q}\Gamma'$. This equation is called a discrete-time Lyapunov equation.

15.9 State Estimation for the Basic State-Variable Model

Prediction, filtering, and smoothing are three types of mean-squared state estimation that have been developed since 1959. A predicted estimate of a state vector $\mathbf{x}(k)$ uses measurements which occur earlier than t_k and a model to make the transition from the last time point, say t_j, at which a measurement is available, to t_k. The success of prediction depends on the quality of the model. In state estimation we use the state equation model. Without a model, prediction is dubious at best.

A recursive mean-squared state filter is called a Kalman filter, because it was developed by Kalman around 1959 [9]. Although it was originally developed within a community of control theorists, and is regarded as the most widely used result of so-called "modern control theory," it is no longer viewed as a control theory result. It is a result within estimation theory; consequently, we now prefer to view it as a signal processing result. A filtered estimate of state vector $\mathbf{x}(k)$ uses all of the measurements up to and including the one made at time t_k.

A smoothed estimate of state vector $\mathrm{x}(k)$ not only uses measurements which occur earlier than t_k plus the one at t_k, but also uses measurements to the right of t_k. Consequently, smoothing can never be carried out in real time, because we have to collect "future" measurements before we can compute a smoothed estimate. If we don't look too far into the future, then smoothing can be performed subject to a delay of LT seconds, where T is our data sampling time and L is a fixed positive integer that describes how many sample points to the right of t_k are to be used in smoothing.

Depending upon how many future measurements are used and how they are used, it is possible to create three types of smoother: (1) the fixed-interval smoother, $\hat{\mathbf{x}}(k|N)$, $k = 0, 1, \ldots, N - 1$, where N is a fixed positive integer; (2) the fixed-point smoother, $\hat{\mathbf{x}}(k|j)$, $j = k + 1, k + 2, \ldots$, where k is a fixed positive integer; and (3) the fixed-lag smoother, $\hat{\mathbf{x}}(k|k + L)$, $k = 0, 1, \ldots$, where L is a fixed positive integer.

15.9.1 Prediction

A single-stage predicted estimate of $\mathbf{x}(k)$ is denoted $\hat{\mathbf{x}}(k|k - 1)$. It is the mean-squared estimate of $\mathbf{x}(k)$ that uses all the measurements up to and including the one made at time t_{k-1}; hence, a single-stage predicted estimate looks exactly one time point into the future. This estimate is needed by the Kalman filter. From the fundamental theorem of estimation theory, we know that $\hat{\mathbf{x}}(k|k - 1) = \mathbf{E}\{\mathbf{x}(k)|Z(k - 1)\}$ where $\mathbf{Z}(k - 1) = \text{col} \left(\mathbf{z}(1), \mathbf{z}(2), \ldots, \mathbf{z}(k - 1)\right)$, from which it follows that

$$\hat{\mathbf{x}}(k|k - 1) = \Phi(k, k - 1)\hat{\mathbf{x}}(k - 1|k - 1) + \Psi(k, k - 1)\mathbf{u}(k - 1) \qquad (15.26)$$

where $k = 1, 2, \ldots$. Observe that $\hat{\mathbf{x}}(k|k - 1)$ depends on the filtered estimate $\hat{\mathbf{x}}(k - 1|k - 1)$ of the preceding state vector $\mathbf{x}(k - 1)$. Therefore, Equation (15.26) cannot be used until we provide the Kalman filter.

Let $\mathbf{P}(k|k - 1)$ denote the error-covariance matrix that is associated with $\hat{\mathbf{x}}(k|k - 1)$, i.e.,

$$\mathbf{P}(k|k - 1) = \mathbf{E}\left\{[\tilde{\mathbf{x}}(k|k - 1) - \mathbf{m}_{\tilde{\mathbf{x}}}(k|k - 1)]\,[\tilde{\mathbf{x}}(k|k - 1) - \mathbf{m}_{\tilde{\mathbf{x}}}(k|k - 1)]'\right\} ,$$

where $\tilde{\mathbf{x}}(k|k - 1) = \mathbf{x}(k) - \hat{\mathbf{x}}(k|k - 1)$. Additionally, let $\mathbf{P}(k - 1|k - 1)$ denote the error-covariance matrix that is associated with $\hat{\mathbf{x}}(k - 1|k - 1)$, i.e.,

$$\mathbf{P}(k - 1|k - 1) = \mathbf{E}\left\{[\tilde{\mathbf{x}}(k - 1|k - 1) - \mathbf{m}_{\tilde{\mathbf{x}}}(k - 1|k - 1)]\,[\tilde{\mathbf{x}}(k - 1|k - 1) - \mathbf{m}_{\tilde{\mathbf{x}}}(k - 1|k - 1)]'\right\} ,$$

where $\tilde{\mathbf{x}}(k - 1|k - 1) = \mathbf{x}(k - 1) - \hat{\mathbf{x}}(k - 1|k - 1)$. Then

$$\mathbf{P}(k|k - 1) = \Phi(k, k - 1)\mathbf{P}(k - 1|k - 1)\Phi'(k, k - 1) + \Gamma(k, k - 1)\mathbf{Q}(k - 1)\Gamma'(k, k - 1) \qquad (15.27)$$

where $k = 1, 2, \ldots$.

Observe, from (15.26) and (15.27), that $\hat{\mathbf{x}}(0|0)$ and $\mathbf{P}(0|0)$ initialize the single-stage predictor and its error covariance, where $\hat{\mathbf{x}}(0|0) = \mathbf{m}_{\mathbf{x}}(0)$ and $\mathbf{P}(0|0) = \mathbf{P}(0)$.

A more general state predictor is possible, one that looks further than just one step. See ([12] Lesson 16) for its details.

The single-stage predicted estimate of $\mathbf{z}(k + 1)$, $\hat{\mathbf{z}}(k + 1|k)$, is given by $\hat{\mathbf{z}}(k + 1|k) = \mathbf{H}(k + 1)\hat{\mathbf{x}}(k + 1|k)$. The error between $\mathbf{z}(k + 1)$ and $\hat{\mathbf{z}}(k + 1|k)$, is $\tilde{\mathbf{z}}(k + 1|k)$; $\tilde{\mathbf{z}}(k + 1|k)$ is called the innovations process (or, prediction error process, or, measurement residual process), and this process plays a very important role in mean-squared filtering and smoothing. The following representations of the innovations process $\tilde{\mathbf{z}}(k + 1|k)$ are equivalent:

$$\begin{aligned}
\tilde{\mathbf{z}}(k + 1|k) &= \mathbf{z}(k + 1) - \hat{\mathbf{z}}(k + 1|k) = \mathbf{z}(k + 1) - \mathbf{H}(k + 1)\hat{\mathbf{x}}(k + 1|k) \\
&= \mathbf{H}(k + 1)\tilde{\mathbf{x}}(k + 1|k) + \mathbf{v}(k + 1) \qquad (15.28)
\end{aligned}$$

The innovations is a zero-mean Gaussian white noise sequence, with

$$\mathbf{E}\left\{\tilde{\mathbf{z}}(k + 1|k)\tilde{\mathbf{z}}'(k + 1|k)\right\} = \mathbf{H}(k + 1)\mathbf{P}(k + 1|k)\mathbf{H}'(k + 1) + \mathbf{R}(k + 1) \qquad (15.29)$$

The paper by Kailath [7] gives an excellent historical perspective of estimation theory and includes a very good historical account of the innovations process.

15.9.2   Filtering (the Kalman Filter)

The Kalman filter (KF) and its later extensions to nonlinear problems represent the most widely applied by-product of modern control theory. We begin by presenting the KF, which is the mean-squared filtered estimator of $\mathbf{x}(k + 1)$, $\hat{\mathbf{x}}(k + 1|k + 1)$, in predictor-corrector format:

$$\hat{\mathbf{x}}(k + 1|k + 1) = \hat{\mathbf{x}}(k + 1|k) + \mathbf{K}(k + 1)\tilde{\mathbf{z}}(k + 1|k) \tag{15.30}$$

for $k = 0, 1, \ldots$, where $\hat{\mathbf{x}}(0|0) = \mathbf{m}_\mathbf{x}(0)$ and $\tilde{\mathbf{z}}(k + 1|k)$ is the innovations sequence in (15.28) (use the second equality to implement the KF). Kalman gain matrix $\mathbf{K}(k + 1)$ is $n \times m$, and is specified by the set of relations:

$$\begin{aligned}
\mathbf{K}(k + 1) &= \mathbf{P}(k + 1|k)\mathbf{H}'(k + 1) \left[\mathbf{H}(k + 1)\mathbf{P}(k + 1|k)\mathbf{H}'(k + 1) + \mathbf{R}(k + 1) \right]^{-1} & (15.31)\\
\mathbf{P}(k + 1|k) &= \Phi(k + 1, k)\mathbf{P}(k|k)\Phi'(k + 1, k) + \Gamma(k + 1, k)\mathbf{Q}(k)\Gamma'(k + 1, k) & (15.32)
\end{aligned}$$

and

$$\mathbf{P}(k + 1|k + 1) = \left[\mathbf{I} - \mathbf{K}(k + 1)\mathbf{H}(k + 1) \right] \mathbf{P}(k + 1|k) \tag{15.33}$$

for $k = 0, 1, \ldots$, where $\mathbf{I}$ is the $n \times n$ identity matrix, and $\mathbf{P}(0|0) = \mathbf{P}_\mathbf{x}(0)$.

The KF involves feedback and contains within its structure a model of the plant. The feedback nature of the KF manifests itself in *two* different ways: in the calculation of $\hat{\mathbf{x}}(k + 1|k + 1)$ and also in the calculation of the matrix of gains, $\mathbf{K}(k + 1)$. Observe, also, from (15.26) and (15.32), that the predictor equations, which compute $\hat{\mathbf{x}}(k + 1|k)$ and $\mathbf{P}(k + 1|k)$, use information only from the state equation, whereas the corrector equations, which compute $\mathbf{K}(k + 1)$, $\hat{\mathbf{x}}(k + 1|k + 1)$, and $\mathbf{P}(k + 1|k + 1)$, use information only from the measurement equation. Once the gain is computed, then (15.30) represents a time-varying recursive digital filter. This is seen more clearly when (15.26) and (15.28) are substituted into (15.30). The resulting equation can be rewritten as

$$\begin{aligned}
\hat{\mathbf{x}}(k + 1|k + 1) &= \left[\mathbf{I} - \mathbf{K}(k + 1)\mathbf{H}(k + 1) \right] \Phi(k + 1, k)\hat{\mathbf{x}}(k|k) + \mathbf{K}(k + 1)\mathbf{z}(k + 1)\\
&\quad + \left[\mathbf{I} - \mathbf{K}(k + 1)\mathbf{H}(k + 1) \right] \Psi(k + 1, k)\mathbf{u}(k) & (15.34)
\end{aligned}$$

for $k = 0, 1, \ldots$. This is a state equation for state vector $\hat{\mathbf{x}}$, whose time-varying plant matrix is $\left[\mathbf{I} - \mathbf{K}(k + 1)\mathbf{H}(k + 1) \right]\Phi(k + 1, k)$. Equation (15.34) is time-varying even if our basic state-variable model is time-invariant and stationary, because gain matrix $\mathbf{K}(k + 1)$ is still time-varying in that case. It is possible, however, for $\mathbf{K}(k + 1)$ to reach a limiting value (i.e., steady-state value, $\overline{\mathbf{K}}$), in which case (15.34) reduces to a recursive constant coefficient filter. Equation (15.34) is in recursive filter form, in that it relates the filtered estimate of $\mathbf{x}(k + 1)$, $\hat{\mathbf{x}}(k + 1|k + 1)$, to the filtered estimate of $\mathbf{x}(k)$, $\hat{\mathbf{x}}(k|k)$. Using substitutions similar to those in the derivation of (15.34), we can also obtain the following recursive predictor form of the KF:

$$\begin{aligned}
\hat{\mathbf{x}}(k + 1|k) &= \Phi(k + 1, k) \left[\mathbf{I} - \mathbf{K}(k)\mathbf{H}(k) \right] \hat{\mathbf{x}}(k|k - 1)\\
&\quad + \Phi(k + 1, k)\mathbf{K}(k)\mathbf{z}(k) + \Psi(k + 1, k)\mathbf{u}(k) & (15.35)
\end{aligned}$$

Observe that in (15.35) the predicted estimate of $\mathbf{x}(k + 1)$, $\hat{\mathbf{x}}(k + 1|k)$, is related to the predicted estimate of $\mathbf{x}(k)$, $\hat{\mathbf{x}}(k|k - 1)$, and that the time-varying plant matrix in (15.35) is different from the time-varying plant matrix in (15.34).

Embedded within the recursive KF is another set of recursive equations, (15.31) to (15.33). Because $\mathbf{P}(0|0)$ initializes these calculations, these equations must be ordered as follows: $\mathbf{P}(k|k) \to \mathbf{P}(k + 1|k) \to \mathbf{K}(k + 1) \to \mathbf{P}(k + 1|k + 1) \to$, etc. By combining these equations, it is possible to get a matrix equation for $\mathbf{P}(k + 1|k)$ as a function of $\mathbf{P}(k|k - 1)$ or a similar equation for $\mathbf{P}(k + 1|k + 1)$ as a function of $\mathbf{P}(k|k)$. These equations are nonlinear and are known as matrix Riccati equations.

A measure of recursive predictor performance is provided by matrix $\mathbf{P}(k + 1|k)$, and a measure of recursive filter performance is provided by matrix $\mathbf{P}(k + 1|k + 1)$. These covariances can be calculated prior to any processing of real data, using (15.31) to (15.33). These calculations are often referred to as a performance analysis, and $\mathbf{P}(k + 1|k + 1) \neq \mathbf{P}(k + 1|k)$. It is indeed interesting that the KF utilizes a measure of its mean-squared error during its real-time operation.

Because of the equivalence between mean-squared, BLUE, and WLS filtered estimates of our state vector $\mathbf{x}(k)$ in the Gaussian case, we must realize that the KF equations are just a recursive solution to a system of normal equations. Other implementations of the KF that solve the normal equations using stable algorithms from numerical linear algebra (see, e.g., [2]) and involve orthogonal transformations have better numerical properties than (15.30) to (15.33) (see, e.g., [4]).

A recursive BLUE of a random parameter vector θ can be obtained from the KF equations by setting $\mathbf{x}(k) = \theta$, $\Phi(k + 1, k) = \mathbf{I}$, $\Gamma(k + 1, k) = \mathbf{0}$, $\Psi(k + 1, k) = \mathbf{0}$ and $\mathbf{Q}(k) = \mathbf{0}$. Under these conditions we see that $\mathbf{w}(k) = \mathbf{0}$ for all k, and $\mathbf{x}(k + 1) = \mathbf{x}(k)$, which means, of course, that $\mathbf{x}(k)$ is a vector of constants, θ. The KF equations reduce to: $\hat{\theta}(k + 1|k + 1) = \hat{\theta}(k|k) + \mathbf{K}(k + 1)[\mathbf{z}(k + 1) - \mathbf{H}(k + 1)\hat{\theta}(k|k)]$, $\mathbf{P}(k + 1|k) = \mathbf{P}(k|k)$, $\mathbf{K}(k + 1) = \mathbf{P}(k|k)\mathbf{H}'(k + 1)[\mathbf{H}(k + 1)\mathbf{P}(k|k)\mathbf{H}'(k + 1) + \mathbf{R}(k + 1)]^{-1}$, and $\mathbf{P}(k + 1|k + 1) = [\mathbf{I} - \mathbf{K}(k + 1)\mathbf{H}(k + 1)]\mathbf{P}(k|k)$. Note that it is no longer necessary to distinguish between filtered and predicted quantities, because $\hat{\theta}(k + 1|k) = \hat{\theta}(k|k)$ and $\mathbf{P}(k + 1|k) = \mathbf{P}(k|k)$; hence, the notation $\hat{\theta}(k|k)$ can be simplified to $\hat{\theta}(k)$, for example, which is consistent with our earlier notation for the estimate of a vector of constant parameters.

A divergence phenomenon may occur when either the process noise or measurement noise or both are too small. In these cases the Kalman filter may lock onto wrong values for the state, but believes them to be true values; i.e., it "learns" the wrong state too well. A number of different remedies have been proposed for controlling divergence effects, including: (1) adding fictitious process noise, (2) finite-memory filtering, and (3) fading memory filtering. Fading memory filtering seems to be the most successful and popular way to control divergence effects. See [6] or [12] for discussions about these remedies.

For time-invariant and stationary systems, if $\lim_{k \to \infty} \mathbf{P}(k + 1|k) = \mathbf{P}_p$ exists, then $\lim_{k \to \infty} \mathbf{K}(k) = \bar{\mathbf{K}}$ and the Kalman filter becomes a constant coefficient filter. Because $\mathbf{P}(k + 1|k)$ and $\mathbf{P}(k|k)$ are intimately related, then if $\mathbf{P}_p$ exists, $\lim_{k \to \infty} \mathbf{P}(k|k) = \mathbf{P}_f$ also exists. If the basic state-variable model is time-invariant, stationary, and asymptotically stable, then: (a) for any nonnegative symmetric initial condition $\mathbf{P}(0|-1)$, we have $\lim_{k \to \infty} \mathbf{P}(k + 1|k) = \mathbf{P}_p$ with $\mathbf{P}_p$ independent of $\mathbf{P}(0|-1)$ and satisfying the following steady-state algebraic matrix Riccati equation,

$$\mathbf{P}_p = \Phi \mathbf{P}_p \left[\mathbf{I} - \mathbf{H}' \left(\mathbf{H}\mathbf{P}_p\mathbf{H}' + \mathbf{R} \right)^{-1} \mathbf{H}\mathbf{P}_p \right] \Phi' + \Gamma\mathbf{Q}\Gamma'. \tag{15.36}$$

(b) The eigenvalues of the steady-state KF, $\lambda[\Phi - \bar{\mathbf{K}}\mathbf{H}\Phi]$, all lie within the unit circle, so that the filter is asymptotically stable, i.e., $|\lambda[\Phi - \bar{\mathbf{K}}\mathbf{H}\Phi]| < 1$. If the basic state-variable model is time-invariant and stationary, but is not necessarily asymptotically stable (e.g., it may have a pole on the unit circle), the points (a) and (b) still hold as long as the basic state-variable model is completely stabilizable and detectable (e.g., [8]). To design a steady-state KF: (1) Given $(\Phi, \Gamma, \Psi, \mathbf{H}, \mathbf{Q}, \mathbf{R})$, compute $\mathbf{P}_p$, the positive definite solution of (15.36); (2) compute $\bar{\mathbf{K}}$, as $\bar{\mathbf{K}} = \mathbf{P}_p\mathbf{H}'(\mathbf{H}\mathbf{P}_p\mathbf{H}' + \mathbf{R})^{-1}$; and (3) use $\bar{\mathbf{K}}$ in

$$\begin{aligned} \hat{\mathbf{x}}(k + 1|k + 1) &= \Phi\hat{\mathbf{x}}(k|k) + \Psi\mathbf{u}(k) + \bar{\mathbf{K}}\mathbf{z}(k + 1|k) \\ &= \left(\mathbf{I} - \bar{\mathbf{K}}\mathbf{H} \right) \Phi\hat{\mathbf{x}}(k|k) + \bar{\mathbf{K}}\mathbf{z}(k + 1) + \left(\mathbf{I} - \bar{\mathbf{K}}\mathbf{H} \right) \Psi\mathbf{u}(k) \end{aligned} \tag{15.37}$$

Equation (15.37) is a steady-state filter state equation. The main advantage of the steady-state filter is a drastic reduction in on-line computations.

15.9.3 Smoothing

Although there are three types of smoothers, the most useful one for digital signal processing is the fixed-interval smoother, hence, we only discuss it here. The fixed-interval smoother is $\hat{\mathbf{x}}(k|N), k =$

$0, 1, \ldots, N - 1$, where N is a fixed positive integer. The situation here is as follows: with an experiment completed, we have measurements available over the fixed interval $1 \leq k \leq N$. For each time point within this interval we wish to obtain the optimal estimate of the state vector $\mathbf{x}(k)$, which is based on all the available measurement data $\{\mathbf{z}(j), j = 1, 2, \ldots, N\}$. Fixed-interval smoothing is very useful in signal processing situations, where the processing is done after all the data are collected. It cannot be carried out on-line during an experiment like filtering can. Because all the available data are used, we cannot hope to do better (by other forms of smoothing) than by fixed-interval smoothing.

A mean-squared fixed-interval smoothed estimate of $\mathbf{x}(k)$, $\hat{\mathbf{x}}(k|N)$, is

$$\hat{\mathbf{x}}(k|N) = \hat{\mathbf{x}}(k|k-1) + \mathbf{P}(k|k-1)\mathbf{r}(k|N) \tag{15.38}$$

where $k = N - 1, N - 2, \ldots, 1$, and $n \times 1$ vector $\mathbf{r}$ satisfies the backward-recursive equation

$$\mathbf{r}(j|N) = \boldsymbol{\Phi}'_p(j+1, j)\mathbf{r}(j+1|N) + \mathbf{H}'(j)\left[\mathbf{H}(j)\mathbf{P}(j|j-1)\mathbf{H}'(j) + \mathbf{R}(j)\right]^{-1}\tilde{\mathbf{z}}(j|j-1) \tag{15.39}$$

where $\boldsymbol{\Phi}_p(k+1, k) = \boldsymbol{\Phi}(k+1, k)[\mathbf{I} - \mathbf{K}(k)\mathbf{H}(k)]$ and $j = N, N-1, \ldots, 1$, and, $\mathbf{r}(N+1|N) = \mathbf{0}$. The smoothing error-covariance matrix $\mathbf{P}(k|N)$, is

$$\mathbf{P}(k|N) = \mathbf{P}(k|k-1) - \mathbf{P}(k|k-1)\mathbf{S}(k|N)\mathbf{P}(k|k-1) \tag{15.40}$$

where $k = N - 1, N - 2, \ldots, 1$, and $n \times n$ matrix $\mathbf{S}(j|N)$, which is the covariance matrix of $\mathbf{r}(j|N)$, satisfies the backward-recursive equation

$$\begin{aligned}
\mathbf{S}(j|N) = \;& \boldsymbol{\Phi}'_p(j+1, j)\mathbf{S}(j+1|N)\boldsymbol{\Phi}_p(j+1, j) \\
& + \mathbf{H}'(j)\left[\mathbf{H}(j)\mathbf{P}(j|j-1)\mathbf{H}'(j) + \mathbf{R}(j)\right]^{-1}\mathbf{H}(j)
\end{aligned} \tag{15.41}$$

where $j = N, N - 1, \ldots, 1$, and $\mathbf{S}(N + 1|N) = \mathbf{0}$. Observe that fixed-interval smoothing involves a forward pass over the data, using a KF, and then a backward pass over the innovations, using (15.39). The smoothing error-covariance matrix, $\mathbf{P}(k|N)$, can be precomputed; but, it is not used during the computation of $\hat{\mathbf{x}}(k|N)$. This is quite different than the active use of the filtering error-covariance matrix in the KF.

An important application for fixed-interval smoothing is deconvolution. Consider the single-input single-output system

$$z(k) = \sum_{i=1}^{k} \mu(i)h(k-i) + v(k) \quad k = 1, 2, \ldots, N \tag{15.42}$$

where $\mu(j)$ is the system's input, which is assumed to be white, and not necessarily Gaussian, and $h(j)$ is the system's impulse response. Deconvolution is the signal-processing procedure for removing the effects of $h(j)$ and $v(j)$ from the measurements so that we are left with an estimate of $\mu(j)$. In order to obtain a fixed-interval smoothed estimate of $\mu(j)$, we must first convert (15.42) into an equivalent state-variable model. The single-channel state-variable model $\mathbf{x}(k+1) = \boldsymbol{\Phi}\mathbf{x}(k) + \boldsymbol{\gamma}\mu(k)$ and $z(k) = \mathbf{h}'\mathbf{x}(k) + v(k)$ is equivalent to (15.42) when $\mathbf{x}(0) = \mathbf{0}$, $\mu(0) = 0$, $h(0) = 0$, and $h(l) = \mathbf{h}'\boldsymbol{\Phi}^{l-i}\boldsymbol{\gamma}$ $(l = 1, 2, \ldots)$. A two-pass fixed-interval smoother for $\mu(k)$ is $\hat{\mu}(k|N) = q(k)\boldsymbol{\gamma}'\mathbf{r}(k+1|N)$ where $k = N-1, N-2, \ldots, 1$. The smoothing error variance, $\sigma_\mu^2(k|N)$, is $\sigma_\mu^2(k|N) = q(k) - q(k)\boldsymbol{\gamma}'\mathbf{S}(k+1|N)\boldsymbol{\gamma}q(k)$. In these formulas $\mathbf{r}(k|N)$ and $\mathbf{S}(k|N)$ are computed using (15.39) and (15.41), respectively, and $\mathrm{E}\{\mu^2(k)\} = q(k)$.

15.10 Digital Wiener Filtering

The steady-state KF is a recursive digital filter with filter coefficients equal to $h_f(j)$, $j = 0, 1, \ldots$. Quite often $h_f(j) \approx 0$ for $j \geq J$, so that the transfer function of this filter, $H_f(z)$, can be truncated, i.e.,

$H_f(z) \approx h_f(0) + h_f(1)z^{-1} + \ldots + h_f(J)z^{-J}$. The truncated steady-state, KF can then be implemented as a finite-impulse response (FIR) digital filter. There is, however, a more direct way for designing a FIR minimum mean-squared error filter, i.e., a digital Wiener filter (WF).

Consider the scalar measurement case, in which measurement $z(k)$ is to be processed by a digital filter $F(z)$, whose coefficients, $f(0), f(1), \ldots, f(\eta)$, are obtained by minimizing the mean-squared error $I(\mathbf{f}) = \mathbf{E}\{[d(k) - y(k)]^2\} = \mathbf{E}\{e^2(k)\}$, where $y(k) = f(k) * z(k) = \sum_{i=0}^{\eta} f(i)z(k - i)$ and $d(k)$ is a desired filter output signal. Using calculus, it is straightforward to show that the filter coefficients that minimize $I(\mathbf{f})$ satisfy the following discrete-time Wiener-Hopf equations:

$$\sum_{i=0}^{\eta} f(i)\phi_{zz}(i - j) = \phi_{zd}(j) \quad j = 0, 1, \ldots, \eta \tag{15.43}$$

where $\phi_{zd}(i) = \mathbf{E}\{d(k)z(k - i)\}$ and $\phi_{zz}(i - m) = \mathbf{E}\{z(k - i)z(k - m)\}$. Observe that (15.43) are a system of normal equations and can be solved in many different ways, including the Levinson algorithm. The minimum mean-squared error, $I^*(\mathbf{f})$, in general, approaches a nonzero limiting value which is often reached for modest values of filter length η.

To relate this FIR WF to the truncated steady-state KF, we must first assume a signal-plus-noise model for $z(k)$, because a KF uses a system model, i.e., $z(k) = s(k) + v(k) = h(k) * w(k) + v(k)$, where $h(k)$ is the IR of a linear time-invariant system and, as in our basic state-variable model, $w(k)$ and $v(k)$ are mutually uncorrelated (stationary) white noise sequences with variances q and r, respectively. We must also specify an explicit form for "desired signal" $d(k)$. We shall require that $d(k) = s(k) = h(k) * w(k)$, which means that we want the output of the FIR digital WF to be as close as possible to signal $s(k)$. The resulting Wiener-Hopf equations are

$$\sum_{i=0}^{\eta} f(i) \left[\frac{q}{r} \phi_{hh}(j - i) + \delta(j - i) \right] = \frac{q}{r} \phi_{hh}(j), \quad j = 0, 1, \ldots, \eta \tag{15.44}$$

where $\phi_{hh}(i) = \sum_{l=0}^{\infty} h(l)h(l + i)$. The truncated steady-state KF is a FIR digital WF. For a detailed comparison of Kalman and Wiener filters, see ([12] Lesson 19).

To obtain a digital Wiener deconvolution filter, we assume that filter $F(z)$ is an infinite impulse response (IIR) filter, with coefficients $\{f(j), j = 0, \pm 1, \pm 2, \ldots\}$; $d(k) = \mu(k)$ where $\mu(k)$ is a white noise sequence and $\mu(k)$ and $v(k)$ are stationary and uncorrelated. In this case, (15.43) becomes

$$\sum_{i=-\infty}^{\infty} f(i)\phi_{zz}(i - j) = \phi_{z\mu}(j) = qh(-j) \quad j = 0, \pm 1, \pm 2 \ldots \tag{15.45}$$

This system of equations cannot be solved as a linear system of equations, because there are a doubly infinite number of them. Instead, we take the discrete-time Fourier transform of (15.45), i.e., $F(\omega)\Phi_{zz}(\omega) = qH^*(\omega)$, but, from (15.42), $\Phi_{zz}(\omega) = q|H(\omega)|^2 + r$; hence,

$$F(\omega) = \frac{qH^*(\omega)}{q|H(\omega)|^2 + r} \tag{15.46}$$

The inverse Fourier transform of (15.46), or spectral factorization, gives $\{f(j), j = 0, \pm 1, \pm 2, \ldots\}$.

15.11 Linear Prediction in DSP, and Kalman Filtering

A well-studied problem in digital signal processing (e.g., [5]), is the linear prediction problem, in which the structure of the predictor is fixed ahead of time to be a linear transformation of the data. The "forward" linear prediction problem is to predict a future value of stationary discrete-time random sequence

$\{y(k), k = 1, 2, \ldots\}$ using a set of past samples of the sequence. Let $\hat{y}(k)$ denote the predicted value of $y(k)$ that uses M past measurements; i.e.,

$$\hat{y}(k) = \sum_{i=1}^{M} a_{M,i} y(k - i) \tag{15.47}$$

The forward prediction error filter (PEF) coefficients, $a_{M,1}, \ldots, a_{M,M}$, are chosen so that either the mean-squared or least-squared forward prediction error (FPE), $f_M(k)$, is minimized, where $f_M(k) = y(k) - \hat{y}(k)$. Note that in this filter design problem the length of the filter, M, is treated as a design variable, which is why the PEF coefficients are argumented by M. Note, also, that the PEF coefficients do not depend on t_k; i.e., the PEF is a constant coefficient predictor, whereas our mean-squared state-predictor and filter are time-varying digital filters.

Predictor $\hat{y}(k)$ uses a finite window of past measurements: $y(k - 1), y(k - 2), \ldots, y(k - M)$. This window of measurements is different for different values of t_k. This use of measurements is quite different than our use of the measurements in state prediction, filtering, and smoothing. The latter are based on an expanding memory, whereas the former is based on a fixed memory.

Digital signal-processing specialists have invented a related type of linear prediction named backward linear prediction in which the objective is to predict a past value of a stationary discrete-time random sequence using a set of future values of the sequence. Of course, backward linear prediction is not prediction at all; it is smoothing. But the term backward linear prediction is firmly entrenched in the DSP literature. Both forward and backward PEFs have a filter architecture associated with them that is known as a tapped delay line. Remarkably, when the two filter design problems are considered simultaneously, their solutions can be shown to be coupled, and the resulting architecture is called a lattice. The lattice filter is doubly recursive in both time, k, and filter order, M. The tapped delay line is only recursive in time. Changing its filter length leads to a completely new set of filter coefficients. Adding another stage to the lattice filter does not affect the earlier filter coefficients. Consequently, the lattice filter is a very powerful architecture. No such lattice architecture is known for mean-squared state estimators.

In a second approach to the design of the FPE coefficients, the constraint that the FPE coefficients are constant is transformed into the state equations:

$$a_{M,1}(k + 1) = a_{M,1}(k), a_{M,2}(k + 1) = a_{M,2}(k), \ldots, a_{M,M}(k + 1) = a_{M,M}(k)$$

Equation (15.47) then plays the role of the observation equation in our basic state-variable model, and is one in which the observation matrix is time-varying. The resulting mean-squared error design is then referred to as the Kalman filter solution for the PEF coefficients. Of course, we saw above that this solution is a very special case of the KF, the BLUE. In yet a third approach, the PEF coefficients are modeled as:

$$\begin{aligned} a_{M,1}(k + 1) &= a_{M,1}(k) + w_1(k), a_{M,2}(k + 1) \\ &= a_{M,2}(k) + w_2(k), \ldots, a_{M,M}(k + 1) = a_{M,M}(k) + w_M(k) \end{aligned}$$

where $w_i(k)$ are white noises with variances q_i. Equation (15.47) again plays the role of the measurement equation in our basic state-variable model and is one in which the observation matrix is time-varying. The resulting mean-squared error design is now a full-blown KF.

15.12 Iterated Least Squares

Iterated least squares (ILS) is a procedure for estimating parameters in a nonlinear model. Because it can be viewed as the basis for the extended KF, which is described in the next section, we describe ILS briefly here. To keep things simple, we describe ILS for the scalar parameter model $z(k) = f(\theta, k) + v(k)$ where

$k = 1, 2, \ldots, N$. ILS is basically a four-step procedure: (1) Linearize $f(\theta, k)$ about a nominal value of θ, θ^*. Doing this, we obtain the perturbation measurement equation

$$\delta z(k) = F_\theta(k; \theta^*)\delta\theta + v(k) \quad k = 1, 2, \ldots, N \tag{15.48}$$

where $\delta z(k) = z(k) - z^*(k) = z(k) - f(\theta^*, k)$, $\delta\theta = \theta - \theta^*$, and $F_\theta(k; \theta^*) = \partial f(\theta, k)/\partial\theta|_{\theta=\theta^*}$; (2) Concatenate (15.48) for the N values of k and compute $\hat{\delta\theta}_{\text{WLS}}(N)$ using (15.2); (3) Solve the equation $\hat{\delta\theta}_{\text{WLS}}(N) = \hat{\theta}_{\text{WLS}}(N) - \theta^*$ for $\hat{\theta}_{\text{WLS}}(N)$, i.e., $\hat{\theta}_{\text{WLS}}(N) = \theta^* + \hat{\delta\theta}_{\text{WLS}}(N)$; (4) Replace θ^* with $\hat{\theta}_{\text{WLS}}(N)$ and return to step 1. Iterate through these steps until convergence occurs. Let $\hat{\theta}^i_{\text{WLS}}(N)$ and $\hat{\theta}^{i+1}_{\text{WLS}}(N)$ denote estimates of θ obtained at iterations i and $i + 1$, respectively. Convergence of the ILS method occurs when $|\hat{\theta}^{i+1}_{\text{WLS}}(N) - \hat{\theta}^i_{\text{WLS}}(N)| < \varepsilon$ where ε is a prespecified small positive number.

Observe from this four-step procedure that ILS uses the estimate obtained from the linearized model to generate the nominal value of θ about which the nonlinear model is relinearized. Additionally, in each complete cycle of this procedure, we use both the nonlinear and linearized models. The nonlinear model is used to compute $z^*(k)$ and subsequently $\delta z(k)$. The notions of relinearizing about a filter output and using both the nonlinear and linearized models are also at the very heart of the extended KF.

15.13   Extended Kalman Filter

Many real-world systems are continuous-time in nature and are also nonlinear. The extended Kalman filter (EKF) is the heuristic, but very widely used, application of the KF to estimation of the state vector for the following nonlinear dynamical system:

$$\dot{x}(t) = f[\mathbf{x}(t), \mathbf{u}(t), t] + \mathbf{G}(t)\mathbf{w}(t) \tag{15.49}$$

$$\mathbf{z}(t) = \mathbf{h}[\mathbf{x}(t), \mathbf{u}(t), t] + \mathbf{v}(t) \quad t = t_i, \quad i = 1, 2, \ldots \tag{15.50}$$

In this model measurement equation (15.50) is treated as a discrete-time equation, whereas state equation (15.49) is treated as a continuous-time equation; $\dot{x}(t)$ is short for $d\mathbf{x}(t)/dt$; both $\mathbf{f}$ and $\mathbf{h}$ are continuous and continuously differentiable with respect to all elements of $\mathbf{x}$ and $\mathbf{u}$; $\mathbf{w}(t)$ is a zero-mean continuous-time white noise process, with $\mathbf{E}\{\mathbf{w}(t)\mathbf{w}'(\tau)\} = \mathbf{Q}(t)\delta(t - \tau)$; $\mathbf{v}(t_i)$ is a discrete-time zero-mean white noise sequence, with $\mathbf{E}\{\mathbf{v}(t_i)\mathbf{v}'(t_j)\} = \mathbf{R}(t_i)\delta_{ij}$; and, $\mathbf{w}(t)$ and $\mathbf{v}(t_i)$ are mutually uncorrelated at all $t = t_i$, i.e., $\mathbf{E}\{\mathbf{w}(t)\mathbf{v}'(t_i)\} = \mathbf{0}$ for $t = t_i, i = 1, 2, \ldots$.

In order to apply the KF to (15.49) and (15.50) we must linearize and discretize these equations. Linearization is done about a nominal input $\mathbf{u}^*(t)$ and nominal trajectory $\mathbf{x}^*(t)$, whose choices we discuss below. If we are given a nominal input $\mathbf{u}^*(t)$, then $\mathbf{x}^*(t)$ satisfies the nonlinear differential equation.

$$\dot{\mathbf{x}}^*(t) = \mathbf{f}[\mathbf{x}^*(t), \mathbf{u}^*(t), t] \tag{15.51}$$

and associated with $\mathbf{x}^*(t)$ and $\mathbf{u}^*(t)$ is the following nominal measurement, $\mathbf{z}^*(t)$, where

$$\mathbf{z}^*(t) = \mathbf{h}[\mathbf{x}^*(t), \mathbf{u}^*(t), t] \quad t = t_i, \quad i = 1, 2, \ldots \tag{15.52}$$

Equations (15.51) and (15.52) are referred to as the nominal system model. Letting $\delta\mathbf{x}(t) = \mathbf{x}(t) - \mathbf{x}^*(t)$, $\delta\mathbf{u}(t) = \mathbf{u}(t) - \mathbf{u}^*(t)$, and $\delta\mathbf{z}(t) = \mathbf{z}(t) - \mathbf{z}^*(t)$, we have the following linear perturbation state-variable model:

$$\delta\dot{\mathbf{x}}(t) = \mathbf{F_x}[\mathbf{x}^*(t), \mathbf{u}^*(t), t]\delta\mathbf{x}(t) + \mathbf{F_u}[\mathbf{x}^*(t), \mathbf{u}^*(t), t]\delta\mathbf{u}(t) + \mathbf{G}(t)\mathbf{w}(t) \tag{15.53}$$

$$\delta\mathbf{z}(t) = \mathbf{H_x}[\mathbf{x}^*(t), \mathbf{u}^*(t), t]\delta\mathbf{x}(t) + \mathbf{H_u}[\mathbf{x}^*(t), \mathbf{u}^*(t), t]\delta\mathbf{u}(t) + \mathbf{v}(t),$$

$$t = t_i, \quad i = 1, 2, \ldots \tag{15.54}$$

Where $\mathbf{F_X}[\mathbf{x}^*(t), \mathbf{u}^*(t), t]$, for example, is the following time-varying Jacobian matrix,

$$\mathbf{F_X}\left[\mathbf{x}^*(t), \mathbf{u}^*(t), t\right] = \begin{pmatrix} \partial f_1/\partial x_1^* & \cdots & \partial f_1/\partial x_n^* \\ \vdots & \ddots & \vdots \\ \partial f_n/\partial x_1^* & \cdots & \partial f_n/\partial x_n^* \end{pmatrix} \tag{15.55}$$

in which $\partial f_i/\partial x_j^* = \partial f_i[\mathbf{x}(t), \mathbf{u}(t), t]/\partial x_j(t)|_{\mathbf{x}(t)=\mathbf{x}^*(t), \mathbf{u}(t)=\mathbf{u}^*(t)}$.

Starting with (15.53) and (15.54), we obtain the following discretized perturbation state variable model:

$$\delta\mathbf{x}(k+1) = \Phi\left(k+1, k;^*\right)\delta\mathbf{x}(k) + \Psi\left(k+1, k;^*\right)\delta\mathbf{u}(k) + \mathbf{w}_d(k) \tag{15.56}$$

$$\delta\mathbf{z}(k+1) = \mathbf{H_X}\left(k+1;^*\right)\delta\mathbf{x}(k+1) + \mathbf{H_u}\left(k+1;^*\right)\delta\mathbf{u}(k+1) + \mathbf{v}(k+1) \tag{15.57}$$

where the notation $\Phi(k+1, k;^*)$, for example, denotes the fact that this matrix depends on $\mathbf{x}^*(t)$ and $\mathbf{u}^*(t)$. In (15.56), $\Phi(k+1, k;^*) = \Phi(t_{k+1}, t_k;^*)$, where

$$\dot{\Phi}\left(t, \tau;^*\right) = \mathbf{F_X}\left[\mathbf{x}^*(t), \mathbf{u}^*(t), t\right]\Phi\left(t, \tau;^*\right) , \quad \Phi\left(t, t;^*\right) = \mathbf{I} \tag{15.58}$$

Additionally,

$$\Psi\left(k+1, k;^*\right) = \int_{t_k}^{t_{k+1}} \Phi\left(t_{k+1}, \tau;^*\right)\mathbf{F_u}\left[\mathbf{x}^*(\tau), \mathbf{u}^*(\tau), \tau\right]d\tau \tag{15.59}$$

and $\mathbf{w}_d(k)$ is a zero-mean noise sequence that is statistically equivalent to $\int_{t_k}^{t_{k+1}} \Phi(t_{k+1}, \tau)\mathbf{G}(\tau)\mathbf{w}(\tau)d\tau$; hence, its covariance matrix, $\mathbf{Q}_d(k+1, k)$, is

$$\mathrm{E}\left\{\mathbf{w}_d(k)\mathbf{w}_d'(k)\right\} = \mathbf{Q}_d(k+1, k) = \int_{t_k}^{t_{k+1}} \Phi\left(t_{k+1}, \tau\right)\mathbf{G}(\tau)\mathbf{Q}(\tau)\mathbf{G}'(\tau)\Phi'\left(t_{k+1}, \tau\right)d\tau \tag{15.60}$$

Great simplifications of the calculations in (15.58), (15.59), and (15.60) occur if $\mathbf{F}(t)$, $\mathbf{B}(t)$, $\mathbf{G}(t)$, and $\mathbf{Q}(t)$ are approximately constant during the time interval $t \in [t_k, t_{k+1}]$, i.e., if $\mathbf{F}(t) \approx \mathbf{F}_k, \mathbf{B}(t) \approx \mathbf{B}_k, \mathbf{G}(t) \approx \mathbf{G}_k$, and $\mathbf{Q}(t) \approx \mathbf{Q}_k$ for $t \in [t_k, t_{k+1}]$. In this case: $\Phi(k+1, k) = e^{\mathbf{F}_k T}$, $\Psi(k+1, k) \approx \mathbf{B}_k T = \Psi(k)$, and $\mathbf{Q}_d(k+1, k) \approx \mathbf{G}_k\mathbf{Q}_k\mathbf{G}_k'T = \mathbf{Q}_d(k)$ where $T = t_{k+1} - t_k$.

Suppose $\mathbf{x}^*(t)$ is given *a priori*; then we can compute predicted, filtered, or smoothed estimates of $\delta\mathbf{x}(k)$ by applying all of our previously derived state estimators to the discretized perturbation state-variable model in (15.56) and (15.57). We can precompute $\mathbf{x}^*(t)$ by solving the nominal differential equation (15.51). The KF associated with using a precomputed $\mathbf{x}^*(t)$ is known as a relinearized KF. A relinearized KF usually gives poor results, because it relies on an openloop strategy for choosing $\mathbf{x}^*(t)$. When $\mathbf{x}^*(t)$ is precomputed, there is no way of forcing $\mathbf{x}^*(t)$ to remain close to $\mathbf{x}(t)$, and this must be done or else the perturbation state-variable model is invalid.

The relinearized KF is based only on the discretized perturbation state-variable model. It does not use the nonlinear nature of the original system in an active manner. The EKF relinearizes the nonlinear system about each new estimate as it becomes available, i.e., at $k = 0$, the system is linearized about $\hat{\mathbf{x}}(0|0)$. Once $\mathbf{z}(1)$ is processed by the EKF so that $\hat{\mathbf{x}}(1|1)$ is obtained, the system is linearized about $\hat{\mathbf{x}}(1|1)$. By "linearize about $\hat{\mathbf{x}}(1|1)$," we mean $\hat{\mathbf{x}}(1|1)$ is used to calculate all the quantities needed to make the transition from $\hat{\mathbf{x}}(1|1)$ to $\hat{\mathbf{x}}(2|1)$ and subsequently $\hat{\mathbf{x}}(2|2)$. The purpose of relinearizing about the filter's output is to use a better reference trajectory for $\mathbf{x}^*(t)$. Doing this, $\delta\mathbf{x} = \mathbf{x} - \hat{\mathbf{x}}$ will be held as small as possible, so that our linearization assumptions are less likely to be violated than in the case of the relinearized KF.

The EKF is available only in predictor-corrector format [6]. Its prediction equation is obtained by integrating the nominal differential equation for $\mathbf{x}^*(t)$ from t_k to t_{k+1}. Its correction equation is obtained by applying the KF to the discretized perturbation state-variable model. The equations for the EKF are:

$$\hat{\mathbf{x}}(k+1|k) = \hat{\mathbf{x}}(k|k) + \int_{t_k}^{t_{k+1}} \mathbf{f}\left[\hat{\mathbf{x}}(t|t_k), \mathbf{u}^*(t), t\right]dt , \tag{15.61}$$

which must be evaluated by numerical integration formulas that are initialized by $\mathbf{f}[\hat{\mathbf{x}}(t_k|t_k), \mathbf{u}^*(t_k), t_k]$,

$$
\begin{aligned}
\hat{\mathbf{x}}(k+1|k+1) &= \hat{\mathbf{x}}(k+1|k) + \mathbf{K}\left(k+1;^*\right) \\
&\quad \left\{\mathbf{z}(k+1) - \mathbf{h}\left[\hat{\mathbf{x}}(k+1|k), \mathbf{u}^*(k+1), k+1\right]\right. \\
&\quad \left. - \mathbf{H}_{\mathbf{u}}\left(k+1;^*\right)\delta\mathbf{u}(k+1)\right\} \\
\mathbf{K}\left(k+1;^*\right) &= \mathbf{P}\left(k+1|k;^*\right)\mathbf{H}_{\mathbf{x}}'\left(k+1;^*\right) \\
&\quad \left[\mathbf{H}_{\mathbf{x}}\left(k+1;^*\right)\mathbf{P}\left(k+1|k;^*\right)\mathbf{H}_{\mathbf{x}}'\left(k+1;^*\right) + \mathbf{R}(k+1)\right]^{-1} \\
\mathbf{P}\left(k+1|k;^*\right) &= \Phi\left(k+1,k;^*\right)\mathbf{P}\left(k|k;^*\right)\Phi'\left(k+1,k;^*\right) + \mathbf{Q}_d\left(k+1,k;^*\right) \\
\mathbf{P}\left(k+1|k+1;^*\right) &= \left[\mathbf{I} - \mathbf{K}\left(k+1;^*\right)\mathbf{H}_{\mathbf{x}}\left(k+1;^*\right)\right]\mathbf{P}\left(k+1|k;^*\right)
\end{aligned}
$$

$$(15.62)$$
$$(15.63)$$
$$(15.64)$$
$$(15.65)$$

In these equations, $\mathbf{K}(k+1;^*)$, $\mathbf{P}(k+1|k;^*)$, and $\mathbf{P}(k+1|k+1;^*)$ depend on the nominal $\mathbf{x}^*(t)$ that results from prediction, $\hat{\mathbf{x}}(k+1|k)$. For a complete flowchart of the EKF, see Figure 24-2 in [12].

The EKF is very widely used; however, it does not provide an optimal estimate of $\mathbf{x}(k)$. The optimal mean-squared estimate of $\mathbf{x}(k)$ is still $\mathbf{E}\{\mathbf{x}(k)|\mathbf{Z}(k)\}$, regardless of the linear or nonlinear nature of the system's model. The EKF is a first-order approximation of $\mathbf{E}\{\mathbf{x}(k)|\mathbf{Z}(k)\}$ that sometimes works quite well, but cannot be guaranteed to always work well. No convergence results are known for the EKF; hence, the EKF must be viewed as an ad hoc filter. Alternatives to the EKF, which are based on nonlinear filtering, are quite complicated and are rarely used.

The EKF is designed to work well as long as $\delta\mathbf{x}(k)$ is "small." The *iterated EKF* [6] is designed to keep $\delta\mathbf{x}(k)$ as small as possible. The iterated EKF differs from the EKF in that it iterates the correction equation L times until $\|\hat{\mathbf{x}}_L(k+1|k+1) - \hat{\mathbf{x}}_{L-1}(k+1|k+1)\| \le \varepsilon$. Corrector 1 computes $\mathbf{K}(k+1;^*)$, $\mathbf{P}(k+1|k;^*)$, and $\mathbf{P}(k+1|k+1;^*)$ using $\mathbf{x}^* = \hat{\mathbf{x}}(k+1|k)$; corrector 2 computes these quantities using $\mathbf{x}^* = \hat{\mathbf{x}}_1(k+1|k+1)$; corrector 3 computes these quantities using $\mathbf{x}^* = \hat{\mathbf{x}}_2(k+1|k+1)$; etc. Often, just adding one additional corrector (i.e., $L = 2$) leads to substantially better results for $\hat{\mathbf{x}}(k+1|k+1)$ than are obtained using the EKF.

Acknowledgment

The author gratefully acknowledges Prentice-Hall for extending permission to include summaries of material that appeared originally in *Lessons in Estimation Theory for Signal Processing, Communications, and Control* [12].

References

[1] Anderson, B.D.O. and Moore, J.B., *Optimal Filtering*, Prentice-Hall, Englewood Cliffs, NJ, 1979.

[2] Bierman, G.J., *Factorization Methods for Discrete Sequential Estimation*, Academic Press, New York, 1977.

[3] Golub, G.H. and Van Loan, C.F., *Matrix Computations*, 2nd ed., Johns Hopkins Univ. Press, Baltimore, MD, 1989.

[4] Grewal, M.S. and Andrews, A.P., *Kalman Filtering: Theory and Practice*, Prentice-Hall, Englewood Cliffs, NJ, 1993.

[5] Haykin, S., *Adaptive Filter Theory*, 2nd ed., Prentice-Hall, Englewood Cliffs, NJ, 1991.

[6] Jazwinski, A.H., *Stochastic Processes and Filtering Theory*, Academic Press, New York, 1970.

[7] Kailath, T.K., A view of three decades of filtering theory, *IEEE Trans. Inf. Theory*, IT-20: 146–181, 1974.

[8] Kailath, T.K., *Linear Systems*, Prentice-Hall, Englewood Cliffs, NJ, 1980.

[9] Kalman, R.E., A new approach to linear filtering and prediction problems, *Trans. ASME J. Basic Eng. Series D*, 82: 35–46, 1960.

[10] Kashyap, R.L. and Rao, A.R., *Dynamic stochastic Models from Empirical Data,* Academic Press, New York, 1976.

[11] Ljung, L., *System Identification: Theory for the User,* Prentice-Hall, Englewood Cliffs, NJ, 1987.

[12] Mendel, J.M., *Lessons in Estimation Theory for Signal Processing, Communications, and Control,* Prentice-Hall PTR, Englewood Cliffs, NJ, 1995.

Further Information

Recent articles about estimation theory appear in many journals, including the following engineering journals: *AIAA J., Automatica, IEEE Trans. on Aerospace and Electronic Systems, IEEE Trans. on Automatic Control, IEEE Trans. on Information Theory, IEEE Trans. on Signal Processing, Int. J. Adaptive Control and Signal Processing, Int. J. Control,* and *Signal Processing.* Nonengineering journals that also publish articles about estimation theory include: *Annals Inst. Statistical Math., Ann. Math Statistics, Ann. Statistics, Bull. Inst. Internat. Stat.,* and *Sankhya.*

Some engineering conferences that continue to have sessions devoted to aspects of estimation theory, include: American Automatic Control Conference, IEEE Conference on Decision and Control, IEEE International Conference on Acoustics, Speech and Signal Processing, IFAC International Congress, and, some IFAC Workshops.

MATLAB toolboxes that implement some of the algorithms described in this chapter are: *Control Systems, Optimization,* and *System Identification.* See [12], at the end of each lesson, for descriptions of which M-files in these toolboxes are appropriate. Additionally, [12] lists six estimation algorithm M-files that do not appear in any MathWorks toolboxes or in MATLAB. They are **rwlse** — a recursive least-squares algorithm; **kf** — a recursive KF; **kp** — a recursive Kalman predictor; **sof** — a recursive suboptimal filter in which the gain matrix must be prespecified; **sop** — a recursive suboptimal predictor in which the gain matrix must be prespecified; and, **fis** — a fixed-interval smoother.

16

Validation, Testing, and Noise Modeling

Jitendra K. Tugnait
Auburn University

16.1 Introduction

Linear parametric models of stationary random processes, whether signal or noise, have been found to be useful in a wide variety of signal processing tasks such as signal detection, estimation, filtering, and classification, and in a wide variety of applications such as digital communications, automatic control, radar and sonar, and other engineering disciplines and sciences. A general representation of a linear discrete-time stationary signal $x(t)$ is given by

$$x(t) = \sum_{i=0}^{\infty} h(i)\epsilon(t - i) \tag{16.1}$$

where $\{\epsilon(t)\}$ is a zero-mean, i.i.d. (independent and identically distributed) random sequence with finite variance, and $\{h(i), i \geq 0\}$ is the impulse response of the linear system such that $\sum_{i=-\infty}^{\infty} h^2(i) < \infty$. Much effort has been expended on developing approaches to linear model fitting given a single measurement record of the signal (or noisy signal). Parsimonious parametric models such as AR (autoregressive), MA (moving average), ARMA or state-space, as opposed to impulse response modeling, have been popular together with the assumption of Gaussianity of the data.

Define

$$H(q) = \sum_{i=0}^{\infty} h(i)q^{-i} \tag{16.2}$$

where q^{-1} is the backward shift operator (i.e., $q^{-1}x(t) = x(t - 1)$, etc.). If q is replaced with the complex variable z, then $H(z)$ is the Z-transform of $\{h(i)\}$, i.e., it is the system transfer function. Using (16.2),

(16.1) may be rewritten as

$$x(t) = H(q)\epsilon(t).$$ (16.3)

Fitting linear models to the measurement record requires estimation of $H(q)$, or equivalently of $\{h(i)\}$ (without observing $\{\epsilon(t)\}$). Typically $H(q)$ is parameterized by a finite number of parameters, say by the parameter vector $\theta^{(M)}$ of dimension M. For instance, an AR model representation of order M means that

$$H_{AR}(q; \theta^{(M)}) = \frac{1}{1 + \sum_{i=1}^{M} a_i q^{-i}}, \quad \theta^{(M)} = (a_1, a_2, \cdots, a_M)^T.$$ (16.4)

This reduces the number of estimated parameters from a "large" number to M.

In this section several aspects of fitting models such as (16.1) to (16.3) to the given measurement record are considered. These aspects are (see also Fig. 16.1):

- Is the model of the type (16.1) appropriate to the given record? This requires testing for linearity and stationarity of the data.

- Linear Gaussian models have long been dominant both for signals as well as for noise processes. Assumption of Gaussianity allows implementation of statistically efficient parameter estimators such as maximum likelihood estimators. A Gaussian process is completely characterized by its second-order statistics (autocorrelation function or, equivalently, its power spectral density). Since the power spectrum of $\{x(t)\}$ of (16.1) is given by

$$S_{xx}(\omega) = \sigma_\epsilon^2 |H(e^{j\omega})|^2, \quad \sigma_\epsilon^2 = E\{\epsilon^2(t)\},$$ (16.5)

one cannot determine the phase of $H(e^{j\omega})$ independent of $|H(e^{j\omega})|$. Determination of the true phase characteristic is crucial in several applications such as blind equalization of digital communications channels. Use of higher-order statistics allows one to uniquely identify nonminimum-phase parametric models. Higher-order cumulants of Gaussian processes vanish, hence, if the data are stationary Gaussian, a minimum-phase (or maximum-phase) model is the "best" that one can estimate. Therefore, another aspect considered in this section is testing for non-Gaussianity of the given record.

- If the data are Gaussian, one may fit models based solely upon the second-order statistics of the data — else use of higher-order statistics in addition to or in lieu of the second-order statistics is indicated, particularly if the phase of the linear system is crucial. In either case, one typically fits a model $H(q; \theta^{(M)})$ by estimating the M unknown parameters through optimization of some cost function. In practice, (the model order) M is unknown and its choice has a significant impact on the quality of the fitted model. In this section another aspect of the model-fitting problem considered is that of order selection.

- Having fitted a model $H(q; \theta^{(M)})$, one would also like to know how good are the estimated parameters? Typically this is expressed in terms of error bounds or confidence intervals on the fitted parameters and on the corresponding model transfer function.

- Having fitted a model, a final step is that of model falsification. Is the fitted model an appropriate representation of the underlying system? This is referred to variously as model validation, model verification, or model diagnostics.

- Finally, various models of univariate noise pdf (probability density function) are discussed to complete the discussion of model fitting.

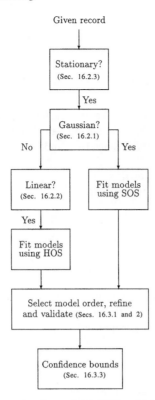

FIGURE 16.1   Section outline (SOS — second-order statistics; HOS — higher-order statistics).

16.2   Gaussianity, Linearity, and Stationarity Tests

Given a zero-mean, stationary random sequence $\{x(t)\}$, its third-order cumulant function $C_{xxx}(i, k)$ is given by [12]

$$C_{xxx}(i, k) := E\{x(t + i)x(t + k)x(t)\}. \tag{16.6}$$

Its bispectrum $B_{xxx}(\omega_1, \omega_2)$ is defined as [12]

$$B_{xxx}(\omega_1, \omega_2) = \sum_{i=-\infty}^{\infty} \sum_{k=-\infty}^{\infty} C_{xxx}(i, k)e^{-j(\omega_1 i + \omega_2 k)}. \tag{16.7}$$

Similarly, its fourth-order cumulant function $C_{xxxx}(i, k, l)$ is given by [12]

$$\begin{aligned}
C_{xxxx}(i, k, l) :=\ & E\{x(t)x(t + i)x(t + k)x(t + l)\} \\
& - E\{x(t)x(t + i)\}E\{x(t + k)x(t + l)\} \\
& - E\{x(t)x(t + k)\}E\{x(t + l)x(t + i)\} \\
& - E\{x(t)x(t + l)\}E\{x(t + k)x(t + i)\}.
\end{aligned} \tag{16.8}$$

Its trispectrum is defined as [12]

$$T_{xxxx}(\omega_1, \omega_2, \omega_3) := \sum_{i=-\infty}^{\infty} \sum_{k=-\infty}^{\infty} \sum_{l=-\infty}^{\infty} C_{xxxx}(i, k, l)e^{-j(\omega_1 i + \omega_2 k + \omega_3 l)}. \tag{16.9}$$

If $\{x(t)\}$ obeys (16.1), then [12]

$$B_{xxx}(\omega_1, \omega_2) = \gamma_{3\epsilon} H(e^{j\omega_1})H(e^{j\omega_2})H^*(e^{j(\omega_1 + \omega_2)}) \tag{16.10}$$

and

$$T_{xxxx}(\omega_1, \omega_2, \omega_3) = \gamma_{4\epsilon} H(e^{j\omega_1}) H(e^{j\omega_2}) H(e^{j\omega_3}) H^*(e^{j(\omega_1+\omega_2+\omega_3)}) \tag{16.11}$$

where

$$\gamma_{3\epsilon} = C_{\epsilon\epsilon\epsilon}(0,0,0) \quad \text{and} \quad \gamma_{4\epsilon} = C_{\epsilon\epsilon\epsilon\epsilon}(0,0,0,0). \tag{16.12}$$

For Gaussian processes, $B_{xxx}(\omega_1, \omega_2) \equiv 0$ and $T_{xxxx}(\omega_1, \omega_2, \omega_3) \equiv 0$; equivalently, $C_{xxx}(i, k) \equiv 0$ and $C_{xxxx}(i, k, l) \equiv 0$. This forms a basis for testing Gaussianity of a given measurement record. When $\{x(t)\}$ is linear (i.e., it obeys (16.1)), then using (16.5) and (16.10),

$$\frac{|B_{xxx}(\omega_1, \omega_2)|^2}{S_{xx}(\omega_1) S_{xx}(\omega_1) S_{xx}(\omega_1 + \omega_2)} = \frac{\gamma_{3\epsilon}}{\sigma_\epsilon^6} = \text{constant} \;\; \forall \, \omega_1, \omega_2, \tag{16.13}$$

and using (16.5) and (16.11),

$$\frac{|T_{xxxx}(\omega_1, \omega_2, \omega_3)|^2}{S_{xx}(\omega_1) S_{xx}(\omega_1) S_{xx}(\omega_3) S_{xx}(\omega_1 + \omega_2 + \omega_3)} = \frac{\gamma_{4\epsilon}}{\sigma_\epsilon^8} = \text{constant} \;\; \forall \, \omega_1, \omega_2, \omega_3. \tag{16.14}$$

The above two relations form a basis for testing linearity of a given measurement record. How the tests are implemented depends upon the statistics of the estimators of the higher-order cumulant spectra as well as that of the power spectra of the given record.

16.2.1 Gaussianity Tests

Suppose that the given zero-mean measurement record is of length N denoted by $\{x(t), \; t = 1, 2, \cdots, N\}$. Suppose that the given sample sequence of length N is divided into K nonoverlapping segments each of size N_B samples so that $N = K N_B$. Let $X^{(i)}(\omega)$ denote the discrete Fourier transform (DFT) of the ith block $\{x(t + (i-1)N_B), \; 1 \le t \le N_B\}$ $(i = 1, 2, \cdots, K)$ given by

$$X^{(i)}(\omega_m) = \sum_{l=0}^{N_B-1} x(l + 1 + (i-1)N_B)\exp(-j\omega_m l) \tag{16.15}$$

where

$$\omega_m = \frac{2\pi}{N_B}m, \quad m = 0, 1, \cdots, N_B - 1. \tag{16.16}$$

Denote the estimate of the bispectrum $B_{xxx}(\omega_m, \omega_n)$ at bifrequency $(\omega_m = \frac{2\pi}{N_B}m, \omega_n = \frac{2\pi}{N_B}n)$ as $\widehat{B}_{xxx}(m, n)$, given by averaging over K blocks

$$\widehat{B}_{xxx}(m, n) = \frac{1}{K} \sum_{i=1}^{K} \left\{ \frac{1}{N_B} X^{(i)}(\omega_m) X^{(i)}(\omega_n) \left[X^{(i)}(\omega_m + \omega_n) \right]^* \right\}, \tag{16.17}$$

where X^* denotes the complex conjugate of X. A principal domain of $\widehat{B}_{xxx}(m, n)$ is the triangular grid

$$D = \left\{ (m, n) \,|\, 0 \le m \le \frac{N_B}{2}, \;\; 0 \le n \le m, \;\; 2m + n \le N_B \right\}. \tag{16.18}$$

Values of $\widehat{B}_{xxx}(m, n)$ outside D can be inferred from that in D.

Select a coarse frequency grid $(\overline{m}, \overline{n})$ in the principal domain D as follows. Let d denote the distance between two adjacent coarse frequency pairs such that $d = 2r + 1$ with r a positive integer. Set $n_0 = 2 + r$ and $\overline{n} = n_0, n_0 + d, \cdots, n_0 + (L_{\overline{n}} - 1)d$ where $L_{\overline{n}} = \lfloor \frac{\lfloor \frac{N_B}{3} \rfloor - 1}{d} \rfloor$. For a given $\overline{n}$, set $m_{0,\overline{n}} = \lfloor \frac{N_B - \overline{n}}{2} \rfloor - r$,

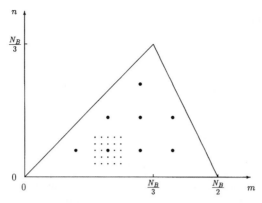

FIGURE 16.2 Coarse and fine grids in the principal domain.

$\overline{m} = \overline{m}_{\overline{n}} = m_{0,\overline{n}}, \; m_{0,\overline{n}} - d, \; \cdots, \; m_{0,\overline{n}} - (L_{\overline{m},\overline{n}} - 1)d$ where $L_{\overline{m},\overline{n}} = \lfloor \frac{m_{0,\overline{n}} - (\overline{n}+r+1)}{d} \rfloor + 1$. Let P denote the number of points on the coarse frequency grid as defined above so that $P = \sum_{\overline{n}=1}^{L_{\overline{n}}} L_{\overline{m},\overline{n}}$. Suppose that $(\overline{m}, \overline{n})$ is a coarse point, then select a fine grid $(\overline{m}, n_{\overline{n}k})$ and $(m_{\overline{m}i}, n_{\overline{n}k})$ consisting of

$$m_{\overline{m}i} = \overline{m} + i, \; |i| \leq r, \quad n_{\overline{n}k} = \overline{n} + k, \; |k| \leq r, \tag{16.19}$$

for some integer $r > 0$ such that $(2r + 1)^2 > P$; see also Fig. 16.2. Order the $L \; (= (2r + 1)^2)$ estimates $\widehat{B}_{xxx}(m_{\overline{m}i}, n_{\overline{n}k})$ on the fine grid around the bifrequency pair $(\overline{m}, \overline{n})$ into an L-vector, which after relabeling, may be denoted as v_{ml}, $l = 1, 2, \cdots, L$, $m = 1, 2, \cdots, P$, where m indexes the coarse grid and l indexes the fine grid. Define P-vectors

$$\overline{\Psi}_i = (v_{1i}, v_{2i}, \cdots, v_{Pi})^T \quad (i = 1, 2, \cdots, L). \tag{16.20}$$

Consider the estimates

$$\overline{\mathcal{M}} = \frac{1}{L} \sum_{i=1}^{L} \overline{\Psi}_i \quad \text{and} \quad \overline{\Sigma} = \frac{1}{L} \sum_{i=1}^{L} \left(\overline{\Psi}_i - \overline{\mathcal{M}} \right) \left(\overline{\Psi}_i - \overline{\mathcal{M}} \right)^{\mathcal{H}}. \tag{16.21}$$

Define

$$\mathcal{F}_G = \frac{2(L - P)}{2P} \overline{\mathcal{M}}^{\mathcal{H}} \overline{\Sigma}^{-1} \overline{\mathcal{M}}. \tag{16.22}$$

If $\{x(t)\}$ is Gaussian, then $\mathcal{F}_G$ is distributed as a central F (Fisher) with $(2P, \; 2(L - P))$ degrees of freedom. A statistical test for testing Gaussianity of $\{x(t)\}$ is to declare it to be a non-Gaussian sequence if $\mathcal{F}_G > T_\alpha$ where T_α is selected to achieve a fixed probability of false alarm $\alpha \; (= Pr\{\mathcal{F}_G > T_\alpha\}$ with $\mathcal{F}_G$ distributed as a central F with $(2P, \; 2(L - P))$ degrees of freedom). If $\mathcal{F}_G \leq T_\alpha$, then either $\{x(t)\}$ is Gaussian or it has zero bispectrum.

The above test is patterned after [3]. It treats the bispectral estimates on the "fine" bifrequency grid as a "data set" from a multivariable Gaussian distribution with unknown covariance matrix. Hinich [4] has simplified the test of [3] by using the known asymptotic expression for the covariance matrix involved, and his test is based upon χ^2 distributions. Notice that $\mathcal{F}_G \leq T_\alpha$ does not necessarily imply that $\{x(t)\}$ is Gaussian; it may result from that fact that $\{x(t)\}$ is non-Gaussian with zero bispectrum. Therefore, a next logical step would be to test for vanishing trispectrum of the record. This has been done in [14] using the approach of [4]; extensions of [3] are too complicated. Computationally simpler tests using "integrated polyspectrum" of the data have been proposed in [6]. The integrated polyspectrum (bispectrum or trispectrum) is computed as cross-power spectrum and it is zero for Gaussian processes. Alternatively, one may test if $C_{xxx}(i, k) \equiv 0$ and $C_{xxxx}(i, k, l) \equiv 0$. This has been done in [8].

Other tests that do not rely on higher-order cumulant spectra of the record may be found in [13].

16.2.2 Linearity Tests

Denote the estimate of the power spectral density $S_{xx}(\omega_m)$ of $\{x(t)\}$ at frequency $\omega_m = \frac{2\pi}{N_B}m$ as $\widehat{S}_{xx}(m)$ given by

$$\widehat{S}_{xx}(m) = \frac{1}{K}\sum_{i=1}^{K}\left\{\frac{1}{N_B}X^{(i)}(\omega_m)\left[X^{(i)}(\omega_m)\right]^*\right\}. \tag{16.23}$$

Consider

$$\widehat{\gamma}_x(m,n) = \frac{|\widehat{B}_{xxx}(m,n)|^2}{\widehat{S}_{xx}(m)\widehat{S}_{xx}(n)\widehat{S}_{xx}(m+n)}. \tag{16.24}$$

It turns out that $\widehat{\gamma}_x(m,n)$ is a consistent estimator of the left side of (16.13), and it is asymptotically distributed as a Gaussian random variable, independent at distinct bifrequencies in the interior of D. These properties have been used by Subba Rao and Gabr [3] to design a test of linearity.

Construct a coarse grid and a fine grid of bifrequencies in D as before. Order the L estimates $\widehat{\gamma}_x(m_{\overline{m}i}, n_{\overline{n}k})$ on the fine grid around the bifrequency pair $(\overline{m}, \overline{n})$ into an L-vector, which after relabeling, may be denoted as β_{ml}, $l = 1, 2, \cdots, L$, $m = 1, 2, \cdots, P$, where m indexes the coarse grid and l indexes the fine grid. Define P-vectors

$$\Psi_i = (\beta_{1i}, \beta_{2i}, \cdots, \beta_{Pi})^T, \quad (i = 1, 2, \cdots, L). \tag{16.25}$$

Consider the estimates

$$\mathcal{M} = \frac{1}{L}\sum_{i=1}^{L}\Psi_i \text{ and } \Sigma = \frac{1}{L}\sum_{i=1}^{L}(\Psi_i - \mathcal{M})(\Psi_i - \mathcal{M})^T. \tag{16.26}$$

Define a $(P-1)\times P$ matrix B whose ijth element B_{ij} is given by $B_{ij} = 1$ if $i = j$; $= -1$ if $j = i+1$; $= 0$ otherwise. Define

$$\mathcal{F}_L = \frac{L-P+1}{P-1}(B\mathcal{M})^T\left(B\Sigma B^T\right)^{-1}B\mathcal{M}. \tag{16.27}$$

If $\{x(t)\}$ is linear, then $\mathcal{F}_L$ is distributed as a central F with $(P-1, L-P+1)$ degrees of freedom. A statistical test for testing linearity of $\{x(t)\}$ is to declare it to be a nonlinear sequence if $\mathcal{F}_L > T_\alpha$ where T_α is selected to achieve a fixed probability of false alarm α $(= Pr\{\mathcal{F}_L > T_\alpha\}$ with $\mathcal{F}_L$ distributed as a central F with $(P-1, L-P+1)$ degrees of freedom). If $\mathcal{F}_L \leq T_\alpha$, then either $\{x(t)\}$ is linear or it has zero bispectrum.

The above test is patterned after [3]. Hinich [4] has "simplified" the test of [3]. Notice that $\mathcal{F}_L \leq T_\alpha$ does not necessarily imply that $\{x(t)\}$ is nonlinear; it may result from that fact that $\{x(t)\}$ is non-Gaussian with zero bispectrum. Therefore, a next logical step would be to test if (16.14) holds true. This has been done in [14] using the approach of [4]; extensions of [3] are too complicated. The approaches of [3] and [4] will fail if the data are noisy. A modification to [3] is presented in [7] when additive Gaussian noise is present. Finally, other tests that do not rely on higher-order cumulant spectra of the record may be found in [13].

16.2.3 Stationarity Tests

Various methods exist for testing whether a given measurement record may be regarded as a sample sequence of a stationary random sequence. A crude yet effective way to test for stationarity is to divide the record into several (at least two) nonoverlapping segments and then test for equivalency (or compatibility) of certain statistical properties (mean, mean-square value, power spectrum, etc.) computed from these segments. More sophisticated tests that do not require *a priori* segmentation of the record are also available.

Consider a record of length N divided into two nonoverlapping segments each of length $N/2$. Let $K N_B = N/2$ and use the estimators such as (16.23) to obtain the estimator $\widehat{S}_{xx}^{(l)}(m)$ of the power spectrum $S_{xx}^{(l)}(\omega_m)$ of the l-th segment ($l = 1, 2$), where ω_m is given by (16.16). Consider the test statistic

$$Y = \frac{2}{N_B - 2} \sqrt{\frac{K}{2}} \sum_{m=1}^{\frac{N_B}{2}-1} \left[ln\, \widehat{S}_{xx}^{(1)}(m) - ln\, \widehat{S}_{xx}^{(2)}(m) \right]. \tag{16.28}$$

Then, asymptotically Y is distributed as zero-mean, unit variance Gaussian if $\{x(t)\}$ is stationary. Therefore, if $|Y| > T_\alpha$, then $\{x(t)\}$ is declared to be nonstationary where the threshold T_α is chosen to achieve a false-alarm probability of α ($= Pr\{|Y| > T_\alpha\}$ with Y distributed as zero-mean, unit variance Gaussian). If $|Y| \leq T_\alpha$, then $\{x(t)\}$ is declared to be stationary. Notice that similar tests based upon higher-order cumulant spectra can also be devised.

The above test is patterned after [10]. More sophisticated tests involving two model comparisons as above but without prior segmentation of the record are available in [11] and references therein. A test utilizing evolutionary power spectrum may be found in [9].

16.3 Order Selection, Model Validation, and Confidence Intervals

As noted earlier, one typically fits a model $H(q; \theta^{(M)})$ to the given data by estimating the M unknown parameters through optimization of some cost function. A fundamental difficulty here is the choice of M. There are two basic philosophical approaches to this problem: one consists of an iterative process of model fitting and diagnostic checking (model validation), and the other utilizes a more "objective" approach of optimizing a cost w.r.t. M (in addition to $\theta^{(M)}$).

16.3.1 Order Selection

Let $f_{\theta^{(M)}}(\mathbf{X})$ denote the probability density function of $\mathbf{X} = [x(1),\ x(2),\ \cdots, x(N)]^T$ parameterized by the parameter vector $\theta^{(M)}$ of dimension M. A popular approach to model order selection in the context of linear Gaussian models is to compute the Akaike information criterion (AIC)

$$AIC(M) = -2\,ln\, f_{\widehat{\theta}^{(M)}}(\mathbf{X}) + 2M \tag{16.29}$$

where $\widehat{\theta}^{(M)}$ maximizes $f_{\theta^{(M)}}(\mathbf{X})$ given the measurement record $\mathbf{X}$. Let $\overline{M}$ denote an upper bound on the true model order. Then the minimum AIC estimate (MAICE), the selected model order, is given by the minimizer of $AIC(M)$ over $M = 1, 2, \cdots, \overline{M}$. Clearly one needs to solve the problem of maximization of $ln\, f_{\theta^{(M)}}(\mathbf{X})$ w.r.t. $\theta^{(M)}$ for each value of $M = 1, 2, \cdots, \overline{M}$. The second term on the right side of (16.29) penalizes overparametrization.

Rissanen's minimum description length (MDL) criterion is given by

$$MDL(M) = -2\,ln\, f_{\widehat{\theta}^{(M)}}(\mathbf{X}) + M\,ln\,N. \tag{16.30}$$

It is known that if $\{x(t)\}$ is a Gaussian AR model, then AIC is an inconsistent estimator of the model order whereas MDL is consistent, i.e., MDL picks the correct model order with probability one as the data length tends to infinity, whereas there is a nonzero probability that AIC will not. Several other variations of these criteria exist [15].

Although the derivation of these order selection criteria is based upon Gaussian distribution, they have frequently been used for non-Gaussian processes with success provided attention is confined to the use of second-order statistics of the data. They may fail if one fits models using higher-order statistics.

16.3.2 Model Validation

Model validation involves testing to see if the fitted model is an appropriate representation of the underlying (true) system. It involves devising appropriate statistical tools to test the validity of the assumptions made in obtaining the fitted model. It is also known as model falsification, model verification, or diagnostic checking. It can also be used as a tool for model order selection. It is an essential part of any model fitting methodology.

Suppose that $\{x(t)\}$ obeys (16.1). Suppose that the fitted model corresponding to the estimated parameter $\widehat{\theta}^{(M)}$ is $H(q; \widehat{\theta}^{(M)})$. Assuming that the true model $H(q)$ is invertible, in the ideal case one should get $\epsilon(t) = H^{-1}(q)x(t)$ where $\{\epsilon(t)\}$ is zero-mean, i.i.d. (or at least white when using second-order statistics). Hence, if the fitted model $H(q; \widehat{\theta}^{(M)})$ is a valid description of the underlying true system, one expects $\epsilon'(t) = H^{-1}(q; \widehat{\theta}^{(M)})x(t)$ to be zero-mean, i.i.d. One of the diagnostic checks then is to test for whiteness or independence of the inverse filtered data (or the residuals or linear innovations, in case second-order statistics are used). If the fitted model is unable to "adequately" capture the underlying true system, one expects $\{\epsilon'(t)\}$ to deviate from i.i.d. distribution. This is one of the most widely used and useful diagnostic checks for model validation.

A test for second-order whiteness of $\{\epsilon'(t)\}$ is as follows [15]. Construct the estimates of the covariance function as

$$\widehat{r}_\epsilon(\tau) = N^{-1} \sum_{t=1}^{N-\tau} \epsilon'(t+\tau)\epsilon'(t) \quad (\tau \geq 0). \tag{16.31}$$

Consider the test statistic

$$R = \frac{N}{\widehat{r}_\epsilon^2(0)} \sum_{i=1}^{m} \widehat{r}_\epsilon^2(i) \tag{16.32}$$

where m is some *a priori* choice of the maximum lag for whiteness testing. If $\{\epsilon'(t)\}$ is zero-mean white, then R is distributed as $\chi^2(m)$ (χ^2 with m degrees of freedom). A statistical test for testing whiteness of $\{\epsilon'(t)\}$ is to declare it to be a nonwhite sequence (hence invalidate the model) if $R > T_\alpha$ where T_α is selected to achieve a fixed probability of false alarm α ($= Pr\{R > T_\alpha\}$ with R distributed as $\chi^2(m)$). If $R \leq T_\alpha$, then $\{\epsilon'(t)\}$ is second-order white, hence the model is validated.

The above procedure only tests for second-order whiteness. In order to test for higher-order whiteness, one needs to examine either the higher-order cumulant functions or the higher-order cumulant spectra (or the integrated polyspectra) of the inverse-filtered data. A statistical test using bispectrum is available in [5]. It is particularly useful if the model fitting is carried out using higher-order statistics. If $\{\epsilon'(t)\}$ is third-order white, then its bispectrum is a constant for all bifrequencies. Let $\widehat{B}_{\epsilon'\epsilon'\epsilon'}(m, n)$ denote the estimate of the bispectrum $B_{\epsilon'\epsilon'\epsilon'}(\omega_m, \omega_n)$ mimicking (16.17). Construct a coarse grid and a fine grid of bifrequencies in D as before. Order the L estimates $\widehat{B}_{\epsilon'\epsilon'\epsilon'}(m_{\overline{m}i}, n_{\overline{n}k})$ on the fine grid around the bifrequency pair $(\overline{m}, \overline{n})$ into an L-vector, which after relabeling may be denoted as μ_{ml}, $l = 1, 2, \cdots, L$, $m = 1, 2, \cdots, P$, where m indexes the coarse grid and l indexes the fine grid. Define P-vectors

$$\widetilde{\Psi}_i = (\mu_{1i}, \mu_{2i}, \cdots, \mu_{Pi})^T, \quad (i = 1, 2, \cdots, L). \tag{16.33}$$

Consider the estimates

$$\widetilde{\mathcal{M}} = \frac{1}{L} \sum_{i=1}^{L} \widetilde{\Psi}_i \text{ and } \widetilde{\Sigma} = \frac{1}{L} \sum_{i=1}^{L} \left(\widetilde{\Psi}_i - \widetilde{\mathcal{M}}\right)\left(\widetilde{\Psi}_i - \widetilde{\mathcal{M}}\right)^{\mathcal{H}}. \tag{16.34}$$

Define a $(P-1) \times P$ matrix $\boldsymbol{B}$ whose ijth element $\boldsymbol{B}_{ij}$ is given by $\boldsymbol{B}_{ij} = 1$ if $i = j$; $= -1$ if $j = i + 1$; $= 0$ otherwise. Define

$$\mathcal{F}_W = \frac{2(L - P + 1)}{2P - 2} \left(\boldsymbol{B}\widetilde{\mathcal{M}}\right)^{\mathcal{H}} \left(\boldsymbol{B}\widetilde{\Sigma}\boldsymbol{B}^T\right)^{-1} \boldsymbol{B}\widetilde{\mathcal{M}}. \tag{16.35}$$

If $\{\epsilon'(t)\}$ is third-order white, then $\mathcal{F}_W$ is distributed as a central F with $(2P - 2, \; 2(L - P + 1))$ degrees of freedom. A statistical test for testing third-order whiteness of $\{\epsilon'(t)\}$ is to declare it to be a nonwhite sequence if $\mathcal{F}_W > T_\alpha$ where T_α is selected to achieve a fixed probability of false alarm $\alpha \; (= Pr\{\mathcal{F}_W > T_\alpha\}$ with $\mathcal{F}_W$ distributed as a central F with $(2P - 2, \; 2(L - P + 1))$ degrees of freedom). If $\mathcal{F}_W \leq T_\alpha$, then either $\{\epsilon'(t)\}$ is third-order white or it has zero bispectrum.

The above model validation test can be used for model order selection. Fix an upper bound on the model orders. For every admissible model order, fit a linear model and test its validity. From among the validated models, select the "smallest" order as the correct order. It is easy to see that this procedure will work only so long as the various candidate orders are nested. Further details may be found in [5] and [15].

16.3.3   Confidence Intervals

Having settled upon a model order estimate M, let $\widehat{\theta}_N^{(M)}$ be the parameter estimator obtained by minimizing a cost function $V_N(\theta^{(M)})$, given a record of length N, such that $V_\infty(\theta) := \lim_{N \to \infty} V_N(\theta)$ exists. For instance, using the notation of the section on order selection, one may take $V_N(\theta^{(M)}) = -N^{-1}ln \, f_{\theta^{(M)}}(X)$. How reliable are these estimates? An assessment of this is provided by confidence intervals.

Under some general technical conditions, it usually follows that asymptotically (i.e., for large N), $\sqrt{N} \left(\widehat{\theta}_N^{(M)} - \theta_0 \right)$ is distributed as a Gaussian random vector with zero-mean and covariance matrix $\mathcal{P}$ where θ_0 denotes the true value of $\theta^{(M)}$. A general expression for $\mathcal{P}$ is given by [15]

$$\mathcal{P} = \left[V_\infty''(\theta_0) \right]^{-1} P_\infty \left[V_\infty''(\theta_0) \right]^{-1} \tag{16.36}$$

where

$$P_\infty = \lim_{N \to \infty} E \left\{ N V_N'^T (\theta_0) V_N'(\theta_0) \right\} \tag{16.37}$$

and V' (a row vector) and V'' (a square matrix) denote the gradient and the Hessian, respectively, of V.

The above result can be used to evaluate the reliability of the parameter estimator. It follows from the above results that

$$\eta_N = N \left(\widehat{\theta}_N^{(M)} - \theta_0 \right)^T \mathcal{P}^{-1} \left(\widehat{\theta}_N^{(M)} - \theta_0 \right) \tag{16.38}$$

is asymptotically $\chi^2(M)$. Define $\chi_\alpha^2(M)$ via $Pr\{y > \chi_\alpha^2(M)\} = \alpha$ where y is distributed as $\chi^2(M)$. For instance, $\chi_{0.05}^2 = 9.49$ so that $Pr\{\eta_N > 9.49\} = 0.05$. The ellipsoid $\eta_N \leq \chi_\alpha^2(M)$ then defines the 95% confidence ellipsoid for the estimate $\widehat{\theta}_N^{(M)}$. It implies that θ_0 will lie with probability 0.95 in this ellipsoid around $\widehat{\theta}_N^{(M)}$.

In practice obtaining expression for $\mathcal{P}$ is not easy; it requires knowledge of θ_0. Typically, one replaces θ_0 with $\widehat{\theta}_N^{(M)}$. If a closed-form expression for $\mathcal{P}$ is not available, it may be approximated by a sample average [16].

16.4   Noise Modeling

As for signal models, Gaussian modeling of noise processes has long been dominant. Typically the central limit theorem is invoked to justify this assumption; thermal noise is indeed Gaussian. Another reason is analytical tractability when the Gaussian assumption is made. Nevertheless, non-Gaussian noise occurs often in practice. For instance, underwater acoustic noise, low-frequency atmospheric noise, radar clutter noise, and urban and man-made radio-frequency noise all are highly non-Gaussian [17]. All these types of noise are impulsive in character, i.e., the noise produces large-magnitude observations more often than predicted by a Gaussian model. This fact has led to development of several models of univariate non-Gaussian noise probability density functions (pdf), all of which have their tails decay at rates lower than the rate of decay of the Gaussian pdf tails. Also, the proposed models are parameterized in such a way as to include Gaussian pdf as a special case.

16.4.1 Generalized Gaussian Noise

A generalized Gaussian pdf is characterized by two constants, variance σ^2, and an exponential decay-rate parameter $k > 0$. It is symmetric and unimodal, given by [17]

$$f_k(x) = \frac{k}{2A(k)\Gamma(1/k)} e^{-[|x|/A(k)]^k} \tag{16.39}$$

where

$$A(k) = \left[\sigma^2 \frac{\Gamma(1/k)}{\Gamma(3/k)}\right]^{1/2} \tag{16.40}$$

and Γ is the gamma function

$$\Gamma(\alpha) := \int_o^\infty x^{\alpha-1} e^{-x} dx. \tag{16.41}$$

When $k = 2$, (16.39) reduces to a Gaussian pdf. For $k < 2$, the tails of f_k decay at a lower rate than for the Gaussian case f_2. The value $k = 1$ leads to the Laplace density (two-sided exponential). It is known that generalized Gaussian density with k around 0.5 can be used to model certain impulsive atmospheric noise [17].

16.4.2 Middleton Class A Noise

Unlike most of the other noise models, the Middleton class A mode is based upon physical modeling considerations rather than an empirical fit to observed data. It is a canonical model based upon the assumption that the noise bandwidth is comparable to, or less than, that of the receiver. The observed noise process is assumed to have two independent components:

$$X(t) = X_G(t) + X_P(t) \tag{16.42}$$

where $X_G(t)$ is a stationary background Gaussian noise component and $X_P(t)$ is the impulsive component. The component $X_P(t)$ is represented by

$$X_P(t) = \sum_i U_i(t, \theta) \tag{16.43}$$

where U_i denotes the ith waveform from an interfering source and θ represents a set of random parameters that describe the scale and structure of the waveform. The arrival time of these independent impulsive events at the receiver is assumed to be Poisson distributed. Under these and some additional assumptions, the class A pdf for the normalized instantaneous amplitude of noise is given by

$$f_A(x) = e^{-A} \sum_{m=0}^\infty \frac{A^m}{m!\sqrt{2\pi\sigma_m^2}} e^{-x^2/(2\sigma_m^2)} \tag{16.44}$$

where

$$\sigma_m^2 = \frac{(m/A) + \Gamma'}{1 + \Gamma'}. \tag{16.45}$$

The parameter A, called the impulsive index, determines how impulsive noise is: a small value of A implies highly impulsive interference (although $A = 0$ degenerates into purely Gaussian $X(t)$). The parameter Γ' is the ratio of power in the Gaussian component of the noise to the power in the Poisson mechanism interference. The term in (16.44) corresponding to $m = 0$ represents the background component of the noise with no impulsive waveform present, whereas the higher-order terms represent the occurrence of m impulsive events overlapping simultaneously at the receiver input.

The class A model has been found to provide very good fits to a variety of noise and interference measurements [17].

16.4.3 Stable Noise Distribution

This is another useful noise distribution model which has a drawback that its variance may not be finite. It is most conveniently described by its characteristic function. A stable univariate probability distribution function (PDF) has characteristic function $\varphi(t)$ of the form [18]

$$\varphi(t) = \exp\left\{jat - \gamma|t|^{\alpha}\left[1 + j\beta\,\text{sgn}(t)\omega(t,\alpha)\right]\right\} \tag{16.46}$$

where

$$\omega(t,\alpha) = \begin{cases} \tan(\alpha\pi/2) & \text{for } \alpha \neq 1 \\ (2/\pi)\log(|t|) & \text{for } \alpha = 1 \end{cases} \tag{16.47}$$

$$\text{sgn}(t) = \begin{cases} 1 & \text{for } t > 0 \\ 0 & \text{for } t = 0 \\ -1 & \text{for } t < 0 \end{cases} \tag{16.48}$$

and

$$-\infty < a < \infty, \quad \gamma > 0, \quad 0 < \alpha \leq 2, \quad -1 \leq \beta \leq 1. \tag{16.49}$$

A stable distribution is completely determined by four parameters: location parameter a, the scale parameter γ, the index of skewness β, and the characteristic exponent α. A stable distribution with characteristic exponent α is called *alpha*$-$ stable.

The characteristic exponent α is a shape parameter and it measures the "thickness" of the tails of the pdf. A small value of α implies longer tails. When $\alpha = 2$, the corresponding stable distribution is Gaussian. When $\alpha = 1$ and $\beta = 0$, then the corresponding stable distribution is Cauchy.

Inverse Fourier transform of $\varphi(t)$ yields the PDF and, therefore, the pdf of noise. No closed-form solution exists in general for the two; however, power series expansion of the pdf is available — details may be found in [18] and references therein.

16.5 Concluding Remarks

In this chapter several fundamental aspects of fitting linear time-invariant parametric (rational transfer function) models to a given measurement record were considered. Before a linear model is fitted, one needs to test for stationarity, linearity, and Gaussianity of the given data. Statistical test for these properties were discussed in the second section. After a model is fitted, one needs to validate the model and assess the reliability of the fitted model parameters. This aspect was discussed in the third section. A cautionary note is appropriate at this point. All of the tests and procedures discussed in this chapter are based upon asymptotic considerations (as record length tends to ∞). In practice, this implies that sufficiently long record length should be available, particularly when higher-order statistics are exploited.

References

[1] Brillinger, D.R., An introduction to polyspectra, *Annals Mathematical Statistics*, 36: 1351-1374, 1965.

[2] Brillinger, D.R., *Time Series, Data Analysis and Theory*, Holt, Rinehart and Winston, New York, 1975.

[3] Subba Rao, T. and Gabr, M.M., A test for linearity of stationary time series, *J. Time Series Analysis*, 1(2): 145-158, 1980.

[4] Hinich, M.J., Testing for Gaussianity and linearity of a stationary time series, *J. Time Series Analysis*, 3(3): 169-176, 1982.

[5] Tugnait, J.K., Linear model validation and order selection using higher-order statistics, *IEEE Trans. Signal Process.*, SP-42: 1728-1736, July, 1994.

[6] Tugnait, J.K., Detection of non-Gaussian signals using integrated polyspectrum, *IEEE Trans. Signal Process.*, SP-42: 3137-3149, Nov., 1994. (Corrections in *IEEE Trans. Signal Process.*, SP-43. Nov., 1995.)

[7] Tugnait, J.K., Testing for linearity of noisy stationary signals, *IEEE Trans.Signal Process.*, SP-42: 2742-2748, Oct., 1994.

[8] Giannakis, G.B. and Tstatsanis, M.K., Time-domain tests for Gaussianity and time-reversibility, *IEEE Trans. Signal Process.*, SP-42: 3460-3472, Dec., 1994.

[9] Priestley, M.B., *Nonlinear and Nonstationary Time Series Analysis*, Academic Press, New York, 1988.

[10] Jenkins, G.M., General considerations in the estimation of spectra, *Technometrics*, 3: 133-166, 1961.

[11] Basseville, M. and Nikiforov, I.V., *Detection of Abrupt Changes*, Prentice-Hall, Englewood Cliffs, NJ, 1993.

[12] Nikias, C.L. and Petropulu, A.P., *Higher-Order Spectra Analysis*, Prentice-Hall, Englewood Cliffs, NJ, 1993.

[13] Tong, H., *Nonlinear Time Series*, Oxford University Press, New York, 1990.

[14] Dalle Molle, J.W. and Hinich, M.J., Tripsectral analysis of stationary time series, *J. Acoust. Soc. Am.*, 97(5), Pt. 1, May, 1995.

[15] Söderström, T. and Stoica, P., *System Identification*, Prentice Hall Int., London, 1989.

[16] Ljung, L., *System Identification: Theory for the User*, Prentice-Hall, Englewood Cliffs, NJ, 1987.

[17] Kassam, S.A., *Signal Detection in Non-Gaussian Noise*, Springer-Verlag, New York, 1988.

[18] Shao, M. and Nikias, C.L., Signal processing with fractional lower order moments: stable processes and their applications, *Proc. IEEE*, 81: 986-1010, July, 1993.

17

Cyclostationary Signal Analysis

Georgios B. Giannakis
University of Virginia

17.1 Introduction

Processes encountered in statistical signal processing, communications, and time series analysis applications are often assumed stationary. The plethora of available algorithms testifies to the need for processing and spectral analysis of stationary signals (see, e.g., [42]). Due to the varying nature of physical phenomena and certain man-made operations, however, time-invariance and the related notion of stationarity are often violated in practice. Hence, study of time-varying systems and nonstationary processes is well motivated.

Research in nonstationary signals and time-varying systems has led both to the development of adaptive algorithms and to several elegant tools, including short-time (or running) Fourier transforms, time-frequency representations such as the Wigner-Ville (a member of Cohen's class of distributions), Loeve's and Karhunen's expansions (leading to the notion of evolutionary spectra), and time-scale representations based on wavelet expansions (see [37, 45] and references therein). Adaptive algorithms derived from stationary models assume slow variations in the underlying system. On the other hand, time-frequency and time-scale representations promise applicability to general nonstationarities and provide useful visual cues for preprocessing. When it comes to nonstationary signal analysis and estimation in the presence of noise, however, they assume availability of multiple independent realizations.

In fact, it is impossible to perform spectral analysis, detection, and estimation tasks on signals involving generally unknown nonstationarities, when only a single data record is available. For instance, consider extracting a deterministic signal $s(n)$ observed in stationary noise $v(n)$, using regression techniques based on nonstationary data $x(n) = s(n) + v(n), n = 0, 1, \ldots, N - 1$. Unless $s(n)$ is finitely parameterized by a $d_{\theta_s} \times 1$ vector $\boldsymbol{\theta}_s$ (with $d_{\theta_s} < N$), the problem is ill-posed because adding a new datum, say $x(n_0)$, adds

0-8493-8572-5/98/$0.00+$.50
© 1998 by CRC Press LLC

a new unknown, $s(n_0)$, to be determined. Thus, only structured nonstationarities can be handled when rapid variations are present; and only for classes of finitely parameterized nonstationary processes can reliable statistical descriptors be computed using a single time series. One such class is that of (wide-sense) cyclostationary processes which are characterized by the periodicity they exhibit in their mean, correlation, or spectral descriptors.

An overview of cyclostationary signal analysis and applications are the main goals of this section. Periodicity is omnipresent in physical as well as manmade processes, and cyclostationary signals occur in various real life problems entailing phenomena and operations of repetitive nature: communications [15], geophysical and atmospheric sciences (hydrology [66], oceanography [14], meteorology [35], and climatology [4]), rotating machinery [43], econometrics [50], and biological systems [48].

In 1961 Gladysev [34] introduced key representations of cyclostationary time series, while in 1969 Hurd's thesis [38] offered an excellent introduction to continuous time cyclostationary processes. Since 1975 [22], Gardner and co-workers have contributed to the theory of continuous-time cyclostationary signals, and especially their applications to communications engineering. Gardner [15] adopts a "non-probabilistic" viewpoint of cyclostationarity (see [19] for an overview and also [36] and [18] for comments on this approach). Responding to a recent interest in digital periodically varying systems and cyclostationary time series, the exposition here is probabilistic and focuses on discrete-time signals and systems, with emphasis on their second-order statistical characterization and their applications to signal processing and communications.

The material in the remaining sections is organized as follows: Section 17.2 provides definitions, properties, and representations of cyclostationary processes, along with their relations with stationary and general classes of nonstationary processes. Testing a time series for cyclostationarity and retrieval of possibly hidden cycles along with single record estimation of cyclic statistics are the subjects of Section 17.3. Typical signal classes and operations inducing cyclostationarity are delineated in Section 17.4 to motivate the key uses and selected applications described in Section 17.5. Finally, Section 17.6 concludes and presents trade-offs, topics not covered, and future directions.

17.2 Definitions, Properties, Representations

Let $x(n)$ be a discrete-index random process (i.e., a time series) with mean $\mu_x(n) := E\{x(n)\}$, and covariance $c_{xx}(n; \tau) := E\{[x(n) - \mu_x(n)][x(n + \tau) - \mu_x(n + \tau)]\}$. For $x(n)$ complex valued, let also $\bar{c}_{xx}(n; \tau) := c_{xx*}(n; \tau)$, where $*$ denotes complex conjugation, and n, τ are in the set of integers $\mathcal{Z}$.

DEFINITION 17.1 Process $x(n)$ is (wide-sense) cyclostationary (CS) iff there exists an integer P such that $\mu_x(n) = \mu_x(n + lP)$, $c_{xx}(n; \tau) = c_{xx}(n + lP; \tau)$, or, $\bar{c}_{xx}(n; \tau) = \bar{c}_{xx}(n + lP; \tau)$, $\forall n, l \in \mathcal{Z}$. The smallest of all such Ps is called the period. Being periodic, they all accept Fourier Series expansions over complex harmonic cycles with the set of cycles defined as: $A_{xx}^c := \{\alpha_k = 2\pi k/P, \ k = 0, \ldots, P - 1\}$; e.g., $c_{xx}(n; \tau)$ and its Fourier coefficients called cyclic correlations are related by:

$$c_{xx}(n; \tau) = \sum_{k=0}^{P-1} C_{xx}\left(\frac{2\pi}{P}k; \tau\right) e^{j\frac{2\pi}{P}kn} \quad \overset{FS}{\longleftrightarrow} \quad C_{xx}\left(\frac{2\pi}{P}k; \tau\right) = \frac{1}{P}\sum_{n=0}^{P-1} c_{xx}(n; \tau) e^{-j\frac{2\pi}{P}kn}.$$

$$(17.1)$$

Strict sense cyclostationarity, or, periodic (non-) stationarity, can also be defined in terms of probability distributions or density functions when these functions vary periodically (in n). But the focus in engineering is on periodically and *almost* periodically correlated[1] time series, since real data are often zero-mean, correlated, and with unknown distributions. Almost periodicity is very common in discrete-time because sampling a continuous-time periodic process will rarely yield a discrete-time periodic signal; e.g., sampling $\cos(\omega_c t + \theta)$ every T_s seconds results in $\cos(\omega_c n T_s + \theta)$ for which an *integer* period exists only if $\omega_c T_s = 2\pi/P$. Because $2\pi/(\omega_c T_s)$ is "almost an integer" period, such signals accept generalized (or limiting) Fourier expansions (see also Eq. (17.2) and [9] for rigorous definitions of almost periodic functions).

DEFINITION 17.2   Process $x(n)$ is (wide-sense) almost cyclostationary (ACS) iff its mean and correlation(s) are almost periodic sequences. For $x(n)$ zero-mean and real, the time-varying and cyclic correlations are defined as the generalized Fourier Series pair:

$$c_{xx}(n; \tau) = \sum_{\alpha_k \in A_{xx}^c} C_{xx}(\alpha_k; \tau) e^{j\alpha_k n} \overset{FS}{\longleftrightarrow}$$

$$C_{xx}(\alpha_k; \tau) = \lim_{N \to \infty} \frac{1}{N} \sum_{n=0}^{N-1} c_{xx}(n; \tau) e^{-j\alpha_k n} . \tag{17.2}$$

The set of cycles, $A_{xx}^c(\tau) := \{\alpha_k : C_{xx}(\alpha_k; \tau) \neq 0, \; -\pi < \alpha_k \leq \pi\}$, must be countable and the limit is assumed to exist at least in the mean-square sense [9, Thm. 1.15].

Definition 17.2 and Eq. (17.2) for ACS, subsume CS Definition 17.1 and Eq. (17.1). Note that the latter require integer period and a finite set of cycles. In the α-domain, ACS signals exhibit lines but not necessarily at harmonically related cycles. The following example will illustrate the cyclic quantities defined thus far:

EXAMPLE 17.1:   Harmonic in multiplicative and additive noise

Let

$$x(n) = s(n) \cos(\omega_0 n) + v(n) , \tag{17.3}$$

where $s(n)$, $v(n)$ are assumed real, stationary, and mutually independent. Such signals appear when communicating through flat-fading channels, and with weather radar or sonar returns when, in addition to sensor noise $v(n)$, backscattering, target scintillation, or fluctuating propagation media give rise to random amplitude variations modeled by $s(n)$ [33]. We will consider two cases:
Case 1: $\mu_s \neq 0$. The mean in (17.3) is $\mu_x(n) = \mu_s \cos(\omega_0 n) + \mu_v$, and the cyclic mean:

$$C_x(\alpha) := \lim_{N \to \infty} \frac{1}{N} \sum_{n=0}^{N-1} \mu_x(n) e^{-j\alpha n} = \frac{\mu_s}{2} [\delta(\alpha - \omega_0) + \delta(\alpha + \omega_0)] + \mu_v \delta(\alpha) , \tag{17.4}$$

where in (17.4) we used the definition of Kronecker's delta

$$\lim_{N \to \infty} \frac{1}{N} \sum_{n=0}^{N-1} e^{j\alpha n} = \delta(\alpha) := \begin{cases} 1 & \alpha = 0 \\ 0 & \text{else} \end{cases} . \tag{17.5}$$

---

[1] The term cyclostationarity is due to Bennet [3]. Cyclostationary processes in economics and atmospheric sciences are also referred to as seasonal time series [50].

Signal $x(n)$ in (17.3) is thus (first-order) cyclostationary with set of cycles $A_x^c = \{\pm\omega_0, 0\}$. If $X_N(\omega) := \sum_{n=0}^{N-1} x(n)\exp(-j\omega n)$, then from (17.4) we find $C_x(\alpha) = \lim_{N\to\infty} N^{-1} E\{X_N(\alpha)\}$; thus, the cyclic mean can be interpreted as an averaged DFT and ω_0 can be retrieved by picking the peak of $|X_N(\omega)|$ for $\omega \neq 0$.

Case 2: $\mu_s = 0$. From (17.3) we find the correlation $c_{xx}(n; \tau) = c_{ss}(\tau)[\cos(2\omega_0 n + \omega_0\tau) + \cos(\omega_0\tau)]/2 + c_{vv}(\tau)$. Because $c_{xx}(n; \tau)$ is periodic in n, $x(n)$ is (second-order) CS with cyclic correlation [c.f. (17.2) and (17.5)]

$$C_{xx}(\alpha; \tau) = \frac{c_{ss}(\tau)}{4}\left[\delta(\alpha + 2\omega_0)e^{j\omega_0\tau} + \delta(\alpha - 2\omega_0)e^{-j\omega_0\tau}\right]$$
$$+ \left[\frac{c_{ss}(\tau)}{2}\cos(\omega_0\tau) + c_{vv}(\tau)\right]\delta(\alpha). \tag{17.6}$$

The set of cycles is $A_{xx}^c(\tau) = \{\pm 2\omega_0, 0\}$ provided that $c_{ss}(\tau) \neq 0$ and $c_{vv}(\tau) \neq 0$. The set $A_{xx}^c(\tau)$ is lag-dependent in the sense that some cycles may disappear while others may appear for different τs. To illustrate the τ-dependence, let $s(n)$ be an MA process of order q. Clearly, $c_{ss}(\tau) = 0$ for $|\tau| > q$, and thus $A_{xx}^c(\tau) = \{0\}$ for $|\tau| > q$.

The CS process in (17.3) is just one example of signals involving products and sums of stationary processes such as $s(n)$ with (almost) periodic deterministic sequences $d(n)$, or, CS processes $x(n)$. For such signals, the following properties are useful:

Property 1 *Finite sums and products of ACS signals are ACS. If $x_i(n)$ is CS with period P_i, then for λ_i constants, $y_1(n) := \sum_{i=1}^{l_1} \lambda_i x_i(n)$ and $y_2(n) := \prod_{i=1}^{l_2} \lambda_i x_i(n)$ are also CS. Unless cycle cancellations occur among $x_i(n)$ components, the period of $y_1(n)$ and $y_2(n)$ equals the least common multiple of the P_is. Similarly, finite sums and products of stationary processes with deterministic (almost) periodic signals are also ACS processes.*

As examples of random–deterministic mixtures, consider

$$x_1(n) = s(n) + d(n) \quad \text{and} \quad x_2(n) = s(n)d(n), \tag{17.7}$$

where $s(n)$ is zero-mean, stationary, and $d(n)$ is deterministic (almost) periodic with Fourier Series coefficients $D(\alpha)$. Time-varying correlations are, respectively,

$$c_{x_1 x_1}(n; \tau) = c_{ss}(\tau) + d(n)d(n + \tau) \quad \text{and} \quad c_{x_2 x_2}(n; \tau) = c_{ss}(\tau)d(n)d(n + \tau). \tag{17.8}$$

Both are (almost) periodic in n, with cyclic correlations

$$C_{x_1 x_1}(\alpha; \tau) = c_{ss}(\tau)\delta(\alpha) + D_2(\alpha; \tau) \quad \text{and} \quad C_{x_2 x_2}(\alpha; \tau) = c_{ss}(\tau)D_2(\alpha; \tau), \tag{17.9}$$

where $D_2(\alpha; \tau) = \sum_\beta D(\beta)D(\alpha - \beta)\exp[j(\alpha - \beta)\tau]$, since the Fourier Series coefficients of the product $d(n)d(n+\tau)$ are given by the convolution of each component's coefficients in the α-domain. To reiterate the dependence on τ, notice that if $d(n)$ is a periodic ± 1 sequence, then $c_{x_2 x_2}(n; 0) = c_{ss}(0)d^2(n) = c_{ss}(0)$, and hence periodicity disappears at $\tau = 0$.

ACS signals appear often in nature with the underlying periodicity hidden, unknown, or inaccessible. In contrast, CS signals are often man-made and arise as a result of, e.g., oversampling (by a known integer factor P) digital communication signals, or by sampling a spatial waveform with P antennas (see also Section 17.4).

Both CS and ACS definitions could also be given in terms of the Fourier Transforms ($\tau \to \omega$) of $c_{xx}(n; \tau)$ and $C_{xx}(\alpha; \tau)$, namely the time-varying and the cyclic spectra which we denote by $S_{xx}(n; \omega)$ and $S_{xx}(\alpha; \omega)$. Suppose $c_{xx}(n; \tau)$ and $C_{xx}(\alpha; \tau)$ are absolutely summable w.r.t. τ for all n in $\mathcal{Z}$ and α_k in $A_{xx}^c(\tau)$. We can then define and relate time-varying and cyclic spectra as follows:

$$S_{xx}(n; \omega) := \sum_{\tau=-\infty}^{\infty} c_{xx}(n; \tau)e^{-j\omega\tau} = \sum_{\alpha_k \in A_{xx}^s} S_{xx}(\alpha_k; \omega)e^{j\alpha_k n} \tag{17.10}$$

$$S_{xx}(\alpha_k; \omega) \quad := \quad \sum_{\tau=-\infty}^{\infty} C_{xx}(\alpha_k; \tau) e^{-j\omega\tau} = \lim_{N \to \infty} \frac{1}{N} \sum_{n=0}^{N-1} S_{xx}(n; \omega) e^{-j\alpha_k n} . \quad (17.11)$$

Absolute summability w.r.t. τ implies vanishing memory as the lag separation increases, and many real life signals satisfy these so called mixing conditions [5, Ch. 2]. Power signals are not absolutely summable, but it is possible to define cyclic spectra equivalently [for real-valued $x(n)$] as

$$S_{xx}(\alpha_k; \omega) := \lim_{N \to \infty} \frac{1}{N} E\{X_N(\omega) X_N(\alpha_k - \omega)\} , \quad X_N(\omega) := \sum_{n=0}^{N-1} x(n) e^{-j\omega n} . \quad (17.12)$$

If $x(n)$ is complex ACS, then one also needs $\bar{S}_{xx}(\alpha_k; \omega) := \lim_{N \to \infty} N^{-1} E\{X_N^*(-\omega) X_N(\alpha_k - \omega)\}$. Both S_{xx} and $\bar{S}_{xx}$ reveal presence of spectral correlation. This must be contrasted to stationary processes whose spectral components, $X_N(\omega_1)$, $X_N(\omega_2)$ are known to be asymptotically uncorrelated unless $|\omega_1 \pm \omega_2| = 0$ (mod 2π) [5, Ch. 4]. Specifically, we have from (17.12) that:

Property 2 *If $x(n)$ is ACS or CS, the N-point Fourier transform $X_N(\omega_1)$ is correlated with $X_N(\omega_2)$ for $|\omega_1 \pm \omega_2| = \alpha_k$ (mod 2π), and $\alpha_k \in A_{xx}^s$.*

Before dwelling further on spectral characterization of ACS processes, it is useful to note the diversity of tools available for processing. Stationary signals are analyzed with time-invariant correlations (lag-domain analysis), or with power spectral densities (frequency-domain analysis). However, CS, ACS, and generally nonstationary signals entail four variables: $(n, \tau, \alpha, \omega)$:=(time, lag, cycle, frequency). Grouping two variables at a time, four domains of analysis become available and their relationship is summarized in Fig. 17.1. Note that pairs $(n; \tau) \leftrightarrow (\alpha; \tau)$, or, $(n; \omega) \leftrightarrow (\alpha; \omega)$, have τ or ω fixed and are Fourier Series pairs; whereas $(n; \tau) \leftrightarrow (n; \omega)$, or, $(\alpha; \tau) \leftrightarrow (\alpha; \omega)$, have n or α fixed and are related by Fourier Transforms. Further insight on the links between stationary and cyclostationary processes is

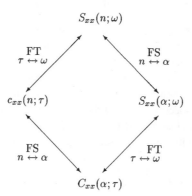

FIGURE 17.1 Four domains for analyzing cyclostationary signals.

gained through the uniform shift (or phase) randomization concept. Let $x(n)$ be CS with period P, and define $y(n) := x(n + \theta)$, where θ is uniformly distributed in $[0, P)$ and independent of $x(n)$. With $c_{yy}(n; \tau) := E_\theta\{E_x[x(n + \theta)x(n + \tau + \theta)]\}$, we find:

$$c_{yy}(n; \tau) = \frac{1}{P} \sum_{p=0}^{P-1} c_{xx}(p; \tau) := C_{xx}(0; \tau) := c_{yy}(\tau) , \quad (17.13)$$

where the first equality follows because θ is uniform and the second uses the CS definition in (17.1). Noting that c_{yy} is not a function of n, we have established (see also [15, 38]):

Property 3 *A CS process $x(n)$ can be mapped to a stationary process $y(n)$ using a shift θ, uniformly distributed over its period, and the transformation $y(n) := x(n + \theta)$.*

Such a mapping is often used with harmonic signals; e.g., $x(n) = A \exp[j(2\pi n/P + \theta)] + v(n)$ is according to Property 2 a CS signal, but can be stationarized by uniform phase randomization. An alternative trick for stationarizing signals which involve complex harmonics is conjugation. Indeed, $c_{xx*}(n; \tau) = A^2 \exp(-j2\pi\tau/P) + c_{vv}(\tau)$ is not a function of n — but why deal with CS or ACS processes if conjugation or phase randomization can render them stationary?

Revisiting Case 2 of Example 17.1 offers a partial answer when the goal is to estimate the frequency ω_0. Phase randomization of $x(n)$ in (17.3) leads to a stationary $y(n)$ with correlation found by substituting $\alpha = 0$ in (17.6). This leads to $c_{yy}(\tau) = (1/2)c_{ss}(\tau)\cos(\omega_0\tau) + c_{vv}(\tau)$, and shows that if $s(n)$ has multiple spectral peaks, or if $s(n)$ is broadband, then multiple peaks or smearing of the spectral peak hamper estimation of ω_0 (in fact, it is impossible to estimate ω_0 from the spectrum of $y(n)$ if $s(n)$ is white). In contrast, picking the peak of $C_{xx}(\alpha; \tau)$ in (17.6) yields ω_0, provided that $\omega_0 \in (0, \pi)$ so that spectral folding is prevented [33]. Equation (17.13) provides a more general answer. Phase randomization restricts a CS process only to one cycle, namely $\alpha = 0$. In other words, the cyclic correlation $C_{xx}(\alpha; \tau)$ contains the "stationarized correlation" $C_{xx}(0; \tau)$ and additional information in cycles $\alpha \neq 0$.

Since CS and ACS processes form a superset of stationary ones, it is useful to know how a stationary process can be viewed as a CS process. Note that if $x(n)$ is stationary, then $c_{xx}(n; \tau) = c_{xx}(\tau)$ and on using (17.2) and (17.5) we find:

$$C_{xx}(\alpha; \tau) = c_{xx}(\tau) \left[\lim_{N \to \infty} \frac{1}{N} \sum_{n=0}^{N-1} e^{-j\alpha n} \right] = c_{xx}(\tau)\delta(\alpha). \tag{17.14}$$

Intuitively, (17.14) is justified if we think that stationarity reflects "zero time-variation" in the correlation $c_{xx}(\tau)$. Formally, (17.14) implies:

Property 4 *Stationary processes can be viewed as ACS or CS with cyclic correlation $C_{xx}(\alpha; \tau) = c_{xx}(\tau)\delta(\alpha)$.*

Separation of information bearing ACS signals from stationary ones (e.g., noise) is desired in many applications and can be achieved based on Property 4 by excluding the cycle $\alpha = 0$.

Next, it is of interest to view CS signals as special cases of general nonstationary processes with 2-D correlation $r_{xx}(n_1, n_2) := E\{x(n_1)x(n_2)\}$, and 2-D spectral densities $S_{xx}(\omega_1, \omega_2) := FT[r_{xx}(n_1, n_2)]$ that are assumed to exist.[2] Two questions arise: What are the implications of periodicity in the (ω_1, ω_2) plane? and how does the cyclic spectra in (17.10) through (17.12) relate to $S_{xx}(\omega_1, \omega_2)$? The answers are summarized in Fig. 17.2, which illustrates that the support of CS processes in the (ω_1, ω_2) plane consists of $2P - 1$ parallel lines (with unity slope) intersecting the axes at equidistant points $2\pi/P$ far apart from each other. More specifically, we have [34]:

Property 5 *A CS process with period P is a special case of a nonstationary (harmonizable) process with 2-D spectral density given by*

$$S_{xx}(\omega_1, \omega_2) = \sum_{k=-(P-1)}^{P-1} S_{xx}\left(\frac{2\pi}{P}k; \omega_1\right) \delta_D\left(\omega_2 - \omega_1 + \frac{2\pi}{P}k\right), \tag{17.15}$$

where δ_D denotes the delta of Dirac.

For stationary processes, only the $k = 0$ term survives in (17.15) and we obtain $S_{xx}(\omega_1, \omega_2) =$

---

[2]Nonstationary processes with Fourier transformable 2-D correlations are called harmonizable processes.

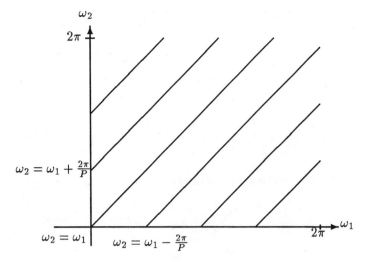

FIGURE 17.2 Support of 2-D spectrum $S_{xx}(\omega_1, \omega_2)$ for CS processes.

$S_{xx}(0; \omega_1)\delta_D(\omega_2 - \omega_1)$; i.e., the spectral mass is concentrated on the diagonal of Fig. 17.2. The well-structured spectral support for CS processes will be used to test for presence of cyclostationarity and estimate the period P. Furthermore, the superposition of lines parallel to the diagonal hints towards representing CS processes as a superposition of stationary processes. Next we will examine two such representations introduced by Gladysev [34] (see also [22, 38, 49], and [56]).

We can uniquely write $n_0 = nP + i$ and express $x(n_0) = x(nP + i)$, where the remainder i takes values $0, 1, \ldots, P - 1$. For each i, define the subprocess $x_i(n) := x(nP + i)$. In multirate processing, the $P \times 1$ vector $\mathbf{x}(n) := [x_0(n) \ldots x_{P-1}(n)]'$ constitutes the so-called polyphase decomposition of $x(n)$ [51, Ch. 12]. As shown in Fig. 17.3, each $x_i(n)$ is formed by downsampling an advanced copy of $x(n)$. On the

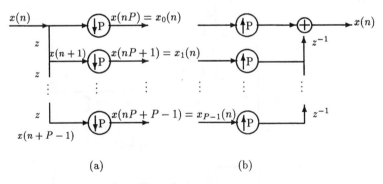

(a)   (b)

FIGURE 17.3 Representation 1: (a) analysis, (b) synthesis.

other hand, combining upsampled and delayed $x_i(n)$s, we can synthesize the CS process as:

$$x(n) = \sum_{i=0}^{P-1} \sum_l x_i(l)\delta(n - i - lP). \tag{17.16}$$

We maintain that subprocesses $\{x_i(n)\}_{i=0}^{P-1}$ are (jointly) stationary, and thus $\mathbf{x}(n)$ is vector stationary. Suppose for simplicity that $E\{x(n)\} = 0$, and start with $E\{x_{i_1}(n)x_{i_2}(n + \tau)\} = E\{x(nP + i_1)x(nP + \tau P + i_2)\} := c_{xx}(i_1 + nP; i_2 - i_1 + \tau P)$. Because $x(n)$ is CS, we can drop nP and c_{xx} becomes independent of n establishing that $x_{i_1}(n)$, $x_{i_2}(n)$ are (jointly) stationary with correlation:

$$c_{x_{i_1} x_{i_2}}(\tau) = c_{xx}(i_1; i_2 - i_1 + \tau P), \quad i_1, i_2 \in [0, P-1]. \tag{17.17}$$

Using (17.17), it can be shown that auto- and cross-spectra of $x_{i_1}(n)$, $x_{i_2}(n)$ can be expressed in terms of the cyclic spectra of $x(n)$ as [56],

$$S_{x_{i_1} x_{i_2}}(\omega) = \frac{1}{P} \sum_{k_1=0}^{P-1} \sum_{k_2=0}^{P-1} S_{xx}\left(\frac{2\pi}{P}k_1; \frac{\omega - 2\pi k_2}{P}\right) e^{j[(\frac{\omega-2\pi k_2}{P})(i_2-i_1)+\frac{2\pi}{P}k_1 i_1]}. \tag{17.18}$$

To invert (17.18), we Fourier transform (17.16) and use (17.12) to obtain [for $x(n)$ real]

$$S_{xx}\left(\frac{2\pi}{P}k; \omega\right) = \sum_{i_1=0}^{P-1} \sum_{i_2=0}^{P-1} S_{x_{i_1} x_{i_2}}(\omega) e^{j\omega(i_2-i_1)} e^{-j\frac{2\pi}{P}k i_2}. \tag{17.19}$$

Based on (17.16) through (17.19), we infer that cyclostationary signals with period P can be analyzed as stationary $P \times 1$ multichannel processes and vice versa. In summary, we have:

Representation 1 (Decimated Components) *CS process $x(n)$ can be represented as a P-variate stationary multichannel process $\mathbf{x}(n)$ with components $x_i(n) = x(nP + i)$, $i = 0, 1, \ldots, P - 1$. Cyclic spectra and stationary auto- and cross-spectra are related as in (17.18) and (17.19).*

An alternative means of decomposing a CS process into stationary components is by splitting the $(-\pi, \pi]$ spectral support of $X_N(\omega)$ into bands each of width $2\pi/P$ [22]. As shown in Fig. 17.4, this can be accomplished by passing modulated copies of $x(n)$ through an ideal low-pass filter $H_0(\omega)$ with spectral support $(-\pi/P, \pi/P]$. The resulting subprocesses $\bar{x}_m(n)$ can be shifted up in frequency and

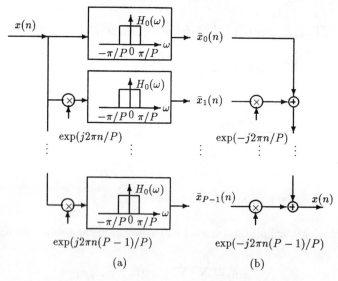

FIGURE 17.4 Representation 2: (a) analysis, (b) synthesis.

recombined to synthesize the CS process as: $x(n) = \sum_{m=0}^{P-1} \bar{x}_m(n) \exp(-j2\pi mn/P)$. Within each band, frequencies are separated by less than $2\pi/P$ and according to Property 2, there is no correlation between spectral components $\bar{X}_{m,N}(\omega_1)$ and $\bar{X}_{m,N}(\omega_2)$; hence, $\bar{x}_m(n)$ components are stationary with auto- and cross-spectra having nonzero support over $-\pi/P < \omega < \pi/P$. They are related with the cyclic spectra as follows:

$$S_{\bar{x}_{m_1}\bar{x}_{m_2}}(\omega) = S_{xx}\left(\frac{2\pi}{P}(m_1 - m_2); \omega + \frac{2\pi}{P}m_1\right) , \quad |\omega| < \frac{\pi}{P} . \tag{17.20}$$

Equation (17.20) suggests that cyclostationary signal analysis is linked with stationary subband processing.

Representation 2 (Subband Components) *CS process $x(n)$ can be represented as a superposition of P stationary narrowband subprocesses according to:* $x(n) = \sum_{m=0}^{P-1} \bar{x}_m(n) \exp(-j2\pi mn/P)$. *Auto- and cross-spectra of $\bar{x}_m(n)$ can be found from the cyclic spectra of $x(n)$ as in (17.20).*

Because ideal low-pass filters cannot be designed, the subband decomposition seems less practical. However, using Representation 1 and exploiting results from uniform DFT filter banks, it is possible using FIR low-pass filters to obtain stationary subband components (see e.g., [51, Ch. 12]). We will not pursue this approach further, but Representation 1 will be used next for estimating time-varying correlations of CS processes based on a single data record.

17.3 Estimation, Time-Frequency Links, Testing

The time-varying and cyclic quantities introduced in (17.1), (17.2), and (17.10) through (17.12), entail ideal expectations (i.e., ensemble averages) and unless reliable estimators can be devised from finite (and often noisy) data records, their usefulness in practice is questionable. For stationary processes with (at least asymptotically) vanishing memory,[3] sample correlations and spectral density estimators converge to their ensembles as the record length $N \to \infty$. Constructing reliable (i.e., consistent) estimators for nonstationary processes, however, is challenging and generally impossible. Indeed, capturing time-variations calls for short observation windows, whereas variance reduction demands long records for sample averages to converge to their ensembles.

Fortunately, ACS and CS signals belong to the class of processes with "well-structured" time-variations that under suitable mixing conditions allow consistent single record estimators. The key is to note that although $c_{xx}(n; \tau)$ and $S_{xx}(n; \omega)$ are *time-varying*, they are expressed in terms of cyclic quantities, $C_{xx}(\alpha_k; \tau)$ and $S_{xx}(\alpha_k; \omega)$, which are *time-invariant*. Indeed, in (17.2) and (17.10) time-variation is assigned to the Fourier basis.

17.3.1 Estimating Cyclic Statistics

First we will consider ACS processes with known cycles α_k. Simpler estimators for CS processes and cycle estimation methods will be discussed later in the section. If $x(n)$ has nonzero mean, we estimate the cyclic mean as in Example 17.1 using the normalized DFT: $\hat{C}_{xx}(\alpha_k) = N^{-1}\sum_{n=0}^{N-1} x(n) \exp(-j\alpha_k n)$. If the set of cycles is finite, we estimate the time-varying mean as: $\hat{c}_{xx}(n) = \sum_{\alpha_k} \hat{C}_{xx}(\alpha_k) \exp(j\alpha_k n)$. Similarly, for zero-mean ACS processes we estimate first cyclic and then time-varying correlations using:

$$\hat{C}_{xx}(\alpha_k; \tau) = \frac{1}{N}\sum_{n=0}^{N-1} x(n)x(n+\tau)e^{-j\alpha_k n} ,$$

$$\hat{c}_{xx}(n; \tau) = \sum_{\alpha_k \in A_{xx}^c(\tau)} \hat{C}_{xx}(\alpha_k; \tau)e^{j\alpha_k n} . \tag{17.21}$$

Note that $\hat{C}_{xx}$ can be computed efficiently using the FFT of the product $x(n)x(n+\tau)$.

---

[3] Well-separated samples of such processes are asymptotically independent. Sufficient (so-called mixing) conditions include absolute summability of cumulants and are satisfied by many real life signals (see [5, 12, Ch. 2]).

For cyclic spectral estimation, two options are available: (1) smoothed cyclic periodograms and (2) smoothed cyclic correlograms. The first is motivated by (17.12) and smooths the cyclic periodogram, $I_{xx}(\alpha; \omega) := N^{-1} X_N(\omega) X_N(\alpha - \omega)$, using a frequency-domain window $W(\omega)$. The second follows (17.2) and Fourier transforms $\hat{C}_{xx}(\alpha; \tau)$ after smoothing it by a lag-window $w(\tau)$ with support $\tau \in [-M, M]$. Either one of the resulting estimates:

$$\hat{S}_{xx}^{(i)}(\alpha; \omega) = \frac{1}{N} \sum_{n=0}^{N-1} W\left(\omega - \frac{2\pi}{N}n\right) I_{xx}\left(\alpha; \frac{2\pi}{N}n\right),$$

$$\hat{S}_{xx}^{(ii)}(\alpha; \omega) = \sum_{\tau=-M}^{M} w(\tau)\hat{C}_{xx}(\alpha; \tau)e^{-j\omega\tau}, \tag{17.22}$$

can be used to obtain time-varying spectral estimates; e.g., using $\hat{S}_{xx}^{(i)}(\alpha; \omega)$, we estimate $S_{xx}(n; \omega)$ as:

$$\hat{S}_{xx}^{(i)}(n; \omega) = \sum_{\alpha_k \in A_{xx}^s} \hat{S}_{xx}^{(i)}(\alpha_k; \omega)e^{j\alpha_k n}. \tag{17.23}$$

Estimates (17.21) through (17.23) apply to ACS (and hence CS) processes with a finite number of known cycles, and rely on the following steps: (1) estimate the time-invariant (or "stationary") quantities by dropping limits and expectations from the corresponding cyclic definitions, and (2) use the cyclic estimates to obtain time-varying estimates relying on the Fourier synthesis Eqs. (17.2) and (17.10). Selection of the windows in (17.22), variance expressions, consistency, and asymptotic normality of the estimators in (17.21) through (17.23) under mixing conditions can be found in [11, 12, 24, 39] and references therein.

When $x(n)$ is CS with known integer period P, estimation of time-varying correlations and spectra becomes easier. Recall that thanks to Representations 1 and 2, not only $c_{xx}(n; \tau)$ and $S_{xx}(n; \omega)$, but the process $x(n)$ itself can be analyzed into P stationary components. Starting with (17.16), it can be shown that $c_{xx}(i; \tau) = c_{x_i x_{i+\tau}}(0)$, where $i = 0, 1, \ldots, P-1$ and subscript $i + \tau$ is understood mod(P). Because the subprocesses $x_i(n)$ and $x_{i+\tau}(n)$ are stationary, their cross-covariances can be estimated consistently using sample averaging; hence, the time-varying correlation can be estimated as:

$$\hat{c}_{xx}(i; \tau) = \hat{c}_{x_i x_{i+\tau}}(0) = \frac{1}{[N/P]} \sum_{n=0}^{[N/P]-1} x(nP + i)x(nP + i + \tau), \tag{17.24}$$

where the integer part $[N/P]$ denotes the number of samples per subprocess $x_i(n)$, and the last equality follows from the definition of $x_i(n)$ in Representation 1. Similarly, the time-varying periodogram can be estimated using: $I_{xx}(n; \omega) = P^{-1} \sum_{k=0}^{P-1} X_P(\omega) X_P(2\pi k/P - \omega) \exp(-j2\pi kn/P)$, and then smoothed to obtain a consistent estimate of $S_{xx}(n; \omega)$.

17.3.2 Links with Time-Frequency Representations

Consistency (and hence reliability) of single record estimates is a notable difference between cyclostationary and time-frequency signal analyses. Short-time Fourier transforms, the Wigner-Ville, and derivative representations are valuable exploratory (and especially graphical) tools for analyzing nonstationary signals. They promise applicability on general nonstationarities, but unless slow variations are present and multiple independent data records are available, their usefulness in estimation tasks is rather limited. In contrast, ACS analysis deals with a specific type of structured variation, namely (almost) periodicity, but allows for rapid variations and consistent single record sample estimates. Intuitively speaking, cyclostationarity provides within a single record, multiple periods that can be viewed as "multiple realizations."

Interestingly, for ACS processes there is a close relationship between the normalized asymmetric ambiguity function $A(\alpha; \tau)$ [37], and the sample cyclic correlation in (17.21):

$$N\hat{C}_{xx}(\alpha; \tau) = A(\alpha; \tau) := \sum_{n=0}^{N-1} x(n)x(n + \tau)e^{-j\alpha n} . \qquad (17.25)$$

Similarly, one may associate the Wigner-Ville with the time-varying periodogram $I_{xx}(n; \omega) = \sum_{\tau=-(N-1)}^{N-1} x(n) \, x(n + \tau) \exp(-j\omega\tau)$. In fact, the aforementioned equivalences and the consistency results of [12] establish that ambiguity and Wigner-Ville processing of ACS signals is reliable even when only a single data record is available. The following example uses a chirp signal to stress this point and shows how some of our sample estimates can be extended to complex processes.

EXAMPLE 17.2: Chirp in multiplicative and additive noise

Consider $x(n) = s(n) \exp(j\omega_0 n^2) + v(n)$, where $s(n)$, $v(n)$, are zero mean, stationary, and mutually independent; $c_{xx}(n; \tau)$ is nonperiodic for almost every ω_0, and hence $x(n)$ is not (second-order) ACS. Even when $E\{s(n)\} \neq 0$, $E\{x(n)\}$ is also nonperiodic, implying that $x(n)$ is not first-order ACS either. However,

$$\tilde{c}_{xx*}(n; \tau) \quad := \quad c_{xx*}(n + \tau; -2\tau) := E\{x(n + \tau)x^*(n - \tau)\}$$
$$= \quad c_{ss}(2\tau) \exp(j4\omega_0 \tau n) + c_{vv*}(2\tau) , \qquad (17.26)$$

exhibits (almost) periodicity and its cyclic correlation is given by: $\tilde{C}_{xx*}(\alpha; \tau) = c_{ss}(\tau)\delta(\alpha - 4\omega_0\tau) + c_{vv*}(2\tau)\delta(\alpha)$. Assuming $c_{ss}(\tau) \neq 0$, the latter allows evaluation of ω_0 by picking the peak of the sample cyclic correlation magnitude evaluated at, e.g., $\tau = 1$, as follows:

$$\hat{\omega}_0 \quad = \quad -\frac{1}{4}\arg \ \max_{\alpha \neq 0} \ |\hat{\tilde{C}}_{xx*}(\alpha; 1)| \ ,$$

$$\hat{\tilde{C}}_{xx*}(\alpha; \tau) \quad = \quad \frac{1}{N}\sum_{n=0}^{N-1} x(n + \tau)x^*(n - \tau)e^{-j\alpha n} . \qquad (17.27)$$

The $\hat{\tilde{C}}_{xx*}(\alpha; \tau)$ estimate in (17.27) is nothing but the symmetric ambiguity function. Because $x(n)$ is ACS, $\hat{\tilde{C}}_{xx*}$ can be shown to be consistent. This provides yet one more reason for the success of time-frequency representations with chirp signals. Interestingly, (17.27) shows that exploitation of cyclostationarity allows not only for additive noise tolerance [by avoiding the $\alpha = 0$ cycle in (17.27)], but also permits parameter estimation of chirps modulated by stationary multiplicative noise $s(n)$.

17.3.3 Testing for Cyclostationarity

In certain applications involving man-made (e.g., communication) signals, presence of cyclostationarity and knowledge of the cycles is assured by design (e.g., baud rates or oversampling factors). In other cases, however, only a time series $\{x(n)\}_{n=0}^{N-1}$ is given and two questions arise: How does one detect cyclostationarity, and if $x(n)$ is confirmed to be CS of a certain order, how does one estimate the cycles present? The former is addressed by testing hypotheses of nonzero $\hat{C}_x(\alpha_k)$, $\hat{C}_{xx}(\alpha_k; \tau)$ or $\hat{S}_{xx}(\alpha_k; \omega)$ over a fine cycle-frequency grid obtained by sufficient zero-padding prior to taking the FFT.

Specifically, to test whether $x(n)$ exhibits cyclostationarity in $\{\hat{C}_{xx}(\alpha; \tau_l)\}_{l=1}^{L}$ for at least one lag, we form the $(2L + 1) \times 1$ vector $\hat{\mathbf{c}}_{xx}(\alpha) := [\hat{C}_{xx}^R(\alpha; \tau_1) \ldots \hat{C}_{xx}^R(\alpha; \tau_L); \hat{C}_{xx}^I(\alpha; \tau_1) \ldots \hat{C}_{xx}^I(\alpha; \tau_L)]'$ where superscript $R(I)$ denotes real (imaginary) part. Similarly, we define the ensemble vector $\mathbf{c}_{xx}(\alpha)$ and the error $\mathbf{e}_{xx}(\alpha) := \hat{\mathbf{c}}_{xx}(\alpha) - \mathbf{c}_{xx}(\alpha)$. For N large, it is known that $\sqrt{N} \, \mathbf{e}_{xx}(\alpha)$ is Gaussian with pdf $\mathcal{N}(\mathbf{0}, \Sigma_c)$.

An estimate $\hat{\Sigma}_c$ of the asymptotic covariance can be computed from the data [12]. If α is not a cycle for all $\{\tau_l\}_{l=1}^L$, then $\mathbf{c}_{xx}(\alpha) \equiv \mathbf{0}$, $\mathbf{e}_{xx}(\alpha) = \hat{\mathbf{c}}_{xx}(\alpha)$ will have zero mean, and $\hat{D}_{2c}(\alpha) := \hat{\mathbf{c}}'_{xx}(\alpha)\hat{\Sigma}_c^\dagger(\alpha)\hat{\mathbf{c}}_{xx}(\alpha)$ will be central chi-square. For a given false-alarm rate, we find from χ^2 tables a threshold Γ and test [10]

$$H_0: \quad \hat{D}_{xx}^c(\alpha) \geq \Gamma \implies \alpha \in A_{xx}^c \quad \text{vs.} \quad H_1: \quad \hat{D}_{xx}^c(\alpha) < \Gamma \implies \alpha \notin A_{xx}^c. \tag{17.28}$$

Alternate $2D$ contour plots revealing presence of spectral correlation rely on (17.15) and more specifically on its normalized version (coherence or correlation coefficient) estimated as [40]

$$\rho_{xx}(\omega_1, \omega_2) := \frac{\frac{1}{M}\sum_{m=0}^{M-1} | X_N(\omega_1 + \frac{2\pi m}{M})X_N^*(\omega_2 + \frac{2\pi m}{M}) |^2}{\frac{1}{M}\sum_{m=0}^{M-1} | X_N(\omega_1 + \frac{2\pi m}{M}) |^2 \frac{1}{M}\sum_{m=0}^{M-1} | X_N(\omega_2 + \frac{2\pi m}{M}) |^2}. \tag{17.29}$$

Plots of $\rho_{xx}(\omega_1, \omega_2)$ with the empirical thresholds discussed in [40] are valuable tools not only for cycle detection and estimation of CS signals but even for general nonstationary processes exhibiting partial (e.g., "transient" lag- or frequency-dependent) cyclostationarity.

EXAMPLE 17.3: Cyclostationarity test

Consider $x(n) = s_1(n)\cos(\pi n/8) + s_2(n)\cos(\pi n/4)$ with $s_1(n)$, $s_2(n)$, and $v(n)$ zero-mean, Gaussian, and mutually independent. To test for cyclostationarity and retrieve the possible periods present, $N = 2,048$ samples were generated; $s_1(n)$ and $s_2(n)$ were simulated as AR(1) with variances $\sigma_{s_1}^2 = \sigma_{s_2}^2 = 2$, while $v(n)$ was white with variance $\sigma_v^2 = 0.1$. Figure 17.5a shows $|\hat{C}_{xx}(\alpha; 0)|$ peaking at $\alpha = \pm 2(\pi/8), \pm 2(\pi/4), 0$ as expected, while Fig. 17.5b depicts $\rho_{xx}(\omega_1, \omega_2)$ computed as in (17.29) with $M = 64$. The parallel lines in Fig. 17.5b are seen at $|\omega_1 - \omega_2| = 0, \pi/8, \pi/4$ revealing the periods present.

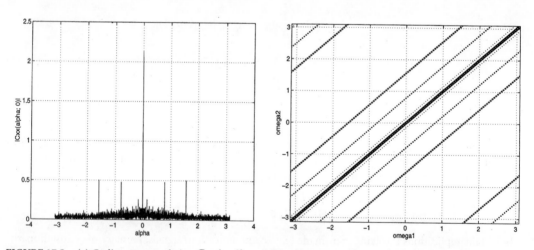

FIGURE 17.5   (a) Cyclic cross-correlation $C_{xx}(\alpha; 0)$, and (b) coherence $\rho_{xx}(\omega_1, \omega_2)$ (Example 17.3).

One can easily verify from (17.11) that $C_{xx}(\alpha; 0) = (2\pi)^{-1}\int_{-\pi}^{\pi} S_{xx}(\alpha; \omega)d\omega$. It also follows from (17.15) that $S_{xx}(\alpha; \omega) = S_{xx}(\omega_1 = \omega, \omega_2 = \omega - \alpha)$; thus, $C_{xx}(\alpha; 0) = (2\pi)^{-1}\int_{-\pi}^{\pi} S_{xx}(\omega, \omega - \alpha)d\omega$, and for each α, we can view Fig. 17.5a as the (normalized) integral (or projection) of Fig. 17.5b along each parallel line [40]. Although $|\hat{C}_{xx}(\alpha; 0)|$ is simpler to compute using the FFT of $x^2(n)$, $\rho_{xx}(\omega_1, \omega_2)$ is generally more informative.

Because cyclostationarity is lag-dependent, as an alternative to $\rho_{xx}(\omega_1, \omega_2)$ one can also plot $|\hat{C}_{xx}(\alpha; \tau)|$ or $|\hat{S}_{xx}(\alpha; \omega)|$ for all τ or ω. Figures 17.6 and 17.7 show perspective and contour plots of $|\hat{C}_{xx}(\alpha; \tau)|$ for

$\tau \in [-31, 31]$ and $|\hat{S}_{xx}(\alpha; \omega)|$ for $\omega \in (-\pi, \pi]$, respectively. Both sets exhibit planes (lines) parallel to the τ-axis and ω-axis, respectively, at cycles $\alpha = \pm 2(\pi/8), \pm 2(\pi/4), 0$, as expected.

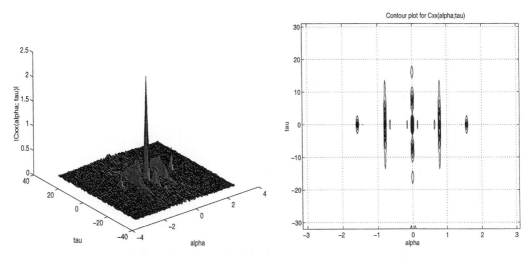

FIGURE 17.6   Cycle detection and estimation (Example 17.3): 3D and contour plots of $\hat{C}_{xx}(\alpha; \tau)$.

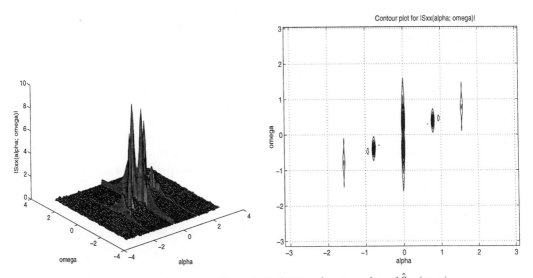

FIGURE 17.7   Cycle detection and estimation (Example 17.3): 3D and contour plots of $\hat{S}_{xx}(\alpha; \omega)$.

17.4   CS Signals and CS-Inducing Operations

We have already seen in Examples 17.1 and 17.2 that amplitude or index transformations of repetitive nature give rise to one class of CS signals. A second category consists of outputs of repetitive (e.g., periodically varying) systems excited by CS or even stationary inputs. Finally, it is possible to have cyclostationarity emerging in the output due to the data acquisition process (e.g., multiple sensors or fractional sampling).

17.4.1 Amplitude Modulation

General examples in this class include signals $x_1(n)$ and $x_2(n)$ of (17.7) or their combinations as de-scribed by Property 1. More specifically, we will focus on communication signals where random (of-ten i.i.d.) information data $w(n)$ are D/A converted with symbol period T_0, to obtain the process: $w_c(t) = \sum_l w(l)\delta_D(t - lT_0)$, which is CS in the continuous variable t. The continuous-time signal $w_c(t)$ is subsequently pulse shaped by the transmit filter $h_c^{(tr)}(t)$, modulated with the carrier $\exp(j\omega_c t)$, and transmitted over the linear time-invariant (LTI) channel $h_c^{(ch)}(t)$. On reception, the carrier is removed and the data are passed through the receive filter $h_c^{(rec)}(t)$ to suppress stationary additive noise. Defining the composite channel $h_c(t) := h_c^{(tr)} \star h_c^{(ch)} \star h_c^{(rec)}(t)$, the continuous time received signal at the baseband is:

$$r_c(t) = e^{j\omega_{ec} t} \sum_l w(l) h_c(t - lT_0 - \epsilon) + v_c(t) , \tag{17.30}$$

where $\epsilon \in (0, T_0)$ is the propagation delay, ω_{ec} denotes the frequency error between transmit-receive carriers, and $v_c(t)$ is AWGN. Signal $r_c(t)$ is CS due to: (1) the periodic carrier offset $e^{j\omega_{ec} t}$, and (2) the cyclostationarity of $w_c(t)$. However, (2) disappears in discrete-time if one samples at the symbol rate because $r(n) := r_c(nT_0)$ becomes

$$r(n) = e^{j\omega_e n} x(n) + v(n) , \quad x(n) := \sum_l w(l) h(n - l) , \quad n \in [0, N - 1] , \tag{17.31}$$

with $\omega_e := \omega_{ec} T_0$, $h(n) := h_c(nT_0 - \epsilon)$, and $v(n) := v_c(nT_0)$.

If $\omega_e = 0$, $x(n)$ (and thus $v(n)$) is stationary, whereas $\omega_e \neq 0$ renders $r(n)$ similar to the ACS signal in Example 17.1. When $w(n)$ is zero-mean, i.i.d., complex symmetric, we have: $E\{w(n)\} \equiv 0$, and $E\{w(n)w(n+\tau)\} \equiv 0$; thus, the cyclic mean and correlations cannot be used to retrieve ω_e. However, peak-picking the cyclic fourth-order correlation [Fourier coefficients of $r^4(n)$] yields $4\omega_e$ uniquely, provided $\omega_e < \pi/4$. If $E\{w^4(n)\} \equiv 0$, higher powers can be used to estimate and recover ω_e.

Having estimated ω_e, we form $\exp(-j\omega_e n)\, r(n)$ in order to demodulate the signal in (17.31). Tra-ditionally, cyclostationarity is removed from the discrete-time information signal, although it may be useful for other purposes (e.g., blind channel estimation) to retain cyclostationarity at the baseband signal $x(n)$. This can be accomplished by multiplying $w(n)$ with a P-periodic sequence $p(n)$ prior to pulse shaping. The noise-free signal in this case is $x(n) = \sum_l p(l)w(l)h(n - l)$, and has correlation, $\bar{c}_{xx}(n; \tau) = \sigma_w^2 \sum_l |p(n - l)|^2 h(l)h^*(l + \tau)$, which is periodic with period P. Cyclic correlations and spectra are given by [28]

$$\bar{C}_{xx}(\alpha; \tau) = \sigma_w^2 P_2(\alpha) \sum_l h(l)h^*(l + \tau)e^{-j\alpha l} ,$$

$$\bar{S}_{xx}(\alpha; \omega) = \sigma_w^2 P_2(\alpha) H^*(-\omega) H(\alpha - \omega) , \tag{17.32}$$

where $P_2(\alpha) := P^{-1} \sum_{m=0}^{P-1} |p(m)|^2 \exp(-j\alpha m)$ and $H(\omega) := \sum_{l=0}^{L} h(l) \exp(-j\omega l)$. As we will see later in this section, cyclostationarity can also be introduced at the transmitter using multirate operations, or at the receiver by fractional sampling. With a CS input, the channel $h(n)$ can be identified using noisy output samples only [28, 64, 65] — an important step towards blind equalization of (e.g., multipath) communication channels.

If $p(n) = 1$ for $n \in [0, P_1) \pmod{P}$ and $p(n) = 0$ for $n \in [P_1, P)$, the CS signal $x(n) = p(n)s(n) + v(n)$ can be used to model systematically missing observations. Periodically, the stationary signal $s(n)$ is observed in noise $v(n)$ for P_1 samples and disappears for the next $P - P_1$ data. Using $C_{xx}(\alpha; \tau) = P_2(\alpha; \tau)c_{ss}(\tau)$, the period P [and thus $P_2(\alpha; \tau)$] can be determined. Subsequently, $c_{ss}(\tau)$ can be retrieved and used for parametric or nonparametric spectral analysis of $s(n)$; see [32] and references therein.

17.4.2 Time Index Modulation

Suppose that a random CS signal $s(n)$ is delayed by D samples and received in zero-mean stationary noise $v(n)$ as: $x(n) = s(n - D) + v(n)$. With $s(n)$ independent of $v(n)$, the cyclic correlation is $C_{xx}(\alpha; \tau) = C_{ss}(\alpha; \tau) \exp(j\alpha D) + \delta(\alpha)c_{vv}(\tau)$ and the delay manifests itself as a phase of a complex exponential. But even when $s(n)$ models a narrowband deterministic signal, the delay appears in the exponent since $s(n - D(n)) \approx s(n) \exp(jD(n))$ [53]. Time-delay estimation of CS signals appears frequently in sonar and radar for range estimation where $D(n) = vn$ and v denotes velocity of propagation. $D(n)$ is also used to model Doppler effects that appear when relative motion is present. Note that with time-varying (e.g., accelerating) motion we have $D(n) = \gamma n^2$ and cyclostationarity appears in the complex correlation as explained in Example 17.2.

Polynomial delays are one form of time scale transformations. Another one is $d(n) = \lambda n + p(n)$, where λ is a constant and $p(n)$ is periodic with period P (e.g., [38]). For stationary $s(n)$, signal $x(n) = s[d(n)]$ is CS because $c_{xx}(n+lP; \tau) = c_{ss}[d(n+lP+\tau) - d(n+lP)] = c_{ss}[\lambda\tau + p(n) - p(n+\tau)] = c_{xx}(n; \tau)$. A special case is the familiar FM model with $d(n) = \omega_c n + h \sin(\omega_0 n)$ where h here denotes the modulation index. The signal and its periodically varying correlation are given by:

$$x(n) = A \cos[\omega_0 n + h \sin(\omega_0 n) + \phi] ,$$

$$c_{xx}(n; \tau) = \frac{A^2}{2} \cos[\omega_0 \tau + h \sin(\omega_0(n + \tau)) - h \sin(\omega_0 n)] . \tag{17.33}$$

In addition to communications, frequency modulated signals appear in sonar and radar when rotating and vibrating objects (e.g., propellers or helicopter blades) induce periodic variations in the phase of incident narrowband waveforms [2, 67].

Delays and scale modulations also appear in 2-D signals. Consider an image frame at time n with the scene displaced relative to time $n = 0$ by $[d_x(n), d_y(n)]$; in spatial and Fourier coordinates we have [8]

$$f(x, y; n) = f_0(x - d_x(n), y - d_y(n)),$$

$$F(\omega_x, \omega_y; n) = F_0(\omega_x, \omega_y)e^{-j\omega_x d_x(n)}e^{-j\omega_y d_y(n)} . \tag{17.34}$$

Images of moving objects having time-varying velocities can be modeled using polynomial displacements, whereas trigonometric $[d_x(n), d_y(n)]$ can be adopted when the motion is circular, or when the imaging sensor (e.g., camera) is vibrating. In either case, $F(\omega_x, \omega_y; n)$ is CS and thus cyclic statistics can be used for motion estimation and compensation [8].

17.4.3 Fractional Sampling and Multivariate/Multirate Processing

Let $\omega_e = 0$ and suppose we oversample (i.e., fractionally sample) (17.30) by a factor P. With $x(n) := r_c(nT_0/P)$, we obtain (see also Fig. 17.8)

$$x(n) = \sum_l w(l)h(n - lP) + v(n) , \tag{17.35}$$

where now $h(n) := h_c(nT_0/P - \epsilon)$, and $v(n) := v_c(nT_0/P)$. Figure 17.8 shows the continuous-time model and the multirate discrete time equivalent of (17.35). With $P = 1$, (17.35) reduces to the stationary part of $r(n)$ in (17.31) but with $P > 1$, $x(n)$ in (17.35) is CS with correlation $c_{xx}(n; \tau) = \sigma_w^2 \sum_l h(n - lP)h^*(n + \tau - lP) + \sigma_v^2 \delta(\tau)$, which can be verified to be periodic with period equal to the oversampling factor P [26, 30, 61]. Cyclic correlations and cyclic spectra are given, respectively, by:

$$\bar{C}_{xx}\left(\frac{2\pi}{P}k; \tau\right) = \frac{\sigma_w^2}{P} \sum_l h(l)h^*(l + \tau)e^{-j\frac{2\pi}{P}kl} + \sigma_v^2 \delta(k)\delta(\tau) \tag{17.36}$$

$$\bar{S}_{xx}\left(\frac{2\pi}{P}k; \omega\right) = \frac{\sigma_w^2}{P} H^*(-\omega)H\left(\frac{2\pi}{P}k - \omega\right) + \sigma_v^2 \delta(k) . \tag{17.37}$$

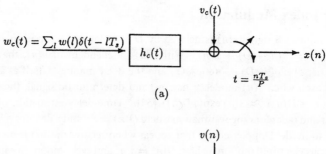

$w_c(t) = \sum_l w(l)\delta(t - lT_s)$ — $h_c(t)$

$v_c(t)$

$t = \frac{nT_s}{P}$

$x(n)$

(a)

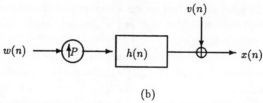

$w(n)$ — $\uparrow P$ — $h(n)$

$v(n)$

$x(n)$

(b)

FIGURE 17.8 (a) Fractionally sampled communications model and (b) multirate equivalent.

Although similar, the order of the FIR channel h in (17.35) is, due to oversampling, P times larger than that of (17.31). Cyclic spectra in (17.32) and (17.37) carry phase information about the underlying H, which is not the case with spectra of stationary processes ($P = 1$). Interestingly, (17.35) can be used also to model spread spectrum and direct sequence code-division multiple access data if $h(n)$ includes also the code [63, 64]. Relying on $\bar{S}_{xx}$ in (17.37), it is possible to identify $h(n)$ based only on output data — a task traditionally accomplished using higher than second order statistics (see e.g., [52]). By avoiding $k = 0$ in (17.36) or (17.37), the resulting cyclic statistics offer a high SNR domain for blind processing in the presence of stationary additive noise of arbitrary color and distribution (c.f., Property 4).

Oversampling by $P > 1$ also allows for estimating the synchronization parameters ω_l and ϵ in (17.31) [25, 54]. Finally, fractional sampling induces cyclostationarity in two-dimensional, linear system outputs [29], as well as in outputs of Volterra-type nonlinear systems [31]. In all these cases, relying on Representation 1 we can view the CS output $x(n)$ as a $P \times 1$ vector output of a multichannel system. Let us focus on 1-D linear channels and evaluate (17.35) at $nP + i$ to obtain the multivariate model

$$x(nP + i) := x_i(n) = \sum_l w(l)h_i(n - l) + v_i(n), \quad i = 0, 1, \ldots, P - 1, \qquad (17.38)$$

where $h_i(n) := h(nP + i)$ denotes the polyphase decomposition (decimated components) of the channel $h(n)$. Figure 17.9 shows how the single-input single output multirate model of Fig. 17.8 can be thought of as a single-input P-output multichannel system. The converse interpretation is equally interesting because it illustrates another CS-inducing operation.

Suppose P sensors (e.g., antennas or cameras) are deployed to receive data from a singe source $w(n)$ propagating through P channels $\{h_i(n)\}_{i=0}^{P-1}$. Using (17.16) we can combine the corresponding sensor data $\{x_i(n)\}_{i=0}^{P-1}$ given by (17.38), in order to create a single channel CS process $x(n)$, identical to the one in (17.35). There is a common feature between fractional sampling and multisensor (i.e., spatial) sampling: they both introduce strict cyclostationarity with known period P.

Strict cyclostationarity is also induced by multirate operators such as upsamplers in synthesis filterbanks, one branch of which corresponds to the multirate diagram of Fig. 17.8(b). We infer that outputs of synthesis filter banks are, in general, CS processes (see also [57]). Analysis filter banks, on the other hand, produce CS outputs when their inputs are also CS, but not if their inputs are stationary. Indeed, downsampling does not affect stationarity, and in contrast to upsamplers, downsamplers do not induce cyclostationarity. Downsamplers can remove cyclostationarity (as verified by Fig. 17.3) and from this point of view, analysis banks can undo CS effects induced by synthesis banks.

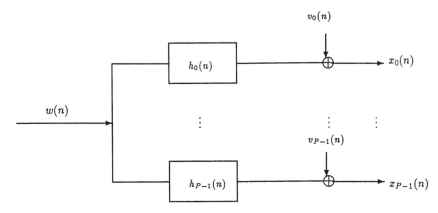

FIGURE 17.9 Multichannel stationary equivalent model of a scalar CS process.

17.4.4 Periodically Varying Systems

Thus far we have dealt with CS signals passing through time-invariant (TI) systems. Here we will focus on (almost) periodically varying (APTV) systems and input-output relationships such as: $x(n) = \sum_l h(n; l) w(n - l)$. Because $h(n; l)$ is APTV, following Definition 2 it accepts a (generalized) Fourier Series expansion $h(n; l) = \sum_\beta H(\beta; l) \exp(j\beta n)$. Coefficients $H(\beta; l)$ are TI, and together with their Fourier Transform are given by

$$
H(\beta; l) := \text{FS}[h(n; l)] = \lim_{N \to \infty} \frac{1}{N} \sum_{n=0}^{N-1} h(n; l) e^{-j\beta n} ,
$$

$$
H(\beta; \omega) := \text{FT}[H(\beta; l)] = \sum_l H(\beta; l) e^{-j\omega l} . \tag{17.39}
$$

In practice, $h(n; l)$ has finite bandwidth and the set of system cycles is finite; i.e., $\beta \in \{\beta_1, \ldots, \beta_Q\}$. Such a finite parametrization could appear, for example, with FIR multipath channels entailing path variations due to Doppler effects present with mobile communicators [62]. Note that when the cycles β are available, knowledge of $h(n; l)$ is equivalent to knowing $H(\beta; l)$ or $H(\beta; \omega)$ in (17.39).

The output correlation of a linear time-varying system is given by

$$
\bar{c}_{xx}(n; \tau) = \sum_{l_1, l_2} h(n; l_1) \, h^*(n + \tau; l_2) \, \bar{c}_{ww}(n - l_1; \tau + l_1 - l_2) . \tag{17.40}
$$

Equation (17.40) shows that if $w(n)$ is ACS, then $x(n)$ is also ACS, regardless of whether h is APTV or TI. More important, if h is APTV, then $x(n)$ is ACS even when $w(n)$ is stationary; i.e., APTV systems are cyclostationarity inducing operators. Similar observations apply to the input-output cross-correlation $\bar{c}_{xw}(n; \tau) := E\{x(n)w^*(n + \tau)\}$, which is given by

$$
\bar{c}_{xw}(n; \tau) = \sum_l h(n; l) \, \bar{c}_{xw}(n - l; l + \tau) . \tag{17.41}
$$

If the n-dependence is dropped from (17.40) and (17.41), one recovers the well-known auto- and cross-correlation expressions of stationary processes passing through linear TI systems. Relying on definitions (17.2), (17.11), and (17.37), the auto- and cross-cyclic correlations and cyclic spectra can be found as

$$\bar{C}_{xx}(\alpha;\tau) = \sum_{l_1,l_2}\sum_{\beta_1,\beta_2} H(\beta_1;l_1)H^*(\beta_2;l_2)e^{-j(\alpha-\beta_1+\beta_2)l_1}e^{-j\beta_2\tau}$$

$$\times \bar{C}_{ww}(\alpha-\beta_1+\beta_2;\tau+l_1-l_2)\,, \tag{17.42}$$

$$\bar{C}_{xw}(\alpha;\tau) = \sum_{\beta}\sum_{l} H(\beta;l)e^{-j(\alpha-\beta)l}\bar{C}_{ww}(\alpha-\beta;l+\tau)\,, \tag{17.43}$$

$$\bar{S}_{xx}(\alpha;\omega) = \sum_{\beta_1,\beta_2} H(\beta_1;\alpha+\beta_2-\beta_1-\omega)H^*(\beta_2;-\omega)\bar{S}_{ww}(\alpha-\beta_1+\beta_2;\omega)\,, \tag{17.44}$$

$$\bar{S}_{xw}(\alpha;\omega) = \sum_{\beta} H(\beta;\alpha-\beta-\omega)\,\bar{S}_{ww}(\alpha-\beta;\omega)\,. \tag{17.45}$$

Simpler expressions are obtained as special cases of (17.42) through (17.45) when $w(n)$ is stationary; e.g., cyclic auto- and cross-spectra reduce to:

$$\bar{S}_{xx}(\alpha;\omega) = \bar{S}_{ww}(\omega)\sum_{\beta} H(\beta;-\omega)H^*(\alpha-\beta;-\omega),$$

$$\bar{S}_{xw}(\alpha;\omega) = \bar{S}_{ww}(\omega)\,H(\alpha;-\omega)\,. \tag{17.46}$$

If $w(n)$ is i.i.d. with variance σ_w^2, then $H(\alpha;\omega)$ can be easily found from (17.46) as $\bar{S}_{xw}(\alpha;-\omega)/\sigma_w^2$. APTV systems and the four domains of characterizing them, namely $h(n;l)$, $H(\beta;l)$, $H(\beta;\omega)$, $H(n;\omega)$, offer diversity similar to that exhibited by ACS statistics. Furthermore, with finite cycles $\{\beta_q\}_{q=1}^Q$, the input-output relation can be rewritten as

$$x(n) = \sum_{q=1}^Q x_q(n) = \sum_{q=1}^Q [\sum_l H(\beta_q;l)\,w(n-l)]e^{j\beta_q n}\,. \tag{17.47}$$

Figure 17.10 depicts (17.47) and illustrates that periodically varying systems can be modeled as a superposition of TI systems weighted by the bases. If separation of the $\{x_q(n)\}_{q=1}^Q$ components is possible, identification and equalization of APTV channels can be accomplished using approaches for multichannel TI systems. In [44], separation is achieved based on fractional sampling or multiple antennas.

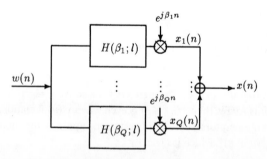

FIGURE 17.10   Multichannel model of a periodically varying system.

17.5 Application Areas

CS signals appear in various applications, but here we will deal with problems where cyclostationarity is exploited for signal extraction, modeling, and system identification. The tools common to all applications are cyclic (cross-)correlations, cyclic (cross-)spectra, or multivariate stationary correlations and spectra which result from the multichannel equivalent stationary processes (recall Representations 1 and 2, and Section 17.4.3). Because these tools are time-invariant, the resulting approaches follow the lines of similar methods developed for applications involving stationary signals.

As a general rule for problems entailing CS signals, one can either map the scalar CS signal model to a multichannel stationary process, or work in the time-invariant domain of cyclic statistics and follow techniques similar to those developed for stationary signals and time-invariant systems. CS signal analysis exploits two extra features not available with scalar stationary signal processing, namely: (1) ability to separate signals on the basis of their cycles and (2) diversity offered by means of cycles. Of course, the cycles must be known or estimated as we discussed in Section 17.3.

Suppose $x(n) = s(n) + v(n)$, where $s(n)$, $v(n)$ are generally CS, and let α be a cycle which is not in $A_{ss}^c(\tau) \cap A_{vv}^c(\tau)$. It then follows for their cyclic correlations and spectra that:

$$
C_{xx}(\alpha; \tau) = \begin{cases} C_{ss}(\alpha; \tau) & \text{if } \alpha \in A_{ss}^c(\tau) \\ C_{vv}(\alpha; \tau) & \text{if } \alpha \in A_{vv}^c(\tau) \end{cases},
$$

$$
S_{xx}(\alpha; \omega) = \begin{cases} S_{ss}(\alpha; \omega) & \text{if } \alpha \in A_{ss}^s(\omega) \\ S_{vv}(\alpha; \omega) & \text{if } \alpha \in A_{vv}^s(\omega) \end{cases}. \tag{17.48}
$$

In words, (17.48) says that signals $s(n)$ and $v(n)$ can be separated in the cyclic correlation or the cyclic spectral domains provided that they possess at least one noncommon cycle. This important property applies to more than two components and is not available with stationary signals because they all have only one cycle, namely $\alpha = 0$, which they share.

More significantly, if $s(n)$ models a CS information bearing signal and $v(n)$ denotes stationary noise, then working in cyclic domains allows for theoretical elimination of the noise, provided that the $\alpha = 0$ cycle is avoided (see also Property 4); i.e.,

$$
C_{xx}(\alpha; \tau) = C_{ss}(\alpha; \tau), \quad \text{and} \quad S_{xx}(\alpha; \omega) = S_{ss}(\alpha; \omega), \quad \text{for } \alpha \neq 0. \tag{17.49}
$$

In practice, noise affects the estimators' variance so that (17.48) and (17.49) hold approximately for sufficiently long data records. Notwithstanding, (17.48), (17.49) and SNR improvement in cyclic domains hold true irrespective of the color and distribution of the CS signals or the stationary noise involved.

EXAMPLE 17.4: Separation based on cycles

Consider the mixture of two modulated signals in noise: $x(n) = s_1(n) \exp[j(\omega_1 n + \varphi_1)] + s_2(n) \exp[j(\omega_2 n + \varphi_2)] + v(n)$, where $s_1(n)$, $s_2(n)$, $v(n)$ are Gaussian zero-mean stationary and mutually uncorrelated. Let $s_1(n)$ be MA(3) with parameters [1, 0.2, 0.3, 0.5] and variance $\sigma_1^2 = 1.38$, $s_2(n)$ be AR(1) with parameters [1, -0.5] and variance $\sigma_2^2 = 2$, and noise $v(n)$ be MA(1) (i.e., colored) with parameters [1, 0.5] and variance $\sigma_v^2 = 1.25$. Frequencies and phases are $(\omega_1, \varphi_1) = (-0.5, 0.6)$, $(\omega_2, \varphi_2) = (1, 1.8)$, and $N = 2,048$ samples are used to compute the correlogram estimates $\hat{S}_{s_1 s_1}(\omega)$, $\hat{S}_{s_2 s_2}(\omega)$, $\hat{S}_{vv}(\omega)$ shown in Figs. 17.11a through c; $\hat{C}_{xx}(\alpha; 0)$ is plotted in Fig. 17.11d and $\hat{S}_{xx}(\alpha; \omega)$ is depicted in Fig. 17.12. The cyclic correlation and cyclic spectrum of $x(n)$ are, respectively:

$$
\begin{aligned}
C_{xx}(\alpha; \tau) =\ & c_{s_1 s_1}(\tau) e^{j(\omega_1 \tau + \varphi_1)} \delta(\alpha - 2\omega_1) \\
& + c_{s_2 s_2}(\tau) e^{j(\omega_2 \tau + \varphi_2)} \delta(\alpha - 2\omega_2) + c_{vv}(\tau) \delta(\alpha),
\end{aligned} \tag{17.50}
$$

$$
\begin{aligned}
S_{xx}(\alpha; \omega) =\ & S_{s_1 s_1}(\omega - \omega_1) e^{j2\varphi_1} \delta(\alpha - 2\omega_1) \\
& + S_{s_2 s_2}(\omega - \omega_2) e^{j2\varphi_2} \delta(\alpha - 2\omega_2) + S_{vv}(\omega) \delta(\alpha).
\end{aligned} \tag{17.51}
$$

As predicted by (17.50), $|C_{xx}(\alpha; 0)| = \sigma_{s_1}^2 \delta(\alpha - 2\omega_1) + \sigma_{s_2}^2 \delta(\alpha - 2\omega_2) + \sigma_v^2 \delta(\alpha)$, which explains the two peaks emerging in Fig. 17.11d at twice the modulating frequencies $(2\omega_1, 2\omega_2) = (-1, 2)$. The third peak at $\alpha = 0$ is due to the stationary noise which can be thought of as being "modulated" by $\exp(j\omega_3 n)$ with $\omega_3 = 0$. Clearly, $2\hat{\omega}_1$, $2\hat{\omega}_2$, $\hat{\sigma}_{s_1}^2$, $\hat{\sigma}_{s_2}^2$, and $\hat{\sigma}_v^2$ can be found from Fig. 17.11d, while the phases at the peaks of $\hat{C}_{xx}(\alpha; 0)$ will yield $\hat{\varphi}_i = \sigma_{s_i}^{-2}\arg[\hat{C}_{xx}(2\hat{\omega}_i; 0)]/2$, $i = 1, 2$. In addition, the correlations of $s_i(n)$ can be retrieved as $\hat{c}_{s_i s_i}(\tau) = \exp[-j(\hat{\omega}_i \tau + 2\hat{\varphi}_i)]\hat{C}_{xx}(2\hat{\omega}_i; \tau)$, $i = 1, 2$.

Separation based on cycles is illustrated in Fig. 17.12, where three distinct slices emerge along the α-axis, each positioned at $\{\alpha_i = 2\omega_i\}_{i=1}^3$, representing the profiles of $\hat{S}_{s_1 s_1}(\omega)$, $\hat{S}_{s_2 s_2}(\omega)$, $\hat{S}_{vv}(\omega)$ shown also in Figs. 17.11a through c.

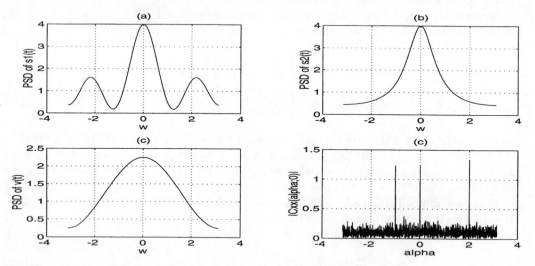

FIGURE 17.11   Spectral densities and cyclic correlation signals in Example 17.4.

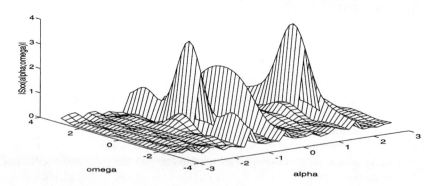

FIGURE 17.12   Cyclic spectrum of $x(n)$ in Example 17.4.

In the ensuing example we will demonstrate how the diversity offered by fractional sampling or by multiple sensors can be exploited for identification of FIR systems when the input is not available. Such a blind scenario appears when estimation and equalization of, e.g., communication channels is to be accomplished without training inputs. Bandwidth efficiencøy and ability to cope with changing multipath environments provide the motivating reasons for blind processing, while fractional sampling or multiple antennas justify the use of cyclic statistics as discussed in Section 17.4.3.

EXAMPLE 17.5: Diversity for channel estimation

Suppose we sample the output of the receiver's filter every $T_0/2$ seconds, to obtain $x(n)$ samples obeying (17.35) with $P = 2$ (see also Fig. 17.8). In the absence of noise, the spectrum of $x(n)$ will be $X_N(\omega) = H(\omega)W_N(2\omega)$. We wish to obtain $H(\omega)$ based only on $X_N(\omega)$ (blind scenario). Note that $W_N(2\omega) = W_N[2(\omega - 2\pi k/2)]$ for any integer k. Considering $k = 1$, we can eliminate the input spectrum $W_N(2\omega)$ from $X_N(\omega)$ and $X_N(\omega - \pi)$, and arrive at [26]

$$H(\omega)\, X_N(\omega - \pi) = H(\omega - \pi)\, X_N(\omega) . \qquad (17.52)$$

With $H(\omega)$ being FIR, the cross-relation (17.52) has turned the output-only identification problem into an input-output problem. The input is $X_N(\omega - \pi) = \text{FT}[(-1)^n x(n)]$, the output is $X_N(\omega)$, and the pole-zero system is $H(\omega)/H(\omega - \pi)$. If the Z-transform $H(z)$ has no zeros on a circle, separated by π, there is no pole-zero cancellation and $H(\omega)$ can be identified uniquely [61], using standard realization (e.g., Padé) methods [42].

Alternatively, with $P = 2$ we can map (17.52) to its one-input two-output time-invariant equivalent model obeying (17.38) with $P = 2$. In the absence of noise, the output spectra are $X_i(\omega) = H_i(\omega)\, W(\omega)$, $i = 0, 1$, from which $W(\omega)$ can be eliminated to arrive at a similar cross-relation [69]

$$H_0(\omega)\, X_1(\omega) = H_1(\omega)\, X_0(\omega) . \qquad (17.53)$$

When oversampling by $P = 2$, $x_0(n)$ $[h_0(n)]$ correspond to the even samples of $x(n)$ $[h(n)]$, whereas $x_1[n]$ $[h_1(n)]$ to the odd ones. Once again, $H_0(\omega)$ and $H_1(\omega)$ can be uniquely recovered using input-output realization methods, provided that they have no common zeros so that cancellations do not occur in (17.53). The desired channel $h(n)$ can be recovered by interleaving $h_0(n)$ with $h_1(n)$.

As explained in Section 17.4.3, oversampling is not the only means of diversity. Even with symbol rate sampling, if multiple (here two) antennas receive a common source through different channels, then $X_i(\omega) = H_i(\omega)\, W(\omega), i = 0, 1$, and thus (17.53) is still applicable.

Interestingly, both (17.52) and (17.53) neither restrict the input to be white (or even random) nor do they assume the channel to be minimum phase as univariate stationary spectral factorization approaches require for blind estimation [52]. The diversity (or overdeterminacy) offered by (17.35) or (17.38) guarantees identifiability provided that no cancellations occur in (17.52) or (17.53) and $W(\omega)$ is nonzero for as many frequencies as the number of channel taps to be estimated [69]. Subspace and least-squares methods are also possible for blind channel estimation and useful when noise is present [26, 47, 60, 69].

In the sequel, we will show how cycle-based separation and diversity can be exploited in selected applications.

17.5.1 CS Signal Extraction

In our first application, a mixture of CS sources with distinct cycles will be recovered using samples collected by an array of sensors.

Application 1: Array Processing

Suppose N_s CS source signals $\{s_l(n)\}_{l=1}^{N_s}$ are received by N_x sensors $\{x_m(n)\}_{m=1}^{N_x}$ in the presence of undesired sources of interference $\{i_m(n)\}_{m=1}^{N_x}$ and stationary noise $\{v_m(n)\}_{m=1}^{N_x}$. The mth sensor samples are: $x_m(n) = \sum_{l=1}^{N_s} \rho_l s_l(n - D_{lm}) + i_m(n) + v_m(n)$, where ρ_l denotes complex gain and D_{lm} the delay experienced by the lth source arriving at the mth sensor relative to the first sensor which is taken as the reference. For uniformly spaced linear arrays $D_{lm} = (m - 1)d \sin\theta_l/v$, where d stands for the sensor spacing, v is the propagation velocity, and θ_l denotes the angle of arrival of the lth source. Assuming that the $s_l(n)$s have a nonzero cycle α not shared by the undesired interferences, we wish to estimate $\boldsymbol{\theta} := [\theta_1 \cdots \theta_{N_s}]$ and subsequently use it to design beamformers that null out the interferences and suppress noise.

For mutually uncorrelated $\{s_l(n), i_m(n), v_m(n)\}$, the time-delay property in Section 17.4.2 yields [68]

$$\bar{C}_{x_m x_m}(\alpha; \tau) = \sum_{l=1}^{N_s} \bar{C}_{s_l s_l}(\alpha; \tau) e^{-j\alpha D_{lm}} + \bar{C}_{i_m i_m}(\alpha; \tau) + \bar{C}_{ww}(\tau)\delta(\alpha) . \qquad (17.54)$$

Choosing a nonzero α not in the interference set of cycles $A^c_{i_m i_m}(\tau)$ and collecting $\{\bar{C}_{x_m x_m}\}_{m=1}^{N_x}$ in an $N_x \times 1$ vector, we arrive at $\bar{\mathbf{c}}_{x_m}(\alpha; \tau) = \mathbf{A}(\alpha; \boldsymbol{\theta})\mathbf{c}_{ss}(\alpha; \tau)$, where the $N_x \times N_s$ matrix $\mathbf{A}(\boldsymbol{\theta})$ is the so-called array manifold containing the propagation parameters. In [68], N_τ lags are used to form the $N_x \times N_\tau$ cyclic correlation matrix

$$\begin{aligned}
\bar{\mathbf{C}}_{xx}(\alpha) &:= [\bar{\mathbf{c}}_{xx}(\alpha; \tau_1) \cdots \bar{\mathbf{c}}_{xx}(\alpha; \tau_{N_\tau})]' = \mathbf{A}(\alpha; \boldsymbol{\theta})\bar{\mathbf{C}}_{ss}(\alpha) , \\
\bar{\mathbf{C}}_{ss}(\alpha) &:= [\bar{\mathbf{c}}_{ss}(\alpha; \tau_1) \cdots \bar{\mathbf{c}}_{ss}(\alpha; \tau_{N_\tau})]' .
\end{aligned} \qquad (17.55)$$

Standard subspace methods can be employed to recover $\boldsymbol{\theta}$ from (17.55). It is worth noting that cycle-based separation of desired from undesired signals and noise is possible for both narrowband and broadband sources [68] (see also [16] for the narrowband case).

With the propagation parameters available, spatio-temporal filtering based on $\bar{\mathbf{C}}_{xx}(\alpha_l; \tau)$ is capable of isolating the source $s_l(n)$ if $\alpha_l \in A^c_{s_l s_l}(\tau)$ and $\alpha_l \notin A^c_{s_k s_k}$ for $k \neq l$. Thus, in addition to interference and noise suppression, cyclic beamformers increase resolution by exploiting known separating cycles. In fact, even sources arriving from the same direction can be separated provided that not all of their cycles are common (see [1, 6, 58] and [16] for detailed algorithms).

In our next application, the desired CS $d(n)$ we wish to extract from noisy data $x(n)$ is known, or at least its (cross-) correlation with $x(n)$ is available.

Application 2: Cyclic Wiener filtering

In a number of real life problems CS data $x(n)$ carry information about a desired CS signal $d(n)$ which may not be available, but the cross-correlation $\bar{c}_{dx}(n; \tau)$ is known or can be estimated otherwise. With reference to Fig. 17.13 we seek a linear (generally time-varying) filter $f(n; k)$ whose output, $\hat{d}(n) = \sum_k f(n; k) x(n - k)$, will come close to the desired $d(n)$ in terms of minimizing $\sigma_e^2(n) = E\{|e(n)|^2\} := E\{|d(n) - \hat{d}(n)|^2\}$. Because both $x(n)$ and $d(n)$ are CS with period P, for $\hat{d}(n)$ to also be CS, filter $f(n; k)$ must be periodically varying with period P; i.e., $f(n; k)$ is equivalent to P time-invariant filters $\{f(n; k)\}_{n=0}^{P-1}$ and accepts a Fourier Series expansion with coefficients $F(\alpha; k)$ defined as in (17.39). Note that $e(n)$ is also CS and $E\{|e(n)|^2\}$ should be minimized for $n = 0, 1, \cdots, P - 1$.

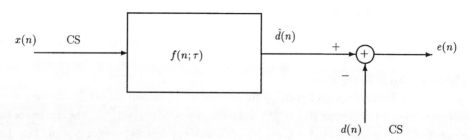

FIGURE 17.13   Cyclic Wiener filtering.

Solving the minimization problem for each n, we arrive at time-varying normal equations

$$\sum_k f(n; k) \bar{c}_{xx}(n - k; k - \tau) = \bar{c}_{dx}(n; -\tau) , \quad n = 0, 1, \ldots, P - 1 , \qquad (17.56)$$

where $\bar{c}_{xx}$ can be estimated consistently from the data as discussed in Section 17.3, and similarly for $\bar{c}_{dx}$ if $d(n)$ is available. Note that with sample estimates, (17.56) could have been reached as a result of minimizing the least-squares error [c.f. (17.24)]: $\hat{\sigma}_e^2(n) = [P/N] \sum_{i=0}^{[N/P]-1} |e(iP + n)|^2$. For each $n \in [0, P-1]$, FIR filters of order K_n can be obtained by concatenating equations such as (17.56) for more than K_n lags τ. As with time-invariant Wiener filters, noncausal and IIR designs are possible for each n in the frequency-domain, $F(n; \omega)$, using nonparametric estimates of the time-varying (cross-)spectra. Depending on $d(n)$, APTV (FIR or IIR) filters can thus be constructed for filtering, prediction, and interpolation or smoothing of CS processes.

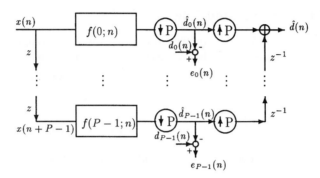

FIGURE 17.14 Multichannel-multirate equivalent of cyclic Wiener filtering.

In Section 17.4.4, we viewed the periodically varying scalar $f(n; k)$ as a time-invariant multichannel filter. Consider the polyphase stationary components $d_i(n)$, $e_i(n)$, and

$$\hat{d}_i(n) := d(nP + i) = \sum_k f(nP + i; k) x(nP + i - k) = \sum_k f(i; k)x(nP + i - k). \quad (17.57)$$

Equation (17.57) allows us to cast the scalar processing in Fig. 17.13 as the filterbank of Fig. 17.14. Because $\sigma_{e_i}^2 = E|e(i)|^2$, for $i = 0, 1, \cdots, P-1$, and $d_i(n)$, $\hat{d}_i(n)$, $e_i(n)$ are stationary, solving for the periodic Wiener filter $f(n; k)$ is equivalent to solving for the P time-invariant Wiener filters $f(i; k)$ in Fig. 17.14. Using the multirate (Noble) identity (e.g., [51, Ch. 12]), one can move the downsamplers before the Wiener filters which now have transfer functions $G(i; \omega) = F(i; \omega/P)$. Such an interchange corresponds to feeding a time-invariant $P \times 1$ vector Wiener filter $\mathbf{g}(k) := [g(0; k) \cdots g(P - 1; k)]'$, with input the $P \times 1$ polyphase component vector $\mathbf{x}(n) := [x(nP)x(nP + 1)...x(nP + P - 1)]'$.

An alternative multichannel interpretation is obtained based on the Fourier Series expansion $f(n; k) = \sum_\alpha F(\alpha; k) \exp(j\alpha n)$. The resulting Wiener processing allows also for APTV filters, which is particularly useful when $d(n)$, $x(n)$, and thus $\hat{d}(n)$, $e(n)$ are ACS processes. Substituting the expansion in the filter output and multiplying by $\exp(i\alpha k) \exp(-i\alpha k) = 1$, we find [22]

$$\hat{d}(n) = \sum_\alpha \sum_k \left[F(\alpha; k)e^{j\alpha k} \right] \left[x(n - k)e^{j\alpha(n-k)} \right] = \sum_\alpha \left\{ \sum_k \tilde{F}(\alpha; k) \, \tilde{x}(n - k) \right\}, \quad (17.58)$$

where $\tilde{F}(\tilde{x})$ are the modulated versions of $F(x)$ shown in the square brackets. For CS processes with period P, the sum over α in (17.58) has finite terms $\{\alpha_i = 2\pi i/P\}_{i=0}^{P-1}$ and shows that scalar cyclic Wiener filtering is equivalent to a superposition of P time-invariant Wiener filters with inputs $\tilde{x}_i(n)$ formed by modulating $x(n)$ with the Fourier bases $\{\exp j(\alpha_i n)\}_{i=1}^{P-1}$ (see also Fig. 17.15).

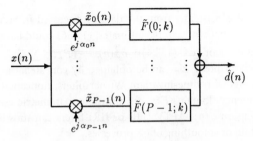

FIGURE 17.15   Multichannel-modulation equivalent of cyclic Wiener filtering.

17.5.2   Identification and Modeling

The need to identify TI and APTV systems (or their inverses for equalization) appears in many applications where input-output or output-only CS data are available. Our first problem in this class deals with identifying pure delay TI systems, $h(n) = \delta(n - D)$, given CS input-output signals observed in correlated noise.

Application 3: Time-delay estimation

We wish to estimate the relative delay D of a CS signal $s(n)$ given data from a pair of sensors

$$x(n) \;=\; s(n) + v_x(n)\,, \qquad y(n) \;=\; s(n - D) + v_y(n)\,. \tag{17.59}$$

Signal $s(n)$ is assumed uncorrelated with $v_x(n)$, $v_y(n)$, but the noises at both sensors are allowed to be colored and correlated with unknown (cross-)spectral characteristics. The time-varying cross-correlation yields the delay (see also [7] and [70] for additional methods relying on cyclic spectra). In addition to suppressing stationary correlated noise, cyclic statistics can also cope with interferences present at both sensors as we show in the following example.

EXAMPLE 17.6:   Time-delay estimation

Consider $x(n) \;=\; w(n)\,\exp[j(-0.5(n)+0.6)] + i(n)\exp[j(n+1.8)] + v_x(n)$, and $y(n) \;=\; w(n - D)\,\exp[j(-0.5(n - D)+0.6)] + i(n - D)\,\exp[j(n - D + 1.8)] + v_y(n)$, with $D = 20$, $v_x(n)$ white, $v_y(n) = v_x \star h(n)$, $h(0) = h(10) = 0.8$ and $h(n) = 0$ for $n \neq 0, 10$.

The magnitude of $\hat{C}_{xy}(\alpha; \tau)$ is computed as in (17.21) with $N = 2,048$ samples and is depicted in Fig. 17.16 (3-D and contour plots). It peaks at the correct delay $D = 20$ at cycles $\alpha = 2(-0.5) = -1$ (due to the signal) and $\alpha = 2(+1) = 2$ (due to the interference). The additional peak at delay 10 occurs at cycle $\alpha = 0$ and reveals the memory introduced in the correlation of $v_y(n)$ due to $h(n)$.

Relying on (17.46), input-output cyclic statistics allow for identification of TI systems, but in certain applications estimation of $h(n)$ or its inverse [call it $g(n)$] is sought based on output data only. In Application 2 we outlined two approaches capable of estimating FIR channels blindly in the absence of noise, even when the input $w(n)$ is not white. If $w(n)$ is white, it follows easily from (17.36) that $\bar{C}_{xx}$ for two cycles k_1, k_2 satisfies [26]

$$\sum_{l=0}^{L} [\, \bar{C}_{xx}\!\left(\frac{2\pi}{P}k_1; \tau + l\right) - e^{j\frac{2\pi}{P}(k_2 - k_1)l}\, \bar{C}_{xx}\!\left(\frac{2\pi}{P}k_2; \tau + l\right)]\, h(l) = 0\,,$$

$$k_1 \neq k_2 \neq 0\,. \tag{17.60}$$

The matrix equation that results from (17.60) for different τs can be solved to obtain $\{h(l)\}_{l=0}^{L}$ within a scale (assuming that the matrix involved is full rank), even when stationary colored noise is present. To fix the scale, we either set $h(0) = 1$, or, $\sum_{l=0}^{L} |h(l)|^2 = 1$. Having estimated $h(l)$, one could find the

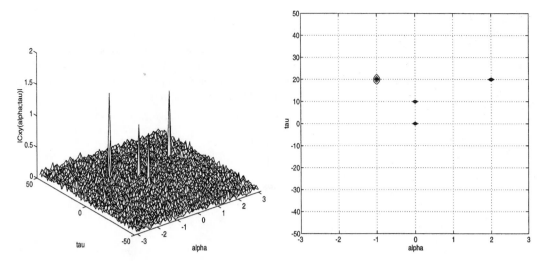

FIGURE 17.16 Cyclic cross-correlation for time-delay estimation.

cross-correlation $\bar{c}_{xw}(n; \tau)$ via (17.35) and use it in (17.56) to obtain FIR minimum mean-square error (MMSE, i.e., Wiener) equalizers for recovering the desired input $d(n) = w(n)$. However, as we will see next, it is possible to construct blind equalizers directly from the data bypassing the channel estimation step.

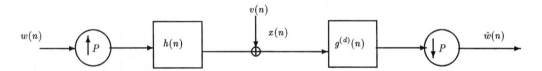

FIGURE 17.17 Cyclic (or multirate) channel-equalizer model.

Application 4: Blind channel equalization

Our setup is described in Fig. 17.8 and the available data satisfy (17.35) with $h(n)$ causal of order L. With reference to Fig. 17.17, we seek a Kth order equalizer, $\{g^{(d)}(n)\}_{n=0}^{K}$, parameterized by the delay d, such that $E\{|w(n-d) - \hat{w}(n)|^2\}$ is minimized. Expressing $\hat{w}(n)$ as $\hat{w}(n) = \sum_k g^{(d)}(k)x(nP - k)$, and using the whiteness of $w(n)$ and the independence between $w(n)$ and $v(n)$, we arrive at:

$$\sum_{k=0}^{K} g^{(d)}(k)\, \bar{c}_{xx}(-k; k - m) \;=\; \sigma_w^2\, h^*(dP - m)$$

$$=\; 0, \quad \text{for } d = 0,\ m > 0. \tag{17.61}$$

Equation (17.61) can be solved for the equalizer coefficients in batch or adaptive forms using recursive least-squares (RLS) or the computationally simpler LMS algorithm suitably modified to compute the cyclic correlation statistics [30]. It turns out that using $\{g^{(0)}(k)\}_{k=0}^{K}$ one can find $\{g^{(d)}(k)\}_{k=0}^{K}$ for $d \in [1, L+K]$, which is important because, in practice, nonzero delay equalizers often achieve lower MSE [30].

Another interesting feature of the overall system in Fig. 17.17 is that in the absence of noise ($v(n) \equiv 0$), the FIR equalizer $\{g^{(d)}(n)\}_{k=0}^{K}$ can equalize the FIR channel $h(n)$ perfectly in the zero-forcing (ZF) sense: $\sum_{k=0}^{K} g^{(d)}(k)\, h(nP - k) = \delta(n - d)$, provided that: (1) the channel $H(z)$ has no equispaced zeros on a circle with each zero separated from the next by $2\pi/P$, and (2) the equalizer has order satisfying:

$K \geq L/(P-1) - 1$. Such a ZF equalizer can be found from the solution of (17.61) provided that conditions (1) and (2) are satisfied. The equalizer obtained is unique when (2) is satisfied as equality, or, when the minimum norm solution is adopted [30]. Recall that with symbol rate sampling ($P = 1$), FIR-ZF equalizers are impossible because the inverse of an FIR $H(z)$ is always the IIR $G(z) := 1/H(z)$. Further with $P = 1$, FIR-MMSE (i.e., Wiener) equalizers cannot be ZF. In [30], it is also shown that under conditions (1) and (2), it is possible to have FIR hybrid MMSE-ZF equalizers.

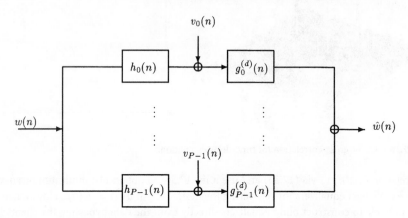

FIGURE 17.18　Multivariate channel-equalizer model.

The FIR channel–FIR equalizer feature can be seen also from the multichannel viewpoint which applies after the CS data $x(n)$ are mapped to the stationary components $\{x_i(n)\}_{i=0}^{P-1}$, or when P sensors collect symbol rate samples as in (17.38). With reference to Fig. 17.18, the channel-equalizer transfer functions satisfy, in the absence of noise, the so-called Bezout's identity: $\sum_{i=0}^{P-1} H_i(z) G_i^{(d)}(z) = z^{-d}$, which is analogous to the condition encountered with perfect reconstruction filterbanks. Given the Lth-order FIR analysis bank (H_i), existence and uniqueness of the Kth-order FIR synthesis filters (G_i) is guaranteed when: (1) $\{H_i(z)\}_{i=0}^{P-1}$ have no common zeros, and (2) $K \geq L/(P-1) - 1$. Next, we illustrate how the blind MMSE equalizer of (17.61) can be used to mitigate intersymbol interference (ISI) introduced by a two-ray multipath channel.

EXAMPLE 17.7:　Direct blind equalization

We generated 16-QAM symbols and passed them through a 7th order FIR channel obtained by sampling at a rate $T_0/2$ the continuous-time channel $h_c(t) = \exp(-j2\pi 0.15)\rho_c(t - 0.25T_0, 0.35) + 0.8 \exp(-j2\pi 0.6)\rho_c(t - T_0, 0.35)$, where $\rho_c(t, 0.35)$ denotes the raised cosine pulse with roll-off factor 0.35 [53, p. 546]. We estimated the time-varying correlations as in (17.24) and solved (17.61) for the equalizer of order $K = 6$ and $d = 0$. At SNR= 25 dB, Fig. 17.19, shows the received and equalized constellations illustrating the ability of the blind equalizer to remove ISI.

In our final application we will be concerned with parameter estimation of APTV systems.

Application 5: Parametric APTV modeling

Seasonal (e.g., atmospheric) time series are often modeled as the CS output of a linear (almost) periodically time varying system $h(n; l)$ with i.i.d. input $w(n)$. Suppose that $x(n)$ obeys an autoregressive [AR(p_n)] model with coefficients $a(n; l)$ which are periodic in n with period P_l. The time series $x(n)$ and

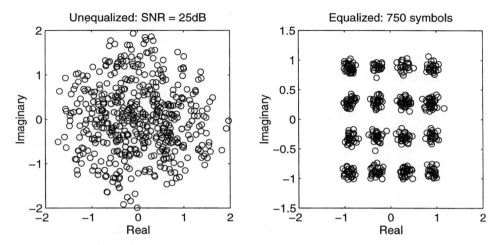

FIGURE 17.19 Before and after equalization (Example 17.7).

its correlation $c_{xx}(n; \tau)$ obey the following periodically varying AR recursions:

$$x(n) \quad + \quad \sum_{l=1}^{p_n} a(n; l)x(n - l) \; = \; w(n),$$

$$c_{xx}(n; \tau) \quad + \quad \sum_{l=1}^{p_n} a(n; l)c_{xx}(n - l; l - \tau) \; = \; \sigma_w^2(n)\delta(\tau). \tag{17.62}$$

The "periodic normal equations" in (17.62) can be solved for each n to estimate the $a(n; l)$ parameters. Relying on Representation 1, [49] showed how PTV-AR modeling algorithms can be used to estimate multivariate AR coefficient matrices. Usage of single channel cyclic (instead of multivariate) statistics for parametric modeling of multichannel stationary time series was motivated on the basis of potential computational savings; see [49] for details and also [55] for cyclic lattice structures. Maximum likelihood estimation of Periodic ARMA models is reported in [66]. PARMA modeling is important for seasonal time series encountered in meteorology, climatology [41], and stratospheric ozone data analysis [4]. Linear methods for estimating periodic MA coefficients along with important TV-MA parameter identifiability issues can be found in [13] using higher than second-order cyclic statistics.

When both input and output CS data are available, it is possible to identify linear periodically time-varying systems $h(n; l)$, even in the presence of correlated stationary input and output noise. Taking advantage of nonzero cycles present in the input and/or the system, one employs auto- and cross-cyclic spectra to identify $H(\beta; \omega)$, the cyclic spectrum of $h(n; l)$, relying on (17.45) or (17.46), when $w(n)$ is stationary.

If the underlying system is time invariant (e.g., a frequency selective communications channel, or a dispersive delay medium), a closed form solution is possible in the frequency domain. With $\beta = 0$, (17.45) yields: $H(\omega) = \bar{S}_{xw}(\alpha; \omega)/\bar{S}_{ww}(\alpha; \omega)$, where $\alpha \in A_{ww}^c$ (see also [17]). For Lth-order FIR system identification a parametric approach in the lag-domain may be preferred because it avoids the trade-offs involved in choosing windows for nonparametric cyclic spectral estimates. One simply solves the following system of linear equations formed by cyclic (cross-) correlations [27]

$$\sum_{l=0}^{L} h(l) \, \bar{C}_{ww}(\alpha; \tau - l) \; = \; \bar{C}_{xw}(\alpha; \tau), \tag{17.63}$$

using batch or adaptive algorithms. If desired, pole-zero models can then be fit in the estimated $\hat{h}(n)$ using Padé or Hankel methods. Estimation of TI systems with correlated input-output disturbances is important not only for open loop identification but also when feedback is present. Therefore, cyclic approaches are also of interest for identification of closed loop systems [27].

17.6 Concluding Remarks

Cyclostationary processes constitute the most common class of nonstationary signals encountered in engineering and time series applications. Cyclostationarity appears in signals and systems exhibiting repetitive variations and allows for separation of components on the basis of their cycles. The diversity offered by such a structured variation can be exploited for suppression of stationary noise with unknown spectral characteristics and for blind parameter estimation using a single data record. Variance of finite sample estimates is affected by noise and increases when the cycles are unknown and have to be estimated prior to applying cyclic signal processing algorithms.

Although our discussion focused on linear systems and second-order statistical descriptors, cyclostationarity appears also with nonlinear systems and certain signals exhibit periodicity in their higher than second-order statistics. The latter are especially useful because in both cases the underlying processes are non-Gaussian and second-order analysis cannot characterize them completely. Cyclostationarity in nonlinear time series of the Volterra type is exploited in [21, 31, 46], whereas sample estimation issues and motivating applications of higher-order cyclostationarity can be found in [11, 12, 23, 59] and references therein.

Topics of current interest and future trends include algorithms for nonlinear signal processing, theoretical performance evaluation, and analysis of cyclostationary point processes. As far as applications, exploitation of cyclostationarity is expected to further improve algorithms in manufacturing problems involving vibrating and rotating components, and will continue to contribute in the design of single- and multi-user digital communication systems especially in the presence of fading and time-varying multipath environments.

Acknowledgments

The author wishes to thank his former and current graduate students for shaping up the content and helping with the preparation of this manuscript. This work was supported by ONR Grant N0014-93-1-0485.

References

[1] Agee, B.G., Schell, S.V., and Gardner, W.A., Spectral self-coherence restoral: a new approach to blind adaptive signal extraction using antenna arrays, *Proc. IEEE*, 78, 753–767, 1990.

[2] Bell, M.R. and Grubbs, R.A., JEM modeling and measurement for radar target identification, *IEEE Trans. on AES*, 29, 73–87, 1993.

[3] Bennet, W.R., Statistics of regenerative digital transmission, *Bell Systems Tech. J.*, 37, 1501–1542, 1958.

[4] Bloomfield, P., Hurd, H.L., and Lund, R.B., Periodic correlation in stratospheric ozone data, *J. Time Series Analysis*, 15, 127–150, 1994.

[5] Brillinger, D.R., *Time Series, Data Analysis and Theory*, McGraw-Hill, New York, 1981.

[6] Castedo, L., Figueiras, V., and Anibal, R., An adaptive beamforming technique based on cyclostationary signal properties, *IEEE Trans. on Signal Processing*, 43, 1637–1650, 1995.

[7] Chen, C.-K. and Gardner, W.A., Signal-selective time-difference-of-arrival estimation for passive location of manmade signal sources in highly-corruptive environments: Part II: algorithms and performance, *IEEE Trans. on Signal Processing*, 40, 1185–1197, 1992.

[8] Chen, W., Giannakis, G.B., and Nandhakumar, N., Spatio-temporal approach for time-varying image motion estimation, *IEEE Transactions on Image Processing,* 10, 1448–1461, 1996.

[9] Corduneanu, C., *Almost Periodic Functions,* Interscience Publishers (John Wiley & Sons), New York, 1968.

[10] Dandawate, A.V. and Giannakis, G.B., Statistical tests for presence of cyclostationarity, *IEEE Trans. on Signal Processing,* 42, 2355–2369, 1994.

[11] Dandawate, A.V. and Giannakis, G.B., Nonparametric polyspectral estimators for kth-order (almost) cyclostationary processes, *IEEE Trans. on Information Theory,* 40, 67–84, 1994.

[12] Dandawate, A.V. and Giannakis, G.B., Asymptotic theory of mixed time averages and kth-order cyclic- moment and cumulant statistics, *IEEE Trans. on Information Theory,* 41, 216–232, 1995.

[13] Dandawate, A.V. and Giannakis, G.B., Modeling (almost) periodic moving average processes using cyclic statistics, *IEEE Trans. on Signal Processing,* 44, 673–684, 1996.

[14] Dragan, Y.P. and Yavorskii, I., The periodic correlation-random field as a model for bidimensional ocean waves, *Peredacha Informatsii,* 51, 15–25, 1982.

[15] Gardner, W.A., *Statistical Spectral Analysis: A Nonprobabilistic Theory,* Prentice-Hall, Englewood Cliffs, NJ, 1988.

[16] Gardner, W.A., Simplification of MUSIC and ESPRIT by exploitation of cyclostationarity, *Proc. IEEE,* 76, 845–847, 1988.

[17] Gardner, W.A., Identification of systems with cyclostationary input and correlated input/output measurement noise, *IEEE Trans. on Automatic Control,* 35, 449–452, 1990.

[18] Gardner, W.A., Two alternative philosophies for estimation of the parameters of time-series, *IEEE Trans. on Information Theory,* 37, 216–218, 1991.

[19] Gardner, W.A., Exploitation of spectral redundancy in cyclostationary signals, *IEEE ASSP Magazine,* 8, 14–36, 1991.

[20] Garder, W.A., Cyclic Wiener filtering: theory and method, *IEEE Trans. on Communications,* 41, 151–163, 1993.

[21] Gardner, W.A. and Archer, T.L., Exploitation of cyclostationarity for identifying the Volterra kernels of nonlinear systems, *IEEE Trans. on Information Theory,* 39, 535–542, 1993.

[22] Gardner, W.A. and Franks, L.E., Characterization of cyclostationary random processes, *IEEE Trans. on Information Theory,* 21, 4–14, 1975.

[23] Gardner, W.A. and Spooner, C.M., The cumulant theory of cyclostationary time-series; foundation, *IEEE Trans. on Signal Processing,* 42, 3387–408, 1994.

[24] Genossar, M.J., Lev-Ari, H., and Kailath, T., Consistent estimation of the cyclic autocorrelation, *IEEE Trans. on Signal Processing,* 42, 595–603, 1994.

[25] Gini, F. and Giannakis, G.B., Frequency offset and timing estimation in slowly-varying fading channels: A cyclostationary approach, *Proc. of 1st IEEE Signal Processing Workshop on Wireless Communications,* 393–396, Paris, France, April 16-18, 1997.

[26] Giannakis, G.B., A linear cyclic correlation approach for blind identification of FIR channels *Proc. of 28th Asilomar Conf. on Signals, Systems, and Computers,* 420–424, Pacific Grove, CA, Oct. 31-Nov. 2, 1994.

[27] Giannakis, G.B., Polyspectral and cyclostationary approaches for identification of closed loop systems, *IEEE Trans. on Auto. Control,* 40, 882–885, 1995.

[28] Giannakis, G.B., Filterbanks for blind channel identification and equalization, *IEEE Signal Processing Letters,* 4, 184–187, June 1997.

[29] Giannakis, G.B. and Chen, W., Blind blur identification and multichannel image restoration using cyclostationarity, *Proc. of IEEE Workshop on Nonlinear Signal and Image Processing,* II, 543–546, June 20-22, 1995, Halkidiki, Greece.

[30] Giannakis, G.B. and Halford, S., Blind fractionally-spaced equalization of noisy FIR channels: direct and adaptive solutions, *IEEE Trans. on Signal Processing,* 1997 (to appear).

[31] Giannakis, G.B. and Serpedin, E., Linear multichannel blind equalizers of nonlinear FIR Volterra channels, *IEEE Trans. on Signal Processing*, 45, 67–81, Jan. 1997.

[32] Giannakis, G.B. and Zhou, G., Parameter estimation of cyclostationary amplitude modulated time series with application to missing observations, *IEEE Trans. on Signal Processing*, 42, 2408–2419, 1994.

[33] Giannakis, G.B. and Zhou, G., Harmonics in multiplicative and additive noise: parameter estimation using cyclic statistics, *IEEE Trans. on Signal Processing*, 43, 2217–2221, 1995.

[34] Gladyšev, E.G., Periodically correlated random sequences, *Soviet Math.*, 2, 385–388, 1961.

[35] Hasselmann, K. and Barnett, T.P., Techniques of linear prediction of systems with periodic statistics, *J. Atmospheric Sci.*, 38, 2275–2283, 1981.

[36] Hinich, M.J., *Statistical Spectral Analysis: Nonprobabilistic Theory*, book review in *SIAM Review*, 33, 677–678, 1991.

[37] Hlawatsch, F. and Boudreaux-Bartels, G.F., Linear and quadratic time-frequency representations, *IEEE Signal Processing Magazine*, 21–67, April 1992.

[38] Hurd, H.L., An Investigation of Periodically Correlated Stochastic Processes, Ph.D. Dissertation, Duke University, Durham, NC, 1969.

[39] Hurd, H.L., Nonparametric time series analysis of periodically correlated processes, *IEEE Trans. on Information Theory*, 350–359, 1989.

[40] Hurd, H.L. and Gerr, N.L., Graphical methods for determining the presence of periodic correlation, *J. Time Series Analysis*, 12, 337–350, 1991.

[41] Jones, R.H. and Brelsford, W.M., Time series with periodic structure, *Biometrika*, 54, 403–408, 1967.

[42] Kay, S.M., *Modern Spectral Estimation — Theory and Application*, Prentice-Hall, Englewood Cliffs, NJ, 1988.

[43] Koenig, D. and Boehme, J., Application of cyclostationarity and time-frequency analysis to engine car diagnostics, *Proc. Intl. Conf. on ASSP*, 149–152, 1994, Adelaide, Australia.

[44] Liu, H., Giannakis, G.B., and Tsatsanis, M.K., Time-Varying System Identification: A Deterministic Blind approach using Antenna Arrays, *Proc. of 30th Conf. on Info. Sciences and Systems*, Princeton University, Princeton, NJ, March 20-22, 1996, 880–884.

[45] Longo, G. and Picinbono, B., Eds., *Time and Frequency Representation of Signals*, Springer-Verlag, New York, 1989.

[46] Marmarelis, V.Z., Practicable identification of nonstationary and nonlinear systems, *IEEE Proc., Part D*, 211–214, 1981.

[47] Moulines, E., Duhamel, P., Cardoso, J.-F., and Mayrargue, S., Subspace Methods for the Blind Identification of Multichannel FIR Filters, *IEEE Trans. on Signal Processing*, 43, 516–525, 1995.

[48] Newton, H.J., Using periodic autoregressions for multiple spectral estimation, *Technometrics*, 24, 109–116, 1982.

[49] Pagano, M., On periodic and multiple autoregressions, *Annal. Stat.*, 6, 1310–1317, 1978.

[50] Parzen, E. and Pagano, M., An approach to modeling seasonally stationary time-series, *J. Econometrics*, North Holland Publishing Company, 9, 137–153, 1979.

[51] Porat, B., *A Course in Digital Signal Processing*, John Wiley & Sons, New York, 1997.

[52] Porat, B. and Friedlander, B., Blind equalization of digital communication channels using high-order moments, *IEEE Trans. on Signal Processing*, 39, 522–526, 1991.

[53] Proakis, J., *Digital Communications*, 3rd ed., McGraw-Hill, New York, 1989.

[54] Riba, J. and Vazquez, G., Bayesian recursive estimation of frequency and timing exploiting the cyclostationarity property, *Signal Processing*, 40, 21–37, 1994.

[55] Sakai, H., Circular lattice filtering using Pagano's method, *IEEE Trans. on Acoust. Speech & Signal Proc.*, 30, 279–287, 1982.

[56] Sakai, H., On the spectral density matrix of a periodic ARMA process, *J. Time Series Analysis*, 12, 73–82, 1991.

[57] Sathe, V.P. and Vaidyanathan, P.P., Effects of multirate systems on the statistical properties of random signals, *IEEE Trans. on Signal Processing*, 131–146, 1993.

[58] Schell, S.V., An overview of sensor array processing for cyclostationary signals, in *Cyclostationarity in Communications and Signal Processing*, Gardner, W.A., Ed., IEEE Press, New York, 1994, 168–239.

[59] Spooner, C.M. and Gardner, W.A., The cumulant theory of cyclostationary time-series: development and applications, *IEEE Trans. on Signal Processing*, 42, 3409–29, 1994.

[60] Tong, L., Xu, G., and Kailath, T., Blind identification and equalization based on second-order statistics: a time domain approach, *IEEE Trans. on Information Theory*, 340–349, 1994.

[61] Tong, L., Xu, G., Hassibi, B., and Kailath, T., Blind channel identification based on second-order statistics: a frequency-domain approach, *IEEE Trans. on Information Theory*, 41, 329–334, 1995.

[62] Tsatsanis, M.K. and Giannakis, G.B., Modeling and equalization of rapidly fading channels, *Intl. J. Adaptive Control and Signal Processing*, 10, 159–176, 1996.

[63] Tsatsanis, M.K. and Giannakis, G.B., Optimal linear receivers for DS-CDMA systems: a signal processing approach, *IEEE Trans. on Signal Processing*, 44, 3044–3055, 1996.

[64] Tsatsanis, M.K. and Giannakis, G.B., Blind estimation of direct sequence spread spectrum signals in multipath, *IEEE Trans. on Signal Processing*, 45, 1241–1252, 1997.

[65] Tsatsanis, M.K. and Giannakis, G.B., Transmitter induced cyclostationarity for blind channel equalization, *IEEE Trans. on Signal Processing*, 45, 1785–1794, 1997.

[66] Vecchia, A.V., Periodic autoregressive-moving average (PARMA) modeling with applications to water resources, *Water Res. Bull.*, 21, 721–730, 1985.

[67] Wilbur, J.-E. and McDonald, R.J., Nonlinear analysis of cyclically correlated spectral spreading in modulated signals, *J. Acoustical Soc. Am.*, 92, 219–230, 1992.

[68] Xu, G. and Kailath, T., Direction-of-arrival estimation via exploitation of cyclostationarity — A combination of temporal and spatial processing, *IEEE Trans. on Signal Processing*, 40, 1775–1786, 1992.

[69] Xu, G., Liu, H., Tong, L., and Kailath, T., A least-squares approach to blind channel identification, *IEEE Trans. on Signal Processing*, 43, 2982–2993, 1995.

[70] Zhou, G. and Giannakis, G.B., Performance analysis of cyclic time-delay estimation algorithms, *Proc. of 29th Conf. on Info. Sciences and Systems*, 780–785, The Johns Hopkins University, Baltimore, MD, March 22-24, 1995.

VI

Adaptive Filtering

Scott C. Douglas
University of Utah

A FILTER IS, IN ITS MOST BASIC SENSE, a device that enhances and/or rejects certain components of a signal. To *adapt* is to change one's characteristics according to some knowledge about one's environment. Taken together, these two terms suggest the goal of an adaptive filter: to alter its selectivity based on the specific characteristics of the signals that are being processed.

In digital signal processing, the term *adaptive filters* refers to a particular set of computational structures and methods for processing digital signals. While many of the most popular techniques used in adaptive filters have been developed and refined within the past forty years, the field of adaptive filters is part of the larger field of optimization theory that has a history dating back to the scientific work of both Galileo and Gauss in the 18th and 19th centuries. Modern developments in adaptive filters began in the 1930s and 1940s with the efforts of Kolmogorov, Wiener, and Levinson to formulate and solve linear estimation tasks. For those who desire an overview of many of the structures, algorithms, analyses, and applications of adaptive filters, the seven chapters in this section provide an excellent introduction to several prominent topics in the field.

Chapter 18 presents an overview of adaptive filters, describing many of the applications for which these systems are used today. This chapter considers basic adaptive filtering concepts while providing an introduction to the popular *least-mean-square (LMS) adaptive filter* that is often used in these applications.

Chapters 19 and 20 focus on the *design* of the LMS adaptive filter from two different viewpoints. In the former chapter, the behavior of the LMS adaptive filter is analyzed within a *statistical* framework that has proven to be quite useful for establishing initial choices of the parameter values of this system. The latter chapter studies the behavior of the LMS adaptive filter from a *deterministic* viewpoint, showing why this system behaves robustly even when modeling errors and finite-precision calculation errors continually perturb the state of this adaptive filter.

Chapter 21 presents the techniques used in another popular class of adaptive systems collectively known as *recursive least-squares (RLS) adaptive filters*. Focusing on the numerical methods that are typically employed in the implementations of these systems, the chapter provides a detailed summary of both conventional and "fast" computational methods for these high-performance systems.

Transform domain adaptive filtering is discussed in Chapter 22. Using the frequency-domain and fast convolution techniques described in this chapter, it is possible both to reduce the computational complexity and to increase the performance of LMS adaptive filters when implemented in block form.

The first five chapters of this section focus almost exclusively on adaptive structures of a finite-impulse response (FIR) form. In Chapter 23, the subtle performance issues surrounding methods for *adaptive infinite-impulse-response (IIR) filters* are carefully described. The most recent technical results concerning the convergence behavior and stability of each major adaptive IIR algorithm class is provided in an easy-to-follow format.

Finally, Chapter 24 presents an important emerging application area for adaptive filters: *blind equalization*. This section indicates how an adaptive filter can be adjusted to produce a desirable input/output characteristic without having an example desired output signal on which to be trained.

While adaptive filters have had a long history, new adaptive filter structures and algorithms are continually being developed. In fact, the range of adaptive filtering algorithms and applications is so great that no one paper, chapter, section, or even book can fully cover the field. Those who desire more information on the topics presented in this section should consult works within the extensive reference lists that appear at the end of each chapter.

18

Introduction to Adaptive Filters

Scott C. Douglas
University of Utah

18.1   What is an Adaptive Filter?

An *adaptive filter* is a computational device that attempts to model the relationship between two signals in real time in an iterative manner. Adaptive filters are often realized either as a set of program instructions running on an arithmetical processing device such as a microprocessor or DSP chip, or as a set of logic operations implemented in a field-programmable gate array (FPGA) or in a semi-custom or custom VLSI integrated circuit. However, ignoring any errors introduced by numerical precision effects in these implementations, the fundamental operation of an adaptive filter can be characterized independently of the specific physical realization that it takes. For this reason, we shall focus on the mathematical forms of adaptive filters as opposed to their specific realizations in software or hardware. Descriptions of adaptive filters as implemented on DSP chips and on a dedicated integrated circuit can be found in [1, 2, 3], and [4], respectively.

An adaptive filter is defined by four aspects:

1. the *signals* being processed by the filter
2. the *structure* that defines how the output signal of the filter is computed from its input signal
3. the *parameters* within this structure that can be iteratively changed to alter the filter's input-output relationship
4. the *adaptive algorithm* that describes how the parameters are adjusted from one time instant to the next

By choosing a particular adaptive filter structure, one specifies the number and type of parameters that can be adjusted. The adaptive algorithm used to update the parameter values of the system can take on a myriad of forms and is often derived as a form of *optimization procedure* that minimizes an *error criterion* that is useful for the task at hand.

In this section, we present the general adaptive filtering problem and introduce the mathematical notation for representing the form and operation of the adaptive filter. We then discuss several different structures that have been proven to be useful in practical applications. We provide an overview of the many and varied applications in which adaptive filters have been successfully used. Finally, we give a simple derivation of the *least-mean-square (LMS) algorithm,* which is perhaps the most popular method for adjusting the coefficients of an adaptive filter, and we discuss some of this algorithm's properties.

As for the mathematical notation used throughout this section, all quantities are assumed to be real-valued. Scalar and vector quantities shall be indicated by lowercase (e.g., x) and uppercase-bold (e.g., $\mathbf{X}$) letters, respectively. We represent scalar and vector sequences or signals as $x(n)$ and $\mathbf{X}(n)$, respectively, where n denotes the discrete time or discrete spatial index, depending on the application. Matrices and indices of vector and matrix elements shall be understood through the context of the discussion.

18.2   The Adaptive Filtering Problem

Figure 18.1 shows a block diagram in which a sample from a digital *input signal* $x(n)$ is fed into a device, called an *adaptive filter,* that computes a corresponding *output signal* sample $y(n)$ at time n. For the moment, the structure of the adaptive filter is not important, except for the fact that it contains adjustable parameters whose values affect how $y(n)$ is computed. The output signal is compared to a second signal $d(n)$, called the *desired response signal,* by subtracting the two samples at time n. This difference signal, given by

$$e(n) = d(n) - y(n) , \tag{18.1}$$

is known as the *error signal.* The error signal is fed into a procedure which alters or *adapts* the parameters of the filter from time n to time $(n + 1)$ in a well-defined manner. This process of adaptation is represented by the oblique arrow that pierces the adaptive filter block in the figure. As the time index n is incremented, it is hoped that the output of the adaptive filter becomes a better and better match to the desired response signal through this adaptation process, such that the magnitude of $e(n)$ decreases over time. In this context, what is meant by "better" is specified by the form of the adaptive algorithm used to adjust the parameters of the adaptive filter.

In the adaptive filtering task, adaptation refers to the method by which the parameters of the system are changed from time index n to time index $(n + 1)$. The number and types of parameters within this system depend on the computational structure chosen for the system. We now discuss different filter structures that have been proven useful for adaptive filtering tasks.

18.3   Filter Structures

In general, any system with a finite number of parameters that affect how $y(n)$ is computed from $x(n)$ could be used for the adaptive filter in Fig. 18.1. Define the *parameter* or *coefficient vector* $\mathbf{W}(n)$ as

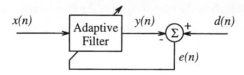

FIGURE 18.1   The general adaptive filtering problem.

$$\mathbf{W}(n) = [w_0(n)\, w_1(n)\, \cdots\, w_{L-1}(n)]^T \tag{18.2}$$

where $\{w_i(n)\}, 0 \le i \le L-1$ are the L parameters of the system at time n. With this definition, we could define a general input-output relationship for the adaptive filter as

$$y(n) = f(\mathbf{W}(n),\, y(n-1),\, y(n-2),\, \ldots,\, y(n-N),\, x(n),\, x(n-1),\, \ldots,\, x(n-M+1)), \tag{18.3}$$

where $f(\cdot)$ represents any well-defined linear or nonlinear function and M and N are positive integers. Implicit in this definition is the fact that the filter is *causal*, such that future values of $x(n)$ are not needed to compute $y(n)$. While noncausal filters can be handled in practice by suitably buffering or storing the input signal samples, we do not consider this possibility.

Although (18.3) is the most general description of an adaptive filter structure, we are interested in determining the best *linear relationship* between the input and desired response signals for many problems. This relationship typically takes the form of a *finite-impulse-response* (FIR) or *infinite-impulse-response* (IIR) filter. Figure 18.2 shows the structure of a direct-form FIR filter, also known as a *tapped-delay-line* or *transversal filter*, where z^{-1} denotes the unit delay element and each $w_i(n)$ is a multiplicative gain within the system. In this case, the parameters in $\mathbf{W}(n)$ correspond to the impulse response values of the filter at time n. We can write the output signal $y(n)$ as

$$y(n) = \sum_{i=0}^{L-1} w_i(n)x(n-i) \tag{18.4}$$

$$= \mathbf{W}^T(n)\mathbf{X}(n), \tag{18.5}$$

where $\mathbf{X}(n) = [x(n)\, x(n-1)\, \cdots\, x(n-L+1)]^T$ denotes the *input signal vector* and $\cdot^T$ denotes vector transpose. Note that this system requires L multiplies and $L-1$ adds to implement, and these computations are easily performed by a processor or circuit so long as L is not too large and the sampling period for the signals is not too short. It also requires a total of $2L$ memory locations to store the L input signal samples and the L coefficient values, respectively.

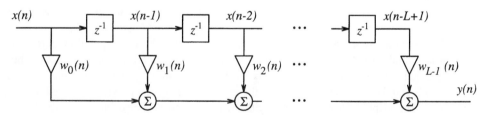

FIGURE 18.2   Structure of an FIR filter.

The structure of a direct-form IIR filter is shown in Fig. 18.3. In this case, the output of the system can be represented mathematically as

$$y(n) = \sum_{i=1}^{N} a_i(n)y(n-i) + \sum_{j=0}^{N} b_j(n)x(n-j), \tag{18.6}$$

although the block diagram does not explicitly represent this system in such a fashion.[1] We could easily

---

[1]The difference between the *direct form II* or *canonical form* structure shown in Fig. 18.3 and the *direct form I* implementation of this system as described by (18.6) is discussed in [5].

write (18.6) using vector notation as

$$y(n) = \mathbf{W}^T(n)\mathbf{U}(n) , \tag{18.7}$$

where the $(2N + 1)$-dimensional vectors $\mathbf{W}(n)$ and $\mathbf{U}(n)$ are defined as

$$\mathbf{W}(n) = [a_1(n)\, a_2(n) \cdots a_N(n)\, b_0(n)\, b_1(n) \cdots b_N(n)]^T \tag{18.8}$$

$$\mathbf{U}(n) = [y(n-1)\, y(n-2) \cdots y(n-N)\, x(n)\, x(n-1) \cdots x(n-N)]^T , \tag{18.9}$$

respectively. Thus, for purposes of computing the output signal $y(n)$, the IIR structure involves a fixed number of multiplies, adds, and memory locations not unlike the direct-form FIR structure.

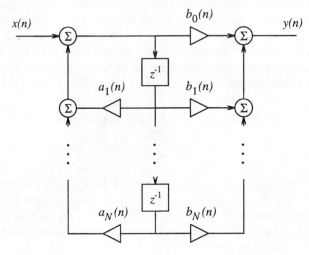

FIGURE 18.3   Structure of an IIR filter.

A third structure that has proven useful for adaptive filtering tasks is the *lattice filter*. A lattice filter is an FIR structure that employs $L - 1$ stages of preprocessing to compute a set of auxiliary signals $\{b_i(n)\}$, $0 \le i \le L - 1$ known as *backward prediction errors*. These signals have the special property that they are *uncorrelated*, and they represent the elements of $\mathbf{X}(n)$ through a *linear transformation*. Thus, the backward prediction errors can be used in place of the delayed input signals in a structure similar to that in Fig. 18.2, and the uncorrelated nature of the prediction errors can provide improved convergence performance of the adaptive filter coefficients with the proper choice of algorithm. Details of the lattice structure and its capabilities are discussed in [6].

A critical issue in the choice of an adaptive filter's structure is its computational complexity. Since the operation of the adaptive filter typically occurs in real time, all of the calculations for the system must occur during one sample time. The structures described above are all useful because $y(n)$ can be computed in a finite amount of time using simple arithmetical operations and finite amounts of memory.

In addition to the linear structures above, one could consider *nonlinear systems* for which the principle of superposition does not hold when the parameter values are fixed. Such systems are useful when the relationship between $d(n)$ and $x(n)$ is not linear in nature. Two such classes of systems are the *Volterra* and *bilinear* filter classes that compute $y(n)$ based on polynomial representations of the input and past output signals. Algorithms for adapting the coefficients of these types of filters are discussed in [7]. In addition, many of the nonlinear models developed in the field of *neural networks*, such as the multilayer perceptron, fit the general form of (18.3), and many of the algorithms used for adjusting the parameters of neural networks are related to the algorithms used for FIR and IIR adaptive filters. For a discussion of neural networks in an engineering context, the reader is referred to [8].

18.4 The Task of an Adaptive Filter

When considering the adaptive filter problem as illustrated in Fig. 18.1 for the first time, a reader is likely to ask, "If we already have the desired response signal, what is the point of trying to match it using an adaptive filter?" In fact, the concept of "matching" $y(n)$ to $d(n)$ with some system obscures the subtlety of the adaptive filtering task. Consider the following issues that pertain to many adaptive filtering problems:

- *In practice, the quantity of interest is not always $d(n)$.* Our desire may be to represent in $y(n)$ a certain component of $d(n)$ that is contained in $x(n)$, or it may be to isolate a component of $d(n)$ within the error $e(n)$ that is *not* contained in $x(n)$. Alternatively, we may be solely interested in the values of the parameters in $\mathbf{W}(n)$ and have no concern about $x(n)$, $y(n)$, or $d(n)$ themselves. Practical examples of each of these scenarios are provided later in this chapter.

- *There are situations in which $d(n)$ is not available at all times.* In such situations, adaptation typically occurs only when $d(n)$ is available. When $d(n)$ is unavailable, we typically use our most-recent parameter estimates to compute $y(n)$ in an attempt to *estimate* the desired response signal $d(n)$.

- *There are real-world situations in which $d(n)$ is never available.* In such cases, one can use additional information about the characteristics of a "hypothetical" $d(n)$, such as its predicted statistical behavior or amplitude characteristics, to form suitable estimates of $d(n)$ from the signals available to the adaptive filter. Such methods are collectively called *blind adaptation algorithms*. The fact that such schemes even work is a tribute both to the ingenuity of the developers of the algorithms and to the technological maturity of the adaptive filtering field.

It should also be recognized that the relationship between $x(n)$ and $d(n)$ can vary with time. In such situations, the adaptive filter attempts to alter its parameter values to follow the changes in this relationship as "encoded" by the two sequences $x(n)$ and $d(n)$. This behavior is commonly referred to as *tracking*.

18.5 Applications of Adaptive Filters

Perhaps the most important driving forces behind the developments in adaptive filters throughout their history have been the wide range of applications in which such systems can be used. We now discuss the forms of these applications in terms of more-general problem classes that describe the assumed relationship between $d(n)$ and $x(n)$. Our discussion illustrates the key issues in selecting an adaptive filter for a particular task. Extensive details concerning the specific issues and problems associated with each problem genre can be found in the references at the end of this chapter.

18.5.1 System Identification

Consider Fig. 18.4, which shows the general problem of *system identification*. In this diagram, the system enclosed by dashed lines is a "black box," meaning that the quantities inside are not observable from the outside. Inside this box is (1) an unknown system which represents a general input-output relationship and (2) the signal $\eta(n)$, called the *observation noise signal* because it corrupts the observations of the signal at the output of the unknown system.

Let $\widehat{d}(n)$ represent the output of the unknown system with $x(n)$ as its input. Then, the desired response signal in this model is

$$d(n) = \widehat{d}(n) + \eta(n) . \tag{18.10}$$

Here, the task of the adaptive filter is to accurately represent the signal $\widehat{d}(n)$ at its output. If $y(n) = \widehat{d}(n)$, then the adaptive filter has accurately modeled or identified the portion of the unknown system that is driven by $x(n)$.

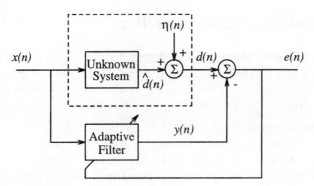

FIGURE 18.4   System identification.

Since the model typically chosen for the adaptive filter is a linear filter, the practical goal of the adaptive filter is to determine the best linear model that describes the input-output relationship of the unknown system. Such a procedure makes the most sense when the unknown system is also a linear model of the same structure as the adaptive filter, as it is possible that $y(n) = \widehat{d}(n)$ for some set of adaptive filter parameters. For ease of discussion, let the unknown system and the adaptive filter both be FIR filters, such that

$$d(n) = \mathbf{W}_{opt}^T(n)\mathbf{X}(n) + \eta(n), \tag{18.11}$$

where $\mathbf{W}_{opt}(n)$ is an optimum set of filter coefficients for the unknown system at time n. In this problem formulation, the ideal adaptation procedure would adjust $\mathbf{W}(n)$ such that $\mathbf{W}(n) = \mathbf{W}_{opt}(n)$ as $n \to \infty$. In practice, the adaptive filter can only adjust $\mathbf{W}(n)$ such that $y(n)$ closely approximates $\widehat{d}(n)$ over time.

The system identification task is at the heart of numerous adaptive filtering applications. We list several of these applications here.

Channel Identification

In communication systems, useful information is transmitted from one point to another across a medium such as an electrical wire, an optical fiber, or a wireless radio link. Nonidealities of the transmission medium or *channel* distort the fidelity of the transmitted signals, making the deciphering of the received information difficult. In cases where the effects of the distortion can be modeled as a linear filter, the resulting "smearing" of the transmitted symbols is known as *inter-symbol interference* (ISI). In such cases, an adaptive filter can be used to model the effects of the channel ISI for purposes of deciphering the received information in an optimal manner. In this problem scenario, the transmitter sends to the receiver a sample sequence $x(n)$ that is known to both the transmitter and receiver. The receiver then attempts to model the received signal $d(n)$ using an adaptive filter whose input is the known transmitted sequence $x(n)$. After a suitable period of adaptation, the parameters of the adaptive filter in $\mathbf{W}(n)$ are fixed and then used in a procedure to decode future signals transmitted across the channel.

Channel identification is typically employed when the fidelity of the transmitted channel is severely compromised or when simpler techniques for sequence detection cannot be used. Techniques for detecting digital signals in communication systems can be found in [9].

Plant Identification

In many control tasks, knowledge of the transfer function of a linear plant is required by the physical controller so that a suitable control signal can be calculated and applied. In such cases, we can characterize the transfer function of the plant by exciting it with a known signal $x(n)$ and then attempting to match the output of the plant $d(n)$ with a linear adaptive filter. After a suitable period of adaptation, the system has been adequately modeled, and the resulting adaptive filter coefficients in $\mathbf{W}(n)$ can be used in a control scheme to enable the overall closed-loop system to behave in the desired manner.

In certain scenarios, continuous updates of the plant transfer function estimate provided by $\mathbf{W}(n)$ are needed to allow the controller to function properly. A discussion of these *adaptive control* schemes and the subtle issues in their use is given in [10, 11].

Echo Cancellation for Long-Distance Transmission

In voice communication across telephone networks, the existence of junction boxes called *hybrids* near either end of the network link hampers the ability of the system to cleanly transmit voice signals. Each hybrid allows voices that are transmitted via separate lines or channels across a long-distance network to be carried locally on a single telephone line, thus lowering the wiring costs of the local network. However, when small impedance mismatches between the long distance lines and the hybrid junctions occur, these hybrids can reflect the transmitted signals back to their sources, and the long transmission times of the long-distance network—about 0.3 s for a trans-oceanic call via a satellite link—turn these reflections into a noticeable echo that makes the understanding of conversation difficult for both callers. The traditional solution to this problem prior to the advent of the adaptive filtering solution was to introduce significant loss into the long-distance network so that echoes would decay to an acceptable level before they became perceptible to the callers. Unfortunately, this solution also reduces the transmission quality of the telephone link and makes the task of connecting long distance calls more difficult.

An adaptive filter can be used to cancel the echoes caused by the hybrids in this situation. Adaptive filters are employed at each of the two hybrids within the network. The input $x(n)$ to each adaptive filter is the speech signal being received prior to the hybrid junction, and the desired response signal $d(n)$ is the signal being sent out from the hybrid across the long-distance connection. The adaptive filter attempts to model the transmission characteristics of the hybrid junction as well as any echoes that appear across the long-distance portion of the network. When the system is properly designed, the error signal $e(n)$ consists almost totally of the local talker's speech signal, which is then transmitted over the network. Such systems were first proposed in the mid-1960s [12] and are commonly used today. For more details on this application, see [13, 14].

Acoustic Echo Cancellation

A related problem to echo cancellation for telephone transmission systems is that of acoustic echo cancellation for conference-style speakerphones. When using a speakerphone, a caller would like to turn up the amplifier gains of both the microphone and the audio loudspeaker in order to transmit and hear the voice signals more clearly. However, the feedback path from the device's loudspeaker to its input microphone causes a distinctive *howling* sound if these gains are too high. In this case, the culprit is the room's response to the voice signal being broadcast by the speaker; in effect, the room acts as an extremely poor hybrid junction, in analogy with the echo cancellation task discussed previously. A simple solution to this problem is to only allow one person to speak at a time, a form of operation called *half-duplex transmission.* However, studies have indicated that half-duplex transmission causes problems with normal conversations, as people typically overlap their phrases with others when conversing.

To maintain *full-duplex transmission,* an acoustic echo canceller is employed in the speakerphone to model the acoustic transmission path from the speaker to the microphone. The input signal $x(n)$ to the acoustic echo canceller is the signal being sent to the speaker, and the desired response signal $d(n)$ is measured at the microphone on the device. Adaptation of the system occurs continually throughout a telephone call to model any physical changes in the room acoustics. Such devices are readily available in the marketplace today. In addition, similar technology can and is used to remove the echo that occurs through the combined radio/room/telephone transmission path when one places a call to a radio or television talk show. Details of the acoustic echo cancellation problem can be found in [14].

Adaptive Noise Cancelling

When collecting measurements of certain signals or processes, physical constraints often limit our ability to cleanly measure the quantities of interest. Typically, a signal of interest is linearly mixed with other extraneous noises in the measurement process, and these extraneous noises introduce unacceptable errors in the measurements. However, if a linearly related *reference* version of any one of the extraneous noises can be cleanly sensed at some other physical location in the system, an adaptive filter can be used to determine the relationship between the noise reference $x(n)$ and the component of this noise that is contained in the measured signal $d(n)$. After adaptively subtracting out this component, what remains in $e(n)$ is the signal of interest. If several extraneous noises corrupt the measurement of interest, several adaptive filters can be used in parallel as long as suitable noise reference signals are available within the system.

Adaptive noise cancelling has been used for several applications. One of the first was a medical application that enabled the electroencephalogram (EEG) of the fetal heartbeat of an unborn child to be cleanly extracted from the much-stronger interfering EEG of the maternal heartbeat signal. Details of this application as well as several others are described in the seminal paper by Widrow and his colleagues [15].

18.5.2 Inverse Modeling

We now consider the general problem of *inverse modeling*, as shown in Fig. 18.5. In this diagram, a *source signal* $s(n)$ is fed into an unknown system that produces the input signal $x(n)$ for the adaptive filter. The output of the adaptive filter is subtracted from a desired response signal that is a delayed version of the source signal, such that

$$d(n) = s(n - \Delta),\qquad (18.12)$$

where Δ is a positive integer value. The goal of the adaptive filter is to adjust its characteristics such that the output signal is an accurate representation of the delayed source signal.

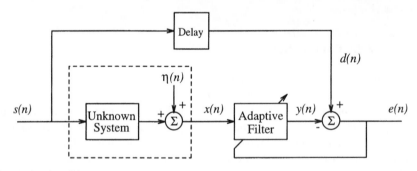

FIGURE 18.5   Inverse modeling.

The inverse modeling task characterizes several adaptive filtering applications, two of which are now described.

Channel Equalization

Channel equalization is an alternative to the technique of channel identification described previously for the decoding of transmitted signals across nonideal communication channels. In both cases, the transmitter sends a sequence $s(n)$ that is known to both the transmitter and receiver. However, in equalization, the received signal is used as the input signal $x(n)$ to an adaptive filter, which adjusts its characteristics so that its output closely matches a delayed version $s(n - \Delta)$ of the known transmitted signal. After a suitable adaptation period, the coefficients of the system either are fixed and used to decode future transmitted messages or are adapted using a crude estimate of the desired response signal that is computed from $y(n)$. This latter mode of operation is known as *decision-directed adaptation.*

Channel equalization was one of the first applications of adaptive filters and is described in the pioneering work of Lucky [16]. Today, it remains as one of the most popular uses of an adaptive filter. Practically every computer telephone modem transmitting at rates of 9600 *baud* (bits per second) or greater contains an adaptive equalizer. Adaptive equalization is also useful for wireless communication systems. Qureshi [17] provides a tutorial on adaptive equalization. A related problem to equalization is *deconvolution,* a problem that appears in the context of geophysical exploration [18]. Equalization is closely related to *linear prediction,* a topic that we shall discuss shortly.

Inverse Plant Modeling

In many control tasks, the frequency and phase characteristics of the plant hamper the convergence behavior and stability of the control system. We can use a system of the form in Fig. 18.5 to compensate for the nonideal characteristics of the plant and as a method for adaptive control. In this case, the signal $s(n)$ is sent at the output of the controller, and the signal $x(n)$ is the signal measured at the output of the plant. The coefficients of the adaptive filter are then adjusted so that the cascade of the plant and adaptive filter can be nearly represented by the pure delay $z^{-\Delta}$. Details of the adaptive algorithms as applied to control tasks in this fashion can be found in [11].

18.5.3 Linear Prediction

A third type of adaptive filtering task is shown in Fig. 18.6. In this system, the input signal $x(n)$ is derived from the desired response signal as

$$x(n) = d(n - \Delta),$$
(18.13)

where Δ is an integer value of delay. In effect, the input signal serves as the desired response signal, and for this reason it is always available. In such cases, the linear adaptive filter attempts to predict future values of the input signal using past samples, giving rise to the name *linear prediction* for this task.

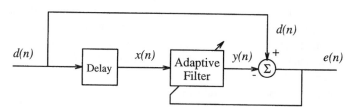

FIGURE 18.6   Linear prediction.

If an estimate of the signal $x(n + \Delta)$ at time n is desired, a copy of the adaptive filter whose input is the current sample $x(n)$ can be employed to compute this quantity. However, linear prediction has a number of uses besides the obvious application of forecasting future events, as described in the following two applications.

Linear Predictive Coding

When transmitting digitized versions of real-world signals such as speech or images, the temporal correlation of the signals is a form of redundancy that can be exploited to code the waveform in a smaller number of bits than are needed for its original representation. In these cases, a linear predictor can be used to model the signal correlations for a short block of data in such a way as to reduce the number of bits needed to represent the signal waveform. Then, essential information about the signal model is transmitted along with the coefficients of the adaptive filter for the given data block. Once received, the signal is synthesized using the filter coefficients and the additional signal information provided for the given block of data.

When applied to speech signals, this method of signal encoding enables the transmission of understandable speech at only 2.4 kb/s, although the reconstructed speech has a distinctly synthetic quality. Predictive coding can be combined with a quantizer to enable higher-quality speech encoding at higher data rates using an *adaptive differential pulse-code modulation* (ADPCM) scheme. In both of these methods, the lattice filter structure plays an important role because of the way in which it parameterizes the physical nature of the vocal tract. Details about the role of the lattice filter in the linear prediction task can be found in [19].

Adaptive Line Enhancement

In some situations, the desired response signal $d(n)$ consists of a sum of a broadband signal and a nearly periodic signal, and it is desired to separate these two signals without specific knowledge about the signals (such as the fundamental frequency of the periodic component).

In these situations, an adaptive filter configured as in Fig. 18.6 can be used. For this application, the delay Δ is chosen to be large enough such that the broadband component in $x(n)$ is uncorrelated with the broadband component in $x(n - \Delta)$. In this case, the broadband signal cannot be removed by the adaptive filter through its operation, and it remains in the error signal $e(n)$ after a suitable period of adaptation. The adaptive filter's output $y(n)$ converges to the narrowband component, which is easily predicted given past samples. The name line enhancement arises because periodic signals are characterized by lines in their frequency spectra, and these spectral lines are enhanced at the output of the adaptive filter.

For a discussion of the adaptive line enhancement task using LMS adaptive filters, the reader is referred to [20].

18.5.4 Feedforward Control

Another problem area combines elements of both the inverse modeling and system identification tasks and typifies the types of problems encountered in the area of adaptive control known as *feedforward control*. Figure 18.7 shows the block diagram for this system, in which the output of the adaptive filter passes through a plant before it is subtracted from the desired response to form the error signal. The plant hampers the operation of the adaptive filter by changing the amplitude and phase characteristics of the adaptive filter's output signal as represented in $e(n)$. Thus, knowledge of the plant is generally required in order to adapt the parameters of the filter properly.

An application that fits this particular problem formulation is *active noise control*, in which unwanted sound energy propagates in air or a fluid into a physical region in space. In such cases, an electroacoustic system employing microphones, speakers, and one or more adaptive filters can be used to create a secondary sound field that interferes with the unwanted sound, reducing its level in the region via destructive interference. Similar techniques can be used to reduce vibrations in solid media. Details of useful algorithms for the active noise and vibration control tasks can be found in [21, 22].

18.6   Gradient-Based Adaptive Algorithms

An adaptive algorithm is a procedure for adjusting the parameters of an adaptive filter to minimize a cost function chosen for the task at hand. In this section, we describe the general form of many adaptive FIR filtering algorithms and present a simple derivation of the LMS adaptive algorithm. In our discussion, we only consider an adaptive FIR filter structure, such that the output signal $y(n)$ is given by (18.5). Such systems are currently more popular than adaptive IIR filters because (1) the input-output stability of the

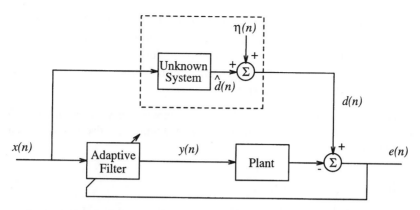

FIGURE 18.7 Feedforward control.

FIR filter structure is guaranteed for any set of fixed coefficients, and (2) the algorithms for adjusting the coefficients of FIR filters are more simple in general than those for adjusting the coefficients of IIR filters.

18.6.1 General Form of Adaptive FIR Algorithms

The general form of an adaptive FIR filtering algorithm is

$$\mathbf{W}(n+1) = \mathbf{W}(n) + \mu(n)\mathbf{G}(e(n), \mathbf{X}(n), \mathbf{\Phi}(n)), \qquad (18.14)$$

where $\mathbf{G}(\cdot)$ is a particular vector-valued nonlinear function, $\mu(n)$ is a *step size* parameter, $e(n)$ and $\mathbf{X}(n)$ are the error signal and input signal vector, respectively, and $\mathbf{\Phi}(n)$ is a vector of states that store pertinent information about the characteristics of the input and error signals and/or the coefficients at previous time instants. In the simplest algorithms, $\mathbf{\Phi}(n)$ is not used, and the only information needed to adjust the coefficients at time n are the error signal, input signal vector, and step size.

The step size is so called because it determines the magnitude of the change or "step" that is taken by the algorithm in iteratively determining a useful coefficient vector. Much research effort has been spent characterizing the role that $\mu(n)$ plays in the performance of adaptive filters in terms of the statistical or frequency characteristics of the input and desired response signals. Often, success or failure of an adaptive filtering application depends on how the value of $\mu(n)$ is chosen or calculated to obtain the best performance from the adaptive filter. The issue of choosing $\mu(n)$ for both stable and accurate convergence of the LMS adaptive filter is addressed in Chapter 19 of this Handbook.

18.6.2 The Mean-Squared Error Cost Function

The form of $\mathbf{G}(\cdot)$ in (18.14) depends on the cost function chosen for the given adaptive filtering task. We now consider one particular cost function that yields a popular adaptive algorithm.

Define the *mean-squared error* (MSE) cost function as

$$J_{MSE}(n) = \frac{1}{2}\int_{-\infty}^{\infty} e^2(n) p_n(e(n)) de(n) \qquad (18.15)$$

$$= \frac{1}{2}E\{e^2(n)\}, \qquad (18.16)$$

where $p_n(e)$ represents the probability density function of the error at time n and $E\{\cdot\}$ is shorthand for the *expectation integral* on the right-hand side of (18.15). The MSE cost function is useful for adaptive FIR filters because

- $J_{MSE}(n)$ has a well-defined minimum with respect to the parameters in $\mathbf{W}(n)$;
- the coefficient values obtained at this minimum are the ones that minimize the power in the error signal $e(n)$, indicating that $y(n)$ has approached $d(n)$; and
- $J_{MSE}(n)$ is a smooth function of each of the parameters in $\mathbf{W}(n)$, such that it is differentiable with respect to each of the parameters in $\mathbf{W}(n)$.

The third point is important in that it enables us to determine both the optimum coefficient values given knowledge of the statistics of $d(n)$ and $x(n)$ as well as a simple iterative procedure for adjusting the parameters of an FIR filter.

18.6.3 The Wiener Solution

For the FIR filter structure, the coefficient values in $\mathbf{W}(n)$ that minimize $J_{MSE}(n)$ are well-defined if the statistics of the input and desired response signals are known. The formulation of this problem for continuous-time signals and the resulting solution was first derived by Wiener [23]. Hence, this optimum coefficient vector $\mathbf{W}_{MSE}(n)$ is often called the *Wiener solution* to the adaptive filtering problem. The extension of Wiener's analysis to the discrete-time case is attributed to Levinson [24].

To determine $\mathbf{W}_{MSE}(n)$, we note that the function $J_{MSE}(n)$ in (18.16) is quadratic in the parameters $\{w_i(n)\}$, and the function is also differentiable. Thus, we can use a result from optimization theory that states that the derivatives of a smooth cost function with respect to each of the parameters is zero at a minimizing point on the cost function error surface. Thus, $\mathbf{W}_{MSE}(n)$ can be found from the solution to the system of equations

$$\frac{\partial J_{MSE}(n)}{\partial w_i(n)} = 0, \qquad 0 \leq i \leq L - 1. \tag{18.17}$$

Taking derivatives of $J_{MSE}(n)$ in (18.16) and noting that $e(n)$ and $y(n)$ are given by (18.1) and (18.5), respectively, we obtain

$$\frac{\partial J_{MSE}(n)}{\partial w_i(n)} = E\left\{e(n)\frac{\partial e(n)}{\partial w_i(n)}\right\} \tag{18.18}$$

$$= -E\left\{e(n)\frac{\partial y(n)}{\partial w_i(n)}\right\} \tag{18.19}$$

$$= -E\{e(n)x(n-i)\} \tag{18.20}$$

$$= -\left(E\{d(n)x(n-i)\} - \sum_{j=0}^{L-1} E\{x(n-i)x(n-j)\}w_j(n)\right). \tag{18.21}$$

where we have used the definitions of $e(n)$ and of $y(n)$ for the FIR filter structure in (18.1) and (18.5), respectively, to expand the last result in (18.21).

By defining the matrix $\mathbf{R}_{XX}(n)$ and vector $\mathbf{P}_{dX}(n)$ as

$$\mathbf{R}_{XX} = E\{\mathbf{X}(n)\mathbf{X}^T(n)\} \text{ and } \mathbf{P}_{dX}(n) = E\{d(n)\mathbf{X}(n)\}, \tag{18.22}$$

respectively, we can combine (18.17) and (18.21) to obtain the system of equations in vector form as

$$\mathbf{R}_{XX}(n)\mathbf{W}_{MSE}(n) - \mathbf{P}_{dX}(n) = \mathbf{0}, \tag{18.23}$$

where $\mathbf{0}$ is the zero vector. Thus, so long as the matrix $\mathbf{R}_{XX}(n)$ is invertible, the optimum Wiener solution vector for this problem is

$$\mathbf{W}_{MSE}(n) = \mathbf{R}_{XX}^{-1}(n)\mathbf{P}_{dX}(n). \tag{18.24}$$

18.6.4 The Method of Steepest Descent

The method of steepest descent is a celebrated optimization procedure for minimizing the value of a cost function $J(n)$ with respect to a set of adjustable parameters $\mathbf{W}(n)$. This procedure adjusts each parameter of the system according to

$$w_i(n+1) = w_i(n) - \mu(n)\frac{\partial J(n)}{\partial w_i(n)} . \tag{18.25}$$

In other words, the ith parameter of the system is altered according to the derivative of the cost function with respect to the ith parameter. Collecting these equations in vector form, we have

$$\mathbf{W}(n+1) = \mathbf{W}(n) - \mu(n)\frac{\partial J(n)}{\partial \mathbf{W}(n)} , \tag{18.26}$$

where $\partial J(n)/\partial \mathbf{W}(n)$ is a vector of derivatives $\partial J(n)/\partial w_i(n)$.

For an FIR adaptive filter that minimizes the MSE cost function, we can use the result in (18.21) to explicitly give the form of the steepest descent procedure in this problem. Substituting these results into (18.25) yields the update equation for $\mathbf{W}(n)$ as

$$\mathbf{W}(n+1) = \mathbf{W}(n) + \mu(n)(\mathbf{P}_{\mathrm{dX}}(n) - \mathbf{R}_{\mathrm{XX}}(n)\mathbf{W}(n)) . \tag{18.27}$$

However, this steepest descent procedure depends on the statistical quantities $E\{d(n)x(n-i)\}$ and $E\{x(n-i)x(n-j)\}$ contained in $\mathbf{P}_{\mathrm{dX}}(n)$ and $\mathbf{R}_{\mathrm{XX}}(n)$, respectively. In practice, we only have measurements of both $d(n)$ and $x(n)$ to be used within the adaptation procedure. While suitable estimates of the statistical quantities needed for (18.27) could be determined from the signals $x(n)$ and $d(n)$, we instead develop an approximate version of the method of steepest descent that depends on the signal values themselves. This procedure is known as the *LMS algorithm*.

18.6.5 The LMS Algorithm

The cost function $J(n)$ chosen for the steepest descent algorithm of (18.25) determines the coefficient solution obtained by the adaptive filter. If the MSE cost function in (18.16) is chosen, the resulting algorithm depends on the statistics of $x(n)$ and $d(n)$ because of the expectation operation that defines this cost function. Since we typically only have measurements of $d(n)$ and of $x(n)$ available to us, we substitute an alternative cost function that depends only on these measurements. One such cost function is the least-squares cost function given by

$$J_{LS}(n) = \sum_{k=0}^{n} \alpha(k)(d(k) - \mathbf{W}^T(n)\mathbf{X}(k))^2 . \tag{18.28}$$

where $\alpha(n)$ is a suitable weighting sequence for the terms within the summation. This cost function, however, is complicated by the fact that it requires numerous computations to calculate its value as well as its derivatives with respect to each $w_i(n)$, although efficient recursive methods for its minimization can be developed. See Chapter 21 for more details on these methods.

Alternatively, we can propose the simplified cost function $J_{LMS}(n)$ given by

$$J_{LMS}(n) = \frac{1}{2}e^2(n) . \tag{18.29}$$

This cost function can be thought of as an *instantaneous estimate* of the MSE cost function, as $J_{MSE}(n) = E\{J_{LMS}(n)\}$. Although it might not appear to be useful, the resulting algorithm obtained when $J_{LMS}(n)$ is used for $J(n)$ in (18.25) is extremely useful for practical applications. Taking derivatives of $J_{LMS}(n)$

with respect to the elements of $\mathbf{W}(n)$ and substituting the result into (18.25), we obtain the LMS adaptive algorithm given by

$$\mathbf{W}(n+1) = \mathbf{W}(n) + \mu(n)e(n)\mathbf{X}(n) . \tag{18.30}$$

Note that this algorithm is of the general form in (18.14). It also requires only multiplications and additions to implement. In fact, the number and type of operations needed for the LMS algorithm is nearly the same as that of the FIR filter structure with fixed coefficient values, which is one of the reasons for the algorithm's popularity.

The behavior of the LMS algorithm has been widely studied, and numerous results concerning its adaptation characteristics under different situations have been developed. For discussions of some of these results, the reader is referred to Chapters 19 and 20 in this Handbook. For now, we indicate its useful behavior by noting that the solution obtained by the LMS algorithm near its convergent point is related to the Wiener solution. In fact, analyses of the LMS algorithm under certain statistical assumptions about the input and desired response signals show that

$$\lim_{n \to \infty} E\{\mathbf{W}(n)\} = \mathbf{W}_{MSE} , \tag{18.31}$$

when the Wiener solution $\mathbf{W}_{MSE}(n)$ is a fixed vector. Moreover, the average behavior of the LMS algorithm is quite similar to that of the steepest descent algorithm in (18.27) that depends explicitly on the statistics of the input and desired response signals. In effect, the iterative nature of the LMS coefficient updates is a form of time-averaging that smooths the errors in the instantaneous gradient calculations to obtain a more reasonable estimate of the true gradient.

18.6.6 Other Stochastic Gradient Algorithms

The LMS algorithm is but one of an entire family of algorithms that are based on instantaneous approximations to steepest descent procedures. Such algorithms are known as *stochastic gradient algorithms* because they use a stochastic version of the gradient of a particular cost function's error surface to adjust the parameters of the filter. As an example, we consider the cost function

$$J_{SA}(n) = |e(n)| , \tag{18.32}$$

where $|\cdot|$ denotes absolute value. Like $J_{LMS}(n)$, this cost function also has a unique minimum at $e(n) = 0$, and it is differentiable everywhere except at $e(n) = 0$. Moreover, it is the instantaneous value of the *mean absolute error* cost function $J_{MAE}(n) = E\{J_{SA}(n)\}$. Taking derivatives of $J_{SA}(n)$ with respect to the coefficients $\{w_i(n)\}$ and substituting the results into (18.25) yields the *sign-error* algorithm as[2]

$$\mathbf{W}(n+1) = \mathbf{W}(n) + \mu(n)\text{sgn}(e(n))\mathbf{X}(n) , \tag{18.33}$$

where

$$\text{sgn}(e) = \begin{cases} 1 & \text{if } e > 0 \\ 0 & \text{if } e = 0 \\ -1 & \text{if } e < 0 \end{cases} . \tag{18.34}$$

This algorithm is also of the general form in (18.14).

The sign error algorithm is a useful adaptive filtering procedure because the terms $\text{sgn}(e(n))x(n-i)$ can be computed easily in dedicated digital hardware. Its convergence properties differ from those of the LMS algorithm, however. Discussions of this and other algorithms based on non-MSE criteria can be found in [25].

---

[2]Here, we have specified $\partial|e|/\partial e = 0$ for $e = 0$, although the derivative of this function does not exist at this point.

18.6.7 Finite-Precision Effects and Other Implementation Issues

In all digital hardware and software implementations of the LMS algorithm in (18.30), the quantities $e(n)$, $d(n)$, and $\{x(n-i)\}$ are represented by finite-precision quantities with a certain number of bits. Small numerical errors are introduced in each of the calculations within the coefficient updates in these situations. The effects of these numerical errors are usually less severe in systems that employ *floating-point arithmetic*, in which all numerical values are represented by both a mantissa and exponent, as compared to systems that employ *fixed-point arithmetic*, in which a mantissa-only numerical representation is used. The effects of the numerical errors introduced in these cases can be characterized; see [26] for a discussion of these issues.

While knowledge of the numerical effects of finite-precision arithmetic are necessary for obtaining the best performance from the LMS adaptive filter, it can be generally stated that the LMS adaptive filter performs robustly in the presence of these numerical errors. In fact, the apparent robustness of the LMS adaptive filter has led to the development of approximate implementations of (18.30) that are more easily implemented in dedicated hardware. The general form of these implementations is

$$w_i(n+1) = w_i(n) + \mu(n)g_1(e(n))g_2(x(n-i)), \tag{18.35}$$

where $g_1(\cdot)$ and $g_2(\cdot)$ are odd-symmetric nonlinearities that are chosen to simplify the implementation of the system. Some of the algorithms described by (18.35) include the *sign-data* $\{g_1(e) = e, g_2(x) = \operatorname{sgn}(x)\}$, *sign-sign* or *zero-forcing* $\{g_1(e) = \operatorname{sgn}(e), g_2(x) = \operatorname{sgn}(x)\}$, and *power-of-two quantized* algorithms, as well as the sign error algorithm introduced previously. A presentation and comparative analysis of the performance of many of these algorithms can be found in [27].

18.6.8 System Identification Example

We now illustrate the actual behavior of the LMS adaptive filter through a system identification example in which the impulse response of a small audio loudspeaker in a room is estimated. A Gaussian-distributed signal with a flat frequency spectrum over the usable frequency range of the loudspeaker is generated and sent through an audio amplifier to the loudspeaker. This same Gaussian signal is sent to a 16-bit analog-to-digital (A/D) converter which samples it at an 8 kHz rate. The sound produced by the loudspeaker propagates to a microphone located several feet away from the loudspeaker, where it is collected and digitized by a second A/D converter also sampling at an 8 kHz rate. Both signals are stored to a computer file for subsequent processing and analysis. The goal of the analysis is to determine the combined impulse response of the loudspeaker/room/microphone sound propagation path. Such information is useful if the loudspeaker and microphone are to be used in the active noise control task described previously, and the general task also resembles that of acoustic echo cancellation for speakerphones.

We process these signals using a computer program that implements the LMS adaptive filter within the MATLAB[3] signal manipulation environment. In this case, we have normalized the powers of both the Gaussian input signal and desired response signal collected at the microphone to unity, and we have highpass-filtered the microphone signal using a filter with transfer function $H(z) = (1 - z^{-1})/(1 - 0.95z^{-1})$ to remove any DC offset in this signal. For this task, we have chosen an $L = 100$-coefficient FIR filter adapted using the LMS algorithm in (18.30) with a fixed step size of $\mu = 0.0005$ to obtain an accurate estimate of the impulse response of the loudspeaker and room. Figure 18.8 shows the convergence of the error signal in this situation. After about 4000 samples (0.5 s), the error signal has been reduced to a power that is about 1/15 (-12 dB) below that of the microphone signal, indicating that the filter has converged. Figure 18.9 shows the coefficients of the adaptive filter at iteration $n = 10000$. The impulse

---

[3]MATLAB is a registered trademark of The MathWorks, Newton, MA.

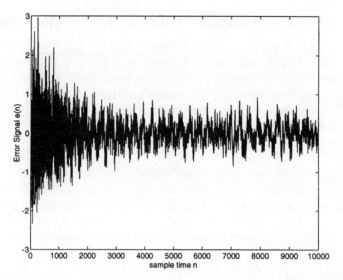

FIGURE 18.8   Convergence of the error signal in the loudspeaker identification experiment.

response of the loudspeaker/room/microphone path consists of a large pulse corresponding to the direct sound propagation path as well as numerous smaller pulses caused by reflections of sounds off walls and other surfaces in the room.

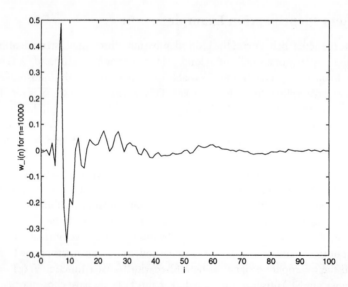

FIGURE 18.9   The adaptive filter coefficients obtained in the loudspeaker identification experiment.

18.7   Conclusions

In this section, we have presented an overview of adaptive filters, emphasizing the applications and basic algorithms that have already proven themselves to be useful in practice. Despite the many contributions in the field, research efforts in adaptive filters continue at a strong pace, and it is likely that new applications for adaptive filters will be developed in the future. To keep abreast of these advances, the reader is urged to consult journals such as the *IEEE Transactions on Signal Processing* as well as the proceedings of yearly conferences and workshops in the signal processing and related fields.

References

[1] Kuo, S. and Chen, C., Implementation of adaptive filters with the TMS320C25 or the TMS320C30, in *Digital Signal Processing Applications with the TMS320 Family,* Papamichalis, P., Ed., Prentice-Hall, Englewood Cliffs, NJ, 1991, 191–271.

[2] Analog Devices, Adaptive Filters, in *ADSP-21000 Family Application Handbook,* vol. 1, Analog Devices, 1994, 157–203.

[3] El-Sharkawy, M., Designing adaptive FIR filters and implementing them on the DSP56002 processor, in *Digital Signal Processing Applications with Motorola's DSP56002 Processor,* Prentice-Hall, Upper Saddle River, NJ, 1996, 319–342.

[4] Borth, D.E., Gerson, I.A., Haug, J.R., and Thompson, C.D., A flexible adaptive FIR filter VLSI IC, *IEEE J. Sel. Areas Commun.,* 6(3), 494–503, April 1988.

[5] Oppenheim, A.V. and Schafer, A.W., *Discrete-Time Signal Processing,* Prentice-Hall, Englewood Cliffs, NJ, 1989.

[6] Friedlander, B., Lattice filters for adaptive processing, *Proc. IEEE,* 70(8), 829–867, Aug. 1982.

[7] Mathews, V.J., Adaptive polynomial filters, *IEEE Signal Processing Mag.,* 8(3), 10–26, July 1991.

[8] Haykin, S., *Neural Networks: A Comprehensive Foundation,* Macmillan, New York, 1994.

[9] Proakis, J.G. and Salehi, M., *Communication Systems Engineering,* Prentice-Hall, Englewood Cliffs, NJ, 1994.

[10] Åström, K.G. and Wittenmark, B., *Adaptive Control,* Addison-Wesley, Reading, MA, 1989.

[11] Widrow, B. and Walach, E., *Adaptive Inverse Control,* Prentice-Hall, Upper Saddle River, NJ, 1996.

[12] Sondhi, M.M., An adaptive echo canceller, *Bell Sys. Tech. J.,* 46, 497–511, March 1967.

[13] Messerschmitt, D.G., Echo cancellation in speech and data transmission, *IEEE J. Sel. Areas Commun.,* SAC-2(2), 283–297, March 1984.

[14] Murano, K., Unagami, S., and Amano, F., Echo cancellation and applications, *IEEE Commun. Mag.,* 28(1), 49–55, Jan. 1990.

[15] Widrow, B., Glover, J.R., Jr., McCool, J.M., Kaunitz, J., Williams, C.S., Hearn, R.H., Zeidler, J.R., Dong, E., Jr., and Goodlin, R.C., Adaptive noise cancelling: principles and applications, *Proc. IEEE,* 63(12), 1692–1716, Dec. 1975.

[16] Lucky, R.W., Techniques for adaptive equalization of digital communication systems, *Bell Sys. Tech. J.,* 45, 255–286, Feb. 1966.

[17] Qureshi, S.U.H., Adaptive equalization, *Proc. IEEE,* 73(9), 1349–1387, Sept. 1985.

[18] Robinson, E.A. and Durrani, T., *Geophysical Signal Processing,* Prentice-Hall, Englewood Cliffs, NJ, 1986.

[19] Makhoul, J., Linear prediction: A tutorial review, *Proc. IEEE,* 63(4), 561–580, April 1975.

[20] Zeidler, J.R., Performance analysis of LMS adaptive prediction filters, *Proc. IEEE,* 78(12), 1781–1806, Dec. 1990.

[21] Kuo, S.M. and Morgan, D.R., *Active Noise Control Systems: Algorithms and DSP Implementations,* John Wiley & Sons, New York, 1996.

[22] Fuller, C.R., Elliott, S.J., and Nelson, P.A., *Active Control of Vibration,* Academic Press, London, 1996.

[23] Wiener, N., *Extrapolation, Interpolation, and Smoothing of Stationary Time Series, with Engineering Applications,* MIT Press, Cambridge, MA, 1949.

[24] Levinson, N., The Wiener RMS (root-mean-square) error criterion in filter design and prediction, *J. Math Phys.*, 25, 261–278, 1947.

[25] Douglas, S.C. and Meng, T.H.-Y., Stochastic gradient adaptation under general error criteria, *IEEE Trans. Signal Processing*, 42(6), 1335–1351, June 1994.

[26] Caraiscos, C. and Liu, B., A roundoff error analysis of the LMS adaptive algorithm, *IEEE Trans. Acoust., Speech, Signal Processing*, ASSP-32(1), 34–41, Feb. 1984.

[27] Duttweiler, D.L., Adaptive filter performance with nonlinearities in the correlation multiplier, *IEEE Trans. Acoust., Speech, Signal Processing*, ASSP-30(4), 578–586, Aug. 1982.

19

Convergence Issues in the LMS Adaptive Filter

Scott C. Douglas
University of Utah

Markus Rupp
Bell Laboratories
Lucent Technologies

19.1 Introduction

In adaptive filtering, the *least-mean-square (LMS) adaptive filter* [1] is the most popular and widely used adaptive system, appearing in numerous commercial and scientific applications. The LMS adaptive filter is described by the equations

$$\mathbf{W}(n+1) = \mathbf{W}(n) + \mu(n)e(n)\mathbf{X}(n) \tag{19.1}$$

$$e(n) = d(n) - \mathbf{W}^T(n)\mathbf{X}(n), \tag{19.2}$$

where $\mathbf{W}(n) = [w_0(n)\; w_1(n)\; \cdots\; w_{L-1}(n)]^T$ is the coefficient vector, $\mathbf{X}(n) = [x(n)\; x(n-1)\; \cdots\; x(n-L+1)]^T$ is the input signal vector, $d(n)$ is the desired signal, $e(n)$ is the error signal, and $\mu(n)$ is the step size.

There are three main reasons why the LMS adaptive filter is so popular. First, it is relatively easy to implement in software and hardware due to its computational simplicity and efficient use of memory. Second, it performs robustly in the presence of numerical errors caused by finite-precision arithmetic. Third, its behavior has been analytically characterized to the point where a user can easily set up the system to obtain adequate performance with only limited knowledge about the input and desired response signals.

0-8493-8572-5/98/$0.00+$.50
© 1998 by CRC Press LLC

Our goal in this chapter is to provide a detailed performance analysis of the LMS adaptive filter so that the user of this system understands how the choice of the step size $\mu(n)$ and filter length L affect the performance of the system through the natures of the input and desired response signals $x(n)$ and $d(n)$, respectively. The organization of this chapter is as follows. We first discuss why analytically characterizing the behavior of the LMS adaptive filter is important from a practical point of view. We then present particular signal models and assumptions that make such analyses tractable. We summarize the analytical results that can be obtained from these models and assumptions, and we discuss the implications of these results for different practical situations. Finally, to overcome some of the limitations of the LMS adaptive filter's behavior, we describe simple extensions of this system that are suggested by the analytical results. In all of our discussions, we assume that the reader is familiar with the adaptive filtering task and the LMS adaptive filter as described in Chapter 18 of this Handbook.

19.2   Characterizing the Performance of Adaptive Filters

There are two practical methods for characterizing the behavior of an adaptive filter. The simplest method of all to understand is *simulation*. In simulation, a set of input and desired response signals are either collected from a physical environment or are generated from a mathematical or statistical model of the physical environment. These signals are then processed by a software program that implements the particular adaptive filter under evaluation. By trial-and-error, important design parameters, such as the step size $\mu(n)$ and filter length L, are selected based on the observed behavior of the system when operating on these example signals. Once these parameters are selected, they are used in an adaptive filter implementation to process additional signals as they are obtained from the physical environment. In the case of a real-time adaptive filter implementation, the design parameters obtained from simulation are encoded within the real-time system to allow it to process signals as they are continuously collected.

While straightforward, simulation has two drawbacks that make it a poor sole choice for characterizing the behavior of an adaptive filter:

- *Selecting design parameters via simulation alone is an iterative and time-consuming process.* Without any other knowledge of the adaptive filter's behavior, the number of trials needed to select the best combination of design parameters is daunting, even for systems as simple as the LMS adaptive filter.
- *The amount of data needed to accurately characterize the behavior of the adaptive filter for all cases of interest may be large.* If real-world signal measurements are used, it may be difficult or costly to collect and store the large amounts of data needed for simulation characterizations. Moreover, once this data is collected or generated, it must be processed by the software program that implements the adaptive filter, which can be time-consuming as well.

For these reasons, we are motivated to develop an *analysis* of the adaptive filter under study. In such an analysis, the input and desired response signals $x(n)$ and $d(n)$ are characterized by certain properties that govern the forms of these signals for the application of interest. Often, these properties are *statistical* in nature, such as the *means* of the signals or the *correlation* between two signals at different time instants. An analytical description of the adaptive filter's behavior is then developed that is based on these signal properties. Once this analytical description is obtained, the design parameters are selected to obtain the best performance of the system as predicted by the analysis. What is considered "best performance" for the adaptive filter can often be specified directly within the analysis, without the need for iterative calculations or extensive simulations.

Usually, both analysis and simulation are employed to select design parameters for adaptive filters, as the simulation results provide a check on the accuracy of the signal models and assumptions that are used within the analysis procedure.

19.3 Analytical Models, Assumptions, and Definitions

The type of analysis that we employ has a long-standing history in the field of adaptive filters [2]– [6]. Our analysis uses *statistical models* for the input and desired response signals, such that any collection of samples from the signals $x(n)$ and $d(n)$ have well-defined joint probability density functions (p.d.f.s). With this model, we can study the *average behavior* of functions of the coefficients $\mathbf{W}(n)$ at each time instant, where "average" implies taking a statistical expectation over the ensemble of possible coefficient values. For example, the mean value of the ith coefficient $w_i(n)$ is defined as

$$E\{w_i(n)\} = \int_{-\infty}^{\infty} w\, p_{w_i}(w, n)dw\,, \tag{19.3}$$

where $p_{w_i}(w, n)$ is the probability distribution of the ith coefficient at time n. The mean value of the coefficient vector at time n is defined as $E\{\mathbf{W}(n)\} = [E\{w_0(n)\}\ E\{w_1(n)\}\ \cdots\ E\{w_{L-1}(n)\}]^T$.

While it is usually difficult to evaluate expectations such as (19.3) directly, we can employ several simplifying *assumptions* and *approximations* that enable the formation of *evolution equations* that describe the behavior of quantities such as $E\{\mathbf{W}(n)\}$ from one time instant to the next. In this way, we can predict the evolutionary behavior of the LMS adaptive filter on average. More importantly, we can study certain characteristics of this behavior, such as the stability of the coefficient updates, the speed of convergence of the system, and the estimation accuracy of the filter in steady-state. Because of their role in the analyses that follow, we now describe these simplifying assumptions and approximations.

19.3.1 System Identification Model for the Desired Response Signal

For our analysis, we assume that the desired response signal is generated from the input signal as

$$d(n) = \mathbf{W}_{opt}^T \mathbf{X}(n) + \eta(n)\,, \tag{19.4}$$

where $\mathbf{W}_{opt} = [w_{0,opt}\ w_{1,opt}\ \cdots\ w_{L-1,opt}]^T$ is a vector of optimum FIR filter coefficients and $\eta(n)$ is a noise signal that is independent of the input signal. Such a model for $d(n)$ is realistic for several important adaptive filtering tasks. For example, in echo cancellation for telephone networks, the optimum coefficient vector $\mathbf{W}_{opt}$ contains the impulse response of the echo path caused by the impedance mismatches at hybrid junctions within the network, and the noise $\eta(n)$ is the near-end source signal [7]. The model is also appropriate in system identification and modeling tasks such as plant identification for adaptive control [8] and channel modeling for communication systems [9]. Moreover, most of the results obtained from this model are independent of the specific impulse response values within $\mathbf{W}_{opt}$, so that general conclusions can be readily drawn.

19.3.2 Statistical Models for the Input Signal

Given the desired response signal model in (19.4), we now consider useful and appropriate statistical models for the input signal $x(n)$. Here, we are motivated by two typically conflicting concerns: (1) the need for signal models that are realistic for several practical situations and (2) the tractability of the analyses that the models allow. We consider two input signal models that have proven useful for predicting the behavior of the LMS adaptive filter.

Independent and Identically Distributed (I.I.D.) Random Processes

In digital communication tasks, an adaptive filter can be used to identify the dispersive characteristics of the unknown channel for purposes of decoding future transmitted sequences [9]. In this application, the transmitted signal is a bit sequence that is usually zero mean with a small number of amplitude levels. For example, a non-return-to-zero (NRZ) binary signal takes on the values of ± 1 with equal probability at each time instant. Moreover, due to the nature of the encoding of the transmitted signal in many cases, any set of L samples of the signal can be assumed to be *independent* and *identically distributed* (i.i.d.). For an i.i.d. random process, the p.d.f. of the samples $\{x(n_1), x(n_2), \ldots, x(n_L)\}$ for any choices of n_i such that $n_i \neq n_j$ is

$$p_X(x(n_1), x(n_2), \ldots, x(n_L)) = p_x(x(n_1)) \, p_x(x(n_2)) \cdots p_x(x(n_L)), \qquad (19.5)$$

where $p_x(\cdot)$ and $p_X(\cdot)$ are the univariate and L-variate probability densities of the associated random variables, respectively.

Zero-mean and statistically independent random variables are also *uncorrelated,* such that

$$E\{x(n_i)x(n_j)\} = 0 \qquad (19.6)$$

for $n_i \neq n_j$, although uncorrelated random variables are not necessarily statistically independent. The input signal model in (19.5) is useful for analyzing the behavior of the LMS adaptive filter, as it allows a particularly simple analysis of this system.

Spherically Invariant Random Processes (SIRPs)

In acoustic echo cancellation for speakerphones, an adaptive filter can be used to electronically isolate the speaker and microphone so that the amplifier gains within the system can be increased [10]. In this application, the input signal to the adaptive filter consists of samples of bandlimited speech. It has been shown in experiments that samples of a bandlimited speech signal taken over a short time period (e.g., 5 ms) have so-called "spherically invariant" statistical properties. *Spherically invariant random processes* (SIRPs) are characterized by multivariate p.d.f.s that depend on a quadratic form of their arguments, given by $X^T(n)R_{XX}^{-1}X(n)$, where

$$\mathbf{R_{XX}} = E\{\mathbf{X}(n)\mathbf{X}^T(n)\} \qquad (19.7)$$

is the L-dimensional *input signal autocorrelation matrix* of the stationary signal $x(n)$. The best-known representative of this class of stationary stochastic processes is the *jointly Gaussian random process* for which the joint p.d.f. of the elements of $\mathbf{X}(n)$ is

$$p_X(x(n), \ldots, x(n-L+1)) = \left((2\pi)^L \det(\mathbf{R_{XX}})\right)^{-1/2} \exp\left(-\frac{1}{2}\mathbf{X}^T(n)\mathbf{R_{XX}^{-1}}\mathbf{X}(n)\right), \qquad (19.8)$$

where $\det(\mathbf{R_{XX}})$ is the *determinant* of the matrix $\mathbf{R_{XX}}$. More generally, SIRPs can be described by a weighted mixture of Gaussian processes as

$$p_X(x(n), \ldots, x(n-L+1)) = \int_0^\infty \left((2\pi|u|)^L \det(\mathbf{\overline{R}_{XX}})\right)^{-1/2}$$
$$\times p_\sigma(u) \exp\left(-\frac{1}{2u^2}\mathbf{X}^T(n)\mathbf{\overline{R}_{XX}^{-1}}\mathbf{X}(n)\right) du, \qquad (19.9)$$

where $\mathbf{\overline{R}_{XX}}$ is the autocorrelation matrix of a zero-mean, unit-variance jointly Gaussian random process. In (19.9), the p.d.f. $p_\sigma(u)$ is a weighting function for the value of u that scales the standard deviation of this process. In other words, any single realization of a SIRP is a Gaussian random process with an autocorrelation matrix $u^2\mathbf{R_{XX}}$. Each realization, however, will have a different variance u^2.

As described, the above SIRP model does not accurately depict the statistical nature of a speech signal. The variance of a speech signal varies widely from phoneme (vowel) to fricative (consonant) utterances, and this burst-like behavior is uncharacteristic of Gaussian signals. The statistics of such behavior can be accurately modeled if a *slowly varying* value for the random variable u in (19.9) is allowed. Figure 19.1 depicts the differences between a *nearly SIRP* and an SIRP. In this system, either the random variable u or a sample from the slowly varying random process $u(n)$ is created and used to scale the magnitude of a sample from an uncorrelated Gaussian random process. Depending on the position of the switch, either an SIRP (upper position) or a nearly SIRP (lower position) is created. The linear filter $F(z)$ is then used to produce the desired autocorrelation function of the SIRP. So long as the value of $u(n)$ changes slowly over time, $\mathbf{R}_{XX}$ for the signal $x(n)$ as produced from this system is approximately the same as would be obtained if the value of $u(n)$ were fixed, except for the amplitude scaling provided by the value of $u(n)$.

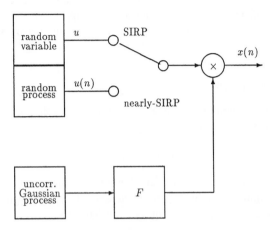

FIGURE 19.1 Generation of SIRPs and nearly SIRPs.

The random process $u(n)$ can be generated by filtering a zero-mean uncorrelated Gaussian process with a narrow-bandwidth lowpass filter. With this choice, the system generates samples from the so-called K_0 p.d.f., also known as the MacDonald function or degenerate Bessel function of the second kind [11]. This density is a reasonable match to that of typical speech sequences, although it does not necessarily generate sequences that sound like speech. Given a short-length speech sequence from a particular speaker, one can also determine the proper $p_\sigma(u)$ needed to generate $u(n)$ as well as the form of the filter $F(z)$ from estimates of the amplitude and correlation statistics of the speech sequence, respectively.

In addition to adaptive filtering, SIRPs are also useful for characterizing the performance of vector quantizers for speech coding. Details about the properties of SIRPs can be found in [12].

19.3.3 The Independence Assumptions

In the LMS adaptive filter, the coefficient vector $\mathbf{W}(n)$ is a complex function of the current and past samples of the input and desired response signals. This fact would appear to foil any attempts to develop equations that describe the evolutionary behavior of the filter coefficients from one time instant to the next. One way to resolve this problem is to make further statistical assumptions about the nature of the input and the desired response signals. We now describe a set of assumptions that have proven to be useful for predicting the behaviors of many types of adaptive filters.

The Independence Assumptions: Elements of the vector $\mathbf{X}(n)$ are statistically independent of the elements of the vector $\mathbf{X}(m)$ if $m \neq n$. In addition, samples from the noise signal $\eta(n)$ are i.i.d. and independent of the input vector sequence $\mathbf{X}(k)$ for all k and n.

A careful study of the structure of the input signal vector indicates that the independence assumptions are never true, as the vector $\mathbf{X}(n)$ shares elements with $\mathbf{X}(n-m)$ if $|m| < L$ and thus cannot be independent of $\mathbf{X}(n-m)$ in this case. Moreover, $\eta(n)$ is not guaranteed to be independent from sample to sample. Even so, numerous analyses and simulations have indicated that these assumptions lead to a reasonably accurate characterization of the behavior of the LMS and other adaptive filter algorithms for small step size values, even in situations where the assumptions are grossly violated. In addition, analyses using the independence assumptions enable a simple characterization of the LMS adaptive filter's behavior and provide reasonable guidelines for selecting the filter length L and step size $\mu(n)$ to obtain good performance from the system.

It has been shown that the independence assumptions lead to a first-order-in-$\mu(n)$ approximation to a more accurate description of the LMS adaptive filter's behavior [13]. For this reason, the analytical results obtained from these assumptions are not particularly accurate when the step size is near the stability limits for adaptation. It is possible to derive an exact statistical analysis of the LMS adaptive filter that does not use the independence assumptions [14], although the exact analysis is quite complex for adaptive filters with more than a few coefficients. From the results in [14], it appears that the analysis obtained from the independence assumptions is most inaccurate for large step sizes and for input signals that exhibit a high degree of statistical correlation.

19.3.4 Useful Definitions

In our analysis, we define the minimum mean-squared error (MSE) solution as the coefficient vector $\mathbf{W}(n)$ that minimizes the mean-squared error criterion given by

$$\xi(n) = E\{e^2(n)\} . \tag{19.10}$$

Since $\xi(n)$ is a function of $\mathbf{W}(n)$, it can be viewed as an *error surface* with a minimum that occurs at the minimum MSE solution. It can be shown for the desired response signal model in (19.4) that the minimum MSE solution is $\mathbf{W}_{opt}$ and can be equivalently defined as

$$\mathbf{W}_{opt} = \mathbf{R}_{XX}^{-1}\mathbf{P}_{dX} , \tag{19.11}$$

where $\mathbf{R}_{XX}$ is as defined in (19.7) and $\mathbf{P}_{dX} = E\{d(n)\mathbf{X}(n)\}$ is the cross-correlation of $d(n)$ and $\mathbf{X}(n)$. When $\mathbf{W}(n) = \mathbf{W}_{opt}$, the value of the *minimum MSE* is given by

$$\xi_{min} = \sigma_\eta^2 , \tag{19.12}$$

where σ_η^2 is the power of the signal $\eta(n)$.

We define the *coefficient error vector* $\mathbf{V}(n) = [v_0(n) \ \cdots \ v_{L-1}(n)]^T$ as

$$\mathbf{V}(n) = \mathbf{W}(n) - \mathbf{W}_{opt} , \tag{19.13}$$

such that $\mathbf{V}(n)$ represents the errors in the estimates of the optimum coefficients at time n. Our study of the LMS algorithm focuses on the statistical characteristics of the coefficient error vector. In particular, we can characterize the approximate evolution of the *coefficient error correlation matrix* $\mathbf{K}(n)$, defined as

$$\mathbf{K}(n) = E\{\mathbf{V}(n)\mathbf{V}^T(n)\} . \tag{19.14}$$

Another quantity that characterizes the performance of the LMS adaptive filter is the *excess mean-squared error (excess MSE)*, defined as

$$\begin{aligned} \xi_{ex}(n) &= \xi(n) - \xi_{min} \\ &= \xi(n) - \sigma_\eta^2 , \end{aligned} \tag{19.15}$$

where $\xi(n)$ is as defined in (19.10). The excess MSE is the power of the additional error in the filter output due to the errors in the filter coefficients. An equivalent measure of the excess MSE in steady-state is the *misadjustment*, defined as

$$M = \lim_{n \to \infty} \frac{\xi_{ex}(n)}{\sigma_\eta^2} , \tag{19.16}$$

such that the quantity $(1 + M)\sigma_\eta^2$ denotes the total MSE in steady-state.

Under the independence assumptions, it can be shown that the excess MSE at any time instant is related to $\mathbf{K}(n)$ as

$$\xi_{ex}(n) = \text{tr}[\mathbf{R}_{XX}\mathbf{K}(n)] , \qquad (19.17)$$

where the *trace* $\text{tr}[\cdot]$ of a matrix is the sum of its diagonal values.

19.4 Analysis of the LMS Adaptive Filter

We now analyze the behavior of the LMS adaptive filter using the assumptions and definitions that we have provided. For the first portion of our analysis, we characterize the mean behavior of the filter coefficients of the LMS algorithm in (19.1) and (19.2). Then, we provide a mean-square analysis of the system that characterizes the natures of $\mathbf{K}(n)$, $\xi_{ex}(n)$, and M in (19.14), (19.15), and (19.16), respectively.

19.4.1 Mean Analysis

By substituting the definition of $d(n)$ from the desired response signal model in (19.4) into the coefficient updates in (19.1) and (19.2), we can express the LMS algorithm in terms of the coefficient error vector in (19.13) as

$$\mathbf{V}(n + 1) = \mathbf{V}(n) - \mu(n)\mathbf{X}(n)\mathbf{X}^T(n)\mathbf{V}(n) + \mu(n)\eta(n)\mathbf{X}(n) . \qquad (19.18)$$

We take expectations of both sides of (19.18), which yields

$$E\{\mathbf{V}(n + 1)\} = E\{\mathbf{V}(n)\} - \mu(n)E\{\mathbf{X}(n)\mathbf{X}^T(n)\mathbf{V}(n)\} + \mu(n)E\{\eta(n)\mathbf{X}(n)\} , \qquad (19.19)$$

in which we have assumed that $\mu(n)$ does not depend on $\mathbf{X}(n)$, $d(n)$, or $\mathbf{W}(n)$.

In many practical cases of interest, either the input signal $x(n)$ and/or the noise signal $\eta(n)$ is zero-mean, such that the last term in (19.19) is zero. Moreover, under the independence assumptions, it can be shown that $\mathbf{V}(n)$ is approximately independent of $\mathbf{X}(n)$, and thus the second expectation on the right-hand side of (19.19) is approximately given by

$$
\begin{aligned}
E\{\mathbf{X}(n)\mathbf{X}^T(n)\mathbf{V}(n)\} &\approx E\{\mathbf{X}(n)\mathbf{X}^T(n)\}E\{\mathbf{V}(n)\} \\
&= \mathbf{R}_{XX}E\{\mathbf{V}(n)\} .
\end{aligned} \qquad (19.20)
$$

Combining these results with (19.19), we obtain

$$E\{\mathbf{V}(n + 1)\} = (\mathbf{I} - \mu(n)\mathbf{R}_{XX}) E\{\mathbf{V}(n)\} . \qquad (19.21)$$

The simple expression in (19.21) describes the evolutionary behavior of the mean values of the errors in the LMS adaptive filter coefficients. Moreover, if the step size $\mu(n)$ is constant, then we can write (19.21) as

$$E\{\mathbf{V}(n)\} = (\mathbf{I} - \mu\mathbf{R}_{XX})^n E\{\mathbf{V}(0)\} , \qquad (19.22)$$

To further simplify this matrix equation, note that $\mathbf{R}_{XX}$ can be described by its *eigenvalue decomposition* as

$$\mathbf{R}_{XX} = \mathbf{Q}\Lambda\mathbf{Q}^T , \qquad (19.23)$$

where $\mathbf{Q}$ is a matrix of the eigenvectors of $\mathbf{R}_{XX}$ and Λ is a diagonal matrix of the eigenvalues $\{\lambda_0, \lambda_1, \ldots, \lambda_{L-1}\}$ of $\mathbf{R}_{XX}$, which are all real valued because of the symmetry of $\mathbf{R}_{XX}$. Through some simple manipulations of (19.22), we can express the $(i + 1)$th element of $E\{\mathbf{W}(n)\}$ as

$$E\{w_i(n)\} = w_{i,opt} + \sum_{j=0}^{L-1} q_{ij}(1 - \mu\lambda_j)^n E\{\tilde{v}_j(0)\} , \qquad (19.24)$$

where q_{ij} is the $(i + 1, j + 1)$th element of the eigenvector matrix $\mathbf{Q}$ and $\widetilde{v}_j(n)$ is the $(j + 1)$th element of the *rotated coefficient error vector* defined as

$$\widetilde{\mathbf{V}}(n) = \mathbf{Q}^T \mathbf{V}(n) . \tag{19.25}$$

From (19.21) and (19.24), we can state several results concerning the mean behaviors of the LMS adaptive filter coefficients:

- *The mean behavior of the LMS adaptive filter as predicted by (19.21) is identical to that of the method of steepest descent for this adaptive filtering task.* Discussed in Chapter 18 of this Handbook, the method of steepest descent is an iterative optimization procedure that requires precise knowledge of the statistics of $x(n)$ and $d(n)$ to operate. That the LMS adaptive filter's average behavior is similar to that of steepest descent was recognized in one of the earliest publications of the LMS adaptive filter [1].

- *The mean value of any LMS adaptive filter coefficient at any time instant consists of the sum of the optimal coefficient value and a weighted sum of exponentially converging and/or diverging terms.* These error terms depend on the elements of the eigenvector matrix $\mathbf{Q}$, the eigenvalues of $\mathbf{R}_{XX}$, and the mean $E\{\mathbf{V}(0)\}$ of the initial coefficient error vector.

- *If all of the eigenvalues $\{\lambda_j\}$ of $\mathbf{R}_{XX}$ are strictly positive and*

$$0 < \mu < \frac{2}{\lambda_j} \tag{19.26}$$

for all $0 < j < L - 1$, then the means of the filter coefficients converge exponentially to their optimum values. This result can be found directly from (19.24) by noting that the quantity $(1 - \mu\lambda_j)^n \to 0$ as $n \to \infty$ if $|1 - \mu\lambda_j| < 1$.

- *The speeds of convergence of the means of the coefficient values depend on the eigenvalues λ_i and the step size μ.* In particular, we can define the time constant τ_j of the jth term within the summation on the right hand side of (19.24) as the approximate number of iterations it takes for this term to reach $(1/e)$th its initial value. For step sizes in the range $0 < \mu \ll 1/\lambda_{max}$ where λ_{max} is the maximum eigenvalue of $\mathbf{R}_{XX}$, this time constant is

$$\tau_j = -\frac{1}{\ln(1 - \mu\lambda_j)} \approx \frac{1}{\mu\lambda_j} . \tag{19.27}$$

Thus, faster convergence is obtained as the step size is increased. However, for step size values greater than $1/\lambda_{max}$, the speeds of convergence can actually decrease. Moreover, the convergence of the system is limited by its *mean-squared behavior,* as we shall indicate shortly.

An Example

Consider the behavior of an $L = 2$-coefficient LMS adaptive filter in which $x(n)$ and $d(n)$ are generated as

$$x(n) = 0.5x(n - 1) + \frac{\sqrt{3}}{2}z(n) \tag{19.28}$$

$$d(n) = x(n) + 0.5x(n - 1) + \eta(n) , \tag{19.29}$$

where $z(n)$ and $\eta(n)$ are zero-mean uncorrelated jointly Gaussian signals with variances of one and 0.01, respectively. It is straightforward to show for these signal statistics that

$$\mathbf{W}_{opt} = \begin{bmatrix} 1 \\ 0.5 \end{bmatrix} \text{ and } \mathbf{R}_{XX} = \begin{bmatrix} 1 & 0.5 \\ 0.5 & 1 \end{bmatrix} . \tag{19.30}$$

Figure 19.2(a) depicts the behavior of the mean analysis equation in (19.24) for these signal statistics, where $\mu(n) = 0.08$ and $\mathbf{W}(0) = [4 \ -0.5]^T$. Each circle on this plot corresponds to the value of $E\{\mathbf{W}(n)\}$ for a particular time instant. Shown on this $\{w_0, \ w_1\}$ plot are the coefficient error axes $\{v_0, \ v_1\}$, the rotated coefficient error axes $\{\tilde{v}_0, \ \tilde{v}_1\}$, and the contours of the excess MSE error surface ξ_{ex} as a function of w_0 and w_1 for values in the set $\{0.1, \ 0.2, \ 0.5, \ 1, \ 2, \ 5, \ 10, \ 20\}$. Starting from the initial coefficient vector $\mathbf{W}(0)$, $E\{\mathbf{W}(n)\}$ converge toward $\mathbf{W}_{opt}$ by reducing the components of the mean coefficient error vector $E\{\mathbf{V}(n)\}$ along the rotated coefficient error axes $\{\tilde{v}_0, \ \tilde{v}_1\}$ according to the exponential weighting factors $(1 - \mu\lambda_0)^n$ and $(1 - \mu\lambda_1)^n$ in (19.24).

For comparison, Fig. 19.2(b) shows five different simulation runs of an LMS adaptive filter operating on Gaussian signals generated according to (19.28) and (19.29), where $\mu(n) = 0.08$ and $\mathbf{W}(0) = [4 \ -0.5]^T$ in each case. Although any single simulation run of the adaptive filter shows a considerably more erratic convergence path than that predicted by (19.24), one observes that the average of these coefficient trajectories roughly follows the same path as that of the analysis.

19.4.2 Mean-Square Analysis

Although (19.24) characterizes the mean behavior of the LMS adaptive filter, it does not indicate the nature of the fluctuations of the filter coefficients about their mean values, as indicated by the actual behavior of the LMS adaptive filter in Fig. 19.2(b). The magnitudes of these fluctuations can be accurately characterized through a *mean-square analysis* of the LMS adaptive filter. Because the coefficient error correlation matrix $\mathbf{K}(n)$ as defined in (19.14) is the basis for our mean-square analysis, we outline methods for determining an evolution equation for this matrix. Then, we derive the forms of this evolution equation for both the i.i.d. and SIRP input signal models described previously, and we summarize the resulting expressions for the steady-state values of the misadjustment and excess MSE in (19.16) and (19.17), respectively, for these different signal types. Finally, several conclusions regarding the mean-square behavior of the LMS adaptive filter are drawn.

Evolution of the Coefficient Error Correlation Matrix

To derive an evolution equation for $\mathbf{K}(n)$, we post-multiply both sides of (19.18) by their respective transposes, which gives

$$
\begin{aligned}
\mathbf{V}(n+1)\mathbf{V}^T(n+1) \ = \ & \left(\mathbf{I} - \mu(n)\mathbf{X}(n)\mathbf{X}^T(n)\right)\mathbf{V}(n)\mathbf{V}^T(n)\left(\mathbf{I} - \mu(n)\mathbf{X}(n)\mathbf{X}^T(n)\right) \\
& + \mu^2(n)\eta^2(n)\mathbf{X}(n)\mathbf{X}^T(n) \\
& + \mu(n)\eta(n)\left(\mathbf{I} - \mu(n)\mathbf{X}(n)\mathbf{X}^T(n)\right)\mathbf{V}(n)\mathbf{X}^T(n) \\
& + \mu(n)\eta(n)\mathbf{X}(n)\mathbf{V}^T(n)\left(\mathbf{I} - \mu(n)\mathbf{X}(n)\mathbf{X}^T(n)\right).
\end{aligned}
\tag{19.31}
$$

Taking expectations of both sides of (19.31), we note that $\eta(n)$ is zero mean and independent of both $\mathbf{X}(n)$ and $\mathbf{V}(n)$ from our models and assumptions, and thus the expectations of the third and fourth terms on the right hand side of (19.31) are zero. Moreover, by using the independence assumptions, it can be shown that

$$
\begin{aligned}
E\{\mathbf{X}(n)\mathbf{X}^T(n)\mathbf{V}(n)\mathbf{V}^T(n)\} \ & \approx \ E\{\mathbf{X}(n)\mathbf{X}^T(n)\}E\{\mathbf{V}(n)\mathbf{V}^T(n)\} \\
& = \ \mathbf{R}_{XX}\mathbf{K}(n).
\end{aligned}
\tag{19.32}
$$

Thus, we obtain from (19.31) the expression

$$
\begin{aligned}
\mathbf{K}(n+1) \ = \ & \mathbf{K}(n) - \mu(n)\left(\mathbf{R}_{XX}\mathbf{K}(n) + \mathbf{K}(n)\mathbf{R}_{XX}\right) \\
& + \mu^2(n)E\{\mathbf{X}(n)\mathbf{X}^T(n)\mathbf{K}(n)\mathbf{X}(n)\mathbf{X}^T(n)\} + \mu^2(n)\sigma_\eta^2\mathbf{R}_{XX},
\end{aligned}
\tag{19.33}
$$

where σ_η^2 is as defined in (19.12).

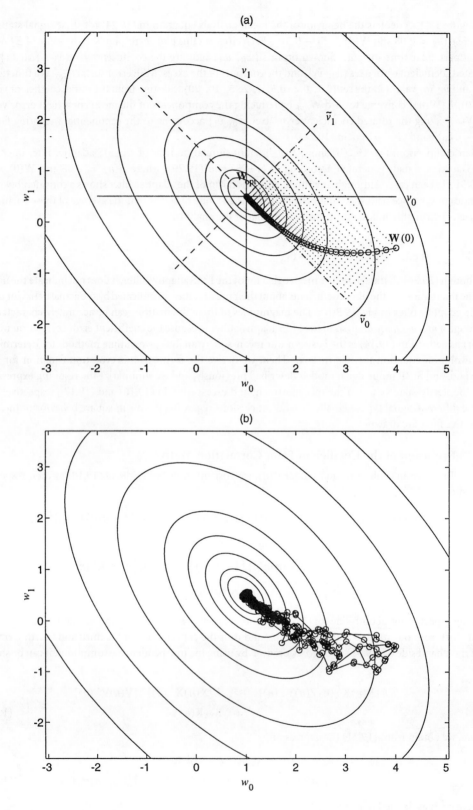

FIGURE 19.2   Comparison of the predicted and actual performances of the LMS adaptive filter in the two-coefficient example: (a) the behavior predicted by the mean analysis, and (b) the actual LMS adaptive filter behavior for five different simulation runs.

At this point, the analysis can be simplified depending on how the third term on the right hand side of (19.33) is evaluated according to the signal models and assumptions.

Analysis for SIRP Input Signals: In this case, the value of $E\{\mathbf{X}(n)\mathbf{X}^T(n)\mathbf{K}(n)\mathbf{X}(n)\mathbf{X}^T(n)\}$ can be expressed as

$$E\{\mathbf{X}(n)\mathbf{X}^T(n)\mathbf{K}(n)\mathbf{X}(n)\mathbf{X}^T(n)\} = m_z^{(2,2)}\left[2\mathbf{R}_{\mathbf{XX}}\mathbf{K}(n)\mathbf{R}_{\mathbf{XX}} + \mathbf{R}_{\mathbf{XX}}\mathrm{tr}\{\mathbf{R}_{\mathbf{XX}}\mathbf{K}(n)\}\right], \qquad (19.34)$$

where the moment term $m_z^{(2,2)}$ is given by

$$m_z^{(2,2)} = E\{z_i^2(n)z_j^2(n)\} \qquad (19.35)$$

for any $0 \leq i \neq j \leq (L-1)$ and

$$z_i(n) = \lambda_i^{1/2}\sum_{l=0}^{L-1}q_{li}x(n-l). \qquad (19.36)$$

If $x(n)$ is a Gaussian random process, then $m_z^{(2,2)} = 1$, and it can be shown that $m_z^{(2,2)} \geq 1$ for SIRPs in general. For more details on these results, see [15].

Analysis for I.I.D. Input Signals: In this case, we can express the (i, j)th element of the matrix $E\{\mathbf{X}(n)\mathbf{X}^T(n)\mathbf{K}(n)\mathbf{X}(n)\mathbf{X}^T(n)\}$ as

$$\left[E\{\mathbf{X}(n)\mathbf{X}^T(n)\mathbf{K}(n)\mathbf{X}(n)\mathbf{X}^T(n)\}\right]_{i,j} = \begin{cases} 2\sigma_x^4[\mathbf{K}(n)]_{i,j}, & \text{if } i \neq j \\ \sigma_x^4\left(\gamma[\mathbf{K}(n)]_{i,i} + \sum_{m=1,\,m\neq i}^{L}[\mathbf{K}(n)]_{m,m}\right), & \text{if } i = j, \end{cases} \qquad (19.37)$$

where $[\mathbf{K}(n)]_{i,j}$ is the (i, j)th element of $\mathbf{K}(n)$,

$$\sigma_x^2 = E\{x^2(n)\}, \text{ and } \gamma = \frac{E\{x^4(n)\}}{\sigma_x^4}, \qquad (19.38)$$

respectively. For details, see [5].

Zeroth-Order Approximation Near Convergence: For small step sizes, it can be shown that the elements of $\mathbf{K}(n)$ are approximately proportional to both the step size and the noise variance σ_η^2 in steady-state. Thus, the magnitudes of the elements in the third term on the right hand side of (19.33) are about a factor of $\mu(n)$ smaller than those of any other terms in this equation at convergence. Such a result suggests that we could set

$$\mu^2(n)E\{\mathbf{X}(n)\mathbf{X}^T(n)\mathbf{K}(n)\mathbf{X}(n)\mathbf{X}^T(n)\} \approx 0 \qquad (19.39)$$

in the steady-state analysis of (19.33) without perturbing the analytical results too much. If this approximation is valid, then the form of (19.33) no longer depends on the form of the amplitude statistics of $x(n)$, as in the case of the mean analysis.

Excess MSE, Mean-Square Stability, and Misadjustment

Given the results in (19.34) through (19.39), we can use the evolution equation for $\mathbf{K}(n)$ in (19.33) to explore the mean-square behavior of the LMS adaptive filter in several ways:

- By studying the structure of (19.33) for different signal types, we can determine conditions on the step size $\mu(n)$ to guarantee the stability of the mean-square analysis equation.
- By setting $\mathbf{K}(n+1) = \mathbf{K}(n)$ and fixing the value of $\mu(n)$, we can solve for the steady-state value of $\mathbf{K}(n)$ at convergence, thereby obtaining a measure of the fluctuations of the coefficients about their optimum solutions.

- Given a value for $V(0)$, we can write a computer program to simulate the behavior of this equation for different signal statistics and step size sequences.

Moreover, once the matrix sequence $\mathbf{K}(n)$ is known, we can obtain the values of the excess MSE and misadjustment from $\mathbf{K}(n)$ by employing the relations in (19.16) and (19.17), respectively.

Table 19.1 summarizes many of the analytical results that can be obtained from a careful study of (19.33). Shown in the table are the conditions on the step size $\mu(n) = \mu$ to guarantee stability, sufficient stability conditions on the step size that can be easily calculated, and the misadjustment in steady-state for the three different methods of evaluating $E\{\mathbf{X}(n)\mathbf{X}^T(n)\mathbf{K}(n)\mathbf{X}(n)\mathbf{X}^T(n)\}$ in (19.34) through (19.39). In the table, the quantity C is defined as

$$C = \sum_{i=0}^{L-1} \frac{\lambda_i}{1 - \mu m_z^{(2,2)}\lambda_i} \, . \tag{19.40}$$

TABLE 19.1   Summary of MSE Analysis Results

Assumption	MSE Stability Conditions	Sufficient Conditions	Misadjustment
I.I.D. input	$0 < \mu < \dfrac{2}{(L-1+\gamma)\sigma_x^2}$	$0 < \mu < \dfrac{2}{(L-1+\gamma)\sigma_x^2}$	$M = \dfrac{\mu\sigma_x^2 L}{2-\mu\sigma_x^2(L-1+\gamma)}$
SIRP input	$0 < \mu < \dfrac{1}{m_z^{(2,2)}\lambda_{max}}$ and $\mu m_z^{(2,2)}C < 2$	$0 < \mu < \dfrac{2}{3Lm_z^{(2,2)}\sigma_x^2}$	$M = \dfrac{\mu C}{2-\mu m_z^{(2,2)}C}$
Approx.	$0 < \mu < \dfrac{1}{\lambda_{max}}$	$0 < \mu < \dfrac{1}{L\sigma_x^2}$	$M = \dfrac{\mu\sigma_x^2 L}{2}$

From these results and others that can be obtained from (19.33), we can infer several facts about the mean-square performance of the LMS adaptive filter:

- *The value of the excess MSE at time n consists of the sum of the steady-state excess MSE, given by $M\sigma_\eta^2$, and a weighted sum of L exponentially converging and/or diverging terms.* Similar to the mean analysis case, these additional terms depend on the elements of the eigenvector matrix $\mathbf{Q}$, the eigenvalues of $\mathbf{R}_{\mathbf{XX}}$, the eigenvalues of $\mathbf{K}(0)$, and the values of $m_z^{(2,2)}$ or γ for the SIRP or i.i.d. input signal models, respectively.

- *For all input signal types, approximate conditions on the fixed step size value to guarantee convergence of the evolution equations for $\mathbf{K}(n)$ are of the form*

$$0 < \mu < \frac{K}{L\sigma_x^2} \, , \tag{19.41}$$

  where σ_x^2 is the input signal power and where the constant K depends weakly on the nature of the input signal statistics and not on the magnitude of the input signal. All of the sufficient stability bounds on μ as shown in Table 19.1 can be put in the form of (19.26). Because of the inaccuracies within the analysis that are caused by the independence assumptions, however, the actual step size chosen for stability of the LMS adaptive filter should be somewhat smaller than these values, and step sizes in the range $0 < \mu(n) < 0.1/(L\sigma_x^2)$ are often chosen in practice.

- *The misadjustment of the LMS adaptive filter increases as the filter length L and step size μ are increased.* Thus, a larger step size causes larger fluctuations of the filter coefficients about their optimum solutions in steady-state.

19.5   Performance Issues

When using the LMS adaptive filter, one must select the filter length L and the step size $\mu(n)$ to obtain the desired performance from the system. In this section, we explore the issues affecting the choices of these parameters using the analytical results for LMS adaptive filter's behavior derived in the last section.

19.5.1   Basic Criteria for Performance

The performance of the LMS adaptive filter can be characterized in three important ways: the *adequacy of the FIR filter model*, the *speed of convergence* of the system, and the *misadjustment* in steady-state.

Adequacy of the FIR Model

The LMS adaptive filter relies on the linearity of the FIR filter model to accurately characterize the relationship between the input and desired response signals. When the relationship between $x(n)$ and $d(n)$ deviates from the linear one given in (19.4), then the performance of the overall system suffers. In general, it is possible to use a nonlinear model in place of the adaptive FIR filter model considered here. Possible nonlinear models include polynomial-based filters such as Volterra and bilinear filters [16] as well as neural network structures [17].

Another source of model inaccuracy is the finite impulse response length of the adaptive FIR filter. It is typically necessary to tune both the length of the filter L and the relative delay between the input and desired response signals so that the input signal values *not* contained in $\mathbf{X}(n)$ are largely uncorrelated with the desired response signal sample $d(n)$. However, such a situation may be impossible to achieve when the relationship between $x(n)$ and $d(n)$ is of an infinite-impulse response (IIR) nature. Adaptive IIR filters can be considered for these situations, although the stability and performance behaviors of these systems are much more difficult to characterize. Adaptive IIR filters are discussed in Chapter 23 of this Handbook.

Speed of Convergence

The rate at which the coefficients approach their optimum values is called the *speed of convergence*. As the analytical results show, there exists no one quantity that characterizes the speed of convergence, as it depends on the initial coefficient values, the amplitude and correlation statistics of the signals, the filter length L, and the step size $\mu(n)$. However, we can make several qualitative statements relating the speed of convergence to both the step size and the filter length. All of these results assume that the desired response signal model in (19.4) is reasonable and that the errors in the filter coefficients are uniformly distributed across the coefficients on average.

- *The speed of convergence increases as the value of the step size is increased, up to step sizes near one-half the maximum value required for stable operation of the system.* This result can be obtained from a careful analysis of (19.33) for different input signal types and correlation statistics. Moreover, by simulating the behavior of (19.33) and (19.34) for typical signal scenarios, it is observed that the speed of convergence of the excess MSE actually *decreases* for large enough step size values. For i.i.d. input signals, the fixed step size providing fastest convergence of the excess MSE is exactly one-half the MSE step size bound as given in Table 19.1 for this type of input signal.

- *The speed of convergence decreases as the length of the filter is increased.* The reasons for this behavior are twofold. First, if the input signal is correlated, the *condition number* of $\mathbf{R}_{\mathrm{XX}}$, defined as the ratio of the largest and smallest eigenvalues of this matrix, generally increases as L is increased for typical real-world input signals. A larger condition number for $\mathbf{R}_{\mathrm{XX}}$ makes it more difficult to choose a good step size to obtain fast convergence of all of the elements of either $E\{\mathbf{V}(n)\}$ or $\mathbf{K}(n)$. Such an effect can be seen in (19.24), as a larger condition number leads to a larger disparity in the values of $(1 - \mu\lambda_j)$ for different j. Second, in the MSE analysis equation of (19.33), the overall magnitude of the expectation term $E\{\mathbf{X}(n)\mathbf{X}^T(n)\mathbf{K}(n)\mathbf{X}(n)\mathbf{X}^T(n)\}$ is larger for larger L, due to the fact that the scalar quantity $\mathbf{X}^T(n)\mathbf{K}(n)\mathbf{X}(n)$ within this expectation increases as the size of the filter is increased. Since this quantity is always positive, it limits the amount that the excess MSE can be decreased at each iteration, and it reduces the maximum step size that is allowed for mean-square convergence as L is increased.

- *The maximum possible speed of convergence is limited by the largest step size that can be chosen for stability for moderately correlated input signals.* In practice, the actual step size needed for stability of the LMS adaptive filter is smaller than one-half the maximum values given in Table 19.1 when the input signal is moderately correlated. This effect is due to the actual statistical relationships between the current coefficient vector $\mathbf{W}(n)$ and the signals $\mathbf{X}(n)$ and $d(n)$, relationships that are neglected via the independence assumptions. Since the convergence speed increases as μ is increased over this allowable step size range, the maximum stable step size provides a practical limit on the speed of convergence of the system.

- *The speed of convergence depends on the desired level of accuracy that is to be obtained by the adaptive filter.* Generally speaking, the speed of convergence of the system decreases as the desired level of misadjustment is decreased. This result is due to the fact that the behavior of the system is dominated by the slower-converging modes of the system as the length of adaptation time is increased. Thus, if the desired level of misadjustment is low, the speed of convergence is dominated by the slower-converging modes, thus limiting the overall convergence speed of the system.

Misadjustment

The misadjustment, defined in (19.16), is the additional fraction of MSE in the filter output above the minimum MSE value σ_η^2 caused by a nonzero adaptation speed. We can draw the following two conclusions regarding this quantity:

- *The misadjustment increases as the step size is increased.*
- *The misadjustment increases as the filter length is increased.*

Both results can be proven by direct study of the analytical results for M in Table 19.1.

19.5.2   Identifying Stationary Systems

We now evaluate the basic criteria for performance to provide qualitative guidance as to how to choose μ and L to identify a stationary system.

Choice of Filter Length

We have seen that as the filter length L is increased, the speed of convergence of the LMS adaptive filter decreases, and the misadjustment in steady-state increases. Therefore, the filter length should be chosen as short as possible but long enough to adequately model the unknown system, as too short a filter model leads to poor modeling performance. In general, there exists an optimal length L for a given μ that exactly balances the penalty for a finite-length filter model with the increase in misadjustment caused by a longer filter length, although the calculation of such a model order requires more information than is typically available in practice. Modeling criteria, such as Akaike's Information Criterion [18] and minimum description length (MDL) [19] could be used in this situation.

Choice of Step Size

We have seen that the speed of convergence increases as the step size is increased, up to values that are roughly within a factor of $1/2$ of the step size stability limits. Thus, if fast convergence is desired, one should choose a large step size according to the limits in Table 19.1. However, we also observe that the misadjustment increases as the step size is increased. Therefore, if highly accurate estimates of the filter coefficients are desired, a small step size should be chosen. This classical tradeoff in convergence speed vs. the level of error in steady state dominates the issue of step size selection in many estimation schemes.

If the user knows that the relationship between $x(n)$ and $d(n)$ is linear and time-invariant, then one possible solution to the above tradeoff is to choose a large step size initially to obtain fast convergence, and then switch to a smaller step size to obtain a more accurate estimate of $\mathbf{W}_{opt}$ near convergence. The point to switch to a smaller step size is roughly when the excess MSE becomes a small fraction (approximately 1/10th) of the minimum MSE of the filter. This method of *gearshifting*, as it is commonly known, is part of a larger class of time-varying step size methods that we shall explore shortly.

Although we have discussed qualitative criteria by which to choose a fixed step size, it is possible to define specific performance criteria by which to choose μ. For one study of this problem for i.i.d. input signals, see [20].

19.5.3   Tracking Time-Varying Systems

Since the LMS adaptive filter continually adjusts its coefficients to approximately minimize the mean-squared error criterion, it can adjust to changes in the relationship between $x(n)$ and $d(n)$. This behavior is commonly referred to as *tracking*. In such situations, it clearly is not desirable to reduce the step size to an extremely small value in steady-state, as the LMS adaptive filter would not be able to follow any changes in the relationship between $x(n)$ and $d(n)$.

To illustrate the issues involved, consider the desired response signal model given by

$$d(n) = \mathbf{W}_{opt}^T(n)\mathbf{X}(n) + \eta(n) \,, \tag{19.42}$$

where the optimum coefficient vector $\mathbf{W}_{opt}(n)$ varies with time according to

$$\mathbf{W}_{opt}(n+1) = \mathbf{W}_{opt}(n) + \mathbf{M}(n) \,, \tag{19.43}$$

and $\{\mathbf{M}(n)\}$ is a sequence of vectors whose elements are all i.i.d. This nonstationary model is similar to others used in other tracking analyses of the LMS adaptive filter [4, 21]; it also enables a simple analysis that is similar to the stationary system identification model discussed earlier.

Applying the independence assumptions, we can analyze the behavior of the LMS adaptive filter for this desired response signal model. For brevity, we only summarize the results of an approximate analysis in which terms of the form $\mu^2 E\{\mathbf{X}(n)\mathbf{X}^T(n)\mathbf{K}(n)\mathbf{X}(n)\mathbf{X}^T(n)\}$ are neglected in the MSE behavioral equations [4]. The misadjustment of the system in steady-state is

$$M_{non} = \frac{L}{2}\left(\mu\sigma_x^2 + \frac{\sigma_m^2}{\mu\sigma_\eta^2}\right) \,, \tag{19.44}$$

where σ_m^2 is the power in any one element of $\mathbf{M}(n)$. Details of a more-accurate tracking analysis can be found in [21].

In this case, the misadjustment is the sum of two terms. The first term is the same as that for the stationary case and is proportional to μ. The second term is the *lag error* and is due to the fact that the LMS coefficients follow or "lag" behind the optimum coefficient values. The lag error is proportional to the speed of variation of the unknown system through σ_m^2 and is inversely proportional to the step size, such that its value increases as the step size is decreased.

In general, there exists an optimum fixed step size that minimizes the misadjustment in steady-state for an LMS adaptive filter that is tracking changes in $\mathbf{W}_{opt}(n)$. For the approximate analysis used to derive (19.44), the resulting step size is

$$\mu_{opt} = \frac{\sigma_m}{\sigma_\eta \sigma_x} . \tag{19.45}$$

As the value of σ_m^2 increases, the level of nonstationarity increases such that a larger step size is required to accurately track changes in the unknown system. Similar conclusions can be drawn from other analyses of the LMS adaptive filter in tracking situations [22].

19.6   Selecting Time-Varying Step Sizes

The analyses of the previous sections enable one to choose a fixed step size μ for the LMS adaptive filter to meet the system's performance requirements when the general characteristics of the input and desired response signals are known. In practice, the exact statistics of $x(n)$ and $d(n)$ are unknown or vary with time. A time-varying step size $\mu(n)$, if properly computed, can provide stable, robust, and accurate convergence behavior for the LMS adaptive filter in these situations. In this section, we consider useful on-line procedures for computing $\mu(n)$ in the LMS adaptive filter to meet these performance requirements.

19.6.1   Normalized Step Sizes

For the LMS adaptive filter to be useful, it must operate in a stable manner so that its coefficient values do not diverge. From the stability results in Table 19.1 and the generalized expression for these stability bounds in (19.41), the upper bound for the step size is inversely proportional to the input signal power σ_x^2 in general. In practice, the input signal power is unknown or varies with time. Moreover, if one were to choose a small fixed step size value to satisfy these stability bounds for the largest anticipated input signal power value, then the convergence speed of the system would be unnecessarily slow during periods when the input signal power is small.

These concerns can be addressed by calculating a *normalized step size* $\mu(n)$ as

$$\mu(n) = \frac{\overline{\mu}}{\delta + L\widehat{\sigma_x^2}(n)} , \tag{19.46}$$

where $\widehat{\sigma_x^2}(n)$ is an *estimate* of the input signal power, $\overline{\mu}$ is a constant somewhat smaller than the value of K required for system stability in (19.41), and δ is a small constant to avoid a divide-by-zero should $\widehat{\sigma_x^2}(n)$ approach zero. To estimate the input signal power, a lowpass filter can be applied to the sequence $x^2(n)$ to track its changing envelope. Typical estimators include

- *exponentially weighted estimate:*

$$\widehat{\sigma_x^2}(n) = (1 - c)\widehat{\sigma_x^2}(n - 1) + cx^2(n) , \tag{19.47}$$

- *sliding-window estimate:*

$$\widehat{\sigma_x^2}(n) = \frac{1}{N} \sum_{i=0}^{N-1} x^2(n - i) , \tag{19.48}$$

where the parameters c, $0 < c \ll 1$ and N, $N \geq L$ control the effective memories of the two estimators, respectively.

The Normalized LMS Adaptive Filter

By choosing a sliding window estimate of length $N = L$, the LMS adaptive filter with $\mu(n)$ in (19.46) becomes

$$\mathbf{W}(n+1) = \mathbf{W}(n) + \frac{\overline{\mu}e(n)}{p(n)}\mathbf{X}(n) \tag{19.49}$$

$$p(n) = \delta + ||\mathbf{X}(n)||^2, \tag{19.50}$$

where $||\mathbf{X}(n)||^2$ is the L_2-norm of the input signal vector. The value of $p(n)$ can be updated recursively as

$$p(n) = p(n-1) + x^2(n) - x^2(n-L), \tag{19.51}$$

where $p(0) = \delta$ and $x(n) = 0$ for $n \leq 0$. The adaptive filter in (19.49) is known as the *normalized LMS (NLMS) adaptive filter*. It has two special properties that make it useful for adaptive filtering tasks:

- *The NLMS adaptive filter is guaranteed to converge for any value of $\overline{\mu}$ in the range*

$$0 < \overline{\mu} < 2, \tag{19.52}$$

  regardless of the statistics of the input signal. Thus, selecting the value of $\overline{\mu}$ for stable behavior of this system is much easier than selecting μ for the LMS adaptive filter.
- *With the proper choice of $\overline{\mu}$, the NLMS adaptive filter can often converge faster than the LMS adaptive filter.* In fact, for noiseless system identification tasks in which $\eta(n)$ in (19.4) is zero, one can obtain $\mathbf{W}_{opt}(n) = \mathbf{W}_{opt}$ after L iterations of (19.49) for $\overline{\mu} = 1$. Moreover, for SIRP input signals, the NLMS adaptive filter provides more uniform convergence of the filter coefficients, making the selection of $\overline{\mu}$ an easier proposition than the selection of μ for the LMS adaptive filter.

A discussion of these and other results on the NLMS adaptive filter can be found in [15, 23]– [25].

19.6.2 Adaptive and Matrix Step Sizes

In addition to stability, the step size controls both the speed of convergence and the misadjustment of the LMS adaptive filter through the statistics of the input and desired response signals. In situations where the statistics of $x(n)$ and/or $d(n)$ are changing, the value of $\mu(n)$ that provides the best performance from the system can change as well. In these situations, it is natural to consider $\mu(n)$ as an adaptive parameter to be optimized along with the coefficient vector $\mathbf{W}(n)$ within the system. While it may seem novel, the idea of computing an *adaptive step size* has a long history in the field of adaptive filters [2]. Numerous such techniques have been proposed in the scientific literature. One such method uses a stochastic gradient procedure to adjust the value of $\mu(n)$ to iteratively minimize the MSE within the LMS adaptive filter. A derivation and performance analysis of this algorithm is given in [26].

In some applications, the task at hand suggests a particular strategy for adjusting the step size $\mu(n)$ to obtain the best performance from an LMS adaptive filter. For example, in echo cancellation for telephone networks, the signal-to-noise ratio of $d(n)$ falls to extremely low values when the near-end talker signal is present, making accurate adaptation during these periods difficult. Such systems typically employ *double-talk detectors*, in which estimates of the statistical characteristics of $x(n)$, $d(n)$, and/or $e(n)$ are used to raise and lower the value of $\mu(n)$ in an appropriate manner. A discussion of this problem and a method for its solution are given in [27].

While our discussion of the LMS adaptive filter has assumed a single step size value for each of the filter coefficients, it is possible to select L *different* step sizes $\mu_i(n)$ for each of the L coefficient updates within the LMS adaptive filter. To select fixed values for each $\mu_i(n) = \mu_i$, these *matrix step size* methods require prior knowledge about the statistics of $x(n)$ and $d(n)$ and/or the approximate values of $\mathbf{W}_{opt}$. It is possible, however, to adapt each $\mu_i(n)$ according to a suitable performance criterion to obtain improved convergence behavior from the overall system. A particularly simple adaptive method for calculating matrix step sizes is provided in [28].

19.6.3 Other Time-Varying Step Size Methods

In situations where the statistics of $x(n)$ and $d(n)$ do not change with time, choosing a variable step size sequence $\mu(n)$ is still desirable, as one can decrease the misadjustment over time to obtain an accurate estimate of the optimum coefficient vector $\mathbf{W}_{opt}$. Such methods have been derived and characterized in a branch of statistical analysis known as *stochastic approximation* [29]. Using this formalism, it is possible to prove under certain assumptions on $x(n)$ and $d(n)$ that the value of $\mathbf{W}(n)$ for the LMS adaptive filter converges to $\mathbf{W}_{opt}$ as $n \to \infty$ if $\mu(n)$ satisfies

$$\sum_{n=0}^{\infty} |\mu(n)| \to \infty \text{ and } \sum_{n=0}^{\infty} \mu^2(n) < \infty ,\tag{19.53}$$

respectively. One step size function satisfying these constraints is

$$\mu(n) = \frac{\mu(0)}{n+1} ,\tag{19.54}$$

where $\mu(0)$ is an initial step size parameter. The gearshifting method described in Section 19.5.2 can be seen as a simple heuristic approximation to (19.54). Moreover, one can derive an optimum step size sequence $\mu_{opt}(n)$ that minimizes the excess MSE at each iteration under certain situations, and the limiting form of the resulting step size values for stationary signals are directly related to (19.54) as well [24].

19.7 Other Analyses of the LMS Adaptive Filter

While the analytical techniques employed in this section are useful for selecting design parameters for the LMS adaptive filter, they are but one method for characterizing the behavior of this system. Other forms of analyses can be used to determine other characteristics of this system, such as the p.d.f.s of the adaptive filter coefficients [30] and the probability of large excursions in the adaptive filter coefficients for different types of input signals [31]. In addition, much research effort has focused on characterizing the stability of the system without extensive assumptions about the signals being processed. One example of such an analysis is given in [32]. Other methods for analyzing the LMS adaptive filter include the method of ordinary differential equations (ODE) [33], the stochastic approximation methods described previously [29], computer-assisted symbolic derivation methods [14], and averaging techniques that are particularly useful for deterministic signals [34].

19.8 Analysis of Other Adaptive Filters

Because of the difficulties in performing multiplications in the first digital hardware implementations of adaptive filters, many of these systems employed nonlinearities in the coefficient update terms to simplify their hardware requirements. An example of one such algorithm is the *sign-error* adaptive filter, in which the coefficient update is

$$\mathbf{W}(n+1) = \mathbf{W}(n) + \mu(n)\mathrm{sgn}(e(n))\mathbf{X}(n) ,\tag{19.55}$$

where the value of $\mathrm{sgn}(e(n))$ is either 1 or -1 depending on whether $e(n)$ is positive or negative, respectively. If $\mu(n)$ is chosen as a power-of-two, this algorithm only requires a comparison and bit shift per coefficient to implement in hardware. Other algorithms employing nonlinearities of the input signal vector $\mathbf{X}(n)$ in the updates are also useful [35].

Many of the analysis techniques developed for the LMS adaptive filter can be applied to algorithms with nonlinearities in the coefficient updates, although such methods require additional assumptions to obtain accurate results. For presentations of two such analyses, see [36, 37]. It should be noted that the performance characteristics and stability properties of these nonlinearly modified versions of the LMS adaptive filter can be quite different from those of the LMS adaptive filter. For example, the sign-error adaptive filter in (19.55) is guaranteed to converge for any fixed positive step size value under fairly loose assumptions on $x(n)$ and $d(n)$ [38].

19.9 Conclusions

In summary, we have described a statistical analysis of the LMS adaptive filter, and through this analysis, suggestions for selecting the design parameters for this system have been provided. While useful, analytical studies of the LMS adaptive filter are but one part of the system design process. As in all design problems, sound engineering judgment, careful analytical studies, computer simulations, and extensive real-world evaluations and testing should be combined when developing an adaptive filtering solution to any particular task.

References

[1] Widrow, B. and Hoff, M.E., Adaptive switching circuits, *IRE WESCON Conv. Rec.*, 4, 96–104, Aug. 1960.

[2] Widrow, B., Adaptive sampled-data systems — a statistical theory of adaptation, *IRE WESCON Conv. Rec.*, 4, 74–85, Aug. 1959.

[3] Senne, K.D., Adaptive linear discrete-time estimation, Ph.D. thesis, Stanford University, Stanford, CA, June 1968.

[4] Widrow, B., McCool, J., Larimore, M.G., and Johnson, C.R., Jr., Stationary and nonstationary learning characteristics of the LMS adaptive filter, *Proc. IEEE*, 64(8), 1151–1162, Aug. 1976.

[5] Gardner, W.A., Learning characteristics of stochastic-gradient-descent algorithms: a general study, analysis, and critique, *Signal Processing*, 6(2), 113–133, April 1984.

[6] Feuer, A. and Weinstein, E., Convergence analysis of LMS filters with uncorrelated data, *IEEE Trans. Acoust., Speech, Signal Processing*, ASSP-331, 222–230, Feb. 1985.

[7] Messerschmitt, D.G., Echo cancellation in speech and data transmission, *IEEE J. Selected Areas Comm.*, 2(2), 283–301, Mar. 1984.

[8] Widrow, B. and Walach, E., *Adaptive Inverse Control*, Prentice-Hall, Upper Saddle River, NJ, 1996.

[9] Proakis, J.G., *Digital Communications*, 3rd ed., McGraw-Hill, New York, 1995.

[10] Murano, K., Unagami, S., and Amano, F., Echo cancellation and applications, *IEEE Commun. Mag.*, 28(1), 49–55, Jan. 1990.

[11] Gradsteyn, I.S. and Ryzhik, I.M., *Table of Integrals, Series and Products*, Academic Press, New York, 1980.

[12] Brehm, H. and Stammler, W., Description and generation of spherically invariant speech-model signals, *Signal Processing*, 12(2), 119–141, Mar. 1987.

[13] Mazo, J.E., On the independence theory of equalizer convergence, *Bell Sys. Tech. J.*, 58(5), 963–993, May-June 1979.

[14] Douglas, S.C. and Pan, W., Exact expectation analysis of the LMS adaptive filter, *IEEE Trans. Signal Processing*, 43(12), 2863–2871, Dec. 1995.

[15] Rupp, M., The behavior of LMS and NLMS algorithms in the presence of spherically invariant processes, *IEEE Trans. Signal Processing*, 41(3), 1149–1160, Mar. 1993.

[16] Mathews, V.J., Adaptive polynomial filters, *IEEE Signal Proc. Mag.*, 8(3), 10–26, July 1991.

[17] Haykin, S., *Neural Networks,* Prentice-Hall, Englewood Cliffs, NJ, 1995.

[18] Akaike, H., A new look at the statistical model identification, *IEEE Trans. Automatic Control,* AC-19(6), 716–723, Dec. 1974.

[19] Rissanen, J., Modelling by shortest data description, *Automatica,* 14(5), 465–471, Sept. 1978.

[20] Bershad, N.J., On the optimum gain parameter in LMS adaptation, *IEEE Trans. Acoust., Speech, Signal Processing,* ASSP-35(7), 1065–1068, July 1987.

[21] Gardner, W.A., Nonstationary learning characteristics of the LMS algorithm, *IEEE Trans. Circ. Syst.,* 34(10), 1199–1207, Oct. 1987.

[22] Farden, D.C., Tracking properties of adaptive signal processing algorithms, *IEEE Trans. Acoust., Speech, Signal Processing,* ASSP-29(3), 439–446, June 1981.

[23] Bitmead, R.R. and Anderson, B.D.O., Performance of adaptive estimation algorithms in dependent random environments, *IEEE Trans. Automatic Control,* AC-25(4), 788–794, Aug. 1980.

[24] Slock, D.T.M., On the convergence behavior of the LMS and the normalized LMS algorithms, *IEEE Trans. Signal Processing,* 41(9), 2811–2825, Sept. 1993.

[25] Douglas, S.C. and Meng, T.H.-Y., Normalized data nonlinearities for LMS adaptation, *IEEE Trans. Signal Processing,* 42(6), 1352–1365, June 1994.

[26] Mathews, V.J. and Xie, Z., A stochastic gradient adaptive filter with gradient adaptive step size, *IEEE Trans. Signal Processing,* 41(6), 2075–2087, June 1993.

[27] Ding, Z., Johnson, C.R., Jr., and Sethares, W.A., Frequency-dependent bursting in adaptive echo cancellation and its prevention using double-talk detectors, *Int. J. Adaptive Contr. Signal Processing,* 4(3), 219–216, May-June 1990.

[28] Harris, R.W., Chabries, D.M., and Bishop, F.A., A variable step (VS) adaptive filter algorithm, *IEEE Trans. Acoust., Speech, Signal Processing,* ASSP-34(2), 309–316, April 1986.

[29] Kushner, H.J. and Clark, D.S., *Stochastic Approximation Methods for Constrained and Unconstrained Systems,* Springer-Verlag, New York, 1978.

[30] Bershad, N.J. and Qu, L.Z., On the probability density function of the LMS adaptive filter weights, *IEEE Trans. Acoust., Speech, Signal Processing,* ASSP-37(1), 43–56, Jan. 1989.

[31] Rupp, M., Bursting in the LMS algorithm, *IEEE Trans. on Signal Processing,* 43(10), 2414–2417, Oct. 1995.

[32] Macchi, O. and Eweda, E., Second-order convergence analysis of stochastic adaptive linear filtering, *IEEE Trans. Automatic Control,* AC-28(1), 76–85, Jan. 1983.

[33] Benveniste, A., Métivier, M. and Priouret, P., *Adaptive Algorithms and Stochastic Approximations,* Springer-Verlag, New York, 1990.

[34] Solo, V. and Kong, X., *Adaptive Signal Processing Algorithms: Stability and Performance,* Prentice-Hall, Englewood Cliffs, NJ, 1995.

[35] Duttweiler, D.L., Adaptive filter performance with nonlinearities in the correlation multiplier, *IEEE Trans. Acoust., Speech, Signal Processing,* ASSP-30(4), 578–586, Aug. 1982.

[36] Bucklew, J.A., Kurtz, T.J., and Sethares, W.A., Weak convergence and local stability properties of fixed step size recursive algorithms, *IEEE Trans. Inform. Theory,* 39(3), 966–978, May 1993.

[37] Douglas, S.C. and Meng, T.H.-Y., Stochastic gradient adaptation under general error criteria, *IEEE Trans. Signal Processing,* 42(6), 1335–1351, June 1994.

[38] Cho, S.H. and Mathews, V.J., Tracking analysis of the sign algorithm in nonstationary environments, *IEEE Trans. Acoust., Speech, Signal Processing,* ASSP-38(12), 2046–2057, Dec. 1990.

20

Robustness Issues in Adaptive Filtering

Ali H. Sayed
University of California, Los Angeles

Markus Rupp
Bell Laboratories
Lucent Technologies

Adaptive filters are systems that adjust themselves to a changing environment. They are designed to meet certain performance specifications and are expected to perform reasonably well under the operating conditions for which they have been designed. In practice, however, factors that may have been ignored or overlooked in the design phase of the system can affect the performance of the adaptive scheme that has been chosen for the system. Such factors include unmodeled dynamics, modeling errors, measurement noise, and quantization errors, among others, and their effect on the performance of an adaptive filter could be critical to the proposed application. Moreover, technological advancements in digital circuit and VLSI design have spurred an increase in the range of new adaptive filtering applications in fields ranging from biomedical engineering to wireless communications. For these new areas, it is increasingly important to design adaptive schemes that are tolerant to unknown or nontraditional factors and effects. The aim of this chapter is to explore and determine the robustness properties of some classical adaptive schemes. Our presentation is meant as an introduction to these issues, and many of the relevant details of specific topics discussed in this section, and alternative points of view, can be found in the references at the end of the chapter.

20.1 Motivation and Example

A classical application of adaptive filtering is that of system identification. The basic problem formulation is depicted in Fig. 20.1, where z^{-1} denotes the unit-time delay operator. The diagram contains two system

0-8493-8572-5/98/$0.00+$.50

blocks: one representing the *unknown plant* or system and the other containing a time-variant tapped-delay-line or finite-impulse-response (FIR) filter structure. The unknown plant represents an arbitrary

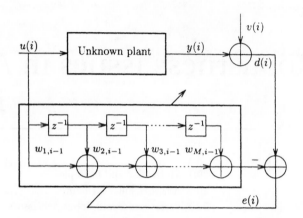

FIGURE 20.1   A system identification example.

relationship between its input and output. This block might implement a pole-zero transfer function, an all-pole or *autoregressive* transfer function, a fixed or time-varying FIR system, a nonlinear mapping, or some other complex system. In any case, it is desired to determine an FIR model for the unknown system of a predetermined impulse response length M, and whose coefficients at time $i - 1$ are denoted by $\{w_{1,i-1}, w_{2,i-1}, \ldots, w_{M,i-1}\}$. The unknown system and the FIR filter are excited by the same input sequence $\{u(i)\}$, where the time origin is at $i = 0$.

If we collect the FIR coefficients into a column vector, say $\mathbf{w}_{i-1} = \mathrm{col}\{w_{1,i-1}, w_{2,i-1}, \ldots, w_{M,i-1}\}$, and define the state vector of the FIR model at time i as $\mathbf{u}_i = \mathrm{col}\{u(i), u(i-1), \ldots, u(i-M+1)\}$, then the output of the FIR filter at time i is the inner product $\mathbf{u}_i^T \mathbf{w}_{i-1}$. In principle, this inner product should be compared with the output $y(i)$ of the unknown plant in order to determine whether or not the FIR output is a good enough approximation for the output of the plant and, therefore, whether or not the current coefficient vector $\mathbf{w}_{i-1}$ should be updated.

In general, however, we do not have direct access to the uncorrupted output $y(i)$ of the plant but rather to a noisy measurement of it, say $d(i) = y(i) + v(i)$. The purpose of an adaptive scheme is to employ the output error sequence $\{e(i) = d(i) - \mathbf{u}_i^T \mathbf{w}_{i-1}\}$, which measures how far $d(i)$ is from $\mathbf{u}_i^T \mathbf{w}_{i-1}$, in order to update the entries of $\mathbf{w}_{i-1}$ and provide a better model, say $\mathbf{w}_i$, for the unknown system. That is, the purpose of the adaptive filter is to employ the available data at time i, $\{d(i), \mathbf{w}_{i-1}, \mathbf{u}_i\}$, in order to update the coefficient vector $\mathbf{w}_{i-1}$ into a presumably better estimate vector $\mathbf{w}_i$.

In this sense, we may regard the adaptive filter as a recursive estimator that tries to come up with a coefficient vector $\mathbf{w}$ that "best" matches the observed data $\{d(i)\}$ in the sense that, for all i, $d(i) \approx \mathbf{u}_i^T \mathbf{w} + v(i)$ to good accuracy. The successive $\mathbf{w}_i$ provide estimates for the unknown and desired $\mathbf{w}$.

20.2 Adaptive Filter Structure

We may reformulate the above adaptive problem in mathematical terms as follows. Let $\{\mathbf{u}_i\}$ be a sequence of *regression vectors* and let $\mathbf{w}$ be an unknown column vector to be estimated or identified. Given noisy measurements $\{d(i)\}$ that are assumed to be related to $\mathbf{u}_i^T \mathbf{w}$ via an additive noise model of the form

$$d(i) = \mathbf{u}_i^T \mathbf{w} + v(i) , \tag{20.1}$$

we wish to employ the given data $\{d(i), \mathbf{u}_i\}$ in order to provide recursive estimates for $\mathbf{w}$ at successive time instants, say $\{\mathbf{w}_0, \mathbf{w}_1, \mathbf{w}_2, \ldots\}$. We refer to these estimates as *weight* estimates since they provide estimates for the coefficients or weights of the tapped-delay model.

Most adaptive schemes perform this task in a recursive manner that fits into the following general description: starting with an initial guess for $\mathbf{w}$, say $\mathbf{w}_{-1}$, iterate according to the learning rule

$$\left(\begin{array}{c} \text{new weight} \\ \text{estimate} \end{array} \right) = \left(\begin{array}{c} \text{old weight} \\ \text{estimate} \end{array} \right) + \left(\begin{array}{c} \text{correction} \\ \text{term} \end{array} \right),$$

where the correction term is usually a function of $\{d(i), \mathbf{u}_i, \text{old weight estimate}\}$. More compactly, we may write $\mathbf{w}_i = \mathbf{w}_{i-1} + f[d(i), \mathbf{u}_i, \mathbf{w}_{i-1}]$, where $\mathbf{w}_i$ denotes an estimate for $\mathbf{w}$ at time i and f denotes a function of the data $\{d(i), \mathbf{u}_i, \mathbf{w}_{i-1}\}$ or of previous values of the data, as in the case where only a filtered version of the error signal $d(i) - \mathbf{u}_i^T \mathbf{w}_{i-1}$ is available. In this context, the well-known least-mean-square (LMS) algorithm has the form

$$\mathbf{w}_i = \mathbf{w}_{i-1} + \mu \cdot \mathbf{u}_i \cdot [d(i) - \mathbf{u}_i^T \cdot \mathbf{w}_{i-1}], \tag{20.2}$$

where μ is known as the *step-size* parameter.

20.3 Performance and Robustness Issues

The performance of an adaptive scheme can be studied from many different points of view. One distinctive methodology that has attracted considerable attention in the adaptive filtering literature is based on stochastic considerations that have become known as the *independence assumptions*. In this context, certain statistical assumptions are made on the natures of the noise signal $\{v(i)\}$ and of the regression vectors $\{\mathbf{u}_i\}$, and conclusions are derived regarding the steady-state behavior of the adaptive filter.

The discussion in this chapter avoids statistical considerations and develops the analysis in a purely deterministic framework that is convenient when prior statistical information is unavailable or when the independence assumptions are unreasonable. The conclusions discussed herein highlight certain features of the adaptive algorithms that hold regardless of any statistical considerations in an adaptive filtering task.

Returning to the data model in (20.1), we see that it assumes the existence of an unknown weight vector $\mathbf{w}$ that describes, along with the regression vectors $\{\mathbf{u}_i\}$, the uncorrupted data $\{y(i)\}$. This assumption may or may not hold.

For example, if the unknown plant in the system identification scenario of Fig. 20.1 is itself an FIR system of length M, then there exists an unknown weight vector $\mathbf{w}$ that satisfies (20.1). In this case, the successive estimates provided by the adaptive filter attempt to identify the unknown weight vector of the plant.

If, on the other hand, the unknown plant of Fig. 20.1 is an autoregressive model of the simple form

$$\frac{1}{1 - cz^{-1}} = 1 + cz^{-1} + c^2 z^{-2} + c^3 z^{-3} + \ldots$$

where $|c| < 1$, then an infinitely long tapped-delay line is necessary to justify a model of the form (20.1). In this case, the first term in the linear regression model (20.1) for a finite order M cannot describe the uncorrupted data $\{y(i)\}$ exactly, and thus modeling errors are inevitable. Such modeling errors can naturally be included in the noise term $v(i)$. Thus, we shall use the term $v(i)$ in (20.1) to account not only for measurement noise but also for modeling errors, unmodeled dynamics, quantization effects, and other kind of disturbances within the system. In many cases, the performance of the adaptive filter depends on how these unknown disturbances affect the weight estimates.

A second source of error in the adaptive system is due to the initial guess $\mathbf{w}_{-1}$ for the weight vector. Due to the iterative nature of our chosen adaptive scheme, it is expected that this initial weight vector plays less of a role in the steady-state performance of the adaptive filter. However, for a finite number of iterations of the adaptive algorithm, both the noise term $v(i)$ and the initial weight error vector $(\mathbf{w} - \mathbf{w}_{-1})$ are disturbances that affect the performance of the adaptive scheme, particularly since the system designer often has little control over them.

The purpose of a robust adaptive filter design, then, is to develop a recursive estimator that minimizes in some well-defined sense the effect of any unknown disturbances on the performance of the filter. For this purpose, we first need to quantify or measure the effect of the disturbances. We address this concern in the following sections.

20.4   Error and Energy Measures

Assuming that the model (20.1) is reasonable, two error quantities come to mind. The first one measures how far the weight estimate $\mathbf{w}_{i-1}$ provided by the adaptive filter is from the true weight vector $\mathbf{w}$ that we are trying to identify. We refer to this quantity as the weight error at time $(i - 1)$, and we denote it by $\tilde{\mathbf{w}}_{i-1} = \mathbf{w} - \mathbf{w}_{i-1}$. The second type of error measures how far the estimate $\mathbf{u}_i^T \mathbf{w}_{i-1}$ is from the uncorrupted output term $\mathbf{u}_i^T \mathbf{w}$. We shall call this the *a priori estimation error*, and we denote it by $e_a(i) = \mathbf{u}_i^T \tilde{\mathbf{w}}_{i-1}$. Similarly, we define an *a posteriori estimation error* as $e_p(i) = \mathbf{u}_i^T \tilde{\mathbf{w}}_i$. Comparing with the definition of the *a priori* error, the *a posteriori* error employs the most recent weight error vector.

Ideally, one would like to make the estimation errors $\{\tilde{\mathbf{w}}_i, e_a(i)\}$ or $\{\tilde{\mathbf{w}}_i, e_p(i)\}$ as small as possible. This objective is hindered by the presence of the disturbances $\{\tilde{\mathbf{w}}_{-1}, v(i)\}$. For this reason, an adaptive filter is said to be *robust* if the effects of the disturbances $\{\tilde{\mathbf{w}}_{-1}, v(i)\}$ on the resulting estimation errors $\{\tilde{\mathbf{w}}_i, e_a(i)\}$ or $\{\tilde{\mathbf{w}}_i, e_p(i)\}$ is small in a well-defined sense. To this end, we can employ one of several measures to denote how "small" these effects are. For our discussion, a quantity known as the *energy* of a signal will be used to quantify these effects. The energy of a sequence $x(i)$ of length N is measured by $\mathcal{E}_x = \sum_{i=0}^{N-1} |x(i)|^2$. A finite energy sequence is one for which $\mathcal{E}_x < \infty$ as $N \to \infty$. Likewise, a finite power sequence is one for which

$$\mathcal{P}_x = \lim_{N \to \infty} \left(\frac{1}{N} \sum_{i=0}^{N-1} |x(i)|^2 \right) < \infty .$$

20.5   Robust Adaptive Filtering

We can now quantify what we mean by robustness in the adaptive filtering context. Let $\mathcal{A}$ denote any adaptive filter that operates causally on the input data $\{d(i), \mathbf{u}_i\}$. A causal adaptive scheme produces a weight vector estimate at time i that depends only on the data available up to and including time i. This adaptive scheme receives as input the data $\{d(i), \mathbf{u}_i\}$ and provides as output the weight vector estimates $\{\mathbf{w}_i\}$. Based on these estimates, we introduce one or more estimation error quantities such as the pair $\{\tilde{\mathbf{w}}_{i-1}, e_a(i)\}$ defined above. Even though these quantities are not explicitly available because $\mathbf{w}$ is unknown, they are of interest to us as their magnitudes determine how well or how poorly a candidate adaptive filtering scheme might perform.

Figure 20.2 indicates the relationship between $\{d(i), \mathbf{u}_i\}$ to $\{\tilde{\mathbf{w}}_{i-1}, e_a(i)\}$ in block diagram form. This schematic representation indicates that an adaptive filter $\mathcal{A}$ operates on $\{d(i), \mathbf{u}_i\}$ and that its performance relies on the sizes of the error quantities $\{\tilde{\mathbf{w}}_{i-1}, e_a(i)\}$, which could be replaced by the error quantities $\{\tilde{\mathbf{w}}_i, e_p(i)\}$ if desired. This representation explicitly denotes the quantities $\{\tilde{\mathbf{w}}_{-1}, v(i)\}$ as disturbances to the adaptive scheme.

In order to measure the effect of the disturbances on the performance of an adaptive scheme, it will be helpful to determine the explicit relationship between the disturbances and the estimation errors that is provided by the adaptive filter. For example, we would like to know what effect the noise terms and the initial weight error guess $\{\tilde{\mathbf{w}}_{-1}, v(i)\}$ would have on the resulting *a priori* estimation errors and the final

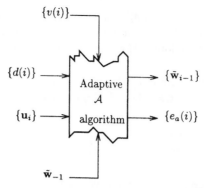

FIGURE 20.2   Input-output map of a generic adaptive scheme.

weight error, $\{e_a(i),\ \tilde{\mathbf{w}}_N\}$, for a given adaptive scheme. Knowing such a relationship, we can then quantify the robustness of the adaptive scheme by determining the degree to which disturbances affect the size of the estimation errors.

We now illustrate how this disturbances-to-estimation-errors relationship can be determined by considering the LMS algorithm in (20.2). Since $d(i) - \mathbf{u}_i^T \mathbf{w}_{i-1} = e_a(i) + v(i)$, we can subtract $\mathbf{w}$ from both sides of (20.2) to obtain the weight-error update equation

$$\tilde{\mathbf{w}}_i = \tilde{\mathbf{w}}_{i-1} - \mu \cdot \mathbf{u}_i \cdot [e_a(i) + v(i)]\,. \qquad (20.3)$$

Assume that we run N steps of the LMS recursion starting with an initial guess $\tilde{\mathbf{w}}_{-1}$. This operation generates the weight error estimates $\{\tilde{\mathbf{w}}_0, \tilde{\mathbf{w}}_1, \dots, \tilde{\mathbf{w}}_N\}$ and the *a priori* estimation errors $\{e_a(0), \dots, e_a(N)\}$.

Define the following two column vectors:

$$\underline{\text{dist}} = \text{col}\left\{ \frac{1}{\sqrt{\mu}} \tilde{\mathbf{w}}_{-1},\ v(0),\ v(1),\ \dots,\ v(N) \right\}\,, \quad \underline{\text{error}} = \text{col}\left\{ e_a(0),\ e_a(1),\ \dots,\ e_a(N),\ \frac{1}{\sqrt{\mu}} \tilde{\mathbf{w}}_N \right\}.$$

The vector $\underline{\text{dist}}$ contains the disturbances that affect the performance of the adaptive filter. The initial weight error vector is scaled by $\mu^{-1/2}$ for convenience. Likewise, the vector $\underline{\text{error}}$ contains the *a priori* estimation errors and the final weight error vector which has also been scaled by $\mu^{-1/2}$. The weight error update relation in (20.3) allows us to relate the entries of both vectors in a straightforward manner. For example,

$$e_a(0) = \mathbf{u}_0^T \tilde{\mathbf{w}}_{-1} = \left(\sqrt{\mu}\, \mathbf{u}_0^T\right)\left(\frac{1}{\sqrt{\mu}}\, \tilde{\mathbf{w}}_{-1}\right),$$

which shows how the first entry of $\underline{\text{error}}$ relates to the first entry of $\underline{\text{dist}}$. Similarly, for $e_a(1) = \mathbf{u}_1^T \tilde{\mathbf{w}}_0$ we obtain

$$e_a(1) = \left(\sqrt{\mu}\, \mathbf{u}_1^T [I - \mu \mathbf{u}_0 \mathbf{u}_0^T]\right) \frac{1}{\sqrt{\mu}} \tilde{\mathbf{w}}_{-1} - \left(\mu \mathbf{u}_1^T \mathbf{u}_0\right) v(0)\,,$$

which relates $e_a(1)$ to the first two entries of the vector $\underline{\text{dist}}$. Continuing in this manner, we can relate $e_a(2)$ to the first three entries of $\underline{\text{dist}}$, $e_a(3)$ to the first four entries of $\underline{\text{dist}}$, and so on.

In general, we can compactly express this relationship as

$$\underbrace{\begin{bmatrix} e_a(0) \\ e_a(1) \\ \vdots \\ \hline e_a(N) \\ \frac{1}{\sqrt{\mu}} \tilde{\mathbf{w}}_N \end{bmatrix}}_{\text{error}} = \underbrace{\begin{bmatrix} \times & & & & & \\ \times & \times & & & O & \\ & & & & & \\ \vdots & & & \ddots & & \\ \times & \times & \times & \times & \times & \times \end{bmatrix}}_{\mathcal{T}} \underbrace{\begin{bmatrix} \frac{1}{\sqrt{\mu}} \tilde{\mathbf{w}}_{-1} \\ \hline v(0) \\ v(1) \\ \vdots \\ v(N) \end{bmatrix}}_{\text{dist}}$$

where the symbol $\times$ is used to denote the entries of the lower triangular mapping $\mathcal{T}$ relating <u>dist</u> to <u>error</u>. The specific values of the entries of $\mathcal{T}$ are not of interest for now, although we have indicated how the expressions for these $\times$ terms can be found. However, the causal nature of the adaptive algorithm requires that $\mathcal{T}$ be of lower triangular form.

Given the above relationship, our objective is to quantify the effect of the disturbances on the estimation errors. Let $\mathcal{E}_d$ and $\mathcal{E}_e$ denote the energies of the vectors <u>dist</u> and <u>error</u>, respectively, such that

$$\mathcal{E}_e = \frac{1}{\mu} \|\tilde{\mathbf{w}}_N\|^2 + \sum_{i=0}^{N} |e_a(i)|^2 \quad \text{and} \quad \mathcal{E}_d = \frac{1}{\mu} \|\tilde{\mathbf{w}}_{-1}\|^2 + \sum_{i=0}^{N} |v(i)|^2 ,$$

where $\| \cdot \|$ denotes the Euclidean norm of a vector. We shall say that the LMS adaptive algorithm is *robust with level γ* if a relation of the form

$$\frac{\mathcal{E}_e}{\mathcal{E}_d} \leq \gamma^2 , \tag{20.4}$$

holds for some positive γ and for *any* nonzero, finite-energy disturbance vector <u>dist</u>. In other words, no matter what the disturbances $\{\tilde{\mathbf{w}}_{-1}, v(i)\}$ are, the energy of the resulting estimation errors will never exceed γ^2 times the energy of the associated disturbances.

The form of the mapping $\mathcal{T}$ affects the value of γ in (20.4) for any particular algorithm. To see this result, recall that for any finite-dimensional matrix A, its maximum singular value, denoted by $\bar{\sigma}(A)$, is defined by $\bar{\sigma}(A) = \max_{x \neq 0} \frac{\|Ax\|}{\|x\|}$. Hence, the square of the maximum singular value, $\bar{\sigma}^2(A)$, measures the maximum energy gain from the vector x to the resulting vector Ax. Therefore, if a relation of the form (20.4) should hold for any nonzero disturbance vector <u>dist</u>, then it means that

$$\max_{\underline{\text{dist}} \neq 0} \frac{\|\mathcal{T} \underline{\text{dist}}\|}{\|\underline{\text{dist}}\|} \leq \gamma .$$

Consequently, the maximum singular value of $\mathcal{T}$ must be bounded by γ. This imposes a condition on the allowable values for γ; its smallest value cannot be smaller than the maximum singular value of the resulting $\mathcal{T}$.

Ideally, we would like the value of γ in (20.4) to be as small as possible. In particular, an algorithm for which the value of γ is 1 would guarantee that the estimation error energy will never exceed the disturbance energy, no matter what the natures of the disturbances are! Such an algorithm would possess a good degree of robustness since it would guarantee that the disturbance energy will never be unnecessarily magnified.

Before continuing our study, we ask and answer the obvious questions that arise at this point:

- *What is the smallest possible value for γ for the LMS algorithm?* It turns out for the LMS algorithm that, under certain conditions on the step-size parameter, the smallest possible value for γ is 1. Thus, $\mathcal{E}_e \leq \mathcal{E}_d$ for the LMS algorithm.

- *Does there exist any other causal adaptive algorithm that would result in a value for γ in (20.4) that is smaller than one?* It can be argued that no such algorithm exists for the model (20.1) and criterion (20.4).

In other words, the LMS algorithm is in fact the most robust adaptive algorithm in the sense defined by (20.4). This result provides a rigorous basis for the excellent robustness properties that the LMS algorithm, and several of its variants, have shown in practical situations. The references at the end of the chapter provide an overview of the published works that have established these conclusions. Here, we only motivate them from first principles. In so doing, we shall also discuss other results (and tools) that can be used in order to impose certain robustness and convergence properties on other classes of adaptive schemes.

20.6 Energy Bounds and Passivity Relations

Consider the LMS recursion in (20.2), with a time-varying step-size $\mu(i)$ for purposes of generality, as given by

$$\mathbf{w}_i = \mathbf{w}_{i-1} + \mu(i) \cdot \mathbf{u}_i \cdot [d(i) - \mathbf{u}_i^T \cdot \mathbf{w}_{i-1}] . \tag{20.5}$$

Subtracting the optimal coefficient vector $\mathbf{w}$ from both sides and squaring the resulting expressions, we obtain

$$\|\tilde{\mathbf{w}}_i\|^2 = \| \tilde{\mathbf{w}}_{i-1} - \mu(i) \cdot \mathbf{u}_i \cdot [e_a(i) + v(i)] \|^2 .$$

Expanding the right-hand side of this relationship and rearranging terms leads to the equality

$$\|\tilde{\mathbf{w}}_i\|^2 - \|\tilde{\mathbf{w}}_{i-1}\|^2 + \mu(i) \cdot |e_a(i)|^2 - \mu(i) \cdot |v(i)|^2 = \mu(i) \cdot |e_a(i) + v(i)|^2 \cdot [\mu(i) \cdot \|\mathbf{u}_i\|^2 - 1] .$$

The right-hand side in the above equality is the product of three terms. Two of these terms, $\mu(i)$ and $|e_a(i) + v(i)|^2$, are nonnegative, whereas the term $(\mu(i) \cdot \|\mathbf{u}_i\|^2 - 1)$ can be positive, negative, or zero depending on the relative magnitudes of $\mu(i)$ and $\|\mathbf{u}_i\|^2$. If we define $\bar{\mu}(i)$ as (assuming nonzero regression vectors):

$$\bar{\mu}(i) = \|\mathbf{u}_i\|^{-2} , \tag{20.6}$$

then the following relations hold:

$$\frac{\|\tilde{\mathbf{w}}_i\|^2 + \mu(i) |e_a(i)|^2}{\|\tilde{\mathbf{w}}_{i-1}\|^2 + \mu(i) |v(i)|^2} \quad \begin{cases} \leq 1 & \text{for} \quad 0 < \mu(i) < \bar{\mu}(i) \\ = 1 & \text{for} \quad \mu(i) = \bar{\mu}(i) \\ \geq 1 & \text{for} \quad \mu(i) > \bar{\mu}(i) \end{cases}$$

The result for $0 < \mu(i) \leq \bar{\mu}(i)$ has a nice interpretation. It states that, no matter what the value of $v(i)$ is and no matter how far $\mathbf{w}_{i-1}$ is from $\mathbf{w}$, the sum of the two energies $\|\tilde{\mathbf{w}}_i\|^2 + \mu(i) \cdot |e_a(i)|^2$ will always be smaller than or equal to the sum of the two disturbance energies $\|\tilde{\mathbf{w}}_{i-1}\|^2 + \mu(i) \cdot |v(i)|^2$. This relationship is a statement of the *passivity* of the algorithm locally in time, as it holds for every time instant. Similar relationships can be developed in terms of the *a posteriori* estimation error.

Since this relationship holds for each time instant i, it also holds over an interval of time such that

$$\frac{\|\tilde{\mathbf{w}}_N\|^2 + \sum_{i=0}^{N} |\bar{e}_a(i)|^2}{\|\tilde{\mathbf{w}}_{-1}\|^2 + \sum_{i=0}^{N} |\bar{v}(i)|^2} \leq 1 , \tag{20.7}$$

where we have introduced the normalized *a priori* residuals and noise signals

$$\bar{e}_a(i) = \sqrt{\mu(i)}\, e_a(i) \quad \text{and} \quad \bar{v}(i) = \sqrt{\mu(i)}\, v(i) ,$$

respectively. Equation (20.7) states that the lower-triangular matrix that maps the normalized noise signals $\{\bar{v}(i)\}_{i=0}^{N}$ and the initial uncertainty $\tilde{\mathbf{w}}_{-1}$ to the normalized *a priori* residuals $\{\bar{e}_a(i)\}_{i=0}^{N}$ and the final weight error $\tilde{\mathbf{w}}_N$ has a maximum singular value that is less than one. Thus, it is a *contraction mapping* for $0 < \mu(i) \leq \bar{\mu}(i)$. For the special case of a constant step-size μ, this is the same mapping $\mathcal{T}$ that we introduced earlier (20.4).

In the above derivation, we have assumed for simplicity of presentation that the denominators of all expressions are nonzero. We can avoid this restriction by working with differences rather than ratios. Let $\Delta_N(\mathbf{w}_{-1}, v(\cdot))$ denote the difference between the numerator and the denominator of (20.7), such that

$$\Delta_N(\mathbf{w}_{-1}, v(\cdot)) = \left\{ \|\tilde{\mathbf{w}}_N\|^2 + \sum_{i=0}^{N} |\bar{e}_a(i)|^2 \right\} - \left\{ \|\tilde{\mathbf{w}}_{-1}\|^2 + \sum_{i=0}^{N} |\bar{v}(i)|^2 \right\} . \tag{20.8}$$

Then, a similar argument that produced (20.7) can be used to show that for any $\{\mathbf{w}_{-1}, v(\cdot)\}$,

$$\Delta_N(\mathbf{w}_{-1}, v(\cdot)) \leq 0 . \tag{20.9}$$

20.7   Min-Max Optimality of Adaptive Gradient Algorithms

The property in (20.7) or (20.9) is valid for any initial guess $\mathbf{w}_{-1}$ and for any noise sequence $v(\cdot)$, so long as the $\mu(i)$ are properly bounded by $\bar{\mu}(i)$. One might then wonder whether the bound in (20.7) is tight or not. In other words, are there choices $\{\mathbf{w}_{-1}, v(\cdot)\}$ for which the ratio in (20.7) can be made arbitrarily close to one or Δ_N in (20.9) arbitrarily close to zero? We now show that there are. We can rewrite the gradient recursion of (20.5) in the equivalent form

$$\mathbf{w}_i = \mathbf{w}_{i-1} + \mu(i) \cdot \mathbf{u}_i \cdot [e_a(i) + v(i)] . \tag{20.10}$$

Envision a noise sequence $v(i)$ that satisfies $v(i) = -e_a(i)$ at each time instant i. Such a sequence may seem unrealistic but is entirely within the realm of our unrestricted model of the unknown disturbances. In this case, the above gradient recursion trivializes to $\mathbf{w}_i = \mathbf{w}_{i-1}$ for all i, thus leading to $\mathbf{w}_N = \mathbf{w}_{-1}$. Thus, Δ_N in (20.8) will be zero for this particular experiment. Therefore,

$$\max_{\{\mathbf{w}_{-1}, v(\cdot)\}} \{\Delta_N(\mathbf{w}_{-1}, v(\cdot))\} = 0 .$$

We now consider the following question: how does the gradient recursion in (20.5) compare with other possible causal recursive algorithms for the update of the weight estimate? Let $\mathcal{A}$ denote any given causal algorithm. Suppose that we initialize algorithm $\mathcal{A}$ with $\mathbf{w}_{-1} = \mathbf{w}$, and suppose the noise sequence is given by $v(i) = -e_a(i)$ for $0 \leq i \leq N$. Then, we have

$$\sum_{i=0}^{N} |\bar{v}(i)|^2 = \sum_{i=0}^{N} |\bar{e}_a(i)|^2 \leq \|\tilde{\mathbf{w}}_N\|^2 + \sum_{i=0}^{N} |\bar{e}_a(i)|^2 ,$$

no matter what the value of $\tilde{\mathbf{w}}_N$ is. This particular choice of initial guess ($\mathbf{w}_{-1} = \mathbf{w}$) and noise sequence $\{v(\cdot)\}$ will always result in a nonnegative value of Δ_N in (20.8), implying for any causal algorithm $\mathcal{A}$ that

$$\max_{\{\mathbf{w}_{-1}, v(\cdot)\}} \{\Delta_N(\mathbf{w}_{-1}, v(\cdot))\} \geq 0 .$$

For the gradient recursion in (20.5), the maximum has to be exactly zero because the global property (20.9) provided us with an inequality in the other direction. Therefore, the algorithm in (20.5) solves the following optimization problem:

$$\min_{Algorithm} \left\{ \max_{\{\mathbf{w}_{-1}, v(\cdot)\}} \Delta_N(\mathbf{w}_{-1}, v(\cdot)) \right\} ,$$

and the optimal value is equal to zero. More details and justification can be found in the references at the end of this chapter, especially connections with so-called H_∞ estimation theory.

As explained before, Δ_N measures the difference between the output energy and the input energy of the algorithm mapping $\mathcal{T}$. The gradient algorithm in (20.5) minimizes the maximum possible difference between these two energies over all disturbances with finite energy. In other words, it minimizes the effect that the worst-possible input disturbances can have on the resulting estimation-error energy.

20.8   Comparison of LMS and RLS Algorithms

To illustrate the ideas in our discussion, we compare the robustness performance of two classical algorithms: the LMS algorithm (20.2) and the recursive least-squares (RLS) algorithm. More details on the example given below can be found in the reference section at the end of the chapter.

Consider the data model in (20.1) where $\mathbf{u}_i$ is a scalar that randomly assumes the values $+1$ and -1 with equal probability. Let $\mathbf{w} = 0.25$, and let $v(i)$ be an uncorrelated Gaussian noise sequence with unit variance. We first employ the LMS recursion in (20.2) and compute the initial 150 estimates $\mathbf{w}_i$, starting with $\mathbf{w}_{-1} = 0$ and using $\mu = 0.97$. Note that μ satisfies the requirement $\mu \leq 1/\|\mathbf{u}_i\|^2 = 1$ for all i. We then evaluate the entries of the resulting mapping $\mathcal{T}$, now denoted by $\mathcal{T}_{lms}$, that we defined in (20.4). We then compute the corresponding $\mathcal{T}_{rls}$ for the recursive-least-squares (RLS) algorithm for these signals, which for this special data model can be expressed as

$$\mathbf{w}_{i+1} = \mathbf{w}_i + \frac{p_i \mathbf{u}_i}{1 + p_i}[d(i) - \mathbf{u}_i^T \mathbf{w}_{i-1}], \quad p_{i+1} = \frac{p_i}{1 + p_i}.$$

The initial condition chosen for p_i is $p_0 = \mu = 0.97$.

Figure 20.3 shows a plot of the 150 singular values of the resulting mappings $\mathcal{T}_{lms}$ and $\mathcal{T}_{rls}$. As predicted from our analysis, the singular values of $\mathcal{T}_{lms}$, indicated by an almost horizontal line at unity, are all bounded by one, whereas the maximum singular value of $\mathcal{T}_{rls}$ is approximately 1.65. This result indicates that the LMS algorithm is indeed more robust than the RLS algorithm, as is predicted by the earlier analysis.

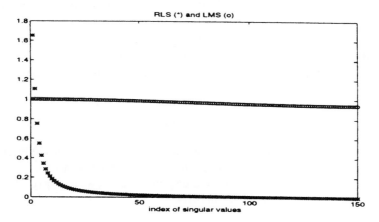

FIGURE 20.3   Singular value plot.

Observe, however, that most of the singular values of $\mathcal{T}_{rls}$ are considerably smaller than one, whereas the singular values of $\mathcal{T}_{lms}$ are clustered around one. This has an interesting interpretation that we explain as follows. An $N \times N$-dimensional matrix A has N singular values $\{\sigma_i\}$ that are equal to the positive square-roots of the eigenvalues of AA^T. For each σ_i, there exists a unit-norm vector x_i such that the energy gain from x_i to Ax_i is equal to σ_i^2, i.e., $\sigma_i = \|Ax_i\|/\|x_i\|$. The vector x_i can be chosen as the ith right singular vector of A. Now, recall that $\mathcal{T}_{lms}$ and $\mathcal{T}_{rls}$ are finite-dimensional matrices that map a disturbance vector <u>dist</u> to the estimation-errors vector <u>error</u>. Considering the plot of the singular values of $\mathcal{T}_{rls}$, we see that if the disturbance vector <u>dist</u> happens to lie in the range space of the right singular vectors associated with the smaller singular values in this plot, then its effect will be significantly attenuated. This fact indicates that while the performance of the LMS algorithm guards against worst-case disturbances, the RLS algorithm is likely to have a better performance than the LMS algorithm on average, as is well-known.

20.9   Time-Domain Feedback Analysis

Robust adaptive filters are designed to induce contractive mappings between sequences of numbers. This fact also has important implications on the convergence performance of a robust adaptive scheme. In the remaining sections of this chapter, we discuss the combined issues of robustness and convergence from a deterministic standpoint. In particular, the following issues are discussed:

- We show that each step of the update equation of the gradient algorithm in (20.5) can be described in terms of an elementary section that possesses a useful feedback structure.

- The feedback structure provides insights into the robust and convergence performance of the adaptive scheme. This is achieved by studying the energy flow through a cascade of elementary sections and by invoking a useful tool from system theory known as the *small gain theorem*.

- The feedback analysis extends to more general update relations. The example considered here is filtered-error LMS algorithm, although the methodology can be extended to other structures such as perceptrons. Details can be found in the references at the end of this chapter.

20.9.1   Time-Domain Analysis

From the update equation in (20.5), $\tilde{w}_i$ satisfies

$$\tilde{w}_i = \tilde{w}_{i-1} - \mu(i) \cdot u_i \cdot [e_a(i) + v(i)] . \tag{20.11}$$

If we multiply both sides of (20.11) by u_i^T from the left, we obtain the following relation among $\{e_p(i), e_a(i), v(i)\}$:

$$e_p(i) = \left(1 - \frac{\mu(i)}{\bar{\mu}(i)}\right) e_a(i) - \frac{\mu(i)}{\bar{\mu}(i)} v(i) , \tag{20.12}$$

where $\bar{\mu}(i)$ is given by (20.6). Using (20.12), (20.5) can be rewritten in the equivalent form

$$\begin{aligned}
w_i &= w_{i-1} + \bar{\mu}(i) \cdot u_i \cdot [e_a(i) - e_p(i)] , \\
&= w_{i-1} + \bar{\mu}(i) \cdot u_i \cdot [e_a(i) + r(i)] ,
\end{aligned} \tag{20.13}$$

where we have defined the signal $r(i) = -e_p(i)$ for convenience. The expression (20.13) shows that (20.5) can be rewritten in terms of a new step-size $\bar{\mu}(i)$ and a modified "noise" term $r(i)$.

Therefore, if we follow arguments similar to those prior to (20.6), we readily conclude that for algorithm (20.5) the following equality holds for *all* $\{\mu(i), v(i)\}$:

$$\frac{\|\tilde{w}_i\|^2 + \bar{\mu}(i) |e_a(i)|^2}{\|\tilde{w}_{i-1}\|^2 + \bar{\mu}(i) |r(i)|^2} = 1 . \tag{20.14}$$

This relation establishes a lossless map (denoted by $\overline{\mathcal{T}}_i$) from the signals $\{\tilde{w}_{i-1}, \sqrt{\bar{\mu}(i)}\ r(i)\}$ to the signals $\{\tilde{w}_i, \sqrt{\bar{\mu}(i)}\ e_a(i)\}$. Correspondingly, using (20.12), the map from the original weighted disturbance $\sqrt{\bar{\mu}(i)}v(i)$ to the weighted estimation error signal $\sqrt{\bar{\mu}(i)}e_a(i)$ can be expressed in terms of the feedback structure shown in Fig. 20.4.

The feedback description provides useful insights into the behavior of the adaptive scheme. Because the map $\overline{\mathcal{T}}_i$ in the feedforward path is lossless or energy-preserving, the design and analysis effort can be concentrated on the terms contained in the feedback path. This feedback block controls:

- How much energy is fed back into the input of each section and whether energy magnification or demagnification may occur (i.e., stability).

- How sensitive the estimation error is to noise and disturbances (i.e., robustness).

- How fast the estimation error energy decays (i.e., convergence rate).

20.9.2 l_2—Stability and the Small Gain Condition

We start by reconsidering the robustness issue. Recall that if the step-sizes $\mu(i)$ are chosen such that $\mu(i) \leq \bar{\mu}(i)$, then robustness is guaranteed in that the ratio of the energies of the signals in (20.7) will be bounded by one.

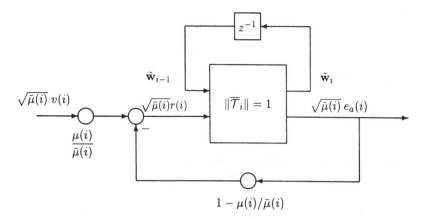

FIGURE 20.4 A time-variant lossless mapping with gain feedback for gradient algorithms. © IEEE 1996. (*Source:* Rupp, M. and Sayed, A.H., A time domain feedback analysis of filtered-error adaptive gradient algorithms, *IEEE Trans. Signal Process.* 44(6): 1428–1439, June 1996. With permission.)

The condition on $\mu(i)$ can be relaxed at the expense of guaranteeing energy ratios that are bounded by some other positive number, say

$$\frac{\text{weighted estimation error energy}}{\text{weighted disturbance energy}} \leq \gamma^2 , \tag{20.15}$$

for some constant γ to be determined. This is still a desirable property because it means that the disturbance energy will be, at most, scaled by a factor of γ. This fact can in turn lead to useful convergence conclusions, as argued later.

In order to guarantee robustness conditions according to (20.15), for some γ, we rely on the observation that feedback configurations of the form shown in Fig. 20.4 can be analyzed using a tool known in system theory as the small gain theorem. In loose terms, this theorem states that the stability of a feedback configuration such as that in Fig. 20.4 is guaranteed if the product of the norms of the feedforward and the feedback mappings are strictly bounded by one. Since the feedforward mapping $\overline{\mathcal{T}}_i$ has a norm (or maximum singular value) of one, the norm of the feedback map needs to be strictly bounded by one for stability of this system.

To illustrate these concepts more fully, consider the feedback structure in Fig. 20.5 that has a lossless mapping $\mathcal{T}$ in its feedforward path and an arbitrary mapping $\mathcal{F}$ in its feedback path. The input/output signals of interest are denoted by $\{x, y, r, v, e\}$. In this system, the signals x, v play the role of the disturbances.

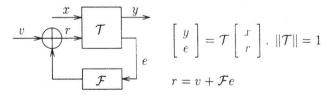

FIGURE 20.5 A feedback structure.

The losslessness of the feedforward path implies conservation of energy such that

$$\|y\|^2 + \|e\|^2 = \|x\|^2 + \|r\|^2 .$$

Consequently, $\|e\| \le \|x\| + \|r\|$. On the other hand, the triangle inequality of norms implies that

$$\|r\| \le \|v\| + \|\mathcal{F}\| \cdot \|e\|,$$

where the notation $\|\mathcal{F}\|$ denotes the maximum singular value of the mapping $\mathcal{F}$. Provided that the small gain condition $\|\mathcal{F}\| < 1$ is satisfied, we have

$$\|e\| \le \frac{1}{1 - \|\mathcal{F}\|} \cdot [\, \|x\| + \|v\| \,]. \tag{20.16}$$

Thus, a contractive $\mathcal{F}$ guarantees a robust map from $\{x, v\}$ to $\{e\}$ with a robustness level that is determined by the factor $1/(1 - \|\mathcal{F}\|)$. In this case, we shall say that the map from $\{x, v\}$ to $\{e\}$ is l_2−stable.

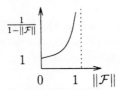

FIGURE 20.6   Plot of the l_2−gain.

A plot of the factor $1/(1 - \|\mathcal{F}\|)$, as a function of $\|\mathcal{F}\|$, is shown in Fig. 20.6. It can be seen that the smaller the value of $\|\mathcal{F}\|$:

- The smaller the effect of $\{x, v\}$ on $\{e\}$.
- The smaller the upper bound on $\|e\|$.

Moreover, we shall argue that smaller values of $\|\mathcal{F}\|$ are associated with faster convergence. Therefore, controlling the norm of $\mathcal{F}$, is important for both the robustness and convergence performance of an adaptive algorithm. In most cases, the feedback filter $\mathcal{F}$ will depend on several quantities, such as the step-sizes $\{\mu(i)\}$ and the data vectors $\{\mathbf{u}_i\}$ (as in Fig. 20.4). It may also depend on error filters and on regression filters that appear in more general adaptive schemes.

Referring to Fig. 20.4, define

$$\eta(N) = \max_{0 \le i \le N} \left| 1 - \frac{\mu(i)}{\bar{\mu}(i)} \right| \quad \text{and} \quad \xi(N) = \max_{0 \le i \le N} \frac{\mu(i)}{\bar{\mu}(i)}.$$

That is, $\eta(N)$ is the maximum absolute value of the gain of the feedback loop over the interval of time $0 \le i \le N$, and $\xi(N)$ is the maximum value of the scaling factor $\mu(i)/\bar{\mu}(i)$ at the input of the feedback interconnection. In this context, the small gain condition requires that $\eta(N) < 1$. This condition is equivalent to choosing the step-size parameter $\mu(i)$ such that $0 < \mu(i) < 2\bar{\mu}(i)$. Under this condition, the general relation (20.16) can be used to deduce either of the following two relationships:

$$\sqrt{\sum_{i=0}^{N} \bar{\mu}(i)\, |e_a(i)|^2} \le \frac{1}{1 - \eta(N)} \left[\|\tilde{\mathbf{w}}_{-1}\| + \xi(N) \sqrt{\sum_{i=0}^{N} \bar{\mu}(i)|v(i)|^2} \right] \tag{20.17}$$

or

$$\sqrt{\sum_{i=0}^{N} \mu(i)\, |e_a(i)|^2} \le \frac{\xi^{1/2}(N)}{1 - \eta(N)} \left[\|\tilde{\mathbf{w}}_{-1}\| + \xi^{1/2}(N) \sqrt{\sum_{i=0}^{N} \mu(i)|v(i)|^2} \right]. \tag{20.18}$$

Note that in either case the upper bound on $\mu(i)$ is now $2\bar{\mu}(i)$ and the robustness level is essentially determined by

$$\frac{1}{1 - \eta(N)} \quad \text{or} \quad \frac{\xi^{1/2}(N)}{1 - \eta(N)},$$

depending on how the estimation errors $\{e_a(i)\}$ and the noise terms $\{v(i)\}$ are normalized [by $\mu(\cdot)$ or $\bar{\mu}(\cdot)$].

20.9.3 Energy Propagation in the Feedback Cascade

By studying the energy flow in the feedback interconnection of Fig. 20.4, we can also obtain some physical insights into the convergence behavior of the gradient recursion (20.5).

Assume that $\mu(i) = \bar{\mu}(i)$, such that the feedback loop of Fig. 20.4 is disconnected. In this situation, there is no energy flowing back into the lower input of the lossless section from its lower output $e_a(\cdot)$. The losslessness of the feedforward path then implies that

$$E_w(i) = E_w(i-1) + E_v(i) - E_e(i),$$

where we have defined the energy terms

$$E_e(i) = \bar{\mu}(i)|e_a(i)|^2, \quad E_v(i) = \bar{\mu}(i)|v(i)|^2, \quad E_w(i) = \|\tilde{w}_i\|^2.$$

In the noiseless case where $v(i) = 0$, the above expression implies that the weight-error energy is a nonincreasing function of time, i.e., $E_w(i) \leq E_w(i-1)$.

However, what happens if $\mu(i) \neq \bar{\mu}(i)$? In this case, the feedback path is active and the convergence speed will be affected because the rate of decrease in the energy of the estimation error will change. Indeed, for $\mu(i) \neq \bar{\mu}(i)$ we obtain for $E_v(i) = 0$

$$E_w(i) = E_w(i-1) - \left(1 - \left|1 - \frac{\mu(i)}{\bar{\mu}(i)}\right|^2\right) E_e(i),$$

where, due to the small gain condition, the coefficient multiplying $E_e(i)$ can be seen to be smaller than 1.

Loosely speaking, this energy argument indicates for $v(i) = 0$ that the smaller the maximum singular value of feedback block $\mathcal{F}$, for a generic feedback interconnection of the form shown in Fig. 20.5, the faster the convergence of the algorithm will be, since less energy is fed back to the input of each section.

20.9.4 A Deterministic Convergence Analysis

The energy argument can be pursued in order to provide sufficient deterministic conditions for the convergence of the weight estimates w_i to the true weight vector w. The argument follows as a consequence of the energy relations (or robustness bounds) (20.17) and (20.18), which essentially establishes that the adaptive gradient algorithm (20.5) maps a finite-energy sequence to another finite-energy sequence.

To clarify this point, we define the quantities

$$\eta = \sup_i \left|1 - \frac{\mu(i)}{\bar{\mu}(i)}\right|, \quad \xi = \sup_i \left[\frac{\mu(i)}{\bar{\mu}(i)}\right],$$

and note that if the step-size parameter $\mu(i)$ is chosen such that $\mu(i)\|u_i\|^2$ is uniformly bounded by 2, then we guarantee $\xi < 2$ and $\eta < 1$. We further note that it follows from the weight-error update relation (20.11) that $\tilde{w}_i$ satisfies

$$\tilde{w}_i = \tilde{w}_{i-1} - \bar{u}_i^T [\bar{e}_a(i) + \bar{v}(i)], \tag{20.19}$$

where we have defined $\bar{u}_i = \sqrt{\mu(i)}\, u_i$ [likewise for $\bar{e}_a(i), \bar{v}(i)$]. The following conclusions can now be established under the stated conditions.

- *Finite noise energy condition.* We assume that the normalized sequence $\{\bar{v}(\cdot) = \sqrt{\mu(i)}\, v(i)\}$ has finite energy, i.e.,

$$\sum_{i=0}^{\infty} \mu(i)|v(i)|^2 < \infty .\qquad(20.20)$$

This in turn implies that $\bar{v}(i) \to 0$ as $i \to \infty$ (but not necessarily $v(i) \to 0$). If the initial weight-error vector is finite, $\|\tilde{\mathbf{w}}_{-1}\| < \infty$, then condition (20.20) along with the energy bound (20.18) [as $N \to \infty$] allows us to conclude that $\sum_{i=0}^{\infty} \mu(i)|e_a(i)|^2 < \infty$. Consequently, $\lim_{i\to\infty} \bar{e}_a(i) \to 0$ (but not necessarily $e_a(i) \to 0$).

- *Persistent excitation condition.* We also assume that the normalized vectors $\{\bar{\mathbf{u}}_i\}$ are persistently exciting. By this we mean that there exists a finite integer $L \geq M$ such that the smallest singular value of

$$\mathrm{col}\,\{\bar{\mathbf{u}}_i^T, \bar{\mathbf{u}}_{i+1}^T, \ldots, \bar{\mathbf{u}}_{i+L}^T\}$$

is uniformly bounded from below by a positive quantity for sufficiently large i. The persistence of excitation condition can be used to further conclude from $\bar{e}_a(i) \to 0$ that $\lim_{i\to\infty} \mathbf{w}_i = \mathbf{w}$.

The above statements can also be used to clarify the behavior of the adaptive algorithm (20.5) in the presence of finite-power (rather than finite-energy) normalized noise sequences $\{\bar{v}(\cdot)\}$, i.e., for $v(\cdot)$ satisfying

$$\lim_{N\to\infty} \frac{1}{N} \sum_{i=0}^{N-1} \mu(i)|v(i)|^2 = P_v < \infty .$$

For this purpose, we divide both sides of (20.18) by $\sqrt{N}$ and take the limit as $N \to \infty$ to conclude that

$$\lim_{N\to\infty} \frac{1}{N} \sum_{i=0}^{N-1} \mu(i)|e_a(i)|^2 \leq \frac{\xi^2 P_v}{(1-\eta)^2} .$$

In other words, a bounded noise power leads to a bounded estimation error power.

20.10   Filtered-Error Gradient Algorithms

The feedback analysis of the former sections can be extended to gradient algorithms that employ filtered versions of the error signal $d(i) - \mathbf{u}_i^T \mathbf{w}_{i-1}$. Such algorithms are useful in applications such as active noise and vibration control and in adaptive IIR filters, where a filtered error signal is more easily observed or measured. Figure 20.7 depicts the context of this problem. The symbol F denotes the filter that operates on $e(i)$. For our discussion, we assume that F is a finite-impulse response filter of order M_F, such that the z-transform of its impulse response is $F(z) = \sum_{j=0}^{M_F-1} f_j z^{-j}$.

For purposes of discussion, we focus on one particular form of adaptive update known as the *filtered-error LMS algorithm*:

$$\mathbf{w}_i = \mathbf{w}_{i-1} + \mu(i) \cdot \mathbf{u}_i \cdot F[d(i) - \mathbf{u}_i^T \mathbf{w}_{i-1}] .\qquad(20.21)$$

Comparing (20.21) with (20.5), the only difference between the two updates is the filter F that acts on the error $d(i) - \mathbf{u}_i^T \mathbf{w}_{i-1}$.

Following the discussion that led to (20.13), it can be verified that (20.21) is equivalent to the following update:

$$\mathbf{w}_i = \mathbf{w}_{i-1} + \bar{\mu}(i) \cdot \mathbf{u}_i \cdot [e_a(i) + r(i)] ,\qquad(20.22)$$

where $\bar{\mu}(i) = 1/\|\mathbf{u}_i\|^2$, $e_a(i) = \mathbf{u}_i^T \tilde{\mathbf{w}}_{i-1}$, and $r(i)$ is defined as

$$\bar{\mu}(i)r(i) = \mu(i)F[v(i)] - \bar{\mu}(i)e_a(i) + \mu(i)F[e_a(i)] .\qquad(20.23)$$

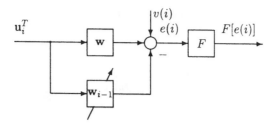

FIGURE 20.7 Structure of filtered-error gradient algorithms. © IEEE 1996. (*Source:* Rupp, M. and Sayed, A.H., A time domain feedback analysis of filtered-error adaptive gradient algorithms, *IEEE Trans. Signal Process.* 44(6): 1428–1439, June 1996. With permission.)

Expression (20.22) is of the same form as (20.13), which implies that the following relation also holds:

$$\frac{\|\tilde{\mathbf{w}}_i\|^2 + \bar{\mu}(i)\,|e_a(i)|^2}{\|\tilde{\mathbf{w}}_{i-1}\|^2 + \bar{\mu}(i)\,|r(i)|^2} = 1 . \qquad (20.24)$$

This establishes that the map $\overline{T}_i$ from the signals $\{\tilde{\mathbf{w}}_{i-1}, \sqrt{\bar{\mu}(i)}r(i)\}$ to the signals $\{\tilde{\mathbf{w}}_i, \sqrt{\bar{\mu}(i)}e_a(i)\}$ is lossless. Moreover, the map from the original disturbance $\sqrt{\bar{\mu}(\cdot)}v(\cdot)$ to the signal $\sqrt{\bar{\mu}(\cdot)}e_a(\cdot)$ can be expressed in terms of a feedback structure, as shown in Fig. 20.8. We remark that the notation $1 - \frac{\mu(i)}{\sqrt{\bar{\mu}(i)}}\,F[\cdot]\,\frac{1}{\sqrt{\bar{\mu}(i)}}$ should be interpreted as follows. We first divide $\sqrt{\bar{\mu}(i)}\,e_a(i)$ by $\sqrt{\bar{\mu}(i)}$ before filtering it by the filter F and then scaling the result by $\mu(i)/\sqrt{\bar{\mu}(i)}$. Similarly, the term $\sqrt{\bar{\mu}(i)}\,v(i)$ is first divided by $\sqrt{\bar{\mu}(i)}$, then filtered by F, and is finally scaled by $\mu(i)/\sqrt{\bar{\mu}(i)}$.

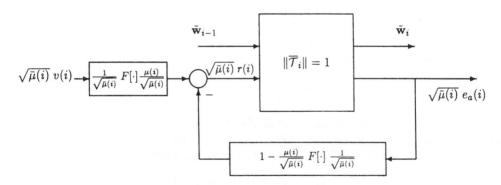

FIGURE 20.8 Filtered-error LMS algorithm as a time-variant lossless mapping with dynamic feedback. © IEEE 1996. (*Source:* Rupp, M. and Sayed, A.H., A time domain feedback analysis of filtered-error adaptive gradient algorithms, *IEEE Trans. Signal Process.* 44(6): 1428–1439, June 1996. With permission.)

The feedback path now contains a dynamic system. The small gain theorem dictates that this system will be robust if the feedback path is a contractive system. For the special case of the *projection* filtered-error LMS algorithm that employs the step-size $\mu(i) = \alpha\,\bar{\mu}(i)$, $\alpha > 0$,

$$\mathbf{w}_i = \mathbf{w}_{i-1} + \alpha\frac{\mathbf{u}_i}{\|\mathbf{u}_i\|^2}\,F[d(i) - \mathbf{u}_i^T\mathbf{w}_{i-1}] , \qquad (20.25)$$

the small-gain condition implies that the following matrix should be strictly contractive:

$$
\mathbf{P}_N = \begin{pmatrix}
1 - \alpha f_0 & & \mathbf{O} \\
-\alpha \dfrac{\sqrt{\bar{\mu}(1)}}{\sqrt{\bar{\mu}(0)}} f_1 & 1 - \alpha f_0 & \\
-\alpha \dfrac{\sqrt{\bar{\mu}(2)}}{\sqrt{\bar{\mu}(0)}} f_2 & -\alpha \dfrac{\sqrt{\bar{\mu}(2)}}{\sqrt{\bar{\mu}(1)}} f_1 & 1 - \alpha f_0 \\
\vdots & & & \ddots
\end{pmatrix} .
$$

Here, the $\{f_i\}$ are the coefficients of the FIR filter F. Since, in practice, the length M_F of this filter is usually much smaller than the length of the regression vector $\mathbf{u}_i$, the energy of the input sequence $\mathbf{u}_i$ does not change very rapidly over the filter length M_F, such that

$$
\bar{\mu}(i) \approx \ldots \approx \bar{\mu}(i - M_F) .
$$

In this case, $\mathbf{P}_N$ becomes

$$
\mathbf{P}_N \approx \mathbf{I} - \alpha \mathbf{F}_N , \tag{20.26}
$$

where $\mathbf{F}_N$ is the lower triangular Toeplitz matrix that describes the convolution of the filter F on an input sequence. This is generally a banded matrix since $M_F \ll M$, as shown below for the special case of $M_F = 3$,

$$
\mathbf{F}_N = \begin{bmatrix}
f_0 & & & \\
f_1 & f_0 & & \\
f_2 & f_1 & f_0 & \\
& f_2 & f_1 & f_0 \\
& & \ddots & \ddots & \ddots
\end{bmatrix} .
$$

In this case, the strict contractivity of $(\mathbf{I} - \alpha \mathbf{F}_N)$ can be guaranteed by choosing the step-size parameter α such that

$$
\max_{\Omega} \left| 1 - \alpha F(e^{j\Omega}) \right| < 1 , \tag{20.27}
$$

where $F(z)$ is the transfer function of the error filter. For better convergence performance, we may choose α by solving the min-max problem

$$
\min_{\alpha} \max_{\Omega} \left| 1 - \alpha F(e^{j\Omega}) \right| . \tag{20.28}
$$

If the resulting minimum is less than one, then the corresponding optimum value of α will result in faster convergence, and it will also guarantee the robustness of the scheme.

We now illustrate these concepts via a simulation example. The error-path filter for this example is

$$
F(z) = 1 - 1.2 \, z^{-1} + 0.72 \, z^{-2} .
$$

We use an FIR filter adapted by the algorithm in (20.25), where the input signal to the adaptive filter consists of a single sinusoid of frequency $\Omega_0 = 1.2/\pi$. In this case, if we assume that the *a priori* error signal is dominated by the frequency component Ω_0, we can solve for the optimum α via the simpler expression [cf. (20.28)] $\min_{\alpha} \left| 1 - \alpha F\left(e^{j\Omega_o}\right) \right|$. The resulting optimum value of α is

$$
\alpha_{opt} = \text{Real} \left\{ \frac{1}{F(e^{-j\Omega_0})} \right\} .
$$

This step size provides the fastest convergence speed. In addition, the stability limits for α can be shown to be $0 < \alpha < 2\alpha_{opt}$ using a similar procedure.

Figure 20.9 shows three convergence curves of the average squared error $\mathrm{Av}[|e(i)|^2] = \frac{1}{50}\sum_{j=1}^{50}|e_j(i)|^2$, as determined from 50 simulation runs of the projection filtered-error LMS algorithm for the choices $\alpha = 0.085, \alpha = 0.15$ and $\alpha = 0.18$, respectively. In this case, we have generated an input sequence of the form $u(i) = \sin[1.2i + \phi]$ for each simulation run, where ϕ is uniformly chosen from the interval $[-\pi, \pi]$ to obtain smoother learning curves after averaging, and the $M = 10$ coefficients of the unknown system were all set to unity. Moreover, the additive noise $v(i)$ corrupting the signal $d(i)$ is uncorrelated Gaussian-distributed with a level that is -40dB below that of the input signal power. The optimal step-size α_{opt} in this case can be calculated to be $\alpha_{opt} = 0.085$ and the stability bounds for the system are $0 < \alpha < 0.17$. As expected, choosing $\alpha = 0.085$ provides the fastest convergence speed for this situation. We also see that for values of α greater than $2\alpha_{opt}$, the error of the system diverges.

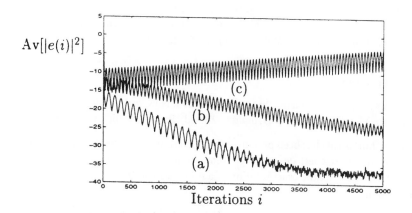

FIGURE 20.9 Convergence behavior for FELMS algorithm with sinusoidal input sequence and various step-sizes $\alpha = 0.085$(a), 0.15 (b), 0.18(c). © IEEE 1996. (*Source:* Rupp, M. and Sayed, A.H., A time domain feedback analysis of filtered-error adaptive gradient algorithms, *IEEE Trans. Signal Process.* 44(6): 1428–1439, June 1996. With permission.)

20.11 References and Concluding Remarks

The intent of this chapter was to highlight certain robustness and convergence issues that arise in the study of adaptive algorithms in the presence of uncertain data. More details, extensions, and related discussions can be found in several of the references indicated in this section. The references are not intended to be complete but rather indicative of the work in the different areas. More complete lists can be found in several of the textbooks mentioned herein.

Detailed discussions on the different forms of adaptive algorithms and their potential applications can be found in:

[1] Haykin, S., *Adaptive Filter Theory*, 3rd ed., Prentice-Hall, Englewood Cliffs, NJ, 1996.
[2] Proakis, J.G., Rader, C.M., Ling, F., and Nikias, C.L., *Advanced Digital Signal Processing*, Macmillan Publishing, New York, 1992.
[3] Widrow, B. and Stearns, S.D., *Adaptive Signal Processing*, Prentice-Hall, Englewood Cliffs, NJ, 1985.
[4] Sayed, A.H. and Kailath, T., A state-space approach to adaptive RLS filtering, *IEEE Signal Processing Magazine*, 11(3), 18–60, July 1994.

The fundamentals of robust or H^∞ design, both in filtering and control applications, can be found in the following references:

[5] Green, M. and Limebeer, D.J.N., *Linear Robust Control*, Prentice-Hall, Englewood Cliffs, NJ, 1995.

[6] Zhou, K., Doyle, J.C., and Glover, K., *Robust and Optimal Control*, Prentice Hall, Englewood Cliffs, NJ, 1996.

[7] Khargonekar, P.P. and Nagpal, K.M., Filtering and smoothing in an $H^\infty-$ setting, *IEEE Trans. on Automatic Control*, 36, 151–166, 1991.

[8] Shaked, U. and Theodor, Y., $H^\infty-$optimal estimation: A tutorial, *Proc. IEEE Conf. Decision and Control*, 2278–2286, Tucson, AZ, Dec. 1992.

[9] Hassibi, B., Sayed, A.H., and Kailath, T., Linear estimation in Krein spaces — Part I: Theory, *IEEE Trans. Automatic Control*, 41(1), 18–33, Jan. 1996.

[10] Hassibi, B., Sayed, A.H., and Kailath, T., Linear estimation in Krein spaces — Part II: Applications, *IEEE Trans. Automatic Control*, 41(1), 34–49, Jan. 1996.

The small gain analysis is a standard tool in linear and nonlinear system theory. More advanced and detailed treatments can be found in:

[11] Khalil, H.K., *Nonlinear Systems*, 2nd ed., Macmillan, New York, 1996.

[12] Vidyasagar, M., *Nonlinear Systems Analysis*, 2nd ed., Prentice Hall, Englewood Cliffs, NJ, 1993.

The LMS algorithm usually has been presented in the literature as an instantaneous-gradient-based approximation for the steepest descent algorithm. Its robustness properties, and the interesting observation that it is in fact the exact solution of a min-max (or H^∞) optimization problem, have been first noted in

[13] Hassibi, B., Sayed, A.H., and Kailath, T., H^∞ optimality of the LMS algorithm, *IEEE Trans. Signal Processing*, 44(2), 267–280, Feb. 1996. See also *Proc. CDC*, 1, 74–79, 1993.

Also, more details on the example comparing the performance of LMS and RLS can be found in the above reference. Extensions of the discussion to the backpropagation algorithm for neural network training, and other related results in adaptive filtering and H_∞ estimation and control can be found in

[14] Hassibi, B., Sayed, A.H., and Kailath, T., LMS and backpropagation are minimax filters, in *Neural Computation and Learning*, Roychowdhurys, V., Siu, K.Y., and Orlitsky, A., Eds., Kluwer Academic Publishers, 1994, 425–447.

[15] Hassibi, B., *Indefinite Metric Spaces in Estimation, Control, and Adaptive Filtering*, Ph.D. Dissertation, Stanford University, August 1996.

[16] Hassibi, B., Sayed, A.H., and Kailath, T., *Indefinite Quadratic Estimation and Control: A Unified Approach to H_2 and H_∞ Theories*, to be published by SIAM, Studies in Applied Mathematics Series, 1997.

Extensions of the feedback analysis to Perceptron training in neural network can be found in

[17] Rupp, M. and Sayed, A.H., Supervised learning of perceptron and output feedback dynamic networks: a feedback analysis via the small gain theorem, *IEEE Trans. Neural Networks*, 8(3), 612–622, May 1997.

A Cauchy-Schwarz argument that further highlights the robustness property of adaptive gradient algorithms, along with other local energy bounds, are given in

[18] Sayed, A.H. and Rupp, M., Error energy bounds for adaptive gradient algorithms, *IEEE Trans. Signal Processing*, 44(8), 1982–1989, Aug. 1996.

[19] Sayed, A.H. and Kailath, T., A state-space approach to adaptive RLS filtering, *IEEE Signal Processing Magazine*, 11(3), 18–60, July 1994.

The time-domain feedback and small gain analyses of adaptive filters, along with extensions to nonlinear settings and connections with Gauss-Newton updates and H^∞ filters, are discussed in

[20] Rupp, M. and Sayed, A.H., A time-domain feedback analysis of filtered-error adaptive gradient algorithms, *IEEE Trans. on Signal Processing*, 44(6), 1428–1439, June 1996.
[21] Rupp, M. and Sayed, A.H., Robustness of Gauss-Newton recursive methods: a deterministic feedback analysis, *Signal Processing*, 50(3), 165–188, June 1996.
[22] Sayed, A.H. and Rupp, M., An l_2−stable feedback structure for nonlinear adaptive filtering and identification, *Automatica*, 33(1), 13–30, 1997.

Discussions of the singular value decomposition and its properties can be found in

[23] Golub, G.H. and Van Loan, C.F., *Matrix Computations*, 3rd ed., The Johns Hopkins University Press, Baltimore, MD, 1996.

21

Recursive Least-Squares Adaptive Filters

Ali H. Sayed
University of California, Los Angeles

Thomas Kailath
Stanford University

The central problem in estimation is to recover, to good accuracy, a set of unobservable parameters from corrupted data. Several optimization criteria have been used for estimation purposes over the years, but the most important, at least in the sense of having had the most applications, are criteria that are based on quadratic cost functions. The most striking among these is the linear least-squares criterion, which was perhaps first developed by Gauss (ca. 1795) in his work on celestial mechanics. Since then, it has enjoyed widespread popularity in many diverse areas as a result of its attractive computational and statistical properties. Among these attractive properties, the most notable are the facts that least-squares solutions:

0-8493-8572-5/98/$0.00+$.50
© 1998 by CRC Press LLC

- can be explicitly evaluated in closed forms;
- can be recursively updated as more input data is made available, and
- are maximum likelihood estimators in the presence of Gaussian measurement noise.

The aim of this chapter is to provide an overview of adaptive filtering algorithms that result when the least-squares criterion is adopted. Over the last several years, a wide variety of algorithms in this class has been derived. They all basically fall into the following main groups (or variations thereof): recursive least-squares (RLS) algorithms and the corresponding fast versions (known as FTF and FAEST), QR and inverse QR algorithms, least-squares lattice (LSL), and QR decomposition-based least-squares lattice (QRD-LSL) algorithms.

Table 21.1 lists these different variants and classifies them into order-recursive and fixed-order algorithms. The acronyms and terminology are not important at this stage and will be explained as the discussion proceeds. Also, the notation $O(M)$ is used to indicate that each iteration of an algorithm requires of the order of M floating point operations (additions and multiplications). In this sense, some algorithms are fast (requiring only $O(M)$), while others are slow (requiring $O(M^2)$). The value of M is the filter order that will be introduced in due time.

TABLE 21.1   Most Common RLS Adaptive Schemes

Adaptive Algorithm	Order Recursive	Fixed Order	Cost per Iteration
RLS		x	$O(M^2)$
QR and Inverse QR		x	$O(M^2)$
FTF, FAEST		x	$O(M)$
LSL	x		$O(M)$
QRD-LSL	x		$O(M)$

It is practically impossible to list here all the relevant references and all the major contributors to the rich field of adaptive RLS filtering. The reader is referred to some of the textbooks listed at the end of this chapter for more comprehensive treatments and bibliographies.

Here we wish to stress that, apart from introducing the reader to the fundamentals of RLS filtering, one of our goals in this exposition is to present the different versions of the RLS algorithm in computationally convenient so-called array forms. In these forms, an algorithm is described as a sequence of elementary operations on arrays of numbers. Usually, a prearray of numbers has to be triangularized by a rotation, or a sequence of elementary rotations, in order to yield a postarray of numbers. The quantities needed to form the next prearray can then be read off from the entries of the postarray, and the procedure can be repeated. The explicit forms of the rotation matrices are not needed in most cases.

Such array descriptions are more truly algorithms in the sense that they operate on sets of numbers and provide other sets of numbers, with no explicit equations involved. The rotations themselves can be implemented in a variety of well-known ways: as a sequence of elementary circular or hyperbolic rotations, in square-root- and/or division-free forms, as Householder transformations, etc. These may differ in computational complexity, numerical behavior, and ease of hardware (VLSI) implementation. But, if preferred, explicit expressions for the rotation matrices can also be written down, thus leading to explicit sets of equations in contrast to the array forms.

For this reason, and although the different RLS algorithms that we consider here have already been derived in many different ways in earlier places in the literature, the derivation and presentation in this chapter are intended to provide an alternative unifying exposition that we hope will help a reader get a deeper appreciation of this class of adaptive algorithms.

Notation

We use small boldface letters to denote column vectors (e.g., $\mathbf{w}$) and capital boldface letters to denote matrices (e.g., $\mathbf{A}$). The symbol $\mathbf{I}_n$ denotes the identity matrix of size $n \times n$, while $\mathbf{0}$ denotes a zero column. The symbol ^{T} denotes transposition. This chapter deals with real-valued data. The case of complex-valued data is essentially identical and is treated in many of the references at the end of this chapter.

Square-Root Factors

A symmetric positive-definite matrix $\mathbf{A}$ is one that satisfies $\mathbf{A} = \mathbf{A}^T$ and $\mathbf{x}^T \mathbf{A} \mathbf{x} > 0$ for all nonzero column vectors $\mathbf{x}$. Any such matrix admits a factorization (also known as eigen-decomposition) of the form $\mathbf{A} = \mathbf{U} \Sigma \mathbf{U}^T$, where $\mathbf{U}$ is an orthogonal matrix, namely a square matrix that satisfies $\mathbf{U} \mathbf{U}^T = \mathbf{U}^T \mathbf{U} = \mathbf{I}$, and Σ is a diagonal matrix with real positive entries. In particular, note that $\mathbf{A} \mathbf{U} = \mathbf{U} \Sigma$, which shows that the columns of $\mathbf{U}$ are the right eigenvectors of $\mathbf{A}$ and the entries of Σ are the corresponding eigenvalues.

Note also that we can write $\mathbf{A} = \mathbf{U} \Sigma^{1/2} (\Sigma^{1/2})^T \mathbf{U}^T$, where $\Sigma^{1/2}$ is a diagonal matrix whose entries are (positive) square-roots of the diagonal entries of Σ. Since $\Sigma^{1/2}$ is diagonal, $(\Sigma^{1/2})^T = \Sigma^{1/2}$. If we introduce the matrix notation $\mathbf{A}^{1/2} = \mathbf{U} \Sigma^{1/2}$, then we can alternatively write $\mathbf{A} = (\mathbf{A}^{1/2})(\mathbf{A}^{1/2})^T$. This can be regarded as a square-root factorization of the positive-definite matrix $\mathbf{A}$. Here, the notation $\mathbf{A}^{1/2}$ is used to denote one such square-root factor, namely the one constructed from the eigen-decomposition of $\mathbf{A}$.

Note, however, that square-root factors are not unique. For example, we may multiply the diagonal entries of $\Sigma^{1/2}$ by $\pm 1's$ and obtain a new square-root factor for Σ and, consequently, a new square-root factor for $\mathbf{A}$.

Also, given any square-root factor $\mathbf{A}^{1/2}$, and any orthogonal matrix Θ (satisfying $\Theta \Theta^T = \mathbf{I}$) we can define a new square-root factor for $\mathbf{A}$ as $\mathbf{A}^{1/2} \Theta$ since

$$(\mathbf{A}^{1/2} \Theta)(\mathbf{A}^{1/2} \Theta)^T = \mathbf{A}^{1/2} (\Theta \Theta^T)(\mathbf{A}^{1/2})^T = \mathbf{A} .$$

Hence, square factors are highly nonunique. We shall employ the notation $\mathbf{A}^{1/2}$ to denote any such square-root factor. They can be made unique, e.g., by insisting that the factors be symmetric or that they be triangular (with positive diagonal elements). In most applications, the triangular form is preferred. For convenience, we also write

$$\left(\mathbf{A}^{1/2}\right)^T = \mathbf{A}^{T/2}, \quad \left(\mathbf{A}^{1/2}\right)^{-1} = \mathbf{A}^{-1/2}, \quad \left(\mathbf{A}^{-1/2}\right)^T = \mathbf{A}^{-T/2} .$$

Thus, note the expressions $\mathbf{A} = \mathbf{A}^{1/2} \mathbf{A}^{T/2}$ and $\mathbf{A}^{-1} = \mathbf{A}^{-T/2} \mathbf{A}^{-1/2}$.

21.1 Array Algorithms

The array form is so important that it will be worthwhile to explain its generic form here.

An array algorithm is described via rotation operations on a prearray of numbers, chosen to obtain a certain zero pattern in a postarray. Schematically, we write

$$
\begin{bmatrix}
x & x & x & x \\
x & x & x & x \\
x & x & x & x \\
x & x & x & x
\end{bmatrix}
\Theta =
\begin{bmatrix}
x & 0 & 0 & 0 \\
x & x & 0 & 0 \\
x & x & x & 0 \\
x & x & x & x
\end{bmatrix},
$$

where Θ is any rotation matrix that triangularizes the prearray. In general, Θ is required to be a J-orthogonal matrix in the sense that it should satisfy the normalization $\Theta J \Theta^T = J$, where J is a given signature matrix with $\pm 1's$ on the diagonal and zeros elsewhere. The orthogonal case corresponds to $J = I$ since then $\Theta \Theta^T = I$.

A rotation Θ that transforms a prearray to triangular form can be achieved in a variety of ways: by using a sequence of elementary Givens and hyperbolic rotations, Householder transformations, or square-root-free versions of such rotations. Here we only explain the elementary forms. The other choices are discussed in some of the references at the end of this chapter.

21.1.1 Elementary Circular Rotations

An elementary 2×2 orthogonal rotation Θ (also known as Givens or circular rotation) takes a row vector $\begin{bmatrix} a & b \end{bmatrix}$ and rotates it to lie along the basis vector $\begin{bmatrix} 1 & 0 \end{bmatrix}$. More precisely, it performs the transformation

$$\begin{bmatrix} a & b \end{bmatrix} \Theta = \begin{bmatrix} \pm\sqrt{|a|^2 + |b|^2} & 0 \end{bmatrix}. \tag{21.1}$$

The quantity $\pm\sqrt{|a|^2 + |b|^2}$ that appears on the right-hand side is consistent with the fact that the pre-array, $\begin{bmatrix} a & b \end{bmatrix}$, and the postarray, $\begin{bmatrix} \pm\sqrt{|a|^2 + |b|^2} & 0 \end{bmatrix}$, must have equal Euclidean norms (since an orthogonal transformation preserves the Euclidean norm of a vector).

An expression for Θ is given by

$$\Theta = \frac{1}{\sqrt{1+\rho^2}} \begin{bmatrix} 1 & -\rho \\ \rho & 1 \end{bmatrix}, \quad \rho = \frac{b}{a}, \quad a \neq 0. \tag{21.2}$$

In the trivial case $a = 0$ we simply choose Θ as the permutation matrix,

$$\Theta = \begin{bmatrix} 0 & 1 \\ 1 & 0 \end{bmatrix}.$$

The orthogonal rotation (21.2) can also be expressed in the alternative form:

$$\Theta = \begin{bmatrix} c & -s \\ s & c \end{bmatrix},$$

where the so-called cosine and sine parameters, c and s, respectively, are defined by

$$c = \frac{1}{\sqrt{1+\rho^2}}, \quad s = \frac{\rho}{\sqrt{1+\rho^2}}.$$

The name *circular rotation* for Θ is justified by its effect on a vector; it rotates the vector along the *circle* of equation $x^2 + y^2 = |a|^2 + |b|^2$, by an angle θ that is determined by the inverse of the above cosine and/or sine parameters, $\theta = \tan^{-1} \rho$, in order to align it with the basis vector $\begin{bmatrix} 1 & 0 \end{bmatrix}$. The trivial case $a = 0$ corresponds to a 90 degrees rotation in an appropriate clockwise (if $b \geq 0$) or counterclockwise (if $b < 0$) direction.

21.1.2 Elementary Hyperbolic Rotations

An elementary 2×2 hyperbolic rotation Θ takes a row vector $\begin{bmatrix} a & b \end{bmatrix}$ and rotates it to lie either along the basis vector $\begin{bmatrix} 1 & 0 \end{bmatrix}$ (if $|a| > |b|$) or along the basis vector $\begin{bmatrix} 0 & 1 \end{bmatrix}$ (if $|a| < |b|$). More precisely, it performs either of the transformations

$$\begin{bmatrix} a & b \end{bmatrix} \Theta = \begin{bmatrix} \pm\sqrt{|a|^2 - |b|^2} & 0 \end{bmatrix} \quad \text{if } |a| > |b|, \tag{21.3}$$

$$\begin{bmatrix} a & b \end{bmatrix} \Theta = \begin{bmatrix} 0 & \pm\sqrt{|b|^2 - |a|^2} \end{bmatrix} \quad \text{if } |a| < |b|. \tag{21.4}$$

The quantity $\sqrt{\pm(|a|^2 - |b|^2)}$ that appears on the right-hand side of the above expressions is consistent with the fact that the prearray, $\begin{bmatrix} a & b \end{bmatrix}$, and the postarrays must have equal hyperbolic "norms." By the hyperbolic "norm" of a row vector $\mathbf{x}^T$ we mean the indefinite quantity $\mathbf{x}^T \mathbf{J} \mathbf{x}$, which can be positive or negative. Here,

$$\mathbf{J} = \begin{bmatrix} 1 & 0 \\ 0 & -1 \end{bmatrix} = (1 \oplus -1) .$$

An expression for a hyperbolic rotation Θ that achieves (21.3) or (21.4) is given by

$$\Theta = \frac{1}{\sqrt{1 - \rho^2}} \begin{bmatrix} 1 & -\rho \\ -\rho & 1 \end{bmatrix} , \tag{21.5}$$

where

$$\rho = \begin{cases} \frac{b}{a} & \text{when } a \neq 0 \text{ and } |a| > |b| \\ \\ \frac{a}{b} & \text{when } b \neq 0 \text{ and } |b| > |a| \end{cases}$$

The hyperbolic rotation (21.5) can also be expressed in the alternative form:

$$\Theta = \begin{bmatrix} ch & -sh \\ -sh & ch \end{bmatrix} ,$$

where the so-called hyperbolic cosine and sine parameters, ch and sh, respectively, are defined by

$$ch = \frac{1}{\sqrt{1 - \rho^2}} \quad , \quad sh = \frac{\rho}{\sqrt{1 - \rho^2}} .$$

The name *hyperbolic rotation* for Θ is again justified by its effect on a vector; it rotates the original vector along the *hyperbola* of equation $x^2 - y^2 = |a|^2 - |b|^2$, by an angle θ determined by the inverse of the above hyperbolic cosine and/or sine parameters, $\theta = \tanh^{-1}[\rho]$, in order to align it with the appropriate basis vector. Note also that the special case $|a| = |b|$ corresponds to a row vector $\begin{bmatrix} a & b \end{bmatrix}$ with zero hyperbolic norm since $|a|^2 - |b|^2 = 0$. It is then easy to see that there does not exist a hyperbolic rotation that will rotate the vector to lie along the direction of one basis vector or the other.

21.1.3 Square-Root-Free and Householder Transformations

We remark that the above expressions for the circular and hyperbolic rotations involve square-root operations. In many situations, it may be desirable to avoid the computation of square-roots because it is usually expensive. For this and other reasons, square-root- and division-free versions of the above elementary rotations have been developed and constitute an attractive alternative.

Therefore one could use orthogonal or $\mathbf{J}$−orthogonal Householder reflections (for given $\mathbf{J}$) to simultaneously annihilate several entries in a row, e.g., to transform $\begin{bmatrix} x & x & x & x \end{bmatrix}$ directly to the form $\begin{bmatrix} x' & 0 & 0 & 0 \end{bmatrix}$. Combinations of rotations and reflections can also be used.

We omit the details here but the idea is clear. There are many different ways in which a prearray of numbers can be rotated into a postarray of numbers.

21.1.4 A Numerical Example

Assume we are given a 2×3 prearray $\mathbf{A}$,

$$\mathbf{A} = \begin{bmatrix} 0.875 & 0.15 & 1.0 \\ 0.675 & 0.35 & 0.5 \end{bmatrix} , \tag{21.6}$$

and wish to triangularize it via a sequence of elementary circular rotations, i.e., reduce $\mathbf{A}$ to the form

$$\mathbf{A\Theta} = \begin{bmatrix} x & 0 & 0 \\ x & x & 0 \end{bmatrix}. \tag{21.7}$$

This can be obtained, among several different possibilities, as follows. We start by annihilating the $(1,3)$ entry of the prearray (21.6) by pivoting with its $(1,1)$ entry. According to expression (21.2), the orthogonal transformation Θ_1 that achieves this result is given by

$$\Theta_1 = \frac{1}{\sqrt{1 + \rho_1^2}} \begin{bmatrix} 1 & -\rho_1 \\ \rho_1 & 1 \end{bmatrix} = \begin{bmatrix} 0.6585 & -0.7526 \\ 0.7526 & 0.6585 \end{bmatrix}, \quad \rho_1 = \frac{1}{0.875}.$$

Applying Θ_1 to the prearray (21.6) leads to (recall that we are only operating on the first and third columns, leaving the second column unchanged):

$$\begin{bmatrix} 0.875 & 0.15 & 1 \\ 0.675 & 0.35 & 0.5 \end{bmatrix} \begin{bmatrix} 0.6585 & 0 & -0.7526 \\ 0 & 1 & 0 \\ 0.7526 & 0 & 0.6585 \end{bmatrix} = \begin{bmatrix} 1.3288 & 0.1500 & 0.0000 \\ 0.8208 & 0.3500 & -0.1788 \end{bmatrix}. \tag{21.8}$$

We now annihilate the $(1,2)$ entry of the resulting matrix in the above equation by pivoting with its $(1,1)$ entry. This requires that we choose

$$\Theta_2 = \frac{1}{\sqrt{1 + \rho_2^2}} \begin{bmatrix} 1 & -\rho_2 \\ \rho_2 & 1 \end{bmatrix} = \begin{bmatrix} 0.9937 & -0.1122 \\ 0.1122 & 0.9937 \end{bmatrix}, \quad \rho_2 = \frac{0.1500}{1.3288}. \tag{21.9}$$

Applying Θ_2 to the matrix on the right-hand side of (21.8) leads to (now we leave the third column unchanged)

$$\begin{bmatrix} 1.3288 & 0.1500 & 0.0000 \\ 0.8208 & 0.3500 & 0.1788 \end{bmatrix} \begin{bmatrix} 0.9937 & -0.1122 & 0 \\ 0.1122 & 0.9937 & 0 \\ 0 & 0 & 1 \end{bmatrix} = \begin{bmatrix} 1.3373 & 0.0000 & 0.0000 \\ 0.8549 & 0.2557 & 0.1788 \end{bmatrix}.$$

$$\tag{21.10}$$

We finally annihilate the $(2,3)$ entry of the resulting matrix in (21.10) by pivoting with its $(2,2)$ entry. In principle this requires that we choose

$$\Theta_3 = \frac{1}{\sqrt{1 + \rho_3^2}} \begin{bmatrix} 1 & -\rho_3 \\ \rho_3 & 1 \end{bmatrix} = \begin{bmatrix} 0.8195 & 0.5731 \\ -0.5731 & 0.8195 \end{bmatrix}, \quad \rho_3 = \frac{0.1788}{-0.2557}, \tag{21.11}$$

and apply it to the matrix on the right-hand side of (21.10), which would then lead to

$$\begin{bmatrix} 1.3373 & 0.0000 & 0.0000 \\ 0.8549 & -0.2557 & 0.1788 \end{bmatrix} \begin{bmatrix} 1 & 0 & 0 \\ 0 & 0.8195 & 0.5731 \\ 0 & -0.5731 & 0.8195 \end{bmatrix} = \begin{bmatrix} 1.3373 & 0.0000 & 0.0000 \\ 0.8549 & -0.3120 & 0.0000 \end{bmatrix}.$$

$$\tag{21.12}$$

Alternatively, this last step could have been implemented without explicitly forming Θ_3. We simply replace the row vector $\begin{bmatrix} -0.2557 & 0.1788 \end{bmatrix}$, which contains the $(2,2)$ and $(2,3)$ entries of the prearray in (21.12), by the row vector $\begin{bmatrix} \pm\sqrt{(-0.2557)^2 + (0.1788)^2} & 0.0000 \end{bmatrix}$, which is equal to $\begin{bmatrix} \pm 0.3120 & 0.0000 \end{bmatrix}$. We choose the positive sign in order to conform with our earlier convention that the diagonal entries of triangular square-root factors are taken to be positive. The resulting postarray is therefore

$$\begin{bmatrix} 1.3373 & 0.0000 & 0.0000 \\ 0.8549 & 0.3120 & 0.0000 \end{bmatrix}. \tag{21.13}$$

We have exhibited a sequence of elementary orthogonal transformations that triangularizes the prearray of numbers (21.6). The combined effect of the sequence of transformations $\{\Theta_1, \Theta_2, \Theta_3\}$ corresponds to the orthogonal rotation Θ required in (21.7). However, note that we do not need to know or to form $\Theta = \Theta_1 \Theta_2 \Theta_3$.

It will become clear throughout our discussion that the different adaptive RLS schemes can be described in array forms, where the necessary operations are elementary rotations as described above. Such array descriptions lend themselves rather directly to parallelizable and modular implementations. Indeed, once a rotation matrix is chosen, then all the rows of the prearray undergo the same rotation transformation and can thus be processed in parallel. Returning to the above example, where we started with the prearray **A**, we see that once the first rotation is determined, both rows of **A** are then transformed by it, and can thus be processed in parallel, and by the same functional (rotation) block, to obtain the desired postarray. The same remark holds for prearrays with multiple rows.

21.2 The Least-Squares Problem

Now that we have explained the generic form of an array algorithm, we return to the main topic of this chapter and formulate the least-squares problem and its regularized version. Once this is done, we shall then proceed to describe the different variants of the recursive least-squares solution in compact array forms.

Let **w** denote a column vector of n unknown parameters that we wish to estimate, and consider a set of $(N + 1)$ noisy measurements $\{d(i)\}$ that are assumed to be linearly related to **w** via the additive noise model

$$d(j) = \mathbf{u}_j^T \mathbf{w} + v(j) ,$$

where the $\{\mathbf{u}_j\}$ are given column vectors. The $(N+1)$ measurements can be grouped together into a single matrix expression:

$$\underbrace{\begin{bmatrix} d(0) \\ d(1) \\ \vdots \\ d(N) \end{bmatrix}}_{\mathbf{d}} = \underbrace{\begin{bmatrix} \mathbf{u}_0^T \\ \mathbf{u}_1^T \\ \vdots \\ \mathbf{u}_N^T \end{bmatrix}}_{\mathbf{A}} \mathbf{w} + \underbrace{\begin{bmatrix} v(0) \\ v(1) \\ \vdots \\ v(N) \end{bmatrix}}_{\mathbf{v}} ,$$

or, more compactly, $\mathbf{d} = \mathbf{A}\mathbf{w} + \mathbf{v}$. Because of the noise component **v**, the observed vector **d** does not lie in the column space of the matrix **A**. The objective of the least-squares problem is to determine the vector in the column space of **A** that is closest to **d** in the least-squares sense.

More specifically, any vector in the range space of **A** can be expressed as a linear combination of its columns, say $\mathbf{A}\hat{\mathbf{w}}$ for some $\hat{\mathbf{w}}$. It is therefore desired to determine the particular $\hat{\mathbf{w}}$ that minimizes the distance between **d** and $\mathbf{A}\hat{\mathbf{w}}$,

$$\min_{\mathbf{w}} \|\mathbf{d} - \mathbf{A}\mathbf{w}\|^2 . \tag{21.14}$$

The resulting $\hat{\mathbf{w}}$ is called the least-squares solution and it provides an estimate for the unknown **w**. The term $\mathbf{A}\hat{\mathbf{w}}$ is called the linear least-squares estimate (l.l.s.e.) of **d**.

The solution to (21.14) *always* exists and it follows from a simple geometric argument. The orthogonal projection of **d** onto the column span of **A** yields a vector $\hat{\mathbf{d}}$ that is the closest to **d** in the least-squares sense. This is because the resulting error vector $(\mathbf{d} - \hat{\mathbf{d}})$ will be orthogonal to the column span of **A**.

In other words, the closest element $\hat{\mathbf{d}}$ to **d** must satisfy the orthogonality condition

$$\mathbf{A}^T (\mathbf{d} - \hat{\mathbf{d}}) = \mathbf{0}.$$

That is, and replacing $\hat{\mathbf{d}}$ by $\mathbf{A}\hat{\mathbf{w}}$, the corresponding $\hat{\mathbf{w}}$ must satisfy

$$\mathbf{A}^T \mathbf{A}\hat{\mathbf{w}} = \mathbf{A}^T \mathbf{d} .$$

These equations always have a solution $\hat{\mathbf{w}}$. But while a solution $\hat{\mathbf{w}}$ may or may not be unique (depending on whether $\mathbf{A}$ is or is not full rank), the resulting estimate $\hat{\mathbf{d}} = \mathbf{A}\hat{\mathbf{w}}$ is always unique no matter which solution $\hat{\mathbf{w}}$ we pick. This is obvious from the geometric argument because the orthogonal projection of $\mathbf{d}$ onto the span of $\mathbf{A}$ is unique.

If $\mathbf{A}$ is assumed to be a full rank matrix then $\mathbf{A}^T\mathbf{A}$ is invertible and we can write

$$\hat{\mathbf{w}} = (\mathbf{A}^T\mathbf{A})^{-1}\mathbf{A}^T\mathbf{d} \,. \tag{21.15}$$

21.2.1 Geometric Interpretation

The quantity $\mathbf{A}\hat{\mathbf{w}}$ provides an estimate for $\mathbf{d}$; it corresponds to the vector in the column span of $\mathbf{A}$ that is closest in Euclidean norm to the given $\mathbf{d}$. In other words,

$$\hat{\mathbf{d}} = \mathbf{A}\left(\mathbf{A}^T\mathbf{A}\right)^{-1}\mathbf{A}^T \cdot \mathbf{d} \triangleq \mathcal{P}_A \cdot \mathbf{d} \,,$$

where $\mathcal{P}_A$ denotes the projector onto the range space of $\mathbf{A}$. Figure 21.1 is a schematic representation of this geometric construction, where $\mathcal{R}(\mathbf{A})$ denotes the column span of $\mathbf{A}$.

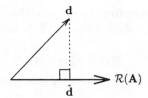

FIGURE 21.1   Geometric interpretation of the least-squares solution.

21.2.2 Statistical Interpretation

The least-squares solution also admits an important statistical interpretation. For this purpose, assume that the noise vector $\mathbf{v}$ is a realization of a vector-valued random variable that is normally distributed with zero mean and identity covariance matrix, written $\mathbf{v} \sim N[\mathbf{0}, \mathbf{I}]$. In this case, the observation vector $\mathbf{d}$ will be a realization of a vector-valued random variable that is also normally distributed with mean $\mathbf{A}\mathbf{w}$ and covariance matrix equal to the identity $\mathbf{I}$. This is because the random vectors are related via the additive model $\mathbf{d} = \mathbf{A}\mathbf{w} + \mathbf{v}$. The probability density function of the observation process $\mathbf{d}$ is then given by

$$\frac{1}{\sqrt{(2\pi)^{(N+1)}}} \cdot \exp\left\{-\frac{1}{2}(\mathbf{d} - \mathbf{A}\mathbf{w})^T(\mathbf{d} - \mathbf{A}\mathbf{w})\right\} \,. \tag{21.16}$$

It follows, in this case, that the least-squares estimator $\hat{\mathbf{w}}$ is also the maximum likelihood (ML) estimator because it maximizes the probability density function over $\mathbf{w}$, given an observation vector $\mathbf{d}$.

21.3 The Regularized Least-Squares Problem

A more general optimization criterion that is often used instead of (21.14) is the following

$$\min_{\mathbf{w}}\left[(\mathbf{w} - \bar{\mathbf{w}})^T \Pi_0^{-1}(\mathbf{w} - \bar{\mathbf{w}}) + \|\mathbf{d} - \mathbf{A}\mathbf{w}\|^2\right] \,. \tag{21.17}$$

This is still a quadratic cost function in the unknown vector $\mathbf{w}$, but it includes the additional term

$$(\mathbf{w} - \bar{\mathbf{w}})^T \, \Pi_0^{-1} (\mathbf{w} - \bar{\mathbf{w}}) \, ,$$

where Π_0 is a given positive-definite (weighting) matrix and $\bar{\mathbf{w}}$ is also a given vector. Choosing $\Pi_0 = \infty \cdot \mathbf{I}$ leads us back to the original expression (21.14).

A motivation for (21.17) is that the freedom in choosing Π_0 allows us to incorporate additional *a priori* knowledge into the statement of the problem. Indeed, different choices for Π_0 would indicate how confident we are about the closeness of the unknown $\mathbf{w}$ to the given vector $\bar{\mathbf{w}}$.

Assume, for example, that we set $\Pi_0 = \epsilon \cdot \mathbf{I}$, where ϵ is a very small positive number. Then the first term in the cost function (21.17) becomes dominant. It is then not hard to see that, in this case, the cost will be minimized if we choose the estimate $\hat{\mathbf{w}}$ close enough to $\bar{\mathbf{w}}$ in order to annihilate the effect of the first term. In simple words, a "small" Π_0 reflects a high confidence that $\bar{\mathbf{w}}$ is a good and close enough guess for $\mathbf{w}$. On the other hand, a "large" Π_0 indicates a high degree of uncertainty in the initial guess $\bar{\mathbf{w}}$.

One way of solving the regularized optimization problem (21.17) is to reduce it to the standard least-squares problem (21.14). This can be achieved by introducing the change of variables $\mathbf{w}' = \mathbf{w} - \bar{\mathbf{w}}$ and $\mathbf{d}' = \mathbf{d} - \mathbf{A}\bar{\mathbf{w}}$. Then (21.17) becomes

$$\min_{\mathbf{w}'} \left[(\mathbf{w}')^T \Pi_0^{-1} \mathbf{w}' + \left\| \mathbf{d}' - \mathbf{A}\mathbf{w}' \right\|^2 \right] \, ,$$

which can be further rewritten in the equivalent form

$$\min_{\mathbf{w}'} \left\| \begin{bmatrix} \mathbf{0} \\ \mathbf{d}' \end{bmatrix} - \begin{bmatrix} \Pi_0^{-1/2} \\ \mathbf{A} \end{bmatrix} \mathbf{w}' \right\|^2 \, .$$

This is now of the same form as our earlier minimization problem (21.14), with the observation vector $\mathbf{d}$ in (21.14) replaced by

$$\begin{bmatrix} \mathbf{0} \\ \mathbf{d}' \end{bmatrix} \, ,$$

and the matrix $\mathbf{A}$ in (21.14) replaced by

$$\begin{bmatrix} \Pi_0^{-1/2} \\ \mathbf{A} \end{bmatrix} \, .$$

21.3.1 Geometric Interpretation

The orthogonality condition can now be used, leading to the equation

$$\begin{bmatrix} \Pi_0^{-1/2} \\ \mathbf{A} \end{bmatrix}^T \left(\begin{bmatrix} \mathbf{0} \\ \mathbf{d}' \end{bmatrix} - \begin{bmatrix} \Pi_0^{-1/2} \\ \mathbf{A} \end{bmatrix} \hat{\mathbf{w}}' \right) = \mathbf{0} \, ,$$

which can be solved for the optimal estimate $\hat{\mathbf{w}}$,

$$\hat{\mathbf{w}} = \bar{\mathbf{w}} + \left[\Pi_0^{-1} + \mathbf{A}^T \mathbf{A} \right]^{-1} \mathbf{A}^T \left[\mathbf{d} - \mathbf{A}\bar{\mathbf{w}} \right] \, . \tag{21.18}$$

Comparing with the earlier expression (21.15), we see that instead of requiring the invertibility of $\mathbf{A}^T \mathbf{A}$, we now require the invertibility of the matrix $\left[\Pi_0^{-1} + \mathbf{A}^T \mathbf{A} \right]$. This is yet another reason in favor of the modified criterion (21.17) because it allows us to relax the full rank condition on $\mathbf{A}$.

TABLE 21.2   Linear Least-Squares Estimation

Optimization / Problem	Solution
$\{\mathbf{w}, \mathbf{d}\}$ $\min_{\mathbf{w}} \|\mathbf{d} - \mathbf{A}\mathbf{w}\|^2$ $\mathbf{A}$ full rank	$\hat{\mathbf{w}} = (\mathbf{A}^T\mathbf{A})^{-1}\mathbf{A}^T\mathbf{d}$
$\{\mathbf{w}, \mathbf{d}, \bar{\mathbf{w}}, \Pi_0\}$ $\min_{\mathbf{w}} \left[(\mathbf{w} - \bar{\mathbf{w}})^T \Pi_0^{-1} (\mathbf{w} - \bar{\mathbf{w}}) + \|\mathbf{d} - \mathbf{A}\mathbf{w}\|^2 \right]$ Π_0 positive-definite	$\hat{\mathbf{w}} = \bar{\mathbf{w}} + \left[\Pi_0^{-1} + \mathbf{A}^T\mathbf{A} \right]^{-1} \mathbf{A}^T [\mathbf{d} - \mathbf{A}\bar{\mathbf{w}}]$ Min. value $= (\mathbf{d} - \mathbf{A}\bar{\mathbf{w}})^T [\mathbf{I} + \mathbf{A}\Pi_0\mathbf{A}^T]^{-1}(\mathbf{d} - \mathbf{A}\bar{\mathbf{w}})$

The solution (21.18) can also be reexpressed as the solution of the following linear system of equations:

$$\underbrace{\left[\Pi_0^{-1} + \mathbf{A}^T\mathbf{A} \right]}_{\Phi}(\hat{\mathbf{w}} - \bar{\mathbf{w}}) = \underbrace{\mathbf{A}^T \left[\mathbf{d} - \mathbf{A}\bar{\mathbf{w}} \right]}_{\mathbf{s}}, \tag{21.19}$$

where we have denoted, for convenience, the coefficient matrix by Φ and the right-hand side by $\mathbf{s}$.

Moreover, it further follows that the value of (21.17) at the minimizing solution (21.18), denoted by $\mathcal{E}_{\min}$, is given by either of the following two expressions:

$$\begin{aligned} \mathcal{E}_{\min} &= \|\mathbf{d} - \mathbf{A}\bar{\mathbf{w}}\|^2 - \mathbf{s}^T (\hat{\mathbf{w}} - \bar{\mathbf{w}}) \\ &= (\mathbf{d} - \mathbf{A}\bar{\mathbf{w}})^T \left[\mathbf{I} + \mathbf{A}\Pi_0\mathbf{A}^T \right]^{-1} (\mathbf{d} - \mathbf{A}\bar{\mathbf{w}}). \end{aligned} \tag{21.20}$$

Expressions (21.19) and (21.20) are often rewritten into the so-called *normal equations*:

$$\begin{bmatrix} \|\mathbf{d} - \mathbf{A}\bar{\mathbf{w}}\|^2 & \mathbf{s}^T \\ \mathbf{s} & \Phi \end{bmatrix} \begin{bmatrix} 1 \\ -(\hat{\mathbf{w}} - \bar{\mathbf{w}}) \end{bmatrix} = \begin{bmatrix} \mathcal{E}_{\min} \\ \mathbf{0} \end{bmatrix}. \tag{21.21}$$

The results of this section are summarized in Table 21.2.

21.3.2   Statistical Interpretation

A statistical interpretation for the regularized problem can be obtained as follows. Given two vector-valued zero-mean random variables $\mathbf{w}$ and $\mathbf{d}$, the minimum-variance unbiased (MVU) estimator of $\mathbf{w}$ given an observation of $\mathbf{d}$ is $\hat{\mathbf{w}} = E(\mathbf{w}|\mathbf{d})$, the conditional expectation of $\mathbf{w}$ given $\mathbf{d}$. If the random variables $(\mathbf{w}, \mathbf{d})$ are jointly Gaussian, then the MVU estimator for $\mathbf{w}$ given $\mathbf{d}$ can be shown to collapse to

$$\hat{\mathbf{w}} = (E\mathbf{w}\mathbf{d}^T) \left(E\mathbf{d}\mathbf{d}^T \right)^{-1} \mathbf{d}. \tag{21.22}$$

Therefore, if $(\mathbf{w}, \mathbf{d})$ are further linearly related, say

$$\mathbf{d} = \mathbf{A}\mathbf{w} + \mathbf{v}, \quad \mathbf{v} \sim N(\mathbf{0}, \mathbf{I}), \quad \mathbf{w} \sim N(\mathbf{0}, \Pi_0) \tag{21.23}$$

with a zero-mean noise vector $\mathbf{v}$ that is uncorrelated with $\mathbf{w}$ ($E\mathbf{w}\mathbf{v}^T = \mathbf{0}$), then the expressions for $(E\mathbf{w}\mathbf{d}^T)$ and $(E\mathbf{d}\mathbf{d}^T)$ can be evaluated as

$$E\mathbf{w}\mathbf{d}^T = E\mathbf{w}(\mathbf{A}\mathbf{w} + \mathbf{v})^T = \Pi_0\mathbf{A}^T, \quad E\mathbf{d}\mathbf{d}^T = \mathbf{A}\Pi_0\mathbf{A}^T + \mathbf{I}.$$

This shows that (21.22) evaluates to

$$\hat{\mathbf{w}} = \Pi_0\mathbf{A}^T (\mathbf{I} + \mathbf{A}\Pi_0\mathbf{A}^T)^{-1}\mathbf{d}. \tag{21.24}$$

By invoking the useful matrix inversion formula (for arbitrary matrices of appropriate dimensions and invertible $\mathbf{E}$ and $\mathbf{C}$):

$$(\mathbf{E} + \mathbf{B}\mathbf{C}\mathbf{D})^{-1} = \mathbf{E}^{-1} - \mathbf{E}^{-1}\mathbf{B}(\mathbf{D}\mathbf{E}^{-1}\mathbf{B} + \mathbf{C}^{-1})^{-1}\mathbf{D}\mathbf{E}^{-1},$$

we can rewrite expression (21.24) in the equivalent form

$$\hat{\mathbf{w}} = (\Pi_0^{-1} + \mathbf{A}^T \mathbf{A})^{-1} \mathbf{A}^T \mathbf{d} \, . \tag{21.25}$$

This expression coincides with the regularized solution (21.18) for $\bar{\mathbf{w}} = \mathbf{0}$ (the case $\bar{\mathbf{w}} \neq \mathbf{0}$ follows from similar arguments by assuming a nonzero mean random variable $\mathbf{w}$).

Therefore, the regularized least-squares solution is the minimum variance unbiased (MVU) estimate of $\mathbf{w}$ given observations $\mathbf{d}$ that are corrupted by additive Gaussian noise as in (21.23).

21.4 The Recursive Least-Squares Problem

The recursive least-squares formulation deals with the problem of updating the solution $\hat{\mathbf{w}}$ of a least-squares problem (regularized or not) when new data are added to the matrix $\mathbf{A}$ and to the vector $\mathbf{d}$. This is in contrast to determining afresh the least-squares solution of the new problem. The distinction will become clear as we proceed in our discussions. In this section, we formulate the recursive least-squares problem as it arises in the context of adaptive filtering.

Consider a sequence of $(N + 1)$ scalar data points, $\{d(j)\}_{j=0}^{N}$, also known as *reference* or *desired* signals, and a sequence of $(N + 1)$ row vectors $\{\mathbf{u}_j^T\}_{j=0}^{N}$, also known as *input* signals. Each input vector $\mathbf{u}_j^T$ is a $1 \times M$ row vector whose individual entries we denote by $\{u_k(j)\}_{k=1}^{M}$, viz.,

$$\mathbf{u}_j^T = \begin{bmatrix} u_1(j) & u_2(j) & \cdots & u_M(j) \end{bmatrix} \, . \tag{21.26}$$

The entries of $\mathbf{u}_j$ can be regarded as the values of M input channels at time j: channels 1 through M.

Consider also a known column vector $\bar{\mathbf{w}}$ and a positive-definite weighting matrix Π_0. The objective is to determine an $M \times 1$ column vector $\mathbf{w}$, also known as the *weight vector*, so as to minimize the weighted error sum:

$$\mathcal{E}(N) = (\mathbf{w} - \bar{\mathbf{w}})^T \left[\lambda^{-(N+1)} \Pi_0 \right]^{-1} (\mathbf{w} - \bar{\mathbf{w}}) + \sum_{j=0}^{N} \lambda^{N-j} \left| d(j) - \mathbf{u}_j^T \mathbf{w} \right|^2 \, , \tag{21.27}$$

where λ is a positive scalar that is less than or equal to one (usually $0 \ll \lambda \leq 1$). It is often called the forgetting factor since past data is exponentially weighted less than the more recent data. The special case $\lambda = 1$ is known as the *growing memory* case, since, as the length N of the data grows, the effect of past data is not attenuated. In contrast, the *exponentially decaying memory* case ($\lambda < 1$) is more suitable for time-variant environments.

Also, and in principle, the factor $\lambda^{-(N+1)}$ that multiplies Π_0 in the error-sum expression (21.27) can be incorporated into the weighting matrix Π_0. But it is left explicit for convenience of exposition.

We further denote the individual entries of the column vector $\mathbf{w}$ by $\{w(j)\}_{j=1}^{M}$,

$$\mathbf{w} = \text{col}\{w(1), w(2), \ldots, w(M)\} \, .$$

A schematic description of the problem is shown in Fig. 21.2. At each time instant j, the inputs of the M channels are linearly combined via the coefficients of the weight vector and the resulting signal is compared with the desired signal $d(j)$. This results in a residual error $e(j) = d(j) - \mathbf{u}_j^T \mathbf{w}$, for every j, and the objective is to find a weight vector $\mathbf{w}$ in order to minimize the (exponentially weighted and regularized) squared-sum of the residual errors over an interval of time, say from $j = 0$ up to $j = N$.

The linear combiner is said to be of order M since it is determined by M coefficients $\{w(j)\}_{j=1}^{M}$.

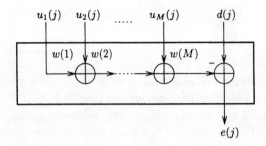

FIGURE 21.2   A linear combiner.

21.4.1   Reducing to the Regularized Form

The expression for the weighted error-sum (21.27) is a special case of the regularized cost function (21.17). To clarify this, we introduce the residual vector $\mathbf{e}_N$, the reference vector $\mathbf{d}_N$, the data matrix $\mathbf{A}_N$, and a diagonal weighting matrix Λ_N,

$$
\mathbf{e}_N \;=\; \underbrace{\begin{bmatrix} d(0) \\ d(1) \\ d(2) \\ \vdots \\ d(N) \end{bmatrix}}_{\mathbf{d}_N} - \underbrace{\begin{bmatrix} u_1(0) & u_2(0) & \ldots & u_M(0) \\ u_1(1) & u_2(1) & \ldots & u_M(1) \\ u_1(2) & u_2(2) & \ldots & u_M(2) \\ \vdots & & & \vdots \\ u_1(N) & u_2(N) & \ldots & u_M(N) \end{bmatrix}}_{\mathbf{A}_N} \mathbf{w} \;,
$$

$$
\Lambda_N^{1/2} \;=\; \begin{bmatrix} \left[\lambda^{\frac{1}{2}}\right]^N & & & & \\ & \left[\lambda^{\frac{1}{2}}\right]^{N-1} & & & \\ & & \ddots & & \\ & & & \left[\lambda^{\frac{1}{2}}\right]^2 & \\ & & & & 1 \end{bmatrix} .
$$

We now use a subscript N to indicate that the above quantities are determined by data that is available up to time N.

With these definitions, we can write $\mathcal{E}(N)$ in the equivalent form

$$
\mathcal{E}(N) = (\mathbf{w} - \bar{\mathbf{w}})^T \left[\lambda^{-(N+1)} \Pi_0 \right]^{-1} (\mathbf{w} - \bar{\mathbf{w}}) + \left\| \Lambda_N^{1/2} \mathbf{e}_N \right\|^2 ,
$$

which is a special case of (21.17) with

$$
\Lambda_N^{1/2} \mathbf{d}_N \text{ and } \Lambda_N^{1/2} \mathbf{A}_N \tag{21.28}
$$

replacing

$$
\mathbf{d}_N \text{ and } \mathbf{A}_N , \tag{21.29}
$$

respectively, and with $\lambda^{-(N+1)} \Pi_0$ replacing Π_0.

We therefore conclude from (21.19) that the optimal solution $\hat{\mathbf{w}}$ of (21.27) is given by

$$
(\hat{\mathbf{w}} - \bar{\mathbf{w}}) = \Phi_N^{-1} \mathbf{s}_N , \tag{21.30}
$$

where we have introduced

$$
\Phi_N \;=\; \left[\lambda^{(N+1)} \Pi_0^{-1} + \mathbf{A}_N^T \Lambda_N \mathbf{A}_N \right] , \tag{21.31}
$$

$$
\mathbf{s}_N \;=\; \mathbf{A}_N^T \Lambda_N \left[\mathbf{d}_N - \mathbf{A}_N \bar{\mathbf{w}} \right] . \tag{21.32}
$$

The coefficient matrix Φ_N is clearly symmetric and positive-definite.

21.4.2 Time Updates

It is straightforward to verify that Φ_N and $\mathbf{s}_N$ so defined satisfy simple time-update relations, viz.,

$$\Phi_{N+1} = \lambda\Phi_N + \mathbf{u}_{N+1}\mathbf{u}_{N+1}^T , \tag{21.33}$$

$$\mathbf{s}_{N+1} = \lambda\mathbf{s}_N + \mathbf{u}_{N+1}\left[d(N+1) - \mathbf{u}_{N+1}^T\bar{\mathbf{w}}\right], \tag{21.34}$$

with initial conditions $\Phi_{-1} = \Pi_0^{-1}$ and $\mathbf{s}_{-1} = \mathbf{0}$. Note that Φ_{N+1} and $\lambda\Phi_N$ differ only by a rank-one matrix.

The solution $\hat{\mathbf{w}}$ obtained by solving (21.30) is the optimal weight estimate based on the available data from time $i = 0$ up to time $i = N$. We shall denote it from now on by $\mathbf{w}_N$,

$$\Phi_N(\mathbf{w}_N - \bar{\mathbf{w}}) = \mathbf{s}_N .$$

The subscript N in $\mathbf{w}_N$ indicates that the data up to, and including, time N were used. This is to differentiate it from the estimate obtained by using a different number of data points.

This notational change is necessary because the main objective of the recursive least-squares (RLS) problem is to show how to update the estimate $\mathbf{w}_N$, which is based on the data up to time N, to the estimate $\mathbf{w}_{N+1}$, which is based on the data up to time $(N+1)$, without the need to solve afresh a new set of linear equations of the form

$$\Phi_{N+1}(\mathbf{w}_{N+1} - \bar{\mathbf{w}}) = \mathbf{s}_{N+1} .$$

Such a recursive update of the weight estimate should be possible since the coefficient matrices $\lambda\Phi_N$ and Φ_{N+1} of the associated linear systems differ only by a rank-one matrix. In fact, a wide variety of algorithms has been devised for this end and our purpose in this chapter is to provide an overview of the different schemes.

Before describing these different variants, we note in passing that it follows from (21.20) that we can express the minimum value of $\mathcal{E}(N)$ in the form:

$$\mathcal{E}_{\min}(N) = \left\|\Lambda_N^{1/2}(\mathbf{d}_N - \mathbf{A}_N\bar{\mathbf{w}})\right\|^2 - \mathbf{s}_N^T(\mathbf{w}_N - \bar{\mathbf{w}}) . \tag{21.35}$$

21.5 The RLS Algorithm

The first recursive solution that we consider is the famed recursive least-squares algorithm, usually referred to as the RLS algorithm. It can be derived as follows.

Let $\mathbf{w}_{i-1}$ be the solution of an optimization problem of the form (21.27) that uses input data up to time $(i-1)$ [that is, for $N = (i-1)$]. Likewise, let $\mathbf{w}_i$ be the solution of the same optimization problem but with input data up to time i [$N = i$].

The recursive least-squares (RLS) algorithm provides a recursive procedure that computes $\mathbf{w}_i$ from $\mathbf{w}_{i-1}$. A classical derivation follows by noting from (21.30) that the new solution $\mathbf{w}_i$ should satisfy

$$\mathbf{w}_i - \bar{\mathbf{w}} = \Phi_i^{-1}\mathbf{s}_i = \left[\lambda\Phi_{i-1} + \mathbf{u}_i\mathbf{u}_i^T\right]^{-1}\left(\lambda\mathbf{s}_{i-1} + \mathbf{u}_i\left[d(i) - \mathbf{u}_i^T\bar{\mathbf{w}}\right]\right) ,$$

where we have also used the time-updates for $\{\Phi_i, \mathbf{s}_i\}$.

Introduce the quantities

$$\mathbf{P}_i = \Phi_i^{-1} , \quad \mathbf{g}_i = \Phi_i^{-1}\mathbf{u}_i . \tag{21.36}$$

Expanding the inverse of $[\lambda\Phi_{i-1} + \mathbf{u}_i\mathbf{u}_i^T]$ by using the matrix inversion formula [stated after (21.24)], and grouping terms, leads after some straightforward algebra to the RLS procedure:

- Initial conditions: $\mathbf{w}_{-1} = \bar{\mathbf{w}}$ and $\mathbf{P}_{-1} = \Pi_0$.
- Repeat for $i \geq 0$:

$$\mathbf{w}_i = \mathbf{w}_{i-1} + \mathbf{g}_i \left[d(i) - \mathbf{u}_i^T \mathbf{w}_{i-1} \right], \tag{21.37}$$

$$\mathbf{g}_i = \frac{\lambda^{-1} \mathbf{P}_{i-1} \mathbf{u}_i}{1 + \lambda^{-1} \mathbf{u}_i^T \mathbf{P}_{i-1} \mathbf{u}_i}, \tag{21.38}$$

$$\mathbf{P}_i = \lambda^{-1} \left[\mathbf{P}_{i-1} - \mathbf{g}_i \mathbf{u}_i^T \mathbf{P}_{i-1} \right]. \tag{21.39}$$

- The computational complexity of the algorithm is $O(M^2)$ per iteration.

21.5.1 Estimation Errors and the Conversion Factor

With the RLS problem we associate two residuals at each time instant i: the *a priori* estimation error $e_a(i)$, defined by

$$e_a(i) = d(i) - \mathbf{u}_i^T \mathbf{w}_{i-1},$$

and the *a posteriori* estimation error $e_p(i)$, defined by

$$e_p(i) = d(i) - \mathbf{u}_i^T \mathbf{w}_i.$$

Comparing the expressions for $e_a(i)$ and $e_p(i)$, we see that the latter employs the most recent weight vector estimate.

If we replace $\mathbf{w}_i$ in the definition for $e_p(i)$ by its update expression (21.37), say

$$e_p(i) = d(i) - \mathbf{u}_i^T (\mathbf{w}_{i-1} + \mathbf{g}_i \left[d(i) - \mathbf{u}_i^T \mathbf{w}_{i-1} \right]),$$

some straightforward algebra will show that we can relate $e_p(i)$ and $e_a(i)$ via a factor $\gamma(i)$ known as the *conversion* factor:

$$e_p(i) = \gamma(i) e_a(i),$$

where $\gamma(i)$ is equal to

$$\gamma(i) = \frac{1}{1 + \lambda^{-1} \mathbf{u}_i^T \mathbf{P}_{i-1} \mathbf{u}_i} = 1 - \mathbf{u}_i^T \mathbf{P}_i \mathbf{u}_i. \tag{21.40}$$

That is, the *a posteriori* error is a scaled version of the *a priori* error. The scaling factor $\gamma(i)$ is defined in terms of $\{\mathbf{u}_i, \mathbf{P}_{i-1}\}$ or $\{\mathbf{u}_i, \mathbf{P}_i\}$. Note that $0 \leq \gamma(i) \leq 1$.

Note further that the expression for $\gamma(i)$ appears in the definition of the so-called *gain* vector $\mathbf{g}_i$ in (21.38) and, hence, we can alternatively rewrite (21.38) and (21.39) in the forms:

$$\mathbf{g}_i = \lambda^{-1} \gamma(i) \mathbf{P}_{i-1} \mathbf{u}_i, \tag{21.41}$$

$$\mathbf{P}_i = \lambda^{-1} \mathbf{P}_{i-1} - \gamma^{-1}(i) \mathbf{g}_i \mathbf{g}_i^T. \tag{21.42}$$

21.5.2 Update of the Minimum Cost

Let $\mathcal{E}_{\min}(i)$ denote the value of the minimum cost of the optimization problem (21.27) with data up to time i. It is given by an expression of the form (21.35) with N replaced by i,

$$\mathcal{E}_{\min}(i) = \left[\sum_{j=0}^{i} \lambda^{i-j} \left\| d(j) - \mathbf{u}_j^T \bar{\mathbf{w}} \right\|^2 \right] - \mathbf{s}_i^T (\mathbf{w}_i - \bar{\mathbf{w}}).$$

Using the RLS update (21.37) for $\mathbf{w}_i$ in terms of $\mathbf{w}_{i-1}$, as well as the time-update (21.34) for $\mathbf{s}_i$ in terms of $\mathbf{s}_{i-1}$, we can derive the following time-update for the minimum cost:

$$\mathcal{E}_{\min}(i) = \lambda\mathcal{E}_{\min}(i-1) + e_p(i)e_a(i) , \tag{21.43}$$

where $\mathcal{E}_{\min}(i-1)$ denotes the value of the minimum cost of the same optimization problem (21.27) but with data up to time $(i-1)$.

21.6 RLS Algorithms in Array Forms

As mentioned in the introduction, we intend to stress the array formulations of the RLS solution due to their intrinsic advantages:

- They are easy to implement as a sequence of elementary rotations on arrays of numbers.
- They are modular and parallelizable.
- They have better numerical properties than the classical RLS description.

21.6.1 Motivation

Note from (21.39) that the RLS solution propagates the variable $\mathbf{P}_i$ as the difference of two quantities. This variable should be positive-definite. But due to roundoff errors, however, the update (21.39) may not guarantee the positive-definiteness of $\mathbf{P}_i$ at all times i. This problem can be ameliorated by using the so-called array formulations. These alternative forms propagate square-root factors of either $\mathbf{P}_i$ or $\mathbf{P}_i^{-1}$, namely, $\mathbf{P}_i^{1/2}$ or $\mathbf{P}_i^{-1/2}$, rather than $\mathbf{P}_i$ itself. By squaring $\mathbf{P}_i^{1/2}$, for example, we can always recover a matrix $\mathbf{P}_i$ that is more likely to be positive-definite than the matrix obtained via (21.39),

$$\mathbf{P}_i = \mathbf{P}_i^{1/2}\mathbf{P}_i^{T/2} .$$

21.6.2 A Very Useful Lemma

The derivation of the array variants of the RLS algorithm relies on a very useful matrix result that encounters applications in many other scenarios as well. For this reason, we not only state the result but also provide one simple proof.

LEMMA 21.1   Given two $n \times m$ $(n \le m)$ matrices $\mathbf{A}$ and $\mathbf{B}$, then $\mathbf{A}\mathbf{A}^T = \mathbf{B}\mathbf{B}^T$ if, and only if, there exists an $m \times m$ orthogonal matrix Θ $(\Theta\Theta^T = \mathbf{I}_m)$ such that $\mathbf{A} = \mathbf{B}\Theta$.

PROOF 21.1   One implication is immediate. If there exists an orthogonal matrix Θ such that $\mathbf{A} = \mathbf{B}\Theta$ then

$$\mathbf{A}\mathbf{A}^T = (\mathbf{B}\Theta)(\mathbf{B}\Theta)^T = \mathbf{B}(\Theta\Theta^T)\mathbf{B}^T = \mathbf{B}\mathbf{B}^T .$$

One proof for the converse implication follows by invoking the singular value decompositions of the matrices $\mathbf{A}$ and $\mathbf{B}$,

$$\mathbf{A} = \mathbf{U}_A \begin{bmatrix} \Sigma_A & \mathbf{0} \end{bmatrix} \mathbf{V}_A^T ,$$
$$\mathbf{B} = \mathbf{U}_B \begin{bmatrix} \Sigma_B & \mathbf{0} \end{bmatrix} \mathbf{V}_B^T ,$$

where $\mathbf{U}_A$ and $\mathbf{U}_B$ are $n \times n$ orthogonal matrices, $\mathbf{V}_A$ and $\mathbf{V}_B$ are $m \times m$ orthogonal matrices, and Σ_A and Σ_B are $n \times n$ diagonal matrices with nonnegative (ordered) entries.

The squares of the diagonal entries of Σ_A (Σ_B) are the eigenvalues of $\mathbf{A}\mathbf{A}^T$ ($\mathbf{B}\mathbf{B}^T$). Moreover, $\mathbf{U}_A$ ($\mathbf{U}_B$) are constructed from an orthonormal basis for the right eigenvectors of $\mathbf{A}\mathbf{A}^T$ ($\mathbf{B}\mathbf{B}^T$).

Hence, it follows from the identity $\mathbf{A}\mathbf{A}^T = \mathbf{B}\mathbf{B}^T$ that we have $\Sigma_A = \Sigma_B$ and we can choose $\mathbf{U}_A = \mathbf{U}_B$. Let $\Theta = \mathbf{V}_B\mathbf{V}_A^T$. We then obtain $\Theta\Theta^T = \mathbf{I}_m$ and $\mathbf{B}\Theta = \mathbf{A}$.

21.6.3 The Inverse QR Algorithm

We now employ the above result to derive an array form of the RLS algorithm that is known as the inverse QR algorithm.

Let $\mathbf{P}_{i-1}^{1/2}$ denote a (preferably lower triangular) square-root factor of $\mathbf{P}_{i-1}$, i.e., any matrix that satisfies

$$\mathbf{P}_{i-1} = \mathbf{P}_{i-1}^{1/2}\ \mathbf{P}_{i-1}^{T/2}.$$

[The triangular square-root factor of a symmetric positive-definite matrix is also known as the Cholesky factor].

Now note that the RLS recursions (21.38) and (21.39) can be expressed in factored form as follows:

$$\begin{bmatrix} 1 & \frac{1}{\sqrt{\lambda}}\mathbf{u}_i^T\mathbf{P}_{i-1}^{1/2} \\ 0 & \frac{1}{\sqrt{\lambda}}\mathbf{P}_{i-1}^{1/2} \end{bmatrix} \begin{bmatrix} 1 & \mathbf{0}^T \\ \frac{1}{\sqrt{\lambda}}\mathbf{P}_{i-1}^{T/2}\mathbf{u}_i & \frac{1}{\sqrt{\lambda}}\mathbf{P}_{i-1}^{T/2} \end{bmatrix}$$

$$= \begin{bmatrix} \gamma^{-1/2}(i) & \mathbf{0}^T \\ \mathbf{g}_i\gamma^{-1/2}(i) & \mathbf{P}_i^{1/2} \end{bmatrix} \begin{bmatrix} \gamma^{-1/2}(i) & \mathbf{g}_i^T\gamma^{-1/2}(i) \\ 0 & \mathbf{P}_i^{T/2} \end{bmatrix}.$$

To verify that this is indeed the case, we simply multiply the factors and compare terms on both sides of the equality.

The point to note is that the above equality fits nicely into the statement of the previous lemma by taking

$$\mathbf{A} = \begin{bmatrix} 1 & \frac{1}{\sqrt{\lambda}}\mathbf{u}_i^T\mathbf{P}_{i-1}^{1/2} \\ 0 & \frac{1}{\sqrt{\lambda}}\mathbf{P}_{i-1}^{1/2} \end{bmatrix} \tag{21.44}$$

and

$$\mathbf{B} = \begin{bmatrix} \gamma^{-1/2}(i) & \mathbf{0}^T \\ \mathbf{g}_i\gamma^{-1/2}(i) & \mathbf{P}_i^{1/2} \end{bmatrix}. \tag{21.45}$$

We therefore conclude that there should exist an orthogonal matrix Θ_i that relates the arrays $\mathbf{A}$ and $\mathbf{B}$ in the form

$$\begin{bmatrix} 1 & \frac{1}{\sqrt{\lambda}}\mathbf{u}_i^T\mathbf{P}_{i-1}^{1/2} \\ 0 & \frac{1}{\sqrt{\lambda}}\mathbf{P}_{i-1}^{1/2} \end{bmatrix} \Theta_i = \begin{bmatrix} \gamma^{-1/2}(i) & \mathbf{0}^T \\ \mathbf{g}_i\gamma^{-1/2}(i) & \mathbf{P}_i^{1/2} \end{bmatrix}.$$

That is, there should exist an orthogonal Θ_i that transforms the prearray $\mathbf{A}$ into the postarray $\mathbf{B}$.

Note that the prearray contains quantities that are available at step i, namely $\{\mathbf{u}_i, \mathbf{P}_{i-1}^{1/2}\}$, while the postarray provides the (normalized) gain vector $\mathbf{g}_i\gamma^{-1/2}(i)$, which is needed to update the weight vector estimate $\mathbf{w}_{i-1}$ into $\mathbf{w}_i$, as well as the square-root factor of the variable $\mathbf{P}_i$, which is needed to form the prearray for the next iteration.

But how do we determine Θ_i? The answer highlights a remarkable property of array algorithms. We do not really need to know or determine Θ_i explicitly!

To clarify this point, we first remark from the expressions (21.44) and (21.45) for the pre and postarrays that Θ_i is an orthogonal matrix that takes an array of numbers of the form (assuming a vector $\mathbf{u}_i$ of dimension $M = 3$)

$$\begin{bmatrix} 1 & x & x & x \\ \hline 0 & x & 0 & 0 \\ 0 & x & x & 0 \\ 0 & x & x & x \end{bmatrix} \tag{21.46}$$

and transforms it to the form

$$\begin{bmatrix} x & 0 & 0 & 0 \\ \hline x & x & 0 & 0 \\ x & x & x & 0 \\ x & x & x & x \end{bmatrix}. \tag{21.47}$$

That is, Θ_i annihilates all the entries of the top row of the prearray (except for the left-most entry).

Now assume we form the prearray $\mathbf{A}$ in (21.44) and choose any Θ_i (say as a sequence of elementary rotations) so as to reduce $\mathbf{A}$ to the triangular form (21.47), that is, in order to annihilate the desired entries in the top row.

Let us denote the resulting entries of the postarray arbitrarily as:

$$\begin{bmatrix} 1 & \frac{1}{\sqrt{\lambda}}\mathbf{u}_i^T \mathbf{P}_{i-1}^{1/2} \\ 0 & \frac{1}{\sqrt{\lambda}}\mathbf{P}_{i-1}^{1/2} \end{bmatrix} \Theta_i = \begin{bmatrix} a & \mathbf{0}^T \\ \mathbf{b} & \mathbf{C} \end{bmatrix}, \tag{21.48}$$

where $\{a, \mathbf{b}, \mathbf{C}\}$ are quantities that we wish to identify [a is a scalar, $\mathbf{b}$ is a column vector, and $\mathbf{C}$ is a lower triangular matrix]. The claim is that by constructing Θ_i in this way (i.e., by simply requiring that it achieves the desired zero pattern in the postarray), the resulting quantities $\{a, \mathbf{b}, \mathbf{C}\}$ will be meaningful and can in fact be identified with the quantities in the postarray $\mathbf{B}$.

To verify that the quantities $\{a, \mathbf{b}, \mathbf{C}\}$ can indeed be identified with $\{\gamma^{-1/2}(i), \mathbf{g}_i\gamma^{-1/2}(i), \mathbf{P}_i^{1/2}\}$, we proceed by squaring both sides of (21.48),

$$\begin{bmatrix} 1 & \frac{1}{\sqrt{\lambda}}\mathbf{u}_i^T \mathbf{P}_{i-1}^{1/2} \\ 0 & \frac{1}{\sqrt{\lambda}}\mathbf{P}_{i-1}^{1/2} \end{bmatrix} \underbrace{\Theta_i\Theta_i^T}_{\mathbf{I}} \begin{bmatrix} 1 & \mathbf{0} \\ \frac{1}{\sqrt{\lambda}}\mathbf{P}_{i-1}^{T/2}\mathbf{u}_i & \frac{1}{\sqrt{\lambda}}\mathbf{P}_{i-1}^{T/2} \end{bmatrix} = \begin{bmatrix} a & \mathbf{0}^T \\ \mathbf{b} & \mathbf{C} \end{bmatrix}\begin{bmatrix} a & \mathbf{b}^T \\ \mathbf{0} & \mathbf{C}^T \end{bmatrix},$$

and comparing terms on both sides of the equality to get the identities:

$$\begin{aligned} a^2 &= 1 + \lambda^{-1}\mathbf{u}_i^T \mathbf{P}_{i-1}\mathbf{u}_i = \gamma^{-1}(i)\,, \\ \mathbf{b}a &= \lambda^{-1}\mathbf{P}_{i-1}\mathbf{u}_i = \mathbf{g}_i\gamma^{-1}(i)\,, \\ \mathbf{C}\mathbf{C}^T &= \lambda^{-1}\mathbf{P}_{i-1} - \mathbf{b}\mathbf{b}^T = \lambda^{-1}\mathbf{P}_{i-1} - \gamma^{-1}(i)\mathbf{g}_i\mathbf{g}_i^T\,. \end{aligned}$$

Hence, as desired, we can make the identifications

$$a = \gamma^{-1/2}(i)\,, \quad \mathbf{b} = \mathbf{g}_i\gamma^{-1/2}(i)\,, \quad \mathbf{C} = \mathbf{P}_i^{1/2}\,.$$

In summary, we have established the validity of an array alternative to the RLS algorithm, known as the inverse QR algorithm (also as square-root RLS). It is listed in Table 21.3. The recursions are known as *inverse* QR since they propagate $\mathbf{P}_i^{1/2}$, which is a square-root factor of the inverse of the coefficient matrix Φ_i.

21.6.4 The QR Algorithm

The RLS recursion (21.39) and the inverse QR recursion of Table 21.3 propagate the variable $\mathbf{P}_i$ or a square-root factor of it. The starting condition for both algorithms is therefore dependent on the weighting matrix Π_0 or its square-root factor $\Pi_0^{1/2}$.

TABLE 21.3   The Inverse QR Algorithm

Initialization. Start with $\mathbf{w}_{-1} = \bar{\mathbf{w}}$ and

$$\mathbf{P}_{-1}^{1/2} = \Pi_0^{1/2}$$

• Repeat for each time instant $i \geq 0$:

$$\begin{bmatrix} 1 & \frac{1}{\sqrt{\lambda}}\mathbf{u}_i^T \mathbf{P}_{i-1}^{1/2} \\ 0 & \frac{1}{\sqrt{\lambda}}\mathbf{P}_{i-1}^{1/2} \end{bmatrix} \Theta_i = \begin{bmatrix} \gamma^{-1/2}(i) & \mathbf{0}^T \\ \mathbf{g}_i \gamma^{-1/2}(i) & \mathbf{P}_i^{1/2} \end{bmatrix}$$

where Θ_i is any orthogonal rotation that
produces the zero pattern in the postarray.

The weight-vector estimate is updated via

$$\mathbf{w}_i = \mathbf{w}_{i-1} + \left[\frac{\mathbf{g}_i}{\gamma^{1/2}(i)}\right]\left[\frac{1}{\gamma^{1/2}(i)}\right]^{-1}\left[d(i) - \mathbf{u}_i^T \mathbf{w}_{i-1}\right]$$

where the quantities $\{\gamma^{-1/2}(i), \mathbf{g}_i \gamma^{-1/2}(i)\}$ are
read from the entries of the postarray.

The computational cost is $O(M^2)$ per iteration.

This situation becomes inconvenient when the initial condition Π_0 assumes relatively large values, say $\Pi_0 = \sigma \mathbf{I}$ with $\sigma \gg 1$. A particular instance arises, for example, when we take $\sigma \to \infty$ in which case the regularized least-squares problem (21.27) reduces to a standard least-squares problem of the form

$$\min_{\mathbf{w}} \left[\mathcal{E}(N) = \sum_{j=0}^{N} \lambda^{N-j} |d(j) - \mathbf{u}_j^T \mathbf{w}|^2 \right]. \tag{21.49}$$

For such problems, it is preferable to propagate the inverse of the variable $\mathbf{P}_i$ rather than $\mathbf{P}_i$ itself. Recall that the inverse of $\mathbf{P}_i$ is Φ_i since we have defined earlier $\mathbf{P}_i = \Phi_i^{-1}$.

The QR algorithm is a recursive procedure that propagates a square-root factor of Φ_i. Its validity can be verified in much the same way as we did for the inverse QR algorithm. We form a prearray of numbers and then choose a sequence of rotations that induces a desired zero pattern in the postarray. Then by squaring and comparing terms on both sides of an equality we can identify the resulting entries of the postarray as meaningful quantities in the RLS context. For this reason, we shall be brief and only highlight the main points.

Let $\Phi_{i-1}^{1/2}$ denote a square-root factor (preferably lower-triangular) of Φ_{i-1}, $\Phi_{i-1} = \Phi_{i-1}^{1/2} \Phi_{i-1}^{T/2}$, and define, for notational convenience, the quantity

$$\mathbf{q}_{i-1} = \Phi_{i-1}^{T/2} \mathbf{w}_{i-1} . \tag{21.50}$$

At time $(i-1)$ we form the prearray of numbers

$$\mathbf{A} = \begin{bmatrix} \sqrt{\lambda}\Phi_{i-1}^{1/2} & \mathbf{u}_i \\ \sqrt{\lambda}\mathbf{q}_{i-1}^T & d(i) \\ \mathbf{0}^T & 1 \end{bmatrix},$$

whose entries have the following pattern (shown for $M = 3$):

$$\mathbf{A} = \begin{bmatrix} x & 0 & 0 & x \\ x & x & 0 & x \\ x & x & x & x \\ x & x & x & x \\ 0 & 0 & 0 & 1 \end{bmatrix}.$$

Now implement an orthogonal transformation Θ_i that reduces $\mathbf{A}$ to the form

$$\mathbf{B} = \begin{bmatrix} x & 0 & 0 & 0 \\ x & x & 0 & 0 \\ x & x & x & 0 \\ \hline x & x & x & x \\ x & x & x & x \end{bmatrix} = \begin{bmatrix} \mathbf{C} & 0 \\ \mathbf{b}^T & a \\ \mathbf{h}^T & f \end{bmatrix},$$

where the quantities $\{\mathbf{C}, \mathbf{b}, \mathbf{h}, a, f\}$ need to be identified. By comparing terms on both sides of the equality

$$\begin{bmatrix} \sqrt{\lambda}\Phi_{i-1}^{1/2} & \mathbf{u}_i \\ \sqrt{\lambda}\mathbf{q}_{i-1}^T & d(i) \\ 0^T & 1 \end{bmatrix} \underbrace{\Theta_i \Theta_i^T}_{\mathbf{I}} \begin{bmatrix} \sqrt{\lambda}\Phi_{i-1}^{1/2} & \mathbf{u}_i \\ \sqrt{\lambda}\mathbf{q}_{i-1}^T & d(i) \\ 0^T & 1 \end{bmatrix}^T = \begin{bmatrix} \mathbf{C} & 0 \\ \mathbf{b}^T & a \\ \mathbf{h}^T & f \end{bmatrix} \begin{bmatrix} \mathbf{C} & 0 \\ \mathbf{b}^T & a \\ \mathbf{h}^T & f \end{bmatrix}^T,$$

we can make the identifications:

$$\mathbf{C} = \Phi_i^{1/2}, \quad \mathbf{b}^T = \mathbf{q}_i^T, \quad \mathbf{h}^T = \mathbf{u}_i^T \Phi_i^{-T/2},$$

$$a = e_a(i)\gamma^{1/2}(i), \quad f = \gamma^{1/2}(i),$$

where $e_a(i) = d(i) - \mathbf{u}_i^T \mathbf{w}_{i-1}$ is the *a priori* estimation error. This derivation establishes the so-called QR algorithm (listed in Table 21.4).

TABLE 21.4 The QR Algorithm

Initialization. Start with $\mathbf{w}_{-1} = \bar{\mathbf{w}}$, $\Phi_{-1}^{1/2} = \Pi_0^{-T/2}$, $\mathbf{q}_{-1} = \Pi_0^{-1/2}\bar{\mathbf{w}}$.

● Repeat for each time instant $i \geq 0$:

$$\begin{bmatrix} \sqrt{\lambda}\Phi_{i-1}^{1/2} & \mathbf{u}_i^T \\ \sqrt{\lambda}\mathbf{q}_{i-1}^T & d(i) \\ 0^T & 1 \end{bmatrix} \Theta_i = \begin{bmatrix} \Phi_i^{1/2} & 0 \\ \mathbf{q}_i^T & e_a(i)\gamma^{1/2}(i) \\ \mathbf{u}_i^T \Phi_i^{-T/2} & \gamma^{1/2}(i) \end{bmatrix}$$

where Θ_i is any orthogonal rotation that produces the zero pattern in the postarray.

The weight-vector estimate can be obtained by solving the triangular linear system of equations

$$\Phi_i^{T/2}\mathbf{w}_i = \mathbf{q}_i$$

where the quantities $\{\Phi_i^{1/2}, \mathbf{q}_i\}$ are available from the entries of the postarray.

The computational complexity is still $O(M^2)$ per iteration.

The QR solution determines the weight-vector estimate $\mathbf{w}_i$ by solving a triangular linear system of equations, e.g., via back-substitution. A major drawback of a back-substitution step is that it involves serial operations and, therefore, does not lend itself to a fully parallelizable implementation.

An alternative procedure for computing the estimate $\mathbf{w}_i$ can be obtained by appending one more block row to the arrays of the QR algorithm, leading to the equations:

$$\begin{bmatrix} \sqrt{\lambda}\Phi_{i-1}^{1/2} & \mathbf{u}_i \\ \sqrt{\lambda}\mathbf{q}_{i-1}^T & d(i) \\ 0^T & 1 \\ \hline \frac{1}{\sqrt{\lambda}}\Phi_{i-1}^{-T/2} & 0 \end{bmatrix} \Theta_i = \begin{bmatrix} \Phi_i^{1/2} & 0 \\ \mathbf{q}_i^T & e_a(i)\gamma^{1/2}(i) \\ \mathbf{u}_i^T \Phi_i^{-T/2} & \gamma^{1/2}(i) \\ \hline \Phi_i^{-T/2} & -\mathbf{g}_i\gamma^{-1/2}(i) \end{bmatrix}. \tag{21.51}$$

In this case, the last row of the postarray provides the gain vector $\mathbf{g}_i$ that can be used to update the weight-vector estimate as follows:

$$\mathbf{w}_i = \mathbf{w}_{i-1} + \left[\frac{\mathbf{g}_i}{\gamma^{1/2}(i)} \right] \left[e_a(i)\gamma^{1/2}(i) \right].$$

Note, however, that the pre- and postarrays now propagate both $\Phi_i^{1/2}$ and its inverse, which may lead to numerical difficulties.

21.7 Fast Transversal Algorithms

The earlier recursive least-squares solutions require $O(M^2)$ floating point operations per iteration, where M is the size of the input vector $\mathbf{u}_i$.

$$\mathbf{u}_i^T = \left[\begin{array}{cccc} u_1(i) & u_2(i) & \ldots & u_M(i) \end{array} \right].$$

It often happens in practice that the entries of $\mathbf{u}_i$ are time-shifted versions of each other. More explicitly, if we denote the value of the first entry of $\mathbf{u}_i$ by $u(i)$ [instead of $u_1(i)$], then $\mathbf{u}_i$ will have the form

$$\mathbf{u}_i^T = \left[\begin{array}{cccc} u(i) & u(i-1) & \ldots & u(i-M+1) \end{array} \right]. \tag{21.52}$$

This has the pictorial representation shown in Fig. 21.3. The term z^{-1} represents a unit-time delay. The structure that takes $u(j)$ as an input and provides the inner product $\sum_{k=1}^{M} u(j+1-k)w(k)$ as an output is known as a transversal or FIR (finite-impulse response) filter.

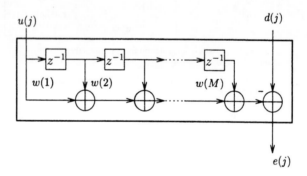

FIGURE 21.3   A linear combiner with shift structure in the input channels.

The shift structure in $\mathbf{u}_i$ can be exploited in order to derive fast variants to the RLS solution that would require $O(M)$ operations per iteration rather than $O(M^2)$. This can be achieved by showing that, in this case, the $M \times M$ variables $\mathbf{P}_i$ that are needed in the RLS recursion (21.39) exhibit certain matrix structure that allows us to replace the RLS recursions by an alternative set of recursions that we now motivate.

21.7.1 The Prewindowed Case

We first assume that no input data is available prior to and including time $i = 0$. That is, $u(i) = 0$ for $i \le 0$. In this case, the values at time 0 of the variables $\{\mathbf{u}_i, \mathbf{g}_i, \gamma(i), \mathbf{P}_i\}$ become:

$$\mathbf{u}_0 = \mathbf{0}, \quad \mathbf{g}_0 = \mathbf{0}, \quad \gamma(0) = 1, \quad \mathbf{P}_0 = \lambda^{-1}\mathbf{P}_{-1} = \lambda^{-1}\Pi_0.$$

It then follows that the following equality holds:

$$\begin{bmatrix} \mathbf{P}_0 & \mathbf{0} \\ \mathbf{0}^T & 0 \end{bmatrix} - \begin{bmatrix} 0 & \mathbf{0}^T \\ \mathbf{0} & \mathbf{P}_{-1} \end{bmatrix} = \begin{bmatrix} \lambda^{-1}\Pi_0 & \mathbf{0} \\ \mathbf{0}^T & 0 \end{bmatrix} - \begin{bmatrix} 0 & \mathbf{0}^T \\ \mathbf{0} & \Pi_0 \end{bmatrix}$$

Note that we have embedded $\mathbf{P}_0$ and $\mathbf{P}_{-1}$ into larger matrices [of size $(M+1) \times (M+1)$ each] by adding one zero row and one zero column. This embedding will allow us to suggest a suitable choice for the initial weighting matrix Π_0 in order to enforce a low-rank difference matrix on the right-hand side of the above expression. In so doing, we guarantee that $(\mathbf{P}_0 \oplus 0)$ can be obtained from $(0 \oplus \mathbf{P}_{-1})$ via a low rank update.

Strikingly enough, the argument will further show that because of the shift structure in the input vectors $\mathbf{u}_i$, if this low-rank property holds for the initial time instant then it also holds for the successive time instants! Consequently, the successive matrices $(\mathbf{P}_i \oplus 0)$ will also be low rank modifications of earlier matrices $(0 \oplus \mathbf{P}_{i-1})$.

In this way, a fast procedure for updating the $\mathbf{P}_i$ can be developed by replacing the propagation of $\mathbf{P}_i$ via (21.39) by a recursion that instead propagates the low rank factors that generate the $\mathbf{P}_i$. We will verify that this procedure also allows us to update the weight vector estimates rapidly (in $O(M)$ operations).

21.7.2 Low-Rank Property

Assume we choose Π_0 in the special diagonal form

$$\Pi_0 = \delta \cdot \text{diagonal} \{\lambda^2, \lambda^3, \dots, \lambda^{M+1}\}, \tag{21.53}$$

where δ is a positive quantity (usually much larger than one, $\delta \gg 1$). In this case, we are led to a rank-two difference of the form

$$\begin{bmatrix} \lambda^{-1}\Pi_0 & \mathbf{0} \\ \mathbf{0}^T & 0 \end{bmatrix} - \begin{bmatrix} 0 & \mathbf{0}^T \\ \mathbf{0} & \Pi_0 \end{bmatrix} = \delta \cdot \lambda \cdot \begin{bmatrix} 1 & & \\ & 0 & \\ & & -\lambda^M \end{bmatrix},$$

which can be factored as

$$\begin{bmatrix} \mathbf{P}_0 & \mathbf{0} \\ \mathbf{0}^T & 0 \end{bmatrix} - \begin{bmatrix} 0 & \mathbf{0}^T \\ \mathbf{0} & \mathbf{P}_{-1} \end{bmatrix} = \lambda \cdot \bar{\mathbf{L}}_0 \mathbf{S}_0 \bar{\mathbf{L}}_0^T, \tag{21.54}$$

where $\bar{\mathbf{L}}_0$ is $(M+1) \times 2$ and $\mathbf{S}_0$ is a 2×2 signature matrix that are given by

$$\bar{\mathbf{L}}_0 = \sqrt{\delta} \cdot \begin{bmatrix} 1 & 0 \\ 0 & 0 \\ 0 & \lambda^{\frac{M}{2}} \end{bmatrix}, \quad \mathbf{S}_0 = \begin{bmatrix} 1 & 0 \\ 0 & -1 \end{bmatrix}.$$

21.7.3 A Fast Array Algorithm

We now argue by induction, and by using the shift property of the input vectors $\mathbf{u}_i$, that if the low-rank property holds at a certain time instant i, say

$$\begin{bmatrix} \mathbf{P}_i & \mathbf{0} \\ \mathbf{0}^T & 0 \end{bmatrix} - \begin{bmatrix} 0 & \mathbf{0}^T \\ \mathbf{0} & \mathbf{P}_{i-1} \end{bmatrix} = \lambda \cdot \bar{\mathbf{L}}_i \mathbf{S}_i \bar{\mathbf{L}}_i^T, \tag{21.55}$$

then three important facts hold:

- The low-rank property also holds at time $i+1$, say

$$\begin{bmatrix} \mathbf{P}_{i+1} & \mathbf{0} \\ \mathbf{0}^T & 0 \end{bmatrix} - \begin{bmatrix} 0 & \mathbf{0}^T \\ \mathbf{0} & \mathbf{P}_i \end{bmatrix} = \lambda \cdot \bar{\mathbf{L}}_{i+1} \mathbf{S}_{i+1} \bar{\mathbf{L}}_{i+1}^T,$$

- There exists an array algorithm that updates $\bar{L}_i$ to $\bar{L}_{i+1}$. Moreover, the algorithm also provides the gain vector $\mathbf{g}_i$ that is needed to update the weight-vector estimate in the RLS solution.
- The signature matrices $\{S_i, S_{i+1}\}$ are equal! That is, all successive low-rank differences have the same signature matrix as the initial difference and, hence,

$$S_i = S_0 = \begin{bmatrix} 1 & 0 \\ 0 & -1 \end{bmatrix} \quad \text{for all } i .$$

To verify these claims, consider (21.55) and form the prearray

$$A = \begin{bmatrix} \gamma^{-1/2}(i) & \begin{bmatrix} u(i+1) & \mathbf{u}_i^T \end{bmatrix}\bar{L}_i \\ \begin{bmatrix} 0 \\ \mathbf{g}_i\gamma^{-1/2}(i) \end{bmatrix} & \bar{L}_i \end{bmatrix} .$$

For $M = 3$, the prearray has the following generic form (recall that $\bar{L}_i$ is $(M+1) \times 2$):

$$A = \begin{bmatrix} x & x & x \\ \hline 0 & x & x \\ x & x & x \\ x & x & x \\ x & x & x \end{bmatrix} .$$

Now let Θ_i be a matrix that satisfies

$$\Theta_i \begin{bmatrix} 1 & & \\ & 1 & \\ & & -1 \end{bmatrix} \Theta_i^T = \begin{bmatrix} 1 & & \\ & 1 & \\ & & -1 \end{bmatrix} = \begin{bmatrix} 1 & \\ & S_i \end{bmatrix} ,$$

and such that it transforms A into the form

$$B = \begin{bmatrix} x & 0 & 0 \\ x & x & x \\ x & x & x \\ x & x & x \\ x & x & x \end{bmatrix} = \begin{bmatrix} a & \mathbf{0}^T \\ \mathbf{b} & C \end{bmatrix} .$$

That is, Θ_i annihilates two entries in the top row of the prearray. This can be achieved by employing a circular rotation that pivots with the left-most entry of the first row and annihilates its second entry. We then employ a hyperbolic rotation that pivots again with the left-most entry and annihilates the last entry of the top row.

The unknown entries $\{a, \mathbf{b}, C\}$ can be identified by resorting to the same technique that we employed earlier during the derivation of the QR and inverse QR algorithms. By comparing entries on both sides of the equality

$$A \begin{bmatrix} 1 & \\ & S_i \end{bmatrix} A^T = \begin{bmatrix} a & \mathbf{0}^T \\ \mathbf{b} & C \end{bmatrix}\begin{bmatrix} 1 & \\ & S_i \end{bmatrix}\begin{bmatrix} a & \mathbf{0}^T \\ \mathbf{b} & C \end{bmatrix}^T$$

we obtain several equalities. For example, by equating the $(1, 1)$ entries we obtain the following relation:

$$\gamma^{-1}(i) + \begin{bmatrix} u(i+1) & \mathbf{u}_i^T \end{bmatrix}\bar{L}_i S_i \bar{L}_i^T \begin{bmatrix} u(i+1) \\ \mathbf{u}_i \end{bmatrix} = a^2 . \tag{21.56}$$

By using (21.55) for $\bar{L}_i S_i \bar{L}_i$ and by noting that we can rewrite the vector $\begin{bmatrix} u(i+1) & \mathbf{u}_i^T \end{bmatrix}$ in two equivalent forms (due to its shift structure):

$$\begin{bmatrix} u(i+1) & \mathbf{u}_i^T \end{bmatrix} = \begin{bmatrix} \mathbf{u}_{i+1}^T & u(i-M+1) \end{bmatrix} , \tag{21.57}$$

we readily conclude that (21.56) collapses to

$$\gamma^{-1}(i) + \lambda^{-1}\mathbf{u}_{i+1}^T\mathbf{P}_i\mathbf{u}_{i+1} - \lambda^{-1}\mathbf{u}_i^T\mathbf{P}_{i-1}\mathbf{u}_i = a^2 .$$

But $\gamma^{-1}(i) = 1 + \lambda^{-1}\mathbf{u}_i^T\mathbf{P}_{i-1}\mathbf{u}_i$. Therefore,

$$a^2 = 1 + \lambda^{-1}\mathbf{u}_{i+1}^T\mathbf{P}_i\mathbf{u}_{i+1} = \gamma^{-1}(i+1),$$

which shows that we can identify a as

$$a = \gamma^{-1/2}(i+1).$$

A similar argument allows us to identify $\mathbf{b}$. By comparing the $(2, 1)$ entries we obtain

$$a\mathbf{b} = \begin{bmatrix} 0 \\ \mathbf{g}_i\gamma^{-1}(i) \end{bmatrix} + \bar{\mathbf{L}}_i\mathbf{S}_i\bar{\mathbf{L}}_i^T \begin{bmatrix} u(i+1) \\ \mathbf{u}_i \end{bmatrix}. \tag{21.58}$$

Again, by (21.55) for $\bar{\mathbf{L}}_i\mathbf{S}_i\bar{\mathbf{L}}_i^T$, (21.57) for the vector $\begin{bmatrix} u(i+1) & \mathbf{u}_i^T \end{bmatrix}$, and by noting from the definition of $\mathbf{g}_i$ that

$$\begin{bmatrix} 0 \\ \mathbf{g}_i\gamma^{-1}(i) \end{bmatrix} = \begin{bmatrix} 0 \\ \lambda^{-1}\mathbf{P}_{i-1}\mathbf{u}_i \end{bmatrix}$$

we obtain

$$\mathbf{b} = \begin{bmatrix} \mathbf{g}_{i+1}\gamma^{-1/2}(i+1) \\ 0 \end{bmatrix}.$$

Finally, for the last term $\mathbf{C}$ we compare the $(2, 2)$ entries to obtain

$$\mathbf{C}\mathbf{S}_i\mathbf{C}^T = \begin{bmatrix} \mathbf{P}_{i+1} & \mathbf{0} \\ \mathbf{0}^T & 0 \end{bmatrix} - \begin{bmatrix} 0 & \mathbf{0}^T \\ \mathbf{0} & \mathbf{P}_i \end{bmatrix}.$$

The difference on the right-hand side is by definition $\lambda\bar{\mathbf{L}}_{i+1}\mathbf{S}_{i+1}\bar{\mathbf{L}}_{i+1}^T$. This shows that we can make the identifications

$$\mathbf{C} = \sqrt{\lambda} \cdot \bar{\mathbf{L}}_{i+1} , \quad \mathbf{S}_{i+1} = \mathbf{S}_i .$$

In summary, we have established the validity of the array algorithm shown in Table 21.5, which minimizes the cost function (21.27) in the prewindowed case and for the special choice of Π_0 in (21.53).

Note that this fast procedure computes the required gain vectors $\mathbf{g}_i$ without explicitly evaluating the matrices $\mathbf{P}_i$. Instead, the low-rank factors $\bar{\mathbf{L}}_i$ are propagated, which explains the lower computational requirements.

21.7.4 The Fast Transversal Filter

The fast algorithm of the last section is an array version of fast RLS algorithms known as FTF (Fast Transversal Filter) and FAEST (Fast A posteriori Error Sequential Technique). In contrast to the above array description, where the transformation Θ_i that updates the data from time i to time $(i + 1)$ is left implicit, the FTF and FAEST algorithms involve explicit sets of equations.

The derivation of these explicit sets of equations can be motivated as follows. Note that the factorization (21.54) is highly nonunique. What is special about (21.54) [and also (21.55)] is that we have forced $\mathbf{S}_0$ to be a signature matrix, i.e., a matrix with $\pm 1's$ on its diagonal. More generally, we can allow for different factorizations with an $\mathbf{S}_0$ that is not restricted to be a signature matrix. Different choices lead to different sets of equations.

More explicitly, assume we factor the difference matrix in (21.55) as

$$\begin{bmatrix} \mathbf{P}_i & \mathbf{0} \\ \mathbf{0}^T & 0 \end{bmatrix} - \begin{bmatrix} 0 & \mathbf{0}^T \\ \mathbf{0} & \mathbf{P}_{i-1} \end{bmatrix} = \lambda \cdot \mathbf{L}_i\mathbf{M}_i\mathbf{L}_i^T , \tag{21.59}$$

TABLE 21.5 A Fast Array Algorithm

Input. Prewindowed data $\{d(j), u(j)\}$ for $j \geq 1$ and Π_0 as in (21.53) in the cost (21.27).

Initialization. Set

$$\mathbf{w}_{-1} = \bar{\mathbf{w}}, \quad \gamma^{-1/2}(0) = 1$$

$$\bar{\mathbf{L}}_0 = \sqrt{\delta} \cdot \begin{bmatrix} 1 & 0 \\ 0 & 0 \\ 0 & \lambda^{\frac{M}{2}} \end{bmatrix}, \quad \mathbf{S}_0 = \begin{bmatrix} 1 & 0 \\ 0 & -1 \end{bmatrix}$$

Repeat for each time instant $i \geq 0$:

$$\begin{bmatrix} \gamma^{-1/2}(i) & \begin{bmatrix} u(i+1) & \mathbf{u}_i^T \end{bmatrix} \bar{\mathbf{L}}_i \\ \begin{bmatrix} 0 \\ \mathbf{g}_i \gamma^{-1/2}(i) \end{bmatrix} & \bar{\mathbf{L}}_i \end{bmatrix} \Theta_i = \begin{bmatrix} \gamma^{-1/2}(i+1) & \mathbf{0}^T \\ \begin{bmatrix} \mathbf{g}_{i+1}\gamma^{-1/2}(i+1) \\ 0 \end{bmatrix} & \sqrt{\lambda}\,\bar{\mathbf{L}}_{i+1} \end{bmatrix}$$

where Θ_i is any $(1 \oplus \mathbf{S}_0)$-orthogonal matrix that produces the zero pattern in the postarray, and $\bar{\mathbf{L}}_i$ is a two-column matrix.

The weight-vector estimate is updated via:

$$\mathbf{w}_i = \mathbf{w}_{i-1} + \begin{bmatrix} \dfrac{\mathbf{g}_i}{\gamma^{1/2}(i)} \end{bmatrix} \begin{bmatrix} \gamma^{-1/2}(i) \end{bmatrix}^{-1} \begin{bmatrix} d(i) - \mathbf{u}_i^T \mathbf{w}_{i-1} \end{bmatrix}$$

The computational cost is $O(M)$ per iteration.

where $\mathbf{Ł}_i$ is an $(M+1) \times 2$ matrix and $\mathbf{M}_i$ is a 2×2 matrix that is not restricted to be a signature matrix. [We already know from the earlier array-based argument that this difference is always low-rank.]

Given the factorization (21.59), it is easy to verify that two successive gain vectors satisfy the relation:

$$\begin{bmatrix} \mathbf{g}_{i+1}\gamma^{-1}(i+1) \\ 0 \end{bmatrix} = \begin{bmatrix} 0 \\ \mathbf{g}_i\gamma^{-1}(i) \end{bmatrix} + \mathbf{Ł}_i \mathbf{M}_i \mathbf{Ł}_i^T \begin{bmatrix} u(i+1) \\ \mathbf{u}_i \end{bmatrix}.$$

This is identical to (21.58) except that $\mathbf{S}_i$ is replaced by $\mathbf{M}_i$ and $\bar{\mathbf{L}}_i$ is replaced by $\mathbf{Ł}_i$. The fast array algorithm of the previous section provides one possibility for enforcing this relation and, hence, of updating $\mathbf{g}_i$ to $\mathbf{g}_{i+1}$ via updates of $\bar{\mathbf{L}}_i$.

The FTF and FAEST algorithms follow by employing one such alternative factorization, where the two columns of the factor $\mathbf{Ł}_i$ turn out to be related to the solution of two fundamental problems in adaptive filter theory: the so-called forward and backward prediction problems. Moreover, the $\mathbf{M}_i$ factor turns out to be diagonal with entries equal to the so-called forward and backward minimum prediction energies. An explicit derivation of the FTF equations can be pursued along these lines. We omit the details and continue to focus on the square-root formulation. We now proceed to discuss order-recursive adaptive filters within this framework.

21.8 Order-Recursive Filters

The RLS algorithms that were derived in the previous sections are all fixed-order solutions of (21.27) in the sense that they recursively evaluate successive weight estimates $\mathbf{w}_i$ that correspond to a fixed-order combiner of order M. This form of computing the minimizing solution $\mathbf{w}_N$ is not convenient from an order-recursive point of view. In other words, assume we pose a new optimization problem of the same form as (21.27) but where the vectors $\{\mathbf{w}, \mathbf{u}_j\}$ are now of order $(M+1)$ rather than M. How do the weight estimates of this new higher-dimensional problem relate to the weight estimates of the lower dimensional problem?

Before addressing this issue any further, it is apparent at this stage that we need to introduce a notational modification in order to keep track of the proper sizes of the variables. Indeed, from now on, we shall explicitly indicate the size of a variable by employing an additional subscript. For example, we shall write $\{\mathbf{w}_M, \mathbf{u}_{M,j}\}$ instead of $\{\mathbf{w}, \mathbf{u}_j\}$ to denote vectors of size M.

Returning to the point raised in the previous paragraph, let $\mathbf{w}_{M+1,N}$ denote the optimal solution of the new optimization problem (with $(M+1)$−dimensional vectors $\{\mathbf{w}_{M+1}, \mathbf{u}_{M+1,j}\}$. The adaptive algorithms of the previous sections give an explicit recursive (time-update) relation between $\mathbf{w}_{M,N}$ and $\mathbf{w}_{M,N-1}$. But they do not provide a recursive (order-update) relation between $\mathbf{w}_{M,N}$ and $\mathbf{w}_{M+1,N}$.

There is an alternative to the FIR implementation of Fig. 21.3 that allows us to easily carry over the information from previous computations for the order M filter. This is the so-called lattice filter.

From now on we assume, for simplicity of presentation, that the weighting matrix Π_0 in (21.27) is very large, i.e., $\Pi_0 \to \infty \mathbf{I}$. This assumption reduces (21.27) to a standard least-squares formulation:

$$\min_{\mathbf{w}_M} \left[\sum_{j=0}^{N} \lambda^{N-i} |d(j) - \mathbf{u}_{M,j}^T \mathbf{w}_M|^2 \right]. \tag{21.60}$$

The order-recursive filters of this section deal with this kind of minimization.

Now suppose that our interest in solving (21.60) is not to explicitly determine the weight estimate $\mathbf{w}_{M,N}$, but rather to determine estimates for the reference signals $\{d(\cdot)\}$, say

$$d_M(N) = \mathbf{u}_{M,N}^T \mathbf{w}_{M,N} = \text{ estimate of } d(N) \text{ of order } M.$$

Likewise, for the higher-order problem,

$$d_{M+1}(N) = \mathbf{u}_{M+1,N}^T \mathbf{w}_{M+1,N} = \text{ estimate of } d(N) \text{ of order } M+1.$$

The resulting estimation errors will be denoted by

$$e_M(N) = d(N) - d_M(N), \quad e_{M+1}(N) = d(N) - d_{M+1}(N).$$

The lattice solution allows us to update $e_M(N)$ to $e_{M+1}(N)$ without explicitly computing the weight estimates $\mathbf{w}_{M,N}$ and $\mathbf{w}_{M+1,N}$.

The discussion that follows relies heavily on the orthogonality property of least-squares solutions and, therefore, serves as a good illustration of the power and significance of this property. It will further motivate the introduction of the forward and backward prediction problems.

21.8.1 Joint Process Estimation

For the sake of illustration, and without loss of generality, the discussion in this section assumes particular values for M and λ, say $M = 3$ and $\lambda = 1$. These assumptions simplify the exposition without affecting the general conclusions. In particular, a nonunity λ can always be incorporated into the discussion by properly normalizing the vectors involved in the derivation [cf. (21.28) and (21.29)] and we will do so later. We continue to assume prewindowed data (i.e., the data is zero for time instants $i \leq 0$).

To begin with, assume we solve the following problem [as suggested by (21.60)]: minimize over $\mathbf{w}_3$ the cost function

$$\left\| \underbrace{\begin{bmatrix} 0 \\ d(1) \\ d(2) \\ \vdots \\ d(N) \end{bmatrix}}_{\mathbf{d}_N} - \underbrace{\begin{bmatrix} 0 & 0 & 0 \\ u(1) & 0 & 0 \\ u(2) & u(1) & 0 \\ \vdots & \vdots & \vdots \\ u(N) & u(N-1) & u(N-2) \end{bmatrix}}_{\mathbf{A}_{3,N}} \begin{bmatrix} w_3(1) \\ w_3(2) \\ w_3(3) \end{bmatrix}_{\mathbf{w}_3} \right\|^2 \tag{21.61}$$

where $\mathbf{d}_N$ denotes the vector of desired signals up to time N, and $\mathbf{A}_{3,N}$ denotes a three-column matrix of input data $\{u(\cdot)\}$, also up to time N.

The optimal solution is denoted by $\mathbf{w}_{3,N}$. The subscript N indicates that it is an estimate based on the data $u(\cdot)$ up to time N. Determining $\mathbf{w}_{3,N}$ corresponds to determining the entries of a 3-dimensional weight vector so as to approximate the column vector $\mathbf{d}_N$ by the linear combination $\mathbf{A}_{3,N}\mathbf{w}_{3,N}$ in the least-squares sense (21.61). We thus say that expression (21.61) defines a third-order estimator for the reference sequence $\{d(\cdot)\}$. The resulting *a posteriori* estimation error vector is denoted by

$$\mathbf{e}_{3,N} = \mathbf{d}_N - \mathbf{A}_{3,N}\mathbf{w}_{3,N} \,,$$

where, for example, the last entry of $\mathbf{e}_{3,N}$ is given by

$$e_3(N) = d(N) - \mathbf{u}_{3,N}^T \mathbf{w}_{3,N} \,,$$

and it denotes the *a posteriori* estimation error in estimating $d(N)$ from a linear combination of the three most recent inputs.

We already know from the orthogonality property of least-squares solutions that the *a posteriori* residual vector $\mathbf{e}_{3,N}$ has to be orthogonal to the data matrix $\mathbf{A}_{3,N}$, viz.,

$$\mathbf{A}_{3,N}^T \mathbf{e}_{3,N} = \mathbf{0}.$$

We also know that the optimal solution $\mathbf{w}_{3,N}$ provides an estimate vector $\mathbf{A}_{3,N}\mathbf{w}_{3,N}$ that is the closest element in the column space of $\mathbf{A}_{3,N}$ to the column vector $\mathbf{d}_N$.

Now assume that we wish to solve the next higher order problem, viz., of order $M = 4$: minimize over $\mathbf{w}_4$ the cost function

$$\left\| \mathbf{d}_N - \mathbf{A}_{4,N}\mathbf{w}_4 \right\|^2 \,, \tag{21.62}$$

where

$$\mathbf{A}_{4,N} = \begin{bmatrix} 0 & 0 & 0 & 0 \\ u(1) & 0 & 0 & 0 \\ u(2) & u(1) & 0 & 0 \\ \vdots & \vdots & \vdots & \vdots \\ u(N-1) & u(N-2) & u(N-3) & u(N-4) \\ u(N) & u(N-1) & u(N-2) & u(N-3) \end{bmatrix}, \quad \mathbf{w}_4 = \begin{bmatrix} w_4(1) \\ w_4(2) \\ w_4(3) \\ w_4(4) \end{bmatrix}.$$

This statement is very close to (21.61) except for an extra column in the data matrix $\mathbf{A}_{4,N}$: the first three columns of $\mathbf{A}_{4,N}$ coincide with those of $\mathbf{A}_{3,N}$, while the last column of $\mathbf{A}_{4,N}$ contains the extra new data that are needed for a fourth-order estimator. More specifically, $\mathbf{A}_{3,N}$ and $\mathbf{A}_{4,N}$ are related as follows:

$$\mathbf{A}_{4,N} = \begin{bmatrix} & 0 \\ & 0 \\ & 0 \\ \mathbf{A}_{3,N} & \vdots \\ & u(N-4) \\ & u(N-3) \end{bmatrix}. \tag{21.63}$$

The problem in (21.62) requires us to linearly combine the four columns of $\mathbf{A}_{4,N}$ in order to compute the fourth-order estimates of $\{0, d(1), d(2), \dots, d(N)\}$. In other words, it requires us to determine the closest element in the column space of $\mathbf{A}_{4,N}$ to the same column vector $\mathbf{d}_N$.

We already know what is the closest element to $\mathbf{d}_N$ in the column space of $\mathbf{A}_{3,N}$, which is a submatrix of $\mathbf{A}_{4,N}$. This suggests that we should try to decompose the column space of $\mathbf{A}_{4,N}$ into two orthogonal subspaces, viz.,

$$\text{Range}(\mathbf{A}_{4,N}) = \text{Range}(\mathbf{A}_{3,N}) \oplus \text{Range}(\mathbf{m}) , \tag{21.64}$$

where $\mathbf{m}$ is a column vector that is orthogonal to $\mathbf{A}_{3,N}$, $\mathbf{A}_{3,N}^T \mathbf{m} = \mathbf{0}$. The notation $\text{Range}(\mathbf{A}_{3,N}) \oplus \text{Range}(\mathbf{m})$ also means that every element in the column space of $\mathbf{A}_{4,N}$ can be expressed as a linear combination of the columns of $\mathbf{A}_{3,N}$ and of $\mathbf{m}$.

The desired decomposition motivates the backward prediction problem.

21.8.2 The Backward Prediction Error Vectors

We continue to assume $\lambda = 1$ and $M = 3$, and we note that the required decomposition can be accomplished by projecting the last column of $\mathbf{A}_{4,N}$ onto the column space of its first three columns (i.e., onto the column space of $\mathbf{A}_{3,N}$) and keeping the residual vector as the desired vector $\mathbf{m}$. This is nothing but a Gram-Schmidt orthogonalization step and it is equivalent to the following minimization problem: minimize over $\mathbf{w}_3^b$

$$\left\| \underbrace{\begin{bmatrix} 0 \\ 0 \\ 0 \\ \vdots \\ u(N-4) \\ u(N-3) \end{bmatrix}}_{\substack{\text{Last column} \\ \text{of } \mathbf{A}_{4,N}}} - \mathbf{A}_{3,N} \underbrace{\begin{bmatrix} w_3^b(1) \\ w_3^b(2) \\ w_3^b(3) \end{bmatrix}}_{\mathbf{w}_3^b} \right\|^2 . \tag{21.65}$$

This is also a special case of (21.60) where we have replaced the sequence

$$\{0, d(1), \ldots, d(N)\}$$

by the sequence

$$\{0, 0, 0, \ldots, u(N-4), u(N-3)\}.$$

We denote the optimal solution by $\mathbf{w}_{3,N}^b$. The subscript N indicates that it is an estimate based on the data $u(\cdot)$ up to time N. Determining $\mathbf{w}_{3,N}^b$ corresponds to determining the entries of a 3-dimensional weight vector so as to approximate the last column of $\mathbf{A}_{4,N}$ by a linear combination of the columns of $\mathbf{A}_{3,N}$, viz., $\mathbf{A}_{3,N} \mathbf{w}_{3,N}^b$, in the least-squares sense.

Note that the entries in every row of the data matrix $\mathbf{A}_{3,N}$ are the three "future" values corresponding to the entry in the last column of $\mathbf{A}_{4,N}$. Hence, the last element of the above linear combination serves as a *backward* prediction of $u(N-3)$ in terms of $\{u(N), u(N-1), u(N-2)\}$. A similar remark holds for the other entries. The superscript ^{b} stands for *backward*.

We thus say that expression (21.65) defines a third-order backward prediction problem. The resulting *a posteriori* backward prediction error vector is denoted by

$$
\mathbf{b}_{3,N} = \begin{bmatrix} 0 \\ 0 \\ 0 \\ \vdots \\ u(N-4) \\ u(N-3) \end{bmatrix} - \mathbf{A}_{3,N}\mathbf{w}_{3,N}^{b} .
$$

In particular, the *last* entry of $\mathbf{b}_{3,N}$ is defined as the a *posteriori* backward prediction error in estimating $u(N-3)$ from a linear combination of the future 3 inputs. It is denoted by $b_3(N)$ and is given by

$$
b_3(N) = u(N-3) - \mathbf{u}_{3,N}^{T}\mathbf{w}_{3,N}^{b} . \tag{21.66}
$$

We further know, from the orthogonality property of least-squares solutions, that the *a posteriori* backward residual vector $\mathbf{b}_{3,N}$ has to be orthogonal to the data matrix $\mathbf{A}_{3,N}$, $\mathbf{A}_{3,N}^{T}\mathbf{b}_{3,N} = \mathbf{0}$, which therefore implies that it can be taken as the **m** column that we mentioned earlier, viz., we can write

$$
\text{Range} (\mathbf{A}_{4,N}) = \text{Range} (\mathbf{A}_{3,N}) \oplus \text{Range} (\mathbf{b}_{3,N}). \tag{21.67}
$$

Our original motivation for introducing the *a posteriori* backward residual vector $\mathbf{b}_{3,N}$ was the desire to solve the fourth-order problem (21.62), not afresh, but in a way so as to exploit the solution of lower order, thus leading to an order-recursive algorithm.

Assume now that we have available the estimation error vectors $\mathbf{e}_{3,N}$ and $\mathbf{b}_{3,N}$, which are both orthogonal to $\mathbf{A}_{3,N}$. Knowing that $\mathbf{b}_{3,N}$ leads to an orthogonal decomposition of the column space of $\mathbf{A}_{4,N}$ as in (21.67), then updating $\mathbf{e}_{3,N}$ into a fourth-order *a posteriori* residual vector $\mathbf{e}_{4,N}$, which has to be orthogonal to $\mathbf{A}_{4,N}$, simply corresponds to projecting the vector $\mathbf{e}_{3,N}$ onto the vector $\mathbf{b}_{3,N}$. More explicitly, it corresponds to determining a scalar coefficient k_3 that solves the optimization problem

$$
\min_{k_3} \left\| \mathbf{e}_{3,N} - k_3\mathbf{b}_{3,N} \right\|^2 . \tag{21.68}
$$

This is a standard least-squares problem and its optimal solution is denoted by

$$
k_3(N) = \frac{1}{\mathbf{b}_{3,N}^{T}\mathbf{b}_{3,N}} \mathbf{b}_{3,N}^{T}\mathbf{e}_{3,N} . \tag{21.69}
$$

We now know how to update $\mathbf{e}_{3,N}$ into $\mathbf{e}_{4,N}$ by projecting $\mathbf{e}_{3,N}$ onto $\mathbf{b}_{3,N}$. In order to be able to proceed with this order update procedure, we still need to know how to order-update the backward residual vector. That is, we need to know how to go from $\mathbf{b}_{3,N}$ to $\mathbf{b}_{4,N}$.

21.8.3 The Forward Prediction Error Vectors

We continue to assume $\lambda = 1$ and $M = 3$. The order-update of the backward residual vector motivates us to introduce the forward prediction problem: minimize over $\mathbf{w}_3^{f}$ the cost function

$$
\left\| \begin{bmatrix} u(1) \\ u(2) \\ u(3) \\ \vdots \\ u(N+1) \end{bmatrix} - \mathbf{A}_{3,N} \underbrace{\begin{bmatrix} w_3^{f}(1) \\ w_3^{f}(2) \\ w_3^{f}(3) \end{bmatrix}}_{\mathbf{w}_3^{f}} \right\|^2 . \tag{21.70}
$$

We denote the optimal solution by $\mathbf{w}_{3,N+1}^f$. The subscript indicates that it is an estimate based on the data $u(\cdot)$ up to time $N + 1$. Determining $\mathbf{w}_{3,N+1}^f$ corresponds to determining the entries of a 3-dimensional weight vector so as to approximate the column vector

$$
\begin{bmatrix}
u(1) \\
u(2) \\
u(3) \\
\vdots \\
u(N+1)
\end{bmatrix}
$$

by a linear combination of the columns of $\mathbf{A}_{3,N}$, viz., $\mathbf{A}_{3,N}\mathbf{w}_{3,N+1}^f$.

Note that the entries of the successive rows of the data matrix $\mathbf{A}_{3,N}$ are the past three inputs relative to the corresponding entries of the column vector. Hence, the last element of the linear combination $\mathbf{A}_{3,N}\mathbf{w}_{3,N+1}^f$ serves as a *forward* prediction of $u(N+1)$ in terms of $\{u(N), u(N-1), u(N-2)\}$. A similar remark holds for the other entries. The superscript f stands for *forward*.

We thus say that expression (21.70) defines a third-order forward prediction problem. The resulting *a posteriori* forward prediction error vector is denoted by

$$
\mathbf{f}_{3,N+1} =
\begin{bmatrix}
u(1) \\
u(2) \\
u(3) \\
\vdots \\
u(N+1)
\end{bmatrix}
- \mathbf{A}_{3,N}\mathbf{w}_{3,N+1}^f .
$$

In particular, the *last* entry of $\mathbf{f}_{3,N+1}$ is defined as the *a posteriori* forward prediction error in estimating $u(N+1)$ from a linear combination of the past three inputs. It is denoted by $f_3(N+1)$ and is given by

$$
f_3(N+1) = u(N+1) - \mathbf{u}_{3,N}\mathbf{w}_{3,N+1}^f . \tag{21.71}
$$

Now assume that we wish to solve the next-higher order problem, viz., of order $M = 4$: minimize over $\mathbf{w}_4^f$ the cost function

$$
\left\|
\begin{bmatrix}
u(1) \\
u(2) \\
u(3) \\
\vdots \\
u(N+1)
\end{bmatrix}
- \mathbf{A}_{4,N}
\begin{bmatrix}
w_4^f(1) \\
w_4^f(2) \\
w_4^f(3) \\
w_4^f(4)
\end{bmatrix}
\right\|^2 . \tag{21.72}
$$

We again observe that this statement is very close to (21.70) except for an extra column in the data matrix $\mathbf{A}_{4,N}$, in precisely the same way as happened with $\mathbf{e}_{4,N}$ and $\mathbf{b}_{3,N}$. We can therefore obtain $\mathbf{f}_{4,N+1}$ by projecting $\mathbf{f}_{3,N+1}$ onto $\mathbf{b}_{3,N}$ and taking the residual vector as $\mathbf{f}_{4,N+1}$,

$$
\min_{k_3^f} \|\mathbf{f}_{3,N+1} - k_3^f \mathbf{b}_{3,N}\|^2 . \tag{21.73}
$$

This is also a standard least-squares problem and we denote its optimal solution by $k_3^f(N+1)$,

$$
k_3^f(N+1) = \frac{\mathbf{b}_{3,N}^T \mathbf{f}_{3,N+1}}{\mathbf{b}_{3,N}^T \mathbf{b}_{3,N}} , \tag{21.74}
$$

with

$$\mathbf{f}_{4,N+1} = \mathbf{f}_{3,N+1} - k_3^f(N+1)\mathbf{b}_{3,N} . \tag{21.75}$$

Similarly, the backward residual vector $\mathbf{b}_{3,N}$ can be updated to $\mathbf{b}_{4,N+1}$ by projecting $\mathbf{b}_{3,N}$ onto $\mathbf{f}_{3,N+1}$,

$$\min_{k_3^b} \left\| \mathbf{b}_{3,N} - k_3^b \mathbf{f}_{3,N+1} \right\|^2 , \tag{21.76}$$

and we get, after denoting the optimal solution by $k_3^b(N+1)$,

$$\mathbf{b}_{4,N+1} = \mathbf{b}_{3,N} - k_3^b(N+1)\mathbf{f}_{3,N+1} , \tag{21.77}$$

where

$$k_3^b(N+1) = \frac{\mathbf{f}_{3,N+1}^T \mathbf{b}_{3,N}}{\mathbf{f}_{3,N+1}^T \mathbf{f}_{3,N+1}} . \tag{21.78}$$

Note the change in the time index as we move from $\mathbf{b}_{3,N}$ to $\mathbf{b}_{4,N+1}$. This is because $\mathbf{b}_{4,N+1}$ is obtained by projecting $\mathbf{b}_{3,N}$ onto $\mathbf{f}_{3,N+1}$, which corresponds to the following definition for $\mathbf{b}_{4,N+1}$,

$$\mathbf{b}_{4,N+1} = \begin{bmatrix} 0 \\ 0 \\ 0 \\ \vdots \\ u(N-4) \\ u(N-3) \end{bmatrix} - \mathbf{A}_{4,N+1} \underbrace{\begin{bmatrix} w_{4,N+1}^b(1) \\ w_{4,N+1}^b(2) \\ w_{4,N+1}^b(3) \\ w_{4,N+1}^b(4) \end{bmatrix}}_{\mathbf{w}_{4,N+1}^b} .$$

Finally, in view of (21.69), the joint process estimation problem involves a recursion of the form

$$\mathbf{e}_{4,N} = \mathbf{e}_{3,N} - k_3(N)\mathbf{b}_{3,N} , \tag{21.79}$$

where

$$k_3(N) = \frac{\mathbf{b}_{3,N}^T \mathbf{e}_{3,N}}{\mathbf{b}_{3,N}^T \mathbf{b}_{3,N}} . \tag{21.80}$$

21.8.4   A Nonunity Forgetting Factor

For a general filter order M and for a nonunity λ, an extension of the above arguments would show that the prediction vectors can be updated as follows:

$$
\begin{aligned}
\mathbf{f}_{M+1,N+1} &= \mathbf{f}_{M,N+1} - k_M^f(N+1)\mathbf{b}_{M,N} , \\
\mathbf{b}_{M+1,N+1} &= \mathbf{b}_{M,N} - k_M^b(N+1)\mathbf{f}_{M,N+1} , \\
\mathbf{e}_{M+1,N} &= \mathbf{e}_{M,N} - k_M(N)\mathbf{b}_{M,N} , \\
k_M^f(N+1) &= \frac{\mathbf{b}_{M,N}^T \Lambda_N \mathbf{f}_{M,N+1}}{\mathbf{b}_{M,N}^T \Lambda_N \mathbf{b}_{M,N}} , \\
k_M^b(N+1) &= \frac{\mathbf{f}_{M,N+1}^T \Lambda_N \mathbf{b}_{M,N}}{\mathbf{f}_{M,N+1}^T \Lambda_N \mathbf{f}_{M,N+1}} , \\
k_M(N) &= \frac{\mathbf{b}_{M,N}^T \Lambda_N \mathbf{e}_{M,N}}{\mathbf{b}_{M,N}^T \Lambda_N \mathbf{b}_{M,N}} ,
\end{aligned}
$$

where

$$\Lambda_N = \text{diag} \{\lambda^N, \lambda^{N-1}, \ldots, \lambda, 1\}.$$

For completeness, we also include the defining relations for the *a priori* and *a posteriori* prediction errors:

$$\beta_M(N) = u(N - M) - \mathbf{u}_{M,N}^T \mathbf{w}_{M,N-1}^b,$$

$$b_M(N) = u(N - M) - \mathbf{u}_{M,N}^T \mathbf{w}_{M,N}^b,$$

$$\alpha_M(N + 1) = u(N + 1) - \mathbf{u}_{M,N}^T \mathbf{w}_{M,N}^f,$$

$$f_M(N + 1) = u(N + 1) - \mathbf{u}_{M,N}^T \mathbf{w}_{M,N+1}^f.$$

Using the definition (21.40) for a conversion factor in a least-squares formulation, it is easy to see that the same factor converts the *a priori* prediction errors to the corresponding *a posteriori* prediction errors. This factor will be denoted by $\gamma_M(N)$.

TABLE 21.6 Useful Relations for the Prediction Problems

Variable	Definition or Relation
A priori forward error	$\alpha_M(N+1) = u(N+1) - \mathbf{u}_{M,N}^T \mathbf{w}_{M,N-1}^f$
A priori backward error	$\beta_M(N) = u(N-M) - \mathbf{u}_{M,N}^T \mathbf{w}_{M,N-1}^b$
A posteriori forward error	$f_M(N+1) = u(N+1) - \mathbf{u}_{M,N}^T \mathbf{w}_{M,N}^f$
A posteriori backward error	$b_M(N) = u(N-M) - \mathbf{u}_{M,N}^T \mathbf{w}_{M,N}^b$
Forward error by conversion	$f_M(N+1) = \alpha_M(N+1)\gamma_M(N)$
Backward error by conversion	$b_M(N) = \beta_M(N)\gamma_M(N)$
Gain vector	$g_{M,N} = \Phi_{M,N}^{-1} \mathbf{u}_{M,N}$
Conversion factor	$\gamma_M(N) = 1 - \mathbf{u}_{M,N}^T \Phi_{M,N}^{-1} \mathbf{u}_{M,N}$
Minimum forward-prediction error energy	$\xi_M^f(N+1) = \lambda\xi_M^f(N) + |\bar{f}_M(N+1)|^2$
Minimum backward-prediction error energy	$\xi_M^b(N+1) = \lambda\xi_M^b(N) + |\bar{b}_M(N+1)|^2$

Table 21.6 summarizes, for ease of reference, the definitions and relations that have been introduced thus far. In particular, the last two lines of the table also provide time-update relations for the minimum costs of the forward and backward prediction problems. These costs are denoted by $\xi_M^f(N+1)$ and $\xi_M^b(N)$ and they are equal to the quantities $\mathbf{f}_{M,N+1}^T \Lambda_N \mathbf{f}_{M,N+1}$ and $\mathbf{b}_{M,N}^T \Lambda_N \mathbf{b}_{M,N}$ that appear in the denominators of some of the earlier expressions. The last two relations of Table 21.6 use the result (21.43) to express the minimum costs in terms of the so-called *angle-normalized* prediction errors:

$$\bar{f}_M(N + 1) = \alpha_M(N + 1)\gamma_M^{1/2}(N), \tag{21.81}$$

$$\bar{b}_M(N) = \beta_M(N)\gamma_M^{1/2}(N). \tag{21.82}$$

We can derive, in different ways, similar update relations for the inner product terms

$$\Delta_M(N + 1) = \mathbf{f}_{M,N+1}^T \Lambda_N \mathbf{b}_{M,N},$$

$$\rho_M(N) = \mathbf{b}_{M,N}^T \Lambda_N \mathbf{e}_{M,N}.$$

One possibility is to note, after some algebra and using the orthogonality principle, that the following relation holds:

$$\Delta_M(N + 1) = \begin{bmatrix} 1 & -(\mathbf{w}_{M,N}^f)^T & 0 \end{bmatrix} \Phi_{M+2,N+1} \begin{bmatrix} 0 \\ -\mathbf{w}_{M,N}^b \\ 1 \end{bmatrix},$$

where

$$\Phi_{M+2,N+1} = \sum_{j=0}^{N+1} \lambda^{N+1-j} \mathbf{u}_{M+2,j} \mathbf{u}_{M+2,j}^T$$

If we now invoke the time-update expression

$$\Phi_{M+2,N+1} = \lambda \Phi_{M+2,N} + \mathbf{u}_{M+2,N+1}^T \mathbf{u}_{M+2,N+1},$$

we conclude that $\Delta_M(N+1)$ satisfies the time-update formula:

$$
\begin{aligned}
\Delta_M(N+1) &= \lambda \Delta_M(N) + \alpha_M(N+1) b_M(N) \\
&= \lambda \Delta_M(N) + \frac{f_M(N+1) b_M(N)}{\gamma_M(N)}.
\end{aligned}
$$

A similar argument for $\rho_M(N)$ shows that it satisfies the time-update relation

$$\rho_M(N) = \lambda \rho_M(N-1) + \frac{e_M(N) b_M(N)}{\gamma_M(N)}.$$

Finally, the orthogonality principle can again be invoked to derive order-update (rather than time-update) relations for $\xi_M^f(N+1)$ and $\xi_M^b(N)$. Indeed, using $\mathbf{f}_{M+1,N+1}^T \Lambda_N \mathbf{b}_{M,N} = 0$ we obtain

$$
\begin{aligned}
\xi_{M+1}^f(N+1) &= \mathbf{f}_{M+1,N+1}^T \Lambda_N \mathbf{f}_{M+1,N+1} = \mathbf{f}_{M+1,N+1}^T \Lambda_N \mathbf{f}_{M,N+1}, \\
&= \xi_M^f(N+1) - \frac{\|\Delta_M(N+1)\|^2}{\xi_M^b(N)}.
\end{aligned}
$$

Likewise,

$$\xi_{M+1}^b(N+1) = \xi_M^b(N) - \frac{\|\Delta_M(N+1)\|^2}{\xi_M^f(N+1)}.$$

Table 21.7 summarizes the order-update relations derived thus far.

21.8.5 The QRD Least-Squares Lattice Filter

There are many variants of adaptive lattice algorithms. In this section we present one such variant in square-root form. Most, if not all, other alternatives can be obtained as special cases. Some alternatives propagate the *a posteriori* prediction errors $\{f_M(N+1), b_M(N)\}$, while others employ the *a priori* prediction errors $\{\alpha_M(N+1), \beta_M(N)\}$. The QRD-LSL algorithm we present here is invariant to the particular choice of *a posteriori* or *a priori* errors because it propagates the *angle normalized* prediction errors that we introduced earlier in (21.81) and (21.82), viz.,

$$
\begin{aligned}
\bar{f}_M(i+1) &= \alpha_M(i+1) \gamma_M^{1/2}(i) = [u(i+1) - \mathbf{u}_{M,i}^T \mathbf{w}_{M,i}^f] \gamma_M^{1/2}(i), \\
\bar{b}_M(i) &= \beta_M(i) \gamma_M^{1/2}(i) = [u(i-M) - \mathbf{u}_{M,i}^T \mathbf{w}_{M,i-1}^b] \gamma_M^{1/2}(i).
\end{aligned}
$$

The QRD-LSL algorithm can be motivated as follows. Assume we form the following two vectors of angle normalized prediction errors:

$$
\bar{\mathbf{f}}_{M,N+1} = \begin{bmatrix} \bar{f}_M(1) \\ \bar{f}_M(2) \\ \vdots \\ \bar{f}_M(N+1) \end{bmatrix}, \quad \bar{\mathbf{b}}_{M,N} = \begin{bmatrix} \bar{b}_M(0) \\ \bar{b}_M(1) \\ \vdots \\ \bar{b}_M(N) \end{bmatrix}. \tag{21.83}
$$

TABLE 21.7   Order-Update Relations

$$\Delta_M(N+1) = \lambda \Delta_M(N) + \frac{f_M(N+1)b_M(N)}{\gamma_M(N)}$$

$$\rho_M(N) = \lambda \rho_M(N-1) + \frac{e_M(N)b_M(N)}{\gamma_M(N)}$$

$$\xi_M^f(N+1) = \lambda \xi_M^f(N) + \frac{|f_M(N+1)|^2}{\gamma_M(N)}$$

$$\xi_M^b(N) = \lambda \xi_M^b(N-1) + \frac{|b_M(N)|^2}{\gamma_M(N)}$$

$$k_M^f(N+1) = \Delta_M(N+1)/\xi_M^b(N)$$

$$k_M^b(N+1) = \Delta_M(N+1)/\xi_M^f(N+1)$$

$$k_M(N) = \rho_M(N)/\xi_M^b(N)$$

$$f_{M+1}(N+1) = f_M(N+1) - k_M^f(N+1)b_M(N)$$

$$b_{M+1}(N+1) = b_M(N) - k_M^b(N+1)f_M(N+1)$$

$$e_{M+1}(N) = e_M(N) - k_M(N)b_M(N)$$

$$\xi_{M+1}^f(N+1) = \xi_M^f(N+1) - \frac{|\Delta_M(N+1)|^2}{\xi_M^b(N)}$$

$$\xi_{M+1}^b(N+1) = \xi_M^b(N) - \frac{|\Delta_M(N+1)|^2}{\xi_M^f(N+1)}$$

We then conclude from the time-updates in Table 21.6 for $\xi_M^f(N+1)$ and $\xi_M^b(N)$ that $\xi_M^f(N+1)$ and $\xi_M^b(N)$ are the (weighted) squared Euclidean norms of the angle normalized vectors $\bar{\mathbf{f}}_M(N+1)$ and $\bar{\mathbf{b}}_M(N)$, respectively. That is, $\xi_M^f(N+1) = \bar{\mathbf{f}}_{M,N+1}^T \Lambda_N \bar{\mathbf{f}}_{M,N+1}$ and $\xi_M^b(N) = \bar{\mathbf{b}}_{M,N}^T \Lambda_N \bar{\mathbf{b}}_{M,N}$. Likewise, it follows from the time-update for $\Delta_M(N+1)$ that it is equal to the inner product of the angle normalized vectors,

$$\Delta_M(N+1) = \bar{\mathbf{b}}_{M,N}^T \Lambda_N \bar{\mathbf{f}}_{M,N+1} . \tag{21.84}$$

Consequently, the coefficients $k_M^f(N+1)$ and $k_M^b(N+1)$ are also equal to the ratios of the inner product of the angle normalized vectors to their energies. But recall that $k_M^f(N+1)$ is the coefficient we need in order to project $\mathbf{f}_{M,N+1}$ onto $\mathbf{b}_{M,N}$. This means that we can alternatively evaluate the same coefficient by posing the problem of projecting $\bar{\mathbf{f}}_{M,N+1}$ onto $\bar{\mathbf{b}}_{M,N}$. In a similar fashion, $k_M^b(N+1)$ can be evaluated alternatively by projecting $\bar{\mathbf{b}}_{M,N}$ onto $\bar{\mathbf{f}}_{M,N+1}$. (The inner products and projections are to be understood here to include the additional weighting by Λ_N.)

We are therefore reduced to two simple projection problems that involve projecting a vector onto another vector (with exponential weighting). But these are special cases of standard least-squares problems. In particular, recall that the QR solution of Table 21.4 solves the problem of projecting a given vector $\mathbf{d}_N$ onto the range space of a data matrix $\mathbf{A}_N$ (whose rows are $\mathbf{u}_j^T$).

In a similar fashion, we can write down the QR solution that would solve the problem of projecting $\bar{\mathbf{f}}_{M,N+1}$ onto $\bar{\mathbf{b}}_{M,N}$. For this purpose, we introduce the scalar variables $q_M^f(N+1)$ and $q_M^b(N+1)$ [recall the earlier notation (21.50)]:

$$q_M^b(N+1) = \frac{\Delta_M(N+1)}{\xi_M^{b/2}(N)} , \quad q_M^f(N+1) = \frac{\Delta_M(N+1)}{\xi_M^{f/2}(N+1)} . \tag{21.85}$$

The QR array that updates the forward prediction errors can now be obtained as follows. Form the 3×2 prearray (this is a special case of the QR array of Table 21.4):

$$\mathbf{A} = \begin{bmatrix} \sqrt{\lambda}\xi_M^{b/2}(N-1) & \bar{b}_M(N) \\ \sqrt{\lambda}q_M^b(N) & \bar{f}_M(N+1) \\ 0 & 1 \end{bmatrix}$$

and choose an orthogonal rotation $\Theta^b_{M,N}$ that reduces it to the form

$$
A\Theta^b_{M,N} = \begin{bmatrix} x & 0 \\ a & b \\ y & c \end{bmatrix}.
$$

That is, it annihilates the second entry in the top row of the prearray. The scalar quantities $\{a, b, c, x, y\}$ can be identified, as before, by squaring and comparing entries of the resulting equality. This step allows us to make the following identifications very immediately:

$$
\begin{aligned}
x &= \xi_M^{b/2}(N), \\
a &= q_M^b(N+1), \\
y &= \bar{b}_M(N)\xi_M^{-b/2}(N), \\
bc &= \gamma_M^{-1/2}(N)f_{M+1}(N+1), \\
b^2 &= |\bar{f}_{M+1}(N+1)|^2,
\end{aligned}
$$

where for the last equality we used the following relation that follows immediately from the last two lines of Table 21.7:

$$
\left\| q_M^b(N+1) \right\|^2 + \left\| \bar{f}_{M+1}(N+1) \right\|^2 = \lambda \left\| q_M^b(N) \right\|^2 + \left\| \bar{f}_M(N+1) \right\|^2
$$

Therefore, $b^2c^2 = \frac{\gamma_{M+1}(N)}{\gamma_M(N)}|\bar{f}_{M+1}(N+1)|^2$ and we can make the identifications:

$$
c = \frac{\gamma_{M+1}^{1/2}(N)}{\gamma_M^{1/2}(N)}, \quad b = \bar{f}_{M+1}(N+1).
$$

A similar argument leads to an array equation for the update of the backward errors. In summary, we obtain the QRD-LSL algorithm (listed in Table 21.8) for the update of the angle-normalized forward and backward prediction errors with prewindowed data that correspond to the minimization problem:

$$
\min_{\mathbf{w}_M} \sum_{j=0}^{N} \lambda^{N-j} |d(j) - \mathbf{u}_{M,j}^T \mathbf{w}_M|^2.
$$

The recursions of the table can be shown to collapse, by squaring and comparing terms on both sides of the resulting equality, to several lattice forms that are available in the literature. We forgo the details here.

21.8.6 The Filtering or Joint Process Array

We now return to the estimation of the sequence $\{d(\cdot)\}$. We argued earlier that if we are given the backward residual vector $\mathbf{b}_{M,N}$ and the estimation residual vector $\mathbf{e}_{M,N}$, then the higher-order estimation residual vector $\mathbf{e}_{M+1,N}$ can be obtained by projecting $\mathbf{e}_{M,N}$ onto $\mathbf{b}_{M,N}$ and using the corresponding residual vector as $\mathbf{e}_{M+1,N}$.

Arguments similar to what we have done in the previous section will readily show that the array for the joint process estimation problem is the following: define the angle-normalized residual

$$
\bar{e}_M(i) = e_M(i)\gamma_M^{-1/2}(i) = [d(i) - \mathbf{u}_{M,i}^T \mathbf{w}_{M,i}]\gamma_M^{-1/2}(i),
$$

as well as the scalar quantity

$$
q_M^d(N) = \frac{\rho_M(N)}{\xi_M^{b/2}(N)}.
$$

TABLE 21.8   The QRD Least-Squares Lattice Algorithm

Input. Prewindowed data $\{d(j), u(j)\}$ for $j \geq 1$.

Initialization. For each $M = 0, 1, 2, \ldots, M_{max}$ set

$$\xi_M^{f/2}(0) = 0, \quad \xi_M^{b/2}(-1) = 0, \quad q_M^b(0) = 0 = q_M^f(0)$$

- For each time instant $N \geq 0$ do:

$$\gamma_0(N) = 1, \quad \bar{f}_0(N) = u(N), \quad \bar{b}_0(N) = u(N)$$

- For each $M = 0, 1, 2, \ldots, M_{max} - 1$ do:

$$\begin{bmatrix} \sqrt{\lambda}\xi_M^{b/2}(N-1) & \bar{b}_M(N) \\ \sqrt{\lambda}q_M^b(N) & \bar{f}_M(N+1) \\ 0 & \gamma_M^{1/2}(N) \end{bmatrix} \Theta_{M,N}^b = \begin{bmatrix} \xi_M^{b/2}(N) & 0 \\ q_M^b(N+1) & \bar{f}_{M+1}(N+1) \\ b_M(N)\xi_M^{-b/2}(N) & \gamma_{M+1}^{1/2}(N) \end{bmatrix}$$

$$\begin{bmatrix} \sqrt{\lambda}\xi_M^{f/2}(N) & \bar{f}_M(N+1) \\ \sqrt{\lambda}q_M^f(N) & \bar{b}_M(N) \end{bmatrix} \Theta_{M,N+1}^f = \begin{bmatrix} \xi_M^{f/2}(N+1) & 0 \\ q_M^f(N+1) & \bar{b}_{M+1}(N+1) \end{bmatrix}$$

The orthogonal matrices $\Theta_{M,N}^b$ and $\Theta_{M,N+1}^f$ are chosen so as to annihilate the (1, 2) entries in the corresponding postarrays.

- end
◇ end

Then the array for the filtering process is what is shown in Table 21.9. Note that it uses precisely the same rotation as the first array in the QRD-LSL algorithm. Hence, the second line in the above array can be included as one more line in the first array of QRD-LSL, thus completing the algorithm to also include the joint-process estimation part.

TABLE 21.9   Array for Joint Process Estimation

Input. Prewindowed data $\{d(j), u(j)\}$ for $j \geq 1$.

Initialization. For each $M = 0, 1, 2, \ldots, M_{max}$ set

$$\xi_M^{b/2}(-1) = 0, \quad q_M^d(-1) = 0, \quad q_M^b(0) = 0$$

- For each time instant $N \geq 0$ do:

$$\gamma_0(N) = 1, \quad \bar{e}_0(N) = d(N), \quad \bar{b}_0(N) = u(N)$$

- For each $M = 0, 1, 2, \ldots, M_{max} - 1$ do:

$$\begin{bmatrix} \sqrt{\lambda}\xi_M^{b/2}(N-1) & \bar{b}_M(N) \\ \sqrt{\lambda}q_M^d(N-1) & \bar{e}_M(N) \end{bmatrix} \Theta_{M,N}^b = \begin{bmatrix} \xi_M^{b/2}(N) & 0 \\ q_M^d(N) & \bar{e}_{M+1}(N) \end{bmatrix}$$

where the orthogonal matrix $\Theta_{M,N}^b$ is the same as in the QRD-LSL algorithm.

- end
◇ end

21.9   Concluding Remarks

The intent of this chapter was to provide an overview of the fundamentals of recursive least-squares estimation, with emphasis on array formulations of the varied algorithms (slow or fast) that are available for this purpose. More details and related discussion can be found in several of the references indicated in this section. The references are not intended to be complete but rather indicative of the work in the different areas. More complete lists can be found in several of the textbooks mentioned herein.

References

Detailed discussions on the different forms of RLS adaptive algorithms and their potential applications can be found in:

[1] Haykin, S., *Adaptive Filter Theory,* 3rd ed., Prentice-Hall, Englewood Cliffs, NJ, 1996.
[2] Proakis, J.G., Rader, C.M., Ling, F., and Nikias, C.L., *Advanced Digital Signal Processing,* Macmillan, New York, 1992.
[3] Honig, M.L. and Messerschmitt, D.G., *Adaptive Filters — Structures, Algorithms and Applications,* Kluwer Academic Publishers, 1984.
[4] Orfanidis, S.J., *Optimum Signal Processing,* 2nd ed., McGraw-Hill, New York, 1988.
[5] Kalouptsidis, N. and Theodoridis, S., *Adaptive System Identification and Signal Processing Algorithms,* Prentice-Hall, Englewood Cliffs, NJ, 1993.

The array formulation that we emphasized in this chapter is motivated by the state-space approach developed in

[6] Sayed, A.H. and Kailath, T., A state-space approach to adaptive RLS filtering, *IEEE Signal Processing Magazine,* 11(3), 18–60, July 1994.

This reference also clarifies the connections between adaptive RLS filtering and Kalman filter theory and treats other forms of lattice filters.

A detailed discussion of the square-root formulation in the context of Kalman filtering can be found in

[7] Morf, M. and Kailath, T. Square root algorithms for least squares estimation, *IEEE Trans. Automatic Control,* AC-20(4), 487–497, Aug. 1975.

Further motivation, and earlier discussion, on lattice algorithms can be found in several places in the literature:

[8] Lee, D.T.L., Morf, M., and Friedlander, B., Recursive least-squares ladder estimation algorithms, *IEEE Trans. Circuits and Systems,* CAS-28(6), 467–481, June 1981.
[9] Friedlander, B., Lattice filters for adaptive processing, *Proc. IEEE,* 70(8), 829–867, Aug. 1982.
[10] Lev-Ari, H., Kailath, T., and Cioffi, J., Least squares adaptive lattice and transversal filters: a unified geometrical theory, *IEEE Trans. Information Theory,* IT-30(2), 222–236, March, 1984.

The fast fixed-order recursive least-squares algorithms (FTF and FAEST) were independently derived in

[11] Carayannis, G., Manolakis, D., and Kalouptsidis, N., A fast sequential algorithm for least squares filtering and prediction, *IEEE Trans. Acoustics, Speech, and Signal Processing,* ASSP-31(6), 1394–1402, Dec. 1983.
[12] Cioffi, J., and Kailath, T., Fast recursive-least-squares transversal filters for adaptive filtering, *IEEE Trans. Acoustics, Speech and Signal Processing,* ASSP-32, 304–337, April 1984.

These algorithms, however, suffer from numerical instability problems. Some variables that are supposed to remain positive or bounded by one may lose this property due to roundoff errors. A treatment of these issues appears in

[13] Slock, D.T.M. and Kailath, T., Numerically stable fast transversal filters for recursive least squares adaptive filtering, *IEEE Trans. Signal Processing,* SP-39(1), 92–114, Jan. 1991.

More discussion on the QRD least-squares lattice filter, including alternative derivations that are based on the QR decomposition of certain data matrices, can be found in the references:

[14] Cioffi, J., The fast adaptive rotor's RLS algorithm, *IEEE Trans. Acoustics, Speech and Signal Processing,* ASSP-38, 631–653, 1990.

[15] Proudler, I.K., McWhirter, J.G., and Shepherd, T.J., Computationally efficient QR decomposition approach to least squares adaptive filtering, *IEE Proc.,* 138(4), 341–353, Aug. 1991.

[16] Regalia, P.A. and Bellanger, M.G., On the duality between fast QR methods and lattice methods in least squares adaptive filtering, *IEEE Trans. Signal Processing,* 39(4), 879–891, April 1991.

[17] Yang, B. and Böhme, J.F., Rotation-based RLS algorithms: unified derivations, numerical properties, and parallel implementations, *IEEE Trans. Signal Processing,* SP-40(5), 1151–1167, May 1992.

More discussion and examples of elementary and square-root free rotations and Householder transformations can be found in:

[18] Golub, G.B. and Van Loan, C.F., *Matrix Computations,* 2nd ed., The Johns Hopkins University Press, Baltimore, MD, 1989.

[19] Rader, C.M. and Steinhardt, A.O., Hyperbolic householder transformations, *IEEE Trans. Acoustics, Speech and Signal Processing,* ASSP-34(6), 1589–1602, Dec. 1986.

[20] Bojanczyk, A.W. and Steinhardt, A.O., Stabilized hyperbolic householder transformations, *IEEE Trans. Acoustics, Speech, and Signal Processing,* ASSP-37(8), 1286–1288, Aug. 1989.

[21] Hsieh, S.F., Liu, K.J.R., and Yao, K., A unified square-root-free approach for QRD-based recursive least-squares estimation, *IEEE Trans. Signal Processing,* SP-41(3), 1405–1409, March 1993.

Fast fixed-order adaptive algorithms that consider different choices of the initial weighting matrix Π_0, and also the case of data that is not necessarily prewindowed, can be found in:

[22] Houacine, A., Regularized fast recursive least squares algorithms for adaptive filtering, *IEEE Trans. Signal Processing,* SP-39(4), 860–870, April 1991.

Gauss' original exposition of the least-squares criterion can be found in:

[23] Gauss, C.F., *Theory of the Motion of Heavenly Bodies,* Dover, New York, 1963 (English translation of *Theoria Motus Corporum Coelestium, 1809*).

Transform Domain Adaptive Filtering

W. Kenneth Jenkins
*University of Illinois,
Urbana-Champaign*

Daniel F. Marshall
MIT Lincoln Laboratory

One of the earliest works on transform domain adaptive filtering was published in 1978 by Dentino et al. [4], in which the concept of adaptive filtering in the frequency domain was proposed. Many publications have since appeared that further develop the theory and expand the current understanding of performance characteristics for this class of adaptive filters. In addition to the discrete Fourier transform (DFT), other orthogonal transforms such as the discrete cosine transform (DCT) and the Walsh Hadamard transform (WHT) can also be used effectively as a means to improve the LMS algorithm without adding too much computational complexity. For this reason, the general term *transform domain adaptive filtering* is used in the following discussion to mean that the input signal is preprocessed by decomposing the input vector into orthogonal components, which are in turn used as inputs to a parallel bank of simpler adaptive subfilters. With an orthogonal transformation, the adaptation takes place in the transform domain, as it is possible to show that the adjustable parameters are indeed related to an equivalent set of time domain filter coefficients by means of the same transformation that is used for the real time processing [5, 14, 17].

A direct form FIR digital filter structure is shown in Fig. 22.1. The direct form requires $N - 1$ delays, N multiplications, and $N - 1$ additions for each output sample that is produced. The amount of hardware (as well as power) required to implement the direct form structure depends on the degree of hardware multiplexing that can be utilized within the speed demands of the application. A fully parallel implementation consisting of N delay registers, N multipliers, and a tree of two-input adders would be needed for very high-frequency applications. At the opposite end of the performance spectrum, a sequential implementation consisting of a length N delay line and a single time multiplexed multiplier and accumulation adder would provide the cheapest (and slowest) implementation. This latter structure would be characteristic of a filter that is implemented in software on one of the many commercially available DSP chips.

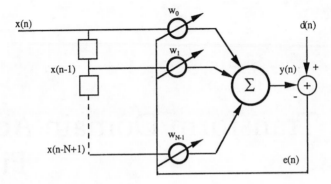

FIGURE 22.1   The direct form adaptive filter structure.

Regardless of the hardware complexity that results from a particular implementation, the computational complexity of the filter is determined by the requirements of the algorithm and, as such, remains invariant with respect to different hardware structures. In particular, the computational complexity of the direct form FIR filter is $O[N]$, since N multiplications and $(N-1)$ additions must be performed at each iteration. When designing an adaptive filter, it seems reasonable to seek an adaptive algorithm whose order of complexity is no greater than the order of complexity of the basic filter structure itself. This goal is achieved by the LMS algorithm, which is the major contributing factor to the enormous success of that algorithm. Extending this principle for 2-D adaptive filters implies that desirable 2-D adaptive algorithms have an order of complexity of $O[N^2]$, since a 2-D FIR direct form filter has $O[N^2]$ complexity inherent in its basic structure [11, 21].

The transform domain adaptive filter is a generalization of the LMS FIR structure, in which a linear transformation is performed on the input signal and each transformed "chanel" is power normalized to improve the convergence rate of the adaptation process. The linear transform is characterized throughout the following discussions as a sliding window operator that consists of a transformation matrix multiplying an input vector [14]. At each iteration n the input vector includes one new input sample $x(n)$, and $N-1$ past input samples $x(n-k), k = 1, \ldots, N-1$. As the window slides forward sample by sample, filtered outputs are produced continuously at each value of the index n.

Since the input transformation is represented by a matrix-vector product, it might appear that the computational complexity of the transform domain filter is at least $O[N^2]$. However, many transformations can be implemented with fast algorithms that have complexities less than $O[N^2]$. For example, the discrete Fourier transform can be implemented by the FFT algorithm, resulting in a complexity of $O[N \log_2 N]$ per iteration. Some transformations can be implemented recursively in a bank of parallel filters, resulting in a net complexity of $O[N]$ per iteration. The main point to be made here is that the complexity of the transform domain filter typically falls between $O[N]$ and $O[N^2]$, with the actual complexity depending on the specific algorithm that is used to compute the sliding window transform operator [17].

22.1   LMS Adaptive Filter Theory

The LMS algorithm is derived as an approximation to the steepest descent optimization strategy. The fact that the field of adaptive signal processing is based on an elementary principle from optimization theory suggests that more advanced adaptive algorithms can be developed by incorporating other results from the field of optimization [22]. This point of view recurs throughout this discussion, as concepts are borrowed from the field of optimization and modified for adaptive filtering as needed. In particular, one of the borrowed ideas that appears later is the quasi-Newton optimization strategy. It will be shown that transform domain adaptive filtering algorithms are closely related to quasi-Newton algorithms, but have computational complexity that is closer to the simple requirements of the LMS algorithm.

For a length N FIR filter with the input expressed as a column vector $\mathbf{x}(n) = [x(n), x(n-1), \ldots, x(n-N+1)]^T$, the filter output $y(n)$ is easily expressed as

$$y(n) = \mathbf{w}^T(n)\mathbf{x}(n) , \tag{22.1}$$

where $\mathbf{w}(n) = [w_0(n), w_1(n), \ldots, w_{N-1}(n)]^T$ is the time varying vector of filter coefficients (tap weights), and the superscript "T" denotes vector transpose. The output error is formed as the difference between the filter output and a training signal $d(n)$, i.e., $e(n) = d(n) - y(n)$. Strategies for obtaining an appropriate $d(n)$ vary from one application to another. In many cases the availability of a suitable training signal determines whether an adaptive filtering solution will be successful in a particular application. The ideal cost function is defined by the mean squared error (MSE) criterion, $E[|e(n)|^2]$. The LMS algorithm is derived by approximating the ideal cost function by the instantaneous squared error, resulting in $J_{\text{LMS}}(n) = |e(n)|^2$. While the LMS seems to make a rather crude approximation at the very beginning, the approximation results in an unbiased estimator. In many applications the LMS algorithm is quite robust and is able to converge rapidly to a small neighborhood of the optimum Wiener solution.

The steepest descent optimization strategy is given by

$$\mathbf{w}(n+1) = \mathbf{w}(n) - \mu \nabla_{E[|e|^2]^{(n)}} , \tag{22.2}$$

where $\nabla_{E[|e|^2]^{(n)}}$ is the gradient of the cost function with respect to the coefficient vector $\mathbf{w}(n)$. When the gradient is formed using the LMS cost function $J_{\text{LMS}}(n) = |e(n)|^2$, the conventional LMS results:

$$
\begin{aligned}
\mathbf{w}(n+1) &= \mathbf{w}(n) + \mu e(n)\mathbf{x}(n) , \\
e(n) &= d(n) - y(n) ,
\end{aligned}
\tag{22.3}
$$

and

$$y(n) = \mathbf{x}(n)^T \mathbf{w}(n) .$$

(Note: Many sources include a "2" before the μ factor in Eq. (22.3) because this factor arises during the derivation of (22.3) from (22.2). In this discussion we assume this factor is absorbed into the μ, so it will not appear explicitly.) Since the LMS algorithm is treated in considerable detail in other sections of this book, we will not present any further derivation or analysis of it here. However, the following observations will be useful when other algorithms are compared to the LMS as a baseline design [2, 3, 6, 8].

1. Assume that all of the signals and filter variables are real-valued. The filter itself requires N multiplications and $N-1$ additions to produce $y(n)$ at each value of n. The coefficient update algorithm requires $2N$ multiplications and N additions, resulting in a total computational burden of $3N$ multiplications and $2N-1$ additions per iteration. Since N is generally much larger than the factor of three, the order of complexity of the LMS algorithm is $O[N]$.

2. The cost function given for the LMS algorithm is a simplified form of the one used for the RLS algorithm. This implies that the LMS algorithm is a simplified version of the RLS algorithm, where averages are replaced by single instantaneous terms.

3. The (power normalized) LMS algorithm is also a simplified form of the transform domain adaptive filter which results by setting the transform matrix equal to the identity matrix.

4. The LMS algorithm is also a simplified form of the Gauss-Newton optimization strategy which introduces second order statistics (the input autocorrelation function) to accelerate the rate of convergence. In order to obtain the LMS algorithm from the Gauss-Newton algorithm, two approximations must be made: (i) The gradient must be approximated by the instantaneous error squared, and (ii) the inverse of the input autocorrelation matrix must be crudely approximated by the identity matrix.

These observations suggest that many of the seemingly distinct adaptive filtering algorithms that appear scattered about in the literature are indeed closely related, and can be considered to be members of a family whose hereditary characteristics have their origins in Gauss-Newton optimization theory [15, 16]. The different members of this family inherit their individual characteristics from approximations that are made on the pure Gauss-Newton algorithm at various stages of their derivations. However, after the individual derivations are complete and each algorithm is packaged in its own algorithmic form, the algorithms look considerably different from one another. Unless a conscious effort is made to reveal their commonality, the fact that they have evolved from common roots may be entirely obscured.

The convergence behavior of the LMS algorithm, as applied to a direct form FIR filter structure, is controlled by the autocorrelation matrix $\mathbf{R}_x$ of the input process, where

$$\mathbf{R}_x \equiv E[\mathbf{x}^*(n)\mathbf{x}^T(n)] . \tag{22.4}$$

(The $*$ in Eq. (22.4) denotes complex conjugate to account for the general case of complex input signals, although throughout most of the following discussions it will be assumed that $x(n)$ and $d(n)$ are both real-valued signals.) The autocorrelation matrix $\mathbf{R}_x$ is usually positive definite, which is one of the conditions necessary to guarantee convergence to the Wiener solution. Another necessary condition for convergence is $0 < \mu < 1/\lambda_{max}$, where λ_{max} is the largest eigenvalue of $\mathbf{R}_x$. It is also well established that the convergence of this algorithm is directly related to the eigenvalue spread of $\mathbf{R}_x$. The eigenvalue spread is measured by the condition number of $\mathbf{R}_x$, defined as $\kappa = \lambda_{max}/\lambda_{min}$, where λ_{min} is the minimum eigenvalue of $\mathbf{R}_x$. Ideal conditioning occurs when $\kappa = 1$ (white noise); as this ratio increases, slower convergence results. The eigenvalue spread (condition number) depends on the spectral distribution of the input signal and can be shown to be related to the maximum and minimum values of the input power spectrum (22.4). From this line of reasoning it becomes clear that white noise is the ideal input signal for rapidly training an LMS adaptive filter. The adaptive process becomes slower and requires more computation for input signals that are more severely colored [6].

Convergence properties are reflected in the geometry of the MSE surface, which is simply the mean squared output error $E[|e(n)|^2]$ expressed as a function of the N adaptive filter coefficients in $(N+1)$-space. An expression for the error surface of the direct form filter is

$$J(\mathbf{z}) \equiv E\left[|e(n)|^2\right] = J_{min} + \mathbf{z}^{*T}\mathbf{R}_x\mathbf{z} , \tag{22.5}$$

with $\mathbf{R}_x$ defined in (22.4) and $\mathbf{z} \equiv \mathbf{w} - \mathbf{w}_{opt}$, where $\mathbf{w}_{opt}$ is the vector of optimum filter coefficients in the sense of minimizing the mean squared error ($\mathbf{w}_{opt}$ is known as the Wiener solution). An example of an error surface for a simple two-tap filter is shown in Fig. 22.2. In this example $x(n)$ was specified to be a colored noise input signal with an autocorrelation matrix

$$\mathbf{R}_x = \begin{bmatrix} 1.0 & 0.9 \\ 0.9 & 1.0 \end{bmatrix} .$$

Figure 22.2 shows three equal-error contours on the three dimensional surface. The term $\mathbf{z}^{*T}\mathbf{R}_x\mathbf{z}$ in Eq. (22.2) is a quadratic form that describes the bowl shape of the FIR error surface. When $\mathbf{R}_x$ is positive definite, the equal-error contours of the surface are hyperellipses (N dimensional ellipses) centered at the origin of the coefficient parameter space. Furthermore, the principle axes of these hyperellipses are the eigenvectors of $\mathbf{R}_x$, and their lengths are proportional to the eigenvalues of $\mathbf{R}_x$. Since the convergence rate of the LMS algorithm is inversely related to the ratio of the maximum to the minimum eigenvalues of $\mathbf{R}_x$, large eccentricity of the equal-error contours implies slow convergence of the adaptive system. In the case of an ideal white noise input, $\mathbf{R}_x$ has a single eigenvalue of multiplicity N, so that the equal-error contours are hyperspheres [8].

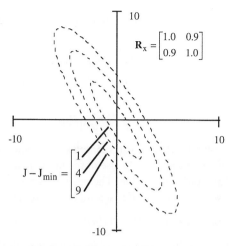

$$R_x = \begin{bmatrix} 1.0 & 0.9 \\ 0.9 & 1.0 \end{bmatrix}$$

$$J - J_{min} = \begin{bmatrix} 1 \\ 4 \\ 9 \end{bmatrix}$$

FIGURE 22.2 Example of an error surface for a simple two-tap filter.

22.2 Orthogonalization and Power Normalization

The transform domain adaptive filter (TDAF) structure is shown in Fig. 22.3. The input $x(n)$ and desired signal $d(n)$ are assumed to be zero mean and jointly stationary. The input to the filter is a vector of N current and past input samples, defined in the previous section and denoted as $\mathbf{x}(n)$. This vector is processed by a unitary transform, such as the DFT. Once the filter order N is fixed, the transform is simply an $N \times N$ matrix $\mathbf{T}$, which is in general complex, with orthonormal rows. The transformed outputs form a vector $\mathbf{v}(n)$ which is given by

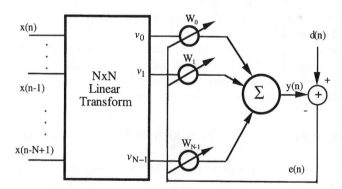

FIGURE 22.3 The transform domain adaptive filter structure

$$\mathbf{v}(n) = [v_0(n), v_1(n), \dots, v_{N-1}(n)]^T = \mathbf{Tx}(n) . \tag{22.6}$$

With an adaptive tap vector defined as

$$\mathbf{W}(n) = [W_0(n), W_1(n), \dots, W_{N-1}(n)]^T , \tag{22.7}$$

the filter output is given by

$$y(n) = \mathbf{W}^T(n)\mathbf{v}(n) = \mathbf{W}^T(n)\mathbf{Tx}(n) . \tag{22.8}$$

The instantaneous output error

$$e(n) = d(n) - y(n) \tag{22.9}$$

is then formed and used to update the adaptive filter taps using a modified form of the LMS algorithm (22.11):

$$
\begin{aligned}
\mathbf{W}(n+1) &= \mathbf{W}(n) + \mu e(n)\Lambda^{-2}\mathbf{v}^*(n) \\
\Lambda^2 &\equiv \mathrm{diag}\left[\sigma_0^2, \sigma_1^2, \ldots, \sigma_{N-1}^2\right]
\end{aligned}
\tag{22.10}
$$

where

$$
\sigma_i^2 = E\left[|v_i(n)|^2\right].
$$

As before, the superscript asterisk in (22.10) indicates complex conjugation to account for the most general case in which the transform is complex. Also, the use of the upper case coefficient vector in Eq. (22.10) denotes that $\mathbf{W}(n)$ is a transform domain variable. The power estimates σ_i^2 can be developed on-line by computing an exponentially weighted average of past samples according to

$$
\sigma_i^2(n) = \alpha\sigma_i^2(n-1) + |v_i(n)|^2, \qquad 0 < \alpha < 1.
\tag{22.11}
$$

If σ_i^2 becomes too small due to an insufficient amount of energy in the i-th channel, the update mechanism becomes ill-conditioned due to a very large effective step size. In some cases the process will become unstable and register overflow will cause the adaptation to catastrophically fail. So the algorithm given by (22.10) should have the update mechanism disabled for the i-th orthogonal channel if σ_i^2 falls below a critical threshold.

Alternatively the transform domain algorithm may be stabilized by adding small positive constants ε to the diagonal elements of Λ^2 according to

$$
\widehat{\Lambda}^2 = \Lambda^2 + \varepsilon\mathbf{I}.
\tag{22.12}
$$

Then $\widehat{\Lambda}^2$ is used in place of Λ^2 in Eq. (22.10). For most input signals $\sigma_i^2 \gg \varepsilon$, and the inclusion of the stabilization factors is transparent to the performance of the algorithm. However, whenever $\sigma_i^2 \approx \varepsilon$, the stabilization terms begins to have a significant effect. Within this operating region the power in the channels will not be uniformly normalized and the convergence rate of the filter will begin to degrade but catatrophic failure will be avoided.

The motivation for using the TDAF adaptive system instead of a simpler LMS based system is to achieve rapid convergence of the filter's coefficients when the input signal is not white, while maintaining a reasonably low computational complexity requirement. In the following section this convergence rate improvement of the TDAF will be explained geometrically.

22.3   Convergence of the Transform Domain Adaptive Filter

In this section the convergence rate improvement of the TDAF is described in terms of the mean squared error surface. From Eqs. (22.4) and (22.6) it is found that $\mathbf{R}_v = \mathbf{T}^*\mathbf{R}_x\mathbf{T}^T$, so that for the transform structure without power normalization Eq. (22.5) becomes

$$
J\mathbf{z} \equiv E\left[|e(n)|^2\right] = J_{\min} + \mathbf{z}^{*T}\left[\mathbf{T}^*\mathbf{R}_x\mathbf{T}^T\right]\mathbf{z}.
\tag{22.13}
$$

The difference between (22.5) and (22.13) is the presence of $\mathbf{T}$ in the quadratic term of (22.13). When $\mathbf{T}$ is a unitary matrix, its presence in (22.13) gives a rotation and/or a reflection of the surface. The eccentricity of the surface is unaffected by the transform, so the convergence rate of the system is unchanged by the transformation alone.

However, the signal power levels at the adaptive coefficients are changed by the transformation. Consider the intersection of the equal-error contours with the rotated axes: letting $\mathbf{z} = [0\cdots z_i\cdots 0]^T$, with z_i in the i-th position, Eq. (22.13) becomes

$$
J(\mathbf{z}) - J_{\min} = \left[\mathbf{T}^*\mathbf{R}_x\mathbf{T}^T\right]_i z_i^2 \approx \sigma_i^2 z_i^2.
\tag{22.14}
$$

If the equal-error contours are hyperspheres (the ideal case), then for a fixed value of the error $J(n)$, (22.14) must give $|z_i| = |z_j|$ for all i and j, since all points on a hypersphere are equidistant from the origin. When the filter input is not white, this will not hold in general. But since the power levels σ_i^2 are easily estimated, the rotated axes can be scaled to have this property. Let $\Lambda^{-1}\hat{\mathbf{z}} = \mathbf{z}$, where Λ is defined in (22.10). Then the error surface of the TDAF, with transform $\mathbf{T}$ and including power normalization, is given by

$$J(\hat{\mathbf{z}}) = J_{min} + \hat{\mathbf{z}}^{*T}\left[\Lambda^{-1}\mathbf{T}^*\mathbf{R_x}\mathbf{T}^T\Lambda^{-1}\right]\hat{\mathbf{z}}. \qquad (22.15)$$

The main diagonal entries of $\Lambda^{-1}\mathbf{T}^*\mathbf{R_x}\mathbf{T}^T\Lambda^{-1}$ are all equal to one, so (22.14) becomes $J(\mathbf{z}) - J_{min} = \hat{z}_i^2$, which has the property described above.

Thus, the action of the TDAF system is to rotate the axes of the filter coefficient space using a unitary rotation matrix $\mathbf{T}$, and then to scale these axes so that the error surface contours become approximately hyperspherical at the points where they can be easily observed, i.e., the points of intersection with the new (rotated) axes. Usually the actual eccentricity of the error surface contours is reduced by this scaling, and faster convergence is obtained.

As a second example, transform domain processing is now added to the previous example, as illustrated in Figs. 22.4 and 22.5. The error surface of Fig. 22.4 was created by using the (arbitrary) transform

$$\mathbf{T} = \begin{bmatrix} 0.866 & 0.500 \\ 0.500 & 0.866 \end{bmatrix},$$

on the error surface shown in Fig. 22.2, which produces clockwise rotation of the ellipsoidal contours so that the major and minor axes more closely align with the coordinate axes than they did without the transform. Power normalization was then applied using the normalization matrix Λ^{-1} as shown in

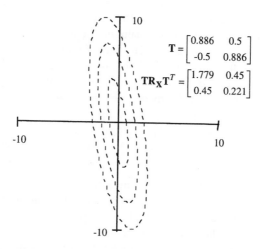

FIGURE 22.4   Error surface for the TDAF with transform $\mathbf{T}$.

Fig. 22.5, which represents the transformed and power normalized error surface. Note that the elliptical contours after transform domain processing are nearly circular in shape, and in fact they would have been perfectly circular if the rotation of Fig. 22.4 had brought the contours into precise alignment with the coordinate axes. Perfect alignment did not occur in this example because $\mathbf{T}$ was not able to perfectly diagonalize the input autocorrelation matrix for this particular $x(n)$. Since $\mathbf{T}$ is a fixed transform in the TDAF structure, it clearly cannot properly diagonalize $\mathbf{R_x}$ for an arbitrary $x(n)$, hence the surface rotation (orthogonalization) will be less than perfect for most input signals. It should be noted here that a well-known conventional algorithm called recursive least squares (RLS) is known to achieve near optimum

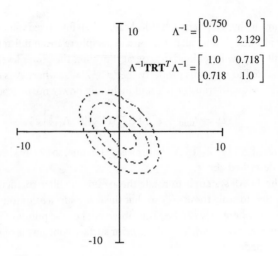

FIGURE 22.5 Error surface with transform and power normalization.

convergence rates by forming an estimate of $\mathbf{R}_x^{-1}$, the inverse of the autocorrelation matrix. This type of algorithm automatically adjusts to whiten any input signal, and it also varies over time if the input signal is a nonstationary process. Unfortunately, the computation required for the RLS algorithm is large and is not easily carried out in real time within the resource limitations of many practical applications. The RLS algorithm falls into the general class of quasi-Newton optimization techniques, which are thoroughly treated in numerous places throughout the literature.

There are two different ways to interpret the mechanism that brings about improved convergence rates achieved through transform domain processing [16]. The first point of view considers the combined operations of orthogonalization and power normalization to be the effective transformation $\Lambda^{-1}T$, an interpretation that is implied by Eq. (22.15). This line of thinking leads to an understanding of the transformed error surfaces as illustrated by example in Figs. 22.4 and 22.5 and leads to the logical conclusion that the faster learning rate is due to the conventional LMS algorithm operating on an improved error surface that has been rendered more properly oriented and more symmetrical via the transformation. While this point of view is useful in understanding the principles of transform domain processing, it is not generally implementable from a practical point of view. This is because for an arbitrary input signal, the power normalization factors that constitute the Λ^{-1} part of the input transformation are not known *a priori*, and must be estimated after $\mathbf{T}$ is used to decompose the input signal into orthogonal channels.

The second point of view interprets the transform domain equations as operating on the *transformed error surface* (without power normalization) with a modified LMS algorithm where the step sizes are adjusted differently in the various channels according to $\mu(n) = \mu\Lambda^{-2}$, where $\mu(n) = \text{diag}[\mu_i(n)]$ is a diagonal matrix that contains the step size for the i-th channel at location (i, i). The dependence of the $\mu_i(n)$'s on the iteration (time) index n acknowledges that the steps sizes are a function of the power normalization factors, which are updated in real time as part of the on-line algorithm. This suggests that the TDAF should be able to track nonstationary input statistics within the limited abilities of the transformation $\mathbf{T}$ to orthogonalize the input and within the accuracy limits of the power normalization factors. Furthermore, when the input signal is white, all of the σ_i^2's are identical and each is equal to the power in the input signal. In this case the TDAF with power normalization becomes the conventional normalized LMS algorithm.

It is straightforward to show mathematically that the above two points of view are indeed compatible [10]. Let $\hat{\mathbf{v}}(n) \equiv \Lambda^{-1}\mathbf{Tx}(n) = \Lambda^{-1}\mathbf{v}(n)$ and let the filter tap vector be denoted $\hat{\mathbf{W}}(n)$ when the matrix $\Lambda^{-1}T$ is treated as the effective transformation. For the resulting filter to have the same response as the filter in

Fig. 22.3 we must have

$$\mathbf{v}^T(n)\mathbf{W} = y(n) = \hat{\mathbf{v}}^T\hat{\mathbf{W}} = \mathbf{v}^T(n)\Lambda^{-1}\hat{\mathbf{W}}, \qquad \forall \mathbf{v}(n) \tag{22.16}$$

which implies that $\mathbf{W} = \Lambda^{-1}\hat{\mathbf{W}}$. It the tap vector $\hat{\mathbf{w}}$ is updated using the LMS algorithm, then

$$
\begin{aligned}
\mathbf{W}(n+1) &= \Lambda^{-1}\hat{\mathbf{W}}(n+1) = \Lambda^{-1}[\hat{\mathbf{W}}(n) + \mu e(n)\hat{\mathbf{v}}^*(n)] \\
&= \Lambda^{-1}\hat{\mathbf{W}}(n) + \mu e(n)\Lambda^{-1}\hat{\mathbf{v}}^*(n) \\
&= \mathbf{W}(n) + \mu e(n)\Lambda^{-2}\mathbf{v}^*(n),
\end{aligned}
\tag{22.17}
$$

which is precisely the algorithm (22.10). This analysis demonstrates that the two interpretations are consistent, and they are, in fact, alternate ways to explain the fundamentals of transform domain processing.

22.4 Discussion and Examples

It is clear from the above development that the power estimates σ_i^2 are the optimum scale factors, as opposed to $|\sigma_i|$ or some other statistic. Also, it is significant to note that no convergence rate improvement can be realized without power normalization. This is the same conclusion that was reached in [6] where the frequency domain LMS algorithm was analyzed with a constant convergence factor. From the error surface description of the TDAF's operation, it is seen that an optimal transform rotates the axes of the hyperellipsoidal equal-error contours into alignment with the coordinate axes. The prescribed power normalization scheme then gives the ideal hyperspherical contours, and the convergence rate becomes the same as if the input were white. The optimal transform is composed of the orthonormal eigenvectors of the input autocorrelation matrix and is known in the literature as the Karhunen-Loe've transform (KLT). The KLT is signal dependent and usually cannot be easily computed in real time. Note that real signals have real KLT's, suggesting the use of real transforms in the TDAF (in contrast to complex transforms such as the DTF).

Since the optimal transform for the TDAF is signal dependent, a universally optimal fixed parameter transform can never be found. It is also clear that once the filter order has been chosen, any unitary matrix of correct dimensions is a possible choice for the transform; there is no need to restrict attention to classes of known transforms. In fact, if a prototype input power spectrum is available, its KLT can be constructed and used. One factor that must be considered in choosing a transform for real-time applications is computational complexity. In this respect, real transforms are superior to complex ones, transforms with fast algorithms are superior to those without, and transforms whose elements are all powers-of-two are attractive since only additions and shifts are needed to compute them. Throughout the literature the discrete Fourier transform (DFT), the discrete cosine transform (DCT), and the Walsh Hadamard transform (WHT) have received considerable attention as possible candidates for use in the TDAF [14]. In spite of the fact that the DFT is a complex transform and not computationally optimal from that point of view, it is often used in practice because of the availability of efficient FFT algorithms.

Figure 22.6 shows learning characteristics for computer-generated TDAF examples using six different orthogonal transforms to decorrelate the input signal. The examples presented are for system identification experiments, where the desired signal was derived by passing the input through an 8-tap FIR filter, which serves as the model system to be identified. Computer-generated white pseudo-noise, uncorrelated with the input signal, was added to the output of the model system, creating a -100 dB noise floor. The filter inputs were generated by filtering white pseudo-noise with a 32-tap linear phase FIR noise-coloring filter to produce an input autocorrelation eigenvalue ratio of 681. Experiments were then performed using the discrete Fourier transform (DFT), the discrete cosine transform (DCT), the Walsh-Hadamard transform (WHT), discrete Hartley transform (DHT), and a specially designed computationally efficient "power-of-2" PO2 transform, as listed in Fig. 22.6. The eigenvalue ratios that result from transform processing with each of these transforms is shown in Fig. 22.6, where it is seen that the PO2 transform with power normalization reduces the input condition number from 681 to 128, resulting in the most effective transform for this particular input coloring. All of the transforms used in this experiment are able to reduce the input condition number and greatly improve convergence rates, although some transforms are seen to be more effective than others for the coloring chosen for these examples.

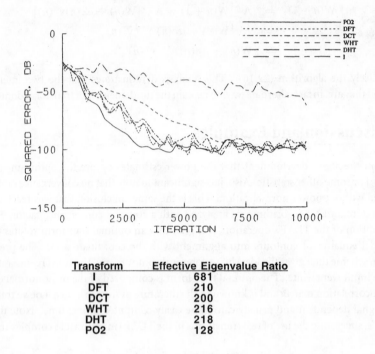

Transform	Effective Eigenvalue Ratio
I	681
DFT	210
DCT	200
WHT	216
DHT	218
PO2	128

FIGURE 22.6 Comparison of (smoothed) learning curves for five different transforms operating on a colored noise input signal with condition number 681.

22.5 Quasi-Newton Adaptive Algorithms

The dependence of the adaptive system's convergence rate on the input power spectrum can be reduced by using second-order statistics via the Gauss-Newton method [9, 10, 21]. The Gauss-Newton algorithm is well known in the field of optimization as one of the basic accelerated search techniques. In recent years it has also appeared in various forms in publications on adaptive filtering. In this section a brief introduction to quasi-Newton adaptive filtering methods is presented. When the quasi-Newton concept is integrated into the LMS algorithm, the resulting adaptive strategy is closely related to the transform domain adaptive filter, but where the transform is computed on-line as an approximation to the Hessian acceleration matrix. For FIR structures it turns out that the Hessian is equivalent to the input autocorrelation matrix inverse, and therefore the quasi-Newton LMS algorithm effectively implements a transform that adjusts to the statistics of the input signal and is capable of tracking slowly varying nonstationary input signals.

The basic Gauss-Newton coefficient update algorithm for an FIR adaptive filter is given by

$$\mathbf{w}(n+1) = \mathbf{w}(n) - \mu \mathbf{H}(n) \nabla_{E[e^2]}(n), \tag{22.18}$$

where $\mathbf{H}(n)$ is the Hessian matrix and $\nabla_{E[e^2]}(n)$ is the gradient of the cost function at iteration n. For an FIR adaptive filter with a stationary input the Hessian is equal to $\mathbf{R}_x^{-1}$. If the gradient is estimated with the instantaneous error squared, as in the LMS algorithm, the result is

$$\mathbf{w}(n+1) = \mathbf{w}(n) + \mu e(n)\widehat{\mathbf{R}}_x^{-1}(n)\mathbf{x}(n) , \tag{22.19}$$

where $\widehat{\mathbf{R}}_x^{-1}(n)$ is an estimate of $\mathbf{R}_x^{-1}$ that varies as a function of the index n. Equation (22.19) characterizes the quasi-Newton LMS algorithm. Note that (22.18) is the starting point for the development of many practical adaptive algorithms that can be obtained by making approximations to one or both of the Hessian and the gradient. Therefore, we typically refer to all such algorithms derived from (22.18) as the family of quasi-Newton algorithms.

The autocorrelation estimate $\widehat{\mathbf{R}}_x(n)$ is constructed from data received up to time step n. It must then be inverted for use in (22.19). This is in general an $O[N^3]$ operation, which must be performed for every iteration of the algorithm. However, the use of certain autocorrelation estimators allows more economical matrix inversion techniques to be applied. Using this approach, the conventional sequential regression algorithm (22.11) and the recursive least squares (RLS) algorithm (22.3) achieve quasi-Newton implementations with a computational requirement of only $O[N^2]$.

The RLS algorithm is probably the best-known member of the class of quasi-Newton algorithms [6]. The drawback that has prevented its widespread use in real-time signal processing is its $O[N^2]$ computational requirement, which is still too high for many applications (and is an order of magnitude higher than the order of complexity of the FIR filter itself). This problem appeared to have been solved by the formulation of $O[N]$ versions of the RLS algorithm. Unfortunately, many of these more efficient forms of the RLS tend to be numerically ill-conditioned. They are often unstable in finite precision implementations, especially in low signal-to-noise applications or where the input signal is highly colored. This behavior is caused by the accumulation of finite precision errors in internal variables of the algorithm and is essentially the same source of numerical instability that occurs in the standard $O[N^2]$ RLS algorithm, although the problem is greater in the $O[N]$ case since these algorithms typically have a larger number of coupled internal recursions. Considerable work has been reported in the literature to stabilize $O[N^2]$ RLS algorithm and to produce a numerically robust $O[N]$ RLS algorithm.

22.5.1   A Fast Quasi-Newton Algorithm

The quasi-Newton algorithms discussed above achieve reduced computation through the use of particular autocorrelation estimators which lend themselves to efficient matrix inversion techniques. This section reviews a particular quasi-Newton algorithm that was developed to provide a numerically robust $O[N]$ algorithm [15]. This particular 1-D algorithm is discussed here simply as a representative algorithm from the quasi-Newton class; numerous variations of the Newton optimization strategy are reported in various places throughout the adaptive filtering literature. The fast quasi-Newton algorithm described below has also been extended successfully to 2-D FIR adaptive filters [11].

To derive the $O[N]$ fast quasi-Newton (FQN) algorithm, an autocorrelation matrix estimate is used which permits the use of more robust and efficient computation techniques. Assuming stationarity, the autocorrelation matrix $\mathbf{R}_x$ has a high degree of structure; it is symmetric and Toeplitz, and thus has only N free parameters, the elements of the first row. This structure can be imposed on the autocorrelation estimate, since this incorporates prior knowledge of the autocorrelation into the estimation process. The estimation problem then becomes that of estimating the N autocorrelation lags $r_i, i = 0, \ldots, N - 1$, which comprise the first row of $\mathbf{R}_x$. The autocorrelation estimate is also required to be positive definite, to ensure the stability of the adaptive update process.

A standard positive semidefinite autocorrelation lag estimator for a block of data is given by

$$\hat{r}_i = \frac{1}{M + 1} \sum_{k=i}^{M} x(k - i)x(k) , \tag{22.20}$$

where $x(k), k = 0, \ldots, M$, is a block of real data samples, and i ranges from 0 to M. However, the preferred form of the estimation equation for use in an adaptive system, from an implementation standpoint, is an exponentially weighted recursion. Thus, (22.20) must be expressed in an exponentially weighted recursive form, without destroying its positive semidefiniteness property. Consider the form of the sum in Eq. (22.20): it is the (deterministic) correlation of the data sequence $x(k), k = 0, \ldots, M$, with itself. Thus, $\hat{r}_i, i = 0, \ldots, M$ is the deterministic autocorrelation sequence of the sequence $x(k)$. (Note that $\hat{r}_i$ must also be defined for $i = -M, \ldots, -1$, according to the requirement that $\hat{r}_i = \hat{r}_{-i}$). In fact, the deterministic autocorrelation for any sequence is positive semidefinite. The goal of exponential weighting, in a general sense, is to weight recent data most heavily and to forget old data by using progressively smaller weighting factors. To construct an exponentially weighted, positive definite autocorrelation estimate, we must weight the data first, then form its deterministic autocorrelation to guarantee positive semidefiniteness. At time step n, the available data are $x(k), k = 0, \ldots, n$. If these samples are exponentially weighted using $\sqrt{\alpha}$, the result is $\alpha^{(n-k)/2} x(k), k = 0, \ldots, n$. Using (22.20) and assuming $n > N - 1$, we then have

$$
\begin{aligned}
\hat{r}_i(n) &= \sum_{k=i}^{n} [\alpha^{(n-k+i)/2} x(k-i)][\alpha^{(n-k)/2} x(k)] \\
&= \alpha \sum_{k=i}^{n-1} \alpha^{(n-1-k)} \alpha^{i/2} x(k-i) x(k) \\
&\quad + \alpha^{i/2} x(n-i) x(n) \\
&= \alpha \hat{r}_i(n-1) + \alpha^{i/2} x(n-i) x(n) \\
&\quad \text{for } i = 0, \ldots, N - 1 .
\end{aligned}
\tag{22.21}
$$

A normalization term is omitted in (22.21), and initialization is ignored. With regard to the latter point, the simplest way to consistently generate $\hat{r}_i(n)$ for $0 \leq n \leq N - 1$ is to assume that $x(n) = 0$ for $n < 0$, set $\hat{r}_i(-1) = 0$ for all i, and then use the above recursion. A small positive constant δ may be added to $\hat{r}_0(n)$ to ensure positive definiteness of the estimated autocorrelation matrix.

With this choice of an autocorrelation matrix estimate, a quasi-Newton algorithm is determined. Thus, the fast quasi-Newton (FQN) algorithm is given by (22.19) and (22.21), where $\widehat{\mathbf{R}}_x(n) \approx \mathbf{R}_x$ is understood to be the Toeplitz symmetric matrix whose first row consists of the autocorrelation lag estimates $\hat{r}_i(n), i = 0, \ldots, N - 1$, generated by (22.21). Because $\widehat{\mathbf{R}}_x(n)$ is Toeplitz, its inverse can be obtained using the Levinson Recursion, leading to an $O[N]$ implementation of this algorithm. The step size μ for the FQN algorithm is given by

$$
\frac{1}{2} \mu^{-1} = \varepsilon + \mathbf{x}^T(n) \widehat{\mathbf{R}}_x^{-i}(n-1) \mathbf{x}(n) .
\tag{22.22}
$$

This step size is used in other quasi-Newton algorithms (22.4), and seems nearly optimal. The parameter ε is intended to be small relative to the average value of $x^T(n) \widehat{\mathbf{R}}_x^{-1}(n-1) x(n)$. Then the normalization term omitted from (22.21), which is a function of α but not of i, cancels out of the coefficient update, since $\widehat{\mathbf{R}}_x^{-1}(n)$ appears in both the numerator and the denominator. Thus, the normalization can be safely ignored.

22.5.2 Examples

The previous examples are used again here to compare the performance of the fast quasi-Newton algorithm with the recursive least squares (RLS), which provides a baseline for performance comparisons. The RLS examples are shown in Fig. 22.7(a) for different values of the exponential forgetting factor α, and the FQN examples are shown in Fig. 22.7(b). Note that the FQN algorithm is somewhat slower to converge due to the fact that the autocorrelation inverse matrix is updated only once every eight samples. In comparison the RLS algorithm converges more quickly, but has a computational complexity of $O[N^3]$ as compared

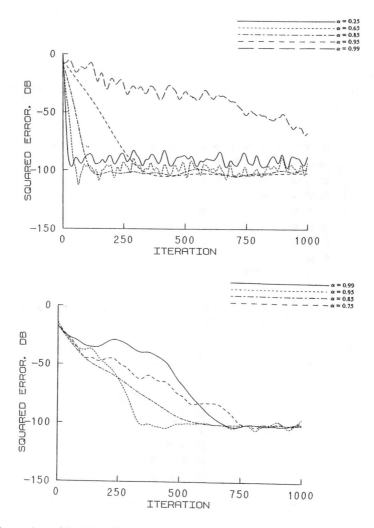

FIGURE 22.7 Comparison of the RLS and FQN performance. Simulated learning curves for (a) the RLS algorithm and (b) the FQN algorithm.

to a complexity of $O[N]$ for the FQN algorithm. But, more important, note that the convergence rate of the FQN algorithm is much faster than any of the transform domain examples shown previously.

22.6 The 2-D Transform Domain Adaptive Filter

Many successful 1-D FIR algorithms have been extended to 2-D filters [7, 10, 19, 21]. Transform domain adaptive algorithms are also well suited to 2-D signal processing. Orthogonal transforms with power normalization can be used to accelerate the convergence of an adaptive filter in the presence of a colored input signal.

The 2-D TDAF structure is shown in Fig. 22.8 with the corresponding (possibly complex) LMS algorithm given as

$$\mathbf{w}_{k+1}(m_1, m_2) = \mathbf{w}_k(m_1, m_2) + \mu \, e(n_1, n_2) \Lambda_u^{-2} \mathbf{u}_k^*(n_1, n_2) \,, \qquad (22.23)$$

where $\mathbf{u}_k(n_1, n_2)$ is the column-ordered vector formed by premultiplying the input column-ordered vector

$\mathbf{x}_k(n_1, n_2)$ by the 2-D unitary transform $\mathbf{T}$, i.e.,

$$\mathbf{u}_k(n_1, n_2) = \mathbf{T}\,\mathbf{x}_k(n_1, n_2)\,. \tag{22.24}$$

Channel normalization results from including $\Lambda_u^2 = \mathrm{diag}[\sigma_u^2(0,0)\sigma_u^2(1,0)\ldots\sigma_u^2(N,N)]$ in (22.23) where $\sigma_u^2(n_1, n_2) \approx E[|\mathbf{u}(n_1, n_2)|^2]$. Ideally, the Karhunen-Loe've transform (KTL) is used to achieve optimal

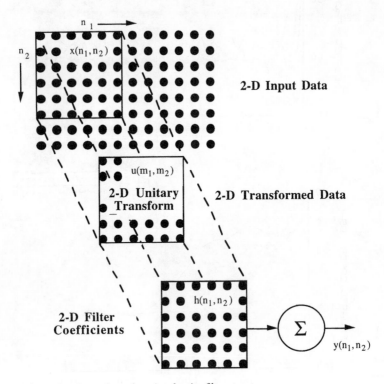

FIGURE 22.8   Two-dimensional transform domain adaptive filter structure.

convergence, but this requires *a priori* knowledge of the input statistical properties. The KLT corresponding to the input autocorrelation matrix $\mathbf{R}_x$ is constructed using as rows of $\mathbf{T}$ the orthonormal eigenvectors of $\mathbf{R}_x$. Therefore, with unitary $\mathbf{Q}_X^H = [\mathbf{q}_1 \ldots \mathbf{q}_M 2]$ and $\Lambda_x = \mathrm{diag}[\lambda_1 \ldots \lambda_M 2]$ $(M = N + 1$ for convenience), the unitary similarity transformation is $\mathbf{R}_x = \mathbf{Q}_x^{-1}\Lambda_x\mathbf{Q}_x$, and the KLT is given by $\mathbf{T} = \mathbf{Q}_x$. However, since the statistical properties of the input process are usually unknown and time varying, the KLT cannot be implemented in practice. Researchers have found that many fixed transforms do provide good orthogonalization for a wide class of input signals. Those include the discrete Fourier transform (DFT or FFT), the discrete cosine transform (DCT), and the Walsh Hadamard transform (WHT). For example, the DFT provides only approximate channel decorrelation since it is well-known that a "sliding" DFT implements a parallel bank of overlapping band-pass filters with center frequencies evenly distributed over the interval $[0, 2\pi]$. Furthermore, the DFT (or FFT) is hampered by the fact that it requires complex arithmetic. It is still a very effective method of orthogonalization which we compare here to the 2-D FQN algorithm.

   The convergence plots in Fig. 22.9 show the comparison between the 2-D FQN, the 2-D TDAF (with the DFT), and the simple 2-D LMS with the same fourth-order low-pass coloring filter. The adaptive filter is second order, and the 2-D FQN algorithm, as expected, outperforms the 2-D TDAF. The 2-D FQN algorithm is effectively attempting to estimate the KLT on-line so that, while not able to perfectly orthogonalize the training signal, it does offer improved convergence over that of the fixed transform algorithm. Similar results appear in Fig. 22.10 with the same coloring filter and a fourth-order adaptive filter.

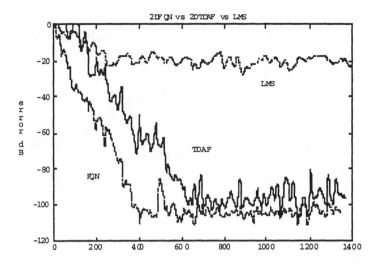

FIGURE 22.9   Convergence plot for 3×3 FIR 2-D LMS, 2-D TDAF, and 2-D FQN adaptive filters in the system identification configuration with low-pass colored inputs.

22.7   Block-Based Adaptive Filters

The block-based LMS (BLMS) algorithm is one of the many efficient adaptive filtering algorithms aimed at increasing convergence speed and reducing the computational complexity. The basic principle of the BLMS algorithm is that the filter coefficients remain unchanged during the processing of each data block and are updated only once per block [1, 20]. The updating equation of the block LMS algorithm for linear adaptive filters is as follows:

$$\mathbf{w}_{k+1} = \mathbf{w}_k + \mu \sum_{m=0}^{L-1} \mathbf{x}(kL + m)e(kL + m) , \tag{22.25}$$

where k refers to the kth block of input data, k-th denotes the index of the incoming data, and L is the block length. Due to the fact that in the BLMS algorithm the computation of the filter output and of the gradient itself are represented by linear convolution and correlation, they can be implemented efficiently using the FFT. Based on the convolutional property of the FFT and the overlap-save method, the BLMS algorithm can be implemented in the frequency domain and the computational complexity can be reduced dramatically. Both the constrained and the unconstrained linear frequency domain LMS algorithms have been developed (22.2 and 22.9). In this chapter we formulate both the constrained and unconstrained FBLMS algorithms and then present performance comparisons between them.

In this section the frequency-domain block LMS (FBLMS) algorithm is summarized. Throughout this chapter, capital letters are used to denote the frequency-domain variables and lowercase letters to denote the time-domain variables. Boldface letters denote vectors and matrices. The basic principle of the FBLMS algorithm is to use the DFT to calculate a block of the filter output and the block gradient. Figure 22.11 shows the block diagram of the frequency domain block LMS algorithm.

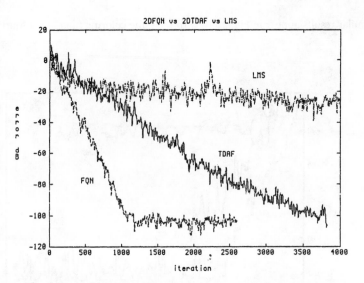

FIGURE 22.10 Convergence plot for 5×5 FIR 2-D LMS, 2-D TDAF, and 2-D FQN adaptive filters in the system identification configuration with low-pass colored inputs.

Because the DFT-based calculation actually implements circular convolution, in order to obtain a result that is equivalent to linear convolution, a data vector $\mathbf{x}_k$ that has a larger dimension than the filter length must be used. Assume the filter length is N, then the formulate a length $2N$ input data vector $\mathbf{x}_k$ as

$$\mathbf{x}_k = [x(2Nk), x(2Nk+1), \cdots, x(2Nk+2N-1)]^T \,, \tag{22.26}$$

where the last N samples in $\mathbf{x}_k$ are taken from the current block of input data and the first N samples are taken from the previous input data block. In order to maintain consistent dimensions, the filter weight vector must also be increased to length $2N$ by appending N zeros to the original filter weights, as indicated below:

$$\mathbf{W}_k = [\overbrace{w_k\,(0)w_k\,(1) \cdots w_k\,(N-1)}^{\mathbf{w}_{k,c}} \overbrace{0 \cdots 0}^{N}]^T \,. \tag{22.27}$$

If $\mathbf{F}$ denotes the $2N \times 2N$ DFT matrix with elements as $F_{mn} = \exp(-j2\pi mn/2N)$, then the output of the adaptive filter in the frequency domain is

$$\mathbf{Y}_k = \mathbf{X}_k \mathbf{W}_k \tag{22.28}$$

where $\mathbf{X}_k = \text{diag}(\mathbf{F}\mathbf{x}_k)$ and $\mathbf{W}_k = \mathbf{F}\mathbf{w}_k$. Because the last N elements of $\mathbf{F}^{-1}\mathbf{Y}_k$ coincide with the result of linear convolution, only these N elements should be used in calculating the error vector. Define the time domain desired signal vector $\mathbf{d}_k$ and the filter output vector as:

$$\mathbf{d}_k = [\overbrace{0 \cdots 0}^{N}\overbrace{d_k\,(0)d_k\,(1) \cdots d_k\,(N-1)}^{\mathbf{d}_{k,c}}]^T \,, \tag{22.29}$$

$$\mathbf{y}_k = [\overbrace{0 \cdots 0}^{N}\overbrace{\text{last } N \text{ elements of } \mathbf{F}^{-1}\mathbf{Y}_k}^{\mathbf{y}_{k,c}}]^T \,. \tag{22.30}$$

Then the error signal in the frequency domain is

$$\mathbf{E}_k = \mathbf{F}(\mathbf{d}_k - \mathbf{y}_k) \,. \tag{22.31}$$

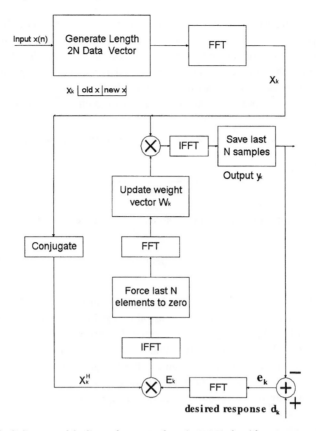

FIGURE 22.11 The block diagram of the linear frequency domain BLMS algorithms.

In order to guarantee that there are only N nonzero terms in the impulse response of the adaptive filter, a gradient constraint should be used to force the last N terms of the gradient in the time domain to be zeros, as shown in Fig. 22.11. The weight vector update equation for the constrained FBLMS algorithm is

$$\mathbf{W}_{k+1} = \mathbf{W}_k + \mu \mathbf{F} \left(\nabla (k)^T, 0 \cdots 0 \right)^T, \tag{22.32}$$

where $\nabla(k) =$ the first N elements of $\mathbf{F}^{-1}\mathbf{X}_k^H \mathbf{E}_k$ and H denotes the complex conjugate transpose. From Fig. 22.11 it can be seen that five forward and inverse FFTs are used in the constrained FBLMS algorithm, where two of them are required to apply the gradient constraint.

In order to reduce the computational complexity, the unconstrained FBLMS algorithm (22.21) has been developed by simply removing the constraint from the constrained FBLMS updating equation as

$$\mathbf{W}_{k+1} = \mathbf{W}_k + \mu \mathbf{X}_k^H \mathbf{E}_k. \tag{22.33}$$

With the constraint removed, the gradient is no longer the result of linear convolution but rather a circular correlation. But this simplified algorithm requires only three FFTs, so the computational complexity is significantly reduced.

22.7.1 Comparison of the Constrained and Unconstrained Frequency Domain Block-LMS Adaptive Algorithms

Some work has been done in analyzing the performance of the constrained and unconstrained FBLMS algorithm. Mansour and Gray [13] proved the almost sure asymptotic exponential convergence of the unconstrained FBLMS algorithm. More thorough analysis of the convergence behavior of the uncon-strained FBLMS algorithm was given in [11]. By formulating the time domain updating equations of the two FBLMS algorithms, it can be shown that the two algorithms are equivalent to time domain adaptive filters with different numbers of taps. It is this effective difference in filter length that results in different convergence performance of the two algorithms. Although the unconstrained FBLMS algorithm corre-sponds to a longer filter, "wrap-around" error prevents it from effectively performing as a higher order filter [12].

In order to guarantee the mean-squared error convergence, the step size parameter μ must satisfy the condition

$$0 < \mu < \frac{2}{\text{Total input power}} = \frac{2}{Tr(\mathbf{R})} . \tag{22.34}$$

Also, the following relationship between the misadjustment and the step size exists

$$\text{Misadjustment} = \frac{\mu N \lambda_{\text{avg}}}{2} \tag{22.35}$$

with $\lambda_{\text{avg}} = \frac{1}{N} \sum_1^N \lambda_i$. For stationary input signals it can be shown easily that $Tr(\mathbf{R}_c) = 2Tr(\mathbf{R}_u)$ where $\mathbf{R}_c$ is the input autocorrelation matrix for the constraint algorithm and $\mathbf{R}_u$ for the unconstrained case. So λ_{avg} is the same for both $\mathbf{R}_c$ and $\mathbf{R}_u$. Combining the above two conditions with the properties of the correlation matrices $\mathbf{R}_c$ and $\mathbf{R}_u$ we can conclude that the unconstrained FBLMS algorithm has a reduced stable range of step size compared to the constrained FBLMS algorithm. If we keep the misadjustment of both algorithms the same, the stepsize of the unconstrained FBLMS algorithm is about one half that of the constrained FBLMS algorithm and its convergence rate is approximately one half that of the constrained FBLMS algorithm.

The unconstrained FBLMS algorithm is equivalent to a length-$2N$ adaptive filter. Is it capable of mod-eling a higher order system than the constrained FBLMS algorithm? The answer is no; the unconstrained FBLMS algorithm works on the circulant data matrix which generates the "wrap-around" error. This error, which is embedded in the gradient of the unconstrained FBLMS algorithm during the learning process, prevents the unconstrained FBLMS algorithm with filter coefficient length $2N$ from accurately modeling an unknown system with greater than N coefficients, although it is capable of achieving more accuracy than the constrained FBLMS algorithm of length N.

22.7.2 Examples and Discussion

Computer simulations are presented here to demonstrate the FBLMS algorithms. In the first example, an 8-tap FIR filter was used to identify an 8-tap lowpass filter using the constrained and the unconstrained FBLMS algorithms. Colored Gaussian noise was used as the input signal. The signal-to-noise ratio of the output signal of the system being identified is 100 dB. During the adaption the best step size μ (as experimentally determined for fastest convergence) for each case was used. Figure 22.12 shows the learning curves of the constrained and unconstrained FBLMS algorithms. From the figure we can see that the constrained FBLMS algorithm converges faster than the unconstrained FBLMS algorithm. In the second example, instead of using the best step size of each algorithm, the best step size of the constrained FBLMS algorithm was used for both the constrained and unconstrained FBLMS algorithms. Figure 22.13 shows the corresponding learning curves of the two algorithms. When the best step size of the constrained FBLMS algorithm is used, the performance of the unconstrained FBLMS algorithm tends to be unstable. The ultimate effect of removing the gradient constraint from the constrained FBLMS algorithm is slower convergence, a smaller range for stable step sizes, doubling of the equivalent filter length, in exchange for a reduction in computational cost.

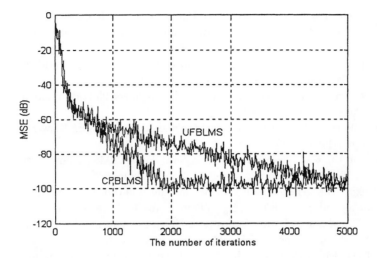

FIGURE 22.12   Learning curves of the length-8 constrained ($\mu = 0.055$) and unconstrained ($\mu = 0.030$) frequency domain BLMS algorithms with colored Gaussian noise input.

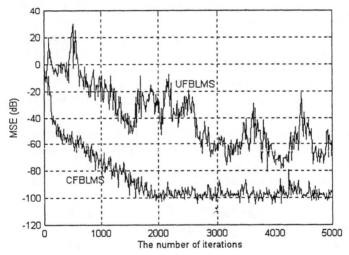

FIGURE 22.13   Learning curves of the length-8 constrained ($\mu = 0.055$) and unconstrained ($\mu = 0.055$) frequency domain BLMS algorithms with colored Gaussian noise input.

References

[1] Clark, G.A., Parker, S.R. and Mitra, S.K., A unified approach to time and frequency realization for FIR adaptive digital filters, *IEEE Trans. Acoust., Speech, Signal Process.*, Vol. ASSP-31(5), 1073–1083, Oct. 1983.

[2] Bellanger, M.G., *Adpative Digital Filters and Signal Analysis,* Marcel Dekker, New York and Basel, 1987.

[3] Cowan, B. and Grant. P., *Adaptive Filters,* Prentice-Hall, Englewood Cliffs, NJ, 1987.

[4] Dentino, M., McCool, J., and Widrow, B., Adaptive filtering in the frequency domain, *Proc. IEEE,* Vol. 66, 1658–1659, Dec. 1978.

[5] Gitlin, R.D. and Magee, F.R., Jr., Self-orthogonalizing adaptive equalization algorithms, *IEEE Trans. Commun.,* Vol. COM-25(7), 666–672, July 1977.

[6]   Haykin, S., *Adaptive Filter Theory*, Prentice-Hall, Englewood Cliffs, NJ, 1991.

[7]   Hadhoud, M.M. and Thomas, D.W., The two-dimensional adaptive LMS (TDLMS) algorithm, *IEEE Trans. Circuits Syst.*, Vol. 35, 485–494, 1988.

[8]   Honig, M.L. and Messerschmidt, D.G., *Adaptive Filters: Structures, Algorithms, and Applications*, Kluwer Academic Press, Boston, 1984.

[9]   Hull, A.W., Orthogonalization techniques for adaptive filters, Ph.D. dissertation, University of Illinois, Urbana-Champaign, 1994.

[10] Jenkins, W.K. et al., *Advanced Concepts in Adaptive Signal Processing*, Kluwer Academic Publishers, Boston, 1996.

[11] Lee, J.C. and Un, C.K., Performance of transform domain LMS adaptive filters, *IEEE Trans. Acoust., Speech, Signal Process.*, Vol. ASSP-34, 499–510, June 1986.

[12] Li and Jenkins, W.K., The comparison of the constrained and unconstrained frequency-domain block LMS algorithms, *IEEE Trans. Signal Process.*, Vol. 44(7), 1813–1816, July 1996.

[13] Mansour, D. and Gray, A.H., Jr., Unconstrained frequency-domain adaptive filters, *IEEE Trans. Acoust., Speech, Signal Process.*, Vol. ASSP-30(5), 726–734, Oct. 1982.

[14] Marshall, D.F., Jenkins, W.K., and Murphy, J.J., The use of orthogonal transforms for improving performance of adaptive filters, *IEEE Trans. Acoust., Speech, Signal Process.*, Vol. ASSP-36(4), 474–484, April 1989.

[15] Marshall, D.F. and Jenkins, W.K., A fast quasi-Newton adaptive filtering algorithm, *IEEE Trans. Acoust., Speech, Signal Process.*, Vol. ASSP-40(7), 1652–1662, July 1992.

[16] Marshall, D.F., Computationally efficient techniques for rapid convergence of adaptive digital filters, Ph.D. dissertation, University of Illinois, Urbana-Champaign, IL, 1988.

[17] Narayan, S.S., Peterson, A.M., and Narasima, M.J., Transform domain LMS algorithm, *IEEE Trans. Acoust., Speech, Signal Process.*, Vol. ASSP-34, 499–510, June 1986.

[18] Regalia, P.A., *Adaptive IIR Filtering in Signal Processing and Control*, Marcel Dekker, New York, 1995.

[19] Shapiro, J.M., Algorithms and systolic architectures for real-time multidimensional adaptive filtering of frequency domain multiplexed video signals, Ph.D. dissertation, M.I.T., Cambridge, MA, 1990.

[20] Shynk, J.J., Frequency-domain and multirate adaptive filtering, *IEEE ASSP Mag.*, 15–37, Jan. 1992.

[21] Strait, J.C., Structures and algorithms for two-dimensional adaptive signal processing, Ph.D. dissertation, University of Illinois at Urbana-Champaign, 1995.

[22] Widrow, B. and Stearns, S.D., *Adaptive Signal Processing*, Prentice-Hall, Englewood Cliffs, NJ, 1985.

23

Adaptive IIR Filters

Geoffrey A. Williamson
Illinois Institute of Technology

23.1 Introduction

In comparison with adaptive finite impulse response (FIR) filters, adaptive infinite impulse response (IIR) filters offer the potential to implement an adaptive filter meeting desired performance levels, as measured by mean-square error, for example, with much less computational complexity. This advantage stems from the enhanced modeling capabilities provided by the pole/zero transfer function of the IIR structure, compared to the "all-zero" form of the FIR structure.

However, adapting an IIR filter brings with it a number of challenges in obtaining stable and optimal behavior of the algorithms used to adjust the filter parameters. Since the 1970s, there has been much active research focused on adaptive IIR filters, but many of these challenges to date have not been completely resolved. As a consequence, adaptive IIR filters are not found in commercial practice in anywhere near the frequency that adaptive FIR filters are. Nonetheless, recent advances in adaptive IIR filter research have provided new results and insights into the behavior of several methods for adapting the filter parameters, and new algorithms have been proposed that address some of the problems and open issues in these systems. Hence, this class of adaptive filter continues to maintain promise as a potentially effective and efficient adaptive filtering option.

In this section, we provide an up-to-date overview of the different approaches to the adaptive IIR filtering problem. Due to the extensive literature on the subject, many readers may wish to peruse several earlier general treatments of the topic. Johnson's 1984 [11] and Shynk's 1989 paper [23] are still current in the sense that a number of open issues cited therein remain open today. More recently, Regalia's 1995 book [19] provides a comprehensive view of the subject.

0-8493-8572-5/98/$0.00+$.50
© 1998 by CRC Press LLC

23.1.1   The System Identification Framework for Adaptive IIR Filtering

The spread of issues associated with adaptive IIR filters is most easily understood if one adopts a system identification perspective to the filtering problem. To this end, consider the diagram presented in Fig. 23.1. Available to the adaptive filter are two external signals: the input signal $x(n)$ and the desired output signal $d(n)$. The adaptive filtering problem is to adjust the parameters of the filter acting on $x(n)$ so that its output $y(n)$ approximates $d(n)$. From the system identification perspective, the task at hand is to adjust the parameters of the filter generating $y(n)$ from $x(n)$ in Fig. 23.1 so that the *filtering operation itself* matches in some sense the *system* generating $d(n)$ from $x(n)$. These two viewpoints are closely related because if the systems are the same, then their outputs will be close. However, by adopting the convention that there is a system generating $d(n)$ from $x(n)$, clearer insights into the behavior and design of adaptive algorithms are obtained. This insight is useful even if the "system" generating $d(n)$ from $x(n)$ has only a statistical and not a physical basis in reality.

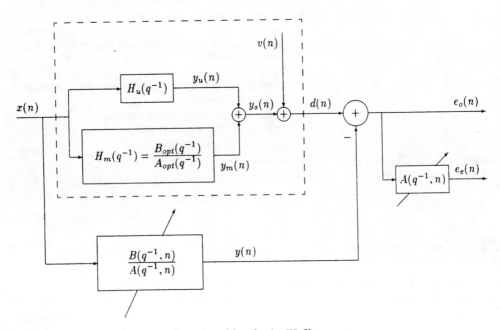

FIGURE 23.1   System identification configuration of the adaptive IIR filter.

The standard adaptive IIR filter is described by

$$y(n)+a_1(n)y(n-1)+\cdots+a_N(n)y(n-N) = b_0(n)x(n)+b_1(n)x(n-1)+\cdots+b_M(n)x(n-M), \quad (23.1)$$

or equivalently

$$\left(1 + a_1(n)q^{-1} + \cdots + a_N(n)q^{-N}\right) y(n) = \left(b_0(n) + b_1(n)q^{-1} + \cdots + b_M(n)q^{-M}\right) x(n). \quad (23.2)$$

As is shown in Fig. 23.1, Eq. (23.2) may be written in shorthand as

$$y(n) = \frac{B(q^{-1}, n)}{A(q^{-1}, n)} x(n), \quad (23.3)$$

where $B(q^{-1}, n)$ and $A(q^{-1}, n)$ are the time-dependent polynomials in the delay operator q^{-1} appearing in (23.2). The parameters that are updated by the adaptive algorithm are the coefficients of these polynomials. Note that the polynomial $A(q^{-1}, n)$ is constrained to be monic, such that $a_0(n) = 1$.

We adopt a rather more general description for the unknown system, assuming that $d(n)$ is generated from the input signal $x(n)$ via some linear time-invariant system $H(q^{-1})$, with the addition of a noise signal $v(n)$ to reflect components in $d(n)$ that are independent of $x(n)$. We further break down $H(q^{-1})$ into a transfer function $H_m(q^{-1})$ that is explicitly modeled by the adaptive filter, and a transfer function $H_u(q^{-1})$ that is unmodeled. In this way, we view $d(n)$ as a sum of three components: the signal $y_m(n)$ that is modeled by the adaptive filter, the signal $y_u(n)$ that is unmodeled but that depends on the input signal, and the signal $v(n)$ that is independent of the input. Hence,

$$d(n) = y_m(n) + y_u(n) + v(n) \tag{23.4}$$
$$= y_s(n) + v(n), \tag{23.5}$$

where $y_s(n) = y_m(n) + y_u(n)$. The modeled component of the system output is viewed as

$$y_m(n) = \frac{B_{\text{opt}}(q^{-1})}{A_{\text{opt}}(q^{-1})} x(n), \tag{23.6}$$

with $B_{\text{opt}}(q^{-1}) = \sum_{i=0}^{M} b_{i,\text{opt}} q^{-i}$ and $A_{\text{opt}}(q^{-1}) = 1 + \sum_{i=i}^{N} a_{i,\text{opt}} q^{-i}$. Note that (23.6) has the same form as (23.3). The parameters $\{a_{i,\text{opt}}\}$ and $\{b_{i,\text{opt}}\}$ are considered to be the optimal values for the adaptive filter parameters, in a manner that we describe shortly.

Figure 23.1 shows two error signals: $e_e(n)$ termed the *equation error* , and $e_o(n)$, termed the *output error*. The parameters of the adaptive filter are usually adjusted so as to minimize some positive function of one or the other of these error signals. However, the figure of merit for judging adaptive filter performance that we will apply throughout this section is the mean-square output error $E\{e_o^2(n)\}$. In most adaptive filtering applications, the desired signal, $d(n)$, is available only during a "training phase" in which the filter parameters are adapted. At the conclusion of the training phase, the filter will be operated to produce the output signal $y(n)$ as shown in the figure, with the difference between the filter output $y(n)$ and the (now unmeasurable) system output $d(n)$ the error. Thus, we adopt the convention that $\{a_{i,\text{opt}}\}$ and $\{b_{i,\text{opt}}\}$ are defined such that when $a_i(n) \equiv a_{i,\text{opt}}$ and $b_i(n) \equiv b_{i,\text{opt}}$, $E\{e_o^2(n)\}$ is minimized, with $A_{\text{opt}}(q^{-1})$ constrained to be stable.

At this point it is convenient to set down some notation and terminology. Define the regressor vectors

$$\mathbf{U}_e(n) = [x(n) \cdots x(n-M) \ -d(n-1) \cdots -d(n-N)]^T, \tag{23.7}$$
$$\mathbf{U}_o(n) = [x(n) \cdots x(n-M) \ -y(n-1) \cdots -y(n-N)]^T, \tag{23.8}$$
$$\mathbf{U}_m(n) = [x(n) \cdots x(n-M) \ -y_m(n-1) \cdots -y_m(n-N)]^T. \tag{23.9}$$

These vectors are the equation error regressor, output error regressor, and modeled system regressor vectors, respectively. Define a noise regressor vector

$$\mathbf{V}(n) = [0 \cdots 0 \ -v(n-1) \cdots -v(n-N)]^T \tag{23.10}$$

with $M+1$ leading zeros corresponding to the $x(n-i)$ values in the preceding regressors. Furthermore, define the parameter vectors

$$\mathbf{W}(n) = [b_0(n) \ b_1(n) \cdots b_M(n) \ a_1(n) \cdots a_N(n)]^T, \tag{23.11}$$
$$\mathbf{W}_{\text{opt}} = [b_{0,\text{opt}} \ b_{1,\text{opt}} \cdots b_{M,\text{opt}} \ a_{1,\text{opt}} \cdots a_{N,\text{opt}}]^T, \tag{23.12}$$
$$\widetilde{\mathbf{W}}(n) = \mathbf{W}_{\text{opt}} - \mathbf{W}(n), \tag{23.13}$$
$$\mathbf{W}_\infty = \lim_{n \to \infty} E\{\mathbf{W}(n)\}. \tag{23.14}$$

We will have occasion to use $\mathbf{W}$ to refer to the adaptive filter parameter vector when the parameters are considered to be held at fixed values. With this notation, we may for instance write $y_m(n) = \mathbf{U}_m^T(n)\mathbf{W}_{\text{opt}}$ and $y(n) = \mathbf{U}_o^T(n)\mathbf{W}(n)$.

The situation in which $y_u(n) \equiv 0$ is referred to as the *sufficient order case*. The situation in which $y_u(n) \not\equiv 0$ is termed the *undermodeled case*.

23.1.2 Algorithms and Performance Issues

A number of different algorithms for the adaptation of the parameter vector $W(n)$ in (23.11) have been suggested. These may be characterized with respect to the form of the error criterion employed by the algorithm. Each algorithm attempts to drive to zero either the equation error, the output error, or some combination or hybrid of these two error criteria. Major algorithm classes that we consider for the equation error approach include the standard least-squares (LS) and least mean-square (LMS) algorithms, which parallel the algorithms used in adaptive FIR filtering. For equation error methods, we also examine the instrumental variables (IV) algorithm, as well as algorithms that constrain the parameters in the denominator of the adaptive filter's transfer function to improve estimation properties. In the output error class, we examine gradient algorithms and hyperstability-based algorithms. Within the equation and output error hybrid algorithm class, we focus predominantly on the Steiglitz-McBride (SM) algorithm, though there are several algorithms that are more straightforward combinations of equation and output error approaches.

In general, we desire that the adaptive filtering algorithm adjusts the parameter vector $\mathbf{W}_n$ so that it converges to $\mathbf{W}_{opt}$, the parameters that minimize the mean-square output error. The major issues for adaptive IIR filtering on which we will focus herein are

1. conditions for the stability and convergence of the algorithm used to adapt $\mathbf{W}(n)$, and
2. the asymptotic value of the adapted parameter vector $\mathbf{W}_\infty$, and its relationship to $\mathbf{W}_{opt}$.

This latter issue relates to the minimum mean-square error achievable by the algorithm, as noted above. Other issues of importance include the convergence speed of the algorithm, its ability to track time variations of the "true" parameter values, and numerical properties, but these will receive less attention here. Of these, convergence speed is of particular concern to practitioners, especially as adaptive IIR filters tend to converge at a far slower rate than their FIR counterparts. However, we emphasize the stability and nature of convergence over the speed because if the algorithm fails to converge or converges to an undesirable solution, the rate at which it does so is of less concern. Furthermore, convergence speed is difficult to characterize for adaptive IIR filters due to a number of factors, including complicated dependencies on algorithm initializations, input signal characteristics, and the relationship between $x(n)$ and $d(n)$.

23.1.3 Some Preliminaries

Unless otherwise indicated, we assume in our discussion that all signals in Fig. 23.1 are stationary, zero mean, random signals with finite variance. In particular, the properties we ascribe to the various algorithms are stated with this assumption and are presumed to be valid. Results that are based on a deterministic framework are similar to those developed here; see [1] for an example.

We shall also make use of the following definitions.

DEFINITION 23.1   A (scalar) signal is persistently exciting (PE) of order L if, with

$$\mathbf{X}(n) = [x(n) \cdots x(n - L + 1)]^T , \tag{23.15}$$

there exist α and β satisfying $0 < \alpha < \beta < \infty$ such that $\alpha I < E\{\mathbf{X}(n)\mathbf{X}^T(n)\} < \beta I$. The (vector) signal $\mathbf{X}(n)$ is then also said to be PE.

If $x(n)$ contains at least $L/2$ distinct sinusoidal components, then $x(n)$ is PE of order L. Any random signal $x(n)$ whose power spectrum is nonzero over a interval of nonzero width will be PE for any value of L in (23.15). Such is the case, for example, if $x(n)$ is uncorrelated or if $x(n)$ is modeled as an AR, MA, or ARMA process driven by uncorrelated noise. PE conditions are required of all adaptive algorithms to ensure good behavior because if there is inadequate excitation to provide information to the algorithm, convergence of the adapted parameters estimates will not necessary follow [22].

DEFINITION 23.2   A transfer function $H(q^{-1})$ is said to be strictly positive real (SPR) if $H(q^{-1})$ is stable and the real part of its frequency response is positive at all frequencies.

An SPR condition will be required to ensure convergence for a few of the algorithms that we discuss. Note that such a condition cannot be guaranteed in practice when $H(q^{-1})$ is an unknown transfer function, or when $H(q^{-1})$ depends on an unknown transfer function.

23.2   The Equation Error Approach

To motivate the equation error approach, consider again Fig. 23.1. Suppose that $y(n)$ in the figure were actually equal to $d(n)$. Then the system relationship $A(q^{-1}, n)y(n) = B(q^{-1}, n)x(n)$ would imply that $A(q^{-1}, n)d(n) = B(q^{-1}, n)x(n)$. But of course this last equation does not hold exactly, and we term its error the "equation error" $e_e(n)$. Hence, we define

$$e_e(n) = A(q^{-1}, n)d(n) - B(q^{-1}, n)x(n) . \tag{23.16}$$

Using the notation developed in (23.7) through (23.14), we find that

$$e_e(n) = d(n) - \mathbf{U}_e^T(n)\mathbf{W}(n) . \tag{23.17}$$

Equation error methods for adaptive IIR filtering typically adjust $\mathbf{W}(n)$ so as to minimize the mean-squared error (MSE) $J_{\text{MSE}}(n) = E\{e_e^2(n)\}$, where $E\{\cdot\}$ denotes statistical expectation, or the exponentially weighted least-squares (LS) error $J_{\text{LS}}(n) = \sum_{k=0}^{n} \lambda^{n-k} e_e^2(k)$.

23.2.1   The LMS and LS Equation Error Algorithms

The equation error $e_e(n)$ of (23.17) is the difference between $d(n)$ and a prediction of $d(n)$ given by $\mathbf{U}_e^T(n)\mathbf{W}(n)$. Noting that $\mathbf{U}_e^T(n)$ does not depend on $\mathbf{W}(n)$, we see that equation error adaptive IIR filtering is a type of linear prediction, and in particular the form of the prediction is identical to that arising in adaptive FIR filtering. One would suspect that many adaptive FIR filter algorithms would then apply directly to adaptive IIR filters with an equation error criterion, and this is in fact the case.

Two adaptive algorithms applicable to equation error adaptive IIR filtering are the LMS algorithm given by

$$\mathbf{W}(n+1) = \mathbf{W}(n) + \mu(n)\mathbf{U}_e(n)e_e(n) , \tag{23.18}$$

and the recursive least-squares (RLS) algorithm given by

$$\mathbf{W}(n+1) = \mathbf{W}(n) + P(n)\mathbf{U}_e(n)e_e(n) , \tag{23.19}$$

$$P(n) = \frac{1}{\lambda}\left(P(n-1) - \frac{P(n-1)\mathbf{U}_e(n)\mathbf{U}_e^T(n)P(n-1)}{\lambda + \mathbf{U}_e^T(n)P(n-1)\mathbf{U}_e(n)} \right) , \tag{23.20}$$

where the above expression for $P(n)$ is a recursive implementation of

$$P(n) = \left(\sum_{k=0}^{n} \lambda^{n-k} \mathbf{U}_e(k)\mathbf{U}_e^T(k) \right)^{-1} . \tag{23.21}$$

Some typical choices for $\mu(n)$ in (23.18) are $\mu(n) \equiv \mu_0$, a constant, or $\mu(n) = \bar{\mu}/(\epsilon + \mathbf{U}_e^T(n)\mathbf{U}_e(n))$, a normalized step size. For convergence of the gradient algorithm in (23.18), μ_0 is chosen in the range

$0 < \mu_0 < 1/((M+1)\sigma_x^2 + N\sigma_d^2)$, where $\sigma_x^2 = E\{x^2(n)\}$ and $\sigma_d^2 = E\{d^2(n)\}$. Typically, values of μ_0 in the range $0 < \mu_0 < 0.1/((M+1)\sigma_x^2 + N\sigma_d^2)$ are chosen. With the normalized step size, we require $0 < \bar{\mu} < 2$ and $\epsilon > 0$ for stability, with typical choices of $\bar{\mu} = 0.1$ and $\epsilon = 0.001$. In 23.20, we require that λ satisfy $0 < \lambda \leq 1$, with λ typically close to or equal to one, and we initialize $P(0) = \gamma I$ with γ a large, positive number. These results are analogous to the FIR filter cases considered in the earlier sections of this chapter.

These algorithms possess nice convergence properties, as we now discuss.

Property 1: *Given that x is PE of order $N + M + 1$, under (23.18) and under (23.19) and (23.20), with algorithm parameters chosen to satisfy the conditions noted above, then $E\{W(n)\}$ converges to a value $\mathbf{W}_\infty$ minimizing $J_{\mathrm{MSE}}(n)$ and $J_{\mathrm{LS}}(n)$, respectively, as $n \to \infty$.*

This property is desirable in that global convergence to parameter values optimal for the equation error cost function is guaranteed, just as with adaptive FIR filters. The convergence result holds whether the filter is operating in the sufficient order case or the undermodeled case. This is an important advantage of the equation error approach over other approaches. The reader is referred to Chapters 19, 20, and 21 for further details on the convergence behaviors of these algorithms and their variations. As in the FIR case, the eigenvalues of the matrix $R = E\{\mathbf{U}_e(n)\mathbf{U}_e^T(n)\}$ determine the rates of convergence for the LMS algorithm. A large eigenvalue disparity in R engenders slow convergence in the LMS algorithm and ill-conditioning, with the attendant numerical instabilities, in the RLS algorithm. For adaptive IIR filters, compared to the FIR case, the presence of $d(n)$ in $\mathbf{U}_e(n)$ tends to increase the eigenvalue disparity, so that slower convergence is typically observed for these algorithms.

Of importance is the value of the convergence points for the LMS and RLS algorithms with respect to the modeling assumptions of the system identification configuration of Fig. 23.1. For simplicity, let us first assume that the adaptive filter is capable of modeling the unknown system exactly; that is, $H_u(q^{-1}) = 0$. One may readily show that the parameter vector $\mathbf{W}$ that minimizes the mean-square equation error (or equivalently the asymptotic least square equation error, given ergodic stationary signals) is

$$\mathbf{W} = E\left\{\mathbf{U}_e(n)\mathbf{U}_e^T(n)\right\}^{-1} E\{\mathbf{U}_e(n)d(n)\} \tag{23.22}$$

$$= \left(E\{\mathbf{U}_m(n)\mathbf{U}_m^T(n)\} + E\left\{\mathbf{V}(n)\mathbf{V}^T(n)\right\}\right)^{-1}$$
$$\left(E\{\mathbf{U}_m(n)y_m(n)\} + E\{\mathbf{V}(n)v(n)\}\right) . \tag{23.23}$$

Clearly, if $v(n) \equiv 0$, the $\mathbf{W}$ so obtained must equal $\mathbf{W}_{\mathrm{opt}}$, so that we have

$$\mathbf{W}_{\mathrm{opt}} = E\left\{\mathbf{U}_m(n)\mathbf{U}_m^T(n)\right\}^{-1} E\{\mathbf{U}_m(n)y_m(n)\} . \tag{23.24}$$

By comparing (23.23) and (23.24), we can easily see that when $v(n) \not\equiv 0, \mathbf{W} \neq \mathbf{W}_{\mathrm{opt}}$. That is, the parameter estimates provided by (23.18) through (23.20) are, in general, *biased* from the desired values, even when the noise term $v(n)$ is uncorrelated.

What effect on adaptive filter performance does this bias impose? Since the parameters that minimize the mean-square *equation* error are not the same as $\mathbf{W}_{\mathrm{opt}}$, the values that minimize the mean-square *output* error, the adaptive filter performance will not be optimal. Situations can arise in which this bias is severe, with correspondingly significant degradation of performance.

Furthermore, a critical issue with regard to the parameter bias is the input-output stability of the resulting IIR filter. Because the equation error is formed as $A(q^{-1})d(n) - B(q^{-1})x(n)$, a difference of two FIR filtered signals, there are no built in constraints to keep the roots of $A(q^{-1})$ within the unit circle in the complex plane. Clearly, if an unstable polynomial results from the adaptation, then the filter output $y(n)$ can grow unboundedly in operational mode, so that the adaptive filter fails. An example of such a situation is given in [25]. An important feature of this example is that the adaptive filter is capable of precisely modeling the unknown system, and that interactions of the noise process within the algorithm are all that is needed to destabilize the resulting model.

Nonetheless, under certain operating conditions, this kind of instability can be shown not to occur, as described in the following.

Property 2: *[18] Consider the adaptive filter depicted in Fig. 23.1, where $y(n)$ is given by (23.2). If $x(n)$ is an autoregressive process of order no more than N, and $v(n)$ is independent of $x(n)$ and of finite variance, then the adaptive filter parameters minimizing the mean-square equation error $E\{e_e^2(n)\}$ are such that $A(q^{-1})$ is stable.*

For instance, if $x(n)$ is an uncorrelated signal, then the convergence point of the equation error algorithms corresponds to a stable filter.

To summarize, for LMS and RLS adaptation in an equation error setting, we have guaranteed global convergence, but bias in the presence of additive noise even in the exact modeling case, and an estimated model guaranteed to be stable only under a limited set of conditions.

23.2.2 Instrumental Variable Algorithms

A number of different approaches to adaptive IIR filtering have been proposed with the intention of mitigating the undesirable biased properties of the LMS- and RLS-based equation error adaptive IIR filters. One such approach, still within the equation error context, is the instrumental variables (IV) method. Observe that the bias problem illustrated above stems from the presence of $v(n)$ in both $\mathbf{U}_e(n)$ and in $e_e(n)$ in the update terms in (23.18) and (23.19), so that second order terms in $v(n)$ then appear in (23.23). This simultaneous presence creates, in expectation, a nonzero, noise-dependent driving term to the adaptation. The IV algorithm approach addresses this by replacing $\mathbf{U}_e(n)$ in these algorithms with a vector $\mathbf{U}_{iv}(n)$ of *instrumental variables* that are independent of $v(n)$. If $\mathbf{U}_{iv}(n)$ remains correlated with $\mathbf{U}_m(n)$, the noiseless regressor, convergence to unbiased filter parameters is possible.

The IV algorithm is given by

$$\mathbf{W}(n+1) = \mathbf{W}(n) + \mu(n)P_{iv}(n)\mathbf{U}_{iv}(n)e_e(n) . \tag{23.25}$$

$$P_{iv}(n) = \frac{1}{\lambda(n)}\left(P_{iv}(n-1) - \frac{P_{iv}(n-1)\mathbf{U}_{iv}(n)\mathbf{U}_e^T(n)P_{iv}(n-1)}{(\lambda(n)/\mu(n)) + \mathbf{U}_e^T(n)P_{iv}(n-1)\mathbf{U}_{iv}(n)}\right) . \tag{23.26}$$

with $\lambda(n) = 1 - \mu(n)$. Common choices for $\lambda(n)$ are to set $\lambda(n) \equiv \lambda_0$, a fixed constant in the range $0 < \lambda < 1$ and usually chosen in the range between 0.9 and 0.99, or to choose $\mu(n) = 1/n$ and $\lambda(n) = 1 - \mu(n)$. As with RLS methods, $P(0) = \gamma I$ with γ a large, positive number. The vector $\mathbf{U}_{iv}(n)$ is typically chosen as

$$\mathbf{U}_{iv}(n) = [x(n) \cdots x(n-M) \quad -z(n-1) \cdots -z(n-N)]^T \tag{23.27}$$

with either

$$z(n) = -x(n-M) \quad \text{or} \quad z(n) = \frac{\bar{B}(q^{-1})}{\bar{A}(q^{-1})}x(n) . \tag{23.28}$$

In the first case, $\mathbf{U}_{iv}(n)$ is then simply an extended regressor in the input $x(n)$, while the second choice may be viewed as a regressor parallel to $\mathbf{U}_m(n)$, with $z(n)$ playing the role of $y_m(n)$. For this choice, one may think of $\bar{A}(q^{-1})$ and $\bar{B}(q^{-1})$ as fixed filters chosen to approximate $A_{opt}(q^{-1})$ and $B_{opt}(q^{-1})$, but the exact choice of $\bar{A}(q^{-1})$ and $\bar{B}(q^{-1})$ is not critical to the qualitative behavior of the algorithm. In both cases, note that $\mathbf{U}_{iv}(n)$ is independent of $v(n)$, since $d(n)$ is not employed in its construction.

The convergence of this algorithm is described by the following property, derived in [15].

Property 3: *In the sufficient order case with $x(n)$ PE of order at least $N+M+1$, the IV algorithm in (23.25) and (23.26) with $\mathbf{U}_{iv}(n)$ chosen according to (23.27) or (23.28) causes $E\{\mathbf{W}(n)\}$ to converge to $\mathbf{W}_\infty = \mathbf{W}_{opt}$.*

There are a few additional technical conditions an $A_{opt}(q^{-1})$, $B_{opt}(q^{-1})$, $\bar{A}(q^{-1})$, and $\bar{B}(q^{-1})$ that are required for the property to hold. These conditions will be satisfied in almost all circumstances; for details, the reader is referred to [15]. This convergence property demonstrates that the IV algorithm does in fact achieve unbiased parameter estimates in the sufficient order case.

In the undermodeled case, little has been said regarding the behavior and performance of the IV algorithm. A convergence point $\mathbf{W}_\infty$ must satisfy $E\{\mathbf{U}_{iv}(n) - (d(n)\mathbf{U}_e^T(n)\mathbf{W}_\infty)\} = 0$, but no characterization of such points exists if N and M are not of sufficient order. Furthermore, it is possible for the IV algorithm to converge to a point such that $1/A(q^{-1})$ is unstable [9].

Notice that (23.25) and (23.26) are similar in form to the RLS algorithm. One may postulate an "LMS-style" IV algorithm as

$$\mathbf{W}(n+1) = \mathbf{W}(n) + \mu(n)\mathbf{U}_{iv}(n)e_e(n), \tag{23.29}$$

which is computationally much simpler than the "RLS-style" IV algorithm of (23.25) and (23.26). However, the guarantee of convergence of the algorithm to $\mathbf{W}_{opt}$ in the sufficient order case for the RLS-style algorithm is now complicated by an additional requirement on $\mathbf{U}_{iv}(n)$ for convergence of the algorithm in (23.29). In particular, all eigenvalues of

$$R_{iv} = E\left\{\mathbf{U}_{iv}(n)\mathbf{U}_e^T(n)\right\} \tag{23.30}$$

must lie strictly in the right half of the complex plane. Since the properties of $\mathbf{U}_e(n)$ depend on the unknown relationship between $x(n)$ and $d(n)$, one is generally unable to guarantee *a priori* satisfaction of such conditions. This situation has parallels with the stability-theory approach to output error algorithms, as discussed later in this section.

Summarizing the IV algorithm properties, we have that in the sufficient order case, the RLS-style IV algorithm is guaranteed to converge to unbiased parameter values. However, an understanding and characterization of its behavior in the undermodeled case is yet incomplete, and the IV algorithm may produce unstable filters.

23.2.3 Equation Error Algorithms with Unit Norm Constraints

A different approach to mitigating the parameter bias in equation error methods arises as follows. Consider modifying the equation error of (23.17) to

$$e_e(n) = a_0(n)d(n) - \mathbf{U}_e^T(n)\mathbf{W}(n). \tag{23.31}$$

In terms of the expression (23.16), this change corresponds to redefining the adaptive filter's denominator polynomial to be

$$A(q^{-1}, n) = a_0(n) + a_1(n)q^{-1} + \cdots + a_N(n)q^{-N}, \tag{23.32}$$

and allowing for adaptation of the new parameter $a_0(n)$. One can view the equation error algorithms that we have already discussed as adapting the coefficients of this version of $A(q^{-1}, n)$, but with a *monic constraint* that imposes $a_0(n) = 1$. Recently, several algorithms have been proposed that consider instead equation error methods with a *unit norm constraint*. In these schemes, one adapts $\mathbf{W}(n)$ and $a_0(n)$ subject to the constraint

$$\sum_{i=0}^{N} a_i^2(n) = 1 \tag{23.33}$$

Note that if $A(q^{-1}, n)$ is defined as in (23.32), then $e_e(n)$ as constructed in Fig. 23.1 is in fact the error $e_e(n)$ given in (23.31).

The effect on the parameter bias stemming from this change from a monic to a unit norm constraint is as follows.

Property 4: *[18] Consider the adaptive filter in Fig. 23.1 with $A(q^{-1}, n)$ given by (23.32), with $v(n)$ an uncorrelated signal and with $H_u(q^{-1}) = 0$ (the sufficient order case). Then the parameter values $\mathbf{W}$ and a_0 that minimize $E\{e_e^2(n)\}$ subject to the unit norm constraint (23.33) satisfy $\mathbf{W}/a_0 = \mathbf{W}_{opt}$.*

That is, the parameter estimates are unbiased in the sufficient order case with uncorrelated output noise. Note that normalizing the coefficients in $\mathbf{W}$ by a_0 recovers the monic character of the denominator for $\mathbf{W}_{opt}$:

$$\frac{B(q^{-1})}{A(q^{-1})} = \frac{b_0 + b_1 q^{-1} + \cdots + b_M q^{-M}}{a_0 + a_1 q^{-1} + \cdots + a_N q^{-N}} \tag{23.34}$$

$$= \frac{(b_0/a_0) + (b_1/a_0)\, q^{-1} + \cdots + (b_M/a_0)\, q^{-M}}{1 + (a_1/a_0)\, q^{-1} + \cdots + (a_N/a_0)\, q^{-N}}. \tag{23.35}$$

In the undermodeled case, we have the following.

Property 5: *[18] Consider the adaptive filter in Fig. 23.1 with $A(q^{-1}, n)$ given by (23.32). If $x(n)$ is an autoregressive process of order no more than N, and $v(n)$ is independent of $x(n)$ and of finite variance, then the parameter values $\mathbf{W}$ and a_0 that minimize $E\{e_e^2(n)\}$ subject to the unit norm constraint (23.33) are such that $A(q^{-1})$ is stable. Furthermore, at those minimizing parameter values, if $x(n)$ is an uncorrelated input, then*

$$E\{e_e^2(n)\} \le \sigma_{N+1}^2 + \sigma_v^2, \tag{23.36}$$

where σ_{N+1} is the $(N+1)^{st}$ Hankel singular value of $H(z)$.

Notice that Property 5 is similar to Property 2, except that we have the added bonus of a bound on the mean-square equation error in terms of the Hankel singular values of $H(q^{-1})$. Note that the $(N+1)^{st}$ Hankel singular value of $H(q^{-1})$ is related to the achievable modeling error in an Nth order, reduced order approximation to $H(q^{-1})$ (see [19, Ch.4] for details). This bound thus indicates that the optimal unit norm constrained equation error filter will in fact do about as well as can be expected with an Nth order filter. However, this adaptive filter will suffer, just as with the equation error approaches with the monic constraint on the denominator, from a possibly unstable denominator if the input $x(n)$ is not an autoregressive process.

An adaptive algorithm for minimizing the mean-square equation error subject to the unit norm constraint can be found in [4]. The algorithm of [4] is formulated as a recursive total least squares algorithm using a two-channel, fast transversal filter implementation. The connection between total least squares and the unit norm constrained equation error adaptive filter implies that the correlation matrices that are embedded within the adaptive algorithm will be more poorly conditioned than the correlation matrices arising in the RLS algorithm. Consequently, convergence will be slower for the unit norm constrained approach than in the standard, monic constraint approach.

More recently, several new algorithms that generalize the above approach to confer unbiasedness in the presence of correlated output noises $v(n)$ have been proposed [5]. These algorithms require knowledge of the statistics of $v(n)$, though versions of the algorithms in which these statistics are estimated on-line are also presented [5]. However, little is known about the transient behaviors or the local stabilities of these adaptive algorithms, particularly in the undermodeled case.

In conclusion, minimizing the equation error cost function with a unit norm constraint on the autoregressive parameter vector provides bias-free estimates in the sufficient order case and a bias level similar to the standard equation error methods in the undermodeled case. Adaptive algorithms for constrained equation error minimization are under development, and their convergence properties are largely unknown.

23.3   The Output Error Approach

We have already noted that the error of merit for adaptive IIR filters is the output error $e_o(n)$. We now describe a class of algorithms that explicitly uses the output error in the parameter updates. We distinguish between two categories within this class: those algorithms that directly attempt to minimize the least squares or mean-square output error, and those formulated using stability theory to enforce convergence to the "true" system parameters. This class of algorithms has the advantage of eliminating the parameter bias that occurs in the equation error approach. However, as we will see, the price paid is that convergence of the algorithms becomes more complicated, and unlike in the equation error methods, global convergence to the desired parameter values is no longer guaranteed.

Critical to the formulation of these output error algorithms is an understanding of the relationship of $\mathbf{W}(n)$ to $e_o(n)$. With reference to Fig. 23.1, we have

$$e_o(n) = d(n) - \frac{B(q^{-1}, n)}{A(q^{-1}, n)} x(n) . \tag{23.37}$$

Using the notation in (23.7) through (23.14) and following a standard derivation [19, Ch.9] shows that

$$y_m(n) - y(n) = \frac{1}{A_{\text{opt}}(q^{-1})} \left[\mathbf{U}_o^T(n) \widetilde{\mathbf{W}}(n) \right] , \tag{23.38}$$

so that

$$e_o(n) = \frac{1}{A_{\text{opt}}(q^{-1})} \left[\mathbf{U}_o^T(n) \widetilde{\mathbf{W}}(n) \right] + y_u(n) + v(n) . \tag{23.39}$$

The expression in (23.39) makes clear two characteristics of $e_o(n)$. First, $e_o(n)$ separates the error due to the modeled component, which is the term based on $\widetilde{\mathbf{W}}(n)$, from the error due to the unmodeled effects in $d(n)$, that is $y_u(n) + v(n)$. Neither $y_u(n)$ nor $v(n)$ appear in the term based on $\widetilde{\mathbf{W}}(n)$. Second, $e_o(n)$ is nonlinear in $\mathbf{W}(n)$, since $\mathbf{U}_o(n)$ depends on $\mathbf{W}(n)$. The first feature leads to the desirable unbiasedness characteristic of output error methods, while the second is a source of difficulty for defining globally convergent algorithms.

23.3.1   Gradient-Descent Algorithms

An output error-based gradient descent algorithm may be defined as follows. Set

$$x_f(n) = \frac{1}{A(q^{-1}, n)} x(n), \quad y_f(n) = \frac{1}{A(q^{-1}, n)} y(n), \tag{23.40}$$

and define

$$\mathbf{U}_{\text{of}}(n) = \left[x_f(n) \cdots x_f(n - M) \; - y_f(n - 1) \cdots - y_f(n - N) \right]^T . \tag{23.41}$$

Then

$$\mathbf{W}(n + 1) = \mathbf{W}(n) + \mu(n) \mathbf{U}_{\text{of}}(n) e_o(n) \tag{23.42}$$

defines an approximate stochastic gradient (SG) algorithm for adapting the parameter vector $\mathbf{W}(n)$. The direction of the update term in (23.42) is opposite to the gradient of $e_o(n)$ with respect to $\mathbf{W}(n)$, assuming that the parameter vector $\mathbf{W}(n)$ varies slowly in time. To see how a gradient descent results in this algorithm, note that the output error may be written as

$$e_o(n) = d(n) - y(n) \tag{23.43}$$

$$= d(n) - \sum_{i=0}^{M} b_i(n) x(n - i) + \sum_{i=1}^{N} a_i(n) y(n - i) \tag{23.44}$$

so that

$$\frac{\partial e_o(n)}{\partial b_i(n)} = -x(n-i) + \sum_{i=1}^{N} a_i(n) \frac{\partial y(n-1)}{\partial b_i(n)} . \tag{23.45}$$

Noting that $\partial e_o(n)/\partial b_i(n) = -\partial y(n)/\partial b_i(n)$, and assuming that the parameter $b_i(n)$ varies slowly enough so that

$$\frac{\partial e(n-i)}{\partial b_i(n)} \approx \frac{\partial e(n-i)}{\partial b_i(n-i)} , \tag{23.46}$$

(23.45) becomes

$$\frac{\partial e(n)}{\partial b_i(n)} \approx -x(n-i) - \sum_{i=1}^{N} a_i(n) \frac{\partial e(n-i)}{\partial b_i(n-i)} . \tag{23.47}$$

This equation can be rearranged to

$$\frac{\partial e(n-i)}{\partial b_i(n)} \approx -A(q^{-1}, n)x(n-i) = -x_f(n-i) . \tag{23.48}$$

The relation

$$\frac{\partial e(n-i)}{\partial a_i(n)} \approx A(q^{-1}, n)y(n-i) = y_f(n-i) \tag{23.49}$$

may be found in a similar fashion. Since the gradient descent algorithm is

$$b_i(n+1) = b_i(n) - \frac{\mu}{2} \frac{\partial e_o^2(n)}{\partial b_i(n)} \tag{23.50}$$

$$\approx b_i(n) + \mu x_f(n-i)e_o(n) \tag{23.51}$$

$$a_i(n+1) = a_i(n) - \frac{\mu}{2} \frac{\partial e_o^2(n)}{\partial a_i(n)} \tag{23.52}$$

$$\approx b_i(n) - \mu y_f(n-i)e_o(n) , \tag{23.53}$$

(23.42) follows.

The step size $\mu(n)$ is typically chosen either as a constant μ_0, or normalized by $\mathbf{U}_{of}(n)$ as

$$\mu(n) = \frac{\bar{\mu}}{1 + \bar{\mu}\mathbf{U}_{of}^T(n)\mathbf{U}_{of}(n)} . \tag{23.54}$$

Due to the nonlinear relationship between the parameters and the output error, selection of values for $\mu(n)$ is less straightforward than in the equation error case. Roughly speaking, one would like that $0 < \mu(n) \le 1/\mathbf{U}_{of}^T(n)\mathbf{U}_{of}(n)$, or more conservatively, $\mu(n) = 0.1/\mathbf{U}_{of}^T(n)\mathbf{U}_{of}(n)$. This suggests setting $\mu_0 = 0.1/E\{\mathbf{U}_{of}^T(n)\mathbf{U}_{of}(n)\}$, given an estimate of the expected value, or $\bar{\mu}$ at about the same value. The behavior of the algorithm using the normalized step size of (23.54) is in general less sensitive to variations in $\bar{\mu}$ than is the unnormalized version with respect to choice of μ_0.

Another alternative to (23.42) is the Gauss-Newton (GN) algorithm given by

$$\mathbf{W}(n+1) = \mathbf{W}(n) + \mu(n)P(n)\mathbf{U}_{of}(n)e_o(n) , \tag{23.55}$$

$$P(n) = \frac{1}{\lambda(n)} \left(P(n-1) - \frac{P(n-1)\mathbf{U}_{of}(n)\mathbf{U}_{of}^T(n)P(n-1)}{(\lambda(n)/\mu(n)) + \mathbf{U}_{of}^T(n)P(n-1)\mathbf{U}_{of}(n)} \right) , \tag{23.56}$$

while setting $\lambda(n) = 1 - \mu(n)$, and $P(0) = \gamma I$ just as for the IV algorithm. Most frequently $\lambda(n)$ is chosen as a constant in the range between 0.9 and 0.99. Another choice is to set $\mu(n) = 1/n$, a decreasing adaptation gain, but when $\mu(n)$ tends to zero, one loses adaptability. The GN algorithm is a descent strategy utilizing approximate second order information, with the matrix $P(n)$ being an approximation of the inverse of the Hessian of $e_o(n)$ with respect to $\mathbf{W}(n)$. Note the similarity of (23.55) and (23.56) to (23.19) and (23.20). In fact, replacing $\mathbf{U}_{of}(n)$ in the GN algorithm with $\mathbf{U}_e(n)$, replacing $e_o(n)$ with $e_e(n)$, and setting $\mu(n) = (1 - \lambda)/(1 - \lambda^n)$, one recovers the RLS algorithm, though the interpretation of $P(n)$ in this form is slightly different. As n gets large, the choice of constant λ and $\mu = 1 - \lambda$ approximates RLS (with $0 < \lambda < 1$).

Precise convergence analyses of these two algorithms are quite involved and rely on a number of technical assumptions. Analyses fall into two categories. One approach treats the step size $\mu(n)$ in (23.42) and in (23.55) and (23.56) as a quantity that tends to zero, satisfying the following properties:

$$(1)\ \mu(n) \to 0, \quad (2)\ \lim_{L \to \infty} \sum_{n=0}^{L} \mu(n) \to \infty, \quad (3)\ \lim_{L \to \infty} \sum_{n=0}^{L} \mu^2(n) < \infty\ ; \quad (23.57)$$

for instance $\mu(n) = 1/n$, as noted above. The ODE analysis of [15] applies in this situation. Assuming a decreasing step size is a necessary technicality to enable convergence of the adapted parameters to their optimum values in a random environment. The second approach allows μ to remain as a fixed, but small, step size, as in [3]. The results describe the probabilistic behavior of $\mathbf{W}(n)$ over finite intervals of time, with the extent of the interval increasing and the degree of variability of $\mathbf{W}(n)$ decreasing as the fixed value of μ becomes smaller.

However, in both cases, the conclusions are essentially the same. The behavior of these algorithms with small enough step size μ is to follow gradient descent of the mean-square output error $E\{e_o^2(n)\}$. We should note that a technical requirement for the analyses to remain valid is that signals within the adaptive filter remain bounded, and to insure this the stability for the polynomial $A(q^{-1}, n)$ must be maintained to ensure this requirement. Therefore, at each iteration of the gradient descent algorithm, one must check the stability of $A(q^{-1}, n)$ and, if it is unstable, prevent the update of the $a_i(n + 1)$ values in $\mathbf{W}(n+1)$, or project $\mathbf{W}(n+1)$ back into the set of parameter vector values whose corresponding $A(q^{-1}, n)$ polynomial is stable [11, 15]. For direct-form adaptive filters, this stability check can be computationally burdensome, but it is necessary as the algorithm often fails to converge without its implementation, especially if $A_{opt}(q^{-1})$ has roots near to the unit circle. Imposing a stability check at each iteration of the algorithm guarantees the following result.

Property 6: *For the SG or GN algorithm with decreasing $\mu(n)$ satisfying (23.57), $\mathbf{W}(n)$ converges to a value locally minimizing the mean-square output error or locks up on a point on the stability boundary where $\mathbf{W}$ represents a marginally stable filter. For the SG or GN algorithm with constant μ that is small enough, $\mathbf{W}(n)$ remains close in probability to a trajectory approaching a value locally minimizing the mean-square output error.*

This property indicates that the value of $\mathbf{W}(n)$ found by these algorithms does in practice approach a local minimum of the mean-square output error surface. A stronger analytic statement of this expected convergence is unfortunately not possible, and in fact, the probability of large deviations of the algorithms from a minimum point becomes large with time with constant μ. As a practical matter, however, one can expect the parameter vector to approach and stay near a minimizing parameter value using these methods.

More problematic, however, is whether effective convergence to a *global* minimum is achieved. A thorough treatment of this issue appears in [16] and [27], with conclusions as follows.

Property 7: *In the sufficient order case ($y_u \equiv 0$) with an uncorrelated input $x(n)$, all minima of $E\{e_o^2(n)\}$ are global minima.*

The same conclusion holds if $x(n)$ is generated as an ARMA process, given satisfaction of certain conditions on the orders of the adaptive filter, the unknown system, and the system generating $x(n)$; see [27] for details. However, in the undermodeled case ($y_u(n) \not\equiv 0$), it is possible for the system to converge to a local but not global minimum. Several examples of this are presented in [16]. Since the insufficient order case will likely be the one encountered in practice, the possibility of convergence to a local but not global minimum will always exist with these gradient descent output error algorithms. It is possible that these local minima will provide a level of mean-square output error much greater than that obtained at the global minimum, so serious performance degradation may result. However, any such minimum must correspond to a stable parametrization of the adaptive filter, in contrast with the equation error methods for which there is no such guarantee in the most general of circumstances.

We have the following summary. The output error gradient descent algorithms converge to a stable filter parametrization that locally minimizes the mean-square output error. These algorithms are unbiased, and reach a global minimum, when $y_u(n) \equiv 0$, but when the true system has been undermodeled, convergence to a local but not global minimum is likely.

23.3.2 Output Error Algorithms Based on Stability Theory

One of the first adaptive IIR filters to be proposed employs the parameter update

$$\mathbf{W}(n+1) = \mathbf{W}(n) + \mu(n)\mathbf{U}_o(n)e_o(n). \tag{23.58}$$

This algorithm, often referred to as a pseudolinear regression, Landau's algorithm, or Feintuch's algorithm, was proposed as an alternative to the gradient descent algorithm of (23.42). This algorithm is similar in form to (23.18), save that the regressor vector and error signal of the output error formulation appear in place of their equation error counterparts. In essence, the update in (23.58) ignores the nonlinear dependence of $e_o(n)$ on $\mathbf{W}(n)$, and takes the form of an algorithm for a linear regression, hence the label "pseudolinear" regression. One advantage of (23.58) over the algorithm in (23.42) is that the former is computationally simpler as it avoids the filtering operations necessary to generate (23.40). An additional requirement is needed for stability of the algorithm, however, as we now discuss.

The algorithm of (23.58) is one possibility among a broad range of algorithms studied in [21], given by

$$\mathbf{W}(n+1) = \mathbf{W}(n) + \mu(n)F(q^{-1},n)[\mathbf{U}_o(n)]G(q^{-1},n)[e_o(n)], \tag{23.59}$$

where $F(q^{-1},n)$ and $G(q^{-1},n)$ are possibly time-varying filters, and where $F(q^{-1},n)$ acting on the vector $\mathbf{U}_o(n)$ denotes an element-by-element filtering operation. We can see that setting $F(q^{-1}) = G(q^{-1}) = 1$ yields (23.58). The convergence of these algorithms have been studied using the theory of hyperstability and the theory of averaging [1]. For this reason, we classify this family as "stability theory based" approaches to adaptive IIR filtering. The method behind (23.59) can be understood by considering the algorithm subclass represented by

$$\mathbf{W}(n+1) = \mathbf{W}(n) + \mu(n)\mathbf{U}_o(n)G(q^{-1})[e_o(n)], \tag{23.60}$$

which is known as the Simplified Hyperstable Adaptive Recursive Filter or SHARF algorithm [12]. A GN-like alternative is

$$\mathbf{W}(n+1) = \mathbf{W}(n) + \mu(n)P(n)\mathbf{U}_o(n)G(q^{-1})[e_o(n)], \tag{23.61}$$

$$P(n) = \frac{1}{\lambda(n)}\left(P(n-1) - \frac{P(n-1)\mathbf{U}_o(n)\mathbf{U}_o^T(n)P(n-1)}{(\lambda(n)/\mu(n)) + \mathbf{U}_o^T(n)P(n-1)\mathbf{U}_o(n)}\right). \tag{23.62}$$

with again $\lambda(n) = 1 - \mu(n)$. Choice of $\mu(n)$, $\lambda(n)$, and $P(0)$ are similar to those for other algorithms.

The averaging analyses applied in [21] to (23.60) and (23.61)–(23.62) obtain the following convergence results, with reference to Definitions 23.1 and 23.2 in the Introduction.

Property 8: *If $G(q^{-1})/A_{\text{opt}}(q^{-1})$ is SPR and $\mathbf{U}_m(n)$ is a bounded, PE vector sequence,[1] then when $y_u(n) = v(n) = 0$, there exists a μ_0 such that $0 < \mu < \mu_0$ implies that (23.60) is locally exponentially stable*

---

[1] The PE condition applies to $\mathbf{U}_m(n)$, rather than $\mathbf{U}_o(n)$, since this is an analysis local to $\mathbf{W} = \mathbf{W}_{\text{opt}}$, where $\mathbf{U}_o(n) = \mathbf{U}_m(n)$.

about $\mathbf{W} = \mathbf{W}_{\text{opt}}$. *It also follows that nonzero, but small,* $y_u(n)$ *and* $v(n)$ *result in a bounded perturbation of* $\mathbf{W}$ *from* $\mathbf{W}_{\text{opt}}$. *If the SPR condition is strengthened to* $(G(q^{-1})/A_{\text{opt}}(q^{-1})) - (1/2)$ *being SPR, then the results apply to* (23.61) *and* (23.62) *with* μ *constant and small enough.*

The essence of the analysis is to describe the average behavior of the parameter error $\widetilde{W}(n)$ under (23.60) by

$$\widetilde{\mathbf{W}}_{\text{avg}}(n+1) = [I - \mu R]\widetilde{\mathbf{W}}_{\text{avg}}(n) + \mu \xi_1(n) + \mu^2 \xi_2(n) . \tag{23.63}$$

In (23.63), the signal $\xi_1(n)$ is a term dependent on $y_u(n)$ and $v(n)$, the signal $\xi_2(n)$ represents the approximation error made in linearizing and averaging the update, and the matrix R is given by

$$R = \text{avg} \left\{ \mathbf{U}_m(n) \left(\frac{G(q^{-1})}{A_{\text{opt}}(q^{-1})} [\mathbf{U}_m(n)] \right)^T \right\} . \tag{23.64}$$

The SPR and PE conditions imply that the eigenvalues of R all have positive real part, so that μ may then be chosen small enough so that the eigenvalues of $I - \mu R$ are all less than one in magnitude. Then $\widetilde{\mathbf{W}}_{\text{avg}}(n+1) = [I - \mu R]\widetilde{\mathbf{W}}_{\text{avg}}(n)$ is exponentially stable, and Property 8 follows. The exponential stability of (23.63) is the property that allows the algorithm to behave robustly in the presence of a number of effects, including that the undermodeling [1].

The above convergence result is local in nature and is the best that can be stated for this class of algorithms in the general case. In the variation proposed for system identification by Landau and interpreted for adaptive filters by Johnson [10], a stronger statement of convergence can be made in the exact modeling case when $y_u(n)$ is zero and assuming that $v(n) = 0$. In that situation, given satisfaction of the SPR and PE conditions, $\mathbf{W}(n)$ can be shown to converge to $\mathbf{W}_{\text{opt}}$. For nonzero $v(n)$, analyses with a vanishing step size $\mu(n) \to \infty$ have established this convergence, again assuming exact modeling, even in the presence of a correlated noise term $v(n)$ [20]. One advantage on this convergence result in comparison to the exact modeling convergence result for the gradient algorithm (23.42) is that the PE condition on the input is less restrictive than the conditions that enable global convergence of the gradient algorithm. Nonetheless, in the undermodeled case, convergence is local in nature, and although the robustness conferred by the local, exponential stability to some extent mitigates this problem, it represents a drawback to the practical application of these techniques.

A further drawback of this technique is the SPR condition that $G(q^{-1})/A_{\text{opt}}(q^{-1})$ must satisfy. The polynomial $A_{\text{opt}}(q^{-1})$ is of course unknown to the adaptive filter designer, presenting difficulties in the selection of $G(q^{-1})$ to ensure that $G(q^{-1})/A_{\text{opt}}(q^{-1})$ is SPR. Recent research into choices of filters $G(q^{-1})$ that render $G(q^{-1})/A_{\text{opt}}(q^{-1})$ SPR for all $A_{\text{opt}}(q^{-1})$ within a set of filters, a form of "robust SPR" result, has begun to address this issue [2], but the problem of selecting $G(q^{-1})$ has not yet been completely resolved.

To summarize, for the SHARF algorithm and its cousins, we have convergence to unbiased parameter values guaranteed in the sufficient order case when there is adequate excitation and an SPR condition is satisfied. Satisfaction of the SPR condition cannot be guaranteed without *a priori* knowledge of the optimal filter, however. In the undermodeled case, no general results can be stated, but as long as the unmodeled component of the optimal filter is small in some sense, the exponential convergence in the sufficient order case implies stable behavior in this situation. Filter order selection to make the unmodeled component small again requires *a priori* knowledge about the optimal filter.

23.4   Equation-Error/Output-Error Hybrids

We have seen that equation error methods enjoy global convergence of their parameters, but suffer from parameter estimation bias, while output error methods enjoy unbiased parameter estimates while suffering from difficulties in their convergence properties. A number of algorithms have been proposed that in a sense strive to combine the best of both of these approaches. The most important of these is the Steiglitz-McBride algorithm, which we consider in detail below. Several other algorithms in this class work by using convex combinations of terms in the equation error and output error parameter updates. Two such algorithms include the Bias Remedy Least Mean-Squares algorithm of [14] and the Composite Regressor Algorithm of [13]. We will not consider these algorithms here; for details, see [14] and [13].

23.4.1 The Steiglitz-McBride Family of Algorithms

The Steiglitz-McBride (SM) algorithm is adapted from an off-line system identification method that iteratively minimizes the squared equation error criterion using prefiltered data. The prefiltering operations are based on the results of the previous iteration in such a way that the algorithm bears a close relationship to an output error approach.

A clear understanding of the algorithm in an adaptive filtering context is best obtained by first considering the original off-line method. Given a finite record of input and output sequences $x(n)$ and $d(n)$, one first forms the equation error according to (23.16). The parameters of $A(q^{-1})$ and $B(q^{-1})$ minimizing the LS criterion for this error are then found, and the minimizing polynomials are labelled as $A^{(0)}(q^{-1})$ and $B^{(0)}(q^{-1})$. The SM method then proceeds iteratively by minimizing the LS criterion for

$$e_e^{(i)}(n) = A(q^{-1})d_f^{(i)}(n) - B(q^{-1})x_f^{(i)}(n) \tag{23.65}$$

to find $A^{(i)}(q^{-1})$ and $B^{(i)}(q^{-1})$, where

$$d_f^{(i)}(n) = \frac{1}{A^{(i-1)}(q^{-1})}d(n); \quad x_f^{(i)}(n) = \frac{1}{A^{(i-1)}(q^{-1})}x(n) . \tag{23.66}$$

Notice that at each iteration, we find $A^{(i)}(q^{-1})$ and $B^{(i)}(q^{-1})$ through equation error minimization, for which we have globally convergent methods as discussed previously.

Let $A^{(\infty)}(q^{-1})$, $B^{(\infty)}(q^{-1})$ denote the polynomials obtained at a convergence point of this algorithm. Then minimizing the LS criterion applied to

$$e_e^{(\infty)}(n) = A(q^{-1})\frac{1}{A^{(\infty)}(q^{-1})}d(n) - B(q^{-1})\frac{1}{A^{(\infty)}(q^{-1})}x(n) \tag{23.67}$$

results again in $A(q^{-1}) = A^{(\infty)}(q^{-1})$ and $B(q^{-1}) = B^{(\infty)}(q^{-1})$ by virtue of this solution being a convergence point, and the error signal at this minimizing choice of parameters is

$$e_e^{(\infty)}(n) = d(n) - \frac{B^{(\infty)}(q^{-1})}{A^{(\infty)}(q^{-1})}x(n) . \tag{23.68}$$

Comparing (23.68) to (23.37), we see that at a convergence point of the SM algorithm, $e_e^{(\infty)}(n) = e_o(n)$, thereby drawing the connection between equation error and output error approaches in the SM approach.[2] Because of this connection, one expects that the parameter bias problem is mitigated, and in fact this is the case, as demonstrated by the following property.

Property 9: *[26] If $y_u(n) \equiv 0$ and $v(n)$ is white noise, then with $x(n)$ PE of order at least $N + M + 1$, $B(q^{-1}) = B_{\mathrm{opt}}(q^{-1})$ and $A(q^{-1}) = A_{\mathrm{opt}}(q^{-1})$ is the only convergence point of the SM algorithm, and this point is locally stable.*

---

[2]Realize, however, that minimizing the square of $e_o^{(\infty)}(n)$ in (23.67) is *not* equivalent to minimizing the squared output error, and in general these two approaches can result in different values for $A(q^{-1})$ and $B(q^{-1})$.

The local stability implies that if the initial denominator estimate $A^{(0)}(q^{-1})$ is close enough to $A_{\text{opt}}(q^{-1})$, then the algorithm converges to the unbiased solution in the uncorrelated noise case.

The on-line variation of the SM algorithm useful for adaptive filtering applications is given as follows. Set $x_f(n)$ as in (23.40) and set

$$d_f(n) = \frac{1}{A(q^{-1}, n+1)} d(n) . \tag{23.69}$$

The $(n + 1)$ index in the above filter is reasonable as only past $d_f(n)$ samples shall appear in the parameter updates at time n. Then by defining the SM regressor vector as

$$\mathbf{U}_{\text{ef}}(n) = \left[x_f(n) \cdots x_f(n - M) \; -d_f(n-1) \cdots - d_f(n-N)\right]^T , \tag{23.70}$$

the algorithm is

$$\mathbf{W}(n + 1) = \mathbf{W}(n) + \mu(n) \mathbf{U}_{\text{ef}}(n) e_o(n) . \tag{23.71}$$

Alternatively, we may employ the GN-style version given by

$$\mathbf{W}(n + 1) = \mathbf{W}(n) + \mu(n) P_{\text{ef}}(n) \mathbf{U}_{\text{ef}}(n) e_o(n) , \tag{23.72}$$

$$P_{\text{ef}}(n) = \frac{1}{\lambda(n)} \left(P_{\text{ef}}(n-1) - \frac{P_{\text{ef}}(n-1) \mathbf{U}_{\text{ef}}(n) \mathbf{U}_{\text{ef}}^T(n) P_{\text{ef}}(n-1)}{(\lambda(n)/\mu(n)) + \mathbf{U}_{\text{ef}}^T(n) P_{\text{ef}}(n-1) \mathbf{U}_{\text{ef}}(n)} \right) , \tag{23.73}$$

with $\lambda(n)$, $\mu(n)$, and $P(0)$ chosen in the same fashion as with the IV and GN algorithms. For these algorithms, the signal $e_o(n)$ is the output error, constructed as shown in Fig. 23.1, a reflection of the connection of the SM and output error approaches noted above. Also, note that $\mathbf{U}_{\text{ef}}(n)$ is a filtered version of the equation error regressor $\mathbf{U}_e(n)$, but with the time index of the filtering operation of (23.69) set to $n + 1$ rather than n, reflecting the derivation of the algorithm from the iterative off-line procedure. This form of the algorithm is only one of several variations; see [8] or [19] for others.

Assuming that one monitors and maintains the stability of the adapted polynomial $A(q^{-1}, n)$, in order that the signals $x_f(n)$ and $d_f(n)$ remain bounded, this algorithm has the following properties [19].

Property 10: *[6] In the sufficient order case where $y_u(n) \equiv 0$ and with $v(n)$ an uncorrelated noise sequence and $x(n)$ PE of order at least $N + M + 1$, the on-line SM algorithm converges to $\mathbf{W}_\infty = \mathbf{W}_{\text{opt}}$ or locks up on the stability boundary.*

Property 11: *[19] In the sufficient order case where $y_u(n) \equiv 0$ with $v(n)$ a correlated sequence, and in the undermodeled case where $y_u(n) \not\equiv 0$, the existence of convergence points $\mathbf{W}_\infty$ of the on-line SM algorithm is not guaranteed, and if these convergence points exist, they are generally biased away from $\mathbf{W}_{\text{opt}}$.*

Property 12: *[19] In the undermodeled case with the order of the adaptive filter numerator and denominator both equal to N, and with $x(n)$ an uncorrelated sequence, then at the convergence points of the on-line SM algorithm, if they exist,*

$$E\{e_o^2(n)\} \leq \sigma_{N+1}^2 + \frac{\max_\omega S_v\left(e^{j\omega}\right) - \sigma_v^2}{\sigma_u^2} + \sigma_v^2 , \tag{23.74}$$

where σ_{N+1} is the $(N + 1)^{\text{st}}$ Hankel singular value of $H(z) = H_m(z) + H_u(z)$ and where $S_v(e^{j\omega})$ is the power spectral density function of $v(n)$.

Note that in the off-line version for either the sufficient order case with correlated $v(n)$ or the undermodeled case, the SM algorithm can possibly converge to a set of parameters yielding an unstable filter. The stability check and projection steps noted above will prevent such convergence in the on-line version, contributing in part to the possibility of non-convergence.

The pessimistic nature of Properties 11 and 12 with regard to existence of convergence points is somewhat unfair in the following sense. In practice, one finds that in most circumstances the SM algorithm does converge, and furthermore that the convergence point is close to the minimum of the mean-square output error surface [7]. Property 12 quantifies this closeness. The $(N + 1)^{st}$ Hankel singular value of a transfer function is an upper bound for the minimum mean-square output error of an Nth order transfer function approximation [19, Ch.4]. Hence, one sees that $\mathbf{W}_\infty$ under the SM algorithm is guaranteed to remain close in this sense to $\mathbf{W}_{opt}$, and this fact remains true regardless of the existence or relative values of local minima on the mean-square output error surface.

The second term in (23.74) describes the effect of the noise term. The fact that $\max_\omega S_v(e^{j\omega}) = \sigma_v^2$ for uncorrelated $v(n)$ shows the disappearance of noise effects in that case. For strongly correlated $v(n)$, the effect of this noise term will increase, as the adaptive filter attempts to model the noise as well as the unknown system, and of course this effect is reduced as the signal-to-noise ratio of $d(n)$ is increased. One sees, then, that with strongly correlated noise, the SM algorithm may produce a significantly biased solution $\mathbf{W}_\infty$.

To summarize, given adequate excitation, the SM algorithm in the sufficient order case converges to unbiased parameter values when the noise $v(n)$ is uncorrelated, and generally converges to biased parameter values when $v(n)$ is correlated. The SM algorithm is not guaranteed to converge in the undermodeled case and, furthermore, if it converges, there is no general guarantee of stability of the resulting filter. However, a bound of the modeling error in these instances quantifies what can be considered as the good performance of the algorithm when it converges.

23.5   Alternate Parametrizations

Thus far we have couched our discussion of adaptive IIR filters in terms of a direct-form implementation of the system. Direct-form implementations suffer from poor finite precision effects both in terms of coefficient quantization and round-off effects in their computations. Furthermore, in output error adaptive IIR filtering, one must check the stability of the adaptive filter's denominator polynomial at each iteration. In direct-form implementations, this stability check is cumbersome and computationally expensive to implement. For these reasons, adaptive IIR filters implemented in alternative realizations such as parallel-form, cascade-form, and lattice-form have been proposed. For these structures, a stability check is easily implemented.

The SG and GN algorithms of (23.42), (23.55), and (23.56), respectively, are easily adapted for many of the alternate parametrizations. The resulting updates are

$$\mathbf{W}(n + 1) = \mathbf{W}(n) + \mu(n)\mathbf{U}_{alt}(n)e_o(n) \tag{23.75}$$

and

$$
\begin{aligned}
\mathbf{W}(n + 1) &= \mathbf{W}(n) + \mu(n)P(n)\mathbf{U}_{alt}(n)e_o(n) & (23.76) \\
P(n) &= \frac{1}{\lambda(n)}\left(P(n-1) - \frac{P(n-1)\mathbf{U}_{alt}(n)\mathbf{U}_{alt}^T(n)P(n-1)}{(\lambda(n)/\mu(n)) + \mathbf{U}_{alt}^T(n)P(n-1)\mathbf{U}_{alt}(n)}\right), & (23.77)
\end{aligned}
$$

respectively, where all signal definitions parallel those for the direct-form algorithms, save that $\mathbf{U}_{alt}(n)$ equals the gradient of the filter output with respect to the adapted parameters. Note that $\mathbf{U}_{of}(n)$ in (23.42), (23.55), and (23.56) is such a gradient for the direct-form implementation. The output gradient $\mathbf{U}_{of}(n)$ for the direct-form was constructed as shown in (23.40) and (23.41). For alternate parametrizations, these output gradients may be constructed as described in [29]. In [29], the implementation for a two-multiplier lattice adaptive IIR filter is shown, but the methodology is applicable to cascade and parallel implementations, resulting for instance in the same algorithm for the parallel-form filter that appears in [24]. We should note that the complexity of the output gradient generation may be an issue; however, implementations for parallel and cascade realizations exist where this complexity is equivalent to that of the direct form. The lattice implementation of [29] presents a sizable computational burden in gradient generation, but the normalized lattice of [17] (see below) can be implemented with the same complexity as the direct form.

The convergence results we have noted for previous output error approaches for the most part apply as well for these alternate realizations. Differences in these results for cascade and parallel form filters stem from the fact that permutations of some of the filter parameters yield equivalent filter transfer functions, but these differences do not affect the convergence results. In general, gradient algorithms for alternate implementations appear to converge more slowly than their direct-form counterparts; however, the reasons for this difference in convergence speed are poorly understood.

Algorithms other than the gradient approach have not been extensively explored for the alternate parametrizations. There may be fundamental limitations in this regard. For direct-form implementations, the signals of the unknown system corresponding to internal states of the adaptive filter are available through the delayed outputs $d(n - i)$, and it is these signals that are used in the equation error-based algorithms. However, the analogous signals for alternate implementations are unavailable, and so equation error methods, as well as the Steiglitz-McBride approach, are a challenge to even devise, let alone implement. Stability theory-based approaches are difficult to formulate, and the results of [28] indicate that simple algorithms of the form of (23.60) would not be stable in a wide set of operating conditions.

One promising alternate structure is the normalized lattice of [17]. The normalized structure is by nature stable, and hence no stability check is necessary in the adaptive algorithm. Furthermore, a clever implementation of the output gradient calculation keeps the computational burden of the SG and GN algorithms for this normalized lattice comparable to direct-form implementations. Convergence rates for this structure appear to be comparable to the direct-form structure as well [19]. While we have noted that SM approaches are, in general, infeasible for alternate parametrizations, it is in fact possible to implement an SM algorithm for the normalized lattice through use of an invertibility property held by stages of the lattice [17]. The convergence results we have noted for SM apply as well to the normalized lattice implementation.

We summarize as follows. Alternate parametrizations of adaptive IIR filters enable the stability of the adaptive system to be easily checked. Convergence results for gradient-based algorithms typically apply to these alternate structures. However, the complexities of the gradient calculations can be large for certain systems, and Gauss-Newton approaches appear to be difficult to implement and stabilize for these systems.

23.6 Conclusions

Adaptive IIR filtering remains an open area of research. The preceding survey has examined a number of different approaches to algorithm design within this field. We have considered equation error algorithm designs, including the well-known LMS and RLS algorithms, but also the IV approach and the more recent equation error algorithms with a unit-norm constraint. Output error algorithm designs that we have treated are gradient descent methods and methods based on stability theory. Somewhere in between these two categories is the Steiglitz-McBride approach to adaptive IIR filtering.

Each of these approaches has certain advantages but also disadvantages. We have evaluated each approach in terms of convergence conditions and also with regard to the nature of the filter parameters to which the algorithm converges. We have taken special interest in whether the algorithm converges to or is biased away from the optimal filter parameters, both in the presence of undermodeling and also measurement noise effects, and a further concern has been the stability of the resulting filter. We have placed less emphasis on convergence speed, as this issue is highly dependent on the particular environment in which the filter is to operate.

Unfortunately, no one algorithm possesses satisfactory properties in all of these regards. Therefore, the choice of algorithm in a given application will depend on which property is most critical in the application setting. Meanwhile, research seeking improvement in adaptive IIR filtering algorithms continues.

References

[1] Anderson, B.D.O. et al., *Stability of Adaptive Systems: Passivity and Averaging Analysis*, MIT Press, Cambridge, MA, 1987.

[2] Anderson, B.D.O. et al., Robust strict positive realness: characterization and construction, *IEEE Trans. Circuits Syst.*, 37(7), 869–876, 1990.

[3] Benveniste, A., Metivier, M. and Priouret, P., *Adaptive Algorithms and Stochastic Approximations*, Springer-Verlag, New York, 1990.

[4] Davila, C.E., An algorithm for efficient, unbiased, equation-error infinite impulse response adaptive filtering, *IEEE Trans. Signal Processing*, 42(5), 1221–1226, 1994.

[5] Douglas, S.C. and Rupp, M., On bias removal and unit-norm constraints in equation-error adaptive IIR filters, *Proc. 30th Ann. Asilomar Conf. Signals, Sys., Comput.*, Pacific Grove, CA, 1996.

[6] Fan, H., Application of Benveniste's convergence results in the study of adaptive IIR filtering algorithms, *IEEE Trans. Inform. Theory*, 34(7), 692–709, 1988.

[7] Fan, H. and Doroslovački, D. On "global convergence" of Steiglitz-McBride adaptive algorithm, *IEEE Trans. Circuits Syst. II*, 40(2), 73–87, 1993.

[8] Fan, H. and Jenkins, W.K., Jr., A new adaptive IIR filter, *IEEE Trans. Circuits Syst.*, 33(10), 939–947, 1986.

[9] Fan, H. and Nayeri, M., On reduced order identification: revisiting On some system identification techniques for adaptive filtering, *IEEE Trans. Circuits Syst.*, 37(9), 1144–1151, 1990.

[10] Johnson, C.R., Jr., A convergence proof for a hyperstable adaptive recursive filter, *IEEE Trans. Inform. Theory*, 25(6), 745–749, 1979.

[11] Johnson, C.R., Jr., Adaptive IIR filtering: current results and open issues, *IEEE Trans. Inform. Theory*, 30(2), 237–250, 1984.

[12] Johnson, C.R., Jr., Larimore, M.G., Treichler, J.R. and Anderson, B.D.O., SHARF convergence properties, *IEEE Trans. Circuits Syst.*, 28(6), 499–510, 1984.

[13] Kenney, J.B. and Rohrs, C.E., The composite regressor algorithm for IIR adaptive systems, *IEEE Trans. Signal Processing*, 41(2), 617–628, 1993.

[14] Lin, J.-N. and Unbehauen, R., Bias-remedy least mean square equation error algorithm for IIR parameter recursive estimation, *IEEE Trans. Signal Processing*, 40(1), 62–69, 1992.

[15] Ljung, L. and Söderström, T., *Theory and Practice of Recursive System Identification*, MIT Press, Cambridge, MA, 1983.

[16] Nayeri, M., Fan, H. and Jenkins, W.K., Jr., Some characteristics of error surfaces for insufficient order adaptive IIR filters, *IEEE Trans. Acoustics, Speech, Signal Processing*, 38(7), 1222–1227, 1990.

[17] Regalia, P.A., Stable and efficient lattice algorithms for adaptive IIR filtering, *IEEE Trans. Signal Processing*, 40(2), 375–388, 1992.

[18] Regalia, P.A., An unbiased equation error identifier and reduced-order approximations, *IEEE Trans. Signal Processing*, 42(6), 1397–1412, 1994.

[19] Regalia, P.A., *Adaptive IIR Filtering in Signal Processing and Control*, Marcel-Dekker, New York, 1995.

[20] Ren, W. and Kumar, P.R., Stochastic parallel model adaptation: theory and applications to active noise canceling, feedforward control, IIR filtering, and identification, *IEEE Trans. Automat. Contr.*, 37(5), 566–578, 1992.

[21] Sethares, W.A., Anderson, B.D.O. and Johnson, C.R., Jr., Adaptive algorithms with filtered regressor and filtered error, *Math. Control Signals Systems*, 2, 381–403, 1988.

[22] Sethares, W.A., Lawrence, D.A., Johnson, Jr., C.R., and Bitmead, R.R., Parameter drift in LMS adaptive filters, *IEEE Trans. Acoustics, Speech, Signal Processing*, 34(8), 868–879, 1986.

[23] Shynk, J.J., Adaptive IIR filtering, *IEEE ASSP Magazine*, 6(2), 4–21, 1989.

[24] Shynk, J.J., Adaptive IIR filtering using parallel-form realizations, *IEEE Trans. Acoustics, Speech, Signal Processing*, 37(4), 519–533, 1989.

[25] Söderström, T. and Stoica, P., On the stability of dynamic models obtained by least-squares identification, *IEEE Trans. Automat. Contr.*, 26(2), 575–577, 1981.

[26] Stoica, P. and Söderström, S., The Steiglitz-McBride identification algorithm revisited – convergence analysis and accuracy aspects, *IEEE Trans. Automat. Contr.*, 26(3), 712–717, 1981.

[27] Söderström, S. and Stoica, P., Some properties of the output error method, *Automatica*, 18(1), 93–99, 1982.

[28] Williamson, G.A., Anderson, B.D.O. and Johnson, C.R., Jr., On the local stability properties of adaptive parameter estimators with composite errors and split algorithms, *IEEE Trans. Automat. Contr.*, 36(4), 463–473, 1991.

[29] Williamson, G.A., Johnson, C.R., Jr., and Anderson, B.D.O., Locally robust identification of linear systems containing unknown gain elements with application to adapted IIR lattice models, *Automatica*, 27(5), 783–798, 1991.

24

Adaptive Filters for Blind Equalization

Zhi Ding
Auburn University

24.1 Introduction

One of the earliest and most successful applications of adaptive filters is adaptive channel equalization in digital communication systems. Using the standard least mean LMS algorithm, an adaptive equalizer is a finite-impulse-response FIR filter whose desired reference signal is a known training sequence sent by the transmitter over the unknown channel. The reliance of an adaptive channel equalizer on a training sequence requires that the transmitter cooperates by (often periodically) resending the training sequence, lowering the effective data rate of the communication link.

In many high-data-rate bandlimited digital communication systems, the transmission of a training sequence is either impractical or very costly in terms of data throughput. Conventional LMS adaptive filters depending on the use of training sequences cannot be used. For this reason, blind adaptive channel equalization algorithms that do not rely on training signals have been developed. Using these "blind" algorithms, individual receivers can begin self-adaptation without transmitter assistance. This ability of blind startup also enables a blind equalizer to self-recover from system breakdowns. This self-recovery ability is critical in broadcast and multicast systems where channel variation often occurs.

0-8493-8572-5/98/$0.00+$.50

In this section, we provide an introduction to the basics of blind adaptive equalization. We describe commonly used blind algorithms, highlight important issues regarding convergence properties of various blind equalizers, outline common initialization tactics, present several open problems, and discuss recent advances in this field.

24.2 Channel Equalization in QAM Data Communication Systems

In data communication, digital signals are transmitted by the sender through an analog channel to the receiver. Nonideal analog media such as telephone cables and radio channels typically distort the transmitted signal.

The problem of blind channel equalization can be described using the simple system diagram shown in Fig. 24.1. The complex baseband model for a typical QAM (quadrature amplitude modulated) data communication system consists of an unknown linear time-invariant (LTI) channel $h(t)$ which represents all the interconnections between the transmitter and the receiver at baseband. The matched filter is also included in the LTI channel model. The baseband-equivalent transmitter generates a sequence of complex-valued random input data $\{a(n)\}$, each element of which belongs to a complex alphabet $\mathcal{A}$ (or constellation) of QAM symbols. The data sequence $\{a(n)\}$ is sent through a baseband-equivalent complex LTI channel whose output $x(t)$ is observed by the receiver. The function of the receiver is to estimate the original data $\{a(n)\}$ from the received signal $x(t)$.

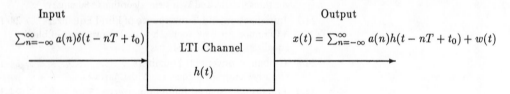

FIGURE 24.1 Baseband representation of a QAM data communication system.

For a causal and complex-valued LTI communication channel with impulse response $h(t)$, the input/output relationship of the QAM system can be written as

$$x(t) = \sum_{n=-\infty}^{\infty} a(n)h(t - nT + t_0) + w(t), \quad a(n) \in \mathcal{A}, \tag{24.1}$$

where T is the symbol (or baud) period. Typically the channel noise $w(t)$ is assumed to be stationary, Gaussian, and independent of the channel input $a(n)$.

In typical communication systems, the matched filter output of the channel is sampled at the known symbol rate $1/T$ assuming perfect timing recovery. For our model, the sampled channel output

$$x(nT) = \sum_{k=-\infty}^{\infty} a(k)h(nT - kT + t_0) + w(nT) \tag{24.2}$$

is a discrete time stationary process. Equation (24.2) relates the channel input to the sampled matched filter output. Using the notations

$$x(n) \overset{\Delta}{=} x(nT), \quad w(n) \overset{\Delta}{=} w(nT), \quad \text{and} \quad h(n) \overset{\Delta}{=} h(nT + t_0), \tag{24.3}$$

the relationship in (24.2) can be written as

$$x(n) = \sum_{k=-\infty}^{\infty} a(k)h(n-k) + w(n) .$$

(24.4)

When the channel is nonideal, its impulse response $h(n)$ is nonzero for $n \neq 0$. Consequently, undesirable signal distortion is introduced as the channel output $x(n)$ depends on multiple symbols in $\{a(n)\}$. This phenomenon, known as *intersymbol* interference (ISI), can severely corrupt the transmitted signal. ISI is usually caused by limited channel bandwidth, multipath, and channel fading in digital communication systems. A simple memoryless decision device acting on $x(n)$ may not be able to recover the original data sequence under strong ISI. Channel equalization has proven to be an effective means of significant ISI removal. A comprehensive tutorial on nonblind adaptive channel equalization by Qureshi [2] contains detailed discussions on various aspects of channel equalization.

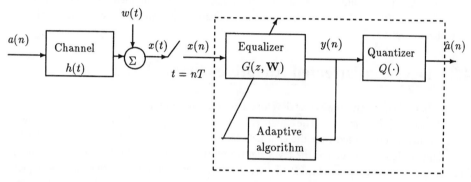

FIGURE 24.2   Adaptive blind equalization system.

Figure 24.2 shows the combined communication system with adaptive equalization. In this system, the equalizer $G(z, \mathbf{W})$ is a linear FIR filter with parameter vector $\mathbf{W}$ designed to remove the distortion caused by channel ISI. The goal of the equalizer is to generate an output signal $y(n)$ that can be quantized to yield a reliable estimate of the channel input data as

$$\hat{a}(n) = Q\left(y(n)\right) = a(n-\delta) ,$$

(24.5)

where δ is a constant integer delay. Typically any constant but finite amount of delay introduced by the combined channel and equalizer is acceptable in communication systems.

The basic task of equalizing a linear channel can be translated to that task of identifying the equivalent discrete channel, defined in z-transform notation as

$$H(z) = \sum_{k=0}^{\infty} h(k)z^{-k} .$$

(24.6)

With this notation, the channel output becomes

$$x(n) = H(z)a(n) + w(n)$$

(24.7)

where $H(z)a(n)$ denotes linear filtering of the sequence $a(n)$ by the channel and $w(n)$ is a white (for a root-raised-cosine matched filter [2]) stationary noise with constant power spectrum N_0. Once the channel has been identified, the equalizer can be constructed according to the minimum mean square error (MMSE) criterion between the desired signal $a(n - \delta)$ and the output $y(n)$ as

$$G_{mmse}(z, \mathbf{W}) = \frac{H^*(z^{-1})z^{-\delta}}{H(z)H^*(z^{-1}) + N_0},$$ (24.8)

where $*$ denotes complex conjugate. Alternatively, if the zero-forcing (ZF) criterion is employed, then the optimum ZF equalizer is

$$G_{zf}(z, \mathbf{W}) = \frac{z^{-\delta}}{H(z)},$$ (24.9)

which causes the combined channel-equalizer response to become a purely δ-sample delay with zero ISI. ZF equalizers tend to perform poorly when the channel noise is significant and when the channels $H(z)$ have zeros near the unit circle.

Both the MMSE equalizer (24.8) and the ZF equalizer (24.9) are of a general infinite impulse response (IIR) form. However, adaptive linear equalizers are usually implemented as FIR filters due to the difficulties inherent in adapting IIR filters. Adaptation is then based on a well-defined criterion such as the MMSE between the ideal IIR and truncated FIR impulse responses or the MMSE between the training signal and the equalizer output.

24.3 Decision-Directed Adaptive Channel Equalizer

Adaptive channel equalization was first developed by Lucky [1] for telephone channels. Figure 24.3 depicts the traditional adaptive equalizer. The equalizer begins adaptation with the assistance of a known training sequence initially transmitted over the channel. Since the training signal is known, standard gradient-based adaptive algorithms such as the LMS algorithm can be used to adjust the equalizer coefficients to minimize the mean square error (MSE) between the equalizer output and the training sequence. It is assumed that the equalizer coefficients are sufficiently close to their optimum values and that much of the ISI is removed by the end of the training period. Once the channel input sequence $\{a(n)\}$ can be accurately recovered from the equalizer output through a memoryless decision device such as a quantizer, the system is switched to the decision-directed mode whereby the adaptive equalizer obtains its reference signal from the decision output.

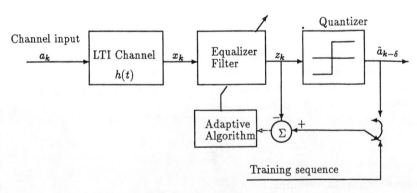

FIGURE 24.3   Decision-directed channel equalization algorithm.

One can construct a blind equalizer by employing decision-directed adaptation without a training sequence. The algorithm minimizes the MSE between the quantizer output

$$\hat{a}(n - \delta) = Q(y(n))$$ (24.10)

and the equalizer output $y(n)$. Naturally, the performance of the decision-directed algorithm depends on the accuracy of the estimate $Q(y(n))$ for the true symbol $a(n - \delta)$. Undesirable convergence to a local minimum with severe residual ISI can occur in this situation such that $Q(y(n))$ and $a(n - \delta)$ differ sufficiently often. Thus, the challenge of blind equalization lies in the design of special adaptive algorithms that eliminate the need for training without compromising the desired convergence to near the optimum MMSE or ZF equalizer coefficients.

24.4 Basic Facts on Blind Adaptive Equalization

In blind equalization, the desired signal or input to the channel is unknown to the receiver, except for its probabilistic or statistical properties over some known alphabet $\mathcal{A}$. As both the channel $h(n)$ and its input $a(n)$ are unknown, the objective of blind equalization is to recover the unknown input sequence based solely on its probabilistic and statistical properties.

The first comprehensive analytical study of the blind equalization problem was presented by Benveniste, Goursat, and Ruget in 1980 [3]. In fact, the very term "blind equalization" can be attributed to Benveniste and Goursat from the title of their 1984 paper [4]. The seminal paper of Benveniste et al. [3] established the connection between the task of blind equalization and the use of higher order statistics of the channel output. Through rigorous analysis, they generalized the original Sato algorithm [5] into a class of algorithms based on non-MSE cost functions. More importantly, the convergence properties of the proposed algorithms were carefully investigated. Based on the work of [3], the following facts about blind equalization are generally noted:

1. Second order statistics of $x(n)$ alone only provide the magnitude information of the linear channel and are insufficient for blind equalization of a mixed phase channel $H(z)$ containing zeros inside and outside the unit circle in the z-plane.

2. A mixed phase linear channel $H(z)$ cannot be identified from its outputs when the input signal is i.i.d. Gaussian, since only second order statistical information is available.

3. Although the exact inverse of a nonminimum phase channel is unstable, a truncated anticausal expansion can be delayed by δ to allow a causal approximation to a ZF equalizer.

4. ZF equalizers cannot be implemented for channels $H(z)$ with zeros on the unit circle.

5. The symmetry of QAM constellations $\mathcal{A} \subset \mathbb{C}$ causes an inherent phase ambiguity in the estimate of the channel input sequence or the unknown channel when input to the channel is uniformly distributed over $\mathcal{A}$. This phase ambiguity can be overcome by differential encoding of the channel input.

Due to the absence of a training signal, it is important to exploit various available information about the input symbol and the channel output to improve the quality of blind equalization. Usually, the following information is available to the receiver for blind equalization:

- The power spectral density (PSD) of the channel output signal $x(t)$, which contains information on the magnitude of the channel transfer function;

- The higher-order statistics (HOS) of the T-sampled channel output $\{x(kT)\}$, which contains information on the phase of the channel transfer function;

- Cyclostationary second and higher order statistics of the channel output signal $x(t)$, which contain additional phase information of the channel; and

- The finite channel input alphabet, which can be used to design quantizers or decision devices with memory to improve the reliability of the channel input estimate.

Naturally in some cases, these information sources are not necessarily independent as they contain overlapping information. Efficient and effective blind equalization schemes are more likely to be designed when all useful information is exploited at the receiver. We now describe various algorithms for blind channel identification and equalization.

24.5   Adaptive Algorithms and Notations

There are basically two different approaches to the problem of blind equalization. The stochastic gradient descent (SGD) approach iteratively minimizes a chosen cost function over all possible choices of the equalizer coefficients, while the statistical approach uses sufficient stationary statistics collected over a block of received data for channel identification or equalization. The latter approach often exploits higher order or cyclostationary statistical information directly. In this discussion, we focus on the the adaptive online equalization methods employing the the gradient descent approach, as these methods are most closely related to other topics in this chapter. Consequently, the design of special, non-MSE cost functions that implicitly exploits the HOS of the channel output is the key issue in our methods and discussions.

For reasons of practicality and ease of adaptation, a linear channel equalizer is typically implemented as an FIR filter $G(z, \mathbf{W})$. Denote the equalizer parameter vector as

$$\mathbf{W} \triangleq [w_0 \ w_1 \ \cdots \ w_m]^T, \quad m < \infty.$$

In addition, define the received signal vector as

$$\mathbf{X}(n) \triangleq [x(n) \ x(n-1) \ \ldots \ x(n-m)]^T. \tag{24.11}$$

The output signal of the linear equalizer is thus

$$\begin{aligned} y(n) &= \mathbf{W}^T \mathbf{X}(n) \\ &= G(z, \mathbf{W})\{x(n)\}, \end{aligned} \tag{24.12}$$

where we have defined the equalizer transfer function as

$$G(z, \mathbf{W}) = \sum_{i=0}^{m} w_i z^{-i}. \tag{24.13}$$

All the ISI is removed by a ZF equalizer if

$$H(z)G(z, \mathbf{W}) = gz^{-\delta}, \quad g \neq 0 \tag{24.14}$$

such that the noiseless equalizer output becomes $y(n) = ga(n - \delta)$, where g is a complex-valued scaling factor. Hence, a ZF equalizer attempts to achieve the inverse of the channel transfer function with a possible gain difference g and/or a constant time delay δ.

Denoting the parameter vector of the equalizer at sample instant n as $\mathbf{W}(n)$, the conventional LMS adaptive equalizer employing a training sequence is given by

$$\mathbf{W}(n+1) = \mathbf{W}(n) + \mu[a(n-\delta) - y(n)]\mathbf{X}(n), \tag{24.15}$$

where $\cdot^*$ denotes complex conjugates and μ is a small positive stepsize. Naturally, this algorithm requires that the channel input $a(n - \delta)$ be available. The equalizer iteratively minimizes the MSE cost function

$$E\left\{|e_n|^2\right\} = E\{|a(n-\delta) - y(n)|^2\}.$$

If the MSE is so small after training that the equalizer output $y(n)$ is a close estimate of the true channel input $a(n-\delta)$, then $Q(y(n))$ can replace $a(n - \delta)$ in a decision-directed algorithm that continues to track modest time-variations in the channel dynamics [2].

In blind equalization, the channel input $a(n - \delta)$ is unavailable, and thus different minimization criteria are explored. The crudest blind equalization algorithm is the decision-directed scheme that updates the adaptive equalizer coefficients as

$$\mathbf{W}(n + 1) = \mathbf{W}(n) + \mu[Q(y(n)) - y(n)]\mathbf{X}^*(n). \tag{24.16}$$

The performance of the decision-directed algorithm depends on how close $\mathbf{W}(n)$ is to its optimum setting $\mathbf{W}_{opt}$ under the MMSE or the ZF criterion. The closer $\mathbf{W}(n)$ is to $\mathbf{W}_{opt}$, the smaller the ISI is and the more accurate the estimate $Q(y(n))$ is to $a(n - \delta)$. Consequently, the algorithm in (24.16) is likely to converge to $\mathbf{W}_{opt}$ if $\mathbf{W}(n)$ is initially close to $\mathbf{W}_{opt}$. The validity of this intuitive argument is shown analytically in [6, 7]. On the other hand, $\mathbf{W}(n)$ can also converge to parameter values that do not remove sufficient ISI from certain initial parameter values $\mathbf{W}(0)$, as $Q(y(n)) \neq a(n - \delta)$ sufficiently often in some cases [6, 7].

The ability of the equalizer to achieve the desired convergence result when it is initialized with sufficiently small ISI accounts for the key role that the decision-directed algorithm plays in channel equalization. In the system of Fig. 24.3, the training session is designed to help $\mathbf{W}(n)$ converge to a parameter vector such that most of the ISI has been removed, from which adaptation can be switched to the decision-directed mode. Without direct training, a blind equalization algorithm is therefore used to provide a good initialization for the decision-directed equalizer because of the decision-directed equalizer's poor convergence behavior under high ISI.

24.6 Mean Cost Functions and Associated Algorithms

Under the zero-forcing criterion, the objective of the blind equalizer is to adjust $\mathbf{W}(n)$ such that (24.14) can be achieved using a suitable rule of self-adaptation. We now describe the general methodology of blind adaptation and introduce several popular algorithms.

Unless otherwise stated, we focus on the blind equalization of pulse-amplitude modulation (PAM) signals, in which the input symbol is uniformly distributed over the following M levels,

$$\{\pm(M - 1)d, \ \pm(M - 3)d, \ \dots, \ \pm 3d, \ \pm d\}, \qquad M \text{ even}. \tag{24.17}$$

We study this particular case because (1) algorithms are often defined only for real signals when first developed [3, 5], and (2) the extension to complex (QAM) systems is generally straightforward [4].

Blind adaptive equalization algorithms are often designed by minimizing special non-MSE cost functions that do not involve the use of the original input $a(n)$ but still reflect the current level of ISI in the equalizer output. Define the *mean cost function* as

$$J(\mathbf{W}) \stackrel{\Delta}{=} E\{\Psi(y(n))\}, \tag{24.18}$$

where $\Psi(\cdot)$ is a scalar function of its argument. The mean cost function $J(\mathbf{W})$ should be specified such that its minimum point $\mathbf{W}$ corresponds to a minimum ISI or MSE condition. Because of the symmetric distribution of $a(n)$ over $\mathcal{A}$ in 24.17, the function Ψ should be even ($\Psi(-x) = \Psi(x)$), so that both $y(n) = a(n - \delta)$ and $y(n) = -a(n - \delta)$ are desired objectives or global minima of the mean cost function.

Using 24.18, the stochastic gradient descent minimization algorithm is easily derived as [3]

$$\begin{aligned}
\mathbf{W}(n + 1) &= \mathbf{W}(n) - \mu \cdot \frac{\partial}{\partial \mathbf{W}(n)} \Psi(y(n)) \\
&= \mathbf{W}(n) - \mu \cdot \Psi'\left(\mathbf{X}^T(n)\mathbf{W}(n)\right) \mathbf{X}^*(n). \tag{24.19}
\end{aligned}$$

Define the first derivative of Ψ as

$$\psi(x) \stackrel{\Delta}{=} \Psi'(x) = \frac{\partial}{\partial x} \Psi(x).$$

The resulting blind equalization algorithm can then be written as

$$\mathbf{W}(n+1) = \mathbf{W}(n) - \mu\psi(\mathbf{X}^T(n)\mathbf{W}(n))\mathbf{X}^*(n) . \tag{24.20}$$

Hence, a blind equalizer can either be defined by its cost function $\Psi(x)$, or equivalently, by the derivative $\psi(x)$ of its cost function, which is also called the *error function* since it replaces the prediction error in the LMS algorithm. Correspondingly, we have the following relationship:

$$\text{Minima of the mean cost } J(\mathbf{W}) \iff \text{Stable equilibria of the algorithm in (24.20) .}$$

The design of the blind equalizer thus translates into the selection of the function Ψ (or ψ) such that local minima of $J(\mathbf{W})$, or equivalently, the locally stable equilibria of the algorithm (24.20) correspond to a significant removal of ISI in the equalizer output.

24.6.1 The Sato Algorithm

The first blind equalizer for multilevel PAM signals was introduced by Sato [5] and is defined by the error function

$$\psi_1(y(n)) = y(n) - R_1 \, \text{sgn}\,(y(n)) , \tag{24.21}$$

where

$$R_1 \triangleq \frac{E|a(n)|^2}{E|a(n)|} .$$

Clearly, the Sato algorithm effectively replaces $a(n-\delta)$ with $R_1\text{sgn}(y(n))$, known as the slicer output. The multilevel PAM signal is viewed as an equivalent binary input signal in this case, so that the error function often has the same sign for adaptation as the LMS error $y(n) - a(n-\delta)$.

24.6.2 BGR Extensions of Sato Algorithm

The Sato algorithm was extended by Benveniste, Goursat, and Ruget [3] who introduced a class of error functions given by

$$\psi_b(y(n)) = \tilde{\psi}(y(n)) - R_b \, \text{sgn}\,(y(n)) , \tag{24.22}$$

where

$$R_b \triangleq \frac{E\{\tilde{\psi}(a(n))a(n)\}}{E|a(n)|} . \tag{24.23}$$

Here, $\tilde{\psi}(x)$ is an odd and twice differentiable function satisfying

$$\tilde{\psi}''(x) \geq 0, \quad \forall x \geq 0 . \tag{24.24}$$

The use of the function $\tilde{\psi}$ generalizes the linear function $\tilde{\psi}(x) = x$ in the Sato algorithm. The class of algorithms satisfying (24.22) and (24.24) are called *BGR algorithms*. They are individually represented by the explicit specification of the $\tilde{\psi}$ function, as with the Sato algorithm.

The generalization of these algorithms to complex signals (QAM) and complex equalizer parameters is straightforward by separating signals into their real and the imaginary as

$$\psi_b(y(n)) = \tilde{\psi}(Re[y(n)]) - R_b\text{sgn}(Re[y(n)]) + j\{\tilde{\psi}(Im[y(n)]) - R_b\text{sgn}(Im[y(n)])\} . \tag{24.25}$$

24.6.3 Constant Modulus or Godard Algorithms

Integrating the Sato error function $\psi_1(x)$ shows that the Sato algorithm has an equivalent cost function

$$\Psi_1(y(n)) = \frac{1}{2}\big(|y(n)| - R_1\big)^2 .$$

This cost function was generalized by Godard into another class of algorithms that are specified by the cost functions [8]

$$\Psi_q(y(n)) = \frac{1}{2q}\big(|y(n)|^q - R_q\big)^2 , \quad q = 1, 2, \ldots \tag{24.26}$$

where

$$R_q \triangleq \frac{E|a(n)|^{2q}}{E|a(n)|^q} .$$

This class of *Godard algorithms* is indexed by the positive integer q. Using the stochastic gradient descent approach, the Godard algorithms given by

$$\mathbf{W}(n + 1) = \mathbf{W}(n) - \mu(|\mathbf{X}(n)^H \mathbf{W}(n)|^q - R_q)|\mathbf{X}(n)^T \mathbf{W}(n)|^{q-2}\mathbf{X}(n)^T \mathbf{W}(n)\mathbf{X}^*(n) . \tag{24.27}$$

The Godard algorithm for the case $q = 2$ was independently developed as the "constant modulus algorithm" (CMA) by Treichler and co-workers [9] using the philosophy of property restoral. For channel input signal that has a constant modulus $|a(n)|^2 = R_2$, the CMA equalizer penalizes output samples $y(n)$ that do not have the desired constant modulus characteristics. The modulus error is simply

$$e(n) = |y(n)|^2 - R_2 ,$$

and the squaring of this error yields the constant modulus cost function that is the identical to the Godard cost function.

This modulus restoral concept has a particular advantage in that it allows the equalizer to be adapted independent of carrier recovery. A carrier frequency offset of Δ_f causes a possible phase rotation of the equalizer output so that

$$y(n) = |y(n)| \exp[j(2\pi \Delta_f n + \phi(n))] .$$

Because the CMA cost function is insensitive to the phase of $y(n)$, the equalizer parameter adaptation can occur independently and simultaneously with the operation of the carrier recovery system. This property also allows CMA to be applied to analog modulation signals with constant amplitude such as those using frequency or phase modulation [9].

24.6.4 Stop-and-Go Algorithms

Given the standard form of the blind equalization algorithm in (24.20), it is apparent that the convergence characteristics of these algorithms are largely determined by the sign of the error signal $\psi(y(n))$. In order for the coefficients of a blind equalizer to converge to the vicinity of the optimum MMSE solution as observed through LMS adaptation, the sign of its error signal should agree with the sign of the LMS prediction error $y(n) - a(n - \delta)$ most of the time. Slow convergence or convergence of the parameters to local minima of the cost function $J(\mathbf{W})$ that do not provide proper equalization can occur if the signs of these two errors differ sufficiently often. In order to improve the convergence properties of blind equalizers, the so-called "stop-and-go" methodology was proposed by Picchi et al. [10]. We now describe its simple concept.

The idea behind the stop-and-go algorithms is to allow adaptation "to go" only when the error function is more likely to have the correct sign for the gradient descent direction. Since there are several criteria for blind equalization, one can expect a more accurate descent direction when more than one of the existing algorithms provide the same sign of the error function. When the error signs differ for a particular output sample, parameter adaptation is "stopped". Consider two algorithms with error functions $\psi_1(y)$ and $\psi_2(y)$. We can devise the following *stop-and-go* algorithm:

$$\mathbf{W}(k+1) = \begin{cases} \mathbf{W}(k) - \mu\psi_1(y(n))\mathbf{X}^*(n), & \text{if } \text{sgn}[\psi_1(y(n))] = \text{sgn}[\psi_2(y(n))]\,; \\ \mathbf{W}(k), & \text{if } \text{sgn}[\psi_1(y(n))] \neq \text{sgn}[\psi_2(y(n))]\,. \end{cases} \tag{24.28}$$

In their work, Picchi and Prati combined only the Sato and the decision-directed algorithms with faster convergence results through the corresponding error function

$$\psi(y(n)) = \frac{1}{2}(y(n) - Q(y(n))) + \frac{1}{2}|y(n) - Q(y(n))|\,\text{sgn}\left(y(n) - R_1\text{sgn}(y(n))\right)\,.$$

However, given the number of existing algorithms, the stop-and-go methodology can include many different combinations of error functions. One that combines Sato and Godard algorithms was tested by Hatzinakos [11].

24.6.5　Shalvi and Weinstein Algorithms

Unlike previously introduced algorithms, the methods of Shalvi-Weinstein [12] are based on higher order statistics of the equalizer output. Define the kurtosis of the equalizer output signal $y(n)$ as

$$K_y \triangleq E\left|y(n)^4\right| - 2E^2\left|y(n)^2\right| - \left|E\left\{y(n)^2\right\}\right|^2\,. \tag{24.29}$$

The Shalvi-Weinstein algorithm maximizes $|K_y|$ subject to the constant power constraint $E|y(n)|^2 = E|a(n)^2|$. Define c_n as the combined channel-equalizer impulse response given by

$$c_n = \sum_{k=0}^{m} h_k w_{n-k}, \qquad -\infty < n < \infty\,. \tag{24.30}$$

Using the fact that $a(n)$ is i.i.d., it can be shown [13] that

$$E\left|y(n)^2\right| = E\left|a(n)^2\right| \sum_{i=-\infty}^{\infty} |c_i|^2 \tag{24.31}$$

$$K_y = K_a \sum |c_n|^4\,, \tag{24.32}$$

where K_a is the kurtosis of the channel input, a quantity that is nonzero for most QAM and PAM signals. Hence, the Shalvi-Weinstein equalizer is equivalent to the following criterion:

$$\text{maximize} \sum_{n=-\infty}^{\infty} |c_n|^4 \quad \text{subject to} \quad \sum_{n=-\infty}^{\infty} |c_n|^2 = 1\,. \tag{24.33}$$

It can be shown [14] that there is a one-to-one correspondence between the minima of the cost function surface searched by this algorithm and those of the Godard algorithm with $q = 2$. However, the methods of adaptation given in [12] can exhibit convergence characteristics different from those of the CMA.

24.6.6　Summary

Over the years, there have been many attempts to derive new algorithms and equalization methods that are more reliable and faster than the existing methods. Nonetheless, the algorithms presented above are still the most commonly used methods in blind equalization due to their computational simplicity and practical effectiveness. In particular, CMA has proven to be useful not only in blind equalization but also in blind array signal processing systems. Because it does not rely on the accuracy of the decision device output nor the knowledge of the channel input signal constellation, CMA is a versatile algorithm that can be used not only for digital communication signals but also for analog signals that do not conform to a finite constellation alphabet.

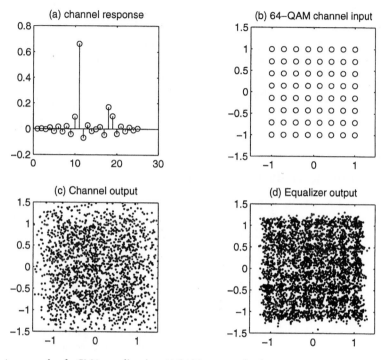

FIGURE 24.4   An example of a CMA equalizer in a 64 QAM communication system.

As a practical example, consider a QAM system in which the channel impulse response is shown in Fig. 24.4(a). This sampled composite channel response results from a continuous time system in which the transmitter and receiver filters both have identical root-raised cosine frequency response with the roll-off factor of 0.13, while the channel between the two filters is nonideal with several nondominant multipaths. The channel input signal is generated from a rectangular 64-QAM constellation as shown in Fig. 24.4(b). The channel output points are shown in Fig. 24.4(c). The channel output signal clearly has significant ISI such that a simple quantizer based on the nearest neighbor principle is likely to make many decision errors. We use a CMA equalizer with 25 parameter taps. The equalizer input is normalized by its power and a stepsize $\mu = 10^{-3}$ is used in the CMA adaptation. After 20,000 iterations, the final equalizer output after parameter convergence is shown in Fig. 24.4(d). The tighter clustering of the equalizer output shows that the decision error rate will be very low so that the equalizer can be switched to the decision-directed or decision-feedback algorithm mode at this point.

24.7   Initialization and Convergence of Blind Equalizers

The success and effectiveness of a QAM blind equalization algorithm clearly hinges on its convergence behavior in practical QAM systems with distortive channels. A desired globally convergent algorithm should only produce stable equilibria that are close to the optimum MMSE or ZF equalizer coefficients. If an equalization algorithm has local equilibria, then the initial equalizer parameter values are critical in determining the final values of parameters at convergence. Due to the analytical difficulty in locating and characterizing these local minima, most analytical studies of blind equalizers focus on the noiseless environment. For noiseless channels, the optimum MMSE and ZF equalizers are identical. The goal in the noiseless system is to remove sufficient ISI so that the open eye condition or errorless decision output, given by

$$Q(y(n)) = a(n - \delta) ,$$

holds.

Although the problem of blind equalization has been studied for over two decades, useful convergence analyses of most blind adaptive algorithms have proven to be difficult to perform. While some recent analytical results have helped to characterize the behavior of several popular algorithms, the overall knowledge of the behaviors of most known effective algorithms is still quite limited. Consequently, practical implementations of blind equalizers still employ heuristic measures to improve their convergence characteristics. We summarize several issues regarding the convergence and initialization of blind equalizers in this subsection.

24.7.1   A Common Analysis Approach

Although many readers may be surprised by the apparent lack of convergence proofs for most blind equalization algorithms, a closer look at the cost functions for these algorithms shows the analytical difficulty of the problem. Specifically, the stable stationary points of the blind algorithm in (24.20) correspond to the local minima of the mean cost function

$$J(\mathbf{W}) = E\left\{ \Psi\left(\sum_{i=0}^{m} w_i x(n-i) \right) \right\} . \tag{24.34}$$

The convergence of the adaptive algorithm is thus determined by the geometry of the error function $J(\mathbf{W})$ over the equalizer parameters $\{w_i\}$. An analysis of the convergence of the algorithm in terms of its parameters $\{w_i\}$ is difficult because the statistical characterization of the channel output signal $x(n)$ is highly dependent on the channel impulse response. For this reason, most blind equalization algorithms have initially been presented with only simulation results and without a rigorous convergence analysis.

Faced with this difficulty, several researchers have studied the global behavior of the equalizer in the combined parameter space c_i of (24.30) since

$$J(\mathbf{W}) = E\left\{ \Psi\left(\sum_{i=-\infty}^{\infty} w_i x(n-i) \right) \right\} = E\left\{ \Psi\left(\sum_{i=-\infty}^{\infty} c_i a(n-i) \right) \right\} . \tag{24.35}$$

Because the probabilistic information of the signal $a(n)$ is completely known, the convergence analysis of the c_i parameters tends to be much simpler than that of the equalizer parameters w_i. The following convergence results are known from these analyses:

- For channel input signals with uniform or sub-Gaussian probability distributions, Sato and BGR algorithms are globally convergent under zero channel noise. The corresponding cost functions only have global minima at parameter settings that result in zero ISI [3].

- For uniform and discrete PAM channel input distributions, undesirable local minima of the Sato and the BGR algorithms exist that do not satisfy the open-eye condition [6, 7, 16].

- For uniform and discrete PAM (or QAM) channel input distributions, the Godard algorithm with $q = 2$ (CMA) and the Shalvi-Weinstein algorithm have no local minima under zero channel noise. Only global minima exist at parameter settings that result in zero ISI [12, 15]. In other words, all minima satisfy the zero-forcing condition

$$c_n^2 = \begin{cases} 1, & n = \delta \\ 0, & n \neq \delta \, . \end{cases} \tag{24.36}$$

24.7.2 Local Convergence of Blind Equalizers

In order for the convergence analysis of the c_i parameters to be valid for the w_i parameters, a one-to-one linear mapping must exist between the two parameter spaces. A cost function of two variables will still have the same number of minima, maxima, and saddle points after a linear one-to-one coordinate change. On the other hand, a mapping that is not one-to-one can turn a nonstationary or saddle point into a local minimum.

If a one-to-one linear mapping exists between the two parameter spaces $\{w_i\}$ and $\{c_i\}$, then a stationary point for the equalizer coefficients w_i must correspond to a stationary point in the c_i parameters. Consequently, the convergence properties in the c_i parameter space will be equivalent to those in the w_i parameter space. However, $c_i = \sum_{k=0}^{m} h_k w_{i-k}$ does not provide a one-to-one mapping. The linear mapping is one-to-one if and only if $c_i = \sum_{k=-\infty}^{\infty} h_k w_{i-k}$, i.e., the equalizer coefficients w_i must exist for $-\infty < i < \infty$.

In this case, the equalizer parameter vector **W** needs to be doubly infinite. Hence, *unless the equalizer has an infinite number of parameters and is infinitely non-causal*, the convergence behavior of the c_i parameters do not completely characterize the behavior of the finite-length equalizer [17].

Undesirable local convergence of the Godard algorithm to a high ISI equalizer was initially thought to be impossible due to some over-zealous interpretations of the global convergence results in the combined c_i space [15]. The local convergence of the Godard ($q = 2$) algorithm or CMA is accurately analyzed by Ding et al. [18], where it is shown that even for noiseless channels whose ISI can be completely eliminated by an FIR equalizer, there can be local convergence of this equalizer to undesirable minima of the cost surface. Furthermore, these equilibria still remain under moderate channel noise. Based on the convergence similarity between the Godard algorithm and the Shalvi-Weinstein algorithm, the local convergence of the Shalvi-Weinstein algorithm to undesirable minima is established in [14]. Using a similar method, Sato and BGR algorithms have also been seen to have additional local minima previously undetected in the combined parameter space [16].

The proof that existing blind equalization algorithms previously thought to be robust can converge to poor solutions demonstrates that rigorous convergence analyses of blind equalizers must be based on the actual equalizer coefficients. Moreover, the undesirable local convergence behavior of existing algorithms indicates the importance of algorithm parameter initialization, which can avoid these local convergent points.

24.7.3 Initialization Issues

In [19], it is shown that local minima of a CMA equalizer cost surface tend to exist near MMSE parameter settings if the delay δ is chosen to be too short or too long. In other words, convergence to local minima is more likely to occur when the equalizer has large tap weights concentrated near either end of the finite equalizer coefficient vector. This type of lopsided parameter weight distribution was also suggested in [15] as being indicative of a local convergence phenomenon. To avoid local convergence to a lopsided tap weight vector, Foschini [15] introduced a tap-centering initialization strategy that requires the gravity center of the equalizer coefficient vector be centered through periodic tap-shifting. A more recent result [14] shows that, by over-parameterization and tap-centering, the Godard algorithm or CMA can effectively reduce the probability of local convergence. This tap-centering method has also been proposed for the Shalvi-Weinstein algorithm [20].

In practice, the tap-centering initialization approach has become an integral part of most blind equalization algorithms. Although a thorough analysis of its effect has not been shown, most reported successful uses of blind equalization algorithms typically rely on tap-centering or center-spike initialization scheme [17]. Although special channels exist that can foil the successful convergence of Sato and BGR algorithms using tap-centering, such channels are atypical [16]. Hence, unless global convergence of the equalizer can be proven, tap-centering is commonly recommended for most blind equalizers.

24.8 Globally Convergent Equalizers

24.8.1 Linearly Constrained Equalizer With Convex Cost

Without a proof of global convergence and a thorough analysis on initialization of existing equalization methods, one can design new and possibly better blind algorithms that can proven to always result in the global minimization of ISI. Here we present one strategy based on highly specialized convex cost functions coupled with a constrained equalizer parameterization designed to avoid ill-convergence.

Recall that the goal of blind equalization is to remove ISI so that the equalizer output is

$$y(n) = ga(n - \delta), \quad g \neq 0 . \tag{24.37}$$

Blind equalization of pulse amplitude modulation (PAM) systems without gain recovery has been proposed in [21]. The idea is to fix the center tap w_0 as a non-zero constant in order to prevent equalizer to the trivial minimum with all zero coefficient values in a convex cost function. For QAM input, a nontrivial extension is shown here.

For the particular equalizer design, assume that the input QAM constellation is square, which resembles the constellation in Fig. 24.4(b). The cost function to be minimized is

$$J(\mathbf{W}) \triangleq \max |Re\{y(n)\}| = \max |Im\{y(n)\}| . \tag{24.38}$$

The convexity of $J(\mathbf{W})$ with respect to the equalizer coefficient vector $\mathbf{W}$ follows from the triangle inequality under the assumption that all input sequences are possible. We constrain the equalizer coefficients $\mathbf{W}(n)$ with the following linear constraint

$$Re\{w_0\} + Im\{w_0\} = 1 , \tag{24.39}$$

where w_0 is the center tap. Due to the linearity of this constraint, the convexity of the cost function (24.38) with respect to both the real and imaginary parts of the equalizer coefficients is maintained, and global convergence is therefore assured.

Because of its convexity, this cost function is unimodal with a unique global minimum for almost all channels. It can then be shown [22] that a doubly infinite noncausal equalizer under the linear constraint is globally convergent to the condition in (24.37).

The linear constraint in (24.39) can be changed to any weighted linear combination of the two terms in (24.39). More general linear constraints on the equalizer coefficients can also be employed [23]. This fact is particularly important for preserving the global convergence property when causal finite-length equalizers are used. This behavior is a direct consequence of convexity, since restricting most of the equalizer taps to zero values as in FIR is a form of linear constraint. Convexity also ensures that one can approximate arbitrarily closely the performance of the ideal nonimplementable double infinite noncausal equalizer with a finite length FIR equalizer. These facts are important since many of the limitations illustrated earlier for convergence analyses of other equalizers can be overcome in this case.

For an actual implementation of this algorithm, a gradient descent method can be derived by using an l_p-norm cost function to approximate (24.38) as

$$J(\mathbf{W}) \approx E|Re\{z_k\}|^p , \tag{24.40}$$

where p is a large integer. As the cost function in (24.40) is strictly convex, linear constraints such as truncation preserve convexity. Simulation examples of this algorithm can be found in [22] and [24].

24.9 Fractionally Spaced Blind Equalizers

A so-called fractionally spaced equalizer (FSE) is obtained from the system in Fig. 24.2 if the channel output is sampled at a rate faster than the baud or symbol rate $1/T$. Recent work on the blind FSE has been motivated by several new results on nonadaptive blind equalization based on second order cyclostationary statistics. In addition to the first noted work by Tong et al. [25], new nonadaptive algorithms are also presented in [26, 27, 28, 29]. Here we only focus on the adaptive framework.

Let p be an integer such that the sampling interval be $\Delta = T/p$. As long as the channel bandwidth is greater than the minimum $1/(2T)$, sampling at higher than $1/T$ can retain channel diversity as shown here.

Let the sequence of sampled channel output be

$$x(k\Delta) = \sum_{n=0}^{\infty} a(n)h(k\Delta - np\Delta + t_0) + w(k\Delta) . \tag{24.41}$$

For notational simplicity, the oversampled channel output $x(k\Delta)$ can be divided into p linearly-independent subsequences

$$x^{(i)}(n) \triangleq x[(np+i)\Delta] = x(nT + i\Delta), \qquad i = 1, \ldots, p . \tag{24.42}$$

Define K as the effective channel length based on

$$\begin{aligned} h_0^{(i)} &\neq 0, \quad \text{for some } 1 \le i \le p \\ h_K^{(i)} &\neq 0, \quad \text{for some } 1 \le i \le p . \end{aligned} \tag{24.43}$$

By denoting the sub-channel transfer function as

$$H_i(z) = \sum_{k=0}^{K} h_k^{(i)} z^{-k} \quad \text{where} \quad h_k^{(i)} \triangleq h(kT + i\Delta + t_0) , \tag{24.44}$$

the p subsequences can be written as

$$x^{(i)}(n) = H_i(z)a(n) + w(nT + i\Delta), \qquad i = 1, \ldots, p . \tag{24.45}$$

Thus, these p subsequences can be viewed as stationary outputs of p discrete FIR channels with a common input sequence $a(n)$ as shown in Fig. 24.5. Naturally, they can also represent physical sub-channels in multisensor receivers.

The vector representation of the FSE is shown in Fig. 24.5. One equalizer filter is provided for each subsequence $x^{(i)}(n)$. In fact, the actual equalizer is a vector of filters

$$G_i(z) = \sum_{k=0}^{m} w_{i,k} z^{-k}, \qquad i = 1, \ldots p . \tag{24.46}$$

The p filter outputs $\{y(n)^{(i)}\}$ are summed to form the stationary equalizer output

$$y(n) = \mathbf{W}^T \mathbf{X}(n) \tag{24.47}$$

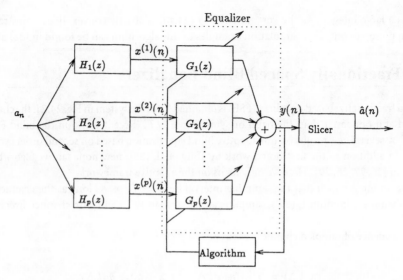

FIGURE 24.5   The structure of blind adaptive FSE.

where

$$\mathbf{W} \triangleq \begin{bmatrix} w_{1,0} & \dots & w_{1,m} & \dots w_{p,0} & \dots & w_{p,m} \end{bmatrix}^T .$$

$$\mathbf{X}(n) \triangleq \begin{bmatrix} x(n)^{(1)} & \dots & x(n-m)^{(1)} & \dots & x(n)^{(p)} & \dots & x(n-m)^{(p)} \end{bmatrix}^T .$$

Given the equalizer output and parameter vector, any T-sampled blind equalization adaptive algorithm can be applied to the FSE via stochastic gradient descent techniques.

Since their first use, adaptive blind equalizers have often been implemented as FSEs. When training data are available, FSEs have the known advantage of suppressing timing phase sensitivity [30]. In fact, a blind FSE has another important advantage: there exists a one-to-one mapping between the combined parameter space and the equalizer parameter space, as shown in [31], under the following *length and zero conditions*:

- *The equalizer length satisfies $(m + 1) \geq K$;*
- *The p discrete sub-channels $\{H_i(z)\}$ do not share any common zeros.*

Note that for T-sampled equalizers (TSE), only one ($p = 1$) sub-channel exists and all zeros are common zeros, and, thus, the length and zero conditions cannot be satisfied. In most practical implementations, p is either 2 or 3. So long as the above conditions hold, the convergence behaviors of blind adaptive FSEs can be characterized completely in the combined parameter space. Based on the work of [12, 15], for QAM channel inputs, there do not exist any algorithm-dependent stable equilibria other than the desired global minima (24.36) for FSEs driven by the Godard ($q = 2$) algorithm (CMA) and the Shalvi-Weinstein algorithms. Thus, the Godard and the Shalvi-Weinstein algorithms are globally convergent for FSEs satisfying these conditions [31].

Notice that global convergence of the Godard FSE is only proven for noiseless channels under the no-common zero condition. There have been recent advances in analyzing the performance of blind equalizers in the presence of Gaussian noise and the existence of common sub-channel zeros. While all possible delays of (24.36) are global minima for noiseless channels, the locations and effects of minima vary when channel noises are present. An analysis by Zeng and Tong shows that for noisy channels, CMA equalizer parameters have minima near the MMSE equilibria [32]. The effects of noise and common zeros was also studied by Fijalkow et al. [33, 34], providing further indications of the robustness of CMA when implemented as an FSE.

24.10   Concluding Remarks

Adaptive channel equalization and blind equalization are among the most successful applications of adaptive filtering. We have introduced the basic concept of blind equalization along with some of the most commonly used blind equalization algorithms. Without the aid of training signals, the key challenge of blind adaptive equalizers lies in the design of special cost functions whose minimization is consistent with the goal of ISI removal. We have also summarized key results on the convergence of blind equalizers. The idea of constrained minimization of a convex cost function to assure global convergence of the blind equalizer was described. Finally, the blind adaptation in fractionally spaced equalizers and multichannel receivers was shown to possess useful convergence properties.

It is important to note that the problem of blind equalization has not been completely solved by any means. In addition to the fact that the convergence behaviors of most algorithms are still unknown, the rates of convergence of typical algorithms such as CMA is quite slow, often needing thousands of iterations to achieve acceptable output. The difficulty of the convergence analysis and the slow rate of convergence of these algorithms have prompted many efforts to modify blind error functions to obtain faster and better algorithms. Furthermore, nonadaptive algorithms that explicitly exploit higher order statistics [35]–[38] and second order cyclostationary statistics [25]–[29] appear to be quite efficient in exploiting small amount of channel output data. A detailed discussion of these methods is beyond the scope of the section. Interested readers may refer to two collected works edited by Haykin [24] and Gardner [39] and the references therein.

References

[1] Lucky, R.W., Techniques for adaptive equalization of digital communication systems, *Bell Syst. Tech. J.*, 45:255–286, Feb., 1966.

[2] Qureshi, S.U.H., Adaptive equalization, *Proc. IEEE*, 73:1349–1387, Sept., 1985.

[3] Benveniste, A., Goursat, M., and Ruget, G., Robust identification of a nonminimum phase system, *IEEE Trans. Auto. Control*, AC-25:385–399, June, 1980.

[4] Benveniste, A. and Goursat, M., Blind equalizers, *IEEE Trans. Commun.*, 32:871–882, Aug., 1982.

[5] Sato, Y., A method of self-recovering equalization for multi-level amplitude modulation, *IEEE Trans. Commun.*, COM-23: 679-682, June, 1975.

[6] Macchi, O. and Eweda, E., Convergence analysis of self-adaptive equalizers, *IEEE Trans. Inform. Theory*, IT-30:162–176, Mar., 1984.

[7] Mazo, J.E., Analysis of decision-directed equalizer convergence, *Bell Syst. Tech. J.*, 59:1857–1876, Dec., 1980.

[8] Godard, D.N., Self-recovering equalization and carrier tracking in two-dimensional data communication systems, *IEEE Trans. Commun.*, COM-28:1867-1875, 1980.

[9] Treichler, J.R. and Agee, B.G., A new approach to multipath correction of constant modulus signals, *IEEE Trans. Acoustics, Speech Signal Process.*, ASSP-31:349-372, 1983.

[10] Picchi, G. and Prati, G., Blind equalization and carrier recovery using a "stop-and-go" decision-directed algorithm, *IEEE Trans. Commun.*, COM-35: 877-887, Sept., 1987.

[11] Hatzinakos, D., Blind equalization using stop-and-go criterion adaptation rules, *Opt. Eng.*, 31:1181-1198, June, 1992.

[12] Shalvi, O. and Weinstein, E., New criteria for blind deconvolution of nonminimum phase systems (channels), *IEEE Trans. Inform. Theory*, IT-36: 312-321, Mar., 1990.

[13] Brillinger, D.R. and Rosenblatt, M., Computation and interpretation of k-th order spectra, in *Spectral Analysis of Time Series*, B. Harris, Ed., Wiley, New York, 1967.

[14] Li, Y. and Ding, Z., Convergence analysis of finite length blind adaptive equalizers, *IEEE Trans. Signal Process.*, pp. 2120-2129, Sept., 1995.

[15] Foschini, G.J., Equalization without altering or detect data, *AT&T Tech. J.*, pp. 1885-1911, Oct., 1985.

[16] Ding, Z., Kennedy, R.A., Anderson, B.D.O., and Johnson, C.R., Jr., Local convergence of the Sato blind equalizer and generalizations under practical constraints, *IEEE Trans. Inform. Theory*, IT-39: 129-144, Jan., 1993.

[17] Ding, Z., Kennedy, R.A., and Johnson, C.R., Jr., On the (non)existence of undesirable equilibria of Godard blind equalizer, *IEEE Trans. Signal Process.*, 40:2425-2432, Oct., 1992.

[18] Ding, Z., Kennedy, R.A., Anderson, B.D.O., and Johnson, C.R., Jr., Ill-convergence of Godard blind equalizers in data communications, *IEEE Trans. Commun.*, 39:1313–1328, Sept., 1991.

[19] Minardi, M.J. and Ingram, M. Ann, Finding misconvergence in blind equalizers and new variance constraint cost functions to mitigate the problem, *Proc. 1996 Int. Conf. Acoustics, Speech, Signal Process.*, pp.1723-1726, May, 1996.

[20] Tugnait, J.K., Shalvi, O., and Weinstein, E., Comments on "new criteria for blind deconvolution of nonminimum phase systems (channels)," *IEEE Trans. Inf. Theory*, IT-38:210-213, Jan., 1992.

[21] Rupprecht, W.T., Adaptive equalization of binary NRZ-signals by means of peak value minimization, in *Proc. 7th Europ. Conf. Circuit Theory and Design*, pp. 352–355, Prague, 1985.

[22] Kennedy, R.A. and Ding, Z., Blind adaptive equalizers for QAM communication systems based on convex cost functions, *Opt. Eng.*, 31:1189-1199, June, 1992.

[23] Yamazaki, K. and Kennedy, R.A., Reformulation of linearly constrained adaptation and its application to blind equalization, *IEEE Trans. Signal Process.*, SP-42:1837-1841, 1994.

[24] Haykin, Simon, Ed., *Blind Deconvolution*, Prentice-Hall, 1994.

[25] Tong, L., Xu, G., and Kailath, T., Blind channel identification and equalization based on second-order statistics: a time-domain approach, *IEEE Trans. Inform. Theory*, IT-40:340-349, Mar., 1994.

[26] Moulines, E. et al., Subspace methods for the blind identification of multichannel FIR filters, *Proc. IEEE ICASSP*, pp. IV:573-576, Adelaide, 1994.

[27] Li, Y. and Ding, Z., ARMA system identification based on second order cyclostationarity, *IEEE Trans. Signal Process.*, pp. 3483-3493, Dec., 1994.

[28] Meriam, K.A., Duhamel, P., Gesbert, D., Loubaton, P., Mayrarague, S., Moulines, E., and Slock, D., Prediction error methods for time-domain blind identification of multichannel FIR filters, *Proc. IEEE Int. Conf. Acoustics, Speech, Signal Process.*, pp. 1968-1971, May, 1995.

[29] Hua, Y., Fast maximum likelihood for blind identification of multiple FIR channels, *IEEE Trans. Signal Process.*, SP-44:661-672, Mar., 1996.

[30] Gitlin, R.D. and Weinstein, S.B., Fractionally spaced equalization: an improved digital transversal equalizer, *Bell Syst. Tech. J.*, 60:275-296, 1981.

[31] Li, Y. and Ding, Z., Global convergence of fractionally spaced Godard adaptive equalizers, *IEEE Trans. Signal Process.*, SP-44:818-826, Apr., 1996.

[32] Zeng, H. and Tong, L., On the performance of CMA in the presence of noise, *Proc. Conf. Inform. Sci. Syst.*, Mar., 1996.

[33] Fijalkow, I., Treichler, J.R., and Johnson, C.R., Jr., Fractionally spaced blind equalization: loss of channel diversity, *Proc. IEEE Int. Conf. Acoustics, Speech, Signal Process.*, pp. 1988-1991, May, 1995.

[34] Touzni, A., Fijalkow, I., and Treichler, J.R., Fractionally spaced CMA under channel noise, *Proc. IEEE Int. Conf. Acoustics, Speech, and Signal Process.*, pp. 2674-2677, May, 1996.

[35] Tugnait, J.K., Identification of linear stochastic systems via second and fourth-order cumulant matching, *IEEE Trans. Inf. Theory*, IT-33:393-407, May, 1987.

[36] Giannakis, G.B. and Mendel, J.M., Identification of nonminimum phase systems using via higher order statistics, *IEEE Trans. Acoustics, Speech, and Signal Process.*, ASSP-37:360-377, 1989.

[37] Hatzinakos, D. and Nikias, C.L., Blind equalization using a tricepstrum based algorithm, *IEEE Trans. Commun.*, 39:669–682, May, 1991.

[38] Shalvi, O. and Weinstein, E., Super-exponential methods for blind deconvolution, *IEEE Trans. Inform. Theory*, IT-39:504-519, Mar., 1993.

[39] Gardner, W.A., Ed., *Cyclostationarity in Communications and Signal Processing*, IEEE Press, 1994.

VII

Inverse Problems and Signal Reconstruction

Richard J. Mammone
Rutgers University

T HERE ARE MANY SITUATIONS where a desired signal cannot be measured directly. The measurement might be degraded by physical limitations of the signal source and/or by the measurement device itself. The acquired signal is thus a transformation of the desired signal. The inversion
of such transformations is the subject of the present chapter. In the following sections we will review several inverse problems and various methods of implementation of the inversion or recovery process. The
methods differ in the ability to deal with the specific limitations present in each application. For example,
the *a priori* constraint of non-negativity is important for image recovery, but not so for adaptive array
processing. The goal of the following sections is to present the basic approaches of inversion and signal
recovery. Each section focuses on a particular application area and describes the appropriate methods for
that area.

The first chapter, 25, is entitled "Signal Recovery from Partial Information" by Christine Podilchuk.
This section reviews the basic problem of signal recovery. The idea of projection onto convex sets (POCs)
is introduced as an elegant solution to the signal recovery problem. The inclusion of linear and non-linear
constraints are addressed. The POCs method is shown to be a subset of the set theoretic approach to signal
estimation. The application of image of restoration is described in detail.

Chapter 26 is entitled "Algorithms for Computed Tomography" by Gabor T. Herman. This section
presents methods to reconstruct the interiors of objects from data collected based on transmitted or emitted
radiation. The problem occurs in a wide range of application areas. The computer algorithms used for
achieving the reconstructions are discussed. The basic techniques of image reconstruction from projections
are classified into "Transform Methods" (including Filtered Backprojection and the Linogram Methods)
and "Series Expansion Methods" (including, in particular, the Algebraic Reconstruction Techniques and the
method of Expectation Maximization). In addition, a performance comparison of the various algorithms
for computed tomography is given.

Chapter 27 is entitled "Robust Speech Processing as an Inverse Problem" by Richard J. Mammone and
Xiaoyu Zhang. The performance of speech and speaker recognition systems is significantly affected by the
acoustic environment. The background noise level, the filtering effects introduced by the microphone and
the communication channel dramatically affect the performance of recognition systems. It is therefore
critical that these speech recognition systems be capable of detecting the ambient acoustic environment
continue and inverse their effects from the speech signal. This is the inverse problem in robust speech
processing that will be addressed in this section. A general approach to solving this inverse problem is
presented based on an affine transform model in the cepstrum domain.

Chapter 28 is entitled "Inverse Problems, Statistical Mechanics and Simulated Annealing" by K.
Venkatesh Prasad. In this section, a computational approach to 3-D coordinate restoration is presented.

The problem is to obtain high-resolution coordinates of 3-D volume-elements (voxels) from observations of their corresponding 2-D picture-elements (pixels). The problem is posed as a combinatorial optimization problem and borrowing from our understanding of statistical mechanics, we show how to adapt the tool of simulated annealing to solve this problem. This method is highly amenable to parallel and distributed processing.

Chapter 29 is entitled "Image Recovery Using the EM Algorithm" by Jun Zhang and Aggelos K. Katsaggelos. In this section, the image recovery/reconstruction problem is formulated as a maximum-likelihood (ML) problem in which the image is recovered by maximizing an appropriately defined likelihood function. These likelihood functions are often highly non-linear and when some of the variables involved are not directly observable, they can only be specified in integral form (i.e., averaging over the "hidden variables"). The EM (expectation-maximization) algorithm is revised and applied to some typical image recovery problems. Examples include image restoration using the Markov random field model and single and multiple channel image restoration with blur identification.

Chapter 30 is entitled "Inverse Problems In Array Processing" by Kevin R. Farrell. Array processing uses multiple sensors to improve signal reception by reducing the effects of interfering signals that originate from different spatial locations. Array processing algorithms are generally implemented via narrowband and broadband arrays, both of which are discussed in this chapter. Two classical approaches, namely sidelobe canceler and Frost beam formers, are reviewed. These algorithms are formulated as an inverse problem and an iterative approach for solving the resulting inverse problem is provided.

Chapter 31 is entitled "Channel Equalization as a Regularized Inverse Problem" by John F. Doherty. In this section, the relationship between communication channel equalization and the inversion of a linear system of equations is examined. A regularized method of inversion is an inversion process in which the noise dominated modes of the restored signal are attenuated. Channel equalization is the process that reduces the effects of a band-limited channel at the receiver of a communication system. A regularized method of channel equalization is presented in this section. Although there are many ways to accomplish this, the method presented uses linear and adaptive filters, which makes the transition to matrix inversion possible.

Chapter 32 is entitled "Inverse Problems in Microphone Arrays" by A.C. Surendran. The response of an acoustic enclosure is, in general, a non-minimum phase function and hence not invertible. In this section, we discuss techniques using microphone arrays that attempt to recover speech signals degraded by the filtering effect of acoustic enclosures by either approximately or exactly "inverting" the room response. The aim of such systems is to force the impulse response of the overall system, after de-reverberation, to be an impulse function. Beamforming and matched-filtering techniques (that approximate this ideal case) and the Diophantine inverse filtering method (a technique that provides an exact inverse) are discussed in detail.

Chapter 33 is entitled "Synthetic Aperture Radar Algorithms" by Clay Stewart and Vic Larson. A synthetic aperture radar (SAR) is a radar sensor that provides azimuth resolution superior to that achievable with its real beam by synthesizing a long aperture by platform motion. This section presents an overview of the basics of SAR phenomenology and the associated algorithms that are used to form the radar image and to enhance it. The section begins with an overview of SAR applications, historical development, fundamental phenomenology, and a survey of modern SAR systems. It also presents examples of SAR imagery. This is followed by a discussion of the basic principles of SAR image formation that begins with side looking radar, progresses to unfocused SAR, and finishes with focused SAR. A discussion of SAR image enhancement techniques, such as the polarimetric whitening filters, follows. Finally, a brief discussion of automatic target detection and classification techniques is offered.

Chapter 34 is entitled "Iterative Image Restoration Algorithms" by Aggelos K. Katsaggelos. In this section, a class of iterative restoration algorithms is presented. Such algorithms provide solutions to the problem of recovering an original signal or image from a noisy and blurred observation of it. This situation is encountered in a number of important applications, ranging from the restoration of images obtained by the Hubble space telescope, to the restoration of compressed images. The successive approximation methods form the basis of the material presented in this section.

The sample of applications and methods described in this chapter are meant to be representative of the large volume of work performed in this field. There is no claim of completeness, any omissions of significant contributors or other errors are solely the responsibility of the section editor, and all praiseworthy contributions are due solely to the chapter authors.

25

Signal Recovery from Partial Information

Christine Podilchuk
Bell Laboratories
Lucent Technologies

25.1 Introduction

Signal recovery has been an active area of research for applications in many different scientific disciplines. A central reason for exploring the feasibility of signal recovery is due to the limitations imposed by a physical device on the amount of data one can record. For example, for diffraction-limited systems, the finite aperture size of the lens constrains the amount of frequency information that can be captured. The image degradation is due to attenuation of high frequency components resulting in a loss of details and other high frequency information. In other words, the finite aperture size of the lens acts like a lowpass filter on the input data. In some cases, the quality of the recorded image data can be improved by building a more costly recording device but many times the required condition for acceptable data quality is physically unrealizable or too costly. Other times signal recovery may be necessary is for the recording of a unique event that cannot be reproduced under more ideal recording conditions.

Some of the earliest work on signal recovery includes the work by Sondhi [1] and Slepian [2] on recovering images from motion blur and Helstrom [3] on least squares restoration. A sampling of some of the signal recovery algorithms applied to different types of problems can be found in [4]–[21]. Further reading includes the other sections in this book, Chapter 53, and the extended list of references provided by all the authors.

The simple signal degradation model described in the next section turns out to be a useful representation for many different problems encountered in practice. Some examples that can be formulated using the general signal recovery paradigm include image restoration, image reconstruction, spectral estimation, and filter design. We distinguish between image restoration, which pertains to image recovery based on a measured distorted version of the original image, and image reconstruction, which refers most commonly to medical imaging where the image is reconstructed from a set of indirect measurements, usually projections. For many of the signal recovery applications, it is desirable to extrapolate a signal outside of a known interval. Extrapolating a signal in the spatial or temporal domain could result in improved spectral resolution and applies to such problems as power spectrum estimation, radio–astronomy, radar target detection, and geophysical exploration. The dual problem, extrapolating the signal in the frequency domain, also known as *superresolution*, results in improved spatial or temporal resolution and is desirable in many image restoration problems. As will be shown later, the standard inverse filtering techniques are not able to resolve the signal estimate beyond the diffraction limit imposed by the physical measuring device.

The observed signal is degraded from the original signal by both the measuring device as well as external conditions. Besides the measured, distorted output signal we may have some additional information about the following: the measuring system and external conditions, such as noise, as well as some *a priori* knowledge about the desired signal to be restored or reconstructed. In order to produce a good estimate of the original signal, we should take advantage of all the available information.

Although the data recovery algorithms described here apply in general to any data type, we derive most of the techniques based on two-dimensional input data for image processing applications. For most cases, it is straightforward to adapt the algorithms to other data types. Examples of data recovery techniques for different inputs are illustrated in the other sections in this book as well as Chapter 53 for image restoration. The material in this section requires some basic knowledge of linear algebra as found in [22].

Section 25.2 presents the signal degradation model and formulates the signal recovery problem. The early attempts of signal recovery based on inverse filtering are presented in Section 25.3. The concept of Projection Onto Convex Sets (POCS) described in Section 25.4 allows us to introduce *a priori* knowledge about the original signal in the form of linear as well as nonlinear constraints into the recovery algorithm. Convex set theoretic formulations allow us to design recovery algorithms that are extremely flexible and powerful. Sections 25.5 and 25.6 present some basic POCS-based algorithms and Section 25.7 presents a POCS-based algorithm for image restoration as well as some results. The sample algorithms presented here are not meant to be exhaustive and the reader is encouraged to read the other sections in this chapter as well as the references for more details.

25.2   Formulation of the Signal Recovery Problem

Signal recovery can be viewed as an estimation process in which operations are performed on an observed signal in order to estimate the ideal signal that would be observed if no degradation was present. In order to design a signal recovery system effectively, it is necessary to characterize the degradation effects of the physical measuring system. The basic idea is to model the signal degradation effects as accurately as possible and perform operations to undo the degradations and obtain a restored signal. When the degradation cannot be modeled sufficiently, even the best recovery algorithms will not yield satisfactory results. For many applications, the degradation system is assumed to be linear and can be modeled as a Fredholm integral equation of the first kind expressed as

$$g(x) = \int_{-\infty}^{+\infty} h(x; a) f(a) da + n(x). \tag{25.1}$$

This is the general case for a one-dimensional signal where f and g are the original and measured signals, respectively, n represents noise, and $h(x; a)$ is the impulse response or the response of the measuring system to an impulse at coordinate a.[1] A block diagram illustrating the general one-dimensional signal degradation system is shown in Fig. 25.1. For image processing applications, we modify this equation to the two-dimensional case, that is,

$$g(x, y) = \int_{-\infty}^{+\infty} \int_{-\infty}^{+\infty} h(x, y; a, b) f(a, b) da db + n(x, y) \tag{25.2}$$

The degradation operator h is commonly referred to as a *point spread function* (PSF) in imaging applications because in optics, h is the measured response of an imaging system to a point of light.

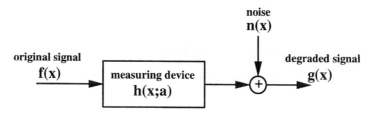

FIGURE 25.1 Block diagram of the signal recovery problem.

The Fourier transform of the point spread function $h(x, y)$ denoted as $\mathcal{H}(w_x, w_y)$ is known as the optical transfer function (OTF) and can be expressed as

$$\mathcal{H}(w_x, w_y) = \frac{\int \int_{-\infty}^{\infty} h(x, y) exp\left\{-i \left(w_x x + w_y y\right)\right\} dx dy}{\int \int_{-\infty}^{\infty} h(x, y) dx dy}. \tag{25.3}$$

The absolute value of the OTF is known as the modulation transfer function (MTF). A commonly used optical image formation system is a circular thin lens. The recovery problem is considered *ill-posed* when a small change in the observed image, g, results in a large change in the solution, f. Most signal recovery problems in practice are ill-posed.

The continuous version of the degradation system for two-dimensional signals formulated in Eq. (25.2) can be expressed in discrete form by replacing the continuous arguments with arrays of samples in two dimensions, that is,

$$g(i, j) = \sum_m \sum_n h(i, j; m, n) f(m, n) + n(i, j). \tag{25.4}$$

It is convenient for image recovery purposes to represent the discrete formulation given in Eq. (25.4) as a system of linear equations expressed as

$$\mathbf{g} = \mathbf{Hf} + \mathbf{n}, \tag{25.5}$$

where $\mathbf{g}$, $\mathbf{f}$, and $\mathbf{n}$ are the lexicographic row-stacked versions of the discretized versions of g, f, and n in Eq. (25.4) and $\mathbf{H}$ is the degradation matrix composed of the PSF.

This section presents an overview of some of the techniques proposed to estimate $\mathbf{f}$ when the recovery problem can be modeled by Eq. (25.5). If there is no external noise or measurement error and the set of equations is consistent, Eq. (25.5) reduces to

$$\mathbf{g} = \mathbf{Hf}. \tag{25.6}$$

It is usually not the case that a practical system can be described by Eq. (25.6). In this section, we will focus on recovery algorithms where an estimate of the distortion operation represented by the matrix $\mathbf{H}$

---

[1]This corresponds to the case of a shift–varying impulse response.

is known. For recovery problems where both the desired signal, **f**, and the degradation operator, **H**, are unknown, refer to other articles in this book.

For most systems, the degradation matrix **H** is highly structured and quite sparse. The additive noise term due to measurement errors and external and internal noise sources is represented by the vector **n**. At first glance, the solution to the signal recovery problem seems to be straightforward — find the inverse of the matrix **H** to solve for the unknown vector **f**. It turns out that the solution is not so simple because in practice the degradation operator is usually ill-conditioned or rank-deficient and the problem of inconsistencies or noise must be addressed. Other problems that may arise include computational complexity due to extremely large problem dimensions especially for image processing applications. The algorithms described here try to address these issues for the general signal recovery problem described by Eq. (25.5).

25.2.1 Prolate Spheroidal Wavefunctions

We introduce the problem of signal recovery by examining a one-dimensional, linear, time-invariant system that can be expressed as

$$g(x) = \int_{-T}^{+T} f(\alpha)h(x - \alpha)d\alpha, \tag{25.7}$$

where $g(x)$ is the observed signal, $f(\alpha)$ is the desired signal of finite support on the interval $(-T, +T)$, and $h(x)$ denotes the degradation operator. Assuming that the degradation operator in this case is an ideal lowpass filter, h can be described mathematically as

$$h(x) = \frac{sin(x)}{x}. \tag{25.8}$$

For this particular case, it is possible to solve for the exact signal $f(x)$ with prolate spheroidal wavefunctions [23]. The key to successfully solving for f lies in the fact that prolate spheroidal wavefunctions are the eigenfunctions of the integral equation expressed by Eq. (25.7) with Eq. (25.8) as the degradation operator. This relationship is expressed as:

$$\int_{-T}^{+T} \psi_n(\alpha)\frac{sin(x - \alpha)}{x - \alpha}d\alpha = \lambda_n\psi_n(x), \ n = 0,1,2,\ldots, \tag{25.9}$$

where $\psi_n(x)$ are the prolate spheroidal wavefunctions and λ_n are the corresponding eigenvalues. A critical feature of prolate spheroidal wavefunctions is that they are complete orthogonal bases in the interval $(-\infty, +\infty)$ as well as the interval $(-T, +T)$, that is,

$$\int_{-\infty}^{+\infty} \psi_n(x)\psi_m(x)dx = \begin{cases} 1, & \text{if } n = m, \\ 0, & \text{if } n \neq m, \end{cases} \tag{25.10}$$

and

$$\int_{-T}^{+T} \psi_n(x)\psi_m(x)dx = \begin{cases} \lambda_n, & \text{if } n = m, \\ 0, & \text{if } n \neq m. \end{cases} \tag{25.11}$$

This allows the functions $g(x)$ and $f(x)$ to be expressed as the series expansion,

$$g(x) = \sum_{n=0}^{\infty} c_n\psi_n(x), \tag{25.12}$$

$$f(x) = \sum_{n=0}^{\infty} d_n\psi_{Ln}(x), \tag{25.13}$$

where $\psi_{Ln}(x)$ are the prolate spheroidal functions truncated to the interval $(-T, T)$. The coefficients c_n and d_n are given by

$$c_n = \int_{-\infty}^{\infty} g(x)\psi_n(x)dx \qquad (25.14)$$

and

$$d_n = \frac{1}{\lambda_n} \int_{-T}^{T} f(x)\psi_n(x)dx. \qquad (25.15)$$

If we substitute the series expansions given by Eqs. (25.12) and (25.13) into Eq. (25.7), we get

$$
\begin{aligned}
g(x) &= \sum_{n=0}^{\infty} c_n \psi_n(x) \\
&= \int_{-T}^{+T} \left[\sum_{n=0}^{\infty} d_n \psi_{Ln}(\alpha) \right] h(x - \alpha)d\alpha \qquad (25.16) \\
&= \sum_{n=0}^{\infty} d_n \left[\int_{-T}^{+T} \psi_n(\alpha)h(x - \alpha)d\alpha \right]. \qquad (25.17)
\end{aligned}
$$

Combining this result with Eq. (25.9),

$$\sum_{n=0}^{\infty} c_n \psi_n(x) = \sum_{n=0}^{\infty} \lambda_n d_n \psi_n(x), \qquad (25.18)$$

where

$$c_n = \lambda_n d_n, \qquad (25.19)$$

and

$$d_n = \frac{c_n}{\lambda_n}. \qquad (25.20)$$

We get an exact solution for the unknown signal $f(x)$ by substituting Eq. (25.20) into Eq. (25.13), that is,

$$f(x) = \sum_{n=0}^{\infty} \frac{c_n}{\lambda_n} \psi_{Ln}(x). \qquad (25.21)$$

Therefore, in theory, it is possible to obtain the exact image $f(x)$ from the diffraction-limited image, $g(x)$, using prolate spheroidal wavefunctions. The difficulties of signal recovery become more apparent when we examine the simple diffraction-limited case in relation to prolate spheroidal wavefunctions as described in Eq. (25.21). The finite aperture size of a diffraction-limited system translates to eigenvalues λ_n which exhibit a unit–step response; that is, the several largest eigenvalues are approximately one followed by a succession of eigenvalues that rapidly fall off to zero. The solution given by Eq. (25.21) will be extremely sensitive to noise for small eigenvalues λ_n. Therefore, for the general problem represented in vector–space by Eq. (25.5), the degradation operator H is ill-conditioned or rank-deficient due to the small or zero-valued eigenvalues, and a simple inverse operation will not yield satisfactory results. Many algorithms have been proposed to find a compromise between exact deblurring and noise amplification. These techniques include Wiener filtering and pseudo-inverse filtering. We begin our overview of signal recovery techniques by examining some of the methods that fall under the category of optimization-based approaches.

25.3   Least Squares Solutions

The earliest attempts toward signal recovery are based on the concept of inverting the degradation operator to restore the desired signal. Because in practical applications the system will often be ill-conditioned, several problems can arise. Specifically, high detail signal information may be masked by observation noise, or a small amount of observation noise may lead to an estimate that contains very large false high frequency components. Another potential problem with such an approach is that for a rank-deficient degradation operator, the zero-valued eigenvalues cannot be inverted. Therefore, the general inverse filtering approach will not be able to resolve the desired signal beyond the diffraction limit imposed by the measuring device. In other words, referring to the vector–space description, the data that has been nulled out by the zero-valued eigenvalues cannot be recovered.

25.3.1   Wiener Filtering

Wiener filtering combines inverse filtering with *a priori* statistical knowledge about the noise and unknown signal [24] in order to deal with the problems associated with an ill-conditioned system.

The impulse response of the restoration filter is chosen to minimize the mean square error as defined by

$$\mathcal{E}_{\mathbf{f}} = E\left\{\left(\mathbf{f} - \hat{\mathbf{f}}\right)^2\right\} \tag{25.22}$$

where $\hat{f}$ denotes the estimate of the ideal signal f and $E\{\cdot\}$ denotes the expected value. The Wiener filter estimate is expressed as

$$\mathbf{H}_{\mathbf{W}}^{-1} = \frac{\mathbf{R_{ff}H}^T}{\mathbf{HR}_{ff}\mathbf{H}^T + \mathbf{R_{nn}}} \tag{25.23}$$

where $\mathbf{R_{ff}}$ and $\mathbf{R_{nn}}$ are the covariance matrices of $\mathbf{f}$ and $\mathbf{n}$, respectively, and $\mathbf{f}$ and $\mathbf{n}$ are assumed to be uncorrelated; that is,

$$\mathbf{R_{ff}} = E\left\{\mathbf{ff}^T\right\}, \tag{25.24}$$

$$\mathbf{R_{nn}} = E\left\{\mathbf{nn}^T\right\}, \tag{25.25}$$

and

$$\mathbf{R_{fn}} = 0. \tag{25.26}$$

The superscript T in the above equations denotes transpose. The Wiener filter can also be expressed in the Fourier domain as

$$\mathcal{H}_W^{-1} = \frac{\mathcal{H}^* S_{ff}}{|\mathcal{H}|^2 S_{ff} + S_{nn}} \tag{25.27}$$

where S denotes the power spectral density, the superscript $*$ denotes the complex conjugate, and $\mathcal{H}$ denotes the Fourier transform of $\mathbf{H}$. Note that when the noise power is zero, the Wiener filter reduces to the inverse filter; that is,

$$\mathcal{H}_W^{-1} = \mathcal{H}^{-1}. \tag{25.28}$$

The Wiener filter approach for signal recovery assumes that the power spectra are known for the input signal and the noise. Also, this approach assumes that finding a least squares solution that optimizes Eq. (25.22) is meaningful. For the case of image processing, it has been shown, specifically in the context of image compression, that the mean square error (mse) does not predict subjective image quality [25]. Many signal processing algorithms are based on the least squares paradigm because the solutions are tractable and, in practice, such approaches have produced some useful results. However, in order to define a more meaningful optimization metric in the design of image processing algorithms, we need to incorporate a human visual model into the algorithm design. In the area of image coding, several coding schemes based on perceptual criteria have been shown to produce improved results over schemes based on maximizing signal–to–noise ratio or minimizing mse [25]. Likewise, the Wiener filtering approach will not necessarily produce an estimate that maximizes perceived image or signal quality. Another limitation of the Wiener filter approach is that the solution will not necessarily be consistent with any *a priori* knowledge about the desired signal characteristics. In addition, the Wiener filter approach does not resolve the desired signal beyond the diffraction limit imposed by the measuring system. For more details on Wiener filtering and the various applications, see other chapters in this book.

25.3.2 The Pseudoinverse Solution

The Wiener filters attempt to minimize the noise amplification obtained in a direct inverse by providing a taper determined by the statistics of the signal and noise process under consideration. In practice, the power spectra of the noise and desired signal might not be known. Here we present what is commonly referred to as the generalized inverse solution. This will be the framework for some of the signal recovery algorithms described later.

The pseudoinverse solution is an optimization approach that seeks to minimize the least squares error as given by

$$\mathcal{E}_\mathbf{n} = \mathbf{n}^T \mathbf{n} = (\mathbf{g} - \mathbf{H}\mathbf{f})^T (\mathbf{g} - \mathbf{H}\mathbf{f}). \tag{25.29}$$

The least squares solution is not unique when the rank of the $M \times N$ matrix $\mathbf{H}$ is $r < N \le M$. In other words, there are many solutions that satisfy Eq. (25.29). However, the Moore-Penrose generalized inverse or pseudoinverse [26] does provide a unique least squares solution based on determining the least squares solution with minimum norm. For a consistent set of equations as described in Eq. (25.6), a solution is sought that minimizes the least squares estimation error; that is,

$$
\begin{aligned}
\mathcal{E}_\mathbf{f} &= \left(\mathbf{f} - \hat{\mathbf{f}}\right)^T (\mathbf{f} - \hat{\mathbf{f}}) \\
&= tr\left\{(\mathbf{f} - \hat{\mathbf{f}})\left(\mathbf{f} - \hat{\mathbf{f}}\right)^T\right\}
\end{aligned}
\tag{25.30}
$$

where $\mathbf{f}$ is the desired signal vector, $\hat{\mathbf{f}}$ is the estimate, and tr denotes the trace [22]. The generalized inverse provides an optimum solution that minimizes the estimation error for a consistent set of equations. Thus, the generalized inverse provides an optimum solution for both the consistent and inconsistent set of equations as defined by the performance functions $\mathcal{E}_\mathbf{f}$ and $\mathcal{E}_\mathbf{n}$, respectively. The generalized inverse solution satisfies the normal equations

$$\mathbf{H}^T \mathbf{g} = \mathbf{H}^T \mathbf{H}\mathbf{f}. \tag{25.31}$$

The generalized inverse solution, also known as the Moore-Penrose generalized inverse, pseudoinverse, or least squares solution with minimum norm is defined as

$$\mathbf{f}^\dagger = \left(\mathbf{H}^T \mathbf{H}\right)^{-1} \mathbf{H}^T \mathbf{g} = \mathbf{H}^\dagger \mathbf{g}, \tag{25.32}$$

where the dagger $\dagger$ denotes the pseudoinverse and the rank of $\mathbf{H}$ is $r = N \le M$.

For the case of an inconsistent set of equations as described in Eq. (25.5), the pseudoinverse solution becomes

$$\mathbf{f}^\dagger = \mathbf{H}^\dagger \mathbf{g} = \mathbf{H}^\dagger \mathbf{H}\mathbf{f} + \mathbf{H}^\dagger \mathbf{n} \tag{25.33}$$

where $\mathbf{f}^\dagger$ is the minimum norm, least squares solution. If the set of equations are overdetermined with rank $r = N < M$, $\mathbf{H}^\dagger \mathbf{H}$ becomes an identity matrix of size N denoted as $\mathbf{I}_N$ and the pseudoinverse solution reduces to

$$
\begin{aligned}
\mathbf{f}^\dagger &= \mathbf{f} + \mathbf{H}^\dagger \mathbf{n} \\
&= \mathbf{f} + \Delta \mathbf{f}.
\end{aligned}
\tag{25.34}
$$

A straightforward result from linear algebra is the bound on the relative error,

$$\frac{\parallel \Delta \mathbf{f} \parallel}{\parallel \mathbf{f} \parallel} \parallel \mathbf{H}^{\dagger} \parallel \parallel \mathbf{H} \parallel \frac{\parallel \mathbf{n} \parallel}{\parallel \mathbf{g} \parallel}, \tag{25.35}$$

where the product $\parallel \mathbf{H}^{\dagger} \parallel \parallel \mathbf{H} \parallel$ is the condition number of $\mathbf{H}$. This quantity determines the relative error in the estimate in terms of the ratio of the vector norm of the noise to the vector norm of the observed image. The condition number of H is defined as

$$\mathcal{C}_H = \parallel \mathbf{H}^{\dagger} \parallel \parallel \mathbf{H} \parallel = \frac{\sigma_1}{\sigma_N} \tag{25.36}$$

where σ_1 and σ_N denote the largest and smallest singular values of the matrix H, respectively. The larger the condition number, the greater the sensitivity to noise perturbations. A matrix with a large condition number, typically greater than 100, results in an ill-conditioned system.

The pseudoinverse solution is best described by diagonalizing the degradation matrix $\mathbf{H}$ using singular value decomposition (SVD) [22]. SVD provides a way to diagonalize any arbitrary $M \times N$ matrix. In this case, we wish to diagonalize $\mathbf{H}$; that is,

$$\mathbf{H} = \mathbf{U} \Sigma \mathbf{V}^T \tag{25.37}$$

where $\mathbf{U}$ is a unitary matrix composed of the orthonormal eigenvectors of $\mathbf{H}^T \mathbf{H}$, $\mathbf{V}$ is a unitary matrix composed of the orthonormal eigenvectors of $\mathbf{H} \mathbf{H}^T$, and Σ is a diagonal matrix composed of the singular values of $\mathbf{H}$. The number of nonzero diagonal terms denotes the rank of $\mathbf{H}$. The degradation matrix can be expressed in series form as

$$\mathbf{H} = \sum_{i=1}^{r} \sigma_i \mathbf{u}_i \mathbf{v}_i^T \tag{25.38}$$

where $\mathbf{u}_i$ and $\mathbf{v}_i$ are the i-th columns of $\mathbf{U}$ and $\mathbf{V}$, respectively and r is the rank of $\mathbf{H}$. From Eqs. (25.37) and (25.38), the pseudoinverse of $\mathbf{H}$ becomes as

$$\mathbf{H}^{\dagger} = \mathbf{V} \Sigma^{\dagger} \mathbf{U}^T = \sum_{i=1}^{r} \sigma_i^{-1} . \mathbf{v}_i \mathbf{u}_i^T \tag{25.39}$$

Therefore, from Eq. (25.39), the pseudoinverse solution can be expressed as

$$\mathbf{f}^{\dagger} = \mathbf{H}^{\dagger} \mathbf{g} = \mathbf{V} \Sigma^{\dagger} \mathbf{U}^T \mathbf{g} \tag{25.40}$$

or

$$\mathbf{f}^{\dagger} = \sum_{i=1}^{r} \sigma_i^{-1} \mathbf{v}_i \mathbf{u}_i^T \mathbf{g} = \sum_{i=1}^{r} \sigma_i^{-1} \left(\mathbf{u}_i^T \mathbf{g} \right) \mathbf{v}_i . \tag{25.41}$$

The series form of the pseudoinverse solution using SVD allows us to solve for the pseudoinverse solution using a sequential restoration algorithm expressed as

$$\mathbf{f}^{\dagger(k+1)} = \mathbf{f}^{\dagger(k)} + \sigma_k^{-1} \left(\mathbf{u}_k^T \mathbf{g} \right) \mathbf{v}_k . \tag{25.42}$$

The iterative approach for finding the pseudoinverse solution is advantageous when dealing with ill-conditioned systems and noise corrupted data. The iterative form can be terminated before the inversion of small singular values resulting in an unstable estimate. This technique becomes quite easy to implement for the case of a circulant degradation matrix $\mathbf{H}$, where the unitary matrices in Eq. (25.37) reduce to the discrete Fourier transform (DFT).

25.3.3 Regularization Techniques

Smoothing and regularization techniques [27, 28, 29] have been proposed in an attempt to overcome the problems associated with inverting ill-conditioned degradation operators for signal recovery. These methods attempt to force smoothness on the solution of a least squares error problem. The problem can be formulated in two different ways. One way of formulating the problem is

minimize:

$$\hat{\mathbf{f}}^T \mathbf{S}\hat{\mathbf{f}} \qquad (25.43)$$

subject to:

$$\left(\mathbf{g} - \mathbf{H}\hat{\mathbf{f}}\right)^T \mathbf{W}(\mathbf{g} - \mathbf{H}\hat{\mathbf{f}}) = e \qquad (25.44)$$

where $\mathbf{S}$ represents a smoothing matrix, $\mathbf{W}$ is an error weighting matrix, and e is a residual scalar estimation error. The error weighting matrix can be chosen as $\mathbf{W} = \mathbf{R}_{nn}^{-1}$. The smoothing matrix is typically composed of the first or second order difference. For this case, we wish to find the stationary point of the Lagrangian expression

$$F(\hat{\mathbf{f}}, \lambda) = \hat{\mathbf{f}}^T \mathbf{S}\hat{\mathbf{f}} + \lambda \left[\left(\mathbf{g} - \mathbf{H}\hat{\mathbf{f}}\right)^T \mathbf{W}(\mathbf{g} - \mathbf{H}\hat{\mathbf{f}}) - e\right]. \qquad (25.45)$$

The solution is found by taking derivatives with respect to $\mathbf{f}$ and λ and setting them equal to zero. The solution for a nonsingular overdetermined set of equations becomes

$$\hat{\mathbf{f}} = \left(\mathbf{H}^T \mathbf{W}\mathbf{H} + \frac{1}{\lambda}\mathbf{S}\right)^{-1} \mathbf{H}^T \mathbf{W}\mathbf{g} \qquad (25.46)$$

where λ is chosen to satisfy the compromise between residual error and smoothness in the estimate.

Alternately, this problem can be formulated as

minimize:

$$\left(\mathbf{g} - \mathbf{H}\hat{\mathbf{f}}\right)^T \mathbf{W}(\mathbf{g} - \mathbf{H}\hat{\mathbf{f}}) \qquad (25.47)$$

subject to:

$$\hat{\mathbf{f}}^T \mathbf{S}\hat{\mathbf{f}} = d \qquad (25.48)$$

where d represents a fixed degree of smoothness. The Lagrangean expression for this formulation becomes

$$G(\hat{\mathbf{f}}, \gamma) = \left(\mathbf{g} - \mathbf{H}\hat{\mathbf{f}}\right)^T \mathbf{W}\left(\mathbf{g} - \mathbf{H}\hat{\mathbf{f}}\right)^T + \gamma \left(\hat{\mathbf{f}}^T \mathbf{S}\hat{\mathbf{f}} - \mathbf{d}\right) \qquad (25.49)$$

and the solution for a nonsingular overdetermined set of equations becomes

$$\hat{\mathbf{f}} = \left(\mathbf{H}^T \mathbf{W}\mathbf{H} + \gamma \mathbf{S}\right)^{-1} \mathbf{H}^T \mathbf{W}\mathbf{g}. \qquad (25.50)$$

Note that for the two problem formulations, the results as given by Eq. (25.46) and Eq. (25.50) are identical if $\gamma = 1/\lambda$. The shortcomings of such a regularization technique is that the smoothing function $\mathbf{S}$ must be estimated and either the degree of smoothness, d, or the degree of error, e, must be known to determine γ or λ.

Constrained restoration techniques have also been developed [30] to overcome the problem of an ill-conditioned system. Linear equality constraints and linear inequality constraints have been enforced to yield one-step solutions similar to those described in this section. All the techniques described thus far attempt to overcome the problem of noise corrupted data and ill-conditioned systems by forcing some sort of taper on the inverse of the degradation operator. The sampling of algorithms discussed thus far fall under the category of optimization techniques where the objective function to be minimized is the least squares error. Recovery algorithms that fall under the category of optimization-based algorithms include maximum likelihood, maximum *a posteriori*, and maximum entropy methods [17].

We now introduce the concept of Projection onto Convex Sets (POCS), which will be the framework for a much broader and more powerful class of signal recovery algorithms.

25.4   Signal Recovery using Projection onto Convex Sets (POCS)

A broad set of recovery algorithms has been proposed to conform to the general framework introduced by the theory of projection onto convex sets (POCS) [31]. The POCS framework enables one to define an iterative recovery algorithm that can incorporate a number of linear as well as nonlinear constraints that satisfy certain properties. The more *a priori* information about the desired signal that one can incorporate into the algorithm, the more effective the algorithm becomes. In [21], POCS is presented as a particular example of a much broader class of algorithms described as Set Theoretic Estimation. The author distinguishes between two basic approaches to a signal estimation or recovery problem: optimization-based approaches and set theoretic approaches. The effectiveness of optimization-based approaches is highly dependent on defining a valid optimization criterion that, in practice, is usually determined by computational tractability rather than how well it models the problem. The optimization-based approaches seek a unique solution based on some predefined optimization criterion. The optimization-based approaches include the least squares techniques of the previous section as well as maximum likelihood (ML), maximum *a posteriori* (MAP), and maximum entropy techniques. Set theoretic estimation is based on the concept of finding a *feasible* solution, that is, a solution that is consistent with all the available *a priori* information. Unlike the optimization-based approaches which seek to find one optimum solution, the set theoretic approaches usually determine one of many possible feasible solutions. Many problems in signal recovery can be approached using the set theoretic paradigm. POCS has been one of the most extensively studied set theoretic approaches in the literature due to its convergence properties and flexibility to handle a wide range of signal characteristics. We limit our discussion here to POCS–based algorithms. The more general case of signal estimation using nonconvex as well as convex sets is covered in [21]. The rest of this section will focus on defining the POCS framework and describing several useful algorithms that fall into this general category.

25.4.1   The POCS Framework

A projection operator onto a closed convex set is an example of a nonlinear mapping that is easily analyzed and contains some very useful properties. Such projection operators minimize error distance and are non-expansive. These are two very important properties of ordinary linear orthogonal projections onto closed linear manifolds (CLMs). The benefit of using POCS for signal restoration is that one can incorporate nonlinear constraints of a certain type into the POCS framework. Linear image restoration algorithms cannot take advantage of *a priori* information based on nonlinear constraints.

The method of POCS depends on the set of solutions that satisfies *a priori* characteristics of the desired signal to lie in a well-defined closed convex set. For such properties, $\mathbf{f}$ is restricted to lie in the region defined by the intersection of all the convex sets, that is,

$$\mathbf{f} \in C_0 = \cap_{i=1}^{l} C_i. \tag{25.51}$$

Here C_i denotes the i-th closed convex set corresponding to the i-th property of $\mathbf{f}$, $C_i \in \mathcal{S}$, and $i \in \mathcal{I}$. The unknown signal $\mathbf{f}$ can be restored by using the corresponding projection operators P_i onto each convex set C_i. A property of closed convex sets is that a projection of a point onto the convex set is unique. This is known as the unique-nearest-neighbor property. The general form of the POCS-based recovery algorithm is expressed as

$$\mathbf{f}^{(k+1)} = P_{i_k} \mathbf{f}^{(k)} \tag{25.52}$$

where k denotes the iteration and i_k denotes a sequence of indices in $\mathcal{I}$. A common technique for iterating through the projections is referred to as *cyclic control* where the projections are applied in a cyclic manner, that is, $i_k = k(modulo l) + 1$. A geometric interpretation of the POCS algorithm for the simple case of two convex sets is illustrated in Fig. (25.2). The original POCS formulation is further generalized by

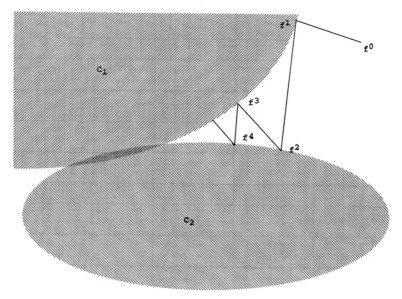

FIGURE 25.2 Geometric interpretation of POCS.

introducing a relaxation parameter expressed as

$$\mathbf{f}^{(k+1)} = \mathbf{f}^{(k)} + \lambda_k (\mathbf{P}_{i_k}(\mathbf{f}^{(k)}) - \mathbf{f}^{(k)}), \tag{25.53}$$

$$0 < \lambda_k < 2$$

where λ_k denotes the relaxation parameter. If $\lambda_k < 1$, the algorithm is said to be *underrelaxed* and if $\lambda_k > 1$, the algorithm is *overrelaxed*. Refer to [31] for further details on the convergence properties of POCS.

Common constraints that apply to many different signals in practice and whose solution space obeys the properties of convex sets are described in [10]. Some examples from [10] include frequency limits, spatial/temporal bounds, nonnegativity, sparseness, intensity or energy bounds, and partial knowledge of the spectral or spatial/temporal components. For further details on commonly used convex sets, see [10]. Most of the commonly used constraints for different signal processing applications fall under the category of convex sets which provide weak convergence. However, in practice, most of the POCS algorithms provide strong convergence.

Many of the commonly used iterative signal restoration techniques are specific examples of the POCS algorithm. The Kaczmarz algorithm [32], Landweber's iteration [33], and the method of alternating projections [9] are all POCS-based algorithms. It is worth noting that the image restoration technique developed independently by Gerchberg and Saxton [4] and Papoulis [5] are also versions of POCS. The algorithm developed by Gerchberg addressed phase retrieval from two images and Papoulis addressed superresolution by iterative methods. The Gerchberg–Papoulis (GP) algorithm is based on applying constraints on the estimate in the signal space and the Fourier space in an iterative fashion until the estimate converges to a solution. For the image restoration problem, the high frequency components of the image are extrapolated by imposing the finite extent of the object in the spatial domain and by imposing the known low frequency components in the frequency domain. The dual problem involves spectral estimation where the signal is extrapolated in the time or spatial domain. The algorithm consists of imposing the known part of the signal in the time domain and imposing a finite bandwidth constraint in the frequency domain. The GP algorithm assumes a space-invariant (or time-invariant) degradation operator.

We now present several signal recovery algorithms that conform to the POCS paradigm which are broadly classified under two categories: row-based and block-based algorithms.

25.5   Row-Based Methods

As early as 1937, Kaczmarz [32] developed an iterative projection technique to solve the inverse problem for a linear set of equations as given by Eq. (25.5). The algorithm takes the following form:

$$\mathbf{f}^{(k+1)} = \mathbf{f}^{(k)} + \lambda_k \frac{g_{i_k} - (\mathbf{h}_{i_k}, \mathbf{f}^{(k)})}{\| \mathbf{h}_{i_k} \|^2} \mathbf{h}_{i_k}. \tag{25.54}$$

The relaxation parameter λ_k is bound by $0 \leq \lambda_k \leq 2$, $\mathbf{h}$ represents a row of the matrix $\mathbf{H}$, i_k denotes a sequence of indices corresponding to a row in $\mathbf{H}$, g_i represents the i-*th* element of the vector g, $(\cdot, \cdot)$ is the standard inner product between two vectors, k denotes the iteration, and $\| \cdot \|$ denotes the Euclidean or $\mathcal{L}_2$ norm of a vector defined as

$$\| g \| = \left(\sum_{i=1}^{N} g_i^2 \right)^{1/2}. \tag{25.55}$$

Kaczmarz proved that Eq. (25.54) converges to the unique solution when the relaxation parameter is unity and H represents a square, nonsingular matrix, that is, H possesses an inverse and under certain conditions, the solution will converge to the minimum norm least squares or pseudoinverse solution. For further reading on the Kaczmarz algorithm and conditions for convergence, see [7, 8, 34, 35].

In general, the order in which one performs the Kaczmarz algorithm on the M existing equations can differ. *Cyclic control*, where the algorithm iterates through the equations in a periodic fashion is described as $i_k = k(modulo M) + 1$ where M is the number of rows in $\mathbf{H}$. *Almost cyclic control* exists when M sequential iterations of the Kaczmarz algorithm yield exactly one operation per equation in any order. *Remotest set control* exists when one performs the operations on the most distant equation first; most distant in the sense that the projection onto the hyperplane represented by the equation is the furthest away. The measure of distance is determined by the norm. This type of control is seldomly used since it requires a measurement dependent on all the equations.

The method of Kaczmarz for $\lambda = 1.0$, can be expressed geometrically as follows. Given $\mathbf{f}^{(k)}$ and the hyperplane $H_{i_k} = \{\mathbf{f} \in R^n \mid (\mathbf{h}_{i_k}, \mathbf{f}) = g_{i_k}\}$, $\mathbf{f}^{(k+1)}$ is the orthogonal projection of $\mathbf{f}^{(k)}$ onto H_{i_k}. This is illustrated in Fig. 25.3. Note that by changing the relaxation parameter, the next iterate can be a point

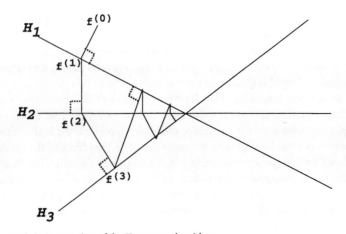

FIGURE 25.3   Geometric interpretation of the Kaczmarz algorithm.

anywhere along the line segment connecting the previous iterate and its orthogonal reflection with respect to the hyperplane.

The technique of Kaczmarz to solve for a set of linear equations has been rediscovered over the years for many different applications where the general problem formulation can be expressed as Eq. (25.5). For this reason, the Kaczmarz algorithm appears as the algebraic reconstruction technique (ART) in the field of medical imaging for computerized tomography (CT) [7], as well as the Widrow-Hoff least mean squares (LMS) algorithm [36] for channel equalization, echo cancellation, system identification, and adaptive array processing.

For the case of solving linear inequalities where Eq. (25.5) is replaced with

$$\mathbf{Hf} \leq \mathbf{g}, \tag{25.56}$$

a method very similar to Kaczmarz's algorithm is developed by Agmon [37] and Motzkin and Schoenberg [38],

$$
\begin{aligned}
\mathbf{f}^{(k+1)} &= \mathbf{f}^{(k)} + c^{(k)} \mathbf{h}_{i_k} \\
c^{(k)} &= \min \left(0, \lambda_k \frac{g_{i_k} - \left(\mathbf{h}_{i_k}, \mathbf{f}^{(k)} \right)}{\| \mathbf{h}_{i_k} \|^2} \right).
\end{aligned}
\tag{25.57}
$$

Once again, the relaxation parameter is defined on the interval $0 \leq \lambda_k \leq 2$. The method of solving linear inequalities by Agmon and Motzkin and Schoenberg is mathematically identical to the perceptron convergence theorem from the theory of learning machines (see [39]).

25.6 Block-Based Methods

A generalization of the Kaczmarz algorithm introduced in the previous section has been suggested by Eggermont [35] which can be described as a block, iterative algorithm. Recall the set of linear equations given by Eq. (25.5) where the dimensions of the problem are redefined so that $H \in \mathcal{R}^{LM \times N}$, $\mathbf{f} \in \mathcal{R}^N$, and $\mathbf{g} \in \mathcal{R}^{LM}$. In order to describe the generalization of the Kaczmarz algorithm, the matrix $\mathbf{H}$ is partitioned into M blocks of length L,

$$\mathbf{H} = \begin{pmatrix} \mathbf{h}_1^T \\ \mathbf{h}_2^T \\ \vdots \\ \mathbf{h}_{LM}^T \end{pmatrix} = \begin{pmatrix} \mathbf{H}_1 \\ \mathbf{H}_2 \\ \vdots \\ \mathbf{H}_M \end{pmatrix} \tag{25.58}$$

and $\mathbf{g}$ is partitioned as

$$\mathbf{g} = \begin{pmatrix} g_1 \\ g_2 \\ \vdots \\ g_{LM} \end{pmatrix} = \begin{pmatrix} \mathbf{G}_1 \\ \mathbf{G}_2 \\ \vdots \\ \mathbf{G}_M \end{pmatrix} \tag{25.59}$$

where $\mathbf{G}_i$, $i = 1, 2, \ldots, M$ is a vector of length L and the subblocks $\mathbf{H}_i$ are of dimension $L \times N$. The generalized group-iterative variation of the Kaczmarz algorithm is expressed as

$$\mathbf{f}^{(k+1)} = \mathbf{f}^{(k)} + \mathbf{H}_{i_k}^T \Sigma_k \left(\mathbf{G}_{i_k} - \mathbf{H}_{i_k} \mathbf{f}^{(k)} \right) \tag{25.60}$$

where $\mathbf{f}^{(0)} \in \mathcal{R}^N$. Eggermont gives details of convergence as well as conditions for convergence to the pseudoinverse solution [35].

A further generalization of Kaczmarz's algorithm led Eggermont [35] to the following form of the general block Kaczmarz algorithm:

$$\mathbf{f}^{(k+1)} = \mathbf{f}^{(k)} + \mathbf{H}_{i_k}^{\dagger} \Lambda_k \left(G_{i_k} - \mathbf{H}_{i_k} x^{(k)} \right) \tag{25.61}$$

where once again $\mathbf{H}_{i_k}^{\dagger}$ denotes the Moore–Penrose inverse of $\mathbf{H}_{i_k}$, Λ_k is the $L \times L$ relaxation matrix, and cyclic control is defined as $i_k = k(moduloM) + 1$.

When the block size L given in Eq. (25.60) is equal to the number of equations M, the algorithm becomes identical to Landweber's iteration [33] for solving Fredholm equations of the first kind; that is,

$$\mathbf{f}^{(k+1)} = \mathbf{f}^{(k)} + \mathbf{H}^T \Sigma_k \left(\mathbf{g} - \mathbf{H}\mathbf{f}^{(k)} \right). \tag{25.62}$$

The resulting Landweber iteration becomes

$$\mathbf{f}^{(k+1)} = \mathbf{H}^T \mathbf{g} + \left(\mathbf{I} - \mathbf{H}^T \mathbf{H} \right) \mathbf{f}^{(k)}. \tag{25.63}$$

Another interesting approach that is similar to the generalized block-Kaczmarz algorithm, with the block size L equal to the number of equations M, is the method of alternating orthogonal projections described by Youla [9] where alternating orthogonal projections are made onto closed linear manifolds (CLM).

The row-based and block-based algorithms described here correspond to a POCS framework where the only *a priori* information incorporated into the algorithm is the original problem formulation as described by Eq. (25.5). At times, the only information we may have is the original measurement $\mathbf{g}$ and an estimate of the degradation operator $\mathbf{H}$ and these algorithms are suited for such applications. However, for most applications, other *a priori* information is known about the desired signal and an effective algorithm should utilize this information.

We now describe a POCS-based algorithm suited for the problem of image restoration where additional *a priori* signal information is incorporated into the algorithm.

25.7 Image Restoration Using POCS

Here we describe an image recovery algorithm [18, 40] that is based on the POCS framework and show some image restoration results [19, 20]. The list of references includes other examples of POCS-based recovery algorithms.

The least squares minimum norm or pseudoinverse solution can be formulated as

$$\mathbf{f}^{\dagger} = \mathbf{H}^{\dagger} \mathbf{H} \mathbf{f} = \mathbf{V} \Lambda \mathbf{V}^T \mathbf{f} \tag{25.64}$$

where the dagger † denotes the pseudoinverse, $\mathbf{V}$ is the unitary matrix found in the diagonalization of $\mathbf{H}$, and Λ is the following diagonal matrix whose first r diagonal terms are equal to one:

$$\Lambda = \begin{pmatrix} 1_1 & 0 & \cdots & & & \\ 0 & 1_2 & 0 & & & \\ & & & \ddots & & \\ & & & & 1_r & \\ & & & & & 0 \end{pmatrix}. \tag{25.65}$$

By defining

$$\mathbf{P} = \mathbf{V} \Lambda \mathbf{V}^T, \tag{25.66}$$

the orthogonal complement to the operator $\mathbf{P}$ is given by the projection operator

$$\mathbf{Q} = \mathbf{I} - \mathbf{P} = \mathbf{V}\Lambda^C\mathbf{V}^T \tag{25.67}$$

where

$$\Lambda^C = \begin{pmatrix} 0 & \cdots & & & & \\ \vdots & \ddots & & & & \\ & & 1 & & & \\ & & & 1_{r+1} & & \\ & & & & \ddots & \end{pmatrix}. \tag{25.68}$$

The diagonal matrix Λ^C contains ones in the last $N - r$ diagonal positions and zeroes elsewhere. The superscript C denotes the complement.

Any arbitrary vector $\mathbf{f}$ can be decomposed as follows:

$$\mathbf{f} = \mathbf{Pf} + \mathbf{Qf} \tag{25.69}$$

where the projection operator $\mathbf{P}$ projects $\mathbf{f}$ onto the range space of the degradation matrix $\mathbf{H}^T\mathbf{H}$ and the orthogonal projection operator $\mathbf{Q}$ projects $\mathbf{f}$ onto the null space of the degradation matrix $\mathbf{H}^T\mathbf{H}$. The component $\mathbf{Pf}$ will be referred to as the "in-band" term and the component $\mathbf{Qf}$ will be referred to as the "out-of-band" term.

In general, the least squares family of solutions to the image restoration problem can be stated as

$$\begin{aligned} \mathbf{f} &= \mathbf{f}_{in-band} + \mathbf{f}_{out-of-band} \\ &= \mathbf{f}^\dagger + K_{r+1}\mathbf{v}_{r+1} + K_{r+2}\mathbf{v}_{r+2} + \ldots + K_N\mathbf{v}_N. \end{aligned} \tag{25.70}$$

The vectors $\mathbf{v}_i$ correspond to the eigenvectors of $\{\sigma_{r+1}^2, \sigma_{r+2}^2, \ldots, \sigma_N^2\}$ for $\mathbf{H}^T\mathbf{H}$; they are the eigenvectors associated with zero valued eigenvalues. The out-of-band solution $K_{r+1}\mathbf{v}_{r+1} + \ldots + K_N\mathbf{v}_N$ must satisfy

$$\mathbf{H}\mathbf{f}_{out-of-band} = 0. \tag{25.71}$$

Adding the terms $\{K_{r+1}\mathbf{v}_{r+1}, K_{r+2}\mathbf{v}_{r+2}, \ldots, K_N\mathbf{v}_N\}$, to the pseudoinverse solution $\mathbf{f}^\dagger$ does not change the $\mathcal{L}_2$ norm of the error since

$$\begin{aligned} \| n \| &= \| \mathbf{g} - \mathbf{H}\mathbf{f} \| \\ &= \| \mathbf{g} - \mathbf{H}\left(\mathbf{f}^\dagger + K_{r+1}\mathbf{v}_{r+1} + \ldots + K_N\mathbf{v}_N\right) \| \\ &= \| \mathbf{g} - \mathbf{H}\mathbf{f}^\dagger - \mathbf{H}K_{r+1}\mathbf{v}_{r+1} - \ldots - \mathbf{H}K_N\mathbf{v}_N \| \\ &= \| \mathbf{g} - \mathbf{H}\mathbf{f}^\dagger \| \end{aligned} \tag{25.72}$$

which is the least squares error. The terms $\mathbf{H}K_{r+1}\mathbf{v}_{r+1}, \ldots, \mathbf{H}K_N\mathbf{v}_N$ are all equal to zero because the vectors $\mathbf{v}_{r+1}, \ldots, \mathbf{v}_N$ are in the null space of $\mathbf{H}$. Therefore, any linear combination of $\mathbf{v}_i$ in the null space of $\mathbf{H}$ can be added to the pseudoinverse solution without affecting the least squares cost function. The pseudoinverse solution, $\mathbf{f}^\dagger$, provides the unique least squares estimate with minimum norm,

$$\min \| \mathbf{f}_{LS} \| = \| \mathbf{f}^\dagger \| \tag{25.73}$$

where $\mathbf{f}_{LS}$ denotes the least squares solution. In practice, it is unlikely that the desired solution is required to possess the minimum norm out of all feasible solutions so that $\mathbf{f}^\dagger$ is not necessarily the optimum solution.

The image restoration algorithm described here provides a framework that allows *a priori* information in the form of signal constraints to be incorporated into the algorithm in order to obtain a better estimate than the least squares minimum norm solution $\mathbf{f}^\dagger$. The constraint operator will be represented by C and can incorporate a variety of linear and nonlinear *a priori* signal characteristics as long as they obey the properties of convex set theory. In the case of image restoration, the constraint operator C includes non-negativity which can be described by

$$(\mathbf{C}_+\mathbf{f})_i = \begin{cases} f_i & f_i \geq 0 \\ 0 & f_i < 0 \end{cases}. \tag{25.74}$$

Concatenating the vectors $\mathbf{v}_i$ in Eq. (25.70) yields

$$\mathbf{f} = \mathbf{f}^\dagger + \mathbf{V}\Lambda^C\mathbf{K} \tag{25.75}$$

where

$$\mathbf{K} = \begin{pmatrix} K_1 \\ K_2 \\ \vdots \\ K_N \end{pmatrix} \tag{25.76}$$

and

$$\mathbf{V}\Lambda^C = \begin{pmatrix} \mathbf{v}_1 & \mathbf{v}_2 & \cdots & \mathbf{v}_N \end{pmatrix} \begin{pmatrix} 0 & & & & \\ & \ddots & & & \\ & & 1_{r+1} & & \\ & & & \ddots & \\ & & & & 1_N \end{pmatrix}. \tag{25.77}$$

We would like to find the solution to the unknown vector K in Eq. (25.75). A reasonable approach is to start with the constrained pseudoinverse solution and solve for $\mathbf{K}$ in a least squares manner; that is,

minimize:

$$\| \mathbf{C}_+\mathbf{f}^\dagger - \left\{ \mathbf{f}^\dagger + \mathbf{V}\Lambda^C\mathbf{K} \right\} \| \tag{25.78}$$

subject to:

$$\mathbf{C}_+\mathbf{f}^\dagger = \mathbf{f}^\dagger + \mathbf{V}\Lambda^C\mathbf{K}. \tag{25.79}$$

The least squares solution becomes

$$\mathbf{C}_+\mathbf{f}^\dagger - \mathbf{f}^\dagger = \mathbf{V}\Lambda^C\mathbf{K}$$
$$\mathbf{K} = \Lambda^C\mathbf{V}^T\left(\mathbf{C}_+\mathbf{f}^\dagger - \mathbf{f}^\dagger\right). \tag{25.80}$$

Since $\Lambda^C\mathbf{V}^T\mathbf{f}^\dagger = 0$, we get

$$\mathbf{K} = \Lambda^C\mathbf{V}^T\mathbf{C}_+\mathbf{f}^\dagger. \tag{25.81}$$

Substituting Eq. (25.81) into Eq. (25.79) yields

$$\mathbf{C}_+\mathbf{f}^\dagger = \mathbf{f}^\dagger + \mathbf{Q}\mathbf{C}_+\mathbf{f}^\dagger + e. \tag{25.82}$$

where e denotes a residual vector. The process of enforcing the overall least squares solution and solving for the out-of-band component to fit the constraints can be implemented in an iterative fashion. The resulting recursion is

$$\mathbf{C}_+ \mathbf{f}^{(k)} = \mathbf{f}^\dagger + \mathbf{Q}\mathbf{C}_+ \mathbf{f}^{(k)} + e^{(k)}. \tag{25.83}$$

By defining

$$\mathbf{f}^{(k+1)} \equiv \mathbf{C}_+ \mathbf{f}^{(k)} - e^{(k)} \tag{25.84}$$

the final iterative algorithm becomes

$$\begin{aligned}
\mathbf{f}^{(0)} &= \mathbf{f}^\dagger \\
\mathbf{f}^{(k+1)} &= \mathbf{f}^\dagger + \mathbf{Q}\mathbf{C}_+ \mathbf{f}^{(k)}. \\
k &= 0, 1, 2, \ldots.
\end{aligned} \tag{25.85}$$

Note that the recursion yields the least squares solution while enforcing the *a priori* constraints through the out-of-band signal component. It is apparent that such an approach will yield a better estimate for the unknown signal $\mathbf{f}$ than the minimum norm least squares solution $\mathbf{f}^\dagger$. Note that this algorithm can easily be generalized to other problems by replacing the non-negativity constraint $\mathbf{C}_+$ with the signal appropriate constraints. In the case when $\mathbf{f}^\dagger$ satisfies all the constraints exactly, the solution to iterative algorithm reduces to the pseudoinverse solution. For more details on this algorithm, convergence issues, and stopping criterion, refer to [18, 20, 40]. By looking at this algorithmic framework from the set theoretic viewpoint described in [21], the original set of solutions is given by all the solutions that satisfy the least squares error criterion. The addition of *a priori* signal constraints attempts to reduce the feasible set of solutions and to provide a better estimate than the pseudoinverse solution.

Finally, we would like to show some image restoration results based on the method described in [19, 20, chap. 4]. The technique is a modification of the Kaczmarz method described here using the theory of POCS. Original, degraded, restored images using the original Kaczmarz algorithm and the restored images using the modified algorithm based on the POCS framework are shown in Fig. (25.4). Similarly, we show the original, degraded and restored images in the frequency domain in Fig. (25.5). The details of the algorithm are found in [19].

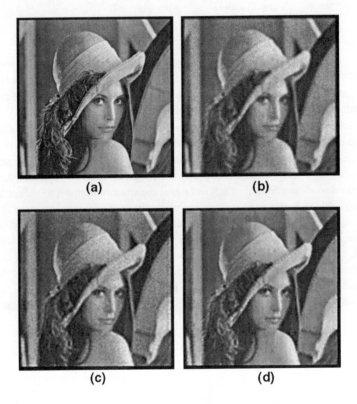

FIGURE 25.4 (a) Original image. (b) Degraded image at 25 dB SNR. (c) Restored image using Kaczmarz iterations. (d) Restored image using the modified Kaczmarz algorithm in a POCS framework. (Courtesy of IEEE: Kuo, S.S. and Mammone, R.J., Image restoration by convex projections using adaptive constraints and the l_1 norm, *IEEE Trans. Signal Process.*, 40, 159–168, 1992.)

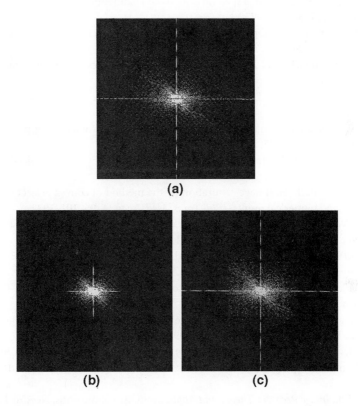

FIGURE 25.5 Spatial frequency response of the (a) original image, (b) degraded image, and (c) restored image using the new algorithm. (Courtesy of IEEE: Kuo, S.S. and Mammone, R.J., Image restoration by convex projections using adaptive constraints and the l_1 norm, *IEEE Trans. Signal Process.*, 40, 159–168, 1992.)

References

[1] Sondhi, M.M., Image restoration: The removal of spatially invariant degradations, *Proc. IEEE*, 60(7), 842–853, July 1972.

[2] Slepian, D., Restoration of photographs blurred by image motion, *Bell Syst. Tech. J.*, XLVI, 2353–2362, 1967.

[3] Helstrom, C.W., Image restoration by the method of least squares, *J. Opt. Soc. Am.*, 57, 297–303, 1967.

[4] Gerchberg, R.W. and Saxton, W.O., A practical algorithm for the determination of phase from image and diffraction plane pictures, *Optik*, 35, 237–246, 1972.

[5] Papoulis, A., A new algorithm in spectral analysis and band–limited extrapolation, *IEEE Trans. Circuits Syst.*, 22, 735–742, 1975.

[6] Hayes, M.H., Lim, J.S., and Oppenheim, A.V., Signal reconstruction from phase or magnitude, *IEEE Trans. Acoust., Speech, Signal Process.*, ASSP-28, 1980.

[7] Lent, A., Herman G.T., and Rowland, S.W., Art: Mathematics and applications, *J. Theoret. Biol.*, 42, 1–32, 1973.

[8] Lent, A., Herman, G.T., and Lutz, P.H., Relaxation methods for image reconstruction, *Comm. ACM*, 21, 152–158, 1978.

[9] Youla, D.C., Generalized image restoration by the method of alternating projections, *IEEE Trans. Circuits Syst.*, CAS-25, 1978.

[10] Youla, D.C. and Webb, H., Image restoration by the method of convex projections: Part I— theory, *IEEE Trans. Med. Imag.*, 1, 81–94, 1982.

[11] Sezan, M.I. and Stark, H., Image restoration by the method of convex projections: Part II— applications and numerical results, *IEEE Trans. Med. Imag.*, 1, 95–101, 1982.

[12] Schafer, R.W., Mersereau, R.M., and Richards, M.A., Constrained iterative restoration algorithms, *Proc. IEEE*, 69(4), 432–449, April 1981.

[13] Civanlar, M.R. and Trussell, H.J., Digital signal restoration using fuzzy sets, *IEEE Trans. Acoustics, Speech, Signal Process.*, 34, 919–936, 1986.

[14] Trussell, H.J. and Civanlar, M.R., The feasible solution in signal restoration, *IEEE Trans. Acoust., Speech, Signal Process.*, 32, 201–212, 1984.

[15] Sezan, M.I. and Trussell, H.J., Prototype image constraints for set–theoretic image restoration, *IEEE Trans. Signal Process.*, 39, 2275–2285, 1991.

[16] Sezan, M.I. and Tekalp, A.M., Adaptive image restoration with artifact suppression using the theory of convex projections, *IEEE Trans. Acoust., Speech, Signal Process.*, 38, 181–185, Jan. 1990.

[17] Stark, H., Ed., *Image Recovery Theory and Applications,* Academic Press, New York, 1987.

[18] Podilchuk, C.I. and Mammone, R.J., Image recovery by convex projections using a least–squares constraint, *J. Opt. Soc. Am. A*, 7, 517–521, March 1990.

[19] Kuo, S.S. and Mammone, R.J., Image restoration by convex projections using adaptive constraints and the l_1 norm, *IEEE Trans. Signal Process.*, 40, 159–168, 1992.

[20] Mammone, R.J., Ed., *Computational Methods of Signal Recovery and Recognition,* John Wiley & Sons, New York, 1992.

[21] Combettes, P.L., The foundations of set theoretic estimation, *Proc. IEEE*, 81, 182–208, 1993.

[22] Noble, B. and Daniel, J.W., *Applied Linear Algebra,* Prentice-Hall, Englewood Cliffs, NJ, 1977.

[23] Landau, H.J. and Miranker, W.L., The recovery of distorted bandlimited signals, *J. Math. Anal. Appl.*, 2, 97–104, 1961.

[24] Wiener, N., On the factorization of matrices, *Commentarii Mathematici Helvetici*, 29, 97–111, 1955.

[25] Jayant, N.S., Johnston, J.D., and Safranek, R.J., Signal compression based on models of human perception, *Proc. IEEE*, October 1993.

[26] Pratt, W.K. and Davarian, F., Fast computational techniques for pseudoinverse and Wiener image restoration, *IEEE Trans. Comp.*, 26, 571–580, 1977.

[27] Twomey, S., On the numerical solution of fredholm integral equations of the first kind by the inversion of the linear system produced by quadrature, *J. Assoc. Comp. Mach.*, 10, 97–101, 1963.

[28] Tikonov, A.N., Regularization of incorrectly posed problems, *Soviet Math.*, 4, 1624–1627, 1963.

[29] Phillips, D.L., A technique for the numerical solution of certain integral equations of the first kind, *J. Assoc. Comput. Mach.*, 9, 84–97, 1964.

[30] Mascarenhas, N.D.A. and Pratt, W.K., Digital image restoration under a regression model, *IEEE Trans. Circuits Syst.*, 22, 252–266, 1975.

[31] Polyak, B.T., Gubin, L.G., and Raik, E.V., The method of projections for finding the common point of convex sets, *U.S.S.R. Comp. Math. Math. Phys.*, 7, 1–24, 1967.

[32] Kaczmarz, S., Angenaherte au flosung von systemen linearer gleichungen, *Bull. Acad. Polon. Sci. Lett.*, A, 355–357, 1937.

[33] Strand, O.N., Theory and methods related to the singular-function expansion and Landweber's iteration for integral equations of the first kind, *SIAM J. Numer. Anal.*, 11, 798–825, 1974.

[34] Tanabe, K., Projection method for solving a singular system of linear equations and its applications, *Numer. Math.*, 17, 203–214, 1971.

[35] Eggermont, P.P.B., Iterative algorithms for large partitioned linear systems with applications to image reconstruction, *Linear Algebra and its Applications*, 40, 37–67, 1981.

[36] Widrow, B. and McCool, J.M., A comparison of adaptive algorithms based on the methods of steepest descent and random search, *IEEE Trans. Antennas Propag.*, 24, 615–637, 1976.

[37] Agmon, S., The relaxation method for linear inequalities, *Can. J. Math.*, 6, 382–392, 1954.

[38] Motzkin, T.S. and Schoenberg, I.J., The relaxation method for linear inequalities, *Can. J. Math.*, 6, 393–404, 1954.

[39] Minsky, M. and Papert, S., *Perceptrons: An Introduction to Computational Geometry*, MIT Press, Cambridge, MA., 1969.

[40] Podilchuk, C.I. and Mammone, R.J., Step size for the general iterative image recovery algorithm, *Opt. Eng.*, 27, 806–811, 1988.

26

Algorithms for Computed Tomography

Gabor T. Herman
University of Pennsylvania

26.1 Introduction

Computed tomography is the process of reconstructing the interiors of objects from data collected based on transmitted or emitted radiation. The problem occurs in a wide range of application areas. Here we discuss the computer algorithms used for achieving the reconstructions.

26.2 The Reconstruction Problem

We want to solve the following general problem. There is a three-dimensional structure whose internal composition is unknown to us. We subject this structure to some kind of radiation, either by transmitting the radiation through the structure or by introducing the emitter of the radiation into the structure. We measure the radiation transmitted through or emitted from the structure at a number of points. *Computed tomography* (CT) is the process of obtaining from these measurements the distribution of the physical parameter(s) inside the structure that have an effect on the measurements. The problem occurs in a wide range of areas, such as x-ray CT, emission tomography, photon migration imaging, and electron microscopic reconstruction; see, e.g., [1, 2]. All of these are inverse problems of various sorts; see, e.g., [3].

Where it is not otherwise stated, we will be discussing the special reconstruction problem of estimating a function of two variables from estimates of its line integrals. As it is quite reasonable for any application, we will assume that the domain of the function is contained in a finite region of the plane. In what follows we will introduce all the needed notation and terminology; in most cases these agree with those used in [1].

Suppose f is a function of the two polar variables r and ϕ. Let $[\mathcal{R}f](\ell, \theta)$ denote the line integral of f along the line that is at distance ℓ from the origin and makes an angle θ with the vertical axis. This operator $\mathcal{R}$ is usually referred to as the *Radon transform*.

0-8493-8572-5/98/$0.00+$.50
© 1998 by CRC Press LLC

The input data to a reconstruction algorithm are estimates (based on physical measurements) of the values of $[\mathcal{R}f](\ell, \theta)$ for a finite number of pairs (ℓ, θ); its output is an estimate, in some sense, of f. More precisely, suppose that the estimates of $[\mathcal{R}f](\ell, \theta)$ are known for I pairs: $(\ell_i, \theta_i), 1 \leq i \leq I$. We use y to denote the I-dimensional column vector (called the *measurement vector*) whose ith component, y_i, is the available estimate of $[\mathcal{R}f](\ell_i, \theta_i)$. The task of a reconstruction algorithm is:

given the data y, **estimate** the function f.

Following [1], reconstruction algorithms are characterized either as transform methods or as series expansion methods. In the following subsections we discuss the underlying ideas of these two approaches and give detailed descriptions of two algorithms from each category.

26.3 Transform Methods

The Radon transform has an inverse, R^{-1}, defined as follows. For a function p of ℓ and θ,

$$\left[R^{-1}p\right](r, \phi) = \frac{1}{2\pi^2} \int_0^\pi \int_{-\infty}^\infty \frac{1}{r\cos(\theta - \phi) - \ell} p_1(\ell, \theta) d\ell d\theta , \tag{26.1}$$

where $p_1(\ell, \theta)$ denotes the partial derivative of p with respect to its first variable ℓ. [Note that it is intrinsically assumed in this definition that p is sufficiently smooth for the existence of the integral in Eq. (26.1)]. It is known [1] that for any function f which satisfies some physically reasonable conditions (such as continuity and boundedness) we have, for all points (r, ϕ),

$$\left[R^{-1}Rf\right](r, \phi) = f(r, \phi). \tag{26.2}$$

Transform methods are numerical procedures that estimate values of the double integral on the right hand side of Eq. (26.1) from given values of $p(\ell_i, \theta_i)$, for $1 \leq i \leq I$. We now discuss two such methods: the widely adopted *filtered backprojection* (FBP) algorithm (called the *convolution* method in [1]) and the more recently developed *linogram* method.

26.4 Filtered Backprojection (FBP)

In this algorithm, the right hand side of Eq. (26.1) is approximated by a two-step process (for derivational details see [1] or, in a more general context, [3]). First, for fixed values of θ, convolutions defined by

$$[p *_Y q](\ell', \theta) = \int_{-\infty}^\infty p(\ell, \theta)q\left(\ell' - \ell, \theta\right) d\ell \tag{26.3}$$

are carried out, using a *convolving function* q (of one variable) whose exact choice will have an important influence on the appearance of the final image. Second, our estimate f^* of f is obtained by *backprojection* as follows:

$$f^*(r, \phi) = \int_0^\pi [p *_Y q](r\cos(\theta - \phi), \theta) d\theta. \tag{26.4}$$

To make explicit the implementation of this for a given measurement vector, let us assume that the data function p is known at points $(nd, m\Delta), -N \leq n \leq N, 0 \leq m \leq M - 1$, and $M\Delta = \pi$. Let us further assume that the function f is to be estimated at points $(r_j, \phi_j), 1 \leq j \leq J$. The computer algorithm operates as follows.

A sequence $f_0, \ldots, f_{M-1}, f_M$ of estimates is produced; the last of these is the output of the algorithm. First we define

$$f_0(r_j, \phi_j) = 0, \tag{26.5}$$

for $1 \leq j \leq J$. Then, for each value of m, $0 \leq m \leq M - 1$, we produce the $(m + 1)$st estimate from the mth estimate by a two-step process:

1. For $-N \leq n' \leq N$, calculate

$$p_c\left(n'd, m\Delta\right) = d \sum_{n=-N}^{N} p\left(nd, m\Delta\right) q\left(\left(n' - n\right)d\right), \qquad (26.6)$$

using the measured values of $p(nd, m\Delta)$ and precalculated values (same for all m) of $q((n' - n)d)$. This is a discretization of Eq. (26.3). One possible (but by no means only) choice is

2. For $1 \leq j \leq J$, we set

$$f_{m+1}\left(r_j, \phi_j\right) = f_m\left(r_j, \phi_j\right) + \Delta p_c\left(r_j \cos\left(m\Delta - \phi_j\right), m\Delta\right). \qquad (26.7)$$

This is a discretization of Eq. (26.4). To do it, we need to interpolate in the first variable of p_c from the values calculated in Eq. (26.6) to obtain the values needed in Eq. (26.7). In practice, once $f_{m+1}(r_j, \phi_j)$ has been calculated, $f_m(r_j, \phi_j)$ is no longer needed and the computer can reuse the same memory location for $f_0(r_j, \phi_j), \ldots, f_{M-1}(r_j, \phi_j), f_M(r_j, \phi_j)$.

In a complete execution of the algorithm, the uses of Eq. (26.6) require $M(2N + 1)$ multiplications and additions, while all the uses of Eq. (26.7) require MJ interpolations and additions. Since J is typically of the order of N^2 and N itself in typical applications is between 100 and 1000, we see that the cost of backprojection is likely to be much more computationally demanding than the cost of convolution. In any case, reconstruction of a typical 512×512 cross-section from data collected by a typical x-ray CT device is not a challenge to state-of-the art computational capabilities; it is routinely done in the order of a second or so and can be done, using a pipeline architecture, in a fraction of a second [4].

26.5   The Linogram Method

The basic result that justifies this method is the well-known *projection theorem* which says that "taking the two-dimensional Fourier transform is the same as taking the Radon transform and then applying the Fourier transform with respect to the first variable" [1]. The method was first proposed in [5] and the reason for the name of the method can be found there. The basic reason for proposing this method is its speed of execution and we return to this below. In the description that follows, we use the approach of [6]. That paper deals with the fully three-dimensional problem; here we simplify it to the two-dimensional case.

For the linogram approach we assume that the data were collected in a special way (that is, at points whose locations will be precisely specified below); if they were collected otherwise, we need to interpolate prior to reconstruction. If the function is to be estimated at an array of points with rectangular coordinates $\{(id, jd) \mid -N \leq i \leq N, -N \leq j \leq N\}$ (this array is assumed to cover the object to be reconstructed), then the data function p needs to be known at points

$$\left(nd_m, \theta_m\right), \ -2N - 1 \leq n \leq 2N + 1, \ -2N - 1 \leq m \leq 2N + 1 \qquad (26.8)$$

and at points

$$\left(nd_m, \frac{\pi}{2} + \theta_m\right), \ -2N - 1 \leq n \leq 2N + 1, \ -2N - 1 \leq m \leq 2N + 1, \qquad (26.9)$$

where

$$\theta_m = \tan^{-1} \frac{2m}{4N + 3} \text{ and } d_m = d \cos \theta_m. \qquad (26.10)$$

The linogram method produces from such data estimates of the function values at the desired points using a multi-stage procedure. We now list these stages, but first point out two facts. One is that the most expensive computation that needs to be used in any of the stages is the taking of *discrete Fourier transforms* (DFTs), which can always be implemented (possibly after some padding by zeros) very efficiently by the use of the *fast Fourier transform* (FFT). The other is that the output of any stage produces estimates of function values at exactly those points where they are needed for the discrete computations of the next stage; there is never any need to interpolate between stages. It is these two facts which indicate why the linogram method is both computationally efficient and accurate. (From the point of view of this handbook, these facts justify the choice of sampling points in Eqs. (26.8) through (26.10); a geometrical interpretation is given in [7].)

1. *Fourier transforming of the data* — For each value of the second variable, we take the DFT of the data with respect to the first variable in Eq. (26.8) and Eq. (26.9). By the projection theorem, this provides us with estimates of the two-dimensional Fourier transform F of the object at points (in a rectangular coordinate system)

$$\left(\frac{k}{(4N+3)d'}, \frac{k}{(4N+3)d} \tan\theta_m\right), \quad -2N-1 \le k \le 2N+1,$$
$$-2N-1 \le m \le 2N+1 \tag{26.11}$$

and at points (also in a rectangular coordinate system)

$$\left(\frac{k}{(4N+3)d} \tan\left(\frac{\pi}{2} + \theta_m\right), \frac{k}{(4N+3)d}\right), \quad -2N-1 \le k \le 2N+1,$$
$$-2N-1 \le m \le 2N+1. \tag{26.12}$$

2. *Windowing* — At this point we may suppress those frequencies which we suspect to be noise-dominated by multiplying with a *window function* (corresponding to the convolving function in FBP).

3. *Separating into two functions* — The sampled Fourier transform F of the object to be reconstructed is written as the sum of two functions, G and H. G has the same values as F at all the points specified in Eq. (26.11) except at the origin and is zero-valued at all other points. H has the same values as F at all the points specified in Eq. (26.12) except at the origin and is zero-valued at all other points. Clearly, except at the origin, $F = G + H$. The idea is that by first taking the two-dimensional inverse Fourier transforms of G and H separately and then adding the results, we get an estimate (except for a *DC* term which has to be estimated separately, see [6]) of f. We only follow what needs to be done with G; the situation with H is analogous.

4. *Chirp z-transforming in the second variable* — Note that the way the θ_m were selected implies that if we fix k, then the sampling in the second variable of Eq. (26.11) is uniform. Furthermore, we know that the value of G is zero outside the sampled region. Hence, for each fixed k, $0 < |k| \le 2N+1$, we can use the chirp z-transform to estimate the inverse DFT in the second variable at points

$$\left(\frac{k}{(4N+3)d}, jd\right), \quad -2N-1 \le k \le 2N+1, -N \le j \le N. \tag{26.13}$$

The chirp z-transform can be implemented using three FFTs, see [7].

5. *Inverse transforming in the first variable* — The inverse Fourier transform of G can now be

estimated at the required points by taking, for every fixed j, the inverse DFT in the first variable of the values at the points of Eq. (26.13).

26.6 Series Expansion Methods

This approach assumes that the function, f, to be reconstructed can be approximated by a linear combination of a finite set of known and fixed basis functions,

$$f(r, \phi) \approx \sum_{j=1}^{J} x_j b_j(r, \phi), \tag{26.14}$$

and that our task is to estimate the unknowns, x_j. If we assume that the measurements depend linearly on the object to be reconstructed (certainly true in the special case of line integrals) and that we know (at least approximately) what the measurements would be if the object to be reconstructed was one of the basis functions (we use $r_{i,j}$ to denote the value of the ith measurement of the jth basis function), then we can conclude [1] that the ith of our measurements of f is approximately

$$\sum_{j=1}^{J} r_{i,j} x_j. \tag{26.15}$$

Our problem is then to estimate the x_j from the measured approximations (for $1 \leq i \leq I$) to Eq. (26.15). The estimate can often be selected as one that satisfies some *optimization criterion*.

To simplify the notation, the image is represented by a J-dimensional *image vector x* (with components x_j) and the data form an I-dimensional measurement vector y. There is an assumed *projection matrix R* (with entries $r_{i,j}$). We let r_i denote the transpose of the ith row of R ($1 \leq i \leq I$) and so the inner product $\langle r_i, x \rangle$ is the same as the expression in Eq. (26.15). Then y is approximately Rx and there may be further information that x belongs to a subset C of $\mathbf{R}^J$, the space of J-dimensional real-valued vectors. In this formulation R, C, and y are known and x is to be estimated. Substituting the estimated values of x_j into Eq. (26.14) will then provide us with an estimate of the function f.

The simplest way of selecting the basis functions is by subdividing the plane into pixels (or space into voxels) and choosing basis functions whose value is 1 inside a specific pixel (or voxel) and is 0 everywhere else. However, there are other choices that may be preferable; for example, [8] uses spherically symmetric basis functions that are not only spatially limited, but also can be chosen to be very smooth. The smoothness of the basis functions then results in smoothness of the reconstructions, while the spherical symmetry allows easy calculation of the $r_{i,j}$. It has been demonstrated [9], for the case of fully three-dimensional *positron emission tomography* (PET) reconstruction, that such basis functions indeed lead to statistically significant improvements in the task-oriented performance of series expansion reconstruction methods.

In many situations only a small proportion of the $r_{i,j}$ are nonzero. (For example, if the basis functions are based on voxels in a $200 \times 200 \times 100$ array and the measurements are approximate line integrals, then the percent of nonzero $r_{i,j}$ is less than 0.01, since a typical line will intersect fewer than 400 voxels.) This makes certain types of iterative methods for estimating the x_j surprisingly efficient. This is because one can make use of a subroutine which, for any i, returns a list of those js for which $r_{i,j}$ is not zero, together with the values of the $r_{i,j}$ [1, 10]. We now discuss two such iterative approaches: the so-called *algebraic reconstruction techniques* (ART) and the use of *expectation maximization* (EM).

26.7 Algebraic Reconstruction Techniques (ART)

The basic version of ART operates as follows [1]. The method cycles through the measurements repeatedly, considering only one measurement at a time. Only those x_j are updated for which the corresponding $r_{i,j}$

for the currently considered measurement i is nonzero and the change made to x_j is proportional to $r_{i,j}$. The factor of proportionality is adjusted so that if Eq. (26.15) is evaluated for the resulting x_j, then it will match exactly the ith measurement. Other variants will use a block of measurements in one iterative step and will update the x_j in different ways to ensure that the iterative process converges according to a chosen estimation criterion.

Here we discuss only one specific optimization criterion and the associated algorithm. (Others can be found, for example, in [1]). Our task is to find the x in $\mathbf{R}^J$ which **minimizes**

$$r^2\|y - R_x\|^2 + \|x - \mu_x\|^2 \tag{26.16}$$

($\|\cdot\|$ indicates the usual Euclidean norm), for a given constant scalar r (called the *regularization parameter*) and a given constant vector μ_x.

The algorithm makes use of an I-dimensional vector u of additional variables, one for each measurement. First we define $u^{(0)}$ to be the I-dimensional zero vector and $x^{(0)}$ to be the J-dimensional zero vector. Then, for $k \geq 0$, we set

$$u^{(k+1)} = u^{(k)} + c^{(k)}e_{i_k},$$

$$x^{(k+1)} = x^{(k)} + rc^{(k)}r_{i_k}, \tag{26.17}$$

where e_i is an I-dimensional vector whose ith component is 1 with all other components being 0 and

$$c^{(k)} = \lambda^{(k)}\frac{r\left(y_{i_k} - \langle r_{i_k}, x^{(k)}\rangle\right) - u_{i_k}^{(k)}}{1 + r^2\|r_{i_k}\|^2}, \tag{26.18}$$

with $i_k = [k(\mathrm{mod}\,I) + 1]$.

THEOREM 26.1   (*see [1] for a proof*). *Let y be any measurement vector, r be any real number, and μ_x be any element of $\mathbf{R}^J$. Then for any real numbers $\lambda^{(k)}$ satisfying*

$$0 < \varepsilon_1 \leq \lambda^{(k)} \leq \varepsilon_2 < 2, \tag{26.19}$$

the sequence $x^{(0)}, x^{(1)}, x^{(2)}, \dots$ determined by the algorithm given above converges to the unique vector x which minimizes Eq. (26.16).

The implementation of this algorithm is hardly more complicated than that of basic ART which is described at the beginning of this subsection. We need an additional sequence of I-dimensional vectors $u^{(k)}$, but in the kth iterative step only one component of $u^{(k)}$ is needed or altered. Since the i_ks are defined in a cyclic order, the components of the vector $u^{(k)}$ (just as the components of the measurement vector y) can be sequentially accessed. (The exact choice of this — often referred to as the *data access ordering* — is very important for fast initial convergence; it is described in [11]. The underlying principle is that in any subsequence of steps, we wish to have the individual actions to be as independent as possible.) We also use, for every integer $k \geq 0$, a positive real number $\lambda^{(k)}$. (These are the so-called *relaxation parameters*. They are free parameters of the algorithm and in practice need to be optimized [11].) The r_i are usually not stored at all, but the location and size of their nonzero elements are calculated as and when needed. Hence, the algorithm described by Eq. (26.17) and Eq. (26.18) shares the storage efficient nature of basic ART and its computational requirements are essentially the same. Assuming, as is reasonable, that the number of nonzero $r_{i,j}$ is of the same order as J, we see that the cost of cycling through the data once using ART is of the order IJ, which is approximately the same as the cost of reconstructing using FBP. (That this is indeed so is confirmed by the timings reported in [12].) An important thing to note about Theorem 26.1 is that there are no restrictions of consistency in its statement. Hence, the algorithm of Eqs. (26.17) and (26.18) will converge to the minimizer of Eq. (26.16) — the so-called *regularized least squares solution* — using the real data collected in any application.

26.8 Expectation Maximization (EM)

We may wish to find the x that maximizes the likelihood of observing the actual measurements, based on the assumption that ith measurement comes from a Poisson distribution whose mean is given by Eq. (26.15). An iterative method to do exactly that, based on the so-called EM (*expectation maximization*) approach, was proposed in [13]. Here we discuss a variant of this approach that was designed for a somewhat more complicated optimization criterion [14], which enforces smoothness of the results where the original maximum likelihood criterion may result in noisy images.

Let $\mathbf{R}_+^J$ denote those elements of $\mathbf{R}^J$ in which all components are non-negative. Our task is to find the x in $\mathbf{R}_+^J$ which **minimizes**

$$\sum_{i=1}^{I} [\langle r_i, x \rangle - y_i \ln\langle r_i, x \rangle] + \frac{\gamma}{2} x^T S x, \tag{26.20}$$

where the $J \times J$ matrix S (with entries denoted by $s_{j,u}$) is a *modified smoothing matrix* [1] which has the following property. (This definition is only applicable if we use pixels to define the basis functions.) Let N denote the set of indexes corresponding to pixels that are not on the border of the digitization. Each such pixel has eight neighbors, let N_j denote the indexes of the pixels associated with the neighbors of the pixel indexed by j. Then

$$x^T S x = \sum_{j \in N} \left(x_j - \frac{1}{8} \sum_{k \in N_j} x_k \right)^2. \tag{26.21}$$

Consider the following rules for obtaining $x^{(k+1)}$ from $x^{(k)}$.

$$p_j^{(k)} = \frac{\sum_{i=1}^{I} r_{i,j}}{9\gamma s_{j,j}} - x_j^{(k)} + \frac{1}{9 s_{j,j}} \sum_{u=1}^{J} s_{j,u} x_u^{(k)}, \tag{26.22}$$

$$q_j^{(k)} = \frac{x_j^{(k)}}{9\gamma s_{j,j}} \sum_{i=1}^{I} \frac{r_{i,j} y_i}{\langle r_i, x^{(k)} \rangle}, \tag{26.23}$$

$$x_j^{(k+1)} = \frac{1}{2} \left(-p_j^{(k)} + \sqrt{\left(p_j^{(k)} \right)^2 + 4 q_j^{(k)}} \right). \tag{26.24}$$

Since the first term of Eq. (26.22) can be precalculated, the execution of Eq. (26.22) requires essentially no more effort than multiplying $x^{(k)}$ with the modified smoothing matrix. As explained in [1], there is a very efficient way of doing this. The execution of Eq. (26.23) requires approximately the same effort as cycling once through the data set using ART; see Eq. (26.18). Algorithmic details of efficient computations of Eq. (26.23) appeared in [15]. Clearly, the execution of Eq. (26.24) requires a trivial amount of computing. Thus, we see that one iterative step of the EM algorithm of Eq. (26.22) to Eq. (26.24) requires, in total, approximately the same computing effort as cycling through the data set once with ART, which costs about the same as a complete reconstruction by FBP. A basic difference between the ART method and the EM method is that the former updates its estimate based on one measurement at a time, while the latter deals with all measurements simultaneously.

THEOREM 26.2 (*see [14] for a proof*). *For any $x^{(0)}$ with only positive components, the sequence $x^{(0)}, x^{(1)}, x^{(2)}, \ldots$ generated by the algorithm of Eqs. (26.22) to (26.24) converges to the minimizer of (26.20) in $\mathbf{R}_+^J$.*

26.9   Comparison of the Performance of Algorithms

We have discussed four very different-looking algorithms and the literature is full of many others, only some of which are surveyed in books such as [1]. Many of the algorithms are available in general purpose image reconstruction software packages, such as SNARK [10]. The novice faced with a problem of reconstruction is justified in being puzzled as to which algorithm to use. Unfortunately, there is not a generally valid answer: the right choice may very well be dependent on the area of application and on the instrument used for gathering the data. Here we make only some general comments regarding the four approaches discussed above, followed by some discussion of the methodologies that are available to comparatively evaluate reconstruction algorithms for a particular application.

Concerning the two transform methods we have discussed, the linogram method is faster than FBP (essentially an $N^2 \log N$ method, rather than an N^3 method as is the FBP) and, when the data are collected according to the geometry expressed by Eqs. (26.8) and (26.9), then the linogram method is likely to be more accurate because it requires no interpolations. However, data are not normally collected this way and the need for an initial interpolation together with the more complicated-looking expressions that need to be implemented for the linogram method may indeed steer some users towards FBP, in spite of its extra computational requirements.

Advantages of series expansion methods over transform methods are their flexibility (no special relationship needs to be assumed between the object to be reconstructed and the measurements taken, such as that the latter are samples of the Radon transform of the former) and the ability to control the type of solution we want by specifying the exact sense in which the image vector is to be estimated from the measurement vector; see Eqs. (26.16) and (26.20). The major disadvantage is that it is computationally much more intensive to find these precise estimators than to numerically evaluate Eq. (26.1). Also, if the model (the basis functions, the projection matrix, and the estimation criterion) is not well chosen, then the resulting estimate may be inferior to that provided by a transform method. The recent literature has demonstrated that usually there are models that make the efficacy of a reconstruction provided by a series expansion method at least as good as that provided by a transform method. To avoid the problem of computational expense, one usually stops the iterative process involved in the optimization long before the method has converged to the mathematically specified estimator. Practical experience indicates that this can be done very efficaciously. For example, as reported in [12], in the area of fully three-dimensional PET, the reconstruction times for FBP are slightly longer than for cycling through the data just once with a version of ART using spherically symmetric basis functions and the accuracy of FBP is significantly worse than what is obtained by this very early iterate produced by ART.

Since the iterative process is, in practice, stopped early, in evaluating the efficacy of the result of a series expansion method one should look at the actual outputs rather than the ideal mathematical optimizer. Reported experiences comparing an optimized version of ART with an optimized version of EM [9, 11] indicate that the former can obtain as good or better reconstructions as the latter, but at a fraction of the computational cost. This computational advantage appears to be due to not trying to make use of all the measurements in each iterative step.

The proliferation of image reconstruction algorithms imposes a need to evaluate the relative performance of these algorithms and to understand the relationship between their attributes (free parameters) and their performance. In a specific application of an algorithm, choices have to be made regarding its parameters (such as the basis functions, the optimization criterion, constraints, relaxation, etc.). Such choices affect the performance of the algorithm and there is a need for an efficient and objective evaluation procedure which enables us to select the best variant of an algorithm for a particular task and to compare the efficacy of different algorithms for that task.

An approach to evaluating an algorithm is to first start with a specification of the task for which the image is to be used and then define a *figure of merit* (FOM) which determines quantitatively how helpful the image is, and hence the reconstruction algorithm, for performing that task. In the numerical observer approach [11, 16, 17], a task-specific FOM is *computed* for each image. Based on the FOMs for all the images produced by two different techniques, we can calculate the statistical significance at which we can reject the null hypothesis that the methods are equally helpful for solving a particular task in favor of the alternative hypothesis that the method with the higher average FOM is more helpful for solving that task. Different imaging techniques can then be rank-ordered on the basis of their average FOMs. It is strongly advised that a reconstruction algorithm should not be selected based on the appearance of a few sample reconstructions, but rather that a study along the lines indicated above be carried out.

In addition to the efficacy of images produced by the various algorithms, one should also be aware of the computational possibilities that exist for executing them. A survey from this point of view can be found in [2].

References

[1] Herman, G.T., *Image Reconstruction from Projections: The Fundamentals of Computerized Tomography,* Academic Press, New York, 1980.

[2] Herman, G.T., Image reconstruction from projections, *J. Real-Time Imag.,* 1, 3–18, 1995.

[3] Herman, G.T., Tuy, H.K., Langenberg, K.J., and Sabatier, P.C., *Basic Methods of Tomography and Inverse Problems,* Adam Hilger, Bristol, England, 1987.

[4] Sanz, J.L.C., Hinkle, E.B., and Jain, A.K., *Radon and Projection Transform-Based Computer Vision,* Springer-Verlag, Berlin, 1988.

[5] Edholm, P. and Herman, G.T., Linograms in image reconstruction from projections, *IEEE Trans. Med. Imag.,* 6, 301–307, 1987.

[6] Herman, G.T., Roberts, D., and Axel, L., Fully three-dimensional reconstruction from data collected on concentric cubes in Fourier space: implementation and a sample application to MRI, *Phys. Med. Biol.,* 37, 673–687, 1992.

[7] Edholm, P., Herman, G.T., and Roberts, D.A., Image reconstruction from linograms: implementation and evaluation, *IEEE Trans. Med. Imag.,* 7, 239–246, 1988.

[8] Lewitt, R.M., Alternatives to voxels for image representation in iterative reconstruction algorithms, *Phys. Med. Biol.,* 37, 705–716, 1992.

[9] Matej, S., Herman, G.T., Narayan, T.K., Furuie, S.S., Lewitt, R.M., and Kinahan, P., Evaluation of task-oriented performance of several fully 3–D PET reconstruction algorithms, *Phys. Med. Biol.,* 39, 355–367, 1994.

[10] Browne, J.A., Herman, G.T., and Odhner, D., SNARK93 — a programming system for image reconstruction from projections, Technical Report MIPG198, Department of Radiology, University of Pennsylvania, Philadelphia, 1993.

[11] Herman, G.T. and Meyer, L.B., Algebraic reconstruction techniques can be made computationally efficient, *IEEE Trans. Med. Imag.,* 12, 600–609, 1993.

[12] Matej, S. and Lewitt, R.M., Efficient 3D grids for image reconstruction using spherically symmetric volume elements, *IEEE Trans. Nucl. Sci.,* 42, 1361–1370, 1995.

[13] Shepp, L.A. and Vardi, Y., Maximum likelihood reconstruction in positron emission tomography, *IEEE Trans. Med. Imag.,* 1, 113–122, 1982.

[14] Herman, G.T., De Pierro, A.R., and Gai, N., On methods for maximum *a posteriori* image reconstruction with a normal prior, *J. Visual Comm. Image Represent.,* 3, 316–324, 1992.

[15] Herman, G.T., Odhner, D., Toennies, K.D., and Zenios, S.A., A parallelized algorithm for image reconstruction from noisy projections, in Coleman, T.F. and Li, Y. Eds., *Large-Scale Numerical Optimization,* SIAM, Philadelphia, PA, 1990, pp. 3–21.

[16] Hanson, K.M., Method of evaluating image-recovery algorithms based on task performance, *J. Opt. Soc. Am. A,* 7, 1294–1304, 1990.

[17] Furuie, S.S., Herman, G.T., Narayan, T.K., Kinahan, P., Karp, J.S., Lewitt, R.M., and Matej, S., A methodology for testing for statistically significant differences between fully 3-D PET reconstruction algorithms, *Phys. Med. Biol.,* 39, 341–354, 1994.

27

Robust Speech Processing as an Inverse Problem

Richard J. Mammone
Rutgers University

Xiaoyu Zhang
Rutgers University

27.1 Introduction

This section addresses the inverse problem in robust speech processing. A problem that speaker and speech recognition systems regularly encounter in the commercialized applications is the dramatic degradation of performance due to the mismatch of the training and operating environments. The mismatch generally results from the diversity of the operating environments. For applications over the telephone network, the operating environments may vary from offices and laboratories to household places and airports. The problem becomes worse when speech is transmitted over the wireless network. Here the system experiences cross-channel interferences in addition to the channel and noise degradations that exist in the regular telephone network. The key issue in robust speech processing is to obtain good performance regardless of the mismatch in the environmental conditions. The inverse problem in this sense refers to the process of modeling the mismatch in the form of a transformation and resolving it via an inverse transformation. In this section, we introduce the method of modeling the mismatch as an affine transformation.

Before getting into the details of the inverse problem in robust speech processing, we would like to give a brief review of the mechanism of speech production, as well as the retrieval of useful information from the speech for the recognition purposes.

0-8493-8572-5/98/$0.00+$.50
© 1998 by CRC Press LLC

27.2 Speech Production and Spectrum-Related Parameterization

The speech signal consists of time-varying acoustic waveforms produced as a result of acoustical excitation of the vocal tract. It is nonstationary in that the vocal tract configuration changes over time. A time-varying digital filter is generally used to describe the vocal tract characteristics. The steady-state system function of the filter is of the form [1, 2]:

$$S(z) = \frac{G}{1 - \sum_{i=1}^{p} a_i z^{-i}} = \frac{G}{\prod_{i=1}^{p} \left(1 - z_i z^{-1}\right)}, \tag{27.1}$$

where p is the order of the system and z_i denote the poles of the transfer function. The time domain representation of this filter is

$$s(n) = \sum_{i=1}^{p} a_i s(n - i) + Gu(n). \tag{27.2}$$

The speech sample $s(n)$ is predicted as a linear combination of previous p samples plus the excitation $Gu(n)$, where G is the gain factor. The factor G is generally ignored in the recognition-type tasks to allow for robustness to variations in the energy of speech signals. This speech production model is often referred to as the linear prediction (LP) model, or the autoregressive model, and the coefficients a_i are called the *predictor coefficients*.

The *cepstrum* of the speech signal $s(n)$ is defined as

$$c(n) = \int_{-\pi}^{\pi} \log \left| S\left(e^{j\omega}\right) \right| e^{j\omega n} \frac{d\omega}{2\pi}. \tag{27.3}$$

It is simply the inverse Fourier transform of the logarithm of the magnitude of the Fourier transform $S(e^{j\omega})$ of the signal $s(n)$.

From the definition of cepstrum in Eq. (27.3), we have

$$\sum_{n=-\infty}^{n=\infty} c(n) e^{-j\omega n} = \log \left| S\left(e^{j\omega}\right) \right| = \left| \log \frac{1}{1 - \sum_{n=1}^{p} a_n e^{-j\omega n}} \right|. \tag{27.4}$$

If we differentiate both sides of the equation with respect to ω and equate the coefficients of like powers of $e^{j\omega}$, the following recursion is obtained:

$$c(n) = \begin{cases} \log G & n = 0 \\ a(n) + \frac{1}{n} \sum_{i=1}^{n-1} i c(i) a(n - i) & n > 0 \end{cases} \tag{27.5}$$

The cepstral coefficients can be calculated using the recursion once the predictor coefficients are solved. The *zeroth* order cepstral coefficient is generally ignored in speech and speaker recognition due to its sensitivity to the gain factor, G.

An alternative solution for the cepstral coefficients is given by

$$c(n) = \frac{1}{q} \sum_{i=1}^{p} z_i^n. \tag{27.6}$$

It is obtained by equating the terms of like powers of z^{-1} in the following equation:

$$\sum_{n=-\infty}^{n=\infty} c(n) z^{-n} = \log \frac{1}{\prod_{n=1}^{p} \left(1 - z_n z^{-1}\right)} = -\sum_{i=1}^{p} \log \left[1 - z_n z^{-1}\right], \tag{27.7}$$

where the logarithm terms can be written as a power series expansion given as

$$\log\left[1 - z_n z^{-1}\right] = \sum_{k=1}^{\infty} \frac{1}{k} z_n^k z^{-k} . \tag{27.8}$$

There are two standard methods of solving for the predictor coefficients, a_i, namely, the *autocorrelation* method and the *covariance* method [3, 4, 5, 6]. Both approaches are based on minimizing the mean square value of the estimation error $e(n)$ as given by

$$e(n) = s(n) - \sum_{i=1}^{p} a_i s(n - i) . \tag{27.9}$$

The two methods differ with respect to the details of numerical implementation. The autocorrelation method assumes that the speech samples are zero outside the processing interval of N samples. This results in a nonzero prediction error, $e(n)$, outside the interval. The covariance method fixes the interval over which the prediction error is computed and has no constraints on the sample values outside the interval. The autocorrelation method is computationally simpler than the covariance approach and assures a stable system where all poles of the transfer function lie within the unit circle. A brief description of the autocorrelation method is given as follows.

The autocorrelation of the signal $s(n)$ is defined as

$$r_s(k) = \sum_{n=0}^{N-1-k} s(n)s(n + k) = s(n) \otimes s(-n) , \tag{27.10}$$

where N is the number of samples in the sequence $s(n)$ and the sign $\otimes$ denotes the convolution operation. The definition of autocorrelation implies that $r_s(k)$ is an even function. The predictor coefficients a_i can therefore be obtained by solving the following set of equations

$$\begin{pmatrix} r_s(0) & r_s(1) & \cdots & r_s(p-1) \\ r_s(1) & r_s(0) & \cdots & r_s(p-2) \\ \vdots & \vdots & \ddots & \vdots \\ r_s(p-1) & r_s(p-2) & \cdots & r_s(0) \end{pmatrix} \begin{pmatrix} a_1 \\ \vdots \\ a_p \end{pmatrix} = \begin{pmatrix} r_s(1) \\ \vdots \\ r_s(p) \end{pmatrix} .$$

Denoting the $p \times p$ Toeplitz autocorrelation matrix on the left hand side by $\mathbf{R}_s$, the predictor coefficient vector by $\mathbf{a}$, and the autocorrelation coefficients by $\mathbf{r}_s$, we have

$$\mathbf{R}_s \mathbf{a} = \mathbf{r}_s . \tag{27.11}$$

The solution for the predictor coefficient vector $\mathbf{a}$ can be solved by the inverse relation

$$\mathbf{a} = \mathbf{R}_s^{-1} \mathbf{r}_s .$$

This equation will be used throughout the analysis in the rest of this article. Since the matrix $\mathbf{R}_s$ is Toeplitz, a computationally efficient algorithm known as Levinson-Durbin recursion can be used to solve for $\mathbf{a}$ [3].

27.3 Template-Based Speech Processing

The template-based matching algorithms for speech processing are generally conducted using the similarity of the vocal tract characteristics inhabited in the spectrum of a particular speech sound. There are two types of speech sounds, namely, *voiced* and *unvoiced* sounds. Figure 27.1 shows the speech waveforms, the spectra, and the spectral envelopes of the voiced and the unvoiced sounds. Voiced sounds such as the vowel /a/ and the nasal sound /n/ are produced by the passage of a quasi-periodic air wave through the vocal tract that creates resonances in the speech waveforms known as *formants*. The quasi-periodic air wave is generated as a result of the vibration of the vocal cord. The fundamental frequency of the vibration is known as the *pitch*. In the case of generating fricative sounds such as /sh/, the vocal tract is excited by random noise, resulting in speech waveforms exhibiting no periodicity, as can be seen in Fig. 27.1. Therefore, the spectral envelopes of voiced sounds constantly exhibit the pitch as well as three to five formants when the sampling rate is 8 kHz, whereas the spectral envelopes of the unvoiced sounds reveal no pitch and formant characteristics. In addition, the formants of different voiced sounds differ with respect to the shape and the location of the center frequencies of the formants. This is due to the unique shape of the vocal tract formed to produce a particular sound. Thus, different sounds can be distinguished based on attributes of the spectral envelope.

The cepstral distance given by

$$d = \sum_{n=-\infty}^{\infty} \left[c(n) - c'(n) \right]^2 \tag{27.12}$$

is one of the metrics for measuring the similarity of two spectra envelopes. The reason is as follows. From the definition of cepstrum, we have

$$\sum_{n=\infty}^{\infty} \left[c(n) - c'(n) \right] e^{j\omega n} = \log |S \left(e^{j\omega} \right)| - \log |S' \left(e^{j\omega} \right)|$$

$$= \log \frac{|S \left(e^{j\omega} \right)|}{|S' \left(e^{j\omega} \right)|} . \tag{27.13}$$

The Fourier transform of the difference between a pair of cepstra is equal to the difference between the corresponding spectra pair. By applying the Parseval's theorem, the cepstral distance can be related to the log spectral distance as

$$d = \sum_{n=\infty}^{\infty} \left[c(n) - c'(n) \right]^2 = \int_{-\pi}^{\pi} \left[\log |S \left(e^{j\omega} \right)| - \log |S' \left(e^{j\omega} \right)| \right]^2 \frac{d\omega}{2\pi} . \tag{27.14}$$

The cepstral distance is usually approximated by the distance between the first few lower order cepstral coefficients, the reason being that the magnitude of the high order cepstral coefficients is small and has a negligible contribution to the cepstral distance.

27.4 Robust Speech Processing

Robust speech processing attempts to maintain the performance of speaker and speech recognition system when variations in the operating environment are encountered. This can be accomplished if the similarity in vocal tract structures of the same sound can be recovered under adverse conditions.

Figure 27.2 illustrates how the deterministic channel and random noise contaminate a speech signal during the recording and transmission of the signal.

First of all, at the front end of the speech acquisition system, additive background noise $N_1(\omega)$ from the speaking environment distorts the speech waveform. Adverse background conditions are also found to put stress on the speech production system and change the characteristics of the vocal tract. It is equivalent to performing a linear filtering of the speech. This problem will be addressed in another chapter and will not be discussed here.

After being sampled and quantized, the speech samples corrupted by the background noise $N_1(\omega)$ are then passed through the transmission channel such as a telephone network to get to the receiver's

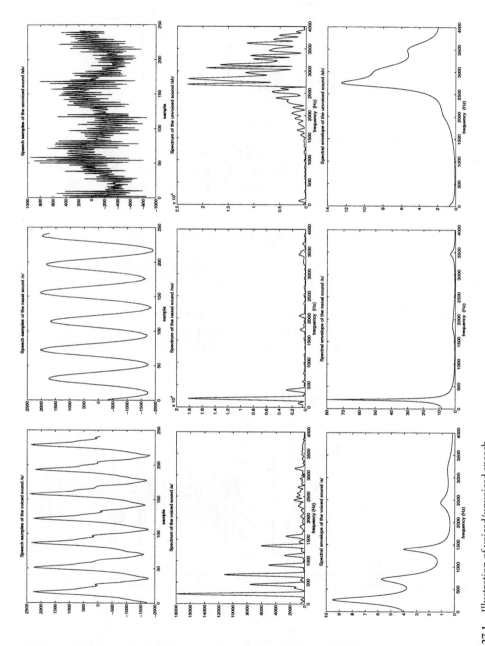

FIGURE 27.1 Illustration of voiced/unvoiced speech.

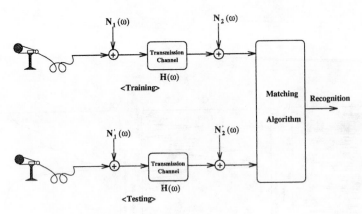

FIGURE 27.2   The speech acquisition system.

site. The transmission channel generally involves two types of degradation sources: the deterministic and convolutional filter with the transfer function $H(\omega)$, and the additive noise denoted by $N_2(\omega)$ in Fig. 27.2.

The signal observed at the output of the system is, therefore,

$$Y(\omega) = H(\omega)\left[X(\omega) + N_1(\omega)\right] + N_2(\omega).\tag{27.15}$$

The spectrum of the output signal is corrupted by both additive and multiplicative interferences. The multiplicative interference due to the linear channel $H(\omega)$ is sometimes referred to as the multiplicative noise.

The various sources of degradation cause distortions of the predictor coefficients and the cepstral coefficients. Fig. 27.4 shows the change of spatial clustering of the cepstral coefficients due to interferences of the linear channel, white noise, and the composite effect of both linear channel and white noise.

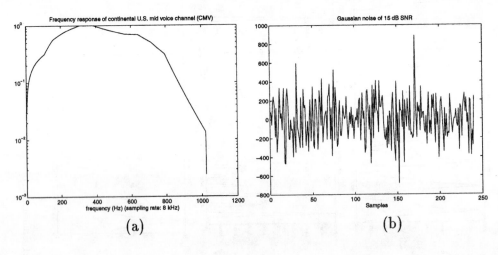

FIGURE 27.3   The simulated environmental interference. (a) Medium voiced channel and (b) Gaussian white noise.

- When the speech is interfered by a linear bandpass channel, the frequency response of which is shown in Fig. 27.3, a translation of the cepstral clusters is observed, as shown in Fig. 27.4(b).

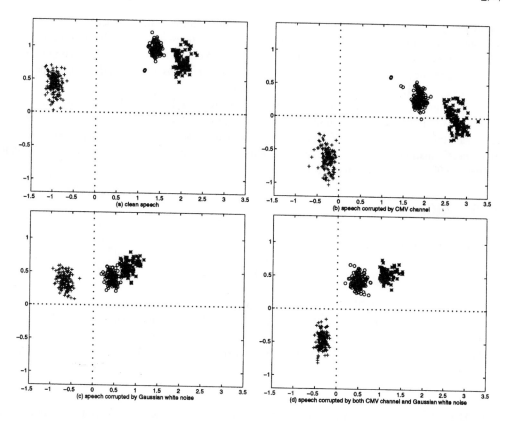

FIGURE 27.4 The spatial distribution of cepstral coefficients under various conditions, "∗" for the sound /a/, "o" for the sound /n/, and "+" for the sound /sh/. (a) Cepstrum of the clean speech; (b) cepstrum of signals filtered by continental U.S. mid-voice channel (CMV); (c) cepstrum of signals with 15 dB SNR, the noise type is additive white Gaussian (AWG); (d) cepstrum of speech corrupted by both CMV channel and AWG noise of 15 dB SNR.

- When the speech is corrupted by Gaussian white noise of 15 dB SNR, a shrinkage of the cepstral vectors results. This is shown in Fig. 27.4(c), where it can be seen that the cepstral clusters move toward the origin.
- When the speech is degraded by both the linear channel and Gaussian white noise, the cepstral vectors are translated and scaled simultaneously.

There are three underlying thoughts behind the various solutions to robust speech processing. The first is to recover the speech signal from the noisy observation by removing an estimate of the noise from the signal. This is also known as the speech enhancement approach. Methods that are executed in the speech sample domain include noise suppression [7] and noise masking [8]. Other speech enhancement methods are carried out in the feature domain, for example, cepstral mean subtraction (CMS) and pole-filtered cepstral mean subtraction (PFCMS). In this category, the key to the problem is to find feature sets that are invariant[1] to the changes of transmission channel and environmental noise. Lifted cepstrum [9] and the adaptive component weighted (ACW) cepstrum [10] are examples of the feature enhancement approach.

---

[1]In practice, it is difficult to find a set of features invariant to the environmental changes. The robust features currently used are mostly less sensitive to environmental changes.

A third category consists of methods for matching the testing features with the models after adaptation of environmental conditions [11, 12, 13, 14]. In this case, the presence of noise in the training and testing environments are tolerable as long as an adaptation algorithm can be found to match the conditions. The adaptations can be performed in either of the following two directions, i.e., adapt the training data to the testing environment, or adapt the testing data to the environment.

The focus of the following discussion will be on viewing the robust speech processing as an inverse problem. We utilize the fact that both deterministic and non-deterministic noise introduce a sound-dependent linear transformation of the predictor coefficients of speech. This can be approximated by an affine transformation in the cepstrum domain. The mismatch can, therefore, be resolved by solving for the inverse affine transformation of the cepstral coefficients.

27.5 Affine Transform

An affine transform $\mathbf{y}$ of a vector $\mathbf{x}$ is defined as

$$\mathbf{y} = \mathbf{A}\mathbf{x} + \mathbf{b}, \quad \text{for } \mathbf{b} \neq \mathbf{0} . \tag{27.16}$$

The matrix, $\mathbf{A}$, represents the linear transformation of the vector, $\mathbf{x}$, and $\mathbf{b}$ is a nonzero vector representing the translation of the vector. Note that the addition of the vector $\mathbf{b}$ to the equation causes the transform to become nonlinear.

The singular value decomposition (SVD) of the matrix, A, can be used to gain some insight into the geometry of an affine transform, i.e.,

$$\mathbf{y} = \mathbf{U}\mathbf{\Sigma}\mathbf{V}^T\mathbf{x} + \mathbf{b} , \tag{27.17}$$

where $\mathbf{U}$ and $\mathbf{V}^T$ are unitary matrices and Σ is a diagonal matrix . The geometric interpretation is thus seen to be that x is rotated by unitary matrix $\mathbf{V}^T$, rescaled by the diagonal matrix Σ, rotated again by the unitary matrix $\mathbf{U}$, and finally translated by the vector $\mathbf{b}$.

27.6 Transformation of Predictor Coefficients

It will be proved in this section that the contamination of a speech signal by a stationary convolutional channel and random white noise is equivalent to a signal dependent linear transformation of the predictor coefficients. The conclusion drawn here will be used in the next section to show that the effect of environmental interference is equivalent to an affine transform in the cepstrum domain.

27.6.1 Deterministic Convolutional Channel as a Linear Transform

When a sample sequence is passed through a convolutional channel of impulse response $h(n)$, the filtered signal $s'(n)$ obtained at the output of the channel is

$$s'(n) = h(n) \otimes s(n) . \tag{27.18}$$

If the power spectra of the signals $s(n)$ and $s'(n)$ are denoted $S_s(\omega)$, and $S_{s'}(\omega)$, respectively, then

$$S_{s'}(\omega) = |H(\omega)|^2 S_s(\omega) . \tag{27.19}$$

Therefore, in the time domain,

$$r_{s'}(k) = [h(n) \otimes h(-n)] \otimes r_s(k) = r_h(k) \otimes r_s(k) , \tag{27.20}$$

where $r_s(k)$ and $r_{s'}(k)$ are the autocorrelation of the input and output signals. The autocorrelation of the impulse response $h(n)$ is denoted $r_h(k)$ and by definition,

$$r_h(k) = h(n) \otimes h(-n) . \tag{27.21}$$

If the impulse response $h(n)$ is assumed to be zero outside the interval $[0, p-1]$, then

$$r_h(k) = 0 \text{ for } |k| > p - 1 . \tag{27.22}$$

Equation (27.20) can therefore be rewritten in matrix form as

$$
\begin{pmatrix}
r_{s'}(0) \\
r_{s'}(1) \\
\vdots \\
r_{s'}(p-1)
\end{pmatrix}
=
\begin{pmatrix}
r_h(0) & r_h(1) & r_h(2) & \cdots & r_h(p-1) \\
r_h(1) & r_h(0) & r_h(1) & \cdots & r_h(p-2) \\
\vdots & \vdots & & \ddots & \vdots \\
r_h(p-1) & r_h(p-2) & r_h(p-3) & \cdots & r_h(0)
\end{pmatrix}
$$

$$
\times
\begin{pmatrix}
r_s(0) \\
r_s(1) \\
\vdots \\
r_s(p-1)
\end{pmatrix}
$$

$$= \mathbf{R}_{h1} \mathbf{r}_s . \tag{27.23}$$

$\mathbf{R}_{h1}$ refers to the autocorrelation matrix of the impulse response of the channel on the right-hand side of the above equation.

The autocorrelation matrix $\mathbf{R}_{s'}$ of the filtered signal $s'(n)$ is then

$$
\mathbf{R}_{s'} =
\begin{pmatrix}
r_{s'}(0) & r_{s'}(1) & r_{s'}(2) & \cdots & r_{s'}(p-1) \\
r_{s'}(1) & r_{s'}(0) & r_{s'}(1) & \cdots & r_{s'}(p-2) \\
\vdots & \vdots & & \ddots & \vdots \\
r_{s'}(p-1) & r_{s'}(p-2) & r_{s'}(p-3) & \cdots & r_{s'}(0)
\end{pmatrix}
$$

$$
=
\begin{pmatrix}
r_h(0) & r_h(1) & r_h(2) & \cdots & r_h(p-1) \\
r_h(1) & r_h(0) & r_h(1) & \cdots & r_h(p-2) \\
\vdots & \vdots & & \ddots & \vdots \\
r_h(p-1) & r_h(p-2) & r_h(p-3) & \cdots & r_h(0)
\end{pmatrix}
$$

$$
\times
\begin{pmatrix}
r_s(0) & r_s(1) & r_s(2) & \cdots & r_s(p-1) \\
r_s(1) & r_s(0) & r_s(1) & \cdots & r_s(p-2) \\
\vdots & \vdots & & \ddots & \vdots \\
r_s(p-1) & r_s(p-2) & r_s(p-3) & \cdots & r_s(0)
\end{pmatrix}
$$

$$= \mathbf{R}_{h1} \mathbf{R}_s . \tag{27.24}$$

Also, the autocorrelation vector $\mathbf{r}_{s'}$ of the filtered signal $s'(n)$ is

$$
\mathbf{r}_{s'} =
\begin{pmatrix}
r_{s'}(1) \\
r_{s'}(2) \\
\vdots \\
r_{s'}(p)
\end{pmatrix}
$$

$$
=
\begin{pmatrix}
r_h(1) & r_h(0) & r_h(1) & \cdots & r_h(p-2) \\
r_h(2) & r_h(1) & r_h(0) & \cdots & r_h(p-3) \\
\vdots & \vdots & & \ddots & \vdots \\
r_h(p) & r_h(p-1) & r_h(p-2) & \cdots & r_h(1)
\end{pmatrix}
$$

$$\times \begin{pmatrix} r_s(1) \\ r_s(2) \\ \vdots \\ r_s(p) \end{pmatrix}$$

$$= \mathbf{R}_{h2}\mathbf{r}_s , \tag{27.25}$$

where $\mathbf{R}_{h2}$ denotes the matrix on the right-hand side.

The predictor coefficients of the output signal $s'(n)$ is thus given by

$$\mathbf{a}_{s'} = \mathbf{R}_{s'}^{-1}\mathbf{r}_{s'} = (\mathbf{R}_{h1}\mathbf{R}_s)^{-1} \times (\mathbf{R}_{h2}\mathbf{r}_s) = \mathbf{R}_s^{-1}\left(\mathbf{R}_{h1}^{-1}\mathbf{R}_{h2}\right)\mathbf{R}_s\mathbf{a} . \tag{27.26}$$

Therefore, the predictor coefficients of a speech signal filtered by a convolutional channel can be obtained via taking a linear transformation of the predictor coefficients of the input speech signal. Note that the transformation in Eq. (27.26) is sound dependent, as the estimates of the autocorrelation matrices assume stationary.

27.6.2 Additive Noise as a Linear Transform

The random noise arising from the background and the fluctuation of the transmission channel is generally assumed to be additive white noise (AWN). The resulted noisy observation of the original speech signal is given by

$$s'(n) = s(n) + e(n) , \tag{27.27}$$

where

$$E[e(n)] = 0 \quad \text{and} \quad E\left[e^2(n)\right] = \sigma^2 , \tag{27.28}$$

and $s'(n)$ results from the original speech signal $s(n)$ being corrupted by the noise $e(n)$.

The autocorrelation of the corrupted speech signal $s'(n)$ is

$$r_{s'}(k) = [s(n) + e(n)] \otimes [s(-n) + e(-n)] = r_s(k) + r_{se}(k) + r_{es}(k) + r_e(k) , \tag{27.29}$$

where $r_s(k)$ and $r_e(k)$ denote the autocorrelation of the signal $s(n)$ and the noise $e(n)$, respectively, and $r_{se}(k)$ and $r_{es}(k)$ represent the cross-correlation of $s(n)$ and $e(n)$. Since

$$\begin{aligned}
r_e(k) &= E\left[\sum_{m=0}^{N-1-k} e(m)e(m+k)\right] = \begin{cases} \sigma^2 & k = 0 \\ 0 & \text{otherwise} \end{cases} \\
r_{se}(k) &= E\left[\sum_{m=0}^{N-1-k} s(m)e(m+k)\right] = \sum_{m=0}^{N-1-k} s(m)E[e(m+k)] = 0, \quad \text{and} \\
r_{es}(k) &= E\left[\sum_{m=0}^{N-1-k} e(m)s(m+k)\right] = \sum_{m=0}^{N-1-k} s(m+k)E[e(m)] = 0 ,
\end{aligned} \tag{27.30}$$

the autocorrelation of the signal $s'(n)$ presented in Eq. (27.29) becomes

$$r_{s'}(k) = \begin{cases} r_s(k) + \sigma^2 & k = 0 \\ r_s(k) & \text{otherwise} \end{cases} . \tag{27.31}$$

Hence, the predictor coefficients as given by Eq. (27.11) are

$$
\begin{aligned}
\mathbf{a}' &= \mathbf{R}_{s'}^{-1}\mathbf{r}_{s'} \\[4pt]
&= \begin{pmatrix}
r_s(0)+\sigma^2 & r_s(1) & r_s(2) & \cdots & r_s(p-1) \\
r_s(1) & r_s(0)+\sigma^2 & r_s(1) & \cdots & r_s(p-2) \\
\vdots & \vdots & & \ddots & \vdots \\
r_s(p-1) & r_s(p-2) & r_s(p-3) & \cdots & r_s(0)+\sigma
\end{pmatrix}^{-1}
\begin{pmatrix}
r_s(1) \\
r_s(2) \\
\vdots \\
r_s(p)
\end{pmatrix} \\[4pt]
&= \left(\mathbf{R}_s+\sigma^2\mathbf{I}\right)^{-1}\mathbf{r}_s = \left(\mathbf{R}_s+\sigma^2\mathbf{I}\right)^{-1}\mathbf{R}_s\mathbf{a}\,.
\end{aligned}
\tag{27.32}
$$

It can be seen from Eq. (27.32) that the addition of AWN noise to the speech is also equivalent to taking a linear transformation of the predictor coefficients. The linear transformation depends on the autocorrelation of the speech and thus in a spectrum-based model, all the spectrally similar predictors will be mapped by a similar linear transform.

The singular value decomposition (SVD) will gain us some insight into what the transformation in Eq. (27.32) actually does. Assume that the Toeplitz autocorrelation matrix of the original speech signal $\mathbf{R}_s$ is decomposed as

$$
\mathbf{R}_s = \mathbf{U}\Lambda\mathbf{U}^T\,,
\tag{27.33}
$$

where $\mathbf{U}$ is a unitary matrix and Λ is a diagonal matrix whose diagonal elements are the eigenvalues of the matrix $\mathbf{R}_s$. Then the autocorrelation matrix of the noise-corrupted signal is

$$
\mathbf{R}_s' = \mathbf{R}_s+\sigma^2\mathbf{I} = \mathbf{U}\left(\Lambda+\sigma^2 I\right)\mathbf{U}^T\,.
\tag{27.34}
$$

Therefore, Eq. (27.32) can be rewritten as

$$
\begin{aligned}
\mathbf{a}' &= \left(\mathbf{U}\left(\Lambda+\sigma^2 I\right)\mathbf{U}^T\right)^{-1}\left(\mathbf{U}\Lambda\mathbf{U}^T\right)\mathbf{a} = \mathbf{U}\left[\left(\Lambda+\sigma^2 I\right)^{-1}\Lambda\right]\mathbf{U}^T \\[4pt]
&= \mathbf{U}\begin{pmatrix}
\dfrac{\lambda_1^2}{\lambda_1^2+\sigma^2} & & & \\
& \dfrac{\lambda_2^2}{\lambda_2^2+\sigma^2} & & \\
& & \ddots & \\
& & & \dfrac{\lambda_n^2}{\lambda_n^2+\sigma^2}
\end{pmatrix}\mathbf{U}^T\,.
\end{aligned}
\tag{27.35}
$$

From the above equation we can see that the norm of the predictor coefficients is reduced when the speech is perturbed by white noise.

27.7 Affine Transform of Cepstral Coefficients

Most speaker and speech recognition systems use a spectrum-based similarity measure to group the vectors, which are normally the LP cepstrum vectors. Thus, we shall investigate the spectrum as to whether or not the cepstral vectors are affinely mapped.

Consider the cepstrum of a speech signal as defined by

$$
c_n = \mathcal{Z}^{-1}\left[\log\frac{1}{A(z)}\right],
\tag{27.36}
$$

where $\frac{1}{A(z)} = \frac{1}{1-\sum_{i=1}^{p} a_i z^{-i}}$ is the transfer function of the linear predictive system. Taking the first order partial derivative of c_n with respect to a_i yields

$$\frac{\partial c_n}{\partial a_i} = \frac{\partial \mathcal{Z}^{-1}\left[\log\left(\frac{1}{1-\sum_{i=1}^{p} a_i z^{-i}}\right)\right]}{\partial a_i} \tag{27.37}$$

$$= \mathcal{Z}^{-1}\left[\frac{\partial \log\left(\frac{1}{1-\sum_{i=1}^{p} a_i z^{-i}}\right)}{\partial a_i}\right] \tag{27.38}$$

$$= -h(n-i), \tag{27.39}$$

where $h(n-i)$ is the nth impulse response delayed by i taps. Therefore, if $\mathbf{c}$ is the vector of the first p cepstral coefficients of the clean speech $s(n)$, then

$$d\mathbf{c} = \mathbf{H}d\mathbf{a} \tag{27.40}$$

where

$$\mathbf{H} = -\begin{pmatrix} h(0) & 0 & \cdots & 0 \\ h(1) & h(0) & \cdots & 0 \\ \vdots & \vdots & \ddots & \vdots \\ h(p-1) & h(p-2) & \cdots & h(0) \end{pmatrix}. \tag{27.41}$$

Note that the impulse response matrix $\mathbf{H}$ would be the same for a group of spectrally similar cepstral vectors. The relationship between a degradation in the predictor coefficients and the corresponding degradation in the cepstral coefficients is given by Eq. (27.40). A degraded set of spectrally similar vectors would undergo the transformation

$$d\mathbf{c}' = \mathbf{H}'d\mathbf{a}', \tag{27.42}$$

where $\mathbf{c}'$ and $\mathbf{a}'$ are the degraded cepstrum and predictor coefficients, respectively. H' is a lower triangular matrix corresponding to the impulse response of the test signal. Since the predictor coefficients satisfy the linear relation $\mathbf{a}' = \mathbf{A}\mathbf{a}$, as shown in Eq. (27.26) and Eq. (27.32), differentiating both sides of the equation yields

$$d\mathbf{a}' = \mathbf{A}d\mathbf{a}. \tag{27.43}$$

If we integrate the above three equations, we have

$$\frac{d\mathbf{c}'}{d\mathbf{c}} = \left(\frac{d\mathbf{c}'}{d\mathbf{a}'}\right)\left(\frac{d\mathbf{a}'}{d\mathbf{a}}\right)\left(\frac{d\mathbf{a}}{d\mathbf{c}}\right) = \mathbf{H}'\mathbf{A}\mathbf{H}^{-1}. \tag{27.44}$$

The degraded cepstrum is then given by

$$\mathbf{c}' = \mathbf{H}'\mathbf{A}\mathbf{H}^{-1}\mathbf{c} + \mathbf{b}_c \tag{27.45}$$

In order to draw the conclusion that there exists an affine transform for the cepstral coefficients, all the variables on the right-hand side of the above equation must be expressed as an explicit function of the training data. However, this is not the case for the matrix $\mathbf{H}'$ in the equation. Since $\mathbf{H}'$ consists of the impulse response of the prediction model of the test data $h'(n)$, we need to represent the impulse response as a function of the training data.

Consider the cases of channel interferences and noise corruption, respectively.

- Assume the training data is of the form

$$s(n) = h_{ch1}(n) \otimes h_{sig}(n) \otimes e(n) = h_{ch1}(n) \otimes s_0(n) , \qquad (27.46)$$

where $e(n)$ represents the innovation sequence, $h_{sig}(n)$ is the impulse response of the all-pole model of the vocal tract, and $h_{ch1}(n)$ is the impulse response of the transmission channel. The convolution of the innovation sequence with the impulse response of the vocal tract yields the clean speech signal $s_0(n)$, the convolution of which with the transmission channel generates the observed sequence $s(n)$.
Similarly, the test data is

$$s'(n) = h_{ch2}(n) \otimes h_{sig}(n) \otimes e(n) = h_{ch2}(n) \otimes s_0(n) , \qquad (27.47)$$

where $h_{ch2}(n)$ is the impulse response of the transmission channel in the operating environment. In practice, the all-pole model is applied to the observation sequence involving channel interference rather than the clean speech signal. The estimated impulse response of the observed signal $h'(n)$ is actually given by

$$h'(n) = h_{ch2}(n) \otimes h_{sig}(n) . \qquad (27.48)$$

The matrix $\mathbf{H}'$ can therefore be written as

$$
\mathbf{H}' = \begin{pmatrix}
h'(0) & 0 & \cdots & 0 \\
h'(1) & h'(0) & \cdots & 0 \\
\vdots & \vdots & \ddots & \vdots \\
h'(p-1) & h'(p-2) & \cdots & h'(0)
\end{pmatrix}
$$

$$
= \begin{pmatrix}
h_{ch2}(0) & 0 & \cdots & 0 \\
h_{ch2}(1) & h_{ch2}(0) & \cdots & 0 \\
\vdots & \vdots & \ddots & \vdots \\
h_{ch2}(p-1) & h_{ch2}(p-2) & \cdots & h_{ch2}(0)
\end{pmatrix}
$$

$$
\begin{pmatrix}
h_{sig}(0) & 0 & \cdots & 0 \\
h_{sig}(1) & h_{sig}(0) & \cdots & 0 \\
\vdots & \vdots & \ddots & \vdots \\
h_{sig}(p-1) & h_{sig}(p-2) & \cdots & h_{sig}(0)
\end{pmatrix} \qquad (27.49)
$$

- When the speech is corrupted by additive noise, the autocorrelation matrix $\mathbf{R}'_s$ can also be written as

$$\mathbf{R}'_s = \mathbf{H}'\mathbf{H}'^T . \qquad (27.50)$$

Equating the right-hand side of Eqs. (27.34) and (27.50) yields

$$\mathbf{H}' = \mathbf{U}\left(\Lambda + \sigma^2\mathbf{I}\right)^{1/2} . \qquad (27.51)$$

where $\mathbf{H}'$ is an explicit function of the training data and the noise.

At this point, we can conclude that the cepstrum coefficients are affinely mapped by mismatches in the noise and channel conditions. Note again that the parameters of the affine mapping are spectrally dependent.

27.8 Parameters of Affine Transform

Assume the knowledge of the correspondence between the set of training cepstral vectors $\{\mathbf{c}_i = (c_{i1}, c_{i2}, \ldots, c_{iq})^T \mid i = 1, 2, \ldots, N\}$ and the set of testing cepstral vectors $\{\mathbf{c}'_i = (c'_{i1}, c'_{i2}, \ldots, c'_{iq})^T \mid i = 1, 2, \ldots, N\}$. Here N is the number of vectors in the vector set and q is the order of the cepstral coefficients. The affine transform holds for the corresponding vectors $\mathbf{c}_i$ and $\mathbf{c}'_i$ in the following way:

$$\mathbf{c}'^T_i = \mathbf{A}\mathbf{c}^T_i + \mathbf{b}$$

$$\Downarrow$$

$$\begin{pmatrix} c'_{i1} \\ \vdots \\ c'_{iq} \end{pmatrix} = \begin{pmatrix} \alpha_{11} & \cdots & \alpha_{1q} \\ \vdots & \ddots & \vdots \\ \alpha_{q1} & \cdots & \alpha_{qq} \end{pmatrix} \begin{pmatrix} c_{i1} \\ \vdots \\ c_{iq} \end{pmatrix} + \begin{pmatrix} b_1 \\ \vdots \\ b_q \end{pmatrix},$$

$$\text{for } i = 1, 2, \ldots, N. \tag{27.52}$$

The entries $\{\alpha_{ij}\}$ and $\{b_j\}$ can be solved in the row by row order since for the jth row of the matrix, i.e., $(\alpha_{j1}, \alpha_{j2}, \ldots, \alpha_{jq})$, there exists a set of equations given by

$$\begin{pmatrix} c'_{1j} \\ \vdots \\ c'_{Nj} \end{pmatrix} = \begin{pmatrix} c_{11} & \cdots & c_{1q} & 1 \\ \vdots & \ddots & \vdots & \vdots \\ c_{N1} & \cdots & c_{Nq} & 1 \end{pmatrix} \begin{pmatrix} \alpha_{j1} \\ \vdots \\ \alpha_{jq} \\ b_j \end{pmatrix},$$

$$\text{for } j = 1, 2, \ldots, q. \tag{27.53}$$

Denoting the vector on the left-hand side of the above equation by γ'_j, the matrix and the vector on the right-hand side by Γ and α_j, respectively, we have

$$\gamma'_j = \Gamma\alpha_j. \tag{27.54}$$

The least squares solution to the above systems of equation is

$$\alpha_j = \begin{pmatrix} \sum_{i=1}^N \mathbf{c}_i\mathbf{c}^T_i & \sum_{i=1}^N \mathbf{c}_i \\ \left(\sum_{i=1}^N \mathbf{c}_i\right)^T & N \end{pmatrix}^{-1} \times \begin{pmatrix} \mathbf{c}_1 & \cdots & \mathbf{c}_N \\ 1 & \cdots & 1 \end{pmatrix} \gamma'_j$$

$$\text{for } j = 1, \ldots, q, \tag{27.55}$$

where

$$\sum_{i=1}^N \mathbf{c}_i\mathbf{c}^T_i = \sum_{i=1}^N \begin{pmatrix} c_{i1} \\ c_{i2} \\ \vdots \\ c_{iq} \end{pmatrix} (c_{i1}, c_{i2}, \ldots, c_{iq}),$$

$$\text{for } i = 1, \ldots, N, \tag{27.56}$$

is the summation of a series of matrices.

The testing vectors can then be adapted to the model by an inverse affine transformation of the form

$$\hat{\mathbf{c}} = \mathbf{A}^{-1}(\mathbf{c}' - \mathbf{b}) \tag{27.57}$$

or vice versa. The adaptation removes the mismatch of environmental conditions due to channel and noise variability.

In the case that the matrix A is diagonal, i.e.,

$$A = \begin{pmatrix} \alpha_{11} & & \\ & \ddots & \\ & & \alpha_{qq} \end{pmatrix},$$ (27.58)

the solutions of α_{ij} in Eq. (27.55) can be simplified as

$$
\begin{aligned}
\alpha_{jj} &= \frac{\sum_{i=1}^{N} \gamma_{ij}'\gamma_{ij} - \sum_{i=1}^{N} \gamma_{ij}' \cdot \sum_{i=1}^{N} \gamma_{ij}/N}{\sum_{i=1}^{N} \gamma_{ij}^2 - \left(\sum_{i=1}^{N} \gamma_{ij}\right)^2/N} \\
&= \frac{E\left[\gamma_j', \gamma_j\right] - E\left[\gamma_j'\right] \cdot E\left[\gamma_j\right]}{E\left[\gamma_j, \gamma_j\right] - E^2\left[\gamma_j\right]} = \frac{COV\left[\gamma_j', \gamma_j\right]}{VAR\left[\gamma_j, \gamma_j\right]}
\end{aligned}
$$ (27.59)

and

$$b_j = \frac{1}{N}\left(\sum_{i=1}^{N} \gamma_{ij}' - \alpha_{jj}\sum_{i=1}^{N} \gamma_{ij}\right) = E\left[\gamma_j'\right] - \alpha_{jj}E\left[\gamma_j\right].$$ (27.60)

Here, $E[\,]$ is the expected value operator, and $VAR[\,]$ and $COV[\,]$ represent the variance and covariance operators, respectively.

As can be seen from Eq. (27.60), the diagonal entries α_{jj} are the weighted covariance of the model and the testing vector, and the value of b_j is equal to the weighted difference between the mean of the training vectors and that of the testing vectors. There are three cases of interest:

1. If the training and operating conditions are matched, then

$$E\left[\gamma_j'\right] = E\left[\gamma_j\right] \text{ and } COV\left[\gamma_j', \gamma_j\right] = VAR\left[\gamma_j, \gamma_j\right].$$ (27.61)

   Therefore,

$$\alpha_{jj} = 1 \text{ and } b_j = 0, \text{ for } j = 1, 2, \ldots, q \quad \Rightarrow \quad \hat{\mathbf{c}} = \mathbf{c}'.$$ (27.62)

   No adaptation is necessary in this case.

2. If the operating environment differs from the training environment due to convolutional distortions, then all the testing vectors are translated by a constant amount as given by

$$\mathbf{c}_i' = \mathbf{c}_i + \mathbf{c}^0,$$ (27.63)

   and

$$E\left[\gamma_j'\right] = E\left[\gamma_j\right] + c0 \text{ and } COV\left[\gamma_j', \gamma_j\right] = VAR\left[\gamma_j, \gamma_j\right].$$ (27.64)

   Therefore,

$$\alpha_{jj} = 1 \text{ and } b_j = c_j^0, \text{ for } j = 1, 2, \ldots, q \quad \Rightarrow \quad \hat{\mathbf{c}} = \mathbf{c}' - \mathbf{b}_c.$$ (27.65)

   This is equivalent to the method of cepstral mean subtraction (CMS) [6].

3. If the mismatch is caused by both channel and random noise, the testing vector is translated as well as shrunk. The shrinkage is measured by α_{jj} and the translation by b_j. The smaller the covariance of the model and the testing data, the greater the scaling of the testing vectors by noise.

The affine matching is similar to matching the *z scores* of the training and testing cepstral vectors. The z score of a set of vectors, c_i, is defined as

$$\mathbf{z}_{ci} = \frac{c_i - \mu_c}{\sigma_c} ,$$

(27.66)

where μ_c is the mean of the vectors, $\mathbf{c}_i$, and σ_c is the variance. Thus, we could form

$$\mathbf{z}_{c_i'} = \sigma_{c'} \left[\frac{\mathbf{c}_i - \mu_c}{\sigma_c} \right] + \mu_{c'} .$$

(27.67)

In the above analysis, we show that the cepstrum domain distortions due to channel and noise interference can be modeled as an affine transformation. The parameters of the affine transformation can be optimally estimated using the least squares method which yields the general result given by Eq. (27.55). In the special case of a similarity transform, we get the result given by Eq. (27.60).

27.9 Correspondence of Cepstral Vectors

While solving for the affine transform parameters, $\mathbf{A}$ and $\mathbf{b}$, we assume to have *a priori* knowledge of the correspondence between the cepstral vectors. A straightforward solution to finding the correspondence is to align the sound units in a speech utterance in terms of the time stamp of the sounds. However, this is generally not realizable in practice due to variations in the content of speech, the identity of a speaker, as well as the rate of speaking. For example, in a speaker recognition system, the text of the testing speech may not be the same as that of the training speech, resulting in a sequence of sounds in a completely different order than the training sequence. Furthermore, even if the text of the speech is the same, the speaking rate may change over time as well as speakers. The time stamp of a particular sound is still not sufficient for lining up corresponding sounds. A valuable solution to the correspondence problem [11] is to use the *expectation-maximization* (EM) algorithm, also known as the Baum-Welch algorithm [15].

The EM algorithm approaches the optimal solution to a system by repeating the procedure of (1) estimating a set of prespecified system parameters and (2) optimizing the system solution based on these parameters. The step of estimating the parameters is known as the expectation step, and the step of optimizing the solution is the maximization step. The second step is usually realized via the maximum-likelihood (ML) method.

With the EM algorithm, the parameters of the affine transform in Eq. (27.52) can be solved at the same time as the correspondence of the cepstral vectors are found. The method can be stated as follows.

- **Expectation**
  Solve for the parameters {$\mathbf{A}$, $\mathbf{b}$} using Eq. (27.55). The vector correspondence is found based on the optimization results obtained in the maximization step.

- **Maximization**
  Compute the *a posteriori* probability $P(\mathbf{c}_j^{ATC}|\mathbf{c}_i)$ and find the optimal matching by maximizing the *a posteriori* probability. This can be formulated as

$$k = \text{argmax}_i P \left(\mathbf{c}_j^{ATC}|\mathbf{c}_i \right) , \quad \text{for all } j .$$

(27.68)

Here, $\mathbf{c}_i^{ATC}$ represents the affine-transformed cepstrum that can be obtained by

$$\mathbf{c}_i^{ATC} = \mathbf{A}^{-1} \left(\mathbf{c}_i' - \mathbf{b}_c \right)$$

(27.69)

Therefore, we have a set of vector pairs denoted by $(\mathbf{c}_k, \mathbf{c}_j')$.

The definition of the *a posteriori* probability is dependent on the models employed by the classifier. In general, for the VQ-based classifiers, the *a posteriori* likelihood probability is defined as a Gaussian given by

$$P\left(\mathbf{c}_j^{ATC}|\mathbf{c}_i\right) = \frac{1}{\sqrt{2\pi}\,\Sigma^{1/2}}e^{\frac{1}{2}\left(\mathbf{c}_j^{ATC}-\mathbf{c}_i\right)^T\Sigma^{-1}\left(\mathbf{c}_j^{ATC}-\mathbf{c}_i\right)}, \qquad (27.70)$$

where Σ is the variance matrix. If we assume that every cepstral coefficient has a unit variance, namely, $\Sigma = \mathbf{I}$, where $\mathbf{I}$ is the identity matrix, then the maximization of the likelihood probability is equivalent to finding the cepstral vector in the VQ codebook that has minimum Euclidean distance to the affine-transformed vector $\mathbf{c}_j^{ATC}$.

References

[1] Flanagan, J.L., *Speech Analysis, Synthesis, and Perception,* Springer-Verlag, 1983.

[2] Fant, G., *Acoustic Theory of Speech Production,* Mouton and Co., Gravenhage, The Netherlands, 1960.

[3] Rabiner, L.R. and Schafer, R.W., *Digital Processing of Speech Signals,* Prentice-Hall, Englewood Cliffs, NJ, 1978.

[4] Atal, B.S., Effectiveness of linear prediction characteristics of the speech wave for automatic speaker identification and verification, *J. Acoust. Soc. Am.,* 55, 1304–1312, 1974.

[5] Atal, B.S., Automatic recognition of speakers from their voices, *Proc. IEEE,* 64, 460–475, April 1976.

[6] Furui, S., Cepstral analysis techniques for automatic speaker verification, *IEEE Trans. Acoust., Speech, Signal Processing,* 29, 254–272, April 1981.

[7] Boll, S.F., Suppression of acoustic noise in speech using spectral subtraction, *IEEE Trans. Acoust., Speech, Signal Processing,* 27, 113–120, April 1979.

[8] Klatt, D.H., A digital filter bank for spectral matching, *ICASSP,* 573–576, 1976.

[9] Juang, B.H., Rabiner, L.R., and Wilpon, J.G., On the use of bandpass liftering in speech recognition, *IEEE Trans. Acoust., Speech, Signal Processing,* 35, 947–954, July 1987.

[10] Assaleh, K.T. and Mammone, R.J., New 1p-derived features for speaker identification, *IEEE Trans. Speech, Audio Processing,* 2, 630–638, October 1994.

[11] Sankar, A. and Lee, C.H., Robust speech recognition based on stochastic matching, *ICASSP,* 1, 121–124, 1995.

[12] Newneyer, L. and Weintraub, M., Probabilistic optimum filtering for robust speech recognition, *Proc. IEEE Intl. Conf. Acoust., Speech, Signal Processing,* 1, 417–420, 1994.

[13] Nadas, A., Nahamoo D., and Picheny, M.A., Adaptive labeling: Normalization of speech by adaptive transformation based on vector quantization, *Proc. IEEE Intl. Conf. Acoust., Speech, Signal Processing,* 521–524, 1988.

[14] Gish, H., Ng, K., and Rohlicek, J.R., Robust mapping of noisy speech parameters for hmm word spotting, *ICASSP,* 2, 109–112, 1992.

[15] Baum, L.E., An inequality and associated maximization technique in statical estimation for probabilistic functions of Markov processes, *Inequalities,* 3, 1–8, 1972.

28

Inverse Problems, Statistical Mechanics and Simulated Annealing

K. Venkatesh Prasad
Ford Motor Company

28.1 Background

The focus of this chapter is on inverse problems — what they are, where they manifest themselves in the realm of digital signal processing (DSP), and how they might be "solved[1]." Inverse problems deal with estimating hidden *causes,* such as a set of transmitted symbols $\{t\}$, given observable *effects* such as a set of received symbols $\{r\}$ and a system (H) responsible for mapping $\{t\}$ into $\{r\}$. Inverse problems are succinctly stated using vector-space notation and take the form of estimating $t \in \mathcal{R}^M$, given:

$$r = Ht , \qquad (28.1)$$

where $r \in \mathcal{R}^N$ and $H \in \mathcal{R}^{M \times N}$ and $\mathcal{R}$ denotes the space of real numbers whose dimensions are specified in the superscript(s). Such problems call for the inversion of H, an operation which may or may not be numerically possible. We will shortly address these issues, but we should note here for completeness that these problems contrast with *direct problems* — where r is to be directly (without matrix inversion) estimated, given H and t.

---

[1] The quotes are used to stress that unique deterministic solutions might not exist for such problems and the observed effects might not continuously track the underlying causes. Formally speaking, this is a result of such problems of being *ill-posed* in the sense of Hadamard [1]. What is typically sought is an optimal solution, such as a minimum norm/minimum energy solution.

28.2   Inverse Problems in DSP

Inverse problems manifest themselves in a broad range of DSP applications in fields as diverse as digital astronomy, electronic communications, geophysics [2], medicine [3], and oceanography. The core of all these problems takes the form shown in Eq. (28.1). This, in fact, is the discrete version of the Fredholm integral equation of the first kind for which, by definition[2], the limits of integration are fixed and the unknown function **f** appears *only* inside the integral. To motivate our discussion, we will describe an application-specific problem, and in the process introduce some of the notations and concepts to be used in the later sections. The inverse problem in the field of electronic communications has to do with estimating **t**, given **r** which is often received with noise, commonly modeled to be additive white Gaussian (AWG) in nature. The communication system and the transmission channel are typically stochastically characterizable and are represented by a linear system matrix (**H**). The problem, therefore, is to solve for **t** in the system of linear equations:

$$\mathbf{r} = \mathbf{H}\mathbf{t} + \mathbf{n} , \qquad\qquad (28.2)$$

where vector **n** denotes AWG noise. Two tempting solutions might come to mind: if matrix **H** is invertible, i.e., $\mathbf{H}^{-1}$ exists, then why not solve for **t** as:

$$\mathbf{t} = \mathbf{H}^{-1}(\mathbf{r} - \mathbf{n}) , \qquad\qquad (28.3)$$

or else why not compute a *minimum-norm* solution such as the *pseudoinverse* solution:

$$\mathbf{t} = \mathbf{H}^{\dagger}(\mathbf{r} - \mathbf{n}) , \qquad\qquad (28.4)$$

where $\mathbf{H}^{\dagger}$ is referred to as the pseudoinverse [5] of **H** and is defined to be $[\mathbf{H}'\mathbf{H}]^{-1}\mathbf{H}'$, where **H**' denotes the transpose of **H**. There are several reasons why neither solution [Eqs. (28.3) or (28.4)] might be viable. One reason is that the dimensions of the system might be extremely large, placing a greater computational load than might be affordable. Another reason is that **H** is often numerically ill-conditioned, implying that inversions or pseudo-inversions might not be reliable even if otherwise reliable numerical inversion procedures, such as Gaussian elimination or singular value decomposition [6, 19], were to be employed. Furthermore, even if preconditioning [6] were possible on the system of linear equations $\mathbf{r} = \mathbf{H}\mathbf{t} + \mathbf{n}$, resulting in a numerical improvement of the coefficients of **H**, there is one even more overbearing hurdle that has often to be dealt with, and this has to do with the fact that such problems are frequently *ill-posed*. In practical terms[3] this means that small changes in the inputs might result in arbitrarily large changes in outputs. For all these reasons the most tempting solution-approaches are often ruled out. As we describe in the next section, inverse problems may be recast as combinatorial optimization problems. We will then show how combinatorial optimization problems may be solved using a powerful tool called *simulated annealing* [7] that has evolved from our understanding of statistical mechanics [8] and the simulation of the annealing (cooling) behavior of physical matter [9].

28.3   Analogies with Statistical Mechanics

Understanding the analogies of inverse problems in DSP to problems in statistical mechanics is valuable to us because we can then draw upon the analytical and computational tools developed over the past century to

---

[2]There exist two classes of integral equations ([4], pg. 865): if the limits of integration are fixed, the equations are referred to as Fredholm integral equations; if one of the limits is a variable the equations are referred to as Volterra integral equations. Further, if the unknown function appears only inside the integral the equation is called "first kind", but if it appears both inside and outside the integral the equation is called "second kind".

[3]For a more complete description see [1].

solve inverse problems in the field of statistical mechanics [8]. The broad analogy is that just as the received symbols **r** in Eq. (28.1) are the observed effects of hidden underlying causes (the transmitted symbols **t**) — the measured temperature and state (solid, liquid, or gaseous) of physical matter are the effects of underlying causes such as the momenta and velocities of the particles that compose the matter. A more specific analogy comes from the reasoning that if the inverse problem were to be treated as a combinatorial optimization problem, where each candidate solution is one possible configuration (or combination of the scalar elements of **t**), then we could use the criterion developed by Metropolis et al. [9] for physical systems to select the optimal configuration. The Metropolis criterion is based on the assumption that candidate configurations have probabilistic distributions of the form originally described by Gibbs [8] to guarantee statistical equilibrium of ensembles of systems. In order to apply Metropolis' selection criterion, we must make one final analogy: we need to treat the combinatorial optimization problem as if it were the outcome of an imaginary physical system in which matter has been brought to boil. When such a physical system is gradually cooled (a process referred to as annealing) then, provided the cooling rate is neither too fast nor too slow, the system will eventually solidify into a minimum energy configuration. As depicted in Fig. 28.1 to solve inverse problems we first recast the problem as a combinatorial optimization problem and then solve this recast problem using *simulated* annealing — a procedure that numerically mimics the annealing of physical systems. In this section we will describe the basic principles of combinatorial optimization, Metropolis' criterion to select or discard potential configurations, and the origins of Gibbs' distribution. We will outline the simulated annealing algorithm in the following section and will follow that with examples of implementation and applications.

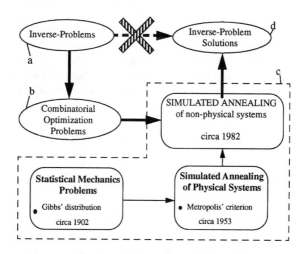

FIGURE 28.1   The direct path ($a \rightarrow d$) to solving the inverse problem is often not viable since it relies on the inversion of a system matrix. An optimal solution, however, may be obtained by an indirect path ($a \rightarrow b \rightarrow c \rightarrow d$) which involves recasting the inverse problem as an equivalent combinatorial optimization problem and then solving this problem using simulated annealing.

28.3.1   Combinatorial Optimization

The optimal solution to the inverse problem [Eq. (28.1)], as explained above, amounts to estimating vector **t**. Under the assumptions enumerated below, the inverse problem can be recast as a combinatorial problem whose solution then yields the desired optimal solution to the inverse problem. The assumptions required are:

1. Each (scalar) element $t(i)$, $1 \leq i \leq M$, of $\mathbf{t} \in \mathcal{R}^M$ can take on only a finite set of finite values. That is $-\infty < t^j(i) < \infty$; $\forall i \& j$, where $t^j(i)$ denotes the jth possible value that the ith element of $\mathbf{t}$ can take, and j is a finite valued index $j \leq J^i < \infty$; $\forall i$. J^i denotes the number of possible values the ith element of $\mathbf{t}$ can take.

2. Let each combination of M scalar values $t(i)$ of $\mathbf{t}$ be referred to as a candidate vector or a feasible configuration $\mathbf{t}_k$, where the index $k \leq K < \infty$. Associated with each candidate vector $\mathbf{t}_k$ we must have a quantifiable measure of *error, cost, or energy* (E_k).

Given the above assumptions, the combinatorial form of the inverse problem may be stated as: out of K possible candidate vectors $\mathbf{t}_k$, $1 \leq k \leq K$, search for the vector $\mathbf{t}_{k_{\text{opt}}}$ with the lowest error $E_{k_{\text{opt}}}$. Although easily stated, the time and computational efficiency with which the solution is obtained hinges on at least two significant factors — the design of the error-function and the choice of the search strategy. The error-function (E_k) must provide a quantifiable measure of *dissimilarity* or *distance*, between a feasible configuration $(\mathbf{t}_k)$ and the true (but unknown) configuration $(\mathbf{t}_{\text{true}})$, i.e.,

$$E\left(\mathbf{t}_k\right) \stackrel{\Delta}{=} d\left(\mathbf{t}_k - \mathbf{t}_{\text{true}}\right), \tag{28.5}$$

where d denotes a distance function. The goal of the combinatorial optimization problem is to efficiently search through the combinatorial space and stop at the optimal, minimum-error (E_{opt}), configuration — $\mathbf{t}_{k_{\text{opt}}}$:

$$E_{\text{opt}} = E\left(\mathbf{t}_{k_{\text{opt}}}\right) = \delta \leq E\left(\mathbf{t}_k\right), \forall k \neq k_{\text{opt}}, \tag{28.6}$$

where k_{opt} denotes the value of index k associated with the optimal configuration. In the ideal case, when $\delta = 0$, from Eq. (28.5), we have that $\mathbf{t}_{k_{\text{opt}}} = \mathbf{t}_{\text{true}}$. In practice, however, owing to a combination of factors such as noise (Eq. 28.2), or the system (Eq. 28.1) being underdetermined, $E_{\text{opt}} = \delta > 0$, implying that $\mathbf{t}_{k_{\text{opt}}} \neq \mathbf{t}_{\text{true}}$, but that $\mathbf{t}_{k_{\text{opt}}}$ is the best possible solution given what is known about the problem and its solutions. In general the error-function must satisfy the requirements of a distance function or metric (adapted from [10], pg. 237):

$$E\left(\mathbf{t}_k\right) = 0 <=> \mathbf{t}_k = \mathbf{t}_{\text{true}}, \tag{28.7a}$$

$$E\left(\mathbf{t}_k\right) = E\left(-\mathbf{t}_k\right) \stackrel{\Delta}{=} d\left(\mathbf{t}_{\text{true}} - \mathbf{t}_k\right), \tag{28.7b}$$

$$E\left(\mathbf{t}_k\right) \leq E\left(-\mathbf{t}_j\right) + d\left(\mathbf{t}_k - \mathbf{t}_j\right), \tag{28.7c}$$

where Eq. (28.7a) follows from Eq. (28.5), and where, like k, index j is defined in the range $(1, K)$ and $K < \infty$. Eq. (28.7a) stated that if the error is zero, t_k is the true configuration. The implication of Eq. (28.7b) is that error is a function of the absolute value of the distance of a configuration from the true configuration. Eq. (28.7c) implies that the triangle inequality law holds.

In designing the error-function, one can classify the sources of error into two distinct categories: The first category of error, denoted by E_k^{signal}, provides a measure of error (or distance) between the observed signal $(\mathbf{r}_k)$ and the estimated signal $(\hat{\mathbf{r}}_k)$ — computed for the current configuration $\mathbf{t}_k$ using Eq. (28.1). The second category, denoted by $E_k^{\text{constraints}}$, accounts for the price to be "paid" when an estimated solution deviates from the constraints we would want to impose on them based on our understanding of the physical world. The physical world, for instance, might suggest that each element of the signal is very probably positive valued. In this case, a negative valued estimate of a signal element will result in an error-value that is proportionate to the magnitude of the signal negativity. This constraint is popularly known as the non-negativity constraint. Another constraint might arise from the assumption that the solution is expected to be smooth [11]:

$$\hat{t}' S \hat{t} = \delta_{\text{smooth}}, \tag{28.8}$$

where S is a smoothing matrix and δ_{smooth} is the degree of smoothness of the signal. The error-function, therefore, takes the following form:

$$
\begin{aligned}
E_k &\overset{\Delta}{=} E_k^{signal} + E_k^{constraints} &\text{where,} \\
E_{signal}^k &\overset{\Delta}{=} \|\mathbf{r}_k - \hat{\mathbf{r}}_k\|_2 &\text{where} \\
\hat{\mathbf{r}}_k &= \mathbf{H} \cdot \mathbf{t_k}, &\text{and} \\
E_k^{constraints} &\overset{\Delta}{=} \sum_{c \in C} (\alpha_c \cdot E_c) ,
\end{aligned}
\tag{28.9}
$$

where $E_{constraints}$ represents the total error from all other factors or constraints that might be imposed on the solution, $\{C\}$ represents the set of constraint indices, and α_c and E_c represent the weight and the error-function, respectively, associated with cth constraint.

28.3.2 The Metropolis Criterion

The core task in solving the combinatorial optimization described above is to search for a configuration $\mathbf{t}_k$ for which the error-function E_K is a minimum. Standard gradient descent methods [6, 12, 13] would have been the natural choice had the E_k been a function with just one minimum (or maximum) value, but this function typically has multiple minimas (or maximas) — gradient descent methods would tend to get locked into a local minimum. The simulated annealing procedure (Fig. 28.2 — discussed in the next section), suggested by Metropolis et al. [9] for the problem of finding stable configurations of interacting atoms and adapted for combinatorial optimization by Kirkpatrick [7], provides a scheme to traverse the surface of the E_k, get out of local minimas, and eventually *cool* into a global minimum. The contribution of Metropolis et al., commonly referred to in the literature as *Metropolis' criterion*, is based on the assumption that the difference in the error of two consecutive feasible configurations (denoted as $\Delta E \overset{\Delta}{=} E_{k+1} - E_k$) takes the form of Gibbs' distribution [Eq. (28.11)]. The criterion states that even if a configuration were to result in increased error, i.e., $\Delta E > 0$, one can select the new configuration if:

$$
random \leq \exp^{\frac{-\Delta E}{kT}} ,
\tag{28.10}
$$

where *random* denotes a random number drawn from a uniform distribution in the range [0,1) and T denotes a the temperature of the physical system.

28.3.3 Gibbs' Distribution

At the turn of the 20th century, Gibbs [8], building upon the work of Clausius, Maxwell, and Boltzmann in statistical mechanics, proposed the probability distribution P:

$$
P = \exp^{\frac{\psi - \epsilon}{\Theta}} ,
\tag{28.11}
$$

where ψ and Θ were constants and ϵ denoted the free energy in a system. This distribution was crafted to satisfy the condition of statistical equilibrium ([8], pg. 32) for ensembles of (thermodynamical) systems:

$$
\sum \left(\frac{dP}{dp_1} \dot{p}_i + \frac{dP}{dq_1} \dot{q}_i \right) = 0 ,
\tag{28.12}
$$

where p_i and q_i represented the generalized momentum and velocity, respectively, of the ith degree of freedom. The negative sign on ϵ in Eq. (28.11) was required to satisfy the condition:

$$
\underbrace{\int \cdots \int}_{\text{all phases}} P dp_1 \cdots dq_n = 1
\tag{28.13}
$$

```
/* SIMULATED ANNEALING */
/* Set initial conditions: */
/* Temperature: T_initial = T_0 */
/* Configuration t_0 = t_initial */
/* Minimum-cost configuration t_opt = t_0 */
while(stopping criterion is not satisfied){
        while(configuration is not in equilibrium){
            Perturb(t_k → t_{k+1});
            ComputeErrorDifference(ΔE_{k+1} = E_{k+1} − E_k);
            if ΔE_{k+1} ≤ 0 then accept else
            if exp(−ΔE_{k+1}/T) > random (0,1] then accept;
            if (accept) then {
            Update(E_opt ← E_{k+1});/* remember the lowest error value */
            Update(t_opt ← t_{k+1})/* remember the lowest error config. */
            } end /* when equilibrium is reached */;
        Cool(T ← T_k)
        k = k + 1
    }end /* when stopping criterion is satisfied */
    return(); /* the global minimum-error configuration */
```

FIGURE 28.2   The outline of the annealing algorithm.

28.4   The Simulated Annealing Procedure

The simulated annealing algorithm as outlined in Fig. 28.2 mimics the annealing (or controlled cooling) of an imaginary physical system. The unknown parameters are treated like particles in a physical system. An initial configuration t_{initial} is chosen along with an initial ("boiling") temperature value (T_{initial}). The choice of T_{initial} is made so as to ensure that a vast majority, say 90%, of configurations are acceptable even if they result in a negative ΔE_k. The initial configuration is perturbed, either by using a random number generator or by sequential selection, to create a second configuration, and ΔE_2 is computed. The Metropolis criterion is applied to decide whether or not to accept the new configuration. After equilibrium is reached, i.e., after $|\Delta E_2| \leq \delta_{\text{equilib}}$, where δ_{equilib} is a small heuristically chosen threshold, the temperature is lowered according to a cooling schedule and the process is repeated until a pre-selected *frozen* temperature is reached. Several different cooling schedules have been proposed in the literature ([18], pg. 59). In one popular schedule [18, 19] each subsequent temperature T_{k+1} is less than the current temperature T_k, by a fixed percentage of T_k, i.e., $T_{k+1} = \beta_k T_k$, where β_k is typically in the range of 0.8 to unity. Based on the behavior of physical systems which attain minimum (free) energy (or global minimum) states when they freeze at the end of an annealing process, the assumption underlying the simulated annealing procedure is that the t_{opt} that is finally attained is also globally minimum.

The results of applying the simulated annealing procedure to the problems of three-dimensional signal restoration [14] is shown in Fig. 28.3. In this problem, a defocused image, vector **r**, of an opaque eight-step staircase object was provided along with the space-varying point-spread-function matrix (**H**), and a well-focused image. The unknown vector **t** represented the intensities of the volume elements (voxels) with the visible voxels taking on positive values and hidden voxels having a value of zero. The vector **t** was lexicographically indexed so that by knowing which elements of **t** were positive, one could reconstruct the three-dimensional structure. Using simulated annealing, and constraints (opacity, non-negativity of intensity, smoothness of intensity and depth, and tight bounds on the voxel intensity values obtained from the well-focused image), the original object was reconstructed.

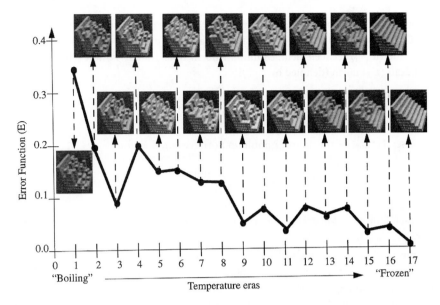

FIGURE 28.3   Three-dimensional signal recovery using simulated annealing. The staircase object shown corresponding to era 17 is recovered from a defocused image by testing a number of feasible configurations and applying the Metropolis criterion to a simulated annealing procedure.

Defining Terms

In the following definitions, as in the preceding discussion, $\mathbf{t} \in \mathcal{R}^M$, $\mathbf{r} \in \mathcal{R}^N$, and $\mathbf{H} \in \mathcal{R}^{M \times N}$.

Combinatorial Optimization: The process of selecting the optimal (lowest-cost) configuration from a large space of candidate or feasible configurations.

Configuration: Any vector $\mathbf{t}$ is a configuration. The term is used in the combinatorial optimization literature.

Cost/energy/error function: The terms *cost, energy, or error function* are frequently used interchangeably in the literature. Cost function is often used in the optimization literature to represent the mapping of a candidate vector into a (scalar) functional whose value is indicative of the optimality of the candidate vector. Energy function is frequently used in electronic communication theory as a pseudonym for the $\mathcal{L}_2$ norm or root-mean-square value of a vector. Error function is typically used to measure a mismatch between an estimated (vector) and its expected value. For purposes of this discussion we use the terms cost, energy, and error function interchangeably.

Gibbs' distribution: The distribution (in reality a probability density function (**pdf**)) in which the η the index of probability (P) is a linear function of energy, i.e., $\eta = \log P = \frac{\psi - \epsilon}{\Theta}$, where ψ and Θ are constants and ϵ represents energy, giving the familiar pdf:

$$P = \exp \frac{\psi - \epsilon}{\Theta}, \qquad (28.14)$$

Inverse problem: Given matrix $\mathbf{H}$ and vector $\mathbf{r}$, find $\mathbf{t}$ that satisfies $\mathbf{r} = \mathbf{Ht}$.

Metropolis' criterion: The criterion first suggested by Metropolis et al. [9] to decide whether or not to accept a configuration that results in an increased error, when trying to search for minimum error configurations in a combinatorial optimization problem.

Minimum-norm: The norm between two vectors is a (scalar) measure of distance (such as the $\mathcal{L}_1$, $\mathcal{L}_2$) (or Euclidean), $\mathcal{L}_\infty$ norms or the Mahalanobis distance ([10], pg. 24), or the Manhattan metric [7]) between them. Minimum-norm, unless otherwise noted, implies minimum Euclidean ($\mathcal{L}_2$) norm (denoted by $\| \cdot \|$):

$$\min_{\text{among all } \mathbf{t}} \|\mathbf{Ht} - \mathbf{r}\|. \tag{28.15}$$

Pseudoinverse: Let $\mathbf{t}_{\text{opt}}$ be the unique minimum norm vector, therefore,

$$\|\mathbf{Ht}_{\text{opt}} - \mathbf{r}\| = \min_{\text{among all } \mathbf{t}} \|\mathbf{Ht} - \mathbf{r}\|. \tag{28.16}$$

The pseudoinverse of matrix $\mathbf{H}$ denoted by $\mathbf{H}^\dagger \in \mathcal{R}^{N \times M}$ is the matrix mapping all $\mathbf{r}$ into its corresponding $\mathbf{t}_{\text{opt}}$.

Statistical mechanics: That branch of mechanics in which the problem is to find the statistical distribution of the parameters of ensembles (large numbers) of systems (each differing not just infinitesimally, but embracing every possible combination of the parameters) at a desired instant in time, given those distributions at the present time. Maxwell, according to Gibbs [8], coined the term "statistical mechanics". This field owes its origin to the desire to explain the laws of thermodynamics as stated by Gibbs ([8], pg. viii): "The laws of thermodynamics, as empirically determined, express the approximate and probable behavior of systems of a great number of particles, or, more precisely, they express the laws of mechanics for such systems as they appear to beings who have not the fineness of perception to enable them to appreciate quantities of the order of magnitude of those which relate to single particles, and who cannot repeat their experiments often enough to obtain any but the most probable results".

References

[1] Hadamard, J., Sur les problèmes aux dérivés partilles et leur signification physique, Bull. 13, Princeton University, 1902.

[2] Frolik, J.L. and Yagle A.E., Reconstruction of multilayered lossy dielectrics from plane-wave impulse responses at 2 angles of incidence, *IEEE Trans. Geosci. Remote Sens.*, 33: 268–279, March, 1995.

[3] Greensite, F., Well-posed formulation of the inverse problem of electrocardiography, *Ann. Biomed. Eng.*, 22 (2): 172–183, 1994.

[4] Arfken, G., *Mathematical Methods for Physicists*, Academic Press, 1985.

[5] Greville, T.N.E., The pseudoinverse of a rectangular or singular matrix and its application to the solution of systems of linear equations, *SIAM Rev.* 1: 38–43, 1959.

[6] Golub, G.H. and Van Loan, C.F., *Matrix Computations*, 2nd ed., The Johns Hopkins University Press, Baltimore, 1989.

[7] Kirkpatrick, S., Optimization by simulated annealing: quantitative studies, *J. Stat. Phys.*, 34(5, 6): 975–986, 1984.

[8] Gibbs, J.W., *Elementary Particles in Statistical Mechanics*, Yale University Press, New Haven, 1902.

[9] Metropolis, N., Rosenbluth, A., Rosenbluth, M., Teller, A., and Teller, E., Equation of state calculations by fast computing machines, *J. Chem. Phys.*, 21: 1087–1092, June, 1953.

[10] Duda, R.O. and Hart, P.E., *Pattern Classification and Scene Analysis*, John Wiley, 1973.

[11] Pratt, W.K., *Digital Image Processing*, John Wiley, New York, 1978.

[12] Luenberger, D.G., *Optimization by Vector Space Methods*, John Wiley & Sons, New York, 1969.

[13] Gill, P.E. and Murray, W., Quasi-Newton methods for linearly constrained optimization, in *Numerical Methods for Constrained Optimization*, Gill, P.E. and Murray, W., Eds., Academic Press, London, 1974.

[14] Prasad, K.V., Mammone, R.J., and Yogeshwar, J., 3-D image restoration using constrained optimization techniques, *Opt. Eng.*, 29: 279–288, April, 1990.

[15] Tikhonov, A.N. and Arsenin, V.Y., *Solutions of Ill-Posed Problems*, V.H. Winston & Sons, Washington, D.C., 1977.

[16] Soumekh, M., Reconnaissance with ultra wideband UHF synthetic aperture radar, *IEEE Acoust. Speech, Signal Process.*, 12: 21–40, July, 1995.

[17] van Laarhoven, P.J.M. and Aarts, E.H.L., *Simulated Annealing: Theory and Applications*, D. Riedel, Dordrecht, Holland, 1987.

[18] Aarts, E. and Korst, J., *Simulated Annealing and Boltzmann Machines*, John Wiley, New York, 1989.

[19] Press, W.H., Flannery, B.D., Teukolsky, S.A., and Vetterling, W.T., *Numerical Recipes in C*, Cambridge University Press, U.K., 1988.

[20] Geman, S. and Geman, D., Stochastic relaxation, Gibbs distributions and the Bayesian restorations of images, *IEEE Trans. Patt. Recog. Mach. Intell.*, PAMI- 6: 721–741, November, 1984.

Further Reading

Inverse problems — The classic by Tikhonov [15] provides a good introduction to the subject matter. For a description of inverse problems related to synthetic aperture radar application see [16].

Statistical mechanics — Gibbs' [8] work is historical treasure.

Vector spaces and optimization — The books by Leunberger [12] and Gill and Murray [13] provide a broad introductory foundation.

Simulated annealing — Two recent books by van Laarhoven and Aarts [17] and Aarts and Korst [18] contain a comprehensive coverage of the theory and application of simulated annealing. A useful simulated annealing algorithm, along with tips for numerical implementation and random number generation, can be found in *Numerical Recipes in C* [19]. An alternative simulated annealing procedure (in which the temperature T is kept constant) can be found in the widely cited work of Geman and Geman [20], applied to image restoration.

29

Image Recovery Using the EM Algorithm

Jun Zhang
University of Wisconsin Milwawkee

Aggelos K. Katsaggelos
Northwestern University

29.1 Introduction

Image recovery constitutes a significant portion of the inverse problems in image processing. Here, by image recovery we refer to two classes of problems, image restoration and image reconstruction. In image restoration, an estimate of the original image is obtained from a blurred and noise-corrupted image. In image reconstruction, an image is generated from measurements of various physical quantities, such as X-ray energy in CT and photon counts in single photon emission tomography (SPECT) and positron emission tomography (PET). Image restoration has been used to restore pictures in remote sensing, astronomy, medical imaging, art history studies, e.g., see [1], and more recently, it has been used to remove picture artifacts due to image compression, e.g., see [2] and [3]. While primarily used in biomedical imaging [4], image reconstruction has also found applications in materials studies [5].

Due to the inherent randomness in the scene and imaging process, images and noise are often best modeled as multidimensional random processes called random fields. Consequently, image recovery becomes the problem of *statistical inference*. This amounts to estimating certain unknown parameters of a probability density function (pdf) or calculating the expectations of certain random fields from the observed image or data. Recently, the *maximum-likelihood estimate* (MLE) has begun to play a central role in image recovery and led to a number of advances [6, 8]. The most significant advantage of the MLE over traditional techniques, such as the Wiener filtering, is perhaps that it can work more autonomously. For example, it can be used to restore an image with unknown blur and noise level by estimating them and the original image *simultaneously* [8, 9]. The traditional Wiener filter and other LMSE (least mean square error) techniques, on the other hand, would require the knowledge of the blur and noise level.

In the MLE, the likelihood function is the pdf evaluated at an observed data sample conditioned on the parameters of interest, e.g., blur filter coefficients and noise level, and the MLE seeks the parameters that maximize the likelihood function, i.e., best explain the observed data. Besides being intuitively appealing, the MLE also has several good asymptotic (large sample) properties [10] such as consistency (the estimate converges to the true parameters as the sample size increases). However, for many nontrivial image recovery problems, the direct evaluation of the MLE can be difficult, if not impossible. This difficulty is due to the fact that likelihood functions are usually highly nonlinear and often cannot be written in closed forms (e.g., they are often integrals of some other pdf's). While the former case would prevent analytic solutions, the latter case could make any numerical procedure impractical.

The EM algorithm, proposed by Dempster, Laird, and Rubin in 1977 [11], is a powerful iterative technique for overcoming these difficulties. Here, EM stands for *expectation-maximization*. The basic idea behind this approach is to introduce an auxiliary function (along with some auxiliary variables) such that it has similar behavior to the likelihood function but is much easier to maximize. By similar behavior, we mean that when the auxiliary function increases, the likelihood function also increases. Intuitively, this is somewhat similar to the use of auxiliary lines for the proofs in elementary geometry.

The EM algorithm was first used by Shepp and Verdi [7] in 1982 in emission tomography (medical imaging). It was first used by Katsaggelos and Lay [8] and Lagendijk et al. [9] for simultaneous image restoration and blur identification around 1989. The work of using the EM algorithm in image recovery has since flourished with impressive results. A recent search on the Compendex data base with key words "EM" and "image" turned up more than 60 journal and conference papers, published over the two and a half year period from January, 1993 to June, 1995.

Despite these successes, however, some fundamental problems in the application of the EM algorithm to image recovery remain. One is convergence. It has been noted that the estimates often do not converge, converge rather slowly, or converge to unsatisfactory solutions (e.g., spiky images) [12, 13]. Another problem is that, for some popular image models such as Markov random fields, the conditional expectation in the E-step of the EM algorithm can often be difficult to calculate [14]. Finally, the EM algorithm is rather general in that the choice of auxiliary variables and the auxiliary function is not unique. Is it possible that one choice is better than another with respect to convergence and expectation calculations [17]?

The purpose of this chapter is to demonstrate the application of the EM algorithm in some typical image recovery problems and survey the latest research work that addresses some of the fundamental problems described above. The chapter is organized as follows. In section 29.2, the EM algorithm is reviewed and demonstrated through a simple example. In section 29.3, recent work in convergence, expectation calculation, and the selection of auxiliary functions is discussed. In section 29.4, more complicated applications are demonstrated, followed by a summary in section 29.5. Most of the examples in this chapter are related to image restoration. This choice is motivated by two considerations — the mathematical formulations for image reconstruction are often similar to that of image restoration and a good account on image reconstruction is available in Snyder and Miller [6].

29.2   The EM Algorithm

Let the observed image or data in an image recovery problem be denoted by $\mathbf{y}$. Suppose that $\mathbf{y}$ can be modeled as a collection of random variables defined over a lattice $\mathbf{S}$ with $\mathbf{y} = \{y_i, i \in \mathbf{S}\}$. For example, $\mathbf{S}$ could be a square lattice of N^2 sites. Suppose that the pdf of $\mathbf{y}$ is $p_{\mathbf{y}}(\mathbf{y}|\theta)$, where θ is a set of parameters. In this chapter, $p(\cdot)$ is a general symbol for pdf and the subscript will be omitted whenever there is no confusion. For example, when $\mathbf{y}$ and $\mathbf{x}$ are two different random fields, their pdf's are represented as $p(\mathbf{y})$ and $p(\mathbf{x})$, respectively.

29.2.1 The Algorithm

Under statistical formulations, image recovery often amounts to seeking an estimate of θ, denoted by $\hat{\theta}$, from an observed $\mathbf{y}$. The MLE approach is to find $\hat{\theta}_{ML}$ such that

$$\hat{\theta}_{ML} = \arg \max_{\theta} p\left(\mathbf{y}|\theta\right) = \arg \max_{\theta} \log p\left(\mathbf{y}|\theta\right), \tag{29.1}$$

where $p(\mathbf{y}|\theta)$, as a function of θ, is called the likelihood. As described previously, a direct solution of (29.1) can be difficult to obtain for many applications. The EM algorithm attempts to overcome this problem by introducing an auxiliary random field $\mathbf{x}$ with pdf $p(\mathbf{x}|\theta)$. Here, $\mathbf{x}$ is somewhat "more informative" [17] than $\mathbf{y}$ in that it is related to $\mathbf{y}$ by a *many-to-one* mapping

$$\mathbf{y} = \mathbf{H}(\mathbf{x}). \tag{29.2}$$

That is, $\mathbf{y}$ can be regarded as a partial observation of $\mathbf{x}$, or *incomplete data*, with $\mathbf{x}$ being the *complete data*.

The EM algorithm attempts to obtain the incomplete data MLE of (29.1) through an iterative procedure. Starting with an initial estimate θ^0, each iteration k consists of two steps:

- *The E-step*: Compute the conditional expectation[1] $\langle \log p(\mathbf{x}|\theta)|\mathbf{y}, \theta^k \rangle$. This leads to a function of θ, denoted by $Q(\theta|\theta^k)$, which is the auxiliary function mentioned previously.
- *M-step*: Find θ^{k+1} from

$$\theta^{k+1} = \arg \max_{\theta} Q\left(\theta|\theta^k\right). \tag{29.3}$$

It has been shown that the EM algorithm is monotonic [11], i.e., $\log p(\mathbf{y}|\theta^k) \geq \log p(\mathbf{y}|\theta^{k+1})$. It has also been shown that under mild regularity conditions, such as that the true θ must lie in the interior of a compact set and that the likelihood functions involved must have continuous derivatives, the estimate of θ from the EM algorithm converges, at least to a local maxima of $p(\mathbf{y}|\theta)$ [20, 21]. Finally, the EM algorithm extends easily to the case in which the MLE is used along with a penalty or a prior on θ. For example, suppose that $q(\theta)$ is a penalty to be minimized. Then, the M-step is modified to maximizing $Q(\theta|\theta^k) - q(\theta)$ with respect to θ.

29.2.2 Example: A Simple MRF

As an illustration of the EM algorithm, we consider a simple image restoration example. Let $\mathbf{S}$ be a two-dimensional square lattice. Suppose that the observed image $\mathbf{y}$ and the original image $\mathbf{u} = \{u_i, i \in \mathbf{S}\}$ are related through

$$\mathbf{y} = \mathbf{u} + \mathbf{w}, \tag{29.4}$$

where $\mathbf{w} = \{u_i, i \in \mathbf{S}\}$ is an i.i.d. additive zero-mean white Gaussian noise with variance σ^2. Suppose that $\mathbf{u}$ is modeled as a random field with an exponential or Gibbs pdf

$$p(\mathbf{u}) = Z^{-1} e^{-\beta E(\mathbf{u})} \tag{29.5}$$

where $E(\mathbf{u})$ is an *energy function* with

$$E(\mathbf{u}) = \frac{1}{2} \sum_{i} \sum_{j \in N_i} \phi\left(u_i, u_j\right) \tag{29.6}$$

---

[1] In this chapter, we use $\langle \cdot \rangle$ rather than $E[\cdot]$ to represent expectations since E is used to denote energy functions of the MRF.

and Z is a normalization factor

$$Z = \sum_{\mathbf{u}} e^{-\beta E(\mathbf{u})} \tag{29.7}$$

called the *partition function* whose evaluation generally involves all possible realizations of $\mathbf{u}$. In the energy function, N_i is a set of neighbors of i (e.g., the nearest four neighbors) and $\phi(\cdot, \cdot)$ is a nonlinear function called the *clique function*. The model for $\mathbf{u}$ is a simple but nontrivial case of the Markov random field (MRF) [22, 23] which, due to its versatility in modeling spatial interactions, has emerged as a powerful model for various image processing and computer vision applications [24].

A restoration that is optimal in the sense of minimum mean square error is

$$\hat{\mathbf{u}} = \langle \mathbf{u} | \mathbf{y} \rangle = \int \mathbf{u} p(\mathbf{u} | \mathbf{y}) \, d\mathbf{u} \,. \tag{29.8}$$

If parameters β and σ^2 are known, the above expectation can be computed, at least approximately (see Conditional Expectation Calculations in section 29.3 for details). To estimate the parameters, now denoted by $\theta = (\beta, \sigma^2)$, one could use the MLE. Since $\mathbf{u}$ and $\mathbf{w}$ are independent,

$$p(\mathbf{y} | \theta) = \int p_{\mathbf{u}}(\mathbf{v} | \theta) p_{\mathbf{w}}(\mathbf{y} - \mathbf{v} | \theta) \, d\mathbf{v} = (p_{\mathbf{u}} * p_{\mathbf{w}})(\mathbf{y} | \theta) \,, \tag{29.9}$$

where $*$ denotes convolution, and we have used some subscripts to avoid ambiguity. Notice that the integration involved in the convolution generally does not have a closed-form expression. Furthermore, for most types of clique functions, Z is a function of β and its evaluation is exponentially complex. Hence, direct MLE does not seem possible.

To try with the EM algorithm, we first need to select the complete data. A natural choice here, for example, is to let

$$\mathbf{x} \quad = \quad (\mathbf{u}, \mathbf{w}) \tag{29.10}$$

$$\mathbf{y} \quad = \quad \mathbf{H}(\mathbf{x}) = \mathbf{H}(\mathbf{u}, \mathbf{w}) = \mathbf{u} + \mathbf{w} \,. \tag{29.11}$$

Clearly, many different $\mathbf{x}$ can lead to the same $\mathbf{y}$. Since $\mathbf{u}$ and $\mathbf{w}$ are independent, $p(\mathbf{x} | \theta)$ can be found easily as

$$p(\mathbf{x} | \theta) = p(\mathbf{u}) p(\mathbf{w}) \,. \tag{29.12}$$

However, as the reader can verify, one encounters difficulty in the derivation of $p(\mathbf{x} | \mathbf{y}, \theta^k)$ which is needed for the conditional expectation of the E-step. Another choice is to let

$$\mathbf{x} \quad = \quad (\mathbf{u}, \mathbf{y}) \tag{29.13}$$

$$\mathbf{y} \quad = \quad H(\mathbf{u}, \mathbf{y}) = \mathbf{y} \tag{29.14}$$

The log likelihood of the complete data is

$$
\begin{aligned}
\log p(\mathbf{x} | \theta) &= \log p(\mathbf{y}, \mathbf{u} | \theta) \\
&= \log p(\mathbf{y} | \mathbf{u}, \theta) p(\mathbf{u} | \theta) \\
&= c - \sum_i \frac{(y_i - u_i)^2}{2\sigma^2} - \log Z(\beta) - \frac{\beta}{2} \sum_i \sum_{j \in N_i} \phi(u_i, u_j) \,,
\end{aligned} \tag{29.15}
$$

where c is a constant. From this we see that in the E-step, we only need to calculate three types of terms, $\langle u_i \rangle$, $\langle u_i^2 \rangle$, and $\langle \phi(u_i, u_j) \rangle$. Here, the expectations are all conditioned on $\mathbf{y}$ and θ^k. To compute these expectations, one needs the conditional pdf $p(\mathbf{u} | \mathbf{y}, \theta^k)$ which is, from Bayes' formula,

$$
\begin{aligned}
p\left(\mathbf{u} | \mathbf{y}, \theta^k\right) &= \frac{p\left(\mathbf{y} | \mathbf{u}, \theta^k\right) p\left(\mathbf{u} | \theta^k\right)}{p\left(\mathbf{y} | \theta^k\right)} \\
&= \left[2\pi\sigma^2\right]^{-\|S\|/2} e^{-\sum_i (y_i - u_i)^2 / 2(\sigma^2)^k} Z^{-1} e^{-\beta^k E(\mathbf{u})} \left[p\left(\mathbf{y} | \theta^k\right)\right]^{-1} \,.
\end{aligned} \tag{29.16}
$$

Here, the superscript k denotes the kth iteration rather than the kth power. Combining all the constants and terms in the exponentials, the above equation becomes that of a Gibbs distribution

$$p\left(\mathbf{u}|\mathbf{y}, \theta^k\right) = Z_1^{-1}\left(\theta^k\right)e^{-E_1\left(\mathbf{u}|\mathbf{y}, \theta^k\right)} \tag{29.17}$$

where the energy function is

$$E_1\left(\mathbf{u}|\mathbf{y}, \theta^k\right) = \sum_i \left[\frac{(y_i - u_i)^2}{2\left(\sigma^2\right)^k} + \frac{\beta^k}{2}\sum_{j \in N_i}\phi\left(u_i, u_j\right)\right]. \tag{29.18}$$

Even with this, the computation of the conditional expectation in the E-step can still be a difficult problem due to the coupling of the u_i and u_j in E_1. This is one of the fundamental problems of the EM algorithm that will be addressed in section 29.3. For the moment, we assume that the E-step can be performed successfully with

$$
\begin{aligned}
Q\left(\theta|\theta^k\right) &= \langle \log p(\mathbf{x}|\theta)|\mathbf{y}, \theta^k\rangle \\
&= c - \sum_i \frac{\langle(y_i - x_i)^2\rangle^k}{2\sigma^2} - \log Z(\beta) - \frac{\beta}{2}\sum_i\sum_{j \in N_i}\langle\phi\left(u_i, u_j\right)\rangle^k,
\end{aligned} \tag{29.19}
$$

where $\langle\cdot\rangle^k$ is an abbreviation for $\langle\cdot|\mathbf{y}, \theta^k\rangle$. In the M-step, the update for θ can be found easily by setting

$$\frac{\partial}{\partial\sigma^2}Q\left(\theta|\theta^k\right) = 0, \qquad \frac{\partial}{\partial\beta}Q\left(\theta|\theta^k\right) = 0. \tag{29.20}$$

From the first of these,

$$\left(\sigma^2\right)^{k+1} = ||\mathbf{S}||^{-1}\sum_i\langle(y_i - u_i)^2\rangle^k \tag{29.21}$$

The solution of the second equation, on the other hand, is generally difficult due to the well-known difficulties of evaluating the partition function $Z(\beta)$ (see also Eq. (29.7)) which needs to be dealt with via specialized approximations [22, 25]. However, as demonstrated by Bouman and Sauer [26], some simple yet important cases exist in which the solution is straightforward. For example, when $\phi(u_i, u_j) = (u_i - u_j)^2$, $Z(\beta)$ can be written as

$$
\begin{aligned}
Z(\beta) &= \int e^{-\frac{\beta}{2}\sum_i\sum_{j \in N_i}(u_i - u_j)^2}d\mathbf{u} \\
&= \beta^{-||\mathbf{S}||/2}\int e^{-\frac{1}{2}\sum_i\sum_{j \in N_i}(v_i - v_j)^2}d\mathbf{v} = \beta^{-||\mathbf{S}||/2}Z(1).
\end{aligned} \tag{29.22}
$$

Here, we have used a change of variable, $v_i = \sqrt{\beta}u_i$. Now, the update of β can be found easily as

$$\beta^{k+1} = ||\mathbf{S}||^{-1}\sum_i\sum_{j \in N_i}\langle\left(u_i - u_j\right)^2\rangle^k. \tag{29.23}$$

This simple technique applies to a wider class of clique functions characterized by $\phi(u_i, u_j) = |u_i - u_j|^r$ with any $r > 0$ [26].

29.3 Some Fundamental Problems

As is in many other areas of signal processing, the power and versatility of the EM algorithm has been demonstrated in a large number of diverse image recovery applications. Previous work, however, has also revealed some of its weaknesses. For example, the conditional expectation of the E-step can be difficult to calculate analytically and too time-consuming to compute numerically, as is in the MRF example in the previous section. To a lesser extent, similar remarks can be made to the M-step. Since the EM algorithm is iterative, convergence can often be a problem. For example, it can be very slow. In some applications, e.g., emission tomography, it could converge to the wrong result — the reconstructed image gets spikier as the number of iterations increases [12, 13]. While some of these problems, such as slow convergence, are common to many numerical algorithms, most of their causes are inherent to the EM algorithm [17, 19].

In previous work, the EM algorithm has mostly been applied in a "natural fashion" (e.g., in terms of selecting incomplete and complete data sets) and the problems mentioned above were dealt with on an ad hoc basis with mixed results. Recently, however, there has been interest in seeking more fundamental solutions [14, 19]. In this section, we briefly describe the solutions to two major problems related to the EM algorithm, namely, the conditional expectation computation in the E-step when the data is modeled as MRF's and fundamental ways of improving convergence.

29.3.1   Conditional Expectation Calculations

When the complete data is an MRF, the conditional expectation of the E-step of the EM algorithm can be difficult to perform. For instance, consider the simple MRF in section 29.2, where it amounts to calculating $\langle u_i \rangle$, $\langle u_i^2 \rangle$, and $\langle \phi(u_i, u_j) \rangle$ and the expectations are taken with respect to $p(\mathbf{u}|\mathbf{y}, \theta^k)$ of Eq. (29.17). For example, we have

$$\langle u_i \rangle = Z_1^{-1} \int u_i e^{-E_1(\mathbf{u})} \, d\mathbf{u} \tag{29.24}$$

Here, for the sake of simplicity, we have omitted the superscript k and the parameters, and this is done in the rest of this section whenever there is no confusion. Since the variables u_i and u_j are coupled in the energy function for all i and j that are neighbors, the pdf and Z_1 cannot be factored into simpler terms, and the integration is exponentially complex, i.e., it involves all possible realizations of $\mathbf{u}$. Hence, some approximation scheme has to be used. One of these is the Monte Carlo simulation. For example, Gibbs samplers [23] and Metropolis techniques [27] have been used to generate samples according to $p(\mathbf{u}|\mathbf{y}, \theta^k)$ [26, 28]. A disadvantage of these is that, generally, hundreds of samples of $\mathbf{u}$ are needed and if the image size is large, this can be computation intensive. Another technique is based on the mean field theory (MFT) of statistical mechanics [25]. This has the advantage of being computationally inexpensive while providing satisfactory results in many practical applications. In this section, we will outline the essentials of this technique.

Let $\mathbf{u}$ be an MRF with pdf

$$p(\mathbf{u}) = Z^{-1} e^{-\beta E(\mathbf{u})} . \tag{29.25}$$

For the sake of simplicity, we assume that the energy function is of the form

$$E(\mathbf{u}) = \sum_i \left[h_i(u_i) + \frac{1}{2} \sum_{j \in N_i} \phi\left(u_i, u_j\right) \right] \tag{29.26}$$

where $h_i(\cdot)$ and $\phi(\cdot, \cdot)$ are some suitable, and possibly nonlinear, functions. The mean field theory attempts to derive a pdf $p_{MF}(\mathbf{u})$ that is an approximation to $p(\mathbf{u})$ and can be factored like an independent pdf.

The MFT used previously can be divided into two classes, the local mean field energy (LMFE) and the ones based on the Gibbs-Bogoliubov-Feynman (GBF) inequality. The LMFE scheme is based on the idea that when calculating the mean of the MRF at a given site, the influence of the random variables at other sites can be approximated by the influence of their means. Hence, if we want to calculate the mean of u_i, a *local energy function* can be constructed by collecting all the terms in (29.26) that are related to u_i and replacing the u_j's by their mean. Hence, for this energy function we have

$$E_i^{MF}(u_i) = h_i(u_i) + \sum_{i \in N_i} \phi\left(u_i, \langle u_j \rangle\right) \tag{29.27}$$

$$p_i^{MF}(u_i) = Z_i^{-1} e^{-\beta E_i^{MF}(u_i)} \tag{29.28}$$

$$p_{MF}(\mathbf{u}) = \prod_i p_i^{MF}(u_i) \tag{29.29}$$

Using this mean field pdf, the expectation of u_i and its functions can be found easily.

Again we use the MRF example from section 29.2.2 as an illustration. Its energy function is (29.18) and for the sake of simplicity, we assume that $\phi(u_i, u_j) = |u_i - u_j|^2$. By the LMFE scheme,

$$E_i^{MF} = \frac{(y_i - u_i)^2}{2\sigma^2} + \sum_{j \in N_i} \beta \left(u_i - \langle u_j \rangle\right)^2 \tag{29.30}$$

which is the energy of a Gaussian. Hence, the mean can be found easily by completing the square in (29.30) with

$$\langle u_i \rangle = \frac{y_i/\sigma^2 + 2\beta \sum_{j \in N_i} \langle u_j \rangle}{1/\sigma^2 + 2\beta \|N_i\|} . \tag{29.31}$$

When $\phi(\cdot, \cdot)$ is some general nonlinear function, numerical integration might be needed. However, compared to (29.24) such integrals are all with respect to one or two variables and are easy to compute.

Compared to the physically motivated scheme above, the GBF is an optimization approach. Suppose that $p_0(\mathbf{u})$ is a pdf which we want to use to approximate another pdf, $p(\mathbf{u})$. According to information theory, e.g., see [29], the *directed-divergence* between p_0 and p is defined as

$$D(p_0\|p) = \langle \log p_0(\mathbf{u}) - \log p(\mathbf{u}) \rangle_0, \tag{29.32}$$

where the subscript 0 indicates that the expectation is taken with respect to p_0, and it satisfies

$$D(p_0\|p) \geq 0 \tag{29.33}$$

with equality holds if and only if $p_0 = p$. When the pdf's are Gibbs distributions, with energy functions E_0 and E and partition functions Z_0 and Z, respectively, the inequality becomes

$$\log Z \geq \log Z_0 - \beta \langle E - E_0 \rangle_0 = \log Z_0 - \beta \langle \Delta E \rangle_0 , \tag{29.34}$$

which is known as the GBF inequality.

Let p_0 be a parametric Gibbs pdf with a set of parameters ω to be determined. Then, one can obtain an optimal p_0 by maximizing the right-hand side of (29.34). As an illustration, consider again the MRF example in section 29.2 with the energy function (29.18) and a quadratic clique function, as we did for the LMFE scheme. To use the GBF, let the energy function of p_0 be defined as

$$E_0(\mathbf{u}) = \sum_i \frac{(u_i - m_i)^2}{2v_i^2} \tag{29.35}$$

where $\{m_i, v_i^2, i \in \mathbf{S}\} = \omega$ is the set of parameters to be determined in the maximization of the GBF. Since this is the energy for an independent Gaussian, Z_0 is just

$$Z_0 = \prod_i \sqrt{2\pi v_i^2}. \tag{29.36}$$

The parameters of p_0 can be obtained by finding an expression for the right-hand side of the GBF inequality, letting its partial derivatives (with respect to the parameters m_i and v_i^2) be zero, and solving for the parameters. Through a somewhat lengthy but straightforward derivation, one can find that [30]

$$m_i = \frac{y_i/\sigma^2 + 2\beta \sum_{j \in N_i} \langle u_j \rangle}{1/\sigma^2 + 2\beta \|N_i\|} . \tag{29.37}$$

Since $m_i = \langle u_i \rangle$, the GBF produces the same result as the LMEF. This, however, is an exception rather than the rule [30] and it is due to the quadratic structures of both energy functions.

We end this section with several remarks. First, compared to the LMFE, the GBF scheme is an optimization scheme, hence more desirable. However, if the energy function of the original pdf is highly nonlinear, the GBF could require the solution of a difficult nonlinear equation in many variables (see e.g., [30]). The LMFE, though not optimal, can always be implemented relatively easily. Secondly, while the MFT techniques are significantly more computation-efficient than the Monte Carlo techniques and provide good results in many applications, no proof exists as yet that the conditional mean computed by the MFT will converge to the true conditional mean. Finally, the performance of the mean field approximations may be improved by using "high-order" models. For example, one simple scheme is to consider LMFE's with a pair of neighboring variables [25, 31]. For the energy function in (29.26), for example, the "second-order" LMFE is

$$E_{i,j}^{MF}(u_i, u_j) = h_i(u_i) + h_i(u_j) + \beta \sum_{i' \in N_i} \phi(u_i, \langle u_{i'} \rangle) + \beta \sum_{j' \in N_j} \phi(u_j, \langle u_{j'} \rangle) \qquad (29.38)$$

and

$$p_{MF}(u_i, u_j) = Z_{MF}^{-1} e^{-\beta E_{i,j}^{MF}(u_i, u_j)}, \qquad (29.39)$$

$$p_{MF}(u_i) = \int p_{MF}(u_i, u_j) \, du_j. \qquad (29.40)$$

Notice that (29.40) is not the same as (29.28) in that the fluctuation of u_j is taken into consideration.

29.3.2   Convergence Problem

Research on the EM algorithm-based image recovery has so far suggested two causes for the convergence problems mentioned previously. The first is whether the random field models used adequately capture the characteristics and constraints of the underlying physical phenomenon. For example, in emission tomography the original EM procedure of Shepp and Verdi tends to produce spikier and spikier images as the number of iteration increases [13]. It was found later that this is due to the assumption that the densities of the radioactive material at different spatial locations are independent. Consequently, various smoothness constraints (density dependence between neighboring locations) have been introduced as penalty functions or priors and the problem has been greatly reduced. Another example is in blind image restoration. It has been found that in order for the EM algorithm to produce reasonable estimate of the blur, various constraints need to be imposed. For instance, symmetry conditions and good initial guesses (e.g., a lowpass filter) are used in [8] and [9]. Since the blur tends to have a smooth impulse response, orthonormal expansion (e.g., the DCT) has also been used to reduce ("compress") the number of parameters in its representation [15].

The second factor that can be quite influential to the convergence of the EM algorithm, noticed earlier by Feder and Weinstein [16], is how the complete data is selected. In their work [18], Fessler and Hero found that for some EM procedures, it is possible to significantly increase the convergence rate by properly defining the complete data. Their idea is based on the observation that the EM algorithm, which is essentially a MLE procedure, often converges faster if the parameters are estimated sequentially in small groups rather than simultaneously. Suppose, for example, that 100 parameters are to be estimated. It is much better to estimate, in each EM cycle, the first 10 while holding the next 90 constant; then estimate the next 10 holding the remaining 80 and the newly updated 10 parameters constant; and so on. This type of algorithm is called the SAGE (Space Alternating Generalized EM) algorithm.

We illustrate this idea through a simple example used by Fessler and Hero [18]. Consider a simple image recovery problem, modeled as

$$\mathbf{y} = \mathbf{A}_1 \theta_1 + \mathbf{A}_2 \theta_2 + \mathbf{n}. \qquad (29.41)$$

Column vectors θ_1 and θ_2 represent two original images or two data sources, A_1 and A_2 are two blur functions represented as matrices, and $\mathbf{n}$ is an additive white Gaussian noise source. In this model, the observed image $\mathbf{y}$ is the noise-corrupted combination of two blurred images (or data sources). A natural choice for the complete data is to view $\mathbf{n}$ as the combination of two smaller noise sources, each associated with one original image, i.e.,

$$\mathbf{x} = \left[\mathbf{A}_1\theta_1 + \mathbf{n_1}, \mathbf{A}_2\theta_2 + \mathbf{n}_2 \right]' . \tag{29.42}$$

where $\mathbf{n}_1$ and $\mathbf{n}_2$ are i.i.d additive white Gaussian noise vectors with covariance matrix $\frac{\sigma^2}{2}\mathbf{I}$ and $'$ denotes transpose. The incomplete data $\mathbf{y}$ can be obtained from $\mathbf{x}$ by

$$\mathbf{y} = [\mathbf{I}, \mathbf{I}]\mathbf{x} . \tag{29.43}$$

Notice that this is a Gaussian problem in that both $\mathbf{x}$ and $\mathbf{y}$ are Gaussian and they are jointly Gaussian as well. From the properties of jointly Gaussian random variables [32], the EM cycle can be found relatively straightforwardly as

$$\theta_1^{k+1} = \theta_1^k + (\mathbf{A}_1'\mathbf{A}_1)^{-1}\mathbf{A}_1'\hat{\epsilon}/2\sigma^2 \tag{29.44}$$
$$\theta_2^{k+1} = \theta_2^k + (\mathbf{A}_2'\mathbf{A}_2)^{-1}\mathbf{A}_2'\hat{\epsilon}/2\sigma^2 \tag{29.45}$$

where

$$\hat{\epsilon} = (\mathbf{y} - \mathbf{A}_1\theta_1^k - \mathbf{A}_2\theta_2^k)/\sigma^2 \tag{29.46}$$

The SAGE algorithm for this simple problem is obtained by defining two smaller "complete data sets",

$$\mathbf{x}_1 = \mathbf{A}_1\theta_1 + \mathbf{n} , \qquad \mathbf{x}_2 = \mathbf{A}_2\theta_2 + \mathbf{n} . \tag{29.47}$$

Notice that now the noise $\mathbf{n}$ is associated "totally" with each smaller complete data set. The incomplete data $\mathbf{y}$ can be obtained from both $\mathbf{x}_1$ and $\mathbf{x}_2$, e.g.,

$$\mathbf{y} = \mathbf{x}_1 + \mathbf{A}_2\theta_2 \tag{29.48}$$

The SAGE algorithm amounts to two sequential and "smaller" EM algorithms. Specifically, corresponding to each classical EM cycle (29.44)-(29.46), the first SAGE cycle is a classical EM cycle with $\mathbf{x}_1$ as the complete data and θ_1 as the parameter set to be updated. The second SAGE cycle is a classical EM cycle with $\mathbf{x}_2$ as the complete data and θ_2 as the parameter set to be updated. The new update of θ_1 is also used. The specific algorithm is

$$\theta_1^{k+1} = \theta_1^k + (\mathbf{A}_1'\mathbf{A}_1)^{-1}\mathbf{A}_1'\hat{\epsilon}_1/2\sigma^2 \tag{29.49}$$
$$\theta_2^{k+1} = \theta_2^k + (\mathbf{A}_2'\mathbf{A}_2)^{-1}\mathbf{A}_2'\hat{\epsilon}_2/2\sigma^2 \tag{29.50}$$

where

$$\hat{\epsilon}_1 = \left(\mathbf{y} - \mathbf{A}_1\theta_1^k - \mathbf{A}_2\theta_2^k\right)/\sigma^2 \tag{29.51}$$

$$\hat{\epsilon}_2 = \left(\mathbf{y} - \mathbf{A}_1\theta_1^{k+1} - \mathbf{A}_2\theta_2^k\right)/\sigma^2 \tag{29.52}$$

We end this subsection with several remarks. First, for a wide class of random field models including the simple one above, Fessler and Hero have shown that the SAGE converges significantly faster than the classical EM [17]. In some applications, e.g., tomography, an acceleration of 5 to 10 times may be achieved. Secondly, just as for the EM algorithm, various constraints on the parameters are often needed and can be imposed easily as penalty functions in the SAGE algorithm. Finally, notice that in (29.41), the original images are treated as parameters (with constraints) rather than as random variables with their own pdfs. It would be of interest to investigate a Bayesian counterpart of the SAGE algorithm.

29.4   Applications

In this section, we describe the application of the EM algorithm to the simultaneous identification of the blur and image model and the restoration of single and multichannel images.

29.4.1   Single Channel Blur Identification and Image Restoration

Most of the work on restoration in the literature was done under the assumption that the blurring process (usually modeled as a linear space-invariant system (LSI) and specified by its point spread function (PSF)) is exactly known (for recent reviews of the restoration work in the literature see [8, 33]). However, this may not be the case in practice since usually we do not have enough knowledge about the mechanism of the degradation process. Therefore, the estimation of the parameters that characterize the degradation operator needs to be based on the available noisy and blurred data.

Problem formulation

The observed image $y(i, j)$ is modeled as the output of a 2D LSI system with PSF $\{d(p, q)\}$. In the following we will use (i, j) to denote a location on the lattice $\mathbf{S}$, instead of a single subscript. The output of the LSI system is corrupted by additive zero-mean Gaussian noise $v(i, j)$ with covariance matrix $\Lambda_{\mathbf{V}}$, which is uncorrelated with the original image $u(i, j)$. That is, the observed image $y(i, j)$ is expressed as

$$y(i, j) = \sum_{(p,q)\in\mathbf{S_D}} d(p, q)u(i - p, j - q) + v(i, j), \tag{29.53}$$

where $\mathbf{S_D}$ is the finite support region of the distortion filter. We assume that the arrays $y(i, j)$, $u(i, j)$, and $v(i, j)$ are of size $N \times N$. By stacking them into $N^2 \times 1$ vectors, Eq. (29.53) can be rewritten in matrix/vector form as [35]

$$\mathbf{y} = \mathbf{Du} + \mathbf{v}, \tag{29.54}$$

where $\mathbf{D}$ is an $N^2 \times N^2$ matrix.

The vector $\mathbf{u}$ is modeled as a zero-mean Gaussian random field. Its pdf is equal to

$$p(\mathbf{u}) = |2\pi \Lambda_{\mathbf{U}}|^{-1/2} \exp\left\{\frac{-1}{2}\mathbf{u}^H \Lambda_{\mathbf{U}}^{-1}\mathbf{u}\right\}, \tag{29.55}$$

where $\Lambda_{\mathbf{U}}$ is the covariance matrix of $\mathbf{u}$, ^{H} denotes the Hermitian (i.e. conjugate transpose) of a matrix and a vector, and $|\cdot|$ denotes the determinant of a matrix. A special case of this representation is when $u(i, j)$ is described by an autoregressive (AR) model. Then $\Lambda_{\mathbf{U}}$ can be parameterized in terms of the AR coefficients and the covariance of the driving noise [38, 57].

Equation (29.53) can be written in the continuous frequency domain according to the convolution theorem. Since the discrete Fourier transform (DFT) will be used in implementing convolution, we assume that Eq. (29.53) represents circular convolution (2D sequences can be padded with zeros in such a way that the result of the linear convolution equals that of the circular convolution, or the observed image can be preprocessed around its boundaries so that Eq. (29.53) is consistent with the circular convolution of $\{d(p, q)\}$ with $\{u(p, q)\}$ [36]). Matrix $\mathbf{D}$ then becomes block circulant [35].

Maximum Likelihood (ML) Parameter Identification

The assumed image and blur models are specified in terms of the deterministic parameters $\theta = \{\Lambda_{\mathbf{U}}, \Lambda_{\mathbf{V}}, \mathbf{D}\}$. Since $\mathbf{u}$ and $\mathbf{v}$ are uncorrelated, the observed image $\mathbf{y}$ is also Gaussian with pdf equal to

$$p(\mathbf{y}/\theta) = |2\pi \left(\mathbf{D}\Lambda_U\mathbf{D}^H + \Lambda_V\right)|^{-1/2}$$

$$\exp\left\{\frac{-1}{2}\mathbf{y}^T \left(\mathbf{D}\Lambda_U\mathbf{D}^H + \Lambda_V\right)^{-1}\mathbf{y}\right\}, \tag{29.56}$$

where the inverse of the matrix $(\mathbf{D}\Lambda_U\mathbf{D}^H + \Lambda_V)$ is assumed to be defined since covariance matrices are symmetric positive definite.

Taking the logarithm of Eq. (29.56) and disregarding constant additive and multiplicative terms, the maximization of the log-likelihood function becomes the minimization of the function $L(\theta)$, given by

$$L(\theta) = \log|\mathbf{D}\Lambda_U\mathbf{D}^H + \Lambda_V| + \left\{\mathbf{y}^T \left(\mathbf{D}\Lambda_U\mathbf{D}^H + \Lambda_V\right)^{-1}\mathbf{y}\right\}. \tag{29.57}$$

By studying the function $L(\theta)$ it is clear that if no structure is imposed on the matrices $\mathbf{D}$, Λ_U, and Λ_V, the number of unknowns involved is very large. With so many unknowns and only one observation (i.e., $\mathbf{y}$), the ML identification problem becomes unmanageable. Furthermore, the estimate of $\{d(p, q)\}$ is not unique, because the ML approach to image and blur identification uses only second order statistics of the blurred image, since all pdfs are assumed to be Gaussian. More specifically, the second order statistics of the blurred image do not contain information about the phase of the blur, which, therefore, is in general undetermined. In order to restrict the set of solutions and hopefully obtain a unique solution, additional information about the unknown parameters needs to be incorporated into the solution process.

The structure we are imposing on Λ_U and Λ_V results from the commonly used assumptions in the field of image restoration [35]. First we assume that the additive noise $\mathbf{v}$ is white, with variance σ_V^2, that is,

$$\Lambda_V = \sigma_V^2\mathbf{I}. \tag{29.58}$$

Further we assume that the random process $\mathbf{u}$ is stationary which results in Λ_U being a block Toeplitz matrix [35]. A block Toeplitz matrix is asymptotically equivalent to a block circulant matrix as the dimension of the matrix becomes large [37]. For average size images, the dimensions of Λ_U are large indeed; therefore, the block circulant approximation is a valid one. Associated with Λ_U are the 2D sequences $\{l_U(p, q)\}$. The matrix $\mathbf{D}$ in Eq. (29.54) was also assumed to be block circulant. Block circulant matrices can be diagonalized with a transformation matrix constructed from discrete Fourier kernels [35]. The diagonal matrices corresponding to Λ_U and $\mathbf{D}$ are denoted respectively by $\mathbf{Q}_U$ and $\mathbf{Q}_D$. They have as elements the raster scanned 2D DFT values of the 2D sequences $\{l_U(p, q)\}$ and $\{d(p, q)\}$, denoted respectively by $S_U(m, n)$ and $\Delta(m, n)$.

Due to the above assumptions Eq. (29.57) can be written in the frequency domain as

$$L(\theta) = \sum_{m=0}^{N-1}\sum_{n=0}^{N-1}$$

$$\left\{\log\left(|\Delta(m, n)|^2 S_U(m, n) + \sigma_V^2\right) + \frac{|Y(m, n)|^2}{|\Delta(m, n)|^2 S_U(m, n) + \sigma_V^2}\right\}, \tag{29.59}$$

where $Y(m, n)$ is the 2D DFT of $y(i, j)$. Equation (29.59) more clearly demonstrates the already mentioned nonuniqueness of the ML blur solution, since only the magnitude of $\Delta(m, n)$ appears in $L(\theta)$. If the blur is zero-phase, as is the case with $\mathbf{D}$ modeling atmospheric turbulence with long exposure times and mild defocussing ($\{d(p, q)\}$ is 2D Gaussian in this case), then a unique solution may be obtained. Nonuniqueness of the estimation of $\{d(p, q)\}$ can in general be avoided by enforcing the solution to satisfy a set of constraints. Most PSFs of practical interest can be assumed to be symmetric, i.e., $d(p, q) = d(-p, -q)$. In this case the phase of the DFT of $\{d(p, q)\}$ is zero or $\pm\pi$. Unfortunately, uniqueness of the ML solution is not always established by the symmetry assumption, due primarily to the phase ambiguity. Therefore, additional constraints may alleviate this ambiguity. Such additional constraints are the following: (1) The PSF coefficients are nonnegative, (2) the support $\mathbf{S_D}$ is finite, and (3) the blurring mechanism preserves energy [35], which results in

$$\sum_{(i,j)\in \mathbf{S_D}} d(i,j) = 1 . \tag{29.60}$$

The EM Iterations for the ML Estimation of θ

The next step to be taken in implementing the EM algorithm is the determination of the mapping **H** in Eq. (29.2). Clearly Eq. (29.54) can be rewritten as

$$\mathbf{y} = \begin{bmatrix} \mathbf{0} & \mathbf{I} \end{bmatrix} \begin{bmatrix} \mathbf{u} \\ \mathbf{y} \end{bmatrix} = \begin{bmatrix} \mathbf{D} & \mathbf{I} \end{bmatrix} \begin{bmatrix} \mathbf{u} \\ \mathbf{v} \end{bmatrix} = \begin{bmatrix} \mathbf{I} & \mathbf{I} \end{bmatrix} \begin{bmatrix} \mathbf{Du} \\ \mathbf{v} \end{bmatrix} , \tag{29.61}$$

where $\mathbf{0}$ and $\mathbf{I}$ represent the $N^2 \times N^2$ zero and identity matrices, respectively. Therefore, according to Eq. (29.61), there are three candidates for representing the complete data $\mathbf{x}$, namely, $\{\mathbf{u}, \mathbf{y}\}$, $\{\mathbf{u}, \mathbf{v}\}$, and $\{\mathbf{Du}, \mathbf{v}\}$. All three cases are analyzed in the following. However, as it will be shown, only the choice of $\{\mathbf{u}, \mathbf{y}\}$ as the complete data fully justifies the term "complete data", since it results in the simultaneous identification of all unknown parameters and the restoration of the image.

For the case when $\mathbf{H}$ in Eq. (29.2) is linear, as are the cases represented by Eq. (29.61), and the data $\mathbf{y}$ is modeled as a zero-mean Gaussian process, as is the case under consideration expressed by Eq. (29.56), the following general result holds for all three choices of the complete data [38, 39, 57].

The E-step of the algorithm results in the computation of $Q(\theta/\theta^k) = \text{constant} - F(\theta/\theta^k)$ where

$$\begin{aligned} F(\theta/\theta^k) &= \log |\Lambda_\mathbf{X}| + tr\left(\Lambda_\mathbf{X}^{-1} \mathbf{C}_{\mathbf{X}|\mathbf{y}}^k\right) \\ &= \log |\Lambda_\mathbf{X}| + tr\left(\Lambda_\mathbf{X}^{-1} \Lambda_{\mathbf{X}|\mathbf{y}}^k\right) + \mu_{\mathbf{X}|\mathbf{y}}^{(k)H} \Lambda_\mathbf{X}^{-1} \mu_{\mathbf{X}|\mathbf{y}}^k , \end{aligned} \tag{29.62}$$

where $\Lambda_\mathbf{X}$ is the covariance of the complete data $\mathbf{x}$ which is also a zero-mean Gaussian process,

$$\begin{aligned} \mathbf{C}_{\mathbf{X}|\mathbf{y}}^k &= \langle \mathbf{xx}^H | \mathbf{y}; \ \theta^k \rangle = \Lambda_{\mathbf{X}|\mathbf{y}}^k + \mu_{\mathbf{X}|\mathbf{y}}^k \mu_{\mathbf{X}|\mathbf{y}}^{(k)H} , \\ \mu_{\mathbf{X}|\mathbf{y}}^k &= \langle \mathbf{x}|\mathbf{y}; \ \theta^k \rangle = \Lambda_{\mathbf{XY}} \Lambda_\mathbf{Y}^{-1} \mathbf{y} = \Lambda_\mathbf{X} \mathbf{H}^H \left(\mathbf{H}\Lambda_{\mathbf{XH}}^H \right)^{-1} \mathbf{y} , \end{aligned} \tag{29.63}$$

and

$$\begin{aligned} \Lambda_{\mathbf{X}|\mathbf{y}} &= \langle \left(\mathbf{x} - \mu_{\mathbf{X}|\mathbf{y}} \right) \left(\mathbf{x} - \mu_{\mathbf{X}|\mathbf{y}} \right)^H | \mathbf{y}; \theta^k \rangle = \Lambda_\mathbf{X} - \Lambda_{\mathbf{XY}} \Lambda_\mathbf{Y}^{-1} \Lambda_{\mathbf{YX}} \\ &= \Lambda_\mathbf{X} - \Lambda_\mathbf{X} \mathbf{H}^H \left(\mathbf{H}\Lambda_\mathbf{X}\mathbf{H}^H \right)^{-1} \mathbf{H}\Lambda_\mathbf{X} . \end{aligned} \tag{29.64}$$

The M-step of the algorithm is described by the following equation

$$\theta^{(k+1)} = arg \left\{ \min_{\{\theta\}} F(\theta/\theta^k) \right\} . \tag{29.65}$$

In our formulation of the identification/restoration problem the original image is not one of the unknown parameters in the set θ. However, as it will be shown in the next section, the restored image will be obtained in the E-step of the iterative algorithm.

{ *u,y*} *as the complete data (CD_uy algorithm)*

Choosing the original and observed images as the complete data, we obtain $\mathbf{H} = [\mathbf{0}\ \mathbf{I}]$ and $\mathbf{x} = [\mathbf{u}^H\ \mathbf{y}^H]^H$. The covariance matrix of $\mathbf{x}$ takes the form

$$\Lambda_{\mathbf{X}} = \langle \mathbf{xx}^H \rangle = \begin{bmatrix} \Lambda_{\mathbf{U}} & \Lambda_{\mathbf{U}} \mathbf{D}^H \\ \mathbf{D}\Lambda_{\mathbf{U}} & \mathbf{D}\Lambda_{\mathbf{U}}\mathbf{D}^H + \Lambda_{\mathbf{V}} \end{bmatrix}, \tag{29.66}$$

and its inverse is equal to [40]

$$\Lambda_{\mathbf{X}}^{-1} = \begin{bmatrix} \Lambda_{\mathbf{U}}^{-1} + \mathbf{D}^H \Lambda_{\mathbf{V}}^{-1}\mathbf{D} & -\mathbf{D}^H \Lambda_{\mathbf{V}}^{-1} \\ -\Lambda_{\mathbf{V}}^{-1}\mathbf{D} & \Lambda_{\mathbf{V}}^{-1} \end{bmatrix}. \tag{29.67}$$

Substituting Eqs. (29.66) and (29.67) into Eqs. (29.62), (29.63), and (29.64), we obtain

$$\begin{aligned} F(\theta/\theta^k) =\ & \log|\Lambda_{\mathbf{U}}| + \log|\Lambda_{\mathbf{V}}| + tr\left\{ \left(\Lambda_{\mathbf{U}}^{-1} + \mathbf{D}^H \Lambda_{\mathbf{V}}^{-1}\mathbf{D} \right) \Lambda_{\mathbf{U}|\mathbf{y}}^k \right\} \\ & + \mu_{\mathbf{U}|\mathbf{y}}^{(k)H}\left(\Lambda_{\mathbf{U}}^{-1} + \mathbf{D}^H \Lambda_{\mathbf{V}}^{-1}\mathbf{D} \right)\mu_{\mathbf{U}|\mathbf{y}}^k \\ & - 2\mathbf{y}^H \Lambda_{\mathbf{V}}^{-1}\mathbf{D}\mu_{\mathbf{U}|\mathbf{y}}^k + \mathbf{y}^H \Lambda_{\mathbf{V}}^{-1}\mathbf{y}, \end{aligned} \tag{29.68}$$

where

$$\mu_{\mathbf{U}|\mathbf{y}}^k = \Lambda_{\mathbf{U}}^k \mathbf{D}^{(k)H}\left(\mathbf{D}^k \Lambda_{\mathbf{U}}^k \mathbf{D}^{(k)H} + \Lambda_{\mathbf{V}}^k \right)^{-1}\mathbf{y}, \tag{29.69}$$

and

$$\Lambda_{\mathbf{U}|\mathbf{y}}^k = \Lambda_{\mathbf{U}}^k - \Lambda_{\mathbf{U}}^k \mathbf{D}^{(k)H}\left(\mathbf{D}^k \Lambda_{\mathbf{U}}^k \mathbf{D}^{(k)H} + \Lambda_{\mathbf{V}}^k \right)^{-1}\mathbf{D}^k \Lambda_{\mathbf{U}}^k. \tag{29.70}$$

Due to the constraints on the unknown parameters described in the subsection Eq. (29.62) can be written in the discrete frequency domain, as follows

$$\begin{aligned} F(\theta/\theta^k) =\ & N^2 \log \sigma_{\mathbf{V}}^2 \\ & + \frac{1}{\sigma_{\mathbf{V}}^2}\sum_{m=0}^{N-1}\sum_{n=0}^{N-1}\left\{ |\Delta(m,n)|^2 \left(S_{\mathbf{U}|\mathbf{y}}^k(m,n) + \frac{1}{N^2}|M_{\mathbf{U}|\mathbf{y}}^k(m,n)|^2 \right)\right. \\ & + \frac{1}{N^2}\left(|Y(m,n)|^2 - 2Re\left[Y^*(m,n)\Delta(m,n)M_{\mathbf{U}|\mathbf{y}}^k(m,n) \right] \right)\Big\} \\ & + \sum_{m=0}^{N-1}\sum_{n=0}^{N-1}\left\{ \log S_{\mathbf{U}}(m,n) + \frac{1}{S_{\mathbf{U}}(m,n)}\left(S_{\mathbf{U}|\mathbf{y}}^k(m,n)\right.\right. \\ & + \frac{1}{N^2}|M_{\mathbf{U}|\mathbf{y}}^k(m,n)|^2 \Big)\Big\} \end{aligned} \tag{29.71}$$

where

$$M_{\mathbf{U}|\mathbf{y}}^k(m,n) = \frac{\Delta^{(k)*}(m,n)S_{\mathbf{U}}^k(m,n)}{|\Delta^k(m,n)|^2 S_{\mathbf{U}}^k(m,n) + \sigma_{\mathbf{V}}^{2(p)}}Y(m,n), \tag{29.72}$$

$$S_{\mathbf{U}|\mathbf{y}}^k(m,n) = \frac{S_{\mathbf{U}}^k(m,n)\sigma_{\mathbf{V}}^{2(k)}}{|\Delta^k(m,n)|^2 S_{\mathbf{U}}^k(m,n) + \sigma_{\mathbf{V}}^{2(k)}}. \tag{29.73}$$

In Eq. (29.71), $Y(m,n)$ is the 2D DFT of the observed image $y(i,j)$ and $M_{\mathbf{U}|\mathbf{y}}^k(m,n)$ is the 2D DFT of the unstacked vector $\mu_{\mathbf{U}|\mathbf{y}}^k$ into an $N \times N$ array. Taking the partial derivatives of $F(\theta/\theta^k)$ with respect to

$S_U(m, n)$ and $\Delta(m, n)$ and setting them equal to zero, we obtain the solutions that minimize $F(\theta/\theta^k)$, which represent $S_U^{(k+1)}(m, n)$ and $\Delta^{(k+1)}(m, n)$. They are equal to

$$S_U^{(k+1)}(m, n) = S_{U|y}^k(m, n) + \frac{1}{N^2}|M_{U|y}^k(m, n)|^2 , \tag{29.74}$$

$$\Delta^{(k+1)}(m, n) = \frac{1}{N^2} \frac{Y(m, n)M_{U|y}^{(k)*}(m, n)}{S_{U|y}^k(m, n) + \frac{1}{N^2}|M_{U|y}^k(m, n)|^2}, \tag{29.75}$$

where $M_{U|y}^k(m, n)$ and $S_{U|y}^k(m, n)$ are computed by Eqs. (29.72) and (29.73). Substituting Eq. (29.75) into Eq. (29.71) and then minimizing $F(\theta/\theta^k)$ with respect to σ_V^2, we obtain

$$\sigma_V^{2(k+1)} = \frac{1}{N^2} \sum_{m=0}^{N-1}\sum_{n=0}^{N-1} \left\{ |\Delta^{(k+1)}(m, n)|^2 \left(S_{U|y}^k(m, n) + \frac{1}{N^2}|M_{U|y}^k(m, n)|^2 \right) \right.$$
$$\left. + \frac{1}{N^2} \left(|Y(m, n)|^2 - 2Re\left[Y^*(m, n)\Delta^{(k+1)}(m, n)M_{U|y}^k(m, n) \right] \right) \right\} . \tag{29.76}$$

According to Eq. (29.72) the restored image (i.e., $M_{U|y}^k(m, n)$) is the output of a Wiener filter, based on the available estimate of θ, with the observed image as input.

{u,v} as the complete data (CD_uv algorithm)

The second choice of the complete data is $\mathbf{x} = [\mathbf{u}^H \ \mathbf{v}^H]^H$, therefore, $\mathbf{H} = [\mathbf{D} \ \mathbf{I}]$. Following similar steps as in the previous case it has been shown that the equations for evaluating the spectrum of the original image are the same as in the previous case, i.e., Eqs. (29.72), (29.73) and (29.74) hold true. The other two unknowns, i.e., the variance of the additive noise and the DFT of the PSF are given by

$$\sigma_V^{2(k+1)} = \frac{1}{N^2} \sum_{m=0}^{N-1}\sum_{n=0}^{N-1} \left(S_{V|y}^k(m, n) + \frac{1}{N}|M_{V|y}^k(m, n)|^2 \right), \tag{29.77}$$

where

$$M_{V|y}^k(m, n) = \frac{\sigma_V^{2(k)}}{|\Delta^k(m, n)|^2 S_U^k(m, n) + \sigma_V^{2(k)}} Y(m, n) , \tag{29.78}$$

$$S_{V|y}^k(m, n) = \frac{|\Delta^k(m, n)|^2 S_U^k(m, n)\sigma_V^{2(k)}}{|\Delta^k(m, n)|^2 S_U^k(m, n) + \sigma_V^{2(k)}} , \tag{29.79}$$

and

$$|\Delta^k(m, n)|^2 = \begin{cases} \dfrac{\frac{1}{N^2}|Y(m,n)|^2 - \sigma_V^{2(k)}}{S_U^k(m,n)} , & \text{if } \frac{1}{N^2}|Y(m, n)|^2 > \sigma_V^{2(k)} \\ 0, & \text{otherwise .} \end{cases} \tag{29.80}$$

From Eq. (29.80) we observe that only the magnitude of $\Delta^k(m, n)$ is available, as was mentioned earlier. A similar observation can be made for Eq. (29.75), according to which the phase of $\Delta(m, n)$ is equal to the phase of $\Delta^0(m, n)$.

In deriving the above expressions the set of unknown parameters θ was divided into two sets $\theta_1 = \{\Lambda_U, \Lambda_V\}$ and $\theta_2 = \{D\}$. $F(\theta_1/\theta^k)$ was then minimized with respect to θ_1, resulting in Eqs. (29.74) and (29.77). The likelihood function in Eq. (29.59) was then minimized directly with respect to $\Delta(m, n)$ assuming knowledge of θ_1^k, resulting in Eq. (29.80). The effect of mixing the optimization procedure into the EM algorithm has not been completely analyzed theoretically. That is, the convergence properties of the EM algorithm do not necessarily hold, although the application of the resulting equations increases the likelihood function. Based on the experimental results, the algorithm derived in this section always converges to a stationary point. Furthermore, the results are comparable to the ones obtained with the **CD_uy** algorithm.

{ Dx,v } as the complete data (CD_Dx,v algorithm)

The third choice of the complete data is $\mathbf{x} = [(\mathbf{D}\mathbf{u})^H, \mathbf{v}^H]^H$. In this case, $\mathbf{D}$ and $\mathbf{x}$ cannot be estimated separately, since various combinations of $\mathbf{D}$ and $\mathbf{u}$ can result in the same $\mathbf{D}\mathbf{u}$. The two quantities $\mathbf{D}$ and $\mathbf{u}$ are lumped into one quantity $\mathbf{t} = \mathbf{D}\mathbf{u}$.

Following similar steps as in the two previous cases it has been shown [38, 39, 57] that the variance of the additive noise is computed according to Eq. (29.77), while the spectrum of the noise-free but blurred image $\mathbf{t}$ by the iterations

$$S_T^{(k+1)}(m, n) = S_{T|y}^k(m, n) + \frac{1}{N^2}|M_{T|y}^k(m, n)|^2, \tag{29.81}$$

where

$$M_{T|y}^k(m, n) = \frac{S_T^k(m, n)}{S_T^k(m, n) + \sigma_V^{2(k)}} Y(m, n), \tag{29.82}$$

and

$$S_{T|y}^k(m, n) = S_T^k(m, n) - \frac{S_T^{(k)2}(m, n)}{S_T^k(m, n) + \sigma_V^{2(k)}} Y(m, n). \tag{29.83}$$

Iterative Wiener Filtering

In this subsection, we deviate somewhat from the original formulation of the identification problem by assuming that the blur function is known. The problem at hand then is the restoration of the noisy-blurred image. Although there are a great number of approaches that can be followed in this case, the Wiener filtering approach represents a commonly used choice. However, in Wiener filtering knowledge of the power spectrum of the original image ($\mathbf{S_U}$) and the additive noise ($\mathbf{S_V}$) is required. A standard assumption is that of ergodicity, i.e., ensemble averages are equal to spatial averages. Even in this case, the estimation of the power spectrum of the original image has to be based on the observed noisy-blurred image, since the original image is not available. Assuming that the noise is white, its variance σ_V^2 needs also to be estimated from the observed image. Approaches, according to which the power spectrum of the original image is computed from images with similar statistical properties, have been suggested in the literature [35]. However, a reasonable idea is to successively use the Wiener-restored image as an improved prototype for updating the unknown S_U and σ_V^2. This idea is precisely implemented by the **CD_uy** algorithm.

More specifically, now that the blur function is known, Eq. (29.75) is removed from the EM iterations. Thus, Eqs. (29.74) and (29.76) are used to estimate $\mathbf{S_U}$ and σ_V^2, respectively, while Eq. (29.72) is used to compute the Wiener-filtered image. The starting point S_U^0 for the Wiener iteration can be chosen to be equal to

$$S_U^0(m, n) = \hat{S}_Y(m, n), \tag{29.84}$$

where $\hat{S}_Y(m, n)$ is an estimate of the power spectral density of the observed image. The value of $\sigma_V^{2(0)}$ can be determined from flat regions in the observed image, since this represents a commonly used approach for estimating the noise variance.

29.4.2 Multi-Channel Image Identification and Restoration

Introduction

We use the term *multi-channel images* to define the multiple image planes (channels) which are typically obtained by an imaging system that measures the same scene using multiple sensors. Multi-channel images exhibit strong between-channel correlations. Representative examples are multispectral images [41], microwave radiometric images [42], and image sequences [43]. In the first case such images are acquired for remote sensing and facilities/military surveillance applications. The channels are the different frequency bands (color images represent a special case of great interest). In the last case the channels are the different time frames after motion compensation. More recent applications of multi-channel filtering theory include the processing of the wavelet decomposed single-channel image [44] and the reconstruction of a high resolution image from multiple low resolution images [45, 46, 47, 48].

Although the problem of single channel image restoration has been thoroughly researched, significantly less work has been done on the problem of multi-channel restoration. The multi-channel formulation of the restoration problem is necessary when cross-channel degradations exist. It can be useful, however, in the case when only within-channel degradations exist, since cross-correlation terms are exploited to achieve better restoration results [49, 50]. The cross-channel degradations may come in the form of channel crosstalks, leakage in detectors, and spectral blurs [51]. Work on restoring multi-channel images is reported in [42, 49, 50, 51, 52, 53, 54, 55], when the within- and cross-channel (where applicable) blurs are known.

29.4.3 Problem Formulation

The degradation process is modeled again as [35]

$$y = Du + v, \tag{29.85}$$

where y, u, and v are the observed (noisy and degraded) image, the original undistorted image, and the noise process, respectively, all of which have been lexicographically ordered, and D the resulting degradation matrix. The noise process is assumed to be white Gaussian, independent of u.

Let P be the number of channels, each of size $N \times N$. If u_i, $i = 0, 1, \ldots, P-1$, represents the i-th channel. Then using the ordering of [56], the multichannel image u can be represented in vector form as

$$u = \left[u_1(0)u_2(0) \ldots u_P(0)u_1(1) \ldots u_P(1) \ldots u_1(N^2 - 1) \ldots u_P(N^2 - 1) \right]^T. \tag{29.86}$$

Defining y and v similarly to that of Eq. (29.86), we can now use the degradation model of Eq. (29.85), recognizing that y, u, and v are of size $PN^2 \times 1$, and D is of size $PN^2 \times PN^2$.

Assuming that the distortion system is linear shift invariant, D is a $PN^2 \times PN^2$ matrix of the form

$$D = \begin{bmatrix} D(0) & D(1) & \cdots & D(N^2 - 1) \\ D(N^2 - 1) & D(0) & \cdots & D(N^2 - 2) \\ \vdots & \vdots & \ddots & \vdots \\ D(1) & D(2) & \cdots & D(0) \end{bmatrix}, \tag{29.87}$$

where the $P \times P$ sub-matrices (sub-blocks) have the form

$$D(m) = \begin{bmatrix} D_{11}(m) & D_{12}(m) & \cdots & D_{1P}(m) \\ D_{21}(m) & D_{22}(m) & \cdots & D_{2P}(m) \\ \vdots & \vdots & \ddots & \vdots \\ D_{P1}(m) & D_{P2}(m) & \cdots & D_{PP}(m) \end{bmatrix}, \quad 0 \le m \le N^2 - 1. \tag{29.88}$$

Note that $D_{ii}(m)$ represents the intrachannel blur, while $D_{ij}(m)$, $i \ne j$ represents the interchannel blur. The matrix D in Eq. (29.87) is circulant at the block level. However, for D to be block-circulant, each of its subblocks $D(m)$ also needs to be circulant, which, in general, is not the case. Matrices of this form are called semiblock circulant (SBC) matrices [56]. The singular values of such matrices can be found with the use of the discrete Fourier transform (DFT) kernels. Equation (29.85) can therefore be written in the vector DFT domain [56].

Similarly, the covariance matrix of the original signal, Λ_U, and the covariance matrix of the noise process, Λ_V, are also semiblock circulant (assuming u and v are stationary). Note that Λ_U is not block-circulant because there is no justification to assume stationarity between channels (i.e., $\Lambda_{U_i U_j}(m) =$

$E[\mathbf{u}_i(m)\mathbf{u}_j(m)^*]$ is not equal to $\Lambda_{U_{i+p}U_{j+p}}(m) = E[\mathbf{u}_{i+p}(m)\mathbf{u}_{j+p}(m)^*]$ [50], where $\Lambda_{U_iU_j}(m)$ is the $(i, j)^{th}$ submatrix of Λ_U). However, Λ_U and Λ_V are semiblock circulant because $\mathbf{u}_i$ and $\mathbf{v}_i$ are assumed to be stationary within each channel.

29.4.4   The E-Step

We follow here similar steps to the ones presented in the previous section. We choose $[\mathbf{u}^H\mathbf{y}^H]^H$ as the complete data. Since the matrices Λ_U, Λ_V, and D, are assumed to be semi-block circulant, the E-step requires the evaluation of

$$F\left(\theta;\theta^k\right) = \sum_{m=0}^{N-1}\sum_{n=0}^{N-1} J(m,n), \tag{29.89}$$

where

$$
\begin{aligned}
J(m,n) &= \log|\Theta_U(m,n)| + \log|\Theta_V(m,n)| + \mathrm{tr}\bigg\{\Big(\Theta_U^{-1}(m,n) \\
&+ \Theta_D^H(m,n)\Theta_V^{-1}(m,n)\Theta_D(m,n)\Big)\Theta_{U|y}^k(m,n)\bigg\} \\
&+ \frac{1}{N^2}\mathrm{tr}\bigg\{\Big[\Theta_U^{-1}(m,n) + \Theta_D^H(m,n)\Theta_V^{-1}(m,n)\Theta_D(m,n)\Big] \\
&\times \mathbf{M}_{U|y}^k(m,n)\mathbf{M}_{U|y}^{(k)H}(m,n)\bigg\} \\
&- \frac{1}{N^2}\Big(\mathbf{Y}^H(m,n)\Theta_V^{-1}(m,n)\Theta_D(m,n)\,\mathbf{M}_{U|y}^k(m,n) \\
&+ \mathbf{M}_{U|y}^{(k)H}(m,n)\Theta_D^H(m,n)\Theta_V^{-1}(m,n)\mathbf{Y}(m,n)\Big) \\
&+ \frac{1}{N^2}\mathbf{Y}^H(m,n)\Theta_V^{-1}(m,n)\mathbf{Y}(m,n). \tag{29.90}
\end{aligned}
$$

The derivation of Eq. (29.90) is presented in detail in [48, 57, 58]. Equation (29.89) is the corresponding equation to Eq. (29.71) for the multichannel case.

In Eq. (29.90), $\Theta_U(m,n)$ is the (m,n)-th component matrix of Θ_U, which is related to Λ_U by a similarity transformation using two-dimensional discrete Fourier kernels [56, 57]. To be more specific, for $P = 3$, the matrix,

$$\Theta_U(m,n) = \begin{bmatrix} S_{11}(m,n) & S_{12}(m,n) & S_{13}(m,n) \\ S_{21}(m,n) & S_{22}(m,n) & S_{23}(m,n) \\ S_{31}(m,n) & S_{32}(m,n) & S_{33}(m,n) \end{bmatrix}, \tag{29.91}$$

consists of all the (m,n)-th component of the power and cross power spectra of the original color image (without loss of generality in the subsequent discussion three-channel examples will be used). It is worthwhile noting here that the power spectra $S_{ii}(m,n)$, $i = 1,2,3$, which are the diagonal entries of $\Theta_U(m,n)$, are real-valued, while the cross power spectra (the off-diagonal entries) are complex. This illustrates one of the main differences between working with multichannel images as opposed to single-channel images. In addition to each frequency component being a $P \times P$ matrix versus a scalar quantity for the single-channel case, the cross power spectra is complex versus being real for the single-channel case. Similarly, the (m,n)-th component of the inverse of the noise spectrum matrix is given by

$$\Theta_V^{-1}(m,n) = \begin{bmatrix} z_{11}(m,n) & z_{12}(m,n) & z_{13}(m,n) \\ z_{21}(m,n) & z_{22}(m,n) & z_{23}(m,n) \\ z_{31}(m,n) & z_{32}(m,n) & z_{33}(m,n) \end{bmatrix}. \tag{29.92}$$

One simplifying assumption that we can make about Eq. (29.92) is that the noise is white within channels and zero across channels. This results in $\Theta_V(m,n)$ being the same diagonal matrix for all (m,n).

$\Theta_{\mathbf{D}}(m, n)$ in Eq. (29.90) is equal to

$$\Theta_{\mathbf{D}}(m, n) = \begin{bmatrix} \Delta_{11}(m, n) & \Delta_{12}(m, n) & \Delta_{13}(m, n) \\ \Delta_{21}(m, n) & \Delta_{22}(m, n) & \Delta_{23}(m, n) \\ \Delta_{31}(m, n) & \Delta_{32}(m, n) & \Delta_{33}(m, n) \end{bmatrix}, \tag{29.93}$$

where $\Delta_{ij}(m, n)$ is the within-channel $(i = j)$ or cross-channel $(i \neq j)$ frequency response of the blur system, and $\mathbf{Y}(m, n)$ is the (m, n)-th component of the DFT of the observed image. $\Theta_{\mathbf{U}|\mathbf{y}}^{k}(m, n)$ and $\mathbf{M}_{\mathbf{U}|\mathbf{y}}^{k}(m, n)$ are the (m, n)-th frequency component matrix and vector of the multichannel counterparts of $\Lambda_{\mathbf{U}|\mathbf{y}}$ and $\mu_{\mathbf{U}|\mathbf{y}}$, respectively, computed by

$$\Theta_{\mathbf{U}|\mathbf{y}}^{k}(m, n) = \Theta_{\mathbf{U}}^{k}(m, n) - \Theta_{\mathbf{U}}^{k}(m, n)\Theta_{\mathbf{D}}^{(k)H}(m, n)\left[\Theta_{\mathbf{V}}^{k}(m, n)\right.$$
$$+ \left.\Theta_{\mathbf{D}}^{k}(m, n)\Theta_{\mathbf{U}}^{k}(m, n)\Theta_{\mathbf{D}}^{(k)H}(m, n)\right]^{-1}\Theta_{\mathbf{D}}^{k}(m, n)\Theta_{\mathbf{U}}^{k}(m, n) \tag{29.94}$$

and

$$\mathbf{M}_{\mathbf{U}|\mathbf{y}}^{k}(m, n) = \Theta_{\mathbf{U}}^{k}(m, n)\Theta_{\mathbf{D}}^{(k)H}(m, n)\left[\Theta_{\mathbf{V}}^{k}(m, n)\right.$$
$$+ \left.\Theta_{\mathbf{D}}^{k}(m, n)\Theta_{\mathbf{U}}^{k}(m, n)\Theta_{\mathbf{D}}^{(k)H}(m, n)\right]^{-1}\mathbf{Y}(m, n). \tag{29.95}$$

29.4.5 The M-Step

The M-step requires the minimization of $J(m, n)$ with respect to $\Theta_{\mathbf{U}}(m, n)$, $\Theta_{\mathbf{V}}(m, n)$ and $\Theta_{\mathbf{D}}(m, n)$. The resulting solutions become $\Theta_{\mathbf{U}}^{(k+1)}(m, n)$, $\Theta_{\mathbf{V}}^{(k+1)}(m, n)$ and $\Theta_{\mathbf{D}}^{(k+1)}(m, n)$, respectively.

The minimization of $J(m, n)$ with respect to $\Theta_{\mathbf{U}}$ is straightforward, since $\Theta_{\mathbf{U}}$ is decoupled from $\Theta_{\mathbf{V}}(\mathbf{m}, \mathbf{n})$ and $\Theta_{\mathbf{D}}$. An equation similar to Eq. (29.74) results. The minimization of $J(m, n)$ with respect to $\Theta_{\mathbf{D}}$ is not as straightforward; $\Theta_{\mathbf{D}}$ is coupled with $\Theta_{\mathbf{V}}$. Therefore, in order to minimize $J(m, n)$ with respect to $\Theta_{\mathbf{D}}$, $\Theta_{\mathbf{V}}$ must be solved first in terms of $\Theta_{\mathbf{D}}$, substituted back into Eq. (29.90), and then minimized with respect to $\Theta_{\mathbf{D}}$.

It is shown in [48, 58] that two conditions must be met in order to obtain explicit equations for the blur. First, the noise spectrum matrix, $\Theta_{\mathbf{V}}(m, n)$, must be a diagonal matrix, which is frequently encountered in practice. Second, all of the blurs must be symmetric, so that there is no phase when working in the discrete frequency domain. The first condition arises from the fact that $\Theta_{\mathbf{V}}(m, n)$ and $\Theta_{\mathbf{D}}(m, n)$ are coupled. The second condition arises from the Cauchy-Riemann theorem, and must be satisfied in order to guarantee the existence of a derivative at every point.

With these conditions, the iterations for $\Delta(m, n)$ and $\sigma_{V}(m, n)$ are derived in [48, 58], which are similar respectively to Eqs. (29.75) and (29.76). Special cases are also analyzed in [48, 58], when the number of unknowns is reduced. For example, if $\Theta_{\mathbf{D}}$ is known, the multichannel Wiener filter results.

29.5   Experimental Results

The effectiveness of both the single channel and multi-channel restoration and identification algorithms is demonstrated experimentally. The red, green, and blue (RGB) channels of the original Lena image used for this experiment are shown in Fig. 29.1. A 5×5 truncated Gaussian blur is used for each channel and Gaussian white noise is added resulting in a blurred signal-to-noise ratio (SNR) of 20 dB. The degraded channels are shown in Fig. 29.2. Three different experiments were performed with the available degraded data. The single channel algorithm of Eqs. (29.74), (29.75), and (29.76) was first run for each of the RGB

FIGURE 29.1   Original RGB Lena.

FIGURE 29.2   Degraded RGB Lena, intra-channel blurs only, 20 dB SNR.

FIGURE 29.3   Restored RGB by the decoupled single channel EM algorithm.

channels independently. The restored images are shown in Fig. 29.3. The corresponding multichannel algorithm was then run, resulting in the restored channels shown in Fig. 29.4. Finally the multichannel Wiener filter was also run, in demonstrating the upper bound of the algorithm's performance, since the blurs are now exactly known. The resulting restored images are shown in Fig. 29.5. The improvement in SNR for the three experiments and for each channel is shown in Table 29.1. According to this table,

TABLE 29.1   Improvement in SNR (dB)

η	Decoupled EM	Multichannel EM	Wiener
Red	1.5573	2.1020	2.3420
Green	1.3814	2.0086	2.3181
Blue	1.1520	1.5148	1.8337

the performance of the algorithm increases from the first to the last experiment. This is to be expected, since in considering the multichannel algorithm over the single channel algorithm the correlation between channels is taken into account, which brings additional information into the problem.

A photographically blurred image is shown next in Fig. 29.6. The restorations of it by the **CD_uy** and **CD_uv** algorithms are shown, respectively, in Figs. 29.7 and 29.8.

FIGURE 29.4   Restored RGB Lena by the multi-channel EM algorithm.

FIGURE 29.5   Restored RGB Lena by the iterative multi-channel Wiener algorithm.

FIGURE 29.6   Photographically blurred image.

FIGURE 29.7   Restored image by the **CD_uy** algorithm.

FIGURE 29.8   Restored image by the **CD_uv** algorithm.

29.5.1   Comments on the Choice of Initial Conditions

The likelihood function which is optimized is highly nonlinear and a number of local minima exist. Although the incorporation of the various constraints, discussed earlier, restricts the set of possible solutions, a number of local minima still exist. Therefore, the final result depends on the initial conditions. Based on our experience in implementing the EM iterations of the previous sections for the single-channel and the multi-channel image restoration cases, the following comments and observations are in order.

It was observed experimentally that the final results are quite insensitive to variations in the values of the noise variance(s) and the original image power spectra. An estimate of the noise variances from flat regions of the noisy and blurred images were used as initial condition. It was observed that using initial estimates of the noise variances larger than the actual ones produced good final results.

The final results are quite sensitive, however, to variations in the values of the PSF. Knowledge of the support of the PSF is quite important. In [38] after convergence of the EM algorithm the estimate of the PSF was truncated, normalized, and used as an initial condition in restarting another iteration cycle.

29.6   Summary and Conclusion

In this chapter, we have described and illustrated how the EM algorithm can be used in image recovery problems. The basic approach can be summarized by the following steps.

1. Select a statistical model for the observed data and formulate the image recovery problem as an MLE problem.
2. If the likelihood function is difficult to optimize directly, the EM algorithm can be used by properly selecting the complete data.
3. Constraints on the parameters or image to be estimated, proper initial conditions, and multiple complete data spaces can be considered to improve the uniqueness and convergence of the estimates.
4. Derive the equations for the E-step and M-step.

We end this chapter with several remarks. We want to emphasize again that the EM algorithm only guarantees convergence to a local optimum. Therefore, the initial conditions are quite critical, as is also discussed in the previous section. Depending on the number of the unknown parameters, one could consider evaluating in a systematic fashion the likelihood function directly at a number of points and use as initial condition the point which results in the largest value of the likelihood function. Improved results can be obtained potentially if the number of the unknown parameters is reduced by parameterizing the unknown functions. For example, separable and nonseparable exponential covariance models are used in [46, 47, 48], and an autoregressive model in [38, 57] to model the original image, and parameterized blur models are discussed in [38]. We want to mention also that the EM algorithm can be implemented in different domains. For example, it is implemented in both spatial and frequency domains, respectively, in sections 29.3 and 29.4. Other domains are also possible by applying proper transforms, e.g., the wavelet transform [59].

References

[1] Jain, A.K., *Fundamentals of Digital Image Processing*, Prentice Hall, Englewood Cliffs, NJ, 1989.

[2] Yang, Y., Galatsanos, N.P., and Katsaggelos, A.K., Regularized image reconstruction to remove blocking artifacts from block discrete cosine transform compressed images, *IEEE Trans. Circuits Syst. Video Technol.*, 3(6): 421–432, December, 1993.

[3] Yang, Y., Galatsanos, N.P., and Katsaggelos, A.K., Projection-based spatially-adaptive reconstruction of block transform compressed images, *IEEE Trans. Image Process.*, 4(7): 896–908, July, 1995.

[4] Parker, A.J., *Image Reconstruction in Radiology*, CRC Press, Boca Raton, FL, 1990.

[5] Russ, J.C., *The Image Processing Handbook*, CRC Press, Boca Raton, FL, 1992.

[6] Snyder, D.L. and Miller, M.I., *Random Processes in Time and Space*, 2nd ed., Springer-Verlag, 1991.

[7] Shepp, L. and Vardi, Y., Maximum-likelihood reconstruction for emission tomography, *IEEE Trans. Med. Imag.*, 1: 113–122, Oct., 1982.

[8] Katsaggelos, A.K., Ed., *Digital Image Restoration*, Springer-Verlag, 1991.

[9] Lagendijk, R.L. and Biemond, J., *Iterative Identification and Restoration of Images*, Kluwer Academic Publishers, 1991.

[10] Cox, D.R and Hinkley, D.V., *Theoretical Statistics*, Chapman and Hall, 1974.

[11] Dempster, A.P., Laird, N.M., and Rubin, D.B., Maximum likelihood from incomplete data via the EM algorithm, *J. Roy. Soc. Statist.*, Series B, 39: 1–38, 1977.

[12] Hebert, T. and Leahy, R., A generalized EM algorithm for 3-D Bayesian reconstruction from Poisson data using Gibbs priors, *IEEE Trans. Med. Imaging*, 8: 194–202, June, 1989.

[13] Green, P.J., On use of the EM algorithm for penalized likelihood estimation, *J. Roy. Soc. Statist.*, Series B, 52: 443–452, 1990.

[14] Zhang, J., The mean field theory in EM procedures for Markov random fields, *IEEE Trans. ASSP*, 40: 2570–2583, October, 1992.

[15] Zhang, J., The mean field theory in EM procedures for blind Markov random field image restoration, *IEEE Trans. Image Process.*, 2: 27–40, Jan., 1993.

[16] Feder, M. and Weinstein, E., Parameter estimation of superimposed signals using the EM algorithm, *IEEE Trans. ASSP*, 36: 477–489, April, 1988.

[17] Fessler, J.A. and Hero, A.O., Space alternating generalized expectation-maximization algorithm, *IEEE Trans. SP*, 42: 2664–2678, Oct., 1994.

[18] Fessler, J.A. and Hero, A.O., Complete data space and generalized EM algorithm, *Proc. ICASSP*, Vol. IV, pp. 1–4, Mineapolis, Minnesota, April 27-30, 1993.

[19] Hero, A.O and Fessler, J.A., Convergence in norm for alternating expectation-maximization (EM) type algorithms, *Statistica Sinica*, 5: 41–54, Jan., 1995.

[20] Wu, J., On the convergence properties of the EM algorithm, *The Annals of Statistics*, 11: 95–103, 1983.

[21] Redner, R.A. and Walker, H.F., Mixture densities, maximum likelihood and the EM algorithm, *SIAM Review*, 26(2): 195–239, 1984.

[22] Besag, J., Spatial interaction and the statistical analysis of lattice systems, *J. Roy. Statist. Soc.*, Series B, 36: 192–226, 1974.

[23] Geman, S. and Geman, D., Stochastic relaxation, Gibbs distribution, and the Bayesian restoration of images, *IEEE Trans. PAMI,* 6: 721–741, Nov., 1984.

[24] Chellappa, R. and Jain, A., Eds., *Markov Random Fields — Theory and Applications,* Academic Press, 1993.

[25] Chandler, D., *Introduction to Modern Statistical Mechanics,* Oxford University Press, 1987.

[26] Bouman, C. and Sauer, K., Maximum likelihood scale estimation for a class of Markov random fields, *Proc. ICASSP,* pp. V537–540, Adelaide, Australia, April, 19-22, 1994.

[27] Metropolis, N., et al., Equation of state calculation by fast computing machines, *J. Chem. Phys.,* 21: 1087–1092, 1953.

[28] Konrad, J. and Dubois, E., Comparison of stochastic and deterministic solution methods in Bayesian estimation of 2D motion, *Image and Visual Computing,* 8(4): 304–317, Nov., 1990.

[29] Cover, T. and Thomas, J., *Elements of Information Theory,* John Wiley & Sons, 1992.

[30] Zhang, J., The application of the Gibbs-Bogoliubov-Feynmann inequality in the mean field theory for Markov random fields, Preprint, 1995.

[31] Wu, C.-H. and Doerschuk, P.C., Cluster expansions for the deterministic computation of Bayesian estimators based on Markov random fields, *IEEE Trans. PAMI,* 17: 275–293, March, 1995.

[32] Anderson, B.D.O. and Moore, J. B., *Optimal Filtering,* Prentice-Hall, Englewood Cliffs, NJ, 1979.

[33] Banham, M.R. and Katsaggelos, A.K., Digital restoration of images, *IEEE Signal Process. Mag.,* 14(2): 24–41, Mar., 1997.

[34] Tekalp, A.M., Kaufman, H., and Woods, J.W., *IEEE Trans. ASSP-***34**: 963–972, 1986.

[35] Andrews, H.C. and Hunt, B.R., *Digital Image Restoration,* Prentice-Hall, Englewood Cliffs, NJ, 1977.

[36] Dudgeon, D.E. and Mersereau, R.M., *Multidimensional Digital Signal Processing,* Prentice-Hall, Englewood Cliffs, NJ, 1984.

[37] Gray, R.M., *IEEE Trans. IT-***18**: 725–730, 1985.

[38] Katsaggelos, A.K. and Lay, K.T., Identification and restoration of images using the expectation maximization algorithm, in *Digital Image Restoration,* Katsaggelos, A.K., Ed., Springer-Verlag, 1991.

[39] Lay, K.T. and Katsaggelos, A.K., Image identification and restoration based on the expectation-maximization algorithm, *Opt. Eng.,* 29: 436–445, May, 1990.

[40] Kailath, T., *Linear Systems,* Prentice-Hall, Englewood Cliffs, NJ, 1980.

[41] Lee, J.B., Woodyatt, A.S., and Berman, M., Enhancement of high spectral resolution remote-sensing data by a noise adjusted principle component transform, *IEEE Trans. Geosci. Remote Sens.,* 28(3): 295–304, 1990.

[42] Chin, R.T., Yeh, C.L., and Olson, W.S., Restoration of multichannel microwave radiometric images, *IEEE Trans. Patt. Anal. Mach. Intell.,* PAMI-7(4): 475–484, July, 1985.

[43] Choi, M.G., Galatsanos, N.P., and Katsaggelos, A.K., Multichannel regularized iterative restoration of image sequences, *J. Visual Commun. Image Represent.,* 7(3): 244–258, Sept., 1996.

[44] Banham, M.R., Galatsanos, N.P., Gonzalez, H., and Katsaggelos, A.K., Multichannel restoration of single channel images using a wavelet-based subband decomposition, *IEEE Trans. Image Process.,* 3(6): 821–833, Nov., 1994.

[45] Tsai, R.Y. and Huang, T.S., Multiframe image restoration and registration, in *Advances in Computer Vision and Registration,* vol. 1, *Image Reconstruction from Incomplete Observations,* Huang, T.S., Ed., pp. 317–339, JAI Press, 1984.

[46] Tom, B.C. and Katsaggelos, A.K., Reconstruction of a high resolution image from multiple degraded mis-registered low resolution images, *Proc. SPIE, Visual Communications and Image Processing,* Chicago, Vol. 2308, pt. 2, pp. 971–981, Sept., 1994.

[47] Tom, B.C., Katsaggelos, A.K., and Galatsanos, N.P., Reconstruction of a high resolution from registration and restoration of low resolution images, *IEEE Proc. Int. Conf. Image Process.,* Austin, Vol. 3, pp. 553–557, Nov., 1994.

[48] Tom, B.C., Reconstruction of a High Resolution Image from Multiple Degraded Mis-Registered Low Resolution Images, Ph.D. thesis, Northwestern University, Dept. of EECS, June, 1995.

[49] Hunt, B.R. and Kübler, O., Karhunen-Loeve multispectral image restoration, part I : theory, *IEEE Trans. Acoust., Speech, Signal Process.,* ASSP-32(3): 592–600, June, 1984.

[50] Galatsanos, N.P. and Chin, R.T., Digital restoration of multichannel images, *IEEE Trans. Acoust., Speech, Signal Process.,* ASSP-37(3): 415–421, March, 1989.

[51] Galatsanos, N.P. and Chin, R.T., Restoration of color images by multichannel Kalman filtering, *IEEE Trans. Signal Process.,* 39(10): 2237–2252, Oct., 1991.

[52] Galatsanos, N.P., Katsaggelos, A.K., Chin, R.T., and Hillery, A.D., Least squares restoration of multichannel images, *IEEE Trans. Signal Process.,* 39: 2222–2236, Oct., 1991.

[53] Tekalp, A.M. and Pavlovic, G., Multichannel image modeling and Kalman filtering for multi-spectral image restoration, *IEEE Trans. Signal Process.,* 19(3): 221–232, March, 1990.

[54] Kang, M.G. and Katsaggelos, A.K., Simultaneous multichannel image restoration and estimation of the regularization parameters, *IEEE Trans. Image Process.,* 6(5) 774–778, May, 1997.

[55] Zhu, W., Galatsanos, N.P., and Katsaggelos, A.K., Regularized multichannel restoration using cross-validation, *Graph. Models Image Process.,* 57(1): 38–54, Jan., 1995.

[56] Katsaggelos, A.K., Lay, K.T., and Galatsanos, N.P., A general framework for frequency domain multichannel signal processing, *IEEE Trans. Image Process.,* 2(3): 417–420, July, 1993.

[57] Lay, K.T., Blur Identification and Image Restoration Using the EM Algorithm, Ph.D. thesis, Northwestern University, Dept. of EECS, Dec., 1991.

[58] Tom, B.C.S., Lay, K.T., and Katsaggelos, A.K., Multi-channel image identification and restoration using the expectation-maximization algorithm, *Optical Engineering,* Special Issue on Visual Communications and Image Processing, 35(1): 241–254, Jan., 1996.

[59] Banham, M.R., Wavelet Based Image Restoration Techniques, Ph.D. thesis, Northwestern University, Dept. of EECS, June, 1994.

30

Inverse Problems in Array Processing

Kevin R. Farrell
T-Netix/SpeakEZ

30.1   Introduction

Signal reception has numerous applications in communications, radar, sonar, and geoscience among others. However, the adverse effects of noise in these applications limit their utility. Hence, the quest for new and improved noise removal techniques is an ongoing research topic of great importance in a vast number of applications of signal reception.

When certain characteristics of noise are known, their effects can be compensated. For example, if the noise is known to have certain spectral characteristics, then a finite impulse response (FIR) or infinite impulse response (IIR) filter can be designed to suppress the noise frequencies. Similarly, if the statistics of the noise are known, then a Weiner filter can be used to alleviate its effects. Finally, if the noise is spatially separated from the desired signal, then multisensor arrays can be used for noise suppression. This last case is discussed in this article.

A multisensor array consists of a set of transducers, i.e., antennas, microphones, hydrophones, seismometers, geophones, etc. that are arranged in a pattern which can take advantage of the spatial location of signals. A two-element television antenna provides a good example. To improve signal reception and/or mitigate the effects of a noise source, the antenna pattern is manually adjusted to steer a low gain component of the antenna pattern towards the noise source. Multisensor arrays typically achieve this adjustment through the use of an array processing algorithm. Most applications of multisensor arrays involve a fixed pattern of transducers, such as a linear array. Antenna pattern adjustments are made by applying weights to the outputs of each transducer. If the noise arrives from a specific non-changing spatial location, then the weights will be fixed. Otherwise, if the noise arrives from random, changing locations then the weights must be adaptive. So, in a military communications application where a communications channel is subject to jamming from random spatial locations, an adaptive array processing algorithm would be the appropriate solution. Commercial applications of microphone arrays include teleconferencing [6] and hearing aids [9].

There are several methods for obtaining the weight update equations in array processing. Most of these are derived from statistically based formulations. The resulting optimal weight vector is then generally expressed in terms of the input autocorrelation matrix. An alternative formulation is to express the array processing problem as a linear system of equations to which iterative matrix inversion techniques can be applied. The matrix inverse formulation will be the focus of this article.

The following section provides a background overview of wave propagation, spatial sampling, and spatial filtering. Next, narrowband and broadband beamforming arrays are described along with the standard algorithms used for these implementations. The narrowband and broadband algorithms are then reformulated in terms of an inverse problem and an iterative technique for solving this system of equations is provided. Finally, several examples are given along with a summary.

30.2 Background Theory

Array processing uses information regarding the spatial locations of signals to aid in interference suppression and signal enhancement. The spatial locations of signals may be determined by the wavefronts that are emanated by the signal sources. Some background theory regarding wave propagation and spatial frequency is necessary to fully understand the interference suppression techniques used within array processing. The following subsections provide this background material.

30.2.1 Wave Propagation

An adaptive array consists of a number of sensors typically configured in a linear pattern that utilizes the spatial characteristics of signals to improve the reception of a desired signal and/or cancellation of undesired signals. The analysis used in this chapter assumes that a linear array is being used, which corresponds to the sensors being configured along a line. Signals may be spatially characterized by their angle of arrival with respect to the array. The angle of arrival of a signal is defined as the angle between the propagation path of the signal and the perpendicular of the array. Consider the wavefront emanating from a point source as is illustrated in Fig. 30.1. Here, the angle of arrival is shown as θ.

Note in Fig. 30.1 that wavefronts emanating from a point source may be characterized by plane waves (i.e., the locus of constant phase form straight lines) when originating from the far field or *Fraunhofer*, region. The far field approximation is valid for signals that satisfy the following condition:

$$s \geq \frac{D^2}{\lambda} \tag{30.1}$$

where s is the distance between the signal and the array, λ is the wavelength of the signal, and D is the length of the array. Wavefronts that originate closer than D^2/λ are considered to be from the near field or *Fresnel*, region. Wavefronts originating from the near field exhibit a convex shape when striking the array sensors. These wavefronts do not create linear phase shifts between consequetive sensors. However, the curvature of the wavefront allows algorithms to determine point source location in addition to direction of arrival [1]. The remainder of this article assumes that all wavefronts arrive from the far field region.

30.2.2 Spatial Sampling

In Fig. 30.1 it can be seen that the signal waveform experiences a time delay between crossing each sensor, assuming that it does not arrive perpendicular to the array. The time delay, τ, of the waveform striking

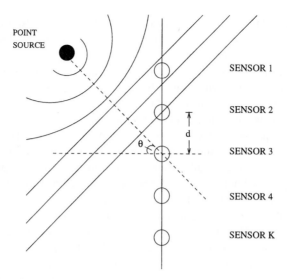

FIGURE 30.1 Propagating wavefront.

the first and then second sensors in Fig. 30.1 may be calculated as

$$\tau = \frac{d}{c} \sin \theta \tag{30.2}$$

where d is the sensor spacing, c is the speed of propagation of the given waveform for a particular medium (i.e., 3×10^8 m/s for electromagnetic waves through air, 1.5×10^3 m/s for sound waves through water, etc.), and θ is the angle of arrival of the wavefront. This time delay corresponds to a shift in phase of the signal as observed by each sensor. The phase shift, ϕ, or electrical angle observed at each sensor due to the angle of arrival of the wavefront may be found as

$$\phi = \frac{2\pi d}{\lambda_o} \sin \theta = \frac{\omega_o d}{c} \sin \theta . \tag{30.3}$$

Here, λ_o is the wavelength of the signal at frequency f_o as defined by

$$\lambda_o = \frac{c}{f_o} . \tag{30.4}$$

Hence, a signal $x(k)$ that crosses the sensor array and exhibits a phase shift ϕ between uniformly spaced, consequetive sensors can be characterized by the vector $\mathbf{x}(k)$, where:

$$\mathbf{x}(k) = x(k) \begin{bmatrix} 1 \\ e^{-j\phi} \\ e^{-2j\phi} \\ \vdots \\ e^{-j(K-1)\phi} \end{bmatrix} . \tag{30.5}$$

Uniform sensor spacing is assumed throughout the remainder of this article.

30.2.3 Spatial Frequency

The angle of arrival of a wavefront defines a quantity known as the *spatial frequency*. Adaptive arrays use information regarding the spatial frequency to suppress undesired signals that originate from different locations than that of the target signal. The spatial frequency is determined from the periodicity that is observed across an array of sensors due to the phase shift of a signal arriving at some angle of arrival.

Signals that arrive perpendicular to the array (known as boresight) create identical waveforms at each sensor. The spatial frequency of such signals is zero. Signals that do not arrive perpendicular to the array will not create waveforms that are identical at each sensor assuming that there is no spatial aliasing due to insufficiently spaced sensors. In general, as the angle increases, so does the spatial frequency. It can also be deduced that retaining signals having an angle of arrival equal to zero degrees while suppressing signals from other directions is equivalent to low pass filtering the spatial frequency. This provides the motivation for conventional or fixed-weight beamforming techniques. Here, the sensor values can be computed via a windowing technique, such as a rectangular, Hamming, etc. to yield a fixed suppression of non-boresight signals. However, adaptive techniques can locate the specific spatial frequency of an interfering signal and position a null in that exact location to achieve greater suppression.

There are two types of beamforming, namely conventional, or "fixed weight", beamforming and adaptive beamforming. A conventional beamformer can be designed using windowing and FIR filter theory. They utilize fixed weights and are appropriate in applications where the spatial locations of noise sources are known and are not changing. Adaptive beamformers make no such assumptions regarding the locations of the signal sources. The weights are adapted to accommodate the changing signal environment.

Arrays that have a visible region of $-90°$ to $+90°$ (i.e., the azimuth range for signal reception) require that the sensor spacing satisfy the relation

$$d \leq \frac{\lambda}{2}. \tag{30.6}$$

The above relation for sensor spacing is analogous to the Nyquist sampling rate for frequency domain analysis. For example, consider a signal that exhibits exactly one period between consequetive sensors. In this case, the output of each sensor would be equivalent, giving the false impression that the signal arrives normal to the array. In terms of the antenna pattern, insufficient sensor spacing results in grating lobes. Grating lobes are lobes other than the main lobe that appear in the visible region and can amplify undesired directional signals.

The spatial frequency characteristics of signals enable numerous enhancement opportunities via array processing algorithms. Array processing algorithms are typically realized through the implementation of narrowband or broadband arrays. These two arrays are discussed in the following sections.

30.3 Narrowband Arrays

Narrowband adaptive arrays are used in applications where signals can be characterized by a single frequency and thus occupy a relatively narrow bandwidth. A signal whose envelope does not change during the time their wavefront is incident on the transducers is considered to be narrowband. A narrowband adaptive array consists of an array of sensors followed by a set of adjustable gains, or weights. The outputs of the weighted sensors are summed to produce the array output. A narrowband array is shown in Fig. 30.2.

The input vector $\mathbf{x}(k)$ consists of the sum of the desired signal $\mathbf{s}(k)$ and noise $\mathbf{n}(k)$ vectors and is defined as

$$\mathbf{x}(k) = \mathbf{s}(k) + \mathbf{n}(k) \tag{30.7}$$

where k denotes the time instant of the input vector. The noise vector $\mathbf{n}(k)$ will generally consist of thermal noise and directional interference. At each time instant, the input vector is multiplied with the weight vector to obtain the array output, which is given as

$$y(k) = \mathbf{x}^T(k)\mathbf{w}, \quad \mathbf{x}, \mathbf{w} \in C^K, \tag{30.8}$$

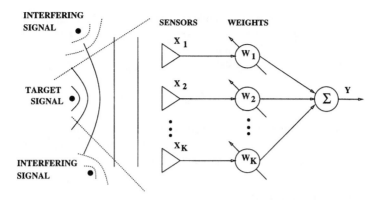

FIGURE 30.2 Narrowband array.

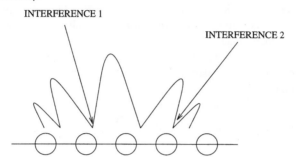

FIGURE 30.3 Sidelobe canceling.

where C^K is the complex space of dimension K. The array output is then passed to the signal processor which uses the previous value of the output and current values of the inputs to determine the adjustment to make to the weights. The weights are then adjusted and multiplied with the new input vector to obtain the next output. The output feedback loop allows the weights to be adjusted adaptively, thus accommodating nonstationary environments.

In Eq. (30.8), it is desired to find a weight vector that will allow the output y to approximately equal the true target signal. For the derivation of the weight update equations, it is necessary to know what *a priori* information is being assumed. One form of *a priori* information could be the spatial location of the target signal, also known as the "look-direction". For example, many array processing algorithms assume that the target signal arrives normal to the array, or else a steering vector is used to make it appear as such. Another form of *a priori* information is to use a signal at the receiving end that is correlated with the input signal, i.e., a pilot signal. Each of these criteria will be considered in the following subsections.

30.3.1 Look-Direction Constraint

One of the first narrowband array algorithms was proposed by Applebaum [2]. This algorithm is known as the *sidelobe canceler* and assumes that the direction of the target signal is known. The algorithm does not attempt to maximize the signal gain, but instead adjusts the sidelobes so that interfering signals coincide with the nulls of the antenna pattern. This concept is illustrated in Fig. 30.3.

Applebaum derived the weight update equation via maximization of the signal to interference plus thermal noise ratio (SINR). As derived in [2], this optimization results in the optimal weight vector as given by Eq. (30.9):

$$\mathbf{w}_{\text{opt}} = \mu R_{xx}^{-1} \mathbf{t} . \tag{30.9}$$

In Eq. (30.9), R_{xx} is the covariance matrix of the input, μ is a constant related to the signal gain, and

$\mathbf{t}$ is a steering vector that corresponds to the angle of arrival of the desired signal. This steering vector is equivalent to the phase shift vector of Eq. (30.5). Note that if the angle of arrival of the desired signal is zero, then the $\mathbf{t}$ vector will simply contain ones.

A discretized implementation of the Applebaum algorithm appears as follows:

$$\mathbf{w}^{(j+1)} = \mathbf{w}^{(j)} + \alpha \left(\mathbf{w}_q - \mathbf{w}^{(j)} \right) - \beta \mathbf{x}(k) y(k) . \tag{30.10}$$

In Eq. (30.10), $\mathbf{w}_q$ represents the quiescent weight vector (i.e., when no interference is present), the superscript j refers to the iteration, α is a gain parameter for the steering vector, and β is a gain parameter controlling the adaptation rate and variance about the steady state solution.

30.3.2   Pilot Signal Constraint

Another form of *a priori* information is to use a pilot signal that is correlated with the target signal. This results in a beamforming algorithm that will concentrate on maintaining a beam directed towards the target signal, as opposed to, or in addition to, positioning the nulls as in the case of the sidelobe canceler. One such adaptive beamforming algorithm was proposed by Widrow [20, 21]. The resulting weight update equation is based on minimizing the quantity $(y(k) - p(k))^2$ where $p(k)$ is the pilot signal. The resulting weight update equation is

$$\mathbf{w}^{(j+1)} = \mathbf{w}^{(j)} + \mu \epsilon(k) \mathbf{x}(k) . \tag{30.11}$$

This corresponds to the least means square (LMS) algorithm, where ϵ is the current error, namely $(y(k) - p(k))$, and μ is a scaling factor.

30.4   Broadband Arrays

Narrowband arrays rely on the assumption that wavefronts normal to the array will create identical waveforms at each sensor and wavefronts arriving at angles not normal to the array will create a linear phase shift at each sensor. Signals that occupy a large bandwidth and do not arrive normal to the array violate this assumption since the phase shift is a function of f_o and varying frequency will cause a varying phase shift. Broadband signals that arrive normal to the array will not be subject to frequency dependent phase shifts at each sensor as will broadband signals that do not arrive normal to the array. This is attributed to the coherent summation of the target signal at each sensor where the phase shift will be a uniform random variable with zero mean. A modified array structure, however, is necessary to compensate the interference waveform inconsistencies that are caused by variations about the center frequency. This can be achieved by having the weight for a sensor being a function of frequency, i.e., a FIR filter, instead of just being a scalar constant as in the narrowband case. Broadband adaptive arrays consist of an array of sensors followed by tapped delay lines, which is the major implementation difference between a broadband and narrowband array. A broadband array is shown in Fig. 30.4.

Consider the transfer functions for a given sensor of the narrowband and broadband arrays, shown by

$$H_{\text{narrow}}(w) = w_1 \tag{30.12}$$

and

$$H_{\text{broad}}(w) = w_1 + w_2 e^{-jwT} + w_3 e^{-2jwT} + \ldots + w_J e^{-j(J-1)wT} . \tag{30.13}$$

The narrowband transfer function has only a single weight that is constant with frequency. However, the broadband transfer function, which is actually a Fourier series expansion, is frequency dependent and allows for choosing a weight vector that may compensate phase variations due to signal bandwidth. This property of tapped delay lines provides the necessary flexibility for processing broadband signals. Note that typically four or five taps will be sufficient to compensate most bandwidth variances [14].

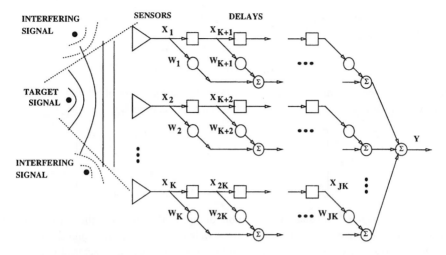

FIGURE 30.4 Broadband array.

The broadband array shown in Fig. 30.4 obtains values at each sensor and then propagates these values through the array at each time interval. Therefore, if the values x_1 through x_K are input at time instant one, then at time instant two, x_{K+1} through x_{2K} will have the values previously held by x_1 through x_K, x_{2K+1} through x_{3K} will have the values previously held by x_{K+1} through x_{2K}, etc. Also, at each time instant, a scalar value y will be calculated as the inner product of the input vector $\mathbf{x}$ and the weight vector $\mathbf{w}$. This array output is calculated as

$$y(k) = \mathbf{x}^T(k)\mathbf{w}, \quad \mathbf{x}, \mathbf{w} \in C^{JK}, \tag{30.14}$$

where C^{JK} is the complex space of dimension JK.

Although not shown in Fig. 30.4, a signal processor exists as in the narrowband array, which uses the previous output and current inputs to determine the adjustments to make to the weight vector $\mathbf{w}$. The output signal y will approach the value of the desired signal as the interfering signals are canceled until it converges to the desired signal in the least squares sense.

Broadband arrays have been analyzed by Widrow [20], Griffiths [10, 12], and Frost [7]. Widrow [20] proposed a LMS algorithm that minimizes the square of the difference between the observed output and the expected output, which was estimated with a pilot signal. This approach assumes that the angle of arrival and a pilot signal are available *a priori*. Griffiths [10] proposed a LMS algorithm that assumes knowledge of the cross-correlation matrix between the input and output data instead of the pilot signal. This method assumes that the angle of arrival and second order signal statistics are known *a priori*. The methods proposed by Widrow and Griffiths are forms of *unconstrained* optimization. Frost [7] proposed a LMS algorithm that assumes *a priori* knowledge of the angle of arrival and the frequency band of interest. The Frost algorithm utilizes a *constrained* optimization technique, which Griffiths later derived an *unconstrained* formulation that utilizes the same constraints [12]. The Frost algorithm will be the focus of this section.

The Frost algorithm implements the look-direction and frequency response constraints as follows. For the broadband array shown in Fig. 30.4, a target signal waveform propagating normal to the array, or steered to appear as such, will create identical waveforms at each sensor. Since the taps in each column, i.e., w_1 through w_K, see the same signal, this array may be collapsed to a single sensor FIR filter. Hence, to constrain the frequency range of the target signal, one just has to constrain the sum of the taps for each column to be equal to the corresponding tap in a FIR filter having J taps and the desired frequency response for the target signal.

These look-direction and frequency response constraints can be implemented by the following optimization problem:

$$\text{minimize} : \mathbf{w}^T \mathbf{R}_{xx} \mathbf{w} \tag{30.15}$$

$$\text{subject to} : \mathbf{C}^T \mathbf{w} = \mathbf{h} \tag{30.16}$$

where $\mathbf{R}_{xx}$ is the covariance matrix of the received signals, $\mathbf{h}$ is the vector of FIR filter coefficients defining the desired frequency response, and $\mathbf{C}^T$ is the constraint matrix given by

$$\mathbf{C}^T = \begin{bmatrix} 11 & \dots & 1 & 00 & \dots & 0 & \dots & 00 & \dots & 0 \\ 00 & \dots & 0 & 11 & \dots & 1 & \dots & 00 & \dots & 0 \\ \vdots & & & & & & & & & \\ 00 & \dots & 0 & 00 & \dots & 0 & \dots & 11 & \dots & 1 \end{bmatrix}.$$

The number of rows in $\mathbf{C}^T$ is equal to the number of taps of the array and the number of ones in each row is equal to the number of sensors. The optimal weight vector $\mathbf{w}_{\text{opt}}$ will minimize the output power of the noise sources subject to the constraint that the sum of each column vector of weights is equal to a coefficient of a FIR filter defining the desired impulse response of the array.

The Frost algorithm [7] is a constrained LMS method derived by solving Eqs. (30.15) and (30.16) via Lagrange Multipliers to obtain an expression for the optimum weight vector, Frost [7] derived the constrained LMS algorithm for broadband array processing using Lagrange multipliers. The function to be minimized may be defined as

$$H(\mathbf{w}) = \frac{1}{2} \mathbf{w}^T \mathbf{R}_{xx} \mathbf{w} + \lambda^T \left(\mathbf{C}^T \mathbf{w} - \mathbf{h} \right) \tag{30.17}$$

where λ is a Lagrange multiplier and F is a vector representative of the desired frequency response. Minimizing the function $H(\mathbf{w})$ with respect to $\mathbf{w}$ will obtain the following optimal weight vector:

$$\mathbf{w}_{\text{opt}} = \mathbf{R}_{xx}^{-1} \mathbf{C} \left(\mathbf{C}^T \mathbf{R}_{xx}^{-1} \mathbf{C} \right)^{-1} \mathbf{h} . \tag{30.18}$$

An iterative implementation of this algorithm was implemented via the following equations:

$$\mathbf{w}^{(j+1)} = \mathbf{P} \left[\mathbf{w}^{(j)} - \mu \mathbf{R}_{xx} \mathbf{w}^{(j)} \right] + \mathbf{C} \left(\mathbf{C}^T \mathbf{C} \right)^{-1} \mathbf{h} \tag{30.19}$$

where μ is a step size parameter and

$$\mathbf{P} = \mathbf{I} - \mathbf{C} \left(\mathbf{C}^T \mathbf{C} \right)^{-1} \mathbf{C}^T$$

$$\mathbf{w}(0) = \mathbf{C} \left(\mathbf{C}^T \mathbf{C} \right)^{-1} \mathbf{h}$$

where $\mathbf{I}$ is the identity matrix and

$$\mathbf{h} = \begin{bmatrix} h_1 & h_2 & \dots & h_J \end{bmatrix}.$$

30.5   Inverse Formulations for Array Processing

The array processing algorithms discussed thus far have all been derived through statistical analysis and/or adaptive filtering techniques. An alternative approach is to view the constraints as equations that can be expressed in a matrix-vector format. This allows for a simple formulation of array processing algorithms to which additional constraints can be easily incorporated. Additionally, this formulation allows for efficient iterative matrix inversion techniques that can be used to adapt the weights in real time.

30.5.1 Narrowband Arrays

Two algorithms were discussed for narrowband arrays, namely, the sidelobe canceler and pilot signal algorithms. We will consider the sidelobe canceler algorithm here. The derivation of the sidelobe canceler is based on the optimization of the SINR and yields an expression for the optimum weight vector as a function of the input autocorrelation matrix. We will use the same constraints as the sidelobe canceler to yield a set of linear equations that can be put in a matrix vector format.

Consider the narrowband array description provided in Section 30.3. In Eq. (30.7), $\mathbf{s}(k)$ is the vector representing the desired signal whose wavefront is normal to the array and $\mathbf{n}(k)$ is the sum of the interfering signals arriving from different directions. A weight vector is desired that will allow the signal vector $\mathbf{s}(k)$ to pass through the array undistorted while nulling any contribution of the noise vector $\mathbf{n}(k)$. An optimal weight vector $\mathbf{w}_{\text{opt}}$ that satisfies these conditions is represented by:

$$\mathbf{w}_{\text{opt}}^T \mathbf{s}(k) = s(k) \tag{30.20}$$

and

$$\mathbf{w}_{\text{opt}}^T \mathbf{n}(k) = 0 \tag{30.21}$$

where $s(k)$ is the scalar value of the desired signal. Since the sidelobe canceler does not have access to $s(k)$, an alternative approach must be taken to implement the condition of Eq. (30.20). One method for finding this constraint is to minimize the expectation of the output power [7]. This expectation can be approximated by the quantity y^2, where $y = \mathbf{x}^T(k)\mathbf{w}$. Minimizing y^2 subject to the look-direction constraint will tend to cancel the noise vector while maintaining the signal vector. This criteria can be represented by the linear equation:

$$\mathbf{x}^T(k)\mathbf{w} = 0. \tag{30.22}$$

Note that Eq. (30.22) implies that the weight vector be orthogonal to the composite input vector as opposed to just the noise component. However, the look-direction constraint imposed by the following equation will maintain the desired signal

$$\begin{bmatrix} 1 & 1 & \dots & 1 \end{bmatrix} \mathbf{w} = 1. \tag{30.23}$$

This equation satisfies the look-direction constraint that a signal arriving perpendicular to the array will have unity gain in the output.

The constraints imposed by Eqs. (30.22) and (30.23) can be expressed in a matrix-vector form as follows:

$$\begin{bmatrix} x_1(k) & x_2(k) & \dots & x_K(k) \\ 1 & 1 & \dots & 1 \end{bmatrix} \mathbf{w} = \begin{bmatrix} 0 \\ 1 \end{bmatrix} \tag{30.24}$$

or, equivalently,

$$\mathbf{A}\mathbf{w} = \mathbf{b}.$$

30.5.2 Broadband Arrays

The broadband array considered in this section will utilize the constraints considered by Frost [7], namely the look-direction and frequency range of the target signal. The linear equations that represent the Frost algorithm are similar to those used for the narrowband formulation derived in the previous section. Once again, the minimization of the cost function in Eq. (30.15) can be achieved by Eq. (30.22), assuming that the target signal arrives normal to the array. The constraint for the desired frequency response in the look direction can also be implemented in a similar fashion to that of the narrowband array in Eq. (30.23). Instead of constraining the sum of the weights to be one, as in the narrowband array, the broadband array implementation will constrain the sum of each *column* of weights to be equal to a corresponding tap value in a FIR filter with the desired frequency response for the target signal.

Hence, the broadband array problem represented by Eqs. (30.15) and (30.16) can be expressed as a linear system of equations by creating a matrix that has the cost function given by Eq. (30.15) augmented with the linear constraint equations given by Eq. (30.16). The problem can now be expressed as:

$$
\begin{bmatrix}
x_1 & .. & x_K & \cdots & x_{(J-1)K+1} & .. & x_{JK} \\
1 & .. & 1 & \cdots & 0 & .. & 0 \\
0 & .. & 0 & \cdots & 0 & .. & 0 \\
\vdots & & & & & & \\
0 & .. & 0 & \cdots & 1 & .. & 1
\end{bmatrix}
\cdot
\begin{bmatrix}
w_1 \\
w_2 \\
\vdots \\
w_{JK}
\end{bmatrix}
=
\begin{bmatrix}
0 \\
h_1 \\
\vdots \\
h_J
\end{bmatrix},
$$

$$\text{or } \mathbf{Aw} = \mathbf{h}' \tag{30.25}$$

where $\mathbf{h}'$ is the vector of FIR filter coefficients augmented with a zero.

30.5.3 Row-Action Projection Method

The matrix-vector formulation for the narrowband beamforming problem, as represented in Eq. (30.24) or the broadband array problem formulated in Eq. (30.25) can now be expressed as an inverse problem. For example, if $\mathbf{A}$ is **n** x **n** and rank$[\mathbf{A}|\mathbf{b}]$ = rank$[\mathbf{A}]$, then a unique solution for $\mathbf{w}$ can be found as

$$\mathbf{w} = \mathbf{A}^{-1}\mathbf{b}. \tag{30.26}$$

If instead, $\mathbf{A}$ is **m** x **n**, then a least squares solution can be implemented as

$$\mathbf{w} = \left(\mathbf{A}^T\mathbf{A}\right)^{-1}\mathbf{A}^T\mathbf{b}. \tag{30.27}$$

Another solution can be obtained by using the Moore-Penrose generalized inverse, or pseudo-inverse, of $\mathbf{A}$ via

$$\mathbf{w}^\dagger = \mathbf{A}^\dagger\mathbf{b} \tag{30.28}$$

where $\mathbf{A}^\dagger$ and $\mathbf{w}^\dagger$ represent the pseudo-inverse of $\mathbf{A}$ and the pseudo-inverse solution for $\mathbf{w}$, respectively.

These methods all provide an immediate solution for the weight vector, $\mathbf{w}$, however, at the expense of requiring a matrix inversion along with any instabilities that may be apparent if the matrix is ill-conditioned. A more convenient approach to solve for the weights is to use an iterative approach. The method that we shall use here is known as the row-action projection (RAP) algorithm. The RAP algorithm is an iterative technique for solving a system of linear equations. The RAP method has found numerous applications in digital signal processing [16] and is applied here to adaptive beamforming.

The RAP method for iteratively solving the system in Eq. (30.24) is given by the update equation:

$$\mathbf{w}^{(j+1)} = \mathbf{w}^{(j)} + \mu \frac{\epsilon_i}{\|\mathbf{a}_i\| \, \|\mathbf{a}_i\|} \mathbf{a}_i^T \tag{30.29}$$

where ϵ_i is the error term for the ith row defined as:

$$\epsilon_i = b_i - \mathbf{a}_i \mathbf{w}^{(k)}. \tag{30.30}$$

In Eqs. (30.29) and (30.30), the superscript j denotes the iteration, the subscript i refers to the row number of the matrix or vector, and μ is a gain parameter, which is known to be stable for values between zero and two. The choice of μ is important for performance characteristics and has the tradeoff that a large μ will provide faster convergence, while a small μ will provide greater accuracy. Also, note that choosing μ between one and two may, in some instances, prevent convergence to the LMS solution.

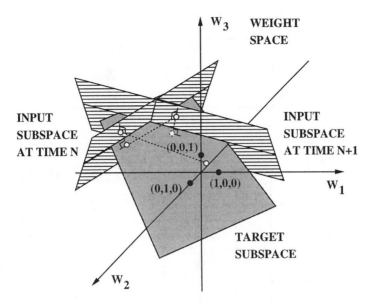

FIGURE 30.5 Orthogonal projections in weight space.

The RAP method operates by creating orthogonal projections in the space defined by the data matrix **A** in Eq. (30.24). A graphical representation of the RAP algorithm, as applied to a three sensor beamforming array, is illustrated in Fig. 30.5.

In Fig. 30.5, the target signal subspace consists of the plane represented by the look-direction constraint, namely $w_1 + w_2 + w_3 = 1$. The input signal subspace, given by $w_1 x_1(k) + w_2 x_2(k) + w_3 x_3(k) = 0$, will consist of a different plane for each discrete time index k. The RAP method first creates an orthogonal projection to the input subspace (i.e., satisfying $\mathbf{w}^T \mathbf{x}(k) = 0$). A projection is then made to the target signal subspace. This procedure will be repeated for the next input subspace, etc. Intuitively, this procedure will find a solution as "orthogonal as possible" to the different input subspaces, which lies in the target signal subspace. Since the RAP method consists of only row operations, it is convenient for parallel implementations. This technique, described by Eqs. (30.24), (30.29), and (30.30), comprises the RAP method for array processing.

30.6 Simulation Results

Several simulations were performed to compare the inverse formulation of the array processing problem to the more traditional adaptive filtering approaches. These simulations compare the inverse formulation to the sidelobe canceler implementation of the narrowband array and to the Frost implementation of the broadband array.

30.6.1 Narrowband Results

The sidelobe canceler application is evaluated with both the Applebaum algorithm and the inverse formulation. Both arrays are simulated for a nine-sensor narrowband array. The RAP algorithm for the inverse formulation uses a gain value of $\mu = 0.001$ and the Applebaum array uses values of $\alpha = 0.25$ and $\beta = 0.01$. The signal environment for the scenario consists of unit amplitude tones whose spectral and spatial characteristics are summarized by Table 30.1. The input spectrum of the narrowband scenario is shown in Fig. 30.6. The input and output spectrums for the inverse formulation and Applebaum algorithm are shown in Figs. 30.6 through 30.8. The inverse formulation and Applebaum algorithms demonstrate similar performance for this example.

TABLE 30.1   Input Scenario for Narrowband
Experiment

Signal	Angle (deg)	Frequency (KHz)
Target signal	0	2.0
Interference 1	28	3.0
Interference 2	41	1.0
Interference 3	72	4.0

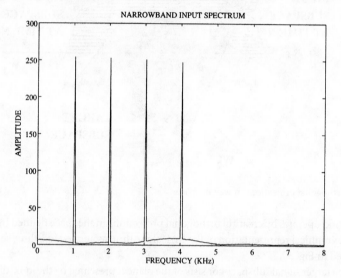

FIGURE 30.6   Narrowband input spectrum.

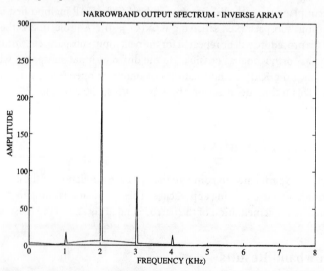

FIGURE 30.7   Output spectrum for inverse formulation.

30.6.2   Broadband Results

The broadband array application is also evaluated with both the inverse formulation and Frost algorithm. The algorithms are both evaluated for a broadband array that consists of nine sensors, each followed by five taps. The signal environment used for the scenario consists of several signals of varying spectral and spatial characteristics as summarized by Table 30.2.

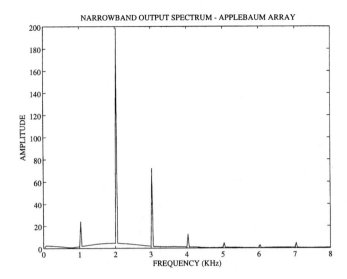

FIGURE 30.8   Output spectrum for Applebaum array.

TABLE 30.2   Input Scenario for Broadband Experiment

Signal	Angle (deg)	Frequency (KHz)
Target signal	0	3.0
Interference 1	27	1.5
Interference 2	41	4.0

The RAP algorithm used for the inverse has a gain value $\mu = 0.5$ and the Frost algorithm uses the gain value $\mu = 0.05$. The **h** vector specifies a low pass frequency response with a passband up to 4 KHz. The input and output signal spectrums are shown in Figs. 30.9 through 30.11. The inverse formulation and Frost algorithms again demonstrate similar performance.

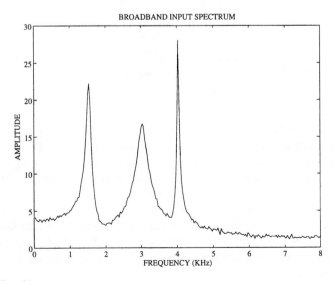

FIGURE 30.9   Broadband input spectrum.

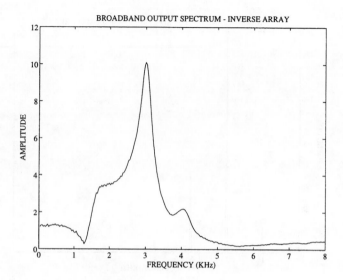

FIGURE 30.10   Output spectrum for inverse array.

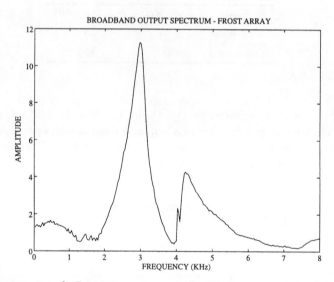

FIGURE 30.11   Output spectrum for Frost array.

The broadband array processing algorithms are also evaluated for a microphone array application [5]. The simulation uses a microphone array with nine equispaced transducers each followed by 13 taps. The microphone spacing is chosen as 4.3 cm and the sampling rate for the speech signals is 16 KHz. The **h** vector contains coefficients for a low pass FIR filter designed with a Hamming window for a passband of 0 to 4 KHz. The signal environment consists of two speech signals. The target signal arrives normal to the array. The interfering signal is applied to the array at uniformly spaced angles ranging from $-90°$ to $+90°$ in unit increments. The interference power is 2.6 dB greater than the desired signal. The resulting interference suppression observed in the array output is illustrated in Fig. 30.12. The maximum interference suppression (i.e., for interference arriving at $\pm90°$) is 11.0 dB for the RAP method and 11.2 dB for the Frost method.

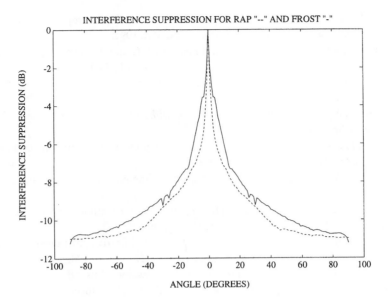

FIGURE 30.12 Interference suppression.

30.7 Summary

This article has formulated the array processing problem as an inverse problem. Inverse formulations for both narrowband and broadband arrays were discussed. Specifically, the sidelobe canceler algorithm for narrowband array processing and Frost algorithm for broadband array processing were analyzed. The inverse formulations provide a flexible, intuitive implementation of the constraints that are used by each algorithm. The inverse formulations were then solved through use of the RAP method. The RAP method is a simple technique for creating orthogonal projections within a space defined by a set of hyperplanes. The RAP method can easily be applied to unconstrained and constrained optimization problems whose solution lies in a convex set (i.e., no local maxima or minima). Many array processing algorithms fall into this category and it has been shown that the RAP method is a viable solution for this application. Since the RAP method only involves row operations, it is also more convenient for parallel processing implementations such as systolic arrays [15].

These algorithms have also been simulated for both narrowband and broadband implementations. The narrowband simulation consisted of a set of tones arriving at different spatial locations. The broadband array was evaluated for a simulation of several signals with differing spatial locations and bandwidths, in addition to a speech enhancement application. For all scenarios, the inverse formulations were found to perform comparable to the traditional approaches.

References

[1] Adugna, E., Speech Enhancement Using Microphone Arrays, Ph.D. thesis, Rutgers University, CAIP Center, New Jersey, June 1994.

[2] Applebaum, S.P., Adaptive arrays, *IEEE Trans. Antennas Propagation,* AP-24, 585–598, 1976.

[3] Censor, Y., Row-action techniques for huge and sparse systems and their applications, *SIAM Review,* 23(4), Oct. 1981.

[4] DeFatta, D., Lucas, J. and Hodgkiss, W., *Digital Signal Processing: A System Design Approach,* John Wiley & Sons, New York, 1988.

[5] Farrell, K.R., Mammone, R.J. and Flanagan, J.L., Beamforming microphone arrays for speech enhancement, in *Proc. ICASSP,* San Francisco, CA, Mar. 1992.

[6] Flanagan, J.L., Johnston, J.D., Zahn, R. and Elko, G.W., Computer-steered microphone arrays for sound transduction in large rooms, *J. Acoustical Soc. Am.,* 78(11), 1508–1518, Nov. 1985.

[7] Frost, O.L., III, An algorithm for linearly constrained adaptive array processing, *Proc. IEEE,* 60(8), 926–935, Aug. 1972.

[8] Giordano, A. and Hsu, F., *Least Square Estimation with Applications to Digital Signal Processing,* John Wiley & Sons, New York, 1985.

[9] Greenberg, J.E. and Zurek, P.M., Evaluation of an adaptive beamforming method for hearing aids, *J. Acoustical Soc. Am.,* 91(3), Mar. 1992.

[10] Griffiths, L.J., A simple adaptive algorithm for real-time processing in antenna arrays, *Proc. IEEE,* 57(10), 1696–1704, Oct. 1969.

[11] Griffiths, L.J., Linearly-constrained adaptive signal processing methods, in *Advanced Algorithms and Architectures for Signal Processing II,* SPIE, 1987, pp. 96–100.

[12] Griffiths, L.J. and Jim, C.W., An alternative approach to linearly constrained adaptive beam-forming, *IEEE Trans. Antennas Propagation,* AP-30(1), 27–34, Jan. 1982.

[13] Haykin, W., *Adaptive Filter Theory,* Prentice-Hall, Englewood Cliffs, NJ, 1991.

[14] Hudson, J.E., *Adaptive Array Principles,* Institute of Electrical Engineers, 1981.

[15] Kung, S.Y., *VLSI Array Processors,* Prentice-Hall, Englewood Cliffs, NJ, 1988.

[16] Mammone, R.J., *Computational Methods of Signal Recovery and Recognition,* John Wiley & Sons, New York, 1992.

[17] Noble, B. and Daniel, J.W., *Applied Linear Algebra,* Prentice-Hall, Englewood Cliffs, NJ, 1988.

[18] Papoulis, A., *Probability, Random Variables, and Stochastic Process,* McGraw-Hill, New York, 1984.

[19] Takao, K., Fujita, M. and Nishi, T., An adaptive antenna array under directional constraint, *IEEE Trans. Antennas Propagation,* AP-24(9), 662–669, Sept. 1976.

[20] Widrow, B., Mantey, P.E. and Goode, B.B., Adaptive antenna systems, *Proc. IEEE,* 55(12), 2143–2158, Dec. 1967.

[21] Widrow, B. and Stearns, S.D., *Adaptive Signal Processing,* Prentice-Hall, Englewood Cliffs, NJ, 1985.

31

Channel Equalization as a Regularized Inverse Problem

John F. Doherty
Pennsylvania State University

31.1 Introduction

In this article we examine the problem of communication channel equalization and how it relates to the inversion of a linear system of equations. Channel equalization is the process by which the effect of a band-limited channel may be diminished, i.e., equalized, at the sink of a communication system. Although there are many ways to accomplish this, we will concentrate on linear filters and adaptive filters. It is through the linear filter approach that the analogy to matrix inversion is possible. Regularized inversion refers to a process in which noise dominated modes of the observed signal are attenuated.

31.2 Discrete-Time Intersymbol Interference Channel Model

Intersymbol interference (ISI) is a phenomenon observed by the equalizer caused by frequency distortion of the transmitted signal. This distortion is usually caused by the frequency selective characteristics of the transmission medium. However, it can also be due to deliberate time dispersion of the transmitted pulse to affect realizable implementations of the transmit filter. In any case, the purpose of the equalizer is to remove deleterious effects of the ISI on symbol detection. The ISI generation mechanism is described next with a description of equalization techniques to follow. The information transmitted by a digital communication system is comprised of a set of discrete symbols. Likewise, the ultimate form of the received information is cast into a discrete form. However, the intermediate components of the digital communications system operate with continuous waveforms which carry the information. The major portions of the communications link are the transmitter pulse shaping filter, the modulator, the channel, the demodulator, and the receiver filter. It will be advantageous to transform the continuous part of the communication system into an equivalent discrete time channel description for simulation purposes. The discrete formulation should be transparent to both the information source and the equalizer when evaluating performance. The equivalent discrete time channel model is attained by combining the transmit filter, $p(t)$, the channel filter, $g(t)$, and the receive filter, $w(t)$, into a single continuous filter, that is,

$$h(t) = w(t) * g(t) * p(t) \tag{31.1}$$

Refer to Fig. 31.1. The effect of the sampler preceding the decision device is to discretize the aggregate filter.

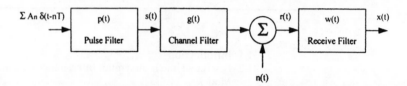

FIGURE 31.1 The signal flow block diagram for the equivalent channel description. The equalizer observes $x(nT)$, a sampled version of the receive filter output $x(t)$.

The equivalent discrete time channel as a means to simulate the performance of digital communications systems was advanced by Proakis [1] and has found subsequent use throughout the communications literature [2, 3].

It has been shown that a bandpass transmitted pulse train has an equivalent low pass representation [1]

$$s(t) = \sum_{n=0}^{\infty} A_n p(t - nT) \tag{31.2}$$

where $\{A_n\}$ is the information bearing symbol set, $p(t)$ is the equivalent low pass transmit pulse waveform, and T is the symbol rate. The observed signal at the input of the receiver is

$$r(t) = \sum_{n=0}^{\infty} A_n \int_{-\infty}^{+\infty} p(t - nT)g(t - nT - \tau)d\tau + n(t) \tag{31.3}$$

where $g(t)$ is the equivalent low pass bandlimited impulse response of the channel and the channel noise, $n(t)$, is modeled as white Gaussian noise. The optimum receiver filter, $w(t)$, is the *matched filter* which is designed to give maximum correlation with the received pulse [4]. The output of the receiver filter, that is, the signal seen by the sampler, can be written as

$$x(t) = \sum_{n=0}^{\infty} A_n h(t - nT) + v(t) \tag{31.4}$$

$$h(t) = \int_{-\infty}^{+\infty} \left[\int_{-\infty}^{+\infty} p(t - nT)g(t - nT - \lambda)d\lambda \right] w(t - \tau)d\tau \tag{31.5}$$

$$v(t) = \int_{-\infty}^{+\infty} n(t)w(t - \tau)d\tau \tag{31.6}$$

where $h(t)$ is the response of the receiver filter to the received pulse, representing the overall impulse response between the transmitter and the sampler, and $v(t) = \int_{-\infty}^{+\infty} n(t)w(t - \tau)d\tau$ is a filtered version of the channel noise. The input to the equalizer is a sampled version of Eq. (31.4), that is, sampling at times $t = kT$ produces

$$x(kT) = \sum_{n=0}^{\infty} A_n h(kt - nT) + v(kT) \tag{31.7}$$

as the input to the discrete time equalizer. By normalizing with respect to the sampling interval and rearranging terms, Eq. (31.7) becomes

$$x_k = \underbrace{h_0 A_k}_{\text{desired symbol}} + \underbrace{\sum_{\substack{n=0 \\ n \neq k}}^{\infty} A_n h_{k-n}}_{\text{intersymbol interference}} + v_k \tag{31.8}$$

31.3 Channel Equalization Filtering

31.3.1 Matrix Formulation of the Equalization Problem

The task of finding the optimum linear equalizer coefficients can be described by casting the problem into a system of linear equations,

$$\begin{bmatrix} \tilde{d}_1 \\ \tilde{d}_2 \\ \vdots \\ \tilde{d}_L \end{bmatrix} = \begin{bmatrix} x_1^T \\ x_2^T \\ \vdots \\ x_L^T \end{bmatrix} c + \begin{bmatrix} e_1 \\ e_2 \\ \vdots \\ e_L \end{bmatrix} \tag{31.9}$$

$$x_k = [x_{k+N-1}, \ldots, x_{k-1}]^T \tag{31.10}$$

where $(\cdot)^T$ denotes the transpose operation. The received sample at time k is x_k, which consists of the channel output corrupted by additive noise. The elements of the $N \times 1$ vector c_k are the coefficients of the equalizer filter at time k. The equalizer is said to be in decision directed mode when $\tilde{d}_k$ is taken as the output of the nonlinear decision device. The equalizer is in training, or reference directed, mode when $\tilde{d}_k$ is explicitly made identical to the transmitted sequence A_k. In either case, e_k is the error between the desired equalizer output, $\tilde{d}_k$, and the actual equalizer output, $x_k^T c$. We will assume that $\tilde{d}_k = A_{k+N}$, then the notation in Eq. (31.9) can be written in the compact form,

$$d = Xc + e \tag{31.11}$$

by defining $d = \left[\tilde{d}_1, \ldots, \tilde{d}_L \right]^T$ and by making the obvious associations with Eq. (31.9). Note that the parameter L determines the number of rows of the time varying matrix X. Therefore, choosing L is analogous to choosing an observation interval for the estimation of the filter coefficients.

31.4 Regularization

We seek a solution for the filter coefficients of the form $c = Yd$, where Y is in some sense an inverse of the data matrix X. The least squares solution requires that

$$Y = \left[X^T X \right]^{-1} X^T \tag{31.12}$$

where $X^\# \triangleq \left[X^T X \right]^{-1} X^T$ represents the *Moore-Penrose (M-P) inverse* of X. If one or more of the eigenvalues of the matrix $X^T X$ is zero, then the Moore-Penrose inverse does not exist.

To investigate the behavior of the inverse, we will decompose the data matrix into the form $X = X_S + X_N$, where X_S is the signal component and X_N is the noise component. Generally, the noise data matrix is full rank and the signal data matrix may be nearly rank deficient from the spectral nulls in the transmission channel. This is illustrated by examining the smallest eigenvalue of $X_S^T X_S$

$$\lambda_{\min} = S_{R\min} + O\left(N^{-k}\right) \tag{31.13}$$

where S_R is the continuous PSD of the received data x_k, $S_{R\min}$ is the minimum value of the PSD, k is the number of non-vanishing derivatives of S_R at $S_{R\min}$, and N is the equalizer filter length. Any spectral loss in the signal caused by the channel is directly translated into a corresponding decrease in the minimum eigenvalue of the received signal. If $\lambda_{\min}$ becomes small, but nonzero, the data correlation matrix $X^T X$ becomes *ill-conditioned* and its inversion becomes sensitive to the noise. The sensitivity is expressed in the quantity

$$\delta \triangleq \frac{\|\tilde{c} - c\|}{\|c\|} \leq \frac{\sigma_n^2}{\lambda_{\min}} + O\left(\sigma_n^4\right) \tag{31.14}$$

where the noiseless least squares filter coefficient vector solution, c, has been perturbed by adding a white noise to the data with variance $\sigma_n^2 \ll 1$, to produce the least squares solution $\tilde{c}$. Substituting Eq. (31.13) into Eq. (31.14) yields

$$\delta \leq \frac{\sigma_n^2}{S_{R\min} + O\left(N^{-k}\right)} + O\left(\sigma_n^4\right) \approx \frac{\sigma_n^2}{S_{R\min}} \tag{31.15}$$

The relation in Eq. (31.15) is an indicator of the potential numerical problems in solving for the equalizer filter coefficients when the data is spectrally deficient.

We see that direct inversion of the data matrix is not recommendable when the channel has severe spectral nulls. This situation is equivalent to stating that the original estimation problem $d = Xc$ is *ill-posed*. That is, the equalizer is asked to reproduce components of the channel input that are unobservable at the channel output or are obscured by noise. Thus, it is reasonable to ascertain the modes of the input dominated by noise and give them little weight, relative to the signal dominated components, when solving for the equalizer filter coefficients. This process of weighting is called *regularization*.

Regularization can be described by relying on a generalization of the M-P inverse that depends on the singular value decomposition (SVD) of the data matrix

$$X = U\Sigma V^T \tag{31.16}$$

where U is an $L \times N$ unitary matrix, V is an $N \times N$ unitary matrix, $\Sigma = \text{diag}\left(\sigma_1, \sigma_2, \ldots, \sigma_N\right)$ is a diagonal matrix of singular values where $\sigma_i \geq 0, \sigma_1 > \sigma_2 > \cdots > \sigma_N$. It is assumed in Eq. (31.16) that $L > N$, which is typical in the equalization problem.

We define the generalized pseudo-inverse of X as

$$X^\dagger = V\Sigma^\dagger U^T \tag{31.17}$$

where $\Sigma^\dagger = \text{diag}\left(\sigma_1^\dagger, \sigma_2^\dagger, \ldots, \sigma_N^\dagger\right)$ and

$$\sigma_i^\dagger = \begin{cases} \sigma_i^{-1} & \sigma_i \neq 0 \\ 0 & \sigma_i = 0 \end{cases} \tag{31.18}$$

The M-P inverse can be reformulated using the SVD as follows

$$X^\# = \left[V\Sigma^2 V^T\right]^{-1} V\Sigma U^T = V\Sigma^{-1} U^T \tag{31.19}$$

Upon examination of Eq. (31.17) and Eq. (31.19), we note that $X^{\#} = X^{\dagger}$ only if all the singular values of X are nonzero, $\sigma_i \neq 0$. Another item to note is that $V \Sigma^2 V^T$ is the eigenvalue decomposition of $X^T X$, which implies that the eigenvalues of $X^T X$ are the squares of the singular values of X.

The generalized pseudo-inverse in Eq. (31.17) provides an eigenvalue spectral weighting given by Eq. (31.18), which differs from the M-P inverse only when one or more of the eigenvalues of $X^T X$ are identically zero. However, this form of regularization is rather restrictive since complete annihilation of the spectral components is rarely encountered in practice. A more likely condition for the eigenvalues of $X^T X$ is that a small band of signal eigen-modes are much smaller in magnitude than the corresponding noise modes. Direct inversion of these eigen-modes, although well-defined mathematically, leads to noise enhancement at the equalizer output and to noise sensitivity in the filter coefficient solution. An alternative to the generalized pseudo-inverse is to use a regularized inverse wherein the eigen-modes are weighted prior to inversion [5]. This approach leads to a trade-off between the noise immunity of the equalizer filter weights and the signal fidelity at the equalizer filter output. To demonstrate this trade-off, let

$$c \ \mathcal{D} \ X^{\dagger} d \tag{31.20}$$

be the least squares solution. Let the regularized inverse be Y_n such that $\lim_{n \to \infty} Y_n = X^{\dagger}$. The regularized estimate for an observation perturbed by a random noise vector, n, is

$$c_n = Y_n \left(d \ \mathcal{C} \ n \right) \tag{31.21}$$

The effects of the regularized inverse and the noise vector are indicated by

$$\| c_n - c \| = \left\| Y_n n + \left(Y_n - X^{\dagger} \right) d \right\| \leq \| Y_n n \| + \left\| Y_n - X^{\dagger} \right\| \| d \| \tag{31.22}$$

The term $\| Y_n n \|$ is the part of the coefficient error due to the noise and is likely to increase as $n \to \infty$. The term $\left\| Y_n - X^{\dagger} \right\|$ represents the contribution due to the regularization error in approximating the pseudo-inverse. This error tends to zero as $n \to \infty$. The trade-off between noise attenuation and regularization error is evident upon inspection of Eq. (31.22), which also points out an idiosyncratic property of the regularization process. At first, the equalizer output error tends to decrease, due to decreasing regularization error, $\left\| Y_n - X^{\dagger} \right\|$. Then, as n increases further, the output error is likely to increase due to the noise amplification component, $\| Y_n n \|$. This behavior leads to the question regarding the best choice for the parameter n. A widely accepted procedure is to use the discrepancy principle, which states that n should satisfy

$$\left\| X c_{n'} - \left(d \ \mathcal{C} \ n \right) \right\| = \| n \| \tag{31.23}$$

Letting $n > n'$ usually results in noise amplification at the equalizer output.

31.5 Discrete-Time Adaptive Filtering

We will next examine three adaptive algorithms in terms of their regularization properties in deriving the equalizer filter. These algorithms are the normalized least mean squares (NLMS) algorithm, the recursive least squares (RLS) algorithm, and the block-iterative NLMS (BINLMS) algorithm. These algorithms are representative of the wider class of adaptive algorithms of which they belong.

31.5.1 Adaptive Algorithm Recapitulation

NLMS

The NLMS algorithm update is given by

$$c_n = c_{n-1} + \mu \left(d_n - x_n^T c_{n-1} \right) \frac{x_n}{\| x_n \|^2} \tag{31.24}$$

for $n = 1, \ldots, L$. This is rewritten as

$$c_n = \left(I - \mu \frac{x_n x_n^T}{\|x_n\|^2} \right) c_{n-1} + \mu \frac{d_n x_n}{\|x_n\|^2} \tag{31.25}$$

Define $P_n \triangleq \left(I - \mu x_n x_n^T / \|x_n\|^2 \right)$ and $p_n \triangleq \mu d_n x_n / \|x_n\|^2$, then Eq. (31.25) becomes

$$c_L = Q c_0 + q \tag{31.26}$$

where

$$Q \triangleq P_L P_{L-1} \cdots P_1 \tag{31.27}$$

and

$$q = [P_L \cdots P_2] p_1 + [P_L \cdots P_3] p_2 + \cdots + P_L p_{L-1} + p_L \tag{31.28}$$

BINLMS

The BINLMS algorithm relies on observing the entire block of filter vectors x_n, $1 \leq n \leq L$, in Eq. (31.9). The BINLMS update procedure is

$$c_{n+1} = c_n + \mu \left(d_j - x_j^T c_n \right) \frac{x_j}{\|x_j\|^2} \tag{31.29}$$

where $j = n \bmod L$. The update in Eq. (31.29) is related to the NLMS update by considering Eq. (31.26). That is, Eq. (31.29) is equivalent to

$$c_{n \cdot L} = Q c_{(n-1) \cdot L} + q \tag{31.30}$$

where L updates of Eq. (31.29) are compacted into a single update in Eq. (31.30). Note that only L updates are possible using Eq. (31.24) compared to an arbitrary number of updates in Eq. (31.29).

RLS

The update procedure for the RLS algorithm is

$$g_n = \frac{\lambda^{-1} Y_{n-1} x_n}{1 + \lambda^{-1} x_n^T Y_{n-1} x_n} \tag{31.31}$$

$$e_n = d_n - c_{n-1}^T x_n \tag{31.32}$$

$$c_n = c_{n-1} + e_n g_n \tag{31.33}$$

$$Y_n = \lambda^{-1} \left[Y_{n-1} - g_n x_n^T Y_{n-1} \right] \tag{31.34}$$

where g_n is called the gain vector, Y_n is the estimate of $\left[X_n^T X_n \right]^{-1}$ using the matrix inversion lemma, and X_n represents the first n rows of X in Eq. (31.9). The forgetting factor $0 < \lambda \ll 1$ allows the RLS algorithm to weight more recent samples providing a tracking capability for time-varying channels. The matrix inversion recursion is initialized with $Y_0 = \delta^{-1} I$, where $0 < \delta \ll 1$. The initialization constant transforms the data correlation matrix into

$$X_n^T \Lambda_n X_n + \lambda^n \delta I \tag{31.35}$$

where $\Lambda_n = \mathrm{diag} \left(1, \lambda, \ldots, \lambda^{n-1} \right)$.

31.5.2   Regularization Properties of Adaptive Algorithms

In this section we examine how each of the adaptive algorithms achieve regularization of the equalizer filter solution. We begin with the BINLMS and will subsequently take the NLMS as a special case. The BINLMS update of Eq. (31.30) is equivalent to

$$c_l = Qc_{l-1} + q \tag{31.36}$$

where an increment in l is equivalent to L increments of n in Eq. (31.29). The recursion in Eq. (31.36) is also equivalent to

$$c_l = B_l d \tag{31.37}$$

where $\lim_{l \to \infty} B_l = X^\dagger$. Let $\hat{\sigma}_{k,l}$ represent the singular values of B_l, then the relationship among the singular values of B_l and the singular values of X is [6]

$$\hat{\sigma}_{k,l} = \begin{cases} \frac{1}{\sigma_k} \left[1 - \left(1 - \frac{\mu}{N} \sigma_k^2 \right)^{l+1} \right] & , \quad \sigma_k \neq 0 \\ \\ 0 & , \quad \sigma_k = 0 \end{cases} \tag{31.38}$$

The regularization property of the BINLMS depends on both μ and l. Since the step size parameter μ is chosen to guarantee convergence, i.e., $0 < \left(1 - \frac{\mu}{N} \sigma_1^2 \right) < 1$, the regularization is primarily controlled by the iteration index l. The regularization behavior of the BINLMS given by Eq. (31.38) is that the signal dominant modes are inverted first, followed by the weaker noise dominant modes, as the index l increases.

The regularization behavior of the NLMS algorithm is directly derived from the BINLMS by setting $l = 1$ in Eq. (31.38). We see that the only control over the regularization for the NLMS algorithm is to decrease the step size μ. However, this leads to a potentially undesirable reduction in the convergence rate of the adaptive equalizer filter.

The RLS algorithm weighting of the singular values is derived upon inspection of Eq. (31.35). The RLS equalizer filter coefficient estimate is

$$c_{LS} = \left[X^T \Lambda_L X + \lambda^L \delta I \right]^{-1} X^T \left(\Lambda_L^{1/2} \right)^T d \tag{31.39}$$

Let $\hat{\sigma}_{LS,k}$ represent the singular values of the effective inverse used in the RLS algorithm, then

$$\hat{\sigma}_{LS,k} = \frac{\sqrt{\lambda_k} \sigma_k}{\lambda_k \sigma_k^2 + \lambda^L \delta} \tag{31.40}$$

There are several points to note about Eq. (31.40). In the absence of the forgetting factor, $\lambda = 1$, and the initialization constant, $\delta = 0$, the RLS algorithm provides the exact inverse of the singular values, as expected. The constant δ prevents the dominator of Eq. (31.40) from getting too small. However, this regularization is lost if $\lambda^L \to 0$, which is the case when the observation interval L becomes large.

The behavior of the regularization functions (31.38) and (31.40) is illustrated in Fig. 31.2.

31.6   Numerical Results

A numerical example of the regularization characteristics of the adaptive equalization algorithms discussed is now presented. A data matrix $X X$ is constructed with dimensions $L = 50$ and $N = 11$, which has the singular value matrix $\Sigma = \text{diag}(1.0, 0.9, \ldots 0.1, 0.0)$. The step size $\mu = 0.2$ is chosen. Since the RLS algorithm computes an estimate of $\left[X^T X \right]^{-1}$, it is sensitive to the eigenvalues of $X^T X$. A graph similar to Fig. 31.2 is produced with the exception that the eigenvalue inverses of $X^T X$ are plotted for the RLS algorithm. These results are shown in Fig. 31.3 using the eigenvalues of X given by $\sigma_i^2 = (1 - (i - 1)/10)^2$ for $1 \leq i \leq 10$ and $\sigma_{11}^2 = 0$. The RLS algorithm exhibits large dynamic range in the eigenvalue inverse using the matrix inversion lemma, which may lead to unstable operation of the adaptive equalizer filter.

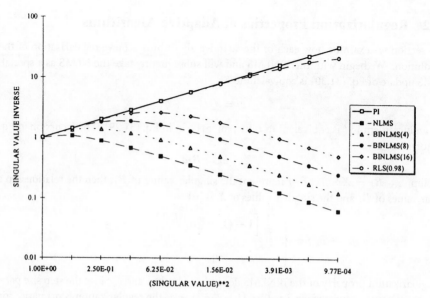

FIGURE 31.2   The regularization functions of the NLMS, BINLMS, and RLS algorithms.

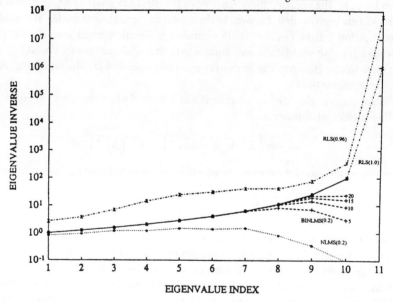

FIGURE 31.3   The regularization behavior of the NLMS, BINLMS, and the RLS adaptive algorithms is shown. The BINLMS curves represent block iterations of 5, 10, 15, and 20. The RLS algorithm uses $\lambda = 1.0$ and $\lambda = 0.96$.

31.7   Conclusion

A short introduction to the basic concepts of regularization analysis are presented in this article. Some further development in the application of this analysis to decision-feedback equalization may be found in [6]. The choice of which adaptive algorithm to use is application-dependent and each one comes with its associated advantages and disadvantages. The LMS-type algorithms are low-complexity solutions that have relatively slow convergence. The RLS-type algorithms have much faster convergence but are typically plagued by stability problems associated with error propagation and unregularized matrix inversion. Circumventing these stability problems tends to lead to more complex algorithm implementation. The

BINLMS algorithm is a trade-off between the convergence speed of the RLS-type algorithms and the stability of the LMS-type algorithms. A disadvantage of the BINLMS algorithm is that instantaneous throughput may be high due to the block-processing required.

References

[1] Proakis, J., *Digital Communications,* 2nd ed., McGraw-Hill, New York, 1989.
[2] Hatzinakos, D. and Nikias, C., Estimation of multipath channel response in frequency selective channels, 7, 12–19, Jan. 1989.
[3] Eleftheriou, E. and Falconer, D., Adaptive equalization techniques for HF channels, SAC-5, 238–247, Feb. 1987.
[4] Wozencraft, J. and Jacobs, I., *Principles of Communication Engineering,* John Wiley & Sons, New York, 1965.
[5] Tikhonov, A. and Arsenin, V., *Solutions to Ill-Posed Problems,* V.H. Winston and Sons, Washington, D.C., 1977.
[6] Doherty, J. and Mammone, R., An adpative algorithm for stable decision-feedback filtering, *IEEE Trans. Circuits Syst. II: Analog and Digital Signal Processing,* 40 CAS-II, Jan. 1993.

32

Inverse Problems in Microphone Arrays

A.C. Surendran
Bell Laboratories
Lucent Technologies

32.1 Introduction: Dereverberation Using Microphone Arrays

An acoustic enclosure usually reduces the intelligibility of the speech transmitted through it because the transmission path is not ideal. Apart from the direct signal from the source, the sound is also reflected off one or more surfaces (usually walls) before reaching the receiver. The resulting signal can be viewed as the output of a convolution in the time domain of the speech signal and the room impulse response. This phenomenon affects the quality of the transmitted sound in important applications such as teleconferencing, cellular telephony, and automatic voice activated systems (speaker and speech recognizers). Room reverberation can be perceptually separated into two broad classes. Early room echoes are manifested as irregularities or "ripples" in the amplitude spectrum. This effect dominates in small rooms, typically offices. Long-term reverberation is typically exhibited as an echo "tail" following the direct sound [1].

If the transfer function $G(z)$ of the system is known, it might be possible to remove the deleterious multi-path effects by inverse filtering the output using a filter $H(z)$ where

$$H(z) = \frac{1}{G(z)} \, . \tag{32.1}$$

Typically $G(z)$ is the transform of the impulse response of the room $g(n)$. In general, the transfer function of a reverberant environment is a non-minimum phase function, i.e., all the zeros of the function do not necessarily lie inside $|z| = 1$. A minimum phase function has a stable causal inverse, while the inverse of a non-minimum phase function is acausal and, in general, infinite in length. In general, $G(z)$ can be expressed as a product of a minimum-phase function and a non-minimum phase function:

$$G(z) = G_{min}(z) \cdot G_{max}(z) . \tag{32.2}$$

Many approaches have been proposed for dereverberating signals. The aim of all the compensation schemes is to bring the impulse response of the system after dereverberation as close as possible to an impulse function. Homomorphic filtering techniques were used to estimate the minimum phase part of $G(z)$ [2, 3]. In [2], the minimum phase component was estimated by zeroing out the cepstrum for negative frequencies. Then the output signal was filtered by the inverse of the minimum phase transfer function. But this technique still did not remove the reverberation contributed by the maximum-phase part of the room response. In [3], the inverse of the maximum-phase part was also estimated from the delayed and truncated version of the acausal inverse. But, the delay can be inordinate and care must be taken to avoid temporal aliasing.

An alternate approach to dereverberation is to calculate, in some form, the least squares estimate of the inverse of the transmission path, i.e., calculate the least squares solution of the equation

$$h(n) * g(n) = d(n) , \tag{32.3}$$

where $d(n)$ is the impulse function and $*$ denotes convolution. Assuming that the system can be modeled by an FIR filter, Eq. (32.3) can be expressed in matrix form as:

$$
\begin{pmatrix}
g(0) & & & \\
g(1) & g(0) & & \\
\vdots & g(1) & \cdots & 0 \\
g(m) & \vdots & \cdots & g(0) \\
0 & g(m) & \cdots & g(1) \\
0 & 0 & \cdots & \vdots \\
& & & g(m)
\end{pmatrix}
\begin{pmatrix}
h(0) \\
h(1) \\
\vdots \\
h(i)
\end{pmatrix}
=
\begin{pmatrix}
1 \\
0 \\
\vdots \\
0
\end{pmatrix} , \tag{32.4}
$$

or,

$$GH = D , \tag{32.5}$$

where D is the unity matrix and G, H and D are matrices of appropriate dimensions as shown in Eq. (32.4). The least squares method finds an approximate solution given by

$$\hat{H}(z) = \left(G^T G\right)^{-1} G^T D . \tag{32.6}$$

Thus, the error vector can be written as

$$
\begin{aligned}
\epsilon &= [D - G\hat{H}] \\
&= [I - G\left(G^T G\right)^{-1} G^T]D \\
&= ED ,
\end{aligned}
$$

where $E = [I - G(G^T G)^{-1} G^T]$. The mean square error or the energy in the error vector is

$$||\epsilon||_2 = ||ED||_2 \le |E|||D||_2 \le \frac{\lambda_{max}}{\lambda_{min}} ||D||_2 , \tag{32.7}$$

where $|E|$ is the norm of E and λ_{max} and λ_{min} are the maximum and minimum eigenvalues of E. The ratio between the maximum and minimum eigenvalues is called the condition number of a matrix and it specifies the noise amplification of the inversion process [4].

Typically, the operation is done on the full-band signal. Sub-band approaches have been proposed in [5, 7, 8]. All these approaches use a single microphone.

The amplitude spectrum of the room response has "ripples" which produce pronounced notches in the signal output spectrum. As the location of the microphone in the room changes, the room response for the same source changes and, as a result, the position of the notches in the amplitude spectrum varies. This property was used to advantage in [1]. In this method, multiple microphones were located in the room. Then, the output of each microphone was divided into multiple bands of equal bandwidth. For each band, by choosing the microphone whose output has the maximum energy, the ripples were reduced. In [9], the signals from all the microphones in each band were first co-phased, and then weighted by a gain calculated from a normalized cross-correlation function calculated based on the outputs of different microphones. Since the reverberation tails are uncorrelated, the cross-correlation-based gain turned off the tail of the signal. These techniques have had modest success in combating reverberation.

In recent years, great progress has been made in the quality, availability, and cost of high performance microphones. Fast digital signal processors that permit complex algorithms to operate in real time have been developed. These advances have enabled the use of large microphone arrays that deploy more sophisticated algorithms for dereverberation. Figure 32.1 shows a generic microphone array system which can "invert" the room acoustics. Different choices of $H_i(z)$ lead to different algorithms, each with their own advantages and disadvantages. In this report, we shall discuss single and multiple beamforming, matched filtering, and Diophantine inverse filtering through multiple input-output (MINT) modeling. In all cases we assume that the source location and the room configuration or, alternatively, the $G_i(z)$s, are known.

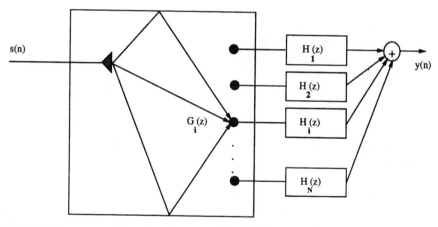

FIGURE 32.1 Modeling a room with a microphone array as a multiple output FIR system.

32.2 Simple Delay-and-Sum Beamformers

Arrays that form a single beam directed towards the source of the sound have been designed and built [11]. In these simple delay-and-sum beamformers, the processing filter has the impulse response

$$h_i(n) = \delta(n - n_i),\tag{32.8}$$

where $n_i = d_i/c$, d_i is the distance of the ith microphone from the source and c is the speed of sound in air. Sound propagation in the room can be modeled by a set of successive reflections off the surfaces (typically the walls) [10]. Figure 32.2 illustrates the impulse response of a single beamformer. The delay at the output of each microphone coheres the sound that arrives at the microphone directly from the source. It can be seen from Fig. 32.2 that in the resulting response, the strength of the coherent pulse is N and there are $N(K-1)$ distributed pulses. So, ideally, the signal-to-reverberant noise ratio (measured as the ratio of undistorted signal power to reverberant noise power) is $N^2/N(K-1)$ [13]. In a highly reverberant room, as the number of images K increases towards infinity, the SNR improvement, $N/K-1$, falls to zero.

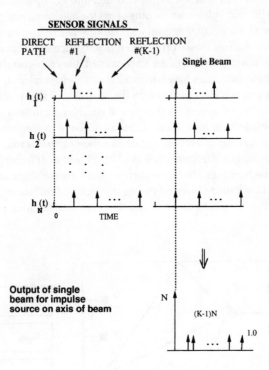

FIGURE 32.2 A single beamformer. (*Source:* Flanagan, J.L., Surendran, A.C., and Jan, E.-E., Spatially selective sound capture for speech and audio processing, *Speech Commun.*, 13: 207–222, 1993. With kind permission of Elsevier Science - NL, Sara Burgerhartstraat 25, 1055 KV Amsterdam, The Netherlands).

The single-beamforming system reported in [11] can automatically determine the direction of the source and rapidly steer the array. But, as the beam is steered away from the broadside, the system exhibits a reduction in spatial discrimination because the beam pattern broadens [12]. Further, beamwidth varies with frequency, so an array has an approximate "useful bandwidth" given by the upper and lower frequencies [12]:

$$f_{upper} = \frac{c}{d|\cos\phi - \cos\phi'|_{\max}}, \tag{32.9}$$

and

$$f_{lower} = \frac{f_{upper}}{N}, \tag{32.10}$$

where c is the speed of sound in air, N is the number of sensors in the array, d is the sensor spacing, ϕ' is the steering angle measured with respect to the axis of the array, and ϕ is the direction of the source.

For example, consider an array with seven microphones and a sensor spacing of 6.5 cm. Further, suppose the desired range of steering is $\pm30°$ from broadside. Then, $|\cos\phi - \cos\phi'|_{\max} = 1.5$ and hence $f_{upper} \approx 3500\,Hz$ and $f_{lower} \approx 500\,Hz$. So, to cover the bandwidth of speech, say from 250 Hz to 7 kHz, three harmonically nested arrays of spacing 3.25, 6.5, and 13 cm can be used. Further, the beamwidth also depends on the frequency of the signal as well as the steering direction. If the beam is steered to an angle ϕ', then the direction of the source for which the beam response falls to half its power is [12]

$$\phi_{3dB} = \cos^{-1}\left\{\cos\phi' \pm \frac{2.8}{N\omega d}\right\}, \tag{32.11}$$

where $\omega = 2\pi f$ and f is the frequency of the signal.

Equation 32.11 shows that the smaller the array, the wider the beam. Since most of the energy of a typical room interfering noise lies at lower frequencies, it would be advantageous to build arrays that have higher directivity (smaller beamwidth) at lower frequencies. This, combined with the fact that the array spacing is larger for lower frequency bands, gives yet another reason to harmonically nest arrays (see Fig. 32.3).

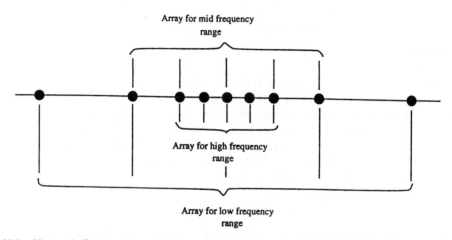

FIGURE 32.3 Harmonically nested array that covers three frequency ranges.

Just as linear one-dimensional arrays display significant fattening of the beams when steered towards the axis of the array, two-dimensional arrays exhibit widening of the beams when steered at angles acute to the plane of the array. Three-dimensional microphone arrays can be constructed [13] that have essentially a constant beamwidth over 4π steradians. Multiple beamforming using three-dimensional arrays of sensors not only provides selectivity in azimuth and elevation but also selectivity in the direction of the beam, i.e., it provides range selectivity.

The performance of single beamformers can degrade severely in the presence of other interfering noise sources, especially if they fall in the direction of the sidelobes. This problem can be mitigated using adaptive arrays. Adaptive arrays are briefly discussed in the next section.

32.2.1 A Brief Look at Adaptive Arrays

Adaptive signal processing techniques can be used to form a beam at the desired source while simultaneously forming a null in the direction of the interfering noise source. Such arrays are called "adaptive arrays". Though adaptive arrays are not effective under conditions of severe reverberation, they are included here because problems in adaptive arrays can be formulated as inverse problems. Hence, we shall discuss adaptive arrays briefly without providing a quantitative analysis of them. Broadband arrays have been analyzed in [14, 15, 16, 17, 18, 19]. In all these methods, the direction of arrival of the signal is assumed to be known.

Let the array have N sensors and M delay taps per sensor. If $X(k) = [x_1(k)\ldots x_i(k)\ldots x_{NM}(k)]^T$ (see Fig. 32.4) is the set of signals observed at the tap points, then $X(k) = S(k) + N(k)$, where $S(k)$

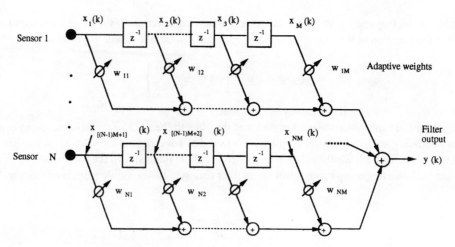

FIGURE 32.4   General form of an adaptive filter.

is the contribution of the desired signal at the tap points and $N(k)$ is the contribution of the unknown interfering noise. The inputs to the sensors, $x_{(jM+1)}(k)$, $j = 0, \ldots, (N-1)$, are the noisy versions of $g(k)$, the actual signal at the source. Now, the filter output $y(k) = W^T X(k)$, where $W^T = [w_{11}, \ldots, w_{1M}, w_{21}, \ldots, w_{2M}, \ldots, w_{N1}, \ldots, w_{NM}]$ is the set of weights at the tap points. The goal of the system is to make the output $y(k)$ as close as possible to the source $g(k)$. One way of doing this is to minimize the error $E\{(g(k) - y(k))^2\}$. The weight W^* that achieves this least mean square (LMS) error is also called the Weiner filter, and is given by

$$W^* = R_{XX}^{-1} C_{gX} \,, \tag{32.12}$$

where R_{XX} is the autocorrelation of $X(k)$ and C_{gX} is the set of cross-correlations between $g(k)$ and each element of $X(k)$. If $g(k)$ and $N(k)$ are uncorrelated, then

$$
\begin{aligned}
C_{gX} &= E\{g(k)X(k)\} = E\{g(k)S(k)\} + E\{g(k)N(k)\} \\
&= E\{g(k)S(k)\}
\end{aligned}
$$

and

$$
\begin{aligned}
R_{XX} &= E\{X(k)X^T(k)\} = E\{(S(k) + N(k))(S(k) + N(k))^T\} \\
&= R_{SS} + R_{NN} \,,
\end{aligned}
$$

where R_{SS} and R_{NN} are the autocorrelation matrices for the signal and noise.

Usually R_{NN} is not known. In such cases, the exact inverse cannot be calculated and an iterative approach to update the weights is needed. In Widrow's approach [15], a known pilot-signal $g(k)$ is injected into the array. Then, the weights are updated using the Widrow-Hopf algorithm that increments the weight vector in the direction of the negative gradient of the error:

$$W^{k+1} = W^k + \mu[g(k) - y(k)]X(k),$$

where W^{k+1} is the weight vector after the kth update and μ is the step size. Griffiths' method also uses the LMS approach, but minimizes the mean square error based on the autocorrelation and the cross-correlation values between the input and the output, rather than the signals themselves. Since the mean square error can be written as

$$E\{(g(k) - y(k))^2\} = R_{gg} - 2C_{gS}^T W + W^T R_{XX} W,$$

where R_{gg} is the auto-correlation matrix of $g(k)$ and C_{gS} is the set of cross-correlation matrix between $g(k)$ and each element of $S(k)$, the weight update can also be done by

$$W^{k+1} = W^k + \mu[C_{gS} - R_{XX}W^k] \tag{32.13}$$
$$= W^k + \mu[C_{gS} - X(k)X^T(k)W^k] \tag{32.14}$$
$$= W^k + \mu[C_{gS} - y(k)X(k)]. \tag{32.15}$$

In the above methods, significant distortion is observed in the primary beam due to null-steering. Constrained LMS techniques which place constraints on the performance of the main lobe can be used to reduce distortion [18, 19]. By specifying the broad-band response and the array beam characteristics as constraints, more robust beams can be formed. The problem now can be formulated as an optimization technique that minimizes the output power of the system. Given that the output power is

$$E\left\{y^2(k)\right\} = E\left\{W^T X(k)X^T(k)W\right\} = W^T R_{XX}W$$
$$= W^T R_{SS}W + W^T R_{NN}W,$$

if W can be chosen such that $W^T R_{NN}W = 0$, the noise can be eliminated. It was proposed [18] that once the array is steered towards the source with appropriate delays, minimizing the output power is equivalent to removing directional interference, since in-phase signals add coherently. In an accurately steered array, the wavefronts arriving from the direction of steering generate identical signals at each sensor. Hence, the array may be collapsed to a single sensor implementation which is equivalent to an FIR filter [18], i.e., the columns of the broadband array sum to an FIR filter. Additional constraints can be placed on this FIR filter. If the weights of the filters can be written as a matrix:

$$\hat{W} = \begin{pmatrix} w_{11} & w_{12} & \cdots & w_{1M} \\ \vdots & \vdots & \vdots & \vdots \\ w_{N1} & w_{N2} & \cdots & w_{NM} \end{pmatrix},$$

then it can be specified that $\sum_{i=1}^{N} w_{ij} = f_j$, $j = 1, \ldots, M$, where f_j, $j = 1, \ldots, M$ are the taps of an FIR filter that provides the desired filter response. Hence, using this method, directional interference can be suppressed by minimizing the output power and spectral interference can be suppressed by constraining the columns of the weight coefficients.

Thus, the problem can be formulated as

$$\text{Minimize:} \quad W^T R_{XX}W \tag{32.16}$$
$$\text{subject to:} \quad C^T W = F, \tag{32.17}$$

where F is the desired FIR filter and

$$C = \begin{pmatrix} 1 & 0 & 0 & \cdots & 0 & 1 & 0 & 0 & \cdots & 0 & \cdots & 1 & 0 & 0 & \cdots & 0 \\ 0 & 1 & 0 & \cdots & 0 & 0 & 1 & 0 & \cdots & 0 & \cdots & 0 & 1 & 0 & \cdots & 0 \\ & & \vdots & & & & & \vdots & & & \vdots & & & & \vdots \\ 0 & 0 & 0 & \cdots & 1 & 0 & 0 & 0 & \cdots & 1 & \cdots & 0 & 0 & 0 & \cdots & 1 \end{pmatrix}. \tag{32.18}$$

C has M rows with NM entries on each row. The first row of C in Eq. 32.18 has ones in positions 1, $(M+1), \ldots, (N-1)*M+1$; the second row has ones in positions 2, $(M+2), \ldots, (N-1)*M+2$, etc. Equation 32.17 can be solved using Lagrange multipliers [18]. This optimization problem can alternatively be posed as an inverse problem.

32.2.2　Constrained Adaptive Beamforming Formulated as an Inverse Problem

Using a similar cost function and the same constraint, the system can be formulated as an inverse problem [19]. The function to be optimized, $W^T R_{XX} W = 0$, can be approximated by $X^T W = 0$. This, combined with the constraint in Eq. 32.17 is written as:

$$
\begin{pmatrix}
x_1 & \cdots & x_M & \cdots & x_{(N-1)*M+1} & \cdots & x_{N*M} \\
1 & \cdots & 0 & \cdots & 1 & \cdots & 0 \\
 & & \vdots & & \vdots & & \vdots \\
0 & \cdots & 1 & \cdots & 0 & \cdots & 1
\end{pmatrix}
*
\begin{pmatrix}
w_{11} \\ \vdots \\ w_{1M} \\ \vdots \\ w_{N1} \\ \vdots \\ w_{NM}
\end{pmatrix}
=
\begin{pmatrix}
0 \\ f_1 \\ \vdots \\ f_M
\end{pmatrix},
\tag{32.19}
$$

$$
AW = F \tag{32.20}
$$

This equation can be solved with any technique that can invert a matrix. There are several problems in solving Eq. 32.20. In general, the equation can be inconsistent. In addition, the system is rank deficient. Further, traditional methods used to solve Eq. 32.20 are not robust to errors such as round-off errors in digital computers, measurement inaccuracies, and noise corruption. In the least squares solution (Eq. 32.6), the noise amplification is dictated by the condition number of the error matrix, i.e., the ratio of the highest and the lowest eigenvalues of E. In the extreme case when $\lambda_{\min} = 0$, the system is rank-deficient. In such cases, the pseudo-inverse solution can be used.

Any matrix A can be written using the singular value decomposition as

$$
A = UDV^T ,
$$

where

$$
D =
\begin{pmatrix}
\sigma_1 & 0 & \cdots & 0 \\
0 & \sigma_2 & \cdots & 0 \\
\vdots & \vdots & \ddots & \vdots \\
0 & 0 & \cdots & \sigma_N
\end{pmatrix},
$$

then,

$$
A^{-1} = VD^{-1}U^T ,
$$

where

$$
D^{-1} =
\begin{pmatrix}
\frac{1}{\sigma_1} & 0 & \cdots & 0 \\
0 & \frac{1}{\sigma_2} & \cdots & 0 \\
\vdots & \vdots & \ddots & \vdots \\
0 & 0 & \cdots & \frac{1}{\sigma_N}
\end{pmatrix}.
$$

σ_i^2, $i = 1, \ldots, N$ are the eigenvalues of AA^T. The matrices U and V are made up of the eigenvectors of AA^T and $A^T A$, respectively.

Extending this definition to rank-deficient matrices, the pseudo-inverse can be written as

$$A^\dagger = VD^\dagger U^T ,$$

where

$$D^\dagger = \begin{pmatrix} \frac{1}{\sigma_1} & 0 & \cdots & & 0 \\ 0 & \frac{1}{\sigma_2} & \cdots & & 0 \\ 0 & 0 & \cdots & \frac{1}{\sigma_r} & \cdots \\ & & & 0 & \\ & & & & 0 \end{pmatrix} ,$$

where r is the rank of the matrix A.

The rank-deficient system has infinite number of solutions. The pseudo-inverse solution can be shown to be the least squares solution with minimum energy. It can also be viewed as the projection of the least squares solution in the range space of A. An iterative technique called the Row Action Projection (RAP) algorithm [4, 19] can be used to solve Eq. 32.20.

Row Action Projection

An effective way to find a solution for Eq. 32.20 is to use the RAP method [4], which has been shown to be effective in providing a fast and stable solution to a system of simultaneous equations. Traditional least squares methods need a block of data to calculate the estimate. Most of these methods demand a lot of memory and processing power. RAP operates on only one row at a time, which makes it a useful sample-by-sample method in adaptive signal processing. Further, the matrix A in Eq. 32.20 is a sparse matrix. RAP has been shown to be effective in solving systems with sparse matrices [4].

For a given system of equations,

$$
\begin{aligned}
a_{01}w_1 + a_{02}w_2 + \ldots + a_{0,NM}w_{NM} &= f_0 \\
a_{11}w_1 + a_{12}w_2 + \ldots + a_{1,NM}w_{NM} &= f_1 \\
\ldots &= \ldots \\
a_{M1}w_1 + a_{M2}w_2 + \ldots + a_{M,NM}w_{NM} &= f_M ,
\end{aligned}
$$

each equation can be viewed as a "hyperplane" in NM dimensional space. If a unique solution exists, then it is at the point of intersection of all the hyperplanes. If the equations are inconsistent or ill-defined, then the solution set is a region in space.

The RAP method defines an iterative method to arrive at a point in the solution set and is as follows: Starting from an initial guess W^0, the algorithm iterates over all the equations by repeatedly projecting the solution on the hyperplanes represented by the equations. At step $i + 1$ the weight vector is updated as:

$$W^{i+1} = W^i + \lambda \frac{e_i}{||\mathbf{a}_p||^2} \mathbf{a}_p \tag{32.21}$$

where $\mathbf{a_p}$ is the pth row of A, λ is the step size, and

$$e_i = f_p - \mathbf{a_p^T W^i} \tag{32.22}$$

is the error at the ith iteration. At the ith iteration, we use the pth row, where $p = i \mod (M + 1)$, i.e., we cycle over all the equations.

The RAP method is a special case of the Projection onto Convex Sets (POCS) algorithm.

The geometrical interpretation of the above algorithm is given in Fig. 32.5. Each equation is modeled as a hyperplane in the solution space. Here, in the figure, it is shown as a line. The initial guess is projected onto the first hyperplane to obtain the second guess. This point is again projected onto the next

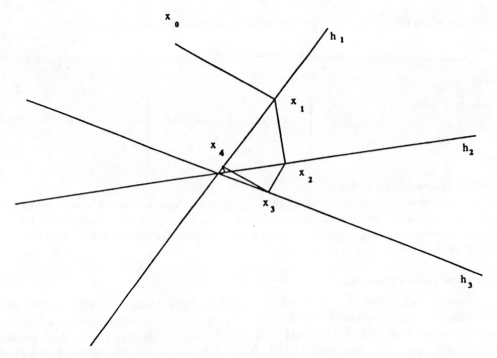

FIGURE 32.5   Geometrical interpretation of RAP.

hyperplane to get the third guess. It can be shown that by repeated projection on to the hyperplanes, the point converges to the solution [4]. λ ($0 \leq \lambda \leq 1$) is called the relaxation parameter. It dictates how far we should proceed along the direction of the estimate. It is also a measure of confidence in the estimate, i.e., if the measurements are noisy, then usually λ is given a small value; if the values are relatively less noisy, then a larger value of λ can be used to speed up convergence. The algorithm is guaranteed to converge to the actual solution (if it exists). If a solution does not exist, then the "guess" is guaranteed to converge to the pseudo-inverse solution. The pseudo-inverse solution is the least squares solution which minimizes the energy in the solution vector. The RAP method provides stable estimates at each iteration. Since the method uses only one row at a time, the system can be made adaptive, i.e., as the source moves around in the room, the system response can be varied.

For a detailed discussion of adaptive arrays, the reader is referred to [20].

32.2.3 Multiple Beamforming

In a highly reverberant environment, many images of the sound source fall along the bore of the beam of a single beamformer. Hence, delay-and-sum single beamformers have limited success in combating reverberation [13]. As shown earlier, the SNR improvement is poor under severe reverberation. Instead of forming a single beam on the source, many beams can be formed, each directed towards the source and its major images [13]. This is called *multiple beamforming*. In a multiple beamformer (Fig. 32.6), the signal-to-reverberant-noise ratio is $\frac{(BN)^2}{BN(K-1)} = \frac{BN}{(K-1)}$. As B, the number of beams, approaches K, the number of images, the SNR approaches N, or the number of microphones. Multiple beamforming, when $B = K$, can be shown to be equivalent to matched filtering.

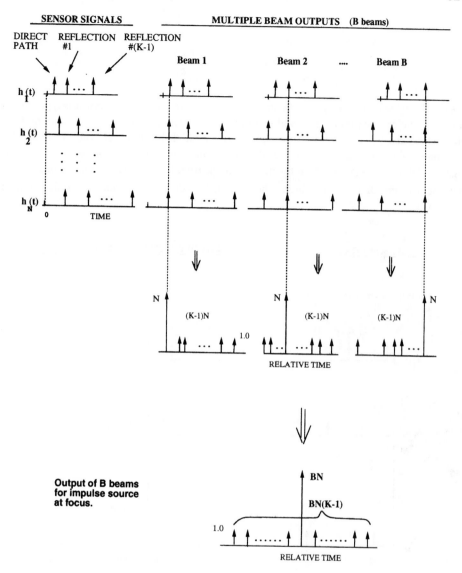

FIGURE 32.6 A multiple beamformer. (*Source:* Flanagan, J.L., Surendran, A.C., and Jan, E.-E., Spatially selective sound capture for speech and audio processing, *Speech Commun.*, 13: 207–222, 1993. With kind permission of Elsevier Science - NL, Sara Burgerhartstraat 25, 1055 KV Amsterdam, The Netherlands).

32.3 Matched Filtering

Matched filtering techniques can be applied to microphone arrays for dereverberation. In this technique, each microphone output is filtered by a causal approximation of the time reverse of the impulse response to that microphone [13]. Thus, if $g_i(n)$ is the impulse response to microphone i, then

$$h_i(n) = g_i(n_0 - n),$$

(32.23)

and

$$H_i(z) = z^{-n_0} G_i\left(\frac{1}{z}\right). \tag{32.24}$$

Since it is desirable for the delay n_0 to be suitably small, the time-reversed response is typically truncated. But careful choice of n_0 leads to a good compromise between delay of the system and high SNR. The matched filter can also be viewed as a special case of a multiple beamformer, when a beam is directed at every image, and when the output of the ith microphone contributing to the beam directed to the jth image is weighted by $\frac{1}{d_{ij}}$, where d_{ij} is the distance of the ith microphone from the jth image. Figure 32.7 shows the principle of a matched filter. The SNR analysis of a matched filter is similar to the multiple beamformer when $B = K$.

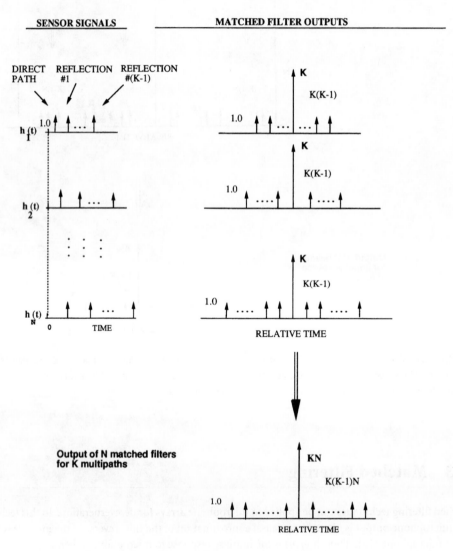

FIGURE 32.7 Principle of a matched filter. (*Source:* Flanagan, J.L., Surendran, A.C., and Jan, E.-E., Spatially selective sound capture for speech and audio processing, *Speech Commun.*, 13: 207–222, 1993. With kind permission of Elsevier Science - NL, Sara Burgerhartstraat 25, 1055 KV Amsterdam, The Netherlands).

Thus, for a source $s(n)$ located at the focal point, the output of the system is

$$o(n) = s(n) * \left\{ \sum_{i=1}^{N} g_i(n) * g_i(n_0 - n) \right\}$$

(32.25)

and the output for a source away from the focus is

$$o(t) = s(t) * \left\{ \sum_{i=1}^{N} g_i'(n) * g_i(n_0 - n) \right\} ,$$

(32.26)

where $g_i'(n)$ is the impulse response for a source located away from the focus. So, additional to mitigating reverberation, matched filters provide *volume selectivity*, i.e., a focal volume of retrieval, which depends on the spatial correlation of the impulse responses $g_i(n)$. Using microphone arrays instead of a single microphone provides not only a smoother frequency response [22], but also a higher SNR improvement, which, even in the worst case, asymptotically approaches N, the number of sensors used [13]. Since each individual matched filter seeks to smooth out the spectral minima due to other matched filters, it is desirable that the matched filters at each microphone be as different as possible. This is a motivation to use a random distribution of sensors [22].

The aim of the matched filter is to maximize the power of the output of the array for a source located at the focus and minimize the power of off-focus sources. This is an important property, which we shall contrast with the exact inverse discussed in the next section.

The power of matched filtering in mitigating reverberation and suppressing interfering noise is demonstrated through examples in Section 32.5. Figure 32.11 shows the response of a matched filter system. It is clear that the matched filter response is similar to, but cannot be exactly, an ideal impulse, i.e., it cannot provide an exact inverse to the room transfer function. Next, we discuss a method that can provide an exact inverse to the room transfer function.

32.4 Diophantine Inverse Filtering Using the Multiple Input-Output (MINT) Model

Miyoshi and Kaneda [23] proposed a novel method to find the exact inverse of a point in a room by using multiple inputs and outputs, each input-output pair modeled by an FIR system. For example, a two-input single-output system is described by the two speaker-to-single-microphone responses, $G_1(z)$ and $G_2(z)$. The inputs need to be pre-processed by the two FIR filters, $H_1(z)$ and $H_2(z)$, such that

$$H_1(z)G_1(z) + H_2(Z)G_2(Z) = 1 .$$

(32.27)

This is a Diophantine equation which has an infinite number of solutions. That is, if $H_1(z)$ and $H_2(z)$ satisfy Eq. 32.27, then

$$\begin{aligned} H_1' &= H_1(z) + G_2(z)K(z) & (32.28) \\ H_2' &= H_2(z) - G_1(z)K(z) , & (32.29) \end{aligned}$$

where $K(z)$ is an arbitrary polynomial, is also a solution for Eq. 32.27. But, if $G_1(z)$ and $G_2(z)$ do not have common zeros in the z-plane, and if the orders of $H_1(z)$ and $H_2(z)$ are less than that of $G_2(z)$ and $G_1(z)$, respectively, by Euclid's theorem, a unique solution is guaranteed to exist [23, 24].

The above system can be used with a microphone array for dereverberation (Fig. 32.1). The problem is to find $H_i(z), i = 1, 2, .., N$ such that

$$G_1(z)H_1(z) + G_2(z)H_2(z) + ... + G_N(z)H_N(z) = 1 .$$

(32.30)

As the number of microphones in the array increases, the chances that all the $G_i(z)$s share a common zero in the z-plane diminishes. This assures that the multiple microphone system yields a unique and exact solution.

In time domain, the previous expression can be written as:

$$d(k) = g_1(k) * h_1(k) + \cdots + g_N(k) * h_N(k), \tag{32.31}$$

where N is the number of microphones. Now,

$$
\begin{pmatrix}
g_1(0) & & & g_N(0) & & \\
g_1(1) & & & g_N(1) & & \\
\vdots & \cdots & 0 & \vdots & \cdots & 0 \\
g_1(m) & \cdots & g_1(0) & \cdots & g_N(l) & \cdots & g_N(0) \\
0 & \cdots & g_1(1) & \cdots & 0 & \cdots & g_N(1) \\
0 & \cdots & \vdots & \cdots & 0 & \cdots & \vdots \\
 & & g_1(m) & & & & g_N(l)
\end{pmatrix}
\begin{pmatrix}
h_1(0) \\
\vdots \\
h_1(i) \\
\vdots \\
h_N(0) \\
\vdots \\
h_N(k)
\end{pmatrix}
=
\begin{pmatrix}
1 \\
0 \\
\vdots \\
0
\end{pmatrix}, \tag{32.32}
$$

$$
(G_1 \quad \cdots \quad G_N)
\begin{pmatrix}
H_1 \\
\vdots \\
H_N
\end{pmatrix}
= D \tag{32.33}
$$

Thus,

$$
\begin{pmatrix}
H_1 \\
\vdots \\
H_N
\end{pmatrix}
= (G_1 \quad \cdots \quad G_N)^{-1} D \tag{32.34}
$$

The RAP algorithm described on page **32**-9 is an effective method to solve Eq. 32.34. In the MINT modeling, even if the different $G_i(z)$s share a common zero, RAP can provide a stable inverse. Even if the data are "noisy", or if the system is ill-conditioned, the algorithm is guaranteed to converge. From computer simulations, it can be shown that the solution converges very fast (see Fig. 32.8). Hence, the system can adapt to the varying conditions without having to recalculate the FIR filters.

Figure 32.8 shows the rate of convergence of the RAP algorithm when the number of microphones in the array is varied. The results suggest that increasing the number of microphones used in the array increases the speed of convergence and also provides more accurate results.

32.5   Results

In this section, computer simulations are presented to demonstrate the effect of matched filtering and the Diophantine inverse filtering method. A room ($20 \times 16 \times 5$ m in size) was simulated using the image model [10]. The source was located at $(14, 9.5, 1.7)$m. 5th order images were assumed and wall reflectivity was assumed to be $\alpha = 0.1$. Sensor spacing was considered to be 40 cm. A large spacing between sensors was chosen to make the impulse responses as dissimilar as possible.

The SNR of the output was calculated using the formula:

$$SNR(dB) = 10 \log_{10} \frac{\sum s(n)^2}{\sum (y(n) - s(n))^2} \tag{32.35}$$

where $s(n)$ is the input speech signal and $y(n)$ is the output speech signal. The two signals are sufficiently staggered to account for the delay in the processing.

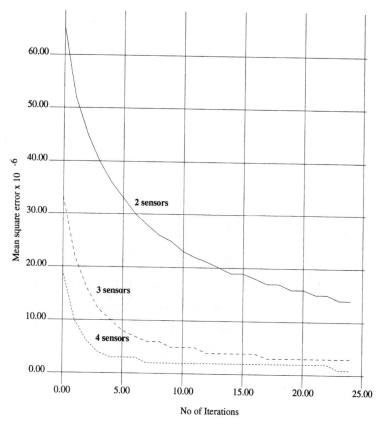

FIGURE 32.8   Rate of convergence of RAP for calculating the exact inverse filters.

The signal-to-noise-ratios were calculated as follows:

No. of mics	SNR
2	15 dB
3	27 dB
4	37 dB

For comparison, the SNR gains of a single beamforming, multiple beamforming, and matched filter linear arrays using five microphones are presented below. The multiple beamformer has one beam directed at each image of the source.

Method	SNR
Single beamformer	-1 dB
Multiple beamformer	11 dB
Matched filter	13 dB

Figure 32.9 shows the impulse response of the room using an unsteered array system consisting of four microphones. Figures 32.10 and 32.11 are the system responses of a single beamformer and the matched filter. The matched filter system response is a much better approximation of an ideal impulse than the single beamformer. But the tail of the response is still significant compared to the exact inverse system (Fig. 32.12) whose final response is very close to an ideal impulse.

For obtaining the same SNR gain, the exact inverse requires a lesser number of microphones than either the matched filter or the multiple beamformer. The Diophantine inverse filtering method does not suffer from the effects of spatial aliasing that may affect traditional beamformers using periodically spaced

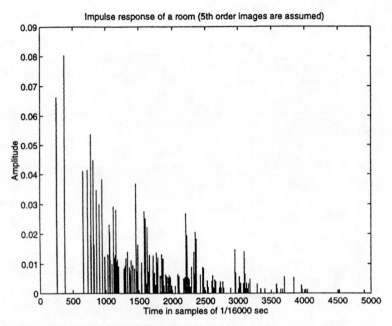

FIGURE 32.9   Impulse response of a room (images up to 5th order are used).

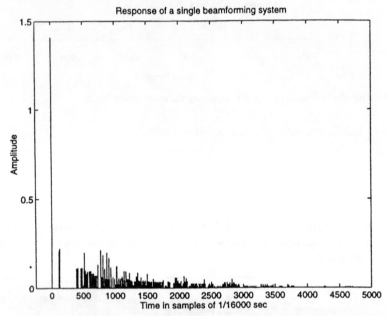

FIGURE 32.10   Response of a single beamformer for a source located on the axis.

microphones. Finding the exact inverse is also more computationally intensive than matched filtering or multiple beamforming.

32.5.1   Speaker Identification

A simple speaker identification experiment was done to test the acoustic fidelity of the exact inverse system. The dimensions of the simulated room, the location of the source and the other conditions was assumed

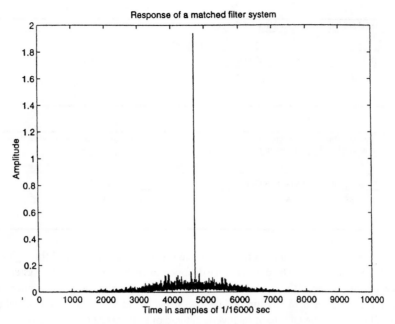

FIGURE 32.11 Response of a matched filtering system for a source located at the focus.

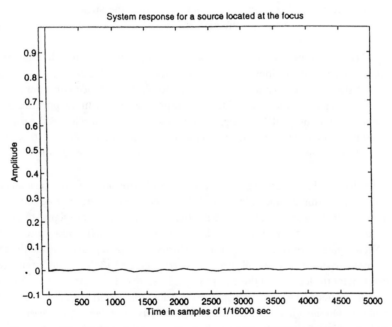

FIGURE 32.12 Response of the Diophantine inverse filtering system (the delay involved is not shown).

to be identical to the experiment reported in the previous section. A part of the TIMIT database with 38 speakers, all from the New England area, was used. Five sentences were used for training and five were used for testing. Twelve cepstral vectors were used and a Learning Vector Quantizer (LVQ) was used for identification [25].

Speaker identification accuracy for the exact inverse system:

Training data	CLS (%)	One mic (%)	Array output (%)
		Testing data	
CLS	91.6	36.3	90
Array output			92.6

Speaker identification accuracy for the exact inverse system when an interfering Gaussian noise source at 15 dB signal-to-competing noise ratio is present:

Training data	CLS (%)	One mic (%)	Array output (%)
		Testing data	
CLS	91.6	14.2	9.5
Array output			49

The identification accuracy when trained and tested on clean speech recorded through a close talking microphone (CLS) was 91.6%. The performance dropped to 36.3% when the same system was tested on a single microphone located at the center of the array. Once the Diophantine inverse filtering was used to clean up the speech, the performance jumped back to 90%. The identification accuracy when the system was trained and tested on the Diophantine inverse filtered output was 92.6%.

But the performance was poor even in the presence of modest interference. When a Gaussian noise source at 15 dB signal-to-competing noise ratio levels was introduced at (3.0,5.0,1.0)m, the performance on the output of the exact inverse filtering system (9.5%) was worse than the single microphone (14.2%). Under matched training and testing conditions, the performance of the exact inverse system was significantly lower (49%).

Recently, speaker identification results were reported on the output of a matched-filtered system [26]. The room dimensions and conditions were similar to the ones in this report and the data sets used for training and testing were the same. The performance under matched conditions for close talking microphone was 94.7% and for the matched filtered output was 88.4%. In the presence of an interfering source producing Gaussian noise at 15 dB signal-to-competing noise ratio levels, the performance when trained on close talking microphone and tested on the matched filtered output was 80%; the performance when trained and tested on the matched filtered output in the presence of noise was approximately 88% [26].

From these results, it is clear that though the exact inverse filtering outperforms the matched filter under clean conditions, it performs significantly poorer when there are interfering noise sources. This can be attributed to the fact that the exact inverse system attempts to maximize the signal-to-reverberant noise ratio (SRNR) for a source at the focus. Though it maximizes the SRNR for a source at the focus and lowers the SRNR for any source located away from the focus, it does not guarantee that the contribution of interfering source to the output power will also be lowered. Figure 32.13 shows the impulse response of the exact inverse system for the location of the interfering noise source. It is clear that the SNR of the source at this location would be poor (the effective response does not look like an ideal impulse). But the signal is effectively amplified. On the other hand, the matched filter *maximizes the output power for a source located at the focus and minimizes the output power for all other sources* thus providing lower signal-to-noise ratio improvement, but higher levels of spatial discrimination.

32.6 Summary

Microphone arrays can be successfully used in "inverting" room acoustics. A simple single beamformer is not effective in combating room reverberation, especially in the presence of interfering noise sources. Adaptive algorithms that project a null in the direction of the interferer can be used, but they introduce

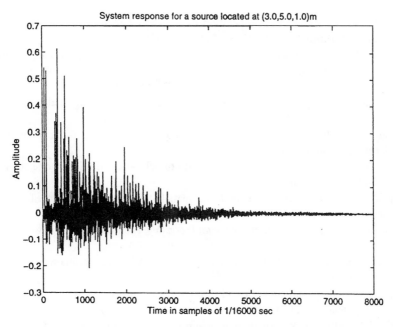

FIGURE 32.13 Response of the Diophantine inverse filtering system for a source located away from the focus.

significant distortion in the main signal. Constrained adaptive arrays mitigate this problem but they are of limited capability in severely reverberant environments. Processing algorithms such as multiple beamforming and matched filtering, combined with three-dimensional array of sensors, though only providing an approximation to the inverse, give robust dereverberant systems that provide selectivity in a spatial volume and thus immunity from interfering noise sources. An exact inverse using Diophantine inverse filtering using the MINT model can be found. Though this method provides a higher signal-to-noise ratio for a source at the focus, it does not provide immunity from noise interference that the matched filtering can offer. Speaker identification results are provided that substantiate the performance analysis of these systems.

References

[1] Flanagan, J.L. and Lummis, R.C., Signal processing to reduce multipath distortions in small rooms, *J. Acoustical Soc. Am.*, 47, 1475–1481, Feb. 1970.

[2] Neely, S. and Allen, J., Invertibility of a room response, *J. Acoustical Soc. Am.*, 66, 165–169, 1979.

[3] Mourjopoulos, J., Clarkson, P.M. and Hammond, J.K., A comparative study of least-squares and homomorphic techniques for the inversion of mixed phase signals, *Proc. IEEE Conf. Acoustics, Speech, Signal Process. '82*, 1858–1861, 1982.

[4] Mammone, R.J., *Computational Methods of Signal Recognition and Recovery*, John Wiley & Sons, New York, 1992.

[5] Mourjopoulos, J. and Hammond, J.K., Modeling and enhancement of reverberant speech using an envelope convolution method, *Proc. IEEE Conf. Acoustics, Speech, Signal Process.*, 1144–1147, 1983.

[6] Stockham, T.G., Cannon, T.M. and Ingebresten, B.R., Blind deconvolution through digital signal processing, *Proc. IEEE*, 63(4), 678–692, 1975.

[7] Langhans, T. and Strube, H.W., Speech enhancement by nonlinear multiband envelope filtering, *Proc. IEEE Conf. Acoustics, Speech, Signal Process.*, 156–159, 1982.

[8] Wang, H. and Itakura, F., Dereverberation of speech signals based on sub-band envelope estimation, *ICIE Trans.,* E 74(11), 3576–3583, Nov. 1991.

[9] Allen, J.B., Berkeley, D.A. and Blauert, J., Multimicrophone signal processing technique to remove room reverberation from speech signals, *J. Acoustical Soc. Am.,* 62, 912–915, Oct., 1977.

[10] Allen, J.B. and Berkeley, D.A., Image method for efficiently simulating small-room acoustics, *J. Acoustical Soc. Am.,* 65(4), 943–950, Apr. 1979.

[11] Flanagan, J.L., Berkeley, D.A., Elko, G.W. and Sondhi, M.M., Autodirective microphone systems, *Acustica,* 73, 58–71, 1991.

[12] Flanagan, J.L., Beamwidth and usable bandwidth of delay-steered microphone arrays, *AT&T Tech. J.,* 64(4), 983–995, Apr. 1985.

[13] Flanagan, J.L., Surendran, A.C. and Jan, E.-E., Spatially selective sound capture for speech and audio processing, *Speech Comm.,* 13, 207–222, 1993.

[14] Widrow, B. and Stearns, S.T., *Adaptive Signal Processing,* Prentice-Hall, Englewood Cliffs, NJ, 1985.

[15] Widrow, B., Mantey, P.E., Griffiths, L.J., and Goode, B.B., Adaptive antenna systems, *Proc. IEEE,* 55, 2143–2159, Dec., 1967.

[16] Griffiths, L.J., A simple adaptive algorithm for real-time processing in antenna arrays, *Proc. IEEE,* 57(10), 1696–1704, Oct. 1969.

[17] Griffiths, L.J. and Jim, C.W., An alternative approach to linearly constrained adaptive beamforming, *IEEE Trans. Antennas Propagation,* AP-30(1), 27–34, Jan. 1982.

[18] Frost III, O.L., An algorithm for linearly constrained adaptive array processing, *Proc. IEEE,* 60(8), 926–935, 1972.

[19] Farrell, K., Mammone, R.J. and Flanagan, J.L., Beamforming microphone arrays for speech enhancement, *Proc. IEEE Conf. Acoustics, Speech, Signal Process. '92,* 1, 285–288, 1992.

[20] *IEEE Trans. Antennas Propagation: Special Issues on Adaptive Arrays,* AP-24, Sept. 1976, 34(3), March 1986.

[21] Applebaum, S.P., Adaptive arrays, *IEEE Trans. Antennas Propagation,* AP-24(5), 585–599, Sept. 1976.

[22] Jan, E.-E. and Flanagan, J.L., Microphone arrays for speech processing, *Intl. Symp. Signals, Syst. Electron.,* San Francisco, CA, 1995.

[23] Miyoshi, M. and Kaneda, Y., Inverse filtering of room acoustics, *IEEE Trans. Acoustics, Speech, Signal Process.,* 36(2), 145–152, Feb., 1988.

[24] Sondhi, M.M., Personal communication.

[25] Surendran, A.C. and Flanagan, J.L., Stable dereverberation using microphone arrays for speaker identification, *J. Acoustical Soc. Am.,* 96(5) 3261, Nov. 1994.

[26] Lin, Q., Jan, E.-E. and Flanagan, J.L., Microphone arrays and speaker identification, *IEEE Trans. Speech Audio Process.,* 2(4), 622–629, Oct., 1994.

33

Synthetic Aperture Radar Algorithms

Clay Stewart
Science Applications International Corporation

Vic Larson
Science Applications International Corporation

33.1 Introduction

A synthetic aperture radar (SAR) is a radar sensor that provides azimuth resolution superior to that achievable with its real beam by synthesizing a long aperture using platform motion. The geometry for the production of the SAR image is shown in Fig. 33.1. The SAR is used to generate an electromagnetic map of the surface of the earth from an airborne or spaceborne platform. This electromagnetic map of the surface contains information that can be used to distinguish different types of objects that make up the surface. The sensor is called a synthetic aperture radar because a synthetic aperture is used to achieve the narrow beamwidth necessary to get a high cross-range resolution. In SAR imagery the two dimensions are range (perpendicular to the sensor) and cross-range (parallel to the sensor). The range resolution is achieved using a high bandwidth pulsed waveform. The cross-range resolution is achieved by making use of the forward motion of the radar platform to synthesize a long aperture giving a narrow beamwidth and high cross-range resolution. The pulse returns collected along this synthetic aperture are coherently combined to create the high cross-range resolution image. A SAR sensor is advantageous compared to an optical sensor because it can operate day and night through clouds, fog, and rain, as well as at very long ranges. At very low nominal operating frequencies, less than 1 GHz, the radar even penetrates foliage and can image objects below the tree canopy. The resolution of a SAR ground map is also not fundamentally limited by the range from the sensor to the ground. If a given resolution is desired at a longer range, the synthetic aperture can simply be made longer to achieve the desired cross-range resolution.

A SAR image may contain "speckle" or coherent noise because it results from coherent processing of the data. This speckle noise is a common characteristic of high frequency SAR imagery and reducing speckle, or building algorithms that minimize speckle, is a major part of processing SAR imagery beyond the image formation stage. Traditional techniques averaged the intensity of adjacent pixels, resulting in a smoother but lower resolution image. Advanced SAR sensors can collect multiple polarimetric and/or frequency channels where each channel contains unique information about the surface. Recent systems have also

0-8493-8572-5/98/$0.00+$.50

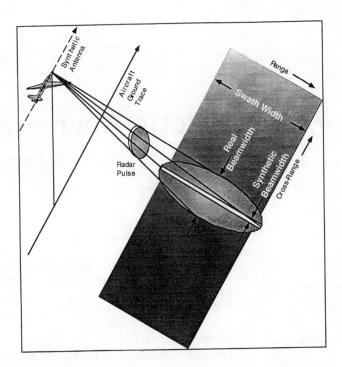

FIGURE 33.1   SAR imaging geometry.

used elevation angle diversity to produce 3-D SAR images using interferometric techniques. In all of these techniques, some sort of averaging is employed to reduce the speckle.

The largest consumers of SAR sensors and products are the defense and intelligence communities. These communities use SAR to locate and target relocatable and fixed objects. Manmade objects, especially ones with sharp corners, have very bright signals in SAR imagery, making these objects particularly easy to locate with a SAR sensor. A technology similar to SAR is inverse synthetic aperture radar (ISAR) which employs motion of the platform to image the target in cross-range. The ISAR data can be collected from a fixed radar platform since the target motion creates the viewing angle diversity necessary to achieve a given cross-range resolution. ISAR systems have been used to image ships, aircraft, and ground vehicles.

In addition to the defense and intelligence applications of SAR, there are several commercial remote sensing applications. Because a SAR sensor can operate day and night and in all weather, it provides the ability to collect data at regular intervals uninterrupted by natural influences. This stable source of ground mapping information is invaluable in tracking agriculture and other natural resources. SAR sensors have also been used to track oil spills (oil-coated water has a different backscatter than natural water), image underground rock formations (at some frequencies the radar will penetrate some soils), track ice conditions in the Arctic, and collect digital terrain elevation data.

Radar is an abbreviation for RAdio Detection And Ranging. Radar was developed in the 1930s and 1940s to detect and track ships and aircraft. These surveillance and tracking radars were designed so that a target was contained in a single resolution cell. The size of the resolution cell was a critical design parameter. Smaller resolution cells allowed one to determine the location of a target more accurately and increased the target-to-clutter ratio, improving the ability to detect a target. In the 1950s it was observed that one could map the ground (an extended target that takes up more than one resolution cell) by mounting the radar on the side of an aircraft and building a surface map from the radar returns. High range resolution was achieved by using a short pulse or high bandwidth waveform. The cross-range resolution was limited by the size of the antenna, with the cross-range resolution roughly proportional to R/L_a where R is the range from the sensor to the ground and L_a is the length of the antenna. The physical length of the antenna was constrained, limiting the resolution. In 1951, Carl Wiley of the Goodyear Aircraft Corporation noted that the reflections from two fixed targets in the antenna beam, but at different angular positions relative to the velocity vector of the platform, could be resolved by frequency analysis of the along track (or cross-range) signal spectrum. Wiley simply observed that each target had different Doppler characteristics because of its relative position to the radar platform and that one could exploit the Doppler to separate the targets. The Doppler effect is, of course, the change in frequency of a signal transmitted or received from a moving platform discovered by Christian J. Doppler in 1853:

$$f_d = v/\lambda$$

where f_d is the Doppler shift, v is the radial velocity between the radar and target, and λ is the radar wavelength. While the Doppler effect had been used in radar processing before the 1950s to separate moving targets from stationary ground clutter, Wiley's contribution was to discover that with a side looking airborne radar (SLAR), Doppler could be used to improve the cross-range spatial resolution of the radar. Other early work on SAR was done independently of Wiley at the University of Illinois and the University of Michigan during the 1950s. The first demonstration of SAR mapping was done in 1953 by the University of Illinois by performing frequency analysis of data collected by a radar operating at a 3-cm wavelength from a C-46 aircraft. Much work has been accomplished perfecting SAR hardware and processing algorithms since the first demonstration. For a much more detailed description of the history of SAR including the development of focused SAR, phase compensation techniques, calibration techniques, and autofocus, see the recent book by Curlander and McDonough [1].

Before offering a brief description of some processing approaches for forming, enhancing, and interpreting SAR imagery, we give two examples of existing SAR systems and their applications. The first system is the Shuttle Imaging Radar (SIR) developed by the NASA Jet Propulsion Laboratory (JPL) and flown on several space shuttle missions. This system was designed for non-military collection of geographic data. The second example is the Advanced Detection Technology Sensor (ADTS) built by the Loral Corporation for the MIT Lincoln Laboratory. The ADTS sensor was designed to demonstrate the capability of a SAR to detect and classify military targets. Table 33.1 contains the basic parameters for the ADTS and SIR SAR systems along with details on several other SAR systems.

Figure 33.2 shows an example image formed from data collected by the SIR SAR. The JPL engineers describe this image as follows:

> This is a radar image of Mount Rainier in Washington state... This image was acquired by the Spaceborne Imaging Radar-C and X-band Synthetic Aperture Radar (SIR-C/X-SAR) aboard the space shuttle Endeavor on its 20th orbit on October 1, 1994. The area shown in the image is approximately 59 kilometers by 60 kilometers (36.5 miles by 37 miles). North is toward the top left of the image, which was composed by assigning red and green colors to the L-band, horizontally transmitted and vertically received, and the L-band, horizontally transmitted and vertically received. Blue indicates the C-band, horizontally transmitted and vertically received. In addition to highlighting topographic slopes facing the space shuttle, SIR-C records rugged areas as brighter and smooth areas as darker. The scene was illuminated by the shuttle's radar from the northwest so that northwest-facing slopes are brighter and southeast-facing slopes are dark. Forested regions are pale green in color; clear cuts and bare ground are bluish or purple; ice is dark green and white. The round cone at the center of the image is the 14,435-foot (4,399-meter) active volcano, Mount Rainier. On the lower slopes is a zone of rock ridges and rubble (purple to reddish) above coniferous forests (in yellow/green). The western boundary of Mount Rainier National Park is seen as a transition from protected, old-growth forest to heavily logged private land, a mosaic of recent clear cuts (bright purple/blue) and partially regrown timber plantations (pale blue).

TABLE 33.1   Example SAR Systems

Platform	Bands polarization	Resolution (m)	Swath width	Interferometry
JPL AIRSAR	C, L, P–Full	4	10–18 km	Cross track L,C Along track L,C
SIR-C/X-SAR	C, L–Full, X - VV	30 × 30	15–90	Multi-pass
ERIM IFSARE	X–HH	2.5 × 0.8	10 km	Cross track
ERIM DCS	X–Full	< 1	1 km	Cross track
MIT LL ADTS	Ka (33 GHz)–Full	0.33	400 m	Multi-pass
NORDEN G11	Ku–VV	1,3	5 km	3 Along track 3 Cross track Phase centers
SRI UWB FOLPEN 2	100–300 MHz, 200–400 MHz, 300–500 MHz, HH	1 × 1	400–600 m	None
LORAL UHF MSAR	500–800 MHz, Full	0.6 × 0.6	280 m	None
NAWC P-3	C, L, X–Full	1.5 × 0.7	5 km	Along track X,C
NAWC P-3 UWB Upgrade	600 MHz–Full tunable over 200–900 MHz	0.33 × 0.66	930 km	None
Tier II+ UAV SAR	X	1 and 0.3	10 km	None

Figure 33.3 is an example image collected by the ADTS system. The ADTS system operates at a nominal frequency of 33 GHz and collects fully polarimetric, 1-ft resolution data. This image was formed using the polarimetric whitening filter (PWF) combination of three polarimetric channels to reduce the speckle noise. The output of the PWF is an estimate of radar backscatter intensity. The image displayed in Fig. 33.3 is based on a false color map which maps low intensity to black followed by green, yellow, and finally white. The color map simply gives the non-color radar sensor output false colors that make the low intensity shadows look black, the grass look green, the trees look yellow, and bright objects look white. This sample image was collected near Stockbridge, New York, and is of a house with an above ground swimming pool and several junked cars in the backyard. The radar is at the top of the image looking down at a 20° depression angle. The scene contains large areas of grass or crops and some foliage. Note the bright returns from the manmade objects, including the circular above-ground swimming pool, and strong corner reflector scattering from some of the cars in the backyard. Also note the relatively strong return from the foliage canopy. At this frequency the radar does not penetrate the foliage canopy. Note the shadows behind the trees where there is no radar illumination.

In this chapter on SAR algorithms, we give a brief introduction to the image formation process in Section 33.2. We review a few simple algorithms for reducing speckle noise in SAR imagery and automatic detection of manmade objects in Section 33.3. We review a few simple automatic object classification algorithms for SAR imagery in Section 33.4. This brief introduction to SAR only contains a few example algorithms. In the Section "Further Reading", we recommend some starting points for further reading on SAR algorithms, and discuss several open issues under current research in the SAR community.

33.2   Image Formation

In this section, we discuss some basic principles of SAR image formation. For more detailed information about SAR image formation, the reader is directed to the references given at the end of this chapter. One fundamental scenario under which SAR data is collected is shown in Fig. 33.1. An aircraft flies in a straight path at a constant velocity and collects radar data at a boresight of 90°. In practice it is impossible for an

FIGURE 33.2 SAR image of Mt. Rainier in Washington State taken from shuttle imaging radar.

aircraft to fly in a perfectly straight line at a constant velocity (at least within a wavelength), so motion (phase) compensation of the received radar signal is needed to account for aircraft perturbations. The radar on the aircraft transmits a short pulsed waveform or uses frequency modulation to achieve high range resolution imaging of the surface. The pulses collected from several positions along the trajectory of the aircraft are coherently combined to synthesize a long synthetic aperture in order to achieve a high cross-range resolution on the surface. In this section, we first discuss SLAR where only range processing is performed. Next, we discuss unfocused SAR where both range and cross-range processing are executed. Finally, we discuss focused SAR where "focusing" is performed in addition to range and cross-range processing to achieve the highest resolution and best image quality. At the end of this section we briefly mention several other important SAR image formation topics such as phase compensation, clutter-lock, autofocus, spotlight SAR, and ISAR. The details of these topics can be found in [1]–[3].

33.2.1 Side-Looking Airborne Radar (SLAR)

SLAR is the earliest radar system for remote surveillance of a surface. These radar systems could only perform range processing to form the 2-D reflectivity map of the surface, so the cross-range resolution is limited by the real antenna beamwidth. These SLAR systems typically operated at high frequencies (microwave or millimeter-wave) to maximize the cross-range resolution. We cover SLAR systems because SLAR performs the same range processing as SAR, and the limitations of a SLAR motivate the need for SAR processing.

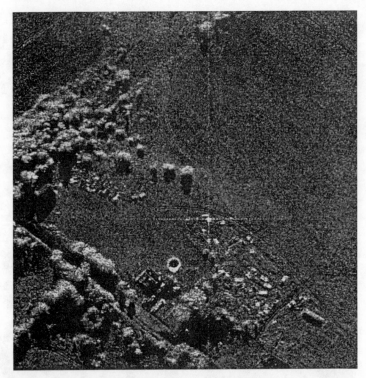

FIGURE 33.3   SAR image near Stockbridge, New York, collected by the ADTS.

The resolution of a SLAR system is limited by the radar pulse width in the range dimension, and the beamwidth and slant range in the cross-range dimension:

$$\delta_r = cT/2 \cos \eta$$
$$\delta_{cr} = R\lambda/L_a$$

where we represent the approximate 3-dB beamwidth of the antenna by λ/L_a, δ_r is the range resolution, δ_{cr} is the cross-range resolution, c is the speed of wave propagation, T is the compressed pulse width, η is the angle between the radar beam and the surface, R is the slant range to the surface, λ is the wavelength, and L_a is the length of the antenna.

The goal is to design the SLAR with a narrow beamwidth, short slant range, and a short pulsewidth to achieve high resolution. In practice, the pulsewidth of the radar is limited by hardware constraints and the amount of "energy on target" required to get sufficient signal-to-noise ratio to obtain a good image. To achieve a high range resolution without a short pulse, frequency modulation can be used to synthesize an effectively short pulse. This process of generating a narrow synthetic pulsewidth is called pulse compression. The approach is to introduce a modulation on the transmitted pulse, and then pass the received signal through a filter matched to the transmit signal modulation. The most common transmit waveforms used for pulse compression are linear FM (or chirp) and phase coded. Some radars use a digital version of linear FM called a stepped frequency waveform.

We illustrate pulse compression with the ideal application of the linear FM waveform. The square pulse is modulated by a linear FM signal, and the resulting transmit signal is

$$s(t) = \begin{cases} \cos\left(\omega_0 t - \frac{1}{2}\mu t^2\right) & |t| \leq T/2 \\ 0 & |t| > T/2 \end{cases}$$

where the bandwidth (frequency deviation) introduced by the linear FM is

$$\Delta f = T\mu/2\pi$$

If this transmit pulse is perfectly reflected from a stationary point target, range losses are ignored, and we shift in time to remove the two-way delay; the received signal is exactly the same as the transmitted signal. The matched filter response for the transmitted signal is

$$h(t) = \left(\frac{2\mu}{\pi}\right)^{1/2} \cos\left(\omega_0 t + \frac{1}{2}\mu t^2\right)$$

The output of the received signal applied to the matched filter is:

$$\Psi(t) = \left(\frac{\mu T^2}{2\pi}\right)^{1/2} \frac{\sin(\mu T t/2)}{(\mu T t/2)} \text{Re}\left[e^{j\left(\omega_0 t + \frac{1}{2}\mu t^2 + \pi/4\right)}\right]$$

This output has a mainlobe that has a 4-dB beamwidth of $1/\Delta f$. The resulting compressed pulse can be significantly narrower than the width of the transmitted pulse with a pulse compression ratio of $T\Delta f$. The range resolution of the radar has been increased by this pulse compression factor and is now given by:

$$\delta_r \approx c/2\Delta f \cos\eta$$

Note that the range resolution in the ideal case is now completely independent of the physical width of the transmitted pulse. Performing range compression against real radar targets that Doppler shift the frequency of the receive signal introduces ambiguities resulting in additional signal processing issues that must be addressed. There is a trade-off between the ability of a radar waveform to resolve a target in range and frequency. The performance of a waveform in range-frequency space is given by its ambiguity. The ambiguity function is the output of the matched filter for the signal for which it is matched and for frequency shifted versions of that signal. The references contain a much more detailed description of ambiguity functions and radar waveform design.

Using pulse compression, a SLAR system can achieve a very high range resolution on the order of 1 ft or less, but the cross-range resolution of the SLAR is limited by the physical beamwidth of the antenna, the operating frequency, and the slant range. This cross-range resolution limitation of SLAR motivates the use of a synthetic array antenna to increase the cross-range resolution.

33.2.2 Unfocused Synthetic Aperture Radar

Figure 33.1 provides a good geometric description of SAR. As with SLAR, the radar platform moves along a straight line collecting radar data from the surface. The SAR system goes one step further than SLAR by coherently combining pulses collected along the flight path to synthesize a long synthetic array. The beamwidth of this synthetic aperture is significantly narrower than the physical beamwidth (real beam) of the real antenna. The ideal synthetic beamwidth of this synthetic aperture is

$$\theta_B = \lambda/2L_\theta$$

The factor of two results from the two-way propagation from the moving platform. The unfocused SAR can be implemented by performing FFT processing in the cross-range dimension for the samples in each range bin. This is simply the conventional beamformer for an array antenna. The difference between SAR and real beam radar is that the aperture samples that comprise the SAR are collected at different times by a moving platform. There are several design constraints on a SAR system, including:

- The speed of the platform and pulse repetition rate (PRF) of the radar must be mutually selected so that the sample points of the synthetic array are separated by less than $\lambda/2$ to avoid grating lobes.

- The PRF must be selected so that the swath width is unambiguously sampled.
- A point on the ground must be visible to the radar real beam across the entire length of the synthetic array. This limits the size of the real beam antenna. This constraint leads to the observation that with SAR, the smaller the real-beam antenna, the better the resolution, whereas with SLAR the larger the real-beam antenna, the better the resolution.
- The SAR assumes that a ground target has an isotropic signal across the collection angle of the radar platform as it flies along the synthetic array.

The resolution of the unfocused SAR is limited because the slant range to a scatterer at a fixed location on the surface changes along the synthetic aperture. If we limit the synthetic aperture to a length so that the range from every array point in the aperture to a fixed surface location differs by less than $\lambda/8$, then the cross-range resolution of the unfocused SAR is limited to:

$$\delta_{cr} = \sqrt{R\lambda/2}$$

33.2.3 Focused Synthetic Aperture Radar

The cross-range limitation of an unfocused SAR can be removed by focusing the data, as in optics. The focusing procedure for the SAR involves adjusting the phase of the received signal for every range sample in the image so that all of the points processed in cross-range through the synthetic beamformer appear to be at the same range. The phase error at each range sample used to form the SAR image is

$$\Delta\phi = \frac{2\pi}{\lambda}\left(\frac{d_n^2}{R}\right) \text{ radiar}$$

where d_n is the cross-range distance from the beam center, R is the slant range to the point on the ground from the beam center, and λ is the wavelength. The range samples can be focused before cross-range processing by removing this phase error from the phase history data. Note that each data point has a different phase correction based on the along-track position of the sensor and the point's range from the sensor.

When focusing is performed, the resulting SAR image resolution is independent of the slant range between the sensor and ground. This can be shown as follows:

$$\delta_{cr} = R\theta_s$$

where,

$$\theta_s \approx \frac{\lambda}{2L_e} \text{ and } L_e \approx \frac{R\lambda}{L_a}$$

therefore,

$$\delta_{cr} \approx L_a/2$$

The effective beamwidth of the synthetic aperture is approximately $\lambda/2L_e$ where the factor of two comes from the two-way propagation of the energy (the exact effective beamwidth depends on the synthetic array taper used to control sidelobes). The length of the effective aperture (L_e) is limited by the fact that a given scatterer on the surface must be in the mainbeam of the real radar beam for every position along the synthetic aperture. The result is that the resolution of the SAR when the data is focused is approximately $L_a/2$.

SAR processing can also be developed by considering the Doppler of the radar signal from the surface as first done by Wiley in 1951. When the real beamwidth of the SAR is small, a point on the surface has an approximately linearly decreasing Doppler frequency as it passes through the main beam of the real SAR beamwidth. This time varying Doppler frequency has been shown to be approximately:

$$f_d(t) = \frac{2v^2|t - t_0|}{\lambda R}$$

where ν is the velocity of the platform and t_0 is the time that the point scatterer is in the center of the main beam. The change in Doppler frequency as the point passes through the main beam is $2\nu^2 T_d/\lambda R$, and T_d is the time that the point is in the main beam. As with linear FM pulse compression, covered in Section 33.2.1, this Doppler signal can be processed through a filter to produce a higher cross-range resolution signal which is limited by the size of the real aperture just as with the synthetic antenna interpretation ($\delta_{cr} = L_a/2$). In a modern SAR system, typically both pulse compression (synthetic range processing) and a synthetic aperture (synthetic cross-range processing) are employed. In most cases, these transformations are separable where the range processing is referred to as "fast time" processing and the cross-range processing is referred to as "slow-time" processing.

A modern SAR system requires several additional signal processing algorithms to achieve high resolution imagery. In practice, the platform does not fly a straight and level path, so the phase of the raw receive signal must be adjusted to account for aircraft perturbations, a procedure called motion compensation. In addition, since it is difficult to exactly estimate the platform parameters necessary to focus the SAR image, an autofocus algorithm is used. This algorithm derives the platform parameters from the raw SAR data to focus the imagery. There is also an interpolation algorithm that converts from polar to rectangular formats for the imagery display. Most modern SAR systems form imagery digitally using either an FFT or a bank of matched filters. Typically, a SAR will operate in either a stripmap or spotlight mode. In the stripmap mode, the SAR antenna is typically pointed perpendicular to the flight path (although it may be squinted slightly to one side). A stripmap SAR keeps its antenna position fixed and collects SAR imagery along a swath to one side of the platform. A spotlight SAR can move its antenna to point at a position on the ground for a longer period of time (thus actually achieving cross-range resolutions even greater than the aperture length over two). Many SAR systems support both stripmap and spotlight modes, using the stripmap mode to cover large areas of the surface in a slightly lower resolution mode, and spotlight modes to perform very high resolution imaging of areas of high interest.

33.3 SAR Image Enhancement

In this section we review a few techniques for removing speckle noise from SAR imagery. Removing the speckle can make it easier to extract information from SAR imagery and improves the visual quality.

Coherent noise or speckle can be a major distortion in high resolution, high frequency SAR imagery. The speckle is caused when the intensity of a resolution cell results from the coherent combination of many wavefronts resulting from randomly oriented clutter surfaces within a resolution cell. These wavefronts can combine constructively or destructively resulting in intensity variations across the image. When the number of wavefronts approaches infinity (i.e., large resolution cell collected by a high frequency radar) the Rayleigh clutter model can be used to represent the speckle under the right statistical assumptions. When the number of wavefronts is less than infinity, the K-distribution and other product models do a better job of theoretically and empirically modeling the clutter.

When the combination of the radar system design and clutter properties results in images that contain large amounts of speckle, it is desirable to perform additional processing to reduce the speckle. One approach for speckle reduction is to noncoherently spatially average adjacent resolution cells, sacrificing resolution for the speckle reduction. This spatial averaging can be performed as a part of the image formation analogous to the Bartlett method of spectral estimation. Another approach for reducing speckle is to average across polarimetric channels if multiple polarimetric channels are available.

The *polarimetric whitening filter* (PWF) reduces the speckle content while preserving the image resolution. The PWF was derived by Novak et al. [5] as a quadratic filter that minimizes a specific speckle metric (defined as the ratio of the clutter standard deviation to its mean). The PWF first whitens the polarimetric data with respect to the clutter's polarimetric covariance, and then noncoherently averages across the polarimetric channels. This whitening filter essentially diagonalizes the covariance matrix of the complex backscatter vector $[HH, HV, VV]^T$, such that the resulting new linear polarization basis

$[HH', HV', VV']^T$ has equal power in each component, where:

$$\begin{bmatrix} HH' \\ HV' \\ VV' \end{bmatrix} = \begin{bmatrix} HH \\ \dfrac{HV}{\sqrt{\varepsilon}} \\ \dfrac{VV - \rho^* \sqrt{\gamma} HH}{\sqrt{\gamma(1-|\rho|^2)}} \end{bmatrix} \tag{33.1}$$

where

$$\varepsilon = \frac{E\left(|HV|^2\right)}{E\left(|HH|^2\right)} , \gamma = \frac{E\left(|W|^2\right)}{E\left(|HH|^2\right)} , \rho = \frac{E\left(HH \cdot W^*\right)}{\sqrt{E\left(|HH|^2\right) \cdot E\left(|W|^2\right)}} \tag{33.2}$$

The polarization scattering matrix (using a linear-polarization basis) can then be expressed as

$$\sum = \sigma_{HH} \begin{bmatrix} 1 & 0 & \rho\sqrt{\gamma} \\ 0 & \varepsilon & 0 \\ \rho^*\sqrt{\gamma} & 0 & \gamma \end{bmatrix} \tag{33.3}$$

The pixel intensity (power) is then derived through non-coherent averaging of the power in each of the new polarization components,

$$Y = |HH|^2 + \left|\frac{HV}{\sqrt{\varepsilon}}\right|^2 + \left|\frac{W - \rho^*\sqrt{\gamma}HH}{\sqrt{\gamma\left(1-|\rho|^2\right)}}\right|^2 \tag{33.4}$$

yielding a minimal speckle image at the original image resolution. Novak et al. [5] have shown that on the ADTS SAR data, the PWF reduces the clutter standard deviation by 2.0 to 2.7 dB compared with the standard deviation of single-polarimetric-channel data. The PWF has a dramatic effect on the visual quality of the SAR imagery and the performance of automatic detection and classification algorithms applied to SAR images. The PWF does not take into account the effect of the speckle reduction operation on target signals. It only minimizes the clutter. There has been recent work on polarimetric speckle reduction filters that both reduce the clutter speckle while preserving the target signal. Fig. 33.4 shows the three polarimetric channels and the resulting PWF image for an ADTS SAR chip of a target-like object.

33.4 Automatic Object Detection and Classification in SAR Imagery

SAR algorithmic tasks of high interest to the defense and intelligence communities include automatic target detection and recognition (ATD/R). Since SAR imagery has very different target and clutter characteristics as compared with visual and infrared imagery, uniquely designed ATD/R algorithms are required for SAR data. In this section, we describe a few basic ATD/R algorithms that have been developed for high resolution, high frequency SAR imagery (10 GHz or above) [6, 7, 8].

Performing target detection and classification against remote sensing imagery and, in particular, SAR imagery is very different from the classical pattern recognition problem. In the classical pattern recognition problem, we have models defining N classes, and the goal is to design a classifier to separate sensor data into one of the N classes. In SAR target classification, the imagery contains regions of diffuse clutter which can be represented to some degree by models, but the imagery also contains a possibly uncountable set of target-like discrete unknown and unmodelable objects. The goal is to reject both the diffuse clutter and

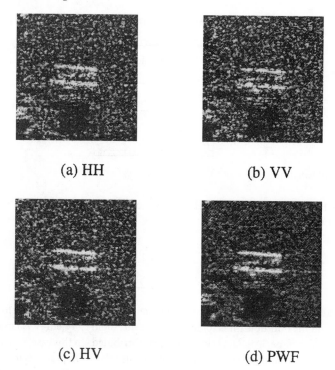

(a) HH   (b) VV

(c) HV   (d) PWF

FIGURE 33.4   Polarimetric processing of SAR data to reduce speckle.

the unknown discrete objects and to classify the target objects. This need to handle the unknown object means that the classifier must have the unknown class as a possible outcome of the classifier. Since the unknown class cannot be modeled, most SAR ATR systems solve the problem by employing a distance metric to compare the sensor data with models for each target of interest, and if the distance is too great, the data is classified as an unknown object.

Another design issue for a SAR ATD/R system is the need to process hundreds of square kilometers of data in near real-time to be of practical benefit. One widely used approach for solving this computational problem is to use a simple focus-of-attention or pre-detection algorithm to reject most of the diffuse clutter and pass only regions of interest (ROI), including all of the targets. These ROIs are then processed through a set of computationally more complicated classifiers which classify objects in the ROIs as one of the targets or as an unknown object.

In high frequency SAR imagery most target signatures have extremely bright peaks caused by physical corners on the target. One effective pre-detection technique involves applying a single pixel detector to find the bright pixels caused by corner reflectors on the targets. Since the background clutter power is unknown and varies across the image, we cannot simply use a thresholding operation to find these bright pixels. One approach for handling the unknown clutter power is to estimate it from clutter samples surrounding a test pixel. This approach for target detection is referred to as a constant false alarm rate (CFAR) detector because with the proper clutter and target models, it can be shown that the output of the detector has a constant false alarm rate in the presence of unknown clutter parameters. Fig. 33.5 depicts one design for a CFAR template. The clutter parameters are estimated using the auxiliary samples along a box with a test sample in the center. This test sample may or may not be on a target. The size of the box containing the auxiliary samples is sized so that the auxiliary samples do not overlap a target when the test sample is on the target. We also need to keep the size of the box containing the auxiliary samples as small as possible, so that we get a good local estimate of the clutter parameters. With these design constraints, a good choice for the CFAR template is just over twice the maximum dimension of the targets of interest.

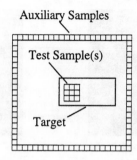

FIGURE 33.5   CFAR template.

One of these CFAR algorithms, first developed by Goldstein [9], is referred to as the two parameter CFAR or the log-*t* test:

$$\frac{\log x - \frac{1}{N}\sum_{i=1}^{N}\log y_i}{\sqrt{\frac{1}{N-1}\sum_{i=1}^{N}\left(\log y_i - \frac{1}{N}\sum_{i=1}^{N}\log y_i\right)^2}} \begin{array}{c} H_1 \\ > \\ < \\ H_0 \end{array} t$$

where x is the test sample, and $y_1, \ldots, y_N$ are the auxiliary samples. This test is performed for every pixel in the SAR scene and the output is thresholded with the threshold t. When N is large, the test statistic is approximately Gaussian if the SAR data is log normally distributed. In this case, Gaussian statistics can be used to determine the threshold for a given probability of false alarm. In practice, it is much more accurate to determine the threshold with a set of training data. This is primarily a corner reflector detector, and the output will almost always get more than one detection per target. In practice, a simple clustering algorithm can be used based on the size of the targets and the expected spacing of targets to get one detection per target and reduce the number of false alarms which are usually also clustered. The two-parameter CFAR test is one example of a simple SAR target detector. Researchers have also developed more sophisticated ordered statistic detectors, multi-polarimetric channel detectors, and feature-based discriminators to get improved SAR target detector performance [6, 7, 8].

This simple pre-detector gets a large number of false alarms (hundreds per square kilometer in single polarimetric channel, one foot resolution imagery) [5]. In order to further reduce the false alarm rate and classify the targets, further processing is necessary on the output of the pre-detector. One widely used approach for performing this classification operation is to apply a linear filter bank classifier to the ROIs identified by the pre-detector. Researchers have developed a large number of approaches for designing these linear filter bank classifiers including spatial matched filters [7], synthetic discriminant functions [7], and vector quantization/learning vector quantization [8]. The simplest approach is to build the spatial matched filters by breaking the target into angle subclasses, and averaging the training signatures in a given angle subclass to represent that subclass. In practice, the templates must be normalized because the absolute energy of a given target signature is unknown. The exact location of a target in the ROI is also unknown, so the matched filter must be applied for every possible spatial position of the target. This is performed more efficiently in the frequency domain as follows:

$$\rho_{ij} = \max\left\{FFT^{-1}\left[FFT\left(\mathbf{t}_{ij}\right)\cdot FFT(\mathbf{x})^*\right]\right\}$$

where $\mathbf{x}$ is a ROI and $\mathbf{t}_{ij}$ is the spatial matched filter representing the ith target and the jth angle subclass of that target. The ρ_{ij} is computed for every angle subclass of every target, and the maximum represents the estimate of the correct target and angle subclass. The output can be thresholded to reject false alarms. In practice the level of the threshold is determined by testing on both target and false alarm data.

In this section, we have reviewed a few basic concepts in SAR ATD/R. For a much more detailed treatment of this topic, consult the references and the recommended further reading given below.

References

[1] Curlander, J.C. and McDonough, R.N., *Synthetic Aperture Radar: Systems and Signal Processing,* John Wiley & Sons, New York, 1991.

[2] Wehner, D.R., *High Resolution Radar,* 2nd ed., Artech House, Boston, MA, 1995.

[3] Stimson, G.W., *Introduction to Airborne Radar,* Hughes Aircraft Company, 1983.

[4] Skolnik, M., *Introduction to Radar Systems,* 2nd ed., McGraw-Hill, New York, 1980.

[5] Novak, L., Burl, M., and Irving, B., Optimal polarimetric processing for enhanced target detection, *IEEE Trans. AES,* 29(1), 234-244, Jan. 1993.

[6] Stewart, C., Moghaddam, B., Hintz, K., and Novak, L., Fractional brownian motion for synthetic aperture radar imagery scene segmentation, *Proc. IEEE,* 81(10), 1511-1522, Oct. 1993.

[7] Novak, L., Owirka, G., and Netishen, C., Radar target identification using spatial matched filters, *Pattern Recognition,* 27(4), 607-617, Apr. 1994.

[8] Stewart, C., Lu, Y.-C., and Larson, V., A neural clustering approach for high resolution radar target classification, *Pattern Recognition,* 27(4), 503-513, Apr. 1994.

[9] Goldstein, G., False-alarm regulation in log-normal and Weibull clutter, *IEEE Trans. AES,* 9, 84-92, 1972.

Further Reading and Open Research Issues

A very brief overview of SAR with a few example algorithms is given here. The items in the reference list give a more detailed treatment of the topics covered in this chapter. SAR is a very active research topic. Articles on SAR algorithms are regularly published in many journals and conferences, including:

Journals

IEEE Transactions on Aerospace and Electronic Systems, IEEE Transactions on Geoscience and Remote Sensing, IEEE Transactions on Antennas and Propagation, IEEE Transactions on Signal Processing, and *IEEE Transactions on Image Processing.*

Conferences

IEEE National Radar Conference, IEEE International Radar Conference, and the International Society for Optical Engineering (SPIE) has several SAR Conferences.

There are numerous open areas of research on SAR signal processing algorithms including:

- Still developing an understanding of the utility and applications of multi-polarimetric, multi-frequency, and 3-D SAR.

- Performance/robustness of model-based image formation not completely understood.

- Performance/robustness of different detection, discrimination, and classification algorithms given radar, clutter, and target parameters not completely understood.

- No fundamental theoretical understanding of performance limitations given radar, clutter, and target parameters (i.e., no Shannon theory).

34

Iterative Image Restoration Algorithms

Aggelos K. Katsaggelos
Northwestern University

34.1 Introduction

In this chapter we consider a class of iterative restoration algorithms. If y is the observed noisy and blurred signal, D the operator describing the degradation system, x the input to the system, and n the noise added to the output signal, the input-output relation is described by [3, 51]

$$y = Dx + n. \tag{34.1}$$

Henceforth, boldface lower-case letters represent vectors and boldface upper-case letters represent a general operator or a matrix. The problem, therefore, to be solved is the inverse problem of recovering x from knowledge of y, D, and n. Although the presentation will refer to and apply to signals of any dimensionality, the restoration of greyscale images is the main application of interest.

There are numerous imaging applications which are described by Eq. (34.1) [3, 5, 28, 36, 52]. D, for example, might represent a model of the turbulent atmosphere in astronomical observations with ground-based telescopes, or a model of the degradation introduced by an out-of-focus imaging device. D might also represent the quantization performed on a signal, or a transformation of it, for reducing the number of bits required to represent the signal (compression application).

The success in solving any recovery problem depends on the amount of the available prior information. This information refers to properties of the original signal, the degradation system (which is in general only partially known), and the noise process. Such prior information can, for example, be represented by the fact that the original signal is a sample of a stochastic field, or that the signal is "smooth," or that the signal takes only nonnegative values. Besides defining the amount of prior information, the ease of incorporating it into the recovery algorithm is equally critical.

After the degradation model is established, the next step is the formulation of a solution approach. This might involve the stochastic modeling of the input signal (and the noise), the determination of the model parameters, and the formulation of a criterion to be optimized. Alternatively it might involve the formulation of a functional to be optimized subject to constraints imposed by the prior information. In the simplest possible case, the degradation equation defines directly the solution approach. For example, if D is a square invertible matrix, and the noise is ignored in Eq. (34.1), $x = D^{-1}y$ is the desired unique solution. In most cases, however, the solution of Eq. (34.1) represents an ill-posed problem [56]. Application of regularization theory transforms it to a well-posed problem which provides meaningful solutions to the original problem.

There are a large number of approaches providing solutions to the image restoration problem. For recent reviews of such approaches refer, for example, to [5, 28]. The intention of this chapter is to concentrate only on a specific type of iterative algorithm, the successive approximation algorithm, and its application to the signal and image restoration problem. The basic form of such an algorithm is presented and analyzed first in detail to introduce the reader to the topic and address the issues involved. More advanced forms of the algorithm are presented in subsequent sections.

34.2 Iterative Recovery Algorithms

Iterative algorithms form an important part of optimization theory and numerical analysis. They date back at least to the Gauss years, but they also represent a topic of active research. A large part of any textbook on optimization theory or numerical analysis deals with iterative optimization techniques or algorithms [43, 44]. In this chapter we review certain iterative algorithms which have been applied to solving specific signal recovery problems in the last 15 to 20 years. We will briefly present some of the more basic algorithms and also review some of the recent advances.

A very comprehensive paper describing the various signal processing inverse problems which can be solved by the successive approximations iterative algorithm is the paper by Schafer et al. [49]. The basic idea behind such an algorithm is that the solution to the problem of recovering a signal which satisfies certain constraints from its degraded observation can be found by the alternate implementation of the degradation and the constraint operator. Problems reported in [49] which can be solved with such an iterative algorithm are the phase-only recovery problem, the magnitude-only recovery problem, the bandlimited extrapolation problem, the image restoration problem, and the filter design problem [10]. Reviews of iterative restoration algorithms are also presented in [7, 25]. There are certain advantages associated with iterative restoration techniques, such as [25, 49]: (1) there is no need to determine or implement the inverse of an operator; (2) knowledge about the solution can be incorporated into the restoration process in a relatively straightforward manner; (3) the solution process can be monitored as it progresses; and (4) the partially restored signal can be utilized in determining unknown parameters pertaining to the solution.

In the following we first present the development and analysis of two simple iterative restoration algorithms. Such algorithms are based on a simpler degradation model, when the degradation is linear and spatially invariant, and the noise is ignored. The description of such algorithms is intended to provide a good understanding of the various issues involved in dealing with iterative algorithms. We then proceed to work with the matrix-vector representation of the degradation model and the iterative algorithms. The degradation systems described now are linear but not necessarily spatially invariant. The relation between the matrix-vector and scalar representation of the degradation equation and the iterative solution is also presented. Various forms of regularized solutions and the resulting iterations are briefly presented. As it will become clear, the basic iteration is the basis for any of the iterations to be presented.

34.3 Spatially Invariant Degradation

34.3.1 Degradation Model

Let us consider the following degradation model

$$y(i, j) = d(i, j) * x(i, j) , \tag{34.2}$$

where $y(i, j)$ and $x(i, j)$ represent, respectively, the observed degraded and original image, $d(i, j)$ the impulse response of the degradation system, and $*$ denotes two-dimensional (2D) convolution. We rewrite Eq. (34.2) as follows

$$\Phi(x(i, j)) = y(i, j) - d(i, j) * x(i, j) = 0. \tag{34.3}$$

The restoration problem, therefore, of finding an estimate of $x(i, j)$ given $y(i, j)$ and $d(i, j)$ becomes the problem of finding a root of $\Phi(x(i, j)) = 0$.

34.3.2 Basic Iterative Restoration Algorithm

The following identity holds for any value of the parameter β

$$x(i, j) = x(i, j) + \beta \Phi(x(i, j)) . \tag{34.4}$$

Equation (34.4) forms the basis of the successive approximation iteration by interpreting $x(i, j)$ on the left-hand side as the solution at the current iteration step and $x(i, j)$ on the right-hand side as the solution at the previous iteration step. That is,

$$
\begin{aligned}
x_0(i, j) &= 0 \\
x_{k+1}(i, j) &= x_k(i, j) + \beta \Phi(x_k(i, j)) \\
&= \beta y(i, j) + (\delta(i, j) - \beta d(i, j)) * x_k(i, j) , \tag{34.5}
\end{aligned}
$$

where $\delta(i, j)$ denotes the discrete delta function and β the relaxation parameter which controls the convergence as well as the rate of convergence of the iteration. Iteration (34.5) is the basis of a large number of iterative recovery algorithms, some of which will be presented in the subsequent sections [1, 14, 17, 31, 32, 38]. This is the reason it will be analyzed in quite some detail. What differentiates the various iterative algorithms is the form of the function $\Phi(x(i, j))$. Perhaps the earliest reference to iteration (34.5) was by Van Cittert [61] in the 1930s. In this case the gain β was equal to one. Jansson et al. [17] modified the Van Cittert algorithm by replacing β with a relaxation parameter that depends on the signal. Also Kawata et al. [31, 32] used Eq. (34.5) for image restoration with a fixed or a varying parameter β.

34.3.3 Convergence

Clearly if a root of $\Phi(x(i, j))$ exists, this root is a *fixed point* of iteration (34.5), that is $x_{k+1}(i, j) = x_k(i, j)$. It is not guaranteed, however, that iteration (34.5) will converge even if Eq. (34.3) has one or more solutions. Let us, therefore, examine under what conditions (sufficient conditions) iteration (34.5) converges. Let us first rewrite it in the discrete frequency domain, by taking the 2D discrete Fourier transform (DFT) of both sides. It should be mentioned here that the arrays involved in iteration (34.5) are appropriately padded with zeros so that the result of 2D circular convolution equals the result of 2D linear convolution in Eq. (34.2). The required padding by zeros determines the size of the 2D DFT. Iteration (34.5) then becomes

$$
\begin{aligned}
X_0(u, v) &= 0 \\
X_{k+1}(u, v) &= \beta Y(u, v) + (1 - \beta D(u, v)) X_k(u, v),
\end{aligned}
\tag{34.6}
$$

where $X_k(u, v)$, $Y(u, v)$, and $D(u, v)$ represent respectively the 2D DFT of $x_k(i, j)$, $y(i, j)$, and $d(i, j)$, and (u, v) the discrete 2D frequency lattice. We express next $X_k(u, v)$ in terms of $X_0(u, v)$. Clearly,

$$
\begin{aligned}
X_1(u, v) &= \beta Y(u, v) \\
X_2(u, v) &= \beta Y(u, v) + (1 - \beta D(u, v)) \beta Y(u, v) \\
&= \sum_{\ell=0}^{1} (1 - \beta D(u, v))^\ell \, \beta Y(u, v) \\
\cdots \quad & \quad \cdots\cdots\cdots \\
X_k(u, v) &= \sum_{\ell=0}^{k-1} (1 - \beta D(u, u))^\ell \, \beta Y(u, v) \\
&= \frac{1 - (1 - \beta D(u, v))^k}{1 - (1 - \beta D(u, v))} \beta Y(u, v) \\
&= (1 - (1 - \beta D(u, v))^k) X(u, v)
\end{aligned}
\tag{34.7}
$$

if $D(u, v) \neq 0$. For $D(u, v) = 0$,

$$
X_k(u, v) = k \cdot \beta Y(u, v) = 0,
\tag{34.8}
$$

since $Y(u, v) = 0$ at the discrete frequencies (u, v) for which $D(u, v) = 0$. Clearly, from Eq. (34.7) if

$$
|1 - \beta D(u, v)| < 1,
\tag{34.9}
$$

then

$$
\lim_{k \to \infty} X_k(u, v) = X(u, v).
\tag{34.10}
$$

Having a closer look at the sufficient condition for convergence, Eq. (34.9), it can be rewritten as

$$
\begin{aligned}
|1 - \beta Re\{D(u, v)\} - \beta Im\{D(u, v)\}|^2 &< 1 \\
\Rightarrow (1 - \beta Re\{D(u, v)\})^2 + (\beta Im\{D(u, v)\})^2 &< 1.
\end{aligned}
\tag{34.11}
$$

Inequality (34.11) defines the region inside a circle of radius $1/\beta$ centered at $c = (1/\beta, 0)$ in the $(Re\{D(u, v)\}, Im\{D(u, v)\})$ domain, as shown in Fig. 34.1. From this figure it is clear that the left half-plane is not included in the region of convergence. That is, even though by decreasing β the size of the region of convergence increases, if the real part of $D(u, v)$ is negative, the sufficient condition for convergence cannot be satisfied. Therefore, for the class of degradations that this is the case, such as the degradation due to motion, iteration (34.5) is not guaranteed to converge.

The following form of (34.11) results when $Im\{D(u, v)\} = 0$, which means that $d(i, j)$ is symmetric

$$
0 < \beta < \frac{2}{D_{\max}(u, v)},
\tag{34.12}
$$

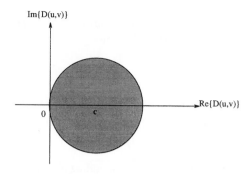

FIGURE 34.1 Geometric interpretation of the sufficient condition for convergence of the basic iteration, where $c = (1/\beta, 0)$.

where $D_{\max}(u, v)$ denotes the maximum value of $D(u, v)$ over all frequencies (u, v). If we now also take into account that $d(i, j)$ is typically normalized, i.e., $\sum_{i,j} d(i, j) = 1$, and represents a low pass degradation, then $D(0, 0) = D_{\max}(u, v) = 1$. In this case (34.11) becomes

$$0 < \beta < 2 . \tag{34.13}$$

From the above analysis, when the sufficient condition for convergence is satisfied, the iteration converges to the original signal. This is also the inverse solution obtained directly from the degradation equation. That is, by rewriting Eq. (34.2) in the discrete frequency domain

$$Y(u, v) = D(u, v) \cdot X(u, v) , \tag{34.14}$$

we obtain, for $D(u, v) \neq 0$,

$$X(u, v) = \frac{Y(u, v)}{D(u, v)} . \tag{34.15}$$

An important point to be made here is that, unlike the iterative solution, the inverse solution (34.15) can be obtained without imposing any requirements on $D(u, v)$. That is, even if Eq. (34.2) or (34.14) has a unique solution, that is, $D(u, v) \neq 0$ for all (u, v), iteration (34.5) may not converge if the sufficient condition for convergence is not satisfied. It is not, therefore, the appropriate iteration to solve the problem. Actually iteration (34.5) may not offer any advantages over the direct implementation of the inverse filter of Eq. (34.15) if no other features of the iterative algorithms are used, as will be explained later. The only possible advantage of iteration (34.5) over Eq. (34.15) is that the noise amplification in the restored image can be controlled by terminating the iteration before convergence, which represents another form of regularization. The effect of noise on the quality of the restoration has been studied experimentally in [47]. An iteration which will converge to the inverse solution of Eq. (34.2) for any $d(i, j)$ is described in the next section.

34.3.4 Reblurring

The degradation Eq. (34.2) can be modified so that the successive approximations iteration converges for a larger class of degradations. That is, the observed data $y(i, j)$ are first filtered (reblurred) by a system with impulse response $d^*(-i, -j)$, where * denotes complex conjugation [33]. The degradation Eq. (34.2), therefore, becomes

$$\tilde{y}(i, j) = y(i, j) * d^*(-i, -j) = d^*(-i, -j) * d(i, j) * x(i, j)$$
$$= \tilde{d}(i, j) * x(i, j) . \tag{34.16}$$

If we follow the same steps as in the previous section substituting $y(i, j)$ by $\tilde{y}(i, j)$ and $d(i, j)$ by $\tilde{d}(i, j)$ the iteration providing a solution to Eq. (34.16) becomes

$$
\begin{aligned}
x_0(i, j) &= 0 \\
x_{k+1}(i, j) &= x_k(i, j) + \beta d^*(-i, -j) * (y(i, j) - d(i, j) * x_k(i, j)) \\
&= \beta d^*(-i, -j) * y(i, j) + (\delta(i, j) \\
&\quad - \beta d^*(-i, -j) * d(i, j)) * x_k(i, j) .
\end{aligned}
\tag{34.17}
$$

Now, the sufficient condition for convergence, corresponding to condition (34.9), becomes

$$
|1 - \beta |D(u, v)|^2| < 1 ,
\tag{34.18}
$$

which can be always satisfied for

$$
0 < \beta < \frac{2}{\max_{u,v} |D(u, v)|^2} .
\tag{34.19}
$$

The presentation so far has followed a rather simple and intuitive path, hopefully demonstrating some of the issues involved in developing and implementing an iterative algorithm. We move next to the matrix-vector formulation of the degradation process and the restoration iteration. We borrow results from numerical analysis in obtaining the convergence results of the previous section but also more general results.

34.4   Matrix-Vector Formulation

What became clear from the previous sections is that in applying the successive approximations iteration the restoration problem to be solved is brought first into the form of finding the root of a function (see Eq. (34.3)). In other words, a solution to the restoration problem is sought which satisfies

$$
\Phi(x) = 0 ,
\tag{34.20}
$$

where $x \in \mathcal{R}^N$ is the vector representation of the signal resulting from the stacking or ordering of the original signal, and $\Phi(x)$ represents a nonlinear in general function. The row-by-row from left-to-right stacking of an image $x(i, j)$ is typically referred to as *lexicographic ordering*.

Then the successive approximations iteration which might provide us with a solution to Eq. (34.20) is given by

$$
\begin{aligned}
x_0 &= 0 \\
x_{k+1} &= x_k + \beta \Phi(x_k) \\
&= \Psi(x_k) .
\end{aligned}
\tag{34.21}
$$

Clearly if x^* is a solution to $\Phi(x) = 0$, i.e., $\Phi(x^*) = 0$, then x^* is also a fixed point to the above iteration since $x_{k+1} = x_k = x^*$. However, as was discussed in the previous section, even if x^* is the unique solution to Eq. (34.20), this does not imply that iteration (34.21) will converge. This again underlines the importance of convergence when dealing with iterative algorithms. The form iteration (34.21) takes for various forms of the function $\Phi(x)$ will be examined in the following sections.

34.4.1   Basic Iteration

From the degradation Eq. (34.1), the simplest possible form $\Phi(x)$ can take, when the noise is ignored, is

$$
\Phi(x) = y - Dx .
\tag{34.22}
$$

Then Eq. (34.21) becomes

$$
\begin{aligned}
x_0 &= 0 \\
x_{k+1} &= x_k + \beta(y - Dx_k) \\
&= \beta y + (I - \beta D)x_k \\
&= \beta y + G_1 x_k,
\end{aligned}
\tag{34.23}
$$

where I is the identity operator.

34.4.2 Least-Squares Iteration

A least-squares approach can be followed in solving Eq. (34.1). That is, a solution is sought which minimizes

$$
M(x) = \| y - Dx \|^2.
\tag{34.24}
$$

A necessary condition for $M(x)$ to have a minimum is that its gradient with respect to x is equal to zero, which results in the *normal equations*

$$
D^T Dx = D^T y
\tag{34.25}
$$

or

$$
\Phi(x) = D^T(y - Dx) = 0,
\tag{34.26}
$$

where ^{T} denotes the transpose of a matrix or vector. Application of iteration (34.21) then results in

$$
\begin{aligned}
x_0 &= 0 \\
x_{k+1} &= x_k + \beta D^T(y - Dx_k) \\
&= \beta D^T y + (I - \beta D^T D)x_k \\
&= \beta D^T y + G_2 x_k.
\end{aligned}
\tag{34.27}
$$

It is mentioned here that the matrix-vector representation of an iteration does not necessarily determine the way the iteration is implemented. In other words, the pointwise version of the iteration may be more efficient from the implementation point of view than the matrix-vector form of the iteration.

34.5 Matrix-Vector and Discrete Frequency Representations

When Eqs. (34.22) and (34.26) are obtained from Eq. (34.2), the resulting iterations (34.23) and (34.27), should be identical to iterations (34.5) and (34.17), respectively, and their frequency domain counterparts. This issue, of representing a matrix-vector equation in the discrete frequency domain is addressed next.

Any matrix can be diagonalized using its singular value decomposition. Finding, in general, the singular values of a matrix with no special structure is a formidable task, given also the size of the matrices involved in image restoration. For example, for a 256×256 image, D is of size 64K$\times$64K. The situation is simplified, however, if the degradation model of Eq. (34.2), which represents a special case of the degradation model of Eq. (34.1), is applicable. In this case, the degradation matrix D is block-circulant [3]. This implies that the singular values of D are the DFT values of $d(i, j)$, and the eigenvectors are the complex exponential basis functions of the DFT. In matrix form, this relationship can be expressed by

$$
D = W \tilde{D} W^{-1},
\tag{34.28}
$$

where $\tilde{D}$ is a diagonal matrix with entries the DFT values of $d(i, j)$ and W the matrix formed by the eigenvectors of D. The product $W^{-1}z$, where z is any vector, provides us with a vector which is formed by lexicographically ordering the DFT values of $z(i, j)$, the unstacked version of z. Substituting D from Eq. (34.28) into iteration (34.23) and premultiplying both sides by W^{-1}, iteration (34.5) results. The same way iteration (34.17) results from iteration (34.27). In this case, *reblurring*, as was named when initially proposed, is nothing else than the least squares solution to the inverse problem. In general, if in a matrix-vector equation all matrices involved are block circulant, a 2D discrete frequency domain equivalent expression can be obtained. Clearly, a matrix-vector representation encompasses a considerably larger class of degradations than the linear spatially-invariant degradation.

34.6   Convergence

In dealing with iterative algorithms, their convergence, as well as their rate of convergence, are very important issues. Some general convergence results will be presented in this section. These results will be presented for general operators, but also equivalent representations in the discrete frequency domain can be obtained if all matrices involved are block circulant.

The *contraction mapping theorem* usually serves as a basis for establishing convergence of iterative algorithms. According to it, iteration (34.21) converges to a unique fixed point x^*, that is, a point such that $\Psi(x^*) = x^*$ for any initial vector if the operator or transformation $\Psi(x)$ is a contraction. This means that for any two vectors z_1 and z_2 in the domain of $\Psi(x)$ the following relation holds

$$\|\Psi(z_1) - \Psi(z_2)\| \le \eta \|z_1 - z_2\|, \qquad (34.29)$$

where η is strictly less than one, and $\|\cdot\|$ denotes any norm. It is mentioned here that condition (34.29) is norm dependent, that is, a mapping may be contractive according to one norm, but not according to another.

34.6.1   Basic Iteration

For iteration (34.23) the sufficient condition for convergence (34.29) results in

$$\|I - \beta D\| < 1, \quad \text{or} \quad \|G_1\| < 1. \qquad (34.30)$$

If the l_2 norm is used, then condition (34.30) is equivalent to the requirement that

$$\max_i |\sigma_i(G_1)| < 1, \qquad (34.31)$$

where $|\sigma_i(G_1)|$ is the absolute value of the i-th singular value of G_1 [54].

The necessary and sufficient condition for iteration (34.23) to converge to a unique fixed point is that

$$\max_i |\lambda_i(G_1)| < 1, \quad \text{or} \quad \max_i |1 - \beta \lambda_i(D)| < 1, \qquad (34.32)$$

where $|\lambda_i(A)|$ represents the magnitude of the i-th eigenvalue of the matrix A. Clearly for a symmetric matrix D conditions (34.30) and (34.32) are equivalent. Conditions (34.29) to (34.32) are used in defining the range of values of β for which convergence of iteration (34.23) is guaranteed.

Of special interest is the case when matrix D is singular (D has at least one zero eigenvalue), since it represents a number of typical distortions of interest (for example, distortions due to motion, defocusing, etc). Then there is no value of β for which conditions (34.31) or (34.32) are satisfied. In this case G_1 is a *nonexpansive mapping* (η in (34.29) is equal to one). Such a mapping may have any number of fixed points (zero to infinitely many). However, a very useful result is obtained if we further restrict the properties of D (this results in no loss of generality, as it will become clear in the following sections). That is, if D is a symmetric, semi-positive definite matrix (all its eigenvalues are nonnegative), then according to *Bialy's theorem* [6], iteration (34.23) will converge to the minimum norm solution of Eq. (34.1), if this solution exists, plus the projection of x_0 onto the null space of D for $0 < \beta < 2 \cdot \|D\|^{-1}$. The theorem provides us with the means of incorporating information about the original signal into the final solution with the use of the initial condition.

Clearly, when D is block circulant the conditions for convergence shown above can be written in the discrete frequency domain. More specifically, conditions (34.31) and (34.9) are identical in this case.

34.6.2 Iteration with Reblurring

The convergence results presented above also holds for iteration (34.27), by replacing G_1 by G_2 in expressions (34.30) to (34.32). If $D^T D$ is singular, according to Bialy's theorem, iteration (34.27) will converge to the minimum norm least squares solution of (34.1), denoted by x^+, for $0 < \beta < 2 \cdot \|D\|^{-2}$, since $D^T y$ is in the range of $D^T D$.

The rate of convergence of iteration (34.27) is linear. If we denote by D^+ the *generalized inverse* of D, that is, $x^+ = D^+ y$, then the rate of convergence of (34.27) is described by the relation [26]

$$\frac{\|x_k - x^+\|}{\|x^+\|} \le c^{k+1} , \tag{34.33}$$

where

$$c = \max\{ |1 - \beta\|D\|^2|, \ |1 - \beta\|D^+\|^{-2}| \} . \tag{34.34}$$

The expression for c in (34.34) will also be used in Section 34.8, where higher order iterative algorithms are presented.

34.7 Use of Constraints

Iterative signal restoration algorithms regained popularity in the 1970s due to the realization that improved solutions can be obtained by incorporating prior knowledge about the solution into the restoration process. For example, we may know in advance that x is bandlimited or space-limited, or we may know on physical grounds that x can only have nonnegative values. A convenient way of expressing such prior knowledge is to define a constraint operator C, such that

$$x = Cx , \tag{34.35}$$

if and only if x satisfies the constraint. In general, C represents the concatenation of constraint operators. With the use of constraints, iteration (34.21) becomes [49]

$$
\begin{aligned}
x_0 &= 0, \\
\tilde{x}_k &= Cx_k, \\
x_{k+1} &= \Psi(\tilde{x}_k) .
\end{aligned}
\tag{34.36}
$$

The already mentioned recent popularity of constrained iterative restoration algorithms is also due to the fact that solutions to a number of recovery problems, such as the bandlimited extrapolation problem [48, 49] and the reconstruction from phase or magnitude problem [49, 57], were provided with the use of algorithms of the form (34.36) by appropriately describing the distortion and constraint operators. These operators are defined in the discrete spatial or frequency domains. A review of the problems which can be solved by an algorithm of the form of (34.36) is presented by Schafer et al. [49].

The contraction mapping theorem can again be used as a basis for establishing convergence of constrained iterative algorithms. The resulting sufficient condition for convergence is that at least one of the operators C and Ψ is contractive while the other is nonexpansive. Usually it is harder to prove convergence and determine the convergence rate of the constrained iterative algorithm, taking also into account that some of the constraint operators are nonlinear, such as the positivity constraint operator.

34.7.1   The Method of Projecting Onto Convex Sets (POCS)

The method of POCS describes an alternative approach in incorporating prior knowledge about the solution into the restoration process. It reappears in the engineering literature in the early 1980s [64], and since then it has been successfully applied to the solution of different restoration problems (from the reconstruction from phase or magnitude [52] to the removal of blocking artifacts [62, 63], for example). According to the method of POCS the incorporation of prior knowledge into the solution can be interpreted as the restriction of the solution to be a member of a closed convex set that is defined as the set of vectors which satisfy a particular property. If the constraint sets have a nonempty intersection, then a solution that belongs to the intersection set can be found by the method of POCS. Indeed, any solution in the intersection set is consistent with the *a priori* constraints and, therefore, it is a feasible solution.

More specifically, let $Q_1, Q_2, \cdots, Q_m$ be closed convex sets in a finite dimensional vector space, with $P_1, P_2, \cdots, P_m$ their respective projectors. Then, the iterative procedure,

$$x_{k+1} = P_1 P_2 \cdots P_m x_k, \tag{34.37}$$

converges to a vector which belongs to the intersection of the sets $Q_i, i = 1, 2, \cdots, m$, for any starting vector x_0. It is interesting to note that the resulting set intersection is also a closed convex set.

Clearly, the application of a projection operator P and the constraint C, discussed in the previous section, express the same idea. Projection operators represent nonexpansive mappings.

34.8   Class of Higher Order Iterative Algorithms

One of the drawbacks of the iterative algorithms presented in the previous sections is their linear rate of convergence. In [26] a unified approach is presented in obtaining a class of iterative algorithms with different rates of convergence, based on a representation of the generalized inverse of a matrix. That is, the algorithm,

$$
\begin{aligned}
x_0 &= \beta D^T y \\
D_0 &= \beta D^T D \\
\Omega_{k+1} &= \sum_{i=0}^{p-1} (I - D_k)^i \\
D_{k+1} &= \Omega_k D_k \\
x_{k+1} &= \Omega_k x_k, \tag{34.38}
\end{aligned}
$$

converges to the minimum norm least squares solution of Eq. (34.1), with $n = 0$. If iteration (34.38) is thought of as corresponding to iteration (34.27), then an iteration similar to (34.38) which corresponds to iteration (34.23) has also been derived [26, 41].

Algorithm (34.38) exhibits a p-th order of convergence. That is, the following relation holds [26]

$$\frac{\|x_k - x^+\|}{\|x^+\|} \le c^{p^k}, \tag{34.39}$$

where the convergence factor c is described by Eq. (34.34).

It is observed that the matrix sequences $\{\Omega_k\}$ and D_k can be computed in advance or *off-line*. When D is block circulant, substantial computational savings result with the use of iteration (34.38) over the linear algorithms. Questions dealing with the best order p of algorithm (34.38) to be used in a given application, as well as comparisons of the trade-off between speed of computation and computational load, are addressed in [26]. One of the drawbacks of the higher order algorithms is that the application of constraints may lead to erroneous results. Combined adaptive or nonadaptive linear and higher order algorithms have been proposed in overcoming this difficulty [11, 26].

34.9 Other Forms of $\Phi(x)$

34.9.1 Ill-Posed Problems and Regularization Theory

The two most basic forms of the function $\Phi(x)$ have only been considered so far. These two forms are represented by Eqs. (34.22) and (34.26), and are meaningful when the noise in Eq. (34.1) is not taken into account. Without ignoring the noise, however, the solution of Eq. (34.1) represents an ill-posed problem. If the image formation process is modeled in a continuous infinite dimensional space, D becomes an integral operator and Eq. (34.1) becomes a Fredholm integral equation of the first kind. Then the solution of Eq. (34.1) is almost always an ill-posed problem [42, 45, 59, 60]. This means that the unique least-squares solution of minimal norm of (34.1) does not depend continuously on the data, or that a bounded perturbation (noise) in the data results in an unbounded perturbation in the solution, or that the generalized inverse of D is unbounded [42]. The integral operator D has a countably infinite number of singular values that can be ordered with their limit approaching zero [42]. Since the finite dimensional discrete problem of image restoration results from the discretization of an ill-posed continuous problem, the matrix D has (in addition to possibly a number of zero singular values) a cluster of very small singular values. Clearly, the finer the discretization (the larger the size of D) the closer the limit of the singular values is approximated. Therefore, although the finite dimensional inverse problem is well posed in the least-squares sense [42], the ill-posedness of the continuous problem translates into an ill-conditioned matrix D.

A regularization method replaces an ill-posed problem by a well-posed problem, whose solution is an acceptable approximation to the solution of the given ill-posed problem [39, 56]. In general, regularization methods aim at providing solutions which preserve the fidelity to the data but also satisfy our prior knowledge about certain properties of the solution. A class of regularization methods associates both the class of admissible solutions and the observation noise with random processes [12]. Another class of regularization methods regards the solution as a deterministic quantity. We give examples of this second class of regularization methods in the following.

34.9.2 Constrained Minimization Regularization Approaches

Most regularization approaches transform the original inverse problem into a constrained optimization problem. That is, a functional needs to be optimized with respect to the original image and possibly other parameters. By using the necessary condition for optimality, the gradient of the functional with respect to the original image is set equal to zero, therefore determining the mathematical form of $\Phi(x)$. The successive approximations iteration becomes in this case a gradient method with a fixed step (determined by β). We briefly mention next the general form of some of the commonly used functionals.

Set Theoretic Formulation

With this approach the problem of solving Eq. (34.1) is replaced by the problem of searching for vectors x which belong to both sets [21, 25, 27]

$$\|Dx - y\| \le \epsilon, \tag{34.40}$$

and

$$\|Cx\| \le E, \tag{34.41}$$

where ϵ is an estimate on the data accuracy (noise norm), E a prescribed constant, and C a high-pass operator. Inequality (34.41) constrains the energy of the signal at high frequencies, therefore requiring that the restored signal is smooth. On the other hand, inequality (34.40) requires that the fidelity to the available data is preserved.

Inequalities (34.40) and (34.41) can be respectively rewritten as [25, 27]

$$\left(x - x^{+}\right)^{T} \frac{D^{T} D}{\epsilon^{2}} \left(x - x^{+}\right) \leq 1 , \tag{34.42}$$

and

$$x^{T} \frac{C^{T} C}{E_{2}} x \leq 1 , \tag{34.43}$$

where $x^{+} = D^{+} y$. That is, each of them represents an N-dimensional ellipsoid, where N is the dimensionality of the vectors involved. The intersection of the two ellipsoids (assuming it is not empty) is also a convex set but not an ellipsoid. The center of one of the ellipsoids which bounds the intersection can be chosen as the solution to the problem [50]. Clearly, even if the intersection is not empty, the center of the bounding ellipsoid may not belong to the intersection, and, therefore, a posterior test is required. The equation the center of one of the bounding ellipsoids is satisfying is given by [25, 27]

$$\Phi(x) = \left(D^{T} D + \alpha C^{T} C\right) x - D^{T} y = 0 , \tag{34.44}$$

where α, the *regularization parameter*, is equal to $(\epsilon/E)^{2}$.

Projection Onto Convex Sets (POCS) Approach

Iteration (34.37) can also be applied in finding a solution which belongs to both ellipsoids (34.42) and (34.43). The respective projections $P_{1}x$ and $P_{2}x$ are defined by [25]

$$P_{1}x = x + \lambda_{1}(I + \lambda_{1} D^{T} D)^{-1} D^{T} (y - Dx) \tag{34.45}$$

$$P_{2}x = [I - \lambda_{2}(I + \lambda_{2} C^{T} C)^{-1} C^{T} C]x, \tag{34.46}$$

where λ_{1} and λ_{2} need to be chosen so that conditions (34.42) and (34.43) are satisfied, respectively. Clearly, a number of other projection operators can be used in (34.37) which force the signal to exhibit certain known *a priori* properties expressed by convex sets.

A Functional Minimization Approach

The determination of the value of the regularization parameter is a critical issue in regularized restoration. A number of approaches for determining its value are presented in [13]. If only one of the parameters ϵ or E in (34.40) and (34.41) is known, a constrained least-squares formulation can be followed [9, 15]. With it, the size of one of the ellipsoids is minimized, subject to the constraint that the solution belongs to the surface of the other ellipsoid (the one defined by the known parameter). Following the Lagrangian approach, which transforms the constrained optimization problem into an unconstrained one, the following functional is minimized

$$M(\alpha, x) = \|Dx - y\|^{2} + \alpha\|Cx\|^{2} . \tag{34.47}$$

The necessary condition for a minimum is that the gradient of $M(\alpha, x)$ is equal to zero. That is, in this case

$$\Phi(x) = \nabla_{x} M(\alpha, x) = \left(D^{T} D + \alpha C^{T} C\right) x - D^{T} y , \tag{34.48}$$

which is identical to (34.44), with the only difference that α now is not known, but needs to be determined.

Spatially Adaptive Iteration

Spatially adaptive image restoration is the next natural step in improving the quality of the restored images. There are various ways to argue the introduction of spatial adaptivity, the most commonly used ones being the nonhomogeneity or nonstationarity of the image field and the properties of the human visual system. In either case, the functional to be minimized takes the form [22, 23, 34]

$$M(\alpha, \boldsymbol{x}) = \|\boldsymbol{D}\boldsymbol{x} - \boldsymbol{y}\|_{\boldsymbol{W}_1}^2 + \alpha \|\boldsymbol{C}\boldsymbol{x}\|_{\boldsymbol{W}_2}^2 , \qquad (34.49)$$

in which case

$$\Phi(\boldsymbol{x}) = \nabla_{\boldsymbol{x}} M(\alpha, \boldsymbol{x}) = \left(\boldsymbol{D}^T \boldsymbol{W}_1^T \boldsymbol{W}_1 \boldsymbol{D} + \alpha \boldsymbol{C}^T \boldsymbol{W}_2^T \boldsymbol{W}_2 \boldsymbol{C} \right) \boldsymbol{x} - \boldsymbol{D}^T \boldsymbol{W}_1 \boldsymbol{y} . \qquad (34.50)$$

The choice of the diagonal weighting matrices $\boldsymbol{W}_1$ and $\boldsymbol{W}_2$ can be justified in various ways. In [16, 22, 23, 25] both matrices are determined by the noise visibility matrix $\boldsymbol{V}$ [2, 46]. That is, $\boldsymbol{W}_1 = \boldsymbol{V}^T \boldsymbol{V}$ and $\boldsymbol{W}_2 = \boldsymbol{I} - \boldsymbol{V}^T \boldsymbol{V}$. The entries of $\boldsymbol{V}$ take values between 0 and 1. They are equal to 0 at the edges (noise is not visible), equal to 1 at the flat regions (noise is visible) and take values in between at the regions with moderate spatial activity. A study of the mapping between the level of spatial activity and the values of the visibility function appears in [11]. The weighting matrices can also be defined by considering the relationship of the restoration approach presented here to the MAP restoration approach [30]. Then, the weighting matrices $\boldsymbol{W}_1$ and $\boldsymbol{W}_2$ contain information about the nonstationarity and/or the nonwhiteness of the high-pass filtered image and noise, respectively.

Robust Functionals

Robust functionals can be employed for the representation of both the noise and the signal statistics. They allow for the efficient suppression of a wide variety of noise processes and permit the reconstruction of sharper edges than their quadratic counterparts. In a robust set-theoretic set-up a solution is sought by minimizing [65]

$$M(\alpha, \boldsymbol{x}) = R_n(\boldsymbol{y} - \boldsymbol{D}\boldsymbol{x}) + \alpha R_x(\boldsymbol{C}\boldsymbol{x}) . \qquad (34.51)$$

$R_n()$ and $R_x()$ are referred to as the residual and stabilizing functionals, respectively, and they are defined in terms of their *kernel functions*. The derivative of the kernel function is called the *influence function*.

$\Phi(\boldsymbol{x})$ in this case equals the gradient of $M(\alpha, \boldsymbol{x})$ in Eq. (34.51). A large number of robust functionals have been proposed in the literature. The properties of potential functions to be used in robust Bayesian estimation are listed in [35]. A robust maximum absolute entropy and a robust minimum absolute-information functionals are introduced in [65]. Clearly since the functionals $R_n()$ and $R_x()$ are typically nonlinear and may not be convex, the convergence analysis of iteration (34.21) or (34.36) is considerably more complicated.

34.9.3 Iteration Adaptive Image Restoration Algorithms

As it has become clear by now there are various pieces of information needed by any regularization algorithms in determining the unknown parameters. In the context of deterministic regularization, the most commonly needed parameter is the regularization parameter. Its determination depends on the noise statistics and the properties of the image. With the set theoretic regularization approach, it is required that the original image is smooth, in which case a bound on the energy of the high-pass filtered image is needed. This bound is proportional to the variance of the image in a stochastic context. In addition, knowledge of the noise variance is also required. In a MAP framework such parameters are called *hyperparameters* [8, 40]. Clearly, such parameters are not typically available and need to be estimated from the available noisy and blurred data. Various techniques for estimating the regularization parameter are discussed, for example, in [13].

In the following we briefly describe a new paradigm we have introduced in the context of iterative image restoration algorithms [18, 19, 20, 29, 30]. According to it, the required information by the deterministic regularization approach is updated at each restoration step, based on the partially restored image.

Spatially Adaptive Algorithm

For the spatially adaptive algorithm we mentioned above, the proposed general form of the weighted smoothing functional whose minimization will result in a restored image is written as

$$
\begin{aligned}
M_w\left(\lambda_w(x), x\right) &= \|y - Dx\|^2_{A(x)} + \lambda_w(x)\|Cx\|^2_{B(x)} \\
&= \|n\|^2_{A(x)} + \lambda_w(x)\|Cx\|^2_{B(x)},
\end{aligned}
\tag{34.52}
$$

where the weighting matrices $A(x)$ and $B(x)$, both functions of the original image, are used to incorporate noise and image characteristics into the restoration process, respectively. The regularization parameter, also a function of x, is defined in such a way as to make the smoothing functional in (34.52) convex with a unique global minimizer.

One of the $\lambda_w(x)$ we have proposed is given by

$$
\lambda_w(x) = \frac{\|y - Dx\|^2_{A(x)}}{(1/\gamma) - \|Cx\|^2_{B(x)}},
\tag{34.53}
$$

where the parameter γ is determined from the convergence and convexity analyses.

The main objective with this approach is to employ an iterative algorithm to estimate the regularization parameter and the proper weighting matrices at the same time with the restored image. The available estimate of the restored image at each iteration step will be used for determining the value of the regularization parameter. That is, the regularization parameter is defined as a function of the original image (and eventually in practice of an estimate of it). Of great importance is the form of this functional, so that the smoothing functional to be minimized preserves its convexity and exhibits a global minimizer. $\lambda_w(x)$ maps a vector x onto the positive real line. Its purpose is as before to control the relative contribution of the error term $\|y - Dx\|^2_{A(x)}$, which enforces "faithfulness" to the data, and the stabilizing functional $\|Cx\|^2_{B(x)}$, which enforces smoothness on the solution. Its dependency, however, on the original image, as well as the available data, is explicitly utilized. This dependency on the other hand is implicitly utilized in the constrained least-squares approach, according to which the minimization of $M_w(\lambda_w(x), x)$ and the determination of the regularization parameter $\lambda_w(x)$ are completely separate steps. The desired properties of $\lambda_w(x)$ and $M_w(\lambda_w(x), x)$ are analyzed in [20]. The relationship of the resulting forms to the hierarchical Bayesian approach towards image restoration and estimation of the regularization parameters is explored in [40].

In this case, therefore, $\Phi(x) = \nabla_x M_w(\lambda_w(x), x)$. The successive approximations iteration after some simplifications takes the form [20, 30]

$$
x_{k+1} = x_k + \left[D^T A(x_k) y - \left(D^T A(x_k) D + \lambda_w(x_k) C^T B(x_k) C \right) x_k \right].
\tag{34.54}
$$

The information required in defining the regularization parameter and the weights for introducing the spatial adaptivity are defined based on the available information about the restored image at the k-th iteration step. Clearly for all this to make sense the convergence of iteration (34.54) has to be guaranteed. Furthermore, convergence to a unique fixed point, which removes the dependency of the final result on the initial conditions, is also desired. These issues are addressed in detail in [20, 30]. A major advantage of the proposed algorithm is that the convexity of the smoothing functional and the convergence of the resulting algorithm are guaranteed regardless of the choice of the weighting matrices. Another advantage of this algorithm is that the proposed adaptive algorithm simultaneously determines the regularization parameter and the desirable weighting matrices based on the restored image at each iteration step and restores the image, without any prior knowledge.

Frequency Adaptive Algorithm

Adaptivity is now introduced into the restoration process by using a constant smoothness constraint, but by assigning a different regularization parameter at each discrete frequency location. We can now "fine-tune" the regularization of each frequency component, thereby achieving improved results and at the same time speeding up the convergence of the iterative algorithm. The regularization parameters are evaluated simultaneously with the restored image based on the partially restored image.

In this algorithm, the following two ellipsoids QE_x and $QE_{x/y}$ are used

$$QE_x = \{x| \ \|Cx\|_R \le E_R\} \tag{34.55}$$

and

$$QE_{x/y} = \{x| \ \|y - Dx\|_P \le \epsilon_P\} , \tag{34.56}$$

where P and R are both block-circulant weighting matrices. Then a solution which belongs to the intersection of QE_x and $QE_{x/y}$ is given by

$$\left(D^T P^T P D + \lambda C^T R^T R C\right) x = D^T P^T P y , \tag{34.57}$$

where $\lambda = (\epsilon_P/E_R)^2$. Let us define $P^T P = B$, $R = PC$ and $\lambda C^T C = A$. Then Eq. (34.57) can be written as

$$B \left(D^T D + A C^T C\right) x = B D^T y , \tag{34.58}$$

since all matrices are block-circulant and they therefore commute. The regularization matrix A is defined based on the set theoretic regularization as

$$A = \|y - Dx\|^2 [\|Cx\|^2 I + \Delta]^{-1} , \tag{34.59}$$

where Δ is a block-circulant matrix used to ensure convergence. B plays the role of the "shaping" matrix [53] for maximizing the speed of convergence at every frequency component as well as for compensating for the near-singular frequency components [19].

With the above formulation, therefore,

$$\Phi(x) = B \left(\left(D^T D + A C^T C\right) x - D^T y\right) , \tag{34.60}$$

and the successive approximations iteration (34.21) becomes

$$x_{k+1} = x_k + B \left[D^T y - \left(D^T D + A_k C^T C\right) x_k\right] , \tag{34.61}$$

where $A_k = \|y - Dx_k\|^2 [\|Cx_k\|^2 I + \Delta_k]^{-1}$. It is mentioned here that iteration (34.61) can also be derived from the regularized equation

$$\left(D^T D + A C^T C\right) x = D^T y , \tag{34.62}$$

using the generalized Landweber's iteration [53]. Since all matrices in iteration (34.61) are block-circulant, the iteration can be written in the discrete frequency domain as

$$X_{k+1}(\underline{p}) = X_k(\underline{p}) + \beta(\underline{p}) \left[D^*(\underline{p})Y(\underline{p}) - \left(|D(\underline{p})|^2 + \lambda_k(\underline{p})|C(\underline{p})|^2\right) X_k(\underline{p})\right] , \tag{34.63}$$

where $\underline{p} = (p_1, p_2)$, $0 \le p_1 \le N - 1$, $0 \le p_2 \le N - 1$, $X_{k+1}(\underline{p})$ and $Y(\underline{p})$ represent the 2D DFT of the unstacked image estimate x_{k+1}, and the noisy-blurred image y and $D(\underline{p})$, $C(\underline{p})$, $\beta(\underline{p})$, and $\lambda_k(\underline{p})$

represent 2D DFTs of the 2D sequences which form the block-circulant matrices $\boldsymbol{D}$, $\boldsymbol{C}$, $\boldsymbol{B}$, and $\boldsymbol{A}_k$, respectively. Since Δ_k is block-circulant $\lambda_k(\underline{p})$ is given by

$$\lambda_k(\underline{p}) = \frac{\sum_m |Y(\underline{m}) - D(\underline{m})X_k(\underline{m})|^2}{\sum_n |C(\underline{n})X_k(\underline{n})|^2 + \delta_k(\underline{p})}, \qquad (34.64)$$

where $\delta_k(\underline{p})$ is the 2D DFT of the sequence which forms Δ_k.

The allowable range of each regularization and control parameter and the convergence analysis of the iterative algorithm are developed in detail in [19]. It is shown that the algorithm has more than two fixed points. The first fixed point is the inverse or generalized inverse solution of Eq. (34.58). The second type of fixed points are regularized approximations to the original image. Since there is more than one solution to iteration (34.63), the determination of the initial condition becomes important. It has been verified experimentally [19] that if a "smooth" image is used for $X_0(\underline{p})$ almost identical fixed points result independently of X_0. The use of spectral filtering functions [53] is also incorporated into the iteration, as shown in [19].

34.10 Discussion

In this chapter we briefly described the application of the successive approximations-based class of iterative algorithms to the problem of restoring a noisy and blurred signal. We analyzed in some detail the simpler forms of the algorithm, while making reference to work which deals with more complicated forms of the algorithms. There are obviously a number of algorithms and issues pertaining to such algorithms which have not been addressed at all. For example, iterative algorithms with a varying relaxation parameter β, such as the steepest descent and conjugate gradient methods, can be applied to the image restoration problem [4, 37]. The number of iterations also represents a means for regularizing the restoration problem [55, 58]. Iterative algorithms which depend on more than one previous restoration steps (multi-step algorithms) have also been considered, primarily for implementation reasons [24].

It is the hope and the expectation of the author that the material presented will form a good introduction to the topic for the engineer or the graduate student who would like to work in this area.

References

[1] Abbiss, J.B., DeMol, C., and Dhadwal, H.S., Regularized iterative and noniterative procedures for object restoration from experimental data, *Opt. Acta*, 107-124, 1983.

[2] Anderson, G.L. and Netravali, A.N., Image restoration based on a subjective criterion, *IEEE Trans. Sys. Man. Cybern.*, SMC-6: 845-853, Dec., 1976.

[3] Andrews, H.C. and Hunt, B.R., *Digital Image Restoration*, Prentice-Hall, Englewood Cliffs, NJ, 1977.

[4] Angel, E.S. and Jain, A.K., Restoration of images degraded by spatially varying point spread functions by a conjugate gradient method, *Appl. Opt.*, 17: 2186-2190, July, 1978.

[5] Banham, M. and Katsaggelos, A.K., Digital image restoration, *Signal Processing Mag.*, 14(2), 24–41, Mar., 1997.

[6] Bialy, H., Iterative Behandlung Linearen Funktionalgleichungen, *Arch. Ration. Mech. Anal.*, 4: 166-176, July, 1959.

[7] Biemond, J., Lagendijk, R.L., and Mersereau, R.M., Iterative methods for image deblurring, *Proc. IEEE*, 78(5): 856-883, May, 1990.

[8] Demoment, G., Image reconstruction and restoration: overview of common estimation structures and problems, *IEEE Trans. Acoust. Speech Signal Process.*, 37(12): 2024-2036, Dec., 1989.

[9] Dines, K.A. and Kak, A.C., Constrained least squares filtering, *IEEE Trans. Acoust. Speech Signal Process.*, ASSP-25: 346-350, 1977.

[10] Dudgeon, D.E. and Mersereau, R.M., *Multidimensional Digital Signal Processing*, Prentice-Hall, Englewood Cliffs, NJ, 1984.

[11] Efstratiadis, S.N. and Katsaggelos, A.K., Adaptive iterative image restoration with reduced computational load, *Opt. Eng.*, 29: 1458-1468, Dec., 1990.

[12] Franklin, J.N., Well-posed stochastic extentions of ill-posed linear problems, *J. Math. Anal.*, 31: 682-716, 1970.

[13] Galatsanos, N.P. and Katsaggelos, A.K., Methods for choosing the regularization parameter and estimating the noise variance in image restoration and their relation, *IEEE Trans. Image Process.*, 1: 322-336, July, 1992.

[14] Huang, T.S., Barker, D.A., and Berger, S.P., Iterative image restoration, *Appl. Opt.*, 14: 1165-1168, May, 1975.

[15] Hunt, B.R., The application of constrained least squares estimation to image restoration by digital computers, *IEEE Trans. Comput.*, C-22: 805-812, Sept., 1973.

[16] Ichioka, Y. and Nakajima, N., Iterative image restoration considering visibility, *J. Opt. Soc. Am.*, 71: 983-988, Aug., 1981.

[17] Jansson, P.A., Hunt R.H., and Pyler, E.K., Resolution enhancement of spectra, *J. Opt. Soc. Am.*, 60: 596-599, May, 1970.

[18] Kang, M.G. and Katsaggelos, A.K., Iterative image restoration with simultaneous estimation of the regularization parameter, *IEEE Trans. Signal Process.*, 40(9): 2329-2334, Sept., 1992.

[19] Kang, M.G. and Katsaggelos, A.K., Frequency domain adaptive iterative image restoration and evaluation of the regularization parameter, *Opt. Eng.*, 33(10): 3222-3232, Oct., 1994.

[20] Kang, M.G. and Katsaggelos, A.K., General choice of the regularization functional in regularized image restoration, *IEEE Trans. Image Process.*, 4(5): 594-602, May, 1995.

[21] Katsaggelos, A.K., Biemond, J., Mersereau, R.M., and Schafer, R.W., A general formulation of constrained iterative restoration algorithms, *Proc. 1985 Int. Conf. Acoust. Speech Signal Process.*, pp. 700-703, Tampa, FL, March, 1985.

[22] Katsaggelos, A.K., Biemond, J., Mersereau, R.M., and Schafer, R.W., Nonstationary iterative image restoration, *Proc. 1985 Int. Conf. Acoust. Speech Signal Process.*, pp. 696-699, Tampa, FL, March, 1985.

[23] Katsaggelos, A.K., A general formulation of adaptive iterative image restoration algorithms, *Proc. 1986 Conf. Inf. Sciences Syst.*, pp. 42-47, Princeton, NJ, March, 1986.

[24] Katsaggelos, A.K. and Kumar, S.P.R., Single and multistep iterative image restoration and VLSI implementation, *Signal Process.*, 16(1): 29-40, Jan., 1989.

[25] Katsaggelos, A.K., Iterative image restoration algorithm, *Opt. Eng.*, 28(7): 735-748, July, 1989.

[26] Katsaggelos, A.K. and Efstratiadis, S.N., A class of iterative signal restoration algorithms, *IEEE Trans. Acoust. Speech Signal Process.*, 38: 778-786, May, 1990 (reprinted in *Digital Image Processing*, R. Chellappa, Ed., IEEE Computer Society Press).

[27] Katsaggelos, A.K., Biemond, J., Mersereau, R.M., and Schafer, R.W., A regularized iterative image restoration algorithm, *IEEE Trans. Signal Process.*, 39(4): 914-929, April, 1991.

[28] Katsaggelos, A.K., Ed., *Digital Image Restoration*, Springer Series in Information Sciences, vol. 23, Springer-Verlag, Heidelberg, 1991.

[29] Katsaggelos, A.K. and Kang, M.G., Iterative evaluation of the regularization parameter in regularized image restoration, *J. Vis. Commun Image Rep.*, special issue on *Image Restoration*, vol. 3, no. 6, pp. 446-455, Dec., 1992.

[30] Katsaggelos, A.K. and Kang, M.G., A spatially adaptive iterative algorithm for the restoration of astronomical images, *Int. J. Imag. Syst. Technol.*, special issue on *Image Reconstruction and Restoration in Astronomy*, vol. 6, no. 4, pp. 305-313, winter, 1995.

[31] Kawata, S., Ichioka, Y., and Suzuki, T., Application of man-machine interactive image processing system to iterative image restoration, *Proc. 4th Int. Conf. Patt. Recog.*, pp. 525-529, Kyoto, 1978.

[32] Kawata, S. and Ichioka, Y., Iterative image restoration for linearly degraded images, I. Basis, *J. Opt. Soc. Am.*, 70: 762-768, July, 1980.

[33] Kawata, S. and Ichioka, Y., Iterative image restoration for linearly degraded images, II. Reblurring procedure, *J. Opt. Soc. Am.*, 70: 768-772, July, 1980.

[34] Lagendijk, R.L., Biemond, J., and Boekee, D.E., Regularized iterative image restoration with ringing reduction, *IEEE Trans. Acoust. Speech Signal Process.*, 36: 1804-1887, Dec., 1988.

[35] Lange, K., Convergence of EM image reconstruction algorithms with Gibbs smoothing, *IEEE Trans. Med. Imag.*, 9(4), Dec., 1990.

[36] Mammone, R.J., *Computational Methods of Signal Recovery and Recognition*, Wiley, 1992.

[37] Marucci, R., Mersereau, R.M., and Schafer, R.W., Constrained iterative deconvolution using a conjugate gradient algorithm, *Proc. 1982 IEEE Int. Conf. Acoust. Speech Signal Process.*, pp. 1845-1848, Paris, France, May, 1982.

[38] Mersereau, R.M. and Schafer, R.W., Comparative study of iterative deconvolution algorithms, *Proc. 1978 IEEE Int. Conf. Acoust. Speech Signal Process.*, pp. 192-195, April, 1978.

[39] Miller, K., Least-squares method for ill-posed problems with a prescribed bound, *SIAM J. Math. Anal.*, 1: 52-74, Feb., 1970.

[40] Molina, R. and Katsaggelos, A.K., The hierarchical approach to image restoration and the iterative evaluation of the regularization parameter, *Proc. 1994 SPIE Conf. Vis. Commun. Image Process.*, pp. 244-251, Chicago, IL, Sept., 1994.

[41] Morris, C.E., Richards M.A., and Hayes, M.H., Fast reconstruction of linearly distorted signals, *IEEE Trans. Acoust. Speech Signal Process.*, 36: 1017-1025, July, 1988.

[42] Nashed, M.Z., Operator theoretic and computational approaches to ill-posed problems with application to antenna theory, *IEEE Trans. Ant. Prop.*, AP-29: 220-231, March, 1981.

[43] Ortega, J.M. and Rheinboldt, W.C., *Iterative Solution of Nonlinear Equations in Several Variables*, Academic Press, New York, 1970.

[44] Ortega, J.M., *Numerical Analysis: A Second Course*, Academic Press, New York, 1972.

[45] Phillips, D.L., A technique for the numerical solution of certain integral equations of the first kind, *Assoc. Comp. Mach.*, 9: pp. 84-97, 1962.

[46] Rajala, S.S. and DeFigueiredo, R.J.P., Adaptive nonlinear image restoration by a modified Kalman filtering approach, *IEEE Trans. Acoust. Speech Signal Process.*, ASSP-29: Oct., 1981.

[47] Richards, M.A., Schafer, R.W., and Mersereau, R.M., An experimental study of the effects of noise on a class of iterative deconvolution algorithms, *Proc. 1979 Int. Conf. Acoust. Speech Signal Process.*, pp. 401-404, April, 1979.

[48] Sanz J.L.C. and Huang, T.S., Iterative Time-Limited Signal Restoration, *IEEE Trans. Acoust. Speech Signal Processing*, ASSP-31: 643-649, June, 1983.

[49] Schafer, R. W., Mersereau, R.M., and Richards, M.A., Constrained iterative restoration algorithms, *Proc. IEEE*, 69: 432-450, April, 1981.

[50] Schweppe, F.C., *Uncertain Dynamic Systems*, Prentice-Hall, 1973.

[51] Sondhi, M.M., Image restoration: the removal of spatially invariant degradations, *Proc. IEEE*, vol. 60, pp. 842-853, July, 1972.

[52] Stark, H., *Image Recovery: Theory and Applications*, Academic Press, New York, 1987.

[53] Strand, O.N., Theory and methods related to the singular-function expansion and Landweber's iteration for integral equations of the first kind, *SIAM J. Numerical Anal.*, 11: 798-825, Sept., 1974.

[54] Strang, G., *Linear Algebra and Its Applications*, 2nd ed., Academic Press, New York, 1980.

[55] Sullivan, B.J. and Katsaggelos, A.K., A new termination rule for linear iterative image restoration algorithms, *Opt. Eng.*, 29: 471-477, May, 1990.

[56] Tikhonov, A.N. and Arsenin, V.Y., *Solution of Ill-Posed Problems*, Winston, Wiley, 1977.

[57] Tom, V.T., Quatieri, T.F., Hayes, M.H., and McClellan, J.M., Convergence of iterative nonexpansive signal reconstruction algorithms, *IEEE Trans. Acoust. Speech Signal Process.*, ASSP-29: 1052-1058, Oct., 1981.

[58] Trussell, H.J., Convergence criteria for iterative restoration methods, *IEEE Trans. Acoust. Speech Signal Process.*, ASSP-31: 129-136, Feb., 1983.

[59] Twomey, S., On the numerical solution of Fredholm integral equations of the first kind by the inversion of the linear system produced by quadrature, *Assoc. Comp. Mach.*, 10: 97-101, 1963.

[60] Twomey, S., The application of numerical filtering of the solution of integral equations encountered in indirect sensing measurements, *J. Franklin Inst.*, 279: 95-109, Feb., 1965.

[61] Van Citttert, P.H., Zum Einfluss der Spaltbreite auf die Intensitatswerteilung in Spektrallinien II, *Z. Physik*, 69: 298-308, 1931.

[62] Yang, Y., Galatsanos N.P., and Katsaggelos, A.K., Regularized image reconstruction from incomplete block discrete cosine transform data, *IEEE Trans. Circuits Syst. Video Technol.*, 3(6): 421-432, Dec., 1993.

[63] Yang, Y., Galatsanos N.P., and Katsaggelos, A.K., Set theoretic spatially-adaptive reconstruction of block transform compressed images, *IEEE Trans. Image Process.*, 4(7): 896-908, July, 1995.

[64] Youla, D.C. and Webb, H., Image reconstruction by the method of convex projections, Pt. 1-Theory, *IEEE Trans. Med. Imag.*, MI-1(2): 81-94, Oct., 1982.

[65] Zervakis, M.E., Katsaggelos, A.K., and Kwon, T.M., A class of robust entropic functionals for image restoration, *IEEE Trans. Image Process.*, 4(6): 752-773, June, 1995.

VIII

Time Frequency and Multirate Signal Processing

Cormac Herley
Hewlett Packard Laboratories
Kambiz Nayebi
Sharif University

A N IMPORTANT PROBLEM IN SIGNAL PROCESSING is the choice of how to represent a signal. It is for this reason that importance is attached to the choice of bases for the *linear* expansion of signals. That is, given a discrete-time signal $x(n)$ how to find $a_i(n)$ and $b_i(n)$ such that we can write

$$x(n) = \sum_i < x(n), a_i(n) > b_i(n). \qquad (VIII.1)$$

If $b_i(n) = a_i(n)$, then (VIII.1) is the familiar orthonormal basis expansion formula [1]. Otherwise, the $b_i(n)$ are a set of biorthogonal functions with the property

$$< b_j(n), a_i(n) > = \delta_{i-j}.$$

The function δ is defined such that $\delta_{i-j} = 0$, unless $i = j$, in which case $\delta_0 = 1$. We shall consider cases where the summation in (VIII.1) is infinite, but restrict our attention to the case where it is finite for the moment; that is, where we have a finite number N of data samples, and so the space is finite dimensional.

We next set up the basic notation used throughout the chapter. Assume that we are operating in C^N, and that we have N basis vectors, the minimum number to span the space. Since the transform is linear, it can be written as a matrix. That is, if the $\mathbf{a}_i^*$ are the rows of a matrix $\mathbf{A}$, then

$$\mathbf{A} \cdot \mathbf{x} = \begin{bmatrix} < x(n), a_0(n) > \\ < x(n), a_1(n) > \\ \vdots \\ < x(n), a_{N-2}(n) > \\ < x(n), a_{N-1}(n) > \end{bmatrix} \tag{VIII.2}$$

and if $\mathbf{b}_i$ are the columns of $\mathbf{B}$ then

$$\mathbf{x} = \mathbf{B} \cdot \mathbf{A} \cdot \mathbf{x}. \tag{VIII.3}$$

Clearly $\mathbf{B} = \mathbf{A}^{-1}$; if $\mathbf{B} = \mathbf{A}^*$ then $\mathbf{A}$ is unitary, $b_i(n) = a_i(n)$ and we have that (VIII.1) is the orthonormal basis expansion.

Clearly the construction of bases is not difficult: any nonsingular $N \times N$ matrix will do for this space. Similarly, to get an orthonormal basis we need merely take the rows of any unitary $N \times N$ matrix, for example the identity $\mathbf{I}_N$. There are many reasons for desiring to carry out such an expansion. Much as Taylor or Fourier series are used in mathematics to simplify solutions to certain problems, the underlying goal is that a cleverly chosen expansion may make a given signal processing task simpler.

A major application is signal compression, where we wish to quantize the input signal in order to transmit it with as few bits as possible, while minimizing the distortion introduced. If the input vector comprises samples of a real signal, then the samples are probably highly correlated, and the identity basis (where the ith vector contains 1 in the ith position and is zero elsewhere) with scalar quantization will end up using many of its bits to transmit information which does not vary much from sample to sample. If we can choose a matrix $\mathbf{A}$ such that the elements of $\mathbf{A} \cdot \mathbf{x}$ are much less correlated than those of $\mathbf{x}$, then the job of efficient quantization becomes a great deal simpler [2]. In fact, the Karhunen-Loève transform, which produces uncorrelated coefficients, is known to be optimal in a mean squared error sense [2].

Since in (VIII.1) the signal is written as a superposition of the basis sequences $b_i(n)$, we can say that if $b_i(n)$ has most of its energy concentrated around time $n = n_0$, then the coefficient $< x(n), a_i(n) >$ measures to some degree the concentration of $x(n)$ at time $n = n_0$. Equally, taking the discrete Fourier transform of (VIII.1)

$$X(k) = \sum_i < x(n), a_i(n) > B_i(k).$$

Thus, if $B_i(k)$ has most of its energy concentrated about frequency $k = k_0$, then $< x(n), a_i(n) >$ measures to some degree the concentration of $X(k)$ at $k = k_0$. This basis function is mostly localized about the point (n_0, k_0) in the discrete-time discrete-frequency plane. Similarly, for each of the basis functions $b_i(n)$ we can find the area of the discrete-time discrete-frequency plane where most of their energy lies. All of the basis functions together will effectively cover the plane, because if any part were not covered there would be a "hole" in the basis, and we would not be able to completely represent all sequences in the space. Similarly the localization areas, or tiles, corresponding to distinct basis functions should not overlap by too much, since this would represent a redundancy in the system.

Choosing a basis can then be loosely thought of as choosing some tiling of the discrete-time discrete-frequency plane. For example, Fig. VIII.1 shows the tiling corresponding to various orthonormal bases in C^{64}. The horizontal axis represents discrete-time, and the vertical axis discrete-frequency. Naturally, each of the diagrams contains 64 tiles, since this is the number of vectors required for a basis, and each tile can be thought of as containing 64 points out of the total of 64^2 in this discrete-time discrete-frequency

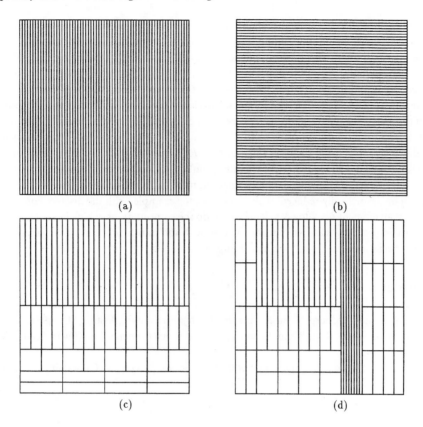

FIGURE VIII.1   Examples of tilings of the discrete-time discrete-frequency plane; time is the horizontal axis, frequency the vertical. (a) The identity transform. (b) Discrete Fourier transform. (c) Finite length discrete wavelet transform. (d) Arbitrary finite length transform.

plane. The first is the identity basis, which has narrow vertical strips as tiles, since the basis sequences $\delta(n + k)$ are perfectly localized in time, but have energy spread equally at all discrete frequencies. That is, the tile is one discrete-time point wide and 64 discrete-frequency points long. The second, shown in Fig. VIII.1(b), corresponds to the discrete Fourier transform basis vectors $e^{j2\pi in/N}$; these of course are perfectly localized at the frequencies $i = 0, 1, \cdots N - 1$, but have equal energy at all times (i.e., 64 points wide, one point long). Figure VIII.1(c) shows the tiling corresponding to a discrete orthogonal wavelet transform (or logarithmic subband coder) operating over a finite length signal. Figure VIII.1(d) shows the tiling corresponding to a discrete orthogonal wavelet packet transform operating over a finite length signal, with arbitrary splits in time and frequency; construction of such schemes is discussed in Section 7.1. In Fig. VIII.1(c) and (d), the tiles have varying shapes but still contain 64 points each.

It should be emphasized that the localization of the energy of a basis function to the area covered by one of the tiles is only approximate. In practice, of course, we will always deal with real signals, and in general we will restrict the basis functions to be real also. When this is so, $\mathbf{B}^* = \mathbf{B}^T$ and the basis is orthonormal provided $\mathbf{A}^T\mathbf{A} = \mathbf{I} = \mathbf{A}\mathbf{A}^T$. Of the bases shown in Fig. VIII.1 only the discrete Fourier transform will be excluded with this restriction. One can, however, consider a real transform which has many properties in common with the DFT, for example the discrete Hartley transform [3].

While the above description was given in terms of finite-dimensional signal spaces, the interpretation of the linear transform as a matrix operation, and the tiling approach remains essentially unchanged in the case of infinite length discrete-time signals. In fact, for bases with the structure we desire, construction in the infinite-dimensional case is easier than in the finite-dimensional case. The modifications necessary for the transition from R^N to $l^2(R)$ are that an infinite number of basis functions is required instead of N, the matrices **A** and **B** become doubly infinite, and the tilings are in the discrete-time continuous-frequency plane (the time axis ranges over Z, the frequency axis goes from 0 to π, assuming real signals).

Good decorrelation is one of the important factors in the construction of bases. If this were the only requirement, we would always use the Karhunen-Loève transform, which is an orthogonal data-dependent transform which produces uncorrelated samples. This is not used in practice, because estimating the coefficients of the matrix **A** can be very difficult. Very significant also, however, is the complexity of calculating the coefficients of the transform using (VIII.2), and of putting the signal back together using (VIII.3). In general, for example, using the basis functions for R^N, evaluating each of the matrix multiplications in (VIII.2) and (VIII.3) will require $O(N^2)$ floating point operations, unless the matrices have some special structure. If, however, **A** is sparse, or can be factored into matrices that are sparse, then the complexity required can be dramatically reduced. This is the case, for example, with the discrete Fourier transform, where there is an efficient $O(N \log N)$ algorithm to do the computations, which has been responsible for its popularity in practice. This will also be the case with the transforms that we consider, **A** and **B** will always have special structure to allow efficient implementation.

References

[1] Gohberg, I. and Goldberg, S., *Basic Operator Theory*, Birkhäuser, Boston, MA, 1981.
[2] Gersho, A. and Gray, R.M., *Vector Quantization and Signal Compression*, Kluwer Academic, Norwell, MA, 1992.
[3] Bracewell, R., *The Fourier Transform and its Applications*, 2nd ed., McGraw-Hill, New York, 1986.

35

Wavelets and Filter Banks

Cormac Herley
Hewlett Packard Laboratories

35.1 Filter Banks and Wavelets

The methods of designing bases that we will employ draw on ideas first used in the construction of multirate filter banks. The idea of such systems is to take an input system and split it into subsequences using banks of filters. This simplest case involves splitting into just two parts using a structure such as that shown in Fig. 35.1. This technique has a long history of use in the area of subband coding: first of speech [1, 2] and more recently of images [3, 4]. In fact, the most successful image coding schemes are based on filter bank expansions [5, 6, 7]. Recent texts on the subject are [8, 9, 10]. We will consider only the two-channel case in this section. If $\hat{X}(z) = X(z)$, then the filter bank has the perfect reconstruction property.

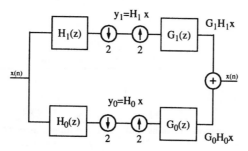

FIGURE 35.1   Maximally decimated two-channel multirate filter bank.

It is easily shown that the output $\hat{X}(z)$ of the overall analysis/synthesis system is given by:

$$
\begin{aligned}
\hat{X}(z) &= \frac{1}{2}[G_0(z)\, G_1(z)]
\begin{bmatrix} H_0(z) & H_0(-z) \\ H_1(z) & H_1(-z) \end{bmatrix}
\begin{bmatrix} X(z) \\ X(-z) \end{bmatrix} \qquad (35.1) \\
&= \frac{1}{2}[H_0(z)G_0(z) + H_1(z)G_1(z)] \cdot X(z) \\
&\quad + \frac{1}{2}[H_0(-z)G_0(z) + H_1(-z)G_1(z)] \cdot X(-z).
\end{aligned}
$$

Call the above 2×2 matrix $\mathbf{H}_m(z)$. This gives that the unique choice for the synthesis filters is

$$\left[\begin{array}{c} G_0(z) \\ G_1(z) \end{array} \right] = \left[\begin{array}{cc} H_0(z) & H_0(-z) \\ H_1(z) & H_1(-z) \end{array} \right]^{-1} \cdot \left[\begin{array}{c} 2 \\ 0 \end{array} \right]$$

$$= \frac{2}{\Delta_m(z)} \left[\begin{array}{c} H_1(-z) \\ -H_0(-z) \end{array} \right], \tag{35.2}$$

where $\Delta_m(z) = \det \mathbf{H}_m(z)$.

If we observe that $\Delta_m(z) = -\Delta_m(-z)$ and define $P(z) = 2 \cdot H_0(z)H_1(-z)/\Delta_m(z) = H_0(z)G_0(z)$, it follows from (35.2) that $G_1(z)H_1(z) = 2 \cdot H_1(z)H_0(-z)/\Delta_m(-z) = P(-z)$. We can then write that the necessary and sufficient condition for perfect reconstruction (35.1) is:

$$P(z) + P(-z) = 2. \tag{35.3}$$

Since this condition plays an important role in what follows, we will refer to any function having this property as *valid*. The implication of this property is that all but one of the even-indexed coefficients of $P(z)$ are zero. That is

$$P(z) + P(-z) = \sum_n (p(n)z^{-n} + p(n)(-z)^{-n})$$

$$= \sum_n 2 \cdot p(2n)z^{-(2n+1)}.$$

For this to satisfy (35.3) requires $p(2n) = \delta_n$; thus, one of the polyphase components of $P(z)$ must be the unit sample. By polyphase components we mean the set of even-indexed samples, and the set of the odd-indexed samples. Such a function is illustrated in Fig. 35.2(a).

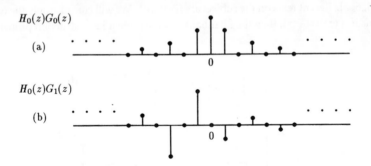

$H_0(z)G_0(z)$

(a)

0

$H_0(z)G_1(z)$

(b)

0

FIGURE 35.2   Zeros of the correlation functions. (a) Autocorrelation $H_0(-z)H_0(z^{-1})$. (b) Cross-correlation $H_0(-z)H_1(z^{-1})$.

Constructing such a function is not difficult. In general, however, we will wish to impose additional constraints on the filter banks. So, $P(z)$ will have to satisfy other constraints in addition to (35.3).

Observe that as a consequence of (35.2) $G_0(z)H_1(z)$, i.e., the cross-correlation of $g_1(n)$ and the time-reversed filter $h_0(-n)$, and $G_1(z)H_0(z)$, the cross-correlation of $g_1(n)$ and $h_0(-n)$, have only odd-indexed coefficients, just as for the function in Fig. 35.2(b), that is:

$$< g_0(n), h_1(2k - n) > = 0, \tag{35.4}$$

$$< g_1(n), h_0(2k - n) > = 0, \tag{35.5}$$

(note the time reversal in the inner product). Define now the matrix H_0 as

$$
H_0 = \begin{bmatrix}
\ddots & \vdots & \vdots & & & \vdots & \vdots & \vdots & \vdots & \\
& h_0(L-1) & h_0(L-2) & \cdots & \cdots & h_0(0) & 0 & 0 & & \\
& 0 & 0 & h_0(L-1) & \cdots & h_0(2) & h_0(1) & h_0(0) & & \ddots \\
& \vdots & \vdots & & & \vdots & \vdots & \vdots & \vdots &
\end{bmatrix} \tag{35.6}
$$

which has as its kth row the elements of the sequence $h_0(2k - n)$. Pre-multiplying by H_0 corresponds to filtering by $H_0(z)$ followed by subsampling by a factor of 2. Also define

$$
G_0^T = \begin{bmatrix}
& \vdots & \vdots & \vdots & \vdots & \vdots & & \vdots & \vdots & \\
\ddots & g_0(0) & g_0(1) & \cdots & \cdots & g_0(L-1) & 0 & 0 & & \\
& 0 & 0 & g_0(0) & \cdots & g_0(L-3) & g_0(L-2) & g_0(L-1) & & \ddots \\
& \vdots & \vdots & \vdots & \vdots & \vdots & & \vdots & \vdots &
\end{bmatrix}, \tag{35.7}
$$

so G_0 has as its kth column the elements of the sequence $g_0(n - 2k)$. Define H_1 by replacing the coefficients of $h_0(n)$ with those of $h_1(n)$ in (35.6) and G_1 by replacing the coefficients of $g_0(n)$ with those of $g_1(n)$ in (35.7).

We find that (35.4) gives that all rows of H_1 are orthogonal to all columns of G_0. Similarly we find, from (35.5), that all of the columns of G_1 are orthogonal to the rows of H_0. So, in matrix notation:

$$
H_0 G_1 = 0 = H_1 G_0. \tag{35.8}
$$

Now $P(z) = G_0(z)H_0(z) = z^{-1}H_0(z)H_1(-z)$ and $P(-z) = G_1(z)H_1(z)$ are both valid and have the form given in Fig. 35.2 (a). Hence, the impulse responses of $g_i(n)$ and $h_i(n)$ are orthogonal with respect to even shifts

$$
< g_i(n), h_i(2l - n) > = \delta_l. \tag{35.9}
$$

In operator notation:

$$
H_0 G_0 = I = H_1 G_1. \tag{35.10}
$$

Since we have a perfect reconstruction system we get:

$$
G_0 H_0 + G_1 H_1 = I. \tag{35.11}
$$

Of course (35.11) indicates that no nonzero vector can lie in the column nullspaces of both G_0 and G_1. Note that (35.10) implies that $G_0 H_0$ and $G_1 H_1$ are each projections (since $G_i H_i G_i H_i = G_i H_i$). They project onto subspaces which are not, in general, orthogonal (since the operators are not self-adjoint). Because of (35.4), (35.5), and (35.9) the analysis/synthesis system is termed biorthogonal. If we interleave the rows of H_0 and H_1, much as was done in the orthogonal case, and form again a block Toeplitz matrix

$$
A = \begin{bmatrix}
& \vdots & \vdots & & & \vdots & \vdots & \vdots & \vdots & \\
& h_0(L-1) & h_0(L-2) & \cdots & \cdots & h_0(0) & 0 & 0 & & \\
\ddots & h_1(L-1) & h_1(L-2) & \cdots & \cdots & h_1(0) & 0 & 0 & & \\
& 0 & 0 & h_0(L-1) & \cdots & h_0(2) & h_0(1) & h_0(0) & & \ddots \\
& 0 & 0 & h_1(L-1) & \cdots & h_1(2) & h_1(1) & h_1(0) & & \\
& \vdots & \vdots & & & \vdots & \vdots & \vdots & \vdots &
\end{bmatrix}, \tag{35.12}
$$

we find that the rows of A form a basis for $l^2(Z)$. If we form B by interleaving the columns of G_0 and G_1, we find

$$
B \cdot A = I.
$$

In the special case where we have a unitary solution, one finds: $\mathbf{G}_0 = \mathbf{H}_0^T$ and $\mathbf{G}_1 = \mathbf{H}_1^T$, and (35.8) gives that we have projections onto subspaces which are mutually orthogonal. The system then simplifies to the orthogonal case, where $\mathbf{B} = \mathbf{A}^{-1} = \mathbf{A}^T$.

A point that we wish to emphasize is that in the conditions for perfect reconstruction, (35.2) and (35.3), the filters $H_0(z)$ and $G_0(z)$ are related via their product $P(z)$. It is the choice of the function $P(z)$ and the factorization taken that determines the properties of the filter bank. We conclude the introduction with a proposition that sums up the foregoing.

PROPOSITION 35.1   To design a two-channel perfect reconstruction filter bank, it is necessary and sufficient to find a $P(z)$ satisfying (35.3), factor it $P(z) = G_0(z)H_0(z)$ and assign the filters as given in (35.2).

35.1.1   Deriving Continuous-Time Bases From Discrete-Time Ones

We have seen that the construction of bases from discrete-time signals can be accomplished easily by using a perfect reconstruction filter bank as the basic building block. This gives us bases that have a certain structure, and for which the analysis and synthesis can be efficiently performed. The design of bases for continuous-time signals appears more difficult. However, it works out that we can mimic many of the ideas used in the discrete-time case, when we go about the construction of continuous-time bases.

In fact, there is a very close correspondence between the discrete-time bases generated by two-channel filter banks, and dyadic wavelet bases. These are continuous-time bases formed by the stretches and translates of a single function, where the stretches are integer powers of two:

$$\{\psi_{jk}(x) = 2^{-j/2}\psi(2^{-j}x - k), \quad j, k, \in Z\} \tag{35.13}$$

This relation has been thoroughly explored in [11, 12].

To be precise, a basis of the form in (35.13) necessarily implies the existence of an underlying two-channel filter bank. Conversely, a two-channel filter bank can be used to generate a basis as in (35.13) provided that the lowpass filter $H_0(z)$ is *regular*. It is not our intention to go into the details of this connection, but the generation of wavelets from filter banks goes briefly as follows:

Considering the logarithmic tree of discrete-time filters in Fig. 35.3, one notices that the lower branch is a cascade of filters $H_0(z)$ followed by subsampling by 2. It is easily shown [12], that the cascade of i blocks of filtering operations, followed by subsampling by 2, is equivalent to a filter $H_0^{(i)}(z)$ with z-transform:

$$H_0^{(i)}(z) = \prod_{l=0}^{i-1} H_0(z^{2^l}), \qquad i = 1, 2 \cdots, \tag{35.14}$$

followed by subsampling by 2^i. We define $H_0^{(0)}(z) = 1$ to initialize the recursion. Now, in addition to the discrete-time filter, consider the function $f^{(i)}(x)$ which is piecewise constant on intervals of length $1/2^i$, and equal to:

$$f^{(i)}(x) = 2^{i/2} \cdot h_0^{(i)}(n), \qquad n/2^i \le x < (n+1)/2^i. \tag{35.15}$$

Note that the normalization by $2^{i/2}$ ensures that if $\sum(h_0^{(i)}(n))^2 = 1$ then $\int (f^{(i)}(x))^2 dx = 1$ as well. Also, it can be checked that $\|h_0^{(i)}\|_2 = 1$ when $\|h_0^{(i-1)}\|_2 = 1$. The relation between the sequence $H_0^{(i)}(z)$ and the function $f^{(i)}(x)$ is clarified in Fig. 35.3, where the first three iterations of each is shown for the simple case of a filter of length 4.

We are going to use the sequence of functions $f^{(i)}(x)$ to converge to the scaling function $\phi(x)$ of a wavelet basis. Hence, a fundamental question is to find out whether and to what the function $f^{(i)}(x)$ converges as $i \to \infty$. First assume that the filter $H_0(z)$ has a zero at the half sampling frequency, or $H_0(e^{j\pi}) = 0$.

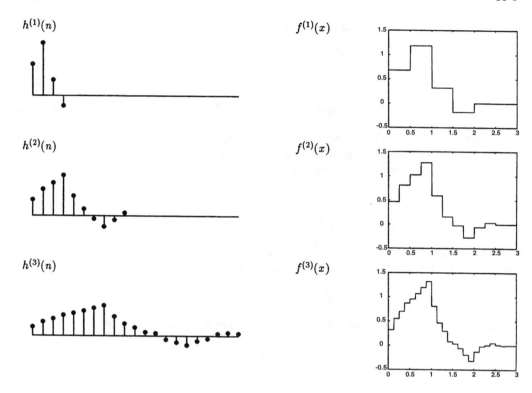

FIGURE 35.3   Iterations of the discrete-time filter (35.14) and the continuous-time function (35.15) for the case of a length-4 filter $H_0(z)$. The length of the filter $H_0^{(i)}(z)$ increases without bound, while the function $f^{(i)}(x)$ actually has bounded support.

This together with the fact that the filter impulse response is orthogonal to its even translates is equivalent to $\sum h_0(n) = H_0(1) = \sqrt{2}$. Define $M_0(z) = 1/\sqrt{2} \cdot H_0(z)$, that is $M_0(1) = 1$. Now factor $M_0(z)$ into its roots at π (there is at least one by assumption) and a remainder polynomial $K(z)$, in the following way:

$$M_0(z) = [(1 + z^{-1})/2]^N K(z).$$

Note that $K(1) = 1$ from the definitions. Now call B the supremum of $|K(z)|$ on the unit circle:

$$B = \sup_{\omega \in [0, 2\pi]} |K(e^{j\omega})|.$$

Then the following result from [11] holds:

PROPOSITION 35.2   [Daubechies 1988] If $B < 2^{N-1}$, and

$$\sum_{n=-\infty}^{\infty} |k(n)|^2 |n|^\epsilon \; < \; \infty, \quad \text{for some } \epsilon > 0, \tag{35.16}$$

then the piecewise constant function $f^{(i)}(x)$ defined in (35.15) converges pointwise to a continuous function $f^{(\infty)}(x)$.

This is a sufficient condition to ensure pointwise convergence to a continuous function, and can be used as a simple test. We shall refer to any filter for which the infinite product converges as *regular*.

If we indeed have convergence, then we define

$$f^{(\infty)}(x) = \phi(x)$$

as the analysis scaling function, and

$$\psi(x) = 2^{-1/2} \sum h_1(n)\phi(2x - n), \tag{35.17}$$

as the analysis wavelet. It can be shown that if the filters $h_0(n)$ and $h_1(n)$ are from a perfect reconstruction filter bank, then (35.13) indeed forms a continuous-time basis.

In a similar way we examine the cascade of i blocks of the synthesis filter $g_0(n)$

$$G_0^{(i)}(z) = \prod_{l=0}^{i-1} G_0(z^{2^l}), \qquad i = 1, 2 \cdots . \tag{35.18}$$

Again, define $G_0^{(0)}(z) = 1$ to initialize the recursion, and normalize $G_0(1) = 1$. From this define a function which is piecewise constant on intervals of length $1/2^i$:

$$\check{f}^{(i)}(x) = 2^{i/2} \cdot g_0^{(i)}(-n), \qquad n/2^i \le x < (n+1)/2^i. \tag{35.19}$$

We call the limit $\check{f}^{(\infty)}(x)$, if it exists, $\check\phi(x)$ the synthesis scaling function, and we find

$$\check\phi(x) = 2^{1/2} \cdot \sum_{n=0}^{L-1} g_0(-n) \cdot \check\phi(2x - n) \tag{35.20}$$

$$\check\psi(x) = 2^{1/2} \cdot \sum_{n=0}^{L-1} g_1(-n) \cdot \check\phi(2x - n). \tag{35.21}$$

The biorthogonality properties of the analysis and synthesis continuous-time functions follow from the corresponding properties of the discrete-time ones. That is, (35.9) leads to

$$< \check\phi(x), \phi(x - k) > = \delta_k. \tag{35.22}$$

and

$$< \check\psi(x), \psi(x - k) > = \delta_k. \tag{35.23}$$

Similarly

$$< \check\phi(x), \psi(x - k) > = 0 \tag{35.24}$$

$$< \check\psi(x), \phi(x - k) > = 0, \tag{35.25}$$

come from (35.4) and (35.5), respectively.

We have shown that the conditions for perfect reconstruction on the filter coefficients lead to functions that have the biorthogonality properties as shown above. Orthogonality across scales is also easily verified:

$$< \check\psi(2^j x), \psi(2^i x - k) > = \delta_{i-j}\delta_k.$$

Thus, the set $\{\psi(2^j x), \check\psi(2^i x - k), i, j, k \in Z\}$ is biorthogonal. That it is complete can be verified as in the orthogonal case [13]. Hence, any function from $L^2(R)$ can be written:

$$f(x) = \sum_j \sum_l < f(x), 2^{-j/2}\psi(2^j x - l) > 2^{-j/2}\check\psi(2^j x - l).$$

Note that $\psi(x)$ and $\check\psi(x)$ play interchangeable roles.

35.1.2 Two-Channel Filter Banks and Wavelets

We have seen that the design of discrete-time bases is not difficult: using two-channel filter banks as the basic building block they can be easily derived. We also know that, using (35.15) and (35.19), we can generate continuous-time bases quite easily as well. If we were just interested in the construction of bases, with no further requirements, we could stop here. However, for applications such as compression, we will often be interested in other properties of the basis functions, for example, whether or not they have any symmetry or finite support, and whether or not the basis is an orthonormal one. We examine these three structural properties for the remainder of this section. Chapter 36 deals with the design of the filters. Chapter 37 deals with time-varying filter banks, where the filters used, or the tree structure employing them, varies over time. Chapter 38 deals with the case of Lapped Transforms, a very important class of multirate filter banks that have achieved considerable success.

From the filter bank point of view, the properties we are most interested in are the following:

- **Orthogonality:**

$$< h_0(n), h_0(n + 2k) > \; = \; \delta_k \; = \; < h_1(n), h_1(n + 2k) >, \qquad (35.26)$$

$$< h_0(n), h_1(n + 2k) > \; = \; 0. \qquad (35.27)$$

- **Linear phase:** $H_0(z)$, $H_1(z)$, $G_0(z)$, and $G_1(z)$ are all linear phase filters.
- **Finite support:** $H_0(z)$, $H_1(z)$, $G_0(z)$, and $G_1(z)$ are all FIR filters.

The reason for our interest is twofold. First, these properties are possibly of value in perfect reconstruction filter banks used in subband coding schemes. For example, orthogonality implies that the quantization noise in the two channels will be independent; linear phase is possibly of interest in very low bit-rate coding of images, and FIR filters have the advantage of having very simple low-complexity implementations. Second, these properties are carried over to the wavelets that are generated. So, if we design a filter bank with a certain set of properties, then the continuous-time basis that it generates will also have these properties.

PROPOSITION 35.3 If the filters belong to an orthogonal filter bank, we shall have

$$< \phi(x), \phi(x + k) > \; = \; \delta_k \; = < \psi(x), \psi(x + k) >,$$

$$< \phi(x), \psi(x + k) > \; = \; 0.$$

PROOF 35.1 From the definition (35.15) $f^{(0)}(x)$ is just the indicator function on the interval $[0, 1)$; so we immediately get orthogonality at the 0th level, that is: $< f^{(0)}(x - l), f^{(0)}(x - k) > \; = \delta_{kl}$. Now we assume orthogonality at the ith level:

$$< f^{(i)}(x - l), f^{(i)}(x - k) > \; = \; \delta_{kl}, \qquad (35.28)$$

and prove that this implies orthogonality at the $(i + 1)$st level:

$$< f^{(i+1)}(x - l), f^{(i+1)}(x - k) > \; = \; 2 \sum_n \sum_m h_0(n) h_0(m)$$

$$< f^{(i)}(2x - 2l - n), f^{(i)}(2x - 2k - m) > \frac{\delta_{n+2l-2k-m}}{2}$$

$$= \; \sum_n h_0(n) h_0(n + 2l - 2k)$$

$$= \; \delta_{kl}.$$

Hence, by induction (35.28) holds for all i. So in the limit $i \to \infty$:

$$< \phi(x - l), \phi(x - k) > = \delta_{kl}. \tag{35.29}$$

The orthogonal case gives considerable simplification, both in the discrete-time and continuous-time cases.

PROPOSITION 35.4   If the filters belong to an FIR filter bank, then $\phi(x)$, $\psi(x)$, $\check{\phi}(x)$, and $\check{\psi}(x)$ will have support on some finite interval.

PROOF 35.2   The filters $H_0^{(i)}(z)$ and $G_0^{(i)}(z)$ defined in (35.14) have respective lengths $(2^i - 1)(L_a - 1) + 1$ and $(2^i - 1)(L_s - 1) + 1$ where L_a and L_s are the lengths of $H_0(z)$ and $G_0(z)$. Hence, $f^{(i)}(x)$ in (35.15) is supported on the interval $[0, L_a - 1)$ and $\check{f}^{(i)}(x)$ on the interval $[0, L_s - 1)$. This holds $\forall i$; hence, in the limit $i \to \infty$ this gives the support of the scaling functions $\phi(x)$ and $\check{\phi}(x)$. That $\psi(x)$ and $\check{\psi}(x)$ have bounded support follow from (35.20) and (35.21).

PROPOSITION 35.5   If the filters belong to a linear phase filter bank, then $\phi(x)$, $\psi(x)$, $\check{\phi}(x)$, and $\check{\psi}(x)$ will be symmetric or antisymmetric.

PROOF 35.3   The filter $H_0^{(i)}(z)$ will have linear phase if $H_0(z)$ does. If $H_0^{(i)}(z)$ has length $(2^i - 1)(L_a - 1) + 1$, the point of symmetry is $(2^i - 1)(L_a - 1)/2$ which need not be an integer. The point of symmetry for $f^{(i)}(x)$ will then be $[(2^i - 1)(L_a - 1) + 1]/2^{i+1}$ or $[(2^i - 1)(L_a - 1) + 2]/2^{i+1}$. In either case, by taking the limit $i \to \infty$ we find that $\phi(x)$ is symmetric about the point $(L_a - 1)/2$ and similarly for the other cases.

Thus having established the relation between wavelets and filter banks we can examine the structure of filter banks in detail, and afterward use them to generate wavelets as described above. It should be emphasized that we are speaking of the two-channel, one-dimensional case. Multidimensional filter banks are a large subject in their own right [8, 10].

35.1.3   Structure of Two-Channel Filter Banks

We saw already that it is the choice of the function $P(z)$ and the factorization taken that determines the properties of the filter bank. In terms of $P(z)$, we give necessary and sufficient conditions for the three properties mentioned above:

- **Orthogonality:** $P(z)$ is an autocorrelation, and $H_0(z)$ and $G_0(z)$ are its spectral factors.
- **Linear phase:** $P(z)$ is linear phase, and $H_0(z)$ and $G_0(z)$ are its linear phase factors.
- **Finite support:** $P(z)$ is FIR, and $H_0(z)$ and $G_0(z)$ are its FIR factors.

Obviously the factorization is not unique in any of the cases above. The FIR case has been examined in detail in [11, 12, 14, 15, 16] and the linear phase case in [12, 15, 17]. In the rest of this paper we will present new results on the orthogonal case, but we shall also review the solutions that explicitly satisfy simultaneous constraints.

PROPOSITION 35.6   To have an orthogonal filter bank it is necessary and sufficient that $P(z)$ be an autocorrelation, and that $H_0(z)$ and $G_0(z)$ be its spectral factors.

PROPOSITION 35.7 To have a linear phase filter bank it is necessary and sufficient that $P(z)$ be a linear phase, and that $H_0(z)$ and $G_0(z)$ be its linear phase factors.

PROPOSITION 35.8 To have an FIR filter bank it is necessary and sufficient that $P(z)$ be FIR, and that $H_0(z)$ and $G_0(z)$ be its FIR factors.

Proofs can be found in [18]. Having seen that the design problem can be considered in terms of $P(z)$ and its factorizations, we consider the three conditions of interest from this point of view.

Orthogonality

In the case where the filter bank is to be orthogonal, we can obtain a complete constructive characterization of the solutions, as given by the following theorem, taken from [18].

THEOREM 35.1 *All orthogonal rational two channel filter banks can be formed as follows:*

1. *Choosing an arbitrary polynomial $R(z)$, form:*

$$P(z) = \frac{2 \cdot R(z)R(z^{-1})}{R(z)R(z^{-1}) + R(-z)R(-z^{-1})},$$

2. *factor as $P(z) = H(z)H(z^{-1})$,*
3. *form the filter $H_0(z) = A_0(z)H(z)$, where $A_0(z)$ is an arbitrary allpass,*
4. *choose $H_1(z) = z^{2k-1}H_0(-z^{-1})A_1(z^2)$, where $A_1(z)$ is again an arbitrary allpass,*
5. *choose $G_0(z) = H_0(z^{-1})$, and $G_1(z) = -H_1(z^{-1})$.*

For a proof, see [18, 19].

EXAMPLE 35.1:

Take $R(z) = (1 + z^{-1})^N$ as above and $N = 7$. It works out that in this case there is a closed form factorization for the filters.

$$P(z) = \frac{(1, 14, 91, 364, 1001, 2002, 3003, 3432, 3003, 2002, 1001, 364, 91, 14, 1) \cdot z^7}{14z^6 + 364z^4 + 2002z^2 + 3432 + 2002z^{-2} + 364z^{-4} + 14z^{-6}}$$

$$= \frac{E(z)E(z^{-1})}{K(z)K(z^{-1})},$$

where

$$\frac{E(z)}{K(z)} = \frac{(1 + 7z^{-1} + 21z^{-2} + 35z^{-3} + 35z^{-4} + 21z^{-5} + 7z^{-6} + z^{-7})}{\sqrt{2} \cdot (1 + 21z^{-2} + 35z^{-4} + 7z^{-6})}.$$

Note that we have used the following shorthand notation to list the coefficients of a causal FIR sequence:

$$\sum_{n=0}^{N-1} a_n z^{-n} = (a_0, a_1, a_2, \cdots a_{N-1}).$$

So, using the description of the filters in Theorem 35.1, with the simplest case $A_0(z) = A_1(z) = 1$ and $k = 0$ we find:

$$H_0(z) = \frac{(1 + 7z^{-1} + 21z^{-2} + 35z^{-3} + 35z^{-4} + 21z^{-5} + 7z^{-6} + z^{-7})}{\sqrt{2} \cdot (1 + 21z^{-2} + 35z^{-4} + 7z^{-6})}$$

$$H_1(z) = z^{-1}\frac{(1 - 7z^1 + 21z^2 - 35z^3 + 35z^4 - 21z^5 + 7z^6 - z^7)}{\sqrt{2}\cdot(1 + 21z^2 + 35z^4 + 7z^6)}$$

$$G_0(z) = H_0(z^{-1}) \qquad G_1(z) = H_1(z^{-1}).$$

In the notation of Proposition 35.2, $B = 8 < 2^6$ so that for this choice of $H_0(z)$ the left-hand side of (35.15) converges to a continuous function. The wavelet, scaling function, and their spectra are shown in Fig. 35.4.

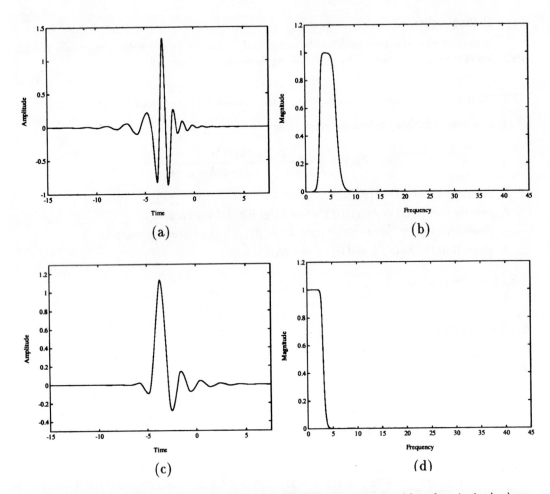

FIGURE 35.4   Example of Butterworth orthogonal wavelet; here $N = 7$, and the closed form factorization has been used. (a) The wavelet. (b) Spectrum of the wavelet. (c) Scaling function. (d) Spectrum of the scaling function.

Finite Impulse Response and Symmetric Solutions

In the case where the filters are to be FIR, we merely require that $P(z)$ be FIR; it is trivially easy to design one. Similarly to have symmetric filters, we merely force $P(z)$ to be symmetric. Obviously any symmetric $P(z)$ which is FIR and satisfies (35.3) can be used to give symmetric FIR filters. We would like, in addition, that the lowpass filters are regular, so that we get symmetric bounded support continuous-time basis functions.

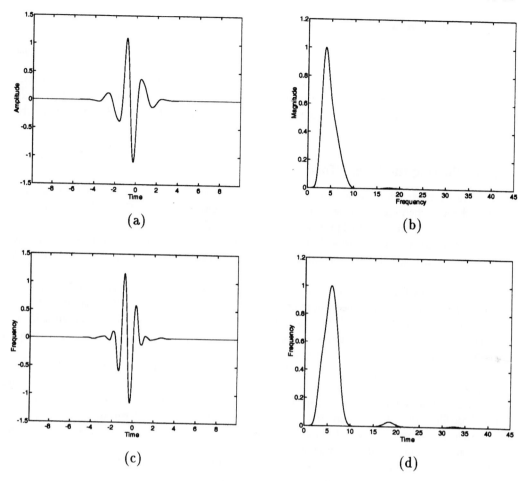

FIGURE 35.5 Biorthogonal wavelets generated by filters of length 18 given in [12]. (a) Analysis wavelet function $\psi(x)$. (b) Spectrum of analysis wavelet. (c) Synthesis wavelet function $\check{\psi}(x)$. (d) Spectrum of synthesis wavelet.

One strategy would be to design a $P(z)$ with the desired properties and then factor to find the filters. Alternatively, we can choose one of the factors, and then find the other necessary to make the product $P(z)$ satisfy (35.3). We will use this approach and, to ensure regularity, choose one factor to be $(1 + z^{-1})^{2N}$. This can be done by solving a linear system of equations [12].

EXAMPLE 35.2:

If we choose $N = 3$ we must find the complement to $(1 + z^{-1})^6$; so we solve the 3 by 3 system found by imposing the constraints on the coefficients of the odd powers of z^{-1} of

$$P(z) = (k_0 + k_1 z^{-1} + k_2 z^{-2} + k_1 z^{-3} + k_0 z^{-4})$$

$$\cdot (1 + 6z^{-1} + 15z^{-2} + 20z^{-3} + 15z^{-4} + 6z^{-5} + z^{-6}) \cdot z^5.$$

So we solve:

$$\begin{pmatrix} 6 & 1 & 0 \\ 20 & 16 & 6 \\ 12 & 30 & 20 \end{pmatrix} \begin{pmatrix} k_0 \\ k_1 \\ k_2 \end{pmatrix} = \begin{pmatrix} 0 \\ 0 \\ 1 \end{pmatrix},$$

giving $\mathbf{k}_6 = (3/2, -9, 19)/128$.

In general, therefore, we solve the system:

$$\mathbf{F}_{2N} \cdot \mathbf{k}_{2N} = \mathbf{e}_{2N}, \tag{35.30}$$

where $\mathbf{F}_{2N}$ is the $N \times N$ matrix, $\mathbf{k}_{2N} = (k_0, \cdots, k_{(k-1)})$, and $\mathbf{e}_{2N}$ is the length k vector $(0, 0, \cdots, 1)$.

Having found the coefficients of $K_{2N}(z)$, we factor it into linear phase components and then regroup these factors of $K_{2N}(z)$ and the $2N$ zeros at $z = -1$ to form two filters: $H_0(z)$ and $H_1(-z)$, both of which are to be regular.

35.1.4 Putting the Pieces Together

An important consideration that is often encountered in the design of wavelets, or of the filter banks that generate them, is the necessity of satisfying competing design constraints. This makes it necessary to clearly understand whether desired properties are mutually exclusive.

Perfect reconstruction solutions, with the constraint that $P(z)$ be rational with real coefficients, must satisfy (35.3). Such general solutions, which do not necessarily have additional properties, were given in [14].

The solutions of set A, where all of the filters involved are FIR, were studied in [14, 15]. Set B contains all orthogonal solutions, and has been the main focus of this paper. A complete characterization of this set was given in Theorem 35.1. A very different characterization, based on lattice structures, is given in [20]. Particular cases of orthogonal solutions were also given in [21]. Set C contains the solutions where all filters are linear phase, first examined in [15].

The earliest examples of perfect reconstruction solutions [22, 23] were orthogonal and FIR; i.e., they were in $A \cap B$. A constructive parametrization of $A \cap B$ was given in [24]. The construction and characterization of examples which converge to wavelets was first done in [11]. Filter banks with FIR linear phase filters (i.e., $A \cap C$) were first given in [15], and also studied in terms of lattices in [17, 25]. The construction of wavelet examples is given in [13] and [12]. Filter banks, which are linear phase and orthogonal, were constructed in Chapter 36 and were presented in [18].

That there exist only trivial solutions which are linear phase, orthogonal and FIR is indicated by the intersection $A \cap B \cap C$; the only solutions are two tap filters [11, 12, 26].

It warrants emphasis that Fig. 35.6 illustrates the filter bank solutions; if the filters are regular, then they will lead to wavelets. Of the dyadic wavelet bases known to the authors, the only ones based on filters where $P(z)$ is not rational are those of Meyer [27], and the only ones where the filter coefficients are complex are those of Lawton [28]. For the case of the Battle-Lemarié wavelets, while the filters themselves are not rational, the $P(z)$ function is; hence, the filters would belong to $B \cap C$ in the figure.

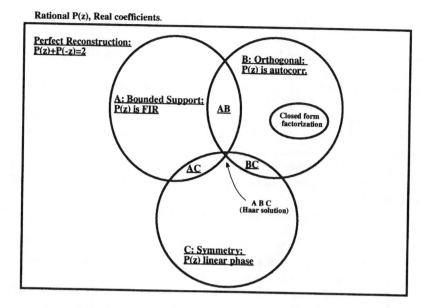

FIGURE 35.6 Two channel perfect reconstruction filter banks. The Venn diagram illustrates which competing constraints can be simultaneously satisfied. The sets A, B, C contain FIR, orthogonal, and linear phase solutions, respectively. Solutions in the intersection $A \cap B$ are examined in [11, 14, 23, 24]; those in the intersection $A \cap C$ are detailed in [12, 13, 15, 17, 25]; solutions in $B \cap C$ are constructed in [18]. The intersection $A \cap B \cap C$ contains only trivial solutions.

References

[1] Croisier, A., Esteban, D. and Galand, C., Perfect channel splitting by use of interpolation, decimation, tree decomposition techniques, in *Int. Conf. on Information Sciences/Systems*, (Patras), 443–446, Aug. 1976.

[2] Crochiere, R.E., Weber, S.A. and Flanagan, J.L., Digital coding of speech in subbands, *Bell System Technical J.*, 55, 1069–1085, Oct. 1976.

[3] Vetterli, M., Multidimensional subband coding: Some theory and algorithms, *Signal Proc.*, 6, 97–112, Feb. 1984.

[4] Woods, J.W. and O'Neil S.D., Subband coding of images, *IEEE Trans. Acoust., Speech, Signal Proc.*, 34(5), 1278–1288, 1986.

[5] Shapiro, J.M., Embedded image coding using zerotrees of wavelet coefficients, *IEEE Trans. on Signal Proc.*, 41, 3445–3462, Dec. 1993.

[6] Said, A. and Pearlman, W.A., An image multiresolution representation for lossless and lossy compression, *IEEE Trans. on Image Proc.*, 5(9), 1303–1310, 1996.

[7] Xiong, Z., Ramchandran, K. and Orchard, M.T., Wavelet packet image coding using space-frequency quantization, *IEEE Trans. on Image Proc.*, submitted, 1996.

[8] Vaidyanathan, P.P., *Multirate Systems and Filter Banks*, Prentice-Hall, Englewood Cliffs, NJ, 1992.

[9] Malvar, H.S., *Signal Processing with Lapped Transforms*, Artech House, 1992.

[10] Vetterli, M. and Kovacevic, J., *Wavelet and Subband Coding*, Prentice-Hall, Englewood Cliffs, NJ, 1995.

[11] Daubechies, I., Orthonormal bases of compactly supported wavelets, *Communications on Pure and Applied Mathematics*, XLI, 909–996, 1988.

[12] Vetterli, M. and Herley, C., Wavelets and filter banks: theory and design, *IEEE Trans. on Signal Proc.*, 40, 2207–2232, Sept. 1992.

[13] Cohen, A., Daubechies, I. and Feauveau, J.-C., Biorthogonal bases of compactly supported wavelets, *Commun. on Pure and Applied Mathematics*, 45, 485–560, 1992.

[14] Smith, M.J.T. and Barnwell III, T.P., Exact reconstruction for tree-structured subband coders, *IEEE Trans. Acoust., Speech, Signal Proc.*, 34, 434–441, June 1986.

[15] Vetterli, M., Filter banks allowing perfect reconstruction, *Signal Proc.*, 10(3), 219–244, 1986.

[16] Vaidyanathan, P.P., Multirate digital filters, filter banks, polyphase networks, and applications: a tutorial, *Proc. IEEE*, 78, 56–93, Jan. 1990.

[17] Nguyen, T.Q. and Vaidyanathan, P.P., Two-channel perfect-reconstruction FIR QMF structures which yield linear-phase analysis and synthesis filters, *IEEE Trans. Acoust., Speech, Signal Proc.*, 37, 676–690, May 1989.

[18] Herley, C. and Vetterli, M., Wavelets and recursive filter banks, *IEEE Trans. on Signal Proc.*, 41, 2536–2556, Aug. 1993.

[19] Herley, C., *Wavelets and Filter Banks*, Ph.D. thesis, Columbia University, New York, April 1993. Available by anonymous ftp to: ftp.ctr.columbia.edu directory: CTR-Research/advent/public/papers/PhD-theses/Herley.

[20] Doğanata, Z. and Vaidyanathan, P.P., Minimal structures for the implementation of digital rational lossless systems, *IEEE Trans. Acoust., Speech, Signal Proc.*, 38, 2058–2074, Dec. 1990.

[21] Smith, M.J.T., IIR analysis/synthesis systems, in *Subband Coding of Images*, Woods, J.W., Ed., Kluwer Academic, Norwell, MA, 1991.

[22] Smith, M.J.T. and Barnwell III, T.P., A procedure for designing exact reconstruction filter banks for tree structured subband coders, in *Proc. IEEE Intl. Conf. ASSP*, San Diego, CA, pp. 27.1.1–27.1.4, March 1984.

[23] Mintzer, F., Filters for distortion-free two-band multirate filter banks, *IEEE Trans. Acoust., Speech, Signal Proc.*, 33, 626–630, June 1985.

[24] Vaidyanathan, P.P. and Hoang, P.-Q., Lattice structures for optimal design and robust implementation of two-band perfect reconstruction QMF banks, *IEEE Trans. Acoust., Speech, Signal Proc.*, 36, 81–94, Jan. 1988.

[25] Vetterli, M. and Le Gall, D., Perfect reconstruction FIR filter banks: some properties and factorizations, *IEEE Trans. Acoust., Speech, Signal Proc.*, 37, 1057–1071, July 1989.

[26] Vaidyanathan, P.P. and Doğanata, Z., The role of lossless systems in modern digital signal processing, *IEEE Trans. Education*, 32, 181–197, Aug. 1989. Special issue on Circuits and Systems.

[27] Meyer, Y., *Ondelettes*, vol. 1 of *Ondelettes et Opérateurs*, Hermann, Paris, 1990.

[28] Lawton, W., Application of complex-valued wavelet transforms to subband decomposition, *IEEE Trans. on Signal Proc.*, submitted, 1992.

36

Filter Bank Design

Joseph Arrowood
Georgia Institute of Technology

Tami Randolph
Georgia Institute of Technology

Mark J.T. Smith
Georgia Institute of Technology

The interest in digital filter banks has grown dramatically over the last few years. Owing to the trend toward lower cost, higher speed microprocessors, digital solutions are becoming attractive for a wide variety of applications. Filter banks allow signals to be decomposed into subbands, often facilitating more efficient and effective processing. They are particularly visible in the areas of image compression, speech coding, and image analysis.

The desired characteristics of a subband decomposition will naturally vary from application to application. Moreover, within any given application, there are a myriad of issues to consider. First, one might consider whether to use FIR or IIR filters. IIR designs can offer computational advantages, while FIR designs can offer greater flexibility in filter characteristics. In this chapter we focus exclusively on FIR design. Second, one might identify the time-frequency or space-frequency representation that is most appropriate. Uniform decompositions and octave-band decompositions are particularly popular at present. At the next level, characteristics of the analysis filters should be defined. This involves imposing specifications on the analysis filter passband deviations, transition bands, and stopband deviations. Alternately or in addition, time domain characteristics may be imposed, such as limits on the step response ripples, and degree of regularity.

One can consider similar constraints for the synthesis filters. For coding applications, the characteristics of the synthesis filters often have a dominant effect on the subjective quality of the output. Finally, one should consider analysis-synthesis characteristics. That is, one has flexibility to specify the overall behavior of the system. In most cases, one views having exact reconstruction as being ideal. Occasionally, however, it may be possible to trade some small loss in reconstruction quality for significant gains in computation, speed, or cost. In addition to specifying the quality of reconstruction, it is generally possible to control the overall delay of the system from end to end. In some applications, such as two-way speech and video coding, latency represents a source of quality degradation. Thus, having explicit control over the analysis-synthesis delay can lead to improvement in quality.

The intelligent design of applications-specific filter banks involves first identifying the relevant parameters and optimizing the system with respect to them. As is typical, the filter bank analysis and reconstruction equations lead to complex tradeoffs among complexity, system delay, filter quality, filter length, and quality of performance. This chapter is devoted to presenting an introduction to filter bank design. Filter bank design has reached a state of maturity in many regards. To cover all of the important

0-8493-8572-5/98/$0.00+$.50
© 1998 by CRC Press LLC

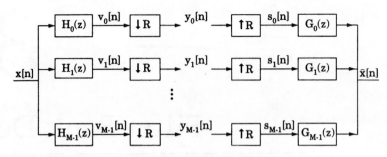

FIGURE 36.1   Block diagram of an M-band analysis-synthesis filter bank.

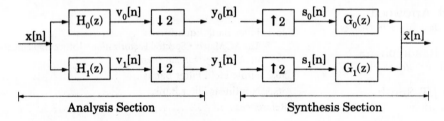

FIGURE 36.2   Two-band analysis-synthesis filter bank.

contributions in any level of detail would be impossible in a single chapter. However, it is possible to gain some insight and appreciation for general design strategies germane to this topic. In addition to discussing design methodologies for linear analysis-synthesis systems, we also consider the design of a couple of new nonlinear classes of filter banks that are currently receiving attention in the literature. This discussion along with the referenced articles should provide a convenient introduction to the design of many useful filter banks.

36.1   Filter Bank Equations

A broad class of linear filter banks can be represented by the block diagram shown in Fig. 36.1. This is a linear time-varying system that decomposes the input into M-subbands, each one of which is decimated by a factor of R. When $R = M$, the system is said to be *critically sampled* or *maximally decimated*. Maximally decimated systems are generally the ones of choice because they can be information preserving, and are not data expansive.

The simplest filter bank of this class is the two-band system, an example of which is shown in Fig. 36.2. Here, there are only two analysis filters: $H_0(z)$, a lowpass filter; and $H_1(z)$, a highpass filter. Similarly, there are two synthesis filters: a lowpass $G_0(z)$, and a highpass $G_1(z)$. Let us consider this two-band filter bank first. In the process, we will develop a design methodology that can be extended to the more complex problem of M-band systems.

Examining the two-band filter bank in Fig. 36.2, we see that the input $x[n]$ is lowpass and highpass filtered, resulting in $v_0[n]$ and $v_1[n]$. These signals are then downsampled by a factor of two, leading to the analysis section outputs, $y_0[n]$ and $y_1[n]$. The downsampling operation is time varying, which implies a non-trivial relationship between $v_k[n]$ and $y_k[n]$ (where $k = 0, 1$). In general, downsampling a signal $v_k[n]$ by an integer factor R is described in the time domain by the equation

$$y_k[n] = v_k[Rn].$$

In the frequency domain, this relationship is given by

$$Y_k\left(e^{j\omega}\right) = \frac{1}{R}\sum_{r=0}^{R-1} V_k\left(e^{j\left(\frac{\omega}{R}+\frac{2\pi r}{R}\right)}\right) .$$

The equivalent equation in the z domain is

$$Y_k(z) = \frac{1}{R}\sum_{r=0}^{R-1} V_k\left(W_R^r z^{\frac{1}{R}}\right)$$

where $W_R^r = e^{-j\frac{2\pi r}{R}}$.

In the synthesis section, the subband signals $y_0[n]$ and $y_1[n]$ are upsampled to give $s_0[n]$ and $s_1[n]$. They are then filtered by the lowpass and highpass filters, $G_0(z)$ and $G_1(z)$, respectively, before being summed together. The upsampling operation (for an arbitrary positive integer R) can be defined by

$$s_k[n] = \begin{cases} y_k[n/R] & \text{for } n = 0, \pm R, \pm 2R, \pm 3R, \ldots \\ 0 & \text{otherwise} \end{cases}$$

in the time domain, and

$$S_k\left(e^{j\omega}\right) = Y_k\left(e^{jR\omega}\right) \quad \text{and} \quad S_k(z) = Y_k\left(z^R\right)$$

in the frequency and z domains, respectively.

Using the expressions for the downsampling and upsampling operations, we can describe the two-band filter bank in terms of z-domain equations. The outputs after analysis filtering are

$$V_k(z) = H_k(z)X(z), \qquad k = 0, 1.$$

After decimation and recognizing that $W_2^1 = -1$, we obtain

$$Y_k(z) = \frac{1}{2}\left[H_k\left(z^{\frac{1}{2}}\right)X\left(z^{\frac{1}{2}}\right) + H_k\left(-z^{\frac{1}{2}}\right)X\left(-z^{\frac{1}{2}}\right)\right], \qquad k = 0, 1. \tag{36.1}$$

Thus, Eq. (36.1) defines completely the input-output relationship for the analysis section in the z domain. In the synthesis section, the subbands are upsampled giving

$$S_k(z) = Y_k(z^2), \qquad k = 0, 1.$$

This implies that

$$S_k(z) = \frac{1}{2}\left(H_k(z)X(z) + H_k(-z)X(-z)\right), \qquad k = 0, 1.$$

Passing $S_k(z)$ through the synthesis filters and then summing yields the reconstructed output

$$\begin{aligned} \hat{X}(z) &= \frac{1}{2}G_0(z)\left[H_0(z)X(z) + H_0(-z)X(-z)\right] \\ &+ \frac{1}{2}G_1(z)\left[H_1(z)X(z) + H_1(-z)X(-z)\right] . \end{aligned} \tag{36.2}$$

For virtually any application for which one can conceive, the synthesis filters should allow the input to be reconstructed exactly or with a minimal amount of distortion. In other words, ideally we want

$$\hat{X}(z) = z^{-n_0}X(z) ,$$

where n_0 is the integer system delay. An intuitive approach to handing this problem is to use the AC-matrix formulation, which we introduce next.

36.1.1　The AC Matrix

The aliasing component matrix (or AC matrix) represents a simple and intuitive idea originally introduced in [6] for handling analysis and reconstruction. The analysis-synthesis equation (36.2) for the two-band case can be expressed as

$$\hat{X}(z) = \frac{1}{2}\left[H_0(z)G_0(z) + H_1(z)G_1(z)\right]X(z)$$
$$+ \frac{1}{2}\left[H_0(-z)G_0(z) + H_1(-z)G_1(z)\right]X(-z).$$

The idea of the AC matrix is to represent the equations in matrix form. For the two-band system, this results in

$$\hat{X}(z) = \frac{1}{2}\left[X(z), X(-z)\right]\underbrace{\begin{bmatrix} H_0(z) & H_1(z) \\ H_0(-z) & H_1(-z) \end{bmatrix}}_{\text{AC matrix}}\begin{bmatrix} G_0(z) \\ G_1(z) \end{bmatrix},$$

where the AC matrix is as shown above. The AC matrix is so designated because it contains the analysis filters and all the associated aliasing components. Exact reconstruction is then obtained when

$$\begin{bmatrix} H_0(z) & H_1(z) \\ H_0(-z) & H_1(-z) \end{bmatrix}\begin{bmatrix} G_0(z) \\ G_1(z) \end{bmatrix} = \begin{bmatrix} T(z) \\ 0 \end{bmatrix}$$

where $T(z)$ is required to be the scaled integer delay $2z^{-n_0}$. The term $T(z)$ is the transfer function of the overall system. The zero term below $T(z)$ determines the amount of aliasing present in the reconstructed signal. Because this term is zero, all aliasing is explicitly removed.

With the equations expressed in matrix form, we can solve for the synthesis filters, which yields

$$\begin{bmatrix} G_0(z) \\ G_1(z) \end{bmatrix} = \frac{1}{H_0(z)H_1(-z) - H_0(-z)H_1(z)}\begin{bmatrix} H_1(-z) & -H_1(z) \\ -H_0(-z) & H_0(z) \end{bmatrix}\begin{bmatrix} T(z) \\ 0 \end{bmatrix}. \qquad (36.3)$$

Often for a variety of reasons, we would like both the analysis and synthesis filters to be FIR. This means the determinant of the AC matrix should be a constant delay. The earliest solution to the FIR filter bank problem was presented by Croisier et al. in 1976 [18]. Their solution was to let

$$H_1(z) = H_0(-z)$$

and

$$G_0(z) = H_0(z)$$
$$G_1(z) = -H_0(-z).$$

This is the quadrature mirror filter (QMF) solution. From the equations in (36.3), it can be seen that this solution cancels all the aliasing and results in a system transfer function

$$T(z) = H_0(z)H_1(-z) - H_0(-z)H_1(z).$$

As it turns out, with careful design $T(z)$ can be made to be close to a constant delay. However, some amount of distortion will always be present. In 1980 Johnston designed a set of optimized QMFs which are now widely used. The coefficient values may be found in several sources [16, 17, 19].

Interestingly, the equations in (36.3) imply that exact reconstruction is possible by forcing the AC-matrix determinant to be a constant delay. The design of such exact reconstruction filters is discussed in the next section.

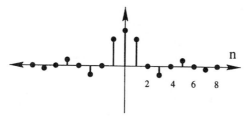

FIGURE 36.3   Example of a zero-phase half-band lowpass filter.

36.1.2 Spectral Factorization

The question at hand is how do we determine $H_0(z)$ and $H_1(z)$ such that $T(z)$ is an integer delay z^{-n_0}. A solution to this problem was introduced in 1984 [7], based on the observation that $H_0(z)H_1(-z)$ is a lowpass filter [which we denote $F_0(z)$] and $H_0(-z)H_1(z)$ is its corresponding frequency shifted highpass filter. A unity transfer function can be constructed by forcing $F_0(z)$ and $F_0(-z)$ to be complementary half-band lowpass and highpass filters. Many fine techniques are available for the design of half-band lowpass filters, such as the Parks-McClellan algorithm, Kaiser window design, Hamming window design, the eigenfilter method, and others. Zero-phase half-band filters have the property that zeros occur in the impulse response at $n = \pm2, \pm4, \pm6, \ldots$, etc. An illustration is shown in Fig. 36.3. Once designed, $F_0(z)$ can be factored into two lowpass filters, $H_0(z)$ and $H_1(-z)$. The design procedure can be summarized as follows.

1. First design a $(2N - 1)$-tap half-band lowpass filter, using the Parks-McClellan algorithm, for example. This can be done by constraining the passband and stopband cutoff frequencies to be $\omega_p = \pi - \omega_s$, and using equal passband and stopband error weightings. The resulting filter will have equal passband and stopband ripples, i.e., $\delta_p = \delta_s = \delta$.
2. Add the value δ to the $f[0]$ (center) tap value. This forces $F(e^{j\omega}) \geq 0$ for all ω.
3. Spectrally factor $F(z)$ into two lowpass filters, $H_0(z)$ and $H_1(-z)$. Generally the best way to factor $F(z)$ is such that $H_1(-z) = H_0(z^{-1})$. Note that the factorization will not be unique and the roots should be split so that if a particular root is assigned to $H_0(z)$, its reciprocal should be given to $H_0(z^{-1})$.

The result of the above procedure is that $H_0(z)$ will be a power complementary, even length, FIR filter that will form the basis for a perfect reconstruction filter bank. Note that since $H_1(z)$ is just a time-reversed, spectrally shifted version of $H_0(z)$,

$$\left|H_0(e^{j\omega})\right| = \left|H_1(-e^{j\omega})\right| .$$

Smith and Barnwell designed and published a set of optimal exact reconstruction filters [1]. The filter coefficients for $H_0(z)$ are given in Table 36.1. The analysis and synthesis filters are obtained from $H_0(z)$ by

$$\begin{aligned}
G_0(z) &= H_0\left(z^{-1}\right) \\
G_1(z) &= H_0(-z) \\
H_1(z) &= H_0\left(-z^{-1}\right) .
\end{aligned}$$

A complete discussion of this approach can be found in many references [1, 6, 7, 25, 27, 28].

TABLE 36.1 CQF (Smith-Barnwell) Filter Bank
Coefficients with 40dB Attenuation

32-Tap filter	16-Tap filter
8.494372478233170D−03	2.193598203004352D−02
−9.617816873474045D−05	1.578616497663704D−03
−8.795047132402801D−03	−6.025449102875281D−02
7.087795490845020D−04	−1.189065962053910D−02
1.220420156035413D−02	0.137537915636625D+00
−1.762639314795336D−03	5.745450056390939D−02
−1.558455903573829D−02	−0.321670296165893D+00
4.082855675060479D−03	−0.528720271545339D+00
1.765222024089335D−02	−0.295779674500919D+00
−8.385219782884901D−03	2.043110845170894D−04
−1.674761388473688D−02	2.906699709446796D−02
1.823906210869841D−02	−3.533486088708146D−02
5.781735813341397D−03	−6.821045322743358D−03
−4.692674090907675D−02	2.606678468264118D−02
5.725005445073179D−02	1.033363491944126D−03
0.354522945953839D+00	−1.435930957477529D−02
0.504811839124518D+00	
0.264955363281817D+00	
−8.329095161140063D−02	
−0.139108747584926D+00	
3.314036080659188D−02	
9.035938422033127D−02	

	8-Tap filter
−1.468791729134721D−02	
−6.103335886707139D−02	
6.606122638753900D−03	3.489755821785150D−02
4.051555088035685D−02	−1.098301946252854D−02
−2.631418173168537D−03	−6.286453934951963D−02
−2.592580476149722D−02	0.223907720892568D+00
9.319532350192227D−04	0.556856993531445D+00
1.535638959916169D−02	0.357976304997285D+00
−1.196832693326184D−04	−2.390027056113145D−02
−1.057032258472372D−02	−7.594096379188282D−02

For the M-channel case shown in Fig. 36.1, where the bands are assumed to be maximally decimated, the same AC-matrix approach can be employed, leading to the equations

$$\hat{X}(z) = \frac{1}{M} \underbrace{\left[X(z), \ldots, X(zW_M^{M-1}) \right]}_{\mathbf{x}^T}$$

$$\underbrace{\begin{bmatrix} H_0(z) & \cdots & H_{M-1}(z) \\ H_0(zW_M^1) & \cdots & H_{M-1}(zW_M^1) \\ \vdots & & \vdots \\ H_0(zW_M^{M-1}) & \cdots & H_{M-1}(zW_M^{M-1}) \end{bmatrix}}_{\mathbf{H}} \underbrace{\begin{bmatrix} G_0(z) \\ G_1(z) \\ \vdots \\ G_{M-1}(z) \end{bmatrix}}_{\mathbf{g}} ,$$

where $W_M = e^{-j\frac{2\pi}{M}}$. This can be rewritten compactly as

$$\hat{X}(z) \frac{1}{M} \mathbf{x}^T(z) \mathbf{H}(z) \mathbf{g}(z) ,$$

where $\mathbf{x}$ is the input vector, $\mathbf{g}$ is the synthesis filter vector, and $\mathbf{H}$ is the AC matrix. However, the AC-matrix determinant for systems with $M > 2$ is typically too intricate for the spectral factorization approach outlined above. An effective approach for handling the design of M-band systems was introduced by Vaidyanathan in [30]. It is based on a lattice implementation structure and is discussed next.

36.1.3 Lattice Implementations

In addition to the direct form structures shown in Figs. 36.1 and 36.2, filter banks can be implemented using lattice structures. For simplicity, consider the two-band case first. An example of a lattice structure

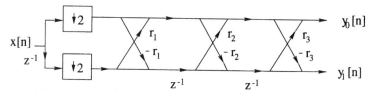

FIGURE 36.4 Flow graph of a two-band lattice structure with three stages.

for a two-band analysis system is shown in Fig. 36.4. It is composed of a cascade of criss-cross elements, each of which has a set of coefficients associated with it. Conveniently, each section, which we denote $\mathbf{R}_m$, can be described by a matrix. For the two-band lattice, these matrices have the form

$$\mathbf{R}_m = \begin{bmatrix} 1 & r_m \\ -r_m & 1 \end{bmatrix}.$$

Interspersed between the coefficient matrices are delay matrices, $\Lambda(z)$, having the form

$$\Lambda(z) = \begin{bmatrix} 1 & 0 \\ 0 & z^{-1} \end{bmatrix}.$$

It can be shown [27] that lattice filters can represent a wide class of exact reconstruction filter banks. Two points regarding lattice filter banks are particularly noteworthy. First, the lattice structure provides an efficient form of implementation. Moreover, the synthesis filter bank is directly related to the analysis bank, since each matrix in the analysis cascade is invertible. Consequently, the synthesis bank consists of the cascade of inverse section matrices. Second, the structure also provides a convenient way to design the filter bank. Each lattice coefficient can be optimized using standard minimization routines to minimize a passband-stopband error cost function for the filters. This approach to design can be used for two-band as well as M-band filter banks [5, 27, 28].

36.1.4 Time-Domain Design

One of the most flexible design approaches is the time domain formulation proposed by Nayebi et al. [3, 8]. This formulation has enabled the discovery of previously unknown classes of filter banks, such as low and variable delay systems [12], time-varying filter banks [4], and block decimation systems [9]. It is attractive because it enables the design of virtually all linear filter banks. The idea underlying this approach is that the conditions for exact reconstruction can be expressed in the time domain in a convenient matrix form. Let us explore this approach in the context of an M-band filter bank. Because of the decimation operations, the overall M-band analysis-synthesis system is periodically time-varying. Thus, we can view an arbitrary maximally decimated M-band system as having M linear time invariant transfer functions associated with it. One can think of the problem as trying to devise M subsampled systems, each one of which exactly reconstructs. This is equivalent to saying that for each impulse input, $\delta[n - i]$, to the analysis-synthesis system, that impulse should appear at the system output at time $n = i + n_0$, where $i = 0, 1, 2, \ldots, M - 1$ and n_0 is the system delay.

This amounts to setting up an overconstrained linear system $\mathbf{AS} = \mathbf{B}$, where the matrix $\mathbf{A}$ is created using the analysis filter coefficients, the matrix $\mathbf{B}$ is the desired response of zeros except at the appropriate delay points (i.e., $\delta[n - n_0]$) and $\mathbf{S}$ is a matrix containing synthesis filter coefficients. Particular linear combinations of analysis and synthesis filter coefficients occur at different points in time for different input impulses. The idea is to make $\mathbf{A}$, $\mathbf{S}$, and $\mathbf{B}$ such that they describe completely all M transfer functions that comprise the periodically time-varying system.

The matrix $\mathbf{A}$ is a matrix of filter coefficients and zeros that effectively describe the decimated convolution operations inherent in the filter bank. For convenience, we express the analysis coefficients as a matrix $\mathbf{h}$,

where

$$
\mathbf{h} = \begin{bmatrix}
h_0[0] & h_1[0] & \cdots & h_{M-1}[0] \\
h_0[1] & h_1[1] & \cdots & h_{M-1}[1] \\
\vdots & \vdots & & \vdots \\
h_0[N-1] & h_1[N-1] & \cdots & h_{M-1}[N-1]
\end{bmatrix}.
$$

The zeros are represented by an $M \times M$ matrix of zeros, denoted $\mathbf{O_M}$. With these terms, we can write the $(2N - M) \times N$ matrix $\mathbf{A}$,

$$
\mathbf{A} = \begin{bmatrix}
\begin{bmatrix} \mathbf{h}[n] \\ \mathbf{O_M} \\ \vdots \\ \mathbf{O_M} \end{bmatrix} &
\begin{bmatrix} \mathbf{O_M} \\ \mathbf{h}[n] \\ \vdots \\ \mathbf{O_M} \end{bmatrix} &
\cdots &
\begin{bmatrix} \mathbf{O_M} \\ \vdots \\ \mathbf{O_M} \\ \mathbf{h}[n] \end{bmatrix}
\end{bmatrix}.
$$

The synthesis filters $\mathbf{S}$ can be expressed most conveniently in terms of the $M \times M$ matrix

$$
\mathbf{Q}_i = \begin{bmatrix}
g_0[i] & g_0[i+1] & \cdots & g_0[i+M-1] \\
g_1[i] & g_1[i+1] & \cdots & g_1[i+M-1] \\
\vdots & \vdots & & \vdots \\
g_{M-1}[i] & g_{M-1}[i+1] & \cdots & g_{M-1}[i+M-1]
\end{bmatrix},
$$

where $i = 0, 1, \ldots, L-1$ and N is assumed to be equal to LM. The synthesis matrix $\mathbf{S}$ is then given by

$$
\mathbf{S} = \begin{bmatrix}
\mathbf{Q}_0 \\
\mathbf{Q}_M \\
\vdots \\
\mathbf{Q}_{iM} \\
\vdots \\
\mathbf{Q}_{(L-1)M}
\end{bmatrix}.
$$

Finally, to achieve exact reconstruction we want the impulse responses associated with each of the M constituent transfer functions in the periodically time-varying system to be an impulse. Therefore, $\mathbf{B}$ is a matrix of zero-element column vectors, each with a single "one" at the location of the particular transfer function group delay. More specifically, the matrix has the form

$$
\mathbf{B} = \begin{bmatrix}
\mathbf{O_M} \\
\mathbf{O_M} \\
\vdots \\
\mathbf{J_M} \\
\vdots \\
\mathbf{O_M} \\
\mathbf{O_M}
\end{bmatrix}
$$

where $\mathbf{J_M}$ is the $M \times M$ antidiagonal identity matrix

$$
\mathbf{J_M} = \begin{bmatrix}
0 & \cdots & 0 & 1 \\
0 & \cdots & 1 & 0 \\
\vdots & & \vdots & \vdots \\
1 & \cdots & 0 & 0
\end{bmatrix}.
$$

It is important to mention here that the location of $\mathbf{J_M}$ within the matrix $\mathbf{B}$ is a system design issue. The case shown here, where it is centered within $\mathbf{B}$, corresponds to an overall system delay of $N - 1$. This is the natural case for systems with N-tap filters. There are many fine points associated with these time domain conditions. For a complete discussion, the reader is referred to [3].

With the reconstruction equations in place, we now turn our attention to the design of the filters. The problem here is that this is an over-constrained system. The matrix $\mathbf{A}$ is of size $(2N - M) \times N$. If we think of the synthesis filter coefficients as the parameters to be solved for, we find $M(2N - M)$ equations and MN unknowns. Clearly, the best we can hope for is to determine $\mathbf{B}$ in an approximate sense. Using least-squares approximation, we let

$$\mathbf{S} = \left(\mathbf{A}^T \mathbf{A}\right)^{-1} \mathbf{B}.$$

Here, it is assumed that $\left(\mathbf{A}^T \mathbf{A}\right)^{-1}$ exists. This is not automatically the case. However, if reasonable lowpass and highpass filters are used as an initial starting point, there is rarely a problem.

This solution gives the best synthesis filter set for a particular analysis set and system delay $N - 1$. The resulting matrix $\mathbf{AS} = \hat{\mathbf{B}}$ will be close to $\mathbf{B}$ but not equal to it in general. The next step in the design is to allow the analysis filter coefficients to vary in an optimization routine to reduce the Frobenius matrix norm, $\left\|\hat{\mathbf{B}} - \mathbf{B}\right\|_F^2$. The locally optimal solution will be,

$$\mathbf{S} = \left(\mathbf{A}^T \mathbf{A}\right)^{-1} \mathbf{B}, \quad \text{such that} \quad \left\|\hat{\mathbf{B}} - \mathbf{B}\right\|_F^2 \text{ is minimized}.$$

Any number of routines may be used to find this minimum. A simple gradient search that updates the analysis filter coefficients will suffice in most cases. Note that, as written, there are no constraints on the analysis filters other than that they provide an invertible $\mathbf{A}^T \mathbf{A}$ matrix. One can easily start imposing constraints relevant to system quality. Most often we find it appropriate to include constraints on the frequency domain characteristics of the individual analysis filters. This can be done conveniently by creating a cost function comprised of the passband and stopband filter errors. For example, in the two-band case, inclusion of such filter frequency constraints gives rise to the overall error function

$$\epsilon = \left\|\hat{\mathbf{B}} - \mathbf{B}\right\|_F^2 + \int_0^{\pi_p} \left|1 - H_1(e^{j\omega})\right|^2 d\omega + \int_{\pi_s}^{\pi} \left|H_0(e^{j\omega})\right|^2 d\omega.$$

This reduces the overall system error of the filter bank while at the same time reducing the stopband errors in analysis filters. Other options in constructing the error function can address control over the step response of the filters, the width of the transition bands, and whether an l_2 norm or an l_∞ norm is used as an optimality criterion.

By properly weighting the reconstruction and frequency response terms in the error function, exact reconstruction can be obtained, if such a solution exists. If an exact reconstruction solution does not exist, the design algorithm will find the locally optimal solution subject to the specified constraints.

Functionality of the Design Formulation

One of the distinct advantages of the time-domain design method is its flexibility. The discussion above assumed that the system delay was $N - 1$ where N is the filter length. For the time-domain formulation, the amount of overall system delay can be thought of as an input to the design algorithm. In other words, one can pre-specify the desired system delay and then find the locally optimal set of analysis and synthesis filters that reduce the cost function while maintaining the specified delay. Control over the system delay is given by the position of $\mathbf{J_M}$ in the matrix $\mathbf{B}$. Placing $\mathbf{J_M}$ at or near the top of $\mathbf{B}$ lowers the system delay while positioning it at or near the bottom increases the system delay. One consideration here is the effect on filter bank quality. Experiments have shown that as the delay moves toward the extremes, the impact of the overconstrained equations is more severe. One is forced to either tolerate poorer frequency response characteristics or perhaps allow a little distortion in the reconstruction.

The cost function allows for an infinite variety of systems to be designed. The algorithm will converge to a filter set that optimizes the cost function as it is given. This provides the freedom to tradeoff among reconstruction error, frequency domain characteristics, and time domain characteristics. To aid in finding a particular locally optimal solution, the cost function can be allowed to be "adaptive". If exact reconstruction is desired, a heavy weighting may be placed on the reconstruction term in the cost function initially, until that term goes to zero. Then the cost function can be adjusted with new weightings that address reducing the error associated with the remaining distortion components.

This time domain formulation has been used to design an unprecedented variety of filter banks, including the first block decimation systems, the first time-varying systems, the first low delay systems, cosine modulated filter banks, nonuniform band filter banks, and many others [3, 4, 9, 10, 11]. One of the most important in this list is cosine modulated filter banks because they can be implemented very efficiently by using FFT-class algorithms. Cosine modulated filter banks may be designed in a variety of ways. Excellent discussions on this topic are given by Malvar [20, 24], Vaidyanathan [21, 27], Vetterli [23], and many others.

Linear filter banks have proven to be effective in many applications. Perhaps their most widespread use is in the area of coding. Subband coders for speech audio, image, and video signals tend to work very well. However, at low bit rates, distortions can be detected. Thus, there is interest in designing filter banks that are less prone to producing annoying distortions in these cases. Other nonlinear classes of filter banks can be considered that display different forms of distortion at low bit rates. In the remainder of this chapter, we discuss the design of two nonlinear filter banks that are presently being studied.

36.2   Finite Field Filter Banks

A new and interesting variant of the classical analysis-synthesis system can be achieved by imposing the explicit constraint that the discrete amplitude range of the subbands is confined. For conventional filter banks, we assume the input signal has a finite number of amplitude values. For instance, in the case of subband image coding, the input will typically contain 8 bits or 256 amplitude levels. However, the subband outputs may contain millions of possible amplitude values. For a coding application, we can think of the input as having a small alphabet (e.g., 256 unique values), and the analysis filter output as having a large alphabet (millions). Conceivably, one might be able to improve coding performance in some situations by designing a filter bank that constrains the output alphabet to be small. With this as motivation, we consider the problem of designing exact reconstruction filter banks with this constraint, an idea originally introduced by Vaidyanathan [37].

To begin our discussion, consider an input image with an alphabet size N (e.g., 256 gray levels). The output is expanded to an alphabet size of $M \times N$ after subband filtering. The value of M is governed by the length and coefficient values of the filter. M can be very large. The design task of interest here is to construct a filter bank where M is very small, ideally unity. In other words, we are constraining the system to operate in a finite field of our choosing, e.g., GF(N). In order to meet this finite field condition, an operational change is needed. Specifically, the finite field filter bank should operate in an integer field. Consequently, the filters used should be perfect reconstruction filters with integer coefficients. This modification makes it possible to perform wrap-around arithmetic. Wrap-around arithmetic restricts outputs to a finite field by performing all operation modulo N.

The design of a finite field filter bank is relatively simple. The image is passed through analysis filters using wrap-around arithmetic. This means that every operation is either modulo-N addition or modulo-N multiplication. Hence, the subband outputs will have an integer alphabet of size N. To reconstruct, the image is passed through the synthesis filters using the same wrap-around arithmetic within the same finite integer field. The bands are then combined using modulo-N addition. As it turns out, the resulting signal will not match the original. However, the signal can be corrected by applying a mapping based on the gain of the filter banks, M, and the dynamic range, N. Let us assume that the input is an image with

N' discrete levels, and that all operations have been performed modulo N. Each value of the output image is found in set **B** and can be mapped into set **A**, where

$$\mathbf{A} = \{0, 1, 2, ..., N' - 1\} \qquad \mathbf{B} = ((M \times \mathbf{A}))_N .$$

The resulting output image $\widehat{x}$ will be, under certain conditions, an exact reconstruction of the input image x.

There are two conditions that must be satisfied in order to obtain exact reconstruction. First, the subband output alphabet size N must be equal to or greater than the input alphabet size N'. This is a necessary condition in order to unambiguously resolve all values of the input. Second, the system gain M is constrained in relation to the subband output size N. The system gain is governed by the analysis and synthesis filters in the following way:

$$M = \left(\sum_n |h_0[n]| \times \sum_n |g_0[n]| \right) + \left(\sum_n |h_1[n]| \times \sum_n |g_1[n]| \right)$$

where $h_0[n]$ and $h_1[n]$ are the analysis filters and $g_0[n]$ and $g_1[n]$ are the synthesis filters. The relation between M and N is crucial in obtaining perfect reconstruction. These two numbers must be relatively prime. That is, M and N can have no common factors. For example, if M is two, any odd value of N would be valid. Ideally, we might want $N = N'$. However, to satisfy the last condition M is determined by the system and N is adjusted slightly up from N'. It is typically easier to adjust N.

To illustrate the differences in outputs obtained from conventional and finite field filter banks, consider the following comparison. For a conventional two-band system with two-tap Haar analysis filters, an input of

$$x = 0, 0, 0, 4, 2, 3, 0, 1, 2, 0, 0, \ldots$$

will yield the outputs

$$y_0 = 0, 4, 5, 1, 2, \ldots$$
$$y_1 = 0, 4, 1, 1, -2, \ldots .$$

However, for the equivalent finite field system (like the one shown in Fig. 36.5), the outputs are noticeably different. For the finite field case, all operations are performed modulo N. Thus, for the same input the outputs produced are

$$y_0 = 0, 4, 0, 1, 2, \ldots$$
$$y_1 = 0, 4, 1, 1, 3 \ldots .$$

Notice that the alphabet here is confined to the integers 0, 1, 2, 3, 4 because we have set $N = 5$. For the reconstruction, the outputs shown in the figure will be

$$\widehat{x}_0 = 0, 4, 4, 0, 0, 1, 1, 2, 2, \ldots$$
$$\widehat{x}_1 = 0, 1, 4, 4, 1, 4, 1, 2, 3, \ldots .$$

Adding these together, modulo N gives

$$\widehat{x}_p = 0, 0, 3, 4, 1, 0, 2, 4, 0 \ldots .$$

Now unscrambling them in the post-mapping step shown in the figure gives

$$\widehat{x} = 0, 0, 4, 2, 3, 0, 1, 2, 0 \ldots = x .$$

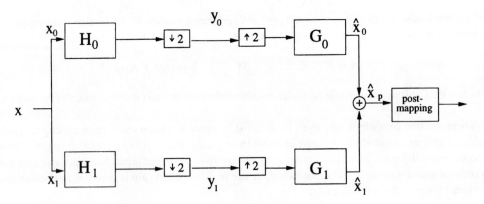

FIGURE 36.5   Block diagram of a two-band finite field filter bank.

It is interesting to compare the analysis section outputs of finite field and conventional filter banks for the two-band case. The lower band output of a conventional filter bank has a dynamic range that is usually much greater than the dynamic range of the input. The values in the lower band tend to have a Gaussian distribution over the range. By constraining the alphabet size, the first-order entropy can be reduced. The amount of the reduction depends on the size of M. The higher band in the conventional filter bank has a dynamic range that might be larger than N; however, the values are clustered around zero. When modulo operations are performed, the negative values go to a high value so not much overlap is obtained. Therefore, the alphabet constraint has little or no affect on the higher bands. The finite field filter bank reduces the overall first-order entropy because the entropy is reduced in the lower band. The degree by which the entropy is reduced is greatly dependent on the image and the filter gains.

How do finite field filter banks affect input images with different dynamic ranges? This effect is dependent on the same two components that have previously been discussed, the system gain M, and the subband output size N. Let us assume the subband output range N is set equal to the input image range N'. Now we can examine the effects of different system gains given N. For example, if the image is binary ($N = 2$), the system gain must be odd. Examining the decomposition of such an image, we can see that it appears very noisy. This is because the dynamic range of the system is small and the gain is large. The image is essentially wrapping around on itself so many times it is difficult to observe the original image in the bands. In a case where $N > 2$, a filter with a smaller gain is more realizable. For example, if $N = 255$, we can choose a system gain of 2. In this decomposition (Fig. 36.6), the lower band image is not what we are accustomed to observing in a conventional decomposition. This case does have a lower first-order entropy than its conventional counterpart.

Finite field filter banks are still in their early phases of study. As a result of the constraints, filter quality is limited. Thus, the net gains achievable in an application could be favorable or unfavorable. One must pay careful attention to the subband output size, filter length, and coefficient values during the design of the filter bank. Nonetheless, it seems that finite field filter banks are potentially attractive in some applications.

36.3 Nonlinear Filter Banks

One of the driving forces for research in filter banks is image coding for low bit rate applications. Presently, subband image coders represent the best approach known for image compression. As with any coder, at low rates distortions occur. Subband coders based on conventional linear filter banks suffer from ringing effects due to the Gibbs phenomenon. These ringing effects occur around edges or high contrast regions. One way to eliminate ringing is to use nonlinear filter banks. There are pros and cons regarding the utility of nonlinear filter banks. However, the design of the systems is rather new and interesting.

Nonlinear filter banks can be constructed within a general two-band framework. A nonlinear filter

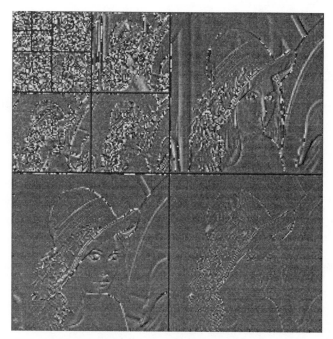

FIGURE 36.6   A four-level octave band decomposition using finite field filter banks.

may be placed in the highpass analysis and in the lowpass synthesis block of the systems. The condition for exact reconstruction will be discussed later. What type of nonlinear filter is an open question. While there are many candidates, the constraints of the overall system restrict the design of filters in terms of type and degrees of freedom in optimization. The most widely used nonlinear filter is the rank-order filter. In this discussion, we consider rank-order filters, more specifically, median filters. The performance of such filters is determined by the rank used and the region of support. The popular N-point median filter has a rank of $(N + 1)/2$, where N is assumed to be odd. Egger et al. [31] suggested a simple two-band nonlinear filter bank that upholds the exact reconstruction property. The lowpass channel consists of direct downsampling, while the highpass channel involves a median filter (differencing) operation to achieve a highpass representation for the other channel. Because straight downsampling and median filtering are involve, there is an inherent finite field constraining property built in to the system. Although these features seem attractive, the system is severely limited by its lack of filtering power. Most notably, the lowpass channel has massive aliasing since no filtering is performed. For many applications, aliasing of this type is not desirable. This problem can be addressed somewhat by using the modified filter bank introduced by Florencio and Schafer [32]. In the two-band system of Florencio and Schafer shown in Fig. 36.7, each channel can be expressed as a filtered combination of the input. This structure can be recognized as a classical polyphase implementation for a two-band filter bank. Here, however, we allow the polyphase filters f_{ij} and g_{ij} to be nonlinear filters. Thus,

$$
\begin{aligned}
y_0[n] &= f_{00}(x_0[n]) + f_{01}(x_1[n]) \\
y_1[n] &= f_{10}(x_0[n]) + f_{11}(x_1[n])
\end{aligned}
$$

where $f_{ij}(\cdot)$ are the linear or nonlinear polyphase analysis filters. To reconstruct the signal, the output can be expressed as a filtered combination of the channels,

$$
\widehat{x}_0 = g_{00}(y_0) + g_{01}(y_1) \qquad \widehat{x}_1 = g_{10}(y_0) + g_{11}(y_1)
$$

where $g_{ij}(\cdot)$ are the linear or nonlinear polyphase synthesis filters. The perfect reconstruction conditions are based on these different classes or structures. The Type I structure consists of $f_{00}(\cdot) = f_{11}(\cdot) = I$

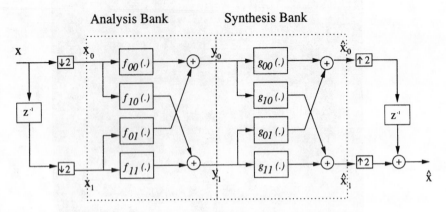

FIGURE 36.7   A two-band polyphase nonlinear filter bank.

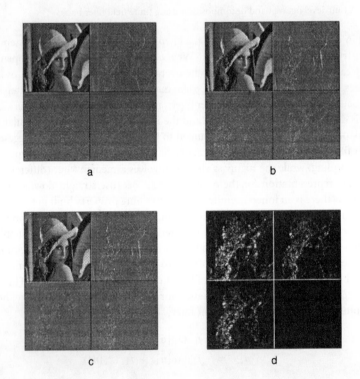

FIGURE 36.8   Comparison of outputs from one linear and three nonlinear filter banks. (a) A four-band linear decomposition using four-tap QMFs. (b) A four-band nonlinear decomposition using the method of Egger and Li. (c) A four-band nonlinear decomposition using the two-stage method of Florencio and Schafer. (d) The residual image obtained from subtracting the nonlinear decomposition result in (b) from the result in (c).

(identity), and either $f_{10}(\cdot) = 0$ or $f_{01}(\cdot) = 0$. The other is any causal transformation. To obtain perfect reconstruction, $g_{00}(\cdot) = g_{11}(\cdot) = I$, $g_{10}(\cdot) = f_{10}(\cdot)$, and $g_{01}(\cdot) = -f_{01}(\cdot)$. The Type II structure consists of $f_{10}(\cdot) = f_{01}(\cdot) = 0$ and both $f_{00}(\cdot)$ and $f_{11}(\cdot)$ being invertible functions. To obtain perfect reconstruction $g_{01}(\cdot) = g_{10}(\cdot) = 0$, $g_{00}(\cdot) = f_{00}^{-1}(\cdot)$, and $g_{11}(\cdot) = f_{11}^{-1}(\cdot)$. The Type III structure consists of $f_{10}(\cdot) = f_{01}(\cdot) = I$ and $f_{00}(\cdot) = f_{11}(\cdot) = 0$. To obtain perfect reconstruction, $g_{01}(\cdot) = g_{10}(\cdot) = I$ and $g_{00}(\cdot) = g_{11}(\cdot) = 0$.

Similar to linear filter banks, this nonlinear filter bank achieves an overall reduction in first-order entropy. Since perfect reconstruction is achieved in the two-band decomposition, perfect reconstruction can be maintained when used in tree structured systems for compression applications. After quantization, coding, and reconstruction, different features will be affected in different ways. The main advantage of nonlinear filtering is that the edges associated with high contrast features are preserved well, and no "ringing" occurs. However, because of the nature of the sampling in the lower band, texture regions are distorted. Using cascaded sections is a way to help preserve the texture. As it turns out, sections can be cascaded in a way that preserves exact reconstruction. For example, let the first stage of the filter contain $f_{01}(\cdot) = 0$, with $f_{10}(\cdot)$ being a four-point median filter. Let the second stage be defined by $f_{10}(\cdot) = 0$ with $f_{01}(\cdot)$ being a four-point median filter with a 0.5 gain (to maintain the dynamic range of the input). The resulting two bands are similar to the bands of the comparable linear case but have the advantages of a nonlinear system. Most notably, the lower band of the nonlinear case has a reduction in higher frequencies, very similar to the linear case. These differences are illustrated in Fig. 36.8 for a four-band decomposition of an image. A conventional QMF decomposition is shown in (*a*). Next to it in (*b*) and (*c*) are the nonlinear decompositions obtained using the Egger and Li approach, and the two-stage approach of Florencio and Schafer, respectively. All show similarities. However, more energy is contained in the high frequency subbands of the nonlinear results. In comparing carefully the two nonlinear results, we can observe that the two-stage approach of Florencio and Schafer has less aliasing in the lowest band and more closely follows the linear result. The difference image between the two nonlinear results is given in (*d*).

It is clear that there are many possibilities for constructing nonlinear filter banks. What is less obvious at this point is the impact of these systems in practical situations. Given that development related to these filter banks is only in the formative stages, only time will tell. Regardless of whether conventional or nonlinear filter banks are ultimately employed, the variety of design options and design techniques offer many useful solutions to engineering problems. More in-depth discussions on applications can be found in the references.

References

[1] Smith, M. and Barnwell, T., The design of digital filters for exact reconstruction in subband coding, *Trans. on Acoustics, Speech, and Signal Proc.*, ASSP-34(3), 434–441, June 1986.

[2] Smith, M. and Barnwell, T., A new filter bank theory for time-frequency representation, *Trans. on Acoustics, Speech, and Signal Proc.*, ASSP-35(3), 314–327, March 1987.

[3] Nayebi, K., Barnwell, T. and Smith, M., Time domain filter bank analysis: A new design theory, *IEEE Trans. on Signal Processing*, 40(6), 1412–1429, June 1992.

[4] Nayebi, K., Barnwell, T. and Smith, M., Analysis-synthesis systems based on time varying filter banks, *Intl. Conf. on Acoustics, Speech, and Signal Processing*, IV, 617–620, March 1992.

[5] Schuller, G. and Smith, M.J.T., A new framework for modulated perfect reconstruction filter banks, *IEEE Trans. Signal Processing*, August 1996.

[6] Smith, M. and Barnwell, T., A unifying framework for maximally decimated analysis/synthesis systems, *Proc. Intl. Conf. on Acoustics, Speech, and Signal Proc.*, 521–524, March 1985.

[7] Smith, M. and Barnwell, T., A procedure for designing exact reconstruction filter banks for tree-structured subband coders, *Proc. Intl. Conf. on Acoustics, Speech, and Signal Proc.*, 27.1.1–27.1.4, March 1984.

[8] Nayebi, K., Barnwell, T. and Smith, M., Time domain conditions for exact reconstruction in analysis/synthesis systems based on maximally decimated filter banks, *19th Southeastern Symposium on System Theory*, 498–502, March 1987. Analysis/Synthesis

[9] Nayebi, K., Barnwell, T. and Smith, M., Block decimated analysis-synthesis filter banks, *IEEE Intl. Symposium on Circuits and Systems*, San Diego, 947–950, May 1992.

[10] Nayebi, K., Barnwell, T. and Smith, M., Design and implementation of computationally efficient modulated filter banks, *Proc. Intl. Symposium on Circuits and Systems*, Singapore, 650–653, June 12-14, 1991.

[11] Nayebi, K., Barnwell, T.P. and Smith, M.J.T., Design of perfect reconstruction nonuniform band filter banks, *Proc. Intl. Conf. on ASSP*, 1781–1784, May 1991.

[12] Nayebi, K., Barnwell, T.P. and Smith, M.J.T., Design of low delay FIR analysis-synthesis filter bank systems, *Proc. Conf. on Information Sciences and Systems*, March 1991.

[13] Nayebi, K., Barnwell, T. and Smith, M., Time-domain view of filter banks and wavelets, *Asilomar Conference on Signals, Systems and Computers*, Nov. 2-6, 1991.

[14] Mersereau, R.M. and Smith, M.J.T., *Digital Filtering: A Computer Laboratory Textbook*, John Wiley & Sons, New York, 1993.

[15] Akansu, A. and Smith, M., Eds., *Subband and Wavelet Transforms: Design and Applications*, Kluwer Academic Publishers, 1995.

[16] Smith, M. and Docef, A., *A Study Guide to Digital Image Processing*, Scientific Publishers, Riverdale, GA, 1997.

[17] Johnston, J., A filter family designed for use in quadrature mirror filter banks, *Proc. IEEE Intl. Conf. Acoustics, Speech, Signal Processing*, Denver, CO, April 1980.

[18] Croisier, A., Esteban, D. and Galand, C., Perfect channel splitting by use of interpolation/decimation/tree decomposition techniques, *Conf. on Information Sciences and Systems*, 1976.

[19] Crochiere, R.E. and Rabiner, L.R., *Multirate Digital Signal Processing*, Prentice-Hall, Englewood Cliffs, NJ, 1983.

[20] Malvar, H.S., *Signal Processing with Lapped Transforms*, Artech House, 1991.

[21] Koilpillai, R. and Vaidyanathan, P., New results on cosine modulated FIR filter banks satisfying perfect reconstruction, *Proc. IEEE Intl. Conf. Acoustics, Speech, Signal Processing*, 1991.

[22] Rothweiler, J., Polyphase quadrature mirror filters — a new sub-band coding technique, *Proc. IEEE Intl. Conf. Acoustics, Speech, Signal Processing*, 1983.

[23] Nussbaumer, H.J. and Vetterli, M., Computationally efficient QMF filter banks, *Proc. IEEE Intl. Conf. Acoustics, Speech, Signal Processing*, 1984.

[24] Malvar, H., Modulated QMF filter banks with perfect reconstruction, *Electronics Lett.*, 26(13), 906–907, June 1990.

[25] Mintzer, F., Filters for distortion-free two-band multirate filter banks, *IEEE Trans. on Acoustics, Speech, and Signal Processing*, ASSP-33, 626–630, June 1985.

[26] Akansu, Ali N. and Haddad, R.A., *Multiresolution Signal Decomposition*, Academic Press, 1992.

[27] Vaidyanathan, P.P., *Multirate Systems and Filterbanks*, Prentice-Hall, Englewood Cliffs, NJ, 1993.

[28] Vetterli, M. and Kovacevic, J., *Wavelets and Subband Coding*, Prentice-Hall, Englewood Cliffs, NJ, 1995.

[29] Fleige, N.J., *Multirate Digital Signal Processing*, John Wiley & Sons, New York, 1993.

[30] Vaidyanathan, P.P., Quadrature mirror filter banks, M-band extensions and perfect reconstruction techniques, *IEEE Trans. on Acoustics, Speech, and Signal Processing,* July 1987.

[31] Egger, O. and Li, W., Very low bit rate image coding using morphological operators and adaptive decompositions, *ICIP '94,* 2, 326–330, Nov. 1994.

[32] Florencio, D.A.F. and Schafer, R.W., Perfect reconstructing nonlinear filter banks, *ICASSP '96,* 1996.

[33] Florencio, D.A.F. and Schafer, R.W., A non-expansive pyramidal morphological image coder, *ICIP '94,* 2, 331–334, Nov. 1994.

[34] Sun, F.-K. and Maragos, P., Experiments on image compression using morphological pyramids, *VCIP '89,* 1303–1312, 1989.

[35] Toet, A., A morphological pyramidal image decomposition, *Patter Recog. Lett.,* 9, 255–261, May 1989.

[36] Bruekers, F.A.M.L. and van den Enden, A.W.M., New networks for perfect inversion and perfect reconstruction, *IEEE J. Selected Areas Comm.,* 10, 130–137, Jan. 1992.

[37] Vaidyanathan, P.P., Unitary and paraunitary systems in finite fields, *Proc. 1990 IEEE Intl. Symp. Circuits Syst.,* New Orleans, LA, 1189–1192, 1990.

[38] Tewfik, A.H., Hosur, S. and Sowelam, S., Recent progress in the application of wavelet in surveillance systems, *Optic. Eng.,* 33, 2509–2519, Aug. 1994.

[39] Swanson, M. and Tewfik, A.H., A binary wavelet decomposition of binary images, *IEEE Trans. Image Processing,* 5, 1637–1650, Dec. 1996.

[40] Flornes, K., Grossman, A., Hoschneider, M. and Torresani, B., Wavelets on finite fields, preprint, Nov. 1993.

37

Time-Varying Analysis-Synthesis Filter Banks

Iraj Sodagar
David Sarnoff Research Center

37.1 Introduction

Time-frequency representations (TFR) combine the time-domain and frequency-domain representations into a single framework to obtain the notion of time-frequency. TFR offer the time localization vs. frequency localization tradeoff between two extreme cases of time-domain and frequency-domain representations. The short-time Fourier transform (STFT) [1, 2, 3, 4, 5] and the Gabor transform [6] are the classical examples of linear time-frequency transforms which use time-shifted and frequency-shifted basis functions.

In conventional time-frequency transforms, the underlying basis functions are fixed in time and define a specific tiling of the time-frequency plane. The term *time-frequency tile* of a particular basis function is meant to designate the region in the plane that contains most of that function's energy. The short-time Fourier transform and the wavelet transform are just two of many possible tilings of the time-frequency plane. These two are illustrated in Fig. 37.1(a) and (b), respectively. In these figures, the rectangular representation for a tile is purely symbolic, since no function can have compact support in both time and frequency. Other arbitrary tilings of the time-frequency plane are possible such as the example shown in Fig. 37.1(c). In the discrete domain, linear time-frequency transforms can be implemented in the form of filter bank structures.

It is well known that the time-frequency energy distribution of signals often changes with time. Thus, in this sense, the conventional linear time-frequency transform paradigm is fundamentally mismatched to many signals of interest. A more flexible and accurate approach is obtained if the basis functions of the transform are allowed to adapt to the signal properties. An example of such a time-varying tiling is shown in Figure 37.1(d). In this scenario, the time-frequency tiling of the transform can be changed from good frequency localization to good time localization and vice versa. Time-varying filter banks provide such flexible and adaptive time-frequency tilings.

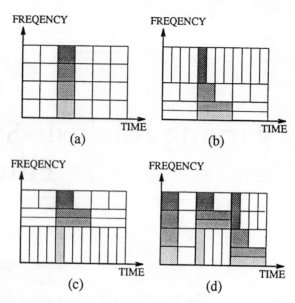

FIGURE 37.1   The time-frequency tiling for different time-frequency transforms: (a) The STFT, (b) the wavelet transform, (c) an example of general tiling, and (d) an example of the time-varying tiling.

The concept of time varying (or adaptive) filter banks was originally introduced in [7] by Nayebi et al. The ideas underlying their method were later developed and extended to a more general case in which it was also shown that the number of frequency bands could also be made adaptive [8, 9, 10, 11]. De Queiroz and Rao [12] reported time-varying extended lapped transforms and Herley et al. [13, 14, 15] introduced another time-domain approach for designing time-varying lossless filter banks. Arrowood and Smith [16] demonstrated a method for switching between filter banks using lattice structures. In [17], the authors presented yet another formulation for designing time-varying filter banks using a different factorization of the paraunitary transform. Chen and Vaidyanathan [18] reported a noncausal approach to time-varying filter banks by using time-reversed filters. Phoong and Vaidyanathan [19] studied time-varying paraunitary filter banks using polyphase approach. In [11, 20, 21, 22], the post filtering technique for designing time-varying filter bank was reported. The design of multidimensional time-varying filter bank was addressed in [23, 24]. In this article, we introduce the notion of the time-varying filter banks and briefly discuss some design methods.

37.2   Analysis of Time-Varying Filter Banks

Time-varying filter banks are analysis-synthesis systems in which the analysis filters, the synthesis filters, the number of bands, the decimation rates, and the frequency coverage of the bands are changed (in part or in total) in time, as is shown in Fig. 37.2. By carefully adapting the analysis section to the temporal properties of the input signal, better performance can be achieved in processing the signal. In the absence of processing errors, the reconstructed output $\hat{x}(n)$ should closely approximate a delayed version of the original signal $x(n)$. When $\hat{x}(n - \Delta) = x(n)$ for some integer constant, Δ, then we say that the filter bank is perfectly reconstructing (PR). The intent of the design is to choose the time-varying analysis and synthesis filters along with the time-varying down/up samplers so that the system requirements are met subject to the constraint that the analysis-synthesis filter bank be PR at all times.

One general method for analysis of time-varying filter banks is the time-domain formulation reported in [10, 22]. In this method, the time-varying impulse response of the entire filter bank is derived in terms of the analysis and synthesis filter coefficients.

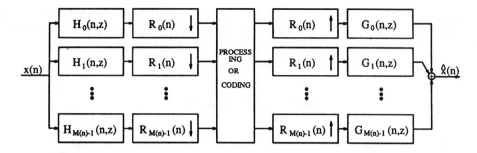

FIGURE 37.2 The time-varying filter bank structure with time-varying filters and time-dependent down/up samplers.

Figure (37.3) shows the diagram of a time-varying filter bank. In this figure, the filter bank is divided into three stages: the analysis filters, the down/up samplers, and the synthesis filters. The signals $x(n)$ and $\hat{x}(n)$ are the filter bank input and output at time n, respectively. The outputs of the analysis filters are shown by $\mathbf{v}(n) = [v_0(n), v_1(n), \ldots, v_{M(n)-1}(n)]^T$, where $v_i(n)$ is the output of the ith analysis filter at time n. The outputs of the down/up samplers at time n is called $\mathbf{w}(n) = [w_0(n), w_1(n), \ldots, w_{M(n)-1}(n)]^T$.

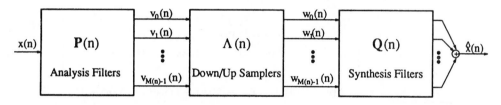

FIGURE 37.3 Time-varying filter bank as a cascade of analysis filters, down/up samplers, and synthesis filters.

The input/output relation of the analysis filters can be expressed by

$$\mathbf{v}(n) = \mathbf{P}(n)\mathbf{x}_N(n) . \tag{37.1}$$

$\mathbf{P}(n)$ is an $M(n) \times N(n)$ matrix whose mth row is comprised of the coefficients of the mth analysis filter at time n and $\mathbf{x}_N(n)$ is the input vector of length $N(n)$ at time n:

$$\mathbf{x}_N(n) = [x(n), x(n-1), x(n-2), \ldots, x(n-N(n)+1)]^T . \tag{37.2}$$

The input/output function of down/up samplers can be expressed in the form

$$\mathbf{w}(n) = \Lambda(n)\mathbf{v}(n) \tag{37.3}$$

where $\Lambda(n)$ is a diagonal matrix of size $M(n) \times M(n)$. The mth diagonal element of $\Lambda(n)$, at time n, is 1 if the input and output of the mth down/up sampler are identical, otherwise it is zero.

To write the input/output relationship of the synthesis filters, $\mathbf{Q}(n)$ is defined as

$$\mathbf{Q}(n) = \begin{bmatrix} g_0(n,0) & g_0(n,1) & g_0(n,2) & \cdots & g_0(n,N(n)-1) \\ g_1(n,0) & g_1(n,1) & g_1(n,2) & \cdots & g_1(n,N(n)-1) \\ g_2(n,0) & g_2(n,1) & g_2(n,2) & \cdots & g_2(n,N(n)-1) \\ \vdots & \vdots & \vdots & \vdots & \vdots \\ g_{M(n)-1}(n,0) & g_{M(n)-1}(n,1) & g_{M(n)-1}(n,2) & \cdots & g_{M(n)-1}(n,N(n)-1) \end{bmatrix}$$

$$= \begin{bmatrix} \mathbf{q}_0(n) & \mathbf{q}_1(n) & \mathbf{q}_2(n) & \cdots & \mathbf{q}_{N(n)-1}(n) \end{bmatrix} \tag{37.4}$$

where $\mathbf{q}_i(n) = [g_0(n, i), g_1(n, i), g_2(n, i), \ldots, g_{M(n)-1}(n, i)]^T$, is a vector of length $M(n)$ and $g_i(n, j)$ denotes the jth coefficient of the ith synthesis filter. At time n, the mth synthesis filter is convolved with vector $[w_m(n), w_m(n-1), \ldots, w_m(n-N(n)+1)]^T$ and all outputs are added together. Using Eq. (37.4), the output of the filter bank at time n can be written as:

$$\hat{x}(n) = \sum_{i=0}^{N(n)-1} \mathbf{q}_i^T(n)\, \mathbf{w}(n-i) \, . \tag{37.5}$$

If $\mathbf{s}(n)$ and $\hat{\mathbf{w}}(n)$ are defined as

$$\mathbf{s}(n) = \left[\mathbf{q}_0^T(n), \mathbf{q}_1^T(n), \mathbf{q}_2^T, \ldots, \mathbf{q}_{N(n)-1}^T(n) \right]^T \tag{37.6}$$

$$\hat{\mathbf{w}}(n) = \left[\mathbf{w}^T(n), \mathbf{w}^T(n-1), \mathbf{w}^T(n-2), \ldots, \mathbf{w}^T(n-N(n)+1) \right]^T , \tag{37.7}$$

then Eq. (37.5) can be written in the form of one inner product,

$$\hat{x}(n) = \mathbf{s}^T(n)\hat{\mathbf{w}}(n) \tag{37.8}$$

where $\mathbf{s}(n)$ and $\hat{\mathbf{w}}(n)$ are vectors of length $N(n)M(n)$. Using Eqs. (37.1), (37.3), (37.7), and (37.8), the input/output function of the filter bank can be written as:

$$\hat{x}(n) = \mathbf{s}^T(n) \begin{bmatrix} \Lambda(n)\, \mathbf{P}(n)\, \mathbf{x}_N(n) \\ \Lambda(n-1)\, \mathbf{P}(n-1)\, \mathbf{x}_N(n-1) \\ \Lambda(n-2)\, \mathbf{P}(n-2)\, \mathbf{x}_N(n-2) \\ \vdots \\ \Lambda(n-N(n)+1)\, \mathbf{P}(n-N(n)+1)\, \mathbf{x}_N(n-N(n)+1) \end{bmatrix} . \tag{37.9}$$

As the last $N(n) - 1$ elements of vector $\mathbf{x}_N(n-i)$ are identical to the first $N(n) - 1$ elements of vector $\mathbf{x}_N(n-i-1)$, the latter equation can be expressed by

$$\hat{x}(n) = \mathbf{s}^T(n) \begin{bmatrix} \big[& \Lambda(n)\, \mathbf{P}(n) & \big]\, \mathbf{O} \ldots \ldots \ldots \ldots \mathbf{O} \\ \mathbf{O}\, \big[& \Lambda(n-1)\, \mathbf{P}(n-1) & \big]\, \mathbf{O} \ldots \ldots \ldots \mathbf{O} \\ \mathbf{O}\, \mathbf{O}\, \big[& \Lambda(n-2)\, \mathbf{P}(n-2) & \big]\, \mathbf{O} \ldots \ldots \ldots \mathbf{O} \\ & \ddots & \\ \mathbf{O} \ldots \ldots \ldots \ldots \ldots \mathbf{O}\, \big[\; \Lambda(n-N(n)+1)\, \mathbf{P}(n-N(n)+1)\; \big] \end{bmatrix}$$
$$\begin{bmatrix} x(n) \\ x(n-1) \\ x(n-2) \\ \vdots \\ x(n-2N(n)+1) \end{bmatrix} \tag{37.10}$$

where $\mathbf{O}$ is the zero column vector with length $M(n)$. Thus, the input/output function of a time-varying filter bank can be expressed in the form of

$$\hat{x}(n) = \mathbf{z}^T(n)\mathbf{x}_I(n) \tag{37.11}$$

where $\mathbf{x}_I(n) = [x(n), x(n-1), \ldots, x(n-I+1)]^T$ and $I(n) = 2N(n) - 1$ and $\mathbf{z}(n)$ is the time-varying impulse response vector of the filter bank at time n:

$$\mathbf{z}(n) = \mathbf{A}(n)\, \mathbf{s}(n). \tag{37.12}$$

The matrix $\mathbf{A}(n)$ is the $[2N(n) - 1] \times [N(n) M(n)]$ matrix

$$
\mathbf{A}(n) =
\begin{bmatrix}
\begin{bmatrix} \mathbf{P}(n)^T \Lambda(n) \\ \mathbf{0}^T \\ \mathbf{0}^T \\ \vdots \\ \mathbf{0}^T \end{bmatrix} &
\begin{bmatrix} \mathbf{0}^T \\ \mathbf{P}(n-1)^T \Lambda(n-1) \\ \mathbf{0}^T \\ \vdots \\ \mathbf{0}^T \end{bmatrix} &
\ddots &
\begin{bmatrix} \mathbf{0}^T \\ \mathbf{0}^T \\ \vdots \\ \mathbf{0}^T \\ \big[\mathbf{P}(n-N(n)+1)^T \Lambda(n-N(n)+1) \big] \end{bmatrix}
\end{bmatrix}.
$$

$$(37.13)$$

For a perfect reconstruction filter bank with a delay of Δ, it is necessary and sufficient that all elements but the $(\Delta + 1)$th in $\mathbf{z}(n)$ be equal to zero at all times. The $(\Delta + 1)$th entry of $\mathbf{z}(n)$ must be equal to one. If the ideal impulse response is $\mathbf{b}(n)$, the filter bank is PR if and only if

$$
\mathbf{A}(n)\,\mathbf{s}(n) = \mathbf{b}(n) \qquad \text{for all } n. \tag{37.14}
$$

37.3 Direct Switching of Filter Banks

Changing from one arbitrary filter bank to another independently designed filter bank without using any intermediate filters is called *direct switching*. Direct switching is the simplest switching scheme and does not require additional steps in switching between two filter banks. But such switching will result in a substantial amount of reconstruction distortion during the transition period. This is because during the transition, none of the synthesis filters satisfies the exact reconstruction conditions. Figure (37.4) shows an example of a direct switching filter bank. Figure (37.5) shows the time-varying impulse response of the above system around the transition periods. In this figure, $z(n, m)$ is the response of the system at time n to the unit input at time m. For a PR system, $z(n, m)$ has a height of 1 along the diagonal and 0 everywhere else in the (m, n)-plane. As is shown, the time-varying filter bank is PR before and after but not during the transition periods. In this case, each switching operation generates a distortion with an 8-sample duration. One way to reduce the distortion is to switch the synthesis filters with an appropriate delay with respect to the analysis switching time. This delay may reduce the output distortion, but it can not eliminate it.

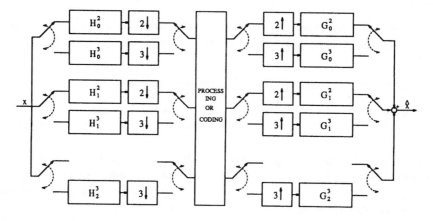

FIGURE 37.4 Block diagram of a time-varying analysis/synthesis filter bank that switches between a two- and three-band decomposition.

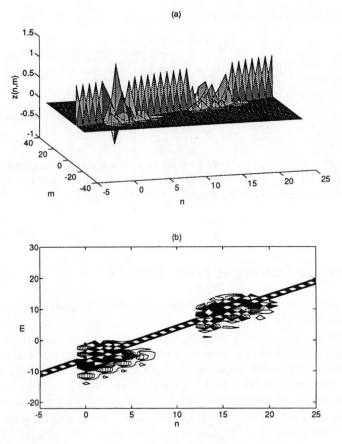

FIGURE 37.5 The time-varying impulse response for direct switching between the two- and the three-band system. The filter bank is switched from the two-band to the three-band at time $n = 0$ and switched back at time $n = 13$. (a) Surface plot, (b) contour plot.

37.4 Time-Varying Filter Bank Design Techniques

The basic time-varying filter bank design methods are summarized in Table 37.1. These techniques can be divided into two major approaches which are briefly described in the following sections.

37.4.1 Approach I: Intermediate Analysis-Synthesis (IAS)

In the first approach, both analysis and synthesis filters are allowed to change during the transition period to maintain perfect reconstruction. We refer to this approach as the intermediate analysis-synthesis (IAS) approach.

In [16], the authors have chosen to start with the lattice implementation of time-invariant two-band filter banks, originally proposed by Vaidyanathan [25] for time-invariant case. Consider the lattice structure shown in Fig. 37.6. Figure 37.6(a) represents a lossless two-band analysis filter bank, consisting of $J + 1$ lattice stages. The corresponding synthesis filter bank is shown in Fig. 37.6(b). As is shown, for each stage in the analysis filter bank, there exists a corresponding stage in the synthesis filter bank with similar, but inverse functionality. As long as each two corresponding lattice stages in the analysis and synthesis sections are PR, the overall system is PR. To switch one filter bank to another, the lattice stages of the analysis section are changed from one set to another. If the corresponding lattice stages of the synthesis

TABLE 37.1   Comparison of Time-Varying Filter Bank Different Designing Methods

		Intermediate analysis	Changing freq. resolution	Filter bank requirement	Computational complexity
Intermediate analysis synthesis (IAS)	Arrowood Smith	Yes	Indirect	Lattice structures	Low
	de Queiroz Rao	Yes	Indirect	ELT	Low
	Gopinath Burrus	Yes	Indirect	Paraunitary	Low
	Herley et. al	Yes	Direct	Paraunitary	Low
	Chen Vaidyanathan	Yes	Direct	Noncausal synthesis	Low
Instantaneous transform switching (ITS)	Least square synthesis	No	Direct	General (not PR)	Low
	Redesigning analysis	No	Direct	General	High
	Post filtering	No	Direct	General	Low

section are also changed according to the changes of the analysis section, the PR property will hold during transition. Due to the existence of delay elements, any change in the analysis section must be followed with the corresponding change in the synthesis section, but with an appropriate delay. For example, the parameter α_j of the analysis and synthesis filter banks can be changed instantaneously. But any change in parameter α_{j-1} in the analysis filter bank must be followed with the similar change in the synthesis filter bank after one sample delay. Because of such delays, switching between two PR filter banks can occur only by going through a transition period in which both analysis and synthesis filter banks are changing in time.

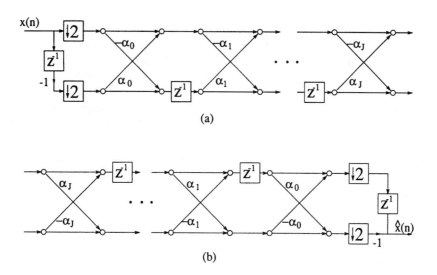

FIGURE 37.6   The block diagram of a two-band paraunitary filter bank in lattice form: (a) analysis lattice, (b) synthesis lattice.

In [12, 26], the design of time-varying extended lapped transform (ELT) [27, 28] was reported. The extended lapped transform is a cosine-modulated filter bank with an additional constraint on the filter lengths. Here, the design procedure is based on factorization of the time-domain transform matrix into permutation and rotation matrices. As the ELT is paraunitary, the inverse transform can be obtained by reversing the order of the matrix multiplication. Since any orthogonal transform is a succession of plane rotations, any changes in these rotation angles result in changing the filter bank without losing the orthogonality property. The authors derived a general frame work for M-band ELT transforms compared to the two-band case approach in [16]. This method parallels the lattice technique [16] except with the mild modification of imposing the additional ELT constraints. In [17], the authors presented yet another formulation for designing time-varying filter banks. In this paper, a different factorization of the paraunitary transform has been shown which is not based on plane rotations unlike the ones in [12, 26]. Using this factorization, a paraunitary filter bank can be implemented in the form of some cascade structures. Again, to switch one filter bank to another, the corresponding structures in the analysis and synthesis filter bank are changed similarly but with an appropriate delay. If the orthogonality property in each cascade structure is maintained, the time-varying filter bank remains PR. This formulation is very similar to the ones in [12, 16, 26], but represent a more general form of factorization. In fact, all above procedures consider similar frameworks of structures that inherently guarantee the exact reconstruction.

Herley et al. [13, 14, 15, 29] introduced a time-domain method for designing time-varying paraunitary filter banks. In this approach, the time-invariant analysis transforms do not overlap. As a simple example, consider the case of switching between two paraunitary time-invariant filter banks. The analysis transform around the transition period can be written as

$$
\mathbf{T} = \begin{bmatrix} \begin{bmatrix} \mathbf{P}_1 \end{bmatrix} & & \\ & \begin{bmatrix} \mathbf{P}_T \end{bmatrix} & \\ & & \begin{bmatrix} \mathbf{P}_2 \end{bmatrix} \end{bmatrix} . \tag{37.15}
$$

The matrices $\mathbf{P}_1$ and $\mathbf{P}_2$ represent paraunitray transforms and therefore are unitary matrices. Their nonzero columns also do not overlap with each other. The matrix $\mathbf{P}_T$ represents the analysis filter bank during the transition period. In order to find this filter bank, the matrix $\mathbf{P}_T$ is initially replaced with a zero matrix. Then, the null space of the transform $\mathbf{T}$ is found. Any matrix that spans this subspace can be a candidate vector for $\mathbf{P}_T$. By choosing enough independent vectors of this null space and applying the Gram-Schimidt procedure to them, an orthogonal transform can be selected for $\mathbf{P}_T$. This method has also been applied to time-varying modulated lapped transforms [24] and two-dimensional time-varying paraunitary filter banks [30].

The basic property of all above procedures is the use of intermediate analysis transforms in the transition period. The characteristics of these analysis transforms are not easy to control and typically the intermediate filters are not well-behaved.

37.4.2 Approach II: Instantaneous Transform Switching (ITS)

In the second approach, the analysis filters are switched instantaneously and time-varying synthesis filters are used in the transition period. We refer to this approach as the instantaneous transform switching (ITS) approach. In the ITS approach, the analysis filter bank may be switched to another set of analysis filters arbitrarily. This means that the basis vectors and the tiling of the time-frequency plane can be changed instantaneously. To achieve PR at each time in the transition period, a new synthesis section is designed to ensure proper reconstruction.

In the least squares (LS) method [10], for any given set of analysis filters, a LS solution of Eq. (37.14) can be used to obtain the "best" synthesis filters of the corresponding system (in $L2$ norm):

$$\mathbf{s}(n)^{LS} = \left(\mathbf{A}(n)^T \mathbf{A}(n)\right)^{-1} \mathbf{A}(n)^T \mathbf{b}(n) \tag{37.16}$$

The advantage of the LS approach is that there is no limitation on the number of analysis filter banks that can be used in the system. The disadvantage of the LS method is that it does not achieve PR. However, experiments have shown that the reconstruction is significantly improved in this method compared to direct switching [10].

In the LS solution, $\mathbf{b}(n)$ is projected onto the column space of $\mathbf{A}(n)$. For PR, the projection error should be zero. Thus, to obtain time-varying PR filter banks, the reconstruction error, $||\mathbf{A}(n)\mathbf{s}(n) - \mathbf{b}(n)||^2$, can be brought to zero with an optimization procedure. The optimization operates on the analysis filter coefficients and modifies the range space of $\mathbf{A}(n)$ until $\mathbf{b}(n) \in range(\mathbf{A}(n))$. Although the $\mathbf{s}(n)$'s at different states are independent of each other, since the $\mathbf{A}(n)$'s have some common elements, optimization procedures should be applied to all analysis sections at the same time. This method is referred to as "redesigning analysis" [10].

The last ITS method, post filtering, uses conventional filter banks with time-varying coefficients followed by a time-varying post filter. The post filter provides exact reconstruction during transition periods, while it operates as a constant delay elsewhere. Assume at time n_0 the time-varying filter bank is switched from the first filter bank to the second. If the length of the transition period is L samples, the output of the filter bank in the interval $[n_0, n_0 + L - 1]$ is distorted because of switching. The post filter removes this distortion. The block diagram of such a system is shown in Fig. (37.7). In this figure, $\mathbf{z}(n)$ and $\mathbf{y}(n)$ are

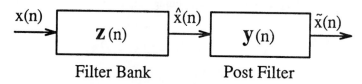

FIGURE 37.7   The block diagram of time-varying filter bank and post filter.

the analysis/synthesis filter bank and post filter impulse responses, respectively. If the delays of the filter bank and the post filter are denoted Δ and Θ, respectively, we can write

$$\hat{x}(n) = \begin{cases} \text{Distorted} & \text{if } n_0 \le n < n_0 + L \\ x(n - \Delta) & \text{otherwise}. \end{cases} \tag{37.17}$$

The desired output of the post filter is

$$\tilde{x}(n) = x(n - \Theta - \Delta) \tag{37.18}$$

The input/output relation of the time-varying filter bank during the transition period can be written as

$$\hat{x}(n) = \mathbf{z}^T(n)\, \mathbf{x}_I(n) \tag{37.19}$$

where $\mathbf{x}_I(n)$ is the input vector at time n:

$$\mathbf{x}_I(n) = [x(n), x(n-1), x(n-2), \dots, x(n-I+1)]^T .$$

$\mathbf{z}(n)$ is a vector of length I and represents the time-varying impulse response of the filter bank at time n. If the transition impulse response matrix is defined to be

$$\mathbf{Z} = \begin{bmatrix} \begin{bmatrix} \mathbf{z}(n_0 + L - 1) \\ O \\ O \\ \vdots \\ O \end{bmatrix} & \begin{bmatrix} O \\ \mathbf{z}(n_0 + L - 2) \\ O \\ \vdots \\ O \end{bmatrix} & \begin{matrix} O \\ O \\ \ddots \\ O \\ \begin{bmatrix} \mathbf{z}(n_0) \end{bmatrix} \end{matrix} \end{bmatrix}, \tag{37.20}$$

then the input/output relation of the filter bank in the transition period can be described as

$$\hat{\mathbf{x}}_L (n_0 + L - 1) = \mathbf{Z}^T \mathbf{x}_K (n_0 + L - 1) \tag{37.21}$$

where $\mathbf{Z}$ is a $K \times L$ matrix and $K = I + L - 1$. In Eq. (37.21), the $I - \Delta - 1$ samples before and Δ samples after the transition period are used to evaluate the output. The above intervals are called the tail and head of the transition period, respectively. Since the first and second filter banks are PR, the tail and head samples are exactly reconstructed. We write $\mathbf{x}_K (n_0 + L - 1)$ as the concatenation of three vectors:

$$\mathbf{x}_K (n_0 + L - 1) = \begin{bmatrix} \mathbf{x}_a \\ \mathbf{x}_t \\ \mathbf{x}_b \end{bmatrix}, \tag{37.22}$$

where $\mathbf{x}_a$ and $\mathbf{x}_b$ are the input signals in the head and tail regions while $\mathbf{x}_t$ represents the input samples which are distorted during the transition period. Using this notation, Eq. (37.21) can be written as

$$\hat{\mathbf{x}}_L (n_0 + L - 1) = \mathbf{Z}_a^T \mathbf{x}_a + \mathbf{Z}_t^T \mathbf{x}_t + \mathbf{Z}_b^T \mathbf{x}_b \tag{37.23}$$

where

$$\mathbf{Z} = \begin{bmatrix} \mathbf{Z}_a \\ \mathbf{Z}_t \\ \mathbf{Z}_b \end{bmatrix}. \tag{37.24}$$

By replacing vectors $\mathbf{x}_b$ and $\mathbf{x}_a$ with their corresponding output vectors $\hat{\mathbf{x}}_a$ and $\hat{\mathbf{x}}_b$, $\mathbf{x}_t$ of Eq. (37.23) can be written as

$$\begin{aligned} \mathbf{x}_t &= (\mathbf{Z}_t^T)^{-1} (\hat{\mathbf{x}}_t - \mathbf{Z}_a^T \hat{\mathbf{x}}_a - \mathbf{Z}_b^T \hat{\mathbf{x}}_b) \\ &= \mathbf{Y}^T \hat{\mathbf{x}}_K. \end{aligned} \tag{37.25}$$

Equation (37.25) describes the post filter input-output relationship during the transition region. In this equation, $\mathbf{Y}$ is the time-varying post filter impulse response which is defined as

$$\mathbf{Y} = \begin{bmatrix} -\mathbf{Z}_a \mathbf{Z}_t^{-1} \\ \mathbf{Z}_t^{-1} \\ -\mathbf{Z}_b \mathbf{Z}_t^{-1} \end{bmatrix}. \tag{37.26}$$

From Eq. (37.25), it is obvious that the condition for causal post filtering is

$$\Theta \geq L + \Delta - 1. \tag{37.27}$$

The post filter exists if $\mathbf{Z}_t$ has an inverse. It can be shown that the transition response matrix $\mathbf{Z}_t$, can be described by a matrix, product of the form

$$\mathbf{Z}_t = \Psi_L \, \mathbf{S} \tag{37.28}$$

where Ψ_L is the analysis transform applied to those input samples that are distorted during the transition period and $\mathbf{S}$ contains the synthesis filters during the transition period. In order for $\mathbf{Z}_t$ to be invertible, it is necessary (but not sufficient) that Ψ_L and $\mathbf{S}$ be full rank matrices. The analysis sections are defined by the required properties of the first and second filter banks and Ψ_L is fixed. Therefore, a filter bank is switchable to another filter bank if the corresponding Ψ_L is a full rank matrix. In this case, by proper design of the synthesis section, both $\mathbf{S}$ and $\mathbf{Z}_t$ will be full rank. Two methods to obtain proper synthesis filters are shown in [20, 22].

37.5 Conclusion

In this article, we briefly review some analysis and design methods of time-varying filter banks.

Time-varying filter banks can provide a more flexible and accurate approach in which the basis functions of the time-frequency transform are allowed to adapt to the signal properties.

A simple form of time-varying filter bank is achieved by changing the filters of an analysis-synthesis system among a number of choices. Even if all the analysis and synthesis filters are PR sets, exact reconstruction will not normally be achieved during the transition periods. To eliminate all distortion during a transition period, new time-varying analysis and/or synthesis sections are required for the transition periods.

Two different approaches for the design were discussed here. In the first approach, both analysis and synthesis filters are allowed to change during the transition period to maintain PR and so it is called the intermediate analysis-synthesis (IAS) approach. In the second approach, the analysis filters are switched instantaneously and time-varying synthesis filters are used in the transition period. This approach is known as the instantaneous transform switching (ITS) approach.

In the IAS approach, both analysis and synthesis filters can change during the transitions rather than only the synthesis filters in ITS approach. That implies that maintaining PR conditions is easier in the IAS approach. Note that the analysis filters in the transition periods are designed only to satisfy PR conditions and they do not usually meet the desired time and frequency characteristics.

In the ITS approach, only synthesis filters are allowed to be time-varying in the transition periods. These methods have the advantage of providing instantaneous switching between the analysis transforms compared to IAS methods. But they have different drawbacks: the LS method does not satisfy PR conditions at all times, the redesigning analysis method requires jointly optimization of the time-invariant analysis section, and finally the post filtering method has the drawback of additional computational complexity required for post filtering.

The analysis and design methods of the time-varying filter bank have been developed to design adaptive time-frequency transforms. These adaptive transforms have many potential applications in areas such as time-frequency representation, subband image and video coding, and speech and audio coding. But since the developments of the time-varying filter bank theory is very new, its applications have not been investigated yet.

References

[1] Allen, J.B., Short-term spectral analysis, synthesis, and modification by discrete fourier transform, *IEEE Trans. Acoustics, Speech, Signal Processing*, 25, 235–238, June 1977.

[2] Allen, J.B. and Rabiner, L.R., A unified approach to STFT analysis and synthesis, *Proc. IEEE*, 65, 1558–1564, Nov. 1977.

[3] Rabiner, L.R. and Schafer, R.W., *Digital Processing of Speech Signals*, Prentice-Hall, Englewood Cliffs, NJ, 1978.

[4] Portnoff, M.R., Time-frequency representation of digital signals and systems based on short-time fourier analysis, *IEEE Trans. Acoustics, Speech, Signal Processing*, 55–69, Feb. 1980.

[5] Nawab, S.N. and Quatieri, T.F., *Short-Time Fourier Transform, Chapter in Advanced Topics in Signal Processing*, Prentice-Hall, Englewood Cliffs, NJ, 1988.

[6] Gabor, D., Theory of communication, *J. IEE (London)*, 93(III), 429–457, Nov. 1946.

[7] Nayebi, K., Barnwell, T.P., and Smith, M.J.T., Analysis-synthesis systems with time-varying filter bank structures, *Proc. Intl. Conf. Acoustics, Speech, Signal Processing*, Mar. 1991.

[8] Nayebi, K., Sodagar, I., and Barnwell, T.P., The wavelet transform and time-varying tiling of the time-frequency plane, *IEEE-SP Intl. Symp. Time-Frequency and Time-Scale Analysis*, Oct. 1992.

[9] Sodagar, I., Nayebi, K., and Barnwell, T.P., A class of time-varying wavelet transforms, *Proc. Intl. Conf. Acoustics, Speech, Signal Processing*, April 1993.

[10] Sodagar, I., Nayebi, K., Barnwell, T.P., and Smith, M.J.T., Time-varying filter banks and wavelets, *IEEE Trans. Signal Processing*, Nov. 1994.

[11] Sodagar, I., Analysis and Design of Time-Varying Filter Banks, Ph.D. thesis, Georgia Institute of Technology, Atlanta, GA, Dec. 1994.

[12] de Queiroz, R.L. and Rao, K.R., Adaptive extended lapped transforms, *Proc. Intl. Conf. Acoustics, Speech, Signal Processing*, April 1993.

[13] Herley, C., Kovacevic, J., Ramchandran, K., and Vetterli, M., Arbitrary orthogonal tilings of the time-frequency plane, *IEEE-SP Intl. Symp. Time-Frequency and Time-Scale Analysis*, Oct. 1992.

[14] Herley, C. and Vetterli, M., Orthogonal time-varying filter banks and wavelets, *Proc. Intl. Symp. Circuits Syst.*, Apr. 1993.

[15] Herley, C., Wavelets and Filter Banks, Ph.D. thesis, Columbia University, New York, 1993.

[16] Arrowood, J.L. and Smith, M.J.T., Exact reconstruction analysis/synthesis filter banks with time-varying filters, *Proc. Intl. Conf. Acoustics, Speech, Signal Processing*, Apr. 1993.

[17] Gopinath, R.A., Factorization approach to time-varying filter banks and wavelets, *Proc. Intl. Conf. Acoustics, Speech, Signal Processing*, Apr. 1994.

[18] Chen, T. and Vaidyanathan, P.P., Time-reversed inversion for time-varying filter banks, *Proc. 27th Asilomar Conf. on Signals, Systems, and Computers*, 1993.

[19] Phoong, S. and Vaidyanathan, P.P., On the study of lossless time-varying filter banks, *Proc. 29th Asilomar Conf. on Signals, Systems, and Computers*, 1995.

[20] Sodagar, I., Nayebi, K., Barnwell, T.P., and Smith, M.J.T., A new approach to time-varying FIR filter banks, *Proc. 27th Asilomar Conf. on Signals, Systems, and Computers*, 1993.

[21] Sodagar, I., Nayebi, K., Barnwell, T.P., and Smith, M.J.T., A novel structure for time-varying FIR filter banks, *Proc. Intl. Conf. Acoustics, Speech, and Signal Processing*, 1994.

[22] Sodagar, I., Nayebi, K., and Barnwell, T.P., Time-varying analysis-synthesis systems based on filter banks and post filtering, *IEEE Trans. Signal Processing*, Oct. 1995.

[23] Sodagar, I., Nayebi, K., Barnwell, T.P., and Smith, M.J.T., Perfect reconstruction multidimensional filter banks with time-varying basis functions, *Proc. 27th Asilomar Conf. on Signals, Systems, and Computers*, 1993.

[24] Kovacevic, J. and Vetterli, M., Time-varying modulated lapped transforms, *Proc. 27th Asilomar Conf. on Signals, Systems, and Computers*, 1993.

[25] Vaidyanathan, P.P., Theory and design of M channel maximally decimated QMF with arbitrary M, having perfect reconstruction property, *IEEE Trans. Acoustics, Speech, and Signal Processing*, Apr. 1987.

[26] de Queiroz, R.L. and Rao, K.R., Time-varying lapped transforms and wavelet packets, *IEEE Trans. Signal Processing*, 3293–3305, Dec. 1993.

[27] Malvar, H.S. and Staelin, D.H., The LOT: Transform coding without blocking effects, *IEEE Trans. Acoustics, Speech, and Signal Processing*, 553–559, Apr. 1989.

[28] Malvar, H.S., The lapped transforms for efficient transform/subband coding, *IEEE Trans. Acoustics, Speech, and Signal Processing*, 553–559, Apr. 1989.

[29] Herley, C. and Vetterli, M., Orthogonal time-varying filter banks and wavelet packets, *IEEE Trans. Signal Processing*, 2650–2664, Oct. 1994.

[30] Herley, C. and Vetterli, M., Spatially varying two-dimensional filter banks, *Proc. 27th Asilomar Conf. on Signals, Systems, and Computers*, 1993.

38

Lapped Transforms

Ricardo L. de Queiroz
Advanced Color Imaging,
Xerox Corporation

38.1 Introduction

The idea of a lapped transform (LT) maintaining orthogonality and non-expansion of the samples was developed in the early 1980s at MIT by a group of researchers unhappy with the blocking artifacts so common in traditional block transform coding of images. The idea was to extend the basis function beyond the block boundaries, creating an overlap, in order to eliminate the blocking effect. This idea was not new, but the new ingredient to overlapping blocks would be the fact that the number of transform coefficients would be the same as if there was no overlap, and that the transform would maintain orthogonality. Cassereau [1] introduced the lapped orthogonal transform (LOT), and Malvar [5, 6, 7] gave the LOT its design strategy and a fast algorithm. The equivalence between an LOT and a multirate filter bank was later pointed out by Malvar [9]. Based on cosine modulated filter banks [15], modulated lapped transforms were designed [8, 25]. Modulated transforms were generalized for an arbitrary overlap later creating the class of extended lapped transforms (ELT) [10]– [13]. Recently a new class of LTs with symmetric bases was developed yielding the class of generalized LOTs (GenLOT) [17, 19, 20]. As we mentioned, filter banks and LTs are the same, although studied independently in the past. We, however, refer to LTs for paraunitary uniform FIR filter banks with fast implementation algorithms based on special factorizations of the basis functions.

We assume a one-dimensional input sequence $x(n)$ which is transformed into several coefficients $y_i(n)$, where $y_i(n)$ would belong to the ith subband. We also will use the discrete cosine transform [23] and another cosine transform variation, which we abbreviate as DCT and DCT-IV (DCT type 4), respectively [23].

38.2 Orthogonal Block Transforms

In traditional block-transform processing, such as in image and audio coding, the signal is divided into blocks of M samples, and each block is processed independently [2, 3, 12, 14, 22, 23, 24]. Let the samples

0-8493-8572-5/98/$0.00+$.50

in the mth block be denoted as

$$\mathbf{x}_m^T = [x_0(m), x_1(m), \ldots, x_{M-1}(m)] , \tag{38.1}$$

for $x_k(m) = x(mM + k)$ and let the corresponding transform vector be

$$\mathbf{y}_m^T = [y_0(m), y_1(m), \ldots, y_{M-1}(m)] . \tag{38.2}$$

For a real unitary transform $\mathbf{A}$, $\mathbf{A}^T = \mathbf{A}^{-1}$. The forward and inverse transforms for the mth block are

$$\mathbf{y}_m = \mathbf{A}\mathbf{x}_m , \tag{38.3}$$

and

$$\mathbf{x}_m = \mathbf{A}^T \mathbf{y}_m . \tag{38.4}$$

The rows of $\mathbf{A}$, denoted $\mathbf{a}_n^T$ ($0 \leq n \leq M - 1$), are called the basis vectors because they form an orthogonal basis for the M-tuples over the real field [24]. The transform vector coefficients $[y_0(m), y_1(m), \ldots, y_{M-1}(m)]$ represent the corresponding weights of vector $\mathbf{x}_m$ with respect to this basis.

If the input signal is represented by vector $\mathbf{x}$ while the subbands are grouped into blocks in vector $\mathbf{y}$, we can represent the transform $\mathbf{T}$ which operates over the entire signal as a block diagonal matrix:

$$\mathbf{T} = \text{diag} \{\ldots, \mathbf{A}, \mathbf{A}, \mathbf{A}, \ldots\} , \tag{38.5}$$

where, of course, $\mathbf{T}$ is an orthogonal matrix.

38.2.1 Orthogonal Lapped Transforms

For lapped transforms [12], the basis vectors can have length L, such that $L > M$, extending across traditional block boundaries. Thus, the transform matrix is no longer square and most of the equations valid for block transforms do not apply to an LT. We will concentrate our efforts on *orthogonal* LTs [12] and consider $L = NM$, where N is the overlap factor. Note that N, M, and hence L are all integers. As in the case of block transforms, we define the transform matrix as containing the orthonormal basis vectors as its rows. A lapped transform matrix $\mathbf{P}$ of dimensions $M \times L$ can be divided into square $M \times M$ submatrices $\mathbf{P}_i$ ($i = 0, 1, \ldots, N - 1$) as

$$\mathbf{P} = [\mathbf{P}_0\ \mathbf{P}_1\ \cdots\ \mathbf{P}_{N-1}] . \tag{38.6}$$

The orthogonality property does not hold because $\mathbf{P}$ is no longer a square matrix and it is replaced by other properties which we will discuss later.

If we divide the signal into blocks, each of size M, we would have vectors $\mathbf{x}_m$ and $\mathbf{y}_m$ such as in 38.1 and 38.2. These blocks are not used by LTs in a straightforward manner. The actual vector which is transformed by the matrix $\mathbf{P}$ has to have L samples and, at block number m, it is composed of the samples of $\mathbf{x}_m$ plus $L - M$ samples. These samples are chosen by picking $(L - M)/2$ samples at each side of the block $\mathbf{x}_m$, as shown in Fig. 38.1, for $N = 2$. However, the number of transform coefficients at each step is M, and, in this respect, there is no change in the way we represent the transform-domain blocks $\mathbf{y}_m$. The input vector of length L is denoted as $\mathbf{v}_m$, which is centered around the block $\mathbf{x}_m$, and is defined as

$$\mathbf{v}_m^T = \left[x\left(mM - (N - 1)\frac{M}{2}\right) \cdots x\left(mM + (N + 1)\frac{M}{2} - 1\right)\right] . \tag{38.7}$$

Then, we have

$$\mathbf{y}_m = \mathbf{P}\mathbf{v}_m . \tag{38.8}$$

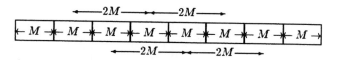

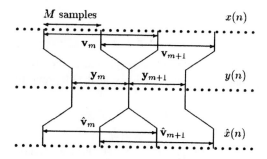

FIGURE 38.1 The signal samples are divided into blocks of M samples. The lapped transform uses neighboring block samples, as in this example for $N = 2$, i.e., $L = 2M$, yielding an overlap of $(L - M)/2 = M/2$ samples on either side of a block.

The inverse transform is not direct as in the case of block transforms, i.e., with the knowledge of $\mathbf{y}_m$ we do not know the samples in the support region of $\mathbf{v}_m$, and neither in the support region of $\mathbf{x}_m$. We can reconstruct a vector $\hat{\mathbf{v}}_m$ from $\mathbf{y}_m$, as

$$\hat{\mathbf{v}}_m = \mathbf{P}^T \mathbf{y}_m \ . \tag{38.9}$$

where $\hat{\mathbf{v}}_m \neq \mathbf{v}_m$. To reconstruct the original sequence, it is necessary to accumulate the results of the vectors $\hat{\mathbf{v}}_m$, in a sense that a particular sample $x(n)$ will be reconstructed from the sum of the contributions it receives from all $\hat{\mathbf{v}}_m$, such that $x(n)$ was included in the region of support of the corresponding $\mathbf{v}_m$. This additional complication comes from the fact that $\mathbf{P}$ is not a square matrix [12]. However, the whole analysis-synthesis system (applied to the entire input vector) is orthogonal, assuring the PR property using 38.9.

We can also describe the process using a sliding rectangular window applied over the samples of $x(n)$. As an M-sample, block $\mathbf{y}_m$ is computed using $\mathbf{v}_m$, $\mathbf{y}_{m+1}$ is computed from $\mathbf{v}_{m+1}$ which is obtained by shifting the window to the right by M samples, as shown in Fig. 38.2.

FIGURE 38.2 Illustration of a lapped transform with $N = 2$ applied to signal $x(n)$, yielding transform domain signal $y(n)$. The input L-tuple as vector $\mathbf{v}_m$ is obtained by a sliding window advancing M samples, generating $\mathbf{y}_m$. This sliding is also valid for the synthesis side.

As the reader may have noticed, the region of support of all vectors $\mathbf{v}_m$ is greater than the region of support of the input vector. Hence, a special treatment has to be given to the transform at the borders. We will discuss this fact later and assume infinite-length signals until then, or assume the length is very large and the borders of the signal are far enough from the region to which we are focusing our attention.

If we denote by $\mathbf{x}$ the input vector and by $\mathbf{y}$ the transform-domain vector, we can be consistent with our notation of transform matrices by defining a matrix $\mathbf{T}$ such that $\mathbf{y} = \mathbf{Tx}$ and $\hat{\mathbf{x}} = \mathbf{T}^T \mathbf{y}$. In this case, we have

$$T = \begin{bmatrix} \ddots & & & \\ & P & & \\ & & P & \\ & & & P \\ & & & & \ddots \end{bmatrix}. \tag{38.10}$$

where the displacement of the matrices P obeys the following

$$T = \begin{bmatrix} \ddots & \ddots & & \ddots & & \\ & P_0 & P_1 & \cdots & P_{N-1} & \\ & & P_0 & P_1 & \cdots & P_{N-1} \\ & & \ddots & \ddots & & \ddots \end{bmatrix}. \tag{38.11}$$

T has as many block-rows as transform operations over each vector v_m.

Let the rows of P be denoted by $1 \times L$ vectors p_i^T ($0 \le i \le M - 1$), so that $P^T = [p_0, \cdots, p_{M-1}]$. In an analogy to the block transform case, we have

$$y_i(m) = p_i^T v_m. \tag{38.12}$$

The vectors p_i are the basis vectors of the lapped transform. They form an orthogonal basis for an M-dimensional subspace (there are only M vectors) of the L-tuples over the real field.

Assuming that the entire input and output signals are represented by the vectors x and y, respectively, and that the signals have infinite length, then, from 38.10, we have

$$y = Tx \tag{38.13}$$

and, if T is orthogonal,

$$x = T^T y. \tag{38.14}$$

The conditions for orthogonality of the LT are expressed as the orthogonality of T. Therefore, the following equations are equivalent in a sense that they state the PR property along with the orthogonality of the LT.

$$\sum_{i=0}^{N-1-l} P_i P_{i+l}^T = \sum_{i=0}^{N-1-l} P_i^T P_{i+l} = \delta(l) I_M. \tag{38.15}$$

$$TT^T = T^T T = I_\infty \tag{38.16}$$

It is worthwhile to reaffirm that orthogonal LTs are a uniform maximally decimated FIR filter bank. Assume the filters in such a filter bank have L-tap impulse responses $f_i(n)$ and $g_i(n)$ ($0 \le i \le M - 1, 0 \le n \le L - 1$), for the analysis and synthesis filters, respectively. If the filters originally have a length smaller than L, one can pad the impulse response with 0s until $L = NM$. In other words, we force the basis vectors to have a common length which is an integer multiple of the block size. Assume the entries of P are denoted by $\{p_{ij}\}$. One can translate the notation from LTs to filter banks by using

$$p_{kn} = f_k(L - 1 - n) = g_k(n) \tag{38.17}$$

38.3 Useful Transforms

38.3.1 Extended Lapped Transform (ELT)

Cosine modulated filter banks are filter banks based on a low-pass prototype filter modulating a cosine sequence. By a proper choice of the phase of the cosine sequence, Malvar developed the modulated lapped transform (MLT) [8], which led to the so-called extended lapped transforms (ELT) [10, 11, 12, 13]. The ELT allows several overlapping factors N, generating a family of LTs with good filter frequency response and fast implementation algorithm.

In the ELTs, the filter length L is basically an even multiple of the block size M, as $L = NM = 2kM$. The MLT-ELT class is defined by

$$p_{k,n} = h(n) \cos\left[\left(k + \frac{1}{2}\right)\left(\left(n - \frac{L-1}{2}\right)\frac{\pi}{M} + (N+1)\frac{\pi}{2}\right)\right] \tag{38.18}$$

for $k = 0, 1 \ldots, M-1$ and $n = 0, 1, \ldots, L-1$. $h(n)$ is a symmetric window modulating the cosine sequence and the impulse response of a low-pass prototype (with cutoff frequency at $\pi/2M$) which is translated in frequency to M different frequency slots in order to construct the uniform filter bank. The ELTs have as their major plus a fast implementation algorithm, which is depicted in Fig. 38.3 in an example for $M = 8$. The free parameters in the design of an ELT are the coefficients of the prototype filter. Such degrees of freedom are translated in the fast algorithm as rotation angles.

For the case $N = 4$ there is a useful parameterized design [11, 12, 13]. In this design, we have:

$$\theta_{k0} = -\frac{\pi}{2} + \mu_{M/2+k} \tag{38.19}$$

$$\theta_{k1} = -\frac{\pi}{2} + \mu_{M/2-1-k} \tag{38.20}$$

where

$$\mu_i = \left[\left(\frac{1-\gamma}{2M}\right)(2k+1) + \gamma\right] \tag{38.21}$$

and γ is a control parameter, for $0 \leq k \leq (M/2) - 1$. γ controls the trade-off between the attenuation and transition region of the prototype filter. For $N = 4$, the relation between angles and $h(n)$ is:

$$h(k) = \cos(\theta_{k0})\cos(\theta_{k1}) \tag{38.22}$$

$$h(M-1-k) = \cos(\theta_{k0})\sin(\theta_{k1}) \tag{38.23}$$

$$h(M+k) = \sin(\theta_{k0})\cos(\theta_{k1}) \tag{38.24}$$

$$h(2M-1-k) = -\sin(\theta_{k0})\sin(\theta_{k1}) \tag{38.25}$$

for $k = 0, 1, \ldots, M/2 - 1$. See [12] for optimized angles for ELTs. Further details on ELTs can be found in [10, 11, 12, 13, 17].

38.3.2 Generalized Linear-Phase Lapped Orthogonal Transform (GenLOT)

The generalized linear-phase lapped orthogonal transform (GenLOT) is also a useful family of LTs possessing symmetric bases (linear-phase filters). The use of linear-phase filters is a popular requirement in image processing applications. Let

$$\mathbf{W} = \frac{1}{\sqrt{2}}\begin{bmatrix} \mathbf{I}_{M/2} & \mathbf{I}_{M/2} \\ \mathbf{I}_{M/2} & -\mathbf{I}_{M/2} \end{bmatrix} \quad \text{and} \quad \mathbf{\Psi}_i = \begin{bmatrix} \mathbf{U}_i & \mathbf{0}_{M/2} \\ \mathbf{0}_{M/2} & \mathbf{V}_i \end{bmatrix}, \tag{38.26}$$

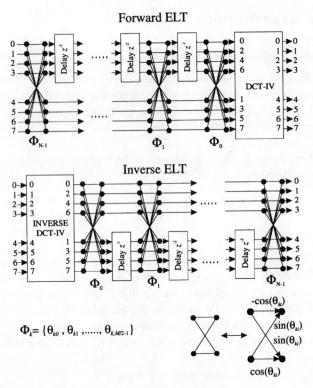

FIGURE 38.3   Implementation flow-graph for the ELT with $M = 8$.

where $\mathbf{U}_i$ and $\mathbf{V}_i$ can be any $M/2 \times M/2$ orthogonal matrices. Let the transform matrix $\mathbf{P}$ for the GenLOT be constructed interactively. Let $\mathbf{P}^{(i)}$ be the partial reconstruction of $\mathbf{P}$ after including up to the ith stage. We start by setting $\mathbf{P}^{(0)} = \mathbf{E}_0$ where $\mathbf{E}_0$ is an orthogonal matrix with symmetric rows. The recursion is given by:

$$\mathbf{P}^{(i)} = \Psi_i \mathbf{W} \mathbf{Z} \begin{bmatrix} \mathbf{W}\mathbf{P}^{(i-1)} & \mathbf{0}_M \\ \mathbf{0}_M & \mathbf{W}\mathbf{P}^{(i-1)} \end{bmatrix} \qquad (38.27)$$

where

$$\mathbf{Z} = \begin{bmatrix} \mathbf{0}_{M/2} & \mathbf{0}_{M/2} & \mathbf{I}_{M/2} & \mathbf{0}_{M/2} \\ \mathbf{0}_{M/2} & \mathbf{I}_{M/2} & \mathbf{0}_{M/2} & \mathbf{0}_{M/2} \end{bmatrix}. \qquad (38.28)$$

At the final stage we set $\mathbf{P} = \mathbf{P}^{(N-1)}$. $\mathbf{E}_0$ is usually the DCT while the other factors ($\mathbf{U}_i$ and $\mathbf{V}_i$) are found through optimization routines. More details on GenLOTs and their design can be found in [17, 19, 20]. The implementation flow-graph of a GenLOT with $M = 8$ is shown in Fig. 38.4.

38.4 Remarks

We hope this introductory work is helpful in understanding the basic concepts of lapped transforms. Filter banks are covered in other parts of this book. An excellent book by Vaidyanathan [28] has a thorough coverage of such subject. The interrelations of filter banks and LTs are well covered by Malvar [12] and Queiroz [17]. For image processing and coding, it is necessary to process finite-length signals. As we discussed, such an issue is not so straightforward in a general case. Algorithms to implement LTs over finite-length signals are discussed in [7, 12, 16, 17, 18, 21]. These algorithms can be general or specific. The

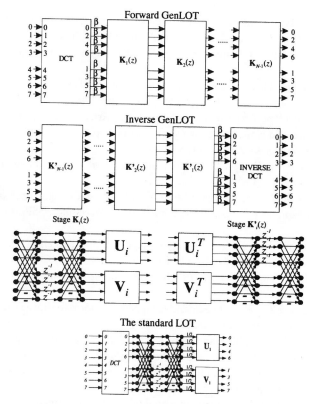

FIGURE 38.4 Implementation flow-graph for the GenLOT with $M = 8$, where $\beta = 2^{N-1}$.

specific algorithms are generally targeted to a particular LT invariantly seeking a very fast implementation. In general, Malvar's book [12] is an excellent reference for lapped transforms and their related topics.

References

[1] Cassereau, P., *A New Class of Optimal Unitary Transforms for Image Processing*, Master's Thesis, MIT, Cambridge, MA, May 1985.

[2] Clarke, R.J., *Transform Coding of Images*, Academic Press, Orlando, FL, 1985.

[3] Jayant, N.S. and Noll, P., *Digital Coding of Waveforms*, Prentice-Hall, Englewood Cliffs, NJ, 1984.

[4] Jozawa, H. and Watanabe, H., Intrafield/interfield adaptive lapped transform for compatible HDTV coding, *4th International Workshop on HDTV and Beyond*, Torino, Italy, Sept. 4-6, 1991.

[5] Malvar, H.S., *Optimal pre- and post-filtering in noisy sampled-data systems*, Ph.D. Dissertation, MIT, Cambridge, MA, Aug. 1986.

[6] Malvar, H.S., Reduction of blocking effects in image coding with a lapped orthogonal transform, *Proc. of Intl. Conf. on Acoust., Speech, Signal Processing*, Glasgow, Scotland, pp. 781-784, Apr. 1988.

[7] Malvar, H.S. and Staelin, D.H., The LOT: transform coding without blocking effects, *IEEE Trans. Acoust., Speech, Signal Processing*, ASSP-37, 553–559, Apr. 1989.

[8] Malvar, H.S., Lapped transforms for efficient transform/subband coding, *IEEE Trans. Acoust., Speech, Signal Processing*, ASSP-38, 969–978, June 1990.

[9] Malvar, H.S., The LOT: a link between block transform coding and multirate filter banks, *Proc. Intl. Symp. Circuits and Systems,* Espoo, Finland, pp. 835–838, June 1988.

[10] Malvar, H.S., Modulated QMF filter banks with perfect reconstruction, *Elect. Letters,* 26, 906-907, June 1990.

[11] Malvar, H.S., Extended lapped transform: fast algorithms and applications, *Proc. of Intl. Conf. on Acoust., Speech, Signal Processing,* Toronto, Canada, pp. 1797–1800, 1991.

[12] Malvar, H.S., *Signal Processing with Lapped Transforms,* Artech House, Norwood, MA, 1992.

[13] Malvar, H.S., Extended lapped transforms: properties, applications and fast algorithms, *IEEE Trans. Signal Processing,* 40, 2703–2714, Nov. 1992.

[14] Pennebaker, W.B. and Mitchell, J.L., *JPEG: Still Image Compression Standard,* Van Nostrand Reinhold, New York, 1993.

[15] Princen, J.P. and Bradley, A.B., Analysis/synthesis filter bank design based on time domain aliasing cancellation, *IEEE Trans. Acoust., Speech, Signal Processing,* ASSP-34, 1153–1161, Oct. 1986.

[16] de Queiroz, R.L. and Rao, K.R., Time-varying lapped transforms and wavelet packets, *IEEE Trans. on Signal Processing,* 41, 3293–3305, Dec. 1993.

[17] de Queiroz, R.L., *On Lapped Transforms,* Ph.D Dissertation, University of Texas at Arlington, August 1994.

[18] de Queiroz, R.L. and Rao, K.R., The extended lapped transform for image coding, *IEEE Trans. on Image Processing,* 4, 828–832, June, 1995.

[19] de Queiroz, R.L., Nguyen, T.Q. and Rao, K.R., GENLOT: generalized linear-phase lapped orthogonal transforms, *IEEE Trans. Signal Processing,* 44, 497–507, Apr. 1996.

[20] de Queiroz, R.L., Nguyen, T.Q. and Rao, K.R., The generalized lapped orthogonal transforms, *Electron. Lett.,* 30, 107, Jan. 1994.

[21] de Queiroz, R.L. and Rao, K.R., On orthogonal transforms of images using paraunitary filter banks, *J. Visual Commun. Image Representation,* 6(2), 142–153, June 1995.

[22] Rabbani, M. and Jones, P.W., *Digital Image Compression Techniques,* SPIE Optical Engineering Press, Bellingham, WA, 1991.

[23] Rao, K.R. and Yip, P., *Discrete Cosine Transform : Algorithms, Advantages, Applications,* Academic Press, San Diego, CA, 1990.

[24] Rao, K.R., Ed., *Discrete Transforms and Their Applications,* Van Nostrand Reinhold, New York, 1985.

[25] Schiller, H., Overlapping block transform for image coding preserving equal number of samples and coefficients, *Proc. SPIE, Visual Communications and Image Processing,* 1001, 834–839, 1988.

[26] Soman, A.K., Vaidyanathan, P.P. and Nguyen, T.Q., Linear-phase paraunitary filter banks: theory, factorizations and applications, *IEEE Trans. on Signal Processing,* 41, 3480–3496, Dec. 1993.

[27] Temerinac, M. and Edler, B., A unified approach to lapped orthogonal transforms, *IEEE Trans. Image Processing,* 1, 111–116, Jan. 1992.

[28] Vaidyanathan, P.P., *Multirate Systems and Filter Banks,* Prentice-Hall, Englewood Cliffs, NJ, 1993.

[29] Young, R.W. and Kingsbury, N.G., Frequency domain estimation using a complex lapped transform, *IEEE Trans. Image Processing,* 2, 2–17, Jan. 1993.

Digital Audio Communications

Nikil Jayant
Bell Laboratories, Lucent Technologies

A S WE ENTER THE 21ST CENTURY, digital audio communications will have become nearly as prevalent as digital speech communications. In particular, new technologies for audio storage and transmission will make available music and wideband signals in a flexible variety of standard formats.

The fundamental underpinning for these technologies is audio compression based on perceptually-tuned shaping of the quantization noise. The next chapter in this section describes psychoacoustics knowledge that suggests the general principles of *perceptual audio coding*. Succeeding chapters in this section are devoted to descriptions of established examples of *perceptual audio coders*. These include MPEG standards, and coders developed by Dolby, Sony, and Bell Laboratories.

The dimensions of coder performance are quality, bit rate, delay, and complexity. The quality vs. bit rate tradeoffs are particularly important.

Audio Quality

The three parameters of digital audio quality are *signal bandwidth, fidelity* and *spatial realism.*

Compact-disk (CD) signals have a bandwidth of 20–20,000 Hz, while traditional telephone speech has a bandwidth of 200–3400 Hz. Intermediate bandwidths characterize various grades of wideband speech and audio, including roughly defined ranges of quality referred to as AM radio and FM radio quality (bandwidths on the order of 7–10 and 12–15 kHz, respectively).

In the context of digital coding, fidelity refers to the level of perceptibility of quantization or reconstruction noise. The highest level of fidelity is one where the noise is imperceptible in formal listening tests. Lower levels of fidelity are acceptable in some applications if they are not annoying, although in general it is good practice to sacrifice some bandwidth in the interest of greater fidelity, for a given bit rate in coding. Five-point scales of signal fidelity are common in both speech and audio coding.

Spatial realism is generally provided by increasing the number of coded (and reproduced) spatial channels. Common formats are 1-channel *(mono),* 2-channel *(stereo),* 5-channel (3 front, 2 rear), 5.1-channel (5-channel plus subwoofer) and 8-channel (6 front, 2 rear). For given constraints on bandwidth and fidelity, the required bit rate in coding increases as a function of the number of channels; but the increase is slower than linear, because of the presence of interchannel redundancy. The notion of perceptual coding originally developed for exploiting the perceptual irrelevancies of a single-channel audio signal extends also to the methods used in exploiting interchannel redundancy.

Bit Rate

The CD-stereo signal has a digital representation rate of 1406 kilobits per second (kb/s). Current technology for perceptual audio coding reproduces CD-stereo with perfect fidelity at bit rates as low as 128 kb/s, depending on the input signal. CD-like reproduction is possible at bit rates as low as 64 kb/s for stereo. Single-channel reproduction of FM-radio-like music is possible at 32 kb/s. Single-channel reproduction of AM-radio-like music and wideband speech is possible at rates approaching 16 kb/s for all but the most demanding signals. Techniques for so-called "pseudo-stereo" can provide additional enhancement of digital single-channel audio.

Applications of Digital Audio

The capabilities of audio compression have combined with increasingly affordable implementations on platforms for digital signal processing (DSP), native signal processing (NSP) in a computer's (native) processor, and application-specific integrated circuits (ASICs) to create revolutionary applications of digital audio. International and national standards have contributed immensely to this revolution. Some of these standards only specify the bit-stream syntax and decoder, leaving room for future, sometimes proprietary, enhancements of the encoding algorithm.

The domains of applications include *transmission* (for example, digital audio broadcasting), *storage* (for example, the minidisk and the digital versatile disk, DVD), and *networking* (music preview, distribution, and publishing). The networking applications will make digital audio communications as commonplace as digital telephony.

The Future of Digital Audio

Remarkable as the capabilities and applications mentioned above are, there are even greater challenges and opportunities for the practitioners of digital audio technology. It is unlikely that we have reached or even approached the fundamental limits of performance in terms of audio quality at a given bit rate. Newer capabilities in this technology (in terms of audio fidelity, bandwidth, and spatial realism) will continue to lead to newer classes of applications in audio communications. New technologies for embedded coding and universal coding will create interesting new options for digital networking and seamless communication of speech and music signals. Finally, co-designs of audio processing with image and video processing will lead to currently unavailable capabilities for multimedia networking games, computer agents, and personal communication services. These scenarios will call upon our best capabilities in signal compression as well as advances in the sister disciplines of signal synthesis and recognition by machine.

39

Auditory Psychophysics for Coding Applications

Joseph L. Hall
Bell Laboratories
Lucent Technologies

In this chapter we review properties of auditory perception that are relevant to the design of coders for acoustic signals. The chapter begins with a general definition of a perceptual coder, then considers what the "ideal" psychophysical model would consist of and what use a coder could be expected to make of this model. We then present some basic definitions and concepts. The chapter continues with a review of relevant psychophysical data, including results on threshold, just-noticeable differences, masking, and loudness. Finally, we attempt to summarize the present state of the art, the capabilities and limitations of present-day perceptual coders for audio and speech, and what areas most need work.

39.1 Introduction

A coded signal differs in some respect from the original signal. One task in designing a coder is to minimize some measure of this difference under the constraints imposed by bit rate, complexity, or cost. What is the appropriate measure of difference? The most straightforward approach is to minimize some physical measure of the difference between original and coded signal. The designer might attempt to minimize RMS difference between the original and coded waveform, or perhaps the difference between original and coded power spectra on a frame-by-frame basis. However, if the purpose of the coder is to encode acoustic signals that are eventually to be listened to[1] by people, these physical measures do not directly address the appropriate issue. For signals that are to be listened to by people, the "best" coder is the one

---

[1] Perceptual coding is not limited to speech and audio. It can be applied also to image and video [16]. In this paper we consider only coders for acoustic signals.

0-8493-8572-5/98/$0.00+$.50

that sounds the best. There is a very clear distinction between *physical* and *perceptual* measures of a signal (frequency vs. pitch, intensity vs. loudness, for example). A perceptual coder can be defined as a coder that minimizes some measure of the difference between original and coded signal so as to minimize the perceptual impact of the coding noise. We can define the best coder given a particular set of constraints as the one in which the coding noise is least objectionable.

It follows that the designer of a perceptual coder needs some way to determine the perceptual quality of a coded signal. "Perceptual quality" is a poorly defined concept, and it will be seen that in some sense it cannot be uniquely defined. We can, however, attempt to provide a partial answer to the question of how it can be determined. We can present something of what is known about human auditory perception from psychophysical listening experiments and show how these phenomena relate to the design of a coder.

One requirement for successful design of a perceptual coder is a satisfactory model for the signal-dependent sensitivity of the auditory system. Present-day models are incomplete, but we can attempt to specify what the properties of a complete model would be. One possible specification is that, for any given waveform (the signal), it accurately predicts the loudness, as a function of pitch and of time, of any added waveform (the noise). If we had such a complete model, then we would in principle be able to build a transparent coder, defined as one in which the coded signal is indistinguishable from the original signal, or at least we would be able to determine whether or not a given coder was transparent. It is relatively simple to design a psychophysical listening experiment to determine whether the coding noise is audible, or equivalently, whether the subject can distinguish between original and coded signal. Any subject with normal hearing could be expected to give similar results to this experiment. While present-day models are far from complete, we can at least describe the properties of a complete model.

There is a second requirement that is more difficult to satisfy. This is the need to be able to determine which of two coded samples, each of which has audible coding noise, is preferable. While a satisfactory model for the signal-dependent sensitivity of the auditory system is in principle sufficient for the design of a transparent coder, the question of how to build the best nontransparent coder does not have a unique answer. Often, design constraints preclude building a transparent coder. Even the best coder built under these constraints will result in audible coding noise, and it is under some conditions impossible to specify uniquely how best to distribute this noise. One listener may prefer the more intelligible version, while another may prefer the more natural sounding version. The preferences of even a single listener might very well depend on the application. In the absence of any better criterion, we can attempt to minimize the loudness of the coding noise, but it must be understood that this is an incomplete solution.

Our purpose in this paper is to present something of what is known about human auditory perception in a form that may be useful to the designer of a perceptual coder. We do not attempt to answer the question of how this knowledge is to be utilized, how to build a coder. Present-day perceptual coders for the most part utilize a *feedforward* paradigm: analysis of the signal to be coded produces specifications for allowable coding noise. Perhaps a more general method is a *feedback* paradigm, in which the perceptual model somehow makes possible a decision as to which of two coded signals is "better". This decision process can then be iterated to arrive at some optimum solution. It will be seen that for proper exploitation of some aspects of auditory perception the feedforward paradigm may be inadequate and the potentially more time-consuming feedback paradigm may be required. How this is to be done is part of the challenge facing the designer.

39.2 Definitions

In this section we define some fundamental terms and concepts and clarify the distinction between physical and perceptual measures.

39.2.1 Loudness

When we increase the intensity of a stimulus its loudness increases, but that does not mean that intensity and loudness are the same thing. *Intensity* is a physical measure. We can measure the intensity of a signal with an appropriate measuring instrument, and if the measuring instrument is standardized and calibrated correctly anyone else anywhere in the world can measure the same signal and get the same result. *Loudness* is *perceptual magnitude*. It can be defined as "that attribute of auditory sensation in terms of which sounds can be ordered on a scale extending from quiet to loud" ([23], p.47). We cannot measure it directly. All we can do is ask questions of a subject and from the responses attempt to infer something about loudness. Furthermore, we have no guarantee that a particular stimulus will be as loud for one subject as for another. The best we can do is assume that, for a particular stimulus, loudness judgments for one group of normal-hearing people will be similar to loudness judgments for another group.

There are two commonly used measures of loudness. One is *loudness level* (unit *phon*) and the other is *loudness* (unit *sone*). These two measures differ in what they describe and how they are obtained. The phon is defined as the intensity, in dB SPL, of an equally loud 1-kHz tone. The sone is defined in terms of subjectively measured loudness ratios. A stimulus half as loud as a one-sone stimulus has a loudness of 0.5 sones, a stimulus ten times as loud has a loudness of 10 sones, etc. A 1-kHz tone at 40 dB SPL is arbitrarily defined to have a loudness of one sone.

The argument can be made that loudness matching, the procedure used to obtain the phon scale, is a less subjective procedure than loudness scaling, the procedure used to obtain the sone scale. This argument would lead to the conclusion that the phon is the more objective of the two measures and that the sone is more subject to individual variability. This argument breaks down on two counts: first, for dissimilar stimuli even the supposedly straightforward loudness-matching task is subject to large and poorly understood order and bias effects that can only be described as subjective. While loudness matching of two equal-frequency tone bursts generally gives stable and repeatable results, the task becomes more difficult when the frequencies of the two tone bursts differ. Loudness matching between two dissimilar stimuli, as for example between a pure tone and a multicomponent complex signal, is even more difficult and yields less stable results. Loudness-matching experiments have to be designed carefully, and results from these experiments have to be interpreted with caution. Second, it is possible to measure loudness in sones, at least approximately, by means of a loudness-matching procedure. Fletcher [6] states that under some conditions loudness adds. Binaural presentation of a stimulus results in loudness doubling; and two equally-loud stimuli, far enough apart in frequency that they do not mask each other, are twice as loud as one. If loudness additivity holds, then it follows that the sone scale can be generated by matching loudness of a test stimulus to binaural stimuli or to pairs of tones. This approach must be treated with caution. As Fletcher states, "However, this method [scaling] is related more directly to the scale we are seeking (the sone scale) than the two preceding ones (binaural or monaural loudness additivity)" ([6], p. 278). The loudness additivity approach relies on the assumption that loudness summation is perfect, and there is some more recent evidence [28, 33] that loudness summation, at least for binaural vs. monaural presentation, is not perfect.

39.2.2 Pitch

The American Standards Association defines pitch as "that attribute of auditory sensation in which sounds may be ordered on a musical scale". Pitch bears much the same relationship to frequency as loudness does to intensity: frequency is an objective physical measure, while pitch is a subjective perceptual measure. Just as there is not a one-to-one relationship between intensity and loudness, so also there is not a one-to-one relationship between frequency and pitch. Under some conditions, for example, loudness can be shown to decrease with decreasing frequency with intensity held constant, and pitch can be shown to decrease with increasing intensity with frequency held constant ([40], p. 409).

39.2.3   Threshold of Hearing

Since the concept of threshold is basic to much of what follows, it is worthwhile at this point to discuss it in some detail. It will be seen that thresholds are determined not only by the stimulus and the observer but also by the method of measurement. While this discussion is phrased in terms of threshold of hearing, much of what follows applies as well to differential thresholds (just-noticeable differences) discussed in the next subsection.

By the simplest definition, the threshold of hearing (equivalently, auditory threshold) is the lowest intensity that the listener can hear. This definition is inadequate because we cannot directly measure the listener's perception. A first-order correction, therefore, is that the threshold of hearing is the lowest intensity that elicits from the listener the response that the sound is audible. Given this definition, we can present a stimulus to the listener and ask whether he or she can hear it. If we do this, we soon find that identical stimuli do not always elicit identical responses. In general, the probability of a positive response increases with increasing stimulus intensity and can be described by a *psychometric function* such as that shown for a hypothetical experiment in Fig. 39.1. Here the stimulus intensity (in dB) appears on the abscissa and the probability $P(C)$ of a positive response appears on the ordinate. The yes-no experiment could be described by a psychometric function that ranges from zero to one, and threshold could be defined as the stimulus intensity that elicits a positive response in 50% of the trials.

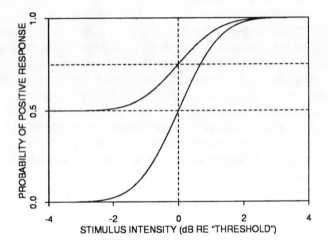

FIGURE 39.1   Idealized psychometric functions for hypothetical yes-no experiment (zero to one) and for hypothetical two-interval forced-choice experiment (0.5 to one).

A difficulty with the simple yes-no experiment is that we have no control over the subject's *criterion level.* The subject may be using a strict criterion ("yes" only if the signal is definitely present) or a lax criterion ("yes" if the signal might be present). The subject can respond correctly either by a positive response in the presence of a stimulus (*hit*) or by a negative response in the absence of a stimulus (*correct rejection*). Similarly the subject can respond incorrectly either by a negative response in the presence of a stimulus (*miss*) or by a positive response in the absence of a stimulus (*false alarm*). Unless the experimenter is willing to use an elaborate and time-consuming procedure that involves assigning rewards to correct responses and penalties to incorrect responses, the criterion level is uncontrolled.

The field of psychophysics that deals with this complication is called *detection theory.* The field of psychophysical detection theory is highly developed [12] and a complete description is far beyond the scope of this paper. Very briefly, the subject's response is considered to be based on an internal *decision variable,* a random variable drawn from a distribution with mean and standard deviation that depend on the stimulus. If we assume that the decision variable is normally distributed with a fixed standard deviation σ and a mean that depends only on stimulus intensity, then we can define an *index of sensitivity* d' for a given stimulus intensity as the difference between m_0 (the mean in the absence of the stimulus) and m_s (the mean in the presence of the stimulus), divided by σ. An *ideal observer* (a hypothetical subject who does the best possible job for the task at hand) gives a positive response if and only if the decision variable exceeds an internal criterion level. An increase in criterion level decreases the probability of a false alarm and increases the probability of a miss.

A simple and satisfactory way to deal with the problem of uncontrolled criterion level is to use a *criterion-free* experimental paradigm. The simplest is perhaps the two-interval forced choice (2IFC) paradigm, in which the stimulus is presented at random in one of two observation intervals. The subject's task is to determine which of the two intervals contained the stimulus. The ideal observer selects the interval that elicits the larger decision variable, and criterion level is no longer a factor. Now the subject has a 50% chance of choosing the correct interval even in the absence of any stimulus, so the psychometric function goes from 0.5 to 1.0 as shown in Fig. 39.1. A reasonable definition of threshold is $P(C) = 0.75$, halfway between the chance level of 0.5 and one. If the decision variable is normally distributed with a fixed standard deviation, it can be shown that this definition of threshold corresponds to a d' of 0.95.

The number of intervals can be increased beyond two. In this case, the ideal observer responds correctly if the decision variable for the interval containing the stimulus is larger than the largest of the N-1 decision variables for the intervals not containing the stimulus. A common practice is, for an N-interval forced choice paradigm (NIFC), to define threshold as the point halfway between the chance level of 1/N and one. This is a perfectly acceptable practice so long as it is recognized that the measured threshold is influenced by the number of alternatives. For a 3IFC paradigm this definition of threshold corresponds to a d' of 1.12 and for a 4IFC paradigm it corresponds to a d' of 1.24.

39.2.4 Differential Threshold

The differential threshold is conceptually similar to the auditory threshold discussed above, and many of the same comments apply. The differential threshold, or just-noticeable difference (JND), is the amount by which some attribute of a signal has to change in order for the observer to be able to detect the change. A tone burst, for example, can be specified in terms of frequency, intensity, and duration, and a differential threshold for any of these three attributes can be defined and measured.

The first attempt to provide a quantitative description of differential thresholds was provided by the German physiologist E. H. Weber in the first half of the 19th century. According to *Weber's law*, the just-noticeable difference ΔI is proportional to the stimulus intensity I, or $\Delta I / I = K$, where the constant of proportionality $\Delta I / I$ is known as the *Weber fraction*. This was supposed to be a general description of sensitivity to changes of intensity for a variety of sensory modalities, not limited just to hearing, and it has since been applied to perception of nonintensive variables such as frequency. It was recognized at an early stage that this law breaks down at near-threshold intensities, and in the latter half of the 19th century the German physicist G. T. Fechner suggested the modification that is now known as the *modified Weber law*, $\Delta I / (I + I_0) = K$, where I_0 is a constant. While Weber's law provides a reasonable first-order description of intensity and frequency discrimination in hearing, in general it does not hold exactly, as will be seen below.

As with the threshold of hearing, the differential threshold can be measured in different ways, and the result depends to some extent on how it is measured. The simplest method is a same-different paradigm, in which two stimuli are presented and the subject's task is to judge whether or not they are the same. This method suffers from the same drawback as the yes-no paradigm for auditory threshold: we do not have control over the subject's criterion level.

If the physical attribute being measured is simply related to some perceptual attribute, then the differential threshold can be measured by requiring the subject to judge which of two stimuli has more of that perceptual attribute. A just-noticeable difference for frequency, for example, could be measured by requiring the subject to judge which of two stimuli is of higher pitch; or a just noticeable difference for intensity could be measured by requiring the subject to judge which of two stimuli is louder. As with the 2IFC paradigm discussed above for auditory threshold, this method removes the problem of uncontrolled criterion level.

There are more general methods that do not assume a knowledge of the relationship between the physical attribute being measured and a perceptual attribute. The most useful, perhaps, is the N-interval forced choice method: N stimuli are presented, one of which differs from the other N-1 along the dimension being measured. The subject's task is to specify which one of the N stimuli is different from the other N-1.

Note that there is a close parallel between the differential threshold and the auditory threshold described in the previous subsection. The auditory threshold can be regarded as a special case of the just-noticeable difference for intensity, where the question is by how much the intensity has to differ from zero in order to be detectable.

39.2.5   Masked Threshold

The *masked threshold* of a signal is defined as the threshold of that signal (the *probe*) in the presence of another signal (the *masker*). A related term is *masking*, which is the elevation of threshold of the probe by the masker: it is the difference between masked and absolute threshold. More generally, the reduction of loudness of a supra-threshold signal is also referred to as masking. It will be seen that masking can appear in many forms, depending on spectral and temporal relationships between probe and masker.

Many of the comments that applied to measurement of absolute and differential thresholds also apply to measurement of masked threshold. The simplest method is to present masker plus probe and ask the subject whether or not the probe is present. Once again there is a problem with criterion level. Another method is to present stimuli in two intervals and ask the subject which one contains the probe. This method can give useful results but can, under some conditions, give misleading results. Suppose, for example, that the probe and masker are both pure tones at 1 kHz, but that the two signals are 180° out of phase. As the intensity of the probe is increased from zero, the intensity of the composite signal will first decrease, then increase. The two signals, masker alone and masker plus probe, may be easily distinguishable, but in the absence of additional information the subject has no way of telling which is which.

A more robust method for measuring masked threshold is the N-interval forced choice method described above, in which the subject specifies which of the N stimuli differs from the other N-1. Subjective percepts in masking experiments can be quite complex and can differ from one observer to another. In the N-interval forced choice method the observer has the freedom to base judgments on whatever attribute is most easily detected, and it is not necessary to instruct the observer what to listen for.

Note that the differential threshold for intensity can be regarded as a special case of the masked threshold in which the probe is an intensity-scaled version of the masker.

A note on terminology: suppose two signals, $x_1(t)$ and $[x_1(t) + x_2(t)]$ are just distinguishable. If $x_2(t)$ is a scaled version of $x_1(t)$, then we are dealing with intensity discrimination. If $x_1(t)$ and $x_2(t)$ are two different signals, then we are dealing with masking, with $x_1(t)$ the masker and $x_2(t)$ the probe. In either case, the difference can be described in several ways. These ways include (1) the intensity increment between $x_1(t)$ and $[x_1(t)+x_2(t)]$, ΔI; (2) the intensity increment relative to $x_1(t)$, $\Delta I / I$; (3) the intensity ratio between $x_1(t)$ and $[x_1(t)+x_2(t)]$, $(I+\Delta I)/I$; (4) the intensity increment in dB, $10 \times \log_{10}(\Delta I/I)$; and (5) the intensity ratio in dB, $10 \times \log_{10}[(I + \Delta I)/I]$. These ways are equivalent in that they show the same information, although for a particular application one way may be preferable to another for presentation purposes. Another measure that is often used, particularly in the design of perceptual coders, is the intensity of the probe $x_2(t)$. This measure is subject to misinterpretation and must be used with caution. Depending on the coherence between $x_1(t)$ and $x_2(t)$, a given probe intensity can result in a wide range of intensity increments ΔI. The resulting ambiguity has been responsible for some confusion.

39.2.6   Critical Bands and Peripheral Auditory Filters

The concepts of *critical bands* and *peripheral auditory* filters are central to much of the auditory modeling work that is used in present-day perceptual coders. Scharf, in a classic review article [33], defines the empirical critical bandwidth as "that bandwidth at which subjective responses rather abruptly change". Simply put, for some psychophysical tasks the auditory system behaves as if it consisted of a bank of bandpass filters (the critical bands) followed by energy detectors. Examples of critical-band behavior that are particularly relevant for the designer of a coder include the relationship between bandwidth and loudness (Fig. 39.5) and the relationship between bandwidth and masking (Fig. 39.10). Another example of critical-band behavior is phase sensitivity: in experiments measuring the detectability of amplitude and of frequency modulation, the auditory system appears to be sensitive to the relative phase of the components of a complex sound only so long as the components are within a critical band [9, 45].

The concept of the critical band was introduced more than a half-century ago by Fletcher [6], and since that time it has been studied extensively. Fletcher's pioneering contribution is ably documented by Allen [1], and Scharf's 1970 review article [33] gives references to some later work. More recently, Moore and his co-workers have made extensive measurements of peripheral auditory filters [24].

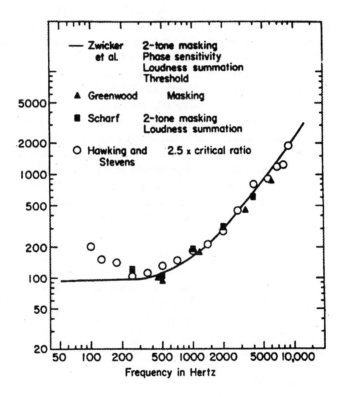

FIGURE 39.2   Empirical critical bandwidth. (*Source:* Scharf, B., Critical bands, ch. 5 in *Foundations of Modern Auditory Theory,* Vol. 1, Tobias, J.V., ed., Academic Press, NY, 1970. With permission).

The value of critical bandwidths has been the subject of some discussion, because of questions of definition and method of measurement. Figure 39.2 ([31], Fig. 1) shows critical bandwidth as a function of frequency for Scharf's empirical definition (the bandwidth at which subjective responses undergo some sort of change). Results from several experiments are superimposed here, and they are in substantial agreement with each other. Moore and Glasberg [26] argue that the bandwidths shown in Fig. 39.2 are determined not only by the bandwidth of peripheral auditory filters but also by changes in processing efficiency. By their argument, the bandwidth of peripheral auditory filters is somewhat smaller than the values shown in Fig. 39.2 at frequencies above 1 kHz and substantially smaller, by as much as an octave, at lower frequencies.

39.3 Summary of Relevant Psychophysical Data

In Section 39.2, we introduced some basic concepts and definitions. In this section, we review some relevant psychophysical results. There are several excellent books and book chapters that have been written on this subject, and we have neither the space nor the inclination to duplicate material found in these other sources. Our attempt here is to make the reader aware of some relevant results and to refer him or her to sources where more extensive treatments may be found.

39.3.1 Loudness

Loudness Level and Frequency

For pure tones, loudness depends on both intensity and frequency. Figure 39.3 (modified from [37], p. 124) shows loudness level contours. The curves are labeled in phons and, in parentheses, sones. These curves have been remeasured many times since, with some variation in the results, but the basic conclusions remain unchanged. The most sensitive region is around 2-3 kHz. The low-frequency slope of the loudness level contours is flatter at high loudness levels than at low. It follows that loudness level grows more rapidly with intensity at low frequencies than at high. The 38- and 48-phon contours are (by definition) separated by 10 dB at 1 kHz, but they are only about 5 dB apart at 100 Hz.

This figure also shows contours that specify the dynamic range of hearing. Tones below the 8-phon contour are inaudible, and tones above the dotted line are uncomfortable. The dynamic range of hearing, the distance between these two contours, is greatest around 2 to 3 kHz and decreases at lower and higher frequencies. In practice, the useful dynamic range is substantially less. We know today that extended exposure to sounds at much lower levels than the dotted line in Fig. 39.3 can result in temporary or permanent damage to the ear. It has been suggested that extended exposure to sounds as low as 70 to 75 dB(A) may produce permanent high-frequency threshold shifts in some individuals [39].

Loudness and Intensity

Figure 39.4 (modified from [32], Fig. 5) shows *loudness growth functions,* the relationship between stimulus intensity in dB SPL and loudness in sones, for tones of different frequencies. As can be seen in Fig. 39.4, the loudness growth function depends on frequency. Above about 40 dB SPL for a 1-kHz tone the relationship is approximately described by the power law $L(I) = (I/I_0)^{1/3}$, so that if the intensity I is increased by 9 dB the loudness L is approximately doubled.[2] The relationship between loudness and intensity has been modeled extensively [1, 6, 46].

Loudness and Bandwidth

The loudness of a complex sound of fixed intensity, whether a tone complex or a band of noise, depends on its bandwidth, as is shown in Fig. 39.5 ([48], Fig. 3). For sounds well above threshold, the

---

[2]This power-law relationship between physical and perceptual measures of a stimulus was studied in great detail by S. S. Stevens. This relationship is now commonly referred to as *Stevens' Law.* Stevens measured exponents for many sensory modalities, ranging from a low of 0.33 for loudness and brightness to a high of 3.5 for electric shock produced by a 60-Hz electric current delivered to the skin.

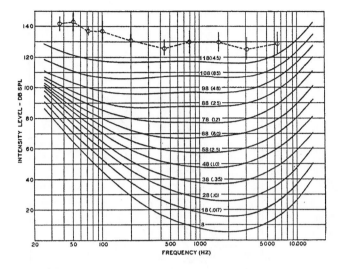

FIGURE 39.3   Loudness level contours. Parameters: phons (sones). The bottom curve (8 phons) is at the threshold of hearing. The dotted line shows Wegel's 1932 results for "threshold of feeling". This line is many dB above levels that are known today to produce permanent damage to the auditory system. (Modified from Stevens, S.S. and Davis, H.W., *Hearing*, John Wiley & Sons, New York, 1938).

loudness remains more or less constant so long as the bandwidth is less than a critical band. If the bandwidth is greater than a critical band, the loudness increases with increasing bandwidth. Near threshold the trend is reversed, and the loudness decreases with increasing bandwidth.[3]

These phenomena have been modeled successfully by utilizing the loudness growth functions shown in Fig. 39.4 in a model that calculates total loudness by summing the specific loudness per critical band [49]. The loudness growth function is very steep near threshold, so that dividing the total energy of the signal into two or more critical bands results in a reduction of total loudness. The loudness growth function well above threshold is less steep, so that dividing the total energy of the signal into two or more critical bands results in an increase of total loudness.

Loudness and Duration

Everything we have talked about so far applies to steady-state, long-duration stimuli. These results are reasonably well understood and can be modeled reasonably well by present-day models. However, there is a host of psychophysical data having to do with aspects of temporal structure of the signal that are less well understood and less well modeled. The subject of temporal dynamics of auditory perception is an area where there is a great deal of room for improvement in models for perceptual auditory coders. One example of this subject is the relationship between loudness and duration discussed here. Other examples appear in a later section on temporal aspects of masking.

There is general agreement that, for fixed intensity, loudness increases with duration up to stimulus durations of a few hundred milliseconds. (Other factors, usually discussed under the terms adaptation or fatigue, come into play for longer durations of many seconds or minutes. We will not discuss these factors here.) The duration below which loudness increases with increasing duration is sometimes referred to

---

[3] These data were obtained by comparing the loudness of a single 1-kHz tone and the loudness of a four-tone complex of the specified bandwidth centered at 1 kHz. The systematic difference between results when the tone was adjusted ("T" symbol) and when the complex was adjusted ("C" symbol) is an example of the bias effects mentioned in section 39.2.1 (**Loudness**).

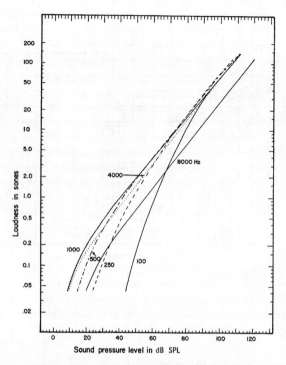

FIGURE 39.4   Loudness growth functions. (Modified from Scharf, B., Loudness, ch. 6 in *Handbook of Perception*, Vol. IV, *Hearing,* Carterette, E.C. and Friedman M.P., eds., Academic Press, New York, 1978. With permission).

as the *critical duration.* Scharf [32] provides an excellent summary of studies of the relationship between loudness and duration. In his survey, he cites values of critical duration ranging from 10 msec to over 500 msec. About half the studies in Scharf's survey show that the total energy (intensity x duration) stays constant below the critical duration for constant loudness, while the remaining studies are about evenly split between total energy increasing and total energy decreasing with increasing duration.

One possible explanation for this confused state of affairs is the inherent difficulty of making loudness matches between dissimilar stimuli, discussed above in Section 39.2.1 (**Loudness**). Two stimuli of different durations differ by more than "loudness", and depending on a variety of poorly-understood experimental or individual factors what appears to be the same experiment may yield different results in different laboratories or with different subjects.

Some support for this explanation comes from the fact that studies of threshold intensity as a function of duration are generally in better agreement with each other than studies of loudness as a function of duration. As discussed above in Section 39.2.3 (**Threshold of Hearing**) measurements of auditory threshold depend to some extent on the method of measurement, but it is still possible to establish an internally-consistent criterion-free measure. The exact results depend to some extent on signal frequency, but there is reasonable agreement among various studies that total energy at threshold remains approximately constant between about 10 msec and 100 msec. (See [41] for a survey of studies of threshold intensity as a function of duration.)

39.3.2   Differential Thresholds

Frequency

Figure 39.6 shows frequency JND as a function of frequency and intensity as measured in the most recent comprehensive study [43]. The frequency JND generally increases with increasing frequency and

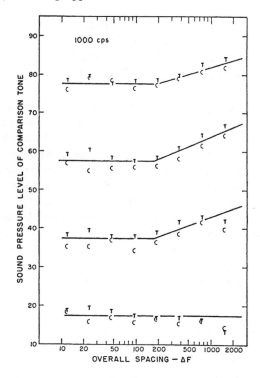

FIGURE 39.5 Loudness vs. bandwidth of tone complex. (Source: Zwicker, E. et al., Critical bandwidth in loudness summation, *J. Acoust. Soc. Am.*, 29: 548-557, 1957. With permission).

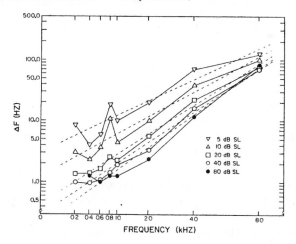

FIGURE 39.6 Frequency JND as a function of frequency and intensity (Modified from Wier, C.C. et al., Frequency discrimination as a function of frequency and sensation level, *J. Acoust. Soc. Am.*, 61: 178-184, 1977. With permission).

decreases with increasing intensity, ranging from about 1 Hz at low frequency and moderate intensity to more than 100 Hz at high frequency and low intensity.

The results shown in Fig. 39.6 are in basic agreement with results from most other studies of frequency JND's with the exception of the earliest comprehensive study, by Shower and Biddulph ([43], p. 180). Shower and Biddulph [35] found a more gradual increase of frequency JND with frequency. As we have noted above, the results obtained in experiments of this nature are strongly influenced by details of the method of measurement. Shower and Biddulph measured detectability of frequency modulation of a pure tone; most other experimenters measured the ability of subjects to correctly identify whether one tone burst was of higher or lower frequency than another. Why this difference in procedure should produce this difference in results, or even whether this difference in procedure is solely responsible for the difference in results, is unclear.

The Weber fraction $\Delta f / f$, where Δf is the frequency JND, is smallest at mid frequencies, in the region from 500 Hz to 2 kHz. It increases somewhat at lower frequencies, and it increases very sharply at high frequencies above about 4 kHz. Wier et al. [43] in their Fig. 1, reproduced here as our Fig. 39.6, plotted $\log \Delta f$ against $\sqrt{f}$. They found that this choice of axes resulted in the closest fit to a straight line. It is not clear that this choice of axes has any theoretical basis; it appears simply to be a choice that happens to work well. There have been extensive attempts to model frequency selectivity. These studies suggest that the auditory system uses the timing of individual nerve impulses at low frequencies, but that at high frequencies above a few kHz this timing information is no longer available and the auditory system relies exclusively on place information from the mechanically tuned inner ear.

Rosenblith and Stevens [30] provide an interesting example of the interaction between method of measurement and observed result. They compared frequency JNDs using two methods. One was an "AX" method, in which the subject judged whether the second of a pair of tone bursts was of higher or lower frequency than the first of the pair. The other was an "ABX" method, in which the subject judged whether the third of three tone bursts, at the same frequency as one of the first two tone bursts, was more similar to the first or to the second burst. They found that frequency JNDs measured using the AX method were approximately half the size of frequency JNDs measured using the ABX method, and they concluded that "... it would be rather imprudent to postulate a "true" DL (difference limen), or to infer the behavior of the peripheral organ from the size of a DL measured under a given set of conditions". They discussed their results in terms of information theory, an active topic at the time, and were unable to reach any definite conclusion. An analysis of their results in terms of detection theory, which at that time was in its infancy, predicts their results almost exactly.[4]

Intensity

The Weber fraction $\Delta I / I$ for pure tones is not constant but decreases slightly as stimulus intensity increases. This change has been termed the *near miss to Weber's law*. In most studies, the Weber fraction has been found to be independent of frequency. An exception is Riesz's study [29], in which the Weber fraction was at a minimum at approximately 2 kHz and increased at higher and lower frequencies.

Typical results are summarized in Fig. 39.7 ([18], Fig. 4). The solid straight line is a good fit to Jesteadt's intensity JND data at frequencies from 200 Hz to 8 kHz. The Weber fraction decreases from about 0.44 at 5 dB SL (decibels above threshold) to about 0.12 at 80 dB SL. These results are in substantial agreement with most other studies with the exception of Riesz's study. Riesz's data are shown in Fig. 39.7 as the curves identified by symbols. There is a larger change of intensity JND with intensity, and the intensity JND depends on frequency.

There is an interesting parallel between the results for intensity JND and the results for frequency JND. In both cases, results from most studies are in agreement with the exception of one study: Shower and Biddulph for frequency JND, and Riesz for intensity JND. In both cases, most studies measured the ability of subjects to correctly identify the difference between two tone bursts. Both of the outlying studies measured, instead, the ability of subjects to identify modulation of a tone: Shower and Biddulph used

---

[4]Assume RV's A, B, and X are drawn independently from normal distributions with means m_A, m_B and m_X, respectively, and equal standard deviations σ. It can be shown that the relevant decision variable in the AX experiment has mean $m_A - m_X$ and standard deviation $\sqrt{2} \times \sigma$, while the relevant decision variable in the ABX experiment has mean $m_A - m_B$ and standard deviation $\sqrt{6} \times \sigma$, a value almost twice as large.

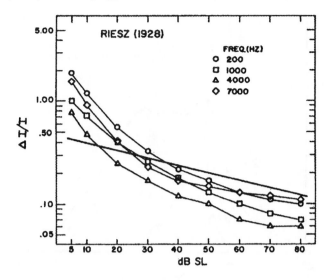

FIGURE 39.7   Summary of intensity JNDs for pure tones. Jesteadt et al. [18] found that the Weber fraction $\Delta I/I$ was independent of frequency (straight line). Riesz [29], using a different procedure, found a dependence (connected points). (Source: Jesteadt, W. et al., Intensity Discrimination as a function of frequency and sensation level, *J. Acoust. Soc. Am.*, 61: 169-177, 1977. With permission).

frequency modulation and Riesz used amplitude modulation. It appears that a modulated continuous tone may give different results than a pair of tone bursts. Whether this is a real effect, and, if it is, whether it is due to stimulus artifact or to properties of the auditory system, is unclear. The subject merits further investigation.

The Weber fraction for wideband noise appears to be independent of intensity. Miller [21] measured detectability of intensity increments in wide-band noise and found that the Weber fraction $\Delta I/I$ was approximately constant at 0.099 above 30 dB SL. It increased below 30 dB SL, which led Miller to revive Fechner's modification of Weber's law as discussed above in Section 39.2.4 (**Differential Threshold**).

39.3.3   Masking

No aspect of auditory psychophysics is more relevant to the design of perceptual auditory coders than masking, since the basic objective is to use the masking properties of speech to hide the coding noise. It will be seen that while we can use present-day knowledge of masking to great advantage, there is still much to be learned about properties of masking if we are to fully exploit it. Since some of the major unresolved problems in modeling masking are related to the relative bandwidth of masker and probe, our approach here is to present masking in terms of this relative bandwidth.

Tone Probe, Tone Masker

At one time, perhaps because of the demonstrated power of the Fourier transform in the analysis of linear time-invariant systems, the sine wave was considered to be the "natural" signal to be used in studies of human hearing. Much of the earliest work on masking dealt with the masking of one tone by another [42]. Typical results are shown in Fig. 39.8 ([3], Fig. 1). Similar results appear in Wegel and Lane [42]. The abscissa is probe frequency and the ordinate is masking in dB, the elevation of masked over absolute threshold (15 dB SPL for 400-Hz tone). Three curves are shown, for 400-Hz maskers at 40, 60, and 80 dB SPL.

Masking is greatest for probe frequencies slightly above or below the masker frequency of 400 Hz. Maximum probe-to-masker ratios are −19 dB for an 80 dB SPL masker (probe intensity elevated 46 dB

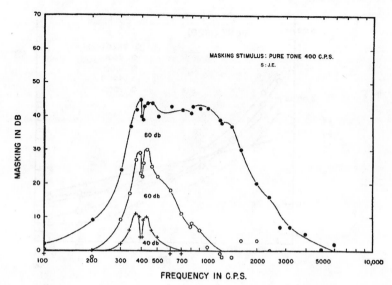

FIGURE 39.8   Masking of tones by a 400-Hz tone at 40, 60, and 80 dB SPL. (Source: Egan, J.P. and Hake, H.W., On the masking pattern of a simple auditory stimulus, *J. Acoust. Soc. Am.*, 22: 622-630, 1950).

above the absolute threshold of 15 dB SPL), -15 dB for a 60 dB SPL masker, and -14 dB for a 40 dB SPL masker.

Masking decreases as probe frequency gets closer to 400 Hz. The probe frequencies closest to 400 Hz are 397 and 403 Hz, and at these frequencies the threshold probe-to-masker ratio is -26 dB for an 80 dB SPL masker, -23 dB for a 60 dB SPL masker, and -21 dB for a 40 dB SPL masker.

Masking also decreases as probe frequency gets further away from masker frequency. For the 40 dB SPL masker this selectivity is nearly symmetric in log frequency, but as the masker intensity increases the masking becomes more and more asymmetric so that the 400-Hz masker produces much more masking at higher frequencies than at lower.

The irregularities seen near probe frequencies of 400, 800, and 1200 Hz are the result of interactions between masker and probe. When masker and probe frequencies are close, beating results. Even when their frequencies are far apart, nonlinear effects in the auditory system result in complex interactions. These irregularities provided incentive to use narrow bands of noise, rather than pure tones, as maskers.

Tone Probe, Noise Masker

Fletcher and Munson [8] were among the first to use bands of noise as maskers. Figure 39.9 ([3], Fig. 2) shows typical results. The conditions are similar to those for Fig. 39.8 except that now the masker is a band of noise 90 Hz wide centered at 410 Hz. The maximum probe-to-masker ratios occur for probe frequencies slightly above the center frequency of the masker, and they are much greater than they were for the tone maskers shown in Fig. 39.8. Maximum probe-to-masker ratios are -4 dB for an 80 dB SPL masker and -3 dB for 60 and 40 dB SPL maskers. The frequency selectivity and upward spread of masking seen in Fig. 39.8 appear in Fig. 39.9 as well, but the irregularities seen at harmonics of the masker frequency are greatly reduced.

An important effect that occurs in connection with masking of a tone probe by a band of noise is the relationship between masker bandwidth and amount of masking. This relationship can be presented in many ways, but the results can be described to a reasonable degree of accuracy by saying that noise energy within a narrow band of frequencies surrounding the probe contributes to masking while noise energy

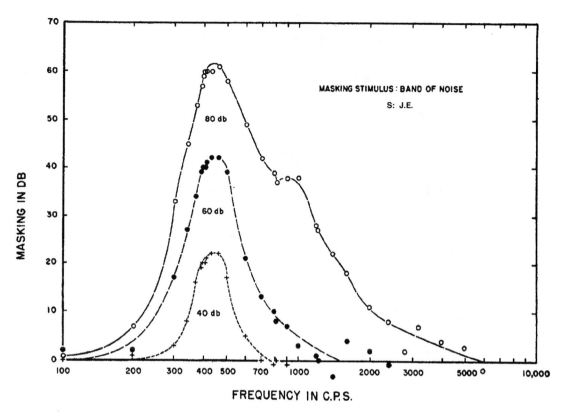

FIGURE 39.9 Masking of tones by a 90-Hz wide band of noise centered at 410 Hz at 40, 60, and 80 dB SPL (Source: Egan, J.P. and Hake, H.W., On the masking pattern of a simple auditory stimulus, *J. Acoust. Soc. Am.*, 22: 622-630, 1950).

outside this band of frequencies does not. This is one manifestation of the *critical band* described in Section 39.2.6 (**Critical Bands and Peripheral Auditory Filters**).

Figure 39.10 ([2], Fig. 6) shows results from a series of experiments designed to determine the widths of critical bands. We are most concerned here with the closed symbols and the associated solid and dotted straight lines. These show an expanded and elaborated repeat of a test Fletcher reported in 1940 to measure the width of critical bands, and the results shown here are similar to Fletcher's results ([7], Fig. 124). The closed symbols show threshold level of probe signals at frequencies ranging from 500 Hz to 8 kHz in dB relative to the intensity of a masking band of noise centered at the frequency of the test signal and with the bandwidth shown on the abscissa. The intensity of the masking noise is 60 dB SPL per 1/3 octave. Note that for narrow-band maskers the probe-to-masker ratio is nearly independent of bandwidth, while for wide-band maskers the probe-to-masker ratio decreases at approximately 3 dB per doubling of bandwidth. This result indicates that above a certain bandwidth, approximated in this figure as the intersection of the asymptotic narrow-band horizontal line and the asymptotic wide-band sloping lines, noise energy outside of this band does not contribute to masking.

The results shown in Fig. 39.10 are from only one of many studies of masking of pure tones by noise bands of varying bandwidths that lead to similar conclusions. The list includes Feldtkeller and Zwicker [5] and Greenwood [13]. Scharf [33] provides additional references.

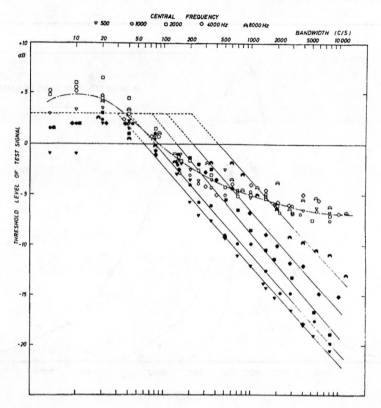

FIGURE 39.10　Threshold level of probe signals from 500 Hz to 8 kHz relative to overall level of noise masker at bandwidth shown on the abscissa. (Modified from Bos, C.E. and de Boer, E., Masking and discrimination, *J. Acoust. Soc. Am.*, 39: 708-715, 1966. With permission).

Noise Probe, Tone or Noise Masker

Masking of bands of noise, either by tone or noise maskers, has received relatively little attention. This is unfortunate for the designer who is concerned with masking wide-band coding noise. Masking of noise by tones is touched on in Zwicker [47], but the earliest study that gives actual data points appears to be Hellman [15]. The threshold probe-to-masker ratios for a noise probe approximately one critical band wide were −21 dB for a 60 dB SPL masker and −28 dB for a 90 dB SPL masker. Threshold probe-to-masker ratios for an octave-band probe were −55 dB for 1-kHz maskers at 80 and 100 dB SPL. A 1-kHz masker at 90 dB SPL produced practically no masking of a wide-band probe.

Hall [34] measured threshold intensity for noise bursts one-half, one, and two critical bands wide with various center frequencies in the presence of 80 dB SPL pure-tone maskers ranging from an octave below to an octave above the center frequency. Figure 39.11 shows results for a critical-band 1-kHz probe. The threshold probe-to-masker ratio for a 1-kHz masker is −24 dB, in agreement with Hellman's results, and the figure shows the same upward spread of masking that appears in Figs. 39.8 and 39.9. (Note that in Figs. 39.8 and 39.9 the masker is fixed and the abscissa is probe frequency, while in Fig. 39.11 the probe is fixed and the abscissa is masker frequency.) A tone below 1 kHz produces more masking than a tone above 1 kHz.

Masking of noise by noise is confounded by the question of phase relationships between probe and masker. If masker and probe are identical in bandwidth and phase, then as we saw in Section 39.2.5 (**Masked Threshold**) the masked threshold becomes identical to the differential threshold. Miller's [21]

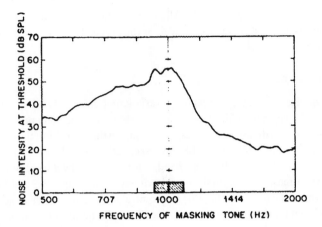

FIGURE 39.11   Threshold intensity for a 923-1083 Hz band of noise masked by an 80-dB SPL tone at the frequency shown on the abscissa. (Source: Schroeder, M.R. et al., Optimizing digital speech coders by exploiting masking properties of the human ear, *J. Acoust. Soc. Am.*, 66: 1647-1652, 1979. With permission).

Weber fraction $\Delta I / I$ of 0.099 for intensity discrimination of wide-band noise, phrased in terms of intensity of the just-detectable increment, leads to a probe-to-masker ratio of -26.3 dB.

More recently, Hall [14] measured threshold intensity for various combinations of probe and masker bandwidths. These experiments differ from earlier experiments in that phase relationships between probe and masker were controlled: all stimuli were generated by adding together equal-amplitude random phase sinusoidal components, and components common to probe and masker had identical phase. Results for one subject are shown in Fig. 39.12. Masker bandwidth appears on the abscissa, and the parameter is probe bandwidth: $A \Rightarrow 0$ Hz, $B \Rightarrow 4$ Hz, $C \Rightarrow 16$ Hz, and $D \Rightarrow 64$ Hz. All stimuli were centered at 1 kHz and the overall intensity of the masker was 70 dB SPL.

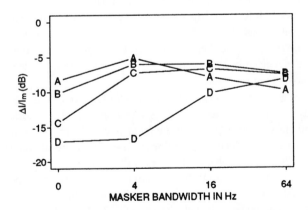

FIGURE 39.12   Intensity increment between masker alone and masker plus just-detectable probe. Probe bandwidth 0 Hz (A); 4 Hz (B); 16 Hz (C); 64 Hz (D). Frequency components common to probe and masker have identical phase. (Source: Hall, J.L., Asymmetry of masking revisited: generalization of masker and probe bandwidth, *J. Acoust. Soc. Am.*, 101: 1023–1033, 1997. With permission).

This figure differs from Figs. 39.8 through 39.11 in that the vertical scale shows intensity increment between masker alone and masker plus just-detectable probe rather than intensity of the just-detectable probe, and the results look quite different. For all probe bandwidths shown, the intensity increment varies only slightly so long as the masker is at least as wide as the probe. The intensity increment decreases when the probe is wider than the masker.

Asymmetry of Masking

Inspection of Figs. 39.8– 39.11 reveals large variation of threshold probe-to-masker intensity ratios depending on the relative bandwidth of probe and masker. Tone maskers produce threshold probe-to-masker ratios of −14 to −26 dB for tone probes, depending on the intensity of the masker and the frequency of the probe (Fig. 39.8), and threshold probe-to-masker ratios of −21 to −28 dB for critical-band noise probes ([15]; also Fig. 39.11). On the other hand, a tone masked by a band of noise is audible only at much higher probe-to-masker ratios, in the neighborhood of 0 dB (Figs. 39.9 and 39.10). This *asymmetry of masking* (the term is due to Hellman, [15]) is of central importance in the design of perceptual coders because of the different masking properties of noise-like and tone-like portions of the coded signal [19]. Current perceptual models do not handle this asymmetry well, so it is a subject we must examine closely.

The logical conclusion to be drawn from the numbers in the preceding paragraph at first appears to be that a band of noise is a better masker than a tone, for both noise and tone probes. In fact, the correct conclusion may be completely different. It can be argued that so long as the masker bandwidth is at least as wide as the probe bandwidth, tones or bands of noise are equally effective maskers and the psychophysical data can be described satisfactorily by current energy-based perceptual models, properly applied. It is only when the bandwidth of the probe is greater than the bandwidth of the masker that energy-based models break down and some criterion other than average energy must be applied.

Figure 39.13 shows Egan and Hake's results for 80 dB SPL tone and noise maskers superimposed on each other. Results for the tone masker are shown as a solid curve and results for the noise masker are shown as a dashed curve. (These curves are not identical to the corresponding curves in Figs. 39.8 and 39.9: They are average results from five subjects, while Figs. 39.8 and 39.9 were for a single subject.) The maximum amount of masking produced by the band of noise is 61 dB, while the tone masker produces only 37 dB of masking for a 397-Hz probe.

The difference between tone and noise maskers may be more apparent than real, and for masking of a tone the auditory system may be similarly affected by tone and noise maskers. What is plotted in this figure is the elevation in threshold intensity of the probe tone by the masker, but the discrimination the subject makes is in fact between masker alone and masker plus probe. As was discussed above in Section 39.2.5 (**Masked Threshold**), since coherence between tone probe and masker depends on the bandwidth of the masker, a probe tone of a given intensity can produce a much greater change in intensity of probe plus masker for a tone masker than for a noise masker.

The stimulus in the Egan and Hake experiment with a 400-Hz masker and a 397-Hz probe is identical to the stimulus Riesz used to measure intensity JND (see Section 39.3.2 **Differential Thresholds: Intensity**, above). As Egan and Hake observe "... When the frequency of the masked stimulus is 397 or 403 c.p.s. [Hz], the amount of masking is evidently determined by the value of the differential threshold for intensity at 400 c.p.s." ([3], p. 624). Specifically, for the results shown in Fig. 39.13, the threshold intensity of a 397-Hz tone is 52 dB SPL. This leads to a Weber fraction $\Delta I / I$ (power at envelope maximum minus power at envelope minimum, divided by power at envelope minimum) of 0.17, which is only slightly higher than values obtained by Riesz and by Jesteadt et al. shown in Fig. 39.7.

The situation with noise masker is more difficult to analyze because of the random nature of the masker. The effective intensity increment between masker alone and masker plus probe depends on the phase relationship between probe and 400-Hz component of the masker, which are uncontrolled in the Egan and Hake experiment, and also on the effective time constant and bandwidth of the analyzing auditory filter, which are unknown. However, for the experiment shown in Fig. 39.12 the maskers were computer-generated repeatable stimuli, so that the intensity of masker plus probe could be computed. The results shown in Fig. 39.12 lead to a Weber fraction $\Delta I / I$ of 0.15 for tone masked by tone and 0.10 for tone

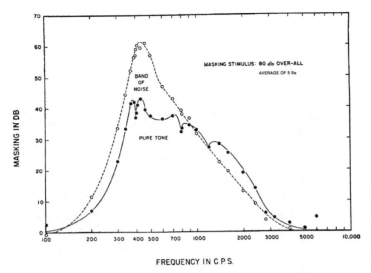

FIGURE 39.13   Masking produced by a 400-Hz masker at 80 dB SPL and a 90-Hz wide band of noise centered at 410 Hz. (Source: Egan, J.P. and Hake, H.W., On the masking pattern of a simple auditory stimulus, *J. Acoust. Soc. Am.*, 22: 622-630, 1950).

masked by 64-Hz wide noise. Results are similar for noise masked by noise, so long as the masker is at least as wide as the probe. Weber fractions for the 64-Hz wide masker in Fig. 39.12 range from 0.18 for a 4-Hz wide probe to 0.15 for a 64-Hz wide probe.

Our understanding of the factors leading to the results shown in Fig. 39.12 is obviously very limited, but these results appear to be consistent with the view that to a first-order approximation the relevant variable for masking is the Weber fraction $\Delta I/I$, the intensity of masker plus probe relative to the intensity of the masker, so long as the masker is at least as wide as the probe. This is true for both tone and noise maskers. Because of changes in coherence between probe and masker as masker bandwidth changes, the corresponding probe intensity at threshold can be much lower for a tone masker than for a probe masker, as is shown in Fig. 39.13.

The asymmetry that Hellman was primarily concerned with in her 1972 paper is the striking difference between the threshold of a band of noise masked by a tone and of a tone masked by a band of noise. It appears that this is a completely different effect than the asymmetry shown in Fig. 39.13 and one that cannot be accounted for by current energy-based models of masking. The difference between the −5 to +5 dB threshold probe-to-masker ratios seen in Figs. 39.9 and 39.10 for tones masked by noise and the −21 to −28 dB threshold probe-to-masker ratios for noise masked by tone reported by Hellman and seen in Fig. 39.11 is due in part to the random nature of the noise masker and to the change in coherence between masker and probe that we have already discussed. Even when these factors are controlled, as in Fig. 39.12, decrease of masker bandwidth for a 64-Hz wide band of noise results in a decrease of threshold intensity increment. (The situation is complicated by the possibility of off-frequency listening. As we have already seen, neither a tone nor a noise masker masks remote frequencies effectively. The 64-Hz band is narrow enough that off-frequency listening is not a factor.) These and similar results lead to the conclusion that present-day models operating on signal power are inadequate and that some envelope-based measure, such as the envelope maximum or ratio of envelope maximum to minimum, must be considered [10, 11, 38].

Temporal Aspects of Masking

Up until now, we have discussed masking effects with simultaneous masker and probe. In order to be able to deal effectively with a dynamically varying signal such as speech, we need to consider nonsimultaneous masking as well. When the probe follows the masker, the effect is referred to as *forward masking*.

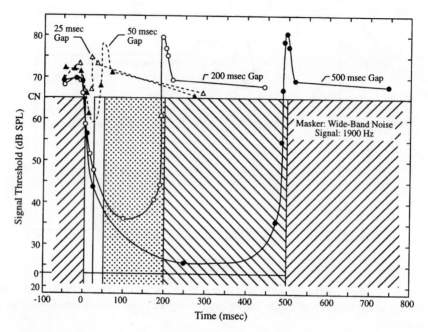

FIGURE 39.14   Masking of tone by ongoing wide-band noise with silent interval of 25, 50, 200, or 500 msec. This figure shows simultaneous, forward, backward, forward fringe, and backward fringe masking. (Source: Elliott, L.L., Masking of tones before, during, and after brief silent periods in noise, *J. Acoust. Soc. Am.*, 45: 1277-1279, 1969. With permission).

When the masker follows the probe, it is referred to as *backward masking*. Effects have also been measured with a brief probe near the beginning or the end of a longer-duration masker. These effects have been referred to as *forward* or *backward fringe masking*, respectively ([44], p. 162).

The various kinds of simultaneous and non-simultaneous masking are nicely illustrated in Fig. 39.14 ([4], Fig. 1). The masker was wideband noise at an overall level of 70 dB SPL and the probe was a brief 1.9-kHz tone burst. The masker was on continuously except for a silent interval of 25, 50, 200, or 500 msec. beginning at the 0-msec point on the abscissa. The four sets of data points show thresholds for probes presented at various times relative to the gap for the four gap durations. Probe thresholds in silence and in continuous noise are indicated on the ordinate by the symbols "**Q**" and "**CN**".

Forward masking appears as the gradual drop of probe threshold over a duration of more than 100 msec following the cessation of the masker. Backward masking appears as the abrupt increase of masking, over a duration of a few tens of msec, immediately before the reintroduction of the masker. Forward fringe masking appears as the more than 10-dB overshoot of masking immediately following the reintroduction of the masker, and backward fringe masking appears as the smaller overshoot immediately preceding the cessation of the masker. Backward masking is an important effect for the designer of coders for acoustic signals because of its relationship to audibility of pre-echo. It is a puzzling effect, because it is caused by a masker that begins only after the probe has been presented. Stimulus-related electrophysiological events can be recorded in the cortex several tens of msec after presentation of the stimulus, so there may be some physiological basis for backward masking. It is an unstable effect, and there is some evidence that backward masking decreases with practice [20], ([23], p. 119).

Forward masking is a more robust effect, and it has been studied extensively. It is a complex function of stimulus parameters, and we do not have a comprehensive model that predicts amount of forward masking as a function of frequency, intensity, and time course of masker and of probe. The following two examples illustrate some of its complexity.

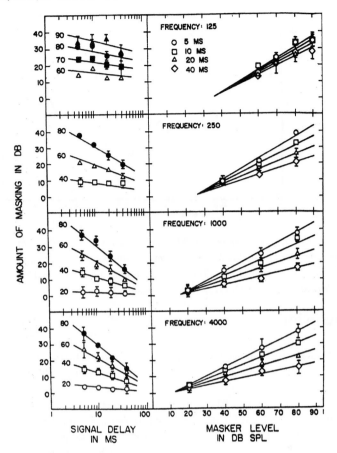

FIGURE 39.15   Forward masking with identical masker and probe frequencies, as a function of frequency, delay, and masker level (Source: Jesteadt, W. et al., Forwarding masking as a function of frequency, masker level, and signal delay, *J. Acoust. Soc. Am.*, 71: 950-962, 1982. With permission).

Figure 39.15 ([17], Fig. 1) is from a study of the effects of masker frequency and intensity on forward masking. Masker and probe were of the same frequency. The left and right columns show the same data, plotted on the left against probe delay with masker intensity as a parameter and plotted on the right against masker intensity with probe delay as a parameter. The amount of masking depends in an orderly way on masker frequency, masker intensity, and probe delay. Jesteadt et al. were able to fit these data with a single equation with three free constants. This equation, with minor modification, was later found to give a satisfactory fit to data obtained with forward masking by wide-band noise [25].

Striking effects can be observed when probe and masker frequencies differ. Figure 39.16 (modified from [22], Fig. 8) superimposes simultaneous (open symbols) and forward (filled symbols) masking curves for a 6-kHz probe at 36 dB SPL, 10 dB above the absolute threshold of 26 dB SPL. Rather than showing the amount of masking for a fixed masker, this figure shows masker level, as a function of masker frequency, sufficient to just mask the probe. It is clear that simultaneous and forward masking differ, and that the difference depends on the relative frequency of masker and probe. Results such as those shown in Fig. 39.16 are of interest to the field of auditory physiology because of similarities between forward masking results and frequency selectivity of primary auditory neurons.

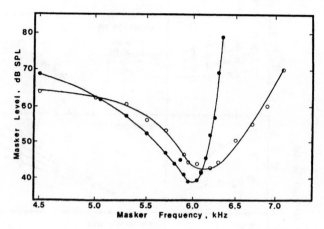

FIGURE 39.16   Simultaneous (open symbols) and forward (closed symbols) masking of a 6-kHz probe tone at 36 dB SPL. Masker frequency appears on the abscissa, and masker intensity just sufficient to mask the probe appears on the abscissa. (Modified from Moore, B.C.J., Psychophysical tuning curves measured in simultaneous and forward masking, *J. Acoust. Soc. Am.,* 63: 524-532, 1978. With permission).

39.4   Conclusions

Notwithstanding the successes obtained to date with perceptual coders for speech and audio [16, 19, 27, 36], there is still a great deal of room for further advancement. The most widely applied perceptual models today apply an energy-based criterion to some critical-band transformation of the signal and arrive at a prediction of acceptable coding noise. These models are essentially refinements of models first described by Fletcher and his co-workers and further developed by Zwicker and others [34]. These models do a good job describing masking and loudness for steady-state bands of noise, but they are less satisfactory for other signals. We can identify two areas in which there seem to be great room for improvement. One of these areas presents a challenge jointly to the designer of coders and to the auditory psychophysicist, and the other area presents a challenge primarily to the auditory psychophysicist.

One area for additional research has to do with asymmetry of masking. Noise is a more effective masker than tones, and this difference is not handled well by present-day perceptual models. Present-day coders first compute a measure of tonality of the signal and then use this measure empirically to obtain an estimate of masking. This empirical approach has been applied successfully to a variety of signals, but it is possible that an approach that is less empirical and more based on a comprehensive model of auditory perception would be more robust.

As discussed in Section 39.3.3 (**Masking:** Asymmetry of Masking), there is evidence that there are two separate factors contributing to this asymmetry of masking. The difference between noise and tone maskers for narrow-band coding noise appears to result from problems with signal definition rather than a difference in processing by the auditory system, and it may be that an effective way of dealing with it will result not from an improved understanding of auditory perception but rather from changes in the coder. A feedforward prediction of acceptable coding noise based on the energy of the signal does not take into account phase relationships between signal and noise. What may be required is a feedback, analysis-by-synthesis approach, in which a direct comparison is made between the original signal and the proposed coded signal. This approach would require a more complex encoder but leave the decoder complexity unchanged [27]. The difference between narrow-band and wide-band coding noise, on the other hand, appears to call for a basic change in models of auditory perception. For largely historical reasons, the idea of signal energy as a perceptual measure is deeply ingrained in present-day perceptual models. There is increasing realization that under some conditions signal energy is not the relevant measure but that some envelope-based measure may be required.

A second area in which additional research may prove fruitful is in the area of temporal aspects of masking. As is discussed in Section 39.3.3 (**Masking:** Temporal Aspects of Masking), the situation with time-varying signal and noise is more complex than the steady-state situation. There is an extensive body of psychophysical data on various aspects of nonsimultaneous masking, but we are still lacking a satisfactory comprehensive perceptual model. As is the case with asymmetry of masking, present-day coders deal with this problem at an empirical level, in some cases very effectively. However, as with asymmetry of masking, an approach based on fundamental properties of auditory perception would perhaps be better able to deal with a wide variety of signals.

References

[1] Allen, J.B., Harvey Fletcher's role in the creation of communication acoustics, *J. Acoust. Soc. Am.*, 99: 1825-1839, 1996.

[2] Bos, C.E. and de Boer, E., Masking and discrimination, *J. Acoust. Soc. Am.*, 39: 708-715, 1966.

[3] Egan, J.P. and Hake, H.W., On the masking pattern of a simple auditory stimulus, *J. Acoust. Soc. Am.*, 22: 622-630, 1950.

[4] Elliott, L.L., Masking of tones before, during, and after brief silent periods in noise, *J. Acoust. Soc. Am.*, 45: 1277-1279, 1969.

[5] Feldtkeller, R. and Zwicker, E., *Das Ohr als Nachrichtenempfänger*, S. Hirzel, Stuttgart, 1956.

[6] Fletcher, H., Loudness, masking, and their relation to the hearing process and the problem of noise measurement, *J. Acoust. Soc. Am.*, 9: 275-293, 1938.

[7] Fletcher, H., *Speech and Hearing in Communication, ASA Edition*, Allen, J.B., Ed., American Institute of Physics, New York, 1995.

[8] Fletcher, H. and Munson, W.A., Relation between loudness and masking, *J. Acoust. Soc. Am.*, 9: 1-10, 1937.

[9] Goldstein, J.L., Auditory spectral filtering and monaural phase perception, *J. Acoust. Soc. Am.*, 41: 458-479, 1967.

[10] Goldstein, J.L., Comparison of peak and energy detection for auditory masking of tones by narrow-band noise, *J. Acoust. Soc. Am.*, 98(A): 2907, 1995.

[11] Goldstein, J.L. and Hall, J.L., Peak detection for auditory sound discrimination, *J. Acoust. Soc. Am.*, 97(A): 3330, 1995.

[12] Green, D.M. and Swets, J.A., *Signal Detection Theory and Psychophysics*, John Wiley & Sons, New York, 1966.

[13] Greenwood, D.D., Auditory masking and the critical band, *J. Acoust. Soc. Am.*, 33: 484-502, 1961.

[14] Hall, J.L., Asymmetry of masking revisited: generalization of masker and probe bandwidth, *J. Acoust. Soc. Am.*, 101: 1023–1033, 1997.

[15] Hellman, R.P., Asymmetry of masking between noise and tone, *Perception and Psychophsyics*, 11: 241-246, 1972.

[16] Jayant, N., Johnston, J., and Safranek, R., Signal compression based on models of human perception, *Proc. IEEE*, 81: 1385-1422, 1993.

[17] Jesteadt, W., Bacon, S.P., and Lehman, J.R., Forward masking as a function of frequency, masker level, and signal delay, *J. Acoust. Soc. Am.*, 71: 950-962, 1982.

[18] Jesteadt, W., Wier, C.C., and Green, D.M., Intensity discrimination as a function of frequency and sensation level, *J. Acoust. Soc. Am.*, 61: 169-177, 1977.

[19] Johnston, J.D., Audio coding with filter banks, in *Subband and Wavelet Transforms, Design and Applications*, ch. 9, Akansu, A.N. and Smith, M.J.T., Eds., Kluwer Academic, Boston, 1966a.

[20] Johnston, J.D., Personal communication, 1996b.

[21] Miller, G.A., Sensitivity to changes in the intensity of white noise and its relation to masking and loudness, *J. Acoust. Soc. Am.*, 19: 609-619, 1947.

[22] Moore, B.C.J., Psychophysical tuning curves measured in simultaneous and forward masking, *J. Acoust. Soc. Am.*, 63: 524-532, 1978.

[23] Moore, B.C.J., *An Introduction to the Psychology of Hearing*, Academic Press, London, 1989.

[24] Moore, B.C.J., *Frequency Selectivity in Hearing*, Academic Press, London, 1986.

[25] Moore, B.C.J. and Glasberg, B.R., Growth of forward masking for sinusoidal and noise maskers as a function of signal delay: implications for suppression in noise, *J. Acoust. Soc. Am.*, 73: 1249-1259, 1983a.

[26] Moore, B.C.J. and Glasberg, B.R., Suggested formulae for calculating auditory-filter bandwidths and excitation patterns, *J. Acoust. Soc. Am.*, 74: 750-757, 1983b.

[27] Noll, P., MPEG/Audio coding standards.

[28] Reynolds, G.S. and Stevens, S.S., Binaural summation of loudness, *J. Acoust. Soc. Am.*, 32: 1337-1344, 1960.

[29] Riesz, R.R., Differential intensity sensitivity of the ear for pure tones, *Phys. Rev.*, 31: 867-875, 1928.

[30] Rosenblith, W.A. and Stevens, K.N., On the DL for frequency, *J. Acoust. Soc. Am.*, 25: 980-985, 1953.

[31] Scharf, B., Critical bands, in *Foundations of Modern Auditory Theory*, Vol. 1, ch. 5, Tobias, J.V., Ed., Academic Press, New York, 1970.

[32] Scharf, B., Loudness, in *Handbook of Perception, Vol. IV, Hearing*, ch. 6, Carterette, E.C. and Friedman, M.P., Eds., Academic Press, New York, 1978.

[33] Scharf, B. and Fishkin, D., Binaural summation of loudness: reconsidered, *J. Exp. Psychol.*, 86: 374-379, 1970.

[34] Schroeder, M.R., Atal, B.S., and Hall, J.L., Optimizing digital speech coders by exploiting masking properties of the human ear, *J. Acoust. Soc. Am.*, 66: 1647-1652, 1979.

[35] Shower, E.G. and Biddulph, R., Differential pitch sensitivity of the ear, *J. Acoust. Soc. Am.*, 3: 275-287, 1931.

[36] Sinha, D., Johnston, J.D., Dorward, S., and Quackenbush, S.R., The perceptual audio coder (PAC).

[37] Stevens, S.S. and Davis, H.W., *Hearing*, John Wiley & Sons, New York, 1938.

[38] Strickland, E.A. and Viemeister, N.F., Cues for discrimination of envelopes, *J. Acoust. Soc. Am.*, 99: 3638-3646, 1996.

[39] Von Gierke, H.E. and Ward, W.D., Criteria for noise and vibration exposure, in *Handbook of Acoustical Measurements and Noise Control*, 3rd ed., ch. 26, Harris, C.M., Ed., McGraw-Hill, New York, 1991.

[40] Ward, W.D., Musical perception, in *Foundations of Modern Auditory Theory*, Vol. 1, ch. 11, Tobias, J.V., Ed., Academic Press, New York, 1970.

[41] Watson, C.S. and Gengel, R.W., Signal duration and signal frequency in relation to auditory sensitivity, *J. Acoust. Soc. Am.*, 46: 989-997, 1969.

[42] Wegel, R.L. and Lane, C.E., The auditory masking of one pure tone by another and its probable relation to the dynamics of the inner ear, *Phys. Rev.*, 23: 266-285, 1924.

[43] Wier, C.C., Jesteadt, W., and Green, D.M., Frequency discrimination as a function of frequency and sensation level, *J. Acoust. Soc. Am.*, 61: 178-184, 1977.

[44] Yost, W.A., *Fundamentals of Hearing, An Introduction*, 3rd ed., Academic Press, New York, 1994.

[45] Zwicker, E., Die Grenzen der Hörbarkeit der Amplitudenmodulation und der Frequenzmodulation eines Tones, *Acustica* 2: 125-133, 1952.

[46] Zwicker, E., Über psychologische und methodische Grundlagen der Lautheit, *Acustica* 8: 237-258, 1958.

[47] Zwicker, E., Über die Lautheit von ungedrosselten und gedrosselten Schallen, *Acustica* 13: 194-211, 1963.

[48] Zwicker, E., Flottorp, G., and Stevens, S.S., Critical bandwidth in loudness summation, *J. Acoust. Soc. Am.*, 29: 548-557, 1957.

[49] Zwicker, E. and Scharf, B., A model of loudness summation, *Psychol. Rev.*, 16: 3-26, 1965.

40

MPEG Digital Audio Coding Standards

Peter Noll
Technical University of Berlin

40.1   Introduction

PCM Bit Rates

Typical audio signal classes are telephone speech, wideband speech, and wideband audio, all of which differ in bandwidth, dynamic range, and in listener expectation of offered quality. The quality of telephone-bandwidth speech is acceptable for telephony and for some videotelephony and video-conferencing services. Higher bandwidths (7 kHz for wideband speech) may be necessary to improve the intelligibility and naturalness of speech. Wideband (high fidelity) audio representation including multichannel audio needs bandwidths of at least 15 kHz.

The conventional digital format for these signals is PCM, with sampling rates and amplitude resolutions (PCM bits per sample) as given in Table 40.1.

The *compact disc* (CD) is today's *de facto standard* of digital audio *representation*. On a CD with its 44.1 kHz sampling rate the resulting stereo net bit rate is $2 \times 44.1 \times 16 \times 1000 \equiv 1.41$ Mb/s (see Table 40.2). However, the CD needs a significant overhead for a runlength-limited line code, which maps 8 information bits into 14 bits, for synchronization and for error correction, resulting in a 49-bit representation of each 16-bit audio sample. Hence, the total stereo bit rate is $1.41 \times 49/16 = 4.32$Mb/s. Table 40.2 compares bit rates of the compact disc and the *digital audio tape* (DAT).

0-8493-8572-5/98/$0.00+$.50
© 1998 by CRC Press LLC

TABLE 40.1   Basic Parameters for Three Classes of Acoustic Signals

	Frequency range in Hz	Sampling rate in kHz	PCM bits per sample	PCM bit rate in kb/s
Telephone speech	300 - 3,400^a	8	8	64
Wideband speech	50 - 7,000	16	8	128
Wideband audio (stereo)	10 - 20,000	48^b	2 × 16	2 × 768

^{a} Bandwidth in Europe; 200 to 3200 Hz in the U.S.
^{b} Other sampling rates: 44.1 kHz, 32 kHz.

TABLE 40.2   CD and DAT Bit Rates

Storage device	Audio rate (Mb/s)	Overhead (Mb/s)	Total bit rate (Mb/s)
Compact disc (CD)	1.41	2.91	4.32
Digital audio tape (DAT)	1.41	1.05	2.46

Note: Stereophonic signals, sampled at 44.1 kHz; DAT supports also sampling rates of 32 kHz and 48 kHz.

For archiving and processing of audio signals, sampling rates of at least 2 × 44.1 kHz and amplitude resolutions of up to 24 b per sample are under discussion. Lossless coding is an important topic in order not to compromise audio quality in any way [1]. The digital versatile disk (DVD) with its capacity of 4.7 GB is the appropriate storage medium for such applications.

Bit Rate Reduction

Although high bit rate channels and networks become more easily accessible, low bit rate coding of audio signals has retained its importance. The main motivations for low bit rate coding are the need to minimize transmission costs or to provide cost-efficient storage, the demand to transmit over channels of limited capacity such as mobile radio channels, and to support variable-rate coding in packet-oriented networks.

Basic requirements in the design of low bit rate audio coders are first, to retain a high quality of the reconstructed signal with robustness to variations in spectra and levels. In the case of stereophonic and multichannel signals spatial integrity is an additional dimension of quality. Second, robustness against random and bursty channel bit errors and packet losses is required. Third, low complexity and power consumption of the codecs are of high relevance. For example, in broadcast and playback applications, the complexity and power consumption of audio decoders used must be low, whereas constraints on encoder complexity are more relaxed. Additional network-related requirements are low encoder/decoder delays, robustness against errors introduced by cascading codecs, and a graceful degradation of quality with increasing bit error rates in mobile radio and broadcast applications. Finally, in professional applications, the coded bit streams must allow editing, fading, mixing, and dynamic range compression [1].

We have seen rapid progress in bit rate compression techniques for speech and audio signals [2]–[7]. Linear prediction, subband coding, transform coding, as well as various forms of vector quantization and entropy coding techniques have been used to design efficient coding algorithms which can achieve substantially more compression than was thought possible only a few years ago. Recent results in speech and audio coding indicate that an excellent coding quality can be obtained with bit rates of 1 b per sample for speech and wideband speech and 2 b per sample for audio. Expectations over the next decade are that the rates can be reduced by a factor of four. Such reductions shall be based mainly on employing sophisticated forms of adaptive noise shaping controlled by psychoacoustic criteria. In storage and ATM-based applications additional savings are possible by employing variable-rate coding with its potential to offer a time-independent constant-quality performance.

Compressed digital audio representations can be made less sensitive to channel impairments than analog ones if source and channel coding are implemented appropriately. Bandwidth expansion has often been mentioned as a disadvantage of digital coding and transmission, but with today's data compression and multilevel signaling techniques, channel bandwidths can be reduced actually, compared with analog systems. In broadcast systems, the reduced bandwidth requirements, together with the error robustness of the coding algorithms, will allow an efficient use of available radio and TV channels as well as "taboo" channels currently left vacant because of interference problems.

MPEG Standardization Activities

Of particular importance for digital audio is the standardization work within the International Organization for Standardization (ISO/IEC), intended to provide international standards for audiovisual coding. ISO has set up a Working Group WG 11 to develop such standards for a wide range of communications-based and storage-based applications. This group is called MPEG, an acronym for *Moving Pictures Experts Group.*

MPEG's initial effort was the MPEG Phase 1 (MPEG-1) coding standards *IS 11172* supporting bit rates of around 1.2 Mb/s for video (with video quality comparable to that of today's analog video cassette recorders) and 256 kb/s for two-channel audio (with audio quality comparable to that of today's compact discs) [8].

The more recent MPEG-2 standard *IS 13818* provides standards for high quality video (including High Definition TV) in bit rate ranges from 3 to 15 Mb/s and above. It provides also new audio features including low bit rate digital audio and multichannel audio [9].

Finally, the current MPEG-4 work addresses standardization of audiovisual coding for applications ranging from mobile access low complexity multimedia terminals to high quality multichannel sound systems. MPEG-4 will allow for interactivity and universal accessibility, and will provide a high degree of flexibility and extensibility [10].

MPEG-1, MPEG-2, and MPEG-4 standardization work will be described in Sections 40.3 to 40.5 of this paper. *Web information about MPEG* is available at different addresses. The official MPEG Web site offers crash courses in MPEG and ISO, an overview of current activities, MPEG requirements, workplans, and information about documents and standards [11]. Links lead to collections of frequently asked questions, listings of MPEG, multimedia, or digital video related products, MPEG/Audio resources, software, audio test bitstreams, etc.

40.2 Key Technologies in Audio Coding

First proposals to reduce wideband audio coding rates have followed those for speech coding. Differences between audio and speech signals are manifold; however, audio coding implies higher sampling rates, better amplitude resolution, higher dynamic range, larger variations in power density spectra, stereophonic and multichannel audio signal presentations, and, finally, higher listener expectation of quality. Indeed, the high quality of the CD with its 16-b per sample PCM format has made digital audio popular.

Speech and audio coding are similar in that in both cases quality is based on the properties of human auditory perception. On the other hand, speech can be coded very efficiently because a *speech production model* is available, whereas nothing similar exists for audio signals.

Modest reductions in audio bit rates have been obtained by instantaneous companding (e.g., a conversion of uniform 14-bit PCM into a 11-bit nonuniform PCM presentation) or by forward-adaptive PCM (block companding) as employed in various forms of *near-instantaneously companded audio multiplex* (NICAM) coding [ITU-R, Rec. 660]. For example, the British Broadcasting Corporation (BBC) has used the NICAM 728 coding format for digital transmission of sound in several European broadcast television networks; it uses 32-kHz sampling with 14-bit initial quantization followed by a compression to a 10-bit format on the basis of 1-ms blocks resulting in a total stereo bit rate of 728 kb/s [12]. Such adaptive PCM schemes can solve the problem of providing a sufficient dynamic range for audio coding but they are not efficient compression schemes because they do not exploit statistical dependencies between samples and do not sufficiently remove signal irrelevancies.

Bit rate reductions by fairly simple means are achieved in the interactive CD (CD-i) which supports 16-bit PCM at a sampling rate of 44.1 kHz and allows for three levels of adaptive differential PCM (ADPCM) with switched prediction and noise shaping. For each block there is a multiple choice of fixed predictors from which to choose. The supported bandwidths and b/sample-resolutions are 37.8 kHz/8 bit, 37.8 kHz/4 bit, and 18.9 kHz/4 bit.

In recent audio coding algorithms *four key technologies* play an important role: perceptual coding, frequency domain coding, window switching, and dynamic bit allocation. These will be covered next.

40.2.1   Auditory Masking and Perceptual Coding

Auditory Masking

The inner ear performs short-term critical band analyses where frequency-to-place transformations occur along the basilar membrane. The power spectra are not represented on a linear frequency scale but on limited frequency bands called *critical bands*. The auditory system can roughly be described as a bandpass filterbank, consisting of strongly overlapping bandpass filters with bandwidths in the order of 50 to 100 Hz for signals below 500 Hz and up to 5000 Hz for signals at high frequencies. Twenty-five critical bands covering frequencies of up to 20 kHz have to be taken into account.

Simultaneous masking is a frequency domain phenomenon where a low-level signal (the maskee) can be made inaudible (masked) by a simultaneously occurring stronger signal (the masker), if masker and maskee are close enough to each other in frequency [13]. Such masking is greatest in the critical band in which the masker is located, and it is effective to a lesser degree in neighboring bands. A *masking threshold* can be measured below which the low-level signal will not be audible. This masked signal can consist of low-level signal contributions, quantization noise, aliasing distortion, or transmission errors. The masking threshold, in the context of source coding also known as *threshold of just noticeable distortion* (JND) [14], varies with time. It depends on the sound pressure level (SPL), the frequency of the masker, and on characteristics of masker and maskee. Take the example of the masking threshold for the SPL = 60 dB narrowband masker in Fig. 40.1: around 1 kHz the four maskees will be masked as long as their individual sound pressure levels are below the masking threshold. The slope of the masking threshold is steeper towards lower frequencies, i.e., higher frequencies are more easily masked. It should be noted that the distance between masker and masking threshold is smaller in noise-masking-tone experiments than in tone-masking-noise experiments, i.e., noise is a better masker than a tone. In MPEG coders both thresholds play a role in computing the masking threshold.

Without a masker, a signal is inaudible if its sound pressure level is below the *threshold in quiet* which depends on frequency and covers a dynamic range of more than 60 dB as shown in the lower curve of Figure 40.1.

The qualitative sketch of Fig. 40.2 gives a few more details about the masking threshold: a critical band, tones below this threshold (darker area) are masked. The distance between the level of the masker and the masking threshold is called *signal-to-mask ratio (SMR)*. Its maximum value is at the left border of the critical band (point *A* in Fig. 40.2), its minimum value occurs in the frequency range of the masker and is around 6 dB in noise-masks-tone experiments. Assume a m-bit quantization of an audio signal. Within a critical band the quantization noise will not be audible as long as its signal-to-noise ratio SNR is higher than its SMR. Noise *and* signal contributions *outside* the particular critical band will also be masked, although to a lesser degree, if their SPL is below the masking threshold.

Defining SNR(m) as the signal-to-noise ratio resulting from an m-bit quantization, the perceivable distortion in a given subband is measured by the *noise-to-mask ratio*

$$\text{NMR (m)} = \text{SMR} - \text{SNR (m) (in dB)}.$$

The noise-to-mask ratio NMR(m) describes the difference in dB between the signal-to-*mask* ratio and the signal-to-*noise* ratio to be expected from an m-bit quantization. The NMR value is also the difference

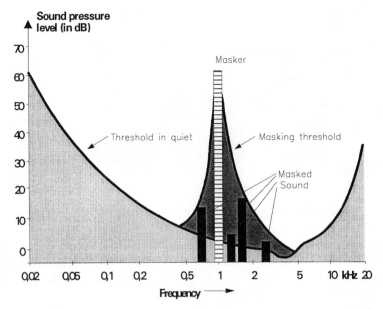

FIGURE 40.1   Threshold in quiet and masking threshold. Acoustical events in the shaded areas will not be audible.

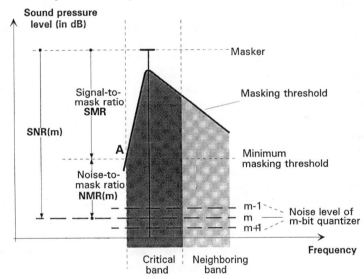

FIGURE 40.2   Masking threshold and signal-to-mask ratio (SMR). Acoustical events in the shaded areas will not be audible.

(in dB) between the level of quantization noise and the level where a distortion may just become audible in a given subband. Within a critical band, coding noise will not be audible as long as NMR(m) is negative.

We have just described masking by only one masker. If the source signal consists of many simultaneous maskers, each has its own masking threshold, and a *global masking threshold* can be computed that describes the threshold of just noticeable distortions as a function of frequency.

In addition to simultaneous masking, the time domain phenomenon of *temporal masking* plays an important role in human auditory perception. It may occur when two sounds appear within a small interval of time. Depending on the individual sound pressure levels, the stronger sound may mask the weaker one, even if the maskee precedes the masker (Fig. 40.3)!

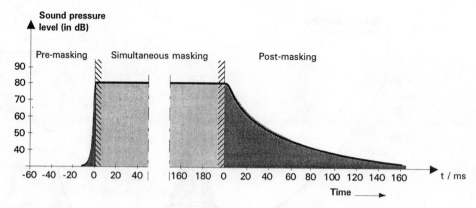

FIGURE 40.3   Temporal masking. Acoustical events in the shaded areas will not be audible.

Temporal masking can help to mask pre-echoes caused by the spreading of a sudden large quantization error over the actual coding block. The duration within which *pre-masking* applies is significantly less than one tenth of that of the *post-masking* which is in the order of 50 to 200 ms. Both pre- and postmasking are being exploited in MPEG/Audio coding algorithms.

Perceptual Coding

Digital coding at high bit rates is dominantly waveform-preserving, i.e., the amplitude-vs.-time waveform of the decoded signal approximates that of the input signal. The difference signal between input and output waveform is then the basic error criterion of coder design. Waveform coding principles have been covered in detail in [2]. At lower bit rates, facts about the production and perception of audio signals have to be included in coder design, and the error criterion has to be in favor of an output signal that is useful to the human receiver rather than favoring an output signal that follows and preserves the input waveform. Basically, an efficient source coding algorithm will (1) remove redundant components of the source signal by exploiting correlations between its samples and (2) remove components that are irrelevant to the ear. Irrelevancy manifests itself as unnecessary amplitude or frequency resolution; portions of the source signal that are masked do not need to be transmitted.

The dependence of human auditory perception on frequency and the accompanying perceptual tolerance of errors can (and should) directly influence encoder designs; *noise-shaping techniques* can emphasize coding noise in frequency bands where that noise perceptually is not important. To this end, the noise shifting must be dynamically adapted to the actual short-term input spectrum in accordance with the signal-to-mask ratio which can be done in different ways. However, frequency weightings based on linear filtering, as typical in speech coding, cannot make full use of results from psychoacoustics. Therefore, in wideband audio coding, noise-shaping parameters are dynamically controlled in a more efficient way to exploit simultaneous masking and temporal masking.

Figure 40.4 depicts the structure of a *perception-based coder* that exploits auditory masking. The encoding process is controlled by the SMR vs. frequency curve from which the needed amplitude resolution (and hence the bit allocation and rate) in each frequency band is derived. The SMR is typically determined from a high resolution, say, a 1024-point FFT-based spectral analysis of the audio block to be coded. Principally, any coding scheme can be used that can be dynamically controlled by such perceptual information. Frequency domain coders (see next section) are of particular interest because they offer a direct method for noise shaping. If the frequency resolution of these coders is high enough, the SMR can be derived directly from the subband samples or transform coefficients without running a FFT-based spectral analysis in parallel [15, 16].

If the necessary bit rate for a complete masking of distortion is available, the coding scheme will be perceptually transparent, i.e., the decoded signal is then subjectively indistinguishable from the source

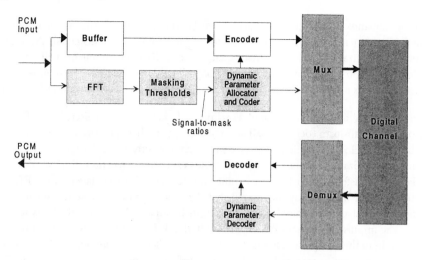

FIGURE 40.4   Block diagram of perception-based coders.

signal. In practical designs, we cannot go to the limits of just noticeable distortion because postprocessing of the acoustic signal by the end-user and multiple encoding/decoding processes in transmission links have to be considered. Moreover, our current knowledge about auditory masking is very limited. Generalizations of masking results, derived for simple and stationary maskers and for limited bandwidths, may be appropriate for most source signals, but may fail for others. Therefore, as an additional requirement, we need a sufficient safety margin in practical designs of such perception-based coders. It should be noted that the MPEG/Audio coding standard is open for better encoder-located psychoacoustic models because such models are not normative elements of the standard (see Section 40.3).

40.2.2 Frequency Domain Coding

As one example of dynamic noise-shaping, quantization noise feedback can be used in predictive schemes [17, 18]. However, frequency domain coders with dynamic allocations of bits (and hence of quantization noise contributions) to subbands or transform coefficients offer an easier and more accurate way to control the quantization noise [2, 15].

In all frequency domain coders, redundancy (the non-flat short-term spectral characteristics of the source signal) and irrelevancy (signals below the psychoacoustical thresholds) are exploited to reduce the transmitted data rate with respect to PCM. This is achieved by splitting the source spectrum into frequency bands to generate nearly uncorrelated spectral components, and by quantizing these separately. Two coding categories exist, *transform coding* (TC) and *subband coding* (SBC). The differentiation between these two categories is mainly due to historical reasons. Both use an analysis filterbank in the encoder to decompose the input signal into subsampled spectral components. The spectral components are called subband samples if the filterbank has low frequency resolution, otherwise they are called spectral lines or transform coefficients. These spectral components are recombined in the decoder via synthesis filterbanks.

In *subband coding*, the source signal is fed into an analysis filterbank consisting of M bandpass filters which are contiguous in frequency so that the set of subband signals can be recombined additively to produce the original signal or a close version thereof. Each filter output is critically decimated (i.e., sampled at twice the nominal bandwidth) by a factor equal to M, the number of bandpass filters. This decimation results in an aggregate number of subband samples that equals that in the source signal. In the receiver, the sampling rate of each subband is increased to that of the source signal by filling in the appropriate number of zero samples. Interpolated subband signals appear at the bandpass outputs of the synthesis filterbank. The sampling processes may introduce aliasing distortion due to the overlapping nature of the subbands. If perfect filters, such as two-band quadrature mirror filters or polyphase filters, are applied, aliasing terms will cancel and the sum of the bandpass outputs equals the source signal in the absence of quantization [19]–[22]. With quantization, aliasing components will not cancel ideally; nevertheless, the errors will be inaudible in MPEG/Audio coding if a sufficient number of bits is used. However, these errors may reduce the original dynamic range of 20 bits to around 18 bits [16].

In *transform coding*, a block of input samples is linearly transformed via a discrete transform into a set of near-uncorrelated transform coefficients. These coefficients are then quantized and transmitted in digital form to the decoder. In the decoder, an inverse transform maps the signal back into the time domain. In the absence of quantization errors, the synthesis yields exact reconstruction. Typical transforms are the Discrete Fourier Transform or the Discrete Cosine Transform (DCT), calculated via an FFT, and modified versions thereof. We have already mentioned that the decoder-based inverse transform can be viewed as the synthesis filterbank, the impulse responses of its bandpass filters equal the basis sequences of the transform. The impulse responses of the analysis filterbank are just the time-reversed versions thereof. The finite lengths of these impulse responses may cause so-called block boundary effects. State-of-the-art transform coders employ a *modified DCT* (MDCT) filterbank as proposed by Princen and Bradley [21]. The MDCT is typically based on a 50% overlap between successive analysis blocks. Without quantization they are free from block boundary effects, have a higher transform coding gain than the DCT, and their basis functions correspond to better bandpass responses. In the presence of quantization, block boundary effects are deemphasized due to the doubling of the filter impulse responses resulting from the overlap.

Hybrid filterbanks, i.e., combinations of discrete transform and filterbank implementations, have frequently been used in speech and audio coding [23, 24]. One of the advantages is that different frequency resolutions can be provided at different frequencies in a flexible way and with low complexity. A high spectral resolution can be obtained in an efficient way by using a cascade of a filterbank (with its short delays) and a linear MDCT transform that splits each subband sequence further in frequency content to achieve a high frequency resolution. MPEG-1/Audio coders use a subband approach in layers I and II, and a hybrid filterbank in layer III.

40.2.3 Window Switching

A crucial part in frequency domain coding of audio signals is the appearance of *pre-echoes*, similar to copying effects on analog tapes. Consider the case that a silent period is followed by a percussive sound, such as from castanets or triangles, within the same coding block. Such an onset ("attack") will cause comparably large instantaneous quantization errors. In TC, the inverse transform in the *decoding* process will distribute such errors over the block; similarly, in SBC, the decoder bandpass filters will spread such errors. In both mappings pre-echoes can become distinctively audible, especially at low bit rates with comparably high error contributions. Pre-echoes can be masked by the time domain effect of pre-masking if the time spread is of short length (in the order of a few milliseconds). Therefore, they can be reduced or avoided by using blocks of short lengths. However, a larger percentage of the total bit rate is typically required for the transmission of side information if the blocks are shorter. A solution to this problem is to switch between block sizes of different lengths as proposed by Edler (*window switching*) [25], typical block sizes are between $N = 64$ and $N = 1024$. The small blocks are only used to control pre-echo artifacts during nonstationary periods of the signal, otherwise the coder switches back to long blocks. It is clear that the block size selection has to be based on an analysis of the characteristics of the actual audio coding block. Figure 40.5 demonstrates the effect in transform coding: if the block size is $N = 1024$ [Fig. 40.5(b)] pre-echoes are clearly (visible and) audible whereas a block size of 256 will reduce these effects because they are limited to the block where the signal attack and the corresponding quantization errors occur [Fig. 40.5(c)]. In addition, pre-masking can become effective.

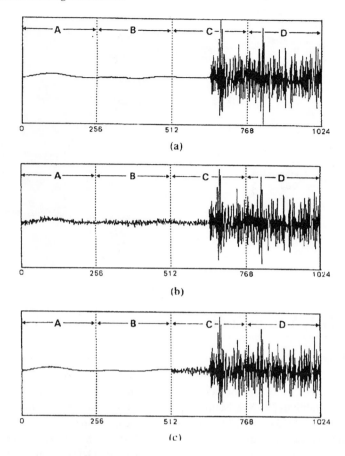

FIGURE 40.5 Window switching. (a) Source signal, (b) reconstructed signal with block size N = 1024, and (c) reconstructed signal with block size N = 256. (*Source:* Iwadare, M., Sugiyama, A., Hazu, F., Hirano, A., and Nishitani, T., IEEE J. Sel. Areas Commun., 10(1), 138-144, Jan. 1992.)

40.2.4 Dynamic Bit Allocation

Frequency domain coding significantly gains in performance if the number of bits assigned to each of the quantizers of the transform coefficients is adapted to short-term spectrum of the audio coding block on a block-by-block basis. In the mid-1970s, Zelinski and Noll introduced *dynamic bit allocation* and demonstrated significant SNR-based and subjective improvements with their adaptive transform coding (ATC, see Fig. 40.6 [15, 27]). They proposed a DCT mapping and a dynamic bit allocation algorithm which used the DCT transform coefficients to compute a DCT-based short-term spectral envelope. Parameters of this spectrum were coded and transmitted. From these parameters, the short-term spectrum was estimated using linear interpolation in the log-domain. This estimate was then used to calculate the optimum number of bits for each transform coefficient, both in the encoder and decoder.

That ATC had a number of shortcomings, such as block boundary effects, pre-echoes, marginal exploitation of masking, and insufficient quality at low bit rates. Despite these shortcomings, we find many of the features of the conventional ATC in more recent frequency domain coders.

MPEG/Audio coding algorithms, described in detail in the next section, make use of the above key technologies.

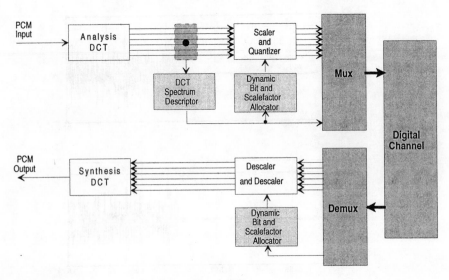

FIGURE 40.6   Conventional adaptive transform coding (ATC).

40.3   MPEG-1/Audio Coding

The MPEG-1/Audio coding standard [8], [28]–[30] is about to become a universal standard in many application areas with totally different requirements in the fields of consumer electronics, professional audio processing, telecommunications, and broadcasting [31]. The standard combines features of *MUSICAM* and *ASPEC coding algotithms* [32, 33]. Main steps of development towards the MPEG-1/Audio standard have been described in [30, 34]. The MPEG-1/Audio standard represents the state of the art in audio coding. Its subjective quality is equivalent to CD quality (16-bit PCM) at stereo rates given in Table 40.3 for many types of music. Because of its high dynamic range, MPEG-1/audio has potential to exceed the quality of a CD [31, 35].

TABLE 40.3   Approximate MPEG-1 Bit Rates for Transparent
Representations of Audio Signals and Corresponding Compression
Factors (Compared to CD Bit Rate)

MPEG-1 audio coding	Approximate stereo bit rates for transparent quality	Compression factor
Layer I	384 kb/s	4
Layer II	192 kb/s	8
Layer III	128 kb/s[a]	12

[a] Average bit rate; variable bit rate coding assumed.

40.3.1   The Basics

Structure

The basic structure follows that of perception-based coders (see Fig. 40.4). In the first step, the audio signal is converted into spectral components via an analysis filterbank; layers I and II make use of a subband filterbank, layer III employs a hybrid filterbank. Each spectral component is quantized and coded with the goal to keep the quantization noise below the masking threshold. The number of bits for each subband and a scalefactor are determined on a block-by-block basis, each block has 12 (layer I) or 36 (layers II and III) subband samples (see Section 40.2). The number of quantizer bits is obtained from a dynamic bit allocation algorithm (layers I and II) that is controlled by a *psychoacoustic model* (see below). The subband codewords, scalefactor, and bit allocation information are multiplexed into one bitstream, together with a header and optional ancillary data. In the decoder, the synthesis filterbank reconstructs a block of 32 audio output samples from the demultiplexed bitstream.

MPEG-1/Audio supports sampling rates of 32, 44.1, and 48 kHz and bit rates between 32 kb/s (mono) and 448 kb/s, 384 kb/s, and 320 kb/s (stereo; layers I, II, and III, respectively). Lower sampling rates (16, 22.05, and 24 kHz) have been defined in MPEG-2 for better audio quality at bit rates at, or below, 64 kb/s per channel [9]. The corresponding maximum audio bandwidths are 7.5, 10.3, and 11.25 kHz. The syntax, semantics, and coding techniques of MPEG-1 are maintained except for a small number of parameters.

Layers and Operating Modes

The standard consists of three layers I, II, and III of increasing complexity, delay, and subjective performance. From a hardware and software standpoint, the higher layers incorporate the main building blocks of the lower layers (Fig. 40.7). A standard *full MPEG-1/Audio decoder* is able to decode bit streams of all three layers. The standard also supports MPEG-1/Audio *layer X decoders* ($X =$ I, II, or III). Usually, a layer II decoder will be able to decode bitstreams of layers I and II, a layer III decoder will be able to decode bitstreams of all three layers.

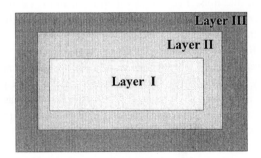

FIGURE 40.7   Hierarchy of layers I, II, and III of MPEG-1/Audio.

Stereo Redundancy Coding

MPEG-1/Audio supports four *modes:* mono, stereo, dual with two separate channels (useful for bilingual programs), and joint stereo. In the optimal joint stereo mode, interchannel dependencies are exploited to reduce the overall bit rate by using an irrelevancy reducing technique called *intensity stereo*. It is known that above 2 kHz and within each critical band, the human auditory system bases its perception of stereo imaging more on the temporal envelope of the audio than on its temporal fine structure. Therefore, the MPEG audio compression algorithm supports a stereo redundancy coding mode called *intensity stereo coding* which reduces the total bit rate without violating the spatial integrity of the stereophonic signal.

In intensity stereo mode, the encoder codes some upper-frequency subband outputs with a single sum signal L + R (or some linear combination thereof) instead of sending independent left (L) and right (R) subband signals. The decoder reconstructs the left and right channels based only on the single L + R signal and on independent left and right channel scalefactors. Hence, the spectral shape of the left and right outputs is the same within each intensity-coded subband but the magnitudes are different [36]. The optional joint stereo mode will only be effective if the required bit rate exceeds the available bit rate, and it will only be applied to subbands corresponding to frequencies of around 2 kHz and above.

Layer III has an additional option: in the mono/stereo (M/S) mode the left and right channel signals are encoded as middle (L + R) and side (L − R) channels. This latter mode can be combined with the joint stereo mode.

Psychoacoustic Models

We have already mentioned that the adaptive bit allocation algorithm is controlled by a psychoacoustic model. This model computes SMR taking into a account the short-term spectrum of the audio block to be coded and knowledge about noise masking. The model is only needed in the encoder which makes the decoder less complex; this asymmetry is a desirable feature for audio playback and audio broadcasting applications.

The normative part of the standard describes the decoder and the meaning of the encoded bitstream, but the encoder is not standardized thus leaving room for an evolutionary improvement of the encoder. In particular, *different psychoacoustic models can be used* ranging from very simple (or none at all) to very complex ones based on quality and implementability requirements. Information about the short-term spectrum can be derived in various ways, for example, as an accurate estimate from an FFT-based spectral analysis of the audio input samples or, less accurate, directly from the spectral components as in the conventional ATC [15]; see also Fig. 40.6. Encoders can also be optimized for a certain application. All these encoders can be used with complete compatibility with all existing MPEG-1/Audio decoders.

The informative part of the standard gives two examples of FFT-based models; see also [8, 30, 37]. Both models identify, in different ways, tonal and non-tonal spectral components and use the corresponding results of tone-masks-noise and noise-masks-tone experiments in the calculation of the global masking thresholds. Details are given in the standard, experimental results for both psychoacoustic models are described in [37]. In the informative part of the standard a 512-point FFT is proposed for layer I, and a 1024-point FFT for layers II and III. In both models, the audio input samples are Hann-weighted. *Model 1*, which may be used for layers I and II, computes for each masker its individual masking threshold, taking into account its frequency position, power, and tonality information. The global masking threshold is obtained as the sum of all individual masking thresholds and the absolute masking threshold. The SMR is then the ratio of the maximum signal level within a given subband and the minimum value of the global masking threshold in that given subband (see Fig. 40.2).

Model 2, which may be used for all layers, is more complex: tonality is assumed when a simple prediction indicates a high prediction gain, the masking thresholds are calculated in the cochlea domain, i.e., properties of the inner ear are taken into account in more detail, and, finally, in case of potential pre-echoes the global masking threshold is adjusted appropriately.

40.3.2   Layers I and II

MPEG layer I and II coders have very similar structures. The layer II coder achieves a better performance, mainly because the overall scalefactor side information is reduced exploiting redundancies between the scalefactors. Additionally, a slightly finer quantization is provided.

Filterbank

Layer I and II coders map the digital audio input into 32 subbands via equally spaced bandpass filters (Figs. 40.8 and 40.9). A polyphase filter structure is used for the frequency mapping; its filters have 512 coefficients. Polyphase structures are computationally very efficient because a DCT can be used in the filtering process, and they are of moderate complexity and low delay. On the negative side, the filters are equally spaced, and therefore the frequency bands do not correspond well to the critical band partition (see Section 40.2.1). At 48-kHz sampling rate, each band has a width of $24000/32 = 750$ Hz; hence, at low frequencies, a single subband covers a number of adjacent critical bands. The subband signals are resampled (critically decimated) at a rate of 1500 Hz. The impulse response of subband k, $h_{\text{sub}(k)}(n)$, is obtained by multiplication of the impulse response of a single *prototype lowpass filter*, $h(n)$, by a modulating function which shifts the lowpass response to the appropriate subband frequency range:

$$h_{\text{sub}(k)}(n) = h(n) \cos \left[\frac{(2k+1)\pi n}{2M} + \varphi(k) \right] ;$$
$$M = 32 ; \ k = 0, 1, \ldots, 31 ; \ n = 0, 1, \ldots, 511$$

The prototype lowpass filter has a 3-dB bandwidth of $750/2 = 375$ Hz, and the center frequencies are at odd multiples thereof (all values at 48 kHz sampling rate). The subsampled filter outputs exhibit a significant overlap. However, the design of the prototype filter and the inclusion of appropriate phase shifts in the cosine terms result in an aliasing cancellation at the output of the decoder synthesis filterbank. Details about the coefficients of the prototype filter and the phase shifts $\varphi(k)$ are given in the ISO/MPEG standard. Details about an efficient implementation of the filterbank can be found in [16] and [37], and, again, in the standardization documents.

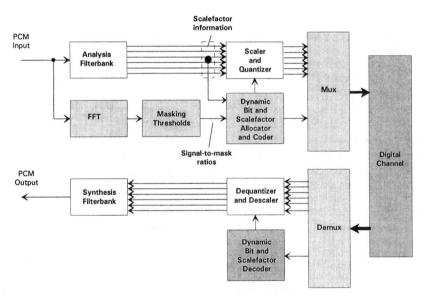

FIGURE 40.8   Structure of MPEG-1/Audio encoder and decoder, layers I and II.

Quantization

The number of quantizer levels for each spectral component is obtained from a dynamic bit allocation rule that is controlled by a psychoacoustic model. The bit allocation algorithm selects one uniform midtread quantizer out of a set of available quantizers such that both the bit rate requirement and the masking requirement are met. The iterative procedure minimizes the NMR in each subband. It starts with the number of bits for the samples and scalefactors set to zero. In each iteration step, the quantizer SNR(m) is increased for the one subband quantizer producing the largest value of the NMR at the quantizer output. (The increase is obtained by allocating one more bit). For that purpose, NMR(m) = SMR − SNR(m) is calculated as the difference (in dB) between the actual quantization noise level and the minimum global masking threshold. The standard provides tables with estimates for the quantizer SNR(m) for a given m.

Block companding is used in the quantization process, i.e., blocks of decimated samples are formed and divided by a *scalefactor* such that the sample of largest magnitude is unity. In *layer I* blocks of 12 decimated and scaled samples are formed in each subband (and for the left and right channel) and there is one bit

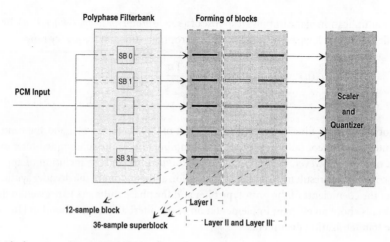

FIGURE 40.9 Block companding in MPEG-1/Audio coders.

allocation for each block. At 48-kHz sampling rate, 12 subband samples correspond to 8 ms of audio. There are 32 blocks, each with 12 decimated samples, representing $32 \times 12 = 384$ audio samples.

In *layer II* in each subband a 36-sample *superblock* is formed of three consecutive blocks of 12 decimated samples corresponding to 24 ms of audio at 48 kHz sampling rate. There is one bit allocation for each 36-sample superblock. All 32 superblocks, each with 36 decimated samples, represent, altogether, $32 \times 36 = 1152$ audio samples. As in layer I, a scalefactor is computed for each 12-sample block. A redundancy reduction technique is used for the transmission of the scalefactors: depending on the significance of the changes between the three consecutive scalefactors, one, two, or all three scalefactors are transmitted, together with a 2-bit *scalefactor select information*. Compared with layer I, the bit rate for the scalefactors is reduced by around 50% [30]. Figure 40.9 indicates the block companding structure.

The scaled and quantized spectral subband components are transmitted to the receiver together with scalefactor, scalefactor select (layer II), and bit allocation information. Quantization with block companding provides a very large dynamic range of more than 120 dB. For example, in layer II uniform midtread quantizers are available with 3, 5, 7, 9, 15, 31, . . . , 65535 levels for subbands of low index (low frequencies). In the mid and high frequency region, the number of levels is reduced significantly. For subbands of index 23 to 26 there are only quantizers with 3, 5, and 65535 (!) levels available. The 16-bit quantizers prevent overload effects. Subbands of index 27 to 31 are not transmitted at all. In order to reduce the bit rate, the codewords of three successive subband samples resulting from quantizing with 3-, 5, and 9-step quantizers are assigned one common codeword. The savings in bit rate is about 40% [30].

Figure 40.10 shows the time-dependence of the assigned number of quantizer bits in all subbands for a layer II encoded high quality speech signal. Note, for example, that quantizers with ten or more bits resolution are only employed in the lowest subbands, and that no bits have been assigned for frequencies above 18 kHz (subbands of index 24 to 31).

Decoding

The decoding is straightforward: the subband sequences are reconstructed on the basis of blocks of 12 subband samples taking into account the decoded scalefactor and bit allocation information. If a subband has no bits allocated to it, the samples in that subband are set to zero. Each time the subband samples of all 32 subbands have been calculated, they are applied to the *synthesis filterbank*, and 32 consecutive 16-bit PCM format audio samples are calculated. If available, as in bidirectional communications or in recorder systems, the encoder (analysis) filterbank can be used in a reverse mode in the decoding process.

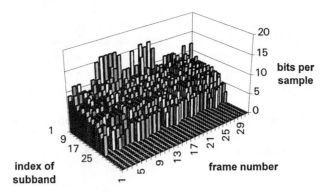

FIGURE 40.10   Time-dependence of assigned number of quantizer bits in all subbands for a layer II encoded high quality speech signal.

40.3.3 Layer III

Layer III of the MPEG-1/Audio coding standard introduces many new features (see Fig. 40.11), in particular a switched hybrid filterbank. In addition, it employs an analysis-by-synthesis approach, an advanced pre-echo control, and nonuniform quantization with entropy coding. A buffer technique, called *bit reservoir*, leads to further savings in bit rate. Layer III is the only layer that provides mandatory decoder support for variable bit rate coding [38].

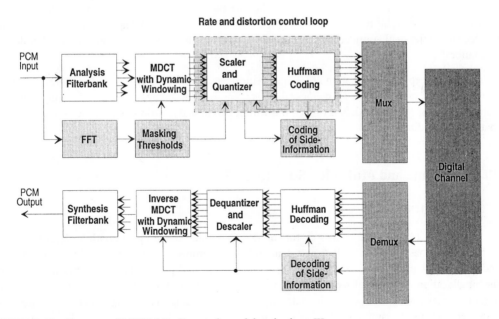

FIGURE 40.11   Structure of MPEG-1/Audio encoder and decoder, layer III.

Switched Hybrid Filterbank

In order to achieve a higher frequency resolution closer to critical band partitions, the 32 subband signals are subdivided further in frequency content by applying, to each of the subbands, a 6- or 18-point modified DCT block transform, with 50% overlap; hence, the windows contain, respectively, 12 or 36 subband samples. The maximum number of frequency components is $32 \times 18 = 576$ each representing a bandwidth of only $24000/576 = 41.67$ Hz. Because the 18-point block transform provides better frequency resolution, it is normally applied, whereas the 6-point block transform provides better time resolution and is applied in case of expected pre-echoes (see Section 40.2.3). In principle, a pre-echo is assumed, when an instantaneous demand for a high number of bits occurs. Depending on the nature of potential, all pre-echoes or a smaller number of transforms are switched. Two special MDCT windows, a start window and a stop window, are needed in case of transitions between short and long blocks and vice versa to maintain the time domain alias cancellation feature of the MDCT [22, 25, 37]. Figure 40.12 shows a typical sequence of windows.

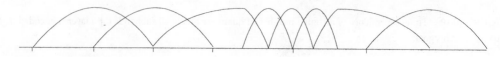

FIGURE 40.12 Typical sequence of windows in adaptive window switching.

Quantization and Coding

The MDCT output samples are nonuniformly quantized thus providing both smaller mean-squared errors and masking because larger errors can be tolerated if the samples to be quantized are large. Huffman coding, based on 32 code tables, and additional run-length coding are applied to represent the quantizer indices in an efficient way. The encoder maps the variable wordlength codewords of the Huffman code tables into a constant bit rate by monitoring the state of a bit reservoir. The bit reservoir ensures that the decoder buffer neither underflows nor overflows when the bitstream is presented to the decoder at a constant rate.

In order to keep the quantization noise in all critical bands below the global masking threshold (noise allocation) an *iterative analysis-by-synthesis method* is employed whereby the process of scaling, quantization, and coding of spectral data is carried out within two nested iteration loops. The decoding follows that of the encoding process.

40.3.4 Frame and Multiplex Structure

Frame Structure

Figure 40.13 shows the frame structure of MPEG-1/Audio coded signals, both for layer I and layer II. Each frame has a header; its first part contains 12 synchronisation bits, 20 bit system information, and an optional 16-bit cyclic redundancy check code. Its second part contains side information about the bit allocation and the scalefactors (and, in layer II, scalefactor information). As main information, a frame carries a total of 32×12 subband samples (corresponding to 384 PCM audio input sample — equivalent to 8 ms at a sampling rate of 48 kHz) in layer I, and a total of 32×36 subband samples in layer II (corresponding to 1152 PCM audio input samples — equivalent to 24 ms at a sampling rate of 48 kHz). Note that the layer I and II frames are autonomous: each frame contains all information necessary for decoding. Therefore, each frame can be decoded independently from previous frames, it defines an entry point for audio storage and audio editing applications. Please note that the lengths of the frames are not fixed, due to (1) the length of the main information field, which depends on bit-rate and sampling frequency, (2) the side information field which varies in layer II, and (3) the ancillary data field, the length of which is not specified.

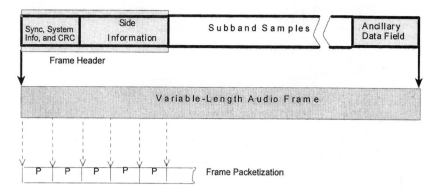

FIGURE 40.13 MPEG-1 frame structure and packetization. Layer I: 384 subband samples; layer II: 1152 subband samples; packets P: 4-byte header; 184-byte payload field (see also Fig. 40.14).

Multiplex Structure

We have already mentioned that the systems part of the MPEG-1 coding standard IS 11172 defines a *packet structure for multiplexing* audio, video, and ancillary data bitstreams in one stream. The variable-length MPEG frames are broken down into packets. The packet structure uses 188-byte packets consisting of a 4-byte header followed by 184 bytes of payload (see Fig. 40.14). The header includes a sync byte, a 13-bit field called packet identifier to inform the decoder about the type of data, and additional information. For example, a *1-bit payload unit start indicator* indicates if the payload starts with a frame header. No predetermined mix of audio, video, and ancillary data bitstreams is required, the mix may change dynamically, and services can be provided in a very flexible way. If additional header information is required, such as for periodic synchronization of audio and video timing, a variable-length *adaptation header* can be used as part of the 184-byte payload field.

Although the lengths of the frames are not fixed, the interval between frame headers is constant (within a byte) throughout the use of padding bytes. The MPEG systems specification describes how MPEG-compressed audio and video data streams are to be multiplexed together to form a single data stream. The terminology and the fundamental principles of the systems layer are described in [39].

40.3.5 Subjective Quality

The standardization process included extensive subjective tests and objective evaluations of parameters such as complexity and overall delay. The MPEG (and equivalent ITU-R) listening tests were carried out under very similar and carefully defined conditions with around 60 experienced listeners, approximately 10 test sequences were used, and the sessions were performed in stereo with both loudspeakers and headphones. In order to detect even small impairments, the 5-point ITU-R impairment scale was used in all experiments. Details are given in [40] and [41]. *Critical test items were chosen in the tests to evaluate the coders by their worst case (not average) performance.* The subjective evaluations, which have been based on triple stimulus/hidden reference/double blind tests, have shown very similar and stable evaluation results. In these tests the subject is offered three signals, A,B, and C (triple stimulus). A is always the unprocessed source signal (the reference). B and C, or C and B, are the reference and the system under test (hidden reference). The selection is neither known to the subjects nor to the conductors(s) of the test (double blind test). The subjects have to decide if B or C is the reference and have to grade the remaining one.

The MPEG-1/Audio coding standard has shown an excellent performance for all layers at the rates given in Table 40.3. It should be mentioned again that the standard leaves room for encoder-based improvements by using better psychoacoustic models. Indeed, many improvements have been achieved since the first subjective results had been carried out in 1991.

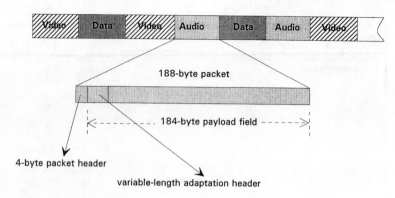

FIGURE 40.14   MPEG packet delivery.

40.4   MPEG-2/Audio Multichannel Coding

A logical further step in digital audio is the definition of a multichannel audio representation system to create a convincing, lifelike soundfield both for audio-only applications and for audiovisual systems, including video conferencing, videophony, multimedia services, and electronic cinema. Multichannel systems can also provide multilingual channels and additional channels for visually impaired (a verbal description of the visual scene) and for hearing impaired (dialog with enhanced intelligibility). ITU-R has recommended a five-channel loudspeaker configuration, referred to as 3/2-stereo, with a left and a right channel (L and R), an additional center channel C, two side/rear surround channels (LS and RS) augmenting the L and R channels, see Fig. 40.15 [ITU-R Rec. 775]. Such a configuration offers an improved realism of auditory ambience with a stable frontal sound image and a large listening area.

Multichannel digital audio systems support p/q presentations with p front and q back channels, and also provide the possibilities of transmitting two independent stereophonic programs and/or a number of commentary or multilingual channels. Typical combinations of channels include.

- 1 channel   1/0-configuration: centre (mono)

- 2 channels   2/0-configuration: left, right (stereophonic)

- 3 channels   3/0-configuration: left, right, centre

- 4 channels: 3/1-configuration   left, right, centre, mono-surround

- 5 channels: 3/2-configuration: left, right, centre, surround left, surround right

FIGURE 40.15   3/2 Multichannel loudspeaker configuration.

ITU-R Recommendation 775 provides a set of downward mixing equations if the number of loud-speakers is to be reduced (*downward compatibility*). An additional *low frequency enhancement (LFE-or subwoofer-) channel* is particularly useful for HDTV applications, it can be added, optionally, to any of the configurations. The LFE channel extends the low frequency content between 15 and 120 Hz in terms of both frequency and level.

One or more loudspeakers can be positioned freely in the listening room to reproduce this LFE signal. (Film industry uses a similar system for their digital sound systems).[1]

In order to reduce the overall bit rate of multichannel audio coding systems, redundancies and irrelevancy, such as interchannel dependencies and interchannel masking effects, respectively, may be exploited. In addition, stereophonic-irrelevant components of the multichannel signal, which do not contribute to the localization of sound sources, may be identified and reproduced in a monophonic format to further reduce bit rates. State-of-the-art multichannel coding algorithms make use of such effects. A careful design is needed, otherwise such joint coding may produce artifacts.

40.4.1   MPEG-2/Audio Multichannel Coding

The second phase of MPEG, labeled MPEG-2, includes in its audio part two multichannel audio coding standards, one of which is forward- and backward-compatible with MPEG-1/Audio [8], [42]–[45]. *Forward compatibility* means that an MPEG-2 multichannel decoder is able to properly decode MPEG-1 mono or stereophonic signals, *backward compatibility* (BC) means that existing MPEG-1 stereo decoders, which only handle two-channel audio, is able to reproduce a meaningful basic 2/0 stereo signal from a MPEG-2 multichannel bit stream so as to serve the need of users with simple mono or stereo equipment. *Non-backward compatible* (NBC) multichannel coders will not be able to feed a meaningful bit stream into a MPEG-1 stereo decoder. On the other hand, NBC codecs have more freedom in producing a high quality reproduction of audio signals.

With backward compatibility, it is possible to introduce multichannel audio at any time in a smooth way without making existing two-channel stereo decoders obsolete. An important example is the European Digital Audio Broadcast system, which will require MPEG-1 stereo decoders in the first generation but may offer multichannel audio at a later point.

40.4.2   Backward-Compatible (BC) MPEG-2/Audio Coding

BC implies the use of compatibility matrices. A down-mix of the five channels ("matrixing") delivers a correct basic 2/0 stereo signal, consisting of a left and a right channel, LO and RO, respectively. A typical set of equations is

$$LO = \alpha \, (L + \beta^{'} C + \delta^{'} LS)$$
$$\alpha = \tfrac{1}{1+\sqrt{2}} \; ; \beta = \delta = \sqrt{2}$$
$$RO = \alpha \, (R + \beta^{'} C + \delta^{'} RS)$$

Other choices are possible, including $LO = L$ and $RO = R$. The factors α, β, and δ attenuate the signals to avoid overload when calculating the compatible stereo signal (LO, RO). The signals LO and RO are transmitted in MPEG-1 format in transmission channels $T1$ and $T2$. Channels $T3, T4$, and $T5$ together form the *multichannel extension signal* (Fig. 40.16). They have to be chosen such that the decoder can recompute the complete 3/2-stereo multichannel signal. Interchannel redundancies and masking effects are taken into account to find the best choice. A simple example is $T3 = C, T4 = LS$, and $T5 = RS$. In MPEG-2 the matrixing can be done in a very flexible and even time-dependent way.

---

[1]A 3/2-configuration with five high-quality full-range channels plus a subwoofer channel is often called a 5.1 system.

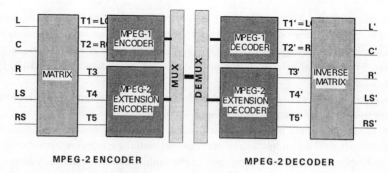

MPEG-2 ENCODER   **MPEG-2 DECODER**

FIGURE 40.16   Compatibility of MPEG-2 multichannel audio bit streams.

BC is achieved by transmitting the channels LO and RO in the subband-sample section of the MPEG-1 audio frame and all multichannel extension signals $T3$, $T4$, and $T5$ in the first part of the MPEG-1/Audio frame reserved for ancillary data. This ancillary data field is ignored by MPEG-1 decoders (see Fig. 40.17). The length of the ancillary data field is not specified in the standard. If the decoder is of type MPEG-1, it uses the 2/0-format front left and right down-mix signals, LO' and RO', directly (see Fig. 40.18). If the decoder is of type MPEG-2, it recomputes the complete 3/2-stereo multichannel signal with its components L', R', C', LS', and RS' via "dematrixing" of LO', RO', $T3'$, $T4'$, and $T5'$ (see Fig. 40.16).

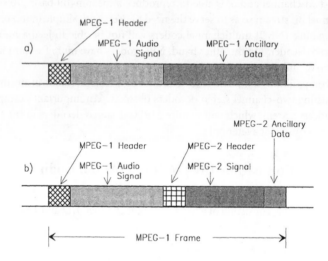

FIGURE 40.17   Data format of MPEG audio bit streams. a.)   MPEG-1 audio frame; b.)   MPEG-2 audio frame, compatible with MPEG-1 format.

Matrixing is obviously necessary to provide BC; however, if used in connection with perceptual coding, "unmasking" of quantization noise may appear [46]. It may be caused in the dematrixing process when sum and difference signals are formed. In certain situations, such a masking sum or difference signal component can disappear in a specific channel. Since this component was supposed to mask the quantization noise in that channel, this noise may become audible. Note that the masking signal will still be present in the multichannel representation but it will appear on a different loudspeaker. Measures against "unmasking" effects have been described in [47].

MPEG-1 decoders have a bit rate limitation (384 kb/s in layer II). In order to overcome this limitation, the MPEG-2 standard allows for a second bit stream, the extension part, to provide compatible multichannel audio at higher rates. Figure 40.19 shows the structure of the bit stream with extension.

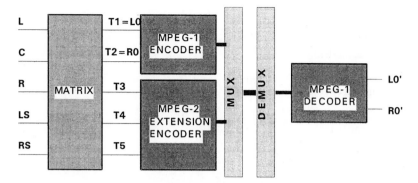

FIGURE 40.18 MPEG-1 stereo decoding of MPEG-2 multichannel bit stream.

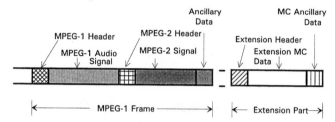

FIGURE 40.19 Data format of MPEG-2 audio bit stream with extension part.

40.4.3 Advanced/MPEG-2/Audio Coding (AAC)

A second standard within MPEG-2 supports applications that do not request compatibility with the existing MPEG-1 stereo format. Therefore, matrixing and dematrixing are not necessary and the corresponding potential artifacts disappear (see Fig. 40.20). The advanced multichannel coding mode will have the sampling rates, audio bandwidth, and channel configurations of MPEG-2/Audio, but shall be capable of operating at bit rates from 32kb/s up to a bit rate sufficient for high quality audio.

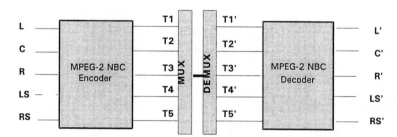

FIGURE 40.20 Non-backward-compatible MPEG-2 multichannel audio coding (advanced audio coding).

The last two years have seen extensive activities to optimize and standardize a MPEG-2 AAC algorithm. Many companies around the world contributed advanced audio coding algorithms in a collaborative effort to come up with a flexible high quality coding standard [44]. The MPEG-2 AAC standard employs high resolution filter banks, prediction techniques, and Huffman coding.

Modules

The MPEG-2 AAC standard is based on recent evaluations and definitions of basic modules each having been selected from a number of proposals. The self-contained modules include:

- optional preprocessing
- time-to-frequency mapping (filterbank)
- psychoacoustic modeling
- prediction
- quantization and coding
- noiseless coding
- bit stream formatter

Profiles

In order to serve different needs, the standard will offer three profiles:

1. main profile
2. low complexity profile
3. sampling-rate-scaleable profile

For example, in its main profile, the filter bank is a modified discrete cosine transform of blocklength 2048 or 256, it allows for a frequency resolution of 23.43 Hz and a time resolution of 2.6 ms (both at a sampling rate of 48 kHz). In the case of the long blocklength, the window shape can vary dynamically as a function of the signal; a temporal noise shaping tool is offered to control the time dependence of the quantization noise; time domain prediction with second order backward-adaptive linear predictors reduces the bit rate for coding subsequent subband samples in a given subband; iterative non-uniform quantization and noiseless coding are applied.

The low complexity profile does not employ temporal noise shaping and time domain prediction, whereas in the sampling-rate-scaleable profile a preprocessing module is added that allows for samplig rates of 6, 12, 18, and 24 kHz. The default configurations of MPEG-2 AAC include 1.0, 2.0, and 5.1 (mono, stereo, and five channel with LFE-channel). However, 16 configurations can be defined in the encoder. A detailed description of the MPEG-2 AAC multichannel standard can be found in the literature [44].

The above listed selected modules define the MPEG-2/AAC standard which became International Standard in April 1997 as an extension to MPEG-2 (*ISO/MPEG 13818 - 7*). The standard offers high quality at lowest possible bit rates between 320 and 384 kb/s for five channels, it will find many applications, both for consumer and professional use.

40.4.4   Simulcast Transmission

If bit rates are not of high concern, a *simulcast transmission* may be employed where a full MPEG-1 bitstream is multiplexed with the full MPEG-2 AAC bit stream in order to support BC without matrixing techniques (Fig. 40.21).

40.4.5   Subjective Tests

First subjective tests, independently run at German Telekom and BBC (UK) under the umbrella of the MPEG-2 standardization process had shown a satisfactory average performance of NBC and BC coders. The tests had been carried out with experienced listeners and critical test items at low bit rates (320 and 384 kb/s). However, all codecs showed deviations from transparency for some of the test items [48, 49]. Very recently [50], extensive formal subjective tests have been carried out to compare MPEG-2 AAC

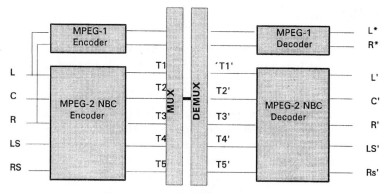

FIGURE 40.21   BC MPEG-2 multichannel audio coding (simulcast mode).

coders, operating, respectively, at 256 and 320 kb/s, and a BC MPEG-2 layer II coder,[2] operating at 640 kb/s. All coders showed a very good performance, with a slight advantage of the 320 kb/s MPEG-2 AAC coder compared with the 640 kb/s MPEG-2 layer II BC coder. The performances of those coders are indistinguishable from the original in the sense of the EBU definition of *indistinguishable quality* [51].

40.5   MPEG-4/Audio Coding

Activities within MPEG-4 aim at proposals for a broad field of applications including multimedia. MPEG-4 will offer higher compression rates, and it will merge the whole range of audio from high fidelity audio coding and speech coding down to synthetic speech and synthetic audio. In order to represent, integrate, and exchange pieces of audio-visual information, MPEG-4 offers standard tools which can be combined to satisfy specific user requirements [52]. A number of such configurations may be standardized. A syntactic description will be used to convey to a decoder the choice of tools made by the encoder. This description can also be used to describe new algorithms and download their configuration to the decoding processor for execution. The current toolset supports audio and speech compression at monophonic bit rates ranging from 2 to 64 kb/s. Three *core coders* are used:

1. a parametric coding scheme for low bit rate speech coding
2. an analysis-by-synthesis coding scheme for medium bit rates (6 to 16 kb/s)
3. a subband/transform-based coding scheme for higher bit rates.

These three coding schemes have been integrated into a so-called verification model that describes the operations both of encoders and decoders, and that is used to carry out simulations and optimizations. In the end, the verification model will be the embodiment of the standard [52]. Let us also note that MPEG-4 will offer new functionalities such as time scale changes, pitch control, edibility, database access, and scalability, which allows extraction from the transmitted bitstream of a subset sufficient to generate audio signals with lower bandwidth and/or lower quality depending on channel capacity or decoder complexity. MPEG-4 will become an international standard in November 1998.

---

[2]A 1995 version of this latter coder was used, therefore its test results do not reflect any subsequent enhancements.

40.6   Applications

MPEG/Audio compression technologies will play an important role in consumer electronics, professional audio, telecommunications, broadcasting, and multimedia. A few, but typical application fields are described in the following.

Main applications will be based on delivering digital audio signals over terrestrial and satellite-based digital *broadcast and transmission systems* such as subscriber lines, program exchange links, cellular mobile radio networks, cable-TV networks, local area networks, etc. [53]. For example, in narrowband *Integrated Services Digital Networks* (ISDN) customers have physical access to one or two 64-kb/s B channels and one 16-kb/s D channel (which supports signaling but can also carry user information). Other configurations are possible including $p \times 64$ kb/s ($p = 1, 2, 3, \ldots$) services. ISDN rates offer useful channels for a practical distribution of stereophonic and multichannel audio signals.

Because ISDN is a bidirectional service, it also provides upstream paths for future on-demand and interactive audiovisual *just-in-time audio* services. The backbone of digital telecommunication networks will be broadband (B-) ISDN with its cell-oriented structure. Cell delays and cell losses are sources of distortions to be taken into account in designs of digital audio systems [54].

Lower bit rates than those given by the 16-bit PCM format are mandatory if audio signals are to be stored efficiently on *storage media*—although the upcoming digital versatile disk (DVD) with its capacity of 4.7 GB relieves the pressure for extreme compression factors. In the field of digital storage on digital audio tape and (re-writeable) disks, a number of MPEG-based consumer products have recently reached the audio market. Of these products, Philips *Digital Compact Cassette* (DCC) essentially makes use of layer I of the MPEG-1/Audio coder employing its 384 kb/s stereo rate; its audio coding algorithm is called PASC (*Precision Audio Subband Coding*) [16]. The DCC encoder obtains an estimate of the short-term spectrum directly from the 32 subbands.

In the movie theater world, a 7.1-channel configuration is becoming popular due to an improved front-back stability of the stereo image and an improved impression of spaciousness. A scalable 7.1-channel reproduction is applied in the digital video disc (DVD). It is based on the MPEG-1 and MPEG-2 standards by down-mixing the 7-channel signal into a 5-channel signal, and a subsequent down-mixing of the latter one into a 2-channel signal [55]. The 2-channel signal, three contributions from the 5-channel signal, and two contributions from the 7-channel signal can then be transmitted or stored. The decoder uses the 2-channel signal directly, or it employs matrixing to reconstruct 5- or 7-channel signals. Other formats are possible, such as storing a 5-channel signal and an additional stereo signal in simulcast mode, without down-mixing the stereo signal from the multichannel signal.

A further example is solid state audio playback systems (e.g., for announcements) with the compressed data stored on chip-based memory cards or smart cards. One example is NEC's prototype *Silicon Audio Player* which uses a one-chip MPEG-1/Audio layer II decoder and offers 24 min of stereo at its recommended stereo bit rate of 192 kb/s [56].

A number of decisions concerning the introduction of *digital audio broadcast* (DAB) and digital video broadcast (DVB) services have been made recently. In Europe, a project group named Eureka 147 has worked out a DAB system able to cope with the problems of digital broadcasting [57]–[59]. ITU-R has recommended the MPEG-1/Audio coding standard after it had made extensive subjective tests. Layer II of this standard is used for program emission, the Layer III version is recommended for commentary links at low rates. The sampling rate is 48 kHz in all cases, the ancillary data field is used for program associated data (PAD information). The DAB system has a significant bit rate overhead for error correction based on punctured convolutional codes in order to support *source-adapted channel coding*, i.e., an unequal error protection that is in accordance with the sensitivity of individual bits or a group of bits to channel errors [60]. Additionally, error concealment techniques will be applied to provide a *graceful degradation* in case of severe errors. In the U.S. a standard has not yet been defined. Simulcasting analog and digital versions of the same audio program in the FM terrestrial band (88 to 108 MHz) is an important issue (whereas the European solution is based on new channels) [61].

As examples of satellite-based digital broadcasting, we mention the Hughes DirecTV satellite subscription television system and ADR (Astra Digital Radio) both of which make use of MPEG-1 layer II. As a further example, the Eutelsat SaRa system will be based on layer III coding.

Advanced digital TV systems provide HDTV delivery to the public by terrestrial broadcasting and a variety of alternate media and offer full-motion high resolution video and high quality multichannel surround audio. The overall bit rate may be transmitted within the bandwidth of an analog UHF television channel. The U.S. *Grand Alliance HDTV* system and the European *Digital Video Broadcast (DVB)* system both make use of the MPEG-2 video compression system and of the MPEG-2 transport layer which uses a flexible ATM-like packet protocol with headers/descriptors for multiplexing audio and video bit streams in one stream with the necessary information to keep the streams synchronized when decoding. The systems differ in the way the audio signal is compressed: the Grand Alliance system will use Dolby's AC-3 transform coding technique [62]–[64], whereas the DVB system will use the MPEG-2/Audio algorithm.

40.7 Conclusions

Low bit rate digital audio is applied in many different fields, such as consumer electronics, professional audio processing, telecommunications, and broadcasting. Perceptual coding in the frequency domain has paved the way to high compression rates in audio coding. ISO/MPEG-1/Audio coding with its three layers has been widely accepted as an international standard. Software encoders, single DSP chip implementations, and computer extensions are available from a number of suppliers.

In the area of broadcasting and mobile radio systems, services are moving to portable and handheld devices, and new, third generation mobile communication networks are evolving. Coders for these networks must not only operate at low bit rates but must be stable in burst-error and packet- (cell-) loss environments. Error concealment techniques will play a significant role. Due to the lack of available bandwidth, traditional channel coding techniques may not be able to sufficiently improve the reliability of the channel.

MPEG/Audio coders are controlled by psychoacoustic models which may be improved thus leaving room for an evolutionary improvement of codecs. In the future, we will see new solutions for encoding. A better understanding of binaural perception and of stereo presentation will lead to new proposals.

Digital multichannel audio improves stereophonic images and will be of importance both for audio-only and multimedia applications. MPEG-2/audio offers both BC and NBC coding schemes to serve different needs. Ongoing research will result in enhanced multichannel representations by making better use of interchannel correlations and interchannel masking effects to bring the bit rates further down. We can also expect solutions for special presentations for people with impairments of hearing or vision which can make use of the multichannel configurations in various ways.

Emerging activities of the ISO/MPEG expert group aim at proposals for audio coding which will offer higher compression rates, and which will merge the whole range of audio from high fidelity audio coding and speech coding down to synthetic speech and synthetic audio (ISO/IEC MPEG-4). Because the basic audio quality will be more important than compatibility with existing or upcoming standards, this activity will open the door for completely new solutions.

References

[1] Bruekers, A.A.M.L. et al., Lossless coding for DVD audio, 101th Audio Engineering Society Convention, Los Angeles, Preprint 4358, 1996.

[2] Jayant, N.S. and Noll, P., *Digital coding of waveforms: Principles and Applications to Speech and Video*, Prentice-Hall, Englewood Cliffs, NJ, 1984.

[3] Spanias, A.S., Speech coding: A tutorial review, *Proc. IEEE*, 82(10), 1541–1582, Oct.94.

[4] Jayant, N.S., Johnston, J.D. and Shoham, Y., Coding of wideband speech, *Speech Commun.*, 11, 127–138, 1992.

[5] Gersho, A., Advances in speech and audio compression, *Proc. IEEE*, 82(6), 900–918, 1994.

[6] Noll, P., Wideband speech and audio coding, *IEEE Commun. Mag.*, 31(11), 34–44, 1993.

[7] Noll, P., Digital audio coding for visual communications, *Proc. IEEE*, 83(6), June 1995.

[8] ISO/IEC JTC1/SC29, Information technology—Coding of moving pictures and associated audio for digital storage media at up to about 1.5 Mbit/s–IS 11172 (Part 3, Audio), 1992.

[9] ISO/IEC JTC1/SC29, Information technology—Generic coding of moving pictures and associated audio information–IS 13818 (Part 3, Audio), 1994.

[10] ISO/MPEG, Doc. N0821, Proposal Package Description - Revision 1.0, Nov. 1994.

[11] WWW — official MPEG home page: address http://drogo.cselt.stet.it/mpeg/. Important link: http:/www.vol.it/MPEG/

[12] Hathaway, G.T., A NICAM digital stereophonic encoder, in *Audiovisual Telecommunications* Nigthingale, N.D. Ed., Chapman & Hall, 1992, 71 - 84.

[13] Zwicker, E. and Feldtkeller, R., *Das Ohr als Nachrichtenempfänger,* S. Hirzel Verlag, Stuttgart, 1967.

[14] Jayant, N.S., Johnston, J.D. and Safranek, R., Signal compression based on models of human perception, *Proc. IEEE*, 81(10), 1385–1422, 1993.

[15] Zelinski, R. and Noll, P., Adaptive transform coding of speech signals, *IEEE Trans. on Acoustics, Speech, and Signal Proc.*, ASSP-25, 299–309, Aug. 1977.

[16] Hoogendorn, A., Digital compact cassette, *Proc. IEEE*, 82(10), 1479–1489, Oct. 1994.

[17] Noll, P., On predictive quantizing schemes, *Bell System Tech. J.*, 57, 1499–1532, 1978.

[18] Makhoul, J. and Berouti, M., Adaptive noise spectral shaping and entropy coding in predictive coding of speech. *IEEE Trans. on Acoustics, Speech, and Signal Processing*, 27(1), 63–73, Feb. 1979.

[19] Esteban, D. and Galand, C., Application of quadrature mirror filters to split band voice coding schemes, *Proc. ICASSP*, 191–195, 1987.

[20] Rothweiler, J.H., Polyphase quadrature filters, a new subband coding technique, *Proc. Intl. Conf. ICASSP'83*, 1280–1283, 1983.

[21] Princen, J. and Bradley, A., Analysis/synthesis filterbank design based on time domain aliasing cancellation, *IEEE Trans. on Acoust. Speech, and Signal Process.*, ASSP-34, 1153–1161, 1986.

[22] Malvar, H.S., *Signal Processing with Lapped Transforms,* Artech House, 1992.

[23] Yeoh, F.S. and Xydeas, C.S., Split-band coding of speech signals using a transform technique, *Proc. ICC*, 3, 1183–1187, 1984.

[24] Granzow, W., Noll, P. and Volmary, C., Frequency-domain coding of speech signals, (in German), NTG-Fachbericht No. 94, VDE-Verlag, Berlin, 150–155, 1986.

[25] Edler, B., Coding of audio signals with overlapping block transform and adaptive window functions, (in German), *Frequenz*, 43, 252–256, 1989.

[26] Iwadare, M., Sugiyama, A., Hazu, F., Hirano, A. and Nishitani, T., A 128 kb/s hi-fi audio CODEC based on adaptive transform coding with adaptive block size, *IEEE J. on Sel. Areas in Commun.*, 10(1), 138–144, Jan. 1992.

[27] Zelinski, R. and Noll, P., Adaptive Blockquantisierung von Sprachsignalen, Technical Report No. 181, Heinrich-Hertz-Institut für Nachrichtentechnik, Berlin, 1975.

[28] van der Waal, R.G., Brandenburg, K. and Stoll, G., Current and future standardization of high-quality digital audio coding in MPEG, *Proc. IEEE ASSP Workshop on Applications of Signal Processing to Audio and Acoustics*, New Paltz, NY, 1993.

[29] Noll, P. and Pan, D., ISO/MPEG audio coding, *Intl. J. High Speed Electronics and Systems*, 1997.

[30] Brandenburg, K. and Stoll, G., The ISO/MPEG-audio codec: A generic standard for coding of high quality digital audio, *J. Audio Eng. Soc. (AES)*, 42(10), 780–792, Oct. 1994.

[31] van de Kerkhof, L.M. and Cugnini, A.G., The ISO/MPEG audio coding standard, *Widescreen Review*, 1994.

[32] Dehery, Y.F., Stoll, G. and Kerkhof, L.v.d., MUSICAM source coding for digital sound, 17th International Television Symposium, Montreux, Record 612–617, june 1991.

[33] Brandenburg, K., Herre, J., Johnston, J.D., Mahieux, Y. and Schroeder, E.F., ASPEC: Adaptive spectral perceptual entropy coding of high quality music signals, *90th. Audio Engineering Society-Convention, Paris,* Preprint 3011, 1991.

[34] Musmann, H.G., The ISO audio coding standard, *Proc. IEEE Globecom*, Dec. 1990.

[35] van der Waal, R.G., Oomen, A.W.J. and Griffiths, F.A., Performance comparison of CD, noise-shaped CD and DCC, in *Proc. 96th Audio Engineering Society Convention*, Amsterdam, Preprint 3845, 1994.

[36] Herre, J., Brandenburg, K. and Lederer, D., Intensity stereo coding, *96th Audio Engineering Society Convention*, Amsterdam, Preprint no. 3799, 1994.

[37] Pan, D., A tutorial on MPEG/audio compression, *IEEE Trans. on Multimedia*, 2(2), 60–74, 1995.

[38] Brandenburg, K. et al., Variable data-rate recording on a PC using MPEG-audio layer III, *5th Audio Engineering Society Convention*, New York, 1993.

[39] Sarginson, P.A., MPEG-2: Overview of the system layer, *BBC Research and Development Report*, BBC RD 1996/2, 1996.

[40] Ryden, T., Grewin, C. and Bergman, S., The SR report on the MPEG audio subjective listening tests in Stockholm April/May 1991, ISO/IEC JTC1/SC29/WG 11: Doc.-No. MPEG 91/010, May 1991.

[41] Fuchs, H., Report on the MPEG/audio subjective listening tests in Hannover, ISO/IEC JTC1/SC29/WG 11: Doc.-No. MPEG 91/331, Nov. 1991.

[42] Stoll, G. et al., Extension of ISO/MPEG-audio layer II to multi-channel coding: The future standard for broadcasting, telecommunication, and multimedia application, *94th Audio Engineering Society Convention*, Berlin, Preprint no. 3550, 1993.

[43] Grill, B. et al., Improved MPEG-2 audio multi-channel encoding, *96th Audio Engineering Society Convention*, Amsterdam, Preprint 3865, 1994.

[44] Bosi, M. et al., ISO/IEC MPEG-2 advanced audio coding, *101th Audio Engineering Society Convention*, Los Angeles, Preprint 4382, 1996.

[45] Johnston J.D. et al., NBC-audio - stereo and multichannel coding methods, *101th Audio Engineering Society Convention*, Los Angeles, Preprint 4383, 1996.

[46] Ten Kate, W.R.Th. et al., Matrixing of bit rate reduced audio signals, *Proc. Int. Conf. on Acoustics, Speech, and Signal Processing* (ICASSP'92), 2, II-205–II-208, 1992.

[47] ten Kate, W.R.Th., Compatibility matrixing of multi-channel bit-rate-reduced audio signals, *96th Audio Engineering Society Convention*, Preprint 3792, Amsterdam, 1994.

[48] Feige, F. and Kirby, D., Report on the MPEG/audio multichannel formal subjective listening tests, ISO/IEC JTC1/SC29/WG 11: Doc. N 0685, March 1994.

[49] Meares, D. and Kirby, D., Brief subjective listening tests on MPEG-2 backwards compatible multichannel audio codecs, ISO/IEC JTC1/SC29/WG 11: Aug. 1994.

[50] ISO/IEC/JTC1/SC29, Report on the formal subjective listening tests of MPEG-2 NBC multi-channel audio coding, Document N1371, Oct. 1996.

[51] ITU-R Document TG 10-2/3, Oct. 1991.

[52] /IEC/JTC1/SC29, Description of MPEG-4, Document N1410, Oct. 1996.

[53] Burpee, D.S. and Shumate, P.W., Emerging residential broadband telecommunications, *Proc. IEEE*, 82(4), 604–614, 1994.

[54] Jayant, N.S., High quality networking of audio-visual information, *IEEE Commun. Mag.*, 84–95, 1993.

[55] Ten Kate, W.R.Th., Akagiri, K., van de Kerkhof, L.M. and Kohut, M. J., Scalability in MPEG audio compression. From stereo via 5.1-channel surround sound to 7.1-channel augmented sound fields, *100th Audio Engineering Society Convention*, Copenhagen, 1996, Preprint 4196.

[56] Sugiyama, A. et al., A new implementation of the silicon audio player based on an MPEG/Audio decoder LSI, Technical Report DSP94-99 (1994-12) of the IEICE, 39–45, 1994.

[57] Lau, A. and Williams, W.F., Service planning for terrestrial digital audio broadcasting, *EBU Technical Review*, 4–25, 1992.

[58] Plenge, G., DAB—A new sound broadcasting systems: status of the development—routes to its introduction, *EBU Review*, April 1991.

[59] ETSI, European Telecommunication Standard, Draft prETS 300 401, Jan. 1994.

[60] Weck, Ch., The error protection of DAB, *Audio Engineering Society-Conference "DAB - The Future of Radio"*, London, May 1995.

[61] Jurgen, R.D., Broadcasting with digital audio, *IEEE Spectrum*, 52–59, March 1996.

[62] Todd, C. et al., AC-3: Flexible perceptual coding for audio transmission and storage, *96th Audio Engineering Society Convention*, Amsterdam, Preprint 3796, 1994.

[63] Hopkins, R., Choosing an American digital HDTV terrestrial broadcasting system, *Proc. IEEE*, 82(4), 554–563, 1994.

[64] The grand alliance, *IEEE Spectrum* 36–45, April 1995.

41

Digital Audio Coding: Dolby AC-3

Grant A. Davidson
Dolby Laboratories, Inc.

41.1 Overview

In order to more efficiently transmit or store high-quality audio signals, it is often desirable to reduce the amount of information required to represent them. In the case of digital audio signals, the amount of binary information needed to accurately reproduce the original pulse code modulation (PCM) samples may be reduced by applying compression algorithm. A primary goal of audio compression algorithms is to maximally reduce the amount of digital information (bit-rate) required for conveyance of an audio signal while rendering differences between the original and decoded signals inaudible.

Digital audio compression is useful wherever there is an economic benefit realized by reducing the bit-rate. Typical applications are in satellite or terrestrial audio broadcasting, delivery of audio over electrical or optical cables, or storage of audio on magnetic, optical, semiconductor, or other storage media. One application which has received considerable attention in the United States is digital television (DTV). Audio and video compression are both necessary in DTV to meet the requirement that one high-definition DTV channel fit within the 6 MHz transmission bandwidth occupied by one preexisting NTSC (analog) channel. In December 1996, the United States Federal Communications Commission adopted the ATSC standard for DTV which is consistent with a consensus agreement developed by a broad cross-section of parties, including the broadcasting and computer industries. The audio technology used in the ATSC digital audio compression standard [1] is Dolby AC-3.

Dolby AC-3 is an audio compression technology capable of encoding a range of audio channel formats into a bit stream ranging from 32 kb/s to 640 kb/s. AC-3 technology is primarily targeted toward delivery of multiple discrete channels intended for simultaneous presentation to consumers. Channel formats range from 1 to 5.1 channels, and may include a number of associated audio services. The 5.1 channel format consists of five full bandwidth (20 kHz) channels plus an optional low frequency effects (lfe or subwoofer) channel.

A typical application of the algorithm is shown in Fig. 41.1. In this example, a 5.1 channel audio program is converted from a PCM representation requiring more than 5 Mbps (6 channels × 48 kHz × 18 bits = 5.184 Mbps) into a 384 kbps serial bit stream by the AC-3 encoder. Satellite transmission equipment converts this bit stream to an RF transmission which is directed to a satellite transponder. The amount of bandwidth and power required by the transmission has been reduced by more than a factor of 13 by the AC-3 digital compression. The signal received from the satellite is demodulated back into the 384 kbps serial bit stream, and decoded by the AC-3 decoder. The result is the original 5.1 channel audio program.

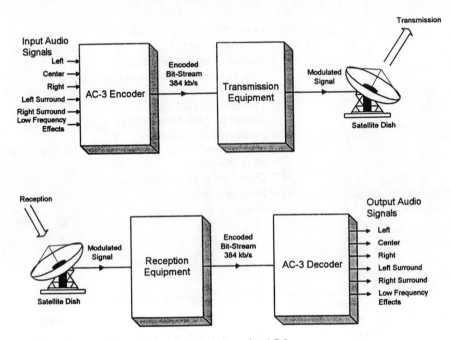

FIGURE 41.1   Example application of satellite transmission using AC-3.

There are a diverse set of requirements for a coder intended for widespread application. While the most critical members of the audience may be anticipated to have complete 6-speaker multi-channel reproduction systems, most of the audience may be listening in mono or stereo, and still others will have three front channels only. Some of the audience may have matrix-based (e.g., Dolby Surround) multi-channel reproduction equipment without discrete channel inputs, thus requiring a dual-channel matrix-encoded output from the AC-3 decoder. Most of the audience welcomes a restricted dynamic range reproduction, while a few in the audience will wish to experience the full dynamic range of the original signal. The visually and hearing impaired wish to be served. All of these and other diverse needs were considered early in the AC-3 design process. Solutions to these requirements have been incorporated from the beginning, leading to a self-contained and efficient system.

As an example, one of the more important listener features built-in to AC-3 is dynamic range compression. This feature allows the program provider to implement subjectively pleasing dynamic range reduction for most of the intended audience, while allowing individual members of the audience the option to experience more (or all) of the original dynamic range. At the discretion of the program originator, the encoder computes dynamic range control values and places them into the AC-3 bit stream. The compression is actually applied in the decoder, so the encoded audio has full dynamic range. It is permissible (under listener control) for the decoder to fully or partially apply the dynamic range control values. In this case, some of the dynamic range will be limited. It is also permissible (again under listener control) for the decoder to ignore the control words, and hence reproduce full-range audio. By default, AC-3 decoders will apply the compression intended by the program provider.

Other user features include decoder downmixing to fewer channels than were present in the bit stream, dialog normalization, and Dolby Surround compatibility. A complete description of these features and the rest of the ATSC Digital Audio Compression Standard is contained in [1].

AC-3 achieves high coding gain (the ratio of the encoder input bit-rate to the encoder output bit-rate) by quantizing a frequency domain representation of the audio signal. A block diagram of this process is shown in Fig. 41.2. The first step in the encoding process is to transform the representation of audio from a sequence of PCM signal sample blocks into a sequence of frequency coefficient blocks. This is done in the analysis filter bank as follows. Signal sample blocks of length 512 are multiplied by a set of window coefficients and then transformed into the frequency domain. Each sample block is overlapped by 256 samples with the two adjoining blocks. Due to the overlap, every PCM input sample is represented in two adjacent transformed blocks. The frequency domain representation includes decimation by an extra factor of two so that each frequency block contains only 256 coefficients. The individual frequency coefficients are then converted into a binary exponential notation as a binary exponent and a mantissa. The set of exponents is encoded into a coarse representation of the signal spectrum which is referred to as the spectral envelope. This spectral envelope is processed by a bit allocation routine to calculate the amplitude resolution required for encoding each individual mantissa. The spectral envelope and the quantized mantissas for 6 audio blocks (1536 audio samples) are formatted into one AC-3 synchronization frame. The AC-3 bit stream is a sequence of consecutive AC-3 frames.

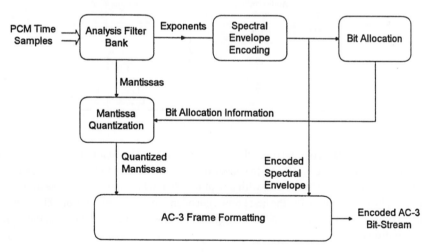

FIGURE 41.2 The AC-3 Encoder.

The decoding process is essentially a mirror-inverse of the encoding process. The decoder, shown in Fig. 41.3, must synchronize to the encoded bit stream, check for errors, and deformat the various types of data such as the encoded spectral envelope and the quantized mantissas. The spectral envelope is decoded to reproduce the exponents. The bit allocation routine is run and the results used to unpack and dequantize the mantissas. The exponents and mantissas are recombined into frequency coefficients, which are then transformed back into the time domain to produce decoded PCM time samples. Figs. 41.2 and 41.3 present a somewhat simplified, high-level view of an AC-3 encoder and decoder.

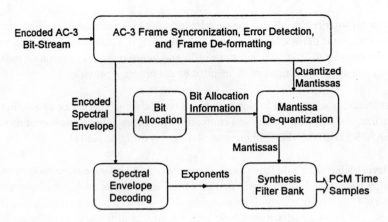

FIGURE 41.3   The AC-3 Decoder.

Table 41.1 presents the different channel formats that are accommodated by AC-3. The three-bit control variable acmod is embedded in the bit stream to convey the encoder channel configuration to the decoder. If acmod is '000', then two completely independent program channels (dual mono) are encoded into the bit stream (referenced as Ch1, Ch2). The traditional mono and stereo formats are denoted when acmod equals '001' and '010', respectively. If acmod is greater than '100', the bit stream format includes one or more surround channels. The optional lfe channel is enabled/disabled by a separate control bit called lfeon.

TABLE 41.1   AC-3 Audio Coding Modes

acmod	Audio coding mode	Number of full bandwidth channels	Channel array ordering
'000'	1 + 1	2	Ch1, Ch2
'001'	1/0	1	C
'010'	2/0	2	L, R
'011'	3/0	3	L, C, R
'100'	2/1	3	L, R, S
'101'	3/1	4	L, C, R, S
'110'	2/2	4	L, R, SL, SR
'111'	3/2	5	L, C, R, SL, SR

Table 41.2 presents the different bit-rates that are accommodated by AC-3. The six-bit control variable frmsizecod is embedded in the bit stream to convey the encoder bit-rate to the decoder. In principle, it is possible to use the bit-rates in Table 41.2 with any of the channel formats from Table 41.1. However, in high-quality applications employing the best known encoder, the typical bit-rate for 2 channels is 192 kb/s, and for 5.1 channels is 384 kb/s. As AC-3 encoding technologies mature in the future, these bit-rates can be expected to drop farther.

TABLE 41.2   AC-3 Audio Coding Bit-Rates

frmsizecod	Nominal bit-rate (kb/sec)	frmsizecod	Nominal bit-rate (kb/sec)	frmsizecod	Nominal bit-rate (kb/sec)
0	32	14	112	28	384
2	40	16	128	30	448
4	48	18	160	32	512
6	56	20	192	34	576
8	64	22	224	36	640
10	80	24	256		
12	96	26	320		

41.2   Bit Stream Syntax

An AC-3 serial coded audio bit stream is composed of a contiguous sequence of synchronization frames. A synchronization frame is defined as the minimum-length bit stream unit which can be decoded independently of any other bit stream information. Each synchronization frame represents a time interval corresponding to 1536 samples of digital audio (for example, 32 ms at a sampling rate of 48 kHz). All of the synchronization codes, preamble, coded audio, error correction, and auxiliary information associated with this time interval is completely contained within the boundaries of one audio frame.

Figure 41.4 presents the various bit stream elements within each synchronization frame. The five different components are: SI (Synchronization Information), BSI (Bit Stream Information), AB (Audio Block), AUX (Auxiliary Data Field), and CRC (Cyclic Redundancy Code). The SI and CRC fields are of fixed-length, while the length of the other four depends upon programming parameters such as the number of encoded audio channels, the audio coding mode, and the number of optionally-conveyed listener features. The length of the AUX field is adjusted by the encoder such that the CRC element falls on the last 16-bit word of the frame. A summary of the bit stream elements and their purpose is provided in Table 41.3.

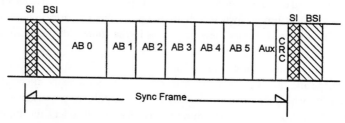

FIGURE 41.4   AC-3 synchronization frame.

TABLE 41.3   AC-3 Bit Stream Elements

Bit stream element	Purpose	Length (bits)
SI	Synchronization information — Header at the beginning of each frame containing information needed to acquire and maintain bit stream synchronization.	40
BSI	Bit stream information — Preamble following SI containing parameters describing the coded audio service, e.g., number of input channels (acmod), dynamic compression control word (dynrng), and program time codes (timecod1, timecod2).	Variable
AB	Audio block — Coded information pertaining to 256 quantized samples of audio from all input channels. There are six audio blocks per AC-3 synchronization frame.	Variable
Aux	Auxiliary data field — Block used to convey additional information not already defined in the AC-3 bit stream syntax.	Variable
CRC	Frame error detection field — Error check field containing a CRC word for error detection. An additional CRC word is located in the SI header, the use of which is optional.	17

The number of bits in a synchronization frame (frame length) is a function of sampling rate and total bit-rate. In a conventional encoding scenario, these two parameters are fixed, resulting in synchronization frames of constant length. However, AC-3 also supports variable-rate audio applications, as will be discussed shortly.

Each Audio Block contains coded information for 256 samples from each input channel. Within one synchronization frame, the AC-3 encoder can change the relative size of the six Audio Blocks depending on audio signal bit demand. This feature is particularly useful when the audio signal is non-stationary over the 1536-sample synchronization frame. Audio Blocks containing signals with a high bit demand can be weighted more heavily than others in the distribution of the available bits (bit pool) for one frame. This feature provides one mechanism for local variation of bit-rate while keeping the overall bit-rate fixed.

In applications such as digital audio storage, an improvement in audio quality can often be achieved by varying the bit-rate on a long-term basis (more than one synchronization frame). This can also be realized in AC-3 by adjusting the bit-rate of different synchronization frames on a signal-dependent basis. In regions where the audio signal is less bit-demanding (for example, during quiet passages), the frame bit-rate (frmsizecod) is reduced. As the audio signal becomes more demanding, the frame bit-rate is increased so that coding distortion remains inaudible. Frame-to-frame bit-rate changes selected by the encoder are automatically tracked by the decoder.

41.3 Analysis/Synthesis Filterbank

The design of an analysis/synthesis filterbank is fundamental to any frequency-domain audio coding system. The frequency and time resolution of the filterbank play critical roles in determining the achievable coding gain. Of significant importance as well are the properties of critical sampling and overlap-add reconstruction. This section discusses these properties in the context of the AC-3 multichannel audio coding system.

Of the many considerations involved in filterbank design, two of the most important for audio coding are the window shape and the impulse response length. The window shape affects the ability to resolve frequency components which are in close proximity, and the impulse response length affects the ability to resolve signal events which are short in time duration. For transform coders, the impulse response length is determined by the transform block length.

A long transform length is most suitable for input signals whose spectrum remains stationary, or varies only slowly with time. A long transform length provides greater frequency resolution, and hence improved coding performance for such signals. On the other hand, a shorter transform length, possessing greater time resolution, is more effective for coding signals that change rapidly in time. The best of both cases can be obtained by dynamically adjusting the frequency/time resolution of the transform depending upon spectral and temporal characteristics of the signal being coded. This behavior is very similar to that known to occur in human hearing, and is embodied in AC-3.

The transform selected for use in AC-3 is based on a 512-point Modified Discrete Cosine Transform (MDCT) [2]. In the encoder, the input PCM block for each successive transform is constructed by taking 256 samples from the last half of the previous audio block and concatenating 256 new samples from the current block. Each PCM block is therefore overlapped by 50% with its two neighbors. In the decoder, each inverse transform produces 512 new PCM samples, which are subsequently windowed, 50% overlapped, and added together with the previous block. This approach has the desirable property of crossfade reconstruction, which reduces waveform discontinuities (and audible distortion) at block boundaries.

41.3.1 Window Design

To achieve perfect-reconstruction with a unity-gain MDCT transform filterbank, the shape of the analysis and synthesis windows must satisfy two design constraints. First of all, the analysis/synthesis windows for two overlapping transform blocks must be related by:

$$a_i(n + N/2)s_i(n + N/2) + a_{i+1}(n)s_{i+1}(n) = 1, \qquad n = 0, \ldots, N/2 - 1 \qquad (41.1)$$

where $a_i(n)$ is the analysis window, $s_i(n)$ is the synthesis window, n is the sample number, N is the transform block length, and i is the transform block index. This is the well-known condition that the analysis/synthesis windows must add so that the result is flat [3]. The second design constraint is:

$$a_i(N/2 - n - 1)s_i(n) - a_i(n)s_i(N/2 - n - 1) = 0, \qquad n = 0, \ldots, N/2 - 1 \qquad (41.2)$$

This constraint must be satisfied so that the time-domain alias distortion introduced by the forward transform is completely canceled during synthesis.

To design the window used in AC-3, a convolution technique was employed which guarantees that the resultant window satisfies Eq. (41.1). Equation (41.2) is then satisfied by choosing the analysis and synthesis windows to be equal. The procedure consists of convolving an appropriately chosen symmetric kernel window with a rectangular window. The window obtained by taking the square root of the result satisfies Eq. (41.1). Tradeoffs between the width of the window main-lobe and the ultimate rejection can be made simply by choosing different kernel windows. This method provides a means for transforming a kernel window having desirable spectral analysis properties (such as in [4]) into one satisfying the MDCT window design constraints.

The window generation technique is based on the following equation:

$$a_i(n) = s_i(n) = \sqrt{\frac{\sum\limits_{j=L}^{M}[w(j)r(n-j)]}{\sum\limits_{j=0}^{K}[w(j)]}} \qquad \text{for } n = 0, \ldots, N-1, \text{ where} \qquad (41.3)$$

$$L = \begin{cases} 0 & 0 \leq n < N - K \\ n - N + K + 1 & N - K \leq n < N \end{cases}$$

$$M = \begin{cases} n & 0 \leq n < K \\ K & K \leq n < N \end{cases}$$

In this equation, $w(n)$ is the kernel window of length $K + 1$, $r(n)$ is a rectangular window of length $N - K$, N is the transform sample block length, and K is the width of the (non-flat) transition region in the resulting window (note that K must satisfy $0 \leq K \leq N/2$). The rectangular window is defined as:

$$r(n) = \begin{cases} 0 & 0 \leq n < (N/2 - K)/2 \text{ and } (3N/2 - K)/2 \leq n < N - K \\ 1 & (N/2 - K)/2 \leq n < (3N/2 - K)/2 \end{cases} \qquad (41.4)$$

The rectangular window is defined to contain $(N/2 - K)/2$ zeros, followed by $N/2$ unity samples, followed by another $(N/2 - K)/2$ zeros. The AC-3 window uses $K = N/2$, implying the transition region length is one-half the total window length.

The Kaiser-Bessel window is used as the kernel in designing the AC-3 analysis/synthesis windows because of its near-optimal transition band slope and good ultimate rejection characteristic. A scalar parameter α in the Kaiser-Bessel window definition can be adjusted to vary this ratio. The AC-3 window uses $\alpha = 5$.

The selection of the Kaiser-Bessel window function and alpha factor used for the AC-3 algorithm is determined by considering the shape of masking template curves. A useful criterion is to use a filter response which is at or below the worst-case combination of all masking templates [5]. Such a filter response is advantageous in reducing the number of bits required for a given level of audio quality. When the filter response is at or below the worst-case combination of all masking templates, the number of bits assigned to transform coefficients adjacent to each tonal component is reduced.

41.3.2   Transform Equations

The transform employed in AC-3 is an extension of the oddly-stacked TDAC (OTDAC) filter bank reported by Princen and Bradley [2]. The extension involves the capability to switch transform block length from $N = 512$ to 256 for audio signals with rapid amplitude changes. As originally formulated by Princen, the filter bank operates with a time-invariant block-length, and therefore has constant time/frequency resolution. An adaptive time/frequency resolution transform can be implemented by changing the time offset of the transform basis functions during short blocks. The time offset is selected to preserve critical sampling and perfect reconstruction before, during, and following transform length changes.

Prior to transforming the audio signal from time to frequency dimension, the encoder performs an analysis of the spectral and/or temporal nature of the input signal and selects the appropriate block length. A one-bit code per channel per Audio Block is embedded in the bit stream which conveys length information: (blksw = 0 or 1 for 512 or 256 samples, respectively). The decoder uses this information to deformat the bit stream, reconstruct the mantissa data, and apply the appropriate inverse transform equations.

Transforming a long block (512 samples) produces 256 unique transform coefficients. Short blocks are constructed starting with 512 windowed audio samples and splitting them into two abutting subblocks of length 256. Each subblock is transformed independently, producing 128 unique non-zero transform coefficients. Hence, the total number of transform coefficients produced in the short-block mode is identical to that produced in long-block mode, but with doubly improved temporal resolution. Transform coefficients from the two subblocks are interleaved together on a coefficient-by-coefficient basis. This block is quantized and transmitted identically to a single long block.

A similar, mirror image procedure is applied in the decoder. Quantized transform coefficients for the two short transforms arrive in the decoder interleaved in frequency. The decoder processes the interleaved sequences identically to long-block sequences, except during the inverse transformation as described below.

A definition of the AC-3 forward transform equation for long and short blocks is:

$$X(k) = 1/N \sum_{n=0}^{N-1} x(n) \cos((2\pi/N)(k + 1/2)(n + n_0)), \quad k = 0, 1, \ldots, N - 1, \tag{41.5}$$

where n is the sample index, k is the frequency index, $x(n)$ is the windowed sequence of N audio samples, and $X(k)$ is the resulting sequence of transform coefficients.

The corresponding inverse transform equation for long and short blocks is:

$$y(n) = \sum_{k=0}^{N-1} X(k) \cos((2\pi/N)(k + 1/2)(n + n_0)), \quad n = 0, 1, \ldots, N - 1 \tag{41.6}$$

Parameter n_0 represents a time offset of the modulator basis vectors used in the transform kernel. For long blocks, and for the second of each short block pair, $n_0 = 257/2$. For the first short block, $n_0 = 1/2$.

When $x(n)$ in Eq. (41.5) is real, $X(k)$ is odd-symmetric for the MDCT. Therefore, only $N/2$ unique non-zero transform coefficients are generated for each new block of N samples. Accordingly, some information is lost during the transform, which ultimately leads to an alias component in $y(n)$. However, with an appropriate choice of n_0, and in the absence of transform coefficient quantization, the aliasing is completely canceled during the window/overlap/add procedure following the inverse transform. Hence, the AC-3 filterbank has the properties of critical sampling and perfect reconstruction. A fundamental advantage of this approach is that 50% frame overlap is achieved without increasing the required bit-rate. Any non-zero overlap used with conventional transforms (such as the DFT or standard DCT) precludes critical sampling, generally resulting in a higher bit-rate for the same level of subjective quality.

Several memory and computation-efficient techniques are available for implementing the AC-3 forward and inverse transforms (for example, see [6]). The most efficient ones can be derived by rewriting

Eqs. (41.5) and (41.6) in the form of an N-point DFT and IDFT, respectively, combined with two complex vector multiplies. The DFT and IDFT can be efficiently computed using an FFT and IFFT, respectively. Two properties further reduce the fast transform length. First, the input signal is real, and second, the N-length sequence $y(n)$ contains only $N/2$ unique samples. When these two properties are combined, the result is an $N/4$-point complex FFT or IFFT. The AC-3 decoder filter bank computation rate is about 13 multiply-accumulate operations per sample per channel, including the window/overlap/add. This computation rate remains virtually unchanged during block length changes.

41.4 Spectral Envelope

The most basic form of audio information conveyed by an AC-3 bit stream consists of quantized frequency coefficients. The coefficients are delivered in floating-point form, whereby each consists of an exponent and a mantissa. The exponents from one audio block provide an estimate of the overall spectral content as a function of frequency. This representation is often termed a spectral envelope. This section describes spectral envelope coding strategies in AC-3, and explores an important relationship between exponent coding and mantissa bit allocation.

Due to the inherent variety of audio spectra within one frame, the AC-3 spectral envelope coding scheme contains significant degrees of freedom. In essence, the six spectral envelopes contained in one frame represent a two-dimensional signal, varying in time (block index) and frequency. AC-3 spectral envelope coding provides for variable coarseness of representation in both dimensions. In the frequency dimension, either one, two, or four mantissas can be shared by one floating-point exponent. In the time dimension, any two or more consecutive audio blocks from one frame can share common set of exponents.

The concepts of spectral envelope coding and bit allocation are closely linked in AC-3. More specifically, the effectiveness with which mantissa bits are utilized can depend greatly upon the encoder's choice of spectral envelope coding. To see this, note that the dominant contributors to the total bit-rate for a frame are the audio exponents and mantissas. Sharing exponents in either the time or frequency dimension, or both, reduces the total cost of exponent transmission for one frame. More liberal use of exponent sharing therefore frees more bits for mantissa quantization. Conversely, retransmitting exponents increases the total cost of exponent transmission for one frame relative to mantissa quantization. Furthermore, the block positions at which exponents are retransmitted can significantly alter the effectiveness of mantissa bit assignments among the various audio blocks. As will be seen later in Section 41.6, bit assignments are derived in part from the coded spectral envelope. In summary, the encoder decisions regarding when to use frequency or time exponent sharing, and when to retransmit exponents depend upon signal conditions. Collectively, these decisions are called exponent strategy.

For short-term stationary signals, the signal spectrum remains substantially invariant from block-to-block. In this case, the AC-3 encoder transmits exponents once in audio block 0, and then typically reuses them for blocks 1-5. The resulting bit allocation would generally be identical for all 6 blocks, which is appropriate for these signal conditions.

For short-term non-stationary signals, the signal spectrum changes significantly from block-to-block. In this case, the AC-3 encoder transmits exponents in block 0 and typically in one or more other blocks as well. In this case, exponent retransmission produces a time trajectory of coded spectral envelopes which better matches dynamics of the original signal. Ultimately, this results in a quality improvement if the cost of exponent retransmission is less than the benefit of redistributing mantissa bits among blocks.

Exponent strategy decisions can be based, for example, on a cost-benefit analysis for each frame. The objective of such an analysis would be to minimize a cost-benefit ratio by considering encoding parameters such as total available bit-rate, audibility of quantization noise (noise-to-mask ratio), exponent coding mode for each audio block (reuse, D15, D25, or D45), channel coupling on/off, and reconstructed audio bandwidth.

The block(s) at which bit assignment updates occur is governed by several different parameters, but primarily by the exponent strategy fields. AC-3 bit streams contain coded exponents for up to five independent channels, and for the coupling and low frequency effects channels (when enabled). The respective exponent strategy fields are called chexpstr[ch], cplexpstr, and lfeexpstr. Bit allocation updates are triggered if the state of any one or more strategy flags is D15, D25, or D45; however, updates can be triggered in between shared exponent block boundaries as well.

Exponents are 5-bit values which indicate the number of leading zeros in the binary representation of a frequency coefficient. For the D15 exponent strategy, the unsigned integer exponent $e(i)$ represents a scale factor for the ith mantissa, equal to $2^{-e(i)}$. Frequency coefficients are normalized in the encoder by multiplying by $2^{e(i)}$, and denormalized in the decoder by multiplying by $2^{-e(i)}$. Exponent values are allowed to range from 0 (for the largest value coefficients with no leading zeros) to 24. Exponents for coefficients which have more than 24 leading zeros are fixed at 24, and the corresponding mantissas are allowed to have leading zeros. Exponents require 5 bits in order to represent all allowed values.

AC-3 exponent transmission employs differential coding, in which the exponents for a channel are differentially coded across frequency. The first exponent of a full bandwidth or lfe channel is always sent as a 4-bit absolute value, ranging from 0-15. The value indicates the number of leading zeros of the first (DC term) transform coefficient. Successive exponents (ascending in frequency) are sent as differential values which must be added to the prior exponent value in order to form the next absolute value.

The differential exponents are combined into groups in the audio block. The grouping is done by one of three methods, D15, D25, or D45. The number of grouped differential exponents placed in the audio block for a particular channel depends on the exponent strategy and on the frequency bandwidth information for that channel. The number of exponents in each group depends only on the exponent strategy.

Exponent strategy information for every channel is included in every AC-3 audio block. Information is never shared across frames, so block 0 will always contain a strategy indication for each channel.

The three exponent strategies provide a tradeoff between bit-rate required for exponents, and their frequency resolution. The overall exponent bit-rate for a frame depends on the exponent strategy, the number of blocks over which the exponents are shared, and the audio signal bandwidth. Table 41.4 presents the per-coefficient bit-rate required to transmit the spectral envelope for each strategy, and for each block share interval. The D15 mode provides the finest frequency resolution (one exponent per frequency coefficient), while the D45 mode consumes the lowest per-coefficient bit-rate.

TABLE 41.4 Exponent Bit-Rate for Different Exponent Strategies

Exponent strategy	Share interval (number of audio blocks)					
	1	2	3	4	5	6
D15	2.33	1.17	0.78	0.58	0.47	0.39
D25	1.17	0.58	0.39	0.29	0.23	0.19
D45	0.58	0.29	0.19	0.15	0.12	0.10

Note: Bits per frequency coefficient

The absolute exponents found in the bit stream at the beginning of the differentially coded exponent sets are sent as 4-bit values which have been limited in either range or resolution in order to save one bit. For full bandwidth and lfe channels, the initial 4-bit absolute exponent represents a value from 0 to 15. Exponent values larger than 15 are limited to a value of 15. For the coupled channel, the 5-bit absolute exponent is limited to even values, and the least significant bit is not transmitted. The resolution has been limited to valid values of 0, 2, 4,..., 24. Each differential exponent can take on one of five values: $-2, -1, 0, +1, +2$. This allows deltas of up to ± 2 (± 12 *dB*) between exponents. These five values are mapped into the values 0, 1, 2, 3, 4 before being grouped, as shown in Table 41.5.

TABLE 41.5 Mapping of
Differential Exponent Values

Differential exponent	Mapped value, M_i
+2	4
+1	3
0	2
−1	1
−2	0

In D15 mode, the above mapping is applied to each individual differential exponent for coding into the bit stream. In D25 mode, each pair of differential exponents is represented by a single mapped value in the bit stream. In this mode, the differential exponent is used once to compute an exponent that is shared between two consecutive frequency coefficients.

The D45 mode is similar to D25 mode except that quadruplets of differential exponents are represented by a single mapped value. Again, the differential exponent is used once to compute an exponent that is shared between four consecutive frequency coefficients.

For all modes, sets of three adjoining (in frequency) mapped values (M_1, M_2 and M_3) are grouped together and coded as a 7-bit unsigned integer I according to the following relation:

$$I = 25M_1 + 5M_2 + M_3 . \tag{41.7}$$

Following the exponent strategy fields in the bit stream is a set of 6-bit channel bandwidth codes, chbwcod[ch]. These are only present for independent channels (not in coupling) that have new exponents in the current block. The channel bandwidth code defines the end mantissa bin number for that channel according to the following:

$$\text{endmant[ch]} = ((\text{chbwcod[ch]} + 12)3) + 37 \tag{41.8}$$

Exponent strategy for each full bandwidth channel and the lfe channel can be updated independently. Lfe channel exponents are restricted only to reuse or D15 mode. If a full bandwidth channel is in coupling, exponents up to the start coupling frequency are transmitted. If the full bandwidth channel is not in coupling, exponents up to the channel bandwidth code are transmitted. If coupling is on for any full bandwidth channel, a separate and independent set of exponents is transmitted for the coupling channel. Coupling start and end frequencies are transmitted as 4-bit indices.

41.5 Multichannel Coding

In the context of AC-3, multichannel audio is defined as two or more full bandwidth channels that are intended for simultaneous presentation to a listener. Multichannel audio coding offers new opportunities in bit-rate reduction beyond those commonly employed in monophonic coders. The goal of multichannel coding is to compress an audio program by exploiting redundancy between the channels and irrelevancy in the signal while preserving both sound clarity and spatial characteristics of the original program. AC-3 achieves this goal by preserving listener cues which affect perceived directionality of hearing (localization).

The motivation for multichannel audio coding is provided by an understanding of how the ear extracts directional information from an incident sound wave. Hearing research suggests that the auditory system does not evaluate every detail of the complicated interaural signal differences, but rather derives what information is needed from definite, easily recognizable attributes [7]. For example, localization of signals are generally distinguished by:

1. Interaural time differences (ITD)
2. Interaural level differences (ILD)

The ITD cues are caused by the difference between the time of arrival of a sound at both ears. ILD cues are sound pressure level differences caused by a different acoustic transfer functions from the acoustic source to the two ears. Most authors agree ITD is the most important attribute of the audio signal relating to the formation of lateral displacements [7]. For tones below about 800 Hz, perceived lateral displacement is approximately linear as a function of the difference in the time of arrival for the two ears, up to an ITD of 600 μsec. Full lateral displacement is obtained with an ITD of approximately 630 μsec. At any given time, the auditory event corresponding to the shorter time of arrival is dominant.

The ear is able to evaluate spectral components of the ear input signals individually with respect to ITD. Lateral displacement of the auditory event attainable for pure tones is most perceptible below 800 Hz. However, for some non-tonal signals above 800 Hz, such as narrowband noise, the ear is still able to detect ITDs. In this case, the interaural temporal displacement of the energy envelope of the signal is generally regarded as the criterion involved. Experiments have indicated that the ITD of the signal's fine temporal structure contributes negligibly to localization; instead, the ear evaluates only the energy envelopes. The processing occurs individually in each of a multiplicity of spectral bands. The spectrum is dissected to a degree determined by the finite spectral resolution of the inner ear. Then, the envelopes of the separate spectral components are evaluated individually. These experimental results form the basis for the use of channel coupling in AC-3.

41.5.1 Channel Coupling

Channel coupling is a method for reducing the bit-rate of multichannel programs by summing two or more correlated channel spectra in the encoder. Frequency coefficients for the single combined (coupled) channel are transmitted in place of the individual channel spectra, together with additional side information. The side information consists of a unique set of coupling coefficients for each channel. In the decoder, frequency coefficients for each output channel are computed by multiplying the coupled channel frequency coefficients by the coupling coefficients for that channel. Coupling coefficients are computed in the encoder in a manner which preserves the short-time energy envelope of the original signals, thereby preserving spatialization cues used by the listener.

Coupling is active only above the coupling start frequency; below this boundary, frequency coefficients are coded independently. The coupling start frequency can be changed from one audio block to the next, and coupling can be disabled if desired. Any combination of two or more full bandwidth channels can be coupled; each channel has an associated channel-in-coupling bit to indicate if it has been included in the coupling channel.

Channel coupling is intended for use only when independent channel coding at the given bit-rate and desired audio bandwidth would result in audible artifacts. As the audio bit-rate is lowered with a fixed bandwidth of 20 kHz, a point is eventually reached where audible coding errors will occur for critical signals. In these circumstances, channel coupling reduces the need for encoders to take more drastic measures to eliminate artifacts, such as lowering the audio bandwidth.

A diagram depicting the encoder coupling procedure for the case of three input channels is depicted in Fig. 41.5. The coupling channel is formed as the vector summation of frequency coefficients from all channels in coupling. An optional signal-dependent phase adjustment is applied to the frequency coefficients prior to summation so that phase cancellation does not occur. For each input channel in coupling, the AC-3 encoder then calculates the power of the original signal and the coupled signal. The power summation is performed individually on a number of bands. For the simplified case of Fig. 41.5, there are two such bands. In a typical application the number of bands is 14, but can vary between 1 and 18. Next, the power ratio between the original signal and the coupled channel is computed for each input channel and each band. Denoted a coupling coordinate, these ratios are quantized and transmitted to the decoder.

To reconstruct the spectral coefficients corresponding to one channel's worth of transform data, quantized spectral coefficients representing the uncoupled portion of the transform block are prepended to a

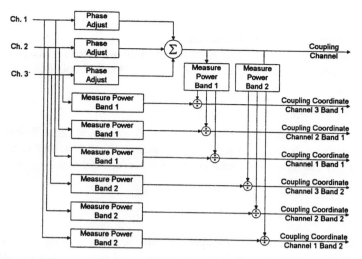

FIGURE 41.5 Block diagram of encoder coupling for three input channels.

set of scaled coupling channel spectral coefficients. The scaled coupling channel coefficients are generated for each channel by multiplying the coupling coordinates for each band by the received coupling channel coefficients, as shown in Fig. 41.6.

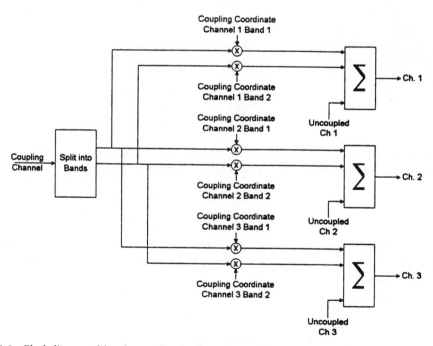

FIGURE 41.6 Block diagram of decoder coupling for three input channels.

Coupling parameters, such as the coupling start/end frequencies and which channels are in coupling, are always transmitted in block 0. They are also optionally transmitted in blocks 1 through 5. Typically only channels with similar spectral shapes are coupled. Level differences between channels are accounted for by the coupling coefficients. It is noteworthy that if in-band spectral differences between coupled input channels are due to level only, the original input channels can still be recovered exactly in the decoder, in the absence of frequency coefficient quantization.

The coupling coefficient dynamic range is -132 to $+18$ dB, with step sizes varying between 0.28 and 0.53 dB. The lowest coupling start frequency is 3.42 kHz at a 48 kHz sampling frequency.

41.5.2 Rematrixing

Rematrixing in AC-3 is a channel combining technique in which sum and difference signals of highly correlated channels are coded rather than the original channels themselves. That is, rather than code and pack left and right (L and R) in a two channel coder, the encoder constructs:

$$
\begin{aligned}
L' &= (L + R)/2 \\
R' &= (L - R)/2 .
\end{aligned}
\tag{41.9}
$$

The usual quantization and data packing operations are then performed on L' and R'.

In the decoder, the original L and R signals are reconstructed using the inverse equations:

$$
\begin{aligned}
L &= L' + R' \\
R &= L' - R' .
\end{aligned}
\tag{41.10}
$$

Clearly, if the original stereo signal were identical in both channels (i.e., two-channel mono), L' is identical to L and R, and R' is identically zero. Therefore, the R' channel can be coded with very few bits, increasing accuracy in the more important L' channel. Rematrixing is only applicable in the 2/0 encoding mode (acmod = '010').

Rematrixing is particularly important when conveying Dolby Surround encoded programs. Consider again a two channel mono source signal. A Dolby Pro Logic decoder will steer all in-phase information to the center channel, and all out-of-phase information to the surround channel. Without rematrixing, the Pro Logic decoder will receive the signals:

$$
\begin{aligned}
Q(L) &= L + n_1 \\
Q(R) &= R + n_2
\end{aligned}
\tag{41.11}
$$

where n_1 and n_2 are uncorrelated quantization noise sequences added by the process of bit-rate reduction. The Pro Logic decoder will then construct the center and surround channels as:

$$
\begin{aligned}
C &= ((L + n_1) + (R + n_2))/2 \\
S &= ((L + n_1) - (R + n_2))/2
\end{aligned}
\tag{41.12}
$$

In the case of the center channel, n_1 and n_2 add, but remain masked by, the dominant $L + R$ signal. In the surround channel, however, $L - R$ cancels to zero, and the surround speakers reproduce the difference in the quantization noise sequences $(n_1 - n_2)$.

If rematrixing is active, the left and right channels will be reproduced as:

$$
\begin{aligned}
L &= (L' + n_3) + (R' + n_4) \cong L + n_3 \\
R &= (L' + n_3) - (R' + n_4) \cong R + n_3 ,
\end{aligned}
\tag{41.13}
$$

where the approximation is made since the quantization noise $n_3 \gg n_4$. More importantly, the center and surround channels will be more faithfully reproduced as:

$$
\begin{aligned}
C &\cong ((L + n_3) + (R + n_3))/2 = (L + R)/2 + n_3 \\
S &\cong ((L + n_3) - (R + n_3))/2 = (L - R)/2 .
\end{aligned}
\tag{41.14}
$$

In this case, the quantization noise in the surround channel is much lower in level.

In AC-3, rematrixing is performed independently in separate frequency bands. There are two to four contiguous bands with boundary locations dependent on coupling information. At a sampling rate of 48 kHz, bands 0 and 1 start at 1.17 and 2.30 kHz, respectively. Bands 2 and 3 start at 3.42 and 5.67 kHz. If coupling is not in use, band 3 stops at a frequency given by the channel bandwidth code. The band boundaries scale proportionally for other sampling frequencies.

Rematrixing is never used in the coupling channel. If coupling and rematrixing are simultaneously in use, the highest rematrixing band ends at the coupling start frequency.

41.6   Parametric Bit Allocation

The process of distributing a finite number of bits B to a block of M frequency bands so as to minimize a suitable distortion criterion is called bit allocation. The result is a bit assignment $b(k)$, $k = 0, 1, ...,$ $M - 1$, which defines the word length of the frequency coefficient(s) transmitted in the kth band. The bit assignment is performed subject to the constraint:

$$\sum_{k=0}^{M-1} b(k) = B .$$

(41.15)

B is determined from the transmission channel capacity, expressed in bits/sec, the block length, and other parameters as well. Performance gains may be realized by allowing B to vary from block to block depending on signal characteristics.

41.6.1   Bit Allocation Strategies

In applications such as digital audio broadcasting and high definition television, one encoder typically distributes programs to many decoders. In these situations, it is advantageous to make the encoder as flexible as possible. If quality improvements are possible even after the decoder design is standardized, the useful life of the coding algorithm can be extended. The bit allocation strategy is a natural candidate for improvement because it plays a crucial role in determining the ultimate quality achievable by a coding algorithm.

One approach to achieving flexibility is to use a forward-adaptive bit allocation strategy, in which the bit assignment $b(k)$ for all bands is explicitly conveyed in the bit stream as side information. A second strategy is termed backward-adaptive allocation, in which $b(k)$ is computed in the encoder and then recomputed in the decoder. Since the computation is based upon quantized information which is transmitted to the decoder anyway, the side information to convey $b(k)$ is not required. The bits that are saved can be used to encode the frequency coefficients themselves.

AC-3 employs an alternative approach [8]. Called parametric bit allocation, the technique combines the advantages of forward and backward-adaptive strategies. It employs a hearing model combining elements from [9] with new features based on more recent psychoacoustic experiments. The term "parametric" refers to the notion that the model is defined by several key variables that influence the masking curve shape and amplitude, and hence the bit assignment. A difference between AC-3 and previous coders is that both the encoder and decoder contain the model, eliminating the need to transmit $b(k)$ explicitly. Only the essential model parameters (psychoacoustic features) are conveyed to the decoder. These parameters can be transmitted with significantly fewer bits than $b(k)$ itself. Furthermore, an improvement path is provided since the specific parameter values are selected by the encoder.

Equally significant, the parametric approach provides latitude for the encoder to adjust the time and frequency resolution of $b(k)$. Bit allocation updates are always present in block 0, and are optionally transmitted in blocks 1 through 5. The frequency resolution of $b(k)$ can be adjusted from 2.7 to 10.7

bands/kHz (one bit assignment every 94 to 375 Hz). The AC-3 encoder typically makes these adjustments in accordance with spectral and temporal changes in the audio signal itself, in a manner similar to the human ear.

Secondary strategy information which also affects block-to-block changes in bit assignment is the bit allocation information, coupling strategy, and SNR offset. This information is always transmitted in block 0, and is optionally transmitted in each subsequent audio block. The presence/absence of the information is controlled by the respective "exists" bits baie, cplstre, and snroffste. The exists bits are set to 1 in blocks 1 to 5 only when a change in strategy results in better audio quality than would be obtained by reusing parameters from the preceding block.

Signal conditions may arise in which the masking curve, and therefore $b(k)$, cannot be sufficiently optimized using the built-in parametric model. Therefore, AC-3 encoders contain a provision for adjusting the masking curve in accordance with an independent psychoacoustic analysis. This is accomplished by transmitting additional bit stream codes, designated as deltas, which convey differences between the two masking curves.

41.6.2   Spreading Function Shape

In Schroeder's model for computing a masking threshold [9], one of the key variables influencing the degree of masking of one spectral component by another is the shape of the spreading function. If the spreading function is a unit impulse, the excitation function (and therefore the masking curve) will be identical in shape to the input signal spectrum. This corresponds to the case where no masking whatsoever is assumed, with the result that all frequency coefficients receive the same bit assignment, and the quantization noise spectrum will conform in shape to the input signal spectrum. As the spreading function is broadened, progressively greater degrees of masking are modeled. This yields a noise spectrum which, in general, contains peaks and valleys that are aligned with features of the input signal spectrum, but broadened in character. As the spreading function is flattened further, eventually a limit is reached where the noise spectrum will be white, corresponding to the minimum mean-square error bit assignment. This noise shaping behavior is identical to the concept of [10, 11].

Since the spreading function strongly influences the level and extent of assumed masking, a parametric description is provided in AC-3. The parametric spreading function is one mechanism available to an AC-3 encoder for making compatible adjustments to the masking model.

The range of spreading function parameter variation was obtained by distilling a family of prototype functions from the available masking data [8]. The variation in shape of the four composite masking curves for 0.5, 1, 2, and 4 kHz tones is shown in Fig. 41.7. We have approximated the envelope of upward masking for each composite curve by two linear segments. The spreading function is defined as the point-by-point maximum of the two segments across frequency. In the example shown in Fig. 41.7, the composite curves can be reasonably approximated by choosing an appropriate slope and vertical offset for each linear segment. Hence, four parameters are transmitted in the bit stream to define the spreading function shape.

41.6.3   Algorithm Description

The AC-3 bit allocation strategy places the majority of computation in the encoder. For example, the encoder can trade off reconstructed signal bandwidth vs. quantization noise power, change the degree of upward masking of one spectral component by another, modify the bit assignment as a function of acoustic signal level, and adaptively control total harmonic distortion. The bit assignment can also be adjusted according to an arbitrary second masking model, as outlined later in the section describing "Delta Bit Allocation". The encoder iteratively converges on an optimal solution. On the other hand, the decoder makes only one pass through the received parameters and exponent data, and is therefore considerably simpler. The bit assignment is reconstructed in the decoder using only basic two's complement operators: add, compare, arithmetic left/right shift, and table lookup.

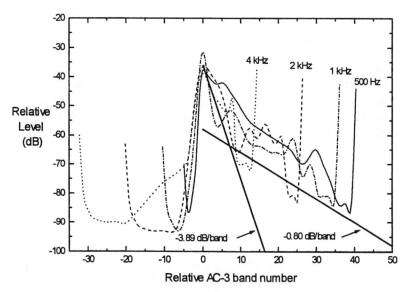

FIGURE 41.7 Comparison between 500 Hz to 4 kHz masking templates and the two slope spreading function.

Frequency Banding

The first step in the masking curve computation is to convert a block of power spectrum samples, taken at equidistant frequency intervals, into a Bark spectrum. This is accomplished by subdividing the power spectrum into multiple frequency bands and then integrating the spectrum samples within each band. The bands are non-uniform in width and derived from the critical-bandwidths defined by Zwicker [12]. The psychoacoustic basis for this procedure is that each critical band corresponds to a fixed distance along the basilar membrane, and therefore to a constant number of auditory nerve fibers.

The summation of the power spectrum samples during linear to Bark frequency conversion requires a linear summation. However, the logarithm of those quantities is most readily available in AC-3. Therefore, a log-adder is employed. The log-addition of two quantities $\log(a)$ and $\log(b)$ is computed using the relation:

$$\log(a + b) = \max(\log(a), \log(b)) + \log(1 + e^d) \tag{41.16}$$

where e is the logarithm base, and

$$d = |\log(a) - \log(b)| . \tag{41.17}$$

The second term on the right side of the equation is implemented as a subtraction $\log(a) - \log(b)$, followed by absolute value and a table lookup. The contents of the table at address d is: $\log(1 + e^d)$. Therefore, the complete log-addition is performed with only add, compare, and table lookup instructions.

Masking Convolution

The technique for modeling masking effects developed by Schroeder specifies convolution. At every frequency point, masking contributions from all other spectral components are weighted and summed. The output of a linear recursive (e.g., IIR) filter may also be viewed as a weighted summation of input samples. Therefore, the convolution of the spreading function with the *linear*-amplitude critical band density in Schroeder's model can be approximated by applying a time-varying linear recursive filter to the spectral components. Upward masking is modeled by filtering the frequency samples from low to high frequency. Downward masking is modeled by filtering input samples in the reverse order. The filter order and coefficients are determined from the desired spreading function. To compute the excitation function using the *logarithmic*-amplitude critical band density used in AC-3, the linear recursive filter is replaced with an equivalent filter which processes logarithmic spectral samples.

The conversion of the linear recursive filter to an equivalent log-domain filter is straightforward. By writing out the difference equation for an IIR filter and taking the logarithm of both sides of the equation, an expression relating the log excitation function with the log power spectrum can be derived. The multiplies and additions of the IIR filter are replaced with additions and log-additions, respectively, in the log domain filter. To implement the two-slope spreading function, two filters are connected in parallel. Each filter implements the characteristic of one of the segments, and the overall excitation value $E(k)$ at band k is computed as the larger of the two filter output samples. The log-domain equations for computing $E(k)$ are:

$$
\begin{aligned}
x_0 &= (x_0 - d_0) \oplus (P(k) - g_0) \\
x_1 &= (x_1 - d_1) \oplus (P(k) - g_1) \\
E(k) &= \max(x_0, x_1)
\end{aligned}
\tag{41.18}
$$

where $P(k)$ is the log-amplitude power spectrum, d_0 and d_1 are the dB spreading function decay values for the first and second segment, respectively, g_0 and g_1 are the dB offsets of the two spreading function segments, and the $\oplus$ symbol denotes log-addition as defined previously. For each of 50 bands, the value of two accumulators x_0 and x_1 is computed by performing a log-addition of the previous accumulator value, decayed by d_0 or d_1, and the current power spectrum value scaled by the gain g_0 or g_1. In AC-3, log-addition is replaced by a maximum operator to reduce computation. This is also more conservative in that additive masking is not assumed.

Compensation for Decoder Selectivity

One basis for determining the bit assignments $b(k)$ in a perception-based allocation strategy is to compute the difference between the signal spectrum and the predicted masking curve. An implicit assumption of this technique is that quantization noise in one particular band is independent of bit assignments in neighboring bands. This is not always a reasonable assumption because the finite frequency selectivity and the high degree of overlap between bands in the decoder filter bank cause localized spreading of the error spectrum (leakage from one band into neighboring bands). The effect is predominant at low frequencies where the slope of the masking curve can equal or exceed the slope of the filter bank transition skirts. Hence, under some conditions, a basis other than the difference between the signal spectrum and masking curve is warranted.

As discussed in [8], decoder selectivity compensation has been found to improve subjective coding performance at low frequencies. Accordingly, the AC-3 masking model employs a straightforward, recursive algorithm for applying compensation from 0 to 2.3 kHz. Although the compensation is a filter bank response correction, not a psychoacoustic effect of human hearing, it can be incorporated into the computation of the excitation curve.

Parameter Variation

The masking computation primarily represents the backward-adaptive portion of the bit allocation strategy. However a number of parameters defining the masking model are transmitted in the compressed data stream. These represent part of the forward-adaptive portion of the bit allocation strategy. As discussed earlier, the shape of the prototype spreading function is controlled by four parameters, where a pair of parameters correspond to each segment. The first linear segment slope is adjustable between -2.95 to -5.77 dB per band, with offsets ranging from -6 to -48 dB. The second segment slope can be adjusted between -0.70 to -0.98 dB per band, with offsets ranging from -49 to -63 dB. The syntax of AC-3 allows the first segment to be controlled independently for each channel. Parameters for the second segment are common to all channels. There are 512 unique spreading function shapes available to an AC-3 encoder.

Delta Bit Allocation

Another forward-adaptive component of the bit allocation strategy is a parametric adjustment that is optionally made to the masking curve computed by the masking model. This adjustment is conveyed to the decoder with the delta bit allocation. For each channel, the encoder can specify nearly arbitrary adjustments to the computed masking curve at a certain cost in bit-rate that otherwise would be used directly to code audio data. Delta bit allocation is used by the encoder to specify a masking curve, and hence a bit assignment, that cannot be generated by the parametric model alone. This feature is useful, for example, if future research points to masking behavior which cannot be simulated by the existing model. In this case, benefits of any new research can be added to an AC-3 encoder by augmenting the parametric model with the improved one.

Determination of the desired delta bit allocation function is straightforward. In an encoder, both the standard AC-3 masking model and the improved one are run in parallel to determine two masking curves. The desired delta bit allocation function is equal to the difference between the masking curves. It may be advantageous to only approximate the desired difference to reduce the required data expenditure. Note that an encoder should first exhaust the flexibility granted by the non-delta parameters before committing any bits to delta bit allocation. The other parameters are less expensive in terms of bit-rate as they must be transmitted periodically in any case.

The delta function is constrained to have a "stair step" shape. Each tread of the stair step corresponds to the masking level adjustment for an integral number of adjoining one-half Bark bands. Taken together, the stair steps comprise a number of non-overlapping, variable-length segments. The segments are run-length coded for efficient transmission.

41.7 Quantization and Coding

All mantissas are quantized to a fixed level of precision indicated by the corresponding bit allocation pointer (bap). Mantissas quantized to 15 or fewer levels use symmetric quantization. Mantissas quantized to more than 15 levels use asymmetric quantization (a conventional two's complement representation).

Some quantized mantissa values are grouped together and encoded into a common codeword. In the case of the 3-level quantizer, 3 quantized values are grouped together and represented by a 5-bit codeword in the data stream. In the case of the 5-level quantizer, 3 quantized values are grouped and represented by a 7-bit codeword. For the 11-level quantizer, 2 quantized values are grouped and represented by a 7-bit codeword. Groups are filled in the order that the mantissas are processed. If the number of mantissas in an exponent set does not fill an integral number of groups, the groups are shared across exponent sets. The next exponent set in the block continues filling the partial groups.

In the encoder, each frequency coefficient is normalized by applying a left-shift equal to its associated exponent (0 to 24). The mantissa is then quantized to a number of levels indicated by the corresponding bap.

Table 41.6 indicates the assignment between bap number and number of quantizer levels. If a bap equals 0, no bits are sent for the mantissa. For more efficient bit utilization, grouping is used for bap values of 1, 2, and 4 (3, 5, and 11 level quantizers).

For bit allocation pointer values between 6 and 15, inclusive, asymmetric fractional two's complement quantization is used. No grouping is employed for asymmetrically quantized mantissas.

For bap values of 1 through 5, inclusive, the mantissas are represented by coded values. The coded values are converted to standard 2's complement fractional binary words. The number of bits indicated by a mantissa's bap are extracted from the bit stream and right justified. This coded value is treated as a table

TABLE 41.6   Quantizer Levels and Mantissa
Bits vs. bap

bap	Quantizer levels	Mantissa bits (group bits / num in group)
0	0	0
1	3	1.67 (5/3)
2	5	2.33 (7/3)
3	7	3
4	11	3.5 (7/2)
5	15	4
6	32	5
7	64	6
8	128	7
9	256	8
10	512	9
11	1024	10
12	2048	11
13	4096	12
14	16,384	14
15	65,536	16

index and is used to look up the quantized mantissa value. The resulting mantissa value is right shifted by the corresponding exponent to generate the transform coefficient value.

The AC-3 decoder may use random noise (dither) values instead of quantized values when the number of bits allocated to a mantissa is zero (bap = 0). The decoder substitution of random values for the quantized mantissas with bap = 0 is conditional on the value of a bit conveyed in the bit stream (dithflag). There is a separate dithflag bit for each transmitted channel. When dithflag = 1, the random noise value is used. When dithflag = 0, a true zero value is used.

41.8 Error Detection

There are several ways in which the AC-3 data may determine that errors are contained within a frame of data. The decoder may be informed of that fact by the transport system which has delivered the data. Data integrity may be checked using the embedded Cyclic Redundancy Check (CRCs) words. Also, some simple consistency checks on the received data can indicate that errors are present. The decoder strategy when errors are detected is user definable. Possible responses include muting, block repeats, frame repeats, or more elaborate schemes based on waveform interpolation to "fill in" missing PCM samples. The amount of error checking performed, and the behavior in the presence of errors are not specified in the AC-3 ATSC standard, but are left to the application and implementation.

Each AC-3 frame contains two 16-bit CRC words. As discussed in Section 41.2, crc1 is the second 16-bit word of the frame, immediately following the synchronization word. crc2 is the last 16-bit word of the frame, immediately preceding the synchronization word of the following frame. crc1 applies to the first 5/8 of the frame, not including the synchronization word. crc2 provides coverage for the last 3/8 of the frame as well as for the entire frame (not including the synchronization word). Decoding of CRC word(s) allows errors to be detected.

The following generator polynomial is used to generate both of the 16-bit CRC words in the encoder:

$$x^{16} + x^{15} + x^2 + 1 .$$
(41.19)

The CRC calculation may be implemented by one of several standard techniques. A convenient hardware implementation is a linear feedback shift register. Details of this technique are presented in [1].

References

[1] United States Advanced Television Systems Committee, ATSC Digital Audio Compression Standard (AC-3), Document A/52, December 20, 1995.

[2] Princen, J.P., Johnson, A.W. and Bradley, A.B., Subband/transform coding using filter bank designs based on time domain aliasing cancellation, *IEEE Intl. Conf. on Acoustics, Speech, and Signal Proc.*, 2161–2164, Dallas, 1987.

[3] Crochiere, R.E. and Rabiner, L.R., *Multirate Digital Signal Processing*, Prentice-Hall, Englewood Cliffs, NJ, 1983, 356–358.

[4] Harris, F.J., On the use of windows for harmonic analysis of the discrete fourier transform, *Proc. IEEE*, 66, 51–83, Jan. 1975.

[5] Fielder, L.D., Bosi, M., Davidson, G., Davis, M., Todd, C. and Vernon, S., AC-2 and AC-3: Low-complexity transform-based audio coding, AES Publication *Collected Papers on Digital Audio Bit Rate Reduction*, Neil Gilchrist and Christer Grewin, Eds., 54–72, 1996.

[6] Sevic, D. and Popovic, M., A new efficient implementation of the oddly-stacked Princen-Bradley filter bank, *IEEE Signal Proc. Lett.*, 1(11), Nov. 1994.

[7] Blauert, J., *Spacial Hearing*, The MIT Press, Cambridge, MA, 1974.

[8] Davidson, G., Parametric bit allocation in a perceptual audio coder, presented at the 97th Convention of the Audio Engineering Society, Preprint 3921, November 1994.

[9] Schroeder, M., Atal, B. and Hall, J., Optimizing digital speech coders by exploiting masking properties of the human ear, *J. Acoustical Soc. Am.*, 66(6), Dec. 1979.

[10] Tribolet, J. and Crochiere, R., Frequency domain coding of speech, *IEEE Trans. Acoustics, Speech, and Signal Proc.*, ASSP-27(5), Oct. 1979.

[11] Crochiere, R. and Tribolet, J., Frequency domain techniques for speech coding, *J. Acoustical Soc. Am.*, 66(6), Dec. 1979.

[12] Zwicker, E., Subdivision of the audible frequency range into critical bands (Frequenzgruppen), *J. Acoustical Soc. Am.*, 33, 248, Feb. 1961.

42

The Perceptual Audio Coder (PAC)

Deepen Sinha
Bell Laboratories
Lucent Technologies

James D. Johnston
AT&T Research Labs

Sean Dorward
Bell Laboratories
Lucent Technologies

Schuyler R. Quackenbush
AT&T Research Labs

PAC is a perceptual audio coder that is flexible in format and bitrate, and provides high-quality audio compression over a variety of formats from 16 kb/s for a monophonic channel to 1024 kb/s for a 5.1 format with four or six auxiliary audio channels, and provisions for an ancillary (fixed rate) and auxiliary (variable rate) side data channel. In all of its forms it provides efficient compression of high-quality audio. For stereo audio signals, it provides near compact disk (CD) quality at about 56 to 64 kb/s, with transparent coding at bit rates approaching 128 kb/s.

PAC has been tested both internally and externally by various organizations. In the 1993 ISO-MPEG-2 5-channel test, PAC demonstrated the best decoded audio signal quality available from any algorithm at 320 kb/s, far outperforming all algorithms, including the layer II and layer III backward compatible algorithms. PAC is the audio coder in most of the submissions to the U.S. Digital Audio Radio (DAR) standardization project, at bit rates of 160 kb/s or 128 kb/s for two-channel audio compression. It has been adapted by various vendors for the delivery of high quality music over the Internet as well as ISDN links. Over the years PAC has evolved considerably. In this paper we present an overview for the PAC algorithm including some recently introduced features such as the use of a signal adaptive switched filterbank for efficient encoding of non-stationary signals.

42.1 Introduction

With the overwhelming success of the compact disc (CD) in the consumer audio marketplace, the public's notion of "high quality audio" has become synonymous with "compact disc quality". The CD represents

0-8493-8572-5/98/$0.00+$.50
© 1998 by CRC Press LLC

stereo audio at a data rate of 1.4112 Mbps (mega bits per second). Despite continued growth in the capacity of storage and transmission systems, many new audio and multi-media applications require a lower data rate.

In compression of audio material, human perception plays a key role. The reason for this is that source coding, a method used very successfully in speech signal compression, does not work nearly as well for music. Recent U.S. and international audio standards work (HDTV, DAB, MPEG-1, MPEG-2, CCIR) therefore has centered on a class of audio compression algorithms known as *perceptual coders*. Rather than minimizing analytic measures of distortion, such as signal-to-noise ratio, perceptual coders attempt to minimize perceived distortion. Implicit in this approach is the idea that signal fidelity perceived by humans is a better quality measure than "fidelity" computed by traditional distortion measures. Perceptual coders define "compact disc quality" to mean "listener indistinguishable from compact disc audio" rather than "two channel of 16-bit audio sampled at 44.1 kHz".

PAC, the Perceptual Audio Coder [10], employs source coding techniques to remove signal redundancy and perceptual coding techniques to remove signal irrelevancy. Combined, these methods yield a high compression ratio while ensuring maximal quality in the decoded signals. The result is a high quality, high compression ratio coding algorithm for audio signals. PAC provides a 20 Hz to 20 kHz signal bandwidth and codes monophonic, stereophonic, and multichannel audio. Even for the most difficult audio material it achieves approximately ten to one compression while rendering the compression effects inaudible. Significantly higher level of compression, e.g., 22 to 1, is achieved with only a little loss in quality.

The PAC algorithm has its roots in a study done by Johnston [7, 8] on the perceptual entropy (PE) vs. the statistical entropy of music. Exploiting the fact that the perceptual entropy (the entropy of that portion of the music signal above the masking threshold) was less than the statistical entropy resulted in the perceptual transform coder (PXFM) [8, 16]. This algorithm used a 2048 point real FFT with 1/16 overlap, which gave good frequency resolution (for redundancy removal) but had some coding loss due to the window overlap.

The next-generation algorithm was ASPEC [2], which used the modified discrete-cosine transform (MDCT) filterbank [15] instead of the FFT, and a more elaborate bit allocation and buffer control mechanism as a means of generating constant-rate output. The MDCT is a critically sampled filterbank, and so does not suffer the 1/16 overlap loss that the PXFM coder did. In addition, ASPEC employed an adaptive window size of 1024 or 256 to control noise spreading resulting from quantization. However, its frequency resolution was half that of PXFM's resulting in some loss in the coding efficiency (c.f., Section 42.3).

PAC as first proposed in [10] is a third-generation algorithm learning from ASPEC and PXFM-Stereo [9]. In its current form, it uses a long transform window size of 2048 for better redundancy removal together with window switching for noise spreading control. It adds composite stereo coding in a flexible and easily controlled form, and introduces improvements in noiseless compression and threshold calculation methods as well. Additional threshold calculations are made for stereo signals to eliminate the problem of binaural noise unmasking.

PAC supports encoders of varying complexity and quality. Broadly speaking, PAC consists of a core codec augmented by various enhancement. The full capability algorithm is sometimes also referred to as *Enhanced* PAC (or EPAC). EPAC is easily configurable to (de)activate some or all of the enhancements depending on the computational budget. It also provides a built-in scheduling mechanism so that some of the enhancements are automatically turned on or off based on averaged short term computational requirement.

One of the major enhancements in the EPAC codec is geared towards improving the quality at lower bit rates of signals with sharp attacks (e.g., castanets, triangles, drums, etc.). Distortion of attacks is a particularly noticeable artifact at lower bit rates. In EPAC, a signal adaptive switched filterbank which switches between a MDCT and a wavelet transform is employed for analysis and synthesis [18]. Wavelet transform offer natural advantages for the encoding of transient signals and the switched filterbank scheme allows EPAC to merge this advantage with the advantages of MDCT for stationary audio segments.

Real-time PAC encoder and decoder hardware have been provided to standards bodies, as well as business partners. Software implementation of real time decoder algorithm is available on PCs and workstations, as well as low cost general-purpose DSPs, making it suitable for mass-market applications. The decoder typically consumes only a fraction of the CPU processing time (even on a 486-PC). Sophisticated encoders run on current workstations and RISC-PCs; simpler real-time encoders that provide moderate compression or quality are realizable on correspondingly less inexpensive hardware.

In the remainder of this paper we present a detailed overview of the various elements of PACs, its applications, audio quality, and complexity issues. The organization of the chapter is as follows. In Section 42.2, some of applications of PAC and its performance on formalized audio quality evaluation tests is discussed. In Section 42.3, we begin with a look at the defining blocks of a perceptual coding scheme followed by the description of the PAC structure and its key components (i.e., filterbank, perceptual model, stereo threshold, noise allocation, etc.). In this context we also describe the switched MDCT/wavelet filterbank scheme employed in the EPAC codec. Section 42.4 focuses on the multichannel version of PAC. Discussions on bitstream formation and decoder complexity are presented in Sections 42.5 and 42.6, respectively, followed by concluding remarks in Section 42.7.

42.2 Applications and Test Results

In the most recent test of audio quality [4] PAC was shown to be the best available audio quality choice [4] for audio compression applications concerning 5-channel audio. This test evaluated both backward compatible audio coders (MPEG Layer II, MPEG Layer III) and non-backward compatible coders, including PAC. The results of these tests showed that PAC's performance far exceeded that of the next best coder in the test.

Among the emerging applications of PAC audio compression technology, the Internet offers one of the best opportunities. High quality audio on demand is increasingly popular and promises both to make existing Internet services more compelling as well as open avenues for new services. Since most Internet users connect to the network using as low bandwidth modem (14.4 to 28.8 kb/s) or at best an ISDN link, high quality low bit rate compression is essential to make audio streaming (i.e., real time playback) applications feasible. PAC is particularly suitable for such applications as it offers near CD quality stereo sound at the ISDN rates and the audio quality continues to be reasonably good for bit rates as low as 12 to 16 kb/s. PAC is therefore finding increasing acceptance in the Internet world.

Another application currently in the process of standardization is digital audio radio (DAR). In the U.S. this may have one of several realizations: a terrestrial broadcast in the existing FM band, with the digital audio available as an adjunct to the FM signal and transmitted either coincident with the analog FM, or in an adjacent transmission slot; alternatively, it can be a direct broadcast via satellite (DBS), providing a commercial music service in an entirely new transmission band. In each of the above potential services, AT&T and Lucent Technologies have entered or partnered with other companies or agencies, providing PAC audio compression at a stereo coding rate of 128 to 160 kb/s as the audio compression algorithm proposed for that service.

Some other applications where PAC has been shown to be the best audio compression quality choice is compression of the audio portion of television services, such as high-definition television (HDTV) or advanced television (ATV).

Still other potential applications of PAC that require compression but are broadcast over wired channels or dedicated networks are DAR, HDTV or ATV delivered via cable TV networks, public switched ISDN, or local area networks. In the last case, one might even envision an "entertainment bus" for the home that broadcasts audio, video, and control information to all rooms in a home.

Another application that entails transmitting information from databases of compressed audio are network-based music servers using LAN or ISDN. This would permit anyone with a networked decoder to have a "virtual music catalog" equal to the size of the music server. Considering only compression, one could envision a "CD on a chip", in which an artist's CD is compressed and stored in a semiconductor ROM and the music is played back by inserting it into a robust, low-power palm-sized music player. Audio compression is also important for read-only applications such as multi-media (audio plus video/stills/text) on CD-ROM or on a PC's hard drive. In each case, video or image data compete with audio for the limited storage available and all signals must be compressed.

Finally, there are applications in which point-to-point transmission requires compression. One is radio station studio to transmitter links, in which the studio and the final transmitter amplifier and antenna may be some distance apart. The on-air audio signal might be compressed and carried to the transmitter via a small number of ISDN B-channels. Another application is the creation of a "virtual studio" for music production. In this case, collaborating artists and studio engineers may each be in different studio, perhaps very far apart, but seamlessly connected via audio compression links running over ISDN.

42.3 Perceptual Coding

PAC, as already mentioned, is a "Perceptual Coder" [6], as opposed to a source modelling coder. For typical examples of source, perceptual, and combined source and perceptual coding, see Figs. 42.1, 42.2, and 42.3. Figure 42.1 shows typical block diagrams of source coders, here exemplified by DPCM, ADPCM, LPC, and transform coding [5]. Figure 42.2 illustrates a basic perceptual coder. Figure 42.3 shows a combined source and perceptual coder.

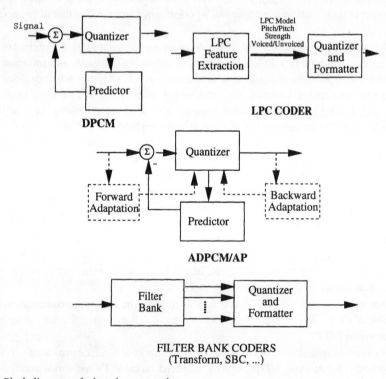

FIGURE 42.1 Block diagrams of selected source-coders.

"Source model" coding describes a method that eliminates redundancies in the source material in the process of reducing the bit rate of the coded signal. A source coder can be either lossless, providing perfect reconstruction of the input signal or lossy. Lossless source coders remove no information from the signal;

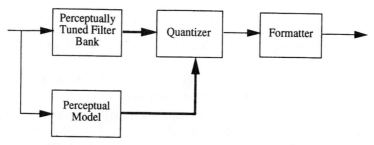

FIGURE 42.2 Block diagrams of a simple perceptual coder.

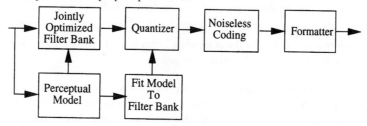

FIGURE 42.3 Block diagrams of an integrated source-perceptual coder.

they remove redundancy in the encoder and restore it in the decoder. Lossy coders remove information from (add noise to) the signal; however, they can maintain a constant compression ratio regardless of the information present in a signal. In practice, most source coders used for audio signals are quite lossy [3].

The particular blocks in source coders, e.g., Fig. 42.1, may vary substantially, as shown in [5], but generally include one or more of the following.

- Explicit source model, for example an LPC model.
- Implicit source model, for example DCPM with a fixed predictor.
- Filterbank, in other words a method of isolating the energy in the signal.
- Transform, which also isolates (or "diagonalizes") the energy in the signal.

All of these methods serve to identify and potentially remove redundancies in the source signal. In addition, some coders may use sophisticated quantizers and information-theoretic compression techniques to efficiently encode the data, and most if not all coders use a bitstream formatter in order to provide data organization. Typical compression methods do not rely on information-theoretic coding alone; explicit source models and filterbanks provide superior source modeling for audio signals.

All perceptual coders are lossy. Rather than exploit mathematical properties of the signal or attempt to understand the producer, perceptual coders model the listener, and attempt to remove irrelevant (unde-tectable) parts of the signal. In some sense, one could refer to it as a "destination" rather than "source" coder. Typically, a perceptual coder will have a lower SNR than an equivalent rate source coder, but will provide superior perceived quality to the listener.

The perceptual coder shown in Fig. 42.2 has the following functional blocks.

- Filterbank — Converts the input signal into a form suitable for perceptual processing.
- Perceptual model — Determines the irrelevancies in the signal, generating a perceptual thresh-old.
- Quantization — Applies the perceptual threshold to the output of the filterbank, thereby removing the irrelevancies discovered by the perceptual model.
- Bit stream former — Converts the quantized output and any necessary side information into a form suitable for transmission or storage.

The combined source and perceptual coder shown in Fig. 42.3 has the following functional blocks.

- Filterbank — Converts the input signal into a form that extracts redundancies and is suitable for perceptual processing.
- Perceptual model — Determines the irrelevancies in the signal, generates a perceptual threshold, and relates the perceptual threshold to the filterbank structure.
- Fitting of perceptual model to filtering domain — Converts the outputs of the perceptual model into a form relevant to the filter bank.
- Quantization – Applies the perceptual threshold to the output of the filterbank, thereby removing the irrelevancies discovered by the perceptual model.
- Information-theoretic compression — Removes redundancy from the output of the quantizer.
- Bit stream former — Converts the compressed output and any necessary side information into a form suitable for transmission or storage.

Most coders referred to as perceptual coders are combined source and perceptual coders. Combining a filterbank with a perceptual model provides not only a means of removing perceptual irrelevancy, but also, by means of the filterbank, provides signal diagonalization, ergo source coding gain. A combined coder may have the same block diagram as a purely perceptual coder; however, the choice of filterbank and quantizer will be different. PAC is a combined coder, removing both irrelevancy and redundancy from audio signals to provide efficient compression.

42.3.1 PAC Structure

Figure 42.4 shows a more detailed block diagram of the monophonic PAC algorithm, and illustrates the flow of data between the algorithmic blocks. There are five basic parts.

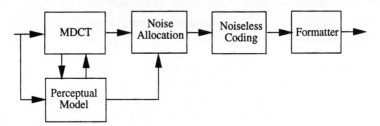

FIGURE 42.4 Block diagram of monophonic PAC encoder.

1. Analysis filterbank — The filterbank converts the time domain audio signal to the short-term frequency domain. Each block is selectably coded by 1024 or 128 uniformly spaced frequency bands, depending on the characteristics of the input signal. PAC's filterbank is used for source coding and cochlear modeling (i.e., perceptual coding).

2. Perceptual model — The perceptual model takes the time domain signal and the output of the filterbank and calculates a frequency domain threshold of masking. A threshold of masking is a frequency dependent calculation of the maximum noise that can be added to the audio material without perceptibly altering it. Threshold values are of the same time and frequency resolution as the filterbank.

3. Noise allocation — Noise is added to the signal in the process of quantizing the filter bank outputs. As mentioned above, the perceptual threshold is expressed as a noise level for each filterbank frequency; quantizers are adjusted such that the perceptual thresholds are met or

exceeded in a perceptually gentle fashion. While it is always possible to meet the perceptual threshold in a unlimited rate coder, coding at high compression ratios requires both overcoding (adding less noise to the signal than the perceptual threshold requires) and undercoding (adding more noise to the signal than the perceptual threshold requires). PAC's noise allocation allows for some time buffering, smoothing local peaks and troughs in the bitrate demand.

4. Noiseless compression — Many of the quantized frequency coefficients produced by the noise allocator are zero; the rest have a non-uniform distribution. Information-theoretic methods are employed to provide an efficient representation of the quantized coefficients.

5. Bitstream former — Forms the bitstream, adds any transport layer, and encodes the entire set of information for transmission or storage.

As an example, Fig. 42.5 shows the perceptual threshold and spectrum for a typical (trumpet) signal. The staircase curve is the calculated perceptual threshold, and the varying curve is the short-term spectrum of the trumpet signal. Note that a great deal of the signal is below the perceptual threshold, and therefore redundant. This part of the signal is what we discard in the perceptual coder.

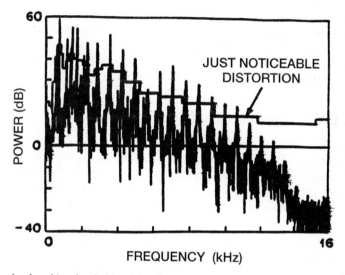

FIGURE 42.5   Example of masking threshold and signal spectrum.

42.3.2   The PAC Filterbank

The filterbank normally used in PAC is referred to as the modified discrete cosine transform (MDCT) [15]. It may be viewed as a modulated, maximally decimated perfect reconstruction filterbank. The subband filters in a MDCT filterbank are linear phase FIR filters with impulse responses twice as long as the number of subbands in the filterbank. Equivalently, MDCT is a lapped orthogonal transform with a 50% overlap between two consecutive transform blocks; i.e., the number of transform coefficients is equal to one half the block length. Various efficient forms of this algorithm are detailed in [11]. Previously, Ferreira [10] has created an alternate form of this filterbank where the decimation is done by dropping the imaginary part of an odd-frequency FFT, yielding and odd-frequency FFT and an MDCT from the same calculations.

In an audio coder it is quite important to appropriately choose the frequency resolution of the filterbank. During the development of the PAC algorithm, a detailed study of the effect of filterbank resolution for a variety of signals was examined. Two important considerations in perceptual coding, i..e, coding gain and non-stationarity within a block, were examined as a function of block length. In general the coding gain increases with the block length indicating a better signal representation for redundancy removal. However, increasing non-stationarity within a block forces the use of more conservative perceptual masking thresholds to ensure the masking of quantization noise at all times. This reduces the realizable or net coding gain. It was found that for a vast majority of music samples the realizable coding gain peaks at the frequency resolution of about 1024 lines or subbands, i.e., a window of 2048 points (this is true for sampling rates in the range of 32 to 48 kHz). PAC therefore employs a 1024 line MDCT as the normal "long" block representation for the audio signal.

In general, some variation in the time frequency resolution of the filterbank is necessary to adapt to the changes in the statistics of the signal. Using a high frequency resolution filterbank to encode a signal segment with a sharp attack leads to significant coding inefficiencies or *pre-echo* conditions. Pre-echos occur when quantization errors are spread over the block by the reconstruction filter. Since pre-masking by an attack in the audio signal lasts for only about 1 msec (or even less for stereo signals), these reconstruction errors are potentially audible as pre-echos unless significant readjustments in the perceptual thresholds are made resulting in coding inefficiencies.

PAC offers two strategies for matching the filterbank resolution to the signal appropriately. A lower computational complexity version is offered in the form of *window switching* approach whereby the MDCT filterbank is switched to a lower 128 line spectral resolution in the presence of attacks. This approach is quite adequate for the encoding of attacks at moderate to higher bit rates (96 kbps or higher for a stereo pair). Another strategy offered as an enhancement in the EPAC codec is the switched MDCT/wavelet filterbank scheme mentioned earlier. The advantages of using such a scheme as well as its functional details are presented below.

42.3.3   The EPAC Filterbank and Structure

The disadvantage of the window switching approach is that the resulting time resolution is uniformly higher for all frequencies. In other words, one is forced to increase the time resolution at the lower frequencies to increase it to the necessary extent at higher frequencies. The inefficient coding of lower frequencies becomes increasingly burdensome at lower bit rates, i.e., 64 kbps and lower. An ideal filterbank for sharp attacks is a non-uniform structure whose subband matches the critical band scale. Moreover, it is desirable that the high frequency filters in the bank be proportionately shorter. This is achieved in EPAC by employing a high spectral resolution *MDCT* for stationary portions of the signal and switching to a non-uniform (tree structured) wavelet filterbank (*WFB*) during non-stationarities.

WFBs are quite attractive for the encoding of attacks [17]. Besides the fact that wavelet representation of such signals is more compact than the representation derived from a high resolution *MDCT*, wavelet filters have desirable temporal characteristics. In a WFB, the high frequency filters (with a suitable moment condition as discussed below) typically have a compact impulse response. This prevents excessive time spreading of quantization errors during synthesis.

The overview of an encoder based on the switched filterbank idea is illustrated in Fig. 42.6. This structure entails the design of a suitable WFB which is discussed next.

The WFB in EPAC consists of a tree structured wavelet filterbank which approximates the critical band scale. The tree structure has the natural advantage that the effective support (in time) of the subband filters is progressively smaller with increasing center frequency. This is because the critical bands are wider at higher frequency so fewer cascading stages are required in the tree to achieve the desired frequency resolution. Additionally, proper design of the prototype filters used in the tree decomposition ensures (see below) that the high frequency filters in particular are compactly localized in time.

The decomposition tree is based on sets of prototype filterbanks. These provide two or more bands of split and are chosen to provide enough flexibility to design a tree structure that approximates the critical band partition closely. The three filterbanks were designed by optimizing parametrized para-unitary filterbanks using standard optimization tools and an optimization criterion based on weighted stopband

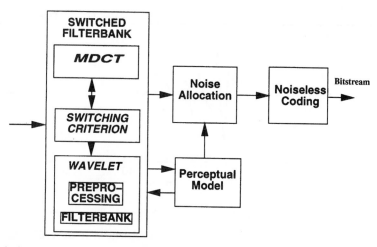

FIGURE 42.6   Block diagram of the switched filterbank audio encoder.

energy [20]. In this design, the *moment* condition plays an important role in achieving desirable temporal characteristics for the high frequency filters. An M band para-unitary filterbank with subband filters $\{H_i\}_{i=1}^{i=M}$ is said to satisfy a Pth order moment condition if $H_i(e^{jw})$ for $i = 2, 3, \ldots M$ has a Pth order zero at $\omega = 0$ [20]. For a given support for the filters, K, requiring $P > 1$ in the design yields filters for which the "effective" support decreases with increasing P. In the other words, most of the energy is concentrated in an interval $K' < K$ and K' is smaller for higher P (for a similar stopband error criterion). The improvement in the temporal response of the filters occurs at the cost of an increased transition band in the magnitude response. However, requiring at least a few vanishing moments yields filters with attractive characteristics.

The impulse response of a high frequency *wavelet* filter (in a 4-band split) is illustrated in Fig. 42.7. For comparison, the impulse response of a filter from a modulated filterbank with similar frequency characteristics is also shown. It is obvious that the *wavelet* filter offers superior localization in time.

Switching Mechanism

The MDCT is a lapped orthogonal transform. Therefore, switching to a wavelet filterbank requires orthogonalization in the overlap region. While it is straightforward to set up a general orthogonalization problem, the resulting transform matrix is inefficient computationally. The orthogonalization algorithm can be simplified by noting that a MDCT operation over a block of $2 * N$ samples is equivalent to a symmetry operation on the windowed data (i.e., outer $N/2$ samples from either end of the window are folded into the inner $N/2$ samples) followed by an N point orthogonal block transform Q over these N samples. Perfect reconstruction is ensured irrespective of the choice of a particular block orthogonal transform Q. Therefore, Q may be chosen to be a DCT for one block and a wavelet transform matrix for the subsequent or any other block. The problem with this approach is that the symmetry operation extends the wavelet filter (or its translates) in time and also introduces discontinuities in these filters. Thus, it impairs the temporal as well as frequency characteristics of the wavelet filters. In the present encoder, this impairment is mitigated by the following two steps: (1) start and stop windows are employed to switch between $MDCT$ and WFB (this is similar to the window switching scheme in PAC), and (2) the effective overlap between the transition and wavelet windows is reduced by the application of a new family of *smooth* windows [19]. The resulting switching sequence is illustrated in Fig. 42.8.

The next design issue in the switched filterbank scheme is the design of a $N \times N$ orthogonal matrix Q^{WFB} based on the prototype filters and the chosen tree structure. To avoid circular convolutions, we employ transition filters at the edge of the blocks. Given a subband filter, c_k, of length K a total of $K_1 = (K/M) - 1$ transition filters are needed at the two ends of the block. The number at a particular

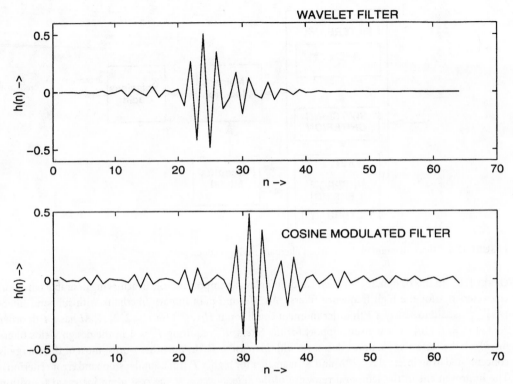

FIGURE 42.7 High frequency wavelet and cosine-modulated filters.

end is determined by the rank of a $K \times (K_1 + 1)$ matrix formed by the translations of c_k. The transition filters are designed through optimization in a subspace constrained by the pre-determined rows of Q^{WFB}.

42.3.4 Perceptual Modeling

Current versions of PAC utilize several perceptual models. Simplest is the monophonic model which calculates an estimated JND in frequency for a single channel. Others add MS (i.e., sum and difference) thresholds and noise-imaging protected thresholds for pairs of channels as well, as "global thresholds" for multiple channels. In this section we discuss the calculation of monophonic thresholds, MS thresholds, and noise-imaging protected thresholds.

Monophonic Perceptual Model

The perceptual model in PAC is similar in method to the model shown as "Psychoacoustic Model II" in the MPEG-1 audio standard annexes [14]. The following steps are used to calculate the masking threshold of a signal.

- Calculate the power spectrum of the signal in 1/3 critical band partitions.
- Calculate the tonal or noiselike nature of the signal in the same partitions, called the tonality measure.
- Calculate the spread of masking energy, based on the tonality measure and the power spectrum.
- Calculate the time domain effects on the masking energy in each partition.
- Relate the masking energy to the filterbank outputs.

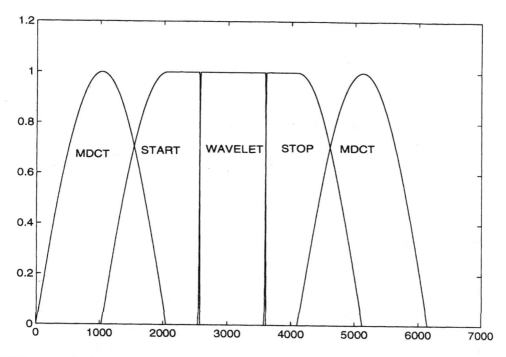

FIGURE 42.8   A filterbank switching sequence.

Application of Masking to the Filterbank

Since PAC uses the same filterbank for perceptual modeling and source coding, converting masking energy into terms meaningful to the filterbank is straightforward. However, the noise allocator quantizes filterbank coefficients in fixed blocks, called *coder bands,* which differ from the 1/3 critical band partitions used in perceptual modeling. Specifically, 49 coder bands are used for the 1024-line filterbank, and 14 for the 128-line filterbank. Perceptual thresholds are mapped to coder bands by using the minimum threshold that overlaps the band.

In EPAC additional processing is necessary to apply the threshold to the WFB. The thresholds for the quantization of wavelet coefficients are based on an estimate of time-varying *spread* energy in each of the subbands and a tonality measure as estimated above. The spread energy is computed by considering the spread of masking across frequency as well as time. In other words, an inter-frequency as well as a temporal spreading function is employed. The shape of these spreading functions may be derived from the cochlear filters [1]. The temporal spread of masking is frequency dependent and is roughly determined by the (inverse of) bandwidth of the cochlear filter at that frequency. A fixed temporal spreading function for a range of frequencies (wavelet subbands) is employed. The coefficients in a subband are grouped in a *coder* band as above and one threshold value per coderband is used in quantization. The coderband span ranges from 10 msec in the lowest frequency subband to about 2.5 msec in the highest frequency subband.

Stereo Threshold Calculation

Experiments have demonstrated that the monaural perceptual model does not extend trivially to the binaural case. Specifically, even if one signal is masked by both the L (left) and R (right) signals individually, it may not be masked when the L and R signals are presented binaurally. For further details, see the discussion of Binary Masking Level Difference (BLMD) in [12].

In stereo PAC Fig. 42.9, we used a model of BLMD in several ways, all based on the calculation of the M (mono, $L + R$) and S (stereo, $L - R$) thresholds in addition to the independent L and R thresholds.

To compute the *M* and *S* thresholds, the following steps are added after the computation of the masking energy.

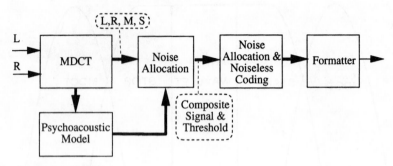

FIGURE 42.9   Stereo PAC block diagram.

- Calculate the spread of masking energy for the other channel, assuming a tonal signal and adding BMLD protection.
- Choose the more restrictive, or smaller, masking energy.

For the *L* and *R* thresholds, the following step is added after the computation of the masking energy.

- Calculation of the spread of masking energy for the other channel. If the two masking energies are similar, add BMLD protection to both.

These four thresholds are used for the calculation of quantization, rate, and so on. An example set of spectra and thresholds for a vocal signal are shown in Fig. 42.10. In this figure, compare the threshold values and energy values in the *S* (or "Difference") signal. As is clear, even with the BMLD protection, most of the *S* signal can be coded as zero, resulting in substantial coding gain. Because the signal is more efficiently coded as MS even at low frequencies where the BLMD protection is in effect, that protection can be greatly reduced for the more energetic *M* channel because the noise will image in the same location as the signal, and not create an unmasking condition for the *M* signal, even at low frequencies. This provides increases in both audio quality and compression rate.

42.3.5   MS vs. LR Switching

In PAC, unlike the MPEG Layer III codec [13] MS decisions are made independently for each group of frequencies. For instance, the coder may alternate coding each group as MS or LR, if that proves most efficient. Each of the L, R, M, and S filterbank coefficients are quantized using the appropriate thresholds, and the number of bits required to transmit coefficients is computed. For each group of frequencies, the more efficient of LR or MS is chosen; this information is encoded with a Huffman codebook and transmitted as part of the bitstream.

42.3.6   Noise Allocation

Compression is achieved by quantizing the filter bank outputs into small integers. Each coder band's threshold is mapped onto 1 of 128 exponentially distributed quantizer step sizes, which is used to quantize the filter bank outputs for that coder band.

PAC controls the instantaneous rate of transmission by adjusting the thresholds according to an equal-loudness calculation. Thresholds are adjusted so that the compression ratio is met, plus or minus a small

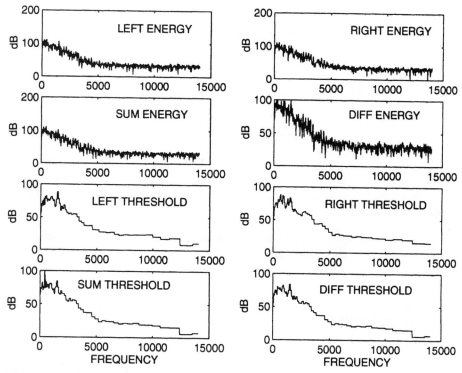

FIGURE 42.10   Examples of stereo PAC thresholds.

amount to allow for short term irregularities in demand. This noise allocation system is iterative, using a single estimator that represents the absolute loudness of the noise relative to the perceptual threshold. Noise allocation is made across all frequencies for all channels, regardless of stereo coding decision: ergo the bits are allocated in a perceptually effective sense between L, R, M, and S, without regard to any measure of how many bits are assigned to L, R, M, and S.

42.3.7   Noiseless Compression

After the quantizers and quantized coefficients for a block are determined, information-theoretic methods are employed to yield an efficient representation.

Coefficients for each coder band are encoded using one of eight Huffman codebooks. One of the tables encodes only zeros; the rest encode coefficients with increasing absolute value. Each codebook encodes groups of two or four coefficients, with the exception of the zero codebook which encodes all of the coefficients in the band. See Table 42.1 for details. In this table, LAV refers to the largest absolute value in a given codebook, and dimension refers to the number of quantized outputs that are coded together in one codeword. Two codebooks are special, and require further mention. The zero codebook is of indeterminate size, it indicates that all quantized values that the zero codebook applies to are in fact zero, no further information is transmitted about those values. Codebook seven is also a special codebook. It is of size -16:16 by -16:16, but the entry of absolute value 16 is not a data value, it is, rather, an escape indicator. For each escape indicator sent in codebook seven (there can be zero, one, or two per codeword), there is an additional escape word sent immediately after the Huffman codeword. This additional codeword, which is generated by rule, transmits the value of the escaped codeword. This generation by rule is a process that has no bounds; therefore, any quantized value can be transmitted by the use of an escape sequence.

Communicating the codebook used for each band constitutes a significant overhead; therefore, similar

TABLE 42.1   PAC Huffman
Codebooks

Codebook	LAV	Dimension
0	0	*
1	1	4
2	1	4
3	2	4
4	4	2
5	7	2
6	12	2
7	ESC	2

codebooks are grouped together in *sections,* with only one codebook transmitted and used for encoding each section.

Since the possible quantizers are precomputed, the indices of the quantizers are encoded rather than the quantizer values. Quantizer indices for coder bands which have only zero coefficients are discarded; the rest are differentially encoded, and the differences are Huffman encoded.

42.4 Multichannel PAC

The multichannel perceptual audio coder (MPAC) extends the stereo PAC algorithm to the coding of multiple audio channels. In general, the MPAC algorithm is software configurable to operate in 2, 4, 5, and 5.1 channel mode. In this document we will describe the MPAC algorithm as it is applied to a 5-channel system consisting of the five full bandwidth channels: Left (L), Right (R), Center (C), Left Surround (Ls), and Right Surround (Rs).

The MPAC 5-channel audio coding algorithm is illustrated in Fig. 42.11. Below we describe the various modules, concentrating in particular on the ones that are different from the stereo algorithm.

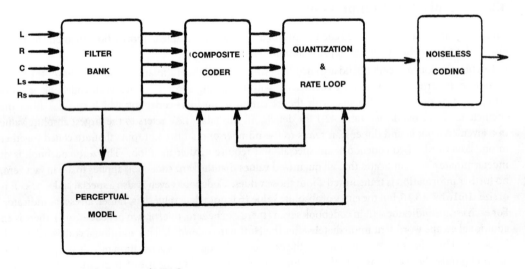

FIGURE 42.11   Block diagram of MPAC.

42.4.1 Filterbank and Psychoacoustic Model

Like the stereo coder, MPAC employs a MDCT filterbank with two possible resolutions, i.e., the usual long block which has 1024 uniformly spaced frequency outputs and a short bank which has 128 uniformly spaced frequency bins. A window switching algorithm, as described above, is used to switch to a short block in the presence of strong non-stationarities in the signal. In the 5-channel setup it desirable to be able to switch the resolution independently for various subsets of channels. For example, one possible, scenario is to apply the window switching algorithm to the front channels (L, R, and C) independently of the surround channels (Ls and Rs). However, this somewhat inhibits the possibilities for composite coding (see below) among the channels. Therefore, one needs to examine the relative gain of independent window switching vs. the gain from a higher level of composite coding. In the present implementation different filterbank resolutions for the front and surround channels are allowed.

The individual masking threshold for the five channels are computed using the PAC psycho-acoustic model described above. In addition, the front pair LR and the surround pair Ls/Rs are used to generate two pairs of MS thresholds (c.f., Section "Stereo Threshold Calculation"). The five channels are coded with their individual thresholds excepting in the case where joint stereo coding is being used (either for the front or the surround pair), in which case the appropriate MS thresholds are used. In addition to the five individual and four stereo thresholds, a joint (or "global") threshold based on all channels is also computed. The computation and role of the global threshold will be discussed later in this section.

42.4.2 The Composite Coding Methods

The MPAC algorithm extends the MS coding of the stereo algorithm to a more elaborate composite coding scheme. Like the MS coding algorithm, the MPAC algorithm uses adaptive composite coding in both time and frequency: the composite coding mode is chosen separately for each of the coder bands at every analysis instance. This selection is based on a "perceptual entropy" criterion and attempts to minimize the bit rate requirement as well as exercise some control over noise localization. The coding scheme uses two complementary sets of inter-channel combinations as described below:

- MS coding for the front and surround pair
- Inter-channel prediction

MS coding is a basis transformation operation and is therefore performed with the uncoded samples of the corresponding pair of channels. The resulting M or S channel is then coded using its own threshold (which is computed separately from the individual channel threshold). Inter-channel prediction, on the other hand, is performed using the quantized samples of the predicting channel. This is done to prevent the propagation of quantization errors (or "cross-talk"). The predicted value for each channel is subtracted from the channel samples and the resulting difference is encoded using the original channel threshold. It may be noted that the two sets of channel combinations are nested so that either, both, or none may be employed for a particular coder band. The coder currently employs the following possibilities for inter-channel prediction.

For the Front Channels (L, R & C): Front L and R channels are coded as LR or MS. In addition, one of the following two possibilities for inter-channel prediction may be used.

1. Center predicts LR (or M if MS coding mode is on).
2. Front M channel predicts the center.

For the Surround Channels (Ls and Rs): Ls and Rs channels are coded as Ls/Rs or Ms/Ss (where Ms and Ss are, respectively, the surround M and surround S). In addition, one or both of following two modes of interchannel prediction may be employed:

1. Front L, R, M channels predict Ls/Rs or Ms.

2. Center channel predicts Ls/Rs or Ms.

In the present implementation, the predictor coefficients in all of the above inter-channel prediction equations are all fixed to either zero or one.

Note that the possibility of completely independent coding is implicit in the above description, i.e., the possibility of turning off any possible prediction is always included. Furthermore, any of these conditions may be independently used in any of the 49 coder bands (long filter band length) or in the 14 coder bands (short filter band length), for each block of filterbank output. Also note that for the short filterbank where the outputs are grouped into 8 groups of 128 (each group of 128 has 14 bands), each of these 8 groups has independently calculated composite coding.

The decisions for composite coding are based primarily on the "perceptual entropy" criterion; i.e., the composite coding mode is chosen to minimize the bit requirement for the perceptual coding of the filterbank outputs from the five channels. The decision for MS coding (for the front and surround pair) is also governed in part by noise localization considerations. As a consequence, the MPAC coding algorithm ensures that signal and noise images are localized at the same place in the front and rear planes. The advantage of this coding scheme is that the quantization noise usually remains masked not only in a listening room environment but also during headphone reproduction of a stereo downmix of the five coded channels (i.e., when two downmixed channels of the form $Lc = L + \alpha C + \beta Ls$, and $Rc = R + \alpha C + \beta Rs$ are produced and fed to a headphone).

The method used for composite coding is still in the experimental phase and subject to refinements/modifications in future.

42.4.3   Use of a Global Masking Threshold

In addition to the five individual thresholds and the four MS thresholds, the MPAC coder also makes use of a global threshold to take advantage of masking across the various channels. This is done when the bit demand is consistently high so that the bit reservoir is close to depletion. The global threshold is taken to be the maximum of five individual thresholds minus a "safety margin". This global threshold is phased in gradually when the bit reservoir is really low (e.g., less than 20%) and in that case it is used as a lower limit for the individual thresholds.

The reason that global threshold is useful is because results in [12] indicate that if the listener is more than a "critical distance" away from the speakers, then the spectrum at either of listener's ear may be well approximated by the sum of power spectrums due to individual speakers.

The computation of a global threshold also involves a safety margin. This safety margin is frequency dependent and is larger for the lower frequencies and smaller for higher frequencies. The safety margin changes with the bit reservoir state.

42.5   Bitstream Formatter

PAC is a block processing algorithm; each block corresponds to 1024 input samples from each channel, regardless of the number of channels. The encoded filter bank outputs, codebook sections, quantizers, and channel combination information for one 1024-sample chunk or eight 128-sample chunks are packed into one *frame*.

Depending on the application, various extra information is added to first frame or to every frame. When storing information on a reliable media, such as a hard disk, one header indicating version, sample rate, number of channels, and encoded rate is placed at the beginning of the compressed music. For extremely unreliable transmission channels, like DAR, a header is added to each frame. This header contains synchronization, error recovery, sample rate, number of channels, an the transmission bit rate.

42.6 Decoder Complexity

The PAC decoder is of approximately equal complexity to other decoders currently known in the art. Its memory requirements are approximately

- 1100 words each for MDCT and WFB workspace
- 512 words per channel for MDCT memory
- (optional) 1024 words per channel for error mitigation
- 1024 samples per channel for output buffer
- 12 Kbytes ROM for codebooks

The calculation requirements for the PAC decoder are slightly more than doing a 512-point complex FFT per 1024 samples per channel. On an Intel 486 based platform, the decoder executes in real time using up approximately 30 to 40.

42.7 Conclusions

PAC has been tested both internally and externally by various organizations. In the 1993 ISO-MPEG-2 5-channel test, PAC demonstrated the best decoded audio signal quality available from any algorithm at 320 kb/s, far outperforming all algorithms, including the backward compatible algorithms. PAC is the audio coder in three of the submissions to the U.S. DAR project, at bit rates of 160 kb/s or 128 kb/s for two-channel audio compression.

PAC presents innovations in the stereo switching algorithm, the psychoacoustic model, filterbank, the noise-allocation method, and the noiseless compression technique. The combination provides either better quality or lower bit rates than techniques currently on the market.

In summary, PAC offers a single encoding solution that efficiently codes signals from AM bandwidth (5 to 10 kHz) to full CD bandwidth, over dynamic ranges that match the best available analog to digital convertors, from one monophonic channel to a maximum of 16 front, 7 back, 7 auxiliary, and at least 1 effects channel. It operates from 16 kb/s up to a maximum of more than 1000 kb/s for the multiple-channel case. It is currently implemented in 2-channel hardware encoder and decoder, and 5-channel software encoder and hardware decoder. Versions of the bitstream that include an explicit transport layer provide very good robustness in the face of burst-error channels, and methods of mitigating the effects of lost audio data.

In the future, we will continue to improve PAC. Some specific improvements that are already in motion are the improvement of the psychoacoustic threshold for unusual signals, reduction of the overhead in the bitstream at low bit rates, improvements of the filterbanks for higher coding efficiency, and the application of vector quantization techniques.

References

[1] Allen, J.B., Ed., *The ASA Edition of Speech Hearing in Communication,* Acoustical Society of America, Woodbury, New York, 1995.
[2] Brandenburg, K. and Johnston, J.D., ASPEC: Adaptive spectral entropy coding of high quality music signals, *AES 90th Convention,* 1991.
[3] G722. *The G722 CCITT Standard for Audio Transmission.*
[4] ISO-II, *Report on the MPEG/Audio Multichannel Formal Subjective Listening Tests,* ISO/MPEG document MPEG94/063. ISO/MPEG-II Audio Committee, 1994.
[5] Jayant, N.S. and Noll, P., *Digital Coding of Waveforms, Principles and Applications to Speech and Video,* Prentice-Hall, Englewoods Cliffs, NJ, 1984.

[6] Jayant, N.S., Johnston, J., and Safranek, R.J., Signal compression based on models of human perception, *Proc. IEEE*, 81(10), 1993.

[7] Johnston, J.D., Estimation of perceptual entropy using noise masking criteria, *ICASSP-88 Conf. Record*, 1988.

[8] Johnston, J.D., Transform coding of audio signals using perceptual noise criteria, *IEEE J. Sepected Areas in Commun.*, Feb. 1988.

[9] Johnston, J.D., Perceptual coding of wideband stereo signals, *ICASSP-89 Conf. Record*, 1989.

[10] Johnston, J.D. and Ferreira, A. J., Sum-difference stereo transform coding, *ICASSP-92 Conf. Record*, II-569 – II-572, 1992.

[11] Malvar, H.S., *Signal Processing with Lapped Transforms*, Artech House, Norwood, MA, 1992.

[12] Moore, B.C.J., *An Introduction to the Psychology of Hearing*, Academic Press, New York, 1989.

[13] MPEG, *ISO-MPEG-1/Audio Standard*.

[14] Mussmann, H.G., The ISO audio coding standard, *Proc. IEEE-Globecom.*, 1990.

[15] Princen, J.P. and Bradlen, A.B., Analysis/synthesis filter bank design based on time domain aliasing cancellation, *IEEE Trans. ASSP*, 34(5), 1986.

[16] Quackenbush, S.R., Ordentlich, E., and Snyder, J.H., Hardware implementation of a 128-kbps monophonic audio coder, in *1989 IEEE ASSP Workshop on Applications of Signal Processing to Audio and Acoustics*, 1989.

[17] Sinha, D. and Tewfik, A. H., Low bit rate transparent audio compression using adapted wavelets, *IEEE Trans. Signal Processing*, 41(12), 3463-3479, Dec. 1993.

[18] Sinha, D. and Johnston, J.D., Audio compression at low bit rates using a signal adaptive switched filterbank, in *Proc. IEEE Intl. Conf. on Acoust. Speech and Signal Proc.*, II-1053, May 1996.

[19] Sinha, D., *A New Family of Smooth Windows*, in preparation.

[20] Vaidyanathan, P.P., Multirate digital filters, filter banks, polyphase networks, and applications: A tutorial, *Proc. IEEE*, 78(1), 56-92, Jan. 1990.

43

Sony Systems

Kenzo Akagiri
Sony Corporation
(Tokyo)

M.Katakura
Sony Corporation
(Kanagawa)

H. Yamauchi
Sony Corporation
(Kanagawa)

E. Saito
Sony Corporation
(Kanagawa)

M. Kohut
Sony Corporation
(California)

Masayuki Nishiguchi
Sony Corporation
(Tokyo)

K. Tsutsui
Sony Corporation
(Tokyo)

43.1 Introduction

Kenzo Akagiri

In digital signal processing, manipulating of the signal is defined as an essentially mathematical procedure, while the AD and DA converters, the front end and the final stage devices of the processing, include analog factor/limitation. Therefore, the performance of the devices determines the degradation from the theoretical performance defined by the format of the system.

Until the 1970s, AD and DA converters with around 16-bit resolution, which were fabricated by module or hybrid technology, were very expensive devices for industry applications. At the beginning of the 1980s, the CD (compact disk) player, the first mass-production digital audio product, was introduced, and required low cost and monolithic type DA converters with 16-bit resolution. The two-step dual slope method [1] and the DEM (Dynamic Element Matching) [2] method were used in the first generation DA converters for CD players. These were methods which relieved the accuracy and matching requirements of the elements to guarantee conversion accuracy by circuit technology. Introducing new ideas on circuit and trimming, like segment decode and laser trimming of the thin film fabricated on monolithic silicon die, for example, classical circuit topologies using binary weighted current source were also used. For AD conversion at same generation, successive approximation topology and the two-step dual slope method were also used.

In the mid-1980s, introductions of the oversampling and the noise shaping technology to the AD and DA converters for audio applications were investigated [3]. The converters using the technologies are the most popular devices for recent audio applications, especially as DA converters.

43.2 Oversampling AD and DA Conversion Principle

M. Katakura

43.2.1 Concept

The concept of the oversampling AD and DA conversion, DS or SD modulation, was known in the 1950s; however, the device technology to fabricate actual devices was impracticable until the 1980s [4].

The oversampling AD and DA conversion is characterized by the following three technologies.

1. oversampling
2. noise shaping
3. fewer bit quantizer (converters used one bit quantizer called the DS or SD type)

It is well known that the quantization noise shown in the next equation is determined by only quantization step D and distributed in bandwidth limited by Nyquist frequency (2/fs), and the spectrum is almost similar to white noise when the step size is smaller than the signal level.

$$V_n = \Delta / \sqrt{12} \tag{43.1}$$

As shown in Fig. 43.1, the oversampling expands a capacity of the quantization noise cavity on the frequency axis and reduces the noise density in the audio band, and the noise shaping moves it to out of the band. Figure 43.2 is first-order noise shaping to show the principle of the noise shaping, in which the quantizer is represented by the adder fed an input $U(n)$ and a quantization noise $Q(n)$. $Y(n)$ and $U(n)$, the output and input signals of the quantizer, respectively, are given as follows:

$$Y_{(n)} \;\; = \;\; U_{(n)} + Q_{(n)} \tag{43.2}$$

$$U_{(n)} \;\; = \;\; X_{(n)} + Q_{(n-1)} \tag{43.3}$$

As a result, the output $Y(n)$ is

$$Y_{(n)} = X_{(n)} + \left\{ Q_{(n)} - Q_{(n-1)} \right\} \tag{43.4}$$

The quantization noise in output $Y(n)$, which is a differentiation of the original quantization noise $Q(n)$ and $Q(n-1)$ shifted a time step, has high frequency boosted spectrum. Equation (43.4) is written as follows using z

$$Y_{(z)} = X_{(z)} + Q_{(z)} \left(1 - Z^{-1} \right) \tag{43.5}$$

The oversampling conversion using one bit quantizer is called DS or SD AD/DA converters. Regarding one bit quantizer, a mismatch of the elements does not affect differential error; in other words, it has no non-linear error. Assume output swing of the quantizer is $\pm$ D, quantization noise $Q(z)$ is white noise, and the magnitude $|Q(Wt)|$ is D2/3, which corresponds to four times in power of Eq. (43.1) since the step

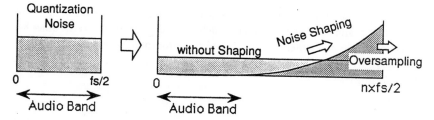

FIGURE 43.1 Quantization noise of the oversampling conversion.

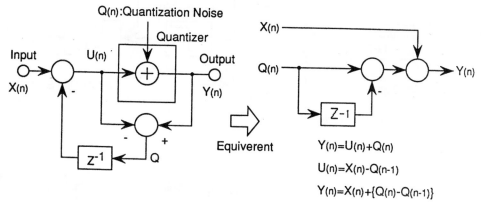

FIGURE 43.2 First-order noise shaping.

size is twice that. Define q which is $2p \cdot f$ max $/fs$, where f max and fs are the audio bandwidth and the sampling frequency, respectively, then the in-band noise in Eq. (43.5) becomes

$$
\begin{aligned}
\bar{N}^2 &= |Q_{(\omega T)}|^2 \frac{1}{2\pi} \int_{-\theta}^{\theta} |H_{(\omega T)}|^2 \, d_{(\omega T)} \\
&= \frac{\Delta^2}{3} \frac{1}{2\pi} \int_{-\theta}^{\theta} \left|1 - e^{-j\omega T}\right|^2 d_{(\omega T)} \\
&= \frac{\Delta^2}{3} \frac{2}{\pi} (\theta - \sin\theta) \\
&= \frac{\Delta^2}{9\pi} \theta^3
\end{aligned}
\tag{43.6}
$$

The oversampling conversion has the following remarkable advantages compared with traditional methods.

1. It is easy to realize "good" one bit converters without superior device accuracy and matching.
2. Analog anti-aliasing filters with sharp cutoff characteristics are unnecessary due to oversampling.

Using the oversampling converting technology, requirements for analog parts are relaxed; however, they require large scale digital circuits because interpolation filters in front of the DA conversion, which increase sampling frequency of the input digital signal, and decimation filters after the AD conversion, which reject quantization noise in high frequency and reduce sampling frequency, are required.

Figure 43.3 shows the block diagram of the DA converter including an interpolation filter. Though the scheme of the noise shaper is different from that of Fig. 43.2, the function is equivalent. Figure 43.4 shows the block diagram of the AD converter including a decimation filter. Note that the AD converter is almost the same as with the DA converters regarding the noise shapers; however, the details of the hardware are different depending on whether the block handles analog or digital signal. For example, to handle digital signals the delay units and the adders should use latches and digital adders; on the other hand, to handle analog signals delay units and adders using switched capacitor topology should be used. In the DS type, the quantizer is just reduction data length to one bit for the DA converter, and is a comparator for the AD converter by the same rule.

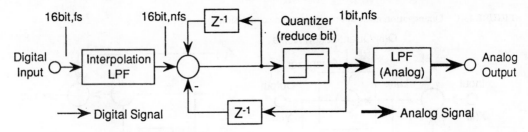

FIGURE 43.3   Oversampling DA converter.

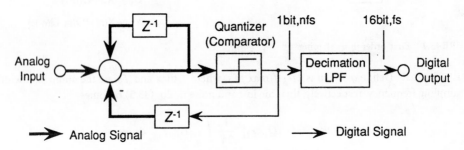

FIGURE 43.4   Oversampling AD converter.

43.2.2   Actual Converters

To achieve resolution of 16 bits or more for digital audio applications, the first-order noise shaping is not acceptable because it requires an extra high oversampling ratio, and the following technologies are actually adopted.

- High-order noise shaping
- Multi-stage (feedforward) noise shaping
- Interpolative conversion

1. High-order noise shaping

Figure 43.5 shows quantization noise spectrum for order of the noise shaping. The third-order noise shaping achieves 16-bit dynamic range using less than an oversampling ratio of 100. Figure 43.6 shows a third-order noise shaping for example of the high order. Order of the noise shaping used is 2 to 5 for audio applications.

In Fig. 43.6 output $Y(z)$ is given

$$Y_{(z)} = X_{(z)} + Q_{(z)} \left(1 - Z^{-1}\right)^3 \qquad (43.7)$$

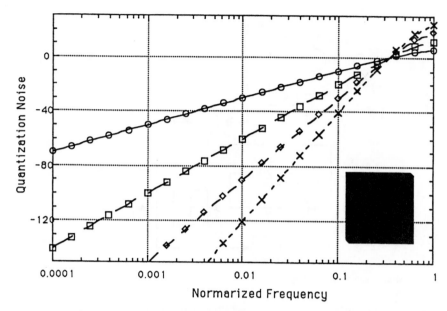

FIGURE 43.5   Quantization noise vs. order of noise shaping.

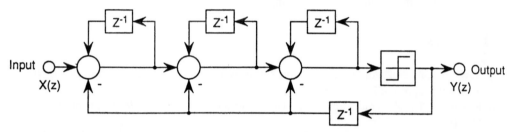

FIGURE 43.6   Third-order noise shaping.

The high-order noise shaping has a stability problem because the phase shift of the open loop in more than a third-order noise shaping exceeds 180°. In order to guarantee the stability, an amplitude limiter at the integrator outputs is used, and modification of the loop transfer function is done, although it degrades the noise shaping performance slightly.

2. Multi-stage (feedforward) noise shaping [5]

Multi-stage (feedforward) noise shaping (called MASH) achieves high-order noise shaping transfer functions using not high-order feedback but feedforward, and is shown in Fig. 43.7. Though two-stage (two-order) is shown in Fig. 43.7, three-stage (three-order) is usually used for audio applications.

3. Interpolative converters [6]

This is a method which uses a few bit resolution converters instead of one bit. The method reduces the oversamplimg ratio and order of the noise shaping to guarantee specified dynamic range and improve the loop stability. Since absolute value of the quantization noise becomes small, it is relatively easy to guarantee noise level; however, linearity of large signal conditions affects the linearity error of the AD/DA converters used in the noise shaping loop.

Oversampling conversion has become a major technique in digital audio application, and one of the distinctions is that it does not inherently zero cross distort. For recent device technology, it is not so difficult to guarantee 18-bit accuracy. Thus far, the available maximum dynamic range is slightly less than 20 bit (120 dB) without noise weighting (wide band) due to analog limitation. On the other hand, converters

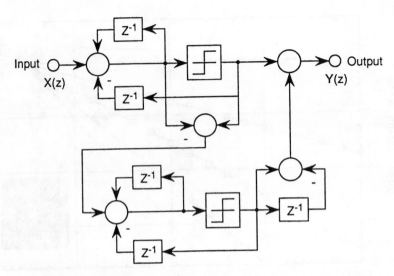

FIGURE 43.7 Multi-stage noise shaping.

with 20-bit or more resolution have been reported [7] and are expected to improve sound quality in very small signal levels from the standpoint of hearing.

References

[1] Kayanuma, A. et al., An integrated 16-bit A/D converter for PCM audio systems, *ISSCC Dig. Tech. Papers*, pp. 56-57, Feb., 1981.
[2] Plassche, R. J. et al., A monolithic 14-bit D/A converter, *IEEE J. Solid State Circuits*, SC-14:552-556, 1979.
[3] Naus, P. J. A. et al., A CMOS stereo 16-bit D/A converter for digital audio, *IEEE J. Solid State Circuits*, SC-22:390-395, June, 1987.
[4] Hauser, M. W., Overview of oversampling A/D Converters, Audio Engineering Society Preprint #2973, 1990.
[5] Matsuya, Y. et al., A 16-bit oversampling A-to-D conversion technology using triple-integration noise shaping, *IEEE J. Solid State Circuits*, SC-22:921-929, Dec., 1987.
[6] Schouwenaars, H. J. et al., An oversampling multibit CMOS D/A converter for digital audio with 115 dB dynamic range, *IEEE J. Solid State Circuits*, SC-26:1775-1780, Dec., 1991.
[7] Maruyama, Y. et al., A 20-bit stereo oversampling D-to-A converter, *IEEE Trans. Consumer Electron.*, 39:274-276, Aug., 1993.

43.3   The SDDS System for Digitizing Film Sound

H. Yamauchi, E. Saito, and M. Kohut

43.3.1   Film Format

There are three basic concepts for developing the SDDS format. They can

1. Provide sound quality similar to CD sound quality. We adapt ATRAC (Adaptive TRansform Acoustic Coding) to obtain good sound quality equivalent to that of CDs. ATRAC is the compression method used

in the mini disc (MD) which has been in sale since 1992. ATRAC enables one record digital sound data by compressing about 1/5 of the original sound.

2. Provide enough numbers of sound channels with good surround effects. We have eight discrete channel systems and six channels to the screen in the front and two channels in the rear as surround speakers shown in Fig. 43.8. We have discrete channel systems, making a good channel separation which provides superior surround effects even in a large theater with no sound defects.

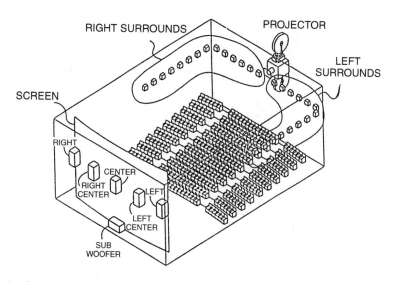

FIGURE 43.8   Speaker arrangement in theater.

3. Be compatible with the current widespread analogue sound system. There are limited spaces between the sprockets, picture frame, and in the external portion of the sprocket hole where the digital sound could be recorded because the analogue sound track is left as usual. As in the cinema scope format, it may be difficult to obtain enough space between picture frames. Because the signal for recording and playback would become intermittent between sprockets, special techniques would be required to process such signals. As shown in Fig. 43.9, we therefore establish track P and track S on a film external portion where continuous recordings are possible and where space can be obtained in the digital sound recording region on the SDDS format.

Data bits are recorded on the film with black and white dot patterns. The size of a bit is decided to overcome the effects caused by film scratch and is able to correct errors. In order to obtain the certainty of reading data, we set a guard band area to the horizontal and track direction.

Now, the method to record digital sound data on these two tracks is to separate eight channels and record four channels each in track P and in track S. A redundant data is also recorded about 18 frames later on the opposite track. By this method, it makes it possible to obtain the equivalent data from track S if any error occurs on track P and the correction is unable to be made, or vice versa. This is called the "Digital Backup System".

Figure 43.10 shows the block structure for the SDDS format. A data compression block of the ATRAC system has 512 bit sound data per film block. A vertical sync region is set at the head of the film block. A film block ID is recorded in this region to reproduce the sound data and picture frame with the right timing and to prevent the "lip sync" offset from discordance; for example, the time accordance between

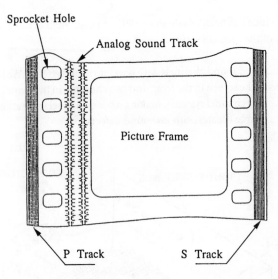

FIGURE 43.9   SDDS track designation.

an actor's/actress' lip movement and his/her voice. Also, a horizontal sync is set on the left-hand side of the film block and is referred to correctly detect the head of the data in reading with the line sensor.

43.3.2   Playback System for Digital Sound

The digital playback sound system for the SDDS system consists of a reader unit, DFP-R2000, and a decoder unit, DFP-D2000 as shown in Fig. 43.11. The reader unit is set between the supply reel and the projector.

The principle of digital sound reading for the reader unit DFP-R2000 is described in Fig. 43.12. The LED light source is derived from the optical fiber and it scans the data portion recorded on track P and track S of the film. Transparent lights through the film give an image formation on the line sensor through the lens. These optical systems are designed to have the appropriate structures which can hardly be affected by scratches on the film. The output of a sensor signal is transmitted to the decoder after the signal processing such as the wave form equalization is made.

The block diagram of the decoder unit DFP-D2000 is shown in Fig. 43.13. The unit consists of EQ, DEC, DSP, and APR blocks.

In the EQ, signals become digital signals after being equalized. Then the digital signals are transmitted to the DEC together with the regenerated clock signal.

In the DEC, jitters elimination and lip sync control are done by the time base collector circuit, and errors caused by scratches and dust on the film are corrected by the strong error correction algorithm. Also in the DEC, signals for track P and track S which have been compressed by the ATRAC system are decoded. This data is transmitted to the DSP as a linear PCM signal.

In the DSP, the sound field of the theater is adjusted and concealment modes are controlled. A CPU is installed in the DSP to control the entire decoder, and control the front panel display and reception and transmission of external control data.

Finally in the APR, 10 channels of digital filter including monitors, D/A converter, and line amplifier are installed. Also, it is possible to directly bypass an analogue input signal by relay as necessary. This bypass is prepared to cope with analogue sound if digital sound would not play back.

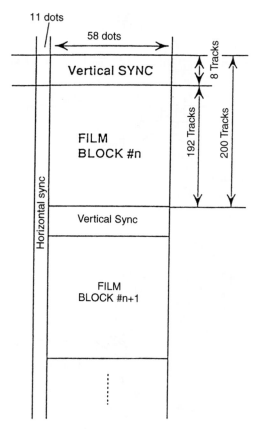

FIGURE 43.10   Data block configuration.

43.3.3   The SDDS Error Correction Technique

The SDDS system adapts the "Reed Solomon" code for error correction. An error correction technique is essential for maintaining high sound quality and high picture quality for digital recording and playback systems, such as CD, MD, digital VTR, etc. Such C1 parity + C2 parity data necessary for error correction are added and recorded in advance to cope with cases when the correct data are not able to be obtained. It enables recovery of the correct data by using this additional data even if a reading error occurs.

If the error rate is 10^{-4} (1 bit for every 10,000 bits), the error rate for C1 parity after correction would normally be 10^{-11}. In other words, an error would occur only once every 1.3 years if a film were showed 24 hours a day. Errors will be extremely close to "zero" by using C2 parity erasure correction. A strong error correction capability is installed in the SDDS digital sound playback system against random errors.

Other errors besides random errors are

- errors caused by a scratch in the film running direction
- errors caused by dust on the film
- errors caused by splice points of films
- errors caused by defocusing during printing or playback

These are considered burst errors which occur consistently. Scratch errors in particular will increase more and more every time the film is shown. SDDS has the capability of dealing with such burst errors. Therefore, in spite of the scratch on the film width direction, error correction towards the film length

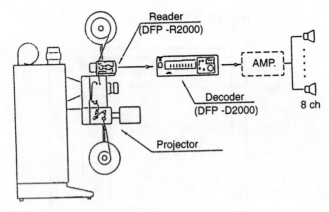

FIGURE 43.11 Playback system.

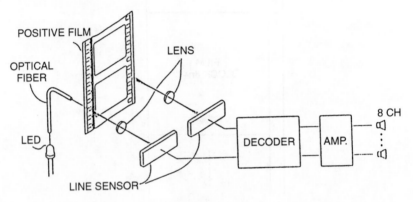

FIGURE 43.12 Optical reader concept.

would be possible up to 1.27 mm and in spite of the scratch on the film running direction, error correction towards the film width would be possible up to 336 μ m.

43.3.4 Features of the SDDS System

The specification characteristics for the SDDS player are shown in Table 43.1. It is not easy to obtain high fidelity in audio data compression compared to the linear recording system of CDs with regard to a sound quality. By adapting a system with high compression efficiency and making use of the human hearing characteristics, we were able to maintain a sound quality equivalent to CDs by adapting the ATRAC system which restrains deterioration to the minimum.

TABLE 43.1   SDDS Player System Electrical Specifications

Item	Specification
Sampling frequency	44.1 KHz
Dynamic range	Over 90 dB
Channel	Max 8 cH
Frequency band	20 Hz - 20 KHz
	$+/-1.0$dB
K.F	$< 0.7\%$
Crosstalk	< -80 dB
Reference output level	-10 dB / Unbalanced
	$+4$ dB / Balanced
Head room	> 20 dB / Balanced

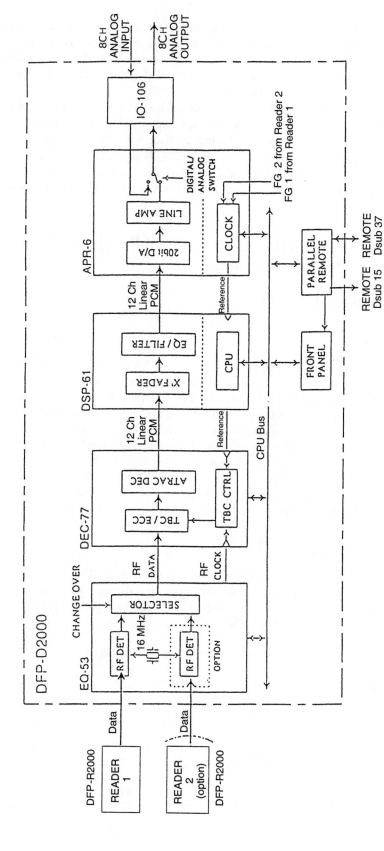

FIGURE 43.13 Overall block diagram.

One of the biggest features of the SDDS is the adaption of a digital backup system. This is a counter-measure system to make up for the damage to the splicing parts of the digital data or the parts of data missing by using the opposite side of the track with a digital data recorded on the backup channel. By this system, it would be possible to obtain an equivalent quality. Next, when finally the film is worn out, the system switches over to an analogue playback signal.

This system also has a digital room EQ function. This supplies 28 bands of graphic EQ with 1/3 octave characteristics and a high and low pass filter. Moreover, a simple operation to control the sound field in the theater will become possible by using a graphic user interface panel of an external personal computer.

Such control usually took hours, but it can be completed in about 30 min with this SDDS player. The stability of its features, reproducibility, and reliability of digitizing is well appreciated.

Furthermore, the SDDS player carries a backup function and a reset function for setting parameters by using memories.

43.4   Switched Predictive Coding of Audio Signals for the CD-I and CD-ROM XA Format

Masayuki Nishiguchi

43.4.1   Abstract

An audio bit rate reduction system for the CD-I and CD-ROM XA format based on switched predictive coding algorithm is described. The principal feature of the system is that the coder provides multiple prediction error filters, each of which has fixed coefficients. The prediction error filter that best matches the input signal is selected every 28 samples (1 block). A first-order and two kinds of second-order prediction error filters are used for signals in the low and middle frequencies, and the straight PCM is used for high-frequency signals. The system also uses near-instantaneous companding to expand the dynamic range. A noise-shaping filter is incorporated in the quantization stage, and its frequency response is varied to minimize the energy of the output noise. With a complexity of less than 8MIPS/channel, audio quality almost the same as CD audio can be achieved at 310 Kbps (8.2 bits/sample), near transparent audio can be achieved at 159 Kbps (4.2 bits/sample), and mid-fidelity audio can be achieved at 80 Kbps (4.2 bits/sample).

43.4.2   Coder Scheme

Figure 43.14 is a block diagram of the encoder and decoder system. The input signal, prediction error, quantization error, encoder output, decoder input, and decoder output are respectively expressed as $x(n), d(n), e(n), \hat{d}(n), \hat{d}'(n)$, and $\hat{x}'(n)$. The z-transforms of the signals are expressed as $X(z), D(z), E(z), \hat{D}(z), \hat{D}'(z)$, and $\hat{X}'(z)$. The encoder response can then be expressed as

$$\hat{D}(z) = G \cdot X(z) \cdot \{1 - P(z)\} + E(z) \cdot \{1 - R(z)\}\,, \tag{43.8}$$

and the decoder response as

$$\hat{X}'(z) = \frac{G^{-1} \cdot \hat{D}'(z)}{1 - P(z)}\,, \tag{43.9}$$

Assuming that there is no channel error, we can write $\hat{D}'(z) = \hat{D}(z)$. Using Eq. (43.8) and (43.9), we can write the decoder output in terms of the encoder input as

$$\hat{X}'(z) = X(z) + G^{-1} \cdot E(z) \cdot \frac{1 - R(z)}{1 - P(z)}\,. \tag{43.10}$$

where

$$P(z) = \sum_{k=1}^{P} \alpha_k \cdot z^{-k} \quad \text{and} \quad R(z) = \sum_{k=1}^{R} \beta_k \cdot z^{-k} \tag{43.11}$$

Here α_k and β_k are, respectively, the coefficients of predictor $P(z)$ and $R(z)$. Equation (43.10) shows the encoder-decoder performance characteristics of the system. It shows that the quantization error $E(z)$ is reduced by the extent of the noise-reduction effect G^{-1}. The distribution of the noise spectrum that appears at the decoder output is

$$N(z) = E(z) \cdot \frac{1 - R(z)}{1 - P(z)} . \tag{43.12}$$

$R(z)$ can be varied according to the spectral shape of the input signal in order to have a maximum masking effect, but we have set $R(z) = P(z)$ to keep from coloring the quantization noise.

G can be regarded as the normalization factor for the peak prediction error (over 28 residual words) from the chosen prediction error filter. The value of G changes according to the frequency response of the prediction gain:

$$G \propto \frac{|X(z)|}{|D(z)|} . \tag{43.13}$$

This is also proportional to the inverse of the prediction error filter, $1/|1 - P(z)|$. So, in order to maximize G, it is necessary to change the frequency response of the prediction error filter $1 - P(z)$ according to the frequency distribution of the input signals.

Selection of the Optimum Filter

Several different strategies of selecting filters are possible in the CD-I/CD-ROM XA format, but the simplest way for the encoder to choose which predictor is most suitable is the following:

- The predictor adaptation section compares the peak value of the prediction errors (over 28 words) from each prediction error filter $1 - P(z)$ and selects the filter that generates the minimum peak.
- The group of prediction errors chosen is then gain controlled (normalized by its maximum value) and noise shaping is executed at the same time.

As a result, a high SNR is obtained by using a first-order and two kinds of second-order prediction error filters for signals with the low and middle frequencies and by using the straight PCM for high-frequency signals.

Coder Parameters

This system provides three bit rates for the CD-I/CD-ROM XA format, and data encoded at any bit rate can be decoded by a single decoder. The following sections explain how the parameters used in the decoder and the encoder change according to the level of sound quality. Table 43.2 lists the parameters for each level.

Level A

We can obtain the highest quality audio sound with Level A, which uses only two prediction error filters. Either the straight PCM or the first-order differential PCM is selected. The transfer functions of the prediction error filters are as follows:

$$H(z) = 1 \tag{43.14}$$

and

$$H(z) = 1 - 0.975z^{-1} , \tag{43.15}$$

where $H(z) = 1 - P(z)$.

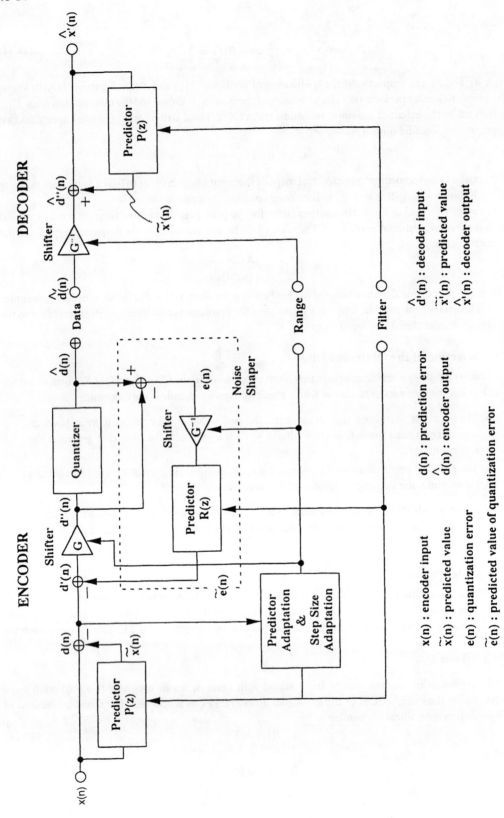

FIGURE 43.14 Block diagram of the bit rate reduction system.

x(n) : encoder input   d(n) : prediction error

$\tilde{x}$(n) : predicted value $\hat{d}$(n) : encoder output

e(n) : quantization error

$\tilde{e}$(n) : predicted value of quantization error

$\hat{d}'$(n) : decoder input

$\tilde{x}'$(n) : predicted value

$\hat{x}'$(n) : decoder output

TABLE 43.2 The Parameters for Each Level

	Level A	Level B	Level C
Sampling frequency (KHz)	37.8	37.8	18.9
Residual word length (bits per sample)	8	4	4
Block length (Number of samples)	28	28	28
Range data (bits per block)	4	4	4
Range values	0-8	0-12	0-12
Filter data (bits per block)	1	2	2
Number of prediction error filters used	2	3	4
Average of bits used per sample (bits per sample)	8.18 $= (8 \times 28 + 4 + 1)/28$	4.21 $= (4 \times 28 + 4 + 1)/28$	4.21 $= (4 \times 28 + 4 + 1)/28$
Bit rate (Kbps)	309	159	80

Level B

The bit rate at Level B is half as high as that at Level A. By using this level, we can obtain high-fidelity audio sound from most high-quality sources. This level uses three filters: the straight PCM, the first-order differential PCM, or the second-order differential PCM-1 is selected. The transfer functions of the first two filters are the same as in Level A, and that for the second-order differential PCM-1 mode is:

$$H(z) = 1 - 1.796875z^{-1} + 0.8125z^{-2} . \tag{43.16}$$

Level C

We can obtain mid-fidelity audio sound at Level C, and a monoaural audio program 16 hours long can be recorded on a single CD. Four filters are used for this level. The transfer function of the first three filters are the same as in Level B. The transfer function of the second-order differential PCM-2 mode, used only at this level, is

$$H(z) = 1 - 1.53125z^{-1} + 0.859375z^{-2} . \tag{43.17}$$

At all levels, the noise-shaping filter and the inverse-prediction-error filter in the decoder have the same coefficients as the prediction error filter in the encoder.

43.4.3 Applications

The simple structure and low complexity of this CD-I/CD-ROM XA audio compression algorithm make it suitable for applications with PCs, workstations, and video games.

References

[1] Nishiguchi, M., Akagiri, K. and Suzuki. T., A new audio bit-rate reduction system for the CD-I format, Preprint 81st AES Convention, Nov. 1986.

[2] Rabiner, L.R. and Schafer, R.W., *Digital Processing of Speech Signals,* Prentice-Hall, Englewood Cliffs, NJ, 1978.

[3] Oppenhein, A.V. and Schafer, R.W., *Digital Signal Processing,* Prentice-Hall, Englewood Cliffs, NJ, 1975.

43.5   ATRAC (Adaptive Transform Acoustic Coding) and ATRAC 2

K. Tsutsui

43.5.1   ATRAC

ATRAC is a coding system designed to meet the following criteria for the MiniDisc system:

- Compression of 16-bit 44.1-kHz audio (705.6 kbps) into 146 kbps with minimal reduction in sound quality.
- Simple hardware implementation suitable for portable players and recorders.

Block diagrams of the encoder and decoder structures are shown in Figs. 43.15 and 43.16, respectively. The time-frequency analysis block of the encoder decomposes the input signal into spectral coefficients grouped into 52 block floating units (BFUs). The bit allocation block divides the available bits among the BFUs adaptively based on the psychoacoustics. The spectrum quantization block normalizes spectral coefficients with the scale factor given to each BFU, and then quantizes each of them to the specified word length. These processes are performed in every sound unit, a block consisting of 512 samples per channel.

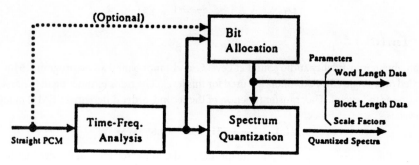

FIGURE 43.15   ATRAC encoder.

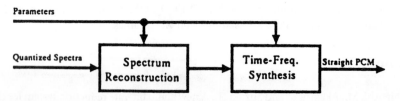

FIGURE 43.16   ATRAC decoder.

In order to generate the BFUs, the time-frequency analysis block first divides the input signal into three subbands. And then, each of these subbands is transformed into the frequency domain by modified discrete cosine transform (MDCT), producing a set of spectral coefficients. Finally, these spectral coefficients are nonuniformly grouped into BFUs. The subband decomposition is performed using cascaded 48-tap quadrature mirror filters (QMFs). The input signal is divided into upper and lower frequency bands by the first QMF, and then, the lower-frequency band is divided again by a second QMF. While the output samples of each filter are decimated by two, the aliasing caused by the subband decomposition is cancelled during reconstruction, due to the use of QMFs. MDCT block length is adaptively determined based on the signal characteristics in each band. There are two block-length modes: long mode (11.6 msec for $f_s = 44.1$ kHz) and short mode (1.45 ms in the high frequency band, 2.9 ms in the others). Normally, long mode is chosen, as this provides good frequency resolution. However, problems occur during attack portions of the signal since the quantization noise is spread over the entire block and the initial quantization noise is not masked by simultaneous masking. In order to prevent this degradation known as pre-echo, ATRAC switches to short mode when it detects an attack signal. In this case, as the noise before the attack exists only for a very short period of time, it is masked by backward masking. The window form is symmetric for both long and short modes, and the window form in the non-zero-nor-one region of the long mode is the same as that of the short mode. Although this window form is somewhat disadvantageous to the separability of the spectrum, it brings the following merits:

- The transform mode can be determined based only on the existence of an attack signal in the current sound unit, and hence, no extra buffer is required in the encoder.
- A smaller size of buffer memory is required to store the overlapped samples for the next sound unit in the encoder and decoder.

The mapping structure of ATRAC is summarized in Fig. 43.18.

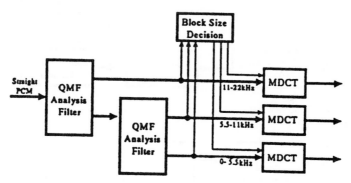

FIGURE 43.17   ATRAC time-frequency analysis.

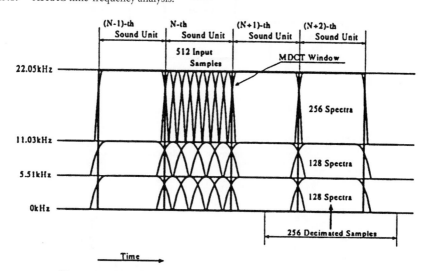

FIGURE 43.18   ATRAC mapping structure.

43.5.2 ATRAC2

The ATRAC2 system, taking advantage of the progress in LSI technologies, allows audio signals of 16 bits per sample with a sampling frequency of 44.1 kHz (705.6 kbps) to be compressed to 64 kbps, sacrificing almost no audio quality. It was designed focusing on efficient coding of tonal signals, as the human ear is very sensitive to distortions in such signals.

Block diagrams of the encoder and decoder structures are shown in Figs. 43.19 and 43.20. The encoder extracts psychoacoustically important tone components from the input signal spectra in order to encode them separately from the other less important spectrum data in an efficient way. A tone component is a group of consecutive spectral coefficients and is defined with several parameters including its location and width data. The remaining spectral coefficients are grouped into 32 non-uniform BFUs. Both the tone components and the remaining spectral coefficients may be encoded with Huffman coding, which is shown in Table 43.3 and for which simple decoding with a look-up table is practical due to its small size. Although the quantization step number is limited to 63, high S/N ratio can be obtained by repeatedly extracting tone components from the same frequency range.

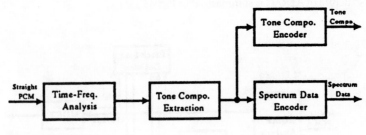

FIGURE 43.19   ATRAC2 encoder.

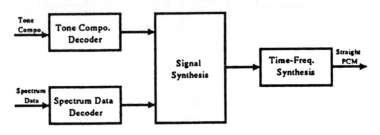

FIGURE 43.20   ATRAC2 decoder.

TABLE 43.3   Huffman Code Table

ID	Quantization step number	Dimension (spectr. num.)	Maximum code length	Look-up table size
0	1	—	—	—
1	3	2	5	32
2	5	1	3	8
3	7	1	4	16
4	9	1	5	32
5	15	1	6	64
6	31	1	7	128
7	63	1	8	256

Note: Total = 536

The mapping structure of ATRAC2 is shown in Fig. 43.21. The frequency resolution is twice that

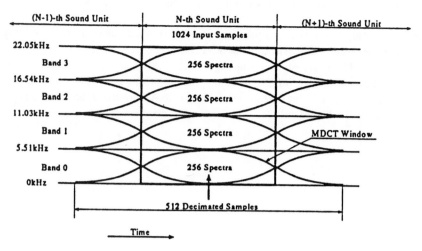

FIGURE 43.21   ATRAC2 mapping structure.

of ATRAC, and in order to secure the frequency separability, ATRAC2 performs a signal analysis using a combination of a 96-tap polyphase quadrature filter (PQF) and a fixed-length 50%-overlap MDCT whose forward and backward window forms are different from each other. ATRAC2 prevents pre-echo by amplifying the signal preceding an attack adaptively before transforming it into spectral coefficients in the encoder and restoring it to the original level after the inverse transform in the decoder. This technique, called gain control, simplifies the spectral structure of the system.

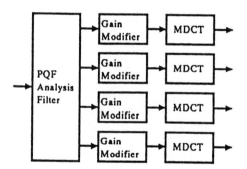

FIGURE 43.22   ATRAC2 time-frequency analysis.

The subband decomposition realizes frequency scalability; decoders with smaller complexity can be constructed by simply decoding only lower-band data. Use of PQF lowers the computational complexity.

References

[1] Kayanuma, A. et al., An integrated 16 bit A/D converter for PCM audio systems, *ISSCC Dig. Tech. Papers,* pp. 56-57, Feb. 1981.
[2] Plassche, R.J. et al., A monolithic 14 bit D/A converter., *IEEE J. Solid State Circuits,* SC-14, 552-556, 1979.

[3] Naus, P.J.A. et al., A CMOS stereo 16 bit D/A converter for digital audio, *IEEE J. Solid State Circuits,* SC-22, 390-395, June 1987.

[4] Hauser, M.W., Overview of oversampling A/D converters, an audio Engineering Society Preprint #2973, 1990.

[5] Matsuya, Y. et al., A 16-bit oversampling A to D conversion technology using triple-integration noise shaping, *IEEE J. Solid State Circuits,* SC-22, 921-929, Dec. 1987.

[6] Schouwenaars, H.J. et al., An oversampling multibit CMOS D/A converter for digital audio with 115dB dynamic range, *IEEE J. Solid State Circuits,* SC-26, 1775-1780, Dec. 1991.

[7] Maruyama, Y. et al., A 20-bit stereo oversampling D to A converter, *IEEE Trans. on Consumer Electronics,* 39, 274-276, Aug. 1993.

Speech Processing

Richard V. Cox
AT&T Labs — Research
Lawrence R. Rabiner
AT&T Labs — Research

W ITH THE ADVENT OF CHEAP, HIGH SPEED PROCESSORS, and with the ever-decreasing
cost of memory, the cost of speech processing has been driven down to the point where it
can be (and has been) embedded in almost any system, from a low cost consumer product
(e.g., solid-state digital answering machines, voice controlled telephones, etc.), to a desktop application
(e.g., voice dictation of a first draft quality manuscript), to an application embedded in a voice or data
network (e.g., voice dialing, packet telephony, voice browser for the Internet, etc.). It is the purpose of
this section of the Handbook to provide discussions of several of the key technologies in speech processing
and to illustrate how the technologies are implemented using special-purpose DSP processor chips or via
standard software packages running on more conventional processors.

The broad area of speech processing can be broken down into several individual areas according to both
applications and technology. These include:

1. **Speech Production Models and their Digital Implementations** (see Chapter 44 by Sondhi and
   Schroeter). In order to understand how the characteristics of a speech signal can be exploited in
   the different application areas, it is necessary to understand the properties and constraints of the
   human vocal apparatus (to understand how speech is generated by humans). It is also necessary
   to understand the way in which models can be built that simulate speech production as well as the
   ways in which they can be implemented as digital systems, since such models form the basis for
   almost all practical speech processing systems.

2. **Speech Coding** (see Chapter 45 by Cox). Speech coding is the process of compressing the information in a speech signal so as to either transit it or store it economically over a channel whose
   bandwidth is significantly smaller than that of the uncompressed signal. Speech coding is used as the
   basis for most modern voice messaging and voice mail systems, for voice response systems, for digital
   cellular and for satellite transmission of speech, for packet telephony, for ISDN teleconferencing,
   and for digital answering machines and digital voice encryption machines.

3. **Text-to-Speech Synthesis** (see Chapter 46 by Sproat and Olive). Speech synthesis is the process of
   creating a synthetic replica of a speech signal so as to transmit a message from a machine to a person,
   with the purpose of conveying the information in the message. Speech synthesis is often called "text-
   to-speech" or TTS, to convey the idea that, in general, the input to the system is ordinary ASCII text,
   and the output of the system is ordinary speech. The goal of most speech synthesis systems is to
   provide a broad range of capability for having a machine speak information (stored in the machine)
   to a user. Key aspects of synthesis systems are the intelligibility and the naturalness of the resulting
   speech. The major applications of speech synthesis include acting as a voice server for text-based
   information services (e.g., stock prices, sports scores, flight information); providing a means for
   reading e-mail, or the text portions of FAX messages over ordinary phone lines; providing a means
   for previewing text stored in documents (e.g., document drafts, Internet files); and finally as a voice
   readout for handheld devices, (e.g., phrase book translators, dictionaries, etc.)

4. **Speech Recognition by Machine** (see Chapter 47 by Rabiner and Juang). Speech recognition is
   the process of extracting the message information in a speech signal so as to control the action of
   a machine in response to spoken commands. In a sense, speech recognition is the complementary
   process to speech synthesis, and together they constitute the building blocks of a voice dialogue
   system with a machine. There are many factors which influence the type of speech recognition

system that is used for different applications, including the mode of speaking to the machine (e.g., single commands, digit sequences, fluent sentences), the size and complexity of the vocabulary which the machine understands, the task which the machine is asked to accomplish, the environment in which the recognition system must run, and finally the cost of the system. Although there is a wide range of applications of speech recognition systems, the most generic systems are simple "command-and-control" systems (with menu-like interfaces), and the most advanced systems support full voice dialogues for dictation, forms entry, catalog ordering, reservation services, etc.

5. **Speaker Verification** (see Chapter 48 by Furui and Rosenberg). Speaker verification is the process of verifying the claimed identity of a speaker for the purpose of restricting access to information (e.g., personal or private records), networks (computer, PBX), or physical premises. The basic problem of speaker verification is to decide whether or not an unknown speech sample was spoken by the individual whose identity was claimed. A key aspect of any speaker verification system is to accept the true speaker as often as possible while rejecting the impostor as often as possible. Since these are inherently conflicting goals, all practical systems arrive at some compromise between levels of these two types of system errors. The major area of application for speaker verification is in access control to information, credit, banking, machines, computer networks, private branch exchanges (PBX's), and even premises. The concept of a "voice lock" that prevents access until the appropriate speech by the authorized individual(s) (e.g., "Open Sesame") is "heard" by the system is made a reality using speaker verification technology.

6. **DSP Implementations of Speech Processing** (see Chapter 49 by Baudendistel). Until a few years ago, almost all speech processing systems were implemented on low-cost DSP fixed-point processors because of their high efficiency in realizing the computational aspects of the various signal processing algorithms. A key problem in the realization of any digital system in integer DSP code is how to map an algorithm efficiently (in both time and space) which is typically running in floating point C code on a workstation to integer C code that takes advantage of the unique characteristics of different DSP chips. Furthermore, because of the rate of change of technology, it is essential that the conversion to DSP code occur rapidly (e.g., on the order of 3-person months) or else by the time a given algorithm is mapped to a specific DSP processor, a new (faster, cheaper) generation of DSP chips will have evolved, obsoleting the entire process.

7. **Software Tools for Speech Research and Development** (see Chapter 50 by Shore). The field of speech processing has become a complex one, where an investigator needs a broad range of tools to record, digitize, display, manipulate, process, store, format, analyze, and listen to speech in its different file forms and manifestations. Although it is conceivable that an individual could create a suite of software tools for an individual application, that process would be highly inefficient and would undoubtedly result in tools which were significantly less powerful than those developed in the commercial sector, such as the Entropic Signal Processing System, MATLAB, Waves, Interactive Laboratory System (ILS), or the commercial packages for TTS and speech recognition such as the Hidden Markov Model Toolkit (HTK).

The material presented in this section should provide the reader with a framework for understanding the signal processing aspects of speech processing and some pointers into the literature for further investigation of this fascinating and rapidly evolving field.

44

Speech Production Models and Their Digital Implementations

M. Mohan Sondhi
Bell Laboratories
Lucent Technologies

Juergen Schroeter
AT&T Labs — Research

44.1 Introduction

The characteristics of a speech signal that are exploited for various applications of speech signal processing to be discussed later in this section on speech processing (e.g., coding, recognition, etc.) arise from the properties and constraints of the human vocal apparatus. It is, therefore, useful in the design of such applications to have some familiarity with the process of speech generation by humans. In this chapter we will introduce the reader to (1) the basic physical phenomena involved in speech production, (2) the simplified models used to quantify these phenomena, and (3) the digital implementations of these models.

44.1.1 Speech Sounds

Speech is produced by acoustically exciting a time-varying cavity — the vocal tract, which is the region of the mouth cavity bounded by the vocal cords and the lips. The various speech sounds are produced by adjusting both the type of excitation as well as the shape of the vocal tract.

There are several ways of classifying speech sounds [1]. One way is to classify them on the basis of the type of excitation used in producing them:

- **Voiced** sounds are produced by exciting the tract by quasi-periodic puffs of air produced by the vibration of the vocal cords in the larynx. The vibrating cords modulate the air stream from the lungs at a rate which may be as low as 60 times per second for some males to as high as 400 or 500 times per second for children. All vowels are produced in this manner. So are laterals, of which **l** is the only exemplar in English.

- **Nasal** sounds such as **m, n, ng,** and nasalized vowels (as in the French word **bon**) are also voiced. However, part or all of the airflow is diverted into the nasal tract by opening the velum.

- **Plosive** sounds are produced by exciting the tract by a sudden release of pressure. The plosives **p, t, k** are voiceless, while **b, d, g** are voiced. The vocal cords start vibrating before the release for the voiced plosives.

- **Fricatives** are produced by exciting the tract by turbulent flow created by air flow through a narrow constriction. The sounds **f, s, sh** belong to this category.

- **Voiced fricatives** are produced by exciting the tract simultaneously by turbulence and by vocal cord vibration. Examples are **v, z,** and **zh** (as in **pleasure**).

- **Affricates** are sounds that begin as a stop and are released as a fricative. In English, **ch** as in **check** is a voiceless affricate and **j** as in **John** is a voiced affricate.

In addition to controlling the type of excitation, the shape of the vocal tract is also adjusted by manipulating the tongue, lips, and lower jaw. The shape determines the frequency response of the vocal tract. The frequency response at any given frequency is defined to be the amplitude and phase at the lips in response to a sinusoidal excitation of unit amplitude and zero phase at the source. The frequency response, in general, shows concentration of energy in the neighborhood of certain frequencies, called **formant frequencies**.

For vowel sounds, three or four resonances can usually be distinguished clearly in the frequency range 0 to 4 kHz. (On average, over 99% of the energy in a speech signal is in this frequency range.) The configuration of these resonance frequencies is what distinguishes different vowels from each other.

For fricatives and plosives, the resonances are not as prominent. However, there are characteristic broad frequency regions where the energy is concentrated.

For nasal sounds, besides formants there are anti-resonances, or zeros in the frequency response. These zeros are the result of the coupling of the wave motion in the vocal and nasal tracts. We will discuss how they arise in a later section.

44.1.2 Speech Displays

We close this section with a description of the various ways of displaying properties of a speech signal. The three common displays are (1) the **pressure waveform**, (2) the **spectrogram**, and (3) the **power spectrum**. These are illustrated for a typical speech signal in Figs. 44.1a–c.

Figure 44.1a shows about half a second of a speech signal produced by a male speaker. What is shown is the **pressure waveform** (i.e., pressure as a function of time) as picked up by a microphone placed a few centimeters from the lips. The sharp click produced at a plosive, the noise-like character of a fricative, and the quasi-periodic waveform of a vowel are all clearly discernible.

Figure 44.1b shows another useful display of the same speech signal. Such a display is known as a **spectrogram** [2]. Here the x-axis is time. But the y-axis is frequency and the darkness indicates the intensity at a given frequency at a given time. [The intensity at a time t and frequency f is just the power in the signal averaged over a small region of the time-frequency plane centered at the point (t, f)]. The dark bands seen in the vowel region are the **formants**. Note how the energy is much more diffusely spread out in frequency during a plosive or fricative.

Finally, Fig. 44.1c shows a third representation of the same signal. It is called the **power spectrum**. Here the power is plotted as a function of frequency, for a short segment of speech surrounding a specified time instant. A logarithmic scale is used for power and a linear scale for frequency. In this particular plot, the power is computed as the average over a window of duration 20 msec. As indicated in the figure, this spectrum was computed in a voiced portion of the speech signal. The regularly spaced peaks — the fine structure — in the spectrum are the harmonics of the fundamental frequency. The spacing is seen to be about 100 Hz, which checks with the time period of the wave seen in the pressure waveform in Fig. 44.1a.

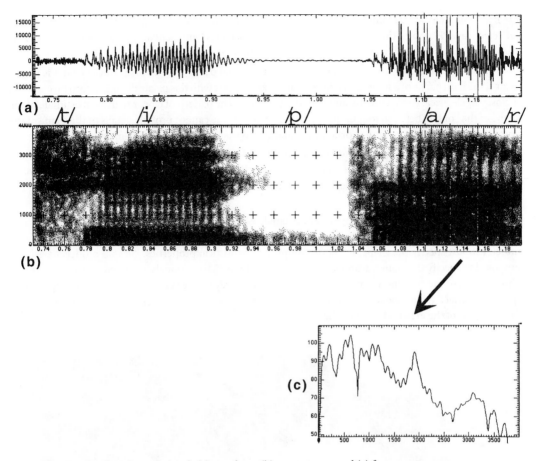

FIGURE 44.1   Display of speech signal: (a) waveform, (b) spectrogram, and (c) frequency response.

The peaks in the envelope of the harmonic peaks are the formants. These occur at about 650, 1100, 1900, and 3200 Hz, which checks with the positions of the formants seen in the spectrogram of the same signal displayed in Fig. 44.1b.

44.2   Geometry of the Vocal and Nasal Tracts

Much of our knowledge of the dimensions and shapes of the vocal tract is derived from a study of x-ray photographs and x-ray movies of the vocal tract taken while subjects utter various specific speech sounds or connected speech [3]. In order to keep x-ray dosage to a minimum, only one view is photographed, and this is invariably the side view (a view of the mid-sagittal plane). Information about the cross-dimensions is inferred from static vocal tracts using frontal X rays, dental molds, etc.

More recently, Magnetic Resonance Imaging (MRI) [4] has also been used to image the vocal and nasal tracts. The images obtained by this technique are excellent and provide three-dimensional reconstructions of the vocal tract. However, at present MRI is not capable of providing images at a rate fast enough for studying vocal tracts in motion.

Other techniques have also been used to study vocal tract shapes. These include:

(1) ultrasound imaging [5]. This provides information concerning the shape of the tongue but not about the shape of the vocal cavity.

(2) Acoustical probing of the vocal tract [6]. In this technique, a known acoustic wave is applied at

the lips. The shape of the time-varying vocal cavity can be inferred from the shape of the time-varying reflected wave. However, this technique has thus far not achieved sufficient accuracy. Also, it requires the vocal tract to be somewhat constrained while the measurements are made.

(3) Electropalatography [7]. In this technique, an artificial palate with an array of electrodes is placed against the hard palate of a subject. As the tongue makes contact with this palate during speech production, it closes an electrical connection to some of the electrodes. The pattern of closures gives an estimate of the shape of the contact between tongue and palate. This technique cannot provide details of the shape of the vocal cavity, although it yields important information on the production of consonants.

(4) Finally, the movement of the tongue and lips has also been studied by tracking the positions of tiny coils attached to them [8]. The motion of the coils is tracked by the currents induced in them as they move in externally applied electromagnetic fields. Again, this technique cannot provide a detailed shape of the vocal tract.

Figure 44.2 shows an x-ray photograph of a female vocal tract uttering the vowel sound /u/. It is seen that the vocal tract has a very complicated shape, and without some simplifications it would be very difficult to just specify the shape, let alone compute its acoustical properties. Several models have been proposed to specify the main features of the vocal tract shape. These models are based on studies of x-ray photographs of the type shown in Fig. 44.2, as well as on x-ray movies taken of subjects uttering various speech materials. Such models are called **articulatory models** because they specify the shape in terms of the positions of the **articulators** (i.e., the tongue, lips, jaw, and velum).

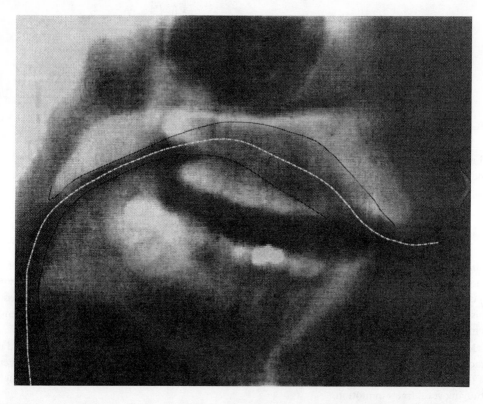

FIGURE 44.2   X-ray side view of a female vocal tract. The tongue, lips, and palate have been outlined to improve visibility. (*Source:* Modified from a single frame from "Laval Film 55," Side 2 of Munhall, K.G., Vatikiotis-Bateson, E., Tohkura, Y., X-ray film data-base for speech research, ATR Technical Report Tr-H-116, 12/28/94, ATR Human Information Processing Research Laboratories, Kyoto, Japan. With permission from Dr. Claude Rochette, Departement de Radiologie de l'Hotel-Dieu de Quebec, Quebec, Canada.)

Figure 44.3 shows such an idealization, similar to one proposed by Coker [9], of the shape of the vocal tract in the mid-sagittal plane. In this model, a fixed shape is used for the palate, and the shape of the vocal cavity is adjusted by specifying the positions of the articulators. The coordinates used to describe

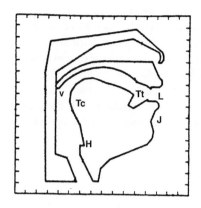

H = **HYTOID POSITION**

J = **ANGLE OF JAW OPENING**

L = **LIP PROTRUSION AND ELEVATION**

Tc= **TONGUE CENTER**

Tt= **POSITION OF TONGUE TIP**

V = **VELUM OPENING**

FIGURE 44.3 An idealized articulatory model similar to that of Coker [9].

the shape are labeled in the figure. They are the position of the tongue center, the radius of the tongue body, the position of the tongue tip, the jaw opening, the lip opening and protrusion, the position of the hyoid, and the opening of the velum. The cross-dimensions (i.e., perpendicular to the sagittal plane) are estimated from static vocal tracts. These dimensions are assumed fixed during speech production. In this manner, the three-dimensional shape of the vocal tract is modeled.

Whenever the velum is open, the nasal cavity is coupled to the vocal tract, and its dimensions must also be specified. The nasal cavity is assumed to have a fixed shape which is estimated from static measurements.

44.3   Acoustical Properties of the Vocal and Nasal Tracts

Exact computation of the acoustical properties of the vocal (and nasal) tract is difficult even for the idealized models described in the previous section. Fortunately, considerable further simplification can be made without affecting most of the salient properties of speech signals generated by such a model. Almost without exception, three assumptions are made to keep the problem tractable. These assumptions are justifiable for frequencies below about 4 kHz [10, 11].

44.3.1   Simplifying Assumptions

1. It is assumed that the vocal tract can be **"straightened out"** in such a way that a center line drawn through the tract (shown dotted in Fig. 44.3) becomes a straight line. In this way, the tract is converted to a straight tube with a variable cross-section.

2. Wave propagation in the straightened tract is assumed to be **planar.** This means that if we consider any plane perpendicular to the axis of the tract, then every quantity associated with the acoustic wave (e.g., pressure, density, etc.) is independent of position in the plane.

3. The third assumption that is invariably made is that wave propagation in the vocal tract is **linear.** Nonlinear effects appear when the ratio of particle velocity to sound velocity (the **Mach number**) becomes large. For wave propagation in the vocal tract the Mach number is usually less than .02, so that nonlinearity of the wave is negligible. There are, however, two exceptions to this. The flow in the **glottis** (i.e., the space between the vocal folds), and that

in the narrow constrictions used to produce fricative sounds, is nonlinear. We will show later how these special cases are handled in current speech production models.

We ought to point out that some computations have been made without the first two assumptions, and wave phenomena studied in two or three dimensions [12]. Recently there has been some interest in removing the third assumption as well [13]. This involves the solution of the so called **Navier-Stokes equation** in the complicated three-dimensional geometry of the vocal tract. Such analyses require very large amounts of high speed computations making it difficult to use them in speech production models. Computational cost and speed, however, are not the only limiting factors. An even more basic barrier is that it is difficult to specify accurately the complicated time-varying shape of the vocal tract. It is, therefore, unlikely that such computations can be used directly in a speech production model. These computations should, however, provide accurate data on the basis of which simpler, more tractable, approximations may be abstracted.

44.3.2 Wave Propagation in the Vocal Tract

In view of the assumptions discussed above, the propagation of waves in the vocal tract can be considered in the simplified setting depicted in Fig. 44.4. As shown there, the vocal tract is represented as a variable area tube of length L with its axis taken to be the x-axis. The glottis is located at $x = 0$ and the lips at $x = L$, and the tube has a cross-sectional area $A(x)$ which is a function of the distance x from the glottis. Strictly speaking, of course, the area is time-varying. However, in normal speech the temporal variation

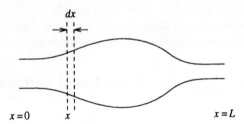

FIGURE 44.4   The vocal tract as a variable area tube.

in the area is very slow in comparison with the propagation phenomena that we are considering. So, the cross-sectional area may be represented by a succession of stationary shapes.

We are interested in the spatial and temporal variation of two interrelated quantities in the acoustic wave: the pressure $p(x, t)$ and the volume velocity $u(x, t)$. The latter is $A(x)v(x, t)$, where v is the **particle velocity**. For the assumption of linearity to be valid, the pressure p in the acoustic wave is assumed to be small compared to the equilibrium pressure P_0, and the particle velocity v is assumed to be small compared to the velocity of sound, c. Two equations can be written down that relate $p(x, t)$ and $u(x, t)$: the equation of motion and the equation of continuity [14]. A combination of these equations will give us the basic equation of wave propagation in the variable area tube. Let us derive these equations first for the case when the walls of the tube are rigid and there are no losses due to viscous friction, thermal conduction, etc.

44.3.3 The Lossless Case

The **equation of motion** is just a statement of Newton's second law. Consider the thin slice of air between the planes at x and $x + dx$ shown in Fig. 44.4. By equating the net force acting on it due to the pressure gradient to the rate of change of momentum one gets

$$\frac{\partial p}{\partial x} = -\frac{\rho}{A} \frac{\partial u}{\partial t} \tag{44.1}$$

(To simplify notation, we will not always explicitly show the dependence of quantities on x and t.)

The **equation of continuity** expresses conservation of mass. Consider the slice of tube between x and $x + dx$ shown in Fig. 44.4. By balancing the net flow of air out of this region with a corresponding decrease in the density of air we get

$$\frac{\partial u}{\partial x} = -\frac{A}{\rho} \frac{\partial \delta}{\partial t} . \tag{44.2}$$

where $\delta(x, t)$ is the fluctuation in density superposed on the equilibrium density ρ. The density is related to pressure by the gas law. It can be shown that pressure fluctuations in an acoustic wave follow the adiabatic law, so that $p = (\gamma P/\rho)\delta$, where γ is the ratio of specific heats at constant pressure and constant volume. Also, $(\gamma P/\rho) = c^2$, where c is the velocity of sound. Substituting this into Eq. (44.2) gives

$$\frac{\partial u}{\partial x} = -\frac{A}{\rho c^2} \frac{\partial p}{\partial t} \tag{44.3}$$

Equations (44.1) and (44.3) are the two relations between p and u that we set out to derive. From these equations it is possible to eliminate u by subtracting $\frac{\partial}{\partial t}$ of Eq. (44.3) from $\frac{\partial}{\partial x}$ of Eq. (44.1). This gives

$$\frac{\partial}{\partial x} A \frac{\partial p}{\partial x} = \frac{A}{c^2} \frac{\partial^2 p}{\partial t^2} . \tag{44.4}$$

Equation (44.4) is known in the literature as **Webster's horn equation** [15]. It was first derived for computations of wave propagation in horns, hence the name. By eliminating p from Eqs. (44.1) and (44.3), one can also derive a single equation in u.

It is useful to write Eqs. (44.1), (44.3), and (44.4) in the frequency domain by taking Laplace transforms. Defining $P(x, s)$ and $U(x, s)$ as the Laplace transforms of $p(x, t)$ and $u(x, t)$, respectively, and remembering that $\frac{\partial}{\partial t} \to s$, we get:

$$\frac{dP}{dx} = -\frac{\rho s}{A} U \tag{44.1a}$$

$$\frac{dU}{dx} = -\frac{sA}{\rho c^2} P \tag{44.3a}$$

and

$$\frac{d}{dx} A \frac{dP}{dx} = \frac{s^2}{c^2} A P \tag{44.4a}$$

It is important to note that in deriving these equations we have retained only first order terms in the fluctuating quantities p and u. Inclusion of higher order terms gives rise to nonlinear equations of propagation. By and large these terms are quite negligible for wave propagation in the vocal tract. However, there is one second order term, neglected in Eq. (44.1), which becomes important in the description of flow through the narrow constriction of the glottis. In deriving Eq. (44.1) we neglected the fact that the slice of air to which the force is applied is moving away with the velocity v. When this effect is correctly taken into account, it turns out that there is an additional term $\rho v \frac{\partial v}{\partial x}$ appearing on the left hand side of that equation. The corrected form of Eq. (44.1) is

$$\frac{\partial}{\partial x}\left[p + \frac{\rho}{2}(u/A)^2\right] = -\rho\frac{d}{dt}\left[\frac{u}{A}\right].$$ (44.5)

The quantity $\frac{\rho}{2}(u/A)^2$ has the dimensions of pressure, and is known as the **Bernoulli pressure**. We will have occasion to use Eq. (44.5) when we discuss the motion of the vocal cords in the section on sources of excitation.

44.3.4 Inclusion of Losses

The equations derived in the previous section can be used to approximately derive the acoustical properties of the vocal tract. However, their accuracy can be considerably increased by including terms that approximately take account of the effect of viscous friction, thermal conduction, and yielding walls [16]. It is most convenient to introduce these effects in the frequency domain.

The effect of viscous friction can be approximated by modifying the equation of motion, Eq. (44.1a) as follows:

$$\frac{dP}{dx} = -\frac{\rho s}{A}U - R(x,s)U.$$ (44.6)

Recall that Eq. (44.1a) states that the force applied per unit area equals the rate of change of momentum per unit area. The added term in Eq. (44.6) represents the viscous drag which reduces the force available to accelerate the air. The assumption that the drag is proportional to velocity can be approximately validated. The dependence of R on x and s can be modeled in various ways [16].

The effect of thermal conduction and yielding walls can be approximated by modifying the equation of continuity as follows:

$$\rho\frac{dU}{dx} = -\frac{A}{c^2}sP - Y(x,s)P$$ (44.7)

Recall that the left hand side of Eq. (44.3a) represents net outflow of air in the longitudinal direction, which is balanced by an appropriate decrease in the density of air. The term added in Eq. (44.7) represents net outward volume velocity into the walls of the vocal tract. This velocity arises from (1) a temperature gradient perpendicular to the walls which is due to the thermal conduction by the walls, and (2) due to the yielding of the walls. Both these effects can be accounted for by appropriate choice of the function $Y(x,s)$, provided the walls can be assumed to be **locally reacting**. By that we mean that the motion of the wall at any point depends on the pressure at that point alone. Models for the function $Y(x,s)$ may be found in [16].

Finally, the lossy equivalent of Eq. (44.4a) is

$$\frac{d}{dx}\frac{A}{\rho s + AR}\frac{dP}{dx} = \left(\frac{As}{\rho c^2} + Y\right)P.$$ (44.8)

44.3.5 Chain Matrices

All properties of linear wave propagation in the vocal tract can be derived from Eqs. (44.1a), (44.3a), (44.4a) or the corresponding Eqs. (44.6), (44.7), and (44.8) for the lossy tract. The most convenient way to derive these properties is in terms of **chain matrices**, which we now introduce.

Since Eq. (44.8) is a second order linear ordinary differential equation, its general solution can be written as a linear combination of two independent solutions, say $\phi(x,s)$ and $\Psi(x,s)$. Thus

$$P(x,s) = a\phi(x,s) + b\Psi(x,s)$$ (44.9)

where a and b are, in general, functions of s. Hence, the pressure at the input of the tube ($x = 0$) and at the output ($x = L$) are linear combinations of a and b. The volume velocity corresponding to the pressure given in Eq. (44.9) is obtained from Eq. (44.6) to be

$$U(x, s) = -\frac{A}{\rho s + AR}[ad\phi/dx + bd\Psi/dx].$$

(44.10)

Thus, the input and output volume velocities are seen to be linear combinations of a and b. Eliminating the parameters a and b from these relationships shows that the input pressure and volume velocity are linear combinations of the corresponding output quantities. Thus, the relationship between the input and output quantities may be represented in terms of a 2×2 matrix as follows:

$$\begin{bmatrix} P_{in} \\ U_{in} \end{bmatrix} = \begin{bmatrix} k_{11} & k_{12} \\ k_{21} & k_{22} \end{bmatrix}\begin{bmatrix} P_{out} \\ U_{out} \end{bmatrix}$$

(44.11)

$$= \mathbf{K}\begin{bmatrix} P_{out} \\ U_{out} \end{bmatrix}.$$

The matrix $\mathbf{K}$ is called a **chain matrix** or ABCD matrix [17]. Its entries depend on the values of ϕ and Ψ at $x = 0$ and $x = L$. For an arbitrarily specified area function $A(x)$ the functions ϕ and ψ are hard to find. However, for a **uniform** tube, i.e., a tube for which the area and the losses are independent of x, the solutions are very easy. For a uniform tube, Eq. (44.8) becomes

$$\frac{d^2 P}{dx^2} = \sigma^2 P$$

(44.12)

where σ is a function of s given by

$$\sigma^2 = (\rho s + AR)\left(\frac{s}{\rho c^2} + \frac{Y}{A}\right).$$

Two independent solutions of Eq. (44.12) are well known to be $\cosh(\sigma x)$ and $\sinh(\sigma x)$, and a bit of algebra shows that the chain matrix for this case is

$$\mathbf{K} = \begin{bmatrix} \cosh(\sigma L) & (1/\beta)\sinh(\sigma L) \\ \beta \sinh(\sigma L) & \cosh(\sigma L) \end{bmatrix}$$

(44.13)

where

$$\beta = \sqrt{\left[Y + \frac{As}{\rho c^2}\right]/\left[R + \frac{\rho s}{A}\right]}.$$

For an arbitrary tract, one can utilize the simplicity of the chain matrix of a uniform tube by approximating the tract as a concatenation of N uniform sections of length $\Delta = L/N$. Now the output quantities of the ith section become the input quantities for the $i + 1$st section. Therefore, if $\mathbf{K}_i$ is the chain matrix for the ith section, then the chain matrix for the variable-area tract is approximated by

$$\mathbf{K} = \mathbf{K}_1 \mathbf{K}_2 \cdots \mathbf{K}_N.$$

(44.14)

This method can, of course, be used to relate the input-output quantities for any portion of the tract, not just the entire vocal tract. Later we shall need to find the input-output relations for various sections of the tract, for example, the tract from the glottis to the velum for nasal sounds, from the narrowest constriction to the lips for fricative sounds, etc.

As stated above, all linear properties of the vocal tract can be derived in terms of the entries of the chain matrix. Let us give several examples.

Let us associate the input with the glottal end, and the output with the lip end of the tract. Suppose the tract is terminated by the radiation impedance Z_R at the lips. Then, by definition, $P_{out} = Z_R U_{out}$. Substituting this in Eq. (44.11) gives

$$\begin{bmatrix} P_{in}/U_{out} \\ U_{in}/U_{out} \end{bmatrix} = \begin{bmatrix} k_{11} & k_{12} \\ k_{21} & k_{22} \end{bmatrix} \begin{bmatrix} Z_R \\ 1 \end{bmatrix}. \tag{44.15}$$

From Eq. (44.15) it follows that

$$\frac{U_{out}}{U_{in}} = \frac{1}{k_{21} Z_R + k_{22}}. \tag{44.16a}$$

Equation (44.16a) gives the **transfer function** relating the output volume velocity to the input volume velocity. Multiplying this by Z_R gives the transfer function relating output pressure to the input volume velocity. Other transfer functions relating output pressure or volume velocity to input pressure may be similarly derived.

Relationships between pressure and volume velocity at a single point may also be derived. For example,

$$\frac{P_{in}}{U_{in}} = \frac{k_{11} Z_R + k_{12}}{k_{21} Z_R + k_{22}} \tag{44.16b}$$

gives the **input impedance** of the vocal tract as seen at the glottis, when the lips are terminated by the radiation impedance.

Also, **formant frequencies**, which we mentioned in the Introduction, can be computed from the transfer function of Eq. (44.16a). They are just the values of s at which the denominator on the right-hand side becomes zero. For a lossy vocal tract, the zeros are complex and have the form $s_n = -\alpha_n + j\omega_n$, $n = 1, 2, \cdots$. Then ω_n is the frequency (in rad/s) of the nth formant, and α_n is its half bandwidth.

Finally, the chain matrix formulation also leads to **linear prediction coefficients** (LPC), which are the most commonly used representation of speech signals today. Strictly speaking, the representation is valid for speech signals for which the excitation source is at the glottis (i.e., voiced or aspirated speech sounds). Modifications are required when the source of excitation is at an interior point.

To derive the LPC formulation, we will assume the vocal tract to be lossless, and the radiation impedance at the lips to be zero. From Eq. (44.16a) we see that to compute the output volume velocity from the input volume velocity, we need only the k_{22} element of the chain matrix for the entire vocal tract. This chain matrix is obtained by a concatenation of matrices as shown in Eq. (44.14). The individual matrices K_i are derived from Eq. (44.13), with $N = L/\Delta$. In the lossless case, R and Y are zero, so $\sigma = s/c$ and $\beta = A/\rho c$. Also, if we define $z = e^{2s\Delta/c}$, then the matrix K_i becomes

$$K_i = z^{N/2} \begin{bmatrix} \frac{1}{2}\left(1 + z^{-1}\right) & \frac{A_i}{2\rho c}\left(1 - z^{-1}\right) \\ \frac{\rho c}{2 A_i}\left(1 - z^{-1}\right) & \frac{1}{2}\left(1 + z^{-1}\right) \end{bmatrix}. \tag{44.17}$$

Clearly, therefore, k_{22} is $z^{N/2}$ times an Nth degree polynomial in z^{-1}. Hence, Eq. (44.16a) can be written as

$$\sum_{k=0}^{N} a_k z^{-k} U_{out} = z^{-N/2} U_{in}. \tag{44.18}$$

where a_k are the coefficients of the polynomial. The frequency domain factor $z = e^{-2s\Delta/c}$ represents a delay of $2\Delta/cs$. Thus, the time domain equivalent of Eq. (44.18) is

$$\sum_{k=0}^{N} a_k u_{\text{out}}(t - 2k\Delta/c) = u_{\text{in}}(t - N\Delta/c).$$ (44.19)

Now $u_{\text{out}}(t)$ is the volume velocity in the speech signal, so we will call it $s(t)$ for brevity. Similarly, since $u_{\text{in}}(t)$ is the input signal at the glottis, we will call it $g(t)$. To get the time-sampled version of Eq. (44.19) we set $t = 2n\Delta/c$ and define $s(2n\Delta/c) = s_n$ and $g((2n - N)\Delta/c) = g_n$. Then Eq. (44.19) becomes

$$\sum_{k=0}^{N} a_k s_{n-k} = \varepsilon_n.$$ (44.20)

Equation (44.20) is the LPC representation of a speech signal.

44.3.6 Nasal Coupling

Nasal sounds are produced by opening the velum and thereby coupling the nasal cavity to the vocal tract. In nasal consonants, the vocal tract itself is closed at some point between the velum and the lips, and all the airflow is diverted into the nostrils. In nasal vowels the vocal tract remains open. (Nasal vowels are common in French and several other languages. They are not nominally phonemes of English. However, some nasalization of vowels commonly occurs in English speech.)

In terms of chain matrices, the nasal coupling can be handled without too much additional effort. As far as its acoustical properties are concerned, the nasal cavity can be treated exactly like the vocal tract, with the added simplification that its shape may be regarded as fixed. The common assumption is that the nostrils are symmetric, in which case the cross-sectional areas of the two nostrils can be added and the nose replaced by a single, fixed, variable-area tube.

The description of the computations is easier to follow with the aid of the block diagram shown in Fig. 44.5. From a knowledge of the area functions and losses for the vocal and nasal tracts three chain matrices $\mathbf{K}_{gv}$, $\mathbf{K}_{vt}$, and $\mathbf{K}_{vn}$ are first computed. These represent, respectively, the matrices from glottis to

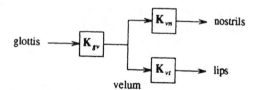

FIGURE 44.5 Chain matrices for synthesizing nasal sounds.

velum, velum to tract closure (or velum to lips, in case of a nasal vowel), and velum to nostrils.

From $\mathbf{K}_{vn}$ with some assumed impedance termination at the nostrils, the input impedance of the nostrils at the velum may be computed as indicated in Eq. (44.16b). Similarly, $\mathbf{K}_{vt}$ gives the input impedance at the velum, of the vocal tract looking toward the lips. At the velum, these two impedances are combined in parallel to give a total impedance, say Z_v. With this as termination, the velocity to velocity transfer function, T_{gv}, from glottis to velum can be computed from $\mathbf{K}_{gv}$ as shown in Eq. (44.16a). For a given volume velocity at the glottis, U_g, the volume velocity at the velum is $U_v = T_{gv}U_g$, and the pressure at the velum is $P_v = Z_vU_v$. Once P_v and U_v are known, the volume velocity and/or pressure at the nostrils and lips can be computed by inverting the matrices $\mathbf{K}_{vn}$ and $\mathbf{K}_{vt}$.

44.4   Sources of Excitation

As mentioned earlier, speech sounds may be classified by type of excitation: periodic, turbulent, or transient. All of these types of excitation are created by converting the potential energy stored in the lungs due to excess pressure into sound energy in the audible frequency range of 20 Hz to 20 kHz.

The lungs of a young adult male may have a maximum usable volume ("vital capacity") of about 5 l. While reading aloud the pressure in the lungs is typically in the range of 6 to 15 cm of water (6000 to 15000 Pa). Vocal cord vibrations can be sustained with a pressure as low as .2 cm of water. At the other extreme, a pressure as high as 195 cm of water has been recorded for a trumpet player. Typical average airflow for normal speech is about 0.1 l/s. It may peak as high as 5 l/s during rapid inhales in singing.

Periodic excitation originates mainly at the vibrating vocal folds, turbulent excitation originates primarily downstream of the narrowest constriction in the vocal tract, and transient excitations occur whenever a complete closure of the vocal pathway is suddenly released. In the following, we will explore these three types of excitation in some detail. The interested reader is referred to [18] for more information.

44.4.1   Periodic Excitation

Many of the acoustic and perceptual features of an individual's voice are believed to be due to specific characteristics of the quasi-periodic excitation signal provided by the vocal folds. These, in turn, depend on the morphology of the voice organ, the **larynx**. The anatomy of the larynx is quite complicated, and descriptions of it may be found in the literature [19]. From an engineering point of view, however, it suffices to note that the larynx is the structure that houses the **vocal folds** whose vibration provides the periodic excitation. The space between the vocal folds, called the **glottis**, varies with the motion of the vocal folds, and thus modulates the flow of air through them. As late as 1950 Husson postulated that each movement of the folds is in fact induced by individual nerve signals sent from the brain (the Neurochronaxis hypothesis) [20]. We now know that the larynx is a self-oscillating acousto-mechanical oscillator. This oscillator is controlled by several groups of tiny muscles also housed in the larynx. Some of these muscles control the rest position of the folds, others control their tension, and still others control their shape. During breathing and production of fricatives, for example, the folds are pulled apart (abducted) to allow free flow of air. To produce voiced speech, the vocal folds are brought close together (adducted). When brought close enough together, they go into a spontaneous periodic oscillation. These oscillations are driven by Bernoulli pressure (the same mechanism that keeps airplanes aloft) created by the airflow through the glottis. If the opening of the glottis is small enough, the Bernoulli pressure due to the rapid flow of air is large enough to pull the folds toward each other, eventually closing the glottis. This, of course, stops the flow and the laryngeal muscles pull the folds apart. This sequence repeats itself until the folds are pulled far enough away, or if the lung pressure becomes too low. We will discuss this oscillation in greater detail later in this section.

Besides the laryngeal muscles, the lung pressure and the acoustic load of the vocal tract also affect the oscillation of the vocal folds.

The larynx also houses many mechanoreceptors that signal to the brain the vibrational state of the vocal folds. These signals help control pitch, loudness, and voice timbre.

Figure 44.6 shows stylized snapshots taken from the side and above the vibrating folds. The view from above can be obtained on live subjects with high speed (or stroboscopic) photography, using a laryngeal mirror or a fiber optic bundle for illumination and viewing. The view from the side is the result of studies on excised (mostly animal) larynges. From studies such as these, we know that, during glottal vibration, the folds carry a mechanical wave that starts at the tracheal (lower) end of the folds and moves upwards to the pharyngeal (upper) end. Consequently, the edge of the folds that faces the vocal tract usually lags behind the edge of the folds that faces the lungs. This phenomenon is called **vertical phasing**. Higher eigenmodes of these mechanical waves have been observed and have been modeled.

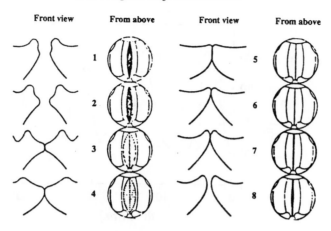

FIGURE 44.6   One cycle of vocal fold oscillation seen from the front and from above. (After Schönhärl, E., 1960 [25]. With permission of Georg Thieme Verlag, Stuttgart, Germany.)

Figure 44.7 shows typical acoustic flow waveforms, called **flow glottograms**, and their first time derivatives. In a normal glottogram, the **closed phase** of the glottal cycle is characterized by zero flow. Often, however, the closure is not complete. Also, in some cases, although the folds close completely, there is a parallel path — a chink — which stays open all the time.

In the **open phase** the flow gradually builds up, reaches a peak, and then falls sharply. The asymmetry is due to the inertia of the airflow in the vocal tract and the sub-glottal cavities. The amplitude of the fundamental frequency is governed mainly by the peak of the flow while the amplitudes of the higher harmonics is governed mainly by the (negative) peak rate of change of flow, which occurs just before closure.

Voice Qualities

Depending on the adjustment of the various parameters mentioned above, the glottis can produce a variety of phonations (i.e., excitations for voiced speech), resulting in different perceptual voice qualities. Some perceptual qualities vary continuously whereas others are essentially categorical (i.e., they change abruptly when some parameters cross a threshold).

Voice timbre is an important continuously variable quality which may be given various labels ranging from "mellow" to "pressed". The spectral slope of the glottal waveform is the main physical correlate of this perceptual quality. On the other hand, nasality and aspiration may be regarded as categorical qualities.

The physical properties that distinguish a "male" voice from a "female" voice are still not well understood, although many distinguishing features are known. Besides the obvious cue of fundamental frequency, the perceptual quality of "breathiness" seems to be important for producing a female-sounding voice. It occurs when the glottis does not close completely during the glottal cycle. This results in a more sinusoidal movement of the folds which makes the amplitude of the fundamental frequency much larger compared to those of the higher harmonics. The presence of leakage in the abducted glottis also increases the damping of the lower formants, thus increasing their bandwidths. Also, the continuous airflow through the leaking glottis gives rise to increased levels of glottal noise (aspiration noise) that masks the higher harmonics of the glottal spectrum. Finally, in glottograms of female voices, the open phase is a larger proportion of the glottal cycle (about 80%) than in glottograms of male voices (about 60%). The points of closure are also smoother for female voices, which results in lower high frequency energy relative to the fundamental.

Finally, the individuality of a voice (which allows us to recognize the speaker) appears to be dependent largely on the exact relationships between the amplitudes of the first few harmonics.

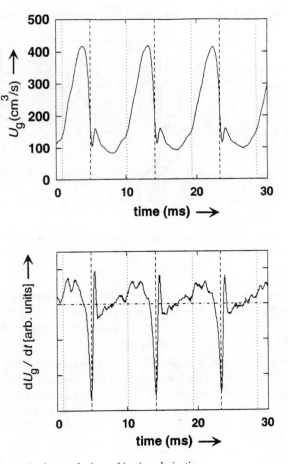

FIGURE 44.7　Example of glottal volume velocity and its time derivative.

Models of the Glottis

A study of the mechanical and acoustical properties of the larynx is still an area of active interdisciplinary research. Modeling in the mechanical and acoustical domains requires making simplifying assumptions about the tissue movements and the fluid mechanics of the airflow. Depending on the degree to which the models incorporate physiological knowledge, one can distinguish three categories of glottal models:

Parametrization of glottal flow is the "black-box" approach to glottal modeling. The glottal flow wave or its first time derivative is parametrized in segments by analytical functions. It seems doubtful that any simple model of this kind can match all kinds of speakers and speaking styles. Examples of speech sounds that are difficult to parametrize in this way are nasal and mixed-excitation sounds (i.e., sounds with an added fricative component) and "simple" high-pitch female vowels.

Parametrization of glottal area is more realistic. In this model, the area of the glottal opening is parametrized in segments, but the airflow is computed from the propagation equations, and includes its interaction with the acoustic loads of the vocal tract and the subglottal structures. Such a model is capable of reproducing much more of the detail and individuality of the glottal wave than the black box approach. Problems are still to be expected for mixed glottal/fricative sounds unless the tract model includes an accurate mechanism for frication (see the section on turbulent excitation below).

In a complete, self-oscillating model of the glottis described below, the amplitude of the glottal opening as well as the instants of glottal closure are automatically derived, and depend in a complicated manner on the laryngeal parameters, lung pressure, and the past history of the flow. The area-driven model has the disadvantage that amplitude and instants of closure must be specified as side information. However, the ability to specify the points of glottal closure can, in fact, be an advantage in some applications; for example, when the model is used to mimic a given speech signal.

Self-oscillating physiological models of the glottis attempt to model the complete interaction of the airflow and the vocal folds which results in periodic excitation. The input to a model of this type is slowly varying physical parameters such as lung pressure, tension of the folds, pre-phonatory glottal shape, etc. Of the many models of this type that have been proposed, the one most often used is the 2-mass model of Ishizaka and Flanagan (I&F). In the following we will briefly review this model.

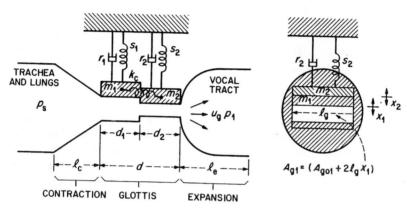

FIGURE 44.8 The two-mass model of Ishizaka and Flanagan [21].

The I&F two-mass model is depicted in Fig. 44.8. As shown there, the thickness of the vocal folds that separates the trachea from the vocal tract is divided into two parts of length d_1 and d_2, respectively, where the subscript 1 refers to the part closest to the trachea and 2 refers to the part closest to the vocal tract. These portions of the vocal folds are represented by damped spring-mass systems coupled to each other. The division into two portions is a refinement of an earlier version that represented the folds by a single spring-mass system. By using two sections the model comes closer to reality and exhibits the phenomenon of vertical phasing mentioned earlier.

In order to simulate tissue, all the springs and dampers are chosen to be nonlinear. Before discussing the choice of these nonlinear elements, let us first consider the relationship between the airflow and the pressure variations from the lungs to the vocal tract.

Airflow in the Glottis

The dimensions d_1 and d_2 are very small — about 1.5 mm each. This is a very small fraction of the wavelength even at the highest frequencies of interest. (The wavelength of a sound wave in air at 100 kHz is about 3 mm!). Therefore we may assume the flow through the glottis to be incompressible. With this assumption the equation of continuity, Eq. (44.2), merely states that the volume velocity is the same everywhere in the glottis. We will call this volume velocity u_g. The relationship of this velocity to the pressure is governed by the equation of motion. Since the particle velocity in the glottis can be very large, we need to consider the nonlinear version given in Eq. (44.5). Also, since the cross-section of the glottis is very small, viscous drag cannot be neglected. So we will include a term representing viscous drag proportional to the velocity. With this addition, Eq. (44.5) becomes:

$$\frac{\partial}{\partial x}\left[p + \frac{\rho}{2}\left(u_g/A\right)^2\right] = -\rho\frac{\partial}{\partial t}\left(\frac{u_g}{A}\right) - R_v\left(u_g/A\right) . \tag{44.21}$$

The drag coefficient R_v can be estimated for simple geometries. In the present application a rectangular aperture is appropriate. If the length of the aperture is l, its width (corresponding to the opening between the folds) is w and its depth in the direction of flow is d, then $R_v = \frac{12\mu d}{lw^3}$, where μ is the coefficient of shear viscosity. The pressure distribution is obtained by repeated use of Eq. (44.21), using the appropriate value of A (and hence of R_v) in the different parts of the glottis. In this manner, the pressure at any point in the glottis may be determined in terms of the volume velocity, u_g, the lung pressure, P_s, and the pressure at the input to the vocal tract, p_1.

The detailed derivation of the pressure distribution is given in [21]. The derivation shows that the total pressure drop across the glottis, $P_s - p_1$, is related to the glottal volume velocity, u_g, by an equation of the form

$$P_s - p_1 = Ru_g + \frac{d}{dt}(Lu_g) + \frac{\rho}{2}\left(u_g/\alpha\right)^2 . \tag{44.22}$$

With the analogy of pressure to voltage and volume velocity to current, the quantity R is analogous to resistance and L to inductance. The term in u_g^2 may be regarded as u_g times a current-dependent resistance. The quantity α has the dimensions of an area.

Models of Vocal Fold Tissue

When the pressure distribution derived above is coupled to the mechanical properties of the vocal folds, we get a self-oscillating system with properties quite similar to those of a real larynx. The mechanical properties of the vocal folds have been modeled in many ways with varying degrees of complexity ranging from a single spring-mass system to a distributed parameter flexible tube. In the following, by way of example, we will summarize only the original 1972 I&F model.

Returning to Fig. 44.8, we observe that the mechanical properties of the folds are represented by the masses m_1 and m_2, the (nonlinear) springs s_1 and s_2, the coupling spring k_c, and the nonlinear dampers r_1 and r_2. The opening in each section of the glottis is assumed to have a rectangular shape with length l_g. The widths of the two sections are $2x_j$, $j = 1, 2$. Assuming a symmetrical glottis, the cross-sectional areas of the two sections are

$$A_{gj} = A_{g0j} + 2l_g x_j, \quad j = 1, 2, \tag{44.23}$$

where A_{g01} and A_{g02} are the areas at rest. From this equation, we compute the lateral displacements $x_{j\,\min}$, $j = 1, 2$ at which the two folds touch each other in each section to be $x_{j\,\min} = -A_{g0j}/(2l_g)$. Displacements more negative than these indicate a collision of the folds. The springs s_1 and s_2 are assumed to have restoring forces of the form $ax + bx^3$, where the constants a and b take on different values for the two sections and for the colliding and non-colliding conditions.

The dampers r_1 and r_2 are assumed to be linear, but with different values in the colliding and non-colliding cases. The coupling spring k_c is assumed to be linear. With these choices, the coupled equations of motion for the two masses are:

$$m_1\frac{d^2x_1}{dt^2} + r_1\frac{dx_1}{dt} \quad + \quad f_{s1}(x_1) + k_c(x_1 - x_2) = F_1 , \tag{44.24a}$$

and

$$m_2\frac{d^2x_2}{dt^2} + r_2\frac{dx_2}{dt} \quad + \quad f_{s2}(x_2) + k_c(x_2 - x_1) = F_2 . \tag{44.24b}$$

Here f_{s1} and f_{s2} are the cubic nonlinear springs. The parameters of these springs as well as the damping constants r_1 and r_2 change when the folds go from a colliding state to a non-colliding state and vice versa. The driving forces F_1 and F_2 are proportional to the average acoustic pressures in the two sections of the glottis. Whenever a section is closed (due to the collision of its sides) the corresponding driving force is zero. Note that it is these forces that provide the feedback of the acoustic pressures to the mechanical system. This feedback is ignored in the area-driven models of the glottis.

We close this section with an example of ongoing research in glottal modeling. In the introduction to this section we had stated that breathiness of a voice is considered important for producing a natural-sounding synthetic female voice. Breathiness results from incomplete closures of the folds. We had also stated that incomplete glottal closures due to abducted folds lead to a steep spectral roll-off of the glottal excitation and a strong fundamental. However, practical experience shows that many voices show clear evidence for breathiness but do not show a steep spectral roll-off, and have relatively weak fundamentals instead. How can this mystery be solved? It has been suggested that the glottal "chink" mentioned in the discussion of Fig. 44.7 might be the answer. Many high-speed videos of the vocal folds show evidence of a separate leakage path in the "posterior commissure" (where the folds join) which stays open all the time. Analysis of such a permanently open path produces the stated effect [22].

44.4.2 Turbulent Excitation

Turbulent airflow shows highly irregular fluctuations of particle velocity and pressure. These fluctuations are audible as broadband noise. Turbulent excitation occurs mainly at two locations in the vocal tract: near the glottis and at constriction(s) between the glottis and the lips. Turbulent excitation at a constriction downstream of the glottis produces fricative sounds or voiced fricatives depending on whether or not voicing is simultaneously present. Also, stressed versions of the vowel i, and liquids l and r are usually accompanied by turbulent flow. Measurements and models for turbulent excitation are even more difficult to establish than for the periodic excitation produced by the glottis because, usually, no vibrating surfaces are involved. Because of the lack of a comprehensive model, much confusion exists over the proper sub-classification of fricatives. The simplest model for turbulent excitation is a "nozzle" (narrow orifice) releasing air into free space. Experimental work has shown that half (or more) of the noise power generated by a jet of air originates within the so-called mixing region that starts at the nozzle outlet and extends as far as a distance four times the diameter of the orifice. The noise source is therefore distributed. Several scaling relations hold between the acoustic output and the nozzle geometry. One of these scaling properties is the so-called **Reynolds number**, Re, that characterizes the amount of turbulence generated as the air from the jet mixes with the ambient air downstream from the orifice:

$$Re = \frac{u}{A}\frac{x}{\nu} \, . \tag{44.25}$$

Here u is the volume velocity, A is the area of the orifice (hence, u/A is the particle velocity), x is a characteristic dimension of the orifice (the width for a rectangular orifice), and $\nu = \mu/\rho$ is the kinematic viscosity of air. Beyond a critical value of the Reynolds number, Re_{crit} (which is about 1200 for the case of a free jet), the flow becomes fully turbulent; below this value, the flow is partly turbulent and becomes fully laminar at very low velocities. Another scaling equation defines the so-called **Strouhal number**, S, that relates the frequency F_{max} of the (usually broad) peak in the power spectrum of the generated noise to the width of the orifice and the velocity:

$$S = F_{max}\frac{x}{u/A} \, . \tag{44.26}$$

For the case of a free jet, the Strouhal number S is 0.15. Within the jet, higher frequencies are generated closer to the orifice and lower frequencies further away.

Distributed sources of turbulence can be modeled by expanding them in terms of monopoles (i.e., pulsating spheres), dipoles (two pulsating spheres in opposite phase), quadrupoles (two dipoles in opposite phase), and higher-order representations. The total power generated by a monopole source in free space is proportional to the fourth power of the particle velocity of the flow, that of a dipole source obeys a $(u/A)^6$ power law, and that of a quadrupole source obeys a $(u/A)^8$ power law. Thus, the low order sources are more important at low flow rates, while the reverse is the case at high flow rates. In a duct, however, the exponents of the power laws decrease by 2, that is, a dipole source's noise power is proportional to $(u/A)^4$, etc.

Thus far, we have summarized noise generation in a free jet or air. A much stronger noise source is created when a jet of air hits an obstacle. Depending on the angle between the surface of the obstacle and the direction of flow, the surface roughness, and the obstacle geometry, the noise generated can be up to 20 dB higher than that generated by the same jet in free space. Because of the spatially concentrated source, modeling obstacle noise is easier than modeling the noise in a free jet. Experiments reveal that obstacle noise can be approximated by a dipole source located at the obstacle.

The above theoretical findings qualitatively explain the observed phenomenon that the fricatives *th* and *f* (and the corresponding voiced *dh* and *v*) are weak compared to the fricatives *s* and *sh*. The teeth (upper for *s* and lower for *sh*) provide the obstacle on which the jet impinges to produce the higher noise levels. A fricative of intermediate strength results from a **distributed** obstacle (the "wall" case) when the jet is forced along the roof of the mouth as for the sound *y*.

In a synthesizer, dipole noise sources can be implemented as series pressure sources. One possible implementation is to make the source pressure proportional to $Re^2 - Re_{crit}^2$ for $Re > Re_{crit}$ and zero otherwise [11]. Another option [23] is to relate the noise source power to the Bernoulli pressure $B = .5\rho(u/A)^2$. Since the power of a dipole source located at the teeth (and radiating into free space) is $(u/A)^6$, it is also proportional to B^3, and the noise source pressure $p_n \propto B^{3/2}$. On the other hand, for wall sources located further away from the lips, we need multiple (distributed) dipole sources with source pressures proportional either to $Re^2 - Re_{crit}^2$ or to B. In either case, the source should have a broadband spectrum with a peak at a frequency given by Eq. (44.26).

When a noise source is located at some point inside the tract, its effect on the acoustic output at the lips is computed in terms of two chain matrices — the matrix $\mathbf{K}_F$ from the glottis to the noise source, and the matrix $\mathbf{K}_L$ from the noise source to the lips. For fricative sounds, the glottis is wide open, so the termination impedance at the glottis end may be assumed to be zero. With this termination, the impedance at the noise source looking toward the glottis is computed from $\mathbf{K}_F$ as explained in the section on chain matrices. Call this impedance Z_1. Similarly, a knowledge of the radiation impedance at the lips and the matrix $\mathbf{K}_L$ allows us to compute the input impedance Z_2 looking toward the lips. The volume velocity at the source is then just $P_n/(Z_1 + Z_2)$ where P_n is the pressure generated by the noise source. The transfer function obtained from Eq. (44.16a) for the matrix $\mathbf{K}_L$ then gives the volume velocity at the lips.

It can be shown that the series noise source P_n excites all formants of the entire tract (i.e., the ones we would see if the source were at the glottis). However, the spectrum of fricative noise usually has a high pass character. This can be understood qualitatively by the following considerations.

When the tract has a very narrow constriction, the front and back cavities are essentially decoupled, and the formants of the tract are the formants of the back cavity plus those of the front cavity. If now the noise source is just downstream of the constriction, the formants of the back cavity are only slightly excited because the impedance Z_1 also has poles at those frequencies. Since the back cavity is usually much longer than the front cavity for fricatives, the lower formants are missing in the velocity at the lips. This gives it a high pass character.

44.4.3   Transient Excitation

Transient excitation of the vocal tract occurs whenever pressure is built up behind a total closure of the tract and suddenly released. This sudden release produces a step-function of input pressure at the point of release. The output velocity is therefore proportional to the integral of the impulse response of the tract from the point of release to the lips. In the frequency domain, this is just P_r/s times the transfer function, where P_r is the step change in pressure. Hence, the velocity at the lips may be computed in the same way as in the case of turbulent excitation, with P_n replaced by P_r/s. In practice, this step excitation is usually followed by the generation of fricative noise for a short period after release when the constriction is still narrow enough. Sometimes, if the glottis is also being constricted (e.g., to start voicing) some aspiration might also result.

44.5 Digital Implementations

The models of the various parts of the human speech production apparatus which we have described above can be assembled to produce fluent speech. Here we will consider how a digital implementation of this process may be carried out. Basically, the standard theory of sampling in the time and frequency domains is used to convert the continuous signals considered above to sampled signals, and the samples are represented digitally to the desired number of bits per sample.

44.5.1 Specification of Parameters

The parameters that drive the synthesizer need to be specified about every 20 ms. (The assumed quasi-stationarity is valid over durations of this size.)

Two sets of parameters are needed — the parameters that specify the shape of the vocal tract and those that control the glottis. The vocal tract parameters implicitly control nasality (by specifying the opening area of the velum) and also frication (by specifying the size of the narrowest constriction).

44.5.2 Synthesis

The vocal tract is approximated by a concatenation of about 20 uniform sections. The cross-sectional areas of these sections is either specified directly, or computed from a specification of articulatory parameters as shown in Fig. 44.3. The chain matrix for each section is computed at an adequate sampling rate in the frequency domain to avoid time-aliasing of the corresponding time functions. (Computation of the chain matrices requires a specification of the losses also. Several models exist which assign the losses in terms of the cross-sectional area [11, 16]).

The chain matrices for the individual sections are combined to derive the matrices for various portions of the tract, as appropriate for the particular speech sound being synthesized. For voiced sounds, the matrices for the sections from the glottis to the lips are sequentially multiplied to give the matrix from the glottis to the lips. From the $\mathbf{k}_{11}$, $\mathbf{k}_{12}$, $\mathbf{k}_{21}$, $\mathbf{k}_{22}$ components of this matrix, the transfer function $\frac{U_{\text{out}}}{U_{\text{in}}}$ and the input impedance are obtained as in Eqs. (44.16a) and (44.16b). Knowing the radiation impedance Z_R at the lips we can compute the transfer function for output pressure, $H = \frac{U_{\text{out}}}{U_{\text{in}}} Z_R$. The inverse FFT of the transfer function H and the input impedance Z_{in} give the corresponding time functions $h(n)$ and $z_{\text{in}}(n)$, respectively. These functions are computed every 20 ms, and the intermediate values are obtained by linear interpolation.

For the current time sampling instant n, the current pressure $p_1(n)$ at the input to the vocal tract is then computed by convolving z_{in} with the past values of the glottal volume velocity u_g. With p_1 known, the pressure difference $P_s - p_1$ on the left hand side of Eq. (44.22) is known. Equation (44.18) is discretized by using a backward difference for the time derivative. Thus, a new value of the glottal volume velocity is derived. This, together with the current values of the displacements of the vocal folds, gives us new values for the driving forces F_1 and F_2 for the coupled oscillator Eqs. (44.24a) and (44.24b). The coupled oscillator equations are also discretized by backward differences for time derivatives. Thus, the new values of the driving forces give new values for the displacements of the vocal folds. The new value of volume velocity also gives a new value for p_1, and the computational cycle repeats, to give successive samples of p_1, u_g, and the vocal fold displacements.

The glottal volume velocity obtained in this way, is convolved with the impulse response $h(n)$ to produce voiced speech.

If the speech sound calls for frication, the chain matrix of the tract is derived as the product of two matrices — from the glottis to the narrowest constriction and from the constriction to the lips, as discussed in the section on turbulent excitation. This enables us to compute the volume velocity at the constriction, and thus introduce a noise source on the basis of the Reynolds number.

Finally, to produce nasal sounds, the chain matrix for the nasal tract is also computed, and the output at the nostrils computed as discussed in the section on chain matrices. If the lips are open, the output from the lips is also computed and added to the output from the nostrils to give the total speech signal. Details of the synthesis procedure may be found in [24].

References

[1] Edwards, H.T., *Applied Phonetics: The Sounds of American English*, Singular Publishing Group, San Diego, 1992, Chap. 3.

[2] Olive, J.P., Greenwood, A., and Coleman, J., *Acoustics of American English Speech*, Springer Verlag, New York, 1993.

[3] Fant, G., *Acoustic Theory of Speech Production*, Mouton Book Co., Gravenhage, 1960, Chap. 2.1, 93-95.

[4] Baer, T., Gore, J.C., Gracco, L.C., and Nye, P.W., Analysis of vocal tract shape and dimensions using magnetic resonance imaging: Vowels, *J. Acoust. Soc. Am.*, 90 (2),799-828, Aug 1991.

[5] Stone, M., A three-dimensional model of tongue movement based on ultrasound and microbeam data, *J. Acoust. Soc. Am.*, 87 (5), 2207-2217, May 1990.

[6] Sondhi, M.M. and Resnick, J.R., The inverse problem for the vocal tract: Numerical methods, acoustical experiments, and speech synthesis, *J. Acoust. Soc. Am.*, 73 (3), 985-1002, March 1983.

[7] Hardcastle, W.J., Jones, W., Knight, C., Trudgeon, A., and Calder, G., New developments in electropalatography: A state of the art report, *Clinical Linguistics and Phonetics*, 3, 1-38, 1989.

[8] Perkell, J.S., Cohen, M.H., Svirsky, M.A., Mathies, M.L., Garabieta, I., and Jackson, M.T.T., Electromagnetic midsagittal articulometer systems for transducing speech articulatory movements, *J. Acoust. Soc. Am.*, 92 (6), 3078-3096, Dec 1992.

[9] Coker, C.H., A model of articulatory dynamics and control, *Proc. IEEE*, 64 (4), 452-460, April 1976.

[10] Sondhi, M.M., Resonances of a bent vocal tract, *J. Acoust. Soc. Am.*, 79 (4), 1113-1116, April 1986.

[11] Flanagan, J.L., *Speech Analysis, Synthesis and Perception*, 2nd ed., Springer Verlag, New York, 1972, Chap. 3.

[12] Lu, C., Nakai, T., and Suzuki, H., Three-dimensional FEM simulation of the effects of the vocal tract shape on the transfer function, *Intl. Conf. on Spoken Lang. Processing*, Banff, Alberta, 1, 771-774, 1992.

[13] Richard, G., Liu, M., Sinder, D., Duncan, H., Lin, O., Flanagan, J.L., Levinson, S.E., Davis, D.W. and Slimon, S., Numerical simulations of fluid flow in the vocal tract, *Proc. Eurospeech '95, European Speech Comm. Assoc.*, Madrid, Spain, 18-21, Sept. 1995.

[14] Morse, P.M., *Vibration and Sound*, McGraw Hill, New York, 1948, Chap. 6.

[15] Pierce, A.D., *Acoustics*, 2nd ed., McGraw-Hill, 360, 1981.

[16] Sondhi, M.M., Model for wave propagation in a lossy vocal tract, *J. Acoust. Soc. Am.*, 55 (5), 1070-1075, May 1974.

[17] Siebert, W. McC., *Circuits, Signals and Systems*, MIT Press/McGraw-Hill, pp. 97, 1986.

[18] Sundberg, J., *The Science of the Singing Voice*, Northern Illinois University Press, DeKalb, IL, 1987.

[19] Zemlin, W.R., *Speech and Hearing Science, Anatomy, and Physiology*, Prentice-Hall, Englewood Cliffs, NJ, 1968.

[20] Husson, R., Etude des phénomenes physiologiques et acoustiques fondamentaux de la voix cantée, Disp edit Rev Scientifique, 1-91, 1950. For a discussion see Diehl, C.F., *Introduction to the anatomy and physiology of the speech mechanisms,* Charles C Thomas, Springfield, IL, 110-111, 1968.

[21] Ishizaka, K. and Flanagan, J.L., Synthesis of voiced sounds from a two-mass model of the vocal cords, *Bell System Tech. J.,* 51 (6), 1233-1268, July-Aug. 1972.

[22] Cranen, B. and Schroeter, J., Modeling a leaky glottis, *J. Phonetics,* 23, 165-177, 1995.

[23] Stevens, K.N., Airflow and turbulence noise for fricative and stop consonants: Static considerations, *J. Acoust. Soc. Am.,* 50 (4), 1180-1192, 1971.

[24] Sondhi, M.M. and Schroeter, J., A hybrid time-frequency domain articulatory speech synthesizer, *IEEE Trans. on Acous., Speech, and Sig. Proc.,* ASSP-35 (7), 955-967, July 1987.

[25] Schönhärl, E., *Die Stroboskopie in der praktischen Laryngologie,* Georg Thieme Verlag, Stuttgart, Germany, 1960.

45

Speech Coding

Richard V. Cox
AT&T Labs — Research

45.1 Introduction

Digital speech coding is used in a wide variety of everyday applications that the ordinary person takes for granted, such as network telephony or telephone answering machines. By speech coding we mean a method for reducing the amount of information needed to represent a speech signal for transmission or storage applications. For most applications this means using a lossy compression algorithm because a small amount of perceptible degradation is acceptable. This section reviews some of the applications, the basic attributes of speech coders, methods currently used for coding, and some of the most important speech coding standards.

45.1.1 Examples of Applications

Digital speech transmission is used in network telephony. The speech coding used is just sample-by-sample quantization. The transmission rate for most calls is fixed at 64 kilobits per second (kb/s). The speech is sampled at 8000 Hz (8 kHz) and a logarithmic 8-bit quantizer is used to represent each sample as one of 256 possible output values. International calls over transoceanic cables or satellites are often reduced in bit rate to 32 kb/s in order to boost the capacity of this relatively expensive equipment. Digital wireless transmission has already begun. In North America, Europe, and Japan there are digital cellular phone systems already in operation with bit rates ranging from 6.7 to 13 kb/s for the speech coders. Secure telephony has existed since World War II, based on the first vocoder. (Vocoder is a contraction of the words voice coder.) Secure telephony involves first converting the speech to a digital form, then digitally encrypting it and then transmitting it. At the receiver, it is decrypted, decoded, and reconverted back to analog. Current videotelephony is accomplished through digital transmission of both the speech and the video signals. An emerging use of speech coders is for simultaneous voice and data. In these applications, users exchange data (text, images, FAX, or any other form of digital information) while carrying on a conversation.

All of the above examples involve real-time conversations. Today we use speech coders for many storage applications that make our lives easier. For example, voice mail systems and telephone answering machines allow us to leave messages for others. The called party can retrieve the message when they wish, even from halfway around the world. The same storage technology can be used to broadcast announcements to many different individuals. Another emerging use of speech coding is multimedia. Most forms of multimedia involve only one-way communications, so we include them with storage applications. Multimedia documents on computers can have snippets of speech as an integral part. Capabilities currently exist to allow users to make voice annotations onto documents stored on a personal computer (PC) or workstation.

45.1.2 Speech Coder Attributes

Speech coders have attributes that can be placed in four groups: *bit rate, quality, complexity,* and *delay*. For a given application, some of these attributes are pre-determined while tradeoffs can be made among the others. For example, the communications channel may set a limit on bit rate, or cost considerations may limit complexity. Quality can usually be improved by increasing bit rate or complexity, and sometimes by increasing delay. In the following sections, we discuss these attributes.

Primarily we will be discussing *telephone bandwidth speech*. This is a slightly nebulous term. In the telephone network, speech is first bandpass filtered from roughly 200 to 3200Hz. This is often referred to as 3 kHz speech. Speech is sampled at 8 kHz in the telephone network. The usual telephone bandwidth filter rolls off to about 35 dB by 4 kHz in order to eliminate the aliasing artifacts caused by sampling.

There is a second bandwidth of interest. It is referred to as *wideband speech*. The sampling rate is doubled to 16 kHz. The lowpass filter is assumed to begin rolling off at 7 kHz. At the low end, the speech is assumed to be uncontaminated by line noise and only the DC component needs to be filtered out. Thus, the highpass filter cutoff frequency is 50 Hz. When we refer to wideband speech, we mean speech with a bandwidth of 50 to 7000 Hz and a sampling rate of 16 kHz. This is also referred to as 7 kHz speech.

Bit Rate

Bit rate tells us the degree of compression that the coder achieves. Telephone bandwidth speech is sampled at 8 kHz and digitized with an 8-bit logarithmic quantizer, resulting in a bit rate of 64 kb/s. For telephone bandwidth speech coders, we measure the degree of compression by how much the bit rate is lowered from 64 kb/s. International telephone network standards currently exist for coders operating from 64 kb/s down to 5.3 kb/s. The speech coders for regional cellular standards span the range from 13 to 3.45 kb/s and those for secure telephony span the range from 16 kb/s to 800 b/s. Finally, there are proprietary speech coders that are in common use which span the entire range.

Speech coders need not have a constant bit rate. Considerable compression can be gained by not transmitting speech during the silence intervals of a conversation. Nor is it necessary to keep the bit rate fixed during the talkspurts of a conversation.

Delay

The communication delay of the coder is more important for transmission than for storage applications. In real-time conversations, a large communication delay can impose an awkward protocol on talkers. Large communication delays of 300 ms or greater are particularly objectionable to users even if there are no echoes.

Most low bit rate speech coders are block coders. They encode a block of speech, also known as a frame, at a time. Speech coding delay can be allocated as follows. First, there is algorithmic delay. Some coders have an amount of *look-ahead* or other inherent delays in addition to their frame size. The sum of frame size and other inherent delays constitutes algorithmic delay. The coder requires computation. The amount of time required for this is called processing delay. It is dependent on the speed of the processor used. Other delays in a complete system are the multiplexing delay and the transmission delay.

Complexity

The degree of complexity is a determining factor in both the cost and power consumption of a speech coder. Cost is almost always a factor in the selection of a speech coder for a given application. With the advent of wireless and portable communications, power consumption has also become an important factor. Simple scalar quantizers, such as linear or logarithmic PCM, are necessary in any coding system and have the lowest possible complexity.

More complex speech coders are first simulated on host processors, then implemented on DSP chips and may later be implemented on special purpose VLSI devices. Speed and random access memory (RAM) are the two most important contributing factors of complexity. The faster the chip or the greater the chip size, the greater the cost. In fact, complexity is a determining factor for both cost and power consumption. Generally 1 word of RAM takes up as much on-chip area as 4 to 6 words of read only memory (ROM). Most speech coders are implemented on fixed point DSP chips, so one way to compare the complexity of coders is to measure their speed and memory requirements when efficiently implemented on commercially available fixed point DSP chips.

DSP chips are available in both 16-bit fixed point and 32-bit floating point. 16-bit DSP chips are generally preferred for dedicated speech coder implementations because the chips are usually less expensive and consume less power than implementations based on floating point DSPs. A disadvantage of fixed-point DSP chips is that the speech coding algorithm must be implemented using 16-bit arithmetic. As part of the implementation process, a representation must be selected for each and every variable. Some can be represented in a fixed format, some in block floating point, and still others may require double precision. As VLSI technology has advanced, fixed point DSP chips contain a richer set of instructions to handle the data manipulations required to implement representations such as block floating point. The advantage of floating point DSP chips is that implementing speech coders is much quicker. Their arithmetic precision is about the same as that of a high level language simulation, so the steps of determining the representation of each and every variable and how these representations affect performance can be omitted.

Quality

The attribute of quality has many dimensions. Ultimately quality is determined by how the speech sounds to a listener. Some of the factors that affect the performance of a coder are whether the input speech is clean or noisy, whether the bit stream has been corrupted by errors, and whether multiple encodings have taken place.

Speech coder quality ratings are determined by means of subjective listening tests. The listening is done in a quiet booth and may use specified telephone handsets, headphones, or loudspeakers. The speech material is presented to the listeners at specified levels and is originally prepared to have particular frequency characteristics. The most often used test is the absolute category rating (ACR) test. Subjects hear pairs of sentences and are asked to give one of the following ratings: *excellent, good, fair, poor,* or *bad.* A typical test contains a variety of different talkers and a number of different coders or reference conditions. The data resulting from this test can be analyzed in many ways. The simplest way is to assign a numerical ranking to each response, giving a 5 to the best possible rating, 4 to the next best, down to a 1 for the worst rating, then computing the mean rating for each of the conditions under test. This is a referred to as a mean opinion score (MOS) and the ACR test is often referred to as a MOS test.

There are many other dimensions to quality besides those pertaining to noiseless channels. Bit error sensitivity is another aspect of quality. For some low bit rate applications such as secure telephones over 2.4 or 4.8 kb/s modems, it might be reasonable to expect the distribution of bit errors to be random and coders should be made robust for low random bit error rates up to 1 to 2%. For radio channels, such as in digital cellular telephony, provision is made for additional bits to be used for channel coding to protect the information bearing bits. Errors are more likely to occur in bursts and the speech coder requires a mechanism to recover from an entire lost frame. This is referred to as frame erasure concealment, another aspect of quality for cellular speech coders.

For the purposes of conserving bandwidth, voice activity detectors are sometimes used with speech coders. During non-speech intervals, the speech coder bit stream is discontinued. At the receiver "comfort noise" is injected to simulate the background acoustic noise at the encoder. This method is used for some cellular systems and also in digital speech interpolation (DSI) systems to increase the effective number of channels or circuits. Most international phone calls carried on undersea cables or satellites use DSI systems. There is some impact on quality when these techniques are used. Subjective testing can determine the degree of degradation.

45.2 Useful Models for Speech and Hearing

45.2.1 The LPC Speech Production Model

Human speech is produced in the vocal tract by a combination of the vocal cords in the glottis interacting with the articulators of the vocal tract. The vocal tract can be approximated as a tube of varying diameter. The shape of the tube gives rise to resonant frequencies called formants. Over the years, the most successful speech coding techniques have been based on linear prediction coding (LPC). The LPC model is derived from a mathematical approximation to the vocal tract representation as a variable diameter tube. The essential element of LPC is the linear prediction filter. This is an all pole filter which predicts the value of the next sample based on a linear combination of previous samples.

Let x_n be the speech sample value at sampling instant n. The object is to find a set of prediction coefficients $\{a_i\}$ such that the prediction error for a frame of size M is minimized:

$$\varepsilon = \sum_{m=0}^{M-1} \left(\sum_{i=1}^{I} a_i x_{n+m-i} + x_{n+m} \right)^2 \tag{45.1}$$

where I is the order of the linear prediction model. The prediction value for x_n is given by

$$\tilde{x}_n = -\sum_{i=1}^{I} a_i x_{n-i} \tag{45.2}$$

The prediction error signal $\{e_n\}$ is also referred to as the *residual* signal. In z-transform notation we can write

$$A(z) = 1 + \sum_{i=1}^{I} a_i z^{-i} \tag{45.3}$$

$1/A(z)$ is referred to as the *LPC synthesis* filter and (ironically) $A(z)$ is referred to as the *LPC inverse* filter.

LPC analysis is carried out as a block process on a frame of speech. The most often used techniques are referred to as the autocorrelation and the autocovariance methods [1]–[3]. Both methods involve inverting matrices containing correlation statistics of the speech signal. If the poles of the LPC filter are close to the unit circle, then these matrices become more ill-conditioned, which means that the techniques used for inversion are more sensitive to errors caused by finite numerical precision. Various techniques for dealing with this aspect of LPC analysis include windows for the data [1, 2], windows for the correlation statistics [4], and bandwidth expansion of the LPC coefficients.

For forward adaptive coders, the LPC information must also be quantized and transmitted or stored. Direct quantization of LPC coefficients is not efficient. A small quantization error in a single coefficient can render the entire LPC filter unstable. Even if the filter is stable, sufficient precision is required and too many bits will be needed. Instead, it is better to transform the LPC coefficients to another domain in which stability is more easily determined and fewer bits are required for representing the quantization levels.

The first such domain to be considered is the reflection coefficient [5]. Reflection coefficients are computed as a byproduct of LPC analysis. One of their properties is that all reflection coefficients must have magnitudes less than 1, making stability easily verified. Direct quantization of reflection coefficients is still not efficient because the sensitivity of the LPC filter to errors is much greater when reflection coefficients are nearly 1 or −1. More efficient quantizers have been designed by transforming the individual reflection coefficients with a nonlinearity that makes the error sensitivity more uniform. Two such nonlinear functions are the inverse sine function, $\arcsin(k_i)$, and the logarithm of the area ratio, $\log \frac{1+k_i}{1-k_i}$.

A second domain that has attracted even greater interest recently is the line spectral frequency (LSF) domain [6]. The transformation is given as follows. We first use $A(z)$ to define two polynomials:

$$P(z) = A(z) + z^{-(I+1)} A\left(z^{-1}\right) \tag{45.4a}$$

$$Q(z) = A(z) - z^{-(I+1)} A\left(z^{-1}\right) \tag{45.4b}$$

These polynomials can be shown to have two useful properties: all zeroes of $P(z)$ and $Q(z)$ lie on the unit circle and they are interlaced with each other. Thus, stability is easily checked by assuring both the interlaced property and that no two zeroes are too close together. A second property is that the frequencies tend to be clustered near the formant frequencies; the closer together two LSFs are, the sharper the formant. LSFs have attracted more interest recently because they typically result in quantizers having either better representations or using fewer bits than reflection coefficient quantizers.

The simplest quantizers are scalar quantizers [8]. Each of the values (in whatever domain is being used to represent the LPC coefficients) is represented by one of the possible quantizer levels. The individual values are quantized independently of each other. There may also be additional redundancy between successive frames, especially during stationary speech. In such cases, values may be quantized differentially between frames.

A more efficient, but also more complex, method of quantization is called vector quantization [9]. In this technique, the complete set of values is quantized jointly. The actual set of values is compared against all sets in the codebook using a distance metric. The set that is nearest is selected. In practice, an exhaustive codebook search is too complex. For example, a 10-bit codebook has 1024 entries. This seems like a practical limit for most codebooks, but does not give sufficient performance for typical 10th order LPC. A 20-bit codebook would give increased performance, but would contain over 1 million vectors. This is both too much storage and too much computational complexity to be practical. Instead of using large codebooks, product codes are used. In one technique, an initial codebook is used, then the remaining error vector is quantized by a second stage codebook. In the second technique, the vector is sub-divided and each sub-vector is quantized using its own codebook. Both of these techniques lose efficiency compared to a full-search vector quantizer, but represent a good means for reducing computational complexity and codebook size for bit rate or quality.

45.2.2 Models of Human Perception for Speech Coding

Our ears have a limited dynamic range that depends on both the level and the frequency content of the input signal. The typical bandpass telephone filter has a stopband of only about 35 dB. Also, the logarithmic quantizer characteristics specified by CCITT Rec. G.711 result in a signal-to-quantization noise ratio of about 35 dB. Is this a coincidence? Of course not! If a signal maintains an SNR of about 35 dB or greater for telephone bandwidth, then most humans will perceive little or no noise.

Conceptually, the masking property tells us that we can permit greater amounts of noise in and near the formant regions and that noise will be most audible in the spectral valleys. If we use a coder that produces a white noise characteristic, then the noise spectrum is flat. The white noise would probably be audible in all but the formant regions.

In modern speech coders, an additional linear filter is added to weight the difference between the original speech signal and the synthesized signal. The object is to minimize the error in a space whose metric is like that of the human auditory system. If the LPC filter information is available, it constitutes the best available estimate of the speech spectrum. It can be used to form the basis for this "perceptual weighting filter" [10]. The perceptual weighting filter is given by

$$W(z) = \frac{1 - A(z/\gamma_1)}{1 - A(z/\gamma_2)} \quad 0 < \gamma_2 < \gamma_1 < 1 \tag{45.5}$$

The perceptual weighting filter de-emphasizes the importance of noise in the formant region and emphasizes its importance in spectral valleys. The quantization noise will have a spectral shape that is similar to that of the LPC spectral estimate, making it easier to mask.

The adaptive postfilter is an additional linear filter that is combined with the synthesis filter to reduce noise in the spectral valleys [11]. Once again the LPC synthesis filter is available as the estimate of the speech spectrum. As in the perceptual weighting filter, the synthesis filter is modified. This idea was later further extended to include a long-term (pitch) filter. A tilt-compensation filter was added to correct for the low pass characteristic that causes a muffled sound. A gain control strategy helped prevent any segments from being either too loud or too soft. Adaptive postfilters are now included as a part of many standards.

45.3 Types of Speech Coders

This part of the section describes a variety of speech coders that are widely used. They are divided into two categories: waveform-following coders and model-based coders. Waveform-following coders have the property that if there were no quantization error, the original speech signal would be exactly reproduced. Model-based coders are based on parametric models of speech production. Only the values of the parameters are quantized. If there were no quantization error, the reproduced signal would not be the original speech.

45.3.1 Model-Based Speech Coders

LPC Vocoders

A block diagram of the LPC vocoder is shown in Fig. 45.1. LPC analysis is performed on a frame of speech and the LPC information is quantized and transmitted. A voiced/unvoiced determination is made. The decision may be based on either the original speech or the LPC residual signal, but it will always be based on the degree of periodicity of the signal. If the frame is classified as unvoiced, the excitation signal is white noise. If the frame is voiced, the pitch period is transmitted and the excitation signal is a periodic pulse train. In either case, the amplitude of the output signal is selected such that its power matches that of the original speech. For more information on the LPC vocoder, the reader is referred to [12].

Multiband Excitation (MBE) Coders

Figure 45.2 is a block diagram of a multiband sinusoidal excitation coder. The basic premise of these coders is that the speech waveform can be modeled as a combination of harmonically related sinusoidal waveforms and narrowband noise. Within a given bandwidth, the speech is classified as periodic or aperiodic. Harmonically related sinusoids are used to generate the periodic components and white noise

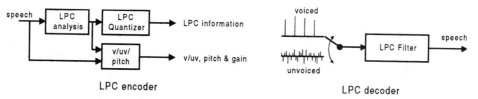

FIGURE 45.1   Block diagram of LPC vocoder.

is used to generate the aperiodic components. Rather than transmitting a single voiced/unvoiced decision, a frame consists of a number of voiced/unvoiced decisions corresponding to the different bands. In addition, the spectral shape and gain must be transmitted to the receiver. LPC may or may not be used to quantize the spectral shape. Most often the analysis of the encoder is performed via fast Fourier transform (FFT). Synthesis at the decoder is usually performed by a number of parallel sinusoid and white noise generators. MBE coders are model-based because they do not transmit the phase of the sinusoids, nor do they attempt to capture anything more than the energy of the aperiodic components. For more information the reader is referred to [13]–[16].

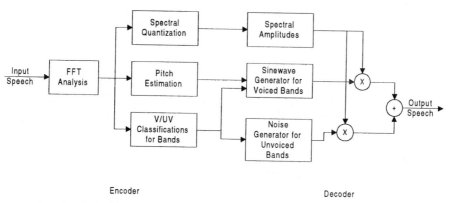

FIGURE 45.2   Block diagram of multiband excitation coder.

Waveform Interpolation Coders

Figure 45.3 is a block diagram of a waveform interpolation coder. In this coder, the speech is assumed to be composed of a slowly evolving periodic waveform (SEW) and a rapidly evolving noise-like waveform (REW). A frame is analyzed first to extract a "characteristic waveform". The evolution of these waveforms is filtered to separate the REW from the SEW. REW updates are made several times more often than SEW updates. The LPC, the pitch, the spectra of the SEW and REW, and the overall energy are all transmitted independently. At the receiver a parametric representation of the SEW and REW information is constructed, summed, and passed through the LPC synthesis filter to produce output speech. For more information the reader is referred to [17, 18].

45.3.2   Time Domain Waveform-Following Speech Coders

All of the time domain waveform coders described in this section include a prediction filter. We begin with the simplest.

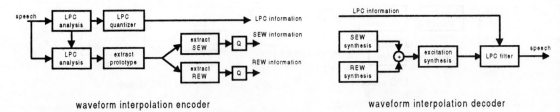

FIGURE 45.3   Block diagram of waveform interpolation coder.

Adaptive Differential Pulse Code Modulation (ADPCM)

Adaptive differential pulse code modulation (ADPCM) [19] is based on sample-by-sample quantization of the prediction error. A simple block diagram is shown in Fig. 45.4. Two parts of the coder may be adaptive: the quantizer step-size and/or the prediction filter. ITU Recommendations G.726 and G.727 adapt both. The adaptation may be either forward or backward adaptive. In a backward adaptive system, the adaptation is based only on the previously quantized sample values and the quantizer codewords. At the receiver, the backward adaptive parameter values must be recomputed. An important feature of such adaptation schemes is that they must use predictors that include a *leakage factor* that allows the effects of erroneous values caused by channel errors to die out over time. In a forward adaptive system, the adapted values are quantized and transmitted. This additional "side information" uses bit rate, but can improve quality. Additionally, it does not require recomputation at the decoder.

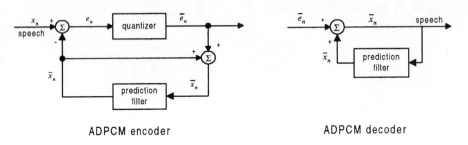

FIGURE 45.4   ADPCM encoder and decoder block diagrams.

Delta Modulation Coders

In delta modulation coders [20], the quantizer is just the sign bit. The quantization step size is adaptive. Not all the adaptation schemes used for ADPCM will work for delta modulation because the quantization is so coarse. The quality of delta modulation coders tends to be proportional to their sampling clock: the greater the sampling clock, the greater the correlation between successive samples, and the finer the quantization step size that can be used. The block diagram for delta modulation is the same as that of ADPCM.

Adaptive Predictive Coding

The better the performance of the prediction filter, the lower the bit rate needed to encode a speech signal. This is the basis of the adaptive predictive coder [21] shown in Fig. 45.5. A forward adaptive higher order linear prediction filter is used. The speech is quantized on a frame-by-frame basis. In this way the bit rate for the excitation can be reduced compared to an equivalent quality ADPCM coder.

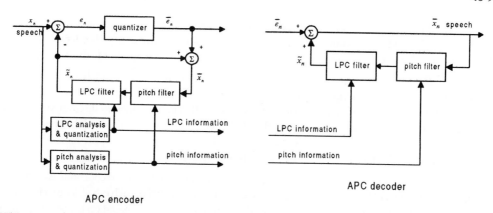

FIGURE 45.5   Adaptive predictive coding encoder and decoder.

Linear Prediction Analysis-by-Synthesis Speech Coders

Figure 45.6 shows a typical linear prediction analysis-by-synthesis speech coder [22]. Like APC, these are frame-by-frame coders. They begin with an LPC analysis. Typically the LPC information is forward adaptive, but there are exceptions. LPAS coders borrow the concept from ADPCM of having a locally available decoder. The difference between the quantized output signal and the original signal is passed through a perceptual weighting filter. Possible excitation signals are considered and the best (minimum mean square error in the perceptual domain) is selected. The long-term prediction filter removes long-term correlation (the pitch structure) in the signal. If pitch structure is present in the coder, the parameters for the long-term predictor are determined first. The most commonly used system is the adaptive codebook, where samples from previous excitation sequences are stored. The pitch period and gain that result in the greatest reduction of perceptual error are selected, quantized, and transmitted. The fixed codebook excitation is next considered and, again, the excitation vector that most reduces the perceptual error energy is selected and its index and gain are transmitted. A variety of different possible fixed excitation codebooks and their corresponding names have been created for coders that fall into this class. Our enumeration touches only the highlights.

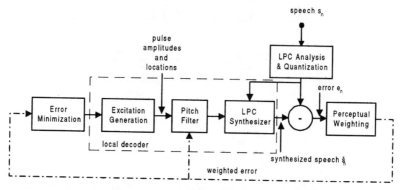

FIGURE 45.6   Linear prediction analysis-by-synthesis coder.

Multipulse Linear Predictive Coding (MPLPC) assumes that the speech frame is sub-divided into smaller sub-frames. After determining the adaptive codebook contribution, the fixed codebook consists of a number of pulses. Typically the number of pulses is about one-tenth the number of samples in a sub-frame. The pulse that makes the greatest contribution to reducing the error is selected first, then the pulse making the next largest contribution, etc. Once the requisite number of pulses have been selected, determination of the pulses is complete. For each pulse, its location and amplitude must be transmitted.

Codebook Excited Linear Predictive Coding (CELP) assumes that the fixed codebook is composed of vectors. This is similar in nature to the *Vector Excitation Coder* (VXC). In the first CELP coder, the codebooks were composed of Gaussian random numbers. It was subsequently discovered that center-clipping these random number codebooks resulted in better quality speech. This had the effect of making the codebook look more like a collection of multipulse LPC excitation vectors. One means for reducing the fixed codebook search is if the codebook consists of overlapping vectors.

Vector Sum Excitation Linear Predictive Coding (VSELP) assumes that the fixed codebook is composed of a weighted sum of a set of basis vectors. The basis vectors are orthogonal to each other. The weights on any basis vector are always either -1 or $+1$. A fast search technique is possible based on using a pseudo-Gray code method of exploration. VSELP was used for several first or second generation digital cellular phone standards [23].

45.3.3 Frequency Domain Waveform-Following Speech Coders

Sub-Band Coders

Figure 45.7 shows the structures of a typical sub-band encoder and decoder [19, 24]. The concept behind sub-band coding is quite simple: divide the speech signal into a number of frequency bands and quantize each band separately. In this way the quantization noise is kept within the band. Typically quadrature mirror or wavelet filterbanks are used. These have the properties that (1) in the absence of quantization error all aliasing caused by decimation in the analysis filterbank is canceled in the synthesis filterbank and (2) the bands can be critically sampled, i.e., the number of frequency domain samples is the same as the number of time domain samples. The effectiveness of these coders depends largely on the sophistication of the quantization algorithm. Generally, algorithms that dynamically allocate the bits according to the current spectral characteristics of the speech give the best performance.

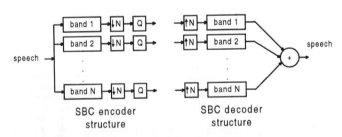

FIGURE 45.7   Sub-band coder.

Adaptive Transform Coders

Adaptive transform coding (ATC) can be viewed as a further extension to sub-band coding [19, 24]. The filterbank structure of SBC is replaced with a transform such as the FFT, the discrete cosine transform (DCT), wavelet transform or other transform-filterbank. They provide a higher resolution analysis than the sub-band filterbanks. This allows the coder to exploit the pitch harmonic structure of the spectrum. As in the case of SBC, the ATC coders that use sophisticated quantization techniques that dynamically allocate the bits usually give the best performance. Most recently, work has combined transform coding with LPC and time-domain pitch analysis [25]. The residual signal is coded using ATC.

45.4 Current Standards

This part of the section is divided into descriptions of current speech coder standards and activities. The subsections contain information on speech coders that have been or will soon be standardized. We begin first by briefly describing the standards organizations who formulate speech coding standards and the processes they follow in making these standards.

The International Telecommunications Union (ITU) is an agency of the United Nations Economic, Scientific and Cultural Organization (UNESCO) charged with all aspects of standardization in telecommunications and radio networks. Its headquarters are in Geneva, Switzerland. The ITU Telecommunications Standardization Sector (ITU-T) formulates standards related to both wireline and wireless telecommunications. The ITU Radio Standardization Sector (ITU-R) handles standardization related to radio issues. There is also a third branch, the ITU – Telecommunications Standards Bureau (ITU-B) is the bureaucracy handling all of the paperwork. Speech coding standards are handled jointly by Study Groups 16 and 12 within the ITU-T. Other Study Groups may originate requests for speech coders for specific applications. The speech coding experts are found in SG16. The experts on speech performance are found in SG12. When a new standard is being formulated, SG16 draws up a list of requirements based on the intended applications. SG12 and other interested bodies may review the requirements before they are finalized. SG12 then creates a test plan and enlists the help of subjective testing laboratories to measure the quality of the speech coders under the various test conditions. The process of standardization can be time consuming and take between 2 to 6 years.

Three different standards bodies make regional cellular standards, including those for the speech coders. In Europe, the parent body is the European Telecommunications Standards Institute (ETSI). ETSI is an organization that is composed mainly of telecommunications equipment manufacturers. In North America, the parent body is the American National Standards Institute (ANSI). The body charged with making digital cellular standards is the Telecommunications Industry Association (TIA). In Japan, the body charged with making digital cellular standards is the Research and Development Center for Radio Systems (RCR).

There are also speech coding standards for satellite, emergencies, and secure telephony. Some of these standards were promulgated by government bodies, while others were promulgated by private organizations.

Each of these standards organizations works according to its own rules and regulations. However, there is a set of common threads among all of the organizations. These are the standards making process. Creating a standard is a long process, not to be undertaken lightly. First, a consensus must be reached that a standard is needed. In most cases this is obvious. Second, the terms of reference need to be created. This becomes the governing document for the entire effort. If defines the intended applications. Based on these applications, requirements can be set on the attributes of the speech coder: quality, complexity, bit rate, and delay. The requirements will later determine the test program that is needed to ascertain whether any candidates are suitable.

Finally, the members of the group need to define a schedule for doing the work. There needs to be an initial period to allow proponents to design coders that are likely to meet the requirements. A deadline is set for submissions. The services of one or more subjective test labs need to be secured and a test plan needs to be defined. A host lab is also needed to process all of the data that will be used in the Selection Test. Some criteria are needed for determining how to make the selection. Based on the selection, a draft standard needs to be written. Only after the standard is fully specified can manufacturers begin to produce implementations of the standard.

45.4.1 Current ITU Waveform Signal Coders

Table 45.1 describes current ITU speech coding recommendations that are based on sample-by-sample scalar quantization. Three of these coders operate in the time domain on the original sampled signal while the fourth is based on a two-band sub-band coder for wideband speech.

TABLE 45.1   ITU Waveform Speech Coders

Standard body	ITU	ITU	ITU	ITU
Number	G.711	G.726	G.727	G.722
Year	1972	1990	1990	1988
Type of coder	Companded PCM	ADPCM	ADPCM	SBC/ADPCM
Bit rate	64 kb/s	16–40 kb/s	16–40 kb/s	48, 56 64 kb/s
Quality	Toll	≤ Toll	≤ Toll	Commentary

Complexity
  MIPS
  RAM
	≪ 1
	1 byte
		1
		< 50 bytes
			1
			< bytes
				10
				1 K words

Delay Frame size	0.125 ms	0.125 ms	0.125 ms	1.5 ms
Specification type Fixed point	Bit exact	Bit exact	Bit exact	Bit exact

The CCITT standardized two 64 kb/s companded PCM coders in 1972. North America and Japan use μ-law PCM. The rest of the world uses A-law PCM. Both coders use 8 bits to represent the signal. Their effective signal-to-noise ratio is about 35 dB. The tables for both of the G.711 quantizer characteristics are contained in [19]. Both coders are considered equivalent in overall quality. A tandem encoding with either coder is considered equivalent to dropping the least significant bit (which is equivalent to reducing the bit rate to 56 kb/s). Both coders are extremely sensitive to bit errors in the most significant bits. Their complexity is very low.

32 kb/s ADPCM was first standardized by the ITU in 1984 [26]–[28]. Its primary application was intended to be digital circuit multiplication equipment (DCME). In combination with digital speech interpolation, a 5:1 increase in the capacity of undersea cables and satellite links was realized for voice conversations. An additional reason for its creation was that such links often encountered the problem of having μ-law PCM at one end and A-law at the other. G.726 can accept either μ-law or A-law PCM as inputs or outputs. Perhaps its most unique feature is a property called synchronous tandeming. If a circuit involves two ADPCM codings with a μ-law or A-law encoding in-between, no additional degradation occurs because of the second encoding. The second bit stream will be identical to the first! In 1986 the Recommendation was revised to eliminate the all-zeroes codeword and so that certain low rate modem signals would be passed satisfactorily. In 1988 extensions for 24 and 40 kb/s were added and in 1990 the 16 kb/s rate was added. All of these additional rates were added for use in digital circuit multiplication equipment applications.

G.727 includes the same rates as G.726, but all of the quantizers have an even number of levels. The 2-bit quantizer is embedded in the 3-bit quantizer, which is embedded in the 4-bit quantizer, which is embedded in the 5-bit quantizer. The is needed for Packet Circuit Multiplex Equipment (PCME) where the least significant bits in the packet can be discarded when there is an overload condition.

Recommendation G.722 is a wideband speech coding standard. Its principal applications are teleconferences and videoteleconferences [29]. The wider bandwidth (50 – 7000 Hz) is more natural sounding and less fatiguing than telephone bandwidth (200 – 3200 Hz). The wider bandwidth increases the intelligibility of the speech, especially for fricative sounds such as /f/ and /s/, which are difficult to distinguish for telephone bandwidth. The G.722 coder is a two-band sub-band coder with ADPCM coding in both bands. The ADPCM is similar in structure to that of the G.727 recommendation. The upper band uses an

ADPCM coder with a 2-bit adaptive quantizer. The lower band uses an ADPCM coder with an embedded 4-5-6 bit adaptive quantizer. This makes the rates of 48, 56, and 64 kb/s all possible. A 24-tap quadrature mirror filter is used to efficiently split the signal.

45.4.2 ITU Linear Prediction Analysis-by-Synthesis Speech Coders

Table 45.2 describes three current analysis-by-synthesis speech coder recommendations of the ITU. All three are block coders based on extensions of the original multipulse LPC speech coder.

TABLE 45.2 ITU Linear Prediction Analysis-By-Synthesis Speech Coders

Standard body	ITU	ITU	ITU
Number	G.728	G.729	G.723.1
Year	1992 and 1994	1995	1995
Type of coder	LD-CELP	CS-ACELP	MPC-MLQ and ACELP
Bit rate	16 kb/s	8 kb/s	6.3 & 5.3 kb/s
Quality	Toll	Toll	$\leq$ Toll
Complexity			
MIPS	30	$\leq$ 22	$\leq$ 16
RAM	2 K	< 2.5 K	2.2 K
Delay			
Frame size	0.625 ms	10 ms	30 ms
Look ahead	0	5 ms	7.5 ms
Specification type			
Floating point	Algorithm exact	None	None
Fixed point	Bit exact	Bit exact C	Bit exact C

G.728 Low-Delay CELP (LD-CELP) [30] is a backward adaptive CELP coder whose quality is equivalent to that of 32 kb/s ADPCM. It was initially specified as a floating point CELP coder that required implementers to follow exactly the algorithm specified in the recommendation. A set of test vectors for verifying correct implementation was created. Subsequently, a bit exact fixed point specification was requested and completed in 1994. The performance of G.728 has been extensively tested by SG12. It gives robust performance for signals with background noise or music. It is very robust to random bit errors, more so than previous ITU standards G.711, G.726, G.727, and the newer standards described below. In addition to passing low bit rate modem signals as high as 2400 bps, it passes all network signaling tones.

In response to a request from CCIR Task Group 8/1 for a speech coder for wireless networks as envisioned in the Future Public Land Mobile Telecommunication Service (FPLMTS), the ITU initiated a work program for a toll quality 8 kb/s speech coder which resulted in G.729. It is a forward adaptive CELP coder with a 10-ms frame size that uses algebraic CELP (ACELP) excitation.

The work program for G.723.1 was initiated in 1993 by the ITU as part of a group of standards to specify a low bit rate videophone for use on the public switched toll networks (PSTN) carried over a high speed modem. Other standards in this group include the video coder, modem, and data multiplexing scheme. A dual rate coder was selected. The two rates differ primarily by their excitation scheme. The higher rate used Multipulse LPC with Maximum Likelihood Quantization (MPC-MLQ) while the lower rate used ACELP. G.723.1 and G.729 are the first ITU coders to be specified by a bit exact fixed point ANSI C code simulation of the encoder and decoder.

45.4.3 Digital Cellular Speech Coding Standards

Table 45.3 describes the first and second generation of speech coders to be standardized for digital cellular telephony. The first generation coders provided adequate quality. Two of the second generation coders are so-called half-rate coders that have been introduced in order to double the capacity of the rapidly growing digital cellular industry. Another generation of coders will soon follow them in order to bring the voice quality of digital cellular service up to that of current wireline network telephony.

TABLE 45.3   Digital Cellular Telephony Speech Coders

Standard body	CEPT	ETSI	TIA	TIA	RCR	RCR
Standard name	GSM	GSM 1/2 Rate	IS-54	IS-96	PDC	PDC 1/2 Rate
Type of coder	RPE-LTP	VSELP	VSELP	CELP	VSELP	PSI-CELP
Date	1987	1994	1989	1993	1990	1993
Bit rate	13 kb/s	5.6 kb/s	7.95 kb/s	0.8 to 8.5	6.7 kb/s	3.45 kb/s
Quality	< toll	= GSM	= GSM	< GSM	< GSM	= PDC
Est. complexity						
MIPS	4.5	30	20	20	20	50
RAM	1K	4K	2K	2K	2K	4K
Delay						
Frame size	20 ms	20 ms	20 ms	20 ms	20 ms	40 ms
Look ahead	0	5 ms	5 ms	5 ms	5 ms	10 ms
Specification type	Bit	Bit	Bit	Bit	Bit	Bit
fixed point	exact	exact C	stream	stream	stream	stream

The RPE-LTP coder [33] was standardized by the Group Special Mobile (GSM) of CEPT in 1987 for pan-European digital cellular telephony. RPE-LTP stands for Regular Pulse Excitation with Long-Term Predictor. The GSM full-rate channel supports 22.8 kb/s. The additional 9.8 kb/s is used for channel coding to protect the coder from bit errors in the radio channel. Voice activity detection and discontinuous transmission are included as part of this standard. In addition to digital cellular telephony, this coder has since been used for other applications, such as messaging, because of its low complexity.

The GSM half-rate coder was standardized by ETSI (an off-shoot of CEPT) in order to double the capacity of the GSM cellular system. The coder is a 5.6 kb/s VSELP coder [23]. A greater percentage of the channel bits are used for error protection because the half-rate channel has less frequency diversity than the full-rate system. The overall performance was measured to be similar to that of RPE-LTP, except for certain signals with background noise.

Vector Sum Excitation Liner Prediction Coding (VSELP) was standardized by the Telecommunications Industry Association (TIA) for time division multiple access (TDMA) digital cellular telephony in North America as a part of Interim Standard 54 (IS-54). It was selected on the basis of subjective listening tests in 1989. The quality of this coder and RPE-LTP are somewhat different in the character of their distortion, but they usually receive about the same MOS in subjective listening tests. IS-54 does not have a bit exact specification. Implementations need only conform to the bit stream specification. The TIA does have a qualification procedure, IS-85, to verify whether the performance of an implementation is good enough to be used for digital cellular [34]. In addition, Motorola provided a floating point C program for their version of the coder, which implementers may use as a guideline.

The IS-96 coder [35] was standardized by the TIA for code division multiple access (CDMA) digital cellular telephony in North America. It is a part of IS-96 and is used in the system specified by IS-95. CDMA system capacity is its most attractive feature. When there is no speech, the rate of the channels is reduced. IS-96 is a variable rate CELP coder which uses digital speech interpolation to achieve this rate reduction. It runs at 8.5 kb/s during most of a talk spurt. When there is no speech on the channel, it drops down to just 0.8 kb/s. At this rate, it is just supplying statistics about the background noise. These two rates are the ones most often used during operation of IS96, although the coder does transition through the two intermediate rates of 2 and 4 kb/s. The validation procedure for this coder is similar to that of IS-85.

The Personal Digital Cellular (PDC) full-rate speech coder was standardized by the Research and Development Center for Radio Systems (RCR) for TDMA digital cellular telephone service in Japan as RCR STD-27B. The coder is very similar to IS-54 VSELP. The principal difference is that instead of two vector sum excitation codebooks, there is only one.

The PDC half-rate coder [37] was standardized by RCR to double the capacity of the Japanese TDMA PDC system. Pitch synchronous innovation CELP (PSI-CELP) uses fixed codebooks that are modified as a function of the pitch in order to improve the speech quality for such a low rate coder. If the pitch period is less than the frame size, then all vectors in the fixed codebook for that frame are made periodic. It has a background noise pre-processor as part of the standard. When it senses that the background noise exceeds a certain threshold, the pre-processor attempts to improve the quality of the speech. To date, this coder appears to be the most complex yet standardized.

45.4.4 Secure Voice Standards

Table 45.4 presents information about three secure voice standards. Two are existing standards, while the third describes a standard that the U.S. government hopes to promulgate in 1996.

TABLE 45.4 Secure Telephony Speech Coding Standards

Standard body	U.S. Dept. of Defense	U.S. Dept. of Defense	U.S. Dept. of Defense
Standard number	FS-1015	FS-1016	?
Type of coder	LPC vocoder	CELP	Model-based
Year	1984	1991	1996
Bit rate	2.4 kb/s	4.8 kb/s	2.4 kb/s
Quality	high DRT	< IS-54	= FS-1016
Complexity			
MIPS	20	19	41[a]
RAM	2K	1.5K	Unknown
Delay			
Frame size	22.5 ms	30 ms	22.5
Look ahead	90 ms	7.5 ms	23
Specification type	Bit stream	Bit stream	Bit stream

[a] Actual goal is 40 MIPS floating point or 80 MIPS fixed point.

FS1015 [12] is a U.S. Federal Standard 2.4 kb/s LPC vocoder that was created over a long period of time beginning in the late 1970s. It was standardized by the U.S. Department of Defense (DoD) and later the North Atlantic Treaty Organization (NATO) before becoming a U.S. Federal Standard in 1984. It was always intended for secure voice terminals. It does not produce natural sounding speech, but over the years its intelligibility has been greatly improved through a series of changes to both its encoder and decoder. Remarkably, these changes never required changes to the bit stream. Presently the intelligibility of FS1015 for clean input speech having telephone bandwidth is almost equivalent to that of the source material as measured by the diagnostic rhyme test (DRT). Most recently an 800 bps vector quantized version of FS1015 has been standardized by NATO [39].

FS1016 [40] is the result of a project undertaken by DoD to increase the naturalness of the secure telephone unit III (STU-3) by the introduction of 4.8 kb/s modem technology. DoD surveyed available 4.8 kb/s speech coder technology in 1988 and 1989. It selected a CELP-based coder having a so-called ternary codebook, meaning that all excitation amplitudes are +1, −1, or 0 before scaling by the gain for that sub-frame. This allows an easier codebook search. FS1016 definitely preserves far more of the naturalness of the original speech than FS1015, but the speech still contains many artifacts and the quality is substantially below that of the cellular coders such as GSM of IS54. Both FS-1015 and FS-1016 have bit stream specifications, but there are C code simulations of them available from the government.

The next coder to be standardized by DoD is a new 2.4 kb/s coder to replace both FS1015 and FS1016. A 3-year project was initiated in 1993 which should culminate in a new standard in 1997. Subjective testing was done in 1993 and 1994 on software versions of potential coders and a realtime hardware evaluation took place in 1995 and 1996 to select a best candidate. The Mixed Excitation Linear Prediction (MELP) coder was selected [41]–[43]. The need for this coder is due to the lack of a sufficient number of satellite channels at 4.8 kb/s. The quality target for this coder is to match or exceed the quality and intelligibility of FS1016 for most scenarios. Many of the scenarios include severe background noise and noisy channel conditions. At 2.4 kb/s, there is not enough bit rate available for explicit channel coding, so the speech coder itself must be designed to be robust for the channel conditions. The noisy background conditions have proven to be difficult for vocoders making voiced/unvoiced classification decisions, whether the decisions are made for all bands or for individual bands.

45.4.5   Performance

Figure 45.8 is included to give an impression of the relative performance for clean speech of most of the standard coders that were included above. There has never been a single subjective test which included all of the above coders. Figure 45.8 is based on the relative performances of these coders across a number of tests that have been reported. In the case of coders that are not yet standards, their performance is projected and shown as a circle. The vertical axis of Fig. 45.8 gives the approximate single encoding quality for clean input speech. The horizontal axis is a logarithmic scale of bit rate. Figure 45.8 only includes telephone bandwidth speech coders. The 7-kHz speech coders have been omitted. Figure 45.9 compares the complexity as measured in MIPS and RAM for a fixed point DSP implementation for most of the same standard coders. The horizontal axis is in RAM and the vertical axis is in MIPS.

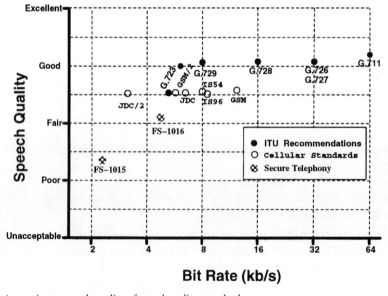

FIGURE 45.8   Approximate speech quality of speech coding standards.

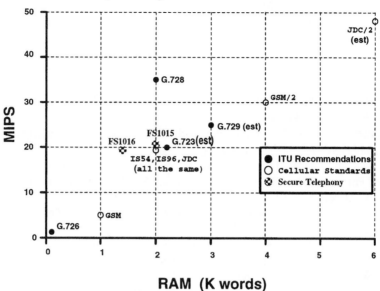

FIGURE 45.9 Approximate complexity of speech coding standards.

References

[1] Markel, J.D. and Gray, Jr., A.H., *Linear Prediction of Speech,* Springer-Verlag, Berlin, 1976.

[2] Rabiner, L.R. and Schafer, R.W., *Digital Processing of Speech Signals,* Prentice-Hall, Englewood Cliffs, NJ, 1978.

[3] LeRoux, J. and Gueguen, C., A fixed point computation of partial correlation coefficients, *IEEE Trans. ASSP,* ASSP-27, 257–259, 1979.

[4] Tohkura, Y., Itakura, F., and Hashimoto, S., Spectral smoothing technique in PARCOR speech analysis/synthesis, *IEEE Trans. ASSP,* 27, 257–259, 1978.

[5] Viswanathan, R. and Makhoul, J., Quantization properties of transmission parameters in linear predictive systems, *IEEE Trans. ASSP,* 23, 309–321, 1975.

[6] Sugamura, N. and Itakura, F., Speech analysis and synthesis methods developed at ECL in NTT — from LPC to LSP, *Speech Commun.,* 5, 199–215, 1986.

[7] Soong, F. and Juang, B.-H., Optimal quantization of LSP parameters, *IEEE Trans. Speech and Audio Processing,* 1, 15–24, 1993.

[8] Lloyd, S.P., Least squares quantization in PCM, *IEEE Trans. Inform. Theory,* 28, 129–137, 1982.

[9] Gersho, A. and Gray, R.M., *Vector Quantization and Signal Compression,* Kluwer-Academic Publishers, Dordrecht, Holland, 1991.

[10] Schroeder, M.R., Atal, B.S., and Hall, J.L., Optimizing digital speech coders by exploiting masking properties of the human ear, *J. Acoustical Soc. Am.,* 66, 1647–1652, Dec. 1979.

[11] Chen, J.-H. and Gersho, A., Adaptive postfiltering for quality enhancement of coded speech, *IEEE Trans. on Speech and Audio Processing,* 3, 59–71, 1995.

[12] Tremain, T., The Government Standard Linear Predictive Coding Algorithm: LPC-10, *Speech Technol.,* 40–49, Apr. 1982. Federal Standard 1015 is available from the U.S. government, as is C source code.

[13] McAulay, R.J. and Quatieri, T.F., Speech analysis/synthesis based on a sinusoidal representation, *IEEE Trans. ASSP*, 34, 744–754, 1986.

[14] McAulay, R.J. and Quatieri, T.F., Low-rate speech coding based on the sinusoidal model, in *Advances in Acoustics and Speech Processing*, Sondhi, M. and Furui, S., Eds., Marcel-Dekker, New York, 1992, 165–207.

[15] Griffin, D.W. and Lim, J.S., Multiband excitation vocoder, *IEEE Trans. ASSP*, 36, 1223–1235, 1988.

[16] Hardwick, J.C. and Lim, J.S., The application of the IMBE speech coder to mobile communications, *Proc. ICASSP '91*, 249–252, 1991.

[17] Kleijn, W.B. and Haagen, J., Transformation and decomposition of the speech signal for coding, *IEEE Signal Processing Lett.*, 136–138, 1994.

[18] Kleijn, W.B. and Haagen, J., A general waveform interpolation structure for speech coding, in *Signal Processing VII*, Holt, M.J.J., Grant, P.M. and Sandham, W.A., Eds., Kluwer Academic Publishers, Dordrecht, Holland, 1994.

[19] Jayant, N.S. and Noll, P., *Digital Coding of Waveforms*, Prentice-Hall, Englewood Cliffs, NJ, 1984, 232–233.

[20] Steele, R., *Delta Modulation Systems*, Halsted Press, New York, 1975.

[21] Atal, B.S., Predictive coding of speech at low bit rates, *IEEE Trans. Comm.*, 30, 600–614, 1982.

[22] Gersho, A., Advances in speech and audio compression, *Proc. IEEE*, 82, 900–918, 1994.

[23] Gerson, I.A. and Jasiuk, M.A., Techniques for improving the performance of CELP-type speech coders, *IEEE JSAC*, 10, 858–865, 1992.

[24] Crochiere, R.E. and Tribolet, J., Frequency domain coding of speech, *Proc. IEEE Trans. ASSP*, 1979.

[25] Lefebvre, R., Salami, R., Laflamme, C., and Adoul, J.-P., High quality coding of wideband audio signals using transform coded excitation (TCX), *Proc. ICASSP '94*, I-193–196, Apr. 1994.

[26] Petr, D.W., 32 kb/s ADPCM-DLQ coding for network applications, *Proc. IEEE GLOBECOM '82*, A8.3-1-A8.3-5, 1982.

[27] Daumer, W.R., Maitre, X., Mermelstein, P., and Tokizawa, I., Overview of the 32 kb/s ADPCM algorithm, *Proc. IEEE GLOBECOM '84*, 774–777, 1984.

[28] Taka, M., Maruta, R., and LeGuyader, A., Synchronous tandem algorithm for 32 kb/s ADPCM, *Proc. IEEE GLOBECOM '84*, 791–795, 1984.

[29] Taka, M. and Maitre, X., CCITT standardizing activities in speech coding, *Proc. ICASSP '86*, 817–820, 1986.

[30] Chen, J.-H., Cox, R.V., Lin, Y.-C., Jayant, N., and Melchner, M.J., A low-delay CELP coder for the CCITT 16 kb/s speech coding standard, *IEEE JSAC*, 10, 830–849, 1992.

[31] Johansen, F.T., A non bit-exact approach for implementation verification of the CCITT LD-CELP speech coder, *Speech Commun.*, 12, 103–112, 1993.

[32] South, C.R., Rugelbak, J., Usai, P., Kitawaki, N., Irii, H., Rosenberger, J., Cavanaugh, J.R., Adesanya, C.A., Pascal, D., Gleiss, N., and Barnes, G.J., Subjective performance assessment of CCITT's 16 kbit/s speech coding algorithm, *Speech Commun.*, 12, 113–134, 1993.

[33] Vary, P., Hellwig, K., Hofmann, R., Sluyter, R.J., Galand, C., and Russo, M., Speech codec for the European mobile radio system, *Proc. ICASSP '88*, 227–230, 1988.

[34] TIA/EIA Interim Standard 85, Recommended minimum performance standards for full rate speech codes, May 1992.

[35] DeJaco, A., Gardner, W., Jacobs, P., and Lee, C., QCELP: the North American CDMA digital cellular variable rate speech coding standard, *Proc. IEEE Workshop on Speech Coding for Telecommunications*, 5-6, 1993.

[36] TIA/EIA Interim Standard 125, Recommended minimum performance for digital cellular wideband spread spectrum speech service option 1, Aug. 1994.

[37] Miki, T., 5.6 kb/s PSI-CELP for digital cellular mobile radio, *Proc. First International Workshop on Mobile Multimedia Communications,* Tokyo, Japan, Dec. 7-10, 1993.

[38] Ohya, T., Suda, H., and Miki, T., 5.6 kb/s PSI-CELP of the half-rate PDC speech coding standard, *Proc. IEEE Vehicular Technol. Conf.,* 1680–1684, June 1994.

[39] Nouy, B., de la Noue, P., and Goudezeune, G., NATO stanag 4479, a standard for an 800 bps vocoder and redundancy protection in HF-ECCM system, *Proc. ICASSP '95,* 480–483, May 1995.

[40] Campbell, J.P., Welch, V.C., and Tremain, T.E., The new 4800 bps voice coding standard, *Proc. Military Speech Tech. 89,* 64–70, Nov 1989. Copies of Federal Standard 1016 are available from the U.S. Government, as is C source code.

[41] McCree, A., Truong, K., George, E., Barnwell, T., and Viswanathan, V., A 2.4 kbit/s MELP coder candidate for the new U.S. federal standard, *Proc. ICASSP '96,* May 1996.

[42] Kohler, M., A comparison of the new 2400 bps MELP federal standard with other standard coders, *Proc. ICASSP'97,* pp. 1587–1590, April 1997.

[43] Supplee, L., Cohn, R., Collura, J., and McCree, A., MELP: the new federal standard at 2400 bps, *Proc. ICASSP'97,* pp. 1591–1594, April 1997.

46

Text-to-Speech Synthesis

Richard Sproat
Bell Laboratories
Lucent Technologies

Joseph Olive
Bell Laboratories
Lucent Technologies

46.1 Introduction

Text-to-speech synthesis has had a long history, one that can be traced back at least to Dudley's "Voder", developed at Bell Laboratories and demonstrated at the 1939 World's Fair [1]. Practical systems for automatically generating speech parameters from a linguistic representation (such as a phoneme string) were not available until the 1960s, and systems for converting from ordinary text into speech were first completed in the 1970s, with MITalk being the best-known such system [2]. Many projects in text-to-speech conversion have been initiated in the intervening years, and papers on many of these systems have been published.[1]

It is tempting to think of the problem of converting written text into speech as "speech recognition in reverse": current speech recognition systems are generally deemed successful if they can convert speech input into the sequence of words that was uttered by the speaker, so one might imagine that a text-to-speech (TTS) synthesizer would start with the words in the text, convert each word one-by-one into speech (being careful to pronounce each word correctly), and concatenate the result together. However, when one considers what literate native speakers of a language must do when they read a text aloud, it quickly becomes clear that things are much more complicated than this simplistic view suggests. Pronouncing words correctly is only part of the problem faced by human readers: in order to sound natural and to sound as if they understand what they are reading, they must also appropriately emphasize (accent) some words, and deemphasize others; they must "chunk" the sentence into meaningful (intonational) phrases; they must pick an appropriate F0 (fundamental frequency) contour; they must control certain aspects of their voice quality; they must know that a word should be pronounced longer if it appears in some positions in the sentence than if it appears in others because *segmental durations* are affected by various factors, including phrasal position.

---

[1]For example, [3] gives an overview of recent Dutch efforts in this area. Audio examples of several current projects on TTS can be found at the WWW URL http://www.cs.bham.ac.uk/~jpi/synth/museum.html.

What makes reading such a difficult task is that *all* writing systems systematically fail to specify many kinds of information that are important in speech. While the written form of a sentence (usually) completely specifies the words that are present, it will only partly specify the intonational phrases (typically with some form of punctuation), will usually not indicate which words to accent or deaccent, and hardly ever give information on segmental duration, voice quality, or intonation. (One might think that a question mark "?" indicates that a sentence should be pronounced with a rising intonation: generally, though, a question mark merely indicates that a sentence is a question, leaving it up to the reader to judge whether this question should be rendered with a rising intonation.) The orthographies of some languages — e.g., Chinese, Japanese, and Thai — fail to give information on where word boundaries are, so that even this needs to be figured out by the reader.[2] Humans are able to perform these tasks because, in addition to being knowledgeable about the grammar of their language, they also (usually) understand the content of the text that they are reading, and can thus appropriately manipulate various extragrammatical "affective" factors, such as appropriate use of intonation and voice quality.

The task of a TTS system is thus a complex one that involves mimicking what human readers do. But a machine is hobbled by the fact that it generally "knows" the grammatical facts of the language only imperfectly, and generally can be said to "understand" nothing of what it is reading. TTS algorithms thus have to do the best they can making use, where possible, of purely grammatical information to decide on such things as accentuation, phrasing, and intonation — and coming up with a reasonable "middle ground" analysis for aspects of the output that are more dependent on actual understanding.

It is natural to divide the TTS problem into two broad subproblems. The first of these is the conversion of text — an imperfect representation of language, as we have seen — into some form of linguistic representation that includes information on the phonemes (sounds) to be produced, their duration, the locations of any pauses, and the F0 contour to be used. The second — the actual synthesis of speech — takes this information and converts it into a speech waveform. Each of these main tasks naturally breaks down into further subtasks, some of which have been alluded to. The first part, text and linguistic analysis, may be broken down as follows:

- *Text preprocessing*: including end-of-sentence detection, "text normalization" (expansion of numerals and abbreviations), and limited grammatical analysis, such as grammatical part-of-speech assignment.

- *Accent assignment*: the assignment of levels of prominence to various words in the sentence.

- *Word pronunciation*: including the pronunciation of names and the disambiguation of homographs.[3]

- *Intonational phrasing*: the breaking of (usually long) stretches of text into one or more intonational units.

- *Segmental durations*: the determination, on the basis of linguistic information computed thus far, of appropriate durations for phonemes in the input.

- *F0 contour computation*.

Speech synthesis breaks down into two parts:

- The selection and concatenation of appropriate concatenative units given the phoneme string.

- The synthesis of a speech waveform given the units, plus a model of the glottal source.

---

[2]Even in English, single *orthographic* words, e.g., *AT&T*, can actually represent multiple words — *A T and T*.

[3]A *homograph* is a single written word that represents two or more different lexical entries, often having different pronunciations: an example would be *bass*, which could be the word for a musical range — with pronunciation /beɪs/ — or a fish — with pronunciation /bæs/. We transcribe pronunciations using the International Phonetic Association's (IPA) symbol set. Symbols used in this chapter are defined in Table 46.1.

46.2 Text Analysis and Linguistic Analysis

46.2.1 Text Preprocessing

The input to TTS systems is text encoded using an electronic coding scheme appropriate for the language, such as ASCII, JIS (Japanese), or Big-5 (Chinese). One of the first tasks facing a TTS system is that of dividing the input into reasonable chunks, the most obvious chunk being the sentence. In some writing systems there is a designated symbol used for marking the end of a declarative sentence and for nothing else — in Chinese, for example, a small circle is used — and in such languages end-of-sentence detection is generally not a problem. For English and other languages we are not so fortunate because a period, in addition to its use as a sentence delimiter, is also used, for example, to mark abbreviations: if one sees the period in *Mr.*, one would not (normally) want to analyze this as an end-of-sentence marker. Thus, before one concludes that a period does in fact mark the end of a sentence, one needs to eliminate some other possible analyses. In a typical TTS system, text analysis would include an abbreviation-expansion module; this module is invoked to check for common abbreviations which might allow one to eliminate one or more possible periods from further consideration. For example, if a preprocessor for English encounters the string *Mr.* in an appropriate context (e.g., followed by a capitalized word), it will expand it as *mister* and remove the period.

Of course, abbreviation expansion itself is not trivial, since many abbreviations are ambiguous. For example, is *St.* to be expanded as *Street* or *Saint*? Is *Dr.*, *Doctor* or *Drive*? Such cases can be disambiguated via a series of heuristics. For *St.*, for example, the system might first check to see if the abbreviation is followed by a capitalized word (i.e., a potential name), in which case it would be expanded as *Saint*; otherwise, if it is preceded by a capitalized word, a number, or an alphanumeric (*49th*), it would be expanded as *Street*. Another problem that must be dealt with is the conversion of numbers into words: *232* should usually be expanded as *two hundred thirty two*, whereas if the same sequence occurs as part of *232-3142* — a likely telephone number — it would normally be read *two three two*.

In languages like English, tokenization into words can to a large extent be done on the basis of white space. In contrast, in many Asian languages, including Chinese, the situation is not so simple because spaces are never used to delimit words. For the purposes of text analysis it is therefore generally necessary to "reconstruct" word boundary information. A minimal requirement for word segmentation is an on-line dictionary that enumerates the wordforms of the language. This is not enough on its own, however, since there are many words that will not be found in the dictionary; among these are personal names, foreign names in transliteration, and morphological derivatives of words that do not occur in the dictionary. It is therefore necessary to build models of these non-dictionary words; see [4] for further discussion.

In addition to lexical analysis, the text-analysis portion of a TTS system will typically perform syntactic analysis of various kinds. One commonly performed analysis is grammatical part-of-speech assignment, as information on the part of speech of words can be useful for accentuation and phrasing, among other things. Thus, in a sentence like *they can can cans*, it is useful for accentuation purposes to know that the first *can* is a *function word* — an auxiliary verb, whereas the second and third are *content words* — respectively a verb and a noun. There are a number of part-of-speech algorithms available, perhaps the best known being the stochastic method of [5], which computes the most likely analysis of a sequence of words, maximizing the product of the *lexical probabilities* of the parts-of-speech in the sentence (i.e., the possible parts of speech of each word and their probabilities), and the *n-gram probabilities* (probabilities of n-grams of parts of speech), which provide a model of the context.

46.2.2 Accentuation

In languages like English, various words in a sentence are associated with *accents*, which are usually manifested as upward or downward movements of fundamental frequency. Usually, not every word in the sentence bears an accent, however, and the decision on which words should be accented and which should be unaccented is one of the problems that must be addressed as part of text analysis. It is common in prosodic analysis to distinguish three levels of *prominence*. Two are accented and unaccented, as just described, and the third is *cliticized*. Cliticized words are unaccented but in addition have lost their word stress, so that they tend to be durationally short: in effect, they behave like unstressed affixes, even though they are written as separate words.

A good first step in assigning accents is to make the accentual determination on the basis of broad lexical categories or parts of speech. Content words — nouns, verbs, adjectives, and perhaps adverbs, tend in general to be accented; function words, including auxiliary verbs and prepositions tend to be deaccented; short function words tend to be cliticized. But accenting has a wider function than merely communicating lexical category distinctions between words. In English, one important set of constructions where accenting is more complicated than what might be inferred from the above discussion are complex noun phrases — basically, a noun preceded by one or more adjectival or nominal modifiers. In a "discourse-neutral" context, some constructions are accented on the final word (*Madison* **Avenue**), some on the penultimate (**Wall** *Street, kitchen* **towel** *rack*), and some on an even earlier word (**sump** *pump factory*). The assignment of accent to complex noun phrases depends on complex lexical and semantic factors; see [6].

Accenting is not only sensitive to syntactic structure and semantics, but also to properties of the discourse. One straightforward effect is *contrast*, as in the example *I didn't ask for* **cherry** *pie, I asked for* **apple** *pie*. For most speakers, the "discourse neutral" accent would be on *pie*, but in this example there is a clear intention to contrast the ingredients in the pies, and *pie* is thus deaccented to effect the contrast between *cherry* and *apple*. See [7] for a discussion of how these kind of effects are handled in a TTS system for English. Note, while humanlike accenting capabilities are possible in many cases, there are still some intractable problems. For example, just as one would often deaccent a word that had been previously mentioned, so would one often deaccent a word if a supercategory of that word had been mentioned: *My son wants a Labrador, but I'm* **allergic** *to dogs*. Handling such cases in any general way is beyond the capabilities of current TTS systems.

46.2.3 Word Pronunciation

The next stage of analysis involves computing pronunciations for the words in the input, given the orthographic representation of those words. The simplest approach is to have a set of "letter-to-sound" rules that simply map sequences of graphemes into sequences of phonemes, along with possible diacritic information, such as stress placement. This approach is naturally best suited to languages where there is a relatively simple relation between orthography and phonology: languages such as Spanish or Finnish fall into this category. However, languages like English manifestly do not, so it has generally been recognized that a highly accurate word pronunciation module must contain a pronouncing dictionary that, at the very least, records words whose pronunciation could not be predicted on the basis of general rules. However, having a dictionary that is merely a list of words presents us with familiar problems of coverage: many text words occur that are not to be found in the dictionary, including morphological derivatives from known words, or previously unseen personal names.

For morphological derivatives, standard techniques for morphological analysis [2, 8] can be applied to achieve a morphological decomposition for a word. The pronunciation of the whole can then, in general, be computed from the (presumably known) pronunciation of the morphological parts, applying appropriate phonological rules of the language. For novel personal names, additional mechanisms may be necessary since novel names cannot always be related *morphologically* to previously seen ones. One such additional method involves computing the pronunciation of a new name by analogy with the pronunciation of a similar name [9, 10]. For example, imagine that we have the name *Califano* in our dictionary and that we know its pronunciation: then we could compute the pronunciation of a hypothetical name *Balifano* by noting that both names share the "suffix" *alifano*. The pronunciation of *Balifano* can then be computed by removing the phoneme /k/, corresponding to the letter *C* in *Califano*, and replacing it with the phoneme /b/.

There are some word forms that are inherently ambiguous in pronunciation, and for which a word pronunciation module as just described can only return a set of possible pronunciations, from which one must then be chosen. A straightforward example is the word *Chevy*, which is most commonly pronounced /ʃ'ɛvi/, but is /tʃ'ɛvi/ in the name *Chevy Chase*, so in this case one could succeed by simply storing the bigram *Chevy Chase*. But n-gram models do not solve all cases of homograph disambiguation. So,

TABLE 46.1 IPA Symbols Used in this Chapter

IPA Symbol	Phonetic value
æ	a as in c*a*t
b	b as in *b*ass
eʲ	a as in b*a*ke
ɛ	e as in b*e*t
ə	a as in d*a*ta
i	i as in lingu*i*ni
ɪ	i as in *i*n
s	s as in *s*oggy
ʃ	sh as in *sh*ip
t	t as in *t*est
tʃ	ch as in *ch*ase
ð	th as in *th*e
r	r as in a*r*e
v	v as in *v*oodoo
w	w as in *w*e
'	primary stress

the word *bass*, is most likely to be pronounced /bæs/ in a "fishy" context like *he was fishing for bass*, but /beʲs/ in a musical context like *he plays bass*. What defines the context as being musical or "fishy" is not characterizable in terms of n-grams, but rather relates to the occurrence of certain words (e.g., *fish*, *lake*, *boat* vs. *play*, *sing*, *orchestra*) in a wider context. A method proposed by Yarowsky [11, 12] allows for both local (n-gram) context and wide context to be used in homograph disambiguation, and excellent results have been achieved using this approach.[4]

46.2.4 Intonational Phrasing

In reading a long sentence, speakers will typically break the sentence up into several phrases, each of which can be said to "stand alone" as an intonational unit. If punctuation is used liberally so that there are relatively few words between the commas, semicolons, or periods, then a reasonable guess at an appropriate phrasing would be simply to break the sentence at the punctuation marks (though this is not always appropriate [13]). The real problem comes when long stretches occur without punctuation; in

---

[4]Clearly the above-described method for homograph disambiguation can also be applied to other formally similar problems in TTS, such as whether *St.* to be expanded as *Saint* or *Street*, or *747* is to be read as a number *seven hundred and forty seven* or the name of an aircraft *seven forty seven*.

such cases, human readers would normally break the string of words into phrases, and the problem then arises of where to place these breaks.

The simplest approach is to have a list of words, typically function words, that are likely indicators of good places to break [1]. One has to use some caution, however, because while a particular function word such as *and* may coincide with a plausible phrase break in some cases (*He got out of the car and walked towards the house*), in other examples it might coincide with a particularly *poor* place to break as in *I was forced to sit through a dog and pony show that lasted most of Wednesday afternoon*. Other approaches to intonational phrasing have been proposed in the literature, including methods that depend on syntactic parsers of various degrees of sophistication [13, 14]. An alternative approach, described in [15], uses a decision tree model [16, 17] that is trained on a corpus of text annotated with prosodic phrase-boundary information.

46.2.5   Segmental Durations

Having computed which phonemes are to be produced by the synthesizer, it is necessary to decide how long to make each one. In this section we briefly describe the methods used for computing segmental durations: the reader is referred to [18] for an extended discussion of this topic.

What duration to assign to a phonemic segment depends on many factors, including:

- The identity of the segment in question. For example, in many dialects of English, the vowel /æ/ has a longer *intrinsic duration* than the vowel /ɪ/.
- The stress of the syllable of which the segment is a member. For example, vowels in stressed syllables tend to be longer than vowels in unstressed syllables.
- Whether the syllable of which the segment is a member bears an accent. Accented syllables tend to be longer than otherwise identical unaccented syllables.
- The quality of the surrounding segments. For example, a vowel preceding a voiced consonant in the same syllable tends to be longer than the same vowel preceding a voiceless consonant.
- The position of the segment in the phrase: elements close to the ends of phrases tend to be longer than elements more internal to the phrase.

Various approaches have been taken to modeling segmental durations in TTS systems. One method involves *duration rules*, which are rules of the form "if the segment is X and it is in phrase-final position, then lengthen X by n milliseconds" [19, 20]. In rule-based systems of this kind, it is not unusual for the duration of a given segment to be rewritten several times as the conditions for the application of the various rules are considered. The rule-based approach can be formalized explicitly in terms of the second approach — *duration models* — which are mathematical expressions that prescribe how the various conditioning factors are to be used in computing the duration of a segment [19]; the successive application of the rules can, in effect, be "compiled" into a single mathematical expression that implements the combined effect of the rules. As argued in [18], all extant duration models can be viewed as instances of a more general *sum-of-products* model, where the duration of a segment is predicted by a formula of the general form:

$$DUR(\mathbf{f}) = \sum_{i \in T} \prod_{j \in I_i} S_{i,j}(f_j) \qquad (46.1)$$

Here the duration assigned to a feature vector — $DUR(\mathbf{f})$ — is computed by scaling each factor f_j in the ith product term by a factor scale $S_{i,j}$; computing the product of all scaled factors within each product term; and then summing over all i product terms. Rather than deciding *a priori* on a particular sums-of-products model (or set of such models) within the space of all possible models, one approach taken to segmental duration is to use exploratory data analysis to arrive at models whose predictions show a good fit to durations from a corpus of labeled speech [18].

More specifically, we start with a text corpus that (ideally) has a good coverage both of various phonemes and of the factors (and their combinations) that are deemed likely to be relevant for duration. A native speaker of the language reads this text and the speech is segmented and labeled. Using the text-analysis modules of TTS, with some possible hand correction, we automatically compute the sequence of phonemes, and the feature vectors (including features on stress, accent, phrasal position, etc.) associated with each phoneme. Given the feature vectors, various sums-of-products models are compared and their predictions of the values of the observed segmental durations are evaluated. In general, different specific duration models may be better suited to different sets of conditions than others: for example, in the English duration system, intervocalic consonants are associated with a different sums-of-products model than consonants that occur in clusters. In the actual implementation of segmental duration predictions, a decision tree is used to determine, on the basis of contextual factors appropriate to the segment at hand, what particular sums-of-products model to use; this model is then used to compute the duration of the segment.

Designing a corpus with good coverage of relevant factors is a non-trivial task in itself: the basic problem is to provide a set that has maximal coverage with the minimal amount of text to be read by a speaker, and analyzed. The method that we use involves starting with a large corpus of text in a language and automatically predicting the phonemic segments along with their features (again, using text analysis components for the language). A *greedy* algorithm is then applied to arrive at a minimal set of sentences that have good (ideally total) coverage of the desired feature vectors.

46.2.6   Intonation

Having computed linguistic information such as the sequence of segments to be produced, their duration, the prominence of the various words, and the locations of prosodic boundaries, the next thing that a TTS system needs to compute is an intonation contour. There are almost as many models of intonation implemented in TTS systems as there are TTS systems, and we do not have the space to review these different approaches here. Suffice it to say that most intonation models that have actually been incorporated into working TTS systems can be classified into one of three "schools":

- The *Fujisaki* school [21, 22, inter alia]. An intonation contour for a phrase is computed from a phrase impulse and some number of accent impulses. These impulses are convolved with a smoothing function to produce phrase and accent curves, which are then summed to produce the final contour.

- The *Dutch* school [23]. Intonation contours are represented as sequences of connected line segments which are chosen so as to perceptually closely approximate real (smooth) intonation contours.

- The *autosegmental/metrical* school [24, 25, inter alia]. Intonation contours are represented abstractly as sequences of high and low targets.

The computation of an intonation contour from a phonological representation can be illustrated by considering the Bell Labs English TTS system, which currently uses a version of the Pierrehumbert autosegmental model [26, 27, 28]. As the first stage in the computation of an intonation contour, a tone-timing function sets up nominal times for each accent in the sentence. Separate routines are called for initial boundary tones, final boundary tones, pitch accents and phrase accents. Roughly, initial boundary tones are aligned with the silence that is placed at the beginning of each minor phrase, whereas final boundary tones are aligned with the final vowel of each minor phrase. Phrase accents are aligned after the final word accent of the minor phrase, if there is one; otherwise at the end of the first vowel of the first word, or else at the end of the first phoneme. Finally, accents on words are aligned with their associated syllables using a complex set of contextual factors. These nominal accent times are then converted into actual F0/time pairs, by another function. F0 values are computed dependent on the prominence of the accent (either determined automatically, or else definable by the user), and various phrasal parameters from the intonation model, as well as the particular type of accent involved. Finally, an *F0 contour* is produced by interpolating the computed pitch/time pairs, and smoothing via convolution with a rectangular window.

46.3   Speech Synthesis

Once the text has been transformed into phonemes, and their associated durations and a fundamental frequency contour have been computed, the system is ready to compute the speech parameters for synthesis.

There are two independent variables in the choice of parametric computation in a TTS system. One variable is the choice between a *rule-based* scheme for the computation of the parameters on the one hand, and a *concatenative scheme* involving concatenation of short segments of previously uttered speech on the other. The second variable is the actual parametric representation chosen: possible choices include articulatory parameters, formants, LPC (linear predictive coding), spectral parameters, or time domain parameters. In a concatenative scheme, any parametric representation that permits independent control of loudness, F0, voicing, timing, and possibly spectral manipulations is appropriate. Rule-based systems are more restrictive of the choice of parameters since such schemes rely both on our understanding of the relation between the parameters and the acoustic signals they represent, and on our ability to compute the dynamics of the parameters as they move from one sound to another. Thus far only articulatory parameters and formants have been used in rule-based systems. The best-known examples of a formant-based synthesizer are the Klatt synthesizer and its commercial offshoot DECtalk.

Rule-based systems are space-efficient because they eliminate the need to store speech segments. Rule-based approaches also make it easier, in principle, to implement new speaker characteristics for different voices, as well as different phone inventories for new dialects and languages. However, since the dynamics of the parameters are very difficult to model it requires a great deal more effort to produce a rule-based system than it does to produce a concatenative system of comparable quality. Given the right choice of units, a concatenative scheme is able to store the dynamics of the speech signal and thus produce high quality synthetic sound. The choice of the exact parameters depends on what the designer values in such a system. Waveform representations — such as PSOLA [29] — have a high sound quality, but they are limiting in terms of the ability to alter the sound, and thus far, no one has been able to change the spectral parameters in a time domain system. Articulatory parameters or formants, on the other hand, can be successfully manipulated. However, the speech quality produced by using these parameters is somewhat degraded because there are no reliable methods to extract these parameters and even in a plain coding application (analysis and resynthesis without manipulations) these methods produce degradation of the speech signal.

Other systems use a concatenative approach. In this approach, parametrized short speech segments of natural speech are connected to form a representation of the synthetic speech. The majority of the natural speech segments are merely transitions between pairs of phonemes. However, due to the large contextual variation of some phonemes, some segments consisting of three or more phoneme elements are often necessary; such elements consist of the transition from the first phoneme to the second, and a transition from the penultimate phoneme to the last, but the intermediate phonemes are stored completely. For the Bell Labs English system, there are approximately 2900 different speech elements — also called *dyads* — in the acoustic inventory, and these elements are sufficient to make up all the legal phoneme combinations for English.

The concatenative approach to speech synthesis requires that speech samples be stored in some parametric representation that will be suitable for connecting the segments and changing the signal's characteristics of loudness, F0, and spectrum. One method for changing the characteristics of natural speech is to analyze the speech in terms of a source/filter model, as diagramed in Fig. 46.1.

This model of speech synthesis has a variety of independent input controls. Starting at the left side of the figure, we show two possible source generators: a noise generator and a simple pulse generator. The noise generator has no controlling input whereas the pulse generator is controlled by the F0 parameter;

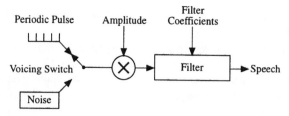

FIGURE 46.1 Source-filter model for speech synthesis.

the F0 parameter specifies the distance between any two pulses thus controlling the F0 of the periodic source. These inputs are selected by a switch which is controlled by a voicing flag. It is also possible to have a mixer to control the relative contribution of the noise and pulse source, and to insert a glottal pulse with additional controls for the shape of the glottal source in place of the simple pulse generator. To the right of the switch, we have a multiplier that multiplies the source by an amplitude parameter. This serves as the loudness control for the system. The signal from the multiplier is fed into a filter controlled by the filter coefficients which are varied slowly to shape the speech spectrum.

The source/filter model can be used to replicate naturally spoken speech when the parameters are obtained by analysis of natural speech. Speech can be parametrized by an amplitude control, voiced/voiceless flag, F0, and filter coefficients at a small interval (on the order of 5 msec. to 15 msec.) The loudness control is determined from the power of the speech at the time frame of the analysis. F0 extraction algorithms determine the voicing of the speech as well as the fundamental frequency. The filter parameters can be determined by various analysis techniques. The parameters obtained from the analysis can be used to drive the source filter model to reproduce the analyzed speech. However, these parameters can also be varied independently to change the speech. The ability to alter the analysis parameters is crucial to a concatenative approach, where the spectral parameters have to be smoothed and interpolated whenever two elements from different utterances are connected, or when the duration of the speech has to be altered. Of course, the fundamental frequency of the original speech is completely discarded and replaced during synthesis by a rule-generated F0, as described earlier.

46.4 The Future of TTS

Using current methods such as those outlined in this chapter, it is possible to produce speech output that is of high intelligibility and reasonable naturalness, given unrestricted input text. (See [30] for a discussion of methods for evaluating TTS systems.) However, there is still much work to be done in all areas of the problem, including: improving voice quality, and allowing for greater user control over aspects of voice quality; producing better models of intonation to allow for more natural-sounding F0 contours; and improving linguistic analysis so that more accurate information on contextually appropriate word pronunciation, accenting, and phrasing can be computed automatically. This latter area — linguistic analysis — is particularly crucial: most high-quality TTS systems allow for user control of the output speech by means of various "escape sequences", which can be inserted into the input text. By use of such escape sequences, it is possible to produce highly appropriate and natural-sounding output. What is still lacking in many cases are natural-language analysis techniques that can mimic what a human annotator is able to do.

References

[1] Klatt, D., Review of text-to-speech conversion for English, *J. Acoustical Soc. Am.*, 82, 737–793, 1987.

[2] Allen, J., Hunnicutt, M.S. and Klatt, D., *From Text to Speech*, Cambridge University Press, Cambridge, 1987.

[3] van Heuven, V. and Pols, L., *Analysis and Synthesis of Speech: Strategic Research towards High-Quality Text-to-Speech Generation*, Mouton de Gruyter, Berlin, 1993.

[4] Sproat, R., Shih, C., Gale, W. and Chang, N., A stochastic finite-state word-segmentation algorithm for Chinese, in *Association for Computational Linguistics, Proceedings of 32nd Annual Meeting*, 66–73, 1994.

[5] Church, K., A stochastic parts program and noun phrase parser for unrestricted text, in *Proceedings of the Second Conference on Applied Natural Language Processing*, Association for Computational Linguistics, Morristown, NJ, 136-143, 1988.

[6] Sproat, R., English noun-phrase accent prediction for text-to-speech, *Computer Speech and Language*, 8, 79–94, 1994.

[7] Hirschberg, J., Pitch accent in context: Predicting intonational prominence from text, *Artificial Intelligence*, 63, 305–340, 1993.

[8] Koskenniemi, K., Two-Level Morphology: A General Computational Model for Word-Form Recognition and Production, Ph.D. thesis, University of Helsinki, Helsinki, 1983.

[9] Coker, C., Church, K. and Liberman, M., Morphology and rhyming: Two powerful alternatives to letter-to-sound rules for speech synthesis, in *Proceedings of the ESCA Workshop on Speech Synthesis*, Bailly, G. and Benoit, C., Eds., 83–86, 1990.

[10] Golding, A., Pronouncing Names by a Combination of Case-Based and Rule-Based Reasoning, Ph.D. thesis, Stanford University, 1991.

[11] Yarowsky, D., Homograph disambiguation in speech synthesis, in *Proceedings of the Second ESCA/IEEE Workshop on Speech Synthesis*, 1994.

[12] Sproat, R., Hirschberg, J. and Yarowsky, D., A corpus-based synthesizer, in *Proceedings of the International Conference on Spoken Language Processing*, Banff, ICSLP, 563–566, Oct. 1992.

[13] O'Shaughnessy, D., Parsing with a small dictionary for applications such as text to speech, *Computational Linguistics*, 15, 97–108, 1989.

[14] Bachenko, J. and Fitzpatrick, E., A computational grammar of discourse-neutral prosodic phrasing in English, *Computational Linguistics*, 16, 155–170, 1990.

[15] Wang, M. and Hirschberg, J., Automatic classification of intonational phrase boundaries, *Computer Speech and Language*, 6, 175–196, 1992.

[16] Breiman, L., Friedman, J.H., Olshen, R.A. and Stone, C.J., *Classification and Regression Trees*, Wadsworth & Brooks, Pacific Grove, CA, 1984.

[17] Riley, M., Some applications of tree-based modelling to speech and language, in *Proceedings of the Speech and Natural Language Workshop*, DARPA, Morgan Kaufmann, Cape Cod, MA, 339–352, Oct. 1989.

[18] van Santen, J., Assignment of segmental duration in text-to-speech synthesis, *Computer Speech and Language*, 8, 95–128, 1994.

[19] Klatt, D., Linguistic uses of segmental duration in English: acoustic and perceptual evidence, *J. Acoustic Soc. Am.*, 59, 1209–1221, 1976.

[20] Syrdal, A.K., Improved duration rules for text-to-speech synthesis, *J. Acoustic Soc. Am.*, 85, S1(Q4), 1989.

[21] Fujisaki, H., Dynamic characteristics of voice fundamental frequency in speech and singing, in *The Production of Speech*, MacNeilage, P. Ed., Springer, New York, 39–55, 1983.

[22] Möbius, B., *Ein quantitatives Modell der deutschen Intonation*, Niemeyer, Tübingen, 1993.

[23] 't Hart, J., Collier, R. and Cohen, A., *A Perceptual Study of Intonation: An Experimental-Phonetic Approach to Speech Melody*, Cambridge University Press, Cambridge, 1990.

[24] Pierrehumbert, J.B., The Phonology and Phonetics of English Intonation, Ph.D. thesis, Massachusetts Institute of Technology, Sept. 1980.

[25] Ladd, D.R., *The Structure of Intonational Meaning,* Indiana University Press, Bloomington, Ind., 1980.

[26] Liberman, M. and Pierrehumbert, J., Intonational invariants under changes in pitch range and length, in *Language Sound Structure,* Aronoff, M. and Oehrle, R., Eds., MIT Press, Cambridge, 1984.

[27] Anderson, M., Pierrehumbert, J. and Liberman, M., Synthesis by rule of English intonation patterns, in *Proceedings of the International Conference on Acoustics, Speech, and Signal Processing,* vol.1, ICASSP, San Diego, 2.8.1–2.8.4, 1984.

[28] Silverman, K., Utterance-internal prosodic boundaries, in *Proceedings of the Second Australian International Conference on Speech Science and Technology,* Sydney, Australia, 86–91, 1988.

[29] Charpentier, F. and Moulines, E., Pitch-synchronous waveform processing techniques for text-to-speech synthesis using diphones, *Speech Commun.,* 9(5/6), 453–467, 1990.

[30] van Santen, J., Perceptual experiments for diagnostic testing of text-to-speech systems, *Computer Speech and Language,* 7, 49–100, 1993.

47

Speech Recognition by Machine

Lawrence R. Rabiner
AT&T Labs — Research

B. H. Juang
Bell Laboratories
Lucent Technologies

47.1 Introduction

Over the past several decades a need has arisen to enable humans to communicate with machines in order to control their actions or to obtain information. Initial attempts at providing human-machine communications led to the development of the keyboard, the mouse, the trackball, the touch screen, and the joy stick. However, none of these communication devices provides the richness or the ease of use of speech which has been the most natural form of communication between humans for tens of centuries. Hence, a need has arisen to provide a voice interface between humans and machines. This need has been met, to a limited extent, by speech processing systems which enable a machine to speak (speech synthesis systems) and which enable a machine to understand (speech recognition systems) human speech. We concentrate on speech recognition systems in this section.

Speech recognition by machine refers to the capability of a machine to convert human speech to a textual form, providing a transcription or interpretation of everything the human speaks while the machine is listening. This capability is required for tasks in which the human is controlling the actions of the machine using only limited speaking capability, e.g., while speaking simple commands or sequences of words from a limited vocabulary (e.g., digit sequences for a telephone number). In the more general case, usually referred to as speech understanding, the machine need only recognize a limited subset of the user input speech, namely the speech that specifies enough about the action requested so that the machine can either respond appropriately, or initiate some action in response to what was understood.

Speech recognition systems have been deployed in applications ranging from control of desktop computers, to telecommunication services, to business services, and have achieved varying degrees of success and commercialization.

In this section we discuss a range of issues involved in the design and implementation of speech recognition systems.

47.2   Characterization of Speech Recognition Systems

A number of issues define the technology of speech recognition systems. These include:

1. The manner in which a user speaks to the machine. There are generally three modes of speaking, including:
    - isolated word (or phrase) mode in which the user speaks individual words (or phrases) drawn from a specified vocabulary;
    - connected word mode in which the user speaks fluent speech consisting entirely of words from a specified vocabulary (e.g., telephone numbers);
    - continuous speech mode in which the user can speak fluently from a large (often unlimited) vocabulary.

2. The size of the recognition vocabulary, including:
    - small vocabulary systems which provide recognition capability for up to 100 words;
    - medium vocabulary systems which provide recognition capability for from 100 to 1000 words;
    - large vocabulary systems which provide recognition capability for over 1000 words.

3. The knowledge of the user's speech patterns, including:
    - speaker dependent systems which have been custom tailored to each individual talker;
    - speaker independent systems which work on broad populations of talkers, most of which the system has never encountered or adapted to;
    - speaker adaptive systems which customize their knowledge to each individual user over time while the system is in use.

4. The amount of acoustic and lexical knowledge used in the system, including:
    - simple acoustic systems which have no linguistic knowledge;
    - systems which integrate acoustic and linguistic knowledge, where the linguistic knowledge is generally represented via syntactical and semantic constraints on the output of the recognition system.

5. The degree of dialogue between the human and the machine, including:
    - one-way (passive) communication in which each user spoken input is acted upon;
    - system-driven dialog systems in which the system is the sole initiator of a dialog, requesting information from the user via verbal input;
    - natural dialogue systems in which the machine conducts a conversation with the speaker, solicits inputs, acts in response to user inputs, or even tries to clarify ambiguity in the conversation.

47.3   Sources of Variability of Speech

Speech recognition by machine is inherently difficult because of the variability in the signal. Sources of this variability include:

1. Within-speaker variability in maintaining consistent pronunciation and use of words and phrases.
2. Across-speaker variability due to physiological differences (e.g., different vocal tract lengths) regional accents, foreign languages, etc.
3. Transducer variability while speaking over different microphones/telephone handsets.
4. Variability introduced by the transmission system (the media through which speech is transmitted, telecommunication networks, cellular phones, etc.).
5. Variability in the speaking environment, including extraneous conversations and acoustic background events (e.g., noise, door slams).

47.4   Approaches to ASR by Machine

47.4.1   The Acoustic-Phonetic Approach [1]

The earliest approaches to speech recognition were based on finding speech sounds and providing appropriate labels to these sounds. This is the basis of the acoustic-phonetic approach which postulates that there exist finite, distinctive phonetic units (phonemes) in spoken language, and that these units are broadly characterized by a set of acoustic properties that are manifest in the speech signal over time. Even though the acoustic properties of phonetic units are highly variable, both with speakers and with neighboring sounds (the so-called coarticulation), it is assumed in the "acoustic-phonetic approach" that the rules governing the variability are straightforward and can be readily learned (by a machine). The first step in the acoustic-phonetic approach is a segmentation and labeling phase in which the speech signal is segmented into stable acoustic regions, followed by attaching one or more phonetic labels to each segmented region, resulting in a phoneme lattice characterization of the speech (see Fig. 47.1). The second step attempts to determine a valid word (or string of words) from the phonetic label sequences produced in the first step. In the validation process, linguistic constraints of the task (i.e., the vocabulary, the syntax, and other semantic rules) are invoked in order to access the lexicon for word decoding based on the phoneme lattice. The acoustic-phonetic approach has not been widely used in most commercial applications.

47.4.2   "Pattern-Matching" Approach [2]

The "pattern-matching approach" involves two essential steps, namely, pattern training and pattern comparison. The essential feature of this approach is that it uses a well- formulated mathematical framework, and establishes consistent speech pattern representations for reliable pattern comparison from a set of labeled training samples via a formal training algorithm. A speech pattern representation can be in the form of a speech template or a statistical model, and can be applied to a sound (smaller than a word), a word, or a phrase. In the pattern-comparison stage of the approach, a direct comparison is made between the unknown speech (the speech to be recognized) with each possible pattern learned in the training stage, in order to determine the identity of the unknown according to the goodness of match of the patterns. The pattern matching approach has become the predominant method of speech recognition in the last decade and we shall elaborate on it in subsequent sections.

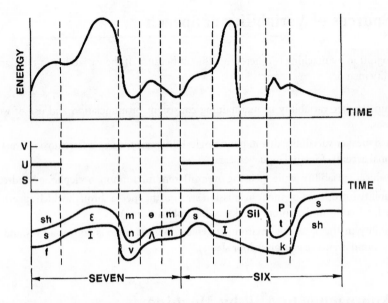

FIGURE 47.1   Segmentation and labeling for word sequence"seven-six".

47.4.3 Artificial Intelligence Approach [3, 4]

The "artificial intelligence approach" attempts to mechanize the recognition procedure according to the way a person applies intelligence in visualizing, analyzing, and characterizing speech based on a set of measured acoustic features. Among the techniques used within this class of methods are use of an expert system (e.g., a neural network) which integrates phonemic, lexical, syntactic, semantic, and even pragmatic knowledge for segmentation and labeling, and uses tools such as artificial neural networks for learning the relationships among phonetic events. The focus in this approach has been mostly in the representation of knowledge and integration of knowledge sources. This method has not been used widely in commercial systems.

47.5   Speech Recognition by Pattern Matching

Figure 47.2 is a block diagram that depicts the pattern matching framework. The speech signal is first analyzed and a feature representation is obtained for comparison with either stored reference templates or statistical models in the pattern matching block. A decision scheme determines the word or phonetic class of the unknown speech based on the matching scores with respect to the stored reference patterns.

There are two types of reference patterns that can be used with the model of Fig. 47.2. The first type,

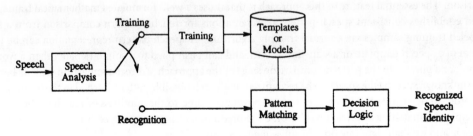

FIGURE 47.2   Block diagram of pattern-recognition speech recognizer.

called a nonparametric reference pattern [5] (or often a template), is a pattern created from one or more spoken tokens (exemplars) of the sound associated with the pattern. The second type, called a statistical reference model, is created as a statistical characterization (via a fixed type of model) of the behavior of a collection of tokens of the sound associated with the pattern. The hidden Markov model [6] is an example of the statistical model.

The model of Fig. 47.2 has been used (either explicitly or implicitly) for almost all commercial and industrial speech recognition systems for the following reasons:

1. It is invariant to different speech vocabularies, user sets, feature sets, pattern matching algorithms, and decision rules.
2. It is easy to implement in software (and hardware).
3. It works well in practice.

We now discuss the elements of the pattern recognition model and show how it has been used in isolated word, connected word, and continuous speech recognition systems.

47.5.1 Speech Analysis

The purpose of the speech analysis block is to transform the speech waveform into a parsimonious representation which characterizes the time varying properties of the speech. The transformation is normally done on successive and possibly overlapped short intervals 10 to 30 msec in duration (i.e., short-time analysis) due to the time-varying nature of speech. The representation [7] could be spectral parameters, such as the output from a filter bank, a discrete Fourier transform (DFT), or a linear predictive coding (LPC) analysis, or they could be temporal parameters, such as the locations of various zero or level crossing times in the speech signal.

Empirical knowledge gained over decades of psychoacoustic studies suggests that the power spectrum has the necessary acoustic information for high accuracy sound identity. Studies in psychoacoustics also suggest that our auditory perception of sound power and loudness involves compression, leading to the use of the logarithmic power spectrum and the cepstrum [8], which is the Fourier transform of the log-spectrum. The low order cepstral coefficients (up to 10 to 20) provide a parsimonious representation of the short-time speech segment which is usually sufficient for phonetic identification.

The cepstral parameters are often augmented by the so-called delta cepstrum [9] which characterizes dynamic aspects of the time-varying speech process.

47.5.2 Pattern Training

Pattern training is the method by which representative sound patterns (for the unit being trained) are converted into reference patterns for use by the pattern matching algorithm. There are several ways in which pattern training can be performed, including:

1. Casual training in which a single sound pattern is used directly to create either a template or a crude statistical model (due to the paucity of data).
2. Robust training in which several (typically 2 to 4) versions of the sound pattern (usually extracted from the speech of a single talker) are used to create a single merged template or statistical model.
3. Clustering training in which a large number of versions of the sound pattern (extracted from a wide range of talkers) is used to create one or more templates or a reliable statistical model of the sound pattern.

In order to better understand how and why statistical models are so broadly used in speech recognition, we now formally define an important class of statistical models, namely the hidden Markov model (HMM) [6].

The HMM

The HMM is a statistical characterization of both the dynamics (time varying nature) and statics (the spectral characterization of sounds) of speech during speaking of a sub-word unit, a word, or even a phrase. The basic premise of the HMM is that a Markov chain can be used to describe the probabilistic nature of the temporal sequence of sounds in speech, i.e., the phonemes in the speech, via a probabilistic state sequence. The states in the sequence are not observed with certainty because the correspondence between linguistic sounds and the speech waveform is probabilistic in nature; hence the concept of a hidden model. Instead, the states manifest themselves through the second component of the HMM which is a set of output distributions governing the production of the speech features in each state (the spectral characterization of the sounds). In other words, the output distributions (which are observed) represent the local statistical knowledge of the speech pattern within the state, and the Markov chain characterizes, through a set of state transition probabilities, how these sound processes evolve from one sound to another. Integrated together, the HMM is particularly well suited for modeling speech processes.

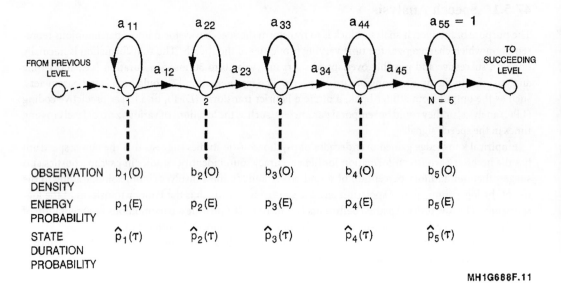

MH1G688F.11

FIGURE 47.3 Characterization of a word (or phrase, or subword) using a N(5) state, left-to-right, HMM, with continuous observation densities in each state of the model.

An example of an HMM of a speech pattern is shown in Fig. 47.3. The model has five states (corresponding to five distinct "sounds" or "phonemes" within the speech), and the state (corresponding to the sound being spoken) proceeds from left-to-right (as time progresses). Within each state (assumed to represent a stable acoustical distribution) the spectral features of the speech signal are characterized by a mixture Gaussian density of spectral features (called the observation density), along with an energy distribution, and a state duration probability. The states represent the changing temporal nature of the speech signal; hence indirectly they represent the speech sounds within the pattern.

The training problem for HMMs consists of estimating the parameters of the statistical distributions within each state (e.g., means, variances, mixture gains, etc.), along with the state transition probabilities for the composite HMM. Well-established techniques (e.g., the Baum-Welch method [10] or the segmental K-means method [11]) have been defined for doing this pattern training efficiently.

47.5.3 Pattern Matching

Pattern matching refers to the process of assessing the similarity between two speech patterns, one of which represents the unknown speech and one of which represents the reference pattern (derived from the training process) of each element that can be recognized. When the reference pattern is a "typical" utterance template, pattern matching produces a gross similarity (or dissimilarity) score. When the reference pattern consists of a probabilistic model, such as an HMM, the process of pattern matching is equivalent to using the statistical knowledge contained in the probabilistic model to assess the likelihood of the speech (which led to the model) being realized as the unknown pattern.

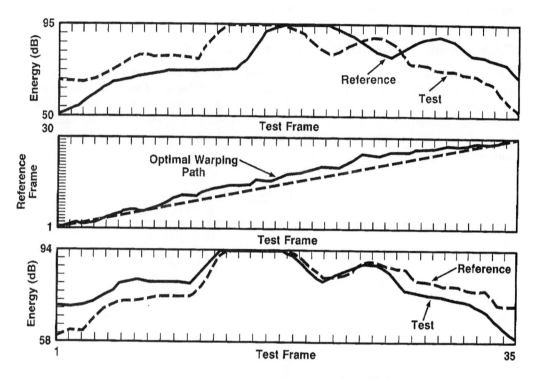

FIGURE 47.4 Results of time aligning two versions of the word "seven", showing linear alignment of the two utterances (top panel); optimal time-alignment path (middle panel); and nonlinearly aligned patterns (lower panel).

A major problem in comparing speech patterns is due to speaking rate variations. HMMs provide an implicit time normalization as part of the process for measuring likelihood. However, for template approaches, explicit time normalization is required. Figure 47.4 demonstrates the effect of explicit time normalization between two patterns representing isolated word utterances. The top panel of the figure shows the log energy contour of the two patterns (for the spoken word "seven") — one called the reference (known) pattern and the other called the test (or unknown input) pattern. It can be seen that the inherent duration of the two patterns, 30 and 35 frames (where each frame is a 15-ms segment of speech), is different and that linear alignment is grossly inadequate for internally aligning events within the two patterns (compare the locations of the vowel peaks in the two patterns). A basic principle of time alignment is to nonuniformly warp the time scale so as to achieve the best possible matching score between the two patterns (regardless of whether the two patterns are of the same word identity or not). This can be accomplished by a dynamic programming procedure, often called dynamic time warping (DTW) [12] when applied to speech template matching. The "optimal" nonlinear alignment result of dynamic time warping is shown at the bottom of Fig. 47.4 in contrast to the linear alignment of the patterns at the top. It is clear that the nonlinear alignment provides a more realistic measure of similarity between the patterns.

47.5.4   Decision Strategy

The decision strategy takes all the matching scores (from the unknown pattern to each of the stored reference patterns) into account, finds the "closest" match, and decides if the quality of the match is good enough to make a recognition decision. If not, the user is asked to provide another token of the speech (e.g., the word or phrase) for another recognition attempt. This is necessary because often the user may speak words that are incorrect in some sense (e.g., hesitation, incorrectly spoken word, etc.) or simply outside of the vocabulary of the recognition system.

47.5.5   Results of Isolated Word Recognition

Using the pattern recognition model of Fig. 47.2, and using either the non-parametric template approach or the statistical HMM method to derive reference patterns, a wide variety of tests of the recognizer have been performed on telephone speech with isolated word inputs in both speaker-dependent (SD) and speaker–independent (SI) modes. Vocabulary sizes have ranged from as few as 10 words (i.e., the digits zero–nine) to as many as 1109 words. Table 47.1 gives a summary of recognizer performance under the conditions described above.

TABLE 47.1   Performance of Isolated Word Recognizers

	Vocabulary	Mode	Word error rate (%)
10	Digits	SI	0.1
		SD	0.0
39	Alphadigits	SI	7.0
		SD	4.5
129	Airline terms	SI	2.9
		SD	1.0
1109	Basic English	SD	4.3

47.6   Connected Word Recognition

The systems we have been describing in previous sections have all been isolated word recognition systems. In this section we consider extensions of the basic processing methods described in previous sections in order to handle recognition of sequences of words, the so-called connected word recognition system.

The basic approach to connected word recognition is shown in Fig. 47.5. Assume we are given a fluently spoken sequence of words, represented by the (unknown) test pattern T, and we are also given a set of V reference patterns, $\{R_1, R_2, \ldots, R_V\}$ each representing one of the words in the vocabulary. The connected word recognition problem consists of finding the concatenated reference pattern, R^S, which best matches the test pattern, in the sense that the overall similarity between T and R^S is maximum over all sequence lengths and over all combinations of vocabulary words.

There are several problems associated with solving the connected word recognition problem, as formulated above. First of all, we do not know how many words were spoken; hence, we have to consider solutions with a range on the number of words in the utterance. Second, we do not know nor can we reliably find word boundaries within the test pattern. Hence, we cannot use word boundary information to segment the problem into simple "word-matching" recognition problems. Finally, since the combinatorics of trying

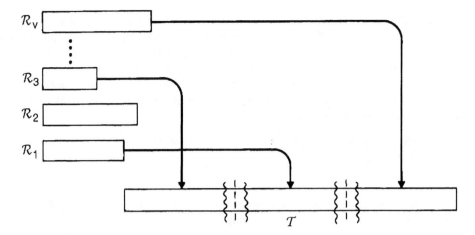

FIGURE 47.5 Illustration of the problem of matching a connected word string, spoken fluently, using whole word patterns concatenated together to provide the best match.

to solve the problem exhaustively (by trying to match every possible string) are exponential in nature, we need to devise efficient algorithms to solve this problem. Such efficient algorithms have been developed and they solve the connected word recognition problem by iteratively building up time-aligned matches between sequences of reference patterns and the unknown test pattern, one frame at a time [13, 14, 15].

47.6.1 Performance of Connected Word Recognizers

Typical recognition performance for connected word recognizers is given in Table 47.2 for a range of vocabularies, and for a range of associated tasks. In the next section we will see how we exploit linguistic constraints of the task to improve recognition accuracy for word strings beyond the level one would expect on the basis of word error rates of the system.

TABLE 47.2 Performance of Connected Word Recognizers

Vocabulary	Mode	Word error rate (%)	Task	String error rate (%)
10 Digits	SD	0.1	Variable length	0.4
	SI	0.2	digit strings (1–7 digits)	0.8
26 Letters of the alphabet	SD	10.0	Name retrieval	4.0
	SI	10.0	from directory of 1700 names	10.0
129 Airline terms	SD	0.1	Sentences in a	1.0
	SI	3.0	grammar	10.0

47.7 Continuous Speech Recognition

The techniques used in connected word recognition systems cannot be extended to the problem of continuous speech recognition for several reasons. First of all, as the size of the vocabulary of the recognizer grows, it becomes impractical to train patterns for each individual word in the vocabulary. Hence, continuous speech recognizers generally use sub-word speech units as the basic patterns to be trained, and use a lexicon to define the structure of word patterns in terms of the sub-word units. Second, the words spoken during continuous speech generally have a syntax associated with the word order, i.e., they are spoken according to a grammar. In order to achieve good recognition performance, account must be taken of the word grammar so as to constrain the set of possible recognized sentences. Finally, the spoken sentence often must make sense according to a semantic model of the task which the recognizer is asked to perform. Again, by explicitly including these semantic constraints on the spoken sentence, as part of the recognition process, performance of the system improves.

Based on the discussion above, there are three distinct new problems associated with continuous speech recognition [16], namely:

1. Choice of sub-word unit used to represent the sounds of speech, and methods of creating appropriate acoustic models for these sub-word units;
2. Choice of a representation of words in the recognition vocabulary, in terms of the sub-word units;
3. Choice of a method for integrating syntactic (and possibly semantic) information into the recognition process so as to properly constrain the sentences that are allowed by the system.

47.7.1   Sub-Word Speech Units and Acoustic Modeling

For the basic sub-word speech recognition unit, one could consider a range of linguistic units, including syllables, half syllables, dyads, dyphones, or phonemes. The most common choice is a simple phoneme set, which for English comprises about 40 to 50 units, depending on fine choices as to what constitutes a unique phoneme. Since the number of phonemes is limited, it is usually straightforward to collect sufficient speech training data for reliable estimation of statistical models of the phonemes. The resulting set of sub-word speech models are usually referred to as "context independent" phone-like units (CI-PLU) since each unit is trained independently of the context of neighboring units. The problem with using such CI-PLU models is that phonemes are highly variable according to different contexts, and therefore using models which cannot represent this variability properly leads to inferior speech recognition performance.

A straightforward way to improve the modeling of phonemes is to augment the CI-PLU set with phoneme models that are context dependent. In this manner, a target phoneme is modeled differently depending on the phonemes that precede and follow it. By using such context dependent PLUs (in addition to the CI-PLUs) the "resolution" of the acoustic models is increased, and the performance of the recognition system improves.

47.7.2   Word Modeling From Sub-Word Units

Once the base set of sub-word units is chosen, one can use standard lexical modeling techniques to represent words in terms of these units. The key problem here is variability of word pronunciation across talkers with different regional accents. Hence, for each word in the recognition vocabulary, the lexicon contains a baseform (or standard) pronunciation of the word, as well as alternative pronunciations, as appropriate.

The lexicon used in most recognition systems is extracted from a standard pronouncing dictionary, and each word pronunciation is represented as a linear sequence of phonemes. This lexical definition is basically data independent because no speech or text data are used to derive the pronunciation. Hence the lexical variability of a word in speech is characterized only indirectly through the sub-word unit models. To improve lexical modeling capability, the use of (multiple) pronunciation networks has been proposed [17].

47.7.3 Language Modeling Within the Recognizer

In order to determine the best match to a spoken sentence, a continuous speech recognition system has to evaluate both an acoustic match score (corresponding to the "local" acoustic matches of the words in the sentence) and a language match score (corresponding to the match of the words to the grammar and syntax of the task). The acoustic matching score is readily determined using dynamic programming methods much like those used in connected word recognition systems. The language match scores are computed according to a production model of the syntax and the semantics. Most often the language model is represented as a finite state network (FSN) for which the language score is computed according to arc scores along the best decoded path (according to an integrated model where acoustic and language modeling are combined) in the network. Other models of language include word pair models as well as N-gram word probabilities.

47.7.4 Performance of Continuous Speech Recognizers

Table 47.3 illustrates current capabilities in continuous speech recognition, for three distinct tasks, namely database access (Resource Management), natural language queries (ATIS) for air travel reservations, and read text from a set of business publications (NAB).

TABLE 47.3   Performance of Continuous Speech Recognition Systems

Task	Syntax	Mode	Vocabulary	Word error rate (%)
Resource management (DARPA)	Finite state grammar (perplexity = 60)	SI fluent input	1,000 Words	4.4
Air travel information system (DARPA)	Backoff trigram (perplexity = 18)	SI natural language	2,500 Words	3.6
North American business (DARPA)	Backoff 5-gram (perplexity = 173)	SI fluent input	60,000 Words	10.8

47.8   Speech Recognition System Issues

This section discusses some key issues in building "real world" speech recognition systems.

47.8.1   Robust Speech Recognition [18]

Robust speech recognition refers to the problem of designing an ASR system that works equally well in various unknown or adverse operating environments. Robustness is important because the performance of existing ASR systems, whose designs are predicated on known or clean environments, often degrades rapidly under field conditions.

There are basically four types of sound degradation, namely, noise, distortion, articulation effects, and pronunciation variations. Noise is an inevitable component of the acoustic environment and is normally considered additive with the speech. Distortion refers to modification to the spectral characteristics of the signal by the room, the transducer (microphone), the channel (e.g., transmission), etc. Articulation effects result from the factors that affect a talker's speaking manner when responding to a machine rather than a human. One well-known phenomenon is the Lombard effect which is related to the changes in articulation when the talker speaks in a noisy environment. Finally, different speakers will pronounce a word differently depending on the regional accent. These conditions are often not known *a priori* when the recognizer is trained in the laboratory and are often detrimental to the recognizer performance.

There are essentially two broad categories of techniques that have been proposed for dealing with adverse conditions. These are invariant methods and adaptive methods, respectively.

Invariant methods use speech features (or the associated similarity measures) that are invariant under a wide range of conditions, e.g., liftering and RASTA [19] (which suppress speech features that are more susceptible to signal variabilities), the short-time modified coherence (SMC) [20] (which has a built-in noise averaging advantage), and the Ensemble Interval Histogram (EIH) [21] (which mimics the human auditory mechanism). Robust distortion measures include the group-delay measure [22] and a family of distortion measures based on the projection operator [23] which were shown to be effective in conditions involving additive noise.

Adaptive methods differ from invariant methods in the way the characteristics of the operating environment are taken into account. Invariant methods assume no explicit knowledge of the signal environment, while adaptive methods attempt to estimate the adverse condition and adjust the signal (or the reference models) accordingly in order to achieve reliable matching results.

When channel or transducer distortions are the major factor, it is convenient to assume that the linear distortion effect appears as an additive signal bias in the cepstral domain. This distortion model leads to the method of cepstral mean subtraction and, more generally, signal bias removal [24] which makes a maximum likelihood estimate of the bias due to distortion and subtracts the estimated bias from the cepstral features before pattern matching is performed.

47.8.2   Speaker Adaptation [25]

Given sufficient training data, a SD recognition system usually performs better than a SI system for the same task. Many systems are designed for SI applications, however, due to the fact that it is often difficult to collect speaker-specific training data that would be adequate for reliable performance. One way to bridge the performance gap is to apply the method of speaker adaptation which uses a very limited amount of speaker-specific data to modify the model parameters of a SI recognition system in order to achieve a recognition accuracy, approaching that of a well-trained SD system.

47.8.3   Keyword Spotting [26] and Utterance Verification [27]

An automatic speech recognition system needs to have both high accuracy and a user-friendly interface in order to be acceptable to the users. One major component in a friendly user-interface is to allow the user to speak naturally and spontaneously without imposing a rigid speaking format. In a typical spontaneously spoken utterance, however, we usually observe various kinds of disfluency, such as hesitation and extraneous sounds such as um and ah and false starts, and unanticipated ambient noise, such as mouth clicks and lip smacks, etc. In the conventional paradigm, which formulates speech recognition as decoding of an unknown utterance into a contiguous sequence of phonetic units, the task is equivalent to designing an unlimited vocabulary continuous speech recognition and understanding system which is, unfortunately, beyond reach with today's technology.

One alternative to the above approach, particularly when implementing domain-specific services, is to focus on a finite set of vocabulary words most relevant to the intended task and design the system using the technology of keyword spotting and, more generally, utterance verification (UV). With UV incorporated into the speech recognition system, the user is allowed to speak spontaneously so long as the keywords appear somewhere in the spoken utterance. The system then detects and identifies the in-vocabulary words (i.e., keywords), while rejecting all other superfluous acoustic events in the utterance (which include out-of-vocabulary words, invalid inputs — any form of disfluency as well as lack of keywords — and ambient sounds). In such cases, no critical constraints are imposed on the users' speaking format, making the user interface natural and effective.

47.8.4 Barge-In

In human-human conversation, talkers often interrupt each other during speaking. This is called "barge-in". For human-machine interactions, in which machine prompts are often routine messages or instructions, the capability of allowing talkers to "barge in" becomes an important enabling technology for a natural human-machine interface.

Two key technologies are integrated in the implementation of "barge-in", namely, an echo canceler (to remove the spoken message from the machine to the recognizer) and a partial rejection mechanism.

With "barge-in", the recognizer needs to be activated and listen starting from the beginning of the system prompt. An echo canceler, with a proper double talk detector, is used to cancel the system prompt while attempting to detect if the near-end signal from the talker (i.e., speech to be recognized) is present. The tentatively detected signal is then passed through the recognizer with rejection thresholds to produce the partial recognition results. The rejection technique is critical because extraneous input is very likely to be present, both from the ambient background and from the talker (breathing, lip smacks, etc.), during the long period when the recognizer is activated.

47.9   Practical Issues in Speech Recognition

As progress is made in fundamental recognition technologies, we need to examine carefully the key attributes that a recognition machine must possess in order for it to be useful. These include: high recognition performance in terms of speed and accuracy, ease of use, and low cost. A recognizer must be able to deliver high recognition accuracy without excessive delay. A system that does not provide high performance often adds to users' frustration and may even be considered counterproductive. A recognition system must also be easy to use. The more naturally a system interacts with the user (e.g., does not require words in a sentence to be spoken in isolation), the higher the perceived effectiveness. Finally, the recognition system must be low cost to be competitive with alternative technologies such as keyboard or mouse devices in computer interface applications.

47.10   ASR Applications

Speech recognition has been successfully applied in a range of systems. We categorize these applications into five broad classes.

1. Office or business system. Typical applications include data entry onto forms, database management and control, keyboard enhancement, and dictation. Examples of voice-activated dictation machines include the Tangora system [28] and the Dragon Dictate system [29].

2. Manufacturing. ASR is used to provide "eyes-free, hands-free" monitoring of manufacturing processes (e.g., parts inspection) for quality control.

3. Telephone or telecommunications. Applications include automation of operator assisted services (the Voice Recognition Call Processing system by AT&T to automate operator service routing according to call types), inbound and outbound telemarketing, information services (the ANSER system by NTT for limited home banking services, the stock price quotation system by Bell Northern Research, Universal Card services by Conversant/AT&T for account information retrieval), voice dialing by name/number (AT&T VoiceLine, 800 Voice Calling services, Conversant FlexWord, etc.), directory assistance call completion, catalog ordering, and telephone calling feature enhancements (AT&T VIP — Voice Interactive Phone for easy activation of advanced calling features such as call waiting, call forwarding, etc. by voice rather than by keying in the code sequences).

4. Medical. The application is primarily in voice creation and editing of specialized medical reports (e.g., Kurzweil's system).

5. Other. This category includes voice controlled and operated toys and games, aids for the handicapped and voice control of non-essential functions in moving vehicles (such as climate control and the audio system).

References

[1] Hemdal, J.F. and Hughes, G.W., A feature based computer recognition program for the modeling of vowel perception, in *Models for the Perception of Speech and Visual Form*, Wathen-Dunn, W. Ed. MIT Press, Cambridge, MA.

[2] Itakura, F., Minimum prediction residual principle applied to speech recognition, *IEEE Trans. Acoustics, Speech, and Signal Processing*, ASSP-23,57–72, Feb. 1975.

[3] Lesser, V.R., Fennell, R.D., Erman, L.D. and Reddy D.R., Organization of the Hearsay-II Speech Understanding System, *IEEE Trans. Acoustics, Speech, and Signal Processing*, ASSP-23(1),11–23, 1975.

[4] Lippmann, R., An introduction to computing with neural networks, *IEEE ASSP Magazine*, 4(2),4–22, Apr. 1987.

[5] Rabiner, L.R. and Levinson, S.E., Isolated and connected word recognition — theory and selected applications, *IEEE Trans. Commun.*, COM-29(5),621–659, May 1981.

[6] Rabiner, L.R., A tutorial on hidden Markov models and selected applications in speech recognition, *Proc. IEEE*, 77(2),257–286, Feb. 1989.

[7] Rabiner, L.R. and Juang, B.H., *Fundamentals of Speech Recognition*, Prentice-Hall, Englewood Cliffs, NJ, 1993.

[8] Davis, S.B. and Mermelstein, P., Comparison of parametric representations for monosyllabic word recognition in continuously spoken sentences, *IEEE Trans. Acoustics, Speech, and Signal Processing*, ASSP-28(4),357–366, Aug. 1980.

[9] Furui, S., Speaker independent isolated word recognition using dynamic features of speech spectrum, *IEEE Trans. Acoustics, Speech, and Signal Processing*, ASSP-34(1),52–59, Feb. 1986.

[10] Baum, L.E., Petrie, T., Soules, G. and Weiss, N., A maximization technique occurring in the statistical analysis of probabilistic functions of Markov chains, *Ann. Math. Stat.*, 41(1),164–171, 1970.

[11] Juang, B.H. and Rabiner, L.R., The segmental k-means algorithm for estimating parameters of hidden Markov models, *IEEE Trans. Acoustics, Speech, and Signal Processing*, ASSP-30(9),1639–1641, Sept. 1990.

[12] Sakoe, H. and Chiba, S., Dynamic programming optimization for spoken word recognition, *IEEE Trans. Acoustics, Speech, and Signal Processing*, ASSP-26(1),43–49, Feb. 1978.

[13] Sakoe, H., Two-level DP matching — a dynamic programming-based pattern matching algorithm for connected word recognition, *IEEE Trans. Acoustics, Speech, and Signal Processing*, ASSP-27(6),588–595, Dec. 1979.

[14] Myers, C.S. and Rabiner, L.R., A level building dynamic time warping algorithm for connected word recognition, *IEEE Trans. Acoustics, Speech, and Signal Processing*, ASSP-29(3),351–363, June 1981.

[15] Bridle, J.S., Brown, M.D. and Chamberlain, R.M., An algorithm for connected word recognition, *Proc. ICASSP-82*, 899–902, May 1982.

[16] Lee, C.H., Rabiner, L.R. and Pieraccini, R., Speaker independent continuous speech recognition using continuous density hidden Markov models, in *Proc. NATO-ASI, Speech Recognition and Understanding: Recent Advances, Trends and Applications*, Laface, P. and DeMori, R., Eds., Springer-Verlag, Cetraro, Italy, 1992, 135–163.

[17] Riley, M.D., A statistical model for generating pronunciation networks, *Proc. ICASSP-91*, 2, 737–740, 1991.

[18] Juang, B.H., Speech recognition in adverse environments, *Computer Speech and Language*, 5, 275–294, 1991.

[19] Hermansky, H. et al., RASTA-PLP speech analysis technique, *Proc. ICASSP-29*, 121–124, 1992.

[20] Mansour, D. and Juang, B.H., The short-time modified coherence representation and noisy speech recognition, *IEEE Trans. Acoustics, Speech, and Signal Processing*, ASSP-37(6), 795–804, June 1989.

[21] Ghitza, O., Auditory nerve representation as a front-end for speech recognition in a noisy environment, *Comp. Speech Lang.*, 1(2), 109–130, Dec. 1986.

[22] Itakura, F. and Umezaki, T., Distance measure for speech recognition based on the smoothed group delay spectrum, *Proc. ICASSP-87*, 1257–1260, Apr. 1987.

[23] Mansour, D. and Juang, B.H., A family of distortion measures based upon projection operation for robust speech recognition, *Proc. ICASSP-88*, Apr. 1988. Also in *IEEE Trans.*, ASSP-37(11), 1659–1671, Nov. 1989.

[24] Rahim, M.G. and Juang, B.H., Signal bias removal for robust telephone speech recognition in adverse environments, *Proc. ICASSP-94*, Apr. 1994.

[25] Lee, C.-H., Lin, C.-H. and Juang, B.H., A study on speaker adaptation of the parameters of continuous density hidden Markov models, *IEEE Trans. Acoustics, Speech, and Signal Processing*, ASSP-39(4), 806–814, Apr. 1991.

[26] Wilpon, J.G., Rabiner, L.R., Lee, C.-H. and Goldman, E., Automatic recognition of keywords in unconstrained speech using hidden Markov models, *IEEE Trans. Acoustics, Speech, and Signal Processing*, 38(11), 1870–1878, Nov. 1990.

[27] Rahim, M., Lee, C.-H. and Juang, B.H., Robust utterance verification for connected digit recognition, *Proc. ICASSP-95*, WA02.02, May 1995.

[28] Jelinek, F., The development of an experimental discrete dictation recognizer, *IEEE Proc.*, 73(11), 1616–1624, Nov. 1985.

[29] Baker, J.M., Large vocabulary speech recognition prototype, *Proc. DARPA Speech and Natural Language Workshop*, 414–415, June 1990.

48

Speaker Verification

Sadaoki Furui
*Tokyo Institute
of Technology*

Aaron E. Rosenberg
AT&T Labs — Research

48.1 Introduction

Speaker recognition is the process of automatically extracting personal identity information by analysis of spoken utterances. In this section, speaker recognition is taken to be a general process whereas speaker identification and speaker verification refer to specific tasks or decision modes associated with this process. Speaker identification refers to the task of determining who is speaking and speaker verification is the task of validating a speaker's claimed identity.

Many applications have been considered for automatic speaker recognition. These include secure access control by voice, customizing services or information to individuals by voice, indexing or labeling speakers

0-8493-8572-5/98/$0.00+$.50
© 1998 by CRC Press LLC

in recorded conversations or dialogues, surveillance, and criminal and forensic investigations involving recorded voice samples. Currently, the most frequently mentioned application is access control. Access control applications include voice dialing, banking transactions over a telephone network, telephone shopping, database access services, information and reservation services, voice mail, and remote access to computers. Speaker recognition technology, as such, is expected to create new services and make our daily lives more convenient. Another potentially important application of speaker recognition technology is its use for forensic purposes [24].

For access control and other important applications, speaker recognition operates in a speaker verification task decision mode. For this reason the section is entitled speaker verification. However, the term speaker recognition is used frequently in this section when referring to general processes.

This section is not intended to be a comprehensive review of speaker recognition technology. Rather, it is intended to give an overview of recent advances and the problems that must be solved in the future. The reader is referred to papers by Doddington [4], Furui [10, 11, 12, 13], O'Shaughnessy [39], and Rosenberg and Soong [48] for more general reviews.

48.2   Personal Identity Characteristics

A universal human faculty is the ability to distinguish one person from another by personal identity characteristics. The most prominent of these characteristics are facial and vocal features. Organized, scientific efforts to make use of personal identifying characteristics for security and forensic purposes began about 100 years ago. The most successful such effort was fingerprint classification which has gained widespread use in forensic investigations.

Today, there is a rapidly growing technology based on biometrics, the measurement of human physiological or behavioral characteristics, for the purpose of identifying individuals or verifying the claimed or asserted identity of an individual [34]. The goal of these technological efforts is to produce completely automated systems for personal identity identification or verification that are convenient to use and offer high performance and reliability. Some of the personal identity characteristics which have received serious attention are blood typing, DNA analysis, hand shape, retinal and iris patterns, and signatures, in addition to fingerprints, facial features, and voice characteristics. In general, characteristics that are subject to the least amount of contamination or distortion and variability provide the greatest accuracy and reliability. Difficulties arise, for example, with smudged fingerprints, inconsistent signature handwriting, recording and channel distortions, and inconsistent speaking behavior for voice characteristics. Indeed, behavioral characteristics, intrinsic to signature and voice features, although potentially an important source of identifying information, are also subject to large amounts of variability from one sample to another.

The demand for effective biometric techniques for personal identity verification comes from forensic and security applications. For security applications, especially, there is a great need for techniques that are not intrusive, that are convenient and efficient, and are fully automated. For these reasons, techniques such as signature verification or speaker verification are attractive even if they are subject to more sources of variability than other techniques. Speaker verification, in addition, is particularly useful for remote access, since voice characteristics are easily recorded and transmitted over telephone lines.

48.3   Vocal Personal Identity Characteristics

Both physiology and behavior underly personal identity characteristics of the voice. Physiological correlates are associated with the size and configuration of the components of the vocal tract (see Fig. 48.1).

For example, variations in the size of vocal tract cavities are associated with characteristic variations in the spectral distributions in the speech signal for different speech sounds. The most prominent of these spectral features are the characteristic resonances associated with voiced speech sounds known as formants [6]. Vocal cord variations are associated with the average pitch or fundamental frequency of

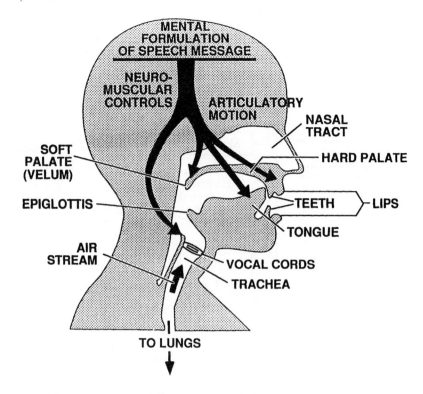

FIGURE 48.1 Simplified diagram of the human vocal tract showing how speech sounds are generated. The size and shape of the articulators differ from person to person.

voiced speech sounds. Variations in the velum and nasal cavities are associated with characteristic variations in the spectrum of nasalized speech sounds. Atypical anatomical variations, in the configuration of the teeth or the structure of the palate are associated with atypical speech sounds such as lisps or abnormal nasality.

Behavioral correlates of speaker identity in the speech signal are more difficult to specify. "Low level" behavioral characteristics are associated with individuality in articulating speech sounds, characteristic pitch contours, rhythm, timing, etc. Characteristics of speech that have to do with individual speech sounds, or phones, are referred to as "segmental", while those that pertain to speech phenomena over a sequence of phones are referred to as "suprasegmental". Phonetic or articulatory suprasegmental "settings" distinguishing speakers have been identified which are associated with characteristic "breathy", nasal, and other voice qualities [38]. "High-level" speaker behavioral characteristics refer to individual choice of words and phrases and other aspects of speaking styles.

48.4 Basic Elements of a Speaker Recognition System

The basic elements of a speaker recognition system are shown in Fig. 48.2. An input utterance from an unknown speaker is analyzed to extract speaker characteristic features. The measured features are compared with prototype features obtained from known speaker models.

Speaker recognition systems can operate in either an identification decision mode (Fig. 48.2(a)) or verification decision mode (Fig. 48.2(b)). The fundamental difference between these two modes is the number of decision alternatives.

In the identification mode, a speech sample from an unknown speaker is analyzed and compared with models of known speakers. The unknown speaker is identified as the speaker whose model best matches

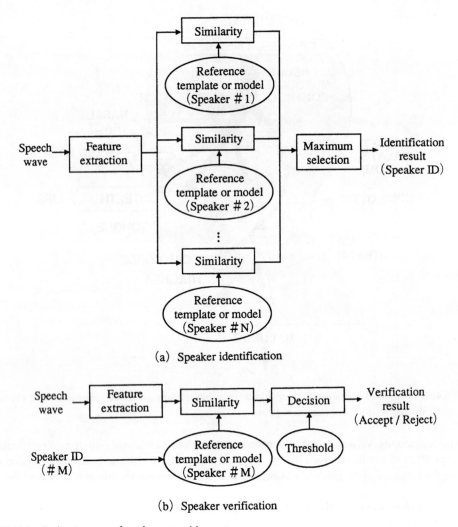

(a) Speaker identification

(b) Speaker verification

FIGURE 48.2   Basic structures of speaker recognition systems.

the input speech sample. In the "closed set" identification mode, the number of decision alternatives is equal to the size of the population. In the "open set" identification mode, a reference model for the unknown speaker may not exist. In this case, an additional alternative, "the unknown does not match any of the models", is required.

In the verification decision mode, an identity claim is made by or asserted for the unknown speaker. The unknown speaker's speech sample is compared with the model for the speaker whose identity is claimed. If the match is good enough, as indicated by passing a threshold test, the identity claim is verified. In the verification mode there are two decision alternatives, accept or reject the identity claim, regardless of the size of the population. Verification can be considered as a special case of the "open set" identification mode in which the known population size is one.

Crucial to the operation of a speaker recognition system is the establishment and maintenance of speaker models. One or more enrollment sessions are required in which training utterances are obtained from known speakers. Features are extracted from the training utterances and compiled into models. In addition, if the system operates in the "open set" or verification decision mode, decision thresholds must also be set. Many speaker recognition systems include an updating facility in which test utterances are used to adapt speaker models and decision thresholds.

A list of terms commonly found in the speaker recognition literature can be found at the end of this chapter. In the remaining sections of the chapter, the following subjects are treated: how speaker characteristic features are extracted from speech signals, how these features are used to represent speakers, how speaker models are constructed and maintained, how speech utterances from unknown speakers are compared with speaker models and scored to make speaker recognition decisions, and how speaker verification performance is measured. The chapter concludes with a discussion of outstanding issues in speaker recognition.

48.5 Extracting Speaker Information from the Speech Signal

Explicit measurements of speaker characteristics in the speech signal are often difficult to carry out. Segmenting, labeling, and measuring specific segmental speech events that characterize speakers, such as nasalized speech sounds, is difficult because of variable speech behavior and variable and distorted recording and transmission conditions. Overall qualities, such as breathiness, are difficult to correlate with specific speech signal measurements and are subject to variability in the same way as segmental speech events.

Even though voice characteristics are difficult to specify and measure explicitly, most characteristics are captured implicitly in the kinds of speech measurements that can be performed relatively easily. Such measurements as short-time and long-time spectral energy, overall energy, and fundamental frequency are relatively easy to obtain. They can often resolve differences in speaker characteristics surpassing human discriminability. Although subject to distortion and variability, features based on these analysis tools form the basis for most automatic speaker recognition systems.

The most important analysis tool is short-time spectral analysis. It is no coincidence that short-time spectral analysis also forms the basis for most speech recognition systems [42]. Short-time spectral analysis not only resolves the characteristics that differentiate one speech sound from another, but also many of the characteristics already mentioned that differentiate one speaker from another. There are two principal modes of short-time spectral analysis: filter bank analysis and linear predictive coding (LPC) analysis.

In filter bank analysis, the speech signal is passed through a bank of bandpass filters covering the available range of frequencies associated with the signal. Typically, this range is 200 to 3,000 Hz for telephone band speech and 50 to 8,000 Hz for wide band speech. A typical filter bank for wide band speech contains 16 bandpass filters spaced uniformly 500 Hz apart. The output of each filter is usually implemented as a windowed, short-time Fourier transform [using fast Fourier transform (FFT) techniques] at the center frequency of the filter. The speech is typically windowed using a 10 to 30 ms Hamming window. Instead of uniformly spacing the bandpass filters, a nonuniform spacing is often carried out reflecting perceptual criteria that allot approximately equal perceptual contributions for each such filter. Such mel scale or bark scale filters [42] provide a spacing linear in frequency below 1000 Hz and logarithmic above.

LPC-based spectral analysis is widely used for speech and speaker recognition. The LPC model of the speech signal specifies that a speech sample at time t, $s(t)$, can be represented as a linear sum of the p previous samples plus an excitation term, as follows:

$$s(t) = a_1 s(t-1) + a_2 s(t-2) + \cdots + a_p s(t-p) + Gu(t) \tag{48.1}$$

The LPC coefficients, a_i, are computed by solving a set of linear equations resulting from the minimization of the mean-squared error between the signal at time t and the linearly predicted estimate of the signal. Two generally used methods for solving the equations, the autocorrelation method and the covariance method, are described in Rabiner and Juang [42].

The LPC representation is computationally efficient and easily convertible to other types of spectral representations. While the computational advantage is less important today than it was for early digital implementations of speech and speaker recognition systems, LPC analysis competes well with other spectral analysis techniques and continues to be widely used.

An important spectral representation for speech and speaker recognition is the cepstrum. The cepstrum is the (inverse) Fourier transform of the log of the signal spectrum. Thus, the log spectrum can be represented as a Fourier series expansion in terms of a set of cepstral coefficients c_n

$$\log S(\omega) = \sum_{n=-\infty}^{\infty} c_n e^{-nj\omega} \tag{48.2}$$

The cepstrum can be calculated from the filter-bank spectrum or from LPC coefficients by a recursion formula [42]. In the latter case it is known as the LPC cepstrum indicating that it is based on an all-pole representation of the speech signal. The cepstrum has many interesting properties. Since the cepstrum represents the log of the signal spectrum, signals that can be represented as the cascade of two effects which are products in the spectral domain are additive in the cepstral domain. Also, pitch harmonics, which produce prominent ripples in the spectral envelope, are associated with high order cepstral coefficients. Thus, the set of cepstral coefficients truncated, for example, at order 12 to 24 can be used to reconstruct a relatively smooth version of the speech spectrum. The spectral envelope obtained is associated with vocal tract resonances and does not have the variable, oscillatory effects of the pitch excitation. It is considered that one of the reasons that cepstral representation has been found to be more effective than other representations for speech and speaker recognition is this property of separability of source and tract. Since the excitation function is considered to have speaker dependent characteristics, it may seem contradictory that a representation which largely removes these effects works well for speaker recognition. However, in short-time spectral analysis the effects of the source spectrum are highly variable so that they are not especially effective in providing consistent representations of the source spectrum.

Other spectral features such as PARCOR coefficients, log area ratio coefficients, LSP (line spectral pair coefficients), have been used for both speech and speaker recognition [42]. Generally speaking, however, the cepstral representation is most widely used and is usually associated with better speaker recognition performance than other representations.

Cruder measures of spectral energy, such as waveform zero-crossing or level-crossing measurements have also been used for speech and speaker recognition in the interest of saving computation with some success.

Additional features have been proposed for speaker recognition which are not used often or considered to be marginally useful for speech recognition. For example, pitch and energy features, particularly when measured as a function of time over a sufficiently long utterance, have been shown to be useful for speaker recognition [27]. Such time sequences or "contours" are thought to represent characteristic speaking inflections and rhythms associated with individual speaking behavior. Pitch and energy measurements have an advantage over short-time spectral measurements in that they are more robust to many different kinds of transmission and recording variations and distortions since they are not sensitive to spectral amplitude variability. However, since speaking behavior can be highly variable due to both voluntary and involuntary activity, pitch and energy can acquire more variability than short-time spectral features and are more susceptible to imitation.

The time course of feature measurements, as represented by so-called feature contours, provides valuable speaker characterizing information. This is because such contours provide overall, suprasegmental information characterizing speaking behavior and also because they contain information on a more local, segmental time scale describing transitions from one speech sound to another. This latter kind of information can be obtained explicitly by measuring the local trajectory in time of a measured feature at each analysis frame. Such measurements can be obtained by averaging successive differences of the feature in a window around each analysis frame, or by fitting a polynomial in time to the successive feature measurements in the window. The window size is typically 5 to 9 analysis frames. The polynomial fit provides a less noisy estimate of the trajectory than averaging successive differences. The order of the polynomial is typically 1 or 2, and the polynomial coefficients are called delta- and delta-delta-feature coefficients. It has been shown in experiments that such dynamic feature measurements are fairly uncorrelated with the original static feature measurements and provide improved speech and speaker recognition performance [9].

48.6 Feature Similarity Measurements

Much of the originality and distinctiveness in the design of a speaker recognition system is found in how features are combined and compared with reference models. Underlying this design is the basic representation of features in some space and the formation of a distance or distortion measurement to use when one set of features is compared with another. The distortion measure can be used to partition the feature vectors representing a speaker's utterances into regions representative of the most prominent speech sounds for that speaker, as in the vector quantization (VQ) codebook representation (Section 48.9.2). It can be used to segment utterances into speech sound units. And it can be used to score an unknown speaker's utterances against a known speaker's utterance models.

A general approach for calculating a distance between two feature vectors is to make use of a distance metric from the family of L_p norm distances d_p, such as the absolute value of the difference between the feature vectors

$$d_1 = \sum_{i=1}^{D} |f_i - f_i'| \tag{48.3}$$

or the Euclidean distance

$$d_2 = \sum_{i=1}^{D} \left(f_i - f_i'\right)^2 \tag{48.4}$$

where f_i, f_i', $i = 1, 2, \ldots, D$ are the coefficients of two feature vectors f and f'. The feature vectors, for example, could comprise filter-bank outputs or cepstral coefficients described in the previous section. (It is not common, however, to use filter bank outputs directly, as previously mentioned, because of the variability associated with these features due to harmonics from the pitch excitation.)

For example, a weighted Euclidean distance distortion measure for cepstral features of the form

$$d_{cw}^2 = \sum_{i=1}^{D} \left(w_i \left(c_i - c_i'\right)\right)^2 \tag{48.5}$$

where

$$w_i = 1/\sigma_i \tag{48.6}$$

and σ_i^2 is an estimate of the variance of the ith coefficient has been shown to provide good performance for both speech and speaker recognition. A still more general formulation is the Mahalanobis distance formulation which accounts for interactions between coefficients with a full covariance matrix.

An alternate approach to comparing vectors in a feature space with a distortion measurement is to establish a probabilistic formulation of the feature space. It is assumed that the feature vectors in a subspace associated with, for example, a particular speech sound for a particular speaker, can be specified by some probability distribution. A common assumption is that the feature vector is a random variable x whose probability distribution is Gaussian

$$p(x|\lambda) = \frac{1}{(2\pi)^{D/2} |\Sigma|^{1/2}} \exp\left[-\frac{1}{2} (x - \mu)^T \Sigma^{-1} (x - \mu)\right] \tag{48.7}$$

where λ represents the parameters of the distribution, which are the mean vector μ and covariance matrix Σ.

When x is a feature vector sample, $p(x|\lambda)$ is referred to as the likelihood of x with respect to λ. Suppose there is a population of n speakers each modeled by a Gaussian distribution of feature vectors, $\lambda_i, i = 1, 2, \ldots, n$. In the maximum likelihood formulation, a sample x is associated with speaker I if

$$p(x|\lambda_I) > p(x|\lambda_i), \text{ for all } i \neq I \qquad (48.8)$$

where $p(x|\lambda i)$ is the likelihood of the test vector x for speaker model λ_i. It is common to use log likelihoods to evaluate Gaussian models. From Eq. (48.7)

$$L(x|\lambda_i) = \log p(x|\lambda_i) = -\frac{D}{2} \log 2\pi - \frac{1}{2} \log |\Sigma_i| - \frac{1}{2}(x - \mu_i)^T \Sigma_i^{-1}(x - \mu_i) \qquad (48.9)$$

It can be seen from Eq. (48.9) that, using log likelihoods, the maximum likelihood classifier is equivalent to the minimum distance classifier using a Mahalanobis distance formulation.

A more general probabilistic formulation is the Gaussian mixture distribution of a feature vector x

$$p(x|\lambda) = \sum_{i=1}^{M} w_i b_i(x) \qquad (48.10)$$

where $b_i(x)$ is the Gaussian probability density function with mean μ_i and covariance Σ_i, w_i is the weight associated with the ith component, and M is the number of Gaussian components in the mixture. The weights w_i are constrained so that $\sum_{i=1}^{n} w_i = 1$. The model parameters λ are

$$\lambda = \{\mu_i, \Sigma_i, w_i, i = 1, 2, \ldots, M\} \qquad (48.11)$$

The Gaussian mixture probability function is capable of approximating a wide variety of smooth, continuous, probability functions.

48.7 Units of Speech for Representing Speakers

An important consideration in the design of a speaker recognition system is the choice of a speech unit to model a speaker's utterances. The choice of units includes phonetic or linguistic units such as whole sentences or phrases, words, syllables, and phone-like units. It also includes acoustic units such as subword segments, segmented from utterances and labeled on the basis of acoustic rather than phonetic criteria. Some speaker recognition systems model speakers directly from single feature vectors rather than through an intermediate speech unit representation. Such systems usually operate in a text independent mode (see Sections 48.8 and 48.9) and seek to obtain a general model of a speaker's utterances from a usually large number of training feature vectors. Direct models might include long-time averages, VQ codebooks, segment and matrix quantization codebooks, or Gaussian mixture models of the feature vectors.

Most speech recognizers of moderate to large vocabulary are based on subword units such as phones so that large numbers of utterances transcribed as sequences of phones can be represented as concatenations of phone models. For speaker recognition, there is no absolute need to represent utterances in terms of phones or other phonetically based units because there is no absolute need to account for the linguistic or phonetic content of utterances in order to build speaker recognition models. Generally speaking, systems in which phonetic representations are used are more complex than other representations because they require phonetic transcriptions for both training and testing utterances and because they require accurate and reliable segmentations of utterances in terms of these units. The case in which phonetic representations are required for speaker recognition is the same as for speech recognition: where there is a need to represent utterances as concatenations of smaller units. Speaker recognition systems based on subword units have been described by Rosenberg et al. [46] and Matsui and Furui [31].

48.8   Input Modes

Speaker recognition systems typically operate in one of two input modes: text dependent or text independent. In the text-dependent mode, speakers must provide utterances of the same text for both training and recognition trials. In the text-independent mode, speakers are not constrained to provide specific texts in recognition trials. Since the text-dependent mode can directly exploit the voice individuality associated with each phoneme or syllable, it generally achieves higher recognition performance than the text-independent mode.

48.8.1   Text-Dependent (Fixed Passwords)

The structure of a system using fixed passwords is rather simple; input speech is time aligned with reference templates or models created by using training utterances for the passwords. If the fixed passwords are different from speaker to speaker, the difference can also be used as additional individual information. This helps to increase performance.

48.8.2   Text Independent (No Specified Passwords)

There are several applications in which predetermined passwords cannot be used. In addition, human beings can recognize speakers irrespective of the content of the utterance. Therefore, text-independent methods have recently been actively investigated. Another advantage of text-independent recognition is that it can be done sequentially, until a desired significance level is reached, without the annoyance of having to repeat passwords again and again.

48.8.3   Text Dependent (Randomly Prompted Passwords)

Both text-dependent and independent methods have a potentially serious problem. Namely, these systems can be defeated because someone who plays back the recorded voice of a registered speaker uttering key words or sentences into the microphone could be accepted as the registered speaker. To cope with this problem, there are methods in which a small set of words, such as digits, are used as key words and each user is prompted to utter a given sequence of key words that is randomly chosen every time the system is used [20, 47].

Recently, a text-prompted speaker recognition method was proposed in which password sentences are completely changed every time [31, 33]. The system accepts the input utterance only when it judges that the registered speaker uttered the prompted sentence. Because the vocabulary is unlimited, prospective impostors cannot know in advance the sentence they will be prompted to say. This method cannot only accurately recognize speakers, but can also reject utterances whose text differs from the prompted text, even if it is uttered by a registered speaker. Thus, a recorded and played-back voice can be correctly rejected.

48.9   Representations

48.9.1   Representations That Preserve Temporal Characteristics

The most common approach to automatic speaker recognition in the text-dependent mode uses representations that preserve temporal characteristics. Each speaker is represented by a sequence of feature vectors (generally, short-term spectral feature vectors), analyzed for each test word or phrase. This approach is usually based on template matching techniques in which the time axes of an input speech sample and each reference template of registered speakers are aligned, and the similarity between them accumulated from the beginning to the end of the utterance is calculated.

Trial-to-trial timing variations of utterances of the same talker, both local and overall, can be normalized by aligning the analyzed feature vector sequence of a test utterance to the template feature vector sequence using a dynamic programming (DP) time warping algorithm or DTW [11, 42]. Since the sequence of phonetic events is the same for training and testing, there is an overall similarity among these sequences of feature vectors. Ideally the intra-speaker differences are significantly smaller than the inter-speaker differences.

Figure 48.3 shows an example of a typical structure of the DTW-based system [9]. Initially, 10 LPC cepstral coefficients are extracted every 10 ms from a short sentence of speech. The spectral equalization technique, which is described in Section 48.12.1, is applied to each cepstral coefficient to compensate for transmission distortion and intraspeaker variability. In addition to the normalized cepstral coefficients, delta-cepstral and delta-delta-cepstral coefficients (polynomial expansion coefficients) are extracted every 10 ms. The time function of the set of parameters is brought into time registration with the reference template in order to calculate the distance between them. The overall distance is then compared with a threshold for the verification decision.

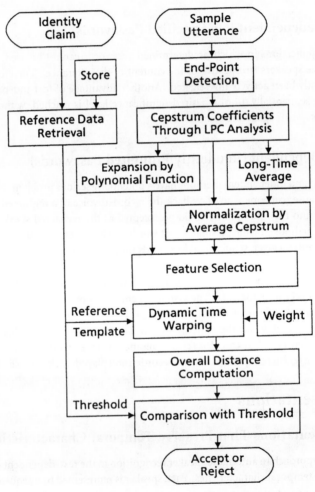

FIGURE 48.3   Typical structure of the DTW-based text-dependent speaker verification system.

Another approach using representations that preserve temporal characteristics is based on the HMM (hidden Markov model) technique [42]. In this approach, a reference model for each speaker is represented by an HMM instead of directly using a time series of feature vectors. An HMM can efficiently model statistical variation in spectral features. Therefore, HMM-based methods have achieved significantly better recognition accuracies than the DTW-based methods [36, 47, 53].

48.9.2 Representations That do not Preserve Temporal Characteristics

In a text-independent system, the words or phrases used in recognition trials generally cannot be predicted. Therefore, it is impossible to model or match speech events at the level of words or phrases. Classical text-independent speaker recognition techniques are based on measurements for which the time dimension is collapsed. Recently text-independent speaker verification techniques based on short duration speech events have been studied. The new approaches extract and measure salient acoustic and phonetic events. The bases for these approaches lie in statistical techniques for extracting and modeling reduced sets of optimally representative feature vectors or feature vector sequences or segments. These techniques fall under the related categories of vector quantization (VQ), matrix and segment quantization, probabilistic mixture models, and HMM.

A set of short-term training feature vectors of a speaker can be used directly to represent the essential characteristics of that speaker. However, such a direct representation is impractical when the number of training vectors is large, since the memory and amount of computation required become prohibitively large. Therefore, efficient ways of compressing the training data have been tried using VQ techniques.

In this method, VQ codebooks consisting of a small number of representative feature vectors are used as an efficient means of characterizing speaker-specific features [25, 29, 45, 52]. A speaker-specific codebook is generated by clustering the training feature vectors of each speaker. In the recognition stage, an input utterance is vector-quantized using the codebook of each reference speaker, and the VQ distortion accumulated over the entire input utterance is used in making the recognition decision.

In contrast with the memoryless VQ-based method, source coding algorithms with memory have also been studied using a segment (matrix) quantization technique [22]. The advantage of a segment quantization codebook over a VQ codebook representation is its characterization of the sequential nature of speech events. Higgins and Wohlford [19] proposed a segment modeling procedure for constructing a set of representative time normalized segments, which they called "filler templates". The procedure, a combination of K-means clustering and dynamic programming time alignment, provides a way to handle temporal variation.

On a longer time scale, temporal variation in speech signal parameters can be represented by stochastic Markovian transitions between states. Poritz [41] proposed using a five-state ergodic HMM (i.e., all possible transitions between states are allowed) to classify speech segments into one of the broad phonetic categories corresponding to the HMM states. A linear predictive HMM was used to characterize the output probability function. Poritz characterized the automatically obtained categories as strong voicing, silence, nasal/liquid, stop burst/post silence, and frication.

Savic and Gupta [50] also used a five-state ergodic linear predictive HMM for broad phonetic categorization. After identifying frames belonging to particular phonetic categories, feature selection was performed. In the training phase, reference templates are generated and verification thresholds are computed for each phonetic category. In the verification phase, after the phonetic categorization, a comparison with the reference template for each particular category provides a verification score for that category. The final verification score is a weighted linear combination of the scores for each category. The weights are chosen to reflect the effectiveness of particular categories of phonemes in discriminating between speakers and are adjusted to maximize the verification performance.

The performances of speaker recognition based on a VQ-based method and that using discrete/continuous ergodic HMM-based methods have been compared, in particular from the viewpoint of robustness against utterance variations [30]. It was shown that a continuous ergodic HMM method is far superior to a discrete ergodic HMM method, and that a continuous ergodic HMM method is as robust as a VQ-based method when enough training data is available. However, when little data is available, the VQ-based method is more robust than a continuous HMM method. It was also shown that the information on transitions between different states is ineffective for text-independent speaker recognition, so the speaker recognition rates using a continuous ergodic HMM are strongly correlated with the total number of mixtures, irrespective of the number of states.

Rose and Reynolds [44] investigated a technique based on maximum likelihood estimation of a Gaussian mixture model representation of speaker identity. This method corresponds to the single-state continuous ergodic HMM. Gaussian mixtures are noted for their robustness as a parametric model and for their ability to form smooth estimates of rather arbitrary underlying densities.

Traditionally, long-term sample statistics of various spectral features, e.g., the mean and variance of spectral components averaged over a series of utterances have been used for speaker recognition [7, 28]. However, long-term spectral averages are extreme condensations of the spectral characteristics of a speaker's utterances and, as such, lack the discriminating power obtained in the sequence of short-term spectral features used as models in text-dependent systems. Moreover, recognition based on long-term spectral averages tends to be less tolerant of recording and transmission variations since many of these variations are themselves associated with long-term spectral averages.

Studies on the use of statistical dynamic features have also been reported. Montacie et al. [35] used a multivariate auto-regression (MAR) model to characterize speakers, and reported good speaker recognition results. Griffin et al. [18] studied distance measures for the MAR-based method, and reported that the identification and verification rates were almost the same as those obtained by an HMM-based method. In these experiments, the MAR model was applied to the time series of cepstral vectors. It was also reported that the optimum order of the MAR model was 2 or 3, and that distance normalization was essential to obtain good results in speaker verification.

Speaker recognition based on feed-forward neural net models has been investigated [40]. Each registered speaker has a personalized neural net that is trained to be activated only by that speaker's utterances. It is assumed that including speech from many people in the training data of each net enables direct modeling of the differences between the registered person's speech and an impostor's speech. It has been found that while the net architecture and the amount of training utterances strongly affect the recognition performance, it is comparable to the performance of the VQ approach based on personalized codebooks.

As an expansion of the VQ-based method, a connectionist approach has also been developed based on the learning vector quantization (LVQ) algorithm [2].

48.10 Optimizing Criteria for Model Construction

The establishment of effective speaker models is fundamental for good performing speaker recognition. In the previous section, we described different kinds of representations for speaker models. In this section, we describe some of the techniques for optimizing model representations.

Statistical and discriminative training techniques are based on optimizing criteria for constructing models. Typical criteria for optimizing the model parameters include likelihood maximization, *a posteriori* probability maximization, linear discriminant analysis (LDA), and discriminative error minimization.

The maximum likelihood (ML) approach is widely used in statistical model parameter estimation, such as for HMM parameter training [42]. Although ML estimation has good asymptotic properties, it often requires a large amount of training data to achieve reliable results.

Linear discriminant analysis techniques have been used in a speaker verification system reported by Netsch and Doddington [37]. A set of LDA weights applied to word-level feature vectors is found by maximizing the ratio of between-speaker to within speaker covariances obtained from pooled customer and impostor training data.

In contrast to conventional ML training, which estimates a model based only on training utterances from the same speaker, discriminative training takes into account the models of other competing speakers and formulates the optimization criterion so that speaker separation is enhanced. In the minimum classification error/generalized probabilistic descent (MCE/GPD) method [23], the optimum solution is obtained with a steepest descent algorithm minimizing recognition error rate for the training data. Unlike the statistical framework, this method does not require estimating the probability distributions, which usually cannot be reliably obtained. However, discriminative training methods require a sufficient amount of representative reference speaker training data, which is often difficult to obtain, to be effective. This method has been applied to speaker recognition with good results [26].

Neural nets are capable of discriminative training. Various investigations have been conducted to cope with training problems, such as overtuning to training data. A typical implementation is the neural tree network (NTN) classifier [5]. In this system each speaker is represented by a VQ codebook and an NTN classifier. The NTN classifier is trained on both customer and impostor training data.

48.11 Model Training and Updating

Trial-to-trial variations have a major impact on the performance of speaker recognition systems. Variations arise from the speaker himself/herself, from differences in recording and transmission conditions, and from noise. Speakers cannot repeat an utterance precisely the same way from trial to trial. It has been found that tokens of the same utterance recorded in one session are much more highly correlated than tokens recorded in separate sessions. There are also long-term trends in voices [7, 8].

There are two approaches for dealing with variability. One, discussed in this section, is to construct and update models to accommodate variability. Another, discussed in the next section, is to condition or normalize the acoustic features or the recognition scores to manage some sources of variability.

Training difficulties are closely related to training conditions. The key training conditions include the number of training sessions, the number of tokens, and transmission channel and recording conditions. Tokens of the same utterance recorded in one session are much more highly correlated than tokens recorded in separate sessions. Therefore, wherever it is practicable, it is desirable to collect training utterances for each speaker in multiple sessions to accommodate trial-to-trial variability. For example, Gish and Schmidt [17] report a text-independent speaker identification system in which multiple models of a speaker are constructed from multiple session training utterances.

It is inconvenient to request speakers to utter training tokens at many sessions before being allowed to use a speaker recognition system. It is possible, however, to compensate for small amounts of training data collected in a small number of enrollment sessions, often only one, by updating models with utterances collected in recognition sessions. Updating is especially important for speaker verification systems used for access control, where it can be expected that user trials will take place periodically over long periods of time in which trial-to-trial variations are likely. Updating models in this way incorporates into the models the effects of trial-to-trial variations we have mentioned. Rosenberg and Soong [45] reported significant improvements in performance in a text independent speaker verification system based on VQ speaker models in which the VQ codebooks were updated with test utterance data. A hazard associated with updating models using test session data is the possibility of adapting a customer model with impostor data.

48.12 Signal Feature and Score Normalization Techniques

Some sources of variability can be managed by normalization techniques applied to signal features or the scores. For example, as noted in Section 48.9.1, it is possible to adjust for trial-to-trial timing variations by aligning test utterances with model parameters using DTW or Viterbi alignment techniques.

48.12.1 Signal Feature Normalization

A typical normalization technique in the parameter domain, spectral equalization, also called "blind equalization" or "blind deconvolution", has been shown to be effective in reducing linear channel effects and long-term spectral variation [1, 9]. This method is especially effective for text-dependent speaker recognition applications using sufficiently long utterances. In this method, cepstral coefficients are averaged over the duration of an entire utterance, and the averaged values are subtracted from the cepstral coefficients of each frame. This method can compensate fairly well for additive variation in the log spectral domain. However, it unavoidably removes some text-dependent and speaker specific features, and is therefore inappropriate for short utterances in speaker recognition applications.

Gish [15] demonstrated that by simply prefiltering the speech transmitted over different telephone lines with a fixed filter, text-independent speaker recognition performance can be significantly improved. Gish et al. [14, 16] have also proposed using multi-variate Gaussian probability density functions to model channels statistically. This can be achieved if enough training samples of channels to be modeled are available. It was shown that time derivatives (short-time spectral dynamic features) of cepstral coefficients (delta-cepstral coefficients) are resistant to linear channel mismatch between training and testing [51].

48.12.2 Likelihood and Normalized Scores

Likelihood measures (see Section 48.6) are commonly used in speaker recognition systems based on statistical models, such as HMMs, to compare test utterances with models. Since likelihood values are highly subject to inter-session variability, it is essential to normalize these variations.

Higgins et al. [20] proposed a normalization method that uses a likelihood ratio. The likelihood ratio is defined as the ratio of the conditional probability of the observed measurements of the utterance given the claimed identity to the conditional probability of the observed measurements given the speaker is an impostor. A mathematical expression in terms of log likelihoods is given as

$$\log l(x) = \log p\left(x | S = S_c\right) - \log p\left(x | S \neq S_c\right) \tag{48.12}$$

Generally, a positive value of $\log l$ indicates a valid claim, whereas a negative value indicates an impostor. The second term of the right hand side of Eq. (48.12) is called the normalization term. Some proposals for calculating the normalization term are described.

The density at point x for all speakers other than the true speaker S can be dominated by the density for the nearest reference speaker, if we assume that the set of reference speakers is representative of all speakers. We can, therefore, arrive at the decision criterion

$$\log l(x) = \log p\left(x | S = S_c\right) - \max_{S \in Ref, S \neq S_c} \log p(x | S) \tag{48.13}$$

This shows that likelihood ratio normalization is approximately equal to optimal scoring in Bayes' sense. However, this decision criterion is unrealistic for two reasons. First, in order to choose the nearest reference speaker, conditional probabilities must be calculated for all the reference speakers, which involves a high computational cost. Second, the maximum conditional probability value is rather variable from speaker to speaker, depending on how close the nearest speaker is in the reference set.

48.12.3 Cohort or Speaker Background Models

A set of speakers, "cohort speakers", has been chosen for calculating the normalization term of Eq. (48.12). Higgins et al. proposed the use of speakers that are representative of the population near the claimed speaker:

$$\log l(x) = \log p\left(x | S = S_c\right) - \log \sum_{S \in Cohort, S \neq S_c} p(x | S) \tag{48.14}$$

Experimental results show that this normalization method improves speaker separability and reduces the need for speaker-dependent or text-dependent thresholding, compared with scoring using only the model of the claimed speaker. Another experiment in which the size of the cohort speaker set was varied from 1 to 5 showed that speaker verification performance increases as a function of the cohort size, and that the use of normalization significantly compensates for the degradation obtained by comparing verification utterances recorded using an electret microphone with models constructed from training utterances recorded with a carbon button microphone [49].

This method using speakers that are representative of the population near the claimed speaker is expected to increase the selectivity of the algorithm against voices similar to the claimed speaker. However, this method has a serious problem in that it is vulnerable to attack by impostors of the opposite gender. Since the cohorts generally model only same-gender speakers, the probability of opposite-gender impostor speech is not well modeled, and the likelihood ratio is based on the tails of distributions giving rise to unreliable values. Another way of choosing the cohort speaker set is to use speakers who are typical of the general population. Reynolds [43] reported that a randomly selected, gender-balanced background speaker population outperformed a population near the claimed speaker.

Matsui and Furui [31] proposed a normalization method based on *a posteriori* probability:

$$\log l(x) = \log p\,(x|S = S_c) - \log \sum_{S \in Ref} p(x|S) \tag{48.15}$$

The difference between the normalization method based on the likelihood ratio and that based on *a posteriori* probability is in whether or not the claimed speaker is included in the speaker set for normalization; the cohort speaker set in the likelihood-ratio-based method does not include the claimed speaker, whereas the normalization term for the *a posteriori*-probability-based method is calculated using all the reference speakers, including the claimed speaker. Matsui and Furui approximated the summation in Eq. (48.15) by the summation over a small set of speakers having relatively high likelihood values. Experimental results indicate that the two normalization methods are almost equally effective.

Carey and Parris [3] proposed a method in which the normalization term is approximated by the likelihood for a world model representing the population in general. This method has the advantage that the computational cost for calculating the normalization term is much smaller than in the original method since it does not need to sum the likelihood values for cohort speakers. Matsui and Furui [32] recently proposed a new method based on tied-mixture HMMs in which the world model is made as a pooled mixture model representing the parameter distribution for all the registered speakers. This model is created by averaging the mixture-weighting factors of each registered speaker calculated using speaker-independent mixture distributions. Therefore, the pooled model can be easily updated when a new speaker is added as a registered speaker. In addition, this method has been shown to give much better results than either of the original normalization methods.

Since these normalization methods neglect the absolute deviation between the claimed speaker's model and the input speech, they cannot differentiate highly dissimilar speakers. Higgins et al. [20] reported that a multilayer network decision algorithm makes effective use of the relative and absolute scores obtained from the matching algorithm.

48.13 Decision Process

48.13.1 Specifying Decision Thresholds and Measuring Performance

A "tight" decision threshold makes it difficult for impostors to be falsely accepted by the system. However, it increases the possibility of rejecting legitimate users (customers). Conversely, a "loose" threshold enables customers to be consistently accepted, while also falsely accepting impostors. To set the threshold at a desired level of customer acceptance and impostor rejection, the distribution of customer and impostor scores must be known. In practice, samples of impostor and customer scores of a reasonable size that will provide adequate estimates of distributions are not readily available. A satisfactory empirical procedure for setting the threshold is to assign a relatively loose initial threshold and then allow it to adapt by setting it to the average, or some other statistic, of recent trial scores, plus some margin that allows a reasonable rate of customer acceptance. For the first few verification trials, the threshold may be so loose that it does not adequately protect against impostor attempts. To prevent impostor acceptance during initial trials, they may be carried out as part of an extended enrollment.

48.13.2   ROC Curves

Measuring the false rejection and false acceptance rates for a given threshold condition is an incomplete description of system performance. A general description can be obtained by varying the threshold over a sufficiently large range and tabulating the resulting false rejection and false acceptance rates. A tabulation of this kind can be summarized in a receiver operating characteristic (ROC) curve, first used in psychophysics. An ROC curve, shown as the probability of correct acceptance vs. the probability of incorrect (false) acceptance is shown in Figure 48.4 [11].

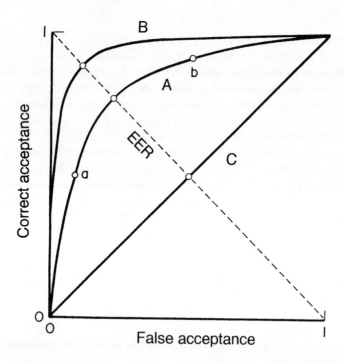

FIGURE 48.4   Receiver operating characteristic (ROC) curves; performance examples of three speaker recognition systems: A, B, and C.

The figure exemplifies the curves for three systems: A, B, and C. Clearly, the performance of curve B is consistently superior to that of curve A, and C corresponds to the limiting case of purely chance performance. Position a in the figure corresponds to the case in which a strict decision criterion is employed, and position b corresponds to a case involving a lax criterion.

The point-by-point knowledge of the ROC curve provides a threshold-independent description of all possible functioning conditions of the system. For example, if a false rejection rate is specified, the corresponding false acceptance rate is obtained as the intersection of the ROC curve with the vertical straight line indicating the false rejection.

Equal-error rate is a commonly accepted summary of system performance. It corresponds to a threshold at which the rate of false acceptance is equal to the rate of false rejection. The equal-error rate point corresponds to the intersection of the ROC curve with the straight line of 45 degrees, indicated in the figure.

48.13.3 Adaptive Thresholds

An issue related to model updating is the selection of a strategy for updating thresholds. A threshold updating strategy must be specified that tolerates trial-to-trial variations while, at the same time, ensures the desired level of performance.

48.13.4 Sequential Decisions (Multi-Attempt Trials)

In either the verification or identification mode, an additional threshold test can be applied to determine whether the match is good enough to accept the decision or whether the decision should be deferred to a new trial.

48.14 Outstanding Issues

There are many outstanding issues and problems in the area of speaker recognition. The most pressing issues, providing challenges for implementing practical and uniformly reliable systems for speaker verification, are rooted in problems associated with variability and insufficient data. As described earlier, variability is associated with trial-to-trial variations in recording and transmission conditions and speaking behavior. The most serious variations occur between enrollment sessions and subsequent test sessions resulting in models that are mismatched to test conditions. Most applications require reliable system operation under a variety of environmental and channel conditions and require that variations in speaking behavior will be tolerated. Insufficient data refers to the unavailability of sufficient amounts of data to provide representative models and accurate decision thresholds. Insufficient data is a serious and common problem because most applications require systems that operate with the smallest practicable amounts of training data recorded in the fewest number of enrollment sessions, preferably one. The challenge is to find techniques that compensate for these deficiencies. A number of techniques have been mentioned which provide partial solutions, such as cepstral subtraction techniques for channel normalization and spectral subtraction for noise removal. An especially effective technique for combating both variability and insufficient data is updating models with data extracted from test utterances. Studies have shown that model adaptation, properly implemented, can improve verification performance significantly with a small number of updates. It is difficult, however, for model adaptation to respond to large, precipitous changes. Moreover, adaptation provides for the possibility that customer models might be updated and possibly captured by impostors. Another effective tool for making speaker verification more robust is the use of likelihood ratio scoring. An utterance recorded in conditions mismatched to the conditions of enrollment will experience degraded scores for both the customer reference model and the cohort or background model so that the ratio of these two scores remains relatively stable. Ongoing research is directed towards constructing efficient and effective background models for which likelihood ratio scores that behave in this manner can be reliably obtained.

A desirable feature for a practical speaker verification system is reasonably uniform performance across a population of speakers. Unfortunately, it is typical to observe in a speaker verification experiment a substantial discrepancy between the best performing individuals, the "sheep", and the worst, the "goats". This additional problem in variability has been widely observed, but there are virtually no studies focusing on its origin. Speakers with no observable speech pathologies, and for whom apparently good reference models have been obtained, are often observed to be "goats". It is possible that such speakers exhibit large amounts of trial-to-trial variability, beyond the ability of the system to provide adequate compensation.

Finally, there are fundamental research issues which require additional study to promote further advances in speaker recognition technology. First, and most important, is the selection of effective features for speaker discrimination and the specification of robust, efficient acoustic measurements for representing these features. Currently, as we have described, the most effective speaker recognition features are short-time spectral features, the same features used for speech recognition. These features are mainly correlated with segmental speech phenomena and have been shown to be capable of resolving very fine spectral differences, possibly exceeding human perceptual resolving ability. Suprasegmental features, such as pitch and energy, are generally acknowledged to be less effective for speaker recognition. However, it may be that suprasegmental features are not being measured or used effectively since human listeners make effective use of such features in their speaker recognition judgments.

Perhaps the single most fundamental speaker recognition research issue is the intrinsic discriminability of speakers. A related issue is whether intrinsic discriminability should be calibrated by the ability of listeners to discriminate speakers. It is not at all clear that the intrinsic discriminability of speakers is the same order as the discriminability that can be obtained using other personal identification characteristics, such as fingerprints and facial features. Speakers' voices differ on the basis of physiological and behavioral characteristics. But it is not clear precisely which characteristics are significant, what acoustic measurements are correlated with specific features, and how close features of different speakers must be to be acoustically and perceptually indistinguishable. Fundamental research on these questions will provide answers for developing better speaker recognition technology.

Defining Terms

Registered speaker: A speaker who belongs to the list of known (registered) users for a given speaker recognition system. Alternative terms: reference speaker, customer.

Genuine speaker: A speaker whose real identity is in accordance with the claimed identity. Alternative terms: true speaker, correct speaker.

Impostor: In the context of speaker identification, a speaker who does not belong to the set of registered speakers. In the context of speaker verification, a speaker whose real identity is different from his/her claimed identity.

Acceptance: A decision outcome which involves a positive response to a speaker (or speaker class) verification task.

Rejection: A decision outcome which involves refusal to assign a registered identity (or class) in the context of open-set speaker identification or speaker verification.

Misclassification: Erroneous identity assignment to a registered speaker in speaker identification.

False rejection: Erroneous rejection of a genuine speaker in open-set speaker identification or speaker verification.

False acceptance: Erroneous acceptance of an impostor in open-set identification or speaker verification.

***A posteriori* equal error threshold:** A decision threshold which is set *a posteriori* on the test data so that the false rejection rate and false acceptance rate become equal. Although this method cannot be put into actual practice, it is the most common constraint because it is a simple way to summarize the overall performance of the system into a single figure.

A priori threshold: A decision threshold which is set beforehand usually based on estimates from a set of training data.

References

[1] Atal, B.S., Effectiveness of linear prediction characteristics of the speech wave for automatic speaker identification and verification, *J. Acoust. Soc. Am.*, 55 (6), 1304–1312, 1974.

[2] Bennani, Y., Fogelman Soulie, F. and Gallinari, P., A connectionist approach for automatic speaker identification, *Proc. IEEE Intl. Conf. Acoust., Speech, Signal Processing*, 265–268, 1990.

[3] Carey, M.J. and Parris, E.S., Speaker verification using connected words, *Proc. Inst. Acoustics*, 14 (6), 95–100, 1992.

[4] Doddington, G.R., Speaker recognition-identifying people by their voices, *Proc. IEEE*, 73 (11), 1651–1664, 1985.

[5] Farrel, K.R., Mammone, R.J. and Assaleh, K.T., Speaker recognition using neural networks and conventional classifiers, *IEEE Trans. On Speech and Audio Processing*, 2 (1), 194–205, 1993.

[6] Flanagan, J.L. *Speech Analysis, Synthesis and Perception*, Springer-Verlag, New York, 1972.

[7] Furui, S., Itakura, F. and Saito, S., Talker recognition by longtime averaged speech spectrum, *Trans. IECE*, 55-A, 1 (10), 549-556, 1972.

[8] Furui, S., An analysis of long-term variation of feature parameters of speech and its application to talker recognition, *Trans. IECE*, 57-A, 12, 880–887, 1974.

[9] Furui, S., Cepstral analysis technique for automatic speaker verification, *IEEE Trans. Acoust., Speech, Signal Processing*, 29 (2), 254–272, 1981.

[10] Furui, S., Research on individuality features in speech waves and automatic speaker recognition techniques, *Speech Commun.*, 5 (2), 183–197, 1986.

[11] Furui, S., *Digital Speech Processing, Synthesis, and Recognition*, Marcel Dekker, New York, 1989.

[12] Furui, S., Speaker-dependent-feature extraction, recognition and processing techniques, *Speech Commun.*, 10 (5-6), 505–520, 1991.

[13] Furui, S., An overview of speaker recognition technology, ESCA Workshop on Automatic Speaker Recognition, Identification and Verification, 1–9, 1994.

[14] Gish, H., Krasner, M., Russell, W. and Wolf, J., Methods and experiments for text-independent speaker recognition over telephone channels, *Proc. IEEE Intl. Conf. Acoust., Speech, Signal Processing*, 865–8, 1986.

[15] Gish, H., Robust discrimination in automatic speaker identification, *Proc. IEEE Intl. Conf. Acoust., Speech, Signal Processing*, 289–292, 1990.

[16] Gish, H., Karnofsky, K., Krasner, K., Roucos, S., Schwartz, R. and Wolf, J., Investigation of text-independent speaker identification over telephone channels, *Proc. IEEE Intl. Conf. Acoust., Speech, Signal Processing*, 379–382, 1985.

[17] Gish, H. and Schmidt, M., Text-independent speaker identification, *IEEE Signal Processing Magazine*, 11(4), 18–32, 1994.

[18] Griffin, C., Matsui, T. and Furui, S., Distance measures for text-independent speaker recognition based on MAR model, *Proc. IEEE Intl. Conf. Acoust., Speech, Signal Processing*, Adelaide, I-309–312, 1994.

[19] Higgins, A.L. and Wohlford, R.E., A new method of text-independent speaker recognition, *Proc. IEEE Intl. Conf. Acoust., Speech, Signal Processing*, 869–872, 1986.

[20] Higgins, A.L., Bahler, L. and Porter, J., Speaker verification using randomized phrase prompting, *Digital Signal Processing*, 1, 89–106, 1991.

[21] Juang, B.-H., Rabiner, L.R. and Wilpon, J.G., On the use of bandpass liftering in speech recognition, *IEEE Trans. Acoust., Speech and Signal Processing*, ASSP-35, 947–954, 1987.

[22] Juang, B.-H. and Soong, F.K., Speaker recognition based on source coding approaches, *Proc. IEEE Intl. Conf. Acoust., Speech, Signal Processing*, 613–616, 1990.

[23] Juang, B.-H. and Katagiri, S., Discriminative learning for minimum error classification, *IEEE Trans. on Signal Processing*, 40, 3043–3054, 1992.

[24] Kunzel, H.J., Current approaches to forensic speaker recognition, ESCA Workshop on Automatic Speaker Recognition, Identification and Verification, 135–141, 1994.

[25] Li, K.-P. and Wrench Jr., E.H., An approach to text-independent speaker recognition with short utterances, *Proc. IEEE Intl. Conf. Acoust., Speech, Signal Processing*, 555–558, 1983.

[26] Liu, C.-S., Lee, C.-H., Chou, W., Juang, B.-H. and Rosenberg, A.E., A study on minimum error discriminative training for speaker recognition, *J. Acoust. Soc. Am.*, 97(1), 637–648, 1995.

[27] Lummis, R.C., Speaker verification by computer using speech intensity for temporal registration, *IEEE Trans. on Audio and Electroacoustics*, AU-21, 80–89, 1973.

[28] Markel, J.D., Oshika, B.T. and Gray, A.H., Long-term feature averaging for speaker recognition, *IEEE Trans. Acoust., Speech, Signal Processing*, ASSP-25(4), 330–337, 1977.

[29] Matsui, T. and Furui, S., Text-independent speaker recognition using vocal tract and pitch information, *Proc. ICSLP 90*, 1, 137–140, 1990 International Conference on Spoken Language Processing, Kobe, Japan.

[30] Matsui, T. and Furui, S., Comparison of text-independent speaker recognition methods using VQ-distortion and discrete/continuous HMMs, *Proc. IEEE Intl. Conf. Acoust., Speech, Signal Processing*, II, 157-160, 1992.

[31] Matsui, T. and Furui, S., Concatenated phoneme models for text-variable speaker recognition, *Proc. IEEE Intl. Conf. Acoust., Speech, Signal Processing*, II, 391–394, 1993.

[32] Matsui, T. and Furui, S., Similarity normalization method for speaker verification based on *a posteriori* probability, ESCA Workshop on Automatic Speaker Recognition, Identification and Verification, 59–62, 1994.

[33] Matsui, T. and Furui, S., Speaker adaptation of tied-mixture-based phoneme models for text-prompted speaker recognition, *Proc. IEEE Intl. Conf. Acoust., Speech, Signal Processing*, I, 125–128, 1994.

[34] Miller, B., Vital signs of identity, *IEEE Spectrum*, 22–30, Feb. 1994.

[35] Montacie, C. et al., Cinematic techniques for speech processing: temporal decomposition and multivariate linear prediction, *Proc. IEEE Intl. Conf. Acoust., Speech, Signal Processing*, I, 153-156, 1992.

[36] Naik, J.M., Netsch, L.P. and Doddington, G.R., Speaker verification over long distance telephone lines, *Proc. IEEE Intl. Conf. Acoust., Speech, Signal Processing*, 524–527, 1989.

[37] Netsch, L.P. and Doddington, G.R., Speaker verification using temporal decorrelation post-processing, *Proc. IEEE Intl. Conf. Acoust., Speech, Signal Processing*, II, 181-184, 1992.

[38] Nolan, F., *The Phonetic Bases of Speaker Recognition*, Cambridge University Press, Cambridge, 1983.

[39] O' Shaughnessy, D., Speaker recognition, *IEEE ASSP Magazine*, 3(4), 4–17, 1986.

[40] Oglesby, J. and Mason, J.S., Optimization of neural models for speaker identification, *Proc. IEEE Intl. Conf. Acoust., Speech, Signal Processing*, 261–264, 1990.

[41] Poritz, A.B., Linear predictive hidden Markov models and the speech signal, *Proc. IEEE Intl. Conf. Acoust., Speech, Signal Processing*, 1291–1294, 1982.

[42] Rabiner, L.R. and Juang, B.-H., *Fundamentals of Speech Recognition*, Prentice-Hall, Englewood Cliffs, NJ, 1993.

[43] Reynolds, D., Speaker identification and verification using Gaussian mixture speaker models, ESCA Workshop on Automatic Speaker Recognition, Identification and Verification, 27–30, 1994.

[44] Rose, R. and Reynolds, R.A., Text independent speaker identification using automatic acoustic segmentation, *Proc. IEEE Intl. Conf. Acoust., Speech, Signal Processing*, 293–296, 1990.

[45] Rosenberg, A.E. and Soong, F.K., Evaluation of a vector quantization talker recognition system in text independent and text dependent modes, *Computer Speech and Language*, 2, 143–157, 1987.

[46] Rosenberg, A.E., Lee, C.-H., Soong, F. K. and McGee, M.A., Experiments in automatic talker verification using sub-word unit hidden Markov models, *Proc. ICSLP 90,* 1, 141–144, 1990 International Conference on Spoken Language Processing, Kobe, Japan.

[47] Rosenberg, A.E., Lee, C.-H. and Gokcen, S., Connected word talker verification using whole word hidden Markov models, *Proc. IEEE Intl. Conf. Acoust., Speech, Signal Processing,* Toronto, 381–384, 1991.

[48] Rosenberg, A.E. and Soong, F.K., Recent research in automatic speaker recognition, in *Advances in Speech Signal Processing,* Furui, S. and Sondhi, M.M., Eds., Marcel Dekker, New York, 1991, 701-737.

[49] Rosenberg, A.E., Delong, J., Lee, C.-H., Juang, B.-H., and Soong, F.K., The use of cohort normalized scores for speaker verification, *Proc. Intl. Conf. Spoken Language Processing,* Banff, 599–602, 1992.

[50] Savic, M. and Gupta, S.K., Variable parameter speaker verification system based on hidden Markov modeling, *Proc. IEEE Intl. Conf. Acoust., Speech, Signal Processing,* 281–284, 1990.

[51] Soong, F.K. and Rosenberg, A.E., On the use of instantaneous and transitional spectral information in speaker recognition, *IEEE Trans. Acoust., Speech, Signal Processing,* ASSP-36(6), 871–879, 1988.

[52] Soong, F.K., Rosenberg, A.E., Juang, B.-H. and Rabiner, L.R., A vector quantization approach to speaker recognition, *AT&T Technical J.,* (66), 14–26, 1987.

[53] Zheng, Y.-C. and Yuan, B.-Z., Text-dependent speaker identification using circular hidden Markov models, *Proc. IEEE Intl. Conf. Acoust., Speech, Signal Processing,* 580–582, 1988.

49

DSP Implementations of Speech Processing

Kurt Baudendistel
Momentum Data Systems

Implementations of digital speech processing algorithms in software can be distinguished from those resulting from general-purpose algorithms basically in the *type of arithmetic* and the *algorithmic constructs* used in their realization. In addition, many speech processing algorithms are realized with Programmable Digital Signal Processors (PDSPs) as the software development target—this leads to important considerations in the languages and paradigms used to realize the algorithms.

Although they are important topics in their own right, this section does not discuss the historical development of PDSPs to explain why these devices provide the architectural features that they do, and it does not provide a primer on PDSP architectures, either in general or in specific. Brief synopses of these topics are presented in the text, however, where they are appropriate.

49.1 Software Development Targets

PDSPs were developed as specialized microprocessors in the late 1970s in response to the needs of speech processing algorithms, and the vast majority of these devices have remained to this day basically audio-rate and, hence, speech processing devices [1]–[5]. These processors present a unique venue in which to both examine and implement speech processing algorithms since even a cursory examination of the device architectures quickly reveals the strong synergy between PDSP features and speech processing algorithms. As a result, specialized but restricted software development skills are necessary to realize speech processing algorithms on these devices.

Within the context of speech processing application realization, PDSPs which provide *fixed-point* data processing capabilities in hardware rather than floating-point capabilities are significantly more important. The simple reason for this is that fixed-point PDSPs are significantly less expensive than floating-point PDSPs but still provide the required computational capabilities for this class of applications. The fixed-point hardware capabilities are used to realize various *types of arithmetic* or *data abstractions* for the infinite-precision mathematical constructs used in algorithms. These abstractions include the well-known *integer arithmetic*, as well as various forms of *fixed-point arithmetic* that use fixed shifts to control the scale of values within a computation and *block floating-point arithmetic* which performs run-time scale manipulations under programmer control.

General-purpose microprocessors also present an implementation medium that is well suited to many speech processing operations, although not one so well tailored to the task as that provided by PDSPs. All of the algorithmic structures presented here can be realized via microprocessors, and in fact many software libraries have been specifically designed to allow such realization [6]–[7].

49.2   Software Development Paradigms

As with general-purpose algorithms, a single software development paradigm cannot be described under which speech processing algorithms are always implemented. A small set of such paradigms do exist, however, and they are distinguished by just a few salient features.

Imperative vs. Applicative Language

Imperative programming languages specify a program as a sequence of commands to be performed in the order given. All of the familiar high-level programming languages, such as C, C++, or FORTRAN, as well as the assembly languages of most PDSPs, are imperative.

Applicative programming languages, on the other hand, describe a program via a collection of relationships that must be maintained between variables. Applicative languages intended to be programmed directly by the user such as Silage, SIGNAL, LUCID/Lustre, and Esterel, as well as the assembly language of data-flow PDSPs, can all be used to specify speech processing algorithms in a non-imperative manner, but their use to date in real applications is quite limited. And, although usually described as a hardware-description language and used as an intermediate language generated by other tools rather than directly by programmers, from the point of view of this discussion VHDL is an applicative language that can be used to describe speech processing algorithms directly.

Graphical programming environments such as Ptolemy, GOSPL, COSSAP, and SPW also provide an applicative "language" in which speech processing algorithms can be described. However, these environments universally rely on atomic elements that are programmed with a separate paradigm, usually an imperative one.

Most speech processing applications are implemented using imperative languages, and this programming model will be used here. Note, however, that the important distinguishing features of speech processing algorithms, arithmetic and algorithmic constructs, are applicable within any programming paradigm.

High Level Language vs. Assembly Language

Given that an imperative programming paradigm is to be used, the choice of a High Level Language (HLL) or assembly language as an implementation vehicle seems very straightforward [8]–[9]. The common wisdom holds that (1) assembly language should be chosen where execution speed is of the essence, in realizing "signal-processing kernels", since HLL compilers cannot produce object code of the same efficiency as can be obtained with hand-coded assembly language. However, (2) a high-level language should be used otherwise, in the realization of "control code", since this allows effective software development and the use of a *top-down* code development strategy.

In PDSP implementations, however, this sensible arrangement is often not possible. The reason for this is that use of a HLL compiler and run-time system makes untoward demands, relatively speaking, on an embedded system where resources such as registers, memory, and instruction cycles are quite scarce. In particular:

1. The settings in the control registers of the processor are often different between signal-processing and control code, and the device is more often than not " in the wrong mode".

2. The run-time memory organization demanded by a high-level language, typically including a stack on which automatic variables are to be allocated but lacking memory bank control, is one that most system designers are not willing to provide.

3. The standard function-call mechanism of a high-level language does not fit well with the customized register usage demanded in embedded systems programming.

Thus, more often than not, HLL programming is not currently utilized in PDSP systems. This will change, however, as PDSP HLL compilers become more sophisticated and as PDSP architectures become more "microprocessor-like".

Specialized vs. Standard High Level Languages

Specialized languages are often developed as dialects of standard high level programming languages by the authors of compilers. DSP/C, for example, is an extension of the C language that contains special vector and signal processing operations [10]. While they appear to be quite useful for target code development for speech processing applications, the lack of general support means that these languages are not often used for either algorithm or target code development.

Extensible languages, on the other hand, allow "dialects" of standard programming languages to be created by the end-user. C++ and Ada allow the construction of specialized arithmetic support via class and generic constructs, respectively. While these languages are quite useful for algorithm development, they generally cannot produce efficient realizations of the kind desired in target code for speech processing applications.

More often than not, when standard high level languages are used, they are simply augmented by libraries of operations. The Signal Processing Toolbox for Matlab^{TM} and the Basic Operators for the C language used in standard speech codecs are good examples of these [6]–[7].

Block vs. Single-Sample Processing

Speech coding applications lend themselves quite well to *block processing,* where individual time-domain signal samples are buffered into vectors or *frames* [11]. This is often done for algorithmic reasons, as in LPC analysis, but significant performance gains can be realized by choosing this processing structure as well, when this is possible.[1]

Buffered data can be processed much more efficiently than single samples with typical PDSP architectures because the overhead associated with data transfer and instruction pipelining in these devices can be amortized over the entire vector rather than occurring for each sample. For example, the ubiquitous multiply-accumulate operation can be performed in a single instruction cycle by most PDSPs, but only *within the instruction execution pipeline,* meaning that overhead of several instruction cycles are required to set up for this level of performance. In single-sample processing, this instruction execution rate cannot be achieved.

Frames can be processed *in toto* or divided into *subframes* that are to be processed individually. This technique provides algorithmic flexibility without sacrificing the significant performance enhancement to be achieved with block processing.

Static vs. Dynamic Run-Time Operation

Two disparate philosophies on the operation of any real-time software system are particularly evident in speech processing implementations. *Static* and *dynamic* here indicate that run-time resource requirements, outlined in Table 49.1, can be computed and known at compile-time or only at run-time,

---

[1] Not all algorithms can use block processing—modems and other signaling systems with very low delay requirements cannot. This technique is generally useful, however, for speech processing applications.

respectively. Of course, some mix of these two philosophies can be found in any system, but the emphasis will usually be placed on one or the other.

TABLE 49.1 Static vs. Dynamic Operation

Resource	Static operation	Dynamic operation
Memory allocation	Global	Stack/heap
$\longrightarrow$ Address computation	Fixed	Stack-relative dynamic
Vector size	Fixed	Data dependent
Execution time	Fixed	Data dependent
$\longrightarrow$ Branch paths	Time-equivalent	Time-disparate
$\longrightarrow$ Wait-state insertion[a]	Must be computed	Can be ignored
Data transfer	Polling possible	DMA required
$\longrightarrow$ Fifo buffers	Not necessary	Required
$\longrightarrow$ Fifo overflow	Impossible	Possible
Operating system	Not typical	Typical

[a] Wait states may be inserted by an interlocked pipeline.

Exact vs. Approximate Arithmetic

The terms *exact* and *approximate* here refer to the concern on the part of the programmer as to whether the results produced by a given arithmetic operation are fully specified by the programmer in a *bit-exact* manner, or whether the best, approximate numerical performance that can be produced by a particular processor is acceptable [12]. For example, IEEE floating-point arithmetic is exact, while machine-dependent floating-point formats can be considered approximate from the point of view of a programmer porting code to that architecture from another. The most important form of exact arithmetic for speech processing applications is that provided by the Basic Operators, which are used in the C language specification provided as part of modern speech coding standards [6]–[7]. As a general rule, the integral and fractional fixed-point arithmetic forms, discussed in Section 49.4, can be considered exact and approximate, respectively.

Approximate arithmetic is much simpler to specify than exact arithmetic, but it is harder to evaluate. In the former case, implementation details are left up to the target architecture, but if the numerical performance of a particular realization does not meet some criteria, gross changes are required in the source code. The problem here is that the criteria are not defined as part of the source code and must be supplied elsewhere. Exact arithmetic, on the other hand, requires excruciating detail in the specification of the algorithm from the outset, but no evaluation of the realization is required since this realization must adhere to the specification.

Approximate arithmetic is the form promoted by the C language where, for example, the data type int does not define the precision of the integer or the results of operations that overflow.[2] It is also the form preferred by software developers working in a native code development environment.[3] Exact arithmetic, on the other hand, is preferred by developers who produce standards and who work in cross-code development environments because it eases the task of porting the algorithm from one environment to the other. Care must be exercised in this case, however, as any cross-development introduces inherent biases into an implementation that may be difficult or impractical to realize on a particular target processor [7].

---

[2] This is not to say that the C language cannot be *used* to realize exact arithmetic, which it often is through the machine-dependent declaration of data types such as int16 and int32, but rather that the language was not *designed* for use with exact arithmetic.

[3] *Native* indicates that code for a particular processor is developed on that processor, while *cross* indicates that the host and target processors are different.

It is well-known that a trade-off always exists between numerical and execution performance, as discussed in Section 49.4. It is not so well-known, however, that approximate arithmetic will always allow an equivalent or better balance to be struck in this trade-off than exact arithmetic. This is because the excruciating detail provided as part of an exact specification supplies not a minimum numerical requirement, but an exact one. In the case where a particular architecture can provide more precision than is specified, extra code must be inserted to remove that precision, resulting in less efficient execution performance. And, precisely because of this, an exact specification is in fact always targeted to a particular PDSP or microprocessor architecture—no algorithm can be specified in an exact manner and be truly portable or architecturally neutral.

49.3 Assembly Language Basics

Assembly languages for PDSPs are closely matched with the PDSP architecture for which they are designed, but they all share common elements [1]–[5]. In particular, multiple processing units must be programmed at the same time:

- adder
- multiplier
- fixed-point logic, such as shifter(s), rounding logic, saturation logic, etc.
- address generation unit
- program memory, for instruction fetch or data fetch
- data memories, perhaps multiple

In some cases, these units operate by default. For example, instruction fetches occur each machine cycle unless program memory is otherwise used. And in other cases, these units are utilized in combination. For example, (1) the DSP56000 multiply-accumulate instructions and (2) all address generation and memory fetch operations are indivisible and not pipelined. In all other cases, however, these processing units must be programmed within the *instruction execution pipeline* in which the outputs of one processing unit are connected directly to the inputs of another.

Coding Paradigms

Distinct coding paradigms are required by the architectures of various PDSPs, basically determined by the pipeline of that device, in order to perform this programming [13, 14]. Several assembly language forms are presented by PDSPs to realize these coding paradigms:

Data stationary coding specifies ultimately the *data* that is operated on by an instruction, but not the *time* at which the operation takes place—the latter is implicit in the form of the instruction. For example, the AT&T DSP32 instruction

$$*r0++ \ = \ a0 \ = \ *r1++ \ + \ *r2++ \tag{49.1}$$

specifies the locations in memory from which the addends should be read and to which the sum should be written, but it is implicit that the sum will be written to memory in the third instruction cycle following this one.

Because of such delays, *illegal* and *erroneous* instruction combinations can be written that cause conflicts in the use of data from both memory and registers—the former can be detected by the assembler, but the latter will simply produce data manipulations different from those intended by programmer.

Time stationary coding specifies the operations that should occur at the *time* that this instruction is executed, while the *data* to be used is whatever is present in the "pipeline registers" at this time. For example, the AT&T DSP16 instruction

$$a1 \; = \; a0 \; + \; y \qquad y \; = \; *r0++ \qquad\qquad (49.2)$$

specifies that a sum should occur at this time between the named registers and that a memory read should occur to the **y** register in parallel. No illegal or erroneous instruction combinations are possible in this case.

Interlocked coding solves the instruction combination problems of data stationary coding by automatically introducing extra machine cycles or *wait states* to ensure that conflicts do not occur. While this is convenient for the programmer, it does not produce more efficient execution than pure data stationary coding—on the contrary, it encourages programmers to be less savvy about their product.

Data flow coding is appropriate for machines that realize an applicative paradigm directly, such as the Hughes DFSP or the NEC μPD7281.

It must be pointed out that a mixture of these coding paradigms is often used in real PDSPs for control of different processing units. For example, the AT&T DSP16, while ostensibly a time-stationary device, utilizes a form of interlocking to allow multiple accesses to the same memory bank in a single instruction cycle [1].

Assembly Languages Forms

Within the four coding paradigms presented above, several assembly language forms can be utilized. First, either an infix form as given in Eq. 49.1 or the traditional assembly language prefix form using instruction mnemonics, as shown in Eq. 49.3 for the Motorola DSP56000, can be used:

$$\mathtt{clr \quad a} \qquad\qquad (49.3)$$

Second, the instruction may consist of a single field, as in Eq. 49.1 or Eq. 49.3, or it may contain multiple fields to be executed in parallel, as in Eq. 49.2 or Eq. 49.4:

$$\mathtt{mac \quad x0, \; y0, a \quad x:(r0)+, x0 \quad y:(r4)+, y0} \qquad\qquad (49.4)$$

Note, however, that even within the multiple fields more than one operation is specified—in both Eq. 49.2 and Eq. 49.4 address register updates are specified along with the memory move. Pure *horizontal microcode*, in which a dedicated field in each instruction word controls a particular processing unit, is used in only a few modern PDSP architectures, but the multiple-field instructions are similar.

Additionally, all PDSPs contain "mode registers" which control operation of particular elements of the device. For example, the auc register of the AT&T DSP16 controls the multipler-shift, and thus the type of arithmetic realized by this processor's p=x*y instruction. Such mode registers, while prevalent and powerful in extending the effective instruction encoding space of a PDSP, are quite difficult to manage in large programming systems, especially in the design of function libraries.

49.4 Arithmetic

The most fundamental problem encountered during the implementation of speech processing algorithms is that the algorithm must be realized (1) using the finite-precision arithmetic capabilities of real processors rather than the infinite-precision available in mathematic formulae (2) under typically severe cost constraints in terms of the processing capabilities of the target system [15]–[17]. Any arbitrary level of arithmetic performance can be achieved by any processor, but the cost of this performance in terms of machine cycles can be prohibitive, and so an engineering trade-off is required.

Finite-precision arithmetic effects can be broadly classified as *representational* and *operational errors*:

- The bit pattern used to represent a finite-precision value can be of many forms, but all restrict the *range* of values over which a representation can be provided as well as the *precision* or number of bits used for the representation of a given value. No forms of arithmetic allow values outside the range to be represented, but some invoke an exception handler when such is requested. This is not appropriate in most speech processing systems, however, and in this case a finite-precision representation must be provided to approximate this value.

  The difference between an infinite-precision value and its finite-precision representation is the representational error, and there are two sources of such error: *truncation error* results from finite precision and *overflow error* results from range violations.

- Finite-precision operators used to transform values can also introduce error. In the case of simple arithmetic operators, this is equivalent to representational error, but it is often useful to conceptualize more complicated operators, such as an FIR or IIR filter, and to characterize the error introduced by that entity.

The engineering trade-off thus becomes an exercise in balancing the *numerical performance* of a realization of an algorithm in terms of truncation error and overflow error under the considerations introduced by possibly wide variance in input signal strengths or *dynamic range*, against implementation cost constraints in terms of target processor choice and available machine cycles on that processor. Because of the importance of this trade-off, it is important to examine different *types of arithmetic* and to evaluate the numerical performance and implementation cost of each type.

For example, floating-point arithmetic produces adequate numerical performance for most speech processing applications. However, the cost of floating-point processors is often prohibitive in dollar terms, and the cost of realizing floating-point arithmetic on a less expensive, fixed-point processor is prohibitive in terms of machine cycles. For this reason, some other type of arithmetic is often a better choice even though it may be numerically inferior and much harder to implement.

Regardless of the type of arithmetic chosen, however, it will be used in speech processing applications as a proxy or *abstraction* for the real-valued, infinite-precision arithmetic of mathematics. An important aspect that must be considered in evaluating finite-precision arithmetic types, then, is the effectiveness of the abstraction they provide for real-valued arithmetic. For example, all arithmetic needed for speech processing applications can be provided by integers, but determining what bit pattern to use to represent π or how to add two values of different scales can be quite difficult with this data abstraction.

Arithmetic Errors as Noise

Considering that most numerical values used in a speech processing algorithm are *signals*, in that they take on distinct values at distinct sample points, the difference between a finite-precision realization and the infinite-precision mathematical model on which it is based can be considered an *error* or *noise signal* that is injected into an algorithm at the point at which that arithmetic is used, as illustrated in Fig. 49.1. Given this model for the error as simply a noise source, finite-precision arithmetic effects can be analyzed in a manner similar to that used for other noise sources in a signal processing system.

An important corollary to this fact is that speech processing algorithms should be, and typically are, designed to be robust in the presence of arithmetic noise, just as they are designed to be robust in the presence of other noise sources.

The model for the noise that is injected at each point is a function of the type of arithmetic used in that operation, however. This noise model is an important element in understanding the motivation for using various types of arithmetic, and is presented where appropriate in the sections that follow.

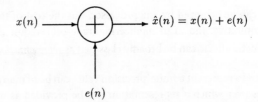

$$x(n):\text{infinite-precision signal}$$
$$e(n):\text{error signal}$$
$$\hat{x}(n):\text{finite-precision signal}$$

FIGURE 49.1 Noise model of arithmetic error.

Floating Point

A *floating-point number* consists of a sign bit, a mantissa, and an exponent, and it presents a well-known model for realizing an approximation to real-valued arithmetic, where the value of the number V is given by

$$V = M \cdot \beta^E \tag{49.5}$$

with β the radix of the representation, usually 2, and M and E the effective values of the signed mantissa and the exponent, respectively. A wide variety of floating-point formats exist, especially for PDSPs, of which the IEEE 754 Floating-Point Standard is the most widely utilized for general-purpose processors. These different formats are distinguished chiefly by the precision of the exponent and the mantissa, and the behavior of the arithmetic at the limits of the representable range.

Floating-point arithmetic is usually used only in applications in which such arithmetic capabilities are provided in hardware by the processor—it is not often simulated via software by a processor that provides only fixed-point arithmetic capabilities, but rather another, similar data abstraction is used. While quite powerful and easy-to-use, floating-point arithmetic is actually of little practical value in the realization of speech processing algorithms.

Block Floating Point

A *block floating-point representation* of a vector of length N of numbers $\bar{v}$ consists of a single signed, 2's complement integer of precision B_e representing the exponent e for the block computed as[4]

$$e = \max_{i \in N} \lceil \log_2 (|v_i|) \rceil \tag{49.6}$$

along with an array of N signed, 2s complement fractions of precision $B_m = b_m + 1$ representing the mantissas m_i to which the exponent can be applied to yield the represented values $\hat{v}_i$ as

$$\hat{v}_i = m_i \cdot 2^{b_m - e} \tag{49.7}$$

The precision of the exponent B_e and of the mantissas B_m are almost always chosen as the word length of the target machine, yielding a *single-precision block floating-point vector*.

Arithmetic on the exponent and mantissas in a block floating-point representation are controlled separately, since significant savings in computation can often be supplied directly by the programmer. For

---

[4]This is a simplified exponent definition used for purposes of illustration. The actual value used in any particular implementation will be machine dependent, but this is of no consequence except as regards the point at which exponent overflow or underflow occurs, rare occurrences in most systems.

example, if a block floating-point vector is to be computed as the result of a correlation, it is known that the zeroeth lag will produce the value with the largest magnitude, and so the exponent for the vector can be immediately determined. In the absence of such direct support from the programmer, block floating-point computations require either (1) that high precision results be saved in a temporary buffer to be scaled after all values have been computed and the maximum exponent found or (2) that all results be computed twice—once to determine the exponent and a second time to compute the mantissas.

An array of length L of block floating-point vectors of length N can be constructed, yielding a construct consisting of L exponents and $L \cdot N$ mantissa values. This *segmented block floating-point representation* allows better representation of values over a wide dynamic range than is available with a single exponent. It is also quite suited to applications in which a segment of values is known to be of one scale that can be quite different from that of neighboring segments.

In the limit with $N = 1$, (segmented) block floating-point yields the *scalar (segmented) block floating-point representation* which is quite like the well-known (vector) floating-point representation, except that normalization occurs only on demand. This is an appropriate representation to use for quantities of large dynamic range in speech processing applications realized on fixed-point processors where true floating-point would be prohibitively expensive.

Fixed Point

A *fixed-point number* consists of a field of $B = b + 1$ data bits that is interpreted as a binary, 2's complement number relative to a scale factor or *size* that is multiplied by the field to yield a value. The two basic forms of fixed-point numbers are the *integral* and *fractional* forms, in which the *justification* of the data bits within the field determines how the value of a bit pattern is interpreted:

Justification	Field	Size	Value	Range
Right	Integer i	Stepsize Δ	$\Delta \cdot i$	$\left[-\Delta \cdot 2^b, \Delta \cdot 2^b\right)$
Left	Fraction f	Fieldsize ϕ	$\phi \cdot f$	$[-\phi, \phi)$

Regardless of the representation, note that the stepsize Δ and fieldsize ϕ are always related as $\phi = \Delta \cdot 2^b$ for quantities of precision $B = b + 1$.

Among other possible fixed-point representations, *center-justified* or *mixed numbers* are quite rare in speech processing applications, and all other common representations are easily derived from the integral and fractional forms.

Given the basic machine word length or precision, usually 16 or 24 bits, fixed-point PDSPs universally provide signed, single-precision multiplication producing a double-precision product, along with double-precision addition, which allows numerically efficient computation of a sum-of-products. *Multiple-precision operations* of greater precision, discussed below, must be simulated in software.

The additive operators (addition, subtraction, negation, and absolute value) are equivalent for any fixed-point representation, with the caveat that only numbers of the same type can be combined with the binary additive operators. That is, only numbers of the same precision, form, and size can be added together directly—other combinations require conversion of one or both quantities to another, possibly a third, type before the operation can take place. Given this equivalency, it can be seen that it is the kind of multiplication, controlled by the shift that occurs at the output of the multiplier and the input to the ALU in all processors, that determines the type of arithmetic realized by a device, as shown in Table 49.2.

Fixed-point PDSPs abound with shifters—at the ALU inputs, the multiplier output, accumulator outputs, and perhaps within an independent barrel shifter. Because of a dearth of instruction encoding space, however, these are often fixed or controlled from mode registers rather than instructions or general registers, as discussed in Section 49.3.

The kind of multiplication realized by a processor also defines the kinds of data abstractions that are most useful given that machine architecture:

TABLE 49.2 Multiplier-Shift Determines
Processor Type

Processor type	Shift[a]
Integral	0
Fractional	1
Biquadratic[b]	2
Summation[b]	-N

[a] This value is a relative one—the value zero could just
as easily have been assigned to the fractional
machine.
[b] These names derive from the use of this type of
arithmetic in second-order IIR filter sections and
long summations, respectively.

Q-notation is a natural extension of integer notation that is useful for right-justified arithmetic. A
B-bit Qn fixed-point number is defined to have a binary point to the right of bit n, where bit
0 is the Least Significant Bit (LSB), yielding a stepsize $\Delta = 2^B$ a range $[-2^{B-n-1}, 2^{B-n-1})$.
Multiplication is defined as producing a product with a precision and Q-value that are the
sums of those of the multiplicands, respectively:

$$B_{x \star y} = B_x + B_y \tag{49.8}$$

$$n_{x \star y} = n_x + n_y \tag{49.9}$$

When precision is increased or reduced, it is naturally done on the left of a right-justified
quantity, as with an integer. This seemingly simple operation is catastrophic when Q-notation
is used to model real-valued arithmetic, however, since it produces overflow. Thus, precision
must not be omitted at any point when Q-notation is in use—the term "a Qn number" should
always be qualified as "a B-bit, Qn number".

Scaled fractions are a natural extension of fractions that are useful for left-justified arithmetic. A
$b + 1$−bit fractional number of fieldsize ϕ has a range $[-\phi, \phi)$, and multiplication is defined
as producing a product with a precision and fieldsize that are the sum and product of those
of the multiplicands, respectively:

$$b_{x \star y} = b_x + b_y \tag{49.10}$$

$$\phi_{x \star y} = \phi_x \cdot \phi_y \tag{49.11}$$

With this notation, biquadratic quantities can be seen to be simply scaled-fractions of fieldsize
2.0.

Precision is much less important for scaled-fractions than for Q-values. This is because
increasing or reducing the precision of a left-justified quantity naturally occurs on the right,
which simply raises or lowers the accuracy of the representation. Thus, while important
as regards numerical performance, precision is not required in describing a quantity as "a
scaled-fractional of fieldsize ϕ".

It should be pointed out that use of a right-justified data abstraction on a left-justified machine, or vice
versa, is quite difficult.

Reduction describes the common response to overflow in fixed-point additive operations, where a sum
is simply allowed to "wrap around" in the 2's complement representation:

$$x + y \equiv \operatorname{sgn}(x + y) \cdot [(|x + y| + \phi) \bmod 2\phi - \phi] \tag{49.12}$$

Saturation describes an alternate response to overflow where the result is set to the maximum representable value of the appropriate sign:

$$\mathtt{x+y} \equiv \begin{cases} \phi - \Delta & x+y \geq \phi \\ x+y & -\phi \geq x+y < \phi \\ -\phi & x+y < -\phi \end{cases} \tag{49.13}$$

The bit patterns that result from saturation are 0x7f...f and 0x80...0 in the cases of positive and negative overflow, respectively. Fixed-point PDSPs typically provide hardware to realize saturation because it gives a significant boost to the numerical performance of many speech processing algorithms in the presence of overflow. In most cases, when reduction arithmetic is in use *no* overflow can be tolerated, even in extremely unlikely situations, while *some* overflow can be tolerated with saturation arithmetic in most algorithms.

General-purpose microprocessors traditionally provide only a single overflow-detection bit. Fixed-point PDSPs, on the other hand, typically provide $N > 1$ overflow bits for each register that can be the destination of an additive operation in the ALU, usually termed *accumulators*. This feature allows summations of up to 2^N terms to be performed while the result can be saturated correctly if overflow does occur. The overflow bits are alternately called *secondary overflow bits, guard bits,* or *extension words* by different manufacturers.

For summations involving more than 2^N terms, it is often useful to determine if overflow occurred during the summation, even though enough information to saturate the result is not available—this capability is also required for the support of block floating-point operations. *Sticky* or *permanent* overflow bits are set when overflow occurs, but they are only cleared under programmer control, allowing such overflow detection. And, it is sometimes useful to provide such permanent overflow detection at a saturation value other than the range, as noted in Section 49.5.

Another option in the case of summations involving more than 2^N terms is to scale the inputs to the summation and then perform saturation at the end of the summation during a *rescaling* operation. As with all scaling operations, however, this one trades off overflow error for truncation error, and it may introduce unacceptable noise levels. For example, in the case of a summation of K i.i.d. Gaussian random variables, prescaling introduces a $3\lceil \log_2 K \rceil$ dB SNR degradation relative to an unscaled summation.

The nature of fixed-point PDSPs as single-precision multiply/double-precision add machines means that conversions between single- and double-precision quantities is quite common. *Extension* from single- to double-precision always takes place on the right for fixed-point quantities, except in the rare cases where integers are involved, and the extension is always with zeroes. Conversion from double- to single-precision, however, can be performed by *truncation* where the extra bits are simply removed,

$$(\mathtt{x} \ \& \ ((\ -1 \ ) << B) \tag{49.14}$$

where B is the basic machine precision, or by *rounding*:

$$((\mathtt{x} \ +(1 \ << B - 1)) \ \& \ ((\ -1 \ ) << B) \tag{49.15}$$

Fixed-point PDSPs typically provide hardware to realize rounding because it gives a significant boost to the numerical performance of many speech processing algorithms. In most applications, it can be safely assumed that the low bits of the 2s complement value that are removed as part of a conversion operation are neither deterministic nor correlated and that they represent values that are uniformly distributed over the range $[0, \Delta)$. In this case, rounding produces errors that statistically are approximately zero-mean, while truncation produces errors with mean $\mu \approx \frac{1}{2}\Delta$, and this *bias error* can be significant in many situations.

Multiple-precision operations can be simulated in software in many ways, but usually one or more of the following formats is used to represent them:

Native format represents double- and higher-precision numbers as simply the appropriate bit pattern broken into multiple machine words. High-precision additive operations can be directly realized in this format using a carry flag, but multiplication of such quantities requires unsigned multiplication capabilities, which are lacking in most PDSPs and many general-purpose processors.

Double precision format (DPF) allows double-precision fractional multiplication, with double-precision inputs and double-precision output, to be realized using signed multiplier capabilities by representing a double-precision value as the concatenation of the high-order word with the low-order word logically right-shifted by one bit.

Double round format (DRF) allows double-precision multiplication, both integral and fractional, to be realized using signed multiplier capabilities by representing a double-precision value as the concatenation of the high-order word that would result from rounding the double-precision quantity to single-precision, with the original low-order word.

These representations are illustrated in Table 49.3.

TABLE 49.3   Double-Precision Formats

Format	High word	Low word
Native	0x89AB	0xCDEF
DPF	0x89AB	0x66F7
DRF	0x89AC	0xCDEF

49.5   Algorithmic Constructs

The second major distinction between implementations of digital speech processing algorithms and general-purpose algorithms concerns the algorithmic constructs used in their realization, and the most important of these are discussed below.

Delay Lines

Delay lines, which allow the storage of sample values from one operational cycle to the next, are an important component of speech processing systems, and they can be realized in a variety of ways with PDSPs:

Registers, including implicit pipeline registers, can be used to effectively realize short delays, including the one- and two-tap delays required in IIR filters.

Modulo addressing causes an address register to "wrap around" within a defined range to the start of a buffer when an attempt is made to increment that register past the end of the defined range. A delay line can be realized using modulo addressing by utilizing the location containing the expired data at a given step for the new data and by bumping the address register accordingly. Most PDSPs do provide modulo-addressing capabilities, but often in only a limited manner. For example, strides greater than one or negative strides may not be supported, and the buffer may require a certain alignment in memory.

Writeback causes a delay line element to be written back to memory at a new location after it is read and used in a computation. While this technique is quite powerful as regards the rearrangement of data in memory, it is quite expensive in terms of memory bandwidth requirements.

These techniques are most useful for fixed delays, but equivalent methods can be used to realize variable delays.

Transforms

Modern PDSPs provide specialized support for transforms, and inverse transforms as well, especially the radix-2 FFT. This can include the ability to

- Compute both a sum and a difference on the same data in parallel.
- Compute addresses using *reverse-carry addition,* where the carry propagates to the right rather than to the left as in ordinary addition. This allows straightforward computation of the bit-reversed addresses needed to unscramble the results of many transform calculations.
- Detect overflow in fixed-point computations at a point other than the saturation point. This can be used to predict that overflow is likely to occur at the current transform stage based on the output of the previous stage *before computation of the current stage begins.* With this capability, the data can be scaled as part of the current processing stage if and only if it is necessary, efficiently producing an optimally scaled transform output.

Vector Structure Organization

Vectors of atomic components are always laid out simply as an array of the elements. When the components are not atomic, however, as with segmented block floating-point or complex quantities, an alternative is to organize the vector as two arrays: an exponent array and a mantissa array for segmented block floating-point quantities, or a real array and an imaginary array for complex quantities.

The choice of *interleaved* or *separate* arrays, as these two techniques are known, is a trade-off between resource demands, in terms of the number of address registers needed to access a single element, vs. flexibility, in terms of the order of access and stride control that is possible.

Zipping

Zipping is a generic term that is used to refer to the process of performing a sequence of multiply-accumulate operations on input arrays to realize the signal processing tasks of scaling, windowing, convolution, auto- and cross-correlation, and FIR filtering. The only real difference between these conceptually distinct tasks is (1) the choice of data or constant input arrays and (2) the order of access within these arrays.

PDSPs are designed to implement this operation, above all others, efficiently—their performance here is what distinguishes them most from general-purpose and RISC microprocessors. Regardless of the coding paradigm used,[5] all PDSPs allow in a single instruction cycle the following:

- two memory accesses, either data-constant or data-data
- two address register updates
- a single-precision multiply
- a double-precision accumulate

Programming contortions are often required to achieve this throughput in the face of processor limitations and memory access penalties, but the holy grail single-cycle operation is always attainable.

---

[5] Coding paradigms are discussed in Section 49.3.

Mathematical Functions

As in general-purpose programming, higher level mathematical functions can be realized within speech processing applications in one of three ways:

Bitwise computation can be used to build an exact representation one bit at a time. This technique is often used to implement single-precision division and square root functions, and some PDSPs even include special iterative instructions to accomplish these operations in a single cycle per output bit. For example, unsigned division can be realized for the Motorola 56000 as follows:

```
and  #$fe,ccr  ;  Clear quotient sign bit
rep  #24       ;  Form 24 bit quotient,          (49.16)
div  x0,a      ;  ... one bit at a time.
```

Approximate computation, such as Newton's method, is often used to produce a double-precision result from a single-precision estimate.

Table lookup is often used, along with linear interpolation between sample points, especially for trigonometric, logarithmic, and inverse functions. Several PDSPs even include the necessary tables in ROM.

Looping Constructs

Loop counting can be done with general registers, and this is required in deeply nested loops, but hardware support is often provided by PDSPs for low- and zero-overhead loops. *Low-overhead loops* utilize a special counter register to realize a branching construct similar to the well-known decrement-and-branch instruction of the Motorola 68000 microprocessor. They are "low overhead" in that the cost of the loop is typically only that of the branch instruction per iteration—separate increment (or decrement) and test instructions are not needed. *Zero-overhead loops* go one step further and eliminate even the cost of the branch instruction per iteration. They do this via special-purpose hardware to perform the program counter manipulations normally handled in the branch instruction. There is an overhead cost at the start of the loop, but the cost per iteration is truly zero.

Loop reversal is an important concept that often allows more efficient coding of speech processing constructs. In its simplest form, a loop counter is run backward to allow more efficient counting of iterations, or an address register is run backward to allow it to be reused without having to reinitialize it. In both of these cases, the reversal of the counter or address register is only possible when there are no dependencies from one loop iteration to the next. More powerful, however, is to perform memory access via a temporary register to allow loops that need to run in one direction for algorithmic reasons to be coded in the opposite direction. This technique can be used to exploit the pipelined nature of PDSPs to great effect.

References

[1] AT&T Microelectronics, *DSP1610 Digital Signal Processor Information Manual,* 1992.
[2] Motorola, Inc., *DSP65000 Digital Signal Processor User's Manual,* 1990.
[3] Texas Instruments, Inc., *TMS320C25 User's Guide,* 1986.
[4] Analog Devices, Inc., *ADSP-2100 User's Guide,* 1988.
[5] NEC, Corp., *NEC μ PD7720 User's Manual,* 1984.
[6] *Draft Recommendation G.723 – Dual Rate Speech Coder for Multimedia Telecommunication Transmitting at 5.3 & 6.3 kbit/s,* International Telecommunication Union Telecommunications Standardization Sector (ITU) Study Group 15, 1995.
[7] *Draft Recommendation G.729 – Coding of Speech at 8 kbit/s using Conjugate-Structure Algebraic-Code-Excited Linear Prediction (CS-ACELP),* ITU Study Group 15, 1995.

[8] Chassaing, C., *Digital Signal Processing with C and the TMS 320C30*, John Wiley & Sons, New York, 1992.

[9] Baudendistel, K., Code generation for the AT&T DSP32, in *Proc. ICASSP-90*, 1073-76, Apr. 1990.

[10] Leary, K. and Waddington, W., DSP/C: A standard high level language for DSP and numeric processing, in *Proc. ICASSP-90*, 1065-68, Apr. 1990.

[11] Sridharan, S. and Dickman, G., Block floating-point implementation of digital filters using the DSP56000, *Microprocessors and Microsystems*, 12, 299–308, July/Aug. 1988.

[12] Baudendistel, K., Compiler Development for Fixed-Point Processors, Ph.D. thesis, Georgia Institute of Technology, 1992.

[13] Madisetti, V. K., *VLSI Digital Signal Processors, An Introduction to Rapid Prototyping and Design Synthesis*, Butterworth-Heinemann, 1995.

[14] Lee, E.A., Programmable DSP architectures: Parts I & II, *IEEE ASSP Magazine*, 5 & 6, 4–19 & 4–14, Oct. 1988 & Jan. 1989.

[15] Oppenheim, A.V. and Schafer, R.W., *Digital Signal Processing*, Prentice-Hall, Englewood Cliffs, NJ, 1975.

[16] Jackson, L., Roundoff-noise analysis for fixed-point digital filters realized in cascade or parallel form, *IEEE Trans. Audio and Electroacoustics*, AU-18, 102-22, June 1970.

[17] Parks, T.W. and Burrus, C.S., *Digital Filter Design*, John Wiley & Sons, New York, 1987.

50

Software Tools for Speech Research and Development

John Shore
Entropic Research Laboratory, Inc.

50.1 Introduction

Experts in every field of study depend on specialized tools. In the case of speech research and development, the dominant tools today are computer programs. In this article, we present an overview of key technical approaches and features that are prevalent today.

0-8493-8572-5/98/$0.00+$.50
© 1998 by CRC Press LLC

We restrict the discussion to software intended to support R&D, as opposed to software for commercial applications of speech processing. For example, we ignore DSP programming (which is discussed in the previous article). Also, we concentrate on software intended to support the specialities of speech analysis, coding, synthesis, and recognition, since these are the main subjects of this chapter. However, much of what we have to say applies as well to the needs of those in such closely related areas as psycho-acoustics, clinical voice analysis, sound and vibration, etc.

We do not attempt to survey available software packages, as the result would likely be obsolete by the time this book is printed. The examples mentioned are illustrative, and not intended to provide a thorough or balanced review. Our aim is to provide sufficient background so that readers can assess their needs and understand the differences among available tools. Up-to-date surveys are readily available online (see Section 50.13).

In general, there are three common uses of speech R&D software:

- *Teaching*, e.g., homework assignments for a basic course in speech processing
- *Interactive, free-form exploration*, e.g., designing a filter and evaluating its effects on a speech processing system
- *Batch experiments*, e.g., training and testing speech coders or speech recognizers using a large database

The relative importance of various features differs among these uses. For example, in conducting batch experiments, it is important that large signals can be handled, and that complicated algorithms execute efficiently. For teaching, on the other hand, these features are less important than simplicity, quick experimentation, and ease-of-use. Because of practical limitations, such differences in priority mean that no one software package today can meet all needs.

To explain the variation among current approaches, we identify a number of distinguishing characteristics. These characteristics are not independent (i.e., there is considerable overlap), but they do help to present the overall view.

For simplicity, we will refer to any particular speech R&D software as "the speech software".

50.2   Historical Highlights

Early or significant examples of speech R&D software include "Visible Speech" [5], *MITSYN* [1], and Lloyd Rice's *WAVE* program of the mid 1970s (not to be confused with David Talkin's *waves* [8]).

The first general, commercial system that achieved widespread acceptance was the Interactive Laboratory System (*ILS*) from Signal Technology Incorporated, which was popular in the late 1970s and early 1980s. Using the terminology defined below, *ILS* is compute-oriented software with an operating-system-based environment. The first popular, display-oriented, workspace-based speech software was David Shipman's LISP-machine application called *Spire* [6].

50.3   The User's Environment
(OS-Based vs. Workspace-Based)

In some cases, the user sees the speech software as an extension of the computer's operating system. We call this "operating-system-based" (or OS-based); an example is the Entropic Signal Processing System (*ESPS*) [7].

In other cases, the software provides its own operating environment. We call this "workspace- based" (from the term used in implementations of the programming language *APL*); an example is *MATLAB*^{TM} (from The Mathworks).

50.3.1 Operating-System-Based Environment

In this approach, signals are represented as files under the native operating system (e.g., Unix, DOS), and the software consists of a set of programs that can be invoked separately to process or display signals in various ways. Thus, the user sees the software as an extension of an already-familiar operating system. Because signals are represented as files, the speech software inherits file manipulation capabilities from the operating system. Under Unix, for example, signals can be copied and moved respectively using the *cp* and *mv* programs, and they can be organized as directory trees in the Unix hierarchical file system (including NFS).

Similarly, the speech software inherits extension capabilities inherent in the operating system. Under Unix, for example, extensions can be created using shell scripts in various languages (*sh, csh, Tcl, perl*, etc.), as well as such facilities as pipes and remote execution. OS-based speech software packages are often called command-line packages because usage typically involves providing a sequence of commands to some type of shell.

50.3.2 Workspace-Based Environment

In this approach, the user interacts with a single application program that takes over from the operating system. Signals, which may or may not correspond to files, are typically represented as variables in some kind of virtual space. Various commands are available to process or display the signals. Such a workspace is often analogous to a personal blackboard.

Workspace-based systems usually offer means for saving the current workspace contents and for loading previously saved workspaces.

An extension mechanism is typically provided by a command interpreter for a simple language that includes the available operations and a means for encapsulating and invoking command sequences (e.g., in a function or procedure definition). In effect, the speech software provides its own shell to the user.

50.4 Compute-Oriented vs. Display-Oriented

This distinction concerns whether the speech software emphasizes computation or visualization or both.

50.4.1 Compute-Oriented Software

If there is a large number of signal processing operations relative to the number of signal display operations, we say that the software is compute-oriented. Such software typically can be operated without a display device and the user thinks of it primarily as a computation package that supports such functions as spectral analysis, filtering, linear prediction, quantization, analysis/synthesis, pattern classification, Hidden Markov Model (HMM) training, speech recognition, etc.

Compute-oriented software can be either OS-based or workspace based. Examples include *ESPS, MATLAB*^{TM}, and the Hidden Markov Model Toolkit (*HTK*) (from Cambridge University and Entropic).

50.4.2 Display-Oriented Software

In contrast, display-oriented speech software is not intended to and often cannot operate without a display device. The primary purpose is to support visual inspection of waveforms, spectrograms, and other parametric representations. The user typically interacts with the software using a mouse or other pointing device to initiate display operations such as scrolling, zooming, enlarging, etc.

While the software may also provide computations that can be performed on displayed signals (or marked segments of displayed signals), the user thinks of the software as supporting visualization more than computation. An example is the *waves* program [8].

50.4.3   Hybrid Compute/Display-Oriented Software

Hybrid compute/display software combines the best of both. Interactions are typically by means of a display device, but computational capabilities are rich. The computational capabilities may be built-in to workspace-based speech software, or may be OS-based but accessible from the display program. Examples include the Computerized Speech Lab (*CSL*) from Kay Elemetrics Corp., and the combination of *ESPS* and *waves*.

50.5   Compiled vs. Interpreted

Here we distinguish according to whether the bulk of the signal processing or display code (whether written by developers or users) is interpreted or compiled.

50.5.1   Interpreted Software

The interpreter language may be specially designed for the software (e.g., S-PLUS from Statistical Sciences, Inc., and $MATLAB^{TM}$), or may be an existing, general purpose language (e.g., LISP is used in *N!Power* from Signal Technology, Inc.).

Compared to compiler languages, interpreter languages tend to be simpler and easier to learn. Furthermore, it is usually easier and faster to write and test programs under an interpreter. The disadvantage, relative to compiled languages, is that the resulting programs can be quite slow to run. As a result, interpreted speech software is usually better suited for teaching and interactive exploration than for batch experiments.

50.5.2   Compiled Software

Compared to interpreted languages, compiled languages (e.g., FORTRAN, C, C++) tend to be more complicated and harder to learn. Compared to interpreted programs, compiled programs are slower to write and test, but considerably faster to run. As a result, compiled speech software is usually better suited for batch experiments than for teaching.

50.5.3   Hybrid Interpreted/Compiled Software

Some interpreters make it possible to create new language commands with an underlying implementation that is compiled. This allows a hybrid approach that can combine the best of both.

Some languages provide a hybrid approach in which the source code is pre-compiled quickly into intermediate code that is then (usually!) interpreted. *Java* is a good example.

If compiled speech software is OS-based, signal processing scripts can typically be written in an interpretive language (e.g., a *sh* script containing a sequence of calls to *ESPS* programs). Thus, hybrid systems can also be based on compiled software.

50.5.4   Computation vs. Display

The distinction between compiled and interpreted languages is relevant mostly to the computational aspects of the speech software. However, the distinction can apply as well to display software, since some display programs are compiled (e.g., using Motif) while others exploit interpreters (e.g., *Tcl/Tk, Java*).

50.6   Specifying Operations Among Signals

Here we are concerned with the means by which users specify what operations are to be done and on what signals. This consideration is relevant to how speech software can be extended with user-defined operations (see Section 50.7), but is an issue even in software that is not extensible.

The main distinction is between a text-based interface and a visual ("point-and-click") interface. Visual interfaces tend to be less general but easier to use.

50.6.1   Text-Based Interfaces

Traditional interfaces for specifying computations are based on a textual-representation in the form of scripts and programs. For OS-based speech software, operations are typically specified by typing the name of a command (with possible options) directly to a shell. One can also enter a sequence of such commands into a text editor when preparing a script.

This style of specifying operations also is available for workspace-based speech software that is based on a command interpreter. In this case, the text comprises legal commands and programs in the interpreter language.

Both OS-based and workspace-based speech software may also permit the specification of operations using source code in a high-level language (e.g., C) that gets compiled.

50.6.2   Visual ("Point-and-Click") Interfaces

The point-and-click approach has become the ubiquitous user-interface of the 1990s. Operations and operands (signals) are specified by using a mouse or other pointing device to interact with on-screen graphical user-interface (GUI) controls such as buttons and menus. The interface may also have a text-based component to allow the direct entry of parameter values or formulas relating signals.

Visual Interfaces for Display-Oriented Software

In display-oriented software, the signals on which operations are to be performed are visible as waveforms or other directly representative graphics.

A typical user-interaction proceeds as follows: A relevant signal is specified by a mouse-click operation (if a signal segment is involved, it is selected by a click-and-drag operation or by a pair of mouse-click operations). The operation to be performed is then specified by mouse click operations on screen buttons, pull-down menus, or pop-up menus.

This style works very well for unary operations (e.g., compute and display the spectrogram of a given signal segment), and moderately well for binary operations (e.g., add two signals). But it is awkward for operations that have more than two inputs. It is also awkward for specifying chained calculations, especially if you want to repeat the calculations for a new set of signals.

One solution to these problems is provided by a "calculator-style" interface that looks and acts like a familiar arithmetic calculator (except the operands are signal names and the operations are signal processing operations).

Another solution is the "spreadsheet-style" interface. The analogy with spreadsheets is tight. Imagine a spreadsheet in which the cells are replaced by images (waveforms, spectrograms, etc.) connected logically by formulas. For example, one cell might show a test signal, a second might show the results of filtering it, and a third might show a spectrogram of a portion of the filtered signal. This exemplifies a spreadsheet-style interface for speech software.

A spreadsheet-style interface provides some means for specifying the "formulas" that relate the various "cells". This formula interface might itself be implemented in a point-and-click fashion, or it might permit direct entry of formulas in some interpretive language. Speech software with a spreadsheet-style interface will maintain consistency among the visible signals. Thus, if one of the signals is edited or replaced, the other signal graphics change correspondingly, according to the underlying formulas.

DADisp (from DSP Development Corporation) is an example of a spreadsheet-style interface.

Visual Interfaces for Compute-Oriented Software

In a visual interface for display-oriented software, the focus is on the signals themselves. In a visual interface for compute-oriented software, on the other hand, the focus is on the operations. Operations among signals typically are represented as icons with one or more input and output lines that interconnect the operations. In effect, the representation of a signal is reduced to a straight line indicating its relationship (input or output) with respect to operations. Such visual interfaces are often called block-diagram interfaces. In effect, a block-diagram interface provides a visual representation of the computation chain. Various point-and-click means are provided to support the user in creating, examining, and modifying block diagrams.

Ptolomy [4] and *N!Power* are examples of systems that provide a block-diagram interface.

Limitations of Visual Interfaces

Although much in vogue, visual interfaces are inherently limited as a means for specifying signal computations.

For example, the analogy between spreadsheets and spreadsheet-style speech software continues. For simple signal computations, the spreadsheet-style interface can be very useful; computations are simple to set up and informative when operating. For complicated computations, however, the spreadsheet-style interface inherits all of the worst features of spreadsheet programming. It is difficult to encapsulate common sub-calculations, and it is difficult to organize the "program" so that the computational structure is self-evident. The result is that spreadsheet-style programs are hard to write, hard to read, and error-prone.

In this respect, block-diagram interfaces do a better job since their main focus is on the underlying computation rather than on the signals themselves. Thus, screen "real-estate" is devoted to the computation rather than to the signal graphics. However, as the complexity of computations grows, the geometric and visual approach eventually becomes unwieldy. When was the last time you used a flowchart to design or document a program?

It follows that visual interfaces for specifying computations tend to be best suited for teaching and interactive exploration.

50.6.3   Parametric Control of Operations

Speech processing operations often are based on complicated algorithms with numerous parameters. Consequently, the means for specifying parameters is an important issue for speech software.

The simplest form of parametric control is provided by command-line options on command-line programs. This is convenient, but can be cumbersome if there are many parameters. A common alternative is to read parameter values from parameter files that are prepared in advance. Typically, command-line values can be used to override values in the parameter file. A third input source for parameter values is directly from the user in response to prompts issued by the program.

Some systems offer the flexibility of a hierarchy of inputs for parameter values, for example:

- default values
- values from a global parameter file read by all programs
- values from a program-specific parameter file
- values from the command line
- values from the user in response to run-time prompts

In some situations, it is helpful if a current default value is replaced by the most recent input from a given parameter source. We refer to this property as "parameter persistence".

50.7   Extensibility (Closed vs. Open Systems)

Speech software is "closed" if there is no provision for the user to extend it. There is a fixed set of operations available to process and display signals. What you get is all you get.

OS-based systems are always extensible to a degree because they inherit scripting capabilities from the OS, which permits the creation of new commands. They may also provide programming libraries so that the user can write and compile new programs and use them as commands.

Workspace-based systems may be extensible if they are based on an interpreter whose programming language includes the concept of an encapsulated procedure. If so, then users can write scripts that define new commands. Some systems also allow the interpreter to be extended with commands that are implemented by underlying code in C or some other compiled language.

In general, for speech software to be extensible, it must be possible to specify operations (see Section 50.6) and also to re-use the resulting specifications in other contexts. A block-diagram interface is extensible, for example, if a given diagram can be reduced to an icon that is available for use as a single block in another diagram.

For speech software with visual interfaces, extensibility considerations also include the ability to specify new GUI controls (visible menus and buttons), the ability to tie arbitrary internal and external computations to GUI controls, and the ability to define new display methods for new signal types.

In general, extended commands may behave differently from the built-in commands provided with the speech software. For example, built-in commands may share a common user interface that is difficult to implement in an independent script or program (such a common interface might provide standard parameters for debug control, standard processing of parameter files, etc.).

If user-defined scripts, programs, and GUI components are indistinguishable from built-in facilities, we say that the speech software provides seamless extensibility.

50.8   Consistency Maintenance

A speech processing chain involves signals, operations, and parameter sets. An important consideration for speech software is whether or not consistency is maintained among all of these. Thus, for example, if one input signal is replaced with another, are all intermediate and output signals recalculated automatically? Consistency maintenance is primarily an issue for speech software with visual interfaces, namely whether or not the software guarantees that all aspects of the visible displays are consistent with each other.

Spreadsheet-style interfaces (for display-oriented software) and block-diagram interfaces (for compute-oriented software) usually provide consistency maintenance.

50.9   Other Characteristics of Common Approaches

50.9.1   Memory-based vs. File-based

"Memory-based" speech software carries out all of its processing and display operations on signals that are stored entirely within memory, regardless of whether or not the signals also have an external representation as a disk file. This approach has obvious limitations with respect to signal size, but it simplifies programming and yields fast operation. Thus, memory-based software is well-suited for teaching and the interactive exploration of small samples.

In "file-based" speech software, on the other hand, signals are represented and manipulated as disk files. The software partially buffers portions of the signal in memory as required for processing and display operations. Although programming can be more complicated, the advantage is that there are no inherent limitations on signal size. The file-based approach is, therefore, well-suited for large scale experiments.

50.9.2   Documentation of Processing History

Modern speech processing involves complicated algorithms with many processing steps and operating parameters. As a result, it is often important to be able to reconstruct exactly how a given signal was produced. Speech software can help here by creating appropriate records as signal and parameter files are processed.

The most common method for recording this information about a given signal is to put it in the same file as the signal. Most modern speech software uses a file format that includes a "file header" that is used for this purpose. Most systems store at least some information in the header, e.g., the sampling rate of the signal. Others, such as *ESPS,* attempt to store all relevant information. In this approach, the header of a signal file produced by any program includes the program name, values of processing parameters, and the names and headers of all source files. The header is a recursive structure, so that the headers of the source files themselves contain the names and headers of files that were prior sources. Thus, a signal file header contains the headers of all source files in the processing chain. It follows that files contain a complete history of the origin of the data in the file and all the intermediate processing steps. The importance of record keeping grows with the complexity of computation chains and the extent of available parametric control.

50.9.3   Personalization

There is considerable variation in the extent to which speech software can be customized to suit personal requirements and tastes. Some systems cannot be personalized at all; they start out the same way, every time. But most systems store personal preferences and use them again next time. Savable preferences may include color selections, button layout, button semantics, menu contents, currently loaded signals, visible windows, window arrangement, and default parameter sets for speech processing operations.

At the extreme, some systems can save a complete "snapshot" that permits exact resumption. This is particularly important for the interactive study of complicated signal configurations across repeated software sessions.

50.9.4   Real-Time Performance

Software is generally described as "real-time" if it is able to keep up with relevant, changing inputs. In the case of speech software, this usually means that the software can keep up with input speech.

Even this definition is not particularly meaningful unless the input speech is itself coming from a human speaker and digitized in real-time. Otherwise, the real-issue is whether or not the software is fast enough to keep up with interactive use.

For example, if one is testing speech recognition software by directly speaking into the computer, real-time performance is important. It is less important, on the other hand, if the test procedure involves running batch scripts on a database of speech files.

If the speech software is designed to take input directly from devices (or pipes, in the case of Unix), then the issue becomes one of CPU speed.

50.9.5   Source Availability

It is unfortunate but true that the best documentation for a given speech processing command is often the source code. Thus, the availability of source code may be an important factor for this reason alone. Typically, this is more important when the software is used in advanced R&D applications. Sources also are needed if users have requirements to port the speech software to additional platforms. Source availability may also be important for extensibility, since it may not be possible to extend the speech software without the sources.

If the speech software is interpreter-based, sources of interest will include the sources for any built-in operations that are implemented as interpreter scripts.

50.9.6 Hardware Requirements

Speech software may require the installation of special purpose hardware. There are two main reasons for such requirements: to accelerate particular computations (e.g., spectrograms), and to provide speech I/O with A/D and D/A converters.

Such hardware has several disadvantages. It adds to the system cost, and it decreases the overall reliability of the system. It may also constrain system software upgrades; for example, the extra hardware may use special device drivers that do not survive OS upgrades. Special purpose hardware used to be common, but is less so now owing to the continuing increase in CPU speeds and the prevalence of built-in audio I/O. It is still important, however, when maximum speed and high-quality audio I/O are important. *CSL* is a good example of an integrated hardware/software approach.

50.9.7 Cross-Platform Compatibility

If your hardware platform may change or your site has a variety of platforms, then it is important to consider whether the speech software is available across a variety of platforms. Source availability (Section 50.9.5) is relevant here.

If you intend to run the speech software on several platforms that have different underlying numeric representations (a byte order difference being most likely), then it is important to know whether the file formats and signal I/O software support transparent data exchange.

50.9.8 Degree of Specialization

Some speech software is intended for general purpose work in speech (e.g., *ESPS/waves*, *MATLAB*^TM). Other software is intended for more specialized usage. Some of the areas where specialized software tools may be relevant include linguistics, recognition, synthesis, coding, psycho-acoustics, clinical-voice, music, multi-media, sound and vibration, etc. Two examples are *HTK* for recognition, and *Delta* (from Eloquent Technology) for synthesis.

50.9.9 Support for Speech Input and Output

In the past, built-in speech I/O hardware was uncommon in workstations and PCs, so speech software typically supported speech I/O by means of add-on hardware supplied with the software or available from other third parties. This provided the desired capability, albeit with the disadvantages mentioned earlier (see Section 50.9.6).

Today most workstations and PCs have built-in audio support that can be used directly by the speech software. This avoids the disadvantages of add-on hardware, but the resulting A/D-D/A quality can be too noisy or otherwise inadequate for use in speech R&D (the built-in audio is typically designed for more mundane requirements). There are various reasons why special-purpose hardware may still be needed, including:

- need for more than two channels
- need for very high sampling rates
- compatibility with special hardware (e.g., DAT tape)

50.10   File Formats (Data Import/Export)

Signal file formats are fundamentally important because they determine how easy it is for independent programs to read and write the files (interoperability). Furthermore, the format determines whether files can contain all of the information that a program might need to operate on the file's primary data (e.g., can the file contain the sampling frequency in addition to a waveform itself?).

The best way to design speech file formats is hotly debated, but the clear trend has been towards "self-describing" file formats that include information about the names, data types, and layout of all data in the file. (For example, this permits programs to retrieve data by name.)

There are many popular file formats, and various programs are available for converting among them (e.g., SOX). For speech sampled data, the most important file format is Sphere (from NIST), which is used in the speech databases available from the Linguistic Data Consortium (LDC). Sphere supports several data compression formats in a variety of standard and specialized formats.

Sphere works well for sampled data files, but is limited for more general speech data files. A general purpose, public-domain format (*Esignal*) has recently been made available by Entropic.

50.11   Speech Databases

Numerous databases (or corpora) of speech are available from various sources. For a current list, see the comp.speech Frequently Asked Questions (FAQ) (see Section 50.13). The largest supplier of speech data is the Linguistic Data Consortium, which publishes a large number of CDs containing speech and linguistic data.

50.12   Summary of Characteristics and Uses

In Section 50.1, we mentioned that the three most common uses for speech software are teaching, interactive exploration, and batch experiments. And at various points during the discussion of speech software characteristics, we mentioned their relative importance for the different classes of software uses. We attempt to summarize this in Table 50.1, where the symbol "•" indicates that a characteristic is particularly useful or important.

It is important not to take Table 50.1 too seriously. As we mentioned at the outset, the various distinguishing characteristics discussed in this section are not independent (i.e., there is considerable overlap). Furthermore, the three classes of software use are broad and not always easily distinguishable; i.e., the importance of particular software characteristics depends a lot on the details of intended use. Nevertheless, Table 50.1 is a reasonable starting point for evaluating particular software in the context of intended use.

50.13   Sources for Finding Out What is Currently Available

The best single online source of general information is the Internet news group comp.speech, and in particular its FAQ (see http://svr-www.eng.cam.ac.uk/comp.speech/). Use this as a starting point.

Here are some other WWW sites that (at this writing) contain speech software information or pointers to other sites:

```
http://svr-www.eng.cam.ac.uk
http://mambo.ucsc.edu/psl/speech.html
http://www.bdti.com/faq/dsp_faq.html
http://www.ldc.upenn.edu
http://www.entropic.com
```

TABLE 50.1 Relative Importance of Software Characteristics

	Teaching	Interactive exploration	Batch experiments
OS-based (50.3.1)			•
Workspace-based (50.3.2)	•		
Compute-oriented (50.4.1)			•
Display-oriented (50.4.2)	•	•	
Compiled (50.5.2)			•
Interpreted (50.5.1)		•	
Text-based interface (50.6.1)		•	•
Visual interface (50.6.2)	•	•	
Memory-based (50.9.1)	•	•	
File-based (50.9.1)			•
Parametric control (50.6.3)	•	•	•
Consistency maintenance (50.8.0)	•	•	
History documentation (50.9.2)		•	•
Extensibility (50.7.0)		•	•
Personalization (50.9.3)		•	•
Real-time performance (50.9.4)	•	•	
Source availability (50.9.5)		•	•
Cross-platform compatibility (50.9.7)	•	•	•
Support for speech I/O (50.9.9)	•	•	

50.14 Future Trends

From the user's viewpoint, speech software will continue to become easier to use, with a heavier reliance on visual interfaces with consistency maintenance.

Calculator, spreadsheet, and block-diagram interfaces will become more common, but will not eliminate text-based (programming, scripting) interfaces for specifying computations and system extensions.

Software will become more open. Seamless extensibility will be more common, and extensions will be easier. GUI extensions as well as computation extensions will be supported.

There will be less of a distinction between compute-oriented and display-oriented speech software. Hybrid compute/display-oriented software will dominate.

Visualization will become more important and more sophisticated, particularly for multidimensional data. "Movies" will be used to show arbitrary 3D data. Sound will be used to represent an arbitrary dimension. Various methods will be available to project N-dimensional data into 2- or 3-space. (This will be used, for example, to show aspects of vector quantization or HMM clustering.)

Public-domain file formats will dominate proprietary formats. Networked computers will be used for parallel computation if available. *Tcl/Tk* and *Java* will grow in popularity as a base for graphical data displays and user interfaces.

References

[1] Henke, W.L., Speech and audio computer-aided examination and analysis facility, Quarterly Progress Rep. No. 95, MIT Research Laboratory for Electronics, 1969, 69–73.
[2] Henke, W.L., MITSYN — An interactive dialogue language for time signal processing, MIT Research Laboratory for Electronics, report RLE TM-1, 1975.

[3] Kopec, G., The integrated signal processing system ISP, *IEEE Trans. on Acoustics, Speech, and Signal Processing,* ASSP-32(4), 842-851, Aug. 1984.

[4] Pino, J.L., Ha, S., Lee, E.A. and Buck, J.T., Software synthesis for DSP using ptolemy, *J. VLSI Signal Processing,* 9(1), 7-21, Jan. 1995.

[5] Potter, R.K., Kopp, G.A. and Green, H.C., *Visible Speech,* D. Van Nostrand Company, New York, 1946.

[6] Shipman, D., SpireX: Statistical analysis in the SPIRE acoustic-phonetic workstation, *Proc. ICASSP,* Boston, 1983.

[7] Shore, J., Interactive signal processing with UNIX, *Speech Technol.,* 3, March/April 1988.

[8] Talkin, D., Looking at speech, *Speech Technol.,* 4, April/May 1989.

Image and Video Processing

Jan Biemond
Delft University of Technology
Russell M. Mersereau
Georgia Institute of Technology

I MAGE AND VIDEO SIGNAL PROCESSING is quite different from other forms of signal processing
for a variety of reasons. The most obvious difference lies in the fact that these signals are two or three
dimensional. This means that some familiar techniques used for processing one-dimensional signals,
for example, those that require factorization of polynomials, have to be abandoned. Other techniques for
filtering, sampling, and transform computation have to be modified. Even more compromises have to be
made, however, because of the signals' size. Images and sequences of images can be huge. For example,
processing sequences of color images each of which contains 780 rows and 1024 columns at a frame rate of 30
frames per second requires a data rate of 72 megabytes per second. Successful image processing techniques
reward careful attention to problem requirements, algorithmic complexity, and machine architecture. The
past decade has been particularly exciting as each new wave of faster computing hardware has opened the
door to new applications. This is a trend that will likely continue for some time.

The following chapters, written by experts in their fields, highlight the state-of-the-art in several aspects
of image and video processing. The range of topics is quite broad. While it includes some discussions of
techniques that go back more than a decade, the emphasis is on current practice. There is some danger
in this, because the field is changing very rapidly, but, on the other hand, many of the concepts on which
these current techniques are based should be around for some time.

Chapter 51 is a very long and thorough discussion of image processing fundamentals. For a novice to the
field, this material is important for a complete understanding. It discusses the basics of how images differ
from other types of signals and how the limitations of cameras, displays, and the human visual system affect
the kinds of processing that can be done. It also defines the basic theory of multidimensional digital signal
processing, particularly with respect to how linear and nonlinear filtering, transform computation, and
sampling are generalized from the one-dimensional case. Other topics treated include statistical models
for images, models for recording distortions, histogram-based methods for image processing, and image
segmentation.

Probably the most visible image processing is occurring in the development of standards for image and
video compression. JPEG, MPEG, and digital television are all highly visible success stories. Chapter 52
looks at methods for still image compression including JPEG, wavelet, and fractal coders. Image compression is successful because image samples are spatially correlated with their neighbors. Operators such as
the discrete cosine transform (DCT) largely remove this correlation and capture the essence of an image
block in a few parameters that can be quantized and transmitted. The transform domain also enables these
coders to exploit limitations in the human visual system. Chapters 55 and 56 extend these approaches to
video and television compression, respectively. Video compression achieves significant additional compression gains by exploiting the temporal redundancy that is present in video sequences. This is done by
using simple models for modeling object motion within a scene, using these models to predict the current
frame, and then encoding only the model parameters and the quantized prediction errors.

Images are often distorted when they are recorded. This might be caused by out-of-focus optics, motion
blur, camera noise, or coding errors. Chapter 53 looks at methods for image and video restoration.
This is the most mathematically based area of image processing, and it is also one of the areas with the

longest history. It has applications in the analysis of astronomical images, in forensic imaging, and in the production of high-quality stills from video sequences.

Chapter 54 looks at methods for motion estimation and video scan conversion. Motion estimation is a key technique for removing temporal redundancy in image sequences and, as a result, it is a key component in all of the video compression standards. It is also, however, a highly time-consuming numerically ill-posed operation. As a result it continues to be highly studied, particularly with respect to more sophisticated motion models. A related problem is the problem of scanning format conversion. This is a major issue in television systems where both interlaced and progressively scanned images are encountered.

Chapter 57 explores stereoscopic and multiview image processing. Traditional image processing assumes that only one camera is present. As a result depth information in a three-dimensional scene is lost. When explicit depth information is needed, multiple cameras can be used. Differences in the displacement of objects in the left and right images can be converted to depth measurements. Mammals do this naturally with their two eyes. Stereoscopic image processing techniques are becoming increasingly used in problems of computer vision and computer graphics. This chapter discusses the state-of-the-art in this emerging area.

The final two chapters in this section, Chapters 58 and 59, look at software and hardware systems for doing image processing. Chapter 58 provides an overview of a representative set of image software packages that embody the core capabilities required by many image processing applications. It also provides a list of Internet addresses for a number of image databases. Chapter 59 provides an overview of VLSI architectures for implementing many of the video compression standards.

51

Image Processing Fundamentals

Ian T. Young
Delft University of Technology,
The Netherlands

Jan J. Gerbrands
Delft University of Technology,
The Netherlands

Lucas J. van Vliet
Delft University of Technology,
The Netherlands

51.1 Introduction

Modern digital technology has made it possible to manipulate multidimensional signals with systems that range from simple digital circuits to advanced parallel computers. The goal of this manipulation can be

0-8493-8572-5/98/$0.00+$.50

divided into three categories:

- Image Processing   *image in → image out*
- Image Analysis   *image in → measurements out*
- Image Understanding   *image in → high-level description out*

In this section we will focus on the fundamental concepts of *image processing*. Space does not permit us to make more than a few introductory remarks about *image analysis*. *Image understanding* requires an approach that differs fundamentally from the theme of this handbook, *Digital Signal Processing*. Further, we will restrict ourselves to two-dimensional (2D) image processing although most of the concepts and techniques that are to be described can be extended easily to three or more dimensions.

We begin with certain basic definitions. An image defined in the "real world" is considered to be a function of two real variables, for example, $a(x, y)$ with a as the amplitude (e.g., brightness) of the image at the *real* coordinate position (x, y). An image may be considered to contain sub-images sometimes referred to as *regions-of-interest, ROIs,* or simply *regions*. This concept reflects the fact that images frequently contain collections of objects each of which can be the basis for a region. In a sophisticated image processing system it should be possible to apply specific image processing operations to selected regions. Thus, one part of an image (region) might be processed to suppress motion blur while another part might be processed to improve color rendition.

The amplitudes of a given image will almost always be either real numbers or integer numbers. The latter is usually a result of a quantization process that converts a continuous range (say, between 0 and 100%) to a discrete number of levels. In certain image-forming processes, however, the signal may involve photon counting which implies that the amplitude would be inherently quantized. In other image forming procedures, such as magnetic resonance imaging, the direct physical measurement yields a complex number in the form of a real magnitude and a real phase. For the remainder of this introduction we will consider amplitudes as reals or integers unless otherwise indicated.

51.2 Digital Image Definitions

A digital image $a[m, n]$ described in a 2D discrete space is derived from an analog image $a(x, y)$ in a 2D continuous space through a *sampling* process that is frequently referred to as digitization. The mathematics of that sampling process will be described in section 51.5. For now we will look at some basic definitions associated with the digital image. The effect of digitization is shown in Fig. 51.1.

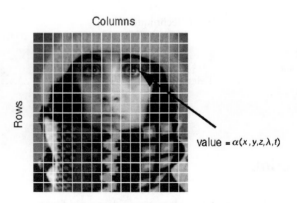

FIGURE 51.1 Digitization of a continuous image. The pixel at coordinates $[m = 10, n = 3]$ has the integer brightness value 110.

The 2D continuous image $a(x, y)$ is divided into N *rows* and M *columns.* The intersection of a row and a column is termed a *pixel.* The value assigned to the integer coordinates $[m, n]$ with $\{m = 0, 1, 2, \ldots, M - 1\}$ and $\{n = 0, 1, 2, \ldots, N - 1\}$ is $a[m, n]$. In fact, in most cases $a(x, y)$ — which we might consider to be the physical signal that impinges on the face of a 2D sensor — is actually a function of many variables including depth (z), color (λ), and time (t). Unless otherwise stated, we will consider the case of 2D, monochromatic, static images in this chapter.

The image shown in Fig. 51.1 has been divided into $N = 16$ rows and $M = 16$ columns. The value assigned to every pixel is the average brightness in the pixel rounded to the nearest integer value. The process of representing the amplitude of the 2D signal at a given coordinate as an integer value with L different gray levels is usually referred to as amplitude quantization or simply *quantization.*

51.2.1 Common Values

There are standard values for the various parameters encountered in digital image processing. These values can be caused by video standards, algorithmic requirements, or the desire to keep digital circuitry simple. Table 51.1 gives some commonly encountered values.

TABLE 51.1
Common Values of Digital Image Parameters

Parameter	Symbol	Typical Values
Rows	N	256,512,525,625,1024,1035
Columns	M	256,512,768,1024,1320
Gray levels	L	2,64,256,1024,4096,16384

Quite frequently we see cases of $M = N = 2^K$ where $\{K = 8, 9, 10\}$. This can be motivated by digital circuitry or by the use of certain algorithms such as the (fast) Fourier transform (see section 51.3.3).

The number of distinct gray levels is usually a power of 2, that is, $L = 2^B$ where B is the number of bits in the binary representation of the brightness levels. When $B > 1$, we speak of a *gray-level image;* when $B = 1$, we speak of a *binary image.* In a binary image there are just two gray levels which can be referred to, for example, as "black" and "white" or "0" and "1".

51.2.2 Characteristics of Image Operations

There is a variety of ways to classify and characterize image operations. The reason for doing so is to understand what type of results we might expect to achieve with a given type of operation or what might be the computational burden associated with a given operation.

Types of Operations

The types of operations that can be applied to digital images to transform an input image $a[m, n]$ into an output image $b[m, n]$ (or another representation) can be classified into three categories as shown in Table 51.2.

This is shown graphically in Fig. 51.2.

Types of Neighborhoods

Neighborhood operations play a key role in modern digital image processing. It is therefore important to understand how images can be sampled and how that relates to the various neighborhoods that can be used to process an image.

TABLE 51.2   Types of Image Operations

Operation	Characterization	Generic Complexity / Pixel
• *Point*	- the output value at a specific coordinate is dependent only on the input value at that same coordinate.	*constant*
• *Local*	- the output value at a specific coordinate is dependent on the input values in the *neighborhood* of that same coordinate.	P^2
• *Global*	- the output value at a specific coordinate is dependent on all the values in the input image.	N^2

Note: Image size $= N \times N$; neighborhood size $= P \times P$. Note that the complexity is specified in operations *per pixel*.

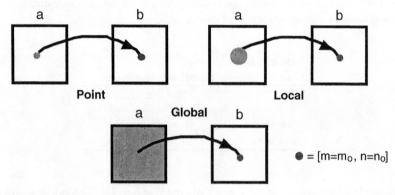

FIGURE 51.2   Illustration of various types of image operations.

• Rectangular sampling — In most cases, images are sampled by laying a rectangular grid over an image as illustrated in Fig. 51.1. This results in the type of sampling shown in Fig. 51.3(a) and 51.3(b).

• Hexagonal sampling — An alternative sampling scheme is shown in Fig. 51.3(c) and is termed hexagonal sampling.

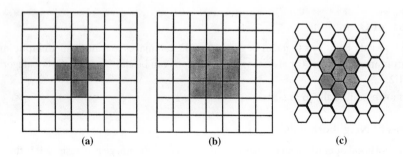

FIGURE 51.3   (a) Rectangular sampling 4-connected; (b) rectangular sampling 8-connected; (c) hexagonal sampling 6-connected.

Both sampling schemes have been studied extensively and both represent a possible periodic tiling of the continuous image space. We will restrict our attention, however, to only rectangular sampling as it remains, due to hardware and software considerations, the method of choice.

Local operations produce an output pixel value $b[m = m_0, n = n_0]$ based on the pixel values in the *neighborhood* of $a[m = m_0, n = n_0]$. Some of the most common neighborhoods are the 4-connected neighborhood and the 8-connected neighborhood in the case of rectangular sampling and the 6-connected neighborhood in the case of hexagonal sampling illustrated in Fig. 51.3.

51.2.3 Video Parameters

We do not propose to describe the processing of dynamically changing images in this introduction. It is appropriate — given that many static images are derived from video cameras and frame grabbers — to mention the standards that are associated with the three standard video schemes currently in worldwide use — NTSC, PAL, and SECAM. This information is summarized in Table 51.3.

TABLE 51.3   Standard Video Parameters

Property	NTSC	Standard PAL	SECAM
images/second	29.97	25	25
ms/image	33.37	40.0	40.0
lines/image	525	625	625
(horiz./vert.) = aspect ratio	4:3	4:3	4:3
interlace	2:1	2:1	2:1
μs /line	63.56	64.00	64.00

In an interlaced image, the odd numbered lines (1, 3, 5, . . .) are scanned in half of the allotted time (e.g., 20 ms in PAL) and the even numbered lines (2, 4, 6, . . .) are scanned in the remaining half. The image display must be coordinated with this scanning format. (See section 51.8.2.) The reason for interlacing the scan lines of a video image is to reduce the perception of flicker in a displayed image. If one is planning to use images that have been scanned from an interlaced video source, it is important to know if the two half-images have been appropriately "shuffled" by the digitization hardware or if that should be implemented in software. Further, the analysis of moving objects requires special care with interlaced video to avoid "zigzag" edges.

The number of rows (N) from a video source generally corresponds one-to-one with lines in the video image. The number of columns, however, depends on the nature of the electronics that is used to digitize the image. Different frame grabbers for the same video camera might produce $M = 384, 512,$ or 768 columns (pixels) per line.

51.3 Tools

Certain tools are central to the processing of digital images. These include mathematical tools such as *convolution, Fourier analysis,* and *statistical* descriptions, and manipulative tools such as *chain codes* and *run codes.* We will present these tools without any specific motivation. The motivation will follow in later sections.

51.3.1 Convolution

There are several possible notations to indicate the convolution of two (multidimensional) signals to produce an output signal. The most common are:

$$c = a \otimes b = a * b \tag{51.1}$$

We shall use the first form, $c = a \otimes b$, with the following formal definitions.

In 2D continuous space:

$$c(x, y) = a(x, y) \otimes b(x, y) = \int_{-\infty}^{+\infty} \int_{-\infty}^{+\infty} a(\chi, \zeta) b(x - \chi, y - \zeta) d\chi d\zeta \tag{51.2}$$

In 2D discrete space:

$$c[m, n] = a[m, n] \otimes b[m, n] = \sum_{j=-\infty}^{+\infty} \sum_{k=-\infty}^{+\infty} a[j, k] b[m - j, n - k] \tag{51.3}$$

51.3.2 Properties of Convolution

There are a number of important mathematical properties associated with convolution.

- Convolution is *commutative*.

$$c = a \otimes b = b \otimes a \tag{51.4}$$

- Convolution is *associative*.

$$c = a \otimes (b \otimes d) = (a \otimes b) \otimes d = a \otimes b \otimes d \tag{51.5}$$

- Convolution is *distributive*.

$$c = a \otimes (b + d) = (a \otimes b) + (a \otimes d) \tag{51.6}$$

where a, b, c, and d are all images, either continuous or discrete.

51.3.3 Fourier Transforms

The Fourier transform produces another representation of a signal, specifically a representation as a weighted sum of complex exponentials. Because of Euler's formula:

$$e^{jq} = \cos(q) + j \sin(q) \tag{51.7}$$

where $j^2 = -1$, we can say that the Fourier transform produces a representation of a (2D) signal as a weighted sum of sines and cosines. The defining formulas for the forward Fourier and the inverse Fourier transforms are as follows. Given an image a and its Fourier transform A, the forward transform goes from the spatial domain (either continuous or discrete) to the frequency domain which is always continuous.

$$\textit{Forward -} \quad A = \mathcal{F}\{a\} \tag{51.8}$$

The inverse Fourier transform goes from the frequency domain back to the spatial domain

$$\textit{Inverse -} \quad a = \mathcal{F}^{-1}\{A\} \tag{51.9}$$

The Fourier transform is a unique and invertible operation so that:

$$a = \mathcal{F}^{-1}\left\{\mathcal{F}\{a\}\right\} \quad \text{and} \quad A = \mathcal{F}\left\{\mathcal{F}^{-1}\{A\}\right\} \tag{51.10}$$

The specific formulas for transforming back and forth between the spatial domain and the frequency domain are given below.

In 2D continuous space:

$$Forward\text{ -} \quad A(u, v) = \int_{-\infty}^{+\infty} \int_{-\infty}^{+\infty} a(x, y) e^{-j(ux+vy)} dx dy \tag{51.11}$$

$$Inverse\text{ -} \quad a(x, y) = \frac{1}{4\pi^2} \int_{-\infty}^{+\infty} \int_{-\infty}^{+\infty} A(u, v) e^{+j(ux+vy)} du dv \tag{51.12}$$

In 2D discrete space:

$$Forward\text{ -} \quad A(\Omega, \Psi) = \sum_{m=-\infty}^{+\infty} \sum_{n=-\infty}^{+\infty} a[m, n] e^{-j(\Omega m + \Psi n)} \tag{51.13}$$

$$Inverse\text{ -} \quad a[m, n] = \frac{1}{4\pi^2} \int_{-\pi}^{+\pi} \int_{-\pi}^{+\pi} A(\Omega, \Psi) e^{+j(\Omega m + \Psi n)} d\Omega d\Psi \tag{51.14}$$

51.3.4 Properties of Fourier Transforms

There are a variety of properties associated with the Fourier transform and the inverse Fourier transform. The following are some of the most relevant for digital image processing.

• The Fourier transform is, in general, a complex function of the real frequency variables. As such, the transform can be written in terms of its magnitude and phase.

$$A(u, v) = |A(u, v)| e^{j\varphi(u,v)} \qquad A(\Omega, \Psi) = |A(\Omega, \Psi)| e^{j\varphi(\Omega, \Psi)} \tag{51.15}$$

• A 2D signal can also be complex and thus written in terms of its magnitude and phase.

$$a(x, y) = |a(x, y)| e^{j\vartheta(x,y)} \qquad a[m, n] = |a[m, n]| e^{j\vartheta[m,n]} \tag{51.16}$$

• If a 2D signal is real, then the Fourier transform has certain symmetries.

$$A(u, v) = A^*(-u, -v) \qquad A(\Omega, \Psi) = A^*(-\Omega, -\Psi) \tag{51.17}$$

The symbol (∗) indicates complex conjugation. For real signals Eq. (51.17) leads directly to:

$$|A(u, v)| = |A(-u, -v)| \qquad \varphi(u, v) = -\varphi(-u, -v)$$

$$|A(\Omega, \Psi)| = |A(-\Omega, -\Psi)| \qquad \varphi(\Omega, \Psi) = -\varphi(-\Omega, -\Psi) \tag{51.18}$$

• If a 2D signal is real and even, then the Fourier transform is real and even.

$$A(u, v) = A(-u, -v) \qquad A(\Omega, \Psi) = A(-\Omega, -\Psi) \tag{51.19}$$

• The Fourier and the inverse Fourier transforms are linear operations.

$$\mathcal{F}\{w_1 a + w_2 b\} = \mathcal{F}\{w_1 a\} + \mathcal{F}\{w_2 b\} = w_1 A + w_2 B$$

$$\mathcal{F}^{-1}\{w_1 A + w_2 B\} = \mathcal{F}^{-1}\{w_1 A\} + \mathcal{F}^{-1}\{w_2 B\} = w_1 a + w_2 b \tag{51.20}$$

where a and b are 2D signals (images) and w_1 and w_2 are arbitrary, complex constants.

• The Fourier transform in discrete space, $A(\Omega, \Psi)$, is periodic in both Ω and Ψ. Both periods are 2π.

$$A(\Omega + 2\pi j, \Psi + 2\pi k) = A(\Omega, \Psi) \qquad j, k \text{ integers} \tag{51.21}$$

• The energy, E, in a signal can be measured either in the spatial domain or the frequency domain. For a signal with finite energy:

Parseval's theorem (2D continuous space):

$$E = \int_{-\infty}^{+\infty} \int_{-\infty}^{+\infty} |a(x, y)|^2 \, dxdy = \frac{1}{4\pi^2} \int_{-\infty}^{+\infty} \int_{-\infty}^{+\infty} |A(u, v)|^2 \, dudv \tag{51.22}$$

Parseval's theorem (2D discrete space):

$$E = \sum_{m=-\infty}^{+\infty} \sum_{n=-\infty}^{+\infty} |a[m, n]|^2 = \frac{1}{4\pi^2} \int_{-\pi}^{+\pi} \int_{-\pi}^{+\pi} |A(\Omega, \Psi)|^2 \, d\Omega d\Psi \tag{51.23}$$

This "signal energy" is not to be confused with the physical energy in the phenomenon that produced the signal. If, for example, the value $a[m, n]$ represents a photon count, then the *physical* energy is proportional to the amplitude, a, and not the square of the amplitude. This is generally the case in video imaging.

• Given three, multi-dimensional signals a, b, and c and their Fourier transforms A, B, and C:

$$c = a \otimes b \quad \overset{\mathcal{F}}{\leftrightarrow} \quad C = A \bullet B$$

and

$$c = a \bullet b \quad \overset{\mathcal{F}}{\leftrightarrow} \quad C = \frac{1}{4\pi^2} A \otimes B \tag{51.24}$$

In words, convolution in the spatial domain is equivalent to multiplication in the Fourier (frequency) domain and vice-versa. This is a central result which provides not only a methodology for the implementation of a convolution but also insight into how two signals interact with each other — under convolution — to produce a third signal. We shall make extensive use of this result later.

• If a two-dimensional signal $a(x, y)$ is scaled in its spatial coordinates then:

$$\text{If } a(x, y) \quad \rightarrow \quad a\left(M_x \bullet x, M_y \bullet y\right)$$

$$\text{Then } A(u, v) \quad \rightarrow \quad A\left(u / M_x, v / M_y\right) / \left|M_x \bullet M_y\right| \tag{51.25}$$

• If a two-dimensional signal $a(x, y)$ has Fourier spectrum $A(u, v)$ then:

$$A(u = 0, v = 0) = \int_{-\infty}^{+\infty} \int_{-\infty}^{+\infty} a(x, y) dxdy$$

$$a(x = 0, y = 0) = \frac{1}{4\pi^2} \int_{-\infty}^{+\infty} \int_{-\infty}^{+\infty} A(u, v) dxdy \tag{51.26}$$

• If a two-dimensional signal $a(x, y)$ has Fourier spectrum $A(u, v)$ then:

$$\frac{\partial a(x, y)}{\partial x} \overset{\mathcal{F}}{\leftrightarrow} ju A(u, v) \qquad \frac{\partial a(x, y)}{\partial y} \overset{\mathcal{F}}{\leftrightarrow} jv A(u, v)$$

$$\frac{\partial^2 a(x, y)}{\partial x^2} \overset{\mathcal{F}}{\leftrightarrow} -u^2 A(u, v) \qquad \frac{\partial^2 a(x, y)}{\partial y^2} \overset{\mathcal{F}}{\leftrightarrow} -v^2 A(u, v) \tag{51.27}$$

Importance of Phase and Magnitude

Equation (51.15) indicates that the Fourier transform of an image can be complex. This is illustrated below in Fig. 51.4(a-c). Figure 51.4(a) shows the original image $a[m, n]$, Fig. 51.4(b) the magnitude in a scaled form as $\log(|A(\Omega, \Psi)|)$, and Fig. 51.4(c) the phase $\varphi(\Omega, \Psi)$.

(a) (b) (c)

FIGURE 51.4   (a) Original; (b) $\log(|A(\Omega, \Psi)|)$; (c) $\varphi(\Omega, \Psi)$.

Both the magnitude and the phase functions are necessary for the complete reconstruction of an image from its Fourier transform. Figure 51.5(a) shows what happens when Fig. 51.4(a) is restored solely on the basis of the magnitude information and Fig. 51.5(b) shows what happens when Fig. 51.4(a) is restored solely on the basis of the phase information.

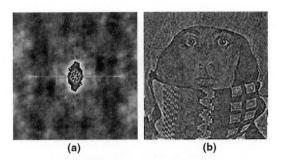

(a) (b)

FIGURE 51.5   (a) $\varphi(\Omega, \Psi) = 0$ and (b) $|A(\Omega, \Psi)| = constant$.

Neither the magnitude information nor the phase information is sufficient to restore the image. The magnitude-only image, Fig. 51.5(a), is unrecognizable and has severe dynamic range problems. The phase-only image, Fig. 51.5(b), is barely recognizable, that is, severely degraded in quality.

Circularly Symmetric Signals

An arbitrary 2D signal $a(x, y)$ can always be written in a polar coordinate system as $a(r, \theta)$. When the 2D signal exhibits a circular symmetry this means that:

$$a(x, y) = a(r, \theta) = a(r) \tag{51.28}$$

where $r^2 = x^2 + y^2$ and $\tan \theta = y/x$. As a number of physical systems, such as lenses, exhibit circular symmetry, it is useful to be able to compute an appropriate Fourier representation.

The Fourier transform $A(u, v)$ can be written in polar coordinates $A(\omega_r, \xi)$ and then, for a circularly symmetric signal, rewritten as a *Hankel transform:*

$$A(u, v) = \mathcal{F}\{a(x, y)\} = 2\pi \int_0^\infty a(r) J_0(\omega_r r) r \, dr = A(\omega_r) \tag{51.29}$$

where $\omega_r^2 = u^2 + v^2$ and $\tan \xi = v/u$ and $J_0(\bullet)$ is a Bessel function of the first kind of order zero.

The inverse *Hankel transform* is given by:

$$a(r) = \frac{1}{2\pi} \int_0^\infty A(\omega_r) J_0(\omega_r r) \omega_r d\omega_r \tag{51.30}$$

The Fourier transform of a circularly symmetric 2D signal is a function of only the radial frequency, ω_r. The dependence on the angular frequency, ξ, has vanished. Further, if $a(x, y) = a(r)$ is real, then it is automatically even due to the circular symmetry. According to Eq. (51.19), $A(\omega_r)$ will then be real and even.

Examples of 2D Signals and Transforms

Table 51.4 shows some basic and useful signals and their 2D Fourier transforms. In using the table entries in the remainder of this chapter, we will refer to a spatial domain term as the *point spread function (PSF)* or the *2D impulse response* and its Fourier transforms as the *optical transfer function (OTF)* or simply *transfer function*. Two standard signals used in this table are $u(\bullet)$, the unit step function, and $J_1(\bullet)$, the Bessel function of the first kind. Circularly symmetric signals are treated as functions of r as in Eq. (51.28).

51.3.5 Statistics

In image processing, it is quite common to use simple statistical descriptions of images and sub-images. The notion of a statistic is intimately connected to the concept of a probability distribution, generally the distribution of signal amplitudes. For a given region — which could conceivably be an entire image — we can define the probability *distribution* function of the brightnesses in that region and the probability *density* function of the brightnesses in that region. We will assume in the discussion that follows that we are dealing with a digitized image $a[m, n]$.

Probability Distribution Function of the Brightnesses

The probability distribution function, $P(a)$, is the probability that a brightness chosen from the region is less than or equal to a given brightness value a. As a increases from $-\infty$ to $+\infty$, $P(a)$ increases from 0 to 1. $P(a)$ is monotonic, nondecreasing in a and thus $dP/da \geq 0$.

Probability Density Function of the Brightnesses

The probability that a brightness in a region falls between a and $a + \Delta a$, given the probability distribution function $P(a)$, can be expressed as $p(a)\Delta a$ where $p(a)$ is the probability density function:

$$p(a)\Delta a = \left(\frac{dP(a)}{da}\right)\Delta a \tag{51.31}$$

Because of the monotonic, nondecreasing character of $P(a)$ we have that:

$$p(a) \geq 0 \quad \text{and} \quad \int_{-\infty}^{+\infty} p(a) da = 1 \tag{51.32}$$

For an image with quantized (integer) brightness amplitudes, the interpretation of Δa is the width of a brightness interval. We assume constant width intervals. The brightness probability *density* function is

TABLE 51.4 2D Images and their Fourier Transforms

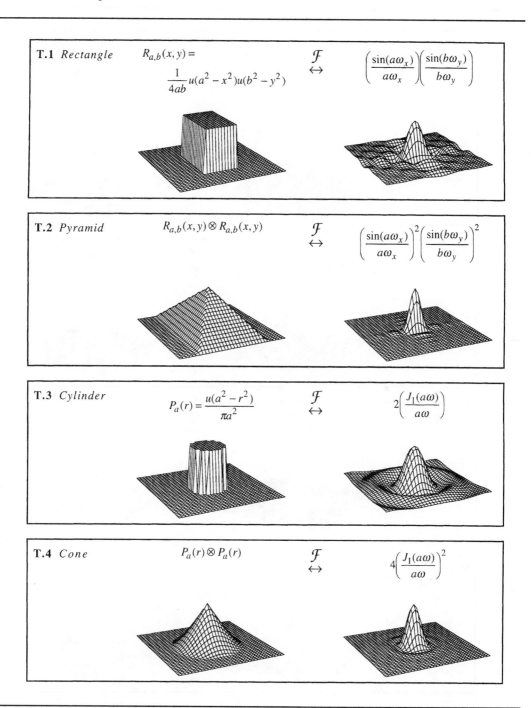

T.1 *Rectangle* $R_{a,b}(x,y) =$

$$\frac{1}{4ab}u(a^2 - x^2)u(b^2 - y^2)$$

$\mathcal{F} \atop \leftrightarrow$

$$\left(\frac{\sin(a\omega_x)}{a\omega_x}\right)\left(\frac{\sin(b\omega_y)}{b\omega_y}\right)$$

T.2 *Pyramid* $R_{a,b}(x,y) \otimes R_{a,b}(x,y)$

$\mathcal{F} \atop \leftrightarrow$

$$\left(\frac{\sin(a\omega_x)}{a\omega_x}\right)^2\left(\frac{\sin(b\omega_y)}{b\omega_y}\right)^2$$

T.3 *Cylinder* $P_a(r) = \dfrac{u(a^2 - r^2)}{\pi a^2}$

$\mathcal{F} \atop \leftrightarrow$

$$2\left(\frac{J_1(a\omega)}{a\omega}\right)$$

T.4 *Cone* $P_a(r) \otimes P_a(r)$

$\mathcal{F} \atop \leftrightarrow$

$$4\left(\frac{J_1(a\omega)}{a\omega}\right)^2$$

TABLE 51.4   2D Images and their Fourier Transforms *(continued)*

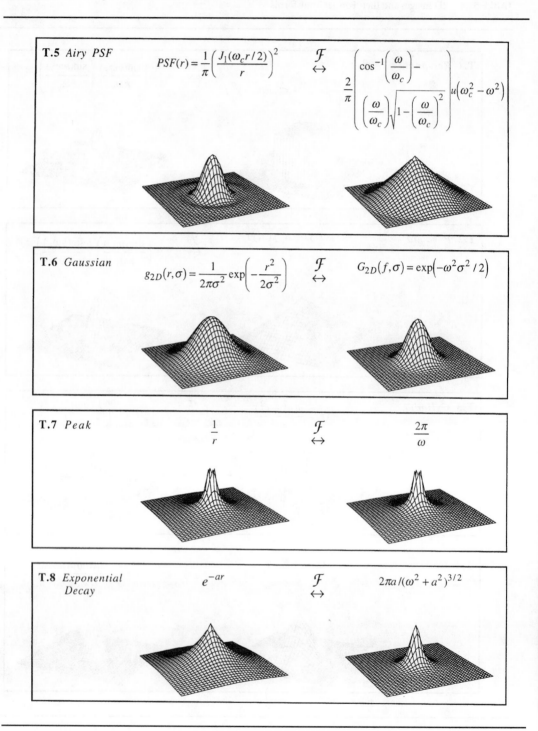

T.5 *Airy PSF*

$$PSF(r) = \frac{1}{\pi}\left(\frac{J_1(\omega_c r/2)}{r}\right)^2 \quad \overset{\mathcal{F}}{\leftrightarrow} \quad \frac{2}{\pi}\left(\cos^{-1}\left(\frac{\omega}{\omega_c}\right) - \left(\frac{\omega}{\omega_c}\right)\sqrt{1-\left(\frac{\omega}{\omega_c}\right)^2}\right)u\left(\omega_c^2 - \omega^2\right)$$

T.6 *Gaussian*

$$g_{2D}(r,\sigma) = \frac{1}{2\pi\sigma^2}\exp\left(-\frac{r^2}{2\sigma^2}\right) \quad \overset{\mathcal{F}}{\leftrightarrow} \quad G_{2D}(f,\sigma) = \exp\left(-\omega^2\sigma^2/2\right)$$

T.7 *Peak*

$$\frac{1}{r} \quad \overset{\mathcal{F}}{\leftrightarrow} \quad \frac{2\pi}{\omega}$$

T.8 *Exponential Decay*

$$e^{-ar} \quad \overset{\mathcal{F}}{\leftrightarrow} \quad 2\pi a/(\omega^2 + a^2)^{3/2}$$

frequently estimated by counting the number of times that each brightness occurs in the region to generate a *histogram, h[a]*. The histogram can then be normalized so that the total area under the histogram is 1 [Eq. (51.32)]. Said another way, the $p[a]$ for a region is the normalized count of the number of pixels, Λ, in a region that have quantized brightness a:

$$p[a] = \frac{1}{\Lambda}h[a] \quad \text{with} \quad \Lambda = \sum_a h[a] \tag{51.33}$$

The brightness probability *distribution* function for the image shown in Fig. 51.4(a) is shown in Fig. 51.6(a). The (unnormalized) brightness histogram of Fig. 51.4(a), which is proportional to the estimated brightness probability density function, is shown in Fig. 51.6(b). The height in this histogram corresponds to the number of pixels with a given brightness.

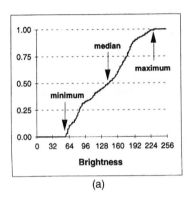

(a)

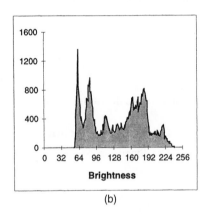

(b)

FIGURE 51.6   (a) Brightness distribution function of Fig. 51.4(a) with *minimum, median,* and *maximum* indicated. See text for explanation. (b) Brightness histogram of Fig. 51.4(a).

Both the distribution function and the histogram as measured from a region are a statistical description of that region. It should be emphasized that both $P[a]$ and $p[a]$ should be viewed as *estimates* of true distributions when they are computed from a specific region. That is, we view an image and a specific region as one realization of the various random processes involved in the formation of that image and that region. In the same context, the statistics defined below must be viewed as estimates of the underlying parameters.

Average

The average brightness of a region is defined as the *sample mean* of the pixel brightnesses within that region. The average, m_a, of the brightnesses over the Λ pixels within a region ($\Re$) is given by:

$$m_a = \frac{1}{\Lambda} \sum_{(m,n)\in\Re} a[m,n] \tag{51.34}$$

Alternatively, we can use a formulation based on the (unnormalized) brightness histogram, $h(a) = \Lambda \bullet p(a)$, with discrete brightness values a, This gives:

$$m_a = \frac{1}{\Lambda} \sum_a a \bullet h[a] \tag{51.35}$$

The average brightness, m_a, is an estimate of the mean brightness, μ_a, of the underlying brightness probability distribution.

Standard Deviation

The *unbiased estimate* of the standard deviation, s_a, of the brightness within a region ($\Re$) with Λ pixels is called the *sample standard deviation* and is given by:

$$s_a = \sqrt{\frac{1}{\Lambda - 1} \sum_{m,n \in \Re} (a[m,n] - m_a)^2}$$

$$= \sqrt{\frac{\sum\limits_{m,n \in \Re} a^2[m,n] - \Lambda m_a^2}{\Lambda - 1}} \tag{51.36}$$

Using the histogram formulation gives:

$$s_a = \sqrt{\frac{\left(\sum\limits_a a^2 \bullet h[a]\right) - \Lambda \bullet m_a^2}{\Lambda - 1}} \tag{51.37}$$

The standard deviation, s_a, is an estimate of σ_a of the underlying brightness probability distribution.

Coefficient-of-Variation

The dimensionless *coefficient-of-variation*, CV, is defined as:

$$CV = \frac{s_a}{m_a} \times 100\% \tag{51.38}$$

Percentiles

The percentile, $p\%$, of an *unquantized* brightness distribution is defined as that value of the brightness a such that:

$$P(a) = p\%$$

or equivalently

$$\int_{-\infty}^{a} p(\alpha)d\alpha = p\% \tag{51.39}$$

Three special cases are frequently used in digital image processing.

- 0%   the *minimum* value in the region
- 50%   the *median* value in the region
- 100%   the *maximum* value in the region

All three of these values can be determined from Fig. 51.6(a).

Mode

The mode of the distribution is the most frequent brightness value. There is no guarantee that a mode exists or that it is unique.

Signal-to-Noise Ratio

The signal-to-noise ratio, SNR, can have several definitions. The noise is characterized by its standard deviation, s_n. The characterization of the signal can differ. If the signal is known to lie between two boundaries, $a_{\min} \leq a \leq a_{\max}$, then the SNR is defined as:

Bounded signal -

$$SNR = 20 \log_{10} \left(\frac{a_{\max} - a_{\min}}{s_n} \right) dB \qquad (51.40)$$

If the signal is not bounded but has a statistical distribution, then two other definitions are known:

Stochastic signal -

$$S \& N \text{ inter-dependent } SNR = 20 \log_{10} \left(\frac{m_a}{s_n} \right) dB \qquad (51.41)$$

$$S \& N \text{ independent } SNR = 20 \log_{10} \left(\frac{s_a}{s_n} \right) dB \qquad (51.42)$$

where m_a and s_a are defined above.

The various statistics are given in Table 51.5 for the image and the region shown in Fig. 51.7.

FIGURE 51.7 Region is the interior of the circle.

TABLE 51.5 Statistics from Fig.51.7

Statistic	Image	ROI
Average	137.7	219.3
Standard deviation	49.5	4.0
Minimum	56	202
Median	141	220
Maximum	241	226
Mode	62	220
SNR (db)	NA	33.3

A SNR calculation for the *entire* image based on Eq. (51.40) is not directly available. The variations in the image brightnesses that lead to the large value of $s (= 49.5)$ are not, in general, due to noise but to the variation in local information. With the help of the region, there is a way to estimate the SNR. We can use the $s_\Re (= 4.0)$ and the dynamic range, $a_{\max} - a_{\min}$, for the image ($= 241 - 56$) to calculate a global $SNR (= 33.3 \text{ dB})$. The underlying assumptions are that (1) the signal is approximately constant in that region and the variation in the region is, therefore, due to noise, and that (2) the noise is the same over the entire image with a standard deviation given by $s_n = s_\Re$.

51.3.6 Contour Representations

When dealing with a region or object, several compact representations are available that can facilitate manipulation of and measurements on the object. In each case we assume that we begin with an image representation of the object as shown in Fig. 51.8(a) and (b). Several techniques exist to represent the region or object by describing its contour.

Chain Code

This representation is based on the work of Freeman. We follow the contour in a clockwise manner and keep track of the directions as we go from one contour pixel to the next. For the standard implementation of the chain code, we consider a contour pixel to be an object pixel that has a background (nonobject) pixel as one or more of its 4-connected neighbors. See Figs. 51.3(a) and 51.8(c).

The codes associated with eight possible directions are the chain codes and, with x as the current contour

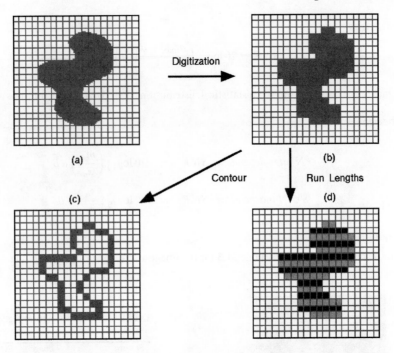

FIGURE 51.8 Region (shaded) as it is transformed from (a) continuous to (b) discrete form and then considered as a (c) contour or (d) run lengths illustrated in alternating colors.

pixel position, the codes are generally defined as:

$$
Chain\ codes = \begin{array}{ccc} 3 & 2 & 1 \\ 4 & x & 0 \\ 5 & 6 & 7 \end{array} \tag{51.43}
$$

Chain Code Properties

• Even codes {0, 2, 4, 6} correspond to horizontal and vertical directions: odd codes {1, 3, 5, 7} correspond to the diagonal directions.

• Each code can be considered as the angular direction, in multiples of 45°, that we must move to go from one contour pixel to the next.

• The absolute coordinates $[m, n]$ of the first contour pixel (e.g., top, leftmost) together with the chain code of the contour represent a complete description of the discrete region contour.

• When there is a change between two consecutive chain codes, then the contour has changed direction. This point is defined as a *corner*.

"Crack" Code

An alternative to the chain code for contour encoding is to use neither the contour pixels associated with the object nor the contour pixels associated with background but rather the line, the "crack", in between. This is illustrated with an enlargement of a portion of Fig. 51.8 in Fig. 51.9.

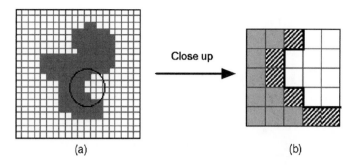

FIGURE 51.9 (a) Object including part to be studied. (b) Contour pixels as used in the chain code are diagonally shaded. The "crack" is shown with the thick black line.

The "crack" code can be viewed as a chain code with four possible directions instead of eight.

$$\text{Crack codes} = \begin{array}{ccc} & 1 & \\ 2 & x & 0 \\ & 3 & \end{array} \tag{51.44}$$

The chain code for the enlarged section of Fig. 51.9(b), from top to bottom, is {5, 6, 7, 7, 0}. The crack code is {3, 2, 3, 3, 0, 3, 0, 0}.

Run Codes

A third representation is based on coding the consecutive pixels along a row — a run — that belongs to an object by giving the starting position of the run and the ending position of the run. Such runs are illustrated in Fig. 51.8(d). There are a number of alternatives for the precise definition of the positions. Which alternative should be used depends on the application and thus will not be discussed here.

51.4 Perception

Many image processing applications are intended to produce images that are to be viewed by human observers (as opposed to, say, automated industrial inspection.) It is, therefore, important to understand the characteristics and limitations of the human visual system — to understand the "receiver" of the 2D signals. At the outset it is important to realize that (1) the human visual system is not well understood, (2) no objective measure exists for judging the quality of an image that corresponds to human assessment to image quality, and (3) the "typical" human observer does not exist. Nevertheless, research in perceptual psychology has provided some important insights into the visual system.

51.4.1 Brightness Sensitivity

There are several ways to describe the sensitivity of the human visual system. To begin, let us assume that a homogeneous region in an image has an intensity as a function of wavelength (color) given by $I(\lambda)$. Further, let us assume that $I(\lambda) = I_o$, a constant.

Wavelength Sensitivity

The perceived intensity as a function of λ, the spectral sensitivity, for the "typical observer" is shown in Fig. 51.10.

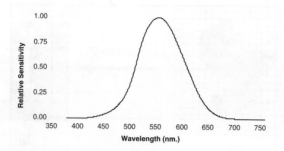

FIGURE 51.10 Spectral sensitivity of the "typical" human observer.

Stimulus Sensitivity

If the constant intensity (brightness) I_o is allowed to vary, then, to a good approximation, the visual response, R, is proportional to the logarithm of the intensity. This is known as the Weber-Fechner law:

$$R = \log (I_o) \tag{51.45}$$

The implications of this are easy to illustrate. Equal *perceived* steps brightness, $\Delta R = k$, require that the physical brightness (the stimulus) increases exponentially. This is illustrated in Fig. 51.11(a) and (b).

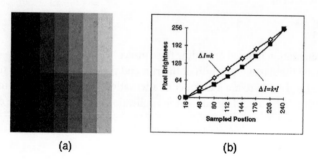

(a)   (b)

FIGURE 51.11 (a) (Top) brightness step $\Delta I = k$, (bottom) brightness step $\Delta I = k \bullet I$. (b) Actual brightnesses plus interpolated values.

A horizontal line through the top portion of Fig. 51.11(a) shows a linear increase in objective brightness (Fig. 51.11(b)) but a logarithmic increase in subjective brightness. A horizontal line through the bottom portion of Fig. 51.11(a) shows an exponential increase in objective brightness Fig. 51.11(b) but a linear increase in subjective brightness.

The *Mach band effect* is visible in Fig. 51.11(a). Although the physical brightness is constant across each vertical stripe, the human observer perceives an "undershoot" and "overshoot" in brightness at what is physically a step edge. Thus, just before the step, we see a slight decrease in brightness compared to the true physical value. After the step we see a slight overshoot in brightness compared to the true physical value. The total effect is one of increased, local, *perceived* contrast at a step edge in brightness.

51.4.2 Spatial Frequency Sensitivity

If the constant intensity (brightness) I_o is replaced by a sinusoidal grating with increasing spatial frequency (Fig. 51.12(a)), it is possible to determine the spatial frequency sensitivity. The result is shown in Fig. 51.12(b).

 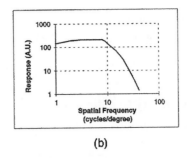

         (a)   (b)

FIGURE 51.12   (a) Sinusoidal test grating and (b) spatial frequency sensitivity.

To translate these data into common terms, consider an "ideal" computer monitor at a viewing distance of 50 cm. The spatial frequency that will give maximum response is at 10 cycles per degree. (See Fig. 51.12(b)). The one degree at 50 cm translates to $50 \tan(1°) = 0.87$ cm on the computer screen. Thus, the spatial frequency of maximum response $f_{max} = 10$ cycles/0.87 cm $= 11.46$ cycles/cm at this viewing distance. Translating this into a general formula gives:

$$f_{max} = \frac{10}{d \bullet \tan(1°)} = \frac{572.9}{d} \text{ cycles / cm} \tag{51.46}$$

where $d = $ viewing distance measured in centimeters.

51.4.3 Color Sensitivity

Human color perception is an exceedingly complex topic. As such we can only present a brief introduction here. The physical perception of color is based on three color pigments in the retina.

Standard Observer

Based on psychophysical measurements, standard curves have been adopted by the CIE (Commission Internationale de l'Eclairage) as the sensitivity curves for the "typical" observer for the three "pigments" $\bar{x}(\lambda)$, $\bar{y}(\lambda)$, and $\bar{z}(\lambda)$. These are shown in Fig. 51.13. These are not the *actual* pigment absorption characteristics found in the "standard" human retina but rather sensitivity curves derived from actual data.

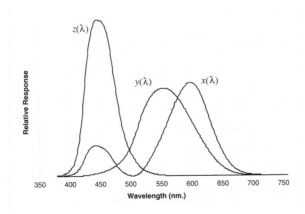

FIGURE 51.13   Standard observer color pigment sensitivity curves.

For an arbitrary homogeneous region in an image that has an intensity as a function of wavelength (color) given by $I(\lambda)$, the three pigment responses are called the *tristimulus values*:

$$X = \int_0^\infty I(\lambda)\bar{x}(\lambda)d\lambda \quad Y = \int_0^\infty I(\lambda)\bar{y}(\lambda)d\lambda \quad Z = \int_0^\infty I(\lambda)\bar{z}(\lambda)d\lambda \qquad (51.47)$$

CIE Chromaticity Coordinates

The *chromaticity coordinates,* which describe the perceived color information, are defined as:

$$x = \frac{X}{X + Y + Z} \qquad y = \frac{Y}{X + Y + Z} \qquad z = 1 - (x + y) \qquad (51.48)$$

The red chromaticity coordinate is given by x and the green chromaticity coordinate by y. The tristimulus values are linear in $I(\lambda)$, and thus the absolute intensity information has been lost in the calculation of the chromaticity coordinates $\{x, y\}$. All color distributions, $I(\lambda)$, that appear to an observer as having the same color will have the same chromaticity coordinates.

If we use a tunable source of pure color (such as dye laser), then the intensity can be modeled as $I(\lambda) = \delta(\lambda - \lambda_0)$ with $\delta(\bullet)$ as the impulse function. The collection of chromaticity coordinates $\{x, y\}$ that will be generated by varying λ_0 gives the *CIE chromaticity triangle* as shown in Fig. 51.14.

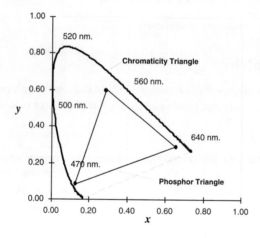

FIGURE 51.14 Chromaticity diagram containing the *CIE chromaticity triangle* associated with pure spectral colors and the triangle associated with CRT phosphors.

Pure spectral colors are along the boundary of the chromaticity triangle. All other colors are inside the triangle. The chromaticity coordinates for some standard sources are given in Table 51.6.

TABLE 51.6

Chromaticity Coordinates for Standard Sources

Source	x	y
Fluorescent lamp @ 4800 °K	0.35	0.37
Sun @ 6000 °K	0.32	0.33
Red phosphor (europium yttrium vanadate)	0.68	0.32
Green phosphor (zinc cadmium sulfide)	0.28	0.60
Blue phosphor (zinc sulfide)	0.15	0.07

The description of color on the basis of chromaticity coordinates not only permits an analysis of color but provides a synthesis technique as well. Using a mixture of two color sources, it is possible to generate any of the colors along the line connecting their respective chromaticity coordinates. Since we cannot have a negative number of photons, this means the mixing coefficients must be positive. Using three color sources such as the red, green, and blue phosphors on CRT monitors leads to the set of colors defined by the *interior* of the "phosphor triangle" shown in Fig. 51.14.

The formulas for converting from the tristimulus values (X, Y, Z) to the well-known CRT colors (R, G, B) and back are given by:

$$
\begin{bmatrix} R \\ G \\ B \end{bmatrix} = \begin{bmatrix} 1.9107 & -0.5326 & -0.2883 \\ -0.9843 & 1.9984 & -0.0283 \\ 0.0583 & -0.1185 & 0.8986 \end{bmatrix} \bullet \begin{bmatrix} X \\ Y \\ Z \end{bmatrix} \tag{51.49}
$$

and

$$
\begin{bmatrix} X \\ Y \\ Z \end{bmatrix} = \begin{bmatrix} 0.6067 & 0.1736 & 0.2001 \\ 0.2988 & 0.5868 & 0.1143 \\ 0.0000 & 0.0661 & 1.1149 \end{bmatrix} \bullet \begin{bmatrix} R \\ G \\ B \end{bmatrix} \tag{51.50}
$$

As long as the position of a desired color (X, Y, Z) is inside the phosphor triangle in Fig. 51.14, the values R, G, and B as computed by Eq. (51.49) will be positive and therefore can be used to drive a CRT monitor.

It is incorrect to assume that a small displacement anywhere in the chromaticity diagram (Fig. 51.14) will produce a proportionally small change in the *perceived* color. An empirically derived chromaticity space where this property is approximated is the (u', v') space:

$$
u' = \frac{4x}{-2x + 12y + 3} \qquad v' = \frac{9y}{-2x + 12y + 3}
$$

and

$$
x = \frac{9u'}{6u' - 16v' + 12} \qquad y = \frac{4v'}{6u' - 16v' + 12} \tag{51.51}
$$

Small changes almost anywhere in the (u', v') chromaticity space produce equally small changes in the perceived colors.

51.4.4 Optical Illusions

The description of the human visual system presented above is couched in standard engineering terms. This could lead one to conclude that there is sufficient knowledge of the human visual system to permit modeling the visual system with standard system analysis techniques. Two simple examples of optical illusions, shown in Fig. 51.15, illustrate that this system approach would be a gross oversimplification. Such models should only be used with extreme care.

The left illusion induces the illusion of gray values in the eye that the brain "knows" do not exist. Further, there is a sense of dynamic change in the image due, in part, to the saccadic movements of the eye. The right illusion, Kanizsa's triangle, shows enhanced contrast and false contours, neither of which can be explained by the system-oriented aspects of visual perception described above.

51.5 Image Sampling

Converting from a continuous image $a(x, y)$ to its digital representation $b[m, n]$ requires the process of sampling. In the ideal sampling system, $a(x, y)$ is multiplied by an ideal 2D impulse train:

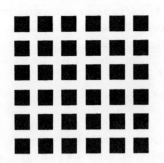 

FIGURE 51.15 Optical illusions.

$$
\begin{aligned}
b_{\text{ideal}}[m, n] &= a(x, y) \bullet \sum_{m=-\infty}^{+\infty} \sum_{n=-\infty}^{+\infty} \delta\left(x - mX_o, y - nY_o\right) \\
&= \sum_{m=-\infty}^{+\infty} \sum_{n=-\infty}^{+\infty} a\left(mX_o, nY_o\right) \delta\left(x - mX_o, y - nY_o\right)
\end{aligned}
\tag{51.52}
$$

where X_o and Y_o are the sampling distances or intervals and $\delta(\bullet, \bullet)$ is the ideal impulse function. (At some point, of course, the impulse function $\delta(x, y)$ is converted to the discrete impulse function $\delta[m, n]$.) *Square sampling* implies that $X_o = Y_o$. Sampling with an impulse function corresponds to sampling with an infinitesimally small point. This, however, does not correspond to the usual situation as illustrated in Fig. 51.1. To take the effects of a *finite* sampling aperture $p(x, y)$ into account, we can modify the sampling model as follows:

$$
b[m, n] = (a(x, y) \otimes p(x, y)) \bullet \sum_{m=-\infty}^{+\infty} \sum_{n=-\infty}^{+\infty} \delta\left(x - mX_o, y - nY_o\right)
\tag{51.53}
$$

The combined effect of the aperture and sampling are best understood by examining the Fourier domain representation.

$$
B(\Omega, \Psi) = \frac{1}{4\pi^2} \sum_{m=-\infty}^{+\infty} \sum_{n=-\infty}^{+\infty} A\left(\Omega - m\Omega_s, \Psi - n\Psi_s\right) \bullet P\left(\Omega - m\Omega_s, \Psi - n\Psi_s\right)
\tag{51.54}
$$

where $\Omega_s = 2\pi/X_o$ is the sampling frequency in the x direction and $\Psi_s = 2\pi/Y_o$ is the sampling frequency in the y direction. The aperture $p(x, y)$ is frequently square, circular, or Gaussian with the associated $P(\Omega, \Psi)$ (see Table 51.4). The periodic nature of the spectrum, described in Eq. (51.21) is clear from Eq. (51.54).

51.5.1 Sampling Density for Image Processing

To prevent the possible *aliasing* (overlapping) of spectral terms that is inherent in Eq. (51.54), two conditions must hold:

- *Bandlimited $A(u, v)$* -

$$
|A(u, v)| \equiv 0 \quad \text{for } |u| > u_c \quad \text{and } |v| > v_c
\tag{51.55}
$$

- *Nyquist sampling frequency* -

$$
\Omega_s > 2 \bullet u_c \quad \text{and} \quad \Psi_s > 2 \bullet v_c
\tag{51.56}
$$

where u_c and v_c are the *cutoff frequencies* in the x and y direction, respectively. Images that are acquired through lenses that are circularly symmetric, aberration-free, and diffraction-limited will, in general, be bandlimited. The lens acts as a lowpass filter with a cutoff frequency in the frequency domain [Eq. (51.11)] given by:

$$u_c = v_c = \frac{2NA}{\lambda} \tag{51.57}$$

where NA is the numerical aperture of the lens and λ is the shortest wavelength of light used with the lens. If the lens does not meet one or more of these assumptions, then it will still be bandlimited but at lower cutoff frequencies than those given in Eq. (51.57). When working with the F-number (F) of the optics instead of the NA and in air (with *index of refraction* $= 1.0$), Eq. (51.57) becomes:

$$u_c = v_c = \frac{2}{\lambda} \left(\frac{1}{\sqrt{4F^2 + 1}} \right) \tag{51.58}$$

Sampling Aperture

The aperture $p(x, y)$ described above will have only a marginal effect on the final signal if the two conditions, Eqs. (51.56) and (51.57), are satisfied. Given, for example, the distance between samples X_o equals Y_o and a sampling aperture that is not wider than X_o, the effect on the overall spectrum — due to the $A(u, v)P(u, v)$ behavior implied by Eq. (51.53) — is illustrated in Fig. 51.16 for square and Gaussian apertures.

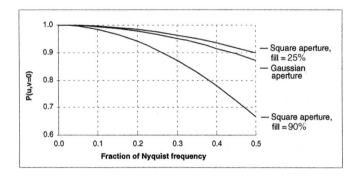

FIGURE 51.16 Aperture spectra $P(u, v = 0)$ for frequencies up to half the Nyquist frequency. For explanation of "fill" see text.

The spectra are evaluated along one axis of the 2D Fourier transform. The Gaussian aperture in Fig. 51.16 has a width such that the sampling interval X_o contains $\pm 3\sigma$ (99.7%) of the Gaussian. The rectangular apertures have a width such that one occupies 95% of the sampling interval and the other occupies 50% of the sampling interval. The 95% width translates to a *fill factor* of 90% and the 50% width to a *fill factor* of 25%. The *fill factor* is discussed in section 51.7.5.

51.5.2 Sampling Density for Image Analysis

The "rules" for choosing the sampling density when the goal is image analysis — as opposed to image processing — are different. The fundamental difference is that the digitization of objects in an image into a collection of pixels introduces a form of spatial quantization noise that is not bandlimited. This leads to the following results for the choice of sampling density when one is interested in the measurement of area and (perimeter) length.

Sampling for Area Measurements

Assuming square sampling, $X_o = Y_o$ and the unbiased algorithm for estimating area which involves simple pixel counting, the CV [see Eq. (51.38)] of the area measurement is related to the sampling density by:

$$2D: \quad \lim_{S \to \infty} CV(S) = k_2 S^{-3/2} \qquad 3D: \quad \lim_{S \to \infty} CV(S) = k_3 S^{-2} \tag{51.59}$$

and in D dimensions:

$$\lim_{S \to \infty} CV(S) = k_D S^{-(D+1)/2} \tag{51.60}$$

where S is the number of samples *per object diameter.* In 2D, the measurement is area; in 3D, volume; and in D-dimensions, hypervolume.

Sampling for Length Measurements

Again assuming square sampling and algorithms for estimating length based on the Freeman chain-code representation (see section 51.3.6), the CV of the length measurement is related to the sampling density *per unit length* as shown in Fig. 51.17.

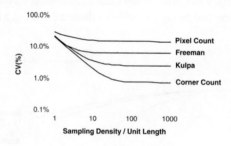

FIGURE 51.17 CV of length measurement for various algorithms.

The curves in Fig. 51.17 were developed in the context of straight lines but similar results have been found for curves and closed contours. The specific formulas for length estimation use a chain code representation of a line and are based on a linear combination of three numbers:

$$L = \alpha \bullet N_e + \beta \bullet N_0 + \gamma \bullet N_c \tag{51.61}$$

where N_e is the number of even chain codes, N_0 the number of odd chain codes, and N_c the number of corners. The specific formulas are given in Table 51.7.

TABLE 51.7 Length Estimation Formulas Based on Chain Code Counts (N_e, N_0, $\mathbf{N_C}$)

Formula	α	β	γ
		Coefficients	
Pixel count	1	1	0
Freeman	1	$\sqrt{2}$	0
Kulpa	0.9481	$0.9481 \bullet \sqrt{2}$	0
Corner count	0.980	1.406	−0.091

Conclusions on Sampling

If one is interested in image processing, one should choose a sampling density based on classical signal theory, that is, the Nyquist sampling theory. If one is interested in image analysis, one should choose a sampling density based on the desired measurement accuracy (*bias*) and precision (*CV*). In a case of uncertainty, one should choose the higher of the two sampling densities (frequencies).

51.6 Noise

Images acquired through modern sensors may be contaminated by a variety of noise sources. By noise we refer to stochastic variations as opposed to deterministic distortions such as shading or lack of focus. We will assume for this section that we are dealing with images formed from light using modern electro-optics. In particular we will assume the use of modern, charge-coupled device (CCD) cameras where photons produce electrons that are commonly referred to as photoelectrons. Nevertheless, most of the observations we shall make about noise and its various sources hold equally well for other imaging modalities.

While modern technology has made it possible to reduce the noise levels associated with various electro-optical devices to almost negligible levels, one noise source can never be eliminated and thus forms the limiting case when all other noise sources are "eliminated".

51.6.1 Photon Noise

When the physical signal that we observe is based on light, then the quantum nature of light plays a significant role. A single photon at $\lambda = 500$ nm carries an energy of $E = h\nu = hc/\lambda = 3.97 \times 10^{-19}$ Joules. Modern CCD cameras are sensitive enough to be able to count individual photons. (Camera sensitivity will be discussed in section 51.7.2). The noise problem arises from the fundamentally statistical nature of photon production. We cannot assume that, in a given pixel for two consecutive but independent observation intervals of length T, the same number of photons will be counted. Photon production is governed by the laws of quantum physics which restrict us to talking about an average number of photons within a given observation window. The probability distribution for p photons in an observation window of length T seconds is known to be Poisson:

$$P(p|\rho, T) = \frac{(\rho T)^p \, e^{-\rho T}}{p!} \tag{51.62}$$

where ρ is the rate or intensity parameter measured in photons per second. It is critical to understand that even if there were no other noise sources in the imaging chain, the statistical fluctuations associated with photon counting over a finite time interval T would still lead to a finite signal-to-noise ratio (SNR). If we use the appropriate formula for the SNR [Eq. (51.41)], then due to the fact that the average value and the standard deviation are given by:

Poisson process -

$$\begin{aligned} \text{average} &= \rho T \\ \sigma &= \sqrt{\rho T} \end{aligned} \tag{51.63}$$

we have for the SNR:

Photon noise -

$$SNR = 10 \log_{10}(\rho T) \quad dB \tag{51.64}$$

The three traditional assumptions about the relationship between signal and noise do not hold for photon noise:

- photon noise is not independent of the signal;
- photon noise is not Gaussian; and

- photon noise is not additive.

For very bright signals, where ρT exceeds 10^5, the noise fluctuations due to photon statistics can be ignored if the sensor has a sufficiently high saturation level. This will be discussed further in section 51.7.3 and, in particular, Eq. (51.73).

51.6.2   Thermal Noise

An additional, stochastic source of electrons in a CCD well is thermal energy. Electrons can be freed from the CCD material itself through thermal vibration and then, trapped in the CCD well, be indistinguishable from "true" photoelectrons. By cooling the CCD chip, it is possible to reduce significantly the number of "thermal electrons" that give rise to thermal noise or *dark current*. As the integration time T increases, the number of thermal electrons increases. The probability distribution of thermal electrons is also a Poisson process where the rate parameter is an increasing function of temperature. There are alternative techniques (to cooling) for suppressing dark current and these usually involve estimating the *average* dark current for the given integration time and then subtracting this value from the CCD pixel values before the A/D converter. While this does reduce the dark current *average*, it does not reduce the dark current *standard deviation* and it also reduces the possible dynamic range of the signal.

51.6.3   On-Chip Electronic Noise

This noise originates in the process of reading the signal from the sensor, in this case through the field effect transistor (FET) of a CCD chip. The general form of the power spectral density of readout noise is:

$$\text{Readout noise - } S_{nn}(\omega) \propto \begin{cases} \omega^{-\beta} & \omega < \omega_{\min} & \beta > 0 \\ k & \omega_{\min} < \omega < \omega_{\max} \\ \omega^{\alpha} & \omega > \omega_{\max} & \alpha > 0 \end{cases} \tag{51.65}$$

where α and β are constants and ω is the (radial) frequency at which the signal is transferred from the CCD chip to the "outside world". At very low readout rates ($\omega < \omega_{\min}$) the noise has a $1/f$ character. Readout noise can be reduced to manageable levels by appropriate readout rates and proper electronics. At very low signal levels [see Eq. (51.64)], however, readout noise can still become a significant component in the overall SNR.

51.6.4   KTC Noise

Noise associated with the gate capacitor of an FET is termed *KTC noise* and can be nonnegligible. The output RMS value of this noise voltage is given by:

KTC noise (voltage) -

$$\sigma_{KTC} = \sqrt{\frac{kT}{C}} \tag{51.66}$$

where C is the FET gate switch capacitance, k is Boltzmann's constant, and T is the absolute temperature of the CCD chip measured in K. Using the relationships $Q = C \bullet V = N_{e-} \bullet e^-$, the ouput RMS value of the KTC noise expressed in terms of the number of photoelectrons (N_{e-}) is given by:

KTC noise (electrons) -

$$\sigma_{N_e} = \frac{\sqrt{kTC}}{e^-} \tag{51.67}$$

where e^- is the electron charge. For $C = 0.5$ pF and $T = 233$ K, this gives $N_{e-} = 252$ electrons. This value is a "one time" noise per pixel that occurs during signal readout and is thus independent of the integration time (see sections 51.6.1 and 51.7.7). Proper electronic design that makes use, for example, of correlated double sampling and dual-slope integration can almost completely eliminate KTC noise.

51.6.5 Amplifier Noise

The standard model for this type of noise is additive, Gaussian, and independent of the signal. In modern well-designed electronics, amplifier noise is generally negligible. The most common exception to this is in color cameras where more amplification is used in the blue color channel than in the green channel or red channel leading to more noise in the blue channel. (See also section 51.7.6.)

51.6.6 Quantization Noise

Quantization noise is inherent in the amplitude quantization process and occurs in the analog-to-digital converter, ADC. The noise is additive and independent of the signal when the number of levels $L \geq 16$. This is equivalent to $B \geq 4$ bits. (See section 51.2.1). For a signal that has been converted to electrical form and thus has a minimum and maximum electrical value, Eq. (51.40) is the appropriate formula for determining the SNR. If the ADC is adjusted so that 0 corresponds to the minimum electrical value and $2^B - 1$ corresponds to the maximum electrical value then:

Quantization noise -

$$SNR = 6B + 11 \; dB \tag{51.68}$$

For $B \geq 8$ bits, this means a $SNR \geq 59$ dB. Quantization noise can usually be ignored as the total SNR of a complete system is typically dominated by the smallest SNR. In CCD cameras, this is photon noise.

51.7 Cameras

The cameras and recording media available for modern digital image processing applications are changing at a significant pace. To dwell too long in this section on one major type of camera, such as the CCD camera, and to ignore developments in areas such as charge injection device (CID) cameras and CMOS cameras, is to run the risk of obsolescence. Nevertheless, the techniques that are used to characterize the CCD camera remain "universal" and the presentation that follows is given in the context of modern CCD technology for purposes of illustration.

51.7.1 Linearity

It is generally desirable that the relationship between the input physical signal (e.g., photons) and the output signal (e.g., voltage) be linear. Formally this means [as in Eq. (51.20)] that if we have two images, a and b, and two arbitrary complex constants, w_1 and w_2, and a linear camera response, then:

$$c = \mathcal{R} \left\{ w_1 a + w_2 b \right\} = w_1 \mathcal{R}\{a\} + w_2 \mathcal{R}\{b\} \tag{51.69}$$

where $\mathcal{R}\{\bullet\}$ is the camera response and c is the camera output. In practice, the relationship between input a and output c is frequently given by:

$$c = \text{gain} \bullet a^\gamma + \text{offset} \tag{51.70}$$

where γ is the *gamma* of the recording medium. For a truly linear recording system we must have $\gamma = 1$ and *offset* $= 0$. Unfortunately, the offset is almost never zero and thus we must compensate for this if the intention is to extract intensity measurements. Compensation techniques are discussed in section 51.10.1.

Typical values of γ that may be encountered are listed in Table 51.8. Modern cameras often have the ability to switch electronically between various values of γ.

TABLE 51.8   Comparison of γ of Various Sensors

Sensor	Surface	γ	Possible advantages
CCD chip	Silicon	1.0	Linear
Vidicon tube	$Sb_2 S_3$	0.6	Compresses dynamic range $\rightarrow$ high contrast scenes
Film	Silver halide	< 1.0	Compresses dynamic range $\rightarrow$ high contrast scenes
Film	Silver halide	> 1.0	Expands dynamic range $\rightarrow$ low contrast scenes

51.7.2   Sensitivity

There are two ways to describe the sensitivity of a camera. First, we can determine the minimum number of detectable photoelectrons. This can be termed the *absolute* sensitivity. Second, we can describe the number of photoelectrons necessary to change from one digital brightness level to the next, that is, to change one *analog-to-digital unit* (ADU). This can be termed the *relative* sensitivity.

Absolute Sensitivity

To determine the absolute sensitivity we need a characterization of the camera in terms of its noise. If the noise has a σ of, say, 100 photoelectrons, then to ensure detectability of a signal we could then say that, at the 3σ level, the minimum detectable signal (or absolute sensitivity) would be 300 photoelectrons. If all the noise sources listed in section 51.6, with the exception of photon noise, can be reduced to negligible levels, this means that an absolute sensitivity of less than 10 photoelectrons is achievable with modern technology.

Relative Sensitivity

The definition of relative sensitivity, S, given above when coupled to the linear case, Eq. (51.70) with $\gamma = 1$, leads immediately to the result:

$$S = 1 / gain = gain^{-1} \tag{51.71}$$

The measurement of the *sensitivity* or *gain* can be performed in two distinct ways.

- If following Eq. (51.70), the input signal a can be precisely controlled by either "shutter" time or intensity (through neutral density filters), then the gain can be estimated by estimating the slope of the resulting straight-line curve. To translate this into the desired units, however, a standard source must be used that emits a known number of photons onto the camera sensor and the quantum efficiency (η) of the sensor must be known. The quantum efficiency refers to how many photoelectrons are produced — on the average — per photon at a given wavelength. In general $0 \leq \eta(\lambda) \leq 1$.
- If, however, the limiting effect of the camera is only the photon (Poisson) noise (see section 51.6.1), then an easy-to-implement, alternative technique is available to determine the sensitivity. Using Eqs. (51.63), (51.70), and (51.71) and after compensating for the *offset* (see section 51.10.1), the *sensitivity* measured from an image c is given by:

$$S = \frac{E\{c\}}{Var\,\{c\}} = \frac{m_c}{s_c^2} \tag{51.72}$$

where m_c and s_c are defined in Eqs. (51.34) and (51.36).

Measured data for five modern (1995) CCD camera configurations are given in Table 51.9.

The extraordinary sensitivity of modern CCD cameras is clear from these data. In a scientific-grade CCD camera (C-1), only 8 photoelectrons (approximately 16 photons) separate two gray levels in the digital representation of the image. For a considerably less expensive video camera (C-5), only about 110 photoelectrons (approximately 220 photons) separate two gray levels.

TABLE 51.9 Sensitivity Measurements

Camera label	Pixels	Pixel size ($\mu m \times \mu m$)	Temp. (K)	S (e^-/ADU)	Bits
C-1	1320 × 1035	6.8 × 6.8	231	7.9	12
C-2	578 × 385	22.0 × 22.0	227	9.7	16
C-3	1320 × 1035	6.8 × 6.8	293	48.1	10
C-4	576 × 384	23.0 × 23.0	238	90.9	12
C-5	756 × 581	11.0 × 5.5	300	109.2	8

Note: The lower the value of S, the more sensitive the camera is.

51.7.3 SNR

As described in section 51.6 in modern camera systems the noise is frequently limited by:

- amplifier noise in the case of color cameras;
- thermal noise which, itself, is limited by the chip temperature K and the exposure time T; and/or
- photon noise, which is limited by the photon production rate ρ and the exposure time T.

Thermal Noise (Dark Current)

Using cooling techniques based on Peltier cooling elements, it is straightforward to achieve chip temperatures of 230 to 250 K. This leads to low thermal electron production rates. As a measure of the thermal noise, we can look at the number of seconds necessary to produce a sufficient number of thermal electrons to go from one brightness level to the next, an ADU, in the absence of photoelectrons. This last condition — the absence of photoelectrons — is the reason for the name *dark current*. Measured data for the five cameras described above are given in Table 51.10.

TABLE 51.10 Thermal Noise Characteristics

Camera label	Temp. (K)	Dark current (*seconds/ADU*)
C-1	231	526.3
C-2	227	0.2
C-3	293	8.3
C-4	238	2.4
C-5	300	23.3

The *video* camera (C-5) has on-chip dark current suppression (see section 51.6.2). Operating at room temperature this camera requires more than 20 seconds to produce one ADU change due to thermal noise. This means at the conventional video frame and integration rates of 25 to 30 images per second (see Table 51.3), the thermal noise is negligible.

Photon Noise

From Eq. (51.64) we see that it should be possible to increase the SNR by increasing the integration time of our image and thus "capturing" more photons. The pixels in CCD cameras have, however, a finite well capacity. This finite capacity, C, means that the maximum SNR for a CCD camera per pixel is given by:

Capacity-limited photon noise -

$$SNR = 10 \log_{10}(C) \, dB \tag{51.73}$$

Theoretical as well as measured data for the five cameras described above are given in Table 51.11.

TABLE 51.11   Photon Noise Characteristics

Camera label	C #e−	Theor. SNR (dB)	Meas. SNR (dB)	Pixel size (μm × μm)	Well depth (#e − /μm²)
C-1	32,000	45	45	6.8 × 6.8	692
C-2	340,000	55	55	22.0 × 22.0	702
C-3	32,000	45	43	6.8 × 6.8	692
C-4	400,000	56	52	23.0 × 23.0	756
C-5	40,000	46	43	11.0 × 5.5	661

Note that for certain cameras, the measured SNR achieves the theoretical maximum indicating that the SNR is, indeed, photon and well capacity limited. Further, the curves of SNR vs. T (integration time) are consistent with Eqs. (51.64) and (51.73). (Data not shown.) It can also be seen that, as a consequence of CCD technology, the "depth" of a CCD pixel well is constant at about $0.7\ ke^-/\mu m^2$.

51.7.4   Shading

Virtually all imaging systems produce shading. By this we mean that if the physical input image $a(x, y) = constant$, then the digital version of the image will not be constant. The source of the shading might be outside the camera, such as in the scene illumination, or the result of the camera itself where a *gain* and *offset* might vary from pixel to pixel. The model for shading is given by:

$$c[m, n] = \text{gain}[m, n] \bullet a[m, n] + \text{offset}[m, n] \tag{51.74}$$

where $a[m, n]$ is the digital image that would have been recorded if there were no shading in the image, that is, $a[m, n] = constant$. Techniques for reducing or removing the effects of shading are discussed in section 51.10.1.

51.7.5   Pixel Form

While the pixels shown in Fig. 51.1 appear to be square and to "cover" the continuous image, it is important to know the geometry for a given camera/digitizer system. In Fig. 51.18 we define possible parameters associated with a camera and digitizer and the effect they have on the pixel.

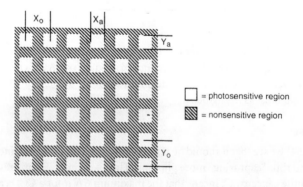

FIGURE 51.18   Pixel form parameters.

The parameters X_o and Y_o are the spacing between the pixel centers and represent the sampling distances from Eq. (51.52). The parameters X_a and Y_a are the dimensions of that portion of the camera's surface that is sensitive to light. As mentioned in section 51.2.3 different video digitizers (frame grabbers) can have different values for X_o while they have a common value for Y_o.

Square Pixels

As mentioned in section 51.5, square sampling implies that $X_o = Y_o$ or alternatively $X_o/Y_o = 1$. It is not uncommon, however, to find frame grabbers where $X_o/Y_o = 1.1$ or $X_o/Y_o = 4/3$. (This latter format matches the format of commercial television. See Table 51.3). The risk associated with nonsquare pixels is that isotropic objects scanned with nonsquare pixels might appear isotropic on a camera-compatible monitor but analysis of the objects (such as length-to-width ratio) will yield nonisotropic results. This is illustrated in Fig. 51.19.

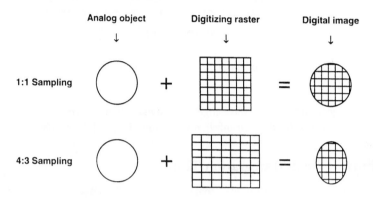

FIGURE 51.19 Effect of nonsquare pixels.

The ratio X_o/Y_o can be determined for any specific camera/digitizer system by using a calibration test chart with known distances in the horizontal and vertical direction. These are straightforward to make with modern laser printers. The test chart can then be scanned and the sampling distances X_o and Y_o determined.

Fill Factor

In modern CCD cameras it is possible that a portion of the camera surface is not sensitive to light and is instead used for the CCD electronics or to prevent *blooming*. Blooming occurs when a CCD well is filled (see Table 51.11) and additional photoelectrons spill over into adjacent CCD wells. Antiblooming regions between the active CCD sites can be used to prevent this. This means, of course, that a fraction of the incoming photons are lost as they strike the nonsensitive portion of the CCD chip. The fraction of the surface that is sensitive to light is termed the *fill factor* and is given by:

$$\text{fill factor} = \frac{X_a \bullet Y_a}{X_o \bullet Y_o} \times 100\% \tag{51.75}$$

The larger the *fill factor*, the more light will be captured by the chip up to the maximum of 100%. This helps improve the SNR. As a tradeoff, however, larger values of the fill factor mean more spatial smoothing due to the aperture effect described in section 51.5.1. This is illustrated in Fig. 51.16.

51.7.6 Spectral Sensitivity

Sensors, such as those found in cameras and film, are not equally sensitive to all wavelengths of light. The spectral sensitivity for the CCD sensor is given in Fig. 51.20.

The high sensitivity of silicon in the infra-red means that for applications where a CCD (or other silicon-based) camera is to be used as a source of images for digital image processing and analysis, consideration should be given to using an IR blocking filter. This filter blocks wavelengths above 750 nm and thus prevents

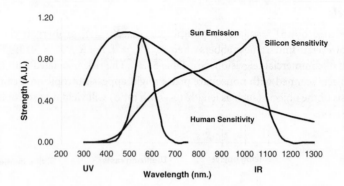

FIGURE 51.20   Spectral characteristics of silicon, the sun, and the human visual system. UV = ultraviolet and IR = infra-red.

"fogging" of the image from the longer wavelengths found in sunlight. Alternatively, a CCD-based camera can make an excellent sensor for the near infrared wavelength range of 750 to 1000 nm.

51.7.7   Shutter Speeds (Integration Time)

The length of time that an image is exposed — that photons are collected — may be varied in some cameras or may vary on the basis of video formats (see Table 51.3). For reasons that have to do with the parameters of photography, this exposure time is usually termed *shutter speed* although integration time would be a more appropriate description.

Video Cameras

Values of the shutter speed as low as 500 ns are available with commercially available CCD *video* cameras, although the more conventional speeds for video are 33.37 ms (NTSC) and 40.0 ms (PAL, SECAM). Values as high as 30 s may also be achieved with certain video cameras although this means sacrificing a continuous stream of video images that contain signal in favor of a single integrated image among a stream of otherwise empty images. Subsequent digitizing hardware must be capable of handling this situation.

Scientific Cameras

Again, values as low as 500 ns are possible and, with cooling techniques based on Peltier-cooling or liquid nitrogen cooling, integration times in excess of one hour are readily achieved.

51.7.8   Readout Rate

The rate at which data is read from the sensor chip is termed the *readout rate*. The readout rate for standard video cameras depends on the parameters of the frame grabber as well as the camera. For standard video — see section 51.2.3 — the readout rate is given by:

$$R = \left(\frac{images}{sec} \right) \bullet \left(\frac{lines}{image} \right) \bullet \left(\frac{pixels}{line} \right) \tag{51.76}$$

While the appropriate unit for describing the readout rate should be *pixels/second*, the term Hz is frequently found in the literature and in camera specifications; we shall therefore use the latter unit. As illustration, readout rates for a video camera with square pixels are given in Table 51.12 (see also section 51.7.5).

TABLE 51.12	Video Camera Readout Rates		
Format	*lines/sec*	*pixels/line*	$R(MHz)$
NTSC	15,750	$(4/3)*525$	≈ 11.0
PAL/SECAM	15,625	$(4/3)*625$	≈ 13.0

Note that the values in Table 51.12 are approximate. Exact values for square-pixel systems require exact knowledge of the way the video digitizer (frame grabber) samples each video line.

The readout rates used in video cameras frequently mean that the electronic noise described in section 51.6.3 occurs in the region of the noise spectrum [Eq. (51.65)] described by $\omega > \omega_{max}$ where the noise power increases with increasing frequency. Readout noise can thus be significant in video cameras.

Scientific cameras frequently use a slower readout rate in order to reduce the readout noise. Typical values of readout rate for scientific cameras, such as those described in Tables 51.9, 51.10, and 51.11 are 20 kHz, 500 kHz, and 1 to 8 MHz.

51.8 Displays

The displays used for image processing — particularly the display systems used with computers — have a number of characteristics that help determine the quality of the final image.

51.8.1 Refresh Rate

The *refresh rate* is defined as the number of complete images that are written to the screen per second. For standard video, the refresh rate is fixed at the values given in Table 51.3, either 29.97 or 25 images/s. For computer displays, the refresh rate can vary with common values being 67 images/s and 75 images/s. At values above 60 images/s, visual flicker is negligible at virtually all illumination levels.

51.8.2 Interlacing

To prevent the appearance of visual flicker at refresh rates below 60 images/s, the display can be interlaced as described in section 51.2.3. Standard interlace for video systems is 2:1. Since interlacing is not necessary at refresh rates above 60 images/s, an interlace of 1:1 is used with such systems. In other words, lines are drawn in an ordinary sequential fashion: $1, 2, 3, 4 \ldots, N$.

51.8.3 Resolution

The pixels stored in computer memory, although they are derived from regions of finite area in the original scene (see sections 51.5.1 and 51.7.5), may be thought of as mathematical points having no physical extent. When displayed, the space between the points must be filled in. This generally happens as a result of the finite spot size of a cathode-ray tube (CRT). The brightness profile of a CRT spot is approximately Gaussian and the number of spots that can be resolved on the display depends on the quality of the system. It is relatively straightforward to obtain display systems with a resolution of 72 spots per inch (28.3 spots per cm.) This number corresponds to standard printing conventions. If printing is not a consideration, then higher resolutions, in excess of 30 spots per cm, are attainable.

51.9 Algorithms

In this section we will describe operations that are fundamental to digital image processing. These operations can be divided into four categories: operations based on the image histogram, on simple mathematics, on convolution, and on mathematical morphology. Further, these operations can also be described in terms of their implementation as a point operation, a local operation, or a global operation as described in section 51.2.2.

51.9.1 Histogram-Based Operations

An important class of point operations is based on the manipulation of an image histogram or a *region* histogram. The most important examples are described below.

Contrast Stretching

Frequently, an image is scanned in such a way that the resulting brightness values do not make full use of the available dynamic range. This can be easily observed in the histogram of the brightness values shown in Fig. 51.6. By stretching the histogram over the available dynamic range, we attempt to correct this situation. If the image is intended to go from brightness 0 to brightness $2^B - 1$ (see section 51.2.1), then one generally maps the 0% value (or *minimum* as defined in section 51.3.5) to the value 0 and the 100% value (or *maximum*) to the value $2^B - 1$. The appropriate transformation is given by:

$$b[m, n] = \left(2^B - 1\right) \bullet \frac{a[m, n] - \text{minimum}}{\text{maximum} - \text{minimum}} \tag{51.77}$$

This formula, however, can be somewhat sensitive to outliers and a less sensitive and more general version is given by:

$$b[m, n] = \begin{cases} 0 & a[m, n] \leq p_{\text{low}}\% \\ \left(2^B - 1\right) \bullet \frac{a[m,n] - p_{\text{low}}\%}{p_{\text{high}}\% - p_{\text{low}}\%} & p_{\text{low}}\% < a[m, n] < p_{\text{high}}\% \\ \left(2^B - 1\right) & a[m, n] \geq p_{\text{high}}\% \end{cases} \tag{51.78}$$

In this second version, one might choose the 1% and 99% values for $p_{\text{low}}\%$ and $p_{\text{high}}\%$, respectively, instead of the 0% and 100% values represented by Eq. (51.77). It is also possible to apply the contrast-stretching operation on a regional basis using the histogram from a region to determine the appropriate limits for the algorithm. Note that in Eqs. (51.77) and (51.78) it is possible to suppress the term $2^B - 1$ and simply normalize the brightness range to $0 \leq b[m, n] \leq 1$. This means representing the final pixel brightnesses as reals instead of integers, but modern computer speeds and RAM capacities make this quite feasible.

Equalization

When one wishes to compare two or more images on a specific basis, such as texture, it is common to first normalize their histograms to a "standard" histogram. This can be especially useful when the images have been acquired under different circumstances. The most common histogram normalization techniques is *histogram equalization* where one attempts to change the histogram through the use of a function $b = f(a)$ into a histogram that is constant for all brightness values. This would correspond to a brightness distribution where all values are equally probable. Unfortunately, for an arbitrary image, one can only approximate this result.

For a "suitable" function $f(\bullet)$ the relation between the input probability density function, the output probability density function, and the function $f(\bullet)$ is given by:

$$p_b(b)db = p_a(a)da \quad \Rightarrow \quad df = \frac{p_a(a)da}{p_b(b)} \tag{51.79}$$

From Eq. (51.79) we see that "suitable" means that $f(\bullet)$ is differentiable and that $df/da \geq 0$. For histogram equalization, we desire that $p_b(b) = $ constant and this means that:

$$f(a) = \left(2^B - 1\right) \bullet P(a) \tag{51.80}$$

where $P(a)$ is the probability distribution function defined in section 51.3.5 and illustrated in Fig. 51.6(a). In other words, the *quantized* probability distribution function normalized from 0 to $2^B - 1$ *is* the look-up table required for histogram equalization. Figures 51.21(a-c) illustrate the effect of contrast stretching and histogram equalization on a standard image. The histogram equalization procedure can also be applied on a regional basis.

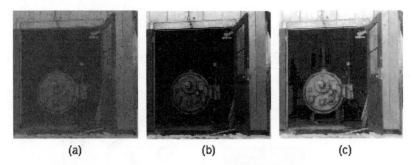

FIGURE 51.21 (a) Original, (b) contrast stretched, and (c) histogram equalized.

Other Histogram-Based Operations

The histogram derived from a local region can also be used to drive local filters that are to be applied to that region. Examples include *minimum* filtering, *median* filtering, and *maximum* filtering. The concepts minimum, median, and maximum were introduced in Fig. 51.6. The filters based on these concepts will be presented formally in sections 51.9.4 and 51.9.6.

51.9.2 Mathematics-Based Operations

In this section we distinguish between binary arithmetic and ordinary arithmetic. In the binary case there are two brightness values "0" and "1". In the ordinary case we begin with 2^B brightness values or levels but the processing of the image can easily generate many more levels. For this reason, many *software* systems provide 16- or 32-bit representations for pixel brightnesses in order to avoid problems with arithmetic overflow.

Binary Operations

Operations based on binary (Boolean) arithmetic form the basis for a powerful set of tools that will be described here and extended in section 51.9.6 mathematical morphology. The operations described below are point operations and thus admit a variety of efficient implementations including simple look-up tables. The standard notation for the basic set of binary operations is:

$$
\begin{array}{lll}
NOT & c = \bar{a} & \\
OR & c = a + b & \\
AND & c = a \bullet b & \\
XOR & c = a \oplus b = a \bullet \bar{b} + \bar{a} \bullet b & \\
SUB & c = a \backslash b = a - b = a \bullet \bar{b} &
\end{array}
\tag{51.81}
$$

The implication is that each operation is applied on a pixel-by-pixel basis. For example, $c[m, n] = a[m, n] \bullet \bar{b}[m, n] \ \forall m, n$. The definition of each operation is:

NOT	
a	
0	1
1	0

OR	b	
a	0	1
0	0	1
1	1	1

AND	b	
a	0	1
0	0	0
1	0	1

↑   ↑
input   output

XOR	b	
a	0	1
0	0	1
1	1	0

SUB	b	
a	0	1
0	0	0
1	1	0

(51.82)

These operations are illustrated in Fig. 51.22 where the binary value "1" is shown in black and the value "0" in white.

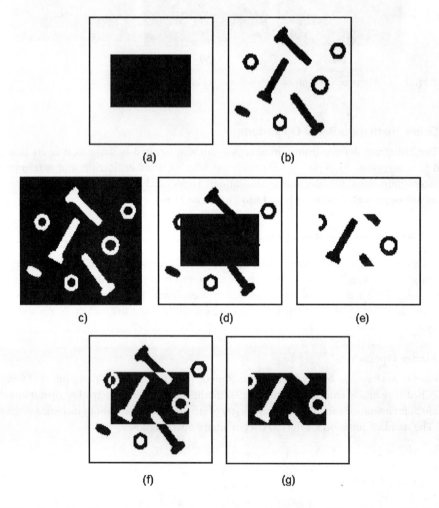

FIGURE 51.22 Examples of the various binary point operations. (a) Image a; (b) Image b; (c) NOT$(b) = \bar{b}$; (d) OR$(a, b) = a + b$; (e) AND$(a, b) = a \bullet b$; (f) XOR$(a, b) = a \oplus b$; and (g) SUB$(a, b) = a \backslash b$.

The SUB($\bullet$) operation can be particularly useful when the image a represents a region-of-interest that we want to analyze systematically and the image b represents objects that, having been analyzed, can now be discarded, that is subtracted, from the region.

Arithmetic-Based Operations

The gray-value point operations that form the basis for image processing are based on ordinary mathematics and include:

Operation	Definition	Preferred Data Type
ADD	$c = a + b$	Integer
SUB	$c = a - b$	Integer
MUL	$c = a \bullet b$	Integer or floating point
DIV	$c = a/b$	Floating point
LOG	$c = \log(a)$	Floating point
EXP	$c = \exp(a)$	Floating point
SQRT	$c = \mathrm{sqrt}(a)$	Floating point
TRIG.	$c = \sin/\cos/\tan(a)$	Floating point
INVERT	$c = (2^B - 1) - a$	Integer

$$(51.83)$$

51.9.3   Convolution-Based Operations

Convolution, the mathematical, *local* operation defined in section 51.3.1, is central to modern image processing. The basic idea is that a window of some finite size and shape — the *support* — is scanned across the image. The output pixel value is the weighted sum of the input pixels within the window where the weights are the values of the filter assigned to every pixel of the window itself. The window with its weights is called the *convolution kernel.* This leads directly to the following variation on Eq. (51.3). If the filter $h[j, k]$ is zero outside the (rectangular) window $\{j = 0, 1, \ldots, J - 1; \ k = 0, 1, \ldots, K - 1\}$, then using Eq. (51.4), the convolution can be written as the following finite sum:

$$c[m, n] = a[m, n] \otimes h[m, n] = \sum_{j=0}^{J-1} \sum_{k=0}^{K-1} h[j, k] a[m - j, n - k] \tag{51.84}$$

This equation can be viewed as more than just a pragmatic mechanism for smoothing or sharpening an image. Further, while Eq. (51.84) illustrates the local character of this operation, Eqs. (51.10) and (51.24) suggest that the operation can be implemented through the use of the Fourier domain which requires a global operation, the Fourier transform. Both of these aspects will be discussed below.

Background

In a variety of image-forming systems, an appropriate model for the transformation of the physical signal $a(x, y)$ into an electronic signal $c(x, y)$ is the convolution of the input signal with the impulse response of the sensor system. This system might consist of both an optical as well as an electrical sub-system. If each of these systems can be treated as a linear, shift-invariant (LSI) system, then the convolution model is appropriate. The definitions of these two possible system properties are given below:

Linearity -
$$\text{If } a_1 \rightarrow c_1 \text{ and } a_2 \rightarrow c_2$$
$$\text{Then } w_1 \bullet a_1 + w_2 \bullet a_2 \rightarrow w_1 \bullet c_1 + w_2 \bullet c_2 \tag{51.85}$$

Shift-Invariance -
$$\text{If } a(x, y) \rightarrow c(x, y)$$
$$\text{Then } a(x - x_o, y - y_o) \rightarrow c(x - x_o, y - y_o) \tag{51.86}$$

where w_1 and w_2 are arbitrary complex constants and x_o y_o are coordinates corresponding to arbitrary spatial translations.

Two remarks are appropriate at this point. First, linearity implies (by choosing $w_1 = w_2 = 0$) that "zero in" gives "zero out". The offset described in Eq. (51.70) means that such camera signals are not the output of a linear system and thus (strictly speaking) the convolution result is not applicable. Fortunately, it is straightforward to correct for this nonlinear effect. (See section 51.10.1).

Second, optical lenses with a magnification, M, other than $1\times$ are not shift invariant; a translation of 1 unit in the input image $a(x, y)$ produces a translation of M units in the output image $c(x, y)$. Due to the Fourier property described in Eq. (51.25), this case can still be handled by linear system theory.

If an impulse point of light $\delta(x, y)$ is imaged through an LSI system, then the impulse response of that system is called the *point spread function (PSF)*. The output image then becomes the convolution of the input image with the PSF. The Fourier transform of the PSF is called the *optical transfer function (OTF)*. For optical systems that are circularly symmetric, aberration-free, and diffraction-limited, the PSF is given by the Airy disk shown in Table 51.4-T.5. The OTF of the Airy disk is also presented in Table 51.4-T.5.

If the convolution window is not the diffraction-limited PSF of the lens but rather the effect of defocusing a lens, then an appropriate model for $h(x, y)$ is a pill box of radius a as described in Table 51.4-T.3. The effect on a test pattern is illustrated in Fig. 51.23.

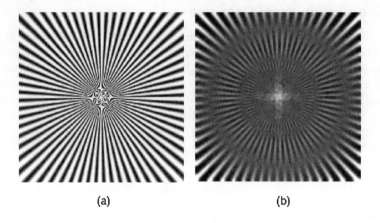

(a)   (b)

FIGURE 51.23   Convolution of test pattern with a pill box of radius $a = 4.5$ pixels. (a) Test pattern; (b) defocused image.

The effect of the defocusing is more than just simple blurring or smoothing. The almost periodic negative lobes in the transfer function in Table 51.4-T.3 produce a $180°$ phase shift in which black turns to white and vice-versa. The phase shift is clearly visible in Fig. 51.23(b).

Convolution in the Spatial Domain

In describing filters based on convolution, we will use the following convention. Given a filter $h[j, k]$ of dimensions $J \times K$, we will consider the coordinate $[j = 0, k = 0]$ to be in the center of the filter matrix, **h**. This is illustrated in Fig. 51.24. The "center" is well defined when J and K are odd: for the case where they are even, we will use the approximations $(J/2, K/2)$ for the "center" of the matrix.

When we examine the convolution sum [Eq. (51.84)] closely, several issues become evident.

- Evaluation of formula (51.84) for $m = n = 0$ while rewriting the limits of the convolution sum based on the "centering" of $h[j, k]$ shows that values of $a[j, k]$ can be required that are

$$\mathbf{h} = \begin{bmatrix} h\left[-\left(J-\frac{1}{2}\right),-\left(K-\frac{1}{2}\right)\right] & \cdots & & h\left[0,-\left(K-\frac{1}{2}\right)\right] & & \cdots & & h\left[\left(J-\frac{1}{2}\right),-\left(K-\frac{1}{2}\right)\right] \\ & \ddots & \vdots & \vdots & \vdots & \ddots & \\ \vdots & \cdots & h[-1,-1] & h[0,-1] & h[1,-1] & \cdots & \vdots \\ h\left[-\left(J-\frac{1}{2}\right),0\right] & \cdots & h[-1,0] & h[0,0] & h[1,0] & \cdots & h\left[\left(J-\frac{1}{2}\right),0\right] \\ & \cdots & h[-1,1] & h[0,1] & h[1,+1] & \cdots & \\ \vdots & & \vdots & \vdots & \vdots & \ddots & \vdots \\ h\left[-\left(J-\frac{1}{2}\right),\left(K-\frac{1}{2}\right)\right] & \cdots & & h\left[0,\left(K-\frac{1}{2}\right)\right] & & \cdots & & h\left[\left(J-\frac{1}{2}\right),\left(K-\frac{1}{2}\right)\right] \end{bmatrix}$$

FIGURE 51.24 Coordinate system for describing $h[j,k]$.

outside the image boundaries:

$$c[0,0] = \sum_{j=-J_0}^{+J_0} \sum_{k=-K_0}^{+K_0} h[j,k]a[-j,-k] \qquad J_0 = \frac{(J-1)}{2}, \quad K_0 = \frac{(K-1)}{2} \qquad (51.87)$$

The question arises — what values should we assign to the image $a[m,n]$ for $m < 0$, $m \geq M$, $n < 0$, and $n \geq N$? There is no "answer" to this question. There are only alternatives among which we are free to choose assuming we understand the possible consequences of our choice. The standard alternatives are (a) extend the images with a constant (possibly zero) brightness value, (b) extend the image periodically, (c) extend the image by mirroring it at its boundaries, or (d) extend the values at the boundaries indefinitely. These alternatives are illustrated in Fig. 51.25.

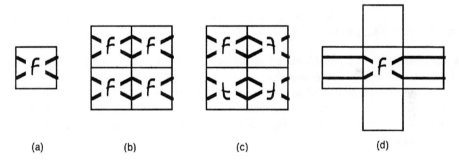

(a)   (b)   (c)   (d)

FIGURE 51.25 Examples of various alternatives to extend an image outside its formal boundaries. See text for explanation.

- When the convolution sum is written in the standard form [Eq. (51.3)] for an image $a[m,n]$ of size $M \times N$:

$$c[m,n] = \sum_{j=0}^{M-1} \sum_{k=0}^{N-1} a[j,k]h[m-j,n-k] \qquad (51.88)$$

we see that the convolution kernel $h[j,k]$ is mirrored around $j = k = 0$ to produce $h[-j,-k]$ before it is translated by $[m,n]$ as indicated in Eq. (51.88). While some convolution kernels in common use are symmetric in this respect, $h[j,k] = h[-j,-k]$, many are not. (See section 51.9.5). Care must therefore be taken in the implementation of filters with respect to the mirroring requirements.

- The computational complexity for a $K \times K$ convolution kernel implemented in the spatial domain on an image of $N \times N$ is $O(K^2)$ where the complexity is measured *per pixel* on the basis of the number of multiplies-and-adds (MADDs).

- The value computed by a convolution that begins with integer brightnesses for $a[m, n]$ may produce a rational number or a floating point number in the result $c[m, n]$. Working exclusively with integer brightness values will, therefore, cause roundoff errors.
- Inspection of Eq. (51.84) reveals another possibility for efficient implementation of convolution. If the convolution kernel $h[j, k]$ is *separable*, that is, if the kernel can be written as:

$$h[j, k] = h_{\text{row}}[k] \bullet h_{\text{col}}[j] \tag{51.89}$$

then the filtering can be performed as follows:

$$c[m, n] = \sum_{j=0}^{J-1} \left\{ \sum_{k=0}^{K-1} h_{\text{row}}[k] a[m-j, n-k] \right\} h_{\text{col}}[j] \tag{51.90}$$

This means that instead of applying one two-dimensional filter, it is possible to apply two one-dimensional filters, the first one in the k direction and the second one in j direction. For an $N \times N$ image this, in general, reduces the computational complexity per pixel from $O(J \bullet K)$ to $O(J + K)$.

An alternative way of writing separability is to note that the convolution kernel Fig. 51.24 is a matrix **h** and, if separable, **h** can be written as:

$$
\begin{aligned}
[\mathbf{h}] &= [\mathbf{h_{col}}] \bullet [\mathbf{h_{row}}]^t \\
(J \times K) &= (J \times 1) \bullet (1 \times K)
\end{aligned} \tag{51.91}
$$

where "t"denotes the matrix transpose operation. In other words, **h** can be expressed as the *outer product* of a column vector $[\mathbf{h_{col}}]$ and a row vector $[\mathbf{h_{row}}]$.

- For certain filters it is possible to find an *incremental implementation* for a convolution. As the convolution window moves over the image [see Eq. (51.88)], the leftmost column of image data under the window is shifted out as a new column of image data is shifted in from the right. Efficient algorithms can take advantage of this and, when combined with separable filters as described above, this can lead to algorithms where the computational complexity per pixel is $O(constant)$.

Convolution in the Frequency Domain

In section 51.3.4 we indicated that there was an alternative method to implement the filtering of images through convolution. Based on Eq. (51.24), it appears possible to achieve the same result as in Eq. (51.84) by the following sequence of operations:

$$
\begin{aligned}
&(i) \quad \text{Compute } A(\Omega, \Psi) = \mathcal{F}\{a[m, n]\} \\
&(ii) \quad \text{Multiply } A(\Omega, \Psi) \text{by the } \textit{precomputed } H(\Omega, \Psi) = \mathcal{F}\{h[m, n]\} \\
&(iii) \quad \text{Compute the result } c[m, n] = \mathcal{F}^{-1}\{A(\Omega, \Psi) \bullet H(\Omega, \Psi)\}
\end{aligned} \tag{51.92}
$$

- While it might seem that the "recipe" given above in Eq. (51.92) circumvents the problems associated with direct convolution in the spatial domain — specifically, determining values for the image outside the boundaries of the image — the Fourier domain approach, in fact, simply "assumes" that the image is repeated periodically outside its boundaries as illustrated in Fig. 51.25(b). This phenomenon is referred to as *circular convolution*.

  If circular convolution is not acceptable, then the other possibilities illustrated in Fig. 51.25 can be realized by embedding the image $a[m, n]$ and the filter $H(\Omega, \Psi)$ in larger matrices with the desired image extension mechanism for $a[m, n]$ being explicitly implemented.

- The computational complexity per pixel of the Fourier approach for an image of $N \times N$ and for a convolution kernel of $K \times K$ is $O(\log N)$ *complex* MADDs *independent of K*. Here we assume that $N > K$ and that N is a highly composite number such as a power of two. (See also section 51.2.1). This latter assumption permits use of the computationally efficient fast Fourier transform (FFT) algorithm. Surprisingly then, the indirect route described by Eq. (51.92) can be faster than the direct route given in Eq. (51.84). This requires, in general, that $K^2 \gg \log N$. The range of K and N for which this holds depends on the specifics of the implementation. For the machine on which this manuscript is being written and the specific image processing package that is being used, for an image of $N = 256$, the Fourier approach is faster than the convolution approach when $K \geq 15$. (It should be noted that in this comparison the direct convolution involves only integer arithmetic while the Fourier domain approach requires complex floating point arithmetic.)

51.9.4 Smoothing Operations

These algorithms are applied in order to reduce noise and/or to prepare images for further processing such as segmentation. We distinguish between linear and nonlinear algorithms where the former are amenable to analysis in the Fourier domain and the latter are not. We also distinguish between implementations based on a rectangular support for the filter and implementations based on a circular support for the filter.

Linear Filters

Several filtering algorithms will be presented together with the most useful supports.

Uniform filter – The output image is based on a local averaging of the input filter where all of the values within the filter support have the same weight. In the continuous spatial domain (x, y) the PSF and transfer function are given in Table 51.4-T.1 for the rectangular case and in Table 51.4-T.3 for the circular (pill box) case. For the discrete spatial domain $[m, n]$, the filter values are the samples of the continuous domain case. Examples for the rectangular case $(J = K = 5)$ and the circular case $(R = 2.5)$ are shown in Fig. 51.26.

$$h_{rect}[j,k] = \frac{1}{25}\begin{bmatrix} 1 & 1 & 1 & 1 & 1 \\ 1 & 1 & 1 & 1 & 1 \\ 1 & 1 & 1 & 1 & 1 \\ 1 & 1 & 1 & 1 & 1 \\ 1 & 1 & 1 & 1 & 1 \end{bmatrix} \qquad h_{circ}[j,k] = \frac{1}{21}\begin{bmatrix} 0 & 1 & 1 & 1 & 0 \\ 1 & 1 & 1 & 1 & 1 \\ 1 & 1 & 1 & 1 & 1 \\ 1 & 1 & 1 & 1 & 1 \\ 0 & 1 & 1 & 1 & 0 \end{bmatrix}$$

(a) (b)

FIGURE 51.26 Uniform filters for image smoothing. (a) Rectangular filter $(J = K = 5)$; (b) circular filter $(R = 2.5)$.

Note that in both cases the filter is normalized so that $\sum h[j, k] = 1$. This is done so that if the input $a[m, n]$ is a constant, then the output image $c[m, n]$ is the same constant. The justification can be found in the Fourier transform property described in Eq. (51.26). As can be seen from Table 51.4, both of these filters have transfer functions that have negative lobes and can, therefore, lead to phase reversal as seen in Fig. 51.23. The square implementation of the filter is separable and incremental; the circular implementation is incremental.

Triangular filter – The output image is based on a local averaging of the input filter where the values within the filter support have differing weights. In general, the filter can be seen as the convolution of two (identical) uniform filters either rectangular or circular, and this has direct consequences for the computational complexity. (See Table 51.13.) In the continuous spatial domain, the PSF and transfer function are given in Table 51.4-T.2 for the rectangular support case and in Table 51.4-T.4 for the circular (pill box) support case. As seen in Table 51.4, the transfer functions of these filters do not have negative lobes and thus do not exhibit phase reversal.

Examples for the rectangular support case ($J = K = 5$) and the circular support case ($R = 2.5$) are shown in Fig. 51.27. The filter is again normalized so that $\sum h[j, k] = 1$.

$$
h_{rect}[j,k] = \frac{1}{81}
\begin{bmatrix}
1 & 2 & 3 & 2 & 1 \\
2 & 4 & 6 & 4 & 2 \\
3 & 6 & 9 & 6 & 3 \\
2 & 4 & 6 & 4 & 2 \\
1 & 2 & 3 & 2 & 1
\end{bmatrix}
\qquad
h_{circ}[j,k] = \frac{1}{25}
\begin{bmatrix}
0 & 0 & 1 & 0 & 0 \\
0 & 2 & 2 & 2 & 0 \\
1 & 2 & 5 & 2 & 1 \\
0 & 2 & 2 & 2 & 0 \\
0 & 0 & 1 & 0 & 0
\end{bmatrix}
$$

(a)   (b)

FIGURE 51.27   Triangular filters for image smoothing. (a) Pyramidal filter ($J = K = 5$); (b) Cone filter ($R = 2.5$).

Gaussian filter – The use of the Gaussian kernel for smoothing has become extremely popular. This has to do with certain properties of the Gaussian (e.g., the central limit theorem, minimum space-bandwidth product) as well as several application areas such as edge finding and scale space analysis. The PSF and transfer function for the continuous space Gaussian are given in Table 51.4-T.6. The Gaussian filter is separable:

$$
\begin{aligned}
h(x, y) &= g_{2D}(x, y) = \left(\frac{1}{\sqrt{2\pi}\sigma} e^{-\left(x^2/2\sigma^2\right)} \right) \bullet \left(\frac{1}{\sqrt{2\pi}\sigma} e^{-\left(y^2/2\sigma^2\right)} \right) \\
&= g_{1D}(x) \bullet g_{1D}(y)
\end{aligned}
\tag{51.93}
$$

There are four distinct ways to implement the Gaussian:

1. Convolution using a finite number of samples (N_o) of the Gaussian as the convolution kernel. It is common to choose $N_o = \lceil 3\sigma \rceil$ or $\lceil 5\sigma \rceil$.

$$
g_{1D}[n] = \begin{cases} \frac{1}{\sqrt{2\pi}\sigma} e^{-\left(n^2/2\sigma^2\right)} & |n| \le N_o \\ 0 & |n| > N_o \end{cases}
\tag{51.94}
$$

2. Repetitive convolution using a uniform filter as the convolution kernel.

$$
g_{1D}[n] \approx u[n] \otimes u[n] \otimes u[n]
$$

$$
u[n] = \begin{cases} 1/(2N_o + 1) & |n| \le N_o \\ 0 & |n| > N_o \end{cases}
\tag{51.95}
$$

The actual implementation (in each dimension) is usually of the form:

$$c[n] = ((a[n] \otimes u[n]) \otimes u[n]) \otimes u[n] \tag{51.96}$$

This implementation makes use of the approximation afforded by the central limit theorem. For a desired σ with Eq. (51.96), we use $N_o = \lceil \sigma \rceil$ although this severely restricts our choice of σ's to integer values.

3. Multiplication in the frequency domain. As the Fourier transform of a Gaussian *is* a Gaussian (see Table 51.4-T.6), this means that it is straightforward to prepare a filter $H(\Omega, \Psi) = G_{2D}(\Omega, \Psi)$ for use with Eq. (51.92). To avoid truncation effects in the frequency domain due to the infinite extent of the Gaussian, it is important to choose a σ that is sufficiently large. Choosing $\sigma > k/\pi$ where $k = 3$ or 4 will usually be sufficient.

4. Use of a recursive filter implementation. A recursive filter has an infinite impulse response and thus an infinite support. The separable Gaussian filter can be implemented by applying the following recipe in each dimension when $\sigma \geq 0.5$.

  (i)   Choose the σ based on the desired goal of the filtering;
  (ii)   Determine the parameter q based on Eq. (51.98);
  (iii)   Use Eq. (51.99) to determine the filter coefficients $\{b_0, b_1, b_2, b_3, B\}$;   (51.97)
  (iv)   Apply the forward difference equation. Eq. (51.100);
  (v)   Apply the backward difference equation. Eq. (51.101).

The relation between the desired σ and q is given by:

$$q = \begin{cases} .98711\sigma - 0.96330 & \sigma \geq 2.5 \\ 3.97156 - 4.14554\sqrt{1 - .26891\,\sigma} & 0.5 \leq \sigma \leq 2.5 \end{cases} \tag{51.98}$$

The *filter coefficients* $\{b_0, b_1, b_2, b_3, B\}$ are defined by:

$$
\begin{aligned}
b_0 &= 1.57825 + (2.44413\,q) + \left(1.4281\,q^2\right) + \left(0.422205\,q^3\right) \\
b_1 &= (2.44413\,q) + \left(2.85619\,q^2\right) + \left(1.26661\,q^3\right) \\
b_2 &= -\left(1.4281\,q^2\right) - \left(1.26661\,q^3\right) \\
b_3 &= 0.422205\,q^3 \\
B &= 1 - (b_1 + b_2 + b_3)/b_0
\end{aligned}
\tag{51.99}
$$

The one-dimensional *forward difference equation* takes an input row (or column) $a[n]$ and produces an intermediate output result $w[n]$ given by:

$$w[n] = Ba[n] + (b_1 w[n-1] + b_2 w[n-2] + b_3 w[n-3])\,/b_0 \tag{51.100}$$

The one-dimensional *backward difference equation* takes the intermediate result $w[n]$ and produces the output $c[n]$ given by:

$$c[n] = Bw[n] + (b_1 c[n+1] + b_2 c[n+2] + b_3 c[n+3])\,/b_0 \tag{51.101}$$

The forward equation is applied from $n = 0$ *up to* $n = N - 1$ while the backward equation is applied from $n = N - 1$ *down* to $n = 0$.

The relative performance of these various implementations of the Gaussian filter can be described as follows. Using the *root-square error* $\sqrt{\sum_{n=-\infty}^{+\infty} |g[n|\sigma] - h[n]|^2}$ between a true, infinite-extent Gaussian, $g[n|\sigma]$, and an approximated Gaussian, $h[n]$, as a measure of accuracy, the various algorithms described above give the results shown in Fig. 51.28(a). The relative speed of the various algorithms is shown in Fig. 51.28(b).

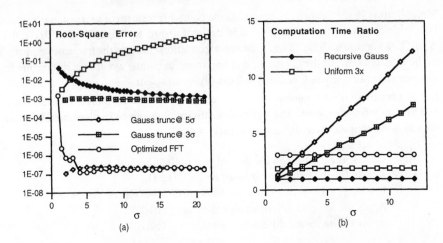

FIGURE 51.28 Comparison of various Gaussian algorithms with $N = 256$. The legend is spread across both graphs. (a) Accuracy comparison; (b) speed comparison.

The root-square error measure is extremely conservative and, thus, all filters, with the exception of "Uniform 3×" for large σ, are sufficiently accurate. The recursive implementation is the fastest independent of σ: the other implementations can be significantly slower. The FFT implementation, for example, is 3.1 times slower for $N = 256$. Further, the FFT requires that N be a highly composite number.

    Other – The Fourier domain approach offers the opportunity to implement a variety of smoothing algorithms. The smoothing filters will then be *lowpass filters*. In general, it is desirable to use a lowpass filter that has zero phase so as not to produce phase distortion when filtering the image. The importance of phase was illustrated in Figs. 51.5 and 51.23. When the frequency domain characteristics can be represented in an analytic form, then this can lead to relatively straightforward implementations of $H(\Omega, \Psi)$. Possible candidates include the lowpass filters "Airy" and "Exponential Decay" found in Table 51.4-T.5 and Table 51.4-T.8, respectively.

Nonlinear Filters

    A variety of smoothing filters have been developed that are not linear. While they cannot, in general, be submitted to Fourier analysis, their properties and domains of application have been studied extensively.

    Median filter – The median statistic was described in section 51.3.5. A median filter is based on moving a window over an image (as in a convolution) and computing the output pixel as the median value of the brightnesses within the input window. If the window is $J \times K$ in size we can order the $J \bullet K$ pixels in brightness value from smallest to largest. If $J \bullet K$ is odd, then the median will be the $(J \bullet K + 1)/2$ entry in the list of ordered brightnesses. Note that the value selected will be exactly equal to one of the existing brightnesses so that no roundoff error will be involved if we want to work exclusively with integer brightness values. The algorithm as it is described above has a generic complexity per pixel of $O(J \bullet K \bullet \log(J \bullet K))$. Fortunately, a fast algorithm (due to Huang et al.) exists that reduces the complexity to $O(K)$ assuming $J \geq K$.

A useful variation on the theme of the median filter is the *percentile filter*. Here the center pixel in the window is replaced not by the 50% (median) brightness value but rather by the $p\%$ brightness value where $p\%$ ranges from 0% (the *minimum filter*) to 100% (the *maximum filter*). Values other than $(p = 50)\%$ do not, in general, correspond to smoothing filters.

Kuwahara filter – Edges play an important role in our perception of images (see Fig. 51.15) as well as in the analysis of images. As such, it is important to be able to smooth images without disturbing the sharpness and, if possible, the position of edges. A filter that accomplishes this goal is termed an *edge-preserving filter* and one particular example is the Kuwahara filter. Although this filter can be implemented for a variety of different window shapes, the algorithm will be described for a square window of size $J = K = 4L + 1$ where L is an integer. The window is partitioned into four regions, as shown in Fig. 51.29.

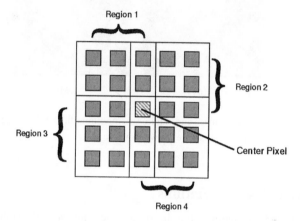

FIGURE 51.29 Four square regions defined for the Kuwahara filter. In this example, $L = 1$ and thus $J = K = 5$. Each region is $[(J + 1)/2] \times [(K + 1)/2]$.

In each of the four regions ($i = 1, 2, 3, 4$), the mean brightness, m_i in Eq. (51.34), and the *variance$_i$*, s_i^2 in Eq. (51.36), are measured. The output value of the center pixel in the window is the mean value of that region that has the smallest variance.

Summary of Smoothing Algorithms

Table 51.13 summarizes the various properties of the smoothing algorithms presented above. The filter size is assumed to be bounded by a rectangle of $J \times K$ where, without loss of generality, $J \geq K$. The image size is $N \times N$. Examples of the effect of various smoothing algorithms are shown in Fig. 51.30.

TABLE 51.13 Characteristics of Smoothing Filters

Algorithm	Domain	Type	Support	Separable/incremental	Complexity/pixel
Uniform	Space	Linear	Square	Y/Y	$O(constant)$
Uniform	Space	Linear	Circular	N/Y	$O(K)$
Triangle	Space	Linear	Square	Y/N	$O(constant)^a$
Triangle	Space	Linear	Circular	N/N	$O(K)^a$
Gaussian	Space	Linear	∞^a	Y/N	$O(constant)^a$
Median	Space	Non-Linear	Square	N/Y	$O(K)^a$
Kuwahara	Space	Non-Linear	Square^a	N/N	$O(J \bullet K)$
Other	Frequency	Linear	—	—/—	$O(\log N)$

^{a} See text for additional explanation.

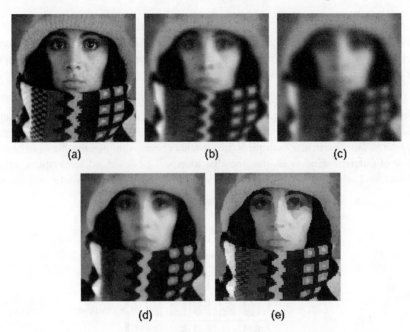

FIGURE 51.30   Illustration of various linear and nonlinear smoothing filters: (a) Original; (b) Uniform 5×5; (c) Gaussian ($\sigma = 2.5$); (d) Median 5×5; and (e) Kuwahara 5×5.

51.9.5   Derivative-Based Operations

Just as smoothing is a fundamental operation in image processing, so is the ability to take one or more spatial derivatives of the image. The fundamental problem is that, according to the mathematical definition of a derivative, this cannot be done. A digitized image is not a continuous function $a(x, y)$ of the spatial variables but rather a discrete function $a[m, n]$ of the integer spatial coordinates. As a result, the algorithms we will present can only be seen as *approximations* to the true spatial derivatives of the original spatially continuous image.

Further, as we can see from the Fourier property in Eq. (51.27), taking a derivative multiplies the signal spectrum by either u or v. This means that high frequency noise will be emphasized in the resulting image. The general solution to this problem is to combine the derivative operation with one that suppresses high frequency noise, in short, smoothing in combination with the desired derivative operation.

First Derivatives

As an image is a function of two (or more) variables, it is necessary to define the direction in which the derivative is taken. For the two-dimensional case, we have the horizontal direction, the vertical direction, or an arbitrary direction that can be considered as a combination of the two. If we use $\mathbf{h_x}$ to denote a horizontal derivative filter (matrix), $\mathbf{h_y}$ to denote a vertical derivative filter (matrix), and $\mathbf{h}_\theta$ to denote the arbitrary angle derivative filter (matrix), then:

$$[\mathbf{h}_\theta] = \cos\theta \bullet [\mathbf{h_x}] + \sin\theta \bullet \left[\mathbf{h_y}\right] \tag{51.102}$$

Gradient filters – It is also possible to generate a vector derivative description as the *gradient*, $\nabla a[m, n]$, of an image:

$$\nabla a = \frac{\partial a}{\partial x}\vec{i}_x + \frac{\partial a}{\partial y}\vec{i}_y = (h_x \otimes a)\,\vec{i}_x + \left(h_y \otimes a\right)\vec{i}_y \tag{51.103}$$

where $\vec{i}_x$ and $\vec{i}_y$ are unit vectors in the horizontal and vertical direction, respectively. This leads to two descriptions:

Gradient magnitude -

$$|\nabla_a| = \sqrt{(h_x \otimes a)^2 + (h_y \otimes a)^2} \tag{51.104}$$

and

Gradient direction -

$$\psi\,(\nabla a) = \arctan\left\{(h_y \otimes a)\,/\,(h_x \otimes a)\right\} \tag{51.105}$$

The gradient magnitude is sometimes approximated by:

Approx. gradient magnitude -

$$|\nabla a| \cong |h_x \otimes a| + \left|h_y \otimes a\right| \tag{51.106}$$

The final results of these calculations depend strongly on the choices of $\mathbf{h_X}$ and $\mathbf{h_Y}$. A number of possible choices for $(\mathbf{h_X}, \mathbf{h_Y})$ will now be described.

Basic derivative filters   – These filters are specified by:

$$(i)\ [\mathbf{h_x}] = [\mathbf{h_y}]^t = [1\ -1]$$

$$(ii)\ [\mathbf{h_x}] = [\mathbf{h_y}]^t = [1\ \ 0\ -1] \tag{51.107}$$

where "^{t}" denotes matrix transpose. These two filters differ significantly in their Fourier magnitude and Fourier phase characteristics. For the frequency range $0 \leq \Omega \leq \pi$, these are given by:

$$(i)\quad [\mathbf{h}] = [1\ -1]\quad \overset{\mathcal{F}}{\leftrightarrow}\quad |H(\Omega)| = 2\,|\sin(\Omega/2)|\,;\ \ \varphi(\Omega) = (\pi - \Omega)/2$$

$$(ii)\quad [\mathbf{h}] = [1\ \ 0\ -1]\quad \overset{\mathcal{F}}{\leftrightarrow}\quad |H(\Omega)| = 2\,|\sin(\Omega)|\,;\ \ \varphi(\Omega) = \pi/2 \tag{51.108}$$

The second form (ii) gives suppression of high frequency terms ($\Omega \approx \pi$) while the first form (i) does not. The first form leads to a phase shift; the second form does not.

Prewitt gradient filters   – These filters are specified by:

$$[\mathbf{h_x}] = \frac{1}{3}\begin{bmatrix} 1 & 0 & -1 \\ 1 & 0 & -1 \\ 1 & 0 & -1 \end{bmatrix} = \frac{1}{3}\begin{bmatrix} 1 \\ 1 \\ 1 \end{bmatrix} \bullet [1\ \ 0\ -1]$$

$$[\mathbf{h_y}] = \frac{1}{3}\begin{bmatrix} 1 & 1 & 1 \\ 0 & 0 & 0 \\ -1 & -1 & -1 \end{bmatrix} = \frac{1}{3}\begin{bmatrix} 1 \\ 0 \\ -1 \end{bmatrix} \bullet [1\ \ 1\ \ 1] \tag{51.109}$$

Both $\mathbf{h_x}$ and $\mathbf{h_y}$ are separable. Beyond the computational implications are the implications for the analysis of the filter. Each filter takes the derivative in one direction using Eq. (51.107 ii) and smoothes in the orthogonal direction using a one-dimensional version of a *uniform* filter as described in section 51.9.4.

Sobel gradient filters   – These filters are specified by:

$$[\mathbf{h_x}] = \frac{1}{4}\begin{bmatrix} 1 & 0 & -1 \\ 2 & 0 & -2 \\ 1 & 0 & -1 \end{bmatrix} = \frac{1}{4}\begin{bmatrix} 1 \\ 2 \\ 1 \end{bmatrix} \bullet [1\ \ 0\ -1]$$

$$[\mathbf{h_y}] = \frac{1}{4}\begin{bmatrix} 1 & 2 & 1 \\ 0 & 0 & 0 \\ -1 & -2 & -1 \end{bmatrix} = \frac{1}{4}\begin{bmatrix} 1 \\ 0 \\ -1 \end{bmatrix} \bullet [1\ \ 2\ \ 1] \tag{51.110}$$

Again, $\mathbf{h_x}$ and $\mathbf{h_y}$ are separable. Each filter takes the derivative in one direction using Eq. (51.107 *ii*) and smoothes in the orthogonal direction using a one-dimensional version of a *triangular* filter as described in section 51.9.4.

Alternative gradient filters   – The variety of techniques available from one-dimensional signal processing for the design of digital filters offers us powerful tools for designing one-dimensional versions of $\mathbf{h_x}$ and $\mathbf{h_y}$. Using the Parks-McClellan filter design algorithm, for example, we can choose the frequency bands where we want the derivative to be taken and the frequency bands where we want the noise to be suppressed. The algorithm will then produce a real, odd filter with a minimum length that meets the specifications.

As an example, if we want a filter that has derivative characteristics in a passband (with weight 1.0) in the frequency range $0.0 \leq \Omega \leq 0.3\pi$ and a stopband (with weight 3.0) in the range $0.32\pi \leq \Omega \leq \pi$, then the algorithm produces the following optimized seven sample filter:

$$[\mathbf{h}_x] = [\mathbf{h}_y]^t = \frac{1}{16348} \begin{bmatrix} -3571 & 8212 & -15580 & 0 & 15580 & -8212 & 3571 \end{bmatrix} \qquad (51.111)$$

The gradient can then be calculated as in Eq. (51.103).

Gaussian gradient filters   – In modern digital image processing, one of the most common techniques is to use a Gaussian filter (see section 51.9.4) to accomplish the required smoothing and one of the derivatives listed in Eq. (51.107). Thus, we might first apply the recursive Gaussian in Eq. (51.97) followed by Eq. (51.107 *ii*) to achieve the desired, smoothed derivative filters $\mathbf{h_x}$ and $\mathbf{h_y}$. Further, for computational efficiency, we can combine these two steps as:

$$w[n] \quad = \quad \left(\frac{B}{2}\right)(a[n+1] - a[n-1]) + (b_1 w[n-1] + b_2 w[n-2] + b_3 w[n-3])/b_0$$

$$c[n] \quad = \quad Bw[n] + (b_1 c[n+1] + b_2 c[n+2] + b_3 c[n+3])/b_0 \qquad (51.112)$$

where the various coefficients are defined in Eq. (51.99). The first (forward) equation is applied from $n = 0$ *up* to $n = N - 1$ while the second (backward) equation is applied from $n = N - 1$ *down* to $n = 0$.

Summary   – Examples of the effect of various *derivative algorithms* on a noisy version of Fig. 51.30(a) $(SNR) = 29$ dB) are shown in Figs. 51.31(a-c). The effect of various *magnitude gradient algorithms* on Fig. 51.30(a) are shown in Figs. 51.32(a-c). After processing, all images are contrast stretched as in Eq. (51.77) for display purposes.

The magnitude gradient takes on large values where there are strong edges in the image. Appropriate choice of σ in the Gaussian-based derivative (Fig. 51.31(c)) or gradient (Fig. 51.32(c)) permits computation of virtually any of the other forms — simple, Prewitt, Sobel, etc. In that sense, the Gaussian derivative represents a superset of derivative filters.

Second Derivatives

It is, of course, possible to compute higher-order derivatives of functions of two variables. In image processing, as we shall see in sections 51.10.2 and 51.10.3, the second derivatives or Laplacian play an important role. The Laplacian is defined as:

$$\nabla^2 a = \frac{\partial^2 a}{\partial x^2} + \frac{\partial^2 a}{\partial y^2} = (h_{2x} \otimes a) + \left(h_{2y} \otimes a\right) \qquad (51.113)$$

where $\mathbf{h_{2x}}$ and $\mathbf{h_{2y}}$ are second derivative filters. In the frequency domain, we have for the Laplacian filter [from Eq. (51.27)]:

$$\nabla^2 a = \overset{\mathcal{F}}{\longleftrightarrow} \; - \left(u^2 + v^2\right) A(u, v) \qquad (51.114)$$

The transfer function of a Laplacian corresponds to a parabola $H(u, v) = -(u^2 + v^2)$.

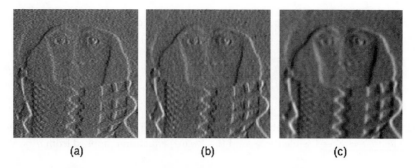

FIGURE 51.31 Application of various algorithms for $\mathbf{h_x}$ — the horizontal derivative. (a) Simple Derivative — Eq. (51.107)ii; (b) Sobel — Eq. (51.110); (c) Gaussian ($\sigma = 1.5$) and Eq. (51.107)ii.

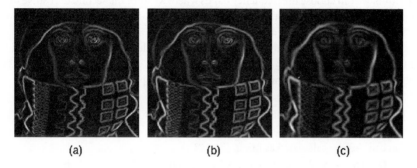

FIGURE 51.32 Various algorithms for the magnitude gradient, $|\nabla$al. (a) Simple Derivative — Eq. (51.107)ii; (b) Sobel — Eq. (51.110); (c) Gaussian ($\sigma = 1.5$) and Eq. (51.107)ii.

Basic second derivative filter – This filter is specified by:

$$[\mathbf{h}_{2x}] = [\mathbf{h}_{2y}]^t = [1 \quad -2 \quad 1] \tag{51.115}$$

and the frequency spectrum of this filter, in each direction, is given by:

$$H(\Omega) = \mathcal{F}\{1 \quad -2 \quad 1\} = -2(1 - \cos\Omega) \tag{51.116}$$

over the frequency range $-\pi \leq \Omega \leq \pi$. The two, one-dimensional filters can be used in the manner suggested by Eq. (51.113) or combined into one, two-dimensional filter as:

$$[\mathbf{h}] = \begin{bmatrix} 0 & 1 & 0 \\ 1 & -4 & 1 \\ 0 & 1 & 0 \end{bmatrix} \tag{51.117}$$

and used as in Eq. (51.84).

Frequency domain Laplacian – This filter is the implementation of the general recipe given in Eq. (51.92) and for the Laplacian filter takes the form:

$$c[m, n] = \mathcal{F}^{-1}\left\{-\left(\Omega^2 + \Psi^2\right) A(\Omega, \Psi)\right\} \tag{51.118}$$

Gaussian second derivative filter – This is the straightforward extension of the Gaussian first derivative filter described above and can be applied independently in each dimension. We first apply Gaussian smoothing with a σ chosen on the basis of the problem specification. We then apply the desired second derivative filter Eq. (51.115) or Eq. (51.118). Again, there is the choice among the various Gaussian smoothing algorithms.

For efficiency, we can use the recursive implementation and combine the two steps — smoothing and derivative operation — as follows:

$$w[n] \;=\; B(a[n] - a[n-1]) + (b_1 w[n-1] + b_2 w[n-2] + b_3 w[n-3])/b_0$$

$$c[n] \;=\; B(w[n+1] - w[n]) + (b_1 c[n+1] + b_2 c[n+2] + b_3 c[n+3])/b_0 \qquad (51.119)$$

where the various coefficients are defined in Eq. (51.99). Again, the first (forward) equation is applied from $n = 0$ *up* to $n = N - 1$ while the second (backward) equation is applied from $n = N - 1$ *down* to $n = 0$.

    Alternative Laplacian filters   – Again one-dimensional digital filter design techniques offer us powerful methods to create filters that are optimized for a specific problem. Using the Parks-McClellan design algorithm, we can choose the frequency bands where we want the second derivative to be taken and the frequency bands where we want the noise to be suppressed. The algorithm will then produce a real, even filter with a minimum length that meets the specifications.

    As an example, if we want a filter that has second derivative characteristics in a passband (with weight 1.0) in the frequency range $0.0 \le \Omega \le 0.3\pi$ and a stopband (with weight 3.0) in the range $0.32\pi \le \Omega \le \pi$, then the algorithm produces the following optimized seven sample filter:

$$[\mathbf{h}_x] = [\mathbf{h}_y]^t = \frac{1}{11043} \, [-3448 \quad 10145 \quad 1495 \; - \; 16383 \quad 1495 \quad 10145 \; - \; 3448] \qquad (51.120)$$

The Laplacian can then be calculated as in Eq. (51.113).

    SDGD filter   – A filter that is especially useful in edge finding and object measurement is the *Second-Derivative-in-the-Gradient-Direction (SDGD)* filter. This filter uses five partial derivatives:

$$
\begin{aligned}
A_{xx} &= \frac{\partial^2 a}{\partial x^2} \quad A_{xy} = \frac{\partial^2 a}{\partial x \partial y} \quad A_x = \frac{\partial a}{\partial x} \\[2mm]
A_{yx} &= \frac{\partial^2 a}{\partial x \partial y} \quad A_{yy} = \frac{\partial^2 a}{\partial y^2} \quad A_y = \frac{\partial a}{\partial y}
\end{aligned}
\qquad (51.121)
$$

Note that $A_{xy} = A_{yx}$, which accounts for the five derivatives.

    This $SDGD$ combines the different partial derivatives as follows:

$$SDGD(a) = \frac{A_{xx} A_x^2 + 2 A_{xy} A_x A_y + A_{yy} A_y^2}{A_x^2 + A_y^2} \qquad (51.122)$$

As one might expect, the large number of derivatives involved in this filter implies that noise suppression is important and that Gaussian derivative filters — both first and second order — are highly recommended, if not required. It is also necessary that the first and second derivative filters have essentially the same passbands and stopbands. This means that if the first derivative filter h_{1x} is given by $[1 \quad 0 \quad -1]$ Eq. (51.107 *ii*) then the second derivative filter should be given by $h_{1x} \otimes h_{1x} = h_{2x} = [1 \quad 0 \quad -2 \quad 0 \quad 1]$.

    Summary   – The effects of the various second derivative filters are illustrated in Figs. 51.33(a-e). All images were contrast stretched for display purposes using Eq. (51.78) and the parameters 1% and 99%.

Other Filters

    An infinite number of filters, both linear and nonlinear, are possible for image processing. It is, therefore, impossible to describe more than the basic types in this section. The description of others can be found be in the reference literature (see section 51.11) as well as in the applications literature. It is important to use a small consistent set of test images that are relevant to the application area to understand the effect of a given filter or class of filters. The effect of filters on images can be frequently understood

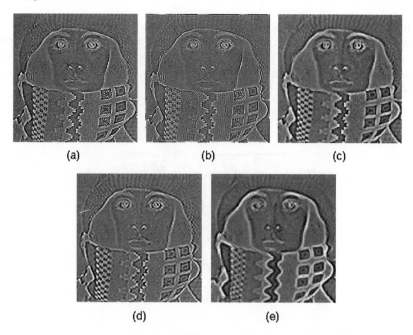

FIGURE 51.33 Various algorithms for the Laplacian and Laplacian-related filters. (a) Laplacian — Eq. (51.117); (b) Fourier parabola — Eq. (51.118); (c) Gaussian ($\sigma = 1.0$) and Eq. (51.117); (d) "Designer" — Eq. (51.120); and (e) SDGD ($\sigma = 1.0$) — Eq. (51.122).

by the use of images that have pronounced regions of varying sizes to visualize the effect on edges or by the use of test patterns such as sinusoidal sweeps to visualize the effects in the frequency domain. The former have been used previously (Figs. 51.21, 51.23 and 51.30 to 51.33), and the latter are demonstrated in Fig. 51.34.

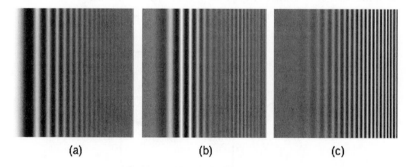

FIGURE 51.34 Various convolution algorithms applied to sinusoidal test image. (a) Lowpass filter, (b) bandpass filter, and (c) highpass filter.

51.9.6 Morphology-Based Operations

In section 51.1, we defined an image as an (amplitude) function of two, real (coordinate) variables $a(x, y)$ or two discrete variables $a[m, n]$. An alternative definition of an image can be based on the notion that an image consists of a set (or collection) of either continuous or discrete coordinates. In a sense, the set corresponds to the points or pixels that belong to the objects in the image. This is illustrated in Fig. 51.35 which contains two objects or sets A and B. Note that the coordinate system is required. For the moment, we will consider the pixel values to be binary as discussed in section 51.2.1 and 51.9.2. Further, we shall restrict our discussion to discrete space (Z^2). More general discussions can be found in Giardina and Dougherty [4], Gonzales and Woods [5], and Heijmans [7].

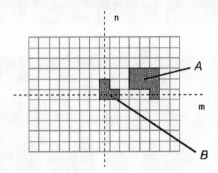

FIGURE 51.35   A binary image containing two objects sets A and B.

The object A consists of those pixels α that share some common property:
  Object -

$$A = \{\alpha | \text{property}(a) == \text{TRUE}\} \tag{51.123}$$

As an example, object B in Fig. 51.35 consists of $\{[0, 0], [1, 0], [0, 1]\}$.

The background of A is given by A^c (the *complement* of A) which is defined as those elements that are not in A:
  Background -

$$A^c = \{\alpha | \alpha \notin A\} \tag{51.124}$$

In Fig. 51.3, we introduced the concept of neighborhood connectivity. We now observe that if an object A is defined on the basis of C-connectivity ($C = 4, 6,$ or 8) then the background A^c has a connectivity given by $12 - C$. The necessity for this is illustrated for the Cartesian grid in Fig. 51.36.

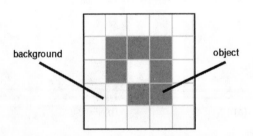

FIGURE 51.36   A binary image requiring careful definition of object and background connectivity.

Fundamental Definitions

The fundamental operations associated with an object are the standard set operations *union, intersection,* and *complement* $\{\cup, \cap, ^c\}$ plus *translation*:

  Translation   – Given a vector **x** and a set A, the *translation,* $A + \mathbf{x}$, is defined as:

$$A + \mathbf{x} = \{\alpha + \mathbf{x} | \alpha \in A\} \tag{51.125}$$

Note that, since we are dealing with a digital image composed of pixels at integer coordinate positions (Z^2), this implies restrictions on the allowable translation vectors $\mathbf{x}$.

The basic *Minkowski set operations* — addition and subtraction — can now be defined. First we note that the individual elements that comprise $\mathbf{B}$ are not only pixels but also *vectors* as they have a clear coordinate position with respect to $[0, 0]$. Given two sets $\mathbf{A}$ and $\mathbf{B}$:

$$\text{Minkowski addition -}\quad \mathbf{A} \oplus \mathbf{B} = \bigcup_{\beta \in \mathbf{B}} (\mathbf{A} + \beta) \tag{51.126}$$

$$\text{Minkowski subtraction -}\quad \mathbf{A} \ominus \mathbf{B} = \bigcap_{\beta \in \mathbf{B}} (\mathbf{A} + \beta) \tag{51.127}$$

Dilation and Erosion

From these two Minkowski operations, we define the fundamental mathematical morphology operations *dilation* and *erosion*:

$$\text{Dilation -}\quad D(\mathbf{A}, \mathbf{B}) = \mathbf{A} \oplus \mathbf{B} = \bigcup_{\beta \in \mathbf{B}} (\mathbf{A} + \beta) \tag{51.128}$$

$$\text{Erosion -}\quad E(\mathbf{A}, \mathbf{B}) = \mathbf{A} \ominus (-\mathbf{B}) = \bigcap_{\beta \in \mathbf{B}} (\mathbf{A} - \beta) \tag{51.129}$$

where $-\mathbf{B} = \{-\beta | \beta \in \mathbf{B}\}$. These two operations are illustrated in Fig. 51.37 for the objects defined in Fig. 51.35.

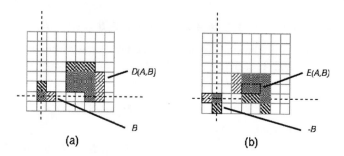

FIGURE 51.37 A binary image containing two object sets $\mathbf{A}$ and $\mathbf{B}$. The three pixels in $\mathbf{B}$ are "color-coded" as is their effect in the result. (a) Dilation $D(\mathbf{A}, \mathbf{B})$ and (b) Erosion $E(\mathbf{A}, \mathbf{B})$.

While either set $\mathbf{A}$ or $\mathbf{B}$ can be thought of as an "image", $\mathbf{A}$ is usually considered as the image and $\mathbf{B}$ is called a *structuring element*. *The structuring element is to mathematical morphology what the convolution kernel is to linear filter theory.*

Dilation, in general, causes objects to dilate or grow in size; *erosion* causes objects to shrink. The amount and the way that they grow or shrink depend on the choice of the structuring element. Dilating or eroding without specifying the structural element makes no more sense than trying to lowpass filter an image without specifying the filter. The two most common structuring elements (given a Cartesian grid) are the 4-connected and 8-connected sets, N_4 and N_8. They are illustrated in Fig. 51.38.

Dilation and *erosion* have the following properties:

$$\text{Commutative -}\quad D(\mathbf{A}, \mathbf{B}) = \mathbf{A} \oplus \mathbf{B} = \mathbf{B} \oplus \mathbf{A} = D(\mathbf{B}, \mathbf{A}) \tag{51.130}$$

$$\text{Noncommutative -}\quad E(\mathbf{A}, \mathbf{B}) \neq E(\mathbf{B}, \mathbf{A}) \tag{51.131}$$

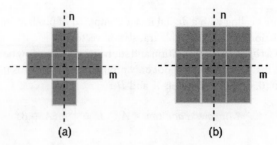

FIGURE 51.38 The standard structuring elements N_4 and N_8. (a) N_4 and (b) N_8.

$$\text{Associative - } A \oplus (B \oplus C) = (A \oplus B) \oplus C \tag{51.132}$$

$$\text{Translation Invariance - } A \oplus (B + \mathbf{x}) = (A \oplus B) + \mathbf{x} \tag{51.133}$$

$$D^c(A, B) = E(A^c, -B)$$

$$\text{Duality - } \tag{51.134}$$

$$E^c(A, B) = D(A^c, -B)$$

With A as an object and A^c as the background, Eq. (51.134) says that the *dilation* of an object is equivalent to the *erosion* of the background. Likewise, the *erosion* of the object is equivalent to the *dilation* of the background.

Except for special cases:

$$\text{Noninverses - } D(E(A, B), B) \neq A \neq E(D(A, B), B) \tag{51.135}$$

Erosion has the following translation property:

$$\text{Translation Invariance - } A \ominus (B + \mathbf{x}) = (A + \mathbf{x}) \ominus B = (A \ominus B) + \mathbf{x} \tag{51.136}$$

Dilation and *erosion* have the following important properties. For any arbitrary structuring element B and two image objects A_1 and A_2 such that $A_1 \subset A_2 (A_1$ is a proper subset of A_2):

$$D(A_1, B) \subset D(A_2, B)$$

$$\text{Increasing in } A - \tag{51.137}$$

$$E(A_1, B) \subset E(A_2, B)$$

For two structuring elements B_1 and B_2 such that $B_1 \subset B_2$:

$$\text{Decreasing in } B - \quad E(A, B_1) \supset E(A, B_2) \tag{51.138}$$

The *decomposition theorems* below make it possible to find efficient implementations for morphological filters.

$$\text{Dilation - } A \oplus (B \cup C) = (A \oplus B) \cup (A \oplus C) = (B \cup C) \oplus A \tag{51.139}$$

$$\text{Erosion - } A \ominus (B \cup C) = (A \ominus B) \cap (A \ominus C) \tag{51.140}$$

$$\text{Erosion - } (A \ominus B) \ominus C = A \ominus (B \oplus C) \tag{51.141}$$

$$\text{Multiple Dilations - } nB = \underbrace{(B \oplus B \oplus B \oplus \cdots \oplus B)}_{n \text{ times}} \tag{51.142}$$

An important decomposition theorem is due to Vincent. First, we require some definitions. A *convex* set (in R^2) is one for which the straight line joining any two points in the set consists of points that are also in the set. Care must obviously be taken when applying this definition to discrete pixels as the concept of a "straight line" must be interpreted appropriately in Z^2. A set is *bounded* if each of its elements has a finite magnitude, in this case distance to the origin of the coordinate system. A set is *symmetric* if $B = -B$. The sets N_4 and N_8 in Fig. 51.38 are examples of convex, bounded, symmetric sets.

Vincent's theorem, when applied to an image consisting of discrete pixels, states that for a bounded, symmetric structuring element B that contains no holes and contains its own center, $[0, 0] \in B$:

$$D(A, B) = A \oplus B = A \cup (\partial A \oplus B) \tag{51.143}$$

where ∂A is the contour of the object. That is, ∂A is the set of pixels that have a background pixel as a neighbor. The implication of this theorem is that it is not necessary to process all the pixels in an object in order to compute a *dilation* or [using Eq. (51.134)] an *erosion*. We only have to process the boundary pixels. This also holds for all operations that can be derived from *dilations* or *erosions*. The processing of boundary pixels instead of object pixels means that, except for pathological images, computational complexity can be reduced from $O(N^2)$ to $O(N)$ for an $N \times N$ image. A number of "fast" algorithms can be found in the literature that are based on this result. The simplest dilation and erosion algorithms are frequently described as follows.

Dilation – Take each binary object pixel (with value "1") and set all background pixels (with value "0") that are C-connected to that object pixel to the value "1".

Erosion – Take each binary object pixel (with value "1") that is C-connected to a background pixel and set the object pixel value to "0".

Comparison of these two procedures to Eq. (51.143) where $B = N_{C=4}$ or $N_{C=8}$ shows that they are equivalent to the formal definitions for dilation and erosion. The procedure is illustrated for *dilation* in Fig. 51.39.

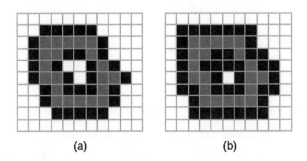

(a)   (b)

FIGURE 51.39 Illustration of *dilation*. Original object pixels are in gray; pixels added through *dilation* are in black. (a) $B = N_4$ and (b) $B = N_8$.

Boolean Convolution

An arbitrary *binary* image object (or structuring element) A can be represented as

$$A \leftrightarrow \sum_{k=-\infty}^{+\infty} \sum_{j=-\infty}^{+\infty} a[j, k] \bullet \delta[m - j, n - k] \tag{51.144}$$

where $\sum$ and $\bullet$ are the Boolean operations OR and AND as defined in Eqs. (51.81) and (51.82), $a[j, k]$ is a *characteristic function* that takes on the Boolean values "1" and "0" as follows:

$$a[j, k] = \begin{cases} 1 & a \in A \\ 0 & a \notin A \end{cases} \tag{51.145}$$

and $\delta[m, n]$ is a Boolean version of the Dirac delta function that takes on the Boolean values "1" and "0" as follows:

$$\delta[j, k] = \begin{cases} 1 & j = k = 0 \\ 0 & \text{otherwise} \end{cases} \tag{51.146}$$

Dilation for binary images can therefore be written as:

$$D(A, B) = \sum_{k=-\infty}^{+\infty} \sum_{j=-\infty}^{+\infty} a[j, k] \bullet b[m - j, n - k] = a \otimes b \tag{51.147}$$

which, because Boolean OR and AND are commutative, can also be written as

$$D(A, B) = \sum_{k=-\infty}^{+\infty} \sum_{j=-\infty}^{+\infty} a[m - j, n - k] \bullet b[j, k] = b \otimes a = D(B, A) \tag{51.148}$$

Using De Morgan's theorem:

$$\overline{(a + b)} = \bar{a} \bullet \bar{b} \text{ and } \overline{(a \bullet b)} = \bar{a} + \bar{b} \tag{51.149}$$

on Eq. (51.148) together with Eq. (51.134), *erosion* can be written as:

$$E(A, B) = \prod_{k=-\infty}^{+\infty} \prod_{j=-\infty}^{+\infty} \left(a[m - j, n - k] + \bar{b}[-j, -k] \right) \tag{51.150}$$

Thus, *dilation* and *erosion* on binary images can be viewed as a form of convolution over a Boolean algebra.

In section 51.9.3 we saw that, when convolution is employed, an appropriate choice of the boundary conditions for an image is essential. Dilation and erosion — being a Boolean convolution — are no exception. The two most common choices are that either everything outside the binary image is "0" or everything outside the binary image is "1".

Opening and Closing

We can combine *dilation* and *erosion* to build two important higher order operations:

$$\text{Opening -} \ O(A, B) = A \circ B = D(E(A, B), B) \tag{51.151}$$

$$\text{Closing -} \ C(A, B) = A \bullet B = E(D(A, -B), -B) \tag{51.152}$$

The *opening* and *closing* have the following properties:

$$\text{Duality -} \quad \begin{array}{l} C^c(A, B) = O(A^c, B) \\ O^c(A, B) = C(A^c, B) \end{array} \tag{51.153}$$

$$\text{Translation -} \quad \begin{array}{l} O(A + \mathbf{x}, B) = O(A, B) + \mathbf{x} \\ C(A + \mathbf{x}, B) = C(A, B) + \mathbf{x} \end{array} \tag{51.154}$$

For the *opening* with structuring element B and images A, A_1, and A_2, where A_1 is a subimage of $A_2 (A_1 \subseteq A_2)$:

$$\text{Antiextensivity -} \qquad O(A, B) \subseteq A \tag{51.155}$$

$$\text{Increasing monotonicity -} \quad O(A_1, B) \subseteq O(A_2, B) \tag{51.156}$$

$$\text{Idempotence -} \qquad O(O(A, B), B) = O(A, B) \tag{51.157}$$

For the *closing* with structuring element B and images A, A_1, and A_2, where A_1 is a subimage of $A_2 (A_1 \subseteq A_2)$:

$$\text{Extensivity -} \quad A \subseteq C(A, B) \quad\quad (51.158)$$

$$\text{Increasing monotonicity -} \quad C(A_1, B) \subseteq C(A_2, B) \quad\quad (51.159)$$

$$\text{Idempotence -} \quad C(C(A, B), B) = C(A, B) \quad\quad (51.160)$$

The two properties given by Eqs. (51.155) and (51.158) are so important to mathematical morphology that they can be considered as the reason for defining *erosion* with $-B$ instead of B in Eq. (51.129).

Hit-and-Miss Operation

The *hit-or-miss operator* was defined by Serra but we shall refer to it as the *hit-and-miss* operator defined as follows. Given an image A and two structuring elements B_1 and B_2, the set definition and Boolean definition are:

Hit-and-Miss -

$$\text{HitMiss}(A, B_1, B_2) = \begin{cases} E(A, B_1) \cap E^c \left(A^c, B_2\right) \\[2mm] E(A, B_1) \bullet \overline{E\left(\bar{A}, B_2\right)} \\[2mm] E(A, B_1) - E\left(\bar{A}, B_2\right) \end{cases} \quad\quad (51.161)$$

where B_1 and B_2 are bounded, disjoint structuring elements. (Note the use of the notation from Eq. (51.81).) Two sets are *disjoint* if $B_1 \cap B_2 = \varnothing$, the empty set. In an important sense the *hit-and-miss operator* is the morphological equivalent of *template matching,* a well-known technique for matching patterns based on cross-correlation. Here, we have a template B_1 for the object and a template B_2 for the background.

Summary of the Basic Operations

The results of the application of these basic operations on a test image are illustrated in Fig. 51.40. In this figure, the various structuring elements used in the processing are defined. The value "$-$" indicates a "don't care". All three structuring elements are symmetric.

$$B = N_8 = \begin{bmatrix} 1 & 1 & 1 \\ 1 & 1 & 1 \\ 1 & 1 & 1 \end{bmatrix} \quad\quad B_1 = \begin{bmatrix} - & - & - \\ - & 1 & - \\ - & - & - \end{bmatrix} \quad\quad B_2 = \begin{bmatrix} - & 1 & - \\ 1 & - & 1 \\ - & 1 & - \end{bmatrix}$$

$$\text{(a)} \quad\quad\quad\quad\quad \text{(b)} \quad\quad\quad\quad\quad \text{(c)}$$

FIGURE 51.40 Structuring elements B, B_1, and B_2 that are 3×3 and symmetric.

The results of processing are shown in Fig. 51.41 where the binary value "1" is shown in black and the value "0" in white.

The *opening* operation can separate objects that are connected in a binary image. The *closing* operation can fill in small holes. Both operations generate a certain amount of smoothing on an object contour given a "smooth" structuring element. The *opening* smoothes from the inside of the object contour and the *closing* smoothes from the outside of the object contour. The *hit-and-miss* example has found the 4-connected contour pixels. An alternative method to find the contour is simply to use the relation:

$$\text{4-connected contour-} \quad \partial A = A - E(A, N_8) \quad\quad (51.162)$$

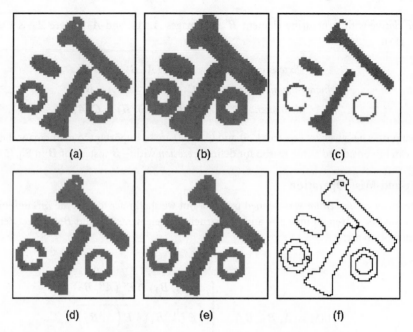

FIGURE 51.41 Examples of various mathematical morphology operations. (a) Image A; (b) *dilation* with $2B$; (c) *erosion* with $2B$; (d) *opening* with $2B$; (e) *closing* with $2B$; and (f) hit-and-miss with B_1 and B_2.

or

$$\text{8-connected contour -} \quad \partial A = A - E(A, N_4) \tag{51.163}$$

Skeleton

The informal definition of a skeleton is a line representation of an object that is:

    (i)   one-pixel thick,
    (ii) through the "middle" of the object, and   (51.164)
    (iii) preserves the topology of the object.

These are not always realizable. Fig. 51.42 shows why this is the case.

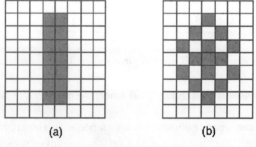

FIGURE 51.42 Counterexamples to the three requirements.

In the first example, Fig. 51.42(a), it is not possible to generate a line that is one pixel thick and in the center of an object while generating a path that reflects the simplicity of the object. In Fig. 51.42(b) it is not possible to remove a pixel from the 8-connected object and simultaneously preserve the topology —

the notion of connectedness — of the object. Nevertheless, there are a variety of techniques that attempt to achieve this goal and to produce a *skeleton*.

A basic formulation is based on the work of Lantuéjoul. The *skeleton subset* $S_k(A)$ is defined as:

$$\text{Skeleton subsets - } S_k(A) = E(A, kB) - [E(A, kB) \circ B] \quad k = 0, 1, \ldots K \tag{51.165}$$

where K is the largest value of k before the set $S_k(A)$ becomes empty. [From Eq. (51.156), $E(A, kB) \circ B \subseteq E(A, kB)$]. The structuring element B is chosen (in Z^2) to approximate a circular disc, that is, convex, bounded, and symmetric. The *skeleton* is then the union of the skeleton subsets:

$$\text{Skeleton - } S(A) = \bigcup_{k=0}^{K} S_k(A) \tag{51.166}$$

An elegant side effect of this formulation is that the original object can be reconstructed given knowledge of the skeleton subsets $S_k(A)$, the structuring element B, and K:

$$\text{Reconstruction - } A = \bigcup_{k=0}^{K} (S_k(A) \oplus kB) \tag{51.167}$$

This formulation for the skeleton, however, does not preserve the topology, a requirement described in Eq. (51.164).

An alternative point of view is to implement a *thinning*, an erosion that reduces the thickness of an object without permitting it to vanish. A general thinning algorithm is based on the *hit-and-miss* operation:

$$\text{Thinning - } \text{Thin } (A, B_1, B_2) = A - \text{HitMiss } (A, B_1, B_2) \tag{51.168}$$

Depending on the choice of B_1 and B_2, a large variety of thinning algorithms — and through repeated application skeletonizing algorithms — can be implemented.

A quite practical implementation can be described in another way. If we restrict ourselves to a 3×3 neighborhood, similar to the structuring element $B = N_8$ in Fig. 51.40(a), then we can view the thinning operation as a window that repeatedly scans over the (binary) image and sets the center pixel to "0" under certain conditions. The center pixel is *not* changed to "0" if and only if:

$$
\begin{aligned}
&(i) &&\text{an isolated pixel is found [e.g., Fig.51.43(a)],} \\
&(ii) &&\text{removing a pixel would change the connectivity [e.g., Fig.51.43(b)],} \\
&(iii) &&\text{removing a pixel would shorten a line [e.g., Fig.51.43(c)].}
\end{aligned}
\tag{51.169}
$$

As pixels are (potentially) removed in each iteration, the process is called a *conditional erosion*. Three test cases of Eq. (51.169) are illustrated in Fig. 51.43. In general, all possible rotations and variations have to be checked. As there are only 512 possible combinations for a 3×3 window on a binary image, this can be done easily with the use of a lookup table.

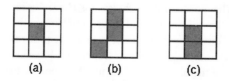

(a) (b) (c)

FIGURE 51.43 Test conditions for *conditional erosion* of the center pixel. (a) Isolated pixel, (b) connectivity pixel, and (c) end pixel.

If only condition (i) is used, then each object will be reduced to a single pixel. This is useful if we wish to count the number of objects in an image. If only condition (ii) is used, then holes in the objects will be found. If conditions ($i + ii$) are used, each object will be reduced to either a single pixel if it does not contain a hole or to closed rings if it does contain holes. If conditions ($i + ii + iii$) are used, then the "complete skeleton" will be generated as an approximation to Eq. (51.164). Illustrations of these various possibilities are given in Figs. 51.44(a) and (b).

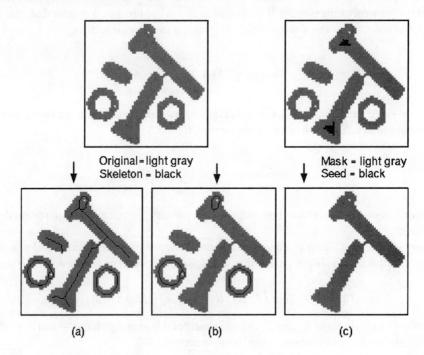

FIGURE 51.44 Examples of *skeleton* and *propagation*. (a) *Skeleton* with end pixels, *condition* Eq. (51.169)*i+ii+iii*; (b) *skeleton* without end pixels, *condition* Eq. (51.169)*i+ii*; (c) *propagation* with N_8.

Propagation

It is convenient to be able to reconstruct an image that has "survived" several erosions or to fill an object that is defined, for example, by a boundary. The formal mechanism for this has several names, including *region-filling, reconstruction,* and *propagation*. The formal definition is given by the following algorithm. We start with a *seed image* $S^{(0)}$, a *mask image* A, and a structuring element B. We then use dilations of S with structuring element B and masked by A in an iterative procedure as follows:

$$Iteration\ k -\quad S^{(k)} = \left[S^{k-1} \oplus B\right] \cap A\ \ \text{until}\ \ S^{(k)} = S^{(k-1)} \tag{51.170}$$

With each iteration, the seed image grows (through dilation) but *within* the set (object) defined by A; S propagates to fill A. The most common choices for B are N_4 or N_8. Several remarks are central to the use of propagation. First, in a straightforward implementation, as suggested by Eq. (51.170), the computational costs are extremely high. Each iteration requires $O(N^2)$ operations for an $N \times N$ image, and with the required number of iterations this can lead to a complexity of $O(N^3)$. Fortunately, a recursive implementation of the algorithm exists in which one or two passes through the image are usually sufficient, meaning a complexity of $O(N^2)$. Second, although we have not paid much attention to the issue of object/background connectivity until now (see Fig. 51.36), it is essential that the connectivity implied by B be matched to be connectivity associated with the boundary definition of A [see Eqs. (51.162) and (51.163)]. Finally, as mentioned earlier, it is important to make the correct choice ("0" or "1") for the boundary condition of the image. The choice depends on the application.

Summary of Skeleton and Propagation

The application of these two operations on a test image is illustrated in Fig. 51.44. In (a) and (b) of the figure the skeleton operation is shown with the end pixel condition [Eq. (51.169) $i + ii + iii$] and without the end pixel condition [Eq. (51.169) $i + ii$]. The propagation operation is illustrated in Fig. 51.44(c). The original image, shown in light gray, was eroded by $E(A, 6N_8)$ to produce the *seed* image shown in black. The original was then used as the *mask* image to produce the final result. The border value in both images was "0".

Several techniques based on the use of *skeleton* and *propagation* operations in combination with other mathematical morphology operations will be given in section 51.10.

Gray-Value Morphological Processing

The techniques of morphological filtering can be extended to gray-level images. To simplify matters, we will restrict our presentation to structuring elements, B, that comprise a finite number of pixels and are convex and bounded. Now, however, the structuring element has gray values associated with every coordinate position as does the image A.

Gray-level dilation, $D_G(\bullet)$, is given by:

$$\text{Dilation -} \quad D_G(A, B) = \max_{[j,k] \in B} \{a[m - j, n - k] + b[j, k]\} \tag{51.171}$$

For a given output coordinate $[m, n]$, the structuring element is summed with a shifted version of the image and the maximum encountered over all shifts within the $J \times K$ *domain* of B is used as the result. Should the shifting require values of the image A that are outside the $M \times N$ domain of A, then a decision must be made as to which model for image extension, as described in section 51.9.3, should be used.

Gray-level erosion, $E_G(\bullet)$, is given by:

$$\text{Erosion -} \quad E_G(A, B) = \min_{[j,k] \in B} \{a[m + j, n + k] - b[j, k]\} \tag{51.172}$$

The duality between *gray-level erosion* and *gray-level dilation* — the gray-level counterpart of Eq. (51.134) — is somewhat more complex than in the binary case:

$$\text{Duality -} \quad E_G(A, B) = -D_G(-\tilde{A}, B) \tag{51.173}$$

where "$-\tilde{A}$" means that $a[j, k] \rightarrow -a[-j, -k]$.

The definitions of higher order operations such as *gray-level opening* and *gray-level closing* are:

$$\text{Opening -} \quad O_G(A, B) = D_G(E_G(A, B), B) \tag{51.174}$$

$$\text{Closing -} \quad C_G(A, B) = -O_G(-A, -B) \tag{51.175}$$

The important properties that were discussed earlier such as idempotence, translation invariance, increasing in A, and so forth are also applicable to gray level morphological processing. The details can be found in Giardina and Dougherty [4].

In many situations the seeming complexity of gray level morphological processing is significantly reduced through the use of symmetric structuring elements where $b[j, k] = b[-j, -k]$. The most common of these is based on the use of $\boldsymbol{B} = constant = 0$. For this important case and using again the domain $[j, k] \in \boldsymbol{B}$, the definitions above reduce to:

$$Dilation \text{ - } \quad D_G(\boldsymbol{A}, \boldsymbol{B}) = \max_{[j,k]\in B} \{a[m - j, n - k]\} = \max_{B}(\boldsymbol{A}) \qquad (51.176)$$

$$Erosion \text{ - } \quad E_G(\boldsymbol{A}, \boldsymbol{B}) = \min_{[j,k]\in B} \{a[m - j, n - k]\} = \min_{B}(\boldsymbol{A}) \qquad (51.177)$$

$$Opening \text{ - } \quad O_G(\boldsymbol{A}, \boldsymbol{B}) = \max_{B} \left(\min_{B}(\boldsymbol{A}) \right) \qquad (51.178)$$

$$Closing \text{ - } \quad C_G(\boldsymbol{A}, \boldsymbol{B}) = \min_{B} \left(\max_{B}(\boldsymbol{A}) \right) \qquad (51.179)$$

The remarkable conclusion is that the *maximum filter* and the *minimum filter*, introduced in section 51.9.4, are gray-level dilation and gray-level erosion for the specific structuring element given by the shape of the filter window with the gray value "0" *inside* the window. Examples of these operations on a simple one-dimensional signal are shown in Fig. 51.45.

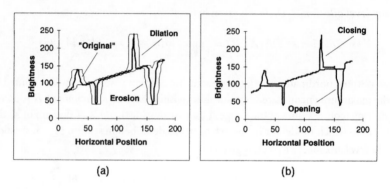

(a)   (b)

FIGURE 51.45 Morphological filtering of gray-level data. (a) Effect of 15×1 *dilation* and *erosion*; (b) effect of 15×1 *opening* and *closing*.

For a rectangular window, $J \times K$, the two-dimensional maximum or minimum filter is separable into two one-dimensional windows. Further, a one-dimensional maximum or minimum filter can be written in incremental form (see section 51.9.3). This means that gray-level dilations and erosions have a computational complexity per pixel that is $O(constant)$, that is, independent of J and K (see also Table 51.13).

The operations defined above can be used to produce morphological algorithms for smoothing, gradient determination and a version of the Laplacian. All are constructed from the primitives for *gray-level dilation* and *gray-level erosion* and in all cases the *maximum* and *minimum* filters are taken over the domain $[j, k] \in \boldsymbol{B}$.

Morphological Smoothing

This algorithm is based on the observation that a *gray-level opening* smoothes a gray-value image from above the brightness surface given by the function $a[m, n]$ and the *gray-level closing* smoothes from below. We use a structuring element $\boldsymbol{B}$ based on Eqs. (51.176) and (51.177).

$$MorphSmooth(\boldsymbol{A}, \boldsymbol{B}) \quad = \quad C_G \left(O_G(\boldsymbol{A}, \boldsymbol{B}) \boldsymbol{B} \right)$$

$$= \quad \min(\max(\max(\min(\boldsymbol{A})))) \tag{51.180}$$

Note that we have supressed the notation for the structuring element $\boldsymbol{B}$ under the *max* and *min* operations to keep the notation simple. Its use, however, is understood.

Morphological Gradient

For linear filters the gradient filter yields a vector representation [Eq. (51.103)] with a magnitude [Eq. (51.104)] and direction [Eq. (51.105)]. The version presented here generates a morphological estimate of the *gradient magnitude*:

$$Gradient(\boldsymbol{A}, \boldsymbol{B}) \quad = \quad \frac{1}{2} \left(D_G(\boldsymbol{A}, \boldsymbol{B}) - E_G(\boldsymbol{A}, \boldsymbol{B}) \right)$$

$$= \quad \frac{1}{2} \left(\max(\boldsymbol{A}) - \min(\boldsymbol{A}) \right) \tag{51.181}$$

Morphological Laplacian

The morphologically based Laplacian filter is defined by:

$$Laplacian(\boldsymbol{A}, \boldsymbol{B}) \quad = \quad \frac{1}{2} \left((D_G(\boldsymbol{A}, \boldsymbol{B}) - \boldsymbol{A}) - (\boldsymbol{A} - E_G(\boldsymbol{A}, \boldsymbol{B})) \right)$$

$$= \quad \frac{1}{2} \left(D_G(\boldsymbol{A}, \boldsymbol{B}) + E_G(\boldsymbol{A}, \boldsymbol{B}) - 2\boldsymbol{A} \right)$$

$$= \quad \frac{1}{2} \left(\max(\boldsymbol{A}) + \min(\boldsymbol{A}) - 2\boldsymbol{A} \right) \tag{51.182}$$

Summary of Morphological Filters

The effect of these filters is illustrated in Fig. 51.46. All images were processed with a 3×3 structuring element as described in Eqs. (51.176) through (51.182). Figure 51.46(e) was contrast stretched for display purposes using Eq. (51.78) and the parameters 1% and 99%. Figures 51.46(c),(d), and (e) should be compared to Figs. 51.30, 51.32, and 51.33.

51.10 Techniques

The algorithms presented in section 51.9 can be used to build techniques to solve specific image processing problems. Without presuming to present the solution to all processing problems, the following examples are of general interest and can be used as models for solving related problems.

51.10.1 Shading Correction

The method by which images are produced — the interaction between objects in real space, the illumination, and the camera — frequently leads to situations where the image exhibits significant shading across the field of view. In some cases, the image might be bright in the center and decrease in brightness as one goes to the edge of the field of view. In other cases, the image might be darker on the left side and lighter on the right side. The shading might be caused by nonuniform illumination, nonuniform camera sensitivity, or even dirt and dust on glass (lens) surfaces. In general, this shading effect is undesirable. Eliminating it is frequently necessary for subsequent processing and especially when image analysis or image understanding is the final goal.

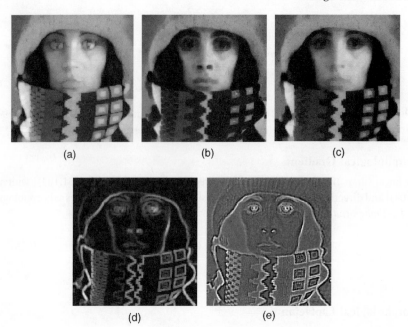

FIGURE 51.46 Examples of gray-level morphological filters. (a) Dilation; (b) Erosion; (c) Smoothing; (d) Gradient; and (e) Laplacian.

Model of Shading

In general, we begin with a model for the shading effect. The illumination $I_{ill}(x, y)$ usually interacts in a multiplicative with the object $a(x, y)$ to produce the image $b(x, y)$:

$$b(x, y) = I_{ill}(x, y) \bullet a(x, y) \tag{51.183}$$

with the object representing various imaging modalities such as:

$$a(x, y) = \begin{cases} r(x, y) & \text{reflectance model} \\ 10^{-OD(x,y)} & \text{absorption model} \\ c(x, y) & \text{fluorescence model} \end{cases} \tag{51.184}$$

where at position (x, y), $r(x, y)$ is the *reflectance*, $OD(x, y)$ is the *optical density*, and $c(x, y)$ is the concentration of fluorescent material. Parenthetically, we note that the fluorescence model only holds for low concentrations. The camera may then contribute *gain* and *offset* terms, as in Eq. (51.74), so that:

$$c[m, n] = gain[m, n] \bullet b[m, n] + offset[m, n]$$

Total shading -

$$= gain[m, n] \bullet I_{ill}[m, n] \bullet a[m, n] + offset[m, n] \tag{51.185}$$

In general, we assume that $I_{ill}[m, n]$ is slowly varying compared to $a[m, n]$.

Estimate of Shading

We distinguish between two cases for the determination of $a[m, n]$ starting from $c[m, n]$. In both cases we intend to estimate the shading terms $\{gain[m, n] \bullet I_{ill}[m, n]\}$ and $\{offset[m, n]\}$. While in the first case we assume that we have only the recorded image $c[m, n]$ with which to work, in the second case we assume that we can record two, additional, calibration images.

A *posteriori estimate* – In this case, we attempt to extract the shading estimate from $c[m, n]$. The most common possibilities are the following.

Lowpass filtering – We compute a smoothed version of $c[m, n]$ where the smoothing is large compared to the size of the objects in the image. This smoothed version is intended to be an estimate of the background of the image. We then subtract the smoothed version from $c[m, n]$ and then restore the desired DC value. In formula:

$$Lowpass \text{-} \hat{a}[m, n] = c[m, n] - LowPass\{c[m, n]\} + constant \tag{51.186}$$

where $\hat{a}[m, n]$ is the estimate of $a[m, n]$. Choosing the appropriate lowpass filter means knowing the appropriate spatial frequencies in the Fourier domain where the shading terms dominate.

Homomorphic filtering – We note that if the $offset[m, n] = 0$, then $c[m, n]$ consists solely of multiplicative terms. Further, the term $\{gain[m, n] \bullet I_{ill}[m, n]\}$ is slowly varying while $a[m, n]$ presumably is not. We therefore take the logarithm of $c[m, n]$ to produce two terms, one of which is low frequency and one of which is high frequency. We suppress the shading by high pass filtering the logarithm of $c[m, n]$ and then take the exponent (inverse logarithm) to restore the image. This procedure is based on *homomorphic filtering* as developed by Oppenheim and Stockham. In formula:

$$(i) \quad c[m, n] = gain[m, n] \bullet I_{ill}[m, n] \bullet a[m, n]$$

$$(ii) \quad \ln\{c[m, n]\} = \ln\left\{\underbrace{gain[m, n] \bullet I_{ill}[m, n]}_{\text{slowly varying}}\right\} + \ln\left\{\underbrace{a[m, n]}_{\text{rapidly varying}}\right\} \tag{51.187}$$

$$(iii) \quad HighPass\{\ln\{c[m, n]\}\} \approx \ln\{a[m, n]\}$$

$$(iv) \quad \hat{a}[m, n] = \exp\left\{HighPass\{\ln\{c[m, n]\}\}\right\}$$

Morphological filtering – We again compute a smoothed version of $c[m, n]$ where the smoothing is large compared to the size of the objects in the image but this time using morphological smoothing as in Eq. (51.180). This smoothed version is the estimate of the background of the image. We then subtract the smoothed version from $c[m, n]$ and then restore the desired DC value. In formula:

$$\hat{a}[m, n] = c[m, n] - MorphSmooth\{c[m, n]\} + constant \tag{51.188}$$

Choosing the appropriate morphological filter window means knowing (or estimating) the size of the largest objects of interest.

A *priori estimate* – If it is possible to record test (calibration) images through the camera's system, then the most appropriate technique for the removal of shading effects is to record two images — $BLACK[m, n]$ and $WHITE[m, n]$. The $BLACK$ image is generated by covering the lens leading to $b[m, n] = 0$ which in turn leads to $BLACK[m, n] = offset[m, n]$. The $WHITE$ image is generated by using $a[m, n] = 1$ which gives $WHITE[m, n] = gain[m, n] \bullet I_{ill}[m, n] + offset[m, n]$. The correction then becomes:

$$\hat{a}[m, n] = constant \bullet \frac{c[m, n] - BLACK[m, n]}{WHITE[m, n] - BLACK[m, n]} \tag{51.189}$$

The *constant* term is chosen to produce the desired dynamic range.

The effects of these various techniques on the data from Fig. 51.45 are shown in Fig. 51.47. The shading is a simple, linear ramp increasing from left to right; the objects consist of Gaussian peaks of varying widths.

In summary, if it is possible to obtain $BLACK$ and $WHITE$ calibration images, then Eq. (51.189) is to be preferred. If this is not possible, then one of the other algorithms will be necessary.

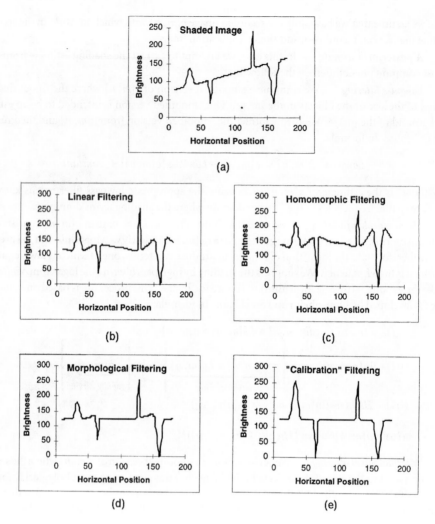

FIGURE 51.47 Comparison of various shading correction algorithms. The final result (e) is identical to the original (not shown). (a) Shaded; (b) correction with lowpass filtering; (c) correction with logarithmic filtering; (d) correction with max/min filtering; and (e) correction with test images.

51.10.2 Basic Enhancement and Restoration Techniques

The process of image acquisition frequently leads (inadvertently) to image degradation. Due to mechanical problems, out-of-focus blur, motion, inappropriate illumination, and noise, the quality of the digitized image can be inferior to the original. The goal of *enhancement* is — starting from a recorded image $c[m, n]$ — to produce the most visually pleasing image $\hat{a}[m, n]$. The goal of restoration is — starting from a recorded image $c[m, n]$ — to produce the best possible estimate $\hat{a}[m, n]$ of the original image $a[m, n]$. The goal of enhancement is beauty; the goal of restoration is truth.

The measure of success in restoration is usually an error measure between the original $a[m, n]$ and the estimate $\hat{a}[m, n]$: $\mathcal{E}\{\hat{a}[m, n], a[m, n]\}$. *No mathematical error function is known that corresponds to human perceptual assessment of error.* The mean-square error function is commonly used because:

1. It is easy to compute;
2. It is differentiable, implying that a minimum can be sought;

3. It corresponds to "signal energy" in the total error; and

4. It has nice properties *vis à vis* Parseval's theorem, Eqs. (51.22) and (51.23).

The *mean-square error* is defined by:

$$\mathcal{E}\{\hat{a}, a\} = \frac{1}{MN} \sum_{m=0}^{M-1} \sum_{n=0}^{N-1} |\hat{a}[m,n] - a[m,n]|^2 \tag{51.190}$$

In some techniques, an error measure will not be necessary; in others it will be essential for evaluation and comparative purposes.

Unsharp Masking

A well-known technique from photography to improve the visual quality of an image is to enhance the edges of the image. The technique is called *unsharp masking*. Edge enhancement means first isolating the edges in an image, amplifying them, and then adding them back into the image. Examination of Fig. 51.33 shows that the Laplacian is a mechanism for isolating the gray level edges. This leads immediately to the technique:

$$\hat{a}[m,n] = a[m,n] - \left(k \bullet \nabla^2 a[m,n] \right) \tag{51.191}$$

The term k is the amplifying term and $k > 0$. The effect of this technique is shown in Fig. 51.48.

FIGURE 51.48 Edge enhanced compared to original. (Left) Original, (right) Laplacian-enhanced.

The Laplacian used to produce Fig. 51.48 is given by Eq. (51.120) and the amplification term $k = 1$.

Noise Suppression

The techniques available to suppress noise can be divided into those techniques that are based on temporal information and those that are based on spatial information. By temporal information we mean that a sequence of images $\{a_p[m,n] | p = 1, 2, \ldots, P\}$ is available that contains *exactly* the same objects and that differs only in the sense of independent noise realizations. If this is the case and if the noise is additive, then simple averaging of the sequence:

$$\text{Temporal averaging-} \quad \hat{a}[m,n] = \frac{1}{P} \sum_{p=1}^{P} a_p[m,n] \tag{51.192}$$

will produce a result where the mean value of each pixel will be unchanged. For each pixel, however, the standard deviation will decrease from σ to $\sigma/\sqrt{P}$.

If temporal averaging is not possible, then spatial averaging can be used to decrease the noise. This generally occurs, however, at a cost to image sharpness. Four obvious choices for spatial averaging are the smoothing algorithms that have been described in section 51.9.4 — Gaussian filtering [Eq. (51.93)], median filtering, Kuwahara filtering, and morphological smoothing [Eq. (51.180)].

Within the class of linear filters, the optimal filter for restoration in the presence of noise is given by the *Wiener filter*. The word "optimal" is used here in the sense of minimum mean-square error (*mse*). Because the square root operation is monotonic increasing, the optimal filter also minimizes the root mean-square error (*rms*). The Wiener filter is characterized in the Fourier domain, and for additive noise that is independent of the signal it is given by:

$$H_W(u, v) = \frac{S_{aa}(u, v)}{S_{aa}(u, v) + S_{nn}(u, v)} \tag{51.193}$$

where $S_{aa}(u, v)$ is the power spectral density of an ensemble of random images $\{a[m, n]\}$ and $S_{nn}(u, v)$ is the power spectral density of the random noise. If we have a single image, then $S_{aa}(u, v) = |A(u, v)|^2$. In practice it is unlikely that the power spectral density of the uncontaminated image will be available. Because many images have a similar power spectral density that can be modeled by Table 51.4-T.8, that model can be used as an estimate of $S_{aa}(u, v)$.

A comparison of the five different techniques described above is shown in Fig. 51.49. The Wiener filter was constructed directly from Eq. (51.193) because the image spectrum and the noise spectrum were known. The parameters for the other filters were determined choosing that value (either σ or window size) that led to the minimum *rms*.

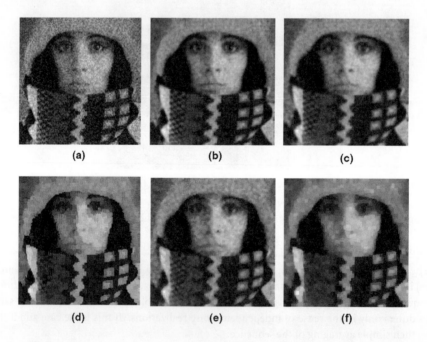

    (a)   (b)   (c)

    (d)   (e)   (f)

FIGURE 51.49 Noise suppression using various filtering techniques. (a) Noisy image (*SNR* = 20 dB) *rms* = 25.7; (b) Wiener filter *rms* = 20.2; (c) Gauss filter (σ = 1.0) *rms* = 21.1; (d) Kuwahara filter (5 × 5) *rms* = 22.4; (e) median filter 3 × 3 *rms* = 22.6; and (f) morphological smoothing (3 × 3) *rms* = 26.2.

The root mean-square errors (*rms*) associated with the various filters are shown in Fig. 51.49. For this specific comparison, the Wiener filter generates a lower error than any of the other procedures that are examined here. The two linear procedures, Wiener filtering and Gaussian filtering, performed slightly better than the three nonlinear alternatives.

Distortion Suppression

The model presented above — an image distorted solely by noise — is not, in general, sophisticated enough to describe the true nature of distortion in a digital image. A more realistic model includes not only the noise but also a model for the distortion induced by lenses, finite apertures, possible motion of the camera and/or an object, and so forth. One frequently used model is of an image $a[m, n]$ distorted by a linear, shift-invariant system $h_o[m, n]$ (such as a lens) and then contaminated by noise $\kappa[m, n]$. Various aspects of $h_o[m, n]$ and $\kappa[m, n]$ have been discussed in earlier sections. The most common combination of these is the additive model:

$$c[m, n] = (a[m, n] \otimes h_o[m, n]) + \kappa[m, n] \tag{51.194}$$

The restoration procedure that is based on linear filtering coupled to a minimum mean-square error criterion again produces a Wiener filter:

$$H_W(u, v) = \frac{H_o^*(u, v) S_{aa}(u, v)}{|H_o(u, v)|^2 \, S_{aa}(u, v) + S_{nn}(u, v)}$$

$$= \frac{H_o^*(u, v)}{|H_o(u, v)|^2 + (S_{nn}(u, v)/S_{aa}(u, v))} \tag{51.195}$$

Once again $S_{aa}(u, v)$ is the power spectral density of an image, $S_{nn}(u, v)$ is the power spectral density of the noise, and $H_o(u, v) = \mathcal{F}\{h_o[m, n]\}$. Examination of this formula for some extreme cases can be useful. For those frequencies where $S_{aa}(u, v) \gg S_{nn}(u, v)$, where the signal spectrum dominates the noise spectrum, the Wiener filter is given by $1/H_o(u, v)$, the *inverse filter* solution. For those frequencies where $S_{aa}(u, v) \ll S_{nn}(u, v)$, where the noise spectrum dominates the signal spectrum, the Wiener filter is proportional to $H_o^*(u, v)$, the *matched filter* solution. For those frequencies where $H_o(u, v) = 0$, the Wiener filter $H_W(u, v) = 0$ preventing overflow.

The Wiener filter is a solution to the restoration problem based on the hypothesized use of a linear filter and the minimum mean-square (or *rms*) error criterion. In the example below, the image $a[m, n]$ was distorted by a bandpass filter and then white noise was added to achieve an $SNR = 30 \, dB$. The results are shown in Fig. 51.50.

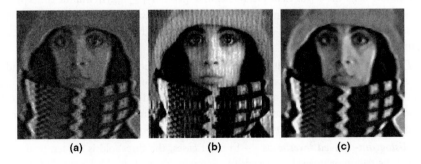

(a)   (b)   (c)

FIGURE 51.50   Noise *and* distortion suppression using the Wiener filter, Eq.(51.195) and the median filter. (a) Distorted, noisy image; (b) Wiener filter, $rms = 108.4$; (c) Median filter (3×3), $rms = 40.9$.

The *rms* after Wiener filtering but before contrast stretching was 108.4; after contrast stretching with Eq. (51.77), the final result as shown in Fig. 51.50(b) has a mean-square error of 27.8. Using a 3×3

median filter as shown in Fig. 51.50(c) leads to a *rms* error of 40.9 before contrast stretching and 35.1 after contrast stretching. Although the Wiener filter gives the minimum *rms* error over the set of all *linear* filters, the *nonlinear* median filter gives a lower *rms* error. The operation *contrast stretching* is itself a nonlinear operation. The "visual quality" of the median filtering result is comparable to the Wiener filtering result. This is due in part to periodic artifacts introduced by the linear filter which are visible in Fig. 51.50(b).

51.10.3 Segmentation

In the analysis of the objects in images, it is essential that we can distinguish between the objects of interest and "the rest". This latter group is also referred to as the background. The techniques that are used to find the objects of interest are usually referred to as *segmentation techniques* — segmenting the foreground from background. In this section we will discuss two of the most common techniques — *thresholding* and *edge finding* — and we will present techniques for improving the quality of the segmentation result. It is important to understand that:

- there is no universally applicable segmentation technique that will work for all images, and,
- no segmentation technique is perfect.

Thresholding

This technique is based on a simple concept. A parameter θ called the *brightness threshold* is chosen and applied to the image $a[m, n]$ as follows:

$$
\begin{aligned}
&\text{If } a[m, n] \geq \theta \quad a[m, n] = object = 1 \\
&\text{Else} \qquad\qquad\quad a[m, n] = background \quad = 0
\end{aligned}
\tag{51.196}
$$

This version of the algorithm assumes that we are interested in light objects on a dark background. For dark objects on a light background we would use:

$$
\begin{aligned}
&\text{If } a[m, n] < \theta \quad a[m, n] = object = 1 \\
&\text{Else} \qquad\qquad\quad a[m, n] = background \quad = 0
\end{aligned}
\tag{51.197}
$$

The output is the label "object" or "background" which, due to its dichotomous nature, can be represented as a Boolean variable "1" or "0". In principle, the test condition could be based on some property other than simple brightness [for example, *If (Redness* $\{a[m, n]\} \geq \theta_{red})$], but the concept is clear.

The central question in thresholding then becomes: How do we choose the threshold θ? While there is no universal procedure for threshold selection that is guaranteed to work on all images, there is a variety of alternatives.

Fixed threshold – One alternative is to use a threshold that is chosen independently of the image data. If it is known that one is dealing with very high-contrast images where the objects are very dark and the background is homogeneous (section 51.10.1) and very light, then a constant threshold of 128 on a scale of 0 to 255 might be sufficiently accurate. By accuracy we mean that the number of falsely classified pixels should be kept to a minimum.

Histogram-derived thresholds – In most cases, the threshold is chosen from the brightness histogram of the region or image that we wish to segment (see sections 51.3.5 and 51.9.1). An image and its associated brightness histogram are shown in Fig. 51.51.

A variety of techniques has been devised to automatically choose a threshold starting from the gray-value histogram, $\{h[b]|b = 0, 1, \ldots, 2^B - 1\}$. Some of the most common ones are presented below. Many of these algorithms can benefit from a smoothing of the raw histogram data to remove small fluctuations, but the smoothing algorithm must not shift the peak positions. This translates into a zero-phase smoothing

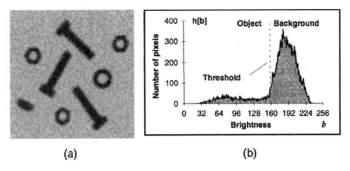

(a)   (b)

FIGURE 51.51 Pixels below the threshold ($a[m, n] < \theta$) will be labeled as object pixels: those above the threshold will be labeled as background pixels. (a) Image to be *thresholded* and (b) brightness histogram of the image.

algorithm given below where typical values for W are 3 or 5:

$$h_{smooth}[b] = \frac{1}{W} \sum_{w=-(W-1)/2}^{(W-1)/2} h_{raw}[b-w] \quad W \text{ odd} \tag{51.198}$$

Isodata algorithm – This iterative technique for choosing a threshold was developed by Ridler and Calvard. The histogram is initially segmented into two parts using a starting threshold value such as $\theta_0 = 2^{B-1}$, half the maximum dynamic range. The sample mean ($m_{f,0}$) of the gray values associated with the foreground pixels and the sample mean ($m_{b,0}$) of the gray values associated with the background pixels are computed. A new threshold value θ_1 is now computed as the average of these two sample means. The process is repeated, based on the new threshold, until the threshold value does not change any more. In formula:

$$\theta_k = \left(m_{f,k-1} + m_{b,k-1}\right)/2 \quad \text{until } \theta_k = \theta_{k-1} \tag{51.199}$$

Background-symmetry algorithm – This technique assumes a distinct and dominant peak for the background that is symmetric about its maximum. The technique can benefit from smoothing as described above [Eq. (51.198)]. The maximum peak ($maxp$) is found by searching for the maximum value in the histogram. The algorithm then searches on the *nonobject pixel side* of that maximum to find a $p\%$ point as in Eq. (51.39).

In Fig. 51.51(b), where the object pixels are located to the *left* of the background peak at brightness 183, this means searching to the right of that peak to locate, as an example, the 95% value. At this brightness value, 5% of the pixels lie to the *right* of (are above) that value. This occurs at brightness 216 in Fig. 51.51(b). Because of the assumed symmetry, we use as a threshold a displacement to the *left* of the maximum that is equal to the displacement to the right where the $p\%$ is found. For Fig. 51.51(b) this means a threshold value given by $183 - (216 - 183) = 150$. In formula:

$$\theta = maxp - \left(p\% - maxp\right) \tag{51.200}$$

This technique can be adapted easily to the case where we have light objects on a dark, dominant background. Further, it can be used if the object peak dominates and we have reason to assume that the brightness distribution around the object peak is symmetric. An additional variation on this symmetry theme is to use an estimate of the sample standard deviation [s in Eq. (51.37)] based on one side of the dominant peak and then use a threshold based on $\theta = maxp \pm 1.96s$ (at the 5% level) or $\theta = maxp \pm 2.57s$ (at the 1% level). The choice of "+" or "−" depends on which direction from $maxp$ is being defined as the object/background threshold. Should the distributions be approximately Gaussian around $maxp$, then the values 1.96 and 2.57 will, in fact, correspond to the 5% and 1% level.

Triangle algorithm – This technique due to Zack is illustrated in Fig. 51.52. A line is constructed between the maximum of the histogram at brightness b_{max} and the lowest value $b_{min} = (p = 0)\%$ in the image. The distance **d** between the line and the histogram $h[b]$ is computed for all values of b from $b = b_{min}$ to $b = b_{max}$. The brightness value b_o where the distance between $h[b_o]$ and the line is maximal is the threshold value, that is, $\theta = b_o$. This technique is particularly effective when the object pixels produce a weak peak in the histogram.

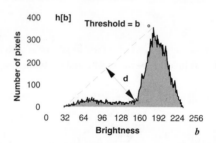

FIGURE 51.52 The triangle algorithm is based on finding the value of b that gives the maximum distance **d**.

The three procedures described above give the values $\theta = 139$ for the Isodata algorithm, $\theta = 150$ for the background symmetry algorithm at the 5% level, and $\theta = 152$ for the triangle algorithm for the image in Fig. 51.51(a).

Thresholding does not have to be applied to entire images but can be used on a region-by-region basis. Chow and Kaneko developed a variation in which the $M \times N$ image is divided into nonoverlapping regions. In each region, a threshold is calculated and the resulting threshold values are put together (interpolated) to form a thresholding surface for the entire image. The regions should be of "reasonable" size so that there are a sufficient number of pixels in each region to make an estimate of the histogram and the threshold. The utility of this procedure — like so many others — depends on the application at hand.

Edge Finding

Thresholding produces a segmentation that yields all the pixels that, in principle, belong to the object or objects of interest in an image. An alternative to this is to find those pixels that belong to the borders of the objects. Techniques that are directed to this goal are termed *edge finding techniques*. From our discussion in section 51.9.6 on mathematical morphology, specifically Eqs. (51.162), (51.163), and (51.170), we see that there is an intimate relationship between edges and regions.

Gradient-based procedure – The central challenge to edge finding techniques is to find procedures that produce *closed* contours around the objects of interest. For objects of particularly high *SNR*, this can be achieved by calculating the gradient and then using a suitable threshold. This is illustrated in Fig. 51.53.

While the technique works well for the 30-dB image in Fig. 51.53(a), it fails to provide an accurate determination of those pixels associated with the object edges for the 20-dB image in Fig. 51.53(b). A variety of smoothing techniques as described in section 51.9.4 and in Eq. (51.180) can be used to reduce the noise effects before the gradient operator is applied.

Zero-crossing based procedure – A more modern view to handling the problem of edges in noisy images is to use the zero crossings generated in the Laplacian of an image (section 51.9.5). The rationale starts from the model of an ideal edge, a step function, that has been blurred by an *OTF* such as Table 51.4.T.3 (out-of-focus), T.5 (diffraction-limited), or T.6 (general model) to produce the result shown in Fig. 51.54.

The edge location is, according to the model, at that place in the image where the Laplacian changes

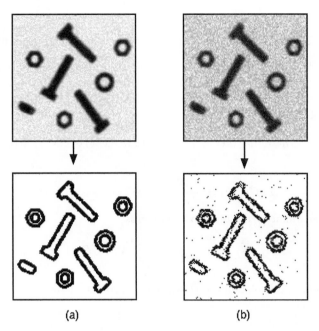

FIGURE 51.53   Edge finding based on the Sobel gradient, Eq. (51.110), combined with the Isodata thresholding algorithm Eq. (51.199). (a) $SNR = 30$ dB and (b) $SNR = 20$ dB.

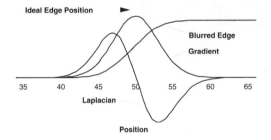

FIGURE 51.54   Edge finding based on the zero crossing as determined by the second derivative, the Laplacian. The curves are not to scale.

sign, the zero crossing. As the Laplacian operation involves a second derivative, this means a potential enhancement of noise in the image at high spatial frequencies; see Eq. (51.114). To prevent enhanced noise from dominating the search for zero crossings, a smoothing is necessary.

The appropriate smoothing filter from among the many possibilities described in section 51.9.4 should have, according to Canny, the following properties:

- In the frequency domain, (u, v) or (Ω, Ψ), the filter should be as narrow as possible to provide suppression of high frequency noise, and;
- In the spatial domain, (x, y) or $[m, n]$, the filter should be as narrow as possible to provide good localization of the edge. A too wide filter generates uncertainty as to precisely where, within the filter width, the edge is located.

The smoothing filter that simultaneously satisfies both these properties — minimum bandwidth and minimum spatial width — is the Gaussian filter described in section 51.9.4. This means that the image should be smoothed with a Gaussian of an appropriate σ followed by application of the Laplacian. In formula:

$$ZeroCrossing\{a(x, y)\} = \left\{(x, y)|\nabla^2\{g_{2D}(x, y) \otimes a(x, y)\} = 0\right\} \tag{51.201}$$

where $g_{2D}(x, y)$ is defined in Eq. (51.93). The derivative operation is linear and shift-invariant as defined in Eqs. (51.85) and (51.86). This means that the order of the operators can be exchanged [Eq. (51.4)] or combined into one single filter [Eq. (51.5)]. This second approach leads to the Marr-Hildreth formulation of the "Laplacian-of-Gaussians" (*LoG*) filter:

$$ZeroCrossing\{a(x, y)\} = \{(x, y)|LoG(x, y) \otimes a(x, y) = 0\} \tag{51.202}$$

where

$$LoG(x, y) = \frac{x^2 + y^2}{\sigma^4}g_{2D}(x, y) - \frac{2}{\sigma^2}g_{2D}(x, y) \tag{51.203}$$

Given the circular symmetry, this can also be written as:

$$LoG(r) = \left(\frac{r^2 - 2\sigma^2}{2\pi\sigma^6}\right)e^{-\left(r^2/2\sigma^2\right)} \tag{51.204}$$

This two-dimensional convolution kernel, which is sometimes referred to as a "Mexican hat filter", is illustrated in Fig. 51.55.

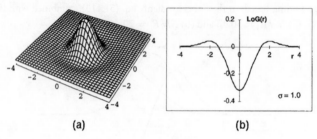

(a)   **(b)**

FIGURE 51.55   *LoG* filter with $\sigma = 1.0$. (a) $-LoG(x, y)$ and (b) $LoG(r)$.

PLUS-based procedure   – Among the zero crossing procedures for edge detection, perhaps the most accurate is the *PLUS* filter as developed by Verbeek and Van Vliet. The filter is defined, using Eqs. (51.121) and (51.122) as:

$$
\begin{aligned}
PLUS(a) &= SDGD(a) + Laplace(a) \\
&= \left(\frac{A_{xx}A_x^2 + 2A_{xy}A_xA_y + A_{yy}A_y^2}{A_x^2 + A_y^2}\right) + \left(A_{xx} + A_{yy}\right) \tag{51.205}
\end{aligned}
$$

Neither the derivation of the *PLUS*'s properties nor an evaluation of its accuracy are within the scope of this section. Suffice it to say that, for positively curved edges in gray value images, the Laplacian-based zero crossing procedure *overestimates* the position of the edge and the *SDGD*-based procedure *underestimates* the position. This is true in both two-dimensional and three-dimensional images with an error on the order of $(\sigma/R)^2$ where R is the radius of curvature of the edge. The *PLUS* operator has an error on the order of $(\sigma/R)^4$ if the image is sampled at, at least, $3\times$ the usual Nyquist sampling frequency as in Eq. (51.56) *or* if we choose $\sigma \geq 2.7$ and sample at the usual Nyquist frequency.

All of the methods based on zero crossings in the Laplacian must be able to distinguish between zero *crossings* and zero *values*. While the former represent edge positions, the latter can be generated by regions that are no more complex than bilinear surfaces, that is, $a(x, y) = a_0 + a_1 \bullet x + a_2 \bullet y + a_3 \bullet x \bullet y$. To distinguish between these two situations, we first find the zero crossing positions and label them as "1" and all other pixels as "0". We then multiply the resulting image by a measure of the *edge strength* at each pixel. There are various measures for the edge strength that are all based on the gradient as described in section 51.9.5 and Eq. (51.181). This last possibility, use of a morphological gradient as an edge strength measure, was first described by Lee, Haralick, and Shapiro and is particularly effective. After multiplication the image is then thresholded (as above) to produce the final result. The procedure is shown in Fig. 51.56.

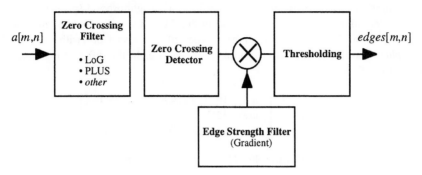

FIGURE 51.56   General strategy for edges based on zero crossings.

The results of these two edge finding techniques based on zero crossings, *LoG* filtering and *PLUS* filtering, are shown in Fig. 51.57 for images with a 20-dB *SNR*.

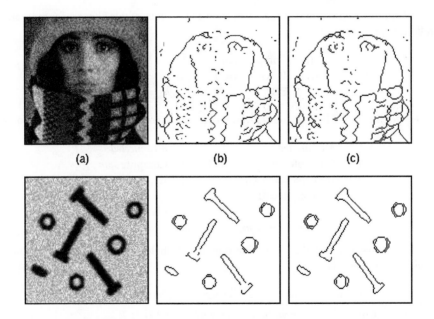

FIGURE 51.57   Edge finding using zero crossing algorithms *LoG* and *PLUS*. In both algorithms $\sigma = 1.5$. (a) Image $SNR = 20$ dB; (b) *LoG* filter; and (c) *PLUS* filter.

Edge finding techniques provide, as the name suggests, an image that contains a collection of edge pixels. Should the edge pixels correspond to objects, as opposed to say simple lines in the image, then a region-filling technique such as Eq. (51.170) may be required to provide the complete objects.

Binary Mathematical Morphology

The various algorithms that we have described for mathematical morphology in section 51.9.6 can be put together to form powerful techniques for the processing of binary images and gray level images. As binary images frequently result from segmentation processes on gray level images, the morphological processing of the binary result permits the improvement of the segmentation result.

Salt-or-pepper filtering – Segmentation procedures frequently result in isolated "1" pixels in a "0" neighborhood (salt) or isolated "0" pixels in a "1" neighborhood (pepper). The appropriate neighborhood definition must be chosen such as in Fig. 51.3. Using the lookup table formulation for Boolean operations in a 3×3 neighborhood that was described in association with Fig. 51.43, *salt filtering* and *pepper filtering* are straightforward to implement. We weight the different positions in the 3×3 neighborhood as follows:

$$\text{Weights} = \begin{bmatrix} w_4 = 16 & w_3 = 8 & w_2 = 4 \\ w_5 = 32 & w_0 = 1 & w_1 = 2 \\ w_6 = 64 & w_7 = 128 & w_8 = 256 \end{bmatrix} \tag{51.206}$$

For a 3×3 window in $a[m, n]$ with values "0" or "1" we then compute:

$$\begin{aligned} sum \quad = \quad & w_0 a[m, n] + w_1 a[m + 1, n] + w_2 a[m + 1, n - 1] + \\ & w_3 a[m, n - 1] + w_4 a[m - 1, n - 1] + w_5 a[m - 1, n] + \\ & w_6 a[m - 1, n + 1] + w_7 a[m, n + 1] + w_8 a[m + 1, n - 1] \end{aligned} \tag{51.207}$$

The result, *sum*, is a number bounded by $0 \leq sum \leq 511$.

Salt filter – The 4-connected and 8-connected versions of this filter are the same and are given by the following procedure:

$$\begin{aligned} &(i) \quad \text{Compute } sum \\ &(ii) \quad \textbf{If } ((sum == 1)) \quad c[m, n] = 0 \\ &\qquad \textbf{Else} \qquad\qquad\qquad c[m, n] = a[m, n] \end{aligned} \tag{51.208}$$

Pepper filter – The 4-connected and 8-connected versions of this filter are the following procedures:

	4-connected		*8-connected*
(i)	Compute *sum*	(i)	Compute *sum*
(ii)	**If** $((sum == 170)$	(ii)	**If** $((sum == 510)$
	$c[m, n] = 1$		$c[m, n] = 1$
	Else		**Else**
	$c[m, n] = a[m, n]$		$c[m, n] = a[m, n]$

(51.209)

Isolate objects with holes – To find objects with holes, we can use the following procedure which is illustrated in Fig. 51.58.

(i) *Segment* image to produce binary mask representation

(ii) Compute *skeleton* without end pixels — Eq. (51.169)   (51.210)

(iii) Use *salt* filter to remove single skeleton pixels

(iv) *Propagate* remaining skeleton pixels into original binary mask — Eq. (51.170)

The binary objects are shown in gray and the skeletons, after application of the salt filter, are shown as a black overlay on the binary objects. Note that this procedure uses no parameters other than the fundamental choice of connectivity; it is free from "magic numbers". In the example shown in Fig. 51.58, the 8-connected definition was used as well as the structuring elements $\boldsymbol{B} = \boldsymbol{N}_8$.

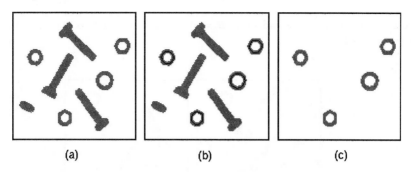

FIGURE 51.58 Isolation of objects with holes using morphological operations. (a) Binary image; (b) skeleton after salt filter; and (c) objects with holes.

Filling holes in objects – To fill holes in objects, we use the following procedure which is illustrated in Fig. 51.59.

$$(i) \quad \textit{Segment} \text{ image to produce binary representation of objects}$$
$$(ii) \quad \text{Compute } \textit{complement} \text{ of binary image as a } \textit{mask image}$$
$$(iii) \quad \text{Generate a } \textit{seed image} \text{ as the border of the image} \qquad (51.211)$$
$$(iv) \quad \textit{Propagate} \text{ the } \textit{seed} \text{ into the } \textit{mask} \text{ — Eq. (51.170)}$$
$$(v) \quad \textit{Complement} \text{ result of propagation to produce final result.}$$

The *mask image* is illustrated in *gray* in Fig. 51.59(a) and the *seed image* is shown in *black* in that same illustration. When the object pixels are specified with a connectivity of $C = 8$, then the propagation into the mask (background) image should be performed with a connectivity of $C = 4$, that is, dilations with the structuring element $\boldsymbol{B} = \boldsymbol{N}_4$. This procedure is also free of "magic numbers".

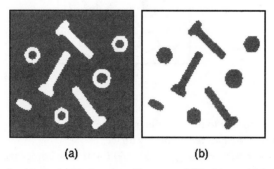

FIGURE 51.59 Filling holes in objects. (a) Mask and seed images and (b) objects with holes filled.

Removing border-touching objects – Objects that are connected to the image border are not suitable for analysis. To eliminate them we can use a series of morphological operations that are illustrated in Fig. 51.60.

$$(i) \quad \textit{Segment} \text{ image to produce binary } \textit{mask image} \text{ of objects}$$
$$(ii) \quad \text{Generate a } \textit{seed image} \text{ as the border of the image} \qquad (51.212)$$
$$(iii) \quad \textit{Propagate} \text{ the } \textit{seed} \text{ into the } \textit{mask} \text{ — Eq. (51.170)}$$
$$(iv) \quad \text{Compute } XOR \text{ of the propagation result and the } \textit{mask image} \text{ as final result.}$$

The *mask image* is illustrated in *gray* in Fig. 51.60(a) and the *seed image* is shown in *black* in that same illustration. If the structuring element used in the propagation is $\boldsymbol{B} = \boldsymbol{N}_4$, then objects are removed that are 4-connected with the image boundary. If $B = N_8$ is used, then objects that 8-connected with the boundary are removed.

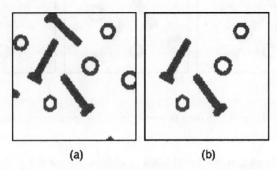

(a)   **(b)**

FIGURE 51.60   Removing objects touching borders. (a) Mask and seed images and (b) remaining objects.

  Exo-skeleton – The *exo-skeleton* of a set of objects is the skeleton of the background that contains the objects. The exo-skeleton produces a partition of the image into regions each of which contains one object. The actual skeletonization [Eq. (51.169)] is performed without the preservation of end pixels and with the border set to "0". The procedure is described below and the result is illustrated in Fig. 51.61.

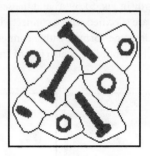

FIGURE 51.61   Exo-skeleton.

   (*i*) *Segment* image to produce binary image
   (*ii*) Compute *complement* of binary image   (51.213)
   (*iii*) Compute *skeleton* using Eq. (51.169) $i + ii$ with border set to "0".

  Touching objects – Segmentation procedures frequently have difficulty separating slightly touching, yet distinct, objects. The following procedure provides a mechanism to separate these objects and makes minimal use of "magic numbers". The exo-skeleton produces a partition of the image into regions each of which contains one object. The actual skeletonization is performed without the preservation of end pixels and with the border set to "0". The procedure is illustrated in Fig. 51.62.

   (*i*) *Segment* image to produce binary image
   (*ii*) Compute a "small number" of *erosions* with $B = N_4$
   (*iii*) Compute *exo-skeleton* of eroded result   (51.214)
   (*iv*) Complement *exo-skeleton* result
   (*v*) Compute *AND* of original binary image and the complemented exo-skeleton.

The *eroded binary image* is illustrated in *gray* in Fig. 51.62(a) and the *exo-skeleton image* is shown in *black* in that same illustration. An enlarged section of the final result is shown in Fig. 51.62(b) and the separation is easily seen. This procedure involves choosing a small, minimum number of erosions, but the number is not critical as long as it initiates a coarse separation of the desired objects. The actual separation is performed by the exo-skeleton which, itself, is free of "magic numbers". If the exo-skeleton is 8-connected, then the background separating the objects will be 8-connected. The objects themselves will be disconnected according to the 4-connected criterion. (See section 51.9.6 and Fig. 51.36.)

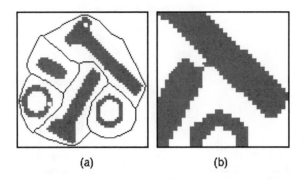

(a)   (b)

FIGURE 51.62   Separation of touching objects. (a) Eroded and exo-skeleton images and (b) objects separated (detail).

Gray-Value Mathematical Morphology

As we have seen is section 51.10.1, gray-value morphological processing techniques can be used for practical problems such as shading correction. In this section, several other techniques will be presented.

Top-hat transform   – The isolation of gray-value objects that are convex can be accomplished with the *top-hat transform* as developed by Meyer. Depending on whether we are dealing with light objects on a dark background or dark objects on a light background, the transform is defined as:

$$\text{Light objects - } TopHat(A, B) = A - (A \circ B) = A - \max_{B}\left(\min_{B}(A)\right) \tag{51.215}$$

$$\text{Dark objects - } TopHat(A, B) = (A \bullet B) - A = \min_{B}\left(\max_{B}(A)\right) - A \tag{51.216}$$

where the structuring element B is chosen to be bigger than the objects in question and, if possible, to have a convex shape. Because of the properties given in Eqs. (51.155) and (51.158), $TopHat(A, B) \geq 0$. An example of this technique is shown in Fig. 51.63.

The original image including shading is processed by a 15×1 structuring element as described in Eqs. (51.215) and (51.216) to produce the desired result. Note that the transform for dark objects has been defined in such a way as to yield "positive" objects as opposed to "negative" objects. Other definitions are, of course, possible.

Thresholding   – A simple estimate of a locally varying threshold surface can be derived from morphological processing as follows:

$$\text{Threshold surface - } \theta[m, n] = \frac{1}{2}(\max(A) + \min(A)) \tag{51.217}$$

Once again, we suppress the notation for the structuring element B under the *max* and *min* operations to keep the notation simple. Its use, however, is understood.

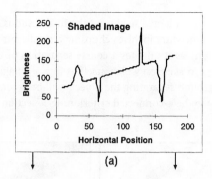

(a)

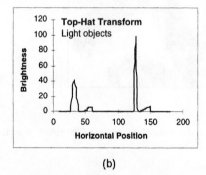

(b)

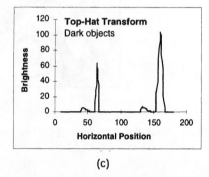

(c)

FIGURE 51.63   Top-hat transforms. (a) Original; (b) light object transform; and (c) dark object transform.

       Local contrast stretching   – Using morphological operations, we can implement a technique for *local contrast stretching*. That is, the amount of stretching that will be applied in a neighborhood will be controlled by the original contrast in that neighborhood. The morphological gradient defined in Eq. (51.181) may also be seen as related to a measure of the local contrast in the window defined by the structuring element B:

$$LocalContrast\,(A, B) = \max(A) - \min(A) \tag{51.218}$$

The procedure for local contrast stretching is given by:

$$c[m, n] = \text{scale} \; \bullet \; \frac{A - \min(A)}{\max(A) - \min(A)} \tag{51.219}$$

The *max* and *min* operations are taken over the structuring element B. The effect of this procedure is illustrated in Fig. 51.64. It is clear that this *local* operation is an extended version of the *point* operation for contrast stretching presented in Eq. (51.77).

Using standard test images (as we have seen in so many examples in this chapter) illustrates the power of this local morphological filtering approach.

51.11   Acknowledgments

This work was partially supported by the Netherlands Organization for Scientific Research (NWO) Grant 900-538-040, the Foundation for Technical Sciences (STW) Project 2987, the ASCI PostDoc program, and the Rolling Grants program of the Foundation for Fundamental Research in Matter (FOM). Images presented above were processed using *TCL-Image* and *SCIL-Image* (both from the TNO-TPD, Stieltjesweg 1, Delft, The Netherlands) and Adobe *Photoshop*™.

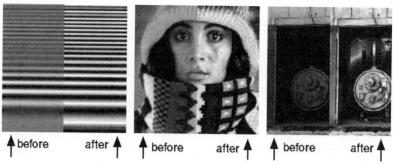

FIGURE 51.64   Local contrast stretching.

References

[1] Castleman, K.R., *Digital Image Processing,* 2nd ed., Prentice-Hall, Englewood Cliffs, NJ, 1996.

[2] Russ, J.C., *The Image Processing Handbook,* 2nd ed., CRC Press, Boca Raton, FL, 1995.

[3] Dudgeon, D.E. and Mersereau, R.M., *Multidimensional Digital Signal Processing,* Prentice-Hall, Englewood Cliffs, NJ, 1984.

[4] Giardina, C.R. and Dougherty, E.R., *Morphological Methods in Image and Signal Processing,* Prentice-Hall, Englewood Cliffs, NJ, 1988.

[5] Gonzalez, R.C. and Woods, R.E., *Digital Image Processing,* Addison-Wesley, Reading, MA, 1992.

[6] Goodman, J.W., *Introduction to Fourier Optics,* 2nd ed., McGraw-Hill, New York, 1996.

[7] Heijmans, H.J.A.M., *Morphological Image Operators,* Academic Press, Boston, 1994.

[8] Hunt, R.W.G., *The Reproduction of Colour in Photography, Printing and Television,* 4th ed., Fountain Press, Tolworth, England, 1987.

[9] Oppenheim, A.V., Willsky, A.S., and Young, I.T., *Systems and Signals,* Prentice-Hall, Englewood Cliffs, NJ, 1983.

[10] Papoulis, A., *Systems and Transforms with Applications in Optics,* McGraw-Hill, New York, 1968.

52

Still Image Compression[1]

Tor A. Ramstad
*Norwegian University of Science
and Technology (NTNU)*

52.1 Introduction

Digital representation of images is important for digital transmission and storage on different media such as magnetic or laser disks. However, pictorial material requires vast amounts of bits if represented through direct quantization. As an example, an SVGA color image requires $3 \times 600 \times 800$ bytes $= 1, 44$ Mbytes when each color component is quantized using 1 byte per pixel, the amount of bytes that can be stored on one standard 3.5-inch diskette. It is therefore evident that *compression* (often called *coding*) is necessary for reducing the amount of data [33].

In this chapter we address three fundamental questions concerning image compression:

- Why is image compression possible?
- What are the theoretical coding limits?
- Which practical compression methods can be devised?

---

[1] Parts of this manuscript are based on Ramtad, T.A., Aase, S.O., and Husøy, J.H., *Subband Compression of Images — Principles and Examples,* Elsevier Science Publishers BV, North Holland, 1995. Permission to use the material is given by ELSEVIER Science Publishers BV.

The first two questions concern statistical and structural properties of the image material and human visual perception. Even if we were able to answer these questions accurately, the methodology for image compression (third question) does not follow thereof. That is, the practical coding algorithms must be found otherwise. The bulk of the chapter will review image coding principles and present some of the best proposed still image coding methods.

The prevailing technique for image coding is *transform coding*. This is part of the JPEG (Joint Picture Expert Group) standard [14] as well as a part of all the existing video coding standards (H.261, H.263, MPEG-1, MPEG-2) [15, 16, 17, 18]. Another closely related technique, *subband coding*, is in some respects better, but has not yet been recognized by the standardization bodies. A third technique, *differential coding*, has not been successful for still image coding, but is often used to code the lowpass-lowpass band in subband coders, and is an integral part of hybrid video coders for removal of temporal redundancy. *Vector quantization* (VQ) is the ultimate technique if there were no complexity constraints. Because all practical systems must have limited complexity, VQ is usually used as a component in a multi-component coding scheme. Finally, *fractal* or *attraclor coding* is based on an idea far from other methods, but it is, nevertheless, strongly related to vector quantization.

For natural images, no exact digital representation exists because the quantization, which is an integral part of digital representations, is a lossy technique. Lossy techniques will always add noise, but the noise level and its characteristics can be controlled and depend on the number of bits per pixel as well as the performance of the method employed. *Lossless* techniques will be discussed as a component in other coding methods.

52.1.1 Signal Chain

We assume a model where the input signal is properly bandlimited and digitized by an appropriate *analog-to-digital converter*. All subsequent processing in the encoder will be digital. The decoder is also digital up to the digital-to-analog converter, which is followed by a lowpass reconstruction filter.

Under idealized conditions, the interconnection of the signal chain excluding the compression unit will be assumed to be noise-free. (In reality, the analog-to-digital conversion will render a noise power which can be approximated by $\Delta^2/12$, where Δ is the quantizer interval. This interval depends on the number of bits, and we assume that it is so high that the contribution to the overall noise from this process is negligible). The performance of the coding chain can then be assessed from the difference between the input and output of the digital compression unit disregarding the analog part.

Still images must be sampled on some two-dimensional grid. Several schemes are viable choices, and there are good reasons for selecting nonrectangular grids. However, to simplify, rectangular sampling will be considered only, and all filtering will be based on separable operations, first performed on the rows and subsequently on the columns of the image. The theory is therefore presented for one-dimensional models, only.

52.1.2 Compressibility of Images

There are two reasons why images can be compressed:

- All meaningful images exhibit some form of internal structure, often expressed through statistical dependencies between pixels. We call this property *signal redundancy*.
- The human visual system is not perfect. This means that certain degradations cannot be perceived by human observers. The degree of allowable noise is called *irrelevancy* or *visual redundancy*. If we furthermore accept visual degradation, we can exploit what might be termed *tolerance*.

In this section we make some speculations about the compression potential resulting from redundancy and irrelevancy.

The two fundamental concepts in evaluating a coding scheme are *distortion*, which measures quality in the compressed signal, and *rate*, which measures how costly it is to transmit or store a signal.

Distortion is a measure of the deviation between the encoded/decoded signal and the original signal. Usually, distortion is measured by a single number for a given coder and bit rate. There are numerous ways of mapping an error signal onto a single number. Moreover, it is hard to conceive that a single number could mimic the quality assessment performed by a human observer. An easy-to-use and well-known error measure is the *mean square error (mse)*. The visual correctness of this measure is poor. The human visual system is sensitive to errors in shapes and deterministic patterns, but not so much in stochastic textures. The *mse* defined over the entire image can, therefore, be entirely erroneous in the visual sense. Still, *mse* is the prevailing error measure, and it can be argued that it reflects well small changes due to optimization in a given coder structure, but poor as for the comparison between different models that create different noise characteristics.

Rate is defined as *bits per pixel* and is connected to the information content in a signal, which can be measured by *entropy*.

A Lower Bound for Lossless Coding

To define image entropy, we introduce the set **S** containing all possible images of a certain size and call the number of images in the set N_S. To exemplify, assume the image set under consideration has dimension 512×512 pixels and each pixel is represented by 8 bits. The number of different images that exist in this set is $2^{512 \times 512 \times 8}$, an overwhelming number!

Given the probability P_i of each image in the set **S**, where $i \in N_S$ is the index pointing to the different images, the source entropy is given by

$$H = - \sum_{i \in N_S} P_i \log_2 P_i .$$

(52.1)

The entropy is a lower bound for the rate in lossless coding of the digital images.

A Lower Bound for Visually Lossless Coding

In order to incorporate perceptual redundancies, it is observed that all the images in the given set cannot be distinguished visually. We therefore introduce *visual entropy* as an abstract measure which incorporates distortion.

We now partition the image set into disjoint subsets, S_i, in which all the different images have similar appearance. One image from each subset is chosen as the *representation* image. The collection of these N_R representation images constitutes a subset $\mathcal{R}$, that is a set spanning all distinguishable images in the original set.

Assume that image $i \in \mathcal{R}$ appears with probability $\hat{P}_i$. Then the *visual entropy* is defined by

$$H_V = - \sum_{i \in N_R} \hat{P}_i \log_2 \hat{P}_i .$$

(52.2)

The minimum attainable bit rate is lower bounded by this number for image coders without visual degradation.

52.1.3 The Ideal Coding System

Theoretically, we can approach the visual entropy limit using an unrealistic *vector quantizer* (VQ), in conjunction with an ideal *entropy coder*. The principle of such an optimal coding scheme is described next.

The set of representation images is stored in what is usually called a *codebook*. The encoder and decoder have similar copies of this codebook. In the encoding process, the image to be coded is compared to all the vectors in the codebook applying the visually correct distortion measure. The codebook member with the closest resemblance to the sample image is used as the *coding approximation*. The corresponding codebook index (address) is entropy coded and transmitted to the decoder. The decoder looks up the image located at the address given by the transmitted index.

Obviously, the above method is unrealistic. The complexity is beyond any practical limit both in terms of storage and computational requirement. Also, the correct visual distortion measure is not presently known. We should therefore only view the indicated coding strategy as the limit for any coding scheme.

52.1.4 Coding with Reduced Complexity

In practical coding methods, there are basically two ways of avoiding the extreme complexity of ideal VQ. In the first method, the encoder operates on small image blocks rather than on the complete image. This is obviously suboptimal because the method cannot profit from the redundancy offered by large structures in an image. But the larger the blocks, the better the method. The second strategy is very different and applies some preprocessing on the image prior to quantization. The aim is to remove statistical dependencies among the image pixels, thus avoiding representation of the same information more than once. Both techniques are exploited in practical coders, either separately or in combination.

A typical image encoder incorporating preprocessing is shown in Fig. 52.1.

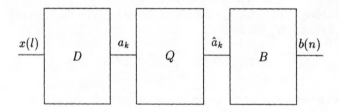

FIGURE 52.1   Generic encoder structure block diagram. D = decomposition unit, Q = quantizer, B = coder for minimum bit-representation.

The first block (D) decomposes the signal into a set of coefficients. The coefficients are subsequently quantized (in Q), and are finally coded to a minimum bit representation (in B). This model is correct for frequency domain coders, but in *closed loop differential coders* (*DPCM*), the decomposition and quantization is performed in the same block, as will be demonstrated later. Usually the decomposition is exact. In *fractal* coding, the decomposition is replaced by approximate modeling.

Let us consider the *decoder* and introduce a *series expansion* as a unifying description of the different image representation methods:

$$\hat{x}(l) = \sum_k \hat{a}_k \phi_k(l) . \tag{52.3}$$

The formula represents the recombination of signal components. Here $\{\hat{a}_k\}$ are the coefficients (the parameters in the representation), and $\{\phi_k(l)\}$ are the *basis functions*. A major distinction between coding methods is their set of basis functions, as will be demonstrated in the next section.

The complete decoder consists of three major parts as shown in Fig. 52.2. The first block (I) receives the bit representation which it partitions into entities representing the different coder parameters and decodes them. The second block (Q^{-1}) is a dequantizer which maps the code to the parametric approximation. The third block (R) reconstructs the signal from the parameters using the series representation.

The second important distinction between compression structures is the coding of the series expansion coefficients in terms of bits. This is dealt with in section 52.3.

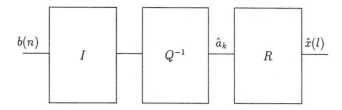

FIGURE 52.2 Block diagram of generic decoder structure. I = bit-representation decoder, Q^{-1} = inverse quantizer, R = signal reconstruction unit.

52.2 Signal Decomposition

As introduced in the previous section, series expansion can be viewed as a common tool to describe signal decomposition. The choice of basis functions will distinguish different coders and influence such features as *coding gain* and the types of distortions present in the decoded image for low bit rate coding. Possible classes of basis functions are:

1. Block-oriented basis functions.

   - The basis functions can cover the whole signal length L. L linearly independent basis functions will make a complete representation.

   - Blocks of size $N \leq L$ can be decomposed individually. *Transform coders* operate in this way. If the blocks are small, the decomposition can catch fast transients. On the other hand, regions with constant features, such as smooth areas or textures, require long basis functions to fully exploit the correlation.

2. Overlapping basis functions:
   The length of the basis functions and the degree of overlap are important parameters. The issue of reversibility of the system becomes nontrivial.

   - In *differential coding*, one basis function is used over and over again, shifted by one sample relative to the previous function. In this case, the basis function usually varies slowly according to some adaptation criterion with respect to the local signal statistics.

   - In subband coding using a uniform filter bank, N distinct basis functions are used. These are repeated over and over with a shift between each group by N samples. The length of the basis functions is usually several times larger than the shifts accommodating for handling fast transients as well as long-term correlations if the basis functions taper off at both ends.

   - The basis functions may be finite (FIR filters) or semi-infinite (IIR filters).

Both time domain and frequency domain properties of the basis functions are indicators of the coder performance. It can be argued that decomposition, whether it is performed by a transform or a filter bank, represents a spectral decomposition. Coding gain is obtained if the different output channels are *decorrelated*. It is therefore desirable that the frequency responses of the different basis functions are localized and separate in frequency. At the same time, they must cover the whole frequency band in order to make a complete representation.

The desire to have highly localized basis functions to handle transients, with localized Fourier transforms to obtain good coding gain, are contradictory requirements due to the *Heisenberg uncertainty relation* [33] between a function and its Fourier transform. The selection of the basis functions must be a compromise between these conflicting requirements.

52.2.1 Decomposition by Transforms

When nonoverlapping block transforms are used, the Karhunen-Loève transform decorrelates, in a statistical sense, the signal within each block completely. It is composed of the eigenvectors of the correlation matrix of the signal. This means that one either has to know the signal statistics in advance or estimate the correlation matrix from the image itself.

Mathematically the eigenvalue equation is given by

$$\mathbf{R}_{xx}\mathbf{h}_n = \lambda_n \mathbf{h}_n \ . \tag{52.4}$$

If the eigenvectors are column vectors, the KLT matrix is composed of the eigenvectors $\mathbf{h}_n$, $n = 0, 1, \cdots, N-1$, as its rows:

$$\mathbf{K} = [\mathbf{h}_0 \mathbf{h}_1 \ldots \mathbf{h}_{N-1}]^T \ . \tag{52.5}$$

The decomposition is performed as

$$\mathbf{y} = \mathbf{K}\mathbf{x} \ . \tag{52.6}$$

The *eigenvalues* are equal to the power of each transform coefficient.

In practice, the so-called *Cosine Transform* (of type II) is usually used because it is a fixed transform and it is close to the KLT when the signal can be described as a first-order autoregressive process with correlation coefficient close to 1.

The cosine transform of length N in one dimension is given by:

$$y(k) = \sqrt{\frac{2}{N}}\alpha(k) \sum_{n=0}^{N-1} x(n) \cos \frac{(2n+1)k\pi}{2N}, \quad k = 0, 1, \cdots, N-1 \ , \tag{52.7}$$

where

$$\alpha(0) = \frac{1}{\sqrt{2}} \quad \text{and} \quad \alpha(k) = 1 \text{ for } k \neq 0 \ . \tag{52.8}$$

The inverse transform is similar except that the scaling factor $\alpha(k)$ is inside the summation.

Many other transforms have been suggested in the literature (DFT, Hadamard Transform, Sine Transform, etc.), but none of these seem to have any significance today.

52.2.2 Decomposition by Filter Banks

Uniform analysis and synthesis filter banks are shown in Fig. 52.3.

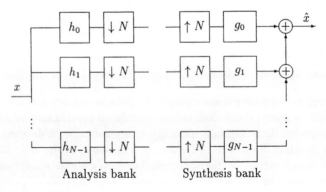

FIGURE 52.3   Subband coder system.

In the analysis filter bank the input signal is split in contiguous and slightly overlapping frequency bands denoted *subbands*. An ideal frequency partitioning is shown in Fig. 52.4.

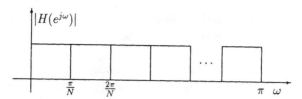

FIGURE 52.4   Ideal frequency partitioning in the analysis channel filters in a subband coder.

If the analysis filter bank was able to *decorrelate* the signal completely, the output signal would be white. For all practical signals, complete decorrelation requires an infinite number of channels.

In the encoder the symbol $\downarrow N$ indicates decimation by a factor of N. By performing this decimation in each of the N channels, the total number of samples is conserved from the system input to decimator outputs. With the channel arrangement in Fig. 52.4, the decimation also serves as a *demodulator*. All channels will have a baseband representation in the frequency range $[0, \pi/N]$ after decimation.

The synthesis filter bank, as shown in Fig. 52.3, consists of N branches with interpolators indicated by $\uparrow N$ and bandpass filters arranged as the filters in Fig. 52.4.

The reconstruction formula constitutes the following series expansion of the output signal:

$$\hat{x}(l) = \sum_{n=0}^{N-1} \sum_{k=-\infty}^{\infty} e_n(k) g_n(l - kN) , \tag{52.9}$$

where $\{e_n(k), \ n = 0, 1, \ldots, N - 1, \ k = -\infty, \ldots, -1, 0, 1, \ldots, \infty\}$ are the expansion coefficients representing the quantized subband signals and $\{g_n(k), \ n = 0, 1, \ldots, N\}$ are the basis functions, which are implemented as unit sample responses of bandpass filters.

Filter Bank Structures

Through the last two decades, an extensive literature on filter banks and filter bank structures has evolved. Perfect reconstruction (PR) is often considered desirable in subband coding systems. It is not a trivial task to design such systems due to the downsampling required to maintain a minimum sampling rate. PR filter banks are often called identity systems. Certain filter bank structures inherently guarantee PR.

It is beyond the scope of this chapter to give a comprehensive treatment of filter banks. We shall only present different alternative solutions at an overview level, and in detail discuss an important two-channel system with inherent perfect reconstruction properties.

We can distinguish between different filter banks based on several properties. In the following, five classifications are discussed.

1. FIR vs. IIR filters — Although IIR filters have an attractive complexity, their inherent long unit sample response and nonlinear phase are obstacles in image coding. The unit sample response length influences the *ringing problem*, which is a main source of objectionable distortion in subband coders. The nonlinear phase makes the *edge mirroring technique* [30] for efficient coding of images near their borders impossible.

2. Uniform vs. nonuniform filter banks — This issue concerns the spectrum partioning in frequency subbands. Currently it is the general conception that nonuniform filter banks perform better than uniform filter banks. There are two reasons for that. The first reason is that our visual system also performs a nonuniform partioning, and the coder should mimic the type of receptor for which it is designed. The second reason is that the filter bank should be able to cope with slowly varying signals (correlation over a large region) as well as transients that are short and represent high frequency signals. Ideally, the filter banks should be adaptive (and good examples of adaptive filter banks have been demonstrated in the literature [2, 11]), but

without adaptivity one filter bank has to be a good compromise between the two extreme cases cited above. Nonuniform filter banks can give the best tradeoff in terms of space-frequency resolution.

3. Parallel vs. tree-structured filter banks — The parallel filter banks are the most general, but tree-structured filter banks enjoy a large popularity, especially for octave band (dyadic frequency partitioning) filter banks as they are easily constructed and implemented. The popular subclass of filter banks denoted *wavelet filter banks* or *wavelet transforms* belong to this class. For octave band partioning, the tree-structured filter banks are as general as the parallel filter banks when perfect reconstruction is required [4].

4. Linear phase vs. nonlinear phase filters — There is no general consensus about the optimality of linear phase. In fact, the traditional wavelet transforms cannot be made linear phase. There are, however, three indications that linear phase should be chosen. (1) The noise in the reconstructed image will be antisymmetrical around edges with nonlinear phase filters. This does not appear to be visually pleasing. (2) The mirror extension technique [30] cannot be used for nonlinear phase filters. (3) Practical coding gain optimizations have given better results for linear than nonlinear phase filters.

5. Unitary vs. nonunitary systems — A unitary filter bank has the same analysis and synthesis filters (except for a reversal of the unit sample responses in the synthesis filters with respect to the analysis filters to make the overall phase linear). Because the analysis and synthesis filters play different roles, it seems plausible that they, in fact, should not be equal. Also, the gain can be larger, as demonstrated in section 52.2.3, for nonunitary filter banks as long as straightforward scalar quantization is performed on the subbands.

Several other issues could be taken into consideration when optimizing a filter bank. These are, among others, the actual frequency partitioning including the number of bands, the length of the individual filters, and other design criteria than coding gain to alleviate coding artifacts, especially at low rates. As an example of the last requirement, it is important that the different phases in the reconstruction process generate the same noise; in other words, the noise should be stationary rather than cyclo-stationary. This may be guaranteed through requirements on the norms of the unit sample responses of the polyphase components [4].

The Two-Channel Lattice Structure

A versatile perfect reconstruction system can be built from two-channel substructures based on lattice filters [36]. The analysis filter bank is shown in Fig. 52.5. It consists of delay-free blocks given in matrix forms as

$$\eta = \begin{bmatrix} a & b \\ c & d \end{bmatrix},$$ (52.10)

and single delays in the lower branch between each block. At the input, the signal is multiplexed into the two branches, which also constitutes the decimation in the analysis system.

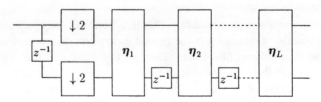

FIGURE 52.5   Multistage two-channel lattice analysis lattice filter bank.

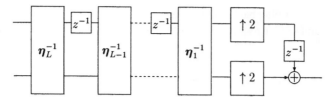

FIGURE 52.6 Multistage two-channel polyphase synthesis lattice filter bank.

A similar synthesis filter structure is shown in Fig. 52.6. In this case, the lattices are given by the inverse of the matrix in Eq. 52.10:

$$\eta^{-1} = \frac{1}{ad - bc} \begin{bmatrix} d & -b \\ -c & a \end{bmatrix}, \qquad (52.11)$$

and the delays are in the upper branches. It is not hard to realize that the two systems are inverse systems provided $ad - bc \neq 0$, except for a system delay.

As the structure can be extended as much as wanted, the flexibility is good. The filters can be made unitary or they can have a linear phase. In the unitary case, the coefficients are related through $a = d = \cos\phi$ and $b = -c = \sin\phi$, whereas in the linear phase case, the coefficients are $a = d = 1$ and $b = c$. In the linear phase case, the last block (η_L) must be a *Hadamard* transform.

Tree Structured Filter Banks

In tree-structured filter banks, the signal is first split in two channels. The resulting outputs are input to a second stage with further separation. This process can go on as indicated in Fig. 52.7 for a system where at every stage the outputs are split further until the required resolution has been obtained.

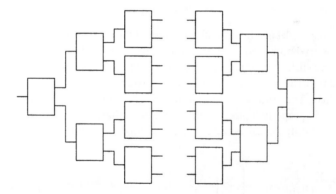

FIGURE 52.7 Left: Tree structured analysis filter bank consisting of filter blocks where the signal is split in two and decimated by a factor of two to obtain critical sampling. Right: Corresponding synthesis filter bank for recombination and interpolation of the signals.

Tree-structured systems have a rather high flexibility. Nonuniform filter banks are obtained by splitting only some of the outputs at each stage. To guarantee perfect reconstruction, each stage in the synthesis filter bank (Fig. 52.7) must reconstruct the input signal to the corresponding analysis filter.

52.2.3 Optimal Transforms/Filter Banks

The *gain* in subband and transform coders depends on the detailed construction of the filter bank as well as the quantization scheme.

Assume that the analysis filter bank unit sample responses are given by $\{h_n(k),\ n = 0, 1, \ldots, N - 1\}$. The corresponding unit sample responses of the synthesis filters are required to have unit norm:

$$\sum_{k=0}^{L-1} g_n^2(k) = 1 .$$

The coding gain of a subband coder is defined as the ratio between the noise using scalar quantization (PCM) and the subband coder noise incorporating optimal bit-allocation as explained in section 52.3:

$$G_{SBC} = \left[\prod_{n=0}^{N-1} \frac{\sigma_{x_n}^2}{\sigma_x^2} \right]^{-1/N} \tag{52.12}$$

Here σ_x^2 is the variance of the input signal while $\{\sigma_{x_n}^2, n = 0, 1 \ldots, N - 1\}$ are the subband variances given by

$$\sigma_{x_n}^2 = \sum_{l=-\infty}^{\infty} R_{xx}(l) \sum_{j=-\infty}^{\infty} h_n(j) h_n(l + j) \tag{52.13}$$

$$= \int_{-\pi}^{\pi} S_{xx}(e^{j\omega}) |H_n(e^{j\omega})|^2 \frac{d\omega}{2\pi} . \tag{52.14}$$

The subband variances depend both on the filters and the second order spectral information of the input signal.

For images, the gain is often estimated assuming that the image can be modeled as a first order Markov source (also called an AR(1) process) characterized by

$$R_{xx}(l) = \sigma_x^2\ 0.95^{|l|} . \tag{52.15}$$

(Strictly speaking, the model is valid only after removal of the image average).

We consider the maximum gain using this model for three special cases. The first is the transform coder performance, which is an important reference as all image and video coding standards are based on transform coding. The second is for unitary filter banks, for which optimality is reached by using ideal brick-wall filters. The third case is for nonunitary filter banks, often denoted *biorthogonal* when the perfect reconstruction property is guaranteed. In the nonunitary case, *halfwhitening* is obtained within each band. Mathematically this can be seen from the optimal magnitude response for the filter in channel n:

$$|H_n(e^{j\omega})| = \begin{cases} c_2 \left[\frac{S_{xx}(e^{j\omega})}{\sigma_x^2} \right]^{-1/4} & \text{for } \omega \in \pm[\frac{\pi n}{N}, \frac{\pi(n+1)}{N}] \\ 0 & \text{otherwise,} \end{cases} \tag{52.16}$$

where c_2 is a constant that can be selected for correct gain in each band.

The inverse operation must be performed in the synthesis filter to make completely flat responses within each band.

In Fig. 52.8, we give optimal coding gains as a function of the number of channels.

52.2.4 Decomposition by Differential Coding

In *closed-loop differential* coding, the generic encoder structure (Fig. 52.1) is not valid as the quantizer is placed inside a feedback loop. The decoder, however, behaves according to the generic decoder structure. Basic block diagrams of a closed-loop differential encoder and the corresponding decoder are shown in Figs. 52.9(a) and (b), respectively.

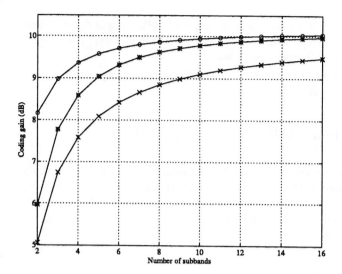

FIGURE 52.8 Maximum coding gain as function of the number of channels for different one-dimensional coders operating on a first order Markov source with one-delay correlation $\rho = 0.95$. Lower curve: Cosine transform. Middle curve: Unitary filter bank. Upper curve: Unconstrained filter bank. Nonunitary case.

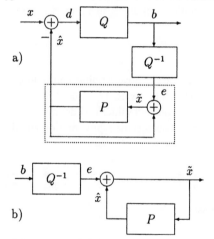

FIGURE 52.9 (a) DPCM encoder. (b) DPCM decoder.

In the encoder, the input signal x is represented by the bit-stream b. Q is the quantizer and Q^{-1} the dequantizer, but $QQ^{-1} \neq 1$, except for the case of infinite resolution in the quantizer. The signal d, which is quantized and transmitted by some binary code, is the difference between the input signal and a predicted value of the input signal based on previous outputs and a prediction filter with transfer function $G(z) = 1/(1 - P(z))$. Notice that the decoder is a substructure of the encoder, and that $\tilde{x} = x$ in the limiting case of infinite quantizer resolution. The last property guarantees exact representation when discarding quantization.

Introducing the inverse z-transform of $G(z)$ as $g(l)$, the reconstruction is performed on the dequantized values as

$$\tilde{x}(l) = \sum_{k=0}^{\infty} e(k)g(l - k) . \tag{52.17}$$

The output is thus a linear combination of unit sample responses excited by the sample amplitudes at different times and, can be viewed as a series expansion of the output signal. In this case, the basis functions are generated by shifts of a single basis function [the unit sample response $g(l)$] and the coefficients represent the coded difference signal $e(n)$.

With an adaptive filter the basis function will vary slowly, depending on some spectral modification derived from the incoming samples.

52.3 Quantization and Coding Strategies

Quantization is the means of providing approximations to signals and signal parameters by a finite number of representation levels. This process is nonreversible and thus always introduces noise. The representation levels constitute a finite alphabet which is usually represented by binary symbols, or *bits*. The mapping from symbols in a finite alphabet to bits is not unique. Some important techniques for quantization and coding will be reviewed next.

52.3.1 Scalar Quantization

The simplest quantizer is the *scalar* quantizer. It can be optimized to match the *probability density function* (*pdf*) of the input signal.

A *scalar* quantizer maps a continuous variable x to a finite set according to the rule

$$x \in R_i \qquad \implies \qquad Q[x] = y_i , \tag{52.18}$$

where $R_i = (x_i, x_{i+1}), i = 1, \ldots, L$, are nonoverlapping, contiguous intervals covering the real line, and $(\cdot, \cdot)$ denotes open, half open, or closed intervals. $\{y_i, \ i = 1, 2, \ldots, L\}$ are referred to as *representation levels* or *reconstruction values*. The associated values $\{x_i\}$ defining the partition are referred to as *decision levels* or *decision thresholds*. Fig. 52.10 depicts the representation and decision levels.

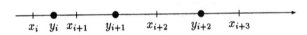

FIGURE 52.10   Quantization notation.

In a *uniform* quantizer, all intervals are of the same length and the representation levels are the midpoints in each interval. Furthermore, in a *uniform threshold* quantizer, the decision levels form a uniform partitioning of the real line, while the representation levels are the *centroids* (see below) in each decision interval. Strictly speaking, uniform quantizers consist of an infinite number of intervals. In practice, the number of intervals is adapted to the dynamic range of the signal. All other quantizers are *non-uniform*.

The optimization task is to minimize the average *distortion* between the original samples and the appropriate representation levels given the number of levels. This is the so-called pdf-optimized quantizer. Allowing for variable rate per symbol, the *entropy constrained quantizer* can be used. These schemes are described in the following two subsections.

The Lloyd-Max Quantizer

The Lloyd-Max quantizer is a scalar quantizer where the 1st order signal pdf is exploited to increase the quantizer performance. It is therefore often referred to as a *pdf-optimized quantizer*. Each signal sample is quantized using the same number of bits. The optimization is done by minimizing the total distortion of a quantizer with a given number L of representation levels. For an input signal X with pdf $p_X(x)$, the average mean square distortion is

$$D = \sum_{i=1}^{L} \int_{x_i}^{x_{i+1}} (x - y_i)^2 p_X(x) dx . \tag{52.19}$$

Minimization of D leads to the following implicit expressions connecting the decision and representation levels:

$$x_{k,opt} = \frac{1}{2}(y_{k,opt} + y_{k-1,opt}), \qquad k = 1, \ldots, L-1 \tag{52.20}$$

$$x_{0,opt} = -\infty \tag{52.21}$$

$$x_{L,opt} = \infty \tag{52.22}$$

$$y_{k,opt} = \frac{\int_{x_{k,opt}}^{x_{k+1,opt}} x p_X(x) dx}{\int_{x_{k,opt}}^{x_{k+1,opt}} p_X(x) dx}, \qquad k = 0, \ldots, L-1 . \tag{52.23}$$

Equation 52.20 indicates that the decision levels should be the midpoints between neighboring representation levels, while Eq. 52.23 requires that the optimal representation levels are the *centroids* of the pdf in the appropriate interval.

The equations can be solved iteratively [21]. For high bit rates it is possible to derive approximate formulas assuming that the signal pdf is flat within each quantization interval [21].

In most practical situations the pdf is not known, and the optimization is based on a training set. This will be discussed in section 52.3.2.

Entropy Constrained Quantization

When minimizing the total distortion for a fixed number of possible representation levels, we have tacitly assumed that every signal sample is coded using the same number of bits: $\log_2 L$ bits/sample. If we allow for a variable number of bits for coding each sample, a further rate-distortion advantage is gained. The Lloyd-Max solution is then no longer optimal. A new optimization is needed, leading to the *entropy constrained quantizer.*

At high bit rates, the optimum is reached when using a uniform quantizer with an infinite number of levels. At low bit rates, uniform quantizers perform close to optimum provided the representation levels are selected as the centroids according to Eq. 52.23. The performance of the entropy constrained quantizer is significantly better than the performance of the Lloyd-Max quantizer [21].

A standard algorithm for assigning codewords of variable length to the representation levels was given by Huffman [12]. The Huffman code will minimize the average rate for a given set of probabilities and the resulting average bit rate will be close to the entropy bound. Even closer performance to the bound is obtained by *arithmetic coders* [32].

At high bit rates, scalar quantization on *statistically independent* samples renders a bit rate which is at least 0.255 bits/sample higher than the *rate distortion bound* irrespective of the signal pdf. Huffman coding of the quantizer output typically gives a somewhat higher rate.

52.3.2 Vector Quantization

Simultaneous quantization of several samples is referred to as *vector quantization* (VQ) [9], as mentioned in the introductory section. VQ is a generalization of scalar quantization:

A vector quantizer maps a continuous N-dimensional vector $\mathbf{x}$ to a discrete-valued N-dimensional vector according to the rule

$$\mathbf{x} \in C_i \qquad \Longrightarrow \qquad Q[\mathbf{x}] = \mathbf{y}_i , \tag{52.24}$$

where C_i is an N-dimensional cell. The L possible cells are nonoverlapping and contiguous and fill the entire geometric space. The vectors $\{\mathbf{y}_i\}$ correspond to the representation levels in a scalar quantizer. In a VQ setting the collection of representation levels is referred to as the *codebook*. The cells C_i, also called *Voronoi regions*, correspond to the decision regions, and can be thought of as solid polygons in the N-dimensional space.

In the scalar case, it is trivial to test if a signal sample belongs to a given interval. In VQ an indirect approach is utilized via a *fidelity criterion or distortion measure* $d(\cdot, \cdot)$:

$$Q[\mathbf{x}] = \mathbf{y}_i \iff d(\mathbf{x}, \mathbf{y}_i) \le d(\mathbf{x}, \mathbf{y}_j), \qquad j = 0, \ldots, L - 1. \tag{52.25}$$

When the best match, $\mathbf{y}_i$, has been found, the *index i* identifies that vector and is therefore coded as an efficient representation of the vector. The receiver can then reconstruct the vector $\mathbf{y}_i$ by looking up the contents of cell number i in a copy of the codebook. Thus, the bit rate in bits per sample in this scheme is $\log_2 L/N$ when using straightforward bit-representation for i. A block diagram of vector quantization is shown in Fig. 52.11.

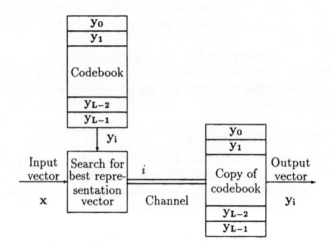

FIGURE 52.11   Vector quantization procedure.

In the previous section we stated that scalar entropy coding was sub-optimal, even for sources producing independent samples. The reason for the sub-optimal performance of the entropy constrained quantizer is a phenomenon called *sphere packing*. In addition to obtaining good sphere packing, a VQ scheme also exploits both correlation and higher order statistical dependencies of a signal. The higher order statistical dependency can be thought of as "a preference for certain vectors". Excellent examples of sphere packing and higher order statistical dependencies can be found in [28].

In principle, the codebook design is based on the N-dimensional pdf. But as the pdf is usually not known, the codebook is optimized from a training data set. This set consists of a large number of vectors that are representative for the signal source. A sub-optimal codebook can then be designed using an iterative algorithm, for example the *K-means* or *LBG* algorithm [25].

Multistage Vector Quantization

To alleviate the complexity problems of vector quantization, several methods have been suggested. They all introduce some structure into the codebook which makes fast search possible. Some systems also reduce storage requirements, like the one we present in this subsection. The obtainable performance is always reduced, but the performance in an implementable coder can be improved.

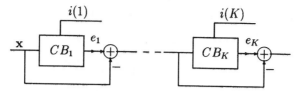

FIGURE 52.12 *K*-stage VQ encoder structure showing the successive approximation of the signal vector.

Fig. 52.12 illustrates the encoder structure.

The first block in the encoder makes a rough approximation to the input vector by selecting the codebook vector which, upon scaling by e_1, is closest in some distortion measure. Then this approximation is subtracted from the input signal. In the second stage, the difference signal is approximated by a vector from the second codebook scaled by e_2. This procedure continues in K stages, and can be thought of as a successive approximation to the input vector. The indices $\{i(k), \ k = 1, 2, \cdots, K\}$ are transmitted as part of the code for the particular vector under consideration.

Compared to unstructured VQ, this method is suboptimal but has a much lower complexity than the optimal case due to the small codebooks that can be used.

A special case is the *mean-gain-shape* VQ [9], where one stage only is kept, but in addition the mean is represented separately.

In all multistage VQs, the code consists of the codebook address and codes for the quantized versions of the scaling coefficients.

52.3.3 Efficient Use of Bit-Resources

Assume we have a signal that can be split in classes with different statistics. As an example, after applying signal decomposition, the different transform coefficients typically have different variances. Assume also that we have a pool of bits to be used for representing a collection of signal vectors from the different classes, or we try to minimize the number of bits to be used after all signals have been quantized. These two situations are described below.

Bit Allocation

Assume that a signal consists of N components $\{x_i, \ i = 1, 2, \cdots, N\}$ forming a vector $\mathbf{x}$ where the variance of component number i is equal to $\sigma_{x_i}^2$ and all components are zero mean.

We want to quantize the vector $\mathbf{x}$ using scalar quantization on each of the components and minimize the total distortion with the only constraint that the total number of bits to be used for the whole vector be fixed and equal to B. Denoting the quantized signal components $Q_i(x_i)$, the average distortion per component can be written as

$$D_{DS} = \frac{1}{N} \sum_{i=1}^{N} E[x_i - Q_i(x_i)]^2 = \frac{1}{N} \sum_{i=1}^{N} D_i , \tag{52.26}$$

where $E[\cdot]$ is the expectation operator, and the subscript DS stands for *decomposed source*.

The bit-constraint is given by

$$B = \sum_{i=1}^{N} b_i , \tag{52.27}$$

where b_i is the number of bits used to quantize component number i.

Minimizing D_{DS} with Eq. 52.27 as a constraint, we obtain the following bit assignment

$$b_j = \frac{B}{N} + \frac{1}{2} \log_2 \frac{\sigma_{x_j}^2}{\left[\prod_{n=1}^{N} \sigma_{x_n}^2 \right]^{1/N}} . \tag{52.28}$$

This formula will in general render noninteger and even negative values of the bit count. So-called "greedy" algorithms can be used to avoid this problem.

To evaluate the coder performance, we use *coding gain*. It is defined as the distortion advantage of the component-wise quantization over a direct scalar quantization at the same rate. For the example at hand, the coding gain is found to be

$$G_{DS} = \frac{\frac{1}{N} \sum_{j=1}^{N} \sigma_{x_j}^2}{(\prod_{j=1}^{N} \sigma_{n_j}^2)^{1/N}} . \tag{52.29}$$

The gain is equal to the ratio between the arithmetic mean and the geometric mean of the component variances. The minimum value of the variance ratio is equal to 1 when all the component variances are equal. Otherwise, the gain is larger than one. Using the optimal bit allocation, the noise contribution is equal in all components.

If we assume that the different components are obtained by passing the signal through a bank of bandpass filters, then the variance from one band is given by the integral of the power spectral density over that band. If the process is non-white, the variances are more different the more colored the original spectrum is. The maximum possible gain is obtained when the number of bands tends to infinity [21]. Then the gain is equal to the maximum gain of a differential coder which again is inversely proportional to the *spectral flatness measure* [21] given by

$$\gamma_x^2 = \frac{\exp[\int_{-\pi}^{\pi} \ln S_{xx}(e^{j\omega}) \frac{d\omega}{2\pi}]}{\int_{-\pi}^{\pi} S_{xx}(e^{j\omega}) \frac{d\omega}{2\pi}}, \tag{52.30}$$

where $S_{xx}(e^{j\omega})$ is the spectral density of the input signal. In both subband coding and differential coding, the complexity of the systems must approach infinity to reach the coding gain limit.

To be able to apply bit allocation dynamically to non-stationary sources, the decoder must receive information about the local bit allocation. This can be done either by transmitting the bit allocation table, or the variances from which the bit allocation was derived. For real images where the statistics vary rapidly, the cost of transmitting the side information may become costly, especially for low rate coders.

Rate Allocation

Assume we have the same signal collection as above. This time we want to minimize the number of bits to be used after the signal components have been quantized. The first order entropy of the decomposed source will be selected as the measure for the obtainable minimum bit-rate when scalar representation is specified.

To simplify, assume all signal components are Gaussian. The entropy of a Gaussian source with zero mean and variance σ_x^2 and statistically independent samples quantized by a uniform quantizer with quantization interval Δ can, for high rates, be approximated by

$$H_G(X) = \frac{1}{2} \log_2(2\pi e(\sigma_x/\Delta)^2) . \tag{52.31}$$

The rate difference [24] between direct scalar quantization of the signal collection using one entropy coder and the rate when using an adapted entropy coder for each component is

$$\Delta H = H_{PCM} - H_{DS} = \frac{1}{2} \log_2 \frac{\sigma_x^2}{[\prod_{i=1}^{N} \sigma_{x_i}^2]^{1/N}} , \tag{52.32}$$

provided the decomposition is power conserving, meaning that

$$\sigma_x^2 = \sum_{i=1}^{N} \sigma_{x_i}^2 . \tag{52.33}$$

The coding gain in Eq. 52.29 and the rate gain in Eq. 52.32 are equivalent for Gaussian sources.

In order to exploit this result in conjunction with signal decomposition, we can view each output component as a stationary source, each with different signal statistics. The variances will depend on the spectrum of the input signal. From Eq. 52.32 and Eq. 52.33 we see that the rate difference is larger the more different the channel variances are.

To obtain the rate gain indicated by Eq. 52.32, different *Huffman* or *arithmetic* coders [9] adapted to the rate given in Eq. 52.31 must be employed. In practice, a pool of such coders should be generated and stored. During encoding, the closest fitting coder is chosen for each block of components. An index indicating which coder was used is transmitted as side information to enable the decoder to reinterpret the received code.

52.4 Frequency Domain Coders

In this section we present the JPEG standard and some of the best subband coders that have been presented in the literature.

52.4.1 The JPEG Standard

The JPEG coder [37] is the only internationally standardized still image coding method. Presently there is an international effort to bring forth a new, improved standard under the title JPEG2000.

The principle can be sketched as follows: First, the image is decomposed using a two-dimensional cosine transform of size 8×8. Then, the transform coefficients are arranged in an 8×8 matrix as given in Fig. 52.13, where i and j are the horizontal and vertical frequency indices, respectively. A vector is formed by a scanning sequence which is chosen to make large amplitudes, on average, appear first, and smaller amplitudes at the end of the scan. In this arrangement, the samples at the end of the scan string approach zero. The scan vector is quantized in a non-uniform scalar quantizer with characteristics as depicted in Fig. 52.14.

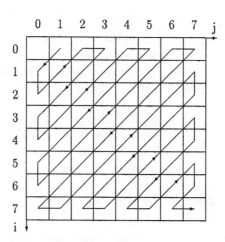

FIGURE 52.13   Zig-zag scanning of the coefficient matrix.

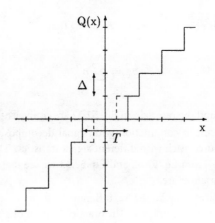

FIGURE 52.14   Non-uniform quantizer characteristic obtained by combining a midtread uniform quantizer and a thresholder. Δ is the quantization interval and T is the threshold.

Due to the thresholder, many of the trailing coefficients in the scan vector are set to zero. Often the zero values appear in clusters. This property is exploited by using *runlength coding*, which basically amounts to finding zero-runs. After runlength coding, each run is represented by a *number pair* (a, r) where the number a is the amplitude and r is the length of the run. Finally, the number pair is entropy coded using the Huffman method, or arithmetic coding.

The thresholding will increase the distortion and lower the entropy both with and without decomposition, although not necessarily with the same amounts.

As can be observed from Fig. 52.13, the coefficient in position (0,0) is not part of the string. This coefficient represents the block average. After collecting all block averages in one image, this image is coded using a DPCM scheme [37].

Coding results for three images are given in Fig. 52.16.

52.4.2   Improved Coders: State-of-the-Art

Many coders that outperform JPEG have been presented in the scientific literature. Most of these are based on subband decomposition (or the special case: wavelet decomposition). Subband coders have a higher potential coding gain by using filter banks rather than transforms, and thus exploiting correlations over larger image areas. Figure 52.8 shows the theoretical gain for a stochastic image model. Visually, subband coders can avoid the blocking-effects experienced in transform coders at low bit-rates. This property is due to the overlap in basis functions in subband coders. On the other hand, Gibb's phenomenon is more prevalent in subband coders and can cause severe ringing in homogeneous areas close to edges. The detailed choice and optimization of the filter bank will strongly influence the visual performance of subband coders. The other factor which decides the coding quality is the detailed quantization of the subband signals. The final bit-representation method does not effect the quality, only the rate for a given quality.

Depending on the bit-representation, the total rate can be preset for some coders, and will depend on some quality factor specified for other coders. Even though it would be desirable to preset the visual quality in a coder, this is a challenging task, which has not yet been satisfactorily solved.

In the following we present four subband coders with different coding schemes and different filter banks.

Subband Coder Based on Entropy Coder Allocation [24]

This coder uses an 8×8 uniform filter bank optimized for reducing blocking and ringing artifacts, plus maximizing the coding gain [1]. The lowpass-lowpass band is quantized using a fixed rate DPCM coder with a third-order two-dimensional predictor. The other subband signals are segmented into blocks of size 4 × 4, and each block is classified based on the block power. Depending on the block power, each block is allocated a corresponding entropy coder (implemented as an arithmetic coder). The entropy coders have been preoptimized by minimizing the first-order entropy given the number of available entropy coders (See section 52.3.3). This number is selected to balance the amount of side information necessary in the decoder to identify the correct entropy decoder and the gain by using more entropy coders. Depending on the bit-rate, the number of entropy coders is typically 3 to 5. In the presented results, three arithmetic coders are used. Conditional arithmetic coding has been used to represent the side information efficiently.

Coding results are presented in Fig. 52.16 under the name "Lervik".

Zero-Tree Coding

Shapiro [35] introduced a method that exploits some dependency between pixels in corresponding location in the bands of an octave band filter bank. The basic assumed dependencies are illustrated in Fig. 52.15. The low-pass band is coded separately. Starting in any location in any of the other three

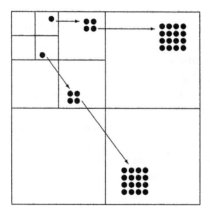

FIGURE 52.15 Zero-tree arrangement in an octave-band decomposed image.

bands of same size, any pixel will have an increasing number of *descendants* as one passes down the tree representing information from the same location in the original image. The number of corresponding pixels increases by a factor of four from one level to the next. When used in a coding context, the tree is terminated at any zero-valued pixel (obtained after quantization using some threshold) after which all subsequent pixels are assumed to be zero as well. Due to the growth by a factor of four between levels, many samples can be discarded this way.

What is the underlying mechanism that makes this technique work so well? On one hand, the image spectrum falls off rapidly as a function of frequency for most images. This means that there is a tendency to have many zeros when approaching the leaves of the tree. Our visual system is furthermore more tolerant to high frequency errors. This should be compared to the zig-zag scan in the JPEG coder. On the other hand, viewed from a pure statistical angle, the subbands are uncorrelated if the filter bank has done what is required from it! However, the statistical argument is based on the assumption of "local ergodicity", which means that statistical parameters derived locally from the data have the same mean values everywhere. With real images composed of objects with edges, textures, etc. these assumptions do not hold. The "activity" in the subbands tends to appear in the same locations. This is typical at edges. One can look at these connections as energy correlations among the subbands. The zero-tree method will efficiently cope with these types of phenomena.

Shapiro furthermore combined the zero-tree representation with bit-plane coding. Said [34] went one step further and introduced what he calls *set partitioning*. The resulting algorithm is simple and fast, and is embedded in the sense that the bit-stream can be cut off at any point in time in the decoder, and the obtained approximation is optimal using that number of bits. The subbands are obtained using the 9/7 biorthogonal spline filters [38].

Coding results from Said's coder are shown in Fig. 52.16 and marked "Said".

Pyramid VQ and Improved Filter Bank

This coder is based on bit-allocation, or rather, allocation of vector quantizers of different sizes. This implies that the coder is fixed rate, that is, we can preset the total number of bits for an image. It is assumed that the subband signals have a Laplacian distribution, which makes it possible to apply *pyramid* vector quantizers [6]. These are suboptimal compared to trained codebook vector quantizers, but significantly better than scalar quantizers without increasing the complexity too much.

The signal decomposition in the encoder is performed using an 8×8 channel uniform filter bank [1], followed by an octave-band filter bank of three stages operating on the resulting lowpass-lowpass band. The uniform filter bank is nonunitary and optimized for coding gain. The building blocks of the octave band filter bank have been carefully selected from all available perfect reconstruction, two-channel filter systems with limited FIR filter orders.

Coding results from this coder are shown in Figs. 52.16 and are marked "Balasingham".

Trellis Coded Quantization

Joshi [22] has presented what is presently the "state-of-the-art" coder. Being based on trellis coded quantization [29], the encoder is more complex than the other coders presented. Furthermore, it does not have the embedded character of Said's coder.

The filter bank employed has 22 subbands. This is obtained by first employing a 4×4 uniform filter bank, followed by a further split of the resulting lowpass-lowpass band using a two-stage octave band filter bank. All filters in the curves shown in the next section are 9/7 biorthogonal spline filters [38].

The encoding of the subbands is performed in several stages:

- Separate classification of signal blocks in each band.
- Rate allocation among all blocks.
- Individual arithmetic coding of the trellis-coded quantized signals in each class.

The trellis coded quantization [7] is a method that can reach the rate distortion bound in the same way as vector quantization. It uses search methods in the encoder, which adds to its complexity. The decoder is much simpler.

Coding results from this coder are shown in Fig. 52.16 and are marked "Joshi".

Frequency Domain Coding Results

The five coders presented above are compared in this section. All of them are simulated using the three images "Lenna", "Barbara", and "Goldhill" of size 512×512. These three images have quite different contents in terms of spectrum, textures, edges, and so on. Fig. 52.16 shows the PSNR as a function of bit-rate for the five coders. The PSNR is defined as

$$\text{PSNR} = 10 \log_{10} \left(\frac{255^2}{\frac{1}{NM} \sum_{n=1}^{N} \sum_{m=1}^{M} (x(n,m) - \hat{x}(n,m))^2} \right). \tag{52.34}$$

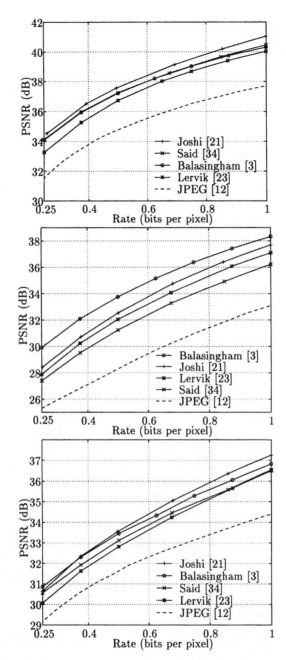

FIGURE 52.16   Coding results. Top: "Lenna", middle: "Barbara", bottom: "Goldhill".

As is observed, the coding quality among the coders varies when exposed to such different stimuli. The exception is that all subband coders are superior to JPEG, which was expected from the use of better decomposition as well as more clever quantization and coding strategies. Joshi's coder is best for "Lenna" and "Goldhill" at high rates. Balasingham's coder is, however, better for "Barbara" and for "Goldhill" at low rates. These results are interpreted as follows. The Joshi coder uses the most elaborate quantization/coding scheme, but the Balasingham coder applies a better filter bank in two respects. First, it has better high frequency resolution, which explains that the "Barbara" image, with a relatively high frequency content, gives a better result for the latter coder. Second, the improved low frequency resolution of this filter bank also implies better coding at low rates for "Goldhill".

From the results above, it is also observed that the Joshi coder performs well for images with a lowpass character such as the "Lenna" image, especially at low rates. In these cases there are many "zeros" to be represented, and the zero-tree coding can typically cope well with zero-representations.

A combination of several of the aforementioned coders, picking up their best components, would probably render an improved system.

52.5 Fractal Coding

This section is placed towards the end of the chapter because fractal coding deviates in many respects from the generic coder on the one hand, but on the other hand can be compared to vector quantization.

A good overview of the field can be found in [8].

Fractal coding (also called attractor coding) is based on *Banach's fixed point theorem* and exploits *self-similarity* or *partial self-similarity* among different scales of a given image. A nonlinear transform gives the fractal image representation. Iterative operations using this transform starting from any initial image will converge to the image approximation, called the *attractor*. The success of such a scheme will rest upon the compactness, in terms of bits, of the description of the nonlinear transform. A classical example of self-similarity is Michael Barnsley's fern, where each branch is a small copy of the complete fern. Even the branches are composed of small copies of itself. A very compact description can be found for the class of images exhibiting self similarity. In fact, the fern can be described by 24 numbers, according to Barnsley.

Self-similarity is a dependency among image elements (possibly objects) that is not described by correlation, but can be called affine correlation.

There is an enormous potential for image compression if images really have the self-similarity property. However, there seems to be no reason to believe that global self-similarity exists in any complex image created, e.g., by photographing natural or man-made scenes. The less requiring notion of partial self-similarity among image blocks of different scales has proven to be fruitful [19].

In this section we will, in fact, present a practical fractal coder exploiting partial self-similarity among different scales, which can be directly compared to mean-gain-shape vector quantization (MGSVQ). The difference between the two systems is that the vector quantizer uses an optimized codebook based on data from a large collection of different images, whereas the fractal coder uses a *self codebook*, in the sense that the codebook is generated from the image itself and implicitly and approximately transmitted to the receiver as part of the image code. The question is then, "Is the 'adaptive' nature of the fractal codebook better than the statistically optimized codebook of standard vector quantization?"

We will also comment on other models and give a brief status of fractal compression techniques.

52.5.1 Mathematical Background

The code of an image in the language of fractal coding is given as the bit-representation of a nonlinear transform T. The transform defines what is called the *collage* x_c of the image. The collage is found by

$$x_c = Tx ,$$

where x is the original image.

The collage is the object we try to make resemble the image as closely as possible in the encoder through minimization of the distortion function

$$D = d(x, x_c) . \tag{52.35}$$

Usually the distortion function is chosen as the Euclidean distance between the two vectors. The decoder cannot reconstruct the collage as it depends on the knowledge of the original image, and not only the transform T. We therefore have to accept reconstruction of the image with less accuracy.

The reconstruction algorithm is based on *Banach's fixed point theorem*: If a transform T is *contractive* or *eventually contractive* [26], the fixed point theorem states that the transform then has a unique *attractor* or *fixed point* given by

$$x_T = T x_T , \qquad (52.36)$$

and that the fixed point can be approached by iteration from any starting vector according to

$$x_T = \lim_{n \to \infty} T^n y; \quad \forall y \in X , \qquad (52.37)$$

where X is a normed linear space.

The similarity between the collage and the attractor is indicated from an extended version of the collage theorem [27]:

Given an original image x and its collage Tx where $\|x - Tx\| \le \epsilon$, then

$$\|x - x_T\| \le \frac{1 - s_1^K}{(1 - s_1)(1 - s_K)} \epsilon \qquad (52.38)$$

where s_1 and s_K are the Lipschitz constants of T and T^K, respectively, provided $|s_1| < 1$ and $|s_K| < 1$.

Provided the collage is a good approximation of the original image and the Lipschitz constants are small enough, there will also be similarity between the original image and the attractor.

In the special case of *fractal block coding*, a given image block (usually called a *domain block*) is supposed to resemble another block (usually called a *range block*) after some *affine transformation*. The transformation that is most commonly used moves the image block to a different position while shrinking the block, rotating it or shuffling the pixels, and adding what we denote a *fixed term*, which could be some predefined function with possible parameters to be decided in the encoding process. In most natural images it is not difficult to find affine similarity, e.g., in the form of objects situated at different distances and positions in relation to the camera. In standard block coding methods, only local statistical dependencies can be utilized. The inclusion of *affine redundancies* should therefore offer some extra advantage.

In this formalism we do not see much resemblance with VQ. However, the similarities and differences between fractal coding and VQ were pointed out already in the original work by Jacquin [20]. We shall, in the following section, present a specific model that enforces further similarity to VQ.

52.5.2 Mean-Gain-Shape Attractor Coding

It has been proven [31] that in all cases where each domain block is a union of range blocks, the decoding algorithms for sampled images where the nonlinear part (fixed term) of the transform is orthogonal to the image transformed by the linear part, full convergence is reached after a finite and small number of iterations. In one special case there are no iterations at all [31], and then $x_T = Tx$. We shall discuss only this important case here because it has an important application potential due to its simplicity in the decoder, but, more importantly, we can more clearly demonstrate the similarity to VQ.

Codebook Formation

In the encoder two tasks have to be performed, the codebook formation and the codebook search, to find the best representation of the transform T with as few bits as possible.

First the image is split in non-overlapping blocks of size $L \times L$ so that the complete image is covered. The codebook construction goes as follows:

- Calculate the mean value m in each block.

- Quantize the mean values, resulting in the approximation $\hat{m}$, and transmit their code to the receiver.

  These values will serve two purposes:

  1. They are the additive, nonlinear terms in the block transform.
  2. They are the building elements for the codebook.

All the following steps must be performed both in the encoder and the decoder.

- Organize the quantized mean values as an image so that it becomes a block averaged and downsampled version of the original image.
- Pick blocks of size $L \times L$ in the obtained image. Overlap between blocks is possible.
- Remove the mean values from each block. The resulting blocks constitute part of the codebook.
- Generate new codebook vectors by a predetermined set of mathematical operations (mainly pixel shuffling).

With the procedure given, the codebook is explicitly known in the decoder, because the mean values also act as the nonlinear part of the affine transforms. The codebook vectors are orthogonal to the nonlinear term due to the mean value removal.

Observe also that the block decimation in the encoder must now be chosen as $L \times L$, which is also the size of the blocks to be coded.

The Encoder

The actual encoding is similar to traditional product code VQ.

In our particular case, the image block in position (k, l) is modeled as

$$\hat{x}_{k,l} = \hat{m}_{k,l} + \hat{\alpha}_{k,l}\rho^{(i)} , \tag{52.39}$$

where $\hat{m}_{k,l}$ is the quantized mean value of the block, $\rho^{(i)}$ is codebook vector number i, and $\hat{\alpha}_{k,l}$ is a quantized scaling factor.

To optimize the parameters, we minimize the Euclidean distance between the image block and the given approximation,

$$d = \|x_{k,l} - \hat{x}_{k,l}\| . \tag{52.40}$$

This minimization is equivalent to the maximization of

$$P^{(i)} = \frac{\langle x_{k,l}, \rho^{(i)}\rangle^2}{\|\rho^{(i)}\|^2} , \tag{52.41}$$

where $\langle u, v \rangle$ denotes the inner product between u and v over one block. If vector number j maximizes P, then the scaling factor can be calculated as

$$\alpha_{k,l} = \frac{\langle x_{k,l}, \rho^{(j)}\rangle}{\|\rho^{(j)}\|^2} . \tag{52.42}$$

The Decoder

In the decoder, the codebook can be regenerated, as previously described, from the mean values. The decoder reconstructs each block according to Eq. 52.39 using the transmitted, quantized parameters. In the particular case given above, the following procedure is followed:

Denote by c an image composed of subblocks of size $L \times L$ which contains the correct mean values. The decoding is then performed by

$$x_1 = Tc = Ac + c , \tag{52.43}$$

where A is the linear part of the transform. The operation of A can be described blockwise.

- It takes a block from c of size $L^2 \times L^2$,
- shrinks it to size $L \times L$ after averaging over subblocks of size $L \times L$,
- subtracts from the resulting block its mean value,
- performs the prescribed pixel shuffling,
- multiplies by the scaling coefficient,
- and finally inserts the resulting block in the correct position.

Notice that x_1 has the correct mean value due to c, and because Ac does not contribute to the block mean values. Another observation is that each block of size $L \times L$ is mapped to one pixel.

The algorithm just described is equivalent to the VQ decoding given earlier.

The iterative algorithm indicated by Banach's fixed point theorem can be used also in this case. The above described algorithm is the first iteration. In the next iteration we get

$$x_2 = Ax_1 + c = A(Ac) + Ac + c . \tag{52.44}$$

But $A(Ac) = 0$ because A and Ac are orthogonal, therefore $x_2 = x_1$. The iteration can, of course, be continued without changing the result. Note also that $Ac = Ax$, where x is the original image!

We will stress the important fact that as the attractor and the collage are equivalent in the noniterative case, we have direct control of the attractor, unlike any other fractal coding method.

Experimental Comparisons with the Performance of MSGVQ

It is difficult to conclude from theory alone as to the performance of the attractor coder model. Experiments indicate, however, that for this particular fractal coder the performance is always worse than for the VQ with optimized codebook for all images tested [23]. The adaptivity of the *self codebook*, does not seem to outcompete the VQ codebook which is optimal in a statistical sense.

52.5.3 Discussion

The above model is severely constrained through the required relation between the block size $(L \times L)$ and the decimation factor (also $L \times L$). Better coding results are obtained by using smaller decimation factors, typically 2×2.

Even with small decimation factors, no pure fractal coding technique has, in general, been shown to outperform vector quantization of similar complexity.

However, fractal methods have potential in hybrid block coding. It can efficiently represent edges and other deterministic structures where a shrunken version of another block is likely to resemble the block we are trying to represent. For instance, edges tend to be edges also after decimation. On the other hand, many textures can be hard to represent, as the decimation process requires that another texture with different frequency contents be present in the image to make a good approximation.

Using several block coding methods, where for each block the best method in a distortion-rate sense is selected, has been proven to give good coding performance [5, 10].

On the practical side, the fractal encoders have a very high complexity. Several methods have been suggested to alleviate this problem. These methods include limited search regions in the vicinity of the block to be coded, clustering of codebook vectors, and hierarchical search at different resolutions.

The iteration-free decoder is one of the fastest decoders obtainable for any coding method.

52.6 Color Coding

Any color image can be split in three color components and thereafter coded individually for each component. If this is done on the RGB (Red, Green, Blue) components, the bit rate tends to be approximately three times as high as for black and white images.

However, there are many other ways of decomposing the colors. The most used representations split the image in a *luminance* component and two *chrominance* components. Examples are so-called YUV and YIQ representations. One rationale for doing this kind of splitting is that the human visual system has different resolution for luminance and chrominance. The chrominance sampling can therefore be performed at a lower resolution, from two to eight times lower resolution depending on the desired quality and the interpolation method used to reconstruct the image. A second rationale is that the RGB components in most images are strongly correlated and therefore direct coding of the RGB components results in repeated coding of the same information. The luminance/chrominance representations try to decorrelate the components.

The transform between RGB and the luminance and chrominance components (YIQ) used in NTSC is given by

$$\begin{bmatrix} Y \\ I \\ Q \end{bmatrix} = \begin{bmatrix} 0.299 & 0.587 & 0.114 \\ 0.596 & -0.274 & -0.322 \\ 0.058 & -0.523 & 0.896 \end{bmatrix} \begin{bmatrix} R \\ G \\ B \end{bmatrix}. \tag{52.45}$$

There are only minor differences between the suggested color transforms. It is also possible to design the optimal decomposition based on the Karhunen-Loève transform. The method could be made adaptive by deriving a new transform for each image based on an estimated color correlation matrix.

We shall not go further into the color coding problem, but state that it is possible to represent color by adding 10 to 20% to the luminance component bit rate.

References

[1] Aase, S.O., *Image Subband Coding Artifacts: Analysis and Remedies*, Ph.D. thesis, The Norwegian Institute of Technology, Norway, March 1993.

[2] Arrowwood, Jr., J.L. and Smith, M.J.T., Exact reconstruction analysis/synthesis filter banks with time varying filters, in *Proc. Int. Conf. on Acoustics, Speech, and Signal Proc. (ICASSP)*, Minneapolis, MN, 3, 233–236, April 1993.

[3] Balasingham, I., Fuldseth, A. and Ramstad, T. A., On optimal tiling of the spectrum in subband image compression, in *Proc. Int. Conf. on Image Processing (ICIP)*, 1997.

[4] Balasingham, I. and Ramstad, T.A., On the optimality of tree-structured filter banks in subband image compression, *IEEE Trans. Signal Processing*, 1997, (submitted).

[5] Barthel, K.U., Schüttemeyer, J., Voyé T. and Noll, P., A new image coding technique unifying fractal and transform coding, in *Proc. Int. Conf. on Image Processing (ICIP)*, Nov. 1994.

[6] Fischer, T.R., A pyramid vector quantizer, *IEEE Trans. Inform. Theory*, IT-32:568–583, July 1986.

[7] Fischer, T.R. and Mercellin, M.W., Joint trellis coded quantization/modulation, *IEEE Trans. Commun.*, 39(2):172–176, Feb. 1991.

[8] Fisher, Y. (Ed.), *Fractal Image Compression. Theory and Applications*, Springer-Verlag, 1995.

[9] Gersho, A. and Gray, R.M., *Vector Quantization and Signal Compression*, Kluwer Academic Publishers, Boston, MA, 1992.

[10] Gharavi-Alkhansari, M., Fractal image coding using rate-distortion optimized matching pursuit, in *Proc. SPIE's Visual Communications and Image Processing*, 2727,1386–1393, March 1996.

[11] Herley, C., Kovacevic, J., Ramchandran, K. and Vetterli, M., Tilings of the time-frequency plane: Construction of arbitrary orthogonal bases and fast tiling transforms, *IEEE Trans. Signal Processing*, 41(12),3341–3359, Dec. 1993.

[12] Huffman, D.A., A method for the construction of minimum redundancy codes, *Proc. IRE*, 40(9),1098–1101, Sept. 1952.

[13] Hung, A.C., *PVRG-JPEG Codec 1.2.1*, Portable Video Research Group, Stanford University, Boston, MA, 1993.

[14] ISO/IEC IS 10918-1, *Digital Compression and Coding of Continuous-Tone Still Images, Part 1: Requirements and Guidelines,* JPEG.

[15] ISO/IEC IS 11172, *Information Technology-Coding of Moving Pictures and Associated Audio for Digital Storage Up to about 1.5 Mbit/s,* MPEG-1.

[16] ISO/IEC IS 13818, *Information Technology – Generic Coding of Moving Pictures and Associated Audio Information,* MPEG-2.

[17] ITU-T (CCITT), *Video Codec for Audiovisual Services at p × 64 kbit/s,* Geneva, Italy, Aug. 1990, Recommendation H.261.

[18] ITU-T (CCITT), *Video Coding for Low Bitrate Communication,* May, 1996. Draft Recommendation H.263.

[19] Jacquin, A., Fractal image coding: A review, *Proc. IEEE,* 81(10):1451–1465, Oct. 1993.

[20] Jacquin, A., Fractal image coding based on a theory of iterated contractive transformations, in *Proc. SPIE's Visual Communications and Image Processing,* 227–239, Oct. 1990.

[21] Jayant, N.S. and Noll, P., *Digital Coding of Waveforms, Principles and Applications to Speech and Video,* Prentice-Hall, Englewood Cliffs, NJ, 1984.

[22] Joshi, R.L., *Subband Image Coding Using Classification and Trellis Coded Quantization,* Ph.D. thesis, Washington State University, Aug. 1996.

[23] Lepsøy, S., *Attractor Image Compression – Fast Algorithms and Comparisons to Related Techniques,* Ph.D. thesis, The Norwegian Institute of Technology, Norway, June 1993.

[24] Lervik, J.M., *Subband Image Communication over Digital Transparent and Analog Waveform Channels,* Ph.D. thesis, Norwegian University of Science and Technology, Dec. 1996.

[25] Linde, Y., Buzo, A. and Gray, R.M., An algorithm for vector quantizer design, *IEEE Trans. Commun.,* COM-28(1),84–95, Jan. 1980.

[26] Luenbereger, D.G., *Optimization by Vector Space Methods,* John Wiley & Sons, New York, 1979.

[27] Lundheim, L., *Fractal Signal Modelling for Source Coding,* Ph.D. thesis, The Norwegian Institute of Technology, Norway, Sept. 1992.

[28] Makhoul, J., Roucos, S. and Gish, H., Vector quantization in speech coding, in *Proc. IEEE,* 1551–1587, Nov. 1985.

[29] Marcellin, M.W. and Fischer, T.R., Trellis coded quantization of memoryless and Gauss-Markov sources, *IEEE Trans. Commun.,* 38(1):82–93, Jan. 1990.

[30] Martucci, S., Signal extension and noncausal filtering for subband coding of images, in *Proc. SPIE's Visual Communications and Image Processing,* 137–148, Nov. 1991.

[31] Øien, G.E., *L2-Optimal Attractor Image Coding with Fast Decoder Convergence,* Ph.D. thesis, The Norwegian Institute of Technology, Norway, June 1993.

[32] Popat, K., Scalar quantization with arithmetic coding, M.Sc. thesis, Massachusetts Institute of Technology, Cambridge, MA, June 1990.

[33] Ramstad, T.A., Aase, S.O. and Husøy, J.H., *Subband Compression of Images — Principles and Examples,* Elsevier Science Publishers BV, North Holland, 1995.

[34] Said, A. and Pearlman, W. A., A new, fast, and efficient image codec based on set partitioning in hierarchical trees, *IEEE Trans. Circuits, Syst. for Video Technol.,* 6(3):243–250, June 1996.

[35] Shapiro, J.M., Embedded image coding using zerotrees of wavelets coefficients, *IEEE Trans. Signal Processing,* 41,3445–3462, Dec. 1993.

[36] Vaidyanathan, P.P., *Multirate Systems and Filter Banks,* Prentice-Hall, Englewood Cliffs, NJ, 1993.

[37] Wallace, G.K., Overview of the JPEG (ISO/CCITT) still image compression standard, in *Proc. SPIE's Visual Communications and Image Processing,* 1989.

[38] Antonini, M., Barland, M., Mathieu, P., and Daubechies, I., Image coding using wavelet transform, *IEEE Trans. Image Processing,* 1, 205–220, Apr. 1992.

53

Image and Video Restoration

A. Murat Tekalp
University of Rochester

53.1 Introduction

Digital images and video, acquired by still cameras, consumer camcorders, or even broadcast-quality video cameras, are usually degraded by some amount of blur and noise. In addition, most electronic cameras have limited spatial resolution determined by the characteristics of the sensor array. Common causes of blur are out-of-focus, relative motion, and atmospheric turbulence. Noise sources include film grain, thermal, electronic, and quantization noise. Further, many image sensors and media have known nonlinear input-output characteristics which can be represented as point nonlinearities. The goal of image and video (image sequence) restoration is to estimate each image (frame or field) as it would appear without any degradations, by first modeling the degradation process, and then applying an inverse procedure. This is distinct from image enhancement techniques which are designed to manipulate an image in order to produce more pleasing results to an observer without making use of particular degradation models. On the other hand, superresolution refers to estimating an image at a resolution higher than that of the imaging sensor. Image sequence filtering (restoration and superresolution) becomes especially important when still images from video are desired. This is because the blur and noise can become rather objectionable when observing a "freeze-frame", although they may not be visible to the human eye at the usual frame rates. Since many video signals encountered in practice are interlaced, we address the cases of both progressive and interlaced video.

0-8493-8572-5/98/$0.00+$.50

The problem of image restoration has sparked widespread interest in the signal processing community over the past 20 or 30 years. Because image restoration is essentially an ill-posed inverse problem which is also frequently encountered in various other disciplines such as geophysics, astronomy, medical imaging, and computer vision, the literature that is related to image restoration is abundant. A concise discussion of early results can be found in the books by Andrews and Hunt [1] and Gonzalez and Woods [2]. More recent developments are summarized in the book by Katsaggelos [3], and review papers by Meinel [4], Demoment [5], Sezan and Tekalp [6], and Kaufman and Tekalp [7]. Most recently, printing high-quality still images from video sources has become an important application for multi-frame restoration and superresolution methods. An in-depth coverage of video filtering methods can be found in the book *Digital Video Processing* by Tekalp [8]. This chapter summarizes key results in digital image and video restoration.

53.2 Modeling

Every image restoration/superresolution algorithm is based on an observation model, which relates the observed degraded image(s) to the desired "ideal" image, and possibly a regularization model, which conveys the available *a priori* information about the ideal image. The success of image restoration and/or superresolution depends on how good the assumed mathematical models fit the actual application.

53.2.1 Intra-Frame Observation Model

Let the observed and ideal images be sampled on the same 2-D lattice Λ. Then, the observed blurred and noisy image can be modeled as

$$g = s(Df) + v \tag{53.1}$$

where g, f, and v denote vectors representing lexicographical ordering of the samples of the observed image, ideal image, and a particular realization of the additive (random) noise process, respectively. The operator D is called the blur operator. The response of the image sensor to light intensity is represented by the memoryless mapping $s(\cdot)$, which is, in general, nonlinear. (This nonlinearity has often been ignored in the literature for algorithm development.)

The blur may be space-invariant or space-variant. For space-invariant blurs, D becomes a convolution operator, which has block-Toeplitz structure; and Eq. (53.1) can be expressed, in scalar form, as

$$g(n_1, n_2) = s\left(\sum_{(m_1, m_2) \in \mathcal{S}_d} d(m_1, m_2) f(n_1 - m_1, n_2 - m_2) \right) + v(n_1, n_2) \tag{53.2}$$

where $d(m_1, m_2)$ and $\mathcal{S}_d$ denote the kernel and support of the operator D, respectively. The kernel $d(m_1, m_2)$ is the impulse response of the blurring system, often called the point spread function (PSF). In case of space-variant blurs, the operator D does not have a particular structure; and the observation equation can be expressed as a superposition summation

$$g(n_1, n_2) = s\left(\sum_{(m_1, m_2) \in \mathcal{S}_d(n_1, n_2)} d(n_1, n_2; m_1, m_2) f(m_1, m_2) \right) + v(n_1, n_2) \tag{53.3}$$

where $\mathcal{S}_d(n_1, n_2)$ denotes the support of the PSF at the pixel location (n_1, n_2).

The noise is usually approximated by a zero-mean, white Gaussian random field which is additive and independent of the image signal. In fact, it has been generally accepted that more sophisticated noise models do not, in general, lead to significantly improved restorations.

53.2.2 Multispectral Observation Model

Multispectral images refer to image data with multiple spectral bands that exhibit inter-band correlations. An important class of multispectral images are color images with three spectral bands. Suppose we have K spectral bands, each blurred by possibly a different PSF. Then, the vector-matrix model (53.1) can be extended to multispectral modeling as

$$g = \mathcal{D}f + v \tag{53.4}$$

where

$$g \doteq \begin{bmatrix} g_1 \\ \vdots \\ g_K \end{bmatrix}, \quad f \doteq \begin{bmatrix} f_1 \\ \vdots \\ f_K \end{bmatrix}, \quad v \doteq \begin{bmatrix} v_1 \\ \vdots \\ v_K \end{bmatrix}$$

denote $N^2 K \times 1$ vectors representing the multispectral observed, ideal, and noise data, respectively, stacked as composite vectors, and

$$\mathcal{D} \doteq \begin{bmatrix} D_{11} & \cdots & D_{1K} \\ \vdots & \ddots & \vdots \\ D_{K1} & \cdots & D_{KK} \end{bmatrix}$$

is an $N^2 K \times N^2 K$ matrix representing the multispectral blur operator. In most applications, $\mathcal{D}$ is block diagonal, indicating no inter-band blurring.

53.2.3 Multiframe Observation Model

Suppose a sequence of blurred and noisy images $g_k(n_1, n_2)$, $k = 1, \ldots, L$, corresponding to multiple shots (from different angles) of a static scene sampled on a 2-D lattice or frames (fields) of video sampled (at different times) on a 3-D progressive (interlaced) lattice, is available. Then, we may be able to estimate a higher-resolution "ideal" still image $f(m_1, m_2)$ (corresponding to one of the observed frames) sampled on a lattice, which has a higher sampling density than that of the input lattice. The main distinction between the multispectral and multiframe observation models is that here the observed images are subject to sub-pixel shifts (motion), possibly space-varying, which makes high-resolution reconstruction possible. In the case of video, we may also model blurring due to motion within the aperture time to further sharpen images.

To this effect, each observed image (frame or field) can be related to the desired high-resolution ideal still-image through the superposition summation [8]

$$g_k(n_1, n_2) = s \left(\sum_{(m_1, m_2) \in S_d(n_1, n_2; k)} d_k(n_1, n_2; m_1, m_2) f(m_1, m_2) \right) + v_k(n_1, n_2) \tag{53.5}$$

where the support of the summation over the high-resolution grid (m_1, m_2) at a particular observed pixel $(n_1, n_2; k)$ depends on the motion trajectory connecting the pixel $(n_1, n_2; k)$ to the ideal image, the size of the support of the low-resolution sensor PSF $h_a(x_1, x_2)$ with respect to the high resolution grid, and whether there is additional optical (out-of-focus, motion, etc.) blur. Because the relative positions of low- and high-resolution pixels in general vary by spatial coordinates, the discrete sensor PSF is space-varying. The support of the space-varying PSF is indicated by the shaded area in Fig. 53.1, where the rectangle depicted by solid lines shows the support of a low-resolution pixel over the high-resolution sensor array. The shaded region corresponds to the area swept by the low-resolution pixel due to motion during the aperture time [8].

Note that the model (53.5) is invalid in case of occlusion. That is, each observed pixel $(n_1, n_2; k)$ can be expressed as a linear combination of several desired high-resolution pixels (m_1, m_2), provided that $(n_1, n_2; k)$ is connected to (m_1, m_2) by a motion trajectory. We assume that occlusion regions can be detected *a priori* using a proper motion estimation/segmentation algorithm.

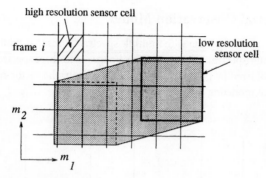

high resolution sensor cell

frame *i*

low resolution
sensor cell

m_2

m_1

FIGURE 53.1 Illustration of the discrete system PSF.

53.2.4 Regularization Models

Restoration is an ill-posed problem which can be regularized by modeling certain aspects of the desired "ideal" image. Images can be modeled as either 2-D deterministic sequences or random fields. *A priori* information about the ideal image can then be used to define hard or soft constraints on the solution. In the deterministic case, images are usually assumed to be members of an appropriate Hilbert space, such as a Euclidean space with the usual inner product and norm. For example, in the context of set theoretic restoration, the solution can be restricted to be a member of a set consisting of all images satisfying a certain smoothness criterion [9]. On the other hand, constrained least squares (CLS) and Tikhonov-Miller regularization use quadratic functionals to impose smoothness constraints in an optimization framework.

In the random case, models have been developed for the pdf of the ideal image in the context of maximum *a posteriori* (MAP) image restoration. For example, Trussell and Hunt [10] have proposed a Gaussian distribution with space-varying mean and stationary covariance as a model for the pdf of the image. Geman and Geman [11] proposed a Gibbs distribution to model the pdf of the image. Alternatively, if the image is assumed to be a realization of a homogeneous Gauss-Markov random process, then it can be statistically modeled through an autoregressive (AR) difference equation [12]

$$f(n_1, n_2) = \sum_{(m_1, m_2) \in \mathcal{S}_c} c(m_1, m_2) f(n_1 - m_1, n_2 - m_2) + w(n_1, n_2) \tag{53.6}$$

where $\{c(m_1, m_2) : (m_1, m_2) \in \mathcal{S}_c\}$ denote the model coefficients, $\mathcal{S}_c$ is the model support (which may be causal, semi-causal, or non-causal), and $w(n_1, n_2)$ represents the modeling error which is Gaussian distributed. The model coefficients can be determined such that the modeling error has minimum variance [12]. Extensions of (53.6) to inhomogeneous Gauss-Markov fields was proposed by Jeng and Woods [13].

53.3 Model Parameter Estimation

In this section, we discuss methods for estimating the parameters that are involved in the observation and regularization models for subsequent use in the restoration algorithms.

53.3.1 Blur Identification

Blur identification refers to estimation of both the support and parameters of the PSF $\{d(n_1, n_2) : (n_1, n_2) \in \mathcal{S}_d\}$. It is a crucial element of image restoration because the quality of restored images is highly sensitive to errors in the PSF [14]. An early approach to blur identification has been based on the assumption that the original scene contains an ideal point source, and that its spread (hence the PSF) can be determined from the observed image. Rosenfeld and Kak [15] show that the PSF can also be determined from an ideal line source. These approaches are of limited use in practice because a scene, in general, does not contain an ideal point or line source and the observation noise may not allow the measurement of a useful spread.

Models for certain types of PSF can be derived using principles of optics, if the source of the blur is known [7]. For example, out-of-focus and motion blur PSF can be parameterized with a few parameters. Further, they are completely characterized by their zeros in the frequency-domain. Power spectrum and cepstrum (Fourier transform of the logarithm of the power spectrum) analysis methods have been successfully applied in many cases to identify the location of these zero-crossings [16, 17]. Alternatively, Chang et al. [18] proposed a bispectrum analysis method, which is motivated by the fact that bispectrum is not affected, in principle, by the observation noise. However, the bispectral method requires much more data than the method based on the power spectrum. Note that PSFs, which do not have zero crossings in the frequency domain (e.g., Gaussian PSF modeling atmospheric turbulence), cannot be identified by these techniques.

Yet another approach for blur identification is the maximum likelihood (ML) estimation approach. The ML approach aims to find those parameter values (including, in principle, the observation noise variance) that have most likely resulted in the observed image(s). Different implementations of the ML image and blur identification are discussed under a unifying framework [19]. Pavlović and Tekalp [20] propose a practical method to find the ML estimates of the parameters of a PSF based on a continuous domain image formation model.

In multi-frame image restoration, blur identification using more than one frame at a time becomes possible. For example, the PSF of a possibly space-varying motion blur can be computed at each pixel from an estimate of the frame-to-frame motion vector at that pixel, provided that the shutter speed of the camera is known [21].

53.3.2 Estimation of Regularization Parameters

Regularization model parameters aim to strike a balance between the fidelity of the restored image to the observed data and its smoothness. Various methods exist to identify regularization parameters, such as parametric pdf models, parametric smoothness constraints, and AR image models. Some restoration methods require the knowledge of the power spectrum of the ideal image, which can be estimated, for example, from an AR model of the image. The AR parameters can, in turn, be estimated from the observed image by a least squares [22] or an ML technique [63]. On the other hand, non-parametric spectral estimation is also possible through the application of periodogram-based methods to a prototype image [69, 23]. In the context of maximum *a posteriori* (MAP) methods, the *a priori* pdf is often modeled by a parametric pdf, such as a Gaussian [10] or a Gibbsian [11]. Standard methods for estimating these parameters do not exist. Methods for estimating the regularization parameter in the CLS, Tikhonov-Miller, and related formulations are discussed in [24].

53.3.3 Estimation of the Noise Variance

Almost all restoration algorithms assume that the observation noise is a zero-mean, white random process that is uncorrelated with the image. Then, the noise field is completely characterized by its variance, which is commonly estimated by the sample variance computed over a low-contrast local region of the observed image. As we will see in the following section, the noise variance plays an important role in defining constraints used in some of the restoration algorithms.

53.4 Intra-Frame Restoration

We start by first looking at some basic regularized restoration strategies, in the case of an LSI blur model with no pointwise nonlinearity. The effect of the nonlinear mapping $s(.)$ is discussed in Section 53.4.2. Methods that allow PSFs with a random components are summarized in Section 53.4.3. Adaptive restoration for ringing suppression and blind restoration are covered in Sections 53.4.4 and 53.4.5, respectively. Restoration of multispectral images and space-varying blurred images are addressed in Sections 53.4.6 and 53.4.7, respectively.

53.4.1 Basic Regularized Restoration Methods

When the mapping $s(.)$ is ignored, it is evident from Eq. (53.1) that image restoration reduces to solving a set of simultaneous linear equations. If the matrix D is nonsingular (i.e., D^{-1} exists) and the vector g lies in the column space of D (i.e., there is no observation noise), then there exists a unique solution which can be found by direct inversion (also known as inverse filtering). In practice, however, we almost always have an underdetermined (due to boundary truncation problem [14]) and inconsistent (due to observation noise) set of equations. In this case, we resort to a minimum-norm least-squares solution. A least squares (LS) solution (not unique when the columns of D are linearly dependent) minimizes the norm-square of the residual

$$J_{LS}(f) \doteq \|g - Df\|^2 \tag{53.7}$$

LS solution(s) with the minimum norm (energy) is (are) generally known as pseudo-inverse solution(s) (PIS).

Restoration by pseudo-inversion is often ill-posed owing to the presence of observation noise [14]. This follows because the pseudo-inverse operator usually has some very large eigenvalues. For example, a typical blur transfer function has zeros; and thus, its pseudo-inverse attains very large magnitudes near these singularities as well as at high frequencies. This results in excessive amplification at these frequencies in the sensor noise. Regularized inversion techniques attempt to roll-off the transfer function of the pseudo-inverse filter at these frequencies to limit noise amplification. It follows that the regularized inverse deviates from the pseudo-inverse at these frequencies which leads to other types of artifacts, generally known as regularization artifacts [14]. Various strategies for regularized inversion (and how to achieve the right amount of regularization) are discussed in the following.

Singular-Value Decomposition Method

The pseudo-inverse D^+ can be computed using the singular value decomposition (SVD) [1]

$$D^+ = \sum_{i=0}^{R} \lambda_i^{-1/2} z_i u_i^T \tag{53.8}$$

where λ_i denote the singular values, z_i and u_i are the eigenvectors of $D^T D$ and DD^T, respectively, and R is the rank of D. Clearly, reciprocation of zero singular-values is avoided since the summation runs to R, the rank of D. Under the assumption that D is block-circulant (corresponding to a circular convolution), the PIS computed through Eq. (53.8) is equivalent to the frequency domain pseudo-inverse filtering

$$D^+(u, v) = \begin{cases} 1/D(u, v) & \text{if } D(u, v) \neq 0 \\ 0 & \text{if } D(u, v) = 0 \end{cases} \tag{53.9}$$

where $D(u, v)$ denotes the frequency response of the blur. This is because a block-circulant matrix can be diagonalized by a 2-D discrete Fourier transformation (DFT) [2].

Regularization of the PIS can then be achieved by truncating the singular value expansion (53.8) to eliminate all terms corresponding to small λ_i (which are responsible for the noise amplification) at the expense of reduced resolution. Truncation strategies are generally ad-hoc in the absence of additional information.

Iterative Methods (Landweber Iterations)

Several image restoration algorithms are based on variations of the so-called Landweber iterations [25, 26, 27, 28, 31, 32]

$$f_{k+1} = f_k + RD^T (g - Df_k) \tag{53.10}$$

where R is a matrix that controls the rate of convergence of the iterations. There is no general way to select the best C matrix. If the system (53.1) is nonsingular and consistent (hardly ever the case), the iterations (53.10) will converge to the solution. If, on the other hand, (53.1) is underdetermined and/or inconsistent, then (53.10) converges to a minimum-norm least squares solution (PIS). The theory of this and other closely related algorithms are discussed by Sanz and Huang [26] and Tom et al. [27]. Kawata and Ichioka [28] are among the first to apply the Landweber-type iterations to image restoration, which they refer to as "reblurring" method.

Landweber-type iterative restoration methods can be regularized by appropriately terminating the iterations before convergence, since the closer we are to the pseudo-inverse, the more noise amplification we have. A termination rule can be defined on the basis of the norm of the residual image signal [29]. Alternatively, soft and/or hard constraints can be incorporated into iterations to achieve regularization. The constrained iterations can be written as [30, 31]

$$f_{k+1} = C \left[f_k + RD^T (g - Df_k) \right] \tag{53.11}$$

where C is a nonexpansive constraint operator, i.e., $||C(f_1) - C(f_2)|| \leq ||f_1 - f_2||$, to guarantee the convergence of the iterations. Application of Eq. (53.11) to image restoration has been extensively studied (see [31, 32] and the references therein).

Constrained Least Squares Method

Regularized image restoration can be formulated as a constrained optimization problem, where a functional $||Q(f)||^2$ of the image is minimized subject to the constraint $||g - Df||^2 = \sigma^2$. Here σ^2 is a constant, which is usually set equal to the variance of the observation noise. The constrained least squares (CLS) estimate minimizes the Lagrangian [34]

$$J_{CLS}(f) = ||Q(f)||^2 + \alpha \left(||g - Df||^2 - \sigma^2 \right) \tag{53.12}$$

where α is the Lagrange multiplier. The operator Q is chosen such that the minimization of Eq. (53.12) enforces some desired property of the ideal image. For instance, if Q is selected as the Laplacian operator, smoothness of the restored image is enforced. The CLS estimate can be expressed, by taking the derivative of Eq. (53.12) and setting it equal to zero, as [1]

$$\hat{f} = \left(D^H D + \gamma Q^H Q \right)^{-1} D^H g \tag{53.13}$$

where ^{H} stands for Hermitian (i.e., complex-conjugate and transpose). The parameter $\gamma = \frac{1}{\alpha}$ (the regularization parameter) must be such that the constraint $||g - Df||^2 = \sigma^2$ is satisfied. It is often computed iteratively [2]. A sufficient condition for the uniqueness of the CLS solution is that Q^{-1} exists. For space-invariant blurs, the CLS solution can be expressed in the frequency domain as [34]

$$\hat{F}(u, v) = \frac{D^*(u, v)}{|D(u, v)|^2 + \gamma |L(u, v)|^2} G(u, v) \tag{53.14}$$

where * denotes complex conjugation. A closely related regularization method is the Tikhonov-Miller (T-M) regularization [33, 35]. T-M regularization has been applied to image restoration [31, 32, 36]. Recently, neural network structures implementing the CLS or T-M image restoration have also been proposed [37, 38].

Linear Minimum Mean Square Error Method

The linear minimum mean square error (LMMSE) method finds the linear estimate which minimizes the mean square error between the estimate and ideal image, using up to second order statistics of the ideal image. Assuming that the ideal image can be modeled by a zero-mean homogeneous random field and the blur is space-invariant, the LMMSE (Wiener) estimate, in the frequency domain, is given by [8]

$$\hat{F}(u, v) = \frac{D^*(u, v)}{|D(u, v)|^2 + \sigma_v^2/|P(u, v)|^2} G(u, v) \tag{53.15}$$

where σ_v^2 is the variance of the observation noise (assumed white) and $|P(u, v)|^2$ stands for the power spectrum of the ideal image. The power spectrum of the ideal image is usually estimated from a prototype. It can be easily seen that the CLS estimate (53.14) reduces to the Wiener estimate by setting $|L(u, v)|^2 = \sigma_v^2/|P(u, v)|^2$ and $\gamma = 1$.

A Kalman filter determines the causal (up to a fixed lag) LMMSE estimate recursively. It is based on a state-space representation of the image and observation models. In the first step of Kalman filtering, a prediction of the present state is formed using an autoregressive (AR) image model and the previous state of the system. In the second step, the predictions are updated on the basis of the observed image data to form the estimate of the present state. Woods and Ingle [39] applied 2-D reduced-update Kalman filter (RUKF) to image restoration, where the update is limited to only those state variables in a neighborhood of the present pixel. The main assumption here is that a pixel is insignificantly correlated with pixels outside a certain neighborhood about itself. More recently, a reduced-order model Kalman filtering (ROMKF), where the state vector is truncated to a size that is on the order of the image model support has been proposed [40]. Other Kalman filtering formulations, including higher-dimensional state-space models to reduce the effective size of the state vector, have been reviewed in [7]. The complexity of higher-dimensional state-space model based formulations, however, limits their practical use.

Maximum *A posteriori* Probability Method

The maximum *a posteriori* probability (MAP) restoration maximizes the *a posteriori* probability density function (pdf) $p(f|g)$, i.e., the likelihood of a realization of f being the ideal image given the observed data g. Through the application of the Bayes rule, we have

$$p(f|g) \propto p(g|f)p(f) \tag{53.16}$$

where $p(g|f)$ is the conditional pdf of g given f (related to the pdf of the noise process) and $p(f)$ is the *a priori* pdf of the ideal image. We usually assume that the observation noise is Gaussian, leading to

$$p(g|f) = \frac{1}{(2\pi)^{N/2} |R_v|^{1/2}} \exp\left\{-1/2 \left(g - Df\right)^T R_v^{-1} \left(g - Df\right)\right\} \tag{53.17}$$

where R_v denotes the covariance matrix of the noise process. Unlike the LMMSE method, the MAP method uses complete pdf information. However, if both the image and noise are assumed to be homogeneous Gaussian random fields, the MAP estimate reduces to the LMMSE estimate, under a linear observation model.

Trussell and Hunt [10] used non-stationary *a priori* pdf models, and proposed a modified form of the Picard iteration to solve the nonlinear maximization problem. They suggested using the variance of the residual signal as a criterion for convergence. Geman and Geman [11] proposed using a Gibbs random field model for the *a priori* pdf of the ideal image. They used simulated annealing procedures to maximize Eq. (53.16). It should be noted that the MAP procedures usually require significantly more computation compared to, for example, the CLS or Wiener solutions.

Maximum Entropy Method

A number of maximum entropy (ME) approaches have been discussed in the literature, which vary in the way that the ME principle is implemented. A common feature of all these approaches, however, is their computational complexity. Maximizing the entropy enforces smoothness of the restored image. (In the absence of constraints, the entropy is highest for a constant-valued image). One important aspect of the ME approach is that the nonnegativity constraint is implicitly imposed on the solution because the entropy is defined in terms of the logarithm of the intensity.

Frieden was the first to apply the ME principle to image restoration [41]. In his formulation, the sum of the entropy of the image and noise, given by

$$J_{ME1}(f) = -\sum_i f(i) \ln f(i) - \sum_i n(i) \ln n(i) \tag{53.18}$$

is maximized subject to the constraints

$$n = g - Df \tag{53.19}$$

$$\sum_i f(i) = K \doteq \sum_i g(i) \tag{53.20}$$

which enforce fidelity to the data and a constant sum of pixel intensities. This approach requires the solution of a system of nonlinear equations. The number of equations and unknowns are on the order of the number of pixels in the image. The formulation proposed by Gull and Daniell [42] can be viewed as another form of Tikhonov regularization (or constrained least squares formulation), where the entropy of the image

$$J_{ME2}(f) = -\sum_i f(i) \ln f(i) \tag{53.21}$$

is the regularization functional. It is maximized subject to the following usual constraints

$$||g - Df||^2 = \sigma_v^2 \tag{53.22}$$

$$\sum_i f(i) = K \doteq \sum_i g(i) \tag{53.23}$$

on the restored image. The optimization problem is solved using an ascent algorithm. Trussell [43] showed that in the case of a prior distribution defined in terms of the image entropy, the MAP solution is identical to the solution obtained by this ME formulation. Other ME formulations were also proposed [44, 45]. Note that all ME methods are nonlinear in nature.

Set-Theoretic Methods

In set-theoretic methods, first a number of "constraint sets" are defined such that their members are consistent with the observations and/or some *a priori* information about the ideal image. A set-theoretic estimate of the ideal image is then defined as a feasible solution satisfying all constraints, i.e., any member of the intersection of the constraint sets. Note that set-theoretic methods are, in general, nonlinear.

Set-theoretic methods vary according to the mathematical properties of the constraint sets. In the method of projections onto convex sets (POCS), the constraint sets C_i are closed and convex in an appropriate Hilbert space $\mathcal{H}$. Given the sets $C_i, i = 1, \ldots, M$, and their respective projection operators P_i, a feasible solution is found by performing successive projections as

$$f_{k+1} = P_M P_{M-1} \ldots P_1 f_k; \quad k = 0, 1, \ldots \tag{53.24}$$

where f_0 is the initial estimate (a point in $\mathcal{H}$). The projection operators are usually found by solving constrained optimization problems. In finite-dimensional problems (which is the case for digital image restoration), the iterations converge to a feasible solution in the intersection set [46, 47, 48]. It should be noted that the convergence point is affected by the choice of the initialization. However, as the size of the intersection set becomes smaller, the differences between the convergence points obtained by different initializations become smaller. Trussell and Civanlar [49] applied POCS to image restoration. For examples of convex constraint sets that are used in image restoration, see [23]. A relationship between the POCS and Landweber iterations were developed in [10].

A special case of POCS is the Gerchberg-Papoulis type algorithms where the constraint sets are either linear subspaces or linear varieties [50]. Extensions of POCS to the case of nonintersecting sets [51] and nonconvex sets [52] have been discussed in the literature. Another extension is the method of fuzzy sets (FS), where the constraints are defined in terms of FS. More precisely, the constraints are reflected in the membership functions defining the FS. In this case, a feasible solution is defined as one that has a high grade of membership (e.g., above a certain threshold) in the intersection set. The method of FS has also been applied to image restoration [53].

53.4.2 Restoration of Images Recorded by Nonlinear Sensors

Image sensors and media may have nonlinear characteristics that can be modeled by a pointwise (memoryless) nonlinearity $s(.)$. Common examples are photographic film and paper, where the nonlinear relationship between the exposure (intensity) and the silver density deposited on the film or paper is specified by a "$d - \log e$" curve. The modeling of sensor nonlinearities was first addressed by Andrews and Hunt [1]. However, it was not generally recognized that results obtained by taking the sensor nonlinearity into account may be far more superior to those obtained by ignoring the sensor nonlinearity, until the experimental work of Tekalp and Pavlović [54, 55].

Except for the MAP approach, none of the algorithms discussed above are equipped to handle sensor nonlinearity in a straightforward fashion. A simple approach would be to expand the observation model with $s(.)$ into its Taylor series about the mean of the observed image and obtain an approximate (linearized) model, which can be used with any of the above methods [1]. However, the results do not show significant improvement over those obtained by ignoring the nonlinearity. The MAP method is capable of taking the sensor nonlinearity into account directly. A modified Picard iteration was proposed in [10], assuming both the image and noise are Gaussian distributed, which is given by

$$\hat{f}_{k+1} = \bar{f}_k + R_f D^T S_b R_n^{-1} \left[g - s\left(D f_k \right) \right] \tag{53.25}$$

where $\bar{f}$ denotes non-stationary image mean, R_f and R_n are the correlation matrices of the ideal image and noise, respectively, and S_b is a diagonal matrix consisting of the derivatives of $s(.)$ evaluated at $b = Df$. It is the matrix S_b that maps the difference $[g - s(Df_k)]$ from the observation domain to the intensity domain.

An alternative approach, which is computationally less demanding, transforms the observed density domain image to the exposure domain [54]. There is a convolutional relationship between the ideal and blurred images in the exposure domain. However, the additive noise in the density domain manifests itself as multiplicative noise in the exposure domain. To this effect, Tekalp and Pavlović [54] derive an LMMSE deconvolution filter in the presence of multiplicative noise under certain assumptions. Their results show that accounting for the sensor nonlinearity may dramatically improve restoration results [54, 55].

53.4.3   Restoration of Images Degraded by Random Blurs

Basic regularized restoration methods (reviewed in Section 53.4.1) assume that the blur PSF is a deterministic function. A more realistic model may be

$$D = \bar{D} + \Delta D \tag{53.26}$$

where $\bar{D}$ is the deterministic part (known or estimated) of the blur operator and ΔD stands for the random component. Random component may represent inherent random fluctuations in the PSF, for instance due to atmospheric turbulence or random relative motion, or it may model the PSF estimation error.

A naive approach would be to employ the expected value of the blur operator in one of the restoration algorithms discussed above. The resulting restoration, however, may be unsatisfactory. Slepian [56] derived the LMMSE estimate, which explicitly incorporated the random component of the PSF. The resulting Wiener filter requires the *a priori* knowledge of the second order statistics of the blur process. Ward et al. [57, 58] also proposed LMMSE estimators. Combettes and Trussell [59] addressed restoration of random blurs within the framework of POCS, where fluctuations in the PSF are reflected in the bounds defining the residual constraint sets. The method of total least squares (TLS) has been used in the mathematics literature to solve a set of linear equations with uncertainties in the system matrix. The TLS method amounts to finding the minimum perturbations on D and g to make the system of equations consistent. A variation of this principle has been applied to image restoration with random PSF by Mesarovic et al. [60]. Various authors have shown that modeling the uncertainty in the PSF (by means of a random component) reduces ringing artifacts that are due to using erroneous PSF estimates.

53.4.4   Adaptive Restoration for Ringing Reduction

Linear space-invariant (LSI) restoration methods introduce disturbing ringing artifacts which originate around sharp edges and image borders [36]. A quantitative analysis of the origins and characteristics of ringing and other restoration artifacts was given by Tekalp and Sezan [14]. Suppression of ringing may be possible by means of adaptive filtering, which tracks edges or image statistics such as local mean and variance.

Iterative and set-theoretic methods are well-suited for adaptive image restoration with ringing reduction. Lagendijk et al. [36] have extended Miller regularization to adaptive restoration by defining the solution in a weighted Hilbert space, in terms of norms weighted by space-variant weights. Later, Sezan and Tekalp [9] extended the method of POCS to the space-variant case by introducing a region-based bound on the signal energy. In both methods, the weights and/or the regions were identified from the degraded image. Recently, Sezan and Trussell [23] have developed constraints based on prototype images for set-theoretic image restoration with artifact reduction.

Kalman filtering can also be extended to adaptive image restoration. For a typical image, the homogeneity assumption will hold only over small regions. Rajala and de Figueiredo [61] used an off-line visibility function to segment the image according to the local spatial activity of the picture being restored. Later, a rapid edge adaptive filter based on multiple image models to account for edges with various orientations was developed by Tekalp et al. [62]. Jeng and Woods [13] developed inhomogeneous Gauss-Markov field models for adaptive filtering, and maximum entropy methods were used for ringing reduction [45]. Results show a significant reduction in ringing artifacts in comparison to LSI restoration.

53.4.5   Blind Restoration (Deconvolution)

Blind restoration refers to methods that do not require prior identification of the blur and regularization model parameters. Two examples are simultaneous identification and restoration of noisy blurred images [63] and image recovery from Fourier phase information [64]. Lagendijk et al. [63] applied the E-M algorithm to blind image restoration, which alternates between ML parameter identification and minimum mean square error image restoration. Chen et al. [64] employed the POCS method to estimate the Fourier magnitude of the ideal image from the Fourier phase of the observed blurred image by assuming a zero-phase blur PSF so that the Fourier phase of the observed image is undistorted. Both methods require the PSF to be real and symmetric.

53.4.6 Restoration of Multispectral Images

A trivial solution to multispectral image restoration, when there is no inter-band blurring, may be to ignore the spectral correlations among different bands and restore each band independently, using one of the algorithms discussed above. However, algorithms that are optimal for single-band imagery may no longer be so when applied to individual spectral bands. For example, restoration of the red, green, and blue bands of a color image independently usually results in objectionable color shift artifacts.

To this effect, Hunt and Kubler [65] proposed employing the Karhunen-Loeve (KL) transform to decorrelate the spectral bands so that an independent-band processing approach can be applied. However, because the KL transform is image dependent, they then recommended using the NTSC YIQ transformation as a suboptimum but easy-to-use alternative. Experimental evidence shows that the visual quality of restorations obtained in the KL, YIQ, or another luminance-chrominance domain are quite similar [65]. In fact, restoration of only the luminance channel suffices in most cases. This method applies only when there is no inter-band blurring. Further, one should realize that the observation noise becomes correlated with the image under a non-orthogonal transformation. Thus, filtering based on the assumption that the image and noise are uncorrelated is not theoretically founded in the YIQ domain.

Recent efforts in multispectral image restoration are concentrated on making total use of the inherent correlations between the bands [66, 67]. Applying the CLS filter expression (53.13) to the observation model (53.4) with $Q^H Q = \mathcal{R}_f^{-1} \mathcal{R}_v$, we obtain the multispectral Wiener estimate $\hat{f}$, given by [68]

$$\hat{f} = \left(\mathcal{D}^T \mathcal{D} + \mathcal{R}_f^{-1} \mathcal{R}_v \right)^{-1} \mathcal{D}^T g \tag{53.27}$$

where

$$\mathcal{R}_f \doteq \begin{bmatrix} \mathbf{R}_{f;11} & \cdots & \mathbf{R}_{f;1K} \\ \vdots & \ddots & \vdots \\ \mathbf{R}_{f;K1} & \cdots & \mathbf{R}_{f;KK} \end{bmatrix}, \text{ and } \mathcal{R}_v \doteq \begin{bmatrix} \mathbf{R}_{v;11} & \cdots & \mathbf{R}_{v;1K} \\ \vdots & \ddots & \vdots \\ \mathbf{R}_{v;K1} & \cdots & \mathbf{R}_{v;KK} \end{bmatrix}$$

Here $\mathbf{R}_{f;ij} \doteq \mathcal{E}\{f_i f_j^T\}$ and $\mathbf{R}_{v;ij} \doteq \mathcal{E}\{v_i v_j^T\}$, $i, j = 1, 2, \ldots, K$ denote the inter-band, cross-correlation matrices. Note that if $\mathbf{R}_{f;ij} = 0$ for $i \neq j$, $i, j = 1, 2, \ldots, K$, then the multiframe estimate becomes equivalent to stacking the K single-frame estimates obtained independently.

Direct computation of $\hat{f}$ through Eq. (53.27) requires inversion of a $N^2 L \times N^2 L$ matrix. Because the blur PSF is not necessarily the same in each band and the inter-band correlations are not shift-invariant, the matrices $\mathcal{D}$, $\mathcal{R}_f$, and $\mathcal{R}_v$ are not block-Toeplitz; thus, a 3-D DFT would not diagonalize them. However, assuming LSI blurs, each $\mathbf{D}_k$ is block Toeplitz. Furthermore, assuming each image and noise band are wide-sense stationary, $\mathbf{R}_{f;ij}$ and $\mathbf{R}_{v;ij}$ are also block-Toeplitz. Approximating the block-Toeplitz submatrices $\mathbf{D}_i$, $\mathbf{R}_{f;ij}$, and $\mathbf{R}_{v;ij}$ by block-circulant ones, each submatrix can be diagonalized by a separate 2-D DFT operation so that we only need to invert a block matrix with diagonal sub-blocks. Galatsanos and Chin [66] proposed a method that successively partitions the matrix to be inverted and recursively computes the inverse of these partitions. Later Ozkan et al. [68] has shown that the desired inverse can be computed by inverting N^2 submatrices, each $K \times K$, in parallel. The resulting numerically stable filter was called the cross-correlated multiframe (CCMF) Wiener filter.

The multispectral Wiener filter requires the knowledge of the correlation matrices $\mathcal{R}_f$ and $\mathcal{R}_v$. If we assume that the noise is white and spectrally uncorrelated, the matrix $\mathcal{R}_v$ is diagonal with all diagonal entries equal to σ_v^2. Estimation of the multispectral correlation matrix $\mathcal{R}_f$ can be performed by either the periodogram method or 3-D AR modeling [68]. Sezan and Trussell [69] show that the multispectral Wiener filter is highly sensitive to the cross-power spectral estimates, which contain phase information. Other multispectral restoration methods include Kalman filtering approach of Tekalp and Pavlović [67], least squares approaches of Ohyama et al. [70] and Galatsanos et al. [71], and set-theoretic approach of Sezan and Trussell [23, 69] who proposed multispectral image constraints.

53.4.7 Restoration of Space-Varying Blurred Images

In principle, all basic regularization methods apply to the restoration of space-varying blurred images. However, because Fourier transforms cannot be utilized to simplify large matrix operations (such as inversion or singular value decomposition) when the blur is space-varying, implementation of some of these algorithms may be computationally formidable. There exist three distinct approaches to attack the space-variant restoration problem: (1) sectioning, (2) coordinate transformation, and (3) direct approaches.

The main assumption in sectioning is that the blur is approximately space-invariant over small regions. Therefore, a space-varying blurred image can be restored by applying the well-known space-invariant techniques to local image regions. Trussell and Hunt [73] propose using iterative MAP restoration within rectangular, overlapping regions. Later, Trussell and Fogel proposed using a modified Landweber iteration [21]. A major drawback of sectioning methods is generation of artifacts at the region boundaries. Overlapping the contiguous regions somewhat reduces these artifacts, but does not completely suppress them.

Most space-varying PSF vary continuously from pixel to pixel (e.g., relative motion with acceleration) violating the basic premise of the sectioning methods. To this effect, Robbins et al. [74] and then Sawchuck [75] proposed a coordinate transformation (CTR) method such that the blur PSF in the transformed coordinates is space-invariant. Then, the transformed image can be restored by a space-invariant filter and then transformed back to obtain the final restored image. However, the statistical properties of the image and noise processes are affected by the CTR, which should be taken into account in restoration filter design. The results reported in [74] and [75] have been obtained by inverse filtering; and thus, this statistical issue was of no concern. Also note that the CTR method is applicable to a limited class of space-varying blurs. For instance, blurring due to depth of field is not amenable to CTR.

The lack of generality of sectioning and CTR methods motivates direct approaches. Iterative schemes, Kalman filtering, and set-theoretic methods can be applied to restoration of space-varying blurs in a computationally feasible manner. Angel and Jain [76] propose solving the superposition Eq. (53.3) iteratively using a conjugate gradient method. Application of constrained iterative methods was discussed in [30]. More recently, Ozkan et al. [72] developed a robust POCS algorithm for space-varying image restoration, where they defined a closed, convex constraint set for each observed blurred image pixel (n_1, n_2), given by:

$$C_{n_1, n_2} = \left\{ \boldsymbol{y} : |r^{(\boldsymbol{y})}(n_1, n_2)| \leq \delta_0 \right\} \tag{53.28}$$

and

$$r^{(\boldsymbol{y})}(n_1, n_2) \doteq g(n_1, n_2) - \sum_{(m_1, m_2) \in \mathcal{S}_d(n_1, n_2)} d(n_1, n_2; m_1, m_2) \, y(m_1, m_2) \tag{53.29}$$

is the residual at pixel (n_1, n_2) associated with $\boldsymbol{y}$, which denotes an arbitrary member of the set. The quantity δ_0 is an *a priori* bound reflecting the statistical confidence with which the actual image is a member of the set C_{n_1, n_2}. Since $r^{(\boldsymbol{f})}(n_1, n_2) = v(n_1, n_2)$, the bound δ_0 is determined from the statistics of the noise process so that the ideal image is a member of the set within a certain statistical confidence. The collection of bounded residual constraints over all pixels (n_1, n_2) enforces the estimate to be consistent with the observed image.

The projection of an arbitrary $x(i_1, i_2)$ onto each C_{n_1, n_2} is defined as:

$$\mathbf{P}_{n_1, n_2}\left[x(i_1, i_2)\right] =$$

$$\begin{cases} x(i_1, i_2) + \dfrac{r^{(\boldsymbol{x})}(n_1, n_2) - \delta_0}{\sum_{o_1}\sum_{o_2} h^2(n_1, n_2; o_1, o_2)}\, h(n_1, n_2; i_1, i_2) & \text{if } r^{(\boldsymbol{x})}(n_1, n_2) > \delta_0 \\[4mm] x(i_1, i_2) & \text{if } -\delta_0 \leq r^{(\boldsymbol{x})}(n_1, n_2) \leq \delta_0 \\[4mm] x(i_1, i_2) + \dfrac{r^{(\boldsymbol{x})}(n_1, n_2) + \delta_0}{\sum_{o_1}\sum_{o_2} h^2(n_1, n_2; o_1, o_2)}\, h(n_1, n_2; i_1, i_2) & \text{if } r^{(\boldsymbol{x})}(n_1, n_2) < -\delta_0 \end{cases} \tag{53.30}$$

The algorithm starts with an arbitrary $x(i_1, i_2)$, and successively projects onto each C_{n_1, n_2}. This is repeated until convergence [72]. Additional constraints, such as bounded energy, amplitude, and limited support, can be utilized to improve the results.

53.5 Multiframe Restoration and Superresolution

Multiframe restoration refers to estimating the ideal image on a lattice that is identical with the observation lattice, whereas superresolution refers to estimating it on a lattice that has a higher sampling density than the observation lattice. They both employ the multiframe observation model (53.5), which establishes a relation between the ideal image and observations at more than one instance. Several authors eluded that the sequential nature of video sources can be statistically modeled by means of temporal correlations [68, 71]. Multichannel filters similar to those described for multispectral restoration were thus proposed for multiframe restoration. Here, we only review motion-compensated (MC) restoration and superresolution methods, because they are more effective.

53.5.1 Multiframe Restoration

The sequential nature of images in a video source can be used to better estimate the PSF parameters, regularization terms, and the restored image. For example, the extent of a motion blur can be estimated from interframe motion vectors, provided that the aperture time is known. The first MC approach was the motion-compensated multiframe Wiener filter (MCMF) proposed by Ozkan et al. [68] who considered the case of frame-to-frame global translations. Then, the auto power spectra of all frames are the same and the cross spectra are related by a phase factor which can be estimated from the motion information. Given the motion vectors (one for each frame) and the auto power spectrum of the reference frame, they derived a closed-form solution, given by

$$\hat{F}_k(u, v) = \frac{S_{f;k}(u, v)\displaystyle\sum_{i=1}^{N} S_{f;i}^*(u, v) D_i^*(u, v) G_i(u, v)}{\displaystyle\sum_{i=1}^{N} |S_{f;i}(u, v) D_i(u, v)|^2 + \sigma_v^2}, \tag{53.31}$$

where k is the index of the ideal frame to be restored, N is the number of available frames, and $P_{f;ki}(u, v) = S_{f;k}(u, v)S_{f;i}^*(u, v)$ denotes the cross power spectrum between the frames k and i in factored form. The fact that such a factorization exists was shown in [68] for the case of global translational motion. The MCMF yields the biggest improvement when the blur PSF changes from frame-to-frame. This is because the summation in the denominator may not be zero at any frequency, even though each term $D_i(u, v)$ may have zeros at certain frequencies. The case of space-varying blurs may be considered as a special case of the last section which covers superresolution with space-varying restoration.

53.5.2 Superresolution

When the interframe motion is subpixel, each frame, in fact, contains some "new" information that can be utilized to achieve superresolution. Superresolution refers to high-resolution image expansion, which aims to remove aliasing artifacts, blurring due to sensor PSF, and optical blurring given the observation model (53.5). Provided that enough frames with subpixel motion are available, the observation model becomes invertible. It can be easily seen, however, that superresolution from a single observed image is ill-posed because we have more unknowns than equations, and there exist infinitely many expanded images that are consistent with the model (53.5). Therefore, single-frame nonlinear interpolation (also called image expansion and digital zooming) methods for improved definition image expansion employ additional regularization criteria, such as edge-preserving smoothness constraints [77, 78]. (It is well-known that no new high-frequency information can be generated by LSI interpolation techniques, including ideal band-limited interpolation, hence the need for nonlinear methods.)

Several early motion-compensated methods are in the form of two-stage interpolation-restoration algorithms [79, 80]. They are based on the premise that pixels from all observed frames can be mapped back onto a desired frame, based on estimated motion trajectories, to obtain an upsampled reference frame. However, unless we assume global translational motion, the upsampled reference frame is nonuniformly sampled. In order to obtain a uniformly spaced upsampled image, interpolation onto a uniform sampling grid needs to be performed. Image restoration is subsequently applied to the upsampled image to remove the effect of the sensor blur. However, these methods do not use an accurate image formation model, and cannot remove aliasing artifacts.

Motion-compensated (multiframe) superresolution methods that are based on the model (53.5) can be classified as those that aim to eliminate (1) aliasing only, (2) aliasing and LSI blurs, and (3) aliasing and space-varying blurs. In addition, some of these methods are designed for global translational motion only, while others can handle space-varying motion fields with occlusion. Multiframe superresolution was first introduced by Tsai and Huang [81] who exploited the relationship between the continuous and discrete Fourier transforms of the undersampled frames to remove aliasing errors, in the special case of global motion. Their formulation has been extended by Kim et. al. [82] to take into account noise and blur in the low-resolution images, by posing the problem in the least squares sense. A further refinement by Kim and Su [83] allowed blurs that are different for each frame of low-resolution data, by using a Tikhonov regularization. However, the resulting algorithm did not treat the formation of blur due to motion or sensor size, and suffers from convergence problems.

Inspection of the model (53.5) suggests that the superresolution problem can be stated in the spatio-temporal domain as the solution of a set of simultaneous linear equations. Suppose that the desired high-resolution frames are $M \times M$, and we have L low-resolution observations, each $N \times N$. Then, from Eq. (53.5), we can set up at most $L \times N \times N$ equations in M^2 unknowns to reconstruct a particular high-resolution frame. These equations are linearly independent provided that all displacements between the successive frames are at subpixel amounts. (Clearly, the number of equations will be reduced by the number of occlusion labels encountered along the respective motion trajectories.) In general, it is desirable to set up an overdetermined system of equations, i.e., $L > R^2 = M^2/N^2$, to obtain a more robust solution in the presence of observation noise. Because the impulse response coefficients $h_{ik}(n_1, n_2; m_1, m_2)$ are spatially varying, and hence the system matrix is not block-Toeplitz, fast methods to solve them are not available. Stark and Oskui [86] proposed a POCS method to compute a high resolution image from observations obtained by translating and/or rotating an image with respect to a CCD array. Irani and Peleg [84, 85] employed iterative methods. Patti et al. [87] extended the POCS formulation to include sensor noise and space-varying blurs. Bayesian approaches were also employed for superresolution [88]. The extension of the POCS method with space-varying blurs is explained in the following.

53.5.3   Superresolution with Space-Varying Restoration

The POCS method described here addresses the most general form of the superresolution problem based on the model (53.5). The formulation is quite similar to the POCS approach presented for intraframe restoration of space-varying blurred images. In this case, we define a different closed, convex set for each observed low-resolution pixel (n_1, n_2, k) (which can be connected to the desired frame i by a motion trajectory) as

$$C_{n_1,n_2;i,k} = \left\{ x_i(m_1, m_2) : |r_k^{(\mathbf{X}_i)}(n_1, n_2)| \leq \delta_0 \right\}, \quad 0 \leq n_1, n_2 \leq N-1, \quad k = 1, \ldots, L \qquad (53.32)$$

where

$$r_k^{(\mathbf{X}_i)}(n_1, n_2) \doteq g_k(n_1, n_2) - \sum_{m_1=0}^{M-1} \sum_{m_2=0}^{M-1} x_i(m_1, m_2) h_{ik}(m_1, m_2; n_1, n_2)$$

and δ_0 represents the confidence that we have in the observation and is set equal to $c\sigma_v$, where σ_v is the standard deviation of the noise and $c \geq 0$ is determined by an appropriate statistical confidence bound. These sets define high-resolution images that are consistent with the observed low-resolution frames within a confidence bound that is proportional to the variance of the observation noise. The projection operator which projects onto $C_{n_1,n_2;i,k}$ can be deduced from Eq. (53.30) [8]. Additional constraints, such as amplitude and/or finite support constraints, can be utilized to improve the results. Excellent reconstructions have been reported using this procedure [68, 87].

A few observations about the POCS method are in order: (1) While certain similarities exist between the POCS iterations and the Landweber-type iterations [79, 84, 85], the POCS method can adapt to the amount of the observation noise, while the latter generally cannot. (2) The POCS method finds a feasible solution, that is, a solution consistent with all available low-resolution observations. Clearly, the more observations (more frames with reliable motion estimation) we have, the better the high-resolution reconstructed image $\hat{s}_i(m_1, m_2)$ will be. In general, it is desirable that $L > M^2/N^2$. Note, however, that the POCS method generates a reconstructed image with any number L of available frames. The number L is just an indicator of how large the feasible set of solutions will be. Of course, the size of the feasible set can be further reduced by employing other closed, convex constraints in the form of statistical or structural image models.

53.6   Conclusion

At present, factors that limit the success of digital image restoration technology include lack of reliable (1) methods for blur identification, especially identification of space-variant blurs, (2) methods to identify imaging system nonlinearities, and (3) methods to deal with the presence of artifacts in restored images. Our experience with the restoration of real-life blurred images indicates that the choice of a particular regularization strategy (filter) has a small effect on the quality of the restored images as long as the parameters of the degradation model, i.e., the blur PSF and the SNR, and any imaging system nonlinearity is properly compensated. Proper compensation of system nonlinearities also plays a significant role in blur identification.

References

[1] Andrews, H.C. and Hunt, B.R., *Digital Image Restoration*, Prentice-Hall, Englewood Cliffs, NJ, 1977.
[2] Gonzales, R.C. and Woods, R.E., *Digital Image Processing*, Addison-Wesley, MA, 1992.
[3] Katsaggelos, A.K., Ed., *Digital Image Restoration*, Springer-Verlag, Berlin, 1991.

[4] Meinel, E.S., Origins of linear and nonlinear recursive restoration algorithms, *J. Opt. Soc. Am.,* A-3(6), 787–799, 1986.

[5] Demoment, G., Image reconstruction and restoration: Overview of common estimation structures and problems, *IEEE Trans. Acoust. Speech Sign. Proc.,* 37, 2024-2036, 1989.

[6] Sezan, M.I. and Tekalp, A.M., Survey of recent developments in digital image restoration, *Optical Eng.,* 29, 393–404, 1990.

[7] Kaufman, H. and Tekalp, A.M., Survey of estimation techniques in image restoration, *IEEE Control Systems Magazine,* 11, 16–24, 1991.

[8] Tekalp, A.M., *Digital Video Processing,* Prentice-Hall, Englewood Cliffs, NJ, 1995.

[9] Sezan, M.I. and Tekalp, A.M., Adaptive image restoration with artifact suppression using the theory of convex projections, *IEEE Trans. Acoust. Speech Sig. Proc.,* 38(1), 181-185, 1990.

[10] Trussell, H.J. and Hunt, B.R., Improved methods of maximum a posteriori restoration, *IEEE Trans. Comput.,* C-27(1), 57–62, 1979.

[11] Geman, S. and Geman, D., Stochastic relaxation, Gibbs distributions, and the Bayesian restoration of images, *IEEE Trans. Pattern Anal. Machine Intell.,* 6(6), 721–741, 1984.

[12] Jain, A.K., Advances in mathematical models for image processing, *Proc. IEEE* 69(5), 502–528, 1981.

[13] Jeng, F.C. and Woods, J.W., Compound Gauss-Markov random fields for image restoration, *IEEE Trans. Sign. Proc.,* SP-39(3), 683–697, 1991.

[14] Tekalp, A.M. and Sezan, M.I., Quantitative analysis of artifacts in linear space-invariant image restoration, *Multidim. Syst. and Signal Proc.,* 1(1), 143–177, 1990.

[15] Rosenfeld, A. and Kak, A.C., *Digital Picture Processing,* Academic, New York, 1982.

[16] Gennery, D.B., Determination of optical transfer function by inspection of frequency-domain plot, *J. Opt. Soc. Am.,* 63(12), 1571–1577, 1973.

[17] Cannon, M., Blind deconvolution of spatially invariant image blurs with phase, *IEEE Trans. Acoust. Speech Sig. Proc.,* ASSP-24(1), 58–63, 1976.

[18] Chang, M.M., Tekalp, A.M. and Erdem, A.T., Blur identification using the bispectrum, *IEEE Trans. on Sign. Proc.,* ASSP-39(10), 2323–2325, 1991.

[19] Lagendijk, R.L., Tekalp, A.M. and Biemond, J., Maximum likelihood image and blur identification: A unifying approach, *Opt. Eng.,* 29(5), 422–435, 1990.

[20] Pavlović, G. and Tekalp, A.M., Maximum likelihood parametric blur identification based on a continuous spatial domain model, *IEEE Trans. Image Proc.,* 1(4), 496–504, 1992.

[21] Trussell, H.J. and Fogel, S., Identification and restoration of spatially variant motion blurs in sequential images, *IEEE Trans. Image Proc.,* 1(1), 123–126, 1992.

[22] Kaufman, H., Woods, J.W., Dravida, S. and Tekalp, A.M., Estimation and Identification of Two-Dimensional Images, *IEEE Trans. Aut. Cont.,* 28, 745–756, 1983.

[23] Sezan, M.I. and Trussell, H.J., Prototype image constraints for set-theoretic image restoration, *IEEE Trans. Sign. Proc.,* 39(10), 2275–2285, 1991.

[24] Galatasanos, N.P. and Katsaggelos, A.K., Methods for choosing the regularization parameter and estimating the noise variance in image restoration and their relation, *IEEE Trans. Image Proc.,* 1(3), 322–336, 1992.

[25] Trussell, H.J. and Civanlar, M.R., The Landweber iteration and projection onto convex sets, *IEEE Trans. Acoust. Speech Sig. Proc.,* ASSP-33(6), 1632–1634, 1985.

[26] Sanz, J.L.C. and Huang, T.S., Unified Hilbert space approach to iterative least-squares linear signal restoration, *J. Opt. Soc. Am.,* 73(11), 1455–1465, 1983.

[27] Tom, V.T., Quatieri, T.F., Hayes, M.H. and McClellan, J.H., Convergence of iterative nonexpansive signal reconstruction algorithms, *IEEE Trans. Acoust. Speech Sig. Proc.,* ASSP-29(5), 1052–1058, 1981.

[28] Kawata, S. and Ichioka, Y., Iterative image restoration for linearly degraded images. II. Reblurring, *J. Opt. Soc. Am.,* 70, 768–772, 1980.

[29] Trussell, H.J., Convergence criteria for iterative restoration methods, *IEEE Trans. Acoust. Speech Sig. Proc.*, ASSP-31(1), 129–136, 1983.

[30] Schafer, R.W., Mersereau, R.M. and Richards, M.A., Constrained iterative restoration algorithms, *Proc. IEEE*, 69(4), 432–450, 1981.

[31] Biemond, J., Lagendijk, R.L. and Mersereau, R.M., Iterative methods for image deblurring, *Proc. IEEE*, 78(5), 856–883, 1990.

[32] Katsaggelos, A.K., Iterative image restoration algorithms, *Opt. Eng.*, 28(7), 735–748, 1989.

[33] Tikhonov, A.N. and Arsenin, V.Y., *Solutions of Ill-Posed Problems*, V. H. Winston and Sons, Washington, D.C., 1977.

[34] Hunt, B.R., The application of constrained least squares estimation to image restoration by digital computer, *IEEE Trans. Comput.*, C-22(9), 805–812, 1973.

[35] Miller, K., Least squares method for ill-posed problems with a prescribed bound, *SIAM J. Math. Anal.*, 1, 52–74, 1970.

[36] Lagendijk, R.L., Biemond, J. and Boekee, D.E., Regularized iterative image restoration with ringing reduction, *IEEE Trans. Acoust. Speech Sig. Proc.*, 36(12), 1874–1888, 1988.

[37] Zhou, Y.T., Chellappa, R., Vaid, A. and Jenkins, B.K., Image restoration using a neural network, *IEEE Trans. Acoust. Speech Sig. Proc.*, ASSP-36(7), 1141-1151, 1988.

[38] Yeh, S.J., Stark H. and Sezan, M.I., Hopfield-type neural networks: their set-theoretic formulations as associative memories, classifiers, and their application to image restoration, in *Digital Image Restoration*, Katsaggelos, A. Ed., Springer Verlag, Berlin, 1991.

[39] Woods, J.W. and Ingle, V.K., Kalman filtering in two-dimensions-further results, *IEEE Trans. Acoust. Speech Sig. Proc.*, ASSP-29, 188–197, 1981.

[40] Angwin, D.L and Kaufman, H., Image restoration using reduced order models, *Sig. Processing*, 16, 21–28, 1988.

[41] Frieden, B.R., Restoring with maximum likelihood and maximum entropy, *J. Opt. Soc. Am.*, 62(4), 511–518, 1972.

[42] Gull, S.F. and Daniell, G.J., Image reconstruction from incomplete and noisy data, *Nature*, 272, 686–690, 1978.

[43] Trussell, H.J., The relationship between image restoration by the maximum *a posteriori* method and a maximum entropy method, *IEEE Trans. Acoust. Speech Sig. Proc.*, ASSP-28(1), 114–117, 1980.

[44] Burch, S.F., Gull, S.F. and Skilling, J., Image restoration by a powerful maximum entropy method, *Comp. Vis. Graph. Image Proc.*, 23, 113–128, 1983.

[45] Gonsalves, R.A. and Kao, H.-M., Entropy-based algorithm for reducing artifacts in image restoration, *Opt. Eng.*, 26(7), 617–622, 1987.

[46] Youla, D.C. and Webb, H., Image restoration by the method of convex projections: part 1 - theory, *IEEE Trans. Med. Imaging*, MI-1, 81–94, 1982.

[47] Sezan, M.I., An overview of convex projections theory and its applications to image recovery problems, *Ultramicroscopy*, 40, 55–67, 1992.

[48] Combettes, P.L., The foundations of set-theoretic estimation, *Proc. IEEE*, 81(2), 182–208, 1993.

[49] Trussell, H.J. and Civanlar, M.R., Feasible solution in signal restoration, *IEEE Trans. Acoust. Speech Sig. Proc.*, ASSP-32(4), 201-212, 1984.

[50] Youla, D.C., Generalized image restoration by the method of alternating orthogonal projections, *IEEE Trans. Circuits Syst.*, CAS-25(9), 694–702, 1978.

[51] Youla, D.C. and Velasco, V., Extensions of a result on the synthesis of signals in the presence of inconsistent constraints, *IEEE Trans. Circuits Syst.*, CAS-33(4), 465–467, 1986.

[52] Stark, H., Ed., *Image Recovery: Theory and Application*, Academic, Florida, 1987.

[53] Civanlar, M.R. and Trussell, H.J., Digital image restoration using fuzzy sets, *IEEE Trans. Acoust. Speech Sign. Proc.*, ASSP-34(8), 919-936, 1986.

[54] Tekalp, A.M. and Pavlović, G., Image restoration with multiplicative noise: Incorporating the sensor nonlinearity, *IEEE Trans. Sign. Proc.*, SP-39, 2132–2136, 1991.

[55] Tekalp, A.M. and Pavlović, G., Digital restoration of images scanned from photographic paper, *J. Electronic Imaging*, 2, 19–27, 1993.

[56] Slepian, D., Linear least squares filtering of distorted images, *J. Opt. Soc. Am.*, 57(7), 918–922, 1967.

[57] Ward, R.K. and Saleh, B.E.A., Deblurring random blur, *IEEE Trans. Acoust. Speech Sig. Proc.*, ASSP-35(10), 1494–1498, 1987.

[58] Quan, L. and Ward, R.K., Restoration of randomly blurred images by the Wiener filter, *IEEE Trans. Acoust. Speech Sig. Proc.*, ASSP-37(4), 589–592, 1989.

[59] Combettes, P.L. and Trussell, H.J., Methods for digital restoration of signals degraded by a stochastic impulse response, *IEEE Trans. Acoust. Speech Sig. Proc.*, ASSP-37(3), 393–401, 1989.

[60] Mesarovic, V.Z. Galatsanos, N.P., and Katsaggelos, A.K. Regularized constrained total least squares image restoration, *IEEE Trans. Image Proc.*, 4(8), 1096-1108, 1995.

[61] Rajala, S.A. and DeFigueiredo, R.P., Adaptive nonlinear image restoration by a modified Kalman filtering approach, *IEEE Trans. Acoust. Speech Sig. Proc.*, ASSP-29(5), 1033–1042, 1981.

[62] Tekalp, A.M., Kaufman, H. and Woods, J., Edge-adaptive Kalman filtering for image restoration with ringing suppression, *IEEE Trans. Acoust. Speech Sig. Proc.*, ASSP-37(6), 892-899, 1989.

[63] Lagendijk, R.L., Biemond, J. and Boekee, D.E., Identification and restoration of noisy blurred images using the expectation-maximization algorithm, *IEEE Trans. Acoust. Speech Sign. Proc.*, ASSP-38, 1180-1191, 1990.

[64] Chen, C.T., Sezan, M.I. and Tekalp, A.M., Effects of constraints, initialization, and finite-word length in blind deblurring of images by convex projections, *Proc. IEEE ICASSP'87*, Dallas, TX, 1201-1204, 1987.

[65] Hunt, B.R. and Kubler, O., Karhunen-Loeve multispectral image restoration, Part I: Theory, *IEEE Trans. Acoust. Speech Sig. Proc.*, ASSP-32(6), 592–599, 1984.

[66] Galatsanos, N.P. and Chin, R.T., Digital restoration of multi-channel images, *IEEE Trans. Acoust. Speech Sig. Proc.*, ASSP-37(3), 415–421, 1989.

[67] Tekalp, A.M. and Pavlović, G., Multichannel image modeling and Kalman filtering for multi-spectral image restoration, *Signal Process.*, 19, 221-232, 1990.

[68] Ozkan, M.K., Erdem, A.T., Sezan, M.I. and Tekalp, A.M., Efficient multiframe Wiener restoration of blurred and noisy image sequences, *IEEE Trans. Image Proc.*, 1(4), 453–476, 1992.

[69] Sezan, M.I. and Trussell, H.J., Use of *a priori* knowledge in multispectral image restoration, *Proc. IEEE ICASSP'89*, Glasgow, Scotland, 1429–1432, 1989.

[70] Ohyama, N., Yachida, M., Badique, E., Tsujiuchi, J. and Honda, T., Least-squares filter for color image restoration, *J. Opt. Soc. Am.*, 5, 19–24, 1988.

[71] Galatsanos, N.P., Katsaggelos, A.K., Chin, R.T. and Hillery, A.D., Least squares restoration of multichannel images, *IEEE Trans. Sign. Proc.*, SP-39(10), 2222–2236, 1991.

[72] Ozkan, M.K., Tekalp, A.M. and Sezan, M.I., POCS-based restoration of space-varying blurred images, *IEEE Trans. Image Proc.*, 3(3), 450–454, 1994.

[73] Trussell, H.J. and Hunt, B.R., Image restoration of space-variant blurs by sectioned methods, *IEEE Trans. Acoust. Speech Sig. Proc.*, ASSP-26(6) 608–609, 1978.

[74] Robbins, G.M. and Huang, T.S., Inverse filtering for linear shift-variant imaging systems, *Proc. IEEE*, 60(7), 1972.

[75] Sawchuck, A.A., Space-variant image restoration by coordinate transformations, *J. Opt. Soc. Am.*, 64(2), 138–144, 1974.

[76] Angel, E.S. and Jain, A.K., Restoration of images degraded by spatially varying point spread functions by a conjugate gradient method, *Appl. Opt.*, 17, 2186–2190, 1978.

[77] Wang, Y. and Mitra, S.K., Motion/pattern adaptive interpolation of interlaced video sequences, *Proc. IEEE ICASSP'91*, Toronto, Canada, 2829–2832, 1991.

[78] Schultz, R.R. and Stevenson, R.L., A Bayesian approach to image expansion for improved definition, *IEEE Trans. Image Proc.,* 3(3), 233–242, 1994.

[79] Komatsu, T., Igarashi, T., Aizawa, K. and Saito, T., Very high-resolution imaging scheme with multiple different aperture cameras, *Signal Proc.: Image Comm.,* 5, 511–526, 1993.

[80] Ur, H. and Gross, D., Improved resolution from subpixel shifted pictures, *CVGIP: Graphical Models and Image Processing,* 54(3), 181–186, 1992.

[81] Tsai, R.Y. and Huang, T.S., Multiframe image restoration and registration, in *Advances in Computer Vision and Image Processing,* vol. 1, Huang, T.S. Ed., Jai Press, Greenwich, CT, 1984, 317–339.

[82] Kim, S.P., Bose, N.K. and Valenzuela, H.M., Recursive reconstruction of high-resolution image from noisy undersampled frames, *IEEE Trans. Acoust., Speech and Sign. Proc.,* ASSP-38(6), 1013–1027, 1990.

[83] Kim, S.P. and Su, W.-Y., Recursive high-resolution reconstruction of blurred multiframe images, *IEEE Trans. Image Proc.,* 2(4), 534–539, 1993.

[84] Irani, M. and Peleg, S., Improving resolution by image registration, *CVGIP: Graphical Models and Image Proc.,* 53, 231–239, 1991.

[85] Irani, M. and Peleg, S., Motion analysis for image enhancement: Resolution, occlusion and transparency, *J. Vis. Comm. Image Rep.,* 4, 324–335, 1993.

[86] Stark, H. and Oskoui, P., High-resolution image recovery from image plane arrays using convex projections, *J. Opt. Soc. Am., A* 6, 1715–1726, 1989.

[87] Patti, A., Sezan, M.I. and Tekalp, A.M., Superresolution video reconstruction with arbitrary sampling lattices and nonzero aperture time, *IEEE Trans. Image Process.,* 6(8), 1064–1076, 1997.

[88] Schultz, R.R. and Stevenson, R.L., Extraction of high-resolution frames from video sequences, *IEEE Trans. Image Process.,* 5(6), 996–1011, 1996.

54

Video Scanning Format Conversion and Motion Estimation

Gerard de Haan
Philips Research Laboratories

54.1 Introduction

The scanning format of a video signal is a major determinant of general picture quality. Specifically, it determines such aspects as stationary and dynamic resolution, motion portrayal, aliasing, scanning structure visibility, and flicker. Various formats have been designed and standardized to strike a particular balance between quality, cost, transmission capacity, and compatibility with other standards.

The field of video scanning format conversion is concerned with the translation of video signals from one format into another. It consists of two basic parts: temporal interpolation and spatial interpolation. A particular case is de-interlacing, which poses an inseparable spatio-temporal interpolation problem.

Vertical and temporal interpolation cause practical and fundamental difficulties in achieving high-quality scanning format conversion. This is because the conditions of the sampling theorem are generally not met in video signals. If they were satisfied, standard conversions of arbitrary accuracy would be possible using suitable linear filters.

The earlier conversion methods neglected the fundamental problems and, consequently, negatively influenced the resolution and the motion portrayal. More recent algorithms apply motion vectors to predict the position of moving objects at unregistered temporal instances to improve the quality of the picture at the output format. A so-called *motion estimator* extracts these vectors from the input signal. The motion vectors partly solve the fundamental problems, but the demands on the motion estimator for scanning format conversion are severe.

0-8493-8572-5/98/$0.00+$.50

In this section we shall first briefly indicate why we can expect that the importance of scanning format conversion will grow. Then we discuss in more detail the fundamental problems of temporal interpolation of video signals. Next we provide a concise overview of the basic methods in scanning format conversion, focused on temporal sampling rate conversion and de-interlacing. Finally, we give an overview of motion estimation algorithms, which are crucial in the more advanced scanning format convertors.

54.2 Conversion vs. Standardization

Scanning formats have been designed in the past to strike a particular compromise between quality, cost, transmission capacity, and compatibility with other standards. There were three main formats in use a decade ago: 50 Hz interlaced, 60 Hz interlaced, and 24 (or 25) Hz progressive (film). With the arrival of video-conferencing, HDTV, workstations, and PCs, many new video formats have appeared. These include low end formats such as CIF and QCIF with smaller picture size and lower frame rates, progressive and interlaced HDTV formats at 50 Hz and 60 Hz, and other video formats used on computer workstations and enhanced television displays with field rates up to 100 Hz. It will be clear that the problem of scanning format conversion is of a growing importance, despite many attempts to globally standardize video formats.

54.3 Problems with Linear Sampling Rate Conversion Applied to Video Signals

High-quality scanning format conversion is difficult to achieve, as the conditions of the sampling theorem are generally not met in video signals. The solution of Sample Rate Conversion (SRC) for systems satisfying the conditions of the sampling theory is well known for arbitrary sampling ratios [1].

Figure 54.1 illustrates the procedure for a ratio of 2. To arrive at the double output sampling rate, in a first step, zero-valued samples are inserted between every input pair of samples. In a second step, a

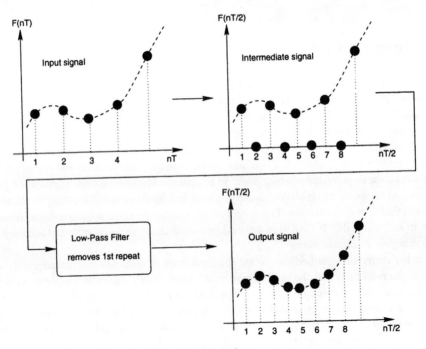

FIGURE 54.1   Consecutive steps in upsampling with a factor of two.

low-pass filter (LPF) at the output rate is applied to remove the first repeat spectrum from the input data. In case of a temporal SRC, the interpolating LPF has to be a temporal LPF, i.e., a filter including picture delays. Though feasible, this makes it a fairly expensive filter.

A more complicated, though still not fundamental, problem occurs at the signal acquisition stage. Since scenes do occur with almost unlimited spatial and/or temporal bandwidth, the sampling theorem requires that this signal be low-pass filtered prior to the scanning process. Interlaced scanning, as commonly applied, even demands two-dimensional prefiltering in the vertical-temporal frequency plane. In a video system, it is the camera that samples the scene in a vertical and temporal sense; therefore, the prefilter has to be realized in the optical path. Although there are considerable practical problems achieving this filtering, it would apparently bring down the problem of temporal interpolation of video images to the common sampling rate conversion problem. The next section will show, however, that in addition to the practical problems there is a fundamental problem as well.

54.3.1 Temporal Interpolation

Considering the eye's sine-wave temporal frequency response for full brightness potential and full field display [2], as shown in Fig. 54.2, temporal prefiltering with a bandwidth of 75 Hz at first sight seems sufficient. The fundamental problem now is that the relation shown in Fig. 54.2 holds for temporal

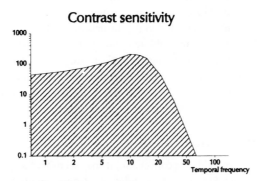

FIGURE 54.2   The contrast sensitivity of the human observer (y-axis) for large areas of uniform brightness, as a function of the temporal frequency (x-axis).

frequencies as they occur at the retina of the observer. These frequencies, however, equal the frequencies at the display only if the eye is stationary with respect to this display. Particularly with the eye tracking objects moving on the screen, this assumption is no longer valid. For a tracking observer very high temporal frequencies on the screen can be transformed to much lower frequencies or even DC at the retina. Consequently, suppression of these frequencies, with an interpolating lowpass filter, results in excessive blurring of moving objects as will be discussed next.

Figure 54.3 shows, in a time-discrete representation, a simple object, a square, moving with a constant velocity. Again, in this example, we consider up-sampling with a factor of two. Therefore, the true position of the object is available at every second temporal position only (e.g., the odd numbered samples). The "tracking observer" views along the motion trajectory, represented with a line in the illustration, which results in a stationary image of the object on the retina. If the output field sampling frequency exceeds the

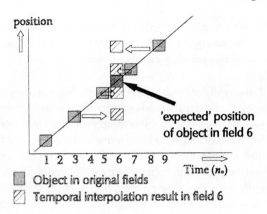

FIGURE 54.3 The effect of temporal interpolation for an object tracking observer. The field numbers are counted at the output field rate.

cutoff temporal frequency of the human visual system,[1] the viewer will have the illusion that the object is continuously present.

Therefore, the object is actually seen at a position corresponding with the motion trajectory. If now, e.g., in the 6th output field, the object is interpolated according to SRC theory, weighted copies of the object from surrounding fields resulting from the interpolating LPF are displayed. Figure 54.3 illustrates the case of a symmetrical transversal lowpass filter. In this situation, the viewer sees the object at the correct position but also various attenuated and displaced copies (the impulse response of the interpolating temporal filter) of the object in a neighborhood. The attenuation depends on the coefficients of the interpolating filter, and the distance between the copies is related to the displacement of the moving object in a field period. For the object-tracking observer, therefore, the temporal LPF is transformed into a spatial LPF. For an object velocity of one pixel per field period (one pel/field), its frequency characteristic equals the temporal frequency characteristic of the interpolating LPF.[2] 1 pel/field is a slow motion, as in broadcast picture material; velocities in a range exceeding 16 pel/field do occur. Thus, the spatial blur caused by the SRC process becomes unacceptable even for moderate object velocities.

54.3.2 Vertical Interpolation and Interlaced Scanning

Much similar to the situation of field rate conversion, it may seem that sequential scan conversion is an up-sampling problem for which SRC-theory provides an adequate solution. However, straightforward, one-dimensional, up-sampling in the vertical frequency domain is incorrect as the data is clearly sub-Nyquist sampled due to interlace.

If, more correctly, the sequential scan conversion is considered as a two-dimensional up-sampling problem in the vertical-temporal frequency domain, we arrive at a discussion similar to the one in Section 54.3.1: the problem cannot be solved as we do not know the temporal frequency at the retina of a movement-tracking observer. It is possible to disregard this problem and to perform a two-dimensional SRC, implicitly assuming a stationary viewer and prefiltered information. Such systems were described and have been implemented for studio applications. With the older image pick-up tubes the results can be satisfactory, as these devices have a poor dynamic resolution. When modern (CCD-)cameras are used, however, the limitations of the assumptions become obvious.

---

[1] Actually the picture update frequency may be even as low as 16 Hz, to guarantee smooth perceived motion (see, e.g., [3]). The higher display rates are merely necessary to prevent the annoying large area flicker.
[2] It is assumed here that both filters are normalized to their respective sampling frequency.

54.4　Alternatives for Sampling Rate Conversion Theory

With the problem of linear interpolation of video signals clarified, we will discuss alternative algorithms developed over time. These algorithms fall into two categories. A first category simplifies the interpolation filter prescribed by SRC-theory, considering that a completely correct solution is impossible anyway. The resulting **"simple algorithms"** are more attractive for hardware realization than the method from which they are derived and under certain conditions can perform quite similarly. The second category includes the most **"advanced algorithms"** for scanning format conversion. These methods can be characterized by their common attempt to interpolate the 3-D image data in the direction in which the correlation is highest. The difference between the various options lies mainly in the number of possible directions, and dimensions, which are considered. The implementation can show various linear interpolation filters controlled by one or more detectors, or a multi-dimensional nonlinear filter that has an inherent edge adaptivity. As this description allows a large number of algorithms, we will illustrate it with some important examples.

54.4.1　Simple Algorithms

SRC-theory in the temporal and vertical frequency domain is not applicable due to the missing prefilter in common video systems. A sophisticated linear interpolation filter therefore makes little sense. Any interpolating (spatio-)temporal low-pass filter will suppress original temporal frequency components as well as aliased signal components, as they occupy, by definition, the same spectrum. As the first effect is desired and the second not, the transfer function of the filter strikes a compromise between alias and blurring. Repetition of the most recent sample in this sense is optimal for the dynamic resolution and worst for alias. A strong temporal low-pass filter suppresses much (not necessarily all) alias and yields a poor dynamic resolution. The annoyance of the temporal alias depends on the input and output picture frequency, and particularly their difference. In the easiest case, both frequencies are high and their difference 50 Hz or more. In the worst case, input and output picture rate are low and their difference in the order of 10 Hz. In case of an annoying beat frequency, an interpolating LPF usually improves picture quality, otherwise the best compromise is closer to repetition of the most recent sample.

54.4.2　Advanced Algorithms

As indicated before, these methods are characterized by their common attempt to interpolate the 3-D image data in the direction in which the correlation is highest. To this end they either have an explicit or implicit detector to find this direction. In case of (1-D) temporal interpolation the explicit detector is usually called a **motion detector,** for 2-D spatial interpolation it is called an **edge detector,** while the most advanced device estimating the optimal spatio-temporal (3-D) interpolation direction is usually called a **motion estimator**. The interpolation filter can be recursive or transversal, and can have any number of taps, but a transversal filter with one or two taps is the most common choice. For a two taps FIR approach we can write the interpolated video signal F_{int}, in picture n, at spatial position $\underline{x} = (x, \ y)^T$ as a function of the input video signal $F(\underline{x}, \ n)$:

$$F_{\text{int}}(\underline{x}, \ n) = 0.5 \left[F\left(\underline{x} + \begin{pmatrix} \delta_1 \\ \delta_2 \end{pmatrix}, \ n + \delta_3 \right) + F\left(\underline{x} - \begin{pmatrix} \delta_1 \\ \delta_2 \end{pmatrix}, \ n - \delta_3 \right) \right] \qquad (54.1)$$

In this terminology a motion detector controls δ_3, an edge detector δ_1, and δ_2, while a motion estimator can be applied to determine δ_1, δ_2, and δ_3.

Algorithms with a Motion Detector

To detect motion, the difference between two successive pictures is calculated. It is too simple, however, to expect this signal to become zero in a picture part without moving objects. The common problems with the detection are noise and alias. Additional problems occurring in some systems are color subcarriers causing non-stationarities in colored regions, interlace causing nonstationarities in vertically detailed picture parts, and timing jitter of the sampling clock which is particularly harmful in detailed areas.

All these problems imply that the output of the motion detector usually is not a binary, but rather a multi-level signal, indicating the probability of motion. Usual (but not always valid) assumptions made to improve the detector are:

1. Noise is small and signal is large.
2. The spectrum part around the color carrier carries no motion information.
3. Low-frequency energy in the signal is larger than in the noise and alias.
4. Moving objects are large compared to a pixel.

The general structure of the motion detector resulting from these assumptions is depicted in Figure 54.4. As can be seen, the difference signal is first low-pass (and carrier reject) filtered to profit from (54.2)

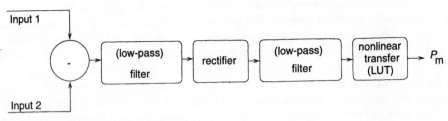

FIGURE 54.4   General structure of a motion detector.

and (54.3). It also makes the detector less "nervous" for timing jitter in detailed areas. After the rectification another low-pass filter improves the consistency of the motion signal, based on assumption (54.4). Finally, the nonlinear (but monotonous) transfer function in the last block translates the signal in a probability figure for the motion P_m, using (54.1). This last function may have to be adapted to the expected noise level. Low-pass filters are not necessarily linear. More than one detector can be used, working on more than just two pictures in the neighborhood of the current image, and a logical or linear combination of their outputs may lead to a more reliable indication of motion.

The motion detector (MD) is applied to switch or fade between two processing modes, one of which is optimal for stationary and the other for moving image parts. Examples are:

- De-interlacing. The MD fades between intra-field interpolation (line-averaging, or edge dependent spatial interpolation) and inter-field interpolation (repetition of the previous field, averaging of neighboring fields, etc.).
- Field rate doubling on interlaced video: The MD fades between repetition of fields (best dynamic resolution without motion compensation for moving picture parts) and repetition of frames (best spatial resolution in stationary image parts).

To slightly elaborate on the first example of de-interlacing, we define the interpolated pixel $X_m(\underline{x}, n)$ in a moving picture part as:

$$X_m\left(\underline{x},\, n\right) = 0.5 \left[F\left(\underline{x} - \begin{pmatrix} 0 \\ 1 \end{pmatrix},\, n\right) + F\left(\underline{x} + \begin{pmatrix} 0 \\ 1 \end{pmatrix},\, n\right) \right] \tag{54.2}$$

while for stationary picture parts the interpolated pixel $X_s(\underline{x}, n)$ is taken as:

$$X_s\left(\underline{x}, n\right) = F\left(\underline{x}, n-1\right) \tag{54.3}$$

and taking the probability of motion P_m, from the motion detector into account, the output is given by:

$$F_{\text{int}}\left(\underline{x}, n\right) = P_m X_m\left(\underline{x}, n\right) + (1 - P(m))X_s\left(\underline{x}, n\right) \tag{54.4}$$

In most practical cases the output P_m has a nonlinear relation with the actual probability.

Algorithms with an Edge Detector

To detect the orientation of a spatial edge, usually the differences between pairs of spatially neighboring pixels are calculated. Again it is a bit unrealistic to expect that a zero difference is a reliable indication of a spatial direction in which the signal is stationary. The same problems (noise, alias, carriers, timing-jitter) occur as with motion detection. The edge detector (ED) is applied to switch or fade between at least two but usually more processing modes, each of them optimal for interpolation of a certain orientation of the spatial edge. Examples are:

- De-interlacing. The ED fades between vertical line-averaging and diagonal averaging ($+/-45°$, or even more angles).
- Up-conversion to a higher resolution format. A simple bi-linear interpolation filter is applied with its coefficients adapted to the output of the edge detector.

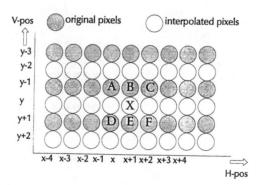

FIGURE 54.5 Identification of pixels as applied for direction dependent spatial interpolation.

In Fig. 54.5, X is the pixel to be interpolated for the sequential scan conversion and the result applying pixels in a neighborhood (A, B, C, D, E and F) is either $X_a, X_b,$ or X_c, where:

$$X_a = 0.5[A+F] = 0.5\left[F\left(\underline{x} - \begin{pmatrix} 1 \\ 1 \end{pmatrix}, n\right) + F\left(\underline{x} + \begin{pmatrix} 1 \\ 1 \end{pmatrix}, n\right)\right] \tag{54.5}$$

and:

$$X_b = 0.5[B+E] = 0.5\left[F\left(\underline{x} - \begin{pmatrix} 0 \\ 1 \end{pmatrix}, n\right) + F\left(\underline{x} + \begin{pmatrix} 0 \\ 1 \end{pmatrix}, n\right)\right] \tag{54.6}$$

and:

$$X_c = 0.5[C+D] = 0.5\left[F\left(\underline{x} + \begin{pmatrix} +1 \\ -1 \end{pmatrix}, n\right) + F\left(\underline{x} + \begin{pmatrix} -1 \\ +1 \end{pmatrix}, n\right)\right] \tag{54.7}$$

The selection of X_a, X_b, or X_c to the interpolated output F_{int} is controlled by a luminance gradient indication calculated from the same neighborhood:

$$F_{int}(\underline{x}, n) = \begin{cases} X_a, & (|A - F| < |C - D| \wedge |A - F| < |B - E|) \\ X_b, & (|B - E| \leq |A - F| \wedge |B - E| \leq |C - D|) \\ X_c, & (|C - D| < |A - F| \wedge |C - D| < |B - E|) \end{cases} \qquad (54.8)$$

In this example, the gradient is calculated on the same pixels that are used in the interpolation step. This is not necessarily the case. Similar to the earlier described motion detector, it is advantageous to filter the video signal prior to and/or after the rectification in Eq. (54.8). Also the decision, i.e., the optimal interpolation angle, can be low-pass filtered to improve the consistency of the interpolation angle. Finally, the edge dependent interpolation can be combined with (motion adaptive or motion compensated) temporal interpolation to improve the interpolation quality of near horizontal edges.

Implicit Detection in Nonlinear Interpolation Filters

Many nonlinear interpolation methods have been described. Most popular is the class of order statistical filters. Combinations with linear (bandsplitting) filters are known, optimizing the interpolation for individual spectrum parts. We will limit ourselves to some basic examples here.

An illustration of a basic inherently adapting filter is shown in Figure 54.6. The line to be interpolated

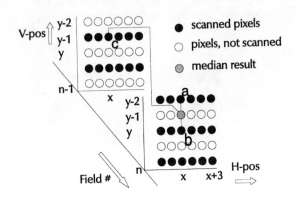

FIGURE 54.6 Sequential scan conversion with three-tap vertical-temporal median filtering. The thin lines show which pixels are input for the median filter.

is found as the median of the spatially neighboring lines (a and b) and the corresponding line (c) from the previous field:

$$F_{int}(\underline{x}, n) = \text{median}\,[a, b, c] =$$

$$\text{median}\left[F\left(\underline{x} + \begin{pmatrix} 0 \\ 1 \end{pmatrix}, n\right), F\left(\underline{x} - \begin{pmatrix} 0 \\ 1 \end{pmatrix}, n\right), F(\underline{x}, n-1) \right] \qquad (54.9)$$

with:

$$\text{median}\,(X, Y, Z) = \begin{cases} X, & (Y \leq X \leq Z \vee Z \leq X \leq Y) \\ Y, & (X < Y \leq Z \vee Z \leq Y < X) \\ Z, & (\text{otherwise}) \end{cases} \qquad (54.10)$$

The inherent adaptation to edges is understood as follows: In case of a temporal edge (i.e., motion) larger than the spatial edge (i.e., vertical detail), the difference between a and b is relatively small compared to their difference with c. Therefore, an intra-field interpolation results (a or b is copied). In case of a non-moving vertical edge, the difference between a and b will be relatively large compared to the difference between c and a or b. In this case, the inter-field interpolation (c is copied) is most likely.

It is possible to combine edge detectors with non-linear filters, e.g., a so-called weighted median filter. In a weighted median filter, the (integer) weight given to a sample indicates the number of times its value is included in the input of the filter to the ranking stage. An increase of this weight increases the chance this sample value is selected as the median. It therefore provides a method, using the output of an edge detector with uncertainties, to statistically improve the performance of the interpolation.

We will again use Fig. 54.5 to identify the location of the pixels used in the interpolation. The output value for the pixel position indicated with X results as:

$$F_{int}(\underline{x}, n) = \text{median}\left[A, B, C, D, E, F, \alpha \cdot X^{-1}, \beta \cdot \frac{B+E}{2}\right], \quad (\alpha, \beta \in \mathbb{N}) \tag{54.11}$$

with:

$$X^{-1} = F(\underline{x}, n-1), \quad A = F\left(\underline{x} - \begin{pmatrix} 1 \\ 1 \end{pmatrix}, n\right), \quad B = F\left(\underline{x} - \begin{pmatrix} 0 \\ 1 \end{pmatrix}, n\right), \dots\dots\dots \tag{54.12}$$

as illustrated in Fig. 54.5. The weighting (α and β) implies that an assumed "important" pixel is fed more than once to the median calculating circuit:

$$\alpha \cdot A = \frac{A, A, A \dots\dots\dots A, A}{\alpha \text{ times}} \tag{54.13}$$

The combination arises if a motion detector is used to control the weighting factors of the pixel from the previous field and that of the value found by line averaging. A large value of α increases the probability of field insertion, while a large β causes an increased probability of line averaging.

Although the examples in this section are limited to de-interlacing, it should be noted that proposals exist for field rate conversion as well.

Algorithms with a Motion Estimator

The idea to interpolate picture content in the direction in which it is most correlated can be extended to a three-dimensional case. This results in an interpolation along the motion trajectory. Figure 54.7 defines the motion trajectory as the line that connects identical picture parts in a sequence of pictures. The projection of this motion trajectory between two successive pictures on the image plane, called the motion vector, is also shown in this figure. Not all temporal information changes can be described adequately as object velocities: e.g., fades and concealed or obscured background. Nevertheless, this method has the strongest physical background, as due to their inertia it always takes time for objects to completely disappear, or change geometry, resulting in a strong correlation of successive images after compensation for motion. This is in contrast to spatial (edge adaptive) interpolation for which there is a statistical but no physical background.

Knowledge of motion vectors allows us to interpolate image data at any temporal instance between two successive pictures. The most common form uses motion compensated averaging according to:

$$\begin{aligned} F_{int}(\underline{x}, n+\alpha) = \ & 1/2 \cdot \big[\, F\left(\underline{x} - \alpha\underline{D}(\underline{x}, n), n\right) \\ & + F\left(\underline{x} + (1-\alpha)\underline{D}(\underline{x}, n), n+1\right)\big], \quad (0 \le \alpha \le 1) \end{aligned} \tag{54.14}$$

where $\underline{D}(\underline{x}, n)$ is the object displacement at position $\underline{x} = (x, y)^T$ estimated between fields n and $n + 1$, while α determines the temporal instance for which the interpolated data has to be valid. However, all previously mentioned interpolation methods that involve a temporal component can be used as a basis

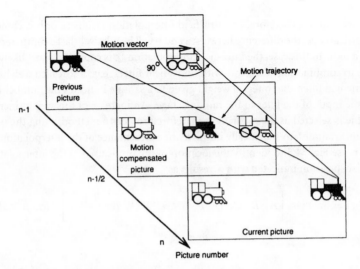

FIGURE 54.7   Identical picture parts of successive images lie on the *motion trajectory*. Its projection in the image plane is the *motion vector*.

of a motion compensated interpolation. So linear, nonlinear, motion adaptive, edge adaptive, and inherently adapting interpolation methods can be upgraded toward their motion-compensated counterparts. Furthermore, bandsplitting can be used to sophisticate the interpolation.

We will not elaborate further on these methods as they follow straightforward from the earlier text. We will make an exception, however, for temporal interpolation on interlaced signals, as this poses non-trivial problems even with knowledge of local motion.

Motion Compensated De-Interlacing

In general, the pixels required for the motion compensated interpolation do not exist in the time discrete input signal, e.g., due to non-integer velocities. In the horizontal domain this problem can be solved with linear SRC-theory, but not in the vertical domain. Three solutions for this problem have been proposed:

1. Application of a generalized sampling theory (GST).
2. Straight extension of the motion vector into earlier pictures until it points (almost) to an existing pixel.
3. Recursive de-interlacing of the signal.

The implication of GST is that it is possible to perfectly reconstruct a signal sampled at $1/n$ times the Nyquist rate if n independent sets of samples describe the signal. The de-interlacing problem is a specific case for which $n = 2$. The required two sets are the current field and the motion compensated previous field, respectively. If the two do not coincide, i.e., the object does not have an odd integer vertical motion vector component, the independency constraint is fulfilled, and the problem can theoretically be solved. Practical problems are:

a. The velocity can have an odd vertical component.
b. Perfect reconstruction requires the use of pixels from many lines, for which the velocity need not be constant.
c. For nearly odd integer valued vertical velocities, noise may be enhanced.

Solution 2 is valid only if we assume the velocity constant over a larger temporal interval. This is a rather severe limitation which makes the method practically useless. Solution 3 is based on the assumption that it is possible at some time to have a perfectly de-interlaced picture in a memory. Once this is true, the picture is used to de-interlace the next input field. With motion compensation, this solution can be perfect as the de-interlaced picture in the memory allows the use of SRC-theory also in the vertical domain. If this new de-interlaced field is written in the memory, it can be used to de-interlace the next incoming field. Limitations of this method are:

a. Propagation of motion vector and interpolation errors.
b. Even a perfectly de-interlaced picture can contain alias in the vertical domain in the common case of a camera without an optical prefilter.

In practice, problem a is the more serious one, particularly for nearly odd vertical velocities.

Although there are restrictions, motion compensated interpolation techniques for field rate up-conversion and de-interlacing provide the most advanced option. However, they require nontrivial algorithms to measure object displacements between consecutive images. These motion estimation methods therefore shall be discussed more extensively in the next section.

54.5 Motion Estimation

This section provides an overview of motion estimation algorithms developed over time. The estimators applicable for scanning format conversion require additional constraints which are discussed in the last part of this section.

54.5.1 Pel-Recursive Estimators

The category of pel-recursive motion estimators can be derived from iterative methods that use a previously calculated motion vector $\underline{D}^{i-1}$ to find the result vector $\underline{D}^i$ according to:

$$\underline{D}^i = \underline{D}^{i-1} + \text{update} \tag{54.15}$$

Several algorithms based on iteration can be found in the literature. A common form applies iterative minimization of the squared value of the displaced frame difference (DFD) along the steepest gradient of the luminance function:

$$\underline{D}^i = \underline{D}^{i-1} - \frac{1}{2} \cdot \alpha \cdot \left(\begin{array}{c} \delta/\delta D_x^{i-1} \\ \delta/\delta D_y^{i-1} \end{array} \right) DFD^2 \left(\underline{x}, \underline{D}^{i-1}, n \right) \tag{54.16}$$

where the DFD is defined as:

$$DFD \left(\underline{x}, \underline{D}^{i-1}, n \right) = F \left(\underline{x}, n \right) - F \left(\underline{x} - \underline{D}^{i-1}, n - 1 \right) \tag{54.17}$$

and:

$$\underline{D}^i = \left(\begin{array}{c} D_x^i \\ D_y^i \end{array} \right) \tag{54.18}$$

As before, n stands for the field or picture number. The constant α is positive and determines the speed of convergence and the accuracy of the estimate. The value of α is limited to a maximum, since instability or a noisy estimation result can occur for higher values. Equation (54.16) can be rewritten as:

$$\underline{D}^i = \underline{D}^{i-1} - \alpha \cdot DFD \left(\underline{x}, \underline{D}^{i-1}, n \right) \cdot \left(\begin{array}{c} \delta/\delta x \\ \delta/\delta y \end{array} \right) F \left(\underline{x} - \underline{D}^{i-1}, n \right) \tag{54.19}$$

The method is known as "steepest descent algorithm". The updating process can be stopped after a fixed number of iterations, at the moment the update term falls under a threshold, or in case slow convergence or even divergence is detected. Rather than iterating the estimation process in a fixed position of the picture, the estimated result from a previously scanned position in the same picture can be used as the prediction for the present location. We shall then speak of a spatial recursive process, and if for every pixel an update is calculated, the name "pel-recursive motion estimation" is commonly used. The spatial prediction can be based on either a single previously calculated result, in which case the convergence shall be one-dimensional, or on a number of earlier calculated vectors. In case more than one vector is used, the design can select the best according to a criterion before or after updating, e.g., the smallest DFD or a weighted average can be calculated The coefficients that determine the weighting can be based on statistical properties of the vector field. Depending on the choice of the relative positions in the picture from which prediction vectors are taken, a one- or two-dimensional convergence can result. In the case of temporal recursion, a further refinement can be obtained by motion compensating the prediction values from the preceding field before weighting them with the values from the present field.

The algorithm can be improved by calculating the update term from a group of pixels rather than from only one pixel. This is then referred to as "gradient summed error algorithm":

$$\underline{D}^i = \underline{D}^{i-1} - \alpha \cdot \sum_{\underline{x} \in \text{ group}} \left[DFD\left(\underline{x},\, \underline{D}^{i-1},\, n\right) \cdot \left(\begin{array}{c} \delta/\delta x \\ \delta/\delta y \end{array} \right) F\left(\underline{x} - \underline{D}^{i-1},\, n\right) \right] \qquad (54.20)$$

Again the group can extend into a one-, two-, or three-dimensional neighborhood. Weighted averaging is an option and weights can be adapted to image statistics. In case of gradients taken from a temporally neighboring position, motion compensation can be applied prior to weighting with the spatial neighboring gradients.

Simplifications of the algorithm are possible. Particularly the prevention of multiplication is useful, and possible, e.g., by only using the sign of the gradient to determine the direction of the update with a fixed length.

In the literature, many variants of the steepest descent or gradient summed error algorithm are described, which mainly differ from the above-mentioned algorithms in that the convergence speed determining constant α is substituted by variables to adapt the estimator to local picture statistics.

54.5.2 Block-Matching Algorithm

In block-matching motion estimation algorithms, a displacement vector is assigned to the center $\underline{X}$ of a block of pixel positions $B(\underline{X})$ in the current field n by searching a similar block within a search area $SA(\underline{X})$, also centered at $\underline{X}$, but in the previous field $n - 1$. The similar block has a center that is shifted with respect to $\underline{X}$ over the displacement vector $\underline{D}(\underline{X},\, n)$. To find $\underline{D}(\underline{X},\, n)$, a number of candidate vectors $\underline{C}$ are evaluated applying an error measure $\in (\underline{C},\, \underline{X},\, n)$ to quantify block similarity. Figure 54.8 illustrates the procedure.

More formally, $CS^{\max}$ is defined as the set of candidate vectors $\underline{C}$, describing all possible displacements (integer on the pixel grid) with respect to $\underline{X}$ within the search area $SA(\underline{X})$ in the previous image:

$$CS^{\max} = \left\{ \underline{C} \mid -N \leq C_x \leq +N,\ -M \leq C_y \leq +M \right\} \qquad (54.21)$$

where N and M are constants limiting $SA(\underline{X})$. Furthermore, a block $B(\underline{X})$ centered at $\underline{X}$ and of size $X \times Y$ consisting of pixel positions $\underline{x}$ in the present field n, is now considered:

$$B(\underline{X}) = \left\{ \underline{x} \mid X_x - X/2 \leq x \leq X_x + X/2 \wedge X_y - Y/2 \leq y \leq X_y + Y/2 \right\} \qquad (54.22)$$

The displacement vector $\underline{D}(\underline{X},\, n)$ resulting from the block-matching process is a candidate vector $\underline{C}$ which yields the minimum value of an error function $\in (\underline{C},\, \underline{X},\, n)$:

$$\underline{D}(\underline{X},\, n) \in \left\{ \underline{C} \in CS^{\max} \mid \in (\underline{C},\, \underline{X},\, n) \leq \in (\underline{F},\, \underline{X},\, n)\, \forall \underline{F} \in CS^{\max} \right\} \qquad (54.23)$$

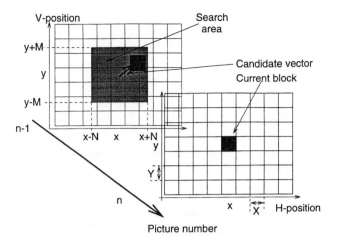

FIGURE 54.8 Block of size $X \times Y$ in current field n and trial block in search area $SA(\underline{X})$ in previous field $n - 1$, shifted over candidate vector $\underline{C}$.

If the vector $\underline{D}(\underline{X}, n)$ with the smallest matching error is assigned to all pixel positions $\underline{x}$ in the block $B(\underline{X})$:

$\forall \underline{x} \in B(\underline{X})$:

$$\underline{D}(\underline{x}, n) \in \left\{ \underline{C} \in CS^{\max} | \in (\underline{C}, \underline{X}, n) \leq \in (\underline{F}, \underline{X}, n) \forall \underline{F} \in CS^{\max} \right\} \tag{54.24}$$

rather than to the center pixel only, a large reduction of computations is achieved. As an implication, consecutive blocks $B(\underline{X})$ are not overlapping.

The error value for a given candidate vector $\underline{C}$ is a function (COST) of the luminance values of the pixels in the current block and those of the shifted block from a previous field, summed over the block $B(\underline{X})$:

$$\in (\underline{C}, \underline{X}, n) = \sum_{\underline{x} \in B(\underline{X})} \text{COST} \left[F(\underline{x}, n), F(\underline{x} - \underline{C}, n - p) \right] \tag{54.25}$$

A common choice for p is either 1 or 2, depending on whether the signal is interlaced or not.

Although the COST function itself can be rather straightforward and simple to implement, the high repetition factor for this calculation creates a huge burden. To save calculational effort in block-matching motion estimation algorithms, several methods have been published. The usual ingredients are:

1. The use of a simpler COST function.
2. Estimation on sub-sampled picture material.
3. Design of a clever search strategy, preventing that all possible vectors need to be checked.

Concerning option 1, there is almost general consensus. The most popular choice thus far for the error function is the Summed Absolute Difference (SAD) criterion:

$$\in (\underline{C}, \underline{X}, n) = SAD(\underline{C}, \underline{X}, n) = \sum_{\underline{x} \in B(\underline{X})} |F(\underline{x}, n), -F(\underline{x} - \underline{C}, n - p)| \tag{54.26}$$

Most important alternatives are the Mean Square Error (MSE), and the Normalized Cross Correlation Function (NCCF) criterion. The simpler error functions that have been designed will not be discussed here, as the economizing hardly ever justifies the performance loss.

Option 2 is straightforward and has little negative effect on the performance with sub-sampling factors up to four. Option 3 is the most effective, and will be dealt with separately in the next section.

54.5.3 Search Strategies

Sub-Sampled Full Search

In the most straightforward search strategy for all candidate vectors $\underline{C}$ in the search area, the matching error for a block $B(\underline{X})$ of pixel positions is calculated. The method is referred to as full search, exhaustive search, or brute force block-matching. To economize the calculational effort, the matching errors of only half of the possible result vectors $\underline{D}(\underline{X},\ n)$ can be calculated in a first step, using a first candidate set CS^1 which is a subset of CS^{max}. Figure 54.9 illustrates this option, further showing the candidate vectors in the second step of the algorithm.

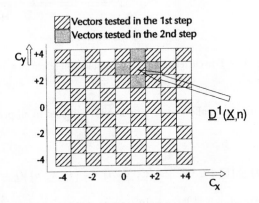

FIGURE 54.9 Candidate vectors tested in the second step around $\underline{D}^1(\underline{X},\ n)$ for sub-sampled full search block-matching. The grid shown is the pixel grid.

N-Step Search

The idea to adapt the search area from coarse to fine is not limited to a two-step process. As illustrated in Fig. 54.10, the first step of a three-step block-matcher performs a search on a coarse grid consisting of only nine candidate vectors in the entire search area. The second step includes a finer search, with eight candidate vectors around the best matching vector of the first step, and finally in the third step a search on

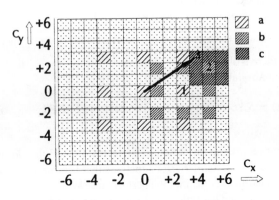

FIGURE 54.10 Illustration of the three-step search. Vectors resulting from the steps are indicated with the step number. The candidates in each step are shaded as in a, b, and c, respectively.

the full resolution grid is performed, with another eight candidates around the best vector of the second step. Note that a search range of $+/-6$ pels is assumed; other search areas require modifications, either resulting in less accurate vectors or in more consecutive steps. Generalizations to N-steps block-matching are obvious.

Related is the 2-D logarithmic, or cross-search, method that checks five vectors per step, one in the middle, four symmetrically around it (two with a different x-component and two with a different y-component). Again, four vectors are checked around the result, and the distance between the candidates is halved when the best matching vector is the middle one. Hence, the number of consecutive steps depends on the resulting vector, which is a drawback, as the hardware has to be designed for the worst case situation and cannot profit from a low average number of steps.

One-at-a-Time-Search

Yet a further reduction of candidate vectors can be realized if the two-dimensional optimization problem is split into two separate one-dimensional optimizations. The candidate set, for step i of the algorithm, $CS^i(\underline{X}, n)$, is adapted during the process, as in the previously discussed algorithms, but contains only three candidate vectors $\underline{C}(\underline{X}, n)$. Departing from vector $\underline{0}$, this method performs a search for the minimum error along the x-axis of the search area:

$$CS^i_x(\underline{X}, n) = \left\{ \underline{C} \mid \underline{C} = \underline{D}^{i-1}(\underline{X}, n) + \underline{U}, \ U_x = 0 \vee \pm 1, \ U_y = 0 \right\} \tag{54.27}$$

The procedure is repeated N times until $\underline{D}^N(\underline{X}, n) = \underline{D}^{N-1}(\underline{X}, n)$. From this minimum a search is started parallel to the y-axis and repeated M times until $\underline{D}^{M+N}(\underline{X}, n) = \underline{D}^{D+N-1}(\underline{X}, n)$.

In its simplest form, shown in Fig. 54.11, the process stops at this minimum and the estimated motion vector $\underline{D}(\underline{x}, n) = \underline{D}^{M+N}(\underline{X}, n)$ for all pixel positions $\underline{x}$ in $B(\underline{X})$. It is possible, however, to refine the result by repeating the OTS procedure, departing with every iteration from the previous result $\underline{D}^{M+N}(\underline{X}, n)$.

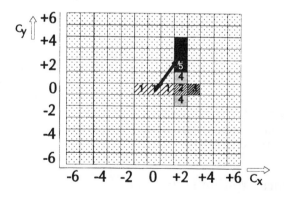

FIGURE 54.11 One-at-a-time search (OTS) block-matching. The new candidate vectors that have to be evaluated for a number of successive steps is indicated.

A problem of all efficient search techniques is the risk of converging to a local rather than the global minimum of the match error function. The coarser the initial grid of candidate vectors is, the higher this risk. It can be reduced by prefiltering the video information prior to the motion estimation, but this introduces inaccuracies in detailed picture parts. If the prefiltering and the block size are adapted separately for every step in the search procedure, we arrive at the hierarchical block-matching algorithms, dealt with in the next subsection.

54.6 Motion Estimation and Scanning Format Conversion

In situations where motion vectors are generated for temporal interpolation of pictures, it is important that the vectors represent the real velocities of objects, or the "true-motion" as it is called, in the picture. None of the described motion estimators is guaranteed to yield true motion vectors. They generate a vector that yields the "best match" or the minimal displaced frame difference and often even only the *local* best match, or the *local* minimum of the *DFD*.

To improve this relation between estimated displacement vectors and actual object velocity, methods have been designed which modify either the algorithm or the displacement vectors. The common solution is based on the observation that the velocity field does not usually contain many fine details. In other words, the motion vector field is spatially consistent: large areas (objects, background) with identical vectors usually exist. Object inertia further causes velocity fields to be temporally consistent. To improve consistency, a number of methods have been proposed. Two classes can be distinguished, combinations of which are possible:

- Methods, that perform a *post-processing* on the output vector field to improve the consistency.
- Methods in which a smoothness constraint is *integrated* in the estimator.

Postprocessing can be straightforward, applying basically low-pass filtering to improve the spatial and/or temporal consistency or smoothness of the vector field generated by any motion estimation algorithm. Often the filter is a nonlinear one; the median particularly is popular as it is edge preserving. More sophisticated methods in this class merely use the output vector field of the estimator to initialize a simulated annealing or genetic optimization algorithm using a new cost function, usually including smoothness constraints.

Integrated solutions can be expected to realize a better performance than the straightforward representatives of the first class at a lower expense than the sophisticated processing methods. The constraint can either be explicit, e.g., by adding a "discontinuity penalty" to the error criterion of a block-matcher:

$$\varrho\left(\underline{C}, \underline{X}, n\right) = \sum_{\underline{x} \in B(\underline{X})} \left| F\left(\underline{x}, n\right), -F\left(\underline{x} - \underline{C}, n - 1\right) \right|$$

$$+ \alpha \cdot \left\| \underline{D}\left(\underline{X}, -\begin{pmatrix} X \\ 0 \end{pmatrix}, n\right) - \underline{C} \right\| + \beta \cdot \left\| \underline{D}\left(\underline{X}, -\begin{pmatrix} 0 \\ Y \end{pmatrix}, n\right) - \underline{C} \right\| \quad (54.28)$$

(where the values of α and β determine the smoothness and it is proposed to adapt their value in the neighborhood of edges in the image), or implicit through *hierarchy* or *recursion*, which will be discussed separately. Again, both classes can be combined.

54.6.1 Hierarchical Motion Estimation

Hierarchical motion estimators realize a consistent velocity field by initializing local estimators with a global estimate, often in more than two steps. In sub-band coding terminology, a resolution pyramid is built and coarse vectors are estimated on the low frequency band. The result is used as a prediction for a more accurate estimate at the next sub-band, which contains higher frequencies, etc. At the top of the pyramid, the signal is strongly prefiltered and sub-sampled. The bandwidth of the filter increases and the sub-sampling factors decrease, going down in the hierarchy, until the full resolution is reached on the lowest hierarchical level.

The value of the motion vector in field n at hierarchical level l, $\underline{D}^{i-1}(\underline{X}, n, l)$, and using logarithmic search, is found as:

$$\underline{D}^{i-1}\left(\underline{X},\ n,\ l\right)\ \in$$

$$
\left(
\begin{array}{l}
\left\{\underline{D}^{N}\left(\underline{X},\ n, l-1\right)\right\}, \hspace{3cm} (i=1) \\[1em]
\left\{\underline{C}\in C S^{i-1}\left(\underline{X},\ n,\ l\right) \mid \varrho\left(\underline{C},\underline{X},\ n\right)\leq \varrho\left(\underline{F},\underline{X},\ n\right),\forall\underline{F}\in C S^{i-1}\left(\underline{X},\ n,\ l\right)\right\}, \hspace{0.6cm} (i>1)
\end{array}
\right.
\tag{54.29}
$$

where the search area is defined as:

$$
C S^{i}\left(\underline{X},\ n,\ l\right)\ =\ \left\{\underline{C}\mid\underline{C}=\underline{D}^{i-1}\left(\underline{X},\ n,\ l\right)+\underline{U},\ U_x=0\vee\pm 2^{N-i}\wedge U_y=0\vee 2^{N-i}\right\},
$$
$$
i=1\ldots N,\quad l=1\ldots L
\tag{54.30}
$$

$\underline{D}^{N}(\underline{X},\ n,\ l-1)$ is the result vector for the block at position $\underline{X}$ in field n in the last (Nth) step of the logarithmic search, at one higher ($l-1$) hierarchical level.

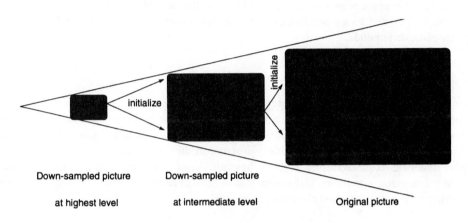

Down-sampled picture   **Down-sampled picture**

at highest level   **at intermediate level**   **Original picture**

FIGURE 54.12   Hierarchical block-matching. Results from an estimation process at a down-sampled image are used to initialize the next estimation process on a higher resolution image.

The method is also referred to as multi-resolution, or multi-grid motion estimation. The initial block size can be the total image, which prevents limitation of the consistency to parts of the picture. The inverted approach has also been published, performing block-matching on initially small blocks, which are grown to larger sizes until the minimum of the match error is considered clearly distinct. Combinations with other than the logarithmic search strategy are possible, and the hierarchical method is not limited to block-matching algorithms either.

Phase Plane Correlation

An important variant of a two step hierarchical motion estimation is a method called phase plane correlation (PPC). This algorithm is an extension of earlier Fourier techniques for motion estimation, which were capable of generating global displacement vectors only. In the PPC algorithm a two-level hierarchy is proposed.

In the first hierarchical level, on fairly large blocks (typically 64 by 64), a limited number of candidate vectors, usually less than 10, is generated, which are fed to the second level. Here one of these candidate vectors is assigned as the resulting vector to a much smaller area (typically 1 by 1 up to 8 by 8, is reported) inside the large initial block.

The name of the method refers to the procedure used for generating the candidate vectors in the first step. For the block in the current field n, the Discrete Fourier Transform (DFT) of the luminance function $F(\underline{x},\ n)$ will be notated as $G(\underline{f},\ n)$. The so-called phase difference matrix $P D(\underline{x},\ n)$ is calculated according to:

$$\text{PD}\left(\underline{x},\, n\right) = \mathcal{F}^{-1}\left(\frac{G\left(\underline{f},\, n\right) \cdot G^*\left(\underline{f},\, n-p\right)}{\mid G\left(\underline{f},\, n\right) \mid \cdot \mid G\left(\underline{f},\, n-p\right) \mid}\right) \tag{54.31}$$

The resulting matrix or "correlation surface" exhibits peaks corresponding to the relative displacement of the information in the two blocks. The Fourier transformation reduces the computational complexity, and enables simple filtering in the frequency domain. Most important is the significantly increased sharpness of the correlation peaks by normalizing each frequency component prior to the reverse transformation. A "peak hunting" algorithm is applied to find the largest peaks in the phase difference matrix, which correspond to the best matching candidate vectors. Sub-pixel accuracy better than a tenth of a pixel can be achieved by fitting a quadratic curve through the elements in this matrix. For interlaced video signals, $p = 2$ is the common choice.

The peaks in the phase plane can be applied to identify the most likely candidate vectors $\underline{C}(x,\, n)$ for a consecutive block-matching algorithm, evaluating all candidates in each sub-block in the area to which the phase plane corresponds.

54.6.2 Recursive Search Block-Matching

Rather than calculating promising candidate vectors for a block-matching algorithm on a lower resolution level or in the frequency domain on larger blocks, the recursive search block-matcher takes spatial and/or temporal "prediction vectors" from a 3-D neighborhood. This implicitly assumes spatial and/or temporal consistency. If the assumption is false, this consistency in the vector field results anyway, as there are no other candidate vectors available. As far as the predictions are concerned, there is a strong similarity with the pel-recursive algorithms, and the various options described there are globally valid here too. Figure 54.13 illustrates a proposed choice of predictions.

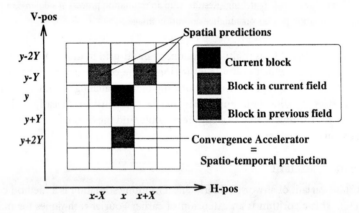

FIGURE 54.13   Relative position of current block and blocks from which prediction vectors can be taken in a recursive search block-matcher.

The most common updating process involves a single, or a very few, update vectors added to either of the prediction vectors. It was suggested, for example, to apply a candidate set $CS(\underline{X},\, n)$:

$$CS\left(\underline{X},\, n\right) =$$

$$\left\{\underline{C} \in CS^{\max} \mid \underline{C} = \underline{D}\left(\underline{X} - \begin{pmatrix} X \\ Y \end{pmatrix},\, n\right) + \underline{U}_a\left(\underline{X},\, n\right) \vee \underline{C} = \underline{D}\left(\underline{X} - \begin{pmatrix} -X \\ Y \end{pmatrix},\, n\right) + \underline{U}_b\left(\underline{X},\, n\right)\right\}$$

$$\cup \left\{ \underline{D}\left(\underline{X} - \begin{pmatrix} X \\ Y \end{pmatrix}, n \right), \underline{D}\left(\underline{X} - \begin{pmatrix} -X \\ Y \end{pmatrix}, n \right), \underline{C} = \underline{D}\left(\underline{X} + \begin{pmatrix} 0 \\ 2Y \end{pmatrix}, n - 1 \right) \right\} \quad (54.32)$$

where the update vectors $\underline{U}_a(\underline{X}, n)$ and $\underline{U}_b(\underline{X}, n)$ may be alternatingly available, and taken from a limited fixed integer update set, such as:

$$US_i = \left\{ \begin{pmatrix} 0 \\ 0 \end{pmatrix}, \begin{pmatrix} 0 \\ 1 \end{pmatrix}, \begin{pmatrix} 0 \\ -1 \end{pmatrix}, \begin{pmatrix} 0 \\ 2 \end{pmatrix}, \begin{pmatrix} 0 \\ -2 \end{pmatrix}, \right.$$
$$\left. \begin{pmatrix} 1 \\ 0 \end{pmatrix}, \begin{pmatrix} -1 \\ 0 \end{pmatrix}, \begin{pmatrix} 3 \\ 0 \end{pmatrix}, \begin{pmatrix} -3 \\ 0 \end{pmatrix} \right\} \quad (54.33)$$

Result vectors can have sub-pixel accuracy, if the update set (also) contains fractional update values. Quarter pel resolution, for example, is realized with adding:

$$US_f = \left\{ \begin{pmatrix} 0 \\ 0.25 \end{pmatrix}, \begin{pmatrix} 0 \\ -0.25 \end{pmatrix}, \begin{pmatrix} 0.25 \\ 0 \end{pmatrix}, \begin{pmatrix} -0.25 \\ 0 \end{pmatrix} \right\} \quad (54.34)$$

The method is very efficient and realizes, due to the inherent smoothness constraint, very coherent and close to true-motion vector fields, most suitable for scanning format conversion.

References

[1] Engstrom, E.W., A study of television image characteristics. Part Two. Determination of frame frequency for television in terms of flicker characteristics, *Proc. of the I.R.E.*, 23 (4), 295-310, 1935.

[2] van den Enden, A.W.M. and Verhoeckx, N.A.M., *Discrete-Time Signal Processing*, Prentice-Hall, Englewood Cliffs, NJ.

[3] Zworykin, V.K. and Morton, G.A., *Television*, 2nd ed., John Wiley & Sons, New York, 1954.

55

Video Sequence Compression

Osama Al-Shaykh
University of California,
Berkeley

Ralph Neff
University of California,
Berkeley

David Taubman
Hewlett Packard

Avideh Zakhor
University of California,
Berkeley

The image and video processing literature is rich with video compression algorithms. This chapter overviews the basic blocks of most video compression systems, discusses some important features required by many applications, e.g., scalability and error resilience, and reviews the existing video compression standards such as H.261, H.263, MPEG-1, MPEG-2, and MPEG-4.

55.1 Introduction

Video sources produce data at very high bit rates. In many applications, the available bandwidth is usually very limited. For example, the bit rate produced by a 30 frame/s color common intermediate format (CIF) (352×288) video source is 73 Mbits/s. In order to transmit such a sequence over a 64 Kbits/s channel (e.g., ISDN line), we need to compress the video sequence by a factor of 1140. A simple approach is to subsample the sequence in time and space. For example, if we subsample both chroma components by 2 in each dimension, i.e., 4:2:0 format, and the whole sequence temporally by 4, the bit rate becomes 9.1 Mbits/s. However, to transmit the video over a 64 kbits/s channel, it is necessary to compress the subsampled sequence by another factor of 143. To achieve such high compression ratios, we must tolerate some distortion in the subsampled frames.

Compression can be either lossless (reversible) or lossy (irreversible). A compression algorithm is lossless if the signal can be reconstructed from the compressed information; otherwise it is lossy. The compression performance of any lossy algorithm is usually described in terms of its *rate-distortion* curve, which represents the potential trade-off between the bit rate and the distortion associated with the lossy representation. The primary goal of any lossy compression algorithm is to optimize the rate-distortion curve over some range of rates or levels of distortion. For video applications, rate is usually expressed in terms of bits per second. The distortion is usually expressed in terms of the peak-signal-to-noise ratio (PSNR) per frame or, in some cases, measures that try to quantify the subjective nature of the distortion.

In addition to good compression performance, many other properties may be important or even critical to the applicability of a given compression algorithm. Such properties include robustness to errors in the compressed bit stream, low complexity encoders and decoders, low latency requirements, and scalability. Developing scalable video compression algorithms has attracted considerable attention in recent years. Generally speaking, scalability refers to the potential to effectively decompress subsets of the compressed bit stream in order to satisfy some practical constraint, e.g., display resolution, decoder computational complexity, and bit rate limitations.

The demand for compatible video encoders and decoders has resulted in the development of different video compression standards. The international standards organization (ISO) has developed MPEG-1 to store video on compact discs, MPEG-2 for digital television, and MPEG-4 for a wide range of applications including multimedia. The international telecommunication union (ITU) has developed H.261 for video conferencing and H.263 for video telephony.

All existing video compression standards are hybrid systems. That is, the compression is achieved in two main stages. The first stage, motion compensation and estimation, predicts each frame from its neighboring frames, compresses the prediction parameters, and produces the prediction error frame. The second stage codes the prediction error. All existing standards use block-based discrete cosine transform (DCT) to code the residual error. In addition to DCT, others non-block-based coders, e.g., wavelets and matching pursuit, can be used.

In this chapter, we will provide an overview of hybrid video coding systems. In Section 55.2, we discuss the main parts of a hybrid video coder. This includes motion compensation, signal decompositions and transformations, quantization, and entropy coding. We compare various transformations such as DCT, subband, and matching pursuit. In Section 55.3, we discuss scalability and error resilience in video compression systems. We also describe a non-hybrid video coder that provides scalable bit-streams [28]. Finally, in Section 55.4, we review the key video compression standards: H.261, H.263, MPEG 1, MPEG 2, and MPEG 4.

55.2 Motion Compensated Video Coding

Virtually all video compression systems identify and reduce four basic types of video data redundancy: inter-frame (temporal) redundancy, interpixel redundancy, psychovisual redundancy, and coding redundancy. Figure 55.1 shows a typical diagram of a hybrid video compression system. First the current frame is predicted from previously decoded frames by estimating the motion of blocks or objects, thus reducing the inter-frame redundancy. Afterwards to reduce the interpixel redundancy, the residual error after frame prediction is transformed to another format or domain such that the energy of the new signal is concentrated in few components and these components are as uncorrelated as possible. The transformed signal is then quantized according to the desired compression performance (subjective or objective). The quantized transform coefficients are then mapped to codewords that reduce the coding redundancy. The rest of this section will discuss the blocks of the hybrid system in more detail.

55.2.1 Motion Estimation and Compensation

Neighboring frames in typical video sequences are highly correlated. This inter-frame (temporal) redundancy can be significantly reduced to produce a more compressible sequence by predicting each frame from its neighbors. Motion compensation is a nonlinear predictive technique in which the feedback loop contains both the inverse transformation and the inverse quantization blocks, as shown in Fig. 55.1.

Most motion compensation techniques divide the frame into regions, e.g., blocks. Each region is then predicted from the neighboring frames. The displacement of the block or region, d, is not fixed and must

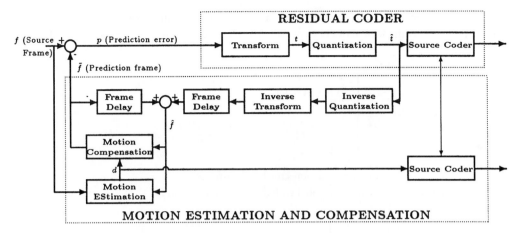

FIGURE 55.1 Motion compensated coding of video.

be encoded as side information in the bit stream. In some cases, different prediction models are used to predict regions, e.g., affine transformations. These prediction parameters should also be encoded in the bit stream.

To minimize the amount of side information, which must be included in the bit stream, and to simplify the encoding process, motion estimation is usually block based. That is, every pixel $\vec{i}$ in a given rectangular block is assigned the same motion vector, d. Block-based motion estimation is an integral part of all existing video compression standards.

55.2.2   Transformations

Most image and video compression schemes apply a transformation to the raw pixels or to the residual error resulting from motion compensation before quantizing and coding the resulting coefficients. The function of the transformation is to represent the signal in a few uncorrelated components. The most common transformations are linear transformations, i.e., the multi-dimensional sequence of input pixel values, $f[\vec{i}]$, is represented in terms of the transform coefficients, $t[\vec{k}]$, via

$$f[\vec{i}] = \sum_{\vec{k}} t[\vec{k}] w_{\vec{k}}[\vec{i}] \tag{55.1}$$

for some $w_{\vec{k}}[\vec{i}]$. The input image is thus represented as a linear combination of basis vectors, $w_{\vec{k}}$. It is important to note that the basis vectors need not be orthogonal. They only need to form an over-complete set (matching pursuits), a complete set (DCT and some subband decompositions), or very close to complete (some subband decompositions). This is important since the coder should be able to code a variety of signals. The remainder of the section discusses and compares DCT, subband decompositions, and matching pursuits.

The DCT

There are two properties desirable in a unitary transform for image compression: the energy should be packed into a few transform coefficients, and the coefficients should be as uncorrelated as possible. The optimum transform under these two constraints is the Karhunen-Loéve transform (KLT) where the eigenvectors of the covariance matrix of the image are the vectors of the transform [10]. Although the KLT is optimal under these two constraints, it is data-dependent, and is expensive to compute. The discrete cosine transform (DCT) performs very close to KLT especially when the input is a first order Markov process [10].

The DCT is a block-based transform. That is, the signal is divided into blocks, which are independently transformed using orthonormal discrete cosines. The DCT coefficients of a one-dimensional signal, f, are computed via

$$t^{\text{DCT}}[Nb + k] = \frac{1}{\sqrt{N}} \begin{cases} \sum_{i=0}^{N-1} f[Nb + i], & k = 0 \\ \sum_{i=0}^{N-1} \sqrt{2} f[Nb + i] \cos \frac{(2i + 1)k\pi}{2N}, & 1 \le k < N \end{cases} \qquad \forall b \qquad (55.2)$$

where N is the size of the block and b denotes the block number.

The orthonormal basis vectors associated with the one-dimensional DCT transformation of Eq. (55.2) are

$$w_k^{\text{DCT}}[i] = \frac{1}{\sqrt{N}} \begin{cases} 1, & k = 0, 0 \le i < N \\ \sqrt{2} \cos \frac{(2i+1)k\pi}{2N}, & 1 \le k < N, 0 \le i < N \end{cases} \qquad (55.3)$$

Figure 55.2(a) shows these basis vectors for $N = 8$.

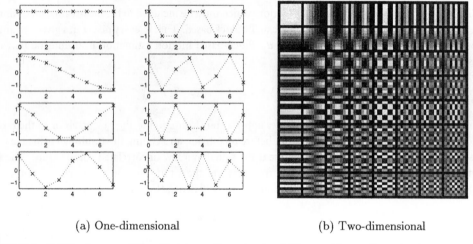

(a) One-dimensional   (b) Two-dimensional

FIGURE 55.2   DCT basis vectors ($N = 8$): (a) one-dimensional and (b) separable two-dimensional.

The one-dimensional DCT described above is usually separably extended to two dimensions for image compression applications. In this case, the two-dimensional basis vectors are formed by the tensor product of one-dimensional DCT basis vectors and are given by

$$w_{\vec{k}}^{\text{DCT}}[\vec{i}] = w_{k_1,k_2}^{\text{DCT}}[i_1, i_2] \stackrel{\Delta}{=} w_{k_1}^{\text{DCT}}[i_1] \cdot w_{k_2}^{\text{DCT}}[i_2]; \qquad 0 \le k_1, k_2, i_1, i_2 < N$$

Figure 55.2(b) shows the two-dimensional basis vectors for $N = 8$.

The DCT is the most common transform in video compression. It is used in the JPEG still image compression standard, and all existing video compression standards. This is because it performs reasonably well at different bit rates. Moreover, there are fast algorithms and special hardware chips to compute the DCT efficiently.

The major objection to the DCT in image or video compression applications is that the non-overlapping blocks of basis vectors, $w_{\vec{k}}$, are responsible for distinctly "blocky" artifacts in the decompressed frames, especially at low bit rates. This is due to the quantization of the transform coefficients of a block independent from neighboring blocks. Overlapped DCT representation addresses this problem [15]; however, the common solution is to post-process the frame by smoothing the block boundaries [18, 22].

Due to bit rate restrictions, some blocks are only represented by one or a small number of coarsely quantized transform coefficients, hence the decompressed block will only consist of these basis vectors. This will cause artifacts commonly known as ringing and mosquito noise.

Figure 55.8(b) shows frame 250 of the 15 frame/s CIF COAST-GUARD sequence coded at 112 Kbits/s using a DCT hybrid video coder.[1] This figure provides a good illustration of the "blocking" artifacts.

Subband Decomposition

The basic idea of subband decomposition is to split the frequency spectrum of the image into (disjoint) subbands. This is efficient when the image spectrum is not flat and is concentrated in a few subbands, which is usually the case. Moreover, we can quantize the subbands differently according to their visual importance.

As for the DCT, we begin our discussion of subband decomposition by considering only a one-dimensional source sequence, $f[i]$. Figure 55.3 provides a general illustration of an N-band one-dimensional subband system. We refer to the subband decomposition itself as *analysis* and to the inverse

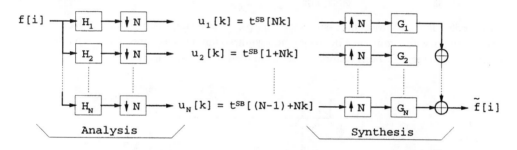

FIGURE 55.3 1D, N-band subband analysis and synthesis block diagrams. (*Source:* Taubman, D., Chang, E., and Zakhor, A., Directionality and scalability in subband image and video compression, in *Image Technology: Advances in Image Processing, Multimedia, and Machine Vision*, Jorge L.C. Sanz, Ed., Springer-Verlag, New York, 1996. With permission).

transformation as *synthesis*. The transformation coefficients of bands $1, 2, \ldots, N$ are denoted by the sequences $u_1[k], u_2[k], \ldots, u_N[k]$, respectively. For notational convenience and consistency with the DCT formulation above, we write $t^{SB}[\cdot]$ for the sequence of all subband coefficients, arranged according to $t^{SB}[(\beta - 1) + Nk] = u_\beta[k]$, where $1 \leq \beta \leq N$ is the subband number. These coefficients are generated by filtering the input sequence with filters $H_1, \ldots, H_N$ and downsampling the filtered sequences by a factor of N, as depicted in Fig. 55.3. In subband synthesis, the coefficients for each band are upsampled, interpolated with the synthesis filters, $G_1, \ldots, G_N$, and the results summed to form a reconstructed sequence, $\tilde{f}[i]$, as depicted in Fig. 55.3.

If the reconstructed sequence, $\tilde{f}[i]$, and the source sequence, $f[i]$, are identical, then the subband system is referred to as perfect reconstruction (PR) and the corresponding basis set is a complete basis set. Although perfect reconstruction is a desirable property, near perfect reconstruction (NPR), for which subband synthesis is only approximately the inverse of subband analysis, is often sufficient in practice.

---

[1]It is coded using H.263 [3], which is an ITU standard.

This is because distortion introduced by quantization of the subband coefficients, $t^{SB}[k]$, usually dwarfs that introduced by an imperfect synthesis system.

The filters, $H_1, \ldots, H_N$, are usually designed to have band-pass frequency responses, as indicated in Fig. 55.4, so that the coefficients $u_\beta[k]$ for each subband, $1 \leq \beta \leq N$, represent different spectral components of the source sequence.

FIGURE 55.4 Typical analysis filter magnitude responses. (*Source:* Taubman, D., Chang, E., and Zakhor, A., Directionality and scalability in subband image and video compression, in *Image Technology: Advances in Image Processing, Multimedia, and Machine Vision,* Jorge L.C. Sanz, Ed., Springer-Verlag, New York, 1996. With permission).

The basis vectors for subband decomposition are the N-translates of the impulse responses, $g_1[i], \ldots, g_N[i]$, of synthesis filters $G_1, \ldots, G_N$. Specifically, denoting the kth basis vector associated with subband β by $w^{SB}_{Nk+\beta-1}$, we have

$$w^{SB}_{Nk+\beta-1}[i] = g_\beta[i - Nk] \qquad (55.4)$$

Figure 55.5 illustrates five of the basis vectors for a particularly simple, yet useful, two-band PR subband decomposition, with symmetric FIR analysis and synthesis impulse responses. As shown in Fig. 55.5 and in contrast with the DCT basis vectors, the subband basis vectors overlap.

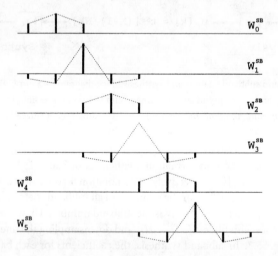

FIGURE 55.5 Subband basis vectors with $N = 2$, $h_1[-2\ldots 2] = \sqrt{2} \cdot (-\frac{1}{8}, \frac{1}{4}, \frac{3}{4}, \frac{1}{4}, -\frac{1}{8})$, $h_2[-2\ldots 0] = \sqrt{2} \cdot (-\frac{1}{4}, \frac{1}{2}, -\frac{1}{4})$, $g_1[-1\ldots 1] = \sqrt{2} \cdot (\frac{1}{4}, \frac{1}{2}, \frac{1}{4})$, and $g_2[-1\ldots 3] = \sqrt{2} \cdot (-\frac{1}{8}, -\frac{1}{4}, \frac{3}{4}, -\frac{1}{4}, -\frac{1}{8})$. h_i and g_i are the impulse responses of the H_i *(analysis)* and G_i *(synthesis)* filters, respectively. (*Source:* Taubman, D., Chang, E., and Zakhor, A., Directionality and scalability in subband image and video compression, in *Image Technology: Advances in Image Processing, Multimedia, and Machine Vision,* Jorge L.C. Sanz, Ed., Springer-Verlag, New York, 1996. With permission).

As for the DCT, one-dimensional subband decompositions may be separably extended to higher dimensions. By this we mean that a one-dimensional subband decomposition is first applied along one dimension of an image or video sequence. Any or all of the resulting subbands are then further decomposed into subbands along another dimension and so on. Figure 55.6 depicts a separable two-dimensional subband system. For video compression applications, the prediction error is sometimes decomposed into subbands of equal size.

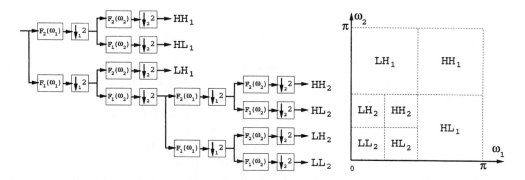

FIGURE 55.6 Separable spatial subband pyramid. Two level analysis system configuration and subband passbands shown. (*Source:* Taubman, D., Chang, E., and Zakhor, A., Directionality and scalability in subband image and video compression, in *Image Technology: Advances in Image Processing, Multimedia, and Machine Vision,* Jorge L.C. Sanz, Ed., Springer-Verlag, New York, 1996. With permission).

Two-dimensional subband decompositions have the advantage that they do not suffer from the disturbing blocking artifacts exhibited by the DCT at high compression ratios. Instead, the most noticeable quantization-induced distortion tends to be 'ringing' or 'rippling' artifacts, which become most bothersome in the vicinity of image edges. Figures 55.11(c) and 55.8(c) clearly show this effect. Figure 55.11 shows frame 210 of the PING-PONG sequence compressed using a scalable, three-dimensional subband coder [28] at 1.5 Mbits/s, 300 Kbits/s, and 60 Kbits/s. As the bit rate decreases, we notice loss of detail and introduction of more ringing noise. Figure 55.8(c) shows frame 250 of the COAST-GUARD sequence compressed at 112 Kbits/s using a zerotree scalable coder [16]. The edges of the trees and the boat are affected by ringing noise.

Matching Pursuit

Representing a signal using an over-complete basis set implies that there is more than one representation for the signal. For coding purposes, we are interested in representing the signal with the fewest basis vectors. This is an NP-complete problem [14]. Different approaches have been investigated to find or approximate the solution. Matching pursuits is a multistage algorithm, which in each stage finds the basis vector that minimizes the mean-squared-error [14].

Suppose we want to represent a signal $f[i]$ using basis vectors from an over-complete dictionary (basis set) $\mathcal{G}$. Individual dictionary vectors can be denoted as:

$$w_\gamma[i] \in \mathcal{G}. \tag{55.5}$$

Here γ is an indexing parameter associated with a particular dictionary element. The decomposition begins by choosing γ to maximize the absolute value of the following inner product:

$$t = < f[i], w_\gamma[i] >, \tag{55.6}$$

where t is the transform (expansion) coefficient. A residual signal is computed as:

$$R[i] = f[i] - t\, w_\gamma[i]. \tag{55.7}$$

This residual signal is then expanded in the same way as the original signal. The procedure continues iteratively until either a set number of expansion coefficients are generated or some energy threshold for the residual is reached. Each stage k yields a dictionary structure specified by γ_k, an expansion coefficient $t[k]$, and a residual R_k, which is passed on to the next stage. After a total of M stages, the signal can be approximated by a linear function of the dictionary elements:

$$\hat{f}[i] = \sum_{k=1}^{M} t[k]\, w_{\gamma_k}[i]. \tag{55.8}$$

The above technique has useful signal representation properties. For example, the dictionary element chosen at each stage is the element that provides the greatest reduction in mean square error between the true signal $f[i]$ and the coded signal $\hat{f}[i]$. In this sense, the signal structures are coded in order of importance, which is desirable in situations where the bit budget is limited. For image and video coding applications, this means that the most visible features tend to be coded first. Weaker image features are coded later, if at all. It is even possible to control which types of image features are coded well by choosing dictionary functions to match the shape, scale, or frequency of the desired features.

An interesting feature of the matching pursuit technique is that it places very few restrictions on the dictionary set. The original Mallat and Zhang paper considers both Gabor and wave-packet function dictionaries, but such structure is not required by the algorithm itself [14]. Mallat and Zhang showed that if the dictionary set is at least complete, then $\hat{f}[i]$ will eventually converge to $f[i]$, though the rate of convergence is not guaranteed [14]. Convergence speed and thus coding efficiency are strongly related to the choice of dictionary set. However, true dictionary optimization can be difficult because there are so few restrictions. Any collection of arbitrarily sized and shaped functions can be used with matching pursuits, as long as completeness is satisfied.

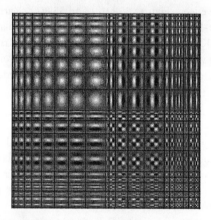

FIGURE 55.7 Separable two-dimensional 20×20 Gabor dictionary.

Bergeaud and Mallat used the matching pursuit technique to represent and process images [1]. Neff and Zakhor have used the matching pursuit technique to code the motion prediction error signal [20]. Their coder divides each motion residual into blocks and measures the energy of each block. The center of the block with the largest energy value is adopted as an initial estimate for the inner product search. A dictionary of Gabor basis vectors, shown in Fig. 55.7, is then exhaustively matched to an $S \times S$ window around the initial estimate. The exhaustive search can be thought of as follows. Each $N \times N$ dictionary structure is centered at each location in the search window, and the inner product between the structure and the corresponding $N \times N$ region of image data is computed. The largest inner-product is then quantized. The location, basis vector index, and quantized inner product are then coded together.

Video sequences coded using matching pursuit do not suffer from either blocking or ringing artifacts, because the basis vectors are only coded when they are well-matched to the residual signal. As bit rate decreases, the distortion introduced by matching pursuit coding takes the form of a gradually increasing blurriness (or loss of detail). Since matching pursuits involves exhaustive search, it is more complex than DCT approaches, especially at high bit rates.

Figure 55.8(d) shows frame 250 of the 15 frame/s CIF COAST-GUARD sequence coded at 112 Kbits/s using the matching pursuit video coder described by Neff and Zakhor [20]. This frame does not suffer from the blocky artifacts, which affect the DCT coders as shown in Fig. 55.8(b). Moreover, it does not suffer from the ringing noise, which affects the subband coders as shown in Figs. 55.8(c) and 55.11(c).

55.2.3 Discussion

Figure 55.8 shows frame 250 of the 15 frame/s CIF COAST-GUARD sequence coded at 112 Kbits/s using DCT, subband, and matching pursuit coders. The DCT coded frame suffers from blocking artifacts. The subband coded frame suffers from ringing artifact.

Figure 55.9 compares the PSNR performance of the matching pursuit coder [20] to a DCT (H.263) coder [3] and a zerotree subband coder [16] when coding the COAST-GUARD sequence at 112 Kbits/s. The matching pursuit coder [20] in this example has consistently higher PSNR than the H.263 [3] and the zerotree subband [16] coders. Table 55.1 shows the average luminance PSNRs for different sequences at different bit rates. In all examples mentioned in Table 55.1, the matching pursuit coder has higher average PSNR than the DCT coder. The subband coder has the lowest average PSNR.

TABLE 55.1   The Average Luminance PSNR of Different Sequences at Different Bit Rates When Coding Using a DCT Coder (H.263) [3], Zero-Tree Subband Coder (ZTS) [16], and Matching Pursuit Coder (MP) [20]

		Rate		PSNR (dB)		
Sequence	Format	Bit	Frame	DCT	ZTS	MP
Container-ship	QCIF	10 K	7.5	29.43	28.01	31.10
Hall-Monitor	QCIF	10 K	7.5	30.04	28.44	31.27
Mother-Daughter	QCIF	10 K	7.5	32.50	31.07	32.78
Container-ship	QCIF	24 K	10.0	32.77	30.44	34.26
Silent-Voice	QCIF	24 K	10.0	30.89	29.41	31.71
Mother-Daughter	QCIF	24 K	10.0	35.17	33.77	35.55
Coast-Guard	QCIF	48 K	10.0	29.00	27.65	29.82
News	CIF	48 K	7.5	30.95	29.97	31.96

55.2.4 Quantization

Motion compensation and residual error decomposition reduce the redundancy in the video signal. However, to achieve low bit rates, we must tolerate some distortion in the video sequence. This is because we need to map the residual and motion information to a fewer collection of codewords to meet the bit rate requirements.

Quantization, in a general sense, is the mapping of vectors (or scalars) of an information source into a finite collection of codewords for storage or transmission [8]. This involves two processes: encoding and decoding. The encoder blocks the source $\{t[i]\}$ into vectors of length n, and maps each vector $T^n \in \mathcal{T}^n$ into a codeword c taken from a finite set of codewords $\mathcal{C}$. The decoder maps the codeword c into a reproduction vector $Y^n \in \mathcal{Y}^n$ where $\mathcal{Y}$ is a reproduction alphabet. If $n = 1$, it is called *scalar quantization*. Otherwise, it is called *vector quantization*.

(a) Original

(b) DCT

(c) Subband

(d) Matching pursuit

FIGURE 55.8 Frame 250 of COAST-GUARD sequence, original shown in (a), coded at 112 Kbits/s using (b) DCT based coder (H.263) [3], (c) zerotree subband coder [16], and (d) matching pursuit coder [20]. Blocking artifacts can be noticed on the DCT coded frame. Ringing artifacts can be noticed on the subband coded frame.

The problem of optimum mean squared scalar quantization for a given reproduction alphabet size was independently solved by Lloyd [13] and Max [17]. They found that if t is a real scalar random variable with continuous probability density function $p_t(t)$, then the quantization thresholds are

$$\hat{t}_k = \frac{r_k + r_{k-1}}{2}, \tag{55.9}$$

which is the geometric mean of the interval $(r^{k-1}, r^k]$, where

$$r_k = \frac{\displaystyle\int_{\hat{t}_k}^{\hat{t}_{k+1}} x p_x(x)\,dx}{\displaystyle\int_{\hat{t}_k}^{\hat{t}_{k+1}} p_x(x)\,dx} \tag{55.10}$$

are the reconstruction levels. Iterative numerical methods are required to solve for the reconstruction and quantization levels.

The simplest scalar quantizer is the uniform quantizer for which the reconstruction intervals are of equal length. The uniform quantizer is optimal when the coefficients have a uniform distribution. Moreover, due to its simplicity and good general performance, it is commonly used in coding systems.

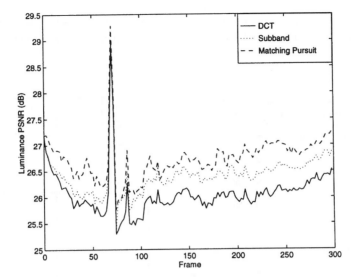

FIGURE 55.9   Frame-by-frame distortion of the luminance component of the COAST-GUARD sequence, reconstructed from 112 Kbits/s H.263 bit stream (solid line) [3], a zerotree subband bit-stream (dotted line) [16], and from a matching pursuit bit stream (dashed line) [20]. Consistently, the matching pursuit coder had the highest PSNR while the DCT coder had the lowest PSNR.

A fundamental result of Shannon's rate distortion theory is that better performance can be achieved by coding vectors instead of scalars, even if the source is memoryless [8, 19]. Linde et al. [12] generalized the Lloyd-Max algorithm to vector quantization. Vector quantization exploits spatial redundancy in images, a function also served by the transformation block of Fig. 55.1, so it is sometimes applied directly to the image or video pixels [19].

Memory can be incorporated into scalar quantization by predicting the current sample from the previous samples and quantizing the residual error, e.g., linear predictive coding.

The human visual system is sensitive to some frequency bands more than others. So, humans tolerate more losses in some bands and less in others. In practice, the DCT coefficients corresponding to a particular frequency are grouped together to form a band, or in the case of subband decomposition, the bands are simply the subband channels. Different quantizers are then applied to each band according to its visual importance.

55.2.5   Coding of Quantized Symbols

The simplest method to code quantized symbols is to assign a fixed number of bits per symbol. For an alphabet of L symbols, this approach requires $\lceil \log_2 L \rceil$ bits per symbol. This method, however, does not exploit the coding redundancy in the symbols. Coding redundancy is eliminated by minimizing the average number of bits per symbol. This is achieved by giving fewer bits to more frequent symbols and more bits to less frequent symbols. Huffman [9] or arithmetic coding [21] schemes are usually used for this purpose.

In image and video coding, a significant number of the transform coefficients are zeros. Moreover, the "significant" DCT transform coefficients (low frequency coefficients) of a block can be predicted from the neighboring blocks resulting in a larger number of zero coefficients. To code the zero coefficients, run-length is performed on a reordered version of the transform coefficients. Figure 55.10(a) shows a commonly used zigzag scan to code 8×8 block DCT coefficients. Figure 55.10(b) shows a scan used to code subband coefficients commonly known as zero-tree coding [24]. The basic idea behind zero-tree coding is that if a coefficient in a lower frequency band (coarse scale) is zero or insignificant, then all the coefficients of the same orientation at higher frequencies (finer scales) are very likely to be zero or insignificant [16, 24]. Thus, the subband coefficients are organized in a data structure design based on this observation.

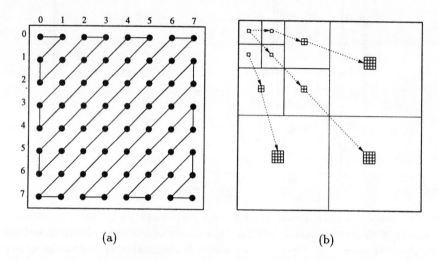

(a)   (b)

FIGURE 55.10 (a) A common scan for an 8 × 8 block DCT. (b) A common scan for subband decompositions (zero-tree).

55.3 Desirable Features

Some video applications require the encoder to provide more than good compression performance. For example, it is desirable to have scalable video compression schemes so that different users with different bandwidth, resolution, or computational capabilities can decode from the same bit-stream. Cellular applications require the coder to provide a bit-stream that is robust when transmission errors occur. Other features include object-based manipulation of the bit-stream and the ability to perform content search. This section addresses two important desired features, namely scalability and error resilience.

55.3.1 Scalability

Developing scalable video compression algorithms has attracted considerable attention in recent years. Scalable compression refers to encoding a sequence in such a way so that subsets of the encoded bit-stream correspond to compressed versions of the sequence at different rates and resolutions. Scalable compression is useful in today's heterogeneous networking environment in which different users have different rate, resolution, display, and computational capabilities.

In rate scalability, appropriate subsets are extracted in order to trade distortion for bit rate at a fixed display resolution. Resolution-scalability, on the other hand, means that extracted subsets represent the image or video sequence at different resolutions. Rate- and resolution-scalability usually also provide a means of scaling the computational demands of the decoder. Resolution-scalability is best thought of as a property of the transformation block of Fig. 55.1. Both the DCT and subband transformations may be used to provide resolution-scalability. Rate-scalability, however, is best thought of as a property of the quantization and coding blocks.

Hybrid video coders can achieve scalability using multi-layer schemes. For example, in a two layer rate-scalable coder, the first layer codes the video at a low bit rate, while the second layer codes the residual error based on the source material and what has been coded thus far. These layers are usually called the base and enhancement layers. Such schemes, however, do not support fully scalable video, i.e., they can only provide a few levels of scalability, e.g., a few rates. The bottleneck is motion compensation, which is a nonlinear feedback predictor. To understand this, observe that the storage block of Fig. 55.1 is a memory element, storing values $\tilde{f}[\vec{i}]$ or $\tilde{t}[\vec{k}]$, recovered during decoding, until they are required for prediction. In scalable compression algorithms, the value of $\tilde{f}[\vec{i}]$ or $\tilde{t}[\vec{k}]$, obtained during decoding, depends on constraints, which may be imposed after the bit-stream has been generated. For example, if the algorithm is to permit rate scalability, then the value of $\tilde{f}[\vec{i}]$ or $\tilde{t}[\vec{k}]$ obtained by decoding a low rate subset of the bit-stream can be expected to be a poorer approximation to $f[\vec{i}]$ or $t[\vec{k}]$, respectively, than the value obtained by decoding from a higher rate subset of the bit-stream. This ambiguity presents a difficulty for the compression algorithm, which must select a particular value for $\tilde{f}[\vec{i}]$ or $\tilde{t}[\vec{k}]$ to serve as a prediction reference.

This inherent non-scalability of motion compensation is particularly problematic for video compression where scalability and motion compensation are both highly desirable features. As a solution, Taubman and Zakhor [28, 29] used three-dimensional subband decompositions to code video. They first compensated for the camera pan motion, then used three-dimensional subband decomposition. The coefficients in each subband are then quantized by a layered quantizer in order to generate a fully scalable video with fine granularity of bit rates. Temporal filtering, however, introduces significant overall latency, a critical parameter for interactive video compression applications. To reduce this effect, it is possible to use a 2-tap temporal filter, which results in one frame of delay.

As a visual demonstration of the quality tradeoff inherent to rate-scalable video compression, Fig. 55.11 shows frame 210 of the PING-PONG video sequence, decompressed at bit rates of 1.5 Mbits/s, 300 kbits/s, and 60 kbits/s for monochrome display using the scalable coder developed by Taubman and Zakhor [28]. As the bit rate decreases, the frame is less detailed and suffers more from ringing noise, i.e., the visual quality decreases. Figure 55.12 shows the PSNR characteristics of the scalable coder and MPEG-1 coder as a function of bit rate. The curve corresponding to the scalable coder corresponds to one encoded bit-stream decoded at arbitrary bit rates, while the three points for the MPEG-1 coder correspond to three different encoded bit-streams encoded and decoded at these different rates. As seen the scalable codec offers a fine granularity of available bit rates with little or no loss in PSNR as compared to MPEG-1 codec.

Real time software only implementation of scalable video codec has also received a great deal of attention over the past few years. Tan et al. [27] have recently proposed a real-time software only implementation of the modified version of the algorithm in [28] by replacing the arithmetic coding with block coding. The resulting scalable coder is symmetric in encoding and decoding complexity and can encode up to 17 frames/s for rates as high as 1 Mbits/s on a 171 MHz Ultra-Sparc workstation.

55.3.2   Error Resilience

When transmitting video over noisy channels, it is important for bit-streams to be robust to transmission errors. It is also important, in case of errors, for the error to be limited to a small region and not to propagate to other areas. If the coder is using fixed-length codes, the error will be limited to the region of the bit-stream where it occurred and the rest of the bit-stream will not be affected. Unfortunately, fixed-length codes do not provide good compression performance, especially since the histogram of the transform coefficients has a significant peak around low frequency.

In order to achieve such features when using variable length codes, the bit-stream is usually partitioned into segments that can be independently decoded. Thus, if a segment is lost, only that region of the video is affected. A segment is usually a small part of a frame. If an error occurs, the decoder should have enough information to know the beginning and the end of a segment. Therefore, synchronization codes are added to the beginning and end of each segment. Moreover, to limit the error to a smaller part of the segment,

(a) 1.5 Mbits / s   (b) 300 Kbits /s

(c) 60 Kbits/ s

FIGURE 55.11   Frame 210 of PING-PONG sequence decoded from scalable bit stream at (a) 1.5 Mbits/s, (b) 300 Kbits/s, and (c) 60 Kbits/s [28]. (*Source:* Taubman, D., Chang, E., and Zakhor, A., Directionality and scalability in subband image and video compression, in *Image Technology: Advances in Image Processing, Multimedia, and Machine Vision,* Jorge L.C. Sanz, Ed., Springer-Verlag, New York, 1996. With permission).

reversible variable length codes may be used [26]. So, if an error occurs, the decoder will advance to the next synchronization code and can decode in the backward direction till the error is reached.

As is evident, there is a tradeoff between good compression performance and error resilience. In order to reduce the cost of error resilient codes, some approaches jointly optimize the source and channel codes [6, 23].

55.4   Standards

In this section we review the major video compression standards. Essentially, these schemes are based on the building blocks introduced in Section 55.2. All these standards use the DCT. Table 55.2 summarizes the basic characteristics and functionalities supported by existing standards. Sections 55.4.2, 55.4.3, and 55.4.5 outline the Motion Picture Experts Group (MPEG) standards for video compression. Sections 55.4.1 and 55.4.4 review the CCITT H.261 and H.263 standards for digital video communications. This section lists the standards according to their chronological order in order to provide an understanding of the progress of the video compression standardization process.

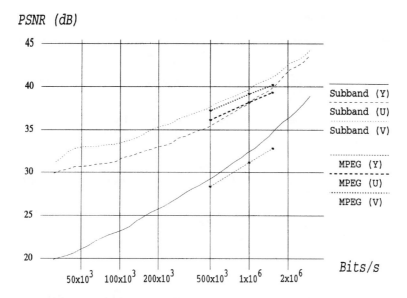

FIGURE 55.12 Rate-distortion curves for PING-PONG sequence. Overall PSNR values for Y, U, and V components for the codec in [28] are plotted against the bit rate limit imposed on the rate-scalable bit stream prior to decompression. MPEG-1 distortion values are also plotted as connected dots for reference. (*Source:* Taubman, D., Chang, E., and Zakhor, A., Directionality and scalability in subband image and video compression, in *Image Technology: Advances in Image Processing, Multimedia, and Machine Vision,* Jorge L.C. Sanz, Ed., Springer-Verlag, New York, 1996. With permission).

55.4.1 H.261

Recommendation H.261 of the CCITT Study Group XV was adopted in December 1990 [2] as a video compression standard to be used for video conferencing applications. The bit rates supported by H.261 are $p \times 64$ Kbits/s, where p is in the range 1 to 30. H.261 supports two source formats: CIF (352×288 luminance and 176×144 chrominance) and QCIF (176×144 luminance and 88×72 chrominance). The chrominance components are subsampled by two in both the vertical and horizontal directions.

The transformation used in H.261 is the 8×8 block-DCT. Thus, there are four luminance (Y) DCT blocks for each pair of U and V chrominance DCT blocks. These six DCT blocks are collectively referred to as a *macro-block*. The macro-blocks are grouped together to construct a group of blocks (GOB), which relates to 11×3 region of macro-blocks. Each macro-block may individually be specified as intra-coded or inter-coded. The Intra-coded blocks are coded independently of the previous frame and so do not conform to the model of Fig. 55.1. They are used when successive frames are not related, such as during scene changes, and to avoid excessive propagation of the effects of communication errors. Inter-coded blocks use the motion compensation predictive feedback loop of Fig. 55.1 to improve compression performance. The motion estimation scheme is based on 16×16 pixel blocks. Each macro-block is predicted from the previous frame and is assigned exactly one motion vector with one pixel accuracy.

The data for each frame consists of a picture header that includes a start code, a temporal reference for the current coded picture, and the source format. The picture header is followed by the GOB layer. The data of each GOB has a header that includes a start code to indicate the beginning of a GOB, the GOB number to indicate the position of the GOB, and all information necessary to code each GOB independently. This will limit the loss if an error occurs during the transmission of a GOB. The header of the GOB is followed by the motion data, then followed by the block information.

TABLE 55.2 Summary of the Functionalities and Characteristics of the Existing Standards

	ITU		ISO		
Attribute	H.261	H.263	MPEG-1	MPEG-2	MPEG-4
Applications	Video-conferencing	Video-phone	CD storage	Broadcast	Wide range (multimedia)
Bit rate	64K - 1M	$< 64K$	1.0 - 1.5M	2 - 10M	5K - 4M
Material	Progressive	Progressive	Progressive, interlaced	Progressive, interlaced	Progressive, interlaced
Object shape	Rectangular	Arbitrary (simple)	Rectangular	Rectangular	Arbitrary
Residual Coding					
Transform	8 × 8 DCT	8 × 8 DCT	8 × 8 DCT	8 × 8 DCT	8 × 8 DCT
Quantizer	Uniform	Uniform	Weighted uniform	Weighted uniform	Weighted uniform
Motion Compensation					
Type	Block	Block	Block	Block	Block, sprites
Block size	16 × 16	16 × 16, 8 × 8	16 × 16	16 × 16	16 × 16, 8 × 8
Prediction type	Forward	Forward, backward	Forward, backward	Forward, backward	Forward, backward
Accuracy	One pixel	Half pixel	Half pixel	Half pixel	Half pixel
Loop filter	Yes	No	No	No	No
Scalability					
Temporal	No	Yes	Yes	Yes	Yes
Spatial	No	Yes	No	Yes	Yes
Bit rate	No	Yes	No	Yes	Yes
Object	No	No	No	No	Yes

55.4.2 MPEG-1

The first (MPEG) video compression standard [7], MPEG-1, is intended primarily for progressive video at 30 frames/s. The targeted bit rate is in the range 1.0 to 1.5 Mbits/s. MPEG-1 was designed to store video on compact discs. Such applications require MPEG-1 to support random access to the material on the disc, fast forward and backward searches, reverse playback, and audio visual synchronization. MPEG-1 is also a hybrid coder that is based on the 8 × 8 block DCT and 16 × 16 motion compensated macro-blocks with half pixel accuracy.

The most significant departure from H.261 in MPEG-1 is the introduction of the concept of bi-directional prediction, together with that of group of pictures (GOP). These concepts may be understood with the aid of Fig. 55.13. Each GOP commences with an intra-coded picture (frame), denoted I in the figure. The motion compensated predictive feedback loop of Fig. 55.1 is used to compress the subsequent inter-coded frames, marked P. Finally, the bi-directionally predicted frames, marked B in Fig. 55.13, are coded using motion compensated prediction based on both previous and successive I or P frames. Bidirectional prediction conforms essentially to the model of Fig. 55.1, except that the prediction signal is given by

$$a\,\tilde{f}[\vec{i} - \vec{d}_i^f] + b\,\tilde{f}[\vec{i} - \vec{d}_i^b]$$

In this notation, $\tilde{f}$ is a reconstructed frame, $\vec{d}_i^f$ (h_i^f, v_i^f, n^f), where (h_i^f, v_i^f) is a forward motion vector describing the motion from the previous I or P frame, and n^f is the frame distance to this previous I or P frame. Similarly, $\vec{d}_i^b = (h_i^b, v_i^b, -n^b)$, where (h_i^b, v_i^b) is a backward motion vector describing the motion to the next I or P frame, and n_b is the temporal distance to that frame. The weights a and b are given either by

$$\begin{matrix} a = 1 \\ b = 0 \end{matrix} , \quad \begin{matrix} a = 0 \\ b = 1 \end{matrix} , \quad or \quad \begin{matrix} a = n_b/(n_f + n_b) \\ b = n_f/(n_f + n_b) \end{matrix}$$

corresponding to forward, backward, and average prediction, respectively. Each bi-directionally predicted macro-block is independently assigned one of these three prediction strategies.

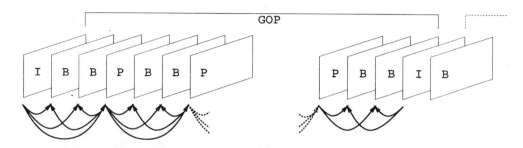

FIGURE 55.13 MPEG's group of pictures (GOP). Arrows represent direction of prediction. (*Source:* Taubman, D., Chang, E., and Zakhor, A., Directionality and scalability in subband image and video compression, in *Image Technology: Advances in Image Processing, Multimedia, and Machine Vision*, Jorge L.C. Sanz, Ed., Springer-Verlag, New York, 1996. With permission).

An MPEG-1 decoder can reconstruct the I and P frames without the need to decode the B frames. This is a form of temporal scalability and is the only form of scalability supported by MPEG-1.

55.4.3 MPEG-2

The second MPEG standard, MPEG-2, targets 60 fields/s interlaced television; however, it also supports progressive video. The targeted bit rate is between 2 Mbits/s and 10 Mbits/s. MPEG supports frames sizes up to $2^{14} - 1$ in each direction; however, the most popular formats are CCIR 601 (720 × 480), CIF (352 × 288), and SIF (352 × 240). The chrominance can be sampled in either the 4:2:0 (half as many samples in the horizontal and vertical directions), 4:2:2 (half as many samples in the horizontal direction only), or 4:4:4 (full chrominance size) formats.

MPEG-2 supports scalability by offering four tools: data partitioning, signal-to-noise-ratio (SNR) scalability, spatial scalability, and temporal scalability. Data partitioning can be used when two channels are available. The bit-stream is partitioned into two streams according to their importance. The most important stream is transmitted in the more reliable channel for better error resilience performance. SNR (rate), spatial, and temporal scalable bit-streams are achieved through the definition of a two-layer coder. The sequence is encoded into two bit-streams called lower and enhancement layer bit-streams. The lower bit-stream can be encoded independently from the enhancement layer using an MPEG-2 basic encoder. The enhancement layer is combined with the lower layer to get a higher quality sequence. The MPEG-2 standard supports hybrid scalabilities by combining these tools.

55.4.4 H.263

The international telecommunication union recommended H.263 standard to be used for video telephony (video coding for narrow telecommunications channels) [3]. Although, the bit rates specified are smaller than 64 Kbits/s, H.263 is also suitable for higher bit rates. H.263 supports three source formats: CIF (352 × 288 luminance and 176 × 144 chrominance), QCIF (176 × 144 luminance and 88 × 72 chrominance), and sub-QCIF (128 × 96 luminance and 64 × 48 chrominance).

The transformation used in H.263 is the 8 × 8 block-DCT. As in H.261, a macro-block consists of four luminance and two chrominance blocks. The motion estimation scheme is based on 16 × 16 and 8 × 8 pixel blocks. It alternates between them according to the residual error in order to achieve better performance.

Each inter-coded macro-block is assigned one or four motion vectors with half pixel accuracy. Motion estimation is done in both forward and backward directions.

H.263 provides a scalable bit-stream in the same fashion MPEG-2 does. This includes temporal, spatial, and rate (SNR) scalabilities. Moreover, H.263 has been extended to support coding of video objects of arbitrary shape. The objects are segmented and then coded the same way rectangular objects are coded with slight modification at the boundaries of the object. The shape information is embedded in the chrominance part of the stream by assigning the least used color to the parts outside the object in the rectangular frame. The decoder uses the color information to detect the object in the decoded stream.

55.4.5 MPEG-4

The moving picture expert group is developing a video standard that targets a wide range of applications including Internet multimedia, interactive video games, video-conferencing, video-phones, multimedia storage, wireless multimedia, and broadcasting applications. Such a wide range of applications needs a large range of bit rates, thus MPEG-4 supports a bit rate range of 5 Kbits/s to 4 Mbits/s. In order to support multimedia applications effectively, MPEG-4 supports synthetic and natural image and video in both progressive and interlaced formats. It is also required to provide object-based scalabilities (temporal, spatial, and rate) and object-based bit-stream manipulation, editing, and access [5, 25]. Since it is also intended to be used in wireless communications, it should be robust to high error rates. The standard is expected to be finalized in 1998.

Acknowledgment

The authors would like to acknowledge support from AFOSR grants F49620-93-1-0370 and F49620-94-1-0359, ONR grant N00014-92-J-1732, Tektronix, HP, SUN Microsystems, Philips, and Rockwell. Thanks to Iraj Sodagar of David Sarnoff Research Center for providing the zerotree coded video sequence.

References

[1] Bergeaud, F. and Mallat, S., Matching pursuit of images, *Proc. IEEE-SP Intl. Symp. on Time-Frequency and Time-Scale Analysis*, 330–333, Oct. 1994.

[2] *CCITT Recommendation H.261, Video codec for audio visual services at $p \times 64$ kbit/s*, 1990.

[3] *CCITT Recommendation H.263, Video codec for audio visual services at $p \times 64$ kbit/s*, 1995.

[4] Chao, T.-H., Lau, B. and Miceli, W.J., Optical implementation of a matching pursuit for image representation, *Optical Eng.*, 33(2), 2303–2309, July 1994.

[5] Chiarilione, L., MPEG and multimedia communications, *IEEE Trans. Circuits and Systems for Video Technology*, 7(1), 5–18, Feb. 1997.

[6] Cheung, G. and Zakhor, A., Joint source/channel coding of scalable video over noisy channels, *Proc. IEEE Intl. Conf. on Image Processing*, 3, 767–770, 1996.

[7] *Committee Draft of Standard ISO11172, Coding of Moving Pictures and Associated Audio*, ISO/MPEG 90/176, Dec. 1990.

[8] Gray, R., Vector quantization, *IEEE Acoustics, Speech, and Signal Processing Magazine*, 4–29, April 1984.

[9] Huffman, D., A method for the construction of minimal redundancy codes, *Proc. IRE*, 1098–1101, Sept. 1952.

[10] Jain, A.K., *Fundamentals of Digital Image Processing*, Prentice-Hall, Englewood Cliffs, NJ, 1989.

[11] Jayant, N. and Noll, P., *Digital Coding of Waveforms*, Prentice-Hall, Englewood Cliffs, NJ, 1984.

[12] Linde, Y., Buzo, A. and Gray, R.M., An algorithm for vector quantizer design, *IEEE Trans. Communications*, COM-28(1), 84–95, Jan. 1980.

[13] Lloyd, S.P., Least squares optimization in PCM, *IEEE Trans. Information Theory* (reproduction of a paper presented at the Institute of Mathematical Statistics meeting in Atlantic City, NJ, September 10-13, 1957), IT-28(2), 129–137, Mar. 1982.

[14] Mallat, S. and Zhang, Z., Matching pursuits with time-frequency dictionaries, *IEEE Trans. Signal Processing*, 41(12), 3397–3415, Dec. 1993.

[15] Malvar, H.S., *Signal Processing with Lapped Transforms*, Artech House, 1992.

[16] Martucci, S.A., Sodagar, I., Chiang, T. and Zhang, Y.-Q., A zerotree wavelet coder, *IEEE Trans. Circuits and Systems for Video Technology*, 7(1), 109–118, Feb. 1997.

[17] Max, J., Quantization for minimum distortion, *IRE Trans. Information Theory*, IT-16(2), 7-12, Mar. 1960.

[18] Minami, S. and Zakhor, A., An optimization approach for removing blocking effects in transform coding, *IEEE Trans. Circuits and Systems for Video Technology*, 5(2), 74–82, April 1995.

[19] Nasrabadi, N.M. and King, R.A., Image coding using vector quantization: a review, *IEEE Trans. Commun.*, 36(8), 957–971, Aug. 1988.

[20] Neff, R. and Zakhor, A., Very low bit rate video coding based on matching pursuits, *IEEE Trans. Circuits and Systems for Video Technology*, 7(1), 158–171, Feb. 1997.

[21] Rissanen, J. and Langdon, G., Arithmetic coding, *IBM J. Res. Dev.*, 23(2), 149–162, Mar. 1979.

[22] Rosenholtz, R. and Zakhor, A., Iterative procedures for reduction of blocking effects in transform image coding, *IEEE Trans. Circuits and Systems for Video Technology*, 2, 91–95, Mar. 1992.

[23] Ruf, M.J. and Modestino, J.W., Rate-distortion performance for joint source channel coding of images, *Proc. IEEE Intl. Conf. on Image Processing*, 2, 77–80, 1995.

[24] Shapiro, J.M., Embedded image coding using zerotrees of wavelet coefficients, *IEEE Trans. Signal Processing*, 41(12), 3445–3462, Dec. 1993.

[25] Sikora, T., The MPEG-4 video standard verification model, *IEEE Trans. Circuits and Systems for Video Technology*, 7(1), 19–31, Feb. 1997.

[26] Takishima, Y., Wada, M. and Murakami, H., Reversible variable length codes, *IEEE Trans. Commun.*, 43(2-4), 158–162, Feb.-April 1995.

[27] Tan, W., Chang, E. and Zakhor, A., Real time software implementation of scalable video codec, *IEEE Intl. Conf. on Image Processing*, 1, 17–20, 1996.

[28] Taubman, D. and Zakhor, A., Multirate 3-D subband coding of video, *IEEE Trans. Image Processing*, 3(5), 572–588, Sept. 1994.

[29] Taubman, D. and Zakhor, A., A common framework for rate and distortion based scaling of highly scalable compressed video, *IEEE Trans. Circuits and Systems for Video Technology*, 6(4), 329–354, Aug. 1996.

[30] Vetterli, M. and Kalker, T., Matching pursuit for compression and application to motion compensated video coding, *Proc. IEEE Intl. Conf. on Image Processing*, 1, 725–729, Nov. 1994.

[31] Woods, J., Ed., *Subband Image Coding*, Kluwer Academic Publishers, 1991.

[32] Taubman, D., Chang, E., and Zakhor, A., Directionality and scalability in subband image and video compression, in *Image Technology: Advances in Image Processing, Multimedia, and Machine Vision*, Jorge L.C. Sanz, Ed., Springer-Verlag, New York, 1996.

56

Digital Television

Kou-Hu Tzou
Hyundai Network Systems

56.1 Introduction

Digital television is being widely adopted for various applications ranging from high-end applications, such as studio recording, to consumer applications, such as digital cable TV and digital DBS (Direct Broadcasting Satellite) TV. For example, several digital video tape recording standards, using component format (D1 and D5), composite format (D2 and D3), or compressed component formats (Digital Betacam) are commonly used by broadcasters and TV studios [1]. These standards preserve the best possible picture quality at the expense of high data rates, ranging from approximately 150 to 300 Mbps. When captured in a digital format, the picture quality can be free from degradation during multiple generations of recording and playback, which is extremely attractive to studio editing. However, transmission of these high data-rate signals may be hindered due to lack of transmission media with an adequate bandwidth. Although it is possible, the associated transmission cost will be very high. The bit rate requirement for high definition television (HDTV) is even more demanding, which may exceed 1 Gbps in an uncompressed form. Therefore, data compression is essential for economical transmission of digital TV/HDTV.

Before motion-compensated DCT coding technology became mature in recent years, transmission of high-quality digital television used to be carried out at 45 Mbps using DPCM techniques. Today, by incorporating advanced motion-compensated DCT coding, comparable picture quality can be achieved at about one-third of the rate required by DPCM-coded video. For entertainment applications, the requirement on picture quality can be relaxed a little bit to allow more TV channels to fit into the same bandwidth. It is generally agreed that 3 to 4 Mbps for movie-originated or low-activity interlaced video (talk shows, etc.) materials is acceptable, and 6-8 Mbps for high-activity interlaced video (sports, etc.) is acceptable. The targeted bit rate for HDTV transmission is usually around 20 Mbps, which is chosen to match the available digital bandwidth of terrestrial broadcast channels allocated for conventional TV signals.

56.2 EDTV/HDTV Standards

The concept of HDTV system and efficient transmission format was originally explored by researches at NHK (Japan Broadcasting Corp.) more than 20 years ago [2] in order to offer superior picture quality while conserving bandwidth. Main HDTV features, including more scan lines, higher horizontal resolution, wider aspect ratio, better color representation, and higher frame rate, were identified. With these new features, HDTV is geared to offer picture quality close to that of 35-mm prints. However, the transmission of such a signal will require a very wide bandwidth. During the last 20 years, intensive research efforts have been engaged toward video coding to reduce bandwidth.

Currently there are two dominant HDTV production formats being used worldwide; one is the 1125-line/60-Hz system primarily used in Japan and the U.S. and the other is the 1250-line/50-Hz system primarily used in Europe. The main scanned raster characteristics of these two formats are listed in Table 56.1. The nominal bandwidth of the luminance component is about 30 MHz (in some cases, 20 MHz was quoted). Roughly speaking, the HDTV signal can carry about six times as much information as a conventional TV signal.

TABLE 56.1   Main Scanned Raster Characteristics of the 1125-line/60-Hz System and the 1250-line/50-Hz System

Format	Total scan lines per frame	Active lines per frame	Scanning format	Aspect ratio	Field rate
1	1125	1035	2:1 interlaced	16:9	60.00/59.94
2	1250	1152	2:1 interlaced	16:9	50.00

Development of HDTV transmission techniques in the early days was focused on bandwidth-compatible approaches that use the same analog bandwidth as a conventional TV signal. In some cases, in order to conserve bandwidth or to offer compatibility with an existing conventional signal or display, a compromised system—Enhanced or Extended Definition TV—was developed instead. The EDTV signal does not offer the picture quality and resolution required for an HDTV signal; however, it enhances the picture quality/resolution of conventional TV.

56.2.1 MUSE System

The most well-known early development in HDTV coding is the MUSE (Multiple Sub-Nyquist Sampling Encoding) system at NHK [3, 4]. The main concept of the MUSE system is adaptive spatial-temporal subsampling. Since human eyes have better spatial sensitivity for stationary or slow-moving scenes, the full spatial resolution is preserved while the temporal resolution is reduced for these scenes in the MUSE system. For fast moving scenes, the spatial sensitivity of human eyes declines so that reducing the spatial resolution will not significantly affect perceived picture quality. The MUSE signal is intended for analog transmission with a baseband bandwidth of 8.1 MHz, which can be fitted into a satellite transponder for a conventional analog TV signal. However, it should be noted that most signal processing employed in the MUSE system is in the digital domain. The MUSE coding technique was later modified to reduce bandwidth requirement for transmission over 6-MHz terrestrial broadcasting channels (Narrow-MUSE) [5]. Currently, MUSE-based HDTV programming is being broadcast regularly through a DBS in Japan.

56.2.2   HD-MAC System

A development similar to the MUSE was initiated in Europe as well. The system, HD-MAC (High-Definition Multiplexed Analog Component), is also based on the concept of adaptive spatial-temporal subsampling. Depending on the amount of motion, each block, consisting of 8 × 8 pixels, is classified into either the 20-, 40-, or 80-ms mode [6]. For a fast-moving block (the 20-ms mode), it is transmitted at the full temporal resolution, but at 1/4 spatial resolution. For a stationary or slow-moving block (the 80-ms mode), it is transmitted at full spatial resolution, but at 1/4 temporal resolution (25/4 frames/sec). For the 40-ms block, it is transmitted at half spatial and half temporal resolutions. The mode associated with each block is transmitted as side information through a digital channel at a bit rate nearly 1 Mbps. The subsampling process of the HD-MAC system is illustrated in Fig. 56.1, where the numbers indicate the corresponding fields of transmitted pixels and the "·" indicates a pixel not transmitted.

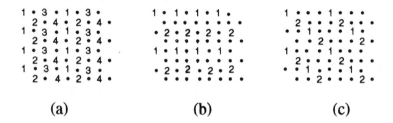

FIGURE 56.1   Adaptive spatial-temporal subsampling of the HD-MAC system. (a) The 80-ms mode for stationary to very-slow moving scenes, (b) the 40-ms mode for medium-speed moving scenes, and (c) the 20-ms mode for fast moving scene.

56.2.3   HDTV in North America

HDTV development in North America started much later than that in Japan and Europe. The Advisory Committee on Advanced Television Services (ACATS) was formed in 1987 to advise Federal Communications Commission (FCC) on the facts and circumstances regarding advanced television systems for terrestrial broadcasting. The proposed systems in early days were all intended for analog transmission [7]. However, the direction of U.S. HDTV development took a 180-degree turn in 1990 since General Instrument (GI) entered the U.S. HDTV race by submitting an all-digital HDTV system proposal to the FCC. The final contender in the U.S. HDTV race consisted of one analog system (Narrow-MUSE) and four digital systems, which all employed motion compensated DCT coding. Extensive testings on the five proposed systems were conducted in 1991 and 1992 and the testing concluded that there are major advantages in the performance of the digital HDTV systems and only the digital system shall be considered as the standard. However, none of these four digital systems was ready to be selected as the standard without implementing improvements.

With the encouragement from ACATS, the four U.S. HDTV proponents formed the Grand Alliance (GA) to combine their efforts for developing a better system. Two HDTV scan formats were adopted by the GA. The main parameters are shown in Table 56.2. The lower-resolution format, 1280 × 720, is only used for progressive source materials while the high-resolution format, 1920 × 1080, can be used for both progressive and interlaced source materials. The digital formats of GA HDTV are carefully designed to accommodate the square-pixel feature, which provides better interoperability with digital video/graphics in the computer environment. Since the main structure of MPEG-2 system and video coding standards were settled at that time and the MPEG-2 video coding standard provides extension to accommodate HDTV formats, the GA adopted MPEG-2 system and video coding (Main Profile (MP) at High Level

(HL)) standards for the U.S. HDTV, instead of creating another standard [8]. However, the GA HDTV adopted the AC-3 audio compression standard [9] instead of the MPEG-2 Layer 1 and Layer 2 audio coding.

TABLE 56.2 Main Scanned Raster Characteristics of the GA HDTV Input Signals

Active samples/line	Active lines per frame	Scanning format	Aspect ratio	Frame rate
1280	720	1:1 progressive	16:9 square pixels	60.00/59.94 30/29.97 24/23.976
1920	1080	1:1 progressive	16:9 square pixels	30/29.97 24/23.976
1920	1080	2:1 interlaced	16:9 square pixels	30/29.97

56.2.4 EDTV

EDTV refers to the TV signal that offers quality between the conventional TV and HDTV. Usually, EDTV has the same number of scan lines as the conventional TV, but offers better horizontal resolution. Though it is not a required feature, most EDTV systems offer a wide aspect ratio. When the compatibility with a conventional TV signal is of concern, the additional information (more horizontal details, side panels, etc.) required by the EDTV signal is embedded in the unused spatial-temporal spectrum (called *spectrum holes*) of the conventional TV signal and can be transmitted in either an analog or digital form [10, 11]. When the compatibility with the conventional TV is not required, EDTV can use the component format to avoid the artifacts caused by mixing of chrominance and luminance signals in the composite format. For example, several MAC (Multiplexed Analog Component) systems for analog transmission were adopted in Europe for DBS and cable TV applications [12, 13]. Usually, these signals offer better horizontal resolution and better color fidelity. There were many fully digital TV systems developed in the past. These systems that used adequate spatial resolution and higher bit rates were likely to achieve superior quality to the conventional TV and were qualified as EDTV [14]. Nevertheless, an efficient EDTV system is already embedded in the MPEG-2 video coding standard. Within the context of the standard, the 16:9 aspect ratio and horizontal and vertical resolutions exceeding the conventional TV can be specified in the "Sequence Header". When coded with adequate bit rates, the resulting signal can be qualified as EDTV.

56.3 Hybrid Analog/Digital Systems

Today, existing conventional TV sets and other home video equipment represent a massive investment by consumers. The introduction of any new video system that is not compatible with the existing system may face strong resistance in initial acceptance and may take a long time to penetrate households. One way to circumvent this problem during the transition period is to "simulcast" a program in both formats. The redundant conventional TV, being simulcast in a separate channel, can be phased out gradually when most households are able to receive the EDTV or HDTV signal. Intuitively, a more bandwidth efficient approach may be achieved if the transmitted conventional TV signal can be incorporated as a baseline signal and only the enhancement signal is transmitted in an additional channel (called "augmentation channel"). In order to facilitate the compatibility, an analog conventional TV signal has to be transmitted to allow conventional TV sets to receive the signal. On the other hand, digital video compression techniques may be employed to code the enhancement signals in order to accomplish the best compression efficiency. Such systems belong to the category of hybrid analog/digital system. A generic system structure for the hybrid analog/digital approach is shown in Fig. 56.2. Due to the interlacing processing used in TV standards, there are some unused holes in the spatial-temporal spectrum [15], which can be used to carry partial enhancement components as shown in Fig. 56.2.

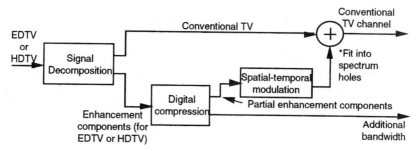

FIGURE 56.2   A generic hybrid analog digital HDTV coding system.

The Advanced Compatible Television System II (ACTV-II), developed by the consortium of NBC, RCA, and the David Sarnoff Research Center during the U.S. ATV standardization process, is an example of a hybrid system. The ACTV-II signal uses a 6-MHz channel to carry an NTSC compatible ACTV-I signal and uses an additional 6-MHz channel to carry the enhancement signal. The ACTV-I consists of a main signal, which is fully compatible with the conventional NTSC signal, and enhancement components (luminance horizontal details, luminance vertical-temporal details, and side-panel details of the wide-screen signal), which are transmitted in 3-D spectrum holes of the NTSC signal. The differences between the input HDTV signal and the ACTV-I signal are digitally coded using 4-band subband coding. The digitally coded video difference signal and digital audio signal require a total bandwidth of 20 Mbps and are expected to fit into the 6-MHz bandwidth by using the 16-QAM modulation. The enhancement components of the ACTV-I signal are digitally processed (time expansion and compression) and transmitted in an analog format. Nevertheless, they could be digitally compressed and transmitted, which would result in a hybrid analog/digital ACTV-I signal. For users with conventional TV sets, conventional TV pictures (4:3 aspect ratio) will be displayed. For users with an ACTV-I decoder and a wide screen (16:9) TV monitor, the wide-screen EDTV can be viewed by receiving the signal from the main channel. For those who have an ACTV-II decoder and an HDTV monitor, the HDTV picture can be received by using signals from both the main channel and the associated augmentation channel.

The HDS/NA system developed by Philips Laboratories is another example of hybrid analog/digital system where the augmentation signal is carried in a 3-MHz channel [16]. The augmentation signal consists of side panels to convert the aspect ratio from 4:3 to 16:9, and high-resolution spatial components. The side panels from two consecutive frames are combined into one frame of panels and are intraframe compressed by using DCT coding with a block size of 16 × 16 pixels. Both the horizontal and vertical high-resolution components are also compressed by intraframe DCT coding with some modifications to take into account the characteristics of these signals. The augmentation signals result in a total bit rate of 6 Mbps, which is expected to fit into a 3-MHz channel using modulation schemes with efficiency of 2 bits/Hz. However, the HDS/NA system was later modified into an analog simulcast system, HDS/NA-6, which occupies only a 6-MHz bandwidth and is intended to be transmitted simultaneously with a conventional TV in a *taboo* channel.

The augmentation-based hybrid analog/digital approach may be more efficient than the simulcast approach when both conventional TV and HDTV receivers have to be accommodated at the same time. However, for the augmentation-based approach, the reconstruction of the HDTV signal relies on the availability of the conventional TV signal, which implies that the main channel carrying the conventional TV signal can never be eliminated. Due to the inefficient use of bandwidth by the conventional analog TV signal, the overall bandwidth efficiency of the hybrid analog/digital approach is inferior to that of the fully digital-based simulcast approach. Furthermore, the system complexity of the hybrid approach is likely to be higher than that of the fully digital approach because it requires both analog and digital types of processing.

56.4   Error Protection and Concealment

Video coding results in a very compact representation of digital video by removing its redundancy, which leaves the compressed data very vulnerable to transmission errors. Usually, a single transmission error will only affect a single pixel for uncompressed data. However, due to the coding process employed, such as DCT transform and motion-compensated inter-field/frame prediction, a single transmission error may affect a whole block or blocks in consecutive frames. Furthermore, variable length coding is extensively used in most video coding systems, which is even more susceptible to transmission errors. For variable-length coded data, a single bit error may cause the decoder to lose track of codeword boundaries and results in decoding errors in subsequent data. Generally speaking, a single transmission error may result in noticeable picture impairment if no error concealment is applied.

56.4.1   FEC

The first effort to protect the compressed digital video in an environment susceptible to transmission errors should be to reduce transmission errors by employing forward error correction (FEC) coding. FEC adds redundancy, just opposite to data compression, in order to protect the underlying data from transmission errors. One trivial FEC example is to transmit each bit repeatedly, say three times. A single bit error in each three transmitted bits can be easily corrected by a majority-vote circuit. There are many known FEC techniques which can achieve much better protection without devoting too much bandwidth to redundancy. Today, two types of FEC codes are popularly used for digital transmission over various media. One is Reed-Solomon (RS) code, which belongs to the class of block codes. The other is the convolutional code, which usually operates on continuous data.

The RS code appends a number of redundant bytes to a block of data to achieve error correction. Usually $2n$ redundant bytes can correct up to n byte errors. When a higher level protection is required, more redundant bytes can be attached or alternatively the redundant bytes can be added to shorter data blocks. For digital transmission using the MPEG-2 transport format, in order to maintain the structure of the MPEG-2 transport packets, the (204,188) RS code has been particularly chosen by many standards, which appends 16 redundant bytes to each MPEG-2 transport packet. On the other hand, the U.S. GA-HDTV chose the (207,187) RS code, where the RS redundancy computation is based on the 187-byte data block with the sync byte excluded.

The convolutional code is a powerful FEC code, which generates m output bits for every n input bits. The code rate, r, is defined as $r = n/m$. The output bits are not only determined by the current input bits, but also depend on previous input bits. The depth of the previous input data affecting the output is called the constraint length, k. The output stream of the convolutional code is the result of a generator function convolved with the input stream. Viterbi decoding is an efficient algorithm to decode convolutionally coded data. The complexity of the Viterbi algorithm is proportional to 2^k. Therefore, longer constraint length results in higher decoding complexity. However, longer constraint length also improves FEC performance. A lower rate convolutional code provides more protection at the expense of higher redundancy. For a $r = 1/2$ and $k = 7$ convolutional code, a BER of 10^{-2} can be reduced to below 10^{-5}.

In order to maintain nearly error-free transmission, a very low BER has to be achieved. For example, if an average error-free interval of two hours needs to be achieved for a 6-Mbps compressed bit stream, the required BER is 2.3×10^{-11}. For some transmission media that have limited carrier-to-noise ratio, such a low BER may not be achievable using the RS code or convolutional code alone. However, an extremely powerful coding can be accomplished by concatenating the RS code and the convolutional code, where the RS code (called outer code) is used toward the source or sink side and the convolutional code (called inner code) is used toward the channel side. An interleaver to spread bursts of errors is usually used between the inner and outer code in order to improve error correction capability. The interleaver needs to be carefully designed so that the locations of the sync byte in the ATM packets remain unchanged through the interleaver. A block diagram of the concatenated RS code and convolutional code is shown in Fig.56.3. Some simulations showed that satisfactory performance can be achieved by using the concatenate codes for digital video transmission over the satellite link [17]. In [17], the overall BER is about 2^{-11}, which corresponds to a BER of about $2 \cdot 10^{-4}$ using the convolutional code only.

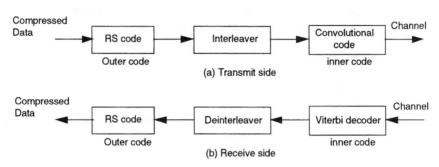

FIGURE 56.3   Block diagram of concatenated RS code and convolutional code.

56.4.2   Error Detection and Confinement

While FEC techniques can improve BER significantly, there are still chances that errors may occur. As mentioned earlier, a single bit error may cause catastrophic effects on compressed digital video if precaution is not exercised. To avoid the infinite error propagation, one needs to identify the occurrences of errors and to confine the errors during decoding. Due to the use of variable length coding, a single bit error in the compressed bit stream may cause the decoder to lose track of codeword boundaries. Even though the decoder may regain code synchronization later, the number of decoded data may be more or less than the actual number of samples transmitted, which will affect proper display of the remaining samples. To avoid error propagation, compressed data need to be organized into smaller self-contained data units with *unique words* to identify the beginning or boundaries of the data unit. In case transmission errors occur in preceding data units, the current data unit can still be properly decoded. In the MPEG-2 video coding standard, the *slice* is the smallest self-contained data unit, which has a unique 32-bit *slice_start_code* and information regarding its location within a picture [18]. Therefore, a transmission error in one *slice* will not affect the proper decoding of subsequent *slices*. However, for inter-field/frame coded pictures, the artifacts in the error-contaminated slice will still propagate to subsequent pictures, which use this slice as reference. Error concealment is a technique to mitigate artifacts caused by transmission errors in the reconstructed picture.

56.4.3   Error Concealment

For DCT-based video coding, some analytic work was conducted in [19] to derive an optimal reconstruction method based on received blocks with missing DCT coefficients. The solution consists of three linear interpolation in the spatial, temporal, and frequency domains from the boundary data, reconstructed reference block, and received DCT block, respectively. When the complete block is missing, the optimal solution becomes a linear combination of a block replaced by the corresponding block in the previous frame and a spatially interpolated block from boundary pixels. This method needs to go through an iterative process to restore damaged data when consecutive blocks are corrupted by errors. The above concealment technique was further improved in [20, 21] by incorporating an adaptive spatial-temporal interpolation scheme and a multi-directional spatial interpolation scheme.

When a temporal concealment scheme is used, the picture quality in the moving area can be improved by incorporating a motion compensation technique The motion vector for a missing or corrupted macroblock can be estimated from the motion vectors of surrounding macroblocks. For example, the motion vector can be estimated based on the averaged motion vector from the macroblocks above and below the underlying block, as suggested in the MPEG-2 video standard. However, when the neighboring reference macroblocks are intra-coded, there are no motion vectors associated with these macroblocks. The MPEG-2 video coding standard allows transmission of the *"concealment_motion_vectors"* associated with intra-coded macroblocks, which can be used to estimate the motion vector for the missing or corrupted macroblock.

56.4.4 Scalable Coding for Error Concealment

When the requirement of error-free transmission cannot be met, it may be useful to provide different protection of underlying data according to the visual importance of the compressed data. This will be useful for transmission media which have different delivery priorities or provide different levels of FEC protection for underlying data. The data that can be used to reconstruct basic pictures are usually treated as high-priority data while the data used to enhance the pictures are treated as low-priority data. For these visually important data, high redundancy is used to offer more protection (or high priority in a cell-based transport system). Therefore, the high-priority data can always be reliably delivered. On the other hand, any errors in the low-priority data will only result in minor degradation. Therefore, if any error is detected in the low-priority data, the affected data can be discarded without significantly degrading the picture quality. Nevertheless, if concealment techniques by spatial-temporal interpolation as described above can be applied to affected areas, this will further improve picture quality. The scalable source coding processes the underlying signal in a hierarchical fashion according to the spatial resolution, temporal resolution, or picture signal-to-noise ratio, and organizes the compressed data into layers so that a lower-level data set can be used to reconstruct a basic video sequence and the quality can be improved by adding higher levels. Many coding systems can offer the scalable coding feature if the underlying data is carefully partitioned [22, 23]. The MPEG-2 video coding standard also offers scalable extension to accommodate spatial, temporal, and SNR scalability.

56.5 Terrestrial Broadcasting

In conventional analog TV standards, in order to allow low-cost TV receivers to acquire the carrier and subcarrier frequencies easily, the transmitted analog signals always contain these two frequencies in high strength, which are the potential cause for co-channel and adjacent-channel interferences. This problem becomes more prominent in the terrestrial broadcasting environment, where the transmitter of an undesired signal (adjacent channel) may be much closer than that of a desired signal. The strong undesired signal may interfere with the desired weak signal. Therefore, some of the terrestrial broadcasting channels (*taboo* channels) are prohibited in the same coverage area in order to reduce the potential interference. In digital TV transmission, the power spectrum of the signal is widespread over the allocated spectrum, which substantially reduces the potential interference. On the other hand, the bandwidth efficiency of digital coding may significantly increase the capacity of terrestrial broadcasting. Therefore, digital video coding is a very attractive alternative to solving the channel congestion problem in major cities.

56.5.1 Multipath Interference

One notorious impairment of the terrestrial broadcasting channel is the multipath interference, which manifests as the ghost effect in received pictures. For digital transmission, the multi-path interference will cause signal distortion and degrade system performance. An effective way to cope with multipath interference is to use adaptive equalization, which can restore the impaired signal by using a known training data sequence. The GA HDTV system for terrestrial broadcasting adopted this method to overcome the multipath problem [9]. A very different approach—Coded Orthogonal Frequency Division Multiplexing (COFDM)— has been advocated in Europe for terrestrial broadcasting [10]. The COFDM technology employs multiple carriers to transport parallel data so that the data rate for each carrier is very low. The COFDM system is carefully designed to ensure that the symbol duration for each carrier is longer than the multipath delay. Consequently, the effect of multipath interference will be significantly reduced. The carrier spacing of the COFDM system is carefully arranged so that each subcarrier is orthogonal to the other subcarriers, which achieves high spectrum efficiency. A performance simulation of COFDM for terrestrial broadcasting was reported in [24], which indicated that COFDM is a viable alternative to digital transmission of 20 Mbps in a 6-MHz terrestrial channel.

56.5.2 Multi-Resolution Transmission

In terrestrial broadcasting, the carrier-to-noise ratio (CNR) of the received signal decreases gradually when the distance between a receiver and the transmitter increases. In an analog transmission system, the picture quality usually degrades gracefully when the CNR decreases. In a digital transmission system, a lower CNR will result in a higher BER and the decoded picture contaminated by errors may become unusable when the BER exceeds a certain threshold. A technique to extend the coverage area of terrestrial broadcasting is to use scalable source coding in conjunction with multiresolution (MR) channel coding [25, 26]. In MR modulation, the constellation of the modulated signal is carefully organized in a hierarchical fashion so that a low-density modulation can be derived from the constellation with high protection while a high-density modulation can be achieved by further demodulation of the received signal. An example of MR modulation using QAM (Quadrature Amplitude Modulation) is shown in Fig. 56.4, where the nonuniform constellation represents 4-QAM/16-QAM MR modulation. The scalable source coding processes the underlying signal in a hierarchical fashion according to the spatial resolution, temporal resolution, or picture signal-to-noise ratio, and organizes the compressed data in layers so that a lower-level data set can be used to reconstruct a basic video sequence and the quality can be improved by adding more levels. The MPEG-2 video coding standard also offers scalable extension to accommodate spatial, temporal, and SNR scalability. In light of the fact that MPEG-2-based systems are being widely used for digital satellite TV broadcasting and being adopted by the Digital Audio-Visual Interactive Council (DAVIC) for the set-top box standard, MPEG-2-based scalable coding in conjunction with the MR modulation likely will be used to offer graceful degradation in terrestrial broadcasting if it is desired.

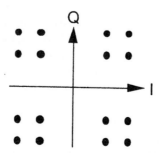

FIGURE 56.4   The constellation of a MR modulation for a 4-QAM/16-QAM.

56.6   Satellite Transmission

Satellite video broadcasting provides an effective way for point-to-multipoint video distribution. It has been widely used in video distribution to cable headends and to satellite TVRO (TV Receive Only) users for years. Due to recent development in high-powered Kuband satellite transponders, satellite video broadcasting to small home antennas becomes feasible. The cost of consumer satellite receive systems, including receive dish antenna/LNB and Integrated Receiver/Decoder (IRD) falls below U.S. $600 today and is expected to decline gradually. Furthermore, due to advances in digital video compression technology, the capacity of the satellite transponder has been increased substantially. Today, digital TV with 100 or more channels per satellite is being broadcast in North America.

In an analog satellite transmission system, the baseband video signal is FM modulated and transmitted from an uplink site to geo-stationary satellite. The signal is received by the satellite and retransmitted downward at a different frequency. At a receive site, the signal is received by the receive antenna, block frequency converted to a lower frequency band, and carried through a coax cable to an indoor IRD unit. A simplified system is shown in Fig. 56.5.

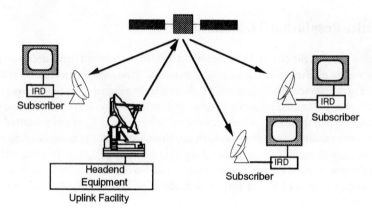

FIGURE 56.5   A satellite video transmission system.

Due to constraint on the power limit, the satellite transmitters are normally operated in the saturated mode, which introduces system nonlinearity and causes waveform distortion. The available signal-to-noise ratio for satellite channels is usually much lower than that for cable channels. In order to overcome the nonlinearity as well as to improve the signal-to-noise ratio, the FM technique is always used for analog TV transmission over satellites. For Ku-band applications, a 27-MHz bandwidth is normally allocated to carry one analog TV signal.

For digital transmission over satellite, the QPSK (Quadrature Phase Shift Keying) modulation is the most popular technique. The QAM (Quadrature Amplitude Modulation) technique, which requires a linear system response, is not suitable for satellite applications. In the North American region, the Ku-band DBS uses the 12/14 GHz frequency band (14 GHz for uplink and 12 GHz for downlink), which allows the subscribers to use a smaller dish antenna. However, the Ku-band link is more susceptible to rain fading than the C-band link and, therefore, more margin for rain fading is required for the Ku-band link. Due to the typical low signal-to-noise ratio available for satellite links, powerful coding techniques are required in order to achieve a high-quality link. For an MPEG-2 video stream at 10 Mbit/s, an average of 1-h error-free transmission will require a BER of 2.778×10^{-11}.

Satellite link has been notorious for its nonlinearily and relatively low carrier-to-noise ratio. Due to the nonlinearity, any amplitude modulation technique is discouraged in satellite environment. Without forward error correction coding, typical satellite links can only achieve a BER around 10^{-2} to 10^{-5}.

This BER is far from the targeted quality of service for compressed digital video. As discussed earlier, concatenated inner and outer codes are very effective for satellite communications, which can reduce the BER to below 10^{-12} from 10^{-4}.

Recently, European Broadcasting Union (EBU) launched a project intended to set a standard for digital video transmission over satellite, cable, and Satellite Master Antenna TV (SMATV) channels. A draft standard [27] was published by EBU/European Telecommunications Standards Institute (ETSI). This draft specifies a powerful error correction scheme based on concatenation of convolutional and Reed-Solomon (RS) codes as shown in Fig. 56.3. The convolutional code can be configured to operate at different rates, including 1/2, 2/3, 3/4, 5/6, 7/8, and 1 to optimize the performance for transponder power and bandwidth. At the receive end, Viterbi decoding with soft-decision is often used to decode the convolutional code. By using the convolutional code alone, a BER between 10^{-3} and 10^{-8} may be achieved for typical satellite links. However, this is still not adequate for real-time digital video applications.

In order to further improve the BER performance, an outer code using the Reed-Solomon code is applied to correct errors remaining uncorrected by the convolutional code. Channel errors generated at the output of Viterbi decoder tend to occur in bursts. The Reed-Solomon code operates on byte-oriented data and is effective in correcting burst errors. To improve the effectiveness of the RS code, an interleaver is usually used between the convolutional code and the RS code. By using the (204,188) RS code and a convolutional interleaver of depth 12, the BER of $2 \cdot 10^{-4}$ for the convolutional code can be improved to around 10^{-11}. A recent report [28] showed that a BER around 10^{-11} can be achieved for typical high-powered DBS with bit rates ranging from 23 to 41 Mbit/s by using concatenate convolutional and RS codes.

56.7 ATM Transmission of Video

ATM is a cell-based transport technology that multiplexes fixed-length cells from a variety of sources to a variety of remote locations. Each ATM cell consists of a 5-byte header and 48-byte payload. The routing, flow control and payload type information is carried in the header, which is then protected by a 1-byte error correction code. However, unlike the packet data communication, ATM is a connection-oriented protocol. Connections, either permanent, semi-temporary, or permanent, between ATM users are established before data exchanges commence. The header information in each cell determines to which port at an ATM switch the cell should be routed. This substantially reduces the processing complexity required in a switching equipment. The flexibility of ATM technology allows both constant rate and variable rate services to be easily offered through the network. Also, it allows multimedia services, such as video, voice, and data of different characteristics to be multiplexed into a single stream and delivered to customers.

56.7.1 ATM Adaptation Layer for Digital Video

In order to carry data units other than the 48-octets payload size in ATM cells, an adaptation layer is needed. The ATM Adaptation Layer (AAL) provides for segmentation and reassembly (SAR) of higher-layer data units and detection of errors in transmission. Five AALs are specified in ITU-T Recommendation, I.363. AAL1 is intended for constant bit rate services while AAL2 is intended for variable bit rate services with a required timing relationship between the source and destination. AAL3/4 is intended for variable bit rate services that require bursty bandwidth. AAL5 is a simple and efficient adaptation layer intended to reduce the complexity and overhead of AAL3/4. Both AAL1 and AAL5 have been seriously considered as a candidate for real-time digital video applications. However, the AAL5 was adopted by the ATM Forum as the standard for Audiovisual Multimedia Services (AMS) [29].

The standard process of ATM is undertaken by several international standard bodies such as ATM Forum, and International Telecommunication Union–Transmission (ITU-T) Study Groups (SG) 9, 13, and 15. For digital television transmission, the MPEG-2 transport standard seems to be the sole format being considered. MPEG-2 transport standard relies on frequent and low-jitter delivery of transport stream (TS) packets containing PCR (Presentation Clock References) to recover the 27-MHz clock at the receiving end. There are several key parameters in designing an AAL for digital video, which include packaging efficiency, complexity, error handling capability and performance, and PCR jitter. When AAL1 is employed, each MPEG-2 TS packet is mapped into 4 ATM cells as shown in Fig. 56.6(a). A 1-byte AAL1 header is inserted into the first payload byte of each ATM cell. The AAL1 header contains a sequence number field and a sequence number protection field. The AAL1 uses the synchronous residual time stamp (SRTS) method to support source clock recovery.

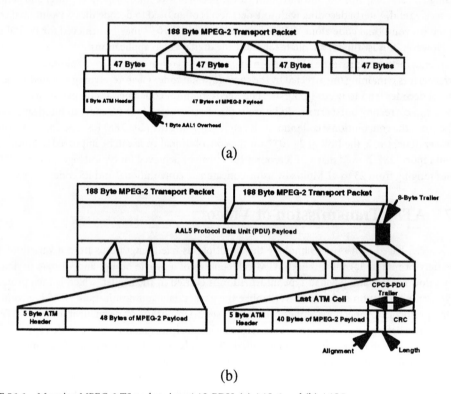

(a)

(b)

FIGURE 56.6   Mapping MPEG-2 TS packets into AAL PDU. (a) AAL-1 and (b) AAL5.

The AAL5 specified in [29] maps N MPEG-2 Single Program TS (SPTS) packets into an AAL5-SDU (service data unit) unless there are fewer than N TS packets left in the sequence. In the case when there are fewer than N packets left in the SPTS, the last AAL5-SDU contains all the remaining packets. The default value for N is 2, which results in a default SDU size of 376 bytes. This default SDU along with an 8-byte trail fits nicely into the payloads of 8 ATM cells, as shown in Fig. 56.6(b). The trailer contains a 2-byte alignment field, a 2-byte length indicator field, and a 4-byte CRC field. For constant bit rate transmission, the MPEG-2 SPTS is considered as a constant packet rate (CPR) stream of information, which implies that the interarrival time between packets of the MPEG-2 TS is constant. In order to ensure satisfactory timing recovery, the time interval of the last byte containing the PCR should be constant.

The AAL5 is meant for both constant bit rate and variable bit rate applications while the AAL1 is mainly intended for constant bit rate applications. The AAL5 contains a 4-byte CRC field and a 2-byte length indicator field to check the payload integrity. On the other hand, the AAL1 only offers sequence integrity to detect lost cells. The most attractive factor of AAL5 is the wide support of major service and equipment vendors.

56.7.2   Cell Loss Protection

In the ATM environment, cells may be corrupted due to transmission errors or lost due to traffic congestion. The transmission bit error rate usually is very small for fiber-based systems. However, the cell loss due to congestion seems to be unavoidable in order to increase link utilization efficiency. Depending on how compressed data is mapped into ATM cells, the loss of a single cell may corrupt a number of cells. In the ATM header, there is 1-bit information to indicate the delivery priority of the underlying payload. This priority bit can be used to cope with the cell loss issue. To take advantage of the priority bit, the coding systems will have to separate the compressed data into high- and low-priority layers and to pack the data into cells with a corresponding priority indicator. When network congestion occurs, these cells labeled with low priority are subject to discarding at the switch. Since the low-priority cells carry visually less important information, the impairments in the reconstructed low-priority data will be less objectionable. Some two-layer coding techniques were proposed for MPEG-2 video and have shown significant improvement over a single-layer coding under cell loss circumstance [20, 21].

References

[1] Strachan, D. and Conrad, R., Serial video basics, *SMPTE J.,* 254-257, Aug. 1994.

[2] Fujii, T. et al., Film simulation for high definition TV picture and subjective test of picture quality, *NHK Technical Report,* 18, No. 11, 1975. Some papers related to HDTV camera and display also appeared in the same issue.

[3] Nonomiya, Y., MUSE coding system for HDTV broadcast, *Proc. 1st Intl. HDTV Signal Processing Workshop,* Torino, Italy, 1986.

[4] Nonomiya, Y. et al., HDTV broadcasting and transmission system—MUSE, *Proc. 2nd Intl. HDTV Signal Processing Workshop,* Torino, Italy, 1986.

[5] Nishizawa, et al., HDTV and transmission system—MUSE and its family, *Proc. 1988 Intl. Broadcasting Conf.,* 37-40, 1988.

[6] Vreeswijk, F.W.P. et al., An HD-MAC coding system, *Proc. 2nd Intl. HDTV Signal Processing Workshop,* Torino, Italy, 1988.

[7] Hopkins, R., Advanced televisions systems, *IEEE Trans. Consumer Electronics,* 34(1), 1-15, Feb. 1988.

[8] United States Advanced Television Systems Committee, *Digital Television Standard for HDTV Transmission,* Doc. A/53, April 12, 1995.

[9] United States Advanced Television Systems Committee, *Digital Audio Compression* (AC-3), Doc. A/52, 1994.

[10] Isnardi, M. et al., Decoding issues in the ACTV system, *IEEE Trans. Consumer Electronics,* 34(1), 111-120, Feb. 1988.

[11] Kawai, K. et al., A wide screen EDTV, *IEEE Trans. Consumer Electronics,* 35(3), 133-141, Aug. 1989.

[12] Gardiner, P.N., The UK D-MAC/packet standard for DBS, *IEEE Trans. Consumer Electronics,* 34(1), 128-136, Feb. 1988.

[13] Garault, T. et al., A digital MAC decoder for the display of a 16/9 aspect ratio picture on a conventional TV receiver, *IEEE Trans. Consumer Electronics,* 34(1), 137-146, Feb. 1988.

[14] Jalali, A. et al., A component CODEC and line multiplexer, *IEEE Trans. Consumer Electronics,* 34(1), 156-165, Feb. 1988.

[15] Fukinuki, T. and Hirano, Y., Extended definition TV fully compatible with existing standards, *IEEE Trans. Commun.,* COM-32, 948-953, Aug. 1984.

[16] Tsinberg, M., Compatible introduction of HDTV: The HDS/NA system, *Proc. 3rd Intl. HDTV Signal Processing Workshop,* Torino, Italy, 1989.

[17] Cominetti, M. and Morello, A., Direct-to-home digital multi-programme television by satellite, *Proc. Intl. Broadcasting Convention,* 358-365, Sept. 16-20, 1994.

[18] ISO/IEC IS 13818-2/ITU-T Recommendation H.262, Information technology— generic coding of moving picture and associated audio—Part 2: Video, ISO/IEC, May 10, 1994.

[19] Zhu, Q.-F., Wang, Y. and Shaw, L., Coding and cell-loss recovery in DCT-based packet video, *IEEE Trans. Circuits and Systems for Video Technology,* 3(3), 238-247, June 1993.

[20] Sun, H. and Zdepski, J., Adaptive error concealment algorithm for MPEG compressed video, *Proc. SPIE, Visual Comm. and Image Proc.,* 1818, 814-824, Nov. 1992.

[21] Kwok, W. and Sun, H., Multi-directional interpolation for spatial error concealment, *IEEE Trans. Consumer Elec.,* 39(3), 455-460, Aug. 1993.

[22] Yu, Y. and Anastassiou, D., High quality two layer video coding using MPEG-2 syntax, *Proc. 6th Intl. Workshop on Packet Video,* A4.1-4, Portland, Oregon, Sept. 26-27, 1994.

[23] Chan, S.K. et al., Layer transmission of MPEG-2 video in ATM environment, *Proc. 6th Intl. Workshop on Packet Video,* D1.1-4, Portland, Oregon, Sept. 26-27, 1994.

[24] Wu, Y. and Zou, W., Performance simulation of COFDM for TV broadcasting application, *SMPTE J.,* 258-265, May 1995.

[25] Schreiber, W.F., Advanced television systems for terrestrial broadcasting: some problems and some proposed solutions, *Proc. IEEE,* 83(6), 958-981, June 1995.

[26] deBot, P.G.M., Multiresolution transmission over the AWGN Channel, Technical Reports, Philips Labs., Eindhoven, The Netherlands, June 1992.

[27] EBU/ETSI JTC, *Draft Digital Broadcasting System for Television, Sound and Data Services; Framing Structure, Channel Coding and Modulation for 11/12 GHz Satellite Services,* Draft prETS 300 421, June 1994.

[28] Cominetti, M. and Morello, A., Direct-to-home digital multi-programme television by satellite, *Proc. Intl. Broadcasting Conv.,* 358-365, June 1994.

[29] ATM Forum, *Audiovisual Multimedia Services: Video on Demand Implementation Agreement 1.0,* ATMF/95-0012R6, Oct. 1995.

57

Stereoscopic Image Processing[1]

Reginald L. Lagendijk
Delft University of Technology

Ruggero E.H. Franich
AEA Technology,
Culham Laboratory

Emile A. Hendriks
Delft University of Technology

57.1 Introduction

Static images and dynamic image sequences are the projection of time-varying three-dimensional real world scenes onto a two-dimensional plane. As a result of this planar projection, depth information of objects in the scene is generally lost. Only by *cues* such as shadow, relative size and sharpness, interposition, perspective factors, and object motion, can we form an impression of the depth organization of the real world scene.

In a wide variety of image processing applications, explicit depth information is required in addition to the scene's gray value information (representing intensities, color, densities, etc.) [2, 4, 7]. Examples of such applications are found in 3-D vision (robot vision, photogrammetry, remote sensing systems); in medical imaging (computer tomography, magnetic resonance imaging, microsurgery); in remote handling of objects, for instance in inaccessible industrial plants or in space exploration; and in visual communications aiming at virtual presence (conferencing, education, virtual travel and shopping, virtual reality). In each of these cases, depth information is essential for accurate image analysis or for enhancing the realism. In remote sensing the terrain's elevation needs to be accurately determined for map production, in remote handling an operator needs to have precise knowledge of the three-dimensional organization of the area to avoid collisions and misplacements, and in visual communications the quality and ease of information exchange significantly benefits from the high degree of realism provided by scenes with depth.

Depth in real world scenes can be explicitly measured by a number of range sensing devices such as by laser range sensors, structured light, or ultrasound. Often it is, however, undesirable or unnecessary to have separate systems for acquiring the intensity and the depth information because of the relative low resolution of the range sensing devices and because of the question of how to fuse information from different types of sensors.

---

[1]This work was supported in part by the European Union under the RACE-II project DISTIMA and the ACTS project PANORAMA.

An often used alternative to acquire depth information is to record the real world scene from different perspective viewpoints. In this way, multiple images or (preferably time-synchronized) image sequences are obtained that implicitly contain the scene's depth information. In the case that multiple views of a single scene are taken without any specific relation between the spatial positions of the viewpoints, such recordings are called *multiview images*. Generally speaking, when recordings are obtained from an increasing number of different viewpoints, the 3-D surfaces and/or interior structures of the real world scene can be reconstructed more accurately. The terms *stereoscopic image* and *stereoscopic image sequence* are reserved for the special case that two perspective viewpoints are recorded or computed such that they can be viewed by a human observer to produce the effect of natural depth perception (see Fig. 57.1). Therefore, the two views are required to be recorded under specific constraints such as the cameras' separation, convergence angle, and alignment [8]. Stereoscopic images are not truly 3-D images since they merely contain information about the 2-D projected real world surfaces plus the depth information at the perspective viewpoints. They are, therefore, sometimes called 2.5-D images.

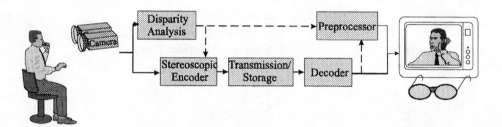

FIGURE 57.1 Illustration of system for stereoscopic image (sequence) recording, processing, transmission, and display.

In the broadest meaning of the word, a digital stereoscopic system contains the following components: stereoscopic camera setup, depth analysis of the digitized and recorded views, compression, transmission or storage, decompression, preprocessing prior to display, and, finally, the stereoscopic display system. The emphasis here is on the image processing components of this stereoscopic system; that is, depth analysis, compression, and preprocessing prior to the stereoscopic display. Nonetheless, we first briefly review the perceptual basis for stereoscopic systems and techniques for stereoscopic recording and display in Section 57.2. The issue of depth or *disparity analysis* of stereoscopic images is discussed in Section 57.3, followed by the application of compression techniques to stereoscopic images in Section 57.4. Finally, Section 57.5 considers the issue of stereoscopic image interpolation as a preprocessing step required for multiviewpoint stereoscopic display systems.

57.2 Acquisition and Display of Stereoscopic Images

The human perception of depth is brought about by the hardly understood brain process of fusing two planar images obtained from slightly different perspective viewpoints. Due to the different viewpoint of each eye, a small horizontal shift exists, called *disparity*, between corresponding image points in the left and right view images on the retinas. In stereoscopic vision, the objects to which the eyes are focused and accommodated have zero disparity, while objects to the front and to the back have negative and positive disparity, respectively, as is illustrated in Figure 57.2. The differences in disparity are interpreted by the brain as differences in depth ΔZ.

In order to be able to perceive depth using recorded images, a stereoscopic camera is required which consists of two cameras that capture two different, horizontally shifted perspective viewpoints. This results in a shift (or disparity) of objects in the recorded scene between the left and the right view depending on

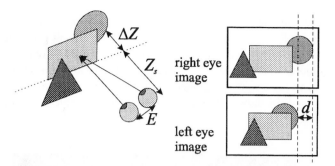

FIGURE 57.2 Stereoscopic vision, resulting in different disparities depending on depth.

their depth. In most cases, the interaxial separation or baseline B between the two lenses of the stereoscopic camera is in the same order as the eye distance E (6 to 8 cm). In a simple camera model, the optical axes are assumed to be parallel. The depth Z and disparity d are then related as follows:

$$d = \lambda \frac{B}{\lambda - Z}, \tag{57.1}$$

where λ is the focal length of the cameras. Fig. 57.3(a) illustrates this relation for a camera with $B = 0.1$ m and $\lambda = 0.05$ m. A more complicated camera model takes into account the convergence of the camera

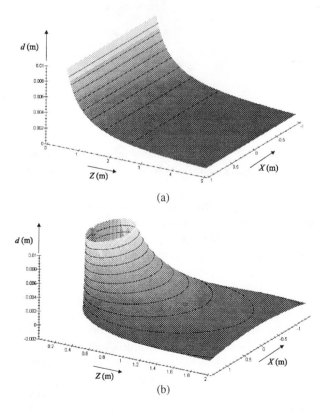

FIGURE 57.3 (a) Disparity as a function of depth for a sample parallel camera configuration; (b) disparity for a sample converging camera configuration.

axes with angle β. The resulting relation between depth and disparity, which is a much more elaborate expression in this case, is illustrated in Fig. 57.3(b) for the same camera parameters and $\beta = 1°$. It shows that, in this case, the disparity is not only dependent on the depth Z of an object, but also on the horizontal object position X. Furthermore, a converging camera configuration also leads to small vertical disparity components, which are, however, often ignored in subsequent processing of the stereoscopic data. Figures. 57.4(a) and (b) show as an example a pair of stereoscopic images encountered in video communications.

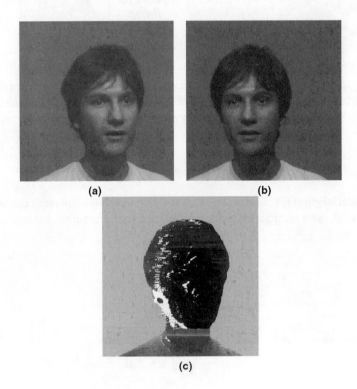

FIGURE 57.4   The left (a) and right (b) view image from a stereoscopic image pair. (c) Disparity field in the stereoscopic image pair represented as gray values (black is foreground, gray is background, white is occlusion).

When recording stereoscopic image sequences, the camera setup should be such that, when displaying the stereoscopic images, the resulting shifts between corresponding points in the left and right view images on the display screen allow for comfortable viewing. If the observer is at a distance Z_s from the screen, then the observed depth Z_{obs} and displayed disparity d are related as:

$$Z_{\mathrm{obs}} = Z_s \frac{E}{E - d} . \tag{57.2}$$

In the case that the camera position and focusing are changing dynamically, as is the case, for instance, in stereoscopic television production where the stereoscopic camera may be zooming, the camera geometry is controlled by a set of production rules. If the recorded images are to be used for multiviewpoint stereoscopic display, a larger interaxial lens separation needs to be used, sometimes even up to 1 m. In any case, the camera setup should be geometrically calibrated such that the two cameras capture the same part of the real world scene. Furthermore, the two cameras and A/D converters need to be electronically calibrated to avoid unbalances in gray value of corresponding points in the left and right view image.

The stereoscopic image pair should be presented such that each perspective viewpoint is seen only by one of the eyes. Most practical state-of-the-art systems require viewers to wear special viewing glasses [6]. In a *time-parallel* display system, the left and right view images are presented simultaneously to the viewer. The views are separated by passive viewing glasses such as red-green viewing glasses requiring the left and right view to be displayed in red and green, respectively, or polarized viewing glasses requiring different polarization of the two views. In a *time-sequential* stereoscopic display, the left and right view images are multiplexed in time and displayed at a double field rate, for instance 100 or 120 Hz. The views are separated by means of the active synchronized shuttered glasses that open and close the left and right eyeglasses depending on the viewpoint being shown. Alternatively, lenticular display screens can be used to create spatial interference patterns such that the left and right view images are projected directly into the viewer's eyes. This avoids the need of wearing viewing glasses.

57.3 Disparity Estimation

The key difference between planar and stereoscopic images and image sequences is that the latter implicitly contains depth information in the form of disparity between the left and right view images. Not only is the presence of disparity information essential to the ability of humans to perceive depth, disparity can also be exploited for automated depth segmentation of real world scenes, and for compression and interpolation of stereoscopic images or image sequences [1].

To be able to exploit disparity information in a stereoscopic pair in image processing applications, the relation between the contents of the left view image and the right view image has to be established, yielding the *disparity (vector) field*. The disparity field indicates for each point in the left view image the relative shift of the corresponding point in the right view image and vice versa. Since some parts of one view image may not be visible in the alternate view image due to occlusion, not all points in the image pair can be assigned a disparity vector.

Disparity estimation is essentially a correspondence problem. The correspondence between the two images can be determined by either matching features or by operating on or matching of small patches of gray values. Feature matching requires as a preprocessing step the extraction of appropriate features from the images, such as object edges and corners. After obtaining the features, the correspondence problem is first solved for the spatial locations at which the features occur, from which next the full disparity field can be deduced by, for instance, interpolation or segmentation procedures. Feature-based disparity estimation is especially useful in the analysis of scenes for robot vision applications [4, 11].

Disparity field estimation by operating directly on the image gray value information is not unlike the problem of motion estimation [11, 12]. The first difference is that disparity vectors are approximately horizontally oriented. Deviations from the horizontal orientation are caused by the convergence of the camera axes and by differences between the camera optics. Usually vertical disparity components are either ignored or rectified. A second difference is that disparity vectors can take on a much larger range of values within a single image pair. Furthermore, the disparity field may have large discontinuities associated with objects neighboring in the planar projection but having a very much different depth. In those regions of the stereoscopic image pair where one finds large discontinuities in the disparity field due to abrupt depth changes, large regions of occlusion will be present. Estimation methods for disparity fields must therefore be able not only to find the correspondence between information in the left and right view images, but must also be able to detect and handle discontinuities and occlusions [1].

Most disparity estimation algorithms used in stereoscopic communications rely on matching small patches of gray values from one view to the gray values in the alternate view. The matching of this small patch is not carried out in the entire alternate image, but only within a relatively small search region to limit the computational complexity. Standard methods typically use a rectangular match block of relatively small size (e.g., 8×8 pixels), as illustrated in Fig. 57.5. The relative horizontal shift between a match block and the block within the search region of the alternate image that results in the smallest value of a criterion function used is then assigned as disparity vector to the center of that match block. Often used criterion

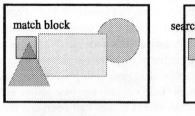 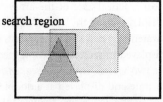

left view image   right view image

FIGURE 57.5   Block matching disparity estimation procedure by comparing a match block from the left image to the blocks within a horizontally oriented search region in the right image.

functions are the sum of squares and the sum of the absolutes values of the differences between the gray values in the match block and the block being considered in the search region [3, 12].

The above procedure is carried out for all pixels, first matching the blocks from the left view image to the right view image, then vice versa. From the combination of the two resulting disparity fields and the values of the criterion function, the final disparity field is computed, and occluding areas in the stereoscopic image pair are detected. For instance, one way of detecting occlusions is a local abrupt increase of the criterion function, indicating that no acceptable correspondence between the two image pairs could be found locally. Fig. 57.4(c) illustrates the result of a disparity estimation process as an image in which different gray values correspond to different disparities (and thus depth), and in which "white" indicates occluding regions that can be seen in the left view image but that cannot be seen in the right view image.

More advanced versions of the above *block matching disparity estimator* use hierarchical or recursive approaches to improve the consistency or smoothness of the resulting disparity field, or are based on the optical flow model often used in motion estimation. Other approaches use preprocessing steps to determine the dominant disparity values that are then used as candidate solutions during the actual estimation procedure. Finally, most recent approaches use advanced Markov random field models for the disparity field and/or they make use of more complicated cost functions such as the *disparity space image*. These approaches typically require exhaustive optimization procedures but they have the potential of accurately estimating large discontinuities and of precisely detecting the presence of occluding regions [1].

In image analysis problems, disparity estimation is often considered in combination with the segmentation of the stereoscopic image pair. Joint disparity estimation and texture segmentation methods partition the image pair into spatially homogeneous regions of approximately equal depth. Disparity estimation in image sequences is typically carried out independently on successive frame pairs. Nevertheless, the need for temporal consistency of successive disparity fields often requires temporal dependencies to be exploited by postprocessing of the disparity fields. If an image sequence is recorded as an interlaced video signal, disparity estimation should be carried out on the individual fields instead of frames to avoid confusion between motion displacements and disparity.

57.4   Compression of Stereoscopic Images

Compression of digital images and image sequences is necessary to limit the required transmission bandwidth or storage capacity [3, 5]. One of the compression principles underlying the JPEG and MPEG standards is to avoid transmitting or storing gray value information that is predictable from the signal's spatial or temporal past, i.e., information that is redundant. In both JPEG and MPEG, this principle is exploited by a spatial DPCM system, while in MPEG motion-compensated temporal prediction is also used to exploit temporal redundancies.

When dealing with stereoscopic image pairs, a third dimension of redundancy appears, namely the mutual predictability of the two perspective views [9]. Although the left and right view images are not identical, gray value information in, for instance, the left view image is highly predictable from the right view image if the horizontal shift of corresponding points, i.e., the disparity, is taken into account. Thus, instead of transmitting or storing both views of a stereoscopic image pair, only the right view image is retained, together with the disparity field. Since the construction of the left view image from the right view is not perfect due to errors in the estimated disparity field and due to presence of occluding areas and perspective differences, some information of the *disparity-compensated prediction error* of the left view (i.e., the difference between the predicted gray values and the actual gray values in the left view image) also needs to be retained. Figure 57.6 shows the *disparity-compensated prediction* and the *disparity-compensated prediction error* of the left view image from Fig. 57.4(a) using the right view image in Fig. 57.4(b) and the disparity field in Fig. 57.4(c). In most cases, the sum of the bit rates needed for coding the disparity vector field and the disparity-compensated prediction error is much smaller than the bit rate needed for the left view image when compressed without disparity compensation.

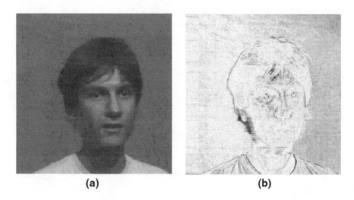

(a) **(b)**

FIGURE 57.6 (a) Disparity-compensated prediction and (b) disparity-compensated prediction error of the left view image (scaled for maximal visibility) in Fig. 57.4. Black areas indicate a large error.

In image sequence, left view images can be compressed efficiently by carrying out motion-compensated prediction from previous left view images, by disparity-compensated prediction from the corresponding right view image, or by a combination of the two by choosing for motion-compensation or disparity-compensation on a block-by-block basis, as illustrated in Fig. 57.7(a). Basically this is a direct extension of the MPEG compression standard with an additional prediction mode for the left view image sequence. The effect of this additional (disparity-compensated) prediction mode is that the variance of the prediction error of the left view image sequence is further decreased [see Fig. 57.7(b)], meaning that more compression of the left view sequence is possible than when independently compressing the two views of the stereoscopic sequence. Figure 57.8 schematically shows the architecture of a disparity- and motion-compensated encoder for stereoscopic video.

57.5 Intermediate Viewpoint Interpolation

The system illustrated in Fig. 57.1 assumes that the stereoscopic image captured by the cameras is directly displayed at the receiver's end. One of the shortcomings of such a two-channel stereoscopic system is that shape and depth distortion occur when the stereoscopic images are viewed from an off-center position. Furthermore, since the cameras are in a fixed position, the viewer's (horizontal) movements do not provide additional information about, for instance, objects that are partly occluded. The lack of this "look around" capability especially is a limiting factor in the truly realistic visualization of a recorded real world scene.

In a multi-channel or *multiview stereoscopic system,* multiple viewpoints of the same real world scene

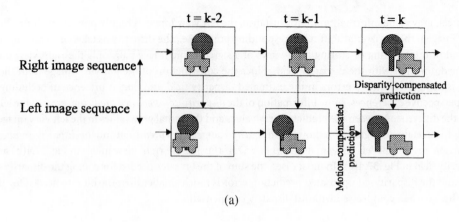

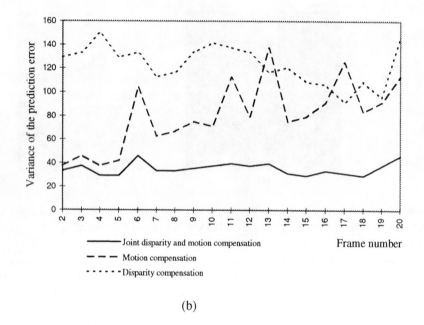

(b)

FIGURE 57.7   (a) Principle of joint disparity- and motion-compensated prediction for the left view of a stereoscopic image sequence; (b) variance of the prediction error of the left view image sequence when using motion-compensation, disparity-compensation, or joint motion-disparity compensation on a block-by-block basis.

are available. The stereoscopic display then shows only those two perspective views which correspond as well as possible with the viewer's position. To this end some form of tracking the viewer's position is necessary. The additional viewpoints could be obtained by installing more cameras at a wide range of possible viewpoints. On grounds of complexity and costs the number of cameras will typically be limited to three to five, meaning that not all possible positions of the viewer are covered in this way. If, because of the viewer's position, a view of the scene is needed from an unavailable camera position, a *virtual camera* or *intermediate viewpoint* must be constructed from the available camera viewpoints (see Fig. 57.9).

The construction of intermediate viewpoints is an interpolation problem, which has much in common with the problem of video standards conversion [11]. In its most simple form, the interpolated viewpoint is merely a weighted average between the images from the nearest two camera viewpoints, which are called the key images. Such a straightforward averaging ignores the presence of disparity between the key images, yielding a highly blurred and essentially useless result [see Fig. 57.10(a)]. If, however, the disparity vector

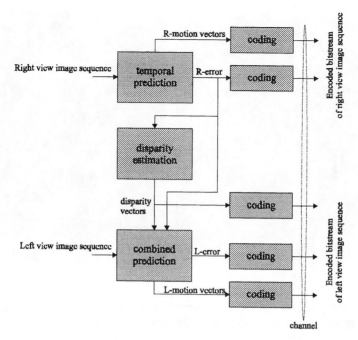

FIGURE 57.8 Architecture of a disparity- and motion-compensated encoder for stereoscopic video.

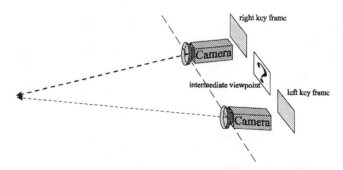

FIGURE 57.9 Multiview stereoscopic system with interpolated intermediate viewpoint (virtual camera).

field between the two key images has been estimated and the areas of occlusions are known, the interpolation can be carried out along the *disparity axis,* such that the disparity information in the interpolated image corresponds exactly to the virtual camera position. For the points where a correspondence exists between the two key images, this construction process is called *disparity-compensated interpolation,* while for the occluding regions *extrapolation* has to be carried out from the key images [10]. Figure 57.10(b) illustrates the result of intermediate viewpoint interpolation on the stereoscopic image pair in Figs. 57.4(a) and (b).

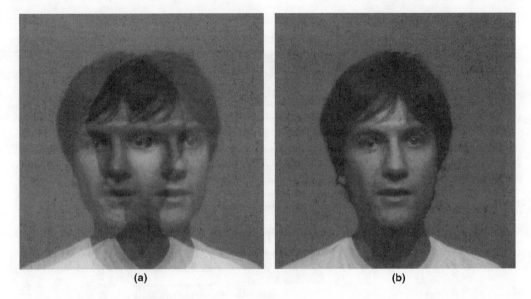

(a)   (b)

FIGURE 57.10   Interpolation of an intermediate viewpoint image of the stereoscopic pair in Fig. 57.4: (a) without and (b) with taking into account the disparity information between the key frames.

References

[1] *Proceedings of the 1995 International Workshop on Stereoscopic and Three Dimensional Imaging,* Efstratiadis, S. et al., Eds., Santorini, Greece, 1995.

[2] Dhond, U.R. and Aggerwal, J.K., Structure from stereo, *IEEE Trans. on System, Man and Cybernetics,* 19(6), 1489-1509, 1989.

[3] Hang, H.-M and Woods, J.W., *Handbook of Visual Communications,* Academic Press, San Diego, CA, 1995.

[4] Horn, B.K.P., *Robot Vision,* MIT Press, Cambridge, 1986.

[5] Jayant, N.S. and Noll, P., *Digital Coding of Waveforms,* Prentice-Hall, London, 1984.

[6] Lipton, L., *The Crystal Eyes Handbook,* StereoGraphics Corporation, 1991.

[7] Marr, D., *Vision,* Freeman, San Francisco, 1982.

[8] Pastoor, S., 3-D television: A survey of recent research results on subjective requirements, *Signal Processing: Image Communications,* 4(1), 21-32, 1991.

[9] Perkins, M.G., Data compression of stereopairs, *IEEE Trans. Commun.,* 40(4), 684-696, 1992.

[10] Skerjanc, R. and Liu, J., A three camera approach for calculating disparity and synthesizing intermediate pictures, *Signal Processing: Image Communications,* 4(1), 55-64, 1991.

[11] Tekalp, A.M., *Digital Video Processing,* Prentice-Hall, Upper Saddle River, NJ, 1995.

[12] Tziritas, G. and Labit, C., *Motion Analysis for Image Sequence Coding,* Elsevier, Amsterdam, 1994.

58

A Survey of Image Processing Software and Image Databases

Stanley J. Reeves
Auburn University

Image processing has moved into the mainstream, not only of the engineering world, but of society in general. Personal computers are now capable of handling large graphics and images with ease, and fast networks and modems transfer images in a fraction of the time required just a few years ago. Image manipulation software is a common item on PCs, and CD-ROMs filled with images and multimedia databases are standard fare in the realm of electronic publishing. Furthermore, the development of areas such as data compression, neural networks and pattern recognition, computer vision, and multimedia systems have all contributed to the use of and interest in image processing. Likewise, the growth of image processing as an engineering discipline has fueled interest in these other areas. As a result of this symbiotic growth, image processing has increasingly become a standard tool in the repertoire of the engineer.

Because of the popularity of image processing, a large array of tools has emerged for accomplishing various image processing tasks. In addition, a variety of image databases has been created to address the needs of various specialty areas. In this article, we will survey some of the tools available for accomplishing basic image processing tasks and indicate where they may be obtained. Furthermore, we will describe and provide pointers to some of the most generally useful images and image databases. The goal is to identify a basic collection of images and software that will be of use to the nonspecialist. It should also be of use to the specialist who needs a general tool in an area outside his or her specialty.

58.1 Image Processing Software

Image processing has become such a broad area that it is sometimes difficult to distinguish what might be considered an image processing package from other software systems. The boundaries among the areas of computer graphics, data visualization, and image processing have become blurred. Furthermore, to discuss or even to list all the image processing software available would require many pages and would not be particularly useful to the nonspecialist. Therefore, we emphasize a representative set of image processing software packages that embody core capabilities in scientific image processing applications. Core capabilities, in our view, include the following:

- **Image utilities:** These include display, manipulation, and file conversion. Images come in such a variety of formats that a package for converting images from one format to another is

0-8493-8572-5/98/$0.00+$.50

essential. Furthermore, basic display and manipulation (cropping, rotating, etc.) are essential for almost any image processing task. The ability to edit images using cut-and-paste, draw, and annotate operations is also useful in many cases.

- **Image filtering and transformation:** These are necessary capabilities for most scientific applications of image processing. Convolution, median filtering, FFTs, morphological operations, scaling, and other image functions form the core of many scientific image processing algorithms.

- **Image compression:** Anyone who works with images long enough will learn that they require a large amount of storage space. A number of standard image compression utilities are available for storing images in compressed form and for retrieving compressed images from image databases.

- **Image analysis:** Scientific image processing applications often have the goal of deriving information from an image. Simple image analysis tools such as edge detection and segmentation are powerful methods for gleaning important visual information.

- **Programming and data analysis environment:** While many image processing packages have a wide variety of functions, a whole new level of utility and flexibility arises when the image processing functions are built around a programming and/or data analysis environment. Programming environments allow for tailoring image processing techniques to the specific task, developing new algorithms, and interfacing image processing tasks with other scientific data analysis and numerical computational techniques.

Other capabilities include higher-level object recognition and other computer vision tasks, visualization and rendering techniques, computed imaging such as medical image reconstruction, and morphing and other special effects of the digital darkroom and the film industry. These areas require highly specialized software and/or very specialized skills to apply the methods and are not likely to be part of the image processing world of the nonspecialist.

The packages to be discussed here encompass as a group all of the core image processing capabilities mentioned above. Because these packages offer such a wide variety and mix of functions, they defy simple categorization. We have chosen to group the packages into three categories: general image utilities, specialized utilities, and programming/analysis environments. Keep in mind, however, that the distinctions among these groups is blurry at best. We have chosen to emphasize packages that are freely distributable and available on the Internet because these can be obtained and used with a minimum of expense and hassle.

58.1.1 General Image Utilities

netpbm

pbmplus is a set of tools that allows the user to convert to and from a large number of common image formats. The package has its own intermediate formats so that the conversion routines can be written to convert to or from one of these formats. The user can then convert to and from any combination of formats by going through one of the intermediate formats. Functions are also provided to convert from different color resolutions, such as from color to grayscale. Several other functions do basic image manipulation such as cropping, rotating, and smoothing. The source is available from ftp://ftp.wustl.edu/graphics/graphics/packages/NetPBM/.

xv

xv is an X11 utility that combines several important image handling functions. It can display images in a wide variety of display formats, including binary, 8-bit, and 24-bit. It allows the user to manipulate the colormap both in RGB and HSV space. It crops, resizes, smooths, rotates, detects edges, and produces other special effects. In addition, it reads and writes a large variety of image formats, so it can serve as a format conversion utility. Until recently, xv has been freely distributable. The latest version, however, is shareware and requires a small fee to become a registered user. The source is available from http://www.trilon.com/xv.

NCSA Image

NCSA Image is available in versions for the Mac, DOS, and Unix (X11). The Unix version is called ximage. ximage allows the user to display color images. It can also display the actual data in the form of a spreadsheet. A number of other display options are available. Like xv, it allows for manipulation of the colormap in a variety of ways. In addition, the user may display multiple images as an animated sequence, either from disk or server memory. The functionality of NCSA Image is augmented by other programs available from NCSA, including DataSlice for visualization tasks and Reformat for converting image formats. The source is available from NCSA by ftp at ftp://ftp.ncsa.uiuc.edu/Visualization/Image/.

ImageMagick

ImageMagick is an X11 package for display and interactive image manipulation. It reads and writes a large number of standard formats, does standard operations such as cropping and rotating as well as more specialized editing operations such as cutting, pasting, color filling, annotating, and drawing. Separate utilities are provided for grabbing images from a display, for converting, combining, resizing, blurring, adding borders, and doing many other operations. The source is available by ftp from ftp://ftp.x.org/contrib/applications/ImageMagick/.

NIH Image

NIH Image is available only in a Macintosh version. However, the popularity of NIH Image among Mac users and the breadth of features justify inclusion of the package in this survey. It reads/writes a small number of image formats, acquires images using compatible frame grabbers, and displays. It allows image manipulation such as flipping, rotating, and resizing; and editing such as drawing and annotating. It has a number of built-in enhancement and filtering functions: contrast enhancement, smoothing, sharpening, median filtering, and convolution. It supports a number of analysis operations such as edge detection and measurement of area, mean, centroid, and perimeter of user-defined regions of interest. It also performs automated particle analysis. In addition, the user can animate a set of images. NIH Image has a Pascal-like macro capability and the ability to add precompiled plug-in modules. The source is available from NIH by ftp at ftp://zippy.nimh.nih.gov/pub/nih-image/.

LaboImage

LaboImage is an X11 package for mouse- and menu-driven interactive image processing. It reads/writes a special format as well as Sun raster format and displays grayscale and RGB and provides dithering. Basic filtering operations are possible, as well as enhancement tasks such as background subtraction and histogram equalization. It computes various measures such as histograms, image statistics, and image power. Region outlining and object counting can be done as well. Images can be modified interactively at the pixel level, and an expert system is available for region segmentation. LaboImage has a macro capability for combining operations. LaboImage can be obtained from http://cuiwww.unige.ch/ftp/sgaico/research/geneve/vision/labo.html.

Paint Shop Pro

Paint Shop Pro is a Windows-based package for creating, displaying, and manipulating images. It has a large number of image editing features, including painting, photo retouching, and color enhancement. It reads and writes a large number of formats. It includes several standard image processing filters and geometrical transformations. It can be obtained from http://www.jasc.com/psp.html. It is shareware and costs $69.

58.1.2   Specialized Image Utilities

Compression

JPEG is a standard for image compression developed by the Joint Photographic Experts Group. Free, portable C code that implements JPEG compression and decompression has been developed by the Independent JPEG Group, a volunteer organization. It is available from ftp://ftp.uu.net/graphics/jpeg. The downloadable package contains source and documentation. The code converts between JPEG and several other common image formats. A lossless JPEG implementation can be obtained from ftp://ftp.cs.cornell.edu/pub/multimed/.

A fractal image compression program is available from ftp://inls.ucsd.edu/pub/young-fractal/. The package contains source for both compression and decompression. A number of other fractal compression programs are also available and can be found in the sci.fractal FAQ at ftp://rtfm.mit.edu/pub/usenet/news.answers/sci/fractals-faq.

JBIG is a standard for binary image compression developed by the Joint Binary Images Group. A JBIG coder/decoder can be obtained from ftp://nic.funet.fi/pub/graphics/misc/test-images/.

MPEG is a standard for video/audio compression developed by the Moving Pictures Experts Group. A set of MPEG tools is available from ftp://mm-ftp.cs.berkeley.edu/pub/multimedia/mpeg/. These tools allow for encoding, decoding (playing), and analyzing the MPEG data.

H.261 and H.263 are standards for video compression for videophone applications. An H.261 coder/decoder is available from ftp://havefun.stanford.edu/pub/p64/. An H.263 video coder/decoder is available from http://www.fou.telenor.no/brukere/DVC/h263_software/.

Computer Vision

Vista is an X11-based image processing environment specifically designed for computer vision applications. It allows a variety of display and manipulation options. It has a library that lets the user easily create applications with menus, mouse interaction, and display options. Vista defines a very flexible data format that represents a variety of images, collections of images, or other objects. It also has the ability to add new objects or new image attributes without changing existing software or data files. It does edge detection and linking, optical flow estimation and camera calibration, and viewing of images and edge vectors. Vista includes routines for common image processing operations such as convolution, FFTs, simple enhancement tasks, scaling, cropping, and rotating. Vista is available from http://www.cs.ubc.ca/nest/lci/vista/vista.html.

58.1.3   Programming/Analysis Environments

Khoros

Khoros is a comprehensive software development and data analysis environment. It allows the user to perform a large variety of image and signal processing and visualization tasks. A graphical programming environment called Cantata allows the user to construct programs visually using a data flowgraph approach. It has a user interface design tool with automatic code generation for writing customized applications. Software objects (programs) are accessible from the command line, from within Cantata, and in libraries.

A large set of standard numerical and statistical algorithms are available within Khoros. Common image processing operations such as FFTs, convolution, median filtering, and morphological operators are available. In addition, a variety of image display and geometrical manipulation programs, animation, and colormap editing are included. Khoros has a very general data model that allows for images of up to five dimensions.

Khoros is free-access software — it is available for downloading free of charge but cannot be distributed without a license. It can be obtained from Khoral Research, Inc., at ftp://ftp.khoral.com/pub/. Note that the Khoros distribution is quite large and requires significant disk space.

MATLAB

MATLAB is a general numerical analysis and visualization environment. Matrices are the underlying data structure in MATLAB, and this structure lends itself well to image processing applications. All data in MATLAB is represented as double-precision, which makes the calculations more precise and interaction more convenient. However, it may also mean that MATLAB uses more memory and processing time than necessary.

A large number of numerical algorithms and visualization options are available with the standard package. The Image Processing Toolbox provides a great deal of added functionality for image processing applications. It reads/writes several of the most common image formats; does convolution, FFTs, median filtering, histogram equalization, morphological operations, two-dimensional filter design, general non-linear filtering, colormap manipulation, and basic geometrical manipulation. It also allows for a variety of display options, including surface warping and movies.

MATLAB is an interactive environment, which makes interactive image processing and manipulation convenient. One can also add functionality by creating scripts or functions that use MATLAB's functions and other user-added functions. Additionally, one can add functions that have been written in C or Fortran. Conversely, C or Fortran programs can call MATLAB and MATLAB library functions. MATLAB is commercial software. More information on MATLAB and how to obtain it can be found through the homepage of The Mathworks, Inc., at http://www.mathworks.com/.

PV-Wave

PV-Wave is a general graphical/visualization and numerical analysis environment. It can handle images of arbitrary dimensionality — 2-D, 3-D, and so on. The user can specify the data type of each data structure, which allows for flexibility but may be inconvenient for interactive work.

PV-Wave contains a large collection of visualization and rendering options, including colormap manipulation, volume rendering, and animation. In addition, the IMSL library is available through PV-Wave. Basic image processing operations such as convolution, FFTs, median filtering, morphological operations, and contrast enhancement are included.

Like MATLAB, PV-Wave is an interactive environment. One can create scripts or functions from the PV-Wave language to add functionality. It can also call C or Fortran functions. PV-Wave can be invoked from within C or Fortran too. PV-Wave is commercial software. More information on PV-Wave can be found through the homepage of Visual Numerics, Inc., at http://www.vni.com/.

58.2 Image Databases

A huge number of image databases and archives are available on the Internet now, and more are continually being added. These databases serve various purposes. For the practicing engineer, the primary value of an image database is for developing, testing, evaluating, or comparing image processing and manipulation algorithms. Standard images provide a benchmark for comparing various algorithms. Furthermore, standard test images can be selected so that their characteristics are particularly suited to demonstrating the strengths and weaknesses of particular types of image processing techniques. In some areas of image processing no real standards exist, although de facto standards have arisen. In the discussion that follows, we provide pointers to some standard images, some de facto standards, and a few other databases that might provide images of value to algorithm work in image processing. We have deliberately steered away from images whose copyright is known to prohibit use for research purposes. However, some of the images in the list have certain copyright restrictions. Be sure to check any auxiliary information provided with the images before assuming that they are public domain.

The images listed are in a variety of formats and may require conversion using one of the packages discussed previously such as netpbm. We list the databases according to two categories: (1) form and (2) content. By form, we mean that the images are organized according to the form of the image — color, stereo, sequence, etc. By content, we mean that the images are grouped according to the image content — faces, fingerprints, etc.

58.2.1   Images by Form

Binary Images

A set of standard CCITT fax test images has been made available for testing compression schemes. These are binary images that have come from scanning actual documents. They can be found at ftp://nic.funet.fi/pub/graphics/misc/test-images/ under ccitt[1-8].pbm.gz.

Grayscale Images

A collection of grayscale images can be obtained from ftp://ipl.rpi.edu/pub/image/still/canon/gray/. A compilation of de facto standard images can be found at
http://www.sys.uea.ac.uk/Research/ResGroups/SIP/images_ftp/index.html.
Note that the Lena image is copyrighted and should not be used in publications.

Color Images

A set of test images that were used by the JPEG committee in the development of the JPEG algorithm are available from ftp://ipl.rpi.edu/pub/image/still/jpeg/bgr/.
These are 24-bit RGB images. Other 24-bit color images can be found at
ftp://ipl.rpi.edu/pub/image/still/canon/bgr/.
A set of miscellaneous images in JPEG and Kodak CD format can be found at
http://www.kodak.com/digitalImages/samples/samples.shtml.

Image Sequences

Image sequences may be intended for study of computer vision applications or video coding. A huge set of sequences for computer vision applications are archived at
http://www.ius.cs.cmu.edu/idb/.
A set of sequences commonly used for video coding applications can be found at
ftp://ipl.rpi.edu/pub/image/sequence/.

Stereo Image Pairs

Stereo image pairs are available from http://www.ius.cs.cmu.edu/idb/.

Texture Images

A large set of texture images can be found at
http://www-white.media.mit.edu/vismod/imagery/VisionTexture/vistex.html.
These images include textures from various angles and under different lighting conditions.

Face Images

The USENIX FACES database contains hundreds of face images in various formats. The database is archived at ftp://ftp.uu.net/published/usenix/faces/.

Fingerprint Images

Fingerprint images can be obtained from ftp://sequoyah.ncsl.nist.gov/pub/databases/data/.

Medical Images

A variety of medical images are available over the Internet. An excellent collection of CT, MRI, and cryosection images of the human body has been made available by the National Library of Medicine's The Visual Human Project. Samples of these images can be acquired at
http://www.nlm.nih.gov/research/visible/visible_human.html.
A collection of over 3500 images that cover an entire human body is available via ftp and on tape by signing a license agreement.

MRI and CT volume images are available from
ftp://omicron.cs.unc.edu/pub/projects/softlab.v/CHVRTD/.
PET images and other modalities can be found in
gopher://gopher.austin.unimelb.edu.au/11/images/petimages.

Astronomical Images

A collection of astronomical images can be found at
https//www.univ-rennesl.fr/ASTRO/astro.english.html.
Hubble telescope imagery can be obtained from http://archive.stsci.edu/archive.html.

Range Images

Range images are available from
http://www.eecs.wsu.edu/~irl/3DDB/RID/, along with a list of other sources of range imagery, and also from http://marathon.csee.usf.edu/range/DataBase.html.

The tools and databases discussed here should provide a convenient set of capabilities for the nonspecialist. The capabilities that are readily available are not static, however. Image processing will continue to become more and more mainstream, so we expect to see the development of image processing tools representing greater variety and sophistication. The advent of the World Wide Web will also stimulate further development and publishing of image databases on the Internet. Therefore, image processing capabilities will continue to grow and will be more readily available. The items discussed here are only a small sample of what will be available as time goes on.

59

VLSI Architectures for Image Communications

P. Pirsch
*Laboratorium für
Informationstechnologie,
University of Hannover*

W. Gehrke
Philips Semiconductors

59.1   Introduction

Video processing has been a rapidly evolving field for telecommunications, computer, and media industries. In particular, for real time video compression applications a growing economical significance is expected for the next years. Besides digital TV broadcasting and videophone, services such as multimedia education, teleshopping, or video mail will become audiovisual mass applications.

To facilitate worldwide interchange of digitally encoded audiovisual data, there is a demand for international standards, defining coding methods, and transmission formats. International standardization committees have been working on the specification of several compression schemes. The Joint Photographic Experts Group (JPEG) of the International Standards Organization (ISO) has specified an algorithm for compression of still images [4]. The ITU proposed the H.261 standard for video telephony and video conference [1]. The Motion Pictures Experts Group (MPEG) of ISO has completed its first standard MPEG-1, which will be used for interactive video and provides a picture quality comparable to VCR quality [2]. MPEG made substantial progress for the second phase of standards MPEG-2, which will provide audiovisual quality of both broadcast TV and HDTV [3]. Besides the availability of international standards, the successful introduction of the named services depends on the availability of VLSI components, supporting a cost efficient implementation of video compression applications. In the following, we give a short overview of recent coding schemes and discuss implementation alternatives. Furthermore, the efficiency estimation of architectural alternatives is discussed and implementation examples of dedicated and programmable architectures are presented.

0-8493-8572-5/98/$0.00+$.50

59.2 Recent Coding Schemes

Recent video coding standards are based on a hybrid coding scheme that combines transform coding and predictive coding techniques. An overview of these hybrid encoding schemes is depicted in Fig. 59.1.

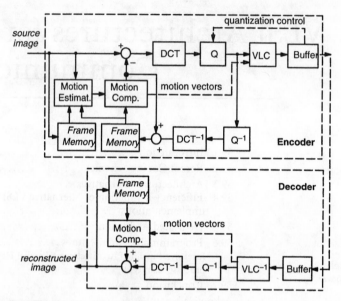

FIGURE 59.1 Hybrid encoding and decoding scheme.

The encoding scheme consists of the tasks motion estimation, typically based on blockmatching algorithms, computation of the prediction error, discrete cosine transform (DCT), quantization (Q), variable length coding (VLC), inverse quantization (Q^{-1}), and inverse discrete cosine transform (IDCT or DCT-1). The reconstructed image data are stored in an image memory for further predictions. The decoder performs the tasks variable length decoding (VLC^{-1}), inverse quantization, and motion compensated reconstruction.

Generally, video processing algorithms can be classified in terms of regularity of computation and data access. This classification leads to three classes of algorithms:

- *Low-Level Algorithms* — These algorithms are based on a predefined sequence of operations and a predefined amount of data at the input and output. The processing sequence of low-level algorithms is predefined and does not depend on the values of data processed. Typical examples of low-level algorithms are block matching or transforms such as the DCT.

- *Medium-Level Algorithms* — The sequence and number of operations of medium-level algorithms depend on the data. Typically, the amount of input data is predefined, whereas the amount of output data varies according to the input data values. With respect to hybrid coding schemes, examples for these algorithms are quantization, inverse quantization, or variable length coding.

- *High-Level Algorithms* — High-level algorithms are associated with a variable amount of input and output data and a data-dependent sequence of operations. As for medium-level algorithms, the sequence of operations is highly data dependent. Control tasks of the hybrid coding scheme can be assigned to this class.

Since hybrid coding schemes are applied for different video source rates, the required absolute processing power varies in the range from a few hundred MOPS (Mega Operations Per Second) for video signals in QCIF format to several GOPS (Giga Operations Per Second) for processing of TV or HDTV signals. Nevertheless, the relative computational power of each algorithmic class is nearly independent of the processed video format. In case of hybrid coding applications, approximately 90% of the overall processing power is required for low-level algorithms. The amount of medium-level tasks is about 7% and nearly 3% is required for high-level algorithms.

59.3 Architectural Alternatives

In terms of a VLSI implementation of hybrid coding applications, two major requirements can be identified. First, the high computational power requirements have to be provided by the hardware. Second, low manufacturing cost of video processing components is essential for the economic success of an architecture. Additionally, implementation size and architectural flexibility have to be taken into account.

Implementations of video processing applications can either be based on standard processors from workstations or PCs or on specialized video signal processors. The major advantage of standard processors is their availability. Application of these architectures for implementation of video processing hardware does not require the time consuming design of new VLSI components. The disadvantage of this implementation strategy is the insufficient processing power of recent standard processors. Video processing applications would still require the implementation of cost intensive multiprocessor systems to meet the computational requirements. To achieve compact implementations, video processing hardware has to be based on video signal processors, adapted to the requirements of the envisaged application field.

Basically, two architectural approaches for the implementations of specialized video processing components can be distinguished. *Dedicated architectures* aim at an efficient implementation of one specific algorithm or application. Due to the restriction of the application field, the architecture of dedicated components can be optimized by an intensive adaptation of the architecture to the requirements of the envisaged application, e.g., arithmetic operations that have to be supported, processing power, or communication bandwidth. Thus, this strategy will generally lead to compact implementations. The major disadvantage of dedicated architecture is the associated low flexibility. Dedicated components can only be applied for one or a few applications. In contrast to dedicated approaches with limited functionality, *programmable architectures* enable the processing of different algorithms under software control. The particular advantage of programmable architectures is the increased flexibility. Changes of architectural requirements, e.g., due to changes of algorithms or an extension of the aimed application field, can be handled by software changes. Thus, a generally cost-intensive redesign of the hardware can be avoided. Moreover, since programmable architectures cover a wider range of applications, they can be used for low-volume applications, where the design of function specific VLSI chips is not an economical solution.

For both architectural approaches, the computational requirements of video processing applications demand for the exploitation of the algorithm-inherent independence of basic arithmetic operations to be performed. Independent operations can be processed concurrently, which enables the decrease of processing time and thus an increased through-put rate. For the architectural implementation of concurrency, two basic strategies can be distinguished: pipelining and parallel processing.

In case of pipelining several tasks, operations or parts of operations are processed in subsequent steps in different hardware modules. Depending on the selected granularity level for the implementation of pipelining, intermediate data of each step are stored in registers, register chains, FIFOs, or dual-port memories. Assuming a processing time of T_P for a non-pipelined processor module and $T_{D,IM}$ for the delay of intermediate memories, we get in the ideal case the following estimation for the throughput-rate $R_{T,\text{Pipe}}$ of a pipelined architecture applying N_{Pipe} pipeline stages:

$$R_{T,\text{Pipe}} = \frac{1}{\frac{T_P}{N_{\text{Pipe}}} + T_{D,IM}} = \frac{N_{\text{Pipe}}}{T_P + N_{\text{Pipe}} \cdot T_{D,IM}} \qquad (59.1)$$

From this follows that the major limiting factor for the maximum applicable degree of pipelining is the access delay of these intermediate memories.

The alternative to pipelining is the implementation of parallel units, processing independent data concurrently. Parallel processing can be applied on operation level as well as on task level. Assuming the ideal case, this strategy leads to a linear increase of processing power and we get:

$$R_{T,\text{Par}} = \frac{N_{\text{Par}}}{T_P} \tag{59.2}$$

where N_{Par} = number of parallel units.

Generally, both alternatives are applied for the implementation of high-performance video processing components. In the following sections, the exploitation of algorithmic properties and the application of architectural concurrency is discussed considering the hybrid coding schemes.

59.4 Efficiency Estimation of Alternative VLSI Implementations

Basically, architectural efficiency can be defined by the ratio of performance over cost. To achieve a figure of merit for architectural efficiency we assume in the following that performance of a VLSI architecture can be expressed by the achieved throughput rate R_T and the cost is equivalent to the required silicon area A_{Si} for the implementation of the architecture:

$$E = \frac{R_T}{A_{Si}} \tag{59.3}$$

Besides the architecture, efficiency mainly depends on the applied semiconductor technology and the design-style (semi-custom, full-custom). Therefore, a realistic efficiency estimation has to consider the gains provided by the progress in semiconductor technology. A sensible way is the normalization of the architectural parameters according to a reference technology. In the following we assume a reference process with a grid length $\lambda_0 = 1.0$ micron. For normalization of silicon area, the following equation can be applied:

$$A_{Si,0} = A_{Si} \left(\frac{\lambda_0}{\lambda} \right)^2 \tag{59.4}$$

where the index 0 is used for the system with reference gate length λ_0.

According to [7] the normalization of throughput can be performed by:

$$R_{T,0} = R_T \left(\frac{\lambda}{\lambda_0} \right)^{1.6} \tag{59.5}$$

From Eqs. (59.3), (59.4), and (59.5), the normalization for the architectural efficiency can be derived:

$$E_0 = \frac{R_{T,0}}{A_{Si,0}} = \frac{R_T}{A_{Si}} \left(\frac{\lambda}{\lambda_0} \right)^{3.6} \tag{59.6}$$

E can be used for the selection of the best architectural approach out of several alternatives. Moreover, assuming a constant efficiency for a specific architectural approach leads to a linear relationship of throughput rate and silicon area and this relationship can be applied for the estimation of the required silicon area for a specific application. Due to the power of 3.6 in Equ. (59.6), the chosen semiconductor technology for implementation of a specific application has a significant impact on the architectural efficiency.

In the following, examples of dedicated and programmable architectures for video processing applications are presented. Additionally, the discussed efficiency measure is applied to achieve a figure of merit for silicon area estimation.

59.5 Dedicated Architectures

Due to their algorithmic regularity and the high processing power required for the discrete cosine transform and motion estimation, these algorithms are the first candidates for a dedicated implementation. As typical examples, alternatives for a dedicated implementation of these algorithms are discussed in the following.

The discrete cosine transform (DCT) is a real-valued frequency transform similar to the Discrete Fourier transform (DFT). When applied to an image block of size L × L, the two dimensional DCT (2D-DCT) can be expressed as follows:

$$Y_{k,l} = \sum_{i=0}^{L-1} \sum_{j=0}^{L-1} x_{i,j} \cdot C_{i,k} \cdot C_{j,l} \qquad (59.7)$$

$$\text{where} \quad C_{n,m} = \begin{cases} \frac{1}{\sqrt{2}} & \text{for } m = 0 \\ \cos\left[\frac{(2n+1)m\pi}{2L}\right] & \text{otherwise} \end{cases}$$

with

(i, j) = coordinates of the pixels in the initial block
(k, l) = coordinates of the coefficients in the transformed block
$x_{i,j}$ = value of the pixel in the initial block
$Y_{k,l}$ = value of the coefficient in the transformed block

Computing a 2D DCT of size L × L directly according to Eq. (59.7) requires L^4 multiplications and L^4 additions.

The required processing power for the implementation of the DCT can be reduced by the exploitation of the arithmetic properties of the algorithm. The two-dimensional DCT can be separated into two one-dimensional DCTs according to Eq. (59.8)

$$Y_{k,l} = \sum_{i=0}^{L-1} C_{i,k} \cdot \left[\sum_{j=0}^{L-1} x_{i,j} \cdot C_{j,l} \right] \qquad (59.8)$$

The implementation of the separated DCT requires $2L^3$ multiplications and $2L^3$ additions. As an example, the DCT implementation according to [9] is depicted in Fig. 59.2. This architecture is based on two

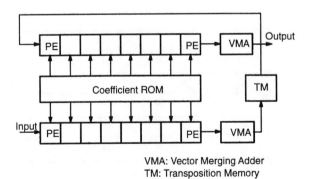

VMA: Vector Merging Adder
TM: Transposition Memory

FIGURE 59.2 Separated DCT implementation according to [9].

one-dimensional processing arrays. Since this architecture is based on a pipelined multiplier/accumulator implementation in carry-save technique, vector merging adders are located at the output of each array. The results of the 1D-DCT have to be reordered for the second 1D-DCT stage. For this purpose, a transposition memory is used. Since both one-dimensional processor arrays require identical DCT coefficients, these coefficients are stored in a common ROM.

Moving from a mathematical definition to an algorithm that can minimize the number of calculations required is a problem of particular interest in the case of transforms such as the DCT. The 1D-DCT can also be expressed by the matrix-vector product :

$$[\mathbf{Y}] = [\mathbf{C}][\mathbf{X}] \tag{59.9}$$

where $[\mathbf{C}]$ is an $L \times L$ matrix and $[\mathbf{X}]$ and $[\mathbf{Y}]$ 8-point input and output vectors. As an example, with $\theta = p/16$, the 8-points DCT matrix can be computed as denoted in Eq. (59.10)

$$
\begin{bmatrix} Y_0 \\ Y_1 \\ Y_2 \\ Y_3 \\ Y_4 \\ Y_5 \\ Y_6 \\ Y_7 \end{bmatrix} =
\begin{bmatrix}
\cos 4\theta & \cos 4\theta & \cos 4\theta & \cos 4\theta & \cos 4\theta & \cos 4\theta & \cos 4\theta & \cos 4\theta \\
\cos \theta & \cos 3\theta & \cos 5\theta & \cos 7\theta & -\cos 7\theta & -\cos 5\theta & -\cos 3\theta & -\cos \theta \\
\cos 2\theta & \cos 6\theta & -\cos 6\theta & -\cos 2\theta & -\cos 2\theta & -\cos 6\theta & \cos 6\theta & \cos 2\theta \\
\cos 3\theta & -\cos 7\theta & -\cos \theta & -\cos 5\theta & \cos 5\theta & \cos \theta & \cos 7\theta & -\cos 3\theta \\
\cos 4\theta & -\cos 4\theta & -\cos 4\theta & \cos 4\theta & \cos 4\theta & -\cos 4\theta & -\cos 4\theta & \cos 4\theta \\
\cos 5\theta & -\cos \theta & \cos 7\theta & \cos 3\theta & -\cos 3\theta & -\cos 7\theta & \cos \theta & -\cos 5\theta \\
\cos 6\theta & -\cos 2\theta & \cos 2\theta & -\cos 6\theta & -\cos 6\theta & \cos 2\theta & -\cos 2\theta & \cos 6\theta \\
\cos 7\theta & -\cos 5\theta & \cos 3\theta & -\cos \theta & \cos \theta & -\cos 3\theta & \cos 5\theta & -\cos 7\theta
\end{bmatrix}
\begin{bmatrix} x_0 \\ x_1 \\ x_2 \\ x_3 \\ x_4 \\ x_5 \\ x_6 \\ x_7 \end{bmatrix}
$$

$$\tag{59.10}$$

$$
\begin{bmatrix} Y_0 \\ Y_2 \\ Y_4 \\ Y_6 \end{bmatrix} =
\begin{bmatrix}
\cos 4\theta & \cos 4\theta & \cos 4\theta & \cos 4\theta \\
\cos 2\theta & \cos 6\theta & -\cos 6\theta & -\cos 2\theta \\
\cos 4\theta & -\cos 4\theta & -\cos 4\theta & \cos 4\theta \\
\cos 6\theta & -\cos 2\theta & \cos 2\theta & -\cos 6\theta
\end{bmatrix}
\begin{bmatrix} x_0 + x_7 \\ x_1 + x_6 \\ x_2 + x_5 \\ x_3 + x_4 \end{bmatrix}
\tag{59.11}
$$

$$
\begin{bmatrix} Y_1 \\ Y_3 \\ Y_5 \\ Y_7 \end{bmatrix} =
\begin{bmatrix}
\cos \theta & \cos 3\theta & \cos 5\theta & \cos 7\theta \\
\cos 3\theta & -\cos 7\theta & -\cos \theta & -\cos 5\theta \\
\cos 5\theta & -\cos \theta & \cos 7\theta & \cos 3\theta \\
\cos 7\theta & -\cos 5\theta & \cos 3\theta & -\cos \theta
\end{bmatrix}
\begin{bmatrix} x_0 + x_7 \\ x_1 + x_6 \\ x_2 + x_5 \\ x_3 + x_4 \end{bmatrix}
\tag{59.12}
$$

More generally, the matrices in Eqs. (59.11) and (59.12) can be decomposed in a number of simpler matrices, the composition of which can be expressed as a flowgraph. Many fast algorithms have been proposed. Figure 59.3 illustrates the flowgraph of the B.G. Lee's algorithms, which is commonly used [10]. Several implementations using fast flow-graphs have been reported [11, 12].

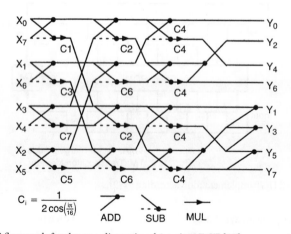

FIGURE 59.3   Lee FDCT flowgraph for the one-dimensional 8-points DCT [10].

Another approach that has been extensively used is based on the technique of distributed arithmetic. Distributed arithmetic is an efficient way to compute the DCT totally or partially as scalar products. To illustrate the approach, let us compute a scalar product between two length-M vectors C and X :

$$Y = \sum_{i=0}^{M-1} c_i \cdot x_i \text{ with } x_i = -x_{i,0} + \sum_{j=1}^{B-1} x_{i,j} \cdot 2^{-j} \tag{59.13}$$

where $\{c_i\}$ are N-bit constants and $\{x_i\}$ are coded in B bits in 2s complement. Then Eq. (59.13) can be rewritten as :

$$Y = \sum_{j=0}^{B-1} C_j \cdot 2^{-j} \text{ with } C_{j \neq 0} = \sum_{i=0}^{M-1} c_i x_{i,j} \text{ and } C_0 = -\sum_{i=0}^{M-1} c_i x_{i,0} \tag{59.14}$$

The change of summing order in i and j characterizes the distributed arithmetic scheme in which the initial multiplications are distributed to another computation pattern. Since the term C_j has only 2^M possible values (which depend on the $x_{i,j}$ values), it is possible to store these 2^M possible values in a ROM. An input set of M bits $\{x_{0,j}, x_{1,j}, x_{2,j}, \ldots, x_{M-1,j}\}$ is used as an address, allowing retrieval of the C_j value. These intermediate results are accumulated in B clock cycles, for producing one Y value. Figure 59.4 shows a typical architecture for the computation of a M input inner product. The inverter and the MUX are used for inverting the final output of the ROM in order to compute C_0.

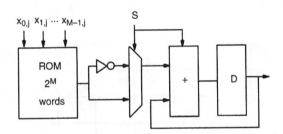

FIGURE 59.4 Architecture of a M input inner product using distributed arithmetic.

Figure 59.5 illustrates two typical uses of distributed arithmetic for computing a DCT. Figure 59.5(a) implements the scalar products described by the matrix of Eq. (59.10). Figure 59.5(b) takes advantage of a first stage of additions and substractions and the scalar products described by the matrices of Eq. (59.11) and Eq. (59.12).

Properties of several dedicated DCT implementations have been reported in [6]. Figure 59.6 shows the silicon area as a function of the throughput rate for selected design examples. The design parameters are normalized to a fictive 1.0 μm CMOS process according to the discussed normalization strategy. As a figure of merit, a linear relationship of throughput rate and required silicon area can be derived:

$$\alpha_{T,0} \approx 0.5 \text{ mm}^2 \text{ / Mpel/s} \tag{59.15}$$

Equation (59.15) can be applied for the silicon area estimation of DCT circuits. For example, assuming TV signals according to the CCIR-601 format and a frame rate of 25Hz, the source rate equals 20.7 Mpel/s. As a figure of merit from Eq. (59.15) a normalized silicon area of about 10.4 mm^2 can be derived. For HDTV signals the video source rate equals 110.6 Mpel/s and approximately 55.3 mm^2 silicon area is required for the implementation of the DCT. Assuming an economically sensible maximum chip size of about 100 mm^2 to 150 mm^2, we can conclude that the implementation of the DCT does not necessarily require the realization of a dedicated DCT chip and the DCT core can be combined with several other on-chip modules that perform additional tasks of the video coding scheme.

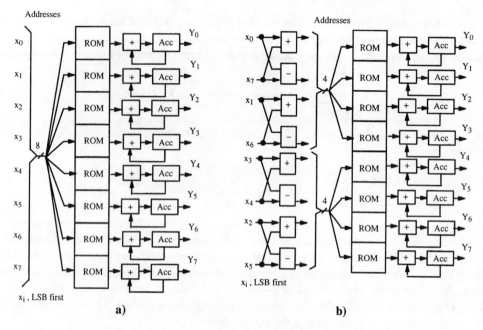

FIGURE 59.5 Architecture of an 8-point one-dimensional DCT using distributed arithmetic. (a) Pure distributed arithmetic. (b) Mixed D.A.: first stage of flowgraph decomposition products of 8 points followed by 2 times 4 scalar products of 4 points.

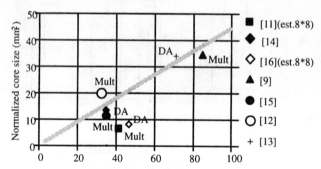

FIGURE 59.6 Normalized silicon area and throughput for dedicated DCT circuits.

For motion estimation several techniques have been proposed in the past. Today, the most important technique for motion estimation is block matching, introduced by [21]. Block matching is based on the matching of blocks between the current and a reference image. This can be done by a full (or exhaustive) search within a search window, but several other approaches have been reported in order to reduce the computation requirements by using an "intelligent" or "directed" search [17, 18, 19, 23, 25, 26, 27].

In case of an exhaustive search block matching algorithm, a block of size N × N pels of the current image (reference block, denoted X) is matched with all the blocks located within a search window (candidate blocks, denoted Y) The maximum displacement will be denoted by w. The matching criterium generally consists in computing the mean absolute difference (MAD) between the blocks. Let $x(i, j)$ be the pixels of the reference block and $y(i, j)$ the pixels of the candidate block. The matching distance (or distortion) D is computed according to Eq. (59.16). The indexes m and n indicate the position of the candidate block within the search window. The distortion D is computed for all the $(2w + 1)^2$ possible positions of the candidate block within the search window [Eq. (59.16)] and the block corresponding to the minimum distortion is used for prediction. The position of this block within the search window is represented by the motion vector **v** (59.17).

$$D(m, n) = \sum_{i=0}^{N-1} \sum_{j=0}^{N-1} |x(i, j) - y(i + m, \ j + n)| \tag{59.16}$$

$$\mathbf{v} = \begin{bmatrix} m \\ n \end{bmatrix} \Big|_{D_{\min}} \tag{59.17}$$

The operations involved for computing $D(m, n)$ and DMIN are associative. Thus, the order for exploring the index spaces (i, j) and (m, n) are arbitrary and the block matching algorithm can be described by several different dependence graphs. As an example, Fig. 59.7 shows a possible dependence graph (DG) for $w = 1$ and $N = 4$. In this figure, AD denotes an absolute difference and an addition, M denotes a minimum value computation.

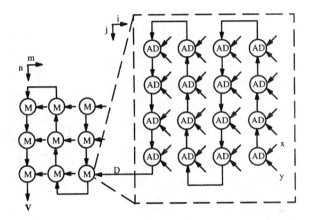

FIGURE 59.7   Dependence graphs of the block matching algorithm. The computation of **v** (X, Y) and $D(m, n)$ are performed by 2D linear DGs.

The dependence graph for computing $D(m, n)$ is directly mapped into a 2-D array of processing elements (PE), while the dependence graph for computing $v(X, Y)$ is mapped into time (59.8). In other words, block matching is performed by a sequential exploration of the search area, while the computation of each distortion is performed in parallel. Each of the AD nodes of the DG is implemented by an AD processing element (AD-PE). The AD-PE stores the value of $x(i, j)$ and receives the value of $y(m+i, n+j)$ corresponding to the current position of the reference block in the search window. It performs the subtraction and the absolute value computation, and adds the result to the partial result coming from the upper PE. The partial results are added on columns and a linear array of adders performs the horizontal summation of the row sums, and computes $D(m, n)$. For each position (n, m) of the reference block, the M-PE checks if the distortion $D(m, n)$ is smaller than the previous smaller distortion value, and, in this case, updates the register which keeps the previous smaller distortion value.

To transform this naive architecture into a realistic implementation, two problems must be solved: (1) a reduction of the cycle time and (2) the I/O management.

1. The architecture of Fig. 59.8 implicitly supposes that the computation of D(m,n) can be done combinatorially in one cycle time. While this is theoretically possible, the resulting cycle time would be very large and would increase as 2N. Thus, a pipeline scheme is generally added.

2. This architecture also supposes that each of the AD-PE receives a new value of $y(m+i, n+j)$ at each clock cycle.

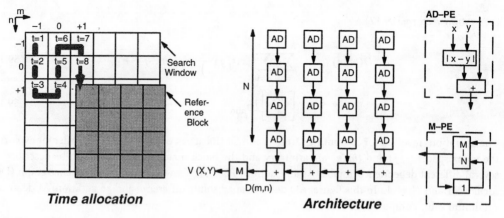

FIGURE 59.8   Principle of the 2-D block-based architecture.

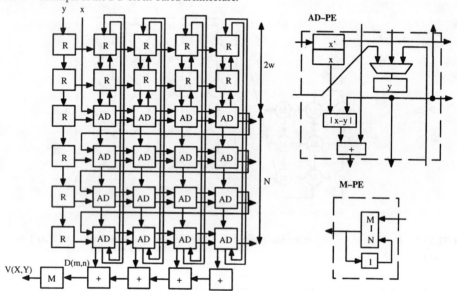

FIGURE 59.9   Practical implementation of the 2-D block-based architecture.

Since transmitting the N^2 values from an external memory is clearly impossible, advantage must be taken from the fact that these values belong to the search window. A portion of the search window of size $N*(2w + N)$ is stored in the circuit, in a 2-D bank of shift registers able to shift in the up, down, and right direction. Each of the AD-PEs has one of these registers and can, at each cycle, obtain the value of $y(m + i, n + j)$ that it needs. To update this register bank, a new column of $2w + N$ pixels of the search area is serially entered in the circuit and is inserted in the bank of registers. A mechanism must also be provided for loading a new reference with a low I/O overhead: a double buffering of $x(i, j)$ is required, with the pixels $x'(i, j)$ of a new reference block serially loaded during the computation of the current reference block (Fig. 59.9).

Figure 59.10 shows the normalized computational rate vs. normalized chip area for block matching circuits. Since one MAD operation consists of three basic ALU operations (SUB, ABS, ADD), for a 1.0 micron CMOS process, we can derive from this figure that:

$$\alpha_{T,0} \approx 30 \text{ mm}^2 + 1.9 \text{ mm}^2 / \text{GOPS} \tag{59.18}$$

The first term of this expression indicates that the block matching algorithm requires a large storage area

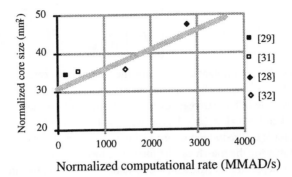

FIGURE 59.10   Normalized silicon area and computational rate for dedicated motion estimation architectures.

(storage of parts of the actual and previous frame), which cannot be reduced even when the throughput is reduced. The second term corresponds to the linear dependency on computation throughput. The second term has the same amount as that determined for the DCT for GADDS because the three types of operations for the matching require approximately the same expense of additions.

From equation Eq. (59.18), the silicon area required for the dedicated implementation of the exhaustive search block matching strategy for a displacement of $\pm w$ pels can be derived by:

$$\alpha_{T,0} \approx 0.0057 \cdot (2w + 1)^2 \cdot R_S + 30 \text{ mm}^2 \tag{59.19}$$

According to Eq. (59.19), a dedicated implementation of exhaustive search block matching for telecommunication applications based on a source rate of $R_S = 1.01$ Mpel/s (CIF format, 10 Hz frame rate) and a maximum displacement of $w = 15$, the required silicon area can be estimated to 35.5 mm². For TV ($R_S = 10.4$ Mpel/s) the silicon area for $w = 31$ can be estimated to 265 mm². Estimating the required silicon area for HDTV signals and $w = 31$ leads to 1280 mm² for the fictive 1.0 μm CMOS process. From this follows that the implementation for TV and HDTV applications will require the realization of a dedicated block matching chip. Assuming a recent 0.5 μm semiconductor processes the core size estimation leads to about 22 mm² for TV signals and 106 mm² for HDTV signals.

To reduce the high computational complexity required for exhaustive search block matching, two strategies can be applied:

1. Decrease of the number of candidate blocks.
2. Decrease of the pels per block by subsampling of the image data.

Typically, (1) is implemented by search strategies in successive steps. As an example, a modified scheme according to the original proposal of [25] will be discussed. In this scheme, the best match $\mathbf{v}_{s-1}$ in the previous step $s - 1$ is improved in the present step s by comparison with displacements $\pm\Delta_s$. The displacement vector $\mathbf{v}_s$ for each step s is calculated according to

$$D_s(m_s, n_s) = \sum_{i=0}^{N-1} \sum_{j=0}^{N-1} |x(i, j) - y(i + m_s + q \cdot \Delta_s, j + n_s + q \cdot \Delta_s)|$$

with $q \in \{-1, 0, 1\}$

$$\begin{bmatrix} m_s \\ n_s \end{bmatrix} = \mathbf{v}_{s-1} \qquad \text{for } s > 0$$

$$\begin{bmatrix} m_s \\ n_s \end{bmatrix} = \begin{bmatrix} 0 \\ 0 \end{bmatrix} \qquad \text{for } s = 0$$

and

$$\mathbf{v}_s = \left[\begin{array}{c} m_s \\ n_s \end{array} \right] |_{D_s,\min} \tag{59.20}$$

Δ_s depends on the maximum displacement w and the number of search steps N_s. Typically, when $w = 2^k - 1$, N_s is set to $k = \log_2(w + 1)$ and $\Delta_s = 2^{k-s+1}$. For example, for $w = 15$, four steps with $\Delta_s = 8, 4, 2, 1$ are performed. This strategy reduces the number of candidate blocks from $(2w + 1)^2$ in case of exhaustive search to $1 + 8 * \log_2(w + 1)$, e.g., for $w = 15$ the number of candidate blocks is reduced from 961 to 33 which leads to a reduction of processing power by a factor of 29. For large block sizes N, the number of operations for the match can be further reduced by combining the search strategy with subsampling in the first steps. Architectures for block matching based on hierarchical search strategies are presented in [20, 22, 24, 30].

59.6 Programmable Architectures

According to the three ways for architectural optimization, adaptation, pipelining, and parallel processing, three architectural classes for the implementation of video signal processors can be distinguished:

- *Intensive Pipelined Architectures* — These architectures are typically scalar architectures that achieve high clock frequencies of several hundreds of MHz due to the exploitation of pipelining.
- *Parallel Data Paths* — These architectures exploit data distribution for the increase of computational power. Several parallel data paths are implemented on one processor die, which leads in the ideal case to a linear increase of supported computational power. The number of parallel data paths is limited by the semiconductor process, since an increase of silicon area leads to a decrease of hardware yield.

- *Coprocessor Architectures* — Coprocessors are known from general processor designs and are often used for specific tasks, e.g., floating point operations. The idea of the adaptation to specific tasks and increase of computational power without an increase of the required semiconductor area has been applied by several designs. Due to their high regularity and the high processing power requirements, low-level tasks are the most promising candidates for an adapted implementation. The main disadvantage of this architectural approach is the decrease of flexibility by an increase of adaptation.

59.6.1 Intensive Pipelined Architectures

Applying pipelining for the increase of clock frequency leads to an increased latency of the circuit. For algorithms that require a data dependent control flow, this fact might limit the performance gain. Additionally, increasing arithmetic processing power leads to an increase of data access rate. Generally, the required data access rate cannot be provided by external memories. The gap between provided external and required internal data access rate increases for processor architectures with high clock frequency. To provide the high data access rate, the amount of internal memory which provides a low access time has to be increased for high-performance signal processors. Moreover, it is unfeasible to apply pipelining to speed-up on-chip memory. Thus, the minimum memory access time is another limiting factor for the maximum degree of pipelining. At least speed optimization is a time consuming task of the design process, which has to be performed for every new technology generation.

Examples for video processors with high clock frequency are the S-VSP [39] and the VSP3 [40]. Due to intensive pipelining, an internal clock frequency of up to 300 MHz can be achieved. The VSP3 consists of two parallel data paths, the Pipelined Arithmetic Logic Unit (PAU) and Pipelined Convolution Unit (PCU) (Fig. 59.11). The relatively large on-chip data memory of size 114 kbit is split into seven blocks, six data memories and one FIFO memory for external data exchange. Each of the six data memories is provided with an address generation unit (AGU), which provides the addressing modes "block", "DCT", and "zig-zag". Controlling is performed by a Sequence Control Unit (SCU) which involves a 1024x32bit instruction memory. A Host Interface Unit (HIU) and a Timing Control Unit (TCU) for the derivation of the internal clock frequency are integrated onto the VSP3 core.

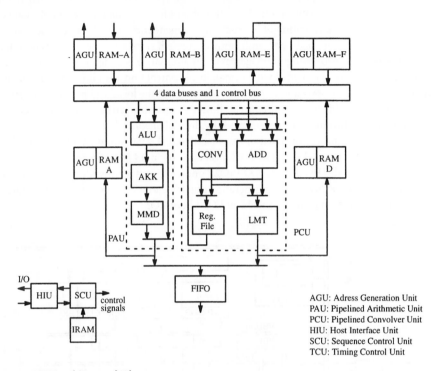

FIGURE 59.11 VSP3 architecture [40].

The entire VSP3 core consists of 1.27 million transistors, implemented based on a 0.5 micron BiCMOS technology on a 16.5 x 17.0-mm^2 die. The VSP3 performs the processing of the CCITT-H.261 tasks (neglecting Huffman coding) for one macroblock in 45 μs. Since realtime processing of 30Hz-CIF signals requires a processing time of less than 85 μs for one macroblock, a H.261 coder can be implemented based on one VSP3.

59.6.2 Parallel Data Paths

In the previous section, pipelining was presented as a strategy for processing power enhancement. Applying pipelining leads to a subdivision of a logic operation into sub-operations, which are processed in parallel with increased processing speed. An alternative to pipelining is the distribution of data among functional units. Applying this strategy leads to an implementation of parallel data paths. Typically, each data path is connected to an on-chip memory which provides the access distributed image segments.

Generally, two types of controlling strategies for parallel data paths can be distinguished. An MIMD concept provides a private control unit for each data path, whereas SIMD-based controlling provides a single common controller for parallel data paths. Compared to SIMD, the advantage of MIMD is a greater flexibility and a higher performance for complex algorithms with highly data dependent control flow. On the other hand, MIMD requires a significantly increased silicon area. Additionally, the access rate to the program memory is increased, since several controllers have to be provided with program data. Moreover, a software-based synchronization of the data paths is more complex. In case of an SIMD concept synchronization is performed implicitly by the hardware.

Since actual hybrid coding schemes require a large amount of processing power for tasks that require a data independent control flow, a single control unit for the parallel data path provides sufficient processor performance. The controlling strategy has to provide the execution of algorithms that require a data dependent control flow, e.g., quantization. A simple concept for the implementation of a data dependent control flow is to disable the execution of instruction in dependence of the local data path status. In this case, the data path utilization might be significantly decreased, since several of the parallel data path idle while others perform the processing of image data. An alternative is a hierarchical controlling concept. In this case, each data path is provided with a small local control unit with limited functionality and the global controller initiates the execution of control sequences of the local data path controllers. To reduce the required chip area for this controlling concept, the local controller can be reduced to a small instruction memory. Addressing of this memory is performed by the global control unit.

An example of a video processor based on parallel identical data path with a hierarchical controlling concept is the IDSP [42] (Fig. 59.12). The IDSP processor includes four pipelined data processing units (DPU0-DPU3), three parallel I/O ports (PIO0-PIO2), one 8×16-bit register file, five dual-ported memory blocks of size 512×16-bit each, an address generation unit for the data memories, and a program sequencer with 512×32-bit instruction memory and 32×32-bit boot ROM.

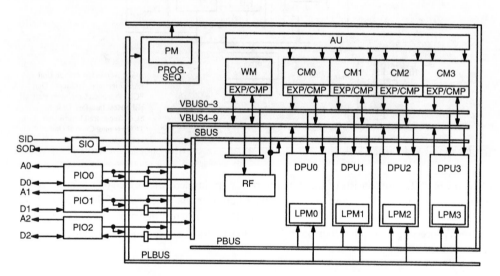

FIGURE 59.12   IDSP architecture [42].

The data processing units consist of a three-stage pipeline structure based on a ALU, multiplier, and an accumulator. This data path structure is well suited for L1 and L2 norm calculations and convolution-like algorithms. The four parallel data paths support a peak computational power of 300 MOPS at a typical clock frequency of 25 MHz. The data required for parallel processing are supplied by four cache memories (CM0-CM3) and a work memory (WM). Address generation for these memories is performed by an address generation unit (AU) which supports address sequences such as block scan, bit reverse, and butterfly. The three parallel I/O units contain a data I/O port, an address generation unit, and a DMA control processor (DMAC).

The IDSP integrates 910,000 transistors in 15.2×15.2 mm^2 using an 0.8 micron BiCMOS technology. For a full-CIF H.261 video codec four IDSP are required.

Another example of an SIMD-based video signal processor architecture based on identical parallel data paths is the HiPAR-DSP [44] (Fig. 59.13). The processor core consists of 16 RISC data paths, controlled by a common VLIW instruction word. The data paths contain a multiplier/accumulator unit, a shift/round unit, an ALU, and a 16×16bit register file. Each data path is connected to a private data cache. To support the characteristic data access pattern of several image processing tasks efficiently, a shared memory with parallel data access is integrated on-chip and provides parallel and conflict-free access to the data stored in this memory. The supported access patterns are "matrix", "vector" and "scalar". Data exchange with external devices is supported by an on chip DMA unit and a hypercube interface.

At present, a prototype of the HiPAR-DSP, based on four parallel data paths, is implemented. This chip will be manufactured in a 0.6 micron CMOS technology and will require a silicon area of about 180 mm^2. One processor chip is sufficient for realtime decoding of video signals, according to MPEG-2 Main Profile at Main Level. For encoding an external motion estimator is required.

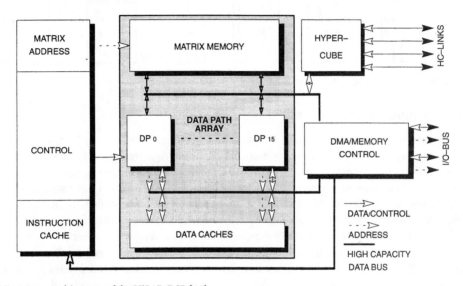

FIGURE 59.13   Architecture of the HiPAR-DSP [44].

In contrast to SIMD-based HiPAR-DSP architecture, the TMS320C80 (MVP) is based on an MIMD approach [43]. The MVP consists of four parallel processors (PP) and one master processor (Fig. 59.14). The processors are connected to 50-kbyte on-chip data memory via a global crossbar interconnection network. A DMA controller provides the data transfer to an external data memory and video I/O is supported by an on-chip video interface.

The master processor is a general-purpose RISC processor with an integral IEEE-compatible floating-point unit (FPU). The processor has a 32-bit instruction word and can load or store 8-, 16-, 32-, and 64-bit data sizes. The master processor includes a 32×32-bit general purpose register file. The master processor is intended to operate as the main supervisor and distributor of tasks within the chip and is also responsible for the communication with external processors. Due to the integrated FPU, the master processor will perform tasks such as audio signal processing and 3-D graphics transformation.

The parallel processors architecture has been designed to perform typical DSP algorithms, e.g., filtering, DCT, and to support bit and pixel manipulations for graphics applications. The parallel processors contain two address units, a program flow control unit, and a data unit with 32-bit ALU, 16×16-bit multiplier, and a barrel rotator.

The MVP has been designed using a 0.5 micron CMOS technology. Due to the supported flexibility, about four million transistors on a chip area of 324 mm^2 are required. A computational power of 2 GOPS is supported: A single MVP is able to encode CIF-30Hz video signals according to the MPEG-1 standard.

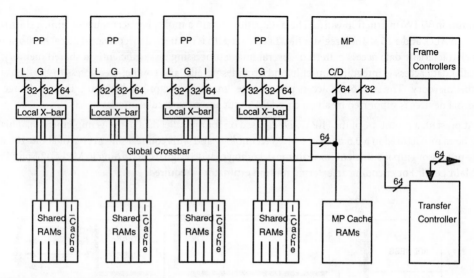

FIGURE 59.14   TMS320C80 (MVP) [43].

59.6.3   Coprocessor Concept

Most programmable architectures for video processing applications achieve an increase of processing power by an adaptation of the architecture to the algorithmic requirements. A feasible approach is the combination of a flexible programmable processor module with one or more adapted modules. This approach leads to an increase of processing power for specific algorithms and leads a significant decrease of required silicon area. The decrease of silicon area is caused by two effects. At first, the implementation of the required arithmetic operations can be optimized, which leads to an area reduction. Second, dedicated modules require significantly less hardware expense for module controlling, e.g., for program memory.

Typically, computation intensive tasks, such as DCT, block matching, or variable length coding, are candidates for an adapted or even dedicated implementation. Besides the adaptation to one specific task, mapping of several different tasks onto one adapted processor module might be advantageous. For example, mapping successive tasks, such as DCT, quantization, inverse quantization, IDCT, onto the same module reduces the internal communication overhead.

Coprocessor architectures that are based on highly adapted coprocessors achieve high computational power on a small chip area. The main disadvantage of these architectures is the limited flexibility. Changes of the envisaged applications might lead to an unbalanced utilization of the processor modules and therefore to a limitation of the effective processing power of the chip.

Applying the coprocessor concept opens up a variety of feasible architecture approaches, which differ in achievable processing power and flexibility of the architecture. In the following several architectures are presented, which clarify the wide variety of sensible approaches for video compression based on a coprocessor concept. Most of these architectures aim at an efficient implementation of hybrid coding schemes. As a consequence, these architectures are based on highly adapted coprocessors.

A chip set for video coding has been proposed in [8]. This chip set consists of four devices: two encoder options (the AVP1300E and AVP1400E), the AVP1400D decoder, and the AVP1400C system controller. The AVP1300E has been designed for H.261 and MPEG-1 frame-based encoding. Full MPEG-1 encoding (I-frame, P-frame, and B-frame) is supported by the AVP1400E. In the following, the architecture of the encoder chips is presented in more detail.

The AVP1300E combines function oriented modules, mask programmable modules, and user programmable modules (Fig. 59.15). It consists of a dedicated motion estimator for exhaustive search block matching with a search area of $+/- 15$ pels. The variable length encoder unit contains an ALU, a register array, a coefficient RAM, and a table ROM. Instructions for the VLE unit are stored in a program ROM.

Special instructions for conditional switching, run-length coding, and variable-to-fixed-length conversion are supported. The remaining tasks of the encoder loop, i.e., DCT/IDCT, quantization, and inverse quantization, are performed in two modules called SIMD processor and quantization processor (QP). The SIMD processor consists of six parallel processors each with ALU, multiplier-accumulator units. Program information for this module is again stored in a ROM memory. The QP's instructions are stored in a 1024 × 28-bit RAM. This module contains 16-bit ALU, a multiplier, and a register file of size 144 × 16-bit. Data communication with external DRAMs is supported by a memory management unit (MMAFC). Additionally, the processor scheduling is performed by a global controller (GC).

Due to the adaptation of the architecture to specific tasks of the hybrid coding scheme, a single chip of size 132 mm^2 (at 0.9 micron CMOS technology) supports the encoding of CIF-30Hz video signals according to the H.261 standard, including the computation intensive exhaustive search motion estimation strategy. An overview of the complete chipset is given in [33].

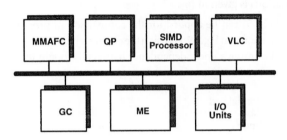

FIGURE 59.15   AVP encoder architecture [8].

The AxPe640V [37] is another typical example of the coprocessor approach (Fig. 59.16). To provide high flexibility for a broad range of video processing algorithms, the two processor modules are fully user programmable. A scalar RISC core supports the processing of tasks with data dependent control flow, whereas the typically more computation intensive low level tasks with data independent control flow can be executed by a parallel SIMD module.

The RISC core functions as a master processor for global control and for processing of tasks such as variable length encoding and quantization. To improve the performance for typical video coding schemes, the data path of the RISC core has been adapted to the requirements of quantization and variable length coding, by an extension of the basic instruction set. A program RAM of size is placed on-chip and can be loaded from an external PROM during start-up. The SIMD oriented arithmetic processing unit (APU) contains four parallel datapaths with a subtracter-complementer-multiplier pipeline. The intermediate results of the arithmetic pipelines are fed into a multi-operand accumulator with shift/limit circuitry. The results of the APU can be stored in the internal local memory or read out to the external data output bus.

Since both RISC core and APU include a private program RAM and address generation units, these processor modules are able to work in parallel on different tasks. This MIMD-like concept enables an execution of two tasks in parallel, e.g., DCT and quantization.

The AxPe640V is currently available in a 66-MHz version, designed in a 0.8 micron CMOS technology. A QCIF-10Hz H.261 codec can be realized with a single chip. To achieve higher computation power several AxPe640V can be combined to a multiprocessor system. For example, three AxPe640V are required for an implementation of a CIF-10Hz codec.

The examples presented above clarify the wide range of architectural approaches for the VLSI implementation of video coding schemes. The applied strategies are influenced by several demands, especially the desired flexibility of the architecture and maximum cost for realization and manufacturing. Due to the high computational requirements of real time video coding, most of the presented architectures apply a coprocessor concept with flexible programmable modules in combination with modules that are more

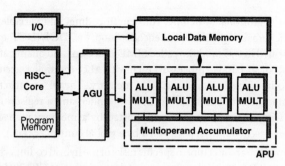

FIGURE 59.16   AxPe640V architecture [37].

or less adapted to specific tasks of the hybrid coding scheme. An overview of programmable architectures for video coding applications is given in [6].

Equations (59.4) and (59.5) can be applied for the comparison of programmable architectures. The result of this comparison is shown in Fig. 59.17, using the coding scheme according to ITU recommendation H.261 as a benchmark. Assuming a linear dependency between throughput rate and silicon area, a linear

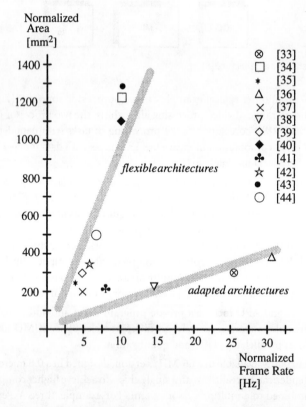

FIGURE 59.17   Normalized silicon area and throughput (frame rate) for adapted and flexible programmable architectures for a H.261 codec.

relationship corresponds to constant architectural efficiency, indicated by the two grey lines in Fig. 59.17. According to these lines, two groups of architectural classes can be identified. The first group consists of adapted architectures, optimized for hybrid coding applications. The architectures contain one or more adapted modules for computation intensive tasks, such as DCT or block matching. It is obvious that the application field of these architectures is limited to a small range of applications. This limitation is avoided by the members of the second group of architectures. Most of these architectures do not contain function specific circuitry for specific tasks of the hybrid coding scheme. Thus, they can be applied for wider variety of applications without a significant loss of sustained computational power. On the other hand, these architectures are associated with a decreased architectural efficiency compared to the first group of proposed architectures: Adapted architectures achieve an efficiency gain of about 6 to 7.

For a typical video phone application a frame rate of 10 Hz can be assumed. For this application the required normalized silicon area of about 130 mm^2 is required for adapted programmable approaches and approximately 950 mm^2 are required for flexible programmable architectures. For a rough estimation of the required silicon area for an MPEG-2 decoder, we assume that the algorithmic complexity of an MPEG-2 decoder for CCIR-601 signals is about half the complexity of an H.261 codec. Additionally, it has to be taken into account that the number of pixels per frame is about 5.3 times larger for CCIR signals than for CIF signals. From this the normalized implementation size of an MPEG-2 decoder for CCIR-601 signals and a frame rate of 25 Hz can be estimated to 870 mm^2 for an adapted architecture and 6333 mm^2 for flexible programmable architecture. Scaling these figures according to the defined scaling rules, a silicon area of about 71 mm^2 and 520 mm^2 can be estimated for an implementation based on an 0.5 μm CMOS process. Thus, the realization of video coding hardware for TV or HDTV based on flexible programmable processors still requires several monolithic components.

59.7 Conclusion

The properties of recent hybrid coding schemes in terms of VLSI implementation have been presented. Architectural alternatives for the dedicated realization of the DCT and block matching have been discussed. Architectures of programmable video signal processors have been presented and compared in terms of architectural efficiency. It has been shown that adapted circuits achieve a six to seven times higher efficiency than flexible programmable circuits. This efficiency gap might decrease for future coding schemes associated with a higher amount of medium- and high-level algorithms. Due to their flexibility, programmable architectures will become more and more attractive for future VLSI implementations of video compression schemes.

Acknowledgment

Figures 59.1–59.12 and 59.14–59.17 are reprinted from Pirsch, P., Demassieux, N., and Gehrke, W., VLSI architectures for video compression — a survey, *Proc. IEEE*, 83(2), 220–246, Feb., 1995, and used with permission from IEEE.

References

[1] ITU-T Recommendation H.261, Video codec for audiovisual services at px64 kbit/s, 1990.
[2] ISO-IEC IS 11172, Coding of moving pictures and associated audio for digital storage media at up to about 1.5 Mbit/s, 1993.
[3] ISO-IEC IS 13818, Generic coding of moving pictures and associated audio, 1994.
[4] ISO-IEC IS 10918, Digital compression and coding of continuous–tone still images, 1992.
[5] ISO/IEC JTC1/SC29/WG11, MPEG-4 functionalities, Nov. 1994.
[6] Pirsch, P., Demassieux, N. and Gehrke, W., VLSI architectures for video compression — a survey, *Proc. IEEE*, 83(2), 220–246, Feb. 1995.

[7] Bakoglu, H.B., Circuits interconnections and packaging for VLSI, Addison Wesley, Reading, MA, 1987.

[8] Rao, S.K., Matthew, M.H. et. al., A real-time P*64/ MPEG video encoder chip, *Proc. IEEE Intl. Solid State Circuits Conf.*, 32–35, 1993.

[9] Totzek, U., Matthiesen, F., Wohlleben, S. and Noll, T.G., CMOS VLSI implementation of the 2D-DCT with linear processor arrays, *Proc. Intl. Conf. on Acoustics Speech and Signal Processing*, V3.3, 1990.

[10] Lee, B.G., A new algorithm to compute the discrete cosine transform, *IEEE Trans. Acoustics, Speech and Signal Processing*, 32(6), 1243–1245, Dec. 1984.

[11] Artieri, A., Macoviak, E., Jutand, F. and Demassieux, N., A VLSI one chip for real time two-dimensional discrete cosine transform, *Proc. IEEE Intl. Symp. on Circuits and Systems*, Helsinki, 1988.

[12] Jain, P.C., Schlenk, W. and Riegel, M., VLSI implementation of two-dimensional DCT processor in real-time for video codec, *IEEE Trans. Consumer Electron.*, 38(3), Aug. 1992.

[13] Chau, K.K., Wang, I.F. and Eldridge, C.K., VLSI implementation of 2D-DCT in a compiler, *Proc. IEEE ICASSP*, 1233–1236, Toronto, Canada, 1991.

[14] Carlach, J.C., Penard, P. and Sicre, J.L., TCAD: a 27 MHz 8x8 Discrete Cosine Transform Chip, *Proc. Intl. Conf. Acoustics Speech and Signal Processing*, V2.3, 1989.

[15] Kim, S.P. and Pan, D.K., Highly modular and concurrent 2-D DCT chip, *Proc. IEEE Intl. Symp. on Circuits and Systems*, 1992.

[16] Sun, M.T., Chen, T.C. and Gottlieb, A.M., VLSI implementation of a 16×16 discrete cosine transform, *IEEE Trans. Circuits and Systems*, 36(4), April 1989.

[17] Bierling, M., Displacement estimation by hierarchical block-matching, *Proc. SPIE Visual Comm. Image Proc.*, 1001, 942–951, 1988.

[18] Chow, K H. and Liou, M.L., Genetic motion search for video compression, *Proc. IEEE Visual Sig. Proc. and Comm.*, Melbourne, Australia, 167–170, Sept. 1993.

[19] Ghanbari, M., The cross-search algorithm for motion estimation, *IEEE Trans. Commun.*, COM 38(7), 950–953, July 1990.

[20] Gupta, G. et. al., VLSI architecture for hierarchical block matching, *Proc. IEEE Intl. Symp. on Circuits and Systems*, 4, 215–218, 1994.

[21] Jain, J.R. and Jain, A.K., Displacement measurement and its application in interframe image coding, *IEEE Trans. Commun.*, COM 29(12), 1799–1808, Dec. 1981.

[22] Jong, H.M. et al., Parallel architectures of 3-step search block-matching algorithms for video coding, *Proc. IEEE Intl. Symp. on Circuits and Systems*, 3, 209–212, 1994.

[23] Kappagantula, S. and Rao, K.R., Motion compensated interframe image prediction, *IEEE Trans. Commun.*, COM 33(9), 1011–1015, Sept. 1985.

[24] Kim, H.C. et al., A pipelined systolic array architecture for the hierarchical block-matching algorithm, *Proc. IEEE Intl. Symp. on Circuits and Systems*, 3, 221–224, 1994.

[25] Koga, T., Iinuma, K., Hirano, A., Iijima, Y. and Ishiguro, T., Motion compensated interframe coding for video conferencing, *Proc. Nat. Telecom. Conf.*, New Orleans, G5.3.1–5.3.5, Nov. 29-Dec. 3, 1981.

[26] Puri, A., Hang, H.M. and Schilling, D.L., An efficient block-matching algorithm for motion compensated coding, *Proc. IEEE ICASSP*, 25.4.1–25.4.4, 1987.

[27] Srinivasan, R. and Rao, K.R., Predictive coding based on efficient motion estimation, *IEEE Trans. Commun.*, COM 33(8), 888–896, Aug. 85.

[28] Colavin, O., Artieri, A., Naviner, J.F. and Pacalet, R., A dedicated circuit for real-time motion estimation, *EuroASIC*, 1991.

[29] Dianysian, R. et al., Bit-serial architecture for real-time motion compensation, *Proc. SPIE Visual Communications and Image Processing*, 1988.

[30] Komarek, T. et al., Array architectures for block-matching algorithms, *IEEE Trans. on Circuits and Systems*, 36(10), Oct. 1989.

[31] Yang, K.M. et al., A family of VLSI designs for the motion compensation block-matching algorithms, *IEEE Trans. on Circuits and Systems*, 36(10), Oct. 1989.

[32] Ruetz, P., Tong, P., Bailey, D., Luthi, D.A. and Ang, P.H., A high-performance full-motion video compression chip set, *IEEE Trans. Circuits and Systems for Video Technol.*, 2(2), 111–122, June 1992.

[33] Ackland, B., The role of VLSI in multimedia, *IEEE J. Solid-State Circuits*, 29(4), 1886–1893, April 1994.

[34] Akari, T. et al., Video DSP architecture for MPEG2 codec, *Proc. ICASSP '94*, 2, 417–420, 1994, IEEE Press.

[35] Aono, K. et al., A video digital signal processor with a vector-pipeline architecture, *IEEE J. Solid-State Circuits*, 27(12), 1886–1893, Dec. 1992.

[36] Bailey, D. et. al., Programmable vision processor/controller, *IEEE MICRO*, 12(5), 33–39, Oct. 1992.

[37] Gaedke, K., Jeschke, H. and Pirsch, P., A VLSI based MIMD architecture of a multiprocessor system of real-time video processing applications, *J. VLSI Signal Processing*, 5, 159–169, April 1993.

[38] Gehrke, W., Hoffer, R. and Pirsch, P., A hierarchical multiprocessor architecture based on heterogeneous processors for video coding applications, *Proc. ICASSP '94*, 2, 1994.

[39] Goto, J. et al., 250-MHz BiCMOS super-high-speed video signal processor (S-VSP) ULSI, *IEEE J. Solid-State Circuits*, 26(12), 1876–1884, 1991.

[40] Inoue, T. et al., A 300-MHz BiCMOS video signal processor, *IEEE J. Solid-State Circuits*, 28(12), 1321–1329, Dec. 1993.

[41] Micke, T., Müller, D. and Heiß, R., ISDN-bildtelefon auf der grundlage eines array-prozessor-IC, mikroelektronik, vde-verlag, 5(3), 116–119, May/June 1991. (In German.)

[42] Yamauchi, H. et al., Architecture and implementation of a highly parallel single chip video DSP, *IEEE Trans. on Circuits and Systems for Videotechnology*, 2(2), 207–220, June 1992.

[43] Guttag, K., The multiprocessor video processor, MVP, *Proc. IEEE Hot Chips V*, Stanford, CA, Aug. 1993.

[44] Kneip, J. Rönner, K. and Pirsch, P., A single chip highly parallel architecture for image processing applications, *Proc. SPIE Visual Communications and Image Processing*, 2308(3), 1753–1764, Sept. 1994.

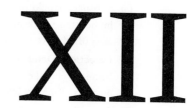

Sensor Array Processing

Mostafa Kaveh
University of Minnesota

A SENSOR ARRAY SYSTEM consists of a number of spatially-distributed elements, such as dipoles, hydrophones, geophones or microphones, followed by receivers and a processor. The array samples propagating wavefields in time and space. The receivers and the processor vary in mode of implementation and complexity according to the types of signals encountered, desired operation, and the adaptability of the array. For example, the array may be narrowband or wideband and the processor may be for determining the directions of the sources of signals or for beamforming to reject interfering signals and to enhance the quality of the desired signal in a communication system. The broad range of applications and the multifaceted nature of technical challenges for modern array signal processing have provided a fertile ground for contributions by and collaborations among researchers and practitioners from many disciplines, particularly those from the signal processing, statistics, and numerical linear algebra communities.

The following chapters present a sampling of the latest theory, algorithms, and applications related to array signal processing. The range of topics and algorithms include some which have been in use for more than a decade as well as some which are results of active current research. The sections on applications give examples of current areas of significant research and development.

Modern array signal processing often requires the use of the formalism of complex variables in modeling received signals and noise. Chapter 60 provides an introduction to complex random processes which are useful for bandpass communication systems and arrays. A classical use for arrays of sensors is to exploit the differences in the location (direction) of sources of transmitted signals to perform spatial filtering. Such techniques are reviewed in Chapter 61.

Another common use of arrays is the estimation of informative parameters about the wavefields impinging on the sensors. The most common parameter of interest is the direction of arrival (DOA) of a wave. Subspace techniques have been advanced as means of estimating the DOAs of sources, which are very close to each other, with high accuracy. The large number of developments in such techniques is reflected in the topics covered in Chapters 62 to 66. Chapter 62 gives a general overview of subspace processing for direction finding, while Chapter 63 discusses a particular type of subspace algorithm which is extended to sensing of azimuth and elevation angles with planar arrays. Most estimators assume knowledge of the needed statistical characteristics of the measurement noise. This requirement is relaxed in the approach given in Chapter 64. Chapter 65 extends the capabilities of traditional sensors to those which can measure the complete electric and magnetic field components and provides estimators which exploit such information. When signal sources move, or when computational requirements for real-time processing prohibit batch estimation of the subspaces, computationally efficient adaptive subspace updating techniques are called for. Chapter 66 presents many of the recent techniques which have been developed for this purpose. Before subspace methods are used for estimating the parameters of the waves received by an array, it is necessary to determine the number of sources which generate the waves. This aspect of the problem, often termed detection, is discussed in Chapter 67.

An important area of application for arrays is in the field of communications, particularly as it pertains to emerging mobile and cellular systems. Chapter 68 gives an overview of a number of techniques for improving the reception of signals in mobile systems, while Chapter 69 considers problems which arise in beamforming in the presence of multipath signals—a common occurrence in mobile communications. Chapter 70 discusses radar systems which employ sensor arrays, thereby providing the opportunity for space-time signal processing for improved resolution and target detection.

60

Complex Random Variables and Stochastic Processes

Daniel R. Fuhrmann
Washington University

60.1 Introduction

Much of modern digital signal processing is concerned with the extraction of information from signals which are noisy, or which behave randomly while still revealing some attribute or parameter of a system or environment under observation. The term in popular use now for this kind of computation is *statistical signal processing*, and much of this Handbook is devoted to this very subject. Statistical signal processing is classical statistical inference applied to problems of interest to electrical engineers, with the added twist that answers are often required in "real time", perhaps seconds or less. Thus, computational algorithms are often studied hand-in-hand with statistics.

One thing that separates the phenomena electrical engineers study from that of agronomists, economists, or biologists, is that the data they process are very often *complex*; that is, the data points come in pairs of the form $x + jy$, where x is called the *real part*, y the *imaginary part*, and $j = \sqrt{-1}$. Complex numbers are entirely a human intellectual creation: there are no complex physical measurable quantities such as time, voltage, current, money, employment, crop yield, drug efficacy, or anything else. However, it is possible to attribute to physical phenomena an underlying mathematical model that associates complex causes with real results. Paradoxically, the introduction of a complex-number-based theory can often simplify mathematical models.

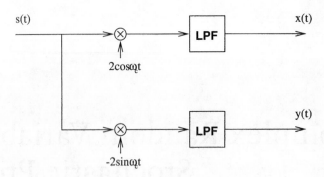

FIGURE 60.1 Quadrature demodulator.

Beyond their use in the development of analytical models, complex numbers often appear as actual data in some information processing systems. For representation and computation purposes, a complex number is nothing more than an ordered pair of real numbers. One just mentally attaches the "j" to one of the two numbers, then carries out the arithmetic or signal processing that this interpretation of the data implies.

One of the most well-known systems in electrical engineering that generates complex data from real measurements is the quadrature, or IQ, demodulator, shown in Fig. 60.1. The theory behind this system is as follows. A real bandpass signal, with bandwidth small compared to its center frequency, has the form

$$s(t) = A(t)\cos(\omega_c t + \phi(t)) \tag{60.1}$$

where ω_c is the center frequency, and $A(t)$ and $\phi(t)$ are the amplitude and angle modulation, respectively. By viewing $A(t)$ and $\phi(t)$ together as the polar coordinates for a complex function $g(t)$, i.e.,

$$g(t) = A(t)e^{j\phi(t)}, \tag{60.2}$$

we imagine that there is an underlying *complex modulation* driving the generation of $s(t)$, and thus

$$s(t) = \text{Re}\{g(t)e^{j\omega_c t}\}. \tag{60.3}$$

Again, $s(t)$ is physically measurable, while $g(t)$ is a mathematical creation. However, the introduction of $g(t)$ does much to simplify and unify the theory of bandpass communication. It is often the case that information to be transmitted via an electronic communication channel can be mapped directly into the magnitude and phase, or the real and imaginary parts, of $g(t)$. Likewise, it is possible to *demodulate* $s(t)$, and thus "retrieve" the complex function $g(t)$ and the information it represents. This is the purpose of the quadrature demodulator shown in Fig. 60.1. In Section 60.2 we will examine in some detail the operation of this demodulator, but for now note that it has one real input and two real outputs, which are *interpreted* as the real and imaginary parts of an information-bearing complex signal.

Any application of statistical inference requires the development of a probabilistic model for the received or measured data. This means that we imagine the data to be a "realization" of a multivariate random variable, or a stochastic process, which is governed by some underlying probability space of which we have incomplete knowledge. Thus, the purpose of this section is to give an introduction to probabilistic models for complex data. The topics covered are 2nd-order stochastic processes and their complex representations, the multivariate complex Gaussian distribution, and related distributions which appear in statistical tests. Special attention will be paid to a particular class of random variables, called *circular* complex random variables. Circularity is a type of symmetry in the distributions of the real and imaginary parts of complex random variables and stochastic processes, which can be physically motivated in many applications and is almost always assumed in the statistical signal processing literature. Complex representations for signals and the assumption of circularity are particularly useful in the processing of data or signals from an array of sensors, such as radar antennas. The reader will find them used throughout this chapter of the Handbook.

60.2 Complex Envelope Representations of Real Bandpass Stochastic Processes

60.2.1 Representations of Deterministic Signals

The motivation for using complex numbers to represent real phenomena, such as radar or communication signals, may be best understood by first considering the complex envelope of a real deterministic finite-energy signal.

Let $s(t)$ be a real signal with a well-defined Fourier transform $S(\omega)$. We say that $s(t)$ is *bandlimited* if the support of $S(\omega)$ is finite, that is,

$$S(\omega) \quad = \quad 0 \qquad \omega \notin B$$
$$\neq \quad 0 \qquad \omega \in B \tag{60.4}$$

where B is the *frequency band* of the signal, usually a finite union of intervals on the ω-axis such as

$$B = [-\omega_2, -\omega_1] \cup [\omega_1, \omega_2] . \tag{60.5}$$

The Fourier transform of such a signal is illustrated in Fig. 60.2.

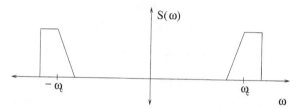

FIGURE 60.2 Fourier transform of a bandpass signal.

Since $s(t)$ is real, the Fourier transform $S(\omega)$ exhibits conjugate symmetry, i.e., $S(-\omega) = S^*(\omega)$. This implies that knowledge of $S(\omega)$, for $\omega \geq 0$ only, is sufficient to uniquely identify $s(t)$.

The complex envelope of $s(t)$, which we denote $g(t)$, is a frequency-shifted version of the complex signal whose Fourier transform is $S(\omega)$ for positive ω, and 0 for negative ω. It is found by the operation indicated graphically by the diagram in Fig. 60.3, which could be written

$$g(t) = \text{LPF}\{2s(t)e^{-j\omega_c t}\} . \tag{60.6}$$

ω_c is the center frequency of the band B, and "LPF" represents an ideal lowpass filter whose bandwidth is greater than half the bandwidth of $s(t)$, but much less than $2\omega_c$. The Fourier transform of $g(t)$ is given

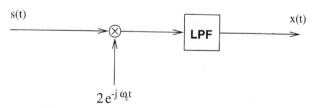

FIGURE 60.3 Quadrature demodulator.

by

$$G(\omega) = 2S(\omega - \omega_c) \qquad |\omega| < BW \tag{60.7}$$
$$= 0 \qquad \text{otherwise}.$$

The Fourier transform of $g(t)$, for $s(t)$ as given in Fig. 60.2, is shown in Fig. 60.4.

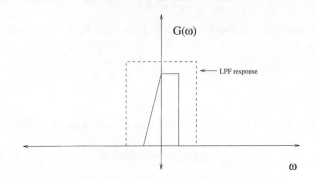

FIGURE 60.4   Fourier transform of the complex representation.

The inverse operation which gives $s(t)$ from $g(t)$ is

$$s(t) = \text{Re}\{g(t)e^{j\omega_c t}\}. \tag{60.8}$$

Our interest in $g(t)$ stems from the information it represents. Real bandpass processes can be written in the form

$$s(t) = A(t)\cos(\omega_c t + \phi(t)) \tag{60.9}$$

where $A(t)$ and $\phi(t)$ are slowly varying functions relative to the *unmodulated carrier* $\cos(\omega_c t)$, and carry information about the signal source. From the complex envelope representation (60.3), we know that

$$g(t) = A(t)e^{j\phi(t)} \tag{60.10}$$

and hence $g(t)$, in its polar form, is a direct representation of the information-bearing part of the signal.

In what follows we will outline a basic theory of complex representations for real stochastic processes, instead of the deterministic signals discussed above. We will consider representations of second-order stochastic processes, those with finite variances and correlations and well-defined spectral properties. Two classes of signals will be treated separately: those with finite energy (such as radar signals) and those with finite power (such as radio communication signals).

60.2.2   Finite-Energy Second-Order Stochastic Processes

Let $\mathbf{x}(t)$ be a real, second-order stochastic process, with the defining property

$$E\{\mathbf{x}^2(t)\} < \infty, \qquad \text{all } t. \tag{60.11}$$

Furthermore, let $\mathbf{x}(t)$ be finite-energy, by which we mean

$$\int_{-\infty}^{\infty} E\{\mathbf{x}^2(t)\}dt < \infty. \tag{60.12}$$

The *autocorrelation function* for $\mathbf{x}(t)$ is defined as

$$R_{\mathbf{xx}}(t_1, t_2) = E\{\mathbf{x}(t_1)\mathbf{x}(t_2)\}, \tag{60.13}$$

and from (60.11) and the Cauchy-Schwartz inequality we know that $R_{\mathbf{xx}}$ is finite for all t_1, t_2.

The *bi-frequency energy spectral density function* is

$$S_{xx}(\omega_1, \omega_2) = \int_{-\infty}^{\infty} \int_{-\infty}^{\infty} R_{xx}(t_1, t_2) e^{-j\omega_1 t_1} e^{+j\omega_2 t_2} dt_1 dt_2 . \tag{60.14}$$

It is assumed that $S_{xx}(\omega_1, \omega_2)$ exists and is well defined. In an advanced treatment of stochastic processes (e.g., Loeve [1]) it can be shown that $S_{xx}(\omega_1, \omega_2)$ exists if and only if the Fourier transform of $x(t)$ exists with probability 1; in this case, the process is said to be *harmonizable*.

If $x(t)$ is the input to a linear time-invariant system $\mathbf{H}$, and $y(t)$ is the output process, as shown in Fig. 60.5, then $y(t)$ is also a second-order finite-energy stochastic process. The bi-frequency energy

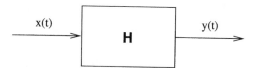

FIGURE 60.5 LTI system with stochastic input and output.

spectral density of $y(t)$ is

$$S_{yy}(\omega_1, \omega_2) = H(\omega_1) H^*(\omega_2) S_{xx}(\omega_1, \omega_2) . \tag{60.15}$$

This last result aids in a natural interpretation of the function $S_{xx}(\omega, \omega)$, which we denote as the *energy spectral density*. For any process, the total energy E_x is given by

$$E_x = \frac{1}{2\pi} \int_{-\infty}^{\infty} S_{xx}(\omega, \omega) d\omega . \tag{60.16}$$

If we pass $x(t)$ through an ideal filter whose frequency response is 1 in the band B and 0 elsewhere, then the total energy in the output process is

$$E_y = \frac{1}{2\pi} \int_B S_{xx}(\omega, \omega) d\omega . \tag{60.17}$$

This says that the energy in the stochastic process $x(t)$ can be partitioned into different frequency bands, and the energy in each band is found by integrating $S_{xx}(\omega, \omega)$ over the band.

We can define a *bandpass* stochastic process, with band B, as one that passes undistorted through an ideal filter $\mathbf{H}$ whose frequency response is 1 within the frequency band and 0 elsewhere. More precisely, if $x(t)$ is the input to an ideal filter $\mathbf{H}$, and the output process $y(t)$ is equivalent to $x(t)$ in the mean-square sense, that is

$$E\{(x(t) - y(t))^2\} = 0 \qquad \text{all } t , \tag{60.18}$$

then we say that $x(t)$ is a bandpass process with frequency band equal to the passband of $\mathbf{H}$. This is equivalent to saying that the integral of $S_{xx}(\omega_1, \omega_2)$ outside of the region $\omega_1, \omega_2 \in B$ is 0.

60.2.3 Second-Order Complex Stochastic Processes

A *complex* stochastic process $z(t)$ is one given by

$$z(t) = x(t) + jy(t) \tag{60.19}$$

where the real and imaginary parts, $x(t)$ and $y(t)$, respectively, are any two stochastic processes defined on a common probability space. A finite-energy, second-order complex stochastic process is one in which $x(t)$ and $y(t)$ are both finite-energy, second-order processes, and thus have all the properties given above.

Furthermore, because the two processes have a joint distribution, we can define the *cross-correlation function*

$$R_{xy}(t_1, t_2) = E\{\mathbf{x}(t_1)\mathbf{y}(t_2)\} \,. \tag{60.20}$$

By far the most widely used class of second-order complex processes in signal processing is the class of *circular* complex processes. A circular complex stochastic process is one with the following two defining properties:

$$R_{xx}(t_1, t_2) = R_{yy}(t_1, t_2) \tag{60.21}$$

and

$$R_{xy}(t_1, t_2) = -R_{yx}(t_1, t_2) \qquad \text{all } t_1, t_2 \,. \tag{60.22}$$

From Eqs. (60.21) and (60.22) we have that

$$E\{\mathbf{z}(t_1)\mathbf{z}^*(t_2)\} = 2R_{xx}(t_1, t_2) + 2j R_{yx}(t_1, t_2) \tag{60.23}$$

and furthermore

$$E\{\mathbf{z}(t_1)\mathbf{z}(t_2)\} = 0 \tag{60.24}$$

for all t_1, t_2. This implies that all of the joint second-order statistics for the complex process $\mathbf{z}(t)$ are represented in the function

$$R_{zz}(t_1, t_2) = E\{\mathbf{z}(t_1)\mathbf{z}^*(t_2)\} \tag{60.25}$$

which we define unambiguously as the *autocorrelation function* for $\mathbf{z}(t)$. Likewise, the *bi-frequency spectral density function* for $\mathbf{z}(t)$ is given by

$$S_{zz}(\omega_1, \omega_2) = \int_{-\infty}^{\infty} \int_{-\infty}^{\infty} R_{zz}(t_1, t_2) e^{-j\omega_1 t_1} e^{+j\omega_2 t_2} dt_1 dt_2 \,. \tag{60.26}$$

The functions $R_{zz}(t_1, t_2)$ and $S_{zz}(\omega_1, \omega_2)$ exhibit Hermitian symmetry, i.e.,

$$R_{zz}(t_1, t_2) = R_{zz}^*(t_2, t_1) \tag{60.27}$$

and

$$S_{zz}(\omega_1, \omega_2) = S_{zz}^*(\omega_2, \omega_1) \,. \tag{60.28}$$

However, there is no requirement that $S_{zz}(\omega_1, \omega_2)$ exhibit the conjugate symmetry for positive and negative frequencies, given in Eq. (60.6), as is the case for real stochastic processes.

Other properties of real second-order stochastic processes given above carry over to complex processes. Namely, if H is a linear time-invariant system with arbitrary complex impulse response $h(t)$, frequency response $H(\omega)$, and complex input $\mathbf{z}(t)$, then the complex output $\mathbf{w}(t)$ satisfies

$$S_{ww}(\omega_1, \omega_2) = H(\omega_1)H^*(\omega_2)S_{zz}(\omega_1, \omega_2) \,. \tag{60.29}$$

A bandpass circular complex stochastic process is one with finite spectral support in some arbitrary frequency band B.

Complex stochastic processes undergo a frequency translation when multiplied by a deterministic complex exponential. If $\mathbf{z}(t)$ is circular, then

$$\mathbf{w}(t) = e^{j\omega_c t}\mathbf{z}(t) \tag{60.30}$$

is also circular, and has bi-frequency energy spectral density function

$$S_{ww}(\omega_1, \omega_2) = S_{zz}(\omega_1 - \omega_c, \omega_2 - \omega_c) \,. \tag{60.31}$$

60.2.4 Complex Representations of Finite-Energy Second-Order Stochastic Processes

Let $s(t)$ be a bandpass finite-energy second-order stochastic process, as defined in Section 60.2.2. The complex representation of $s(t)$ is found by the same down-conversion and filtering operation described for deterministic signals:

$$g(t) = \text{LPF}\{2s(t)e^{-j\omega_c t}\} . \tag{60.32}$$

The lowpass filter in Eq. (60.32) is an ideal filter that passes the baseband components of the frequency-shifted signal, and attenuates the components centered at frequency $-2\omega_c$.

The inverse operation for Eq. (60.32) is given by

$$\hat{s}(t) = \text{Re}\{g(t)e^{j\omega_c t}\} . \tag{60.33}$$

Because the operation in Eq. (60.32) involves the integral of a stochastic process, which we define using mean-square stochastic convergence, we cannot say that $s(t)$ is identically equal to $\hat{s}(t)$ in the manner that we do for deterministic signals. However, it can be shown that $s(t)$ and $\hat{s}(t)$ are equivalent in the mean-square sense, that is,

$$E\{(s(t) - \hat{s}(t))^2\} = 0 \qquad \text{all } t . \tag{60.34}$$

With this interpretation, we say that $g(t)$ is the unique complex envelope representation for $s(t)$.

The assumption of circularity of the complex representation is widespread in many signal processing applications. There is an equivalent condition which can be placed on the real bandpass signal that guarantees its complex representation has this circularity property. This condition can be found indirectly by starting with a circular $g(t)$ and looking at the $s(t)$ which results.

Let $g(t)$ be an arbitrary lowpass circular complex finite-energy second-order stochastic process. The frequency-shifted version of this process is

$$p(t) = g(t)e^{+j\omega_c t} \tag{60.35}$$

and the real part of this is

$$s(t) = \frac{1}{2}(p(t) + p^*(t)) . \tag{60.36}$$

By the definition of circularity, $p(t)$ and $p^*(t)$ are orthogonal processes ($E\{p(t_1)(p^*(t_2))^* = 0\}$) and from this we have

$$
\begin{aligned}
S_{ss}(\omega_1, \omega_2) &= \frac{1}{4}(S_{pp}(\omega_1, \omega_2) + S_{p^*p^*}(\omega_1, \omega_2) \\
&= \frac{1}{4}(S_{gg}(\omega_1 - \omega_c, \omega_2 - \omega_c) + S_{gg}^*(-\omega_1 - \omega_c, -\omega_2 - \omega_c)) .
\end{aligned}
\tag{60.37}
$$

Since $g(t)$ is a baseband signal, the first term in Eq. (60.37) has spectral support in the first quadrant in the (ω_1, ω_2) plane, where both ω_1 and ω_2 are positive, and the second term has spectral support only for both frequencies negative. This situation is illustrated in Fig. 60.6.

It has been shown that a necessary condition for $s(t)$ to have a circular complex envelope representation is that it have spectral support only in the first and third quadrants of the (ω_1, ω_2) plane. This condition is also sufficient: if $g(t)$ is *not* circular, then the $s(t)$ which results from the operation in Eq. (60.33) will have non-zero spectral components in the second and fourth quadrants of the (ω_1, ω_2) plane, and this contradicts the mean-square equivalence of $s(t)$ and $\hat{s}(t)$.

An interesting class of processes with spectral support only in the first and third quadrants is the class of processes whose autocorrelation function is separable in the following way:

$$R_{ss}(t_1, t_2) = R_1(t_1 - t_2)R_2\left(\frac{t_1 + t_2}{2}\right) . \tag{60.38}$$

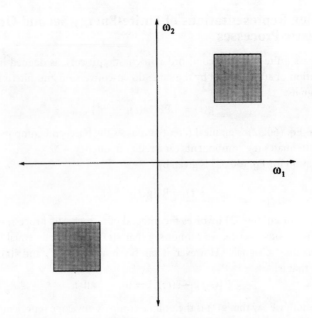

FIGURE 60.6   Spectral support for bandpass process with circular complex representation.

For these processes, the bi-frequency energy spectral density separates in a like manner:

$$S_{ss}(\omega_1, \omega_2) = S_1(\omega_1 - \omega_2)S_2\left(\frac{\omega_1 + \omega_2}{2}\right). \qquad (60.39)$$

In fact, S_1 is the Fourier transform of R_2 and vice versa. If S_1 is a lowpass function, and S_2 is a bandpass function, then the resulting product has spectral support illustrated in Fig. 60.7.

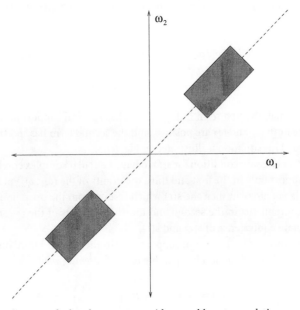

FIGURE 60.7   Spectral support for bandpass process with separable autocorrelation.

The assumption of circularity in the complex representation can often be physically motivated. For example, in a radar system, if the reflected electromagnetic wave undergoes a phase shift, or if the reflector position cannot be resolved to less than a wavelength, or if the reflection is due to a sum of reflections at slightly different path lengths, then the absolute phase of the return signal is considered random and uniformly distributed. Usually it is not the absolute phase of the received signal which is of interest; rather, it is the *relative phase* of the signal value at two different points in time, or of two different signals at the same instance in time. In many radar systems, particularly those used for direction-of-arrival estimation or delay-Doppler imaging, this relative phase is central to the signal processing objective.

60.2.5   Finite-Power Stochastic Processes

The second major class of second-order processes we wish to consider is the class of finite power signals. A finite-power signal $x(t)$ as one whose mean-square value exists, as in Eq. (60.4), but whose total energy, as defined in Eq. (60.12), is infinite. Furthermore, we require that the time-averaged mean-square value, given by

$$P_x = \lim_{T \to \infty} \frac{1}{2T} \int_{-T}^{T} R_{xx}(t, t)dt \ , \tag{60.40}$$

exist and be finite. P_x is called the *power* of the process $x(t)$.

The most commonly invoked stochastic process of this type in communications and signal processing is the *wide-sense-stationary* process, one whose autocorrelation function $R_{xx}(t_1, t_2)$ is a function of the time difference $t_1 - t_2$ only. In this case, the mean-square value is constant and is equal to the average power. Such a process is used to model a communication signal that transmits for a long period of time, and for which the beginning and end of transmission are considered unimportant.

A wide-sense-stationary (w.s.s.) process may be considered to be the limiting case of a particular type of finite-energy process, namely a process with separable autocorrelation as described by Eqs. (60.38) and (60.39). If in Eq. (60.38) the function $R_2 \left(\frac{t_1 + t_2}{2} \right)$ is equal to a constant, then the process is w.s.s. with second-order properties determined by the function $R_1(t_1 - t_2)$. The bi-frequency energy spectral density function is

$$S_{xx}(\omega_1, \omega_2) = 2\pi \delta(\omega_1 - \omega_2) S_2 \left(\frac{\omega_1 + \omega_2}{2} \right) \tag{60.41}$$

where

$$S_2(\omega) = \int_{-\infty}^{\infty} R_1(\tau) e^{-j\omega\tau} d\tau \ . \tag{60.42}$$

This last pair of equations motivates us to describe the second-order properties of $x(t)$ with functions of one argument instead of two, namely the autocorrelation function $R_{xx}(\tau)$ and its Fourier transform $S_{xx}(\omega)$, known as the *power spectral density*. From basic Fourier transform properties we have

$$P_x = \frac{1}{2\pi} \int_{-\infty}^{\infty} S_{xx}(\omega) d\omega \ . \tag{60.43}$$

If w.s.s. $x(t)$ is the input to a linear time-invariant system with frequency response $H(\omega)$ and output $y(t)$, then it is not difficult to show that

1. $y(t)$ is wide-sense-stationary, and
2. $S_{yy}(\omega) = |H(\omega)|^2 S_{xx}(\omega)$.

These last results, combined with Eq. (60.43), lead to a natural interpretation of the power spectral density function. If $x(t)$ is the input to an ideal bandpass filter with passband B, then the total power of the filter output is

$$P_y = \frac{1}{2\pi} \int_B S_x(\omega) d\omega \ . \tag{60.44}$$

This shows how the total power in the process $x(t)$ can be attributed to components in different spectral bands.

60.2.6 Complex Wide-Sense-Stationary Processes

Two real stochastic processes $x(t)$ and $y(t)$, defined on a common probability space, are said to be jointly wide-sense-stationary if:

1. Both $x(t)$ and $y(t)$ are w.s.s., and
2. The cross-correlation $R_{xy}(t_1, t_2) = E\{x(t_1)y(t_2)\}$ is a function of $t_1 - t_2$ only.

For jointly w.s.s. processes, the cross-correlation function is normally written with a single argument, e.g., $R_{xy}(\tau)$, with $\tau = t_1 - t_2$. From the definition we see that

$$R_{xy}(\tau) = R_{yx}(-\tau) . \tag{60.45}$$

A complex wide-sense-stationary stochastic process $z(t)$ is one that can be written

$$z(t) = x(t) + jy(t) \tag{60.46}$$

where $x(t)$ and $y(t)$ are jointly wide-sense stationary. A *circular* complex w.s.s. process is one in which

$$R_{xx}(\tau) = R_{yy}(\tau) \tag{60.47}$$

and

$$R_{xy}(\tau) = -R_{yx}(\tau) \qquad \text{all } \tau . \tag{60.48}$$

The reader is cautioned not to confuse the meanings of Eqs. (60.45) and (60.48).

For circular complex w.s.s. processes, it is easy to show that

$$E\{z(t_1)z(t_2)\} = 0 \tag{60.49}$$

for all t_1, t_2, and therefore the function

$$
\begin{aligned}
R_{zz}(t_1, t_2) &= E\{z(t_1)z^*(t_2)\} \\
&= 2R_{xx}(t_1, t_2) + 2j R_{yx}(t_1, t_2)
\end{aligned}
\tag{60.50}
$$

defines all the second-order properties of $z(t)$. All the quantities involved in Eq. (60.50) are functions of $\tau = t_1 - t_2$ only, and thus the single-argument function $R_{zz}(\tau)$ is defined as the autocorrelation function for $z(t)$.

The power spectral density for $z(t)$ is

$$S_{zz}(\omega) = \int_{-\infty}^{\infty} R_{zz}(\tau)e^{-j\omega\tau} d\tau . \tag{60.51}$$

$R_{zz}(\tau)$ exhibits conjugate symmetry ($R_{zz}(\tau) = R_{zz}^*(-\tau)$); $S_{zz}(\omega)$ is non-negative but otherwise has no symmetry constraints.

If $z(t)$ is the input to a complex linear time-invariant system with frequency response $H(\omega)$, then the output process $w(t)$ is wide-sense-stationarity with power spectral density

$$S_{ww}(\omega) = |H(\omega)|^2 S_{zz}(\omega) . \tag{60.52}$$

A bandpass w.s.s. process is one with finite (possible asymmetric) support in frequency.

If $z(t)$ is a circular w.s.s. process, then

$$w(t) = e^{j\omega_c t} z(t) \tag{60.53}$$

is also circular, and has power spectral density

$$S_{ww}(\omega) = S_{zz}(\omega - \omega_c) . \tag{60.54}$$

60.2.7   Complex Representations of Real Wide-Sense-Stationary Signals

Let $\mathbf{s}(t)$ be a real bandpass w.s.s. stochastic process. The complex representation for $\mathbf{s}(t)$ is given by the now-familiar expression

$$\mathbf{g}(t) = \text{LPF}\{2\mathbf{s}(t)e^{-j\omega_c t}\} \tag{60.55}$$

with inverse relationship

$$\hat{\mathbf{s}}(t) = \text{Re}\{\mathbf{g}(t)e^{j\omega_c t}\} . \tag{60.56}$$

In Eqs. (60.55) and (60.56), ω_c is the center frequency for the passband of $\mathbf{s}(t)$, and the lowpass filter has bandwidth greater than that of $\mathbf{s}(t)$ but much less than $2\omega_c$. $\mathbf{s}(t)$ and $\hat{\mathbf{s}}(t)$ are equivalent in the mean-square sense, implying that $\mathbf{g}(t)$ is the unique complex envelope representation for $\mathbf{s}(t)$.

For arbitrary real w.s.s. $\mathbf{s}(t)$, the circularity of the complex representation comes without any additional conditions like the ones imposed for finite-energy signals. If w.s.s. $\mathbf{s}(t)$ is the input to a quadrature demodulator, then the output signals $\mathbf{x}(t)$ and $\mathbf{y}(t)$ are jointly w.s.s., and the complex process

$$\mathbf{g}(t) = \mathbf{x}(t) + j\mathbf{y}(t) \tag{60.57}$$

is circular. There are various ways of showing this, with the simplest probably being a proof by contradiction. If $\mathbf{g}(t)$ is a complex process that is *not* circular, then the process $\text{Re}\{\mathbf{g}(t)e^{j\omega_c t}\}$ can be shown to have an autocorrelation function with nonzero terms which are a function of $t_1 + t_2$, and thus it cannot be w.s.s.

Communication signals are often modeled as w.s.s. stochastic processes. The stationarity results from the fact that the carrier phase, as seen at the receiver, is unknown and considered random, due to lack of knowledge about the transmitter and path length. This in turn leads to a circularity assumption on the complex modulation.

In many communication and surveillance systems, the quadrature demodulator is an actual electronic subsystem which generates a pair of signals interpreted directly as a complex representation of a bandpass signal. Often these signals are sampled, providing complex digital data for further digital signal processing. In array signal processing, there are multiple such receivers, one behind each sensor or antenna in a multi-sensor system. Data from an array of receivers is then modeled as a *vector* of complex random variables. In the next section, we consider multivariate distributions for such complex data.

60.3   The Multivariate Complex Gaussian Density Function

The discussions of Section 60.2 centered on the second-order (correlation) properties of real and complex stochastic processes, but to this point nothing has been said about joint probability distributions for these processes. In this section, we consider the distribution of samples from a complex process in which the real and imaginary parts are Gaussian distributed. The key concept of this section is that the assumption of circularity on a complex stochastic process (or any collection of complex random variables) leads to a compact form of the density function which can be written directly as a function of a complex argument z rather than its real and imaginary parts.

From a data processing point-of-view, a collection of N complex numbers is simply a collection of $2N$ real numbers, with a certain mathematical significance attached to the N numbers we call the "real parts" and the other N numbers we call the "imaginary parts". Likewise, a collection of N complex random variables is really just a collection of $2N$ real random variables with some joint distribution in $\mathbb{R}^{2N}$. Because these random variables have an interpretation as real and imaginary parts of some complex numbers, and because the $2N$-dimensional distribution may have certain symmetries such as those resulting from circularity, it is often natural and intuitive to express joint densities and distributions using a notation which makes explicit the complex nature of the quantities involved. In this section we develop such a density for the case where the random variables have a Gaussian distribution and are samples of a circular complex stochastic process.

Let z_i, $i = 1..N$ be a collection of complex numbers that we wish to model probabilistically. Write

$$\mathbf{z}_i = \mathbf{x}_i + j\mathbf{y}_i \tag{60.58}$$

and consider the vector of numbers $[\mathbf{x}_1, \mathbf{y}_1, .., \mathbf{x}_N, \mathbf{y}_N]^T$ as a set of $2N$ random variables with a distribution over $\mathbb{R}^{2N}$. Suppose further that the vector $[\mathbf{x}_1, \mathbf{y}_1, .., \mathbf{x}_N, \mathbf{y}_N]^T$ is subject to the usual multivariate Gaussian distribution with $2N \times 1$ mean vector μ and $2N \times 2N$ covariance matrix $\mathbf{R}$. For compactness, denote the entire random vector with the symbol $\mathbf{x}$. The density function is

$$f_{\mathbf{x}}(x) = (2\pi)^{\frac{-2N}{2}} (\det \mathbf{R})^{\frac{-1}{2}} e^{-\frac{x^T \mathbf{R}^{-1} x}{2}} . \tag{60.59}$$

We seek a way of expressing the density function of Eq. (60.59) directly in terms of the complex variable z, i.e., a density of the form $f_{\mathbf{z}}(z)$. In so doing it is important to keep in mind what such a density represents. $f_{\mathbf{z}}(z)$ will be a non-negative real-valued function $f : \mathbb{C}^N \to \mathbb{R}^+$, with the property that

$$\int_{\mathbb{C}^N} f_{\mathbf{z}}(z) dz = 1 . \tag{60.60}$$

The probability that $\mathbf{z} \in A$, where A is some subset of $\mathbb{C}^N$, is given by

$$P(A) = \int_A f_{\mathbf{z}}(z) dz . \tag{60.61}$$

The differential element dz is understood to be

$$dz = dx_1 dy_1 dx_2 dy_2 .. dx_N dy_N . \tag{60.62}$$

The most general form of the complex multivariate Gaussian density is in fact given by Eq. (60.59), and further simplification requires further assumptions. Circularity of the underlying complex process is one such key assumption, and it is now imposed. To keep the following development simple, it is assumed that the mean vector μ is 0. The results for nonzero μ are not difficult to obtain by extension.

Consider the four real random variables $\mathbf{x}_i$, $\mathbf{y}_i$, $\mathbf{x}_k$, $\mathbf{y}_k$. If these numbers represent the samples of a circular complex stochastic process, then we can express the 4×4 covariance as

$$E \left\{ \begin{bmatrix} \mathbf{x}_i \\ \mathbf{y}_i \\ \mathbf{x}_k \\ \mathbf{y}_k \end{bmatrix} \begin{bmatrix} \mathbf{x}_i \mathbf{y}_i \mathbf{x}_k \mathbf{y}_k \end{bmatrix} \right\} = \frac{1}{2} \begin{bmatrix} \alpha_{ii} & 0 & | & \alpha_{ik} & -\beta_{ik} \\ 0 & \alpha_{ii} & | & \beta_{ik} & \alpha_{ik} \\ - & - & - & - & - \\ \alpha_{ki} & -\beta_{ki} & | & \alpha_{kk} & 0 \\ \beta_{ki} & \alpha_{ki} & | & 0 & \alpha_{kk} \end{bmatrix} \tag{60.63}$$

where

$$\alpha_{ik} = 2E\{\mathbf{x}_i \mathbf{x}_k\} = 2E\{\mathbf{y}_i \mathbf{y}_k\} \tag{60.64}$$

and

$$\beta_{ik} = -2E\{\mathbf{x}_i \mathbf{y}_k\} = +2E\{\mathbf{x}_k \mathbf{y}_i\} . \tag{60.65}$$

Extending this to the full $2N \times 2N$ covariance matrix $\mathbf{R}$, we have

$$
\mathbf{R} = \frac{1}{2}
\left[
\begin{array}{ccc|cc|ccc|cc}
\alpha_{11} & 0 & | & \alpha_{12} & -\beta_{12} & | & \cdot & \cdot & \cdot & | & \alpha_{1N} & -\beta_{1N} \\
0 & \alpha_{11} & | & \beta_{12} & \alpha_{12} & | & \cdot & \cdot & \cdot & | & \beta_{1N} & \alpha_{1N} \\
- & - & & - & - & & & & & & \\
\alpha_{21} & -\beta_{21} & | & \alpha_{22} & 0 & | & \cdot & \cdot & \cdot & | & \alpha_{2N} & -\beta_{2N} \\
\beta_{21} & \alpha_{21} & | & 0 & \alpha_{22} & | & \cdot & \cdot & \cdot & | & \beta_{2N} & \alpha_{2N} \\
- & - & & - & - & & & & & & \\
\cdot & \cdot & | & \cdot & \cdot & | & \cdot & & & | & \cdot & \cdot \\
\cdot & \cdot & | & \cdot & \cdot & | & & \cdot & & | & \cdot & \cdot \\
\cdot & \cdot & | & \cdot & \cdot & | & & & \cdot & | & \cdot & \cdot \\
- & - & & - & - & & & & & & \\
\alpha_{N1} & -\beta_{N1} & | & \alpha_{N2} & -\beta_{N2} & | & \cdot & \cdot & \cdot & | & \alpha_{NN} & 0 \\
\beta_{N1} & \alpha_{N1} & | & \beta_{N2} & \alpha_{N2} & | & \cdot & \cdot & \cdot & | & 0 & \alpha_{NN}
\end{array}
\right].
\tag{60.66}
$$

The key thing to notice about the matrix in Eq. (60.66) is that, because of its special structure, it is completely specified by N^2 real quantities: one for each of the 2×2 diagonal blocks, and two for each of the 2×2 upper off-diagonal blocks. This is in contrast to the $N(2N + 1)$ free parameters one finds in an unconstrained $2N \times 2N$ real Hermitian matrix.

Consider now the complex random variables z_i and z_k. We have that

$$
\begin{aligned}
E\{z_i z_i^*\} &= E\{(x_i + jy_i)(x_i - jy_i)\} \\
&= E\{x_i^2 + y_i^2\} = \alpha_{ii}
\end{aligned}
\tag{60.67}
$$

and

$$
\begin{aligned}
E\{z_i z_k^*\} &= E\{(x_i + jy_i)(x_k - jy_k)\} \\
&= E\{x_i x_k + y_i y_k - jx_k y_i + jx_i y_k\} \\
&= \alpha_{ik} + j\beta_{ik}.
\end{aligned}
\tag{60.68}
$$

Similarly

$$
E\{z_k z_i^*\} = \alpha_{ik} - j\beta_{ik}
\tag{60.69}
$$

and

$$
E\{z_k z_k^*\} = \alpha_{kk}.
\tag{60.70}
$$

Using Eqs. (60.66) through (60.70), it is possible to write the following $N \times N$ complex Hermitian matrix:

$$
E\{\mathbf{z}\mathbf{z}^H\} =
\left[
\begin{array}{ccccccc}
\alpha_{11} & | & \alpha_{12} + j\beta_{12} & | & \cdot\cdot & | & \alpha_{1N} + j\beta_{1N} \\
- - & - & - - - & - & - - - & - & - - - \\
\alpha_{21} + j\beta_{21} & | & \alpha_{22} & | & \cdots & | & \alpha_{2N} + j\beta_{2N} \\
- - - & - & - - & - & - - - & - & - - - \\
\cdot & | & \cdot & | & \cdot & | & \cdot \\
\cdot & | & \cdot & | & \cdot & | & \cdot \\
\cdot & | & \cdot & | & \cdot & | & \cdot \\
- - - & - & - - & - & - - - & - & - - - \\
\alpha_{N1} + j\beta_{N1} & | & \alpha_{N2} + j\beta_{N2} & | & \cdot\cdot & | & \alpha_{NN}
\end{array}
\right].
\tag{60.71}
$$

Note that this complex matrix has exactly the same N^2 free parameters as did the $2N \times 2N$ real matrix $\mathbf{R}$ in Eq. (60.66), and thus it tells us everything there is to know about the joint distribution of the real and imaginary components of $\mathbf{z}$. Under the symmetry constraints imposed on $\mathbf{R}$, we can define

$$
\mathbf{C} = E\{\mathbf{z}\mathbf{z}^H\}
\tag{60.72}
$$

and call this matrix the covariance matrix for $\mathbf{z}$. In the 0-mean Gaussian case, this matrix parameter uniquely identifies the multivariate distribution for $\mathbf{z}$.

The derivation of the density function $f_{\mathbf{z}}(z)$ rests on a set of relationships between the $2N \times 1$ real vector $\mathbf{x}$, and its $N \times 1$ complex counterpart $\mathbf{z}$. We say that $\mathbf{x}$ and $\mathbf{z}$ are *isomorphic* to one another, and denote this with the symbol

$$\mathbf{z} \approx \mathbf{x} . \tag{60.73}$$

Likewise we say that the $2N \times 2N$ real matrix $\mathbf{R}$, given in Eq. (60.66), and the $N \times N$ complex matrix $\mathbf{C}$, given in Eq. (60.71) are isomorphic to one another, or

$$\mathbf{C} \approx \mathbf{R} . \tag{60.74}$$

The development of the complex Gaussian density function $f_{\mathbf{z}}(z)$ is based on three claims based on these isomorphisms.

Proposition 1. If $\mathbf{z} \approx \mathbf{x}$, and $\mathbf{R} \approx \mathbf{C}$, then

$$\mathbf{x}^{T}(2\mathbf{R})\mathbf{x} = \mathbf{z}^{H}\mathbf{C}\mathbf{z} . \tag{60.75}$$

Proposition 2. If $\mathbf{R} \approx \mathbf{C}$, then

$$\frac{1}{4}\mathbf{R}^{-1} \approx \mathbf{C}^{-1} . \tag{60.76}$$

Proposition 3. If $\mathbf{R} \approx \mathbf{C}$, then

$$\det \mathbf{R} = |\det \mathbf{C}|^{2}\left(\frac{1}{2}\right)^{2N} . \tag{60.77}$$

The density function $f_{\mathbf{z}}(z)$ is found by substituting the results from Propositions 1 through 3 directly into the density function $f_{\mathbf{x}}(x)$. This is possible because the mapping from $\mathbf{z}$ to $\mathbf{x}$ is one-to-one and onto, and the Jacobian is 1 [see Eq. (60.62)]. We have

$$
\begin{aligned}
f_{\mathbf{z}}(z) &= (2\pi)^{\frac{-2N}{2}}(\det \mathbf{R})^{\frac{-1}{2}}e^{-\frac{\mathbf{x}^{T}\mathbf{R}^{-1}\mathbf{x}}{2}} \tag{60.78}\\
&= \left(\frac{1}{2}\right)^{-N}(2\pi)^{-N}(\det \mathbf{C})^{-1}e^{-\mathbf{z}^{H}\mathbf{C}^{-1}\mathbf{z}} .\\
&= \pi^{-N}(\det \mathbf{C})^{-1}e^{-\mathbf{z}^{H}\mathbf{C}^{-1}\mathbf{z}} . \tag{60.79}
\end{aligned}
$$

At this point it is straightforward to introduce a non-zero mean μ, which is the complex vector isomorphic to the mean of the real random vector $\mathbf{x}$. The resulting density is

$$f_{\mathbf{z}}(z) = \pi^{-N}(\det \mathbf{C})^{-1}e^{-(z-\mu)^{H}\mathbf{C}^{-1}(z-\mu)} . \tag{60.80}$$

The density function in Eq. (60.80) is commonly referred to as the *complex Gaussian density function*, although in truth one could be more general and have an arbitrary $2N$-dimension Gaussian distribution on the real and imaginary components of $\mathbf{z}$. It is important to recognize that the use of Eq. (60.80) implies those symmetries in the real covariance of $\mathbf{x}$ implied by circularity of the underlying complex process. This symmetry is expressed by some authors in the equation

$$E\{\mathbf{z}\mathbf{z}^{T}\} = 0 \tag{60.81}$$

where the superscript "T" indicates transposition without complex conjugation. This comes directly from Eqs. (60.24) and (60.49).

For many, the functional form of the complex Gaussian density in Eq. (60.80) is actually simpler and cleaner than its N-dimensional real counterpart, due to elimination of the various factors of 2 which complicate it. This density is the starting point for virtually all of the multivariate analysis of complex data seen in the current signal and array processing literature.

60.4 Related Distributions

In many problems of interest in statistical signal processing, the raw data may be complex and subject to a complex Gaussian distribution described in the density function in Eq. (60.80). The processing may take the form of the computation of a test statistic for use in a hypothesis test. The density functions for these test statistics are then used to determine probabilities of false alarm and/or detection. Thus, it is worthwhile to study certain distributions that are closely related to the complex Gaussian in this way.

In this section we will describe and give the functional form for four densities related to the complex Gaussian: the complex χ^2, the complex F, the complex β, and the complex t. Only the "central" versions of these distributions will be given, i.e., those based on 0-mean Gaussian data. The central distributions are usually associated with the null hypothesis in a detection problem and are used to compute probabilities of false alarm. The non-central densities, used in computing probabilities of detection, do not exist in closed form but can be easily tabulated.

60.4.1 Complex Chi-Squared Distribution

One very common type of detection problem in radar problems is the "signal present" vs. "signal absent" decision problem. Often under the "signal absent" hypothesis, the data is zero-mean complex Gaussian, with known covariance, whereas under the "signal present" hypothesis the mean is non-zero, but perhaps unknown or subject to some uncertainty. A common test under these circumstances is to compute the sum of squared magnitudes of the data points (after pre-whitening, if appropriate) and compare this to a threshold. The resulting test statistic has a χ^2-*squared* distribution.

Let $z_1..z_N$ be N complex Gaussian random variables, independent and identically distributed with mean 0 and variance 1 (meaning that the covariance matrix for the z vector is I). Define the real non-negative random variable q according to

$$q = \sum_i^N |z_i|^2 .$$
(60.82)

Then the density function for q is given by

$$f_q(q) = \frac{1}{(N-1)!} q^{N-1} e^{-q} U(q) .$$
(60.83)

To establish this result, show that the density function for $|z_i|^2$ is a simple exponential. Equation (60.83) is the N-fold convolution of this exponential density function with itself.

We often say that q is χ^2 with N complex degrees of freedom. A "complex degree of freedom" is like two real degrees of freedom. Note, however, that Eq. (60.83) is not the usual χ^2 density function with $2N$ degrees of freedom. Each of the real variables going into the computation of q has variance $\frac{1}{2}$, not 1. $f_q(q)$ is a gamma density with an integer parameter N, and, like the complex Gaussian density in Eq. (60.60), it is cleaner and simpler than its real counterpart.

60.4.2 Complex F Distribution

In some "signal present" vs. "signal absent" problems, the variance or covariance of the noise is not known under the null hypothesis, and must be estimated from some auxiliary data. Then the test statistic becomes the ratio of the sum of square magnitudes of the test data to the sum of square magnitudes of the auxiliary data. The resulting test statistic is subject to a particular form of the F-distribution.

Let q_1 and q_2 be two independent random variables subject to the χ^2 distribution with N and M complex degrees of freedom, respectively. Define the real, nonnegative random variable f according to

$$f = \frac{q_1}{q_2} .$$
(60.84)

The density function for **f** is

$$f_{\mathbf{f}}(f) = \frac{(N + M - 1)!}{(N - 1)!(M - 1)!} \frac{f^{N-1}}{(1 + f)^{N+M}} U(f) . \tag{60.85}$$

We say that **f** is subject to an F-distribution with N and M complex degrees of freedom.

60.4.3 Complex Beta Distribution

An F-distributed random variable can be transformed in such a way that the resulting density has finite support. The random variable **b**, defined by

$$\mathbf{b} = \frac{1}{(1 + \mathbf{f})} , \tag{60.86}$$

where **f** is an F-distributed random variable, has this property. The density function is given by

$$f_{\mathbf{b}}(b) = \frac{(N + M - 1)!}{(N - 1)!(M - 1)!} b^{M-1} (1 - b)^{N-1} \tag{60.87}$$

on the interval $0 \le b \le 1$, and is 0 elsewhere.

The random variable **b** is said to be beta-distributed, with N and M complex degrees of freedom.

60.4.4 Complex Student-t Distribution

In the "signal present" vs. "signal absent" problem, if the signal is known exactly (including phase) then the optimal detector is a pre-whitener followed by a matched filter. The resulting test statistic is complex Gaussian, and the detector partitions the complex plane into two half-planes which become the decision regions for the two hypotheses. Now it may be that the signal is known, but the variance of the noise is not. In this case, the Gaussian test statistic must be scaled by an estimate of the standard deviation, obtained as before from zero-mean auxiliary data. In this case the test statistic is said to have a complex t (or Student-t) distribution. Of the four distributions discussed in this section, this is the only one in which the random variables themselves are complex: the χ^2, F, and β distributions all describe real random variables functionally dependent on complex Gaussians.

Let **z** and **q** be independent scalar random variables. **z** is complex Gaussian with mean 0 and variance 1, and **q** is χ^2 with N complex degrees of freedom. Define the random variable **t** according to

$$\mathbf{t} = \frac{\mathbf{z}}{\sqrt{\mathbf{q}/N}} . \tag{60.88}$$

The density of **t** is then given by

$$f_{\mathbf{t}}(t) = \frac{1}{\pi \left(1 + \frac{|t|^2}{N}\right)^{N+1}} . \tag{60.89}$$

This density is said to be "heavy-tailed" relative to the Gaussian, and this is a result in the uncertainty in the estimate of the standard deviation. Note that as $N \to \infty$, the denominator Eq. (60.88) approaches 1 (i.e., the estimate of the standard deviation approaches truth) and thus $f_{\mathbf{t}}(t)$ approaches the Gaussian density $\pi^{-1} e^{-|t|^2}$ as expected.

60.5 Conclusion

In this chapter we have outlined a basic theory of complex random variables and stochastic processes as they most often appear in statistical signal and array processing problems. The properties of complex representations for real bandpass signals were emphasized, since this is the most common application in electrical engineering where complex data appear. Models for both finite-energy signals, such as radar pulses, and finite-power signals, such as communication signals, were developed. The key notion of circularity of complex stochastic processes was explored, along with the conditions that a real stochastic process must satisfy in order for it to have a circular complex representation. The complex multivariate Gaussian distribution was developed, again building on the circularity of the underlying complex stochastic process. Finally, related distributions which often appear in statistical inference problems with complex Gaussian data were introduced.

The general topic of random variables and stochastic processes is fundamental to modern signal processing, and many good textbooks are available. Those by Papoulis [2], Leon-Garcia [3], and Melsa and Sage [4] are recommended. The original short paper deriving the complex multivariate Gaussian density function is by Wooding [5]; another derivation and related statistical analysis is given in Goodman [6], whose name is more often cited in connection with complex random variables. The monograph by Miller [7] has a mathematical flavor, and covers complex stochastic processes, stochastic differential equations, parameter estimation, and least-squares problems. The paper by Neeser and Massey [8] treats circular (which they call "proper") complex stochastic processes and their application in information theory. There is a good discussion of complex random variables in Kay [9], which includes Cramer-Rao lower bounds and optimization of functions of complex variables. Kelly and Forsythe [10] is an advanced treatment of inference problems for complex multivariate data, and contains a number of appendices with valuable background information, including one on distributions related to the complex Gaussian.

References

[1] Loeve, M., *Probability Theory,* D. Van Nostrand Company, New York, 1963.

[2] Papoulis, A., *Probability, Random Variables, and Stochastic Processes,* 3rd ed., McGraw-Hill, New York, 1991.

[3] Leon-Garcia, A., *Probability and Random Processes for Electrical Engineering,* 2nd ed., Addison-Wesley, Reading, MA, 1994.

[4] Melsa, J. and Sage, A., *An Introduction to Probability and Stochastic Processes,* Prentice-Hall, Englewood Cliffs, NJ, 1973.

[5] Wooding, R., The multivariate distribution of complex normal variables, *Biometrika,* 43, 212-215, 1956.

[6] Goodman, N., Statistical analysis based on a certain multivariate complex Gaussian distribution, *Ann. Math. Stat.,* 34, 152-177, 1963.

[7] Miller, K., *Complex Stochastic Processes,* Addison-Wesley, Reading, MA, 1974.

[8] Neeser, F. and Massey, J., Proper complex random processes with applications to information theory, *IEEE Trans. Information Theory,* 39(4), 1293-1302, July 1993.

[9] Kay, S., *Fundamentals of Statistical Signal Processing: Estimation Theory,* Prentice-Hall, Englewood Cliffs, NJ, 1993.

[10] Kelly, E. and Forsythe, K., Adaptive Detection and Parameter Estimation for Multidimensional Signal Models, MIT Lincoln Laboratory Technical Report 848, April 1989.

61

Beamforming Techniques for Spatial Filtering

Barry Van Veen
University of Wisconsin

Kevin M. Buckley
Villanova University

61.1 Introduction

Systems designed to receive spatially propagating signals often encounter the presence of interference signals. If the desired signal and interferers occupy the same temporal frequency band, then temporal filtering cannot be used to separate signal from interference. However, desired and interfering signals often originate from different spatial locations. This spatial separation can be exploited to separate signal from interference using a spatial filter at the receiver.

A beamformer is a processor used in conjunction with an array of sensors to provide a versatile form of spatial filtering. The term beamforming derives from the fact that early spatial filters were designed to form pencil beams (see polar plot in Fig. 61.5(c)) in order to receive a signal radiating from a specific location and attenuate signals from other locations. "Forming beams" seems to indicate radiation of energy; however, beamforming is applicable to either radiation or reception of energy. In this section we discuss formation of beams for reception, providing an overview of beamforming from a signal processing perspective. Data independent, statistically optimum, adaptive, and partially adaptive beamforming are discussed.

0-8493-8572-5/98/$0.00+$.50
© 1998 by CRC Press LLC

Implementing a temporal filter requires processing of data collected over a temporal aperture. Similarly, implementing a spatial filter requires processing of data collected over a spatial aperture. A single sensor such as an antenna, sonar transducer, or microphone collects impinging energy over a continuous aperture, providing spatial filtering by summing coherently waves that are in phase across the aperture while destructively combining waves that are not. An array of sensors provides a discrete sampling across its aperture. When the spatial sampling is discrete, the processor that performs the spatial filtering is termed a beamformer. Typically a beamformer linearly combines the spatially sampled time series from each sensor to obtain a scalar output time series in the same manner that an FIR filter linearly combines temporally sampled data. Two principal advantages of spatial sampling with an array of sensors are discussed below.

Spatial discrimination capability depends on the size of the spatial aperture; as the aperture increases, discrimination improves. The absolute aperture size is not important, rather its size in wavelengths is the critical parameter. A single physical antenna (continuous spatial aperture) capable of providing the requisite discrimination is often practical for high frequency signals because the wavelength is short. However, when low frequency signals are of interest, an array of sensors can often synthesize a much larger spatial aperture than that practical with a single physical antenna.

A second very significant advantage of using an array of sensors, relevant at any wavelength, is the spatial filtering versatility offered by discrete sampling. In many application areas, it is necessary to change the spatial filtering function in real time to maintain effective suppression of interfering signals. This change is easily implemented in a discretely sampled system by changing the way in which the beamformer linearly combines the sensor data. Changing the spatial filtering function of a continuous aperture antenna is impractical.

This section begins with the definition of basic terminology, notation, and concepts. Succeeding sections cover data-independent, statistically optimum, adaptive, and partially adaptive beamforming. We then conclude with a summary.

Throughout this section we use methods and techniques from FIR filtering to provide insight into various aspects of spatial filtering with beamformer. However, in some ways beamforming differs significantly from FIR filtering. For example, in beamforming a source of energy has several parameters that can be of interest: range, azimuth and elevation angles, polarization, and temporal frequency content. Different signals are often mutually correlated as a result of multipath propagation. The spatial sampling is often nonuniform and multidimensional. Uncertainty must often be included in characterization of individual sensor response and location, motivating development of robust beamforming techniques. These differences indicate that beamforming represents a more general problem than FIR filtering and, as a result, more general design procedures and processing structures are common.

61.2   Basic Terminology and Concepts

In this section we introduce terminology and concepts employed throughout. We begin by defining the beamforming operation and discussing spatial filtering. Next we introduce second order statistics of the array data, developing representations for the covariance of the data received at the array and discussing distinctions between narrowband and broadband beamforming. Last, we define various types of beamformers.

61.2.1   Beamforming and Spatial Filtering

Figure 61.1 depicts two beamformers. The first, which samples the propagating wave field in space, is typically used for processing narrowband signals. The output at time k, $y(k)$, is given by a linear combination of the data at the J sensors at time k :

$$y(k) = \sum_{l=1}^{J} w_l^* x_l(k) \tag{61.1}$$

where $*$ represents complex conjugate. It is conventional to multiply the data by conjugates of the weights to simplify notation. We assume throughout that the data and weights are complex since in many applications a quadrature receiver is used at each sensor to generate in phase and quadrature (I and Q) data. Each sensor is assumed to have any necessary receiver electronics and an A/D converter if beamforming is performed digitally.

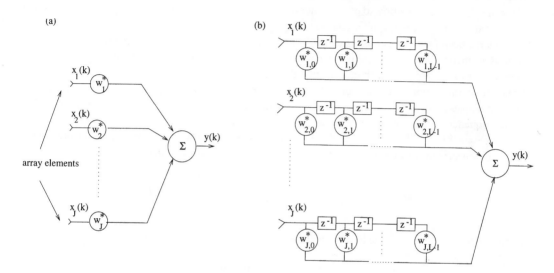

FIGURE 61.1   A beamformer forms a linear combination of the sensor outputs. In (a), sensor outputs are multiplied by complex weights and summed. This beamformer is typically used with narrowband signals. A common broadband beamformer is illustrated in (b).

The second beamformer in Fig. 61.1 samples the propagating wave field in both space and time and is often used when signals of significant frequency extent (broadband) are of interest. The output in this case can be expressed as

$$y(k) = \sum_{l=1}^{J} \sum_{p=0}^{K-1} w_{l,p}^* x_l(k - p) \tag{61.2}$$

where $K - 1$ is the number of delays in each of the J sensor channels. If the signal at each sensor is viewed as an input, then a beamformer represents a multi-input single output system.

It is convenient to develop notation that permits us to treat both beamformers in Fig. 61.1 simultaneously. Note that Eqs. (61.1) and (61.2) can be written as

$$y(k) = \mathbf{w}^H \mathbf{x}(k) \tag{61.3}$$

by appropriately defining a weight vector $\mathbf{w}$ and data vector $\mathbf{x}(k)$. We use lower and upper case boldface to denote vector and matrix quantities, respectively, and let superscript H represent Hermitian (complex conjugate) transpose. Vectors are assumed to be column vectors. Assume that $\mathbf{w}$ and $\mathbf{x}(k)$ are N dimensional; this implies that $N = KJ$ when referring to Eq. (61.2) and $N = J$ when referring to Eq. (61.1). Except for Section 61.5 on adaptive algorithms, we will drop the time index and assume that its presence is understood throughout the remainder of the paper. Thus, Eq. (61.3) is written as $y = \mathbf{w}^H \mathbf{x}$. Many of the techniques described in this section are applicable to continuous time as well as discrete time beamforming.

The frequency response of an FIR filter with tap weights w_p^*, $1 \leq p \leq J$ and a tap delay of T seconds is given by

$$r(\omega) = \sum_{p=1}^{J} w_p^* e^{-j\omega T(p-1)} . \tag{61.4}$$

Alternatively

$$r(\omega) = \mathbf{w}^H \mathbf{d}(\omega) \tag{61.5}$$

where $\mathbf{w}^H = [w_1^* \ w_2^* \ ... w_J^*]$ and $\mathbf{d}(\omega) = [1 \ e^{j\omega T} \ e^{j\omega 2T} \ ... e^{j\omega(J-1)T}]^H$. $r(\omega)$ represents the response of the filter[1] to a complex sinusoid of frequency ω and $\mathbf{d}(\omega)$ is a vector describing the phase of the complex sinusoid at each tap in the FIR filter relative to the tap associated with w_1.

Similarly, beamformer response is defined as the amplitude and phase presented to a complex plane wave as a function of location and frequency. Location is, in general, a three dimensional quantity, but often we are only concerned with one- or two-dimensional direction of arrival (DOA). Throughout the remainder of the section we do not consider range. Figure 61.2 illustrates the manner in which an array of sensors samples a spatially propagating signal. Assume that the signal is a complex plane wave with DOA θ and frequency ω. For convenience let the phase be zero at the first sensor. This implies $x_1(k) = e^{j\omega k}$ and $x_l(k) = e^{j\omega[k - \Delta_l(\theta)]}$, $2 \leq l \leq J$. $\Delta_l(\theta)$ represents the time delay due to propagation from the first to the lth sensor. Substitution into Eq. (61.2) results in the beamformer output

$$y(k) = e^{j\omega k} \sum_{l=1}^{J} \sum_{p=0}^{K-1} w_{l,p}^* e^{-j\omega[\Delta_l(\theta) + p]} = e^{j\omega k} \, r(\theta \omega) \tag{61.6}$$

where $\Delta_1(\theta) = 0$. $r(\theta, \omega)$ is the beamformer response and can be expressed in vector form as

$$r(\theta, \omega) = \mathbf{w}^H \mathbf{d}(\theta, \omega) . \tag{61.7}$$

The elements of $\mathbf{d}(\theta, \omega)$ correspond to the complex exponentials $e^{j\omega[\Delta_l(\theta) + p]}$. In general it can be expressed as

$$\mathbf{d}(\theta, \omega) = [1 \ e^{j\omega\tau_2(\theta)} \ e^{j\omega\tau_3(\theta)} \ ... e^{j\omega\tau_N(\theta)}]^H. \tag{61.8}$$

where the $\tau_i(\theta), 2 \leq i \leq N$ are the time delays due to propagation and any tap delays from the zero phase reference to the point at which the ith weight is applied. We refer to $\mathbf{d}(\theta, \omega)$ as the array response vector. It is also known as the steering vector, direction vector, or array manifold vector. Nonideal sensor characteristics can be incorporated into $\mathbf{d}(\theta, \omega)$ by multiplying each phase shift by a function $a_i(\theta, \omega)$, which describes the associated sensor response as a function of frequency and direction.

The *beampattern* is defined as the magnitude squared of $r(\theta, \omega)$. Note that each weight in $\mathbf{w}$ affects both the temporal and spatial response of the beamformer. Historically, use of FIR filters has been viewed as providing frequency dependent weights in each channel. This interpretation is somewhat incomplete since the coefficients in each filter also influence the spatial filtering characteristics of the beamformer. As a multi-input single output system, the spatial and temporal filtering that occurs is a result of mutual interaction between spatial and temporal sampling.

The correspondence between FIR filtering and beamforming is closest when the beamformer operates at a single temporal frequency ω_o and the array geometry is linear and equi-spaced as illustrated in

---

[1]An FIR filter is by definition linear, so an input sinusoid produces at the output a sinusoid of the same frequency. The magnitude and argument of $r(\omega)$ are, respectively, the magnitude and phase responses.

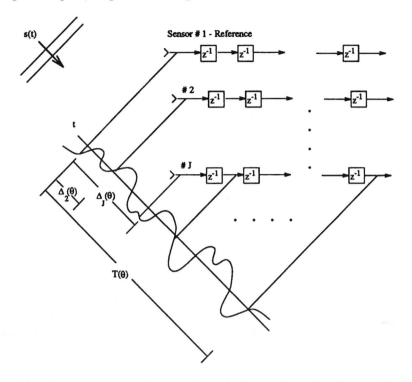

FIGURE 61.2 An array with attached delay lines provides a spatial/temporal sampling of propagating sources. This figure illustrates this sampling of a signal propagating in plane waves from a source located at DOA θ. With J sensors and K samples per sensor, at any instant in time the propagating source signal is sampled at JK nonuniformly spaced points. $T(\theta)$, the time duration from the first sample of the first sensor to the last sample of the last sensor, is termed the temporal aperture of the observation of the source at θ. As notation suggests, temporal aperture will be a function of DOA θ. Plane wave propagation implies that at any time k a propagating signal, received anywhere on a planar front perpendicular to a line drawn from the source to a point on the plane, has equal intensity. Propagation of the signal between two points in space is then characterized as pure delay. In this figure, $\Delta_l(\theta)$ represents the time delay due to plane wave propagation from the 1st (reference) to the lth sensor.

Fig. 61.3. Letting the sensor spacing be d, propagation velocity be c, and θ represent DOA relative to broadside (perpendicular to the array), we have $\tau_i(\theta) = (i - 1)(d/c)\sin\theta$. In this case we identify the relationship between temporal frequency ω in $\mathbf{d}(\omega)$ (FIR filter) and direction θ in $\mathbf{d}(\theta, \omega_o)$ (beamformer) as $\omega = \omega_o(d/c)\sin\theta$. Thus, temporal frequency in an FIR filter corresponds to the sine of direction in a narrowband linear equi-spaced beamformer. Complete interchange of beamforming and FIR filtering methods is possible for this special case provided the mapping between frequency and direction is accounted for.

The vector notation introduced in (61.3) suggests a vector space interpretation of beamforming. This point of view is useful both in beamformer design and analysis. We use it here in consideration of spatial sampling and array geometry. The weight vector $\mathbf{w}$ and the array response vectors $\mathbf{d}(\theta, \omega)$ are vectors in an N-dimensional vector space. The angles between $\mathbf{w}$ and $\mathbf{d}(\theta, \omega)$ determine the response $r(\theta, \omega)$. For example, if for some (θ, ω) the angle between $\mathbf{w}$ and $\mathbf{d}(\theta, \omega)$ $90°$ (i.e., if $\mathbf{w}$ is orthogonal to $\mathbf{d}(\theta, \omega)$), then the response is zero. If the angle is close to $0°$, then the response magnitude will be relatively large. The ability to discriminate between sources at different locations and/or frequencies, say (θ_1, ω_1) and (θ_2, ω_2), is determined by the angle between their array response vectors, $\mathbf{d}(\theta_1, \omega_1)$ and $\mathbf{d}(\theta_2, \omega_2)$.

The general effects of spatial sampling are similar to temporal sampling. Spatial aliasing corresponds to an ambiguity in source locations. The implication is that sources at different locations have the same

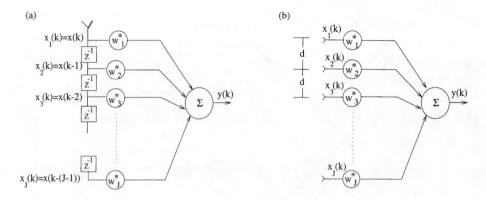

FIGURE 61.3   The analogy between an equi-spaced omni-directional narrowband line array and a single-channel FIR filter is illustrated in this figure.

array response vector, e.g., for narrowband sources $\mathbf{d}(\theta_1, \omega_o)$ and $\mathbf{d}(\theta_2, \omega_o)$. This can occur if the sensors are spaced too far apart. If the sensors are too close together, spatial discrimination suffers as a result of the smaller than necessary aperture; array response vectors are not well dispersed in the N dimensional vector space. Another type of ambiguity occurs with broadband signals when a source at one location and frequency cannot be distinguished from a source at a different location and frequency, i.e., $\mathbf{d}(\theta_1, \omega_1) = \mathbf{d}(\theta_2, \omega_2)$. For example, this occurs in a linear equi-spaced array whenever $\omega_1 sin\theta_1 = \omega_2 sin\theta_2$. (The addition of temporal samples at one sensor prevents this particular ambiguity.)

A primary focus of this section is on designing response via weight selection; however, (61.7) indicates that response is also a function of array geometry (and sensor characteristics if the ideal omnidirectional sensor model is invalid). In contrast with single channel filtering where A/D converters provide a uniform sampling in time, there is no compelling reason to space sensors regularly. Sensor locations provide additional degrees of freedom in designing a desired response and can be selected so that over the range of (θ, ω) of interest the array response vectors are unambiguous and well dispersed in the N dimensional vector space. Utilization of these degrees of freedom can become very complicated due to the multidimensional nature of spatial sampling and the nonlinear relationship between $r(\theta, \omega)$ and sensor locations.

61.2.2   Second Order Statistics

Evaluation of beamformer performance usually involves power or variance, so the second order statistics of the data play an important role. We assume the data received at the sensors are zero mean throughout this section. The variance or expected power of the beamformer output is given by $E\{|y|^2\} = \mathbf{w}^H E\{\mathbf{x}\, \mathbf{x}^H\}\mathbf{w}$. If the data are wide sense stationary, then $\mathbf{R}_x = E\{\mathbf{x}\, \mathbf{x}^H\}$, the data covariance matrix, is independent of time. Although we often encounter nonstationary data, the wide sense stationary assumption is used in developing statistically optimal beamformers and in evaluating steady state performance.

Suppose $\mathbf{x}$ represents samples from a uniformly sampled time series having a power spectral density $S(\omega)$ and no energy outside of the spectral band $[\omega_a, \omega_b]$. $\mathbf{R}_x$ can be expressed in terms of the power spectral density of the data using the Fourier transform relationship as

$$\mathbf{R}_x = \frac{1}{2\pi} \int_{\omega_a}^{\omega_b} S(\omega)\, \mathbf{d}(\omega)\, \mathbf{d}^H(\omega)\, d\omega \tag{61.9}$$

with $\mathbf{d}(\omega)$ as defined for (61.5). Now assume the array data $\mathbf{x}$ is due to a source located at direction θ. In like manner to the time series case we can obtain the covariance matrix of the array data as

$$\mathbf{R}_x = \frac{1}{2\pi} \int_{\omega_a}^{\omega_b} S(\omega)\, \mathbf{d}(\theta, \omega)\, \mathbf{d}^H(\theta, \omega)\, d\omega \tag{61.10}$$

A source is said to be narrowband of frequency ω_o if $\mathbf{R}_x$ can be represented as the rank one outer product

$$\mathbf{R}_x = \sigma_s^2 \, \mathbf{d}(\theta, \omega_o) \, \mathbf{d}^H(\theta, \omega_o) \tag{61.11}$$

where σ_s^2 is the source variance or power.

The conditions under which a source can be considered narrowband depend on both the source bandwidth and the time over which the source is observed. To illustrate this, consider observing an amplitude modulated sinusoid or the output of a narrowband filter driven by white noise on an oscilloscope. If the signal bandwidth is small relative to the center frequency (i.e., if it has small fractional bandwidth), and the time intervals over which the signal is observed are short relative to the inverse of the signal bandwidth, then each observed waveform has the shape of a sinusoid. Note that as the observation time interval is increased, the bandwidth must decrease for the signal to remain sinusoidal in appearance. It turns out, based on statistical arguments, that the observation time bandwidth product (TBWP) is the fundamental parameter that determines whether a source can be viewed as narrowband (see Buckley [2]).

An array provides an effective temporal aperture over which a source is observed. Figure 61.2 illustrates this temporal aperture $T(\theta)$ for a source arriving from direction θ. Clearly the TBWP is dependent on the source DOA. An array is considered narrowband if the observation TBWP is much less than one for all possible source directions.

Narrowband beamforming is conceptually simpler than broadband since one can ignore the temporal frequency variable. This fact, coupled with interest in temporal frequency analysis for some applications, has motivated implementation of broadband beamformers with a narrowband decomposition structure, as illustrated in Fig. 61.4. The narrowband decomposition is often performed by taking a discrete Fourier transform (DFT) of the data in each sensor channel using an FFT algorithm. The data across the array at each frequency of interest are processed by their own beamformer. This is usually termed frequency domain beamforming. The frequency domain beamformer outputs can be made equivalent to the DFT of the broadband beamformer output depicted in Fig. 61.1(b) with proper selection of beamformer weights and careful data partitioning.

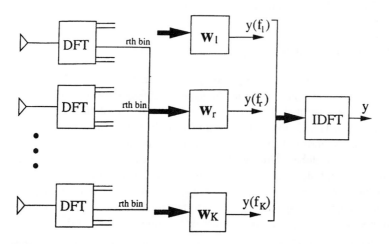

FIGURE 61.4 Beamforming is sometimes performed in the frequency domain when broadband signals are of interest. This figure illustrates transformation of the data at each sensor into the frequency domain. Weighted combinations of data at each frequency (bin) are performed. An inverse discrete Fourier transform produces the output time series.

61.2.3 Beamformer Classification

Beamformers can be classified as either data independent or statistically optimum, depending on how the weights are chosen. The weights in a data independent beamformer do not depend on the array data and are chosen to present a specified response for all signal/interference scenarios. The weights in a statistically optimum beamformer are chosen based on the statistics of the array data to "optimize" the array response. In general, the statistically optimum beamformer places nulls in the directions of interfering sources in an attempt to maximize the signal-to-noise ratio at the beamformer output. A comparison between data independent and statistically optimum beamformers is illustrated in Fig. 61.5.

The next four sections cover data independent, statistically optimum, adaptive, and partially adaptive beamforming. Data independent beamformer design techniques are often used in statistically optimum beamforming (e.g., constraint design in linearly constrained minimum variance beamforming). The statistics of the array data are not usually known and may change over time so adaptive algorithms are typically employed to determine the weights. The adaptive algorithm is designed so the beamformer response converges to a statistically optimum solution. Partially adaptive beamformers reduce the adaptive algorithm computational load at the expense of a loss (designed to be small) in statistical optimality.

61.3 Data Independent Beamforming

The weights in a data independent beamformer are designed so the beamformer response approximates a desired response independent of the array data or data statistics. This design objective — approximating a desired response — is the same as that for classical FIR filter design (see, for example, Parks and Burrus [8]). We shall exploit the analogies between beamforming and FIR filtering where possible in developing an understanding of the design problem. We also discuss aspects of the design problem specific to beamforming.

The first part of this section discusses forming beams in a classical sense, i.e., approximating a desired response of unity at a point of direction and zero elsewhere. Methods for designing beamformers having more general forms of desired response are presented in the second part.

61.3.1 Classical Beamforming

Consider the problem of separating a single complex frequency component from other frequency components using the J tap FIR filter illustrated in Fig. 61.3. If frequency ω_o is of interest, then the desired frequency response is unity at ω_o and zero elsewhere. A common solution to this problem is to choose $\mathbf{w}$ as the vector $\mathbf{d}(\omega_o)$. This choice can be shown to be optimal in terms of minimizing the squared error between the actual response and desired response. The actual response is characterized by a main lobe (or beam) and many sidelobes. Since $\mathbf{w} = \mathbf{d}(\omega_o)$, each element of $\mathbf{w}$ has unit magnitude. Tapering or windowing the amplitudes of the elements of $\mathbf{w}$ permits trading of main lobe or beam width against sidelobe levels to form the response into a desired shape. Let $\mathbf{T}$ be a J by J diagonal matrix with the real-valued taper weights as diagonal elements. The tapered FIR filter weight vector is given by $\mathbf{T}\,\mathbf{d}(\omega_o)$. A detailed comparison of a large number of tapering functions is given in [5].

In spatial filtering one is often interested in receiving a signal arriving from a known location point θ_o. Assuming the signal is narrowband (frequency ω_o), a common choice for the beamformer weight vector is the array response vector $\mathbf{d}(\theta_o, \omega_o)$. The resulting array and beamformer is termed a phased array because the output of each sensor is phase shifted prior to summation. Figure 61.5(b) depicts the magnitude of the actual response when $\mathbf{w} = \mathbf{T}\mathbf{d}(\theta_o, \omega_o)$, where $\mathbf{T}$ implements a common Dolph-Chebyshev tapering function. As in the FIR filter discussed above, beam width and sidelobe levels are the important characteristics of the response. Amplitude tapering can be used to control the shape of the response, i.e., to form the beam. The equivalence of the narrowband linear equi-spaced array and FIR filter (see Fig. 61.3) implies that the same techniques for choosing taper functions are applicable to either problem. Methods for choosing tapering weights also exist for more general array configurations.

61.3.2 General Data Independent Response Design

The methods discussed in this section apply to design of beamformers that approximate an arbitrary desired response. This is of interest in several different applications. For example, we may wish to receive any signal arriving from a range of directions, in which case the desired response is unity over the entire range. As another example, we may know that there is a strong source of interference arriving from a certain range of directions, in which case the desired response is zero in this range. These two examples are analogous to bandpass and bandstop FIR filtering. Although we are no longer "forming beams", it is conventional to refer to this type of spatial filter as a beamformer.

Consider choosing $\mathbf{w}$ so the actual response $r(\theta, \omega) = \mathbf{w}^H \mathbf{d}(\theta, \omega)$ approximates desired response $r_d(\theta, \omega)$. Ad hoc techniques similar to those employed in FIR filter design can be used for selecting $\mathbf{w}$. Alternatively, formal optimization design methods can be employed (see, for example, Parks and Burrus [8]). Here, to illustrate the general optimization design approach, we only consider choosing $\mathbf{w}$ to minimize the weighted averaged square of the difference between desired and actual response.

Consider minimizing the squared error between the actual and desired response at P points (θ_i, ω_i), $1 < i < P$. If $P > N$, then we obtain the overdetermined least squares problem

$$\min_{\mathbf{w}} \ |\mathbf{A}^H \mathbf{w} - \mathbf{r}_d|^2 \tag{61.12}$$

where

$$\mathbf{A} = [\mathbf{d}(\theta_1, \omega_1), \mathbf{d}(\theta_2, \omega_2)...\mathbf{d}(\theta_P, \omega_P)] ; \tag{61.13}$$

$$\mathbf{r}_d = [r_d(\theta_1, \omega_1), r_d(\theta_2, \omega_2)...r_d(\theta_P, \omega_P)]^H . \tag{61.14}$$

Provided $\mathbf{AA}^H$ is invertible (i.e., $\mathbf{A}$ is full rank), then the solution to Eq. (61.12) is given as

$$\mathbf{w} = \mathbf{A}^+ \mathbf{r}_d \tag{61.15}$$

where $\mathbf{A}^+ = (\mathbf{AA}^H)^{-1}\mathbf{A}$ is the pseudo-inverse of $\mathbf{A}$.

A note of caution is in order at this point. The white noise gain of a beamformer is defined as the output power due to unit variance white noise at the sensors. Thus, the norm squared of the weight vector, $\mathbf{w}^H \mathbf{w}$, represents the white noise gain. If the white noise gain is large, then the accuracy by which $\mathbf{w}$ approximates the desired response is a moot point because the beamformer output will have a poor SNR due to white noise contributions. If $\mathbf{A}$ is ill-conditioned, then $\mathbf{w}$ can have a very large norm and still approximate the desired response. The matrix $\mathbf{A}$ is ill-conditioned when the effective numerical dimension of the space spanned by the $\mathbf{d}(\theta_i, \omega_i)$, $1 \leq i \leq P$, is less than N. For example, if only one source direction is sampled, then the numerical rank of $\mathbf{A}$ is approximately given by the TBWP for that direction. Low rank approximates of $\mathbf{A}$ and $\mathbf{A}^+$ should be used whenever the numerical rank is less than N. This ensures that the norm of $\mathbf{w}$ will not be unnecessarily large.

Specific directions and frequencies can be emphasized in Eq. (61.12) by selection of the sample points (θ_i, ω_i) and/or unequally weighting of the error at each (θ_i, ω_i). Parks and Burrus [8] discuss this in the context of FIR filtering.

61.4 Statistically Optimum Beamforming

In statistically optimum beamforming, the weights are chosen based on the statistics of the data received at the array. Loosely speaking, the goal is to "optimize" the beamformer response so the output contains minimal contributions due to noise and interfering signals. We discuss several different criteria for

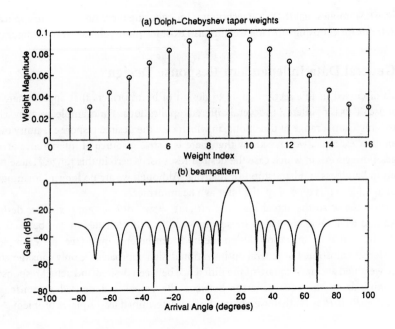

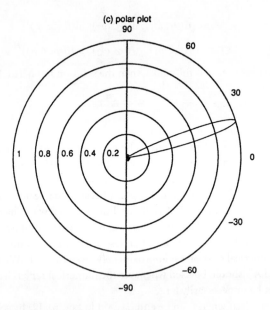

FIGURE 61.5 Beamformers come in both data independent and statistically optimum varieties. In (a) through (e) of this figure we consider an equi-spaced narrowband array of 16 sensors spaced at one-half wavelength. In (a), (b), and (c) the magnitude of the weights, the beampattern, and the beampattern, in polar coordinates are shown, respectively, for a Dolph-Chebyshev beamformer with -30 dB sidelobes. In (d) and (e) beampatterns are shown of statistically optimum beamformers which were designed to minimize output power subject to a constraint that the response be unity for an arrival angle of 18°. Energy is assumed to arrive at the array from several interference sources. In (d) several interferers are located between −20° and −23°, each with power of 30 dB relative to the uncorrelated noise power at a single sensor. Deep nulls are formed in the interferer directions. The interferers in (e) are located between 20° and 23°, again with relative power of 30 dB. Again deep nulls are formed at the interferer directions; however, the sidelobe levels are significantly higher at other directions. (f) depicts the broadband LCMV beamformer magnitude response at eight frequencies on the normalized frequency interval $[2\pi/5, 4\pi/5]$ when two interferers arrive from directions −5.75° and −17.5° in the presence of white noise.

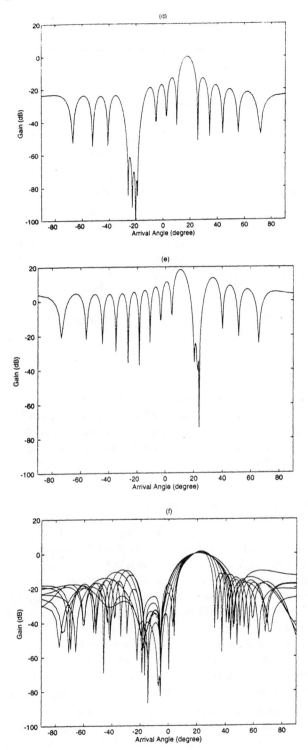

FIGURE 61.5 *(continued)* The interferers have a white spectrum on $[2\pi/5, 4\pi/5]$ and have powers of 40 dB and 30 dB relative to the white noise, respectively. The constraints are designed to present a unit gain and linear phase over $[2\pi/5, 4\pi/5]$ at a DOA of $18°$. The array is linear equi-spaced with 16 sensors spaced at one-half wavelength for frequency $4\pi/5$ and five tap FIR filters are used in each sensor channel.

TABLE 61.1   Summary of Optimum Beamformers

Type	MSC	Reference signal	Max SNR	LCMV
Definitions	x_a — auxiliary data y_m — primary data $r_{ma} = E\{x_a y_m^*\}$ $R_a = E\{x_a x_a^H\}$ output: $y = y_m - w_a^H x_a$	x — array data y_d — desired signal $r_{xd} = E\{x y_d^*\}$ $R_x = E\{x x^H\}$ output: $y = w^H x$	$x = s + x$ — array data s — signal component n — noise component $R_s = E\{s s^H\}$ $R_n = E\{n n^H\}$ output: $y = w^H x$	x — array data C — constraint matrix f — response vector $R_x = E\{x x^H\}$ output: $y = w^H x$
Criterion	$\min_{w_a} E\{\|y_m - w_a^H x_a\|^2\}$	$\min_{w} E\{\|y - y_d\|^2\}$	$\max_{w} \dfrac{w^H R_s w}{w^H R_n w}$	$\min_{w} \{w^H R_x w\} s.t. C^H w = f$
Optimum weights	$w_a = R_a^{-1} r_{ma}$	$w_a = R_x^{-1} r_{rd}$	$R_n^{-1} R_s w = \lambda_{max} w$	$w = R_x^{-1} C [C^H R_x^{-1} C]^{-1} f$
Advantages	Simple	Direction of desired signal can be unknown	True maximization of SNR	Flexible and general constraints
Disadvantages	Requires absence of desired signal from auxiliary channels for weight determination	Must generate reference signal	Must know R_s and R_n Solve generalized eigen-problem for weights	Computation of constrained weight vector
References	Applebaum [1976]	Widrow [1967]	Monzingo and Miller [1980]	Frost [1972]

choosing statistically optimum beamformer weights. Table 61.1 summarizes these different approaches. Where possible, equations describing the criteria and weights are confined to Table 61.1. Throughout the section we assume that the data is wide-sense stationary and that its second order statistics are known. Determination of weights when the data statistics are unknown or time varying is discussed in the following section on adaptive algorithms.

61.4.1   Multiple Sidelobe Canceller

The multiple sidelobe canceller (MSC) is perhaps the earliest statistically optimum beamformer. An MSC consists of a "main channel" and one or more "auxiliary channels" as depicted in Fig. 61.6(a). The main channel can be either a single high gain antenna or a data independent beamformer (see Section 61.3). It has a highly directional response, which is pointed in the desired signal direction. Interfering signals are assumed to enter through the main channel sidelobes. The auxiliary channels also receive the interfering signals. The goal is to choose the auxiliary channel weights to cancel the main channel interference component. This implies that the responses to interferers of the main channel and linear combination of auxiliary channels must be identical. The overall system then has a response of zero as illustrated in Fig. 61.6(b). In general, requiring zero response to all interfering signals is either not possible or can result in significant white noise gain. Thus, the weights are usually chosen to trade off interference suppression for white noise gain by minimizing the expected value of the total output power as indicated in Table 61.1.

Choosing the weights to minimize output power can cause cancellation of the desired signal because it also contributes to total output power. In fact, as the desired signal gets stronger it contributes to a larger fraction of the total output power and the percentage cancellation increases. Clearly this is an undesirable effect. The MSC is very effective in applications where the desired signal is very weak (relative to the interference), since the optimum weights will not pay any attention to it, or when the desired signal is known to be absent during certain time periods. The weights can then be adapted in the absence of the desired signal and frozen when it is present.

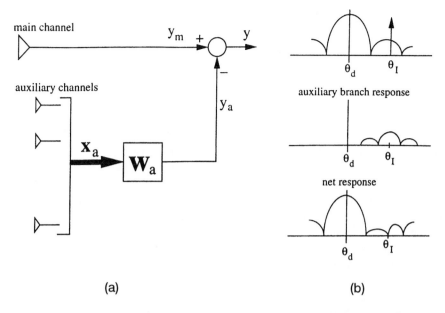

main channel

auxiliary channels

auxiliary branch response

net response

(a)

(b)

FIGURE 61.6 The multiple sidelobe canceller (MSC) consists of a main channel and several auxiliary channels as illustrated in (a). The auxiliary channel weights are chosen to "cancel" interference entering through sidelobes of the main channel. (b) Depicts the main channel, auxiliary branch, and overall system response when an interferer arrives from direction θ_I.

61.4.2 Use of a Reference Signal

If the desired signal were known, then the weights could be chosen to minimize the error between the beamformer output and the desired signal. Of course, knowledge of the desired signal eliminates the need for beamforming. However, for some applications, enough may be known about the desired signal to generate a signal that closely represents it. This signal is called a reference signal. As indicated in Table 61.1, the weights are chosen to minimize the mean square error between the beamformer output and the reference signal.

The weight vector depends on the cross covariance between the unknown desired signal present in $\mathbf{x}$ and the reference signal. Acceptable performance is obtained provided this approximates the covariance of the unknown desired signal with itself. For example, if the desired signal is amplitude modulated, then acceptable performance is often obtained by setting the reference signal equal to the carrier. It is also assumed that the reference signal is uncorrelated with interfering signals in $\mathbf{x}$. The fact that the direction of the desired signal does not need to be known is a distinguishing feature of the reference signal approach. For this reason it is sometimes termed "blind" beamforming. Other closely related blind beamforming techniques choose weights by exploiting properties of the desired signal such as constant modulus, cyclostationarity, or third and higher order statistics.

61.4.3 Maximization of Signal-to-Noise Ratio

Here the weights are chosen to directly maximize the signal-to-noise ratio (SNR) as indicated in Table 61.1. A general solution for the weights requires knowledge of both the desired signal, $\mathbf{R}_s$, and noise, $\mathbf{R}_n$, covariance matrices. The attainability of this knowledge depends on the application. For example, in an active radar system $\mathbf{R}_n$ can be estimated during the time that no signal is being transmitted and $\mathbf{R}_s$ can be obtained from knowledge of the transmitted pulse and direction of interest. If the signal component is narrowband, of frequency ω, and direction θ, then $\mathbf{R}_s = \sigma^2 \mathbf{d}(\theta, \omega) \mathbf{d}^H(\theta, \omega)$ from the results in

Section 61.2. In this case, the weights are obtained as

$$\mathbf{w} = \alpha \mathbf{R}_n^{-1} \mathbf{d}(\theta, \omega) \tag{61.16}$$

where the α is some non-zero complex constant. Substitution of Eq. (61.16) into the SNR expression shows that the SNR is independent of the value chosen for α.

61.4.4   Linearly Constrained Minimum Variance Beamforming

In many applications none of the above approaches is satisfactory. The desired signal may be of unknown strength and may always be present, resulting in signal cancellation with the MSC and preventing estimation of signal and noise covariance matrices in the maximum SNR processor. Lack of knowledge about the desired signal may prevent utilization of the reference signal approach. These limitations can be overcome through the application of linear constraints to the weight vector. Use of linear constraints is a very general approach that permits extensive control over the adapted response of the beamformer. In this section we illustrate how linear constraints can be employed to control beamformer response, discuss the optimum linearly constrained beamforming problem, and present the generalized sidelobe canceller structure.

The basic idea behind linearly constrained minimum variance (LCMV) beamforming is to constrain the response of the beamformer so signals from the direction of interest are passed with specified gain and phase. The weights are chosen to minimize output variance or power subject to the response constraint. This has the effect of preserving the desired signal while minimizing contributions to the output due to interfering signals and noise arriving from directions other than the direction of interest. The analogous FIR filter has the weights chosen to minimize the filter output power subject to the constraint that the filter response to signals of frequency ω_o be unity.

In Section 61.2 we saw that the beamformer response to a source at angle θ and temporal frequency ω is given by $\mathbf{w}^H \mathbf{d}(\theta, \omega)$. Thus, by linearly constraining the weights to satisfy $\mathbf{w}^H \mathbf{d}(\theta, \omega) = g$ where g is a complex constant, we ensure that any signal from angle θ and frequency ω is passed to the output with response g. Minimization of contributions to the output from interference (signals not arriving from θ with frequency ω) is accomplished by choosing the weights to minimize the output power or variance $E\{|y|^2\} = \mathbf{w}^H \mathbf{R}_x \mathbf{w}$. The LCMV problem for choosing the weights is thus written

$$\min_{\mathbf{w}} \quad \mathbf{w}^H \mathbf{R}_x \mathbf{w} \quad \text{subject to} \quad \mathbf{d}^H(\theta, \omega)\mathbf{w} = g^* . \tag{61.17}$$

The method of Lagrange multipliers can be used to solve Eq. (61.17) resulting in

$$\mathbf{w} = g^* \frac{\mathbf{R}_x^{-1} \mathbf{d}(\theta, \omega)}{\mathbf{d}^H(\theta, \omega)\mathbf{R}_x^{-1}\mathbf{d}(\theta, \omega)} . \tag{61.18}$$

Note that, in practice, the presence of uncorrelated noise will ensure that $\mathbf{R}_x$ is invertible. If $g = 1$, then Eq. (61.18) is often termed the minimum variance distortionless response (MVDR) beamformer. It can be shown that Eq. (61.18) is equivalent to the maximum SNR solution given in Eq. (61.16) by substituting $\sigma^2 \mathbf{d}(\theta, \omega)\mathbf{d}^H(\theta, \omega) + \mathbf{R}_n$ for $\mathbf{R}_x$ in Eq. (61.18) and applying the matrix inversion lemma.

The single linear constraint in Eq. (61.17) is easily generalized to multiple linear constraints for added control over the beampattern. For example, if there is fixed interference source at a known direction ϕ, then it may be desirable to force zero gain in that direction in addition to maintaining the response g to the desired signal. This is expressed as

$$\begin{bmatrix} \mathbf{d}^H(\theta, \omega) \\ \mathbf{d}^H(\phi, \omega) \end{bmatrix} \mathbf{w} = \begin{bmatrix} g^* \\ 0 \end{bmatrix} . \tag{61.19}$$

If there are $L < N$ linear constraints on $\mathbf{w}$, we write them in the form $\mathbf{C}^H \mathbf{w} = \mathbf{f}$ where the N by L matrix $\mathbf{C}$ and L dimensional vector $\mathbf{f}$ are termed the constraint matrix and response vector. The constraints are assumed to be linearly independent so **C** has rank L. The LCMV problem and solution with this more general constraint equation are given in Table 61.1.

Several different philosophies can be employed for choosing the constraint matrix and response vector. Specifically point, derivative, and eigenvector constraint approaches are popular. Each linear constraint uses one degree of freedom in the weight vector so with L constraints there are only $N - L$ degrees of freedom available for minimizing variance. See Van Veen and Buckley [11] or Van Veen [12] for a more in-depth discussion on this topic.

Generalized Sidelobe Canceller. The generalized sidelobe canceller (GSC) represents an alternative formulation of the LCMV problem, which provides insight, is useful for analysis, and can simplify LCMV beamformer implementation. It also illustrates the relationship between MSC and LCMV beamforming. Essentially, the GSC is a mechanism for changing a constrained minimization problem into unconstrained form.

Suppose we decompose the weight vector **w** into two orthogonal components $\mathbf{w}_o$ and $-\mathbf{v}$ (i.e., $\mathbf{w} = \mathbf{w}_o - \mathbf{v}$) that lie in the range and null spaces of **C**, respectively. The range and null spaces of a matrix span the entire space so this decomposition can be used to represent any **w**. Since $\mathbf{C}^H \mathbf{v} = \mathbf{0}$, we must have

$$\mathbf{w}_o = \mathbf{C}(\mathbf{C}^H \mathbf{C})^{-1}\mathbf{f} \tag{61.20}$$

if **w** is to satisfy the constraints. Equation (61.20) is the minimum L_2 norm solution to the underdetermined equivalent of Eq. (61.12). The vector **v** is a linear combination of the columns of an N by M ($M = N - L$) matrix $\mathbf{C}_n$ (i.e., $\mathbf{v} = \mathbf{C}_n \mathbf{w}_M$) provided the columns of $\mathbf{C}_n$ form a basis for the null space of **C**. $\mathbf{C}_n$ can be obtained from **C** using any of several orthogonalization procedures such as Gram-Schmidt, QR decomposition, or singular value decomposition. The weight vector $\mathbf{w} = \mathbf{w}_o - \mathbf{C}_n \mathbf{w}_M$ is depicted in block diagram form in Fig. 61.7. The choice for $\mathbf{w}_o$ and $\mathbf{C}_n$ implies that **w** satisfies the constraints independent of $\mathbf{w}_M$ and reduces the LCMV problem to the unconstrained problem

$$\min_{\mathbf{w}_M} \ [\mathbf{w}_o - \mathbf{C}_n \mathbf{w}_M]^H \mathbf{R}_x [\mathbf{w}_o - \mathbf{C}_n \mathbf{w}_M] . \tag{61.21}$$

The solution is

$$\mathbf{w}_M = (\mathbf{C}_n^H \mathbf{R}_x \mathbf{C}_n)^{-1} \mathbf{C}_n^H \mathbf{R}_x \mathbf{w}_o . \tag{61.22}$$

The primary implementation advantages of this alternate but equivalent formulation stem from the facts that the weights $\mathbf{w}_M$ are unconstrained and a data independent beamformer $\mathbf{w}_o$ is implemented as an integral part of the optimum beamformer. The unconstrained nature of the adaptive weights permits much simpler adaptive algorithms to be employed and the data independent beamformer is useful in situations where adaptive signal cancellation occurs (see Section 61.4.5).

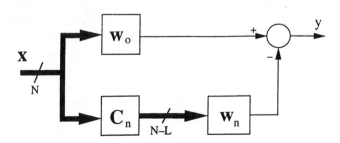

FIGURE 61.7 The generalized sidelobe canceller (GSC) represents an implementation of the LCMV beamformer in which the adaptive weights are unconstrained. It consists of a preprocessor composed of a fixed beamformer $\mathbf{w}_o$ and a blocking matrix $\mathbf{C}_n$, and a standard adaptive filter with unconstrained weight vector $\mathbf{w}_M$.

As an example, assume the constraints are as given in Eq. (61.17). Equation (61.20) implies $\mathbf{w}_o = g^*\mathbf{d}(\theta, \omega)/[\mathbf{d}^H(\theta, \omega)\mathbf{d}(\theta, \omega)]$. $\mathbf{C}_n$ satisfies $\mathbf{d}^H(\theta, \omega)\mathbf{C}_n = \mathbf{0}$ so each column $[\mathbf{C}_n]_i$; $1 < i < N - L$, can be viewed as a data independent beamformer with a null in direction θ at frequency ω: $\mathbf{d}^H(\theta, \omega)[\mathbf{C}_n]_j = 0$. Thus, a signal of frequency ω and direction θ arriving at the array will be blocked or nulled by the matrix $\mathbf{C}_n$. In general, if the constraints are designed to present a specified response to signals from a set of directions and frequencies, then the columns of $\mathbf{C}_n$ will block those directions and frequencies. This characteristic has led to the term "blocking matrix" for describing $\mathbf{C}_n$. These signals are only processed by $\mathbf{w}_o$ and since $\mathbf{w}_o$ satisfies the constraints, they are presented with the desired response independent of $\mathbf{w}_M$. Signals from directions and frequencies over which the response is not constrained will pass through the upper branch in Fig. 61.7 with some response determined by $\mathbf{w}_o$. The lower branch chooses $\mathbf{w}_M$ to estimate the signals at the output of $\mathbf{w}_o$ as a linear combination of the data at the output of the blocking matrix. This is similar to the operation of the MSC, in which weights are applied to the output of auxiliary sensors in order to estimate the primary channel output (see Fig. 61.6).

61.4.5 Signal Cancellation in Statistically Optimum Beamforming

Optimum beamforming requires some knowledge of the desired signal characteristics, either its statistics (for maximum SNR or reference signal methods), its direction (for the MSC), or its response vector $\mathbf{d}(\theta, \omega)$ (for the LCMV beamformer). If the required knowledge is inaccurate, the optimum beamformer will attenuate the desired signal as if it were interference. Cancellation of the desired signal is often significant, especially if the SNR of the desired signal is large. Several approaches have been suggested to reduce this degradation (e.g., Cox et al. [3]).

A second cause of signal cancellation is correlation between the desired signal and one or more interference signals. This can result either from multipath propagation of a desired signal or from smart (correlated) jamming. When interference and desired signals are uncorrelated, the beamformer attenuates interferers to minimize output power. However, with a correlated interferer the beamformer minimizes output power by processing the interfering signal in such a way as to cancel the desired signal. If the interferer is partially correlated with the desired signal, then the beamformer will cancel the portion of the desired signal that is correlated with the interferer. Methods for reducing signal cancellation due to correlated interference have been suggested (e.g., Widrow et al. [13], Shan and Kailath [10]).

61.5 Adaptive Algorithms for Beamforming

The optimum beamformer weight vector equations listed in Table 61.1 require knowledge of second order statistics. These statistics are usually not known, but with the assumption of ergodicity, they (and therefore the optimum weights) can be estimated from available data. Statistics may also change over time, e.g., due to moving interferers. To solve these problems, weights are typically determined by adaptive algorithms.

There are two basic adaptive approaches: (1) block adaptation, where statistics are estimated from a temporal block of array data and used in an optimum weight equation; and (2) continuous adaptation, where the weights are adjusted as the data is sampled such that the resulting weight vector sequence converges to the optimum solution. If a nonstationary environment is anticipated, block adaptation can be used, provided that the weights are recomputed periodically. Continuous adaptation is usually preferred when statistics are time-varying or, for computational reasons, when the number of adaptive weights M is moderate to large; values of $M > 50$ are common.

Among notable adaptive algorithms proposed for beamforming are the Howells-Applebaum adaptive loop developed in the late 1950s and reported by Howells [7] and Applebaum [1], and the Frost LCMV algorithm [4]. Rather than recapitulating adaptive algorithms for each optimum beamformer listed in Table 61.1, we take a unifying approach using the standard adaptive filter configuration illustrated on the right side of Fig. 61.7.

In Fig. 61.7 the weight vector $\mathbf{w}_M$ is chosen to estimate the desired signal y_d as linear combination of the elements of the data vector $\mathbf{u}$. We select $\mathbf{w}_M$ to minimize the MSE

$$J(\mathbf{w}_M) = E\{|y_d - \mathbf{w}_M^H \mathbf{u}|^2\} = \sigma_d^2 - \mathbf{w}_M^H \mathbf{r}_{ud} - \mathbf{r}_{ud}^H \mathbf{w}_M + \mathbf{w}_M^H \mathbf{R}_u \mathbf{w}_M , \tag{61.23}$$

where $\sigma_d^2 = E\{|y_d|^2\}$, $\mathbf{r}_{ud} = E\{\mathbf{u}\, y_d^*\}$ and $\mathbf{R}_u = E\{\mathbf{u}\, \mathbf{u}^H\}$. $J(\mathbf{w}_M)$ is minimized by

$$\mathbf{w}_{opt} = \mathbf{R}_u^{-1} \mathbf{r}_{ud} . \tag{61.24}$$

Comparison of (61.23) and the criteria listed in Table 61.1 indicates that this standard adaptive filter problem is equivalent to both the MSC beamformer problem (with $y_d = y_m$ and $\mathbf{u} = \mathbf{x}_a$) and the reference signal beamformer problem (with $\mathbf{u} = \mathbf{x}$). The LCMV problem is apparently different. However closer examination of Fig. 61.7 and Eqs. (61.22), and (61.24) reveals that the standard adaptive filter problem is equivalent to the LCMV problem implemented with the GSC structure. Setting $\mathbf{u} = \mathbf{C}_n^H \mathbf{x}$ and $y_d = \mathbf{w}_o^H \mathbf{x}$ implies $\mathbf{R}_u = \mathbf{C}_n^H \mathbf{R}_x \mathbf{C}_n$ and $\mathbf{r}_{ud} = \mathbf{C}_n^H \mathbf{R}_x \mathbf{w}_o$. The maximum SNR beamformer cannot in general be represented by Fig. 61.7 and Eq. (61.24). However, it was noted after (61.18) that if the desired signal is narrowband, then the maximum SNR and the LCMV beamformers are equivalent.

The block adaptation approach solves (61.24) using estimates of $\mathbf{R}_u$ and $\mathbf{r}_{ud}$ formed from K samples of $\mathbf{u}$ and y_d: $\mathbf{u}(k)$, $y_d(k)$; $0 < k < K - 1$. The most common are the sample covariance matrix

$$\hat{\mathbf{R}}_u = \frac{1}{K} \sum_{k=0}^{K-1} \mathbf{u}(k) \mathbf{u}^H(k) \tag{61.25}$$

and sample cross-covariance vector

$$\hat{\mathbf{r}}_{ud} = \frac{1}{K} \sum_{k=0}^{K-1} \mathbf{u}(k) y_d^*(k) . \tag{61.26}$$

Performance analysis and guidelines for selecting the block size K are provided in Reed et al. [9].

Continuous adaptation algorithms are easily developed in terms of Fig. 61.7 and Eq. (61.23). Note that $J(\mathbf{w}_M)$ is a quadratic error surface. Since the quadratic surface's "Hessian" $\mathbf{R}_u$ is the covariance matrix of noisy data, it is positive definite. This implies that the error surface is a "bowl". The shape of the bowl is determined by the eigenstructure of $\mathbf{R}_u$. The optimum weight vector $\mathbf{w}_{opt}$ corresponds to the bottom of the bowl.

One approach to adaptive filtering is to envision a point on the error surface that corresponds to the present weight vector $\mathbf{w}_M(k)$. We select a new weight vector $\mathbf{w}_M(k + 1)$ so as to descend on the error surface. The gradient vector

$$\nabla_{\mathbf{w}_M(k)} = \left. \frac{\partial}{\partial \mathbf{w}_M} J(\mathbf{w}_M) \right|_{\mathbf{w}_M = \mathbf{w}_m(k)} = -2\mathbf{r}_{ud} + 2\mathbf{R}_u \mathbf{w}_M(k) \tag{61.27}$$

tells us the direction in which to adjust the weight vector. Steepest descent, i.e., adjustment in the negative gradient direction, leads to the popular least mean-square (LMS) adaptive algorithm. The LMS algorithm replaces $\nabla_{\mathbf{w}_M(k)}$ with the instantaneous gradient estimate $\hat{\nabla}_{\mathbf{w}_M(k)} = -2[\mathbf{u}(k) y_d^*(k) - \mathbf{u}(k) \mathbf{u}^H(k) \mathbf{w}_M(k)]$. Denoting $y(k) = y_d(k) - \mathbf{w}_M^H \mathbf{u}(k)$, we have

$$\mathbf{w}_M(k + 1) = \mathbf{w}_M(k) + \mu\, \mathbf{u}(k)\, y^*(k) . \tag{61.28}$$

The gain constant μ controls convergence characteristics of the random vector sequence $\mathbf{w}_M(k)$. Table 61.2 provides guidelines for its selection.

The primary virtue of the LMS algorithm is its simplicity. Its performance is acceptable in many applications; however, its convergence characteristics depend on the shape of the error surface and therefore

TABLE 61.2 Comparison of the LMS and RLS Weight Adaptation Algorithms

Algorithm	LMS	RLS
Initialization	$\mathbf{w}_M(0) = 0$ $\mathbf{y}(0) = \mathbf{y}_d(0)$ $0 < \mu < \frac{1}{\text{Trace}[\mathbf{R}_u]}$	$\mathbf{w}_M(0) = 0$ $\mathbf{P}(0) = \delta^{-1}\mathbf{I}$ δ small, $\mathbf{I}$ identity matrix
Update Equations	$\mathbf{w}_M(k) = \mathbf{w}_M(k-1) + \mu\mathbf{u}(k-1)y^*(k-1)$ $y(k) = y_d(k) - \mathbf{w}_M^H(k)\mathbf{u}(k)$	$\mathbf{v}(k) = \mathbf{P}(k-1)\mathbf{u}(k)$ $\mathbf{k}(k) = \frac{\lambda^{-1}\mathbf{v}(k)}{1 + \lambda^{-1}\mathbf{u}^H(k)\mathbf{v}(k)}$ $\alpha(k) = y_d(k) - \mathbf{w}_M^H(k-1)\mathbf{u}(k)$ $\mathbf{w}_M(k) = \mathbf{w}_M(k-1) + \mathbf{k}(k)\alpha^*(k)$ $\mathbf{P}(k) = \lambda^{-1}\mathbf{P}(k-1) - \lambda^{-1}\mathbf{k}(k)\mathbf{v}^H(k)$
Multiplies per update	$2M$	$4M^2 + 4M + 2$
Performance Characteristics	Under certain conditions, convergence of $\mathbf{w}_M(k)$ to the statistically optimum weight vector $\mathbf{w}_{opt}$ in the mean-square sense is guaranteed if μ is chosen as indicated above. The convergence rate is governed by the eigenvalue spread of $\mathbf{R}_u$. For large eigenvalue spread, convergence can be very slow.	The $\mathbf{w}_M(k)$ represents the least squares solution at each instant k and are optimum in a deterministic sense. Convergence to the statistically optimum weight vector $\mathbf{w}_{opt}$ is often faster than that obtained using the LMS algorithm because it is independent of the eigenvalue spread of $\mathbf{R}_u$.

the eigenstructure of $\mathbf{R}_u$. When the eigenvalues are widely spread, convergence can be slow and other adaptive algorithms with better convergence characteristics should be considered. Alternative procedures for searching the error surface have been proposed in addition to algorithms based on least squares and Kalman filtering. Roughly speaking, these algorithms trade-off computational requirements with speed of convergence to $\mathbf{w}_{opt}$. We refer you to texts on adaptive filtering for detailed descriptions and analysis (Widrow and Stearns [14], Haykin [6], and others).

One alternative to LMS is the exponentially weighted recursive least squares (RLS) algorithm. At the Kth time step, $\mathbf{w}_M(K)$ is chosen to minimize a weighted sum of past squared errors

$$\min_{\mathbf{w}_M(K)} \sum_{k=0}^{K} \lambda^{K-k} |y_d(k) - \mathbf{w}_M^H(K)\mathbf{u}(k)|^2 . \tag{61.29}$$

λ is a positive constant less than one which determines how quickly previous data are deemphasized. The RLS algorithm is obtained from (61.29) by expanding the magnitude squared and applying the matrix inversion lemma. Table 61.2 summarizes both the LMS and RLS algorithms.

61.6 Interference Cancellation and Partially Adaptive Beamforming

The computational requirements of each update in adaptive algorithms are proportional to either the weight vector dimension M (e.g., LMS) or dimension squared M^2 (e.g., RLS). If M is large, this requirement is quite severe and for practical real time implementation it is often necessary to reduce M. Furthermore, the rate at which an adaptive algorithm converges to the optimum solution may be very slow for large M. Adaptive algorithm convergence properties can be improved by reducing M.

The concept of "degrees of freedom" is much more relevant to this discussion than the number of weights. The expression degrees of freedom refers to the number of unconstrained or "free" weights in an implementation. For example, an LCMV beamformer with L constraints on N weights has $N - L$ degrees of freedom; the GSC implementation separates these as the unconstrained weight vector $\mathbf{w}_M$. There are M degrees of freedom in the structure of Fig. 61.7. A fully adaptive beamformer uses all available degrees of freedom and a partially adaptive beamformer uses a reduced set of degrees of freedom. Reducing degrees of freedom lowers computational requirements and often improves adaptive response time. However, there is a performance penalty associated with reducing degrees of freedom. A partially adaptive beamformer cannot generally converge to the same optimum solution as the fully adaptive beamformer. The goal of partially adaptive beamformer design is to reduce degrees of freedom without significant degradation in performance.

The discussion in this section is general, applying to different types of beamformers although we borrow much of the notation from the GSC. We assume the beamformer is described by the adaptive structure of Fig. 61.7 where the desired signal y_d is obtained as $y_d = \mathbf{w}_o^H \mathbf{x}$ and the data vector $\mathbf{u}$ as $\mathbf{u} = \mathbf{T}^H \mathbf{x}$. Thus, the beamformer output is $y = \mathbf{w}^H \mathbf{x}$ where $\mathbf{w} = \mathbf{w}_o - \mathbf{T}\mathbf{w}_M$. In order to distinguish between fully and partially adaptive implementations, we decompose $\mathbf{T}$ into a product of two matrices $\mathbf{C}_n \mathbf{T}_M$. The definition of $\mathbf{C}_n$ depends on the particular beamformer and $\mathbf{T}_M$ represents the mapping which reduces degrees of freedom. The MSC and GSC are obtained as special cases of this representation. In the MSC $\mathbf{w}_o$ is an N vector that selects the primary sensor, $\mathbf{C}_n$ is an N by $N - 1$ matrix that selects the $N - 1$ possible auxiliary sensors from the complete set of N sensors, and $\mathbf{T}_M$ is an $N - 1$ by M matrix that selects the M auxiliary sensors actually utilized. In terms of the GSC, $\mathbf{w}_o$ and $\mathbf{C}_n$ are defined as in Section 61.4.4 and $\mathbf{T}_M$ is an $N - L$ by M matrix that reduces degrees of freedom ($M < N - L$).

The goal of partially adaptive beamformer design is to choose $\mathbf{T}_M$ (or $\mathbf{T}$) such that good interference cancellation properties are retained even though M is small. To see that this is possible in principle, consider the problem of simultaneously cancelling two narrowband sources from direction θ_1 and θ_2 at frequency ω_o. Perfect cancellation of these sources requires $\mathbf{w}^H \mathbf{d}(\theta_1, \omega_o) = 0$ and $\mathbf{w}^H \mathbf{d}(\theta_2, \omega_o) = 0$ so we must choose $\mathbf{w}_M$ to satisfy

$$\mathbf{w}_M^H [\mathbf{T}^H \mathbf{d}(\theta_1, \omega_o) \quad \mathbf{T}^H \mathbf{d}(\theta_2, \omega_o)] = [g_1, g_2] \tag{61.30}$$

where $g_i = \mathbf{w}_o^H \mathbf{d}(\theta_i, \omega_o)$ is the response of the $\mathbf{w}_o$ branch to the ith interferer. Assuming $\mathbf{T}^H \mathbf{d}(\theta_1, \omega_o)$ and $\mathbf{T}^H \mathbf{d}(\theta_2, \omega_o)$ are linearly independent and nonzero, and provided $M \geq 2$, then at least one $\mathbf{w}_M$ exists that satisfies (61.30). Extending this reasoning, we see that $\mathbf{w}_M$ can be chosen to cancel M narrowband interferers (assuming the $\mathbf{T}^H \mathbf{d}(\theta_i, \omega_o)$ are linearly independent and nonzero), independent of $\mathbf{T}$. Total cancellation occurs if $\mathbf{w}_M$ is chosen so the response of $\mathbf{T}\,\mathbf{w}_M$ perfectly matches the $\mathbf{w}_o$ branch response to the interferers. In general, M narrowband interferers can be cancelled using M adaptive degrees of freedom with relatively mild restrictions on $\mathbf{T}$.

No such rule exists in the broadband case. Here complete cancellation of a single interferer requires choosing $\mathbf{T}\,\mathbf{w}_M$ so that the response of the adaptive branch, $\mathbf{w}_M^H \mathbf{T}^H \mathbf{d}(\theta_1, \omega)$, matches the response of the $\mathbf{w}_o$ branch, $\mathbf{w}_o^H \mathbf{d}(\theta_1, \omega)$, over the entire frequency band of the interferer. In this case, the degree of cancellation depends on how well these two responses match and is critically dependent on the interferer direction, frequency content, and $\mathbf{T}$. Good cancellation can be obtained in some situations when $M = 1$, while in others even large values of M result in poor cancellation.

A variety of intuitive and optimization-based techniques have been proposed for designing $\mathbf{T}_M$ that acheive good interference cancellation with relatively small degrees of freedom. See Van Veen and Buckley [11] and Van Veen [12] for further review and discussion.

61.7 Summary

A beamformer forms a scalar output signal as a weighted combination of the data received at an array of sensors. The weights determine the spatial filtering characteristics of the beamformer and enable separation of signals having overlapping frequency content if they originate from different locations. The weights in a data independent beamformer are chosen to provide a fixed response independent to the received data. Statistically optimum beamformers select the weights to optimize the beamformer response based on the statistics of the data. The data statistics are often unknown and may change with time so adaptive algorithms are used to obtain weights that converge to the statistically optimum solution. Computational and response time considerations dictate the use of partially adaptive beamformers with arrays composed of large numbers of sensors.

61.8 Defining Terms

Beamformer: A device used in conjunction with an array of sensors to separate signals and interference on the basis of their spatial characteristics. The beamformer output is usually given by a weighted combination of the sensor outputs.

Array response vector: Vector describing the amplitude and phase relationships between propagating wave components at each sensor as a function of spatial direction and temporal frequency. Forms the basis for determining the beamformer response.

Beampattern: The magnitude squared of the beamformer's spatial filtering response as a function of spatial direction and possibly temporal frequency.

Data independent, statistically optimum, adaptive, and partially adaptive beamformers: The weights in a data independent beamformer are chosen independent of the statistics of the data. A statistically optimum beamformer chooses its weights to optimize some statistical function of the beamformer output, such as signal-to-noise ratio. An adaptive beamformer adjusts its weights in response to the data to accomodate unknown or time varying statistics. A partially adaptive beamformer uses only a subset of the available adaptive degrees of freedom to reduce the computational burden or improve the adaptive convergence rate.

Multiple sidelobe canceller: Adaptive beamformer structure in which the data received at low gain auxiliary sensors is used to adaptively cancel the interference arriving in the mainlobe or sidelobes of a spatially high gain sensor.

Linearly constrained minimum variance (LCMV) beamformer: Beamformer in which the weights are chosen to minimize the output power subject to a linear response constraint. The constraint preserves the signal of interest while power minimization optimally attenuates noise and interference.

Minimum variance distortionless response (MVDR) beamformer: A form of LCMV beamformer employing a single constraint designed to pass a signal of given direction and frequency with unit gain.

Generalized sidelobe canceller: Structure for implementing LCMV beamformers that separates the constrained and unconstrained components of the adaptive weight vector. The unconstrained components adaptively cancel interference that leaks through the sidelobes of a data independent beamformer designed to satisfy the constraints.

References

[1] Applebaum, S.P., Adaptive arrays, Syracuse University Research Corp., Report SURC SPL TR 66-001, Aug. 1966 (reprinted in *IEEE Trans. on AP*, AP-24, 585–598, Sept. 1976).

[2] Buckley, K.M., Spatial/spectral filtering with linearly-constrained minimum variance beamformers, *IEEE Trans. on ASSP*, ASSP-35, 249–266, Mar. 1987.

[3] Cox, H., Zeskind, R.M., and Owen, M.M., Robust adaptive beamforming, *IEEE Trans. on ASSP*, ASSP-35, 1365–1375, Oct. 1987.

[4] Frost III, O.L., An algorithm for linearly constrained adaptive array processing, *Proc. IEEE*, 60, 926–935, Aug. 1972.

[5] Harris, F.J., On the use of windows for harmonic analysis with the discrete Fourier transform, *Proc. IEEE*, 66, 51–83, Jan. 1978.

[6] Haykin, S., *Adaptive Filter Theory*, 3rd ed., Prentice-Hall, Englewood Cliffs, NJ, 1996.

[7] Howells, P.W., Explorations in fixed and adaptive resolution at GE and SURC, *IEEE Trans. on AP*, AP-24, 575–584, Sept. 1976.

[8] Parks, T.W. and Burrus, C.S., *Digital Filter Design*, Wiley-Interscience, New York, 1987.

[9] Reed, I.S., Mallett, J.D., and Brennen, L.E., Rapid convergence rate in adaptive arrays, *IEEE Trans. on AES*, AES-10, 853–863, Nov. 1974.

[10] Shan, T. and Kailath, T., Adaptive beamforming for coherent signals and interference, *IEEE Trans. on ASSP*, ASSP-33, 527–536, June 1985.

[11] Van Veen, B. and Buckley, K., Beamforming: a versatile approach to spatial filtering, *IEEE ASSP Magazine*, 5(2), 4–24, Apr. 1988.

[12] Van Veen, B., Minimum Variance Beamforming, in *Adaptive Radar Detection and Estimation*, Haykin, S. and Steinhardt, A., Eds., John Wiley & Sons, New York, Chap. 4, 161–236, 1992.

[13] Widrow, B., Duvall, K.M., Gooch, R.P., and Newman, W.C., Signal cancellation phenomena in adaptive arrays: causes and cures, *IEEE Trans. on AP*, AP30, 469–478, May 1982.

[14] Widrow, B. and Stearns, S., *Adaptive Signal Processing*, Prentice-Hall, Englewood Cliffs, NJ, 1985.

Further Reading

For further information, we refer the reader to the following books,

[1] Compton, Jr., R.T., *Adaptive Antennas: Concepts and Performance*, Prentice-Hall, Englewood Cliffs, NJ, 1988.

[2] Haykin, S., Ed., *Array Signal Processing*, Prentice-Hall, Englewood Cliffs, NJ, 1985.

[3] Johnson, D. and Dudgeon, D., *Array Signal Processing: Concepts and Techniques*, Prentice-Hall, Englewood Cliffs, NJ, 1993.

[4] Monzingo, R. and Miller, T., *Introduction to Adaptive Arrays*, John Wiley & Sons, New York, 1980.

a tutorial article,

[5] Gabriel, W.F., Adaptive arrays: an introduction, *Proc. IEEE*, 64, 239–272, Aug. 1976.

and bibliography

[6] Marr, J., A selected bibliography on adaptive antenna arrays, *IEEE Trans. on AES*, AES-22, 781–798, Nov. 1986.

Several special journal issues have been devoted to beamforming — *IEEE Transactions on Antennas and Propagation*, September 1976 and March 1986, and the *Journal of Ocean Engineering* 1987. Papers devoted to beamforming are often found in the *IEEE Transactions on: Antennas and Propagation, Signal Processing, Aerospace and Electronic Systems*, and in the *Journal of the Acoustical Society of America*.

62

Subspace-Based Direction Finding Methods

Egemen Gönen
Globalstar

Jerry M. Mendel
University of Southern California, Los Angeles

62.1   Introduction

Estimating bearings of multiple narrowband signals from measurements collected by an array of sensors has been a very active research problem for the last two decades. Typical applications of this problem are radar, communication, and underwater acoustics. Many algorithms have been proposed to solve the bearing estimation problem. One of the first techniques that appeared was beamforming which has a resolution limited by the array structure. Spectral estimation techniques were also applied to the problem. However, these techniques fail to resolve closely spaced arrival angles for low signal-to-noise ratios. Another approach is the maximum-likelihood (ML) solution. This approach has been well documented in the literature. In the stochastic ML method [29], the signals are assumed to be Gaussian whereas they are regarded as arbitrary and deterministic in the deterministic ML method [37]. The sensor noise is modeled as Gaussian in both methods, which is a reasonable assumption due to the central limit theorem. The stochastic ML estimates of the bearings achieve the Cramer-Rao bound (CRB). On the other hand, this does not hold for deterministic ML estimates [32]. The common problem with the ML methods in general is the necessity of solving a nonlinear multidimensional optimization problem which has a high computational cost and for which there is no guarantee of global convergence. *Subspace-based* (or, super-resolution) approaches have attracted much attention, after the work of [29], due to their computational simplicity as compared to the ML approach, and their possibility of overcoming the Rayleigh bound on the resolution power of classical direction finding methods. Subspace-based direction finding methods are summarized in this section.

62.2   Formulation of the Problem

Consider an array of M antenna elements receiving a set of plane waves emitted by P ($P < M$) sources in the far field of the array. We assume a narrow-band propagation model, i.e., the signal envelopes do not change during the time it takes for the wavefronts to travel from one sensor to another. Suppose that the signals have a common frequency of f_0; then, the wavelength $\lambda = c/f_0$ where c is the speed of propagation. The received M-vector $\mathbf{r}(t)$ at time t is

$$\mathbf{r}(t) = \mathbf{A}\mathbf{s}(t) + \mathbf{n}(t) \tag{62.1}$$

where $\mathbf{s}(t) = [s_1(t), \cdots, s_P(t)]^T$ is the P-vector of sources; $\mathbf{A} = [\mathbf{a}(\theta_1), \cdots, \mathbf{a}(\theta_P)]$ is the $M \times P$ steering matrix in which $\mathbf{a}(\theta_i)$, the ith steering vector, is the response of the array to the ith source arriving from θ_i; and, $\mathbf{n}(t) = [n_1(t), \cdots, n_M(t)]^T$ is an additive noise process.

We assume: (1) the source signals may be statistically independent, partially correlated, or completely correlated (i.e., coherent); the distributions are unknown; (2) the array may have an arbitrary shape and response; and, (3) the noise process is independent of the sources, zero-mean, and it may be either partially white or colored; its distribution is unknown. These assumptions will be relaxed, as required by specific methods, as we proceed.

The direction finding problem is to estimate the bearings [i.e., directions of arrival (DOA)] $\{\theta_i\}_{i=1}^P$ of the sources from the snapshots $\mathbf{r}(t), t = 1, \cdots, N$.

In applications, the Rayleigh criterion sets a bound on the resolution power of classical direction finding methods. In the next sections we summarize some of the so-called super-resolution direction finding methods which may overcome the Rayleigh bound. We divide these methods into two classes, those that use second-order and those that use second- and higher-order statistics.

62.3   Second-Order Statistics-Based Methods

The second-order methods use the sample estimate of the array spatial covariance matrix $\mathbf{R} = E\{\mathbf{r}(t)\mathbf{r}(t)^H\} = \mathbf{A}\mathbf{R}_s\mathbf{A}^H + \mathbf{R}_n$, where $\mathbf{R}_s = E\{\mathbf{s}(t)\mathbf{s}(t)^H\}$ is the $P \times P$ signal covariance matrix and $\mathbf{R}_n = E\{\mathbf{n}(t)\mathbf{n}(t)^H\}$ is the $M \times M$ noise covariance matrix. For the time being, let us assume that the noise is spatially white, i.e., $\mathbf{R}_n = \sigma^2\mathbf{I}$. If the noise is colored and its covariance matrix is known or can be estimated, the measurements can be "whitened" by multiplying the measurements from the left by the matrix $\Lambda^{-1/2}\mathbf{E}_n^H$ obtained by the orthogonal eigendecomposition $\mathbf{R}_n = \mathbf{E}_n\Lambda\mathbf{E}_n^H$. The array spatial covariance matrix is estimated as $\hat{\mathbf{R}} = \sum_{t=1}^N \mathbf{r}(t)\mathbf{r}(t)^H/N$.

Some spectral estimation approaches to the direction finding problem are based on optimization. Consider the *minimum variance* algorithm, for example. The received signal is processed by a beamforming vector $\mathbf{w}_o$ which is designed such that the output power is minimized subject to the constraint that a signal from a desired direction is passed to the output with unit gain. Solving this optimization problem, we obtain the array output power as a function of the arrival angle θ as

$$P_{mv}(\theta) = \frac{1}{\mathbf{a}^H(\theta)\mathbf{R}^{-1}\mathbf{a}(\theta)}.$$

The arrival angles are obtained by scanning the range $[-90°, 90°]$ of θ and locating the peaks of $P_{mv}(\theta)$. At low signal-to-noise ratios the conventional methods, such as minimum variance, fail to resolve closely spaced arrival angles. The resolution of conventional methods are limited by signal-to-noise ratio even if exact $\mathbf{R}$ is used, whereas in subspace methods, there is no resolution limit; hence, the latter are also referred to as *super-resolution* methods. The limit comes from the sample estimate of $\mathbf{R}$.

The subspace-based methods exploit the eigendecomposition of the estimated array covariance matrix $\hat{\mathbf{R}}$. To see the implications of the eigendecomposition of $\hat{\mathbf{R}}$, let us first state the properties of $\mathbf{R}$: (1) If the source signals are independent or partially correlated, $rank(\mathbf{R}_s) = P$. If there are coherent sources,

$rank(\mathbf{R}_s) < P$. In the methods explained in Sections 62.3.1 and 62.3.2, except for the WSF method (see Search-Based Methods), it will be assumed that there are no coherent sources. The coherent signals case is described in Section 62.3.3. (2) If the columns of $\mathbf{A}$ are independent, which is generally true when the source bearings are different, then $\mathbf{A}$ is of full-rank P. (3) Properties 1 and 2 imply $rank(\mathbf{AR}_s\mathbf{A}^H) = P$; therefore, $\mathbf{AR}_s\mathbf{A}^H$ must have P nonzero eigenvalues and $M - P$ zero eigenvalues. Let the eigendecomposition of $\mathbf{AR}_s\mathbf{A}^H$ be $\mathbf{AR}_s\mathbf{A}^H = \sum_{i=1}^{M} \alpha_i \mathbf{e}_i \mathbf{e}_i^H$; then $\alpha_1 \geq \alpha_2 \geq \cdots \geq \alpha_P \geq \alpha_{P+1} = \cdots = \alpha_M = 0$ are the rank-ordered eigenvalues, and $\{\mathbf{e}_i\}_{i=1}^{M}$ are the corresponding eigenvectors. (4) Because $\mathbf{R}_n = \sigma^2 \mathbf{I}$, the eigenvectors of $\mathbf{R}$ are the same as those of $\mathbf{AR}_s\mathbf{A}^H$, and its eigenvalues are $\lambda_i = \alpha_i + \sigma^2$, if $1 \leq i \leq P$, or $\lambda_i = \sigma^2$, if $P+1 \leq i \leq M$. The eigenvectors can be partitioned into two sets: $\mathbf{E}_s \overset{\triangle}{=} [\mathbf{e}_1, \cdots, \mathbf{e}_P]$ forms the *signal subspace*, whereas $\mathbf{E}_n \overset{\triangle}{=} [\mathbf{e}_{P+1}, \cdots, \mathbf{e}_M]$ forms the *noise subspace*. These subspaces are orthogonal. The signal eigenvalues $\Lambda_s \overset{\triangle}{=} diag\{\lambda_1, \cdots, \lambda_P\}$, and the noise eigenvalues $\Lambda_n \overset{\triangle}{=} diag\{\lambda_{P+1}, \cdots, \lambda_M\}$. (5) The eigenvectors corresponding to zero eigenvalues satisfy $\mathbf{AR}_s\mathbf{A}^H \mathbf{e}_i = \mathbf{0}, i = P + 1, \cdots, M$; hence, $\mathbf{A}^H \mathbf{e}_i = \mathbf{0}, i = P + 1, \cdots, M$, because $\mathbf{A}$ and $\mathbf{R}_s$ are full rank. This last equation means that *steering vectors are orthogonal to noise subspace eigenvectors*. It further implies that because of the orthogonality of signal and noise subspaces, *spans of signal eigenvectors and steering vectors are equal*. Consequently there exists a nonsingular $P \times P$ matrix $\mathbf{T}$ such that $\mathbf{E}_s = \mathbf{AT}$.

Alternatively, the signal and noise subspaces can also be obtained by performing a singular value decomposition directly on the received data without having to calculate the array covariance matrix. Li and Vaccaro [17] state that the properties of the bearing estimates do not depend on which method is used; however, singular value decomposition must then deal with a data matrix that increases in size as the new snapshots are received. In the sequel, we assume that the array covariance matrix is estimated from the data and an eigendecomposition is performed on the estimated covariance matrix.

The eigenvalue decomposition of the spatial array covariance matrix, and the eigenvector partitionment into signal and noise subspaces, leads to a number of subspace-based direction finding methods. The signal subspace contains information about where the signals are whereas the noise subspace informs us where they are not. Use of either subspace results in better resolution performance than conventional methods. In practice, the performance of the subspace-based methods is limited fundamentally by the accuracy of separating the two subspaces when the measurements are noisy [18]. These methods can be broadly classified into signal subspace and noise subspace methods. A summary of direction-finding methods based on both approaches follows next.

62.3.1 Signal Subspace Methods

In these methods, only the signal subspace information is retained. Their rationale is that by discarding the noise subspace we effectively enhance the SNR because the contribution of the noise power to the covariance matrix is eliminated. Signal subspace methods are divided into search-based and algebraic methods, which are explained next.

Search-Based Methods

In search-based methods, it is assumed that the response of the array to a single source, *the array manifold* $\mathbf{a}(\theta)$, is either known analytically as a function of arrival angle, or is obtained through the calibration of the array. For example, for an M-element uniform linear array, the array response to a signal from angle θ is analytically known and is given by

$$\mathbf{a}(\theta) = \left[1, e^{-j2\pi \frac{d}{\lambda}\sin(\theta)}, \cdots, e^{-j2\pi(M-1)\frac{d}{\lambda}\sin(\theta)}\right]^T$$

where d is the separation between the elements, and λ is the wavelength.

In search-based methods to follow (except for the subspace fitting methods), which are spatial versions of widely known power spectral density estimators, the estimated array covariance matrix is approximated by its signal subspace eigenvectors, or its *principal components*, as $\hat{\mathbf{R}} \approx \sum_{i=1}^{P} \lambda_i \mathbf{e}_i \mathbf{e}_i{}^H$. Then the arrival angles are estimated by locating the peaks of a function, $S(\theta)$ $(-90° \leq \theta \leq 90°)$, which depends on the particular method. Some of these methods and the associated function $S(\theta)$ are summarized in the following [13, 18, 20]:

Correlogram method: In this method, $S(\theta) = \mathbf{a}(\theta)^H \hat{\mathbf{R}} \mathbf{a}(\theta)$. The resolution obtained from the Correlogram method is lower than that obtained from the MV and AR methods.

Minimum variance (MV) [1] method: In this method, $S(\theta) = 1/\mathbf{a}(\theta)^H \hat{\mathbf{R}}^{-1} \mathbf{a}(\theta)$. The MV method is known to have a higher resolution than the correlogram method, but lower resolution and variance than the AR method.

Autoregressive (AR) method: In this method, $S(\theta) = 1/|\mathbf{u}^T \hat{\mathbf{R}}^{-1} \mathbf{a}(\theta)|^2$ where $\mathbf{u} = [1, 0, \cdots, 0]^T$. This method is known to have a better resolution than the previous ones.

Subspace fitting (SSF) and weighted subspace fitting (WSF) methods: In Section 62.2 we saw that the spans of signal eigenvectors and steering vectors are equal; therefore, bearings can be solved from the best least-squares fit of the two spanning sets when the array is calibrated [35]. In the Subspace Fitting Method the criterion $[\hat{\theta}, \hat{\mathbf{T}}] = argmin \ ||\mathbf{E}_s \mathbf{W}^{1/2} - \mathbf{A}(\theta)\mathbf{T}||^2$ is used, where $||.||$ denotes the Frobenius norm, $\mathbf{W}$ is a positive definite weighting matrix, $\mathbf{E}_s$ is the matrix of signal subspace eigenvectors, and the notation for the steering matrix is changed to show its dependence on the bearing vector θ. This criterion can be minimized directly with respect to $\mathbf{T}$, and the result for $\mathbf{T}$ can then be substituted back into it, so that

$$\hat{\theta} = argmin \ Tr\{(\mathbf{I} - \mathbf{A}(\theta)\mathbf{A}(\theta)^{\#})\mathbf{E}_s \mathbf{W} \mathbf{E}_s^H\},$$

where $\mathbf{A}^{\#} = (\mathbf{A}^H \mathbf{A})^{-1} \mathbf{A}^H$.

Viberg and Ottersten have shown that a class of direction finding algorithms can be approximated by this subspace fitting formulation for appropriate choices of the weighting matrix $\mathbf{W}$. For example, for the deterministic ML method $\mathbf{W} = \Lambda_s - \sigma^2 \mathbf{I}$, which is implemented using the empirical values of the signal eigenvalues, Λ_s, and the noise eigenvalue σ^2. TLS-ESPRIT, which is explained in the next section, can also be formulated in a similar but more involved way. Viberg and Ottersten have also derived an optimal Weighted Subspace Fitting (WSF) Method, which yields the smallest estimation error variance among the class of subspace fitting methods. In WSF, $\mathbf{W} = (\Lambda_s - \sigma^2 \mathbf{I})^2 \Lambda_s^{-1}$. The WSF method works regardless of the source covariance (including coherence) and has been shown to have the same asymptotic properties as the stochastic ML method; hence, it is asymptotically efficient for Gaussian signals (i.e., it achieves the stochastic CRB). Its behavior in the finite sample case may be different from the asymptotic case [34]. Viberg and Ottersten have also shown that the asymptotic properties of the WSF estimates are identical for both cases of Gaussian and non-Gaussian sources. They have also developed a consistent detection method for arbitrary signal correlation, and an algorithm for minimizing the WSF criterion. They do point out several practical implementation problems of their method, such as the need for accurate calibrations of the array manifold and knowledge of the derivative of the steering vectors w.r.t θ. For nonlinear and nonuniform arrays, multidimensional search methods are required for SSF, hence it is computationally expensive.

Algebraic Methods

Algebraic methods do not require a search procedure and yield DOA estimates directly.

ESPRIT (Estimation of Signal Parameters via Rotational Invariance Techniques) [23]: The ESPRIT algorithm requires "translationally invariant" arrays, i.e., an array with its *identical copy* displaced in space. The geometry and response of the arrays do not have to be known; only the measurements from these arrays and the displacement between the identical arrays are required. The computational complexity of ESPRIT is less than that of the search-based methods.

Let $\mathbf{r}^1(t)$ and $\mathbf{r}^2(t)$ be the measurements from these arrays. Due to the displacement of the arrays the following holds:

$$\mathbf{r}^1(t) = \mathbf{A}\mathbf{s}(t) + \mathbf{n}_1(t) \ \text{and} \ \mathbf{r}^2(t) = \mathbf{A}\Phi\mathbf{s}(t) + \mathbf{n}_2(t),$$

where $\Phi = diag\{e^{-j2\pi\frac{d}{\lambda}\sin\theta_1}, \cdots, e^{-j2\pi\frac{d}{\lambda}\sin\theta_P}\}$ in which d is the separation between the identical arrays, and the angles $\{\theta_i\}_{i=1}^P$ are measured with respect to the normal to the displacement vector between the identical arrays. Note that the auto covariance of $\mathbf{r}^1(t)$, $\mathbf{R}^{11}$, and the cross covariance between $\mathbf{r}^1(t)$ and $\mathbf{r}^2(t)$, $\mathbf{R}^{21}$, are given by

$$\mathbf{R}^{11} = \mathbf{ADA}^H + \mathbf{R}_{n_1}$$

and

$$\mathbf{R}^{21} = \mathbf{A\Phi DA}^H + \mathbf{R}_{n_2 n_1},$$

where $\mathbf{D}$ is the covariance matrix of the sources, and $\mathbf{R}_{n_1}$ and $\mathbf{R}_{n_2 n_1}$ are the noise auto- and cross-covariance matrices.

The ESPRIT algorithm solves for Φ, which then gives the bearing estimates. Although the subspace separation concept is not used in ESPRIT, its LS and TLS versions are based on a signal subspace formulation. The LS and TLS versions are more complicated, but are more accurate than the original ESPRIT, and are summarized in the next subsection. Here we summarize the original ESPRIT:

(1) Estimate the autocovariance of $\mathbf{r}^1(t)$ and cross covariance between $\mathbf{r}^1(t)$ and $\mathbf{r}^2(t)$, as

$$\mathbf{R}^{11} = \frac{1}{N}\sum_{t=1}^N \mathbf{r}^1(t)\mathbf{r}^1(t)^H$$

and

$$\mathbf{R}^{21} = \frac{1}{N}\sum_{t=1}^N \mathbf{r}^2(t)\mathbf{r}^1(t)^H.$$

(2) Calculate $\hat{\mathbf{R}}^{11} = \mathbf{R}^{11} - \mathbf{R}_{n_1}$ and $\hat{\mathbf{R}}^{21} = \mathbf{R}^{21} - \mathbf{R}_{n_2 n_1}$ where $\mathbf{R}_{n_1}$ and $\mathbf{R}_{n_2 n_1}$ are the estimated noise covariance matrices. (3) Find the singular values λ_i of the matrix pencil $\hat{\mathbf{R}}^{11} - \lambda_i\hat{\mathbf{R}}^{21}$, $i = 1, \cdots, P$. (4) The bearings, θ_i $(i = 1, \cdots, P)$, are readily obtained by solving the equation

$$\lambda_i = e^{j2\pi\frac{d}{\lambda}\sin\theta_i}$$

for θ_i. In the above steps, it is assumed that the noise is spatially and temporally white or the covariance matrices $\mathbf{R}_{n_1}$ and $\mathbf{R}_{n_2 n_1}$ are known.

LS and TLS ESPRIT [28]: (1) Follow Steps 1 and 2 of ESPRIT; (2) stack $\hat{\mathbf{R}}^{11}$ and $\hat{\mathbf{R}}^{21}$ into a $2M \times M$ matrix $\mathbf{R}$, as $\mathbf{R} \triangleq \left[\hat{\mathbf{R}}^{11T}\ \hat{\mathbf{R}}^{21T}\right]^T$, and perform an SVD of $\mathbf{R}$, keeping the first $2M \times P$ submatrix of the left singular vectors of $\mathbf{R}$. Let this submatrix be $\mathbf{E}_s$; (3) partition $\mathbf{E}_s$ into two $M \times P$ matrices $\mathbf{E}_{s1}$ and $\mathbf{E}_{s2}$ such that

$$\mathbf{E}_s = \left[\mathbf{E}_{s1}^T\ \mathbf{E}_{s2}^T\right]^T.$$

(4) For LS-ESPRIT, calculate the eigendecomposition of $(\mathbf{E}_{s1}^H\mathbf{E}_{s1})^{-1}\mathbf{E}_{s1}^H\mathbf{E}_{s2}$. The eigenvalue matrix gives

$$\Phi = diag\{e^{-j2\pi \frac{d}{\lambda}\sin\theta_1}, \cdots, e^{-j2\pi \frac{d}{\lambda}\sin\theta_P}\}$$

from which the arrival angles are readily obtained. For TLS-ESPRIT, proceed as follows: (5) Perform an SVD of the $M \times 2P$ matrix $[\mathbf{E}_{s1}, \mathbf{E}_{s2}]$, and stack the last P right singular vectors of $[\mathbf{E}_{s1}, \mathbf{E}_{s2}]$ into a $2P \times P$ matrix denoted $\mathbf{F}$; (6) Partition $\mathbf{F}$ as

$$\mathbf{F} \triangleq \left[\mathbf{F}_x{}^T \ \mathbf{F}_y{}^T\right]^T$$

where $\mathbf{F}_x$ and $\mathbf{F}_y$ are $P \times P$; (7) Perform the eigendecomposition of $-\mathbf{F}_x\mathbf{F}_y^{-1}$. The eigenvalue matrix gives

$$\Phi = diag\{e^{-j2\pi \frac{d}{\lambda}\sin\theta_1}, \cdots, e^{-j2\pi \frac{d}{\lambda}\sin\theta_P}\}$$

from which the arrival angles are readily obtained.

Different versions of ESPRIT have different statistical properties. The Toeplitz Approximation Method (TAM) [16], in which the array measurement model is represented as a state-variable model, although different in implementation from LS-ESPRIT, is equivalent to LS-ESPRIT; hence, it has the same error variance as LS-ESPRIT.

Generalized Eigenvalues Utilizing Signal Subspace Eigenvectors (GEESE) [24]: (1) Follow Steps 1 through 3 of TLS ESPRIT. (2) Find the singular values λ_i of the pencil

$$\mathbf{E}_{s1} - \lambda_i \mathbf{E}_{s2}, i = 1, \cdots, P;$$

(3) The bearings, θ_i ($i = 1, \cdots, P$), are readily obtained from

$$\lambda_i = e^{j2\pi \frac{d}{\lambda}\sin\theta_i}.$$

The GEESE method is claimed to be better than ESPRIT [24].

62.3.2 Noise Subspace Methods

These methods, in which only the noise subspace information is retained, are based on the property that the steering vectors are orthogonal to any linear combination of the noise subspace eigenvectors. Noise subspace methods are also divided into search-based and algebraic methods, which are explained next.

Search-Based Methods

In search-based methods, the array manifold is assumed to be known, and the arrival angles are estimated by locating the peaks of the function $S(\theta) = 1/\mathbf{a}(\theta)^H \mathbf{N}\mathbf{a}(\theta)$ where $\mathbf{N}$ is a matrix formed using the noise space eigenvectors.

Pisarenko method: In this method, $\mathbf{N} = \mathbf{e}_M\mathbf{e}_M{}^H$, where $\mathbf{e}_M$ is the eigenvector corresponding to the minimum eigenvalue of $\mathbf{R}$. If the minimum eigenvalue is repeated, any unit-norm vector which is a linear combination of the eigenvectors corresponding to the minimum eigenvalue can be used as $\mathbf{e}_M$. The basis of this method is that when the search angle θ corresponds to an actual arrival angle, the denominator of $S(\theta)$ in the Pisarenko method, $|\mathbf{a}(\theta)^H \mathbf{e}_M|^2$, becomes small due to orthogonality of steering vectors and noise subspace eigenvectors; hence, $S(\theta)$ will peak at an arrival angle.

MUSIC (Multiple Signal Classification) [29] method: In this method, $\mathbf{N} = \sum_{i=P+1}^{M} \mathbf{e}_i\mathbf{e}_i{}^H$. The idea is similar to that of the Pisarenko method; the inner product $|\mathbf{a}(\theta)^H \sum_{i=P+1}^{M} \mathbf{e}_i|^2$ is small when θ is an actual arrival angle. An obvious signal-subspace formulation of MUSIC is also possible. The MUSIC spectrum is equivalent to the MV method using the exact covariance matrix when SNR is infinite, and therefore performs better than the MV method.

Asymptotic properties of MUSIC are well established [32, 33], e.g., MUSIC is known to have the same asymptotic variance as the deterministic ML method for uncorrelated sources. It is shown by Xu and Buckley [38] that although, asymptotically, bias is insignificant compared to standard deviation, it is an important factor limiting the performance for resolving closely spaced sources when they are correlated.

In order to overcome the problems due to finite sample effects and source correlation, a multidimensional (MD) version of MUSIC has been proposed [29, 28]; however, this approach involves a computationally involved search, as in the ML method. MD MUSIC can be interpreted as a norm minimization problem, as shown in [8]; using this interpretation, strong consistency of MD MUSIC has been demonstrated. An optimally weighted version of MD MUSIC, which outperforms the deterministic ML method, has also been proposed in [35].

Eigenvector (EV) method: In this method,

$$\mathbf{N} = \sum_{i=P+1}^{M} \frac{1}{\lambda_i} \mathbf{e}_i \mathbf{e}_i^H.$$

The only difference between the EV method and MUSIC is the use of inverse eigenvalue (the λ_i are the noise subspace eigenvalues of $\mathbf{R}$) weighting in EV and unity weighting in MUSIC, which causes EV to yield fewer spurious peaks than MUSIC [13]. The EV Method is also claimed to shape the noise spectrum better than MUSIC.

Method of direction estimation (MODE): MODE is equivalent to WSF when there are no coherent sources. Viberg and Ottersten [35] claim that, for coherent sources, only WSF is asymptotically efficient. A minimum norm interpretation and proof of strong consistency of MODE for ergodic and stationary signals, has also been reported [8]. The norm measure used in that work involves the source covariance matrix. By contrasting this norm with the Frobenius norm that is used in MD MUSIC, Ephraim et al. relate MODE and MD MUSIC.

Minimum-norm [15] method: In this method, the matrix $\mathbf{N}$ is obtained as follows [12]:

1. Form $\mathbf{E}_n = [\mathbf{e}_{P+1}, \cdots, \mathbf{e}_M]$;
2. partition $\mathbf{E}_n$ as $\mathbf{E}_n = \begin{bmatrix} \mathbf{c} & \mathbf{C}^T \end{bmatrix}^T$, to establish $\mathbf{c}$ and $\mathbf{C}$;
3. compute $\mathbf{d} = \begin{bmatrix} 1 & ((\mathbf{c}^H\mathbf{c})^{-1}\mathbf{C}^*\mathbf{c})^T \end{bmatrix}^T$, and, finally, $\mathbf{N} = \mathbf{d}\mathbf{d}^H$.

For two closely spaced, equal power signals, the Minimum Norm Method has been shown to have a lower SNR threshold (i.e., the minimum SNR required to separate the two sources) than MUSIC [14]. [17] derive and compare the mean-squared errors of the DOA estimates from Minimum Norm and MUSIC algorithms due to finite sample effects, calibration errors, and noise modeling errors for the case of finite samples and high SNR. They show that mean-squared errors for DOA estimates produced by the MUSIC algorithm are always lower than the corresponding mean-squared errors for the Minimum Norm algorithm.

Algebraic Methods

When the array is uniform linear, so that

$$\mathbf{a}(\theta) = \begin{bmatrix} 1, e^{-j2\pi \frac{d}{\lambda} \sin(\theta)}, \cdots, e^{-j2\pi(M-1)\frac{d}{\lambda} \sin(\theta)} \end{bmatrix}^T,$$

the search in $S(\theta) = 1/\mathbf{a}(\theta)^H \mathbf{N}\mathbf{a}(\theta)$ for the peaks can be replaced by a root-finding procedure which yields the arrival angles. So doing results in better resolution than the search-based alternative because the root-finding procedure can give distinct roots corresponding to each source whereas the search function may not have distinct maxima for closely spaced sources. In addition, the computational complexity of algebraic methods is lower than that of the search-based ones. The algebraic version of MUSIC (Root-MUSIC) is given next; for algebraic versions of Pisarenko, EV, and Minimum-Norm, the matrix $\mathbf{N}$ in Root-Music is replaced by the corresponding $\mathbf{N}$ in each of these methods.

Root-MUSIC Method: In Root-MUSIC, the array is required to be uniform linear, and the search procedure in MUSIC is converted into the following root-finding approach:

1. Form the $M \times M$ matrix $\mathbf{N} = \sum_{i=P+1}^{M} \mathbf{e}_i \mathbf{e}_i^{H}$.

2. Form a polynomial $p(z)$ of degree $2M - 1$ which has for its ith coefficient $c_i = tr_i[\mathbf{N}]$, where tr_i denotes the trace of the ith diagonal, and $i = -(M - 1), \cdots, 0, \cdots, M - 1$. Note that tr_0 denotes the main diagonal, tr_1 denotes the first super-diagonal, and tr_{-1} denotes the first sub-diagonal.

3. The roots of $p(z)$ exhibit inverse symmetry with respect to the unit circle in the z-plane. Express $p(z)$ as the product of two polynomials $p(z) = h(z)h^*(z^{-1})$.

4. Find the roots z_i ($i = 1, \cdots, M$) of $h(z)$. The angles of roots that are very close to (or, ideally on) the unit circle yield the direction of arrival estimates, as

$$\theta_i = \sin^{-1}(\frac{\lambda}{2\pi d} \angle z_i), \quad \text{where } i = 1, \cdots, P.$$

The Root-MUSIC algorithm has been shown to have better resolution power than MUSIC [27]; however, as mentioned previously, Root-MUSIC is restricted to uniform linear arrays. Steps (2) through (4) make use of this knowledge. Li and Vaccaro show that algebraic versions of the MUSIC and Minimum Norm algorithms have the same mean-squared errors as their search-based versions for finite samples and high SNR case. The advantages of Root-MUSIC over search-based MUSIC is increased resolution of closely spaced sources and reduced computations.

62.3.3   Spatial Smoothing [9, 31]

When there are coherent (completely correlated) sources, $rank(\mathbf{R}_s)$, and consequently $rank(\mathbf{R})$, is less than P, and hence the above described subspace methods fail. If the array is uniform linear, then by applying the spatial smoothing method, described below, a new rank-P matrix is obtained which can be used in place of $\mathbf{R}$ in any of the subspace methods described earlier.

Spatial smoothing starts by dividing the M-vector $\mathbf{r}(t)$ of the ULA into $K = M - S + 1$ overlapping subvectors of size S, $\mathbf{r}_{S,k}^{f}$ ($k = 1, \cdots, K$), with elements $\{r_k, \cdots, r_{k+S-1}\}$, and $\mathbf{r}_{S,k}^{b}$ ($k = 1, \cdots, K$), with elements $\{r_{M-k+1}^{*}, \cdots, r_{M-S-k+2}^{*}\}$. Then, a forward and backward spatially smoothed matrix $\mathbf{R}^{fb}$ is calculated as

$$\mathbf{R}^{fb} = \sum_{t=1}^{N} \sum_{k=1}^{K} (\mathbf{r}_{S,k}^{f}(t)\mathbf{r}_{S,k}^{f}{}^{H}(t) + \mathbf{r}_{S,k}^{b}(t)\mathbf{r}_{S,k}^{b}{}^{H}(t))/KN.$$

The rank of $\mathbf{R}^{fb}$ is P if there are at most $2M/3$ coherent sources. S must be selected such that

$$P_c + 1 \leq S \leq M - P_c/2 + 1$$

in which P_c is the number of coherent sources. Then, any subspace-based method can be applied to $\mathbf{R}^{fb}$ to determine the directions of arrival. It is also possible to do spatial smoothing based only on $\mathbf{r}_{S,k}^{f}$ or $\mathbf{r}_{S,k}^{b}$, but in this case at most $M/2$ coherent sources can be handled.

62.3.4   Discussion

The application of all the subspace-based methods requires exact knowledge of the number of signals, in order to separate the signal and noise subspaces. The number of signals can be estimated from the data using either the Akaike Information Criterion (AIC) [36] or Minimum Descriptive Length (MDL) [37] methods. The effect of underestimating the number of sources is analyzed by [26], whereas the case of overestimating the number of signals can be treated as a special case of the analysis in [32].

The second-order methods described above have the following disadvantages:

1. Except for ESPRIT (which requires a special array structure), all of the above methods require calibration of the array which means that the response of the array for every possible combination of the source parameters should be measured and stored; or, analytical knowledge of the array response is required. However, at any time, the antenna response can be different from when it was last calibrated due to environmental effects such as weather conditions for radar, or water waves for sonar. Even if the analytical response of the array elements is known, it may be impossible to know or track the precise locations of the elements in some applications (e.g., towed array). Consequently, these methods are sensitive to errors and perturbations in the array response. In addition, physically identical sensors may not respond identically in practice due to lack of synchronization or imbalances in the associated electronic circuitry.

2. In deriving the above methods, it was assumed that the noise covariance structure is known; however, it is often unrealistic to assume that the noise statistics are known due to several reasons. In practice, the noise is not isolated; it is often observed along with the signals. Moreover, as [33] state, there are noise phenomena effects that cannot be modeled accurately, e.g., channel crosstalk, reverberation, near-field, wide-band, and distributed sources.

3. None of the methods in Sections 62.3.1 and 62.3.2, except for the WSF method and other multidimensional search-based approaches, which are computationally very expensive, work when there are coherent (completely correlated) sources. Only if the array is uniform linear, can the spatial smoothing method in Section 62.3.3 be used. On the other hand, higher-order statistics of the received signals can be exploited to develop direction finding methods which have less restrictive requirements.

62.4 Higher-Order Statistics-Based Methods

The higher-order statistical direction finding methods use the spatial cumulant matrices of the array. They require that the source signals be non-Gaussian so that their higher than second order statistics convey extra information. Most communication signals (e.g., QAM) are *complex circular* (a signal is complex circular if its real and imaginary parts are independent and symmetrically distributed with equal variances) and hence their third-order cumulants vanish; therefore, even-order cumulants are used, and usually fourth-order cumulants are employed. The fourth-order cumulant of the source signals must be nonzero in order to use these methods. One important feature of cumulant-based methods is that they can suppress Gaussian noise regardless of its coloring. Consequently, the requirement of having to estimate the noise covariance, as in second-order statistical processing methods, is avoided in cumulant-based methods. It is also possible to suppress non-Gaussian noise [6], and, when properly applied, cumulants extend the aperture of an array [5, 30], which means that more sources than sensors can be detected. As in the second-order statistics-based methods, it is assumed that the number of sources is known or is estimated from the data.

The fourth-order moments of the signal $s(t)$ are

$$E\{s_i s_j{}^* s_k s_l{}^*\} \quad 1 \leq i, j, k, l \leq P$$

and the fourth-order cumulants are defined as

$$\begin{aligned}
c_{4,s}(i, j, k, l) &\triangleq cum(s_i, s_j{}^*, s_k, s_l{}^*) \\
&= E\{s_i s_j{}^* s_k s_l{}^*\} - E\{s_i s_j{}^*\} E\{s_k s_l{}^*\} \\
&\quad - E\{s_i s_l{}^*\} E\{s_k s_j{}^*\} - E\{s_i s_j\} E\{s_k{}^* s_l{}^*\},
\end{aligned}$$

where $1 \leq i, j, k, l \leq P$. Note that two arguments in the above fourth-order moments and cumulants are conjugated and the other two are unconjugated. For circularly symmetric signals, which is often the case in communication applications, the last term in $c_{4,s}(i, j, k, l)$ is zero.

In practice, sample estimates of the cumulants are used in place of the theoretical cumulants, and these sample estimates are obtained from the received signal vector $\mathbf{r}(t)$ $(t = 1, \cdots, N)$, as:

$$\hat{c}_{4,r}(i, j, k, l) = \sum_{t=1}^{N} r_i(t) r_j^*(t) r_k(t) r_l^*(t)/N$$

$$- \sum_{t=1}^{N} r_i(t) r_j^*(t) \sum_{t=1}^{N} r_k(t) r_l^*(t)/N^2$$

$$- \sum_{t=1}^{N} r_i(t) r_l^*(t) \sum_{t=1}^{N} r_k(t) r_j^*(t)/N^2,$$

where $1 \leq i, j, k, l \leq M$. Note that the last term in $c_{4,r}(i, j, k, l)$ is zero and, therefore, it is omitted.

Higher-order statistical subspace methods use fourth-order spatial cumulant matrices of the array output, which can be obtained in a number of ways by suitably selecting the arguments i, j, k, l of $c_{4,r}(i, j, k, l)$. Existing methods for the selection of the cumulant matrix, and their associated processing schemes are summarized next.

Pan-Nikias [22] and Cardoso-Moulines [2] method: In this method, the array needs to be calibrated, or its response must be known in analytical form. The source signals are assumed to be independent or partially correlated (i.e, there are no coherent signals). The method is as follows:

1. An estimate of an $M \times M$ fourth-order cumulant matrix $\mathbf{C}$ is obtained from the data. The following two selections for $\mathbf{C}$ are possible [22, 2]:

$$c_{ij} = c_{4,r}(i, j, j, j) \, 1 \leq i, j \leq M,$$

or

$$c_{ij} = \sum_{m=1}^{M} c_{4,r}(i, j, m, m) 1 \leq i, j \leq M.$$

Using cumulant properties [19], and (62.1), and a_{ij} for the ijth element of $\mathbf{A}$, it is easy to verify that

$$c_{4,r}(i, j, j, j) = \sum_{p=1}^{P} a_{ip} \sum_{q,r,s=1}^{P} a_{jq}^* a_{jr} a_{js}^* c_{4,s}(p, q, r, s)$$

which, in matrix format, is $\mathbf{C} = \mathbf{AB}$ where $\mathbf{A}$ is the steering matrix and $\mathbf{B}$ is a $P \times M$ matrix with elements

$$b_{ij} = \sum_{q,r,s=1}^{P} a_{jq}^* a_{jr} a_{js}^* c_{4,s}(i, q, r, s).$$

Similarly,

$$\sum_{m=1}^{M} c_{4,r}(i, j, m, m) = \sum_{p,q=1}^{P} a_{ip} \left(\sum_{r,s=1}^{P} \sum_{m=1}^{M} a_{mr} a_{ms}^* c_{4,s}(p, q, r, s) \right) a_{jq}^*, 1 \leq i, j \leq M$$

which, in matrix form, can be expressed as $\mathbf{C} = \mathbf{ADA}^H$, where $\mathbf{D}$ is a $P \times P$ matrix with elements

$$d_{ij} = \sum_{r,s=1}^{P} \sum_{m=1}^{M} a_{mr} a_{ms}^* c_{4,s}(i, j, r, s).$$

Note that additive Gaussian noise is suppressed in both C matrices because higher than second-order statistics of a Gaussian process are zero.

2. The P left singular vectors of $\mathbf{C} = \mathbf{AB}$, corresponding to nonzero singular values or the P eigenvectors of $\mathbf{C} = \mathbf{ADA}^H$ corresponding to nonzero eigenvalues form the signal subspace. The orthogonal complement of the signal subspace gives the noise subspace. Any of the Section 62.3 covariance-based search and algebraic DF methods (except for the EV method and ESPRIT) can now be applied (in exactly the same way as described in Section 62.3) either by replacing the signal and noise subspace eigenvectors and eigenvalues of the array covariance matrix by the corresponding subspace eigenvectors and eigenvalues of $\mathbf{ADA}^H$, or by the corresponding subspace singular vectors and singular values of $\mathbf{AB}$. A cumulant-based analog of the EV method does not exist because the eigenvalues and singular values of $\mathbf{ADA}^H$ and $\mathbf{AB}$ corresponding to the noise subspace are theoretically zero. The cumulant-based analog of ESPRIT is explained later.

The same assumptions and restrictions for the covariance-based methods apply to their analogs in the cumulant domain. The advantage of using the cumulant-based analogs of these methods is that there is no need to know or estimate the noise-covariance matrix.

The asymptotic covariance of the DOA estimates obtained by MUSIC based on the above fourth-order cumulant matrices are derived in [2] for the case of Gaussian measurement noise with arbitrary spatial covariance, and are compared to the asymptotic covariance of the DOA estimates from the covariance-based MUSIC algorithm. Cardoso and Moulines show that covariance- and fourth-order cumulant-based MUSIC have similar performance for the high SNR case, and as SNR decreases below a certain SNR threshold, the variances of the fourth-order cumulant-based MUSIC DOA estimates increase with the fourth power of the reciprocal of the SNR, whereas the variances of covariance-based MUSIC DOA estimates increase with the square of the reciprocal of the SNR. They also observe that for high SNR and uncorrelated sources, the covariance-based MUSIC DOA estimates are uncorrelated, and the asymptotic variance of any particular source depends only on the power of that source (i.e., it is independent of the powers of the other sources). They observe, on the other hand, that DOA estimates from cumulant-based MUSIC, for the same case, are correlated, and the variance of the DOA estimate of a weak source increases in the presence of strong sources. This observation limits the use of cumulant-based MUSIC when the sources have a high dynamic range, even for the case of high SNR. Cardoso and Moulines state that this problem may be alleviated when the source of interest has a large fourth-order cumulant.

Porat and Friedlander [25] method: In this method, the array also needs to be calibrated, or its response is required in analytical form. The model used in this method divides the sources into groups that are partially correlated (but not coherent) within each group, but are statistically independent across the groups, i.e.,

$$\mathbf{r}(t) = \sum_{g=1}^{G} \mathbf{A}_g \mathbf{s}_g + \mathbf{n}(t)$$

where G is the number of groups each having p_g sources ($\sum_{g=1}^{G} p_g = P$). In this model, the p_g sources in the gth group are partially correlated, and they are received from different directions. The method is as follows:

1. Estimate the fourth-order cumulant matrix, $\mathbf{C}_r$, of $\mathbf{r}(t) \otimes \mathbf{r}(t)^*$ where $\otimes$ denotes the Kronecker product. It can be verified that

$$\mathbf{C}_r = \sum_{g=1}^{G} (\mathbf{A}_g \otimes \mathbf{A}^*_g) \mathbf{C}_{s_g} (\mathbf{A}_g \otimes \mathbf{A}^*_g)^H$$

where $\mathbf{C}_{s_g}$ is the fourth-order cumulant matrix of $\mathbf{s}_g$. The rank of $\mathbf{C}_r$ is $\sum_{g=1}^{G} p_g^2$, and since $\mathbf{C}_r$ is $M^2 \times M^2$, it has $M^2 - \sum_{g=1}^{G} p_g{}^2$ zero eigenvalues which correspond to the noise subspace. The other eigenvalues correspond to the signal subspace.

2. Compute the SVD of $\mathbf{C}_r$ and identify the signal and noise subspace singular vectors. Now, second-order subspace-based search methods can be applied, using the signal or noise subspaces, by replacing the array response vector $\mathbf{a}(\theta)$ by $\mathbf{a}(\theta) \otimes \mathbf{a}^*(\theta)$.

The eigendecomposition in this method has computational complexity $O(M^6)$ due to the Kronecker product, whereas the second-order statistics-based methods (e.g., MUSIC) have complexity $O(M^3)$.

Chiang-Nikias [4] method: This method uses the ESPRIT algorithm and requires an array with its entire identical copy displaced in space by distance d; however, no calibration of the array is required. The signals

$$\mathbf{r}^1(t) = \mathbf{A}s(t) + \mathbf{n}_1(t)$$

and

$$\mathbf{r}^2(t) = \mathbf{A}\Phi s(t) + \mathbf{n}_2(t).$$

Two $M \times M$ matrices $\mathbf{C}^1$ and $\mathbf{C}^2$ are generated as follows:

$$c^1{}_{ij} = cum(r^1{}_i, r^1{}_j{}^*, r^1{}_k, r^1{}_k{}^*), 1 \le i, j, k \le M$$

and

$$c^2{}_{ij} = cum(r^2{}_i, r^1{}_j{}^*, r^1{}_k, r^1{}_k{}^*) 1 \le i, j, k \le M.$$

It can be shown that $\mathbf{C}^1 = \mathbf{A}\mathbf{E}\mathbf{A}^H$ and $\mathbf{C}^2 = \mathbf{A}\Phi\mathbf{E}\mathbf{A}^H$, where

$$\Phi = diag\{e^{-j2\pi \frac{d}{\lambda} \sin\theta_1}, \cdots, e^{-j2\pi \frac{d}{\lambda} \sin\theta_P}\}$$

in which d is the separation between the identical arrays, and $\mathbf{E}$ is a $P \times P$ matrix with elements

$$e_{ij} = \sum_{q,r=1}^{P} a_{kq} a_{kr}{}^* c_{4,s}(i, q, r, j).$$

Note that these equations are in the same form as those for covariance-based ESPRIT (the noise cumulants do not appear in $\mathbf{C}^1$ and $\mathbf{C}^2$ because the fourth-order cumulants of Gaussian noises are zero); therefore, any version of ESPRIT or GEESE can be used to solve for Φ by replacing $\mathbf{R}^{11}$ and $\mathbf{R}^{21}$ by $\mathbf{C}^1$ and $\mathbf{C}^2$, respectively.

Virtual cross correlation computer (VC^3) **[5]:** In VC^3, the source signals are assumed to be statistically independent. The idea of VC^3 can be demonstrated as follows: Suppose we have three identical sensors as in Fig. 62.1, where $r_1(t)$, $r_2(t)$, and $r_3(t)$ are measurements, and $\vec{d}_1$, $\vec{d}_2$, and $\vec{d}_3$ ($\vec{d}_3 = \vec{d}_1 + \vec{d}_2$) are the vectors joining these sensors. Let the response of each sensor to a signal from θ be $a(\theta)$. A *virtual* sensor is one at which no measurement is actually made. Suppose that we wish to compute the correlation between the virtual sensor $v_1(t)$ and $r_2(t)$, which (using the plane wave assumption) is

$$E\{r_2^*(t) v_1(t)\} = \sum_{p=1}^{P} |a(\theta_p)|^2 \sigma_p{}^2 e^{-j\vec{k}_p . \vec{d}_3}.$$

Consider the following cumulant

$$\begin{aligned}
cum(r_2^*(t), r_1(t), r_2^*(t), r_3(t)) &= \sum_{p=1}^{P} |a(\theta_p)|^4 \gamma_p e^{-j\vec{k}_p . \vec{d}_1} e^{-j\vec{k}_p . \vec{d}_2} \\
&= \sum_{p=1}^{P} |a(\theta_p)|^4 \gamma_p e^{-j\vec{k}_p . \vec{d}_3}.
\end{aligned}$$

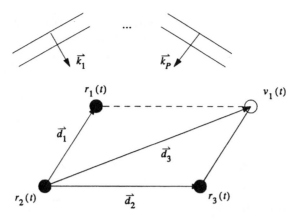

FIGURE 62.1   Demonstration of VC^3.

This cumulant carries the same angular information as the cross correlation $E\{r_2^*(t)v_1(t)\}$, but for sources having different powers.

The fact that we are interested only in the directional information carried by correlations between the sensors therefore let us interpret a cross correlation as a vector (e.g., $\vec{d}_3$), and a fourth-order cumulant as the addition of two vectors (e.g., $\vec{d}_1 + \vec{d}_2$). This interpretation leads to the idea of decomposing the computation of a cross correlation into that of computing a cumulant. Doing this means that the directional information that would be obtained from the cross correlation between nonexisting sensors (or between an actual sensor and a nonexisting sensor) at certain virtual locations in the space can be obtained from a suitably defined cumulant that uses the real sensor measurements.

One advantage of virtual cross correlation computation is that it is possible to obtain a larger aperture than would be obtained by using only second-order statistics. This means that more sources than sensors can be detected using cumulants. For example, given an M element uniform linear array, VC^3 lets its aperture be extended from M to $2M - 1$ sensors, so that $2M - 2$ targets can be detected (rather than $M - 1$) just by using the array covariance matrix obtained by VC^3 in any of the subspace-based search methods explained earlier. This use of VC^3 requires the array to be calibrated. Another advantage of VC^3 is a fault tolerance capability. If sensors at certain locations in a given array fail to operate properly, these sensors can be replaced using VC^3.

Virtual ESPRIT (VESPA) [5]: For VESPA, the array only needs two identical sensors; the rest of the array may have arbitrary and unknown geometry and response. The sources are assumed to be statistically independent. VESPA uses the ESPRIT solution applied to cumulant matrices. By choosing a suitable pair of cumulants in VESPA, the need for a copy of the entire array, as required in ESPRIT, is totally eliminated. VESPA preserves the computational advantage of ESPRIT over search-based algorithms. An example array configuration is given in Fig. 62.2.

Without loss of generality, let the signals received by the identical sensor pair be r_1 and r_2. The sensors r_1 and r_2 are collectively referred to as the *guiding sensor pair*. The VESPA algorithm is

1. Two $M \times M$ matrices, $\mathbf{C}^1$ and $\mathbf{C}^2$, are generated as follows:

$$c^1{}_{ij} = cum(r_1, r_1{}^*, r_i, r_j{}^*), 1 \le i, j \le M$$

$$c^2{}_{ij} = cum(r_2, r_1{}^*, r_i, r_j{}^*), 1 \le i, j \le M.$$

It can be shown that these relations can be expressed as $\mathbf{C}^1 = \mathbf{AFA}^H$ and $\mathbf{C}^2 = \mathbf{A\Phi FA}^H$, where the $P \times P$ matrix

$$\mathbf{F} \overset{\triangle}{=} diag\{\gamma_{4,s_1}|a_{11}|^2, \cdots, \gamma_{4,sp}|a_{1P}|^2\}, \{\gamma_{4,sp}\}_{p=1}^{P},$$

and Φ has been defined before.

2. Note that these equations are in the same form as ESPRIT and Chiang and Nikias's ESPRIT-like method; however, as opposed to these methods, there is no need for an identical copy of the array; only an identical response sensor pair is necessary for VESPA. Consequently, any version of ESPRIT or GEESE can be used to solve for Φ by replacing $\mathbf{R}^{11}$ and $\mathbf{R}^{21}$ by $\mathbf{C}^1$ and $\mathbf{C}^2$, respectively.

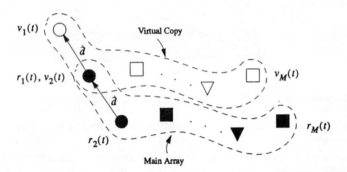

FIGURE 62.2   The main array and its virtual copy.

Note, also, that there exists a very close link between VC^3 and VESPA. Although the way we chose $\mathbf{C}^1$ and $\mathbf{C}^2$ above seems to be not very obvious, there is a unique geometric interpretation to it. According to VC^3, as far as the bearing information is concerned, $\mathbf{C}^1$ is equivalent to the autocorrelation matrix of the array, and $\mathbf{C}^2$ is equivalent to the cross-correlation matrix between the array and its virtual copy (which is created by displacing the array by the vector that connects the second and the first sensors).

If the noise component of the signal received by one of the guiding sensor pair elements is independent of the noises at the other sensors, VESPA suppresses the noise regardless of its distribution [6]. In practice, the noise does affect the standard deviations of results obtained from VESPA.

An iterative version of VESPA has also been developed for cases where the source powers have a high dynamic range [11]. Iterative VESPA has the same hardware requirements and assumptions as in VESPA.

Extended VESPA [10]: When there are coherent (or completely correlated) sources, all of the above second- and higher-order statistics methods, except for the WSF method and other multidimensional search-based approaches, fail. For the WSF and other multidimensional methods, however, the array must be calibrated accurately and the computational load is expensive. The coherent signals case arises in practice when there are multipaths. Porat and Friedlander present a modified version of their algorithm to handle the case of coherent signals; however, their method is not practical because it requires selection of a highly redundant subset of fourth-order cumulants that contains $O(N^4)$ elements, and no guidelines exist for its selection and 2nd-, 4th-, 6th-, and 8th-order moments of the data are required. If the array is *uniform linear*, coherence can be handled using spatial smoothing as a preprocessor to the usual second- or higher-order [3, 39] methods; however, the array aperture is reduced. Extended VESPA can handle coherence and provides increased aperture. Additionally, the array does not have to be completely uniform linear or calibrated; however, a uniform linear subarray is still needed. An example array configuration is shown in Figure 62.3.

Consider a scenario in which there are G statistically independent narrowband sources, $\{u_g(t)\}_{i=1}^G$. These source signals undergo multipath propagation, and each produces p_i coherent wavefronts

$$\{s_{1,1}, \cdots, s_{1,p_1}, \cdots, s_{G,1}, \cdots, s_{G,p_G}\} \ (\sum_{i=1}^{G} p_i = P)$$

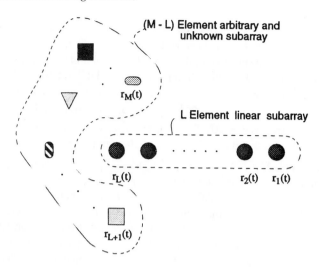

FIGURE 62.3 An example array configuration. There are M sensors, L of which are uniform linearly positioned; $r_1(t)$ and $r_2(t)$ are identical guiding sensors. Linear subarray elements are separated by Δ.

that impinge on an M element sensor array from directions

$$\{\theta_{1,1}, \cdots, \theta_{1,p_1}, \cdots, \theta_{G,1}, \cdots, \theta_{G,p_G}\},$$

where $\theta_{m,p}$ represents the angle-of-arrival of the wavefront $s_{g,p}$ that is the pth coherent signal in the gth group. The collection of p_i coherent wavefronts, which are scaled and delayed replicas of the ith source, are referred to as the ith *group*. The wavefronts are represented by the P-vector $\mathbf{s}(t)$. The problem is to estimate the DOAs $\{\theta_{1,1}, \cdots, \theta_{1,p_1}, \cdots, \theta_{G,1}, \cdots, \theta_{G,p_G}\}$.

When the multipath delays are insignificant compared to the bit durations of signals, then the signals received from different paths differ by only amplitude and phase shifts, thus the coherence among the received wavefronts can be expressed by the following equation:

$$\mathbf{s}(t) = \begin{bmatrix} \mathbf{s}_1(t) \\ \mathbf{s}_2(t) \\ \vdots \\ \mathbf{s}_G(t) \end{bmatrix} = \begin{bmatrix} \mathbf{c}_1 & \mathbf{0} & \cdots & \mathbf{0} \\ \mathbf{0} & \mathbf{c}_2 & \cdots & \mathbf{0} \\ \vdots & \vdots & \ddots & \vdots \\ \mathbf{0} & \mathbf{0} & \cdots & \mathbf{c}_G \end{bmatrix} \begin{bmatrix} u_1(t) \\ u_2(t) \\ \vdots \\ u_G(t) \end{bmatrix} \stackrel{\triangle}{=} \mathbf{Q}u(t) \qquad (62.2)$$

where $\mathbf{s}_i(t)$ is a $p_i \times 1$ signal vector representing the coherent wavefronts from the ith independent source $u_i(t)$, $\mathbf{c}_i$ is a $p_i \times 1$ complex attenuation vector for the ith source ($1 \leq i \leq G$), and $\mathbf{Q}$ is $P \times G$. The elements of $\mathbf{c}_i$ account for the attenuation and phase differences among the multipaths due to different arrival times. The received signal can then be written in terms of the independent sources as follows:

$$\mathbf{r}(t) = \mathbf{As}(t) + \mathbf{n(t)} = \mathbf{AQu}(t) + \mathbf{n(t)} = \mathbf{Bu}(t) + \mathbf{n(t)} \qquad (62.3)$$

where $\mathbf{B} \stackrel{\triangle}{=} \mathbf{AQ}$. The columns of $M \times G$ matrix $\mathbf{B}$ are known as the *generalized steering vectors*.

Extended VESPA has three major steps:

Step 1: Use Step (1) of VESPA by choosing $r_1(t)$ and $r_2(t)$ as any two sensor measurements. In this case $\mathbf{C}^1 = \mathbf{BGB}^H$ and $\mathbf{C}^2 = \mathbf{BCGB}^H$, where

$$\mathbf{G} \stackrel{\triangle}{=} diag(\gamma_{4,u_1}|b_{11}|^2, \cdots, \gamma_{4,u_G}|b_{1G}|^2), \{\gamma_{4,u_g}\}_{g=1}^G$$

$$\mathbf{C} \stackrel{\triangle}{=} diag(\frac{b_{21}}{b_{11}}, \cdots, \frac{b_{2G}}{b_{1G}}).$$

Due to the coherence, the DOAs cannot be obtained at this step from just C^1 and C^2 because the columns of B depend on a vector of DOAs (all those within a group). In the independent sources case, the columns of A depend only on a single DOA. Fortunately, the columns of B can be solved for as follows: (1.1) Follow Steps 2 through 5 of TLS ESPRIT by replacing R^{11} and R^{21} by C^1 and C^2, respectively, and using appropriate matrix dimensions; (1.2) determine the eigenvectors and eigenvalues of $-F_x F_y^{-1}$; Let the eigenvector and eigenvalue matrices of $-F_x F_y^{-1}$ be E and D, respectively; and, (1.3) obtain an estimate of B to within a diagonal matrix, as $B = \left(U_{11} E + U_{12} E D^{-1} \right)/2$, for use in Step 2.

Step 2: Partition the matrices B and A as $B = [b_1, \cdots, b_G]$ and $A = [A_1, \cdots, A_G]$, where the steering vector for the ith group b_i is $M \times 1$, $A_i \triangleq [a(\theta_{i,1}), \cdots, a(\theta_{i,p_i})]$ is $M \times p_i$, and $\theta_{i,m}$ is the angle-of-arrival of the mth source in the ith coherent group $(1 \leq m \leq p_i)$. Using the fact that the ith column of Q has p_i nonzero elements, express B as $B = AQ = [A_1 c_1, \cdots, A_G c_G]$; therefore, the ith column of B, b_i, is $b_i = A_i c_i$ where $i = 1, \cdots, G$. Now, the problem of solving for the steering vectors is transformed into the problem of solving for the steering vectors from *each* coherent group *separately*. To solve this new problem, each generalized steering vector b_i can be interpreted as a received signal for an array illuminated by p_i coherent signals having a steering matrix A_i, and covariance matrix $c_i c_i^H$. The DOAs could then be solved for by using a second-order-statistics-based high-resolution method such as MUSIC, if the array was calibrated, and the rank of $c_i c_i^H$ was p_i; however, the array is not calibrated and $rank(c_i c_i^H) = 1$. The solution is to keep the portion of each b_i that corresponds to the uniform linear part of the array, $b_{L,i}$, and to then apply the Section 62.3.3 spatial smoothing technique to a pseudocovariance matrix $b_{L,i} b_{L,i}^H$ for $i = 1, \cdots, G$. Doing this *restores* the rank of $c_i c_i^H$ to p_i. In the Section 62.3.3 spatial smoothing technique, we must replace $r(t)$ by $b_{L,i}$ and set $N = 1$.

The conditions on the length of the linear subarray and the parameter S under which the rank of $b_{S,i} b_{S,i}^H$ is restored to p_i are [11]: (a) $L \geq 3p_i/2$, which means that the linear subarray must have at least $3p_{max}/2$ elements, where p_{max} is the maximum number of multipaths in anyone of the G groups; and (b) given L and p_{max}, the parameter S must be selected such that $p_{max} + 1 \leq S \leq L - p_{max}/2 + 1$.

Step 3: Apply any second-order-statistics-based subspace technique (e.g., root-MUSIC, etc.) to R_i^{fb} $(i = 1, \cdots, G)$ to estimate DOAs of up to $2L/3$ coherent signals in each group.

Note that the matrices C and G in C^1 and C^2 are not used; however, if the received signals are independent, choosing $r_1(t)$ and $r_2(t)$ from the linear subarray lets DOA estimates be obtained from C in Step 1 because, in that case,

$$C = diag\{e^{-j2\pi \frac{d}{\lambda} \sin \theta_1}, \cdots, e^{-j2\pi \frac{d}{\lambda} \sin \theta_P}\};$$

hence, extended VESPA can also be applied to the case of independent sources.

62.4.1 Discussion

One advantage of using higher-order statistics-based methods over second-order methods is that the covariance matrix of the noise is not needed when the noise is Gaussian. The fact that higher-order statistics have more arguments than covariances leads to more practical algorithms that have less restrictions on the array structure (for instance, the requirement of maintaining identical arrays for ESPRIT is reduced to only maintaining two identical sensors for VESPA). Another advantage is more sources than sensors can be detected, i.e., the array aperture is increased when higher-order statistics are properly applied; or, depending on the array geometry, unreliable sensor measurements can be replaced by using the VC^3 idea. One disadvantage of using higher-order statistics-based methods is that sample estimates of higher-order statistics require longer data lengths than covariances; hence, computational complexity is increased. In their recent study, Cardoso and Moulines [2] present a comparative performance analysis of second- and fourth-order statistics-based MUSIC methods. Their results indicate that dynamic range of the sources may be a factor limiting the performance of the fourth-order statistics-based MUSIC. A comprehensive performance analysis of the above higher-order statistical methods is still lacking; therefore, a detailed comparison of these methods remains as a very important research topic.

62.5 Flowchart Comparison of Subspace-Based Methods

Clearly, there are many subspace-based direction finding methods. In order to see the forest from the trees, to know when to use a second-order or a higher-order statistics-based method, we present Figs. 62.4 through 62.9. These figures provide a comprehensive summary of the existing subspace-based methods for direction finding and constitute guidelines to selection of a proper direction-finding method for a given application.

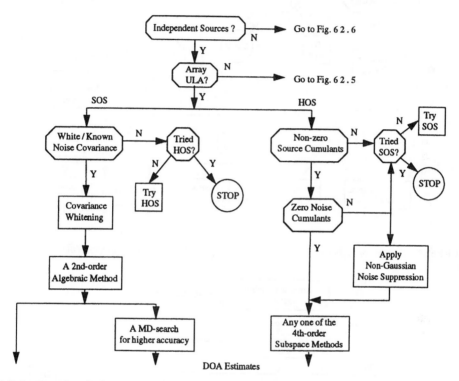

FIGURE 62.4 Second- or higher-order statistics-based subspace DF algorithm. Independent sources and ULA.

Note that: Fig. 62.4 depicts independent sources and ULA, Fig. 62.5 depicts independent sources and NL/mixed array, Fig. 62.6 depicts coherent and correlated sources and ULA, and Fig. 62.7 depicts coherent and correlated sources and NL/mixed array.

All four figures show two paths: SOS (second-order statistics) and HOS (higher-order statistics). Each path terminates in one or more method boxes, each of which may contain a multitude of methods. Figures 62.8 and 62.9 summarize the pros and cons of all the methods we have considered in this chapter.

Using Fig. 62.4 through 62.9, it is possible for a potential user of a subspace-based direction finding method to decide which method(s) is (are) most likely to give best results for his/her application.

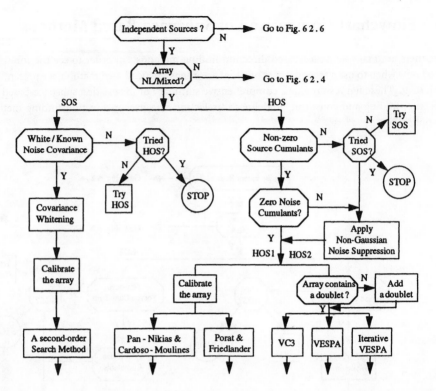

FIGURE 62.5   Second- or higher-order statistics-based subspace DF algorithm. Independent sources and NL/mixed array.

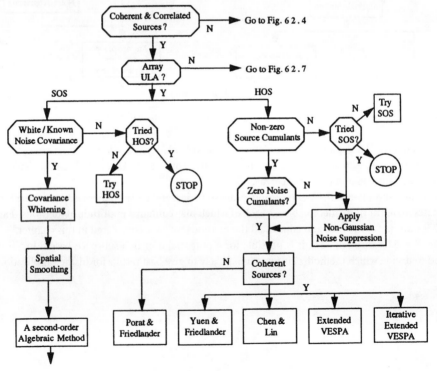

FIGURE 62.6   Second- or higher-order statistics-based subspace DF algorithms. Coherent and correlated sources and ULA.

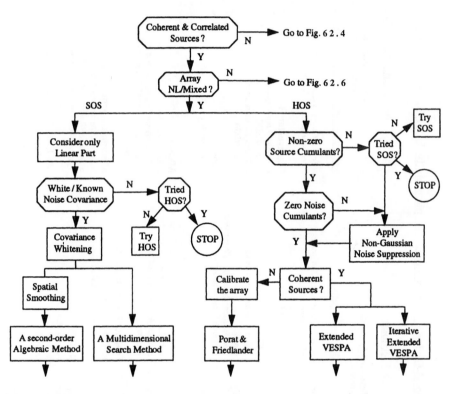

FIGURE 62.7 Second- or higher-order statistics-based subspace DF algorithms. Coherent and correlated sources and NL/mixed array.

Second-Order Statistics based Subspace Methods for Direction Finding

Signal Subspace Methods

> SNR is enhanced effectively by retaining the signal subspace only

Search Based Methods

> Select if array is calibrated or response is known analytically

Correlogram
> Lower resolution than MV and AR

Minimum Variance (MV)
> Narrower mainlobe and smoother sidelobes than conventional beamformers
> Higher resolution than Correlogram
> Lower resolution than AR
> Lower variance than AR

Autoregressive (AR)
> Higher resolution than MV and Correlogram

Subspace Fitting (SF)
> Weighted SF works regardless of source correlation, and has the same asymptotic properties as the stochastic ML method, i.e., it achieves CRB.
> Requires accurate calibration of the manifold and its derivative with respect to arrival angle

Algebraic Methods

> Select if the array is ULA or its identical copy exists
> Computationally simpler than search-based methods.

ESPRIT
> Select if the array has an identical copy
> Computationally simple as compared to search based methods
> Sensitive to perturbations in the sensor response and array geometry
> LS and TLS versions are best. They have the same asymptotic performance, but TLS converges faster and is better than LS for low SNR and short data lengths

Toeplitz Approximation Method (TAM)
> Equivalent to LS-ESPRIT

GEESE
> Better than ESPRIT

Noise Subspace Methods

> Methods are based on the orthogonality of steering vectors and noise subspace eigenvectors

Search Based Methods

> Select if array is calibrated or response is known analytically

Eigenvector (EV)
> Produces fewer spurious peaks than MUSIC
> Shapes the noise spectrum better than MUSIC

Pisarenko
> Performance with short data is poor

MUSIC
> Better than MV
> Same asymptotic performance as the deterministic ML for uncorrelated sources

Minimum Norm
> Select if the array is ULA
> Lower SNR threshold than MUSIC for resolution of closely spaced sources

Method of Direction Estimation (MODE)
> Consistent for ergodic and stationary signals

Algebraic Methods

> Select if the array is ULA
> Algebraic versions of EV, Pisarenko, MUSIC and Minimum Norm are possible
> Better resolution than search-based versions

Root MUSIC
> Lower SNR threshold than MUSIC for resolution of closely spaced sources
> Simple root-finding procedure

FIGURE 62.8   Pros and cons of all the methods considered.

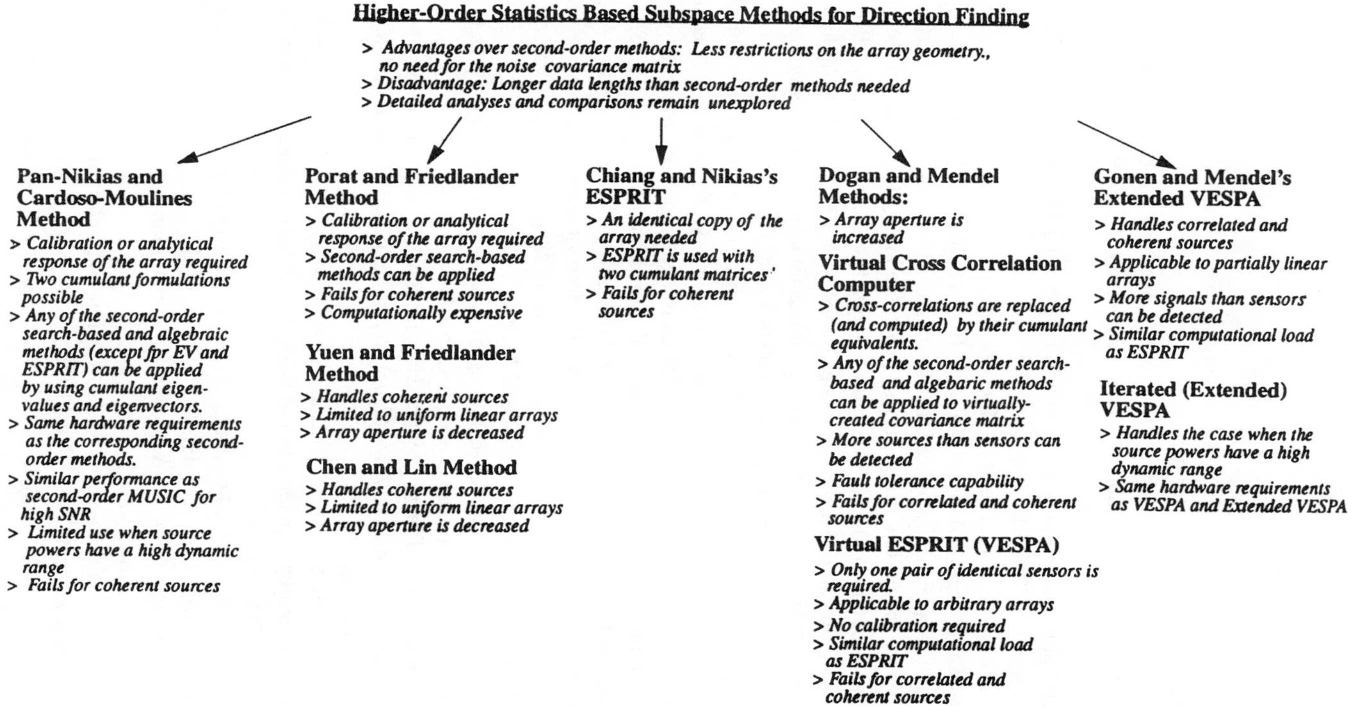

Higher-Order Statistics Based Subspace Methods for Direction Finding

> Advantages over second-order methods: Less restrictions on the array geometry,, no need for the noise covariance matrix
> Disadvantage: Longer data lengths than second-order methods needed
> Detailed analyses and comparisons remain unexplored

Pan-Nikias and Cardoso-Moulines Method

> Calibration or analytical response of the array required
> Two cumulant formulations possible
> Any of the second-order search-based and algebraic methods (except for EV and ESPRIT) can be applied by using cumulant eigenvalues and eigenvectors.
> Same hardware requirements as the corresponding second-order methods.
> Similar performance as second-order MUSIC for high SNR
> Limited use when source powers have a high dynamic range
> Fails for coherent sources

Porat and Friedlander Method

> Calibration or analytical response of the array required
> Second-order search-based methods can be applied
> Fails for coherent sources
> Computationally expensive

Yuen and Friedlander Method

> Handles coherent sources
> Limited to uniform linear arrays
> Array aperture is decreased

Chen and Lin Method

> Handles coherent sources
> Limited to uniform linear arrays
> Array aperture is decreased

Chiang and Nikias's ESPRIT

> An identical copy of the array needed
> ESPRIT is used with two cumulant matrices'
> Fails for coherent sources

Dogan and Mendel Methods:

> Array aperture is increased

Virtual Cross Correlation Computer

> Cross-correlations are replaced (and computed) by their cumulant equivalents.
> Any of the second-order search-based and algebraic methods can be applied to virtually-created covariance matrix
> More sources than sensors can be detected
> Fault tolerance capability
> Fails for correlated and coherent sources

Virtual ESPRIT (VESPA)

> Only one pair of identical sensors is required.
> Applicable to arbitrary arrays
> No calibration required
> Similar computational load as ESPRIT
> Fails for correlated and coherent sources

Gonen and Mendel's Extended VESPA

> Handles correlated and coherent sources
> Applicable to partially linear arrays
> More signals than sensors can be detected
> Similar computational load as ESPRIT

Iterated (Extended) VESPA

> Handles the case when the source powers have a high dynamic range
> Same hardware requirements as VESPA and Extended VESPA

FIGURE 62.9 Pros and cons of all the methods considered.

Acknowledgments

The authors would like to thank Profs. A. Paulraj, V.U. Reddy, and M. Kaveh for reviewing the manuscript.

References

[1] Capon, J., High-resolution frequency-wavenumber spectral analysis, *Proc. IEEE,* 57(8), 1408–1418, Aug. 1969.

[2] Cardoso, J.-F. and Moulines, E., Asymptotic performance analysis of direction-finding algorithms based on fourth-order cumulants, *IEEE Trans. on Signal Processing,* 43(1), 214–224, Jan. 1995.

[3] Chen, Y.H. and Lin, Y.S., A modified cumulant matrix for DOA estimation, *IEEE Trans. on Signal Processing,* 42, 3287–3291, Nov. 1994.

[4] Chiang, H.H. and Nikias, C.L., The ESPRIT algorithm with higher-order statistics, *Proc. Workshop on Higher-Order Spectral Analysis,* Vail, CO, 163–168, June 28-30, 1989.

[5] Dogan, M.C. and Mendel, J.M., Applications of cumulants to array processing, Part I: Aperture extension and array calibration, *IEEE Trans. on Signal Processing,* 43(5), 1200–1216, May 1995.

[6] Dogan, M.C. and Mendel, J.M., Applications of cumulants to array processing, Part II: Non-Gaussian noise suppression, *IEEE Trans. on Signal Processing,* 43(7), 1661–1676, July 1995.

[7] Dogan, M.C. and Mendel, J.M., Method and apparatus for signal analysis employing a virtual cross-correlation computer, U.S. Patent No. 5,459,668, Oct. 17, 1995.

[8] Ephraim, T., Merhav, N. and Van Trees, H.L., Min-norm interpretations and consistency of MUSIC, MODE and ML, *IEEE Trans. on Signal Processing,* 43(12), 2937–2941, Dec. 1995.

[9] Evans, J.E., Johnson, J.R. and Sun, D.F., High resolution angular spectrum estimation techniques for terrain scattering analysis and angle of arrival estimation, in *Proc. First ASSP Workshop Spectral Estimation,* Communication Research Laboratory, McMaster University, Aug. 1981.

[10] Gönen, E., Dogan, M.C. and Mendel, J.M., Applications of cumulants to array processing: direction finding in coherent signal environment, *Proc. of 28th Asilomar Conference on Signals, Systems, and Computers,* Asilomar, CA, 633–637, 1994.

[11] Gönen, E., Cumulants and subspace techniques for array signal processing, Ph.D. thesis, University of Southern California, Los Angeles, CA, Dec. 1996.

[12] Haykin, S.S., *Adaptive Filter Theory,* Prentice-Hall, Englewood Cliffs, NJ, 1991.

[13] Johnson, D.H. and Dudgeon, D.E., *Array Signal Processing: Concepts and Techniques,* Prentice-Hall, Englewood Cliffs, NJ, 1993.

[14] Kaveh, M. and Barabell, A.J., The statistical performance of the MUSIC and the Minimum-Norm algorithms in resolving plane waves in noise, *IEEE Trans. on Acoustics, Speech and Signal Processing,* 34, 331–341, Apr. 1986.

[15] Kumaresan, R. and Tufts, D.W., Estimating the angles of arrival multiple plane waves, *IEEE Trans. on Aerosp. Electron. Syst.,* AES-19, 134-139, Jan. 1983.

[16] Kung, S.Y., Lo, C.K. and Foka, R., A Toeplitz approximation approach to coherent source direction finding, *Proc. ICASSP,* 1986.

[17] Li, F. and Vaccaro, R.J., Unified analysis for DOA estimation algorithms in array signal processing, *Signal Processing,* 25(2), 147–169, Nov. 1991.

[18] Marple, S.L., *Digital Spectral Analysis with Applications,* Prentice-Hall, Englewood Cliffs, NJ, 1987.

[19] Mendel, J.M., Tutorial on higher-order statistics (spectra) in signal processing and system theory: theoretical results and some applications, *Proc. IEEE,* 79(3), 278–305, March 1991.

[20] Nikias, C.L. and Petropulu, A.P., *Higher-Order Spectra Analysis: A Nonlinear Signal Processing Framework,* Prentice-Hall, Englewood Cliffs, NJ, 1993.

[21] Ottersten, B., Viberg, M. and Kailath, T., Performance analysis of total least squares ESPRIT algorithm, *IEEE Trans. on Signal Processing*, 39(5), 1122–1135, May 1991.

[22] Pan, R. and Nikias, C.L., Harmonic decomposition methods in cumulant domains, *Proc. ICASSP'88*, New York, 2356–2359, 1988.

[23] Paulraj, A., Roy, R. and Kailath, T., Estimation of signal parameters via rotational invariance techniques-ESPRIT, *Proc. 19th Asilomar Conf. on Signals, Systems, and Computers*, Asilomar, CA, Nov. 1985.

[24] Pillai, S.U., *Array Signal Processing*, Springer-Verlag, New York, 1989.

[25] Porat, B. and Friedlander, B., Direction finding algorithms based on high-order statistics, *IEEE Trans. on Signal Processing*, 39(9), 2016–2023, Sept. 1991.

[26] Radich, B.M. and Buckley, K., The effect of source number underestimation on MUSIC location estimates, 42(1), 233–235, Jan. 1994.

[27] Rao, D.V.B. and Hari, K.V.S., Performance analysis of Root-MUSIC, *IEEE Trans. on Acoustics, Speech, Signal Processing*, ASSP-37, 1939–1949, Dec. 1989.

[28] Roy, R.H., ESPRIT-Estimation of signal parameters via rotational invariance techniques, Ph.D. dissertation, Stanford University, Stanford, CA, 1987.

[29] Schmidt, R.O., A signal subspace approach to multiple emitter location and spectral estimation, Ph.D. dissertation, Stanford University, Stanford, CA, Nov. 1981.

[30] Shamsunder, S. and Giannakis, G.B., Detection and parameter estimation of multiple non-Gaussian sources via higher order statistics, *IEEE Trans. on Signal Processing*, 42, 1145–1155, May 1994.

[31] Shan, T.J., Wax, M. and Kailath, T., On spatial smoothing for direction-of-arrival estimation of coherent signals, *IEEE Trans. on Acoustics, Speech, Signal Processing*, ASSP-33(2), 806–811, Aug. 1985.

[32] Stoica, P. and Nehorai, A., MUSIC, maximum likelihood and Cramer-Rao bound: Further results and comparisons, *IEEE Trans. on Signal Processing*, 38, 2140–2150, Dec. 1990.

[33] Swindlehurst, A.L. and Kailath, T., A performance analysis of subspace-based methods in the presence of model errors, Part 1: the MUSIC algorithm, *IEEE Trans. on Signal Processing*, 40(7), 1758–1774, July 1992.

[34] Viberg, M., Ottersten, B. and Kailath, T., Detection and estimation in sensor arrays using weighted subspace fitting, *IEEE Trans. on Signal Processing*, 39(11), 2436–2448, Nov. 1991.

[35] Viberg, M. and Ottersten, B., Sensor array processing based on subspace fitting, *IEEE Trans. on Signal Processing*, 39(5), 1110–1120, May 1991.

[36] Wax, M. and Kailath, T., Detection of signals by information theoretic criteria, *IEEE Trans. on Acoustics, Speech, Signal Processing*, ASSP-33(2), 387–392, Apr. 1985.

[37] Wax, M., Detection and estimation of superimposed signals, Ph.D. dissertation, Stanford University, Stanford, CA, Mar. 1985.

[38] Xu, X.-L. and Buckley, K., Bias and variance of direction-of-arrival estimates from MUSIC, MIN-NORM and FINE, *IEEE Trans. on Signal Processing*, 42(7), 1812–1816, July 1994.

[39] Yuen, N. and Friedlander, B., DOA estimation in multipath based on fourth-order cumulants, in *Proc. IEEE Signal Processing ATHOS Workshop on Higher-Order Statistics*, 71–75, June 1995.

63

ESPRIT and Closed-Form 2-D Angle Estimation with Planar Arrays

Martin Haardt
Siemens AG
Mobile Radio Networks

Michael D. Zoltowski
Purdue University

Cherian P. Mathews
University of West Florida

Javier Ramos
Polytechnic University of Madrid

63.1 Introduction

Estimating the directions of arrival (DOAs) of propagating plane waves is a requirement in a variety of applications including radar, mobile communications, sonar, and seismology. Due to its simplicity and high-resolution capability, *ESPRIT (Estimation of Signal Parameters via Rotational Invariance Techniques)* [18] has become one of the most popular signal subspace-based DOA or spatial frequency estimation schemes. ESPRIT is explicitly premised on a point source model for the sources and is restricted to use with array geometries that exhibit so-called invariances [18]. However, this requirement is not very restrictive as many of the common array geometries used in practice exhibit these invariances, or their output may be transformed to effect these invariances.

ESPRIT may be viewed as a complement to the MUSIC algorithm, the forerunner of all signal subspace-based DOA methods, in that it is based on properties of the signal eigenvectors whereas MUSIC is based on properties of the noise eigenvectors. This chapter concentrates solely on the use of ESPRIT to estimate the DOAs of plane waves incident upon an antenna array. It should be noted, though, that ESPRIT may be used in the dual problem of estimating the frequencies of sinusoids embedded in a time series [18]. In this application, ESPRIT is more generally applicable than MUSIC as it can handle damped sinusoids and provides estimates of the damping factors as well as the constituent frequencies. The standard ESPRIT

0-8493-8572-5/98/$0.00+$.50
© 1998 by CRC Press LLC

algorithm for one-dimensional (1-D) arrays is reviewed in Section 63.2. There are three primary steps in any ESPRIT-type algorithm:

1. **Signal Subspace Estimation** computation of a basis for the estimated signal subspace,

2. **Solution of the Invariance Equation** solution of an (in general) overdetermined system of equations, the so-called invariance equation, derived from the basis matrix estimated in Step 1, and

3. **Spatial Frequency Estimation** computation of the eigenvalues of the solution of the invariance equation formed in Step 2.

Many antenna arrays used in practice have geometries that possess some form of symmetry. For example, a linear array of equi-spaced identical antennas is symmetric about the center of the linear aperture it occupies. In Section 63.3.1, an efficient implementation of ESPRIT is presented that exploits the symmetry present in so-called centro-symmetric arrays to formulate the three steps of ESPRIT in terms of real-valued computations, despite the fact that the input to the algorithm needs to be the complex analytic signal output from each antenna. This reduces the computational complexity significantly. A reduced dimension beamspace version of ESPRIT is developed in Section 63.3.2. Advantages to working in beamspace include reduced computational complexity [3], decreased sensitivity to array imperfections [1], and lower SNR resolution thresholds [11].

With a 1-D array, one can only estimate the angle of each incident plane wave relative to the array axis. For source localization purposes, this only places the source on a cone whose axis of symmetry is the array axis. The use of a 2-D or planar array enables one to passively estimate the 2-D arrival angles of each emitting source. The remainder of the chapter presents ESPRIT-based techniques for use in conjunction with circular and rectangular arrays that provide estimates of the azimuth and elevation angle of each incident signal. As in the 1-D case, the symmetries present in these array geometries are exploited to formulate the three primary steps of ESPRIT in terms of real-valued computations.

63.1.1 Notation

Throughout this chapter, column vectors and matrices are denoted by lower case and upper case boldfaced letters, respectively. For any positive integer p, $\boldsymbol{I}_p$ is the $p \times p$ identity matrix and $\boldsymbol{\Pi}_p$ the $p \times p$ exchange matrix with ones on its antidiagonal and zeros elsewhere,

$$\boldsymbol{\Pi}_p = \begin{bmatrix} & & 1 \\ & 1 & \\ & \cdot & \\ 1 & & \end{bmatrix} \in \mathbb{R}^{p \times p}. \tag{63.1}$$

Pre-multiplication of a matrix by $\boldsymbol{\Pi}_p$ will reverse the order of its rows, while post-multiplication of a matrix by $\boldsymbol{\Pi}_p$ reverses the order of its columns. Furthermore, the superscripts $(\cdot)^H$ and $(\cdot)^T$ denote complex conjugate transposition and transposition without complex conjugation, respectively. Complex conjugation by itself is denoted by an overbar $\overline{(\cdot)}$, such that $X^H = \overline{X}^T$. A diagonal matrix $\boldsymbol{\Phi}$ with the diagonal elements $\phi_1, \phi_2, \ldots, \phi_d$ may be written as

$$\boldsymbol{\Phi} = \text{diag}\,\{\phi_i\}_{i=1}^d = \begin{bmatrix} \phi_1 & & & \\ & \phi_2 & & \\ & & \cdot & \\ & & & \phi_d \end{bmatrix} \in \mathbb{C}^{d \times d}.$$

Moreover, matrices $\boldsymbol{Q} \in \mathbb{C}^{p \times q}$ satisfying

$$\boldsymbol{\Pi}_p \overline{\boldsymbol{Q}} = \boldsymbol{Q} \tag{63.2}$$

will be called left $\boldsymbol{\Pi}$-real [10]. Often left $\boldsymbol{\Pi}$-real matrices are also called conjugate centro-symmetric [24].

63.2 The Standard ESPRIT Algorithm

The algorithm ESPRIT [18] must be used in conjunction with an M-element sensor array composed of m pairs of pairwise identical, but displaced, sensors (doublets) as depicted in Fig. 63.1. If the subarrays do not overlap, i.e., if they do not share any elements, $M = 2m$, but in general $M \leq 2m$ since overlapping subarrays are allowed, cf. Fig. 63.2. Let Δ denote the distance between the two subarrays. Incident on both subarrays

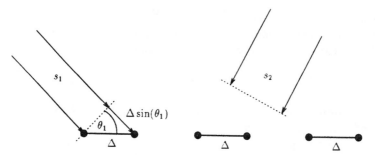

FIGURE 63.1 Planar array composed of $m = 3$ pairwise identical, but displaced, sensors (doublets).

are d narrowband noncoherent[1] planar wavefronts with distinct directions of arrival (DOAs) θ_i, $1 \leq i \leq d$, relative to the displacement between the two subarrays.[2] Their complex pre-envelope at an arbitrary reference point may be expressed as $s_i(t) = \alpha_i(t)e^{j(2\pi f_c t + \beta_i(t))}$, where f_c denotes the common carrier frequency of the d wavefronts. Without loss of generality, we assume that the reference point is the array centroid. The signals are called *narrowband* if their amplitudes $\alpha_i(t)$ and phases $\beta_i(t)$ vary slowly with respect to the propagation time across the array τ, i.e., if

$$\alpha_i(t - \tau) \approx \alpha_i(t) \quad \text{and} \quad \beta_i(t - \tau) \approx \beta_i(t). \tag{63.3}$$

In other words, the narrowband assumption allows the time-delay of the signals across the array τ to be modeled as a simple phase shift of the carrier frequency, such that

$$s_i(t - \tau) \approx \alpha_i(t)e^{j(2\pi f_c(t-\tau) + \beta_i(t))} = e^{-j2\pi f_c \tau}s_i(t).$$

Figure 63.1 shows that the propagation delay of a plane wave signal between the two identical sensors of a doublet equals $\tau_i = \frac{\Delta \sin \theta_i}{c}$, where c denotes the signal propagation velocity. Due to the narrowband assumption (63.3), this propagation delay τ_i corresponds to the multiplication of the complex envelope signal by the complex exponential $e^{j\mu_i}$, referred to as the phase factor, such that

$$s_i(t - \tau_i) = e^{-j\frac{2\pi f_c}{c}\Delta \sin \theta_i}s_i(t) = e^{j\mu_i}s_i(t), \tag{63.4}$$

where the *spatial frequencies* μ_i are given by $\mu_i = -\frac{2\pi}{\lambda}\Delta \sin \theta_i$. Here, $\lambda = \frac{c}{f_c}$ denotes the common wavelength of the signals. We also assume that there is a one-to-one correspondence between the spatial frequencies $-\pi < \mu_i < \pi$ and the range of possible DOAs. Thus, the maximum range is achieved for $\Delta \leq \lambda/2$. In this case, the DOAs are restricted to the interval $-90° < \theta_i < 90°$ to avoid ambiguities.

---

[1] This restriction can be modified later as Unitary ESPRIT can estimate the directions of arrival of two coherent wavefronts due to an inherent forward-backward averaging effect. Two wavefronts are called *coherent* if their cross-correlation coefficient has magnitude one. The directions of arrival of *more than two* coherent wavefronts can be estimated by using spatial smoothing as a preprocessing step.
[2] $\theta_k = 0$ corresponds to the direction perpendicular to Δ.

In the sequel, the d impinging signals $s_i(t)$, $1 \le i \le d$, are combined to a column vector $s(t)$. Then the noise-corrupted measurements taken at the M sensors at time t obey the linear model

$$x(t) = \begin{bmatrix} a(\mu_1) & a(\mu_2) & \cdots & a(\mu_d) \end{bmatrix} \begin{bmatrix} s_1(t) \\ s_2(t) \\ \vdots \\ s_d(t) \end{bmatrix} + n(t) = As(t) + n(t) \in \mathbb{C}^M, \qquad (63.5)$$

where the columns of the array steering matrix $A \in \mathbb{C}^{M \times d}$, the array response or array steering vectors $a(\mu_i)$, are functions of the unknown spatial frequencies μ_i, $1 \le i \le d$. For example, for a uniform linear array (ULA) of M identical omnidirectional antennas,

$$a(\mu_i) = e^{-j\left(\frac{M-1}{2}\right)\mu_i} \begin{bmatrix} 1 & e^{j\mu_i} & e^{j2\mu_i} & \cdots & e^{j(M-1)\mu_i} \end{bmatrix}^T, \quad 1 \le i \le d.$$

Moreover, the additive noise vector $n(t)$ is taken from a zero-mean, spatially uncorrelated random process with variance σ_N^2, which is also uncorrelated with the signals. Since every row of A corresponds to an element of the sensor array, a particular subarray configuration can be described by two selection matrices, each choosing m elements of $x(t) \in \mathbb{C}^M$, where m, $d \le m < M$, is the number of elements in each subarray. Figure 63.2, for example, displays the appropriate subarray choices for three centro-symmetric arrays of $M = 6$ identical sensors.

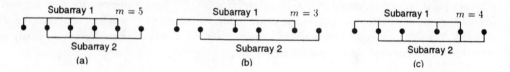

FIGURE 63.2   Three centro-symmetric line arrays of $M = 6$ identical sensors and the corresponding subarrays required for ESPRIT-type algorithms.

In case of a ULA with maximum overlap, cf. Figure 63.2 (a), J_1 picks the first $m = M - 1$ rows of A, while J_2 selects the last $m = M - 1$ rows of the array steering matrix. In this case, the corresponding selection matrices are given by

$$J_1 = \begin{bmatrix} 1 & 0 & 0 & \cdots & 0 & 0 \\ 0 & 1 & 0 & \cdots & 0 & 0 \\ \vdots & \vdots & \vdots & \ddots & \vdots & \vdots \\ 0 & 0 & 0 & \cdots & 1 & 0 \end{bmatrix} \in \mathbb{R}^{m \times M} \quad \text{and} \quad J_2 = \begin{bmatrix} 0 & 1 & 0 & \cdots & 0 & 0 \\ 0 & 0 & 1 & \cdots & 0 & 0 \\ \vdots & \vdots & \vdots & \ddots & \vdots & \vdots \\ 0 & 0 & 0 & \cdots & 0 & 1 \end{bmatrix} \in \mathbb{R}^{m \times M}.$$

Notice that J_1 and J_2 are centro-symmetric with respect to one another, i.e., they obey $J_2 = \Pi_m J_1 \Pi_M$. This property holds for all centro-symmetric arrays and plays a key role in the derivation of Unitary ESPRIT [7]. Since we have two identical, but physically displaced subarrays, Eq. (63.4) indicates that an array steering vector of the *second* subarray $J_2 a(\mu_i)$ is just a scaled version of the corresponding array steering vector of the *first* subarray $J_1 a(\mu_i)$, namely

$$J_1 a(\mu_i) e^{j\mu_i} = J_2 a(\mu_i), \quad 1 \le i \le d. \qquad (63.6)$$

This *shift invariance property* of all d array steering vectors $a(\mu_i)$ may be expressed in compact form as

$$J_1 A \Phi = J_2 A, \quad \text{where} \quad \Phi = \text{diag}\{e^{j\mu_i}\}_{i=1}^{d} \qquad (63.7)$$

is the unitary diagonal $d \times d$ matrix of the phase factors. All ESPRIT-type algorithms are based on this invariance property of the array steering matrix A, where A is assumed to have full column rank d.

Let X denote an $M \times N$ complex data matrix composed of N snapshots $x(t_n), 1 \leq n \leq N$,

$$
\begin{aligned}
X &= \begin{bmatrix} x(t_1) & x(t_2) & \cdots & x(t_N) \end{bmatrix} \\
&= A \begin{bmatrix} s(t_1) & s(t_2) & \cdots & s(t_N) \end{bmatrix} + \begin{bmatrix} n(t_1) & n(t_2) & \cdots & n(t_N) \end{bmatrix} \\
&= A \cdot S + N \in \mathbb{C}^{M \times N}.
\end{aligned}
\tag{63.8}
$$

The starting point is a singular value decomposition (SVD) of the noise-corrupted data matrix X (direct data approach). Assume that $U_s \in \mathbb{C}^{M \times d}$ contains the d left singular vectors corresponding to the d largest singular values of X. Alternatively, U_s can be obtained via an eigendecomposition of the (scaled) sample covariance matrix XX^H (covariance approach). Then, $U_s \in \mathbb{C}^{M \times d}$ contains the d eigenvectors corresponding to the d largest eigenvalues of XX^H.

Asymptotically, i.e., as the number of snapshots N becomes infinitely large, the range space of U_s is the d-dimensional range space of the array steering matrix A referred to as the *signal subspace*. Therefore, there exists a nonsingular $d \times d$ matrix T such that $A \approx U_s T$. Let us express the shift-invariance property (63.7) in terms of the matrix U_s that spans the estimated signal subspace,

$$
J_1 U_s T \Phi \approx J_2 U_s T \iff J_1 U_s \Psi \approx J_2 U_s, \quad \text{where} \quad \Psi = T \Phi T^{-1}
$$

is a nonsingular $d \times d$ matrix. Since Φ in Eq. (63.7) is diagonal, $T \Phi T^{-1}$ is in the form of an eigenvalue decomposition. This implies that $e^{j\mu_i}, 1 \leq i \leq d$, are the eigenvalues of Ψ. These observations form the basis for the subsequent steps of the algorithm. By applying the two selection matrices to the signal subspace matrix, the following (in general) overdetermined set of equations is formed,

$$
J_1 U_s \Psi \approx J_2 U_s \in \mathbb{C}^{m \times d}.
\tag{63.9}
$$

This set of equations, the so-called invariance equation, is usually solved in the least squares (LS) or total least squares (TLS) sense. Notice, however, that Eq. (63.9) is highly structured if overlapping subarray configurations are used. Structured least squares (SLS) is a new algorithm to solve the invariance equation by preserving its structure [8]. Formally, SLS was derived as a linearized iterative solution of a nonlinear optimization problem. If SLS is initialized with the LS solution of the invariance equation, only one "iteration", i.e., the solution of one linear system of equations, is required to achieve a significant improvement of the estimation accuracy [8].

Then an eigendecomposition of the resulting solution $\Psi \in \mathbb{C}^{d \times d}$ may be expressed as

$$
\Psi = T \Phi T^{-1} \quad \text{with} \quad \Phi = \text{diag} \{\phi_i\}_{i=1}^{d}.
\tag{63.10}
$$

The eigenvalues ϕ_i, i.e., the diagonal elements of Φ, represent estimates of the phase factors $e^{j\mu_i}$. Notice that the ϕ_i are not guaranteed to be on the unit circle. Notwithstanding, estimates of the spatial frequencies μ_i and the corresponding DOAs θ_i are obtained via the relationships,

$$
\mu_i = \arg(\phi_i) \quad \text{and} \quad \theta_i = -\frac{\lambda}{2\pi\Delta} \arcsin(\mu_i), \quad 1 \leq i \leq d.
\tag{63.11}
$$

To end this section, a brief summary of the standard ESPRIT algorithm is given in Table 63.1.

63.3 1-D Unitary ESPRIT

In contrast to the standard ESPRIT algorithm, Unitary ESPRIT is efficiently formulated in terms of real-valued computations throughout [7]. It is applicable to centro-symmetric array configurations that possess

TABLE 63.1   Summary of the Standard ESPRIT Algorithm

1. *Signal Subspace Estimation:* Compute $U_s \in \mathbb{C}^{M \times d}$ as the d dominant left singular vectors of $X \in \mathbb{C}^{M \times N}$.

2. *Solution of the Invariance Equation:* Solve

$$\underbrace{J_1 U_s}_{\mathbb{C}^{m \times d}} \Psi \approx \underbrace{J_2 U_s}_{\mathbb{C}^{m \times d}}$$

   by means of LS, TLS, or SLS.

3. *Spatial Frequency Estimation:* Calculate the eigenvalues of the resulting complex-valued solution

$$\Psi = T \, \Phi \, T^{-1} \in \mathbb{C}^{d \times d} \quad \text{with} \quad \Phi = \text{diag} \{\phi_i\}_{i=1}^d$$

   - $\mu_i = \arg(\phi_i), \qquad 1 \le i \le d$

the discussed invariance structure, cf. Figs. 63.1 and 63.2. A sensor array is called *centro-symmetric* [23] if its element locations are symmetric with respect to the centroid. If the sensor elements have identical radiation characteristics, the array steering matrix of a centro-symmetric array satisfies

$$\Pi_M \overline{A} = A, \tag{63.12}$$

since the array centroid is chosen as the phase reference.

63.3.1 1-D Unitary ESPRIT in Element Space

Before presenting an efficient element space implementation of Unitary ESPRIT, let us define the sparse unitary matrices

$$Q_{2n} = \frac{1}{\sqrt{2}} \begin{bmatrix} I_n & j I_n \\ \Pi_n & -j \Pi_n \end{bmatrix} \quad \text{and} \quad Q_{2n+1} = \frac{1}{\sqrt{2}} \begin{bmatrix} I_n & 0 & j I_n \\ 0^T & \sqrt{2} & 0^T \\ \Pi_n & 0 & -j \Pi_n \end{bmatrix}. \tag{63.13}$$

They are left Π-real matrices of even and odd order, respectively.

Since Unitary ESPRIT involves forward-backward averaging, it can efficiently be formulated in terms of real-valued computations throughout, due to a one-to-one mapping between centro-Hermitian and real matrices [10]. The forward-backward averaged sample covariance matrix is centro-Hermitian and can, therefore, be transformed into a real-valued matrix of the same size, cf. [12], [15], and [7]. A real-valued square-root factor of this transformed sample covariance matrix is given by

$$\mathcal{T}(X) = Q_M^H \begin{bmatrix} X & \Pi_M \overline{X} \Pi_N \end{bmatrix} Q_{2N} \in \mathbb{R}^{M \times 2N}, \tag{63.14}$$

where Q_M and Q_{2N} were defined in Eq. (63.13).[3] If M is even, an efficient computation of $\mathcal{T}(X)$ from the complex-valued data matrix X only requires $M \times 2N$ real additions and no multiplication [7]. Instead of computing a complex-valued SVD as in the standard ESPRIT case, the signal subspace estimate is obtained via a real-valued SVD of $\mathcal{T}(X)$ (direct data approach). Let $E_s \in \mathbb{R}^{M \times d}$ contain the d left singular vectors corresponding to the d largest singular values of $\mathcal{T}(X)$.[4] Then the columns of

$$U_s = Q_M E_s \tag{63.15}$$

---

[3] The results of this chapter also hold if Q_M and Q_{2N} denote arbitrary left Π-real matrices that are also unitary.
[4] Alternatively, E_s can be obtained through a real-valued eigendecomposition of $\mathcal{T}(X) \mathcal{T}(X)^H$ (covariance approach).

span the estimated signal subspace, and spatial frequency estimates could be obtained from the eigenvalues of the complex-valued matrix $\mathbf{\Psi}$ that solves Eq. (63.9). These complex-valued computations, however, are not required because the transformed array steering matrix

$$\mathbf{D} = \mathbf{Q}_M^H \mathbf{A} = [\ \mathbf{d}(\mu_1) \ \ \mathbf{d}(\mu_2) \ \ \cdots \ \ \mathbf{d}(\mu_d) \ ] \in \mathbb{R}^{M \times d} \tag{63.16}$$

satisfies the following shift invariance property

$$\mathbf{K}_1 \mathbf{D} \mathbf{\Omega} = \mathbf{K}_2 \mathbf{D}, \quad \text{where} \quad \mathbf{\Omega} = \text{diag}\left\{ \tan\left(\frac{\mu_i}{2}\right) \right\}_{i=1}^d \tag{63.17}$$

and the transformed selection matrices $\mathbf{K}_1$ and $\mathbf{K}_2$ are given by

$$\mathbf{K}_1 = 2 \cdot \text{Re}\{\mathbf{Q}_m^H \mathbf{J}_2 \, \mathbf{Q}_M\} \quad \text{and} \quad \mathbf{K}_2 = 2 \cdot \text{Im}\{\mathbf{Q}_m^H \mathbf{J}_2 \, \mathbf{Q}_M\}. \tag{63.18}$$

Here, Re $\{\cdot\}$ and Im $\{\cdot\}$ denote the real and the imaginary part, respectively. Notice that Eq. (63.17) is similar to Eq. (63.7) except for the fact that all matrices in Eq. (63.17) are real-valued.

Let us take a closer look at the transformed selection matrices defined in Eq. (63.18). If $\mathbf{J}_2$ is sparse, $\mathbf{K}_1$ and $\mathbf{K}_2$ are also sparse. This is illustrated by the following example. For the ULA with $M = 6$ sensors and maximum overlap sketched in Fig. 63.2 (a), $\mathbf{J}_2$ is given by

$$\mathbf{J}_2 = \begin{bmatrix} 0 & 1 & 0 & 0 & 0 & 0 \\ 0 & 0 & 1 & 0 & 0 & 0 \\ 0 & 0 & 0 & 1 & 0 & 0 \\ 0 & 0 & 0 & 0 & 1 & 0 \\ 0 & 0 & 0 & 0 & 0 & 1 \end{bmatrix} \in \mathbb{R}^{5 \times 6}.$$

According to Eq. (63.18), straightforward calculations yield the transformed selection matrices

$$\mathbf{K}_1 = \begin{bmatrix} 1 & 1 & 0 & 0 & 0 & 0 \\ 0 & 1 & 1 & 0 & 0 & 0 \\ 0 & 0 & \sqrt{2} & 0 & 0 & 0 \\ 0 & 0 & 0 & 1 & 1 & 0 \\ 0 & 0 & 0 & 0 & 1 & 1 \end{bmatrix} \quad \text{and} \quad \mathbf{K}_2 = \begin{bmatrix} 0 & 0 & 0 & -1 & 1 & 0 \\ 0 & 0 & 0 & 0 & -1 & 1 \\ 0 & 0 & 0 & 0 & 0 & -\sqrt{2} \\ 1 & -1 & 0 & 0 & 0 & 0 \\ 0 & 1 & -1 & 0 & 0 & 0 \end{bmatrix}.$$

In this case, applying $\mathbf{K}_1$ or $\mathbf{K}_2$ to $\mathbf{E}_s$ only requires $(m-1)d$ real additions and d real multiplications.

Asymptotically, the real-valued matrices $\mathbf{E}_s$ and $\mathbf{D}$ span the same d-dimensional subspace, i.e., there is a nonsingular matrix $\mathbf{T} \in \mathbb{R}^{d \times d}$ such that $\mathbf{D} \approx \mathbf{E}_s \mathbf{T}$. Substituting this into Eq. (63.17) yields the real-valued invariance equation

$$\mathbf{K}_1 \mathbf{E}_s \, \mathbf{\Upsilon} \approx \mathbf{K}_2 \mathbf{E}_s \in \mathbb{R}^{m \times d}, \quad \text{where} \quad \mathbf{\Upsilon} = \mathbf{T} \, \mathbf{\Omega} \, \mathbf{T}^{-1}. \tag{63.19}$$

Thus, the eigenvalues of the solution $\mathbf{\Upsilon} \in \mathbb{R}^{d \times d}$ to the matrix equation above are

$$\omega_i = \tan\left(\frac{\mu_i}{2}\right) = \frac{1}{j} \frac{e^{j\mu_i} - 1}{e^{j\mu_i} + 1}, \qquad 1 \leq i \leq d. \tag{63.20}$$

This reveals a spatial frequency warping identical to the temporal frequency warping incurred in designing a digital filter from an analog filter via the bilinear transformation. Consider $\Delta = \frac{\lambda}{2}$ so that $\mu_i = -\frac{2\pi}{\lambda} \Delta \sin \theta_i = -\pi \sin \theta_i$. In this case, there is a one-to-one mapping between $-1 < \sin \theta_i < 1$, corresponding to the range of possible values for the DOAs $-90° < \theta_i < 90°$, and $-\infty < \omega_i < \infty$.

Note that the fact that the eigenvalues of a real matrix have to either be real-valued or occur in complex conjugate pairs gives rise to an ad-hoc *reliability test*. That is, if the final step of the algorithm yields a complex conjugate pair of eigenvalues, then either the SNR is too low, not enough snapshots have been averaged, or two corresponding signal arrivals have not been resolved. In the latter case, taking the tangent inverse of the real part of the eigenvalues can sometimes provide a rough estimate of the direction of arrival of the two closely spaced signals. In general, though, if the algorithm yields one or more complex-conjugate pairs of eigenvalues in the final stage, the estimates should be viewed as unreliable.

The element space implementation of 1-D Unitary ESPRIT is summarized in Table 63.2.

TABLE 63.2   Summary of 1-D Unitary ESPRIT in Element Space

1. *Signal Subspace Estimation:* Compute $E_s \in \mathbb{R}^{M \times d}$ as the d dominant left singular vectors of
$$\mathcal{T}(X) \in \mathbb{R}^{M \times 2N}.$$

2. *Solution of the Invariance Equation:* Then solve
$$\underbrace{K_1 E_s}_{\mathbb{R}^{m \times d}} \Upsilon \approx \underbrace{K_2 E_s}_{\mathbb{R}^{m \times d}}$$

by means of LS, TLS, or SLS.

3. *Spatial Frequency Estimation:* Calculate the eigenvalues of the resulting real-valued solution
$$\Upsilon = T \Omega T^{-1} \in \mathbb{R}^{d \times d} \quad \text{with} \quad \Omega = \text{diag}\{\omega_i\}_{i=1}^{d}$$

- $\mu_i = 2 \arctan(\omega_i), \qquad 1 \leq i \leq d$

63.3.2 1-D Unitary ESPRIT in DFT Beamspace

Reduced dimension processing in beamspace, yielding reduced computational complexity, is an option when one has *a priori* information on the general angular locations of the incident signals, as in a radar application, for example. In the case of a uniform linear array (ULA), transformation from element space to DFT beamspace may be effected by pre-multiplying the data by those rows of the DFT matrix that form beams encompassing the sector of interest. (Each row of the DFT matrix forms a beam pointed to a different angle.) If there is no *a priori* information, one may examine the DFT spectrum and apply Unitary ESPRIT in DFT beamspace to a small set of DFT values around each spectral peak above a particular threshold. In a more general setting, Unitary ESPRIT in DFT beamspace can simply be applied via parallel processing to each of a number of sets of successive DFT values corresponding to overlapping sectors.

Note, though, that in the development to follow, we will initially employ all M DFT beams for the sake of notational simplicity. Without loss of generality, we consider an omnidirectional ULA. Let $W_M^H \in \mathbb{C}^{M \times M}$ be the scaled M-point DFT matrix with its M rows given by

$$w_k^H = e^{j\left(\frac{M-1}{2}\right)k\frac{2\pi}{M}} \begin{bmatrix} 1 & e^{-jk\frac{2\pi}{M}} & e^{-j2k\frac{2\pi}{M}} & \cdots & e^{-j(M-1)k\frac{2\pi}{M}} \end{bmatrix}, \qquad 0 \leq k \leq (M-1). \tag{63.21}$$

Notice that W_M is left Π-real or column conjugate symmetric, i.e., $\Pi_M \overline{W}_M = W_M$. Thus, as pointed out for D in Eq. (63.16), the transformed steering matrix of the ULA

$$B = W_M^H A = \begin{bmatrix} b(\mu_1) & b(\mu_2) & \cdots & b(\mu_d) \end{bmatrix} \in \mathbb{R}^{M \times d} \tag{63.22}$$

is real-valued. It has been shown in [24] that B satisfies a shift invariance property which is similar to Eq. (63.17), namely

$$\Gamma_1 B \Omega = \Gamma_2 B, \quad \text{where} \quad \Omega = \text{diag}\left\{\tan\left(\frac{\mu_i}{2}\right)\right\}_{i=1}^{d}. \tag{63.23}$$

Here, the selection matrices $\mathbf{\Gamma}_1$ and $\mathbf{\Gamma}_2$ of size $M \times M$ are defined as

$$
\mathbf{\Gamma}_1 = \begin{bmatrix}
1 & \cos\left(\frac{\pi}{M}\right) & 0 & 0 & \cdots & 0 & 0 \\
0 & \cos\left(\frac{\pi}{M}\right) & \cos\left(\frac{2\pi}{M}\right) & 0 & \cdots & 0 & 0 \\
0 & 0 & \cos\left(\frac{2\pi}{M}\right) & \cos\left(\frac{3\pi}{M}\right) & \cdots & 0 & 0 \\
\vdots & \vdots & \vdots & \vdots & \ddots & \vdots & \vdots \\
0 & 0 & 0 & 0 & \cdots & \cos\left((M-2)\frac{\pi}{M}\right) & \cos\left((M-1)\frac{\pi}{M}\right) \\
(-1)^M & 0 & 0 & 0 & \cdots & 0 & \cos\left((M-1)\frac{\pi}{M}\right)
\end{bmatrix}
\tag{63.24}
$$

$$
\mathbf{\Gamma}_2 = \begin{bmatrix}
0 & \sin\left(\frac{\pi}{M}\right) & 0 & 0 & \cdots & 0 & 0 \\
0 & \sin\left(\frac{\pi}{M}\right) & \sin\left(\frac{2\pi}{M}\right) & 0 & \cdots & 0 & 0 \\
0 & 0 & \sin\left(\frac{2\pi}{M}\right) & \sin\left(\frac{3\pi}{M}\right) & \cdots & 0 & 0 \\
\vdots & \vdots & \vdots & \vdots & \ddots & \vdots & \vdots \\
0 & 0 & 0 & 0 & \cdots & \sin\left((M-2)\frac{\pi}{M}\right) & \sin\left((M-1)\frac{\pi}{M}\right) \\
0 & 0 & 0 & 0 & \cdots & 0 & \sin\left((M-1)\frac{\pi}{M}\right)
\end{bmatrix}.
\tag{63.25}
$$

As an alternative to Eq. (63.14), another real-valued square-root factor of the transformed sample covariance matrix is given by

$$
\begin{bmatrix} \text{Re}\{\mathbf{Y}\} & \text{Im}\{\mathbf{Y}\} \end{bmatrix} \in \mathbb{R}^{M \times 2N}, \quad \text{where} \quad \mathbf{Y} = \mathbf{W}_M^H \mathbf{X} \in \mathbb{C}^{M \times N}.
\tag{63.26}
$$

The matrix $\mathbf{Y}$ can efficiently be computed via an FFT, which exploits the Vandermonde form of the rows of the DFT matrix, followed by an appropriate scaling, cf. Eq. (63.21). Let the columns of $\mathbf{E}_s \in \mathbb{R}^{M \times d}$ contain the d left singular vectors corresponding to the d largest singular values of Eq. (63.26). Asymptotically, the real-valued matrices $\mathbf{E}_s$ and $\mathbf{B}$ span the same d-dimensional subspace, i.e., there is a nonsingular matrix $\mathbf{T} \in \mathbb{R}^{d \times d}$, such that $\mathbf{B} \approx \mathbf{E}_s \mathbf{T}$. Substituting this into Eq. (63.23), yields the real-valued invariance equation

$$
\mathbf{\Gamma}_1 \mathbf{E}_s \mathbf{\Upsilon} \approx \mathbf{\Gamma}_2 \mathbf{E}_s \in \mathbb{R}^{M \times d}, \quad \text{where} \quad \mathbf{\Upsilon} = \mathbf{T} \mathbf{\Omega} \mathbf{T}^{-1}.
\tag{63.27}
$$

Thus, the eigenvalues of the solution $\mathbf{\Upsilon} \in \mathbb{R}^{d \times d}$ to the matrix equation above are also given by Eq. (63.20).

It is a crucial observation that one row of the matrix equation (63.23) relates *two successive components* of the transformed array steering vectors $\mathbf{b}(\mu_i)$, cf. (63.24) and (63.25). This insight enables us to apply only $B \ll M$ successive rows of $\mathbf{W}_M^H$ (instead of all M rows) to the data matrix $\mathbf{X}$ in Eq. (63.26). To stress the reduced number of rows, we call the resulting beamforming matrix $\mathbf{W}_B^H \in \mathbb{C}^{B \times M}$. The number of its rows, B, depends on the width of the sector of interest and may be substantially less than the number of sensors M. Thereby, the SVD of Eq. (63.26) and, therefore, also $\mathbf{E}_s \in \mathbb{R}^{B \times d}$ and the invariance equation (63.27) will have a reduced dimensionality. Employing the appropriate subblocks of $\mathbf{\Gamma}_1$ and $\mathbf{\Gamma}_2$ as selection matrices, the algorithm is the same as the one described previously except for its reduced dimensionality. In the sequel, the resulting selection matrices of size $(B-1) \times B$ will be called $\mathbf{\Gamma}_1^{(B)}$ and $\mathbf{\Gamma}_2^{(B)}$. The whole algorithm that operates in a B-dimensional DFT beamspace is summarized in Table 63.3.

Consider, for example, a ULA of $M = 8$ sensors. The structure of the corresponding selection matrices $\mathbf{\Gamma}_1$ and $\mathbf{\Gamma}_2$ is sketched in Fig. 63.3. Here, the symbol $\times$ denotes entries of both selection matrices that might be nonzero, cf. (63.24) and (63.25). If one employed rows 4, 5, and 6 of $\mathbf{W}_8^H$ to form $B = 3$ beams in estimating the DOAs of two closely spaced signal arrivals, as in the low-angle radar tracking scheme described by Zoltowski and Lee [26], the corresponding 2×3 subblock of the selection matrices $\mathbf{\Gamma}_1$ and $\mathbf{\Gamma}_2$ is shaded in Fig. 63.3 (a).[5] Notice that the first and the last (Mth) row of $\mathbf{W}_M^H$ steer beams that are also

---

[5] Here, the first row of $\mathbf{\Gamma}_1^{(3)}$ and $\mathbf{\Gamma}_2^{(3)}$ combines beams 4 and 5, while the second row of $\mathbf{\Gamma}_1^{(3)}$ and $\mathbf{\Gamma}_2^{(3)}$ combines beams 5 and 6.

TABLE 63.3   Summary of 1-D Unitary ESPRIT in DFT Beamspace

0. Transformation to Beamspace: $Y = W_B^H X \in \mathbb{C}^{B \times N}$

1. Signal Subspace Estimation: Compute $E_s \in \mathbb{R}^{B \times d}$ as the d dominant left singular vectors of
$[\ \mathrm{Re}\{Y\} \quad \mathrm{Im}\{Y\} \ ] \in \mathbb{R}^{B \times 2N}$.

2. Solution of the Invariance Equation: Solve

$$\underbrace{\Gamma_1^{(B)} E_s}_{\mathbb{R}^{(B-1) \times d}} \ \Upsilon \approx \underbrace{\Gamma_2^{(B)} E_s}_{\mathbb{R}^{(B-1) \times d}}$$

by means of LS, TLS, or SLS.

3. Spatial Frequency Estimation: Calculate the eigenvalues of the resulting real-valued solution

$$\Upsilon = T \, \Omega \, T^{-1} \in \mathbb{R}^{d \times d} \quad \text{with} \quad \Omega = \mathrm{diag}\, \{\omega_i\}_{i=1}^{d}$$

- $\mu_i = 2 \arctan (\omega_i), \qquad 1 \le i \le d$

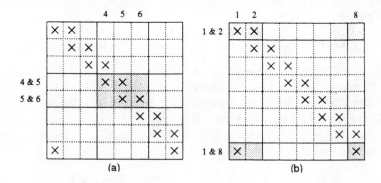

FIGURE 63.3   Structure of the selection matrices Γ_1 and Γ_2 for a ULA of $M = 8$ sensors. The symbol $\times$ denotes entries of both selection matrices that might be nonzero. The shaded areas illustrate how to choose the appropriate subblocks of the selection matrices for reduced dimension processing, i.e., how to form $\Gamma_1^{(B)}$ and $\Gamma_2^{(B)}$, if only $B = 3$ successive rows of W_8^H are applied to the data matrix X. Here, the following two examples are used: (a) rows 4, 5, and 6.   (b) rows 8, 1, and 2.

physically adjacent to one another (the wrap-around property of the DFT). If, for example, one employed rows 8, 1, and 2 of W_8^H to form $B = 3$ beams in estimating the DOAs of two closely spaced signal arrivals, the corresponding subblocks of the selection matrices Γ_1 and Γ_2 are shaded in Fig. 63.3 (b).[6]

63.4   UCA-ESPRIT for Circular Ring Arrays

UCA-ESPRIT [15, 16, 17] is a 2-D angle estimation algorithm developed for use with uniform circular arrays (UCAs). The algorithm provides automatically paired azimuth and elevation angle estimates of far-field signals incident on the UCA via a closed-form procedure. The rotational symmetry of the UCA makes it desirable for a variety of applications where one needs to discriminate in both azimuth and elevation, as opposed to just conical angle of arrival which is all the ULA can discriminate on. For example, UCAs

---

[6]Here, the first row of $\Gamma_1^{(3)}$ and $\Gamma_2^{(3)}$ combines beams 1 and 2, while the second row of $\Gamma_1^{(3)}$ and $\Gamma_2^{(3)}$ combines beams 1 and 8.

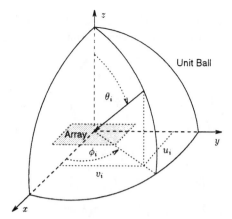

FIGURE 63.4   Definitions of azimuth $(-180° < \phi_i \le 180°)$ and elevation $(0° \le \theta_i \le 90°)$. The direction cosines u_i and v_i are the rectangular coordinates of the projection of the corresponding point on the unit ball onto the equatorial plane.

are commonly employed as part of an anti-jam spatial filter for GPS receivers. Some experimental UCA based systems are described in [4]. The development of a closed-form 2-D angle estimation technique for a UCA provides further motivation for the use of a UCA in a given application.

Consider an M element UCA in which the array elements are uniformly distributed over the circumference of a circle of radius R. We will assume that the array is located in the x-y plane, with its center at the origin of the coordinate system. The elevation angles θ_i and azimuth angles ϕ_i of the d impinging sources are defined in Fig. 63.4, as are the direction cosines u_i and v_i, $1 \le i \le d$. UCA-ESPRIT is premised on phase mode excitation-based beamforming. The maximum phase mode (integer valued) excitable by a given UCA is

$$K \approx \frac{2\pi R}{\lambda},$$

where λ is the common (carrier) wavelength of the incident signals. Phase mode excitation-based beamforming requires $M > 2K$ array elements ($M = 2K + 3$ is usually adequate). UCA-ESPRIT can resolve a maximum of $d_{max} = K - 1$ sources. As an example, if the array radius is $r = \lambda$, $K = 6$ (the largest integer smaller than 2π) and at least $M = 15$ array elements are needed. UCA-ESPRIT can resolve five sources in conjunction with this UCA.

UCA-ESPRIT operates in a $K' = 2K + 1$ dimensional beamspace. It employs a $K' \times M$ beamforming matrix to transform from element space to beamspace. After this transformation, the algorithm has the same three basic steps of any ESPRIT-type algorithm: (1) the computation of a basis for the signal subspace, (2) the solution to an (in general) overdetermined system of equations derived from the matrix of vectors spanning the signal subspace, and (3) the computation of the eigenvalues of the solution to the system of equations formed in Step (2). As illustrated in Fig. 63.6, the ith eigenvalue obtained in the final step is ideally of the form $\xi_i = \sin \theta_i \, e^{j\phi_i}$, where ϕ_i and θ_i are the azimuth and elevation angles of the ith source. Note that

$$\xi_i = \sin \theta_i \, e^{j\phi_i} = u_i + jv_i, \quad 1 \le i \le d,$$

where u_i and v_i are the direction cosines of the ith source relative to the x- and y-axis, respectively, as indicated in Fig. 63.4.

The formulation of UCA-ESPRIT is based on the special structure of the resulting K'-dimensional beamspace manifold. The following vector and matrix definitions are needed to summarize the algorithm in Table 63.4.

$$v_k^H = \frac{1}{M} \begin{bmatrix} 1 & e^{jk\frac{2\pi}{M}} & e^{j2k\frac{2\pi}{M}} & \cdots & e^{j(M-1)k\frac{2\pi}{M}} \end{bmatrix} \tag{63.28}$$

$$V = \sqrt{M} \begin{bmatrix} v_{-K} & \cdots & v_{-1} & v_0 & v_1 & \cdots & v_K \end{bmatrix} \in \mathbb{C}^{M \times K'}$$

$$C_v = \operatorname{diag}\left\{ j^k \right\}_{k=-K}^{K} \in \mathbb{C}^{K' \times K'}$$

$$F_r^H = Q_{K'}^T C_v V^H \in \mathbb{C}^{K' \times M} \tag{63.29}$$

$$C_o = \operatorname{diag}\left\{ \operatorname{sign}(k)^{-k} \right\}_{k=-K}^{K} \in \mathbb{R}^{K' \times K'}$$

$$D = \operatorname{diag}\left\{ (-1)^{|k|} \right\}_{k=-(K-2)}^{K} \in \mathbb{R}^{(K'-2) \times (K'-2)}$$

$$\Gamma = \frac{\lambda}{\pi r} \cdot \operatorname{diag}\{k\}_{k=-(K-1)}^{(K-1)} \in \mathbb{R}^{(K'-2) \times (K'-2)}$$

Note that the columns of the matrix V consist of the DFT weight vectors v_k defined in Eq. (63.28). The beamforming matrix F_r^H in Eq. (63.29) synthesizes a real-valued beamspace manifold and facilitates signal subspace estimation via a real-valued SVD or eigendecomposition. Recall that the sparse left Π-real matrix $Q_{K'} \in \mathbb{C}^{K' \times K'}$ has been defined in Eq. (63.13). The complete UCA-ESPRIT algorithm is summarized in Table 63.4.

TABLE 63.4 Summary of UCA-ESPRIT

---

0. Transformation to Beamspace: $\quad Y = F_r^H X \in \mathbb{C}^{K' \times N}$

1. Signal Subspace Estimation: Compute $E_s \in \mathbb{R}^{K' \times d}$ as the d dominant left singular vectors of $[\ \operatorname{Re}\{Y\} \quad \operatorname{Im}\{Y\}\ ] \in \mathbb{R}^{K' \times 2N}$.

2. Solution of the Invariance Equation:

- Compute $E_u = C_o \overline{Q}_{K'} E_s$. Form the matrix E_{-1} that consists of all but the last two rows of E_u. Similarly form the matrix E_0 that consists of all but the first and last rows of E_u.

- Compute $\underline{\Psi} \in \mathbb{C}^{2d \times d}$, the least squares solution to the system

$$\begin{bmatrix} E_{-1} & D\Pi_{(K'-2)}\overline{E}_{-1} \end{bmatrix} \underline{\Psi} = \Gamma E_0 \in \mathbb{C}^{(K'-2) \times d}.$$

Recall that the overbar denotes complex conjugation. Form Ψ by extracting the upper $d \times d$ block from $\underline{\Psi}$. Note that Ψ can be computed efficiently by solving a *real-valued* system of $2d$ equations (see [17]).

3. Spatial Frequency Estimation: Compute the eigenvalues ξ_i, $1 \le i \le d$, of $\Psi \in \mathbb{C}^{d \times d}$. The estimates of the elevation and azimuth angles of the ith source are

$$\theta_i = \arcsin(|\xi_i|) \quad \text{and} \quad \phi_i = \arg(\xi_i),$$

respectively. If direction cosine estimates are desired, we have

$$u_i = \operatorname{Re}\{\xi_i\} \quad \text{and} \quad v_i = \operatorname{Im}\{\xi_i\}.$$

Again, ξ_i can be efficiently computed via a *real-valued* EVD (see [17]).

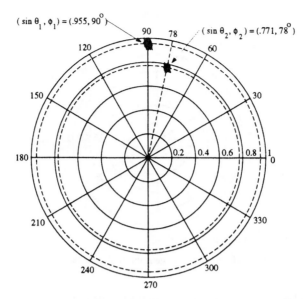

FIGURE 63.5 Plot of the UCA-ESPRIT eigenvalues $\xi_1 = \sin\theta_1 e^{j\phi_1}$ and $\xi_2 = \sin\theta_2 e^{j\phi_2}$ for 200 trials.

63.4.1 Results of Computer Simulations

Simulations were conducted with a UCA of radius $R = \lambda$, with $K = 6$ and $M = 19$ (performance close to that reported below can be expected even if $M = 15$ elements are employed). The simulation employed two sources with arrival angles given by $(\theta_1, \phi_1) = (72.73°, 90°)$ and $(\theta_2, \phi_2) = (50.44°, 78°)$. The sources were highly correlated, with the correlation coefficient referred to the center of the array being $0.9e^{j\frac{\pi}{4}}$. The signal-to-noise ratio (SNR) was 10 dB (per array element) for each source. The number of snapshots was $N = 64$, and arrival angle estimates were obtained for 200 independent trials. Figure 63.5 depicts the results of the simulation. Here, the UCA-ESPRIT eigenvalues ξ_i are denoted by the symbol $\times$.[7] The results from all 200 trials are superimposed in the figure. The eigenvalues are seen to be clustered around the expected locations (the dashed circles indicate the true elevation angles).

63.5 FCA-ESPRIT for Filled Circular Arrays

The use of a circular ring array and the attendant use of UCA-ESPRIT is ideal for applications where the array aperture is not very large as on the top of a mobile communications unit. For much larger array apertures as in phased array surveillance radars, too much of the aperture is devoid of elements so that a lot of the signal energy impinging on the aperture is not intercepted. As an example, each of the four panels comprising either the SPY-1A or SPY-1B radars of the AEGIS series is composed of 4400 identical elements regularly spaced on a flat panel over a circular aperture [19]. The sampling lattice is hexagonal. Recent prototype arrays for satellite-based communications have also employed the filled circular array geometry [2].

This section presents an algorithm similar to UCA-ESPRIT that provides the same closed-form 2-D angle estimation capability for a *Filled Circular Array* (FCA). Similar to UCA-ESPRIT, the far field pattern arising from the sampled excitation is approximated by the far field pattern arising from the continuous excitation from which the sampled excitation is derived through sampling. (Note, Steinberg [20] shows

---

[7]The horizontal axis represents $\mathrm{Re}\{\xi_i\}$, and the vertical axis represents $\mathrm{Im}\{\xi_i\}$.

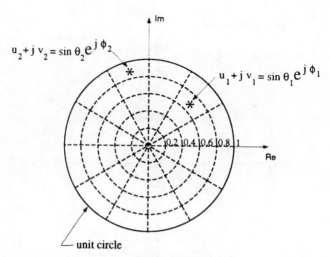

FIGURE 63.6 Illustrating the form of signal roots (eigenvalues) obtained with UCA-ESPRIT or FCA-ESPRIT.

that the array pattern for a ULA of N elements with interelement spacing d is nearly identical to the far field pattern for a continuous linear aperture of length $(N + 1)d$, except near the fringes of the visible region.) That is, it is assumed that the interelement spacings have been chosen so that aliasing effects are negligible as in the generation of phase modes with a single ring array. It can be shown that this is the case for any sampling lattice as long as the inter-sensor spacings is roughly half a wavelength or less on the average and that the sources of interest are at least 20° in elevation above the plane of the array, i.e., we require that the elevation angle of the ith source satisfies $0 \le \theta_i \le 70°$. In practice, many phased arrays only provide reliable coverage for $0 \le \theta_i \le 60°$ (plus or minus 60° away from boresite) due to a reduced aperture effect and the fact that the gain of each individual antenna has a significant roll-off at elevation angles near the horizon, i.e., the plane of the array. FCA-ESPRIT has been successfully applied to rectangular, hexagonal, polar raster, and random sampling lattices.

The key to the development of UCA-ESPRIT was phase-mode (DFT) excitation and exploitation of a recurrence relationship that Bessel functions satisfy. In the case of a filled circular array, the same type of processing is facilitated by the use of a phase-mode dependent aperture taper derived from an integral relationship that Bessel functions satisfy.

Consider an M element FCA where the array elements are distributed over a circular aperture of radius R. We assume that the array is centered at the origin of the coordinate system and contained in the x-y plane. The ith element is located at a radial distance r_i from the origin and at an angle γ_i relative to the x-axis measured counter-clockwise in the x-y plane. In contrast to a UCA, $0 \le r_i \le R$, i.e., the elements lie within, rather than on, a circle of radius R. The beamforming weight vectors employed in FCA-ESPRIT are

$$
\boldsymbol{w}_m = \frac{1}{M}
\begin{bmatrix}
A_1 \left(\frac{r_1}{R}\right)^{|m|} e^{-jm\gamma_1} \\
\vdots \\
A_i \left(\frac{r_i}{R}\right)^{|m|} e^{-jm\gamma_i} \\
\vdots \\
A_M \left(\frac{r_M}{R}\right)^{|m|} e^{-jm\gamma_M}
\end{bmatrix},
\tag{63.30}
$$

where m ranges from $-K$ to K with $K \approx \frac{2\pi R}{\lambda}$. Here A_i is proportional to the area surrounding the ith array element. A_i is a constant (and can be omitted) for hexagonal and rectangular lattices and proportional to the radius ($A_i = r_i$) for a polar raster. The transformation from element space to beamspace is effected through pre-multiplication by the beamforming matrix

$$W = \sqrt{M} \left[\; w_{-K} \quad \cdots \quad w_{-1} \quad w_0 \quad w_1 \quad \cdots \quad w_K \; \right] \in \mathbb{C}^{M \times K'} \; (K' = 2K + 1). \tag{63.31}$$

The following matrix definitions are needed to summarize FCA-ESPRIT.

$$B = WC \in \mathbb{C}^{M \times K'} \tag{63.32}$$

$$C = \text{diag} \left\{ \text{sign}(k) \cdot j^k \right\}_{k=-K}^{K} \in \mathbb{C}^{K' \times K'}$$

$$B_r = BF\bar{Q}_{K'} \in \mathbb{C}^{M \times K'}$$

$$F = \text{diag} \left([(-1)^{-M-1}, \cdots, (-1)^{-2}, 1, 1, \cdots, 1] \right) \in \mathbb{R}^{K' \times K'}$$

$$\Gamma = \frac{\lambda}{\pi R} \text{diag}([\overbrace{-M, \cdots, -3}^{M-1}, -2, 0, \overbrace{2, \cdots, M}^{M-1}]) \in \mathbb{R}^{(K'-2) \times (K'-2)}$$

$$C_1 = \text{diag}([\overbrace{1, \cdots, 1}^{M-2}, -1, -1, \overbrace{1, \cdots, 1}^{M-1}]) \in \mathbb{R}^{(K'-2) \times (K'-2)}$$

The whole algorithm is summarized in Table 63.5. The beamforming matrix B_r^H synthesizes a real-valued manifold that facilitates signal subspace estimation via a real-valued SVD or eigenvalue decomposition in the first step. As in UCA-ESPRIT, the eigenvalues of Ψ computed in the final step are asymptotically of the form $\sin(\theta_i)e^{j\phi_i}$, where θ_i and ϕ_i are the elevation and azimuth angles of the ith source, respectively.

TABLE 63.5 Summary of FCA-ESPRIT

0. Transformation to Beamspace: $Y = B_r^H X$

1. Signal Subspace Estimation: Compute $E_s \in \mathbb{R}^{K' \times d}$ as the d dominant left singular vector of
$[\; \text{Re}\{Y\} \quad \text{Im}\{Y\} \;] \in \mathbb{R}^{K' \times 2N}$.

2. Solution of the Invariance Equation:

- Compute $E_u = FQ_{K'}E_s$. Form the matrices E_{-1}, E_0, and E_1 that consist of all but the last two, first and last, and first two rows, respectively.

- Compute $\underline{\Psi} \in \mathbb{C}^{2d \times d}$, the least squares solution to the system

$$[\; E_{-1} \quad C_1 E_1 \;] \underline{\Psi} = \Gamma E_0 \in \mathbb{C}^{(K'-2) \times d}.$$

Form Ψ by extracting the upper $d \times d$ block from $\underline{\Psi}$.

3. Spatial Frequency Estimation: Compute the eigenvalues ξ_i, $1 \le i \le d$, of $\Psi \in \mathbb{C}^{d \times d}$. The estimates of the elevation and azimuth angles of the ith source are

$$\theta_i = \arcsin(|\xi_i|) \quad \text{and} \quad \phi_i = \arg(\xi_i),$$

respectively.

63.5.1 Computer Simulation

As an example, a simulation involving a *random filled array* is presented. The element locations are depicted in Fig. 63.7. The outer radius is $R = 5\lambda$ and the average distance between elements is $\lambda/4$. Two plane waves of equal power were incident upon the array. The Signal to Noise Ratio (SNR) per antenna per signal was 0 dB. One signal arrived at 10° elevation and 40° azimuth, while the other arrived at 30° elevation and 60°

azimuth. Figure 63.8 shows the results of 32 independent trials of FCA-ESPRIT overlaid; each execution of the algorithm (with a different realization of the noise) produced two eigenvalues. The eigenvalues are observed to be clustered around the expected locations (the dashed circles indicate the true elevation angles).

FIGURE 63.7   Random filled array.

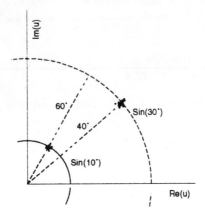

FIGURE 63.8   Plot of the FCA-ESPRIT eigenvalues from 32 independent trials.

63.6   2-D Unitary ESPRIT

For uniform circular arrays and filled circular arrays, UCA-ESPRIT and FCA-ESPRIT provide closed-form, automatically paired 2-D angle estimates as long as the direction cosine pair of each signal arrival is unique. In this section, we develop 2-D Unitary ESPRIT, a closed-form 2-D angle estimation algorithm that achieves automatic pairing in a similar fashion. It is applicable to 2-D centro-symmetric array configurations with a dual invariance structure such as uniform rectangular arrays (URAs). In the derivations of UCA-ESPRIT and FCA-ESPRIT it was necessary to approximate the sampled aperture pattern by the continuous aperture pattern. Such an approximation is not required in the development of 2-D Unitary ESPRIT.

Apart from the 2-D extension presented here, Unitary ESPRIT has also been extended to the R-dimensional case to solve the R-dimensional harmonic retrieval problem, where $R \geq 3$. R-D Unitary ESPRIT is a closed-form algorithm to estimate several undamped R-dimensional modes (or frequencies) along with their correct pairing. In [6], automatic pairing of the R-dimensional frequency estimates is achieved through a new simultaneous Schur decomposition of R real-valued, non-symmetric matrices that reveals their "average eigenstructure". Like its 1-D and 2-D counterparts, R-D Unitary ESPRIT inherently includes forward-backward averaging and is efficiently formulated in terms of real-valued computations throughout. In the array processing context, a three-dimensional extension of Unitary ESPRIT can be used to estimate the 2-D arrival angles and carrier frequencies of several impinging wavefronts simultaneously.

63.6.1 2-D Array Geometry

Consider a 2-D centro-symmetric sensor array of M elements lying in the x-y plane (Fig. 63.4). Assume that the array also exhibits a *dual* invariance, i.e., two identical subarrays of m_x elements are displaced by Δ_x along the x-axis, and another pair of identical subarrays, consisting of m_y elements each, is displaced by Δ_y along the y-axis. Notice that the four subarrays can overlap and m_x is not required to equal m_y. Such array configurations include uniform rectangular arrays (URAs), uniform rectangular frame arrays (URFAs), i.e., URAs without some of their center elements, and cross arrays consisting of two orthogonal linear arrays with a common phase center as shown in Fig. 63.9.[8]

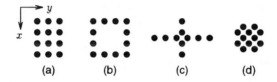

(a) (b) (c) (d)

FIGURE 63.9 Centro-symmetric array configurations with a dual invariance structure: (a) URA with $M = 12$, $m_x = 9, m_y = 8$. (b) URFA with $M = 12, m_x = m_y = 6$. (c) Cross array with $M = 10, m_x = 3, m_y = 5$. (d) $M = 12, m_x = m_y = 7$.

Incident on the array are d narrowband planar wavefronts with wavelength λ, azimuth ϕ_i, and elevation θ_i, $1 \leq i \leq d$. Let

$$u_i = \cos \phi_i \sin \theta_i \quad \text{and} \quad v_i = \sin \phi_i \sin \theta_i, \quad 1 \leq i \leq d,$$

denote the direction cosines of the ith source relative to the x- and y-axes, respectively. These definitions are illustrated in Fig. 63.4. The fact that $\xi_i = u_i + jv_i = \sin \theta_i \, e^{j\phi_i}$ yields a simple formula to determine azimuth ϕ_i and elevation θ_i from the corresponding direction cosines u_i and v_i, namely

$$\phi_i = \arg(\xi_i) \quad \text{and} \quad \theta_i = \arcsin(|\xi_i|), \quad \text{with} \quad \xi_i = u_i + jv_i, \quad 1 \leq i \leq d. \tag{63.33}$$

Similar to the 1-D case, the data matrix X is an $M \times N$ matrix composed of N snapshots $x(t_n)$, $1 \leq n \leq N$, of data as columns. Referring to Fig. 63.10 for a URA of $M = 4 \times 4 = 16$ sensors as an illustrative example, the antenna element outputs are stacked columnwise. Specifically, the first element of $x(t_n)$ is the output of the antenna in the upper left corner. Then sequentially progress downwards along the positive x-axis such that the fourth element of $x(t_n)$ is the output of the antenna in the bottom left corner. The fifth element of $x(t_n)$ is the output of the antenna at the top of the second column; the eighth element of

---

[8]In the examples of Fig. 63.9, all values of m_x and m_y correspond to selection matrices with maximum overlap in both directions. For a URA of $M = M_x \cdot M_y$ elements, cf. Fig. 63.9 (a), this assumption implies $m_x = (M_x - 1) M_y$ and $m_y = M_x (M_y - 1)$.

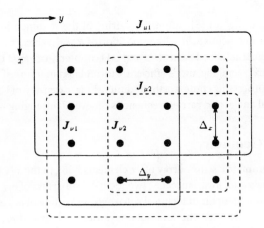

FIGURE 63.10   Subarray selection for a URA of $M = 4 \cdot 4 = 16$ sensor elements (maximum overlap in both directions: $m_x = m_y = 12$).

$x(t_n)$ is the output of the antenna at the bottom of the second column, etc. This forms a 16×1 vector at each sampling instant t_n.

Similar to the 1-D case, the array measurements may be expressed as $x(t) = As(t) + n(t) \in \mathbb{C}^M$. Due to the centro-symmetry of the array, the steering matrix $A \in \mathbb{C}^{M \times d}$ satisfies Eq. (63.12). The goal is to construct two pairs of selection matrices that are centro-symmetric with respect to each other, i.e.,

$$J_{\mu 2} = \Pi_{m_x} J_{\mu 1} \Pi_M \quad \text{and} \quad J_{v2} = \Pi_{m_y} J_{v1} \Pi_M, \tag{63.34}$$

and cause the array steering matrix A to satisfy the following two invariance properties,

$$J_{\mu 1} A \Phi_\mu = J_{\mu 2} A \quad \text{and} \quad J_{v1} A \Phi_v = J_{v2} A, \tag{63.35}$$

where the diagonal matrices

$$\Phi_\mu = \text{diag} \left\{ e^{j\mu_i} \right\}_{i=1}^d \quad \text{and} \quad \Phi_v = \text{diag} \left\{ e^{j v_i} \right\}_{i=1}^d \tag{63.36}$$

are unitary and contain the desired 2-D angle information. Here $\mu_i = \frac{2\pi}{\lambda} \Delta_x u_i$ and $v_i = \frac{2\pi}{\lambda} \Delta_y v_i$ are the spatial frequencies in x- and y-direction, respectively.

Figure 63.10 visualizes a possible choice of the selection matrices for a URA of $M = 4 \times 4 = 16$ sensor elements. Given the stacking procedure described above and the 1-D selection matrices for a ULA of 4 elements

$$J_1^{(4)} = \begin{bmatrix} 1 & 0 & 0 & 0 \\ 0 & 1 & 0 & 0 \\ 0 & 0 & 1 & 0 \end{bmatrix} \quad \text{and} \quad J_2^{(4)} = \begin{bmatrix} 0 & 1 & 0 & 0 \\ 0 & 0 & 1 & 0 \\ 0 & 0 & 0 & 1 \end{bmatrix},$$

the appropriate selection matrices corresponding to maximum overlap are

$$
J_{\mu 1} = I_{M_y} \otimes J_1^{(M_x)} =
\begin{bmatrix}
1 & 0 & 0 & 0 & 0 & 0 & 0 & 0 & 0 & 0 & 0 & 0 & 0 & 0 & 0 & 0 \\
0 & 1 & 0 & 0 & 0 & 0 & 0 & 0 & 0 & 0 & 0 & 0 & 0 & 0 & 0 & 0 \\
0 & 0 & 1 & 0 & 0 & 0 & 0 & 0 & 0 & 0 & 0 & 0 & 0 & 0 & 0 & 0 \\
0 & 0 & 0 & 0 & 1 & 0 & 0 & 0 & 0 & 0 & 0 & 0 & 0 & 0 & 0 & 0 \\
0 & 0 & 0 & 0 & 0 & 1 & 0 & 0 & 0 & 0 & 0 & 0 & 0 & 0 & 0 & 0 \\
0 & 0 & 0 & 0 & 0 & 0 & 1 & 0 & 0 & 0 & 0 & 0 & 0 & 0 & 0 & 0 \\
0 & 0 & 0 & 0 & 0 & 0 & 0 & 0 & 1 & 0 & 0 & 0 & 0 & 0 & 0 & 0 \\
0 & 0 & 0 & 0 & 0 & 0 & 0 & 0 & 0 & 1 & 0 & 0 & 0 & 0 & 0 & 0 \\
0 & 0 & 0 & 0 & 0 & 0 & 0 & 0 & 0 & 0 & 1 & 0 & 0 & 0 & 0 & 0 \\
0 & 0 & 0 & 0 & 0 & 0 & 0 & 0 & 0 & 0 & 0 & 0 & 1 & 0 & 0 & 0 \\
0 & 0 & 0 & 0 & 0 & 0 & 0 & 0 & 0 & 0 & 0 & 0 & 0 & 1 & 0 & 0 \\
0 & 0 & 0 & 0 & 0 & 0 & 0 & 0 & 0 & 0 & 0 & 0 & 0 & 0 & 1 & 0
\end{bmatrix}
\in \mathbb{R}^{12 \times 16}
$$

$$
J_{\mu 2} = I_{M_y} \otimes J_2^{(M_x)} =
\begin{bmatrix}
0 & 1 & 0 & 0 & 0 & 0 & 0 & 0 & 0 & 0 & 0 & 0 & 0 & 0 & 0 & 0 \\
0 & 0 & 1 & 0 & 0 & 0 & 0 & 0 & 0 & 0 & 0 & 0 & 0 & 0 & 0 & 0 \\
0 & 0 & 0 & 1 & 0 & 0 & 0 & 0 & 0 & 0 & 0 & 0 & 0 & 0 & 0 & 0 \\
0 & 0 & 0 & 0 & 0 & 1 & 0 & 0 & 0 & 0 & 0 & 0 & 0 & 0 & 0 & 0 \\
0 & 0 & 0 & 0 & 0 & 0 & 1 & 0 & 0 & 0 & 0 & 0 & 0 & 0 & 0 & 0 \\
0 & 0 & 0 & 0 & 0 & 0 & 0 & 1 & 0 & 0 & 0 & 0 & 0 & 0 & 0 & 0 \\
0 & 0 & 0 & 0 & 0 & 0 & 0 & 0 & 0 & 1 & 0 & 0 & 0 & 0 & 0 & 0 \\
0 & 0 & 0 & 0 & 0 & 0 & 0 & 0 & 0 & 0 & 1 & 0 & 0 & 0 & 0 & 0 \\
0 & 0 & 0 & 0 & 0 & 0 & 0 & 0 & 0 & 0 & 0 & 1 & 0 & 0 & 0 & 0 \\
0 & 0 & 0 & 0 & 0 & 0 & 0 & 0 & 0 & 0 & 0 & 0 & 0 & 1 & 0 & 0 \\
0 & 0 & 0 & 0 & 0 & 0 & 0 & 0 & 0 & 0 & 0 & 0 & 0 & 0 & 1 & 0 \\
0 & 0 & 0 & 0 & 0 & 0 & 0 & 0 & 0 & 0 & 0 & 0 & 0 & 0 & 0 & 1
\end{bmatrix}
\in \mathbb{R}^{12 \times 16}
$$

$$
J_{\nu 1} = J_1^{(M_y)} \otimes I_{M_x} =
\begin{bmatrix}
1 & 0 & 0 & 0 & 0 & 0 & 0 & 0 & 0 & 0 & 0 & 0 & 0 & 0 & 0 & 0 \\
0 & 1 & 0 & 0 & 0 & 0 & 0 & 0 & 0 & 0 & 0 & 0 & 0 & 0 & 0 & 0 \\
0 & 0 & 1 & 0 & 0 & 0 & 0 & 0 & 0 & 0 & 0 & 0 & 0 & 0 & 0 & 0 \\
0 & 0 & 0 & 1 & 0 & 0 & 0 & 0 & 0 & 0 & 0 & 0 & 0 & 0 & 0 & 0 \\
0 & 0 & 0 & 0 & 1 & 0 & 0 & 0 & 0 & 0 & 0 & 0 & 0 & 0 & 0 & 0 \\
0 & 0 & 0 & 0 & 0 & 1 & 0 & 0 & 0 & 0 & 0 & 0 & 0 & 0 & 0 & 0 \\
0 & 0 & 0 & 0 & 0 & 0 & 1 & 0 & 0 & 0 & 0 & 0 & 0 & 0 & 0 & 0 \\
0 & 0 & 0 & 0 & 0 & 0 & 0 & 1 & 0 & 0 & 0 & 0 & 0 & 0 & 0 & 0 \\
0 & 0 & 0 & 0 & 0 & 0 & 0 & 0 & 1 & 0 & 0 & 0 & 0 & 0 & 0 & 0 \\
0 & 0 & 0 & 0 & 0 & 0 & 0 & 0 & 0 & 1 & 0 & 0 & 0 & 0 & 0 & 0 \\
0 & 0 & 0 & 0 & 0 & 0 & 0 & 0 & 0 & 0 & 1 & 0 & 0 & 0 & 0 & 0 \\
0 & 0 & 0 & 0 & 0 & 0 & 0 & 0 & 0 & 0 & 0 & 1 & 0 & 0 & 0 & 0
\end{bmatrix}
\in \mathbb{R}^{12 \times 16}
$$

$$J_{v2} = J_2^{(M_y)} \otimes I_{M_x} = \begin{bmatrix}
0 & 0 & 0 & 0 & 1 & 0 & 0 & 0 & 0 & 0 & 0 & 0 & 0 & 0 & 0 & 0 \\
0 & 0 & 0 & 0 & 0 & 1 & 0 & 0 & 0 & 0 & 0 & 0 & 0 & 0 & 0 & 0 \\
0 & 0 & 0 & 0 & 0 & 0 & 1 & 0 & 0 & 0 & 0 & 0 & 0 & 0 & 0 & 0 \\
0 & 0 & 0 & 0 & 0 & 0 & 0 & 1 & 0 & 0 & 0 & 0 & 0 & 0 & 0 & 0 \\
0 & 0 & 0 & 0 & 0 & 0 & 0 & 0 & 1 & 0 & 0 & 0 & 0 & 0 & 0 & 0 \\
0 & 0 & 0 & 0 & 0 & 0 & 0 & 0 & 0 & 1 & 0 & 0 & 0 & 0 & 0 & 0 \\
0 & 0 & 0 & 0 & 0 & 0 & 0 & 0 & 0 & 0 & 1 & 0 & 0 & 0 & 0 & 0 \\
0 & 0 & 0 & 0 & 0 & 0 & 0 & 0 & 0 & 0 & 0 & 1 & 0 & 0 & 0 & 0 \\
0 & 0 & 0 & 0 & 0 & 0 & 0 & 0 & 0 & 0 & 0 & 0 & 1 & 0 & 0 & 0 \\
0 & 0 & 0 & 0 & 0 & 0 & 0 & 0 & 0 & 0 & 0 & 0 & 0 & 1 & 0 & 0 \\
0 & 0 & 0 & 0 & 0 & 0 & 0 & 0 & 0 & 0 & 0 & 0 & 0 & 0 & 1 & 0 \\
0 & 0 & 0 & 0 & 0 & 0 & 0 & 0 & 0 & 0 & 0 & 0 & 0 & 0 & 0 & 1
\end{bmatrix} \in \mathbb{R}^{12 \times 16},$$

where $M_x = M_y = 4$. Notice, however, that it is not required to compute all four selection matrices explicitly, since they are related via Eq. (63.34). In fact, to be able to compute the four transformed selection matrices for 2-D Unitary ESPRIT, it is sufficient to specify $J_{\mu 2}$ and J_{v2}, cf. (63.38) and (63.39).

63.6.2 2-D Unitary ESPRIT in Element Space

Similar to Eq. (63.16) in the 1-D case, let us define the transformed 2-D array steering matrix as $D = Q_M^H A$. Based on the two invariance properties of the 2-D array steering matrix A in Eq. (63.35), it is a straightforward 2-D extension of the derivation of 1-D Unitary ESPRIT to show that the transformed array steering matrix D satisfies

$$K_{\mu 1} D \cdot \Omega_\mu = K_{\mu 2} D \quad \text{and} \quad K_{v1} D \cdot \Omega_v = K_{v2} D, \tag{63.37}$$

where the two pairs of transformed selection matrices are defined as

$$K_{\mu 1} = 2 \cdot \text{Re}\{Q_{m_x}^H J_{\mu 2} Q_M\} \qquad K_{\mu 2} = 2 \cdot \text{Im}\{Q_{m_x}^H J_{\mu 2} Q_M\} \tag{63.38}$$

$$K_{v1} = 2 \cdot \text{Re}\{Q_{m_y}^H J_{v2} Q_M\} \qquad K_{v2} = 2 \cdot \text{Im}\{Q_{m_y}^H J_{v2} Q_M\} \tag{63.39}$$

and the real-valued diagonal matrices

$$\Omega_\mu = \text{diag}\left\{\tan\left(\frac{\mu_i}{2}\right)\right\}_{i=1}^d \quad \text{and} \quad \Omega_v = \text{diag}\left\{\tan\left(\frac{v_i}{2}\right)\right\}_{i=1}^d \tag{63.40}$$

contain the desired (spatial) frequency information.

Given the noise-corrupted data matrix X, a real-valued matrix E_s, spanning the dominant subspace of $T(X)$, is obtained as described in Section 63.3.1 for the 1-D case. Asymptotically or without additive noise, E_s and D span the same d-dimensional subspace, i.e., there is a nonsingular matrix T of size $d \times d$ such that $D \approx E_s T$. Substituting this relationship into Eq. (63.37) yields two *real-valued* invariance equations

$$K_{\mu 1} E_s \Upsilon_\mu \approx K_{\mu 2} E_s \in \mathbb{R}^{m_x \times d} \quad \text{and} \quad K_{v1} E_s \Upsilon_v \approx K_{v2} E_s \in \mathbb{R}^{m_y \times d}, \tag{63.41}$$

where $\Upsilon_\mu = T \Omega_\mu T^{-1} \in \mathbb{R}^{d \times d}$ and $\Upsilon_v = T \Omega_v T^{-1} \in \mathbb{R}^{d \times d}$. Thus, Υ_μ and Υ_v are related with the diagonal matrices Ω_μ and Ω_v via eigenvalue preserving similarity transformations. Moreover, the real-valued matrices Υ_μ and Υ_v share the *same set of eigenvectors*. As in the 1-D case, the two real-valued invariance equations (63.41) can be solved independently via LS, TLS, or SLS [9]. As an alternative, they may be solved jointly via 2-D SLS, which is a 2-D extension of structured least squares (SLS) [8].

63.6.3   Automatic Pairing of the 2-D Frequency Estimates

Asymptotically or without additive noise, the real-valued eigenvalues of the solutions $\mathbf{\Upsilon}_\mu \in \mathbb{R}^{d \times d}$ and $\mathbf{\Upsilon}_\nu \in \mathbb{R}^{d \times d}$ to the invariance equations above are given by $\tan(\mu_i/2)$ and $\tan(\nu_i/2)$, respectively. If theses eigenvalues were calculated independently, it would be quite difficult to pair the resulting two distinct sets of frequency estimates. Notice that one can choose a real-valued eigenvector matrix $\mathbf{T}$ such that all matrices that appear in the spectral decompositions of $\mathbf{\Upsilon}_\mu = \mathbf{T}\,\mathbf{\Omega}_\mu\,\mathbf{T}^{-1}$ and $\mathbf{\Upsilon}_\nu = \mathbf{T}\,\mathbf{\Omega}_\nu\,\mathbf{T}^{-1}$ are real-valued. Moreover, the subspace spanned by the columns of $\mathbf{T} \in \mathbb{R}^{d \times d}$ is unique. These observations are critical to achieve automatic pairing of the spatial frequencies μ_i and ν_i, $1 \le i \le d$.

With additive noise and a finite number of snapshots N, however, the real-valued matrices $\mathbf{\Upsilon}_\mu$ and $\mathbf{\Upsilon}_\nu$ do not exactly share the same set of eigenvectors. To determine an approximation of the set of common eigenvectors from one of these matrices is, obviously, not the best solution, since this strategy would rely on an arbitrary choice and would also discard information contained in the other matrix. Moreover, $\mathbf{\Upsilon}_\mu$ and $\mathbf{\Upsilon}_\nu$ might have some degenerate (multiple) eigenvalues, while both of them have well determined common eigenvectors $\mathbf{T}$ (for $N \to \infty$ or $\sigma_N^2 \to 0$). 2-D Unitary ESPRIT circumvents these difficulties and achieves automatic pairing of the spatial frequency estimates μ_i and ν_i by computing the eigenvalues of the "complexified" matrix $\mathbf{\Upsilon}_\mu + j\mathbf{\Upsilon}_\nu$ since this complex-valued matrix may be spectrally decomposed as

$$\mathbf{\Upsilon}_\mu + j\mathbf{\Upsilon}_\nu = \mathbf{T}\left(\mathbf{\Omega}_\mu + j\mathbf{\Omega}_\nu\right)\mathbf{T}^{-1}. \tag{63.42}$$

Here, automatically paired estimates of $\mathbf{\Omega}_\mu$ and $\mathbf{\Omega}_\nu$ in Eq. (63.40) are given by the real and imaginary parts of the complex eigenvalues of $\mathbf{\Upsilon}_\mu + j\mathbf{\Upsilon}_\nu$. The maximum number of sources 2-D Unitary ESPRIT can handle is the minimum of m_x and m_y, assuming that at least $d/2$ snapshots are available. If only a single snapshot is available (or more than two sources are highly correlated), one can extract $d/2$ or more identical subarrays out of the overall array to get the effect of multiple snapshots (spatial smoothing), thereby decreasing the maximum number of sources that can be handled. A brief summary of the described element space implementation of 2-D Unitary ESPRIT is given in Table 63.6.

TABLE 63.6   Summary of 2-D Unitary ESPRIT in Element Space

1. *Signal Subspace Estimation:* Compute $\mathbf{E}_s \in \mathbb{R}^{M \times d}$ as the d dominant left singular vectors of $\mathcal{T}(\mathbf{X}) \in \mathbb{R}^{M \times 2N}$.

2. *Solution of the Invariance Equations:* Solve

$$\underbrace{\mathbf{K}_{\mu 1}\mathbf{E}_s}_{\mathbb{R}^{m_x \times d}}\mathbf{\Upsilon}_\mu \approx \underbrace{\mathbf{K}_{\mu 2}\mathbf{E}_s}_{\mathbb{R}^{m_x \times d}} \quad \text{and} \quad \underbrace{\mathbf{K}_{\nu 1}\mathbf{E}_s}_{\mathbb{R}^{m_y \times d}}\mathbf{\Upsilon}_\nu \approx \underbrace{\mathbf{K}_{\nu 2}\mathbf{E}_s}_{\mathbb{R}^{m_y \times d}}$$

by means of LS, TLS, SLS, or 2-D SLS.

3. *Spatial Frequency Estimation:* Calculate the eigenvalues of the complex-valued $d \times d$ matrix

$$\mathbf{\Upsilon}_\mu + j\mathbf{\Upsilon}_\nu = \mathbf{T}\,\mathbf{\Lambda}\,\mathbf{T}^{-1} \quad \text{with} \quad \mathbf{\Lambda} = \text{diag}\{\lambda_i\}_{i=1}^{d}$$

   - $\mu_i = 2\arctan\left(\text{Re}\{\lambda_i\}\right), \quad 1 \le i \le d$
   - $\nu_i = 2\arctan\left(\text{Im}\{\lambda_i\}\right), \quad 1 \le i \le d$

It is instructive to examine a very simple numerical example. Consider a uniform rectangular array (URA) of $M = 2 \times 2 = 4$ sensor elements, i.e., $M_x = M_y = 2$. Effecting maximum overlap, we have $m_x = m_y = 2$. For the sake of simplicity, assume that the true covariance matrix of the noise-corrupted measurements

$$R_{xx} = E\{x(t)x^H(t)\} = AR_{ss}A^H + \sigma_N^2 I_4 = \begin{bmatrix} 3 & 0 & 1-j & -1+j \\ 0 & 3 & 1-j & 1-j \\ 1+j & 1+j & 3 & 0 \\ -1-j & 1+j & 0 & 3 \end{bmatrix}$$

is known. Here, $R_{ss} = E\{s(t)s^H(t)\} \in \mathbb{C}^{d \times d}$ denotes the unknown signal covariance matrix. Furthermore, the measurement vector $x(t)$ is defined as

$$x(t) = \begin{bmatrix} x_{11}(t) & x_{12}(t) & x_{21}(t) & x_{22}(t) \end{bmatrix}^T. \tag{63.43}$$

In this example, we have to use a covariance approach instead of the direct data approach summarized in Table 63.6, since the array measurements $x(t)$ themselves are not known. To this end, we will compute the eigendecomposition of the real part of the transformed covariance matrix as, for instance, discussed in [25]. According to Eq. (63.13), the left Π-real transformation matrices Q_M and $Q_{m_x} = Q_{m_y}$ take the form

$$Q_4 = \frac{1}{\sqrt{2}} \begin{bmatrix} 1 & 0 & j & 0 \\ 0 & 1 & 0 & j \\ 0 & 1 & 0 & -j \\ 1 & 0 & -j & 0 \end{bmatrix} \quad \text{and} \quad Q_2 = \frac{1}{\sqrt{2}} \begin{bmatrix} 1 & j \\ 1 & -j \end{bmatrix},$$

respectively. Therefore, we have

$$R_Q = \text{Re}\left\{ Q_4^H R_{xx} Q_4 \right\} = Q_4^H R_{xx} Q_4 = \begin{bmatrix} 2 & 1 & 1 & -1 \\ 1 & 4 & -1 & -1 \\ 1 & -1 & 4 & -1 \\ -1 & -1 & -1 & 2 \end{bmatrix}. \tag{63.44}$$

The eigenvalues of R_Q are given by $\varrho_1 = 5, \varrho_2 = 5, \varrho_3 = 1$, and $\varrho_4 = 1$. Clearly, ϱ_1 and ϱ_2 are the dominant eigenvalues, and the variance of the additive noise is identified as $\sigma_N^2 = \varrho_3 = \varrho_4 = 1$. Therefore, there are $d = 2$ impinging wavefronts. The columns of

$$E_s = \begin{bmatrix} 1 & 0 \\ 1 & 1 \\ 1 & -1 \\ -1 & 0 \end{bmatrix}$$

contain eigenvectors of R_Q corresponding to the $d = 2$ largest eigenvalues ϱ_1 and ϱ_2. The four selection matrices

$$J_{\mu 1} = \begin{bmatrix} 1 & 0 & 0 & 0 \\ 0 & 0 & 1 & 0 \end{bmatrix}, \quad J_{\mu 2} = \begin{bmatrix} 0 & 1 & 0 & 0 \\ 0 & 0 & 0 & 1 \end{bmatrix},$$

$$J_{\nu 1} = \begin{bmatrix} 1 & 0 & 0 & 0 \\ 0 & 1 & 0 & 0 \end{bmatrix}, \quad J_{\nu 2} = \begin{bmatrix} 0 & 0 & 1 & 0 \\ 0 & 0 & 0 & 1 \end{bmatrix},$$

are constructed in accordance with Eq. (63.43), cf. Fig. 63.10, yielding

$$K_{\mu 1} = \begin{bmatrix} 1 & 1 & 0 & 0 \\ 0 & 0 & 1 & 1 \end{bmatrix}, \quad K_{\mu 2} = \begin{bmatrix} 0 & 0 & -1 & 1 \\ 1 & -1 & 0 & 0 \end{bmatrix},$$

$$K_{\nu 1} = \begin{bmatrix} 1 & 1 & 0 & 0 \\ 0 & 0 & 1 & -1 \end{bmatrix}, \quad K_{\nu 2} = \begin{bmatrix} 0 & 0 & -1 & -1 \\ 1 & -1 & 0 & 0 \end{bmatrix},$$

according to Eq. (63.38) and Eq. (63.39). With these definitions, the invariance equations (63.41) turn out to be

$$\begin{bmatrix} 2 & 1 \\ 0 & -1 \end{bmatrix} \Upsilon_\mu \approx \begin{bmatrix} -2 & 1 \\ 0 & -1 \end{bmatrix} \quad \text{and} \quad \begin{bmatrix} 2 & 1 \\ 2 & -1 \end{bmatrix} \Upsilon_\nu \approx \begin{bmatrix} 0 & 1 \\ 0 & -1 \end{bmatrix}.$$

Solving these matrix equations, we get

$$\Upsilon_\mu = \begin{bmatrix} -1 & 0 \\ 0 & 1 \end{bmatrix} \quad \text{and} \quad \Upsilon_\nu = \begin{bmatrix} 0 & 0 \\ 0 & 1 \end{bmatrix}.$$

Finally, the eigenvalues of the "complexified" 2×2 matrix $\Upsilon_\mu + j\Upsilon_\nu$ are observed to be $\lambda_1 = -1$ and $\lambda_2 = 1 + j$, corresponding to the spatial frequencies

$$\mu_1 = -\frac{\pi}{2}, \quad \nu_1 = 0 \quad \text{and} \quad \mu_2 = \frac{\pi}{2}, \quad \nu_2 = \frac{\pi}{2}.$$

If we assume that $\Delta_x = \Delta_y = \lambda/2$, the direction cosines are given by $u_i = \mu_i/\pi$ and $v_i = \nu_i/\pi$, $i = 1, 2$. According to Eq. (63.33), the corresponding azimuth and elevation angles can be calculated as

$$\phi_1 = 180°, \quad \theta_1 = 30°, \quad \text{and} \quad \phi_2 = 45°, \quad \theta_2 = 45°.$$

63.6.4 2-D Unitary ESPRIT in DFT Beamspace

Here, we will restrict the presentation of 2-D Unitary ESPRIT in DFT beamspace to uniform rectangular arrays (URAs) of $M = M_x \cdot M_y$ identical sensors, cf. Fig. 63.10.[9] Without loss of generality, assume that the M sensors are omnidirectional and that the centroid of the URA is chosen as the phase reference.

Let us form B_x out of M_x beams in x-direction and B_y out of M_y beams in y-direction, yielding a total of $B = B_x \cdot B_y$ beams. Then the corresponding scaled DFT-matrices $W_{B_x}^H \in \mathbb{C}^{B_x \times M_x}$ and $W_{B_y}^H \in \mathbb{C}^{B_y \times M_y}$ are formed as discussed in Section 63.3.2. Now, viewing the array output at a given snapshot as an $M_x \times M_y$ matrix, premultiply this matrix by $W_{B_x}^H$ and postmultiply it by $\overline{W}_{B_y}$.[10] Then apply the vec$\{\cdot\}$-operator, and place the resulting $B \times 1$ vector ($B = B_x \cdot B_y$) as a column of a matrix $Y \in \mathbb{C}^{B \times N}$. The vec$\{\cdot\}$-operator maps a $B_x \times B_y$ matrix to a $B \times 1$ vector by stacking the columns of the matrix. Note that if X denotes the $M \times N$ complex-valued element space data matrix, it is easy to show that the relationship between Y and X may be expressed as $Y = (W_{B_y}^H \otimes W_{B_x}^H)X$ [24]. Here, the symbol $\otimes$ denotes the Kronecker matrix product [5].

Let the columns of $E_s \in \mathbb{R}^{B \times d}$ contain the d left singular vectors of

$$\begin{bmatrix} \text{Re}\{Y\} & \text{Im}\{Y\} \end{bmatrix} \in \mathbb{R}^{B \times 2N} \tag{63.45}$$

corresponding to its d largest singular values. To set up two invariance equations similar to Eq. (63.41), but with a reduced dimensionality, let us define the selection matrices

$$\Gamma_{\mu 1} = I_{B_y} \otimes \Gamma_1^{(B_x)} \quad \text{and} \quad \Gamma_{\mu 2} = I_{B_y} \otimes \Gamma_2^{(B_x)} \tag{63.46}$$

of size $b_x \times B$ for the x-direction ($b_x = (B_x - 1) \cdot B_y$) and

$$\Gamma_{\nu 1} = \Gamma_1^{(B_y)} \otimes I_{B_x} \quad \text{and} \quad \Gamma_{\nu 2} = \Gamma_2^{(B_y)} \otimes I_{B_x} \tag{63.47}$$

---

[9]In [24], we have also described how to use 2-D Unitary ESPRIT in DFT beamspace for cross arrays as depicted in Fig. 63.9 (c).

[10]This can be achieved via a 2-D FFT with appropriate scaling.

of size $b_y \times B$ for the y-direction ($b_y = B_x \cdot (B_y - 1)$). Then $\boldsymbol{\Upsilon}_\mu \in \mathbb{R}^{d \times d}$ and $\boldsymbol{\Upsilon}_\nu \in \mathbb{R}^{d \times d}$ can be calculated as the LS, TLS, SLS, or 2-D SLS solution of

$$\boldsymbol{\Gamma}_{\mu 1} \boldsymbol{E}_s \boldsymbol{\Upsilon}_\mu \approx \boldsymbol{\Gamma}_{\mu 2} \boldsymbol{E}_s \in \mathbb{R}^{b_x \times d} \quad \text{and} \quad \boldsymbol{\Gamma}_{\nu 1} \boldsymbol{E}_s \boldsymbol{\Upsilon}_\nu \approx \boldsymbol{\Gamma}_{\nu 2} \boldsymbol{E}_s \in \mathbb{R}^{b_y \times d}, \tag{63.48}$$

respectively. Finally, the desired *automatically paired* spatial frequency estimates μ_i and ν_i, $1 \leq i \leq d$, are obtained from the real and imaginary part of the eigenvalues of the "complexified" matrix $\boldsymbol{\Upsilon}_\mu + j \boldsymbol{\Upsilon}_\nu$ as discussed in Section 63.6.2. Here, the maximum number of sources we can handle is given by the minimum of b_x and b_y, assuming that at least $d/2$ snapshots are available. A summary of 2-D Unitary ESPRIT in DFT beamspace is presented in Table 63.7.

TABLE 63.7 Summary of 2-D Unitary ESPRIT in DFT Beamspace

---

0. Transformation to Beamspace: Compute a 2-D DFT (with appropriate scaling) of the $M_x \times M_y$ matrix of array outputs at each snapshot, apply the vec{$\cdot$}-operator, and place the result as a column of $\boldsymbol{Y}$ $\Longrightarrow$

$\boldsymbol{Y} = \left(\boldsymbol{W}_{B_y}^H \otimes \boldsymbol{W}_{B_x}^H \right) \boldsymbol{X} \in \mathbb{C}^{B \times N} (B = B_x \cdot B_y)$.

1. Signal Subspace Estimation: Compute $\boldsymbol{E}_s \in \mathbb{R}^{B \times d}$ as the d dominant left singular vectors of
$[\ \text{Re}\{\boldsymbol{Y}\} \quad \text{Im}\{\boldsymbol{Y}\}\ ] \in \mathbb{R}^{B \times 2N}$.

2. Solution of the Invariance Equations: Solve

$$\underbrace{\boldsymbol{\Gamma}_{\mu 1} \boldsymbol{E}_s \boldsymbol{\Upsilon}_\mu}_{\mathbb{R}^{b_x \times d}} \approx \underbrace{\boldsymbol{\Gamma}_{\mu 2} \boldsymbol{E}_s}_{\mathbb{R}^{b_x \times d}} \quad \text{and} \quad \underbrace{\boldsymbol{\Gamma}_{\nu 1} \boldsymbol{E}_s \boldsymbol{\Upsilon}_\nu}_{\mathbb{R}^{b_y \times d}} \approx \underbrace{\boldsymbol{\Gamma}_{\nu 2} \boldsymbol{E}_s}_{\mathbb{R}^{b_y \times d}}$$
$$b_x = (B_x - 1) \cdot B_y \qquad b_y = B_x \cdot (B_y - 1)$$

by means of LS, TLS, SLS, or 2-D SLS.

3. Spatial Frequency Estimation: Calculate the eigenvalues of the complex-valued $d \times d$ matrix

$$\boldsymbol{\Upsilon}_\mu + j \boldsymbol{\Upsilon}_\nu = \boldsymbol{T} \boldsymbol{\Lambda} \boldsymbol{T}^{-1} \quad \text{with} \quad \boldsymbol{\Lambda} = \text{diag}\{\lambda_i\}_{i=1}^d$$

- $\mu_i = 2 \arctan(\text{Re}\{\lambda_i\}), \qquad 1 \leq i \leq d$
- $\nu_i = 2 \arctan(\text{Im}\{\lambda_i\}), \qquad 1 \leq i \leq d$

63.6.5 Simulation Results

Simulations were conducted employing a URA of 8×8 elements, i.e., $M_x = M_y = 8$, with $\Delta_x = \Delta_y = \lambda/2$. The source scenario consisted of $d = 3$ equi-powered, uncorrelated sources located at $(u_1, v_1) = (0, 0)$, $(u_2, v_2) = (1/8, 0)$, and $(u_3, v_3) = (0, 1/8)$, where u_i and v_i are the direction cosines of the ith source relative to the x- and y-axes, respectively. Notice that sources 1 and 2 have the same v-coordinates, while sources 2 and 3 have the same u-coordinates. A given trial run at a given SNR level (per source per element) involved $N = 64$ snapshots. The noise was *i.i.d.* from element to element and from snapshot to snapshot. The RMS error defined as

$$\text{RMSE}_i = \sqrt{\text{E}\{(\hat{u}_i - u_i)^2\} + \text{E}\{(\hat{v}_i - v_i)^2\}}, \quad i = 1, 2, 3, \tag{63.49}$$

was employed as the performance metric. Let $(\hat{u}_{i_k}, \hat{v}_{i_k})$ denote the coordinate estimates of the ith source obtained at the kth run. Sample performance statistics were computed from $K = 1000$ independent trials as

$$\widehat{\text{RMSE}}_i = \sqrt{\frac{1}{K} \sum_{k=1}^{T} \{(\hat{u}_{i_k} - u_i)^2 + (\hat{v}_{i_k} - v_i)^2\}}, \quad i = 1, 2, 3. \tag{63.50}$$

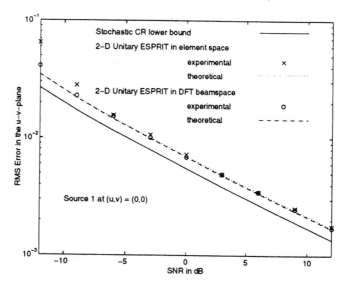

FIGURE 63.11 RMS error of source 1 at $(u_1, v_1) = (0, 0)$ in the u-v plane as a function of the SNR (8×8 sensors, $N = 64$, 1000 trial runs).

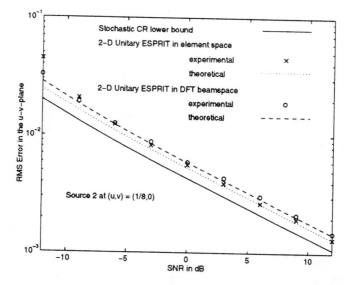

FIGURE 63.12 RMS error of source 2 at $(u_2, v_2) = (1/8, 0)$ in the u-v plane as a function of the SNR (8×8 sensors, $N = 64$, 1000 trial runs).

2-D Unitary ESPRIT in DFT beamspace was implemented with a set of $B = 9$ beams centered at $(u, v) = (0, 0)$, using $B_x = 3$ out of $M_x = 8$ in x-direction (rows 8, 1, and 2 of W_8^H) and also $B_y = 3$ out of $M_y = 8$ in y-direction (again, rows 8, 1, and 2 of W_8^H). Thus, the corresponding subblocks of the selection matrices $\Gamma_1 \in \mathbb{R}^{8 \times 8}$ and $\Gamma_2 \in \mathbb{R}^{8 \times 8}$, used to form $\Gamma_1^{(B_x)}$ and $\Gamma_2^{(B_x)}$ in Eq. (63.46) and also used to form $\Gamma_1^{(B_y)}$ and $\Gamma_2^{(B_y)}$ in Eq. (63.47), are shaded in Fig. 63.3 (b). The bias of 2-D Unitary ESPRIT in element space and DFT beamspace was found to be negligible, facilitating comparison with the Cramér-Rao (CR) lower bound [15]. The resulting performance curves are plotted in Figs. 63.11, 63.12, and 63.13. We have also included theoretical performance predictions of both implementations based on an asymptotic performance analysis [13, 14]. Observe that the empirical RMSEs closely follow the theoretical

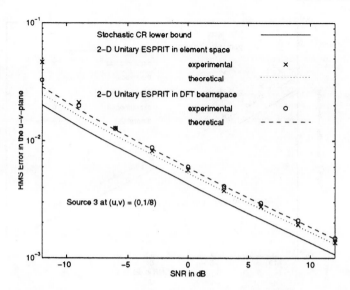

FIGURE 63.13   RMS error of source 3 at $(u_3, v_3) = (0, 1/8)$ in the u-v plane as a function of the SNR (8×8 sensors, $N = 64$, 1000 trial runs).

predictions, except for deviations at low SNRs. The performance of the DFT beamspace implementation is comparable to that of the element space implementation. However, the former requires significantly less computations than the latter, since it operates in a $B = B_x \cdot B_y = 9$ dimensional beamspace as opposed to an $M = M_x \cdot M_y = 64$ dimensional element space.

For SNRs lower than -9 dB, the DFT beamspace version outperformed the element space version of 2-D Unitary ESPRIT. This is due to fact that the DFT beamspace version exploits *a priori* information on the source locations by forming beams pointed in the general directions of the sources.

References

[1] Bienvenu, G. and Kopp, L., Decreasing high resolution method sensitivity by conventional beamforming preprocessing, *Proc. IEEE Int. Conf. Acoust., Speech, Signal Processing*, 33.2.1–33.2.4, San Diego, CA, Mar. 1984.

[2] Brennan, P.V., A low cost phased array antenna for land-mobile satcom applications, *IEEE Proceedings-H*, 138, 131–136, Apr. 1991.

[3] Buckley, K.M. and Xu, X.L., Spatial-spectrum estimation in a location sector, *IEEE Trans. Acoust., Speech, Signal Processing*, ASSP-38, 1842–1852, Nov. 1990.

[4] Davies, D.E.N., *The Handbook of Antenna Design*, vol. 2, Peter Peregrinus, London, U.K., 1983, chap. 12.

[5] Graham, A., *Kronecker Products and Matrix Calculus: With Applications*, Ellis Horwood, Chichester, U.K., 1981.

[6] Haardt, M., Hüper, K., Moore, J.B. and Nossek, J.A., Simultaneous Schur decomposition of several matrices to achieve automatic pairing in multidimensional harmonic retrieval problems, in *Signal Processing VIII: Theories and Applications (Proc. of EUSIPCO-96)*, Trieste, Italy, Sept. 1996, European Association for Signal Processing.

[7] Haardt, M. and Nossek, J.A., Unitary ESPRIT: How to obtain increased estimation accuracy with a reduced computational burden, *IEEE Trans. Signal Processing*, 43, 1232–1242, May 1995.

[8] Haardt, M. and Nossek, J.A., Structured least squares to improve the performance of ESPRIT-type high-resolution techniques, in *Proc. IEEE Int. Conf. Acoust., Speech, Signal Processing*, V, 2805–2808, Atlanta, GA, May 1996.

[9] Haardt, M., Zoltowski, M.D., Mathews, C.P. and Nossek, J.A., 2D Unitary ESPRIT for efficient 2D parameter estimation, in *Proc. IEEE Int. Conf. Acoust., Speech, Signal Processing*, 3, 2096–2099, Detroit, MI, May 1995.

[10] Lee, A., Centrohermitian and skew-centrohermitian matrices, *Linear Algebra and its Applications*, 29, 205–210, 1980.

[11] Lee, H.B. and Wengrovitz, M.S., Resolution threshold of beamspace MUSIC for two closely spaced emitters, *IEEE Trans. Acoust., Speech, Signal Processing*, ASSP-38, 1545–1559, Sept. 1990.

[12] Linebarger, D.A., DeGroat, R.D. and Dowling, E.M., Efficient direction finding methods employing forward/backward averaging, *IEEE Trans. Signal Processing*, 42, 2136–2145, Aug. 1994.

[13] Mathews, C.P., Haardt, M. and Zoltowski, M.D., Implementation and performance analysis of 2D DFT Beamspace ESPRIT, in *Proc. 29th Asilomar Conf. on Signals, Systems, and Computers*, 1, 726–730, Pacific Grove, CA, Nov. 1995, IEEE Computer Society Press.

[14] Mathews, C.P., Haardt, M. and Zoltowski, M.D., Performance analysis of closed-form, ESPRIT based 2-D angle estimator for rectangular arrays, *IEEE Signal Processing Letters*, 3, 124–126, Apr. 1996.

[15] Mathews, C.P. and Zoltowski, M.D., Eigenstructure techniques for 2-D angle estimation with uniform circular arrays, *IEEE Trans. Signal Processing*, 42, 2395–2407, Sept. 1994.

[16] Mathews, C.P. and Zoltowski, M.D., Performance analysis of the UCA-ESPRIT algorithm for circular ring arrays, *IEEE Trans. Signal Processing*, 42, 2535–2539, Sept. 1994.

[17] Mathews, C.P. and Zoltowski, M.D., Closed-form 2D angle estimation with circular arrays/apertures via phase mode exitation and ESPRIT, in *Advances in Spectrum Analysis and Array Processing*, Haykin, S., Ed., vol. III, 171–218, Prentice-Hall, Englewood Cliffs, NJ, 1995.

[18] Roy, R. and Kailath, T., ESPRIT — Estimation of signal parameters via rotational invariance techniques, *IEEE Trans. Acoust., Speech, Signal Processing*, ASSP-37, 984–995, July 1989.

[19] Sensi, J., *Aspects of Modern Radar*, Artech House, 1988.

[20] Steinberg, B.D., Introduction to periodic array synthesis, *Principle of Aperture and Array System Design*, John Wiley & Sons, New York, chap. 6, 98–99, 1976.

[21] Swindlehurst, A.L. and Kailath, T., Azimuth/elevation direction finding using regular array geometries, *IEEE Trans. Aerospace and Electronic Systems*, 29, 145–156, Jan. 1993.

[22] Swindlehurst, A.L., Ottersten, B., Roy, R. and Kailath, T., Multiple invariance ESPRIT, *IEEE Trans. Signal Processing*, 40, 867–881, Apr. 1992.

[23] Xu, G., Roy, R.H. and Kailath, T., Detection of number of sources via exploitation of centro-symmetry property, *IEEE Trans. Signal Processing*, 42, 102–112, Jan. 1994.

[24] Zoltowski, M.D., Haardt, M. and Mathews, C.P., Closed-form 2D angle estimation with rectangular arrays in element space or beamspace via Unitary ESPRIT, *IEEE Trans. Signal Processing*, 44, 316–328, Feb. 1996.

[25] Zoltowski, M.D., Kautz, G.M. and Silverstein, S.D., Beamspace root-MUSIC, *IEEE Trans. Signal Processing*, 41, 344–364, Jan. 1993.

[26] Zoltowski, M.D. and Lee, T., Maximum likelihood based sensor array signal processing in the beamspace domain for low-angle radar tracking, *IEEE Trans. Signal Processing*, 39, 656–671, Mar. 1991.

[27] Zoltowski, M.D. and Stavrinides, D., Sensor array signal processing via a Procrustes rotations based eigenanalysis of the ESPRIT data pencil, *IEEE Trans. Acoust., Speech, Signal Processing*, ASSP-37, 832–861, June 1989.

64

A Unified Instrumental Variable Approach to Direction Finding in Colored Noise Fields[1]

P. Stoica
Uppsala University

M. Viberg
Chalmers University of Technology

M. Wong
McMaster University

Q. Wu
CELWAVE

The main goal herein is to describe and analyze, in a unifying manner, the *spatial* and *temporal* IV-SSF approaches recently proposed for array signal processing in colored noise fields. (The acronym IV-SSF stands for "Instrumental Variable - Signal Subspace Fitting"). Despite the generality of the approach taken herein, our analysis technique is simpler than those used in previous more specialized publications. We derive a general, optimally-weighted (optimal, for short), IV-SSF direction estimator and show that this estimator encompasses the UNCLE estimator of Wong and Wu, which is a spatial IV-SSF method, and the temporal IV-SSF estimator of Viberg, Stoica and Ottersten. The latter two estimators have seemingly different forms (among others, the first of them makes use of four weights, whereas the second one uses three weights "only"), and hence their asymptotic equivalence shown in this paper comes as a surprising unifying result. We hope that the present paper, along with the original works aforementioned, will stimulate the interest in the IV-SSF approach to array signal processing, which is sufficiently flexible to handle colored noise fields, coherent signals and indeed also situations were only some of the sensors in the array are calibrated.

[1]This work was supported in part by the Swedish Research Council for Engineering Sciences (TFR).

64.1   Introduction

Most parametric methods for Direction-Of-Arrival (DOA) estimation require knowledge of the spatial (sensor-to-sensor) color of the background noise. If this information is unavailable, a serious degradation of the quality of the estimates can result, particularly at low Signal-to-Noise Ratio (SNR) [1, 2, 3]. A number of methods have been proposed over the recent years to alleviate the sensitivity to the noise color. If a parametric model of the covariance matrix of the noise is available, the parameters of the noise model can be estimated along with those of the interesting signals [4, 5, 6, 7]. Such an approach is expected to perform well in situations where the noise can be accurately modeled with relatively few parameters. An alternative approach, which does not require a precise model of the noise, is based on the principle of Instrumental Variables (IV). See [8, 9] for thorough treatments of IV methods (IVM) in the context of identification of linear time-invariant dynamical systems. A brief introduction is given in the appendix of this chapter. Computationally simple IVMs for array signal processing appeared in [10, 11]. These methods perform poorly in difficult scenarios involving closely spaced DOAs and correlated signals.

More recently, the combined Instrumental Variable Signal Subspace Fitting (IV-SSF) technique has been proposed as a promising alternative to array signal processing in spatially colored noise fields [12, 13, 14, 15]. The IV-SSF approach has a number of appealing advantages over other DOA estimation methods. These advantages include:

- IV-SSF can handle noises with arbitrary spatial correlation, under minor restrictions on the signals or the array. In addition, estimation of a noise model is avoided, which leads to statistical robustness and computational simplicity.
- The IV-SSF approach is applicable to both non-coherent and coherent signal scenarios.
- The spatial IV-SSF technique can make use of the information contained in the output of a completely uncalibrated subarray under certain weak conditions, which other methods cannot.

Depending on the type of "instrumental variables" used, two classes of IV methods have appeared in the literature:

1. *Spatial IVM*, for which the instrumental variables are derived from the output of a (possibly uncalibrated) subarray the noise of which is uncorrelated with the noise in the main calibrated subarray under consideration (see [12, 13]).
2. *Temporal IVM*, which obtains instrumental variables from the delayed versions of the array output, under the assumption that the temporal-correlation length of the noise field is shorter than that of the signals (see [11, 14]).

The previous literature on IV-SSF has treated and analyzed the above two classes of spatial and temporal methods separately, ignoring their common basis. In this contribution, we reveal the common roots of these two classes of DOA estimation methods and study them under the same umbrella. Additionally, we establish the statistical properties of a general (either spatial or temporal) weighted IV-SSF method and present the optimal weights that minimize the variance of the DOA estimation errors. In particular, we point out that the optimal four-weight spatial IV-SSF of [12, 13] (called UNCLE there, and arrived at by using canonical correlation decomposition ideas) and the optimal three-weight temporal IV-SSF of [14] are asymptotically equivalent when used under the same conditions. This asymptotic equivalence property, which is a main result of the present section, is believed to be important as it shows the close ties that exist between two seemingly different DOA estimators.

This section is organized as follows. In Section 64.2 the data model and technical assumptions are introduced. Next, in Section 64.3 the IV-SSF method is presented in a fairly general setting. In Section 64.4, the statistical performance of the method is presented along with the optimal choices of certain user-specified quantities. The data requirements and the optimal IV-SSF (UNCLE) algorithm are summarized in Section 64.5. The anxious reader may wish to jump directly to this point to investigate the usefulness of the algorithm in a specific application. In Section 64.6, some numerical examples and computer simulations are presented to illustrate the performance. The conclusions are given in Section 64.7. In the appendix we give a brief introduction to IV methods. The reader who is not familiar with IV might be helped by reading the appendix before the rest of the paper. Background material on the subspace-based approach to DOA estimation can be found in Chapter 62 of this Handbook.

64.2   Problem Formulation

Consider a scenario in which n narrowband plane waves, generated by point sources, impinge on an array comprising m calibrated sensors. Assume, for simplicity, that the n sources and the array are situated in the same plane. Let $a(\theta)$ denote the complex array response to a unit-amplitude signal with DOA parameter equal to θ. Under these assumptions, the output of the array, $y(t) \in C^{m \times 1}$, can be described by the following well-known equation [16, 17]:

$$y(t) = Ax(t) + e(t) \tag{64.1}$$

where $x(t) \in C^{n \times 1}$ denotes the signal vector, $e(t) \in C^{m \times 1}$ is a noise term, and

$$A = [a(\theta_1) \cdots a(\theta_n)] \tag{64.2}$$

Hereafter, θ_k denotes the kth DOA parameter.

The following assumptions on the quantities in the array equation, (64.1), are considered to hold throughout this section:

A1. The signal vector $x(t)$ is a normally distributed random variable with zero mean and a possibly singular covariance. The signals may be temporally correlated; in fact the temporal IV-SSF approach relies on the assumption that the signals exhibit some form of temporal correlation (see below for details).

A2. The noise $e(t)$ is a random vector that is temporally white, uncorrelated with the signals and circularly symmetric normally distributed with zero mean and unknown covariance matrix[2] $Q > 0$,

$$E[e(t)e^*(s)] = Q\,\delta_{t,s}\,; \quad E[e(t)e^T(s)] = O \tag{64.3}$$

A3. The manifold vectors $\{a(\theta)\}$, corresponding to any set of m different values of θ, are linearly independent.

Note that assumption A1 above allows for coherent signals, and that in A2 the noise field is allowed to be arbitrarily spatially correlated with an unknown covariance matrix. Assumption A3 is a well-known condition that, under a weak restriction on m, guarantees DOA parameter identifiability in the case Q is known (to within a multiplicative constant) [18]. When Q is completely unknown, DOA identifiability can only be achieved if further assumptions are made on the scenario under consideration. The following assumption is typical of the IV-SSF approach:

A4. There exists a vector $z(t) \in C^{\bar{m} \times 1}$, which is normally distributed and satisfies

$$\begin{aligned}
E[z(t)e^*(s)] &= O \quad \text{for } t \leq s & (64.4)\\
E[z(t)e^T(s)] &= O \quad \text{for all } t, s & (64.5)
\end{aligned}$$

---

[2] Henceforth, the superscript "*" denotes the conjugate transpose; whereas the transpose is designated by a superscript "T". The notation $A \geq B$, for two Hermitian matrices A and B, is used to mean that $(A - B)$ is a nonnegative definite matrix. Also, O denotes a zero matrix of suitable dimension.

Furthermore, denote

$$\Gamma = E[z(t)x^*(t)] \qquad (\bar{m} \times n) \qquad (64.6)$$
$$\bar{n} = \text{rank}(\Gamma) \leq \bar{m}. \qquad (64.7)$$

It is assumed that no row of Γ is identically zero and that the inequality

$$\bar{n} > 2n - m \qquad (64.8)$$

holds (note that a rank-one Γ matrix can satisfy the condition (64.8) if m is large enough, and hence the condition in question is rather weak). Owing to its (partial) uncorrelatedness with $\{e(t)\}$, the vector $\{z(t)\}$ can be used to eliminate the noise from the array output equation (64.1), and for this reason $\{z(t)\}$ is called an IV vector. Below, we briefly describe three possible ways to derive an IV vector from the available data measured with an array of sensors (for more details on this aspect, the reader should consult [12, 13, 14]).

EXAMPLE 64.1: Spatial IV

Assume that the n signals, which impinge on the main (sub)array under consideration, are also received by another (sub)array that is sufficiently distanced from the main one so that the noise vectors in the two subarrays are uncorrelated with one another. Then $z(t)$ can be made from the outputs of the sensors in the second subarray (note that those sensors need not be calibrated) [12, 13, 15].

EXAMPLE 64.2: Temporal IV

When a second subarray, as described above, is not available but the signals are temporally correlated, one can obtain an IV vector by delaying the output vector: $z(t) = [y^T(t-1) \ y^T(t-2) \ \cdots \ ]^T$. Clearly, such a vector $z(t)$ satisfies (64.4) and (64.5), and it also satisfies (64.8) under weak conditions on the signal temporal correlation. This construction of an IV vector can be readily extended to cases where $e(t)$ is temporally correlated, provided that the signal temporal correlation length is longer than that corresponding to the noise [11, 14].

In a sense, the above examples are both special cases of the following more general situation:

EXAMPLE 64.3: Reference Signal

In many systems a reference or pilot signal [19, 20] $z(t)$ (scalar or vector) is available. If the reference signal is sufficiently correlated with all signals of interest (in the sense of (64.8)) and uncorrelated with the noise, it can be used as an IV. Note that all signals that are not correlated with the reference will be treated as noise. Reference signals are commonly available in communication applications, for example a PN-code in spread spectrum communication [20] or a training signal used for synchronization and/or equalizer training [21]. A closely related possibility is utilization of cyclo-stationarity (or self-coherence), a property that is exhibited by many man-made signals. The reference signal(s) can then consist, for example, of sinusoids of different frequencies [22, 23]. In these techniques, the data is usually pre-processed by computing the auto-covariance function (or a higher-order statistic) before correlating with the reference signal.

The problem considered in this section concerns the estimation of the DOA vector

$$\boldsymbol{\theta} = [\theta_1, \cdots, \theta_n]^T \qquad (64.9)$$

given N snapshots of the array output and of the IV vector, $\{y(t), z(t)\}_{t=1}^N$. The number of signals, n, and the rank of the covariance matrix Γ, $\bar{n}$, are assumed to be given (for the estimation of these integer-valued parameters by means of IV/SSF-based methods, we refer to [24, 25]).

64.3   The IV-SSF Approach

Let

$$\hat{R} = \hat{W}_L \left[\frac{1}{N} \sum_{t=1}^{N} z(t) y^*(t) \right] \hat{W}_R \qquad (\bar{m} \times m) \tag{64.10}$$

where $\hat{W}_L$ and $\hat{W}_R$ are two nonsingular Hermitian weighting matrices which are possibly data-dependent (as indicated by the fact that they are roofed). Under the assumptions made, as $N \rightarrow \infty$, $\hat{R}$ converges to the matrix:

$$R = W_L E[z(t) y^*(t)] W_R = W_L \Gamma A^* W_R \tag{64.11}$$

where W_L and W_R are the limiting weighting matrices (assumed to be bounded and nonsingular). Owing to assumptions *A2* and *A3*,

$$\text{rank}(R) = \bar{n} \tag{64.12}$$

Hence, the Singular Value Decomposition (SVD) [26] of R can be written as

$$R = [U \; ?] \begin{bmatrix} \Lambda & O \\ O & O \end{bmatrix} \begin{bmatrix} S^* \\ ? \end{bmatrix} = U \Lambda S^* \tag{64.13}$$

where $U^*U = S^*S = I$, $\Lambda \in \mathcal{R}^{\bar{n} \times \bar{n}}$ is diagonal and nonsingular, and where the question marks stand for blocks that are of no importance for the present discussion.

The following key equality is obtained by comparing the two expressions for R in Eqs. (64.11) and (64.13) above:

$$S = W_R AC \tag{64.14}$$

where $C \overset{\triangle}{=} \Gamma^* W_L U \Lambda^{-1} \in \mathcal{C}^{n \times \bar{n}}$ has full column rank. For a given S, the true DOA vector can be obtained as the unique solution to Eq. (64.14) under the parameter identifiability condition (64.8) (see, e.g., [18]). In the more realistic case when S is unknown, one can make use of Eq. (64.14) to estimate the DOA vector in the following steps.

The IV step — Compute the pre- and post-weighted sample covariance matrix $\hat{R}$ in Eq. (64.10), along with its SVD:

$$\hat{R} = \begin{bmatrix} \hat{U} & ? \end{bmatrix} \begin{bmatrix} \hat{\Lambda} & O \\ O & ? \end{bmatrix} \begin{bmatrix} \hat{S}^* \\ ? \end{bmatrix} \tag{64.15}$$

where $\hat{\Lambda}$ contains the $\bar{n}$ largest singular values. Note that $\hat{U}$, $\hat{\Lambda}$, and $\hat{S}$ are consistent estimates of U, Λ, and S in the SVD of R.

The SSF step — Compute the DOA estimate as the minimizing argument of the following signal subspace fitting criterion:

$$\min_{\theta} \{ \min_{C} [\text{vec}(\hat{S} - \hat{W}_R AC)]^* \hat{V} [\text{vec}(\hat{S} - \hat{W}_R AC)] \} \tag{64.16}$$

where $\hat{V}$ is a positive definite weighting matrix, and "vec" is the vectorization operator[3]. Alternatively, one can estimate the DOA instead by minimizing the following criterion:

$$\min_{\theta} \{ [\text{vec}(B^* \hat{W}_R^{-1} \hat{S})]^* \hat{W} [\text{vec}(B^* \hat{W}_R^{-1} \hat{S})] \} \tag{64.17}$$

---

[3]If x_k is the kth column of a matrix X, then $\text{vec}(X) = [x_1^T \; x_2^T \; \cdots \;]^T$.

where $\hat{W}$ is a positive definite weight, and $B \in \mathcal{C}^{m \times (m-n)}$ is a matrix whose columns form a basis of the null-space of A^* (hence, $B^*A = 0$ and rank $(B) = m - n$). The alternative fitting criterion above is obtained from the simple observation that Eq. (64.14) along with the definition of B imply that

$$B^* W_R^{-1} S = 0 \tag{64.18}$$

It can be shown [27] that *the classes of DOA estimates derived from Eqs. (64.16) and (64.17), respectively, are asymptotically equivalent.* More exactly, for any $\hat{V}$ in Eq. (64.16) one can choose $\hat{W}$ in Eq. (64.17) so that the DOA estimates obtained by minimizing Eq. (64.16) and, respectively, Eq. (64.17) have the same asymptotic distribution and vice-versa.

In view of the previous result, in an asymptotical analysis it suffices to consider only one of the two criteria above. In the following, we focus on Eq. (64.17). Compared with Eq. (64.16), the criterion (64.17) has the advantage that it depends on the DOA only. On the other hand, for a general array there is no known closed-form parameterization of B in terms of θ. However, as shown in the following, this is no drawback because the optimally weighted criterion (which is the one to be used in applications) is an explicit function of θ.

64.4   The Optimal IV-SSF Method

In what follows, we deal with the essential problem of choosing the weights $\hat{W}$, $\hat{W}_R$, and $\hat{W}_L$ in the IV-SSF criterion (64.17) so as to maximize the DOA estimation accuracy. First, we optimize the accuracy with respect to $\hat{W}$, and then with respect to $\hat{W}_R$ and $\hat{W}_L$.

Optimal Selection of $\hat{W}$

Define

$$g(\theta) = \text{vec}\,(B^* \hat{W}_R^{-1} \hat{S}) \tag{64.19}$$

and observe that the criterion function in Eq. (64.17) can be written as,

$$g^*(\theta) \hat{W} g(\theta) \tag{64.20}$$

In [27] it is shown that $g(\theta)$ (evaluated at the true DOA vector) has, asymptotically in N, a circularly symmetric normal distribution with zero mean and the following covariance:

$$G(\theta) = \frac{1}{N}[(W_L U \Lambda^{-1})^* R_z (W_L U \Lambda^{-1})]^T \otimes [B^* R_y B] \tag{64.21}$$

where $\otimes$ denotes the Kronecker matrix product [28]; and where, for a stationary signal $s(t)$, we use the notation

$$R_s = \text{E}\,[s(t)s^*(t)]\,. \tag{64.22}$$

Then, it follows from the ABC (Asymptotically Best Consistent) theory of parameter estimation[4] that the minimum variance estimate, in the class of estimates under discussion, is given by the minimizing argument of the criterion in Eq. (64.20) with $\hat{W} = \hat{G}^{-1}(\theta)$, that is

$$f(\theta) = g^*(\theta) \hat{G}^{-1}(\theta) g(\theta) \tag{64.23}$$

---

[4]For details on the ABC theory, which is an extension of the classical BLUE (Best Linear Unbiased Estimation) / Markov theory of linear regression to a class of nonlinear regressions with asymptotically vanishing residuals, the reader is referred to [9, 29].

where

$$\hat{G}(\theta) = \frac{1}{N}[(\hat{W}_L \hat{U} \hat{\Lambda}^{-1})^* \hat{R}_z (\hat{W}_L \hat{U} \hat{\Lambda}^{-1})]^T \otimes [B^* \hat{R}_y B] \tag{64.24}$$

and where $\hat{R}_z$ and $\hat{R}_y$ are the usual sample estimates of R_z and R_y. Furthermore, it is easily shown that the minimum variance estimate, obtained by minimizing Eq. (64.23), is asymptotically normally distributed with mean equal to the true parameter vector and the following covariance matrix:

$$H = \frac{1}{2} \{\mathrm{Re}\,[J^* G^{-1}(\theta) J]\}^{-1} \tag{64.25}$$

where

$$J = \lim_{N \to \infty} \frac{\partial g(\theta)}{\partial \theta}. \tag{64.26}$$

The following more explicit formula for H is derived in [27]:

$$H = \frac{1}{2N} \left(\mathrm{Re} \left\{ \left[D^* R_y^{-1/2} \Pi^{\perp}_{R_y^{-1/2} A} R_y^{-1/2} D \right] \odot \Omega^T \right\} \right)^{-1} \tag{64.27}$$

where $\odot$ denotes the Hadamard-Schur matrix product (elementwise multiplication) and

$$\Omega = \Gamma^* W_L U (U^* W_L R_z W_L U)^{-1} U^* W_L \Gamma. \tag{64.28}$$

Furthermore, the notation $Y^{-1/2}$ is used for a Hermitian (for notational convenience) square root of the inverse of a positive definite matrix Y, the matrix D is made from the direction vector derivatives,

$$D = [d_1 \quad \cdots \quad d_n]; \quad d_k = \frac{\partial a(\theta_k)}{\partial \theta_k}$$

and, for a full column-rank matrix X, $\Pi^{\perp}_X$ defines the orthogonal projection onto the nullspace of X^* as

$$\Pi^{\perp}_X = I - \Pi_X; \quad \Pi_X = X(X^* X)^{-1} X^*. \tag{64.29}$$

To summarize, for fixed $\hat{W}_R$ and $\hat{W}_L$, the statistically optimal selection of $\hat{W}$ leads to DOA estimates with an asymptotic normal distribution with mean equal to the true DOA vector and covariance matrix given by Eq. (64.27).

Optimal Selection of $\hat{W}_R$ and $\hat{W}_L$

The optimal weights $\hat{W}_R$ and $\hat{W}_L$ are, by definition, those that minimize the limiting covariance matrix H of the DOA estimation errors. In the expression (64.27) of H, only Ω depends on W_R and W_L (the dependence on W_R is implicit, via U). Since the matrix Γ has rank $\bar{n}$, it can be factorized as follows:

$$\Gamma = \Gamma_1 \Gamma_2^* \tag{64.30}$$

where both $\Gamma_1 \in C^{\bar{m} \times \bar{n}}$ and $\Gamma_2 \in C^{n \times \bar{n}}$ have full column rank. Insertion of Eq. (64.30) into the equality $W_L \Gamma A^* W_R = U \Lambda S^*$ yields the following equation, after a simple manipulation,

$$W_L \Gamma_1 T = U \tag{64.31}$$

where $T = \Gamma_2^* A^* W_R S \Lambda^{-1} \in C^{\bar{n} \times \bar{n}}$ is a nonsingular transformation matrix. By using Eq. (64.31) in Eq. (64.28), we obtain:

$$\Omega = \Gamma_2 (\Gamma_1^* W_L^2 \Gamma_1)(\Gamma_1^* W_L^2 R_z W_L^2 \Gamma_1)^{-1}(\Gamma_1^* W_L^2 \Gamma_1)\Gamma_2^* \tag{64.32}$$

Observe that Ω does not actually depend on W_R. Hence, $\hat{W}_R$ *can be arbitrarily selected, as any nonsingular Hermitian matrix, without affecting the asymptotics of the DOA parameter estimates!*

Concerning the choice of $\hat{W}_L$, it is easily verified that

$$\Omega \leq \Omega \mid_{W_L = R_z^{-1/2}} = \Gamma_2(\Gamma_1^* R_z^{-1} \Gamma_1)\Gamma_2^* = \Gamma^* R_z^{-1} \Gamma \tag{64.33}$$

Indeed,

$$\Gamma^* R_z^{-1} \Gamma - \Omega = \Gamma_2[\Gamma_1^* R_z^{-1} \Gamma_1 - (\Gamma_1^* W_L^2 \Gamma_1)(\Gamma_1^* W_L^2 R_z W_L^2 \Gamma_1)^{-1} \times$$
$$\times (\Gamma_1^* W_L^2 \Gamma_1)]\Gamma_2^* = \Gamma^* R_z^{-1/2} \Pi_{R_z^{1/2} W_L^2 \Gamma_1}^{\perp} R_z^{-1/2} \Gamma \tag{64.34}$$

which is obviously a nonnegative definite matrix. Hence, $W_L = R_z^{-1/2}$ maximizes Ω. Then, it follows from the expression of the matrix H and the properties of the Hadamard-Schur product that this same choice of W_L minimizes H. The conclusion is that *the optimal weight $\hat{W}_L$, which yields the best limiting accuracy, is*

$$\hat{W}_L = \hat{R}_z^{-1/2} \tag{64.35}$$

The (minimum) covariance matrix H, corresponding to the above choice, is given by

$$H_o = \frac{1}{2N} \{ \text{Re} \, [(D^* R_y^{-1/2} \Pi_{R_y^{-1/2} A}^{\perp} R_y^{-1/2} D) \odot (\Gamma^* R_z^{-1} \Gamma)^T] \}^{-1} \tag{64.36}$$

Remark It is worth noting that H_o *monotonically decreases as $\bar{m}$ (the dimension of $z(t)$) increases.* The proof of this claim is similar to the proof of the corresponding result in [9], Complement C8.5. Hence, as could be intuitively expected, one should use all available instruments (spatial and/or temporal) to obtain maximal theoretical accuracy. However, practice has shown that too large a dimension of the IV vector may in fact decrease the empirically observed accuracy. This phenomenon can be explained by the fact that increasing $\bar{m}$ means that a longer data set is necessary for the asymptotic results to be valid.

Optimal IV-SSF Criteria

Fortunately, the criterion, (64.23) and (64.24) can be expressed in a functional form that depends on the indeterminate θ in an explicit way (recall that, for most cases, the dependence of B in Eq. (64.23) on θ is not available in explicit form). By using the following readily verified equality [28],

$$\text{tr}\,(AX^* BY) = [\text{vec}\,(X)]^* [A^T \otimes B][\text{vec}\,(Y)] \tag{64.37}$$

which holds for any conformable matrices A, X, B, and Y, one can write Eq. (64.23) as:[5]

$$f(\theta) = \text{tr}\,\{[(\hat{W}_L \hat{U} \hat{\Lambda}^{-1})^* \hat{R}_z (\hat{W}_L \hat{U} \hat{\Lambda}^{-1})]^{-1} \hat{S}^* \hat{W}_R^{-1} B(B^* \hat{R}_y B)^{-1} B^* \hat{W}_R^{-1} \hat{S}\} \tag{64.38}$$

However, observe that

$$B(B^* \hat{R}_y B)^{-1} B^* = \hat{R}_y^{-1/2} \Pi_{\hat{R}_y^{1/2} B} \hat{R}_y^{-1/2} = \hat{R}_y^{-1/2} \Pi_{\hat{R}_y^{-1/2} A}^{\perp} \hat{R}_y^{-1/2} \tag{64.39}$$

Inserting Eq. (64.39) into Eq. (64.38) yields:

$$f(\theta) = \text{tr}\,[\hat{\Lambda}(\hat{U}^* \hat{W}_L \hat{R}_z \hat{W}_L \hat{U})^{-1} \hat{\Lambda} \hat{S}^* \hat{W}_R^{-1} \hat{R}_y^{-1/2} \Pi_{\hat{R}_y^{-1/2} A}^{\perp} \hat{R}_y^{-1/2} \hat{W}_R^{-1} \hat{S}] \tag{64.40}$$

which is an explicit function of θ. Insertion of the optimal choice of W_L into Eq. (64.40) leads to a further simplification of the criterion as seen below.

---

[5]To within a multiplicative constant.

Owing to the arbitrariness in the choice of $\hat{W}_R$, there exists an infinite class of optimal IV-SSF criteria. In what follows, we consider two members of this class. Let

$$\hat{W}_R = \hat{R}_y^{-1/2} \tag{64.41}$$

Insertion of Eq. (64.41), along with Eq. (64.35), into Eq. (64.40) yields the following criterion function:

$$f_{WW}(\theta) = \text{tr}\left(\Pi^{\perp}_{\hat{R}_y^{-1/2}A} \tilde{S}\tilde{\Lambda}^2\tilde{S}^* \right) \tag{64.42}$$

where $\tilde{S}$ and $\tilde{\Lambda}$ are made from the principal singular right vectors and singular values of the matrix

$$\tilde{R} = \hat{R}_z^{-1/2}\hat{R}_{zy}\hat{R}_y^{-1/2} \tag{64.43}$$

(with $\hat{R}_{zy}$ defined in an obvious way). The function (64.42) is the UNCLE (spatial IV-SSF) criterion of Wong and Wu [12, 13].
Next, choose $\hat{W}_R$ as

$$\hat{W}_R = I \tag{64.44}$$

The corresponding criterion function is

$$f_{VSO}(\theta) = \text{tr}\left(\Pi^{\perp}_{\hat{R}_y^{-1/2}A} \hat{R}_y^{-1/2}\bar{S}\bar{\Lambda}^2\bar{S}^*\hat{R}_y^{-1/2} \right) \tag{64.45}$$

where $\bar{S}$ and $\bar{\Lambda}$ are made from the principal singular pairs of

$$\bar{R} = \hat{R}_z^{-1/2}\hat{R}_{zy} \tag{64.46}$$

The function (64.45) above is recognized as the optimal (temporal) IV-SSF criterion of Viberg et al. [14]. An important consequence of the previous discussion is that the DOA estimation methods of [12, 13] and [14], respectively, which were derived in seemingly unrelated contexts and by means of somewhat different approaches, are in fact asymptotically equivalent when used under the same conditions. These two methods have very similar computational burdens, which can be seen by comparing Eqs. (64.42) and (64.43) with Eqs. (64.45) and (64.46). Also, their finite-sample properties appear to be rather similar, as demonstrated in the simulation examples. Numerical algorithms for the minimization of the type of criterion function associated with the optimal IV-SSF methods are discussed in [17]. Some suggestions are also given in the summary below.

64.5 Algorithm Summary

The estimation method presented in this section is useful for direction finding in the presence of noise of unknown spatial color. The underlying assumptions and the algorithm can be summarized as follows:

      Assumptions — A batch of N samples of the array output $y(t)$, that can accurately be described by the model (64.1) and (64.2) is available. The array is calibrated in the sense that $a(\theta)$ is a known function of its argument θ. In addition, N samples of the IV-vector $z(t)$, fulfilling Eqs. (64.4) through (64.8), are given. In words, the IV vector is uncorrelated with the noise but well correlated with the signal. In practice, $z(t)$ may be taken from a second subarray, a delayed version of $y(t)$, or a reference (pilot) signal. In the former case, the second subarray need not be calibrated.

Algorithm — In the following we summarize the UNCLE version (64.42) of the algorithm. First, compute $\tilde{R}$ from the sample statistics of $y(t)$ and $z(t)$, according to

$$\tilde{R} = \hat{R}_z^{-1/2} \hat{R}_{zy} \hat{R}_y^{-1/2} .$$

From a numerical point of view, this is best done using QR factorization. Next, partition the singular value decomposition of $\tilde{R}$ according to

$$\tilde{R} = \begin{bmatrix} \tilde{U} & ? \end{bmatrix} \begin{bmatrix} \tilde{\Lambda} & O \\ O & ? \end{bmatrix} \begin{bmatrix} \tilde{S}^* \\ ? \end{bmatrix} ,$$

where $\tilde{S}$ contains the $\bar{n}$ principal right singular vectors and the diagonal matrix $\tilde{\Lambda}$ the corresponding singular values. If $\bar{n}$ is unknown, it can be estimated as the number of significant singular values. Finally, compute the DOA estimates as the minimizing arguments of the criterion function

$$f_{WW}(\boldsymbol{\theta}) = \mathrm{tr}\left(\Pi_{\hat{R}_y^{-1/2}A}^{\perp} \tilde{S}\tilde{\Lambda}^2 \tilde{S}^* \right)$$

using $n = \bar{n}$. If the minimum value of the criterion is "large", it is an indication that more than $\bar{n}$ sources are present. In the general case, a numerical search must be performed to find the minimum. The **leastsq** implementation in Matlab^{TM}, which uses the Levenberg-Marquardt or Gauss-Newton techniques [30], is a possible choice. To initialize the search, one can use the alternating projection procedure [31]. In short, a grid search over $f_{WW}(\boldsymbol{\theta})$ is first performed assuming $n = 1$, i.e., using $f_{WW}(\theta_1)$. The resulting DOA estimate $\hat{\theta}_1$ is then "projected out" from the data, and a grid search for the second DOA is performed using the modified criterion $f_2(\theta_2)$. The procedure is repeated until initial estimates are available for all DOAs. The kth modified criterion can be expressed as

$$f_k(\theta_k) = -\frac{a^*(\theta_k)\Pi_{\hat{R}_y^{-1/2}\hat{A}_{k-1}}^{\perp} \tilde{S}\tilde{\Lambda}^2 \tilde{S}^* \Pi_{\hat{R}_y^{-1/2}\hat{A}_{k-1}}^{\perp} a(\theta_k)}{a^*(\theta_k)\Pi_{\hat{R}_y^{-1/2}\hat{A}_{k-1}}^{\perp} a(\theta_k)}$$

where

$$\hat{A}_k = A(\hat{\boldsymbol{\theta}}_k)$$
$$\hat{\boldsymbol{\theta}}_k = [\hat{\theta}_1, \ldots, \hat{\theta}_k]^T .$$

The initial estimate of θ_k is taken as the minimizing argument of $f_k(\theta_k)$. Once all DOAs have been initialized one can, in principle, continue the alternating projection minimization in the same way. However, the procedure usually converges rather slowly and therefore it is recommended instead to switch to a Newton-type search as indicated above. Empirical investigations in [17, 32] using similar subspace fitting criteria, have indicated that this indeed leads to the global minimum with high probability.

64.6 Numerical Examples

This section reports the results of a comparative performance study based on Monte-Carlo simulations. The scenarios are identical to those presented in [33] (spatial IV-SSF) and [14] (temporal IV-SSF). The plots presented below contain theoretical standard deviations of the DOA estimates along with empirically observed RMS (root mean square) errors. The former are obtained from Eq. (64.36), whereas the latter are based on 512 independent noise and signal realizations. The minimizers of Eq. (64.42) (UNCLE) and Eq. (64.45) (IV-SSF) are computed using a modified Gauss-Newton search initialized at the true DOAs

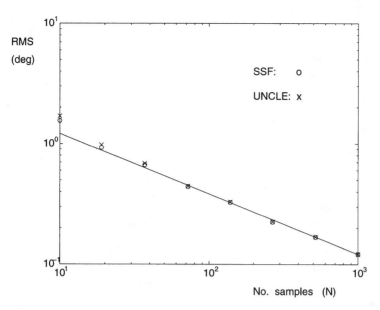

FIGURE 64.1   RMS error of DOA estimate vs. number of snapshots. Spatial IVM. The solid line is the theoretical standard deviation.

(since here we are interested only in the quality of the global optimum). DOA estimates that are more than 5° off the true value are declared failures, and not included in the empirical RMS calculation. If the number of failures exceeds 30%, no RMS value is calculated.

In all scenarios, two planar wavefronts arrive from DOAs 0° and 5° relative to the array broadside. Unless otherwise stated, the emitter signals are zero-mean Gaussian with signal covariance matrix $P = I$. Only the estimation statistics for $\theta_1 = 0°$ are shown in the plots below, the ones for θ_2 being similar.

The array output (both subarrays in the spatial IV scenario) is corrupted by additive zero-mean temporally white Gaussian noise. The noise covariance matrix has klth element

$$Q_{kl} = \sigma^2 \, 0.9^{|k-l|} e^{j\frac{\pi}{2}(k-l)} \, . \tag{64.47}$$

The noise level σ^2 is adjusted to give a desired SNR, defined as $P_{11}/\sigma^2 = P_{22}/\sigma^2$. This noise is reminiscent of a strong signal cluster at the location $\theta = 30°$.

EXAMPLE 64.4:   Spatial IVM

In the first example, a ULA of 16 elements and half-wavelength separation is employed. The first $m = 8$ contiguous sensors form a calibrated subarray, whereas the outputs of the last $\bar{m} = 8$ sensors are used as instrumental variables, and these sensors could therefore be uncalibrated. Letting $\tilde{y}(t)$ denote the 16-element array output, we thus take

$$y(t) = \tilde{y}_{1:8}(t) \quad z(t) = \tilde{y}_{9:16}(t) \, .$$

Both subarray outputs are perturbed by independent additive noise vectors, both having 8×8 covariance matrices given by Eq. (64.47). In this example, the emitter signals are assumed to be temporally white.

In Fig. 64.1, the theoretical and empirical RMS errors are displayed vs. the number of samples. The SNR is fixed at 6 dB.

Figure 64.2 shows the theoretical and empirical RMS errors vs. the SNR. The number of snapshots is here fixed to $N = 100$.

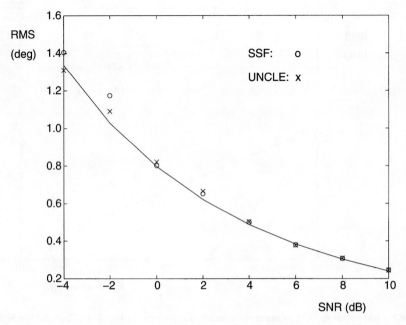

FIGURE 64.2   RMS error of DOA estimate vs SNR. Spatial IVM. The solid line is the theoretical standard deviation.

To demonstrate the applicability to situations involving highly correlated signals, Fig. 64.2 is repeated but using the signal covariance

$$P = \begin{bmatrix} 1 & 1 \\ 1 & 1 \end{bmatrix}$$

The resulting RMS errors are plotted with their theoretical values in Fig. 64.3. By comparing Figs. 64.2 and 64.3, we see that the methods are not insensitive to the signal correlation. However, the observed RMS errors agree well with the theoretically predicted values, and in spatial scenarios this is the best possible RMS performance (the empirical RMS error appears to be lower than the CRB for low SNR; however this is at the price of a notable bias).

In conclusion, no significant performance difference is observed between the two IV-SSF versions. The observed RMS errors of both methods follow the theoretical curves quite closely, even in fairly difficult scenarios involving closely spaced DOAs and highly correlated signals.

EXAMPLE 64.5:   Temporal IVM

In this example, the temporal IV approach is investigated. The array is a 6-element ULA of half wavelength interelement spacing. The real and imaginary parts of both signals are generated as uncorrelated first-order complex AR processes with identical spectra. The poles of the driving AR-processes are 0.6. In this case, $y(t)$ is the array output, whereas the instrumental variable vector is chosen as $z(t) = y(t-1)$.

In Fig. 64.4, we show the theoretical and empirical RMS errors vs. the number of snapshots. The SNR is fixed at 10 dB. Figure 64.5 displays the theoretical and empirical RMS errors vs. the SNR. The number of snapshots is here fixed at $N = 100$.

The figures indicate a slight performance difference among the methods in temporal scenarios, namely when the number of samples is small but the SNR is relatively high. However, no definite conclusions can be drawn regarding this somewhat unexpected phenomenon from our limited simulation study.

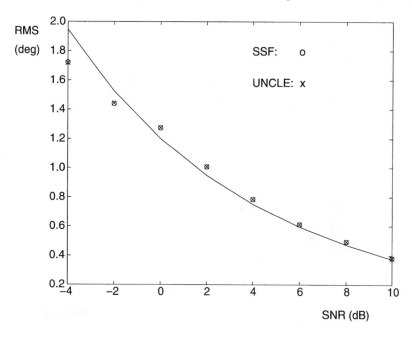

FIGURE 64.3   RMS error of DOA estimate vs. SNR. Spatial IVM. Coherent signals. The solid line is the theoretical standard deviation.

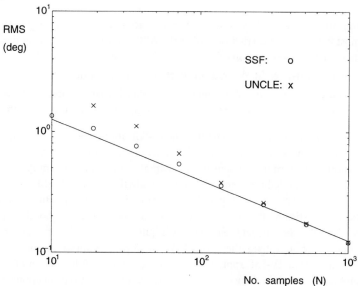

FIGURE 64.4   RMS error of DOA estimate vs. number of snapshots. Temporal IVM. The solid line is the theoretical standard deviation.

64.7   Concluding Remarks

The main points made by the present contribution can be summarized as follows:

1. The spatial and temporal IV-SSF approaches can be treated in a unified manner under general conditions. In fact, a general IV-SSF approach using both spatial and temporal instruments is also possible.

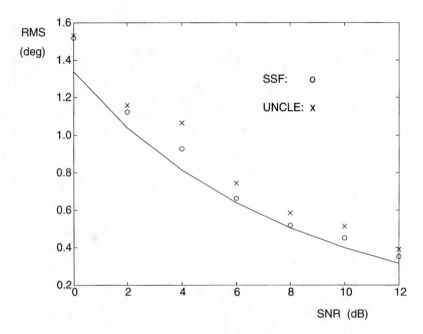

FIGURE 64.5   RMS error of DOA estimate vs. SNR. Temporal IVM. The solid line is the theoretical standard deviation.

2. The optimization of the DOA parameter estimation accuracy, for fixed weights $\hat{W}_L$ and $\hat{W}_R$, can be most conveniently carried out using the ABC theory. The resulting derivations are more concise than those based on other analysis techniques.
3. The column (or post-)weight $\hat{W}_R$ has no effect on the asymptotics.
4. An important corollary of the above-mentioned result is that the optimal IV-SSF methods of [12, 13] and, respectively, [14] are asymptotically equivalent when used on the same data.

In closing this section, we reiterate the fact that the IV-SSF approaches can deal with coherent signals, handle noise fields with general (unknown) spatial correlations, and, in their spatial versions, can make use of outputs from completely uncalibrated sensors. They are also comparatively simple from a computational standpoint, since no noise modelling is required. Additionally, the optimal IV-SSF methods provide highly accurate DOA estimates. More exactly, in spatial IV scenarios these DOA estimation methods can be shown to be asymptotically statistically efficient under weak conditions [33]. In temporal scenarios, they are no longer exactly statistically efficient, yet their accuracy is quite close to the best possible one [14]. All these features and properties should make the optimal IV-SSF approach appealing for practical array signal processing applications. The IV-SSF approach can also be applied, with some modifications, to system identification problems [34] and is hence expected to play a role in that type of application as well.

References

[1] Li, F. and Vaccaro, R.J., Performance degradation of DOA estimators due to unknown noise fields, *IEEE Trans. SP*, SP-40(3), 686–689, March 1992.
[2] Viberg, M., Sensitivity of parametric direction finding to colored noise fields and undermodeling, *Signal Processing*, 34(2), 207–222, Nov. 1993.
[3] Swindlehurst, A. and Kailath, T., A performance analysis of subspace-based methods in the presence of model errors: Part 2 — Multidimensional algorithms, *IEEE Trans. on SP*, SP-41, 2882–2890, Sept. 1993.

[4] Böhme, J.F. and Kraus, D., On least squares methods for direction of arrival estimation in the presence of unknown noise fields, *Proc. ICASSP 88*, 2833–2836, New York, 1988.

[5] Le Cadre, J.P., Parametric methods for spatial signal processing in the presence of unknown colored noise fields, *IEEE Trans. on ASSP*, ASSP-37(7), 965–983, July 1989.

[6] Nagesha, V. and Kay, S., Maximum likelihood estimation for array processing in colored noise, *Proc. ICASSP 93*, 4, 240–243, Minneapolis, MN, 1993.

[7] Ye, H. and DeGroat, R., Maximum likelihood DOA and unknown colored noise estimation with asymptotic Cramér-Rao bounds, *Proc. 27th Asilomar Conf. Sig., Syst., Comput.*, 1391–1395, Pacific Grove, CA, Nov. 1993.

[8] Söderström, T. and Stoica, P., *Instrumental Variable Methods for System Identification*, Springer-Verlag, Berlin, 1983.

[9] Söderström, T. and Stoica, P., *System Identification*, Prentice-Hall, London, U.K., 1989.

[10] Moses, R.L. and Beex, A.A., Instrumental variable adaptive array processing, *IEEE Trans. on AES*, AES-24, 192–202, March 1988.

[11] Stoica, P., Viberg, M. and Ottersten, B., Instrumental variable approach to array processing in spatially correlated noise fields, *IEEE Trans. SP*, SP-42, 121–133, Jan. 1994.

[12] Wu, Q. and Wong, K.M., UN-MUSIC and UN-CLE: An application of generalized canonical correlation analysis to the estimation of the directions of arrival of signals in unknown correlated noise, *IEEE Trans. SP*, 42, 2331–2341, Sept. 1994.

[13] Wu, Q. and Wong, K.M., Estimation of DOA in unknown noise: Performance analysis of UN-MUSIC and UN-CLE, and the optimality of CCD, *IEEE Trans. SP*, 43, 454–468, Feb. 1995.

[14] Viberg, M., Stoica, P. and Ottersten, B., Array processing in correlated noise fields based on instrumental variables and subspace fitting, *IEEE Trans. SP*, 43, 1187–1199, May 1995.

[15] Stoica, P., Viberg, M., Wong, M. and Wu, Q., Maximum-likelihood bearing estimation with partly calibrated arrays in spatially correlated noise fields, *IEEE Trans on SP*, 44, 88–899, Apr. 1996.

[16] Schmidt, R.O., Multiple emitter location and signal parameter estimation, *IEEE Trans. on AP*, 34, 276–280, Mar. 1986.

[17] Ottersten, B., Viberg, M., Stoica, P. and Nehorai, A., Exact and large sample ML techniques for parameter estimation and detection in array processing, in *Radar Array Processing*, Haykin, Litva, and Shepherd, Eds., Springer-Verlag, Berlin, 1993, 99–151.

[18] Wax, M. and Ziskind, I., On unique localization of multiple sources by passive sensor arrays, *IEEE Trans. on ASSP*, ASSP-37(7), 996–1000, July 1989.

[19] Hudson, J.E., *Adaptive Array Principles*, Peter Peregrinus, 1981.

[20] Compton, R.T., Jr., *Adaptive Antennas*, Prentice-Hall, Englewood Cliffs, NJ, 1988.

[21] Lee, W.C.Y., *Mobile Communications Design Fundamentals*, 2nd ed., John Wiley & Sons, New York, 1993.

[22] Agee, B.G., Schell, A.V. and Gardner, W.A., Spectral self-coherence restoral: A new approach to blind adaptive signal extraction using antenna arrays, *Proc. IEEE*, 78, 753–767, Apr. 1990.

[23] Shamsunder, S. and Giannakis, G., Signal selective localization of nonGaussian cyclostationary sources, *IEEE Trans. SP*, 42, 2860–2864, Oct. 1994.

[24] Zhang, Q.T. and Wong, K.M., Information theoretic criteria for the determination of the number of signals in spatially correlated noise, *IEEE Trans. SP*, SP-41(4), 1652–1663, Apr. 1993.

[25] Wu, Q. and Wong, K.M., Determination of the number of signals in unknown noise environments, *IEEE Trans. SP*, 43, 362–365, Jan. 1995.

[26] Golub, G.H. and VanLoan, C.F., *Matrix Computations*, 2nd ed., Johns Hopkins University Press, Baltimore, MD, 1989.

[27] Stoica, P., Viberg, M., Wong, M. and Wu, Q., A unified instrumental variable approach to direction finding in colored noise fields: Report version, Technical Report CTH-TE-32, Chalmers University of Technology, Gothenburg, Sweden, July 1995.

[28] Brewer, J.W., Kronecker products and matrix calculus in system theory, *IEEE Trans. on CAS*, 25(9), 772–781, Sept. 1978.

[29] Porat, B., *Digital Processing of Random Signals*, Prentice-Hall, Englewood Cliffs, NJ, 1993.

[30] Gill, P.E., Murray, W. and Wright, M.H., *Practical Optimization*, Academic Press, London, 1981.

[31] Ziskind, I. and Wax, M., Maximum likelihood localization of multiple sources by alternating projection, *IEEE Trans. on ASSP*, ASSP-36, 1553–1560, Oct. 1988.

[32] Viberg, M., Ottersten, B. and Kailath, T., Detection and estimation in sensor arrays using weighted subspace fitting, *IEEE Trans. SP*, SP-39(11), 2436–2449, Nov. 1991.

[33] Stoica, P., Viberg, M., Wong, M. and Wu, Q., Optimal direction finding with partly calibrated arrays in spatially correlated noise fields, *Proc. 28th Asilomar Conf. Sig., Syst., Comput.*, Pacific Grove, CA, Oct. 1994.

[34] Cedervall, M. and Stoica, P., System identification from noisy measurements by using instrumental variables and subspace fitting, *Proc. ICASSP 95*, 1713–1716, Detroit, MI, May 1995.

[35] Ljung, L., *System Identification: Theory for the User*, Prentice-Hall, Englewood Cliffs, NJ, 1987.

Appendix A: Introduction to IV Methods

In this appendix we give a brief introduction to instrumental variable methods in their original context, which is time series analysis. Let $y(t)$ be a real-valued scalar time series, modeled by the auto-regressive moving average (ARMA) equation

$$y(t) + a_1 y(t-1) + \cdots + a_p y(t-p) = e(t) + b_1 e(t-1) + \cdots + b_q e(t-q). \tag{64.48}$$

Here, $e(t)$ is assumed to be a stationary white noise. Suppose we are given measurements of $y(t)$ for $t = 1, \ldots, N$ and wish to estimate the AR parameters $a_1, \ldots, a_p$. The roots of the AR polynomial $z^p + a_1 z^{p-1} + \cdots + a_p$ are the system poles, and their estimation is of importance, for instance, for stability monitoring. Also, the first step of any "linear" method for ARMA modeling involves finding the AR parameters as the first step. The optimal way to approach the problem requires a non-linear search over the entire parameter set $\{a_k\}_{k=1}^{p}, \{b_k\}_{k=1}^{q}$; using a maximum likelihood or a prediction error criterion [9, 35]. However, in many cases this is computationally prohibitive, and in addition the "noise model" (the MA parameters) is sometimes of less interest per se. In contrast, the IV approach produces estimates of the AR part from a solution of a (possibly overdetermined) linear system of equations as follows: Rewrite Eq. (64.48) as

$$y(t) = \varphi^T(t)\theta + v(t), \tag{64.49}$$

where

$$\varphi(t) = [-y(t-1), \ldots, -y(t-p)]^T \tag{64.50}$$

$$\theta = [a_1, \ldots, a_p]^T \tag{64.51}$$

$$v(t) = e(t) + b_1 e(t-1) + \cdots + b_q e(t-q). \tag{64.52}$$

Note that Eq. (64.49) is a linear regression in the unknown parameter θ. A standard least-squares (LS) estimate is obtained by minimizing the LS criterion

$$V_{LS}(\theta) = \mathrm{E}\left[(y(t) - \varphi^T(t)\theta)^2\right]. \tag{64.53}$$

Equating the derivative of Eq. (64.53) (w.r.t. θ) to zero gives the so-called normal equations

$$\mathrm{E}\left[\varphi(t)\varphi^T(t)\right]\hat{\theta} = \mathrm{E}\left[\varphi(t)y(t)\right]. \tag{64.54}$$

resulting in

$$\hat{\theta} = R_{\varphi\varphi}^{-1} R_{\varphi y} = \left(E\left[\varphi(t)\varphi^T(t)\right] \right)^{-1} E\left[\varphi(t)y(t)\right]. \tag{64.55}$$

Inserting Eq. (64.49) into Eq. (64.55) shows that

$$\hat{\theta} = \theta + R_{\varphi\varphi}^{-1} R_{\varphi v}. \tag{64.56}$$

In case $q = 0$ (i.e., $y(t)$ is an AR process), we have $v(t) = e(t)$. Because $\varphi(t)$ and $e(t)$ are uncorrelated, Eq. (64.56) shows that the LS method produces a consistent estimate of θ. However, when $q > 0$, $\varphi(t)$ and $v(t)$ are in general correlated, implying that the LS method gives biased estimates.

From the above we conclude that the problem with the LS estimate in the ARMA case is that the regression vector $\varphi(t)$ is correlated with the "equation error noise" $v(t)$. An instrumental variable vector $\zeta(t)$ is one that is uncorrelated with $v(t)$, while still "sufficiently correlated" with $\varphi(t)$. The most natural choice in the ARMA case (provided the model orders are known) is

$$\zeta(t) = \varphi(t - q) \tag{64.57}$$

which clearly fulfills both requirements. Now, multiply both sides of the linear regression model (64.49) by $\zeta(t)$ and take expectation, resulting in the "IV normal equations"

$$E\left[\zeta(t)y(t)\right] = E\left[\zeta(t)\varphi^T(t)\right]\theta. \tag{64.58}$$

The IV estimate is obtained simply by solving the linear system of equations (64.58), but with the unknown cross-covariance matrices $R_{\zeta\varphi}$ and $R_{\zeta y}$ replaced by their corresponding estimates using time averaging. Since the latter are consistent, so are the IV estimates of θ. The method is also referred to as the *extended Yule-Walker* approach in the literature. Its finite sample properties may often be improved upon by increasing the dimension of the IV vector, which means that Eq. (64.58) must be solved in an LS sense, and also by appropriately pre-filtering the IV-vector. This is quite similar to the optimal weighting proposed herein.

In order to make the connection to the IV-SSF method more clear, a slightly modified version of Eq. (64.58) is presented. Let us rewrite Eq. (64.58) as follows

$$R_{\zeta\phi} \begin{bmatrix} 1 \\ \theta \end{bmatrix} = 0, \tag{64.59}$$

where

$$R_{\zeta\phi} = E\left\{\zeta(t)\left[y(t), -\varphi^T(t)\right]\right\}. \tag{64.60}$$

The relation (64.59) shows that $R_{\zeta\phi}$ is singular, and that θ can be computed from a suitably normalized vector in its one-dimensional nullspace. However, when $R_{\zeta\phi}$ is estimated using a finite number of data, it will with probability one have full rank. The best (in a least squares sense) low-rank approximation of $R_{\zeta\phi}$ is obtained by truncating its singular value decomposition. A natural estimate of θ can therefore be obtained from the right singular vector of $\hat{R}_{\zeta\phi}$ that corresponds to the minimum singular value. The proposed modification is essentially an IV-SSF version of the extended Yule-Walker method, although the SSF step is trivial because the parameter vector of interest can be computed directly from the estimated subspace.

Turning to the array processing problem, the counterpart of Eq. (64.49) is the (Hermitian transposed) data model (64.1)

$$y^*(t) = x^*(t)A^* + e^*(t). $$

Note that this is a non-linear regression model, owing to the non-linear dependence of A on θ. Also observe that $y(t)$ is a complex vector as opposed to the real scalar $y(t)$ in Eq. (64.49). Similar to Eq. (64.58), the IV normal equations are given by

$$E\left[z(t)y^*(t)\right] = E\left[z(t)x^*(t)\right]A^* \tag{64.61}$$

under the assumption that the IV-vector $z(t)$ is uncorrelated with the noise $e(t)$. Unlike the standard IV problem, the "regressor" $x(t)$ [corresponding to $\varphi(t)$ in Eq. (64.49)] cannot be measured. Thus, it is not possible to get a direct estimate of the "regression variable" A. However, its range space, or at least a subset thereof, can be computed from the principal right singular vectors. In the finite sample case, the performance can be improved by using row and column weighting, which leads to the weighted IV normal equations (64.11). The exact relation involving the principal right singular vectors is Eq. (64.14), and two SSF formulations for revealing θ from the computed signal subspace are given in Eqs. (64.16) and (64.17).

65

Electromagnetic Vector-Sensor Array Processing[1]

Arye Nehorai
The University of Illinois at Chicago

Eytan Paldi
Haifa, Israel

Dedicated to the memory of our physics teacher, Isaac Paldi

65.1 Introduction

This article (see also [1, 2]) considers new methods for multiple electromagnetic source localization using sensors whose output is a *vector* corresponding to the complete electric and magnetic fields at the sensor. These sensors, which will be called *vector sensors*, can consist for example of two orthogonal triads of scalar sensors that measure the electric and magnetic field components. Our approach is in contrast to other articles in this chapter that employ sensor arrays in which the output of each sensor is a scalar corresponding, for example, to a scalar function of the electric field. The main advantage of the vector sensors is that they make use of all available electromagnetic information and hence should outperform the scalar sensor arrays in accuracy of direction of arrival (DOA) estimation. Vector sensors should also

---

[1]This work was supported by the U.S. Air Force Office of Scientific Research under Grant no. F49620-97-1-0481, the Office of Naval Research under Grant no. N00014-96-1-1078, the National Science Foundation under Grant no. MIP-9615590, and the HTI Fellowship.

allow the use of smaller array apertures while improving performance. (Note that we use the term "vector sensor" for a device that measures a complete physical vector quantity.)

Section 65.2 derives the measurement model. The electromagnetic sources considered can originate from two types of transmissions: (1) Single signal transmission (SST), in which a single signal message is transmitted, and (2) dual signal transmission (DST), in which two separate signal messages are transmitted simultaneously (from the same source), see for example [3, 4]. The interest in DST is due to the fact that it makes full use of the two spatial degrees of freedom present in a transverse electromagnetic plane wave. This is particularly important in the wake of increasing demand for economical spectrum usage by existing and emerging modern communication technologies.

Section 65.3 analyzes the minimum attainable variance of unbiased DOA estimators for a general vector sensor array model and multi-electromagnetic sources that are assumed to be stochastic and stationary. A compact expression for the corresponding Cramér-Rao bound (CRB) on the DOA estimation error that extends previous results for the scalar sensor array case in [5] (see also [6]) is presented.

A significant property of the vector sensors is that they enable DOA (azimuth and elevation) estimation of an electromagnetic source with a *single* vector sensor and a single snapshot. This result is explicitly shown by using the CRB expression for this problem in Section 65.4. A bound on the associated normalized mean-square angular error (MSAE, to be defined later) which is invariant to the reference coordinate system is used for an in-depth performance study. Compact expressions for this MSAE bound provide physical insight into the SST and DST source localization problems with a single vector sensor.

The CRB matrix for an SST source in the sensor coordinate frame exhibits some nonintrinsic singularities (i.e., singularities that are not inherent in the physical model while being dependent on the choice of the reference coordinate system) and has complicated entry expressions. Therefore, we introduce a new vector angular error defined in terms of the incoming wave frame. A bound on the normalized asymptotic covariance of the vector angular error (CVAE) is derived. The relationship between the CVAE and MSAE and their bounds is presented. The CVAE matrix bound for the SST source case is shown to be diagonal, easy to interpret, and to have only intrinsic singularities.

We propose a simple algorithm for estimating the source DOA with a single vector sensor, motivated by the Poynting vector. The algorithm is applicable to various types of sources (e.g., wide-band and non-Gaussian); it does not require a minimization of a cost function and can be applied in real time. Statistical performance analysis evaluates the variance of the estimator under mild assumptions and compares it with the MSAE lower bound.

Section 65.5 extends these results to the multi-source multi-vector sensor case, with special attention to the two-source single-vector sensor case. Section 65.6 summarizes the main results and gives some ideas of possible extensions.

The main difference between the topics of this article and other articles on source direction estimation is in our use of vector sensors with *complete* electric and magnetic data. Most papers have dealt with scalar sensors. Other papers that considered estimation of the polarization state and source direction are [7]–[12]. Reference [7] discussed the use of subspace methods to solve this problem using diversely polarized electric sensors. References [8]–[10] devised algorithms for arrays with two dimensional electric measurements. Reference [11] provided performance analysis for arrays with two types of electric sensor polarizations (diversely polarized). An earlier reference, [12], proposed an estimation method using a three-dimensional vector sensor and implemented it with magnetic sensors. All these references used only part of the electromagnetic information at the sensors, thereby reducing the observability of DOAs. In most of them, time delays between distributed sensors played an essential role in the estimation process.

For a plane wave (typically associated with a single source in the far-field) the magnitude of the electric and magnetic fields can be found from each other. Hence, it may be felt that one (complete) field is deducible from the other. However, this is not true when the source direction is unknown. Additionally, the electric and magnetic fields are orthogonal to each other and to the source DOA vector, hence measuring both fields increases significantly the accuracy of the source DOA estimation. This is true in particular for an incoming wave which is nearly linearly polarized, as will be explicitly shown by the CRB (see Table 65.1).

The use of the complete electromagnetic vector data enables source parameter estimation with a single sensor (even with a single snapshot) where time delays are not used at all. In fact, this is shown to be possible for at least two sources. As a result, the derived CRB expressions for this problem are applicable to wide-band sources. The source DOA parameters considered include azimuth and elevation. This section also considers direction estimation to DST sources, as well as the CRB on wave ellipticity and orientation angles (to be defined later) for SST sources using vector sensors, which were first presented in [1, 2]. This is true also for the MSAE and CVAE quality measures and the associated bounds. Their application is not limited to electromagnetic vector sensor processing.

We comment that electromagnetic vector sensors as measuring devices are commercially available and actively researched. EMC Baden Ltd. in Baden, Switzerland, is a company that manufactures them for signals in the 75 Hz to 30 MHz frequency range, and Flam and Russell, Inc. in Horsham, Pennsylvania, makes them for the 2 to 30 MHz frequency band. Lincoln Labs at MIT has performed some preliminary localization tests with vector sensors [13]. Some examples of recent research on sensor development are [14] and [15].

Following the recent impressive progress in the performance of DSP processors, there is a trend to fuse as much data as possible using smart sensors. Vector sensors, which belong to this category of sensors, are expected to find larger use and provide important contribution in improving the performance of DSP in the near future.

65.2 The Measurement Model

This section presents the measurement model for the estimation problems that are considered in the latter parts of the article.

65.2.1 Single-Source Single-Vector Sensor Model

Basic Assumptions

Throughout the article it will be assumed that the wave is traveling in a nonconductive, homogeneous, and isotropic medium. Additionally, the following will be assumed:

A1: Plane wave at the sensor: This is equivalent to a far-field assumption (or maximum wave-length much smaller than the source to sensor distance), a point source assumption (i.e., the source size is much smaller than the source to sensor distance) and a point-like sensor (i.e., the sensor's dimensions are small compared to the minimum wave-length).

A2: Band-limited spectrum: The signal has a spectrum including only frequencies ω satisfying $\omega_{min} \leq |\omega| \leq \omega_{max}$ where $0 < \omega_{min} < \omega_{max} < \infty$. This assumption is satisfied in practice. The lower and upper limits on ω are also needed, respectively, for the far-field and point-like sensor assumptions.

Let $\mathcal{E}(t)$ and $\mathcal{H}(t)$ be the vector phasor representations (or complex envelopes, see e.g., [16, 17] and [1, Appendix A]) of the electric and magnetic fields at the sensor. Also, let u be the unit vector at the sensor pointing towards the source, i.e.,

$$u = \begin{bmatrix} \cos\theta_1 \cos\theta_2 \\ \sin\theta_1 \cos\theta_2 \\ \sin\theta_2 \end{bmatrix} \qquad (65.1)$$

where θ_1 and θ_2 denote, respectively, the azimuth and elevation angles of u, see Fig. 65.1. Thus, $\theta_1 \in [0, 2\pi)$ and $|\theta_2| \leq \pi/2$.

In [1, Appendix A] it is shown that for plane waves Maxwell's equations can be reduced to an equivalent set of two equations without any loss of information. Under the additional assumption of a band-limited

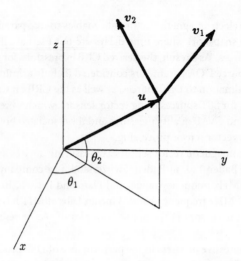

FIGURE 65.1   The orthonormal vector triad $(\boldsymbol{u}, \boldsymbol{v}_1, \boldsymbol{v}_2)$.

signal, these two equations can be written in terms of phasors. The results are summarized in the following theorem.

THEOREM 65.1   *Under assumption A1, Maxwell's equations can be reduced to an equivalent set of two equations. With the additional band-limited spectrum assumption A2, they can be written as:*

$$\boldsymbol{u} \times \mathcal{E}(t) \;=\; -\eta \mathcal{H}(t) \tag{65.2a}$$

$$\boldsymbol{u} \cdot \mathcal{E}(t) \;=\; 0 \tag{65.2b}$$

where η is the intrinsic impedance of the medium and "$\times$" and "$\cdot$" are the cross and inner products of $\mathbb{R}^3$ applied to vectors in $\mathbb{C}^3$. (That is, if $\boldsymbol{v}, \boldsymbol{w} \in \mathbb{C}^3$ then $\boldsymbol{v} \cdot \boldsymbol{w} = \sum_i v_i w_i$. This is different than the usual inner product of $\mathbb{C}^3$).

PROOF 65.1   See [1, Appendix A]. (Note that $\boldsymbol{u} = -\boldsymbol{\kappa}$ where $\boldsymbol{\kappa}$ is the unit vector in the direction of the wave propagation).

Thus, under the plane and band-limited wave assumptions, the vector phasor equations (65.2) provide all the information contained in the original Maxwell equations. This result will be used in the following to construct measurement models in which the Maxwell equations are incorporated entirely.

The Measurement Model

Suppose that a vector sensor measures all six components of the electric and magnetic fields. (It is assumed that the sensor does not influence the electric and magnetic fields). The measurement model is based on the phasor representation of the measured electromagnetic data (with respect to a reference frame) at the sensor. Let $\boldsymbol{y}_E(t)$ be the measured electric field phasor vector at the sensor at time t and $\boldsymbol{e}_E(t)$ its noise component. Then the electric part of the measurement will be

$$\boldsymbol{y}_E(t) = \mathcal{E}(t) + \boldsymbol{e}_E(t) \tag{65.3}$$

Similarly, from Eq. (65.2a), after appropriate scaling, the magnetic part of the measurement will be taken as

$$y_H(t) = u \times \mathcal{E}(t) + e_H(t) \tag{65.4}$$

In addition to Eq. (65.3) and (65.4), we have the constraint (65.2b).

Define the matrix cross product operator that maps a vector $v \in \mathbb{R}^{3 \times 1}$ to $(u \times v) \in \mathbb{R}^{3 \times 1}$ by

$$(u \times) \overset{\triangle}{=} \begin{bmatrix} 0 & -u_z & u_y \\ u_z & 0 & -u_x \\ -u_y & u_x & 0 \end{bmatrix} \tag{65.5}$$

where u_x, u_y, u_z are the x, y, z components of the vector u. With this definition, Eqs. (65.3) and (65.4) can be combined to

$$\begin{bmatrix} y_E(t) \\ y_H(t) \end{bmatrix} = \begin{bmatrix} I_3 \\ (u \times) \end{bmatrix} \mathcal{E}(t) + \begin{bmatrix} e_E(t) \\ e_H(t) \end{bmatrix} \tag{65.6}$$

where I_3 denotes the 3×3 identity matrix. For notational convenience the dimension subscript of the identity matrix will be omitted whenever its value is clear from the context.

The constraint (65.2b) implies that the electric phasor $\mathcal{E}(t)$ can be written

$$\mathcal{E}(t) = V\xi(t) \tag{65.7}$$

where V is a 3×2 matrix whose columns span the orthogonal complement of u and $\xi(t) \in \mathbb{C}^{2 \times 1}$. It is easy to check that the matrix

$$V = \begin{bmatrix} -\sin\theta_1 & -\cos\theta_1\sin\theta_2 \\ \cos\theta_1 & -\sin\theta_1\sin\theta_2 \\ 0 & \cos\theta_2 \end{bmatrix} \tag{65.8}$$

whose columns are orthonormal, satisfies this requirement. We note that since $\|u\|^2 = 1$ the columns of V, denoted by v_1 and v_2, can be constructed, for example, from the partial derivatives of u with respect to θ_1 and θ_2 and post-normalization when needed. Thus,

$$v_1 = \frac{1}{\cos\theta_2} \frac{\partial u}{\partial \theta_1} \tag{65.9a}$$

$$v_2 = u \times v_1 = \frac{\partial u}{\partial \theta_2} \tag{65.9b}$$

and (u, v_1, v_2) is a right orthonormal triad, see Fig. 65.1. (Observe that the two coordinate systems shown in the figure actually have the same origin). The signal $\xi(t)$ fully determines the components of $\mathcal{E}(t)$ in the plane where it lies, namely the plane orthogonal to u spanned by v_1, v_2. This implies that there are two degrees of freedom present in the spatial domain (or the wave's plane), or two independent signals can be transmitted simultaneously.

Combining Eq. (65.6) and Eq. (65.7) we now have

$$\begin{bmatrix} y_E(t) \\ y_H(t) \end{bmatrix} = \begin{bmatrix} I \\ (u \times) \end{bmatrix} V\xi(t) + \begin{bmatrix} e_E(t) \\ e_H(t) \end{bmatrix} \tag{65.10}$$

This system is equivalent to Eq. (65.6) with Eq. (65.2b).

The measured signals in the sensor reference frame can be further related to the original source signal at the transmitter using the following lemma.

LEMMA 65.1   Every vector $\boldsymbol{\xi} = [\xi_1, \xi_2]^T \in \mathbb{C}^{2 \times 1}$ has the representation

$$\boldsymbol{\xi} = \|\boldsymbol{\xi}\| e^{i\varphi} Q\boldsymbol{w} \tag{65.11}$$

where

$$Q = \begin{bmatrix} \cos\theta_3 & \sin\theta_3 \\ -\sin\theta_3 & \cos\theta_3 \end{bmatrix} \tag{65.12a}$$

$$\boldsymbol{w} = \begin{bmatrix} \cos\theta_4 \\ i\sin\theta_4 \end{bmatrix} \tag{65.12b}$$

and where $\varphi \in (-\pi, \pi]$, $\theta_3 \in (-\pi/2, \pi/2]$, $\theta_4 \in [-\pi/4, \pi/4]$. Moreover, $\|\boldsymbol{\xi}\|$, φ, θ_3, θ_4 in Eq. (65.11) are uniquely determined if and only if $\xi_1^2 + \xi_2^2 \neq 0$.

PROOF 65.2   See [1, Appendix B].

The equality $\xi_1^2 + \xi_2^2 = 0$ holds if and only if $|\theta_4| = \pi/4$, corresponding to circular polarization (defined below). Hence, from Lemma 65.1 the representation (65.11), (65.12) is not unique in this case as should be expected, since the orientation angle θ_3 is ambiguous. It should be noted that the representation (65.11), (65.12) is known and was used (see, e.g., [18]) without a proof. However, Lemma 65.1 of existence and uniqueness appears to be new. The existence and uniqueness properties are important to guarantee identifiability of parameters.

The physical interpretations of the quantities in the representation (65.11), (65.12) are as follows.

$\|\boldsymbol{\xi}\| e^{i\varphi}$: Complex envelope of the source signal (including amplitude and phase).

$\boldsymbol{w}$: Normalized overall transfer vector of the source's antenna and medium, i.e., from the source complex envelope signal to the principal axes of the received electric wave.

Q: A rotation matrix that performs the rotation from the principal axes of the incoming electric wave to the $(\boldsymbol{v}_1, \boldsymbol{v}_2)$ coordinates.

Let ω_c be the reference frequency of the signal phasor representation, see [1, Appendix A]. In the narrow-band SST case, the incoming electric wave signal $\text{Re}\{e^{i\omega_c t}\|\boldsymbol{\xi}(t)\|e^{i\varphi(t)}Q\boldsymbol{w}\}$ moves on a quasistationary ellipse whose semi-major and semi-minor axes' lengths are proportional, respectively, to $\cos\theta_4$ and $\sin\theta_4$, see Fig. 65.2 and [19]. The ellipse's eccentricity is thus determined by the magnitude of θ_4. The sign of θ_4 determines the spin sign or direction. More precisely, a positive (negative) θ_4 corresponds to a positive (negative) spin with right-(left) handed rotation with respect to the wave propagation vector $\boldsymbol{\kappa} = -\boldsymbol{u}$. As shown in Fig. 65.2, θ_3 is the rotation angle between the $(\boldsymbol{v}_1, \boldsymbol{v}_2)$ coordinates and the electric ellipse axes $(\tilde{\boldsymbol{v}}_1, \tilde{\boldsymbol{v}}_2)$. The angles θ_3 and θ_4 will be referred to, respectively, as the orientation and ellipticity angles of the received electric wave ellipse. In addition to the electric ellipse, there is also a similar but perpendicular magnetic ellipse.

It should be noted that if the transfer matrix from the source to the sensor is time invariant, then so are θ_3 and θ_4.

The signal $\boldsymbol{\xi}(t)$ can carry information coded in various forms. In the following we discuss briefly both existing forms and some motivated by the above representation.

Single Signal Transmission (SST) Model

Suppose that a single modulated signal is transmitted. Then, using Eq. (65.11), this is a special case of Eq. (65.10) with

$$\boldsymbol{\xi}(t) = Q\boldsymbol{w}s(t) \tag{65.13}$$

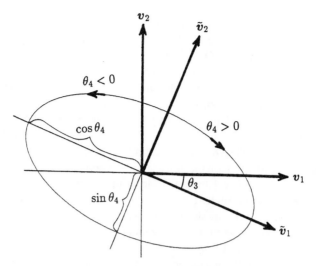

FIGURE 65.2   The electric polarization ellipse.

where $s(t)$ denotes the complex envelope of the (scalar) transmitted signal. Thus, the measurement model is

$$\begin{bmatrix} \boldsymbol{y}_E(t) \\ \boldsymbol{y}_H(t) \end{bmatrix} = \begin{bmatrix} I \\ (\boldsymbol{u}\times) \end{bmatrix} V Q \boldsymbol{w} s(t) + \begin{bmatrix} \boldsymbol{e}_E(t) \\ \boldsymbol{e}_H(t) \end{bmatrix} \tag{65.14}$$

Special cases of this transmission are linear polarization with $\theta_4 = 0$ and circular polarization with $|\theta_4| = \pi/4$.

Recall that since there are two spatial degrees of freedom in a transverse electromagnetic plane wave, one could, in principle, transmit two separate signals simultaneously. Thus, the SST method does not make full use of the two spatial degrees of freedom present in a transverse electromagnetic plane wave.

Dual Signal Transmission (DST) Models

Methods of transmission in which two separate signals are transmitted simultaneously from the same source will be called *dual signal transmissions*. Various DST forms exist, and all of them can be modeled by Eq. (65.10) with $\boldsymbol{\xi}(t)$ being a linear transformation of the two-dimensional source signal vector.

One DST form uses two linearly polarized signals that are spatially and temporally orthogonal with an amplitude or phase modulation (see e.g., [3, 4]). This is a special case of Eq. (65.10), where the signal $\boldsymbol{\xi}(t)$ is written in the form

$$\boldsymbol{\xi}(t) = Q \begin{bmatrix} s_1(t) \\ i s_2(t) \end{bmatrix} \tag{65.15}$$

where $s_1(t)$ and $s_2(t)$ represent the complex envelopes of the transmitted signals. To guarantee unique decoding of the two signals (when θ_3 is unknown) using Lemma 65.1, they have to satisfy $s_1(t) \neq 0$, $s_2(t)/s_1(t) \in (-1, 1)$. (Practically this can be achieved by using a proper electronic antenna adapter that yields a desirable overall transfer matrix.)

Another DST form uses two circularly polarized signals with opposite spins. In this case

$$\boldsymbol{\xi}(t) = Q[\boldsymbol{w}\tilde{s}_1(t) + \overline{\boldsymbol{w}}\tilde{s}_2(t)] \tag{65.16a}$$

$$\boldsymbol{w} = (1/\sqrt{2})[1, \ i]^T \tag{65.16b}$$

where $\overline{\boldsymbol{w}}$ denotes the complex conjugate of $\boldsymbol{w}$. The signals $\tilde{s}_1(t), \tilde{s}_2(t)$ represent the complex envelopes of the transmitted signals. The first term on the r.h.s. of Eqs. (65.16) corresponds to a signal with positive spin and circular polarization ($\theta_4 = \pi/4$), while the second term corresponds to a signal with negative spin and circular polarization ($\theta_4 = -\pi/4$). The uniqueness of Eqs. (65.16) is guaranteed without the conditions needed for the uniqueness of Eq. (65.15).

The above-mentioned DST models can be applied to communication problems. Assuming that u is given, it is possible to measure the signal $\xi(t)$ and recover the original messages as follows. For Eq. (65.15), an existing method resolves the two messages using mechanical orientation of the receiver's antenna (see, e.g., [4]). Alternatively, this can be done electronically using the representation of Lemma 65.1, without the need to know the orientation angle. For Eqs. (65.16), note that $\xi(t) = we^{i\theta_3}\tilde{s}_1(t) + \overline{w}e^{-i\theta_3}\tilde{s}_2(t)$, which implies the uniqueness of Eqs. (65.16) and indicates that the orientation angle has been converted into a phase angle whose sign depends on the spin sign. The original signals can be directly recovered from $\xi(t)$ up to an additive constant phase without knowledge of the orientation angle. In some cases, it is of interest to estimate the orientation angle. Let W be a matrix whose columns are $w, \overline{w}$. For Eqs. (65.16) this can be done using equal calibrating signals and then premultiplying the measurement by W^{-1} and measuring the phase difference between the two components of the result. This can also be used for real time estimation of the angular velocity $d\theta_3/dt$.

In general it can be stated that the advantage of the DST method is that it makes full use of the spatial degrees of freedom of transmission. However, the above DST methods need the knowledge of u and, in addition, may suffer from possible cross polarizations (see, e.g., [3]), multipath effects, and other unknown distortions from the source to the sensor.

The use of the proposed vector sensor can motivate the design of new improved transmission forms. Here we suggest a new dual signal transmission method that uses on line electronic calibration in order to resolve the above problems. Similar to the previous methods it also makes full use of the spatial degrees of freedom in the system. However, it overcomes the need to know u and the overall transfer matrix from source to sensor.

Suppose the transmitted signal is $z(t) \in \mathbb{C}^{2\times 1}$ (this signal is as it appears before reaching the source's antenna). The measured signal is

$$\begin{bmatrix} y_E(t) \\ y_H(t) \end{bmatrix} = C(t)z(t) + \begin{bmatrix} e_E(t) \\ e_H(t) \end{bmatrix} \tag{65.17}$$

where $C(t) \in \mathbb{C}^{6\times 2}$ is the unknown source to sensor transfer matrix that may be slowly varying due to, for example, the source dynamics. To facilitate the identification of $z(t)$, the transmitter can send calibrating signals, for instance, transmit $z_1(t) = [1, 0]^T$ and $z_2(t) = [0, 1]^T$ separately. Since these inputs are in phasor form, this means that actually constant carrier waves are transmitted. Obviously, one can then estimate the columns of $C(t)$ by averaging the received signals, which can be used later for finding the original signal $z(t)$ by using, for example, least-squares estimation. Better estimation performance can be achieved by taking into account *a priori* information about the model.

The use of vector sensors is attractive in communication systems as it doubles the channel capacity (compared with scalar sensors) by making full use of the electromagnetic wave properties. This spatial multiplexing has vast potential for performance improvement in cellular communications.

In future research it would be of interest to develop optimal coding methods (modulation forms) for maximum channel capacity while maintaining acceptable distortions of the decoded signals despite unknown varying channel characteristics. It would also be of interest to design communication systems that utilize entire arrays of vector sensors.

Observe that actually any combination of the variables $\|\xi\|$, φ, θ_3 and θ_4 can be modulated to carry information. A binary signal can be transmitted using the spin sign of the polarization ellipse (sign of θ_4). Lemma 65.1 guarantees the identifiability of these signals from $\xi(t)$.

65.2.2 Multi-Source Multi-Vector Sensor Model

Suppose that waves from n distant electromagnetic sources are impinging on an array of m vector sensors and that assumptions A1 and A2 hold for each source. To extend the model (65.10) to this scenario we need the following additional assumptions, which imply that A1, A2 hold uniformly on the array:

A3: Plane wave across the array: In addition to A1, for each source the array size d_A has to be much smaller than the source to array distance, so that the vector $\boldsymbol{u}$ is approximately independent of the individual sensor positions.

A4: Narrow-band signal assumption: The maximum frequency of $\mathcal{E}(t)$, denoted by ω_m, satisfies $\omega_m d_A/c \ll 1$, where c is the velocity of wave propagation (i.e., the minimum modulating wave-length is much larger than the array size). This implies that $\mathcal{E}(t - \tau) \simeq \mathcal{E}(t)$ for all differential delays τ of the source signals between the sensors.

Note that (under the assumption $\omega_m < \omega_c$) since $\omega_m = \max\{|\omega_{\min} - \omega_c|, |\omega_{\max} - \omega_c|\}$, it follows that A4 is satisfied if $(\omega_{\max} - \omega_{\min})d_A/2c \ll 1$ and ω_c is chosen to be close enough to $(\omega_{\max} + \omega_{\min})/2$.

Let $\boldsymbol{y}_{EH}(t)$ and $\boldsymbol{e}_{EH}(t)$ be the $6m \times 1$ dimensional electromagnetic sensor phasor measurement and noise vectors,

$$\boldsymbol{y}_{EH}(t) \triangleq \left[(\boldsymbol{y}_E^{(1)}(t))^T, (\boldsymbol{y}_H^{(1)}(t))^T, \cdots, (\boldsymbol{y}_E^{(m)}(t))^T, (\boldsymbol{y}_H^{(m)}(t))^T \right]^T \tag{65.18a}$$

$$\boldsymbol{e}_{EH}(t) \triangleq \left[(\boldsymbol{e}_E^{(1)}(t))^T, (\boldsymbol{e}_H^{(1)}(t))^T, \cdots, (\boldsymbol{e}_E^{(m)}(t))^T, (\boldsymbol{e}_H^{(m)}(t))^T \right]^T \tag{65.18b}$$

where $\boldsymbol{y}_E^{(j)}(t)$ and $\boldsymbol{y}_H^{(j)}(t)$ are, respectively, the measured phasor electric and magnetic vector fields at the jth sensor and similarly for the noise components $\boldsymbol{e}_E^{(j)}(t)$ and $\boldsymbol{e}_H^{(j)}(t)$. Then, under assumptions A3 and A4 and from Eq. (65.10), we find that the array measured phasor signal can be written as

$$\boldsymbol{y}_{EH}(t) = \sum_{k=1}^n \boldsymbol{e}_k \otimes \left[\begin{array}{c} I_3 \\ (\boldsymbol{u}_k \times) \end{array} \right] V_k \boldsymbol{\xi}_k(t) + \boldsymbol{e}_{EH}(t) \tag{65.19}$$

where $\otimes$ is the Kronecker product, $\boldsymbol{e}_k$ denotes the kth column of the matrix $E \in \mathbb{C}^{m \times n}$ whose (j, k) entry is

$$E_{jk} = e^{-i\omega_c \tau_{jk}} \tag{65.20}$$

where τ_{jk} is the differential delay of the kth source signal between the jth sensor and the origin of some fixed reference coordinate system (e.g., at one of the sensors). Thus, $\tau_{jk} = -(\boldsymbol{u}_k \cdot \boldsymbol{r}_j)/c$, where $\boldsymbol{u}_k$ is the unit vector in the direction from the array to the kth source and $\boldsymbol{r}_j$ is the position vector of the jth sensor in the reference frame. The rest of the notation in Eq. (65.19) is similar to the single source case, cf. Eqs. (65.1), (65.8), and (65.10). The vector $\boldsymbol{\xi}_k(t)$ can have either the SST or the DST form described above.

Observe that the signal manifold matrix in Eq. (65.19) can be written as the Khatri-Rao product (see, e.g., [20, 21]) of E and a second matrix whose form depends on the source transmission type (i.e., SST or DST), see also later.

65.3 Cramér-Rao Bound for a Vector Sensor Array

65.3.1 Statistical Model

Consider the problem of finding the parameter vector $\boldsymbol{\theta}$ in the following discrete-time vector sensor array model associated with n vector sources and m vector sensors:

$$y(t) = A(\boldsymbol{\theta})x(t) + e(t) \qquad t = 1, 2, \ldots \qquad (65.21)$$

where $y(t) \in \mathbb{C}^{\bar{\mu} \times 1}$ are the vectors of observed sensor outputs (or snapshots), $x(t) \in \mathbb{C}^{\bar{\nu} \times 1}$ are the unknown source signals, and $e(t) \in \mathbb{C}^{\bar{\mu} \times 1}$ are the additive noise vectors. The transfer matrix $A(\boldsymbol{\theta}) \in \mathbb{C}^{\bar{\mu} \times \bar{\nu}}$ and the parameter vector $\boldsymbol{\theta} \in \mathbb{R}^{\bar{q} \times 1}$ are given by

$$A(\boldsymbol{\theta}) \;=\; \left[A_1(\boldsymbol{\theta}^{(1)}) \cdots A_n(\boldsymbol{\theta}^{(n)})\right] \qquad (65.22a)$$

$$\boldsymbol{\theta} \;=\; \left[(\boldsymbol{\theta}^{(1)})^T, \cdots, (\boldsymbol{\theta}^{(n)})^T\right]^T \qquad (65.22b)$$

where $A_k(\boldsymbol{\theta}^{(k)}) \in \mathbb{C}^{\bar{\mu} \times \nu_k}$ and the parameter vector of the kth source $\boldsymbol{\theta}^{(k)} \in \mathbb{R}^{q_k \times 1}$, thus $\bar{\nu} = \sum_{k=1}^{n} \nu_k$ and $\bar{q} = \sum_{k=1}^{n} q_k$. The following notation will also be used:

$$y(t) \;=\; \left[(y^{(1)}(t))^T, \cdots, (y^{(m)}(t))^T\right]^T \qquad (65.23a)$$

$$x(t) \;=\; \left[(x^{(1)}(t))^T, \cdots, (x^{(n)}(t))^T\right]^T \qquad (65.23b)$$

where $y^{(j)}(t) \in \mathbb{C}^{\mu_j \times 1}$ is the vector measurement of the jth sensor, implying $\bar{\mu} = \sum_{j=1}^{m} \mu_j$, and $x^{(k)}(t) \in \mathbb{C}^{\nu_k \times 1}$ is the vector signal of the kth source. Clearly $\bar{\mu}$ and $\bar{\nu}$ correspond, respectively, to the total number of sensor components and source signal components.

The model (65.21) generalizes the commonly used multi-scalar source multi-scalar sensor one (see, e.g., [7, 22]). It will be shown later that the electromagnetic multi-vector source multi-vector sensor data models are special cases of Eq. (65.21) with appropriate choices of matrices.

For notational simplicity, the explicit dependence on $\boldsymbol{\theta}$ and t will be occasionally omitted.

We make the following commonly used assumptions on the model (65.21):

A5: The source signal sequence $\{x(1), x(2), \ldots\}$ is a sample from a temporally uncorrelated stationary (complex) Gaussian process with zero mean and

$$E\,x(t)x^*(s) = P\delta_{t,s}$$
$$E\,x(t)x^T(s) = 0 \qquad \text{(for all } t \text{ and } s).$$

where E is the expectation operator, the superscript "*" denotes the conjugate transpose, and $\delta_{t,s}$ is the Kronecker delta.

A6: The noise $e(t)$ is (complex) Gaussian distributed with zero mean and

$$E\,e(t)e^*(s) = \sigma^2 I \delta_{t,s}$$
$$E\,e(t)e^T(s) = 0 \qquad \text{(for all } t \text{ and } s).$$

It is also assumed that the signals $x(t)$ and the noise $e(s)$ are independent for all t and s.

A7: The matrix A has full rank $\bar{\nu} < \bar{\mu}$ (thus A^*A is p.d.) and a continuous Jacobian $\partial A/\partial \boldsymbol{\theta}$ in some neighborhood of the true $\boldsymbol{\theta}$. The matrix $APA^* + \sigma^2 I$ is assumed to be positive definite, which implies that the probability density functions of the model are well defined in some neighborhood of the true $\boldsymbol{\theta}$, P, σ^2. Additionally, the matrix in braces in Eq. (65.24) below is assumed to be nonsingular.

The unknown parameters in the model (65.21) include the vector $\boldsymbol{\theta}$, the signal covariance matrix P, and the noise variance σ^2. The problem of estimating $\boldsymbol{\theta}$ in (65.21) from N snapshots $\mathbf{y}(1), \ldots, \mathbf{y}(N)$ and the statistical performance of estimation methods are the main concerns of this article.

65.3.2 The Cramér-Rao Bound

Consider the estimation of $\boldsymbol{\theta}$ in the model (65.21) under the above assumptions and with $\boldsymbol{\theta}, P, \sigma^2$ unknown. We have the following theorem.

THEOREM 65.2 *The Cramér-Rao lower bound on the covariance matrix of any (locally) unbiased estimator of the vector $\boldsymbol{\theta}$ in the model (65.21), under assumptions A5 through A7 with $\boldsymbol{\theta}, P, \sigma^2$ unknown and $v_k = v$ for all k, is a positive definite matrix given by*

$$\text{CRB}(\boldsymbol{\theta}) = \frac{\sigma^2}{2N} \left\{ \text{Re}\left[\text{btr}\left((1 \boxtimes U) \odot (D^* \Pi_c D)^{bT} \right) \right] \right\}^{-1} \tag{65.24}$$

where

$$
\begin{align}
U &= P\left(A^*AP + \sigma^2 I\right)^{-1} A^*AP \tag{65.25a} \\
\Pi_c &= I - \Pi \tag{65.25b} \\
\Pi &= A(A^*A)^{-1}A^* \tag{65.25c} \\
D &= \left[D_1^{(1)} \cdots D_{q_1}^{(1)} \quad \cdots \quad D_1^{(n)} \cdots D_{q_n}^{(n)} \right] \tag{65.25d} \\
D_\ell^{(k)} &= \frac{\partial A_k}{\partial \theta_\ell^{(k)}} \tag{65.25e}
\end{align}
$$

and where 1 denotes a $\bar{q} \times \bar{q}$ matrix with all entries equal to one, and the block trace operator btr $(\cdot)$, the block Kronecker product $\boxtimes$, the block Schur-Hadamard product $\odot$, and the block transpose operator bT are as defined in the Appendix with blocks of dimensions $v \times v$, except for the matrix 1 that has blocks of dimensions $q_i \times q_j$.

Furthermore, the CRB in Eq. (65.24) remains the same independently of whether σ^2 is known or unknown.

PROOF 65.3 See [1, Appendix C].

Theorem 65.2 can be extended to include a larger class of unknown sensor noise covariance matrices (see [1, Appendix D]).

65.4 MSAE, CVAE, and Single-Source Single-Vector Sensor Analysis

This section introduces the MSAE and CVAE quality measures and their bounds for source direction and orientation estimation in three-dimensional space. The bounds are applied to analyze the statistical performance of parameter estimation of an electromagnetic source whose covariance is unknown using a single vector sensor. Note that single vector sensor analysis is valid for *wide-band* sources, as assumptions A3 and A4 are not needed.

65.4.1   The MSAE

We define the mean-square angular error which is a quality measure that is useful for gaining physical insight into DOA (azimuth and elevation) estimation and for performance comparisons. The analysis of this subsection is not limited to electromagnetic measurements or to Gaussian data.

The angular error, say δ, corresponding to a direction error Δu in u, can be shown to be $\delta = 2 \arcsin(\|\Delta u\|/2)$. Hence, $\delta^2 = \|\Delta u\|^2 + O(\|\Delta u\|^4)$. Since $\Delta u = \left(\frac{\partial u}{\partial \theta_1}\right) \Delta\theta_1 + \left(\frac{\partial u}{\partial \theta_2}\right) \Delta\theta_2 + O((\Delta\theta_1)^2 + (\Delta\theta_2)^2)$ where $\Delta\theta_1$, $\Delta\theta_2$ are the errors in θ_1 and θ_2, we have

$$\delta^2 = (\cos\theta_2 \cdot \Delta\theta_1)^2 + (\Delta\theta_2)^2 + O(|\Delta\theta_1|^3 + |\Delta\theta_2|^3) \tag{65.26}$$

We introduce the following definitions.

DEFINITION 65.1   A model will be called *regular* if it satisfies any set of sufficient conditions for the CRB to hold (see, e.g., [23, 24]).

DEFINITION 65.2   The *normalized asymptotic mean-square angular error* of a direction estimator will be defined as

$$\text{MSAE} \triangleq \lim_{N\to\infty} \left\{ NE\left(\delta^2\right) \right\} \tag{65.27}$$

whenever this limit exists.

DEFINITION 65.3   A direction estimator will be called *regular* if its errors satisfy $E\left[|\Delta\theta_1|^3 + |\Delta\theta_2|^3\right] = o(1/N)$, the gradient of its bias with respect to θ_1, θ_2 exists and is $o(1)$ as $N \to \infty$, and its MSAE exists. (If $|\theta_2| = \pi/2$ then θ_1 is undefined and we can use the equivalent condition $E\left[\|\Delta u\|^3\right] = o(1/N)$).

Equation (65.26) shows that under the assumptions that the model and estimator are regular we have

$$E(\delta)^2 \geq [\cos^2\theta_2 \cdot \text{CRB}(\theta_1) + \text{CRB}(\theta_2)] + o(1/N) \qquad \text{as } N \to \infty \tag{65.28}$$

where $\text{CRB}(\theta_1)$ and $\text{CRB}(\theta_2)$ are, respectively, the Cramér-Rao bounds for the azimuth and elevation. Using Eq. (65.28) we have the following theorem.

THEOREM 65.3   *For a regular model MSAE of any regular direction estimator is bounded from below by*

$$\text{MSAE}_{CR} \triangleq N[\cos^2\theta_2 \cdot \text{CRB}(\theta_1) + \text{CRB}(\theta_2)] \tag{65.29}$$

Observe that MSAE_{CR} is not a function of N. Additionally, MSAE_{CR} is a tight bound if it is attained by some second order efficient regular estimator (usually the maximum likelihood (ML) estimator, see e.g., [25]). For vector sensor measurements this bound has the desirable property of being invariant to the choice of reference coordinate frame, since the information content in the data is invariant under rotational transformations. This invariance property also holds for the MSAE of an estimator if the estimate is independent of known rotational transformations of the data.

For a regular model, the bound (65.29) can be used for performance analysis of any regular direction (azimuth and elevation) finding algorithm.

It is of interest to note that the bound (65.29) actually holds for finite data, when the estimators of u are unbiased and constrained to be of unit norm, see [26].

65.4.2 DST Source Analysis

Assume that it is desired to estimate the direction to a DST source whose covariance is unknown using a vector sensor. We will first present a statistical model for this problem as a special case of Eq. (65.21) and then investigate in detail the resulting CRB and MSAE.

The measurement model for the DST case is given in Eq. (65.10). Suppose the noise vector of Eq. (65.10) is (complex) Gaussian with zero mean and the following covariances:

$$
E \left[\begin{array}{c} e_E(t) \\ e_H(t) \end{array} \right] \left[\begin{array}{cc} e_E^*(s), & e_H^*(s) \end{array} \right] = \left[\begin{array}{cc} \sigma_E^2 I_3 & 0 \\ 0 & \sigma_H^2 I_3 \end{array} \right] \delta_{t,s}
$$

$$
E \left[\begin{array}{c} e_E(t) \\ e_H(t) \end{array} \right] \left[\begin{array}{cc} e_E^T(s), & e_H^T(s) \end{array} \right] = 0 \qquad \text{(for all } t \text{ and } s\text{)}.
$$

Our assumption that the noise components are statistically independent stems from the fact that they are created separately at *different* sensor components (even if the sensor components belong to a vector sensor). Note that under assumption A1 the measurement includes a source plane wave component and sensor self noise.

To relate the model (65.10) to (65.21), define a scaled measurement $y(t) \triangleq \left[r y_E^T(t), \ y_H^T(t) \right]^T$ where $r \triangleq \sigma_H/\sigma_E$ is assumed to be known. (The results of this section actually hold also when r is unknown as is explained in [1]). The resulting scaled noise vector $e(t) \triangleq \left[r e_E^T(t), \ e_H^T(t) \right]^T$ then satisfies assumption A6 with $\sigma = \sigma_H$. Assume further that the signal $\xi(t)$ satisfies assumption A5 with $x(t) = \xi(t)$. Then, under these assumptions, the scaled version of the DST source (65.10) can be viewed as a special case of Eq. (65.21) with $m = n = 1$ and

$$
A = \left[\begin{array}{c} r V \\ (u \times) V \end{array} \right] \qquad x(t) = \xi(t) \qquad \sigma^2 = \sigma_H^2
$$

$$
\theta = \left[\begin{array}{cc} \theta_1, & \theta_2 \end{array} \right]^T \tag{65.30}
$$

where the unknown parameters are θ, P, σ^2. The parameter vector of interest is θ while P and σ^2 are the so-called nuisance parameters.

The above discussion shows that the CRB expression (65.24) is applicable to the present problem with the special choice of variables in Eq. (65.30), thus $n = 1$ and $\bar{q} = 2$. The computation of the CRB is given in [1]. The result is independent of whether r is known or unknown.

Using the CRB results of [1] we find that MSAE_{CR} for the present DST problem is

$$
\text{MSAE}_{CR}^D = \frac{\left(\sigma_E^2 + \sigma_H^2 \right) \sigma_E^2 \sigma_H^2 \, \text{tr} \, U}{2 \left[\sigma_E^2 \sigma_H^2 \, (\text{tr} \, U)^2 + \left(\sigma_E^2 - \sigma_H^2 \right)^2 \det \left(\text{Re} \, U \right) \right]} \tag{65.31}
$$

Observe that MSAE_{CR}^D is symmetric with respect to σ_E, σ_H, as should be expected from the Maxwell equations. MSAE_{CR}^D is not a function of $\theta_1, \theta_2, \theta_3$, as should be expected since for vector sensor measurements the MSAE bound is by definition invariant to the choice of coordinate system. Note that MSAE_{CR}^D is independent of whether σ_E and σ_H are known or unknown.

65.4.3 SST Source (DST Model) Analysis

Consider the MSAE for a single signal transmission source when the estimation is done under the assumption that the source is of a dual signal transmission type. In this case, the model (65.10) has to be used but with a signal in the form of (65.13). The signal covariance is then

$$
P = \sigma_s^2 Q w (Q w)^* \tag{65.32}
$$

where $\sigma_s^2 = \mathrm{E}\, s^2(t)$ and Q and $\boldsymbol{w}$ are defined in Eq. (65.12). Thus, rank $P = 1$ and P has a unit norm eigenvector $Q\boldsymbol{w}$ with an eigenvalue σ_s^2.

Let

$$\sigma_{\|}^2 \triangleq \frac{\sigma_E^2 \cdot \sigma_H^2}{\sigma_E^2 + \sigma_H^2} \tag{65.33}$$

The variance $\sigma_{\|}^2$ can be viewed as an equivalent noise variance of two measurements with independent noise variances σ_E^2 and σ_H^2. Define ϱ, $\sigma_s^2/\sigma_{\|}^2$, which is an effective SNR.

Using the analysis of U in [1] and expression (65.31) we find that

$$\mathrm{MSAE}_{CR}^S = \frac{(1 + \varrho)(\sigma_E^2 + \sigma_H^2)^2}{2\varrho^2 \left[\sigma_E^2\sigma_H^2 + (\sigma_E^2 - \sigma_H^2)^2 \sin^2 \theta_4 \cos^2 \theta_4\right]}$$

$$= \frac{(1 + \varrho)(1 + r^2)^2}{2\varrho^2 \left[r^2 + (1 - r^2)^2 \sin^2 \theta_4 \cos^2 \theta_4\right]} \tag{65.34}$$

where MSAE_{CR}^S denotes the MSAE_{CR} bound for the SST problem under the DST model. (It will be shown later that the same result also holds under the SST model.) Observe that MSAE_{CR}^S is symmetric with respect to σ_E, σ_H. It is also independent of whether σ_H and σ_E are known or unknown, as can be shown from Theorem 65.2 and [1, Appendix D]. Also, MSAE_{CR}^S is not a function of $\theta_1, \theta_2, \theta_3$, since for vector sensor measurements the MSAE bound is invariant under rotational transformations of the reference coordinate system. On the other hand, MSAE_{CR}^S is influenced by the ellipticity angle θ_4 through the difference in the electric and magnetic noise variances.

Table 65.1 summarizes several special cases of the expression (65.34) for MSAE_{CR}^S. The elliptical polarization column corresponds to an arbitrary polarization angle $\theta_4 \in [-\pi/4, \pi/4]$. The circular and linear polarization columns are obtained, respectively, as special cases of Eq. (65.34) with $|\theta_4| = \pi/4$ and $\theta_4 = 0$. The row of precise (noise-free) electric measurement (with noisy magnetic measurements) is obtained by substituting $\sigma_E^2 = 0$ in (65.34). The row of electric measurement only is obtained by deriving the corresponding CRB and MSAE_{CR}^S. Alternatively, MSAE_{CR}^S can be found for this case by taking the limit of Eq. (65.34) as $\sigma_H^2 \to \infty$.

TABLE 65.1   MSAE Bounds for a Single Signal Transmission Source

	Elliptical	Circular	Linear
General MSAE_{CR}^S	(65.34)	$\dfrac{2(1 + \varrho)}{\varrho^2}$	$\dfrac{(1 + \varrho)(\sigma_E^2 + \sigma_H^2)}{2\varrho\sigma_s^2}$
Precise electric measurement	0	0	$\dfrac{\sigma_H^2}{2\sigma_s^2}$
Electric measurement only	$\dfrac{\sigma_E^2(\sigma_E^2 + \sigma_s^2)}{2\sigma_s^4 \sin^2 \theta_4 \cos^2 \theta_4}$	$\dfrac{2\sigma_E^2(\sigma_E^2 + \sigma_s^2)}{\sigma_s^4}$	∞

Observe from Eq. (65.34) that when $\sigma_H^2 \neq \sigma_E^2$, MSAE_{CR}^S is minimized for circular polarization and maximized for linear polarization. This result is illustrated in Fig. 65.3, which shows the square root of MSAE_{CR}^S as a function of $r = \sigma_H/\sigma_E$ for three types of polarizations ($\theta_4 = 0, \pi/12, \pi/4$). The equivalent signal-to-noise ratio $\mathrm{SNR} = \sigma_s^2/\sigma_{\|}^2$ is kept at one, while the individual electric and magnetic noise variances are varied to give the desired value of r. As r becomes larger or smaller than one, MSAE_{CR}^S increases more significantly for sources with polarization closer to linear.

When the electric (or magnetic) field is measured precisely and the source polarization is circular or elliptical, the MSAE_{CR}^S is zero (i.e., no angular error), while for linearly polarized sources it remains positive.

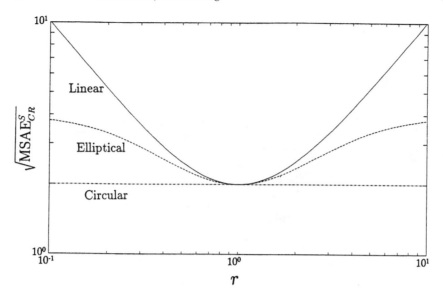

FIGURE 65.3 Effect of change in $r = \sigma_H/\sigma_E$ on MSAE^S_{CR} for three types of polarizations ($\theta_4 = 0, \pi/12, \pi/4$). A single SST source, $\mathrm{SNR} = \sigma^2_s/\sigma^2_\| = 1$.

In the latter case, the contribution to MSAE^S_{CR} stems from the magnetic (or electric) noisy measurement. When only the electric (or magnetic) field is measured, MSAE^S_{CR} increases as the polarization changes from circular to linear. In the linear polarization case, MSAE^S_{CR} tends to infinity. In this case, it is impossible to uniquely identify the source direction $\boldsymbol{u}$ from the electric field only, since $\boldsymbol{u}$ can then be anywhere in the plane orthogonal to the electric field vector.

The immediate conclusion is that as the source becomes closer to being linearly polarized it becomes more important to measure both the electric and magnetic fields to get good direction estimates using a single vector sensor.

These results are illustrated in Fig. 65.4, which shows the square root of MSAE^S_{CR} as a function of σ^2_H and three polarization types ($\theta_4 = 0, \pi/12, \pi/4$). The standard deviations of the signal and electric noise are $\sigma_s = \sigma_E = 1$. The left side of the figure corresponds to (nearly) precise magnetic measurement, while the right side to (nearly) electric measurement only.

65.4.4 SST Source (SST Model) Analysis

Suppose that it is desired to estimate the direction to an SST source whose variance is unknown using a single vector sensor, and the estimation is done under the correct model of an SST source. In the following, the CRB for this problem will be derived and it will be shown that the resulting MSAE bound remains the same as when the estimation was done under the assumption of a DST source. That is, knowledge of the source type does not improve the accuracy of its direction estimate.

To get a statistical model for the SST measurement model (65.14) as a special case of Eq. (65.21), we will make the same assumptions on the noise and use a similar data scaling as in the above DST source case. That will give again equal noise variances in all the sensor coordinates. Assume also that the signal envelope $s(t)$ satisfies assumption A5 with $x(t) = s(t)$ in Eq. (65.14). Then the resulting statistical model becomes a special case of Eq. (65.21) with

$$A = \begin{bmatrix} rV \\ (\boldsymbol{u}\times)V \end{bmatrix} Q\boldsymbol{w} \qquad x(t) = s(t) \qquad \sigma^2 = \sigma^2_H$$

$$\boldsymbol{\theta} = \begin{bmatrix} \theta_1, & \theta_2, & \theta_3, & \theta_4 \end{bmatrix}^T \tag{65.35}$$

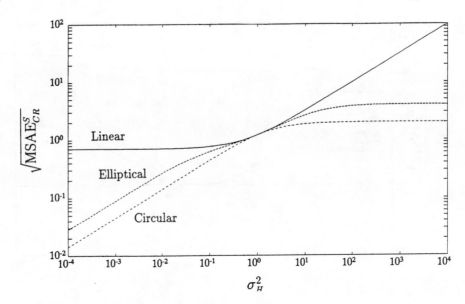

FIGURE 65.4 Effect of change in magnitude of σ_H^2 on MSAE_{CR}^S for three types of polarizations ($\theta_4 = 0, \pi/12, \pi/4$). A single SST source, $\sigma_s = \sigma_E = 1$.

The unknown parameters are $\boldsymbol{\theta}, P, \sigma^2$.

The matrix expression of $\text{CRB}(\boldsymbol{\theta})$ was calculated and its entries are presented in [1, Appendix F]. The results show that the ellipticity angle θ_4 is decoupled from the rest of the parameters and that its variance is not a function of these parameters. Additionally, the parameter vector $\boldsymbol{\theta}$ is decoupled from σ_E and σ_H.

The MSAE bound for an SST source under the SST model was calculated using the analysis of [1]. The result coincides with Eq. (65.34). That is, *the MSAE bound for an SST source is the same under both the SST and the DST models.*

The CRB expression in [1, Appendix F] implies that the CRB variance of the orientation angle θ_3 tends to infinity as the elevation angle θ_2 approaches $\pi/2$ or $-\pi/2$. This singularity is explained by the fact that the orientation angle is a function of the azimuth (through v_1, v_2), and the latter becomes increasingly sensitive to measurement errors as the elevation angle approaches the zenith or nadir. (Note that the azimuth is undefined in the zenith and nadir elevations). However, this singularity is not an intrinsic one, as it depends on the chosen reference system, while the information in the vector measurement does not.

65.4.5 CVAE and SST Source Analysis in the Wave Frame

In order to get performance results intrinsic to the SST estimation problem and thereby solve the singularity problems associated with the above model, we choose an alternative error vector that is invariant under known rotational transformations of the coordinate system. The details of the following analysis appear in [1, Appendix G].

Denote by W the wave frame whose coordinate axes are $(\boldsymbol{u}, \tilde{\boldsymbol{v}}_1, \tilde{\boldsymbol{v}}_2)$ where $\tilde{\boldsymbol{v}}_1$ and $\tilde{\boldsymbol{v}}_2$ correspond, respectively, to the major and minor axes of the source's electric wave ellipse (see Fig. 65.2). For any estimator $\hat{\theta}_i$, $i = 1, 2, 3$ there is an associated estimated wave frame $\widehat{W}$. Define the vector angular error $\boldsymbol{\phi}_{W\widehat{W}}$ which is the vector angle by which $\widehat{W}$ is (right-handed) rotated about W, and by $[\boldsymbol{\phi}_{W\widehat{W}}]_W$ the representation of $\boldsymbol{\phi}_{W\widehat{W}}$ in the coordinate system W (see [1, Appendix G]). The proposed vector angular error will be $[\boldsymbol{\phi}_{W\widehat{W}}]_W$.

Observe that $[\boldsymbol{\phi}_{W\widehat{W}}]_W$ depends, by definition, only on the frames $W, \widehat{W}$. Thus, for an estimator that is independent of known rotations of the data, the estimated wave frame $\widehat{W}$, the vector angular error and its covariance are independent of the sensor frame. We introduce the following definitions.

DEFINITION 65.4 The *normalized asymptotic covariance of the vector angular error* in the wave frame is defined as

$$\text{CVAE} \triangleq \lim_{N \to \infty} \left\{ NE \left([\boldsymbol{\phi}_{w\widehat{w}}]_W [\boldsymbol{\phi}_{w\widehat{w}}]_W^T \right) \right\} \tag{65.36}$$

whenever this limit exists.

DEFINITION 65.5 A direction and orientation estimator will be called *regular* if its errors satisfy $E \sum_{i=1}^{3} |\Delta\theta_i|^3 = o(1/N)$ and the gradient of its bias with respect to $\theta_1, \theta_2, \theta_3$ is $o(1)$ as $N \to \infty$.

Then we have the following theorems.

THEOREM 65.4 *For a regular model the CVAE of any regular direction and orientation estimator, whenever it exists, is bounded from below by*

$$\text{CVAE}_{CR} \triangleq N \cdot K \; \text{CRB}(\theta_1, \theta_2, \theta_3) K^T \tag{65.37}$$

where

$$K = \begin{bmatrix} \sin\theta_2 & 0 & -1 \\ -\cos\theta_2 \sin\theta_3 & -\cos\theta_3 & 0 \\ \cos\theta_2 \cos\theta_3 & -\sin\theta_3 & 0 \end{bmatrix} \tag{65.38}$$

and $\text{CRB}(\theta_1, \theta_2, \theta_3)$ *is the Cramér-Rao submatrix bound for the azimuth, elevation, and orientation angles for the particular model used.*

PROOF 65.4 See [1, Appendix G].

Observe that the result of Theorem 65.4 is obtained using geometrical considerations only. Hence, it is applicable to general direction and orientation estimation problems and is not limited to the SST problem only. It is dependent only on the ability to define a wave frame. For example, one can apply this theorem to a DST source with a wave frame defined by the orientation angle that diagonalizes the source signal covariance matrix. A generalization of this theorem to estimating non-unit vector systems is given in [26].

For vector sensor measurements, CVAE_{CR} has the desirable property of being invariant to the choice of reference coordinate frame. This invariance property also holds for the CVAE of an estimator if the estimate is independent of deterministic rotational transformations of the data. Note that CVAE_{CR} is not a function of N.

THEOREM 65.5 *The MSAE and CVAE of any regular estimator are related through*

$$\text{MSAE} = [\text{CVAE}]_{2,2} + [\text{CVAE}]_{3,3} \tag{65.39}$$

Furthermore, a similar equality holds for a regular model where the MSAE and CVAE in Eq. (65.39) are replaced by their lower bounds MSAE_{CR} *and* CVAE_{CR}.

PROOF 65.5 See [1, Appendix G].

In our case, $\text{CRB}(\theta_1, \theta_2, \theta_3)$ is the 3×3 upper left block entry of the CRB matrix in the sensor frame given in [1, Appendix F]. Substituting this block entry into Eq. (65.37) and denoting the CVAE matrix bound for the SST problem by CVAE_{CR}^S, we have that this matrix is diagonal with nonzero entries given by

$$[\text{CVAE}_{CR}^S]_{1,1} = \frac{(1+\varrho)}{2\varrho^2 \cos^2 2\theta_4} \tag{65.40a}$$

$$[\text{CVAE}_{CR}^S]_{2,2} = \frac{(1+\varrho)(\sigma_E^2 + \sigma_H^2)}{2\varrho^2 [\sigma_H^2 \sin^2 \theta_4 + \sigma_E^2 \cos^2 \theta_4]} \tag{65.40b}$$

$$[\text{CVAE}_{CR}^S]_{3,3} = \frac{(1+\varrho)(\sigma_E^2 + \sigma_H^2)}{2\varrho^2 [\sigma_E^2 \sin^2 \theta_4 + \sigma_H^2 \cos^2 \theta_4]} \tag{65.40c}$$

Some observations on Eqs. (65.40) are summarized in the following:

- Rotation around u: Singular only for a circularly polarized signal.
- Rotation around $\tilde{v}_1$ (electric ellipse's major axis): Singular only for a linearly polarized signal and no magnetic measurement.
- Rotation around $\tilde{v}_2$ (electric ellipse's minor axis): Singular only for a linearly polarized signal and no electric measurement.
- The rotation variances around $\tilde{v}_1$ and $\tilde{v}_2$ are symmetric with respect to the electric and magnetic measurements.
- All the three variances in Eq. (65.40) are bounded from below by $(1+\varrho)/2\varrho^2$ (independent of the wave parameters).

The singular cases above are found by checking when their variances in CVAE_{CR}^S tend to infinity (see, e.g., [25, Theorem 6.3]). The three singular cases above should be expected as the corresponding rotations are unobservable. These singularities are intrinsic to the SST estimation problem and are independent of the reference coordinate system. The symmetry of the variances of the rotations around the major and minor axes of the ellipse with respect to the magnetic and electric measurements should be expected as their axes have a spatial angle difference of $\pi/2$.

The fact that the resulting singularities in the rotational errors are intrinsic (independent of the reference coordinate system) as well as the diagonality of the CVAE_{CR}^S bound matrix with its simple entry expressions indicate that the wave frame is a natural system in which to do the analysis.

65.4.6 A Cross-Product-Based DOA Estimator

We propose a simple algorithm for estimating the DOA of a single electromagnetic source using the measurements of a single vector sensor. The motivation for this algorithm stems from the average cross-product Poynting vector. Observe that $-u$ is the unit vector in the direction of the Poynting vector given by [27],

$$\begin{aligned} S(t) &= E(t) \times H(t) = \text{Re}\left\{ e^{i\omega_c t} \mathcal{E}(t) \right\} \times \text{Re}\left\{ e^{i\omega_c t} \mathcal{H}(t) \right\} \\ &= \tfrac{1}{2}\text{Re}\left\{ \mathcal{E}(t) \times \overline{\mathcal{H}}(t) \right\} + \tfrac{1}{2}\text{Re}\left\{ e^{i2\omega_c t} \mathcal{E}(t) \times \mathcal{H}(t) \right\} \end{aligned}$$

where $\overline{\mathcal{H}}$ denotes the complex conjugate of $\mathcal{H}$. The carrier time average of the Poynting vector is defined as $\langle S \rangle_t \triangleq \tfrac{1}{2}\text{Re}\left\{ \mathcal{E}(t) \times \overline{\mathcal{H}}(t) \right\}$. Note that unlike $\mathcal{E}(t)$ and $\mathcal{H}(t)$ this average is not a function of ω_c. Thus, it has an intrinsic physical meaning.

At this point we can see two possible ways for estimating u:

1. Phasor time averaging of $\langle S \rangle_t$ yielding a vector denoted by $\langle S \rangle$ with the estimated $\boldsymbol{u}$ taken as the unit vector in the direction of $-\langle S \rangle$.

2. Estimation of $\boldsymbol{u}$ by phasor time averaging of the unit vectors in the direction of $\mathrm{Re}\left\{\mathcal{E}(t) \times \overline{\mathcal{H}}(t)\right\}$.

Clearly, the first way is preferable, since then $\boldsymbol{u}$ is estimated after the measurement noise is reduced by the averaging process, while the estimated $\boldsymbol{u}$ in the second way is more sensitive to the measurement noises which may be magnified considerably.

Thus, the proposed algorithm computes

$$\widehat{s} = \frac{1}{N} \sum_{t=1}^{N} \mathrm{Re}\left\{ \boldsymbol{y}_E(t) \times \overline{\boldsymbol{y}}_H(t) \right\} \tag{65.41a}$$

$$\widehat{\boldsymbol{u}} = \widehat{s}/\|\widehat{s}\| \tag{65.41b}$$

This algorithm and some of its variants have been patented [28].

The statistical performance of this estimator $\widehat{\boldsymbol{u}}$ is analyzed in [1, Appendix H] under the previous assumptions on $\boldsymbol{\xi}(t)$, $\boldsymbol{e}_E(t)$, $\boldsymbol{e}_H(t)$, except that the Gaussian assumption is omitted. The results are summarized by the following theorem.

THEOREM 65.6   *The estimator $\widehat{\boldsymbol{u}}$ has the following properties (for both DST and SST sources):*

a) If $\|\boldsymbol{\xi}(t)\|^2$, $\|\boldsymbol{e}_E(t)\|$, $\|\boldsymbol{e}_H(t)\|$ have finite first order moments, then $\widehat{\boldsymbol{u}} \to \boldsymbol{u}$ almost surely.

b) If $\|\boldsymbol{\xi}(t)\|^2$, $\|\boldsymbol{e}_E(t)\|$, $\|\boldsymbol{e}_H(t)\|$ have finite second order moments, then $\sqrt{N}(\widehat{\boldsymbol{u}} - \boldsymbol{u})$ is asymptotically normal.

c) If $\|\boldsymbol{\xi}(t)\|^2$, $\|\boldsymbol{e}_E(t)\|$, $\|\boldsymbol{e}_H(t)\|$ have finite fourth order moments, then the MSAE is

$$\mathrm{MSAE} = \tfrac{1}{2}\varrho^{-1}\left(1 + 4\varrho^{-1}\right)\left(r + r^{-1}\right)^2 \tag{65.42}$$

where $\varrho = \mathrm{tr}\,(P)/\sigma_\parallel^2 = \mathrm{SNR}$.

d) Under the conditions of (c), $N\delta^2$ is asymptotically χ^2 distributed with two degrees of freedom.

PROOF 65.6   See [1, Appendix H].

For the Gaussian SST case, the ratio between the MSAE of this estimator to MSAE_{CR}^S in Eq. (65.34) is

$$\mathrm{eff} \triangleq \frac{\mathrm{MSAE}}{\mathrm{MSAE}_{CR}^S} = \frac{\varrho + 4}{\varrho + 1}\left[1 + (r - r^{-1})^2 \sin^2\theta_4 \cos^2\theta_4\right] \tag{65.43}$$

Hence, this estimator is nearly efficient if the following two conditions are met:

$$\varrho \gg 1 \tag{65.44a}$$

$$r \simeq 1 \quad \text{or} \quad \theta_4 \simeq 0 \tag{65.44b}$$

Figure 65.5 illustrates these results using plots of the efficiency factor (65.43) as a function of the ellipticity angle θ_4 for $\mathrm{SNR} = \varrho = 10$ and three different values of r.

The estimator (65.41) can be improved using a weighted average of cross products between all possible pairs of real and imaginary parts of $\boldsymbol{y}_E(t)$ and $\boldsymbol{y}_H(s)$ taken at arbitrary times t and s. (Note that these cross products have directions nearly parallel to the basic estimator $\widehat{\boldsymbol{u}}$ in Eq. (65.41); however, before averaging,

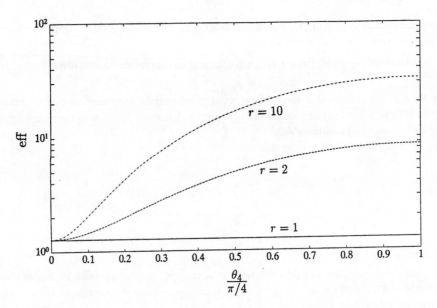

FIGURE 65.5 The efficiency factor (65.43) of the cross-product-based direction estimator as a function of the normalized ellipticity angle for three values of $r = \sigma_H / \sigma_E$. A single source, SNR = 10.

these cross products should be premultiplied by $+1$ or -1 in accordance with the direction of the basic estimator $\hat{u}$). A similar algorithm suitable for real time applications can also be developed in the time domain without preprocessing needed for phasor representation. It can be extended to nonstationary inputs by using a moving average window on the data. It is of interest to find the optimal weights and the performances of these estimators.

The main advantages of the proposed cross-product-based algorithm (65.41) or one of its variants above are

- It can give a direction estimate instantly, i.e., with one time sample.
- It is simple to implement (does not require minimization of a cost function) and can be applied in real time.
- It is equally applicable to sources of various types, including SST, DST, wide-band, and non-Gaussian.
- Its MSAE is nearly optimal in the Gaussian SST case under Eq. (65.44).
- It does not depend on time delays and therefore does not require data synchronization among different sensor components.

65.5 Multi-Source Multi-Vector Sensor Analysis

Consider the case in which it is desired to estimate the directions to multiple electromagnetic sources whose covariance is unknown using an array of vector sensors. The MSAE_{CR} and CVAE_{CR} bound expressions in Eqs. (65.29) and (65.37) are applicable to each of the sources in the multi-source multi-vector sensor scenario. Suppose that the noise vector $e_{EH}(t)$ in Eq. (65.19) is complex white Gaussian with zero mean and diagonal covariance matrix (i.e., noises from different sensors are uncorrelated) and with electric and magnetic variances σ_E^2 and σ_H^2, respectively. Suppose also that $r = \sigma_H / \sigma_E$ is known. Similarly to the single sensor case, multiply the electric measurements in Eq. (65.19) by r to obtain equal noise variances in all the sensor coordintates. The resulting models then become special cases of Eq. (65.21) as follows.

For DST signals, the block columns $A_k \in \mathbb{C}^{6m \times 2}$ and the signals $x(t) \in \mathbb{C}^{2n \times 1}$ are

$$A_k = e_k \otimes \begin{bmatrix} r I_3 \\ (u_k \times) \end{bmatrix} V_k \tag{65.45a}$$

$$x(t) = \begin{bmatrix} \xi_1^T(t), \cdots, \xi_n^T(t) \end{bmatrix}^T \tag{65.45b}$$

The parameter vector of the kth source includes here its azimuth and elevation.

For the SST case, the columns $A_k \in \mathbb{C}^{6m \times 1}$ and the signals $x(t) \in \mathbb{C}^{n \times 1}$ are

$$A_k = e_k \otimes \begin{bmatrix} r I_3 \\ (u_k \times) \end{bmatrix} V_k Q_k w_k \tag{65.46a}$$

$$x(t) = [s_1(t), \cdots, s_n(t)]^T \tag{65.46b}$$

The parameter vector of the kth source includes here its azimuth, elevation, orientation, and ellipticity angles.

The matrices A whose (block) columns are given in Eqs. (65.45a) and (65.46a) are the Khatri-Rao products (see, e.g., [20, 21]) of the two matrices whose (block) columns are the arguments of the Kronecker products in these equations.

Mixed single and dual signal transmissions are also special cases of Eq. (65.21) with appropriate combinations of the above expressions.

65.5.1 Results for Multiple Sources, Single-Vector Sensor

We present several results for the multiple-source model and a single-vector sensor. It is assumed that the signal and noise vectors satisfy, respectively, assumptions A5 and A6. The results are applicable to wide-band sources since a single vector sensor is used and thus A3 and A4 are not needed.

We first present results obtained by numerical evaluation concerning the localization of *two* uncorrelated sources, assuming r is known:

1. When only the electric field is measured, the information matrix is singular.
2. When the electric measurement is precise, the CRB variances are generally nonzero.
3. The MSAE$_{CR}^S$ can increase without bound with decreasing source angular separation for sources with the same ellipticity and spin direction, but remarkably it remains bounded for sources with different ellipticities or opposite spin directions.

Properties 1 and 2 are, in general, different from the single source case. Property 1 shows that it is necessary to include both the electric and magnetic measurements to estimate the direction to more than one source. Property 3 demonstrates the great advantage of using the electromagnetic vector sensor, in that it allows high resolution of sources with different ellipticities or opposite spins. Note that this generally requires a very large aperture using a scalar sensor array.

The above result on the ability to resolve two sources that are different only in their ellipticity or spin direction appears to be new. Note also the analogy to Pauli's "exclusion principle", as in our case two narrow-band SST sources are distinguishable if and only if they have different sets of parameters. The set in our case includes wave-length, direction, ellipticity, and spin sign.

Now we present conditions for identifiability of multiple SST (or polarized) sources and a single vector sensor, which are analytically proven in [29] and [30], assuming the noise variances are known:

1. A single source is always identifiable.
2. Two sources that are not fully correlated are identifiable if they have different DOAs.

3. Two fully correlated sources are identifiable if they have different DOAs and ellipticities.

4. Three sources that are not fully correlated are identifiable if they have different DOAs and ellipticities.

Note that by identifiability we refer to both the DOA and polarization parameters.

Figures 65.6 and 65.7 illustrate the resolution of two uncorrelated equal power SST sources with a single electromagnetic vector sensor. The figures show the square root of the $MSAE_{CR}^S$ of one of the sources for a variety of spin directions, ellipticities, and orientation angles, as a function of the separation angle between the sources. (The $MSAE_{CR}^S$ values of the two sources are found to be equal in all the following cases.) The covariances of the signals and noise are normalized such that $P = I_2, \sigma_E = \sigma_H = 1$. The azimuth angle of the first source and the elevation angles of the two sources are kept constant ($\theta_1^{(1)} = \theta_2^{(1)} = \theta_2^{(2)} = 0$). The second source's azimuth is varied to give the desired separation angle $\Delta\theta_1$, $\theta_1^{(2)}$. In

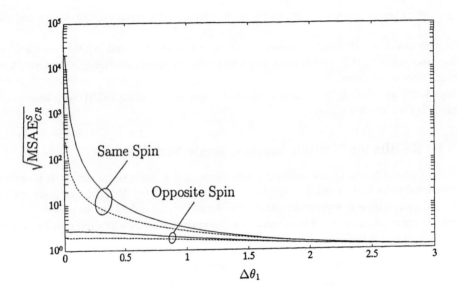

FIGURE 65.6 $MSAE_{CR}^S$ for two uncorrelated equal power SST sources and a single vector sensor as a function of the source angular separation. Upper two curves: Same spin directions ($\theta_4^{(1)} = \theta_4^{(2)} = \pi/12$). Lower two curves: Opposite spin directions ($\theta_4^{(1)} = -\theta_4^{(2)} = \pi/12$). Solid curves: Same orientation angles ($\theta_3^{(1)} = \theta_3^{(2)} = \pi/4$). Dashed curves: Different orientation angles ($\theta_3^{(1)} = -\theta_3^{(2)} = \pi/4$). Remaining parameters are $\theta_1^{(1)} = \theta_2^{(1)} = \theta_2^{(2)} = 0$, $\Delta\theta_1 \triangleq \theta_1^{(2)}$, $P = I_2, \sigma_E = \sigma_H = 1$.

Fig. 65.6, the cases shown are of same spin directions ($\theta_4^{(1)} = \theta_4^{(2)} = \pi/12$) and opposite spin directions ($\theta_4^{(1)} = -\theta_4^{(2)} = \pi/12$), same orientation angles ($\theta_3^{(1)} = \theta_3^{(2)} = \pi/4$) and different orientation angles ($\theta_3^{(1)} = -\theta_3^{(2)} = \pi/4$). The figure shows that the resolution of the two sources with a single vector sensor is remarkably good when the sources have opposite spin directions. In particular, the $MSAE_{CR}^S$ remains bounded even for zero separation angle and equal orientation angles! On the other hand, the resolution is not so significant when the two sources have different orientation angles but equal ellipticity angles (then, for example, the $MSAE_{CR}^S$ tends to infinity for zero separation angle). In Fig. 65.7, the orientation angles of the sources is the same ($\theta_3^{(1)} = \theta_3^{(2)} = \pi/4$), the polarization of the first source is kept linear ($\theta_4^{(1)} = 0$) while the ellipticity angle of the second source is varied ($|\theta_4^{(2)}| = \pi/12, \pi/6, \pi/4$) to illustrate

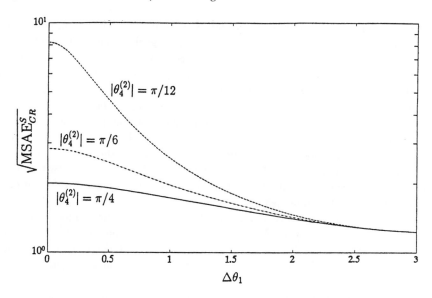

FIGURE 65.7 MSAE_{CR}^{S} for two uncorrelated equal power SST sources and a single vector sensor as a function of the source angular separation. Sources are with the same orientation angles $(\theta_3^{(1)} = \theta_3^{(2)} = \pi/4)$ and different ellipticity angles $(\theta_4^{(1)} = 0$ and $\theta_4^{(2)}$ as shown in the figure). Remaining parameters are as in Fig. 65.6.

the remarkable resolvability due to different ellipticities. It can be seen that the MSAE_{CR}^{S} remains bounded here even for zero separation angle.

Thus, Figs. 65.6 and 65.7 show that with one vector sensor it is possible to resolve extremely well two uncorrelated SST sources that have only different spin directions or different ellipticities (these sources can have the same direction of arrival and the same orientation angle). This demonstrates a great advantage of the vector sensor over scalar sensor arrays, in that the latter require large array apertures to resolve sources with small separation angle.

65.6 Concluding Remarks

An approach has been presented for the localization of electromagnetic sources using vector sensors. We summarize some of the main results of this article and give an outlook to their possible extensions.

Models: New models that include the complete electromagnetic data at each sensor have been introduced. Furthermore, new signal models and vector angular error models in the wave frame have been proposed. The wave frame model provides simple performance expressions that are easy to interpret and have only intrinsic singularities. Extensions of the proposed models may include additional structures for specific applications.

Cramér-Rao bounds and quality measures: A compact expression for the CRB for multi-vector source multi-vector sensor processing has been derived. The derivation gave rise to new block matrix operators. New quality measures in three-dimensional space, such as the MSAE for direction estimation and CVAE for direction and orientation estimation, have been defined. Explicit bounds on the MSAE and CVAE, having the desirable property of being invariant to the choice of the reference coordinate frame, have been derived and can be used for performance analysis. Some generalizations of the bounds appear in [26]. These bounds are not limited to electromagnetic vector sensor processing. Performance comparisons of vector sensor processing with scalar sensor counterparts are of interest.

Identifiablity: The derived bounds and the identifiability analysis of [29] and [30] were used to show that the fusion of magnetic and electric data at a single vector sensor increases the number of identifiable sources (or resolution capacity) in three-dimensional space from one source in the electric data case to up to three sources in the electromagnetic case. For a single signal transmission source, in order to get good direction estimates, the fusion of the complete data becomes more important as the polarization gets closer to linear. Finding the number of identifiable sources per sensor in a general vector sensor array is of interest. Preliminary results on this issue can be found in [29, 31].

Resolution: Source resolution using vector sensors is inherently different from scalar sensors, where the latter case is characterized by the classical Rayleigh principle. For example, it was shown that a single vector sensor can be used to resolve two sources in three-dimensional space. In particular, a vector sensor exhibits remarkable resolvability when the sources have opposite spin directions or different ellipticity angles. This is very different from the scalar sensor array case in which a plane array with large aperture is required to achieve the same goal. Analytical results on source resolution using vector sensor arrays and comparisons with their scalar counterparts are of interest.

Algorithms: A simple algorithm has been proposed and analyzed for finding the direction to a single source using a single vector sensor based on the cross-product operation. It is of interest to analyze the performance of the aforementioned variants of this algorithm and to extend them to more general source scenarios (e.g., larger number of sources). It is also of interest to develop new algorithms for the vector sensor array case.

Communication: The main considerations in communication are transmission of signals over channels with limited bandwidth and their recovery at the sensor. Vector sensors naturally fit these considerations as they have maximum observability to incoming signals and they double the channel capacity (compared with scalar sensors) with DST signals. This has vast potential for performance improvement in cellular communications. Future goals will include development of optimum signal estimation algorithms, communication forms, and coding design with vector-sensor arrays.

Implementations: The proposed methods should be implemented and tested with real data.

Sensor development: The use of complete electromagnetic data seems to be virtually nonexistent in the literature on source localization. It is hoped that the results of this research will motivate the systematic development of high quality electromagnetic sensors that can operate over a broad range of frequencies. Recent references on this topic can be found in [14] and [15].

Extensions: The vector sensor concept can be extended to other areas and open new possibilities. An example of this can be found in [32] and [33] for the acoustic case.

Acknowledgment

The authors are grateful to Professor I.Y. Bar-Itzhack from the Department of Aeronautical Engineering, Technion, Israel, for bringing reference [34] to their attention.

References

[1] Nehorai, A. and Paldi, E., Vector-sensor array processing for electromagnetic source localization, *IEEE Trans. on Signal Processing*, SP-42, 376–398, Feb. 1994.

[2] Nehorai, A. and Paldi, E., Vector sensor processing for electromagnetic source localization, *Proc. 25th Asilomar Conf. Signals, Syst. Comput.*, Pacific Grove, CA, Nov. 1991, 566–572.

[3] Schwartz, M., Bennett, W.R. and Stein, S., *Communication Systems and Techniques*, McGraw-Hill, New York, 1966.

[4] Keiser, B.E., *Broadband Coding, Modulation, and Transmission Engineering*, Prentice-Hall, Englewood Cliffs, New Jersey, 1989.

[5] Stoica, P. and Nehorai, A., Performance study of conditional and unconditional direction-of-arrival estimation, *IEEE Trans. Acoust., Speech, Signal Processing*, ASSP-38, 1783–1795, Oct. 1990.

[6] Ottersten, B., Viberg, M. and Kailath, T., Analysis of subspace fitting and ML techniques for parameter estimation from sensor array data, *IEEE Trans. Signal Processing*, SP-40, 590–600, March 1992.

[7] Schmidt, R.O., A signal subspace approach to multiple emitter location and spectral estimation, Ph.D., Dissertation, Stanford University, Stanford, CA, Nov. 1981.

[8] Ferrara, Jr., E.R. and Parks, T.M., Direction finding with an array of antennas having diverse polarization, *IEEE Trans. Antennas Propagat.*, AP-31, 231–236, March 1983.

[9] Ziskind, I. and Wax, M., Maximum likelihood localization of diversely polarized sources by simulated annealing, *IEEE Trans. Antennas Propagat.*, AP-38, 1111–1114, July 1990.

[10] Li, J. and Compton, R.T., Jr., Angle and polarization estimation using ESPRIT with a polarization sensitive array, *IEEE Trans. Antennas Propagat.*, AP-39, 1376–1383, Sept. 1991.

[11] Weiss, A.J. and Friedlander, B., Performance analysis of diversely polarized antenna arrays, *IEEE Trans. Signal Processing*, SP-39, 1589–1603, July 1991.

[12] Means, J.D., Use of three-dimensional covariance matrix in analyzing the polarization properties of plane waves, *J. Geophys. Res.*, 77, 5551–5559, Oct. 1972.

[13] Hatke, G.F., Performance analysis of the SuperCART antenna array, Project Report No. AST-22, Lincoln Laboratory, Massachusetts Institute of Technology, Lexington, MA, March 1992.

[14] Kanda, M., An electromagnetic near-field sensor for simultaneous electric and magnetic-field measurements, *IEEE Trans. on Electromagnetic Compatibility*, 26(1), 102–110, Aug. 1984.

[15] Kanda, M. and Hill, D., A three-loop method for determining the radiation characteristics of an electrically small source, *IEEE Trans. on Electromagnetic Compatibility*, 34(1), 1–3, Feb. 1992.

[16] Dugundji, J., Envelopes and pre-envelopes of real waveforms, *IRE Trans. Information Theory*, IT-4, 53–57, March 1958.

[17] Rice, S.O., Envelopes of narrow-band signals, *Proc. IEEE*, 70, 692–699, July 1982.

[18] Giuli, D., Polarization diversity in radars, *Proc. IEEE*, 74, 245–269, Feb. 1986.

[19] Born, M. and Wolf, E., Eds., *Principles of Optics*, 6th ed., Pergamon Press, Oxford, 1980 [1st ed., 1959].

[20] Khatri, C.G. and Rao, C.R., Solution to some functional equations and their applications to characterization of probability distribution, *Sankhyā Ser. A*, 30, 167–180, 1968.

[21] Rao, C.R. and Mitra, S.K., *Generalized Inverse of Matrices and its Applications*, John Wiley & Sons, New York, 1971.

[22] Stoica, P. and Nehorai, A., MUSIC, maximum likelihood and Cramér-Rao bound, *IEEE Trans. Acoust., Speech, Signal Processing*, ASSP-37, 720–741, May 1989.

[23] Ibragimov, I.A. and Has'minskii, R.Z., *Statistical Estimation: Asymptotic Theory*, Springer-Verlag, New York, 1981.

[24] Paldi, E. and Nehorai, A., A generalized Cramér-Rao bound, in preparation.

[25] Caines, P.E., *Linear Stochastic Systems*, John Wiley & Sons, New York, 1988.

[26] Nehorai, A. and Hawkes, M., Performance bounds on estimating vector systems, in preparation.

[27] Jackson, J.D., *Classical Electrodynamics*, 2nd ed., John Wiley & Sons, New York, 1975 [1st ed., 1962].

[28] Nehorai, A. and Paldi, E., Method for electromagnetic source localization, U.S. Patent No. 5,315,308, May 24, 1994.

[29] Hochwald, B. and Nehorai, A., Identifiability in array processing models with vector-sensor applications, *IEEE Trans. Signal Process.*, SP-44, 83–95, Jan. 1996.

[30] Ho, K.-C., Tan, K.-C. and Ser, W., An investigation on number of signals whose direction-of-arrival are uniquely determinable with an electromagnetic vector sensor, *Signal Processing*, 47, 41–54, Nov. 1995.

[31] Tan, K.-C., Ho, K.-C. and Nehorai, A., Uniqueness study of measurements obtainable with arrays of electromagnetic vector sensors, *IEEE Trans. Signal Process.*, SP-44, 1036–1039, Apr. 1996.

[32] Nehorai, A. and Paldi, E., Acoustic vector-sensor array processing, *IEEE Trans. on Signal Processing*, SP-42, 2481–2491, Sept. 1994. A short version appeared in *Proc. 26th Asilomar Conf. Signals, Syst. Comput.*, Pacific Grove, CA, Oct. 1992, 192–198.

[33] Hawkes, M. and Nehorai, A., Acoustic vector-sensor beamforming and capon direction estimation, *IEEE Int. Conf. Acoust., Speech, Signal Processing*, Detroit, MI, May 1995, 1673–1676.

[34] Shuster, M.D., A Survey of attitude representations, *J. Astronaut. Sci.*, 41(4), 439–517, Oct.–Dec. 1993.

Appendix A: Definitions of Some Block Matrix Operators

This appendix defines several block matrix operators that are found to be useful in this article. The following notation will be used for a blockwise partitioned matrix A:

$$A = \begin{bmatrix} A_{<11>} & \cdots & A_{<1n>} \\ \vdots & & \vdots \\ A_{<m1>} & \cdots & A_{<mn>} \end{bmatrix} \overset{\Delta}{=} \left[A_{<ij>} \right] \tag{A.1}$$

with the block entries $A_{<ij>}$ of dimensions $\mu_i \times \nu_j$. Define $\bar{\mu} \overset{\Delta}{=} \sum_{i=1}^{m} \mu_i$, $\bar{\nu} \overset{\Delta}{=} \sum_{j=1}^{n} \nu_j$, so A is a $\bar{\mu} \times \bar{\nu}$ matrix. Since the block entries may not be of the same size, this is sometimes called an unbalanced partitioning. The following definitions will be considered.

DEFINITION 65.6   *Block transpose.* Let A be an $m\mu \times n\nu$ blockwise partitioned matrix, with blocks $A_{<ij>}$ of equal dimensions $\mu \times \nu$. Then the block transpose A^{bT} is an $n\mu \times m\nu$ matrix defined through

$$\left(A^{bT} \right)_{<ij>} = A_{<ji>} \tag{A.2}$$

DEFINITION 65.7   *Block Kronecker product.* Let A be a blockwise partitioned matrix of dimension $\bar{\mu} \times \bar{\nu}$, with block entries $A_{<ij>}$ of dimensions $\mu_i \times \nu_j$, and let B be a blockwise partitioned matrix of dimensions $\bar{\eta} \times \bar{\rho}$, with block entries $B_{<ij>}$ of dimensions $\eta_i \times \rho_j$. Also $\bar{\mu} = \sum_{i=1}^{m} \mu_i$, $\bar{\nu} = \sum_{j=1}^{n} \nu_j$, $\bar{\eta} = \sum_{i=1}^{m} \eta_i$, $\bar{\rho} = \sum_{j=1}^{n} \rho_j$. Then the block Kronecker product $A \boxtimes B$ is an $(\sum_{i=1}^{m} \mu_i \eta_i \times \sum_{j=1}^{n} \nu_j \rho_j)$ matrix defined through

$$(A \boxtimes B)_{<ij>} = A_{<ij>} \otimes B_{<ij>} \tag{A.3}$$

i.e., the (i, j) block entry of $A \boxtimes B$ is $A_{<ij>} \otimes B_{<ij>}$ of dimension $\mu_i \eta_i \times \nu_j \rho_j$.

DEFINITION 65.8   *Block Schur-Hadamard product.* Let A be an $m\mu \times n\nu$ matrix consisting of blocks $A_{<ij>}$ of dimensions $\mu \times \nu$, and let B be an $m\nu \times n\eta$ matrix consisting of blocks $B_{<ij>}$ of dimensions $\nu \times \eta$. Then the block Schur-Hadamard product $A \boxdot B$ is an $m\mu \times n\eta$ matrix defined through

$$(A \boxdot B)_{<ij>} = A_{<ij>} B_{<ij>} \tag{A.4}$$

Thus, each block of the product is a *usual* product of a pair of blocks and is of dimension $\mu \times \eta$.

DEFINITION 65.9   *Block trace operator.* Let A be an $m\mu \times n\mu$ matrix consisting of blocks $A_{<ij>}$ of dimensions $\mu \times \mu$. Then the block trace matrix operator $\text{btr}[A]$ is an $m \times n$ matrix defined by

$$(\text{btr}\,[A])_{ij} = \text{tr}\, A_{<ij>} \tag{A.5}$$

66

Subspace Tracking

R.D. DeGroat
The University of Texas at Dallas

E.M. Dowling
The University of Texas at Dallas

D.A. Linebarger
The University of Texas at Dallas

66.1 Introduction

Most high resolution direction-of-arrival (DOA) estimation methods rely on subspace or eigen-based information which can be obtained from the eigenvalue decomposition (EVD) of an estimated correlation matrix, or from the singular value decomposition (SVD) of the corresponding data matrix. However, the expense of directly computing these decompositions is usually prohibitive for real-time processing. Also, because the DOA angles are typically time-varying, repeated computation is necessary to track the angles. This has motivated researchers in recent years to develop low cost eigen and subspace tracking methods. Four basic strategies have been pursued to reduce computation: (1) computing only a few eigencomponents, (2) computing a subspace basis instead of individual eigencomponents, (3) approximating the eigencomponents or basis, and (4) recursively updating the eigencomponents or basis. The most efficient methods usually employ several of these strategies.

In 1990, an extensive survey of SVD tracking methods was published by Comon and Golub [7]. They classified the various algorithms according to complexity and basically two categories emerge: $O(n^2 r)$ and $O(nr^2)$ methods, where n is the snapshot vector size and r is the number of extreme eigenpairs to be tracked. Typically, $r < n$ or $r \ll n$, so the $O(nr^2)$ methods involve significantly fewer computations

0-8493-8572-5/98/$0.00+$.50
© 1998 by CRC Press LLC

than the $O(n^2 r)$ algorithms. However, since 1990, a number of $O(nr)$ algorithms have been developed. This article will primarily focus on recursive subspace and eigen updating methods developed since 1990, especially, the $O(nr^2)$ and $O(nr)$ algorithms.

66.2 Background

66.2.1 EVD vs. SVD

Let $X = [x_1 | x_2 | ... | x_N]$ be an $n \times N$ data matrix where the kth column corresponds to the kth snapshot vector, $x_k \in C^n$. With block processing, the correlation matrix for a zero mean, stationary, ergodic vector process is typically estimated as $R = \frac{1}{N} X X^H$ where the true correlation matrix, $\Phi = E[x_k x_k^H] = E[R]$.

The EVD of the estimated correlation matrix is closely related to the SVD of the corresponding data matrix. The SVD of X is given by $X = USV^H$ where $U \in C^{n \times n}$ and $V \in C^{N \times N}$ are unitary matrices and $S \in C^{n \times N}$ is a diagonal matrix whose nonzero entries are positive. It is easy to see that the left singular vectors of X are the eigenvectors of $X X^H = U S S^T U^H$, and the right singular vectors of X are the eigenvectors of $X^H X = V S^T S V^H$. This is so because $X X^H$ and $X^H X$ are positive definite Hermitian matrices (which have orthogonal eigenvectors and real, positive eigenvalues). Also note that the nonzero singular values of X are the positive square roots of the nonzero eigenvalues of $X X^H$ and $X^H X$. Mathematically, the eigen information contained in the SVD of X or the EVD of $X X^H$ (or $X^H X$) is equivalent, but the dynamic range of the eigenvalues is twice that of the corresponding singular values. With finite precision arithmetic, the greater dynamic range can result in a loss of information. For example, in rank determination, suppose the smallest singular value is ϵ where ϵ is machine precision. The corresponding eigenvalue, ϵ^2, would be considered a machine precision zero and the EVD of $X X^H$ (or $X^H X$) would incorrectly indicate a rank deficiency. Because of the dynamic range issue, it is generally recommended to use the SVD of X (or a square root factor of R). However, because additive sensor noise usually dominates numerical errors, this choice may not be critical in most signal processing applications.

66.2.2 Short Memory Windows for Time Varying Estimation

Ultimately, we are interested in tracking some aspect of the eigenstructure of a time varying correlation (or data) matrix. For simplicity we will focus on time varying estimation of the correlation matrix, realizing that the EVD of R is trivially related to the SVD of X. A time varying estimator must have a short term memory in order to track changes. An example of long memory estimation is an estimator that involves a growing rectangular data window. As time goes on, the estimated quantities depend more and more on the old data, and less and less on the new data. The two most popular short memory approaches to estimating a time varying correlation matrix involve (1) a moving rectangular window and (2) an exponentially faded window. Unfortunately, an unbiased, causal estimate of the true instantaneous correlation matrix at time k, $\Phi_k = E[x_k x_k^H]$, is not possible if averaging is used and the vector process is truly time varying. However, it is usually assumed that the process is varying slowly enough within the effective observation window that the process is approximately stationary and some averaging is desirable. In any event, at time k, a length N moving rectangular data window results in a rank two modification of the correlation matrix estimate, i.e.,

$$R_k^{(rect)} = R_{k-1}^{(rect)} + \frac{1}{N}(x_k x_k^H - x_{k-N} x_{k-N}^H) \tag{66.1}$$

where x_k is the new snapshot vector and x_{k-N} is the oldest vector which is being removed from the estimate. The corresponding data matrix is given by $X_k^{(rect)} = [x_k | x_{k-1} | ... | x_{k-N+1}]$ and $R_k^{(rect)} = \frac{1}{N} X_k^{(rect)} \left(X_k^{(rect)} \right)^H$. Subtracting the rank one matrix from the correlation estimate is referred to as a rank one downdate. Downdating moves all the eigenvalues down (or unchanged). Updating, on the other hand, moves all eigenvalues up (or unchanged). Downdating is potentially ill-conditioned because the smallest eigenvalue can move towards zero.

An exponentially faded data window produces a rank one modification in

$$R_k^{(fade)} = \alpha R_{k-1}^{(fade)} + (1-\alpha)x_k x_k^H \tag{66.2}$$

where α is the fading factor with $0 \leq \alpha \leq 1$. In this case, the data matrix is growing in size, but the older data is de-emphasized with a diagonal weighting matrix,

$X_k^{(fade)} = [x_k | x_{k-1} | ... | x_1] \text{ sqrt}(\text{diag}(1, \alpha, \alpha^2, ..., \alpha^{k-1}))$ and $R_k^{(fade)} = (1-\alpha)X_k^{(fade)} \left(X_k^{(fade)} \right)^H$.

Of course, the two windows could be combined to produce an exponentially faded moving rectangular window, but this kind of hybrid short memory window has not been the subject of much study in the signal processing literature. Similarly, not much attention has been paid to which short memory windowing scheme is most appropriate for a given data model. Since downdating is potentially ill-conditioned, and since two rank one modifications usually involve more computation than one, the exponentially faded window has some advantages over the moving rectangular window. The main advantage of a (short) rectangular window is in tracking sudden changes. Assuming stationarity within the effective observation window, the power in a rectangular window will be equal to the power in an exponentially faded window when

$$N \approx \frac{1}{(1-\alpha)} \quad \text{or equivalently} \quad \alpha \approx 1 - \frac{1}{N} = \frac{N-1}{N}. \tag{66.3}$$

Based on a Fourier analysis of linearly varying frequencies, equal frequency lags occur when [14]

$$N \approx \frac{(1+\alpha)}{(1-\alpha)} \quad \text{or equivalently} \quad \alpha \approx \frac{N-1}{N+1}. \tag{66.4}$$

Either one of these relationships could be used as a rule of thumb for relating the effective observation window of the two most popular short memory windowing schemes.

66.2.3 Classification of Subspace Methods

Eigenstructure estimation can be classified as (1) block or (2) recursive. Block methods simply compute an EVD, SVD, or related decomposition based on a block of data. Recursive methods update the previously computed eigen information using new data as it arrives. We focus on recursive subspace updating methods in this article.

Most subspace tracking algorithms can also be broadly categorized as (1) modified eigen problem (MEP) methods or (2) adaptive (or non-MEP) methods. With short memory windowing, MEP methods are adaptive in the sense that they can track time varying eigen information. However, when we use the word adaptive, we mean that exact eigen information is not computed at each update, but rather, an adaptive method tends to move towards an EVD (or some aspect of an EVD) at each update. For example, gradient-based, perturbation-based, and neural network-based methods are classified as adaptive because on average they move towards an EVD at each update. On the other hand, rank one, rank k, and sphericalized EVD and SVD updates are, by definition, MEP methods because exact eigen information associated with an explicit matrix is computed at each update. Both MEP and adaptive methods are supposed to track the eigen information of the instantaneous, time varying correlation matrix.

66.2.4 Historical Overview of MEP Methods

Many researchers have studied SVD and EVD tracking problems. Golub [19] introduced one of the first eigen-updating schemes, and his ideas were developed and expanded by Bunch and co-workers in [3, 4]. The basic idea is to update the EVD of a symmetric (or Hermitian) matrix when modified by a rank one matrix. The rank-one eigen update was simplified in [37], when Schreiber introduced a transformation that makes the core eigenproblem real. Based on an additive white noise model, Karasalo [21] and Schreiber [37] suggested that the noise subspace be "sphericalized", i.e., replace the noise eigenvalues by their average value so that deflation [4] could be used to significantly reduce computation. By deflating the noise subspace and only tracking the r dominant eigenvectors, the computation is reduced from $O(n^3)$ to $O(nr^2)$ per update. DeGroat reduced computation further by extending this concept to the signal subspace [8]. By sphericalizing and deflating both the signal and the noise subspaces, the cost of tracking the r dimensional signal (or noise) subspace is $O(nr)$ and no iteration is involved. To make eigen updating more practical, DeGroat and Roberts developed stabilization schemes to control the loss of orthogonality due to the buildup of roundoff error [10]. Further work related to eigenvector stabilization is reported in [15, 28, 29, 30]. Recently, a more stable version of Bunch's algorithm was developed by Gu and Eisenstat [20]. In [46], Yu extended rank one eigen updating to rank k updating.

DeGroat showed in [8] that forcing certain subspaces of the correlation matrix to be spherical, i.e., replacing the associated eigenvalues with a fixed or average value, is an easy way to deflate the size of the updating problem and reduce computation. Basically, a spherical subspace (SS) update is a rank one EVD update of a sphericalized correlation matrix. Asymptotic convergence analysis of SS updating is found in [11, 13]. A four level SS update capable of automatic signal subspace rank and size adjustment is described in [9, 11]. The four level and the two level SS updates are the only MEP updates to date that are $O(nr)$ and noniterative. For more details on SS updating, see Section 66.3.6, *Spherical Subspace (SS) Updating: A General Framework for Simplified Updating*.

In [42], Xu and Kailath present a Lanczos based subspace tracking method with an associated detection scheme to track the number of sources. A reference list for systolic implementations of SVD based subspace trackers is contained in [12].

66.2.5 Historical Overview of Adaptive, Non-MEP Methods

Owsley pioneered orthogonal iteration and stochastic-based subspace trackers in [32]. Yang and Kaveh extended Owsley's work in [44] by devising a family of constrained gradient-based algorithms. A highly parallel algorithm, denoted the inflation method, is introduced for the estimation of the noise subspace. The computational complexity of this family of gradient-based methods varies from (approximately) n^2r to $\frac{7}{2}nr$ for the adaptation equation. However, since the eigenvectors are only approximately orthogonal, an additional nr^2 flops may be needed if Gram Schmidt orthogonalization is used. It may be that a partial orthogonalization scheme (see Section 66.3.2 *Controlling Roundoff Error Accumulation and Orthogonality Errors*) can be combined with Yang and Kaveh's methods to improve orthogonality enough to eliminate the $O(nr^2)$ Gram Schmidt computation. Karhunen [22] also extended Owsley's work by developing a stochastic approximation method for subspace computation. Bin Yang [43] used recursive least squares (RLS) methods with a projection approximation approach to develop the projection approximation subspace tracker (PAST) which tracks an arbitrary basis for the signal subspace, and PASTd which uses deflation to track the individual eigencomponents. A multi-vector eigen tracker based on the conjugate gradient method is developed in [18]. Previous conjugate gradient-based methods tracked a single eigenvector only. Orthogonal iteration, lossless adaptive filter, and perturbation-based subspace trackers appear in [40] [36], and [5] respectively. A family of non-EVD subspace trackers is given in [16]. An adaptive subspace method that uses a linear operator, referred to as the Propagator, is given in [26]. Approximate SVD methods that are based on a QR update step followed by a single (or partial) Jacobi sweep to move the triangular factor towards a diagonal form appear in [12, 17, 30]. These methods can be described as approximate SVD methods because they will converge to an SVD if the Jacobi sweeps are repeated.

Subspace estimation methods based on URV or rank revealing QR (RRQR) decompositions are referenced in [6]. These rank revealing decompositions can divide a set of orthonormal vectors into sets that span the signal and noise subspaces. However, a threshold (noise power) level that lies between the largest noise eigenvalue and the smallest signal eigenvalue must be known in advance. In some ways, the URV decomposition can be viewed as an approximate SVD. For example, the transposed QR (TQR) iteration [12] can be used to compute the SVD of a matrix, but if the iteration is stopped before convergence, the resulting decomposition is URV-like.

Artificial neural networks (ANN) have also been used to estimate eigen information [35]. In 1982, Oja [31] was one of the first to develop an eigenvector estimating ANN. Using a Hebbian type learning rule, this ANN adaptively extracts the first principal eigenvector. Much research has been done in this area since 1982. For an overview and a list of references, see [35].

66.3 Issues Relevant to Subspace and Eigen Tracking Methods

66.3.1 Bias Due to Time Varying Nature of Data Model

Because direction-of-arrival (DOA) angles are typically time varying, a range of spatial frequencies is usually included in the effective observation window. Most spatial frequency estimation methods yield frequency estimates that are approximately equal to the effective frequency average in the window. Consequently, the estimates lag the true instantaneous frequency. If the frequency variation is assumed to be linear within the effective observation window, this lag (or bias) can be easily estimated and compensated [14].

66.3.2 Controlling Roundoff Error Accumulation and Orthogonality Errors

Numerical algorithms are generally defined as stable if the roundoff error accumulates in a linear fashion. However, recursive updating algorithms cannot tolerate even a linear buildup of error if large (possibly unbounded) numbers of updates are to be performed. For real time processing, periodic reinitialization is undesirable. Most of the subspace tracking algorithms involve the product of at least k orthogonal matrices by the time the kth update is computed. According to Parlett [33], the error propagated by a product of orthogonal matrices is bounded as

$$|U_k U_k^H - I|_E \leq (k+1)n^{1.5}\epsilon \tag{66.5}$$

where the $n \times n$ matrix $U_k = U_{k-1}Q_k = Q_k Q_{k-1}...Q_1$ is a product of k matrices that are each orthogonal to working accuracy, ϵ is machine precision, and $|.|_E$ denotes the Euclidean matrix norm. Clearly, if k is large enough, the roundoff error accumulation can be significant.

There are really only two sources of error in updating a symmetric or Hermitian EVD: (1) the eigenvalues and (2) the eigenvectors. Of course, the eigenvectors and eigenvalues are interrelated. Errors in one tend to produce errors in the other. At each update, small errors may occur in the EVD update so that the eigenvalues become slowly perturbed and the eigenvectors become slowly nonorthonormal. The solution is to prevent significant errors from ever accumulating in either.

We do not expect the main source of error to be from the eigenvalues. According to Stewart [38], the eigenvalues of a Hermitian matrix are perfectly conditioned, having condition numbers of one. Moreover, it is easy to show that when exponential weighting is used, the accumulated roundoff error is bounded by a constant, assuming no significant errors are introduced by the eigenvectors. By contrast, if exponential windowing is not used, the bound for the accumulated error builds up in a linear fashion. Thus, the fading factor not only fades out old data, but also old roundoff errors that accumulate in the eigenvalues.

Unfortunately, the eigenvectors of a Hermitian matrix are not guaranteed to be well conditioned. An eigenvector will be ill-conditioned if its eigenvalue is closely spaced with other eigenvalues. In this case, small roundoff perturbations to the matrix may cause relatively large errors in the eigenvectors.

The greatest potential for nonorthogonality then is between eigenvectors with adjacent (closely spaced) eigenvalues. This observation led to the development of a partial orthogonalization scheme known as pairwise Gram Schmidt (PGS) [10] which attacks the roundoff error buildup problem at the point of greatest numerical instability — nonorthogonality of adjacent eigenvectors. If the intervening rotations (orthogonal matrix products) inherent in the eigen update are random enough, the adjacent vector PGS can be viewed as a full orthogonalization spread out over time. When PGS is combined with exponential fading, the roundoff accumulation in both the eigenvectors and the eigenvalues is controlled. Although PGS was originally designed to stabilize Bunch's EVD update, it is generally applicable to any EVD, SVD, URV, QR, or orthogonal vector update. Moonen et al. [29] suggested that the bulk of the eigenvector stabilization in the PGS scheme is due to the normalization of the eigenvectors. Simulations seem to indicate that normalization alone stabilizes the eigenvectors almost as well as the PGS scheme, but not to working precision orthogonality. Edelman and Stewart provide some insight into the normalization only approach to maintaining orthogonality [15]. For additional analysis and variations on the basic idea of spreading orthogonalization out over time, see [30] and especially [28].

Many of the $O(nr)$ adaptive subspace methods produce eigenvector estimates that are only approximately orthogonal and normalization alone does not always provide enough stabilization to keep the orthogonality and other error measures small enough. We have found that PGS stabilization can noticeably improve both the subspace estimation performance as well as the DOA (or spatial frequency) estimation performance. For example, without PGS (but with normalization only), we found that Champagne's $O(nr)$ perturbation-based eigen tracker (method PC) [5] sometimes gives spurious MUSIC-based frequency estimates. On the other hand, with PGS, Champagne's PC method produced improved subspace and frequency estimates. The orthogonality error was also significantly reduced. Similar performance boosts could be expected for any subspace or eigen tracking method (especially those that produce eigenvector estimates that are only approximately orthogonal, e.g., PAST and PASTd [43] or Yang and Kaveh's family of gradient based methods [44, 45]). Unfortunately, normalization only and PGS are $O(nr)$. Adding this kind of stabilization to an $O(nr)$ subspace tracking method could double its overall computation.

Other variations on the original PGS idea involve symmetrizing the 2×2 transformation and making the pairwise orthogonalization cyclic [28]. The symmetric transformation assumes that the vector pairs are almost orthgonal so that higher order error terms can be ignored. If this is the case, the symmetric version can provide slightly better results at a somewhat higher computational cost. For methods that involve working precision orthogonal vectors, the original PGS scheme is overkill. Instead of doing PGS orthogonalization on each adjacent vector pair, cyclic PGS orthogonalizes only one pair of vectors per update, but cycles through all possible combinations over time. Thus, cyclic PGS covers all vector pairs without relying on the randomness of intervening rotations. Cyclic PGS spreads the orthogonalization process out in time even more than the adjacent vector PGS method. Moreover, cyclic PGS (or cyclic normalization) involves $O(n)$ flops per update, but there is a small overhead associated with keeping track of the vector pair cycle.

In summary, we can say that stabilization may not be needed for a small number of updates. On the other hand, if an unbounded number of updates is to be performed, some kind of stabilization is recommended. For methods that yield nearly orthogonal vectors at each update, only a small amount of orthogonalization is needed to control the error buildup. In these cases, cyclic PGS may be best. However, for methods that produce vectors that are only approximately orthogonal, a more complete orthogonalization scheme may be appropriate, e.g., a cyclic scheme with two or three vector pairs orthogonalized per update will produce better results than a single pair scheme.

66.3.3 Forward-Backward Averaging

In many subspace tracking problems, forward-backward (FB) averaging can improve subspace as well as DOA (or frequency) estimation performance. Although FB averaging is generally not appropriate for nonstationary processes, it does appear to improve spatial frequency estimation performance if the frequencies vary linearly within the effective observation window. Based on Fourier analysis of linearly varying frequencies, we infer that this is probably due to the fact that the average frequency in the window is identical for both the forward and the backward cases [14]. Consequently, the frequency estimates are reinforced by FB averaging. Besides improved estimation performance, FB averaging can be exploited to reduce computation by as much as 75% [24]. FB averaging can also reduce computer memory requirements because (conjugate symmetric or anti-symmetric) symmetries in the complex eigenvectors of an FB averaged correlation matrix (or the singular vectors of an FB data matrix) can be exposed through appropriate normalization.

66.3.4 Frequency vs. Subspace Estimation Performance

It has recently been shown with asymptotic analysis that a better subspace estimate does not necessarily result in a better MUSIC-based frequency estimate [23]. In subspace tracking simulations, we have also observed that some methods produce better subspace estimates, but the associated MUSIC-based frequency estimates are not always better. Consequently, if DOA estimation is the ultimate goal, subspace estimation performance may not be the best criterion for evaluating subspace tracking methods.

66.3.5 The Difficulty of Testing and Comparing Subspace Tracking Methods

A significant amount of research has been done on subspace and eigen tracking algorithms in the past few years, and much progress has been made in making subspace tracking more efficient. Not surprisingly, all of the methods developed to date have different strengths and weaknesses. Unfortunately, there has not been enough time to thoroughly analyze, study, and evaluate all of the new methods. Over the years, several tests have been devised to "experimentally" compare various methods, e.g., convergence tests [44], response to sudden changes [7], and crossing frequency tracks (where the signal subspace temporarily collapses) [8]. Some methods do well on one test, but not so well on another. It is difficult to objectively compare different subspace tracking methods because optimal operating parameters are usually unknown and therefore unused, and the performance criteria may be ill-defined or contradictory.

66.3.6 Spherical Subspace (SS) Updating — A General Framework for Simplified Updating

Most eigen and subspace tracking algorithms are based directly or indirectly on tracking some aspect of the EVD of a time varying correlation matrix estimate that is recursively updated according to Eq. (66.1) or (66.2). Since Eqs. (66.1) and (66.2) involve rank one and rank two modifications to the correlation matrix, most subspace tracking algorithms explicitly or implicitly involve rank one (or two) modification of the correlation matrix. Since rank two modifications can be computed as two rank one modifications, we will focus on rank one updating.

Basically, spherical subspace (SS) updates are simplified rank one EVD updates. The simplification involves sphericalizing subsets of eigenvalues (i.e., forcing each subset to have the same eigenlevel) so that the sphericalized subspaces can be deflated.

Based on an additive white noise signal model, Karasalo [21] and Schreiber [37] first suggested that the "noise" eigenvalues be replaced by their average value in order to reduce computation by deflation. Using Ljung's ODE-based method for analyzing stochastic recursive algorithms [25], it has recently been shown that, if the noise subspace is sphericalized, the dominant eigenstructure of a correlation matrix asymptotically converges to the true eigenstructure with probability one (under any noise assumption) [11]. It is important to realize that averaging the noise eigenvalues yields a spherical subspace in which the eigenvectors can be arbitrarily oriented as long as they form an orthonormal basis for the subspace. A rank-one modification affects only one component of the sphericalized subspace. Thus, only one of the multiple noise eigenvalues is changed by a rank-one modification. Consequently, making the noise subspace spherical (by averaging the noise eigenvalues, or replacing them with a constant eigenlevel) deflates the eigenproblem to an $(r + 1) \times (r + 1)$ problem, which corresponds to a signal subspace of dimension r, and the single noise component whose power is changed. For details on deflation, see [4].

The analysis in [11] shows that any number of sphericalized eigenlevels can be used to track various subspace spans associated with the correlation matrix. For example, if both the noise and the signal subspaces are sphericalized (i.e., the dominant and subdominant set of eigenvalues is replaced by their respective averages), the problem deflates to a 2×2 eigenproblem that can be solved in closed form, noniteratively. We will call this doubly deflated SS update, SA2 (Signal Averaged, Two Eigenlevels) [8]. In [13] we derived the SA2 algorithm ODE and used a Lyapunov function to show asymptotic convergence to the true subspaces w.p. 1 under a diminishing gain assumption. In fact, the SA2 subspace trajectories can be described with Lie bracket notation and follow an isospectral flow as described by Brockett's ODE [2]. A four level SS update (called SA4) was introduced in [9] to allow for information theoretic source detection (based on the eigenvalues at the boundary of the signal and noise subspaces) and automatic subspace size adjustment. A detailed analysis of SA4 and an SA4 minimum description length (SA4-MDL) detection scheme can be found in [11, 41]. SA4 sphericalizes all the signal eigenvalues except the smallest one, and all the noise eigenvalues except the largest one, resulting in a 4×4 deflated eigenproblem. By tracking the eigenvalues that are on the boundary of the signal and noise subspaces, information theoretic detection schemes can be used to decide if the signal subspace dimension should be increased, decreased, or remain unchanged. Both SA2 and SA4 are $O(nr)$ and noniterative.

The deflated core problem in SS updating can involve any EVD or SVD method that is desired. It can also involve other decompositions, e.g., the URVD [34]. To illustrate the basic idea of SS updating, we will explicitly show how an update is accomplished when only the smallest $(n - r)$ "noise" eigenvalues are sphericalized. This particular SS update is called a Signal Eigenstructure (SE) update because only the dominant r "signal" eigencomponents are tracked. This case is equivalent to that described by Schreiber [37] and an SVD version is given by Karasalo [21].

To simplify and more clearly illustrate the idea SS updating, we drop the normalization factor, $(1 - \alpha)$, and the k subscripts from Eq. (66.2) and use the eigendecomposition of $R = UDU^H$ to expose a simpler underlying structure for a single rank-one update

$$
\begin{align}
\tilde{R} &= \alpha R + xx^H && \text{(66.6)} \\
&= \alpha UDU^H + xx^H && \text{(66.7)} \\
&= U(\alpha D + \beta\beta^H)U^H, && \beta = U^H x && \text{(66.8)} \\
&= UG(\alpha D + \gamma\gamma^T)G^H U^H, && \gamma = G^H \beta && \text{(66.9)} \\
&= UGH(\alpha D + \zeta\zeta^T)H^T G^H U^H, && \zeta = H^T \gamma && \text{(66.10)} \\
&= UGH(Q\tilde{D}Q^T)H^T G^H U^H && \text{(66.11)} \\
&= \tilde{U}\tilde{D}\tilde{U}^H, && \tilde{U} = UGHQ && \text{(66.12)}
\end{align}
$$

where $G = \mathrm{diag}\,(\beta_1/|\beta_1|, ..., \beta_n/|\beta_n|)$ is a diagonal unitary transformation that has the effect of making the matrix inside the parenthesis real [37], H is an embedded Householder transformation that deflates the core problem by zeroing out certain elements of ζ (see the SE case below), and $Q\tilde{D}Q^T$ is the EVD of the simplified, deflated core matrix, $(\alpha D + \zeta\zeta^T)$. In general, H and Q will involve smaller matrices embedded in an $n \times n$ identity matrix. In order to more clearly see the details of deflation, we must concentrate on finding the eigendecomposition of the completely real matrix, $S = (\alpha D + \gamma\gamma^T)$ for a specific case. Let us consider the SE update and assume that the noise eigenvalues contained in the diagonal matrix have been replaced by their average values, $d^{(n)}$, to produce a sphericalized noise subspace. We must then apply block Householder transformations to concentrate all of the power in the new data vector into a single component of the noise subspace. The update is thus deflated to an $(r + 1) \times (r + 1)$ embedded eigenproblem as shown below,

$$
\begin{aligned}
S &= (\alpha D + \gamma\gamma^T) \tag{66.13}\\
&= H(\alpha D + \zeta\zeta^T)H^T, \qquad \zeta = H^T\gamma \tag{66.14}\\
&= \begin{bmatrix} I_r & 0 \\ 0 & H^{(n)}_{n-r} \end{bmatrix}\left(\alpha\begin{bmatrix} D^{(s)}_r & 0 \\ 0 & d^{(n)}I_{n-r} \end{bmatrix}+\zeta\zeta^T\right)\begin{bmatrix} I_r & 0 \\ 0 & H^{(n)}_{n-r} \end{bmatrix}^T \tag{66.15}\\
&= \begin{bmatrix} I_r & 0 \\ 0 & H^{(n)}_{n-r} \end{bmatrix}\left(\begin{bmatrix} Q_{r+1} & 0 \\ 0 & I_{n-r-1} \end{bmatrix}\begin{bmatrix} \tilde{D}^{(s)}_r & 0 & 0 \\ 0 & \tilde{d}^{(n)} & 0 \\ 0 & 0 & \alpha d^{(n)}I_{n-r-1} \end{bmatrix}\right.\\
&\qquad\qquad\left.\times\begin{bmatrix} Q_{r+1} & 0 \\ 0 & I_{n-r-1} \end{bmatrix}^T\right)\begin{bmatrix} I_r & 0 \\ 0 & H^{(n)}_{n-r} \end{bmatrix}^T \tag{66.16}\\
&= H(Q\tilde{D}Q^T)H^T \tag{66.17}
\end{aligned}
$$

where

$$
\zeta^T = (H^T\gamma)^T = [\gamma^{(s)}, |\gamma^{(n)}|, 0_{(n-r-1)\times 1}]^T, \tag{66.18}
$$

$$
H^{(n)}_{n-r} = I_{n-r} - 2\frac{v^{(n)}(v^{(n)})^T}{(v^{(n)})^T v^{(n)}}, \tag{66.19}
$$

$$
H = \begin{bmatrix} I_r & 0 \\ 0 & H^{(n)}_{n-r} \end{bmatrix}, \tag{66.20}
$$

$$
\gamma = \begin{bmatrix} \gamma^{(s)} \\ \gamma^{(n)} \end{bmatrix}\begin{matrix} \}r \\ \}n-r \end{matrix} \tag{66.21}
$$

$$
v^{(n)} = \gamma^{(n)} + |\gamma^{(n)}|\begin{bmatrix} 1 \\ 0_{(n-r-1)\times 1} \end{bmatrix} \tag{66.22}
$$

The superscripts (s) and (n) denote signal and noise subspace, respectively, and the subscripts denote the size of the various block matrices. In the actual implementation of the SE algorithm, the Householder transformations are not explicitly computed, as we will see below. Moreover, it should be stressed that the Householder transformation does not change the span of the noise subspace, but merely "aligns" the subspace so that all of the new data vector, x, that projects into the noise subspace lies in a single component of the noise subspace.

The embedded (deflated) $(r+1)\times(r+1)$ eigenproblem,

$$
E = \left(\begin{bmatrix} D^{(s)} & 0 \\ 0 & d^{(n)} \end{bmatrix}+\begin{bmatrix} \gamma^{(s)} \\ |\gamma^{(n)}| \end{bmatrix}\begin{bmatrix} \gamma^{(s)} \\ |\gamma^{(n)}| \end{bmatrix}^T\right)_{(r+1)\times(r+1)} = Q_{r+1}\tilde{D}_{r+1}Q^T_{r+1} \tag{66.23}
$$

can be solved using any EVD algorithm. Or, an SVD (square root) version can be computed by finding the SVD of

$$
F = \begin{bmatrix} \Sigma^{(s)} & 0 & \gamma^{(s)} \\ 0 & \sigma^{(n)} & |\gamma^{(n)}| \end{bmatrix}_{(r+1)\times(r+2)} = Q_{r+1}\tilde{\Sigma}_{r+1}P^T_{r+1} \tag{66.24}
$$

where $E = FF^T$, $\Sigma^{(s)} = \text{sqrt}(D^{(s)})$, $\sigma^{(n)} = \sqrt{d^{(n)}}$ and $\tilde{\Sigma}_{r+1} = \text{sqrt}(\tilde{D}_{r+1})$. The right singular vectors, P_{r+1}, are generally not needed or explicitly computed in most subspace tracking problems.

The new signal and noise subspaces are thus given by

$$\tilde{U} = [\tilde{U}^{(s)}, \tilde{U}^{(n)}] \tag{66.25}$$

$$= UGHQ \tag{66.26}$$

$$= \left[\underbrace{U^{(s)}G^{(s)}}_{n \times r}, \underbrace{U^{(n)}G^{(n)}H^{(n)}}_{n \times (n-r)} \right] \left[\begin{array}{cc} Q_{r+1} & 0 \\ 0 & I_{n-r-1} \end{array} \right] \tag{66.27}$$

where $U^{(s)}$ and $U^{(n)}$ are the old signal and noise subspaces, G represents the diagonal unitary transformation that makes the rest of the problem real, H is the block Householder transformation that rotates (or more precisely, reflects) the spherical subspaces so that all of the noise power contained in the new data vector can be concentrated into a single component of noise subspace, and Q represents the evolution and interaction of the two subspaces induced by the new data vector. Basically, this update partitions the data space into two subspaces: the signal subspace is not sphericalized and all of its eigencomponents are explicitly tracked whereas the noise subspace is sphericalized and not explicitly tracked (to save computation). Using the properties of the Householder transformation, it can be shown that the single component of the noise subspace that mixes with the signal subspace via Q_{r+1} is given by

$$u^{(n)} = \text{the first column of } U^{(n)}G^{(n)}H^{(n)} \tag{66.28}$$

$$= \frac{U^{(n)}(U^{(n)})^H x}{|U^{(n)}(U^{(n)})^H x|} \tag{66.29}$$

$$= \frac{1}{|\gamma^{(n)}|}(I - U^{(s)}(U^{(s)})^H)x \tag{66.30}$$

$$= \frac{1}{|\gamma^{(n)}|}(x - U^{(s)}\gamma^{(s)}) \tag{66.31}$$

where $u^{(n)}$ is the projection of x into the noise subspace and $|\gamma^{(n)}| = |x - U^{(s)}\gamma^{(s)}|$ is the power of x projected into the noise subspace. Once the eigenvectors of the core $(r+1) \times (r+1)$ problem are found, the signal subspace eigenvectors can be updated as

$$\tilde{U} = \left[\tilde{U}^{(s)}, \tilde{u}^{(n)} \right] \tag{66.32}$$

$$= \left[\underbrace{U^{(s)}G^{(s)}}_{n \times r}, \underbrace{u^{(n)}}_{n \times 1} \right] Q_{r+1} \tag{66.33}$$

where updating the new noise eigenvector is not necessary (if the noise subspace is resphericalized). The complexity of the core eigenproblem is $O(r^3)$ and updating the signal eigenvectors is $O(nr^2)$. Thus, the SE update is $O(nr^2)$.

After an update is accomplished, one of the noise eigencomponents is altered by the embedded eigenproblem. To maintain noise subspace sphericity, the noise eigenvalues must be re-averaged before the next SE update can be accomplished. On the other hand, if the noise eigenvalues are not re-averaged, the SE update eventually reverts to a full eigen update.

A whole family of related SS updates is possible by simple modification of the above described process. For example, to obtain SA2, the H transformation in Eq. (66.20) would be modified by replacing the I_r with an $r \times r$ Householder matrix that deflates the signal subspace. This would make the core eigenproblem 2×2 and the Q matrix an identity with an embedded 2×2 orthogonal matrix.

66.3.7   Initialization of Subspace and Eigen Tracking Algorithms

It is impossible to give generic initialization requirements that would apply to all subspace tracking algorithms, but one feature that is common to many updating methods is a fading factor. For cold start initialization (e.g., starting from nothing) at $k = 0$, initial convergence can often be sped up by ramping up the fading factor, e.g.,

$$\alpha_k = (1 - \frac{1}{k+1})\alpha, \quad k = 0, 1, 2, ... \tag{66.34}$$

where α is the final steady state value for the fading factor.

66.3.8   Detection Schemes for Subspace Tracking

Several subspace tracking methods have detection schemes that were specifically designed for them. Xu and Kailath developed a strongly consistent detection scheme for their Lanczos-based method [42]. DeGroat and Dowling adapted information theoretic criteria for use with SA4 [9] and an asymptotic proof of consistency is given in [11]. Stewart proposed the URV update as a rank revealing method [39]. Bin Yang proposed that the eigenvalue estimates from PASTd be used for information theoretic-based rank estimation [43].

66.4   Summary of Subspace Tracking Methods Developed Since 1990

66.4.1   Modified Eigen Problems

An $O(n^2r)$ fast subspace decomposition (FSD) method based on the Lanczos algorithm and a strongly consistent source detection scheme was introduced by Xu and Kailath [42].

A transposed QR (TQR) iteration-based SVD update was introduced in [12]. To reduce computation to $O(nr^2)$, the noise subspace is sphericalized and deflated. Based on various performance tests, one or two TQR iterations per update yield results that are comparable to the fully converged SVD. Moreover, because the diagonalization process is taking place on a triangular factor, the partially converged, deflated TQR-SVD update is very similar to a deflated URV update [34].

DeGroat and Roberts [10] simplified Bunch's rank one eigen update [4] and proposed a partial orthogonalization scheme, called pair-wise Gram Schmidt (PGS), to stabilize the eigenvectors. Together with exponential fading to stabilize the eigenvalues, the buildup of roundoff error is essentially controlled and machine precision orthogonality is maintained. For a more complete discussion, see Section 66.3.2 *Controlling Roundoff and Orthogonality Error*. Recently, Gu and Eisenstat [20] presented an improved version of Bunch's rank one EVD update. The new algorithm contains a more stable way to compute the eigenvectors. DeGroat and Dowling have also developed a family of sphericalized EVD and SVD updates (see Section 66.3.6 *Spherical Subspace Updating*).

66.4.2   Gradient-Based Eigen Tracking

Jar-Ferr Yang and Hui-Ju Lin [45] proposed a generalized inflation method which extends the gradient-based work of Yang and Kaveh [44]. An $O(nr^2)$ noise sphericalized and deflated conjugate gradient-based eigen tracking method is presented by Fu and Dowling in [18]. This method can be described as an SS update with a conjugate gradient-based eigen tracker at the core.

Bin Yang [43] introduced a projection approximation approach that uses RLS techniques to update the signal subspace. The projection approximation subspace tracker (PAST) algorithm computes an arbitrary basis for the signal subspace in $3nr + O(r^2)$ flops per update. The PASTd algorithm (which uses deflation to track the individual eigenvalues and vectors of the signal subspace) requires $4nr + O(n)$ flops per update. Both methods produce eigenvector estimates that are only approximately orthogonal.

Regalia and Loubaton [36] use an adaptive lossless transfer matrix (multivariable lattice filter) excited by sensor output to achieve a condition of maximum "power splitting" between two groups of output bins. The update equations resemble standard gradient descent algorithms, but they do not properly follow the gradient of the error surface. Nonetheless, the convergence speed may be a strong function of the source spectral and spatial characteristics.

Recently, Marcos and Benidir [26] introduced an adaptive subspace-based method that relies on a linear operator, referred to as the Propagator, which exploits the linear independency of the source steering vectors, and which allows the determination of the signal and noise subspaces without any eigendecomposition of the correlation matrix. Two gradient-based adaptive algorithms are proposed for the estimation of the Propagator, and then the basis of the signal subspace. The overall computational complexity of the adaptive Propagator subspace update is $O(nr^2)$.

A family of three perturbation-based EVD tracking methods (denoted PA, PB, and PC) are presented by Champagne [5]. Each method uses perturbation-based approximations to track the eigencomponents. Progressively more simplifications are used to reduce the complexity from $\frac{1}{2}n^3 + O(n^2)$ for PA to $\frac{1}{2}nr^2 + O(nr)$ for PB to $5nr + O(n)$ for PC. Both the PB and PC methods use a sphericalized noise subspace to reduce computation. Thus, PB and PC can be viewed as SS updates that use perturbation-based approximations to track the deflated core eigenproblem. The PC method achieves greater computational simplifications by assuming well-separated eigenvalues. Some special decompositions are also used to reduce the computation of the PC algorithm. Surprisingly, simulations seem to indicate that the PC method achieves good overall performance even when the eigenvalues are not well separated. Convergence rates are also very good for the PC method. However, we have noticed that occasionally spurious frequency estimates may be obtained with PC-based MUSIC. Ironically, the PC estimated subspaces tend to be closer to the true subspaces than other subspace tracking methods that do not exhibit occasionally spurious frequency estimates. Because PC only tracks approximations of the eigencomponents, the orthogonality error is typically much greater than machine precision orthogonality. Nevertheless, partial orthogonality schemes can be used to improve orthogonality and other measures of performance (see Section 66.3.2 *Controlling Roundoff Error Accumulation and Orthogonality Errors*).

Artificial neural networks (ANN) have been developed to find eigen information, e.g., see [27] and [35] as well as the references contained therein. An ANN consists of many richly interconnected simple and similar processing elements (called artificial neurons) operating in parallel. High computational rates (due to massive parallelism) and robustness (due to local neural connectivity) are two important features of ANNs. Most of the eigenvector estimating ANNs appear under the topic of principal component analysis (PCA). The principal eigenvectors are defined as the eigenvectors associated with the larger eigenvalues.

66.4.3 The URV and Rank Revealing QR (RRQR) Updates

The URV update [39] is based on the URV decomposition (URVD) developed by G.W. Stewart as a two sided generalization of the RRQR methods. The URVD can also be viewed as a generalization of the SVD because U and V are orthogonal matrices and R is an upper triangular matrix. Clearly, the SVD is a special case of the URVD. If $X = URV^H$ is the URVD of X, then the R factor can be rank revealing in the sense that the Euclidean norm of the $n - r$ rightmost columns of R is approximately equal to the Euclidean norm of the $n - r$ smallest singular values of X. Also, the smallest singular value of the first r columns of R is approximately equal to the rth singular value of X. These two conditions effectively partition the corresponding columns of U and V into an r-dimensional dominant subspace and an $(n-r)$-dimensional subdominant subspace that can be used as estimates for the signal and noise subspace spans. The URV update is $O(n^2)$ per update. An RRQR update [that is usually $O(n^2)$ per update] is developed by Bischof and Schroff in [1]. RRQR methods that use the traditional pivoting strategy to maintain a rank revealing structure involve $O(n^3)$ flops per update. An analysis of problems associated with RRQR methods along with a fairly extensive reference list on RRQR methods can be found in [6]. An $O(nr)$ deflated URV update is presented by Rabideau and Steinhardt in [34] (and the references contained therein).

66.4.4 Miscellaneous Methods

Strobach [40] recently introduced a family of low rank or eigensubspace adaptive filters based on orthogonal iteration. The computational complexity ranges from $O(nr^2)$ to $O(nr)$.

A family of Subspace methods Without EigenDEcomposition (SWEDE) has been proposed by Eriksson et al. [16]. With SWEDE, portions of the correlation matrix must be updated for each new snapshot vector at a cost of approximately $12nr$ flops. However, the subspace basis (which is computed from the correlation matrix partitions) need only be computed every time a DOA estimate is needed. Computing the subspace estimate is $O(nr^2)$, so if the subspace is computed every kth update, the overall complexity is $O(nr^2/k) + 12nr$ per update. At high SNR, SWEDE performs almost as well as eigen-based MUSIC.

Key References:

As previously mentioned, Comon and Golub did a nice survey of SVD tracking methods in 1990 [7]. In 1995, Reddy et al. published a selected overview of eigensubspace estimation methods, including ANN approaches [35]. For a study of URV and RRQR methods, see [6]. Partial orthogonalization schemes are studied in [28]. Finally, a special issue of Signal Processing [41] is planned for April 1996 featuring *Subspace Methods for Detection and Estimation*.

TABLE 66.1 Efficient Subspace Tracking Methods Developed Since 1990

Complexity	Subspace or eigen tracking method	Orthog. span
$O(n^2r)$	Fast subspace decomposition (FSD) [42]	Yes[a]
$O(n^2)$	URV update [39]	Yes
	Rank revealing QR [possibly $O(n^3)$] [1]	Yes
	Approximate SVD updates [17, 30]	Yes[a]
	Neural Network Based Updates [35]	No[a]
$O(nr^2)$	Stabilized signal eigenstructure (SE) update[b] [8, 10]	Yes[a]
	Sphericalized transposed QR SVD update[b] [12]	Yes[a]
	Sphericalized conjugate gradient SVD update[b] [18]	Yes[a]
	SWEDE [16]	No
	Gradient-based EVD updates with gram schmidt orthog. [44, 45]	Yes[a]
$O(nr)$	Signal averaged 2-level (SA2) update[b] [8]	Yes
	Signal averaged 4-level (SA4) update[b] [9, 11]	Yes[a]
	Projection approximation subspace tracking (PAST) [43]	No
	PAST with deflation (PASTd) [43]	No[a]
	Sphericalized perturbation based eigen update (PC method)[b] [5]	Yes[a]
	Sphericalized URV update[b] [34]	Yes

Key: $n =$ no. of sensors, $r =$ rank of subspace.
[a] Tracks individual eigencomponents.
[b] Uses sphericalized subspaces.

References

[1] Bischof, C.H. and Shroff, G.M., On updating signal subspaces, *IEEE Trans. on Sig. Proc.*, 40(1), 96–105, Jan. 1992.

[2] Brockett, R.W., Dynamical systems that sort list, diagonalize matrices and solve linear programming problems, *Proc. of the 27th Conf. on Decis. and Cntrl.*, 799–803, 1988.

[3] Bunch, J.R. and Nielsen, C.P., Updating the singular value decomposition, *Numer. Math.*, 31, 111–129, 1978.

[4] Bunch, J.R., Nielsen, C.P. and Sorensen, D.C., Rank-one modification of the symmetric eigenproblem, *Numer. Math.*, 31, 31–48, 1978.

[5] Champagne, B., Adaptive eigendecomposition of data covariance matrices based on first-order perturbations, *IEEE Trans. Sig. Proc.*, SP-42(10), 2758–2770, Oct. 1994.

[6] Chandrasekaran, S. and Ipsen, I.C.F., On rank-revealing factorisations, *SIAM J. Matrix Anal. Appl.*, 15(2), 592–622, April 1994.

[7] Comon, P. and Golub, G.H., Tracking a few extreme singular values and vectors in signal processing, *Proc. IEEE*, 78(8), 1327–1343, Aug. 1990.

[8] DeGroat, R.D., Non-iterative subspace tracking, *IEEE Trans. Sig. Proc.*, SP-40(3), 571–577, Mar. 1992.

[9] DeGroat, R.D. and Dowling, E.M., Spherical subspace tracking: analysis, convergence and detection schemes, in *26th Annual Asilomar Conf. on Signals, Systems, and Computers*, (invited paper) Oct. 1992, 561–565.

[10] DeGroat, R.D. and Roberts, R.A., Efficient, numerically stabilized rank-one eigenstructure updating, *IEEE Trans. ASSP*, ASSP-38(2), 301–316, Feb. 1990.

[11] Dowling, E.M., DeGroat, R.D., Linebarger, D.A. and Ye, H., Sphericalized SVD updating for subspace tracking, in Moonen, M. and De Moor, B., Eds., *SVD and Signal Processing III: Algorithms, Applications and Architectures*, Elsevier, 1995, 227–234.

[12] Dowling, E.M., Ammann, L.P. and DeGroat, R.D., A TQR-iteration based SVD for real time angle and frequency tracking, *IEEE Trans. on Sig. Proc.*, 914–925, April 1994.

[13] Dowling, E.M. and DeGroat, R.D., Adaptation dynamics of the spherical subspace tracker, *IEEE Trans. on Sig. Proc.*, 2599–2602, Oct. 1992.

[14] Dowling, E.M., DeGroat, R.D. and Linebarger, D.A., Efficient, high performance subspace based tracking problems, in *Adv. Sig. Proc. Algs., Archs. and Appls. VI*, SPIE 2563, 253–264, 1995.

[15] Edelman, A. and Stewart, G.W., Scaling for orthogonality, *IEEE Trans. Sig. Proc.*, SP-41(4), 1676–1677, Apr. 1993.

[16] Eriksson, A., Stoica, P. and Soderstrom, T., On-line subspace algorithms for tracking moving sources, *IEEE Trans. on Sig. Proc.*, 42(9), 2319–2330, Sept. 1994.

[17] Ferzali, W. and Proakis, J.G., Adaptive SVD algorithm and applications, in *SVD and Signal Processing II*, Elsevier, 1992, 14–21.

[18] Fu, Z. and Dowling, E.M., Conjugate gradient eigenstructure tracking for adaptive spectral estimation, *IEEE Trans. Sig. Proc.*, 43(5), 1151–1160, May 1995.

[19] Golub, G.H. and VanLoan, C.F., Some modified matrix eigenvalue problems, *SIAM Review*, 15, 318–334, 1973.

[20] Gu, M. and Eisenstat, S.C., A stable and efficient algorithm for the rank-one modification of the symmetric eigenproblem, *SIAM J. Matrix Anal. Appl.*, 15(4), 1266–1276, Oct. 1994.

[21] Karasalo, I., Estimating the covariance matrix by signal subspace averaging, *IEEE Trans. ASSP*, ASSP-34(1), 8–12, Feb. 1986.

[22] Karhunen, J., Adaptive algorithms for estimating eigenvectors of correlation type matrices, in *ICASSP-84*, 14.6.1–14.6.4, 1984.

[23] Linebarger, D.A., DeGroat, R.D., Dowling, E.M., Stoica, P. and Fudge, G., Incorporating a priori information into MUSIC - algorithms and analysis, *Signal Processing*, 46(1), 85–104, 1995.

[24] Linebarger, D.A., DeGroat, R.D. and Dowling, E.M., Efficient direction finding methods employing forward/backward averaging, *IEEE Tr. SP*, 42(8), 2136–2145, Aug. 1994.

[25] Ljung, L., Analysis of recursive stochastic algorithms, *IEEE Trans. on Automatic Control*, AC-22(4), 551–575, Aug. 1977.

[26] Marcos, S. and Benidir, M., An adaptive subspace algorithm for direction finding and tracking, in *Adv. Sig. Proc. Algs., Archs. and Appls. VI*, SPIE 2563, 230–241, 1995.

[27] Mathew, G. and Reddy, V.U., Orthogonal eigensubspace estimation using neural networks, *IEEE Trans. on Sig. Proc.*, 42, 1803–1811, July 1994.

[28] Mathias, R., Analysis of algorithms for orthogonalizing products of unitary matrices, *J. Numerical Linear Algebra with Applic.*, 3(2), 125–145, 1996.

[29] Moonen, M., VanDooren, P. and Vanderwalle, J., A note on efficient, numerically stabilized rank-one eigenstructure updating, *IEEE Trans. Sig. Proc.*, SP-39(8), 1913–1914, Aug. 1991.

[30] Moonen, M., VanDooren, P. and Vanderwalle, J., A singular value decomposition updating algorithm for subspace tracking, *SIAM J. Matrix Anal. Appl.*, 13(4), 1015–1038, Oct. 1992.

[31] Oja, E., A simplified neuron model as a principal component analyzer, *J. Math. Biol.*, 15, 267–273, 1982.

[32] Owsley, N.L., Adaptive data orthogonalization, *ICASSP*, 109–112, 1978.

[33] Parlett, B.N., *The Symmetric Eigenvalue Problem*, Prentice-Hall, Englewood Cliffs, NJ, 1980.

[34] Rabideau, D.J., Subspace invariance: The RO-FST and TQR-SVD adaptive subspace tracking algorithms, *IEEE Trans. SP*, SP-43, 2016–2018, Aug. 1995.

[35] Reddy, V.U., Mathew, G. and Paulraj, A., Some algorithms for eigensubspace estimation, *Digital Signal Processing*, 5, 97–115, 1995.

[36] Regalia, P.A. and Loubaton, P., Rational subspace estimation using adaptive lossless filters, *IEEE Trans. on Sig. Proc.*, 40, 2392–2405, Oct. 1992.

[37] Schreiber, R., Implementation of adaptive array algorithms, *IEEE Trans. ASSP*, ASSP-34, 1038–1045, Oct. 1986.

[38] Stewart, G.W., *Introduction to Matrix Computations*, Academic Press, New York, 1973.

[39] Stewart, G.W., An updating algorithm for subspace tracking, *IEEE Trans. Sig. Proc.*, SP-40(6), 1535–1541, June 1992.

[40] Strobach, P., Fast recursive eigensubspace adaptive filters, in *International Conference on Acoustics, Speech and Sig. Proc.*, 1416–1419, 1995.

[41] Viberg, M. and Stoica, P., Eds., *Signal Processing*, 50(1-2) of *Special Issue on Subspace Methods for Detection and Estimation*, April 1996.

[42] Xu, G., Zha, H., Golub, G. and Kailath, T., Fast and robust algorithms for updating signal subspaces, *IEEE Trans. CAS*, 41(6), 537–549, June 1994.

[43] Yang, B., Projection approximation subspace tracking, *IEEE Trans. SP*, SP-43(1), 95–107, Jan. 1995.

[44] Yang, J.F. and Kaveh, M., Adaptive eigensubspace algorithms for direction or frequency estimation and tracking, *IEEE Trans. ASSP*, ASSP-36(2), 241–251, Feb. 1988.

[45] Yang, J.-F. and Lin, H.-J., Adaptive high-resolution algorithms for tracking nonstationary sources without the estimation of source number, *IEEE Trans. on Sig. Proc.*, 42(3), 563–571, Mar. 1994.

[46] Yu, K.B., Recursive updating the eigenvalue decomposition of a covariance matrix, *IEEE Trans. Sig. Proc.*, SP-39(5), 1136–1145, May 1991.

67

Detection: Determining the Number of Sources

Douglas B. Williams
Georgia Institute of Technology

The processing of signals received by sensor arrays generally can be separated into two problems: (1) detecting the number of sources and (2) isolating and analyzing the signal produced by each source. We make this distinction because many of the algorithms for separating and processing array signals make the assumption that the number of sources is known *a priori* and may give misleading results if the wrong number of sources is used [3]. A good example are the errors produced by many high resolution bearing estimation algorithms (e.g., MUSIC) when the wrong number of sources is assumed. Because, in general, it is easier to determine how many signals are present than to estimate the bearings of those signals, signal detection algorithms typically can correctly determine the number of signals present even when bearing estimation algorithms cannot resolve them. In fact, the capability of an array to resolve two closely spaced sources could be said to be limited by its ability to detect that there are actually two sources present. If we have a reliable method of determining the number of sources, not only can we correctly use high resolution bearing estimation algorithms, but we can also use this knowledge to utilize more effectively the information obtained from the bearing estimation algorithms. If the bearing estimation algorithm gives fewer source directions than we know there are sources, then we know that there is more than one source in at least one of those directions and have thus essentially increased the resolution of the algorithm. If analysis of the information provided by the bearing estimation algorithm indicates more source directions than we know there are sources, then we can safely assume that some of the directions are the results of false alarms and may be ignored, thus decreasing the probability of false alarm for the bearing estimation algorithms. In this section we will present and discuss the more common approaches to determining the number of sources.

67.1 Formulation of the Problem

The basic problem is that of determining how many signal producing sources are being observed by an array of sensors. Although this problem addresses issues in several areas including sonar, radar, communications, and geophysics, one basic formulation can be applied to all these applications. We will give only a basic, brief description of the assumed signal structure, but more detail can be found in references such as the book by Johnson and Dudgeon [3]. We will assume that an array of M sensors observes signals produced by N_s sources. The array is allowed to have an arbitrary geometry. For our discussion here, we will assume that the sensors are omnidirectional. However, this assumption is only for notational convenience as the algorithms to be discussed will work for more general sensor responses.

The output of the mth sensor can be expressed as a linear combination of signals and noise

$$y_m(t) = \sum_{i=1}^{N_s} s_i\,(t - \Delta_i(m)) + n_m(t)\,.$$

The noise observed at the mth sensor is denoted by $n_m(t)$. The propagation delays, $\Delta_i(m)$, are measured with respect to an origin chosen to be at the geometric center of the array. Thus, $s_i(t)$ indicates the ith propagating signal observed at the origin, and $s_i(t - \Delta_i(m))$ is the same signal measured by the mth sensor. For a plane wave in a homogeneous medium, these delays can be found from the dot product between a unit vector in the signal's direction of propagation, $\vec{\zeta}_i^o$, and the sensor's location, $\vec{x}_m$,

$$\Delta_i(m) = \frac{\vec{\zeta}_i^o \cdot \vec{x}_m}{c}\,,$$

where c is the plane wave's speed of propagation.

Most algorithms used to detect the number of sources incident on the array are frequency domain techniques that assume the propagating signals are narrowband about a common center frequency, ω^o. Consequently, after Fourier transforming the measured signals, only one frequency is of interest and the propagation delays become phase shifts

$$Y_m\left(\omega^o\right) = \sum_{i=1}^{N_s} S_i\left(\omega^o\right) e^{-j\omega^o \Delta_i(m)} + N_m\left(\omega^o\right)\,.$$

The detection algorithms then exploit the form of the spatial correlation matrix, $\mathbf{R}$, for the array. The spatial correlation matrix is the $M \times M$ matrix formed by correlating the vector of the Fourier transforms of the sensor outputs at the particular frequency of interest

$$\mathbf{Y} = \left[Y_0\left(\omega^o\right)\quad Y_1\left(\omega^o\right)\quad \cdots \quad Y_{M-1}\left(\omega^o\right)\right]^T\,.$$

If the sources are assumed to be uncorrelated with the noise, then the form of $\mathbf{R}$ is

$$\mathbf{R} = E\left\{\mathbf{YY}'\right\} = \mathbf{K}_n + \mathbf{SCS}'\,,$$

where $\mathbf{K}_n$ is the correlation matrix of the noise, $\mathbf{S}$ is the matrix whose columns correspond to the vector representations of the signals, $\mathbf{S}'$ is the conjugate transpose of $\mathbf{S}$, and $\mathbf{C}$ is the matrix of the correlations between the signals. Thus, the matrix $\mathbf{S}$ has the form

$$\mathbf{S} = \begin{bmatrix} e^{-j\omega^o \Delta_1(0)} & \cdots & e^{-j\omega^o \Delta_{N_s}(0)} \\ \vdots & & \vdots \\ e^{-j\omega^o \Delta_1(M-1)} & \cdots & e^{-j\omega^o \Delta_{N_s}(M-1)} \end{bmatrix}\,.$$

If we assume that the noise is additive, white Gaussian noise with power σ_n^2 and that none of the signals are perfectly coherent with any of the other signals, then $\mathbf{K}_n = \sigma_n^2 \mathbf{I}_m$, $\mathbf{C}$ has full rank, and the form of $\mathbf{R}$ is

$$\mathbf{R} = \sigma_n^2 \mathbf{I}_M + \mathbf{SCS}'\,. \tag{67.1}$$

We will assume that the columns of S are linearly independent when there are fewer sources than sensors, which is the case for most common array geometries and expected source locations. As C is of full rank, if there are fewer sources than sensors, then the rank of SCS' is equal to the number of signals incident on the array or, equivalently, the number of sources. If there are N_s sources, then SCS' is of rank N_s and its N_s eigenvalues in descending order are $\delta_1, \delta_2, \cdots, \delta_{N_s}$. The M eigenvalues of $\sigma_n^2 I_M$ are all equal to σ_n^2, and the eigenvectors are any orthonormal set of length M vectors. So the eigenvectors of R are the N_s eigenvectors of SCS' plus any $M - N_s$ eigenvectors which complete the orthonormal set, and the eigenvalues in descending order are $\sigma_n^2 + \delta_1, \cdots, \sigma_n^2 + \delta_{N_s}, \sigma_n^2, \cdots, \sigma_n^2$. The correlation matrix is generally divided into two parts: the signal-plus-noise subspace formed by the largest eigenvalues ($\sigma_n^2 + \delta_1, \cdots, \sigma_n^2 + \delta_{N_s}$) and their eigenvectors, and the noise subspace formed by the smallest, equal eigenvalues and their eigenvectors. The reason for these labels is obvious as the space spanned by the signal-plus-noise subspace eigenvectors contains the signals and a portion of the noise while the noise subspace contains only that part of the noise that is orthogonal to the signals [3]. If there are fewer sources than sensors, the smallest $M - N_s$ eigenvalues of R are all equal and to determine exactly how many sources there are, we must simply determine how many of the smallest eigenvalues are equal. If there are not fewer sources than sensors ($N_s \geq M$), then none of the smallest eigenvalues are equal. The detection algorithms then assume that only the smallest eigenvalue is in the noise subspace as it is not equal to any of the other eigenvalues. Thus, these algorithms can detect up to $M - 1$ sources and for $N_s \geq M$ will say that there are $M - 1$ sources as this is the greatest detectable number. Unfortunately, all that is usually known is $\widehat{R}$, the sample correlation matrix, which is formed by averaging N samples of the correlation matrix taken from the outputs of the array sensors. As $\widehat{R}$ is formed from only a finite number of samples of R, the smallest $M - N_s$ eigenvalues of $\widehat{R}$ are subject to statistical variations and are unequal with probability one [4]. Thus, solutions to the detection problem have concentrated on statistical tests to determine how many of the eigenvalues of R are equal when only the sample eigenvalues of $\widehat{R}$ are available.

When performing statistical tests on the eigenvalues of the sample correlation matrix to determine the number of sources, certain assumptions must be made about the nature of the signals. In array processing, both deterministic and stochastic signal models are used depending on the application. However, for the purpose of testing the sample eigenvalues, the Fourier transforms of the signals at frequency ω^o; $S_i(\omega^o)$, $i = 1, \ldots, N_s$; are assumed to be zero mean Gaussian random processes that are statistically independent of the noise and have a positive definite correlation matrix C. We also assume that the N samples taken when forming $\widehat{R}$ are statistically independent of each other. With these assumptions, the spatial correlation matrix is still of the same form as in (67.1), except that now we can more easily derive statistical tests on the eigenvalues of $\widehat{R}$.

67.2 Information Theoretic Approaches

We will see that the source detection methods to be described all share common characteristics. However, we will classify them into two groups—information theoretic and decision theoretic approaches—determined by the statistical theories used to derive them. Although the decision theoretic techniques are quite a bit older, we will first present the information theoretic algorithms as they are currently much more commonly used.

67.2.1 AIC and MDL

AIC and MDL are both information theoretic model order determination techniques that can be used to test the eigenvalues of a sample correlation matrix to determine how many of the smallest eigenvalues of the correlation matrix are equal. The AIC and MDL algorithms both consist of minimizing a criterion over the number of signals that are detectable, i.e., $N_s = 0, \ldots, M - 1$. To construct these criteria, a family of probability densities, $f(Y|\theta(N_s))$, $N_s = 0, \ldots, M - 1$, is needed, where θ, which is a function of the number of sources, N_s, is the vector of parameters needed for the model that generated the data Y. The criteria are composed of the negative of the log-likelihood function of the density $f(Y|\hat{\theta}(N_s))$, where $\hat{\theta}(N_s)$ is the maximum likelihood estimate of θ for N_s signals, plus an adjusting term for the model dimension. The adjusting term is needed because the negative log-likelihood function always achieves a minimum for the highest dimension model possible, which in this case is the largest possible number of sources. Therefore, the adjusting term will be a monotonically increasing function of N_s and should be chosen so that the algorithm is able to determine the correct model order.

AIC was introduced by Akaike [1]. Originally, the "IC" stood for information criterion and the "A" designated it as the first such test, but it is now more commonly considered an acronym for the "Akaike Information Criterion." If we have N independent observations of a random variable with probability density $g(Y)$ and a family of models in the form of probability densities $f(Y|\theta)$ where θ is the vector of parameters for the models, then Akaike chose his criterion to minimize

$$I(g; \ f(\cdot|\theta)) = \int g(Y) \ln g(Y) dY - \int g(Y) \ln f(Y|\theta) dY \tag{67.2}$$

which is known as the Kullback-Leibler mean information distance. $\frac{1}{N} AIC(\theta)$ is an estimate of $-E\{\int g(Y) \ln f(Y|\theta) dY\}$ and minimizing $AIC(\theta)$ over the allowable values of θ should minimize (67.2). The expression for $AIC(\theta)$ is

$$AIC(\theta) = -2 \ln \left[f\left(Y|\hat{\theta}\ (N_s)\right)\right] + 2\eta ,$$

where η is the number of independent parameters in θ.

Following AIC, MDL was developed by Schwarz [6] using Bayesian techniques. He assumed that the *a priori* density of the observations comes from a suitable family of densities that possess efficient estimates [7]; they are of the form

$$f(Y|\theta) = \exp(\theta \cdot p(Y) - b(\theta)) .$$

The MDL criterion was then found by choosing the model that is most probable *a posteriori*. This choice is equivalent to selecting the model for which

$$MDL(\theta) = -\ln \left[f\left(Y|\hat{\theta}\ (N_s)\right)\right] + \frac{1}{2}\eta \ln N$$

is minimized. This criterion was independently derived by Rissanen [5] using information theoretic techniques. Rissanen noted that each model can be perceived as encoding the observed data and that the optimum model is the one that yields the minimum code length. Hence, the name MDL comes from "Minimum Description Length".

For the purpose of using AIC and MDL to determine the number of sources, the forms of the log-likelihood function and the adjusting terms have been given by Wax [8]. For N_s signals the parameters that completely parameterize the correlation matrix R are $\{\sigma_n^2, \lambda_1, \cdots, \lambda_{N_s}, v_1, \cdots, v_{N_s}\}$ where λ_i and v_i, $i = 1, ..., N_s$, are the eigenvalues and their respective eigenvectors of the signal-plus-noise subspace of the correlation matrix. As the vector of sensor outputs is a Gaussian random vector with correlation matrix R and all the samples of the sensor outputs are independent, the log-likelihood function of $f(Y|\theta)$ is

$$\ln f\left(Y|\sigma_n^2, \lambda_1, \cdots, \lambda_{N_s}, v_1, \cdots, v_{N_s}\right) = \pi^{-pN} (\det R)^{-N} \exp\left(-N\text{tr}\left(R^{-1}\hat{R}\right)\right)$$

where $\text{tr}(\cdot)$ denotes the trace of the matrix, $\hat{R}$ is the sample correlation matrix, and R is the unique correlation matrix formed from the given parameters. The maximum likelihood estimate of the parameters are [2, 4]

$$
\begin{aligned}
\hat{\mathbf{v}}_i &= \mathbf{u}_i; \ i = 1, \cdots, N_s \\
\hat{\lambda}_i &= l_i; \ i = 1, \cdots, N_s \\
\hat{\sigma}_n^2 &= \frac{1}{M - N_s} \sum_{i=N_s+1}^{M} l_i = \bar{l},
\end{aligned}
\tag{67.3}
$$

where $l_1, \cdots, l_M$ are the eigenvalues in descending order of $\hat{\mathbf{R}}$ and $\mathbf{u}_i$ are the corresponding eigenvectors. Therefore, the log-likelihood function of $f(\mathbf{Y}|\hat{\theta}(N_s))$ is

$$
\ln f(\mathbf{Y}|\bar{l}, l_1, \cdots, l_{N_s}, \mathbf{u}_1, \cdots, \mathbf{u}_{N_s}) = \ln \left[\frac{\prod_{i=N_s+1}^{M} l_i^{1/(M-N_s)}}{\frac{1}{M-N_s} \sum_{i=N_s+1}^{M} l_i} \right]^{(M-N_s)N}.
$$

Remembering that the eigenvalues of a complex correlation matrix are real and that the eigenvectors are complex and orthonormal, the number of degrees of freedom in the parameters of the model is classically chosen to be $\eta = N_s(2M - N_s) + 1$. Noting that any constant term in the criteria which is common to the entire family of models for either AIC or MDL may be ignored, we have the criterion for AIC as

$$
AIC(\hat{N}_s) = -2N \ln \left[\frac{\prod_{i=\hat{N}_s+1}^{M} l_i}{\left[\frac{1}{M-\hat{N}_s} \sum_{i=\hat{N}_s+1}^{M} l_i \right]^{M-\hat{N}_s}} \right] + 2\hat{N}_s(2M - \hat{N}_s); \ \hat{N}_s = 0, \ldots, M-1
$$

and the criterion for MDL as

$$
MDL(\hat{N}_s) = -N \ln \left[\frac{\prod_{i=\hat{N}_s+1}^{M} l_i}{\left[\frac{1}{M-\hat{N}_s} \sum_{i=\hat{N}_s+1}^{M} l_i \right]^{M-\hat{N}_s}} \right] + \frac{1}{2}\hat{N}_s(2M - \hat{N}_s) \ln N; \ \hat{N}_s = 0, \ldots, M-1.
$$

For both of these methods, the estimate of the number of sources is that value of $\hat{N}_s$ which minimizes the criterion. In [9] there is a more thorough discussion concerning determining the number of degrees of freedom and the advantages of choosing instead $\eta = N_s(2M - N_s - 1)$.

In general, MDL is considered to perform better than AIC. Schwarz [6], through his derivation of the MDL criterion, showed that if his assumptions are accepted, then AIC cannot be asymptotically optimal. He also mentioned that MDL tends toward lower-dimensional models than AIC as the model dimension term is multiplied by $\frac{1}{2} \ln N$ in the MDL criterion. Zhao et al. [14] showed that MDL is consistent (the probability of detecting the correct number of sources, i.e., $\Pr(\hat{N}_s = N_s)$, goes to 1 as N goes to infinity), but AIC is not consistent and will tend to overestimate the number of sources as N goes to infinity. Thus, most people in array processing prefer to use MDL over AIC. Interestingly, many statisticians prefer AIC because many of their modeling problems have a very large penalty for underestimating the model order but a relatively mild penalty for overestimating it. Xu and Kaveh [12] have provided a thorough discussion of the asymptotic properties of AIC and MDL, including an examination of their sensitivities to modelling errors and bounds on the probability that AIC will overestimate the number of sources.

67.2.2 EDC

Clearly, the only difference between the implementations of AIC and MDL is the choice of the adjusting term that penalizes for choosing larger model orders. Several people have examined using other adjusting terms to arrive at other criteria. In particular, statisticians at the University of Pittsburgh [13, 14] have developed the Efficient Detection Criterion (EDC) procedure which is actually a family of criteria chosen such that they are all consistent. The general form of these criteria is

$$EDC(\theta) = -\ln\left[f\left(\mathbf{Y}|\hat{\theta}\left(N_s\right)\right)\right] + \eta C_N \,,$$

where C_N can be any function of N such that

$$(1) \qquad \lim_{N\to\infty} C_N/N = 0$$

$$(2) \qquad \lim_{N\to\infty} C_N/\ln(\ln(N)) = \infty \,.$$

Thus, for the array processing source detection problem the EDC procedure chooses the value of $\widehat{N}_s$ that minimizes

$$EDC(\widehat{N}_s) = -N \ln\left[\frac{\prod\limits_{i=\widehat{N}_s+1}^{M} l_i}{\left[\frac{1}{M-\widehat{N}_s}\sum\limits_{i=\widehat{N}_s+1}^{M} l_i\right]^{M-\widehat{N}_s}}\right] + \widehat{N}_s(2M-\widehat{N}_s)C_N; \ \widehat{N}_s = 0, \ldots, M-1 \,.$$

In their analysis of the EDC procedure, Zhao et al. [14] showed that not only are all the EDC criteria consistent for the data assumptions we have made, but under certain conditions they remain consistent even when the data sample vectors used to form the estimate $\widehat{\mathbf{R}}$ are not independent or Gaussian.

The choice of $C_N = \frac{1}{2}\ln(N)$ satisfies the restrictions on C_N and, thus, produces one of the EDC procedures. This particular criterion is identical to MDL and shows that the MDL criterion is included as one of the EDC procedures. Another relatively common choice for C_N is $C_N = \sqrt{N \ln(N)}$.

67.3 Decision Theoretic Approaches

The methods that we term decision theoretic approaches all rely on the statistical theory of hypothesis testing to determine the number of sources. The first of these that we will discuss, the sphericity test, is by far the oldest algorithm for source detection.

67.3.1 The Sphericity Test

Originally, the sphericity test was a hypothesis testing method designed to determine if the correlation (or covariance) matrix, $\mathbf{R}$, of a length M Gaussian random vector is proportional to the identity matrix, $\mathbf{I}_M$, when only $\widehat{\mathbf{R}}$, the sample correlation matrix, is known. If $\mathbf{R} \propto \mathbf{I}_M$, then the contours of equal density for the Gaussian distribution form concentric spheres in M-dimensional space. The sphericity test derives its name from being a test of the sphericity of these contours.

The original sphericity test had two possible hypotheses

$$H_0 : \qquad \mathbf{R} = \sigma_n^2 \mathbf{I}_M$$
$$H_1 : \qquad \mathbf{R} \neq \sigma_n^2 \mathbf{I}_M$$

for some unknown σ_n^2. If we denote the eigenvalues of $\mathbf{R}$ in descending order by $\lambda_1, \lambda_2, \cdots, \lambda_M$, then equivalent hypotheses are

$$
\begin{aligned}
H_0: & \quad \lambda_1 = \lambda_2 = \cdots = \lambda_M \\
H_1: & \quad \lambda_1 > \lambda_M .
\end{aligned}
$$

For the appropriate statistic, $T(\widehat{\mathbf{R}})$, the test is of the form

$$
T(\widehat{\mathbf{R}}) \underset{H_0}{\overset{H_1}{\gtrless}} \gamma
$$

where the threshold, γ, can be set according to the Neyman-Pearson criterion [7]. That is, if the distribution of $T(\widehat{\mathbf{R}})$ is known under the null hypothesis, H_0, then for a given probability of false alarm, P_F, we can choose γ such that

$$
\Pr(T(\widehat{\mathbf{R}}) > \gamma | H_0) = P_F.
$$

Using the alternate form of the hypotheses, $T(\widehat{\mathbf{R}})$ is actually $T(l_1, l_2, \cdots, l_M)$, and the eigenvalues of the sample correlation matrix are a sufficient statistic for the hypothesis test. The correct form of the sphericity test statistic is the generalized likelihood ratio [4]

$$
T(l_1, l_2, \cdots, l_M) = \ln \left[\frac{\left(\frac{1}{M} \sum_{i=1}^{M} l_i \right)^M}{\prod_{i=1}^{M} l_i} \right]
$$

which was also a major component of the information theoretic tests.

For the source detection problem we are interested in testing a subset of the smaller eigenvalues for equality. In order to use the sphericity test, the hypotheses are generally broken down into pairs of hypotheses that can be tested in a series of hypothesis tests. For testing $M - \widehat{N}_s$ eigenvalues for equality, the hypotheses are

$$
\begin{aligned}
H_0: & \quad \lambda_1 \geq \cdots \geq \lambda_{\widehat{N}_s} \geq \lambda_{\widehat{N}_s+1} = \cdots = \lambda_M \\
H_1: & \quad \lambda_1 \geq \cdots \geq \lambda_{\widehat{N}_s} \geq \lambda_{\widehat{N}_s+1} > \lambda_M .
\end{aligned}
$$

We are interested in finding the smallest value of $\widehat{N}_s$ for which H_0 is true, which is done by testing $\widehat{N}_s = 0$, $\widehat{N}_s = 1, \cdots$ until $\widehat{N}_s = M - 2$ or the test does not fail. If the test fails for $\widehat{N}_s = M - 2$, then we consider none of the smallest eigenvalues to be equal and say that there are $M - 1$ sources. If $\widehat{N}_s$ is the smallest value for which H_0 is true, then we say that there are $\widehat{N}_s$ sources. There is also a problem involved in setting the desired P_F. The Neyman-Pearson criterion is not able to determine a threshold for given P_F for the overall detection problem. The best that can be done is to set a P_F for each individual test in the nested series of hypothesis tests using Neyman-Pearson methods. Unfortunately, as the hypothesis tests are obviously not statistically independent and their statistical relationship is not very clear, how this P_F for each test relates to the P_F for the entire series of tests is not known.

To use the sphericity test to detect sources, we need to be able to set accurately the threshold γ according to the desired P_F, which requires knowledge of the distribution of the sphericity test statistic $T(l_{\widehat{N}_s+1}, \cdots, l_M)$ under the null hypothesis. The *exact* form of this distribution is not available in a form that is very useful as it is generally written as an infinite series of Gaussian, chi-squared, or beta distributions [2, 4]. However, if the test statistic is multiplied by a suitable function of the eigenvalues of $\widehat{R}$, then its distribution can be accurately approximated as being chi-squared [10]. Thus, the statistic

$$
2\left((N-1)-\widehat{N}_s-\frac{2\left(M-\widehat{N}_s\right)^2+1}{6\left(M-\widehat{N}_s\right)}+\sum_{i=1}^{\widehat{N}_s}\left(\frac{l_i}{\bar{l}}-1\right)^{-2}\right)\ln\left[\frac{\left(\frac{1}{M-\widehat{N}_s}\sum_{i=\widehat{N}_s+1}^{M}l_i\right)^{M-\widehat{N}_s}}{\prod_{i=\widehat{N}_s+1}^{M}l_i}\right]
$$

is approximately chi-squared distributed with degrees of freedom given by

$$
d=\left(M-\widehat{N}_s\right)^2-1,
$$

where $\bar{l}=\frac{1}{M-\widehat{N}_s}\sum_{i=\widehat{N}_s+1}^{M}l_i$.

Although the performance of the sphericity test is comparable to that of the information theoretic tests, it is not as popular because it requires selection of the P_F and calculation of the test thresholds for each value of $\widehat{N}_s$. However, if the received data does not match the assumed model, the ability to change the test thresholds gives the sphericity test a robustness lacking in the information theoretic methods.

67.3.2 Multiple Hypothesis Testing

The sphericity test relies on a sequence of binary hypothesis tests to determine the number of sources. However, the optimum test for this situation would be to test all hypotheses simultaneously:

$$
\begin{aligned}
H_0: &\quad \lambda_1=\lambda_2=\cdots=\lambda_M\\
H_1: &\quad \lambda_1>\lambda_2=\cdots=\lambda_M\\
H_2: &\quad \lambda_1\geq\lambda_2>\lambda_3=\cdots=\lambda_M\\
&\quad\vdots\\
H_{M-1}: &\quad \lambda_1\geq\lambda_2\geq\cdots\geq\lambda_{M-1}>\lambda_M
\end{aligned}
$$

to determine how many of the smaller eigenvalues are equal. While it is not possible to generalize the sphericity test directly, it is possible to use an approximation to the probability density function (*pdf*) of the eigenvalues to arrive at a suitable test. Using the theory of multiple hypothesis tests, we can derive a test that is similar to AIC and MDL and is implemented in *exactly* the same manner, but is designed to minimize the probability of choosing the wrong number of sources.

To arrive at our statistic, we start with the joint probability density function (*pdf*) of the eigenvalues of the $M\times M$ sample covariance when the $M-\widehat{N}_s$ smallest eigenvalues are known to be equal. We will denote this pdf by $f_{\widehat{N}_s}(l_1,\ldots,l_M|\lambda_1\geq\cdots\geq\lambda_{\widehat{N}_s+1}=\cdots=\lambda_M)$ where the l_i denote the eigenvalues of the sample matrix and the λ_i are the eigenvalues of the true covariance matrix. The asymptotic expression for $f_{\widehat{N}_s}(\cdot)$ is given by Wong et al. [11] for the complex-valued data case as

$$
f_{\widehat{N}_s}(l_1,\ldots,l_M|\lambda_1\geq\cdots\geq\lambda_{\widehat{N}_s+1}=\cdots=\lambda_M)\approx\frac{n^{mn-\frac{\widehat{N}_s}{2}(2M-\widehat{N}_s-1)}\pi^{M(M-1)-\frac{\widehat{N}_s}{2}(2M-\widehat{N}_s-1)}}{\widetilde{\Gamma}_M(n)\widetilde{\Gamma}_{M-\widehat{N}_s}(M-\widehat{N}_s)}
$$

$$
\prod_{i=1}^{M}\lambda_i^{-n}\prod_{i=1}^{M}l_i^{n-M}\exp\left\{-n\sum_{i=1}^{M}\frac{l_i}{\lambda_i}\right\}\prod_{i=\widehat{N}_s+1}^{M}\prod_{i<j}^{M}\left(l_i-l_j\right)^2
$$

$$
\prod_{i=1}^{\widehat{N}_s}\prod_{i<j}^{\widehat{N}_s}\left(\frac{(l_i-l_j)\lambda_i\lambda_j}{\lambda_i-\lambda_j}\right)\prod_{i=1}^{\widehat{N}_s}\prod_{j=\widehat{N}_s+1}^{M}\left(\frac{(l_i-l_j)\lambda_i\lambda_j}{\lambda_i-\lambda_j}\right)
$$

where $n = N - 1$ is one less than the number of samples and $\tilde{\Gamma}_N(\cdot)$ is the multivariate gamma function for complex-valued data [11]. We then form M likelihood ratios by dividing each joint pdf by $f_{M-1}(\cdot)$ to form

$$\Lambda(\widehat{N}_s) = \frac{f_{\widehat{N}_s}\left(l_1, \ldots, l_M \mid \lambda_1 \geq \cdots \geq \lambda_{\widehat{N}_s+1} = \cdots = \lambda_M\right)}{f_{M-1}(l_1, \ldots, l_M \mid \lambda_1 \geq \cdots \geq \lambda_M)}, \quad \widehat{N}_s = 0, \ldots, M-1.$$

Assuming that each value of $\widehat{N}_s$ is equally likely, then multiple hypothesis testing theory tells us that the value of $\widehat{N}_s$ that *maximizes* $\Lambda(\widehat{N}_s)$ is the optimum choice in that it minimizes the probability of choosing the incorrect $\widehat{N}_s$ [7]. Because $\Lambda(\widehat{N}_s)$ in this form requires knowledge of the unknown parameters λ_i, we must use a generalized likelihood ratio test and independently substitute the maximum likelihood estimates of the λ_i [see Eq.(67.3) for these expressions] into both $f_{\widehat{N}_s}(\cdot)$, for which we assume $M - \widehat{N}_s$ equal λ_is, and $f_{M-1}(\cdot)$, for which we assume no equal λ_is, to get our new statistics $\Lambda(\widehat{N}_s)$. After much simplification including dropping terms that are common to $\Lambda(\widehat{N}_s)$ for every allowable value of $\widehat{N}_s$ and then taking the natural logarithm of each $\Lambda(\widehat{N}_s)$, we get the statistic

$$\Lambda\left(\widehat{N}_s\right) = (n - \widehat{N}_s) \ln \left[\frac{\prod_{i=\widehat{N}_s+1}^{M} l_i}{\left(\frac{1}{M - \widehat{N}_s} \sum_{i=\widehat{N}_s+1}^{M} l_i \right)^{M-\widehat{N}_s}} \right] - \frac{1}{2} \widehat{N}_s \left(2M - \widehat{N}_s - 1\right) \ln[n] +$$

$$\ln \left[\frac{\pi^{-\widehat{N}_s\left(2M-\widehat{N}_s-1\right)/2}}{\tilde{\Gamma}_{M-\widehat{N}_s}\left(M - \widehat{N}_s\right)} \right] + \sum_{i=1}^{\widehat{N}_s} \sum_{j=\widehat{N}_s+1}^{M} \ln\left[\frac{l_i - l_j}{l_i - \bar{l}}\right] - \sum_{i=\widehat{N}_s+1}^{M} \sum_{j=i+1}^{M} 2\ln\left[\frac{(l_i l_j)^{1/2}}{l_i - l_j}\right]$$

where $\bar{l} = \frac{1}{M-\widehat{N}_s} \sum_{i=\widehat{N}_s+1}^{M} l_i$.

The terms in the first line of this equation are almost identical to the negative of the MDL criterion, especially when the degrees of freedom recommended in [9] are used. Note that the change in sign is necessary because we are finding the maximum of this criterion, not the minimum. The extra terms on the following line include both the eigenvalues being tested for equality and those not being tested. These extra terms allow this test to outperform the information theoretic techniques, since the use of all the eigenvalues for each value of $\widehat{N}_s$ being tested allows this criterion to be more adaptive.

67.4 For More Information

Most of the original papers on model order determination appeared in the statistical literature in journals such as *The Annals of Statistics* and the *Journal of Multivariate Analysis*. However, almost all of the more recent developments that apply these techniques to the source detection problem have appeared in signal processing journals such as the *IEEE Transactions on Signal Processing*. More advanced topics that have been addressed in the signal processing literature but not discussed here include: detecting coherent (i.e., completely correlated) signals, detecting sources in unknown colored noise, and developing more robust source detection methods.

References

[1] Akaike, H., A new look at the statistical model identification, *IEEE Trans. on Automatic Control*, AC-19, 716–723, Dec. 1974.

[2] Anderson, T.W., *An Introduction to Multivariate Statistical Analysis*, 2nd ed., John Wiley & Sons, New York 1984.

[3] Johnson, D.H. and Dudgeon, D.E., *Array Signal Processing: Concepts and Techniques*, Prentice-Hall, Englewood Cliffs, NJ, 1993.

[4] Muirhead, R.J., *Aspects of Multivariate Statistical Theory*, John Wiley & Sons, New York, 1982.

[5] Rissanen, J., Modeling by shortest data description, *Automatica*, 14, 465–471, Sept. 1978.

[6] Schwarz, G., Estimating the dimension of a model, *Annal. Stat.*, 6, 461–464, Mar. 1978.

[7] Van Trees, H.L., *Detection, Estimation, and Modulation Theory, Part I*, John Wiley & Sons, New York, 1968.

[8] Wax, M. and Kailath, T., Detection of signals by information theoretic criteria, *IEEE Trans. Acoustics, Speech, and Signal Processing*, ASSP-33, 387–392, Apr. 1985.

[9] Williams, D.B., Counting the degrees of freedom when using AIC and MDL to detect signals, *IEEE Trans. Signal Processing*, 42, 3282–3284, Nov. 1994.

[10] Williams, D.B. and Johnson, D.H., Using the sphericity test for source detection with narrow-band passive arrays, *IEEE Trans. Acoustics, Speech, and Signal Processing*, 38, 2008–2014, Nov. 1990.

[11] Wong, K.M., Zhang, Q.-T., Reilly, J.P. and Yip, P.C., On information theoretic criteria for determining the number of signals in high resolution array processing, *IEEE Trans. Acoustics, Speech, and Signal Processing*, 38, 1959–1971, Nov. 1990.

[12] Xu, W. and Kaveh, M., Analysis of the performance and sensitivity of eigendecomposition-based detectors, *IEEE Trans. Signal Processing*, 43, 1413–1426, June 1995.

[13] Yin, Y.Q. and Krishnaiah, P.R., On some nonparametric methods for detection of the number of signals, *IEEE Trans. Acoustics, Speech, and Signal Processing*, ASSP–35, 1533–1538, Nov. 1987.

[14] Zhao, L.C., Krishnaiah, P.R. and Bai, Z.D., On detection of the number of signals in presence of white noise, *J. Multivariate Analysis*, 20, 1–25, Oct. 1986.

68

Array Processing for Mobile Communications

A. Paulraj
Stanford University

C. B. Papadias
Stanford University

68.1 Introduction and Motivation

This chapter reviews the applications of antenna array signal processing to mobile networks. Cellular networks are rapidly growing around the world and a number of emerging technologies are seen to be critical to their improved economics and performance. Among these is the use of multiple antennas and spatial signal processing at the base station. This technology is referred to as Smart Antennas or, more accurately, as Space-Time Processing (STP). STP refers to processing the antenna outputs in both space and time to maximize signal quality.

A cellular architecture is used in a number of mobile/portable communications applications. Cell sizes may range from large macrocells, which serve high speed mobiles, to smaller microcells or very small picocells, which are designed for outdoor and indoor applications. Each of these offers different channel characteristics and, therefore, poses different challenges for STP. Likewise, different service delivery goals such as grade of service and type of service: voice, data, or video, also need specific STP solutions. STP provides three processing leverages. The first is *array gain*. Multiple antennas capture more signal energy, which can be combined to improve the signal-to-noise ratio (SNR). Next is spatial diversity to combat space-selective fading. Finally, STP can reduce *co-channel, adjacent channel,* and inter-symbol interference.

The organization of this chapter is as follows. In Section 68.2, we describe the vector channel model for a base station antenna array. In Section 68.3 we discuss the algorithms for STP. Section 68.4 outlines the applications of STP in cellular networks. Finally, we conclude with a summary in Section 68.5.

68.2   Vector Channel Model

Channel effects in a cellular radio link arise from multipath propagation and user motion. These create special challenges for STP. A thorough understanding of channel characteristics is the key to developing successful STP algorithms. The main features of a mobile wireless channel are described below.

68.2.1   Propagation Loss and Fading

The signal radiated by the mobile loses strength as it travels to the base station. These losses arise from the mean propagation loss and from slow and fast fading. The mean propagation loss comes from square law spreading, absorption by foliage, and the effect of vertical multipath. A number of models exist for characterizing the mean propagation loss [22, 30], which is usually around 40 dB per decade. Slow fading results from shadowing by buildings and natural features and is usually characterized by a log-normal distribution with standard deviation agreed to 8 dB. Fast fading results from multipath scattering in the vicinity of the moving mobile. It is usually Rayleigh distributed. However, if there is a direct path component present, the fading will be Rician distributed.

68.2.2   Multipath Effects

Multipath propagation plays a central role in determining the nature of the channel. By channel we mean the impulse, or frequency response, of the radio channel from the mobile to the output of the antenna array. We refer to it as a *vector* channel, because we have multiple antennas and, therefore, we have a collection of channels. The mobile radiates omnidirectionally in azimuth using a vertical E-field antenna. The transmitted signal then undergoes scattering, reflection, or diffraction before reaching the base station, where it arrives from different paths, each with its own fading, propagation delay, and angle-of-arrival. This multipath propagation, in conjunction with user motion, determines the behavior of the wireless channel. Multipath scattering arises from three sources (see Fig. 68.1). There are scatterers local to the mobile, remote dominant scatterers, and scatterers local to the base. We will now describe these three scattering mechanisms and their effect on the channel.

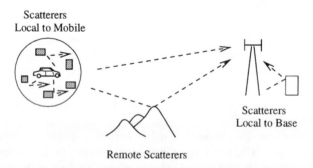

FIGURE 68.1   Multipath propagation has three distinct classes, each of which gives rise to different channel effects.

Scatterers Local to Mobile

Scattering local to the mobile is caused by buildings in the vicinity of the mobile (a few tens of meters). Mobile motion and local scattering give rise to Doppler spread, which causes time-selective fading. For a vertical, polarized E-field antenna, it has been shown [22] that the fading signal has a characteristic *classical spectrum*. For a mobile traveling at 55 MPH, the Doppler spread is about +/- 200

Hz in the 1900 MHz band. This effect results in rapid signal fluctuations also called time-selective fading. While local scattering contributes to Doppler spread, the delay spread will usually be insignificant because of the small scattering radius. Likewise, the angle spread will also be small.

Remote Scatterers

The emerging wavefront from the local scatterers may then travel directly to the base and also be scattered toward the base by remote *dominant scatterers*, giving rise to specular multipath. These remote scatterers can be terrain features or high rise buildings. Remote scattering can cause significant delay and angle spreads. Delay spread causes frequency-selective fading, and the angle spread results in space-selective fading.

Scatterers Local to Base

Once these multiple wavefronts reach the base station, they may be scattered further by local structures such as buildings or other structures that are in the vicinity of the base. Such scattering will be more pronounced for low elevation below-roof-top antennas. The scattering local to the base can cause severe angle spread.

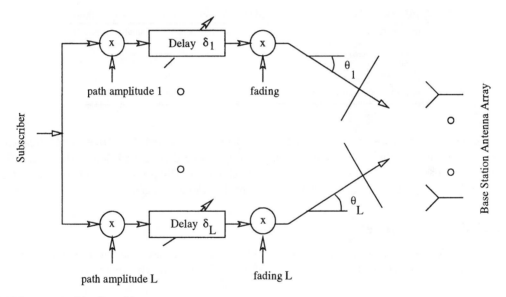

FIGURE 68.2 Multipath model.

68.2.3 Typical Channels

Measurements in macrocells indicate that up to 6 to 12 paths may be present. Typical channel delay, angle, and one-sided Doppler (1800 MHz) spreads are given in Table 68.1.

A multipath channel structure is illustrated in Fig. 68.2. Typical path power and delay statistics can be obtained from the GSM[1] standard. Angle-of-arrival statistics have been less well studied but several results have been reported (see [1, 2, 3]). The resulting channel is shown in Fig. 68.3. We show a frequency

---

[1] Global System for Mobile communications.

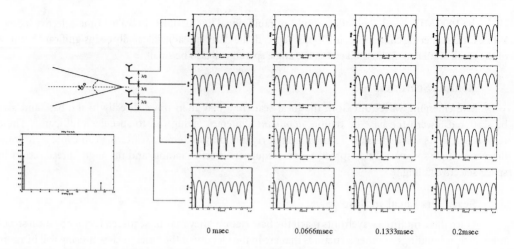

0 msec   0.0666msec   0.1333msec   0.2msec

FIGURE 68.3 Channel frequency response at four different antennas for GSM in a typical hilly terrain channel at 1800 MHz. Mobile speed is 100 KPH. The response is plotted at four time instances spaced 66 μsecs apart.

TABLE 68.1 Typical Delay, Angle and Doppler Spreads in Cellular Applications

Environment	Delay spread (μsec)	Angle spread (deg)	Doppler spread (Hz)
Flat rural (Macro)	0.5	1	190
Urban (Macro)	5	20	120
Hilly (Macro)	20	30	190
Microcell (Mall)	0.3	120	10
Picocell (Indoors)	0.1	360	5

response at each antenna for a GSM system. Since the channel bandwidth is 200 KHz, it is highly frequency-selective in a hilly terrain environment. Also, the large angle spread causes variations of the channel from antenna to antenna. The channel variation in time depends on the Doppler spread. Notice that since GSM uses a short time slot, the channel variation during the time slot is negligible.

68.2.4   Signal Model

We study the case when a single user transmits and is received at a base station with multiple antennas. The noiseless baseband signal $x_i(t)$ received by the base station at the ith element of an m element antenna array is given by

$$x_i(t) = \sum_{l=1}^{L} a_i(\theta_l)\alpha_l^R(t)u(t - \tau_l) \tag{68.1}$$

where L is the number of multipaths, $a_i(\theta_l)$ is the response of the ith element for an lth path from direction θ_l, $\alpha_l^R(t)$ is the complex path fading, τ_l is the path delay, and $u(\cdot)$ is the transmitted signal that depends on the modulation waveform and the information data stream. In the IS-54 TDMA standard, $u(\cdot)$ is a $\pi/4$ shifted DQPSK, gray-coded signal that is modulated using a pulse with square-root raised cosine spectrum with excess bandwidth of 0.35. In GSM, a Gaussian Minimum Shift Keying (GMSK) modulation is used. See [12, 30, 55] for more details. For a linear modulation (e.g., BPSK), we can write

$$u(t) = \sum_{k} g(t - kT)s(k) \tag{68.2}$$

where $g(\cdot)$ is the pulse shaping waveform and $s(k)$ represents the information bits.

In the above model, we have assumed that the inverse signal bandwidth is large compared to the travel time across the array. For example, in GSM the inverse signal bandwidth is 5 μs, whereas the travel time across the array is, at most, a few ns. This is the narrowband assumption in array processing. The signal bandwidth is a sum of the modulation bandwidth and the Doppler spread, with the latter being comparatively negligible. Therefore, the complex envelope of the signal received by different antennas from a given path are identical except for phase and amplitude differences that depend on the path angle-of-arrival, array geometry, and the element pattern. This angle-of-arrival dependent phase and amplitude response at the ith element is $a_i(\theta_l)$ [37].

We collect all the element responses to a path arriving from angle θ_l into an m-dimensional vector, called the *array response vector* defined as

$$\mathbf{a}(\theta_l) = [a_1(\theta_l) \, a_2(\theta_l) \, \ldots \, a_m(\theta_l)]^T$$

where $(\cdot)^T$ denotes matrix transpose.

In array processing literature the array vector $\mathbf{a}(\theta)$ is also known as the *steering vector*. We can rewrite the array output at the base station as

$$\mathbf{x}(t) = \sum_{l=1}^{L} \mathbf{a}(\theta_l)\alpha_l^R(t)u(t - \tau_l) \tag{68.3}$$

where

$$\mathbf{x}(t) = [x_1(t) \, x_2(t) \, \ldots \, x_m(t)]^T$$

and $\mathbf{x}(t)$ and $\mathbf{a}(\theta_l)$ are m-dimensional complex vectors. The fade amplitude $|\alpha^R(t)|$ is Rayleigh or Rician distributed depending on the propagation model.

The channel model described above uses physical path parameters such as path gain, delay, and angle of arrival. When the received signal is sampled at the receiver at symbol (or higher) rate, such a model may be inconvenient to use. For linear modulation schemes, it is more convenient to use a "symbol response" channel model.

Such a discrete-time signal model can be obtained easily as follows. Let the continuous-time output from the receive antenna array $\mathbf{x}(t)$ be sampled at the symbol rate at instants $t = t_o + kT$. The output may be written as

$$\mathbf{x}(k) = \mathbf{H}\mathbf{s}(k) + \mathbf{n}(k) \tag{68.4}$$

where $\mathbf{H}$ is the symbol response channel (a $m \times N$ matrix) that captures the effects of the array response, symbol waveform, and path fading. m is the number of antennas, N is the channel length in symbol periods, and $\mathbf{n}(k)$ is the sampled vector of additive noise. Note that $\mathbf{n}(k)$ may be colored in space and time, as will be shown later. $\mathbf{H}$ is assumed to be time invariant, i.e., α^R is constant. $\mathbf{s}(k)$ is a vector of N consecutive elements of the data sequence and is defined as

$$\mathbf{s}(k) = \begin{bmatrix} s(k) \\ \vdots \\ s(k - N + 1) \end{bmatrix} \tag{68.5}$$

It can be shown [49] that the ijth element of the $\mathbf{H}$ is given by

$$[\mathbf{H}]_{ij} = \sum_{l=1}^{L} a_i(\theta_l)\alpha_l^R g((M_d + \Delta - j)T - \tau_l), \quad i = 1 \ldots, m \; ; \; j = 1, \ldots, N \tag{68.6}$$

where M_d is the maximum path delay and $2\Delta T$ is the duration of the pulse shaping waveform $g(t)$.

68.2.5   Co-Channel Interference

In wireless networks a cellular layout with frequency reuse is exploited to support a large number of geographically dispersed users. In TDMA and FDMA networks, when a co-channel mobile operates in a neighboring cell, co-channel interference (CCI) will be present. The average signal-to-interference power ratio (SIR), also called the protection ratio [24], depends on the reuse factor (K). It is 18.7 dB for reuse $K = 7$ (IS-54), and 13.8 dB for reuse $K = 4$ (GSM). In sectored cells, CCI is significant mainly from cells that lie within the sector beam. The received signal at a base station will therefore be a sum of the desired signal and co-channel interference.

68.2.6   Signal-Plus-Interference Model

The overall signal-plus-interference-and-noise model at the base station antenna array can now be rewritten as

$$x(k) = H_s s_s(k) + \sum_{q=1}^{Q-1} H_q s_q(k) + n(k) \tag{68.7}$$

where H_s and H_q are channels for signal and CCI, respectively, while s_s and s_q are the corresponding data sequences. Note that Eq. (68.7) appears to suggest that the signal and interference are band synchronous. However, this can be relaxed and the time offsets can be absorbed into the channel H_q. In multi-user cases, all the signals are desired and Eq. (68.7) can be rewritten to reflect this situation.

68.2.7   Block Signal Model

It is often convenient to handle signals in blocks. Therefore, we may collect M consecutive snapshots of $x(\cdot)$ corresponding to time instants $k, \ldots, k + M - 1$, (and dropping subscripts for a moment), we get

$$X(k) = HS(k) + N(k) \tag{68.8}$$

where $X(k)$, $S(k)$, and $N(k)$ are defined as

$$
\begin{aligned}
X(k) &= [x(k) \cdots x(k + M - 1)] & (m \times M) \\
S(k) &= [s(k) \cdots s(k + M - 1)] & (N \times M) \\
N(k) &= [n(k) \cdots n(k + M - 1)] & (m \times M)
\end{aligned}
$$

Note that $S(k)$ by definition is constant along the diagonals and is therefore Toeplitz.

68.2.8   Spatial and Temporal Structure

Given the signal model at Eq. (68.8), an important question is whether the unknown channel, H, and data, s, can be determined from the observations X. This leads us to examine the underlying constraints on H and $S(\cdot)$ which we call *structure*.

Spatial Structure

From Eq. (68.6), the jth column of H is given by

$$H_{1:m,j} = \sum_{l=1}^{L} a(\theta_l) \alpha_l^R g((M_d + \Delta - j)T - \tau_l) \tag{68.9}$$

Spatial structure can help determine $\mathbf{a}(\theta_l)$ if the angles of arrival θ_l are known or can be estimated. $\mathbf{a}(\theta_l)$ lies on a *array manifold* $\mathcal{A}$, which is the set of all possible array response vectors indexed by θ.

$$\mathcal{A} = \{\mathbf{a}(\theta)|\theta \in \Theta\} \tag{68.10}$$

where Θ is the set of all possible values of θ. $\mathcal{A}$ includes the effect of array geometry, element patterns, inter-element coupling, scattering from support structures, and objects near the base station.

Temporal Structure

The temporal structure relates to the properties of the signal $u(t)$ and includes modulation format, pulse-shaping function, and symbol constellation. Some typical temporal structures are

- Constant modulus (CM)

In many wireless applications, the transmitted waveform has a constant envelope (e.g., in FM modulation). A typical example of a constant envelope waveform is the GMSK modulation used in the GSM cellular system which has the following general form

$$u(t) = e^{j(\omega t + \phi(t))}$$

where $\phi(t)$ is a Gaussian-filtered phase output of a minimum shift keyed (MSK) signal [40].

- Finite alphabet (FA)

Another important temporal structure in mobile communication signals is the *finite alphabet*. This structure underlies all digitally modulated schemes. The modulated signal is a linear or nonlinear map of an underlying finite alphabet. For example, the IS-54 signal is a $\pi/4$ shifted DQPSK signal given by

$$
\begin{aligned}
u(t) &= \sum_p A_p g(t - pT) + j \sum_p B_p g(t - pT) \\
A_p &= \cos(\phi_p), \quad B_p = \sin(\phi_p), \quad \phi_p = \phi_{p-1} + \Delta\phi_p
\end{aligned}
\tag{68.11}
$$

where $g(\cdot)$ is the pulse shaping function (which is a square root raised cosine function in the case of IS-54), and $\Delta\phi_p$ is chosen from a set of finite phase shifts $\{\frac{5\pi}{4}, \frac{3\pi}{4}, \frac{\pi}{4}, \frac{7\pi}{4}\}$ depending on the data $s(\cdot)$. These finite set of phase shifts represent the FA structure.

- Distance from Gaussianity

The distribution of digitally modulated signals is not Gaussian,[2] and this property can be exploited to estimate the channel from the higher-order moments such as cumulants, see e.g., [15, 33]. Clearly CM signals are non-Gaussian. These higher order statistics (HOS) based methods are usually slower converging than those based on second order statistics.

- Cyclostationarity

Recent theoretical results [14, 28, 39, 44] suggest that exploiting the *cyclostationary* characteristic of the communication signal can lead to second-order statistics based algorithms to identify the channel, $\mathbf{H}$, and therefore a more attractive approach than HOS techniques.

It can be shown [10] that the continuous-time stochastic process $x(t)$ defined in Eq. (68.1) (assuming the fade amplitude α^R is constant) is cyclostationary. Moreover, the discrete sequence $\{x_i\}$ obtained by

---

[2]The distribution may, however, approach Gaussian when constellation shaping is used for spectral efficiency [56].

sampling $x(t)$ at the symbol rate $\frac{1}{T}$ is wide-sense stationary, whereas the sequence obtained by temporal oversampling (i.e., at a rate higher than $1/T$) or spatial oversampling (multiple antenna elements) is cyclostationary. The cyclostationary signal consists of a number of sampling *phases* each of which is stationary. A phase corresponds to a shift in the sampling point in temporal oversampling and different antenna element in spatial oversampling.

The cyclostationary property of sampled communication signals carries important information about the channel phase, which can be exploited in several ways to identify the channel. The cyclostationarity property can also be interpreted as a *finite duration* property. Put simply, this says that the oversampling increases the number of samples in the signal $x(t)$ and phases in the channel, H, but does not change the value of the data for the duration of the symbol period. This allows H to become tall (more rows than columns) and full column rank. Also, the stationarity of the channel makes H Toeplitz (or rather block Toeplitz). Tallness and Toeplitz properties are key to the blind estimation of H.

- The temporal manifold

Just as the array manifold captures spatial wavefront information, the *temporal manifold* captures the temporal pulse-shaping function information [48, 49]. We define the temporal manifold $\mathbf{k}(\tau)$ as the sampled response of a receiver to an incoming pulse with delay τ. Unlike the array manifold, the temporal manifold can be estimated with good accuracy because it depends only on our knowledge of the pulse-shaping function. Table 68.2 summarizes the duality between the array and the temporal manifold.

TABLE 68.2   The Duality Between the Array and the Time Manifold

Manifold	Indexed by:	Characterizes:
Array	Angle θ	Antenna array response
Time	Delay τ	Transmitted pulse shape

The different structures and properties inherent in the signal model are depicted in Fig. 68.4

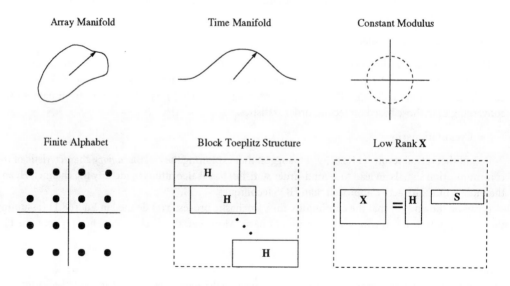

FIGURE 68.4   Space-time structures.

68.3   Algorithms for STP

The history of array signal processing goes back nearly four decades to adaptive antenna combining techniques using phase-lock loops for antenna tracking. An important beginning was made by Howells [21], when he proposed the sidelobe canceller for adaptive nulling, and later Applebaum developed a feedback control algorithm for maximizing SINR. Another significant advance was the LMS algorithm proposed by Widrow [54]. Yet another important milestone was the work of Capon who proposed an adaptive antenna system [8] using a look direction constraint that resulted in the minimum variance distortionless beamformer. Further advances were made by Frost [13] and Griffiths and Jim [17] among several others. See [50] for a review on spatial filtering.

Because of significant delay spread in the channel, array processing in mobile communications can be greatly leveraged by processing the signals in space and in time (STP) to minimize *both* co-channel interference and inter symbol interference while maximizing SNR. See [35] for a review of channel equalization.

We begin with the single-user case where we are only interested in demodulating the signal of interest. We therefore treat interference from other users as unknown additive noise. This is an interference-suppression approach [53]. Later in the section, we will discuss multi-user detection which jointly detects all impinging signals.

68.3.1   Single-User ST-ML and ST-MMSE

The first criterion for optimality in space-time processing is Maximum Likelihood (ML) or is usually referred to as *Maximum Likelihood Sequence Estimation* (MLSE). ST-MLSE seeks to estimate the data sequence that is most likely to have been sent given the received vector signal. Another frequently used criterion is *Minimum Mean Square Error* (MMSE). In ST-MMSE we obtain an estimate of the transmitted signal as a space-time weighted sum of the received signal and seek to minimize the mean square error between the estimate and the true signal at every time instant.

We present ST-MLSE and ST-MMSE in a form that is a space-time extension of the well-known ML and MMSE algorithms.

ST-MLSE

With the channel model described by Eq. (68.8), we assume that the noise **N** is spatially and temporally white and Gaussian, and that there is *no interference*. The MLSE problem can be shown to reduce to finding **S** so as to satisfy the following criterion:

$$\min_{\mathbf{S}} \|\mathbf{X} - \mathbf{HS}\|_F^2 \tag{68.12}$$

where the channel **H** is assumed to be known and $\|.\|_F$ denotes Frobenius norm. This is a generalization of the standard MLSE problem where the channel is now defined in space and time. We can, therefore, use a space-time generalization of the well-known Viterbi algorithm (VA) to carry out the search in Eq. (68.12) efficiently as is shown in Fig. 68.5. See [34] for a discussion on the VA methods.

In the presence of co-channel interference, which is likely to be both spatially and temporally correlated (due to delay spread), the MLSE criterion can be reformulated with a new metric to address the problem. However, the temporal correlation due to delay spread in CCI complicates the implementation of the Viterbi equalizer.

ST-MMSE

In the presence of CCI with delay spread, a ST-MMSE receiver is more attractive. This receiver combines the input in space and time to generate an output that minimizes the error between itself and the desired signal (see Fig. 68.5). Before proceeding further, we need to introduce some preliminaries.

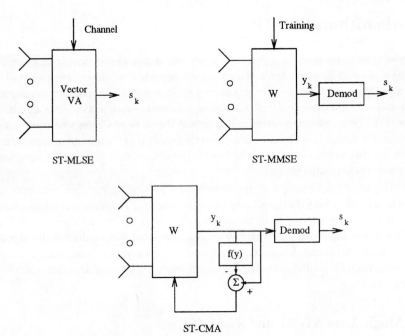

FIGURE 68.5   Different structures for space-time processing.

In a space-time filter (equalizer cum beamformer), W has the following form:

$$W(k) = \begin{bmatrix} w_{11}(k) & \cdots & w_{1M}(k) \\ \vdots & \cdots & \vdots \\ w_{m1}(k) & \cdots & w_{mM}(k) \end{bmatrix} \tag{68.13}$$

In order to obtain a convenient formulation for the space-time filter output, we introduce the quantities $W(k)$ and $X(k)$ as follows:

$$\begin{aligned} X(k) &= vec\,(\mathbf{X}(k)) & (mM \times 1) \\ W(k) &= vec\,(\mathbf{W}(k)) & (mM \times 1) \end{aligned} \tag{68.14}$$

where the operator $vec(\cdot)$ is defined as:

$$vec\,([\mathbf{v}_1 \cdots \mathbf{v}_M]) = \begin{bmatrix} \mathbf{v}_1 \\ \vdots \\ \mathbf{v}_M \end{bmatrix}$$

The scalar equalizer output $y(k)$ can then be written as

$$y(k) = W^H(k)X(k) = Tr(\mathbf{W}^H(k)\mathbf{X}(k)) \tag{68.15}$$

where $(\cdot)^H$ denotes complex conjugate transpose.

The ST-MMSE filter chooses the space-time filter weights to approach the minimum mean square error, i.e.,

$$\min_{W} E \left\| W^H X(k) - s(k - \varsigma) \right\|_2^2 \tag{68.16}$$

where ζ is a delay chosen to center the space-time filter (the choice of this parameter strongly affects performance). The solution to this least-squares (LS) problem follows from the well-known projection theorem

$$E\left(X(k)(X^H(k)W - s^*(k-\zeta))\right) = 0 \tag{68.17}$$

This leads to

$$W = \left\{E\left(X(k)X^H(k)\right)\right\}^{-1} E\left(X(k)s^*(k-\zeta)\right) \tag{68.18}$$

where superscript $*$ denotes complex conjugate. If the interference and noise are independent of the signal, the transmitted bit sequence is white, and $M > N$

$$E\left(X(k)s^*(k-\zeta)\right) = \left[0 \cdots 0 \; vec^T(\mathbf{H}) \; 0 \cdots 0\right]^T = \overline{H} \tag{68.19}$$

where the number of zeros preceding and succeeding $vec^T(\mathbf{H})$ depends on the choice of ζ. Defining the space-time $mN \times mN$ covariance matrix $\mathbf{R}_{XX} = E\left(XX^H\right)$, Eq. (68.18) takes the familiar form

$$W = \mathbf{R}_{XX}^{-1}\overline{H} \tag{68.20}$$

Note that when $M = N, \overline{H} = vec(\mathbf{H})$. A number of techniques are available in order to solve Eq. (68.20), such as the least mean square (LMS) [54] or the recursive least square (RLS) [29]. These have different tradeoffs of computational complexity, tracking capability and steady-state error. See [29] for a discussion. Alternatively, if a block method is used, we can explicitly calculate $\mathbf{R}_{XX}^{-1}$ and then use $\overline{H}$ to find W. This is known as Sample Matrix Inversion (SMI).

The relative performance of ST-MLSE and ST-MMSE schemes is influenced by the dominance of CCI and ISI, and the nature of the channel. When the channel is CCI dominated and contains multipath, a Viterbi equalizer is complicated to implement. In this case, the MMSE approach appears desirable. On the other hand, in an ISI-dominated large delay spread scenario, an MLSE has natural advantages.

Having reviewed the basic approaches to space-time processing, we will now study two important issues: blind vs. non-blind and single vs. multi-user approaches.

Training Signal Methods

In many mobile communications standards such as GSM and IS-54, explicit training signals are inserted inside the TDMA data bursts. These training signals can be used to estimate the channel needed for the MLSE or MMSE receivers.

Let $\mathbf{T}$ be the training sequence arranged in a matrix form (again $\mathbf{T}$ is Toeplitz). Then, during the training burst, the received data is given by

$$X = \mathbf{HT} + \mathbf{N} \tag{68.21}$$

Clearly $\mathbf{H}$ can be estimated using LS

$$\mathbf{H} = X\mathbf{T}^\dagger \tag{68.22}$$

where $\mathbf{T}^\dagger = \mathbf{T}^H\left(\mathbf{TT}^H\right)^{-1}$.

In a ST-MMSE receiver, we need W and this can be computed readily from $\mathbf{H}$ using Eq. (68.20).

Blind Methods

The term "blind" methods (other names are "self-recovering" or "unsupervised"), do not need training signals and rather exploit the temporal structure such as non-Gaussianity; constant modulus (CM); finite alphabet (FA); cyclostationarity; or the spatial structure, such as the array manifold. The performance of blind methods will, of course, be sensitive to the validity of structural properties assumed.

Spatial Structure or DOA-Based Methods   These techniques use DOA estimates as a basis for determining the optimum beamformer. These methods were developed vigorously in the 1980s in military applications for reception of unknown or noise-like signals. The modern era in DOA estimation began with the MUSIC algorithm first proposed in 1979 by Schmidt and independently by Bienvenu and Kopp [6, 36], thus launching the "subspace era" in signal processing. See [52] for a survey. Another class of DOA estimation techniques was launched when Paulraj et al. proposed the ESPRIT algorithm which has striking advantages when compared to MUSIC but needs a special array geometry. See [23] for a survey.

DOA-based methods suffer from serious drawbacks in cellular applications. First, DOA estimation requires an accurate knowledge of the array manifold. This needs expensive calibration support. Next, the number of antennas at cellular base stations vary from four to eight per sector, an insufficient number for the multipath and interference rich cellular environments. Finally, these methods do not exploit the knowledge of the modulation format of the communication signal and the time delay relationship between multipath signals.

A *subspace approach* can be used to estimate the directions-of-arrival of the impinging wavefronts. The signal model is given by

$$\mathbf{x}(t) = \mathbf{A}\mathbf{u}(t) + \mathbf{n}(t) \tag{68.23}$$

where $\mathbf{A}$ is an $m \times Q$ matrix whose columns are the array response vectors for each wavefront (assuming no multipath)

$$\mathbf{A} = [\ \mathbf{a}(\theta_1) \quad \cdots \quad \mathbf{a}(\theta_Q)\ ],$$

$\mathbf{u}(t)$ contains the fading signals from the Q users

$$\mathbf{u}(t) = [\alpha_1(t)u_1(t - \tau_1) \quad \cdots \quad \alpha_Q(t)u_Q(t - \tau_Q)]^T$$

and

$$u_q(t) = \sum_k s_q(k)g(t - kT)$$

The sampled block signal model then takes the following form

$$\mathbf{X} = \mathbf{A}\mathbf{S} + \mathbf{N} \tag{68.24}$$

In the subspace approach, we seek to estimate $\mathbf{A}$ from the array data by exploiting the underlying array manifold structure. When the number of antennas, m, is greater than the number of signals, Q, the signal $\mathbf{x}(t)$ in the absence of noise is confined to a subspace, referred to as the *signal subspace*.

We first estimate this signal subspace from the received data $\mathbf{X}$. We then search for an $m \times Q$ matrix $\mathbf{A}$ whose *columns lie on the array manifold* and whose (column) subspace matches the estimated signal subspace. A good estimate of the signal subspace is given by the first Q dominant eigenvectors of the space-only $m \times m$ covariance matrix $\mathbf{R}_{xx} = E(\mathbf{x}\mathbf{x}^H)$. If $\mathbf{E}_s$ is a matrix of these eigenvectors, then the subspace fitting approach estimates $\mathbf{A}$ to minimize the following criterion

$$\min_{\mathbf{A}} \ \|\mathbf{E}_s - \mathbf{A}\mathbf{Z}\|_F^2$$

where $\mathbf{Z}$ is an arbitrary $Q \times Q$ square matrix.

Once $\mathbf{A}$ is estimated, we have the array vector for the desired signal. The MMSE and ML estimators (assuming no multipath) of $u_q(t)$ are identical [7] and are given by

$$\mathbf{w}_q = \mathbf{R}_{xx}^{-1}\mathbf{a}(\theta_q) \tag{68.25}$$

$\mathbf{w}_q$ is a (space-only) beamformer that has been studied extensively.

When multipath and delay spread is present, the solution in Eq. (68.25) will have a poor performance and improved techniques are needed. If we use a ST-MMSE structure, we can extend the above subspace methods to compute the optimum beamformer given in Eq. (68.20).

Temporal Structure Methods These techniques include a vast range that spans from the well-studied CM and HOS methods to the more recent second order methods that exploit the cyclostationarity of the received signal.

The fading and dynamics of the mobile propagation channel create special problems for blind techniques, and their performance in mobile channels is only recently gaining attention. A widely known class of simple blind algorithms is the so-called Bussgang class that contains, among others, the CM 1-2, CM 2-2, Sato, and Decision-Directed (DD) algorithms. See [11, 19, 20] for a survey of blind algorithms.

Contrary to non-blind techniques, where a training signal drives the recursive algorithms, in the CM approach, we replace the training signals by a modulus corrected version of the output signal. The CM 2-2 minimizes the following cost function

$$\min_{W} J(W) \; = \; E||y(k)|^2 - 1|^2 \qquad\qquad (68.26)$$

where $y(k)$ is the output of the ST filter [see Eq. (68.15)].

The resulting LMS-type algorithm is given by

$$W(k+1) \; = \; W(k) \; - \; \mu \, X^*(k) \, y(k) \, \left(|y(k)|^2 \; - \; 1\right) \qquad\qquad (68.27)$$

$W(k+1)$, under the right conditions, approaches the optimum ST-MMSE solution in Eq. (68.20).

Important performance issues for blind algorithms are speed of convergence, ability to reach the global optimum solution, and capacity to track time varying mobile channels.

Polyphase Methods Following the path-breaking paper by Tong, et al. [44] that presented a blind channel identification method using oversampling and relying only on second order statistics, a number of techniques that exploit cyclostationarity have since dominated the blind-deconvolution litera-ture.

Polyphase methods provide a blind solution by starting with the data

$$X(k) \; = \; HS(k) \; + \; N(k) \qquad\qquad (68.28)$$

or its second order statistics. They then extract **H** and **S** by exploiting the tallness structure (obtained via oversampling) of **H** [28, 29]. See [25] for a tutorial presentation of polyphase techniques.

68.3.2 Multi-User Algorithms

In multi-user (MU) algorithms, we address the problem of extracting multiple co-channel user signals arriving at an antenna array. Such problems occur in channel reuse within cell (RWC) applications or in situations where we attempt to demodulate the interference signal in order to improve interference suppression. The data model is once again

$$X = HS + N \qquad\qquad (68.29)$$

where **H** and **S** are suitably defined to include multiple users and are of dimensions $m \times NQ$ and $NQ \times M$, respectively.

We have several approaches that parallel the single-user case. We begin with the ML and MMSE prototypes and then explain in more detail some recently developed blind techniques.

Multi-User MLSE and MMSE

If the channels for all arriving signals are known, then we can extend the earlier MLSE to jointly demodulate all the user data sequences. Starting with the data model in Eq. (68.29), we can then search for multiple user data sequences that minimize the ML cost function in Eq. (68.12). The multi-user MLSE will have a large number of states in the trellis. Efficient techniques for implementing a Viterbi equalizer need to be developed.

In multi-user MMSE, we usually estimate each user signal separately using the single-user MMSE processor given in Eq. (68.20). In this case, the MMSE treats other user signals as interference with unknown structure. Multi-user techniques either need training signals for all the users or adopt blind methods. The multiple training signals should be designed to have low cross correlation properties so as to minimize cross coupling in the channel estimates.

Multi-User Blind Methods

Once again, the techniques are parallel to the single-user spatial and temporal blind methods. The spatial structure multi-user algorithms are again applicable under the conditions discussed in Section 68.3.1. The approach is identical; we first estimate $\mathbf{A}$ using subspace methods, and the beamformer $\mathbf{w}_q$ for each user follows from Eq. (68.25).

We briefly describe some illustrative algorithms.

Finite Alphabet (FA) Method

This approach exploits the FA property of the digitally modulated signals. Assuming no delay spread and perfect multi-user symbol synchronization, the channel model is given by Eq. (68.24), which we repeat here for convenience:

$$\mathbf{X} = \mathbf{AS} + \mathbf{N} \tag{68.30}$$

where both $\mathbf{A}$ and $\mathbf{S}$ are unknown and the additive noise is assumed to be white and Gaussian. The joint ML criterion for this reduces to the familiar minimization problem

$$\min_{\mathbf{A},\mathbf{S}} \|\mathbf{X} - \mathbf{AS}\|_F^2 \tag{68.31}$$

This is a joint ML problem where both the channel and data are unknown. The FA property allows us to solve Eq. (68.31) and estimate both $\mathbf{A}$ and $\mathbf{S}$. Since the ML criterion is separable with respect to the unknowns, one approach to minimize the cost function in Eq. (68.31) is alternating projections. Starting with an initial estimate of $\mathbf{A}$, we minimize Eq. (68.31) with respect to $\mathbf{S}$, keeping $\mathbf{A}$ fixed. This is a data detection problem. With an estimate of $\mathbf{S}$, an improved estimate of $\mathbf{A}$ can be obtained by minimizing Eq. (68.31) with respect to $\mathbf{A}$, keeping $\mathbf{S}$ fixed. This is a standard least-squares problem. We continue this iterative process until a fixed point is reached. The global solution is a fixed point of the iteration. In order to avoid a computationally expensive search, two suboptimal iterative techniques, ILSP and ILSE [42, 43], can be used to make this minimization tractable.

Note that the joint ML problem can also be formulated for the single user case where we estimate the channel and the data jointly using the FA or other signal properties. Joint ML methods are also known as adaptive ML.

Finite Alphabet – Oversampling (FA-OS) Method

In the presence of delay spread (and unsynchronized symbols), the FA algorithm has to be modified to estimate the space-time channel $\mathbf{H}$ as against the spatial channel $\mathbf{A}$ described earlier. An attractive technique to estimate the temporal channel using polyphase or oversampling method was proposed recently in [47].

We therefore first need to extend the multi-user data model in Eq. (68.30) to incorporate oversampling. Assuming oversampling at P samples per symbol, we define a new $mP \times M$ data matrix $\mathbf{X}$ where each entry is a vector of P data samples per symbol period.

This results once again in the familiar model

$$\mathbf{X} = \mathbf{HS} + \mathbf{N} \tag{68.32}$$

Note that now the dimensions of $\mathbf{X}$, $\mathbf{H}$, and $\mathbf{S}$ are $mP \times M$, $mP \times NQ$, and $NQ \times M$, respectively.

As noted earlier, $\mathbf{H}$ is tall, full column rank and has a block Toeplitz structure.

Once again we can estimate $\mathbf{H}$ and $\mathbf{S}$ using a Joint-ML approach. This reduces to minimizing

$$\min_{\mathbf{H},\mathbf{S}} \|\mathbf{X} - \mathbf{HS}\|_F^2 \tag{68.33}$$

A direct approach to Eq. (68.33) is computationally prohibitive. The approach in [47] breaks up the joint problem into two smaller subproblems. First, the channels are equalized by enforcing the low rank and block-Toeplitz structure of $\mathbf{H}$. This yields the row subspace of $\mathbf{S}$. The FA property can now be enforced to determine the symbols in $\mathbf{S}$.

Multi-User CM

While the FA approach exploits the FA property, CM is another structure for multi-user signal separation. When the channel has no delay spread, we have a standard source separation problem, which can be dealt with the so-called *constant modulus array* [4, 16, 26]. The resulting algorithms are space-only multi-user counterparts of the temporal CM algorithms: instead of combatting the channel ISI in order to retrieve the single-user signal, they try to combat the channel CCI and demodulate the different user signals.

Recent advances in multi-user CM include an analytical CM (ACM) [32, 45], a multi-stage CM algorithm [38], and related approaches [5] and [9]. The extension to a delay spread environment was proposed in [31].

68.3.3 Simulation Example

Figure 68.6 illustrates the effect of STP in a mobile environment. The channel model chosen was a typical urban channel. A four element linear array with $\frac{\lambda}{2}$ spacing was employed. The desired and interference signals arrived from mean directions of 0° and 45°, respectively. An IS-54 channel interface was used. Section (a) shows the received signal constellation for a simple antenna. Note that this implies that the eye is completely closed. Section (b) shows the constellation after STP, using a ST-MMSE equalizer employing training signals. Note the dramatic improvement in the received constellation.

68.4 Applications of Spatial Processing

In this section, we briefly describe three applications of antennas and STP in cellular base stations.

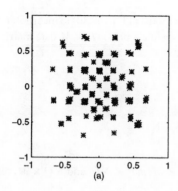 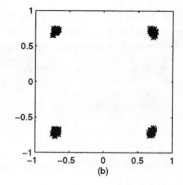

(a)   (b)

FIGURE 68.6   Interference cancellation using STP.

68.4.1   Switched Beam Systems

Switched Beam Systems (SBS) consist of a beamformer in the RF stage that forms multiple (non-adaptive) beams, a "sniffer" that determines the beam that has the best SINR and a switch that is used to select the best or best two beams for the receiver. These systems are used as an appliqué unit, where the existing diversity antennas are replaced by a switch-beam antenna system. The SBS operates by sniffer scanning the beamformer outputs to detect the best two beams which are then switched through to the receiver. In order to reduce the probability of incorrect beam selection, the beam outputs are validated by checking the color code (for example CDVCC in IS-54 or SAT tone in AMPS) prior to determining the best beam.

The main advantage of using SBS is the improvement in cell coverage on the reverse link due to the array gain and improved voice quality due to reduced interference. Since the pre-formed beams are narrower than the sector beamwidth, reduction in interference power is obtained when the desired signal and the interference are separated in angle and fall into different beams. This SINR improvement offers better voice quality and may also allow use of a smaller reuse factor and therefore improve capacity.

The performance of SBS depends on a number of factors, including angle spread of multipath, relative angles-of-arrival of the signals and interference, and the array topology. Performance gains in SBS come from array gain, diversity gain, reduced interference, and trunking efficiency.

68.4.2   Space-Time Filtering

Space-time filtering (STF) applies space-time processing to maximize signal power and minimize inter-symbol and co-channel interference. As is evident from earlier sections, space-time processing will be very different for each air interface standard. In GSM, the slot duration is 0.577 ms with 26 training bits and the symbol period is 3.7 μsecs. Channel equalization is, of course, necessary. The presence of controlled ISI further complicates the equalization problem. It is reasonable to assume that the signal and interference channels are invariant across a slot. In IS-54, the slot duration is 6.66 ms, resulting in substantial variation of the channel over the slot. However, the symbol period is 41.6 μsecs, and therefore channel equalization is usually not needed. Thus, the STF architectures must find some means to track the signal and interference channels across the burst, using blind techniques, as those described earlier in this chapter.

68.4.3   Channel Reuse Within Cell

Reuse within cell (RWC) refers to the reuse of a channel or radio resource *within* a cell by exploiting differences in the channels. This is akin to spectrum reuse in cellular systems, where a channel or a spectrum resource used in one cell is reused in another cell separated by sufficient *distance* such that the co-channel interference is sufficiently small.

When RWC is used in TDMA or FDMA, a cell supports two or more users in a given channel, as against a single user in conventional cells. Antenna arrays and space-time processing is used for joint demodulation of multiple users, assuming such users are sufficiently separated in channels (directions). When two or more users become closely aligned in their channels, they will no longer be separable, and one of the users should be handed off to another frequency or time slot. RWC needs to work on both forward and reverse links, therefore, signal separability must be achieved on both links.

The principal challenge in RWC when used with TDMA or FDMA is to estimate and track the reverse and forward channels to a high degree of accuracy. The problem is further complicated by the near-far problem resulting in power imbalance between users. The ability to estimate and track the reverse link channel depends on angle, delay, and Doppler spreads. The higher these spreads, the higher the sources of channel estimation errors. Therefore, flat rural environments with low angle and delay spreads score over urban and microcells which use antennas below roof top. Also, fixed wireless applications score over mobile applications. In the forward link, we need to once again predict the channel accurately. We can do this by an open loop method, i.e., use the reverse link channel to predict forward channel. Alternatively, we can use feedback from the mobile to estimate the forward channel. For the open loop method, in FDD, angle spread is a source of error and in TDD the Doppler spread is a source of error. Due to these complications, RWC is not a promising technology in most TDMA and FDMA applications.

68.5   Summary

Use of array signal processing or STP is emerging as a powerful tool for improving cellular wireless networks. STP can improve cell coverage, enhance link quality, and increase system capacity. The rapidly varying mobile channel with large multipath delay and angle spreads offer a significant challenge to STP. Effective solutions have to be specific to each air interface and the propagation environment. More work is needed to develop robust STP techniques and to characterize their performance. See [1, 2, 3] for a review of the current state of the art in smart antennas technology.

68.6   References

[1] First workshop on smart antennas in wireless mobile communications, Center for Telecommunications and Information Systems Laboratory, Stanford University, Stanford, CA, June 1994.

[2] Second workshop on smart antennas in wireless mobile communications, Center for Telecommunications and Information Systems Laboratory, Stanford University, Stanford, CA, July 20-21, 1995.

[3] Third workshop on smart antennas in wireless mobile communications, Center for Telecommunications and Information Systems Laboratory, Stanford University, Stanford, CA, July 25-26, 1996.

[4] Agee, B.G., Blind separation and capture of communication signals using multitarget constant modulus beamformer, *Proc. MILCOM'89*, 1989.

[5] Batra, A. and Barry, J.R., Blind cancellation of co-channel interference, *Proc. IEEE Globecom Conference*, 1995.

[6] Bienvenu, G. and Kopp, L., Principe de la goniometri passive adaptative, *Proc. 7ème Colloque GRETSI*, 106/1–106/10, Nice, France, 1979.

[7] Capon, J., High resolution frequency wave number spectrum analysis, *Proc. IEEE*, 57, 1408–1418, 1969.

[8] Capon, J., Greenfield, R.J. and Kolker, R.J., Multidimensional maximum likelihood processing of a large aperture seismic array, *Proc. of IEEE*, 55, 192–211, Feb. 1967.

 [9] Castedo, L., Escudero, C.J. and Dapena, A., A blind signal separation method for multiuser communications, *IEEE Trans. on Signal Processing, Special Issue on Signal Processing for Advanced Communications*, 45(1), Jan. 1997.

[10] Ding, Z., Blind channel identification and equalization using spectral correlation measurements, Part I: frequency-domain analysis, in *Cyclostationarity in Communications and Signal Processing*, Gardner, W.A., Ed., New Jersey, 1994, 417–436.

[11] Duhamel, P., Blind equalization, Tutorial presentation, International Conference on Acoustics, Speech, and Signal Processing, Detroit, MI, May 1995.

[12] Feher, K., *Wireless Digital Communications*, Feher/Prentice-Hall, Upper Saddle River, NJ, 1995.

[13] Frost, O.L., An algorithm for linearly constrained adaptive array processing, *Proc. IEEE*, 60, 926–935, 1972.

[14] Gardner, W.A., Ed., *Cyclostationarity in Communications and Signal Processing*, IEEE Press, New Jersey, 1994.

[15] Giannakis, G.B. and Mendel, J.M., Identification of nonminimum phase systems using higher order statistics, *IEEE Trans. on Acoustics, Speech, and Signal Processing*, 37(3), 360–377, March 1989.

[16] Gooch, R.P. and Lundell, J., The CM array: an adaptive beamformer for constant modulus signals, *Proc. ICASSP'86*, 2523–2526, Tokyo, Japan, 1986.

[17] Griffiths, L.J. and Jim, C.W., An alternative approach to linearly constrained adaptive beamforming, *IEEE Transactions on Antennas Propag.*, AP-30, 27–34, May 1982.

[18] Hansen, L.K. and Xu, G., Geometric properties of the blind digital co-channel communications problem, *Proc. ICASSP'96*, Atlanta, May 1996.

[19] Haykin, S., *Blind Deconvolution*, Prentice-Hall, Englewood Cliffs, NJ, 1994.

[20] Haykin, S., *Adaptive Filter Theory*, 3rd ed., Prentice-Hall, Englewood Cliffs, NJ, 1995.

[21] Howells, P., Intermediate frequency side-lobe canceller, U.S. Patent 3,202,990, Aug. 1965.

[22] Jakes, W.C., *Microwave Mobile Communications*, John Wiley & Sons, New York, 1974.

[23] Krim, H. and Viberg, M., Two decades of array signal processing research: the parametric approach, *IEEE Signal Processing Magazine*, 13(4), 67–94, July 1996.

[24] Lee, W.C., *Mobile Communications – Design Fundamentals*, Howard Sams, Indianapolis, IN, 1986.

[25] Liu, H., Xu, G., Tong, L. and Kailath, T., Recent developments in blind channel equalization: From cyclostationarity to subspaces, *Signal Processing*, 50, 83–99, 1996.

[26] Lundell, J.D. and Widrow, B., Application of the constant modulus adaptive beamformer to constant and nonconstant modulus signals, in *Proc. Asilomar 21st Conference on Signals, Systems, and Computers*, 432–436, Pacific Grove, CA, Nov. 1991.

[27] Matsumoto, T., Nishioka, S. and Hodder, D., Beam selection performance analysis of a switched multi-beam antenna system in mobile communications environments, Stanford, CA, July 1995. Second workshop on Smart Antennas in Wireless Mobile Communications.

[28] Moulines, E., Duhamel, P., Cardoso, J.F. and Mayrargue, S., Subspace methods for the blind identification of multichannel FIR filters, *IEEE Trans. on Signal Processing*, 1995.

[29] Orfanidis, S.J., *Optimal Signal Processing – An Introduction*, Macmillan Publishing Co., New York, 1985.

[30] Pahlavan, K. and Levesque, A.H., *Wireless Information Networks*, John Wiley & Sons, New York, 1995.

[31] Papadias, C.B. and Paulraj, A., A constant modulus algorithm for multi-user signal separation in presence of delay spread using antenna arrays, *IEEE Signal Processing Letters*, 4(6): 178–181, June 1997.

[32] Papadias, C.B. and Slock, D.T.M., Towards globally convergent blind equalization of constant modulus signals: a bilinear approach, in *Proc. VII European Signal Processing Conference*, Edinburgh, Scotland, Sept. 13-16, 1994.

[33] Porat, B. and Friedlander, B., Blind equalization of digital communication channels using higher order moments, *IEEE Trans. Acoust. Speech, Signal Processing,* SP-39(2), 522–526, Feb. 1991.

[34] Proakis, J.G., *Digital Communications,* McGraw-Hill, New York, 1983.

[35] Qureshi, S.U.H., Adaptive equalization, *Proc. IEEE,* 53(12), 1349–1387, Sept. 1985.

[36] Schmidt, R.O., Multiple emitter location and signal parameter estimation, in *Proc. RADC Spectrum Estimation Workshop,* 243–258, Griffiss AFB, NY, 1979.

[37] Schmidt, R.O., *A Signal Subspace Approach to Multiple Emitter Location and Spectral Estimation,* Ph.D. thesis, Stanford University, Stanford, CA, Nov. 1981.

[38] Shynk, J.J. and Gooch, R.P., The constant modulus array for co-channel signal copy and direction finding, *IEEE Trans. on Signal Processing,* 44(3), 652–660, March 1996.

[39] Slock, D.T.M. and Papadias, C.B., Blind fractionally-spaced equalization based on cyclostationarity, in *Proc. Vehicular Technology Conf.,* Stockholm, Sweden, June 1994.

[40] Steele, R., *Mobile Radio Communications,* Pentech Press, 1992.

[41] Swindlehurst, A. and Yang, J., Using least squares to improve blind signal copy performance, *IEEE Signal Processing Letters,* 1(5), 80–82, May 1994.

[42] Talwar, S., Paulraj, A. and Viberg, M., Reception of multiple co-channel digital signals using antenna arrays with applications to PCS, in *Proc. ICC'94,* 700–794, 1994.

[43] Talwar, S., Viberg, M. and Paulraj, A., Blind separation of synchronous co-channel digital signals using an antenna array. Part I. Algorithms, *IEEE Trans. on Signal Processing,* 44(5), 1184–1197, May 1996.

[44] Tong, L., Xu, G. and Kailath, T., Blind identification and equalization of multipath channels: a time domain approach, *IEEE Trans. on Information Theory,* 40(2), 340–349, March 1994.

[45] van der Veen, A. and Paulraj, A., An analytical constant modulus algorithm, *IEEE Trans. on Signal Processing,* 44(5), 1136–1195, May 1996.

[46] van der Veen, A.J., Talwar, S. and Paulraj, A., Blind identification of FIR channels carrying multiple finite alphabet signals, in *Proc. IEEE ICASSP,* 2, 1213–1216, 1995.

[47] van der Veen, A.J., Talwar, S. and Paulraj, A., Blind estimation of multiple digital signals transmitted over FIR channels, *IEEE Sig. Process. Lett.,* 5(2), 99–102, May 1995.

[48] Vanderveen, M.C., Ng, B., Papadias, C.B. and Paulraj, A.J., Joint angle and delay estimation (JADE) for signals in multipath environments, in *30th Asilomar Conference on Signals, Systems, and Computers,* Pacific Grove, CA, 1996.

[49] Vanderveen, M.C., Papadias, C.B. and Paulraj, A.J., Joint angle and delay estimation (JADE) for multipath signals arriving at an antenna array, *IEEE Comm. Lett.,* Jan. 1997.

[50] Van Veen, B.D. and Buckley, K.M., Beamforming: a versatile approach to spatial filtering, *IEEE ASSP Magazine,* 4–24, April 1988.

[51] Verdu, S., Minimum probability of error for asynchronous Gaussian multiple-access channels, *IEEE Trans. Inform. Theory,* 32(1), 85–96, Jan. 1986.

[52] Viberg, M. and Stoica, P., Editorial note, in special issue on subspace methods, Part I: array signal processing and subspace computations, *Sig. Process.,* 50(1,2), 1–3, April 1996.

[53] Wales, S., Technique for co-channel interference suppression in TDMA mobile radio systems, *IEEE Proc. Comm,* 142, 106–114, April 1995.

[54] Widrow, B. and Stearns, S., *Adaptive Signal Processing,* Prentice-Hall, Englewood Cliffs, NJ, 1985.

[55] Yacoub, M.D., *Foundations of Mobile Radio Engineering,* CRC Press, Boca Raton, FL, 1993.

[56] Zervas, E., Proakis, J. and Eyuboglu, V., Effects of constellation shaping on blind equalization, in *Proc. SPIE 1991,* 1565, 1991.

69

Beamforming with Correlated Arrivals in Mobile Communications[1]

Victor A.N. Barroso
Instituto Superior Técnico,
Instituto de Sistemas e Robótica

José M.F. Moura
Carnegie Mellon University

69.1 Introduction

The classical definition of a beamformer basically specifies its goal: to estimate the signal waveform arriving at the array from a given direction. Beamformers are spatial processors that combine the signals impinging on an array of captors. Combining the outputs of the captors forms a narrow beam pointing towards the direction of the source (look direction). This narrow beam can discriminate between sources spatially located at distinct sites. This important property of beamformers is used to design techniques that localize active or passive sources particularly in RADAR/ SONAR systems.

In the last two decades, beamforming methods have had significant theoretical and practical advances. This, together with other technological advances, has broadened the application of sophisticated beamforming techniques to a diversity of areas, including imaging, geophysical and oceanographic exploration, astrophysical exploration, and biomedical. See [19, 20] for an excellent overview of modern beamforming techniques and applications.

Communications is another attractive application area for beamforming. In fact, beamforming has been widely used for directional transmission and reception as well as for sector broadcasting in satellite communications systems. More recently, due to the drastic increase of users in cellular radio systems [10,

---

[1]Initial date of submission of this article September 28, 1995.

15, 18], including indoors and outdoors mobile systems, it is increasingly being recognized that the design of base station and mobile antennas based on beamforming methods improves significantly the system's spectrum efficiency [1, 11]. In turn, this enables accommodating larger numbers of users [3, 16]. The most striking argument in favor of using advanced beamforming techniques such as adaptive or blind beamforming for mobile communications is based on the idea of Space Division Multiple Access (SDMA) schemes. With SDMA, several mobiles share simultaneously the same frequency channel by creating virtual channels in the spatial domain. Another important argument in favor of using beamforming in cellular radio is that beamforming yields flexible signal processing schemes that properly handle multipath effects which are typical in radio communications. Multipath is the term given when the same signal arrives at the destination through different paths. This may arise when signals bounce off obstacles in their path of propagation. At the receiver, these arrivals are correlated. Their recombination causes severe signal distortions and fading. In limiting cases, the power of the received signal can become so small that the reliability of the data communications link is completely lost.

In this section we design a multichannel beamformer to combat multipath effects. The receiver uses a base station antenna array which handles several radio links operating simultaneously at the same carrier frequency, while preserving the reliability of the communications. The approach relies on statistical signal processing methods, yielding a solution that operates in a blind mode with respect to the parameters that specify the propagation channel. This means that, except for a few quantities related to system specifications, e.g., link budget and array geometry, the receiver that we describe here does not assume any prior knowledge about the locations of the sources and of the structures of the ray arrivals, including directions of arrival and correlations. The simulation results show the excellent performance of this multichannel beamformer in SDMA schemes.

The chapter is organized as follows. In Section 69.2 we introduce the beamforming problem (see also [20]), and classical beamformers such as the delay-and-sum beamformer, the minimum output noise power beamformer, and the minimum variance beamformer. We show that these beamformers present severe drawbacks when operating in multipath environments. Section 69.3 presents a solution to the beamforming problem for the case of correlated arrivals. This solution is based on a minimum mean square error (MMSE) approach. We compare the performance of this beamformer with the performance of the beamformers introduced in Section 69.2. We emphasize, in particular, the case of multipath propagation. In this section, we also discuss issues regarding the implementation of the minimum mean square error beamformer. In Section 69.4, we describe a method to implement the minimum mean square error beamformer in the context of a digital mobile communications system. The method operates in a blind mode and strongly exploits the structure of the received multipath data. Since the propagation channel parameters, e.g., angles of arrivals of the multiple paths, are not known, we estimate them with a maximum likelihood approach supported on a finite mixture distribution model of the array data. We maximize the likelihood function with an iterative scheme. We describe an efficient procedure to initialize the iterative algorithm. In general, this procedure converges rapidly to the global maximum of the likelihood function. Section 69.5 presents simulation results obtained with data synthesized by a simple mobile communications simulator. These results confirm the excellent performance of the MMSE beamformer described in the paper.

69.2 Beamforming

Beamforming is an array processing technique for estimating a desired signal waveform impinging on an array of sensors from a given direction. This technique applies to both narrowband and wideband signals. Here, we will consider only the narrowband case.

Let $s(t)$ be the complex envelope of the source radiated signature. Under the farfield assumption the signal at the receiving array is a planar wavefront, see Fig. 69.1. In this case, and according to the model derived in [20], the complex envelope of the signal received at each sensor of a uniform and linear array

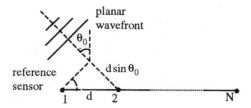

FIGURE 69.1   Source/receiver geometry.

of N omnidirectional sensors is

$$s_n(t) = s(t - \tau_n)e^{-j\omega_0\tau_n} , \tag{69.1}$$

where ω_0 is the carrier frequency and τ_n is the intersensor propagation delay. Let d, c, and θ_0 be, respectively, the distance between sensors, the propagation velocity, and the direction of arrival (DOA). The intersensor delays are then

$$\tau_n = \frac{(n-1)d}{c}\sin\theta_0, \quad n = 1, 2, \ldots, N . \tag{69.2}$$

Because of the narrowband assumption, we can make the simplification

$$s(t - \tau_n) \simeq s(t)$$

in Eq. (69.1). This means that, for the values of τ_n of interest, the source complex envelope $s(t)$ is slowly varying when compared with the carrier $e^{j\omega_0 t}$.

We model each array sensor by a quadrature receiver, its output being given by

$$z_n(t) = s_n(t) + n_n(t), \tag{69.3}$$

where $s_n(t)$ is the complex envelope of the signal component and $n_n(t)$ is a complex additive disturbance, such as sensor noise, ambient noise, or another signal interfering with the desired one. Collecting in a vector $\mathbf{z}(t)$ all the responses of the N sensors of the array to a narrowband source coming at the array from the DOA$= \theta_0$, we get the N-dimensional complex vector

$$\mathbf{z}(t) = \mathbf{a}(\theta_0)s(t) + \mathbf{n}(t) . \tag{69.4}$$

The vector $\mathbf{a}(\theta_0)$ is referred to as the steering vector for the DOA θ_0.

The elements of the steering vector $\mathbf{a}(\theta_0)$ are given by $a_n(\theta_0) = e^{-j\omega_0\tau_n}$, $n = 1, 2, \ldots, N$. The noise vector $\mathbf{n}(t)$ is an N-dimensional complex vector collecting the N sensor noises $n_n(t)$. In general, it includes components correlated with the desired signal as in multipath propagation environments. With multipath, several replicas of the same signal, each one propagating along a different path, arrive at the array with distinct DOAs.

In beamforming, the goal is to estimate the source signal $s(t)$ given $\mathbf{a}(\theta_0)$. The narrowband beamformer is illustrated in Fig. 69.2. The output of the beamformer is

$$y(t) = \mathbf{w}^H\mathbf{z}(t) , \tag{69.5}$$

where $\mathbf{w} = [w_1, w_2, w_3, \ldots, w_N]^T$ is a vector of complex weights. We use the notation $\{\cdot\}^T$ to denote vector and matrix transposition, and $\{\cdot\}^H$ for transposition followed by complex conjugation. The beamformer is completely specified by the vector of weights $\mathbf{w}$.

In the absence of the noise term $\mathbf{n}(t)$ in (69.4), it is readily seen that choosing

$$\mathbf{w} = (1/N)\mathbf{a}(\theta_0) ,$$

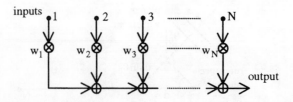

FIGURE 69.2   Narrowband beamformer.

the beamformer output is $y(t) = s(t)$. This corresponds to the simplest implementation of the narrowband beamformer, known as the delay–and–sum (DS) beamformer: it combines coherently the signal replicas received at each sensor after compensating for their corresponding relative delays.

The interpretation of the DS beamformer operation is rather intuitive. However, we may ask ourselves the following question: is DS the best we can do to estimate the desired signal when the disturbance $\mathbf{n}(t)$ is present? To answer the question satisfactorily, we begin by noting that, in the presence of noise, the output of the DS beamformer is

$$y(t) = s(t) + (1/N)\mathbf{a}^H(\theta_0)\mathbf{n}(t) .$$

The influence of the error term on the estimate $y(t)$ of $s(t)$ depends basically on the structure of $\mathbf{n}(t)$. The optimal design of beamformers depends now on the choice of an adequate optimization criterion that takes into account the disturbance vector, with the goal of improving in some sense the quality of the desired estimate. In the sequel, we will consider several cases of practical interest.

69.2.1 Minimum Output Noise Power Beamforming (MNP)

To reduce the effect of the error term at the beamformer output, we formulate the beamforming problem as follows:

find the weight vector $\mathbf{w}$ such that the noise output power $E\left\{\left|\mathbf{w}^H\mathbf{n}(t))\right|^2\right\}$ is minimized subject to the constraint $\mathbf{w}^H\mathbf{a}(\theta_0) = 1$,

where E{·} denotes the statistical average. The cost function is

$$E\left\{\left|\mathbf{w}^H\mathbf{n}(t))\right|^2\right\} = \mathbf{w}^H\mathbf{R_n}\mathbf{w} \tag{69.6}$$

with $\mathbf{R_n}$ the covariance matrix of the disturbance vector $\mathbf{n}(t)$, i.e., $\mathbf{R_n} = E\{\mathbf{n}(t)\mathbf{n}^H(t)\}$. The constraint guarantees that the signal along the look direction θ_0 is not distorted.

The solution to this constrained optimization problem is obtained by Lagrange multipliers techniques. It is given by

$$\mathbf{w} = (\mathbf{a}^H(\theta_0)\mathbf{R_n}^{-1}\mathbf{a}(\theta_0))^{-1}\mathbf{R_n}^{-1}\mathbf{a}(\theta_0) . \tag{69.7}$$

The vector $\mathbf{w}$ in Eq. (69.7) is the gain of the MNP beamformer [20].

When the source signal is uncorrelated with the disturbance,

$$E\{s(t)\mathbf{n}^H(t)\} = \mathbf{0} ,$$

it can be shown that the weight vector (69.7) of the MNP beamformer takes the form

$$\mathbf{w} = (\mathbf{a}^H(\theta_0)\mathbf{R}^{-1}\mathbf{a}(\theta_0))^{-1}\mathbf{R}^{-1}\mathbf{a}(\theta_0) , \tag{69.8}$$

where

$$\mathbf{R} = E\left\{\mathbf{z}(t)\mathbf{z}^H(t)\right\} \tag{69.9}$$

is the covariance matrix of the array data vector **z**. The vector **w** in Eq. (69.8) is the gain of the minimum variance (MV) beamformer [20]. The MV beamformer minimizes the total output power

$$E\left\{\left|\mathbf{w}^H\mathbf{z}(t))\right|^2\right\} = \mathbf{w}^H\mathbf{R}\mathbf{w}$$

subject to $\mathbf{w}^H\mathbf{a}^H(\theta_0) = 1$. The MV beamformer presents an important advantage over the MNP beamformer. While to implement the MNP beamformer we need to know the covariance matrix $\mathbf{R_n}$ of the disturbance vector **n**, in general, to implement the MV beamformer it is sufficient to estimate the array covariance matrix **R** using the available data **z**.

We discuss how to estimate **R**. Let T time samples (snapshots) of the array response vector $\mathbf{z}(t)$ be available. An estimate of **R** is the data sample covariance matrix $\mathbf{R}_s$:

$$\mathbf{R}_s = \frac{1}{T}\sum_{t=1}^{T}\mathbf{z}(t)\mathbf{z}^H(t) . \tag{69.10}$$

Under technical conditions that we will not discuss here, the sample covariance matrix, $\mathbf{R}_s$, converges (in the appropriate sense) to the array covariance matrix **R** when T approaches infinity. This means that, for a large enough number T of snapshots, we can replace **R** in Eq. (69.8) by $\mathbf{R}_s$, without a significant performance degradation.

We provide an alternative interpretation to the MNP beamformer. Using Eq. (69.7) in Eq. (69.5), and taking into account Eq. (69.4), we see that the output of the MNP beamformer has a signal component $s(t)$ and an error term $\mathbf{w}^H\mathbf{n}(t)$ with average power

$$P_o = \mathbf{w}^H E\left\{\mathbf{n}(t)\mathbf{n}^H(t)\right\}\mathbf{w} = \left(\mathbf{a}^H(\theta_0)\mathbf{R_n}^{-1}\mathbf{a}(\theta_0)\right)^{-1} . \tag{69.11}$$

Since the power of the signal is preserved and the MNP beamformer minimizes the power of the noise at its output, the MNP beamformer maximizes the output signal-to-noise ratio (SNR).

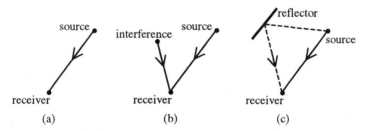

FIGURE 69.3 (a) Single source in white noise; (b) uncorrelated interference; (c) correlated interference.

We will not discuss in detail the behavior of the MNP and MV beamformers. The reader is referred to the work in [2]. We list some of the properties of the MNP and MV beamformers in two scenarios of practical interest.

Case 1: Single Source in White Noise
Here (see Fig. 69.3(a)), we assume that the noise $\mathbf{n}(t)$ is sensor noise. We model it as $\mathbf{n}(t) = \mathbf{u}(t)$ where the components of $\mathbf{u}(t)$ are jointly independent and identically distributed samples of zero mean white noise sequences with variance σ^2, i.e.,

$$\mathbf{R_n} = \mathbf{R_u} = \sigma^2\mathbf{I} ,$$

where $\mathbf{I}$ is the identity matrix. The sensor noise models the thermal noise generated at each receiver and is assumed independent of (thus, uncorrelated with) the source signal. If S is the power of the desired signal, the SNR at each sensor is $SNR_i = S/\sigma^2$. Also, from Eq. (69.7), we conclude that when the additive noise is white, as in this case, the MNP beamformer reduces to the DS beamformer.

Moreover, computing the power at the output of the beamformer with Eq. (69.11) for this particular situation yields $P_o = \sigma^2/N$. This means that, at the output of the beamformer, the signal-to-noise ratio is $SNR_o = N\,SNR_i$.

We conclude, for the case of a single source in white noise, that the DS beamformer is optimum in the sense of maximizing the output signal-to-noise ratio SNR_o. Further, SNR_o increases linearly with the number N of array sensors.

Case 2: Directional Interferences and White Noise

Now, we assume that the disturbance $\mathbf{n}(t)$ is the superposition of possibly several directional interferences and white noise. Without loss of generality, we consider the case of a single interferer:

$$\mathbf{n}(t) = \mathbf{a}(\theta_i)i(t) + \mathbf{u}(t) , \tag{69.12}$$

where $i(t)$ is the signal radiated by the interferer, θ_i is the DOA of the interference signal, and $\mathbf{u}(t)$ is the white noise vector. In general, we assume that $\mathbf{u}(t)$ is uncorrelated with $i(t)$.

Case 2.1: Uncorrelated Arrivals

This is the case where the desired signal and the interference are generated by distinct sources, see Fig. 69.3(b). It is clear that under this assumption, $s(t)$ and $\mathbf{n}(t)$ are uncorrelated. As we emphasized before, this is the situation where the MNP beamformer (69.7) is equivalent to the MV beamformer (69.8).

The covariance of the noise $\mathbf{n}(t)$ is now

$$\mathbf{R_n} = \mathbf{a}(\theta_i)S_i\mathbf{a}^H(\theta_i) + \sigma^2\mathbf{I} ,$$

where S_i is the average power of the interference $i(t)$. At the DOA=θ_i, the beamformer has an amplitude response

$$|\mathbf{w}^H\mathbf{a}(\theta_i)| = \frac{|\beta|}{1 + (1 - |\beta|^2)INR} , \tag{69.13}$$

where

$$INR = S_i/(\sigma^2/N)$$

is the interference-to-noise ratio (INR), and $\beta = (1/N)\mathbf{a}^H(\theta_i)\mathbf{a}(\theta_0)$ measures the spatial coherence between the desired source and the interference.

Well Separated Arrivals

When the signal and interference are well separated, their spatial coherence is small, i.e., $|\beta| \ll 1$. In Eq. (69.13), the denominator is approximately given by $1 + INR$. The net effect is that the beamformer output along the interference direction decreases when INR increases. In other words, the MNP and the MV beamformers direct a beam with gain 1 towards the DOA of the desired signal and *null* the interference. The interference canceling property is reflected on the average power of the beamformer output error which is evaluated to

$$P_o = \frac{\sigma^2}{N}\frac{1}{1 - |\beta|^2\frac{INR}{1+INR}} . \tag{69.14}$$

For large INR and well-separated DOAs, $P_o \simeq (\sigma^2/N)$. This means that the interference contributes little to the estimation error at the output of the beamformer.

Close Arrivals

When the source and the interferer are spatially close, their spatial coherence is large, $|\beta| \simeq 1$, and the output at the interference DOA is $\simeq 1$. This means that the MNP and MV beamformers no longer have the ability to discriminate the two sources.

We conclude from the simple analysis of these two cases that the DOA discrimination capability is strongly related to the *spatial resolution* of the array geometry through the parameter β. In practice, to improve upon the resolution of a linear and uniform array, we increase, when feasible, the number N of sensors. This results in narrower beamwidths which can resolve closer arrivals.

Case 2.2: Correlated Arrivals

This is the case where the interference $i(t)$ is correlated with the desired signal $s(t)$, i.e.,

$$E\{s(t)i^*(t)\} = \rho = |\rho|e^{j\phi_\rho} \neq 0 .$$

We denote complex conjugate by $(\cdot)^*$.

With reference to Fig. 69.3(c), we discuss a simple example where the interference results from a secondary path generated by a reflector (multipath propagation)

$$i(t) = \gamma s(t) . \tag{69.15}$$

The complex parameter, γ, accounts for the relative attenuation and delay of the reflected path. The correlation factor, ρ, between $i(t)$ and $s(t)$ is, in this case, given by

$$\rho = \gamma/|\gamma| .$$

The desired signal and the disturbance vector $\mathbf{n}(t)$ are now correlated and the MV beamformer is no longer equivalent to the MNP beamformer. Recall that the MV beamformer attempts to minimize the total output power under the constraint of a unitary gain at the DOA of the desired source. As the array output vector has a correlated signal component at a different DOA, to minimize the output power may cause the desired signal itself to be strongly attenuated. This is the *signal cancellation* effect, typical of MV beamforming when operating in multipath environments like the one just considered. On the contrary, the behavior of the MNP beamformer is independent of the correlation degree between the desired signal and the disturbance: the MNP beamformer filters out correlated arrivals just as if they were uncorrelated interferences.

To implement the MNP beamformer, besides the DOA of the desired signal, we also need to know the covariance matrix $\mathbf{R_n}$ of the disturbance vector. In general, this covariance is not known *a priori*. It has to be estimated using the available data, and this can be a rather complicated task, not discussed here.

In this section, we discussed the MNP solution to the beamforming problem. The MNP beamformer is optimum in the sense of maximizing the output SNR. When the noise is white, the DS beamformer is recovered as the optimum solution for the single source case. It points a beam towards the source DOA and reduces the sensor noise power by a factor of N, see Fig. 69.4(a). We also saw that, in the more general situation where the disturbance vector includes directional interferences, the MNP beamformer acts like an interference canceler: it points a beam towards the DOA of the desired signal, while nulling the remaining arrivals regardless of their correlation degrees, see Figs. 69.4(b) and (c).

We comment on the adequacy of the MNP solution to the correlated arrivals scenario. By treating the correlated arrivals as an interference, the MNP beamformer neglects the information about the desired signal that may be provided by the reflected path. It is clear that any solution that can combine coherently the information contents of all the correlated arrivals will be more effective in recovering the desired signal from the background noise. This type of solution should behave as a combiner of the outputs of different beams steered towards the DOAs of the correlated replicas of the desired signal, see Fig. 69.4(d). In the following section, we will see how this solution can be designed using a different optimization criterion when solving the beamforming problem.

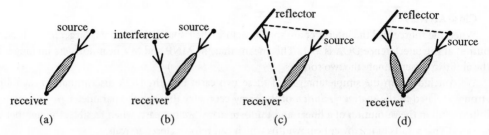

FIGURE 69.4   Artistic representation of alternative solutions to the beamforming problem.

69.3   MMSE Beamformer: Correlated Arrivals

In this section, we study a different beamforming technique that, as we will see, is specially suited to multipath propagation environments. The approach is similar to that used in the previous section. The beamformer is still given by Eq. (69.5), but we choose a different optimization criterion to solve the beamforming problem. We formulate the problem in the following way:

> find the weight vector $\mathbf{w}$ such that the output error power
> $E\left\{\left|\mathbf{w}^H \mathbf{z}(t) - s(t)\right|^2\right\}$ is minimized,

i.e., we want to find the weight vector $\mathbf{w}$ that minimizes the mean square error between the beamformer output $y(t) = \mathbf{w}^H \mathbf{z}(t)$ and the desired signal $s(t)$. The general solution, which we call the Wiener solution, is [2]

$$\mathbf{w} = \mathbf{R}^{-1} \mathbf{r}_{zs}, \tag{69.16}$$

where $\mathbf{r}_{zs} = E\{\mathbf{z}(t)s^*(t)\}$ is the correlation between the array vector response $\mathbf{z}(t)$ and the desired signal $s(t)$. To understand the behavior of the MMSE beamformer (69.16), we address the same alternative configurations considered in Case 2 of the previous section.

Directional Interference and White Noise

In this scenario, the disturbance $\mathbf{n}(t)$ is like in Eq. (69.12), so the received array signal is

$$\mathbf{z}(t) = \mathbf{a}(\theta_0)s(t) + \mathbf{a}(\theta_i)i(t) + \mathbf{u}(t),$$

where, as before, $i(t)$ and $\mathbf{u}(t)$ are the interference and the array sensor noise vector, respectively.

Uncorrelated Arrivals

The signal $s(t)$ and the interference $i(t)$ are uncorrelated. The correlation between the array vector and the desired signal is

$$\mathbf{r}_{zs} = \mathbf{a}(\theta_0)S,$$

where S is the average power of $s(t)$. The MMSE beamformer weight vector in Eq. (69.16) takes the particular form

$$\mathbf{w} = S\mathbf{R}^{-1}\mathbf{a}(\theta_0). \tag{69.17}$$

Comparing (69.17) with (69.8), we conclude that in the present situation, except for a scale factor, the MMSE and the MNP (or MV) beamformers are equivalent. Thus, the MMSE beamformer cancels uncorrelated interferences and directs a beam towards the DOA of the desired signal. However, contrary to what happens with the MNP (or MV) beamformer, the gain of the MMSE beamformer at the look direction is not unity. On the other hand, the MMSE beamformer provides a stronger noise rejection and a smaller output error power than the MNP beamformer. In fact, it can be shown [2] that

$$P_o(\text{MMSE}) \leq \frac{\text{SNR}_o}{1 + \text{SNR}_o} P_o(\text{MNP}), \tag{69.18}$$

where $\mathrm{SNR}_o = S/(\sigma^2/N)$, and $P_o(\mathrm{MNP})$ is the average power of the MNP beamformer output error given by (69.14). Equation (69.18) is particularly significant in low SNR environments.

Correlated Arrivals

We take again the interference to be like in Eq. (69.15), i.e., $i(t) = \gamma s(t)$. It leads to the correlation

$$\mathbf{r}_{zs} = (\mathbf{a}(\theta_0) + \gamma \mathbf{a}(\theta_i))S$$

and to the weight vector

$$\mathbf{w} = S\mathbf{R}^{-1}\mathbf{a}(\theta_0) + S\gamma\mathbf{R}^{-1}\mathbf{a}(\theta_i) \ . \tag{69.19}$$

It is clear from Eq. (69.19) that the MMSE beamformer directs distinct beams towards the DOAs θ_0 and θ_i of the correlated arrivals in order to combine coherently their respective outputs. If $\mathbf{R}$ has contributions of other sources uncorrelated with $s(t)$, then these will be filtered out by both beams. This simple example shows how the MMSE beamformer uses the correlated arrivals to improve the output error power.

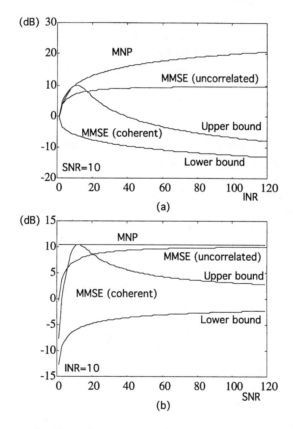

FIGURE 69.5 Output error power. (a) As a function of INR; (b) as a function of SNR. (*Source:* Barroso, V.A.N. and Moura, J.M.F., l_2 and l_1 Beamformers: recursive implementation and performance analysis, *IEEE Trans. on Signal Processing*, SP-42(6), 1323–1334, June 1994.© 1994 IEEE. Used with permission).

To have an idea of how the behavior of the MMSE beamformer compares with that of the MNP beamformer, we represent in Fig. 69.5 the output error power as a function of the INR and the SNR, for the case of spatially close arrivals. We see from Fig. 69.5 that the MMSE beamformer outperforms the

MNP beamformer in all scenarios considered. This is particularly apparent in the limiting case of coherent arrivals, when the correlation between the signal and the interference is such that $|\rho| = 1$. The upper and lower bounds of the output error power shown in the figure are determined by $-1 \leq \cos(\phi_\rho) \leq 1$.

To implement the MMSE beamformer in a multipath propagation environment, we need to know the correlation vector $\mathbf{r}_{zs}$ or, alternatively, the DOAs of all correlated replicas of the desired signal as well as their relative attenuations and propagation delays. In practice, $\mathbf{r}_{zs}$ can only be estimated from the data if we have available a reference signal whose correlation with the array vector is similar to that of the desired signal. This is the basic idea underlying adaptive implementations (least mean squares approach); see, e.g., [8, 9]. In these implementations, a time sequence, which is known by the receiver, is transmitted by the source and used to adapt the beamforming weights until some error threshold is achieved. In many applications, a reference signal is not available or, as in communications, the transmission of a reference signal may represent a significant waste of the channel capacity. In the next section, we explain how to implement the MMSE beamformer without the need of a reference signal. We do this in the context of a wireless digital communication system.

69.4   MMSE Beamformer for Mobile Communications

In this section, we discuss the implementation of the MMSE beamformer in the context of a digital wireless communication system. The users are mobiles transmitting simultaneously in the same frequency channel. This precludes the use of time/frequency methods to discriminate between the source signals. Here, we assume that the multipath propagation delays are smaller than the baud period (symbol time interval). We develop a multiple output or multichannel MMSE beamformer for this application.

Each mobile is assigned to a specific beamforming processing channel that has the following capabilities:

- Combine the multipath arrivals generated by its assigned mobile.
- Cancel all the arrivals generated by other mobiles.

We illustrate the development of the array receiver in the simple context of Fig. 69.6. We consider two mobiles, M_1 and M_2, each one generating a direct path (dm, $m = 1, 2$) and a reflected path (rm, $m = 1, 2$). The propagation channel used by each mobile is characterized by four parameters: two complex numbers,

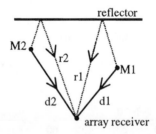

FIGURE 69.6   Example of a mobile communications scenario.

γ_{dm} and γ_{rm}, accounting for the attenuation and the propagation delay in each path, and two DOAs θ_{dm} and θ_{rm}. We model all the parameters that characterize the propagation channels as deterministic unknown parameters. Furthermore, and in contrast with what we assumed in the previous sections, the *DOAs of the desired signals are not known a priori*. The receiver itself has to track the sources while they are moving. In a mobile communications system, we can safely assume that we know the number of mobiles, their transmitted power, and the sensor noise variance. The number of mobiles is determined by some higher layer of the receiver architecture (e.g., handover procedures to establish the connection between the receiver and each mobile), while the power and the noise variance are system parameters specified by the *link budget*.

69.4.1 Model of the Array Output

Let $\mathbf{h}_m, m = 1, 2$, be the single input/multiple output complex transfer function describing, for each source, the channel/receiving array system. Each $\mathbf{h}_m$ is $N \times 1$, where N is the number of array sensors. We write

$$\mathbf{h}_1 = \gamma_{d1}\mathbf{a}(\theta_{d1}) + \gamma_{r1}\mathbf{a}(\theta_{r1})$$

$$\mathbf{h}_2 = \gamma_{d2}\mathbf{a}(\theta_{d2}) + \gamma_{r2}\mathbf{a}(\theta_{r2}) \,,$$

where $\mathbf{a}(\cdot)$ represents the array steering vector. Let $s_1(t)$ and $s_2(t)$ be the baseband signals transmitted by the mobiles. They are assumed to be independent binary sequences taking the values $+1$ and -1 with equal probability. Thus, the signal vector $\mathbf{s}(t) = [s_1(t), s_2(t)]^T$ has zero mean and covariance matrix $\mathbf{S} = \mathbf{I}$. Defining the $(N \times 2)$ matrix $\mathbf{H} = [\mathbf{h}_1, \mathbf{h}_2]$, the array output vector is

$$\mathbf{z}(t) = \mathbf{H}\mathbf{s}(t) + \mathbf{u}(t) \,, \tag{69.20}$$

where $\mathbf{u}(t)$ represents the complex sensor noise vector. The noise $\mathbf{u}(t)$ will be assumed to be zero mean Gaussian with known covariance matrix $\sigma^2\mathbf{I}$. The goal is to estimate the signal vector $\mathbf{s}(t)$ using a MMSE beamformer which, as we saw in Section 69.3, is the appropriate approach for multipath propagation environments. However, to implement the MMSE beamformer (69.16), we need to know the correlation between the array output $\mathbf{z}(t)$ and the signal vector $\mathbf{s}(t)$ which in this case is

$$\mathbf{r}_{zs} = \mathrm{E}\{\mathbf{z}(t)\mathbf{s}^H(t)\} = \mathbf{H} \,.$$

The matrix $\mathbf{H}$ is parameterized by all the unknown θ and γ parameters of our model. Thus, it has to be estimated from the available data. Once we have the estimate of $\mathbf{H}$, we can use it to compute a structured estimate of the array covariance matrix. In this case, where $\mathbf{S} = \mathbf{I}$, this estimate is

$$\mathbf{R} = \mathbf{H}\mathbf{H}^H + \sigma^2\mathbf{I} \,. \tag{69.21}$$

This estimate is better than the sample covariance matrix computed by Eq. (69.10) because it avoids eventual mismatches between the estimate of $\mathbf{H}$ and its actual value embedded in $\mathbf{R}_s$. In the following section, we will describe a method that provides accurate estimates of $\mathbf{H}$.

69.4.2 Maximum Likelihood Estimation of H

Suppose that T is the number of snapshots, i.e., samples, of the array output that are available for estimating the matrix $\mathbf{H}$, while tracking the mobiles. The choice of T should strike a balance between the two following conflicting requirements:

- T should be large enough to guarantee in the appropriate sense a good estimate of $\mathbf{H}$.
- T should be small enough so that we can safely model $\mathbf{H}$ as time invariant.

Fortunately, these two conditions are easily met in the application under study. High data rates usually will be involved in mobile communications. For example, data rates of 2 Mbps or higher are envisaged for the UMTS (Universal Mobile Telecommunications System) and the MBS (Mobile Broadband System), systems presently under development in Europe. These high data rates, the typical velocities of the mobiles (less than 100 Km/h), and the typical mobile/receiver distances ensure that a large amount of data can be collected over a time period during which $\mathbf{H}$ can be considered to be approximately constant. In other words, the geometry is very slowly varying when compared to the transmission data rates.

Under the assumptions used to establish the model of the array output vector, the *likelihood function* is

$$L(\mathbf{H}) = \prod_{t=1}^{T} \sum_{i=1}^{4} \text{Gauss}_i(\mathbf{H}, \mathbf{z}(t)) , \qquad (69.22)$$

where the unnormalized Gauss function is

$$\text{Gauss}_i(\mathbf{H}, \mathbf{z}(t)) = \exp\left(-\frac{|\mathbf{z}(t) - \mathbf{H}\mathbf{s}_i|^2}{\sigma^2}\right) ,$$

and $\{\mathbf{s}_i\}_{i=1}^{4}$ spans the finite alphabet of the sources, i.e., the four possible realizations of $\mathbf{s}(t)$. In Eq. (69.22), we assume that the sampling of the received signals is synchronized with the symbol clock. This is a very strong requirement as time synchronization may be a very difficult problem, specially when the sampling rate equals the data rate. However, the sensitivity of the receiver with respect to timing errors is, in general, efficiently reduced when the sampling rate is made larger than the symbol rate. Although we do not discuss this issue here, the algorithm that we present is easily extended to this case. The value of $\mathbf{H}$ that maximizes the likelihood function $L(\mathbf{H})$ makes the collected data more likely from a probabilistic point of view. The maximization of Eq. (69.22) with respect to $\mathbf{H}$ is not an easy task. Equation (69.22) is strongly non-linear in $\mathbf{H}$ which precludes the analytical solution of the optimization problem. On the other hand, the parameterization of $\mathbf{H}$ in terms of the channel parameters does not help much because optimization over these parameters requires a multidimensional search approach, which may be excessively time consuming.

The EM Algorithm

To maximize Eq. (69.22) with respect to $\mathbf{H}$, we use an algorithm based on the EM (Expectation-Maximization) approach [17]. This approach yields an iterative algorithm that, under some technical conditions that are beyond the scope of this paper, is known to converge to the true maximum likelihood estimate. The EM iteration is

$$\mathbf{H}_{l+1} = \left(\sum_{t=1}^{T} \mathbf{z}(t) \frac{\sum_{i=1}^{4} \text{Gauss}_i(\mathbf{H}_l, \mathbf{z}(t))\mathbf{s}_i^H}{\sum_{i=1}^{4} \text{Gauss}_i(\mathbf{H}_l, \mathbf{z}(t))}\right)$$

$$\left(\sum_{t=1}^{T} \frac{\sum_{i=1}^{4} \text{Gauss}_i(\mathbf{H}_l, \mathbf{z}(t))\mathbf{s}_i \mathbf{s}_i^H}{\sum_{i=1}^{4} \text{Gauss}_i(\mathbf{H}_l, \mathbf{z}(t))}\right)^{-1} . \qquad (69.23)$$

To run this iterative algorithm, we must be aware of two important aspects:

- In general, this algorithm has a very slow convergence rate.
- Due to strong non-linearities, it can easily get trapped in a local maximum.

These two aspects reflect a high sensitivity of the EM algorithm with respect to the initial estimate. Special care has to be taken to specify the initial estimate $\mathbf{H}_0$. Next, we describe how to obtain this initial condition, such that the EM iteration converges in just a few iterations to the global maximum of the likelihood function.

Initialization of the EM Algorithm

In this paragraph, we do not go into the detailed derivation of the initialization procedure. The interested reader is referred to [4]. Here, we describe the basic idea underlying the initialization procedure and present the steps necessary to implement it.

Recall the structure of $\mathbf{H}$ defined in Section 69.4.1. For the simple example that we have been considering, we write

$$\mathbf{H} = \mathbf{A}(\theta)\Gamma , \qquad (69.24)$$

where

$$A(\theta) = [\mathbf{a}(\theta_{d1}), \mathbf{a}(\theta_{r1}), \mathbf{a}(\theta_{d2}), \mathbf{a}(\theta_{r2})] \tag{69.25}$$

is the ($N \times 4$) matrix of the steering vectors associated with each incoming path, θ being the vector of the unknown DOAs. The remaining unknown parameters of our model are collected in

$$\Gamma = \begin{bmatrix} \gamma_{d1} & 0 \\ \gamma_{r1} & 0 \\ 0 & \gamma_{d2} \\ 0 & \gamma_{r2} \end{bmatrix}. \tag{69.26}$$

The problem of estimating the initial condition $\mathbf{H_0}$ is decomposed in two steps: (1) estimation of θ, and (2) estimation of Γ.

Step (1): Estimation of θ

DOA estimation is a well-studied problem in array processing. The classical approach is based on the MV beamformer [6, 12]. The main idea consists in detecting the maxima of the output power, when a set of quantized values of the possible angles of arrival (in the range $[-\pi/2, \pi/2]$ in the case of a linear array) is scanned by varying the look direction specified by the complex weights of the MV beamformer. One of the problems with this approach is that angular resolution is mainly determined by the beamwidth. A narrow beam requires a large number of array sensors. To overcome this limitation, we might use the MUSIC algorithm, a well-known high resolution technique [5, 12, 13] based on the eigenvalue decomposition of the sample covariance matrix. The angular resolution of the MUSIC algorithm improves as the sample covariance matrix approaches the array covariance matrix. However, these two techniques (MV beamformer and MUSIC) fail when the arrivals are correlated as is the case in mobile communications. The spatial smoothing method [12, 14] extends MUSIC to correlated arrivals. The goal of spatial smoothing is to decorrelate the arrivals. It breaks the original array into subarrays. The problem is that the performance of spatial smoothing depends strongly on the correlation degree of the arrivals. To improve the efficiency of spatial smoothing, we have to use a large number of spatial samples, i.e., a larger array.

To circumvent all these difficulties, keeping the array size manageable, and achieving acceptable performance, we use a different approach which is insensitive to the correlation degree of the arrivals. This technique is similar to MUSIC but, rather than relying on the sample covariance matrix (second order statistics), it is based on first order statistics (statistical average) of the array data, see [12] for a detailed derivation of the algorithm. We show how the method is used in the mobile communications problem.

Looking at Eq. (69.20), and recalling the assumptions on $s(t)$ and $\mathbf{u}(t)$, we conclude that $E\{\mathbf{z}(t)\} = \mathbf{0}$. This seems to contradict our goal of using first order statistics of the array data to estimate the DOAs. However, a more careful analysis of the structure of our model shows that the array samples can be partitioned into a number of sets equal to the cardinality of the source's alphabet (4, in the example that is being used). Each of the sets, $\mathbf{Z}_i$, in the partition has samples of the form

$$\mathbf{z}_i(t) = \mathbf{Hs}_i + \mathbf{u}(t), \quad i = 1, \dots, 4. \tag{69.27}$$

The statistical average of these samples is

$$\mathbf{x}_i = \mathbf{Hs}_i, \quad i = 1, \dots, 4. \tag{69.28}$$

This means that the array data is organized in clusters, $\mathbf{Z}_i$, of points centered about each $\mathbf{x}_i$.

To estimate the DOAs, we need to use only one of the $\mathbf{x}_i$'s. Therefore, we are left with finding a scheme to determine one of the sets in the partition of the array data. In a statistical sense, the distance between any two samples of the array data is minimum if the two samples belong to the same partition. In other words, for any pair of time instants, t_l and t_k, $\mathbf{z}(t_l)$ and $\mathbf{z}(t_k)$ belong to the same partition $\mathbf{Z}_i$ if the source signal vectors $\mathbf{s}(t_l)$ and $\mathbf{s}(t_k)$ have the same realization $\mathbf{s}_i$. In this case,

$$\Delta = E\left\{|\mathbf{z}(t_l) - \mathbf{z}(t_k)|^2\right\} = 2N\sigma^2. \tag{69.29}$$

To find one set Z_i of the partition, after choosing randomly one array sample $z(t_l)$, we look for all other samples whose distance to $z(t_l)$ is below some threshold appropriately related with Δ. Once we obtain the set Z_i in the partition, we estimate the statistical average x_i of the points in the set by its sample mean.

At the end of this step, we have calculated the first order statistics and are ready to apply the first order statistics algorithm to estimate the DOAs θ [4]. Having the estimate $\hat{\theta}$ of the DOAs, we estimate the matrix of the steering vectors A in Eq. (69.25) by $\widehat{A} = A(\hat{\theta})$.

Step (2): Estimation of Γ

Recall from Eq. (69.24) that $H = A(\theta)\Gamma$. Step (1) estimated A as $\widehat{A} = A(\hat{\theta})$. We now proceed with the estimation of Γ. To estimate Γ, we use the estimate $\widehat{A} = A(\hat{\theta})$ just obtained, as well as the cluster center x_i on which that estimate was based.

Denote the left pseudoinverse of a matrix A by $A^\dagger$. It is defined by

$$A^\dagger = (A^H A)^{-1} A^H .$$

Recall that the matrix of steering vectors A defined in Eq. (69.25) is a rank 4 ($N \times 4$) matrix.

We now motivate how to estimate Γ. Define the matrix

$$Q = \widehat{A}^\dagger (R - \sigma^2 I) \widehat{A}^{H\dagger} . \tag{69.30}$$

From Eq. (69.21), we get

$$R - \sigma^2 I = HH^H .$$

Substituting this equation in (69.30), we get

$$Q = \widehat{A}^\dagger HH^H \widehat{A}^{H\dagger} .$$

But by (69.24)

$$Q = \widehat{A}^\dagger A\Gamma\Gamma^H A^H \widehat{A}^{H\dagger} . \tag{69.31}$$

Clearly, if $\widehat{A} = A$, then

$$Q = \Gamma\Gamma^H . \tag{69.32}$$

Apparently, to estimate Γ, we should compute Q and then find a factorization of the form (69.32). This is not so easy because, in general, a unique factorization of a Hermitian matrix like Q does not exist. Nevertheless, Q plays a key role in the procedure for estimating Γ. This is essentially based on the singular value decomposition of Q, as we see now. Notice that Q can be computed from the data using the data sample covariance matrix R_s given by Eq. (69.10) instead of the array data covariance matrix R.

Recalling from Eq. (69.26) the structure of Γ, we conclude that Q should be a (4×4) Hermitian matrix of rank 2. This means that Q should only have two nonzero eigenvalues (in fact, because Q is Hermitian, these eigenvalues of Q are real and positive). The singular value factorization of Q is

$$Q = V\Lambda V^H , \tag{69.33}$$

where Λ is a diagonal matrix whose elements are the nonzero singular values, in this case the eigenvalues of Q, and V is a matrix whose columns are the orthonormal eigenvectors associated to these eigenvalues.

From Eq. (69.26) defining Γ, we see that Γ has itself orthogonal columns. In general, the columns of Γ have different norms. Using the eigenvalue decomposition in Eq. (69.33), we conclude that, except for a phase difference, the columns of $V\Lambda^{1/2}$ equal the columns of Γ. Thus, we can conclude that

$$
\begin{aligned}
Q^{1/2} &= V\Lambda^{1/2} & (69.34) \\
&= \Gamma \text{Diag}(e^{j\phi_m}) & (69.35)
\end{aligned}
$$

where $\mathbf{Diag}(\cdot)$ is a diagonal matrix and the ϕ_m, $m = 1, 2$, measure the phase uncertainty of the columns of $\mathbf{Q}^{1/2}$ with respect to the columns of Γ. The last equality in Eq. (69.35) is true only when $\widehat{\mathbf{A}} = \mathbf{A}$.

We now consider how to resolve this phase uncertainty. We show that this uncertainty can be resolved up to a sign difference. This means that, at least theoretically, we can achieve an estimate $\widehat{\Gamma}$ whose columns are colinear with those of Γ. Using Eq. (69.34) and Eq. (69.28), it can be shown after some algebra that

$$\mathbf{f} = \mathbf{Q}^{1/2^\dagger} \widehat{\mathbf{A}}^\dagger \mathbf{x}_i \left(= \mathbf{Diag}(e^{-j\phi_m})\mathbf{s}_i \right) . \tag{69.36}$$

The elements of the vector $\mathbf{f}$ have the form $e^{-j\phi_m} s_{im}$, $m = 1, 2$. Defining $\mathbf{F}$ as the diagonal matrix formed with the elements of $\mathbf{f}$, we define the estimate $\widehat{\Gamma}$ of Γ as

$$\widehat{\Gamma} = \mathbf{Q}^{1/2}\mathbf{F} \left(= \Gamma \mathbf{Diag}(s_i) \right) . \tag{69.37}$$

Recalling that the elements of s_i are $+1$ or -1, we see that, ideally, the columns of Γ are estimated just with a sign uncertainty. In our application, this sign uncertainty is not relevant. To see this, suppose that one of the columns of $\widehat{\Gamma}$ has its sign changed. This sign change is equivalent to an inversion of the data stream generated by one of the mobiles. This inversion of the data stream can be solved using a differential encoding scheme of the transmitted data.

In summary, the algorithm to estimate Γ is as follows:

- use $\mathbf{R}_s$ to compute $\mathbf{Q}$ as defined by Eq. (69.30);
- compute the singular value decomposition of $\mathbf{Q}$ and then form the matrix $\mathbf{Q}^{1/2}$ as given by Eq. (69.34);
- compute $\mathbf{f}$ as given by Eq. (69.36) and use its elements to form the diagonal matrix $\mathbf{F}$;
- compute $\widehat{\Gamma}$ defined in Eq. (69.37).

Using the results of Steps (**1**) and (**2**), we compute the estimate used to initialize the EM iteration as:

$$\mathbf{H}_0 = \mathbf{A}(\widehat{\theta})\widehat{\Gamma} .$$

In practice, it will be shown in the following section by computer simulations that this initial condition enables the EM algorithm to converge to the global maximum of the likelihood function with a small number of iterations. This behavior is maintained even with critical scenarios arising under very specific geometries, see Section 69.5.

The apparent drawback of the proposed initialization procedure is its computational complexity. However, this is not really a problem because we are processing blocks of data with time length T during which the geometry remains, for all practical purposes, unchanged. Also, the changes in geometry between adjacent blocks are small. This means that, except in specific situations, the EM iteration can be initialized using the estimate of $\mathbf{H}$ obtained with the previous block of data. Hence, the algorithm is seldom reinitialized, e.g., when the number of sources present in the cell is changed.

69.5 Experiments

In this section, we present the results obtained with the array receiver designed in the previous section. The context is a cell in a cellular radio mobile communications system. The system is operating at a carrier frequency $f_0 = 1$ GHz and accommodates transmission rates of 1 Mbps with differential encoding binary phase shift keying modulation [7]. All the mobiles present in the cell generate signals that are transmitted simultaneously in the same frequency band. The power emitted by each mobile is assumed to be unity.

The receiving array is a horizontal linear, uniform array, with $N = 19$ identical omnidirectional captors. These captors are separated by half wavelength, i.e., $d = \lambda_0/2$, where $\lambda_0 = c/f_0 = 30$ cm is the center frequency wavelength, yielding an array length of 2.7 m. The geometry of the scenario is similar to that depicted in Fig. 69.6, where the reflector is parallel to the array. The reflector is located at a distance of 50 m from the array. The reflector absorbs half of the impinging power. The two mobiles M_1 and M_2 describe circular trajectories around the receiver, from right to left (counterclockwise). Their velocities are 18 Km/h for M_1 and 19 Km/h for M_2. These are typical speeds for city traffic with slow automobiles circling a rotary. The radii of these trajectories are 30 m and 45 m, respectively. Comparing these distances with the array length, we conclude that we can safely make the planar wavefront assumption; in other words, the received signals across the array are planar wavefronts. Finally, the receiver SNR equals 10 dB for the case of the direct path generated by M_1. This means that all the remaining paths have smaller SNRs.

In Fig. 69.7, we illustrate examples of the geometries for which we ran the multichannel MMSE beam-former. The data synthesized by the simulator was collected during each second. The data blocks have length 1 Mbit per mobile. The matrix **H** is estimated using 3 Kbit (per mobile). In practically all scenarios

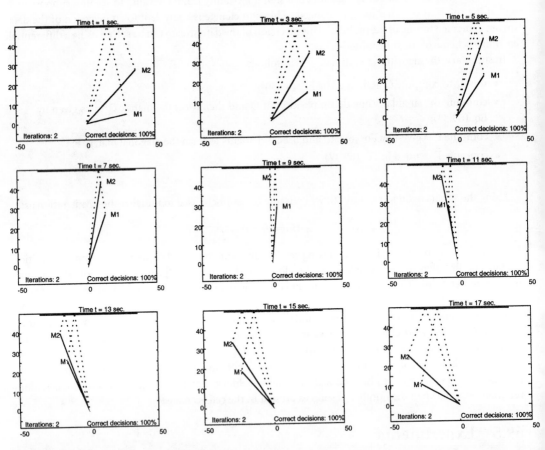

FIGURE 69.7   Simulation scenarios.

that we tested, we observed that the EM algorithm converges in just two iterations. The multichannel MMSE beamforming achieved 100% of correct symbol detections for the outputs of both channels. The initialization procedure described in Section 69.4 was necessary only for the first data block (beginning at time $t = 0$). Except for the first data block, the EM algorithm was initialized with the estimate of **H** obtained with the last adjacent data block. Notice that even for scenarios $t = 11$ and $t = 13$, where the DOAs of the direct paths are very close to each other, the EM algorithm was able to keep tracking of the columns of **H**. In fact, each column of **H**, representing the channel/receiving array transfer function, captures the global structure of the multipath propagation channel used by each mobile, which does not depend explicitly on the DOAs. For illustrative purposes, we show in Fig. 69.8 the amplitude response of

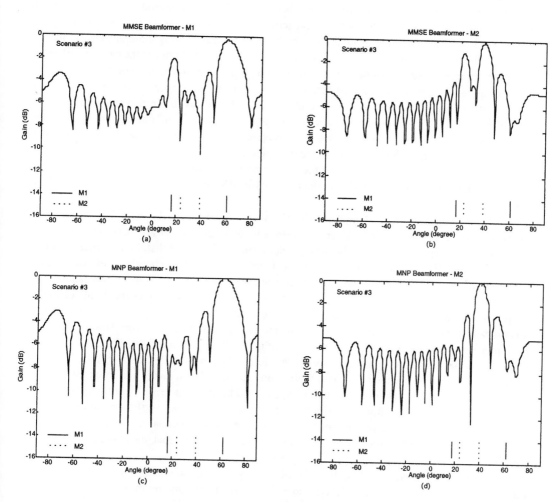

FIGURE 69.8   MMSE and MNP beampatterns: (a), (c)-channel M_1; (b), (d)-channel M_2.

the MMSE and the MNP beamformers as a function of the angle of arrival (beampatterns). The MMSE beampatterns were computed using the weights obtained from the synthesized data, while in the case of the MNP we used the ideal weights. We see that the behavior of both beamformers confirm what was predicted by the simple analysis carried out in Sections 69.2 and 69.3. The MMSE beamformer directs distinct beams towards the arrivals generated by the mobile being tracked and nulls all the remaining arrivals. On the contrary, the MNP beamformer cancels all the incoming wavefronts except the one corresponding to the direct path. The simulations also confirmed the improvement in the output SNR achieved by the MMSE beamformer when compared with the MNP beamformer. In our experiments, a gain close to 3 dB was obtained in almost all scenarios considered.

In this section, we showed the adequacy of the multichannel MMSE beamformer for combating multi-path propagation effects as occurring in mobile communications systems. The simulations illustrated the efficiency of the implementation proposed in Section 69.4.

69.6 Conclusions

In this article, we addressed the problem of designing beamforming arrays specially suited to operate in multipath environments. The problem is motivated in the context of cellular mobile communication systems. This is a field where array processing is being recognized as a powerful technology that has the potential to improve significantly the traffic capacity of such systems. As we saw, beamforming is an efficient technique for handling several mobiles in the same cell transmitting simultaneously in the same frequency channel. Moreover, specific beamformers can be designed to optimize the receiver with respect to distortions introduced by channels, such as multipath propagation and interferences.

We proposed a solution based on the MMSE beamformer. The study presented in Section 69.3 emphasized the relevant properties of this method. In particular, it was shown that the minimum mean square error (MMSE) beamformer has an important advantage over the minimum output noise power beamformer (MNP) introduced in Section 69.2: it combines coherently the correlated signal replicas generated by one source, while canceling the remaining interferences. Recall that the MNP beamformer nulls the secondary multipath arrivals as if they were uncorrelated interferences. Thus, the output SNR of the MMSE beamformer is always better than that of the MNP beamformer, particularly in multipath environments.

We identified the difficulties of implementing the MMSE beamformer. In particular, we pointed out that, except for the case of uncorrelated arrivals, knowledge of the direction of arrival (DOA) of the desired signal is not sufficient to implement the MMSE beamformer. We showed that, in most cases of interest, we need to know the cross-correlation between the desired signal and the array observations. In general, this cross-correlation is not previously known. It is necessary to estimate it from the available data. In other words, it is necessary to perform a blind estimation of the propagation channel. In Section 69.4 we presented a new method for estimating that correlation. It can be applied in situations where the array data have non-zero first order statistics. The method is insensitive to the correlation degree of the incoming wavefronts. This is the case that is relevant in mobile communications. Basically, the method that we described exploits the finite mixture distribution of the array data, which is organized in clusters with non-zero averages. We used a maximum likelihood approach based on the finite mixture model, the global maximum of the likelihood function being obtained by an EM iterative algorithm. As discussed, initialization of the EM algorithm is a critical issue. For a poor initial guess, the algorithm can diverge or, at most, converge to a local maximum. We provided a technique to initialize the EM algorithm. This initialization method enables the EM algorithm to converge fast to the global maximum of the likelihood function.

We tested the multichannel MMSE beamforming receiver using a simulator for a mobile communications system. We generated a simple scenario with moving sources and multiple replicas. Our results illustrated the efficiency of the receiver. In particular, the results confirm that, under the conditions studied which are typical in cellular communications, the reliability of the system is high. This is a consequence of the way we designed the MMSE beamformer which minimizes the power of the error at the output of the receiver. The MMSE beamformer relies on the accuracy of the blind estimation of the radio propagation channel. The algorithm we described can do a good job of estimating the propagating channel as long as the receiver has the ability to properly track the mobiles. In normal operation, the EM algorithm is initialized using the channel estimates obtained with the previous data block. The receiver is globally reinitialized only in very specific situations, for instance when the number of mobiles is changed. This fact simplifies the computational effort associated with the receiver. Our studies showed that the receiver converges so fast that it operates practically in real time.

Acknowledgments

The authors acknowledge discussions with João Xavier and his help with the development of the simulations software.

References

[1] Barroso, V.A.N., Moura, J.M.F. and Xavier, J., Blind array channel division multiple access (AChDMA) for mobile communications, accepted for publication, subject to minor revision, *IEEE Trans. on Signal Processing,* submitted Nov. 1995.

[2] Barroso, V.A.N. and Moura, J.M.F., l_2 and l_1 Beamformers: recursive implementation and performance analysis, *IEEE Trans. on Signal Processing,* SP-42(6), 1323–1334, June 1994.

[3] Barroso, V.A.N., Rendas, M.J. and Gomes, J., Impact of array processing techniques on the design of mobile communication systems, *Proc. MELECON'94,* Turkey, April 1994.

[4] Barroso, V.A.N. and Xavier, J., Blind estimation of multipath channels for the design of array receivers in mobile communications, *Instituto de Sistemas e Robótica, Internal Report,* 1995.

[5] Bienvenu, G. and Kopp, L., Adaptivity to background noise spatial coherence for high resolution passive methods, *Proc. IEEE ICASSP'80,* 307–310, 1980.

[6] Capon, J., High resolution frequency-wavenumber spectrum analysis, *Proc. IEEE,* 57, 1408-1418, Aug. 1969.

[7] Carlson, A.B., *Communication Systems: An Introduction to Signals and Noise in Electrical Communication,* 3rd ed., McGraw-Hill, New York, 1986.

[8] Compton, R.T., Jr., *Adaptive Antennas: Concepts and Performance,* Prentice-Hall, Englewood Cliffs, NJ, 1988.

[9] Haykin, S., *Adaptive Filter Theory,* 2nd ed., Prentice-Hall, Englewood Cliffs, NJ, 1991.

[10] Lee, W.C.Y., *Mobile Cellular Telecommunications Systems,* McGraw-Hill, New York, 1989.

[11] Paulraj, A. and Papadias, C.B., Array processing for mobile communications, Chapter 68 in this book.

[12] Pillai, S.U., *Array Signal Processing,* Springer-Verlag, 1989.

[13] Schmidt, R.O., Multiple emitter location and signal parameter estimation, *Proc. RADC Spectral Estimation Workshop,* 243–258, Oct. 1979.

[14] Shan, T.J., Wax, M. and Kailath, T., On spatial smoothing for estimation of coherent signals, *IEEE Trans. on Acoust., Speech, and Signal Processing,* ASSP-33, 806–811, Aug. 1985.

[15] Steele, R., *Mobile Radio Communications,* IEEE Press-Pentech Press, 1992.

[16] Swales, S., Beach, M., Edwards, D. and McGreehan, J., The performance of enhancement of multibeam adaptive base station antennas for cellular mobile radio, *IEEE Trans. on Vehicular Technology,* 31, Feb. 1990.

[17] Titterington, D.M., Smith, A.F.M. and Makov, U.E., *Statistical Analysis of Finite Mixture Distributions,* John Wiley & Sons, New York, 1985.

[18] Walker, J., *Mobile Information Systems,* Artech House, 1990.

[19] Van Veen, B.D. and Buckley, K.M., Beamforming: A versatile approach to spatial filtering, *IEEE ASSP Magazine,* 4–24, April 1988.

[20] Van Veen, B. and Buckley, K.M., Beamforming techniques for spatial filtering, Chapter 61 in this book.

70

Space-Time Adaptive Processing for Airborne Surveillance Radar

Hong Wang
Syracuse University

Space-Time Adaptive Processing (STAP) is a multi-dimensional filtering technique developed for minimizing the effects of various kinds of interference on target detection with a pulsed airborne surveillance radar. The most common dimensions, or filtering domains, generally include the azimuth angle, elevation angle, polarization angle, doppler frequency, etc. in which the relatively weak target signal to be detected and the interference have certain differences. In the following, the STAP principle will be illustrated for filtering in the joint azimuth angle (space) and doppler frequency (time) domain only.

STAP has been a very active research and development area since the publication of Reed et al.'s seminal paper [1]. With the recently completed Multichannel Airborne Radar Measurement project (MCARM) [2]– [5], STAP has been established as a valuable alternative to the traditional approaches, such as ultra-low sidelobe beamforming and Displaced Phase Center Antenna (DPCA) [6]. Much of STAP research and development efforts have been driven by the needs to make the system affordable, to simplify its front-hardware calibration, and to minimize the system's performance loss in severely nonhomogeneous environments. Figure 70.1 is a general configuration of STAP functional blocks [5, 7] whose principles will be discussed in the following sections.

0-8493-8572-5/98/$0.00+$.50
© 1998 by CRC Press LLC

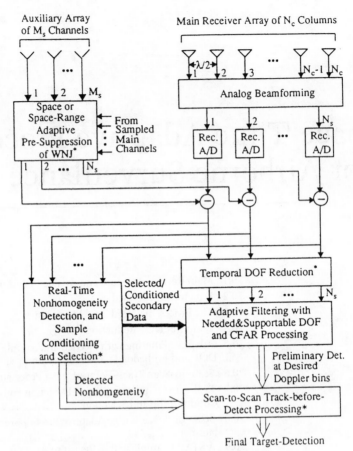

FIGURE 70.1 A general STAP configuration with auxiliary and main arrays.

70.1 Main Receive Aperture and Analog Beamforming

For conceptual clarity, the STAP configuration of Fig. 70.1 separates a possibly integrated aperture into two parts: the main aperture which is most likely shared by the radar transmitter, and an auxiliary array of spatially distributed channels for suppression of Wideband Noise Jammers (WNJ). For convenience of discussion, the main aperture is assumed to have N_c columns of elements, with the column spacing equal to a half wavelength and elements in each column being combined to produce a pre-designed, nonadaptive elevation beam-pattern.

The size of the main aperture in terms of the system's chosen wavelength is an important system parameter, usually determined by the system specifications of the required transmitter power-aperture product as well as azimuth resolution. Typical aperture size spans from a few wavelengths for some short-range radars to over 60 wavelengths for some airborne early warning systems. The analog beamforming network combines the N_c columns of the main aperture to produce N_s receiver channels whose outputs are digitized for further processing. One should note that the earliest STAP approach presented in [1], i.e., the so-called "element space" approach, is a special case of Fig. 70.1 when $N_s = N_c$ is chosen.

The design of the analog beamformer affects

1. the system's overall performance (especially in nonhomogeneous environments),
2. implementation cost,

3. channel calibration burden,
4. system reliability, and
5. controllability of the system's response pattern.

The design principle will be briefly discussed in Section 70.9; and because of the array's element error, column-combiner error, and column mutual-coupling effects, it is quite different from what is available in the adaptive array literature such as [8], where already digitized, perfectly matched channels are generally assumed.

Finally, it should be pointed out that the main aperture and analog beamforming network in Fig. 70.1 may also include nonphased-array hardware, such as the common reflector-feed as well as the hybrid reflector and phased-array feed [9]. Also, subarraying such as [10] is considered as a form of analog beamforming of Fig. 70.1.

70.2 Data to be Processed

Assume that the radar transmits, at each look angle, a sequence of N_t uniformly spaced, phase-coherent RF pulses as shown in Fig. 70.2 for its envelope only. Each of N_s receivers typically consists of a front-end amplifier, down-converter, waveform-matched filter, and A/D converter with a sampling frequency at least equal to the signal bandwidth. Consider the kth sample of radar return over the N_t pulse repetition intervals (PRI) from a single receiver, where the index "k" is commonly called the range index or cell. The total number of range cells, K_0 is approximately equal to the product of the PRI and signal bandwidth. The coherent processing interval (CPI) is the product of the PRI and N_t; and since a fixed PRI can usually be assumed at a given look angle, CPI and N_t are often used interchangeably.

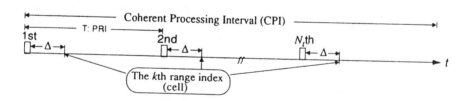

FIGURE 70.2   A sequence of N_t phase-coherent RF pulses (only envelope shown) transmitted at a given angle. The pulse repetition frequency (PRF) is 1/T.

With N_s receiver channels, the data at the kth range cell can be expressed by a matrix X_k, $N_s \times N_t$, for $k = 1, 2, \ldots K_0$. The total amount of data visually forms a "cube" shown in Fig. 70.3, which is the raw data cube to be processed at a given look angle. It is important to note from Fig. 70.3 that the term "time" is associated with the CPI for any given range cell, i.e., across the multiple PRIs, while the term "range" is used within a PRI. Therefore, the meaning of the frequency corresponding to the time is the so-called doppler frequency, describing the rate of the phase-shift progression of a return component with respect to the initial phase of the phase-coherent pulse train. The doppler frequency of a return, e.g., from a moving target, depends on the target velocity and direction as well as the airborne radar's platform velocity and direction, etc.

70.3 The Processing Needs and Major Issues

At a given look angle, the radar is to detect the existence of targets of unknown range and unknown doppler frequency in the presence of various interference. In other words, one can view the processing as a mapping

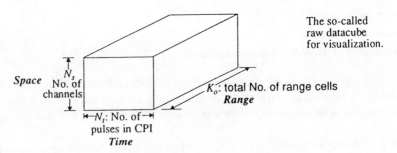

FIGURE 70.3   Raw data at a given look angle and the space, time, and range axes.

from the data cube to a range-doppler plane with sufficient suppression of unwanted components in the data. Like any other filtering, the interference suppression relies on the differences between wanted target components and unwanted interference components in the angle-doppler domain. Figure 70.4 illustrates the spectral distribution of potential interference in the spatial and temporal (doppler) frequency domain before the analog beamforming network, while Fig. 70.5 shows a typical range distribution of interference power. As targets of interest usually have unknown doppler frequencies and unknown distances, detection needs to be carried out at sufficiently dense doppler frequencies along the look angle for each range cell within the system's surveillance volume. For each cell at which target detection is being carried out, some of surrounding cells can be used to produce an estimate of interference statistics (usually up to the second order), i.e., providing "sample support", under the assumption that all cells involved have an identical statistical distribution. Figure 70.4 also shows that, in terms of their spectral differences, traditional wideband noise jammers, whether entering the system through direct path or multipath (terrain scattering/near-field scattering), require spatial nulling only; while clutter and chaff require angle-doppler coupled nulling. Coherent repeater jammers (CRJ) represent a nontraditional threat of a target-like spectral feature with randomized ranges and doppler frequencies, making them more harmful to adaptive systems than to conventional nonadaptive systems [11].

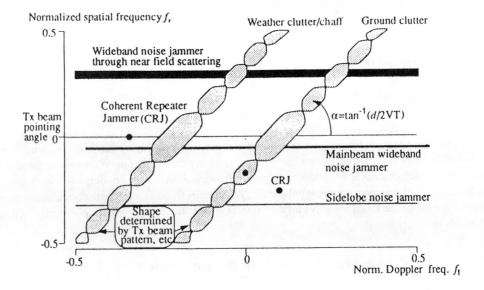

FIGURE 70.4   Illustration of interference spectral distribution for a side-mounted aperture.

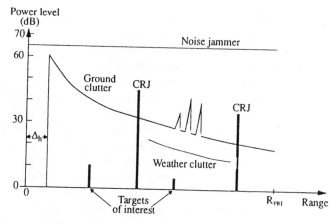

FIGURE 70.5 Illustration of interference-power range distribution, where Δ_h indicates the radar platform height.

Although Fig. 70.5 has already served to indicate that the interference is nonhomogeneous in range, i.e., its statistics vary along the range axis, recent airborne experiments have revealed that its severeness may have long been underestimated, especially over land [3]. Figure 70.6 [5, 7] summarizes the sources

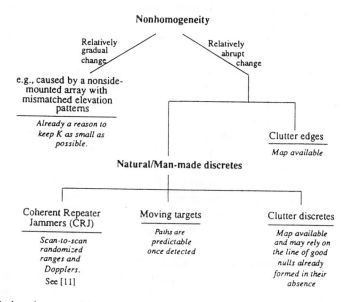

FIGURE 70.6 Typical nonhomogeneities.

of various nonhomogeneity together with their main features. As pointed out in [12], a serious problem associated with any STAP approach is its basic assumption that there is a sufficient amount of sample support for its adaptive learning, which is most often void in real environments even in the absence of any nonhomogeneity type of jammers such as CRJ. Therefore, a crucial issue for the success of STAP in real environments is the development of data-efficient STAP approaches, in conjunction with the selection of reasonably identically distributed samples before estimating interference statistics. To achieve a sufficient level of the data efficiency in nonhomogeneous environments, the three most performance- and cost-effective methods are temporal degrees-of-freedom (DOF) reduction, analog beamforming to control the spatial DOF creation, and pre-suppression of WNJ as shown in Fig. 70.1.

Another crucial issue is the affordability of STAP-based systems. As pointed out in [9], phased-arrays, especially those active ones (i.e., with the so-called T/R modules), remain very expensive despite the 30-year research and development. For multichannel systems, the cost of adding more receivers and A/D converters with a sufficient quality makes the affordability even worse.

Of course, more receiver channels mean more system's available spatial DOF. However, it is often the case in practice that the excessive amount of the DOF, e.g., obtained via one receiver channel for each column of a not-so-small aperture, is not necessary to the system. Ironically, excessive DOF can make the control of the response pattern more difficult, even requiring significant algorithm constraints [8]; and after all, it has to be reduced to a level supportable by the available amount of reasonably identically distributed samples in real environments. An effective solution, as demonstrated in a recent STAP experiment [13], is via the design of the analog beamformer that does not create unnecessary spatial DOF from the beginning — a sharp contrast to the DOF reduction/constraint applied in the spatial domain.

Channel calibration is a problem issue for many STAP approaches. In order to minimize performance degradation, the channels with some STAP approaches must be matched across the signal band, and steering vectors must be known to match the array. Considering the fact that channels generally differ in both elevation and azimuth patterns (magnitude as well as phase) even at a fixed frequency, the calibration difficulty has been underestimated as experienced in recent STAP experiments [5]. It is still commonly wished that the so-called "element-space" approaches, i.e., the special case of $N_s = N_c$ in Fig. 70.1, with an adaptive weight for each error-bearing "element" which hopefully can be modeled by a complex scalar, could solve the calibration problem at a significantly increased system-implementation cost as each element needs a digitized receiver channel. Unfortunately, such a wish can rarely materialize for a system with a practical aperture size operated in nonhomogeneous environments. With a spatial DOF reduction required by these approaches to bring down the number of adaptive weights to a sample-supportable level, the element errors are no longer directly accessible by the adaptive weights, and thus the wishful "embedded robustness" of these element-space STAP approaches is almost gone. In contrast, the MCARM experiment has demonstrated that, by making best use of what has already been excelled in antenna engineering [13], the channel calibration problem associated with STAP can be largely solved at the analog beamforming stage , which will be discussed in Section 70.9.

The above three issues all relate to the question: "What is the minimal spatial and temporal DOF required?" To simplify the answer, it can be assumed first that clutter has no Doppler spectral spread caused by its internal motion during the CPI, i.e., its spectral width cut along the doppler frequency axis of Fig. 70.4 equals to zero. For WNJ components of Fig. 70.4, the required minimal spatial DOF is well established in array processing, and the required minimal temporal DOF is zero as no temporal processing can help suppress these components. The CRJ components appear only in isolated range cells as shown in Fig. 70.5, and thus they should be dealt with by sample conditioning and selection so that the system response does not suffer from their random disturbance. With the STAP configuration of Fig. 70.1, i.e., pre-suppression of WNJ and sample conditioning and selection for CRJ, the only interference components left are those angle-doppler coupled clutter/chaff spectra of Fig. 70.4. It is readily available from the two-dimensional filtering theory [14] that suppression of each of these angle-doppler coupled components only requires one spatial DOF and one temporal DOF of the joint domain processor! In other words, a line of infinitely many nulls can be formed with one spatial DOF and one temporal DOF on top of one angle-doppler coupled interference component under the assumption that there is no clutter internal motion over the CPI. It is also understandable that, when such an assumption is not valid, one only needs to increase the temporal DOF of the processor so that the null width along the doppler axis can be correspondingly increased.

For conceptual clarity, $N_s - 1$ will be called the system's available spatial DOF and $N_t - 1$ the system's available temporal DOF. While the former has a direct impact on the implementation cost, calibration burden, and system reliability, the latter is determined by the CPI length and PRI with little cost impact, etc. Mainly due to the nonhomogeneity-caused sample support problem discussed earlier, the adaptive joint domain processor may have its spatial DOF and temporal DOF, denoted by N_{ps} and N_{pt} respectively, different from the system's availables by what is so-called DOF reduction. However, the spatial DOF reduction should be avoided by establishing the system's available spatial DOF as close to what is needed as possible from the beginning.

70.4 Temporal DOF Reduction

Typically an airborne surveillance radar has N_t anywhere between 8 and 128, depending on the CPI and PRI. With the processor's temporal DOF, N_{pt}, needed for the adjustment of the null width, normally being no more than $2 \sim 4$, huge DOF reduction is usually performed for the reasons of the sample support and better response-pattern control explained in Section 70.3.

An optimized reduction could be found, given N_t, N_{pt}, and the interference statistics which are still unknown at this stage of processing in practice [7]. There are several non-optimized temporal DOF reduction methods available, such as the Doppler-domain (joint domain) localized processing (DDL/JDL) [12, 15, 16] and the PRI-staggered Doppler-decomposed processing (PRI-SDD) [17], which are well behaved and easy to implement. The DDL/JDL principle will be discussed below.

The DDL/JDL consists of unwindowed/untapered DFT of (at least) N_t-point long, operated on each of the N_s receiver outputs. The same $N_{pt} + 1$ most adjacent frequency bins of the DFTs of the N_s receiver outputs form the new data matrix at a given range cell, for detection of a target whose Doppler frequency is equal to the center bin. Figure 70.7 shows an example for $N_s = 3$, $N_t = 8$, and $N_{pt} = 2$. In other words, the DDL/JDL transforms the raw data cube of $N_s \times N_t \times K_0$ into (at least) N_t smaller data cubes, each of $N_s \times (N_{pt} + 1) \times K_0$ for target detection at the center doppler bin.

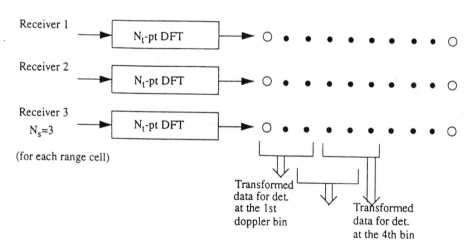

FIGURE 70.7   The DDL/JDL principle for temporal DOF reduction illustrated with $N_s = 3$, $N_t = 8$, and $N_{pt} = 2$.

The DDL/JDL is noticeable for the following features.

1. There is no so-called signal cancellation, as the unwindowed/untapered DFT provides no desired signal components in the adjacent bins (i.e., reference "channel") for the assumed target doppler frequency.

2. The grouping of $N_{pt} + 1$ most adjacent bins gives a high degree of correlation between the interference component at the center bin and those at the surrounding bins — a feature

important to cancellation of any spectrum-distributed interference such as clutter. The cross-spectral algorithm [18] also has this feature.

3. The response pattern can be well controlled as N_{pt} can be kept small — just enough for the needed null-width adjustment; and N_{pt} itself easily can be adjusted to fit different clutter spectral spread due to its internal motion.

4. Obviously the DDL/JDL is suitable for parallel processing.

While the DDL/JDL is a typical transformation-based temporal DOF reduction method, other methods involving the use of DFTs are not necessarily transformation-based. An example is the PRI-SDD [17] which applies time-domain temporal DOF reduction on each doppler component. This explains why the PRI-SDD requires N_{pt} times more DFTs that should be tapered. It also serves as an example that an algorithm classification by the existence of the DFT use may cause a conceptual confusion.

70.5 Adaptive Filtering with Needed and Sample-Supportable DOF and Embedded CFAR Processing

After the above temporal DOF reduction, the dimension of the new data cube to be processed at a given look angle for each doppler bin is $N_s \times (N_{pt} + 1) \times K_0$. Consider a particular range cell at which target detection is being performed. Let $\mathbf{x}$, $N_s(N_{pt} + 1) \times 1$, be the stacked data vector of this range cell, which is usually called the primary data vector. Let $\mathbf{y}_1, \mathbf{y}_2, \ldots, \mathbf{y}_k$, all $N_s(N_{pt} + 1) \times 1$ and usually called the secondary data, be the same-stacked data vectors of the K surrounding range cells, which have been selected and/or conditioned to eliminate any significant nonhomogeneities with respect to the interference contents of the primary data vector. Let $\mathbf{s}$, $N_s(N_{pt} + 1) \times 1$, be the target-signal component of $\mathbf{x}$ with the assumed angle of arrival equal to the look angle and the assumed doppler frequency corresponding to the center doppler bin. In practice, a look-up table of the "steering vector" $\mathbf{s}$ for all look-angles and all doppler bins usually has to be stored in the processor, based on updated system calibration. A class of STAP systems with the steering-vector calibration-free feature has been developed, and an example from [13] will be presented in Section 70.9.

There are two classes of adaptive filtering algorithms: one with a separately designed constant false alarm rate (CFAR) processor, and the other with embedded CFAR processing. The original sample matrix inversion algorithm (SMI) [1] belongs to the former, which is given by

$$\eta_{SMI} = \left| \hat{\mathbf{w}}_{SMI}^H \mathbf{x} \right|^2 \underset{H_0}{\overset{H_1}{\underset{<}{>}}} \eta_0 \tag{70.1}$$

where

$$\mathbf{w}_{SMI} = \hat{\mathbf{R}}^{-1} \mathbf{s}, \tag{70.2}$$

and

$$\hat{\mathbf{R}} = \frac{1}{K} \sum_{k=1}^{K} \mathbf{y}_k \mathbf{y}_k^H \tag{70.3}$$

The SMI performance under the Gaussian noise/interference assumption has been analyzed in detail [1], and in general it is believed that acceptable performance can be expected if the data vectors are independent and identically distributed (iid) with K, the number of the secondary, being at least two times $N_s(N_{pt}+1)$. Detection performance evaluation using a SINR-like measure deserves some care when K is finite, even under the iid assumption [19, 20].

If the output of an adaptive filter, when directly used for threshold detection, produces a probability of false alarm independent of the unknown interference correlation matrix under a set of given conditions, the adaptive filter is said to have an embedded CFAR. Under the iid Gaussian condition, two well-known algorithms with embedded CFAR are the Modified SMI [21] and Kelly's generalized likelihood ratio detector (GLR) [22], both of which are linked to the SMI as shown in Fig. 70.8. The GLR has the following

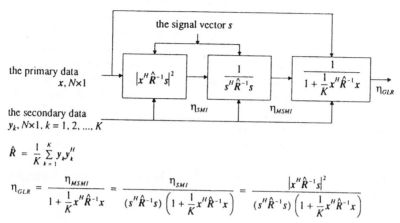

FIGURE 70.8   The link among the SMI, modified SMI (MSMI), and GLR where $N = (N_{ps} + 1)(N_{pt} + 1) \times 1$.

interesting features:

1. $0 < \frac{1}{K}\eta_{GLR} < 1$, which is a necessary condition for robustness in nongaussian interference [23].

2. Invariance with respect to scaling all data or scaling **s**.

3. One cannot express η_{GLR} as $\hat{\mathbf{w}}^H\mathbf{x}$; and with a finite K, an objective definition of its output SINR becomes questionable.

Table 70.1 summarizes the modified SMI and GLR performance, based on [21, 24].

TABLE 70.1   Performance Summary of Modified SMI and GLR

Performance compared	GLR	Modified SMI
Gaussian interference suppression	Similar performance	
Nongaussian interference suppression	More robust	Less robust
Rejection of signals mismatched to the steering vector	Better	Worse

It should be noted that the use of the scan-to-scan track-before-detect processor (SSTBD to be discussed in Section 70.6) does not make the CFAR control any less important because the SSTBD itself is not error-free even with the assumption that almost infinite computing power would be available. Moreover, the initial CFAR thresholding can actually optimize the overall performance, in addition to a dramatic reduction of the computation load of the SSTBD processor. Traditionally, filter and CFAR designs have been carried out separately, which is valid as long as the filter is not data-dependent. Therefore, such a traditional practice becomes questionable for STAP, especially when K is not very large with respect to $N_s(N_{pt} + 1)$, or when some of the secondary data depart from the iid Gaussian assumption that will affect both filtering and CFAR portions. The GLR and Modified SMI start to change the notion that "CFAR is the other guy's job", and their performance has been evaluated in some nongaussian interference [21] as well as in some nonhomogeneities [25]. Finally, it should be pointed out that performance evaluation of STAP algorithms with embedded CFAR by an output SINR-like measure may result in underestimating the effects of some nonhomogeneity such as the CRJ [11].

70.6   Scan-To-Scan Track-Before-Detect Processing

The surveillance volume is usually visited by the radar many times, and the output data collected over multiple scans (i.e., revisits) are correlated and should be further processed together for the updated and improved final target-detection report. For example, random threshold-crossings over multiple scans due to the noise/interference suppression residue can rarely form a meaningful target trajectory and therefore their effect can be deleted from the final report with a certain level of confidence (but not error-free).

For a conventional ground-based radar, scan-to-scan track-before-detect processing (SSTBD) has been well studied and a performance demonstration can be found in [26]. With a STAP-based airborne system, however, much remains to be researched. One crucial issue, coupled with the initial CFAR control, is to answer what is the optimal or near optimal setting of the first CFAR threshold, given an estimate of the current environment including the detected nonhomogeneity. Further discussion of this subject seems out of the scope of this book and still premature.

70.7   Real-Time Nonhomogeneity Detection and Sample Conditioning and Selection

Recent experience with MCARM Flight 5 data has further demonstrated that successful STAP system operation over land heavily relies on the handling of the nonhomogeneity contamination of samples [3, 5], even without intentional nonhomogeneity producing jammers such as CRJ. It is estimated that the total number of reasonably good samples over land may be as few as $10 \sim 20$. Although some system approaches to obtaining more good samples are available, such as multiband signaling [27, 28], it is still essential that a system has the capability of real-time detection of nonhomogeneities, selection of sufficiently good samples to be used as the secondary, and conditioning not-so-good samples in the case of a severe shortage of the good samples. The development of a nonhomogeneity detector can be found in [3], and its integration into the system remains to be a research issue.

Finally, it should be pointed out that the utilization of a sophisticated sample selection scheme makes it nearly unnecessary to look into the so-called training strategy such as sliding window, sliding hole, etc. Also, desensitizing a STAP algorithm via constraints and/or diagonal loading has been found to be less effective than the sample selection [28].

70.8   Space or Space-Range Adaptive Pre-Suppression of Jammers

Wideband noise jammers (WNJ) have a flat or almost flat Doppler spectrum which means that without multipath/terrain-scattering (TS), only spatial nulling is necessary. Although STAP could handle, at least theoretically, the simultaneous suppression of WNJ and clutter simply with an increase of the processor's spatial DOF (N_{ps}), doing so would unnecessarily raise the size of the correlation matrix which, in turn, requires more samples for its estimation. Therefore, spatial adaptive pre-suppression (SAPS) of WNJ, followed by STAP-based clutter suppression, is preferred for systems to be operated in severely nonhomogenous environments. Space-range adaptive processing (SRAP) may become necessary in the presence of multipath/TS to exploit the correlation between the direct path and indirect paths for better suppression of the total WNJ effects on the system performance.

The idea of cascading SAPS and STAP itself is not new, and the original work can be found in [29], with other names such as "two step nulling (TSN)" used in [30]. A key issue in applying this idea is the acquisition of the necessary jammer-only statistics for adaptive suppression, free from strong clutter contamination. Available acquisition methods include the use of clutter-free range-cells for low PRF systems, clutter-free Doppler bins for high PRF systems, or receive-only mode between two CPIs. All of these techniques require jammer data to be collected within a restricted region of the available space-time domain, and may not always be able to generate sufficient jammer-only data. Moreover, fast-changing jamming environments and large-scale PRF hopping can also make these techniques unsuitable. Reference [31] presents a new technique that makes use of frequency sidebands close to, but disjointed from, the radar's mainband, to estimate the jammer-only covariance matrix. Such an idea can be applied to a system with any PRF, and the entire or any appropriate portion of the Range Processing Interval (RPI) could be used to collect jammer data. It should be noted that wideband jammers are designed to sufficiently cover the radar's mainband, making sidebands, of more or less bandwidth, containing their energy always available to the new SAPS technique.

The discussion of the sideband-based STAP can be carried out with different system configurations, which determine the details on the sideband-to-mainband jammer information conversion, as well as the mainband jammer-cancellation signal generation. Reference [31] chooses a single array-based system, while a discussion involving an auxiliary-main array configuration can be found in [7].

70.9 A STAP Example with a Revisit to Analog Beamforming

In the early stage of STAP research, it is always assumed that $N_s = N_c$, i.e., each column consumes a digitized receiver channel, regardless of the size of the aperture. More recent research and experiments have revealed that such an "element-space" set up is only suitable for sufficiently small apertures, and the analog beamforming network has become an important integrated part of STAP-based systems with more practical aperture sizes.

The theoretically optimized analog beamformer design could be carried out for any given N_s, which yields a set of N_s nonrealizable beams once the element error, column-combiner error, and column mutual-coupling effects are factored in. A more practical approach is to select, from what antenna design technology has excelled, those beams that also meet the basic requirements for successful adaptive processing, such as the "signal blocking" requirement developed under the generalized sidelobe canceller [32]. Two examples of proposed analog beamforming methods for STAP applications are (1) multiple shape-identical Fourier beams via the Butler matrix [12], and (2) the sum and difference beams [13]. Both selections have been shown to enable the STAP system to achieve near optimal performance with N_s very close to the theoretical minimum of two for clutter suppression.

In the following, the clutter suppression performance of a STAP with the sum(Σ)-difference(Δ) beams is presented using the MCARM Flight 2 data. The clutter in this case was collected from a rural area in the eastern shore region south of Baltimore, Maryland. A known target signal was injected at a Doppler frequency slightly offset from mainlobe clutter and the results compared for the factored approach (FA-STAP) [16] and $\Sigma\Delta$-STAP. A Modified SMI processor was used in each case to provide a known threshold level based on a false alarm probability of 10^{-6}. As seen in Figs. 70.9 and 70.10, the injected target lies below the detection threshold for FA-STAP , but exceeds the threshold in the case of $\Sigma\Delta$-STAP. This performance was obtained using far fewer samples for covariance estimation in the case of $\Sigma\Delta$-STAP. Also, the $\Sigma\Delta$-STAP uses only 2 receiver channels, while the FA-STAP consumes all 16 channels.

In terms of calibration burden, the $\Sigma\Delta$-STAP uses two different channels to begin with and its corresponding signal (steering) vector easily remains the simplest form as long as the null of the Δ beam is correctly placed (a job in which antenna engineers have excelled already). In that sense, the $\Sigma\Delta$-STAP is both channel calibration-free and steering-vector calibration-free. On the other hand, keeping the 16 channels of FA-STAP calibrated and updating its steering vector look-up table have been a considerable burden during the MCARM experiment [4].

Another significant affordability issue is the applicability of $\Sigma\Delta$-STAP to existing radar systems, both phased array and continuous aperture. Adaptive clutter rejection in the joint angle-doppler domain can be incorporated into existing radar systems by digitizing the difference channel, or making relatively minor antenna modifications to add such a channel. Such a relatively low cost add-on can significantly improve the clutter suppression performance of an existing airborne radar system, whether its original design is based on low sidelobe beamforming or $\Sigma\Delta$-DPCA.

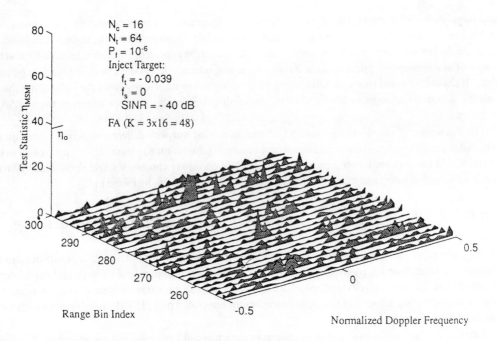

FIGURE 70.9   Range-Doppler plot of MCARM data, factored approach.

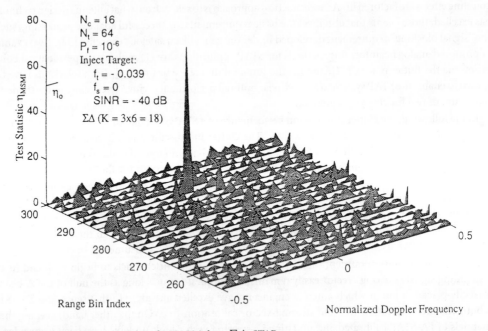

FIGURE 70.10   Range-Doppler plot of MCARM data, $\Sigma\Delta$-STAP.

While the trend is toward more affordable computing hardware, STAP processing still imposes a considerable burden which increases sharply with the order of the adaptive processor and radar bandwidth. In this respect, $\Sigma\Delta$-STAP reduces computational requirements in matrix order N^3 adaptive problems. Moreover, the signal vector characteristic (mostly zero) can be exploited to further reduce test statistic numerical computations.

Finally, it should be pointed out that more than one Δ-beam can be incorporated if needed for clutter suppression [33].

70.10 Summary

Over the 22 years from a theoretical paper [1] to the MCARM experimental system, STAP has been established as a valuable alternative to the traditional airborne surveillance radar design approaches. Initially, STAP was viewed as an expensive technique only for newly designed phased-arrays with many receiver channels; and now it has become much more affordable for both new and some existing systems. Future challenges lie in the area of real system design and integration, to which the MCARM experience is invaluable.

References

[1] Reed, I.S., Mallet, J.D. and Brennan, L.E., Rapid convergence rate in adaptive arrays, *IEEE Trans. on Aerospace and Electronic Systems*, AES-10, 853–863, Nov. 1974.

[2] Little, M.O. and Berry, W.P., Real-time multichannel airborne radar measurements, *Proc. IEEE National Radar Conference*, 138–142, Syracuse, NY, May 13-15, 1997.

[3] Melvin, W.L., Wicks, M.C. and Brown, R.D., Assessment of multichannel airborne radar measurements for analysis and design of space-time processing architectures and algorithms, *Proc. IEEE 1996 National Radar Conference*, 130–135, Ann Arbor, MI, May 13-16, 1996.

[4] Fenner, D.K. and Hoover, Jr., W.F., Test results of a space-time adaptive processing system for airborne early warning radar, *Proc. IEEE 1996 National Radar Conference*, 88–93, Ann Arbor, MI, May 13-16, 1996.

[5] Wang, H., Zhang, Y. and Zhang, Q., Lessons learned from recent STAP experiments, *Proc. CIE International Radar Conference*, Beijing, China, Oct. 8-10, 1996.

[6] Staudaher, F.M., Airborne MTI, *Radar Handbook*, Skolnik, M.I., Ed., McGraw-Hill, New York, 1990, chap. 16.

[7] Wang, H., *Space-Time Processing and Its Radar Applications*, Lecture Notes for ELE891, Syracuse University, Summer 1995.

[8] Tseng, C.Y. and Griffiths, L.J., A unified approach to the design of linear constraints in minimum variance adaptive beamformers, *IEEE Trans. on Antennas and Propagation*, AP-40(12), 1533–1542, Dec. 1992.

[9] Skolnik, M., The radar antenna-Circa 1995, *J. Franklin Inst.*, Elsevier Science Ltd., 332B(5), 503–519, 1995.

[10] Klemm, R., Antenna design for adaptive airborne MTI, *Proc. 1992 IEE Intl. Conf. Radar*, 296–299, Brighton, U.K., Oct. 12-13, 1992.

[11] Wang, H., Zhang, Y. and Wicks, M.C., Performance evaluation of space-time processing adaptive array radar in coherent repeater jamming environments, *Proc. IEEE Long Island Section Adaptive Antenna Syst. Symp.*, 65–69, Melville, NY, Nov. 7-8, 1994.

[12] Wang, H. and Cai, L., On adaptive spatial-temporal processing for airborne surveillance radar systems, *IEEE Trans. on Aerospace and Electronic Systems*, AES-30(3), 660–670, July 1994. Part of this paper is also in *Proc. 25th Annual Conference on Information Sciences and Systems*, 968–975, Baltimore, MD, March 20-22, 1991, and *Proc. CIE 1991 International Conference on Radar*, 365–368, Beijing, China, Oct. 22-24, 1991.

[13] Brown, R.D., Wicks, M.C., Zhang, Y., Zhang, Q. and Wang, H., A space-time adaptive processing approach for improved performance and affordability, *Proc. IEEE 1996 National Radar Conference*, 321–326, Ann Arbor, MI, May 13-16, 1996.

[14] Pendergrass, N.A., Mitra, S.K. and Jury, E.I., Spectral transformations for two-dimensional digital filters, *IEEE Trans. on Circuits and Systems,* CAS-23(1), 26–35, Jan. 1976.

[15] Wang, H. and Cai, L., A localized adaptive MTD processor, *IEEE Trans. on Aerospace and Electronic Systems,* AES-27(3), 532–539, May 1991.

[16] DiPietro, R.C., Extended factored space-time processing for airborne radar systems, *Proc. 26th Asilomar Conference on Signals, Systems, and Computers,* 425–430, Pacific Grove, CA, Nov. 1992.

[17] Brennan, L.E., Piwinski, D.J. and Staudaher, F.M., Comparison of space-time adaptive processing approaches using experimental airborne radar data, *IEEE 1993 National Radar Conference,* 176–181, Lynnfield, MA, April 20-22, 1993.

[18] Goldstein, J.S., Williams, D.B. and Holder, E.J, Cross-spectral subspace selection for rank reduction in partially adaptive sensor array processing, *Proc. IEEE 1994 National Radar Conference,* Atlanta, Georgia, May 29-31, 1994.

[19] Nitzberg, R., Detection loss of the sample matrix inversion technique, *IEEE Trans. on Aerospace and Electronic Systems,* AES-20, 824–827, Nov. 1984.

[20] Khatri, C.G. and Rao, C.R., Effects of estimated noise covariance matrix in optimal signal detection, *IEEE Trans. on Acoustics, Speech, and Signal Processing,* ASSP-35(5), 671–679, May 1987.

[21] Cai, L. and Wang, H., On adaptive filtering with the CFAR feature and its performance sensitivity to non-Gaussian interference, *Proc. of the 24th Annual Conference on Information Sciences and Systems,* 558–563, Princeton, NJ, March 21-23, 1990. Also published in *IEEE Trans. on Aerospace and Electronic Systems,* AES-27(3), 487–491, May 1991.

[22] Kelly, E.J., An adaptive detection algorithm, *IEEE Trans. on Aerospace and Electronic Systems,* AES-22(1), 115–127, March 1986.

[23] Kazakos, D. and Papantoni-Kazakos, P., *Detection and Estimation,* Computer Science Press, New York, 1990.

[24] Robey, F.C. et. al., A CFAR adaptive matched filter detector, *IEEE Trans. on Aerospace and Electronic Systems,* AES-28(1), 208–216, Feb. 1992.

[25] Cai, L. and Wang, H., Further results on adaptive filtering with embedded CFAR, *IEEE Trans. on Aerospace and Electronic Systems,* AES-30(4), 1009–1020, Oct. 1994.

[26] Corbeil, A., Hawkins, L. and Gilgallon, P., Knowledge-based tracking algorithm, Proc. Signal and Data Processing of Small Targets, SPIE Proc. Series, Vol. 1305, Paper 16, 180–192, Orlando, FL, April 16-18, 1990.

[27] Wang, H. and Cai, L., On adaptive multiband signal detection with SMI algorithm, *IEEE Trans. on Aerospace and Electronic Systems,* AES-26, 768–773, Sept. 1990.

[28] Wang, H., Zhang, Y. and Zhang, Q., A view of current status of space-time processing algorithm research, *Proc. IEEE 1995 Intl. Radar Conf.,* 635–640, Alexandria, VA, May 8-11, 1995.

[29] Klemm, R., Adaptive air and spaceborne MTI under jamming conditions, *Proc. 1993 IEEE Natl. Radar Conf.,* 167–172, Boston, MA, April 1993.

[30] Marshall, D.F., A two step adaptive interference nulling algorithm for use with airborne sensor arrays, *Proc. 7th SP Workshop on SSAP,* Quebec City, Canada, June 26-29, 1994.

[31] Rivkin, P., Zhang, Y. and Wang, H., Spatial adaptive pre-suppression of wideband jammers in conjunction with STAP: a sideband approach, *Proc. CIE Intl. Radar Conf.,* 439–443, Beijing, China, Oct. 8-10, 1996.

[32] Griffiths, L.J. and Jim, C.W., An alternative approach to linearly constrained adaptive beamforming, *IEEE Trans. on Antennas and Propagation,* AP-30(1), 27–34, Jan. 1982.

[33] Zhang, Y. and Wang, H., Further results of $\Sigma\Delta$-STAP approach to airborne surveillance radars, *Proc. IEEE National Radar Conference,* 337–342, Syracuse, NY, May 13–15, 1997.

XIII

Nonlinear and Fractal Signal Processing

Alan V. Oppenheim
Massachusetts Institute of Technology
Gregory W. Wornell
Massachusetts Institute of Technology

T RADITIONALLY, SIGNAL PROCESSING as a discipline has relied heavily on a theoretical foundation of linear time-invariant system theory in the development of algorithms for a broad range of applications. In recent years a considerable broadening of this theoretical base has begun to take place. In particular, there has been substantial growth in interest in the use of a variety of nonlinear systems with special properties for diverse applications. Promising new techniques for the synthesis and analysis of such systems continue to emerge. At the same time, there has also been rapid growth in interest in systems that are not constrained to be time-invariant. These may be systems that exhibit temporal fluctuations in their characteristics, or, equally importantly, systems characterized by other invariance properties, such as invariance to scale changes. In the latter case, this gives rise to systems with fractal characteristics.

In some cases, these systems are directly applicable for implementing various kinds of signal processing operations such as signal restoration, enhancement, or encoding, or for modeling certain kinds of distortion encountered in physical environments. In other cases, they serve as mechanisms for generating new classes of signal models for existing and emerging applications. In particular, when autonomous or driven by simpler classes of input signals, they generate rich classes of signals at their outputs. In turn, these new classes of signals give rise to new families of algorithms for efficiently exploiting them in the context of applications.

The spectrum of techniques for nonlinear signal processing is extremely broad, and in this chapter we make no attempt to cover the entire array of exciting new directions being pursued within the community. Rather, we present a very small sampling of several highly promising and interesting ones to suggest the richness of the topic.

A brief overview of the specific chapters comprising this section is as follows.

Chapters 71 and 72 discuss the chaotic behavior of certain nonlinear dynamical systems and suggest ways in which this behavior can be exploited. In particular, Chapter 71 focuses on continuous-time chaotic systems characterized by a special self-synchronization property that makes them potentially attractive for a range of secure communications applications. Chapter 72 describes a family of discrete-time nonlinear dynamical and chaotic systems that are particularly attractive for use in a variety of signal processing applications ranging from signal modeling in power converters to pseudorandom number generation and error-correction coding in signal transmission applications.

Chapter 73 discusses fractal signals which arise out of self-similar system models characterized by scale-invariance. These represent increasingly important models for a range of natural and man-made phenomena in applications involving both signal synthesis and analysis. Multidimensional fractals also arise in the state-space representation of chaotic signals, and the fractal properties in this representation are important in the identification, classification, and characterization of such signals.

Chapter 74 focuses on morphological signal processing, which encompasses an important class of nonlinear filtering techniques together with some powerful associated signal representations. Morphological signal processing is closely related to a number of classes of algorithms including order-statistics filtering, cellular automata methods for signal processing, and others. Morphological algorithms are currently among the most successful and widely used nonlinear signal processing techniques in image processing and vision for such tasks as noise suppression, feature extraction, segmentation, and others.

Chapter 75 discusses the analysis and synthesis of soliton signals and their potential use in communication applications. These signals arise in systems satisfying certain classes of nonlinear wave equations. Because they propagate through those equations without dispersion, there has been longstanding interest in their use as carrier waveforms over fiber-optic channels having the appropriate nonlinear characteristics. As they propagate through these systems, they also exhibit a special type of reduced-energy superposition property that suggests an interesting multiplexing strategy for communications over linear channels.

Finally, Chapter 76 discusses nonlinear representations for stochastic signals in terms of their higher-order statistics. Such representations are particularly important in the processing of non-Gaussian signals for which more traditional second-moment characterizations are often inadequate. The associated tools of higher-order spectral analysis find increasing application in many signal detection, identification, modeling, and equalization contexts, where they have led to new classes of powerful signal processing algorithms.

Again, these articles are only representative examples of the many emerging directions in this active area of research within the signal processing community, and developments in many other important and exciting directions can be found in the community's journal and conference publications.

71

Chaotic Signals and Signal Processing

Alan V. Oppenheim
Massachusetts Institute of Technology

Kevin M. Cuomo
MIT
Lincoln Laboratory

71.1 Introduction

Signals generated by chaotic systems represent a potentially rich class of signals both for detecting and characterizing physical phenomena and in synthesizing new classes of signals for communications, remote sensing, and a variety of other signal processing applications.

In classical signal processing a rich set of tools has evolved for processing signals that are deterministic and predictable such as transient and periodic signals, and for processing signals that are stochastic. Chaotic signals associated with the homogeneous response of certain nonlinear dynamical systems do not fall in either of these classes. While they are deterministic, they are not predictable in any practical sense in that even with the generating dynamics known, estimation of prior or future values from a segment of the signal or from the state at a given time is highly ill-conditioned. In many ways these signals appear to be noise-like and can, of course, be analyzed and processed using classical techniques for stochastic signals. However, they clearly have considerably more structure than can be inferred from and exploited by traditional stochastic modeling techniques.

The basic structure of chaotic signals and the mechanisms through which they are generated are described in a variety of introductory books, e.g., [1, 2] and summarized in [3].

Chaotic signals are of particular interest and importance in experimental physics because of the wide range of physical processes that apparently give rise to chaotic behavior. From the point of view of signal processing, the detection, analysis, and characterization of signals of this type present a significant challenge. In addition, chaotic systems provide a potentially rich mechanism for signal design and generation for a variety of communications and remote sensing applications.

0-8493-8572-5/98/$0.00+$.50

71.2   Modeling and Representation of Chaotic Signals

The state evolution of chaotic dynamical systems is typically described in terms of the nonlinear state equation $\dot{x}(t) = F[x(t)]$ in continuous time or $x[n] = F(x[n-1])$ in discrete time. In a signal processing context, we assume that the observed chaotic signal is a nonlinear function of the state and would typically be a scalar time function. In discrete-time, for example, the observation equation would be $y[n] = G(x[n])$. Frequently the observation $y[n]$ is also distorted by additive noise, multipath effects, fading, etc.

Modeling a chaotic signal can be phrased in terms of determining from clean or distorted observations, a suitable state space and mappings $F(\cdot)$ and $G(\cdot)$ that capture the aspects of interest in the observed signal y. The problem of determining from the observed signal a suitable state space in which to model the dynamics is referred to as the embedding problem. While there is, of course, no unique set of state variables for a system, some choices may be better suited than others. The most commonly used method for constructing a suitable state space for the chaotic signal is the method of delay coordinates in which a state vector is constructed from a vector of successive observations.

It is frequently convenient to view the problem of identifying the map associated with a given chaotic signal in terms of an interpolation problem. Specifically, from a suitably embedded chaotic signal it is possible to extract a codebook consisting of state vectors and the states to which they subsequently evolve after one iteration. This codebook then consists of samples of the function F spaced, in general, non-uniformly throughout state space. A variety of both parametric and nonparametric methods for interpolating the map between the sample points in state space have emerged in the literature, and the topic continues to be of significant research interest. In this section we briefly comment on several of the approaches currently used. These and others are discussed and compared in more detail in [4].

One approach is based on the use of locally linear approximations to F throughout the state space [5, 6]. This approach constitutes a generalization of autoregressive modeling and linear prediction and is easily extended to locally polynomial approximations of higher order. Another approach is based on fitting a global nonlinear function to the samples in state space [7].

A fundamentally rather different approach to the problem of modeling the dynamics of an embedded signal involves the use of hidden Markov models [8, 9, 10]. With this method, the state space is discretized into a large number of states, and a probabilistic mapping is used to characterize transitions between states with each iteration of the map. Furthermore, each state transition spawns a state-dependent random variable as the observation $y[n]$. This framework can be used to simultaneously model both the detailed characteristics of state evolution in the system and the noise inherent in the observed data. While algorithms based on this framework have proved useful in modeling chaotic signals, they can be expensive both in terms of computation and storage requirements due to the large number of discrete states required to adequately capture the dynamics.

While many of the above modeling methods exploit the existence of underlying nonlinear dynamics, they do not explicitly take into account some of the properties peculiar to *chaotic* nonlinear dynamical systems. For this reason, in principle, the algorithms may be useful in modeling a broader class of signals. On the other hand, when the signals of interest are truly chaotic, the special properties of chaotic nonlinear dynamical systems ought to be taken into account, and, in fact, may often be exploited to achieve improved performance. For instance, because the evolution of chaotic systems is acutely sensitive to initial conditions, it is often important that this numerical instability be reflected in the model for the system. One approach to capturing this sensitivity is to require that the reconstructed dynamics exhibit Lyapunov exponents consistent with what might be known about the true dynamics. The sensitivity of state evolution can also be captured using the hidden Markov model framework since the structural uncertainty in the dynamics can be represented in terms of the probabilistic state transactions. In any case, unless sensitivity of the dynamics is taken into account during modeling, detection and estimation algorithms involving chaotic signals often lack robustness.

Another aspect of chaotic systems that can be exploited is that the long term evolution of such systems lies on an attractor whose dimension is not only typically non-integral, but occupies a small fraction of the entire state space. This has a number of important implications both in the modeling of chaotic signals and ultimately in addressing problems of estimation and detection involving these signals. For example, it implies that the nonlinear dynamics can be recovered in the vicinity of the attractor using comparatively less data than would be necessary if the dynamics were required everywhere in state space.

Identifying the attractor, its fractal dimension, and related invariant measures governing, for example, the probability of being in the neighborhood of a particular state on the attractor, are also important aspects of the modeling problem. Furthermore, we can often exploit various ergodicity and mixing properties of chaotic systems. These properties allow us to recover information about the attractor using a single realization of a chaotic signal, and assure us that different time intervals of the signal provide qualitatively similar information about the attractor.

71.3 Estimation and Detection

A variety of problems involving the estimation and detection of chaotic signals arises in potential application contexts. In some scenarios, the chaotic signal is a form of noise or other unwanted interference signal. In this case, we are often interested in detecting, characterizing, discriminating, and extracting known or partially known signals in backgrounds of chaotic noise. In other scenarios, it is the chaotic signal that is of direct interest and which is corrupted by other signals. In these cases we are interested in detecting, discriminating, and extracting known or partially known chaotic signals in backgrounds of other noises or in the presence of other kinds of distortion.

The channel through which either natural or synthesized signals are received can typically be expected to introduce a variety of distortions including additive noise, scattering, multipath effects, etc. There are, of course, classical approaches to signal recovery and characterization in the presence of such distortions for both transient and stochastic signals. When the desired signal in the channel is a chaotic signal, or when the distortion is caused by a chaotic signal, many of the classical techniques will not be effective and do not exploit the particular structure of chaotic signals.

The specific properties of chaotic signals exploited in detection and estimation algorithms depend heavily on the degree of *a priori* knowledge of the signals involved. For example, in distinguishing chaotic signals from other signals, the algorithms may exploit the functional form of the map, the Lyapunov exponents of the dynamics, and/or characteristics of the chaotic attractor such as its structure, shape, fractal dimension and/or invariant measures.

To recover chaotic signals in the presence of additive noise, some of the most effective noise reduction techniques proposed to date take advantage of the nonlinear dependence of the chaotic signal by constructing accurate models for the dynamics. Multipath and other types of convolutional distortion can best be described in terms of an augmented state space system. Convolution or filtering of chaotic signals can change many of the essential characteristics and parameters of chaotic signals. Effects of convolutional distortion and approaches to compensating for it are discussed in [11].

71.4 Use of Chaotic Signals in Communications

Chaotic systems provide a rich mechanism for signal design and generation, with potential applications to communications and signal processing. Because chaotic signals are typically broadband, noise-like, and difficult to predict, they can be used in various contexts in communications. A particularly useful class of chaotic systems are those that possess a self-synchronization property [12, 13, 14]. This property allows two identical chaotic systems to synchronize when the second system (receiver) is driven by the first (transmitter). The well-known Lorenz system is used below to further describe and illustrate the chaotic self-synchronization property.

The Lorenz equations, first introduced by E. N. Lorenz as a simplified model of fluid convection [15], are given by

$$
\begin{aligned}
\dot{x} &= \sigma(y - x) \\
\dot{y} &= rx - y - xz \\
\dot{z} &= xy - bz,
\end{aligned}
\tag{71.1}
$$

where σ, r, and b are positive parameters. In signal processing applications, it is typically of interest to adjust the time scale of the chaotic signals. This is accomplished in a straightforward way by establishing the convention that $\dot{x}$, $\dot{y}$, and $\dot{z}$ denote $dx/d\tau$, $dy/d\tau$, and $dz/d\tau$, respectively, where $\tau = t/T$ is normalized time and T is a time scale factor. It is also convenient to define the normalized frequency $\omega = \Omega T$, where Ω denotes the angular frequency in units of rad/s. The parameter values $T = 400\ \mu sec$, $\sigma = 16$, $r = 45.6$, and $b = 4$ are used for the illustrations in this chapter.

Viewing the Lorenz system (71.1) as a set of transmitter equations, a dynamical receiver system that will synchronize to the transmitter is given by

$$
\begin{aligned}
\dot{x}_r &= \sigma(y_r - x_r) \\
\dot{y}_r &= rx(t) - y_r - x(t)z_r \\
\dot{z}_r &= x(t)y_r - bz_r.
\end{aligned}
\tag{71.2}
$$

In this case, the chaotic signal $x(t)$ from the transmitter is used as the driving input to the receiver system. In Section 71.4.1, an identified equivalence between self-synchronization and asymptotic stability is exploited to show that the synchronization of the transmitter and receiver is global, i.e., the receiver can be initialized in any state and the synchronization still occurs.

71.4.1 Self-Synchronization and Asymptotic Stability

A close relationship exists between the concepts of self-synchronization and asymptotic stability. Specifically, self-synchronization in the Lorenz system is a consequence of globally stable error dynamics. Assuming that the Lorenz transmitter and receiver parameters are identical, a set of equations that govern their error dynamics is given by

$$
\begin{aligned}
\dot{e}_x &= \sigma(e_y - e_x) \\
\dot{e}_y &= -e_y - x(t)e_z \\
\dot{e}_z &= x(t)e_y - be_z.
\end{aligned}
\tag{71.3}
$$

where

$$
\begin{aligned}
e_x(t) &= x(t) - x_r(t) \\
e_y(t) &= y(t) - y_r(t) \\
e_z(t) &= z(t) - z_r(t).
\end{aligned}
$$

A sufficient condition for the error equations to be globally asymptotically stable at the origin can be determined by considering a Lyapunov function of the form

$$
E(\mathbf{e}) = \frac{1}{2}\left(\frac{1}{\sigma}e_x^2 + e_y^2 + e_z^2\right).
$$

Since σ and b in the Lorenz equations are both assumed to be positive, E is positive definite and $\dot{E}$ is negative definite. It then follows from Lyapunov's theorem that $\mathbf{e}(t) \to 0$ as $t \to \infty$. Therefore, synchronization occurs as $t \to \infty$ regardless of the initial conditions imposed on the transmitter and receiver systems.

For practical applications, it is also important to investigate the sensitivity of the synchronization to perturbations of the chaotic drive signal. Numerical experiments are summarized in Section 71.4.2, which demonstrates the robustness and signal recovery properties of the Lorenz system.

71.4.2 Robustness and Signal Recovery in the Lorenz System

When a message or other perturbation is added to the chaotic drive signal, the receiver does not regenerate a perfect replica of the drive; there is always some synchronization error. By subtracting the regenerated drive signal from the received signal, successful message recovery would result if the synchronization error was small relative to the perturbation itself. An interesting property of the Lorenz system is that the synchronization error is *not* small compared to a narrowband perturbation; nevertheless, the message can be recovered because the synchronization error is nearly *coherent* with the message. This section summarizes experimental evidence for this effect; a more detailed explanation has been given in terms of an approximate analytical model [16].

The series of experiments that demonstrate the robustness of synchronization to white noise perturbations and the ability to recover speech perturbations focus on the synchronizing properties of the transmitter Eqs. (71.1) and the corresponding receiver equations,

$$
\begin{aligned}
\dot{x}_r &= \sigma(y_r - x_r) \\
\dot{y}_r &= rs(t) - y_r - s(t)z_r \\
\dot{z}_r &= s(t)y_r - bz_r.
\end{aligned}
\tag{71.4}
$$

Previously, it was stated that with $s(t)$ equal to the transmitter signal $x(t)$, the signals x_r, y_r, and z_r will asymptotically synchronize to x, y, and z, respectively. Below, we examine the synchronization error when a perturbation $p(t)$ is added to $x(t)$, i.e., when $s(t) = x(t) + p(t)$.

First, we consider the case where the perturbation $p(t)$ is Gaussian white noise. In Fig. 71.1, we show the perturbation and error spectra for each of the three state variables vs. normalized frequency ω. Note that at relatively low frequencies, the error in reconstructing $x(t)$ slightly exceeds the perturbation of the drive but that for normalized frequencies above 20 the situation quickly reverses. An analytical model closely predicts and explains this behavior [16]. These figures suggest that the sensitivity of synchronization depends on the spectral characteristics of the perturbation signal. For signals that are bandlimited to the frequency range $0 < \omega < 10$, we would expect that the synchronization errors will be larger than the perturbation itself. This turns out to be the case, although the next experiment suggests there are additional interesting characteristics as well.

In a second experiment, $p(t)$ is a low-level speech signal (for example a message to be transmitted and recovered). The normalizing time parameter is 400 μsec and the speech signal is bandlimited to 4 kHz or equivalently to a normalized frequency ω of 10. Figure 71.2 shows the power spectrum of a representative speech signal and the chaotic signal $x(t)$. The overall chaos-to-perturbation ratio in this experiment is approximately 20 dB.

To recover the speech signal, the regenerated drive signal is subtracted at the receiver from the received signal. In this case, the recovered message is $\hat{p}(t) = p(t) + e_x(t)$. It would be expected that successful message recovery would result if $e_x(t)$ was small relative to the perturbation signal. For the Lorenz system, however, although the synchronization error is not small compared to the perturbation, the message can be recovered because $e_x(t)$ is nearly coherent with the message. This coherence has been confirmed experimentally and an explanation has been developed in terms of an approximate analytical model [16].

71.4.3 Circuit Implementation and Experiments

In Section 71.4.2, we showed that, theoretically, a low-level speech signal could be added to the synchronizing drive signal and approximately recovered at the receiver. These results were based on an analysis of the *exact* Lorenz transmitter and receiver equations. When implementing synchronized chaotic systems in hardware, the limitations of available circuit components result in approximations of the defining equations. The Lorenz transmitter and receiver equations can be implemented relatively easily with standard analog circuits [17, 20, 21]. The resulting system performance is in excellent agreement with numerical and theoretical predictions. Some potential implementation difficulties are avoided by scaling the Lorenz

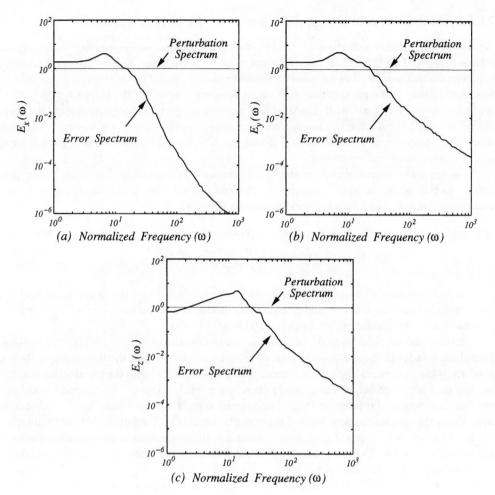

FIGURE 71.1   Power spectra of the error signals: (a) $E_x(\omega)$. (b) $E_y(\omega)$. (c) $E_z(\omega)$.

state variables according to $u = x/10$, $v = y/10$, and $w = z/20$. With this scaling, the Lorenz equations are transformed to

$$
\begin{aligned}
\dot{u} &= \sigma(v - u) \\
\dot{v} &= ru - v - 20uw \\
\dot{w} &= 5uv - bw.
\end{aligned}
\tag{71.5}
$$

For this system, which we refer to as the circuit equations, the state variables all have similar dynamic range and circuit voltages remain well within the range of typical power supply limits. Below, we discuss and demonstrate some applied aspects of the Lorenz circuits.

In Fig. 71.3, we illustrate a communication scenario that is based on chaotic signal masking and recovery [18, 19, 20, 21]. In this figure, a chaotic masking signal $u(t)$ is added to the information-bearing signal $p(t)$ at the transmitter, and at the receiver the masking is removed. By subtracting the regenerated drive signal $u_r(t)$ from the received signal $s(t)$ at the receiver, the recovered message is

$$
\hat{p}(t) = s(t) - u_r(t) = p(t) + [u(t) - u_r(t)].
$$

In this context, $e_u(t)$, the error between $u(t)$ and $u_r(t)$, corresponds directly to the error in the recovered message.

For this experiment, $p(t)$ is a low-level speech signal (the message to be transmitted and recovered). The normalizing time parameter is 400 μsec and the speech signal is bandlimited to 4 kHz or, equivalently, to

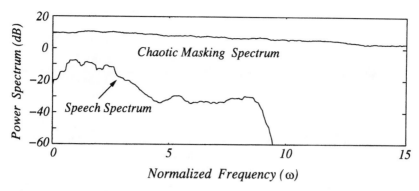

FIGURE 71.2 Power spectra of $x(t)$ and $p(t)$ when the perturbation is a speech signal.

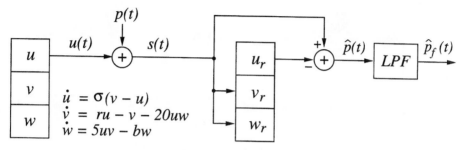

FIGURE 71.3 Chaotic signal masking and recovery system.

a normalized frequency ω of 10. In Fig. 71.4, we show the power spectrum of $p(t)$ and $\hat{p}(t)$, where $\hat{p}(t)$ is obtained from both a simulation and from the circuit. The two spectra for $\hat{p}(t)$ are in excellent agreement, indicating that the circuit performs very well. Because $\hat{p}(t)$ includes considerable energy beyond the bandwidth of the speech, the speech recovery can be improved by lowpass filtering $\hat{p}(t)$. We denote the lowpass filtered version of $\hat{p}(t)$ by $\hat{p}_f(t)$. In Fig. 71.5(a) and (b), we show a comparison of $\hat{p}_f(t)$ from both a simulation and from the circuit, respectively. Clearly, the circuit performs well and, in informal listening tests, the recovered message is of reasonable quality.

Although $\hat{p}_f(t)$ is of reasonable quality in this experiment, the presence of additive channel noise will produce message recovery errors that cannot be completely removed by lowpass filtering; there will always be some error in the recovered message. Because the message and noise are directly added to the synchronizing drive signal, the message-to-noise ratio should be large enough to allow a faithful recovery of the original message. This requires a communication channel that is nearly noise free.

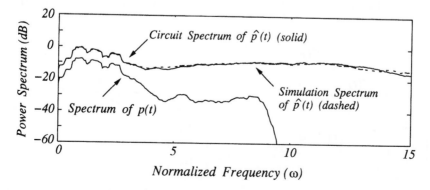

FIGURE 71.4 Power spectra of $p(t)$ and $\hat{p}(t)$ when the perturbation is a speech signal.

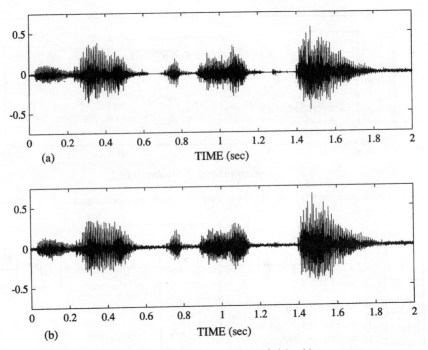

FIGURE 71.5   (a) Recovered speech (simulation). (b) Recovered speech (circuit).

An alternative approach to private communications allows the information-bearing waveform to be exactly recovered at the self-synchronizing receiver(s), even when moderate-level channel noise is present. This approach is referred to as chaotic binary communications [20, 21]. The basic idea behind this technique is to modulate a transmitter parameter with the information-bearing waveform and to transmit the chaotic drive signal. At the receiver, the parameter modulation will produce a synchronization error between the received drive signal and the receiver's regenerated drive signal with an error signal amplitude that depends on the modulation. Using the synchronization error, the modulation can be detected.

This modulation/detection process is illustrated in Fig. 71.6. To illustrate the approach, we use a periodic square-wave for $p(t)$ as shown in Fig. 71.7(a). The square-wave has a repetition frequency of approximately 110 Hz with zero volts representing the zero-bit and one volt representing the one-bit. The square-wave modulates the transmitter parameter b with the zero-bit and one-bit parameters given by $b(0) = 4$ and $b(1) = 4.4$, respectively. The resulting drive signal $u(t)$ is transmitted and the noisy received signal $s(t)$ is used as the driving input to the synchronizing receiver circuit. In Fig. 71.7(b), we show the

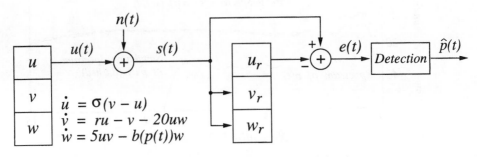

FIGURE 71.6   Communicating binary-valued bit streams with synchronized chaotic systems.

synchronization error power $e^2(t)$. The parameter modulation produces significant synchronization error during a "1" transmission and very little error during a "0" transmission. It is plausible that a detector based on the average synchronization error power, followed by a threshold device, could yield reliable performance. We illustrate in Fig. 71.7(c) that the square-wave modulation can be reliably recovered by lowpass filtering the synchronization error power waveform and applying a threshold test. The threshold device used in this experiment consisted of a simple analog comparator circuit.

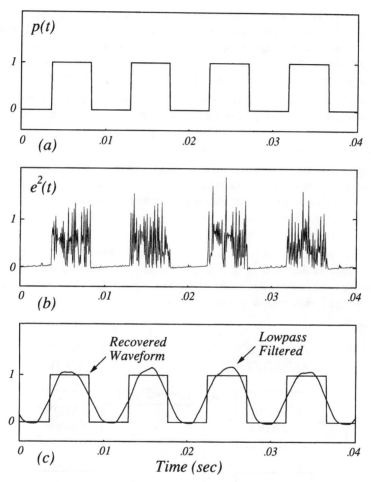

FIGURE 71.7   (a) Binary modulation waveform. (b) Synchronization error power. (c) Recovered binary waveform.

The allowable data rate of this communication technique is, of course, dependent on the synchronization response time of the receiver system. Although we have used a low bit rate to demonstrate the technique, the circuit time scale can be easily adjusted to allow much faster bit rates.

While the results presented above appear encouraging, there are many communication scenarios where it is undesirable to be restricted to the Lorenz system, or for that matter, any other low-dimensional chaotic system. In private communications, for example, the ability to choose from a wide variety of synchronized chaotic systems would be highly advantageous. In the next section, we briefly describe an approach for synthesizing an unlimited number of high-dimensional chaotic systems. The significance of this work lies in the fact that the ability to synthesize high-dimensional chaotic systems further enhances their applicability for practical applications.

71.5   Synthesizing Self-Synchronizing Chaotic Systems

An effective approach to synthesis is based on a systematic four step process. First, an algebraic model is specified for the transmitter and receiver systems. As shown in [22, 23], the chaotic system models can be very general; in [22] the model represents a large class of quadratically nonlinear systems, while in [23] the model allows for an unlimited number of Lorenz oscillators to be mutually coupled via an N-dimensional linear system.

The second step in the synthesis process involves subtracting the receiver equations from the transmitter equations and imposing a global asymptotic stability constraint on the resulting error equations. Using Lyapunov's direct method, sufficient conditions for the error system's global stability are usually straightforward to obtain. The sufficient conditions determine constraints on the free parameters of the transmitter and receiver which guarantee that they possess the global self-synchronization property.

The third step in the synthesis process focuses on the global stability of the transmitter equations. First, a family of ellipsoids in state space is defined and then sufficient conditions are determined which guarantee the existence of a *trapping region*. The trapping region imposes additional constraints on the free parameters of the transmitter and receiver equations.

The final step involves determining sufficient conditions that render all of the transmitter's fixed points unstable. In most cases, this involves numerically integrating the transmitter equations and computing the system's Lyapunov exponents and/or attractor dimension. If stable fixed points exist, the system's bifurcation parameter is adjusted until they all become unstable. Below, we demonstrate the synthesis approach for linear feedback chaotic systems.

Linear feedback chaotic systems (LFBCSs) are composed of a low-dimensional chaotic system and a linear feedback system as illustrated in Fig. 71.8. Because the linear system is N-dimensional, considerable design flexibility is possible with LFBCSs. Another practical property of LFBCSs is that they synchronize via a single drive signal while exhibiting complex dynamics.

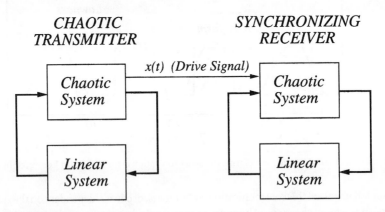

FIGURE 71.8   Linear feedback chaotic systems.

While many types of LFBCSs are possible, two specific cases have been considered in detail: (1) the chaotic Lorenz signal $x(t)$ drives an N-dimensional linear system and the output of the linear system is added to the equation for $\dot{x}$ in the Lorenz system; and (2) the Lorenz signal $z(t)$ drives an N-dimensional linear system and the output of the linear system is added to the equation for $\dot{z}$ in the Lorenz system. In both cases, a complete synthesis procedure was developed.

Below, we summarize the procedure; a complete development is given elsewhere [24].

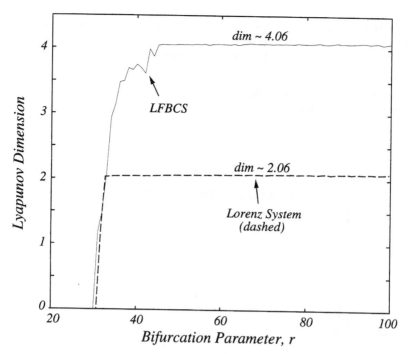

FIGURE 71.9 Lyapunov dimension of a 5-D LFBCS.

Synthesis Procedure

1. Choose any stable A matrix and any $N \times N$ symmetric positive definite matrix Q.
2. Solve $PA + A^T P + Q = 0$ for the positive definite solution P.
3. Choose any vector B and set $C = -B^T P/r$.
4. Choose any D such that $\sigma - D > 0$.

The first step of the procedure is simply the self-synchronization condition; it requires the linear system to be stable. Clearly, many choices for A are possible. The second and third steps are akin to a negative feedback constraint, i.e., the linear feedback tends to stabilize the chaotic system. The last step in the procedure restricts $\sigma - D > 0$ so that the $\dot{x}$ equation of the Lorenz system remains dissipative after feedback is applied.

For the purpose of demonstration, consider the following five-dimensional x-input/x-output LFBCS.

$$
\begin{aligned}
\dot{x} &= \sigma(y - x) + v \\
\dot{y} &= rx - y - xz \\
\dot{z} &= xy - bz \\
\begin{bmatrix} \dot{l}_1 \\ \dot{l}_2 \end{bmatrix} &= \begin{bmatrix} -\frac{1}{2} & 10 \\ -10 & -\frac{1}{2} \end{bmatrix} \begin{bmatrix} l_1 \\ l_2 \end{bmatrix} + \begin{bmatrix} 1 \\ 1 \end{bmatrix} x \\
v &= -\begin{bmatrix} 1 & 1 \end{bmatrix} \begin{bmatrix} l_1 \\ l_2 \end{bmatrix}
\end{aligned}
\tag{71.6}
$$

It can be shown in a straightforward way that the linear system satisfies the synthesis procedure for suitable choices of P, Q, and R. For the numerical demonstrations presented below, the Lorenz parameters chosen are $\sigma = 16$ and $b = 4$; the bifurcation parameter r will be varied.

In Fig. 71.9, we show the computed Lyapunov dimension as r is varied over the range, $20 < r < 100$. This figure demonstrates that the LFBCS achieves a greater Lyapunov dimension than the Lorenz system without feedback. The Lyapunov dimension could be increased by using more states in the linear system. However, numerical experiments suggest that stable linear feedback creates only negative Lyapunov exponents, limiting the dynamical complexity of LFBCSs. Nevertheless, their relative ease of implementation is an attractive practical feature.

In Fig. 71.10, we demonstrate the rapid synchronization between the transmitter and receiver systems. The curve measures the distance in state space between the transmitter and receiver trajectories when the receiver is initialized from the zero state. Synchronization is maintained indefinitely.

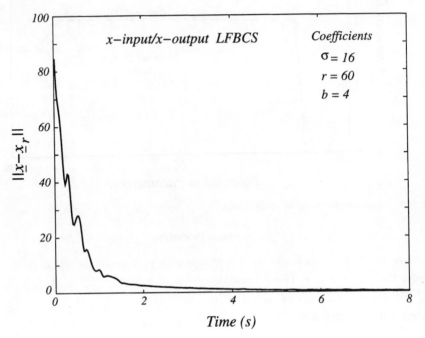

FIGURE 71.10   Self-synchronization in a 5-D LFBCS.

References

[1] Moon, F., *Chaotic Vibrations*, John Wiley & Sons, New York, 1987.

[2] Strogatz, S.H., *Nonlinear Dynamics and Chaos: with Applications to Physics, Biology, Chemistry, and Engineering*, Addison-Wesley, 1994.

[3] Abarbanel, H.D.I., Chaotic signals and physical systems, *Proc. 1992 IEEE ICASSP*, IV, 113–116, 1992.

[4] Sidorowich, J.J., Modeling of chaotic time series for prediction, interpolation and smoothing, *Proc. 1992 IEEE ICASSP*, IV, 121–124, 1992.

[5] Singer, A., Oppenheim, A.V. and Wornell, G., Codebook prediction: A nonlinear signal modeling paradigm, *Proc. 1992 IEEE ICASSP*, V, 325–328, 1992.

[6] Farmer, J.D and Sidorowich, J.J., Predicting chaotic time series, *Phys. Rev. Lett.*, 59, 845, 1987.

[7] Haykin, S. and Leung, H., Chaotic signal processing: First experimental radar results, *Proc. 1992 IEEE ICASSP*, IV, 125–128, 1992.

[8] Meyers, C., Kay, S. and Richard, M., Signal separation for nonlinear dynamical systems, *Proc. 1992 IEEE ICASSP*, IV, 129–132, 1992.

[9] Hsu, C.S., *Cell-to-Cell Mapping*, Springer-Verlag, 1987.

[10] Meyers, C., Singer, A., Shin, B. and Church, E., Modeling chaotic systems with hidden Markov models, *Proc. 1992 IEEE ICASSP*, IV, 565–568, 1992.

[11] Isabelle, S.H., Oppenheim, A.V. and Wornell, G.W., Effects of convolution on chaotic signals, *Proc. 1992 IEEE ICASSP*, IV, 133–136, 1992.

[12] Pecora, L.M. and Carroll, T.L., Synchronization in chaotic systems, *Phys. Rev. Lett.*, 64(8), 821–824, Feb. 1990.

[13] Pecora, L.M. and Carroll, T.L., Driving systems with chaotic signals, *Phys. Rev. A.*, 44, 2374–2383, Aug. 1991.

[14] Carroll, T.L. and Pecora, L.M., Synchronizing chaotic circuits, *IEEE Trans. Circuits Syst.*, 38, 453–456, Apr. 1991.

[15] Lorenz, E.N., Deterministic nonperiodic flow, *J. Atmospheric Sci.*, 20, 130–141, Mar. 1963.

[16] Cuomo, K.M., Oppenheim, A.V. and Strogatz, S.H., Robustness and signal recovery in a synchronized chaotic system, *Int. J. Bifurcation Chaos*, 3(6), 1629–1638, Dec. 1993.

[17] Cuomo, K.M. and Oppenheim, A.V., Synchronized chaotic circuits and systems for communications, *Technical Report 575*, MIT Research Laboratory of Electronics, 1992.

[18] Cuomo, K.M., Oppenheim, A.V. and Isabelle, S.H., Spread spectrum modulation and signal masking using synchronized chaotic systems, *Technical Report 570*, MIT Research Laboratory of Electronics, 1992.

[19] Oppenheim, A.V., Wornell, G.W., Isabelle, S.H. and Cuomo, K.M., Signal processing in the context of chaotic signals, in *Proc. 1992 IEEE ICASSP*, IV, 117–120, 1992.

[20] Cuomo, K.M. and Oppenheim, A.V., Circuit implementation of synchronized chaos with applications to communications, *Phys. Rev. Lett.*, 71(1), 65–68, July 1993.

[21] Cuomo, K.M., Oppenheim, A.V. and Strogatz, S.H., Synchronization of Lorenz-based chaotic circuits with applications to communications, *IEEE Trans. Circuits Syst*, 40(10), 626–633, Oct. 1993.

[22] Cuomo, K.M., Synthesizing self-synchronizing chaotic systems, *Int. J. Bifurcation Chaos*, 3(5), 1327–1337, Oct. 1993.

[23] Cuomo, K.M., Synthesizing self-synchronizing chaotic arrays, *Int. J. Bifurcation Chaos*, 4(3), 727–736, June 1994.

[24] Cuomo, K.M., Analysis and synthesis of self-synchronizing chaotic systems, Ph.D. thesis, Massachusetts Institute of Technology, Feb. 1994.

72

Nonlinear Maps

Steven H. Isabelle
Massachusetts Institute of Technology

Gregory W. Wornell
Massachusetts Institute of Technology

72.1 Introduction

One-dimensional nonlinear systems, although simple in form, are applicable in a surprisingly wide variety of engineering contexts. As models for engineering systems, their richly complex behavior has provided insight into the operation of, for example, analog-to-digital converters [1], nonlinear oscillators [2], and power converters [3]. As realizable systems, they have been proposed as random number generators [4] and as signal generators for communication systems [5, 6]. As analytic tools, they have served as mirrors for the behavior of more complex, higher dimensional systems [7, 8, 9].

Although one-dimensional nonlinear systems are, in general, hard to analyze, certain useful classes of them are relatively well understood. These systems are described by the recursion

$$x[n] = f(x[n-1]) \tag{72.1a}$$
$$y[n] = g(x[n]), \tag{72.1b}$$

initialized by a scalar initial condition $x[0]$, where $f(\cdot)$ and $g(\cdot)$ are real-valued functions that describe the evolution of a nonlinear system and the observation of its state, respectively. The dependence of the sequence $x[n]$ on its initial condition is emphasized by writing $x[n] = f^n(x[0])$ where $f^n(\cdot)$ represents the n-fold composition of $f(\cdot)$ with itself.

Without further restrictions of the form of $f(\cdot)$ and $g(\cdot)$, this class of systems is too large to easily explore. However, systems and signals corresponding to certain "well-behaved" maps $f(\cdot)$ and observation functions $g(\cdot)$ can be rigorously analyzed. Maps of this type often generate chaotic signals—loosely speaking, bounded signals that are neither periodic nor transient—under easily verifiable conditions. These chaotic signals, although completely deterministic, are in many ways analogous to stochastic processes. In fact,

0-8493-8572-5/98/$0.00+$.50
© 1998 by CRC Press LLC

one-dimensional chaotic maps illustrate in a relatively simple setting that the distinction between deterministic and stochastic signals is sometimes artificial and can be profitably emphasized or deemphasized according to the needs of an application. For instance, problems of signal recovery from noisy observations are often best approached with a deterministic emphasis, while certain signal generation problems [10] benefit most from a stochastic treatment.

72.2 Eventually Expanding Maps and Markov Maps

Although signal models of the form [1] have simple, one-dimensional state spaces, they can behave in a variety of complex ways that model a wide range of phenomena. This flexibility comes at a cost, however; without some restrictions on its form, this class of models is too large to be analytically tractable. Two tractable classes of models that appear quite often in applications are eventually expanding maps and Markov maps.

72.2.1 Eventually Expanding Maps

Eventually expanding maps—which have been used to model sigma-delta modulators [11], switching power converters [3], other switched flow systems [12], and signal generators [6, 13]—have three defining features: they are piecewise smooth, they map the unit interval to itself, and they have some iterate with slope that is everywhere greater than unity. Maps with these features generate time series that are chaotic, but on average well behaved. For reference, the formal definition is as follows, where the restriction to the unit interval is convenient but not necessary:

DEFINITION 72.1   A nonsingular map $f : [0, 1] \rightarrow [0, 1]$ is called *eventually expanding* if

1. There is a set of partition points $0 = a_0 < a_1 < \cdots a_N = 1$ such that restricted to each of the intervals $V_i = [a_{i-1}, a_i)$, called partition elements, the map $f(\cdot)$ is monotonic, continuous and differentiable.

2. The function $1/|f'(x)|$ is of bounded variation [14]. (In some definitions, this smoothness condition on the reciprocal of the derivative is replaced with a more restrictive bounded slope condition, i.e., there exists a constant B such that $|f'(x)| < B$ for all x.)

3. There exists a real $\lambda > 1$ and a integer m such that

$$\left| \frac{d}{dx} f^m(x) \right| \geq \lambda$$

wherever the derivative exists. This is the eventually expanding condition.

Every eventually expanding map can be expressed in the form

$$f(x) = \sum_{i=1}^{N} f_i(x) \chi_i(x) \tag{72.2}$$

where each $f_i(\cdot)$ is continuous, monotonic, and differentiable on the interior of the ith partition element and the indicator function $\chi_i(x)$ is defined by

$$\chi_i(x) = \begin{cases} 1 & x \in V_i , \\ 0 & x \notin V_i . \end{cases} \tag{72.3}$$

This class is broad enough to include for example, discontinuous maps and maps with discontinuous or unbounded slope. Eventually expanding maps also include a class that is particularly amenable to analysis—the Markov maps.

Markov maps are analytically tractable and broadly applicable to problems of signal estimation, signal generation, and signal approximation. They are defined as eventually expanding maps that are piecewise-linear and have some extra structure.

DEFINITION 72.2 A map $f : [0, 1] \to [0, 1]$ is an *eventually expanding, piecewise-linear, Markov map* if f is an eventually expanding map with the following additional properties:

1. The map is piecewise-linear, i.e., there is a set of partition points $0 = a_0 < a_1 < \cdots < a_N = 1$ such that restricted to each of the intervals $V_i = [a_{i-1}, a_i)$, called partition elements, the map $f(\cdot)$ is affine, i.e., the functions $f_i(\cdot)$ on the right side of (72.2) are of the form

$$f_i(x) = s_i x + b_i .$$

2. The map has the Markov property that partition points map to partition points, i.e., for each i, $f(a_i) = a_j$ for some j.

Every Markov map can be expressed in the form

$$f(x) = \sum_{i=1}^{N} (s_i x + b_i) \, \chi_i(x) , \qquad (72.4)$$

where $s_i \neq 0$ for all i. Fig. 72.1 shows the Markov map

$$f(x) = \begin{cases} (1-a)x/a + a & 0 \le x \le a \\ (1-x)/(1-a) & a < x \le 1 , \end{cases} \qquad (72.5)$$

which has partition points $\{0, a, 1\}$, and partition elements $V_1 = [0, a)$ and $V_2 = [a, 1)$.

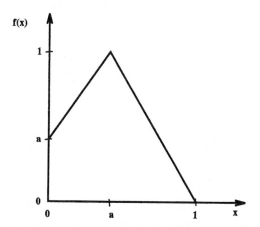

FIGURE 72.1 An example of a piecewise-linear Markov map with two partition elements.

Markov maps generate signals with two useful properties: they are, when suitably quantized, indistinguishable from signals generated by Markov chains; they are close, in a sense, to signals generated by more general eventually expanding maps [15]. These two properties lead to applications of Markov maps for generating random numbers and approximating other signals. The analysis underlying these types of applications depends on signal representations that provide insight into the structure of chaotic signals.

72.3 Signals From Eventually Expanding Maps

There are several general representations for signals generated by eventually expanding maps. Each provides different insights into the structure of these signals and proves useful in different applications. First, and most obviously, a sequence generated by a particular map is completely determined by (and is thus represented by) its initial condition $x[0]$. This representation allows certain signal estimation problems to be recast as problems of estimating the scalar initial condition. Second, and less obviously, the quantized signal $y[n] = g(x[n])$, for $n \geq 0$ generated by (72.1) with $g(\cdot)$ defined by

$$g(x) = i \qquad x \in V_i ,\tag{72.6}$$

uniquely specifies the initial condition $x[0]$ and hence the entire state sequence $x[n]$. Such quantized sequences $y[n]$ are called the symbolic dynamics associated with $f(\cdot)$ [7]. Certain properties of a map, such as the collection of initial conditions leading to periodic points, are most easily described in terms of its symbolic dynamics. Finally, a hybrid representation of $x[n]$ combining the initial condition and symbolic representations

$$H[N] = \{g(x[0]), \ldots, g(x[N]), x[N]\}$$

is often useful.

72.4 Estimating Chaotic Signals in Noise

The hybrid signal representation described in the previous section can be applied to a classical signal processing problem—estimating a signal in white Gaussian noise. For example, suppose the problem is to estimate a chaotic sequence $x[n]$, $n = 0, \ldots, N - 1$ from the noisy observations

$$r[n] = x[n] + w[n], \qquad n = 0, \ldots, N - 1 \tag{72.7}$$

where $w[n]$ is a stationary, zero-mean white Gaussian noise sequence with variance σ_w^2, and $x[n]$ is generated by iterating (72.1) from an unknown initial condition. Because $w[n]$ is white and Gaussian, the maximum likelihood estimation problem is equivalent to the constrained minimum distance problem

$$\begin{array}{c} \text{minimize} \\ x[n] : x[i] = f(x[i-1]) \end{array} \quad \varepsilon[N] = \sum_{k=0}^{N} (r[k] - x[k])^2 \tag{72.8}$$

and to the scalar problem

$$\begin{array}{c} \text{minimize} \\ x[0] \in [0, 1] \end{array} \quad \varepsilon[N] = \sum_{k=0}^{N} \left(r[k] - f^k(x[0]) \right)^2 \tag{72.9}$$

Thus, the maximum-likelihood problem can, in principle, be solved by first estimating the initial condition, then iterating (72.1) to generate the remaining estimates. However, the initial condition is often difficult to estimate directly because the likelihood function (72.9), which is highly irregular with fractal characteristics, is unsuitable for gradient-descent type optimization [16]. Another solution divides the domain of $f(\cdot)$ into subintervals and then solves a dynamic programming problem [17]; however, this solution is, in general, suboptimal and computationally expensive.

Although the maximum likelihood problem described above need not, in general, have a computationally efficient recursive solution, it does have one when, for example, the map $f(\cdot)$ is a symmetric tent map of the form

$$f(x) = \beta - 1 - \beta|x|, \quad x \in [-1, 1] \tag{72.10}$$

with parameter $1 < \beta \leq 2$ [5]. This algorithm solves for the hybrid representation of the initial condition from which an estimate of the entire signal can be determined. The hybrid representation is of the form

$$\mathbf{H}[N] = \{y[0], \ldots, y[N], x[N]\} \,,$$

where each $y[i]$ takes one of two values which, for convenience, we define as $y[i] = \text{sgn}\,(x[i])$. Since each $y[n]$ can independently takes one of two values, there are 2^N feasible solutions to this problem and a direct search for the optimal solution is thus impractical even for moderate values of N.

The resulting algorithm has computational complexity that is linear in the length of the observation, N. This efficiency is the result of a special *separation property*, possessed by the map [10]: given $y[0], \ldots,$ $y[i-1]$ and $y[i+1], \ldots, y[N]$ the estimate of the parameter $y[i]$ is independent of $y[i+1], \ldots, y[N]$. The algorithm is as follows. Denoting by $\hat{\phi}[n|m]$ the ML estimates of any sequence $\phi[n]$ given $r[k]$ for $0 \leq k \leq m$, the ML solution is of the form,

$$\hat{x}[n|n] \quad = \quad \frac{(\beta^2 - 1)\,\beta^{2n} r[n] + (\beta^{2n} - 1)\,\hat{x}[n|n-1]}{\beta^{2(n+1)} - 1} \tag{72.11}$$

$$\hat{y}[n|N] \quad = \quad \text{sgn}\,\hat{x}[n|n] \tag{72.12}$$

$$\hat{x}_{ML}[n|n] \quad = \quad \mathcal{L}_{\beta}(\hat{x}[n|n]) \,, \tag{72.13}$$

where $\hat{x}[n|n-1] = f(\hat{x}[n-1|n-1])$, the initialization is $\hat{x}[0|0] = r[0]$, and the function $\mathcal{L}_{\beta}(\hat{x}[n|n])$, defined by

$$\mathcal{L}_{\beta}(x) = \begin{cases} x & x \in (-1, \beta - 1) \\ -1 & x \leq -1 \\ \beta - 1 & x \geq \beta - 1 \end{cases} \,, \tag{72.14}$$

serves to restrict the ML estimates to the interval $x \in (-1, \beta - 1)$. The smoothed estimates $\hat{x}_{ML}[n|N]$ are obtained by converting the hybrid representation to the initial condition and then iterating the estimated initial condition forward.

72.5 Probabilistic Properties of Chaotic Maps

Almost all waveforms generated by a particular eventually expanding map have the same average behavior [18], in the sense that the time average

$$\bar{h}(x[0]) = \lim_{n \to \infty} \frac{1}{n} \sum_{k=0}^{n-1} h(x[k]) = \lim_{n \to \infty} \frac{1}{n} \sum_{k=0}^{n-1} h\left(f^k(x[0])\right) \tag{72.15}$$

exists and is essentially independent of the initial condition $x[0]$ for sufficiently well-behaved functions $h(\cdot)$. This result, which is reminiscent of results from the theory of stationary stochastic processes [19], forms the basis for a probabilistic interpretation of chaotic signals, which in turn leads to analytic methods for characterizing their time-average behavior.

To explore the link between chaotic and stochastic signals, first consider the stochastic process generated by iterating (72.1) from a random initial condition $x[0]$, with probability density function $p_0(\cdot)$. Denote by $p_n(\cdot)$ the density of the nth iterate $x[n]$. Although, in general, the members of the sequence $p_n(\cdot)$ will differ, there can exist densities, called *invariant densities*, that are time-invariant, i.e.,

$$p_0(\cdot) = p_1(\cdot) = \ldots = p_n(\cdot) \stackrel{\Delta}{=} p(\cdot) \,. \tag{72.16}$$

When the initial condition $x[0]$ is chosen randomly according to an invariant density, the resulting stochastic process is stationary [19] and its ensemble averages depend on the invariant density. Even when the initial condition is not random, invariant densities play an important role in describing the time-average behavior of chaotic signals. This role depends on, among other things, the number of invariant densities that a map possesses.

A general one-dimensional nonlinear map may possess many invariant densities. For example, eventually expanding maps with N partition elements have at least one and at most N invariant densities [20]. However, maps can often be decomposed into collections of maps, each with only one invariant density [19], and little generality is lost by concentrating on maps with only one invariant density. In this special case, the results that relate the invariant density to the average behavior of chaotic signals are more intuitive.

The invariant density, although introduced through the device of a random initial condition, can also be used to study the behavior of individual signals. Individual signals are connected to ensembles of signals, which correspond to random initial conditions, through a classical result due to Birkhoff, which asserts that the time average $\bar{h}(x[0])$ defined by Eq. (72.15) exists whenever $f(\cdot)$ has an invariant density. When the $f(\cdot)$ has only one invariant density, the time average is independent of the initial condition for almost all (with respect to the invariant density $p(\cdot)$) initial conditions and equals

$$\lim_{n\to\infty} \frac{1}{n} \sum_{k=0}^{n-1} h(x[k]) = \lim_{n\to\infty} \frac{1}{n} \sum_{k=0}^{n-1} h\left(f^k(x[0])\right) = \int h(x)p(x)dx . \tag{72.17}$$

where the integral is performed over the domain of $f(\cdot)$ and where $h(\cdot)$ is measurable.

Birkhoff's theorem leads to a relative frequency interpretation of time-averages of chaotic signals. To see this, consider the time-average of the indicator function $\tilde{\chi}_{[s-\epsilon,s+\epsilon]}(x)$, which is zero everywhere but in the interval $[s - \epsilon, s + \epsilon]$ where it is equal to unity. Using Birkhoff's theorem with Eq. (72.17) yields

$$\lim_{n\to\infty} \frac{1}{n} \sum_{k=0}^{n-1} \tilde{\chi}_{[s-\epsilon,s+\epsilon]}(x[k]) = \int \tilde{\chi}_{[s-\epsilon,s+\epsilon]}(x)p(x)dx \tag{72.18}$$

$$= \int_{[s-\epsilon,s+\epsilon]} p(x)dx \tag{72.19}$$

$$\approx 2\epsilon p(s) , \tag{72.20}$$

where Eq. (72.20) follows from Eq. (72.19) when ϵ is small and $p(\cdot)$ is sufficiently smooth. The time-average (72.18) is exactly the fraction of time that the sequence $x[n]$ takes values in the interval $[s-\epsilon, s+\epsilon]$. Thus, from (72.20), the value of the invariant density at any point s is approximately proportional to the relative frequency with which $x[n]$ takes values in a small neighborhood of the point. Motivated by this relative frequency interpretation, the probability that an arbitrary function $h(x[n])$ falls into an arbitrary set A can be defined by

$$Pr\{h(x) \in A\} = \lim_{n\to\infty} \frac{1}{n} \sum_{k=0}^{n-1} \tilde{\chi}_A(h(x[k])) . \tag{72.21}$$

Using this definition of probability, it can be shown that for any Markov map, the symbol sequence $y[n]$ defined in Section 72.3 is indistinguishable from a Markov chain in the sense that

$$Pr\{y[n]|y[n-1], \ldots, y[0]\} = Pr\{y[n]|y[n-1]\} , \tag{72.22}$$

holds for all n [21]. The first order transition probabilities can be shown to be of the form

$$Pr(y[n]|y[n-1]) = \frac{|\mathcal{V}_{y[n]}|}{|s_{y[n]}| |\mathcal{V}_{y[n-1]}|} ,$$

where the s_i are the slopes of the map $f(\cdot)$ as in Eq. (72.4) and $|\mathcal{V}_{y[n]}|$ denotes the length of the interval $\mathcal{V}_{y[n]}$. As an example, consider the asymmetric tent map

$$f(x) = \begin{cases} x/a & 0 \le x \le a \\ (1-x)/(1-a) & a < x \le 1, \end{cases}$$

with parameter in the range $0 < a < 1$ and a quantizer $g(\cdot)$ of the form (72.6). The previous results establish that $y[n] = g(x[n])$ is equivalent to a sample sequence from the Markov chain with transition probability matrix

$$[P]_{ij} = \begin{bmatrix} a & 1-a \\ a & 1-a \end{bmatrix},$$

where $[P]_{ij} = Pr\{y[n] = i | y[n-1] = j\}$. Thus, the symbolic sequence appears to have been generated by independent flips of a biased coin with the probability of heads, say, equal to a. When the parameter takes the value $a = 1/2$, this corresponds to a sequence of independent equally likely bits. Thus, a sequence of Bernoulli random variables can been constructed from a deterministic sequence $x[n]$. Based on this remarkable result, a circuit that generates statistically independent bits for cryptographic applications has been designed [4].

Some of the deeper probabilistic properties of chaotic signals depend on the integral (72.17), which in turn depends on the invariant density. For some maps, invariant densities can be determined explicitly. For example, the tent map (72.10) with $\beta = 2$ has invariant density

$$p(x) = \begin{cases} 1/2 & -1 \le x \le 1 \\ 0 & \text{otherwise} \end{cases}$$

as can be readily verified using elementary results from the theory of derived distributions of functions of random variables [22]. More generally, all Markov maps have invariant densities that are piecewise-constant function of the form

$$\sum_{i=1}^{n} c_i \chi_i(x) \tag{72.23}$$

where c_i are real constants that can be determined from the map's parameters [23]. This makes Markov maps especially amenable to analysis.

72.6 Statistics of Markov Maps

The transition probabilities computed above may be viewed as statistics of the sequence $x[n]$. These statistics, which are important in a variety of applications, have the attractive property that they are defined by integrals having, for Markov maps, readily computable, closed-form solutions. This property holds more generally—Markov maps generate sequences for which a large class of statistics can be determined in closed form. These analytic solutions have two primary advantages over empirical solutions computed by time averaging: they circumvent some of the numerical problems that arise when simulating the long sequences of chaotic data that are necessary to generate reliable averages; and they often provide insight into aspects of chaotic signals, such as dependence on a parameter, that could not be easily determined by empirical averaging.

Statistics that can be readily computed include correlations of the form

$$R_{f;h_0,h_1,\ldots,h_r} [k_1, \ldots, k_r] = \lim_{L\to\infty} \frac{1}{L} \sum_{n=0}^{L-1} h_0(x[n])h_1(x[n+k_1]) \cdots h_r(x[n+k_r]) \tag{72.24}$$

$$= \int h_0(x[n])h_1(x[n+k_1]) \cdots h_r(x[n+k_r]) p(x)\, dx, \tag{72.25}$$

where the $h_i(\cdot)'s$ are suitably well-behaved but otherwise arbitrary functions, the $k_i's$ are nonnegative integers, the sequence $x[n]$ is generated by Eq. (72.1), and $p(\cdot)$ is the invariant density. This class of statistics includes as important special cases the autocorrelation function and all higher-order moments of the time-series. Of primary importance in determining these statistics is a linear transformation called the Frobenius-Perron (FP) operator, which enters into the computation of these correlations in two ways. First, it suggests a method for determining an invariant density. Second, it provides a "change of variables" within the integral that leads to simple expressions for correlation statistics.

The definition of the FP operator can be motivated by using the device of a random initial condition $x[0]$ with density $p_0(x)$ as in Section 72.5. The FP operator describes the time evolution of this initial probability density. More precisely, it relates the initial density to the densities $p_n(\cdot)$ of the random variables $x[n] = f^n(x[0])$ through the equation

$$p_n(x) = P_f^n p_0(x) \tag{72.26}$$

where P_f^n denotes the n-fold self-composition of P_f. This definition of the FP operator, although phrased in terms of its action on probability densities, can be extended to all integrable functions. This extended operator, which is also called the FP operator, is linear and continuous. Its properties are closely related to the statistical structure of signals generated by chaotic maps (see [9] for a thorough discussion of these issues). For example, the evolution equation (72.26) implies that an invariant density of a map is a fixed point of its FP operator, that is, it satisfies

$$p(x) = P_f p(x) . \tag{72.27}$$

This relation can be used to determine explicitly the invariant densities of Markov maps [23], which may in turn be used to compute more general statistics.

Using the change of variables property of the FP operator, the correlation statistic (72.25) can be expressed as the ensemble average

$$R_{f;h_0,h_1,...,h_r} [k_1, ..., k_r] = \tag{72.28}$$

$$\int h_r(x) P_f^{k_r-k_{r-1}} \left\{ h_{r-1}(x) \cdots P_f^{k_2-k_1} \left\{ h_1(x) P_f^{k_1} \{h_0(x)p(x)\} \right\} \cdots \right\} dx . \tag{72.29}$$

Although such integrals are, for general one-dimensional nonlinear maps, difficult to evaluate, closed-form solutions exist when $f(\cdot)$ is a Markov map— a development that depends on an explicit expression for FP operator.

The FP operator of a Markov map has a simple, finite-dimensional matrix representation when it operates on certain piecewise polynomial functions. Any function of the form

$$h(\cdot) = \sum_{i=0}^{K} \sum_{j=1}^{N} a_{ij} x^i \chi_j(x)$$

can be represented by an $N(K+1)$ dimensional coordinate vector with respect to the basis

$$\{\theta_1(x), \theta_2(x), ..., \theta_{N(K+1)}\} \stackrel{\Delta}{=}$$

$$\left\{ \chi_1(x), ..., \chi_N(x), x\chi_1(x), ..., x\chi_N(x), ..., x^K \chi_1(x), ..., x^K \chi_N(x) \right\} . \tag{72.30}$$

The action of the FP operator on any such function can be expressed as a matrix-vector product: when the coordinate vector of $h(x)$ is $\mathbf{h}$, the coordinate vector of $q(x) = P_f h(x)$ is

$$\mathbf{q} = \mathbf{P}_K \mathbf{h} ,$$

where P_k is the square $N(K + 1)$ dimensional, block upper-triangular matrix

$$\mathbf{P}_K = \begin{bmatrix} \mathbf{P}_{00} & \mathbf{P}_{01} & \cdots & \cdots & \mathbf{P}_{0K} \\ 0 & \mathbf{P}_{11} & \mathbf{P}_{12} & \cdots & \mathbf{P}_{1K} \\ \vdots & \vdots & \vdots & \vdots & \vdots \\ 0 & 0 & \cdots & \cdots & \mathbf{P}_{KK} \end{bmatrix}, \tag{72.31}$$

and where each nonzero $N \times N$ block is of the form

$$\mathbf{P}_{ij} = \binom{j}{i} \mathbf{P}_0 \mathbf{B}^{j-i} \mathbf{S}^j \quad \text{for } j \geq i. \tag{72.32}$$

The $N \times N$ matrices $\mathbf{B}$ and $\mathbf{S}$ are diagonal with elements $B_{ii} = -b_i$ and $S_{ii} = 1/s_i$, respectively, while $\mathbf{P}_0 = \mathbf{P}_{00}$ is the $N \times N$ matrix with elements

$$[\mathbf{P}_0]_{ij} = \begin{cases} 1/|s_j| & i \in \mathcal{I}_j, \\ 0 & \text{otherwise.} \end{cases} \tag{72.33}$$

The invariant density of a Markov map, which is needed to compute the correlation statistic (72.25), can be determined as the solution of an eigenvector problem. It can be shown that such invariant densities are piecewise constant functions so that the fixed point equation (72.27) reduces to the matrix expression

$$\mathbf{P}_0 \mathbf{p} = \mathbf{p}.$$

Due to the properties of the matrix $\mathbf{P}_0$, this equation always has a solution that can be chosen to have nonnegative components. It follows that the correlation statistic (72.29) can always be expressed as

$$R_{f;h_0,h_1,\ldots,h_r}[k_1, \ldots, k_r] = \mathbf{g}_1^T \mathbf{M} \mathbf{g}_2 \tag{72.34}$$

where $\mathbf{M}$ is a basis correlation matrix with elements

$$[\mathbf{M}]_{ij} = \int \theta_i(x)\theta_j(x)\, dx. \tag{72.35}$$

and $\mathbf{g}_i$ are the coordinate vectors of the functions

$$g_1(x) = h_r(x) \tag{72.36}$$

$$g_2(x) = P_f^{k_r - k_{r-1}} \left\{ h_{r-1}(x) \cdots P_f^{k_2 - k_1} \left\{ h_1(x) P_f^{k_1} \left\{ h_0(x) p(x) \right\} \right\} \cdots \right\}. \tag{72.37}$$

By the previous discussion, the coordinate vectors $\mathbf{g}_1$ and $\mathbf{g}_2$ can be determined using straightforward matrix-vector operations. Thus, expression (72.34) provides a practical way of exactly computing the integral (72.29), and reveals some important statistical structure of signals generated by Markov maps.

72.7 Power Spectra of Markov Maps

An important statistic in the context of many engineering applications is the power spectrum. The power spectrum associated with a Markov map is defined as the Fourier transform of its autocorrelation sequence

$$R_{xx}[k] = \int x[n]x[n+k]p(x)dx \tag{72.38}$$

which, using Eq. (72.34) can be rewritten in the form

$$R_{xx}[k] = \mathbf{g}_1^T \mathbf{M}_1 \mathbf{P}_1^k \tilde{\mathbf{g}}_2 , \tag{72.39}$$

where $\mathbf{P}_1$ is the matrix representation of the FP operator restricted to the space of piecewise linear functions, and where $\mathbf{g}_1$ is the coordinate vector associated with the function x, and where $\tilde{\mathbf{g}}_2$ is the coordinate vector associated with $\tilde{g}_2(x) = xp(x)$.

The power spectrum is obtained from the Fourier transform of Eq. (72.39), yielding,

$$S_{xx}\left(e^{j\omega}\right) = \mathbf{g}_1^T \mathbf{M}_1 \left(\sum_{k=-\infty}^{+\infty} \mathbf{P}_1^{|k|} e^{-j\omega k} \right) \tilde{\mathbf{g}}_2 . \tag{72.40}$$

This sum can be simplified by examining the eigenvalues of the FP matrix $\mathbf{P}_1$. In general, $\mathbf{P}_1$ has eigenvalues whose magnitude is strictly less than unity, and others with unit-magnitude [9]. Using this fact, Eq. (72.40) can be expressed in the form

$$S_{xx}\left(e^{j\omega}\right) = \mathbf{h}_1^T \mathbf{M} \left(\mathbf{I} - \Gamma_2 e^{-j\omega}\right)^{-1} \left(\mathbf{I} - \Gamma_2^2\right) \left(\mathbf{I} - \Gamma_2 e^{j\omega}\right)^{-1} \tilde{\mathbf{g}}_2 + \sum_{i=1}^{m} C_i \delta\left(\omega - \omega_i\right) , \tag{72.41}$$

where Γ_2 has eigenvalues that are strictly less than one in magnitude, and C_i and ω_i depend on the unit magnitude eigenvalues of $\mathbf{P}_1$.

As Eq. (72.41) reflects, the spectrum of a Markov map is a linear combination of an impulsive component and a rational function. This implies that there are classes of rational spectra that can be generated not only by the usual method of driving white noise through a linear time-invariant filter with a rational system function, but also by iterating deterministic nonlinear dynamics. For this reason it is natural to view chaotic signals corresponding to Markov maps as "chaotic ARMA (autoregressive moving-average) processes". Special cases correspond to the "chaotic white noise" described in [5] and the first order autoregressive processes described in [24].

Consider now a simple example involving the Markov map defined in Eq. (72.5) and shown in Figure 72.1. Using the techniques described above, the invariant density is determined to be the piecewise-constant function

$$p(x) = \begin{cases} 1/(1+a) & 0 \le x \le a \\ 1/\left(1-a^2\right) & a \le x \le 1. \end{cases}$$

Using Eq. (72.41) and a parameter value $a = 8/9$, the rational part of the autocorrelation sequence associated with $f(\cdot)$ is determined to be

$$S_{xx}(z) = -\frac{42632}{459} \frac{36z^{-1} - 145 + 36z}{(9 + 8z)(9 + 8z^{-1})(64z^2 + z + 81)(64z^{-2} + z^{-1} + 81)} . \tag{72.42}$$

The power spectrum corresponding to evaluating Eq. (72.42) on the unit circle $z = e^{j\omega}$ is plotted in Figure 72.2, along with an empirical spectrum computed by periodogram averaging with a window length of 128 on a time series of length 50,000. The solid line corresponds to the analytically obtained expression (72.42), while the circles represent the spectral samples estimated by periodogram averaging.

72.8 Modeling Eventually Expanding Maps with Markov Maps

One approach to studying the statistics of more general eventually expanding maps involves approximation by Markov maps—the statistics of any eventually expanding map can be approximated to arbitrary accuracy by those of some Markov map. This approximation strategy provides a powerful method for analyzing chaotic time series from eventually expanding maps: first approximate the map by a Markov map, then use

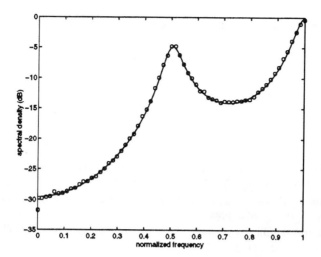

FIGURE 72.2 Comparison of analytically computed power spectrum to empirical power spectrum for the map of Figure 72.1. The solid line indicates the analytically computed spectrum, while the circles indicate the samples of the spectrum estimated by applying periodogram averaging to a time series of length 50,000.

the previously described techniques to determine its statistics. In order for this approach to be useful, an appropriate notion, the approximation quality, and a constructive procedure for generating an approximate map are required.

A sequence of piecewise-linear Markov maps $\hat{f}_i(\cdot)$ with statistics that converge to those of a given eventually expanding map $f(\cdot)$ is said to *statistically converge* to $f(\cdot)$. More formally:

DEFINITION 72.3 Let $f(\cdot)$ be an eventually expanding map with a unique invariant density $p(\cdot)$. A sequence of maps $\{\hat{f}_i(\cdot)\}$ statistically converges to $f(\cdot)$ if each $\hat{f}_i(\cdot)$ has a unique invariant density $p_i(\cdot)$ and

$$R_{\hat{f}_i,h_0,h_1,\ldots,h_r}[k_1,\ldots,k_r] \to R_{f,h_0,h_1,\ldots,h_r}[k_1,\ldots,k_r] \quad \text{as } i \to \infty$$

for any continuous $h_j(\cdot)$ and all finite k_j and finite r.

Any eventually expanding map $f(\cdot)$ is the limit of a sequence of Markov maps that statistically converges and can be constructed in a straightforward manner. The idea is to define a Markov map on an increasingly fine set of partition points that includes the original partition points of $f(\cdot)$. Denote by $\mathcal{Q}$ the set of partition points of $f(\cdot)$, and by $\mathcal{Q}_i$ the set of partition points of the ith map in the sequence of Markov map approximations. The sets of partition points for the increasingly fine approximations are defined recursively via

$$\mathcal{Q}_i = \mathcal{Q}_{i-1} \cup f^{-1}(\mathcal{Q}_{i-1}) . \tag{72.43}$$

In turn, each approximating map $\hat{f}_i(\cdot)$ is defined by specifying its value at the partition points $\mathcal{Q}_i$ by a procedure that ensures that the Markov property holds [15]. At all other points, the map $\hat{f}_i(\cdot)$ is defined by linear interpolation.

Conveniently, if $f(\cdot)$ is an eventually expanding map in the sense of Definition 72.1, then the sequence of piecewise-linear Markov approximations $\hat{f}_i(\cdot)$ obtained by the above procedure statistically converges to $f(\cdot)$, i.e., converges in the sense of Definition 72.3. This means that, for sufficiently large i, the statistics of $\hat{f}_i(\cdot)$ are close to those of $f(\cdot)$. As a practical consequence, the correlation statistics of the eventually expanding map $f(\cdot)$ can be approximated by first determining a Markov map $\hat{f}_k(\cdot)$ that is a good approximation to $f(\cdot)$, and then finding the statistics of Markov map using the techniques described in Section 72.6.

References

[1] Feely, O. and Chua, L.O., Nonlinear dynamics of a class of analog-to-digital converters, *Intl. J. Bifurcation and Chaos in Appl. Sci. Eng.*, 325, June 1992.

[2] Tang, Y.S., Mees, A.I. and Chua, L.O., Synchronization and chaos, *IEEE Trans. Circuits and Systems*, CAS-30(9), 620–626, 1983.

[3] Deane, J.H.B. and Hamill, D.C., Chaotic behavior in a current-mode controlled DC-DC converter, *Electron. Lett.*, 27, 1172–1173, 1991.

[4] Espejo, S., Martin, J.D. and Rodriguez-Vazquez, A., Design of an analog/digital truly random number generator, in *1990 IEEE International Symposium on Circuits and Systems*, 1368–1371, 1990.

[5] Papadopoulos, H.C. and Wornell, G.W., Maximum likelihood estimation of a class of chaotic signals, *IEEE Trans. Inform. Theory*, 41, 312–317, Jan. 1995.

[6] Chen, B. and Wornell, G.W., Efficient channel coding for analog sources using chaotic systems, in *Proc. IEEE GLOBECOM*, Nov. 1996.

[7] Devaney, R., *An Introduction to Chaotic Dynamical Systems*, Addison-Wesley, Reading, MA, 1989.

[8] Collet, P. and Eckmann, J.P., *Iterated Maps on the Interval as Dynamical Systems*, Birkhauser, Boston, MA, 1980.

[9] Lasota, A. and Mackey, M., *Probabilistic Properties of Deterministic Systems*, Cambridge University Press, Cambridge, 1985.

[10] Richard, M.D., *Estimation and Detection with Chaotic Systems*, Ph.D. thesis, M.I.T., Cambridge, MA, Feb. 1994. Also RLE Tech. Rep. No. 581, Feb. 1994.

[11] Risbo, L., On the design of tone-free sigma-delta modulators, *IEEE Trans. Circuits and Systems II*, 42(1), 52–55, 1995.

[12] Chase, C., Serrano, J. and Ramadge, P.J., Periodicity and chaos from switched flow systems: Contrasting examples of discretely controlled continuous systems, *IEEE Trans. Automat. Contr.*, 38, 71–83, 1993.

[13] Chua, L.O., Yao, Y. and Yang, Q., Generating randomness from chaos and constructing chaos with desired randomness, *Intl. J. Circuit Theory and Applications*, 18, 215–240, 1990.

[14] Natanson, I.P., *Theory of Functions of a Real Variable*, Frederick Ungar Publishing, New York, 1961.

[15] Isabelle, S.H., *A Signal Processing Framework for the Analysis and Application of Chaos*, Ph.D. thesis, M.I.T., Cambridge, MA, Feb. 1995. Also RLE Tech. Rep. No. 593, Feb. 1995.

[16] Myers, C., Kay S. and Richard, M., Signal separation for nonlinear dynamical systems, in *Proc. Intl. Conf. Acoust. Speech, Signal Processing*, 1992.

[17] Kay, S. and Nagesha, V., Methods for chaotic signal estimation, *IEEE Trans. Signal Processing*, 43(8), 2013, 1995.

[18] Hofbauer, F. and Keller, G., Ergodic properties of invariant measures for piecewise monotonic transformations, *Math. Z.*, 180, 119–140, 1982.

[19] Peterson, K., *Ergodic Theory*, Cambridge University Press, Cambridge, 1983.

[20] Lasota, A. and Yorke, J.A., On the existence of invariant measures for piecewise monotonic transformations, *Trans. Am. Math. Soc.*, 186, 481–488, Dec. 1973.

[21] Kalman, R., Nonlinear aspects of sampled-data control systems, in *Proc. Symp. Nonlinear Circuit Analysis*, 273–313, Apr. 1956.

[22] Drake, A.W., *Fundamentals of Applied Probability Theory*, McGraw-Hill, New York, 1967.

[23] Boyarsky, A. and Scarowsky, M., On a Class of transformations which have unique absolutely continuous invariant measures, *Trans. Am. Math. Soc.*, 255, 243–262, 1979.

[24] Sakai, H. and Tokumaru, H., Autocorrelations of a certain chaos, *IEEE Trans. Acoust., Speech, Signal Processing*, 28(5), 588–590, 1990.

73

Fractal Signals

Gregory W. Wornell
Massachusetts Institute of Technology

73.1 Introduction

Fractal signal models are important in a wide range of signal processing applications. For example, they are often well-suited to analyzing and processing various forms of natural and man-made phenomena. Likewise, the synthesis of such signals plays an important role in a variety of electronic systems for simulating physical environments. In addition, the generation, detection, and manipulation of signals with fractal characteristics has become of increasing interest in communication and remote-sensing applications.

A defining characteristic of a fractal signal is its invariance to time- or space-dilation. In general, such signals may be one-dimensional (e.g., fractal time series) or multidimensional (e.g., fractal natural terrain models). Moreover, they may be continuous-time or discrete-time in nature, and may be continuous or discrete in amplitude.

73.2 Fractal Random Processes

Most generally, fractal signals are signals having detail or structure on all temporal or spatial scales. The fractal signals of most interest in applications are those in which the structure at different scales is similar. Formally, a zero-mean random process $x(t)$ defined on $-\infty < t < \infty$ is *statistically self-similar* if its statistics are invariant to dilations and compressions of the waveform in time. More specifically, a random process $x(t)$ is statistically self-similar with parameter H if for any real $a > 0$ it obeys the scaling relation $x(t) \overset{\mathcal{P}}{=} a^{-H} x(at)$, where $\overset{\mathcal{P}}{=}$ denotes equality in a statistical sense. For *strict-sense* self-similar processes, this equality is in the sense of all finite-dimensional joint probability distributions. For *wide-sense* self-similar processes, the equality is interpreted in the sense of second-order statistics, i.e., the

$$R_x(t, s) \overset{\triangle}{=} E\left[x(t)x(s)\right] = a^{-2H} R_x(at, as)$$

A sample path of a self-similar process is depicted in Fig. 73.1.

While regular self-similar random processes cannot be stationary, many physical processes exhibiting self-similarity possess some stationary attributes. An important class of models for such phenomena are

0-8493-8572-5/98/$0.00+$.50

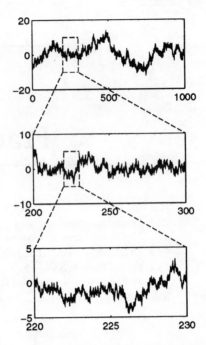

FIGURE 73.1 A sample waveform from a statistically scale-invariant random process, depicted on three different scales.

referred to as "$1/f$ processes". The $1/f$ family of statistically self-similar random processes are empirically defined as processes having measured power spectra obeying a power law relationship of the form

$$S_x(\omega) \sim \frac{\sigma_x^2}{|\omega|^\gamma} \qquad (73.1)$$

for some spectral parameter γ related to H according to $\gamma = 2H + 1$.

Generally, the power law relationship (73.1) extends over several decades of frequency. While data length typically limits access to spectral information at lower frequencies, and data resolution typically limits access to spectral content at higher frequencies, there are many examples of phenomena for which arbitrarily large data records justify a $1/f$ spectrum of the form (73.1) over all accessible frequencies. However, (73.1) is not integrable and hence, strictly speaking, does not constitute a valid power spectrum in the theory of stationary random processes. Nevertheless, a variety of interpretations of such spectra have been developed based on notions of generalized spectra [1, 2, 3].

As a consequence of their inherent self-similarity, the sample paths of $1/f$ processes are typically fractals [4]. The graphs of sample paths of random processes are one-dimensional curves in the plane; this is their "topological dimension". However, fractal random processes have sample paths that are so irregular that their graphs have an "effective" dimension that exceeds their topological dimension of unity. It is this effective dimension that is usually referred to as the "fractal" dimension of the graph. However, it is important to note that the notion of fractal dimension is not uniquely defined. There are several different definitions of fractal dimension from which to choose for a given application—each with subtle but significant differences [5]. Nevertheless, regardless of the particular definition, the fractal dimension D of the graph of a fractal function typically ranges between $D = 1$ and $D = 2$. Larger values of D correspond to functions whose graphs are increasingly rough in appearance and, in an appropriate sense, fill the plane in which the graph resides to a greater extent. For $1/f$ processes, there is an inverse relationship between the fractal dimension D and the self-similarity parameter H of the process: an increase in the parameter H yields a decrease in the dimension D, and vice-versa. This is intuitively reasonable, since an increase in H corresponds to an increase in γ, which, in turn, reflects a redistribution of power from high to low frequencies and leads to sample functions that are increasingly smooth in appearance.

A truly enormous and tremendously varied collection of natural phenomena exhibit $1/f$-type spectral behavior over many decades of frequency. A partial list includes (see, e.g., [4, 6, 7, 8, 9] and the references therein): geophysical, economic, physiological, and biological time series; electromagnetic and resistance fluctuations in media; electronic device noises; frequency variation in clocks and oscillators; variations in music and vehicular traffic; spatial variation in terrestrial features and clouds; and error behavior and traffic patterns in communication networks.

While $\gamma \approx 1$ in many of these examples, more generally $0 \leq \gamma \leq 2$. However, there are many examples of phenomena in which γ lies well outside this range. For $\gamma \geq 1$, the lack of integrability of (73.1) in a neighborhood of the spectral origin reflects the preponderance of low-frequency energy in the corresponding processes. This phenomenon is termed the *infrared catastrophe*. For many physical phenomena, measurements corresponding to very small frequencies show no low-frequency roll off, which is usually understood to reveal an inherent nonstationarity in the underlying process. Such is the case for the Wiener process (regular Brownian motion), for which $\gamma = 2$. For $\gamma \leq 1$, the lack of integrability in the tails of the spectrum reflects a preponderance of high-frequency energy and is termed the *ultraviolet catastrophe*. Such behavior is familiar for generalized Gaussian processes such as stationary white Gaussian noise ($\gamma = 0$) and its usual derivatives. When $\gamma = 1$, both catastrophes are experienced. This process is referred to as "pink" noise, particularly in the audio applications where such noises are often synthesized for use in room equalization.

An important property of $1/f$ processes is their persistent statistical dependence. Indeed, the generalized Fourier pair [10]

$$\frac{|\tau|^{\gamma-1}}{2\Gamma(\gamma)\cos(\gamma\pi/2)} \xleftrightarrow{\mathcal{F}} \frac{1}{|\omega|^{\gamma}} \tag{73.2}$$

valid for $\gamma > 0$ but $\gamma \neq 1, 2, 3, \ldots$, reflects that the autocorrelation $R_x(\tau)$ associated with the spectrum (73.1) for $0 < \gamma < 1$ is characterized by slow decay of the form $R_x(\tau) \sim |\tau|^{\gamma-1}$.

This power law decay in correlation structure distinguishes $1/f$ processes from many traditional models for time series analysis. For example, the well-studied family of autoregressive moving-average (ARMA) models have a correlation structure invariably characterized by *exponential* decay. As a consequence, ARMA models are generally inadequate for capturing long-term dependence in data.

One conceptually important characterization for $1/f$ processes is that based on the effects of bandpass filtering on such processes [11]. This characterization is strongly tied to empirical characterizations of $1/f$ processes, and is particularly useful for engineering applications. With this characterization, a $1/f$ process is formally *defined* as a wide-sense statistically self-similar random process having the property that when filtered by some arbitrary ideal bandpass filter (where $\omega = 0$ and $\omega = \pm\infty$ are strictly not in the passband), the resulting process is wide-sense stationary and has finite variance.

Among a variety of implications of this definition, it follows that such a process also has the property that when filtered by *any* ideal bandpass filter (again such that $\omega = 0$ and $\omega = \pm\infty$ are strictly not in the passband), the result is a wide-sense stationary process with a spectrum that is $\sigma_x^2/|\omega|^\gamma$ within the passband of the filter.

73.2.1 Models and Representations for $1/f$ Processes

A variety of exact and approximate mathematical models for $1/f$ processes are useful in signal processing applications. These include fractional Brownian motion, generalized autoregressive-moving-average, and wavelet-based models.

Fractional Brownian Motion and Fractional Gaussian Noise

Fractional Brownian motion and fractional Gaussian noise have proven to be useful mathematical models for Gaussian $1/f$ behavior. In particular, the fractional Brownian motion framework provides a useful construction for models of $1/f$-type spectral behavior corresponding to spectral exponents in the range $-1 < \gamma < 1$ and $1 < \gamma < 3$; see, e.g., [4, 7]. In addition, it has proven useful for addressing certain classes of signal processing problems; see, e.g., [12, 13, 14, 15].

Fractional Brownian motion is a nonstationary Gaussian self-similar process $x(t)$ with the property that its corresponding self-similar increment process

$$\Delta x(t; \varepsilon) \stackrel{\Delta}{=} \frac{x(t + \varepsilon) - x(t)}{\varepsilon}$$

is stationary for every $\varepsilon > 0$.

A convenient though specialized definition of fractional Brownian motion is given by Barton and Poor [12]:

$$x(t) \stackrel{\Delta}{=} \frac{1}{\Gamma(H + 1/2)} \left[\int_{-\infty}^{0} \left(|t - \tau|^{H-1/2} - |\tau|^{H-1/2} \right) w(\tau) \, d\tau \right.$$
$$\left. + \int_{0}^{t} |t - \tau|^{H-1/2} w(\tau) \, d\tau \right] \tag{73.3}$$

where $0 < H < 1$ is the self-similarity parameter, and where $w(t)$ is a zero-mean, stationary white Gaussian noise process with unit spectral density. When $H = 1/2$, (73.3) specializes to the Wiener process, i.e., classical Brownian motion. Sample functions of fractional Brownian motion have a fractal dimension (in the Hausdorff-Besicovitch sense) given by [4, 5]

$$D = 2 - H.$$

Moreover, the correlation function for fractional Brownian motion is given by

$$R_x(t, s) = E[x(t)x(s)] = \frac{\sigma_H^2}{2} \left(|s|^{2H} + |t|^{2H} - |t - s|^{2H} \right),$$

where

$$\sigma_H^2 = \operatorname{var} x(1) = \Gamma(1 - 2H) \frac{\cos(\pi H)}{\pi H}.$$

The increment process leads to a conceptually useful interpretation of the derivative of fractional Brownian motion: as $\varepsilon \to 0$, fractional Brownian motion has, with $H' = H - 1$, the generalized derivative [12]

$$x'(t) = \frac{d}{dt} x(t) = \lim_{\varepsilon \to 0} \Delta x(t; \varepsilon) = \frac{1}{\Gamma(H' + 1/2)} \int_{-\infty}^{t} |t - \tau|^{H'-1/2} w(\tau) \, d\tau, \tag{73.4}$$

which is termed *fractional Gaussian noise*. This process is stationary and statistically self-similar with parameter H'. Moreover, since (73.4) is equivalent to a convolution, $x'(t)$ can be interpreted as the output of an unstable linear time-invariant system with impulse response

$$v(t) = \frac{1}{\Gamma(H - 1/2)} t^{H-3/2} u(t)$$

driven by $w(t)$. Fractional Brownian motion $x(t)$ is recovered via

$$x(t) = \int_{0}^{t} x'(t) \, dt.$$

The character of the fractional Gaussian noise $x'(t)$ depends strongly on the value of H. This follows from the autocorrelation function for the increments of fractional Brownian motion, viz.,

$$R_{\Delta x}(\tau; \varepsilon) \overset{\triangle}{=} E\left[\Delta x(t; \varepsilon)\Delta x(t - \tau; \varepsilon)\right]$$

$$= \frac{\sigma_H^2 \varepsilon^{2H-2}}{2}\left[\left(\frac{|\tau|}{\varepsilon} + 1\right)^{2H} - 2\left(\frac{|\tau|}{\varepsilon}\right)^{2H} + \left(\frac{|\tau|}{\varepsilon} - 1\right)^{2H}\right],$$

which at large lags ($|\tau| \gg \varepsilon$) takes the form

$$R_{\Delta x}(\tau) \approx \sigma_H^2 H(2H - 1)|\tau|^{2H-2}. \tag{73.5}$$

Since the right side of Eq. (73.5) has the same algebraic sign as $H - 1/2$, for $1/2 < H < 1$ the process $x'(t)$ exhibits long-term dependence, i.e., persistent correlation structure; in this regime, fractional Gaussian noise is stationary with autocorrelation

$$R_{x'}(\tau) = E\left[x'(t)x'(t - \tau)\right] = \sigma_H^2(H' + 1)(2H' + 1)|\tau|^{2H'},$$

and the generalized Fourier pair (73.2) suggests that the corresponding power spectral density can be expressed as $S_{x'}(\omega) = 1/|\omega|^{\gamma'}$, where $\gamma' = 2H' + 1$. In other regimes, for $H = 1/2$ the derivative $x'(t)$ is the usual stationary white Gaussian noise, which has no correlation, while for $0 < H < 1/2$, fractional Gaussian noise exhibits persistent anti-correlation.

A closely related discrete-time fractional Brownian motion framework for modeling $1/f$ behavior has also been extensively developed based on the notion of fractional differencing [16, 17].

ARMA Models for $1/f$ Behavior

Another class of models that has been used for addressing signal processing problems involving $1/f$ processes is based on a generalized autoregressive moving-average framework. These models have been used both in signal modeling and processing applications, as well as in synthesis applications as $1/f$ noise generators and simulators [18, 19, 20].

One such framework is based on a "distribution of time constants" formulation [21, 22]. With this approach, a $1/f$ process is modeled as the weighted superposition of an infinite number of independent random processes, each governed by a distinct characteristic time-constant $1/\alpha > 0$. Each of these random processes has correlation function $R_\alpha(\tau) = e^{-\alpha|\tau|}$ corresponding to a Lorentzian spectra of the form $S_\alpha(\omega) = 2\alpha/(\alpha^2 + \omega^2)$, and can be modeled as the output of a causal LTI filter with system function $\Upsilon_\alpha(s) = \sqrt{2\alpha}/(s + \alpha)$ driven by an independent stationary white noise source. The weighted superposition of a continuum of such processes has an effective spectrum

$$S_x(\omega) = \int_0^\infty S_\alpha(\omega) f(\alpha)\, d\alpha, \tag{73.6}$$

where the weights $f(\alpha)$ correspond to the density of poles or, equivalently, relaxation times. If an un-normalizable, scale-invariant density of the form $f(\alpha) = \alpha^{-\gamma}$ is chosen for $0 < \gamma < 2$, the resulting spectrum (73.6) is $1/f$, i.e., of the form (73.1).

More practically, useful approximate $1/f$ models result from using a countable collection of single time-constant processes in the superposition. With this strategy, poles are uniformly distributed along a logarithmic scale along the negative part of the real axis in the s-plane. The process $x(t)$ synthesized in this manner has a nearly-$1/f$ spectrum in the sense that it has a $1/f$ characteristic with superimposed ripple that is uniform-spaced and of uniform amplitude on a log-log frequency plot. More specifically, when the poles are exponentially spaced according to

$$\alpha_m = \Delta^m, \qquad -\infty < m < \infty, \tag{73.7}$$

for some $1 < \Delta < \infty$, the limiting spectrum

$$S_x(\omega) = \sum_m \frac{\Delta^{(2-\gamma)m}}{\omega^2 + \Delta^{2m}} \tag{73.8}$$

satisfies

$$\frac{\sigma_L^2}{|\omega|^\gamma} \le S_x(\omega) \le \frac{\sigma_U^2}{|\omega|^\gamma} \tag{73.9}$$

for some $0 < \sigma_L^2 \le \sigma_U^2 < \infty$, and has exponenentially spaced ripple such that for all integers k

$$|\omega|^\gamma S_x(\omega) = |\Delta^k \omega|^\gamma S_x(\Delta^k \omega). \tag{73.10}$$

As Δ is chosen closer to unity, the pole spacing decreases, which results in a decrease in both the amplitude and spacing of the spectral ripple on a log-log plot.

The $1/f$ model that results from this discretization may be interpreted as an infinite-order ARMA process, i.e., $x(t)$ may be viewed as the output of a rational LTI system with a countably infinite number of both poles and zeros driven by a stationary white noise source. This implies, among other properties, that the corresponding space descriptions of these models for long-term dependence require infinite numbers of state variables. These processes have been useful in modeling physical $1/f$ phenomena; see, e.g., [23, 24, 25]. And practical signal processing algorithms for them can often be obtained by extending classical tools for processing regular ARMA processes.

The above method focuses on selecting appropriate pole locations for the extended ARMA model. The zero locations, by contrast, are controlled indirectly, and bear a rather complicated relationship to the pole locations. With other extended ARMA models for $1/f$ behavior, both pole and zero locations are explicitly controlled, often with improved approximation characteristics [20]. As an example, [6, 26] describe a construction as filtered white noise where the filter structure consists of a cascade of first-order sections each with a single pole and zero. With a continuum of such sections, exact $1/f$ behavior is obtained. When a countable collection of such sections is used, nearly-$1/f$ behavior is obtained as before. In particular, when stationary white noise is driven through an LTI system with a rational system function

$$\Upsilon(s) = \prod_{m=-\infty}^{\infty} \left[\frac{s + \Delta^{m+\gamma/2}}{s + \Delta^m} \right], \tag{73.11}$$

the output has power spectrum

$$S_x(\omega) \propto \prod_{m=-\infty}^{\infty} \left[\frac{\omega^2 + \Delta^{2m+\gamma}}{\omega^2 + \Delta^{2m}} \right]. \tag{73.12}$$

This nearly-$1/f$ spectrum also satisfies both (73.9) and (73.10). Comparing the spectra (73.12) and (73.8) reveals that the pole placement strategy for both is identical, while the zero placement strategy is distinctly different.

The system function (73.11) associated with this alternative extended ARMA model lends useful insight into the relationship between $1/f$ behavior and the limiting processes corresponding to $\gamma \to 0$ and $\gamma \to 2$. On a logarithmic scale, the poles and zeros of (73.11) are each spaced uniformly along the negative real axis in the s-plane, and to the left of each pole lies a matching zero, so that poles and zeros are alternating along the half-line. However, for certain values of γ, pole-zero cancellation takes place. In particular, as $\gamma \to 2$, the zero pattern shifts left canceling all poles except the limiting pole at $s = 0$. The resulting system is therefore an integrator, characterized by a single state variable, and generates a Wiener process as anticipated. By contrast, as $\gamma \to 0$, the zero pattern shifts right canceling all poles. The resulting system is therefore a multiple of the identity system, requires no state variables, and generates stationary white noise as anticipated.

An additional interpretation is possible in terms of a Bode plot. Stable, rational system functions composed of real poles and zeros are generally only capable of generating transfer functions whose Bode plots have slopes that are integer multiples of $20 \log_{10} 2 \approx 6$ dB/octave. However, a $1/f$ synthesis filter must fall off at $10\gamma \log_{10} 2 \approx 3\gamma$ dB/octave, where $0 < \gamma < 2$ is generally not an integer. With the extended ARMA models, a rational system function with an alternating sequence of poles and zeros is used to generate a stepped approximation to a -3γ dB/octave slope from segments that alternate between slopes of -6 dB/octave and 0 dB/octave.

Wavelet-Based Models for $1/f$ Behavior

Another approach to $1/f$ process modeling is based on the use of wavelet basis expansions. These lead to representations for processes exhibiting $1/f$-type behavior that are useful in a wide range of signal processing applications.

Orthonormal wavelet basis expansions play the role of Karhunen-Loève-type expansions for $1/f$-type processes [11, 27]. More specifically, wavelet basis expansions in terms of uncorrelated random variables constitute very good models for $1/f$-type behavior. For example, when a sufficiently regular orthonormal wavelet basis $\{\psi_n^m(t) = 2^{m/2} \psi(2^m t - n)\}$ is used, expansions of the form

$$x(t) = \sum_m \sum_n x_n^m \psi_n^m(t),$$

where the x_n^m are a collection of mutually uncorrelated, zero-mean random variables with the geometric scale-to-scale variance progression

$$\operatorname{var} x_n^m = \sigma^2 2^{-\gamma m}, \tag{73.13}$$

lead to a nearly-$1/f$ power spectrum of the type obtained via the extended ARMA models. This behavior holds regardless of the choice of wavelet within this class, although the detailed structure of the ripple in the nearly-$1/f$ spectrum can be controlled by judicious choice of the particular wavelet.

More generally, wavelet decompositions of $1/f$-type processes have a decorrelating property. For example, if $x(t)$ is a $1/f$ process, then the coefficients of the expansion of the process in terms of a sufficiently regular wavelet basis, i.e., the

$$x_n^m = \int_{-\infty}^{+\infty} x(t) \, \psi_n^m(t) \, dt$$

are very weakly correlated and obey the scale-to-scale variance progression (73.13). Again, the detailed correlation structure depends on the particular choice of wavelet [3, 11, 28, 29].

This decorrelating property is exploited in many wavelet-based algorithms for processing $1/f$ signals, where the residual correlation among the wavelet coefficients can usually be ignored. In addition, the resulting algorithms typically have very efficient implementations based on the discrete wavelet transform. Examples of robust wavelet-based detection and estimation algorithms for use with $1/f$-type signals are described in [11, 27, 30].

73.3 Deterministic Fractal Signals

While stochastic signals with fractal characteristics are important models in a wide range of engineering applications, deterministic signals with such characteristics have also emerged as potentially important in engineering applications involving signal generation ranging from communications to remote sensing.

Signals $x(t)$ of this type satisfying the deterministic scale-invariance property

$$x(t) = a^{-H} x(at) \tag{73.14}$$

for all $a > 0$, are generally referred to in mathematics as *homogeneous* functions of degree H. Strictly homogeneous functions can be parameterized with only a few constants [31], and constitute a rather limited class of models for signal generation applications. A richer class of homogeneous signal models is obtained by considering waveforms that are required to satisfy (73.14) only for values of a that are integer powers of two, i.e., signals that satisfy the dyadic self-similarity property $x(t) = 2^{-kH} x(2^k t)$ for all integers k.

Homogeneous signals have spectral characteristics analogous to those of $1/f$ processes, and have fractal properties as well. Specifically, although all non-trivial homogeneous signals have infinite energy and many have infinite power, there are classes of such signals with which one can associate a generalized $1/f$-like Fourier transform, and others with which one can associate a generalized $1/f$-like power spectrum. These two classes of homogeneous signals are referred to as energy-dominated and power-dominated, respectively [11, 32]. An example of such a signal is depicted in Fig. 73.2.

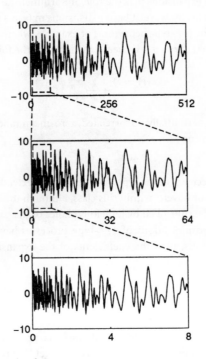

FIGURE 73.2   Dilated homogeneous signal.

Orthonormal wavelet basis expansions provide convenient and efficient representations for these classes of signals. In particular, the wavelet coefficients of such signals are related according to

$$x_n^m = \int_{-\infty}^{+\infty} x(t) \psi_n^m(t) = \beta^{-m/2} q[n],$$

where $q[n]$ is termed a generating sequence and $\beta = 2^{2H+1} = 2^\gamma$. This relationship is depicted in Fig. 73.3, where the self-similarity inherent in these signals is immediately captured in the time-frequency portrait of such signals as represented by their wavelet coefficients. More generally, wavelet expansion naturally lead to "orthonormal self-similar bases" for homogeneous signals [11, 32]. Fast synthesis and analysis algorithms for these signals are based on the discrete wavelet transform.

For some communications applications, the objective is to embed an information sequence into a fractal waveform for transmission over an unreliable communication channel. In this context, it is often natural

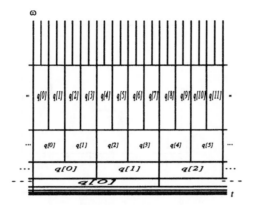

FIGURE 73.3   The time-frequency portrait of a homogeneous signal.

for $q[n]$ to be the information bearing sequence such as a symbol stream to be transmitted, and the corresponding modulation

$$x(t) = \sum_m \sum_n x_n^m \psi_n^m(t)$$

to be the fractal waveform to be transmitted. This encoding, referred to as "fractal modulation" [32] corresponds to an efficient diversity transmission strategy for certain classes of communication channels. Moreover, it can be viewed as a multirate modulation strategy in which data is transmitted simultaneously at multiple rates, and is particularly well-suited to channels having the characteristic that they are "open" for some unknown time interval T, during which they have some unknown bandwidth W and a particular signal-to-noise ratio (SNR). Such a channel model can be used, for example, to capture both characteristics of the transmission medium, such as in the case of meteor-burst channels, the constraints inherent in disparate receivers in broadcast applications, and/or the effects of jamming in military applications.

73.4   Fractal Point Processes

Fractal point processes correspond to event distributions in one or more dimensions having self-similar statistics, and are well-suited to modeling, among other examples, the distribution of stars and galaxies, demographic distributions, the sequence of spikes generated by auditory neural firing in animals, vehicular traffic, and data traffic on packet-switched data communication networks [4, 33, 34, 35, 36].

A point process is said to be self-similar if the associated counting process $N_X(t)$, whose value at time t is the total number of arrivals up to time t, is statistically invariant to temporal dilations and compressions, i.e., $N_X(t) \overset{\mathcal{P}}{=} N_X(at)$ for all $a > 0$, where the notation $\overset{\mathcal{P}}{=}$ again denotes statistical equality in the sense of all finite-dimensional distributions. An example of a sample path for such a counting process is depicted in Fig. 73.4.

Physical fractal point process phenomena generally also possess certain quasi-stationary attributes. For example, empirical measurements of the statistics of the interarrival times $X[n]$, i.e., the time interval between the $(n-1)$st and nth arrivals, are consistent with a *renewal* process. Moreover, the associated interarrival density is a power-law, i.e.,

$$f_X(x) \sim \frac{\sigma_x^2}{x^\gamma} u(x), \tag{73.15}$$

where $u(x)$ is the unit-step function. However, (73.15) is an unnormalizable density, which is a reflection of the fact that a point process cannot, in general, be both self-similar and renewing. This is analogous to the result that a continuous process cannot, in general, be simultaneously self-similar and stationary.

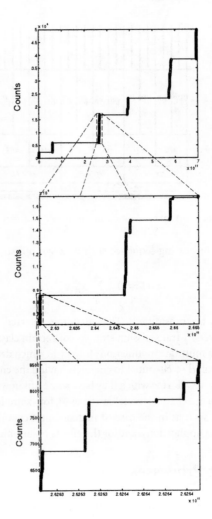

FIGURE 73.4   Dilated fractal renewal process sample path.

However, self-similar processes can possess a milder "conditionally renewing" property [37, 38]. Such processes are referred to as "fractal renewal processes" and have an effectively stationary character. The shape parameter γ in the unnormalizable interarrival density (73.15) is related to the fractal dimension D of the process via [4] $D = \gamma - 1$, and is a measure of the extent to which arrivals "cover" the line.

73.4.1 Multiscale Models

As in the case of continuous fractal processes, multiscale models are both conceptually and practically important representations for discrete fractal processes. As an example, one useful class of multiscale models corresponds to a mixture of simple Poisson processes on different time scales [37]. The construction of such processes involves a collection $\{N_{W_A}(t)\}$ of mutually independent Poisson counting processes such that $N_{W_A}(t) \overset{P}{=} N_{W_0}(e^{-A}t)$. The process $N_{W_0}(t)$ is a prototype whose mean arrival rate we denote by λ, so that the mean arrival rates of the constituent processes are related according to $\lambda_A = e^{-A}\lambda$. A random mixture of this continuum of Poisson processes yields a fractal renewal process when the index choice $A[n]$ for the nth arrival is distributed according to the extended exponential density $f_A(a) \sim \sigma_A^2 e^{-(\gamma-1)a}$. In particular, the first interarrival of the composite process is chosen to be the first arrival of the Poisson process indexed by $A[1]$; the second arrival of the composite process is chosen to be the next arrival in the Poisson process indexed by $A[2]$; and so on.

Useful alternative but equivalent constructions result from exploiting the memoryless property of Poisson processes. For example, interarrival times can be generated according to $X[n] = W_{A[n]}[n]$ or

$$X[n] = e^{A[n]} W_0[n], \tag{73.16}$$

where $W_A[n]$ is the nth interarrival time for the Poisson process indexed by A. The synthesis (73.16) is particularly appealing in that it requires access to only exponential random variables that can be obtained in practice from a single prototype Poisson process. The construction (73.16) also leads to the interpretation of a fractal point process as a Poisson process in which the arrival rate is selected randomly and independently after each arrival (and held constant between consecutive arrivals). Related doubly stochastic process models are described by Johnson et al. [39].

In addition to their use in applications requiring the synthesis of fractal point processes, these multiscale models have also proven useful in signal estimation problems. For these kinds of signal analysis applications, it is frequently convenient to replace the continuum Poisson mixture with a discrete Poisson mixture. Typically, a collection of constituent Poisson counting processes $N_{W_M}(t)$ is used, where M is an integer-valued scale index, and where the mean arrival rates are related according to $\lambda_M = \rho^{-M}\lambda$ for some λ. In this case, the scale selection is governed by an extended geometric probability mass function of the form $p_M(m) \sim \sigma_M^2 \rho^{-(\gamma-1)m}$. This discrete synthesis leads to processes that are approximate fractal renewal processes, in the sense that the interarrival densities follow a power law with a typically small amount of superimposed ripple. A number of efficient algorithms for exploiting such models in the development of robust signal estimation algorithms for use with fractal renewal processes are described in, e.g., [37].

From a broader perspective, the Poisson mixtures can be viewed as a nonlinear multiresolution signal analysis framework that can be generalized to accommodate a broad class of point process phenomena. As such, this framework is the point process counterpart to the linear multiresolution signal analysis framework based on wavelets that is used for a broad class of continuous-valued signals.

73.4.2 Extended Markov Models

An equivalent description of the discrete Poisson mixture model is in terms of an extended Markov model. The associated multiscale pure-birth process, depicted in Fig. 73.5, involves a state space consisting of a set of "superstates," each of which corresponds to fixed number of arrivals (births). Included in a superstate is a set of states corresponding to the scales in the Poisson mixture. Hence, each state is indexed by an ordered pair (i, j), where i is the superstate index and j is the scale index within each superstate.

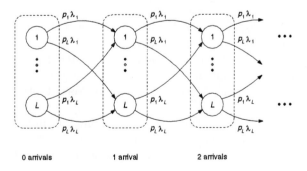

FIGURE 73.5 Multiscale pure-birth process corresponding to Poisson mixture.

The extended Markov model description has proven useful in analyzing the properties of fractal point processes under some fundamental transformations, including superposition and random erasure. These properties, in turn, provide key insight into the behavior of merging and branching traffic at nodes in data communication, vehicular, and other networks. See, e.g., [40].

Other important classes of fractal point process transformations that arise in applications involving queuing. And the extended Markov model also plays an important role in analyzing fractal queues. To address these problems, a multiscale birth-death process model is generally used [40].

References

[1] Mandelbrot, B.B. and Van Ness, H.W., Fractional Brownian motions, fractional noises and applications, *SIAM Rev.,* 10, 422–436, Oct. 1968.

[2] Mandelbrot, B., Some noises with $1/f$ spectrum: A bridge between direct current and white noise, *IEEE Trans. Inform. Theory,* IT-13, 289–298, Apr. 1967.

[3] Flandrin, P., On the spectrum of fractional Brownian motions, *IEEE Trans. Inform. Theory,* IT-35, 197–199, Jan. 1989.

[4] Mandelbrot, B.B., *The Fractal Geometry of Nature,* Freeman, San Francisco, CA, 1982.

[5] Falconer, K., *Fractal Geometry: Mathematical Foundations and Applications,* John Wiley & Sons, New York, 1990.

[6] Keshner, M.S., $1/f$ noise, *Proc. IEEE,* 70, 212–218, Mar. 1982.

[7] Pentland, A.P., Fractal-based description of natural scenes, *IEEE Trans. Pattern Anal. Machine Intell.,* PAMI-6, 661–674, Nov. 1984.

[8] Voss, R.F., $1/f$ (flicker) noise: A brief review, in *Proc. Ann. Symp. Freq. Contr.,* 40–46, 1979.

[9] van der Ziel, A., Unified presentation of $1/f$ noise in electronic devices: Fundamental $1/f$ noise sources, *Proc. IEEE,* 233–258, Mar. 1988.

[10] Champeney, D.C., *A Handbook of Fourier Theorems,* Cambridge University Press, Cambridge, England, 1987.

[11] Wornell, G.W., *Signal Processing with Fractals: A Wavelet-Based Approach,* Prentice-Hall, Upper Saddle River, NJ, 1996.

[12] Barton, R.J. and Poor, V.H., Signal detection in fractional Gaussian noise, *IEEE Trans. Inform. Theory,* IT-34, 943–959, Sept. 1988.

[13] Lundahl, T., Ohley, W.J., Kay, S.M., and Siffert, R., Fractional Brownian motion: A maximum likelihood estimator and its application to image texture, *IEEE Trans. on Medical Imaging,* MI-5, 152–161, Sept. 1986.

[14] Deriche, M. and Tewfik, A.H., Maximum likelihood estimation of the parameters of discrete fractionally differenced Gaussian noise process, *IEEE Trans. Signal Processing,* 41, 2977–2989, Oct. 1993.

[15] Deriche, M. and Tewfik, A.H., Signal modeling with filtered discrete fractional noise processes, *IEEE Trans. Signal Processing,* 41, 2839–2849, Sept. 1993.

[16] Granger, C.W. and Joyeux, R., An introduction to long memory time series models and fractional differencing, *J. Time Series Anal.,* 1 (1), 1980.

[17] Hosking, J.R.M., Fractional differencing, *Biometrika,* 68 (1), 165–176, 1981.

[18] Pellegrini, B., Saletti, R., Neri, B., and Terreni, P., $1/f^{\nu}$ noise generators, in *Noise in Physical Systems and $1/f$ Noise,* D'Amico A. and Mazzetti, P., Eds., North-Holland, Amsterdam, 1986, 425–428.

[19] Corsini, G. and Saletti, R., Design of a digital $1/f^{\nu}$ noise simulator, in *Noise in Physical Systems and $1/f$ Noise,* Van Vliet, C.M., Ed., World Scientific, Singapore, 1987, 82–86.

[20] Saletti, R., A comparison between two methods to generate $1/f^{\gamma}$ noise, *Proc. IEEE,* 74, 1595–1596, Nov. 1986.

[21] Bernamont, J., Fluctuations in the resistance of thin films, *Proc. Phys. Soc.*, 49, 138–139, 1937.

[22] van der Ziel, A., On the noise spectra of semi-conductor noise and of flicker effect, *Physica*, 16 (4), 359–372, 1950.

[23] Machlup, S., Earthquakes, thunderstorms and other $1/f$ noises, in *Noise in Physical Systems*, Meijer, P.H.E., Mountain, R.D., and Soulen, Jr., R.J., Eds., National Bureau of Standards, Washington, DC, Special publ. no. 614, 1981, 157–160.

[24] West, B.J. and Shlesinger, M.F., On the ubiquity of $1/f$ noise, *Int. J. Mod. Phys.*, 3(6), 795–819, 1989.

[25] Montroll, E.W. and Shlesinger, M.F., On $1/f$ noise and other distributions with long tails, *Proc. Natl. Acad. Sci.*, 79, 3380–3383, May 1982.

[26] Oldham, K.B. and Spanier, J., *The Fractional Calculus*, Academic Press, New York, 1974.

[27] Wornell, G.W., Wavelet-based representations for the $1/f$ family of fractal processes, *Proc. IEEE*, 81, 1428–1450, Oct. 1993.

[28] Flandrin, P., Wavelet analysis and synthesis of fractional Brownian motion, *IEEE Trans. Inform. Theory*, IT-38, 910–917, Mar. 1992.

[29] Tewfik, A.H. and Kim, M., Correlation structure of the discrete wavelet coefficients of fractional Brownian motion, *IEEE Trans. Inform. Theory*, IT-38, 904–909, Mar. 1992.

[30] Wornell, G.W. and Oppenheim, A.V., Estimation of fractal signals from noisy measurements using wavelets, *IEEE Trans. Signal Processing*, 40, 611–623, Mar. 1992.

[31] Gel'fand, I.M., Shilov, G.E., Vilenkin, N.Y., and Graev, M.I.,*Generalized Functions*, Academic Press, New York, 1964.

[32] Wornell, G.W. and Oppenheim, A.V., Wavelet-based representations for a class of self-similar signals with application to fractal modulation, *IEEE Trans. Inform. Theory*, 38, 785–800, Mar. 1992.

[33] Schroeder, M., *Fractals, Chaos, Power Laws*, Freeman, W.H., New York, 1991.

[34] Teich, M.C., Johnson, D.H., Kumar, A.R., and Turcott, R.G., Rate fluctuations and fractional power-law noise recorded from cells in the lower auditory pathway of the cat, *Hearing Res.*, 46, 41–52, June 1990.

[35] Leland, W.E., Taqqu, M.S., Willinger, W., and Wilson, D.V., On the self-similar nature of ethernet traffic, *IEEE/ACM Trans. Networking*, 2, 1–15, Feb. 1994.

[36] Paxson, V. and Floyd, S., Wide area traffic: The failure of poisson modeling, *IEEE/ACM Trans. Networking*, 3(3), 226–244, 1995.

[37] Lam, W.M. and Wornell, G.W., Multiscale representation and estimation of fractal point processes, *IEEE Trans. Signal Processing*, 43, 2606–2617, Nov. 1995.

[38] Mandelbrot, B.B., Self-similar error clusters in communication systems and the concept of conditional stationarity, *IEEE Trans. Commun. Technol.*, COM-13, 71–90, Mar. 1965.

[39] Johnson, D.H. and Kumar, A.R., Modeling and analyzing fractal point processes, in *Proc. Int. Conf. Acoust. Speech, Signal Processing*, 1990.

[40] Lam, W.M. and Wornell, G.W., Multiscale analysis of fractal point processes and queues, in *Proc. Int. Conf. Acoust. Speech, Signal Processing*, 1996.

74

Morphological Signal and Image Processing

Petros Maragos
Georgia Institute of Technology

74.1 Introduction

This chapter provides a brief introduction to the theory of morphological signal processing and its applications to image analysis and nonlinear filtering. By "morphological signal processing" we mean a broad and coherent collection of theoretical concepts, mathematical tools for signal analysis, nonlinear signal operators, design methodologies, and applications systems that are based on or related to *mathematical morphology* (MM), a set- and lattice-theoretic methodology for image analysis. MM aims at quantitatively describing the geometrical structure of image objects. Its mathematical origins stem from set theory, lattice algebra, convex analysis, and integral and stochastic geometry. It was initiated mainly by Matheron [42] and Serra [58] in the 1960s. Some of its early signal operations are also found in the work of other researchers who used cellular automata and Boolean/threshold logic to analyze binary image data in the 1950s and 1960s, as surveyed in [49, 54]. MM has formalized these earlier operations and has also added numerous new concepts and image operations. In the 1970s it was extended to gray-level

0-8493-8572-5/98/$0.00+$.50
© 1998 by CRC Press LLC

images [22, 45, 58, 62]. Originally MM was applied to analyzing images from geological or biological spec-
imens. However, its rich theoretical framework, algorithmic efficiency, easy implementability on special
hardware, and suitability for many shape-oriented problems have propelled its widespread diffusion and
adoption by many academic and industry groups in many countries as one among the dominant image
analysis methodologies. Many of these research groups have also extended the theory and applications of
MM. As a result, MM nowadays offers many theoretical and algorithmic tools to and inspires new directions in many research areas from the fields of signal processing, image processing and machine vision,
and pattern recognition.

As the name 'morphology' implies (study/analysis of shape/form), morphological signal processing can
quantify the shape, size, and other aspects of the geometrical structure of signals viewed as image objects,
in a rigorous way that also agrees with human intuition and perception. In contrast, the traditional tools of
linear systems and Fourier analysis are of limited or no use for solving geometry-based problems in image
processing because they do not directly address the fundamental issues of how to quantify shape, size, or
other geometrical structures in signals and may distort important geometrical features in images. Thus,
morphological systems are more suitable than linear systems for shape analysis. Further, they offer simple
and efficient solutions to other nonlinear problems, such as non-Gaussian noise suppression or envelope
estimation. They are also closely related to another class of nonlinear systems, the median, rank, and
stack operators, which also outperform linear systems in non-Gaussian noise suppression and in signal
enhancement with geometric constraints. Actually, rank and stack operators can be represented in terms
of elementary morphological operators. All of the above, coupled with the rich mathematical background
of mathematical morphology, make morphological signal processing a rigorous and efficient framework
to study and solve many problems in image analysis and nonlinear filtering.

74.2   Morphological Operators for Sets and Signals

74.2.1   Boolean Operators and Threshold Logic

Early works in the fields of visual pattern recognition and cellular automata dealt with analysis of binary
digital images using local neighborhood operations of the Boolean type. For example, given a sampled[1]
binary image signal $f[x]$ with values 1 for the image foreground and 0 for the background, typical signal
transformations involving a neighborhood of n samples whose indices are arranged in a window set
$W = \{y_1, y_2, \ldots, y_n\}$ would be

$$\psi_b(f)[x] = b\left(f[x - y_1], \ldots, f[x - y_n]\right)$$

where $b(v_1, \ldots, v_n)$ is a Boolean function of n variables. The mapping $f \mapsto \psi_b(f)$ is a nonlinear system,
called a *Boolean operator*. By varying the Boolean function b, a large variety of Boolean operators can be
obtained; see Table 74.1 where $W = \{-1, 0, 1\}$. For example, choosing a Boolean AND for b would *shrink*
the input image foreground, whereas a Boolean OR would *expand* it.

Two alternative implementations and views of these Boolean operations are (1) *thresholded convolutions*,
where a binary input is linearly convolved with an n-point mask of ones and then the output is thresholded at
1 or n to produce the Boolean OR or AND, respectively, and (2) min / max operations, where the moving
local minima and maxima of the binary input signal produce the same output as Boolean AND/OR,
respectively. In the thresholded convolution interpretation, thresholding at an intermediate level r between
1 and n produces a binary rank operation of the binary input data (inside the moving window). For

---

[1] Signals of a continuous variable $x \in \mathbb{R}^d$ are usually denoted by $f(x)$, whereas for signals with discrete variable
$x \in \mathbb{Z}^d$ we write $f[x]$.

TABLE 74.1 Discrete Set Operators and Their Generating Boolean Function

Set Operator $\Psi(X)$, $X \subseteq \mathbb{Z}$	Boolean function $b(v_1, v_2, v_3)$
Erosion: $X \ominus \{-1, 0, 1\}$	$v_1 v_2 v_3$
Dilation: $X \oplus \{-1, 0, 1\}$	$v_1 + v_2 + v_3$
Median: $X \square_2 \{-1, 0, 1\}$	$v_1 v_2 + v_1 v_3 + v_2 v_3$
Hit-Miss: $X \otimes (\{-1, 1\}, \{0\})$	$v_1 \overline{v_2} v_3$
Opening: $X \circ \{0, 1\}$	$v_1 v_2 + v_2 v_3$
Closing: $X \bullet \{0, 1\}$	$v_2 + v_1 v_3$

example, if $r = (n + 1)/2$, we obtain the binary median filter whose Boolean function expresses the majority voting logic; see the third example of Table 74.1. Of course, numerous other Boolean operators are possible, since there are 2^{2^n} possible Boolean functions of n variables. The main applications of such Boolean signal operations have been in biomedical image processing, character recognition, object detection, and general 2D shape analysis. Detailed accounts and more references of these approaches and applications can be found in [49, 54].

74.2.2 Morphological Set Operators

Among the new important conceptual leaps offered by mathematical morphology was to use *sets* to represent binary image signals and set operations to represent binary image transformations. Specifically, given a binary image, let its foreground be represented by the set X and its background by the set complement X^c. The Boolean OR transformation of X by a (window) set B (local neighborhood of pixels) is mathematically equivalent to the Minkowski set addition $\oplus$, also called **dilation**, of X by B:

$$X \oplus B \equiv \{x + y : x \in X, y \in B\} = \bigcup_{y \in B} X_{+y} \tag{74.1}$$

where $X_{+y} \equiv \{x + y : x \in X\}$ is the *translation* of X along the vector y. Likewise, if $B^r \equiv \{x : -x \in B\}$ denotes the *reflection* of B with respect to the axes' origin, the Boolean AND transformation of X by the reflected B is equivalent to the Minkowski set subtraction [24] $\ominus$, also called **erosion**, of X or B:

$$X \ominus B \equiv \{x : B_{+x} \subseteq X\} = \bigcap_{y \in B} X_{-y} \tag{74.2}$$

In applications, B is usually called a *structuring element* and has a simple geometrical shape and a size smaller than the image set X. As shown in Fig. 74.1, erosion shrinks the original set, whereas dilation expands it.

The erosion (74.2) can also be viewed as Boolean template matching since it gives the center points at which the shifted structuring elements fits inside the image foreground. If we now consider a set A probing the image foreground set X and another set B probing the background X^c, the set of points at which the shifted pair (A, B) fits inside the images is the **hit-miss** transformation of X by (A, B):

$$X \otimes (A, B) \equiv \{x : A_{+x} \subseteq X, B_{+x} \subseteq X^c\} \tag{74.3}$$

In the discrete case, this can be represented by a Boolean product function whose uncomplemented (complemented) variables correspond to points of $A(B)$; see Table 74.1. It has been used extensively for binary feature detection [58] and especially in document image processing [8, 9].

Dilating an eroded set by the same structuring element in general does not recover the original set but only a part of it, its opening. Performing the same series of operations to the set complement yields a set containing the original, its closing. Thus, cascading erosion and dilation gives rise to two new operations, the **opening** $X \circ B \equiv (X \ominus B) \oplus B$ and the **closing** $X \bullet B \equiv (X \oplus B) \ominus B$ of X by B. As shown in

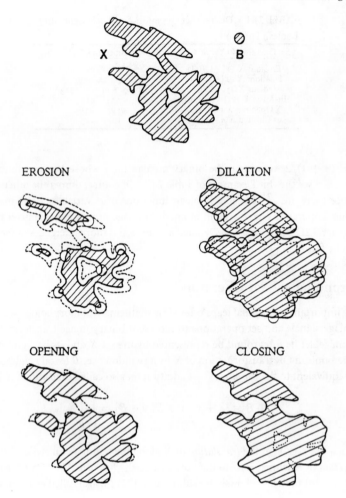

FIGURE 74.1   Erosion, dilation, opening, and closing of X (binary image of an island) by a disk B centered at the origin. The shaded areas correspond to the interior of the sets, the dark solid curve to the boundary of the transformed sets, and the dashed curve to the boundary of the original set X.

Fig. 74.1, the opening suppresses the sharp capes and cuts the narrow isthmuses of X, whereas the closing fills in the thin gulfs and small holes. Thus, if the structuring element B has a regular shape, both opening and closing can be thought of as nonlinear filters which smooth the contours of the input signal.

These set operations make mathematical morphology more general than previous approaches because it unifies and systematizes all previous digital and analog binary image operations, mathematically rigorous and notationally elegant since it is based on set theory, and intuitive since the set formalism is easily connected to mathematical logic. Further, the basic morphological set operators directly relate to the shape and size of binary images in a way that has many common points with human perception about geometry and spatial reasoning.

74.2.3   Morphological Signal Operators and Nonlinear Convolutions

In the 1970s, morphological operators were extended from binary to gray-level images and real-valued signals. Going from sets to functions was made possible by using *set representations of signals* and transforming these input sets via morphological set operations. Thus, consider a signal $f(x)$ defined on the

d-dimensional continuous or discrete domain $\mathbb{D} = \mathbb{R}^d$ or $\mathbb{Z}^d$ and assuming values in $\bar{\mathbb{R}} = \mathbb{R} \cup \{-\infty, \infty\}$. Thresholding the signal at all amplitude values v produces an ensemble of **threshold binary signals**

$$\theta_v(f)(x) \equiv 1 \text{ if } f(x) \geq v, \text{ and } 0 \text{ else,} \tag{74.4}$$

represented by the **threshold sets** [58]

$$\Theta_v(f) \equiv \{x \in \mathbb{D} : f(x) \geq v\}, \quad -\infty < v < +\infty \tag{74.5}$$

The signal can be exactly reconstructed from all its thresholded versions since

$$f(x) = \sup\{v \in \mathbb{R} : x \in \Theta_v(f)\} = \sup\{v \in \mathbb{R} : \theta_v(f)(x) = 1\} \tag{74.6}$$

Transforming each threshold set by a set operator Ψ and viewing the transformed sets as threshold sets of a new signal creates a **flat** signal operator ψ whose output is

$$\psi(f)(x) = \sup\{v \in \mathbb{R} : x \in \Psi[\Theta_v(f)]\} \tag{74.7}$$

Using set dilation and erosion in place of Ψ, the above procedure creates the two most elementary morphological signal operators: the dilation and erosion of a signal $f(x)$ by a set B:

$$(f \oplus B)(x) \quad \equiv \quad \bigvee_{y \in B} f(x - y) \tag{74.8}$$

$$(f \ominus B)(x) \quad \equiv \quad \bigwedge_{y \in B} f(x + y) \tag{74.9}$$

where $\bigvee$ denotes supremum (or maximum for finite B) and $\bigwedge$ denotes infimum (or minimum for finite B). These gray-level morphological operations can also be created from their binary counterparts using concepts from fuzzy sets where set union and intersection becomes maximum and minimum on gray-level images [22, 45]. As Fig. 74.2 shows, flat erosion (dilation) of a function f by a small convex set B reduces (increases) the peaks (valleys) and enlarges the minima (maxima) of the function. The flat opening $f \circ B = (f \ominus B) \oplus B$ of f by B smooths the graph of f from below by cutting down its peaks, whereas the closing $f \bullet B = (f \oplus B) \ominus B$ smoothes it from above by filling up its valleys.

More general morphological operators for gray-level 2D image signals $f(x)$ can be created [62] by representing the surface of f and all the points underneath by a 3D set $U(f) = \{(x, v) : v \leq f(x)\}$, called its *umbra*; then dilating or eroding $U(f)$ by the umbra of another signal g yields the umbras of two new signals, the dilation or erosion of f by g, which can be computed directly by the formulae:

$$(f \oplus g)(x) \quad \equiv \quad \bigvee_{y \in \mathbb{D}} f(x - y) + g(y) \tag{74.10}$$

$$(f \ominus g)(x) \quad \equiv \quad \bigwedge_{y \in \mathbb{D}} f(x + y) - g(y) \tag{74.11}$$

and two supplemental rules for adding and subtracting with infinities: $r \pm s = -\infty$ if $r = -\infty$ or $s = -\infty$, and $+\infty - r = +\infty$ if $r \in \mathbb{R} \cup \{+\infty\}$. These two signal transformations are nonlinear and translation-invariant. Their computational structure closely resembles that of a linear convolution $(f * g)[x] = \sum_y f[x - y]g[y]$ if we correspond the sum of products to the supremum of sums in the dilation. Actually, in the areas of convex analysis [50] and optimization [6], the operation (74.10) has been known as the **supremal convolution**. Similarly, replacing $-g(-x)$ with $g(x)$ in the erosion (74.11) yields the **infimal convolution**

$$(f \square g)(x) \equiv \bigwedge_{y \in \mathbb{D}} f(x - y) + g(y) \tag{74.12}$$

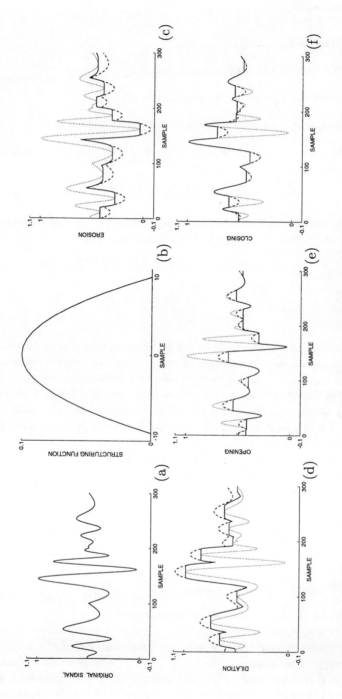

FIGURE 74.2 (a) Original signal f. (b) Structuring function g (a parabolic pulse). (c) Erosion $f \ominus g$ with dashed line and flat erosion $f \ominus B$ with solid line, where the set $B = \{x \in \mathbb{Z} : |x| \leq 10\}$ is the support of g. Dotted line shows original signal f. (d) Dilation $f \oplus g$ (dashed line) and flat dilation $f \oplus B$ (solid line). (e) Opening $f \circ g$ (dashed line) and flat opening $f \circ B$ (solid line). (f) Closing $f \bullet g$ (dashed line) and flat closing $f \bullet B$ (solid line).

The nonlinearity of ⊕ and ⊖ causes some differences between these signal operations and the linear convolutions. A major difference is that serial or parallel interconnections of systems represented by linear convolutions are equivalent to an overall linear convolution, whereas interconnections of dilations and erosions lead to entirely different nonlinear systems. Thus, there is an infinite variety of nonlinear operators created by cascading dilations and erosions or by interconnecting them in parallel via max / min or addition. Two such useful examples are the opening ∘ and closing •:

$$f \circ g \equiv (f \ominus g) \oplus g \tag{74.13}$$

$$f \bullet g \equiv (f \oplus g) \ominus g \tag{74.14}$$

which act as nonlinear smoothers.

Figure 74.2 shows that the four basic morphological transformations of a 1D signal f by a concave even function g with a compact support B have similar effects as the corresponding flat transformations by the set B. Among the few differences, the erosion (dilation) of f by g subtracts from (adds to) f the values of the moving template g during the decrease (increase) of signal peaks (valleys) and the broadening of the local signal minima (maxima) that would incur during erosion (dilation) by B. Similarly, the opening (closing) of f by g cuts the peaks (fills up the valleys) inside which no translated version of $g(-g)$ can fit and replaces these eliminated peaks (valleys) by replicas of $g(-g)$. In contrast, the flat opening or closing by B only cuts the peaks or fills valleys and creates flat plateaus in the output.

The four above morphological operators of dilation, erosion, opening, and closing have a rich collection of algebraic properties, some of which are listed in Tables 74.2 and 74.3, which endow them with a broad range of applications, make them rigorous, and lead to a variety of efficient serial or parallel implementations.

TABLE 74.2 Definitions of Operator Properties

Property	Set operator Ψ	Signal operator ψ
Translation-Invar.	$\Psi(X_{+y}) = \Psi(X)_{+y}$	$\psi[f(x - y) + c] = c + \psi(f)(x - y)$
Shift-Invariant	$\Psi(X_{+y}) = \Psi(X)_{+y}$	$\psi[f(x - y)] = \psi(f)(x - y)$
Increasing	$X \subseteq Y \Longrightarrow \Psi(X) \subseteq \Psi(Y)$	$f \leq g \Longrightarrow \psi(f) \leq \psi(g)$
Extensive	$X \subseteq \Psi(X)$	$f \leq \psi(f)$
Anti-extensive	$\Psi(X) \subseteq X$	$\psi(f) \leq f$
Idempotent	$\Psi(\Psi(X)) = \Psi(X)$	$\psi(\psi(f)) = \psi(f)$

TABLE 74.3 Properties of Basic Morphological Signal Operators

Property	Dilation	Erosion	Opening	Closing
Duality	$f \oplus g = -[(-f) \ominus g^r]$		$f \circ g = -[(-f) \bullet g^r]$	No
Distributivity	$(\vee_i f_i) \oplus g = \vee_i f_i \oplus g$	$(\wedge_i f_i) \ominus g = \wedge_i f_i \ominus g$	No	No
Composition	$(f \oplus g) \oplus h = f \oplus (g \oplus h)$	$(f \ominus g) \ominus h = f \ominus (g \oplus h)$		
Extensive	Yes if $g(0) \geq 0$	No	No	Yes
Anti-Extensive	No	Yes if $g(0) \geq 0$	Yes	No
Commutative	$f \oplus g = g \oplus f$	No	No	No
Increasing	Yes	Yes	Yes	Yes
Translation-Invar.	Yes	Yes	Yes	Yes
Idempotent	No	No	Yes	Yes

74.3 Median, Rank, and Stack Operators

Flat erosion and dilation of a discrete-domain signal $f[x]$ by a finite window $W = \{y_1, \ldots, y_n\} \subseteq \mathbb{Z}^d$ is a moving local minimum or maximum. Replacing min / max with a more general rank leads to rank operators. At each location $x \in \mathbb{Z}^d$, sorting the signal values within the reflected and shifted n-point window $(W^r)_{+x}$ in decreasing order and picking the pth largest value, $p = 1, 2, \ldots, n = $ card (W), yields the output signal from the pth **rank operator:**

$$(f \,\square_p\, W)[x] \equiv p\text{th rank of } (f[x - y_1], \ldots, f[x - y_n]) \tag{74.15}$$

For odd n and $p = (n + 1)/2$ we obtain the **median** operator. If the input signal is binary, the output is also binary since sorting preserves a signal's range. Representing the input binary signal with a set $S \subseteq \mathbb{Z}^d$, the output set produced by the pth **rank set operators** is

$$S \,\square_p\, W \equiv \{x : \text{card}\,((W^r)_{+x} \cap S) \geq p\} \tag{74.16}$$

Thus, computing the output from a set rank operator involves only counting of points and no sorting.

All rank operators *commute with thresholding* [21, 27, 41, 45, 58, 65]; i.e.,

$$\Theta_v \left[f \,\square_p\, W \right] = [\Theta_v(f)] \,\square_p\, W, \quad \forall v\,, \forall p\,. \tag{74.17}$$

This property is also shared by all morphological operators that are finite compositions or maxima/minima of flat dilations and erosions, e.g., openings and closings, by finite structuring elements. All such signal operators ψ that have a corresponding set operator Ψ and commute with thresholding can be alternatively implemented via *threshold superposition* [41, 58] as in (74.7). Namely, to transform a multilevel signal f by ψ is equivalent to decomposing f into all its threshold sets, transforming each set by the corresponding set operator Ψ, and reconstructing the output signal $\psi(f)$ via its thresholded versions. This allows us to study all rank operators and their cascade or parallel (using $\vee, \wedge$) combinations by focusing on their corresponding binary operators. Such representations are much simpler to analyze and they suggest alternative implementations that do not involve numeric comparisons or sorting.

Binary rank operators and all other binary discrete translation-invariant finite-window operators can be described by their generating Boolean function; see Table 74.1. Thus, in synthesizing discrete multilevel signal operators from their binary countparts via threshold superposition all that is needed is knowledge of this Boolean function. Specifically, transforming all the threshold binary signals $\Theta_v(f)[x]$ of an input signal $f[x]$ with an increasing Boolean function $b(u_1, \ldots, u_n)$ (i.e., containing no complemented variables) in place of the set operator Ψ in (74.7) creates a large variety of nonlinear signal operators via threshold superposition, called **stack filters** [41, 70]

$$\phi_b(f)[x] \equiv \sup\{v : b\,(\Theta_v(f)[x - y_1], \ldots, \Theta_v(f)[x - y_n]) = 1\} \tag{74.18}$$

For example, ϕ_b becomes the pth rank operator if b is equal to the sum $\binom{n}{p}$ product terms where each contains one distinct p-point subset from the n variables. In general, the use of Boolean functions facilitates the design of such discrete flat operators with determinable structural properties. Since each increasing Boolean function can be uniquely represented by an irreducible sum (product) of product (sum) terms, and each product (sum) term corresponds to an erosion (dilation), each stack filter can be represented as a finite maximum (minimum) of flat erosions (dilations) [41].

74.4 Universality of Morphological Operators

Dilations or erosions, the basic nonlinear convolutions of morphological signal processing, can be combined in many ways to create more complex morphological operators that can solve a broad variety of problems in image analysis and nonlinear filtering. In addition, they can be implemented using simple and fast software or hardware; examples include various digital [58, 61] and analog, i.e., optical or hybrid optical-electronic implementations [46, 63]. Their wide applicability and ease of implementation poses the question which signal processing systems can be represented by using dilations and erosions as the basic building blocks. Toward this goal, a theory was introduced in [33, 34] that represents a broad class of nonlinear and linear operators as a minimal combination of erosions or dilations. Here we summarize the main results of this theory, in a simplified way, restricting our discussion only to signals with discrete domain $\mathbb{D} = \mathbb{Z}^d$.

Consider a translation-invariant set operator Ψ on the class $\mathcal{P}(\mathbb{D})$ of all subsets of $\mathbb{D}$. Any such Ψ is uniquely characterized by its **kernel** that is defined [42] as the subclass $\mathrm{Ker}(\Psi) \equiv \{X \in \mathcal{P}(\mathbb{D}) : 0 \in \Psi(X)\}$ of input sets, where 0 is the origin of $\mathbb{D}$. If Ψ is also increasing, then it can be represented [42] as the union of erosions by its kernel sets and as the intersection of dilations by the reflected kernel sets of its *dual operator* $\Psi^d(X) \equiv [\Psi(X^c)]^c$. This kernel representation can be extended to signal operators ψ on the class $\mathrm{Fun}(\mathbb{D}, \bar{\mathbb{R}})$ of signals with domain $\mathbb{D}$ and range $\bar{\mathbb{R}}$. The kernel of ψ is defined as the subclass $\mathrm{Ker}(\psi) = \{f \in \mathrm{Fun}(\mathbb{D}, \bar{\mathbb{R}}) : [\psi(f)](0) \geq 0\}$ of input signals. If ψ is translation-invariant and increasing, then it can be represented [33, 34] as the pointwise supremum of erosions by its kernel functions, and as the infimum of dilations by the reflected kernel functions of its dual operator $\psi^d(f) \equiv -\psi(-f)$.

The two previous kernel representations require an infinite number of erosions or dilations to represent a given operator because the kernel contains an infinite number of elements. However, we can find more efficient (requiring less erosions) representations by using only a substructure of the kernel, its basis. The **basis** $\mathrm{Bas}(\cdot)$ of a set (signal) operator is defined [33, 34] as the collection of kernel elements that are *minimal* with respect to the ordering $\subseteq$ ($\leq$).

If a *translation-invariant increasing* set operator Ψ is also *upper semicontinuous*, i.e., obeys a monotonic continuity where $\Psi(\bigcap_n X_n) = \bigcap_n \Psi(X_n)$ for any decreasing set sequence X_n, then Ψ has a nonempty basis and can be represented via erosions only by its basis sets. If the dual Ψ^d is also upper semicontinuous, then its basis sets provide an alternative representation of Ψ via dilations:

$$\Psi(X) = \bigcup_{A \in \mathrm{Bas}(\Psi)} X \ominus A = \bigcap_{B \in \mathrm{Bas}(\Psi^d)} X \oplus B^r \tag{74.19}$$

Similarly, any signal operator ψ that is translation-invariant, increasing, and upper semicontinuous (i.e., $\psi(\wedge_n f_n) = \wedge_n \psi(f_n)$ for any decreasing function sequence f_n) can be represented as the supremum of erosions by its basis functions, and (if ψ^d is upper semicontinuous) as the infimum of dilations by the reflected basis functions of its dual operators:

$$\psi(f) = \bigvee_{g \in \mathrm{Bas}(\psi)} f \ominus g = \bigwedge_{h \in \mathrm{Bas}(\psi^d)} f \oplus h^r \tag{74.20}$$

where $h^r(x) \equiv h(-x)$. Finally, if ϕ is a flat signal operator as in (74.7) that is translation-invariant and commutes with thresholding, then ϕ can be represented as a supremum of erosions by the basis sets of its corresponding set operator Φ:

$$\phi(f) = \bigvee_{A \in \mathrm{Bas}(\Phi)} f \ominus A = \bigwedge_{B \in \mathrm{Bas}(\Phi^d)} f \oplus B^r \tag{74.21}$$

While all the above representations express translation-invariant increasing operators via erosions or dilations, operators that are not necessarily increasing can be represented [4] via operations closely related to hit-miss transformations.

Representing operators that satisfy a few general properties in terms of elementary morphological operations can be applied to more complex morphological systems and various other filters such as linear rank, hybrid linear/rank, and stack filters, as the following examples illustrate.

EXAMPLE 74.1: Morphological Filters

All systems made up of serial or sup/inf combinations of erosions, dilations, opening, and closings admit a basis, which is finite if the system's local definition depends on a finite window. For example, the set opening $\Phi(X) = X \circ A$ has as a basis the set collection $\text{Bas}(\Phi) = \{A_{-a} : a \in A\}$. Consider now 1D discrete-domain signals and let $A = \{-1, 0, 1\}$. Then, the basis of Φ has 3 sets: $G_1 = A_{-1}, G_2 = A, G_3 = A_{+1}$. The basis of the dual operator $\Phi^d(X) = X \bullet A$ has 4 sets: $H_1 = \{0\}, H_2 = \{-2, 1\}, H_3 = \{-1, 2\}, H_4 = \{-1, 1\}$. The flat signal operator corresponding to Φ is the opening $\phi(f) = f \circ A$. Thus, from (74.21), the signal opening can also be realized as a max (min) of local minima (maxima):

$$(f \circ A)[x] = \bigvee_{i=1}^{3} \left\{ \bigwedge_{y \in G_i} f[x+y] \right\} = \bigwedge_{k=1}^{4} \left\{ \bigvee_{y \in H_k} f[x+y] \right\} . \tag{74.22}$$

EXAMPLE 74.2: Linear Filters

A linear shift-invariant filter is translation-invariant and increasing (see Table 74.2 for definitions) if its impulse response is everywhere nonnegative and has area equal to one. Consider the 2-point FIR filter $\psi(f)[x] = af[x] + (1-a)f[x-1]$, where $0 < a < 1$. The basis of ψ consists of all functions $g[x]$ with $g[0] = r \in \mathbb{R}, g[-1] = -ar/(1-a)$, and $g[x] = -\infty$ for $x \neq 0, -1$. Then (74.20) yields

$$af[x] + (1-a)f[x-1] = \bigvee_{r \in \mathbb{R}} \left[\min \left\{ f[x] - r, f[x-1] + \frac{ar}{1-a} \right\} \right], \tag{74.23}$$

which expresses a linear convolution as a supremum of erosions. FIR linear filters have an infinite basis, which is a finite-dimensional vector space.

EXAMPLE 74.3: Median Filters

All rank operators have a finite basis; hence, they can be expressed as a finite max-of-erosions or min-of-dilations. Further, they commute with thresholding, which allows us to focus only on their binary versions. For example, the set median by the window $W = \{-1, 0, 1\}$ has 3 basis sets: $\{-1, 0\}, \{-1, 1\}$, and $\{0, 1\}$. Hence, (74.21) yields

$$\text{median} \left(f[x-1], f[x], f[x+1] \right) = \max \left\{ \begin{array}{l} \min(f[x-1], f[x]), \\ \min[f(x-1), f(x+1)], \\ \min[f(x), f(x+1)] \end{array} \right\} . \tag{74.24}$$

EXAMPLE 74.4: Stack Filters

Stack filters (74.18) are discrete translation-invariant flat operators ϕ_b, locally defined on a finite window W, and are generated by a increasing Boolean function $b(v_1, \ldots, v_n)$, where $n = \text{card}(W)$. This function corresponds to a translation-invariant increasing set operator Φ. For example, consider 1D signals, let $W = \{-2, -1, 0, 1, 2\}$ and

$$b(v_1, \ldots, v_5) = v_1 v_2 v_3 + v_2 v_3 v_4 + v_3 v_4 v_5 = v_3 (v_1 + v_4)(v_2 + v_4)(v_2 + v_5) . \tag{74.25}$$

This function generates via threshold superposition the flat opening $\phi_b(f) = f \circ A, A = \{-1, 0, 1\}$, of (74.22). There is one-to-one correspondence between the three prime implicants of b and the erosions (local min) by the three basis sets of Φ, as well as between the four prime implicates of β and the dilations (local max) by the four basis sets of the dual Φ^d. In general, given b, Φ or ϕ_b is found by replacing Boolean AND/OR with set $\cap/\cup$ or with min / max, respectively. Conversely, given ϕ_b, we can find its generating Boolean function from the basis of its set operator (or directly from its max / min representation if available) [41].

The above examples show the power of the general representation theorems. An interesting applications of these results is the design of morphological systems via their basis [5, 20, 31]. Given the wide applicability of erosions/dilations, their parallelism, and their simple implementations, the previous theorems theoretically support a general purpose vision (software or hardware) module that can perform erosions/dilations, based on which numerous other complex image operations can be built.

74.5 Morphological Operators and Lattice Theory

In the late 1980s and 1990s a new and more general formalization of morphological operators was introduced [59, chaps.1,5-8], [26, 51, 52], which views them as operators on complete lattices. A *complete lattice* is a set $\mathcal{L}$ equipped with a partial ordering $\leq$ such that $(\mathcal{L}, \leq)$ has the algebraic structure of a *partially ordered set (poset)* where the supremum and infimum of any of its subsets exist in $\mathcal{L}$. For any subset $\mathcal{K} \subseteq \mathcal{L}$, its *supremum* $\vee \mathcal{K}$ and *infimum* $\wedge \mathcal{K}$ are defined as the lowest (with respect to $\leq$) upper bound and greatest lower bound of $\mathcal{K}$, respectively. The two main examples of complete lattices used in morphological processing are: (1) the set space $\mathcal{P}(\mathbb{D})$ where the $\vee/\wedge$ lattice operations are the set union/intersection, and (2) the signal space $\mathrm{Fun}(\mathbb{D}, \bar{\mathbb{R}})$ where the $\vee/\wedge$ lattice operations are the supremum/infimum of sets of real numbers. Increasing operators on $\mathcal{L}$ are of great importance because they preserve the partial ordering, and among them four fundamental examples are:

$$\delta \text{ is } \textbf{dilation} \quad \Longleftrightarrow \quad \delta\left(\bigvee_{i \in I} f_i\right) = \bigvee_{i \in I} \delta(f_i) \tag{74.26}$$

$$\varepsilon \text{ is } \textbf{erosion} \quad \Longleftrightarrow \quad \varepsilon\left(\bigwedge_{i \in I} f_i\right) = \bigwedge_{i \in I} \varepsilon(f_i) \tag{74.27}$$

$$\alpha \text{ is } \textbf{opening} \quad \Longleftrightarrow \quad \alpha \text{ is increasing, idempotent, and anti-extensive} \tag{74.28}$$

$$\beta \text{ is } \textbf{closing} \quad \Longleftrightarrow \quad \beta \text{ is increasing, idempotent, and extensive} \tag{74.29}$$

where I is an arbitrary index set.

The above definitions allow broad classes of signal operators to be grouped as lattice dilations, erosions, openings, or closing and their common properties to be studied under the unifying lattice framework. Thus, the translation-invariant morphological dilations, erosions, openings, and closings we saw before are simply special cases of their lattice counterparts. Next, we see some examples and applications of the above general definitions.

EXAMPLE 74.5:   Dilation and Translation-Invariant (DTI) Systems

Consider a signal operator that is shift-invariant and obeys a supremum-of-sums superposition:

$$\mathcal{D}\left[\bigvee_i c_i + f_i(x)\right] = \bigvee_i c_i + \mathcal{D}[f_i(x)] \tag{74.30}$$

Then $\mathcal{D}$ is both a lattice dilation and translation-invariant. We call it a DTI system in analogy to linear time-invariant (LTI) systems that are shift-invariant and obey a linear (sum-of-products) superposition. As an LTI system corresponds in the time-domain to a linear convolution with its impulse response, a DTI system can be represented as a supremal convolution with its upper 'impulse response' $g_\vee(x)$ defined as its output when the input is the upper *zero impulse* $\iota(x)$, defined in Table 74.4. Specifically,

$$\mathcal{D} \text{ is DTI } \Longleftrightarrow \mathcal{D}(f) = f \oplus g_\vee, \quad g_\vee \equiv \mathcal{D}(\iota) \tag{74.31}$$

A similar class is the **erosion and translation-invariant (ETI)** systems ε which are shift-invariant and obey an infimum-of-sums superposition as in (74.30) but with $\vee$ replaced by $\wedge$. Such systems are equivalent to infimal convolutions with their lower impulse response $g_\wedge = \varepsilon(-\iota)$, defined as the system's output due to the lower impulse $-\iota(x)$. Thus, DTI and ETI systems are uniquely determined in the time/spatial domain by their impulse responses, which also control their causality and stability [37].

TABLE 74.4   Examples of Upper Slope Transform

Signal: $f(x)$	Transform: $F_\vee(a)$
$\iota(x - x_0) \equiv 0$ if $x = x_0$, and $-\infty$ else	$-a x_0$
$a_0 x$	$-\iota(a - a_0)$
$\lambda(x) \equiv 0$ if $x \geq 0$, and $-\infty$ else	$-\lambda(a)$
$a_0 x + \lambda(x)$	$-\lambda(a - a_0)$
$\begin{cases} 0, & \|x\| \leq r \\ -\infty, & \|x\| > r \end{cases}$	$r\|a\|$
$-a_0\|x\|, \ a_0 > 0$	$\begin{cases} 0, & \|a\| \leq a_0 \\ +\infty, & \|a\| > a_0 \end{cases}$
$\sqrt{1 - x^2}, \ \|x\| \leq 1$	$\sqrt{1 + a^2}$
$-(\|x\|^p)/p, \ p > 1$	$(\|a\|^q)/q, \ 1/p + 1/q = 1$
$\exp(x)$	$a(1 - \log a)$

EXAMPLE 74.6:   Shift-Varying Dilation

Let $\delta_B(f) = f \oplus B$ be the shift-invariant flat dilation of (74.8). In applying it to nonstationary signals, the need may arise to vary the moving window B by actually having a family of windows $B(x)$, possibly varying at each location x. This creates the new operator

$$\delta_B(f)(x) = \bigvee_{y \in B(x)} f(x - y) \tag{74.32}$$

which is still a lattice dilation, i.e., it distributes over suprema, but it is shift-varying.

EXAMPLE 74.7:   Adjunctions

An operator pair (ε, δ) is called an **adjunction** if $\delta(f) \leq g \Longleftrightarrow f \leq \varepsilon(g)$ for all $f, g \in \mathcal{L}$. Given a dilation δ, there is a unique erosion ε such that (ε, δ) is adjunction, and vice versa. Further, if (ε, δ) is an adjunction, then δ is a dilation, ε is an erosion, $\delta\varepsilon$ is an opening, and $\varepsilon\delta$ is a closing. Thus, from any adjunction we can generate an opening via the composition of its erosion and dilation. If ε and δ are the translation-invariant morphological erosion and dilation in (74.11) and (74.10), then $\delta\varepsilon$ coincides with the translation-invariant morphological opening of (74.13). But there are also numerous other possibilities.

EXAMPLE 74.8:   Radial Opening

If a 2D image f contains 1D objects, e.g., lines, and B is a 2D convex structuring element, then the opening or closing of f by B will eliminate these 1D objects. Another problem arises when f contains large-scale objects with sharp corners that need to be preserved; in such cases opening or closing f by a disk B will round these corners. These two problems could be avoided in some cases if we replace the conventional opening with

$$\alpha(f) = \bigvee_{\theta} f \circ L_{\theta} \qquad (74.33)$$

where the sets L_{θ} are rotated versions of a line segment L at various angles $\theta \in [0, 2\pi)$. The operator α, called **radial opening**, is a lattice opening in the sense of (74.28). It has the effect of preserving an object in f if this object is left unchanged after the opening by L_{θ} in at least one of the possible orientations θ.

EXAMPLE 74.9: Opening by Reconstruction

Consider a set $X = \bigcup_i X_i$ as a union of disjoint connected components X_i and let $M \subseteq X_j$ be a **marker** in the jth component; i.e., M could be a single point or some feature set in X that lies only in X_j. Then, define the **conditional dilation** of M by B within X as

$$\delta_{B|X}(M) \equiv (M \oplus B) \cap X \qquad (74.34)$$

If B is a disk with a radius smaller than the distance between X_j and any of the other components, then by iterating this conditional dilation we can obtain in the limit

$$MR_{B|X}(M) = \lim_{n \to \infty} \underbrace{\left(\delta_{B|X} \cdots (\delta_{B|X}(\delta_{B|X}(M))) \right)}_{n \text{ times}} \qquad (74.35)$$

the whole component X_j. The operator MR is a lattice opening, called **opening by reconstruction**, and its output is called the *morphological reconstruction* of the component from the marker. An example is shown in Fig. 74.3. It can extract large-scale components of the image from knowledge only of a smaller marker inside them.

74.6 Slope Transforms

Fourier transforms are among the most useful linear signal transformations because they enable us to analyze the processing of signals by linear time-invariant (LTI) systems in the frequency domain, which could be more intuitive or easier to implement. Similarly, there exist some nonlinear signal transformations, called slope transforms, which allow the analysis of the dilation and erosion translation-invariant (DTI and ETI) systems in a transform domain, the slope domain. First, we note that the lines $f(x) = ax + b$ are *eigenfunctions* of any DTI system $\mathcal{D}$ or ETI system $\mathcal{E}$ because

$$\mathcal{D}[ax + b] = ax + b + G_{\vee}(a), \quad G_{\vee}(a) \equiv \bigvee_x g_{\vee}(x) - ax$$

$$\mathcal{E}[ax + b] = ax + b + G_{\wedge}(a), \quad G_{\wedge}(a) \equiv \bigwedge_x g_{\wedge}(x) - ax \qquad (74.36)$$

with corresponding eigenvalues $G_{\vee}(a)$ and $G_{\wedge}(a)$, which are called, respectively, the upper and lower **slope response** of the DTI and ETI system. They measure the amount of shift in the intercept of the input lines with slope a and are conceptually similar to the frequency response of LTI systems.

Then, by viewing the slope response as a signal transform with variable the slope $a \in \mathbb{R}$, we define [37] for a 1D signal $f : \mathbb{D} \to \bar{\mathbb{R}}$ its **upper slope transform** $F_{\vee}$ and its **lower slope transform**[2] $F_{\wedge}$ as the

---

[2]In convex analysis [50], to a convex function h there uniquely corresponds its *Fenchel conjugate* $h^*(a) = \bigvee_x ax - h(x)$, which is the negative of the lower slope transform of h.

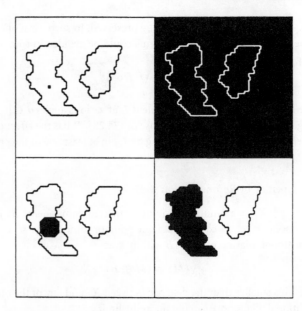

FIGURE 74.3 Let X be the union of the two region boundaries in the top left image, and let M be the single-point marker inside the left region. Top right shows the complement X^c. If $Y_0 = M$ and B is a disk-like set whose radius does not exceed the width of the region boundary, iterating the conditional dilation $Y_i = (Y_{i-1} \oplus B) \cap X^c$, for $i = 1, 2, 3, \ldots$, yields in the limit (reached at $i = 18$ in this case) the interior Y_∞ of the left region via morphological reconstruction, shown in bottom right. (Bottom left shows an intermediate result for $i = 9$.)

functions

$$F_\vee(a) \equiv \bigvee_{x \in D} f(x) - ax \tag{74.37}$$

$$F_\wedge(a) \equiv \bigwedge_{x \in D} f(x) - ax \tag{74.38}$$

Since $f(x) - ax$ is the intercept of a line with slope a passing from the point $(x, f(x))$ on the signal's graph, for each a the upper (lower) slope transform of f is the maximum (minimum) value of this intercept, which occurs when the above line becomes a tangent. Examples of slope transforms are shown in Fig. 74.4. For *differentiable* signals, f, the maximization or minimization of the intercept $f(x) - ax$ can also be done by finding the stationary point(s) x^* such that $df(x^*)/dx = a$. This extreme value of the intercept is the **Legendre transform** of f:

$$F_L(a) \equiv f\left((df/dx)^{-1}(a)\right) - a[(df/dx)^{-1}(a)] \tag{74.39}$$

It is extensively used in mathematical physics. If the signal $f(x)$ is concave or convex and has an invertible derivative, its Legendre transform is single-valued and equal (over the slope regions it is defined) to the upper or lower transform; e.g., see the last three examples in Table 74.4. If f is neither convex nor concave or if it does not have an invertible derivative, its Legendre transform becomes a set $F_L(a) = \{f(x^*) - ax^* : df(x^*)/dx = a\}$ of real numbers for each a. This multivalued Legendre transform, defined and studied in [19] as a 'slope transform', has properties similar to those of the upper/lower slope transform, but there are also some important differences [37].

The upper and lower slope transform have a limitation in that they do not admit an inverse for arbitrary signals. The closest to an '*inverse*' upper slope transform is

$$\hat{f}(x) \equiv \bigwedge_{a \in \mathbb{R}} F_\vee(a) + ax \tag{74.40}$$

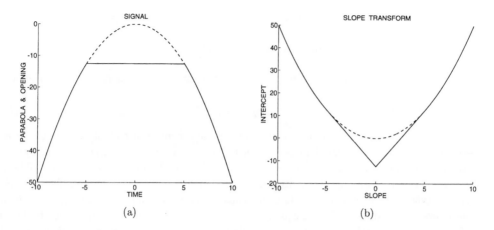

FIGURE 74.4 (a) Original parabola signal $f(x) = -x^2/2$ (in dashed line) and its morphological opening (in solid line) by a flat structuring element $[-5, 5]$. (b) Upper slope transform $F_\vee(a)$ of the parabola (in dashed line) and of its opening (in solid line).

which is equal to f only if f is concave; otherwise, $\hat{f}$ covers f from above by being its smallest concave *upper envelope*. Similarly, the supremum over a of all lines $F_\wedge(a) + ax$ creates the greatest convex *lower envelope* $\check{f}(x)$ of f, which plays the role of an "inverse" lower slope transform and is equal to f only if f is convex. Thus, for arbitrary signals we have $\check{f} \le f \le \hat{f}$.

Tables 74.4 and 74.5 list several examples and properties of the upper slope transform. The most striking is that (dilation) supremal convolution in the time/space domain corresponds to addition in the slope domain. Note the analogy with LTI systems where linearly convolving two signals corresponds to multiplying their Fourier transforms. Very similar properties also hold for the lower slope transform, the only differences being the interchange of suprema with infima, concave with convex, and the supremal $\oplus$ with the infimal convolution $\square$.

TABLE 74.5 Properties of Upper Slope Transform

Signal: $f(x)$	Transform: $F_\vee(a)$
$\vee_i c_i + f_i(x)$	$\vee_i c_i + F_i(a)$
$f(x - x_0)$	$F(a) - ax_0$
$f(x) + a_0 x$	$F(a - a_0)$
$f(rx)$	$F(a/r)$
$f(x) \oplus g(x)$	$F(a) + G(a)$
$\vee_y f(x) + g(x + y)$	$F(-a) + G(a)$
$f(x) \le g(x) \; \forall x$	$F(a) \le G(a) \; \forall a$
$g(x) = \begin{cases} f(x), & \|x\| \le r \\ -\infty, & \|x\| > r \end{cases}$	$G(a) = F(a) \square r\|a\|$

The upper/lower slope transforms for discrete-domain and/or multi-dimensional signals are defined as in the 1D continuous case by replacing the real variable x with an integer and/or multidimensional variable, and their properties are very similar or identical to the ones for signals defined on $\mathbb{R}$. See [37, 38] for details.

One of the most useful applications of LTI systems and Fourier transform is the design of frequency-selective filters. Similarly, it is also possible to design morphological systems that have a slope selectivity. Imagine a DTI system that rejects all line components with slopes in the band $[-a_0, a_0]$ and passes all the rest unchanged. Then its slope response would be

$$G(a) = 0 \text{ if } |a| \leq a_0, \text{ and } +\infty \text{ else} . \tag{74.41}$$

This is an ideal-cutoff **slope bandpass** filter. In the time domain it acts as a supremal convolution with its impulse response

$$g(x) = -a_0|x| \tag{74.42}$$

However, $f \oplus g$ is a non-causal infinite-extent dilation, and hence not realizable. Instead, we could implement it as a cascade of a causal dilation by the half-line $g_1(x) = -a_0 x + \lambda(x)$ followed by an anti-causal dilation by another half-line $g_2(x) = a_0 x + \lambda(-x)$, where $\lambda(x)$ is the zero step defined in Table 74.4. This works because $g = g_1 \oplus g_2$. For a discrete-time signal $f[x]$, this slope-bandpass filtering could be implemented via the recursive max-sum difference equation $f_1[x] = \max(f_1[x] - a_0, f[x])$ run forward in time, followed by another difference equation $f_2[x] = \max(f_2[x + 1] + a_0, f_1[x])$ run backward in time. The final result would be $f_2 = f \oplus g$. Such slope filters are useful for envelope estimation [37].

74.7 Multiscale Morphological Image Analysis

Multiscale signal analysis has recently emerged as a useful framework for many computer vision and signal processing tasks. Examples include: (1) detecting geometrical features or other events at large scales and then refining their location or value at smaller scales, (2) video and audio data compression using multiband frequency analysis, and (3) measurements and modeling of fractal signals. Most of the work in this area has obtained multiscale signal versions via *linear* multiscale smoothing, i.e., convolutions with a Gaussian with a variance proportional to scale [15, 53, 72]. There is, however, a variety of *nonlinear* smoothing filters, including the morphological openings and closings [35, 42, 58] that can provide a multiscale image ensemble and have the advantage over the linear Gaussian smoothers that they do not blur or shift edges, as shown in Fig. 74.5. There we see that the gray-level close-openings by reconstruction are especially useful because they can extract the exact outline of a certain object by locking on it while smoothing out all its surroundings; these nonlinear smoothers have been applied extensively in multiscale image segmentation [56]. The use of morphological operators for multiscale signal analysis is not limited to operations of a smoothing type; e.g., in fractal image analysis, erosion and dilation can provide multiscale distributions of the shrink-expand type from which the fractal dimension can be computed [36].

Overall, many applications of morphological signal processing such as nonlinear smoothing, geometrical feature extraction, skeletonization, size distributions, and segmentation, inherently require or can benefit from performing morphological operations at multiples scales. The required building blocks for a morphological scale-space are the multiscale dilations and erosions. Consider a planar compact convex set $B = \{(x, y) : \|(x, y)\|_p \leq 1\}$ that is the unit ball generated by the L_p norm, $p = 1, 2, \ldots, \infty$. Then the simplest multiscale dilation and erosion of a signal $f(x, y)$ at scales $t > 0$ are the multiscale flat sup/inf convolutions by $t B = \{tz : z \in B\}$

$$\delta(x, y, t) \equiv (f \oplus t B)(x, y) \tag{74.43}$$

$$\varepsilon(x, y, t) \equiv (f \ominus t B)(x, y) \tag{74.44}$$

which apply both to gray-level and binary images.

74.7.1 Binary Multiscale Morphology via Distance Transforms

Viewing the boundaries of multiscale erosions/dilations of a binary image by disks as wavefronts propagating from the original image boundary at uniform unit normal velocity and assigning to each pixel the time t of wavefront arrival creates a distance function, called the distance transform [10]. This transform is a

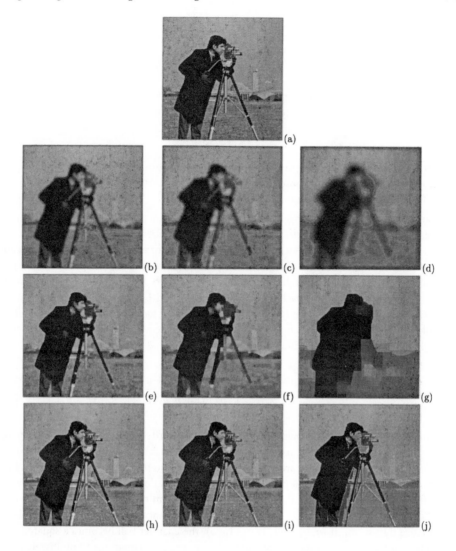

FIGURE 74.5 (a) Original image and its multiscale smoothings via: (b,c,d) Gaussian convolution at scales 2, 4, 16; (e,f,g) close-opening by a square at scales 2, 4, 16; (h,i,j) close-opening by reconstruction at scales 2, 4, 16.

compact way to represent their multiscale dilations and erosions by disks and other polygonial structuring elements whose shape depends on the norm $\| \cdot \|_p$ used to measure distances. Formally, the **distance transform** of the foreground set F of a binary image is defined as

$$D_p(F)(x, y) \equiv \bigwedge_{(v,u) \in F^c} \{\|(x - v, y - u)\|_p\} \tag{74.45}$$

Thresholding the distance transform at various levels $t > 0$ yields the erosions of the foreground F (or the dilation of the background F^c) by the norm-induced ball B at scale t:

$$F \ominus tB = \Theta_t[D_p(F)] \tag{74.46}$$

Another view of the distance transform results from seeing it as the infimal convolution of the $(0, +\infty)$ indicator function of F^c,

$$I_{F^c}(x) \equiv 0 \text{ if } x \in F^c, \text{ and } +\infty \text{ else,} \tag{74.47}$$

with the norm-induced conical structuring function:

$$D_p(F)(x) = I_{F^c}(x) \square \|x\|_p \tag{74.48}$$

Recognizing $g_\wedge(x) = \|x\|_p$ as the lower impulse response of an ETI system with slope response

$$G_\wedge(a) = 0 \text{ if } \|a\|_q \leq 1, \text{ and } -\infty \text{ else}, \tag{74.49}$$

where $1/p + 1/q = 1$, leads to seeing the distance transform as the output of an ideal-cutoff slope-selective filter that rejects all input planes whose slope vector falls outside the unit ball with respect to the $\| \cdot \|_q$ norm, and passes all the rest unchanged.

To obtain isotropic distance propagation, the *Euclidean* distance transform is desirable because it gives multiscale morphology with the *disk* as the structuring element. However, since this has a significant computational complexity, various techniques are used to obtain approximations to the Euclidean distance transform of discrete images at a lower complexity. A general such approach is the use of discrete distances [54] and their generalization via *chamfer metrics* [11]. Given a discrete binary image $f[i, j] \in \{0, +\infty\}$ with 0 marking background/source pixels and $+\infty$ marking foreground/object pixels, its global chamfer distance transform is obtained by propagating local distances within a small neighborhood mask. An efficient method to implement it is a two-pass sequential algorithm [11, 54] where for a 3×3 neighborhood the min-sum difference equation

$$\begin{aligned} u_n[i, j] \quad = \quad \min(\ &u_{n-1}[i, j], u_n[i-1, j] + a, u_n[i, j-1] + a, \\ &u_n[i-1, j-1] + b, u_n[i+1, j-1] + b\ ) \end{aligned} \tag{74.50}$$

is run recursively over the image domain: first ($n = 1$), in a forward scan starting from $u_0 = f$ to obtain u_1, and second ($n = 2$) in a backward scan on u_1 using a reflected mask to obtain u_2, which is the final distance transform. The coefficients a and b are the local distances within the neighborhood mask. The unit ball associated with chamfer metrics is a polygon whose approximation of the disk improves by increasing the size of the mask and optimizing the local distances so as to minimize the error in approximating the true Euclidean distances. In practice, integer-valued local distances are used for faster implementation of the distance transform. If (a, b) is $(1, 1)$ or $(1, \infty)$, the chamfer ball becomes a square or rhombus, respectively, and the chamfer distance transform gives poor approximations to multiscale morphology with disks. The commonly used ($a = 3, b = 4$) chamfer metric gives a maximum absolute error of about 6%, but even better approximations can be found by optimizing a, b.

74.7.2 Multiresolution Morphology

In certain multiscale image analysis tasks, the need also arises to subsample the multiscale image versions and thus create a multiresolution *pyramid* [15, 53]. Such concepts are very similar to the ones encountered in classical signal decimation. Most research in image pyramids has been based on linear smoothers. However, since morphological filters preserve essential shape features, they may be superior in many applications. A theory of morphological decimation and interpolation has been developed in [25] to address these issues which also provides algorithms on reconstructing a signal after morphological smoothing and decimation with quantifiable error. For example, consider a binary discrete image represented by a set X that is smoothed first to $Y = X \circ B$ via opening and then down-sampled to $Y \cap S$ by intersecting it with a periodic sampling set S (satisfying certain conditions). Then the Hausdorff distance between the smoothed signal Y and the interpolation (via dilation) $(Y \cap S) \oplus B$ of its down-sampled version does not exceed the radius of B. These ideas also extend to multilevel signals.

74.8   Differential Equations for Continuous-Scale Morphology

Thus far, most of the multiscale image filtering implementations have been discrete. However, due to the current interest in analog VLSI and neural networks, there is renewed interest in analog computation. Thus, continuous models have been proposed for several computer vision tasks based on **partial differential equations (PDEs)**. In multiscale linear analysis [72] a continuous (in scale t and spatial argument x, y) multiscale signal ensemble

$$\gamma(x, y, t) = f(x, y) * G_t(x, y) \, , \quad G_t(x, y) = \frac{\exp[-(x^2 + y^2)/4t]}{\sqrt{4\pi t}} \tag{74.51}$$

is created by linearly convolving an original signal f with a multiscale Gaussian function G_t whose variance $(2t)$ is proportional to the scale parameter t. The Gaussian multiscale function γ can be generated [28] from the *linear diffusion* equation

$$\frac{\partial \gamma}{\partial t} = \frac{\partial^2 \gamma}{\partial x^2} + \frac{\partial^2 \gamma}{\partial y^2} \tag{74.52}$$

starting from the initial condition $\gamma(x, y, 0) = f(x, y)$.

Motivated by the limitations or inability of linear systems to successfully model several image processing problems, several nonlinear PDE-based approaches have been developed. Among them, some PDEs have been recently developed to model multiscale morphological operators as dynamical systems evolving in scale-space [1, 14, 66].

Consider the multiscale morphological flat dilation and erosion of a 2D image signal $f(x, y)$ by the unit-radius disk at scales $t \geq 0$ as the space-scale functions $\delta(x, y, t)$ and $\varepsilon(x, y, t)$ of (74.43) and (74.44). Then [14] the PDE generating these multiscale flat dilations is

$$\frac{\partial \delta}{\partial t} = \|\nabla \delta\| = \sqrt{\left(\frac{\partial \delta}{\partial x}\right)^2 + \left(\frac{\partial \delta}{\partial y}\right)^2} \tag{74.53}$$

and for the erosions is $\partial \varepsilon / \partial t = -\|\nabla \varepsilon\|$. These morphological PDEs directly apply to binary images because flat dilations/erosions commute with thresholding and hence, when the gray-level image is dilated/eroded, each one of its thresholded versions representing a binary image is simultaneously dilated/eroded by the same element and at the same scale.

In equivalent formulations [10, 57, 66], the boundary of the original binary image is considered as a closed curve and this curve is expanded perpendicularly at constant unit speed. The dilation of the original image with a disk of radius t is the expanded curve at time t. This propagation of the image boundary is a special case of more general curvature-dependent propagation schemes for **curve evolution** studied in [47]. This general curve evolution methodology was applied in [57] to obtain multiscale morphological dilations/erosions of binary images, using an algorithm [47] where the original curve is first embedded in the surface of a 2D continuous function $\Phi_0(x, y)$ as its zero level set and then the evolving 2D curve is obtained as the zero level set of a 2D function $\Phi(x, y, t)$ that evolves from the initial condition $\Phi(x, y, 0) = \Phi_0(x, y)$ according to the PDE $\partial \Phi / \partial t = \|\nabla \Phi\|$. This function evolution PDE makes zero level sets expand at unit normal speed and is identical to the PDE (74.53) for flat dilation by disk. The main steps in its numerical implementations [47] are:

$$\begin{aligned}
\Phi_{i,j}^n &= \text{estimate of } \Phi(i\,\Delta x, j\,\Delta y, n\,\Delta t) \text{ on a grid} \\
D_x^+ &= \left(\Phi_{i+1,j}^n - \Phi_{i,j}^n\right)/\Delta x \, , \quad D_x^- = \left(\Phi_{i,j}^n - \Phi_{i-1,j}^n\right)/\Delta x \\
D_y^+ &= \left(\Phi_{i,j+1}^n - \Phi_{i,j}^n\right)/\Delta y \, , \quad D_y^- = \left(\Phi_{i,j}^n - \Phi_{i,j-1}^n\right)/\Delta y \\
G^2 &= \min^2(0, D_x^-) + \max^2(0, D_x^+) + \min^2(0, D_y^-) + \max^2(0, D_y^+) \\
\Phi_{i,j}^n &= \Phi_{i,j}^{n-1} + G\Delta t \, , \quad n = 1, 2, \ldots, (R/\Delta t)
\end{aligned}$$

where R is the maximum scale (radius) of interest, Δx, Δy are the spatial grid spacings, and Δt is the time (scale) step.

Continuous multiscale morphology using the above curve evolution algorithm for numerically implementing the dilation PDE yields better approximations to disks and avoids the abrupt shape discretization inherent in modeling digital multiscale using discrete polygons [16, 57]. Comparing it to discrete multiscale morphology using chamfer distance transforms, we note that for binary images: (1) the chamfer distance transform is easier to implement and yields similar errors for small scale dilations/erosions; (2) implementing the distance transform via curve evolution is more complex, but at medium and large scales gives a better and very close approximation to Euclidean geometry, i.e., to morphological operations with the disk structuring element. See Fig. 74.6.

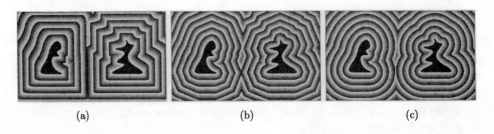

(a)   (b)   (c)

FIGURE 74.6   Distance transforms of a binary image, shown as intensity images modulo 20, obtained using: (a) Metric $\| \cdot \|_\infty$ (chamfer metric with local distances $(1,1)$), (b) chamfer metric with 3×3 neighborhood and local distances $(24,34)/25$, and (c) curve evolution.

74.9   Applications to Image Processing and Vision

There are numerous applications of morphological image operators to image processing and computer vision. Examples of broad application areas include biomedical image processing, automated visual inspection, character and document image processing, remote sensing, nonlinear filtering, multiscale image analysis, feature extraction, motion analysis, segmentation, and shape recognition. Next we shall review a few of these applications to specific problems of image processing and low/mid-level vision.

74.9.1   Noise Suppression

Rank filters and especially medians have been applied mainly to suppress impulse noise or noise whose probability density has heavier tails than the Gaussian for enhancement of image and other signals [2, 12, 27, 64, 65], since they can remove this type of noise without blurring edges, as would be the case for linear filtering. The rank filters have also been used for envelope detection. In their behavior as nonlinear smoothers, as shown in Fig. 74.7, the medians act similarly to an 'open-closing' $(f \circ B) \bullet B$ by a convex set B of diameter about half the diameter of the median window. The open-closing has the advantages over the median that it requires less computation and decomposes the noise suppression task into two independent steps, i.e., suppressing positive spikes via the opening and negative spikes via the closing. Further, cascading open-closings $\beta_t \alpha_t$ at multiple scales $t = 1, \ldots, r$, where $\alpha_t(f) = f \circ tB$ and $\beta_t(f) = f \bullet tB$, generates a class of efficient nonlinear smoothing filters $\beta_r \alpha_r \ldots \beta_2 \alpha_2 \beta_1 \alpha_1$, called **alternating sequential filters**, which smooth progressively from the smallest scale possible up to a maximum scale r and have a broad range of applications [59, 60, 62].

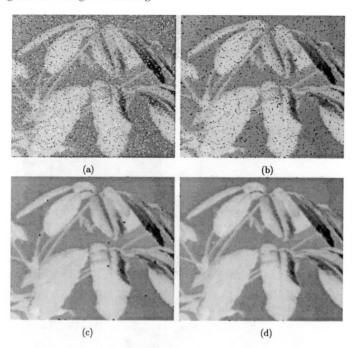

FIGURE 74.7   (a) Noisy image f, corrupted with salt-and-pepper noise of probability 10%. (b) Opening $f \circ B$ of f by a 2×2-pixel square B. (c) Open-closing $(f \circ B) \bullet B$. (d) Median of f by a 3×3-pixel square window.

74.9.2   Feature Extraction

Residuals between a signal and some morphologically transformed versions of it can extract line- or blob-type features or enhance their contrast. An example is the difference between the flat dilation and erosion of an image f by a symmetric disk-like set B whose diameter, diam(B), is very small;

$$\text{edge}\,(f) = \frac{(f \oplus B) - (f \ominus B)}{\text{diam}\,(B)} \tag{74.54}$$

If f is binary, edge (f) extracts its boundary. If f is gray-level, the above residual enhances its edges [7, 58] by yielding an approximation to $\|\nabla f\|$, which is obtained in the limit of (74.54) as diam(B) $\to 0$. See Fig. 74.8. This morphological edge operator can be made more robust for edge detection by first smoothing the input image signal and compares favorably with other gradient approaches based on linear filtering.

Another example involves subtracting the opening of a signal f by a compact convex set B from the input signal yields an output consisting of the signal peaks whose support cannot contain B. This is the *top-hat transformation* [43, 58]

$$\text{peak}\,(f) = f - (f \circ B) \tag{74.55}$$

and can detect bright *blobs*, i.e., regions with significantly brighter intensities relative to the surroundings. Similarly, to detect dark blobs, modeled as intensity valleys, we can use the closing residual operator $f \mapsto (f \bullet B) - f$. See Fig. 74.8. The morphological peak/valley extractors, in addition to their being simple and efficient, have some advantages over curvature-based approaches.

74.9.3   Shape Representation via Skeleton Transforms

There are applications in image processing and vision where a binary shape needs to be summarized down to its thin medial axis and then reconstructed exactly from this axial information. This process, known as *medial axis (or skeleton) transform* has been studied extensively for shape representation and

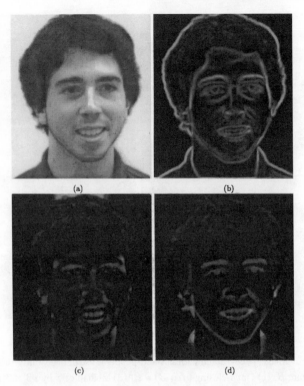

FIGURE 74.8   (a) Image f. (b) Edge enhancement: dilation-erosion residual $f \oplus B - f \ominus B$, where B is a 21-pixel octagon. (c) Peak detection: opening residual $f - f \circ B^{\oplus 3}$. (d) Valley detection: closing residual $f \bullet B^{\oplus 3} - f$.

description [10, 54]. Among many approaches, it can also be obtained via multiscale morphological operators, which offer as a by-product a multiscale representation of the original shape via its skeleton components [39, 58]. Let $X \subseteq \mathbb{Z}^2$ represent the foreground of a finite discrete binary image and let $B \subseteq \mathbb{Z}^2$ be a convex disk-like set at scale 1 and $B^{\oplus n}$ be its multiscale version at scale $n = 1, 2, \ldots$ The nth **skeleton component** of X is the set

$$S_n = (X \ominus B^{\oplus n}) \backslash \left[(X \ominus B^{\oplus n}) \circ B \right] \ , \quad n = 0, 1, \ldots, N \ , \tag{74.56}$$

where $\backslash$ denotes the difference, n is a discrete scale parameter, and $N = \max\{n : X \ominus B^{\oplus n} \neq \emptyset\}$ is the maximum scale. The S_n are disjoint subsets of X, whose union is the **morphological skeleton** of X.

The **morphological skeleton transform** of X is the finite sequence $(S_0, S_1, \ldots, S_N)$. The union of all the S_ns dilated by a n-scale disk reconstructs exactly the original shape; omitting the first k components leads to a smooth partial reconstruction, the opening of X at scale k:

$$X \circ B^{\oplus k} = \bigcup_{k \leq n \leq N} S_n \oplus B^{\oplus n} \ , \quad 0 \leq k \leq N \ . \tag{74.57}$$

Thus, we can view the S_n as '*shape components*', where the small-scale components are associated with the lack of smoothness of the boundary of X, whereas skeleton components of large scale indices n are related to the bulky interior parts of X that are shaped similarly to $B^{\oplus n}$. Figure 74.9 shows a detailed description of the skeletal decomposition and reconstruction of an image.

Several generalizations or modifications of the morphological skeletonization include: using structuring elements different than disks that might result in fewer skeletal points, or removing redundant points from the skeleton [29, 33, 39]; using different structuring elements for each skeletonization step [23, 33]; using

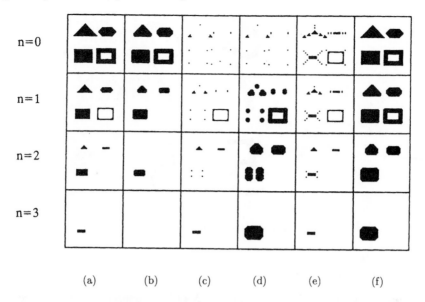

FIGURE 74.9 Morphological skeletonization of a binary image X (top left image) with respect to a 3×3-pixel square structuring element B. (a) Erosions $X \ominus B^{\oplus n}$, $n = 0, 1, 2, 3$. (b) Openings of erosions $(X \ominus B^{\oplus n}) \circ B$. (c) Skeleton subsets S_n. (d) Dilated skeleton subsets $S_n \oplus B^{\oplus n}$. (e) Partial unions of skeleton subsets $\cup_{N \geq k \geq n} S_k$. (f) Partial unions of dilated skeleton subsets $\cup_{N \geq k \geq n} S_k \oplus B^{\oplus k}$.

lattice generalizations of the erosions and openings involved in skeletonization [30]; image representation based on skeleton-like multiscale residuals [23]; and shape decomposition based on residuals between image parts and maximal openings [48]. In addition to its general use for shape analysis, a major application of skeletonization has been binary image coding [13, 30, 39].

74.9.4 Shape Thinning

The skeleton is not necessarily connected; for *connected* skeletons see [3]. Another approach for summarizing a binary shape down to a thin medial axis that is connected but does not necessarily guarantee reconstruction is via thinning. **Morphological thinning** is defined [58] as the difference between the original set X (representing the foreground of a binary image) and a set of feature locations extracted via hit-miss transformations by pairs of foreground-background probing sets (A_i, B_i) designed to detect features that thicken the shape's axis:

$$X \odot \{(A_i, B_i)\}_{i=1}^n \equiv X \setminus \bigcup_{i=1}^n X \otimes (A_i, B_i) \qquad (74.58)$$

Usually each hit-miss by a pair (A_i, B_i) detects a feature at some orientation, and then the difference from the original peels off this feature from X. Since this feature might occur at several orientations, the above thinning operator is applied iteratively by rotating its set of probing elements until there is no further change in the image. Thinning has been applied extensively to character images. Examples are shown in Fig. 74.10, where each thinning iteration used $n = 3$ template pairs (A_i, B_i) for the hit-miss transformations of (74.58) designed in [8].

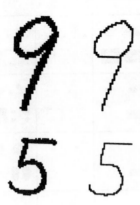

FIGURE 74.10 Left column shows binary images of handwritten characters. Right column shows their thinned version.

74.9.5 Size Distributions

Multiscale openings $X \mapsto X \circ rB$ and closings $X \mapsto X \bullet rB$ of compact sets X in $\mathbb{R}^d$ by convex compact structuring elements rB, parameterized by a scale parameter $r \geq 0$, are called *granulometries* and can unify all sizing (sieving) operations [42]. Because they satisfy a monotonic ordering

$$\ldots X \circ sB \subseteq X \circ rB \subseteq \ldots \subseteq X \subseteq \ldots X \bullet rB \subseteq X \bullet sB \subseteq \ldots , \quad r < s , \qquad (74.59)$$

if we measure the volume (or area) of these sets as a function of scale, this function will also satisfy the same ordering and hence create **size distributions**. Further, taking its derivative leads to a **size density function** (or *size histogram* in the discrete case)

$$h(r) \equiv \begin{cases} -\dfrac{d\text{vol}\,(X \circ rB)}{dr} , & r \geq 0 \\[2ex] \dfrac{d\text{vol}\,(X \bullet |r|B)}{d|r|} , & r < 0 \end{cases} \qquad (74.60)$$

This conveys several types of information useful for shape description and multiscale image analysis. For example, the *boundary roughness* of X relative to B manifests itself as contributions in the lower-size part of the size histogram. Long capes or bulky protruding parts in X that consist of patterns sB show up as isolated impulses in the histogram around positive $r = s$. Finally, the size density can be defined for 'negative' sizes by using closings instead of openings; in this case impulses at negative sizes indicate the existence of prominent intruding *gulfs* or *holes* in X.

   If X is a random set [42], then probabilistic measures of its size distribution have been used extensively in image analysis applications to petrography and biology [58]. All of the above ideas can be extended to gray-level images [35]. One application of gray-level size distributions is texture classification [17].

74.9.6 Fractals

A large variety of natural image objects (e.g., clouds, coastlines, mountains, islands, trees, leaves, etc.) can be modeled with *fractals* [32]. Fractals are mathematical sets with a very high level of geometrical complexity; formally, their Hausdorff dimension is larger than their topological dimension. An important characteristic of fractals to measure for purposes of shape description or classification is their fractal dimension. Among the various methods [32] to estimate the *fractal dimension D* of the surface of a set $F \subseteq \mathbb{R}^3$, the *covering method* is based conceptually on Minkowski's idea of finding the area of irregular sets; dilate them with spheres of radius r, find the volume $V(r)$ of the dilated set, and set its area equal to

$\lim_{r \downarrow 0} A(r)$, where $A(r) = V(r)/2r$. Further, the fractal dimension of F can be found by

$$D = \lim_{r \downarrow 0} \frac{\log[V(r)/r^3]}{\log[(1/r)]} \tag{74.61}$$

The intuitive meaning of D is that $V(r) \approx (\text{constant}) \cdot r^{3-D}$ as $r \downarrow 0$, from which D can be estimated by least-squares fitting a straight line to a log-log plot of $V(r)$.

The theory of morphological operators allows us to find more efficient implementations of the above idea when F is the graph of a 2D function $f(x, y)$. Then, instead of multiscale 3D set dilations of F by spheres, it is computationally more efficient to perform 2D multiscale signal dilations and erosions of f by disks rB and measure the multiscale volumes by

$$V(r) = \int \int [(f \oplus rB)(x, y) - (f \ominus rB)(x, y)]dxdy \tag{74.62}$$

Thus, morphological flat dilations and erosions are used to create a volume-blanket as a layer either covering or being peeled off from the surface of f at various scales. This morphological covering method can also be applied to 1D signals $f(x)$ by replacing volumes with areas and disks rB with horizontal linear segments $[-r, r]$; such a 1D application is shown in Fig. 74.11.

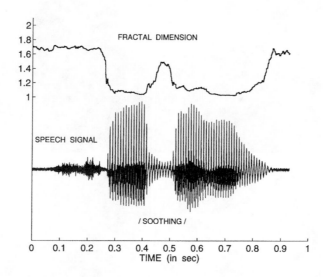

FIGURE 74.11 Speech waveform of the word 'soothing' sampled at 10 kHz and its short-time fractal dimension over 10-ms speech segments, computed every 1 ms and post-smoothed by a 3-point median filter. The short-time fractal dimension increases with the amount of turbulence existing during production of the corresponding sound, having a small value for vowels, medium for weak voiced fricatives, and high for unvoiced fricatives.

74.9.7 Image Segmentation

One of the most powerful and advanced tools of mathematical morphology is the watershed transformation [7] as applied to image segmentation. Let us regard the gray-level image to be segmented as a topographic relief and assume a drop of water falling at a point on it and flowing down along a steep slope path until it is trapped in a local minimum M of the relief. The set of points such that a drop falling on them eventually reaches M is the catchment basin associated with the minimum M. The union of the boundaries of the different catchment basins of the image constitute its **watershed**. Thus, the watershed consists of contours located on crest lines separating adjacent minima.

To ease the segmentation of the original image f, the watershed transformation is usually applied to its gradient magnitude $g = \|\nabla f\|$, which has higher contrast. However, direct computation of the watershed of g usually leads to poor results, i.e., oversegmentation of f, because, even after smoothing f or g, the latter often exhibits far too many minima. One of the best solutions to this problem is to use markers for the regions to be extracted. A **marker** is a small connected component of pixels, a feature, located inside a region. Once the markers have been extracted, the gradient image g is modified via morphological reconstruction so that these markers are imposed as the only minima of the modified function while preserving the highest crest lines of g located between two markers. Then, computing the watershed of the modified g usually provides a good segmentation whose quality depends mainly on the markers and somewhat on g and the initial smoothing of f. An example is shown in Fig. 74.12. The power of this approach as well as its difficulty lies in the choice of the markers. Efficient ways to choose markers as well as fast algorithms for the watershed computation are detailed in [44, 69]. This watershed methodology has already proved to be very useful in various fields of image analysis, ranging from medical imaging to material sciences, remote sensing, and digital elevation models.

FIGURE 74.12   (a) Image f. (b) Edge enhancement (magnitude of gradient) of f. (c) Markers. (d) Watershed.

74.10   Conclusions

This chapter has provided a brief introduction of the theory of morphological signal processing and its applications to image analysis and nonlinear filtering. This methodology nowadays offers a large diversity of theoretical and algorithmic ideas that provide useful tools and inspire new directions in the following research areas from the fields of signal processing, image processing and machine vision, and pattern recognition: nonlinear filtering, nonlinear signal and system representation, image feature extraction, multiscale analysis and geometry-driven diffusion, image segmentation, region-based image coding, motion analysis, automated visual inspection, and detection/estimation in random sets.

Some attractive aspects of morphological signal operators for efficiently solving problems in the above areas include: (1) suitability for geometry-related signal/image analysis problems; (2) unification power because they can be defined both for numerical signals as well as for more abstract data using their lattice generalizations; (3) simplicity of software or hardware implementations of the basic operators; and (4) existence of efficient algorithms for implementing complex morphological systems [68].

Three current research areas where successful future developments may significantly broaden and improve the applicability of morphological signal processing are: *(A) Optimal design of nonlinear systems based on morphological and related signal operators*, where, despite their numerous applications, very few ideas exist for their optimal design. (The current three main approaches are: (1) designing binary systems as a finite union of erosions [20, 31] or hit-miss operations [5] using the morphological representation theory of [33, 34] or [4]; (2) designing stack filters via threshold decomposition and linear programming [18]; (3) gradient-based optimization of morphological/rank filters either via simulated annealing [71] or via a least-mean-square algorithm and adaptive filtering [55].) *(B) The continuous (differential) approach to mathematical morphology via PDEs* and exploitation of its exciting relationships to the physics of wave propagation and eikonal optics [38, 57, 67]. *(C) Development of morphological systems for image pattern recognition* by exploiting the efficiency of morphological operators for shape analysis and their logic-related structure.

Acknowledgment

This chapter was written while the author's research work was supported by the U.S. National Science Foundation under Grant MIP-94-21677.

References

[1] Alvarez, L. and Morel, J.M., Formalization and computational aspects of image analysis, *Acta Numerica*, 1–59, 1994.

[2] Arce, G.R., Gallagher, N.C. and Nodes, T.A., Median filters: Theory for one- and two-dimensional filters, in *Advances in Computer Vision and Image Processing, Vol.2*, Huang, T.S., Ed., JAI Press, Connecticut, 1986.

[3] Arcelli, C., Cordella, L. and Levialdi, S., From local maxima to connected skeletons, *IEEE Trans. Pattern Anal. Mach. Intellig.*, PAMI-3, 134–143, Mar. 1981.

[4] Banon, G.J.F. and Barrera, A.J., Minimal representations for translation-invariant set mappings by mathematical morphology, *SIAM J. Appl. Math.*, 51, 1782–1798, Dec. 1991.

[5] Barrera, A.J., Salas, B.G.P. and Hashimoto, C.R.F., Set operations on closed intervals and their applications to the programming of MMach's, in *Mathematical Morphology and its Application to Image and Signal Processing*, Maragos, P., Schafer, R.W. and Butt, M.A., Eds., Kluwer Acad. Publishing, 1996.

[6] Bellman, R. and Karush, W., On the maximum transform, *J. Math. Anal. Appl.*, 6, 67–74, 1963.

[7] Beucher, S. and Lantuejoul, C., Use of watersheds in contour detection, *Proc. Int'l. Workshop on Image Processing: Real-time Edge & Motion Detection/Estimation*, Rennes, France, 1979.

[8] Bloomberg, D.S., Connectivity-preserving morphological image transformations, in *Visual Communications and Image Processing '91*, Tzou, K.-H. and Koga, T., Eds., Proc. SPIE, vol. 1606, 1991.

[9] Bloomberg, D.S., Multiresolution morphological analysis of document images, in *Visual Communications and Image Processing'92*, Maragos, P., Ed., Proc. SPIE, 1818, 648–662, 1992.

[10] Blum, H., Biological shape and visual science (part I), *J. Theor. Biol.*, 38, 205–287, 1973.

[11] Borgefors, G., Distance transformations in digital images, *Comp. Vision, Graphics, Image Process.*, 34, 344–371, 1986.

[12] Bovik, A.C., Huang, T.S. and Munson, Jr., D.C., A generalization of median filtering using linear combinations of order statistics, *IEEE Trans. Acoust. Speech, Signal Process.*, 31, 1342–1349, Dec. 1983.

[13] Brandt, J.W., Jain, A.K. and Algazi, V.R., Medial axis representation and encoding of scanned documents, *J. Vis. Commun. Image Repres.*, 2, 151–165, June 1991.

[14] Brockett, R.W. and Maragos, P., Evolution equations for continuous-scale morphological filtering, *IEEE Trans. on Signal Processing*, 42, 3377–3386, Dec. 1994.

[15] Burt, P.J. and Adelson, E.H., The Laplacian pyramid as a compact image code, *IEEE Trans. Commun.*, 31, 532–540, Apr. 1983.

[16] Butt, M.A. and Maragos, P., Comparison of multiscale morphology approaches: PDE implemented via curve evolution versus chamfer distance transform in *Mathematical Morphology and its Application to Image and Signal Processing*, Maragos, P., Schafer, R.W. and Butt, M.A., Eds., Kluwer Acad. Publishing, 1996.

[17] Chen, Y. and Dougherty, E.R., Gray-scale morphological granulometric texture classification, *Optical Engineering*, 33, 2713–2722, Aug. 1994.

[18] Coyle, E.J. and Lin, J.H., Stack filters and the mean absolute error criterion, *IEEE Trans. Acoust. Speech Signal Processing*, 36, 1244–1254, Aug. 1988.

[19] Dorst, L. and van der Boomgaard, R., Morphological signal processing and the slope transform, *Signal Processing*, 38, 79–98, July 1994.

[20] Dougherty, E.R., Optimal mean-square N-observation digital morphological filters: I. Optimal binary filters, *CVGIP: Image Understanding*, 55, 36–54, Jan. 1992.

[21] Fitch, J.P., Coyle, E.J. and Gallagher, Jr., N.C., Median filtering by threshold decomposition, *IEEE Trans. Acoust. Speech, Signal Processing*, 32, 1183–1188, Dec. 1984.

[22] Goetcherian, V., From binary to grey tone image processing using fuzzy logic concepts, *Pattern Recognition*, 12, 7–15, 1980.

[23] Goutsias, J. and Shonfeld, D., Morphological representation of discrete and binary images, *IEEE Trans. Signal Processing*, 39, 1369–1379, June 1991.

[24] Hadwiger, H., *Vorlesungen über Inhalt, Oberfläche, und Isoperimetrie*, Springer-Verlag, Berlin, 1957.

[25] Haralick, R.M., Zhuang, X., Lin, C. and Lee, J.S.J., The digital morphological sampling theorem, *IEEE Trans. Acoust. Speech, Signal Process*, 37, 2067–2090, Dec. 1989.

[26] Heijmans, H.J.A.M. and Ronse, C., The algebraic basis of mathematical morphology—part I: Dilations and erosions, *Comput. Vision Graph. Image Process.*, 50, 245–295, 1990.

[27] Justusson, B.I., Median filtering: Statistical properties, in *Two-Dimensional Digital Signal Processing II: Transforms and Median Filters*, Huang, T.S., Ed., Springer-Verlag, NY, 1981.

[28] Koenderink, J.J., The structure of images, *Biol. Cybern.*, 50, 363–370, 1984.

[29] Kresh, R. and Malah, D., Morphological reduction of skeleton redundancy, *Signal Processing*, 38, 143–151, 1994.

[30] Kresh, R., Morphological image representation for coding applications, Ph.D. thesis, Technion, Israel, June 1995.

[31] Loce, R.P. and Dougherty, E.R., Facilitation of optimal binary morphological filter design via structuring element libraries and design constraints, *Optical Engineering*, 31, 1008–1025, May 1992.

[32] Mandelbrot, B.B., *The Fractal Geometry of Nature*, Freeman, San Francisco, CA, 1982.

[33] Maragos, P., A unified theory of translation-invariant systems with applications to morphological analysis and coding of images, Ph.D. thesis, Georgia Institute of Technology, Atlanta, July 1985.

[34] Maragos, P., A representation theory for morphological image and signal processing, *IEEE Trans. Pattern Anal. Mach. Intellig.*, 11, 586–599, June 1989.

[35] Maragos, P., Pattern spectrum and multiscale shape representation, *IEEE Trans. Pattern Anal. Mach. Intellig.*, 11, 701–716, July 1989.

[36] Maragos, P., Fractal signal analysis using mathematical morphology, in *Advances in Electronics and Electron Physics*, vol. 88, Hawkes, P. and Kazan, B., Eds., Academic Press, New York, 1994, 199–246.

[37] Maragos, P., Morphological systems: Slope transforms and max-min difference and differential equations, *Signal Processing*, 38, 57–77, July 1994.

[38] Maragos, P., Differential morphology and image processing, *IEEE Trans. Image Processing*, 5, 922–937, June 1996.

[39] Maragos, P. and Schafer, R.W., Morphological skeleton representation and coding of binary images, *IEEE Trans. Acoust. Speech, Signal Process.*, 34, 1228–1244, Oct. 1986.

[40] Maragos, P. and Schafer, R.W., Morphological filters—Part I: Their set-theoretic analysis and relations to linear shift-invariant filters, *IEEE Trans. Acoust. Speech, Signal Processing*, 35, 1153–1169, Aug. 1987.

[41] Maragos, P. and Schafer, R.W., Morphological filters—Part II: Their relations to median, order-statistic, and stack filters, *IEEE Trans. Acoust. Speech, Signal Process.*, 35, 1170–1184, Aug. 1987; *ibid*, 37, 597, Apr. 1989.

[42] Matheron, G., *Random Sets and Integral Geometry*, John Wiley & Sons, NY, 1975.

[43] Meyer, F., Contrast feature extraction, in *Quantitative Analysis of Microstructures in Materials Science, Biology and Medicine*, Chermant, J.L., Ed., *Special Issues of Practical Metallography*, Riederer-Verlag, Stuttgart, 1978, 374–380.

[44] Meyer, F. and Becheur, S., Morphological segmentation, *J. Vis. Commun. Image Representation*, 1, 21–46, Sept. 1990.

[45] Nakagawa, Y. and Rosenfeld, A., A note on the use of local min and max operations in digital picture processing, *IEEE Trans. Syst., Man, and Cybern.*, 8, 632–635, 1978.

[46] O'Neil, K.S. and Rhodes, W.T., Morphological transformations by hybrid optical-electronic methods, in *Hybrid Image Processing*, Casasent, D. and Tescher, A., Eds., *Proc. SPIE*, 638, 41–44, 1986.

[47] Osher, S. and Sethian, J.A., Fronts propagating with curvature-dependent speed: Algorithms based on Hamilton-Jacobi formulations, *J. Comput. Physics*, 79, 12–49, 1988.

[48] Pitas, I. and Venetsanopoulos, A., Morphological shape decomposition, *IEEE Trans. Pattern Anal. Mach. Intellig.*, 12, 38–45, 1990.

[49] Preston, Jr., K. and Duff, M.J.B., *Modern Cellular Automata*, Plenum Press, New York, 1984.

[50] Rockafellar, R.T., *Convex Analysis*, Princeton University Press, Princeton, NJ, 1972.

[51] Roerdink, J.B.T.M., Mathematical morphology with non-commutative symmetry groups, in *Mathematical Morphology in Image Processing*, Dougherty, E.R., Ed., Marcel Dekker, New York, 1993.

[52] Ronse, C. and Heijmans, H.J.A.M., The algebraic basis of mathematical morphology—part II: Openings and closings, *CVGIP: Image Understanding*, 54, 74–97, 1991.

[53] Rosenfeld, A., Ed., *Multiresolution Image Processing and Analysis*, Springer-Verlag, New York, 1984.

[54] Rosenfeld, A. and Kak, A.C., *Digital Picture Processing*, vols. 1 and 2, Academic Press, New York, 1982.

[55] Salembier, P., Structuring element adaptation for morphological filters, *J. Visual Commun. Image Repres.*, 3, 115–136, June 1992.

[56] Salembier, P. and Serra, J., Morphological multiscale image segmentation, in *Visual Communications and Image Processing'92*, Maragos, P., Ed., *Proc SPIE*, 1818, 620–631, 1992.

[57] Sapiro, G., Kimmel, R., Shaked, D., Kimia, B. and Bruckstein, A., Implementing continuous-scale morphology via curve evolution, *Pattern Recognition*, 26(9), 1363–1372, 1993.

[58] Serra, J., *Image Analysis and Mathematical Morphology*, Academic Press, New York, 1982.

[59] Serra, J., Ed., *Image Analysis and Mathematical Morphology, Vol. 2: Theoretical Advances*, Academic Press, New York, 1988.

[60] Schonfeld, D. and Goutsias, J., Optimal morphological pattern restoration from noisy binary images, *IEEE Trans. Pattern Anal. Machine Intellig.*, 13, 14–29, Jan. 1991.

[61] Sternberg, S.R., Cellular computers and biomedical image processing, in *Biomedical Images and Computers*, Sklansky, J. and Bisconte, J.C., Eds., Springer-Verlag, Berlin, 1982.

[62] Sternberg, S.R., Grayscale Morphology, *Comput. Vision Graph. Image Process.*, 35, 333–355, 1986.

[63] Szoplik, T., Ed., *Selected Papers on Morphological Image Processing: Principles and Optoelectronic Implementations*, SPIE Press, Bellingham, WA, 1996.

[64] Tukey, J.W., *Exploratory Data Analysis*, Addison-Wesley, Reading, MA, 1977.

[65] Tyan, S.G., Median filtering: Deterministic properties, in *Two-Dimensional Digital Signal Processing II: Transforms and Median Filters*, Huang, T.S., Ed., Springer-Verlag, New York, 1981.

[66] van der Boomgaard, R. and Smeulders, A., The morphological structure of images: The differential equations of morphological scale-space, *IEEE Trans. Pattern Anal. Mach. Intellig.*, 16, 1101–1113, Nov. 1994.

[67] Verbeek, P.W. and Verwer, B.J.H., Shading from shape, the eikonal equation solved by grey-weighted distance transform, *Pattern Recogn. Lett.*, 11, 618–690, 1990.

[68] Vincent, L., Morphological algorithms, in *Mathematical Morphology in Image Processing*, Dougherty, E.R., Ed., Marcel Dekker, New York, 1993.

[69] Vincent, L. and Soille, P., Watersheds in digital spaces: An efficient algorithm based on immersion simulations, *IEEE Trans. Pattern Anal. Mach. Intellig.*, 13, 583–598, June 1991.

[70] Wendt, P.D., Coyle, E.J. and Gallagher, N.C., Stack filters, *IEEE Trans. Acoust., Speech, Signal Process.*, 34, 898–911, Aug. 1986.

[71] Wilson, S.S., Training structuring elements in morphological networks, in *Mathematical Morphology in Image Processing*, Dougherty, E.R., Ed., Marcel Dekker, New York, 1993.

[72] Witkin, A.P., Scale-space filtering, in *Proc. Int'l. Joint Conf. Artif. Intellig.*, Karlsruhe, 1983.

75

Signal Processing and Communication with Solitons

Andrew C. Singer
Sanders,
A Lockheed Martin Company

75.1 Introduction

As we increasingly turn to nonlinear models to capture some of the more salient behavior of physical or natural systems that cannot be expressed by linear means, systems that support solitons may be a natural class to explore because they share many of the properties that make LTI systems attractive from an engineering standpoint. Although nonlinear, these systems are solvable through inverse scattering, a technique analogous to the Fourier transform for linear systems [1]. Solitons are eigenfunctions of these systems which satisfy a nonlinear form of superposition. We can therefore decompose complex solutions in terms of a class of signals with simple dynamical structure. Solitons have been observed in a variety of natural phenomena from water and plasma waves [7, 12] to crystal lattice vibrations [2] and energy transport in proteins [7]. Solitons can also be found in a number of man-made media including super-conducting transmission lines [11] and nonlinear circuits [6, 13]. Recently, solitons have become of significant interest for optical telecommunications, where optical pulses have been shown to propagate as solitons for tremendous distances without significant dispersion [4].

We view solitons from a different perspective. Rather than focusing on the propagation of solitons over nonlinear channels, we consider using these nonlinear systems to both generate and process signals for

0-8493-8572-5/98/$0.00+$.50
© 1998 by CRC Press LLC

transmission over traditional linear channels. By using solitons for signal synthesis, the corresponding nonlinear systems become specialized signal processors which are naturally suited to a number of complex signal processing tasks. This section can be viewed as an exploration of the properties of solitons as signals. In the process, we explore the potential application of these signals in a multi-user wireless communication context. One possible benefit of such a strategy is that the soliton signal dynamics provide a mechanism for simultaneously decreasing transmitted signal energy and enhancing communication performance.

75.2 Soliton Systems: The Toda Lattice

The Toda lattice is a conceptually simple mechanical example of a nonlinear system with soliton solutions.[1] It consists of an infinite chain of masses connected with springs satisfying the nonlinear force law $f_n = a(e^{-b(y_n-y_{n-1})} - 1)$ where f_n is the force on the spring between masses with displacements y_n and y_{n-1} from their rest positions. The equations of motion for the lattice are given by

$$m\ddot{y}_n = a\left(e^{-b(y_n-y_{n-1})} - e^{-b(y_{n+1}-y_n)}\right),\tag{75.1}$$

where m is the mass, and a and b are constants. This equation admits pulse-like solutions of the form

$$f_n(t) = \left(\frac{m}{ab}\right)\beta^2\text{sech}^2(\sinh^{-1}(\sqrt{m/ab}\,\beta)n - \beta t),\tag{75.2}$$

which propagate as compressional waves stored as forces in the nonlinear springs. A single right-traveling wave $f_n(t)$ is shown in Fig. 75.1(a).

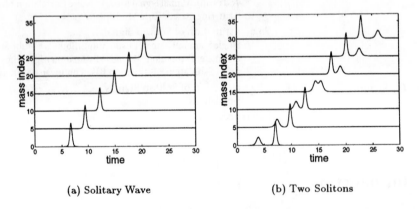

(a) Solitary Wave   (b) Two Solitons

FIGURE 75.1   Propagating wave solutions to the Toda lattice equations. Each trace corresponds to the force $f_n(t)$ stored in the spring between mass n and $n - 1$.

This compressional wave is localized in time, and propagates along the chain maintaining constant shape and velocity. The parameter β appears in both the amplitude and the temporal- and spatial-scales of this one parameter family of solutions giving rise to tall, narrow pulses which propagate faster than small, wide pulses. This type of localized pulse-like solution is what is often referred to as a *solitary wave*.

The study of solitary wave solutions to nonlinear equations dates back to the work of John Scott Russell in 1834 and perhaps the first recorded sighting of a solitary wave. Scott Russell's observations of an unusual

---

[1]A comprehensive treatment of the lattice and its associated soliton theory can be found in the monograph by Toda [18].

water wave in the Union Canal near Edinburgh, Scotland, are interpreted as a solitary wave solution to the Korteweg deVries (KdV) equation [12].[2] In a 1965 paper, Zabusky and Kruskal performed numerical experiments with the KdV equation and noticed that these solitary wave solutions retained their identity upon collision with other solitary waves, which prompted them to coin the term *soliton* implying a particle-like nature. The ability to form solutions to an equation from a superposition of simpler solutions is the type of behavior we would expect for linear wave equations. However, that nonlinear equations such as the KdV or Toda lattice equations permit such a form of superposition is an indication that they belong to a rather remarkable class of nonlinear systems.

An example of this form of soliton superposition is illustrated in Fig. 75.1(b) for two solutions of the form of Eq. (75.2). Note that as a function of time, a smaller, wider soliton appears before a taller, narrower one. However, as viewed by, e.g., the thirtieth mass in the lattice, the larger soliton appears first as a function of time. Since the larger soliton has arrived at this node before the smaller soliton, it has therefore traveled faster. Note that when the larger soliton catches up to the smaller soliton as viewed on the fifteenth node, the combined amplitude of the two solitons is actually less than would be expected for a linear system, which would display a linear superposition of the two amplitudes. Also, the signal shape changes significantly during this nonlinear interaction.

An analytic expression for the two soliton solution for $\beta_1 > \beta_2 > 0$ is given by [6]

$$f_n(t) = \frac{m}{ab} \frac{\beta_1^2 \text{sech}^2(\eta_1) + \beta_2^2 \text{sech}^2(\eta_2) + A\text{sech}^2(\eta_1)\text{sech}^2(\eta_2)}{(\cosh(\phi/2) + \sinh(\phi/2)\tanh(\eta_1)\tanh(\eta_2))^2}, \tag{75.3}$$

where

$$\begin{aligned}
A &= \sinh(\phi/2)\left(\left(\beta_1^2 + \beta_2^2\right)\sinh(\phi/2) + 2\beta_1\beta_2\cosh(\phi/2)\right), \\
\phi &= \ln\left(\frac{\sinh((p_1 - p_2)/2)}{\sinh((p_1 + p_2)/2)}\right),
\end{aligned} \tag{75.4}$$

and $\beta_i = \sqrt{ab/m}\sinh(p_i)$, and $\eta_i = p_i n - \beta_i(t - \delta_i)$. Although Eq. (75.3) appears rather complex, Fig. 75.1(b) illustrates that for large separations, $|\delta_1 - \delta_2|$, $f_n(t)$ essentially reduces to the linear superposition of two solitons with parameters β_1 and β_2. As the relative separation decreases, the multiplicative cross term becomes significant, and the solitons interact nonlinearly. This asymptotic behavior can also be evidenced analytically

$$\begin{aligned}
f_n(t) &= \frac{m}{ab}\beta_1^2 \text{sech}^2(p_1 n - \beta_1(t - \delta_1) \pm \phi/2) \\
&+ \frac{m}{ab}\beta_2^2 \text{sech}^2(p_2 n - \beta_2(t - \delta_2) \mp \phi/2), \quad t \to \pm\infty,
\end{aligned} \tag{75.5}$$

where each component soliton experiences a net displacement ϕ from the nonlinear interaction. The Toda lattice also admits periodic solutions which can be written in terms of Jacobian elliptic functions [18].

An interesting observation can be made when the Toda lattice equations are written in terms of the forces,

$$\frac{d^2}{dt^2}\ln\left(1 + \frac{f_n}{a}\right) = \frac{b}{m}(f_{n+1} - 2f_n + f_{n-1}). \tag{75.6}$$

If the substitution $f_n(t) = \frac{d^2}{dt^2}\ln\phi_n(t)$ is made into Eq. (75.6), then the lattice equations become

$$\frac{m}{ab}\left(\dot{\phi}_n^2 - \phi_n\ddot{\phi}_n\right) = \phi_n^2 - \phi_{n-1}\phi_{n+1}. \tag{75.7}$$

---

[2]A detailed discussion of linear and nonlinear wave theory including KdV can be found in [21].

In view of the Teager energy operator introduced by Kaiser in [8], the left-hand side of Eq. (75.7) is the Teager instantaneous-time energy at the node n, and the right-hand side is the Teager instantaneous-space energy at time t. In this form, we may view solutions to Eq. (75.7) as propagating waveforms that have equal Teager energy as calculated in time and space, a relationship also observed by Kaiser [9].

75.2.1 The Inverse Scattering Transform

Perhaps the most significant discovery in soliton theory was that under a rather general set of conditions, certain nonlinear evolution equations such as KdV or the Toda lattice could be solved analytically. That is, given an initial condition of the system, the solution can be explicitly determined for all time using a technique called inverse scattering. Since much of inverse scattering theory is beyond the scope of this section, we will only present some of the basic elements of the theory and refer the interested reader to [1].

The nonlinear systems that have been solved by inverse scattering belong to a class of systems called conservative Hamiltonian systems. For the nonlinear systems that we discuss in this section, an integral component of their solution via inverse scattering lies in the ability to write the dynamics of the system implicitly in terms of an operator differential equation of the form

$$\frac{dL(t)}{dt} = B(t)L(t) - L(t)B(t), \tag{75.8}$$

where $L(t)$ is a symmetric linear operator, $B(t)$ is an anti-symmetric linear operator, and both $L(t)$ and $B(t)$ depend explicitly on the state of the system.

Using the Toda lattice as an example, the operators L and B would be the symmetric and antisymmetric tridiagonal matrices

$$L = \begin{bmatrix} \ddots & a_{n-1} & \\ a_{n-1} & b_n & a_n \\ & a_n & \ddots \end{bmatrix}, \quad B = \begin{bmatrix} \ddots & -a_{n-1} & \\ a_{n-1} & 0 & -a_n \\ & a_n & \ddots \end{bmatrix}, \tag{75.9}$$

where $a_n = e^{(y_n - y_{n+1})/2}/2$, and $b_n = \dot{y}_n/2$, for mass positions y_n in a solution to Eq. (75.1). Written in this form, the entries of the matrices in Eq. (75.8) yield the following equations

$$\begin{aligned} \dot{a}_n &= a_n(b_n - b_{n+1}), \\ \dot{b}_n &= 2(a_{n-1}^2 - a_n^2). \end{aligned} \tag{75.10}$$

These are equivalent to the Toda lattice equations, Eq. (75.1), in the coordinates a_n and b_n. Lax has shown [10] that when the dynamics of such a system can be written in the form of Eq. (75.8), then the eigenvalues of the operator $L(t)$ are time-invariant, i.e., $\dot{\lambda} = 0$. Although each of the entries of $L(t)$, $a_n(t)$, and $b_n(t)$ evolve with the state of a solution to the Toda lattice, the eigenvalues of $L(t)$ remain constant.

If we assume that the motion on the lattice is confined to lie within a finite region of the lattice, i.e., the lattice is at rest for $|n| \to \infty$, then the spectrum of eigenvalues for the matrix $L(t)$ can be separated into two sets. There is a continuum of eigenvalues $\lambda \in [-1, 1]$ and a discrete set of eigenvalues for which $|\lambda_k| > 1$. When the lattice is at rest, the eigenvalues consist only of the continuum. When there are solitons in the lattice, one discrete eigenvalue will be present for each soliton excited. This separation of eigenvalues of $L(t)$ into discrete and continuous components is common to all of the nonlinear systems solved with inverse scattering.

The inverse scattering method of solution for soliton systems is analogous to methods used to solve linear evolution equations. For example, consider a linear evolution equation for the state $y(x, t)$. Given an initial condition of the system, $y(x, 0)$, a standard technique for solving for $y(x, t)$ employs Fourier methods. By decomposing the initial condition into a superposition of simple harmonic waves, each of the component harmonic waves can be independently propagated. Given the Fourier decomposition of the state at time t, the harmonic waves can then be recombined to produce the state of the system $y(x, t)$. This process is depicted schematically in Fig. 75.2(a).

$$y(x, 0) \xrightarrow{\text{F.T.}} k, Y(k, 0) \qquad\qquad y(x, 0) \xrightarrow{\text{F.S.}} \lambda, \psi(x, 0)$$

$$\Big\downarrow e^{-j\omega(k)t} \qquad\qquad\qquad\qquad\qquad \Big\downarrow e^{j\beta(k)t}$$

$$y(x, t) \xleftarrow{\text{I.F.T.}} k, Y(k, t) \qquad\qquad y(x, t) \xleftarrow{\text{I.S.}} \lambda, \psi(x, t)$$

$$\text{(a) Linear} \qquad\qquad\qquad\qquad \text{(b) Soliton}$$

FIGURE 75.2   Schematic solution to evolution equations.

An outline of the inverse scattering method for soliton systems is similar. Given an initial condition for the nonlinear system, $y(x, 0)$, the eigenvalues λ and eigenfunctions $\psi(x, 0)$ of the linear operator $L(0)$ can be obtained. This step is often called *forward scattering* by analogy to quantum mechanical scattering, and the collection of eigenvalues and eigenfunctions is called the nonlinear spectrum of the system in analogy to the Fourier spectrum of linear systems. To obtain the nonlinear spectrum at a point in time t, all that is needed is the time evolution of the eigenfunctions, since the eigenvalues do not change with time. For these soliton systems, the eigenfunctions evolve simply in time, according to linear differential equations. Given the eigenvalue-eigenfunction decomposition of $L(t)$, through a process called *inverse scattering*, the state of the system $y(x, t)$ can be completely reconstructed. This process is depicted in Fig. 75.2(b) in a similar fashion to the linear solution process.

For a large class of soliton systems, the inverse scattering method generally involves solving either a linear integral equation or a linear discrete-integral equation. Although the equation is linear, finding its solution is often very difficult in practice. However, when the solution is made up of pure solitons, then the integral equation reduces a set of simultaneous linear equations.

Since the discovery of the inverse scattering method for the solution to KdV, there has been a large class of nonlinear wave equations, both continuous and discrete, for which similar solution methods have been obtained. In most cases, solutions to these equations can be constructed from a nonlinear superposition of soliton solutions. For a comprehensive study of inverse scattering and equations solvable by this method, the reader is referred to the text by Ablowitz and Clarkson [1].

75.3   New Electrical Analogs for Soliton Systems

Since soliton theory has its roots in mathematical physics, most of the systems studied in the literature have at least some foundation in physical systems in nature. For example, KdV has been attributed to studies ranging from ion-acoustic waves in plasma [22] to pressure waves in liquid gas bubble mixtures [12]. As a result, the predominant purpose of soliton research has been to explain physical properties of natural systems. In addition, there are several examples of man-made media that have been designed to support soliton solutions and thus exploit their robust propagation. The use of optical fiber solitons for telecommunications and of Josephson junctions for volatile memory cells are two practical examples [11, 12].

Whether its goal has been to explain natural phenomena or to support propagating solitons, this research has largely focused on the properties of propagating solitons through these nonlinear systems. In this section, we will view solitons as signals and consider exploiting some of their rich signal properties in a signal processing or communication context. This perspective is illustrated graphically in Fig. 75.3, where a signal containing two solitons is shown as an input to a soliton system which can either combine or separate the component solitons according to the evolution equations. From the "solitons-as-signals"

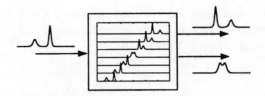

FIGURE 75.3   Two-soliton signal processing by a soliton system.

perspective, the corresponding nonlinear evolution equations can be viewed as special-purpose signal processors that are naturally suited to such signal processing tasks as signal separation or sorting. As we shall see, these systems also form an effective means of generating soliton signals.

75.3.1   Toda Circuit Model of Hirota and Suzuki

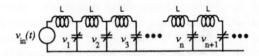

FIGURE 75.4   Nonlinear LC ladder circuit of Hirota and Suzuki.

Motivated by the work of Toda on the exponential lattice, the nonlinear LC ladder network implementation shown in Fig. 75.4 was given by Hirota and Suzuki in [6]. Rather than a direct analogy to the Toda lattice, the authors derived the functional form of the capacitance required for the LC line to be equivalent. The resulting network equations are given by

$$\frac{d^2}{dt^2} \ln\left(1 + \frac{V_n(t)}{V_0}\right) = \frac{1}{LC_0 V_0}(V_{n-1}(t) - 2V_n(t) + V_{n+1}(t)),\qquad(75.11)$$

which is equivalent to the Toda lattice equation for the forces on the nonlinear springs given in Eq. (75.6). The capacitance required in the nonlinear LC ladder is of the form

$$C(V) = \frac{C_0 V_0}{V_0 + V},\qquad(75.12)$$

where V_0 and C_0 are constants representing the bias voltage and the nominal capacitance, respectively. Unfortunately, such a capacitance is rather difficult to construct from standard components.

75.3.2   Diode Ladder Circuit Model for Toda Lattice

In [14], the circuit model shown in Fig. 75.5(a) is presented which accurately matches the Toda lattice and is a direct electrical analog of the nonlinear spring mass system. When the shunt impedance Z_n has the voltage-current relation $\ddot{v}_n(t) = \alpha(i_n(t) - i_{n+1}(t))$, then the governing equations become

$$\frac{d^2 v_n(t)}{dt^2} = \alpha I_s \left(e^{(v_{n-1}(t)-v_n(t))/v_t} - e^{(v_n(t)-v_{n+1}(t))/v_t}\right),\qquad(75.13)$$

or,

$$\frac{d^2}{dt^2} \ln\left(1 + \frac{i_n(t)}{I_s}\right) = \frac{\alpha}{v_t}(i_{n-1}(t) - 2i_n(t) + i_{n+1}(t)),\qquad(75.14)$$

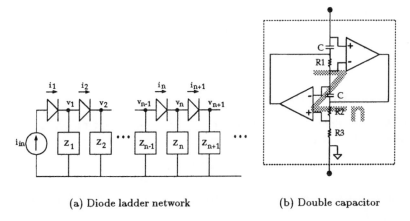

(a) Diode ladder network   (b) Double capacitor

FIGURE 75.5   Diode ladder network in (a), with Z_n realized with a double capacitor as shown in (b).

where $i_1(t) = i_{in}(t)$. These are equivalent to the Toda lattice equations with $a/m = \alpha I_s$ and $b = 1/v_t$. The required shunt impedance is often referred to as a double capacitor, which can be realized using ideal operational amplifiers in the gyrator circuit shown in Fig. 75.5(b), yielding the required impedance of $Z_n = \alpha/s^2 = R_3/R_1 R_2 C^2 s^2$ [13].

This circuit supports a single soliton solution of the form

$$i_n(t) = \beta^2 \mathrm{sech}^2(pn - \beta\tau), \qquad (75.15)$$

where $\beta = \sqrt{I_s}\sinh(p)$, and $\tau = t\sqrt{\alpha/v_t}$. The diode ladder circuit model is very accurate over a large range of soliton wavenumbers, and is significantly more accurate than the LC circuit of Hirota and Suzuki. Shown in Fig. 75.6(a) is an HSPICE simulation with two solitons propagating in the diode ladder circuit.

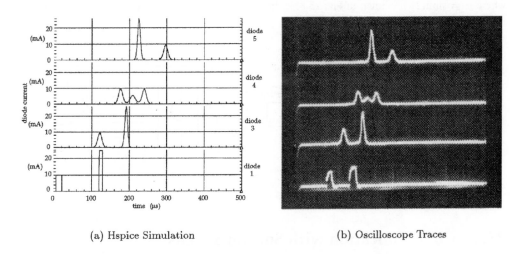

(a) Hspice Simulation   (b) Oscilloscope Traces

FIGURE 75.6   Evolution of a two-soliton signal through the diode lattice. Each horizontal trace shows the current through one of the diodes 1, 3, 4, and 5.

As illustrated in the bottom trace of Fig. 75.6(a), a soliton can be generated by driving the circuit with a square pulse of approximately the same area as the desired soliton. As seen on the third node in the lattice, once the soliton is excited, the non-soliton components rapidly become insignificant.

A two-soliton signal generated by a hardware implementation of this circuit is shown on the oscilloscope traces in Fig 75.6(b). The bottom trace in the figure corresponds to the input current to the circuit, and the remaining traces, from bottom to top, show the current through the third, fourth, and fifth diodes in the lattice.

75.3.3 Circuit Model for Discrete-KdV

The discrete-KdV equation (dKdV), sometimes referred to as the nonlinear ladder equations [1], or the KM system (Kac and vanMoerbeke) [17] is governed by the equation

$$\dot{u}_n(t) = e^{u_{n-1}(t)} - e^{u_{n+1}(t)}. \tag{75.16}$$

In [14], the circuit shown in Fig. 75.7, is shown to be governed by the discrete-KdV equation

$$\dot{v}_n(t) = \frac{I_s}{C} \left(e^{v_{n-1}(t)/v_t} - e^{v_{n+1}(t)/v_t} \right), \tag{75.17}$$

where I_s is the saturation current of the diode, C is the capacitance, and v_t is the thermal voltage. Since this circuit is first order, the state of the system is completely specified by the capacitor voltages.

Rather than processing continuous-time signals as with the Toda lattice system, we can use this system to process discrete-time solitons as specified by v_n. For the purposes of simulation, we consider the periodic dKdV equation by setting $v_{n+1}(t) = v_0(t)$ and initializing the system with the discrete-time signal corresponding to a listing of node capacitor voltages. We can place a multi-soliton solution in the circuit using inverse scattering techniques to construct the initial voltage profile. The single soliton solution to the dKdV system is given by

$$v_n(t) = \ln \left(\frac{\cosh(\gamma\,(n-2) - \beta\,t)\cosh(\gamma\,(n+1) - \beta\,t)}{\cosh(\gamma\,(n-1) - \beta\,t)\cosh(\gamma\,n - \beta\,t)} \right), \tag{75.18}$$

where $\beta = \sinh(2\gamma)$. Shown in Fig. 75.8, is the result of an HSPICE simulation of the circuit with 30 nodes in a loop configuration.

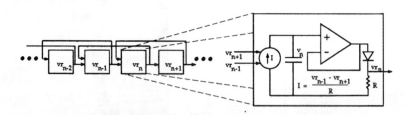

FIGURE 75.7   Circuit model for discrete-KdV.

75.4 Communication with Soliton Signals

Many traditional communication systems use a form of sinusoidal carrier modulation, such as amplitude modulation (AM) or frequency/phase modulation (FM/PM) to transmit a message-bearing signal over a physical channel. The reliance upon sinusoidal signals is due in part to the simplicity with which such signals can be generated and processed using linear systems. More importantly, information contained in sinusoidal signals with different frequencies can easily be separated using linear systems or Fourier techniques. The complex dynamic structure of soliton signals and the ease with which these signals can

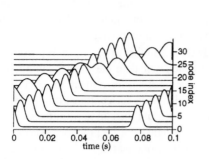
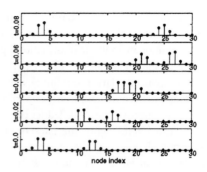

FIGURE 75.8 To the left, the normalized node capacitor voltages, $v_n(t)/v_t$ for each node is shown as a function of time. To the right, the state of the circuit is shown as a function of node index for five different sample times. The bottom trace in the figure corresponds to the initial condition.

be both generated and processed with analog circuitry renders them potentially applicable in the broad context of communication in an analogous manner to sinusoidal signals.

We define a soliton carrier as a signal that is composed of a periodically repeated single soliton solution to a particular nonlinear system. For example, a soliton carrier signal for the Toda lattice is shown in Fig. 75.9. As a Toda lattice soliton carrier is generated, a simple amplitude modulation scheme could be devised by

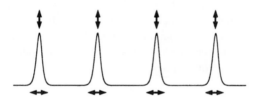

FIGURE 75.9 Modulating the relative amplitude or position of soliton carrier signal for the Toda lattice.

slightly modulating the soliton parameter β, since the amplitude of these solitons is proportional to β^2. Similarly, an analog of FM or pulse-position modulation could be achieved by modulating the relative position of each soliton in a given period, as shown in Fig. 75.9.

As a simple extension, these soliton modulation techniques can be generalized to include multiple solitons in each period and accommodate multiple information-bearing signals, as shown in Fig. 75.10 for a four soliton example using the Toda lattice circuits presented in [14]. In the figure, a signal is

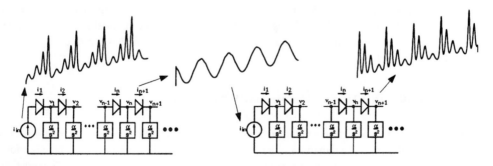

FIGURE 75.10 Multiplexing of a four soliton solution to the Toda lattice.

generated as a periodically repeated train of four solitons of increasing amplitude. The relative amplitudes or positions of each of the component solitons could be independently modulated about their nominal values to accommodate multiple information signals in a single soliton carrier.

The nominal soliton amplitudes can be appropriately chosen so that as this signal is processed by the diode ladder circuit, the larger amplitude solitons propagate faster than the smaller solitons, and each of the solitons can become nonlinearly superimposed as viewed at a given node in the circuit. From an input-output perspective, the diode ladder circuit can be used to make each of the solitons coincidental in time. As indicated in the figure, this packetized soliton carrier could then be transmitted over a wireless communication channel. At the receiver, the multi-soliton signal can be processed with an identical diode ladder circuit which is naturally suited to perform the nonlinear signal separation required to demultiplex the multiple soliton carriers. As the larger amplitude solitons emerge before the smaller, after a given number of nodes, the original multi-soliton carrier re-emerges from the receiver in amplitude-reversed order. At this point, each of the component soliton carriers could be demodulated to recover the individual message signals it contains. Aside from a packetization of the component solitons, we will see that multiplexing the soliton carriers in this fashion can lead to an increased energy efficiency for such carrier modulation schemes, making such techniques particularly attractive for a broad range of portable wireless and power-limited communication applications.

Since the Toda lattice equations are symmetric with respect to time and node index, solitons can propagate in either direction. As a result, a single diode ladder implementation could be used as both a modulator and demodulator simultaneously. Since the forward propagating solitons correspond to positive eigenvalues in the inverse scattering transform and the reverse propagating solitons have negative eigenvalues, the dynamics of the two signals will be completely decoupled.

A technique for modulation of information on soliton carriers was also proposed by Hirota et al. in [15] and [16]. In their work, an amplitude and phase modulation of a two-soliton solution to the Toda lattice were presented as a technique for private communication. Although their signal generation and processing methods relied on an inexact phenomenon known as recurrence, the modulation paradigm they presented is essentially a two-soliton version of the carrier modulation paradigm presented in [14].

75.4.1 Low Energy Signaling

A consequence of some of the conservation laws satisfied by the Toda lattice is a reduction of energy in the transmitted signal for the modulation techniques of this section. In fact, as a function of the relative separation of two solitons, the minimum energy of the transmitted signal is obtained precisely at the point of overlap. This can be shown [14] for the two soliton case by analysis of the form of the equation for the energy in the waveform, $v(t) = f_n(t)$,

$$E = \int_{-\infty}^{\infty} v(t; \delta_1, \delta_2)^2 dt , \tag{75.19}$$

where $v(t; \delta_1, \delta_2)$ is given in Eq. (75.3). In [14] it is proven that E is exactly minimized when $\delta_1 = \delta_2$, i.e., the two solitons are mutually co-located. Significant energy reduction can be achieved for a fairly wide range of separations and amplitudes, indicating that the modulation techniques described here could take advantage of this reduction.

75.5 Noise Dynamics in Soliton Systems

In order to analyze the modulation techniques presented here, accurate models are needed for the effects of random fluctuations on the dynamics of soliton systems. Such disturbances could take the form of additive or convolutional corruption incurred during terrestrial or wired transmission, circuit thermal noise, or modeling errors due to system deviation from the idealized soliton dynamics. A fundamental property of solitons is that they are stable in the presence of a variety of disturbances.

With the development of the inverse scattering framework and the discovery that many soliton systems were conservative Hamiltonian systems, many of the questions regarding the stability of soliton solutions are readily answered. For example, since the eigenvalues of the associated linear operator remain unchanged under the evolution of the dynamics, then any solitons that are initially present in a system must remain present for all time, regardless of their interactions. Similarly, the dynamics of any non-soliton components that are present in the system are uncoupled from the dynamics of the solitons. However, in the communication scenario discussed in [14], soliton waveforms are generated and then propagated over a noisy channel. During transmission, these waveforms are susceptible to additive corruption from the channel. When the waveform is received and processed, the inverse scattering framework can provide useful information about the soliton and noise content of the received waveform.

In this section, we will assume that soliton signals generated in a communication context have been transmitted over an additive white Gaussian noise channel. We can then consider the effects of additive corruption on the processing of soliton signals with their nonlinear evolution equations. Two general approaches are taken to this problem. The first primarily deals with linearized models and investigates the dynamic behavior of the noise component of signals composed of an information bearing soliton signal and additive noise. The second approach is taken in the framework of inverse scattering and is based on some results from random matrix theory. Although the analysis techniques developed here are applicable to a large class of soliton systems, we focus our attention on the Toda lattice as an example.

75.5.1 Toda Lattice Small Signal Model

If a signal that is processed in a Toda lattice receiver contains only a small amplitude noise component, then the dynamics of the receiver can be approximated by a small signal model,

$$\frac{d^2 V_n(t)}{dt^2} = \frac{1}{LC}(V_{n-1}(t) - 2V_n(t) + V_{n+1}(t)),\tag{75.20}$$

when the amplitude of $V_n(t)$ is appropriately small.

If we consider processing signals with an infinite linear lattice and obtain an input-output relationship where a signal is input at the zeroth node and the output is taken as the voltage on the Nth node, it can be shown that the input-output frequency response of the system can be given by

$$H_N(j\omega) = \begin{cases} e^{-2j\sin^{-1}(\omega\sqrt{LC}/2)N}, & |\omega| < 2/\sqrt{LC} \\ e^{[j\pi - 2\cosh^{-1}(\omega\sqrt{LC}/2)]N}, & \text{else}, \end{cases}\tag{75.21}$$

which behaves as a low pass filter, and for $N \gg 1$, approaches

$$|H_N(j\omega)|^2 = \begin{cases} 1, & |\omega| < \omega_c = 2/\sqrt{LC} \\ 0, & \text{else}. \end{cases}\tag{75.22}$$

Our small signal model indicates that in the absence of solitons in the received signal, small amplitude noise will be processed by a low pass filter. If the received signal also contains solitons, then the small signal model of Eq. (75.20) will no longer hold. A linear small signal model can still be used if we linearize Eq. (75.11) about the known soliton signal. Assuming that the solution contains a single soliton in small amplitude noise, $V_n(t) = S_n(t) + v_n(t)$, we can write Eq. (75.11) as an exact equation that is satisfied by the non-soliton component

$$\frac{d^2}{dt^2}\ln\left(1 + \frac{v_n(t)}{1 + S_n(t)}\right) = \frac{1}{LC}(v_{n-1}(t) - 2v_n(t) + v_{n+1}(t)),\tag{75.23}$$

which can be viewed as the fully nonlinear model with a time-varying parameter, $(1 + S_n(t))$. As a result, over short time scales relative to $S_n(t)$, we would expect this model to behave in a similar manner to the small signal model of Eq. (75.20). With $v_n(t) \ll (1 + S_n(t))$, we obtain

$$\frac{d^2}{dt^2} \frac{v_n(t)}{1 + S_n(t)} \approx \frac{1}{LC}(v_{n-1}(t) - 2v_n(t) + v_{n+1}(t)) . \tag{75.24}$$

When the contribution from the soliton is small, Eq. (75.24) reduces to the linear system of Eq. (75.20). We would therefore expect that both before and after a soliton has passed through the lattice, the system essentially low pass filters the noise. However, as the soliton is processed, there will be a time-varying component to the filter.

To confirm the intuition developed through small signal analyses, the fully nonlinear dynamics are shown in Fig. 75.11 in response to a single soliton at 20 dB signal-to-noise ratio. As expected, the response

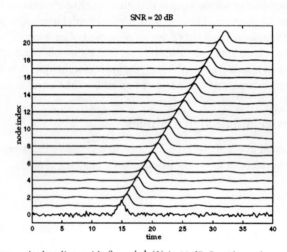

FIGURE 75.11   Response to a single soliton with $\beta = \sinh(1)$ in 20 dB Gaussian noise.

to the lattice is essentially the unperturbed soliton with an additional low pass perturbation. The spectrum of the noise remains essentially flat over the bandwidth of the soliton and is attenuated out of band.

75.5.2   Noise Correlation

The statistical correlation of the system response to the noise component can also be estimated from our linear analyses. Given that the lattice behaves as a lowpass filter, the small amplitude noise $v_n(t)$ is zero mean and has an auto-correlation function given by

$$R_{n,n}(\tau) = E\{v_n(t)v_n(t + \tau)\} \approx N_0 \frac{\sin(\omega_c \tau)}{\pi \tau} , \tag{75.25}$$

and a variance $\sigma_{v_n}^2 \approx N_0 \omega_c / \pi$, for $n \gg 1$.

Although the autocorrelation of the noise at each node is only affected by the magnitude response of Eq. (75.21), the cross-correlation between nodes is also affected by the phase response. The cross-correlation between nodes m and n is given by

$$R_{m,n}(\tau) = R_{m,m}(\tau) * h_{n-m}(-\tau) , \tag{75.26}$$

where $h_m(\tau)$ is the inverse Fourier transform of $H_m(j\omega)$ in Eq. (75.21). Since $h_m(\tau) * h_m(-\tau)$ approaches the impulse response of an ideal low pass filter for $m \gg 1$, we have

$$R_{m,n}(\tau) \approx N_0 \frac{\sin(\omega_c \tau)}{\pi \tau} * h_{n-m}(\tau) . \tag{75.27}$$

For small amplitude noise, the correlation structure can be examined through the linear lattice, which acts as a dispersive low pass filter. A corresponding analysis of the nonlinear system in the presence of solitons is prohibitively complex. However, we can explore the analyses numerically by linearizing the dynamics of the system about the known soliton trajectory.

From our earlier linearized analyses, the linear time-varying small signal model can be viewed over short time scales as a linear time-invariant chain, with a slowly varying parameter. The resulting input-output transfer function can be viewed as a low pass filter with time varying cutoff frequency equal to ω_c when a soliton is far from the node, and to $\omega_0 \sqrt{1 + V_n^0}$ as a soliton passes through. Thus, we would expect the variance of the node voltage to rise from a nominal value as a soliton passes through. This intuition can be verified experimentally by numerically integrating the corresponding Riccati equation for the node covariance and computing the resulting variance of the noise component on each node. Since the lattice was assumed initially at rest, there will be a startup transient, as well as an initial spatial transient at the beginning of the lattice, after which the variance of the noise is amplified from the nominal variance as each soliton passes through, confirming our earlier intuition.

75.5.3 Inverse Scattering-Based Noise Modeling

The inverse scattering transform provides a particularly useful mechanism for exploring the long term behavior of soliton systems. In a similar manner to the use of the Fourier transform for describing the ability of linear processors to extract a signal from a stationary random background, the nonlinear spectrum of a received soliton signal in noise can effectively characterize the ability of the nonlinear system to extract or process the component solitons. In this section, we focus on the effects of random perturbations on the dynamics of solitons in the Toda lattice from the viewpoint of inverse scattering.

As seen in Section 75.2.1, the dynamics of the Toda lattice may be described by the evolution of the matrix

$$L(t) = \begin{bmatrix} \ddots & a_{n-1}(t) & \\ a_{n-1}(t) & b_n(t) & a_n(t) \\ & a_n(t) & \ddots \end{bmatrix} , \tag{75.28}$$

whose eigenvalues outside the range $|\lambda| \leq 1$ give rise to soliton behavior. By considering the effects of small amplitude perturbations to the sequences $a_n(t)$ and $b_n(t)$ on the eigenvalues of $L(t)$, we can observe the effects on the soliton dynamics through the eigenvalues corresponding to solitons.

Following [20], we write the $N \times N$ matrix L as $L = L_0 + D$, where L_0 is the unperturbed symmetric matrix, and D is the symmetric random perturbation. To second order, the eigenvalues are given by

$$\lambda_g = \mu_g + \hat{d}_{gg} - \sum_{i=1,i\neq g}^{N} \frac{\hat{d}_{gi}\hat{d}_{ig}}{\mu_{ig}} , \tag{75.29}$$

where μ_g is the gth eigenvalue of L_0, $\mu_{ig} = \mu_i - \mu_g$, and $\hat{d}_{ij}$ are the elements of the matrix $\widehat{D}$ defined by $\widehat{D} = C^\top D C$, and C is a matrix that diagonalizes L, $C^\top L_0 C = \mathrm{diag}(\mu_1, \ldots, \mu_N)$.

To second order, the means of the eigenvalues are given by

$$E\{\lambda_g\} = \mu_g - \sum_{i=1,i\neq g}^{N} \frac{\hat{d}_{gi}\hat{d}_{ig}}{\mu_{ig}} , \tag{75.30}$$

indicating that the eigenvalues of L are asymptotically (SNR$\to \infty$) unbiased estimates of the eigenvalues of L_0. To first order, $\lambda_g \approx \mu_g - \hat{d}_{gg}$, and $\hat{d}_{gg}$ is a linear combination of the elements of D,

$$\hat{d}_{gg} = \sum_{r=1,s=1}^{N} c_{gr} c_{gs} d_{rs} . \tag{75.31}$$

Therefore, if the elements of D are jointly Gaussian, then to first order, the eigenvalues of L will be jointly Gaussian, distributed about the eigenvalues of L_0.

The variance of the eigenvalues can be shown to be approximately given by

$$\text{Var}(\lambda_g) \approx \frac{\sigma_\beta^2 + 2\sigma_\alpha^2(1 + \cos(4\pi g/N))}{N}, \tag{75.32}$$

to second order, where σ_β^2 and σ_α^2 are the variances of the iid perturbations to b_n and a_n, respectively. This indicates that the eigenvalues of L are consistent estimates of the eigenvalues of L_0.

To first order, when processing small amplitude noise alone, the noise only excites eigenvalues distributed about the continuum, corresponding to non-soliton components. When solitons are processed in small amplitude noise, to first order, there is a small Gaussian perturbation to the soliton eigenvalues as well.

75.6 Estimation of Soliton Signals

In the communication techniques suggested in Section 75.4, the parameters of a multi-soliton carrier are modulated with message-bearing signals and the carrier is then processed with the corresponding nonlinear evolution equation. A potential advantage to transmission of this packetized soliton carrier is a net reduction in the transmitted signal energy. However, during transmission, the multi-soliton carrier signal can be subjected to distortions due to propagation, which we have assumed can be modeled as additive white Gaussian noise (AWGN). In this section, we investigate the ability of a receiver to estimate the parameters of a noisy multi-soliton carrier. In particular, we consider the problems of estimating the scaling parameters and the relative positions of component solitons of multi-soliton solutions, once again focusing on the Toda lattice as an example. For each of these problems, we derive Cramér-Rao lower bounds for the estimation error variance through which several properties of multi-soliton signals can be observed. Using these bounds, we will see that although the net transmitted energy in a multi-soliton signal can be reduced through nonlinear interaction, the estimation performance for the parameters of the component solitons can also be enhanced. However, at the receiver there are inherent difficulties in parameter estimation imposed by this nonlinear coupling. We will see that the Toda lattice can act as a tuned receiver for the component solitons, naturally decoupling them so that the parameters of each soliton can be independently estimated. Based on this strategy, we develop robust algorithms for maximum likelihood parameter estimation. We also extend the analogy of the inverse scattering transform as an analog of the Fourier transform for linear techniques, by developing a maximum likelihood estimation algorithm based on the nonlinear spectrum of the received signal.

75.6.1 Single Soliton Parameter Estimation: Bounds

In our simplified channel model, the received signal $r(t)$ contains a soliton signal $s(t)$ in an additive white Gaussian noise background $n(t)$ with noise power N_0. A bound on the variance of an estimate of the parameter β may be useful in determining the demodulation performance of an AM-like modulation or PAM, where the component soliton wavenumbers are slightly amplitude modulated by a message-bearing waveform. When $s(t)$ contains a single soliton for the Toda lattice, $s(t) = \beta^2 \text{sech}^2(\beta t)$, the variance of any unbiased estimator $\hat{\beta}$ of β must satisfy the Cramér-Rao lower bound (CRB) [19],

$$\text{Var}\left(\hat{\beta}\right) \geq \frac{N_0}{\int_{t_i}^{t_f} \left(\frac{\partial s(t;\beta)}{\partial \beta}\right)^2 dt}, \tag{75.33}$$

where the observation interval is assumed to be $t_i < t < t_f$. For the infinite observation interval, $-\infty < t < \infty$, the CRB (75.33) is given by

$$\text{Var}\left(\hat{\beta}\right) \geq \frac{N_0}{\left(\frac{8}{3} + \frac{4\pi^2}{45}\right)\beta} \approx \frac{N_0}{3.544\beta}. \tag{75.34}$$

A slightly different bound may be useful in determining the demodulation performance of an FM-like modulation or PPM, where the soliton position, or time-delay, is slightly modulated by a message-bearing waveform. The fidelity of the recovered message waveform will be directly affected by the ability of a receiver to estimate the soliton position. When the signal $s(t)$ contains a single soliton $s(t) = \beta^2 \text{sech}^2(\beta(t - \delta))$, where δ is the relative position of the soliton in a period of the carrier, the CRB for $\hat{\delta}$ is given by

$$\text{Var}\left(\hat{\delta}\right) \geq \frac{N_0}{\int_{t_i}^{t_f} 4\beta^6 \text{sech}^4(\beta(t - \delta)) \tanh^2(\beta(t - \delta)) dt} = \frac{N_0}{\left(\frac{16}{15}\right)\beta^5}. \tag{75.35}$$

As a comparison, for estimating the time of arrival of the raised cosine pulse, $\beta^2(1 + \cos(2\pi\beta(t - \delta)))$, the CRB for this more traditional pulse position modulation would be

$$\text{Var}\left(\hat{\delta}\right) \geq \frac{N_0}{\pi^2\beta^5}, \tag{75.36}$$

which has the same dependence on signal amplitude as Eq. (75.35). These bounds can be used for multiple soliton signals if the component solitons are well separated in time.

75.6.2 Multi-Soliton Parameter Estimation: Bounds

When the received signal is a multi-soliton waveform where the component solitons overlap in time, the estimation problem becomes more difficult. It follows that the bounds for estimating the parameters of such signals must also be sensitive to the relative positions of the component solitons.

We will focus our attention on the two-soliton solution to the Toda lattice, given by Eq. (75.3). We are generally interested in estimating the parameters of the multi-soliton carrier for an unknown relative spacing among the solitons present in the carrier signal. Either the relative spacing of the solitons has been modulated and is therefore unknown, or the parameters β_1 and β_2 are slightly modulated and the induced phase shift in the received solitons, ϕ, is unknown. For large separations, $\delta = \delta_1 - \delta_2$, the CRB for estimating the parameters of either of the component solitons will be unaffected by the parameters of the other soliton. As shown in Fig. 75.12, when the component solitons are well separated, the CRB for either β_1 or β_2 approaches the CRB for estimation of a single soliton with that parameter value in the same level of noise. The bounds for estimating β_1, and β_2 are shown in Fig. 75.12 as a function of the relative separation, δ.

Note that both of the bounds are reduced by the nonlinear superposition, indicating that the potential performance of the receiver is *enhanced* by the nonlinear superposition. However, if we let the parameter difference $\beta_2 - \beta_1$ increase, we notice a different character to the bounds. Specifically, we maintain $\beta_1 = \sinh(2)$, and let $\beta_1 = \sinh(1.25)$. The performance of the larger soliton is inhibited by the nonlinear superposition, while the smaller soliton is still enhanced. In fact, the CRB for the smaller soliton becomes lower than that for the larger soliton near the range $\delta = 0$. This phenomenon results from the relative

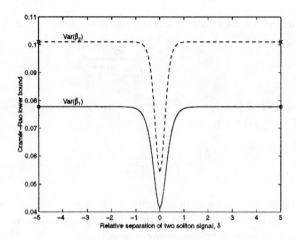

FIGURE 75.12   The Cramér-Rao lower bound for estimating $\beta_1 = \sinh(2)$ and $\beta_2 = \sinh(1.75)$ with all parameters unknown in AWGN with $N_0 = 1$. The bounds are shown as a function of the relative separation, $\delta = \delta_1 - \delta_2$. The CRB for estimating β_1 and β_2 of a single soliton with the same parameter value is indicated with 'o' and '×' marks, respectively.

sensitivity of the signal $s(t)$ to each of the parameters β_1 and β_2. The ability to simultaneously enhance estimation performance while decreasing signal energy is an inherently nonlinear phenomena.

Combining these results with the results of Section 75.4.1, we see that the nonlinear interaction of the component solitons can simultaneously enhance the parameter estimation performance and reduce the net energy of the signal. This property may make superimposed solitons attractive for use in a variety of communication systems.

75.6.3   Estimation Algorithms

In this section we will present and analyze several parameter estimation algorithms for soliton signals. Again, we will focus on the diode ladder circuit implementation of the Toda lattice equations, Eq. (75.14). As motivation, consider the problem of estimating the position, δ, of a single soliton solution $s(t; \delta) = \beta^2 \text{sech}^2(\beta(t - \delta))$, with the parameter β known. This is a classical time-of-arrival estimation problem. For observations $r(t) = s(t) + n(t)$, where $n(t)$ is a stationary white Gaussian process, the maximum likelihood estimate is given by the value of the parameter δ which minimizes the expression

$$\hat{\delta} = \arg \min_{\tau} \int_{t_i}^{t_f} (r(t) - s(t - \tau))^2 dt .$$ (75.37)

Since the replica signals all have the same energy, we can represent the minimization in (75.37) as a maximization of the correlation

$$\hat{\delta} = \arg \min_{\tau} \int_{t_i}^{t_f} r(t)s(t - \tau)dt .$$ (75.38)

It is well known that an efficient way to perform the correlation (75.38) with all of the replica signals $s(t - \tau)$ over the range $\delta_{\min} < \tau < \delta_{\max}$, is through convolution with a matched filter followed by a peak-detector [19].

When the signal $r(t)$ contains a multi-soliton signal, $s(t; \underline{\beta}, \underline{\delta})$, where we wish to estimate the parameter vector $\underline{\delta}$, the estimation problem becomes more involved. If the component solitons are well separated in time, then the maximum likelihood estimator for the positions of each of the component solitons would again involve a matched filter processor followed by a peak-detector for each soliton. If the component solitons are not well separated and are therefore nonlinearly combined, the estimation problems are tightly coupled and should not be performed independently. The estimation problems can be decoupled by preprocessing the signal $r(t)$ with the Toda lattice. By setting $i_{in}(t) = r(t)$, that is the current through the first diode in the diode ladder circuit, then as the signal propagates through the lattice, the component solitons will naturally separate due to their different propagation speeds.

Defining the signal and noise components as viewed on the kth node in the lattice as $s_k(t)$ and $n_k(t)$, respectively, i.e., $i_k(t) = s_k(t) + n_k(t)$, where $n_0(t)$ is the stationary white Gaussian noise process $n(t)$, in Section 75.5, we saw that in the high SNR limit, $n_k(t)$ will be low pass and Gaussian. In this limit, the ML estimator for the positions, δ_i, can again be formulated using matched filters for each of the component solitons. Since the lattice equations are invertible, at least in principle through inverse scattering, then the ML estimate of the parameter $\underline{\delta}$ based on $r(t)$ must be the same as the estimate based on $i_n(t) = T(r(t))$, for any invertible transformation $T(\cdot)$. If the component solitons are well separated as viewed on the Nth node of the lattice, $i_N(t)$, then an ML estimate based on observations of $i_N(t)$ will reduce to the aggregate of ML estimates for each of the separated component solitons on low pass Gaussian noise. For soliton position estimation, this amounts to a bank of matched filters. We can view this estimation procedure as a form of nonlinear matched filtering, whereby first, dynamics matched to the soliton signals are used to perform the necessary signal separation, and then filters matched to the separated signals are used to estimate their arrival time.

75.6.4 Position Estimation

We will focus our attention on the two-soliton signal (75.3). If the component solitons are well-separated as viewed on the Nth node of the Toda lattice, the signal appears to be a linear superposition of two solitons,

$$
\begin{aligned}
i_N(t) &\approx \beta_1^2 \mathrm{sech}^2(\beta_1(t - \delta_1) - p_1 N - \phi/2) \\
&+ \beta_2^2 \mathrm{sech}^2(\beta_2(t - \delta_2) - p_2 N + \phi/2) ,
\end{aligned}
\tag{75.39}
$$

where $\phi/2$ is the time-shift incurred due to the nonlinear interaction. Matched filters can now be used to estimate the time of the arrival of each soliton at the Nth node. We formulate the estimate

$$
\hat{\delta}_1 = \left(t_{N,1}^a - \frac{p_1 N + \phi/2}{\beta_1} \right), \quad
\hat{\delta}_2 = \left(t_{N,2}^a - \frac{p_2 N - \phi/2}{\beta_2} \right) ,
\tag{75.40}
$$

where $t_{N,i}^a$ is the time of arrival of the ith soliton on node N. The performance of this algorithm for a two-soliton signal with $\beta = [\sinh(2), \sinh(1.5)]$ is shown in Fig. 75.13. Note that although the error variance of each estimate appears to be a constant multiple of the CRB, the estimation error variance approaches the CRB in an absolute sense as $N_0 \to 0$.

75.6.5 Estimation Based on Inverse Scattering

The transformation $L(t) = T\{r(t)\}$, where $L(t)$ is the symmetric matrix from the inverse scattering transform, is also invertible in principle. Therefore, an ML estimate based on the matrix $L(t)$ must be the same as an ML estimate based on $r(t)$. We therefore seek to form an estimate of the parameters of the signal $r(t)$ by performing the estimation in the nonlinear spectral domain. This can be accomplished by viewing the Toda lattice as a nonlinear filterbank which projects the signal $r(t)$ onto the spectral components of $L(t)$. This use of the inverse scattering transform is analogous to performing frequency estimation with the Fourier transform.

If $v_n(t)$ evolves according to the Toda lattice equations, then the eigenvalues of the matrix, $L(t)$ are time-invariant, where, $a_n(t) = \frac{1}{2}e^{(v_n(t)-v_{n+1}(t))/2}$, and $b_n = \dot{v}_n(t)/2$. Further, the eigenvalues of $L(t)$ for which $|\lambda_i| > 1$ correspond to soliton solutions, with $\beta_i = \sinh(\cosh^{-1}(\lambda_i)) = \sqrt{\lambda_i^2 - 1}$. The eigenvalues

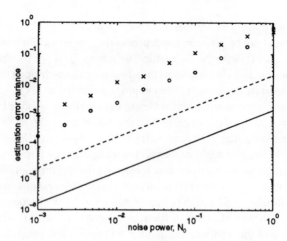

FIGURE 75.13   The CRBs for δ_1 and δ_2 are shown with solid and dashed lines, while the estimation error results of 100 Monte-Carlo trials are indicated with 'o' and '×' marks, respectively.

of $L(t)$ are, to first order, jointly Gaussian and distributed about the true eigenvalues corresponding to the original multi-soliton signal, $s(t)$. Therefore, estimation of the parameters β_i from the eigenvalues of $L(t)$ as described above constitutes a maximum likelihood approach in the high SNR limit.

The parameter estimation algorithm now amounts to an estimation of the eigenvalues of $L(t)$. Note that since $L(t)$ is tridiagonal, very efficient techniques for eigenvalue estimation may be used [3]. The estimate of the parameter β is then found by the relation $\hat{\beta}_i = \sqrt{\lambda_i^2 - 1}$, where $|\lambda_i| > 1$, and the sign of β_i can be recovered from the sign of λ_i. Clearly if there is a pre-specified number of solitons, k, present in the signal, then the k largest eigenvalues would be used for the estimation. If the number k were unknown, then a simultaneous detection and estimation algorithm would be required.

An example of the joint estimation of the parameters of a two-soliton signal is shown in Fig. 75.14(a). The estimation error variance decreases with the noise power at the same exponential rate as the CRB.

To verify that the performance of the estimation algorithm has the same dependence on the relative separation of solitons as indicated in Section 75.6.2, the estimation error variance is also indicated in Fig. 75.14(b) vs. the relative separation, δ. In the figure, the mean-squared parameter estimation error for each of the parameters β_i are shown along with their corresponding CRB. At least empirically, we see that the fidelity of the parameter estimates are indeed enhanced by their nonlinear interaction, even though this corresponds to a signal with lower energy, and therefore lower observational SNR.

75.7   Detection of Soliton Signals

The problem of detecting a single soliton or multiple non-overlapping solitons in AWGN falls within the theory of classical detection. The Bayes optimal detection of a known or multiple known signals in AWGN can be accomplished with matched filter processing. When the signal $r(t)$ contains a multi-soliton signal where the component solitons are not resolved, the detection problem becomes more involved. Specifically, consider a signal comprising a two-soliton solution to the Toda lattice, where we wish to decide which, if any, solitons are present. If the relative positions of the component solitons are known *a priori*, then the detection problem reduces to deciding which among four possible known signals is present,

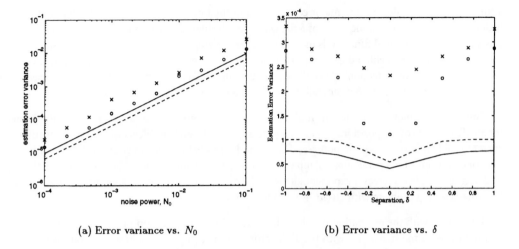

(a) Error variance vs. N_0 (b) Error variance vs. δ

FIGURE 75.14 The estimation error variance for the inverse scattering-based estimates of $\beta_1 = \sinh(2)$, $\beta_2 = \sinh(1.5)$. The bounds for β_1 and β_2 are indicated with solid and dashed lines, respectively. The estimation results for 100 Monte Carlo trials with a diode lattice of $N = 10$ nodes for β_1 and β_2 are indicated by the points labeled 'o' and '×', respectively.

$$
\begin{aligned}
H_0 &: \quad r(t) = n(t), \\
H_1 &: \quad r(t) = s_1(t) + n(t), \\
H_2 &: \quad r(t) = s_2(t) + n(t), \\
H_{12} &: \quad r(t) = s_{12}(t) + n(t),
\end{aligned}
$$

where $s_1(t)$, $s_2(t)$, and $s_{12}(t)$ are soliton one, soliton two, and the multi-soliton signals, respectively. Once again, this problem can be solved with standard Gaussian detection theory.

If the relative positions of the solitons are unknown, as would be the case for a modulated soliton carrier, then the signal $s_{12}(t)$ will vary significantly as a function of the relative separation. Similarly, if the signals are to be transmitted over a soliton channel where different users occupy adjacent soliton wavenumbers, any detection at the receiver would have to be performed with the possibility of another soliton component present at an unknown position. We therefore obtain a composite hypothesis testing problem, whereby under each hypothesis, we have

$$
\begin{aligned}
H_0 &: \quad r(t) = n(t), \\
H_1 &: \quad r(t) = s_1(t; \delta_1) + n(t), \\
H_2 &: \quad r(t) = s_2(t; \delta_2) + n(t), \\
H_{12} &: \quad r(t) = s_{12}(t; \underline{\delta}) + n(t),
\end{aligned}
$$

where $\underline{\delta} = [\delta_1, \delta_2]^{\mathsf{T}}$. The general problem of detection with an unknown parameter, $\underline{\delta}$, can be handled in a number of ways. For example, if the parameter can be modeled as random and the distribution for the parameter were known, $p_{\underline{\delta}}(\underline{\delta})$, along with the distributions $p_{r|\underline{\delta},H}(R|\underline{\delta}, H_i)$ for each hypothesis, then the Bayes or Neyman-Pearson criteria can be used to formulate a likelihood ratio test. Unfortunately, even when the distribution for the parameter $\underline{\delta}$ is known, the likelihood ratios cannot be found in closed form for even the single soliton detection problem.

Another approach that is commonly used in radar processing [5, 19] applies when the distribution of δ does not vary rapidly over a range of possible values while the likelihood function has a sharp peak as a function of δ. In this case, the major contribution to the integral in the averaged likelihood function is due to the region around the value of δ for which the likelihood function is maximum, and therefore this value of the likelihood function is used as if the maximizing value, $\hat{\delta}_{ML}$, were the *actual* value. Since $\hat{\delta}_{ML}$ is the maximum likelihood estimate of δ based on the observation, $r(t)$, such techniques are called "maximum likelihood detection". Also, the term "generalized likelihood ratio test" (GLRT) is used since the hypothesis test amounts to a generalization of the standard likelihood ratio test.

If we plan to employ a GLRT for the multi-soliton detection problem, we are again faced with the need for an ML estimate of the position, $\hat{\underline{\delta}}_{ML}$. A standard approach to such problems would involve turning the current problem into one with hypotheses H_0, H_1, and H_2 as before, and an additional M hypotheses— one for each value of the parameter $\underline{\delta}$ sampled over a range of possible values. The complexity of this type of detection problem increases exponentially with the number of component solitons, N_s, resulting in a hypothesis testing problem with $O((M+1)^{N_s})$ hypotheses.

However, as with the estimation problems in Section 75.6, the detection problems can be decoupled by preprocessing the signal $r(t)$ with the Toda lattice. If the component solitons separate as viewed on the Nth node in the lattice, then the detection problem can be more simply formulated using $i_N(t)$. The invertibility of the lattice equations implies that a Bayes optimal decision based on $r(t)$ must be the same as that based on $i_N(t)$. Since the Bayes optimal decision can be performed based on the likelihood function $\Lambda(r(t))$, and $\Lambda(i_N(t)) = \Lambda(T\{r(t)\}) = \Lambda(r(t))$, the optimal decisions based on $r(t)$ and $i_N(t)$ must be the same for any invertible transformation $T\{\cdot\}$. Although we will be using a GLRT, where the value of $\hat{\underline{\delta}}_{ML}$ is used for the unknown positions of the multi-soliton signal, since the ML estimates based on $r(t)$ and $i_N(t)$ must also be the same, the detection performance of a GLRT using those estimates must also be the same. Since at high SNR, the noise component of the signal $i_N(t)$ can be assumed low pass and Gaussian, the GLRT can be performed by pre-processing $r(t)$ with the Toda lattice equations followed by matched filter processing.

75.7.1   Simulations

To illustrate the algorithm, we consider the hypothesis test between H_0 and H_{12}, where the separation of the two solitons, $\delta_1 - \delta_2$, varies randomly in the interval $[-1/\beta_2, 1/\beta_2]$. The detection processor comprises a Toda lattice of $N = 20$ nodes, with the detection performed based on the signal $i_{10}(t)$. To implement the GLRT, we search over a fixed time interval about the expected arrival time for each soliton. In this manner we obtain a sequence of 1000 Monte Carlo values of the processor output for each hypothesis. A set of Monte Carlo runs has been completed for each of three different levels of the noise power, N_0.

The receiver operating characteristic (ROC) for the soliton with $\beta_2 = \sinh(1.5)$ is shown in Fig. 75.15, where the probability of detection, P_D, for this hypothesis test is shown as a function of the false alarm probability, P_F. For comparison, we also show the ROC that would result from a detection of the soliton alone, at the same noise level and with the time-of-arrival known. The detection index, $d = \sqrt{E/N_0}$, is indicated for each case, where E is the energy in the component soliton. The corresponding results for the larger soliton are qualitatively similar, although the detection indices for that soliton alone, with $\beta_1 = \sinh(2)$, are 5.6, 4, and 3.3, respectively. Therefore, the detection probabilities are considerably higher for a fixed probability of false alarm. Note that the detection performance for the smaller soliton is well modeled by the theoretical performance for detection of the smaller soliton alone. This implies, at least empirically, that the ability to detect the component solitons in a multi-soliton signal appears to be unaffected by the nonlinear coupling with other solitons. Further, although the unknown relative separation results in significant waveform uncertainty and would require a prohibitively complex receiver for standard detection techniques, Bayes optimal performance can still be achieved with a minimal increase in complexity.

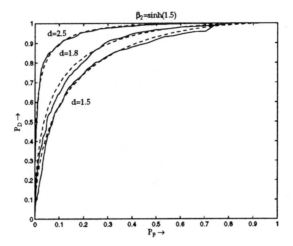

FIGURE 75.15 A set of empirically generated ROCs are shown for the detection of the smaller soliton from a two-soliton signal. For each of the three noise levels, the ROC for detection of the smaller soliton alone is also indicated along with the corresponding detection index, d.

References

[1] Ablowitz, M.J. and Clarkson, A.P., *Solitons, Nonlinear Evolution Equations and Inverse Scattering,* Number 149 in London Mathematical Society Lecture Note Series, Cambridge University Press, Cambridge, Great Britain, 1991.

[2] Fermi, E., Pasta, J.R. and Ulan, S.M., Studies of nonlinear problems, in *Collected Papers of E. Fermi,* vol. II, pp. 977–988, University of Chicago Press, Illinois, 1965.

[3] Golub, G.H. and Van Loan, C.F., *Matrix Computations,* The Johns Hopkins University Press, Baltimore, MD, 1989.

[4] Haus, H.A., Molding light into solitons, *IEEE Spectrum,* 48–53, March 1993.

[5] Helstrom, C.W., *Statistical Theory of Signal Detection,* 2nd ed., Pergamon Press, New York, 1968.

[6] Hirota, R. and Suzuki, K., Theoretical and experimental studies of lattice solitons in nonlinear lumped networks, *Proc. IEEE,* 61(10), 1483–1491, Oct. 1973.

[7] Infeld, E. and Rowlands, R., *Nonlinear Waves, Solitons and Chaos,* Cambridge University Press, New York, 1990.

[8] Kaiser, J.F., On a simple algorithm to calculate the 'energy' of a signal, in *Proc. Int. Conf. Acoust. Speech, Signal Processing,* 381–384, Albuquerque, NM, 1990.

[9] Kaiser, J.F., personal communication, June 1994.

[10] Lax, P.D., Integrals of nonlinear equations of evolution and solitary waves, *Comm. Pure Appl. Math.,* XXI, 467–490, 1968.

[11] Scott, A.C., *Active and Nonlinear Wave Propagation in Electronics,* Wiley-Interscience, New York, 1970.

[12] Scott, A.C., Chu, F.Y.F. and McLaughlin, D., The soliton: A new concept in applied science, *Proc. IEEE,* 61(10), 1443–1483, Oct. 1973.

[13] Singer, A.C., A new circuit for communication using solitons, in *Proc. IEEE Workshop on Nonlinear Signal and Image Processing,* vol. I, 150–153, 1995.

[14] Singer, A.C., Signal Processing and Communication with Solitons, Ph.D. thesis, Massachusetts Institute of Technology, Feb. 1996.

[15] Suzuki, K., Hirota, R. and Yoshikawa, K., Amplitude modulated soliton trains and coding-decoding applications, *Int. J. Electron.*, 34(6), 777–784, 1973.

[16] Suzuki, K., Hirota, R. and Yoshikawa, K., The properties of phase modulated soliton trains, *Japan. J. Appl. Phys.*, 12(3), 361–365, March 1973.

[17] Toda, M., *Theory of Nonlinear Lattices*, Number 20 in Springer Series in Solid-State Science, Springer-Verlag, New York, 1981.

[18] Toda, M., *Nonlinear Waves and Solitons, Mathematics and Its Applications*, Kluwer Academic Publishers, Boston, 1989.

[19] Van Trees, H.L., *Detection, Estimation, and Modulation Theory: Part I Detection, Estimation and Linear Modulation Theory*, John Wiley & Sons, 1968.

[20] vom Scheidt, J. and Purkert, W., *Random Eigenvalue Problems, Probability and Applied Mathematics*, North-Holland, 1983.

[21] Whitham. G.B., *Linear and Nonlinear Waves*, Wiley, New York, 1974.

[22] Zabusky, N.J. and Kruskal, M.D., Interaction of solitons in a collisionless plasma and the recurrence of initial states, *Phys. Rev. Lett.*, 15(6), 240–243, Aug. 1965.

76

Higher-Order Spectral Analysis

Athina P. Petropulu
Drexel University

76.1 Introduction

The past 20 years witnessed an expansion of power spectrum estimation techniques, which have proved essential in many applications, such as communications, sonar, radar, speech/image processing, geophysics, and biomedical signal processing [13, 11, 7]. In power spectrum estimation the process under consideration is treated as a superposition of statistically uncorrelated harmonic components. The distribution of power among these frequency components is the power spectrum. As such, phase relations between frequency components are suppressed. The information in the power spectrum is essentially present in the autocorrelation sequence, which would suffice for the complete statistical description of a Gaussian process of known mean. However, there are applications where one would need to obtain information regarding deviations from the Gaussianity assumption and presence of nonlinearities. In these cases power spectrum is of little help, and one would have to look beyond the power spectrum or autocorrelation domain. Higher-Order Spectra (HOS) (of order greater than 2), which are defined in terms of higher-order cumulants of the data, do contain such information [16]. The third-order spectrum is commonly referred to as bispectrum, the fourth-order one as trispectrum, and in fact, the power spectrum is also a member of the higher-order spectral class; it is the second-order spectrum.

HOS consist of higher-order moment spectra, which are defined for deterministic signals, and cumulant spectra, which are defined for random processes. In general, there are three motivations behind the use of HOS in signal processing: (1) to suppress Gaussian noise of unknown mean and variance; (2) to reconstruct the phase as well as the magnitude response of signals or systems; and (3) to detect and characterize nonlinearities in the data.

The first motivation stems from the property of Gaussian processes to have zero higher-order spectra. Due to this property, HOS are high signal-to-noise ratio domains, in which one can perform detection, parameter estimation, or even signal reconstruction even if the time domain noise is spatially correlated. The same property of cumulant spectra can provide means of detecting and characterizing deviations of the data from the Gaussian model.

The second motivation is based on the ability of cumulant spectra to preserve the Fourier-phase of signals. In the modeling of time series, second-order statistics (autocorrelation) have been heavily used because they are the result of least-squares optimization criteria. However, an accurate phase reconstruction in the autocorrelation domain can be achieved only if the signal is minimum phase. Nonminimum phase signal reconstruction can be achieved only in the HOS domain, due to the HOS ability to preserve phase. Figure 76.1 shows two signals, a nonminimum phase and a minimum phase, with identical magnitude spectra but different phase spectra. Although power spectrum cannot distinguish between the two signals, the bispectrum that uses phase information can.

Being nonlinear functions of the data, HOS are quite natural tools in the analysis of nonlinear systems operating under a random input. General relations for arbitrary stationary random data passing through an arbitrary linear system exist and have been studied extensively. Such expression, however, are not available for nonlinear systems, where each type of nonlinearity must be studied separately. Higher-order correlations between input and output can detect and characterize certain nonlinearities [34], and for this purpose several higher-order spectra-based methods have been developed.

The organization of this chapter is as follows. First the definitions and properties of cumulants and higher-order spectra are introduced. Then two methods for the estimation of HOS from finite length data are outlined and the asymptotic statistics of the obtained estimates are presented. Following that, parametric and nonparametric methods for HOS-based identification of linear systems are described, and the use of HOS in the identification of some particular nonlinear systems is briefly discussed. The chapter concludes with a section on applications of HOS and available software.

76.2 Definitions and Properties of HOS

In this chapter we will consider random one-dimensional processes only. The definitions can be easily extended to the two-dimensional case [15].

The joint moments of order r of the random variables $x_1, \ldots, x_n$ are given by [22]

$$
\begin{aligned}
M\,om\left[x_1^{k_1}, \ldots, x_n^{k_n}\right] &= E\{x_1^{k_1}, \ldots, x_n^{k_n}\} \\
&= (-j)^r \frac{\partial^r \Phi\,(\omega_1, \ldots, \omega_n)}{\partial \omega_1^{k_1} \ldots \partial \omega_n^{k_n}}\Big|_{\omega_1 = \cdots = \omega_n = 0}\,,
\end{aligned}
\tag{76.1}
$$

where $k_1 + \cdots + k_n = r$, and $\Phi()$ is their joint characteristic function. The joint cumulants are defined as

$$
C\,um[x_1^{k_1}, \ldots, x_n^{k_n}] = (-j)^r \frac{\partial^r \ln \Phi\,(\omega_1, \ldots, \omega_n)}{\partial \omega_1^{k_1} \ldots \partial \omega_n^{k_n}}\}|_{\omega_1 = \cdots = \omega_n = 0}\,.
\tag{76.2}
$$

For a stationary discrete time random process $X(k)$, (k denotes discrete time), the *moments* of order n are given by

$$
m_n^x (\tau_1, \tau_2, \ldots, \tau_{n-1}) = E\{X(k)X(k + \tau_1) \cdots X(k + \tau_{n-1})\}\,,
\tag{76.3}
$$

where $E\{.\}$ denotes expectation. The nth order *cumulants* are functions of the moments of order up to n, i.e.,

1st order cumulants:

$$
c_1^x = m_1^x = E\{X(k)\} \quad \text{(mean)}
\tag{76.4}
$$

2nd order cumulants:

$$
c_2^x (\tau_1) = m_2^x (\tau_1) - \left(m_1^x\right)^2 \quad \text{(covariance)}
\tag{76.5}
$$

3rd order cumulants:

$$
c_3^x (\tau_1, \tau_2) = m_3^x (\tau_1, \tau_2) - \left(m_1^x\right)\left[m_2^x (\tau_1) + m_2^x (\tau_2) + m_2^x (\tau_2 - \tau_1)\right] + 2\left(m_1^x\right)^3
\tag{76.6}
$$

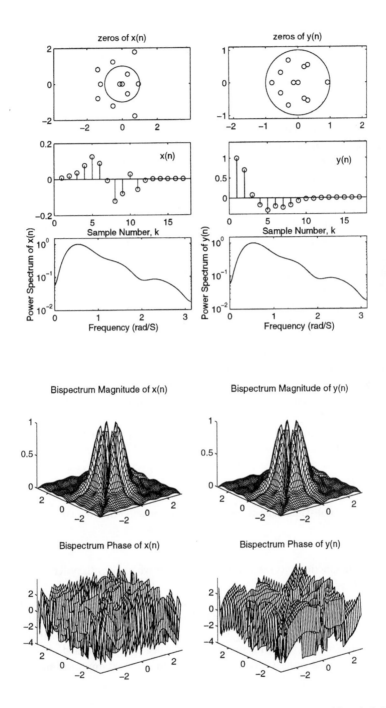

FIGURE 76.1 $x(n)$ is a nonminimum phase signal and $y(n)$ is a minimum phase one. Although their power spectra are identical, their bispectra are different because they contain phase information.

4th order cumulants:

$$
\begin{aligned}
c_4^x(\tau_1, \tau_2, \tau_3) \;=\;& m_4^x(\tau_1, \tau_2, \tau_3) - m_2^x(\tau_1)\,m_2^x(\tau_3 - \tau_2) - m_2^x(\tau_2)\,m_2^x(\tau_3 - \tau_1) \\
& - m_2^x(\tau_3)\,m_2^x(\tau_2 - \tau_1) \\
& - m_1^x\left[m_3^x(\tau_2 - \tau_1, \tau_3 - \tau_1) + m_3^x(\tau_2, \tau_3) + m_3^x(\tau_2, \tau_4) + m_3^x(\tau_1, \tau_2)\right] \\
& + \left(m_1^x\right)^2\left[m_2^x(\tau_1) + m_2^x(\tau_2) + m_2^x(\tau_3) + m_2^x(\tau_3 - \tau_1) + m_2^x(\tau_3 - \tau_2)\right. \\
& + \left. m_1^x(\tau_2 - \tau_1)\right] - 6\left(m_1^x\right)^4
\end{aligned}
\tag{76.7}
$$

where $m_3^x(\tau_1, \tau_2)$ is the 3rd order moment sequence, and m_1^x is the mean. The general relationship between cumulants and moments can be found in [16].

Some important properties of moments and cumulants are summarized next.

[P1] If $X(k)$ is Gaussian, the $c_n^x(\tau_1, \tau_2, \ldots, \tau_{n-1}) = 0$ for $n > 2$. In other words, all the information about a Gaussian process is contained in its first and second-order cumulants. This property can be used to suppress Gaussian noise, or as a measure for non-Gaussianity in time series.

[P2] If $X(k)$ is symmetrically distributed, then $c_3^x(\tau_1, \tau_2) = 0$. Third-order cumulants suppress not only Gaussian processes, but also all symmetrically distributed processes, such as uniform, Laplace, and Bernoulli-Gaussian.

[P3] For cumulants additivity holds. If $X(k) = S(k) + W(k)$, where $S(k)$, $W(k)$ are stationary and statistically independent random processes, then $c_n^x(\tau_1, \tau_2, \ldots, \tau_{n-1}) = c_n^s(\tau_1, \tau_2, \ldots, \tau_{n-1}) + c_n^w(\tau_1, \tau_2, \ldots, \tau_{n-1})$. It is important to note that additivity does not hold for moments.

If $W(k)$ is Gaussian representing noise which corrupts the signal of interest, $S(k)$, then by means of (P2) and (P3), we get that $c_n^x(\tau_1, \tau_2, \ldots, \tau_{n-1}) = c_n^s(\tau_1, \tau_2, \ldots, \tau_{n-1})$, for $n > 2$. In other words, in higher-order cumulant domains the signal of interest propagates noise free. Property (P3) can also provide a measure of statistical dependence of two processes.

[P4] if $X(k)$ has zero mean, then $c_n^x(\tau_1, \ldots, \tau_{n-1}) = m_n^x(\tau_1, \ldots, \tau_{n-1})$, for $n \leq 3$.

Higher-order spectra are defined in terms of either cumulants (e.g., cumulant spectra) or moments (e.g., moment spectra).

Assuming that the nth order cumulant sequence is absolutely summable, the nth order **cumulant spectrum** of $X(k)$, $C_n^x(\omega_1, \omega_2, \ldots, \omega_{n-1})$, exists, and is defined to be the $(n-1)$-dimensional Fourier transform of the nth order cumulant sequence. In general, $C_n^x(\omega_1, \omega_2, \ldots, \omega_{n-1})$ is complex, i.e., it has magnitude and phase. In an analogous manner, **moment spectrum** is the multi-dimensional Fourier transform of the moment sequence.

If $v(k)$ is a stationary non-Gaussian process with zero mean and nth order cumulant sequence

$$
c_n^v(\tau_1, \ldots, \tau_{n-1}) = \gamma_n^v \delta(\tau_1, \ldots, \tau_{n-1}),
\tag{76.8}
$$

where $\delta(.)$ is the delta function, $v(k)$ is said to be nth order white. Its nth order cumulant spectrum is then flat and equal to γ_n^v.

Cumulant spectra are more useful in processing random signals than moment spectra since they posses properties that the moment spectra do not share: (1) the cumulants of the sum of two independent random processes equals the sum of the cumulants of the process; (2) cumulant spectra of order > 2 are zero if the underlying process in Gaussian; (3) cumulants quantify the degree of statistical dependence of time series; and (4) cumulants of higher-order white noise are multidimensional impulses, and the corresponding cumulant spectra are flat.

76.3   HOS Computation from Real Data

The definitions of cumulants presented in the previous section are based on expectation operations, and they assume infinite length data. In practice we always deal with data of finite length; therefore, the cumulants can only be approximated. Two methods for cumulants and spectra estimation are presented next for the third-order case.

Indirect Method :

Let $X(k), k = 1, \ldots, N$ be the available data.

1. Segment the data into K records of M samples each. Let $X^i(k), \ k = 1, \ldots, M$, represent the ith record.

2. Subtract the mean of each record.

3. Estimate the moments of each segments $X^i(k)$ as follows:

$$m_3^{x_i}(\tau_1, \tau_2) = \frac{1}{M} \sum_{l=l_1}^{l_2} X^i(l) X^i(l + \tau_1) X^i(l + \tau_2) ,$$

$$l_1 = max(0, -\tau_1, -\tau_2), \ l_2 = min(M - 1, M - 2),$$
$$|\tau_1| < L, \ |\tau_2| < L, \ i = 1, 2, \ldots, K . \tag{76.9}$$

Since each segment has zero mean, its third-order moments and cumulants are identical, i.e., $c_3^{x_i}(\tau_1, \tau_2) = m_3^{x_i}(\tau_1, \tau_2)$.

4. Compute the average cumulants as:

$$\hat{c}_3^x(\tau_1, \tau_2) = \frac{1}{K} \sum_{i=1}^{K} m_3^{x_i}(\tau_1, \tau_2) \tag{76.10}$$

5. Obtain the third-order spectrum (bispectrum) estimate as

$$\hat{C}_3^x(\omega_1, \omega_2) = \sum_{\tau_1=-L}^{L} \sum_{\tau_2=-L}^{L} \hat{c}_3^x(\tau_1, \tau_2) e^{-j(\omega_1 \tau_1 + \omega_2 \tau_2)} w(\tau_1, \tau_2) , \tag{76.11}$$

where $L < M-1$, and $w(\tau_1, \tau_2)$ is a two-dimensional window of bounded support, introduced to smooth out edge effects. The bandwidth of the final bispectrum estimate is $\Delta = 1/L$.

A complete description of appropriate windows that can be used in (76.11) and their properties can be found in [16]. A good choice of cumulant window is:

$$w(\tau_1, \tau_2) = d(\tau_1) d(\tau_2) d(\tau_1 - \tau_2) , \tag{76.12}$$

where

$$d(\tau) = \begin{cases} \frac{1}{\pi} |\sin \frac{\pi \tau}{L}| + (1 - \frac{|\tau|}{L}) \cos \frac{\pi \tau}{L} & |\tau| \leq L \\ 0 & |\tau| > L \end{cases} \tag{76.13}$$

which is known as the minimum bispectrum bias supremum [17].

Direct Method

Let $X(k), k = 1, \ldots, N$ be the available data.

1. Segment the data into K records of M samples each. Let $X^i(k), \ k = 1, \ldots, M$, represent the ith record.

2. Subtract the mean of each record.

3. Compute the Discrete Fourier Transform $F_x^i(k)$ of each segment, based on M points, i.e.,

$$F_x^i(k) = \sum_{n=0}^{M-1} X^i(n)e^{-j\frac{2\pi}{M}nk}, \quad k = 0, 1, \ldots, M-1, \ i = 1, 2, \ldots, K. \tag{76.14}$$

4. The third-order spectrum of each segment is obtained as

$$C_3^{x_i}(k_1, k_2) = \frac{1}{M} F_x^i(k_1) F_x^i(k_2) F_x^{i\,*}(k_1 + k_2), \quad i = 1, \ldots, K. \tag{76.15}$$

Due to the bispectrum symmetry properties, $C_3^{x_i}(k_1, k_2)$ need to be computed only in the triangular region $0 \le k_2 \le k_1, \ k_1 + k_2 < M/2$.

5. In order to reduce the variance of the estimate additional smoothing over a rectangular window of size $(M_3 \times M_3)$ can be performed around each frequency, assuming that the third-order spectrum is smooth enough, i.e.,

$$\tilde{C}_3^{x_i}(k_1, k_2) = \frac{1}{M_3^2} \sum_{n_1=-M_3/2}^{M_3/2-1} \sum_{n_2=-M_3/2}^{M_3/2-1} C_3^{x_i}(k_1 + n_1, k_2 + n_2). \tag{76.16}$$

6. Finally, the third-order spectrum is given as the average over all third-order spectra, i.e.,

$$\hat{C}_3^x(\omega_1, \omega_2) = \frac{1}{K} \sum_{i=1}^{K} \tilde{C}_3^{x_i}(\omega_1, \omega_2), \quad \omega_i = \frac{2\pi}{M} k_i, \ i = 1, 2. \tag{76.17}$$

The final bandwidth of this bispectrum estimate is $\Delta = M_3/M$, which is the spacing between frequency samples in the bispectrum domain.

For large N, and as long as

$$\Delta \to 0, \ and \ \Delta^2 N \to \infty \tag{76.18}$$

[32], both the direct and the indirect methods produce asymptotically unbiased and consistent bispectrum estimates, with real and imaginary part variances:

$$\text{var}\left(\text{Re}\left[\hat{C}_3^x(\omega_1, \omega_2)\right]\right) = \text{var}\left(\text{Im}\left[\hat{C}_3^x(\omega_1, \omega_2)\right]\right) \tag{76.19}$$

$$= \frac{1}{\Delta^2 N} C_2^x(\omega_1) C_2^x(\omega_2) C_2^x(\omega_1 + \omega_2) = \begin{cases} \frac{VL^2}{MK} C_2^x(\omega_1) C_2^x(\omega_2) C_2^x(\omega_1 + \omega_2) & \text{indirect} \\[2mm] \frac{M}{KM_3^2} C_2^x(\omega_1) C_2^x(\omega_2) C_2^x(\omega_1 + \omega_2) & \text{direct}, \end{cases}$$

where V is the energy of the bispectrum window.

From the above expressions, it becomes apparent that the bispectrum estimate variance can be reduced by increasing the number of records, or reducing the size of the region of support of the window in the cumulant domain (L), or increasing the size of the frequency smoothing window (M_3), etc. The relation between the parameters M, K, L, M_3 should be such that (76.18) is satisfied.

76.4   Linear Processes

Let $x(k)$ be generated by exciting a linear time-invariant (LTI) system with frequency response $H(\omega)$ with a non-Gaussian process $v(k)$. Its nth order spectrum can be written as

$$C_n^x (\omega_1, \omega_2, \ldots, \omega_{n-1}) = C_n^v (\omega_1, \omega_2, \ldots, \omega_{n-1}) H (\omega_1) \cdots H (\omega_{n-1}) H^* (\omega_1 + \cdots + \omega_{n-1}) . \tag{76.20}$$

If $v(k)$ is nth order white then (76.20) becomes

$$C_n^x (\omega_1, \omega_2, \ldots, \omega_{n-1}) = \gamma_n^v H (\omega_1) \cdots H (\omega_{n-1}) H^* (\omega_1 + \cdots + \omega_{n-1}) , \tag{76.21}$$

where γ_n^v is a scalar constant and equals the nth order spectrum of $v(k)$. For a linear non-Gaussian random process $X(k)$, the nth order spectrum can be factorized as in (76.21) for every order n, while for a nonlinear process such a factorization might be valid for some orders only (it is always valid for $n = 2$). If we express $H(\omega) = |H(\omega)| \exp\{j\phi_h(\omega)\}$, then (76.21) can be written as

$$\left| C_n^x (\omega_1, \omega_2, \ldots, \omega_{n-1}) \right| = \gamma_n^v |H (\omega_1)| \cdots |H (\omega_{n-1})| \left| H^* (\omega_1 + \cdots + \omega_{n-1}) \right| , \tag{76.22}$$

and

$$\left| \psi_n^x (\omega_1, \omega_2, \ldots, \omega_{n-1}) \right| = \phi_h (\omega_1) + \cdots + \phi_h (\omega_{n-1}) - \phi_h (\omega_1 + \cdots + \omega_{n-1}) , \tag{76.23}$$

where $\psi_n^x()$ is the phase of the nth order spectrum.

It can be shown easily that the cumulant spectra of successive orders are related as follows:

$$C_n^x (\omega_1, \omega_2, \ldots, 0) = C_{n-1}^x (\omega_1, \omega_2, \ldots, \omega_{n-2}) H(0) \frac{\gamma_n^v}{\gamma_{n-1}^v} . \tag{76.24}$$

As a result, the power spectrum of a Gaussian linear process can be reconstructed from the bispectrum up to a constant term, i.e.,

$$C_3^x (\omega, 0) = C_2^x (\omega) \frac{\gamma_3^v}{\gamma_2^v} . \tag{76.25}$$

To reconstruct the phase $\phi_h(\omega)$ from the bispectral phase $\psi_3^x(\omega_1, \omega_2)$ several algorithms have been suggested. A description of different phase estimation methods can be found in [14] and also in [16].

76.4.1   Nonparametric Methods

Consider $x(k)$ generated as shown in Fig. 76.2. The system transfer function can be written as

$$H(z) = cz^{-r} I \left(z^{-1} \right) O(z) = cz^{-r} \frac{\Pi_i (1 - a_i z^{-1})}{\Pi_i (1 - b_i z^{-1})} \Pi_i (1 - c_i z), \ |a_i|, |b_i|, |c_i| < 1 , \tag{76.26}$$

where $I(z^{-1})$ and $O(z)$ are the minimum and maximum phase parts of $H(z)$, respectively; c is a constant; and r is an integer. The output nth order cumulant equals [2]

$$\begin{aligned}
c_n^x (\tau_1, \ldots, \tau_{n-1}) &= c_n^y (\tau_1, \ldots, \tau_{n-1}) + c_n^w (\tau_1, \ldots, \tau_{n-1}) \\
&= c_n^y (\tau_1, \ldots, \tau_{n-1}) \tag{76.27} \\
&= \gamma_n^v \sum_{k=0}^{\infty} h(k)h (k + \tau_1) \cdots h (k + \tau_{n-1}), \ n \geq 3 \tag{76.28}
\end{aligned}$$

where the noise contribution in (76.27) was zero due to the Gaussianity assumption. The Z-domain equivalent of (76.28) for $n = 3$ is

$$C_3^x (z_1, z_2) = \gamma_3^v H (z_1) H (z_2) H \left(z_1^{-1} z_2^{-1} \right) . \tag{76.29}$$

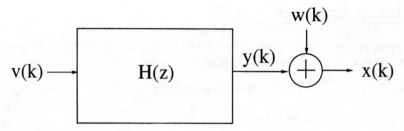

FIGURE 76.2 Single channel model.

Taking the logarithm of $C_3^x(z_1, z_2)$ followed by an inverse 2-D Z-transform we obtain the output bicep-strum $b_x(m, n)$. The bicepstrum of linear processes is nonzero only along the axes ($m = 0$, $n = 0$) and the diagonal $m = n$ [21]. Along these lines the bicepstrum is equal to the complex cepstrum, i.e.,

$$
b_x(m, n) = \begin{cases}
\hat{h}(m) & m \neq 0, \ n = 0 \\
\hat{h}(n) & n \neq 0, \ m = 0 \\
\hat{h}(-n) & m = n, \ m \neq 0 \\
\ln(c\gamma_n^v) & m = n = 0, \\
0 & \text{elsewhere}
\end{cases} \tag{76.30}
$$

where $\hat{h}(n)$ denotes complex cepstrum [20]. From (76.30), the system impulse response $h(k)$ can be reconstructed from $b_x(m, 0)$ (or $b_x(0, m)$, or $b_x(m, m)$), within a constant and a time delay, via inverse cepstrum operations. The minimum and maximum phase parts of $H(z)$ can be reconstructed by applying inverse cepstrum operations on $b_x(m, 0)u(m)$ and $b_x(m, 0)u(-m)$, respectively, where $u(m)$ is the unit step function.

To avoid phase unwrapping with the logarithm of the bispectrum which is complex, the bicepstrum can be estimated using the group delay approach:

$$
b_x(m, n) = \frac{1}{m} F^{-1}\{\frac{F\left[\tau_1 c_3^x(\tau_1, \tau_2)\right]}{C_3^x(\omega_1, \omega_2)}\}, \ m \neq 0 \tag{76.31}
$$

with $b_x(0, n) = b_x(n, 0)$, and $F\{\cdot\}$ and $F^{-1}\{\cdot\}$ denoting 2-D Fourier transform operator and its inverse, respectively.

The cepstrum of the system can also be computed directly from the cumulants of the system output based on the equation [21]:

$$
\sum_{k=1}^{\infty} k\hat{h}(k)\left[c_3^x(m-k, n) - c_3^x(m+k, n+k)\right] + k\hat{h}(-k)\left[c_3^x(m-k, n-k) - c_3^x(m+k, n)\right]
$$

$$
= mc_3^x(m, n) \tag{76.32}
$$

If $H(z)$ has no zeros on the unit circle its cepstrum decays exponentially, thus (76.32) can be truncated to yield an approximate equation. An overdetermined system of truncated equations can be formed for different values of m and n, which can be solved for $\hat{h}(k)$, $k = \ldots, -1, 1, \ldots$. The system response $h(k)$ then can be recovered from its cepstrum via inverse cepstrum operations.

The bicepstrum approach for system reconstruction described above led to estimates with smaller bias and variance than other parametric approaches at the expense of higher computational complexity [21]. The analytic performance evaluation of the bicepstrum approach can be found in [25].

The inverse Z-transform of the logarithm of the trispectrum (fourth-order spectrum), or otherwise tricepstrum, $t_x(m, n, l)$, of linear processes is also zero everywhere except along the axes and the diagonal $m = n = l$. Along these lines it equals the complex cepstrum, thus $h(k)$ can be recovered from slices of the tricepstrum based on inverse cepstrum operations.

For the case of nonlinear processes, the bicepstrum will be nonzero everywhere [4]. The distinctly different structure of the bicepstrum corresponding to linear and nonlinear processes has led to tests of linearity [4].

A new nonparametric method has been recently proposed in [1, 26] in which the cepstrum $\hat{h}(k)$ is obtained as:

$$\hat{h}(-k) = \frac{\hat{p}_n^x\left(k;\ e^{j\beta_1}\right) - \hat{p}_n^x\left(k;\ e^{j\beta_2}\right)}{e^{j(n-2)\beta_1 k} - e^{j(n-2)\beta_2 k}}, \ k \neq 0, \ n > 2 \tag{76.33}$$

where $p_n^x\left(k;\ e^{jb_i}\right)$ is the time domain equivalent of the nth order spectrum slice defined as:

$$P_n^x\left(z;\ e^{j\beta_i}\right) = C_n^x\left(z,\ e^{j\beta_i},\cdots,e^{j\beta_i}\right). \tag{76.34}$$

The denominator of (76.33) is nonzero if

$$|\beta_1 - \beta_2| \neq \frac{2\pi l}{k(n-2)}, \ \text{for every integer } k \text{ and } l. \tag{76.35}$$

This method reconstructs a complex system using two slices of the nth order spectrum. The slices, defined as shown above, can be selected arbitrarily as long as their distance satisfy (76.35). If the system is real, one slices is sufficient for the reconstruction. It should be noted that the cepstra appearing in (76.33) require phase unwrapping. The main advantage of this method is that the freedom to choose the higher-order spectra areas to be used in the reconstruction allows one to avoid regions dominated by noise or finite data length effects. Also, corresponding to different slice pairs various independent representations of the system can be reconstructed. Averaging out these representations can reduce estimation errors [26].

Along the lines of system reconstruction from selected HOS slices, another method has been proposed in [28, 29] where the $\log H(k)$ is obtained as a solution to a linear system of equations. Although logarithimc operation is involved, no phase unwrapping is required and the principal argument can be used instead of real phase. It was also shown that, as long as the grid size and the distance between the slices are coprime, reconstruction is always possible.

76.4.2 Parametric Methods

One of the popular approaches in system identification has been the construction of a white noise driven, linear time invariant model from a given process realization.

Consider the real autoregressive moving average (ARMA) stable process $y(k)$ given by:

$$\sum_{i=0}^{p} a(i)y(k-i) = \sum_{j=0}^{q} b(j)v(k-j) \tag{76.36}$$

$$x(k) = y(k) + w(k) \tag{76.37}$$

where $a(i)$, $b(j)$ represent the AR and MA parameters of the system, $v(k)$ is an independent identically distributed random process, and $w(k)$ represents zero-mean Gaussian noise.

Equations analogous to the Yule-Walker equations can be derived based on third-order cumulants of $x(k)$, i.e.,

$$\sum_{i=0}^{p} a(i) c_3^x (\tau - i, j) = 0, \ \tau > q , \tag{76.38}$$

or

$$\sum_{i=1}^{p} a(i) c_3^x (\tau - i, j) = -c_3^x (\tau, j), \ \tau > q , \tag{76.39}$$

where it was assumed $a(0) = 1$. Concatenating (76.39) for $\tau = q + 1, \ldots, q + M$, $M \geq 0$ and $j = q - p, \ldots, q$, the matrix equation

$$\underline{C} \underline{a} = \underline{c} \tag{76.40}$$

can be formed, where $\underline{C}$ and $\underline{c}$ are a matrix and a vector, respectively, formed by third-order cumulants of the process according to (76.39), and the vector $\underline{a}$ contains the AR parameters. If the AR order p is unknown and (76.40) is formed based on an overestimate of p, the resulting matrix $\underline{C}$ always has rank p. In this case, the AR parameters can be obtained using a low-rank approximation of $\underline{C}$ [5].

Using the estimated AR parameters, $\hat{a}(i)$, $i = 1, \ldots, p$, a pth order filter with transfer function $\hat{A}(z) = 1 + \sum_{i=1}^{p} \hat{a}(i) z^{-1}$ can be constructed. Based on the filtered through $\hat{A}(z)$ process $x(k)$, i.e., $\tilde{x}(k)$, or otherwise known as the residual time series [5], the MA parameters can be estimated via any MA method [15], for example:

$$b(k) = \frac{c_3^{\tilde{x}}(q, k)}{c_3^{\tilde{x}}(q, 0)}, \ k = 0, 1, \ldots, q \tag{76.41}$$

known as the $c(q, k)$ formula [6].

Practical problems associated with the described approach are sensitivity to model order mismatch, and AR estimation errors that propagate in the estimation of the MA parameters. A significant amount of research has been devoted to the ARMA parameter estimation problem. A thorough review of existing ARMA system identification methods can be found in [15, 16]; a more recent method can be found in [24].

76.5　Nonlinear Processes

Despite the fact that progress has been established in developing the theoretical properties of nonlinear models, only a few statistical methods exist for detection and characterization of nonlinearities from a finite set of observations. In this section, we will consider nonlinear Volterra systems excited by Gaussian stationary inputs. Let $y(k)$ be the response of a discrete time invariant pth order Volterra filter whose input is $x(k)$. Then,

$$y(k) = h_0 + \sum_{i} \sum_{\tau_1, \ldots, \tau_i} h_i (\tau_1, \ldots, \tau_i) x (k - \tau_1) \cdots x (k - \tau_i) , \tag{76.42}$$

where $h_i (\tau_1, \ldots, \tau_i)$ are the Volterra kernels of the system, which are symmetric functions of their arguments; for causal systems $h_i (\tau_1, \ldots, \tau_i) = 0$ for any $\tau_i < 0$.

The output of a second-order Volterra system when the input is zero-mean stationary is

$$y(k) = h_0 + \sum_{\tau_1} h_1(\tau_1) x(k - \tau_1) + \sum_{\tau_1} \sum_{\tau_2} h_2 (\tau_1, \tau_2) x (k - \tau_1) x (k - \tau_2) . \tag{76.43}$$

Equation (76.43) can be viewed as a parallel connection of a linear system $h_1(\tau_1)$ and a quadratic system $h_2(\tau_1, \tau_2)$ as illustrated in Fig. 76.3. Let

$$c_2^{xy} (\tau) = E \left\{ x(k + \tau) \left[y(k) - m_1^y \right] \right\} \tag{76.44}$$

be the cross-covariance of input and output, and

$$c_3^{xxy}(\tau_1, \tau_2) = E\left\{x\,(k + \tau_1)\,x\,(k + \tau_2)\left[y(k) - m_1^y\right]\right\} \tag{76.45}$$

be the third-order cross-cumulant sequence of input and output.

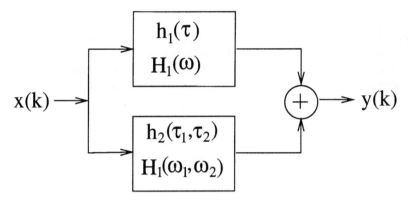

FIGURE 76.3 Second-order Volterra system. Linear and quadratic parts are connected in parallel.

It can be shown that the system's linear part can be identified by

$$H_1(-\omega) = \frac{C_2^{xy}(\omega)}{C_2^{x}(\omega)}, \tag{76.46}$$

and the quadratic part by

$$H_2(-\omega_1, -\omega_2) = \frac{C_3^{xxy}(\omega_1, \omega_2)}{2C_2^{x}(\omega_1)\,C_2^{x}(\omega_2)}, \tag{76.47}$$

where $C_2^{xy}(\omega)$ and $C_3^{xxy}(\omega_1, \omega_2)$ are the Fourier transforms of $c_2^{xy}(\tau)$ and $c_3^{xxy}(\tau_1, \tau_2)$, respectively. It should be noted that the above equations are valid only for Gaussian input signals. More general results assuming non-Gaussian input have been obtained in [9, 27]. Additional results on particular nonlinear systems have been reported in [3, 33].

An interesting phenomenon caused by a second-order nonlinearity is the quadratic phase coupling. There are situations where nonlinear interaction between two harmonic components of a process contribute to the power of the sum and/or difference frequencies. The signal

$$x(k) = A\cos(\lambda_1 k + \theta_1) + B\cos(\lambda_2 k + \theta_2) \tag{76.48}$$

after passing through the quadratic system:

$$z(k) = x(k) + \epsilon x^2(k), \quad \epsilon \neq 0. \tag{76.49}$$

contains cosinusoidal terms in (λ_1, θ_1), (λ_2, θ_2), $(2\lambda_1, 2\theta_1)$, $(2\lambda_2, 2\theta_2)$, $(\lambda_1 + \lambda_2, \theta_1 + \theta_2)$, $(\lambda_1 - \lambda_2, \theta_1 - \theta_2)$. Such a phenomenon that results in phase relations that are the same as the frequency relations is called quadratic phase coupling [12]. Quadratic phase coupling can arise only among harmonically related components. Three frequencies are harmonically related when one of them is the sum or difference of the other two. Sometimes it is important to find out if peaks at harmonically related positions in the power spectrum are in fact phase coupled. Due to phase suppression, the power spectrum is unable to provide an answer to this problem.

As an example, consider the process [30]

$$X(k) = \sum_{i=1}^{6} \cos(\lambda_i k + \phi_i)$$

(76.50)

where $\lambda_1 > \lambda_2 > 0$, $\lambda_4 + \lambda_5 > 0$, $\lambda_3 = \lambda_1 + \lambda_2$, $\lambda_6 = \lambda_4 + \lambda_5$, $\phi_1, \ldots, \phi_5$ are all independent, uniformly distributed random variables over $(0, 2\pi)$, and $\phi_6 = \phi_4 + \phi_5$. Among the six frequencies, $(\lambda_1, \lambda_2, \lambda_3)$ and $(\lambda_4, \lambda_5, \lambda_6)$ are harmonically related, however, only λ_6 is the result of phase coupling between λ_4 and λ_5. The power spectrum of this process consists of six impulses at λ_i, $i = 1, \ldots, 6$ (see Fig. 76.4), offering no indication whether each frequency component is independent or result of frequency

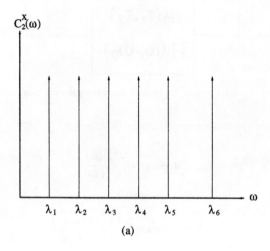

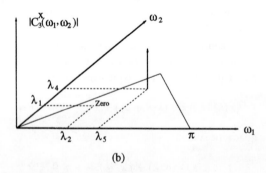

FIGURE 76.4   Quadratic phase coupling. (a) The power spectrum of the process described in Eq. (76.50) cannot determine what frequencies are coupled. (b) The corresponding magnitude bispectrum is zero everywhere in the principle region, except at points corresponding to phase coupled frequencies.

coupling. On the other hand, the bispectrum of $X(k)$, $C_3^x(\omega_1, \omega_2)$ (evaluate in its principal region) is zero everywhere, except at point (λ_4, λ_5) of the (ω_1, ω_2) plane, where it exhibits an impulse (Fig. 76.4(b)). The peak indicates that only λ_4, λ_5 are phase coupled.

The bicoherence index, defined as

$$P_3^x(\omega_1, \omega_2) = \frac{C_3^x(\omega_1, \omega_2)}{\sqrt{C_2^x(\omega_1) C_2^x(\omega_2) C_2^x(\omega_1 + \omega_2)}} \, , \tag{76.51}$$

has been extensively used in practical situations for the detection and quantification of quadratic phase coupling [12]. The value of the bicoherence index at each frequency pair indicates the degree of coupling among the frequencies of that pair. Almost all bispectral estimators can be used in (76.51). However, estimates obtained based on parametric modeling of the bispectrum have been shown to yield superior resolution [30, 31] than the ones obtained with conventional methods.

76.6 Applications/Software Available

Applications of HOS span a wide range of areas [19] such as oceanography (description of wave phenomena), earth sciences (atmospheric pressure, turbulence), crystallography, plasma physics (wave interaction, nonlinear phenomena), mechanical systems (vibration analysis, knock detection), economic time series, biomedical signal analysis (ultrasonic imaging, detection of wave coupling) image processing (texture modeling and characterization, reconstruction, inverse filtering), speech processing (pitch detection, voiced/unvoiced decision), communications (equalization, interference cancellation), array processing (direction of arrival estimation, estimation of number of sources, beamforming, source signal estimation, source classification), harmonic retrieval (frequency estimation), and time delay estimation. Over 500 references can be found in [37]. Additional references can be found in [16, 19, 23].

A software package for signal processing with HOS is the Hi-Spec toolbox, product of Mathworks, Inc. The functions included in Hi-Spec together with a short description are included in Table 76.1.

TABLE 76.1 Functions Included in the Hi-Spec Package

Function name	Description
AR_RCEST	AR parameter estimation based on cumulants
ARMA_QS	ARMA parameter estimation via the Q-slice algorithm
ARMA_RTS	ARMA parameter estimation via the residual time series method
ARMA_SYN	Generates ARMA synthetics
BICEPS	System identification via the bicepstrum approach
BISPEC_D	Bispectrum estimation via the direct method
BISPEC_I	Bispectrum estimation via the indirect method
CUM_EST	Estimates 2nd, 3rd, or 4th order cumulants
CUM_TRUE	Computes the theoretical cumulants of an ARMA model
DOA	Direction-of-arrival estimation
DOA_GEN	Generates synthetics for direction-of-arrival estimation
GL_STAT	Detection statistics for Hinich's Gaussianity and linearity tests
HARM_EST	Estimates frequencies of harmonics in colored noise
HARM_GEN	Generates synthetics for the harmonic retrieval problem
MA_EST	MA parameters estimation
MATUL	System identification via the Matsuoka-Ulrych algorithm
QPC_GEN	Simulation generator for quadratic phase coupling
QPC_TOR	Detects quadratic phase coupling via parametric modeling of bispectrum
RP_IID	Generates samples of an i.i.d. random process
TDE	Estimates time delay between two signals using the parametric cross-cumulant method
TDE_GEN	Synthetics for time delay estimation

Acknowledgments

Most of the material presented in this chapter is based on the book *Higher-Order Spectra Analysis: A Non-Linear Signal Processing Framework* [16]. The author wishes to thank Dr. C.L. Nikias for his valuable input on the organization of the material. She also thanks U.R. Abeyratne for producing the figures in this chapter. Support for this work came from NSF under grant MIP-9553227 and the Whitaker Foundation.

References

[1] Abeyratne, U.R. and Petropulu, A.P., α-Weighted cumulant projections: a novel tool for system identification, *29th Annual Asilomar Conference on Signals, Systems and Computers,* California, Oct. 1995.

[2] Brillinger, D.R. and Rosenblatt, M., Computation and interpretation of kth-order spectra, *Spectral Analysis of Time Series,* B. Harris, Ed., John Wiley & Sons, New York, 1967, 189–232.

[3] Brillinger, D.R., The identification of a particular nonlinear time series system, *Biometrika,* 64(3), 509–515, 1977.

[4] Erdem, A.T. and Tekalp, A.M., Linear bispectrum of signals and identification of nonminimum phase FIR systems driven by colored input, *IEEE Trans. on Signal Processing,* 40, 1469–1479, June 1992.

[5] Giannakis, G.B. and Mendel, J.M., Cumulant-based order determination of non-Gaussian ARMA models, *IEEE Trans. on Acoustics, Speech and Signal Processing,* 38, 1411–1423, 1990.

[6] Giannakis, G.B., Cumulants: a powerful tool in signal processing, *Proc. IEEE,* 75, 1987.

[7] Haykin, S., *Nonlinear Methods of Spectral Analysis,* 2nd ed., Berlin, Germany, Springer-Verlag, 1983.

[8] Hinich, M.J., Testing for gaussianity and linearity of a stationary time series, *J. Time Series Analysis,* 3(3), 169–176, 1982.

[9] Hinich, M.J., Identification of the coefficients in a nonlinear time series of the quadratic type, *J. Economics,* 30, 269–288, 1985.

[10] Huber, P.J., Kleiner, B. et.al., Statistical methods for investigating phase relations in stochastic processes, *IEEE Trans. on Audio and Electroacoustics,* Au-19(1), 78–86, 1976.

[11] Kay, S.M., *Modern Spectral Estimation,* Prentice-Hall, Englewood Cliffs, NJ, 1988.

[12] Kim, Y.C. and Powers, E.J., Digital bispectral analysis of self-excited fluctuation spectral, *Phys. Fluids,* 21(8), 1452–1453, Aug. 1978.

[13] Marple, Jr., S.L., *Digital Spectral Analysis with Applications,* Prentice-Hall, Englewood Cliffs, NJ, 1987.

[14] Matsuoka, T. and Ulrych, T.J., Phase estimation using bispectrum, *Proc. of IEEE,* 72, 1403–1411, Oct., 1984.

[15] Mendel, J.M., Tutorial on higher-order statistics (spectra) in signal processing and system theory: Theoretical results and some applications, *IEEE Proc.,* 79, 278–305, March 1991.

[16] Nikias, C.L. and Petropulu, A.P., *Higher-Order Spectra Analysis: a Nonlinear Signal Processing Framework,* Prentice-Hall, Englewood Cliffs, NJ, 1993.

[17] Nikias, C.L. and Raghuveer, M.R., Bispectrum estimation: a digital signal processing framework, *Proc. IEEE,* 75(7), 869–891, July 1987.

[18] Nikias, C.L. and Chiang, H.-H., Higher-order spectrum estimation via noncausal autoregressive modeling and deconvolution, *IEEE Trans. Acoustics, Speech and Signal Processing,* 36(12), 1911–1913, Dec. 1988.

[19] Nikias, C.L. and Mendel, J.M., Signal processing with higher-order spectra, *IEEE Signal Processing Magazine,* 10–37, July 1993.

[20] Oppenheim, A.V. and Schafer, R.W., Discrete-Time Signal Processing, Prentice-Hall, Englewood Cliffs, NJ., 1989.

[21] Pan, R. and Nikias, C.L., The complex cepstrum of higher order cumulants and nonminimum phase system identification, *IEEE Trans. on Acoust., Speech and Signal Processing,* 36(2), 186–205, Feb. 1988.

[22] Papoulis, A., Probability random variables and stochastic processes, McGraw-Hill, New York, 1984.

[23] Petropulu, A.P., Higher-order spectra in biomedical signal processing, *CRC Press Biomedical Engineering Handbook,* CRC Press, Boca Raton, FL, 1995.

[24] Petropulu, A.P., Noncausal nonminimum phase ARMA modeling of non-Gaussian processes, *IEEE Trans. on Signal Processing,* 43(8), 1946–1954, Aug. 1995.

[25] Petropulu, A.P and Nikias, C.L., The complex cepstrum and bicepstrum: analytic performance evaluation in the presence of Gaussian noise, *IEEE Transactions Acoustics, Speech and Signal Processing, special mini-section on Higher-Order Spectral Analysis,* ASSP-38(7), July 1990.

[26] Petropulu, A.P. and Abeyratne U.R., Signal reconstruction for higher-order spectra slices, *IEEE Trans. on Signal Processing,* Sept. 1997.

[27] Powers, E.J., Ritz, C.K. et.al., Applications of digital polyspectral analysis to nonlinear systems modeling and nonlinear wave phenomena, *Workshop on Higher-Order Spectral Analysis,* Vail, CO, 73–77, June 1989.

[28] Pozidis, H. and Petropulu, A.P., System reconstruction from selected bispectrum slices, *IEEE Signal Processing Workshop on Higher-Order Statistics,* Banff, Alberta, Canada, June 1997.

[29] Pozidis, H. and Petropulu, A.P., System reconstruction using selected regions of the discretized HOS, *IEEE Transactions on Signal Processing,* submitted in 1997.

[30] Raghuveer, M.R. and Nikias, C.L., Bispectrum estimation: A parametric approach, *IEEE Trans. on Acoust., Speech and Signal Processing,* ASSP 33(5), 1213–1230, Oct. 1985.

[31] Raghuveer, M.R. and Nikias, C.L., Bispectrum estimation via AR modeling, *Signal Processing,* 10, 35–48, 1986.

[32] Rao, T. Subba and Gabr, M.M., An introduction to bispectral analysis and bilinear time series models, *Lecture Notes in Statistics,* 24, Springer-Verlag, New York, 1984, 24.

[33] Rozario, N. and Papoulis, A., The identification of certain nonlinear systems by only observing the output, *Workshop on Higher-Order Spectral Analysis,* Vail, CO, 73–77, June 1989.

[34] Schetzen, M., *The Volterra and Wiener Theories on Nonlinear System,* updated edition, Krieger Publishing Company, Malabar, FL, 1989.

[35] Swami, A. and Mendel, J.M., ARMA parameter estimation using only output cumulants, *IEEE Trans. Acoust., Speech and Signal Processing,* 38, 1257–1265, July 1990.

[36] Tick, L.J., The estimation of transfer functions of quadratic systems, *Technometrics,* 3(4), 562–567, Nov. 1961.

[37] United Signals & Systems, Inc., Comprehensive bibliography on higher-order statistics (spectra), Culver City, CA, 1992.

DSP Software and Hardware

Vijay K. Madisetti
Georgia Institute of Technology

T HE PRIMARY TRAITS OF EMBEDDED signal processing systems that distinguish them from general purpose computer systems are their *predictable* reactions to *real-time*[1] stimuli from the environment, their *form-* and *cost*-optimized design, and their compliance with *required* or *specified* modes of response behavior and functionality [1].

---

[1] Real-time indicates behavior related to wall-clock time and does not necessarily imply a quick response.

Other traits that they share with other forms of digital products include the need for reliability, fault-tolerance, and maintainability, to name just a few. An embedded system usually consists of hardware components such as memories, application-specific ICs (ASICs), processors, DSPs, buses, analog-digital interfaces, and also software components that provide control, diagnostic, and application-specific capabilities required of it. In addition, they often contain electromechanical (EM) components such as sensors and transducers and operate in harsh environmental conditions. Unlike general purpose computers they may not allow much flexibility in support of a diverse range of programming applications, and it is not unusual to dedicate such systems to specific application. Embedded systems, thus, range from simple, low-cost sensor/actuator systems consisting of a few tens of lines of code and 8/16-bit processors (CPU) (e.g., bank ATM machines) to sophisticated high-performance signal processing systems consisting of runtime operating system support, tens of x86-class processors, digital signal processing (DSP) chips, interconnection networks, complex sensors, and other interfaces (e.g., radar-based tracking and navigational systems). Their lack of flexibility may be apparent when one considers that an ATM machine cannot be easily programmed to support additional image processing tasks, unless *upgraded* in terms of resources. Finally, embedded systems typically do not support direct *user* interaction in terms of higher order programming languages (HOLs) such as Fortran or C, but allow users to provide inputs that are sensor- or menu-driven. The debug and diagnostic interfaces, however, support HOLs and other lower level software and hardware programmability.

Embedded systems in general may be classified into one of the following four general categories of products. The prices are indicative of the multi-billion dollar marketplace in 1996, and their relative magnitudes are more significant than their actual values. The relationship of the categories to dollar cost is intentional and is an early harbinger of the fact that underlying cost and performance tradeoffs motivate and drive most of the system design and prototyping methodologies.

Commodity DSP Products: High-volume market and valued at less than $ 300 a piece. These include CD players, recorders, VCRs, facsimile and answering machines, telemetry applications, simple signal processing filtering packages, etc., primarily aimed at the highly competitive mass-volume consumer market.

Portable DSP Products: High-volume market and valued at less than $ 800. These include portable and hand-held low-power electronic products for man-machine communications such as DSP boards, digital audio, security systems, modems, camcorders, industrial controllers, scanners, communications equipment, and others.

Cost-Performance DSP Products: High-volume market, and valued at less than $ 3000. These products trade off cost for performance, and include DSP products such as video teleconferencing equipment, laptops, audio, telecommunications switches, high-performance DSP boards and coprocessors, and DSP CAD packages for hardware and software design.

High-Performance Products: Low-to-moderate volume market, and valued at over $8000. These products include high-end workstations with DSP coprocessors, real-time signal processors, real-time database processing systems, digital HDTV, radar signal processor systems, avionics and military systems, sensor and data processing hardware and software systems. This class of products contains a significant amount of software compared to the earlier classes, which often focus on large volume, low-cost, hardware-only solutions.

It may be useful to classify high-performance products further into three categories.

- *Real-time embedded control systems*: These systems are characterized by the following features: interrupt driven, large numerical processing requirements, small databases, tight real-time constraints, well-defined user interface, requirements and design driven by performance requirements. Examples include an aircraft control system, or a control system for a steel plant.

- *Embedded information systems*: These systems are characterized by the following features: transaction-based, moderate numerical/DSP processing, flexible time constraints, complex user interfaces, requirements and design driven by user interface. Examples include accounting and inventory management systems.

- *Command, control, communication, and intelligence (C4I) systems:* These systems are characterized by large numerical processing, large databases, moderate to tight real-time constraints, flexible and complex user interfaces, requirements and design driven by performance and user interface. Examples include missile guidance systems, radar-tracking systems, and inventory and manufacturing control systems.

These four categories of embedded systems can be further distinguished in terms of other metrics such as computing speed (integer or floating point performance), input/output transfer rates, memory capacities, market volume, environmental issues, typical design and development budgets, lifetimes, reliability issues, upgrades, and other lifecycle support costs. Another interesting fact is that the higher the software value in a product, the greater its profitability margin. Recent studies by Andersen Consulting have shown that profit margin pressures are increasing due to increasing semiconductor content in systems' sales' values. In 1985, silicon represented 9.5 percent of a system's value. By 1995, that had shot up to 19.1 percent. The higher the silicon content, the greater the pressure on margins resulting in lower profits. In PCs, integrated circuit components represent 30 to 35 percent of the sales value and the ratio is steadily increasing. More than 50 percent of value of the new network computers (NCs) is expected to be in integrated circuits. In the area of DSPs, we estimate that this ratio is about 20 percent.

In this section, the chapter "Introduction to the TMS320 Family of Digital Signal Processors" by Panos Papamichalis, outlines the programmable DSP families developed by Texas Instruments, the leading organization in this area. In, "Rapid Design and Prototyping of DSP Systems", T. Egolf, M. Pettigrew, J. Debardelaben, R. Hezar, S. Famorzadeh, A. Kavipurapu, M. Khan, L.-R. Dung, K. Balemarthy, N. Desai, Y. Jung, and V. Madisetti, discuss how signal processing systems are designed and integrated using a novel top down design approach developed as part of DARPA's RASSP program.

References

[1] Madisetti, V. K., *VLSI Digital Signal Processors,* IEEE Press, Piscataway, NJ, 1995.

77

Introduction to the TMS320 Family of Digital Signal Processors

Panos Papamichalis
Texas Instruments

This article discusses the architecture and the hardware characteristics of the TMS320 family of Digital Signal Processors. The TMS320 family includes several generations of programmable processors with several devices in each generation. Since the programmable processors are split between fixed-point and floating-point devices, both categories are examined in some detail. The TMS320C25 serves here as a simple example for the fixed-point processor family, while the TMS320C30 is used for the floating-point family.

77.1 Introduction

Since its introduction in 1982 with the TMS32010 processor, the TMS320 family of DSPs has been exceedingly popular. Different members of this family were introduced to address the existing needs for real-time processing, but then, designers capitalized on the features of the devices to create solutions and products in ways never imagined before. In turn, these innovations fed the architectural and hardware configurations of newer generations of devices.

0-8493-8572-5/98/$0.00+$.50

Digital Signal Processing encompasses a variety of applications, such as digital filtering, speech and audio processing, image and video processing, and control. All DSP applications share some common characteristics:

- The algorithms used are mathematically intensive. A typical example is the computation of an FIR filter, implemented as sum-of-products. This operation involves a lot of multiplications combined with additions.
- DSP algorithms must typically run in real time: i.e., the processing of a segment of the arriving signal must be completed before the next segment arrives, or else data will be lost.
- DSP techniques are under constant development. This implies that DSP systems should be flexible to support changes and improvements in the state of the art. As a result, programmable processors have been the preferred way of implementation. In recent times, though, fixed-function devices have also been introduced to address high-volume consumer applications with low-cost requirements.

These needs are addressed in the TMS320 family of DSPs by using appropriate architecture, instruction sets, I/O capabilities, as well as the raw speed of the devices. However, it should be kept in mind that these features do not cover all the aspects describing a DSP device, and especially a programmable one. Availability and quality of software and hardware development tools (such as compilers, assemblers, linker, simulators, hardware emulators, and development systems), application notes, third-party products and support, hot-line support, etc. play an important role on how easy it will be to develop an application on the DSP processor. The TMS320 family has very extensive such support, but its description goes beyond the scope of this article. The interested reader should contact the TI DSP hotline (Tel. 713-274-2320).

For the purposes of this article, two devices have been selected to be highlighted from the Texas Instruments TMS320 family of digital signal processors. One is the TMS320C25, a 16-bit, fixed-point DSP, and the other is the TMS320C30, a 32-bit, floating-point DSP. As a short-hand notation, they will be called 'C25 and 'C30, respectively. The choice was made so that both fixed-point issues are considered.

There have been newer (and more sophisticated) generations added to the TMS320 family but, since the objective of this article is to be more tutorial, they will be discussed as extensions of the 'C25 and the 'C30. Such examples are other members of the 'C2x and the 'C3x generations, as well as the TMS320C5x generation ('C5x for short) of fixed-point devices, and the TMS320C4x ('C4x) of floating-point devices. Customizable and fixed-function extensions of this family of processors will be also discussed.

Texas Instruments, like all vendors of DSP devices, publishes detailed User's Guides that explain at great length the features and the operation of the devices. Each of these User's Guides is a pretty thick book, so it is not possible (or desirable) to repeat all this information here. Instead, the objective of this article is to give an overview of the basic features for each device. If more detail is necessary for an application, the reader is expected to refer to the User's Guides. If the User's Guides are needed, it is very easy to obtain them from Texas Instruments.

77.2 Fixed-Point Devices: TMS320C25 Architecture and Fundamental Features

The Texas Instruments TMS320C25 is a fast, 16-bit, fixed-point digital signal processor. The speed of the device is 10 MHz, which corresponds to a cycle time of 100 ns. Since the majority of the instructions execute in a single cycle, the figure of 100 ns also indicates how long it takes to execute one instruction. Alternatively, we can say that the device can execute 10 million instructions per second (MIPS). The actual signal from the external oscillator or crystal has a frequency four times higher, at 40 MHz. This frequency is then divided on-chip to generate the internal clock with a period of 100 ns. Figure 77.1 shows the relationship between the input clock CLKIN from the external oscillator, and the output clock CLKOUT.

CLKOUT is the same as the clock of the device, and it is related to CLKIN by the equation CLKOUT = CLKIN /4. Note that in Fig. 77.1 the shape of the signal is idealized ignoring rise and fall times.

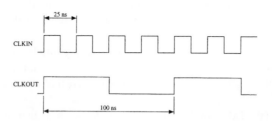

FIGURE 77.1 Clock timing of the TMS320C25. CLKIN = external oscillator; CLKOUT = clock of the device.

Newer versions of the TMS320C25 operate in higher frequencies. For instance, there is a spinoff that has a cycle time of 80 ns, resulting in a 12.5 MIPS operation. There are also slower (and cheaper) versions for applications that do not need this computational power.

Figure 77.2 shows in a simplified form the key features of the TMS320C25. The major parts of the DSP processor are the memory, the Central Processing Unit (CPU), the ports, and the peripherals. Each of these parts will be examined in more detail later. The on-chip memory consists of 544 words of RAM (read/write memory) and 4K words of ROM (read-only memory). In the notation used here, 1K = 1024 words, and 4K = 4 × 1024 = 4096 words. Each word is 16 bits wide and, when some memory size is given, it is measured in 16-bit words, and not in bytes (as is the custom in microprocessors). Of the 544 words of RAM, 256 words can be used as either program or data memory, while the rest is only data memory. All 4K of on-chip ROM is program memory. Overall, the device can address 64K words of data memory and 64K words of program memory. Except for what resides on-chip, the rest of the memory is external, supplied by the designer.

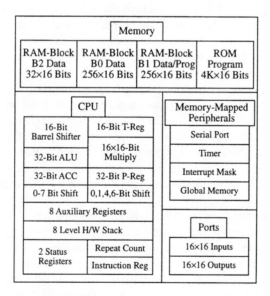

FIGURE 77.2 Key architectural features of the TMS320C25.

The CPU is the heart of the processor. Its most important feature, distinguishing it from the traditional microprocessors, is a hardware multiplier that is capable of performing a 16 × 16 bit multiplication in a single cycle. To preserve higher intermediate accuracy of results, the full 32-bit product is saved in a product register. The other important part of the CPU is the Arithmetic Logic Unit (ALU) that performs additions, subtractions, and logical operations. Again, for increased intermediate accuracy, there is a 32-bit accumulator to handle all the ALU operations.

All the arithmetic and logical functions are accumulator-based. In other words, these operations have two operands, one of which is always the accumulator. The result of the operation is stored in the accumulator.

Because of this approach the form of the instructions is very simple indicating only what the other operand is. This architectural philosophy is very popular but it is not universal. For instance, as is discussed later, the TMS320C30 takes a different approach, where there are several "accumulators" in what is called a register file.

Other components of the TMS320C25 CPU are several shifters to facilitate manipulation of the data and increase the throughput of the device by performing shifting operations in parallel with other functions. As part of the CPU, there are also eight auxiliary registers that can be used as memory pointers or loop counters. There are two status registers, and an 8-deep hardware stack. The stack is used to store the memory address where the program will continue execution after a temporary diversion to a subroutine.

To communicate with external devices, the TMS320C25 has 16 input and 16 output parallel ports. It also has a serial port that can serve the same purpose. The serial port is one of the peripherals that have been implemented on chip. Other peripherals include the interrupt mask, the global memory capability, and a timer. The above components of the TMS320C25 are examined in more detail below.

The device has 68 pins that are designated to perform certain functions, and to communicate with other devices on the same board. The names of the signals and the corresponding definitions appear in Table 77.1. The first column of the table gives the pin names. Note that a bar over the name indicates that the pin is in the active position when it is electrically low. For instance, if the pins take the voltage levels of 0 V and 5 V, a pin indicated with an overbar is asserted when it is set at 0 V. Otherwise, assertion occurs at 5 V. The second column indicates if the pin is used for input to the device or output from the device or both. The third column gives a description of the pin functionality.

Understanding the functionality of the device pins is as important as understanding the internal architecture because it provides the designer with the tools available to communicate with the external world. The DSP device needs to receive data and, often, instructions from the external sources, and send the results back to the external world. Depending on the paths available for such transactions, the design of a program can take very different forms. Within this framework, it is up to the designer to generate implementations that are ingenious and elegant.

The TMS320C25 has its own assembly language to be programmed. This assembly language consists of 133 instructions that perform general-purpose and DSP-specific functions. Familiarity with the instruction set and the device architecture are the two components of efficient program implementation. High-level-language compilers have also been developed that make the writing of programs an easier task. For the TMS320C25, there is a C compiler available. However, there is always a loss of efficiency when programming in high-level languages, and this may not be acceptable in computation-bound real-time systems. Besides, for complete understanding of the device it is necessary to consider the assembly language.

A very important characteristic of the device is its Harvard architecture. In Harvard architecture (see Fig. 77.3), the program and data memory spaces are separated and they are accessed by different buses. One bus accesses the program memory space to fetch the instructions, while another bus is used to bring operands from the data memory space and store the results back to memory. The objective of this approach is to increase the throughput by bringing instructions and data in parallel. An alternate philosophy is the von Neuman architecture. The von Neuman architecture (see Fig. 77.4) uses a single bus and a unified memory space. Unification of the memory space is convenient for partitioning it between program and data, but it presents a bottleneck since both data and program instructions must use the same path and, hence, they must be multiplexed. The Harvard architecture of multiple buses is used in digital signal processors because the increased throughput is of paramount importance in real-time systems.

TABLE 77.1 Names and Functionality of the 68 pins of the TMS320C25

Signals	I/O/Z[a]	Definition
V_{CC}	I	5-V supply pins
V_{SS}	I	Ground pins
X1	O	Output from internal oscillator for crystal
X2/CLKIN	I	Input to internal oscillator from crystal or external clock
CLKOUT1	O	Master clock output (crystal or CLKIN frequency/4)
CLKOUT2	O	A second clock output signal
D15-D0	I/O/Z	16-bit data bus D15 (MSB) through DO (LSB). Multiplexed between program, data, and I/O spaces.
A15-A0	O/Z	16-bit address bus A15 (MSB) through AO (LSB)
$\overline{PS}, \overline{DS}, \overline{IS}$	O/Z	Program, data, and I/O space select signals
R/$\overline{W}$	O/Z	Read/write signal
$\overline{STRB}$	O/Z	Strobe signal
$\overline{RS}$	I	Reset input
$\overline{INT2}$-$\overline{INT0}$	I	External user interrupt inputs
MP/$\overline{MC}$	I	Microprocessor/microcomputer mode select pin
$\overline{MSC}$	O	Microstate complete signal
$\overline{IACK}$	O	Interrupt acknowledge signal
READY	I	Data ready input. Asserted by external logic when using slower devices to indicate that the current bus transaction is complete.
$\overline{BR}$	O	Bus request signal. Asserted when the TMS320C25 requires access to an external global data memory space.
XF	O	External flag output (latched software-programmable signal)
$\overline{HOLD}$	I	Hold input. When asserted. TMS320C25 goes into an idle mode and places the data, address, and control lines in the high impedance state.
$\overline{HOLDA}$	O	Hold acknowledge signal.
$\overline{SYNC}$	I	Synchronization input.
$\overline{BIO}$	I	Branch control input. Polled by BIOZ instruction
DR	I	Serial data receive input
CLKR	I	Clock for receive input for serial port
FSR	I	Frame synchronization pulse for receive input
DX	O/Z	Serial data transmit output
CLKX	I	Clock for transmit output for serial port
FSX	I/O/Z	Frame synchronization pulse for transmit. Configurable as either an input or an output.

[a] I/O/Z denotes input/output/high-impedance state.
Note: The first column is the pin name; the second column indicates if it is an input or an output pin; the third column gives a description of the pin functionality.

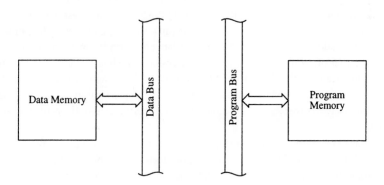

FIGURE 77.3 Simplified block diagram of the Harvard architecture.

The difference of the architectures is important because it influences the programming style. In Harvard architecture, two memory locations can have the same address, as long as one of them is in the data space and the other is in the program space. Hence, when the programmer uses an address label, he has to be alert as to what space he is referring. Another restriction of the Harvard architecture is that the data memory cannot be initialized during loading because loading refers only to placing the program on the memory (and the program memory is separate from the data memory). Data memory can be initialized during execution only. The programmer must incorporate such initialization in his program code. As it

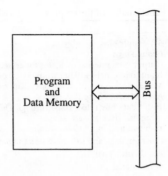

FIGURE 77.4   Simplified block diagram of the von Neuman architecture.

will be seen later, such restrictions have been removed from the TMS320C30 while retaining the convenient feature of multiple buses.

Figure 77.5 shows a functional block diagram of the TMS320C25 architecture. The Harvard architecture of the device is immediately apparent from the separate program and data buses. What is not apparent is that the architecture has been modified to permit communication between the two buses. Through such communication, it is possible to transfer data between the program and memory spaces. Then, the program memory space also can be used to store tables. The transfer takes place by using special instructions such as TBLR (Table Read), TBLW (Table Write), and BLKP (Block transfer from Program memory).

As shown in the block diagram, the program ROM is linked to the program bus, while data RAM blocks B1 and B2 are linked to the data bus. The RAM block B0 can be configured either as program or data memory (using the instructions CNFP and CNFD), and it is multiplexed with both buses. The different segments, such as the multiplier, the ALU, the memories, etc. are examined in more detail below.

77.3   TMS320C25 Memory Organization and Access

Besides the on-chip memory (RAM and ROM), the TMS320C25 can access external memory through the external bus. This bus consists of the 16 address pins A0-A15, and the 16 data pins D0-D15. The address pins carry the address to be accessed, while the data pins carry the instruction word or the operand, depending on whether program or data memory is accessed. The bus can access either program or data memory, the difference indicated by which of the pins PS and DS (with overbars) becomes active. The activation is done automatically when, during the execution, an instruction or a piece of data needs to be fetched. Since the address is 16-bits wide, the maximum memory space is 64K words for program and 64K words for data.

The device starts execution after a reset signal, i.e., after the RS pin is pulled low for a short period of time. The execution always begins at program memory location 0, where there should be an instruction to direct the program execution to the appropriate location. This direction is accomplished by a branch instruction.

B PROG

which loads the program counter with the program memory address that has the label PROG (or any other label you choose). Then, execution continues from the address PROG, where, presumably, a useful program has been placed.

It is clear that the program memory location 0 is very important, and you need to know where it is physically located. The TMS320C25 gives you the flexibility to use as location 0 either the first location of the on-chip ROM, or the first location of the external memory. In the first case, we say that the device operates in the microcomputer mode, while in the second one it is in the microprocessor mode. In the

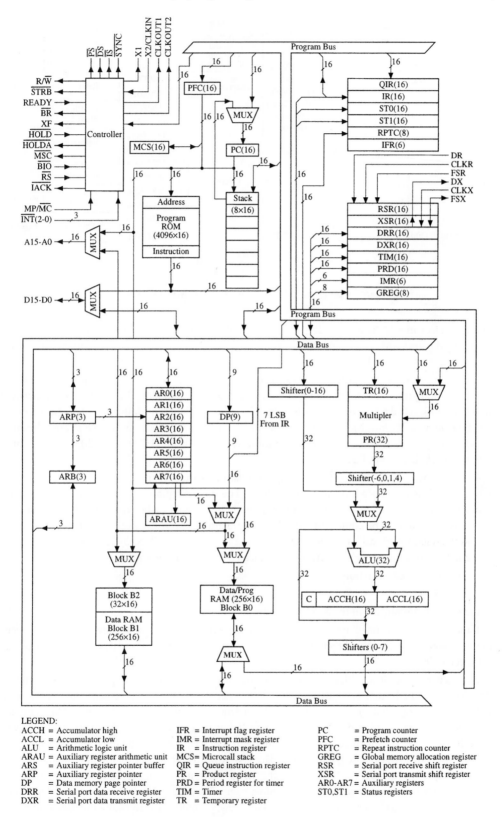

FIGURE 77.5   Functional block diagram of the TMS320C25 architecture.

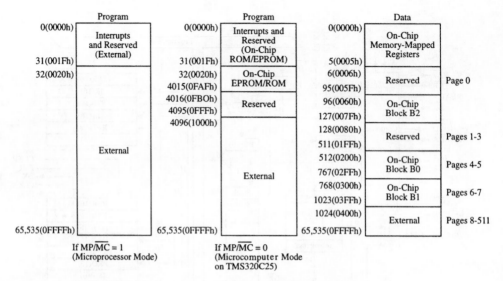

FIGURE 77.6   Memory maps for program and data memory of the TMS320C25.

microprocessor mode, the on-chip ROM is ignored altogether. You can choose between the two modes by pulling the device MP/MC high or low. The microcomputer mode is useful for production purposes, while for laboratory and development work the microprocessor mode is used exclusively.

Figure 77.6 shows the memory configuration of the TMS320C25, where the microprocessor and microcomputer configurations of the program memory are depicted separately. The data memory is partitioned in 512 sections, called pages, of 128 words each. The reason of the partitioning is for addressing purposes, as will be discussed below. Memory boundaries of the 64K memory space are shown in both decimal and hexadecimal notation (hexadecimal notation indicated by an "h" or "H" at the end.) Compare this map with the block diagram in Fig. 77.5.

As mentioned earlier, in two-operand operations, one of the operands resides in the accumulator, and the result is also placed in the accumulator. (The only exceptions is the multiplication operation examined later.) The other operand can either reside in memory or be part of the instruction. In the latter case, the value to be combined with the accumulator is explicitly specified in the instruction, and this addressing mode is called immediate addressing mode. In the TMS320C25 assembly language, the immediate addressing mode instructions are indicated by a "K" at the end of the instruction.

For example, the instruction

ADDK 5

increments the contents of the accumulator by 5.

If the value to be operated upon resides in memory, there are two ways to access it: either by specifying the memory address directly (direct addressing) or by using a register that holds the address of that number (indirect addressing).

As a general rule, it is desirable to describe an instruction as briefly as possible so that the whole description can be held in one 16-bit word. Then, when the program is executed, only one word needs to be fetched before all the information from the instruction is available for execution. This is not always possible and there are two-word instructions as well, but the chip architects always strive to achieve one-word instructions. In the direct addressing mode, full description of a memory address would require a 16-bit word by itself because the memory space is 64K words. To reduce that requirement, the memory space is divided in 512 pages of 128 words each. An instruction using direct addressing contains the 7 bits indicating what word you want to access within a page. The page number (9 bits) is stored in a separate register (actually, part of a register), called the Data Page pointer (DP). You store the page number in the

DP pointer by using the instructions LDP (Load Data Page pointer) or LDPK (Load Data Page pointer immediate).

In the indirect addressing mode, the data memory address is held in a register that acts as a memory pointer. There are eight such registers available, called auxiliary registers, AR0-AR7. The auxiliary registers can also be used for other functions, such as loop counters, etc. To save bits in the instruction, the auxiliary register used as memory pointer is not indicated explicitly, but it is stored in a separate register (actually, part of a register), the auxiliary register pointer (ARP). In other words, there is the concept of the "current register". In an operation using indirect addressing, the contents of the current auxiliary register point to the desired memory location. The current AR is specified by the contents of the ARP as shown in Fig. 77.7. In an instruction, indirect addressing is indicated by an asterisk.

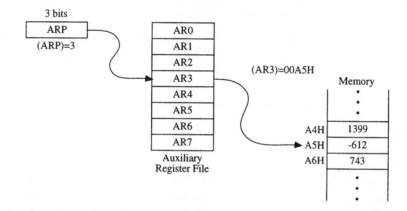

FIGURE 77.7   Example of indirect addressing mode.

A "+" sign at the end of an instruction using indirect addressing means "after the present memory access, increment the contents of the current auxiliary register by 1". This is done in parallel with the load-accumulator operation. The above autoincrementing of the auxiliary register is an optional operation that offers additional flexibility to the programmer. And it is not the only one available. The TMS320C25 has an auxiliary register arithmetic unit (ARAU, see Fig. 77.5) that can execute such operations in parallel with the CPU, and increase the throughput of the device in this way. Table 77.2 summarizes the different operations that can be done while using indirect addressing. As seen from this table, the contents of an auxiliary register can be incremented or decremented by 1, incremented or decremented by the contents of AR0, and incremented or decremented by AR0 in a bit-reversed fashion. The last operation is useful when doing Fast Fourier Transforms. The bit-reversed addressing is implemented by adding AR0 with reverse carry propagation, an operation explained in the TMS320C25 User's Guide. Additionally, it is possible to load at the same time the ARP with a new value, thus saving an extra instruction.

77.4   TMS320C25 Multiplier and ALU

The heart of the TMS320C25 is the CPU consisting, primarily, of the multiplier and the arithmetic logic unit (ALU). The hardware multiplier can perform a 16 bit × 16 bit multiplication in a single machine cycle. This capability is probably the major distinguishing feature of digital signal processors because it permits high throughput in numerically intensive algorithms.

Associated with the multiplier, there are two registers that hold operands and results. The T-register (for temporary register) holds one of the two factors. The other factor comes from a memory location. Again, this construct, with one implied operand residing in the T-register, permits more compact instruction

TABLE 77.2 Operations That Can Be
Performed in Parallel with Indirect Addressing

Notation	Operation
ADD *	No manipulation of AR or ARP
ADD *, Y	Y → ARP
ADD *+	AR(ARP)+1 → AR(ARP)
ADD *+,Y	AR(ARP)+1 → AR(ARP)
	Y → ARP
ADD *-	AR(ARP) - 1 → AR(ARP)
ADD *-,Y	AR(ARP) - 1 → AR(ARP)
	Y → ARP
ADD *0+	AR(ARP) + AR0 → AR(ARP)
ADD *0+,Y	AR(ARP) + AR0 → AR(ARP)
	Y → ARP
ADD *0-	AR(ARP)-AR0 → AR(ARP)
ADD *0-,Y	AR(ARP)-AR0 → AR(ARP)
	Y → ARP
ADD *BR0+	AR(ARP) +rcAR0 → AR(ARP)
ADD *BR0+,Y	AR(ARP) +rcAR0 → AR(ARP)
	Y → ARP
ADD *BR0-	AR(ARP)-rcAR0 → AR(ARP)
ADD *BR0-,Y	AR(ARP)-rcAR0 → AR(ARP)
	Y → ARP

Note: $Y = 0, \ldots, 7$ is the new "current" AR. AR(ARP)
is the AR pointed to by the ARP. BR = bit reversed, rc
= reverse carry.

words. When multiplier and multiplicand (two 16-bit words) are multiplied together, the result is 32-bits long. In traditional microprocessors, this product would have been truncated to 16 bits, and presented as the final result. In DSP applications, though, this product is only an intermediate result in a long stream of multiply-adds, and if truncated at this point, too much computational noise would be introduced to the final result. To preserve higher final accuracy, the full 32-bit result is held in the P-register (for product register). This configuration is shown in Fig. 77.8 which depicts the multiplier and the ALU of the TMS320C25.

Actually, the P-register is viewed as two 16-bit registers concatenated. This viewpoint is convenient if you need to save the product using the instructions SPH (store product high) and SPL (store product low). Otherwise, the product can operate on the accumulator, which is also 32-bits wide. The contents of the product register can be loaded on the accumulator, overwriting whatever was there, using the PAC (product to accumulator) instruction. It can also be added to or subtracted from the accumulator using the instructions APAC or SPAC.

When moving the contents of the T-register to the accumulator, you can shift this number using the built-in shifters. For instance you can shift the result left by 1 or 4 locations (essentially multiplying it by 2 or 16), or you can shift it right by 6 (essentially dividing it by 64). These operations are done automatically, without spending any extra machine cycles, simply by setting the appropriate product mode with SPM instruction. Why would you want to do such shifting? The left shifts have as a main purpose to eliminate any extra sign bits that would appear in computations. The right shift scales down the result and permits accumulation of several products before you start worrying about overflowing the accumulator.

At this point, it is appropriate to discuss the data formats supported on the TMS320C25. This device, as most fixed-point processors, uses two's-complement notation to represent the negative numbers. In two's complement notation, to form the negative of a given number, you take the complement of that number and you add 1. In two's-complement notation, the most significant bit (MSB, the left-most bit) of a positive number is zero, while the MSB of a negative number is one. In the 'C25, the two's complement numbers are sign-extended, which means that, if the absolute value of the number is not large enough to fill all the bits of the word, there will be more than one sign bits.

As seen from Fig. 77.8, the multiplier path is not the only way to access the accumulator. Actually, the ALU and the accumulator support a wealth of arithmetic (ADD, SUB, etc.) and logical (OR, AND, XOR,

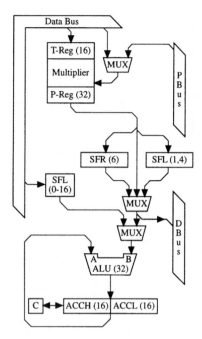

FIGURE 77.8 Diagram of the TMS320C25 multiplier and ALU.

etc.) instructions, in addition to load and store instructions for the accumulator (LAC, ZALH, SACL, SACH, etc.).

An interesting characteristic of the TMS320C25 architecture is the existence of several shifters that can perform such shifts in parallel with other operations. Except for the right shifter at the multiplier, all the other shifters are left shifters. An input shifter to the ALU and the accumulator can shift the input value to the left by up to 16 locations, while output shifters from the accumulator can shift either the high or the low part of the accumulator by up to 7 locations to the left.

A construct that appears very often in mathematical computations is the sum of products. Sums of products appear in the computation of dot products, in matrix multiplication, and in convolution sums for filtering, among other applications. Since it is important to carry out this computation as fast as possible for real-time operation, all digital signal processors have special instructions to speed up this particular function.

The TMS320C25 has the instruction LTA which loads the T-register and, in parallel with that, adds the previous product (which already resides in the P-register) to the accumulator. LTS subtracts the product from the accumulator. Another instruction, LTD, does the same thing as LTA, but it also moves the value that was just loaded on the T-register to the next higher location in memory. This move realizes the delay line that is needed in filtering applications. LTA, when combined with the MPY instruction, can implement very efficiently the sum of products.

For even higher efficiency, there is a MAC instruction that combines LTA and MPY. An additional MACD instruction combines LTD and MPY. The increased efficiency is achieved by using both the data and the program buses to bring in the operands of the multiplication. The data coming from the data bus can be traced in memory by an AR, using indirect addressing. The data coming from the program bus are traced by the program counter (actually, the pre-fetch counter, PFC) and, hence, they must reside in consecutive locations of program memory. To be able to modify the data and then use it in such multiply-add operations, the TMS320C25 permits reconfiguration of block B0 in the on-chip memory. B0 can be configured either as program or as data memory, as shown in Fig. 77.9, using the CNFD and CNFP instructions.

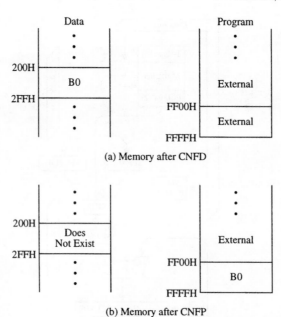

(a) Memory after CNFD

(b) Memory after CNFP

FIGURE 77.9   Partial memory configuration of the TMS320C25 after the CNFD and the CNFP instructions.

77.5   Other Architectural Features of the TMS320C25

The TMS320C25 has many interesting features and capabilities that can be found in the user's guide [1]. Here, we present briefly only the most important of them.

The program counter is a 16-bit register, hidden from the user, which contains the address of the next instruction word to be fetched and executed. Occasionally, the program execution may be redirected, for instance, through a subroutine call. In this case, it is necessary to save the contents of the program counter so that the program flow continues from the correct instruction after the completion of the subroutine call. For this purpose, a hardware stack is provided to save and recover the contents of the program counter.

The hardware stack is a set of eight registers, of which only the top one is accessible to the user. Upon a subroutine call, the address after the subroutine call is pushed on the stack, and it is reinstated in the program counter when the execution returns from the subroutine call. The programmer has control over the stack by using the PUSH, PSHD, POP, and POPD instructions. The PUSH and POP operations push the accumulator on the stack or pop the top of the stack to the accumulator respectively. PSHD and POPD do the same functions but with memory locations instead of the accumulator.

Occasionally the program execution in a processor must be interrupted in order to take care of urgent functions, such as receiving data from external sources. In these cases, a special signal goes to the processor, and an interrupt occurs. The interrupts can be internal or external. During an interrupt, the processor stops execution, wherever it may be, pushes the address of the next instruction on the stack, and starts executing from a predetermined location in memory. The interrupt approach is appropriate when there are functions or devices that need immediate attention. On the TMS320C25, there are several internal and external interrupts, which are prioritized, i.e., when several of the interrupts occur at the same time, the one with the highest priority is executed first. Typically, the memory location where the execution is directed to during an interrupt contains a branch instruction. This branch instruction directs the program execution to an area in the program memory where an interrupt service routine exists. The interrupt service routine will perform the tasks that the interrupt has been designed for, and then return to the execution of the original program.

Besides the external hardware interrupts (for which there are dedicated pins on the device), there are internal interrupts generated by the serial port and the timer. The serial port provides direct communication with serial devices, such as codecs, serial analog-to-digital converters, etc. In these devices, the data are transmitted serially, one bit at a time, and not in parallel, which would require several parallel lines. When 16 bits have been input, the 16-bit word can be retrieved from the register DRR (data receive register). Conversely, to transmit a word, you put it in the DXR (data transmit register). These two registers occupy data memory locations 0 and 1, respectively, and they can be treated like any other memory location.

The timer consists of a period register and a timer register. At the beginning of the operation, the contents of the period register are loaded on the timer register, which is then decremented at every machine cycle. When the value of the timer register reaches zero, it generates a timer interrupt, the period register is loaded again on the timer register, and the whole operation is repeated.

77.6 TMS320C25 Instruction Set

The TMS320C25 has an instruction set consisting of 133 instructions. Some of these assembly language instructions perform general purpose operations, while others are more specific to DSP applications. This section discusses examples of instructions selected from different groups. For a detailed description of each instruction, the reader is referred to the TMS320C25 User's Guide [1].

Each instruction is represented by one or two 16-bit words. Part of the instruction is a unique code identifying the operation to be performed, while the rest of the instruction contains information on the operation. For instance, this additional information determines if direct or indirect addressing is used, if there is a shift of the operand, what is the address of the operand, etc. In the case of two-word instructions, the second word is typically a 16-bit constant or program memory address. As it should be obvious, a two-word instruction takes longer to execute because it has to fetch two words, and it should be avoided if the same operation could be accomplished with a single-word instruction.

For example, if you want to load the accumulator with the contents of the memory location 3FH, shifting it to the left by 8 locations at the same time, you can write the instruction

LAC   3FH,8

The above instruction, when encoded, is represented by the word 283FH. The left-most four bits in this example, i.e., 0010, represent the "opcode" of the instruction. The opcode is the unique identifier of the instruction. The next four bits, 1000, are the shift of the operand. Then there is one bit (zero in this case) to signal that the direct addressing mode is used, and the last 7 bits are the operand address 3Fh (in hexadecimal).

Below, some of the more typical instructions are listed, and the ones that have an important interpretation are discussed. It is a good idea to review carefully the full set of instructions so that you know what tools you have available to implement any particular construct. The instructions are grouped here by functionality.

The accumulator and memory reference instructions involve primarily the ALU and the accumulator. Note that there is a symmetry in the instruction set. The addition instructions have counterparts for subtraction, the direct and indirect-addressing instructions have complementary immediate instructions, and so on.

ABS	Absolute value of accumulator
ADD	Add to accumulator with shift
ADDH	Add to high accumulator
ADDK	Add to accumulator short immediate
AND	Logical AND with accumulator
LAC	Load accumulator with shift
SACH	Store high accumulator with shift
SACL	Store low accumulator with shift
SUB	Subtract from accumulator with shift
SUBC	Subtract conditionally
ZAC	Zero accumulator
ZALH	Zero low accumulator and load high accumulator.

Operations involving the accumulator have versions affecting both the high part and the low part of the accumulator. This capability gives additional flexibility in scaling, logical operations, and double-precision arithmetic.

For example, let location A contain a 16-bit word that you want to scale down dividing by 16, and store the result in B. The following instructions perform this operation:

```
LAC    A,12    ; Load ACC with A shifted by 12 locations
SACH   B       ; Store ACCH to B:B = A/16
```

The auxiliary registers and data page pointer instructions deal with loading, storing, and modifying the auxiliary registers and the data page pointer. Note that the auxiliary registers and the ARP can also be modified during operations using indirect addressing. Since this last approach has the advantage of making the modifications in parallel with other operations, it is the most common method of AR modification.

LAR	Load auxiliary register
LARP	Load auxiliary register pointer
LDP	Load data memory page pointer
MAR	Modify auxiliary register
SAR	Store auxiliary register

The multiplier instructions are more specific to signal processing applications.

APAC	Add P-register to accumulator
LT	Load T-register
LTD	Load T-register, accumulate previous product, and move data
MAC	Multiply and accumulate
MACD	Multiply and accumulate with data move
MPY	Multiply
MPYK	Multiply immediate
PAC	Load accumulator with P-register
SQRA	Square and accumulate

Note that the instructions that perform multiplication and accumulation at the same time do not accumulate the present product but the result of an earlier multiplication. This result is found in the P-register. The square and accumulate function, SQRA, is a special case of the multiplication that appears often enough to prompt the inclusion of this specific instruction.

The branch instructions correspond to the GOTO instruction of high-level languages. They redirect the flow of the execution either unconditionally or depending on some previous result.

B	Branch unconditionally
BANZ	Branch on auxiliary register non zero
BGEZ	Branch if accumulator $>= 0$
CALA	Call with subroutine address in the accumulator
CALL	Call subroutine
RET	Return from subroutine

The CALL and RET instructions go together because the first one pushes the return address on the stack, while the second one pops the address from the stack into the program counter. The BANZ instruction is very helpful in loops where an AR is used as a loop counter. BANZ tests the AR, modifies it, and branches to the indicated address.

The I/O operations are, probably, among the most important in terms of final system configuration, because they help the device interact with the rest of the world. Two instructions that perform that function are the IN and OUT instructions.

BLKD	Block move from data memory to data memory
IN	Input data from port
OUT	Output data to port
TBLR	Table read
TBLW	Table write

The IN and OUT instructions read from or write to the 16 input and the 16 output ports of the TMS320C25. Any transfer of data goes to a specified memory location. The BLKD instruction permits movement of data from one memory location to another without going through the accumulator. To make such a movement effective, though, it is recommended to use BLKD with a repeat instruction, in which case every data move takes only one cycle.

The TBLR and TBLW instructions represent a modification to the Harvard architecture of the device. Using them, data can be moved between the program and the data spaces. In particular, if any tables have been stored in the program memory space they can be moved to data memory before they can be used. That is how the terminology of the instructions originated.

Some other instructions include:

DINT	Disable interrupts
EINT	Enable interrupts
IDLE	Idle until interrupt
RPT	Repeat instruction as specified by data memory value
RPTK	Repeat instruction as specified by immediate value

77.7 Input/Output Operations of the TMS320C25

During program execution on a digital signal processor, the data is moved between the different memory locations, on-chip and off-chip, as well as between the accumulator and the memory locations. This movement is necessary for the execution of the algorithm that is implemented on the processor. However, there is a need to communicate with the external world in order to receive data that will be processed, and return the processed results.

Devices communicate with the external world through their external memory or through the serial and parallel ports. Such a communication can be achieved, for instance, by sharing the external memory. Most often, the communication with the external world takes place through the external parallel or serial ports that the device has. Some devices may have ports of only one kind, serial or parallel, but most modern processors have both types. The two kinds of ports differ in the way in which the bits are read. In a parallel port, there is a physical line (and a processor pin) dedicated to every bit of a word. For example, if the processor reads in words that are 16 bits wide, as is the case with the TMS320C25, it has 16 lines available to read a whole word in a single operation. Typically, the same pins that are used for accessing external memory are also used for I/O.

The TMS320C25 has 16 input and 16 output ports that are accessed with the IN and OUT instructions. These instructions transfer data between memory locations and the I/O port specified.

77.8 Subroutines, Interrupts, and Stack on the TMS320C25

When writing a large program, it is advisable to structure it in a modular fashion. Such modularity is achieved by segmenting the program in small, self-contained tasks that are encoded as separate routines.

Then, the overall program can be simply a sequence of calls to these subroutines, possibly with some "glue" code. Constructing the program as a sequence of subroutines has the advantage that it produces a much more readable algorithm that can greatly help in debugging and maintaining it. Furthermore, each subroutine can be debugged separately, which is far easier than trying to uncover programming errors in a "spaghetti-code" program.

Typically, the subroutine is called during the program execution with an instruction such as

CALL SUBRTN

where SUBRTN is the address where the subroutine begins. In this example, SUBRTN would be the label of the first instruction of the subroutine. The assembler and the linker resolve what the actual value is. Calling a subroutine has the following effects:

- Increments the program counter (PC) by one and pushes its contents on the top of the stack (TOS). The TOS now contains the address of the instruction to be executed after returning from the subroutine.
- Loads the address SUBRTN on the PC.
- Starts execution from where the PC is pointing at (i.e., from location SUBRTN).

At the end of the subroutine execution, a return instruction (RET) will pop the contents of the top of the stack on the program counter, and the program will continue execution from that location.

The stack is a set of memory locations where you can store data, such as the contents of the PC. The difference from regular memory is that the stack keeps track of the location where the most recent data was stored. This location is the TOS. The stack is implemented either in hardware or software.

The TMS320C25 has a hardware stack that is eight locations deep. When a piece of data is put ("pushed") on the stack, everything already there is moved down by one location. Notice that the contents of the last location (bottom of the stack) are lost. Conversely, when a piece of data is retrieved from the stack (it is "popped"), all the other locations are moved up by one location. Pushing and popping always occur at the top of the stack.

The interrupt is a special case of subroutine. The TMS320C25 supports interrupts generated either internally or from external hardware. An interrupt causes a redirection of the program execution in order to accomplish a task. For instance, data may be present at an input port, and the interrupt forces the processor to go and "service" this port (inputting the data). As another example, an external D/A converter may need a sample from the processor, and it uses an interrupt to indicate to the DSP device that it is ready to receive the data. As a result, when the processor is interrupted, it "knows" by the nature of the interrupt that it has to go and do a specific task, and it does just that.

The performance of the designated task is done by the interrupt service routine (ISR). An ISR is like a subroutine with the only difference on the way it is accessed, and in the functions performed upon return. When an interrupt occurs, the program execution is automatically redirected to specific memory locations, associated with each interrupt. As explained earlier, the TMS320C25 continues execution from a specified memory location which, typically, contains a branch instruction to the actual location of the interrupt service routine.

The return from the interrupt service routine, like in a subroutine, pops the top of the stack to the program counter. However, it has the additional effect of re-enabling the interrupts. This is necessary because when an interrupt is serviced, the first thing that happens is that all interrupts are disabled to avoid confusion from additional interrupts. Re-enabling is done explicitly in the TMS320C25 (by using the EINT command).

77.9   Introduction to the TMS320C30 Digital Signal Processor

The Texas Instruments TMS320C30 is a floating-point processor that has some commonalities with the TMS320C25, but that also has a lot of differences. The differences are due more to the fact that the

TMS320C30 is a newer processor than that it is a floating-point processor. The TMS320C30 is a fast, 32-bit, digital signal processor that can handle both fixed-point and floating-point operations. The speed of the device is 16.7 MHz, which corresponds to a cycle time of 60 ns. Since the majority of the instructions execute in a single cycle (after the pipeline is filled), the figure of 60 ns also indicates how long it takes to execute one instruction. Alternatively, we can say that the device can execute 16.7 MIPS. Another figure of merit is based on the fact that the device can perform a floating-point multiplication and addition in a single cycle. Then, it is said that the device has a (maximum) throughput of 33 million floating-point operations per second (MFLOPS).

The actual signal from the external oscillator or crystal has a frequency twice that of the internal device speed, at 33.3 MHz (and period of 30 ns). This frequency is then divided on-chip to generate the internal clock with a period of 60 ns. Newer versions of the TMS320C30 and other members of the 'C3x generation operate in higher frequencies.

Figure 77.10 shows in a simplified form the key features of the TMS320C30. The major parts of the DSP processor are the memory, the CPU, the peripherals, and the direct memory access (DMA) unit. Each of these parts will be examined in more detail later in this article. The on-chip memory consists of 2K words of RAM and 4K words of ROM. There is also a 64-word long program cache. Each word is 32-bits wide and the memory sizes for the TMS320C30 are measured in 32-bit words, and not in bytes. The memory (RAM or ROM) can be used to store either program instructions or data. This presents a departure from the practice of separating the two spaces that the TMS320C25 uses, combining features of a von Neuman architecture with a Harvard architecture. Overall, the device can address 16 M words of memory through two external buses. Except for what resides on-chip, the rest of the memory is external, supplied by the designer.

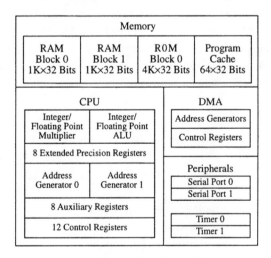

FIGURE 77.10   Key architectural features to the TMS320C30.

The CPU is the heart of the processor. It has a hardware multiplier that is capable of performing a multiplication in a single cycle. The multiplication can be between two 32-bit floating point numbers, or between two integers. To achieve a higher intermediate accuracy of results, the product of two floating-point numbers is saved as a 40-bit result. In integer multiplication, two 24-bit numbers are multiplied together to give a 32-bit result. The other important part of the CPU is the arithmetic logic unit (ALU) that performs additions, subtractions, and logical operations. Again, for increased intermediate accuracy, the ALU can operate on 40-bit long floating-point numbers and generates results that are also 40-bit long.

The 'C30 can handle both integers and floating-point numbers using corresponding instructions. There are three kinds of floating-point numbers, as shown in Fig. 77.11: short, single-precision, and extended-precision. In all three kinds, the number consists of an exponent e, a sign s and a mantissa f. Both the mantissa (part of which is the sign) and the exponent are expressed in two's-complement notation.

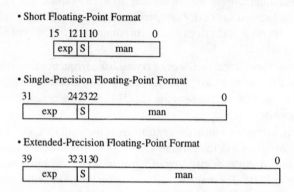

FIGURE 77.11   TMS320C30 floating point formats.

In the short floating-point format, the mantissa consists of 12 bits and the exponent of 4 bits. The short format is used only in immediate operands, where the actual number to operate upon becomes part of the instruction. The single-precision format is the regular format representing the numbers in the TMS320C30, which is a 32-bit device. It has 24 bits for mantissa and 8 bits for exponent. Finally, the extended-precision format is encountered only in the extended-precision registers, to be discussed below. In this case, the exponents is also 8-bits long, but the mantissa is 32 bits, giving extra precision. The mantissa is normalized so that it has a magnitude $|f|$ such that $1.0 =< |f| < 2.0$.

The integer formats supported in the TMS320C30 are shown in Fig. 77.12. Both the short and the single-precision integer formats represent the numbers in two's complement notation. The short format is used in immediate operands, where the actual number to be operated upon is part of the instruction itself.

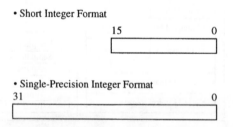

FIGURE 77.12   TMS320C30 integer (fixed-point) formats.

All the arithmetic and logical functions are register-based. In other words, the destination and at least one source operand in every instruction are register file associated with the TMS320C30 CPU. Figure 77.13 shows the components of the register file. There are eight extended-precision registers, R0-R7, that can be used as general purpose accumulators for both integer and floating-point arithmetic. These registers are 40-bits wide. When they are used in floating-point operations, the top 8 bits are the exponent and the bottom 32 bits are the mantissa of the number. When they are used as integers, the bottom 32 bits are the integer, while the top 8 bits are ignored and are left intact.

The eight auxiliary registers, AR0-AR7, are designated to be used as memory pointers or loop counters.

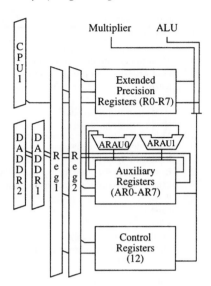

FIGURE 77.13 TMS320C30 register file.

When treated as memory pointers, they are used during the indirect addressing mode, to be examined below. AR0-AR7 can also be used as general-purpose registers but only for integer arithmetic.

Additionally, there are 12 control registers designated for specific purposes. These registers too can be treated as general purpose registers for integer arithmetic if they are not used for their designated purpose. Examples of such control registers are the status register, the stack pointer, the block repeat registers, and the index registers.

To communicate with the external world, the TMS320C30 has two parallel buses, the primary bus and the expansion bus. It also has two serial ports that can serve the same purpose. The serial ports are part of the peripherals that have been implemented on chip. Other peripherals include the direct memory access (DMA) unit, and two timers. These components of the TMS320C30 are examined in more detail in the following.

The device has 181 pins that are designated to perform certain functions, and to communicate with other devices on the same board. The names of the signals and the corresponding definitions appear in Table 77.3. The first column of the table gives the pin names; the second one indicates if the pin is used for input or output; the third column gives a description of the pin functionality. Note that a bar over the name indicates that the pin is in the active position when it is electrically low. The second column indicates if the pin is used for input to the device, output from the device, or both.

The TMS320C30 has its own assembly language consisting of 114 instructions that perform general-purpose and DSP-specific functions. High-level-language compilers have also been developed that make the writing of programs an easier task. The TMS320C30 was designed with a high-level language compiler in mind, and its architecture incorporates some appropriate features. For instance, the presence of the software stack, the register file, and the large memory space were to a large extent motivated by compiler considerations.

The TMS320C30 combines the features of the Harvard and the von Neuman architectures to offer more flexibility. The memory is a unified space where the designer can select the places for loading program instructions or data. This von Neuman feature maximizes the efficient use of the memory. On the other hand, there are multiple buses to access the memory in a Harvard style, as shown in Fig. 77.14. Two of the buses are used for the program, to carry the instruction address and fetch the instruction. Three buses are associated with data: two of those carry data addresses, so that two memory accesses can be done in the same machine cycle. The third bus carries the data. The reason that one bus is sufficient to carry the data is that the device needs only one-half of a machine cycle to fetch an operand from the internal memory.

TABLE 77.3   Names and Functionality of the 181 Pins of the TMS320C30

Signal	I/O	Description
D(31-0)	I/O	32-bit data port of the primary bus
A(23-0)	O	24-bit address port of the primary bus
R/$\overline{W}$	O	Read/write signal for primary bus interface
$\overline{STRB}$	O	External access strobe for the primary bus
$\overline{RDY}$	I	Ready signal
$\overline{HOLD}$	I	Hold signal for primary bus
$\overline{HOLDA}$	O	Hold acknowledge signal for primary bus
XD(31-0)	I/O	32-bit data port of the expansion bus
XA(12-0)	O	13-bit address port of the expansion bus
XR/$\overline{W}$	O	Read/write signal for expansion bus interface
$\overline{MSTRB}$	O	External access strobe for the expansion bus
$\overline{IOSTRB}$	O	External access strobe for the expansion bus
$\overline{XRDY}$	I	Ready signal
$\overline{RESET}$	I	Reset
$\overline{INT}$(3-0)	I	External interrupts
$\overline{IACK}$	O	Interrupt acknowledge signal
MC/$\overline{MP}$	I	Microcomputer/microprocessor mode pin
XF(1-0)	I/O	External flag pins
CLKX(1-0)	I/O	Serial port (1-0) transmit clock
DX(1-0)	O	Data transmit output for port (1-0)
FSX(1-0)	I/O	Frame synchronization pulse for transmit
CLKR(1-0)	I/O	Serial port (1-0) receive clock
DR(1-0)	I	Data receive for serial port (1-0)
FSR(1-0)	I	Frame synchronization pulse for receive
TCLK(1-O)	I/O	Timer (1-0) clock
V_{DD}, etc.	I	12 + 5 V supply pins
V_{SS}, etc.	I	11 ground pins
X1	O	Output pin from internal oscillator for the crystal
X2/CLKIN	I	Input pin to the internal oscillator from the crystal
H1, H3	O	External H1, H3 clock. H1 = H3 = 2 CLKIN
EMU, etc.	I/O	20 reserved and miscellaneous pins

As a result, two data fetches can be accomplished in one cycle over the same bus.

The last two buses are associated with the DMA unit, which transfers data in parallel with and transparently to the CPU. Because of the multiple buses, program instructions and data operands can be moved simultaneously increasing the throughput of the device. Of course, it is conceivable that too many accesses can be attempted to the same memory area, causing access conflicts. However, the TMS320C30 has been designed to resolve such conflicts automatically by inserting the appropriate delays in instruction execution. Hence, the operations always give the correct results.

Figure 77.15 shows a functional block diagram of the TMS320C30 architecture with the buses, the CPU, and the register file. It also points out the peripheral bus with the associated peripherals. Because of the peripheral bus, all the peripherals are memory-mapped, and any operations with them are seen by the programmer as accesses (reads/writes) to the memory.

77.10   TMS320C30 Memory Organization and Access

The TMS320C30 has on-chip 2K words (32-bits wide) of RAM and 4K of ROM. This memory can be accessed twice in a single cycle, a fact that is reflected in the instruction set, which includes three-operand instructions: two of the operands reside in memory, while the third operand is the register where the result is placed.

Besides the on-chip memory, the TMS320C30 can access external memory through two external buses, the primary and the expansion. The primary bus consists of 24 address pins A0-A23, and 32 data pins D0-D31. As the number of address pins suggests, the maximum memory space available is 16M words. Not all of that, though, resides on the primary bus. The primary bus has 16M words minus the on-chip memory, and minus the memory available on the expansion bus.

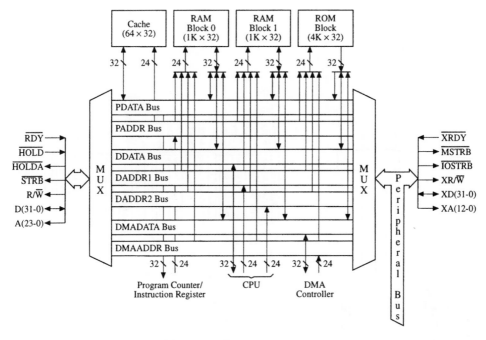

FIGURE 77.14   Internal bus structure of the TMS320C30.

The expansion bus has 13 address pins, XA0-XA12, and 32 data pins, XD0-XD31. The 13 address pins can address 8K words of memory. However, there are two strobes, MSTRB and IOSTRB, that select two different segments of 8K of memory. In other words, the total memory available on the expansion bus is 16K. The differences between the two strobes is in timing. The timing differences can make one of the memory spaces more preferable to the other in certain applications, such as peripheral devices.

As mentioned earlier, the destination operand is always a register in the register file (except for storing a result, where, of course, the destination is a memory location.) The register can also be one of the source operands. It is possible to specify a source operand explicitly and include it in the instruction. This addressing mode is called immediate addressing mode. The immediate constant should be accommodated by a 16-bit wide word, as discussed earlier in the data formats.

For example, if it is desired to increment the (integer) contents of the register R0 by 5, the following instruction can be used:

ADDI   5,R0

To increment the (floating-point) contents of the register R3 by -2.75, you can use the instruction

ADDF   -2.75,R3

If the value to be operated upon resides in memory, there are two ways to access it: either by specifying the memory address directly (direct addressing) or by using an auxiliary register holding that address and, hence, pointing to that number indirectly (indirect addressing). In the direct addressing mode, full description of a memory address would require a 24-bit word because the memory space is 16M words. To reduce that requirement, the memory space is divided in 256 pages of 64K words each. An instruction using direct addressing contains the 16 bits indicating what word you want to access within a page. The page number (8 bits) is stored in one of the control registers, the data page (DP) pointer. The DP pointer can be modified by using either a load instruction or the pseudo-instruction LDP. During assembly time, LDP picks the top 8 bits of a memory address and places them in the DP register.

Of course, if several locations need to be accessed in the same page, you can set the DP pointer only once. Since the majority of the routines written are expected to be less than 64K words long, setting the

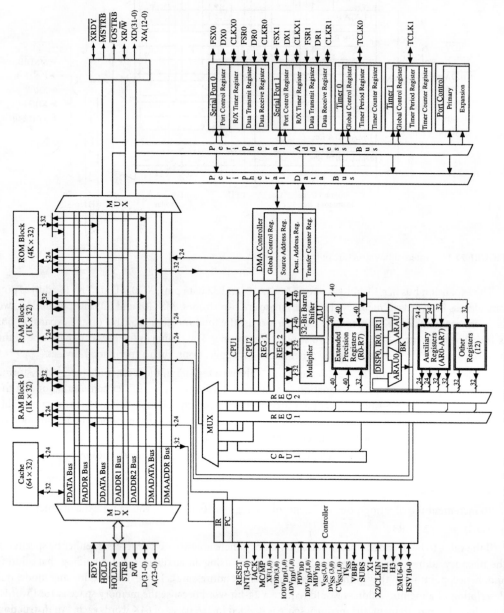

FIGURE 77.15 Functional block diagram of the TMS320C30 architecture.

DP register at the beginning of the program suffices. The exception to that would be placing the code over the boundary of two consecutive pages.

In the indirect addressing mode, the data memory address is held in a register that acts as a memory pointer. There are eight such registers available, AR0-AR7. These registers can also be used for other functions, such as loop counters or general purpose registers. If they are used as memory pointers, they are explicitly specified in the instruction. In an instruction, indirect addressing is indicated by an asterisk preceding the auxiliary register.

For example, the instruction

LDF　　　* AR3++,R0　　　; Load R0 with -612

loads R0 with the contents of the memory location pointed at by AR3.

The "++" sign in the above instruction means "after the present memory access, increment the contents of the current auxiliary register by 1". This is done in parallel with the load-register operation.

The above autoincrementing of the auxiliary register is an optional operation that offers additional flexibility to the programmer, and it is not the only one available. The TMS320C30 has two auxiliary register arithmetic units (ARAU0 and ARAU1) that can execute such operations in parallel with the CPU, and increase the throughput of the device in this way. The primary function of ARAU0 and ARAU1 is to generate the addresses for accessing operands.

Table 77.4 summarizes the different operations that can be done while using indirect addressing. As seen from this table, the contents of an auxiliary register can be incremented or decremented before or after accessing the memory location. In the case of pre-modification, this modification can be permanent or temporary. When an auxiliary register ARn, n = 0-7, is modified, the displacement disp is either a constant (0-255) or the contents of one of the two index registers IR0, IR1 in the register file. If the displacement is missing, a 1 is implied. The auxiliary register contents can be incremented or decremented in a circular fashion, or incremented by the contents of IR0 in a bit-reversed fashion.

TABLE 77.4　Operations That Can Be Performed in Parallel with Indirect Addressing in the TMS320C30

Notation	Operation	Description
*ARn	addr = ARn	Indirect without modification
*+ARn(disp)	addr = ARn + disp	With predisplacement add
*−ARn(disp)	addr = ARn − disp	With predisplacement subtract
*++ARn(disp)	addr = ARn + disp ARn = ARn + disp	With predisplacement add and modify
*−−ARn(disp)	addr = ARn − disp ARn = ARn − disp	With predisplacement subtract and modify
*ARn++(disp)	addr = ARn ARn = ARn + disp	With postdisplacement add and modify
*ARn−−(disp)	addr = ARn ARn = ARn − disp	With postdisplacement subtract and modify
*ARn++(disp)%	addr = ARn ARn = circ(ARn + disp)	With postdisplacement add and circular modify
*ARn−−(disp)%	addr = ARn ARn = circ(ARn − disp)	With postdisplacement subtract and circular modify
*ARn++(IR0)B	addr = ARn ARn = rc(ARn + IR0)	With postdisplacement add and bit-reversed modify

Note: circ = circular modification, B = bit reversed, rc = reverse carry.

The last two kinds of operation have special purposes. Circular addressing is used to create a circular buffer, and it is helpful in filtering applications. Bit-reversed addressing is useful when doing Fast Fourier Transforms. The bit-reversed addressing is implemented by adding IR0 with reverse carry propagation, an operation explained in the TMS320C30 User's Guide.

The TMS320C30 has a software stack that is part of its memory. The software stack is implemented by having one of the control registers, the SP, point to the next available memory location. Whenever a subroutine call occurs, the address to return to after the subroutine completion is pushed on the stack (i.e., it is written on the memory location that SP is pointing at), and SP is incremented by one. Upon return from a subroutine, the SP is decremented by one and the value in that memory location is copied on the program counter.

Since the SP is a regular register, it can be read or written to. As a result, you can specify what part of the memory is used for the stack by initializing SP to the appropriate address. There are specific instructions to push on or pop from the stack any of the registers in the register file: PUSH, POP for integer values, PUSHF, POPF for floating-point numbers. Such instructions can use the stack to pass arguments to subroutines or to save information during an interrupt. In other words, the stack is a convenient scratch-pad that you designate at the beginning, so that you do not have to worry where to store some temporary values.

77.11 Multiplier and ALU of the TMS320C30

The heart of the TMS320C30 is the CPU consisting, primarily, of the multiplier and the ALU. The CPU configuration is shown in Fig. 77.16 which depicts the multiplier and the ALU of the TMS320C30. The hardware multiplier can perform both integer and floating-point multiplications in a single machine cycle.

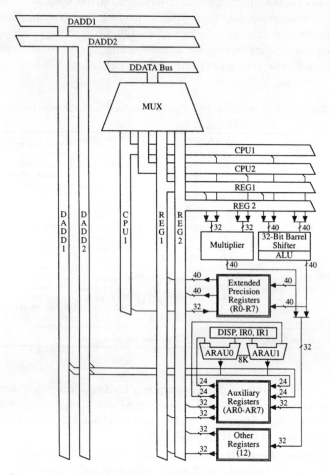

FIGURE 77.16 Central processing unit (CPU) of the TMS320C30.

The inputs to the multiplier come from either the memory or the registers of the register file. The outputs are placed in the register file. When multiplying floating-point numbers, the inputs are 32-bits long (8 bits exponent and 24 bits mantissa), and the result is 40-bits wide directed to one of the extended precision registers. If the input is longer than 32 bits (extended precision) or shorter than 32 bits (short format) it is truncated or extended, respectively, by the device to become a 32-bit number before the operation. Multiplication of integers consists of multiplying two 24-bit numbers to generate a 32-bit result. In this case, the registers used can be any of the registers in the register file.

The other major part of the CPU is the ALU. The ALU can also take inputs from either the memory or the register file and perform arithmetic or logical operations. Operations on floating-point numbers can be done on 40-bit wide inputs (8 bits exponent and 32 bits mantissa) to give also 40-bit results. Integer operations are done on 32-bit numbers. Associated with the ALU, there is a barrel shifter that can perform either a right-shift or a left-shift of a register's contents for any number of locations in a single cycle. The instructions for shifting are ASH (Arithmetic SHift) and LSH (Logical SHift).

77.12 Other Architectural Features of the TMS320C30

The TMS320C30 has many interesting features and capabilities. For a full account, the reader is urged to look them up in the User's Guide [2]. Here, we briefly present only the most important of them so that you have a global view of the device and its salient characteristics.

The TMS320C30 is a very fast device, and it can execute very efficiently instructions from the on-chip memory. Often, though, it is necessary to use external memory for program storage. The existing memory devices either are not as fast as needed, or are quite expensive. To ameliorate this problem, the TMS320C30 has 64 words of program cache on-chip. When executing a program from external memory, every instruction is stored on the cache as it is brought in. Then, if the same instruction needs to be executed again (as is the case for instructions in a loop), it is not fetched from the external memory but from the cache. This approach speeds up the execution, but it also frees the external bus to fetch, for instance, operands. Obviously, the cache is most effective for loops that are shorter than 64 words long, something usual in DSP applications. On the other hand, it does not offer any advantages in the case of straight-line code. However, the structure of DSP problems suggests that the cache is a feature that can be put to good use.

In the instruction set of the 'C30 there is the RPTS (RePeaT Single) instruction

RPTS N

that repeats the following instruction N+1 times. A more generalized repeated mode is implemented by the RPTB (RePeaT Block) instruction that repeats a number of times all the instructions between RPTB and a label that is specified in the block-repeat instruction. The number of repetitions is one more than the number stored in the repeat count register, RC, one of the control registers in the register file.

For example the following instructions are repeated one time more than the number included in the RC.

```
        LDI     63,RC           ; The loop is to be repeated 64 times
                RPTB    LOOP    ; Repeat up to the label LOOP
                LDI     *AR0,R0 ; Load the number on R0
                ADDI    1,R0    ; Increment it by 1
        LOOP    STI     R0,*AR0++ ; Store the result; point to the next
                                ; number; and loop back
```

Besides RC, there are two more control registers used with the block repeat instruction. The repeat-start (RS) contains the beginning of the loop, and the repeat-end (RE) the end of the loop. These registers are initialized automatically by the processor, but they are available to the user in case he needs to save them.

On the TMS320C30, there are several internal and external interrupts, which are prioritized, i.e., when several of the interrupts occur at the same time, the one with the highest priority is executed first.

Besides the reset signal, there are 4 external interrupts, INT0-INT3. Internally, there are the receive and transmit interrupts of the serial ports, and the timer interrupts. There is also an interrupt associated with the DMA. Typically, the memory location where the execution is directed to during an interrupt contains the address where an interrupt service routine starts. The interrupt service routine will perform the tasks for which the interrupt has been designed, and then return to the execution of the original program. All the interrupts (except the reset) are maskable, i.e., they can be ignored by setting the interrupt enable (IE) register to appropriate values. Masking of interrupts, as well as the memory locations where the interrupt addresses are stored, are discussed in the TMS320C30 User's Guide [2].

Each of the two serial ports provides direct communication with serial devices, such as codes, serial analog-to-digital converters, etc. In these devices, the data are transmitted serially, one bit at a time, and not in parallel, which would require several parallel lines. The serial ports have the flexibility to consider the incoming stream as 8-, 16-, 24-, or 32-bit words. Since they are memory-mapped, the programmer goes to certain memory locations to read in or write out the data.

Each of the two timers consists of a period register and a timer register. At the beginning of the operation, the contents of the timer register are incremented at every machine cycle. When the value of the timer register becomes equal to the one in the period register, it generates a timer interrupt, the period register is zeroed out, and the whole operation is repeated.

A very interesting addition to the TMS320C30 architecture is the DMA unit. The DMA can transfer data between memory locations in parallel with the CPU execution. In this way, blocks of data can be transferred transparently, leaving the CPU free to perform computational tasks, and thus increasing the device throughput.

The DMA is controlled by a set of registers, all of which are memory mapped: you can modify these registers by writing to certain memory locations. One register is the source address from where the data is coming. The destination address is where the data is going. The transfer count register specifies how many transfers will take place. A control register determines if the source and the destination addresses are to be incremented, decremented, or left intact after every access. The programmer has several options of synchronizing the DMA data transfers with interrupts or leaving them asynchronous.

77.13   TMS320C30 Instruction Set

The TMS320C30 has an instruction set consisting of 114 instructions. Some of these instructions perform general purpose operations, while others are more specific to DSP applications. The instruction set of the TMS320C30 presents an interesting symmetry that makes programming very easy. Instructions that can be used with integer operands are distinguished from the same instructions for floating-point numbers with the suffix "I" vs. "F". Instructions that take three operands are distinguished from the ones with two operands by using the suffix "3". However, since the assembler permits elimination of the symbol "3", the notation becomes even simpler.

A whole new class of TMS320C30 instructions (as compared to the TMS320C25) are the parallel instructions. Any multiplier or ALU operation can be performed in parallel with a store instruction. Additionally, two stores, two loads, or a multiply and an add/subtract can be performed in parallel. Parallel instructions are indicated by placing two vertical lines in front of the second instruction.

For example, the following instruction adds the contents of *AR3 to R2 and puts the result in R5. At the same time, it stores the previous contents of the R5 into the location *AR0.

```
      ADDF    *AR3++,R2,R5
   ‖  STF     R5,*AR0--
```

Note that the parallel instructions are not really two instructions but one, which is also different from its two components. However, the syntax used helps remembering the instruction mnemonics. One of the most important parallel instructions for DSP applications is the parallel execution of a multiplication with an addition or subtraction. This single-cycle multiply-accumulate is very important in the computation of dot products appearing in vector arithmetic, matrix multiplication, digital filtering, etc.

For example, assume that we want to take the dot product of two vectors having 15 points each. Assume that AR0 points to one vector and AR1 to the other. The dot product can be computed with the following code:

```
    LDF     0.0,R2                      ; Initialize R2=0.0
            LDF 0.0,R0                  ; Initialize R0=0.0
    RPTS 14                             ; Repeat loop (single instruction)
    MPYF *AR0++, *AR1++, R0            ; Multiply two points, and
    ||      ADDF R0,R2                  ; Accumulate previous product
            ADDF R0,R2                  ; Accumulate last product
```

After the operation is completed, R2 holds the dot product.

Before proceeding with the instructions, it is important to understand the working of the device pipeline. At every instant in time, there are 4 execution units operating in parallel in the TMS320C30: the fetch, decode, read, and execute unit, in order of increasing priority. The fetch unit fetches the instruction; the decode unit decodes the instruction and generates the addresses; the read unit reads the operands from the memory or the registers; and the execute unit performs the operation specified in the instruction. Each one of these units takes one cycle to complete. So, an instruction in isolation takes, actually, four cycles to complete. Of course, you never run a single instruction alone.

In the pipeline configuration, as shown in Fig. 77.17, when an instruction is fetched, the previous instruction is decoded. At the same time, the operands of the instruction before that are read, while the third instruction before the present one is executed. So, after the pipeline is full, each instruction takes a single cycle to execute.

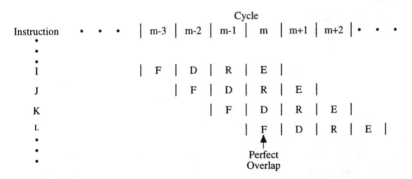

FIGURE 77.17 Pipeline structure of the TMS320C30.

Is it true that all the instructions take a single cycle to execute? No. There are some instructions, like the subroutine calls and the repeat instructions, that need to flush the pipeline before proceeding. The regular branch instructions also need to flush the pipeline. All the other instructions, though, should take one cycle to execute, if there are no pipeline conflicts.

There are a few reasons that can cause pipeline conflicts, and if the programmer is aware of where the conflicts occur, he can take steps to reorganize his code and eliminate them. In this way, the device throughput is maximized. The pipeline conflicts are examined in detail in the User's Guide [2].

The load and store instructions can load a word into a register, store the contents of a register to memory, or manipulate data on the system stack. Note that the instructions with the same functionality that operate on integers or floating-point numbers are presented together in the following selective listing.

LDF, LDI	Load a floating-point or integer value
LDFcond, LDIcond	Load conditionally
POPF, POP	Pop value from stack
PUSHF, PUSH	Push value on stack
STF, STI	Store value to memory

The conditional loads perform the indicated load only if the condition tested is true. The condition tested is, typically, the sign of the last performed operation.

The arithmetic instructions include both multiplier and ALU operations.

ABSF, ABSI	Absolute value
ADDF, ADDI	Add
CMPF, CMPI	Compare values
FIX, FLOAT	Convert between fixed- and floating-point
MPYF, MPYI	Multiply
NEGF, NEGI	Negate
SUBF, SUBI	Subtract
SUBRF,SUBRI	Reverse subtract

The difference between the subtract and the reverse subtract instructions is that the first one subtracts the first operand from the second, while the second one subtracts the second operand from the first.

The logical instructions always operate on integer (or unsigned) operands.

AND	Bitwise logical AND
ANDN	Bitwise logical AND with complement
LSH	Logical shift
NOT	Bitwise logical complement
OR	Bitwise logical OR
XOR	Bitwise exclusive OR

The logical shift differs from an arithmetic shift (which is part of the arithmetic instructions) in that, on a right shift, the logical shift fills the bits to the left with zeros. The arithmetic shift sign-extends the (integer) number.

The program control instructions include the branch instructions (corresponding to GOTO of a high-level languages), and the subroutine call and return instructions.

Bcond[D]	Branch conditionally [with delay]
CALL, CALLcond	Call or call conditionally a subroutine
RETIcond, RETScond	Return from interrupt or subroutine conditionally
RPTB, RPTS	Repeat block or repeat a single instruction

The branch instructions can have an optional "D" at the end to convert them into delayed branches. The delayed branch does the same operation as a regular branch but it takes fewer cycles. A regular branch needs to flush the pipeline before proceeding with the next instruction because it is not known in advance if the branch will be taken or not. As a result, a regular branch costs four machine cycles. If, however, there are three instructions that can be executed no matter if the branch is taken or not, a delayed branch can be used. In a delayed branch, the three instructions following the branch instruction are executed before the branch takes effect. This reduces the effective cost of the delayed branch to one cycle.

77.14   Other Generations and Devices in the TMS320 Family

So far, the discussion in this article has focused on two specific devices of the TMS320 family in order to examine in detail their features. However, the TMS320 family consists of five generations (three fixed-point and two floating-point) of digital signal processors (as well as the latest addition, the TMS320C8x generation, also known as MVP, Multimedia Video Processors). The fixed-point devices are members of the TMS320C1x, TMS320C2x, or TMS320C5x generation, and the floating-point devices belong to the TMS320C3x or TMS320C4x generation.

The TMS320C5x generation is the highest-performance generation of the TI 16-bit fixed-point digital signal processors. The 'C5x performance level is achieved through a faster cycle time, larger on-chip memory space, and systematic integration of more signal-processing functions. As an example, the TMS320C50 (Fig. 77.18) features large on-chip RAM blocks. It is source-code upward-compatible with the first- and second-generation TMS320 devices.

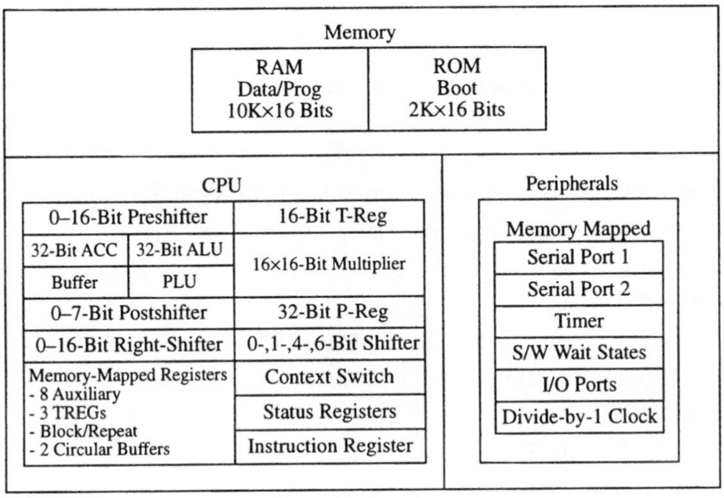

Memory		
RAM Data/Prog 10K×16 Bits	**ROM** Boot 2K×16 Bits	

CPU		Peripherals
0–16-Bit Preshifter	16-Bit T-Reg	Memory Mapped
32-Bit ACC / 32-Bit ALU	16×16-Bit Multiplier	Serial Port 1
Buffer / PLU		Serial Port 2
0–7-Bit Postshifter	32-Bit P-Reg	Timer
0–16-Bit Right-Shifter	0-,1-,4-,6-Bit Shifter	S/W Wait States
Memory-Mapped Registers - 8 Auxiliary	Context Switch	I/O Ports
- 3 TREGs	Status Registers	Divide-by-1 Clock
- Block/Repeat - 2 Circular Buffers	Instruction Register	

FIGURE 77.18 TMS320C50 Block diagram.

Some of the key features of the TMS320C5x generation are listed below. Specific devices that have a particular feature are enclosed in parentheses.

- CPU

  - 25-, 35-, 50-ns single-cycle instruction execution time
  - Single-cycle multiply/accumulate for program code
  - Single-cycle/single-word repeats and block repeats for program code
  - Block memory moves
  - Four-deep pipeline
  - Indexed-addressing mode
  - Bit-reversed/indexed-addressing mode to facilitate FFTs
  - Power-down modes
  - 32-bit ALU, 32-bit accumulator, and 32-bit accumulator buffer
  - Eight auxiliary registers with a dedicated arithmetic unit for indirect addressing
  - 16-bit parallel logic unit (PLU)
  - 16×16-bit parallel multiplier with a 32-bit product capacity
  - 0- to 16-bit right and left barrel-shifters
  - 64-bit incremental data shifter
  - Two indirectly addressed circular data buffers for circular addressing

- Peripherals

  - Eight-level hardware stack

- 11 context-switch registers to shadow the contents of strategic CPU-controlled registers during interrupts
- Full-duplex, synchronous serial port, which directly interfaces to codec
- Time-division multiplexed (TDM) serial port (TMS320C50/C51/C53)
- Interval timer with period and control registers for software stops, starts, and resets
- Concurrent external DMA performance, using extended holds
- On-chip clock generator
- Divide-by-one clock generator (TMS320C50/C51/C53)
- Multiply-by-two clock generator (TMS320C52)

- Memory

  - 10K × 16-bit single cycle on-chip program/data RAM (TMS320C50)
  - 2K × 16-bit single cycle on-chip program/data RAM (TMS320C51)
  - 1K × 16 RAM (TMS320C52)
  - 4K × 16 RAM (TMS320C53)
  - 2K × 16-bit single cycle on-chip boot ROM (TMS320C50)
  - 8K × 16-bit single cycle on-chip boot ROM (TMS320C51)
  - 4K × 16 ROM (TMS320C52)
  - 16K × 16 ROM (TMS320C53)
  - 1056X16-bit dual-access on-chip data/program RAM

- Memory interfaces

  - 16 programmable software wait-state generators for program, data, and I/O memories
  - 224K-word × 16-bit maximum addressable external memory space

Table 77.5 shows the overall TMS320 family. It provides a tabulated overview of each member's memory capacity, number of I/O ports (by type), cycle time, package type, technology, and availability.

Many features are common among these TMS320 processors. When the term TMS320 is used, it refers to all five generations of DSP devices. When referring to a specific member of the TMS320 family (e.g., TMS320C15), the name also implies enhanced-speed in MHz (-14, -25, etc.), erasable/programmable (TMS320E15), low-power (TMS320LC15), and one-time programmable (TMS320P15) versions. Specific features are added to each processor to provide different cost/performance alternatives.

TABLE 77.5 TMS320 Family Overview

Data type	Device	Memory (words) On-chip RAM	ROM	EPROM	Off-chip Dat/Pro	I/O[a] Ser	Par	DMA	Com	On-chip timer	Cycle time (ns)	Package
	TMS320C10[b]	144	1.5K	—	—/4K	—	8 × 16	—	—	—	200	DIP/PLCC
	TMS320C10-14	144	1.5K	—	—/4K	—	8 × 16	—	—	—	280	DIP/PLCC
	TMS320C10-25[b]	144	1.5K	—	—/4K	—	8 × 16	—	—	—	160	DIP/PLCC
	TMS320C14	256	4K	—	—/4K	1	8 × 16	—	—	4	160	PLCC
	TMS320E14[b]	256	—	4K	—/4K	1	7 × 16	—	—	4	160	CERQUAD
	TMS320E14-25[b]	256	—	4K	—/4K	1	7 × 16	—	—	4	167	CERQUAD
	TMS320P14	256	—	4K	—/4K	1	7 × 16	—	—	4	160	PLCC
	TMS320C15[b]	256	4K	—	—/4K	—	8 × 16	—	—	—	200	DIP/PLCC/PQFP
Fixed-point (16-bit word size)	TMS320C15-25[b]	256	4K	—	—/4K	—	8 × 16	—	—	—	160	DIP/PLCC
	TMS320E15[b]	256	—	4K	—/4K	—	8 × 16	—	—	—	200	DIP/CERQUAD
	TMS320E15-25	256	—	4K	—/4K	—	8 × 16	—	—	—	160	DIP/CERQUAD
	TMS320LC15	256	4K	—	—/4K	—	8 × 16	—	—	—	200	DIP/PLCC
	TMS320P15	256	—	4K	—/4K	—	8 × 16	—	—	—	200	DIP/PLCC
	TMS320P15-25	256	—	4K	—/4K	—	8 × 16	—	—	—	160	DIP/PLCC
	TMS320C16	256	8K	—	—/64K	—	8 × 16	—	—	—	114	PQFP
	TMS320LC16	256	8K	—	—/64K	—	8 × 16	—	—	—	250	PQFP
	TMS320C17	256	4K	—	—/—	2	6 × 16	—	—	1	200/160	DIP/PLCC
	TMS320E17	256	—	4K	—/—	2	6 × 16	—	—	1	200/160	DIP

[a] Ser = serial; Par = parallel; DMA = direct memory access (Int = internal; Ext = external); Com = parallel communication ports
[b] A military version is available/planned; contact the nearest TI field sales office for availability.

TABLE 77.5 TMS320 Family Overview (Continued)

Data type	Device	Memory (words)							I/O[a,c]				On-chip timer	Cycle time (ns)	Package type
		On-chip			Off-chip										
		RAM	ROM	EPROM	Dat / Pro		Ser	Par	DMA	Com					
	TMS320LC17	256	4K	—	—/—		2	6 × 16	—	—		1	200	DIP/PLCC	
	TMS320P17	256	—	4K	—/—		2	6 × 16	—	—		1	160	DIP	
	TMS320C25[b]	544	4K	—	64K/ 64K		1	16 × 16	Ext	—		1	100	PGA/PLCC/ PQFP	
Fixed-point (16-bit word size)	TMS320C25-33	544	4K	—	64K/ 64K		1	16 × 16	Ext	—		1	120	PLCC	
	TMS320C25-50[b]	544	4K	—	64K/ 64K		1	16 × 16	Ext	—		1	80	PGA/PLCC	
	TMS320E25	544	—	4K	64K/ 64K		1	16 × 16	Ext	—		1	100	PQFP/PLCC	
	TMS320C26[b]	1.5K	—	—	64K/ 64K		1	16 × 16	Ext	—		1	100	PLCC	
	TMS320C28	544	8K	—	64K/ 64K		1	16 × 16	Ext	—		1	100	PQFP/PLCC	
	TMS320C28-50	544	8K	—	64K/ 64K		1	16 × 16	Ext	—		1	80	PQFP/PLCC	

[a] Ser = serial; Par = parallel; DMA = direct memory access; Int = internal; Ext = external; Com = parallel communication ports.
[b] A military version is available/planned; contact the nearest TI field sales office for availability.
[c] Programmed transcoders (TMS320SS16 and TMS320SA32) are also available.

TABLE 77.5 TMS320 Family Overview (Continued)

Data type	Device	Memory (words) On-chip RAM	ROM	EPROM	Off-chip Dat/Pro	I/O[a,c] Ser	Par	DMA	Com	On-chip timer	Cycle time (ns)	Package
Fixed-point (16-bit word size)	TMS320C50[b]	10K	BL	—	64K/64K	2	$64K \times 16^d$	Ext	—	1	50/35/25/20[e]	PQFP
	TMS320C51	2K	8K	—	64K/64K	2	$64K \times 16^d$	Ext	—	1	50/35/25/20[e]	PQFP/TQFP
	TMS320BC51	2K	BL	—	64K/64K	2	$64K \times 16^d$	Ext	—	1	50/35/25/20[e]	PQFP/TQFP
	TMS320C52	1K	4K	—	64K/64K	1	$64K \times 16^d$	Ext	—	1	50/35/25/20[e]	PQFP/TQFP
	TMS320BC52	1K	BL	—	64K/64K	1	$64K \times 16^d$	Ext	—	1	50/35/25/20[e]	PQFP/TQFP
	TMS320C53	4K	16K	—	64K/64K	2	$64K \times 16^d$	Ext	—	1	50/35/25/20[e]	PQFP/TQFP
	TMS320BC53	4K	BL	—	64K/64K	2	$64K \times 16^d$	Ext	—	1	50/35/25/20[e]	PQFP/TQFP

[a] Ser = serial; Par = parallel; DMA = direct memory access concurrent with CPU operation; Int = internal; Ext = external; Com = parallel communication ports; BL = bootloader.
[b] A military version is available/planned; contact the nearest TI field sales office for availability.
[c] Programmed transcoders (TMS320SS16 and TMS320SA32) are also available.
[d] Sixteen of these parallel I/O ports are memory-mapped.
[e] Planned

TABLE 77.5 TMS320 Family Overview (Continued)

Data type	Device	Memory (words) On-chip RAM	ROM	EPROM	Off-chip Dat/Pro	I/O^{a,c} Ser	Par	DMA	Com	On-chip timer	Cycle time (ns)	Package type
	TMS320C30	2K	4K	—	16M^f	2	16M × 32^g	Int/Ext	—	2(6)^d	60	PGA and PQFP
	TMS320C30-50	2K	4K	—	16M^f	2	16M × 32^g	Int/Ext	—	2(6)^d	40	PGA and PQFP
	TMS320C30-27	2K	4K	—	16M^f	2	16M × 32^g	Int/Ext	—	2(6)^d	74	PGA and PQFP
	TMS320C30-40	2K	4K	—	16M^f	2	16M × 32^g	Int/Ext	—	2(6)^d	50	PGA and PQFP
Floating	TMS320C30-50	2K	4K	—	16M^f	2	16M × 32^g	Int/Ext	—	2(6)^d	40	PGA and PQFP
point	TMS320C31^b	2K	e	—	16M^f	1	16M × 32	Int/Ext	—	2(4)^d	60	PQFP
(32-bit	TMS320LC31	2K	e	—	16M^f	1	16M × 32	Int/Ext	—	2(4)^d	60	PQFP
word	TMS320C31-27	2K	e	—	16M^f	1	16M × 32	Int/Ext	—	2(4)^d	74	PQFP
size)	TMS320C31-40	2K	e	—	16M^f	1	16M × 32	Int/Ext	—	2(4)^d	50	PQFP
	TMS320C31-50	2K	e	—	16M^f	1	16M × 32	Int/Ext	—	2(4)^d	40	PQFP
	TMS320C40	2K	4K^e	—	4G^f	—	4G × 32^g	Int/Ext	6	2	40	PGA
	TMS320C40-40	2K	4K^e	—	4G^f	—	4G × 32^g	Int/Ext	6	2	50	PGA

^{a} Ser = serial; Par = parallel; DMA = direct memory access concurrent with CPU operation; Int = internal; Ext = external; Com = parallel communication ports.
^{b} A military version is available/planned; contact the nearest TI field sales office for availability.
^{c} Programmed transcoders (TMS320SS16 and TMS320SA32) are also available.
^{d} Includes the use of serial port timers.
^{e} Preprogrammed ROM bootloader.
^{f} Single logical memory space for program, data, and I/O; not including on-chip RAM, peripherals, and reserved spaces.
^{g} Dual buses.

References

[1] *TMS320C2x User's Guide,* Texas Instruments, Dallas, TX.
[2] *TMS320C3x User's Guide,* Texas Instruments, Dallas, TX.

78

Rapid Design and Prototyping of DSP Systems

T. Egolf, M. Pettigrew,
J. Debardelaben, R. Hezar,
S. Famorzadeh, A. Kavipurapu,
M. Khan, Lan-Rong Dung,
K. Balemarthy, N. Desai,
Yong-kyu Jung, and
V. Madisetti
Georgia Institute of Technology

The Rapid Prototyping of Application-Specific Signal Processors (RASSP) [1, 2, 3] program of the U.S. Department of Defense (ARPA and Tri-Services) targets a 4X improvement in the design, prototyping, manufacturing, and support processes (relative to current practice). Based on a current practice study (1993) [4], the prototyping time from system requirements definition to production and deployment, of multiboard signal processors, is between 37 and 73 months. Out of this time, 25 to 49 months are devoted to detailed hardware/software (HW/SW) design and integration (with 10 to 24 months devoted to the latter task of integration). With the utilization of a promising top-down hardware-less codesign methodology based on VHDL models of HW/SW components at multiple abstractions, reduction in design time has been shown especially in the area of hardware/software integration [5]. The authors describe a top-down design approach in VHDL starting with the capture of system requirements in an executable form and through successive stages of design refinement, ending with a detailed hardware design. This hardware/software codesign process is based on the RASSP program design methodology called virtual prototyping, wherein VHDL models are used throughout the design process to capture the necessary information to describe the design as it develops through successive refinement and review. Examples are presented to illustrate the information captured at each stage in the process. Links between stages are described to clarify the flow of information from requirements to hardware.

78.1   Introduction

We describe a RASSP-based design methodology for application specific signal processing systems which supports reengineering and upgrading of legacy systems using a virtual prototyping design process. The VHSIC Hardware Description Language (VHDL) [6] is used throughout the process for the following reasons. One, it is an IEEE standard with continual updates and improvements; two, it has the ability to describe systems and circuits at multiple abstraction levels; three, it is suitable for synthesis as well as simulation; and four, it is capable of documenting systems in an executable form throughout the design process.

A *Virtual Prototype* (VP) is defined as an executable requirement or specification of an embedded system and its stimuli describing it in operation at multiple levels of abstraction. *Virtual prototyping* is defined as the top-down design process of creating a virtual prototype for hardware and software cospecification, codesign, cosimulation, and coverification of the embedded system. The proposed top-down design process stages and corresponding VHDL model abstractions are shown in Fig. 78.1. Each stage in the process serves as a starting point for subsequent stages. The testbench developed for requirements capture is used for design verification throughout the process. More refined subsystem, board, and component level testbenches are also developed in-cycle for verification of these elements of the system.

The process begins with requirements definition which includes a description of the general algorithms to be implemented by the system. An algorithm is here defined as a system's signal processing transformations required to meet the requirements of the high level paper specification. The model abstraction created at this stage, the *executable requirement,* is developed as a joint effort between contractor and customer in order to derive a top-level design guideline which captures the customer intent. The executable requirement removes the ambiguity associated with the written specification. It also provides information on the types of signal transformations, data formats, operational modes, interface timing data and control, and implementation constraints. A description of the executable requirement for an MPEG decoder is presented later. Section 78.4 addresses this subject in more detail.

Following the executable requirement, a top-level *executable specification* is developed. This is sometimes referred to as functional level VHDL design. This executable specification contains three general categories of information: (1) the system timing and performance, (2) the refined internal function, and (3) the physical constraints such as size, weight, and power. System timing and performance information include I/O timing constraints, I/O protocols, and system computational latency. Refined internal function information includes algorithm analysis in fixed/floating point, control strategies, functional breakdown, and task execution order. A functional breakdown is developed in terms of primitive signal processing elements which map to processing hardware cells or processor specific software libraries later in the design process. A description of the executable specification of the MPEG decoder is presented later. Section 78.5 investigates this subject in more detail.

The objective of data and control flow modeling is to refine the functional descriptions in the executable specification and capture concurrency information and data dependencies inherent in the algorithm. The intent of the refinement process is to generate multiple implementation independent representations of the algorithm. The implementations capture potential parallelism in the algorithm at a primitive level. The primitives are defined as the set of functions contained in a design library consisting of signal processing

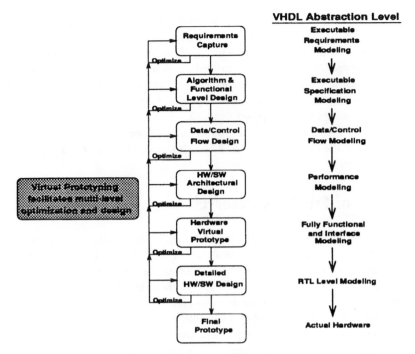

FIGURE 78.1 The VHDL top-down design process.

functions such as Fourier transforms or digital filters at course levels and of adders and multipliers at more fine-grained levels. The control flow can be represented in a number of ways ranging from finite state machines for low level hardware to run-time system controllers with multiple application data flow graphs. Section 78.6 investigates this abstraction model.

After defining the functional blocks, data flow between the blocks, and control flow schedules, hardware-software design trade-offs are explored. This requires architectural design and verification. In support of architecture verification, *performance level modeling* is used. The performance level model captures the time aspects of proposed design architectures such as system throughput, latency, and utilization. The proposed architectures are compared using cost function analysis with system performance and physical design parameter metrics as input. The output of this stage is one or few optimal or nearly optimal system architectural choice(s). In this stage, the interaction between hardware and software is modeled and analyzed. In general, models at this abstraction level are not concerned with the actual data in the system but rather the flow of data through the system. An abstract VHDL data type known as a token captures this flow of data. Examples of performance level models are shown later. Sections 78.7 and 78.8 address architecture selection and architecture verification, respectively.

Following architecture verification using performance level modeling, the structure of the system in terms of processing elements, communications protocols, and input/output requirements is established. Various elements of the defined architecture are refined to create hardware virtual prototypes. *Hardware virtual prototypes* are defined as *software simulatable* models of hardware components, boards, or systems containing sufficient accuracy to guarantee their successful realization in actual hardware. At this abstraction level, fully functional models (FFMs) are utilized. FFMs capture both internal and external (interface) functionality completely. Interface models capturing only the external pin behavior are also used for hardware virtual prototyping. Section 78.9 describes this modeling paradigm.

Application specific component designs are typically done in-cycle and use register transfer level (RTL) model descriptions as input to synthesis tools. The tool then creates gate level descriptions and final layout information. The RTL description is the lowest level contained in the virtual prototyping process and will not be discussed in this paper because existing RTL methodologies are prevalent in the industry.

At least six different hardware/software codesign methodologies have been proposed for rapid prototyping in the past few years. Some of these describe the various process steps without providing specifics for implementation. Others focus more on implementation issues without explicitly considering methodology and process flow. In the next section, we illustrate the features and limitations of these approaches and show how they compare to the proposed approach.

Following the survey, Section 78.3 lays the groundwork necessary to define the elements of the design process. At the end of the paper, Section 78.10 describes the usefulness of this approach for life cycle support and maintenance.

78.2 Survey of Previous Research

The codesign problem has been addressed in recent studies by Thomas et al. [7], Kumar et al. [8], Gupta et al. [9], Kalavade et al. [10, 11], and Ismail et al. [12]. A detailed taxonomy of HW/SW codesign was presented by Gajski et al. [13]. In the taxonomy, the authors describe the desired features of a codesign methodology and show how existing tools and methods try to implement them. However, the authors do not propose a method for implementing their process steps. The features and limitations of the latter approaches are illustrated in Fig. 78.2 [14]. In the table, we show how these approaches compare to the approach presented in this chapter with respect to some desired attributes of a codesign methodology. Previous approaches lack automated architecture selection tools, economic cost models, and the integrated development of test benches throughout the design cycle. Very few approaches allow for true HW/SW cosimulation where application code executes on a simulated version of the target hardware platform.

DSP Codesign Features	TA93	KA93	GD93	KL93 KL94	IJ95	Proposed Method
Executable Functional Specification	✓	✓	✓	✓	✓	✓
Executable Timing Specification		✓	✓			✓
Automated Architecture Selection						✓
Automated Partitioning			✓	✓		✓
Model-Based Performance Estimation		✓	✓			✓
Economic Cost/Profit Estimation Models						✓
HW/SW Cosimulation				✓		✓
Uses IEEE Standard Languages		✓				✓
Integrated Test Bench Generation						✓

FIGURE 78.2   Features and limitations of existing codesign methodologies.

78.3 Infrastructure Criteria for the Design Flow

Four enabling factors must be addressed in the development of a VHDL model infrastructure to support the design flow mentioned in the introduction. These include model verification/validation, interoperability, fidelity, and efficiency.

Verification, as defined by IEEE/ANSI, is the process of evaluating a system or component to determine whether the products of a given development phase satisfy the conditions imposed at the start of that phase. Validation, as defined by IEEE/ANSI, is the process of evaluating a system or component during or at the end of the development process to determine whether it satisfies the specified requirements. The proposed methodology is broken into the design phases represented in Figure 78.1 and uses black- and white-box software testing techniques to verify, via a structured simulation plan, the elements of each stage. In this methodology, the concept of a reference model, defined as the next higher model in the design hierarchy, is used to verify the subsequently more detailed designs. For example, to verify the gate level model after synthesis, the test suite applied to the RTL model is used. To verify the RTL level model, the reference model is the fully functional model. Moving test creation, test application, and test analysis to higher levels of design abstraction, the test description developed by the test engineer is more easily created and understood. The higher functional models are less complex than their gate level equivalents. For system and subsystem verification, which include the integration of multiple component models, higher level models improve the overall simulation time. It has been shown that a processor model at the fully functional level can operate over 1000 times faster than its gate level equivalent while maintaining clock cycle accuracy [5]. Verification also requires efficient techniques for test creation via automation and reuse and requirements compliance capture and test application via structured testbench development.

Interoperability addresses the ability of two models to communicate in the same simulation environment. Interoperability requirements are necessary because models usually developed by multiple design teams and from external vendors must be integrated to verify system functionality. Guidelines and potential standards for all abstraction levels within the design process must be defined when current descriptions do not exist. In the area of fully functional and RTL modeling, current practice is to use IEEE Std. 1164 − 1993 nine-valued logic packages [15]. Performance modeling standards are an ongoing effort of the RASSP program.

Fidelity addresses the problem of defining the information captured by each level of abstraction within the top-down design process. The importance of defining the correct fidelity lies in the fact that information not relevant within a model at a particular stage in the hierarchy requires unnecessary simulation time. Relevant information must be captured efficiently so simulation times improve as one moves toward the top of the design hierarchy. Figure 78.3 describes the RASSP taxonomy [16] for accomplishing this objective. The diagram illustrates how a VHDL model can be described using five resolution axes; temporal, data value, functional, structural, and programming level. Each line is continuous and discrete labels are positioned to illustrate various levels ranging from high to low resolution. A full specification of a model's fidelity requires two charts, one to describe the internal attributes of the model and the second for the external attributes. An "X" through a particular axis implies the model contains no information on the specific resolution. A compressed textual representation of this figure will be used throughout the remainder of the paper. The information is captured in a 5-tuple as follows,

$$\{(\text{Temporal Level}), (\text{Data Value}), (\text{Function}), (\text{Structure}), (\text{Programming Level})\}$$

The temporal axis specifies the time scale of events in the model and is analogous to precision as distinguished from accuracy. At one extreme, for the case of purely functional models, no time is modeled. Examples include Fast Fourier Transform and FIR filtering procedural calls. At the other extreme, time resolutions are specified in gate propagation delays. Between the two extremes, models may be time accurate at the clock level for the case of fully functional processor models, at the instruction cycle level for the case of performance level processor models, or at the system level for the case of application graph switching. In general, higher resolution models require longer simulation times due to the increased number of event transactions.

The data value axis specifies the data resolution used by the model. For high resolution models, data is represented with bit true accuracy and is commonly found in gate level models. At the low end of the spectrum, data is represented by abstract token types where data is represented by enumerated values, for example, *blue*. Performance level modeling uses tokens as its data type. The token only captures the

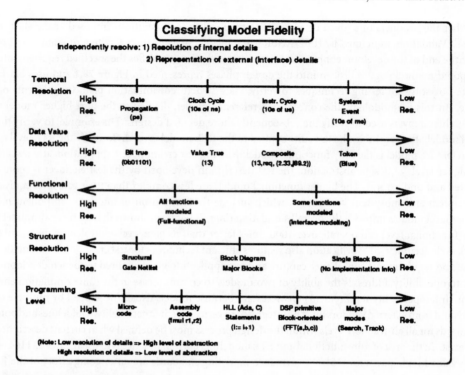

FIGURE 78.3   A model fidelity classification scheme.

control information of the system and no actual data. For the case of no data, the axis would be represented with an "X". At intermediate levels, data is represented with its correct value but at a higher abstraction (i.e., integer or composite types, instead of the actual bits). In general, higher resolutions require more simulation time.

Functional resolution specifies the detail of device functionality captured by the model. At one extreme, no functions are modeled and the model represents the processing functionality as a simple time delay (i.e., no actual calculations are performed). At the high end, all the functions are implemented within the model. As an example, for a processor model, a time delay is used to represent the execution of a specific software task at low resolutions while the actual code is executed on the model for high resolution simulations. As a rule of thumb, the more functions represented, the slower the model executes during simulation.

The structural axis specifies how the model is constructed from its constituent elements. At the low end, the model looks like a black box with inputs and outputs but no detail as to the internal contents. At the high end the internal structure is modeled with very fine detail, typically as a structural net list of lower level components. In the middle, the major blocks are grouped according to related functionality.

The final level of detail needed to specify a model is its programmability. This describes the granularity at which the model interprets software elements of a system. At one extreme, pure hardware is specified and the model does not interpret software, for example, a special purpose FFT processor hard wired for 1024 samples. At the other extreme, the internal micro-code is modeled at the detail of its datapath control. At this resolution, the model captures precisely how the micro-code manipulates the datapath elements. At decreasing resolutions the model has the ability to process assembly code and high level languages as input. At even lower levels, only DSP primitive blocks are modeled. In this case, programming consists of combining functional blocks to define the necessary application. Tools such as MATLAB/Simulink provide examples for this type of model granularity. Finally, models can be programmed at the level of the major modes. In this case, a run-time system is switched between major operating modes of a system by executing alternative application graphs.

Finally, efficiency issues are addressed at each level of abstraction in the design flow. Efficiency will be discussed in coordination with the issues of fidelity where both the model details and information content are related to improving simulation speed.

78.4 The Executable Requirement

The methodology for developing signal processing systems begins with the definition of the system requirement. In the past, common practice was to develop a textual specification of the system. This approach is flawed due to the inherent ambiguity of the written description of a complex system. The new methodology places the requirements in an executable format enforcing a more rigorous description of the system. Thus, VHDL's first application in the development of a signal processing system is an *executable requirement* which may include signal transformations, data format, modes of operation, timing at data and control ports, test capabilities, and implementation constraints [17]. The executable requirement can also define the minimum required unit of development in terms of performance (e.g., SNR, throughput, latency, etc.). By capturing the requirements in an executable form, inconsistencies and missing information in the written specification can also be uncovered during development of the requirements model.

An executable requirement creates an "environment" wherein the surroundings of the signal processing system are simulated. Figure 78.4 illustrates a system model with an accompanying testbench. The testbench generates control and data signals as stimulus to the system model. In addition, the testbench receives output data from the system model. This data is used to verify the correct operation of the system model. The advantages of an executable requirement are varied. First, it serves as a mechanism to define

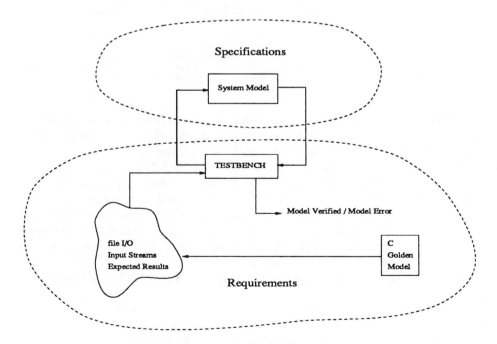

FIGURE 78.4 Illustration of the relation between executable requirements and specifications.

and refine the requirements placed on a system. Also, the VHDL source code along with supporting textual description becomes a critical part of the requirements documentation and life cycle support of the system. In addition, the testbench allows easy examination of different command sequences and data sets. The testbench can also serve as the stimulus for any number of designs. The development of different system models can be tested within a single simulation environment using the same testbench. The requirement is easily adaptable to changes that can occur in lower levels of the design process. Finally, executable requirements are formed at all levels of abstraction and create a documented history of the design process. For example, at the system level, the environment may consist of image data from a camera while at the ASIC level it may be an interface model of another component.

The RASSP program, through the efforts of MIT Lincoln Laboratory, created an executable requirement [18] for a synthetic aperture radar (SAR) algorithm and documented many of the lessons learned in implementing this stage in the top-down design process. Their high level requirements model served as the baseline for the design of two SAR systems developed by separate contractors, Lockheed Sanders and Martin Marietta Advanced Technology Labs. A test bench generation system for capturing high level requirements and automating the creation of VHDL is presented in [19]. In the following sections, we present the details of work done at Georgia Tech in creating an executable requirement and specification for an MPEG-1 decoder.

78.4.1   An Executable Requirements Example: MPEG-1 Decoder

MPEG-1 is a video compression-decompression standard developed under the International Standard Organization originally targeted at CD-ROMs with a data rate of 1.5 Mbits/sec [20]. MPEG-1 is broken into 3 layers: system, video, and audio. Table 78.1 depicts the system clock frequency requirement taken from layer 1 of the MPEG-1 document.[1] The system time is used to control when video frames are decoded and presented via decoder and presentation time stamps contained in the ISO 11172 MPEG-1 bitstream. A VHDL executable rendition of this requirement is illustrated in Table 78.5.

TABLE 78.1   MPEG-1 System Clock Frequency Requirement Example

Layer 1 - System requirement example from ISO 11172 standard	
System clock frequency	The value of the system clock frequency is measured in Hz and shall meet the following constraints: $90,000 - 4.5$ Hz $\leq$ system_clock_frequency $\leq 90,000 + 4.5$ Hz Rate of change of system_clock_frequency $\leq 250 * 10^{-6}$ Hz/s

The testbench of this system uses an MPEG-1 bitstream created from a "golden C model" to ensure correct input. A public-domain C version of an MPEG encoder created at UCal-Berkeley [21] was used as the golden C model to generate the input for the executable requirement. From the testbench, an MPEG bitstream file is read as a series of integers and transmitted to the MPEG decoder model at a constant rate of 174300 Bytes/sec along with a system clock and a control line named *mpeg_go* which activates the decoder. Only 50 lines of VHDL code are required to characterize the top level testbench. This is due to the availability of the golden C MPEG encoder and a shell script which wraps around the output of the golden C MPEG encoder bitstream with system layer information. This script is necessary because there are no *complete* MPEG software codecs in the public domain, i.e., they do not include the system information in the bitstream. Figure 78.6 depicts the process of verification using golden C models. The golden model generates the bitstream sent to the testbench. The testbench reads the bitstream as a series

---

[1] Our efforts at Georgia Tech have only focused on layers 1 and 2 of this standard.

```
- system_time_clk process is a clock process that counts at a rate
- of 90kHz as per MPEG-I requirement. In addition, it is updated by
- the value of the incoming SCR fields read from the ISO11172 stream.
-
system_time_clock : PROCESS(stc_strobe,sys_clk)
        VARIABLE clock_count : INTEGER := 0;
        VARIABLE SCR,system_time_var : bit33;
        CONSTANT clock_divider : INTEGER := 2;
BEGIN
        IF mpeg_go = '1' THEN
        - if stc_strobe is high then update system_time value to latest SCR
        IF (stc_strobe = '1') AND (stc_strobe'EVENT) THEN
                system_time <= system_clock_ref;
                clock_count := 0; - reset counter used for clock downsample
        ELSIF (sys_clk = '1') AND (sys_clk'EVENT) THEN
                clock_count := clock_count + 1;
                IF clock_count MOD clock_divider = 0 THEN
                        system_time_var := system_time + one;
                        system_time <= system_time_var;
                END IF;
        END IF;
        END IF;
END PROCESS system_time_clock;
```

FIGURE 78.5   System clock frequency requirement example translated to VHDL.

of integers. These are in turn sent as data into the VHDL MPEG decoder model driven with appropriate
clock and control lines. The output of the VHDL model is compared with the output of the golden model
(also available from Berkeley) to verify the correct operation of the VHDL decoder. A warning message
alerts the user to the status of the model's integrity.

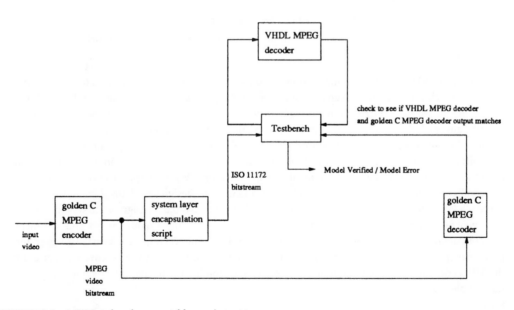

FIGURE 78.6   MPEG-1 decoder executable requirement.

The advantage of the configuration illustrated in Figure 78.6 is its reusability. An obvious example is
MPEG-2 [22], another video compression-decompression standard targeted for the all-digital transmission
of broadcast TV quality video at coded bit rates between 4 and 9 Mbits/sec. The same testbench structure
could be used by replacing the golden C models with their MPEG-2 counterparts. While the system layer
information encapsulation script would have to be changed, the testbench itself remains the same because
the interface between an MPEG-1 decoder and its surrounding environment is identical to the interface
for an MPEG-2 decoder. In general, this testbench configuration could be used for a wide class of video
decoders. The only modifications would be the golden C models and the interface between the VHDL

decoder model and the testbench. This would involve making only minor alterations to the testbench itself.

78.5   The Executable Specification

The executable specification depicted in Fig. 78.4 processes and responds to the outside stimulus, provided by the executable requirement, through its interface. It reflects the particular function and timing of the intended design. Thus, the executable specification describes the behavior of the design and is timing accurate without consideration of the eventual implementation. This allows the user to evaluate the completeness, logical correctness, and algorithmic performance of the system through the test bench. The creation of this formal specification helps identify and correct functional errors at an early stage in the design and reduce total design time [13, 16, 23, 24].

The development of an executable specification is a complex task. Very often, the required functionality of the system is not well-understood. It is through a process of learning, understanding, and defining that a specification is crystallized. To specify system functionality, we decompose it into elements. The relationship between these elements is in terms of their execution order and the data passing between them. The executable specification captures:

- the refined internal functionality of the unit under development (some algorithm parallelism, fixed/floating point bit level accuracies required, control strategies, functional breakdown, task execution order)
- physical constraints of the unit such as size, weight, area, and power
- unit timing and performance information (I/O timing constraints, I/O protocols, computational complexity)

The purpose of VHDL at the executable specification stage is to create a formalization of the elements in a system and their relationships. It can be thought of as the high level design of the unit under development. And although we have restricted our discussion to the system level, the executable specification may describe any level of abstraction (algorithm, system, subsystem, board, device, etc.).

The allure of this approach is based on the user's ability to see what the performance "looks" like. In addition, a stable test mechanism is developed early in the design process (note the complementary relation between the executable requirement and specification). With the specification precisely defined, it becomes easier to integrate the system with other concurrently designed systems. Finally, this executable approach facilitates the re-use of system specifications for the possible redesign of the system.

In general, when considering the entire design process, executable requirements and specifications can potentially cover any of the possible resolutions in the fidelity classification chart. However, for any particular specification or requirement, only a small portion of the chart will be covered. For example, the MPEG decoder presented in this and the previous section has the fidelity information represented by the 5-tuple below,

Internal: {(Clock cycle), (Bit true → Value true), (All), (Major blocks), (X)}
External: {(Clock cycle), (Value true), (Some), (Black box), (X)},

where (Bit true → Value true) means all resolutions between bit true and value true inclusive.

From an internal viewpoint, the timing is at the system clock level, data is represented by bits in some cases and integers in others, the structure is at the major block level, and all the functions are modeled. From an external perspective, the timing is also at the system clock level, the data is represented by a stream of integers, the structure is seen as a single black box fed by the executable requirement and from an external perspective the function is only modeled partially because this does not represent an actual chip interface.

78.5.1   An Executable Specification Example: MPEG-1 Decoder

As an example, an MPEG-1 decoder executable specification developed at Georgia Tech will be examined in detail. Figure 78.7 illustrates how the system functionality was broken into a discrete number of elements. In this diagram each block represents a process and the lines connecting them are signals. Three major areas of functionality were identified from the written specification: memory, control, and the video decoder itself. Two memory blocks, *video_decode_memory* and *system_level_memory* are clearly labeled. The *present_frame_to_decode_file* process contains a frame reorder buffer which holds a frame until its presentation time. All other VHDL processes with the exception of *decode_video_frame_process* are control processes and pertain to the systems layer of the MPEG-1 standard. These processes take the incoming MPEG-1 bitstream and extract system layer information. This information is stored in the *system_level_memory* process where other control processes and the video decoder can access pertinent data. After removing the system layer information from the MPEG-1 bitstream, the remainder is placed in the *video_decode_memory*. This is the input buffer to the video decoder. It should be noted that although MPEG-1 is capable of up to 16 simultaneous video streams multiplexed into the MPEG-1 bitstream only one video stream was selected for simplicity.

The last process, *decode_video_frame_process*, contains all the subroutines necessary to decode the video bitstream from the video buffer (*video_decode_memory*). MPEG video frames are broken into 3 types: (I)ntra, (P)redictive, and (B)idirectional. I frames are coded using block discrete cosine transform (DCT) compression. Thus, the entire frame is broken into 8x8 blocks, transformed with a DCT and the resulting coefficients transmitted. P frames use the previous frame as a prediction of the current frame. The current frame is broken into 16×16 blocks. Each block is compared with a corresponding search window (e.g., 32×32, 48×48) in the previous frame. The 16×16 block within the search window which best matches the current frame block is determined. The motion vector identifies the matching block within the search window and is transmitted to the decoder. B frames are similar to P frames except a previous frame and a future frame are used to estimate the best matching block from either of these frames or an average of the two. It should be noted that this requires the encoder and decoder to store these 2 reference frames.

The functions contained in the *decode_video_frame_process* are shown in Fig. 78.8. In the diagram, there are three main paths representing the procedures or functions in the executable specification which process the I, P, or B frame, respectively. Each box below a path encloses all the procedures executed from within that function. Beneath each path is an estimate of the number of computations required to process each frame type. Comparing the three executable paths in this diagram, one observes the large similarity between each path. Overall, only 25 unique routines are called to process the video frame. By identifying key functions within the video decoding algorithm itself, efficient and reusable code can be created. For instance, the data transmitted from the encoder to the decoder is compressed using a Huffman scheme. The procedures *vlc, advance_bit,* and *extract_n_bits* perform the Huffman decode function and miscellaneous parsing of the MPEG-1 video bitstream. Thus, this set of procedures can be used in each frame type execution path. Reuse of these procedures can be applied in the development of an MPEG-2 decoder executable specification. Since MPEG-2 is structured as a super set of the syntax defined in MPEG-1, there are many procedures that can be utilized with only minor modifications. Other procedures such as *motion_compensate_forward* and *idct* can be reused in a variety of DCT-based video compression algorithms.

The executable specification also allows detailed analysis of the computational complexity on a procedural level. Table 78.2 lists the computational complexity of some of the procedures identified in Fig. 78.8. This breakdown identifies what areas of the algorithm are the most computationally intensive and the numbers were arrived at through a data flow analysis of the VHDL code. Within the MPEG-1 video decoder algorithm, the most intense computational loads occur in the inverse DCT and motion compensation procedures. Thus, such an analysis can alert the user early in the design process to potential design issues.   While parallelism is a logical topic for the data and control flow modeling section, preliminary

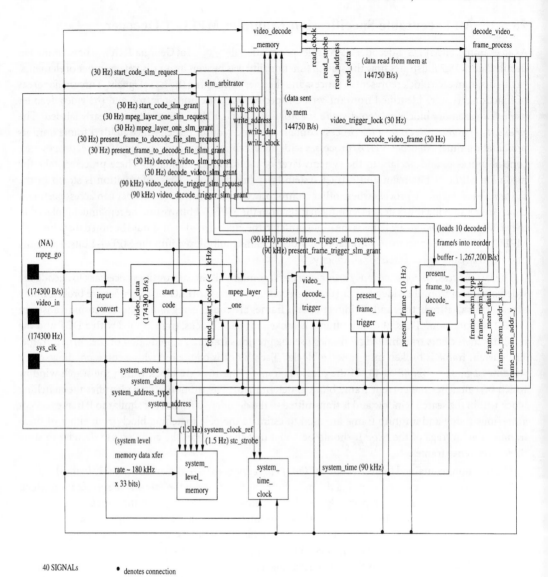

FIGURE 78.7   System functionality breakdown for MPEG-1 decoder.

investigations can be made from the executable specification itself. With the specifications captured in a language, execution order and data passing between procedures are known precisely. This knowledge facilitates the user in extracting potential parallelism from the specification. From the MPEG-1 decoder executable specification, potential parallelism can be seen in several areas. In an I frame, no data dependencies are present between each 8 × 8 block. Therefore, an inverse DCT could potentially be performed on each 8 × 8 block in parallel. In P and B frames, data dependencies occur between consecutive 16 × 16 blocks (called macroblocks) but no data dependencies occur between slices (a grouping of consecutive macroblocks). Thus, parallelism is potentially exploitable at the slice and macroblock level. This information is passed to the data/control flow modeling phase where more detailed analysis of parallelism is done.

It is also possible to delve into implementation requirement issues at the executable specification level. Fixed vs. floating point trade-offs can be examined in detail. The necessary accuracy and resolution

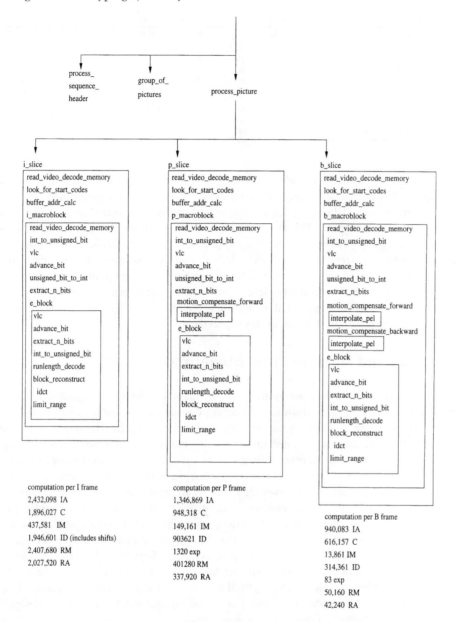

FIGURE 78.8 Description of procedural flow within MPEG-1 decoder executable specification.

required to meet system requirements can be determined through the use of floating and fixed point packages written in VHDL. At Georgia Tech, fixed point packages have been developed. These packages allow the user to experiment with the executable specification and see the effect finite bit accuracy has on the system model. In addition, packages have been developed which implement specific arithmetic architectures such as the ADSP 2100 [25]. This analysis results in additional design requirements being passed to hardware and software developers in later design phases.

Finally, the executable specification allows the explicit capture of internal timing and control flow requirements of the MPEG-1 decoding algorithm itself. The written document is imprecise about the details of how timing considerations for presentation and decoder time stamps will be handled. The control necessary to trigger present and decode video frame events is difficult to articulate in a written

TABLE 78.2 Computational Complexity of Some Specification Procedures

Procedure	Int Adds	Int Div	Comp	Int Mult	exp	Real Add	Real Mult
vlc	—	-	2	-	-	-	-
advance_bit	10	16	9	-	-	-	-
int_to_unsigned_bit	8	16	8	-	-	-	-
extract_n_bits	24	16	20	-	-	-	-
look_for_start_codes	9	16	10	-	-	-	-
runlength_decode	2	-	1	1	-	-	-
block_reconstruct	66	64	258	193	-	-	-
idct	-	-	-	-	-	1024	1216
qmotion_compensate_forward	1422	646	1549	16	-	-	-

form. The most difficult aspects of coding the executable specification for a MPEG-1 decoder were these considerations. The decoder itself hinges on developing a mechanism for robustly determining when to decode or present a frame in the buffer. Events must be triggered using a system time clock which is updated from the input bitstream itself. This task is handled by five processes (*start_code*, *mpeg_layer_one*, *video_decode_trigger*, *present_frame_trigger*, *present_frame_to_decode_file*) grouped around a common memory (*system_level_memory*). This memory was necessary to allow each concurrent process to access timing information extracted from the system layer of the input bitstream. These timing and control considerations had to fit into a larger system timing requirement. For a MPEG-1 decoder, the most critical timing constraints are initial latency and the fixed presentation rate (e.g., 30 frames/sec). All other timing considerations were driven by this requirement.

78.6 Data and Control Flow Modeling

This modeling level captures data and control flow information in the system algorithms. The objective of data flow modeling is to refine the functional descriptions in the executable specification and capture concurrency information and data dependencies inherent in the algorithm. The output of the refinement process is one or a few manually generated implementation independent representations of the algorithm. These multiple implementations capture potential algorithmic parallelism at a primitive level where primitives are defined as that set of functions contained in a design library. The primitives are signal processing functions such as Fast Fourier Transforms or filter routines at coarse-grained levels to adders and multipliers at more fine-grained levels. The breakdown of primitive elements depend on the granularity exploited by the algorithm as well as potential architectural design paradigms to which the algorithm is mapped. For example, if the design paradigm demands architectures using multiple commercial-off-the-shelf (COTS) RISC processors, the primitives consist of signal processing functional block level elements such as FFTs or FIR filters which exist as performance optimized library elements available for the specific processor. For custom computationally intense designs, the data flow of the algorithm may be dissected into lower primitive components such as adders and multipliers using bit-slice architectures. In our design flow, the fidelity captured by data/control flow models is shown below:

Internal: {(X), (Value true → Composite), (All), (X), (Major modes)}
External: {(X), (Value true → Composite), (X), (X), (X)}.

Because the models are purely functional and their major objective is to refine the internal representation of the algorithm, there is no time information captured by its internal or external representation as illustrated by the "*X*". The internal data processed by the model and external data loaded into the model are typically represented by standard data types such as *float* and/or *integer* and in some cases by composite data types such as records or arrays. All internal functionality is represented and is verified using the same data presented to the executable specification. No function is captured via external interfaces since data is input to the model through file input/output. The data processed by the executable specification is also processed by the data/control flow model. No internal or external structural information is captured since the model is implementation independent. Its level of programmability is represented at the application graph level. The applications are major modes of the system under investigation and hence at a low resolution. In general, because the primitive elements can represent adders and/or multipliers, programmability for data/control flow models can resolve to higher resolutions including the microcode level.

The implementation independent representations are compared with the executable specification using the test data supplied by the requirements development phase to verify compliance with the original algorithm design. The representations are then input to the architecture selection phase and, with additional metrics, determine the final architecture of the system.

Signal processing applications inherently follow the data flow execution model. Processing Graph Methodology (PGM) [26] from Naval Research Laboratory was developed specifically to capture signal processing applications. PGM supports specification of full system data flow and its associated control. An application is first captured as a graph, where nodes of the graph represent processing and edges represent queues that hold intermediate data between nodes. The scheduling criteria for each node is based on the state of its corresponding input/output queues. Each queue in the graph can be linked to one node at a time. Associated with each queue is a control block structure containing information such as size, current amount of data, and threshold. A run-time system provides a set of procedures used by each node to check the availability of data from the upstream queue or available space in the downstream queue. Applications consist of one or more graphs, one or more I/O procedures, and a run-time system interfaced with one or more command programs. The PGM graphs serve as the implementation independent representation of the algorithm discussed earlier. An example of a 2-D FFT PGM graph is presented in the next section.

Under the support of the RASSP program, a set of tools is being developed by Management Communications and Control, Inc. (MCCI) and Lockheed Martin Advance Technology Laboratories [27, 28]. The toolset automates the translation of software architecture specifications to design implementations of application and control software for a signal processing system. Hardware/software architectures are presented to the autocoding toolset as PGM application data flow graphs along with a candidate architectures file and graph partition lists. The lists are generated by hardware/software partitioning tools. The proposed partitions are then simulated for performance and verified against the top level specification for correct functionality. The verified partition graphs are then used as inputs to detailed design level autocode tools that generate actual source code. The source code implements the partitions processing specifications using the target processor's math library. It also produces a memory map converting all queues and variables to static buffers. Finally the application graph, with its set of source files, are translated to run-time data structures that are used by the run-time system to create an executable image of the application as distributed tasks on the target processors.

Other tools provide paths from specification to hardware and are briefly mentioned. The Ptolemy [29, 30] design system from the University of California at Berkeley provides a synchronous data flow domain which can be used to perform system level simulations. Silage, another product of UC Berkeley is a data flow modeling language. Data Flow Language (DFL), a commercial version of Silage is used in Mentor Graphics' DSP Station to perform algorithm/architecture tradeoffs. It also provides a path to synthesis as a high-level design entry tool.

78.6.1 Data and Control Flow Example

An example of a small PGM application is presented in Fig. 78.9. The graph represents a two dimensional FFT program implemented in PGM. The graph captures both the functionality and the data flow aspects of the application. The source data is read from a file and represents the I/O processor that would normally provide the input data stream. The data are then distributed to a number of queues serving as inputs to the FFT primitives that perform the operations on the rows of the input stream. The output of the FFT primitives flow to another set of queues that are input to the corner turn graph. Once the data are sorted correctly, they are sent to the input queues of the column FFT primitives. The graph is then executed by the simulator where the functionality, queue sizes, and communication between nodes are examined. This same graph is input to the hardware/software partitioning tools that generate the partition list. Given the partition list and the hardware configuration file, the autocode tool set generates the load image for the target platform.

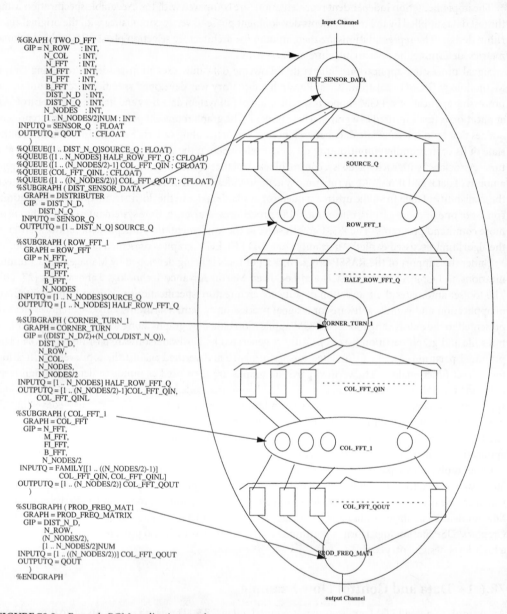

FIGURE 78.9   Example PGM application graph.

78.7   Architectural Design

Signal processing systems are characterized as having high throughput requirements as well as stringent physical constraints. However, due to economic objectives, signal processing systems must also be developed and produced at minimal cost, while meeting time-to-market constraints in order to maximize product profits. Such cost-effective systems can only be produced by applying a high degree of cost emphasis during the early stages of design. Although the conceptual design process typically involves less than 4% of the total prototyping time and cost, it accounts for more than 70% of a system's life cycle cost. Consequently, the goal of the architecture designer is to optimize preliminary architectural design decisions with respect to the dominant system-level cost elements such as acquisition costs, maintenance costs, and time-to-market costs, while satisfying performance and physical constraints.

78.7.1 Cost Models

Current rapid prototyping design methodologies have overlooked an important characteristic of software prototyping. Various parametric studies based on historical project data show that software is difficult to design and test if "slack" margins for hardware CPU and memory resources are overly restrictive [31]. Severe resource constraints may require software developers to interact directly with the operating system and/or hardware in order to optimize the code to meet system requirements. Such constrained architectures particularly increase the integration and test phase because resource constraints usually are not pushed until all software pieces come together. In systems in which most hardware is simply commercial-off-the-shelf (COTS) parts, the time and cost of software prototyping and design can dominate the schedule and budget. If physical constraints permit, the hardware platform can be relaxed to achieve significant reductions in overall development cost and time. This principle of software prototyping is illustrated in Fig. 78.10 [14, 32]. The figure shows how system costs are dominated by software costs, especially at low production volumes, when CPU and memory utilization is high. However, as the utilization of these hardware resources is reduced, the software costs decrease drastically. Most parametric software cost estimation models quantitatively represent this principle. For example, the embedded mode Revised Intermediate COCOMO (REVIC) [33] software development cost and time models can be written as follows:

$$S_C = C_s \left[3.312 \times L^{1.2} \times F_E \times F_M \times \prod_{i=1}^{17} F_i \right] \tag{78.1}$$

$$S_T = 4.376 \left[3.312 \times L^{1.2} \times F_E \times F_M \times \prod_{i=1}^{17} F_i \right]^{0.32} \tag{78.2}$$

S_C refers to the software development cost in dollars. S_T depicts development time in months. C_s is the software labor cost per person-month of effort. L denotes the number of delivered source instructions (thousands) including application code, OS kernel services, control and diagnostics, and support software. The F_is represent additional cost drivers which model the effect of personnel, computer, product, and project attributes on software cost. F_E and F_M are effort adjustment factors which denote the effect of the execution time margin and storage margin on development cost. The relation between these effort adjustment factors and CPU and memory utilization is shown in Table 78.3. Linear interpolation is used to determine the effort multiplier values for utilizations between the given data points displayed in the table.

TABLE 78.3 Execution Time and Main
Storage Constraint Effort Multipliers

Rating	Utilization	F_E	F_M
Nominal	Up to 50%	1.00	1.00
High	70%	1.11	1.06
Very high	85%	1.30	1.21
Extra high	95%	1.66	1.56

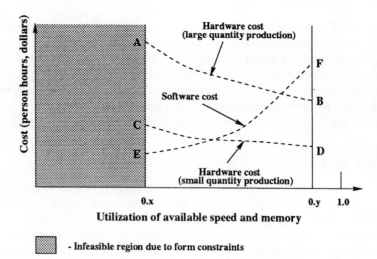

FIGURE 78.10   Hardware/software prototyping costs.

Despite the fact that many signal processing systems are being implemented with purely software solutions due to flexibility and scalability requirements, the combination of high throughput requirements and stringent form factor constraints sometimes necessitate the need for implementing part of the system with dedicated hardware elements such as ASICs or FPGAs. Even though ASICs can provide sizable increases in performance and size efficiency, they come with a heavy development cost penalty which can usually only be justified by high volume production. In order to quantify this effect for trade-off analysis, parametric hardware cost models can be used. For example, the parametric cost model presented in [34, 35] for ASIC design provides the following hardware development time and cost relations for ASIC development:

$$H_C \;=\; C_h \left\{ (1+D)^{YR} \left[A + B(S_h)^H \right] \right\} \tag{78.3}$$

$$H_T \;=\; 3.5 \left\{ (1+D)^{YR} \left[A + B(S_h)^H \right] \right\}^{0.34} \tag{78.4}$$

YR is 1984 minus the year of the bulk of the design effort, C_h is the hardware labor cost per person-month, and A, B, D, and H are parameters of the model. D is the average annual improvement factor; A is the startup manpower; B is a measure of the productivity; and H is a measure of economies/diseconomies of scale. FPGAs provide more flexibility and lower cost penalties than ASICs at the expense of performance and size efficiency. From the FPGA cost model presented in [36], we will assume that FPGA development time and cost can be modeled as roughly one-third of that obtained for a comparable size ASIC. Although rising HW/SW development costs have very detrimental effects on life cycle cost, the effect of an increase in development time can be much more devastating.

Time-to-market costs can often outweigh design, prototyping, and production costs. A recent survey showed that being six months late to market resulted in an average of 33% profit loss. Engineering managers stated that they would rather have a 100% overrun in design and prototyping costs than be three months late to market with a product. Early market entry allows for increased brand name recognition, market share, and product yields. Market research performed by Logic Automation has shown that the demand and potential profits for a new HW/SW product can be modeled by a triangular window of opportunity as shown in Fig. 78.11 [36]. Figure 78.11 illustrates the effect of delivering a product to market late. The shaded region of the triangle signifies the loss of revenue due to late entry in the market. This loss in revenue can be quantitatively stated as follows:

$$R_L = R_0 \times D\,(3W - D)\,/(2W^2) \tag{78.5}$$

R_0 refers to the expected product revenue, D is the delay (months) in delivering a product to market, and W is half the product lifetime (months). Therefore, in order to maximize profits, the product must be on the market by the start of the demand window.

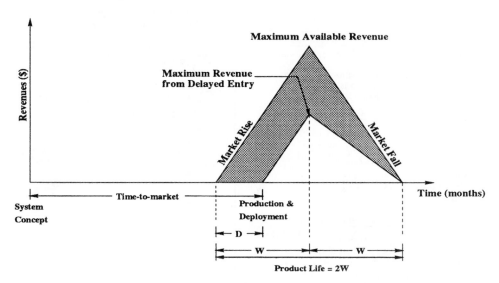

FIGURE 78.11   Time-to-market cost model.

78.7.2   Architectural Design Model

In this section, we present a cost-driven approach to conceptual architecture design. The conceptual architecture design process consists of HW/SW partitioning, architecture selection, and software partitioning. As input, this design stage accepts the application data/control flow graph, system-level performance requirements, form factor constraints, schedule constraints, and HW/SW reuse library parameters. As output, this stage produces an architecture candidate which serves as input to the architectural verification stage. VHDL performance models, described in the next section, are used to verify the architecture candidate. This model is known as the *conceptual prototype* of the system. If the conceptual prototype does not meet performance specifications, updated performance parameters are back annotated to the architecture design process, and the process is repeated.

   The architectural design problem can be modeled as a constrained optimization problem. The objective of cost-effective architecture design is to choose a HW/SW architecture which minimizes total life cycle cost, while maximizing potential product profits subject to performance and form factor constraints. Our approach quantitatively models the architecture design process by formulating it as a non-linear mixed-integer programming problem. In order to provide support for high performance signal processing applications, we assume a distributed memory architecture composed of multiple programmable processors, ASICs, I/O devices, and/or FPGAs connected over a crossbar network. The goal of the architectural design process in this context is to determine the number and type of programmable processors (i860, SHARC, etc.), the memory capacity, the number of dedicated hardware elements (ASIC, FPGA), and to map the data/control flow graph nodes to the architectural elements in a manner that optimally meets design and economic objectives.

   An example mathematical programming formulation that can be used to model the architecture design process is shown in Fig. 78.12. The major decision variables of the model are defined as follows:

Objective: *Minimize software development costs, hardware dev./prod. costs, NRE costs, and revenue losses*

$$\sum_{i=1}^{2} C_h h_i + C_{NRE} \sum_{j=1}^{N} \sum_{i=1}^{N} a_i^{asic} \delta_{ij} + V\left(C^{mcm}\mu\right) +$$
$$V\left(\sum_{i=1}^{M} C_i^{proc} \rho_i\right) + V\left(\sum_{j=1}^{N} \sum_{i=1}^{N} C^{fpga} a_i^{fpga} \beta_{ij}\right) +$$
$$V\left(\sum_{j=1}^{N} \sum_{i=1}^{N} C^{asic} a_i^{asic} \delta_{ij}\right) + R_0 d\,(3W - d)\,/\,(2W^2) +$$
$$C_s s$$

subject to:

Software development cost definition constraint

$$-Zw_{ij} \leq \left(s - 3.312L^{1.2} f_i^E f_j^M \prod_{i=1}^{17} F_i\right) \leq Zw_{ij} \text{ where}$$
$$f_1^E = 1, f_1^M = 1$$
$$f_2^E = 0.55u_p + 0.725,\ f_2^M = 0.3u_m + 0.85$$
$$f_3^E = 1.27u_p + 0.223,\ f_3^M = u_m + 0.36$$
$$f_4^E = 3.6u_p - 1.76,\ f_4^M = 3.5u_m + 1.765$$

Software development time definition constraint

$$4.376s^{0.32} - \left(d_s^+ - d_s^-\right) = T_s$$

ASIC development cost definition constraint

$$h_1 - (1+D)^{YR}\left[A + B\left(\sum_{j=1}^{N} \sum_{i=1}^{N} a_i^{asic} \delta_{ij}\right)^H\right] = 0$$

ASIC development time definition constraint

$$h_c \geq (1+D)^{YR}\left[A + B\left(\sum_{i=1}^{N} a_i^{asic} \delta_{ij}\right)^H\right] \text{ for } j = 1,\dots,N$$
$$3.5h_c^{0.34} - \left(d_{asic}^+ - d_{asic}^-\right) = T_s$$

FPGA development cost definition constraint

$$h_2 - \left\{(1+D)^{YR}\left[A + B\left(\sum_{j=1}^{N} \sum_{i=1}^{N} a_i^{fpga} \beta_{ij}\right)^H\right]\right\}/3 = 0$$

FPGA development time definition constraint

$$h_c' \geq (1+D)^{YR}\left[A + B\left(\sum_{i=1}^{N} a_i^{fpga} \beta_{ij}\right)^H\right] \text{ for } j = 1,\dots,N$$
$$(3.5(h_c')^{0.34})/3 - \left(d_{fpga}^+ - d_{fpga}^-\right) = T_s$$

System-level performance constraint

$$\sum_{k=1}^{N} \sum_{j=1}^{M} \sum_{i=1}^{N} t_{ij}^{proc} \alpha_{ijk} + \sum_{j=1}^{N} \sum_{i=1}^{N} t_{ij}^{asic} \delta_{ij} +$$
$$\sum_{j=1}^{N} \sum_{i=1}^{N} t_{ij}^{fpga} \beta_{ij} - \sum_{i=1}^{M} \rho_i(O_i^P + (O^M(\rho_i - 1)) -$$
$$C_{comm} + \sum_{k=1}^{N} \sum_{j=1}^{N} \sum_{i=1}^{N} \sum_{l=i+1}^{N-1} c_{il}\alpha_{ijk}\alpha_{ljk} +$$
$$\sum_{j=1}^{N} \sum_{i=1}^{N-1} \sum_{k=i+1}^{N} c_{ik}\beta_{ij}\beta_{kj} +$$
$$\sum_{j=1}^{N} \sum_{i=1}^{N-1} \sum_{k=i+1}^{N} c_{ik}\delta_{ij}\delta_{kj} \geq T_{SYS}$$

Hardware/Software partitioning constraint

$$\sum_{k=1}^{N} \sum_{j=1}^{M} \alpha_{ijk} + \sum_{j=1}^{N} (\beta_{ij} + \delta_{ij}) = 1 \text{ for } i = 1,2,\dots,N$$

Time-to-market Constraints

$$d \geq d_s^+, d \geq d_{asic}^+, d \geq d_{fpga}^+$$
$$d \leq W$$

Memory utilization definition constraint

$$u_m \mu = \sum_{k=1}^{N} \sum_{j=1}^{M} \sum_{i=1}^{N} m_i \alpha_{ijk} + O^{MEM}$$

Local memory size definition

$$u_m l_{jk}^m = \sum_{i=1}^{N} m_i \alpha_{ijk} + O^{lm} \text{ for all } j, k$$

Total number of programmable processors definition

$$Zn_{jk} \geq \sum_{i=1}^{N} \alpha_{ijk} \text{ for all } j, k$$
$$\rho_i = \sum_{j=1}^{N} n_{ij} \text{ for } i = 1,2,\dots,M$$

ASIC Yield constraint

$$\sum_{i=1}^{N} a_i^{asic} \delta_{ij} \leq (1/K)\ln(1/Y), \text{ where } j = 1,2,\dots,N$$

FPGA size constraint

$$\sum_{i=1}^{N} a_i^{fpga} \beta_{ij} \leq U^{fpga} \text{ for all } j$$

System-level area and power constraints

$$\sum_{i=1}^{M} a_i^{proc} \rho_i + a_{mcm}\mu + \sum_{j=1}^{N} \sum_{i=1}^{N} a_i^{asic}\delta_{ij} +$$
$$\sum_{j=1}^{N} \sum_{i=1}^{N} a_i^{fpga}\beta_{ij} \leq U_A$$
$$\sum_{i=1}^{M} p_i^{proc}\rho_i + p_{mcm}\mu + \sum_{j=1}^{N} \sum_{i=1}^{N} p_i^{asic}\delta_{ij} +$$
$$\sum_{j=1}^{N} \sum_{i=1}^{N} p_i^{fpga}\beta_{ij} \leq U_P$$

Load Balancing Constraints

$$u_p - \Delta_p - Z(1 - n_{jk}) \leq \left(\sum_{i=1}^{N} t_{ij}^{proc}\alpha_{ijk} + O_j^P\right)/S_j \leq$$
$$u_p + \Delta_p + Z(1 - n_{jk}) \text{ for all } j, k$$

Processor utilization definition constraint

$$u_p\left(\sum_{i=1}^{N} S_i\rho_i\right) =$$
$$\sum_{k=1}^{N} \sum_{j=1}^{M} \sum_{i=1}^{N} t_{ij}^{proc}\alpha_{ijk} + \sum_{i=1}^{M} \rho_i(O_i^P + O^M(\rho_i - 1))$$

Interval Relation

$$w_{ij} = 1 - y_i g_j \text{ for } i, j = 1, 2, 3, 4$$

Memory Utilization Interval Constraints

$$u_m \leq 0.5g_1 + 0.7g_2 + 0.85g_3 + 0.95g_4$$
$$u_m \geq 0.5g_2 + 0.7g_3 + 0.85g_4, \ \sum_{i=1}^{4} g_i = 1$$

Processor Utilization Interval Constraints

$$u_p \leq 0.5y_1 + 0.7y_2 + 0.85y_3 + 0.95y_4$$
$$u_p \geq 0.5y_2 + 0.7y_3 + 0.85y_4$$
$$\text{where } \sum_{i=1}^{4} y_i = 1$$

FIGURE 78.12 Architecture design model.

α_{ijk}	=	1, if DFG task i is implemented on the kth programmable processor of type j, otherwise 0;
β_{ij}	=	1, if DFG task i is implemented on the jth FPGA, otherwise 0;
δ_{ij}	=	1, if DFG task i is implemented on the jth ASIC, otherwise 0;
ρ_i	=	the number of programmable processors of type i in the architecture;
μ	=	the number of DRAM chips in the architecture;
d	=	the delay in delivering the product to market(months);
u_p	=	the overall processor utilization;
u_m	=	the overall memory utilization;
y_i	=	is a binary variable which signifies the processor utilization interval;
g_i	=	is a binary variable which signifies the memory utilization interval;
f_i^E	=	the execution time constraint effort multiplier for processor utilization interval i;
f_i^M	=	the memory constraint effort multiplier for memory utilization interval i;
s	=	the software development effort (person-months);
h_i	=	the hardware development effort for dedicated hardware type i (person-months);
n_{ij}	=	1 if processor j of type i is included in the architecture, otherwise 0;

l_{jk}^m = the local memory allocated to processor k of type j.

Additional model parameters include:

C_{NRE} = the overall non-recurring engineering cost per unit ASIC die area

N = the number of tasks (nodes) in the data/control flow graph; also, the maximum number of computational elements in the architecture

V = the production volume

C_i^{proc} = the procurement cost for a programmable processor of type i

C^{mem} = the procurement cost for per DRAM chip

M = the number of different types of programmable processors

C^{fpga} = the production cost per unit size for an FPGA

C^{asic} = the production cost per unit size for an ASIC

T_S = the amount of time after system concept when the product should be delivered to market

Z = a very large number

Y = the minimum allowable ASIC yield

K = the process defect density

U^{fpga} = the size limit on a single FPGA

U_A = the area limit on the system

U_P = the power limit on the system

S_i = the peak throughput of processor i

t_{ij}^{proc} = the throughput of task i implemented on processor j

t_i^{asic} = the throughput of task i implemented on an asic

t_i^{fpga} = the throughput of task i implemented on an fpga

O_i^P = the processor overhead (scheduling, resource management, dispatching, context switching)(ops/sec)

O^M = multiprocessor overhead factor

T_{sys} = the minimum required system level throughput

m_i = the memory requirement for task i

O^{MEM} = the total memory required for OS services, control and diagnostics, and support software

O^{lm} = the local memory required for OS services, control and diagnostics, and support software

Δ_p = is a constant that defines the desired utilization range on a processor

c_{kl} = the throughput penalty for transferring data between tasks k and l off-chip

C_{comm} = the worst case total throughput penalty due to transferring data between tasks off-chip

We are currently using the GAMS [37] optimization system to solve examples of this form. More specifically, the GAMS/DICOPT non-linear mixed-integer programming package is being employed. DICOPT utilizes non-linear optimization programs such as MINOS or CONOPT and mixed-integer programming packages such as OSL to rigorously solve these problems. Linearization techniques such as those described in [38] are also being applied to the models to improve computational efficiency.

Interestingly, while the rapid prototyping community has largely ignored rigorous integer programming methods for "quick" simplified heuristics, the communications industry (e.g., AT&T, Airlines reservation systems, etc.) routinely uses optimization algorithms with variables numbering in a few tens of thousands and more. The authors feel that complex nonlinear and multiobjective functions cannot be optimized via the "human-in-the-optimization-loop" methods, and any extra effort spent in the conceptual phase of the design process is time well spent.

78.8   Performance Modeling and Architecture Verification

The selection process for possible system architectures was discussed in the previous section. Performance models [39, 40, 41, 42] are used to verify that the architectures adhere to specific time-critical system constraints. The advantages of performance models include:

- They capture the time aspects of the system under development (i.e., throughput, latency, and resource utilization) and present this information for rapid evaluation.
- They verify the performance of proposed architectures found in the architectural design stage.
- They allow for true HW/SW codesign by simulating the behavior of software on performance models of processor hardware. The model of software can take many forms, one being a simple delay which models the performance of a library software primitive executing on a specific processor. For example, an Analog Devices SHARC 2106X chip executes a Fast Fourier Transform (FFT) in shorter time than a Texas Instruments C30 processor. The performance model captures this information through the use of *generic* parameters and when simulated, uses this parameter to determine how long the processor will be utilized while performing the function [40, 41].
- They provide the capability for modeling operating system effects on multiprocessor network architectures [41].
- Performance model development time is shorter when compared with that of fully functional models and hence library population can be done in-cycle.

The fidelity attributes of token-based performance models used for system architecture verification through simulation are listed below:

> Internal: {(Clock cycle → System event), (X), (X), (X), (Assembly → Primitive)}
> External: {(Clock cycle → System event), (X), (X), (Full Structure → Black box), (X)}.

Temporal information for both the internal and external attributes are captured at multiple levels depending on the application modeled. For example, the system event level can capture large blocks of data passing over an interconnect network or the simulating of a large time slice of processing on a single processor. System events occur in the 10s of microseconds to 10s of milliseconds time span and potentially could contain millions of actual clock cycles. The clock cycles, however, are not simulated, only the time events where information is interchanged or processed. At higher resolutions, details may be required to capture how an interconnect network handles data streams from multiple processors at the clock cycle level. The performance models we use for architecture verification fall into these categories and an example is presented later. Performance level models do not capture the function or data values of the system but focus on its time aspects. Internally, a performance model has no structure, but externally, the structure can be represented by any level in the resolution hierarchy. For example, a network architecture model consisting of multiple processors and interconnect ASICs could be described by first instantiating each of the components in a network model, then by connecting them using a performance model of a particular bus or interconnect protocol. At the other extreme, the model may be represented by a black box that outputs tokens based on a specified control input with the internal details represented by abstract behavior. On the RASSP program, the programmability generally was captured at the DSP function primitive level where signal processing procedures (FFT, etc.) are scheduled on performance models of processors. The processor models can, however, be defined for much higher resolutions where the primitives represent assembly level instructions.

   The efficiency of these models is very high because the code is written at the behavioral level of abstraction in VHDL. Signals are used to pass abstract data types known as tokens between component elements. The tokens are record types in VHDL and take the form as shown in Fig. 78.13. All protocol handling information is carried within it. This data type is referred to as *cue* throughout the remainder of the paper. The *cue* data type is used as a virtual packet to pass information between elements in a system architecture. It contains fields for capturing statistical information about how the data is passed through a processor network (priority, collisions, retries, routes), information on the source and destination of packet (src_id, dest_id), packet size (c_size, resp_size), packet identification number (c_id), and transmission status (c_state). There are also user defined fields that allow the performance model designer to implement model specific details (int_user1, int_user2, real_user1, real_user2).

```
-- ucue : the basic unresolved virtual packet type.
TYPE ucue IS
   RECORD
   -- basic information field
   c_id         : integer;
   dest_id      : integer;    -- destination
   src_id       : integer;    -- source
   c_type       : cue_type;   -- packet type
   -- *** TYPE cue_type IS
   -- *** (DATA, READ, WRITE, CONTROL, USER1, USER2);
   -- *** USER1 and USER2 are used to declare some special
   -- *** packet type.
   -- *** e.g. "echo packet" and "idle symbol" in SCI
   c_size       : size_type;
   -- *** TYPE size_type IS RANGE 0 to INTEGER'high
   -- *** UNITS
   -- *** bit_unit;
   -- *** byte       = 8 bit_unit;
   -- *** kbyte      = 1024 byte;
   -- ***   ... ... ... ... ...
   -- ***   ... ... ... ... ...
   -- ***   ... ... ... ... ...
   -- *** END UNITS;

   block_size   : size_type;
   -- *** the block size of response requested by the packet.
   init_time    : TIME;       -- initial time

   -- protocol field
   priority     : INTEGER;
   c_state      : cue_state_type; -- packet state
   -- *** TYPE cue_state_type IS (IDLE,REQ,ACK,BUSY,
   -- *** USERSTATE1, USERSTATE2);
   protocol     : STRING(1 TO 8);
   -- *** the protocol name, e.g. "pt_to_pt"
   collisions   : INTEGER;
   retries      : INTEGER;
   route        : INTEGER;

   -- user-defined field
   int_user1    : INTEGER;
   int_user2    : INTEGER;
   real_user1   : INTEGER;
   bits_user1   : BIT_VECTOR(15 downto 0);
   -- *** used to set some special flags or handshaking signals...
   -- *** e.g. "burst" in HIPPI ...
END RECORD;
```

FIGURE 78.13   The declaration of *cue*. The information of each communication transaction is contained in three fields — basic information field, protocol field, and user-defined field.

Interoperability is determined by the ability of models with a similar external fidelity to communicate. For performance models to achieve this goal, a token format must be defined and standardized. Currently, there are no standards to meet this demand. Protocol converters can be developed to link performance models with alternate token structures but a standard is encouraged. RASSP is pursuing a standard.

78.8.1   A Performance Modeling Example: SCI Networks

A Scalable Coherent Interface (SCI) performance model has been developed as an example. The executable SCI model can serve as an executable specification for the communication protocol. Figure 78.14 illustrates the SCI node interface structure. **Linc** transmits or bypasses packets, and performs the primary SCI protocols. **REC_QUEUE** and **TR_QUEUE** are First-In-First-Out buffers which store receive packets and transmit packets. **Processor** contains a **responder** and a **res_handler**; **responder** generates the response packets for the request packets who ask for responses, and **res_handler** serves as a response packet consumer. The **packet generator** is used to create cues according to the required communication patterns. Connecting the SCI node interface, designers can easily construct an SCI network and create their inputs to evaluate the performance results.

Among the processes of the SCI node, **mux_process** dominates the primary communication protocol. The SCI protocol can be converted to the state diagram shown in Fig. 78.15. Based on the state diagram, the VHDL representation of **mux_process** at the performance level is written as shown in Fig. 78.16. *MUX_Process* is activated by *st_pkt, bf_pkt, tr_pkt,* and *MUX_State* and changes *MUX_State* according to the state diagram.

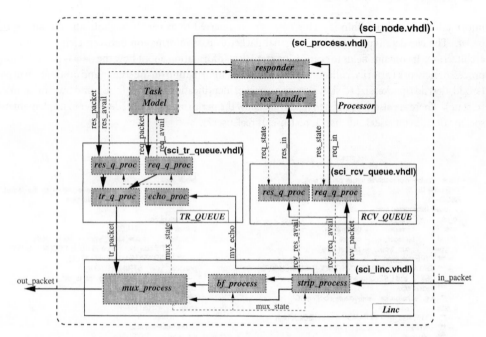

FIGURE 78.14   The SCI node interface performance model.

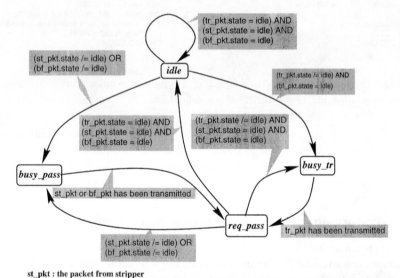

st_pkt : the packet from stripper
bf_pkt: the packet from bypass FIFO
tr_pkt: the packet from transmit queues

FIGURE 78.15   The state diagram of MUX.

78.8.2   Deterministic Performance Analysis for SCI

A key requirement of real-time DSP architectures is determinism. Most designers would like to know the guaranteed worst-case performance rather than the average or peak performance. In order to make the performance determinable, an SCI network must satisfy the following constraints:

Step 1.   *The size of each packet is deterministic.*
Step 2.   *The interprocessor communications are deterministic.*

```
MUX_Process : PROCESS(st_pkt, bf_pkt, tr_pkt, MUX_State)        WHEN busy_tr =>

    VARIABLE temp_pkt    : ucue := init_pkt;                    WHEN busy_pass =>
    VARIABLE time_busy   : TIME;
                                                               WHEN req_pass =>
BEGIN                                                              IF (st_pkt'Event AND st_pkt.state /= idle) THEN
    IF (now = 0 ns) THEN                                              time_busy := TransferTime(st_pkt.size);
        MUX_State <= idle;                                           Out_Token <= st_pkt after WIRE_DELAY+OUT_DELAY;
    END IF;                                                          MUX_State <= busy_pass;
                                                                    MUX_State <= req_pass after time_busy;
    CASE MUX_State IS                                             ELSIF (bf_pkt.state /= idle) THEN
        WHEN idle =>                                                  time_busy := TransferTime(bf_pkt.size);
            IF (tr_pkt.state = idle) THEN                            Out_Token <= bf_pkt after WIRE_DELAY+OUT_DELAY;
                IF (st_pkt'Event AND st_pkt.state /= idle) THEN      MUX_State <= busy_pass;
                    time_busy := TransferTime(st_pkt.size);          MUX_State <= req_pass after time_busy;
                    -- *** calculate the transfer time of the packet ELSIF (tr_pkt.state /= idle) THEN
                    Out_Token <= st_pkt after WIRE_DELAY+OUT_DELAY;  time_busy := TransferTime(tr_pkt.size);
                    MUX_State <= busy_pass;                          Out_Token <= tr_pkt after WIRE_DELAY+OUT_DELAY;
                    MUX_State <= req_pass after time_busy;           Out_Token <= idle_pkt
                ELSE                                                              after time_busy+WIRE_DELAY+OUT_DELAY;
                    IF (bf_pkt.state /= idle) THEN                   MUX_State <= busy_tr;
                        time_busy := TransferTime(bf_pkt.size);      time_busy := time_busy + SCI_BASE_TIME;
                        Out_Token <= bf_pkt after WIRE_DELAY+OUT_DELAY; MUX_State <= req_pass after time_busy;
                        MUX_State <= busy_pass;                   ELSE
                        MUX_State <= req_pass after time_busy;       MUX_State <= idle;
                    END IF;                                       END IF;
                END IF;
            ELSE                                                END CASE;
                time_busy := TransferTime(tr_pkt.size);
                Out_Token <= tr_pkt after WIRE_DELAY+OUT_DELAY; END PROCESS MUX_Process;
                -- insert idle symbol
                Out_Token <= idle_pkt
                            after time_busy+WIRE_DELAY+OUT_DELAY;
                MUX_State <= busy_tr;
                MUX_State <= req_pass after time_busy + SCI_BASE_TIME;
                -- *** SCI_BASE_TIME := 2 ns;
            END IF;
```

FIGURE 78.16 The VHDL process of MUX.

Step 3. *All arrival packets are accepted.* That is to say, packets should not be retransmitted in an SCI ring. The retry packets might prevent sending the fresh packets and make the throughput and latency unpredictable.

DEFINITION 78.1 If a packet is retransmitted from the transmit queues, it is called a retry packet. If a packet is transmitted for the first time, it is called a fresh packet.

A basic SCI network is a unidirectional SCI ring. The maximum number of nodes traversed by a packet is equal to N in an SCI ring, and the worst-case path contains a MUX, a stripper, and $(N - 2)$ links. So, we find the worst case latency , $L_{worst\text{-}case}$, is :

$$L_{worst\text{-}case} = T_{MUX} + T_{wire} + T_{stripper} + (N - 2) \cdot T_{linc} \tag{78.6}$$
$$= (N - 1) \cdot T_{linc} - T_{FIFO} \tag{78.7}$$

where the link delay, T_{linc}, is equal to $T_{MUX} + T_{FIFO} + T_{stripper} + T_{wire}$, T_{MUX} is the MUX delay, T_{wire} is the wire delay between nodes, $T_{stripper}$ is the stripper delay, and T_{FIFO} is the full bypass FIFO delay.

The SCI link bandwidth, BW_{link}, is equal to 1 byte per second per link; the maximum bandwidth of an SCI ring is proportional to the number of nodes:

$$BW_{ring} = N \cdot BW_{link} \text{ (bytes/second)} \tag{78.8}$$

where N is the number of nodes. Now let us consider the bandwidth of an SCI node. Since each link transmits the packets issued by all nodes in the ring, BW_{link} is shared by not only transmitting packets but passing packets, echo packets, and idle symbols.

$$BW_{link} = bw_{transmitting} + bw_{passing} + bw_{echo} + bw_{idle} \tag{78.9}$$

where $bw_{\text{transmitting}}$ is the consumed bandwidth of transmitting packets, bw_{passing} is the consumed bandwidth of passing packets, bw_{echo} is the consumed bandwidth of echo packets, and BW_{idle} is the consumed bandwidth of idle symbols. Assuming that the size of the send packets is fixed, we find $bw_{\text{transmitting}}$ is:

$$bw_{\text{transmitting}} = BW_{\text{link}} \cdot \frac{N_{\text{transmitting}} \cdot D_{\text{packet}}}{D_{\text{link}}} \tag{78.10}$$

$$= BW_{\text{link}} \cdot \frac{N_{\text{transmitting}} \cdot D_{\text{packet}}}{(N_{\text{passing}} + N_{\text{transmitting}}) \cdot (D_{\text{packet}} + 16) + N_{\text{echo}} \cdot 8 + N_{\text{idle}} \cdot 2} \tag{78.11}$$

where D_{packet} is the data size of a transmitting packet, D_{link} is the number of bytes passed through the link, $N_{\text{transmitting}}$ is the number of transmitting packets, N_{passing} is the number of passing packets, N_{echo} is the number of echo packets, and N_{idle} is the number of idle symbols. A transmitting packet consists of an unbroken sequence of data symbols with a 16-byte header that contains address, command, transaction identifier, and status information. The echo packet uses an 8-byte subset of the header while idle symbols require only 2 bytes of overhead. Because each packet is followed by at least an idle symbol, the maximum $bw_{\text{transmitting}}$ is:

$$BW_{\text{transmitting}} = BW_{\text{link}} \cdot \frac{N_{\text{transmitting}} \cdot D_{\text{packet}}}{(N_{\text{passing}} + N_{\text{transmitting}}) \cdot (D_{\text{packet}} + 18) N_{\text{echo}} \cdot 10} \tag{78.12}$$

However, $BW_{\text{transmitting}}$ might be consumed by retry packets; the excessive retry packets will stop sending fresh packets. In general, when the processing rate of arrival packets, $R_{\text{processing}}$, is less than the arrival rate of arrival packets, R_{arrival}, the excessive arrival packets will not be accepted and their retry packets will be transmitted by the sources. This cause for rejecting an arrival packet is the so-called *queue contention*. The number of retry packets will increase with time because retry packets increase the arrival rate. Once $bw_{\text{transmitting}}$ is saturated with fresh packets and retry packets, the transmission of fresh packets is stopped resulting in an increase in the number of retry packets transmitted.

Besides queue contention, incorrect packets cause the rejection of an arrival packet. This indicates a possible component malfunction. No matter what the cause, the retry packets should not exist in a real-time system in that two primary requirements of real-time DSP are data correctness and guaranteed timing behavior.

78.8.3   DSP Design Case: Single Sensor Multiple Processor (SSMP)

Figure 78.17 shows a DSP system with a sensor and N processing elements (PEs). This system is called the *Single Sensor Multiple Processor*. In this system, the sensor uniformly transmits packets to each PE and the sampling rate of the sensor is R_{input}. For the node i, if the arrival rate, $R_{\text{arrival},i}$, is greater than the processing rate, $R_{\text{processing},i}$, receive queue contention will occur and unacceptable arrival packets will be sent again from the sensor node. Retry packets increase the arrival rate and result in more retry packets transmitted by the sensor node. Since the bandwidth of the retry packets and fresh packets is limited by $BW_{\text{transmitting}}$, the sensor will stop reading input data when the bandwidth is saturated. For a real-time DSP, the input data should not be suspended, thus, the following inequality has to be satisfied to avoid the retry packets:

$$\frac{R_{\text{input}}}{N} \leq R_{\text{processing}} \tag{78.13}$$

Because the output link of the sensor node will only transmit the transmitting packets, the maximum transmitting bandwidth is:

$$BW_{\text{transmitting}} = BW_{\text{link}} \cdot \frac{D_{\text{packet}}}{D_{\text{packet}} + 18} \tag{78.14}$$

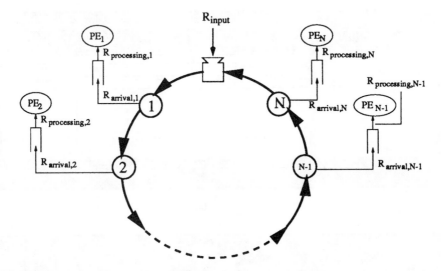

FIGURE 78.17 The SSMP architecture. The sensor uniformly transmits packets to each PE and the sampling rate of sensor is R_{input}; so, the arrival rate of each node is $\frac{R_{input}}{N}$.

and the limitation of R_{input} is:

$$R_{input} \leq BW_{link} \cdot \frac{D_{packet}}{D_{packet} + 18} \tag{78.15}$$

We now assume a SSMP system design with a 10 MBytes/sec sampling rate and five PEs where the computing task of each PE is a 64-point FFT. Since each packet contains 64 32-bit floating-point data, D_{packet} is equal to 256 bytes.

From Eq. (78.13) the processing rate must be greater than 2 MBytes/sec, so the maximum processing time for each packet is equal to 128 μsec. Because an n-point FFT needs $\frac{n}{2} \log_2 n$ butterfly operations and each butterfly needs 10 FLOPs [44], the computing power of each PE should be greater than 15 MFLOPS. From a design library we pick $i860$s to be the processing elements and a single SCI ring to be the communication element in that $i860$ provides 59.63 MFLOPS for 64-point FFT and $BW_{transmitting}$ of a single SCI ring is 934.3 MBytes/sec which satisfies Eq. (78.15). Using 5 $i860$s, the total computing power is equal to 298.15 MFLOPS. The simulation result is shown in Fig. 78.18(a). The result shows that retry packets for $R_{input} = 10$ MBytes/sec do not exist.

As stated earlier, if the processing rate is less than the arrival rate, the retry packets will be generated and the input stream will be stopped. Hence, we changed R_{input} from 10 MBytes/sec to 100 MBytes/sec to test whether the $i860$ can process a sampling rate as high as 100 MBytes/sec. Upon simulating the performance model, we found that the sensor node received an echo packet which asked for retransmitting a packet at 25944 ns in Fig. 78.18(b); thus, we have to substitute another processor with higher MFLOPS for $i860$ to avoid the occurrence of the retry packets.

Under the RASSP program, an additional example where performance level models were used to help define the system architecture can be found in [45].

78.9 Fully Functional and Interface Modeling and Hardware Virtual Prototypes

Fully functional and interface models support the concept of a hardware virtual prototype. The hardware virtual prototype is defined as a software representation of a hardware component, board, or system

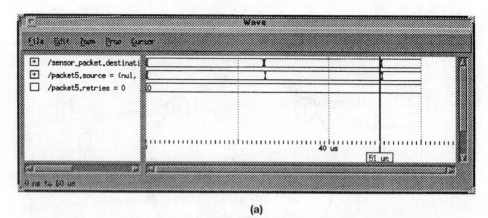

(a)

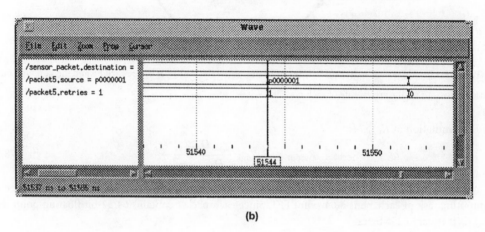

(b)

FIGURE 78.18   The simulation of SSMP using *i*860 and SCI ring. (a) The result of SSMP with $R_{input} = 10$ MBytes/sec. The value of *packet5.retries* shows that there does not exist any request of retry packets. (b) The result of SSMP with $R_{input} = 100$ MBytes/sec. A retry packet is requested at 25944 ns.

containing sufficient accuracy to guarantee its successful hardware system-level realization [5, 47]. The hardware virtual prototype adopts as its main goals (1) verification of the design correctness by eliminating hardware design errors from the in-cycle design loop, (2) decreasing the design process time through first time correctness, (3) allowing concurrent codevelopment of hardware and software, (4) facilitating rapid HW/SW integration, and (5) generation of models to support future system upgrades and maintenance. This model abstraction captures all the documented functionality and interface timing of the unit under development.

Following architectural trade studies, a high level design of the system is determined. This high level design consists of commercial-off-the-shelf (COTS) parts, in-house design library elements, and/or new application specific designs to be done in-cycle. At this level, it is assumed the COTS parts and in-house designs are represented by previously verified fully functional models of the devices. Fully functional models of in-cycle application specific designs serve as high level models useful for system level simulation. They also serve as *golden* models for verification of the synthesizable RTL level representation and define its testbench. For system level simulations, this high level model can improve simulation speed by an order of magnitude while maintaining component interface timing fidelity.

The support infrastructure required for fully functional models is the existence of a library of component elements and appropriate HDL simulation tools. The types of components contained in the library should include models of processors, buses/interconnects, memories, programmable logic, controllers, and medium and large scale integrated circuits. Without sufficient libraries, the development of complex models within the in-cycle design loop can diminish the usefulness of this design philosophy by increasing the design time. The model fidelity used for hardware virtual prototyping can be classified as listed below:

Internal: {(Gate → Clock cycle), (Bit true → Token), (All), (Major blocks), (Micro code → Assembly)}
External: {(Gate → Clock cycle), (Bit true), (All), (Full Structure), (X)}.

Internally and externally, the temporal information of the device should be at least clock cycle accurate. Therefore, internal and external signal events should occur as expected relative to clock edges. For example, if an address line is set to a value after a time of 3 ns from the falling edge of a clock based on the specification for the device, then the model shall capture it. The model shall also contain hooks, via generic parameters, to set the time related parameters. The user selectable generic parameters are placed in a VHDL package and represent the minimum, typical, and maximum setup times for the component being modeled.

Internal data can be represented by any value on the axis, while the interface must be bit true. For example, in the case of an internal 32-bit register, the value could be represented by an integer or a 32-bit vector. Depending on efficiency issues, one or the other choice is selected. The external data resolution must capture the actual hardware pinout footprint and the data on these lines must be bit true. For example, an internally generated address may be in integer format but when it attempts to access external hardware, binary values must be placed on the output pins of the device. The internal and external functionality is represented fully by definition.

Structurally, because the external pins must match those of the actual device, the external resolution is as high as possible and therefore the device can be inserted as a component into a larger system if it satisfies the interoperability constraints. Internally, the structure is composed of high level blocks rather than detailed gates. This improves efficiency because we minimize the signal communication between processes and/or component elements.

Programmability is concerned with the level of software instructions interpreted by the component model. When developing hardware virtual prototypes, the programmable devices are typically general purpose, digital, or video signal processors. In these devices, the internal model executes either microcode or the binary form of assembly instructions and the fidelity of the model captures all the functionality enabling this. This facilitates hardware/software codevelopment and cosimulation. For example, in [5], a processor model of the Intel *i*860 was used to develop and test over 700 lines of Ada code prior to actual hardware prototyping.

An important requirement for fully functional models to support reuse across designs and rapid systems development is the ability to operate in a seamless fashion with models created by other design teams or external vendors. In order to ensure interoperability, the IEEE standard nine value logic package[2] is used for all models. This improves technology insertion for future design system upgrades by allowing segments of the design to be replaced with new designs which follow the same interoperability criteria.

Under the RASSP program, various design efforts utilized this stage in the design process to help achieve first pass success in the design of complex signal processing systems. The Lockheed Sanders team developed an infrared search and track (IRST) system [46, 47] consisting of 192 Intel *i*860 processors using a Mercury RACEWAY network along with custom hardware for data input buffering and distribution and video output handling. The hardware virtual prototype (HVP) served to find a number of errors in the original design both in hardware and software. Control code was developed in Ada and executed on the HVP prior to actual hardware development. Another example where hardware virtual prototypes were used can be found in [48].

---

[2]IEEE 1164-1993 Standard Multi-Value Logic System for VHDL Model Interoperability.

78.9.1   Design Example: I/O Processor for Handling MPEG Data Stream

In this example, we present the design of an I/O processor for the movement of MPEG-1 encoder data from its origin at the output of the encoder to the memory of the decoder. The encoded data obtained from the source is transferred to the VME-bus through a slave interface module which performs the proper handshaking. Upon receiving a request for data (AS low ,WRITE high) and a valid address, the data is presented on the bus in the specified format (the mode of transfer is dictated by the VME signals LWORD,DS0,DS1 and AM[0..5]). The VME DTACK signal is then driven low by the slave indicating that the data is ready on the bus after which the master accepts the data. It repeats this cycle if more data transfer is required, otherwise it releases the bus. In the simulation of the I/O architecture in Fig. 78.19 a Quad-Byte-Block Transfer (QBBT) was done.

The architecture of the I/O processor is described below. The link ports were chosen for the design since they were an existing element in our design library and contain the same functionality as the link ports on the Analog Devices 21060 digital signal processor. The circuit's ASIC controller is designed to interface to the VME bus, buffer data, and distribute it to the link ports. To achieve a fully pipelined design, it contains a 32-bit register buffer both at the input and outputs. The 32-bit data from the VME is read into the input buffer and transferred to the next empty output register. The output registers send the data by unpacking. The unpacking is described as follows: at every rising edge of the clock (LxCLK) a 4-bit nibble of the output register, starting from the LSB, is sent to the link port data line (LxDAT) if the link port acknowledge (LxACK) signal is high. Link ports that are clocked by LxCLK, running at the twice the core processor's clock rate, read the data from the controller ports with the rising edge of the LxCLK signal. When their internal buffers are full they deassert LxACK to stop the data transfer.

Since we have the option of transferring data to the link ports at twice the processor's clock rate, four link ports were devoted to this data transfer to achieve a fully pipelined architecture and maximize utilization of memory bandwidth. With every rising edge of the processor clock (CLK) a new data can be read into the memory. Figure 78.20 shows the pipelined data transfer to the link ports where DATx represents a 4-bit data nibble. As seen from the table, Port0 can start sending the new 32-bit data immediately after it is done with the previous one. Time multiplexing among the ports is done by the use of a token. The token is transferred to the next port circularly with the rising edge of the processor clock. When the data transfer is complete (buffer is empty), each port of the controller deasserts the corresponding LxCLK which disables the data transfer to the link ports. LxCLKs are again clocked when the transfer of a new frame starts. The slave address, the addressing mode, and the data transfer mode require setups for each transfer. The link ports, IOP registers, DMA control units, and multiport memory models were available in our existing library of elements and they were integrated with the VME bus model library element. However, the ASIC controller was designed in-cycle to perform the interface handshaking. In the design of the ASIC, we made use of the existing library elements, i.e., I/O processor link ports, to improve the design time. To verify the performance and correctness of the design, the comparison mechanism we used is shown in Fig. 78.21. The MPEG-1 encoder data is stored in a file prior to being sent over the VME bus via master-slave handshaking. It passes through the controller design and link ports to local memory. The memory then dumps its contents to a file which is compared to the original data. The comparisons are made by reading the files in VHDL and doing a bit by bit evaluation. Any discrepancies are reported to the designer.

The total simulation time required for the transfer of a complete frame of data (28Kbytes) to the memory was approximately 19 min of CPU time and 1 h of wall clock time. These numbers indicate the usefulness of this abstraction level in the design hierarchy. The goal is to prove correctness of design and not simulate algorithm performance. Algorithm simulations at this level would be time prohibitive and must be moved to the performance level of abstraction.

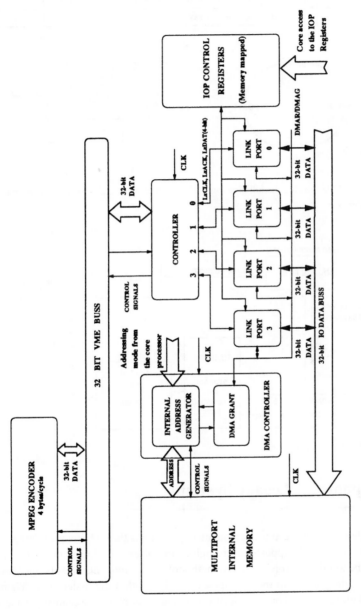

FIGURE 78.19 The system I/O architecture.

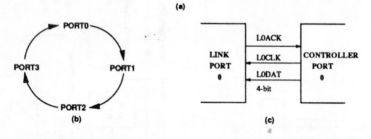

CLK	1		2		3		4		5		6		7		8	
LxCLK	1	2	3	4	5	6	7	8	9	10	11	12	13	14	15	16
PORT0	DAT0	DAT1	DAT2	DAT3	DAT4	DAT5	DAT6	DAT7								
PORT1		DAT0	DAT1	DAT2	DAT3	DAT4	DAT5	DAT6	DAT7							
PORT2			DAT0	DAT1	DAT2	DAT3	DAT4	DAT5	DAT6	DAT7						
PORT3				DAT0	DAT1	DAT2	DAT3	DAT4	DAT5	DAT6	DAT7					

(a)

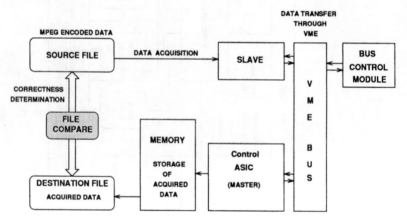

FIGURE 78.20 (a) Table showing the full pipelining, (b) token transfer, and (c) signals between a link port and the associated controller port.

FIGURE 78.21 Data comparison mechanism.

78.10 Support for Legacy Systems

A well-defined design process capturing systems requirements through iterative design refinement improves system life cycle and supports the reengineering of existing legacy systems. With the system captured using the top-down evolving design methodology, components, boards, and/or subsystems can be replaced and redesigned from the appropriate location within the design flow. Figure 78.22 shows examples of possible scenarios. For example, if a system upgrade requires a change in a major system operating mode (e.g., search/track), then the design process can be reentered at the executable requirements or specification stage with the development of an improved algorithm. The remaining system functionality can serve as the environment test bench for the upgrade. If the system upgrade consists of a new processor design to reduce board count or packaging size, then the design flow can be reentered at the hardware virtual prototyping phase using fully functional models. The improved hardware is tested using the previous models of its surrounding environment. If an architectural change using an improved interconnect technology is required, the performance modeling stage is entered. In most cases, only a portion of the entire system is affected, therefore, the remainder serves as a testbench for the upgrade. The test vectors developed in the initial design can be reused to verify the current upgrade.

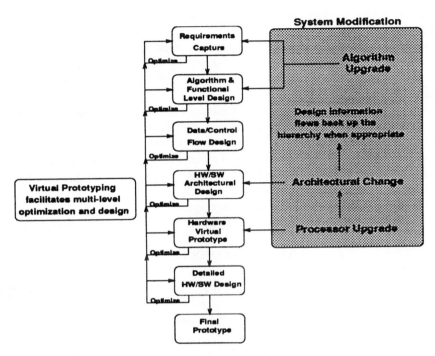

FIGURE 78.22   Reengineering of legacy systems in VHDL.

78.11   Conclusions

In this chapter, we have presented a top-down design process based on the RASSP virtual prototyping methodology. The process starts by capturing the system requirements in an executable form and through successive stages of design refinement, and ends with a detailed hardware design. VHDL models are used throughout the design process to both document the design stages and provide a common language environment for which to perform requirements simulation, architecture verification, and hardware virtual prototyping. The fidelity of the models contain the necessary information to describe the design as it develops through successive refinement and review. Examples were presented to illustrate the information captured at each stage in the process. Links between stages were described to clarify the flow of information from requirements to hardware. Case studies were referenced to point the reader to more detail on how the methodology performs in practice. Tools are being developed by RASSP participants to automate the process at each of the design stages and references are provided for more information.

Acknowledgments

This research was supported in part by DARPA ETO (F33615-94C-1493) as part of the RASSP Program 1994-1997. The authors would like to thank all the RASSP program participants for their effort in creating and demonstrating the usefulness of the methodology and its effectiveness in achieving improvements in the overall design process.

References

[1] Richards, M.A., The rapid prototyping of application specific signal processors (RASSP) program: Overview and accomplishments, *Proceedings 1st Annual RASSP Conference*, pp. 1-8, Arlington, VA, August, 1994.
URL: http://rassp.scra.org/public/confs/1st/papers.html#RASSP P.

[2] Hood, W., Hoffman M., Malley J., et al., RASSP program overview, *Proceedings 2nd Annual RASSP Conference*, pp. 1-18, Arlington, VA, July 24-27, 1995. URL: http://rassp.scra.org/public/confs/2nd/papers.html.

[3] Saultz, J.E., Lockheed Martin advanced technology laboratories RASSP second year overview, *Proceedings 2nd Annual RASSP Conference*, pp. 19-31, Arlington, VA, July 24-27, 1995. URL: http://rassp.scra.org/public/confs/2nd/papers.html#saultz.

[4] Madisetti, V., Corley, J., and Shaw, G., Rapid prototyping of application-specific signal processors: Educator/facilitator current practice (1993) model and challenges, *Proceedings 2nd Annual RASSP Conference*, July 1995.
URL: http://rassp.scra.org/public/confs/2nd/papers.html#current.

[5] Madisetti, V.K. and Egolf, T.W., Virtual prototyping of embedded microcontroller-based DSP systems, *IEEE Micro*, Oct. 1995.

[6] ANSI/IEEE Std 1076 − 1993 IEEE Standard VHDL Language Reference Manual(1 − 55937 − 376 − 8), Order Number [SH16840].

[7] Thomas, D., Adams, J., and Schmit, H., A model and methodology for hardware-software codesign, *IEEE Design & Test of Computers*, pp. 6-15, Sept. 1993.

[8] Kumar, S., Aylor, J., Johnson, B., and Wulf, W., A framework for hardware/software codesign, *Computer*, pp. 39-45, Dec. 1993.

[9] Gupta, R. and De Micheli, G., Hardware-software cosyn thesis for digital systems, *IEEE Design & Test of Computers*, Sept. 1993.

[10] Kalavade, A. and Lee, E., A hardware-software codesign methodology for DSP applications, *IEEE Design & Test of Computers*, pp. 16-28, Sept. 1993.

[11] Kalavade, A. and Lee, E., A global criticality/local phase driven algorithm for the constrained hardware/software partitioning problem, *Proc. of the Third International Workshop on Hardware/Software Codesign*, Sept. 1994.

[12] Ismail, T. and Jerraya, A., Synthesis steps and design models for codesign, *Computer*, pp. 44-52, Feb. 1995.

[13] Gajski, D. and Vahid, F., Specification and design of embedded hardware-software systems, *IEEE Design & Test of Computers*, pp. 53-67, Spring 1995.

[14] DeBardelaben, J. and Madisetti, V., Hardware/software codesign for signal processing systems— A survey and new results, *Proc. of the 29th Annual Asilomar Conference on Signals, Systems, and Computers*, Nov. 1995.

[15] IEEE Std 1164-1993 IEEE Standard Multivalue Logic System for VHDL Model Interoperability (Std_logic_1164) (1 − 55937 − 299 − 0), Order Number [SH16097].

[16] Hein, C., Carpenter, T., Kalutkiewicz, P., and Madisetti, V., RASSP VHDL modeling terminology and taxonomy — Revision 1.0, *Proceedings 2nd Annual RASSP Conference*, pp. 273-281, Arlington, VA, July 24-27, 1995. URL: http://rassp.scra.org/public/confs/2nd/papers.html#taxonomy.

[17] Anderson, A.H. et al., VHDL executable requirements, *Proceedings 1st Annual RASSP Conference*, pp. 87-90, Arlington, VA, August, 1994. URL: http://rassp.scra.org/public/confs/1st/papers.html#VER.

[18] Shaw, G.A. and Anderson A.H., Executable requirements: Opportunities and impediments, *IEEE Proceedings of the International Conference on Acoustics, Speech, and Signal Processing*, pp. 1232-1235, Atlanta, GA. May 7-10, 1996.

[19] Frank, G.A., Armstrong, J.R., and Gray, F.G., Support for model-year upgrades in VHDL test benches, *Proceedings 2nd Annual RASSP Conference,* pp. 211-215, Arlington, VA, July 24-27, 1995. URL: http://rassp.scra.org/public/confs/2nd/papers.html.

[20] ISO/IEC 11172, Information technology—coding of moving picture and associated audio for digital storage media at up to about 1.5 Mbit/s, 1993.

[21] Rowe, L.A., Patel, K. et al., mpeg_encode/mpeg_play, Version 1.0, available via anonymous ftp at ftp://mm-ftp.cs.berkeley.edu/pub/multimedia/mpeg/bmt1r1.tar.gz, Computer Science Department-EECS University of California at Berkeley, May 1995.

[22] ISO/IEC 13818, Coding of moving pictures and associated audio, Nov. 1993.

[23] Tanir, O. et al., A specification-driven architectural design environment, *Computer,* pp. 26-35, June 1995.

[24] Vahid, F. et al., SpecCharts: A VHDL front-end for embedded systems, *IEEE Trans. Computer-Aided Design of Integrated Circuits and Systems,* pp. 694-706, June 1995.

[25] Egolf, T.W., Famorzadeh, S., and Madisetti, V.K., Fixed-point codesign in DSP, *VLSI Signal Processing Workshop, Vol. 8,* Fall, 1994.

[26] Naval Research Laboratory, Processing graph method tutorial, Jan. 8, 1990.

[27] Robbins, C.R., Autocoding in Lockheed Martin ATL-camden RASSP hardware/software code-sign, *Proceedings 2nd Annual RASSP Conference,* pp. 129-133, July 24-27, Arlington, VA, URL: http://rassp.scra.org/public/confs/2nd/papers.html.

[28] Robbins, C.R., Autocoding: An enabling technology for rapid prototyping, *IEEE Proceedings of the International Conference on Acoustics, Speech, and Signal Processing,* pp. 1260-1263, Atlanta, GA., May 7-10, 1996. URL: http://rassp.scra.org/public/confs/2nd/papers.html.

[29] System-Level Design Methodology for Embedded Signal Processors, URL: http://ptolemy.eecs.berkeley. edu/ptolemyrassp.html.

[30] Publications of the DSP Design Group and the Ptolemy Project, URL: http://ptolemy.eecs.berkeley.edu/papers/publications.html/index.html.

[31] Boehm, B., *Software Engineering Economics,* Prentice-Hall, Englewood Cliffs, NJ, 1981.

[32] Madisetti, V. and Egolf, T., Virtual prototyping of embedded microcontroller-based DSP systems, *IEEE Micro,* Oct. 1995.

[33] U.S. Air Force Analysis Agency, *REVIC Software Cost Estimating Model User's Manual Version 9.2,* Dec. 1994.

[34] Fey, C., Custom LSI/VLSI chip design productivity, *IEEE J. Solid-State Circuits,* sc-20(2), April 1985.

[35] Paraskevopoulos, D. and Fey, C., Studies in LSI technology economics III: Design schedules for application-specific integrated circuits, *IEEE J. Solid-State Circuits,* sc-22(2), April 1987.

[36] Liu, J., Detailed model shows FPGAs' true costs, *EDN,* pp. 153-158, May 11, 1995.

[37] Brooke, A., Kendrick, D., and Meeraus, A., *Release 2.25 GAMS: A User's Guide,* Boyd & Fraser, Danvers, MA, 1992.

[38] Oral, M. and Kettani, O., A linearization procedure for quadratic and cubic mixed-integer problems, *Operations Res.,* 40(1), pp. 109–116, 1992.

[39] Rose, F., Steeves, T., and Carpenter, T., VHDL performance modeling, *Proc. 1st Annual RASSP Conf.,* pp. 60-70, Arlington, VA, August 1994. URL: http://rassp.scra.org/public/confs/1st /papers.html#VHDL P.

[40] Hein, C. and Nasoff, D., VHDL-based performance modeling and virtual prototyping, *Proc. 2nd Annual RASSP Conference,* pp. 87-94, Arlington, VA, July 24-27, 1995. URL: http://rassp.scra.org/public/confs/2nd/papers.html.

[41] Steeves, T., Rose, F., Carpenter, T., Shackleton, J., and von der Hoff, O., Evaluating distributed multiprocessor designs, *Proc. 2nd Annual RASSP Conf.,* pp. 95-101, Arlington, VA, July 24-27, 1995. URL: http://rassp.scra.org/public/confs/2nd/papers.html.

[42] Commissariat, H., Gray, F., Armstrong, J., and Frank, G., Developing re-usable performance models for rapid evaluation of computer architectures running DSP algorithms, *Proc. 2nd Annual RASSP Conf.*, pp. 103-108, Arlington, VA, July 24-27, 1995.
URL: http://rassp.scra.org/public/confs/2nd/papers.html.

[43] Athanas, P.M. and Abbott, A. L., Real-time image processing on a custom computing platform, *Computer*, pp. 16-24, Feb. 1995.

[44] Madisetti, V. K., *VLSI Digital Signal Processors: An Introduction to Rapid Prototyping and Design Synthesis*, IEEE Press, Piscataway, NJ, 1995.

[45] Paulson, R.H., Kindling: A RASSP application case study, *Proc. 2nd Annual RASSP Conf.*, pp. 79-85, Arlington, VA, July 24-27, 1995. URL: http://rassp.scra.org/public/confs/ 2nd/papers.html.

[46] Vahey, M. et al., Real time IRST development using RASSP methodology and Process, *Proc. 2nd Annual RASSP Conf.*, pp. 45-51, July 24-27, 1995.
URL: http://rassp.scra.org/public/confs/2nd/papers.html.

[47] Egolf, T., Madisetti, V., Famorzadeh, S., and Kalutkiewicz, P., Experiences with VHDL models of COTS RISC processors in virtual prototyping for complex systems synthesis, *Proc. VHDL Intl. Users' Forum (VIUF)*, Spring 1995, San Diego.

[48] Rundquist, E.A., RASSP benchmark 1: Virtual prototyping of a synthetic aperture radar processor, *Proc. 2nd Annual RASSP Conf.*, pp. 169-175, July 24-27, 1995.
URL: http://rassp.scra.org/public/ confs/2nd/papers.html.

Index

Index